2005 NSTI Nanotechnology Conference and Trade Show

NSTI Nanotech 2005

Volume 2

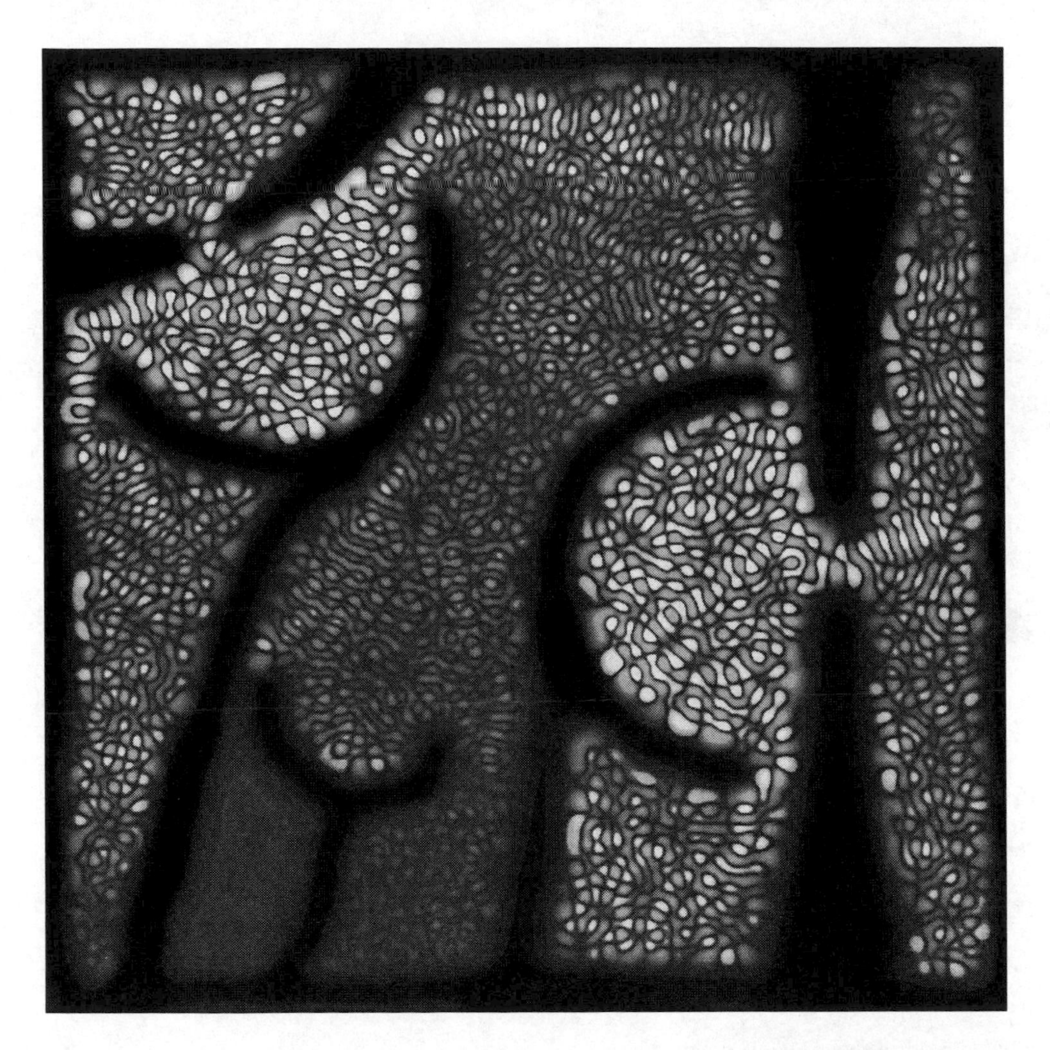

An Interdisciplinary Integrative Forum on
Nanotechnology, Biotechnology and Microtechnology

May 8-12, 2005
Anaheim Marriott & Convention Center
Anaheim, California, U.S.A.
www.nsti.org

Nano Science and Technology Institute
Boston • Geneva • San Francisco

Artwork provided by Prof. Eric J. Heller, http://www-heller.harvard.edu

Waves in a Quantum Resonator System (cover):
Electron propagation through a quantum resonator device. Results follow the launch of two wave packets in a potential landscape that is outlined in the dark regions. The dark regions correspond to high potential barriers. In between these high barriers is a weak random potential. The image is inspired by, and the potential barriers partly taken from, a small quantum resonator device fabricated in the Westervelt group at Harvard, and by Art Gossard at the University of California at Santa Barbara. This is a model for certain two degree of freedom electron gasses which are confined by "gates" (the walls). The waves spill out of the resonant cavities into the surrounding regions, reflecting and diffracting off surfaces and edges that they encounter. The waves correspond to electrons injected into the region and propagating as waves.

2005 NSTI Nanotechnology Conference and Trade Show

NSTI Nanotech 2005

Volume 2

An Interdisciplinary Integrative Forum on
Nanotechnology, Biotechnology and Microtechnology

May 8-12, 2005
Anaheim, California, U.S.A.
www.nsti.org

NSTI Nanotech 2005 Joint Meeting

The 2005 NSTI Nanotechnology Conference and Trade Show includes:
2005 NSTI Bio Nano Conference and Trade Show, Bio Nano 2005
8th International Conference on Modeling and Simulation of Microsystems, MSM 2005
5th International Conference on Computational Nanoscience and Technology, ICCN 2005
2005 Workshop on Compact Modeling, WCM 2005
NSTI Nanotech Ventures 2005

NSTI Nanotech 2005 Proceedings Editors:

Matthew Laudon
mlaudon@nsti.org

Bart Romanowicz
bfr@nsti.org

Nano Science and Technology Institute
Boston • Geneva • San Francisco

Nano Science and Technology Institute,
One Kendall Square, PMB 308
Cambridge, MA 02139
U.S.A.

The papers in this book comprise the proceedings of the 2005 NSTI Nanotechnology Conference and Trade Show, Nanotech 2005, Anaheim, California, May 8-12 2005. They reflect the authors' opinions and, in the interests of timely dissemination, are published as presented and without change. Their inclusion in this publication does not necessarily constitute endorsement by the editors, Nano Science and Technology Institute, or the sponsors. Dedicated to our knowledgeable network of willing and forgiving friends. To Avinash and Nina – for your competing publications, and to Sylvia for all your encouragement – Ed.

Nano Science and Technology Institute Order Number PCP05040392
ISBN 0-9767985-1-4
ISBN 0-9767985-5-7 (www)
ISBN 0-9767985-4-9 (CD-ROM, Vol. 1-3, WCM)

Additional copies may be ordered from:

Nano Science and Technology Institute
Publishing Office
One Kendall Sq., PMB 308
Cambridge, MA 02139, USA
http://www.nsti.org

Printed in the United States of America

Nano Science and Technology Institute
Boston • Geneva • San Francisco

Sponsors

Motorola
National Cancer Institute
Nanotech Japan
Nanophase Technologies
Veeco Instruments
Keithley Instruments
Leica Microsystems
Taiwan Business Alliance Forum
General Electric Company
ChevronTexaco
Olympus Industrial America
Ciphergen Biosystems, Inc.
National Science Foundation
Northern Ireland Nanotechnology
Nanotec NI
Italian Trade Commission
Freescale Semiconductors
Texas Instruments
Agilent Technologies
SEMI
Zygo Corporation
Burns, Doane, Swecker & Mathis, L.L.P.
Accelrys, Inc.
Hielscher
NanoInk, Inc.
Raith
Cantion
piezosystem jena
ALCATEL - Adixen Vacuum Technology
Evans Analytical Group
Edax, Inc.
Carl Zeiss SMT
Jenoptik Mikrotechnik GmbH
LEISTER Technologies, LLC
Intelligent Enclosures
ChemImage
PolyInsight, LLC
First Nano, Inc.
Kinetic Systems, Inc.
HERZAN, LLC
BioForce Nanosciences, Inc.
Australian Government – Invest Australia
Swiss Nanotech
Advance Nanotech, Inc.
Wilmer Cutler Pickering Hale and Dorr L.L.P.
Dorsey & Whitney L.L.P.
IEE Inspec
Finetex Technology
Atomistix
NIST Advanced Technology Program
nanoTITAN
FEI Company
Georgia Technode - NNIN
Nanometrix, Inc.
LioniX BV
Mad City Labs
SUSS MicroTec
Sigma-Aldrich
Bühler AG
NanoWorld AG
Applied Nanotech
Asylum Research
COMSOL
Nanosurf AG
NANOSENSORS
NanoMarkets LLC
Infinitesima, Ltd.
Center for Nanoscale Materials, Argonne National Laboratory
NanoGram Corporation
NanoDynamics, Inc.
mPHASE Technologies, Inc.
PhoeniX
Nanostellar, Inc.
Wyatt Technology Corporation
Particle Sizing Systems, Inc.
HORIBA Jobin Yvon
Seiko Instruments Inc.
Nanotechnology Researchers Network Center of Japan
Foley & Lardner L.L.P.
Malvern Instruments, Inc.
ANSYS, Inc.

Supporting Organizations

National Institutes of Health (NIH)
The Whitaker Foundation
Earth and Sky Radio Series
TIMA-CMP Laboratory, France
Ecos Corporation
Boston University
Greenberg Traurig
The Global Emerging Technology Institute, Ltd. (GETI)
The Electrochemical Society, Inc.
SAGE Crossroads

Table of Contents

Nano Composites

Carbon Nano Structures

Nano Devices and Systems

Surfaces and Films

Micro and Nano Structuring and Assembly

NEMS and MEMS Fabrication

Atomic and Mesoscale Modelling of Nanoscale Phenomena

Catalysis Technologies, Fuel Cells and Physical Chemistry

Inorganic Nanowires and Metallic Nano Structures

Characterization Tools and Microscopy

Nanotechnology: Novel Product Development

Nanotechnology: Societal and Economic Development

NSTI Nanotech 2005 Program Committee

TECHNICAL PROGRAM CO-CHAIRS

Matthew Laudon	*Nano Science and Technology Institute, USA*
Bart Romanowicz	*Nano Science and Technology Institute, USA*

TOPICAL AND REGIONAL SCIENTIFIC ADVISORS AND CHAIRS

Nanotechnology	
Matthew Laudon	*Nano Science and Technology Institute, USA*
Philippe Renaud	*Swiss Federal Institute of Technology, Switzerland*
Mihail Roco	*National Science Foundation, USA*
Wolfgang Windl	*Ohio State University, USA*
Biotechnology	
Srinivas Iyer	*Los Alamos National Laboratory, USA*
Pharmaceutical	
Kurt Krause	*University of Houston, USA*
Microtechnology	
Narayan R. Aluru	*University of Illinois Urbana-Champaign, USA*
Bernard Courtois	*TIMA-CMP, France*
Anantha Krishnan	*Defense Advanced Research Projects Agency, USA*
Bart Romanowicz	*Nano Science and Technology Institute, USA*
Semiconductors	
David K. Ferry	*Arizona State University, USA*
Andreas Wild	*Freescale Semiconductor, France*

SPECIAL SYMPOSIA CHAIRS

Mansoor M. Amiji	*Northeastern University, USA*
Wolfgang S. Bacsa	*Université Paul Sabatier, France*
Fiona Case	*Case Scientific, USA*
David K. Ferry	*Arizona State University, USA*
Elena Gaura	*Coventry University, UK*
Steffen Hardt	*Institute of Microtechnology Mainz, Germany*
Nick Quirke	*Imperial College, London, UK*
Doug Resnick	*Molecular Imprints, USA*
Srinivas Sridhar	*Northeastern University, USA*
Loucas Tsakalakos	*GE Global Research, USA*
Xing Zhou	*Nanyang Technological University, Singapore*

NANOTECHNOLOGY CONFERENCE COMMITTEE

M.P. Anantram	*NASA Ames Research Center, USA*
Phaedon Avouris,	*IBM, USA*
Xavier J. R. Avula	*Washington University, USA*
Wolfgang S. Bacsa	*Université Paul Sabatier, France*
Roberto Car	*Princeton University, USA*
Fiona Case	*Case Scientific, USA*
Franco Cerrina	*University of Wisconsin - Madison, USA*
Murray S. Daw	*Clemson University, USA*
Alex Demkov	*Freescale Semiconductor, USA*
Tejal Desai	*Boston University, USA*
David K. Ferry	*Arizona State University, USA*
Lynn Foster Squire	*Greenberg Traurig L.L.P., USA*
Toshio Fukuda	*Nagoya University, Japan*
Sharon Glotzer	*University of Michigan, USA*
William Goddard	*CalTech, USA*
Gerhard Goldbeck-Wood	*Accelrys, Inc., UK*
Niels Gronbech-Jensen	*UC Davis and Berkeley Laboratory, USA*
Jay T. Groves	*University of California at Berkeley, USA*

NSTI Nanotech 2005 Proceedings Topics

NSTI Nanotech 2005, Vol. 1, ISBN: 0-9767985-0-6:

1. Nanotechnology for Cancer
2. Medical Applications
3. Drug Delivery
4. Bio Nano Particles
5. Bio Nano Analysis and Characterization
6. Bio Nano Materials
7. DNA, Protein, Cells and Tissue Arrays
8. Bio Micro Sensors
9. Computational Biology
10. Micro and Nano Fluidics Design and Phenomena
11. Microfluidics and Lab on Chip
12. Nanostructured Fluids, Soft Materials, and Self Assembly

NSTI Nanotech 2005, Vol. 2, ISBN: 0-9767985-1-4:

1. Nano Particles and Molecules
2. Nano Composites
3. Carbon Nano Structures
4. Nano Devices and Systems
5. Surfaces and Films
6. Micro and Nano Structuring and Assembly
7. NEMS and MEMS Fabrication
8. Atomic and Mesoscale Modelling of Nanoscale Phenomena
9. Catalysis Technologies, Fuel Cells and Physical Chemistry
10. Inorganic Nanowires and Metallic Nano Structures
11. Characterization Tools and Microscopy
12. Nanotechnology: Novel Product Development
13. Nanotechnology: Societal and Economic Development

NSTI Nanotech 2005, Vol. 3, ISBN: 0-9767985-2-2:

1. Nanoscale Device and Process Modeling
2. CNT, Nano and Molecular Electronics
3. Nano Devices and Architectures
4. Nano Scale Electronics Processing
5. Nano Photonics and Optoelectronics
6. Characterization and Parameter Extraction
7. Smart Sensors and Systems
8. MEMS/NEMS Design and Applications
9. MEMS Modeling and Design
10. Computational Methods, Numerics and Software Tools
11. Modules and Circuits

NSTI Nanotech 2005, Workshop on Compact Modeling, ISBN: 0-9767985-3-0:

1. Invited Papers
2. Poster Papers
3. WCM 2004 Invited Papers
4. WCM 2003 Invited Papers
5. A History of MOS Transistor Compact Modeling

Message from the Program Committee

The Nano Science and Technology Institute is proud to present the 2005 NSTI Nanotechnology Conference and Trade Show (Nanotech 2005). The charter of the Nanotech conference, and its numerous sub-conferences, remains the same since its original conception in 1997. The Nanotech provides for a single interdisciplinary integrative community, allowing for core scientific advancements to disseminate into a multitude of industrial sectors and across the breadth of traditional science and technology domains converging under Nanotechnology, Biotechnology and Microtechnology.

The Nanotech Program Committee makes every effort to provide a scientifically outstanding environment, through its review and ranking process. The Nanotech received a total of 1,269 abstracts for the 2005 event, a growth of 193% over the previous year. All abstracts submitted into the Nanotech are reviewed and scored by a minimum of three (3) expert reviewers. Of the 1,269 submitted abstracts, 25% were accepted as oral presentations, 40% were accepted as poster presentations resulting in a 35% rejection rate. In addition to the regular program there are a number of invited sessions, panels, overview presentations and tutorials provided for completeness. The Nanotech program committee thanks the hard work of all of this year's reviewers (116 active reviewers in total). This grassroots self-review process by the nano, bio and micro technology communities is a source of pride for the Nanotech conference. This process is in place to provide for a yearly evolution and technical validation of the Nanotech conference content. We thank the authors for submitting their latest work, making a meeting of this caliber possible.

We hope the reader will find the papers assembled in these proceedings rewarding to read, and that the conference continues to foster further advances in this fascinating and multi-disciplinary field. Although the Nanotech conference makes every effort to be as comprehensive as possible, due to the rapid advancements in science and industry, there will inevitably be under-represented sections of this event. We look to you, as the true science and business leaders of small-technologies, to assist us in identifying the needs of our participants so that we can continue to grow the content of this event to best serve this community.

This year's conference technical program is schedule with special industry focused symposia. These include sessions on the Health Care, Biotechnology, Materials, Surface sciences, Transportation, Personal Care, Energy, Food, Environmental, Electronics, Telecommunications, Computational, Displays & Optics, Design & Modeling, Polymers, and Coatings sectors.

Special symposia this year include:

- Nano Industrial Impact Workshop — Sunday May 8
  One-Day Intensive Program to educate participants in the current state of the art in a range of nanotechnologies and impacted industries.
- National Cancer Institute Symposium — Tuesday May 10
  NSTI is proud to collaborate with the National Cancer Institute (NCI) and Northeastern University in presenting a Special Symposium on Nanotechnology for Cancer Prevention, Diagnosis and Treatment.
- Nanotech Ventures Emerging Company Review — May 9 - 11
  Over 50 early stage nanotech companies will be presenting within the Nanotech 2005 Early Stage Company and Investment Review, representing most every high-tech industrial sector.

We would like to take this opportunity to thank the many individuals who have worked so hard to make this meeting happen, and to welcome the new members of the program committee. Conferences of this scope are possible only because of the continuing interest and support of the community, expressed both by their submission of papers of high quality and by their attendance. The Nanotech 2005 program committee is grateful to all keynote speakers, authors and session chairs for contributing to the success of the event. We are also indebted to the foundations, agencies and companies whose financial contributions made this meeting possible. We encourage your feedback and participation. Additionally, if you have an interest in conference or session organizational assistance, please contact the conference manager and we will attempt to accommodate.

Information concerning next year's conference is posted online at URL: www.nsti.org. We look forward to seeing you again next year, and thank you for your continuing support and participation.

Technical Program Co-chairs, NSTI Nanotech 2005
Matthew Laudon, Nano Science and Technology Institute, USA
Bart Romanowicz, Nano Science and Technology Institute, USA

Structure Determination of Small Metal Clusters by Density-Functional Theory and Comparison with Experimental Far-Infrared Spectra

C. Ratsch*, A. Fielicke**, J. Behler**, M. Scheffler**, G. von Helden**, and G. Meijer**

* UCLA, Department of Mathematics, Los Angeles, CA 90095
** Fritz-Haber-Institut der Max-Planck Gesellschaft, Faradayweg 4-6, D-14195 Berlin, Germany

ABSTRACT

The far-infrared vibrational spectra for charged vanadium clusters as well as charged and neutral niobium clusters have been measured size specifically using far-infrared multiple photon dissociation. The ground state energy and vibrational spectra of a large number of stable and metastable structures for each of these sizes and systems have also been calculated using density-functional theory (DFT). A comparison of the calculated vibrational spectra with those obtained in the experiment allows us to deduce the cluster size specific atomic structures. Our results suggest that sometimes there is one unique atomic structure, while in other cases the experiment might observe several isomers. A comparison of the results for cationic vanadium and cationic niobium and the results for neutral niobium allows us to explore the differences between the different systems, and clusters with different charges.

Keywords: metal clusters, vibrational spectra, atomic structure, density-functional theory

1 INTRODUCTION

Small clusters or nano-particles exhibit properties that are often quite different from those in the bulk phase. For example, small metal clusters have been shown to exhibit unusual magnetic properties [1]. In particular vanadium clusters have been predicted to have large magnetic moments [2]–[6]. Small nanoparticles also play an increasingly important role in catalysis [7], [8]. With new approaches in synthesis, it may become feasible to control the size and possibly also the structure of the nanoparticles, and thus to control their properties. Therefore, it is paramount to gain a better understanding of the atomic structure and properties of small metal clusters.

In this paper we present the details of a combined theoretical and experimental study on the structure determination of metal clusters. We have recently shown that multiple photon dissociation spectroscopy on metal cluster rare gas complexes allows for the determination of cluster size specific far-infrared spectra, and that a comparison with vibrational spectra calculated by DFT can be used to determine the atomic structure [9], [10], because these spectra are typically rather different for different atomic configurations. Moreover, we have shown that a detailed analysis and comparison does even allow for the determination of the atomic structure for different isomers for a certain size [11].

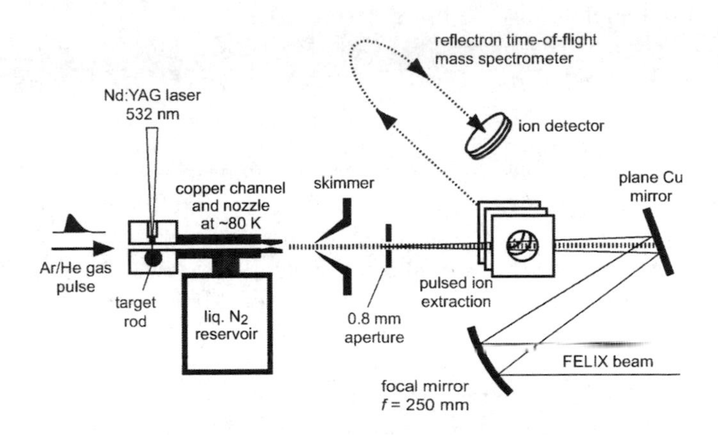

Figure 1: Scheme of the experimental set-up for the IR multiple photon dissociation spectroscopy of the argon complexes of metal cluster cations.

2 THE EXPERIMENT

The experiments are carried out in a molecular beam setup that is coupled to the beam-line of the Free Electron Laser for Infrared eXperiments (FELIX). This laser can produce intense, several μs long pulses of tunable IR radiation in the range of ~40–2000 cm^{-1}. Each pulse consist of a train of ~0.3–3 ps long micropulses of typically ~10 μJ, spaced by 1 ns. The time and intensity profile of the radiation makes FELIX a suitable tool for studying multiple photon excitation processes in molecules or clusters [12]. Some details of the experimental set-up have already been given in Refs. [9], [10]. In short, cationic vanadium clusters V_n^+ are formed in a laser vaporization cluster source (see Fig. 1). Complexes with Ar atoms (V_n^+Ar) are formed after passing through a copper channel that is cooled to about 80 K. The molecular beam containing these complexes is overlapped with the far-IR output of FELIX. Resonant absorption of one or multiple IR photons by the complexes can be followed by evaporation of single or more

Ar atoms from the complex leading to decreases in their abundances.

After normalization for laser power variations and intensity fluctuations stemming from the cluster source, one obtains the absorption spectra of the corresponding $V_n^+Ar_m$ cluster complexes. Although the thus obtained spectra correspond to the spectra of the Ar complexes, the vanadium clusters are the active chromophore and the influence of the Ar atoms is assumed and expected to be negligible. The spectra for cationic and neutral niobium clusters were obtained similarly.

3 DENSITY-FUNCTIONAL THEORY CALCULATIONS

The calculations presented below were obtained within the framework of density-functional theory (DFT). We used the $DMol^3$ code [13], which is an all-electron code that uses numerical atomic orbitals as a basis set. All results presented here were obtained using the generalized gradient approximation (GGA) in the parameterization of Perdew, Burke, and Ernzerhof (PBE) [14] for the exchange-correlation (XC) functional.

We employed the following computational procedure: First, we calculated the ground state energy for a large number of geometries for the clusters of each size. The atomic positions of the structures were always relaxed, and no symmetry constraints are applied. As initial geometries we started with all the structures that have been discussed previously in the literature [15]–[20]. In addition, we tried a large number of different geometries that were derived from previously calculated structures. For example, for many clusters one can describe the geometry as a superposition of smaller building blocks that typically are trigonal, tetragonal, pentagonal, or hexagonal pyramids. Thus, one can often simply add or subtract an atom from a cluster of size n to obtain a start geometry for a cluster of size $n+1$ or $n-1$. In this manner, we typically generated at least 10 or more different geometries for each size. It is important to emphasize that finding the correct initial geometry is a major challenge. As the cluster size increases, the number of possible atomic configurations increases dramatically. Therefore, it is possible that for certain cluster sizes (in particular the larger ones) we did not find the correct atomic structure.

We also tested the different possible spin states for all clusters. Once the structures and spin states that are energetically most favorable were detected, we calculated the vibrational spectrum of these systems. This was done by displacing each atom in each direction, in order to evaluate the $3n$ dimensional force-constant matrix. Then we diagonalized the resulting dynamical matrix. The IR intensities were obtained from the derivative of the dipole moment. For the purpose of an easy comparison to the experimental data, we folded for all results shown the calculated spectra with a Gaussian lineshape function of half width of 2 cm^{-1}. We apply a common scaling factor of 0.87 to all frequencies shown, which was obtained by comparison of the calculated vibrational frequency for V_2 with the experimental value.

All the calculations presented below are for V_n^+ clusters, while the experimental spectra are obtained from the dissociation of $V_nAr^+ \longrightarrow V_n^+ + Ar$. Tests revealed that this is justified, because the rare gas atoms bind only very weakly to the V_n^+ cluster.

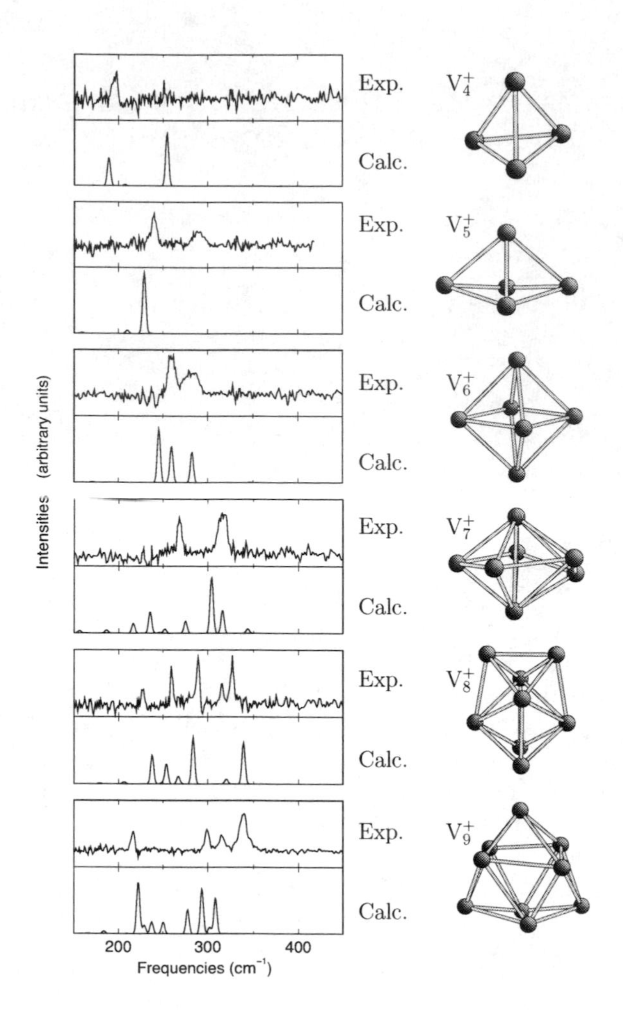

Figure 2: Comparison of the experimentally obtained spectra (Exp.) and calculated spectra (Calc.) for the six clusters V_4^+ to V_9^+. The corresponding atomic structures are shown to the right of each pair of experimental and calculated spectra.

4 RESULTS FOR CATIONIC VANADIUM CLUSTERS

The main results of our calculations and comparisons to the experimental spectra are given in Fig. 2. In this figure we present the experimentally measured spectra for V_4^+ to V_9^+ that were obtained as described above. Underneath the experimental spectra we show calculated spectra for these cluster sizes and on the right side the corresponding structures are shown. Based on similarities between experimental and calculated vibrational spectra and energetical arguments, these structures are suggested to be the structures of the clusters seen in the experiment. More details and further results for V_{10}^+ to V_{15}^+ are given in [10]. For almost all cluster sizes discussed here the lowest energy isomer leads also to the best agreement between calculated and measured spectra. The only exception is V_9^+; here we find two isomers that can be considered (within the accuracy of the calculations) as energetically degenerate, however, best agreement in the spectra is obtained for the isomer that is slightly higher in energy. Most structures are calculated to have the electronic state with the lowest possible multiplicity as the energetically lowest one, e.g. $S = 0$ for odd n and $S = 1/2$ for even n. For V_5^+ and V_9^+ the structures with $S = 0$ and $S = 1$ are close in energy, and the shown spectrum corresponds to the clusters with $S = 1$.

We will now briefly discuss the results presented here and compare them to previous work. For more details, we refer the reader to [10]. We find that a trigonal pyramid is the energetically preferred structure for V_4^+. This is in agreement with the calculations of Wu *et al.* [16]. We note that a planar structure (that has been predicted for neutral clusters) is rather unfavorable. We find that for V_5^+ a (slightly buckled) tetragonal pyramid is the energetically preferred structure, which is in agreement with previous results [16], [20]. We note that a buckled tetragonal pyramid is topologically similar to a distorted trigonal bi-pyramid, which has also been suggested, and that in fact the spectra of these 2 are very similar. For V_5^+ we find that the suggested structure has a spin of $S = 1$, but a spin state that is not the lowest possible spin state has also been reported in optical absorption spectroscopy experiments [21] for V_5^+ The most stable structure for V_6^+ is a tetragonal bi-pyramid. This is in agreement with previous results [15], [16], [18], [20] that all predict this structure. Our calculations for V_7^+ indicate that here a strongly distorted pentagonal bi-pyramid is the energetically most stable structure. The structure appears to be similar to the second lowest structure in Refs. [16], [17]. This result seems to be in contrast to previous DFT calculations [16] that suggested an almost symmetrical pentagonal bi-pyramid. The most stable structure for V_8^+ is the same that has been predicted previously for cationic vanadium [16], [17]. We have discussed the details for this isomer in Ref. [9]. A tricapped trigonal prism with approximate 3-fold symmetry is suggested to be the structure that is present in the experiment for V_9^+. We base this interpretation on the presence of 3 peaks in the frequency range 270 cm^{-1} to 340 cm^{-1}. An alternative structure that in fact might be even slightly lower in energy has only 2 peaks in this region (cf. Ref. [10]).

As mentioned above, we have performed similar studies to determine the structure of larger clusters, e.g. for V_{10}^+ to V_{15}^+. However, for these slightly larger clusters an unambiguous structure determination is not always possible. There are several reasons for this: First, it is possible (and likely) that more than one isomer is present, and that in fact the experimentally observed spectra are a combination of several isomers. But due to space constraints we will not discuss multiple isomers for one size here, and refer the reader to Ref. [10]. An additional complication that arises when larger clusters are considered is the following: The number of possible atomic configuration increases dramatically, and it is very difficult to scan the entire parameter space for larger clusters. It is therefore possible that in particular for the larger clusters we have simply not considered the structure that is actually observed in the experiment. We believe that the development of efficient schemes to test a large number of atomic structure is a major challenge for this type of calculation, and we plan to address this issue in our future work.

5 COMPARISON WITH CATIONIC AND NEUTRAL NIOBIUM CLUSTERS

Following our initial studies on the vanadium cluster cations we measured recently the spectra of cationic as well as neutral niobium clusters using the same experimental technique [11], [22]. A comparison of both metals is interesting, since vanadium and niobium are often similar in their chemical and physical properties. However, the questions that may arise are the following: What are the differences and similarities between cationic vanadium and cationic niobium clusters? What are the differences and similarities between cationic and neutral niobium clusters? Experimentally it has been found that for certain sizes, the spectra for cationic vanadium and niobium clusters are very similar, while for other they are very different. This suggest that the atomic structure of vanadium and niobium clusters are sometimes similar, and very different in other cases. Similarly, for certain sizes the atomic structure for charged and neutral niobium clusters might be very similar, while they might differ more for other sizes. Our initial DFT results indicate that indeed a detailed comparison between calculated and experimentally measured spectra sheds new light on these questions. A detailed description of these results is beyond the scope of this article, and will be reported elsewhere [22].

6 CONCLUSIONS

We have shown that the atomic structure of small metal clusters can be identified by a comparison of experimental far-infrared spectra with vibrational spectra calculated with DFT. Results for V_4^+ to V_9^+ are presented here. Larger clusters are discussed in more details elsewhere [10]. For some cases, a unique atomic structure can be identified. However, for other cluster sizes an unambiguous identification of the atomic structure was not possible. The reason is that several isomers might be present, or that we simply have not found the correct atomic structure. We propose that more efficient computational schemes are required to scan the parameter space of atomic structures. Nevertheless, we believe that the methodology outlined here is useful to identify the atomic structure of small metal clusters, and in fact small nanoparticles in general.

REFERENCES

[1] I.M.L. Billas, J.A. Becker, A. Chatelain, and W.A. de Heer, Phys. Rev. Lett. **71**, 4067 (1993).

[2] F. Liu, S.N. Khanna, and P. Jena, Phys. Rev. B **43**, 8179 (1991).

[3] K. Lee and J. Callaway, Phys. Rev. B **48**, 15358 (1993).

[4] P. Alvarado, J. Dorantes-Dávila, amd H. Dreyssé. Phys. Rev. B **50**, 1039 (1994).

[5] H. Wu, S.R. Desai, and L.-S. Wang, Phys. Rev. Lett. **77**, 2436 (1996).

[6] M. Iseda, T. Nishio, S.Y. Han, H. Yoshida, A. Terasaki, and T. Kondow, J. Chem. Phys. **106**, 2182 (1997).

[7] M. Valden, X. Lai, and D.W. Goodman, Science **281**, 1647 (1998).

[8] A.T. Bell, Science **299**, 1688 (2003).

[9] A. Fielicke, A. Kirilyuk, C. Ratsch, J. Behler, M. Scheffler, G. von Helden, and G. Meijer, Phys. Rev. Lett. **93**, 023401 (2004).

[10] C. Ratsch, A. Fielicke, A. Kirilyuk, J. Behler, M. Scheffler, G. von Helden, and G. Meijer, J. Chem. Phys., in press (2005).

[11] A. Fielicke, C. Ratsch, G. von Helden, and G. Meijer, J. Chem. Phys., in press (2005).

[12] G. von Helden, D. van Heijnsbergen, and G. Meijer, J. Chem. Phys. **107**, 1671 (2003).

[13] B. Delley, J. Chem. Phys. **92**, 508 (1990).

[14] J.P. Perdew, K. Burke, and M. Ernzerhof, Phys. Rev. Lett. **77**, 3865 (1996).

[15] H. Grönbeck and A. Rosén, J. Chem. Phys. **107**, 10620 (1997).

[16] X. Wu and A.K. Ray, J. Chem. Phys. **110**, 2437 (1998).

[17] S. Li, M.M.G. Alemany, and J.R. Chelikowsky, J. Chem. Phys. **121**, 5895 (2004).

[18] H. Sun, Y.-H. Luo, J. Zhao, and G. Wang, Phys. Stat. Sol. (b) **215**, 1127 (1999).

[19] J. Zhao, X. Chen, Q. Sun, F. Liu, G. Wang, and K.D. Lain, Physica B **215**, 377 (1995).

[20] A. Tanedo, T. Shimizu, and Y. Kawazoe, J. Phys. Cond. Matt. **13**, L305 (2001).

[21] S. Minemoto, A. Terasaki, and T. Kondow, J. Electron Spectroscopy **106**, 171 (2000).

[22] A. Fielicke, C. Ratsch, G. von Helden, and G. Meijer, in preparation.

Exploring the Structure-Function Relationship in Multifunctional Nanoparticles

P.S. Casey*, C.J. Rossouw, S. Boskovic, K.A.Lawrence and T.W. Turney

Commonwealth Scientific and Industrial Research Organisation
Division of Manufacturing and Infrastructure Technology
Locked Bag 33, Clayton South, Victoria 3169, Australia
* phil.casey@csiro.au

ABSTRACT

Nanoparticles of ZnO are increasingly being used as a pigment and UV absorber in personal care products (e.g. sunscreens), coatings and paints, predominantly because their absorbance efficiency increases with decreasing particle size. However, whilst protection against UV may be maintained at smaller particle sizes, the rate at which hydroxyl radicals are generated increases, due to the inherent photo-activity of these materials. This paper shows that photoactivity can be reduced (by over an order of magnitude) by doping the material with small amounts metallic cations (<1 atm%). To explore whether or not the dopants have been successfully incorporated within the lattice a new variation of atom location by channelling-enhanced microanalysis (ALCHEMI) has been used. It was found that some dopants were exclusively accommodated on the cation sublattice site, whereas others existed as a foreign phase or decorated the surface. The difference in dopant location may influence consequent photoactivity and other nanoparticle behaviour

KEYWORDS: doped nanoparticle, photoactivity, zinc oxide, ALCHEMI

INTRODUCTION

Nano-particles of ZnO and TiO_2 are increasingly being used as pigments and UV absorbers in personal care products (e.g. sunscreens), coatings and paints, predominantly because their absorbance efficiency increases with decreasing particle size. However, whilst protection against UV may be maintained at smaller particle sizes, the rate at which hydroxyl radicals are generated increases, due to the inherent photoactivity of these materials. Consequently, in the case of personal care products, there may be a negative effect on skin cells due to this increased photocatalytic activity. There is little published data that correlates photoactivity of nanoparticles with biological effects. However, *in vitro* work has shown that supercoiled DNA is indeed damaged in the presence of nano-particulate metal oxides and the rate of unwinding itself can be used as a measure of the photoactivity of the metal oxide [1].

Alteration of the band gap is of increasing interest and may be achieved by lowering the energy gap, relative to the original system, by the introduction of dopant atoms. The majority of published data on the use of dopants to control photoactivity has been with titania [2,3].

In this paper, observations are reported on a series of undoped and doped nanosized ZnO materials designed to quench photoactivity. These were tested for photoactivity, using chemical methods, when exposed to UV. Parameters such as crystallite size, dopant type and level have been considered. Nanoparticle structure, including the location of the dopant, using a new variant of a method related to atom location by channelling-enhanced microanalysis (ALCHEMI) was used [4] and this location explored in relation to particle properties.

EXPERIMENTAL

Characterisation of Materials

A series of undoped and doped zinc oxide was prepared by proprietary methods and characterised chemically and physically. Chemical analysis was performed using Inductively Coupled Plasma Atomic Emission Spectroscopy (ICP-AES) methods. Crystallite phase was determined by using a Bruker ASX-D8 X-Ray Diffractometer using Cu Kα radiation over a 2θ range of 5° to 85° with a step size of 0.02° . Crystallite size was determined by performing a

Rietveld refinement of the diffraction data using Siroquant™ Version 2.5 software.

The method established and implemented for determining the photoactivity of metal oxide particles such as ZnO and TiO_2 was that initially proposed by Dransfield et al. [5]. It is a colorimetric test that follows the photobleaching of the stable radical 1,1-diphenyl-2-picrylhydrazyl (DPPH). Photoactivity is determined by the time it takes the DPPH radical, which is initially purple in colour, to covert to its reduced form, which is yellow. The reciprocal of the time taken for radical decomposition is defined as the Photoactivity Index (PI), with units of min^{-1}

RESULTS

Results of chemical analyses, crystallite size and chemical photoactivity rating for the series of oxides examined are presented in Table 1.

Compound	Crystallite Size (nm)	Time taken for decay of DPPH radical (min)	P I (min^{-1})
ZnO	24	12	0.083
Dopant			
Fe (0.44wt%)	14	30⁺	<0.033
Ni (0.64wt%)	20	100	0.01
Co (0.7 wt%)	16	150⁺	<<.0066
Mn (1.1wt%)	21	150	0.0066
Dopant Level (wt%Mn)			
0.2		30	0.0333
0.3	23.3	38	0.0263
0.6	23	95	0.0105
0.8	18.3	220	0.0045

Table 1. Summary of Chemical analyses, Crystallite Size and Photoactvity Index (PI)

DISCUSSION

Photoactivity is expected to vary in direct proportion with surface area. A decrease in particle size from 80 nm to 40 nm may represent a two- to three-fold increase in surface area per unit weight. A consequent increase in photoactivity of the same order would be expected

It has been found that for undoped zinc oxide, there appear to be two distinct regions of photoactivity behaviour. For crystal sizes of less than 80 nm, photoactivity is linearly dependant on size. However, for larger crystals, photoactivity is independent of size. The increase in photoactivity as particles reduce in size from 80 nm to 40 nm is about 30 %, compared with 2-3 fold increase expected from arguments based on relative surface area.[6].

Whilst experiment shows a monotonic increase in photoactvity with surface area for particles less than 100 nm in size, the increase in photoactivity is less than expected were surface area the dominant controlling factor. Possible reasons for this are (i) other intrinsic or extrinsic material factors, either physical and/or chemical, besides surface area diminish or influence photoactivity (ii) the conditions under which tests were performed affect either the photoactivity or the measurement of photoactivity.

Characterisation suggests that it is unlikely that intrinsic material factors vary significantly between samples of undoped ZnO, and this may be discounted as a significant influence on photoactivity. Extrinsic factors may thus provide a strong influence on photoactivity of which efficient dispersion is likely to be the most significant.

Figure 1 shows that photoactivity of ZnO may be quenched by the incorporation of a dopant material. All dopants used in this study are effective in reducing photoactivity by a significant amount, as shown in Figure 1. The most effective dopants are Mn, Co and Ni, with Fe being less effective. Depending on dopant concentration, photoactivity could be reduced by a factor of 5 or 10 compared with that exhibited by ZnO at a similar crystallite size. Other dopants (not reported here) were tested and displayed less significant changes in photoactivity than Mn, Ni, Co and Fe.

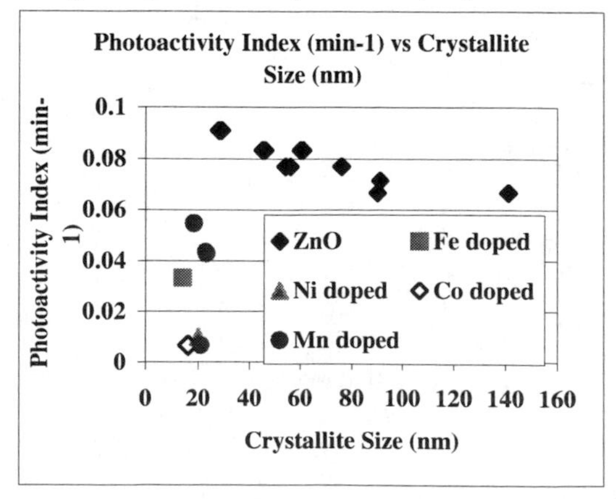

Figure 1 Summary of Photoactivity Index versus Crystallite Size

ALCHEMI of nanoparticles

The question arises as to whether the dopant is in the lattice or merely decorating it and whether differences in photoactivity are correlated with dopant location. In order to ascertain whether or not the dopants are incorporated within the lattice of the nanoparticles, a method related to atom location by channelling-enhanced microanalysis (ALCHEMI) was used [4]. Here a relatively broad but collimated 200 keV electron beam is focused onto a small particle. This particle is then tilted into a low index crystallographic zone axis orientation, and minor adjustments are made to mechanical or electronic tilt such that the beam propagates precisely down the zone axis.

Whilst it should be, in principle, quite straightforward to determine by ALCHEMI whether or not dopant atoms are within the lattice, the analysis of nanoscale particles provides a challenge. The ability to use automated beam rocking is excluded by the necessity to use a small beam in nanoprobe mode. In any event, low count rates and lateral shift incurred whilst electronically rocking the beam excludes, at this stage, automated acquisition of X-ray channelling patterns from nanoscale crystals for analysis [7].

When the nanoparticle is tilted well away from the zone axis orientation, the periodic squeezing of the wavefront onto or between atomic columns within the unit cell is relaxed, and quasi-kinematic diffraction conditions approximate the propagation of the electron wavefront to that of a plane wave. Analogous to Rutherford backscattering (RBS) analysis of charged particles on crystal surfaces, the zone axis orientation is regarded as a 'channelling' orientation, whilst the tilted orientation is regarded as the 'random' orientation.

Thus, in order to reduce the ALCHEMI experiment to a simple determination, we assume that, if incorporated in the lattice, dopant atoms randomly occupy the Zn cation sublattice site. Then, by acquisition of two EDX spectra, one at a 'channelling' orientation and the other at a 'random' orientation, a clear distinction is possible, analogous to RBS analysis, as to whether the introduced cations are incorporated on the Zn sublattice site or not. This provides a useful tool for nanoparticle analysis and crystallographic information not readily accessible by X-ray Rietveld refinement, since the response from each atomic species is individually isolated.

The situation is illustrated schematically in Figure 2. If, on the one hand, Co is incorporated on Zn sublattice sites, the rate of X-ray emission from both Zn and Co will increase in a similar way as the nanoparticle is brought to the zone axis orientation. On the other hand, if Co dopants are in a different phase or form part of a coating on the particle, the emission rate from Co will be invariant with crystal orientation.

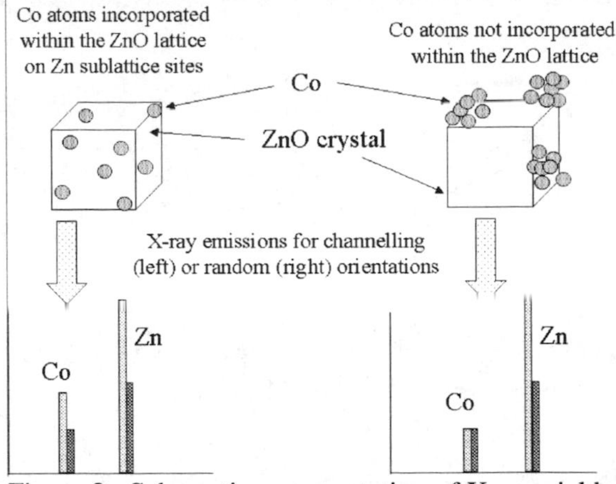

Figure 2 Schematic representation of X-ray yields for Zn and Co with Co in the lattice (left) and Co as a foreign phase (right). The light bar represents a channelling orientation and the dark bar a random orientation.

A comparison of the two EDX spectra is shown in Fig. 3(a) on a log-lin scale. Low energy emissions are from oxygen and Zn L-shell excitations, and a Cu peak is produced by scattered electrons which impinge on the copper grid bars. Fig. 3(b) shows a portion of the two EDX spectra on a linear scale, where the dynamical increase in the Co emissions is more evident.

Concentrations of Co and Mn derived from analyses of different particles under channelling and random diffraction conditions, showed that the apparent Co concentration remains steady within standard deviation for all analyses, indicating it is accommodated on the Zn sublattice site. However results from Mn showed significant variations for channelling and random orientations, as well as differences between different particles. This indicates that a significant portion of Mn atoms exist as a foreign phase or surface coating on the ZnO particles, and that it is not incorporated on the Zn sublattice site.

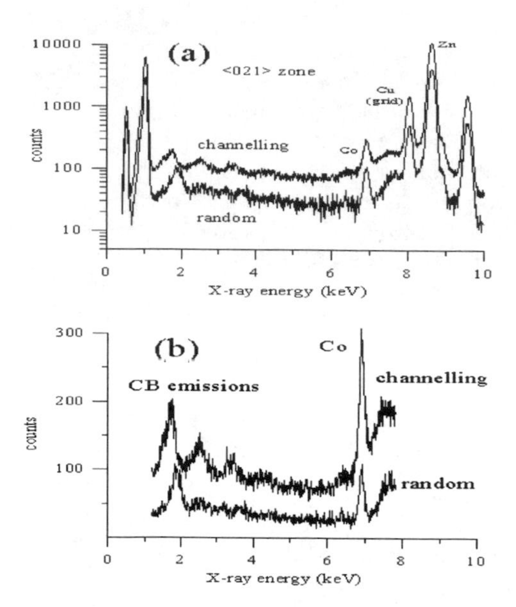

Figure 3 Lin-log plot of full EDX spectra (a), and a linear plot of CB radiation and Co peaks (b).

This data shows reasonable agreement between experiment and theory. Thus this variant form of ALCHEMI, reducing the analysis to two EDX spectra, one acquired under a zone axis channelling configuration and the other under a random orientation, demonstrates that Co is accommodated on Zn sublattice sites, whilst Mn is excluded from the lattice and exists as a foreign phase. This provides a useful tool for nanoparticle analysis and crystallographic information not readily accessible by X-ray Rietveld refinement.

In photoactive nanoparticles, material factors (besides surface area, crystallite size dispersion efficiency etc) will influence the relationship between structure and function. The consequent effect of adding a dopant will be determined by its location (in or on the lattice) as this will affect the mechanism by which photoactivity is quenched. If the dopant is in the lattice the width of the band gap will be determined intrinsically. On the other hand if the dopant decorates the lattice the width of the band gap will not be altered but will effect the fate of radicals generated. The relative contribution that either location makes is unknown for these systems. There is some indication from the preliminary data presented here that the dopant located within the lattice is more effective than that decorating the lattice. For example, the Co dopant appears to be more effective at reducing photoactivity than does the Mn, Ni or Fe and that to achieve a similar level of photoactivity using dopants that do not appear to be in the lattice, increased coverage seems to be required.

CONCLUSION

In general, the photoactivity of nanophase zinc oxide has been found to increase with decreasing particle size but may be controlled by the incorporation of selected dopants. The location of these dopants has been determined by the technique of electron channelling and gives some insight into the method by which the photoactivity of the nanoparticle is controlled together with how preparation conditions may affect the design of additional functionality into the material of interest.

The combination of empirical testing with structural analysis provides a powerful combination to allow the design of nanoparticles with enhanced and/or additional functionality.

REFERENCES

1. Takao Ashikaga, Masayoshi Wada, Hiroshi Kobayashi et al. Effect of photocatalytic activity of TiO_2 on plasmid DNA. Mutation Research 466 (2000) p1-7

2. Sivalingam, G; Nagaveni, K; Hegde, MS, et al. Photocatalytic degradation of various dyes by combustion synthesized nano anataseTiO2. Appl Catal B-Environ, 45 (1): 23-38 Sep 25 2003

3. Hirano, M; Nakahara, C; Ota, K, et al. Photoactivity and phase stability of ZrO2-doped anatase-type TiO2 directly formed as nanometer-sized particles by hydrothermal conditions. J Solid State Chem, 170 (1): 39-47 Jan 2,2003

4. Jones, I.P. Determining the locations of chemical species in ordered compounds: ALCHEMI, Advances in imaging and electron physics 125 (2002) 63-117.

5 Dransfield, G.; Guest, P.J.; Lyth, P.L.; McGarvey, D.J.; Truscott. T.G. Journal of Photochemistry and Photobiology B: Biology 2000, 59, 147.

6. Controlling the Photoactivity of Nanoparticles", Technical Proceedings, Volume 3, Chapter 8, p370-374, NSTI Nanotech 04, Boston, March 2004.

7. Rossouw C.J., Forwood, C.T., Gibson, M.A., and Miller, P.R. Statistical ALCHEMI: General formulation and method with application to TiAl ternary alloys. Phil. Mag. A 74 (1996) 57-76.

Synthesis of Y_2O_3:Eu Phosphor Nanoparticles by Flame Spray Pyrolysis

Xiao Qin*, Yiguang Ju*, Stefan Bernhard** and Nan Yao***

*Dept. of Mech. & Aerospace Engineering, Princeton University, Princeton, NJ 08544
**Dept. of Chemistry, Princeton University, Princeton, NJ 08544
***Princeton Materials Institute, Princeton University, Princeton, NJ 08544

ABSTRACT

Nanoscale europium doped yttrium oxide (Y_2O_3:Eu) particles synthesized by flame spray pyrolysis method and the effect of precursor solution on particle size, morphology and photoluminescence intensity was studied. The structure and morphology of the as-prepared particles were examined using scanning electron microscopy (SEM). The photoluminescence (PL) intensity was measured by a spectrophotometer. Compared with previous report, SEM shows fine and smooth structure for particles prepared from ethanol alcohol as precursor solution. The particle size can be controlled by varying precursor concentration, flame temperature and residence time. Upon excitation with 355 nm UV light, the particles show red emission and the PL intensity of particles using ethanol is 30% stronger than those using water as precursor solution. Annealing particles at 1200°C can convert the monoclinic phase into cubic phase. The concentration quenching limit of the prepared particles was 18% mol Eu.

Keywords: nanoparticles, europium, yttrium oxide, flame spray pyrolysis, photoluminescence, quenching.

1. INTRODUCTION

Nanoparticles have become a research focus in terms of both their fundamental and technological importance, especially in the case of luminescent materials because of a quantum confinement effect which leads to novel optoelectronic properties. It was found that the emission lifetime, luminescence quantum efficiency, and concentration quenching depend strongly on the size in the nanometer range [1-4]. Nanoparticles have great potential for use in applications in electronic, chemical and mechanical industries, as well as technologies which include superconductors, catalyst, drug carriers, sensors, materials, pigment, structural materials, and so on [5].

Europium doped yttrium oxide (Y_2O_3:Eu) is a typical red phosphor that is widely used in optical displays and lighting applications. The better quality of a high resolution display requires smaller size phosphors and nanosized Y_2O_3:Eu phosphor has significant promise in displays due to the quantum efficiency increase of doped nanocrystals [6,7]. Various methods, such as solid state reactions, sol-gel techniques [8], hydroxide precipitation [9], hydrothermal synthesis [10], laser-heated evaporation [11], spray pyrolysis [12], and combustion synthesis [13] were used to prepare nanostructured Y_2O_3:Eu phosphors.

Flame spray pyrolysis (FSP), also called liquid flame spray (LFS), is a promising particle synthesis method because it can employ a wide range of precursors for synthesis of a broad spectrum of functional nanoparticles [14-16]. Using FSP method, Kang et al. [17] prepared nonagglomerated Y_2O_3:Eu phosphor particles of size on the order of 1 μm and had spherical and dense morphology. Their as-prepared particles had a monoclinic phase with small impurities of the cubic phase. Tanner and Wong [18] synthesized Y_2O_3:Eu nanoparticles using preformed sol, conventional spray pyrolysis and flame spray pyrolysis methods and compared the luminescence properties of the powders prepared by these three methods. Chang et al. [19] fabricated cubic nanocrystalline Y_2O_3:Eu phosphors using FSP method without any post-heat treatments.

In flame spray pyrolysis, the precursor composition is a key parameter to control the particle properties. The precursor releases from the droplet and its evaporation, decomposition, and gas phase reaction plays an important role in the formation of the final product. To the best of our knowledge, there's no report on the effect of precursor composition on the synthesis of Y_2O_3:Eu phosphorous nanoparticles in flame spray pyrolysis. Our objective of the present work is to synthesize Y_2O_3:Eu nanoparticles by using FSP method and study the effect of precursor composition on the product properties.

2. EXPERIMENTAL METHOD

Figure 1 shows the schematic of the flame spray pyrolysis system. The system consisted of a spray generator, a coflow burner, a quartz reactor, particle collection filters and a vacuum pump. An ultrasonic spray generator operating at 1.7 MHz was used to generate fine spray droplets which were then carried into the flame by nitrogen gas though a 5.3 mm central pipe. The flame nozzle consisted of three concentric pipes. A Methane and oxygen nonpremixed flame was used and an air coflow was also introduced into the reactor. By varying the flow rate of fuel, oxidant and coflow air, the flame temperature and particle residence time can be controlled. The temperature measurements along the centerline employed uncoated 100 μm diameter R-type wire thermocouples with a junction

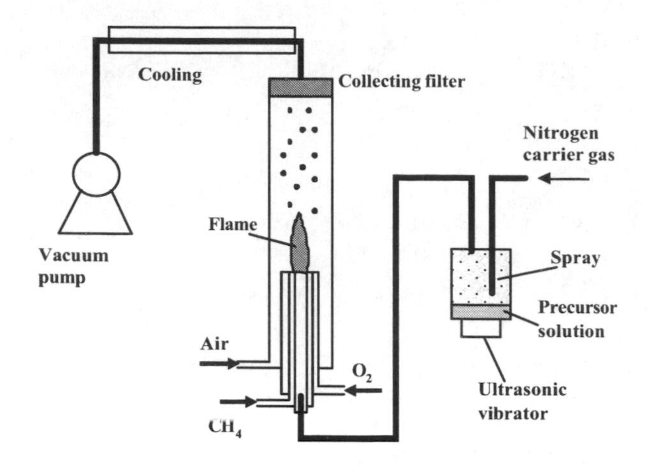

Figure 1: Schematic of experimental setup.

bead diameter of 350±30 μm and were corrected for radiation heat losses.

To study the effect of precursor solution on particle properties, ethanol and water were selected as solvent. The starting precursor solution was prepared by dissolving yttrium and europium nitrate, i.e. $Y(NO_3)_3 \cdot 6H_2O$ and $Eu(NO_3)_3 \cdot 6H_2O$, in pure water or ethanol alcohol. The overall concentration was varied from 0.1 to 0.001 M and the doping concentration of europium varied from 3 to 21 mol% with respect to yttrium. The particles, which were collected using a micron glassfiber filter (Whatmann GF/F) located 30 cm above the flame. The structure and morphology of the as-prepared particles were examined using a Philips XL30 FEG-SEM scanning electron microscope. The photoluminescence intensity of the particles excited by UV light is measured by a spectrophotometer (Fluorolog, Jobin Yvon Group) with a xenon lamp of 150 W.

3. RESULTS AND DISCUSSION

Figure 2 shows SEM photographs of as-prepared Y_2O_3:Eu particles by flame spray pyrolysis using distilled water (Fig. 2a and 2b) and ethanol (Figs. 2c and 2d) as precursor solvent at different overall concentrations. The europium doping concentration was 6 mol% with respect to yttrium for all cases. The effect of precursor solution on particle morphology and size can be clearly seen from these images. The particles made from aqueous solution (c.f. Fig. 2a and 2b) have a fuzzy structure on the surface and broader size distribution. On the other hand, the particles using ethanol as precursor solution (Fig. 2c and 2d) exhibit smoother surface structure and improved homogeneity in distribution. Regardless of the overall concentration and precursor solution type, the particles are generally non-aggregated and have a spherical morphology. Table 1 lists the average particle size, geometric standard deviation calculated from the SEM images at different precursor concentration with a fixed doped Eu concentration of 6 mol% with respect to yttrium. The average size has been determined by measuring the

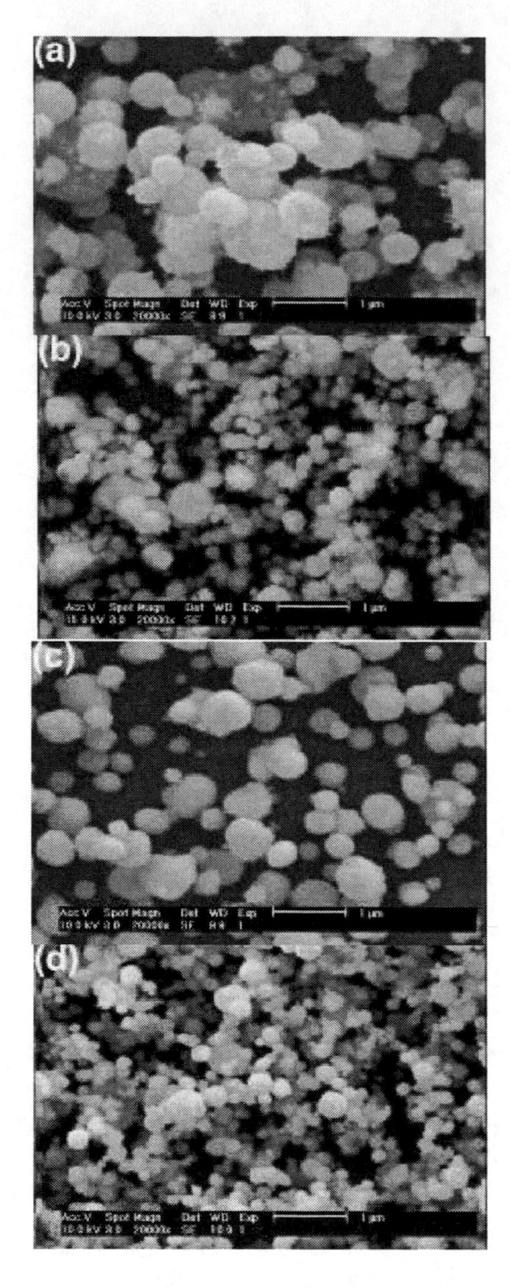

Figure 2: SEM photographs of Y_2O_3:Eu particles prepared from (a) distilled water solution, 0.1 M; (b) distilled water solution, 0.01 M; (c) ethanol solution, 0.1 M; (d) ethanol solution, 0.01 M.

diameters of 500 particles from SEM images. It is seen that particles prepared from ethanol solution has smaller average diameter than those from water solution of the same concentration. The average diameter of the particles varied from 114 nm to 412 nm when the ethanol precursor solution concentration increased from 0.001 M to 0.1 M (c.f. cases 4, 6 and 8) which shows that the particle size can be controlled by changing the overall concentration.

The differences in particle morphology and size distribution when using ethanol and water as precursor solvent arise from the different physical properties between ethanol and water. Ethanol has a lower boiling point and

Table 1: The particle average diameter, geometric standard deviation of the as-prepared particles and flame temperature

case	Precursor concentration (M)	Precursor type	Particle average diameter (nm)	Geometric standard deviation	Flame temperature* (K)
1(Fig.2a)	0.1	Water	535	1.20	1721
2	0.1	Water	498	1.20	2030
3(Fig.2b)	0.01	Water	192	1.31	1720
4(Fig.2c)	0.1	Ethanol	412	1.14	2030
5	0.01	Ethanol	185	1.07	1800
6(Fig.2d)	0.01	Ethanol	198	1.10	2040
7	0.01	Ethanol	214	1.09	2260
8	0.001	Ethanol	114	1.07	2050

*At centerline location of 20 cm above the burner exit.

enthalpy of evaporation (78 °C and 838 kJ/kg) than water (100 °C and 2258 kJ/kg). And more importantly, ethanol is a fuel that directly reacts and release heat in the flame instead of take away heat from the flame when using water. To investigate the effect of precursor solution on flames, the temperature profiles along the centerline for the flames corresponding to Figs. 2a and 2c were measured by thermocouples and shown in Fig. 3. In the measurements, methane, oxygen, nitrogen and coflow air flow rates were kept constant for the two cases. It should be mentioned that near the core of the methane-oxygen flame (<10 cm) the flame temperature is so high that the thermocouple immediately breaks; therefore, only data above 10 cm were taken. Along the centerline, the temperature of the flame using ethanol as precursor solution is consistently 200 K higher than the flame using water. It is known that in flame spray pyrolysis higher flame temperature increases particle sintering and agglomerating, which is not the case in our observation in Fig. 2. This suggests that not only flame temperature but other parameters, such as evaporation, gas phase reaction, nucleation, etc. may also affect particle size and morphology. The faster evaporation of ethanol droplets when passing the preheating zone of the flame produces smaller droplets that determine the final particle size. In the aqueous solution, the presence of water droplets or vapor in the reaction zone results in larger particles. Limaye and Helble [20] observed similar effect of precursor and solvent on the morphology of zirconia nanoparticles produced by combustion aerosol synthesis.

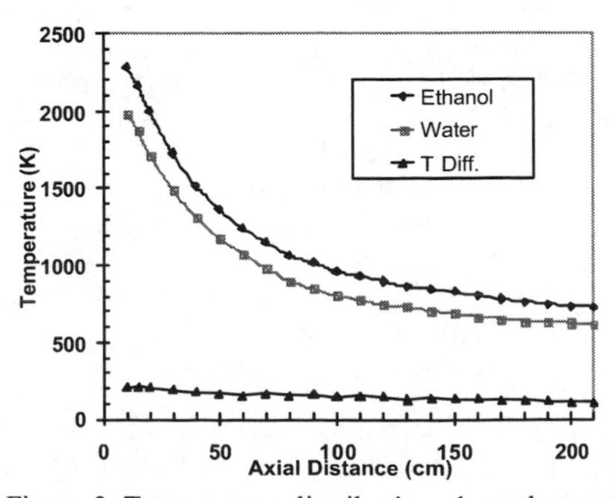

Figure 3: Temperature distribution along the centerline of the flames using different precursor solutions.

By only increasing the methane flow rate of the flame in Fig. 2a, the same temperature distribution as that in Fig. 2b was achieved. The mean particle size decreased to 498 nm from 535 nm (cases 1 and 2 in Table 1), but still larger than that from ethanol flame (412 nm). The increase of flame temperature helped to evaporate the precursor droplet and produced smaller particles. On the other hand, we kept the oxygen, nitrogen and air flow rates constant and adjusted the methane flow rate for the flame in Fig. 2d to achieve three different flame temperatures. As listed in Table 1 (cases 5-7), the temperature at the centerline location of 20 cm above the burner exit is 1800 K, 2040 K, and 2260 K, respectively. The average particle size is 185, 198 and 214 nm, respectively. The flame length for case 7 is about 10 cm longer than in case 5. At higher flame temperatures the sintering and coagulation rates increase and enhance the formation of larger particles. The longer flame length increases the residence time of particles and gives more time for the particles to grow.

Figure 4 shows the photoluminescence spectra of Y_2O_3:Eu nanoparticles exited by UV light at 355 nm The spectra of the particles prepared from water solution (Fig.

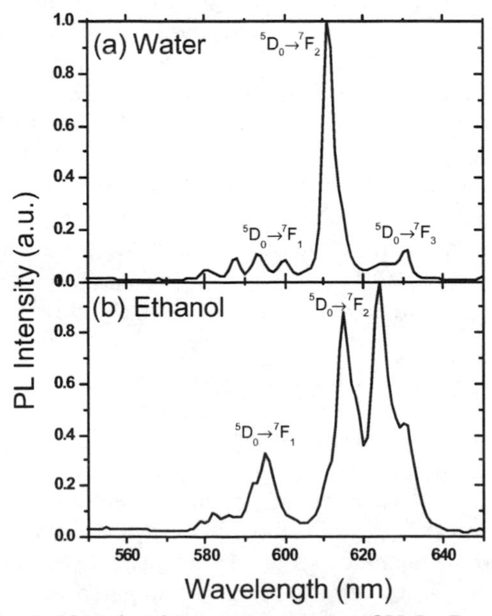

Figure 4: Photoluminescence spectra of Y_2O_3:Eu nanoparticles exited at 355 nm UV light with (a) water (b) ethanol as precursor solution.

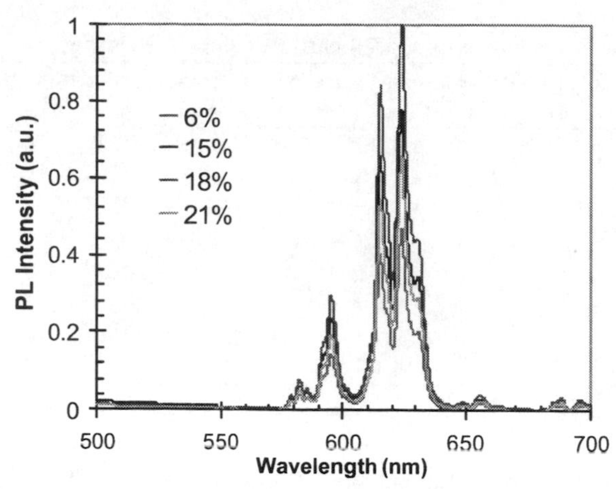

Figure 5: Photoluminescence spectra of Y_2O_3:Eu nanoparticles at different doping concentration (Eu mol% with respect to Y).

4a) shows a typical $Y_2O_3:Eu^{3+}$ emission spectrum, which is described by the well known $^5D_0$? ^7F_J (J = 0, 1, 2 ...) line emissions of the Eu^{3+} ions with the strongest emission for J = 2 at 611 nm. However, the PL spectrum of the particles made from ethanol solution shows a split peak at 615 nm and 624 nm, respectively. This is due to the nonuniformity of the yttrium oxide crystalline structure, which exhibits both monoclinic and cubic phase. By annealing the sample at 1200 °C for 2 hours can transform the monoclinic phase into cubic phase and results in a single peak PL spectrum. The intensity in Fig. 4b is 30% higher than that in Fig. 4a. Further study on the annealing effect is undergoing.

Figure 5 shows the effect of europium doping concentration on the photoluminescence intensity of as-prepared Y_2O_3:Eu nanoparticles using ethanol precursor solution. Different from the reported value of 6% for bulk materials, the quenching concentration is 18% for particles prepared in this study. This is in agreement with the result of Zhang et al. [21] for their 5nm Y_2O_3:Eu crystalline prepared by solid state reaction method. Tao et al. [13] reported a value of 14% and argued that the increase of the quenching concentration could be described as the delay of the energy transfer due to the interface effects of the nanoscale materials. It can be concluded from the previous researches that the quenching concentration increases from 6% to 18 mol% when the crystallite size decrease from 3 µm to 5 nm. XRD examination will be performed on our nanoparticles to obtain the crystallite size in this study.

4. CONCLUSIONS

Europium doped yttrium oxide (Y_2O_3:Eu) phosphor nanoparticles were synthesized by flame spray pyrolysis method and the effect of precursor solution on particle size, morphology and photoluminescence intensity was studied. Compared with previous report, SEM shows fine and smooth structure for particles prepared from ethanol alcohol as precursor solution. The particle size can be controlled by varying precursor concentration, flame temperature and residence time. Upon excitation with 355 nm light, the particles show red emission and the PL intensity of particles using ethanol is 30% stronger than those using water as precursor solution. Annealing particles at 1200°C can convert the monoclinic phase into cubic phase. The concentration quenching limit was also increased to 18% mol Eu. It is expected that these particles will be useful in developing emission field or high resolution displays and be used as markers in biomedical research.

ACKNOWLEDGEMENTS

This work was supported by National Science Foundation grant (0303947).

REFERENCES

[1] C. Suryanarayana. *Int. Mater. Rev.* **40** (1995) 41.

[2] F.E. Kruis, H. Fissan and A. Peled. *J. Aerosol Sci.* **29** (1998) 511.

[3] S. Sun and C.B. Murray. *J. Appl. Phys.* **85** (1999) 4325.

[4] T. Hase, T. Kano, E. Nakazawa and H. Yamamoto. *Adv. Electron. Phys.* **79** (1990) 271.

[5] G. Blasse and B. C. Grabmaier, *Luminescent Materials*, Berlin, Springer, 1994.

[6] R. N. Bharava, D. Gallager, X. Hong and A. Nurmikko. *Phys. Rev. Lett.* **72** (1994) 416.

[7] T. Igarashi, M. Ihara, T. Kusunoki, K. Ohno, T. Isobe and M. Senna. *Appl. Phys. Lett.* **76** (2000) 1549.

[8] C. N. R. Rao. *Mater. Sci. Engng. B* **18** (1993) 1.

[9] K. M. Kinsman, J. Mckittrick, E. Sluzky and K. Hesse. *J. Am. Ceram. Soc.* **77** 11 (1994) 2866.

[10] C. D. Veitch. *J. Mater. Sci.* **26** (1991) 6527.

[11] H. Eilers, B. M. Tissue, *Chem. Phys. Lett.* **251** (1996) 74.

[12] Y. C. Kang, H. S. Roh and S. B. Park. *Adv. Mater.* **12** (2000) 451.

[13] Y. Tao, G. W. Zhao, W. P. Zhang and S. D. Xia. *Mater. Res. Bull.* **32** (1997) 501.

[14] S. E. Pratsinis, *Prog. Energy Combust. Sci.* **24** (1998) 197.

[15] H. K. Kammler, L. Madler and S. E. Pratsinis, *Chem. Eng. Technol.* **24** (2001) 6.

[16] L. Mädler, H. K. Kammler, R. Mueller, and S. E. Pratsinis, *J. Aerosol Sci.,* **33** (2002) 369

[17] Y. C. Kang, D. J. Seo, S. B. Park and H. D. Park, *Jpn. J. Appl. Phys.* **40** (2001) 4083.

[18] P. A. Tanner and K. L. Wong, *J. Phys. Chem. B* **108** (2004) 136.

[19] H. Chang, I. W. Lenggoro, K. Okuyama, and T. O. Kim, *Jpn. J. Appl. Phys.* **43** (2004) 3535.

[20] A. U. Limaye and J. J. Helble, *J. Am. Ceram. Soc.* **86** (2003) 273.

[21] W.-W. Zhang, W.-P. Zhang, P.-B. Xie, M. Y. H.-T. Chen, L. Jing, Y.-S. Zhang, L.-R. Lou, and S.-D. Xia, *J. Colloid Interface. Sci.* **262** (2003) 588.

Synthesis of Polymer-coated Magnetic Nanoparticles

Tania Dey* and Charles J. O'Connor

Advanced Materials Research Institute, 2000 Lakeshore Drive, University of New Orleans, New Orleans, Louisiana 70148

* Corresponding Author. E-mail: taniadey@hotmail.com

ABSTRACT

ATRP (atom transfer radical polymerization) approach was employed to synthesis polymer-coated magnetite nanoparticles with an average diameter of 7.1 nm and of narrow size distribution, followed by various characterization techniques like Transmission Electron Microscopy (TEM), Ultraviolet-Visible spectroscopy (UV-vis), Fourier Transform Infrared spectroscopy (FTIR) and Atomic Force Microscopy (AFM). The challenge was to obtain a thin shell and particles in an unagglomerated state. Several factors like presence/choice of solvent, monomer-to-initiator concentration and structure of initiator were found to play a key role in this study. Attempts have been made to tailor the polymer shells by end-functionalization. This work has an enormous biomedical application potential.

Keywords: magnetite nanoparticles, polystyrene, atom transfer radical polymerization, capping ligand

1 INTRODUCTION

The formation of polymeric shell is essential for biomedical application of magnetic nanoparticles such as targeted drug delivery, magnetic resonance imaging (MRI) contrast enhancement, biosensors, protein and DNA purification, cell separation and so on. In these applications, good control over particle size, size distribution and surface properties are important. Atom transfer radical polymerization (ATRP) is a versatile method which offers several advantages over other polymerization techniques including good control over molecular weight and polydispersity. Lot of work has been done in the area of in-situ preparation of nanoparticle-polymer hybrids using polymer as a matrix to obtain an array of embedded nanoparticles, but to obtain discrete polymer-coated nanoparticle still remains as a challenge. Some ATRP approaches were reported to make core-shell systems including various inorganic cores like SiO_2, Au, $MnFe_2O_4$, Au@SiO_2 and Fe_2O_3 [1-8], but no attempt has been made so far that involves magnetite nanoparticles, which shows strong ferrimagnetic behavior and is less susceptible towards oxidation in comparison to other magnetic transition metals.

We have synthesized spherical magnetite (Fe_3O_4) nanoparticles with an average diameter of 5.6 nm and a narrow size distribution, by co-precipitating iron (II) and iron (III) chloride salts in presence of sodium hydroxide at elevated temperature. Long chain carboxylic acid were found to stabilize these nanoparticles and act as a capping ligand [9]. We have obtained polystyrene coated magnetite nanoparticles of average diameter < 10 nm by a two-step process: firstly modifying the surface of nanoparticles with initiator and then using these initiator-exchanged nanoparticles as a macroinitiator for copper-mediated ATRP reaction. Systems were characterized by TEM, UV-vis, FTIR, AFM and GPC. Choice of initiator, solvent and monomer-to-initiator concentration ratio were found to play a key role in our study. Recent efforts have been made to impart dispersity and stability to magnetite nanoparticles by polymer-coating [10,11]; but polymerization may not be considered as a method of preventing agglomeration only. The potential lies in the tailorability of these polymer shells as it maybe end-functionalized or block-copolymerized.

2 EXPERIMENTAL SECTION

2.1 Instrumentation

Transmission Electron Microscope. The TEM images were recorded on a JEOL Model 2010 at an accelerating voltage of 200 kV. The TEM specimens were made by placing a drop of toluene suspension of nanoparticles on formvar/carbon 200 mesh, copper grid (tedpella.com).
Atomic Force Microscope. AFM images were obtained from a Multimode Scanning Probe Microscope, Digital Instruments (Veeco metrology group).
UV-vis spectrometer. VARIAN Carey 500 scan UV-vis-NIR spectrophotometer was used to measure the absorption of nanoparticle solution taken in a quartz cell.
FT-IR spectrometer. The FT-IR spectra were collected on a Thermo-Nicolet NEXUS 670 FT-IR spectrometer using KBr pellets.

2.2 Synthesis

Synthesis of Magnetite Nanoparticle. 2 mmol of $FeCl_2.4H_2O$ and 4 mmol of $FeCl_3.6H_2O$ were dissolved in 40 g of diethylene glycol (DEG, Alfa Aesar, 99%) in a schlenk flask under protection with argon. Separately, 16 mmol of NaOH (EMD chemicals, GR) was dissolved in 40 g DEG. The solution of NaOH was added to the solution of metal chlorides while stirring at room temperature causing an immediate color change. The temperature of the resulting solution was raised during 1-1.5 h to 210-220 °C and then kept constant for 0.5-1 h. As the solution turned turbid, the reaction was terminated by adding 2.6 mmol of oleic acid (Aldrich, 90%) dissolved in 20 g of DEG. This addition caused immediate precipitation of solid particles. The mixture was cooled at room temperature and then centrifuged. The precipitate was washed with molecularly sieved methanol for several times. The yield of black powder was 0.65 g.

Synthesis of Polystyrene coating. The prepared Fe_3O_4 nanoparticles were dispersed in 1.0 M hexane solution of the initiator, 2-bromo-2-methyl propionic acid (Br-MPA) (Aldrich 98%) and stirred for 72 h at room temperature under protection of argon. The resulting powder was separated using a centrifuge, washed with hexane to remove the excess initiator and was dried under vacuum. The initiator-exchanged nanoparticles (0.055g) were added into 2mL of argon-purged styrene solution (Aldrich, 99+%). CuBr (0.3 mmol, Aldrich, 98%) and 4,4'-dinonyl-2,2'-dipyridyl (1.1 mmol, Aldrich, 97%) were dissolved in 4 mL of xylene (Aldrich, AR) and then the solution was added to the styrene/nanoparticle mixture. The final solution was stirred and kept at 100 °C for 20 h. On completion of reaction, the mixture was 10 times diluted with tetrahydrofuran (THF, Aldrich 99%), centrifuged and washed with methanol.

3 RESULTS AND DISCUSSION

One of the fundamental characterization tools used in this work was TEM. Figure 1 shows the TEM images of oleic acid stabilized well dispersed Fe_3O_4 nanoparticles, as well as the initiator-exchanged and polystyrene-coated Fe_3O_4 nanoparticles. A change in morphology from spherical to faceted ones was observed due to initiator-exchange. A dark center surrounded by a light shell was visible under the microscope upon polymerization. Particle sizes and shell thickness were determined by manually counting over 100 particles. The core/shell nanoparticles showed an average diameter of 7.1 nm, whereas the nanoparticles itself had an average diameter of 5.6 nm, both having a narrow size distribution (as shown in Figure 2).

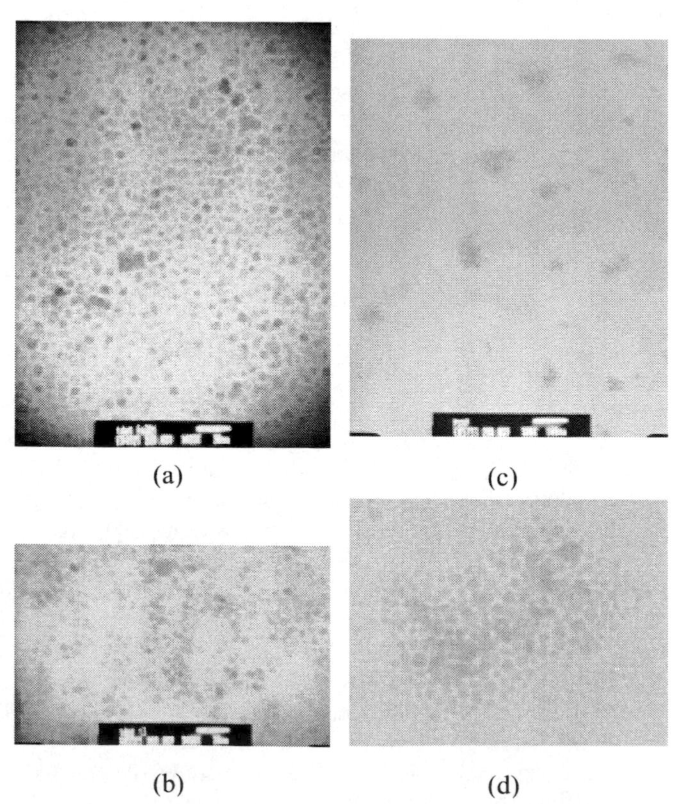

(a) (c)

(b) (d)

Figure 1: TEM images **(a)** Fe_3O_4 nanoparticles **(b)** initiator-exchanged nanoparticles **(c)** poly styrene-coated nanoparticles **(d)** one portion of (c) is enlarged

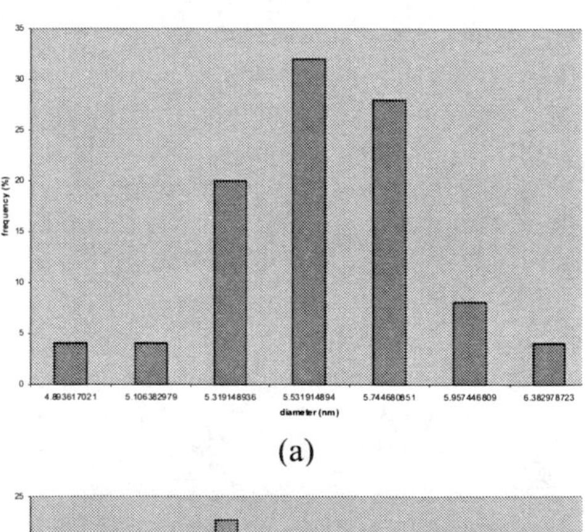

(a)

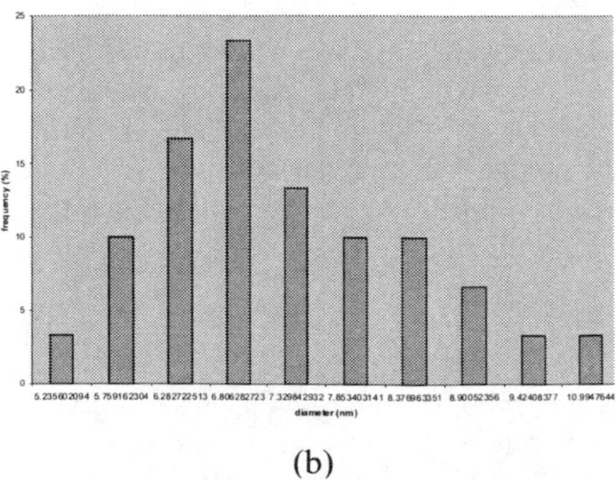

(b)

Figure 2: Size distribution of **(a)** Fe_3O_4 nanoparticles **(b)** polystyrene-coated nanoparticles

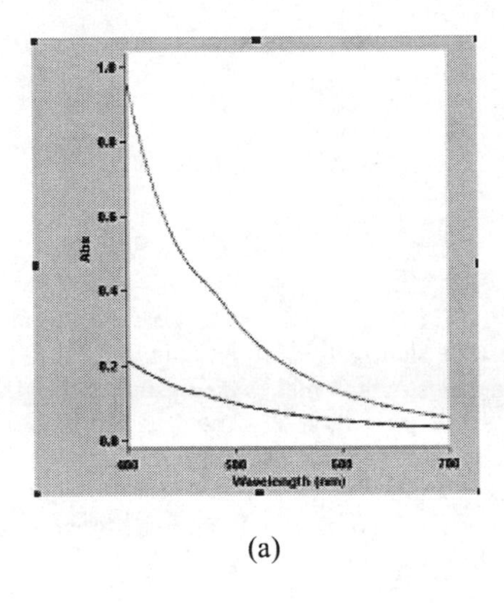

(a)

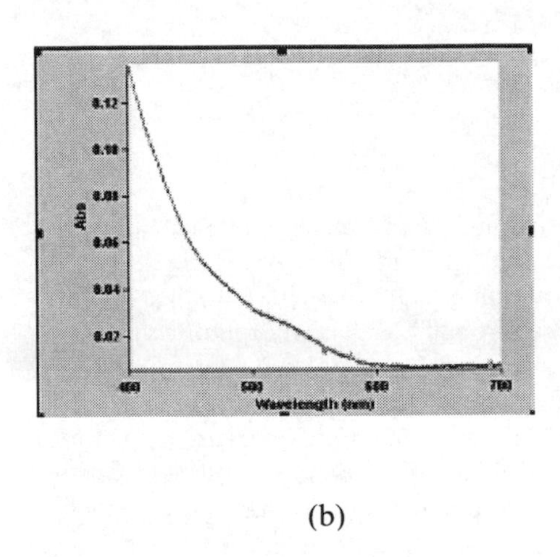

(b)

Figure 3: (a) UV-vis spectra of Fe_3O_4 nanoparticle before and after initiator exchange **(b)** UV-vis spectrum of Fe_3O_4-polystyrene nanoparticle

The clear yellow ethanolic suspension of initiator-exchanged nanoparticle displayed a broad plasmon absorption band at 475 nm indicative of iron oxide, just like the nanoparticles before capping ligand exchange (Figure 3a). A red-shift was observed upon polymerization which may be attributed due to decrease in inter-particle distance caused by increased size, although no precipitation/ aggregation was observed (Figure 3b).

The alkyl C-H vibration bands in the IR spectra (Figure 4a) suggested that there was an equilibrium in the ligand-exchange process. However the occurance of the relatively stronger peaks of CH_3 groups after initiator-exchange was consistent with the fact that CH_3 was the major alkyl group in Br-MPA, while oleic acid predominantly had CH_2 groups. The existence of macroinitiator was also supported by the IR band at 1010 cm^{-1} belonging to Br-MPA. Characteristic peaks of polystyrene at 2700-3500, 1000-1400 and 700 cm^{-1} (Figure 4b) were observed in the core-shell samples indicating polymerization.

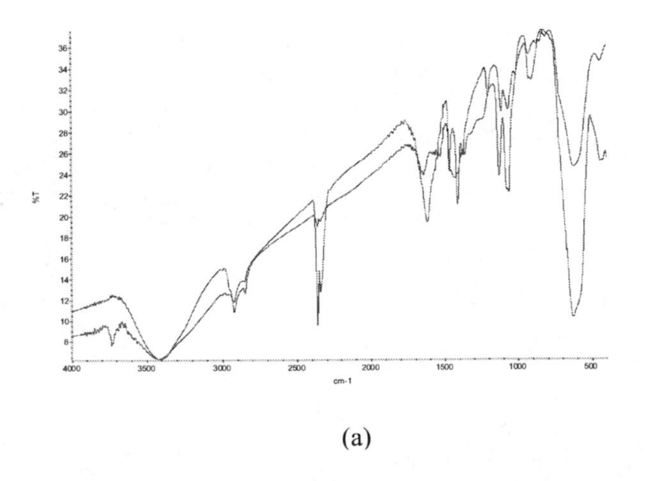

(a)

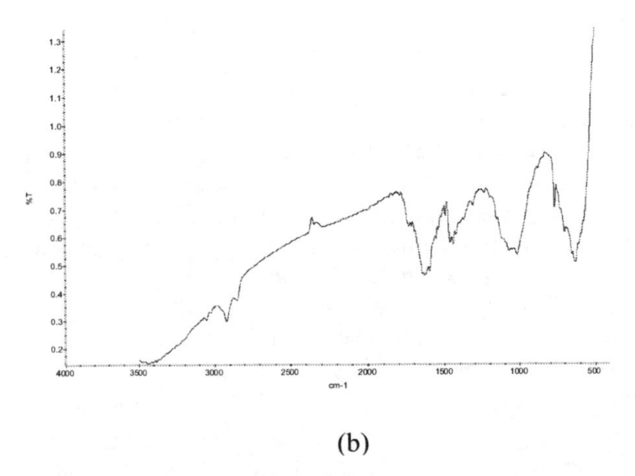

(b)

Figure 4: (a) FTIR spectra of Fe_3O_4 exchange **(b)** FTIR spectrum of polystyrene nanoparticle before and after initiator -coated nanoparticle

The tapping mode AFM measurements (Figure 5) verified the successful grafting of polymer on nanoparticles and was in good agreement with the particle size obtained from TEM. The morphology and volume of hybrid nanoparticles can also be determined by this method, complementing the TEM studies.

Ongoing studies include GPC measurements to obtain molecular weight and polydispersity of the polymeric shells, and the SQUID measurements to determine magnetic surface anisotropy and coercivity.

Figure 5: AFM image of polymer -coated nanoparticles

Presence of a solvent was found to be essential during polymerization to impart good dispersity, unlike the solvent-free approach where the initiator-exchanged as well as the polymer-coated nanoparticles were found to be kind of aggregated. Xylene proved to be the best candidate in this respect. Significant amount of work has been accomplished in varying the initiator structure by incorporating some hydrophobic moiety for better dispersity and choosing appropriate monomer to introduce end-functionalization for better anchoring of biomolecules (to be published elsewhere).

4 CONCLUSION

ATRP approach works very well to produce thin polystyrene shell (~1.5 nm) on magnetic Fe_3O_4 nanoparticles, which makes the system biologically interesting. The core-shell structures can be obtained in a fairly unaggregated state by controlling the presence/choice of solvent, monomer-to-initiator ratio and initiator structure. Polymer shells can be tuned to impart biocompatibility as well as biospecificity.

Acknowledgment. This work was supported by DARPA (Grant No. MDA972-02-1-0001).

REFERENCES

[1] J. Pyun and K. Matyjaszewski, Chem. Mater. *13,* 3436-3448, **2001.**

[2] T. von Werne and T. E. Patten, J. Am. Chem. Soc. *123,* 7497-7505, **2001.**

[3] K. Kamata, Y. Lu and Y. N. Xia, J. Am. Chem. Soc. *125,* 2384-2385, **2003.**

[4] C. R. Vestal and Z. J. Zhang, J. Am. Chem. Soc. *124,* 14312-14313, **2002.**

[5] K. Ohno, K. Koh, Y. Tsujii and T. Fukuda, Macromolecules *35,* 8989-8993, **2002.**

[6] T. K. Mandal, M. S. Fleming and D. R. Walt, Nano Lett. *2*, 3-7, **2002.**

[7] D. A. Savin, J. Pyun, G. D. Patterson, T. Kowalewski and K. Matyjaszewski, J. Polym. Sci. Part B: Polym. Phys. *40,* 2667-2676, **2002.**

[8] Y. Wang, X. Teng, J-S. Wang and H. Yang, Nano Lett. *3*, 789-793, **2003.**

[9] D. Caruntu, Y. Remond, N. H. Chou, M-J. Jun, G. Caruntu, J. He, G. Goloverda, C. O'Connor and V. Kolesnichenko, Inorg. Chem. *41*, 6137-6146, **2002.**

[10] R. Matsuno, K. Yamamoto, H. Otsuka and A. Takahara, Macrocolecules *37*, 2203-2209, **2004.**

[11] R. Matsuno, K. Yamamoto, H. Otsuka and A. Takahara, Chem. Mater. *15*, 3-5, **2003.**

Characterisation of Nanocrystalline PZT Synthesised via a Hydrothermal Method

S. Harada[*] and S. Dunn[**]
Nanotechnology Group, Building 70, Cranfield University,
Cranfield, Bedfordshire, MK43 0AL, [*]s.harada.2003@cranfield.ac.uk
[**]s.c.dunn@cranfield.ac.uk

ABSTRACT

This paper describes the progress made toward characterising lead zirconate titanate (PZT) synthesised via a novel hydrothermal method [1,2]. Low temperatures (≤ 200 °C) and high mineraliser (potassium hydroxide) concentrations were used to produce single-phase, perovskite PZT in a bomb-type reactor. Full phase purity was attained after just 30 minutes of processing at 160 °C. The particles had a wide size distribution, from nanoscale agglomerates with irregular morphology, up to micron-sized cubes. It is believed that the differences in morphology can be attributed to two distinct formation stages.

Keywords: hydrothermal, PZT, nanocrystalline

1 INTRODUCTION

Nanoscale lead zirconate titanate (PZT) has the potential to become a key component in a myriad of emerging future applications. Examples of such technologies include, but are not limited to, non-volatile ferroelectric random access memories and nanoelectronic mechanical systems (NEMS). A sound knowledge of the underlying physical and chemical processes involved in the formation and applications of these particles is clearly essential for successful device development.

The hydrothermal method involves using elevated temperatures and pressures to synthesise inorganic materials from aqueous or non-aqueous solutions. It has several key advantages over other methods when it comes to synthesising PZT. These include the ability to control particle morphology, size and agglomeration whilst retaining high product purity. From a processing point of view, the hydrothermal method is very simple and can also be cost effective [1,2]. In this paper the effects of processing temperature and time on the hydrothermal synthesis of PZT are investigated. Attempts are also made to control the stoichiometry of the resultant powders by altering the precursor ratios.

2 EXPERIMENTAL DETAILS

The precursors for the hydrothermal reaction were lead (II) nitrate ($Pb(NO_3)_2$), zirconyl chloride octahydrate ($ZrOCl_2 \cdot 8H_2O$) and titanium dioxide (TiO_2). All chemicals were reagent grade and obtained from Aldrich, UK or Fisher, UK. The quantities of zirconyl chloride and titanium dioxide were altered to produce the desired PZT stoichiometry ($PbZr_xTi_{1-x}O_3$ where $x = 0.2 \rightarrow 0.7$). A 45 ml Teflon lined autoclave (Parr Instruments, model no. 4744) was filled with the powder precursors and topped up with 25 ml of distilled water. After vigorous stirring, 7 g of potassium hydroxide (KOH) flakes were added slowly to give a 5 M mineraliser concentration. The autoclave was then placed into an oven and heated over a range of temperatures (130 – 200 °C) and times (0.5 – 6.0 hrs). After the hydrothermal treatment the PZT powders were filtered and washed with distilled water and ethanol and then dried at 60 °C for 24 hrs.

The phase composition of the powders was determined using a Siemens D5005 x-ray diffractometer. Qualitative comparisons were made between the resulting patterns and the powder diffraction file for tetragonal $PbZr_{0.52}Ti_{0.48}O_3$ (JCPDS #33-0784). The morphology and size distribution of the particles were analysed using a Sirion S-FEG SEM. Elemental analysis of samples was conducted using an EDAX detector attached to the SEM.

3 RESULTS AND DISCUSSION

3.1 The effects of processing time

The dependence on processing time was tested for 52:48 PZT at 160 °C. Fig. 1 shows XRD patterns of the unreacted precursor material and four PZT powders synthesised over a range of times. The precursor material is a mixture of tetragonal TiO_2 and an amorphous phase. No systematic change in peak intensity occurs after 0.5 hr processing time in the PZT patterns. This suggests that crystallisation is substantially complete within this time period. Any non-systematic variations in peak intensity can be explained by random preferential reflections due to the nature of the x-ray powder method.

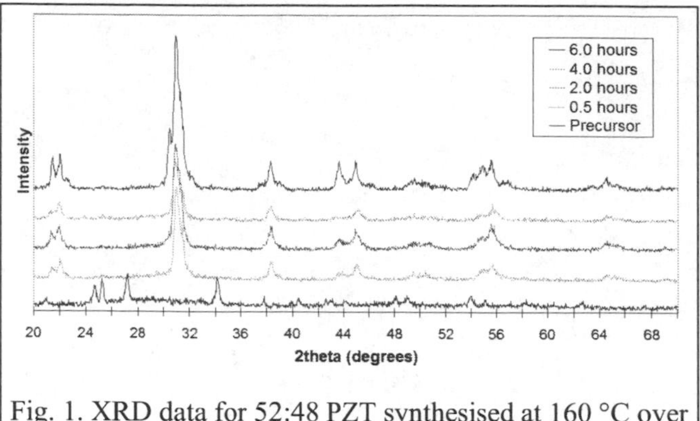

Fig. 1. XRD data for 52:48 PZT synthesised at 160 °C over a range of times

The 0.5 – 6.0 hrs patterns appear to contain a single phase that matches the powder diffraction file for tetragonal $PbZr_{0.52}Ti_{0.48}O_3$ (JCPDS #33-0784). Processing times less than 0.5 hr led to an incomplete reaction of the precursors.

The resultant particle morphology for each processing time is shown in Fig. 2, parts a) – d). The major particles are micron sized and cubic. Smaller, irregularly shaped agglomerates are also present, to varying degrees, in each sample (Fig. 3). Systematic changes in morphology do not appear to take place with increasing processing time – cubic particles, once formed, undergo little change.

Traianidis et al. [1] proposed a two-stage reaction mechanism to account for the existence of two types of particle morphology. In the first stage a Zr-Ti co-precipitate forms, into which lead ions diffuse. This leads to irregularly shaped agglomerates. As the reaction progresses these agglomerates are dissolved leading to the nucleation and growth of cubic PZT particles. A similar process seems to occur in this reaction. Further work is necessary to determine whether the PZT phase has been generated in the irregularly shaped agglomerates. Future experiments, planned to test the ferroelectric properties of nano-sized PZT particles, could verify this.

Fig. 2. SEM images of the PZT powders processed for a) 0.5 hrs, b) 2 hrs, c) 4 hrs and d) 6 hrs

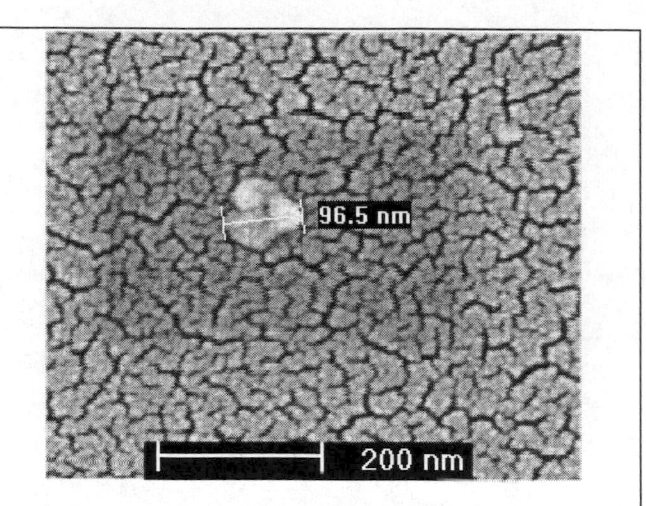

Fig. 3. A nano-sized PZT particle synthesised at 160 °C for 4 hrs

3.2 The effects of processing temperature

Having established that a minimum of 30 minutes processing time at 160 °C was necessary to produce phase-pure PZT, experiments were conducted to reduce processing temperature. Fig. 4 shows the evolution of the PZT phase from 140 – 160 °C. The indexed peaks refer to the accepted values for tetragonal $PbZr_{0.52}Ti_{0.48}O_3$. As the processing temperature increases the TiO_2 phase diminishes until phase purity is reached at 160 °C.

Further evidence for incomplete reaction was found when examining the morphology of the 140 °C powder. EDAX was used to confirm that the small round particles visible in the background of Fig. 5 consisted of unreacted TiO_2; a cubic PZT particle is shown in the foreground. Again, no systematic changes to morphology were evident in the PZT that was produced between 140 – 160 °C.

The breakdown of TiO_2 is a limiting factor for the reaction. In one experiment the mineraliser (KOH) concentration was reduced from 5 M to 0.5 M. After 18 hrs of processing at 200 °C no PZT phase was detectable in the XRD pattern. This indicates that the use of an oxide precursor imposes a minimum limit for processing temperature and mineraliser concentration.

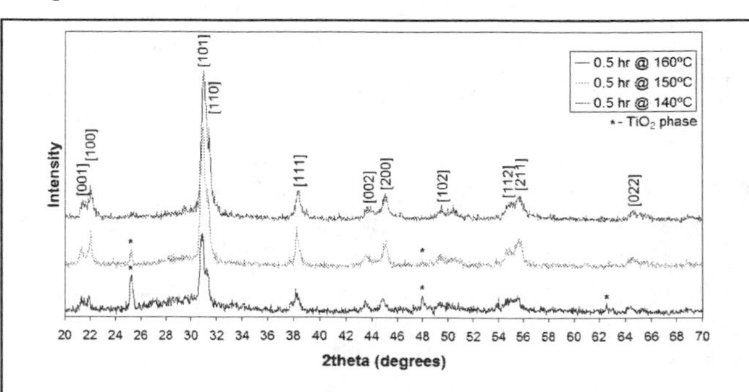

Fig. 4. XRD data for 52:48 PZT showing the evolution of phases with temperature

Fig. 5. SEM image of the PZT powder processed for 0.5 hr at 140 °C

3.3 Altering PZT stoichiometry

Attempts were made to synthesise different PZT stoichiometries by altering the Zr/Ti precursor ratio. Fig. 6 shows XRD data for powders produced using precursor ratios from 20:80 – 70:30. The PZT system has a morphotropic phase boundary (MPB) at approximately Zr/Ti = 52:48. Above this value the powders crystallise with a rhombohedral structure, while those below form a tetragonal structure.

If the synthesis of 70:30 PZT had been successful, we would expect to see a single (100) peak at 2theta ~ 22 degrees as opposed to the split (001) – (100) peaks, characteristic of the tetragonal PZT structure, which are visible in Fig. 6. The clear match of the 70:30 pattern with the peaks expected for 52:48 indicate that the latter has in fact been produced. This also appears to be the case with the attempt to synthesise 20:80 PZT. The presence of a TiO_2 phase in this pattern suggests that the extra titanium has failed to incorporate into the PZT lattice.

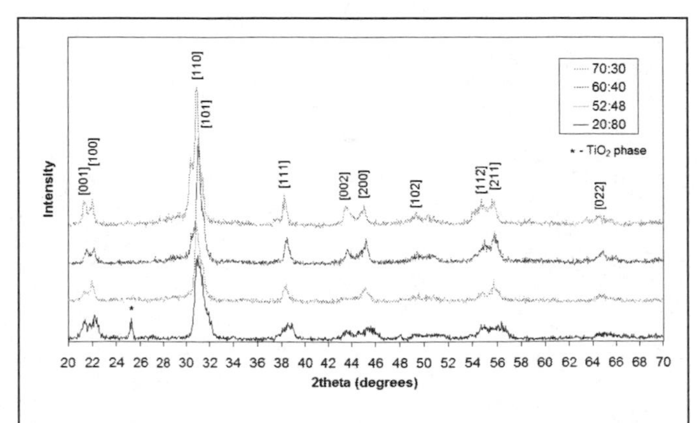

Fig. 6. XRD data for powders synthesized at 160 °C for 4 hrs

Kutty and Balachandran [3] found that the reaction temperature necessary to synthesise PZT was proportional to the Zr/Ti ratio. In other words, the rhombohedral phase required far more energy (340 °C) to form than the tetragonal phase. Attempts to synthesise lead titanate (the tetragonal end member of the PZT system) using this method have been successful, whereas similar attempts to produce lead zirconate have failed, giving credence to the applicability of this theory. However, an alternative explanation may be equally valid. Chen et al. [4] investigated the hydrothermal synthesis of $Zr_xTi_{1-x}O_4$ for x = 0.35 → 0.65. They found that the phases produced were dependent on the pH of the reaction medium. In highly basic solutions (pH = 13), TiO_2 was the only detectable phase when the precursor ratio Zr/Ti < 50:50 whilst $ZrTiO_4$ plus a tetragonal zirconia phase were produced when Zr/Ti > 50:50.

A similar pattern is seen when the results in this paper are compared. Although the zirconium-titanium compound, which forms when the precursors are first mixed, is not $ZrTiO_4$, it seems plausible that its behaviour could be similar, under the same hydrothermal conditions. To begin with, the reaction takes place in a highly basic medium. $PbZr_{0.52}Ti_{0.48}O_3$ plus an excess TiO_2 phase form when the precursor ratio Zr/Ti < 52:48. $PbZr_{0.52}Ti_{0.48}O_3$ forms when the precursor ratio Zr/Ti ≥ 52:48. The lack of a detectable zirconium phase may be due to the differences in the precursors. EDAX analysis has confirmed that a soluble zirconium compound forms when the precursor ratio Zr/Ti > 52:48. This compound gets washed away when the powders are filtered after the hydrothermal process.

4 CONCLUSIONS

Lead zirconate titanate (PZT) powders have been synthesised under a variety of hydrothermal conditions using high mineraliser concentrations, in aqueous solutions. The minimum requirements for phase-pure, $PbZr_{0.52}Ti_{0.48}O_3$ were 0.5 hrs of processing time at 160 °C.

The powders produced in this paper had a wide size distribution and consisted either of nano-scale, irregularly shaped agglomerates or micron-sized cubes. It was thought that particles with the former morphology transformed into the latter through a dissolution-precipitation mechanism.

The reactivity of the TiO_2 precursor was found to limit the formation of PZT. At lower temperatures, below 160 °C, insufficient energy was available to fully breakdown the TiO_2 and as such large quantities remained in the reaction product. These were detected both visually, using a SEM, and as a contaminant phase in the relevant XRD patterns.

Altering the ratio of precursor materials did not lead to changes in the resultant PZT stoichiometry. Two reasons were proposed for this anomaly. The first one relied on the fact that PZT stoichiometries, above the MPB, form at much higher temperatures – double those used in the current work. The second made a comparison with related work regarding the dependence on pH for the hydrothermal synthesis of $Zr_xTi_{1-x}O_4$. It was suggested that the behaviour of the reaction intermediaries in the current work could be similar, under comparable hydrothermal conditions.

REFERENCES

[1] M. Traianidis, C. Courtois and A. Leriche, J. Euro. Ceram. Soc. 20, 2713-2720, 2000

[2] Y. Deng, L. Liu, Y. Cheng, C.W. Nan and S.J. Zhao, Matter. Lett. 57, 11, 1675-1678, 2003

[3] T.R.N. Kutty and R. Balachandran, Mater. Res. Bull. 19, 11, 1479-1488, 1984

[4] D.R. Chen, X.L. Jiao and R.R. Xu, J. Mater. Sci. 34, 6, 1379-1383, 1999

Preparation of Functional LLDPE/LDPE/TiO_2 Membranes

Xin Wang[1, 2], Joong-Hee Lee[1, *], Xu-Yun Wang[2], Chang-Eui Hong[1]

[1]Department of Polymer Science and Technology, Chonbuk National University,

Duckjin-dong 1Ga 664-14, Jeonju, Jeonbuk 561-756, Korea

[2]Qingdao University of Science and Technology, Qingdao 266042, China

ABSTRACT

Nano-scaled TiO_2 is being used in the polymer industry owing to its promising properties as a light catalyst and in UV light shielding. Functional LLDPE/LDPE/TiO_2 membranes were manufactured successfully in this study. To solve the poor dispersion of TiO_2 in polyethylene, TiO_2 was introduced into blow-forming LLDPE/LDPE as a master batch. The nano-TiO_2 had an induced nucleation effect on the crystallization of polyethylene; with the introduction of TiO_2, the size of the crystal spheres decreased, as their numbers increased. TiO_2 was dispersed randomly throughout the membrane, with an average size of about 100 nanometers. This excellent dispersion gave rise to the UV absorption by the membranes . Moreover, the visible light transmittance of the membrane changed slightly with the introduction of TiO_2. In conclusion, the resulting membranes had the desired properties plus a UV absorption function. Such membranes will have many applications in agriculture and food packing owing to the anti-bacterial effect induced by the UV absorption.

Key Words: TiO_2, membranes, crystallization, nucleation, UV adsorption

1. INTRODUCTION

Nano-scaled TiO_2 is being used in the polymer industry owing to its promising properties as a light catalyst and in UV light shielding[1-3]. The anti-bacterium function of TiO_2 comes from its photo-induced reaction. Due to the high oxidation of radicals produced in the reaction, the organic compounds, such as protein and ester, will be attacked, and the remainder will be then degraded. The function of anti-bacterium will be fairly beneficial to the membranes used in agricultural application and food packing.

Linear Low Density Polyethylene (LLDPE) and Low Density Polyethylene (LDPE) are chosen as the matrix of the membranes in our research. Since they are both crystalloid polymers, the transparency of the produced membranes will be great influenced by the degree of crystallinity of polymers. The exhibition of the special functions of TiO_2 in the final products is also determined by its dispersion conditions in the polymer matrix. In this paper, it was presented the preparation of LLDPE/LDPE/TiO_2 membranes and the comprehensive studies focused on the crystallinity of polyethylene and the dispersion of TiO_2.

2. EXPERIMENTS

LDPE (type 150L, Dow Chemical Co LTD, USA) and LLDPE (type 7042, Yangzi Petro-Chem Co LTD, China) were used as the row polymer of membrane. TiO2 (Qingdao Haier-QUST Nano Technology Company, China) possessed an average primary particle size of 80nm and 80% anatase crystal structure. A low molecular weighted polyethylene named as LDPE1 (type 800, Modern Chemical, Korea) was used as the polymer matrix of the master batch containing TiO2.

TiO2 was dried for 8 hr at 100℃, and then subjected to the high speed mixing with LDPE1. The mixture was then extruded in two-screw extruder (ZKS-25 type, Krupp Werner & Pfleiderer Gmbh, Germany) to yield LDPE1/TiO_2 granules. The so-prepared master batch with 20% TiO_2 in weight was blow-formed with LLDPE/LDPE at a single-screw blow-forming system (Laiwu Plastics Machinery Company, China) where the weight ratio of LLDPE to LDPE was 3:1. The dosage of TiO_2 in the final membrane was adjusted by the ratio of master batch to polyethylene.

The melt point and enthalpy of the LLDPE/LDPE membrane were measured on NETZSCH Thermal Analysis DSC Cell with temperature increasing rate of 10K/min from room temperature to 180℃.

The dispersion of TiO_2 in polyethylene was observed in Electron microscope (JEM-2000EX type, Japan Electron Co.). The light absorption of membranes was measured in Ultraviolet radiation-Visible light absorption spectrum instrument (Cary 500, Varian Instrument, U.S.A). The transparency of LLDPE/LDPE membranes was measured in photo-behavior tester (HAZE-GARD puls, BYK Gurder, Germany).

3. RESULTS AND DISCUSSIONS

3.1 Dispersion of TiO_2 and its Nucleation Effect on PE Crystallization

The dispersion of TiO_2 in the final LLDPE/LDPE/TiO_2 membrane was directly observed under an optical microscopy. Rarely could the dark regions be found in the visual field. The microscopic observation was carried out in SEM. Figure 1 indicated a random dispersion of nano-scaled TiO_2 in the polymer matrix, where the average diameter of the dispersed phase was less than 100 nanometers. The complementary application of optical microscopy and SEM confirmed the even and uniform dispersion of TiO_2 in LLDPE/LDPE when introduced in the form of master bitch of low molecular weighted LDPE.

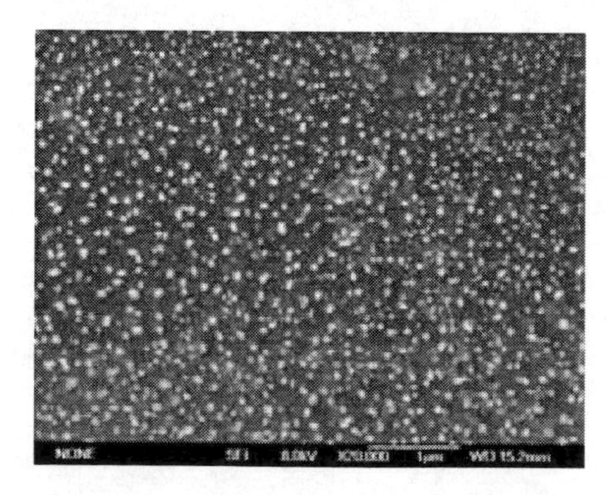

Figure 1. SEM photo of an LLDPE/LDPE membrane with 0.5 wt% TiO_2

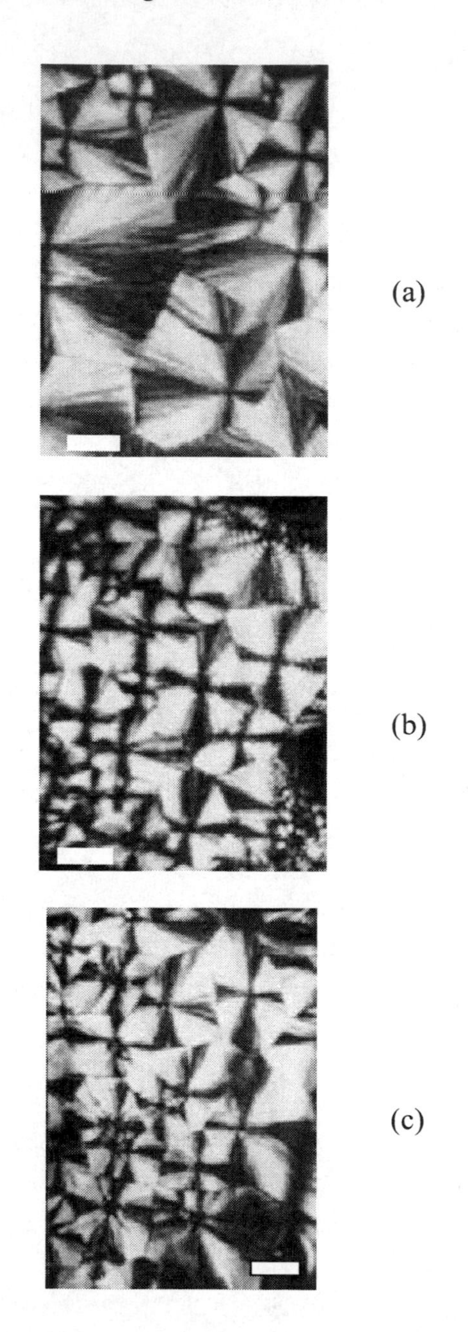

Figure 2. Polarized images of membranes (a) 0, (b) 0.25, and (c) 0.5 wt% TiO_2 (scale bar 100 μm)

LLDPE and LDPE were both crystalloid polymers, and their different molecular structures resulted in the different crystallization behaviors. In general, LLDPE exhibited a greater capacity in crystallization due to its linear structure. Observed in the polarized microscopy, the crystalline of LLDPE was about 380 micrometers in diameter, which was much larger than the 110 micrometers of LDPE. The polarized image of LLDPE/LDPE was shown in (a) of Figure 2, the sizes of the crystalline differed much from each other, most of which ranged from 200 micrometers to 300 micrometers. However the appearance of a minority of crystalline with diameter about 100 micrometers indicated the separated crystallization of LLDPE and LDPE, which was further confirmed by two separated crystallization peaks in DSC curves shown in Figure 3.

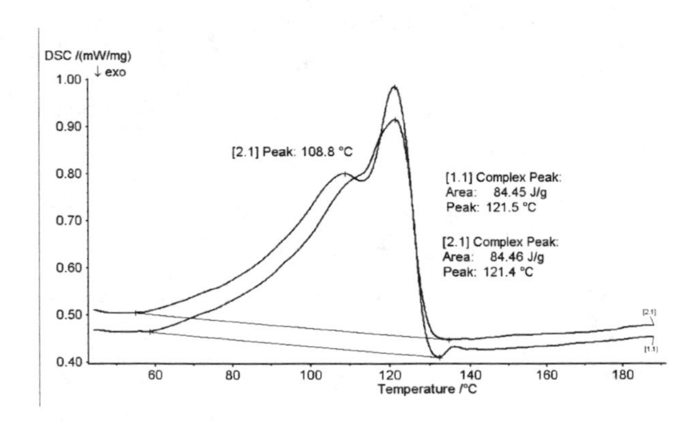

Figure 3. DSC curves of primary LLDPE/LDPE membranes, with [1.1] for LLDPE/LDPE membrane; [2.1] for membrane with 0.5wt% TiO_2

The introduction of TiO_2 into LLDPE/LDPE system changed much the size of crystalline. As shown in （b） and (c) of Figure 2, the diameters of LLDPE/LPDE varied much with each other, however most of crystalline were less than 200um after the introduction of 0.25% TiO_2 in weight. When content of TiO_2 was 0.5% in weight, the polarized microscopic image was similar but with an arising of the number of the crystalline in the visual field. The induced nucleation effect of TiO_2 could be confirmed by the polarized microscopic images.

3.2 UV absorption behaviors and transparency of the membranes

The good dispersion of TiO_2 in the polyethylene matrix provided the probability of an efficient UV absorption of the LLDPE/LDPE membrane. The experimental result shown in Figure 4 indicated the special UV absorption of the membrane after the introduction of 0.5% TiO_2 in weight. The curve (1) of pure LLDPE/LDPE membrane leveled off almost in the whole range of wavelength, while curve(2) of LLDPE/LDPE with 0.5% TiO_2 in weight exhibited an remarkable absorption in the UV region with wavelength ranging from 200 to 400 nanometers. This phenomenon indicated an anti-bacterium function induced by the consequent photochemical reaction.

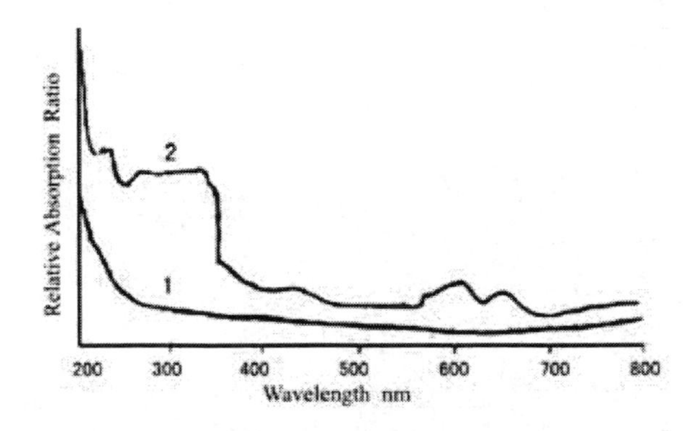

Figure 4. UV absorption spectrum of LLDPE/LDPE membrane s showing the adsorption from 200 to 400 nm; line 1-pure membrane and line 2-membrane with 0.5 wt% TiO_2

Being used as packaging material or agricultural membrane, the transparency was the most important property of LLDPE/LDPE, which, however, might be lessened by the induced nucleation of TiO_2. As shown in Figure 5, the transparency of the membrane decreased with the dosage of TiO_2, however the decline was so slight that could be ignored. This result was a bit surprising, especially accounting for the induced nucleation effect of TiO_2. The DSC determination revealed that the degree of crystallinity changed little with the introduction of TiO_2. As shown in Figure 3, the membranes processed two melting peaks at about 121℃ for LLDPE and 108℃ for LDPE respectively, which proved the crystallization of LLDPE and LDPE occurred independently, while the introduction of TiO_2 led the peak of LDPE more distinguishable. Killer and Hill put forward the concept of liquid-liquid phase separation based on the similar phenomenon observed in HDPE/LDPE system [4], and it was believed that the separation should be regards as a common separation in metastable phase[5], which could be influenced by the molecular weight, the branched chains of the polymers. In our study, the phase separation emerged in the rheological behaviors of LLDPE/LDPE as a big deviation from the pseudo-plastic flow, which would be reported later. Here, the combination of DSC with the polarized microscopic observation revealed the independency in the crystallization of LLDPE/LDPE system.

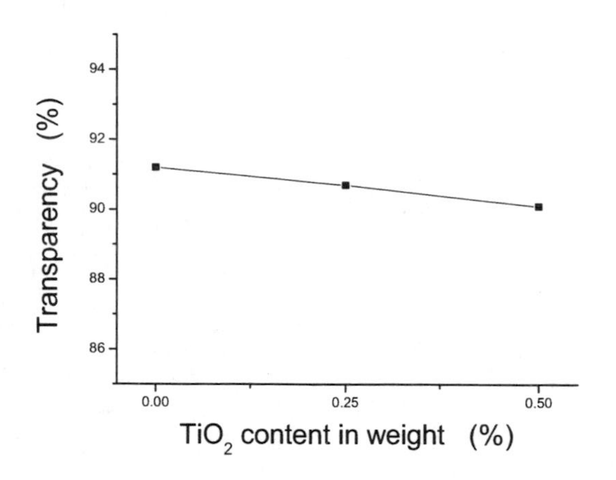

Figure 5. Transparency of LLDPE/LDPE membranes

Though the melt point of polyethylene was little changed, the relative height of these two peaks differed with the introduction of TiO_2, which indicated the influence of TiO_2 on the crystallization of polyethylene. The total areas under the curves, however, were almost the same. So, no loss in transparency of the membrane could be attributed to the crystallinity of polyethylene. The slight decline in transparency could be resulted from the difference between the refraction indexes of polyethylene and TiO_2. After all, the measurement confirmed that the transparency of LLDPE/LDPE membranes could satisfy the request in common applications after the introduction of TiO_2.

In summary, TiO_2 was successfully introduced into LLDPE/LDPE system in the form of master batch, and the membrane manufactured exhibited a special UV absorption while maintaining its transparency. The result indicated a new application of nano-scaled TiO_2 in functional membrane material.

4. CONCLUSIONS

In the present paper, LLDPE/LDPE membrane with special UV absorption was successfully prepared by the introduction of TiO_2. Several points could be concluded as below through the discussion and analysis:

1. The crystallization behavior of polyethylene was great influenced by the introduction of TiO_2 in nano-scale, which could provide a remarkable induced nucleation. As the result, the size of the crystalline of polyethylene decreased while the number of the crystalline increased. However the degree of crystallinity remained almost unchanged, therefore the introduction of TiO_2 changed the transparency of LLDPE/LDPE membranes only a bit.

2. TiO_2 was well dispersed in LLDPE/LDPE membrane with the average diameter less than 100 micrometers after introduced in the form of master batch, which consequently resulted in the remarkable UV adsorption of the membrane.

REFERENCE

1. Martuscelli E Polym.Eng. Sci., 563, 24, 1984
2. Anders Hagfeldt ,Michael Gratzel, Chem. Rev, 95,49, 1995
3. Linsebigler A L. chem. Rev., 95,735,1995
4. Barham P J, Hill M J, Keller A, Rosney C C, J Matter. Sci. Lett., 7,1271,1988
5. Hill M J, Barham P J, Polymer, 38,5595,1997

Vibrating Nanotube–based Nano Powder Production System

I.S. Akhatov, B.Z. Jang, M. Rastgaar Aagaah, D.R. Schmidt, A.S. Mitlyng, M.B. Stewart and M. Mahinfalah

Department of MEAM, North Dakota State University
111 Dolve Hall, Fargo, ND 58105, USA, Iskander.Akhatov@ndsu.edu

ABSTRACT

A nanotube being placed in a vapor atmosphere will provoke condensation of vapor on the tip of nanotube giving rise to a nano droplet formation. Under a resonant mechanical excitation condition the nano droplet may become unstable and lose touch with the tip of the nanotube. This is how nano droplets may be created. Theoretical and experimental research to verify this idea is being conducted. It was shown that the stability of a nano droplet at the end of a carbon nanotube depends on mechanical properties of the carbon nanotube, droplet size, thermal noise, amplitude and frequency of external excitation.

Keywords: carbon nanotube, nanodroplet, ultrasound

1 BASIC IDEA

Ultra-fine particles with a narrow size distribution have enormous application potential in many areas. However, very few of the current nano powder production technologies are capable of providing nano-sized particles with a relatively narrow size distribution. In most cases, this is due to inadequate control over the nucleation, growth and subsequent agglomeration stages of the particle formation process. In this paper, we present a nano powder production method that is expected to provide a good control over the aforementioned three stages of particle formation.

Basic idea of this method is as follows: a nanotube being placed in a vapor atmosphere will provoke condensation of vapor on the tip of the nanotube giving rise to a nano droplet formation. Without intervention, this nano droplet is stabilized at the tip by surface tension and van der Waals forces (Fig.1). Under a resonant mechanical excitation condition the nano droplet may become unstable and lose touch with the tip of the nanotube. This is how nano droplets may be created.

2 BACKGROUND

Since its discovery in 1991 [1] carbon nanotubes attracted research activities in science and engineering devoted to their promising applications [2,3]. Because of their structural perfection, carbon nanotubes possess particularly outstanding physical properties. They are remarkably stiff and strong, conduct electricity, and are projected to conduct heat even better than diamond.

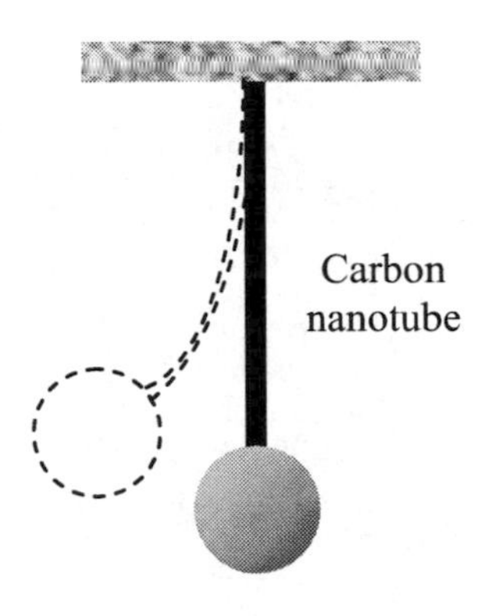

Fig. 1. Schematic of carbon nanotube loaded with a fluid nanodroplet.

Although carbon nanotubes have diameters only several times larger than the length of a bond between carbon atoms, continuum models have been found to describe their mechanical behavior very well. The application of continuum-elasticity methods shows remarkable correspondence with molecular dynamics simulations [4,5]. From continuum mechanics point of view for small deformations, nanotubes may be treated as elastic beams. The equation of motion of such a beam is (see, [6,7]):

$$\rho A \frac{\partial^2 u}{\partial t^2} + EI \frac{\partial^4 u}{\partial x^4} = f(x) \tag{1}$$

where u is the beam displacement due to its bend, ρ is the density of the material in the nanotube wall, A the cross-sectional area of nanotube, E Young's modulus, I the moment of inertia of nanotube cross-section, and $f(x)$ a distributed applied load.

Normally, in elasticity theory, Young's modulus E represents a material property applicable for any type of deformation. It should be noted here that a single value of Young's modulus cannot be uniquely used to describe bending, tension and compression of the beam. The reason is that tension, compression and bending are mainly governed by different bonds between carbon molecules. Hence, E in Eq.(1) denotes the Young's modulus for bending. Moreover, it is expected that tension/compression phenomena will play a minor role in the proposed phenomenon.

The natural frequency of an elastic beam clamped at one end and free at the other is:

$$\omega_n = \frac{\beta_n^2}{l^2}\sqrt{\frac{EI}{\rho A}} \tag{2}$$

where l is the length of the beam, β_n is the root of an equation that is dictated by boundary conditions. Taking into account that

$$A = \pi\left(R_{out}^2 - R_{in}^2\right), \quad I = \frac{\pi}{4}\left(R_{out}^4 - R_{in}^4\right) \tag{3}$$

one obtains for the natural frequency

$$\omega_n = \frac{\beta_n^2}{2l^2}\sqrt{\frac{E\left(R_{out}^2 + R_{in}^2\right)}{\rho}} \tag{4}$$

where R_{out} and R_{in} are outer and inner radii of a carbon nanotube, respectively.

The first experimental measurements of Young's modulus in multiwalled carbon nanotubes were made by Treacy et al. [8]. They took large bundles of carbon nanotubes, and attached them to the edge of a hole in nickel rings for transmission electron microscopy observations. The carbon nanotubes are so small and light that thermal motion of molecules may have substantial impact on their state leading to their intrinsic thermal vibrations. Assuming equipartition of the thermal energy among vibrating modes (Eq. (4)), it was shown that the standard deviation, Δ, of the nanotube tip due to thermally induced vibrations is related to the Young's modulus and temperature through the following equation

$$\Delta^2 = \frac{16 l^3 kT}{\pi E\left(R_{out}^4 - R_{in}^4\right)}\sum_n \beta_n^{-4} \tag{5}$$

where T is the temperature, k is the Boltzmann constant. This mean square vibration amplitude has been measured for different temperatures and the mean value of Young's modulus was found to be of 1.8 TPa with an uncertainty of ± 1.4 TPa.

In [9] an electromechanical excitation was used as a method to initiate the resonance frequencies of multiwalled carbon nanotubes. For tubes of small diameter (less than about 12 nm), they found frequencies consistent with the Young's modulus being in the range of 1 TPa. However, for larger diameters, the bending stiffness was found to decrease.

3 THEORETICAL JUSTIFICATION

To provide theoretical justification for the proposed idea we, *at first*, will estimate if the frequency range we need to work with is appropriate for use and can be achieved in the laboratory experiments. In our estimations we consider two representative examples: multiwalled carbon nanotube and single-walled carbon nanotube.

For the multiwalled carbon nanotube we take Young's modulus of E = 2 TPa and R_{out} = 5 nm, R_{in} = 2 nm. The density of multiwalled carbon nanotube is estimated as ρ = 2,150 kg/m^3 [8]. For the single-walled carbon nanotube, we take an estimated Young's modulus of E = 1 Tpa, R_{out} = 0.75 nm, and R_{in} = 0.41 nm. The thickness of nanotube wall is equal to the interlayer distance in graphite ΔR = 0.34 nm, because each individual single-walled carbon nanotube involves only a single layer of rolled graphene sheet [10]. The density of carbon nanotube is taken to be equal to that of graphite ρ = 2,500 kg/m^3 [11].

One can then use Eq. (4) to calculate resonance frequencies of nanotubes of different lengths. We consider here only the fundamental mode for which β_0 = 1.875 [7]. The results are plotted in Figure 2. It is clear from this figure that the fundamental mode is strongly dependent on the nanotube length. For nanotubes with a length between 1 μm and 1 mm the resonance frequency changes from megahertz to hertz ranges. The resonance frequency of single-walled nanotubes seams to be lower than that of multiwalled ones. In order to work with a frequency of about 20 kHz (the most commonly used frequency in laboratory experiments with ultrasound) one should use multiwalled nanotubes of about 50 μm long or single-walled carbon nanotubes of about 16 μm long. Recent advances in nanotube synthesis [12] have made possible the growth of nanotubes of such lengths.

Being very small, carbon nanotubes will be subject to thermal noise. So, as a *second* step, we would like to estimate the contribution from the vibrations of thermal origin. One can use Eq. (5) to calculate the deflection of the tip. We consider here the superposition of all vibrational modes. The results for room temperature, T = 300 K, are plotted in Figure 3. It is clear from this figure that the intensity of thermal vibrations is strongly dependent on the nanotube length. For nanotubes of length between 1 μm and 1 mm the amplitude of thermal noise vibrations changes from nanometer to millimeter ranges.

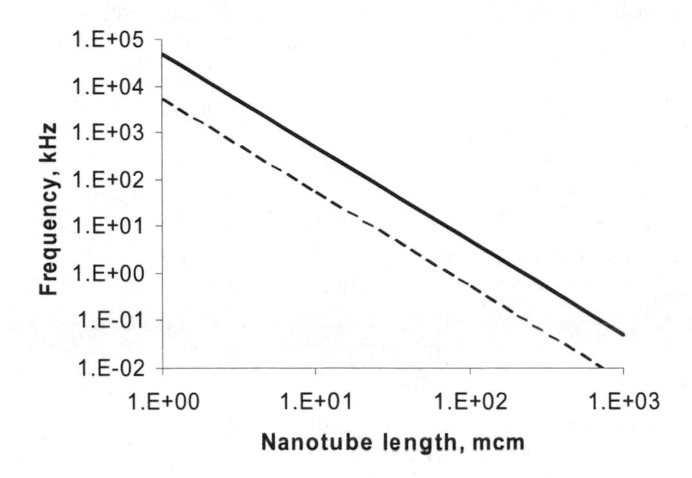

Fig. 2. Resonant frequencies of carbon nanotubes clamped at one end and free at the other. Solid line: multiwalled carbon nanotube (R_{out} = 5 nm, R_{in} = 2 nm, E = 2 TPa, ρ = 2,150 kg/m^3). Dashed line: single-walled carbon nanotube (R_{out} = 0.75 nm, R_{in} = 0.41 nm, E = 1 TPa, ρ = 2,500 kg/m^3).

Thermal vibrations seem to be more pronounced for single-walled nanotubes than for multiwalled ones. At a frequency of about 20 kHz, which corresponds to the multiwalled nanotube of about 50 μm in length and single-walled carbon nanotube of about 16 μm in length, one should expect the amplitudes of thermal vibrations of about 0.4 μm and 5 μm, respectively. This is an indication that one should perform a detailed investigation of the influence of thermal noise on the acoustically induced nanotube oscillations.

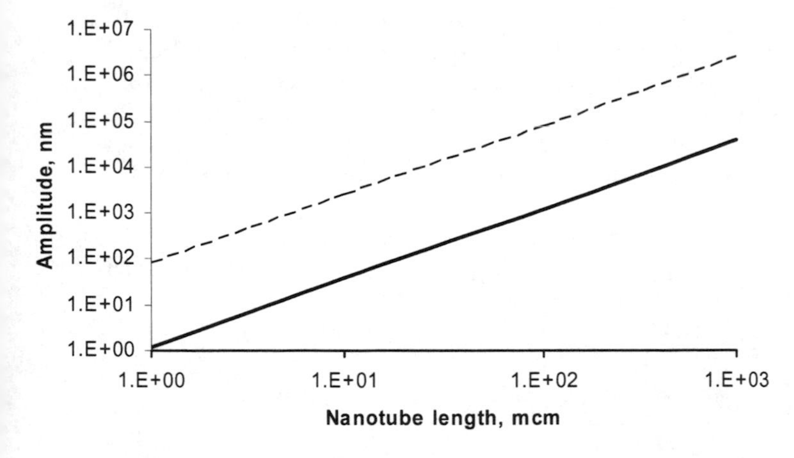

Fig. 3. Thermal vibration amplitudes of carbon nanotubes clamped at one end and free at the other. Solid line: multiwalled carbon nanotube (R_{out} = 5 nm, R_{in} = 2 nm, E = 2 TPa, ρ = 2150 kg/m^3). Dashed line: single-walled carbon nanotube (R_{out} = 0.75 nm, R_{in} = 0.41 nm, E = 1 TPa, ρ = 2,500 kg/m^3).

Although the amplitude of thermal oscillations is substantial, the question arises if those oscillations will play important role in the stability of a nano droplet attached to the tip of nanotube. In other words, will these oscillations have enough intensity to rip nano droplets off the tip? Hence, as the *third* step, we need to estimate the forces that a nano droplet at the tip experiences due to thermal vibrations of the nanotube.

The force that tends to keep the nano droplet at the tip is surface tension force which is estimated as

$$F_\sigma = 2\pi R_{out}\sigma \tag{6}$$

where σ is the surface tension coefficient of the fluid. When the tip of a nanotube performs oscillations with an amplitude Δ and frequency ω_0, it moves with a periodic acceleration of which the amplitude is about $\omega_0^2\Delta$ (here the frequency of the fundamental mode is used for the estimation purposes).

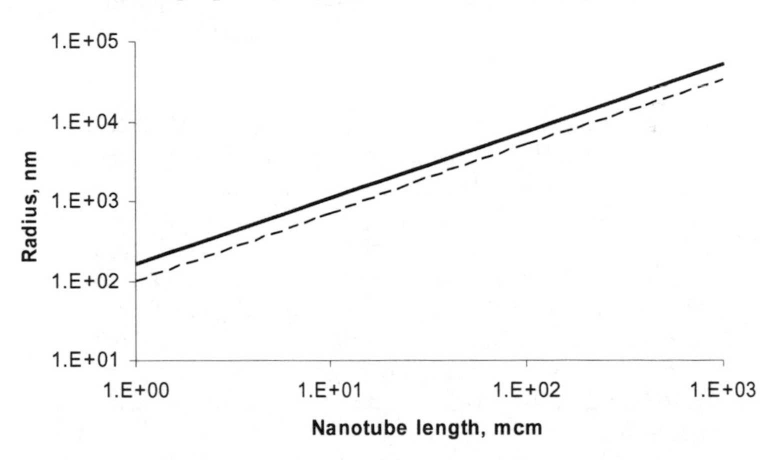

Fig. 4. Critical radii of liquid the droplets attached at the free end of carbon nanotubes. Solid line: multiwalled carbon nanotube (R_{out} = 5 nm, R_{in} = 2 nm, E = 2 TPa, ρ = 2150 kg/m^3). Dashed line: single-walled carbon nanotube (R_{out} = 0.75 nm, R_{in} = 0.41 nm, E = 1 TPa, ρ = 2,500 kg/m^3).

Naturally, the fluid droplet attached to the tip experiences the inertia force of

$$F_i = \frac{4}{3}\pi R_d^3 \rho_d (2\pi f_0)^2 \Delta \tag{7}$$

and the gravity force of

$$F_g = \frac{4}{3}\pi R_d^3 \rho_d g \tag{8}$$

Here R_d and ρ_d is the droplet radius and density of the fluid, g is the gravitational acceleration.

Assuming, for estimation purposes, that all these forces act along the vertical line, we come to the stability condition:

$$F_\sigma > F_i + F_g \quad (9)$$

The Eq. (9) leads to the equation for a critical droplet radius

$$R_d = \left[\frac{3R_{out}\sigma}{2\rho_d\left(4\pi^2 f_0^2 \Delta + g\right)} \right]^{1/3} \quad (10)$$

which means that the droplets of radii larger than R_d will be ripped off from the tip of nanotube.

Having in mind that the resonance frequency f_0 and amplitude of thermal vibrations Δ both depend on the nanotube length, one can estimate the critical droplet radius as a function of the nanotube length. The results are shown in Figure 4. Here for estimation purposes the density and surface tension coefficient of water are used. For nanotubes of length between 1 μm and 1 mm the critical radius changes from hundred nanometers to centimeter ranges. Although thermal vibrations may have strong impact on the amplitude of vibrations of the nanotube, as shown above, they do not have enough intensity to destroy the stability of a nano droplet attached to a nanotube tip. Namely, at a frequency of 20 kHz which corresponds to a multiwalled nanotube of about 50 μm in length and single-walled carbon nanotube of about 16 μm in length, thermal vibrations may rip off the droplets from the tip only if they are relatively large: about 4 μm for the case of a multiwalled nanotube, and 1μm for single-walled nanotube.

That is why we suggest using the acoustically induced nanotube oscillations to create nano droplets. Of course, the resonance properties of a nanotube is expected to be affected by the nano droplet attached to its end, heat exchange between the nanotube and the nano droplet, and the kinetics of condensation, etc. These issues will be addressed in the future research.

4 NANO POWDER PRODUCTION SYSTEM

Herein proposed is a vibrating nano-tube or nano-rod based nano powder production system that promises to deliver a better control over the nano particle size distribution. This system involves a planar array of carbon nanotubes that are suspended from a vibrating plate to form a particle formation zone. A stream of metal vapor carried by a carrier gas is directed to flow into this zone in which the vapor condenses onto the surface of nanotubes. The thermally controlled nanotube surface serves as a preferred site to promote well-controlled heterogeneous nucleation. The growth of nuclei is controllably limited by the size of the nanotube and the vibration frequency and amplitude of the nanotube. The droplets grown to a controllably specified size are continuously shaken off the nanotubes. This step of shaking-off also acts to prevent the condensed particles from aggregating or coalescing among themselves on the nanotube surfaces. The shaken-off droplets are directed to flow through a cooling or passivating zone to further prevent agglomeration. The technological goal is to develop a method of using carbon nanotubes to produce nano particles of controllable sizes and high monodispersity.

Support by NSF is gratefully acknowledged.

REFERENCES

[1] Iijima S., (1991), Helical microtubules of graphitic carbon, *Nature* **354** (6348): 56-58.

[2] Goddard W. A., III *et al.*, Eds. (2003), *Handbook of Nanoscience, Engineering, and Technology*, CRC Press.

[3] Goldstein A. N., (1997), *Handbook of Nanophase Materials*, Marcel Dekker Inc., NY.

[4] Yakobson B.I., Smalley R.E., (1997), Fullerene nanotubes: $C_{1,000,000}$ and beyond, *American Scientist* **85**: 324-337.

[5] Yakobson B.I., Brabec C.J., Bernholc J., (1996), Structural mechanics of carbon nanotubes: From continuum elasticity to atomic fracture, *Journal of Computer-Aided Materials Design* **3**: 173-182.

[6] Timoshenko S., Gere J., (1988), *Theory of Elastic Stability*. McGraw-Hill. New York.

[7] Meirovich L., (1986), *Elements of Vibration Analysis*. McGraw-Hill. New York.

[8] Treacy M.M.J., Ebbesen T.W., Gibson J.M., (1996), Exceptionally high Young's modulus observed for individual carbon nanotubes, *Nature* **381**: 678-680.

[9] Poncharal P., Wang Z.L., Ugarte D., Heer de W.A., (1999), Electrostatic deflections and electromechanical resonances of carbon nanotubes, *Science* **283**: 1513-1516.

[10] Yu M.F., Files B.S., Arepalli S., Ruoff R.S., (2000), Tensile loading of ropes of single-wall carbon nanotubes and their mechanical properties, *Phys. Rev. Lett.* **84**: 5552-5555.

[11] Weast R.C., Selby S.M., Hodgman C.D. (eds), (1965), *Handbook of Chemistry and Physics*. The Chemical Rubber Co. Cleveland.

[12] Journet C. et al., (1997), *Nature* **388**: 756.

Room-Temperature Synthesis and Characterization of Highly Monodisperse Transition Metal-Doped ZnO Nanocrystals

S. P. Singh, O. Perales-Perez, M.S. Tomar, A. Parra-Palomino, A. Ruiz–Mendoza

University of Puerto Rico, Mayagüez, PR 00681-9044, oscar@ge.uprm.edu

ABSTRACT

Recent verifications of intrinsic room-temperature (RT) ferromagnetism in transition metal doped-ZnO have increased its attractiveness as promising material for nano-optoelectronic and spintronics-based devices. A control over dopant speciation and the determination of the size-dependence of the properties at the nanoscale, become then indispensable. We present here the conditions for the room-temperature synthesis in ethanol and characterization of bare, Mn- and Co-doped ZnO nanocrystals. The results evidenced the viability on producing highly monodisperse nanocrystals (5-8nm) with no need for any further thermal treatment. However, the formation of the ZnO structure was delayed when dopant ions co-existed with Zn in starting solutions; well-crystallized doped nanoparticles were produced only after their aging in mother liquors. SQUID measurements on doped nanocrystals evidenced a weak, though noticeable RT-ferromagnetism, or paramagnetism, depending on synthesis conditions.

Keywords: zinc oxide, nanocrystals, room-temperature synthesis, diluted magnetic semiconductors, spintronics

1 INTRODUCTION

The prediction of room temperature ferromagnetism [1, 2] in transition metal doped II-VI nonmagnetic semiconductors leads to a number of magnetic, magneto-optical, magneto-transport and optical properties, which make these structures ideal candidates for spintronic-based devices. Sato et al [3] have suggested the possibility of room temperature ferromagnetism in transition metal-doped ZnO structures, where Mn and Co were found to be suitable dopants. In this regard, ferromagnetism with Curie temperature above 280 K in thin films [4] and large magneto-optic effect [5] has been reported for ZnCoO. Moreover, the recent experimental verifications of the intrinsic ferromagnetism above room temperature in transition metal doped-ZnO nanostructures [6-8] enable these types of materials to be considered for nano-optoelectronic and spintronics-based applications. Accordingly, the control over dopant speciation and the determination of the size-dependence of functional properties at the nanoscale become indispensable.

Nowadays, the production of ZnO-based structures has been attempted through different physical and chemical means. Among them, solution processing routes are based on low-cost, environmental friendly and easily scalable methods. This option also permits the control of the solid formation rate leading to the production of metastable phases, tailored phase compositions and well-tuned particle size at the nanoscale. On these bases, the present work is focused on the synthesis in ethanol solutions of bare and Mn- or Co-doped ZnO nanocrystals at room temperature. The influence of the concentration of dopant species in starting solutions on the stability conditions of the host oxide structure is discussed. Furthermore, the structural, optical and magnetic characterization of the highly monodisperse nanocrystals synthesized by the selected method is also presented.

2 EXPERIMENTAL

2.1 Materials

All reagents were of analytical grade and were used without further purification. Required weighs of the acetate salts of zinc and manganese or cobalt were dissolved in anhydrous ethanol at 65°C to produce a solution with a total concentration of metal ions of 0.01M. A solution of LiOH monohydrate in ethanol was used as the precipitating agent. N-heptane was used to coagulate nanocrystals from their suspensions in ethanol.

2.2 Synthesis of Nanocrystals

Doped ZnO nanocrystals were synthesized by conventional precipitation in ethanol solutions as reported by Spanhel and Anderson for bare ZnO [9]. In our case, the syntheses were carried out under ambient-temperature conditions. When the synthesis of doped-ZnO was attempted, the metal ions solutions were prepared for a specific fraction of dopant coexisting with Zn ions in starting solutions, 'x'. The metals and the hydroxide solutions in ethanol were mixed at room temperature under vigorous stirring for 10 minutes ('contact stage'). In order to follow the progress of the oxide formation, the suspension of nanocrystals were aged in their mother liquors at room temperature and without stirring ('aging stage'). Nanocrystals prepared in this way were recovered by successive coagulation/redispersion cycles using n-heptane/fresh ethanol and submitted for characterization. The present synthesis procedure is simple and permits a very fast nucleation with minimum aggregation of the crystals at a temperature as low as 25°C. Ethanol was selected not only as a suitable solvent but, specifically, because of its de-hydrating property. This de-hydrating capability plays an essential role in the removal of coordinated water from the precursor compound (basic zinc acetate) that promoted the formation of anhydrous oxide structures at room-temperature.

2.3 Products Characterization

Dried powders were submitted for characterization by using XRD, FT-IR, and SQUID techniques. UV/Vis absorbance spectra were obtained from suspensions of nanocrystals in ethanol. HRTEM was used to determine the size and crystallinity of produced crystals.

3 RESULTS AND DISCUSSION

3.1 XRD Analyses

As evidenced by the XRD patterns shown in Figure 1, ZnO having a wurtzite structure was rapidly formed in ethanol even without need of aging (t=0 min). Furthermore, the broadness of the XRD peaks reveals the nanocrystalline nature of the powders. The average crystallite size was estimated to be ~ 6 nm using Scherrer's equation. Although there was not a noticeable sharpening of the ZnO peaks for aged samples, initially clear suspension of nanocrystals turned out whitish with time, which could be attributed to the growth and/or aggregation of ZnO crystals.

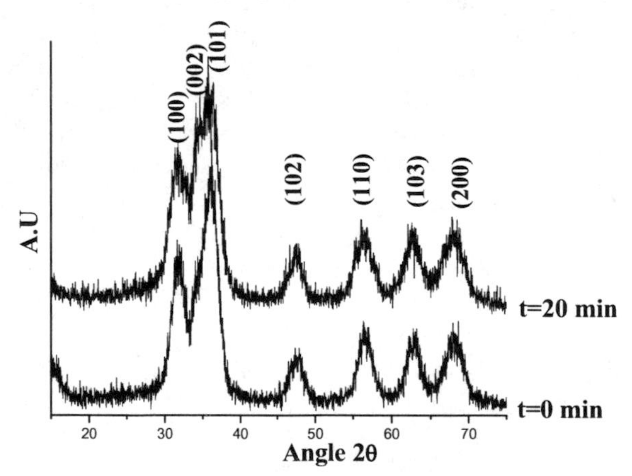

Figure 1: XRD patterns for ZnO powders produced in ethanol after 0 minutes (i. e., no aging) and 20 minutes of aging at 25°C.

XRD analyses also evidenced that the presence of dopant species ('impurities') delayed the formation of the host ZnO structure. Figures 2 and 3 show the XRD patterns for Mn- and Co-doped ZnO powders aged for 30 minutes and 24 hours, respectively, and different fractions of dopant species in starting solutions ('x' values). In both cases it was found that the higher the 'x' values, the more stable the basic zinc acetate intermediate (marked with ▼ in the corresponding figures).

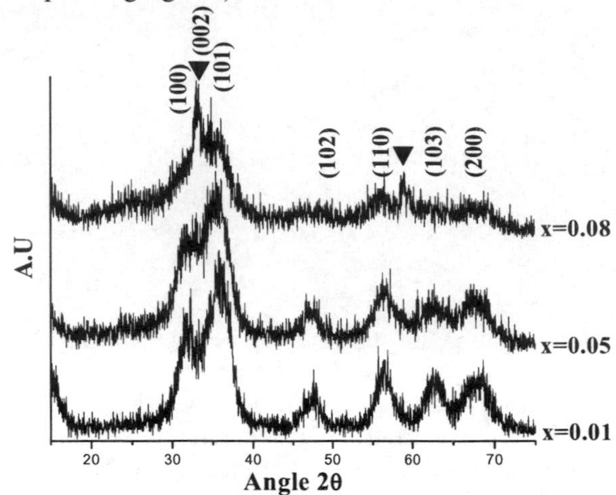

Figure 2: XRD patterns for ZnO powders produced for different fractions of Mn ions in starting solutions, 'x', and 30 minutes of aging at 25°C.

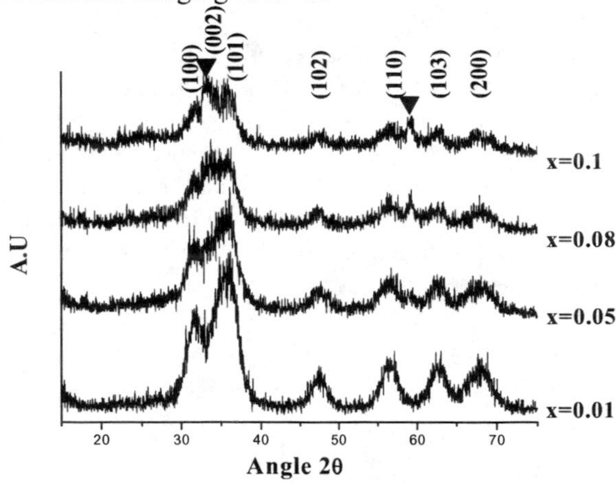

Figure 3: XRD patterns for ZnO powders produced for different fractions of Co ions in starting solutions, 'x', and 24 hours of aging at 25°C.

In doped systems, the complete conversion of the intermediate phase into the host ZnO structure could be achieved after aging of the suspension of nanocrystals. The duration of the aging stage was dependent on the fraction of dopants; longer aging times were required for higher 'x' values. For instance, Co-doped ZnO nanocrystals (x=0.01) had to be aged for 30 minutes to develop the oxide structure. In the Mn-doped ZnO system (x=0.08), the complete conversion of the intermediate into the oxide phase was only possible after 24 hours of aging. Dopant species trying to get incorporated into the ZnO lattice would have generated internal stresses and some lattice distortion. This incorporation will make the oxide structure unstable retarding its formation: *a pure phase always forms faster than an impure one.* Long enough aging times will favor the atomic rearrangement and dehydration conducive to the formation of the oxide structure at room temperature.

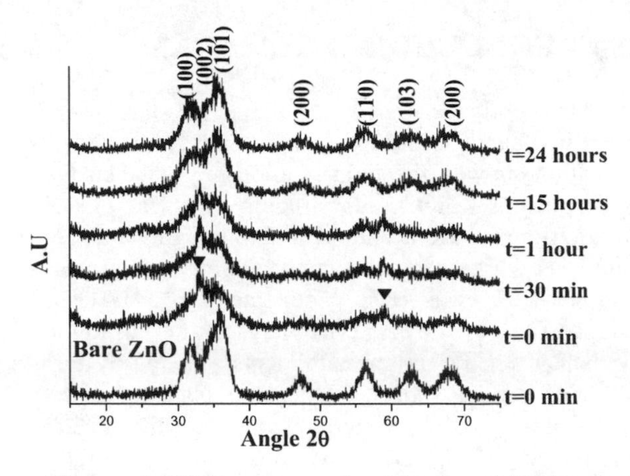

Figure 4: XRD patterns for Mn-doped ZnO powders produced at different aging times ('t' in minutes or hours) and x = 0.08.

3.2 FT-IR Analyses

Figure 5 shows the FT-IR spectra of Mn-ZnO crystals (x=0.08) aged for different times. The systematic increase in the intensity of the band at 525cm^{-1}, assigned to the stretching vibrations of the Zn-O bonds, indicated the favorable effect of aging on the development of the host ZnO structure. In turn, the bands at 1342cm^{-1}, 1420cm^{-1} and 1574cm^{-1} can be assigned to the stretching vibration of C=O, C=C and C-H groups in acetate species. The acetate species were detected even after intensive washing of the solids with ethanol, suggesting their presence as adsorbed species.

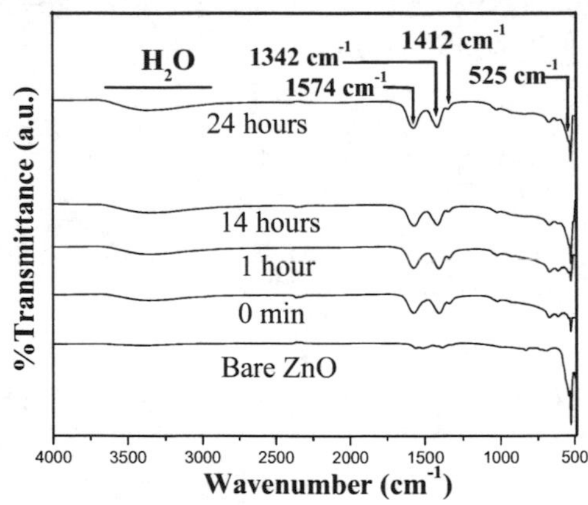

Figure 5: FT-IR spectra for Mn-doped ZnO powders produced at different aging times and x = 0.08. The spectrum for bare ZnO corresponds to a micron-size commercial powder.

3.3 HRTEM Observations

Figure 6 show the high resolution TEM pictures of ZnO nanocrystals after 4 weeks of aging at room temperature. Darker particles show stronger diffraction due to orientation effect. The nanoparticles were highly monodisperse, with sizes between 6 and 8 nm, and well-crystallized. Nanometric sizes of the particles are attributed to the very fast nucleation rate capable to be reached during synthesis in ethanol. Also, adsorbed acetate species must have provided a negative charge to the surface of the nanocrystals that prevented their aggregation.

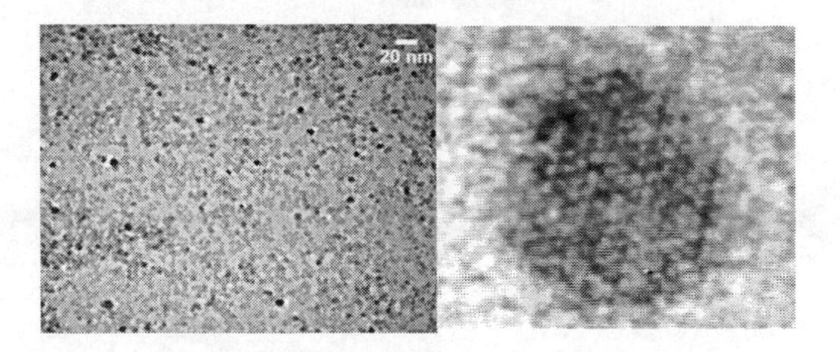

Figure 6: HRTEM pictures of ZnO nanocrystals produced after 4 weeks of aging.

3.4 UV-Vis Measurements

The read shift in absorption peak with aging observed in the room temperature UV-Vis absorption measurements for ZnO nanocrystals, (Figure 7), was attributed to crystal growth. Indicated wavelengths correspond to $\lambda_{1/2}$ values, from which the values of the band gap energy of the inset were estimated. All band gap energy values were above 3.28 eV (for bulk ZnO), evidencing the nanocrystalline nature of aged particles. The end of the growth process is evidenced by the minimum shift in the UV-Vis spectra after 48 hours of aging.

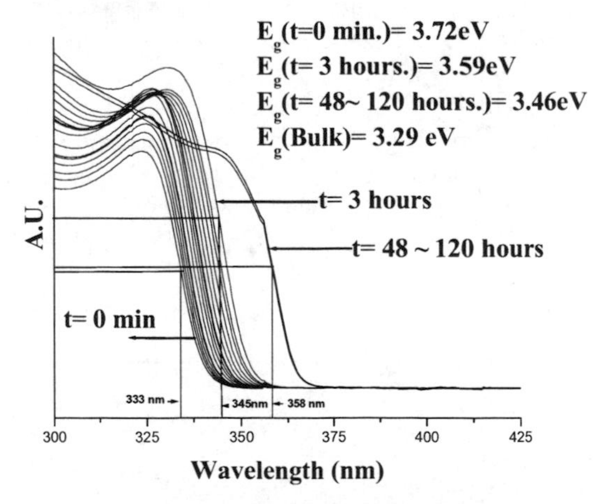

Figure 7: UV/Vis spectra for suspensions of ZnO nanocrystals aged at room temperature.

The crystal growth observed during aging could be stopped by coagulating the nanocrystals with n-heptane and subsequent redispersion in fresh ethanol. To verify the restriction on crystal growth, UV-Vis absorption measurements were carried out on redispersed nanocrystals coagulated after 5 minutes, 1.5 hours and 3 hours of aging (Figure 8). Each curve in this figure corresponds to five observations performed in 15 minutes intervals. The inhibition in crystal growth was evidenced by the invariable position of the absorption peak for each sample.

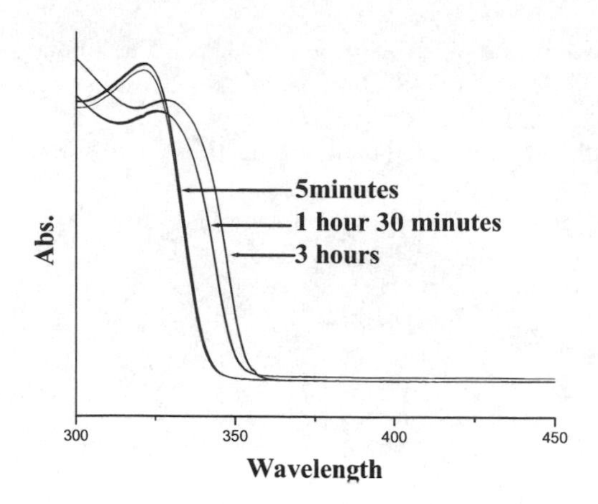

Figure 8: UV/Vis spectra for ZnO nanocrystals coagulated after 5 minutes, 1.5 hours and 3 hours of aging and redispersed in fresh ethanol.

3.5 SQUID Measurements

The results of SQUID analyses suggested the actual incorporation of Mn ions into the otherwise diamagnetic ZnO structure. The Mn-doped ZnO nanocrystals (x=0.05, 24 hours aging) exhibited a room temperature paramagnetic behavior, whereas a very weak coercivity (~15Oe) was observed at 5K. In good agreement with observed behavior no blocking temperature was observed within 2K and 300K in ZFC/FC measurements under an applied field of 100Oe. Similar characteristics were observed in other Mn-ZnO samples synthesized for different 'x' values.

Figure 9 shows the M-H curves at 300K corresponding to as-synthesized Co-doped ZnO nanocrystals (x = 0.01, 30 min aging). The inset is a magnification of the same figure and shows a small, though noticeable, coercivity which suggests a weak ferromagnetism at room temperature. More recent results have evidenced room temperature ferromagnetism in as-synthesized free-standing nanocrystals with a coercivity as high as 150Oe. Detailed work will be published elsewhere.

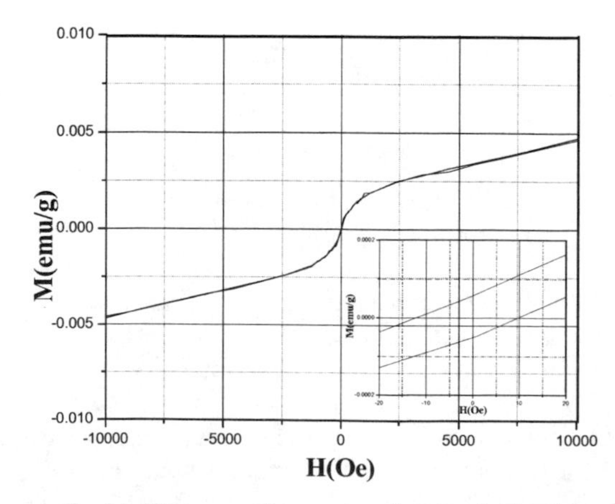

Figure 9: M-H curve for as-synthesized Co-doped ZnO nanocrystals (x = 0.01, 30 min aging). The inset is a magnification of the same figure.

4 CONCLUDING REMARKS

ZnO, Mn- and Co-doped ZnO nanocrystals were synthesized in ethanol solutions at room temperature. XRD characterization showed the tight relationship between the stability of the host ZnO oxide structure, the fraction of dopants in starting solutions and the length of aging at room temperature. The high monodispersity and crystallinity of ZnO nanoparticles was also confirmed by HRTEM observations. The growth of the nanocrystals during aging was revealed by UV-Vis measurements, though a fast coagulation of nanocrystals by n-heptane interrupted the growth process. SQUID analyses suggested the actual incorporation of Mn and Co ions into otherwise diamagnetic ZnO structure. Doped nanocrystals exhibited a predominant paramagnetic behavior at room-temperature, although a small coercivity became evident for the as-synthesized Co-ZnO nanocrystals. Ongoing works are focused on the synthesis of doped ZnO crystals having different sizes and verification of the crystal size dependence of ferromagnetism for a particular composition.

This material is based upon work supported by the National Science Foundation under Grant No. 0351449. Any opinions, findings and conclusions or recommendations expressed in this material are those of the author(s) and do not necessarily reflect the views of the National Science Foundation (NSF). Thanks are also extended to PRSGC-IDEAS-NASA Program for providing funds for the initial part of this research. HRTEM pictures were kindly taken by Paul Voyles at University of Wisconsin-Madison. We also appreciate the contribution from Dr. C. Rinaldi with SQUID measurements.

REFERENCES

[1] T. Dietl, H. Ohno, F. Matsukura, J. Cibert and D. Ferrand, Science 287, 1019, 2000

[2] T. Dietl, H. Ohno and F. Matsukura, Phys. Rev. B 63, 195205, 2001.

[3] K. Sato and H. Katayama-Yoshida, Jpn. J. Appl. Phys. 39, L555, 2000.

[4] K. Ueda, H. Tabata and T. Kawai, Appl.Phys.Lett. 79, 988, 2001.

[5] K. Ando, H. Saito, Z. Jin, T. Fukumura, M. Kawasaki, Y. Matsumoto and H. Koinuma, Appl. Phys. Lett. 78, 2700, 2001.

[6] D.A. Schwartz, N. S. Norberg, Q.P. Nguyen, J.M. Parker and D.R. Gamelin, J. Am. Chem. Soc., 125, 13205, 2003.

[7] P. V. Radonvanovic, N. S. Norberg, K. E. McNally and D. R. Gamelin, J. Am. Chem. Soc. 124, 15192, 2002.

[8] P. V. Radonvanovic and D. R. Gamelin, Phys. Rev. Lett., 91,15, 157202-1, 2003

[9] L. Sphanhel and M. A. Anderson, J. Am. Chem. Soc., 113, 2826, 1991.

Low Temperature Synthesis of Nanosized Metal Oxides by a Supercritical Seed Enhanced Crystallization process (SSEC)

Henrik Jensen[1], Karsten D. Joensen[2], Steen B. Iversen[3], and Erik G. Søgaard[1]

[1]Aalborg University, Department of Chemical Engineering,
Niels Bohrs Vej 8, 6700 Esbjerg, Denmark
[2]JJ X-Ray Systems ApS, Gl. Skovlundevej 54, 2740 Skovlunde Denmark
[3]SCF Technologies A/S, Gl. Køge Landevej 22, Building H, 2500 Valby, Denmark

E-mail: hensen98@aaue.dk, Phone: +45 30569062, Fax: +45 79127677

Summary
A novel low temperature method for the production of homogenous nanosized crystalline materials has been developed. The method is based on a Supercritical Seed Enhanced Crystallization (SSEC) process and has been shown to lower the crystallization temperature for several metal oxides significantly. The novel method is flexible and allows production of homogenous nanocrystalline powders with controlled properties from a wide range of metal oxides. The produced nanopowders have an average particle size of 5-10 nm with a very narrow size distribution. The crystallinity and particle size can be altered by adjusting the process parameters. The novel method allows direct production of dry crystalline nanopowders without the need for additional treatment. The process time is short (0.5 to 4.0 hrs.).

Introduction
The production of nanosized materials is of great interest in both industries as well as in academic research. Despite progress in scaling up and reducing costs, nanoparticles remain relatively expensive materials and the price depends on the production volume, material type, powder characteristics, manufacturing method, and post synthesis processing treatment [1]. The applications for nanosized metal oxides nanoparticles span a wide range of different areas such as UV-protection, plastics, cosmetics, solar cells, dental composites, and photocatalysis. Particle size, specific surface area, crystal phase, crystallinity, and surface properties are considered as key parameters for success in these applications.

In the SSEC process a sol-gel reaction is taking place in a supercritical environment in the proximity of a seeding material [2]. The sol-gel reaction starts with the hydrolysis of the precursor, normally a metal alkoxide ($M(OR)_n$), when it comes into contact with water. The hydrolysis continues simultaneously with the condensation of the hydrolyzed monomers leading to formation of a three dimensional metal oxide network [3]:

$$M(OR)_n + mH_2O \rightarrow M(OR)_{n-m}(OH)_m + mROH$$
$$M(OR)_{n-m}(OH)_m + RO\text{-}M(OR)_{n-m-1}(OH)_m \rightarrow (OR)_{n-m}M\text{-}O\text{-}M(OR)_{n-m-1}(OH)_m + ROH$$

A filling material is introduced in the process to enable the production and collection of nanosized crystalline particles. The filling material acts as seed or catalyst as well as a reservoir for collecting the nanoparticles. The filling material enables homogenous seeding of primary nuclei resulting in a nanosized powder with narrow size distribution.

Supercritical CO_2 is used as solvent in the SSEC process. Using supercritical fluids as solvents in sol-gel processes rather than the traditional alcohols, a significantly lower particle size in the nanometer range can be obtained. This is believed to be due to the higher reaction rate obtained in supercritical media [4].
Producing nanosized metal-oxides by the SSEC process the crystallization temperature is lowered significantly and a reduction of the process time is obtained. Furthermore, the nanopowders produced from the SSEC process do not need a succeeding calcination, which normally causes a reduction in the specific surface area.

In this paper it will be shown that the SSEC process is a promising method for synthesizing nanosized metal-oxides with controllable particle size, crystal phase, and degree of crystallinity without having to resort to costly post-reaction processing.

Materials and Methods

The experimental set-up used in this study is shown on Figure 1 together with a schematic representation of the process.

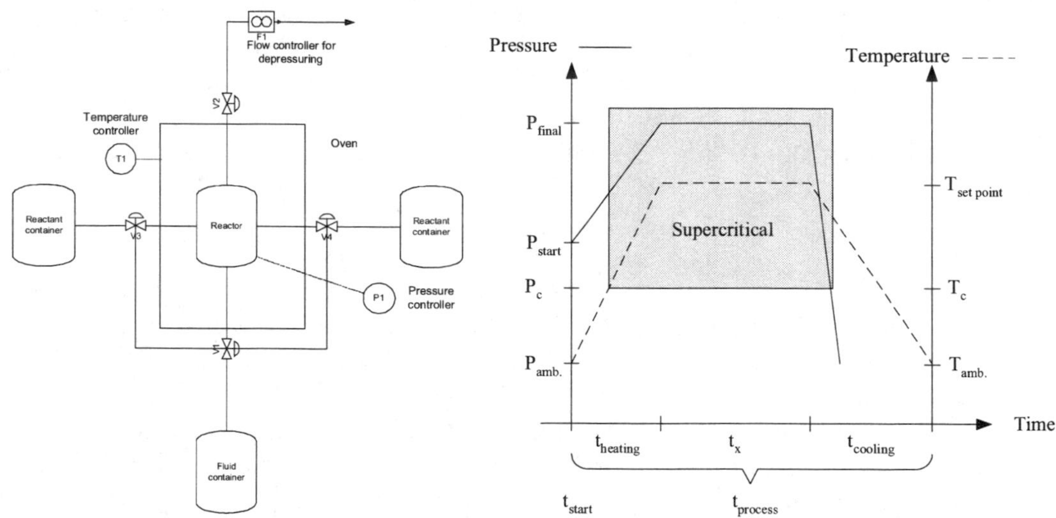

Figure 1 Experimental set-up and schematic representation of the SSEC process.

The central part of the experimental set-up is the reactor which is placed in a pressure safe oven. The experimental set-up can operate in the range from 0-680 bar and from 25-200 °C. The process is occurring in a supercritical environment, which for CO_2 means above 31 °C and 74 bar, which schematically is shown on Figure 1.
The produced powders are characterized by XRD to determine the crystal phase, the absolute crystallinity, and the crystallite size distribution. SAXS is used to determine the particle size distribution of the produced powders. The hard sphere model with an interparticle interference effect is used to model the SAXS data. The characterization methods are described in more details in [5-7].

Results and Discussion

XRD and SAXS spectra obtained from SSEC synthesized TiO_2 are shown in Figure 2. The characterized powder was synthesized with titaniumisopropoxide as precursor and the

NSTI-Nanotech 2005, www.nsti.org, ISBN 0-9767985-1-4 Vol. 2, 2005

experimental conditions were 100 bar and 100 °C. These conditions will be referred to as standard or reference conditions.

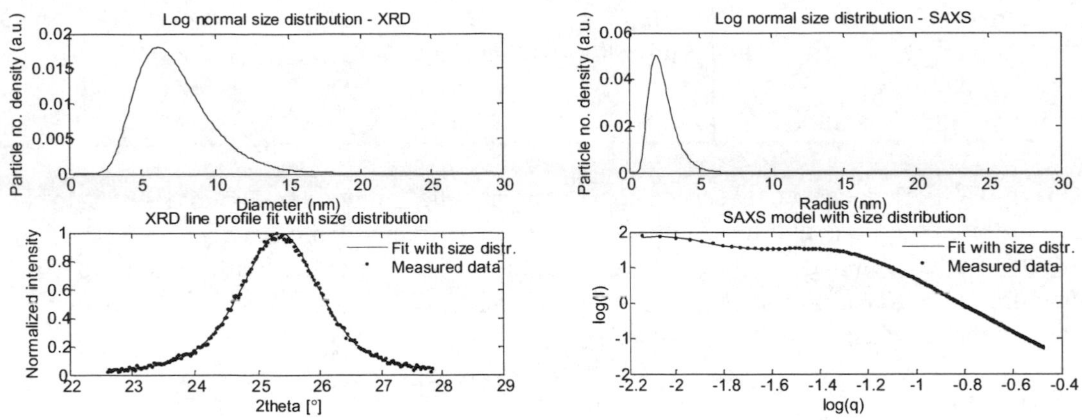

Figure 2 A: XRD spectrum of TiO_2 powder. B: SAXS spectrum of TiO_2 powder

The XRD spectrum in Figure 2A is a close up of the 100 % peak of anatase. Analyzing this peak by a size dependent XRD line profile model shows that TiO_2 prepared at reference conditions has an average crystallite size of 7 nm and a narrow size distribution. The prepared TiO_2 is a pure anatase phase with an absolute crystallinity of 63 % and the remaining part being amorphous titania. The absolute crystallinity is determined with respect to a 100 % crystalline CaF_2 reference.

SAXS analysis of the produced TiO_2 powder shows an average particle size of 4.4 nm. The SAXS analysis reveals that the powder is composed of nanosized primary particles aggregated in larger clusters, which is shown in the SAXS model by the interparticle interference effect.

The properties of the TiO_2 produced at reference conditions are in Table 1 compared to the commercial powder Degussa P25.

Table 1 Measured properties for TiO_2 powders.

	SSEC	Degussa P25
Average crystallite size	6.9 nm	25.7 nm
Average particle size	4.4 nm	18.8 nm
Average particle size (BET)	6.4 nm (236 m^2/g)	30 nm (50 m^2/g)
Crystallinity (anatase)	63 %	~ 100 %[a]

[a] Mixture of anatase (80 %) and rutile (20 %)

The SSEC produced TiO_2 has a pure anatase phase, while Degussa P25 also contains the rutile phase. The crystallite size and particle size of the SSEC produced powder is ~ 4 times smaller than for Degussa P25. Furthermore, in contrary to the normal sol-gel process the SSEC process allows for production of nanopowders without impurities such as residual alkoxy groups [8].

The SSEC process has proven suitable for synthesis of several crystalline metal oxides and hydroxides other than TiO_2. In Table 2 the properties of different metal oxides are shown together with two TiO_2 powders prepared at different temperatures.

Table 2 Measured particle properties for different metal oxides.

	TiO_2	TiO_2	ZrO_2	AlOOH	GeO_2
Temperature	45 °C	100 °C	185 °C	175 °C	185 °C
Pressure	100 bar	100 bar	100 bar	100 bar	100 bar
Crystal phase	Amorphous	Anatase	Tetragonal	Boehmite	Tetragonal
Crystallinity	–	63 %	39 %	94 %	72 %
Average Crystallite Size	–	6.9 nm	6.1 nm	3.7 nm	13.2 nm

From Table 2 it can be observed that changing the temperature from 100 °C to 45 °C an amorphous titania is obtained with a particle size extracted from SAXS of 3.6 nm. Crystalline ZrO_2, AlOOH, and GeO_2 were synthesized at low temperatures with crystallite sizes in the nanorange. Fully amorphous ZrO_2, AlOOH, and GeO_2 are also obtained if the temperature is below the temperatures shown in Table 2.

Conclusion

The results presented in this paper show that nanocrystalline TiO_2, ZrO_2, AlOOH, and GeO_2 can be synthesized by the SSEC process. The crystallization temperatures were lowered significantly compared to traditional sol-gel processes. This alleviates the need for subsequent calcination. The produced homogenous powders consisted of nanocrystallites with an average particle size of approx. 5-10 nm and a very narrow size distribution.

Acknowledgement

The authors would like to thank our colleagues at Department of Chemistry at Aarhus University for help with characterization of the produced nanoparticles.

References

[1] M. N. Rittner. "Market Analysis of Nanostructured Materials", Am. Ceram. Soc. Bull. 81 (3), (2002)

[2] H. Jensen and E. G. Søgaard. "Method for Production of a Product having Sub-Micron Primary Particle Size, Product Produced by the Method and Apparatus for use of the Method." Patent WO2004001278. (2003).

[3] J. Livage, M. Henry and C. Sanchez. "Sol-Gel Chemistry of Transition Metal Oxides", Prog. Solid St. Chemistry 18: 259-341, (1988)

[4] J. Jung and M. Perrut. "Particle design using supercritical fluids: Literature and patent survey", J. Supercrit. Fluids 20: 179–219, (2001)

[5] H. Jensen, K. D. Joensen, J. E. Jørgensen, J. S. Pedersen and E. G. Søgaard. "Characterization of Nanosized Partly Crystalline Photocatalysts", J. Nanopart. Res. 6: 519-526, (2004)

[6] H. Jensen, J. H. Pedersen, K. D. Joensen, J. E. Jørgensen, J. S. Pedersen, S. B. Iversen and E. G. Søgaard. "Determination of Size Distributions in Nanocrystalline Powders by TEM, XRD, and SAXS", In Preparation, (2005)

[7] J. S. Pedersen. "Modelling of Small-Angle Scattering Data from Colloids and Polymer Systems". X-Rays and Light: 391-420, P. Lidner and T. Zemb, Elsevier, (2002).

[8] H. Jensen, A. Soloviev, Z. Lie and E. G. Søgaard. "XPS and FTIR Investigation of the Surface Properties of Different Prepared Titania Nano-Powders", Appl. Surf. Sci. - In Press, (2005)

Surface-Enhanced Raman Scattering in Small Noble-Metal Nanoparticles

V. N. Pustovit and T. V. Shahbazyan

Department of Physics and Computational Center for Molecular Structure and Interactions, Jackson State University, Jackson MS 39217

ABSTRACT

We present a microscopic theory of quantum-size effects in surface-enhanced Raman scattering (SERS) from molecules adsorbed on small metal nanoparticles. In noble-metal nanoparticles, the confining potential has different effect on s-band and d-band electrons. Namely, the spillout of delocalized sp-electrons beyond the classical nanoparticle boundary results in an incomplete embedding of s-electron distribution in the background of localized d-electrons whose density profile follows more closely the classical shape. We demonstrate that a reduction of d-electron screening in the surface layer leads to the enhancement of the surface plasmon local field acting on a molecule located in a close proximity to metal surface. Our numerical calculations of Raman enhancement factor, performed using time-dependent local density approximation, show additional enhancement of the Raman signal which becomes more pronounced for small nanoparticles due to the larger ratio of surface layer to overall nanoparticle size.

Keywords: nanoparticles, Raman scattering, surface plasmon

1 INTRODUCTION

An recent interest in single-molecule surface-enhanced Raman scattering (SERS) stems from the discovery of enormously high (up to 10^{15}) enhancement of Raman spectra from certain (e.g., Rhodamine 6G) molecules fixed at the nanoparticle surfaces in gold and silver colloids [1]–[3]. Major SERS mechanisms include electromagnetic (EM) enhancement by surface plasmon (SP) local field near the metal surface [4]–[8] and chemical enhancement due to dynamical charge transfer between a nanoparticle and a molecule [9]–[12]. Although the origin of this phenomenon has not completely been elucidated so far, the EM enhancement was demonstrated to play the dominant role, especially in dimer systems when the molecule is located in the gap between two closely spaced nanoparticles [10]–[13].

An accurate determination of the SERS signal intensity for molecules located in a close proximity to the nanoparticle surface is a non-trivial issue. The classical approach, used in EM enhancement calculations [4]–[8], is adequate when nanoparticle-molecule or interparticle distances are not very small. For small distances, the quantum-mechanical effects in the electron density distribution can no longer be neglected. These effects are especially important in noble-metal particles where the SP local field is strongly affected by highly-polarizable (core) d-electrons. In the bulk part of the nanoparticle, the (conduction) s-electrons are strongly screened by the localized d-electrons. However, near the nanoparticle boundary, the two electron species have different density profiles. Namely, delocalized s-electrons spill over the classical boundary [14], thus increasing the effective nanoparticle radius, while d-electron density profile mostly retains its classical shape. The incomplete embedding of the conduction electrons in the core electron background [15]–[18] leads to a reduced screening of the s-electron Coulomb potential in the nanoparticle surface layer. The latter has recently been observed as an enhancement of the electron-electron scattering rate in silver nanoparticles [19], [20].

Here we study the role of electron confinement on SERS from molecule adsorbed on the surface of small Ag particles. To this end, we develop a microscopic theory for SERS in noble-metal particle, based on the quantum extension of two-region model [15]–[17], [21], which describes the role of the surface-layer phenomenologically while treating conduction electrons quantum-mechanically within time-dependent density functional theory. We find that the reduction of screening near the surface leads to an additional enhancement of the Raman signal from a molecule located in a close proximity to the nanoparticle. In particular, we address the dependence of SERS on nanoparticle size and show that the interplay of finite-size and screening effects is especially strong for small nanometer-sized particles.

2 QUANTUM TWO-REGION MODEL

We consider SERS from a molecule adsorbed on the surface of Ag spherical particle with radius R. For $R \ll \lambda$, the frequency-dependent potential is determined from Poisson equation,

$$\Phi(\omega, \mathbf{r}) = \phi_0(\mathbf{r}) + e^2 \int d^3r' \frac{\delta N(\omega, \mathbf{r}')}{|\mathbf{r} - \mathbf{r}'|}, \qquad (1)$$

where $\phi_0(\mathbf{r}) = -e\mathbf{E}_i \cdot \mathbf{r}$ is potential of the incident light with electric field amplitude $\mathbf{E}_i = E_i\mathbf{z}$ along the z-axis, and $\delta N(\omega, \mathbf{r})$ is the induced density (hereafter we suppress frequency dependence). There are four contributions to $\delta N(\mathbf{r})$: from valence s-electrons, $\delta N_s(\mathbf{r})$, core d-electrons, $\delta N_d(\mathbf{r})$, dielectric medium, $\delta N_m(\mathbf{r})$, and the molecule, $\delta N_0(\mathbf{r})$. The density profile of delocalized s-electrons is not fully inbedded in the background of localized d-electrons but extends over that of localized d-electrons by $\Delta \sim 1-3$ Å[15]–[17]. Due to localized nature of the d-electron wave-functions, we can adopt a phenomenological description by assuming a uniform bulk-like d-electron ground-state density n_d in the region confined by $R_d < R$ with a sharp step-like edge. Then the induced charge density, $e\,\delta N_d(\mathbf{r}) = -\nabla\cdot\mathbf{P}_d(\mathbf{r})$, is expressed via electric polarization vector vanishing outside of the region $r < R_d$,

$$\mathbf{P}_d(\mathbf{r}) = \frac{\epsilon_d - 1}{4\pi}\,\theta(R_d - r)\mathbf{E}(\mathbf{r}) = -\frac{\epsilon_d - 1}{4\pi e}\,\theta(R_d - r)\nabla\Phi(\mathbf{r}), \tag{2}$$

where $\epsilon_d(\omega)$ is the core dielectric function which can be taken from experiment. Similarly, dielectric medium contribution, which is nonzero for $r > R$, is given by

$$\begin{aligned} e\,\delta N_m(\mathbf{r}) &= -\nabla\cdot\mathbf{P}_m(\mathbf{r}),\\ e\,\mathbf{P}_m(\mathbf{r}) &= -\frac{\epsilon_m - 1}{4\pi}\,\theta(r - R)\nabla\Phi(\mathbf{r}), \end{aligned} \tag{3}$$

where ϵ_m is medium dielectric constant. We represent the molecule by a point dipole with dipole moment $\mathbf{p}_0 = \alpha_0\mathbf{E}$, where α_0 is the molecule polarizability tensor, so that

$$\begin{aligned} e\,\delta N_0(\mathbf{r}) &= -\nabla\cdot\mathbf{P}_0(\mathbf{r}),\\ e\,\mathbf{P}_0(\mathbf{r}) &= \delta(\mathbf{r} - \mathbf{r}_0)\,\mathbf{p}_0 = -\delta(\mathbf{r} - \mathbf{r}_0)\,\alpha_0\nabla\Phi(\mathbf{r}_0), \end{aligned} \tag{4}$$

where $\mathbf{r}_0$ is the vector pointing at the molecule location (we chose origin at the sphere center).

Using Eqs. (2-4), the potential $\Phi(\mathbf{r})$ in Eq. (1) can be expressed in terms of only induced s-electron density, δN_s. Substituting the above expressions into $\delta N = \delta N_s + \delta N_d + \delta N_m + \delta N_0$ in the rhs of Eq. (1) and integrating by parts, we obtain,

$$\begin{aligned} \epsilon(r)\Phi(\mathbf{r}) = \phi_0(\mathbf{r}) &+ e^2\int d^3r'\frac{\delta N_s(\mathbf{r}')}{|\mathbf{r}-\mathbf{r}'|}\\ &+\frac{\epsilon_d - 1}{4\pi}\int d^3r'\nabla'\frac{1}{|\mathbf{r}-\mathbf{r}'|}\cdot\nabla'\theta(R_d - r)\Phi(\mathbf{r}')\\ &+\frac{\epsilon_m - 1}{4\pi}\int d^3r'\nabla'\frac{1}{|\mathbf{r}-\mathbf{r}'|}\cdot\nabla'\theta(r - R)\Phi(\mathbf{r}')\\ &-\nabla_0\frac{1}{|\mathbf{r}-\mathbf{r}_0|}\cdot\alpha_0\nabla_0\Phi(\mathbf{r}_0), \end{aligned} \tag{5}$$

where $\epsilon(r) = \epsilon_d$, 1, and ϵ_m in the intervals $r < R_d$, $R_d < r < R$, and $r > R$, respectively. Since the source term has the form $\phi(\mathbf{r}_0) = \phi(r_0)\cos\theta = -eE_i r\cos\theta$, we expand Φ and δN_s in terms of spherical harmonics and, keeping only the dipole term ($L = 1$), obtain,

$$\begin{aligned} \epsilon(r)\Phi(r) = \phi_0(r) &+ e^2\int dr' r'^2 B(r,r')\delta N_s(r')\\ &-\frac{\epsilon_d - 1}{4\pi}R_d^2\partial_{R_d}B(r,R_d)\Phi(R_d)\\ &+\frac{\epsilon_m - 1}{4\pi}R^2\partial_R B(r,R)\Phi(R)\\ &-\nabla_0[B(r,r_0)\cos\theta_0]\cdot\alpha_0\nabla_0[\Phi(r_0)\cos\theta_0], \end{aligned} \tag{6}$$

where θ_0 is the angle between molecule position and incident light direction (z-axis), and

$$B(r.r') = \frac{4\pi}{3}\left[\frac{r'}{r^2}\,\theta(r - r') + \frac{r}{r'^2}\,\theta(r' - r)\right] \tag{7}$$

is the dipole term of the radial component of the Coulomb potential.

The second terms in rhs of Eq. (6) is the s-electrons contribution to total induced potential, while the third and fourth terms originate from the scattering due to change of dielectric function at $r = R_d$ and $r = R$, respectively. The last term represents the potential of the molecular dipole. The latter depends on the molecule orientation with respect to the nanoparticle surface. In the following we assume averaging over random orientations and replace the polarizability tensor by isotropic α_0.

The values of Φ at the boundaries and at the molecule position can be found by setting $r = R_d, R, r_0$ in Eq. (6). In doing so, the total potential is expressed in terms of only s-electron induced density δN_s. Within TDLDA formalism, the latter can be related back to the potential via

$$\delta N_s(\mathbf{r}) = \int d^3r'\Pi_s(\mathbf{r},\mathbf{r}')\Big[\Phi(\mathbf{r}') + V_x'[n(r')]\delta N_s(\mathbf{r}')\Big], \tag{8}$$

where $\Pi_s(\mathbf{r},\mathbf{r}')$ is the polarization operator for noninteracting s-electrons, $V_x'[n(r')]$ is the (functional) derivative of the exchange-correlation potential (in the endependent-particle approximation) and $n(r)$ is electron ground-state density. The latter is calculated in a standard way using Kohn-Sham equations for jelium model, and then is used as imput in the evaluation of Π_s and V_x'. Eqs. (6) and (8) determine the self-consistent potential in the presence of molecule, nanoparticle, and dielectric medium.

3 CALCULATION OF RAMAN SIGNAL

In the conventional SERS picture, the enhancement of Raman signal from the molecule comes from two sources: far-field of the radiating dipole of the molecule in the local nanopartricle field, and the secondary scattered field of this dipole by the nanoparticle. Accordingly, we present total potential as a sum $\Phi = \phi + \phi^R$, where ϕ^R is the potential of the radiating dipole. Since the molecular polarizability is very small, in the following we will restrict ourselves by the lowest order in α_0, i.e., ϕ is the potential in the absence of molecule and ϕ_R determines the Raman signal in the first order in α_0. Inclusion of higher orders leads to the renormalization of molecular and nanoparticle polarizabilities due to image charges; these effects are not considered here. In the same manner, the induced s-electron density can be decomposed into two contributions, $\delta N_s = \delta n + \delta n^R$, originating from the electric field of incident light and that of the radiating dipole.

Keeping only zero-order terms in Eq. (6), we have

$$\epsilon(r)\phi(r) = \bar{\phi}(r) - \frac{\epsilon_d - 1}{3}\beta(r/R_d)\phi(R_d) + \frac{\epsilon_m - 1}{3}\beta(r/R)\phi(R), \quad (9)$$

where

$$\bar{\phi}(r) = \phi_0(r) + e^2 \int dr' r'^2 B(r,r')\delta n_s(r'), \quad (10)$$

and $\beta(r/R) = \frac{3}{4\pi} R^2 \partial_R B(r,R)$ is given by

$$\beta(x) = x^{-2}\,\theta(x-1) - 2x\theta(1-x). \quad (11)$$

The boundary values of ϕ can be obtained by matching $\phi(r)$ at $r = R_d, R$,

$$\begin{aligned}(\epsilon_d + 2)\phi(R_d) + 2a(\epsilon_m - 1)\phi(R) &= 3\bar{\phi}(R_d)\\ (\epsilon_d - 1)a^2\phi(R_d) + (2\epsilon_m + 1)\phi(R) &= 3\bar{\phi}(R)\end{aligned} \quad (12)$$

where $a = R_d/R$. Substituting $\phi(R_d)$ and $\phi(R)$ back into Eq. (9), we arrive at

$$\epsilon(r)\phi(r) = \bar{\phi}(r) - \beta(r/R_d)\frac{\lambda_d}{\eta}\left[\bar{\phi}(R_d) - 2a\lambda_m\bar{\phi}(R)\right] + \beta(r/R)\frac{\lambda_m}{\eta}\left[\bar{\phi}(R) - a^2\lambda_d\bar{\phi}(R_d)\right], \quad (13)$$

where

$$a = R_d/R, \quad \lambda_d = \frac{\epsilon_d - 1}{\epsilon_d + 2}, \quad \lambda_m = \frac{\epsilon_m - 1}{2\epsilon_m + 1}, \quad \eta = 1 - 2a^3\lambda_d\lambda_m \quad (14)$$

It is convenient to separate out δn_s-independent contribution by writing

$$\phi = \varphi_0 + \delta\varphi_0 + \delta\varphi_s, \quad (15)$$

where $\varphi_0 = \phi_0/\epsilon(r) = -eE_i r/\epsilon(r)$,

$$\delta\varphi_0(r) = \frac{1}{\epsilon(r)}\Big[-\beta(r/R_d)\phi_0(R_d)\,\lambda_d(1-2\lambda_m)/\eta + \beta(r/R)\phi_0(R)\,\lambda_m(1-a^3\lambda_d)/\eta\Big], \quad (16)$$

and

$$\delta\varphi_s(r) = \int dr' r'^2 A(r,r')\delta n_s(r'), \quad (17)$$

with

$$A(r,r') = \frac{e^2}{\epsilon(r)}\Big[B(r,r') - \beta(r/R_d)\Big[B(R_d,r') - 2a\lambda_m B(R,r')\Big]\lambda_d/\eta + \beta(r/R)\Big[B(R,r') - a^2\lambda_d B(R_d,r')\Big]\lambda_m/\eta\Big]. \quad (18)$$

Note that $\delta\varphi_s(r)$ as well as the total potential $\phi(r)$ are continuous at $r = R_d, R$.

Turning to Eq. (8), we use decompositions $\Phi = \phi + \phi^R$ and $\delta N_s = \delta n_s + \delta n_s^R$ to obtain decoupled equations for quantities of zero and first orders in α_0. Then, expanding both parts in spherical harmonics and keeping only the dipole ($L = 1$) terms, we obtain

$$\delta n_s(r) = \int dr' r'^2 \Pi_s(r,r')\Big[\phi(r') + V'_x(r')\delta n_s(r')\Big]. \quad (19)$$

Using Eqs. (15) and (17) then leads to a closed equation for δn_s,

$$\delta n_s(r) = \int dr' r'^2 \Pi_s(r,r')\Big[\varphi_0(r') + \delta\varphi_0(r')\Big] + \int dr' r'^2 \Pi_s(r,r')\Big[\int dr'' r''^2 A(r',r'')\delta n_s(r'') + V'_x(r')\delta n_s(r')\Big], \quad (20)$$

with $A(r,r')$ given by Eq. (18). The effect of d-electrons and dielectric medium is thus encoded in the functions $A(r,r')$ and $\delta\varphi_0(r)$, which reduce to $B(r,r')$ and 0, respectively, for $\epsilon_d = \epsilon_m = 1$.

Turning to the first order in α_0, the equation for ϕ^R takes the form,

$$\epsilon(r)\phi^R(r) = \bar{\phi}^R(r) - \frac{\epsilon_d - 1}{3}\beta(r/R_d)\phi^R(R_d) + \frac{\epsilon_m - 1}{3}\beta(r/R)\phi^R(R), \quad (21)$$

with

$$\bar{\phi}^R(r) = \phi_0^R(r) + e^2\int dr' r'^2 B(r,r')\delta n_s^R(r'), \quad (22)$$

where

$$\phi_0^R(r) = -\alpha_0\nabla_0[B(r,r_0)\cos\theta_0]\cdot\nabla_0[\phi(r_0)\cos\theta_0], \quad (23)$$

is the potential of the molecular dipole in the presence of local field and $\delta n_s^R(r)$ is the induced charge of s-electrons due to molecular potential. The frequency dependence of the Raman field $\phi^R(\omega_s, r)$ is determined by the Stokes shift $\omega_s = \omega - \omega_0$, where ω_0 is the vibrational frequency of the molecule as determined by α_0. The last two terms in Eq. (21) describe potential of molecular dipole scattered from the nanoparticle boundaries at R_d and R. For simplicity, we only consider the case when the molecule is located at the z-axis ($\theta_0 = 0$) so that molecular dipole potential is given by

$$\phi_0^R(r) = -\frac{4\pi\alpha_0}{3r_0^2}\,\beta(r/r_0)\,\frac{\partial\phi(r_0,\omega)}{\partial r_0} = -\chi(\omega)\,eE_i\Big[r\theta(r_0 - r) - \frac{r_0^3}{2r^2}\,\theta(r - r_0)\Big], \quad (24)$$

where

$$\chi(\omega) = \frac{8\pi\alpha_0}{3eE_i r_0^3}\frac{\partial\phi(r_0)}{\partial r_0} \quad (25)$$

Marching ϕ^R at $r = R_d$ and $r = R_d$, we obtain

$$\phi^R = \varphi_0^R + \delta\varphi_0^R + \delta\varphi_s^R = \chi(\omega)\Big[\tilde{\varphi}_0 + \delta\varphi_0 + \delta\tilde{\varphi}_s\Big], \quad (26)$$

where

$$\tilde{\varphi}_0(r) = -\frac{eE_i}{\epsilon(r)}\Big[r\theta(r_0 - r) - \frac{r_0^3}{2r^2}\,\theta(r - r_0)\Big], \quad (27)$$

$\delta\varphi_0$ is given by Eq. (16) with ω_s instead of ω and

$$\delta\tilde{\varphi}_s(r) = \int dr' r'^2 A(r, r')\delta\tilde{n}_s(r'), \quad (28)$$

where $A(r, r')$ is given by Eq. (18), with ω_s instead of ω, and $\delta\tilde{n}_s(r) = \delta n_s^R(r)/\chi(\omega)$ satisfies Eq. (20) but with $\tilde{\varphi}_0(r)$ instead of $\varphi_0(r)$ (and ω_s instead of ω). In the following we consider the case when the molecule is located at the distance of several angstroms from the nanoparticle classical boundary, so the overlap between the molecular orbitals and the s-electron wave function is small. Then we have

$$\int dr' r'^2 \Pi_s(r, r')\tilde{\varphi}_0(r') \simeq \int dr' r'^2 \Pi_s(r, r')\varphi_0(r'), \quad (29)$$

leading to $\delta\tilde{n}_s(r) \simeq \delta n_s(r)$ and, correspondingly, $\delta\tilde{\varphi}_s(r) \simeq \delta\varphi_s(r)$. The Raman signal is determined by the far-field asymptotics of $\phi^R(r)$. From Eqs. (16-18), we find for $r \gg R$

$$\delta\varphi_0(r) = \frac{eE_i}{\epsilon_m r^2}\,\alpha_d, \quad \delta\varphi_s(r) = \frac{eE_i}{\epsilon_m r^2}\,\alpha_s, \quad (30)$$

with

$$\alpha_d(\omega_s) = R^3\Big[1 - (1 + \lambda_m)(1 - a^3\lambda_d)/\eta\Big],$$

$$\alpha_s(\omega_s) = \frac{4\pi}{3eE_0}\Bigg[\int_0^\infty dr' r'^3 \delta n_s(r') - \Big[1 - (1 + \lambda_m)(1 - \lambda_d)/\eta\Big]\int_0^R dr' r'^3 \delta n_s(r') - \Big[1 - (1 + \lambda_m)(1 - a^3\lambda_d)/\eta\Big]R^3\int_R^\infty dr' \delta n_s(r') + \Big[(1 + \lambda_m)\lambda_d/\eta\Big]\int_{R_d}^R dr'(r'^3 - R_d^3)\delta n_s(r')\Bigg], \quad (31)$$

and using Eq. (27) we obtain for the far field

$$\phi^R(r) = \frac{eE_i r_0^3}{2\epsilon_m r^2}\,\chi\,\Big[1 + 2(\alpha_d + \alpha_s)/r_0^3\Big]. \quad (32)$$

Turning to χ, we note that for small molecule-nanopatricle overlap, the local potential at the molecule location can be evaluated using the far-field expressions Eqs. (30). We then obtain

$$\frac{\partial\phi_0}{\partial r_0} = -\frac{eE_i}{\epsilon_m}\Big[1 + 2(\alpha_d + \alpha_s)/r_0^3\Big], \quad (33)$$

and, substituting into Eq. (32), we finally arrive at

$$\phi^R(r) = -\frac{4\pi\alpha_0 eE_i}{3\epsilon_m r^2}\Big[1 + 2g(\omega)\Big]\Big[1 + 2g(\omega_s)\Big], \quad (34)$$

with

$$g = \frac{\alpha}{r_0^3}, \quad \alpha = \alpha_d + \alpha_s. \quad (35)$$

The above expression generalizes the well-known classical result [4], [5] to the case of noble-metal particle with different distributions of d-electron and s-electron densities. The surface-enhanced Raman field retains the same functional dependence on the nanoparticle polarizability, however the latter contains all the information about the surface layer effect.

Finally, the enhancement factor is given by the ratio of Raman to incident field intensities,

$$A(\omega, \omega_s) = \Big|1 + 2g(\omega) + 2g(\omega_s) + 4g(\omega)g(\omega_s)\Big|^2. \quad (36)$$

4 NUMERICAL RESULTS

For large nanoparticle with $R \sim 100$ nm, the classical EM theory provides an enhancement of the Raman signal as large as 10^6-10^7 [6]. In reality, the EM enhancement is inhibited by various factors. In noble-metal particles, the interband transition between d-electron and s-electron bands reduce the SP oscillator strength leading to a weakening of the local fields. For nanoparticle radius below 15 nm, finite-size effects become important. The SP resonance damping comes from the electron scattering at the surface leading to the size-dependent SP resonance width $\gamma_s \simeq v_F/R$. At the resonance frequency, the size-dependence of SERS is quite strong. Indeed, if molecular vibrational frequencies are smaller that the SP width, the enhancement factor decreases as $A \propto R^4$ for small nanoparticles, resulting in several orders of magnitude drop in the Raman signal.

Our main observation is that, in small nanoparticle, the local field enhancement due to reduced screening in

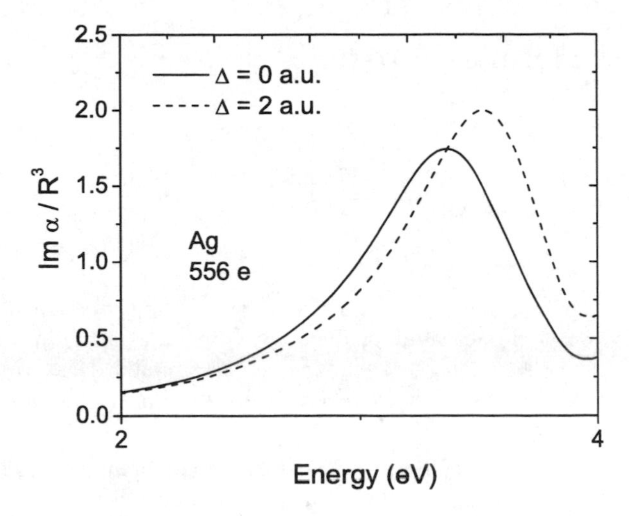

Figure 1: Calculated nanoparticle polarizability for different surface layer thicknesses

.

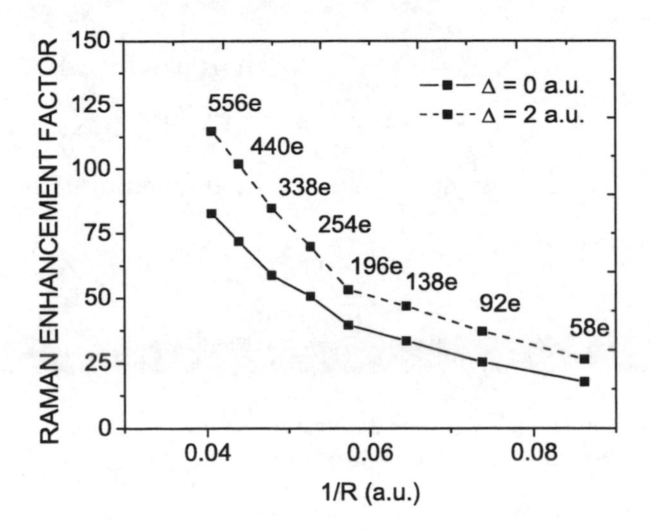

Figure 2: Size-dependence of enhancement factor for different surface layer thicknesses.

the surface layer can provide an additional enhancement of the Raman signal. Although the thickness of the surface layer (0.1-0.3 nm) is small as compared to oveall nanoparticle size [15]–[17], such an enhancement can be condiderable for a molecule located in a close proximity to the surface. In Fig. 1 we show the calculated polarizability with and without surface layer. In the presence of the surface layer, the SP energy experiences a blueshift [15], [16] due an effective decrease in the d-electron dielectric function in the nanoparticle. At the same time, an increase in the peak amplitude, which accompanies the blueshift, indicates a stronger local field at resonance energy acting on a molecule in a close proximity to nanoparticle surface.

In Fig. 2 we show the results of our numerical calculations of SERS with and without surface layer thicknesses, Δ for different nanoparticle sizes. Although the overall magnitude of the enhancemet increases with Δ, a more important effect is its size-dependence. For finite Δ, the enhancement factor descreases more slowly that in the absence of the surface layer: as nanoparticle size decreases, the signals strength ratio *increases*. The reason is that, as the nanoparticle becomes smaller, the fraction of the surface layer increases, and so does the contribution of the unscreened local field into SERS.

This work was supported by NSF under grants DMR-0305557 and NUE-0407108, by NIH under grant 5 SO6 GM008047-31, and by ARL under grant DAAD19-01-2-0014.

REFERENCES

[1] S. Nie and S. R. Emory, Science **275**, 1102 (1997).

[2] K. Kneipp *et al.,* Rev. Lett. 78, 1667 (1997).

[3] K. Kneipp *et al.,* Chem. Rev. **99**, 2957 (1999).

[4] M. Kerker, D.-S. Wang, and H. Chew, Appl. Optics **19**, 4159 (1980).

[5] J. Gersten and A. Nitzan, J. Chem. Phys. **73**, 3023 (1980).

[6] G. S. Schatz and R. P. Van Duyne, in *Handbook of Vibrational Spectroscopy,* edited by J. M. Chalmers and P. R. Griffiths (Wiley, 2002) p. 1.

[7] H. Xu *et al.,* Phys. Rev. Lett. **83**, 4357 (1999).

[8] H. Xu *et al.,* Phys. Rev. B **62**, 4318 (2000).

[9] A. Otto *et al.,* J. Phys. Cond. Matter **4**, 1143 (1992).

[10] M. Michaels, M. Nirmal, and L. E. Brus, J. Am. Chem. Soc. **121**, 9932 (1999).

[11] A. M. Michaels, J. Jiang, and L. E. Brus, J. Phys. Chem. B **104**, 11965 (2000).

[12] Otto, Phys. Phys. Stat. Sol. (a) 4, 1455 (2000).

[13] W. E. Doering and S. Nie, J. Phys. Chem. B **106**, 311 (2002).

[14] W. Ekardt, Phys. Rev. B **31**, 6360 (1985).

[15] A. Liebsch, Phys. Rev. **48**, 11317 (1993).

[16] V. V. Kresin, Phys. Rev. **51**, 1844 (1995).

[17] A. Liebsch and W. L. Schaich, Phys. Rev. **52**, 14219 (1995).

[18] J. Lermé *et al.,* Phys. Rev. Lett. **80**, 5105 (1998).

[19] C. Voisin *et al.,* Phys. Rev. Lett. **85**, 2200 (2000).

[20] C. Lopez-Bastidas, J. A. Maytorena, and A. Liebsch, Phys. Rev. **65**, 035417 (2001).

[21] A. A. Lushnikov, V. V. Maksimenko, and A. J. Simonov, Z. Physik B **27**, 321 (1977).

Vibrational Modes of Metal Nanoshells

A. S. Kirakosyan and T. V. Shahbazyan

Department of Physics and Computational Center for Molecular Structure and Interactions, Jackson State University, Jackson MS 39217

Abstract

We study coherent oscillations of radial breathing modes in metal nanoparticles with a dielectric core. Vibrational modes are impulsively excited by a rapid heating of the particle lattice that occurs after laser excitation, while the energy transfer to a surrounding dielectric leads to a damping of the oscillations. In nanoshells, the presence of two metal surfaces leads to a substantially different energy spectrum of acoustic vibrations. The lowest and first excited modes correspond to in-phase (n=0) and out-of-phase (n=1) contractions of shell-core and shell-matrix interfaces respectively. We calculated the energy spectrum as well as the damping of nanoshell vibrational modes in the presence of surrounding medium, and found that the size-dependences of in-phase and anti-phase modes are different. At the same time, the oscillator strength of the symmetric mode is larger than that in solid nanoparticles leading to stronger oscillations in thin nanoshells.

Keywords: Nanoparticles, surface plasmon, vibrational modes, ultrafast spectroscopy

1 INTRODUCTION

Acoustic vibrational modes in nanoparticles are impulsively excited by a rapid heating of the lattice that takes place after laser excitation. After initial period of rapid expansion, a nanoparticle undergoes radial contractions and expansions around the new equilibrium. The periodic change in nanoparticle volume translates into a modulation in time of the surface plasmon resonance (SPR) energy that dominates nanoparticle optical absorption spectrum. The spectrum of vibrational modes manifests itself via coherent oscillations of differential transmission at SPR energy measured using ultrafast pump-probe spectroscopy [1], [2]. Since the size of laser spot is usually much larger than nanoparticle diameter, the initial expansion is homogeneous so that predominantly the fundamental ($n = 0$) breathing mode, corresponding to oscillations of nanoparticle volume as a whole, is excited. The lowest excited ($n = 1$) mode has weaker oscillator strength ($\approx 1/4$ of that for $n = 0$), and has also been recently observed [3]. When nanoparticle is embedded in a dielectric medium, the oscillations are damped due to the transfer of latice energy to acoustic waves in surrounding dielectric. In solid particles, the size dependences of eigenmodes energy and decay rate are similar – both are inversely proportional to nanoparticle radius [4].

Here we study the vibrartional modes of metal nanoshells. These recently manufactured metal particles with dielectric core [5] attracted much interest due to unique tunability of their optical properties. By varying the shell thickness during the manufacturing process, the SPR can be tuned in a wide energy interval [6]. Recent pump-probe measurements of vibrational modes dynamics in gold nanoshells submerged in water revealed characteristic oscillation pattern of differential transmission. However, the oscillations period and amplitude as well as their damping were significantly larger than those for solid nanoparticles. We perform detailed analysis of energy spectrum of lowest vibrational modes of a nanoshell in a dielectric medium. We find that the modes eigenenergies exhibit a strong dependence on nanoshell aspect ratio, $\kappa = R_1/R_2$, where R_1 and R_2 are inner and outer radii, respectively. Specifically, for thin nanoshells, the fundamental mode energy is consirerably lower than for solid particles while the damping is significantly larger. At the same time, in the thin shell limit, the fundamental mode carries the *entire* oscillator strength which results in an enhanced oscillations amplitude as compared to solid particles. The analysis also reveals two regimes, where the spectrum is dominated by nanoslell geometry or by surrounding medium, with a sharp crossover governed by the interplay between aspect ratio and impendance.

2 SPECTRUM OF VIBRATIONAL MODES FOR A NANOSHELL

We consider radial normal modes of a spherical nanoshell with dielectric core extending up to inner radius R_1 in a dielectric medium over outer radius R_2. The core, shell, and medium are characterized by densities $\rho^{(i)}$ longitudinal and tranverse sound velicities $c_{L,T}^{(i)}$ with $i = c, s, m$, respectively. The radial displacement $u(r)$ is determined from the Helmholtz equation (at zero angular momentum) [7]

$$u'' + \frac{2u'}{r} + k^2 u = 0, \tag{1}$$

where $k = \omega/c_L$ is the wave-vector with the boundary conditions that the displacement u and the radial component of stress tensor,

$$\sigma = \rho\Big[c_L^2 u' + (c_L^2 - 2c_T^2)\,\frac{2u}{r}\Big], \tag{2}$$

are continuous at core/shell and shell/medium interfaces. In the three regions divided by shell boundaries, the solution has the form

$$u^{(c)} \sim \frac{\partial}{\partial r}\frac{\sin k^{(c)} r}{r}, \quad u^{(s)} \sim \frac{\partial}{\partial r}\frac{\sin(k^{(s)} r + \phi)}{r},$$
$$u^{(m)} \sim \frac{\partial}{\partial r}\frac{e^{ik^{(m)} r}}{r}, \tag{3}$$

where ϕ is the phase mismatch. The corresponding eigenenergies are, in general, complex due to energy transfer to outgoing wave in the surrounding medium. Matching $u(r)$ and $\sigma(r)$ at $r = R_1, R_2$, we obtain the following equations for eigenvalues $\xi = kR_2$

$$\frac{\xi^2}{\xi\cot(\xi+\varphi) - 1} + \frac{\eta_m \xi^2}{1 + i\xi/\alpha_m} + \chi_m = 0,$$
$$\frac{\xi^2\kappa^2}{\xi\kappa\cot(\xi\kappa+\varphi) - 1} - \frac{\eta_c \xi^2\kappa^2}{(\xi\kappa/\alpha_c)\cot(\xi\kappa/\alpha_c) - 1} + \chi_c = 0, \tag{4}$$

where $\kappa = R_1/R_2$ is nanoshell aspect ratio and the parameters

$$\alpha_i = c_L^i/c_L^{(s)}, \quad \eta_i = \rho^{(i)}/\rho^{(s)}, \quad \chi_i = 4(\beta_s^2 - \eta_i\delta_i^2)$$
$$\beta_i = c_T^{(i)}/c_L^{(i)}, \quad \delta_i = c_T^{(i)}/c_L^{(s)}. \tag{5}$$

characterize the metal/dielectric interfaces. For the ideal case of free boundary conditions, we have $\alpha_c = \eta_m = \eta_c = 0$ and $\chi_c = \chi_m = 4\beta_s^2$. For a thin nanoshell, $1 - \kappa \ll 1$, we then easily recover the known expression for the fundamental mode

$$\xi_0 = 2\beta_s\sqrt{3 - 4\beta_s^2}. \tag{6}$$

The eigenvalue is purely real since no energy leaks through the interface.

In the realistic case of a nanoshell in a medium, the following simplification occurs. The initial laser pulse causes rapid expansion of both dielectric core and metal shell. However, due to a larger value of the metal thermal expansion coefficient, the shell expands to a greater extend than the core, so at the new equilibrium the core and the shell are, in fact, no longer in contact. In this case, the boundary conditions at he core/shell interface should be taken as stress free. For a thin nanoshell, $1 - \kappa \ll 1$, Eqs. (4) are then reduced to

$$\lambda_c\left[\lambda_m - \lambda_c + \frac{\alpha_m\eta_m\xi^2}{\alpha_m - i\xi}\right] = (1-\kappa)$$
$$\times\left[\left(\lambda_m + \frac{\alpha_m\eta_m\xi^2}{\alpha_m - i\xi}\right)\xi_0^2 - \lambda_c\xi^2\right]. \tag{7}$$

Typically, the metal density of the shell is much large that that of the sorrounding dielectric, i.e., the parameter η_m is small. For $\eta_m \ll 1$, using $\chi_m - \chi_c = -4\eta_m\alpha_m^2\beta_m^2$ and $\chi_m/\chi_c = 1 - \eta_m\beta_m^2$, we obtain

$$x^2 - 1 + \eta_m\beta_m = \frac{\alpha_m\eta_m}{\xi_0(1-\kappa)}\left[\frac{4\alpha_m\beta_m^2}{\xi_0} - \frac{x^2}{\alpha_m/\xi_0 - ix}\right], \tag{8}$$

where $x = \xi/\xi_0$. It can be easily seen that there are two regimes governed by the parameter

$$\lambda = \frac{\alpha_m\eta_m}{\xi_0(1-\kappa)}. \tag{9}$$

For a very thin nanoshell, $\lambda \gg 1$, the lowest eigenvalue is given by

$$\xi \simeq \alpha_m\beta_m - i\frac{\eta_m\alpha_m\beta_m^2}{\xi_0}. \tag{10}$$

In this regime, the energy and damping are comptetely determined by the surrounding medium and do not depend on nanoshell geometry. Note that if the transverse sound speed of the medium is zero (e. g., for water), then both energy and the decay rate vanish. In the opposite case, $\lambda \ll 1$, the solution can be obtained as

$$\omega \approx \frac{c_L\xi_0}{R_2}\left(1 + \frac{\alpha_m^2\eta_m}{2\xi_0^2(1-\kappa)}\left[4\beta_m^2 - \frac{1}{(\alpha_m/\xi_0)^2 + 1}\right]\right),$$
$$\gamma \approx \frac{c_L\alpha_m^2\eta_m}{2\xi_0(1-\kappa)(\alpha_m^2/\xi_0^2 + 1)R_2}. \tag{11}$$

In this case, the spectrum is mostly dominated by nanoshell geometry. Note that in typical experimental situations, the parameter λ is small, and with a good approximation, $\xi \simeq \xi_0$, which is considerably lower that the fundamental mode energy for solid particles, and it depends only weakly on aspect ratio.. At the same time, the damping rate γ is very sensitive to aspect ratio κ and is considerably higher than that for solid particles.

3 DISCUSSION

Here we present the results of our numerical calculations of vibrational mode spectrum for Au nanoshells in water. In Figs. 1 and 2 we show the energies and damping rates for fundamental ($n = 0$) and first excited ($n = 1$) modes. For fundamental mode, the energy decreases with increasing aspect ratio and, for thin nanoshells, is considerably lower that for nanoparticles,

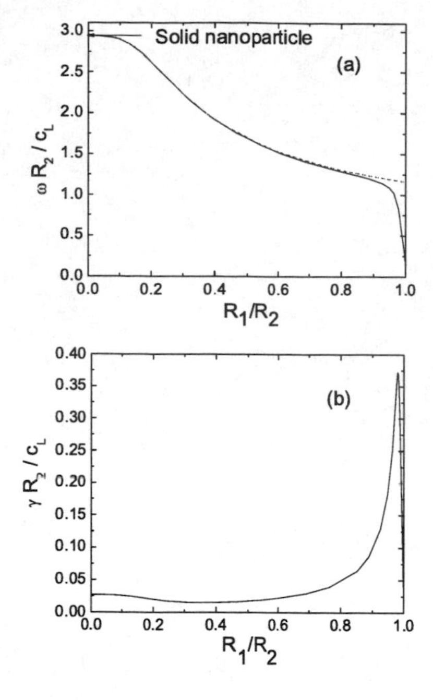

Figure 1: Spectrum for fundamental breathing radial mode in gold nanoshell versus its aspect ratio R_1/R_2. (a) Solid line: eigenfrequency versus R_1/R_2 in the model with free inner boundary and ideal contact between outer shell and matrix. Dashed line: eigenfrequency in the model with free boundaries. (b) Solid line: normalized damping rate versus aspect ratio. Longitudinal sound speed in gold $c_L = 3240$ m/s, transverse sound speed $c_T = 1200$ m/s, the density of gold $\rho = 19700$ kg/m^3. Surrounding matrix is water with $c_L = 1490$ m/s and $c_T = 0$, $\rho = 1000$ kg/m^3.

while the damping rate experiences a rapid increase as nanoshell becomes thiner. The sharp crossover for very thin nanoshells corresponds to the transition between geometry and medium dominated regimes, as discussed above. Note that for water ($c_T = 0$) both energy and damping rate vanish in the thin shell limit. In contrast, for $n = 1$ mode, no such transition takes place, and both energy and damping rate increase with aspect ratio.

Let as now turn to the relative contributions of fundamental and excited modes to the pump-probe signal. Since the intial rapid expansion of nanoshell is homogeneous, the oscillator strength of fundamental (symmetric) mode is expected to be larger than that of $n = 1$ (antisymmetric) mode. The expression for oscillator strength of nth mode has the form

$$C_n = \frac{R_2^{-1} \int r\, U_n(r) dV}{V^{1/2}\Big[\int U_n^2(r) dV\Big]^{1/2}}. \qquad (12)$$

In Fig. 3 we show calculated oscillator strengths for $n = 0$ and $n = 1$ modes versus aspect ratio. In contrast

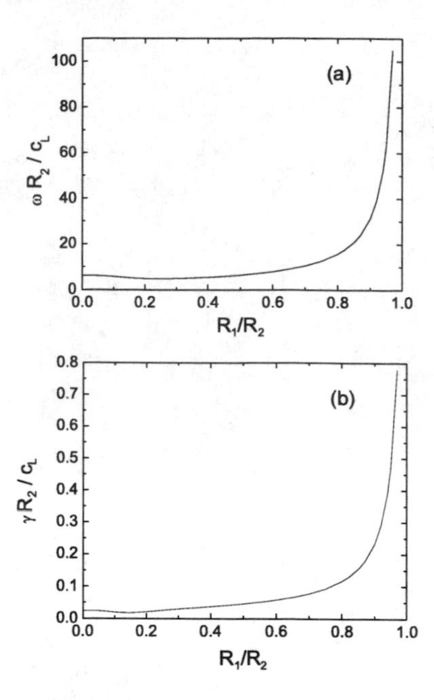

Figure 2: Spectrum for $n = 1$ radial mode in gold nanoshell versus its aspect ratio R_1/R_2. (a) Eigenfrequency calculated with free inner boundary and ideal contact between outer shell and matrix. (b) Normalized damping rate versus aspect ratio.

to solid particles, where the relative strengths of two modes is constant, $C_1/C_0 = 1/4$, here C_1 vanishes in the $\kappa = 1$ limit, while C_0 reaches it maximal value, $C_0 = 1$. Thus, in thin nanoshells, the fundamental mode carries almost entire oscillator strength. As a result, in nanoshells, excitation of the fundamental mode should result in a greater amplitude of oscillations as compared to solid particles while the $n = 1$ should be practically undetectable.

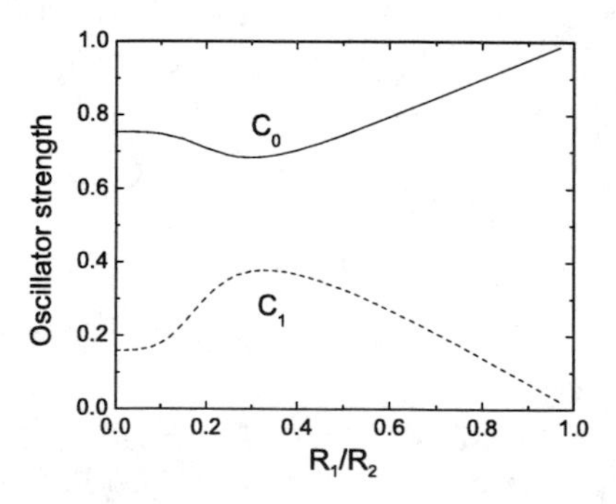

Figure 3: Oscillator strengths for symmetric (solid line) and antisymmetric (dashed line) breathing modes of nanoshell versus nanoshell aspect ratio R_1/R_2. At $R_1 = 0$ oscillator strength coincide with one for nanoparticle.

This work was supported by NSF under grants DMR-0305557 and NUE-0407108, by NIH under grant 5 SO6

GM008047-31, and by ARL under grant DAAD19-01-2-0014. ASK acknowleges support from Armenian State Program "Semiconductor Nanoelectronics" under grant 04-10-30.

REFERENCES

1. N. Del Fatti, C. Voisin, F. Chevy, F. Vallée, and C. Flytzanis, J. Chem. Phys. **110**, 11484 (1999).
2. J.S. Hodak, A.Henglein, G.V. Hartland, J. Chem. Phys. **111**, 8613 (1999).
3. C. Voisin, N. Del Fatti, D. Christofilos, and F. Vallée, J. Chem. Phys. **105**, 2264 (2001).
4. V.A. Dubrovskiy and V.S. Morochnik, Izv. Earth Phys. **17**, 494 (1981).
5. R. D. Averitt, D. Sarkar and N. J. Halas, Phys. Rev. Lett. **78**, 4217 (1997).
6. A. L. Aden and M. Kerker, J. Appl. Phys. **22**, 1242 (1951).
7. L.D. Landau and E.M. Lifshitz, *Theory of Elasticity*, (Addison-Wesley, 1987).

Optical Properties of Silicon Nanocrystals Embedded in SiO_2 Matrix

L. Ding, T. P. Chen and Y. Liu

School of Electrical and Electronic Engineering,
Nanyang Technological University, Singapore 639798
Tel.: (65)67906387; Email: ding0008@ntu.edu.sg

ABSTRACT

From the view point of both science and technology, it would be interesting and useful to examine the optical properties of isolated silicon nanocrystals (*nc*-Si) embedded in SiO_2 matrix. A good understanding to optical properties of the *nc*-Si is definitely important to both the quantum physics and the applications especially in optoelectronic and photonic devices. In this work, spectroscopic ellipsometry (SE), which is a nondestructive method, is employed to determine the optical properties including refractive index, extinction coefficient and absorption coefficient of the *nc*-Si with a mean size of ~ 4.2 nm embedded in a SiO_2 matrix in the photon-energy range of 1.13 - 4.96 eV (or wavelengths of 250 - 1100 nm). It is shown that the optical properties of the *nc*-Si are well described by the four-term F-B (Forouhi-Bloomer) model. A large band gap expansion (~ 0.6 eV) is observed for the *nc*-Si, indicating a significant quantum confinement effect. The bandgap expansion is in accordance with the firs-principle calculation of *nc*-Si optical bandgap based on quantum confinement.

Keywords: silicon nanocrystals (*nc*-Si), spectroscopic ellipsometry (SE), optical constants, dielectric functions, quantum confinement,

INTRODUCTION

Crystalline silicon nanoparticles or nanocrystals (*nc*-Si) embedded in a SiO_2 matrix have received considerable attention in recent years because of their promising applications in optoelectronic devices, memory devices and single electron devices with the main advantage of its full compatibility with silicon technology [1-6] . One promising method of forming silicon nanocrystals in a dielectric film relies on silicon ion implantation into SiO_2 and subsequent high temperature annealing [5-6]. Important advantages of this approach are the precise control for the *nc*-Si depth distribution and being able to form *nc*-Si of smaller size (~ 2 - ~ 5 nm) and a narrow size distribution of *nc*-Si. By this technique, these advantages can be easily achieved by adjusting ion implantation energy and dosage and the annealing conditions. There have been intensive studies on the light emitting mechanism of *nc*-Si formed by this technique, but very few experimental studies have focused on determining the optical properties of isolated Si nanocrystals embedded in SiO_2 matrix. In this study, we present an experimental study on optical properties of Si nanocrystals embedded in a SiO_2 matrix synthesized by this technique. Such a study is obviously important to the fundamental physics as it is concerned with a quantum system of quasiparticles with a size of less than ~ 5 nm isolated by a dielectric matrix, and it is also necessary to device applications especially in Si nanocrystals based optoelectronic and photonic devices.

There have been many reports focusing on the theoretical calculations of optical properties of semiconductor nanocrystals using various methods such as empirical-pseudopotential approach and *ab initio* technique in the recent years [7-9]. In contrast, very few experimental studies of optical properties of *nc*-Si have been reported. Especially, it is difficult to experimentally determine the optical properties of *nc*-Si embedded in a dielectric matrix, because the embedded Si nanocrystals can not be directly investigated by experiment. In our previous studies, we have reported the optical properties of *nc*-Si embedded in a SiO_2 matrix in the photon energy range of 1.1 - 3.1 eV [10, 11]. Nevertheless, there is still lack of a comprehensively experimental study of the optical properties in a wider photon energy range and a proper modeling to the optical properties. In this work, the optical properties including optical constants (refractive index and extinction coefficient), absorption coefficient of the *nc*-Si embedded in a SiO_2 matrix in the photon energy range of 1.13 - 4.96 eV have been determined with spectroscopy ellipsometry (SE). In our opinion, SE can give more complete and detailed information than that obtained from absorbance or reflectance measurements, because SE measurement takes into account polarization of light. The measured values, ellispsometric angles (Ψ and Δ), are related to the ratio of Fresnel reflection coefficients r_p and r_s through the equation $r_p / r_s = \tan(\Psi)\exp(\Delta)$. Dielectric function and optical constants (refractive index and extinction coefficient) can be extracted from the best fitting of Ψ and Δ. The optical constants are well modeled by the Forouhi-Bloomer formulism [12], and the modeling yields the energy bandgap of the *nc*-Si. A strong dielectric suppression and a large bandgap expansion are observed for the *nc*-Si.

EXPERIMENT

The sample under investigation was synthesized by Si^+ implantation with a dose of 1×10^{17} atoms/cm^2 at 100 keV into a 550-nm-thick SiO_2 film on *p*-type Si substrate followed by an annealing at 1000 °C for 30 minutes in nitrogen gas. The average size of *nc*-Si was determined from the broadening of the Bragg peak in XRD spectrum by Scherer's equation

$$D = \frac{0.9\lambda}{\Delta\theta \cos(\theta_B)}, \quad (1)$$

where D is the mean diameter of nanocrystals, λ is the wavelength of the x-ray, θ_B is the Bragg angle and $\Delta\theta$ is the full width of the half maximum (FWHM) of the Bragg peak. Figure 1 shows the XRD measurement of *nc*-Si embedded in SiO_2 matrix and the pseudo-Voigt fit to the data, and the mean *nc*-Si size obtained is ~ 4.2 nm.

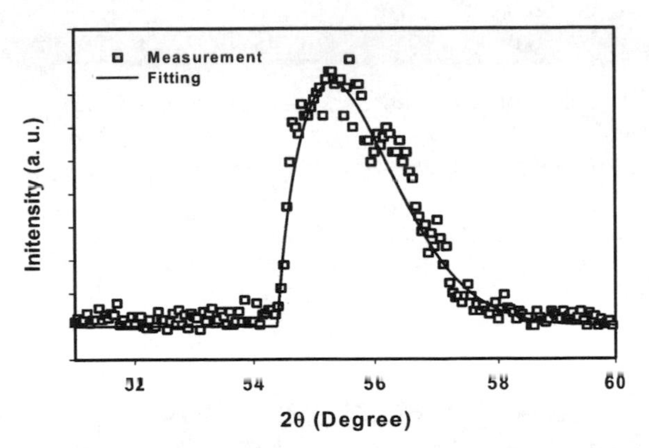

Figure 1: XRD measurement for the *nc*-Si embedded in a SiO_2 matrix and the pseudo-Voigt fit to the data.

In the SE analysis, the thin film system is represented by two layers, namely, the first layer (0≤depth≤250 nm) with *nc*-Si distributing in SiO_2, and the second layer (depth>250 nm) is just a pure SiO_2 layer without *nc*-Si. The *nc*-Si distribution in the SiO_2 thin film is determined from secondary ion mass spectroscopy (SIMS) measurement. Figure 2 shows the best-fit multi-layer model for the SiO_2 film containing silicon nanocrystals and the volume fraction of silicon nanocrystals in SiO_2 calculated from SIMS measurement.

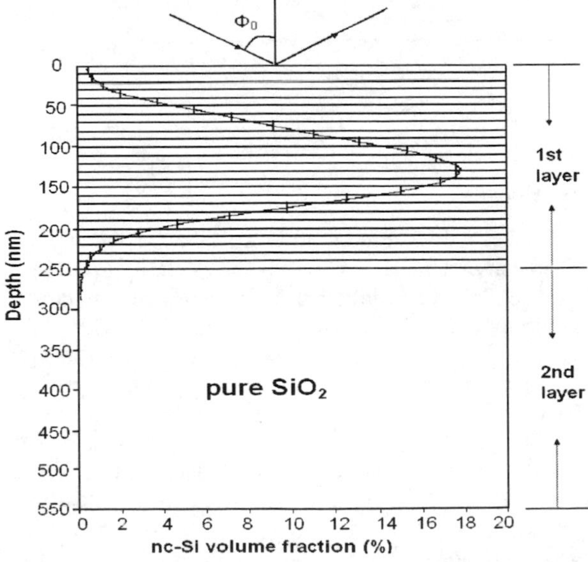

Figure 2: Multi-layer model used in the SE analysis. The *nc*-Si volume fraction is calculated from the SIMS measurement.

As shown in figure 2, in the first layer, the optical properties vary with the depth as the volume fraction of the *nc*-Si varies with depth. In order to model the optical properties of the first layer, it is then divided into 25 sub-layers with equal thickness d_0 = 10 nm. The volume fraction of *nc*-Si in each sub-layer is determined from SIMS measurement, and the *nc*-Si volume fraction is considered to be constant within each sub-layer. Each sub-layer has its own effective dielectric function ε_i (i = 1, 2,... 25) due to its own *nc*-Si volume fraction. Dielectric function of any medium can be described by optical constants through the following equation:

$$\varepsilon(E) = (n(E)^2 - k(E)^2) + i \bullet 2n(E)k(E). \quad (2)$$

As such, it can be optically schematized as an effective medium, in which the SiO_2 is the host matrix while the *nc*-Si is an inclusion embedded in the SiO_2 matrix, represented by the Maxwell-Garnett effective medium approximation (EMA)

$$\frac{\varepsilon_i - \varepsilon_{SiO_2}}{\varepsilon_i + 2\varepsilon_{SiO_2}} = \frac{\varepsilon_{nc-Si} - \varepsilon_{SiO_2}}{\varepsilon_{nc-Si} + 2\varepsilon_{SiO_2}} f_i, \quad (3)$$

where ε_i is the effective complex dielectric function of the ith sub-layer, ε_{SiO_2} is the dielectric function of SiO_2 matrix, ε_{nc-Si} is the dielectric function of the *nc*-Si, and f_i is the volume fraction of *nc*-Si in the ith sub-layer. As the volume fraction (f_i) and ε_{SiO_2} are known, from Eq. (3) the effective complex dielectric function ε_i (and thus the effective complex refractive index N_i) for the ith sub-layer (i=1, 2, ...25) can be expressed in terms of ε_{nc-Si} (or the refractive index and extinction coefficient of *nc*-Si). Therefore, in the SE analysis, the ellipsometric angles (Ψ and Δ) can be expressed as functions of the optical constants of the *nc*-Si, although these functions cannot be displayed with analytical formulas due to their extreme complexity. To avoid solving such complicated equations, a SE spectra fitting is performed to determine the optical constants at each wavelength in the wavelength from 250 to 1100 nm.

In the SE spectra fitting, a proper optical dispersion model should be used to describe the spectra dependence of *nc*-Si optical constants. In this study, the four-term Forouhi-Bloomer (F-B) model is found to be the most suitable one to yield a reasonable fitting. In the four-term F-B model, Optical constants (refractive index n and extinction coefficient k) of silicon nanocrystals given by [12]:

$$k(E) = \left(\sum_{i=1}^{4} \frac{A_i}{E^2 - B_i E + C_i}\right)(E - E_g)^2, \quad (4)$$

$$n(E) = n(\infty) + \sum_{i=1}^{4} \frac{B_{0_i} E + C_{0_i}}{E^2 - B_i E + C_i}, \quad (5)$$

where

$$B_{0_i} = \frac{A_i}{Q_i}\left(-\frac{B_i^2}{2} + E_g B_i - E_g{}^2 + C_i\right), \quad (6)$$

$$C_{0_i} = \frac{A_i}{Q_i}\left((E_g^2 + C_i)\frac{B_i^2}{2} - 2E_g C_i\right), \quad (7)$$

$$Q_i = \frac{1}{2}(4C_i - B_i^2)^{\frac{1}{2}}. \quad (8)$$

Fitting parameters of the model are A_i, B_i, C_i (i = 1, 2, 3, and 4), $n(\infty)$, and energy bandgap E_g. For an efficient fitting, the initial fitting parameters are taken equal to that of bulk crystalline silicon. Figure 3 shows the best spectra fitting, i.e., the comparison of the experimental ellipsometric angles (Ψ and Δ) with the calculated Ψ and Δ in the wavelength range from 250 to 1100 nm.

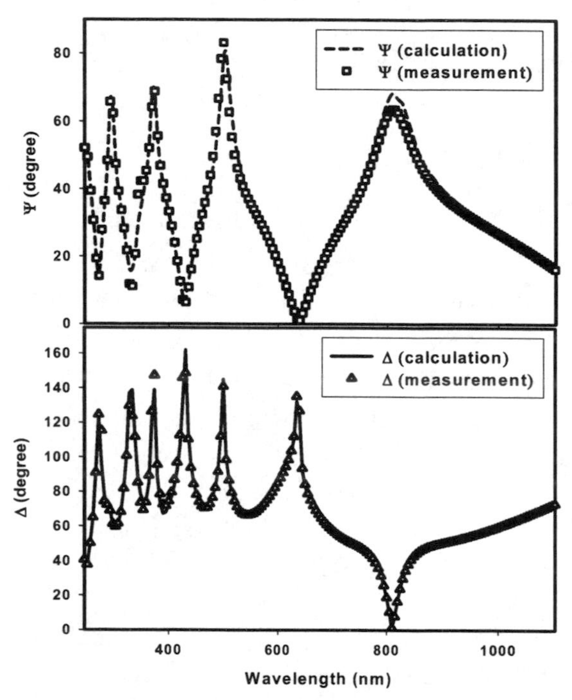

Figure 3: Best spectra fitting of ψ and Δ based on the F-B model with the approach described in the text.

As can be seen in this figure, all the complicated spectra features of both Ψ and Δ are fitted excellently, and the fitting yields reasonable values of all the parameters. This indicates that the above approach is correct, and the four-term F-B dispersion relations can describe the spectral dependence of the optical constants of the *nc*-Si accurately.

RESULTS AND DISCUTIONS

The fitting yields the parameters A_i, B_i, C_i (i = 1, 2, 3, and 4) and $n(\infty)$ of the F-B model and the energy bandgap (E_g) for *nc*-Si, which are shown and compared to their corresponding values of bulk crystalline silicon in table 1. The optical constants of the *nc*-Si are calculated with Eq. (4) and Eq. (5) using the parameters given by table 1. The optical constants of the *nc*-Si are shown in figure 4. The optical constants of bulk crystalline silicon are also included in the two figures for comparison. As can be seen in figure 4, the Si nanocrystals show a significant reduction in the optical constants and as compared to bulk crystalline silicon.

	A_i	B_i	C_i	$n(\infty)$	E_g
Bulk	0.0036	6.881	11.849	2.369	1.12
crystalline	0.014	7.401	13.747		
silicon	0.0683	8.634	18.795		
	0.0496	10.234	26.503		
Silicon	0.0538	7.112	12.718	2.824	1.737
nanocrystals	0.0056	8.016	16.080		
embedded	0.0603	8.030	18.710		
in SiO_2	0.0003	10.323	26.645		

Table 1: Values of the parameters A_i, B_i and C_i (i=1, 2,3,4), $n(\infty)$ and E_g of the F-B model for both bulk crystalline silicon and the *nc*-Si embedded in SiO_2.

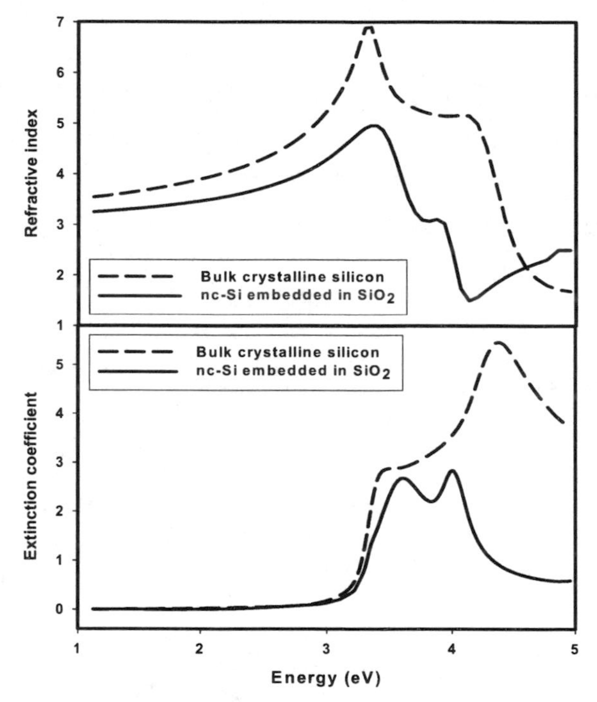

Figure 4: Refractive index (n) and extinction coefficient (k) of the *nc*-Si and bulk crystalline silicon as functions of wavelength.

It has been well known that reduction of the static dielectric constant becomes significant as the size of the quantum confined physical systems, such as quantum dots and wires, approaches the nano-metric range [13-15]. In our study, the static dielectric constant of the *nc*-Si embedded in SiO_2 matrix is 9.7, which is obtained from the calculation with the four-term F-B model by setting the photon energy to zero in Eq. (2). Taking into account the quantum confinement effect and screening effect, Wang *et al.* [7] pointed out that the static dielectric constant of the *nc*-Si as a function of the nanocrystal size could be expressed as follows:

$$\varepsilon_r(D) = 1 + \frac{\varepsilon_r(\infty) - 1}{1 + \left(\frac{6.9}{R}\right)^{1.37}} \quad (9)$$

NSTI-Nanotech 2005, www.nsti.org, ISBN 0-9767985-1-4 Vol. 2, 2005

where $\varepsilon_r(\infty)$ is the static dielectric constant of bulk crystalline silicon, and R is the radius of *nc*-Si with the unit of angstrom. Based on Eq. (9), where $\varepsilon_r(D)$ and $\varepsilon_r(\infty)$ are 9.7 and 11.4 respectively, the diameter of *nc*-Si is found to be ~ 4.5 nm, which is very close to the XRD result (the mean size is ~ 4.2 nm) mentioned above.

Although it is still uncertain that the reduction of dielectric constant is due to the opening of the bandgap, a bandgap expansion is indeed observed in this work for the *nc*-Si. As given in table 1, the bandgap of the *nc*-Si is 1.74 eV, which shows a bandgap expansion of ~ 0.6 eV as compared to the bandgap of bulk crystalline silicon. The bandgap obtained in this work is in very good agreement with the first-principle calculation of the optical gap of silicon nanocrystals in Ref. [16] based on quantum confinement. On the other hand, the size distribution of the *nc*-Si could be estimated from the energy bandgap expansion. Assuming a log-normal size distribution for the nc-Si, the size distribution could be calculated with the following equation proposed by Ranjan *et al.* [17]:

$$\Delta E_g = \frac{3.9}{d_0{}^n}\left(\frac{d_m}{d_0}\right)^{n[2n+5]/3}, \qquad (10)$$

where d_0 (in nm) is the mean size of nanocrystals, d_m is the size for which the maximum occurs in the log-normal distribution, and n = 1.22. For $\Delta E_g = 0.6$ eV and $d_0 = 4.2$ nm, the calculation with Eq. (11) yields $\frac{d_m}{d_0} = 0.96$. This indicates that the *nc*-Si has a very uniform size distribution, being consistent with the general belief that silicon nanocrystals synthesized by ion implantation have a narrow size distribution. A narrow size distribution is very useful to device applications of the *nc*-Si.

SUMMARY

In summary, we have developed a nondestructive approach to study the optical properties of silicon nanocrystals embedded in SiO_2 matrix by spectroscopic ellipsometry based on Maxwell-Garnett effective medium approximation. Thanks to the *nc*-Si depth profile obtained from SIMS measurements and a proper multi-layer model, we have done a complete optical characterization of silicon nanocrystals formed into SiO_2 film in a wide range of 1.13 - 4.96 eV. Optical constants and dielectric functions of *nc*-Si embedded in SiO_2 matrix have been obtained from the best spectra fitting based on the four-term F-B model. It is shown that there are a remarkable dielectric suppression and a large bandgap expansion due to the quantum size effect. The mean size of silicon nanocrystals obtained from the dielectric suppression is consistent with the result of XRD measurement. The bandgap of the *nc*-Si is ~ 1.7 eV, showing a large bandgap expansion of ~ 0.6 eV. The bandgap expansion is in very good agreement with the first-principle calculation of the optical gap of silicon nanocrystals based on quantum confinement effect.

This work has been financially supported by the Ministry of Education Singapore under project NO. ARC 1/04.

REFERENCES

[1] R. A. Rao, R. F. Steimle, M. Sadd, C. T. Swift, B. Hradsky, S. Straub, T. Merchant, M. Stoker, S. G. H. Anderson, M. Rossow *et al.*, Solid-State Electron. **48**, 1463 (2004).
[2] D. N. Kouvatsos, V. Ioannou-Sougleridis, and A. G. Nassiopoulou, Appl. Phys. Lett. **82**, 397 (2003).
[3] S. H. Choi and R. G. Elliman, Appl. Phys. Lett. **75**, 968 (1999).
[4] S. Tiwari, F. Rana, H. Hanafi, A. Hartstein, E. F. Crabbé, and K. Chan, Appl. Phys. Lett. **68**, 1377 (1996).
[5] L. Pavesi, L. Dal Negro, C. Mazzoleni, G. Franzo and F. Priolo, Nature **408**, 440 (2000).
[6] N. Lalic, and J. Linnros, J. Lumin. **80**, 263 (1999).
[7] Lin-Wang Wang and Alex Zunger, Phys. Rev. Lett. **73**, 1039 (1994).
[8] C. Delerue, G. Allan, and M. Lannoo, Phys. Rev. B **48**, 11024 (1993).
[9] Igor Vasiliev, Serdar Ogut, and James R. Chelikowsky, Phys. Rev. Lett. **86**, 1813 (2001).
[10] T. P. Chen, Y. Liu, M. S. Tse, S. Fung, and Gui Dong, J. Appl. Phys. **95**, 8481 (2004).
[11] T. P. Chen, Y. Liu, M. S. Tse, O. K. Tan, P. F. Ho, K. Y. Liu, D. Gui and A. L. K. Tan, Phys. Rev. B **68**, 153301 (2003).
[12] A. R. Forouhi and I. Bloomer, Phys. Rev. B, **38**, 1865 (1988).
[13] D. R. Penn, Phys. Rev. **128**, 2093 (1962).
[14] Raphael Tsu, Davorin Babic, Liderio Ioriatti, Jr., J. Appl. Phys. **82**, 1327 (1997).
[15] Lin–Wang Wang and Alex Zunger, Phys. Rev. B **53**, 9579 (1996).
[16] Serdar Ogut, J. R. Chelikowsky, and S. G. Louie, Phys. Rev. Lett. **79**, 1770 (1997).
[17] V. Ranjan, M. Kapoor, and V. A. Singh, J. Phy. Condensed Matter, **14**, 6647 (2002).

Charge Trapping and Charge Decay in Silicon Nanocrystals

C.Y. Ng[*1], L. Ding[*2] and T.P. Chen[*3]

[*]School of Electrical and Electronic Engineering,
Nanyang Technological University, Singapore 639798, [1]chiyung@pmail.ntu.edu.sg
[2]ding0008@ntu.edu.sg [3]echentp@ntu.edu.sg

ABSTRACT

In this work, we present a study on the charge trapping and charge decay in nc-Si embedded in SiO_2 by electrostatic force microscopy (EFM). The influence of silicon nanocrystals (nc-Si) distribution in the SiO_2 matrix on charge injection (charge spot size) and charge decay (characteristic decay time and decay mechanism) are investigated. The charge decay is found to be affected by the neighboring charge and charge sign for different nc-Si distributions in the SiO_2 matrix. Different dissipation mechanisms of the charges stored in the nc-Si with different distributions inside the SiO_2 matrix are discussed.

Keywords: silicon nanocrystal, charge trapping, charge decay, EFM

1 INTRODUCTION

When the dimensions of a silicon nanocrystal based memory device are scaled down, various quantum effect as well as single electron effects start to play a fundamental role and could affect the memory operation and reliability. In fact, when the area of the memory cell is sufficiently small, small charge fluctuation will change the electrical properties (i.e., flatband voltage) tremendously. For memory device applications, it is necessary to understand the charging and discharging behaviors of the nanocrystals. The distribution of nanocrystal inside the dielectric film will affect the charging and discharging characteristic. As such, it is important to evaluate the charging and discharging characteristics of different nanocrystal distributions inside the dielectric film.

Commonly, monitoring the charging effect for different nanocrystal distributions is done by determining the flatband voltage shift from capacitance-voltage (C-V) measurement [1-3]. However, this technique provides only macroscopic device information as C-V measurement responses to only an average variation over a large area. With the electrostatic force microscopy (EFM) technique, local electrical properties of the dielectric films and the distributions of trapped charges can be obtained in sub-micrometric resolution. In this work, we present a study on the charge trapping and charge decay characteristics and mechanisms of silicon nanocrystal (nc-Si) with different nanocrystal distributions using EFM technique.

2 EXPERIMENTAL SETUP

SiO_2 films were thermally grown to 750 nm on either P-type or N-type (100) oriented Si wafers in dry oxygen at 950 °C. The doping of the Si substrates does not play a role in this study as the substrates are just used for supporting the SiO_2 films. Si^+ ions with a dose of 3×10^{16} cm^{-2} were then implanted to the SiO_2 thin films at 14 keV (sample A) and 10 keV (sample X). Thermal annealing was carried out at 1000 °C in N_2 ambient for 1 hour to induce nc-Si formation. After annealing, about 17 nm SiO_2 was removed in diluted HF solution for sample A. The peak concentration of nc-Si is located very close to the SiO_2 surface for sample A and at the depth of ~ 10 nm underneath the surface for sample X. Transmission electron microscopy (TEM) measurement indicates that the mean size of nc-Si for both samples is ~ 4 nm. EFM studies were performed at room temperature in air with a Veeco/Digital Instrument Dimension 3000 Scanning Probe Microscope.

By using EFM technique, the amount of charge trapped in the nc-Si can be deduced based on the force arising from the Coulomb interactions between the charge of the sample and the charge on the tip. The total charge in nc-Si embedded in SiO_2 dielectric films during charge decay can be mapped with the EFM based on the total force given below [4,5]

$$F(z)=\frac{1}{(z+\frac{d}{\varepsilon})^2}\left(\frac{-\varepsilon_o AV^2}{2}+\frac{dQV}{\varepsilon}-\frac{d^2Q^2}{2\varepsilon^2\varepsilon_o A}\right) \quad (1)$$

where d is the SiO_2 film thickness, A is the area of the charge region, ε is dielectric constant of SiO_2, z is the distance between the tip and the sample, and V is the EFM DC bias. The force is measured from the shift of the resonance frequency of the EFM tip with respect to its nominal frequency. This shift reflects the coulomb force gradient detected by the EFM tip and is written as:

$$\frac{2k\Delta f}{f_o}=\frac{\partial F(z)}{\partial z} \quad (2)$$

where f_o is the nominal frequency, k is the cantilever spring constant estimated from the lever geometry, and Δf is the shift of resonance frequency. From the measured force, the total charge Q is determined with Eq. (1) and (2). Charge was injected to the sample via the contact-mode EFM, while the tapping-mode EFM was used to monitor the discharging of nc-Si.

3 RESULTS AND DISCUSSION

Figure 1 shows the 3D EFM images of initial charge spot for the nc-Si distributing at the SiO_2 surface region (sample A) (a) and nc-Si distributing deep inside the SiO_2 film (sample X) (b) after a charge injection at -4V for 10 s respectively. A wider charge spot is observed for sample A, indicating different charging behaviors between sample X and sample A. For the former, negative charge can be directly trapped in the nc-Si due to the direct contact between EFM tip and nc-Si, while for the latter, negative charge needs to overcome a potential barrier of SiO_2 before charge can be trapped in the nc-Si.

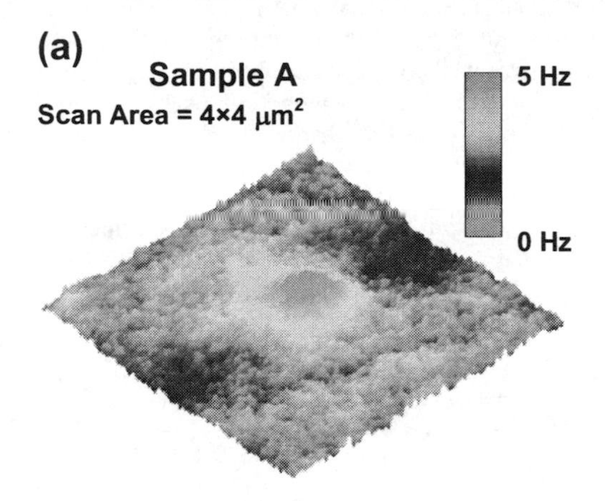

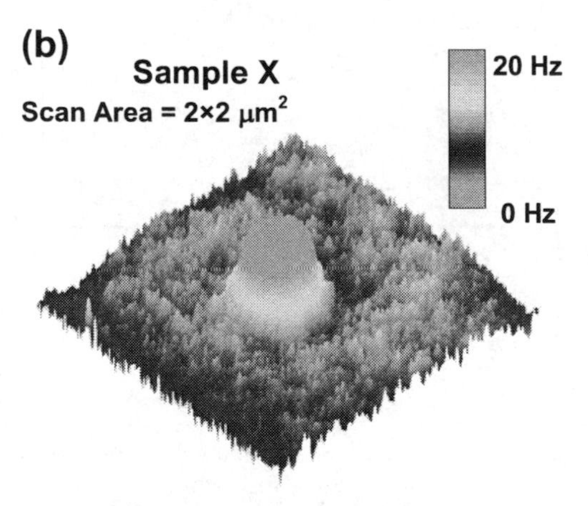

Figure 1: 3D EFM images taken straight after the charge injection for sample A and sample X.

To understand the charge trapping and charge decay in nc-Si, charge decay from the dielectric film is monitored with a series of snap shots of EFM images recorded in real time after constant voltage injection of -4 V for 10 s. Figure 2 shows the charge decay characteristics for the two samples. The initial charges (Q_0) of the stored charges in nc-Si and the characteristic decay time (τ) can be estimated from the curve fitting to the experiment data with $Q(t) = Q_o \exp(-\frac{t}{\tau})$. The initial charge is 1103 electrons for sample A and 84 electrons for sample X. The initial charge of sample A is much larger than that of sample X, indicating that the charge injection into sample A is easier and more efficient. For sample A with most of the nanocrystals distributing in the surface region, lateral charge diffusion occurs easily during the charge injection, leading to a larger charge spot area and larger number of charge. For sample X with most of the nanocrystals distributing in the region at least ~ 5 nm away from the surface, the charge injection is relatively difficult, and thus the charge spot is smaller and the number of trapped charge is less than sample A. A different characteristic decay time τ is observed, indicating a different dissipation mechanism between sample A and sample X. Besides obtaining the initial charge of the two samples, the characteristic decay time is 155 s for sample A and 310 s for sample X. The τ for sample X is 2 times larger than that of sample A is mainly due to the easy charge dissipation on the surface of sample A. At film surface, the charge can easily interact with the surface air and charge can easily dissipate via the air humidity. For sample X, the trapped charge has to overcome a large barrier height of SiO_2 before the charge can either dissipate to the film surface or to the SiO_2 / Si interface. As such, the dominant charge decay mechanism may be due to the trapped charge dissipates to the surrounding uncharged nc-Si.

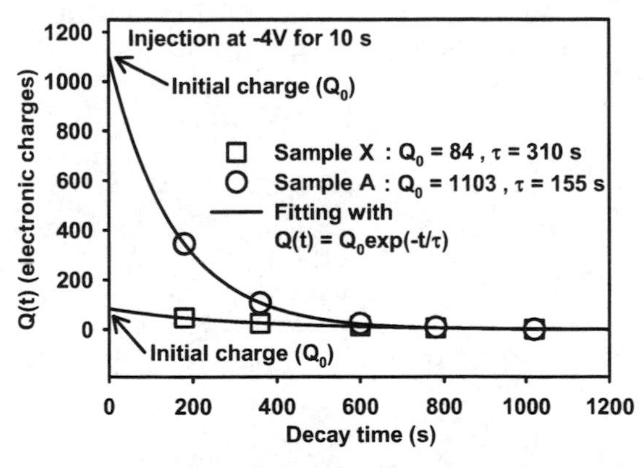

Figure 2: Charge decay characteristics for sample A and sample X.

The influence of neighboring charges is studied with the creation of additional charge spot close to the charge spot under test. To study the influence of neighboring charges on the charge decay, we have created one charged spot, two charged spots. For the cases of 2 charged spots, the charge in one spot has the sign (positive or negative) either the same as or opposite to that of another spot. Such an experimental design can give us a clear picture of the influence of neighboring charges on the charge diffusion.

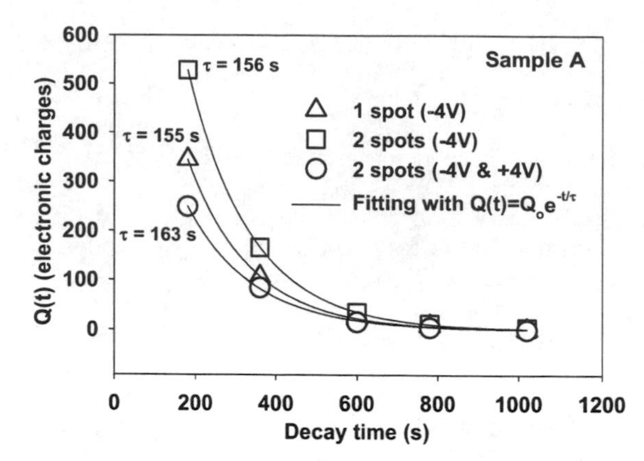

Figure 3: Influence of neighboring charge on the charge decay in sample A [6].

Figure 3 shows the characteristic decay time under the influence of neighboring charge spots on sample A. For the case of 2 spots (-4V) and 2 spots (-4V & +4V), the center of the 2nd charge spot is located ~ 1 μm away from the center of the 1st spot. The time interval between the end of 1st spot charge injection and the start of the 2nd spot charge injection is ~ 2 s. With the existence of neighboring charges, an almost constant τ for sample A is observed. Specially, the surrounding charges with the same charge sign or opposite sign do not lead to a significant change in the decay time. This indicates that lateral charge diffusion is insignificant for sample A. The trapped charge can easily dissipates to the film surface as no lateral charge diffusion and no influence on the neighboring charge sign as well as neighboring charge spot.

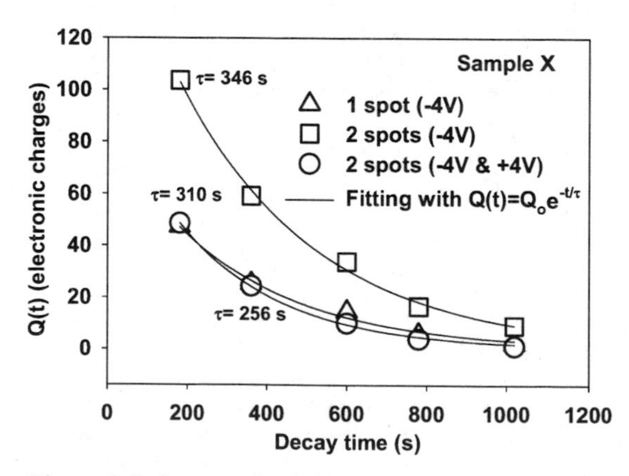

Figure 4: Influence of neighboring charge on the charge decay in sample X [6].

On the other hand, figure 4 shows the characteristic decay time under the influence of neighboring charge spots on sample X. For the case of 2 spots (-4V) and 2 spots (-4V & +4V), the center of the 2nd charge spot is located ~ 0.3 μm away from the center of the 1st spot. With the existence of neighboring charges, a different τ for sample X is observed. For sample X, a positive or negative neighboring charge leads to a shorter or longer characteristic decay time, respectively. During decay, the negative neighboring charge resists the charge dissipation of the negative charge spot under test while the positive neighboring charge accelerates the charge dissipation. The trapped charge will dissipate to the surrounding uncharged nc-Si. This shows that the charge decay is affected by the neighboring charges in sample X while it is not affected in sample A. The difference between the two samples is consistent with the early discussions on the influence of the nc-Si distribution on the charge diffusion. For sample A, as mentioned early there is no lateral charge diffusion during the decay period because the injected charges can easily dissipate on the surface, and therefore the neighboring charges have no significant influence on the charge decay. For sample X, there is a strong lateral charge diffusion during decay. As the neighboring charges can either accelerate or resist the lateral charge diffusion, they have a significant influence on the charge decay.

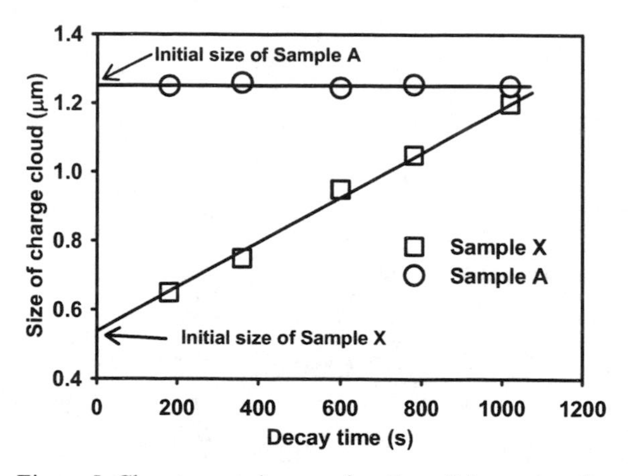

Figure 5: Charge-spot size as a function of decay time for sample A and sample X.

The influence of the nc-Si distribution on the size of charge spot can be clearly seen in Figure 5. The size of charge spot is obtained from pseudo-voigt fitting to the EFM resonance frequency. For sample A, the size of the charge spot doesn't change with the decay time and is ~ 1.25 μm for decay time longer than 3 minutes. For sample X, the size of charge spot increases with the decay time linearly. The linear relationship indicates that the speed of the lateral charge diffusion is constant. The size for sample X is smaller than that for sample A. However, it increases with decay time linearly. In contrast, the size for sample A doesn't change with decay time. For the latter case, there is no lateral charge diffusion as the trapped charges can easily dissipate on the surface. For sample X with most of the nanocrystals distributing in the region away from the SiO_2 surface, the charge injection is relatively difficult, and thus the charge spot is smaller; after the injection, the injected charges diffuse in all directions, leading to the increase in the size of the charge spot with decay time.

4 CONCLUSION

The charge trapping and charge decay in nc-Si embedded in SiO_2 have been studied with electrostatic force microscopy (EFM). The influence of silicon nanocrystals (nc-Si) distribution in the SiO_2 matrix on charge injection and charge decay are investigated and discussed. For nc-Si distributed near the SiO_2 film surface, large charge spot (during charge injection) and shorter characteristic decay time (during charge decay) are observed. In contrast, for nc-Si distributed inside the SiO_2 matrix, smaller charge spot (during charge injection) and longer characteristic decay time (during charge decay) are observed. The observed phenomena are discussed in details.

ACKNOWLEDGEMENTS

This work has been financially supported by the Ministry of Education Singapore under project No. ARC 1/04. The author would like to thank Y. Liu, H.C. Ho and K.H. Lee for sample fabrications.

REFERENCES

[1] C.Y. Ng, T.P. Chen, Y. Liu, M.S. Tse, and D. Gui, Electrochemical and Solid-State Lett. 8, G8, 2005.

[2] C.Y. Ng, T.P. Chen, and M.S. Tse, presented at the 10th International Symposium on Integrated Circuits, Devices and Systems (ISIC-2004), 8-10 September 2004 Singapore, 2004.

[3] Y. Liu, T.P. Chen, C.Y. Ng, and L. Ding, to appear in Proc. Nanotech 2005.

[4] C.Y. Ng, T.P. Chen, H.W. Lau, Y. Liu, M.S. Tse, O.K. Tan, and V.S.W. Lim, Appl. Phys. Lett. 85, 2941, 2004.

[5] C.Y. Ng, T.P. Chen, V.S.W. Lim and M.S. Tse, Proc. ICSE 2004, 160, 2004.

[6] C.Y. Ng, T.P. Chen, V.S.W. Lim, and M.S. Tse, presented at the 10th International Symposium on Integrated Circuits, Devices and Systems (ISIC-2004), 8-10 September 2004 Singapore, 2004.

Synthesis of Antimony Oxide Nano-particles by Thermal Oxidation

C. H. Xu, S. Q. Shi* and Q. Tang

Department of Mechanical Engineering, The Hong Kong Polytechnic University,
Hung Hom, Kowloon, Hong Kong, China, *mmsqshi@polyu.edu.hk

ABSTRACT

Sb_2O_3 nano-particles are synthesized by thermal oxidation in this research. Pure Sb granular is put in the middle of a tube furnace at an air pressure of 1 atm with a flow rate of 4.0 ml/min. The furnace temperature at the position of samples is set at 550°C. Si wafer, glass and Al foil are used as substrates, which are put the downstream of gas flow to collect Sb oxide. The collected Sb oxides on different substrates are examined with field emission scanning electron microscopy, X-ray diffractometer and transmission electron microscopy. Sb oxide obtained in this method shows Sb_2O_3 nano-particles with strong {111} growth texture. The mechanism of synthesis of Sb_2O_3 nano-particles is analyzed, based on the experiment results.

Keywords: anitomony oxide, nano-particles, thermal oxidation, field emission scanning electronic microscope, X-ray diffraction.

1 INTRODUCTION

Antimony (Sb) oxide can be used as catalyst, retardant, fining agent and optical materials [1]. Recently, hydrous Sb oxide was found to have high proton conductivity, which is likely to be a promising humidity-sensing materials [2, 3]. Sb_2O_3 plus ZnO have been used to make films that can be used as gas sensor materials [4].

Nano-particles have increasingly attracted interest over the past decade due to the possibilities of novel or even outstanding properties compared with bulk materials. Semiconductor nanocrystaline materials have been an area of intense investigation [5, 6]. It was reported recently that the synthesis of Sb oxide nano-particles can be done by the chemical method [7] and γ-ray radiation-oxidation route [8], in which $SbCl_3$ and NaOH are used as starting materials. Another reported method of synthesis of Sb oxide nano-particales is vapor condensation, which includes three steps: melting Sb by laser, oxidizing vapor Sb and condensing Sb oxide [9]. However, the mixture of Sb and Sb oxide nano-particles is usually obtained in this method. The experiment equipment is also expensive.

In this paper we report the synthesis of antimony oxide nano-particles by heating metal Sb in solid state in oxidative environment and collecting vapor Sb oxides on the substrates of Si wafer, Al foils, and glass at the downstream of gas flow.

2 EXPERIMENT PROCEDURE

Commercial antimony (Sb) granular with average diameter of 1.5 mm (purity: 99.99% Sb) is used in the present experiment. Sb granulars on an alumina crucible are put in the middle of a tube furnace in compressed air at a pressure of 1 atm with a constant flow rate of 4.0 ml/min for 4 hours. The furnace temperature at the position of samples is set at 550°C. Si (001) wafer, glass and Al foil are used as substrate materials and the substrates are put the downstream of gas flow to collect Sb oxide.

After oxidation, the as-synthesized Sb oxides on the substrates show a white color. The crystal structures of the collected Sb oxides on the substrates are directly examined with a Philips PW3710 X-ray diffractometer, using a 40 kV, 30 mA, Cu Kα X-ray. Morphologies of the collected Sb oxide on the substrates are also characterized by a field emission scanning electron microscopy (JEOL JSM-6335F). The collected Sb oxides on the substrates are dispersed into ethanol. The bottle of the ethanol solution is put in an ultrasonic machine (COLE-Parmer 8890) for 10 min to separate Sb oxides from the substrates. Then the resulting solution is dropped onto a carbon coated TEM copper grid. Transmission electron microscopy (TEM) imaging, electron diffraction and local composition analysis are performed on the samples on the TEM grid using a JEOL 2010F TEM with an energy dispersive (X-ray) spectroscopy (EDS) system, operated at 200 kV.

3 EXPERIMENT RESULTS

3.1 Shape and Size of Nano-particles

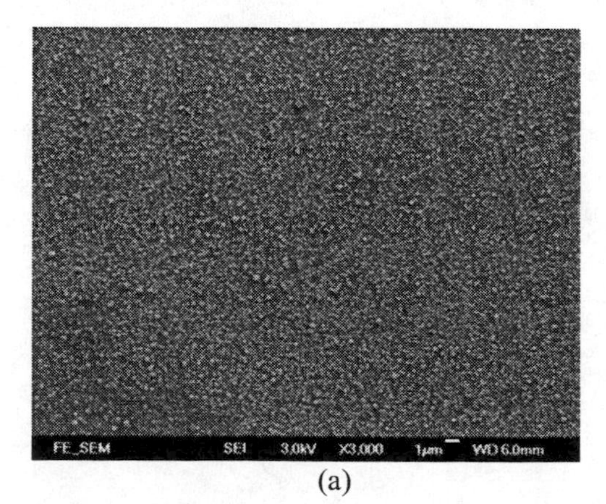

(a)

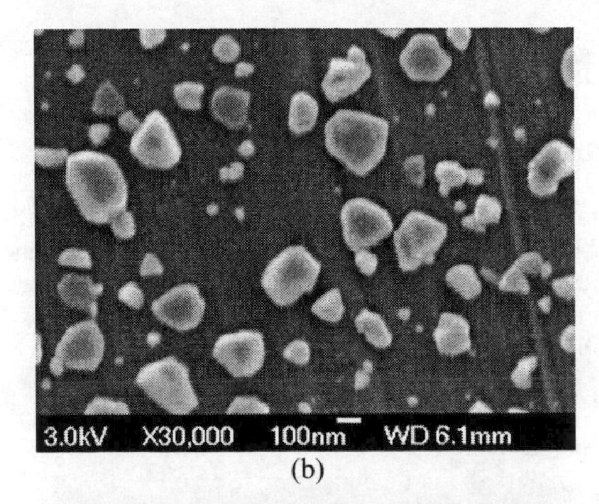

(b)

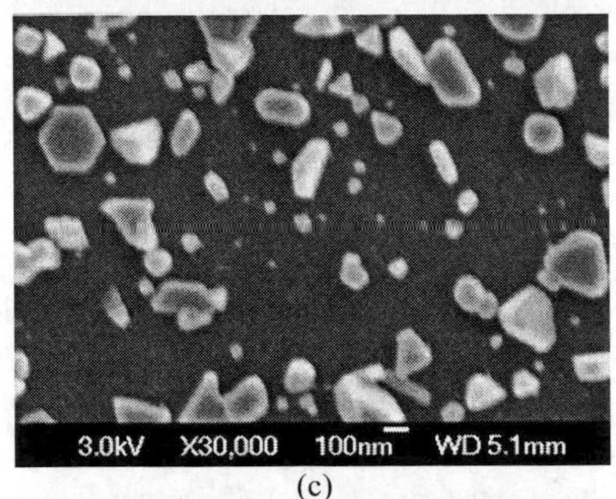

(c)

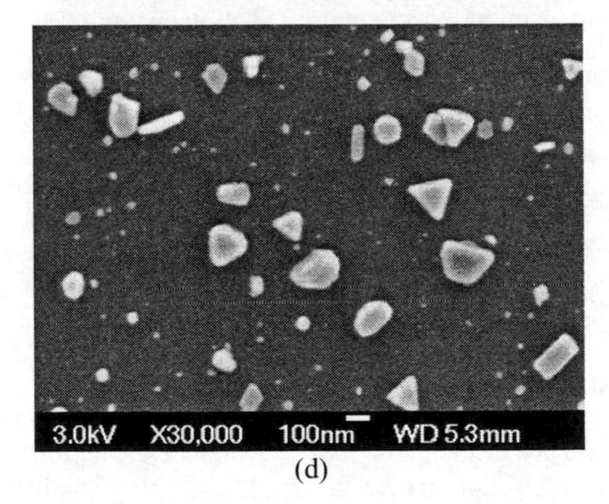

(d)

Figure 1 Morphologies of the collected Sb oxides on the substrate of: (a) Al foil at low magnification, showing uniform particles, (b) Al foil (c) glass and (d) Si (001) wafer at high magnification

The Sb oxide nano-particles uniformly distribute on the substrates. Fig. 1(a) shows the uniform Sb oxide particles on the substrate of Al foil. The shapes and sizes of the collected nano-particles can be seen clearly at high magnifications. Fig 1(b), (c), (d) are the morphologies of the nano-particles on the substrates of Al foil, glass, and Si (001) wafer at high magnifications, respectively. The nano-particles on the different substrate have similar features. The sizes of the nanoparticles are at the range of 10 ~ 200 nm. Large particles show significant crystalline structures. The shapes of these particles include triangle, hexagon, rectangular.

3.2 X-Ray Diffraction

The X-ray diffraction results show that the phase of the collected Sb oxide is Sb_2O_3 with face-centered cubic structure. Only (111), (222) and (444) diffraction peaks occur on XRD spectra, suggesting a strong growth textures of Sb_2O_3 on all three type of substrates. X-ray spectrum of the collected nano-particles on the substrate of glass is shown in Fig. 2(b). An X-ray spectrum for the Sb specimen before oxidation is shown in Fig 2(a). Comparing the two spectra, it can be concluded that no metal Sb is in the collected particles.

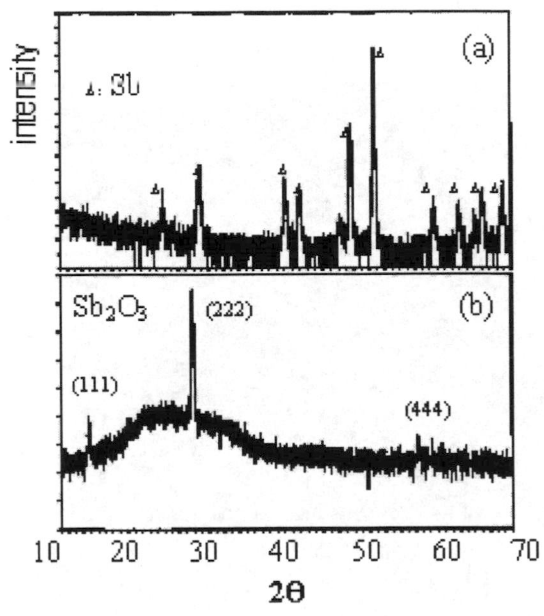

Fig. 2 X-ray spectra (a) pure metal Sb before oxidation and (b) the collected Sb_2O_3 oxides on the glass substrate, showing strong growth texture of (111).

3.3 TEM Results

TEM images in Fig. 3 show Sb oxide nano-particles. Fig. 3(a) shows the morphologies of Sb oxides. The grain size changes from 10 to 150 nm. A large grain tends to be in triangle shape in Fig. 3(a). The corresponding selected area electron diffraction pattern of the triangle grain is illustrated at the right bottom corner. The pattern comes form {440} of Sb_2O_3, which indicates the surface of the triangle Sb oxide is (111) of Sb_2O_3. Fig. 3(b) is a high resolution image taken on the triangle Sb_2O_3 grain in Fig. 3(a). The space of 1.95 A is corresponding for the (440) plane of Sb_2O_3. The Fig. 3(c) is the EDS result of nano-particles in Fig. 3(a), showing Sb and O elements (Cu and C in Fig. 3(c) come from a carbon coated TEM copper grid).

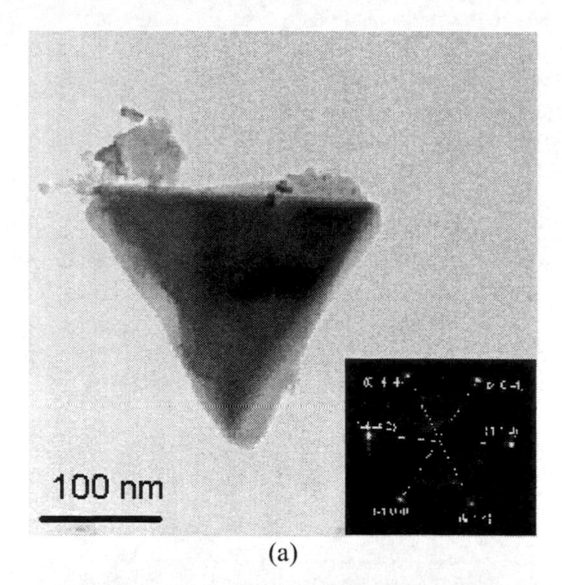

(a)

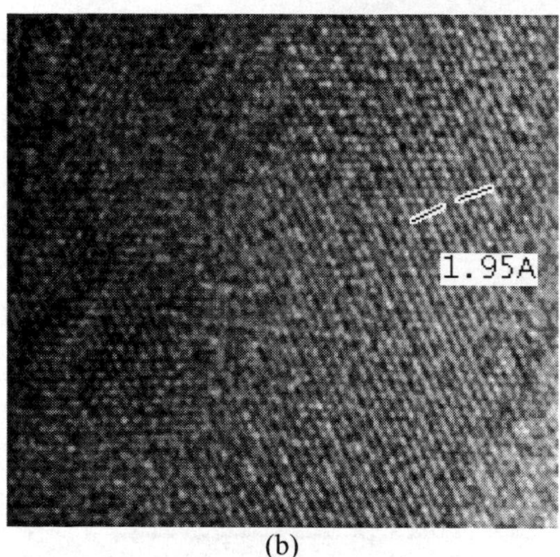

(b)

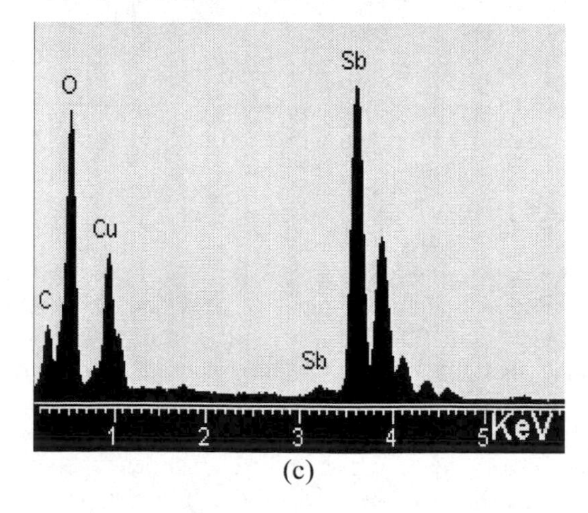

(c)

Fig. 3 TEM images for Sb_2O_3 oxide particles (a) morphology and corresponding selected area electron diffraction of a large triangle grain and (b) a high resolution image taken from the triangle Sb_2O_3, and (c) the EDS result of nano-particles, showing Sb and O

4 DISCUSSIONS

4.1 Phase of Collected Nano-particles

After the oxidation of pure metal Sb in air at 550°C, the collected oxides on the substrates of Si wafer, Al foils and glass at the positions of the downstream of gas flow are Sb_2O_3 with strong {111} growth textures. The size of the nano-oxide particles is in the range of 10 ~ 200 nm. It was reported that melting metal Sb in oxidation environment by laser can be used in syntheses of Sb_2O_3 nano-particales. This technique includes three steps: melting Sb by laser, oxidizing vapor Sb and condensing Sb oxide [9]. The mixture of Sb and Sb oxide nano-particles is usually obtained in this method, due to melting metal during the process. In present experiment, pure Sb_2O_3 nano-oxide is obtained from X-ray spectra in Fig. 2(b). The furnace temperature at the position of Sb metal is set at 550°C, which is less than the melt point of Sb (630°C). According to Sb-O phase diagram [10], the high oxygen phase of Sb_2O_5 can be formed at high oxygen partial pressure, such as in air. The solid Sb_2O_5 oxide decomposes into liquid SbO_2 and O_2 at 525 °C under 1 atm. Compared with metal Sb, Sb oxide moves easily with gas flow, and the temperature of oxidation is higher than the decomposition temperature of Sb_2O_5 oxide. However, the collected oxide on the substrates in the experiment is Sb_2O_3 (not SbO_2). In order to explain the reason, more experiments are needed. We will analyze this in another report in details.

4.2 Growth Texture of Sb_2O_3 Oxide

It can be concluded based on the analysis on X-ray spectra in Fig. 2 that the collected white oxide on the substrates are face-centered cubic Sb_2O_3 with extensive strong {111} growth texture. It can be seen from Fig. 2 that three types of the substrate do not affect the formation of Sb_2O_3 nano-particles much, which can be understood from the crystalline structures of oxide and substrates. The structure of glass substrate is amorphous. The average atomic distance of the glass ($d_{average}$) can be evaluated from the X-ray diffraction pattern in Fig. 2(b). There is widen peak at the position of $2\theta = 15 - 35°$ on the background curve, which stands for the average atomic distance of amorphous material, glass. This distance of $d_{average}$ can be calculated by follow equation [14]

$$d_{average} = \frac{\lambda}{2\sin(\theta)} \tag{1}$$

where λ is the wavelength of diffraction beam, Cu kλ (=1.541 A). The calculation results are listed in Table 1. When one type of material is deposited on a substrate, there are three types of interfaces between a deposition and a substrate, namely coherent, semicoherent and incoherent

[11, 12, 13], depending on the crystalline parameters of the deposition and the substrate. Lattice parameters of Sb_2O_3 and substrates of Al, Si, and glass show big different, as shown in Table 1, so it is easy to form incoherent interface between Sb_2O_3 oxide and the substrates. For this reason, the morphologies of Sb_2O_3 oxides formed on the different substrate are similar.

phase	structure	lattice parameter (A)	Ref.
Sb_2O_3	FCC	a = 11.152	[14]
Al	FCC	a = 4.049	[14]
Si	Diamond cubic	a = 5.428	[14]
Glass	Amorphous	$d_{average}$ = 5.902 – 2.562	

Table 1: Crystalline structures and parameters of materials

The planar energy in a crystal reverses its inter-planar distance *d*. In other word, the plan with the highest density has the lowest energy. {111} planes in face-centered cubic have the highest density [15]. The crystalline structure of Sb_2O_3 oxide is face-centered cubic. In order to remain the lowest energy in the system, the {111} planes dominate on the surface so that the {111} texture Sb_2O_3 oxide forms on different substrates. Recently, three-dimensional molecular dynamics simulation has also indicated that {111} planes in face-centered cubic have the lowest energy [16].

The grains of the collected Sb_2O_3 oxide are in the range of 10 ~ 200 nm, as shown in Fig. 1 and Fig. 3. The large grains show significant crystalline structures, including triangle, hexagon, and rectangular, while small ones tend to be in round shape. When a new phase forms, the formation energy of the new phase includes surface energy and misfit strain energy. The shape of a new phase is determined by maintaining the lowest energy state in the system. For an incoherent Sb_2O_3-substrate interface, the system energy is dominated by the surface energy because the misfit strain energy is small. In order to achieve thc lowest surface energy, the new phase with a small size appears round shape. When the grains of new phase grow larger, the energy difference on various planes becomes significant. The surface of new grains contacted with air is dominated by {111} planes of Sb_2O_3 in order to reduce the total surface energy. Therefore, the large grains show crystalline structures, such as triangle, hexagon and rectangular shapes.

5 CONCLUSIONS

The following conclusions can be drawn:

1. Pure Sb_2O_3 nano-particle oxides can be synthesized by heating metal Sb at 550°C under a pressure of 1 atm in a constant air flow rate and collecting vapor oxides on a substrate at the downstream of gas flow.
2. The size of the Sb_2O_3 nano-particle oxides is in the range of 10 ~ 200 nm.
3. Sb_2O_3 oxides show the strong {111} growth texture, which is independent on the substrates of Al, Si and glass.

ACKNOWLDGEMENTS

This work was funded by Research Grant Council of Hong Kong (B-Q747).

REFERENCES

[1] G. V. Samsonov, "The Oxide Handbook", IFI/Plenum, 320-441,1973

[2] K. Ozawa, Y. Sakka and A. Amamo, Journal of Materials Rcsearch, 13, 830, 1998

[3] D. J. Dzimitrowice, J. B. Goodenough and P. J. Wiseman, Mater. Res. Bull. 17, 971, 1982

[4] B. L. Zhu, C. S. Xie, A. H. Wang, D. W. Zheng, M. L. Hu and W. Y. Wang, Materials Research Bulletin, 39, 409-415, 2004

[5] W. Q. Han, S. S. Fan, Q. Q. Li and Y. D. Hu, Science, 277, 1287, 1997

[6] A. M. Morales and C. M. Liber, Science 279, 208, 1998

[7] Z. L. Zhang, L. Guo, and W. D. Wang, Journal Of Materials Research 16, 803-805, 2001

[8] Y. P. Liu, Y. H. Zhang, M. W. Zhang, W. H. Zhang, Y. T. Qian, L. Yang, C. S. Wang, Z. W. Chen, Materials Science and Engineering B 49, 42-45, 1997

[9] D. W. Zeng, B. L. Zhu, C. S. Xie, W. L. Song and A. H. Wang, Materials science and Engineering A 366, 332-337, 2004

[10] Thaddeus B. Massalski ; editors, Hiroaki Okamoto, P.R. Subramanian, Linda Kacprzak. "Binary alloy phase diagrams", Materilas Park: ASM International, 2912-2913, 1990

[11] J. Mayer, C. P. Flynn, M. Ruhle, Ultramicroscopy, 33, 51, 1990

[12] G. Gutekunst, J. Mayer and M. Ruhle, Philos. Mag. A, 75, 1357, 1997

[13] V. Vitek, G. Gutekunst, J. Mayer, M. Ruhle, Philos. Mag. A, 71, 1219, 1995

[14] D. C. Cullity, "Elements of X-ray Diffraction" Addison-Wesley Publishing Company Inc., 482-485, 1967

[15] Lawrence H. Van Vlack, "Elements of Materials Science and Engineering" Fifth Edition, Addison-Wesley Publishing Company, 63-103, 1985

[16] W. C. Liu, Y. X. Wang, C. H. Woo, and H. C. Huang, Mat. Res. Soc. Symp. Proc. 677, AA7.32.1- AA7.32.5, 2001

Direct Electrochemical Reduction of CuO Nano Particles to Cu Nano Particles in NaCl Solution

Gil-Ho Hwang, Won-Kyu Han, Seok-Jun Hong, Jae-Woong Choi, and Sung-Goon Kang

Div. of Materials Science and Engineering, Hanyang University, Seoul, Korea, 133-791

ABSTRACT

Copper nano particles have been prepared by direct electrochemical reduction from copper oxide(CuO) and the reduction mechanism was investigated. To investigate the reduction mechanism, CuO has been deposited by magnetron sputtering in various Ar/O_2 ratio and cyclic voltammetry(CV) was performed in 0.5M NaCl solution at 300K. The results showed that the oxygen from the CuO was ionized at -0.874V(vs. SCE) and reduced to Cu. To fabricate Cu nano particles, we used CuO nano particles, prepared by a conventional mechanical milling, with a DC rectifier and the electrochemical cell. The characteristics of the films and nano particles were analyzed with XRD, SEM/EDS, and AES.

Keywords: Cu nano particles, electrochemical reduction, cyclic voltammetry

1 INTRODUCTION

Metallic nano particles have received significant attention by researchers due to their unique properties such as color, conductivity, melting point, magnetism, specific heat, and light absorption in comparison with bulky metal[1-5]. Especially, gold and silver nano particles are the most widely researched materials and these nano particles exhibit very useful properties in catalysis and biosensing[3-5]. Since 1990s, copper nano particles have been attracted by researchers due to their applications in catalysis[6]. In other words, copper nano particles show much higher specific catalytic efficiency than bulky copper due to their enormous surface area[7].

Many techniques such as radiation methods[8], micro emulsion techniques[9-11], super critical techniques[12-13], thermal reduction[14], sonochemical reduction[14-15], laser ablation[16], metal vapor synthesis[17], vacuum vapor deposition[18], and chemical reduction[19-20] have been developed to synthesize copper nano particles. Also, it was reported that the new process of electrochemical method for the direct reduction of TiO_2 at 800°C under argon[21] but this process led to the agglomeration of particles(sintering) due to its high processing temperature. Therefore, it is thought that this process is not suitable to produce nano particles.

In our study, we have synthesized Cu nano particles by using a direct electrochemical reduction technique form CuO nano particles without agglomeration of particles.

2 EXPERIMENTAL PROCEDURE

CuO films were prepared by DC magnetron sputtering on the AISI 430 in various Ar/O_2 ratio. The AISI 430 substrates were cleaned with acetone and ethanol before setting in the deposition chamber. The distance between the copper target and the substrate was 60mm. Prior to the copper oxide deposition, the AISI 430 substrate was *in-situ* etched by argon sputtering. The argon flow rate was fixed constantly at 25 sccm and the oxygen flow rate was varied from 0 to 12 sccm. Thus, the base pressure was 5 x 10^{-6} Torr and the working pressure was between 1.1 x 10^{-2} and 1.6 x 10^{-2} Torr. To measure the reduction potential of CuO to Cu, the cyclic voltammetry was carried out by using a conventional three electrodes cell with 0.5M NaCl solution. And CuO, prepared by DC magnetron sputtering, was reduced directly at the potential obtained in the cyclic voltammetry. To fabricate Cu nano particles, we used CuO nano particles, which were prepared by a conventional mechanical milling, with the DC rectifier and the electrochemical cell. The experimental apparatus was represented in Fig. 1. The glass container, filled with CuO nano particles with about 100nm in diameter, was immersed into 0.5M NaCl solution and Pt cathode was dipped into the nano particles. And a voltage of 16V, considering iR drop, was applied for 1h between the cathode and anode. The characteristics of the films and nano particles were analyzed with XRD, SEM/EDS, and AES.

3 RESULTS AND DISCUSSION

3.1 Preparation of CuO thin film

The phases of the films, prepared by reactive sputtering, were strongly dependant on the oxygen flow rate as can be seen in Fig. 2. As soon as the oxygen was introduced into the chamber, Cu_2O was obtained. When the oxygen flow rate was 4 sccm, only Cu_2O was detected. The range of the oxygen flow rate for single Cu_2O is very narrow. Indeed, as the oxygen flow rate went up to 5 sccm, the diffraction peaks of Cu_3O_4 was detected except Cu_2O. When the oxygen flow rate was more than 8 sccm, only CuO phase was obtained.

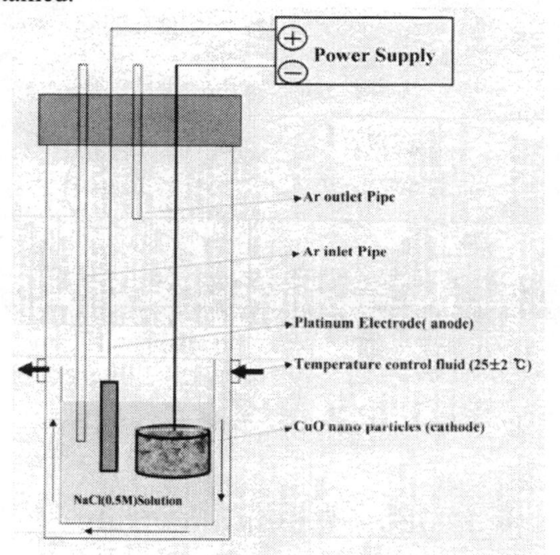

Fig. 1. Experimental apparatus for electrochemical reduction.

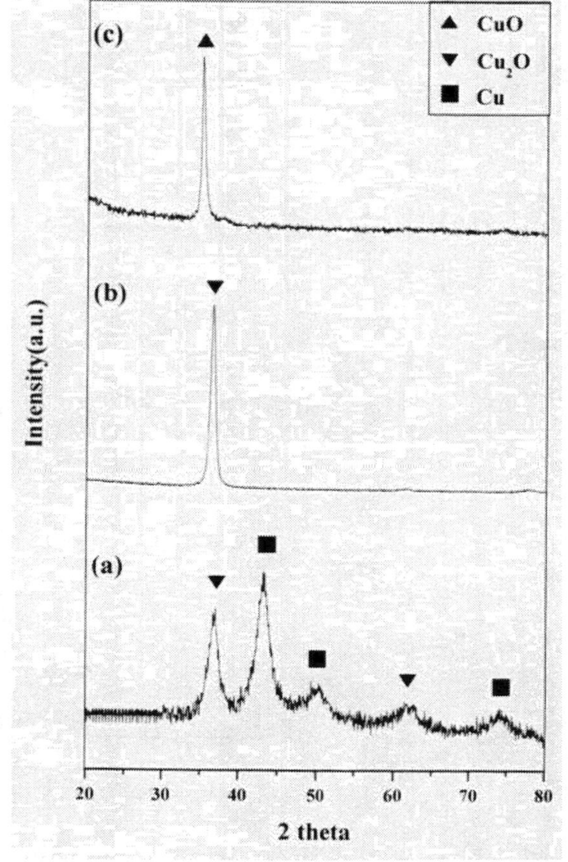

Fig. 2. XRD curves of deposited Cu oxide thin films at various O_2 flow rate(Ar flow rate was fixed at 25 sccm). a) 3 sccm, b) 4 sccm, c) 8 sccm

3.2 Electrochemical reduction of CuO

Fig. 3 shows the cyclic voltammograms(CV) of bulky Cu and CuO thin film in 0.5M NaCl solution at 300K.

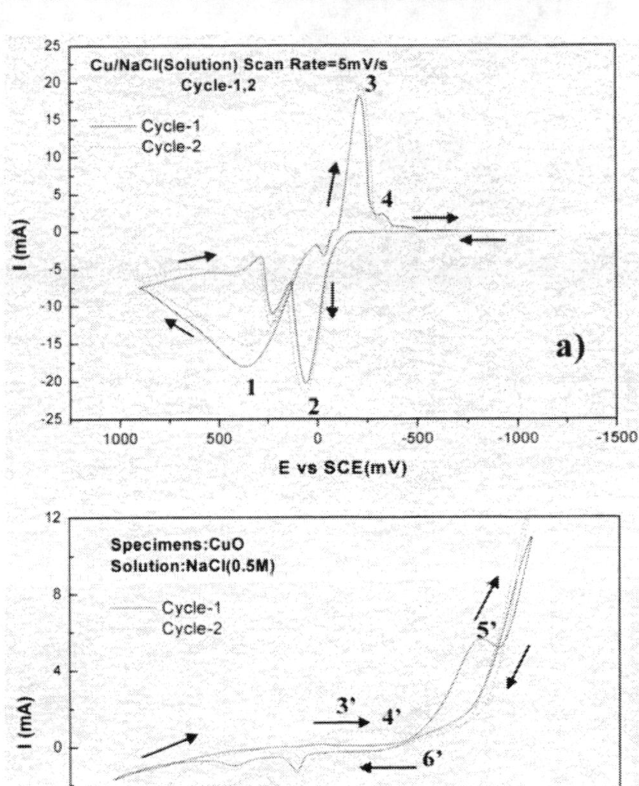

Fig. 3. Cyclic voltammograms of bulky Cu(a) and CuO thin film(b) in 0.5M NaCl at 300K

As can be seen in Fig. 3, the current wave 5' and 6' were observed in only CV curve of CuO thin film. Therefore, it was thought that the current wave 5' and its anodic counter part wave 6' originated from the existence of oxide scale. To ascertain the direct reduction of CuO, CuO thin film was reduced potentio-statically for 1h at the region of wave 5', -0.874V(vs. SCE).

Fig. 4 shows the cross-sectional SEM/EDS of CuO thin film and electrochemically reduced Cu thin film. As shown in Fig. 4, the thin film was about 4 micron in thickness and composed of copper and oxygen. But the thin film reduced potentio-statically was composed of only copper(Fig. 4(b)).

Fig. 5 shows the AES depth profile curves CuO thin film and electrochemically reduced Cu thin film. As can be seen in Fig. 5, the atomic ratio of copper and oxygen in CuO thin film is exactly 1:1. And Fig. 5(b) indicates that the CuO has completely transformed to Cu after electrochemical reduction. Therefore, we can conclude that the CuO thin film was directly reduced to Cu.

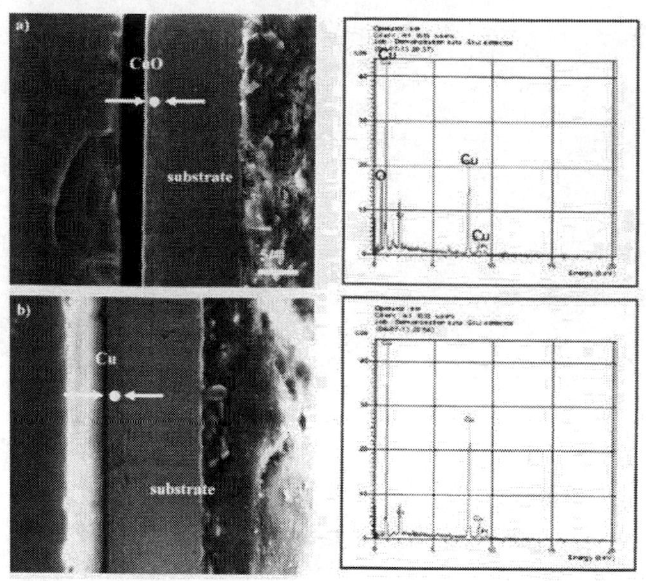

Fig. 4. Cross-sectional SEM/EDS images of CuO thin film(a) and electrochemically reduced Cu thin film(b).

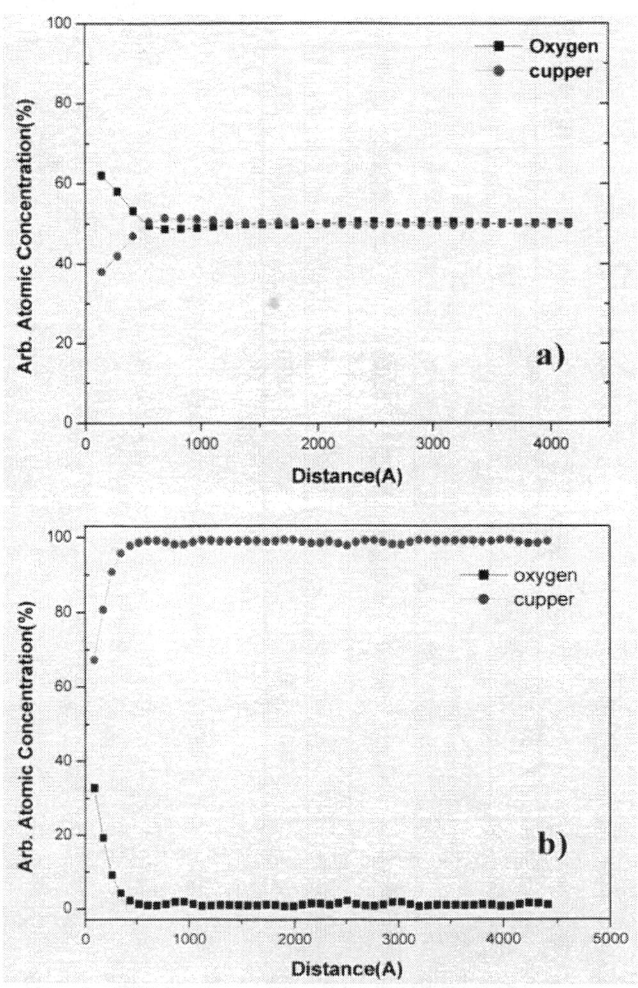

Fig. 5. AES depth profiling of CuO thin film(a) and electrochemically reduced Cu thin film(b).

3.3 Fabrication of Cu nano particles

As mentioned previous section, CuO nano particles were reduced electrochemically for 1h with the apparatus represented in Fig. 1. Fig. 6 shows the XRD curves of CuO nano particles and Cu nano particles obtained by electrochemical reduction. As shown in Fig. 6, the XRD curve of Cu nano particles is composed of Cu and Cu_2O. It was thought that the Cu_2O in Fig. 6(b) was formed during handling Cu nano particles under air. Fig. 7 shows the SEM morphological images of CuO nano particles and Cu nano particles. As can be seen in Fig. 7, the spherical shape and size of CuO nano particles was maintained and the agglomeration of Cu nano particles was not observed.

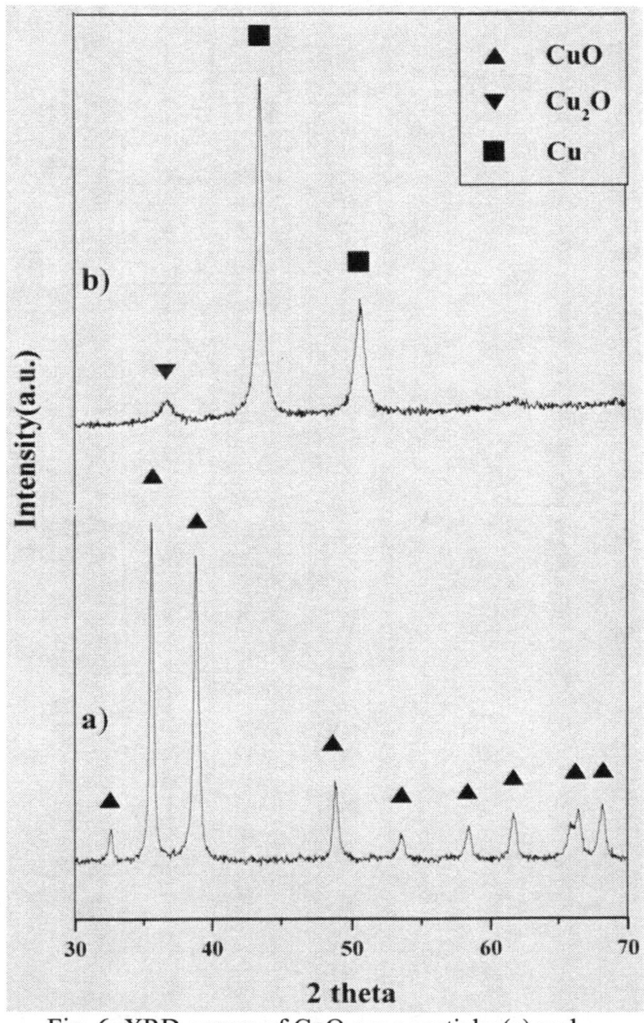

Fig. 6. XRD curves of CuO nano particles(a) and electrochemically reduced Cu nano particles(b).

NSTI-Nanotech 2005, www.nsti.org, ISBN 0-9767985-1-4 Vol. 2, 2005

Fig. 7. SEM morphological images of CuO nano particles(a) and electrochemically reduced Cu nano particles(b).

4 CONCLUSIONS

It was ascertained that CuO thin film was electrochemically reduced to Cu at -0.874V(vs. SCE) by using cyclic voltammetry, XRD, SEM/EDS, and AES. CuO nano particles, prepared by a conventional mechanical milling, were directly reduced to Cu nano particles without agglomeration.

5 REFERENCES

[1] L. D. Zhang, J. M. Mu, Nanoscale Materials and Nanostructures, Science Press, Beijing, 2001.
[2] M. A. El-Sayed, Acc. Chem. Res. 34, 257, 2001.
[3] Y. W. Cao, R. C. Jin, C. A. Mirkin, Science 297, 1535, 2002.
[4] A. D. McFarland, R. P. Van Duyne, Nano Lett. 3, 1057, 2003.
[5] A. M. Yu, Z. J. Liang, J. H. Cho, F. Caruso, Nano Lett. 3, 1203, 2003.
[6] E. R. Savinova, E. N. Savinova, V. N. Parmon, J. Mol. Catal. 48, 231, 1988.
[7] R. Narayanan, M. A. El-Sayed, J. Phys. Chem. B 107, 12416, 2003.
[8] S. S. Joshi, S. F. Patil, V. Iyer, S. Mahumuni, Nanostruct. Mater. 10, 1135, 1998.
[9] I. Lisiecki, M. P. Pileni, J. Am. Chem. Soc. 115, 3887, 1993.
[10] M. P. Pileni, B. W. Ninham, T. Gulik-Krzywicki, J. Tanori, I. Lisiecki, A. Filankembo, Adv. Mater. 11, 1358, 1999.
[11] L. Qi, J. Ma, J. Shen, J. Colloid Interface Sci. 186, 498, 1997.
[12] K. J. Ziegler, R. C. Doty, K. P. Johnston, B. A. Korgel, J. Am. Chem. Soc. 123, 7797, 2001.
[13] H. Ohde, F. Hunt, C. M. Wai, Chem. Mater. 13, 4130, 2001.
[14] N. A. Dhas, C. P. Raj, A. Gedanken, Chem. Mater. 10, 1446, 1998.
[15] R. V. Kumar, Y. Mastai, Y. Diamant, A. Gedanken, J. Mater. Chem. 11, 1209, 2001.
[16] M. S. Yeh, Y. S. Yang, Y. P. Lee, H. F. Lee, Y. H. Yeh, C. S. Yeh, J. Phys. Chem. 103, 6851, 1999
[17] G. Vitulli, M. Bernini, S. Bertozzi, E. Pitzalis, P. Salvadori, S. Coluccia, G. Martra, Chem. Mater. 14, 1183, 2002.
[18] Z. Liu, Y. Bando, Adv. Mater. 15, 303, 2003.
[19] H. H. Huang, F. Q. Yan, Y. M. Kek, C. H. Chew, G. Q. Xu, W. Ji, P. S. Oh, S. H. Tang, Langmuir 13, 172, 1997.
[20] I. Lisiecki, F. Billoudet, M. P. Pileni, J. Phys. Chem. 100, 4160, 1996.
[21] G. Z. Chen, D. J. Fray, T. W. Farthing, Nature 407, 361, 2000.

Environmentally Friendly Pathways for Synthesis of Titanium Dioxide Nano-particles

H. Li, S.G. Sunol and A.K. Sunol,

University of South Florida, Tampa, FL, USA, sunol@eng.usf.edu

ABSTRACT

Titanium dioxide nano-particles were synthesized via the sol-gel method using titanium butoxide as the precursor. Acid catalyzed and aged gels were dried under air to produce xerogels and under supercritical conditions to produce aerogels. Porous nano-particles obtained were characterized using X-ray diffraction, SEM and BET analysis. Aerogel powder was found to have more favorable properties compared to xerogel powder. Aerogel particles were more porous with more profound anatase crystal structure.

Keywords: supercritical carbon dioxide, aerogel, nanostructures, sol-gel, titanium dioxide

1 INTRODUCTION

Titanium dioxide is the most widely used photo-catalyst. It is preferred due to its non-toxicity, availability and relatively high activity. The catalytic properties of titanium dioxide that dictate its activity as photo-catalyst are its surface area, pore size, particle size, and crystallographic structure.

In heterogeneous catalysis, reaction takes place at the active surface of the solid catalyst. For effective use of the catalyst, the solid catalyst needs to provide high active area. Sol-gel process ensures production of high surface area gels. Also, pore size and its distribution are critical in effective use of the solid catalysts. Large pores may be essential for some applications. In other cases, smaller pores may be more desirable because they provide higher surface area. Sol-gel process also provides narrow distribution of pore size. However, during drying of the gel, pores shrink due to capillary action and as a result, pore size of the larger pores decreases and smaller pores collapse resulting also in loss of surface area. In order to prevent the destruction in the pore structure during drying, the solvent can be removed from the wet gel under supercritical conditions [1].

Aerogels are prepared through sol-gel process followed by supercritical drying. Supercritical drying eliminates capillary forces exerted on the walls of the pores and results in highly porous materials. Consequently, aerogel catalysts have desired properties of catalysts such as high surface area [2]. Use of templates during the sol-gel process results in uniform, tailored pore size distribution. Large pore size facilitates easier diffusion of large molecules into the pores and results in effective use of the catalyst. Conventionally, surfactants are removed from the gel by either Soxhlet extraction or by heat treatment. Also, more effective route for removal of the template from the catalyst can be enabled by extraction with supercritical fluids. Supercritical fluid extraction is less time consuming and environmentally friendly compares to solvent extraction. Furthermore, the template can be removed without disturbing the pore structure of the catalyst as opposed to thermal decomposition of the template at high temperatures. [3]

Particle size is another important property of titanium dioxide that needs to be attuned during the synthesis. Sol-gel synthesis results in small nano-particles, where else precipitation results in particles of large size [4].

In photo-catalysis, another issue to be addressed is the crystallographic structure of the metal oxide that is related to its photo-activity. Metal oxides prepared through sol-gel synthesis are basically amorphous. On the other hand, photo-activity of the titanium dioxide is directly related to its crystal structure. Titanium dioxide in the anatase form is found to be the most active photo-catalyst. Altough sol-gel derived titanium dioxide is amorphous, when calcined at relatively low temperatures, it can be transformed into the desired crystalline form.

In the present work, titanium dioxide nano-particles are prepared through sol-gel synthesis and dried using supercritical carbon dioxide. The procedure for supercritical drying is explained and effect of solvent removal step (air drying versus supercritical drying) on the properties of the end material is investigated.

2 EXRERIMENTAL

Titanium dioxide particles were synthesized through sol-gel process. They were dried under air or supercritical conditions to produce xerogel and aerogel samples respectively. After drying they were calcined and characterized in order to determine their pore and crystallographic structure. Preparation of the titanium dioxide particles is summarized in Figure 1.

2.1 Sol-gel Preparation

Titanium butoxide (Supplier: Aldrich) was dissolved in ethanol (Supplier: Aldrich), stirring for ten minutes. Nitric acid (Supplier: Aldrich) was added to the mixture drop-wise. The solution was then mixed with the required amount of distilled water, while stirring continuously at room temperature. Titanium hydroxide gel was formed according to the following hydrolysis reaction:

$$(CH_3\text{-}CH_2\text{-}CH_2\text{-}CH_2O)_4Ti + 4\ H_2O \rightarrow Ti(OH)_4 + 4\ CH_3\text{-}CH_2\text{-}CH_2\text{-}CH_2OH$$

Subsequently, the gel was stored at room temperature for two days to age. The molar ratio of the chemicals $Ti(OBu)_4$: ethanol : H_2O : HNO_3 is 1 : 20 : 6 : 0.8.

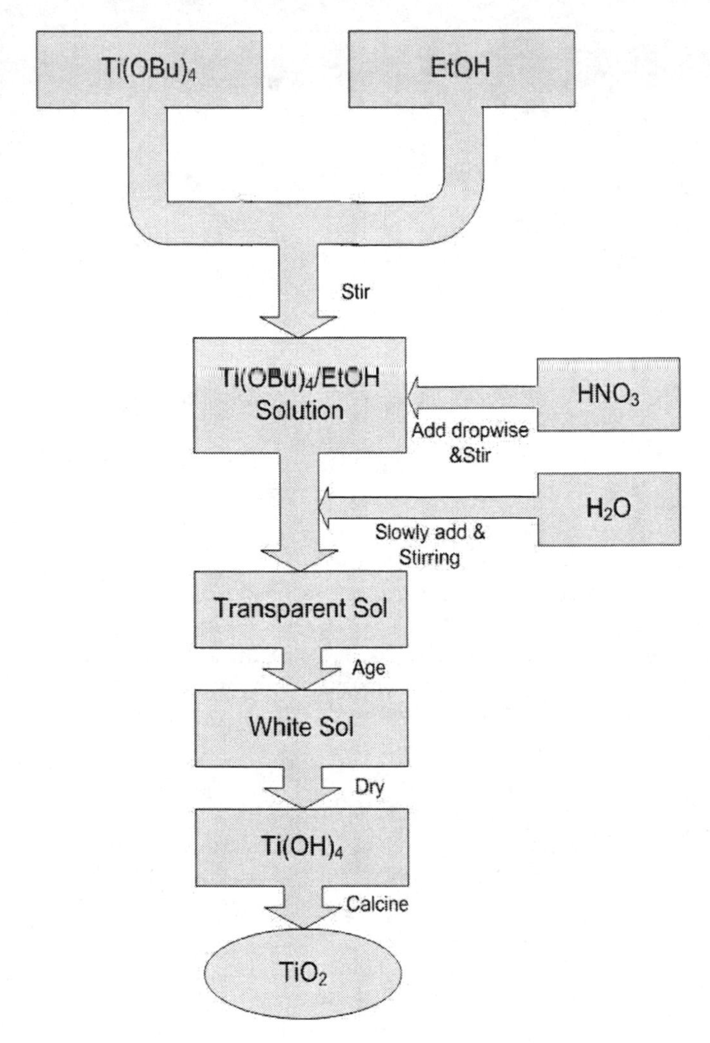

Figure 1. Preparation of titanium dioxide

2.2 Drying of the Xerogels and Aerogels

In order to produce titanium dioxide xerogels, the wet gels were dried under air at atmospheric pressure and 373K for 24 hours. Aerogels were produced through supercritical drying. The supercritical fluid (SCF) apparatus used for supercritical extraction and drying is shown in Figure 2. The SCF extraction and drying apparatus can be divided into three main sections; feeding section, drying section and gas-liquid separation and collection section.

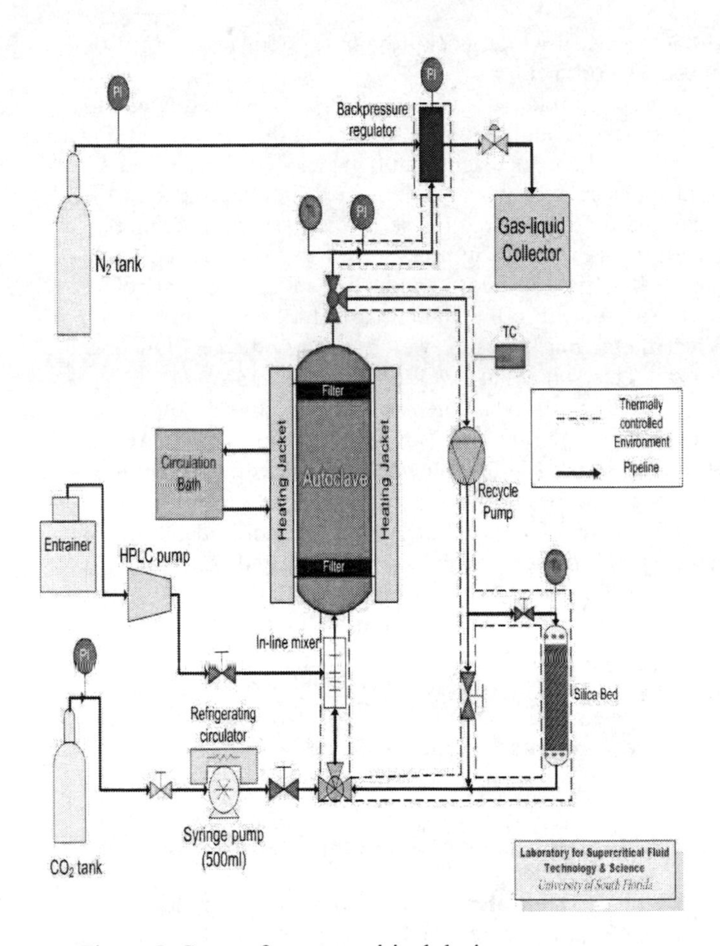

Figure 2. Set-up for supercritical drying

All parts, tubing and *Swagelock*® fittings are made of stainless steel, (SS-316). The tubing used in the main fabrication of the SCF setup is mainly 1/4 inch diameter and in some places 1/8 inch diameter. The feeding section consists of compressed carbon dioxide tank (AIRGAS, CD-50), Pressure regulator (Matheson), Refrigerating circulator (Ecoline RE120, Lauda, Brinkmann), HPLC pump (Waters, model 600E), Entrainer container, Syringe pump (ISCO model 500DX), in-line mixer, magnetic pump (*Ruska*, model 2330-802), and number of plug valves and needle valves. The drying section consists of a 100ml controlled heat-jacketed autoclave (Autoclave Engineer), water circular bath (HAKKE, model B81), several electrical heating tapes (Omega), Pressure gauge (Matheson), Thermocouple Probe (Omega, K-type). The gas-liquid separation and collection consists of Compress Nitrogen tank (AIRGAS, NI200), backpressure regulator, and low-pressure gas-liquid separator and venting system.

The drying procedure was as follows: The alcogel sample was placed into a heat-jacketed autoclave which had been pre-heated to setting temperature at 60°C. The autoclave was then closed tightly and purged with carbon dioxide gas. All the vents and outlet valves were then securely closed and the system pressure was initially raised to 850psi. Meanwhile, backpressure system was turned on and set at 1800 psi.

Carbon dioxide was compressed and cooled down to -4 °C passing through syringe pump, and pressure in the system was kept increasing until it reached the desired set point pressure which was determined and entered into the backpressure regulator at the start of each experimental run. The designed processing pressure for this section was 1800 psi. Carbon dioxide went to an in-line mixer unit just before it was introduced to the autoclave from the bottom section. When the setting pressure was achieved, the feeding flow rate of CO_2 was kept at 10 ml/min for 2 hours.

Recycle pump was then engaged between the input (bottom) and the output (top) of the autoclave. At the beginning of the run, the recycled fluid was passed through the adsorption bed, which contained silica gel particles as the adsorbent. The mixture of carbon dioxide, part of solvents and residual template was evacuated from the SCF experimental setup through the backpressure regulator. Electrical heating element was used to keep the effluent supercritical mixture from getting frozen, hence preventing blocking the system. Fluid leaving the backpressure regulator went to an ambient pressure vessel inside which the effluent mixture separated into gas and liquid mixture phases.

2.3 Calcination of the Xerogels and Aerogels

Samples were calcined at 773K under air-flow in a tubular furnace for three hours.

2.4 Characterization of the Catalysts

Nitrogen adsorption-desorption experiments were carried out in order to study the total surface area, total pore volume, average pore size, and pore size distribution for meso-pores, and the micro porosity of the gels. A Nova 2200 series machine along with Enhanced Data Reduction Software from Quantochrome Corp was used for routine pore structure information. BET method was used in order to determine total surface area. Pore size distribution and average pore size of the meso-pores were determined using BJH method. X-ray powder diffraction (XRD) patterns were collected using CuKα radiation on a Philips Expert System with a 45 KW and 40 mA in order to determine the crystallinity and phase of the samples. Surface images were collected by a Hitachi 800 Scanning Electron Microscope (SEM).

3 RESULTS AND DISCUSSION

Xerogel and aerogel samples were characterized in order to compare their porous and crystallographic structures.

In order to study the crystalline structure, X-Ray diffraction patterns were taken (Figure 3). For both aerogels and xerogels, peaks were observed at 2θ equal to 25.2°, 37.8° and 48.0°. These are diffractions of the anatese form corresponding to (1 0 1), (0 0 4) and (2 0 0) respectively. X-Ray diffraction of the aerogels shows higher intensities than the ones of the xerogels that implies larger crystal size for aerogels.

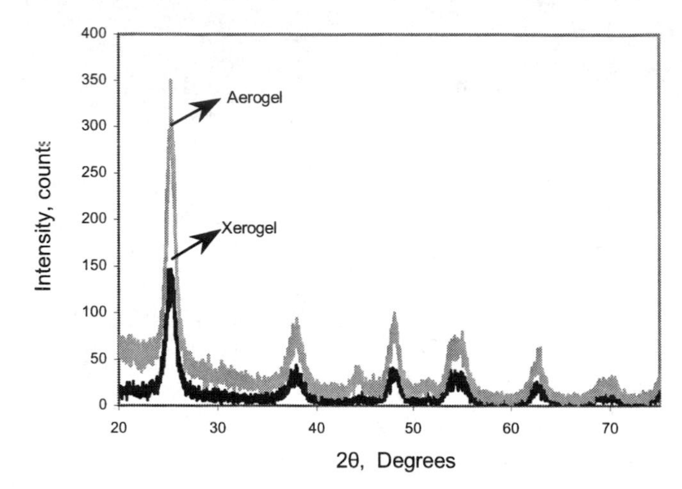

Figure 3: XRD Patterns of aerogels and xerogels.

Comparison of the pore structure of the aerogels and xerogels shows that the aerogels have higher total surface area and pore volume. Also, aerogels have larger average pore diameter than the xerogels. All of these results are indicative of the superior conditions achieved during supercritical drying when compared with evaporation of the solvent.

REFERENCES

[1] A.K. Sunol and S.G. Sunol, "Templating Aerogels for Tunable Nanoporosity" in Dekker Encyclopedia of Nanoscience and Nanotechnology, Marcel Dekker, 3843-3851, 2004.

[2] A.K. Sunol, S.G. Sunol, O. Keskin, O. Guney, "Supercritical fluid aided preparation of aerogels and their characterization", in "Innovations in Supercritical Fluids Science and Technology" Chapter 17, p. 258-268, edited by K. W. Hutchenson and N. R. Foster, ACS Symposium Series 608 (1995).

[3] P. Kluson, P. Kacer, T. Cajthaml, M. Kalaji, J. Mater.Chem., 11, 644-651, 2001.

[4] L. Znaidi, R. Seraphimova, J.F. Bocquet, C. Colbeau-Justin, C. Pommier, Mater. Res. Bulletin., 36, 811-825, 2001.

Effect of Nano-architectural Control on Optical Loss in Electro-optic Polymers

Diyun Huang and Tim Parker

Lumera Corporation
19910 North Creek Parkway
PO Box 3040
Bothell, WA 98011, tim_parker@lumera.com

ABSTRACT

Nonlinear optical chromophores have been used to obtain guest-host and side-chain polymers. The polymers are fabricated with crosslinking groups attached to both the chromophores and the polymer backbones. The guest-host composites often have micro domains that increase optical loss of the material. When the chromophore is covalently linked to the backbone, the micro domains are avoided and the optical loss of the material is decreased by a factor of 2-3 dB/cm. This nano architectural control is crucial to produce polymer electro-optic devices that are commercially viable.

Keywords: electro-optic polymers, optical loss, optical modulators, nonlinear optical chromophores

1 ELECTRO-OPTIC POLYMERS

Polymer electro-optic (EO) devices show great promise for high-speed telecom data transmission, broadband radio frequency (RF) signal distribution, and high-speed data transfer within the next generation of microprocessors running at 20 GHz and above (*i.e.*, optical interconnects). EO polymer devices have the potential to increase performance or enable new applications in several industries including telecom networking, wireless communications, and high-performance computing.

One of the critical device performance parameters is the EO activity of the polymer material. Higher EO activity (r_{33}) leads to more efficient power performance and smaller device size. Dalton and Robinson have used Monte Carlo simulation methods to predict that r_{33} can be more than doubled by sterically modifying the shape of the chromophore in EO polymer materials on the nanoscale.[1] This design concept was used to realize high EO activity in poled thin films to achieve an unprecedented r_{33} of 60 pm/V @ 1300 and reduce the modulator V_π to 0.8 V.[2] Recently, Alex Jen's research group at the University of Washington has pushed performance even further by doubling the electro-optic activity of polymers through a process referred to as "nanoarchitectural control."[3] The approach involves grafting a nonlinear optical chromophore onto a polymer backbone to produce nonlinear optical polymers on the scale of about 100 nm. Further synthetic modification of the material leads to adoption of certain nanoscale geometries that lead to the roughly two-fold increase in electro-optic activity. This work has been a major milestone in EO polymer development, but both the thermal stability and optical loss of the material needs improving to realize commercially viable devices.

2 GUEST-HOST EO POLYMER WITH CROSSLINKING GROUPS

It is well known that increasing the temperature at which the EO polymer forms a rubbery state (*i.e*, the T_g) tends to increase the thermal stability of the EO effect. With this in mind, we studied two high T_g systems to see the effect nanotailoring has on the optical loss performance of EO polymer materials. The first system was a conventional guest-host system where the chromophore (**1**) is included in the polymer matrix (**2**) noncovalently (until thermosetting). The chromophore is based on a π-bridge structure that shows increased thermal and photochemical stability over polyene π-bridges.[4] The chromophores made with this bridge show relatively high $\mu\beta$ while having good processibility. The enhanced photochemical stability is attributed to the presence of alkoxy substituents on the thiophene ring, which prevents hydrogen abstraction and radical formation that may occur when alkyl groups are substituted on the thiophene ring. Guest-host composites of these chromophores can give electro-optic coefficients (r_{33}) over 100 pm/V at 1310 nm in APC.[5]

The backbone polymer is a high Tg polyimide synthesized by grafting a crosslinkable trifluorovinyl ether (TFVE) group[6] on a previously reported hydroxy-functionalized polyimide.[7] The TFVE group has been previously used in crosslinked electro-optic polymers in order to increased chromophore alignment stability.[8] The TFVE group is particularly useful as a crosslinking group in electro-optic polymers because it is fluorinated to reduce optical loss and does not give off condensation products during crosslinking. The crosslinker is also functionalized onto the chromophore to increase compatibility and to incorporate the chromophore into the polymer backbone during thermosetting. It should be noted that attaching the TFVE group to the acceptor increasing the $\mu\beta$ of the chromophore since the TFVE group itself is electron withdrawing and has a dipole moment.

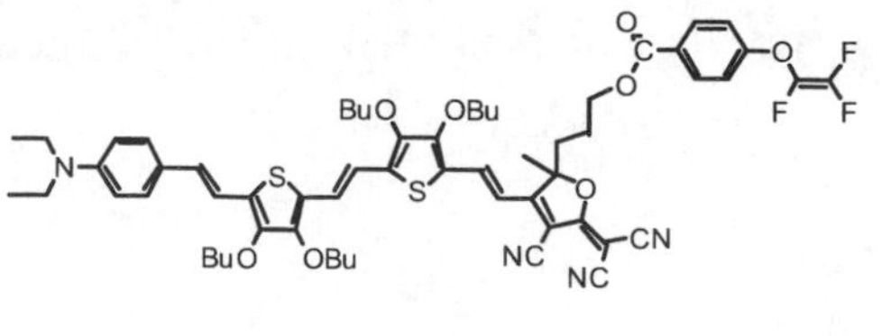

Figure 1: Chemical structures of the polymer (**1**) and the chromophore (**2**) for guest-host composites.

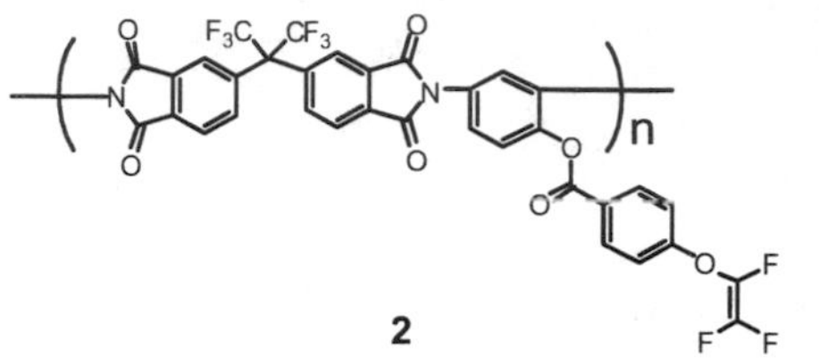

Optical quality thin-films of the guest-host composite were spin deposited on 4-inch diameter silicon wafers. The chromophore (**1**) loading was approximately 20% by weight in the polymer matrix (**2**). The thickness of the film was approximately 3 μm. The films were soft baked at elevated temperatures (80–100 °C) under vacuum for several hours to remove residual spinning solvent. Optical loss measurements were carried out using a prism coupling technique. The average optical loss of five different areas in the thin film was 3.88 dB/cm. This optical loss is relatively high EO polymers materials, and is most likely the results of some degree of phase separation. Phase separation in a fluorinated guest-host composite leads to micro-domains that scatter the light at telecom wavelengths (1.31 and 1.55 μm). A scanning electron micrograph (SEM) of the cross section on of the guest-host composites shows the micro domain formation caused by the phase separation. The globular particles are attributed to aggregates of the chromophore.

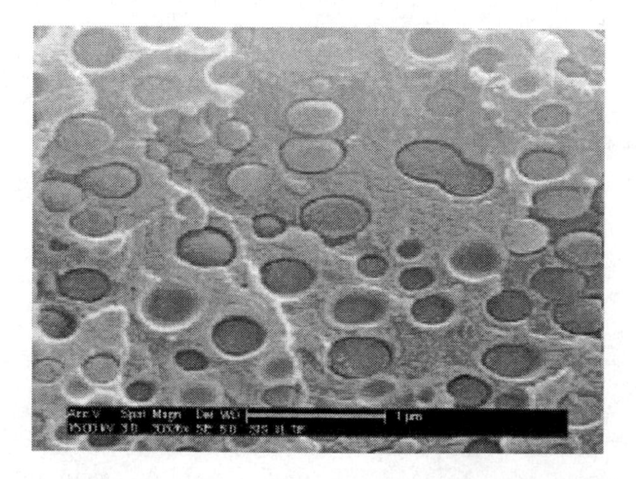

Figure 2: SEM of guest-host composite

3 SIDE-CHAIN EO POLYMER WITH CROSSLINKING GROUPS

The second system was a side-chain polymer (**3**), shown in Figure 3, where the chromophore is covalently attached to the polymer to produce a nonlinear optically active nanostructure. The backbone (**4**) of the polymer used for side-chain attachment and the crosslinking groups were the same as those used in the guest-host composite. The same backbone and crosslinking groups were chosen to obviate any problems associated with structure dependent morphology differences. For the side-chain polymer, a chromophore having a crosslinkable group on the acceptor was grafted onto the polymer backbone followed by the crosslinker. This methodology allows the adjustment of the crosslinker/chromophore ratio to optimize the loading density and processibility of the resulting NLO polymer. After the reaction to form the polymer, unreacted chromophore is removed by repeated precipitation of the polymer. The loading density of the chromophore in the polymer can be determined by monitoring the optical absorption of the chromophore. The refractive index also scales with the amount of chromophore (Figure 4), which is potentially useful as a tunable refractive index system for waveguide fabrication.

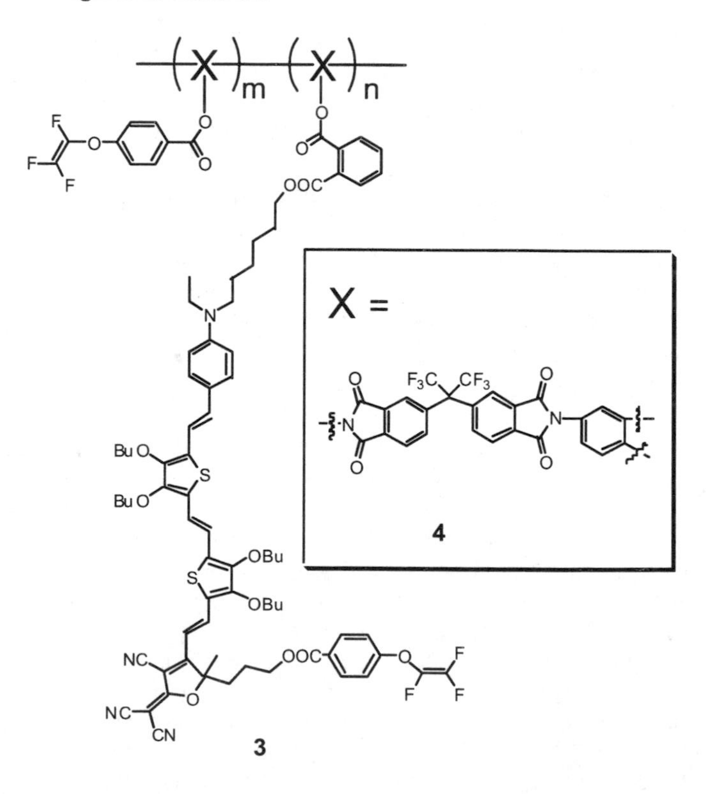

Figure 3: Side-chain polymer (**3**) and the side-chain polymer backbone (**4**)

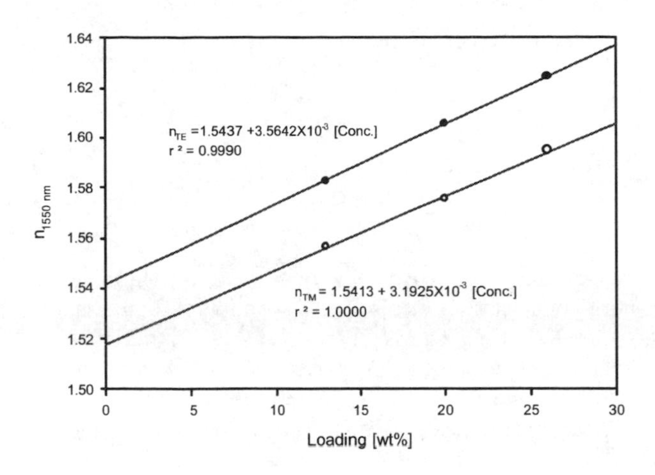

Figure 4: Refractive index vs. chromophore loading in a side-chain polymer.

Optical quality thin films of the side-chain polymer were spin deposited on 4inch diameter silicon wafers. The films were heated at (80-100 °C) under vacuum for the same amount of time as the guest-host composites. The thickness of the films was approximately 3 µm. Optical loss measurements were performed by prism coupling. The average optical loss for five areas of the thin film was 1.9 dB/cm, which is a factor of two improvement over the guest-host system presented here. SEM images of a cross section of the thin film showed no micro domain phase separation in the side-chain polymer, which validates that the chromophore has been included in the nanostructure of the polymer. The nanostructure of the current side-chain polymers is on the order of 100 nm, which is well below the size that may cause significant scattering of µm-wavelength light. This reduction in optical loss is significant especially if one considers the propagation length of a typical Mach-Zehnder modulator. If the propagation length is 3 cm, then a device fabricated form the guest-host system would have a fundamental material loss of around 12 dB, which is more than *twice* that of the 6 dB *total* optical loss (*i.e.*, material loss plus fiber coupling loss) in typical commercial optical modulators made from inorganic material. On the other hand, modulators made from the side-chain materials would have a fundamental material loss of 6 dB, which is near the range of commercially available modulators if < 1 dB/face coupling loss can be achieved.[9] To reduce the total loss of the device, one can either decrease the total device length or decrease the material loss. Further reduction in material optical loss can be expected by increasing the fluorine content of the chromophore and polymer backbone.

4 CONLCUSIONS

Optical loss in electro-optic polymers has been reduced by incorporating chromophores as side-chains in high T_g polyimides. The reduction in optical loss is facilitated by the chromophore's incorporation into the nanostructure of the polymer backbone, which avoids micro domain formation. The optical loss of the material approaches what is necessary to produce a commercially viable electro-optic polymer modulator with a total optical loss (material loss plus fiber coupling loss) of 6 dB. Further studies are being aimed at reducing material loss further by increasing fluorination of the chromophore and polymer backbone as well as decreasing device length.

REFERENCES

[1] Robinson and Dalton *J. Phys. Chem. A* **2000**, *104*, 4785.
[2] Shi *et al., Science* **2000**, *288*, 119.
[3] Luo *et al., J. Phys. Chem B* **2004**, *108*, 8523.
[4] Huang *et al.*, U.S. Patent No. 6,716,995 and 6,750,603.
[5] Ermer *et al., Adv. Funct. Mater.* **2002**, *12*, 605.
[6] Smith *et al., J. Fluorine Chem.* **2000**, *104*, 109.
[7] Lee *et al., J. Polym. Sci: Part A: Polym. Chem.* **1998**, *36*, 301.
[8] Ma *et al., J. Am. Chem. Soc.* **2001**, *123*, 986.
[9] Less than 1 dB of coupling loss in polymer modulators has been reported, for example see: Kim, S. K. *et al., J.. Kor. Phys. Soc.* **2003**, *43*, L645.

Strain and Structure in Nanocrystals

Youri Bae* and Russel E. Caflisch**
* Department of Mathematics,UCLA:**baeri@math.ucla.edu**
** Department of Mathematics,UCLA:**caflisch@math.ucla.edu**

ABSTRACT

Layered nanocrystals consist of a core of one material surrounded by a shell of a second material. We present computation of the atomistic strain energy density in a layered nanocrystal, using an idealized model with a simple cubic lattice and harmonic interatomic potentials. These computations show that there is a critical size r_s^* for the shell thickness r_s at which the energy density has a maximum. This critical size is roughly independent of the geometry and material parameters of the system. Moreover it agrees with the shell thickness at which the quantum yield has a maximum, as observed in several systems.

Keywords: nanocrystals, strain energy, quantum yield.

1 Introduction

The size dependence of electric, optical, and structural properties of nanocrystals has been important issues in nanosciences. The synthesis of an epitaxial core /shell system allows the research on size and shape control. The photostability(hole confinement in the core), electronic accessibility(electron spreading into the shell), and high quantum yield makes these core/shell nanocrystals very attractive for use in optoelectronic devices in [1]. A stringent requirement for the epitaxial growth of several monolayers of one material on the top of another is a low lattice mismatch between the two materials. If this requirement is not met, strain accumulates in the growing layer, and eventually may be released through the formation of misfit dislocations, degrading the optical properties of the system. Moreover, if the strain of shell becomes too great, then it is relieved by irregular growth of shell in [2].

In core/shell epitaxial growth, strain is induced by mismatch between the lattice constants in the core and those in the shell. We use the equilibrium lattice of core as a reference lattice in shell. At the atomistic level, this means that zero displacement corresponds to a homogeneous strain such that the lattice constant has been changed from lattice constant of core to that of shell throughout the material. The derived elastic equation have external force at core/shell and its strength is lattice mismatch.

We examine qualitatively and quantitatively the elastic energy density that arises from the nanocrystal model. We propose that the elastic energy density of nanocrystal is concentrated near interfaces between core/shell and their maximal value as a function of shell thickness has a peak with small shell thickness. We define this shell thickness as critical shell thickness and compare and contrast these results to known photoluminescence quantum yield results from experiments in [1]. Furthermore, we examine the sensitivity of the critical shell thickness to material parameters and the size of core.

2 Atomistic Strain Model

The elastic energy density of the strained substrate is given by the tensor contraction of strain and stress, so that the total elastic energy is the integration of elastic energy density on the whole material. The force balance equations are obtained by setting the variation of the elastic energy with each of the displacements equal to zero. The general approach is not to discretize the force balance equations directly. Instead, we construct a discrete version of the elastic energy density corresponding to finite difference method for continuum elasticity in a cubic lattice. At each point of the grid, the energy contribution only involves terms from the 27 point stencil (nearest and next-nearest neighbors) centered at point itself. If the computational grid is the same as the underlying atomistic lattice, then the discrete version of the energy may be considered purely atomistic.

Here is the discrete elastic energy density

$$E = \sum_{k,p} \alpha_k^p (S_{kk}^p)^2 + \sum_{k \neq l,p,q} \{2\beta_{kl}^{pq}(S_{kl}^{pq})^2 + \gamma_{kl}^{pq} S_{kk}^p S_{ll}^q\}. \tag{1}$$

where p,q is + or - and k,l is 1,2. To make discrete energy density and continuum energy density to be consistent, we can get elastic coefficients from Voight constants

$$(\alpha, \beta, \gamma) = (C_{11}, C_{44}, C_{12})/4. \tag{2}$$

The significant geometric parameters are the core radius r_c and the shell thickness r_s and the lattice mismatch

$$\epsilon = \frac{l_c - l_s}{l_c} \tag{3}$$

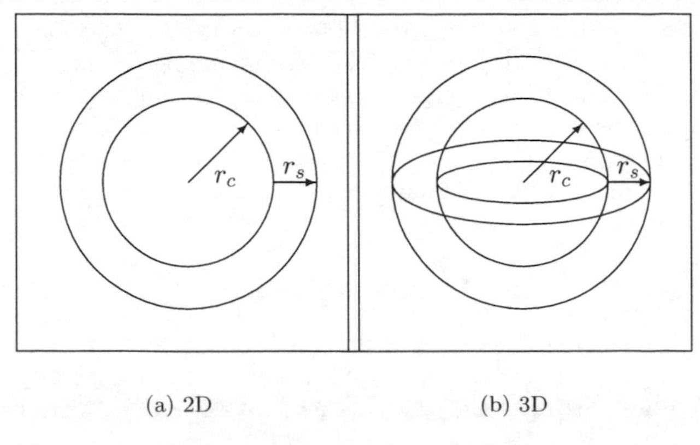

(a) 2D (b) 3D

Figure 1: Basic geometry of core/shell nanocrystal model

for elastic strain energy parameter where l_c and l_s are the lattice constants of core and shell, respectively. The core consists of atoms whose lattice position x (before displacement) satisfies $|x| \leq r_c$, and the shell consists of atoms with $r_c < |x| \leq r_c + r_s$.

We now turn our attention to elastic energy density of core/shell nanocrystal produced by our model. The model in [5] with infinite step train shows the clearly distinct strain fields produced by the surface step and the buried interface step. Our model with consecutive steps on the vicinal and top surfaces gives the possibility of concentration of elastic energy near buried interface steps or surface steps. The experimental results on the irregular growth on shell with large lattice mismatch for thicker nanorods [2] imply that interfacial strain plays a more important role than surface strain. We will verify the significance of interfacial strain by calculating the elastic energy density and showing the concentration of elastic energy fields.

3 Simulation Results

Computational results are presented here from minimization of the total elastic energy (after removing degenerate modes corresponding to translation and rotation), corresponding to balance of all of the forces in the system, for 2D (circular, or equivalently rods of infinite length) and 3D (spherical) nanocrystals. For the harmonic potentials used here, this amounts solving a linear system of equations, in which the forcing terms come from the lattice mismatch ϵ. The simulation results include values of the displacements, the forces and energy density. Graphical results will be presented for the last of these. As a figure of merit for the atomistic strain field in a nanocrystal, we shall use the maximum value E_m of the discrete energy density. The energy at each atom consists of elastic energy and bond energy. The atomic bond can be broken when the elastic energy is larger than bond energy. Therefore, the maximum energy density can be one of the indicators to explain instability with thicker shell.

3.1 Elastic energy density

Figures 2 show the elastic energy density of 3D layered nanocrystals of fixed core size r_c for various values of shell thickness r_s. In the 3D nanocrystal simulation (Figure 2), the shell has thickness values $r_s = 1$, 2 and 7 monolayers, on a core of radius $r_c = 8$ monolayers with the elastic constants $\alpha = 5$, $\beta = 1$ and $\gamma = 3$ and lattice mismatch $\epsilon = 0.04$.

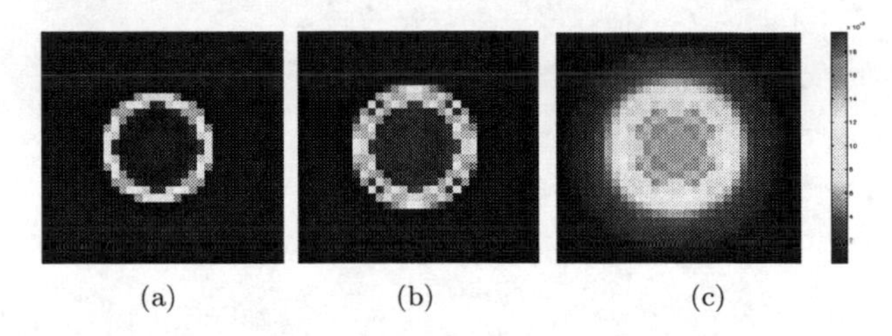

(a) (b) (c)

Figure 2: Elastic energy density on a cross section with maximum energy density for a 3D layered nanocrystals with core size $r_c = 8$ monolayers and with shell thickness r_s of size (a) 1 monolayers, (b) 2 monolayers and (c) 7 monolayers.

Figure 2 show that the energy is concentrated in the region of the shell, along the interface with the core. Moreover, the energy density is more concentrated for thicker shells. In addition the largest values of the energy density are close to the diagonal.

3.2 Critical Thickness

Figure 3 shows the maximum energy density for a layered nanocrystal, as a function of shell thickness r_s, for a fixed value of the core size r_c. Figure 3 shows that the maximum energy density increases with increasing shell thickness r_s up to a critical shell thickness r_s^*. For $r_s > r_s^*$, the maximum energy density is decreasing as a function of r_s. The general similarity between the critical shell thickness in 2D and 3D is indicative of the robustness of this result. The physical core diameter of CdSe/CdS core/shell nanocrystal is ranging from 23Å to 39Å which is equivalent to core radius of 3 monolayers to 6 monolayers, since one full monolayer is approximately 3.5Å [1].

Next we examine the critical shell thickness r_s^* and its dependence on the material and geometric parameters of the nanocrystal, in particular the dependence of r_s^* on the core size r_c, lattice mismatch ϵ and elastic parameters α, β and γ. We find that the critical shell thickness r_s^* does not depend on the lattice mismatch ϵ and has weak sensitivity of critical shell thickness r_s^* on the core radius r_c for 2D and 3D nanocrystals. Additional computations show that the critical shell thickness r_s^* does not depend sensitively on the elastic parameters α, β and γ. All the computational results are shown in [7].

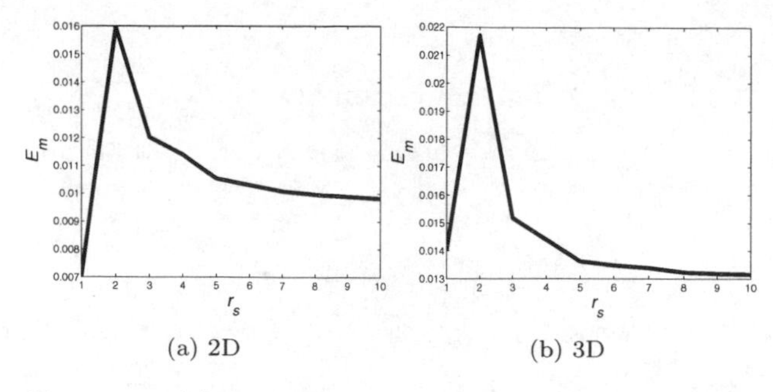

Figure 3: (a)Maximum energy density E_m vs. shell thickness $r_s$for (a) 2D and (b) 3D nanocrystal of core radius $r_c = 8$ monolayers.

4 Discussion

4.1 Step Interactions

Some insight into the existence of a critical shell thickness r_s^* comes from consideration of step interactions. On the outer edge of the shell, the shell atoms will relax towards their equilibrium lattice constant, lowering the strain energy density. This relaxation will be greatest along the diagonal, where the atoms have the smallest number of neighbors. On the other hand, along the core/shell interface, shell atoms near the diagonal have the largest number of core atom neighbors and so they have the largest strain. This maximum is increased by their interaction with the atoms along the diagonal on the outer edge, but that interaction decreases as the shell thickness increases. This indicates a critical thickness.

This is analogous to the strain field produced by a surface step, on an epitaxial surface, interacting with a buried step, on the interface between an epitaxial thin film and the substrate, as studied in [5].

Simplified single step simulations show that the maximum energy E_m peaks when one step on the surface and one on the buried interface are close. This supports the above arguments. More details are in [7].

4.2 Comparison to Quantum Yield

The critical shell thickness, observed in the simulations presented above, correlates closely to the maximum value of the quantum yield from experiments. Photoluminescence quantum yield data come from both CdSe/CdS [1] and InAs/CdSe [3] layered nanocrystals. We compare their quantum yields results with our simulation results and found that both have the peak at the small shell thickness. More computations of the material properties are studied from the simulations in [7].

5 Conclusion

The simulation results presented above are for a highly idealized model of a layered nanocrystal. The robustness of these results with respect to variation of dimension, geometry and material parameters suggests that these results are qualitative and generally applicable. In addition, there is some evidence that the critical shell thickness found in these simulations is related to the maximal values of quantum yield, as found experimentally.

6 Acknowledgement

The authors acknowledge the financial and program support of the Microelectronics Advanced Research Corporation (MARCO) and its Focus Center on Function Engineered NanoArchitectonics (FENA).

REFERENCES

[1] Xiaogang Peng, Michael C. Schlamp, Andreas V. Kadavanich, and A. P. Alivisatos *Epitaxial growth of highly luminescent CdSe/CdS core/shell nanocrystals with photostability and electronic accessibility.* J. Am. Chem. Soc. 119, 7019-7029 (1997).

[2] Liberato Manna, Erik C. Schoer, Liang-Shi Li, and A. Paul Alivisatos *Epitaxial growth and photochemical annealing of graded CdS/ZnS shells on colloidal CdSe nanorods.* J. Am. Chem. Soc. 124, 7136-7145 (2002).

[3] Yun-Wei Cao and Uri Banin *Synthesis and Characterization of InAs/InP and InAs/CdSe Core/Shell Nanocrystals.* Angew. Chem. Int. Ed. 38, No. 24, 3692-3694 (1999).

[4] Taleb Mokari and Uri Banin *Synthesis and Properties of CdSe/ZnS Core/Shell Nanorods.* Chemistry of Materials, 15 (20), 3955-3960 (2003).

[5] A.C. Schindler, M.F. Gyure, G.D. Simms, D.D. Vvedensky, R.E. Caflisch, C. Connell and Erding Luo *Theory of strain relaxation in heteroepitaxial systems.* Phys. Rev. B 67, 075316 (2003).

[6] Alberto Pimpinelli and Jacques Villain *Physics of Crystal Growth.* Cambridge University Press 1998

[7] Youri Bae and Russel E. Caflisch *Strain in Layered Nanocrystals.* In preperation, 2005.

Three-dimensional Structure of Nanocrystals from High-energy X-ray Diffraction and Atomic Pair Distribution Function Analysis

V. Petkov

Department of Physics, Central Michigan University, Mt. Pleasant, MI 48858, USA, petkov@phy.cmich.edu

ABSTRACT

The atomic Pair Distribution Function (PDF) analysis and high-energy x-ray diffraction are introduced as a tool for determining the three-dimensional structure of nanocrystalline materials. The great potential of this experimental approach is demonstrated with results from successful studies on $(Polyaniline)_xV_2O_5nH_2O$ nanocomposites and dendrimer stabilized gold nanoparticles. We find that $(Polyaniline)_xV_2O_5nH_2O$ has a lamellar-type structure wherein polymeric chains are sandwiched between double layers of $V\text{-}O_6$ octahedra. Gold nanoparticles possess a distorted face centered cubic-type structure with the degree of distortion increasing with the decrease in particle's size.

Keywords: x-ray diffraction, nanocomposites, nanoparticles, structure

1 INTRODUCTION

Knowledge of the atomic-scale structure is an important prerequisite to understand and predict the properties of materials. In the case of crystals it is obtained from the positions and intensities of the Bragg peaks in the diffraction patterns. However, materials constructed at the nanoscale lack the translational symmetry and long-range order of perfect crystals. The diffraction patterns of such materials show only a few Bragg peaks, if any, and a pronounced diffuse component. This poses a real challenge to the usual techniques for structure determination. The challenge can be met by employing the so-called atomic Pair Distribution Function (PDF) analysis and high energy x-ray diffraction. This non-traditional experimental approach takes into account both Bragg and diffuse scattering and yields the atomic structure in terms of a small set of parameters such as a unit cell and atomic coordinates [1-4].

The reduced atomic Pair Distribution Function, $G(r)$, is defined as follows:

$$G(r) = 4\pi r[\rho(r) - \rho_0] \quad (1)$$

where $\rho(r)$ and ρ_0 are the local and average atomic number densities, respectively, and r is the radial distance. It peaks at characteristic distances separating pairs of atoms and thus reflects the atomic-scale structure. The PDF $G(r)$ is the Fourier transform of the experimentally observable total structure function, $S(Q)$, i.e.,

$$G(r) = (2/\pi)\int_{Q=o}^{Q_{max}} Q[S(Q)-1]\sin(Qr)dQ \quad (2)$$

where Q is the magnitude of the wave vector ($Q = 4\pi \sin\theta/\lambda$), 2θ is the angle between the incoming and scattered radiation beams and λ is the wavelength of the radiation used. The structure function is related to the coherent part of the total intensity diffracted from the material as follows:

$$S(Q) = 1 + \left[I^{coh}(Q) - \sum c_i |f_i(Q)|^2 \right] / \left|\sum c_i f_i(Q)\right|^2 \quad (3)$$

where $I^{coh}(Q)$ is the coherent scattering intensity per atom in electron units and c_i and f_i are the atomic concentration and x-ray scattering form factor, respectively, for the atomic species of type i [5]. As can be seen from Eqs. 1-3, the PDF is simply another representation of the powder diffraction data. However, exploring the diffraction data in real space is advantageous, especially in the case of materials of limited structural coherence. First, as Eq. 2 implies the *total* scattering, including Bragg scattering as well as diffuse scattering, contributes to the PDF. Therefore both the longer-range atomic structure, responsible for the sharp Bragg-like diffraction features, and the local non-periodic structural imperfections, resulting in the diffuse component of the diffraction pattern, are contained in the PDF. Second, by accessing high values of Q, experimental PDFs with improved real-space resolution can be obtained and hence, quite fine structural features can be revealed. In fact, data at high Q values (Q>10 $Å^{-1}$) are critical to the success of PDF analysis. Third, the PDF is less affected by diffraction optics and experimental factors since these are accounted for when extracting the coherent intensities from the raw diffraction data. This renders the PDF a structure-dependent quantity that gives directly relative positions of atoms in materials and enables convenient testing and refinement of structural models. The great potential of the PDF technique is well demonstrated by the results presented below.

2 EXAMPLES OF PDF STUDIES

2.1 Atomic ordering in (Polyaniline)$_x V_2O_5nH_2O$

In recent years, conducting polymeric nanocomposites have attracted much attention because of their unique and novel properties and a wide variety of potential applications. Of particular interest are nanocomposites based on polyaniline (PANI) which find applications in solar cells, biosensors and color displays [6]. A prime example of the family of PANI-based nanocomposites is (PANI)$_x V_2O_5nH_2O$. Here the combination of the conducting properties of PANI and the inorganic host, $V_2O_5nH_2O$ xerogel, results in a nanomaterial with interesting charge- and ion-transport characteristics [7]. No simulations and theoretical studies on the physical properties of the material are done so far mainly because its nanocrystalline state made it difficult to determine the atomic-scale structure in detail.

The (PANI)$_x V_2O_5nH_2O$ system is composed of two components neither of which exists in a crystalline form. The $V_2O_5nH_2O$ xerogel is nanocrystalline in nature and its structure was unknown ever since the material was discovered in the beginning of the last century. Recently, we determined its atomic-scale structure [4] using the only technique currently available that can do so, namely the PDF technique. Polyaniline is a poorly crystalline material whose three-dimensional structure has never been determined in detail. Here we build upon our previous results and the capabilities of the PDF technique to elucidate the structure of (PANI)$_x V_2O_5nH_2O$. The synthesis of the nanocomposite is described in full detail in ref. [4]. X-ray diffraction (XRD) experiments were carried out at the beamline X7a at the National Synchrotron Light Source at Brookhaven National Laboratory using x-rays of energy 29.09 keV (λ = 0.4257). The higher energy x-rays were used to obtain diffraction data to higher values of the wave vector, Q, which is important for the success of PDF analysis. The experimental diffraction patterns for nanocrystalline (PANI)$_x V_2O_5nH_2O$, crystalline V_2O_5 and pure polyaniline are shown in Fig.1. The latter two materials were used as standards. As can be seen, the XRD pattern of crystalline V_2O_5 exhibits well-defined Bragg peaks up to Q ~ 10 Å^{-1}. The material is obviously a well-ordered crystalline solid. In contrast, the Bragg peaks in the XRD patterns of nanocrystalline (PANI)$_{0.5}V_2O_5nH_2O$ are rather broad and merge into a slowly oscillating diffuse component already at Q values as low as 4-5 Å^{-1}. The diffraction pattern of polyaniline is even less structured: it shows only one sharp feature at low Q values - a behavior typical for severely disordered materials. Such diffraction patterns are very difficult to analyze using ordinary techniques for structure determination. However, when reduced to the corresponding atomic PDFs, they become a structure-sensitive quantity lending itself to structure determination and modeling. The experimental PDF for nanocrystalline (PANI)$_{0.5}V_2O_5nH_2O$ is shown in Fig. 2.

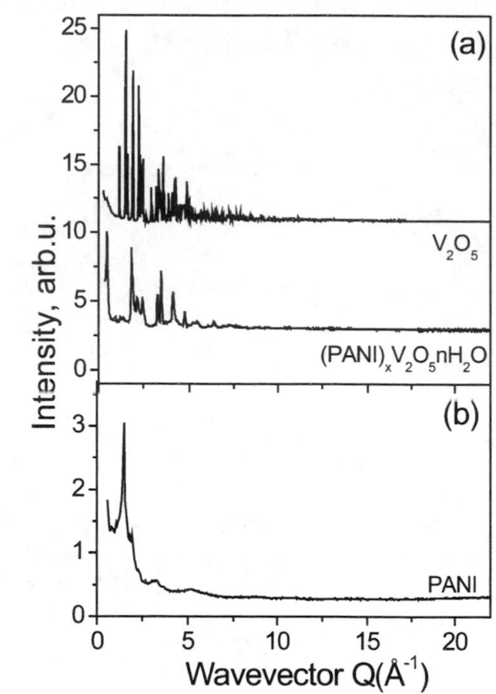

Figure 1. Diffraction patterns of (a) nanocrystalline (PANI)$_x V_2O_5nH_2O$ and crystalline V_2O_5 and (b) polyaniline.

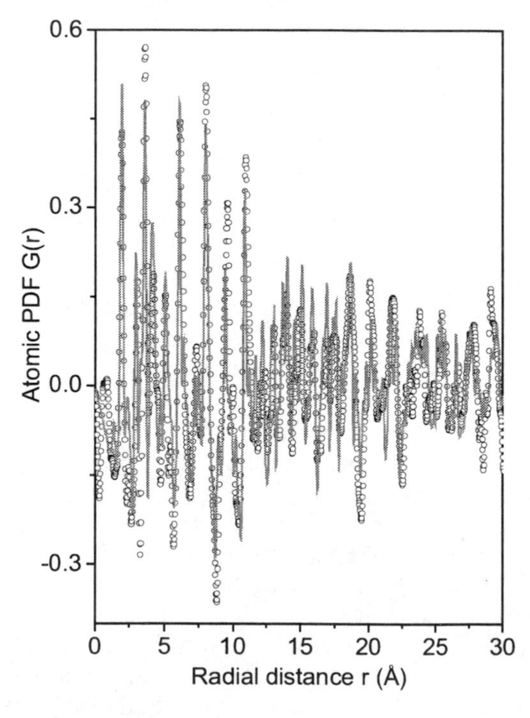

Figure 2. Comparison between the experimental (open circles) and model (solid line, in red) PDFs for (PANI)$_x V_2O_5nH_2O$. The experimental PDF is extracted from the diffraction data shown in Fig. 1. The model PDF is calculated from the atomic configuration shown in Fig. 3.

It shows a series of well-defined peaks each corresponding to a particular set of frequently occurring interatomic distances. Several structural models were tested against the experimental PDF data [8]. The parameters of the models, including unit cells and positions of the atoms in them, were refined as to obtain as good as possible agreement between the experimental and model PDF data. The best model found reproduces the experimental PDF data very well as can be seen in see Fig. 2. The model features the nanocomposite as an assembly of V_2O_5 bilayers enclosing densely packed PANI chains. A fragment of the model is presented in Fig. 3. The parameters of the model, including coordinates of the individual atoms (476 in total) are listed in ref. [8]. The model is periodic and of relatively small size, given the complexity of the nanomaterial studied, thereby allowing the exploration of structure-property relationships in $(PANI)_xV_2O_5nH_2O$. It describes the nanocomposite as an intimate mix of V_2O_5 bilayers and PANI chains, whose nominal planes are perpendicular to the layers, and corroborates the conclusions drawn in previous studies [7]. It is likely that the model misses some subtle details of the atomic structure in the nanocomposite as, for example, the likely presence of NH-O-V hydrogen bonding between the PANI chains and the V_2O_5 matrix and a possible "freezing" of some of the PANI chains inside the constrained interlayer space. Therefore, the model should be considered and used with care bearing in mind that it is still an approximation to the structure of $(PANI)_xV_2O_5nH_2O$. More subtle details as the ones mentioned above may be tested and explored by using structural information provided by local probes and spectroscopic techniques such as EPR, NMR, IR, Raman etc. This structural level of detail goes beyond the scope of the present study. The recent progress in experimental and computational techniques, however, make such an experimental approach feasible, in particular in cases where the more detailed local structural information is known to be critical for improving the properties of a nanocomposite material.

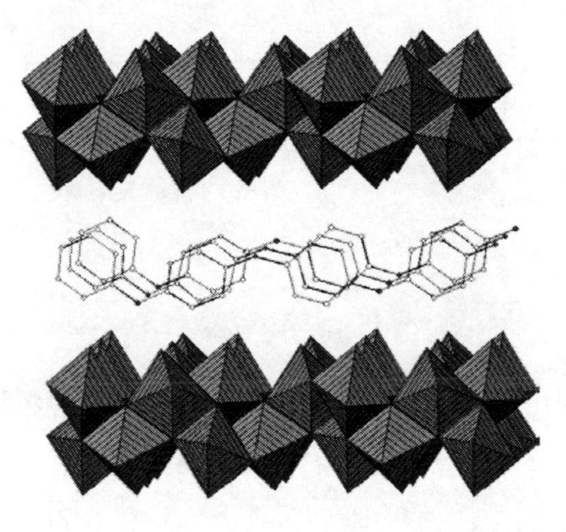

Figure 3. Atomic ordering in $(PANI)_xV_2O_5nH_2O$ as determined by the present PDF studies. The vanadium-oxygen octahedra (in red) are assembled in bilayers and the polyaniline chains occupy the interlayer space.

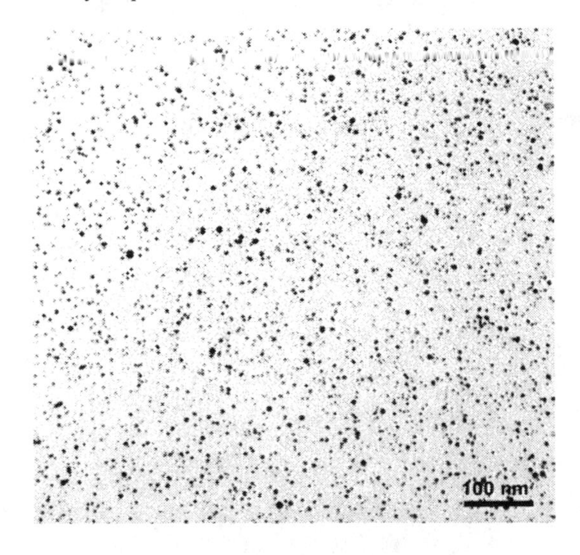

Figure 4. TEM image of dendrimer stabilized 2 nm gold nanoparticles.

2.2 Atomic ordering in dendrimer stabilized gold nanoparticles

The plasmonics of gold nanoparticles is the subject of intense research and has found many uses, ranging from sensors to optical materials [9]. In general, the surface plasmon resonance of gold nanoparticles is controlled by many factors, including their structure and size. We undertook a structure study on dendrimer stabilized gold nanoparticles of sizes 2nm, 12 nm and 30 nm to reveal this interrelation. The particles were made following a procedure described in [10]. An electron microscopy image of the nanoparticles of size 2 nm is shown in Fig. 4. The particles are spherical in shape and well separated from each other thanks to the capping layer of poly(amidoamine) dendrons. X-ray diffraction (XRD) experiments were carried out at the beamline 11IDC at the Advanced Photon Source at the Argonne National Laboratory using X-rays of energy 115.23 keV (λ = 0.1076 Å). The nanoparticles were suspended in water and sealed in glass capillaries. In addition, pure gold (in powder) was measured and used as a standard. The experimental atomic PDFs are shown in Fig. 5. The PDF for bulk gold shows well defined peaks to very long real space distances reflecting the presence of a 3D periodicity and long-range order in this crystalline material. The experimental data are well fit by a model based on the face centered cubic (fcc) structure of crystalline gold as can be seen in Fig. 5. The experimental PDFs for the gold nanoparticles decay to zero at much shorter real space distances reflecting the reduced degree of structural coherence in these nanomaterials. As can be expected this degree diminishes with the particle's size. A comparison between the experimental data and model ones based on the fcc-type structure shows that the atomic ordering in the

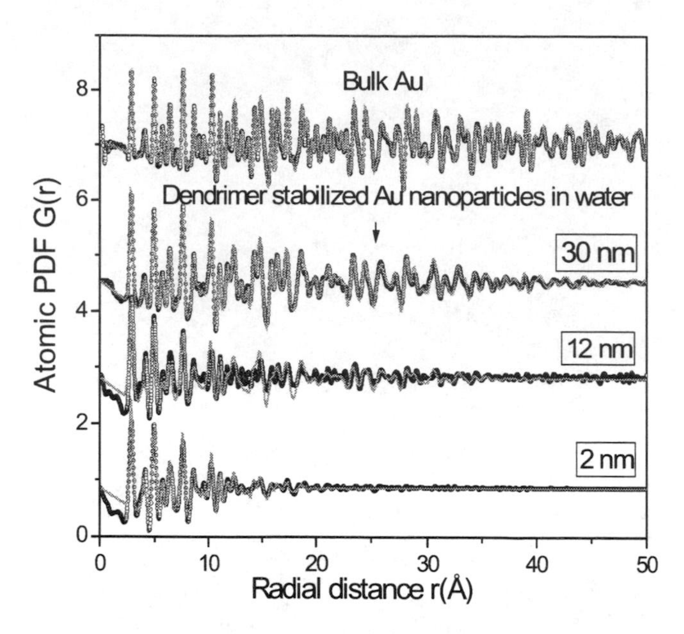

Figure 5. Experimental (circles) atomic PDFs for bulk crystalline gold and dendrimer stabilized gold nanoparticles of different sizes. Model PDFs based on a fcc type structure are shown as solid lines (in red).

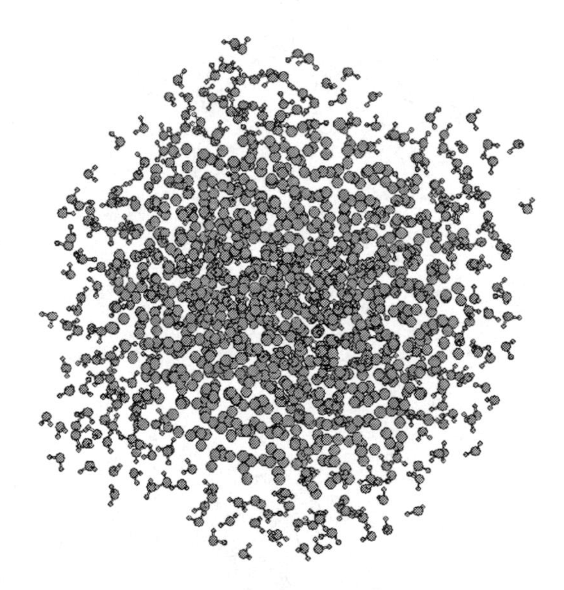

Figure 6. Model of a 2 nm gold nanoparticle (gold atoms are shown as circles in red) surrounded by water (in blue) molecules.

nanoparticles bears resemblance to that occurring crystalline gold (see Fig. 5). To determine the 3D structure of the nanoparticles in detail we generated model atomic configurations by Monte Carlo simulations employing realistic interatomic potentials. A model representing a 2 nm gold nanoparticle in water is shown in Fig. 6. The model reproduces the experimental PDF data quite well and allows the exploration of the structure-sensitive properties of this useful nanomaterial. Modeling studies of 12 nm and 30 nm gold nanoparticles are under way.

3. CONCLUSION

The three-dimensional structures of the scientifically and technologically important $(Polyaniline)_xV_2O_5nH_2O$ nanocomposites and dendrimer stabilized gold nanoparticles have been determined by the PDF technique. The structures are given in terms of a small number of sensible parameters, including atomic coordinates. This development opens up the route to better explaining, predicting, and possibly improving the properties of these materials by using both theoretical and experimental tools. Also, the results of the present study are a direct demonstration of the ability of the PDF technique to yield three-dimensional structural information for materials of limited structural coherence, including nanostructured materials. The technique succeeds because it relies on total scattering data obtained from the material, and as a result, is sensitive to its essential structural features. This allows a convenient testing and refining of structural models. Furthermore, the technique probes the bulk and can provide an important foundation to imaging techniques such as transmission electron and atomic force microscopy, which reveal only structural features projected down one axis or a surface.

ACKNOWLEDGMENTS Thanks are due to Tom Vogt, Yang, Ren, Mercouri Kanatzidis, Donald Tomalia and Baohua Huang for help with the sample preparation and/or the synchrotron radiation experiments. The work was supported by NSF through grant DMR 304391(NIRT) and ARL through grant DAAD19-03-2-0012.

REFERENCES

1. V. Petkov, P. Y. Zavalij, S. Lutta, M. S. Whittingham, V. Parvanov and S. Shastri, *Phys. Rev. B* ***69***. (2004) 085410.
2. V. Petkov, S.J.L. Billinge, T .Vogt, A.S. Ichimura and J.L. Dye, *Phys. Rev. Lett.* **89** (2002) 075502.
3. M. Gateshki, S-J Hwang, D. Hoon Park, Y. Ren, and V. Petkov, *Chem. Mater.,* **16** (2004) 5153.
4. V. Petkov, E. Bozin, S.J.L. Billinge, P. Trikalitis, M. Kanatzidis and T. Vogt,, *J. Am. Chem. Soc.* **124** (2002) 10157.
5. P.H. Klug, and L.A. Alexander, *X-ray Diffraction Procedures for Polycrystalline Materials* (Wiley, New York, 1974).
6. J.M. Zeh, S.J. Liou, C.Y. Lai, C. Wu, C*hem. Mater.*, **13** (2001) 1131; J. Wu , Y.H. Zou; H.L. Li, G.L. Shen, R.Q. Yu, *Sensors and Actuators B* **104** (2005) 43.
7. .M.G. Kanatzidis, C.G. Wu, M. Ho, C.R. Kannefurf, *J. Am. Chem. Soc.*, **111** (1989) 4139.
[8] V. Petkov, V. Parvanov, P. Trikalitis, Ch. Malliakas, T.Vogt and M. Kanatzidis, *J. Am. Chem. Soc.* (2005), submitted.
[9] S. Srivastava, B. L. Frankamp and V.M. Rotello, *Chem. Mater.* **17** (2005) 487.
[10] V. Petkov, Y. Peng, B. Huang and D. Tomalia, in preparation.

Polymeric Nanocomposites Enabled by Controlled Architecture Materials

Terri A. Shefelbine*, Ryan E. Marx**, James M. Nelson*, John W. Longabach**,
Myles L. Brostrom***, Mary I. Bucket***

**3M Corporate Research Process Lab ***3M Corporate Research Analytical Lab
* 3M Dyneon LLC 6744 33rd St. N. Oakdale, MN 55128
Office: 651-737-5099 tshefelbine@mmm.com

ABSTRACT

Clay nanocomposites have been shown to improve barrier properties, increase modulus, increase heat distortion temperature, and improve flame retardancy. To fully utilize these benefits, the clay platelets must be exfoliated and dispersed within the polymer. It is well known that clay will readily exfoliate in some polymeric resins, e.g. nylon. For hydrophobic resins, such as polyolefins, the exfoliation process is much more difficult. 3M is currently developing a portfolio of Controlled Architecture Materials (CAMs) as specialty additives for a variety of applications, including clay nanocomposites. In this paper we demonstrate the ability of CAMs to exfoliate both organically modified and unmodified natural clays in polystyrene (PS) and polypropylene (PP) resins. Complementary characterization techniques of Transmission Electron Microscopy (TEM), x-ray diffraction (XRD), and rheology are used to identify exfoliation. In addition we demonstrate the efficiency of CAMs to exfoliate clay at low additive levels.

Keywords: nanocomposites, functional polymers, olefins, extrusion, clay

1 INTRODUCTION

The benefits in properties gained from nanocomposites are thought to arise from the high number of favorable polymer/filler interactions. In order to achieve these interactions, the clay must be exfoliated. In solution or bulk polymerized monomers exfoliation may be achieved by polymerizing the monomer around the predispersed clay.[1] A second approach involves dissolving the polymer and dispersing the clay in a common solvent, then removing the solvent.[2, 3] In polyolefins neither approach is possible or economical. Instead additives are used during melt processing to exfoliate the clay. Traditional melt additives include random or graft copolymers such as maleic anhydride grafted polyolefins and are mainly utilized with organically modified clays. The main drawback in utilizing these random copolymers in nanocomposites is their inefficient mediation of the exfoliation process, which demands a fairly high level of additive to exfoliate the clay. Block copolymers, on the other hand, have been shown to be much more efficient and can be further designed to augment other physical properties such as impact modification and elongation.[4-6] Furthermore, block copolymers can be tuned to the polymeric resin and the different clays, organically modified or not. Tuning the interaction between filler and polymer allows for optimization of the properties that are dominated by this interface. Exfoliation utilizing this block copolymer mediated approach has been demonstrated in polypropylene (PP), thermoplastic polyolefin (TPO), various polyethylenes (HDPE, LDPE), and polystyrene (PS).

2 EXPERIMENTAL PROCEDURE

2.1 Materials

The base resins used were polypropylene (PP) available from ExxonMobile Corp., Irving, Texas, under the trade designation "Escorene 1024" and polystryene (PS) available from Dow Chemical Co., Midland, Michigan, under the trade designation "Dow Styron 615APR". Montmorillonite clays were used as received from Southern Clay Products, Gonzales, Texas. The clays designated "Cloisite 10A", "Cloisite 20A", and "Cloisite 30B" are organically modified. "Cloisite Na+" does not contain organic modifier.

2.2 Synthetic method

The CAMs additives are made via sequential anionic polymerization in a stirred tubular reactor.[7] A LIST Discotherm B devolatilization system is used to remove the solvent. The polydispersity and molecular weight of each material are characterized by gel permeation chromatography (GPC) analysis and narrow polydispersity polystyrene calibration standards. The composition of each CAM is determined by 1H Nuclear Magnetic Resonance (1H NMR) spectroscopy.

2.3 Nanocomposite Formation by Continuous Twin-Screw Extrusion

Extrusion was carried out using a co-rotating, 25mm twin-screw extruder with 41:1 L/D available under the trade designation "COPERION ZSK-25 WORLD LAB EXTRUDER" from Coperion, Ramsey, New Jersey.

2.4 Film Preparation for XRD and TEM Analysis

Analysis via XRD and TEM was done on 1 mm thick films. To form the films, each compound was placed between untreated polyester liners and pressed between aluminium plates held apart by 1 mm spacers. Each stack was placed in a heated hydraulic press "WABASH MPI MODEL G30H-15-LP" from Wabash MPI, Wabash, Indiana. The press plates were heated to 193 °C. The stack was pressed for 1 minute at 1500 psi (10 MPa). The hot stack was then moved to a low-pressure water-cooled press for 30 seconds.

2.5 X-ray Diffraction (XRD)

Reflection geometry X-ray scattering data were collected using a four-circle diffractometer (available under the trade designation "HUBER (424/511.1)" from Huber Diffraktionstechnik GmbH, D83253 Rimsting, Germany), copper K-alpha radiation, and scintillation detector registry of the scattered radiation. The incident beam was collimated to a circular aperture of 0.70 mm. Scans were conducted in a reflection geometry from 0.5 to 10 degrees (2 theta) using a 0.05 degree step size and 10 second dwell time. A sealed tube X-ray source and X-ray generator settings of 40 kV and 20 mA were used. Data analysis and peak position definition were determined using X-ray diffraction analysis software available under the trade designation "JADE" from MDI, Inc., Livermore, California.

2.6 Transmission Electron Microscopy (TEM)

TEM was performed using a transmission electron microscope operated at 200 kV, available under the trade designation "JEOL 200CX" from JEOL USA, Peabody, Massachusetts. Samples were cryo-ultra microtomed prior to imaging.

2.7 Rheological Characterization

The samples used for rheological characterization were prepared by feeding pelletized nanocomposite samples to a Mini-jector Model 45 injection molder available from Mini-Jector Machinery Corporation, Newbury, Ohio. The temperature of injection was 180 °C with a pressure of 0.48 MPa. The rheological response of the nanocomposites was characterized using a TA AR 2000 rheometer. At a temperature of 200 °C and a frequency of 1 rad/s, three successive strain sweeps from 1-500% strain were applied to the material. The second sweep was performed immediately after the first. Between the second and the third sweeps the sample annealed for an hour at 200 °C with no applied strain. Wagener and Reisinger developed a technique similar technique to characterize exfoliation in polybutylene terephthalate. [8]

3 RESULTS

As described in the experimental section, a variety of resins were compounded with montmorillonite clays to determine whether CAMs improve the efficiency of exfoliation. In Figure 1, a TEM image of 5% of a non-exfoliated organically modified clay in PS is shown along with the corresponding x-ray diffraction pattern. In the TEM image the large dark area in the top center of the image has regularly spaced striations, which correspond to the edge-on view of the layered clay "Cloisite 20A". The three distinct peaks in the XRD pattern arise from the scattering from planes that are 38 Å apart. This spacing is significantly larger than the 24 Å expected for the neat clay and indicates some degree of intercalation of the PS into the clay galleries.

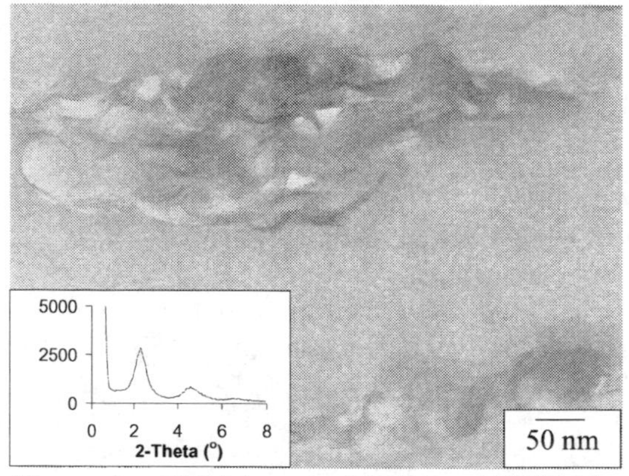

Figure 1. TEM image of 95/5 wt. % PS/20A clay. Inset: XRD pattern of the same sample.

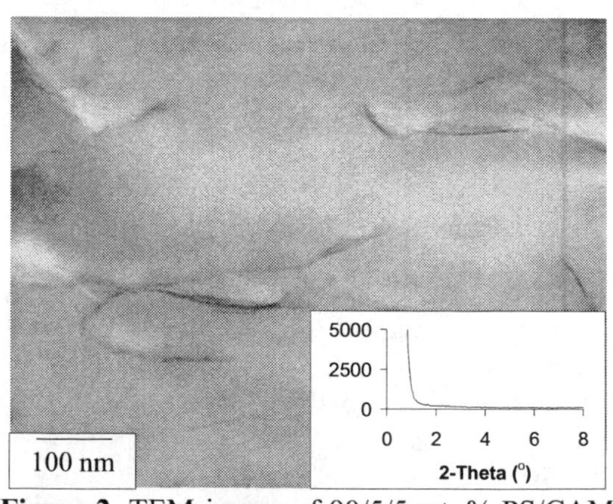

Figure 2. TEM image of 90/5/5 wt. % PS/CAM/20A clay. Inset: XRD pattern of the same sample.

To improve the separation between the clay layers, we processed a compound of 90/5/5 wt.% PS/CAM/Cloisite 20A using the twin-screw extruder as described above. The morphology of this sample is presented in Figure 2. The TEM image shows distinct, separated clay layers on edge that are randomly oriented to each other. The inset is a XRD pattern taken at the same conditions as the one in Figure 1. The lack of features above 1.5° indicates that the long range order observed in the neat clay has been eliminated by exfoliating the clay platelets.

These complementary techniques of x-ray diffraction and TEM clearly show the change in morphology when a PS/clay composite is processed in the presence of a CAM. The amorphous polymer phase in a PS nanocomposite allows one to view the separated clay platelets and their random orientation to each other. For this reason, the PS resin composites were good starting points in our experiments. However, polyolefins are used in significantly higher volume in industry and in a wider variety of applications than polystyrene.

Some of the desired improvements in polyolefin properties could be achieved with nanocomposites, including barrier properties, heat distortion temperature, and mechanical properties. In order to achieve these improvements with nanocomposites, the clay must first be exfoliated. The following data are our results in exfoliating clay with a CAM in a hot melt process for polypropylene.

In Figure 3 we compare the ability of CAMs to exfoliate clay to the ability of random copolymers of maleated polypropylene (MAPP). For formulations made of the same composition of PP/additive/Cloisite 10A (90/5/5 wt.%), the formulation made with a CAM additive has no significant long range order in the XRD pattern. Both the control with no additive and the formulation with MAPP as an additive have significant peaks corresponding to a domain spacing of 18 Å. This is slightly smaller than the published value of 19.2 Å for Cloisite 10A.

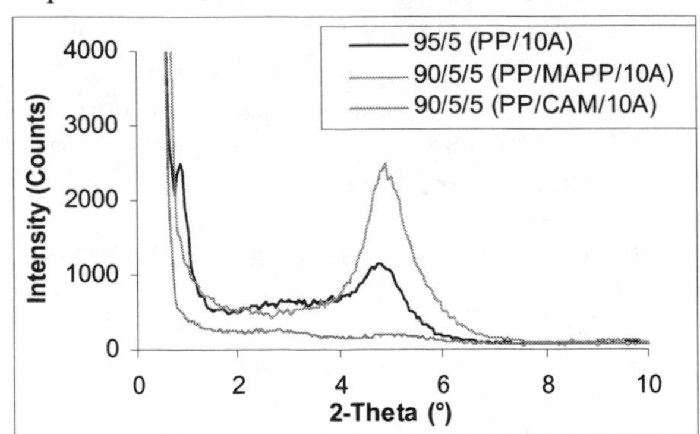

Figure 3. XRD plot of nanocomposites made without additive, with CAM, and with maleated polypropylene.

The lack of long-range order in the formulation with CAM suggests that the clay is exfoliated and randomly oriented by the melt processing method described above. In contrast samples without CAM do not exfoliate, or even intercalate with this processing method.

The efficiency of CAMs at exfoliating clays in PP is further illustrated in Figure 4. In these formulations the clay Cloisite 30B is used at a 5 wt.% level while the CAM loading level is 5, 3, or 1 wt. %. The peak on the PP/30B curve corresponds to 18 Å, which is only slightly smaller than the reported value of 18.5 Å for neat Cloisite 30B. From this image the clay 30B does not exfoliate or intercalate in PP without CAM. The featureless curves for the PP/CAM/30B samples are consistent with exfoliated samples. Remarkably only 1 wt.% CAM in this formulation effectively exfoliates a clay modified with polar organic modifiers. We attribute this efficiency to the carefully designed interaction between CAM and the organic modifier that is possible with our materials.

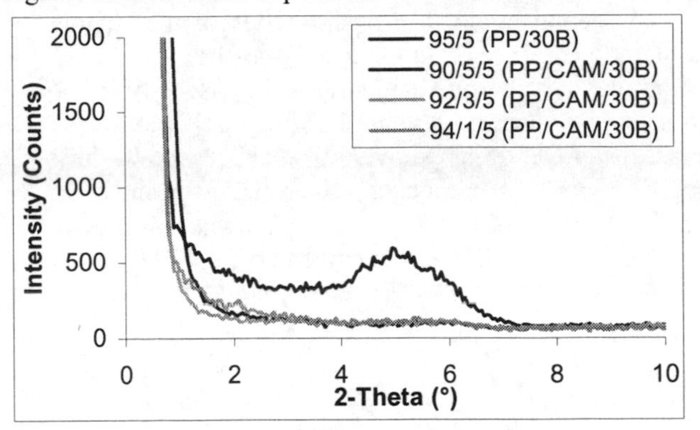

Figure 4. XRD plot of nanocomposites made with various levels of CAM with PP and 30B clay.

While the XRD patterns presented in Figures 3 and 4 indicate exfoliation, we would like to confirm the conclusion that formulations including CAM are exfoliated with a second technique. The TEM method used above in the PS samples, is more difficult to interpret in the case of semicrystalline samples such as PP. Instead we turn to rheology where we examine the hysteresis of the viscosity after strain sweeps.

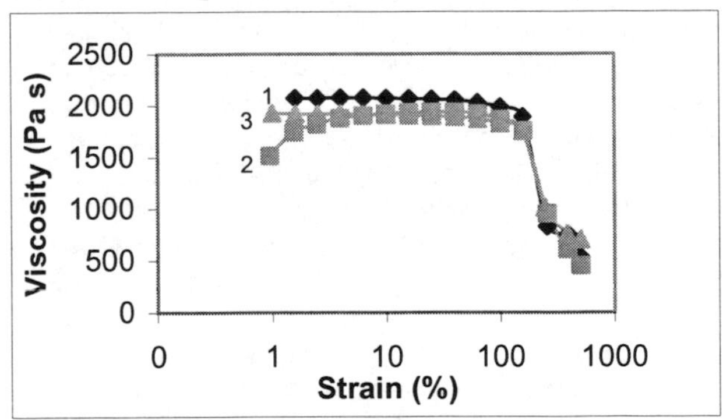

Figure 5. Strain sweeps at 200 °C of 95/5 wt. % PP/10A. (1) is the initial sweep. (2) is the second sweep. (3) is conducted after a 60 min anneal at 200 °C.

In Figure 5, three strain sweeps performed in succession on the sample 95/5 wt.% PP/10A (see Figure 3) are plotted on the same graph. The initial strain sweep shows the highest viscosity and pronounced shear-thinning behavior at strains above 60%. The second sweep, performed immediately after the first finished, shows slightly lower viscosity at low strains, a plateau region at intermediate strains, and pronounced shear-thinning at strains higher than 60%. The sample was annealed at 200 °C for 60 minutes between the second and third sweeps. The viscosity of the third sweep again shows a plateau viscosity up to about 60% strain. This plateau value is slightly lower than the value in the initial strain sweep.

In Figure 6, the sample of 90/5/5 PP/CAM/10A is probed with the strain test. The first strain sweep has a plateau region slightly lower than the sample in Figure 5, with pronounced shear-thinning behavior above strains of 60%. The second strain sweep shows viscosities at low strains that are comparable to the low viscosities seen at the highest strains. At intermediate strains the viscosity in the second sweep recovers to a value less than half that of the initial viscosity. Strains above 60% again show shear thinning behavior. After annealing the sample for one hour, the viscosity again shows a plateau at low strains, with strong shear-thinning behavior at strains above 60%.

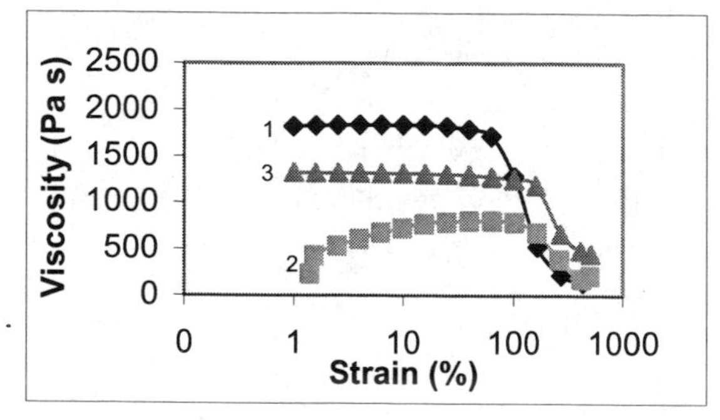

Figure 6. Strain sweeps at 200 °C of 90/5/5 wt. % PP/CAM/10A. (1) is the initial sweep. (2) is the second sweep. (3) is conducted after a 60 min anneal at 200 °C.

The strong hysteresis observed in Figure 6 for the sample containing CAM, which has a XRD pattern consistent with exfoliation, suggests that the platelets are aligned during the first strain sweep. At the beginning of the second strain sweep low viscosity suggests that the platelets are still aligned, but begin to re-randomize over the course of the second sweep. The platelets are re-aligned at high strains in the second sweep, but the one-hour annealing period allows them to re-randomize to a greater extent. This results in an intermediate plateau viscosity at low strains. In contrast the sample present in Figure 5, which has a XRD pattern consistent with non-exfoliated clay (see Figure 3), shows little hysteresis and insignificant drop in viscosity.

The clays presented to this point have all been organically modified with as much as 30% of organic modifier. Some applications will call for pure inorganic clay, and we believe that the physical properties of clay-based nanocomposites will be improved using natural clays. This approach is challenging because natural clays have smaller domain spacing and less flexibility in terms of what chemical functionality the CAM can interact with at the surface. An added complication in determining the exfoliation efficiency of Na+ clay nanocomposites is that the clay has a weak scattering factor yielding weak peaks in XRD. Thus the absence of peaks in a XRD pattern with Na+ Cloisite cannot be used as evidence of exfoliation. In Figure 7 we show a strain hysteresis plot of 90/5/5 wt.% PP/CAM/Na+. As in the case of the exfoliated system shown in Figure 6, the first two strain sweeps show marked deviation. The third shows recovery of the viscosity plateau and increased viscosity compared to the second sweep. As shown in Figure 6, this type of rheological trace is consistent with an exfoliated system.

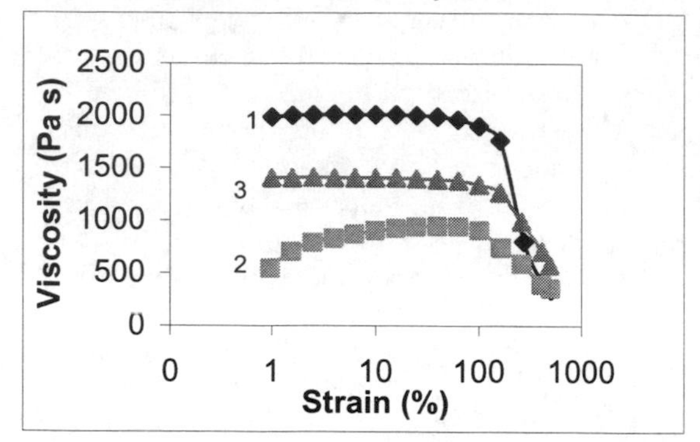

Figure 7. Strain sweeps at 200 °C of 90/5/5 wt. % PP/CAM/Na+. (1) is the initial sweep. (2) is the second sweep. (3) is conducted after a 60 min anneal at 200 °C.

The exfoliation of natural clay in polyolefin by melt processing is a significant advance and is made possible by the flexibility inherent in CAM materials. In addition our ability to characterize Na+ clay nanocomposites by the rheological method described above allows us to explore new applications with different resins. In the future we will continue to probe the physical property enhancements afforded by nanoclay exfoliation in a variety of resins.

REFERENCES

[1] A. Okada, M Kawasumi, A Usuki, Y Kojima, T Kurauchi, O Kamigaito *MRS Symposium Proceedings*, Pittsburg, vol 171; **1990**. 45-50.

[2] DJ Greenland *J. Colloid Science*, vol 18; **1963**. 647-664

[3] P Aranda, E Ruiz-Hitzky *Chem Mater*, vol. 5; **1993**. 1395-1403.

[4] JJ Cernohous, S Papp, NR Granlund, JM Nelson, RE Marx, JG Linert. In US 20040254268 Unitied States, **2004**.

[5] Nelson, J. M.; Marx, R.E.: Hanley, K.J.: Cernohous, J. J. Jones, T.D.: McNerney, J. M.; In *US 20040024130* United States, **2004.**

[6] Nelson, J. M.; Marx, R.E.: Cernohous, J. J.: McNerney, J. M.; In *US 20040023398* United States, **2004**.

[7] Nelson, J. M.; Cernohous, J. J.; Annen, M. J.; McNerney, J. M.;Heldman, B.; Ferguson, R. W.; Higgins, J. A. In *US 6,448,353* United States, **2002**.

[8] R Wagener, TGJ Reisinger *Polymer*, vol. 44; **2003**. 7513-7518.

Effect of Nanoclay Content on Mechanical Behaviour of TGDDM Epoxy Nanocomposites

Nam T.P. Huong*, Alan Crosky*, Ben Qi**, Don Kelly***, David Chongjun Li*

* School of Materials Science and Engineering, University of New South Wales, Australia, a.crosky@unsw.edu.au

** Cooperative Research Centre for Advanced Composite Structures (CRC-ACS), Australia

*** School of Mechanical and Manufacturing Engineering, University of New South Wales, Australia

ABSTRACT

Nanocomposites based on tetraglycidyldiamino diphenylmethane (TGDDM) cured by diethyltoluene diamine (DETDA) reinforced with 1-20 phr of a commercial nanoclay, Nanomer I30E, were prepared. The effect of clay nanoparticle loading on the compression modulus of the nanocomposite was investigated. The level of exfoliation of the nanoclay was measured by small angle X-ray scattering and confirmed by transmission electron microscopy. The modulus of the nanocomposites increased progressively with increasing clay content with the 20 phr nanocomposite showing a 50% increase in modulus compared with the pure resin. The T_g of the cured nanocomposites measured by DSC decreased progressively with increased nanoclay loading.

Keywords: nanocomposites, epoxy, nanoclay, compressive modulus, glass transition temperature.

1 INTRODUCTION

Nanoscale materials have become of considerable interest for particulate reinforcement of polymers. In particular, nanoclay particles are very attractive because of their special structure and properties. Materials made by incorporating nanoclay into epoxy resin matrices have shown a significant improvement in properties, such as mechanical strength, barrier performance, and thermal stability, even at quite low clay loadings. Uniform dispersion of the nanoclay layers in the resin at the nanoscale is however required [1-4]. Two types of nanocomposite reinforced with nanoclay are possible: intercalated or exfoliated [5-8]. The nanoclay loading significantly affects the degree of exfoliation as well as the average distance between the separated nanoclay layers. The improvement in the mechanical properties of epoxy nanocomposites has been explained variously in terms of the ability for both internal and external stress dispersion in the network [9], the homogeneous distribution of nanoclay layers [10, 11], the stable interfacial region close to the nanoclay surface [12], the degree of cross-linking [13], and the stiffness of the particle acting as "network node" [14, 15]. A change in T_g is also observed in nanocomposite reinforced with nanoclay and this has been attributed to factors, such as changes in chain flexibility [14, 16], adsorbed moisture and the residual organics acting as a plasticizer [17, 18], and changes in the degree of cross-linking [19]. The flexibility of the chain is undoubtedly the most important factor influencing T_g. It is a measure of the ability of a chain to rotate about the constituent chain bonds; hence a flexible chain has a low T_g whereas a rigid chain has a high T_g [20].

In this study, the effect of nanoclay content on the mechanical behavior as well as thermal properties of a TGDDM epoxy-clay nanocomposite reinforced with a commercially available montmorillonite clay was investigated.

2 EXPERIMENTAL

2.1 Materials

The epoxy resin used was Araldite® LY568- a tetraglycidyldiamino diphenylmethane (TGDDM) obtained from Ciba Speciality Chemicals. The curing agent was Ethacure 100 supplied by Albemarle Corporation. This is a mixture of two diethyltoluene diamine (DETDA) isomers, containing 74-80% 2,4 isomer and 18-24% 2,6 isomer. The nanoclay used was Nanomer I30E from Nanocor, a montmorillonite clay modified with octadecyl ammonium ions.

2.2 Sample preparation

Oven dried nanoclay in amounts from 1-20 parts per hundred of resin (phr) were dispersed in a volume of neat epoxy resin using a high shear force stirrer (20,000 rpm) with a metal stirring disc at 70°C for 1 hour. Following mixing, the mixture was degassed in a vacuum chamber at 70-80°C for minutes. A stoichiometric concentration of DETDA was then added to the mixture of epoxy and nanoclay, and the mixture again degassed in the vacuum chamber at 80°C. The mixture was finally poured into moulds and cured at 110°C for 1 hour, then at 155°C for 2 hours and then at 195°C for 1 hour.

2.3 Characterization

Small-angle X-ray scattering (SAXS) measurements were carried out with a Philips X'pert X-ray diffraction system using CuKα radiation with a wavelength of 1.5418740Å. An accelerating voltage of 45 kV was used with a current of 40 mA. A scanning range from 0.5 - 7.0° was used with a step size of 0.01°. The time per step was 3 seconds. Transmission electron microscope images were obtained using a Philips CM-200 field emission transmission electron microscope operating at 200 kV. The TEM samples were prepared as thin sections about 50 - 70 nm in thickness using an ultra-microtome (Reichert Ultracut E) equipped with a 45° sharp-edged glass knife. Prior to examination they were mounted on 200-mesh copper grids coated with carbon. Compression samples 6 mm in diameter and 12 mm high were tested on an Instron 1185 Screw at a cross-head speed of 2 mm/min. A Dupont Differential Scanning Calorimeter/Thermal Analyser was used to study the cure reaction and measure the glass transition temperature using ramp speeds of 2°C/min and 20°C/min respectively. The measurements were made using a nitrogen atmosphere at a flow rate of 10ml/min and. The analysis was carried out from 10°C to 240°C.

3 RESULTS AND DISCUSSION

3.1 Effect of surfactant on curing behaviour

The released energy of the curing reaction recorded using DSC is shown for some selected nanoclay additions in Figure 1.

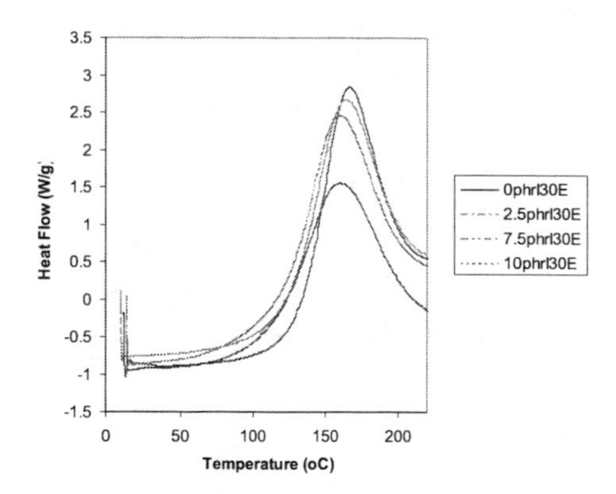

Figure 1: Exothermal curves for curing reaction of TGDDM/DETDA with various I30E nanoclay loadings.

Nanoclay content in nanocomposites (phr)	Onset temperature (°C)	Peak temperature (°C)
0	122.98	166.05
1	119.49	165.49
2.5	111.85	165.15
7.5	93.95	160.58
10	91.55	159.02

Table 1: Peak and onset temperatures of nanocomposites with various nanoclay contents.

With increased nanoclay content, the exothermal peak temperature shifted to the left, indicating that the energy required for the curing reaction of the system decreased progressively with increasing nanoclay content, as shown in Table 1.

The shift in both the exothermal peak and the onset temperatures to lower values is attributed to the catalytic effect of the octadecylamine surfactant on the I30E nanoclay in the curing reaction, leading to a decrease in the energy required for epoxy ring opening polymerization. The catalytic effect becomes stronger with increasing nanoclay content since the surfactant concentration in the resin is increased hence the decrease with increasing clay loading.

3.2 Morphology

The distance between the layers of nanoclay, as measured by SAXS, varied with clay loadings as indicated in Figure 2. For a diffraction angle larger than 0.75°, no peaks were observed with nanoclay contents from 1 to 5 phr, indicating that the d-spacing of the nanoclay layers was greater than 120 Å, consistent with the nanoclay particles being fully exfoliated. However, with further increase in the clay loading the distance between the nanoclay layers decreased, being only 70-80 Å for the composites with 7.5-12.5 phr of nanoclay. These nanocomposites were not considered to be fully exfoliated, and are therefore referred to as pre-exfoliated nanocomposites.

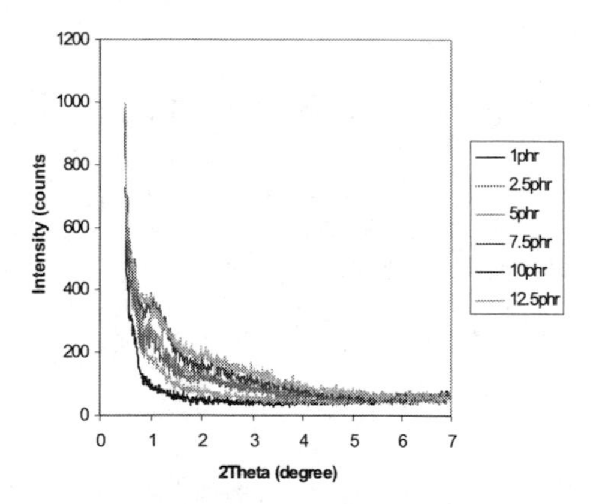

Figure 2: Small angle X-ray diffraction spectra for nanocomposites with varying I30E nanoclay contents.

These results were confirmed by TEM observations as shown in Figure 3. The dark lines define the nanoclay layers. For the 2.5 phr nanoclay composite the spacing of the nanoclay platelets was about 120-125Å, Figure 3(a). This then decreased to 85Å for 7.5 phr of nanoclay, Figure 3(b), 75-80Å for 12.5 phr of nanoclay, Figure 3(c), and 60Å for 20 phr of nanoclay, Figure 3(d). The nanoclay particles were present in groups with almost all the platelets of nanoclay within a group being parallel and the individual groups being well dispersed, indicating that exfoliated nanocomposites had been obtained.

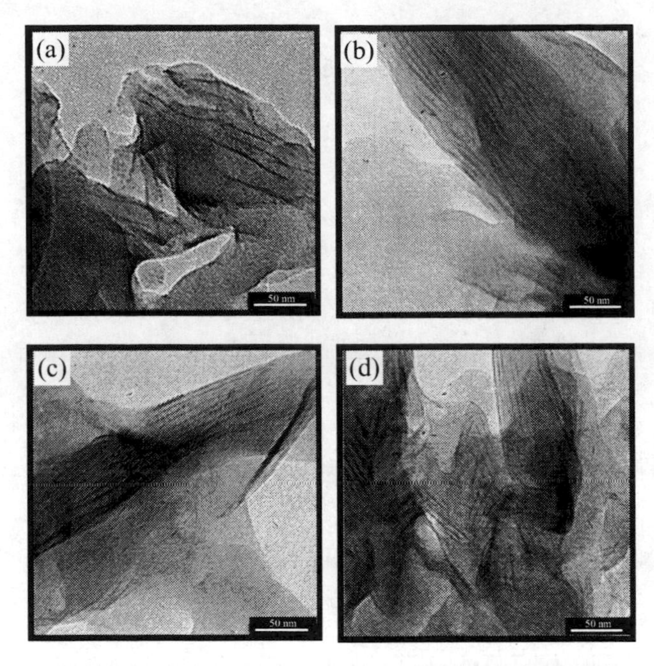

Figure 3: TEM images of nanocomposites with (a) 2.5 phr, (b) 7.5 phr, (c) 12.5 phr and (d) 20 phr of nanoclay.

Just as the presence of the octadecylamine surfactant on the nanoclay surface had a catalytic effect on the curing reaction, it is considered that it also accelerated the cross-linking of the epoxy resin in the spaces between the layers of nanoclay so that the platelets were continuously forced to separate after the curing agent was added, as previously proposed [5, 21]. The surfactant being on the surface of the added clay particles, as well as between the individual layers, would be expected to accelerate the reaction at the surface of the particles also. At low nanoclay loadings this acceleration would be much less pronounced than between the layers because of the comparatively smaller surface area. However with increased nanoclay loading the extrapolymerization reaction would become progressively more accelerated and this may then limit expansion of the nanoclay layers. This would account for the progressive reduction in interlayer spacing observed here.

3.3 Compression modulus

The compressive modulus of the nanocomposites increased progressively with increased nanoclay content, Figure 4. The modulus increased from a value 2352 MPa for the pristine resin to 3030 MPa at a loading of 10 phr of nanoclay eventually attaining a value of 3517 MPa at a loading of 20 phr of nanoclay. These values corresponded to improvements of almost 30% for 10 phr nanoclay and 50% for 20 phr nanoclay additions, Figure 5.

It is evident from Figure 5 that the improvement in modulus varies in an essentially linear manner with nanoclay content. However the values for the two lowest clay loadings seem anomalously high. This may indicate that the better exfoliation produced with the lower clay loadings has provided more effective reinforcement for the volume fraction of clay used. This in turn would suggest that the level of cross linking within the resin-filled galleries between the clay layers was increased in the better exfoliated composites. It is noted that the wider the galleries between the clay layers, the easier would be the access of the hardener to the resin in the galleries. It may be that in the composites with thinner interlayer spacings (ie, those with high clay loadings) insufficient hardener was able to penetrate into the galleries to provide full cross-linking of the resin.

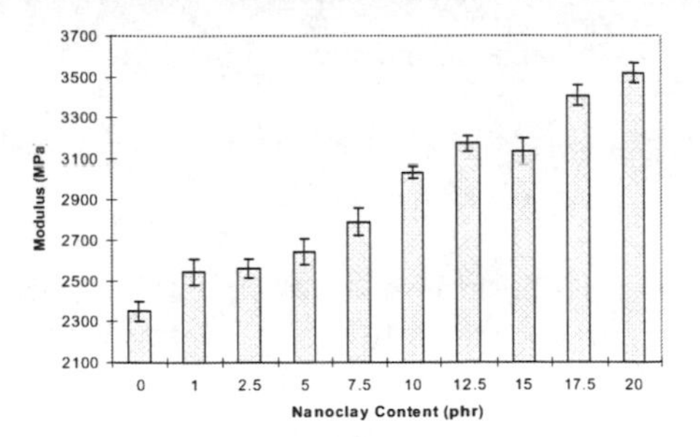

Figure 4: Compressive modulus of nanocomposites based on TGDDM/DETDA/I30E system. Error bars indicate one standard deviation.

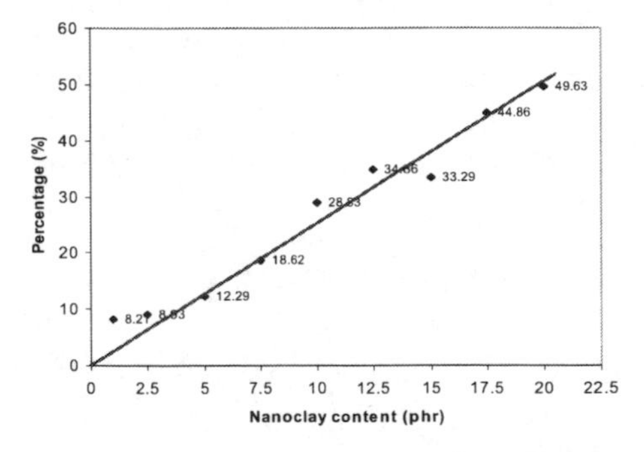

Figure 5: Percentage improvement in modulus for nanocomposites with varying nanoclay contents.

3.4 Thermal properties

An increase in modulus in polymers is usually also accompanied by an increase in the glass transition temperature of the material. However, to the contrary the T_g of the nanocomposites ldecreased progressively with clay content as shown in Figure 6.

The decrease in the T_g of the nanocomposites with increased nanoclay content is considered to be, at least in part, due to the presence of the octadecylamine surfactant. As this was an aliphatic amine with long linear chains, the chains would be very flexible and have low thermal stability [15]. Since the concentration of surfactant in the resin would increase with increased nanoclay content the T_g would also be expected to decrease. A decrease in the level of cross linking in the clay galleries would also contribute to a decrease in T_g.

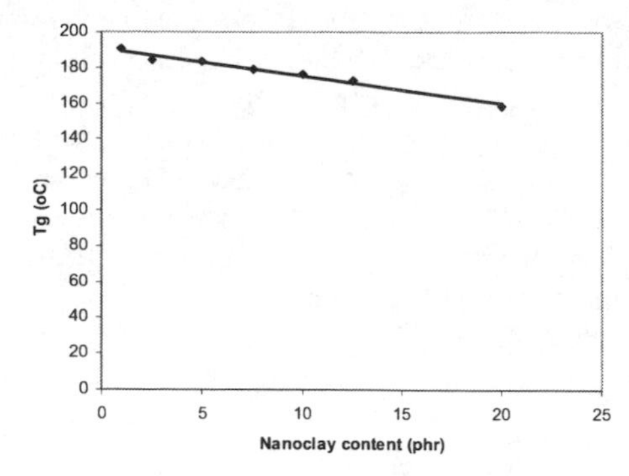

Figure 6: Tg of nanocomposites with varying nanoclay contents.

4 CONCLUSIONS

A strong relationship was observed between the nanoclay content and the properties of nanocomposites based on TGDDM cured by DETDA. The compressive modulus of the nanocomposite increased progressively with increasing nanoclay content, giving an increase of 50% at a nanoclay content of 20 phr. The marked improvement in the modulus is considered to be the result of effective separation of the individual clay layers permitting good penetration of the epoxy resin into the interlayer galleries. While the modulus increased with increasing clay content, the glass transition decreased. The decrease in T_g is attributed, at least in part, to the presence of increasing amounts of surfactant.

ACKNOWLEDGEMENTS

The support of the Cooperative Research Center for Advanced Composite Structures Ltd - Australia is gratefully acknowledged.

REFERENCES

[1] Chenggang Chen, David Curliss, Journal of Applied Polymer Science, 90, 2276-2287, 2003.

[2] Gilman, Jeffrey W., Applied Clay Science, 15, 31-49, 1999.

[3] LeBaron, Peter C., Zhen Wang, and Thomas J. Pinnavaia, Applied Clay Science, 15, 11-29, 1999.

[4] Messersmith, P. B. and E. P. Giannelis, Chemistry of Materials, 6, 1719-1725, 1994.

[5] Chenggang Chen, Tia Benson Tolle, Journal of Polymer Science Part B: Polymer Physics, 42, 3981-3986, 2004.

[6] In-Joo Chin, Thomas Thurn-Albrecht, Ho-Cheol Kim, Thomas P. Russell, Jing Wang, Polymer, 42, 5947-5952, 2001.

[7] Yong Taik Lim, O. Ok Park, Rheol Acta, 40, 220-229, 2001.

[8] Yucai Ke, Jiankun Lü, Xiaosu Yi, Jian Zhao, Zongneng Qi, Journal of Applied Polymer Science, 78, 808-815, 2000.

[9] Carsten Zilg, Ralf Thomann, Jürgen Finter, Rolf Mülhaupt, Macromolecular Materials and Engineering, 280-281, 41-46, 2000.

[10] Bernd Wetzel, Frank Haupert, Klaus Friedrich, Ming Qiu Zhang, Min Zhi Rong, Polymer Engineering and Science, 42, 1919-1927, 2002.

[11] Takahiro Imai, Yoshihiko Hirano, Hisayuki Hirai, Susumu Kojima, Toshio Shimizu, Conference Record of the 2002 IEEE International Symposium on Electrical Insulation. Boston, MA USA. 2002.

[12] Tia Benson Tolle, David P. Anderson, Journal of Applied Polymer Science, 91, 89-100, 2004.

[13] J. Macan, H. Ivankovic, M. Ivankovic, H. J. Mencer, Journal of Applied Polymer Science, 92, 498-505, 2004.

[14] Bernd Wetzel, Frank Haupert, Ming Qiu Zhang, Composites Science and Technology, 63, 2055-2067, 2003.

[15] Jörg Fröhlich, Dietmar Golombowski, Ralf Thomann, Rolf Mülhaupt, Macromolecular Materials and Engineering, 289, 13-19, 2004.

[16] Hiroaki Miyagawa, Michael J. Rich, Lawrence T. Drzal, Journal of Polymer Science Part B: Polymer Physics, 42, 4384-4390, 2004.

[17] Vineeta Nigam, D. K. Setua, G. N. Mathur, Kamal K. Kar, Journal of Applied Polymer Science, 93, 2201-2210, 2004.

[18] Yangayang Sun, Zhuqing Zhang, Kyoung-Sik Moon, C. P. Wong, Journal of Polymer Science Part B: Polymer Physics, 42, 3849-3858, 2004.

[19] Tianxi Liu, Wuiwui Chauhari Tjiu, Yuejin Tong, Chaobin He, Sok Sing Goh, Journal of Applied Polymer Science, 94, 1236-1244, 2004.

[20] Cowie, J. M. G., Polymer: Chemistry and Physics of Modern Materials. Blackie Academic and Professional. 13-327, 1991.

[21] Doil Kong, Chan Eon Park, Chemistry of Materials, 15, 419-424, 2003.

The Impact of Nano-Materials on Coating Technologies

Roger H. Cayton*, Thomas Sawitowski**

*Nanophase Technologies Corporation
Romeoville, IL 60446
** BYK-Chemie
Wesel, Germany 46483

INTRODUCTION

During the past several years, advances in nanomaterials have allowed them to be formulated into numerous applications. The majority of these applications sought performance improvements that were previously unobtainable. Examples of such applications containing nanomaterials that have been commercialized include, scratch/abrasion resistant transparent coatings, sunscreen lotions to provide visible transparent UV protection, polishing slurries to provide pristine surfaces for optics, and environmental catalysts to reduce pollution.

The quest for improved scratch/abrasion resistant coatings is an on going project for many coating formulators. Thousands of scratch resistant coating applications are present in our everyday lives. Examples of these applications include coatings for wood floors, safety glasses, electronic displays, automotive finishes, and polycarbonate panels. Improving the mar, scratch and/or abrasion in these transparent coating applications is a major challenge, particularly with regard to not affecting the other performance attributes of the coating.

Incorporation of inorganic fillers into coatings to improve mechanical properties is well known. Drawbacks associated with this approach can include loss of transparency, reduced coating flexibility, loss of impact resistance, increase in coating viscosity, and appearance defects. To overcome these defects a filler material should impart improved scratch resistance without causing the aforementioned detriments. Nanomaterials, due to their small size and particle morphology, have the potential to overcome many of these detriments.

Maintaining transparency in a coating containing inorganic filler particles is a challenge. Many variables affect the ultimate degree of transparency in a composite material, including film thickness, filler concentration, particle size, particle shape, extent of particle aggregation, homogeneity of the particle dispersion, and the difference in refractive index between the bulk coating and the filler particle.

Silica particles, colloidal or fumed, and clays are among the most widely studied inorganic fillers for improving the scratch/abrasion resistance of transparent coatings. These fillers are attractive from the standpoint that they do not adversely impact the transparency of coatings due to the fact that the refractive indices of these particles (fumed silica = 1.46, bentonite clay = 1.54) closely match those of most resin-based coatings. The drawback to silica-based fillers is that high concentrations of the particles are generally required to show a significant improvement in the scratch/abrasion resistance of a coating, and these high loadings can lead to various other formulation problems associated with viscosity, thixotropy, and film formation.

The use of alumina particles in transparent coatings is much more limited even though alumina is significantly harder than silica-based materials, and as a scratch and abrasion-resistant filler, higher performance at lower loadings is often observed. For alumina particle sizes greater than 100nm, the high refractive index (1.72) results in significant light scattering and a hazy appearance in most clear coatings. Currently, only coatings with a high refractive index, such as the melamine-formaldehyde resins used in laminates, can utilize micron-size alumina particles to gain scratch resistance and still maintain transparency.

NANOPARTICLE PRODUCTION

In order to utilize alumina as scratch-resistant filler in transparent coatings, the particle size must be sufficiently small to overcome its refractive index mismatch. Nanophase Technologies uses plasma processes to produce metal oxide nanoparticles via a bottoms-up method starting from metallic feed. This process allows production of nonporous crystalline metal oxides having primary particle sizes less than 100 nm at economically viable rates with essentially no byproducts or waste streams.

Nanophase Technologies produces three grades of aluminum oxide using the plasma process, two of which are commercial and one that is under development. All grades feature a mixture of γ and δ crystal phases and are spherical in shape, but the grades differ in terms of primary particle size. NanoTek™ alumina has a surface area of 35 m^2/g corresponding to a mean particle size of 48 nm, whereas NanoDur™ alumina has a surface area of 45 m^2/g (37 nm), and the developmental NanoArcTM alumina features a surface area ranging from 80 – 100 m^2/g (17 – 21 nm). A TEM image of NanoDur™ alumina is shown in Figure 1. As is evident in the TEM, the primary particles are spherical and are not "necked" or fused together, rather they are loosely aggregated in the bulk powder. Figure 2 provides a comparison of the particle size distributions among the three plasma alumina grades.

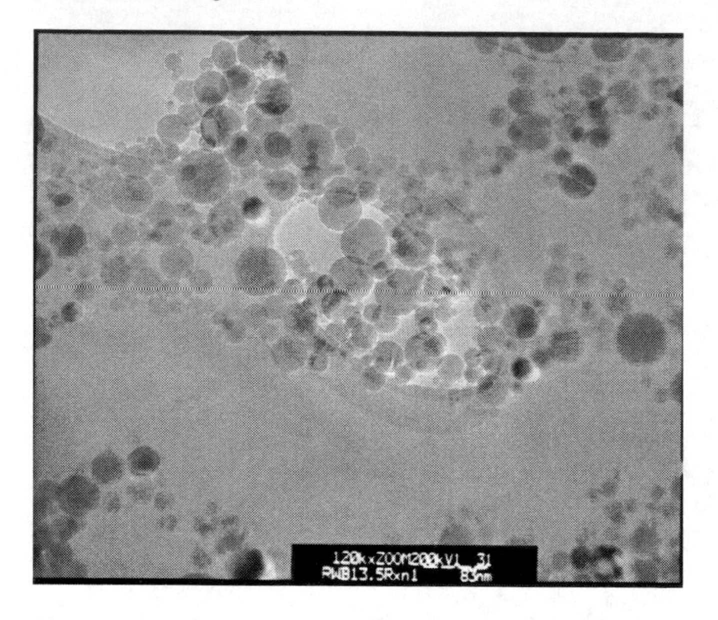

Figure 1: TEM image of NanoDur™ aluminum oxide.

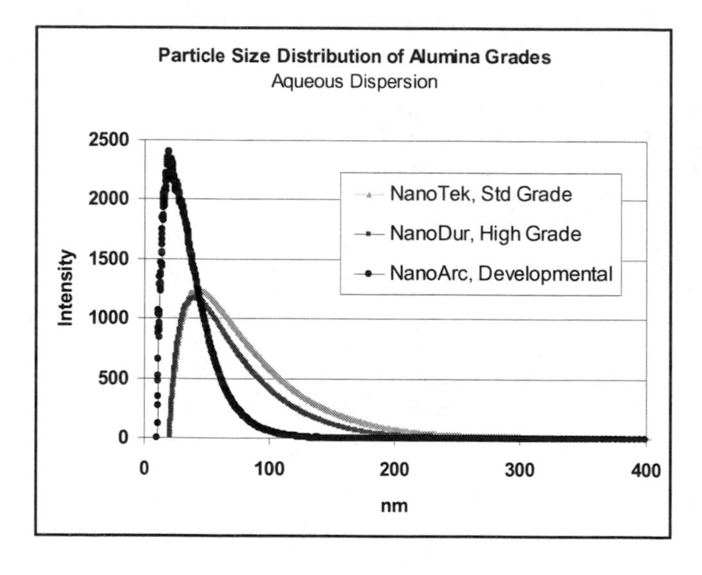

Figure 2: Particle size distributions of aluminum oxide produced by a plasma process.

NANOPARTICLE COATINGS

Nanocomposite coatings are made by embedding nanocrystalline metal oxide particles in a polymer matrix. The benefits accrued by incorporating nanoparticles into a polymer matrix become economically driven when several advantageous properties are obtained simultaneously, compared with conventional materials, i.e., transparency with increased abrasion resistance and toughness.

Nanophase Technologies has developed a variety of commercially scaled, proprietary technologies to modify the surface of nanocrystalline powders as well as processes to disperse the surface treated powders into a range of fluids and organic resins used in the manufacture of nanocomposite coatings.

Nanophase has formed an exclusive partnership with Altana, BYK-Chemie, to develop and market nanoparticles for use in coatings, inks, and plastics. To date the partnership has commercialized two aluminum oxide-based nanoparticle additives for use in UV-curable coatings to reduce scratch and mar.

Figure 3 depicts the improvement in gloss retention of a UV-curable coating after the incorporation of the nano-alumina product, NANOBYK 3600. The improvement in scratch resistance is evident in the images shown in Figure 4 of the control coating and a coating containing the alumina nanoparticles.

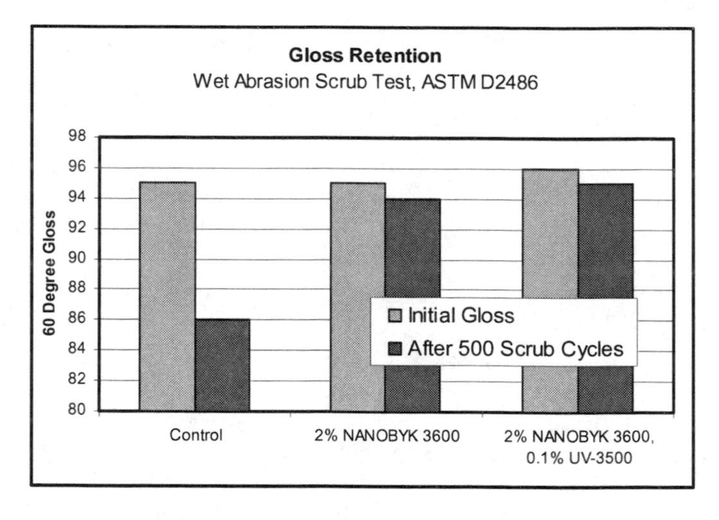

Figure 3: Gloss retention following scrub test of a control coating, and coatings containing alumina nanoparticles, with and without a slip additive.

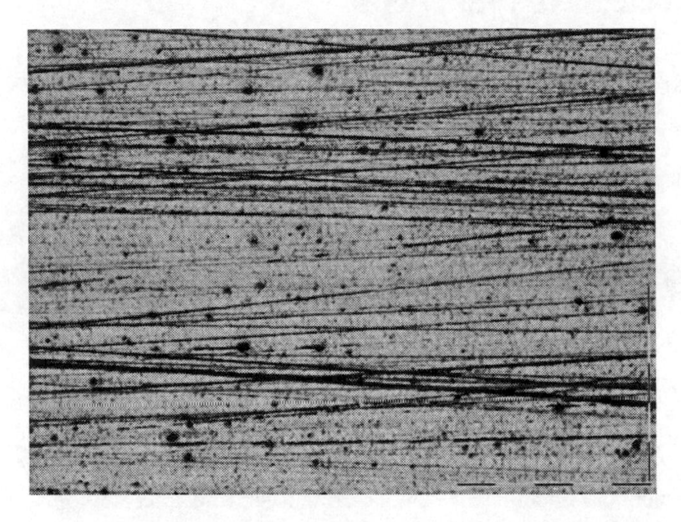

Figure 4A. Microscopic image of control coating following 500 scrub cycles.

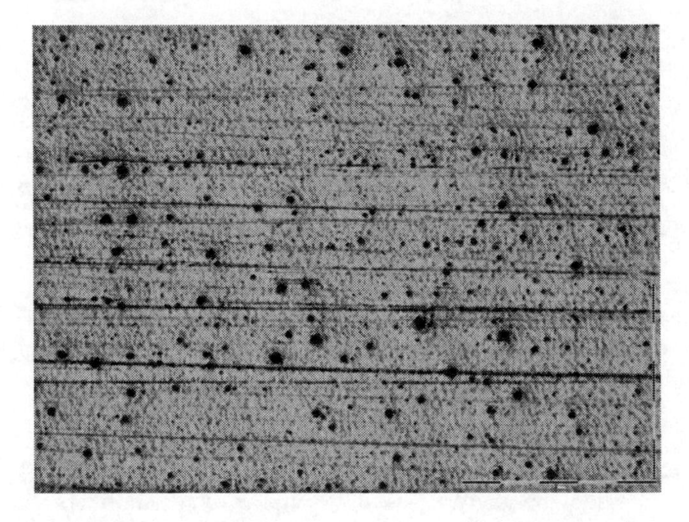

Figure 4B. Microscopic image of a coating containing 2 wt% alumina nanoparticles and 0.1 wt% slip additive following 500 scrub cycles.

Investigation of Thermal and Mechanical Response of Nylon-6 Filaments with the Infusion of Acicular and Spherical Nanoparticles

Hassan Mahfuz*, Mohammad M. Hasan**, Vijaya K. Rangari** and Shaik Jeelani**

*Florida Atlantic University, Boca Raton, FL 33431, USA, hmahfuz@oe.fau.edu
**Tuskegee University's Center for Advanced Material (T-CAM), Tuskegee, AL 36088, USA, mohammad.hasan@tuskegee.edu

ABSTRACT

It has been demonstrated in recent years that infusion of nanoparticles in thermoset as well as in thermoplastic polymers can help improve the chemical, thermal, and mechanical properties significantly. The improvement in properties can be achieved at very low nanoparticles loading (most of the cases ≤ 3% by weight). The shape of nanoparticles can be spherical, layered, irregular and acicular or rod-shaped. It is believed that the properties of the polymer can be maximized by aligning the rod-shaped particles along a preferred direction. The improvement in properties due to alignment of nanoparticles is influenced by two factors; (a) physio-chemical effect on the polymer structure, and (b) load transfer mechanism. In the present study infusion of vapor grown carbon nanofibers (CNFs), and multi-walled carbon nanotubes (MWCNTs) into a thermoplastic polymer, nylon-6 was first investigated. Carbon nanofibers were ≤200nm in diameter and 50-100μm in length. On the other hand MWCNTs were around 10-15 nm in diameter and about 5μm long. These nanoparticles with 1% wt loading were dry-mixed individually with nylon-6 by mechanical means, melted in a single screw extrusion machine, and then extruded through an orifice. Extruded filaments were later stretched and stabilized by sequentially passing them through a set of tension adjusters and a secondary heater. Stabilized filaments of about 80-150 micron dia were finally wound into spools in a filament winder. Two sets of filaments; one with CNFs and the other with MWCNTs were extruded. Individual filaments were then tested under tension. The enhancement in strength and stiffness were phenomenal. In both cases the enhancement in properties was in the 70-100% range. This remarkable increase in strength and stiffness was clearly due to the alignment of acicular reinforcements for MWCNT infused filaments. Although CNFs were relatively weaker than MWCNTs, the improvement with CNF was even more pronounced than MWCNTs. It is believed that since CNFs were larger in diameters, their alignment was much easier during the extrusion process. It is also observed that the improvement in stiffness is at the cost of sacrificing a significant amount of failure strain which is in no way beneficial for nylon. In an attempt to improve upon the breaking strain, spherical particles such as SiO_2 with identical loading were infused with nylon-6. Filament extrusion process was identical with the previous cases. Tensile test results of SiO_2 infused filaments has indicated that with spherical particle infusion a moderate but still significant improvement in strength and stiffness can be achieved without practically any loss of failure strain. Both TGA and DSC results also indicated that with identical loading the acicular nanopartilce infused systems are more thermally stable. Details of fabrication procedures, mechanical and thermal characterizations, and analyses of tension test results with respect to various reinforcements are included in the paper.

Keywords: nylon-6, carbon nanotubes, carbon nanofibers, silicon dioxide

1 INTRODUCTION

Nylon's toughness, low coefficient of friction and good abrasion resistance make it an ideal replacement for a wide variety of materials from metal to rubber. The amide groups of nylon are very polar, and can hydrogen bond with each other. Because of this, and because the nylon backbone is so regular and symmetrical, nylons are often crystalline, and make very good fiber [1, 2].

Extensive work concerning processing, modeling, morphology and thermal and mechanical properties of acicular shaped nano particles, specially multi-walled carbon nanotubes (MWCNTs) [3,4] and carbon nanofibers (CNFs) [5] reinforced filaments and nanocomposites, has been reported by various researchers.

Among several fiber aligning techniques, 'Extrusion' has been found as one of the most widely used fiber-aligning methods[3,4]. However, the end products totally depend on the starting materials and their chemical compositions, their mixing techniques, and type of extruder. In most of these cases, alignment of nanoparticles, specially the ones having high aspect ratios has been considered as a serious problem and decreasing in ductility is also an inevitable issue.

On the other hand, spherical-shaped silicon dioxide nanoparticles possess high thermal stability, have no alignment problems and are an excellent reinforcing agent for polymers [6]. So to get rid of alignment and ductility problem, silicon dioxide might be a better route to reinforce the polymer.

2 EXPERIMENTAL

2.1 Melt Processing

Nylon-6 nanocomposite filaments were synthesized using the following procedure: nanoparticles and nylon-6 were mechanically mixed (1:99 by wt) at 22000 rpm by using a blender with cooling system, then placed into a cylindrical drying chamber for 12 hours to driven out the moisture and finally extruded by a single-screw extruder at a rotational speed of 12 rpm. The temperature profiles of the barrel were 222-232-242 °C from the hopper to the die, respectively. Then the filaments were stabilized by using a set-up of two godet machines and a heater. Finally, the filaments are wound on a spool using *Wayne Desktop Filament Winder* at a winding speed of 70 rpm. The neat nylon-6 filaments were also extruded with the same extrusion condition.

2.2 Testing Procedures

To estimate the size and shape, the neat and nano infused filaments were examined using *JEOL JSM 5800* Scanning Electron Microscope (SEM). The filaments were placed on a carbon double-sided tape and coated with *JEOL Ion beam sputtering* unit before placing into SEM.

The characterization of nanoparticles dispersion and alignment in a filament has been carried out using Field emission scanning electron microscope (FE-SEM, Hitachi S-900). The FE-SEM samples were precisely cut with a carbon coated blade, and polymer was etched by using argon plasma with about 600 mtorr argon pressure and 2.5 W/cm^2 intensity for 1 hour.

In order to obtain the information on the thermal stability of the filaments samples, thermo gravimetric analysis (TGA) tests were carried out by using Hi-Res TGA 2950 TG analyzer under nitrogen gas atmosphere (70 ml/min). The samples were placed in the platinum sample pan, weighed (5 to 15 mg) and heated to 800^0C from room temperature at a heating rate of 10^0C/min.

The differential scanning calorimetry (DSC) tests were carried out non-isothermally by using TA Q1000 instrument under nitrogen gas atmosphere (50 ml/min). The samples were cut into small pieces, weighed (5-15 mg) and put into aluminum crucibles. All samples were held at 5 ^{o}C for 2 min, heated at a rate of 10 ^{o}C/min to 250 ^{o}C and subsequently held for 2 min, then cooled at a rate of 10 ^{o}C/min to 5 ^{o}C. The real time characteristic curves (for both TGA and DSC) were generated by *Universal Analysis 2000 -TA Instruments, Inc.* data acquisition system.

Tensile tests for individual filaments were conducted on a *Zwick Roell tensile* tester equipped with a 20 N Load Cell. The tests were run under displacement control at a crosshead speed of 0.01 1/s strain rate and gage length of 102 mm. At least ten specimens were tested in each category. The stress-strain curves were generated by *Zwick testXpert* software data acquisition system.

3 RESULTS AND DISCUSSION

3.1 Thermal Response

Figure 1 represents the TGA thermogram of neat and nanophased filaments. In the present study the 50% of the total weight loss is considered as the structural stability of the system.

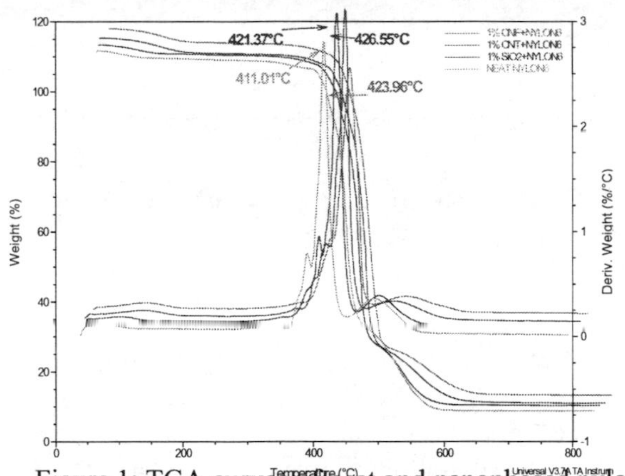

Figure 1: TGA curves of neat and nanophased nylon filaments.

It is clearly shown that the both acicular nanophased samples are more thermally stable (MWCNTs-427 ^{o}C and CNFs-424 ^{o}C) than neat (411 ^{o}C) and spherical SiO_2 (421 ^{o}C). This better thermal improvement indicates better bonding between the matrix and acicular nanoparticles. However, MWCNT nanophased has the highest thermal stability (427 ^{o}C). The reason for this high thermal stability is attributed to slight increase in cross-linking of nylon in the presence of CNT during extrusion process [7].

The DSC measurements (Fig. 2) indicate that the presence of nanoparticles specially both acicular-shaped increases the glass transition temperature (T_g, 62 ^{o}C) of the nylon-6 matrix (neat-49 ^{o}C, SiO_2 nanophased-55 ^{o}C). It is believed that MWCNTs or CNFs restrict the free volume motion of the matrix. The DSC melting peak (T_m) of the composites occurs at a slightly lower temperature (221 ^{o}C) than that of the neat nylon-6 (223 ^{o}C). This may be related to a slight reduction in crystallite size in the presence of nanoparticles [8]. It is also observed that the MWCNT /nylon-6 system has the highest crytallinity (30%) whereas CNF, SiO_2 and neat systems has 27%, 26% and 24% respectively. It might be the cause of high aspect ratio (L/d-500:1), high thermal conductivity and nucleation effect of MWCNTs.

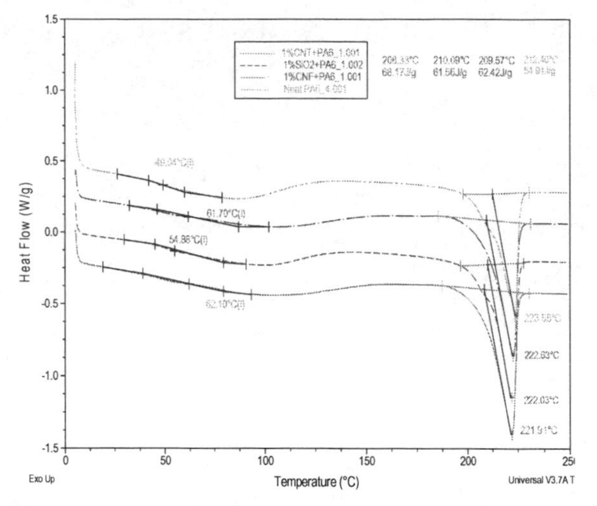

Figure 2: DSC heating curves of neat and nanophased nylon filaments.

All of the nanoparticles cause an increase in the crystallization temperature (Tc) relative to neat nylon-6 (WMCNT & CNF-197 °C, SiO_2-192 °C, Neat-190 °C). To various degrees, nanopartilces may act as nucleating agents causing a higher crystallization than that of the neat nylon-6 and crystallization starts earlier (Fig. 3). However, high

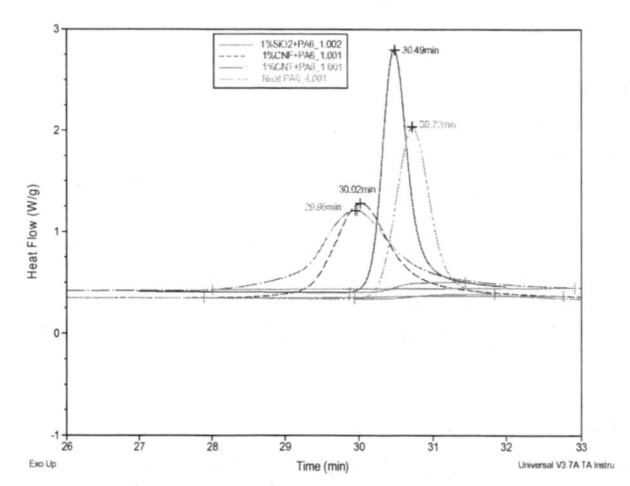

Figure 3: DSC cooling curves of neat and nanophased nylon filaments.

concentrations reduce the rate of crystallization. In some cases, the crystallization of the nanocomposites is slower than the pure extruded polyamide [9]. The combination of a larger number of nucleation sites and limited crystal growth is expected to produce crystals of fine grain size. [10].

3.2 Tensile Response

Stress-strain diagrams for neat and nanophased specimens are shown in Fig. 4. Tensile data in tabular form is shown in Table 1.

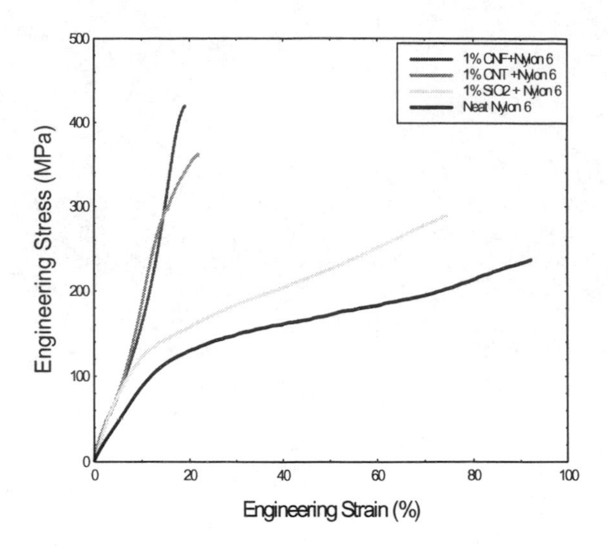

Figure 4: Tensile Stress-Strain curves of various samples.

The enhancement in strength and stiffness is phenomenal for both acicular systems. In both cases the enhancement in properties is in the 70-100% range. This remarkable increase in strength and stiffness is clearly due to the well dispersion and alignment of acicular nanoparticles into filaments (Fig. 5(b)) and higher crystallization with fine grain size. DSC and TGA data, SEM photographs also accord with this phenomenon. The enhancement in strength and stiffness for SiO_2 nanophased system is moderate (36% and 28% respectively). The FE-SEM micrograph (Fig. 5(a)) shows that dry mixing followed by extrusion with high shear force ensures the uniform dispersion of SiO_2 into nylon matrix.

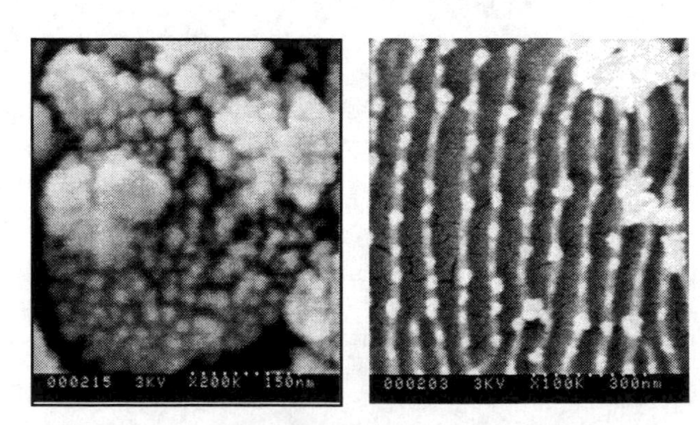

Figure 5: FE-SEM pictures (a) SiO_2/nylon-6 (b) MWCNT/nylon-6

The failure pattern of the MWCNT and CNF reinforced samples represents a different scenario. It is seen that the ultimate failure strain is reduced significantly (23%) whereas for SiO_2 nanophased is about 75%. Masayoshi et al. [11] showed that the deformation of nylon-6 is greatly affected by the crystalline formation. Kanamoto et al. reported [12] that the inter-crystalline network and entanglements of the original samples of polyethylene have a remarkable effect on ductility. Further, they suggested that the interfacial friction and adhesion between crystalline lamellae is the primary mechanism for the deformation. In nylon-6 crystals, the chain sliding and /or slippage are severely restricted by the hydrogen bonding, which induces a high stress on draw.

The embedded MWCNTs within the matrix would have restricted the deformation of the polymer chains as well as chain extension more than SiO_2. The reason why such gain is attributed can be explained in the following way:

When polymers are filled with small particles, the load transfer is dominated by the shearing mechanism between the filler and the matrix. If a perfect bond is assumed between the two constituents, and if fillers within the composite remain highly oriented and its aspect ratio exceeds the critical length at an order of magnitude, the filler can bear 90 percent of the composite load, resulting in high composite strength. Although this model was developed based on the cases of micron size particles, they are still valid for the explanation of the nanoparticle filled composites [13]. The most recent studies related to carbon nanotube reinforced polymeric composites including molecular dynamics simulation and direct experiments also reveal the similar failure mechanism [14, 15].

Material	E-Modulus (GPa)	Gain/ Loss %	Tensile Strength (MPa)	Gain/ Loss %
Neat Nylon 6	1.10±0.10	-	210±10	-
SiO_2-Nylon 6	1.41±0.15	+28.18	285±8	+35.71
CNT-Nylon 6	2.10±0.10	+90.90	371±20	+76.67
CNF-Nylon 6	1.65±0.20	+50.00	410±15	+95.24

Table 1: Comparison of tensile properties of nanophased filaments.

4. SUMMARY

The following are the summary of the above investigation:

1. A comparison study between acicular-shaped and spherical-shaped nanopartilces infused composites has been introduced.
2. SEM studies show that initial dry mixing followed by extrusion under high shear force ensures uniform dispersion and alignment of nanoparticles in a unidirectional manner along the length of the filaments.
3. Filaments fabricated with acicular nanoparticles in this fashion were found to possess higher tensile strength and stiffness by about 70-100% when compared with neat nylon filaments. While the enhancement in modulus was expected, the gain in strength was somewhat surprising and is believed to have been caused by the alignment of nanotubes, and crystal size formation during the extrusion process.
4. It is believed that since CNFs were larger in diameters, their alignment was much easier during the extrusion process and thus CNF-nylon 6 system possesses higher tensile strength compare to MWCNT/nylon-6.
5. Tensile test results of SiO_2 infused filaments indicate that with spherical particle infusion a moderate but still significant improvement in strength and stiffness can be achieved without practically any loss of failure strain.

Acknowledgements

The authors would like to thank the Ship Structures Division of the Office of Naval Research (Grant No. N00014-90-J-11995) and the National Science Foundation (Grant No. HRD –976871) for supporting this research.

REFERENCES

[1] A. M. Hindeleh and D. J. Johnson, Polymer, 19, 27-32, 1978.
[2] T. D. Fornes and D. R.Paul, Polymer 44, 3945-3961, 2003.
[3] H. Mahfuz, A. Adnan, V. Rangari, and S. Jeelani, International Jour. of Nanoscience (IJN), accepted for publication, October 2004.
[4] H. Mahfuz, A. Adnan, V. Rangari, and S. Jeelani, Composites Part A: Applied Science and Manuf., 35, 519-527, 2004.
[5] E. Hammel, X. Tang, M. Trampert, T. Schmitt, K. Mauthner, A. Eder and P. Potschke, Carbon, 42, 1153-1158, 2004.
[6] A. C. C. Esteves, A. M. Barros-Timmons, J. A. Martins, W. Zhang, J. Cruz-Pinto, T. Trindade, Composite : Part B, 36, 51-59, 2005.
[7] J. Bernhilc, D. Brenner, M. Boungiorno, V. Meunier and C. Roland, Annual Review of Materials Research, 32, 347-375, 2002.
[8] J. W. Cho and D. R. Paul, Polymer, 42, 1083, 1094-2001.
[9] G. Jimenez, N. Ogata, H. Kawai and T. Ogihara, Jour Appl Polym Sci, 64(11), 2211–20, 1997.
[10] J. Ma, S. Zhang, Z. Qi, G. Li and Y. Hu, Jour Appl Polym Sci, 83(9), 1978–85, 2002.
[11] M. Ito, K. Mizuochi and T. Kanamoto, Polymer, 39, 4593-4598, 1998.
[12] T. Kanamoto, E. S. Sherman and R. S. Porter, Polymer, 11, 497, 1979.
[13] P. K. Mallick, "Fiber-Reinforced Composites: Materials, Manufacturing and Design," 2nd edition, Marcel Dekker, Inc., New York, USA, 1993.
[14] D. Qian, W. K. Liu and R. S. Ruoff, Composite Science and Technology, 63, 1561-1569, 2003.
[15] O. Meincke, D. Kaempfer, H. Weickmann, C. Friedrich, M. Vathauer and H. Warth, Polymer, 45, 739-748, 2004.

Electromagnetic Containerless Processing of Single-walled Nanotube Reinforced Metal-Matrix Composites

K. Wilson, Q. Zeng, E. Barrera, and Y. Bayazitoglu

Rice University, Mechanical Engineering and Materials Science Dept., MS 321, P.O. Box 1892, 6100 Main St., Houston TX, USA. 77251-5423.
Contact emails: kwilson@rice.edu, qzeng@rice.edu, ebarrera@rice.edu, bayaz@rice.edu

ABSTRACT

Single-walled carbon nanotube (SWNT) reinforced copper (Cu) matrix composites were produced by powder metallurgy (PM) and electromagnetic levitation methods. The nanotubes were first decorated with a nickel coating and then dispersed by ball milling. The optimization of factors such as cold pressing pressure, powder grain size and sintering temperature for full densification were essential in the preform stage of PM in order to produce a sample suitable for levitation. Scanning electron microscopy (SEM) observations have shown that the well-dispersed nanotubes in the composites are not damaged during the high-temperature composite preparation. The preforms were then melted by electromagnetic levitation in an argon environment. SEM observations showed that the nanotubes survived the high-temperature melting. Hardness testing demonstrated that the strength of the composites is increased with increasing volume fraction of SWNTs.

1 INTRODUCTION

Since Iijima's discovery of carbon nanotubes [1], these structures have been recognized as fascinating materials with nanometer dimensions promising exciting new areas of application. It has been demonstrated by both theoretical calculations and experimental measurements that due to their unique structure individual SWNTs and SWNT ropes display remarkable electrical, thermal, and mechanical properties [2]. An ideal-single walled carbon nanotube consists of a rolled graphene layer with a cylindrical hexagonal lattice structure capped by half a fullerene molecule at both ends. Treacy et al. [3] estimated the Young's modulus of individual multi-walled nanaotubes to be in the range of 1 TPa. A number of theoretical studies have suggested that the smallest single-walled nanotubes might have a Young's modulus as high as 5 TPa [4]. Those features combined with high aspect ratios on the order of 1000 or more makes the SWNTs excellent additives for composites [5-8].

2 EXPERIMENTAL PROCEDURE

Cu is widely used in aerospace sciences and numerous industries due to its high electrical conductivity and corrosion resistance. A size range (5-40μm) and spherical morphology has been selected to achieve the best dispersion and flowability of powders during powder metallurgy. The purified single walled carbon nanotubes were supplied by Carbon Nanotechnology, Inc. SWNTs were produced by a High-Pressure CO (HiPCO) process and fabricated into millimeter-sized BuckyPearl pellets. This commercial material contains about 13 wt% Fe catalyst. The pelletized SWNTs were grounded to a finer level. The Fe catalyst was further reduced by refluxing the SWNTs in HCl at 80 °C for 5 hours, and then washed with methanol. These purified SWNTs were nickel coated via the electroless plating method [9-12] and then a mixture of copper powder (5 μm) along with 0-1 wt% of these tubes was mixed by 10 min sonication, followed by 10 hours of ball milling [13,14]. No damage of the SWNTs was observed after the ball milling process by SEM.

The result was a dark powder of uniform dispersion. This well-mixed final sample was slowly cold pressed under a pressure of roughly 100 MPa for 10 minutes [15]. The resulting sample preform was a cylinder 0.25 inches in diameter and 0.25 inches in height weighing roughly 0.75 grams. The samples were placed in a hydrogen furnace for 5 hours at a temperature of 400^0C to help remove any unwanted oxidation that could have formed before or during the cold pressing process. The samples were then sintered in an argon atmosphere at 800^0C for 15 hours [15-17].

SEM was used to analyze the microstructure following the sintering process. After that, the samples were then ready to be levitated and melted using a Radyne EI-40 model radio frequency generator. This generator produces 400-600 A of current at a single frequency of roughly 400 kHz. The large current and frequency create an electromagnetic field strong enough to levitate the copper sample while simultaneously inducing a heating eddy current sufficient enough to melt metals [18]. The samples were melted and mixed in the levitation process in an argon atmosphere for roughly 30 seconds. The surface was then cracked and SEM was again applied to analyze the fractured surface microstructure.

3 RESULTS AND DISCUSSION

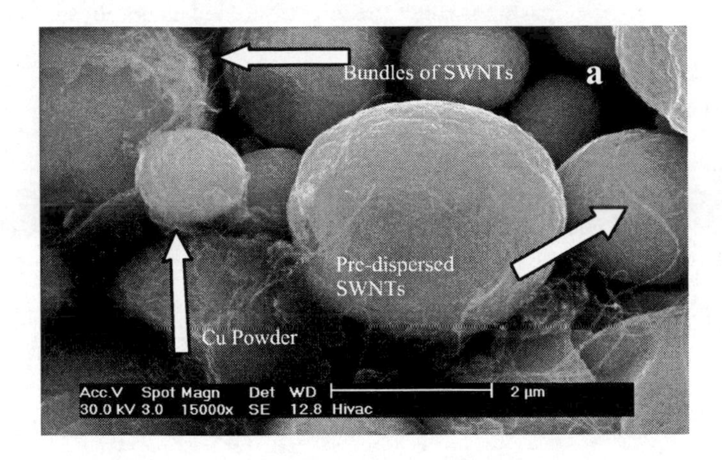

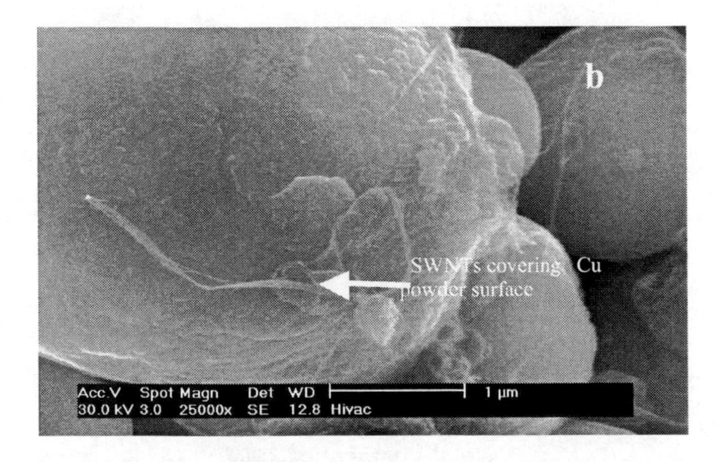

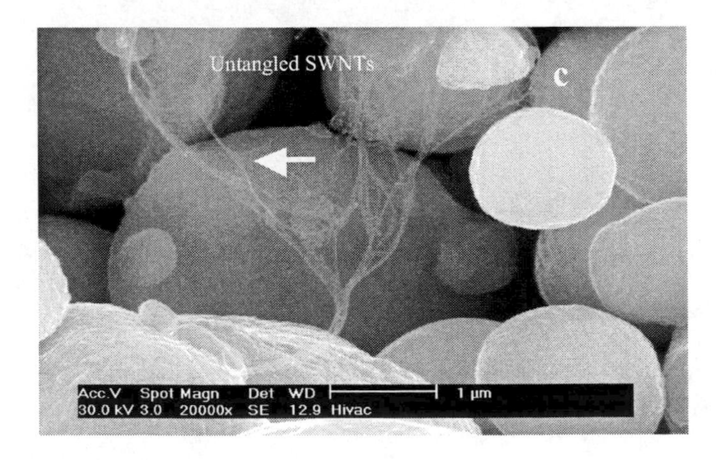

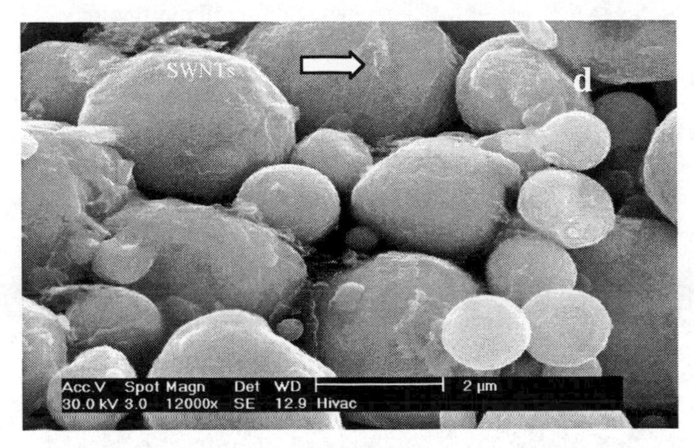

Figure 1(a-d). SEM image of SWNT/Cu particles after ball milling.

The SEM pictures in Fig.1 illustrate the microstructure of the premixed Cu/SWNT sample. It shows that the SWNTs were successfully pre-dispersed among the Cu particles (Fig. 1a) and some SWNTs were already untangled from bundles and attached on the particles surface after 10 hrs ball milling (Fig. 1b-c) without any surface modification or damage. However, some SWNTs were still entangled to each other forming a thin sheet covering the Cu particles surface (Fig. 1d).

Figure 2. SEM image of fractured surface of the SWNTs/Cu after PM process and 800 °C sintering treatment.

Figure 2 shows the microstructure of the composite sample after hydrogen treatment and high-temperature sintering. It is observed by SEM that there is neck-like growth between particles; therefore the distance between copper particle centers decreases. The principal driving force for the shrinkage is surface tension which causes transport of material from surfaces with a small radius of curvature to those with a larger radius of curvature. These forces also cause a decrease in the total surface of the pores and thereby shrinkage of the compact preform. The SWNTs in the SEM image of the composite samples sintered at 800°C are similar to those of the premixed material as shown in Fig. 1. No significant change either in morphology or in microtexture is observed.

Figure 3(a) shows the surface microstructure of the SWNT reinforced copper composite sample after the levitation process. It is noticed that there are SWNTs which still survive even after the high temperature electromagnetic processing. It is observed by SEM that some SWNTs appear on the melted copper composite surface (Fig. 3a). Pulling-out and bridging occurs on the fractographs (Fig. 3b), and the fractures propagate along the direction of the initial cracks. Therefore, by SEM observation, the main fracture mechanism in regard to the SWNTs is pulling-out of the fibers.

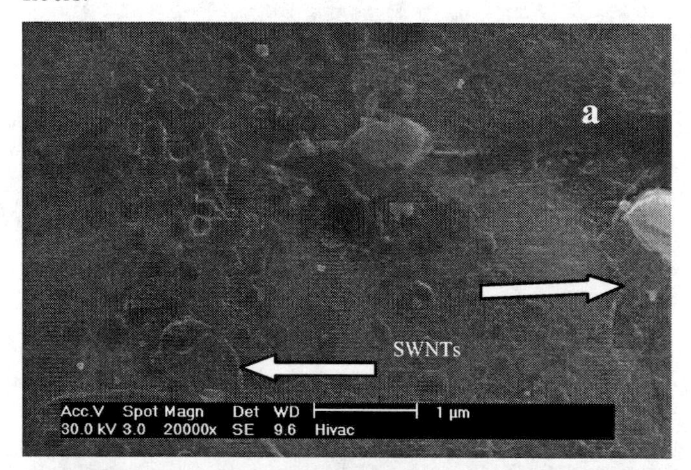

Figure 3. SEM image of SWNT reinforced copper composites (a) the surface of composite microstructure; and (b) the fracture surface of the composite.

Electromagnetic levitation has been discussed in the literature [18-22] in theory and in practice, and for that reason has not been included in this paper. Vickers hardness was used for comparison with [14]. The hardness results of the current study can be seen in Figure 4.

These indicate that increasing hardness is achieved with an increase in percentage of SWNTs. While in other methods adding more SWNTs might lead to an even greater increase in hardness, the electromagnetic levitation process is the factor which dictates how much SWNT reinforcement can be added in the current study. Poor levitation and heating was a result of increasing the SWNT content above 2% wt. The reason is that the nickel-coated nanotubes, because of their surface area to volume ratio, tend to have a negative effect on the internal heat generation and levitating force in the copper. Since nickel is not as electrically conductive as copper, it would follow that the addition of more SWNTs, covered in nickel, would impede the overall electrical properties of the composite. This lowered electrical conductivity causes a significant decrease in the heat generation in the levitation process. It is for this reason that the focus of the present research was on compositions of 1%wt SWNTs and below. Fig. 3 (a) also shows that the Cu powder was completely melted into a relatively smooth surface indicating the SWNTs seen in the figure survived at least the melting point of copper (~1083^{0}C) [23].

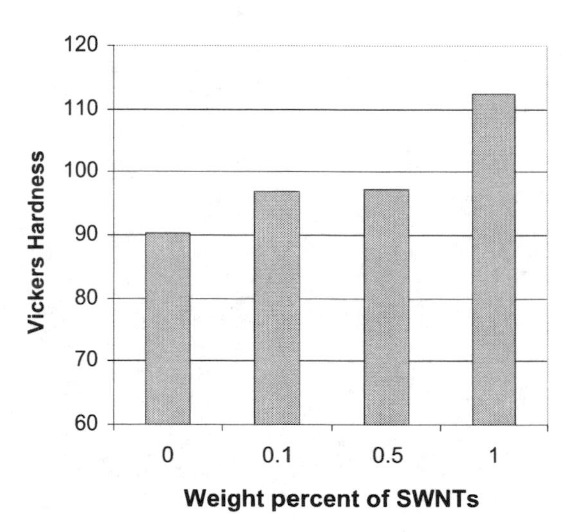

Figure 4. Vickers Hardness vs. Weight percent of SWNTs in the Cu/SWNT matrix showing a 24.6% increase in hardness with the addition of 1 wt% of SWNTs.

4 CONCLUSIONS

The powder metallurgy process along with electroless plating was used to premix nanotubes and copper powder. The powder was then cold pressed, hydrogen treated, sintered, and finally electromagnetically levitated. The result was a copper/SWNT metal-matrix composite with improved hardness. This process is usually capable of producing samples approximately 0.5 – 0.75 grams in mass, based on the frequency used for levitation.

REFERENCES

[1] S. Iijima, Nature, 354, 56, 1991.
[2] P.M. Ajayan, and T.W. Ebbesen, Rep. Prog. Phys., 60, 1025, 1997.

[3] M.M. Treacy, T. W. Ebbesen, J.M. Gibson, Nature, 381, 678, 1996.
[4] B. I. Yakobson, C.J. Barbec, and J. Bernholc, Phys. Rev. Lett. 76, 2511, 1996.
[5] J. Zhu, J.D. Kim, H. Peng, J.L. Margrave, V.N. Khabashesku, and E.V. Barrera, Nano Letters, 3, 1107, 2003.
[6] K. Lozano, S. Yang, and Q. Zeng, J. Appl. Polym. Sci., 93, 155, 2004.
[7] R. Zhong, H.Cong, P. Hou, Carbon, 41, 848, 2003.
[8] P.M. Ajayan, L.S. Schadle, C. Giannaris, A. Rubio, Adv. Mater., 12, 750, 2000.
[9] X. Chen, J. Xia, J. Peng, W. Li, S. Xie, Composites Science and Technology, 60, 301, 2000.
[10] Q. Zeng, Y. Bayazitoglu, J. Zhu, K. Wilson, M. Ashraf Imam, E.V. Barrera, Materials Science Forum, 475-9, 1013, 2005.
[11] L.M. Ang, T.S.A. Hor, G.Q. Xu, C.H. Tung, S. P. Zhao, J.L.S. Wang, Carbon, 38, 363, 2000.
[12] F.Z. Kong, X.B. Zhang, W.Q. Xiong, F. Liu, W.Z. Huang, Y.L. Sun, J.P. Tu, X.W. Chen, Surface and Coatings Technology, 155, 33, 2002.
[13] T. Kuzumaki, K. Miyazawa, H. Ichinose, and K. Ito, J. Mat. Research, 13, 2445, 1998.
[14] D. Shurong, Z. Xiaobin, Transactions of Nonferrous Metals Society of China, 9, 457, 1999.
[15] W. Taubenblat, "Copper Base Powder Metallurgy", Metal Powder Industries Federation, 123-217, 1980.
[16] P. Schwarzkopf, "Powder Metallurgy Its Physics and Production", 47-238, The Macmillan Co., 1947.
[17] F. Lenel, "Powder Metallurgy Principles and Applications," Metal Powder Industries Federation, 99-281, 1980.
[18] E. Okress, Journal of Applied Physics, 23, 545, 1952.
[19] Y. Bayazitoglu, J. of Materials Processing and Manufacturing Science, 3, 117, 1994.
[20] U. Sathuvalli, IEEE Trans. on Magnetics, 32,386, 1996.
[21] Y. Bayazitoglu, J. of Materials Processing and Materials Science, 5, 79, 1996.
[22] E. V. Barrera and Y. Bayazitoglu, "Nanotube Mixing Using Electromagnetic Levitators", Patent Applied, WSM Ref. No.: 11321-P018WOUS
[23] A. Bougrine, N. Dupont-Pavlovsky, A. Naji, J. Ghanbaja, J.F. Mareche, D. Billaud., Carbon, 39, 685, 2001.

Synthesis and Characterization of Perovskite $Ca_{0.5}Sr_{0.5}MnO_3$ Nanoparticles in w/o-Microemulsions

A. López-Trosell* and R. Schomäcker*

* Department of Chemistry, TU-Berlin,TC 8
Strasse des 17. Juni 124, 10623-Berlin, Germany
Lopez.Alejandra@chem.tu-berlin.de and Schomaecker@tu-berlin.de

ABSTRACT

In the present work, w/o-microemulsions were employed to produce nanoparticles of perovskite $(Sr_{0.5}Ca_{0.5}MnO_3)$ with an average diameter of 45 nm. The process was carried out via sol-gel in microemulsions. In order to control particle growth and distribution size, the micellar structure and synthesis conditions were investigated. Therefore, the phase behavior of the microemulsion systems were determined. Samples were characterized by X-ray diffraction (XRD), scanning electron microscopy (SEM), and N_2 adsorption.

Keywords: microemulsions, sol-gel, nanoparticles

1 INTRODUCTION

So-called perovskites are a large family of crystalline ceramics that derive their name from a specific mineral known as perovskite $CaTiO_3$. The unit cell of the cubic aristotype, represented for example by $CaTiO_3$ is type face-centred cubic. The space group is Pm3m. Using the general formula ABO_3, the large cations A occupy the corners of the cube (000), the small cation B the center $(\frac{1}{2}\ \frac{1}{2}\ \frac{1}{2})$, and the anions the center of the faces $(\frac{1}{2}\ \frac{1}{2}\ 0,\ \frac{1}{2}\ 0\ \frac{1}{2},\ 0\ \frac{1}{2}\ \frac{1}{2})$. The ratio of the ionic radii is an important factor for the crystal structure. For non-ideal ratios, distortions from the cubic aristotype are observed [1].

The oxides of perovskite containing Mn ions have been the object of strong interest due to the exhibition of phenomena such as ferroelectricity, extremely high magnetoresistance, high-temperature O_2^--conductivity and high catalytic activity [2],[3],[4]. These properties have been found to be affected by the composition and particle size. More precisely, these properties show an increment when the particle sizes decrease. Therefore, the applied technique to control and to synthesize such materials is very important.

In the past, manganese perovskite have been obtained by conventional solid-stated synthesis. Unfortunately, this technique leads to solids with a larger grain size [5] due to the high-temperature treatment. In comparison with the solid-stated technique, microemulsions offer lower temperature conditions, which favor particle growth and avoid sintering processes. Reverse micelles have been successfully used to synthesize ultrafine particles (1-100 nm as grain diameter). Microemulsions are thermodynamically stable isotropic dispersions of two immiscible liquids consisting in microdomains of one of both liquids in the other, stabilized by an interfacial film of surfactant. These microdomains are used as microreactors, in which chemical reactions can be carried out. Consequently, they present a suitable medium to produce particles with desirable size and morphology. However, in order to control particle growth and distribution size, Schmidt [6], pointed out that not only the micellar structure is important but also the synthesis conditions and the kinetics of the elementary steps, especially the meterial exchange between the droplets.

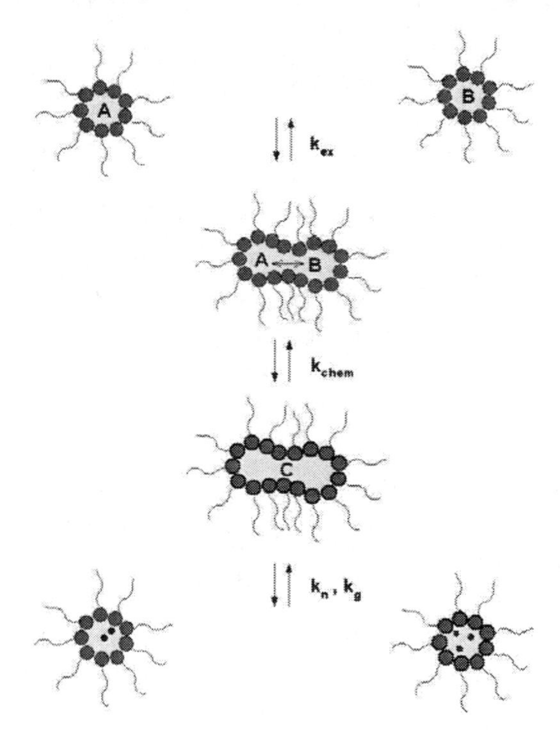

Figure 1: Schematic representation of the mechanism for formation of nanoparticles in w/o-microemulsion proposed by Schmidt. [6]

Several models have been proposed to explain the formation process of ultrafine particles in reverse micelles [7]–[11], [6]. On the one hand, there are some authors such as Tanori and Pileni [12], who suggest that micelles act as confined "nanoreactors", inside which the reaction takes place. They found that the size and form of the particles can be limited to

Table 1: Composition of the microemulsions used for the Perovskite synthesis. Microemulsion A (μE_A) and microemulsion B (μE_B) containing the salts and propionic acid, respectively. Ac denotes the acetate (CH_3COO^-).

	Surfactant	Oil phase	Aqueous phase
μE_A	Marlipal O13/70	cyclohexane	$Ca(Ac)_2$ + $Sr(Ac)_2$ [0.3] mol/l + $Mn(Ac)_2$ [0.6] mol/l
μE_B	Marlipal O13/70	cyclohexane	*propionic acid* [1.2] mol/l
Weight (g)	15	74.38	10.62

colloidal assemblies used as templates.

However, this mechanism seems to be contradicted by experiments that yielded particles larger than the droplets. On the other hand, authors like Towey [9] and Hirai [10] proposed models based on different stages of the precipitation reaction to explain the particle formation.

Figure 1 shows the possible kinetic scheme for the formation of the nanoparticles in reverse microemulsion based on the model proposed by Schmidt [6].

2 EXPERIMENTAL PART

The composition of microemulsions are specified by the weight fraction of oil in the mixture oil water (α) and the weight fraction of surfactant in the ternary mixture (γ), where m_o, m_w and m_s are the weights of oil, water and surfactant, respectively.

$$\alpha = \frac{m_o}{(m_o + m_w)} \quad (1)$$

$$\gamma = \frac{m_s}{(m_o + m_w + m_s)} \quad (2)$$

An important parameter to describe the droplet size is the so-called "water pool", characterized by W_o. This parameter is the water-surfactant molar ratio (equation 3), which is considered to be proportional to the radius *R* of the droplet as was shown by Pileni [13].

$$W_o = \frac{n_w}{n_s} \quad (3)$$

Where n_w is the water mole and n_s is the surfactant mole.

For the nanoparticle synthesis, strontium acetate monohydrate ($Sr(CH_3COO)_2 \cdot H_2O$), calcium acetate monohydrate ($Ca(CH_3COO)_2 \cdot H_2O$) and mangan acetate tetrahydrate ($Mn(CH_3COO)_2 \cdot 4H_2O$) were employed. The acetates were supplied by Fluka with a purity $> 99\%$ and where used without further purifications. The propionic acid was suppled by Aldrich. Marlipal O13/70 (from Sasol) was used as surfactant. The number of carbons in the aliphatic chain as well as the ethoxylation degree are indicated by the label after the name, O13/70. This means that the surfactant has 13 carbons in the lipophilic chain and an average ethoxylation degree of seven. Cyclohexane was employed as oil phase. In order to determine the best microemulsion composition for preparing the precursor of Perovskite, the phase behavior of different systems was studied applying the procedure described in [14].

The synthesis of Perovskite $Ca_{0.5}Sr_{0.5}MnO_3$ was carried out via sol-gel in microemulsion. The reaction was achieved by using propionic acid as gelificant agent. Two microemulsions, "A" and "B", with identical composition but with different aqueous phases were prepared. In the microemulsion "A", the aqueous phase was a solution containing calcium acetate, strontium acetate, and mangan acetate in stoichiometric ratio, whereas the microemulsion "B" was a solution containing the propionic acid. The composition of the microemulsions is shown in table 1.

A semi-batch reactor was used to perform the synthesis of the perovskite. It consists of a tank with four baffles, and can contain 200 ml of microemulsion as maximum. The agitator was a four pitched blade turbine impeller and the feed input was located near to the agitator and above the liquid level. The synthesis was achieved by loading 100 ml of microemulsion "A" to the reactor and adding 100 ml of other one at a constant feed rate and temperature, of 0.30 (ml/s) and 27^oC, respectively. The temperature was controlled using a ultrathermostat K6 supplied by Colora Messtechnik. A peristaltic pump (Besta E100) was utilized for the addition of microemulsion "B" to the reactor.

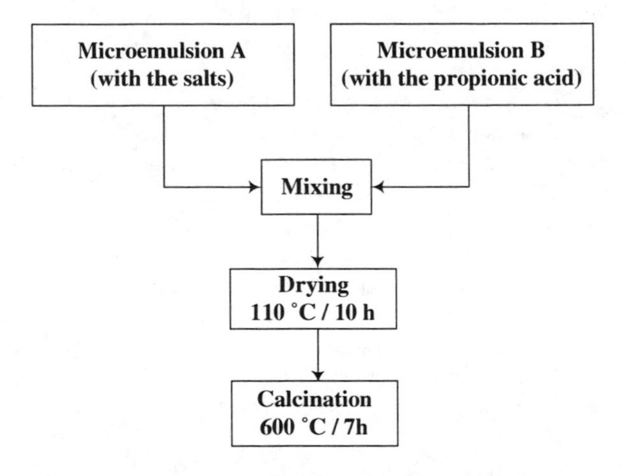

Figure 2: Diagram of the perovskite synthesis.

After addition, the reaction mixture was stirred for another 30 minutes. Subsequently, the solvents were removed by drying at 110^oC for 10 h. Afterwards, the precursor was calcined at different temperatures until the crystalline phase was obtained. The reaction was followed by XRD using a X-ray diffractometer Siemens D500 with Cu anode (radia-

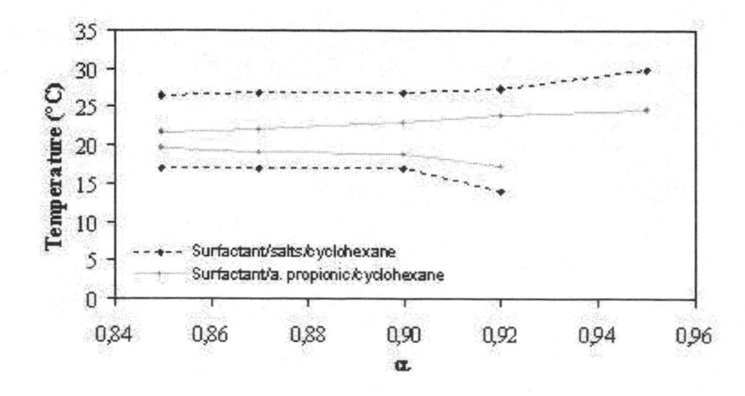

Figure 3: Phase diagram of the microemulsion systems.

tion $K_{\alpha 1}$ of $\lambda = 0.154\ nm$). The crystallized perovskite was analyzed by scanning electron microscopy (SEM) using a Hitachi S-520. The specific surface area was obtained from the N_2 adsorption (BET) using a Micromeritics Gemini with an analysis mode in equilibration, evacuation time of 1 min, equilibrium interval of 5 sec, and saturation pressure of $781.00\ mmHg$. Figure 2 presents a diagram of the employed synthesis process.

3 RESULTS AND DISCUSSIONS

Figure 3 shows the phase behavior of the microemulsion system water-oil-surfactant versus temperature. It is a typical phase diagram at a constant surfactant mass fraction (γ). The one phase region appears as a channel extending from the oil-rich region to the water-rich side of the diagram. In this region, the system is a thermodynamically stable dispersion, which is optically transparent and has a low viscosity.

The cyclohexane was chosen as oil component to prepare the system because its capacity to solubilize water in microemulsions is higher than with other oils such as hexane, octane and iso-octane as found in [6]. The results offer an extensive working range for temperature and composition of the microemulsions.

The X-ray patterns in figure 5 show that the sample reached the crystal structure by heat treatment at 600^oC for seven hours. The average crystall size D_{hkl} was calculated using the Debye-Scherrer's equation 4 with the full width at the half maximum β (in rad.) of the x-ray diffraction peak.

$$D_{hkl} = \frac{K\,\lambda}{\beta\ Cos\theta} \tag{4}$$

In the above equation K is a constant (often taken as 1), λ represents the X-ray wavelength, β and θ are the full width at the half maximum (in rad.) and the Bragg angle, respectively. The results calculated using Debye-Scherrer's equation are in good agreement with those obtained by SEM. Table 2 summarized the particle size obtained by XRD and SEM as well as the employed W_o.

Table 2: Particle diameters obtained by $^*XRD\ (D_{hkl})$ and $^\ddagger SEM\ (D_p)$.

W_o	$*D_{hkl}\ nm$	$\ddagger D_p\ nm$
20	44	46

The SEM analysis (See figure 4) shows that the sample was homogeneous in shape and size. Finally, table 3 lists the specific surface area obtained by N_2 adsorption. This value is in the range of areas reported for some members of the manganite perovskite [4].

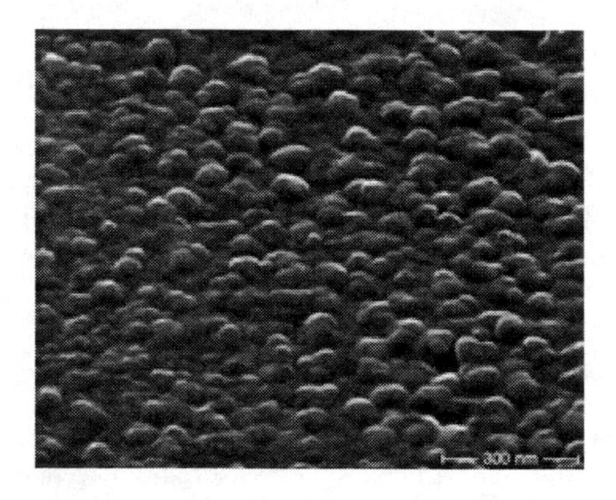

Figure 4: SEM of $Ca_{0.5}Sr_{0.5}MnO_3$ Perovskite.

Table 3: BET surface area (S_{BET}), Volume of pore (V_T) and diameter of pore (D_p) for the perovskite.

$S_{BET}\ (m^2/g)$	$V_T\ (cm^3/g)$	$D_p\ (nm)$
8.3	0.004	1.23

4 CONCLUSIONS

Microemulsion were used to synthesize $Ca_{0.5}Sr_{0.5}MnO_3$. Reverse micelles containg the adequate reactants are a suitable medium to obtain fine particles of perovskite with an av-

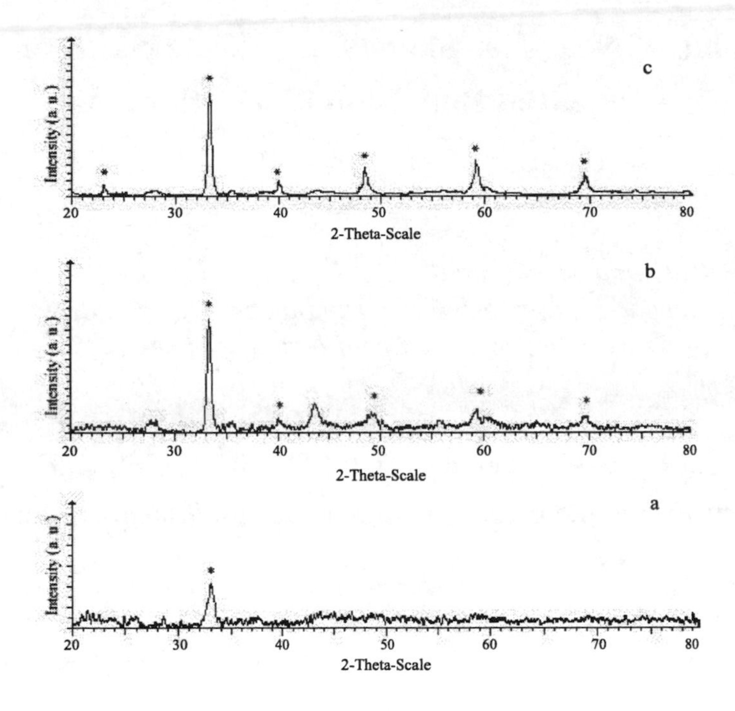

Figure 5: XRD patterns of $Ca_{0.5}Sr_{0.5}MnO_3$ Perovskite at different calcination conditions. (a) 2 h. at 300^oC, (b) 4 h. at 400^oC and (c) 7 h. at 600^oC.

erage diameter of $45\ nm$. The particles were homogeneous in size and shape and exhibited specific surfaces areas in the expected range.

REFERENCES

[1] Chmaissem, O., Dabrowski, B., Mais, J., Brown, D.E., Kruk, R., Prior, P., Jorgensen, J.D.: Relationship between structural parameters and the néll temperature in $sr_{1-x}ca_xmno_3$ $(0 \leq x \leq 1)$ and $sr_{1-y}ba_ymno_3$ $(y \leq 0.2)$. Physical Revew B **64** (2001) 1–9

[2] Galasso, S.F.: Structure, properties and preparation of perovskite type compounds. Pergamon Press: Oxford, Headington Hill Hall, Oxford, 4 and 5. Fitzroy Square, London W.1 (1969)

[3] Chen, C.H., Kruihof, H., Bouwmeester, H.J.M.: Ion conductivity of perovskite $laco_3$ measured by oxygen permeation technique. J. Applied Electrochemistry (1997) 71–75

[4] Song, K.S., Cui, H.X., Kim, S.D., Kang, S.K.: $la_{1-x}m_xmno_3$. Catalysis Today **47** (1999) 155–160

[5] Koch, C.C.: Nanostructured Materials- Processing Properties and Potential Applications. Carl C. Koch., NY (2002)

[6] Schmidt, J., Guesdon, C., Schomaecker, R.: Engineering aspects of preparation of nanocrystalline particles in microemulsion. J. Nanoparticle Reserch **1** (1999) 267–276

[7] Ravet, I., Nagy, J., Derouane, E.: On the mechanism of formation of colloidal monodisperse metal boride particles from reverse micelles composed of ctab-1-hexanol-water. Stud. Surf. Sci. Catal. **31** (1987) 505–516

[8] Osseo-Asare, K., Arriagada, F.: Synthesis of nanosize particles in reversed microemulsions. Ceram. Trans. **12** (1990) 3–16

[9] Towey, T., Khan-Lodhi, A., Robinson, B.: Kinetics and mechanism of formation in $water - aerosol - ot - oil$ microemulsions. J. Chem. Soc. Faraday Trans. **86** (1990) 3757–3762

[10] Hirai, T., Sato, H., Komasawa, I.: Mechamism of formation of titanium dioxide ultrafine particles in reverse micelles by hydrolysis of titanium tetrabutoxide. Ind. Eng. Chem. Res. **32** (1993) 3014–3019

[11] Bandoyopadhyaya, R., Kumar, R., Gandhi, K.: Modeling of precipitation in revese micellar systems. Langmuir **13** (1997) 3610–3620

[12] Tanori, J., Pileni, P.: Control of the shape of copper metallic particles by using a colloidal system as template. Langmuir **13** (1997) 639–646

[13] Pileni, M.P.: Reverse micelle as microreactors. J. Phys. Chem. (1993) 6961–6973

[14] Lade, M., Mays, H., Schmidt, J., Willumeit, R., Schomaecker, R.: On the nanoparticle synthesis in microemulsions: detailed characterization of an applied reaction mixture. Colloids and Surfaces A: Physicochem. Eng. Aspects **163** (2000) 3–15

Plasma Manufacturing of Near-Net-Shape Large Scale Nanocomposite Structures – A Potential Bulk Nanofabrication Tool

P. Georgieva, V. Viswanathan, S. J. Hong, K. Rea, S. C. Kuiry, and S. Seal*

AMPAC and Mechanical Materials Aerospace Engineering
Surface Engineering and Nanotechnology Facility
University of Central Florida, AMPAC,
Eng. 1, Room 381, Orlando, Fl 32816

***Contact: sseal@mail.ucf.edu, 407 882 1119, 823 5277**
http://people.cecs.ucf.edu/sseal/plasma/plasma2.htm

ABSTRACT

We report a brief summary of the fabrication of cost effective, high throughput, near-net shape bulk nanocomponents using plasma spray forming – a potential 3-D engineering tool for large scale assembly of nanostructures. The development of a variety of nano/micro ceramic-ceramic, metal-ceramic, nanotube-ceramic components with retained nanostructure, enhanced oxidation resistance, and high fracture toughness, suitable for a wide range of industrial applications is presented herein. The chemistry and the microstructure of the developed components were studied using state of the art, advanced characterization techniques. Simulation studies were conducted in order to study the proper material selection in the development of advanced bulk nanocomposites through proper design of experiments (DOE).

Keywords: Plasma Spray, Near-Net Shape, Bulk, Large Scale, Nanocomposite

1. TECHNOLOGY NEED

The drive toward developing of materials with small particle size, light weight, and improved mechanical and chemical properties affects a large array of applications in the fields of structural engineering, homeland security, space exploration, and other social issues. In this regard, proper selection methods of consolidation of materials with retained nanostructures are important [1]. For example, Hot Isostatic Pressing (HIP), Laser Direct Consolidation, and Plasma Spray Forming are viewed as the most promising nanomaterial consolidation processes with a few limitations. The disadvantages of the HIP are the size limitation and the complexity of the sample geometry. The inconvenience of the laser process lies in the constraint of material selection. However, one can flow practically all types of materials which can melt in a plasma flame. In this research, we select Plasma Spray Forming (PSF) as one of the most versatile methods of designing large scale nanocomponents with the desired shape and size in minimal time. At present, plasma spraying has advanced from coating technology to materials processing technology with a growing importance in the field of bulk nanocomposite manufacturing [1].

2. PLASMA SPRAYING AND PROCESS CONTROL

Near-net-shape processing using PSF involves simultaneous control of powder melting and then particle acceleration for deposition on a rotating mandrel or substrate with a proper CTE (coefficient of thermal expansion) to release the parts after cooling. The plasma spray gun includes a water cooled copper anode and a tungsten cathode. Due to the applied high voltage an arc is created between them. As a result, the flowing gas (Ar and He) reaches excessive temperatures, dissociates and ionizes to form plasma. Powders are fed into the plasma where they can be melted in a control fashion, accelerated to supersonic speeds, and directed toward a rotating mandrel which is rapidly cooled to form a desired shape and size. Thus, the success of the process lies in the design of the mold mandrel material and the plasma spray parameters, which need to be standardized for each component.

Fig. 1. The Plasma spraying technique in the process of making nanocomposite parts.

3. RETENTION OF NANOSTRUCTURES

The powder particles can be injected internally or externally through the plasma flame to control the final microstructure. The particle size plays an important role. The smaller particles entering into the plasma flame have smaller momenta. Thus, for constant carrier gas rates, smaller size particles have less chance to flow through the hot zone of the plasma. For example, from the micrograph (Fig. 2 b) it is clear that the small nanoparticles reinforcements (Si_3N_4) are partially melted and reside along the grain boundaries of the matrix particles ($MoSi_2$). Therefore, it can be concluded that the nanosized particles contribute to the strengthening by preventing the grain growth. This results in increased fracture toughness of the composite, useful for many applications.

4. A FEW APPLICATIONS OF THE PLAMSA SPRAYED NANOCOMPONENTS AND COATINGS

4.1.1 Defense and Aerospace Applications

Fig 2 shows the successful development the unique near-net-shape nanocomponents with retained nanostructures.

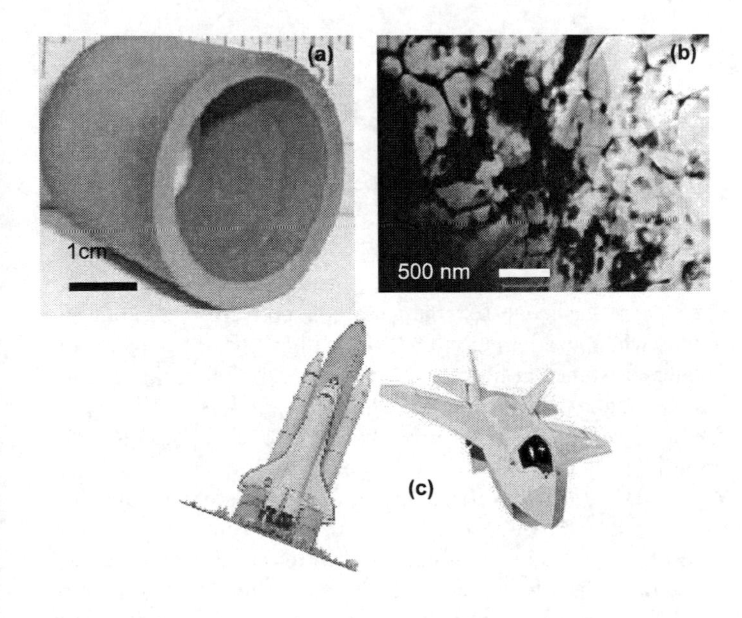

Fig. 2. Plasma sprayed thick nano-$MoSi_2$-Si_3N_4 (a) bulk part and, the correspondent TEM (b) micrograph showing the homogeneous microstructures of the components for applications at specific extreme environments, such as landing gear, high fracture toughness and high temperature resistant materials (c).

4.1.2. Materials Selection Process

Ceramic/Ceramic

The material selection is based on both simulated and experimental results. For example, $MoSi_2$ is a brittle material at low temperature even though it has a high melting point, low density, and excellent high temperature oxidation resistance. Further, it exhibits a poor oxidation resistance of $MoSi_2$ at low temperatures. The problem is due to the "pesting" behavior of $MoSi_2$ [2-5]. The addition of Si_3N_4 is important for increasing the fracture toughness of the composite above 800°C, which is otherwise difficult in silicon-based matrix composites [5]. Therefore, it was assumed that the addition of nanoscale Si_3N_4 to $MoSi_2$ would possibly solve the "pesting" problem and increase the fracture toughness.

Metal/Ceramic

Similarly, Ni/Al_2O_3 (not shown here) is of interest due to the following reasons: alumina has high strength and stiffness, good corrosion resistance, low thermal and electrical conductivity, and low density. Yet, Ni offers several advantages such as high melting point, high thermal expansion coefficient, and relatively low cost. Alumina nanocomposite with Nickel addition is expected to exhibit improved fracture toughness and superior ferromagnetic properties. Therefore, this composite system can be a potential candidate for various applications in diverse environments such as military and aerospace applications.

In the case of $MoSi_2$/Si_3N_4 nanocomposite, the room temperature fracture toughness was calculated to be ~5-7 MPa $m^{1/2}$. It should be noted that this result is almost twice the value reported for a monolithic $MoSi_2$. Therefore, it can be assumed that the enhancement of room temperature fracture toughness of near-net-shape reinforced $MoSi_2$ composites can be attributed to the homogeneously dispersed Si_3N_4 nanoparticles.

It is important to note, that it is rather a difficult task to obtain a homogeneous dispersion of a second phase in the selected matrix, which would lead to enhanced mechanical properties of the composites. Pre-stock powder preparation is very important. To optimize and ensure a proper powder flow through the plasma gun, spray drying is used to make appropriate size agglomerates as part of the pre-stock powder feed preparation.

4.2. Energy Sector: Solution Plasma Spray

In addition to our current thermal plasma spraying possesses we utilize the advantages of SPS to fabricate homogeneous nanostructured porous coating layers or to seal the porosity of the bulk parts. In the plasma flame, the liquid droplets follow a series of chemical reactions including evaporation, droplet disintegration, pyrolysis, and melting. The droplets are deposited on a metallic substrate at high velocity and form a solid thermal resistant coating. The liquid precursors are chosen in such a manner that a direct phase transformation is completed in the flame to achieve the desired material chemistry as end applications. As a result, a unique microstructure is created which can lead to superior toughness and strain resistance to the high thermal shocks often experienced in gas turbine engines (Fig.3).

Fig. 3. A section of turbine blades can be coated using the solution plasma spray for enhanced properties

The as sprayed coatings or functional layers can be applied for insulation of the hot sections of metallic components such as vanes and blades, in combustors and engines, often subjected to extreme conditions and high temperatures.

4.3. In Human Health and Biomedical Applications

The effect of the nanosized materials can be utilized in the improvement of human concerns such as anti-aging and many other therapeutic applications. Most of the present day prostheses, sensor-based systems, and eye and ear implants can be made more efficient with current trends in the nanotechnology revolution. For example, Zirconia has been widely used for biomedical applications; however, in order for ZrO_2 to bond to the bones, the material needs to be stabilized for bio-compatibility with various oxide dopants. By manipulating the plasma parameters, a highly porous structure can be produced, which can be beneficial for the cells to spread, grow, and survive in bodily fluids.

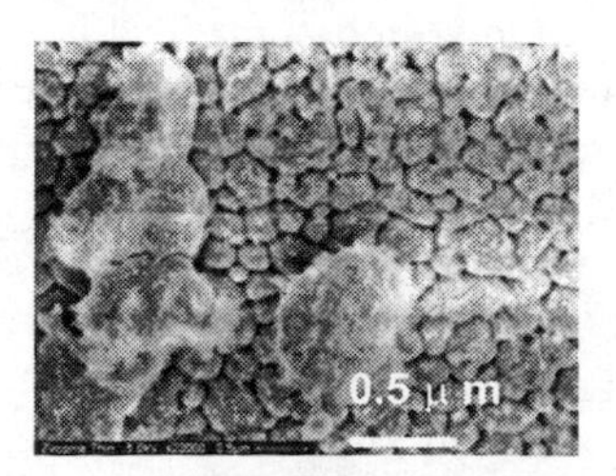

Fig.4. Hip replacements can have an extended life if produced from a biocompatible porous Ceria stabilized Zirconia material using the plasma spraying technique

4.4 Environmental Applications

Many automobile and gas burners technology can use the advantages of the nanomaterials (with high surface area to volume ratio) as components of the catalytic converters due to the presence of an extremely large amount of active sites for reaction. If specific type coatings are applied, harmful emissions such as carbon monoxide, nitrogen oxides, and unburned hydrocarbons could be converted into environmentally friendly by-products such as nitrogen, water and carbon dioxide. Catalytic converter substrates can be produced from the plasma sprayed ceramics. In addition, a washcoat from the Ceria/Zirconia can be applied. For example, due to its unique oxygen storage capacity CeO_2 is often incorporated in the three-way catalysts. CeO_2 acts as an oxygen buffer by storing excess oxygen under oxidizing conditions (oxidation) and releasing it in rich (reduction) conditions. With this process the transformation of (Ce^{3+}) to (Ce^{4+}) and vice versa is advantageous. As a result, CeO_2 oxidizes CO and HC.

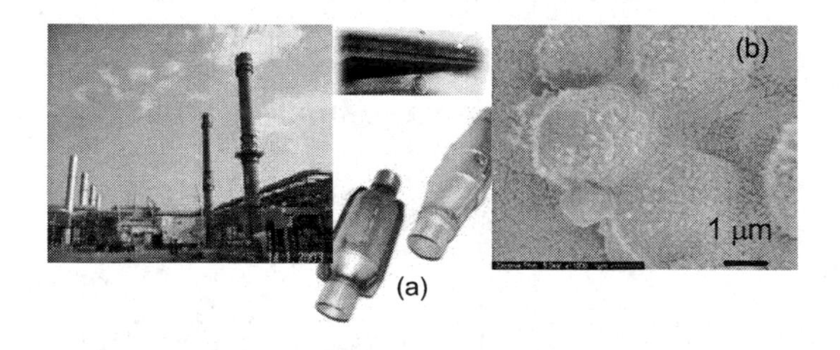

Fig. 5. Catalytic converters (a) are required for the pollution control. A washcoat from Plasma Sprayed Ceria/Zirconia (b) increase the effectiveness of the catalysts

The combination of $CeO_2$with ZrO_2 can be very effectively used as a washcoat for the catalytic converter. Zirconium oxide, when used in conjunction with Ceria, also leads to greater resistance at high temperatures. A problem that might be associated with the poisoning of the catalyst is related to the loss of catalytic activity due to the chemisorption of impurities on the active sites of the catalyst. For example, sulfur may affect the efficiency and oxygen storage capacity of the catalyst. As a solution to this problem, it has been found that Ceria, in combination with either Zirconia or Pd, has a major advantage toward the sulfur reduction reaction [11-13]. Plasma spraying can be effectively utilized for such functional coatings with improved properties.

5. THEORETICAL APPROACH IN MATERIAL DESIGN

Proper design of experiments needs an appropriate material selection methodology. Our group employs a simulation approach for Design of experiments (DOE). The role of the reinforcement in the improvement of the mechanical properties is theoretically studied using the *ab-initio* based Materials Studio software [8].

5.1 Benefit of nanostructures

The effects of nanoparticles size and structure and the stress distributions along the grain boundaries and the interfaces need to be considered. For example, in a nanostructured composite the reinforcement particles are usually dispersed at the grain boundaries or at the triple junctions and in small numbers within the matrix [9]. The nanoparticle reinforcement in the matrix leads to a homogeneous microstructure with enhanced material properties.

The accumulation of a second phase leads to the stabilization of the microstructure and minimizing the grain boundaries free energy. Hence, the increasing of the reinforcement content in the system can lead to the decreasing of the matrix particles and an increasing in the strength of the composite. In conclusion, it is seen that the presence of Si_3N_4 nanoparticles in $MoSi_2$ matrix amplifies the machinabiltiy and ductility of the $MoSi_2$ without sacrificing its properties. Also, the reinforcement nanoparticles serve as obstacles for crack initiation and propagation, and hence contribute to the increased fracture toughness; they can also locally bridge the crack or even porosity and lead to increased capability of energy absorption of the composite.

It should be noted that this paper is presented with only few examples. However, we report both fabrication and characterization of bulk nanopowders to nano/micro alumina, CNT-Vanasil alloy (Al-21%Si) and other plasma manufactured composites with retained nanostructures [1, 14-20].

6. REMARKS

In this report, we tried to convince the research community, that plasma processing is a promising engineering tool toward bulk nanostructured component manufacturing. The proper material selection and the utilization of plasma near-net-shape nanomanufacturing can be of potential interest to a large number of industrial and everyday life applications. Indeed, the theoretical and experimental results need to be combined to explain the mechanisms behind nanostructure retention and enhanced physicochemical properties.

Acknowledgements

The authors are thankful to the funding support under the Office of Naval Research Young Investigator Award Program (Awarded to S. Seal) - ONR 00014-02-1-0591, DURIP for plasma nanomanufacturing facility. We express also a special appreciation for Plasma Process Inc. Huntsville, Alabama for the assistance and design of the bulk nanocomposites.

REFERENCES

1. S. Seal, S. C. Kuiry, P. Georgieva and A. Agarwal, *MRS Bulletin, Jan Issue*, 29(1) 16-22, 2004.
2. J. S. Jayashankar, E. N. Ross. P. D. Eason, M. J. Kaufman, *Mat. Sci. Eng.* A239-240, 485, 1997
3. J. J. Petrovic, *Intermetallics,* 8, 1175, 2000
4. H. Shimizu, M. Yoshinaka, K. Hirota, O. Yamaguchi, *Mater. Research Bulletin,* 37, 1557, 2002
5. S. G. Savio, R. R. Rao, S. K. Ramasesha, *Mater. Letters*, 57, 43, 2002
6. E. Courtrigth, *Mat. Sci Eng.* A261, 1999, 53
7. S. Ogata, and H. Kitagawa, Comput. Mater. Sci. 15, 435, 1999
8. Materials Studio http://www.accelrys.com/sim/
9. Gleiter, *Acta Mater*, 48, 2000
10. X. M. Chen and B. Yang *Mater. Lett.* 33, 237, 1997
11. T. Luo, R. J. Gorte, Applied Catalysis B: Environmental, 53, 2004
12. F. Rohr, S. D. Peter, E. Lox, M. Kogel, A. Sassi, L. Juste, C. Rigaudeau, G. Belot, P. Gelin, M. Primet, Applied Catlysis B: Environmental, 56, 2004, 187
13. T. Luo, J. M. Vohs, R. J. Gorte, Journal of Catalysis, 210, 2002
14. K. E. Rea, A. Agarwal, T. McKechnie and S. Seal, *Microscopy Research Techniques,* 66, xx-xx, 2005
15. K. Balani, T. Laha, A. Agarwal, S. Patil and S. Seal, *Materials and Metallurgical Transactions A.,* 36a (2), 301-309, 2005.
16. T. Laha, A. Agarwal, T. McKechnie and S. Seal, *Materials Science and Engineering A,* 381, 249-258, 2004.
17. S. Wannaparhun, S. C. Kuiry, N. Dahotre and S. Seal, *Scripta Materilia,* 50, 1237-1240, 2004.
18. S. Seal and M. Baraton, *MRS Bulletin, Jan Issue*, 29(1), 9-12, 2004.
19. A. Agarwal, T. Mackenzie, S. Seal, *J of Thermal Spray Technology,* 12(3), 350-359, 2003., *JOM,* Sep Issue, 20-21, 2002.
20. A. Agarwal, T. Mackenzie, S. Seal, *JOM*, Sep issue, 42-44, 2002.

Imation Tera Angstrom™ Technology – Nanotechnology Applications

Dr. Subodh K. Kulkarni

Imation Corporation, Oakdale, MN, USA, skkulkarni@imation.com

ABSTRACT

Imation last year introduced a revolutionary new magnetic tape technology Tera Angstrom™ that will enable high capacity and high transfer rate data cartridges for archiving and backing-up digital data. Tera Angstrom™ technology uses nanometer size magnetic metal particulate (MP) pigments and three key processes to enable these high end data storage products: in-line I-MUF technology to enable separation of these nanometer sized pigments in the dispersion used to coat the substrate; quiescent drying and in-line orientation of the pigments in the direction of the coating; in-line steel calendering using very smooth (Ra < 3 nm) rolls to enable extremely smooth (Ra < 5 nm) magnetic media. This paper describes more details about these unique processes to enable the finished properties of the magnetic media.

Keywords: magnetic media, metal particulate, Tera Angstrom™ technology

1 MAGNETIC TAPE TECHNOLOGY

Magnetic tape has been the dominant technology used in the last 50 years to archive and back-up digital data. Imation (at that time 3M) was the first company to offer magnetic tape products for IBM systems 50 years ago, and Imation continues to be the leader of magnetic technology today, offering the broadest portfolio of removable data storage products across all the segments – Enterprise, Mid-range, Entry level, and Consumer. Magnetic tape technology offers significant advantages over competing technologies, and current trends indicate that this will continue to be the case in foreseeable future. With increasing demand for digital data to be archived, there's great need to keep increasing capacity and transfer rate for data cartridges. This is accomplished with increasing bit and track densities on a given tape, and increasing the amount of tape in a given cartridge.

To enable the increasing bit and track densities, media companies have used increasing coercivity and smaller size acicular shaped metal particulate pigments for every successive generation of tape products. The current high end products (400 GB native capacity) are using acicular magnetic pigments that are ~10 nm in diameter and ~50-70 nm in length. In future, the particle sizes will decrease even further. This is shown in Fig. 1, where each arrow represents a generation of MP technology and the timeline over which products that have been introduced using that technology. With continuous advancements accomplished in the MP technology, we think MP technology will be the most competitive technology for the foreseeable future and 1 TB native capacity cartridges will be achieved using MP technology in the next few years.

Figure 1. Magnetic Tape Technology Migration

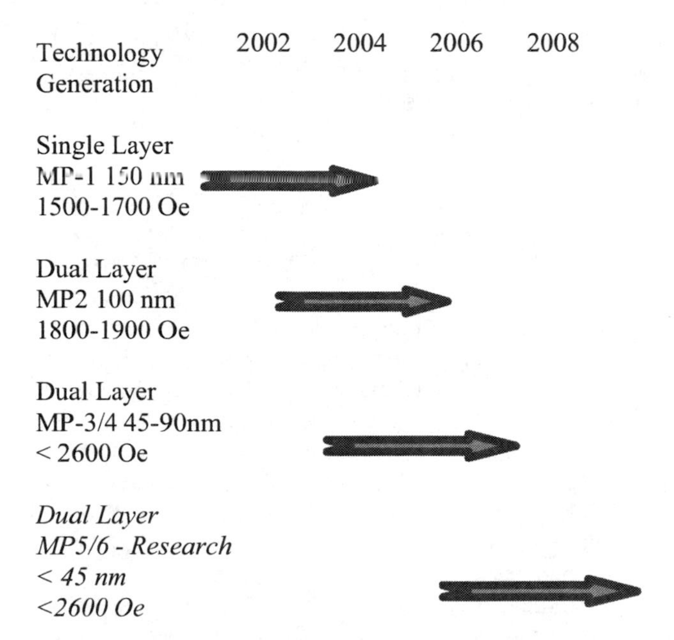

Along with using more sophisticated MP pigments and the formulation chemistry that goes with it, the media processes that are used to enable the media to be produced play a significant role in the end properties of the media. The higher bit and track densities needed for future products necessitate media with high signal to noise ratio (SNR). This is accomplished through a thin magnetic layer on top of a non-magnetic layer or sub-layer coated on a thin flexible plastic substrate with a conductive backside coating. The thicknesses of the substrates and the coatings on the media are depicted in Table 1. Also, in Table 1, are shown the needed smoothness of the magnetic media that is anticipated in future.

Table 1. Media Requirements Migration Path

Parameter	Units	MP 1	MP 2	MP3/4	MP4/5	MP6/7
Mag Layer Caliper	Micro-inches	40	12	8	2-4	<2
Sublayer Caliper	Micro-inches	N/A	70-90	70-90	40-60	40-60
Backside Caliper	Micro-inches	25	20-25	20	10-20	10-20
Substrate Caliper		18 -36ga	24ga	24ga	18 ga	<18 ga
Substrate Material		PEN/PET	PEN/PET	PEN/PET	PEN/PET	TBD
Mag Surface Roughness	Ra nm	6-10	4-6	4-5	3-4	2-3
Bit Density	Kbpi	80 -190	90 -180	180-220	250-300	>300
Data Tracks	in 1/2" wide	350-600	300-400	500-600	1000-2000	>2000
Write Track Width	Microns	17-36	25-30	15	7-10	<7
Read Track Width	Microns	9-20	12-15	7	3-5	<5
Tape Length	Feet	800-2200	2500	1500-2500	2000-3000	3500-5000

In processing of magnetic media suitable for tape products, several steps are used to end with a finished magnetic tape. The process starts with creating dispersions of magnetic, non-magnetic and backside chemistries. To create these fine dispersions, the appropriate pigments, binders and other additives are milled together. These dispersions are coated using thin film coating techniques on a moving web at high speeds and dried using high temperature ovens. After coating and during the process of drying, the magnetic particles are aligned in the direction of the moving substrate to further enhance the electromagnetic properties of the media. After drying the magnetic media is calendered in a stack of rollers to further improve the packing of the magnetic pigment and smoothness of the media, again giving a boost in the electromagnetic performance of the media. Subsequent to calendering, the media is slit into the desired format. This media is further servo-written and then wound into cartridges.

2 IMATION TERA ANGSTROM™ TECHNOLOGY

Imation has developed a revolutionary process to accomplish making the magnetic media that leads to higher smoothness and higher SNR, therefore enable the higher capacity and transfer rates for future products. The features of this process are described in this paper.

The three differentiating elements of Imation's process are:

1. I-MUF technology.
2. Quiescent drying with magnetic coil orientation.
3. In-line calendering using proprietary rollers.

In each of these areas, Imation's process provides an improvement in performance over conventional competitive approaches in the respective areas, and the combination of all the three features provides a significantly improved performance over conventional approach.

Fig. 2 shows the I-MUF schematic and Fig. 3 shows the particle size distribution achieved using conventional milling technologies such as homogenization, compared to Imation's I-MUF technology. I-MUF technology enables elimination of large size particles seen in conventional homogenization, and maximizes the separation of the nano sized MP particles. Fig. 4 shows the schematic and Fig. 5 shows the higher packing density and orientation achieved using Quiescent drying and magnetic coil orientation compared with conventional approach of using impingement drying and magnet orientation. Note the individual nanometer sized MP particles being aligned to the direction of the coating or the direction of the slit media thereby enabling higher EM performance and higher capacity/transfer rates. The effect of in-line calendering with proprietary finished calender rolls is shown in Fig. 6 where it is evident that higher smoothness can be accomplished using the new process. The smoothness of the calendar rolls used in this process is <2.5 nm.

In summary, the new proprietary Tera Angstrom™ process invented and developed by Imation is providing superior magnetic media that will enable future products for archiving and backing-up digital data. The new process has been incorporated in the new coating facility in Weatherford, OK, which is the only facility in the world to have this unique process to make advanced magnetic media. Production has commenced, and the state-of-the-art facility is producing high end tapes for the most advanced products (currently 400 GB native capacity, and in future, higher capacities will be available) used in the industry.

Figure 2. Schematic of I-MUF process

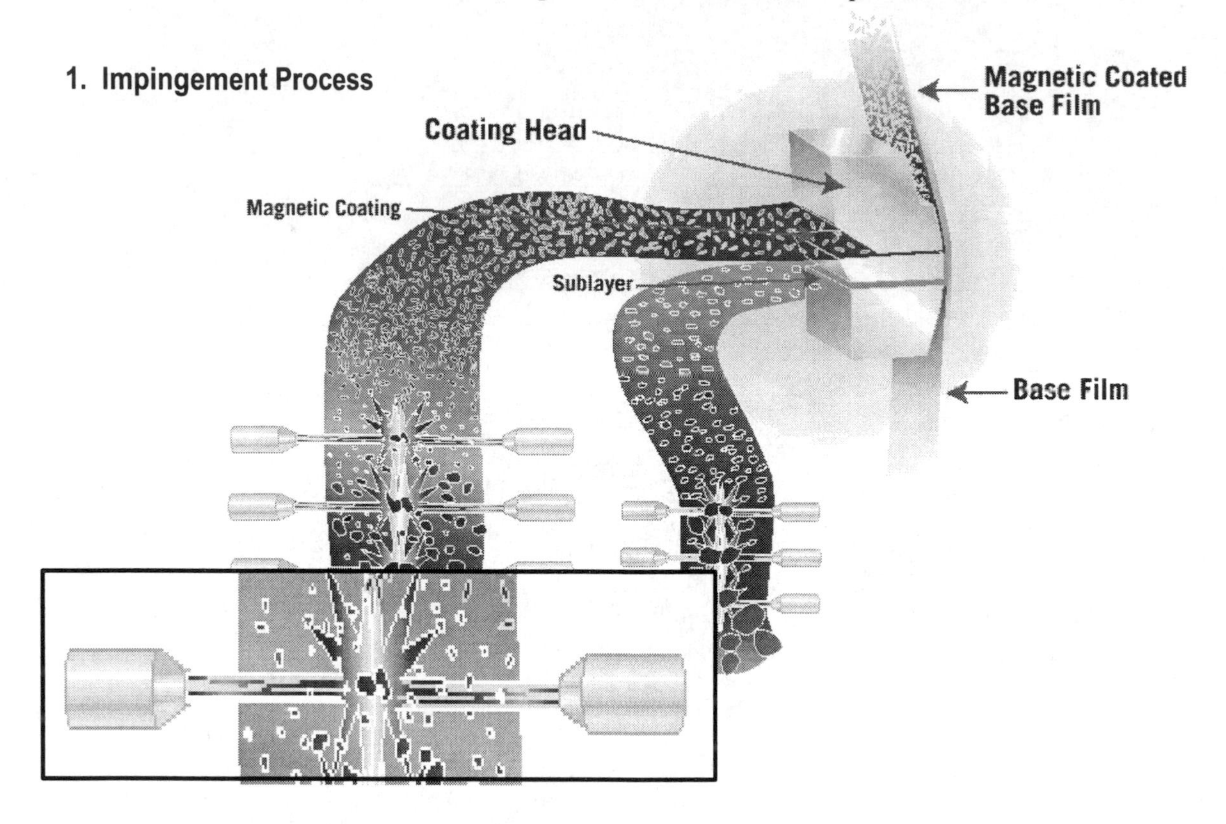

Figure 3. High Pressure I-MUF Impingement Provides More Uniformity and Smaller Particle Size

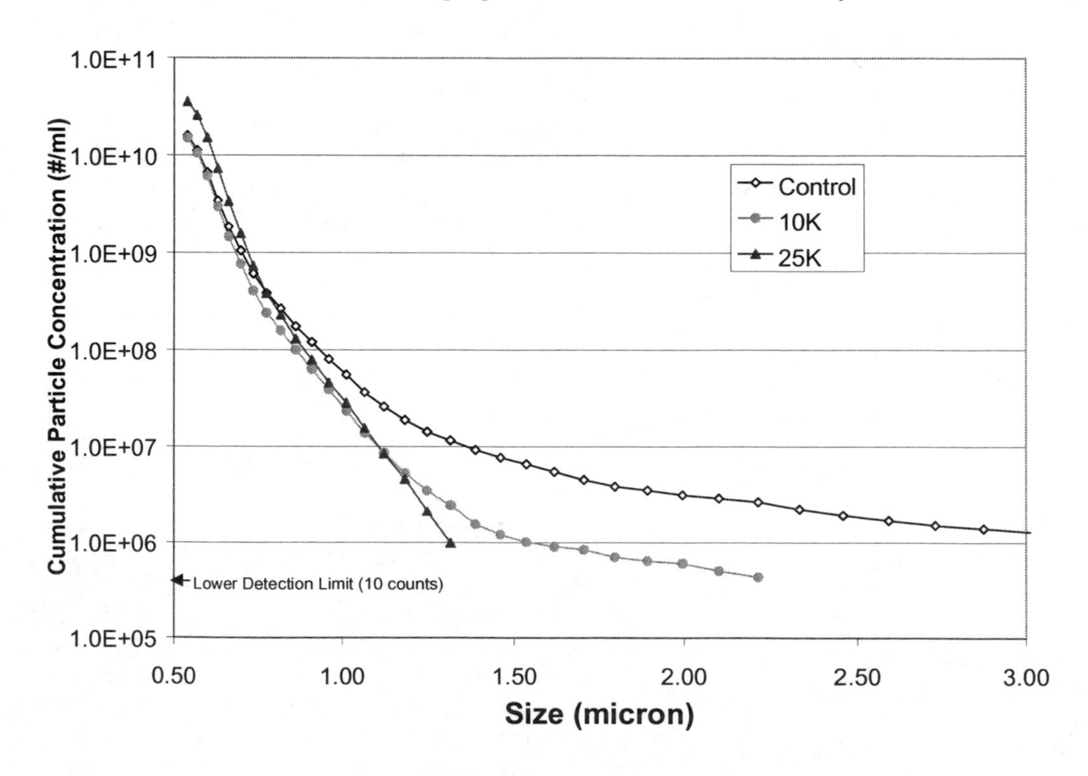

NSTI-Nanotech 2005, www.nsti.org, ISBN 0-9767985-1-4 Vol. 2, 2005

Figure 4. Schematic of quiescent drying and in-line orientation

2. Quiescent Drying Process

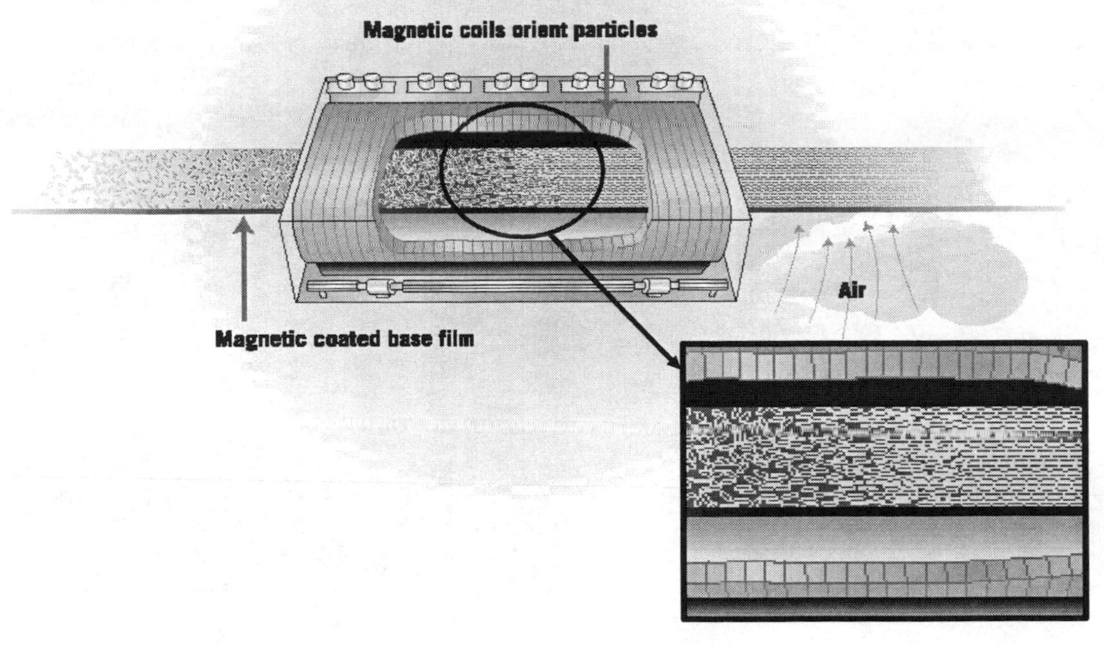

Figure 5. SEM Images at 40,000x. Left: Imation LTO2 sample with Quiescent Drying and Magnetic Coil Orientaion; Right: Competitive LTO2 product

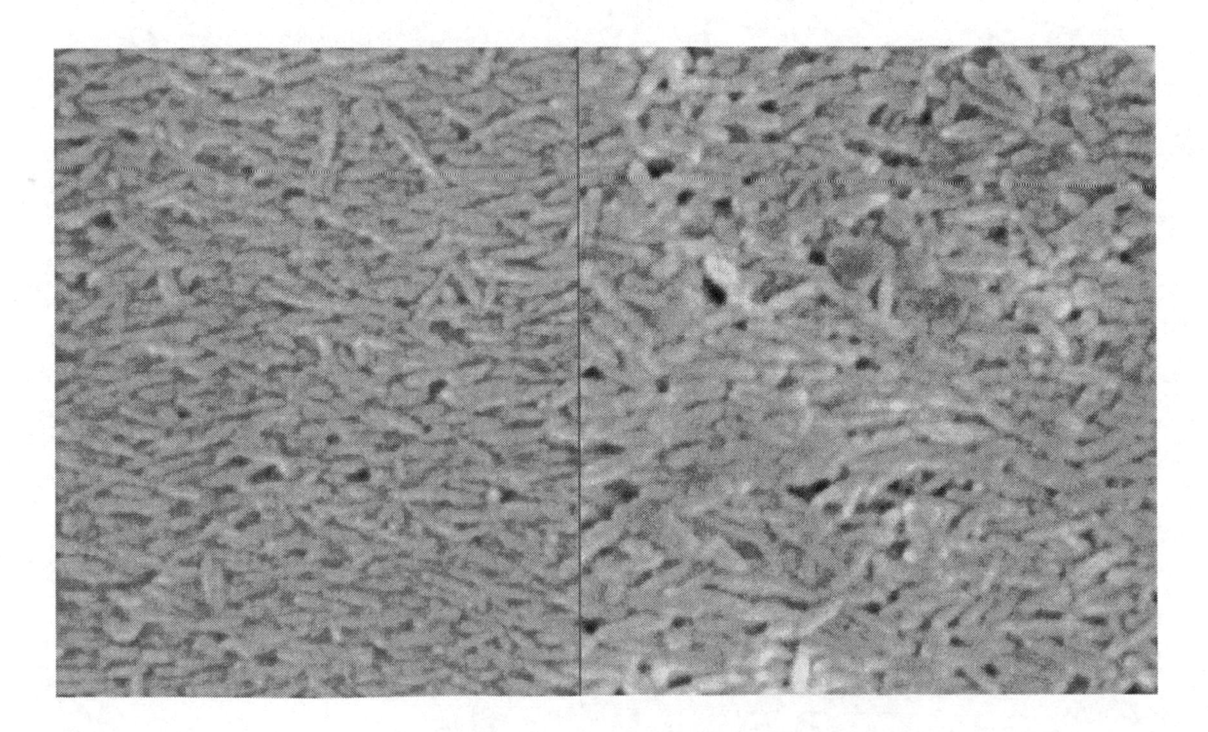

Figure 6. AFM Images of Imation and Competitive Media

AFM Image of Imation LTO2 POR

Ra: 4.0 nm

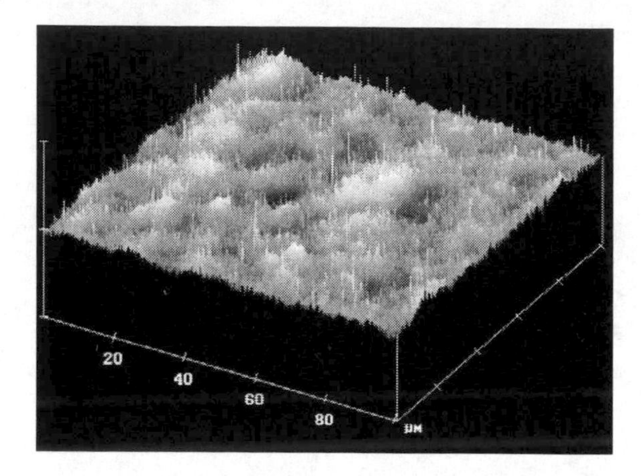

AFM Image of Imation 1 TB target media using Advanced MP Coating Technology in Discovery pre-production facility

Ra: 2.4 nm

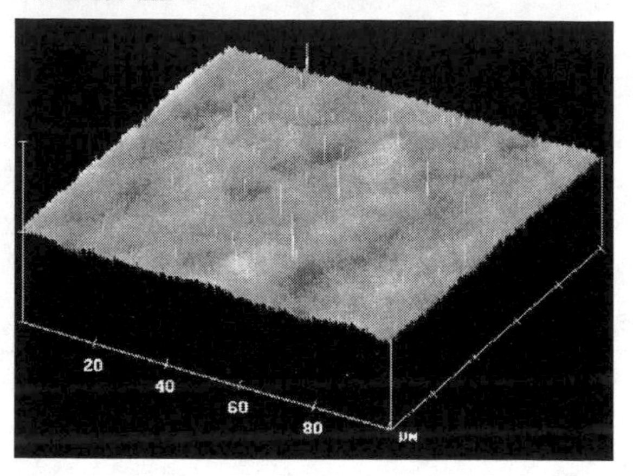

100 µm x 100 µm scans, vertical scale 100 nm/div

Nano-reinforcement of Epoxy Adhesives with POSS

Wei Chian, Chandrakanth Mallampalli and Robb M. Winter*
Department of Chemistry and Chemical Engineering
South Dakota School of Mines & Technology
Rapid City, SD 57701
* Corresponding author. Email address: robb.winter@sdsmt.edu

ABSTRACT

The reinforcement of epoxy adhesives with polyhedral-Oligomeric-SilSesquioxane (POSS) was studied. Shear performance of the nano-filled adhesive was evaluated as a function of filler content using the single-strap bonded joint configuration. It is found that maximal stress and maximum shear strain both increase with the POSS loading. X-ray diffraction (XRD) analyses indicated that aggregation and crystallization of POSS occurred over the curing cycle. Dynamic mechanical analyses (DMA) showed an increased storage modulus with the addition of POSS and decreased loss modulus, hence enhanced network elasticity. The increased chain relaxation enhanced the polymer/aluminum interaction at the interface, resulting in increased interfacial strength and toughness, which was evident from the observed shift in failure mode from adhesive to cohesive for the bonded joints with increased POSS content.

Keywords: epoxy adhesives, mono-epoxide-POSS, nano-reinforcement, bonded joints

INTRODUCTION

The strength and serviceability of adhesive-bonded joints are mainly the result of the play of forces of intermolecular interaction between the adhesive and the substrate. Newly modified adhesives characterized by high adhesion strength, lower curing temperature, high fracture toughness, and other functionalities, such as heat/moisture resistance and noncombustibility is of great practical interest.

The process of achieving high adhesion strength is hindered by a number of phenomena that accompany the formation of adhesive bonded joints. First, an intermediate layer of undercured polymer can be formed when adhesives are used on high-energy substrates due to the absorption of reagents by the solid surface. Second, in the course of setting of the adhesive, its volume decreases due to volatilization of solvents, polymerization or physical structuring. As a result of the adhesion interaction of the adhesive and the substrate, the film can contract only in thickness. The film extends while contraction stresses appear in the substrate. Rapid growth of stresses tends to reduce the length of the film[1]. In addition, thermal stresses caused by differences of the coefficients of linear thermal expansion of the adhesive and the substrate appear in the course of heating or cooling of the adhesive-bonded joint.

Introducing finely dispersed mineral filler with lyophilic or lyophobic surface character into polymer composition (adhesive) was found to increase the polymer-substrate adhesive bond strength, as migration of low molecular weight impurities, including low molecular weight fractions formed during curing, to the polymer-metal substrate boundary decreased as a result of their physiosorption to the filler surface, thus allowing the formation of a less defective boundary layer[1]. However, the internal stresses in the coating grow as the filler content increases, since modified filler simultaneously displays effects of restriction of mobility and polymer structural plasticization in the boundary layer due to the presence of long chain modifier molecules. Increased internal stresses result in reduced strength of the coating-metal substrate adhesive bond. On the other hand, widely used methods for decreasing the internal stresses relate to increasing their rate of relaxation, for example, by decreasing the modulus of elasticity when adding plasticizers to adhesives or when they are combined with elastomers. But the decrease of adhesive rigidity commonly results in decrease of its heat resistance and stress-strain behavior stability, and decrease of strength limits under bending, contraction, and shear due to the lower cohesion strength of the modified adhesive itself [1].

Reinforcement of polymer systems with well-defined nano-sized inorganic clusters, for example Polyhedral-Oligomeric-SilSesquioxane (POSS) has been given considerable attention recently. Functionalized POSS reagents combine a hybrid inorganic-organic composition with nanosized silica-like cage structures having dimensions comparable to those of most polymeric segments or coils[2] and can be covalently bound to the polymer, leading to reinforcement of the system on molecular level. Due to the large surface area, only small amounts are needed to cause significant changes in properties of the matrix. Interestingly, recent study on methacryl-POSS indicates POSS may have the potential to behave like a filler particle or a plasticizing molecule, depending on the degree of dispersion at molecular level[3]. Incorporation of POSS into a wide range of polymers such as polysiloxane[4], polymethacrylate[5], polystyrene[6], polyurethane[7] and epoxy thermosets[2] and has shown improvements in mechanical and rheological properties, including increased glass-transition temperature and modulus, reduced flammability, increased gas permeability and decomposition temperature. Nonetheless, improved mechanical properties in the context of bulk nanocomposites will not necessarily translate into an increase of polymer/substrate adhesion strength and adhesion toughness in the context of adhesive bondline as the presence of solid surface often affects the formation of polymers. In addition, incorporation of multi-functional POSS to polymers has been found to increase the network chemical crosslink density, which may lead to slowing down of the polymer chain relaxation process[2]. In sight of the latter, POSS with monoepoxide functionality (E1-POSS, see Figure 1 for the chemical structure) was used as the reinforcing element in a common two-part epoxy adhesive system in our study with the aim of a strong and tough bondline. Upon cure, the E1-POSS units are expected to covalently bind to the network chain as pendant units without an increase of chemical crosslinks. The weight percentage of nanofiller was kept below 5wt% in consideration of economic viability of future application.

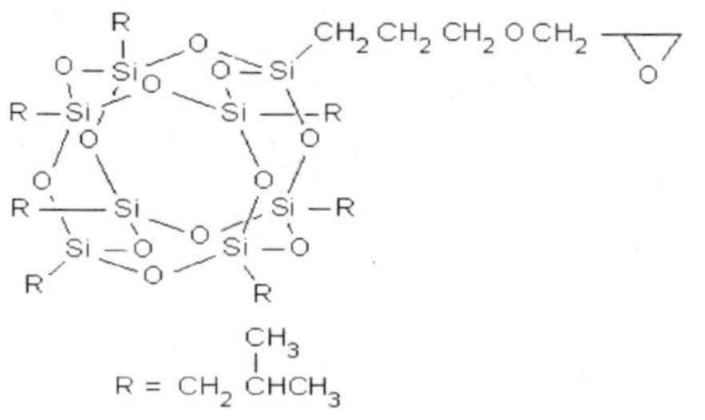

Figure 1. Structure of E1-POSS

EXPERIMENTAL

Mono-epoxide-POSS (PSS-(3-Glycidyl)–Heptaisobutyl substituted, E1-POSS, purchased from Sigma-Aldrich) was added to the epoxy monomer (Araldite GY502 from Vantico, a diglycidyl ether of bisphenol A based resin). A series of mixtures containing different weight percentages of POSS (0-5wt.%) were prepared and heated to 60°C. At this temperature, POSS can fully dissolve in epoxy and form a solution. The solution was allowed to cool. At room temperature, the POSS/epoxy emulsion was mixed with the polyamidoamine curing agent (Aradur 955-1 US from Vantico) at a weight ratio of 100 to 27 (epoxy/amine) to form the adhesive.

Aluminum bars (75mm x 20mm x 3.2mm) were degreased with trichloraethane (Sigma-Aldrich) and etched in Nochromix®/sulfic acid solution for 30 minutes. They were then rinsed with tap water for around 20 minutes and dried at 100°C for 40 minutes. At room temperature, the adhesive was applied on the aluminum bars and single strap bonded joints were made. The joints were allowed to stay 8 hours at room temperature to remove the micro bubbles within the adhesive layer. Four hours curing at 90°C followed. Uniform pressure was applied during the making of joints to keep a consistent thickness of the bondline.

Shear tests of prepared bonded joints were performed on an MTS 810 under displacement control mode (0.2mm/sec). For each adhesive, seven joints were tested. The failure surface of bonded joints (adhesive layer) was subject to X-ray diffraction (XRD) analysis to obtain information about network morphology. Specimen of 17.5mm x 12.7mm x 3.0mm of cured adhesives were made in aluminum foil and dynamic mechanical thermal analyses (DMTA) were performed on a DMA Q800 V3.13 Build 74 under 1Hz 3 point bending mode and a temperature ramp of 2°C/min.

RESULTS AND DISCUSSION

Figures 2 and 3 are plots of static performance of aluminum joints versus the E1-POSS content in the adhesive based on statistical treatment of experimental data. Clearly, the incorporation of E1-POSS nanoparticles into the adhesive layer increased epoxy/aluminum adhesion strength as indicated by the increasing stress at failure in Figure 2. As is shown in Figure 3, addition of 5wt% E1-POSS nearly tripled the adhesive toughness. Visual examination of the failure surfaces shows that cohesive failure was achieved for POSS content higher than 1.0 wt.%, indicating an enhanced polymer/metal interface.

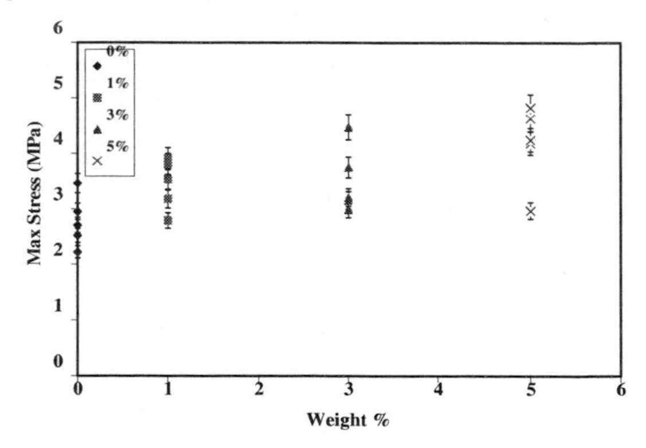

Figure 2. Adhesion Strength vs POSS Content

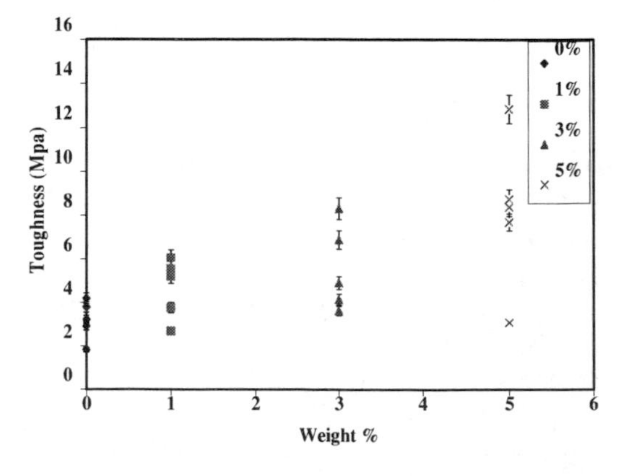

Figure 3. Adhesion Toughness vs POSS Content

Morphology of E1-POSS-modified epoxy network was studied using XRD technique. XRD results of failed adhesive surface are shown as Figure 4 in comparison of the spectrum of pure E1-POSS particles. It appears that the E1-POSS crystal structure was not affected by its incorporation in the epoxy network, as is indicated by the existence of a sharp peak around $2\theta=7.8^{\circ}$ for all E1-POSS concentrations. That is to say, even at very low concentrations, E1-POSS was not dispersed in the network at molecular level. As authors have pointed out, the POSS units in polymers tend to aggregate or even form crystallites [8]. In light of the strain increase of the bondline with the addition of E1-POSS, we infer that a loose transition region of two thermodynamically incompatible phases of formed before POSS crystallization took place. The existence of a loose POSS rich transition region may lower the density of macromolecular packing, which results in decrease of shrinkage during the network formation. This loose transition region in the system may also provide for an increased free volume and lead to higher mobility of the

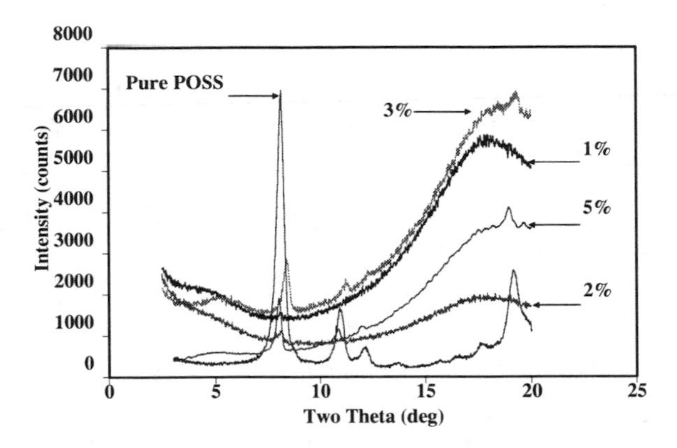

Figure 4. XRD Spectra of E1-POSS/Epoxy

chain segments between the crosslinks of the network. Decrease of internal stress in turn contributes to the formation of a less defective adhesive/metal interface[1]. From the perspective of fracture mechanics, the presence of loose transition region can both initiate and limit the growth of crazes. A large number of small crazes rather than a small number of large cracks are formed and in this case the energy dissipates in a much great volume.

Results of DMTA measurements presented as Figures 5 and 6 show that the bending storage modulus (E') of POSS/epoxy increased with the POSS content for the range 0- 5wt%. As can be seen in Figure 5, the equilibrium rubbery modulus of E1-POSS-modified network at a temperature above breaking of crystalline domains is close to that of the POSS-free network. It can be inferred that the reinforcing effect of POSS disappears at a high temperature and E1-POSS has very insignificant effect on the permanent chemical crosslinking density, based on the theory of rubber elasticity [9]. Namely, reinforcement of the E1-POSS epoxy network is mainly caused by the formation of physical crosslinks between pendant POSS units rather than the increase of chemical crosslinking density.

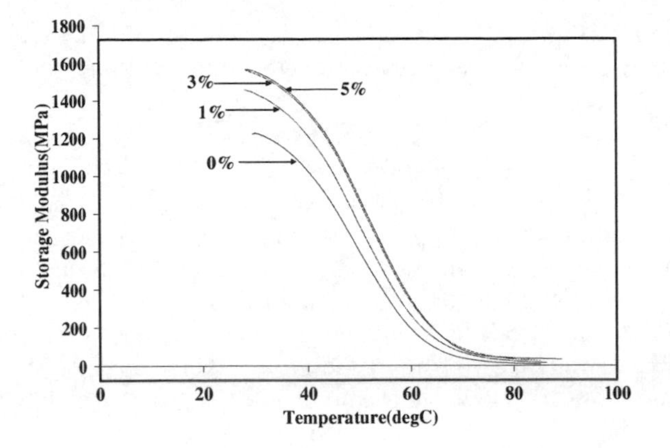

Figure 5. Storage Modulus vs Temperature

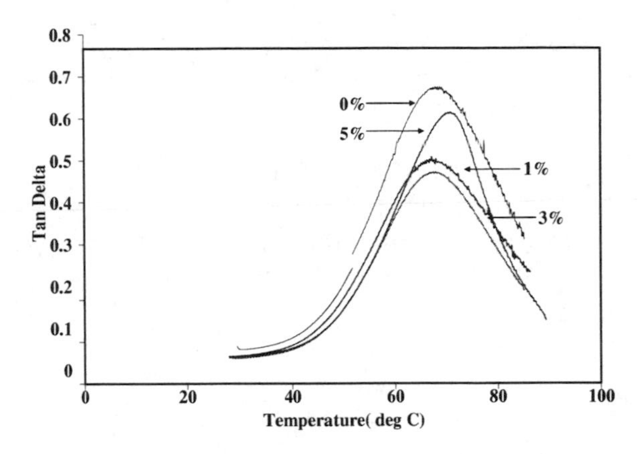

Figure 6. tan δ vs temperature curves

As is shown in Figure 6, the glass transition temperature T_g (defined as the temperature at which tanδ peaks) was not affected by the presence of POSS at low POSS concentrations (<3 wt%). This corresponds to the formation of a loose POSS transition region. As a result, the breadth and amplitude of tanδ peaks both decreased with increasing POSS concentration. Notably, further incorporation of E1-POSS (5 wt%) led to a substantial increase of T_g (5°C), indicating restricted chain segmental mobility and the emergence of a rigid network at high POSS concentrations. This observation also echoes XRD results, which clearly show that the degree of crystallization of E1-POSS increased with the POSS content in the network. Conceivably, an increase of E1-POSS content causes a decrease of the distance between dangling units, hence stronger POSS-POSS interaction. Zheng et al.[8] suggested the existence of critical POSS concentration above which POSS-POSS interactions percolate through the system. As a result of large-scale crystallization, the fraction of mobile free network chains may be considerably reduced. Together with the inertia effect of a large mass POSS molecule, this leads to a decrease of network flexibility and the retardation of chain relaxation process, which contribute negatively to the mechanical performance of the network as a bondline. In addition, tight POSS agglomerates or crystals can act as areas of stress concentrations from the perspective of mechanics.

Therefore, integrating large mass of POSS to epoxy adhesives is not likely to further strengthen the bondline.

CONCLUSIONS

Incorporation of a mono-epoxide POSS in suitable amount can effectively enhance the mechanical performance of epoxy adhesives. Interactions of pendant POSS units covalently bound to the polymer chains form physical crosslinks and lead to increased modulus and increased cohesive strength of the adhesive. When the POSS concentration is low, thermodynamically incompatible components form a loose transition region, which may generate free-volume and promote the chain relaxation process. With the presence of aluminum, interactions between dangling E1-POSS units percolates at a POSS concentration around 4-5wt% and crystallization occurs over a large scale, leading to an increased T_g, which may lead to a decrease of the bondline deformability and limit further enhancement of adhesive performance.

ACKNOWLEGEMENT

Financial support from Scholar/Fellow Program of Camille & Henry Dreyfus Foundation is gratefully acknowledged.

REFERENCES

[1] R.A.Veselovsky and V.N. Kestelman, *Adhesion of Polymers*, McGraw-Hill, **2002**
[2] Lee, A.; Lichtenhan, J. D. *Macormolecules,* **1998**, 31, 4970
[3] Kopesky, E. T.; Cohen R.E.; McKinley, G.H. Conference presentation, *AIChE* Annual Meeting, November, **2004**, Austin, TX
[4] Lichtenhan, J. D.; Vu, N. Q.; Carter, J. A.; Gilman, J.W.; Feher, F.J. *Macromolecules*, **1993**, 26, 2141
[5] Lichtenhan, J. D.; Otonari, Y.A. *Macromolecules* 1995, 28, 8435
[6] Haddad, T. S.; Lichtenhan, J. D.; *Macromolecules* **1996**, 29,7302
[7] Hsiao, B.S.; Mather, P.T.; Chaffee, K.P.; Jeon, H,; White, H.; Rafailovich, M. Lichtenhan, J. D.; Schwab, J. J. *Polym. Mater.Sci.Eng.* 1998, 79, 389
[8] Zheng, L.; Waddon, A.J.; Farris, R.J.; Coughlin, E.B. *Macromolecules* **2002**, 35, 2375
[9] P. J. Flory, *Principles of Polymer Chemistry*, **1953**, Cornell University Press

PREPARATION OF BN-Si_3N_4 NANOSTRUCTURED COMPOSITES

L.F.Krushinskaya, G.N.Makarenko, I.V.Uvarova, I.I.Timofeeva, L.P.Isayeva
Institute for Problems of Materials Science NAS of Ukraine
3, Krzhizhanovsky str, 03142, Kyiv, Ukraine
tel: (+440)424-1533,
e-mail: uvarova@materials.kiev.ua

Abstract
The process of boron silicide nitration in nitrogen and ammonia media has been studied. The composite powders of boron and silicon nitrides were established as the products of nitration. By using the boron silicide compounds instead of boron and silicon powder mixture it can be possible to decrease the temperature of BN and Si_3N_4 formation up to 1200 and 1400 ^{0}C, correspondingly. Without using the catalytic additions, the BN and Si_3N_4 are generated at 1500 and 1700 ^{0}C, correspondingly over a long period of time. Decreasing the temperatures and time of nitration one can increase the dispersity of final products.
The nitration by ammonia occurs more intensive than that by nitrogen and the boron and silicon powder nitrides are more dispersive and have the specific surface area about 20 m^2/g.
The dispersity of BN and Si_3N_4 composites can be increased by preliminary mechanical activation of boron silicide by intensive milling. In this case the temperature of nitration decreases up to 1100 °C and as a result of this the composite powders of boron and silicon nitrides have a size of particles about 50-70 nm.

INTRODUCTION

High-temperature composite materials of the BN-Si_3N_4 system are of great interest due to a wide range of applications such as automobile components, refractory nozzles, sliding gates, break rings for the horizontal continuous casting of steel, friction assemblies operating at high temperatures in aggressive media. In a number of works [1-2], it has been shown that the use of boron nitride as a minor phase increases significantly the fracture toughness and thermal shock resistance, which are sensitive to the grain size of ceramics.

RESULTS AND DISCUSSION

In the present work, the nitration process of boron silicide in a nitrogen and ammonia atmospheres was investigated with the aim of preparing nanocrystalline powders of the BN-Si_3N_4 system. The influences of the nitration temperature, exposure time, nitration atmosphere, compositions and dispersion of initial boron silicide powders on the nitride formation process and nitrogen content in products of nitration were investigated.
As a starting material, boron silicide having a rhombohedral lattice and characterized by a wide region of homogeneity from $B_{2.8}Si$ to B_4Si [3] was chosen. Boron silicide, compared to phases that were expected to form during nitration, is much less stable, which follows from thermodynamic characteristics presented in Table 1.

Table 1. Thermodynamic characteristics of the strengths of the chemical bonds of BN, Si_3N_4 и B_4Si [4,5]

Characteristic	Compound		
	BN	Si_3N_4	B_4Si
Melting point, T_m., K	3240 (under pressure of N_2)	2250 (dissociation at a pressure of N_2=100 kPA)	1600 (decomposes)
Heat of formation, $-\Delta_f H^0$(298 K), kJ/mole	250.5	757.8	86.56
Atomization energy, $\Delta_{at}H^0$(298 K), kJ/mole (kJ/g-atom)	1286.92 643.45*	40285.53 575.50*	2701.16 540.32*

*calculation

The comparative investigation of nitration of boron silicide powders of different dispersion and composition was performed on the initial and mechanically activated powder with the B_4Si $B_{2,8}Si$ compositions. Both powders were prepared by synthesis from elements. The chemical compositions and specific surface areas of the powders are given in Table 2.

Table 2. Characteristics of initial boron silicide powders

Boron silicide	Content, mass %				Specific surface area, m^2/g
	B	Si	$Si_{cв}$	Fe	
B_4Si	62.45	35.9	0.9	-	7.09
$B_{2\cdot8}Si$	52.0	48.1	-	0.9	2.99

The initial B_4Si powder was mechanically activated in a planetary mill in a liquid nitrogen atmosphere. The total time of the operation was 40 min. After mechanical activation, the powder contained 5.12 mass % of iron and o.65 mass % of oxygen; its specific surface was 10.46 m^2/g.
The nitration of the initial boron silicide powders was performed in a resistance furnace in a nitrogen and an ammonia flow in the temperature range 1200—1450°C with isothermal exposures for 1-5 h, and the activated powder was nitrided in the temperature range 1000-1400 °C with exposures for 1-5 h.
The compositions of the reaction products were assessed on the base of the data of the chemical and X-ray phase analyses using standard procedures. The

specific surfaces were determined by the thermal nitrogen desorption method.

From the change of the Gibbs free energy ($\Delta_r G^0(T)$) in the temperature range 1000—2000 K the probabilities of development of nitration reactions of boron silicide.

$3B_4Si_{(c)} + 8N_2 = 12BN_{(c)} + Si_3N_4$, (1)

$3B_4Si_{(c)} + 16NH_3 = 12BN + Si_3N_{4(c)} + 24H_2$, (2)

and, for comparison, the probabilities of development of the nitration reactions of boron and silicon were evaluated

$B_{(c)} + 1/2N_2 = BN_{(c)}$ (3)

$2B_{(c)} + 2NH_3 = 2BN_{(c)} + 3H_2$, (4)

$3Si_{(c)} + 2N_2 = Si_3N_{4(c)}$, (5)

$3Si_{(c)} + 4NH_3 = Si_3N_{4(c)} + 6H_2$ (6)

In the calculation, the following equations were used:

$$\Delta_r G^0(T) = \Delta_r H(298) - T\Delta_r \Phi'(T)$$
$$\Delta_r H(298) = \sum \Delta_f H^0(298)_{prod.} - \sum \Delta_f H^0(298)_{init.}$$
$$\Delta_r \Phi(T) = \sum \Phi'(T)_{prod.} - \sum \Phi'(T)_{init.},$$

where $\Delta_f H^0(298)$ is the standard enthalpy of formation; $\Phi'(T)$ is the reduced isobari-isothermal potential at a temperature T; subscripts: r – reaction, prod. – reaction products, init. – initial substances, c – condensed state.

The thermodynamic characteristics of the components were taken from [4—7].

The results of the thermodynamic calculation are shown in Fig. 1. The results of the experimental investigation of nitration are presented in Table 3. The chemical analysis of the products of nitration of the initial boron silicide powders shows that the intensity of the nitride formation process depends on the ratio of boron to silicon.

The nitration of the $B_{2,8}Si$ powder in a nitrogen flow proceeds somewhat more intensively (despite of its smaller specific surface) than the nitration of the initial B_4Si powder at temperatures to 1450 °C. The nitrogen content in the nitration products of $B_{2,8}Si$ obtained at 1300 и 1400 °C and an exposure for 3 h is higher than that in the nitration products of B_4Si obtained using the same conditions by 4 and 10 mass %, respectively (Table 3). The difference between these powders in the nitrogen content decreases to 1 mass % as the nitration temperature is raised to 1450 °C.

The nitration of the $B_{2,8}Si$ и B_4Si powders proceeds more intensively in ammonia , since the chemical reactivity of dissociated ammonia is larger than that of nitrogen and, moreover, ammonia cleans the surface of particles. These results are substantiated by the thermodynamic calculations (Fig. 1).

According to the X-ray analysis data, boron forms first. It is identified in the nitration products at a temperature of 1200 °C. The formation of silicon nitride begins at temperatures higher by 100-200 °C, which agrees with a significant difference between these compounds in the heat of formation (Table 1). In the whole investigated temperature range, in the reaction products, silicon is present; its amount increases near the temperature of peritectic decomposition of B_4Si and then decreases to traces at the maximal one.

The nitration behavior of the mechanically activated powders during in nitrogen and ammonia is somewhat different. The mechanochemical treatment leads not only to the increase in the dispersion, but also to a rise in the energy of the lattice. As a result, B_4Si becomes more chemically active, and all processes occurring during its nitration shift in the direction of lower temperatures and proceed much more intensively. Moreover, after the treatment of the initial B_4Si powder in the planetary mill, the iron impurity exerts a catalytic influence on the nitration processes.

According to the results of the chemical analysis (Table 3), the nitration of the activated powder begins at a temperature of 1000 °C. Lines of boron nitride were identified in the X-ray diffraction patterns of specimens nitrided at 1100 °C. The formation of silicon nitride begins at 1000 °C. A rise in the temperature to 1400 °C and an increase of the exposure time at the maximal temperature from 1 to 5 h leads to an abrupt increase in the nitrogen contents in the reaction products.

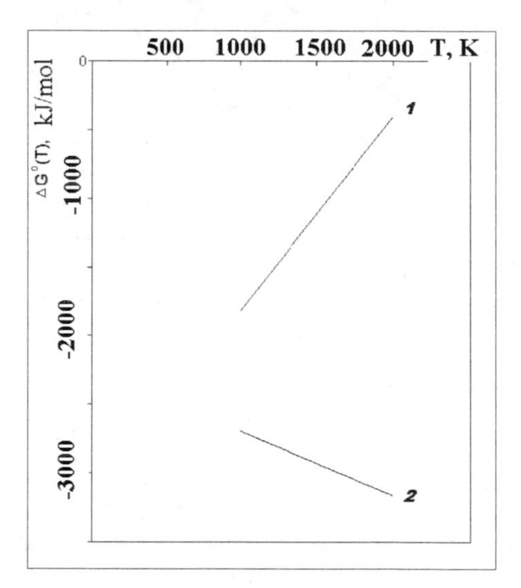

a)

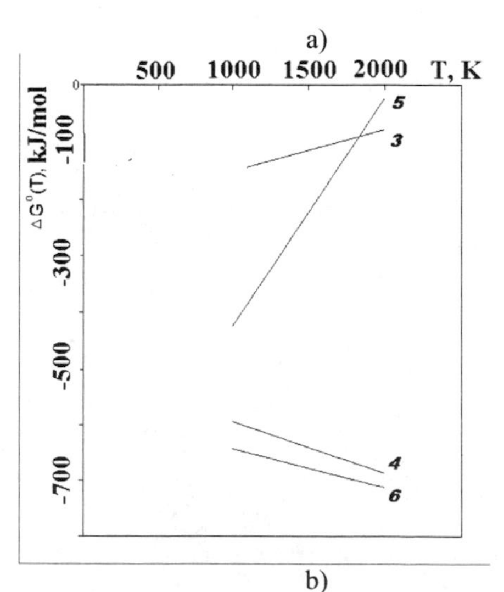

b)

Figure 1. Temperature dependences of Gibbs free energies *a* – reactions *1*, *2*;*b* - reactions *3*, *4*, *5*, 6

Table 3 Data of chemical analysis of nitration products of boron silicide

Boron silicide	Conditions of nitrating			N_2 content mass %	Specific surface area, m^2/g
	Media	T, ° C	Time, h		
Initial powders					
$B_{2,8}Si$	N_2	1200	3	8.22	3.07
		1300	3	19.1	2.63
		1400	3	43.4	7.59
		1450	1	-	6.42
		1450	3	47.4	-
	NH_3	1300	1	19.7	20.31
		1300	3	35.7	16.20
B_4Si	N_2	1300	3	15.7	8.04
		1400	3	33.5	15.13
		1450	1	43.4	15.71
		1450	3	48.7	16.67
		1450	5	49.5	16.92
	NH_3	1300	3	19.5	21.8
		1400	3	48.8	
		1450	1	-	17.06
Mechanically activated powders					
B_4Si	N_2	1000	1	1.77	8.42
		1100	1	5.3	8.81
		1200	1	11.0	10.60
		1300	1	35	20.30
		1400	1	44.4	25.7
		1400	3	45	24.33
		1400	5	44.8	24.85
	NH_3	1300	3	26.9	13.17
		1400	1	42.8	43.56
		1400	3	45.2	31.9

Such an intensive process of formation of boron and silicon nitrides is connected both with the increased activity of the mechanically activated B_4Si [8] and with the catalytic influence of iron [9]. It should be noted that, during the nitration of the mechanically activated boron silicide, silicon nitride first formed in the form of the α-phase, and when the nitration temperature and exposure time were increased, the formation of β-Si_3N_4 was observed; the amount of which rose in comparison with that of α-Si_3N_4. This agrees with data of [8, 10], according to which, in the process of formation of silicon nitride, the α-phase forms in the first minutes of nitration, and then, in the course of the isothermal exposure, annealing of the powders, accompanied by $\alpha \rightarrow \beta$ transition, occurs. In the nitration products of the mechanically activated B_4Si powder, besides the basic phases of BN and Si_3N_4, iron boride and iron silicide phases are present in small amounts. As the nitration temperature is raised, the formed iron borides and silicides interact with nitrogen and partially form BN и Si_3N_4 [11]. As this take place, iron silicide phases transform from higher to lower phases.

The nitrogen contents in the powders obtained by nitration of mechanically activated boron silicide in nitrogen and ammonia differ insignificantly. It is evident that, against the background of the catalytic effect of iron and the increased reactivity of the activated boron silicide, the influence of ammonia does not manifest itself.

To assess the degrees of dispersion of the formed composite powders, the measurement of their specific surface area was carried out. An essential increase in $S_{уд}$, which corresponds to a rise in the dispersion of powders (Fig. 2), occurs after the beginning of their intensive nitration. In this case, the preliminary activation of the initial B_4Si powder exerts a particular influence on the dispersion process, which agrees well with data of [10]. The authors of [10] showed that, in the nitration of silicon powder with a highly defective lattice, the failure of coherence on the Si—Si_3N_4 interface occurs in the certain stage of nitration. This results in the fracture of the formed silicon nitride layer, which facilitates the access of nitrogen to the new-formed surface of silicon. According to electron microscopy data, the consequence of this is the much smaller particle of the formed silicon nitride against that of the initial silicon powder. It is assumed that an analogous situation takes place in the formation of boron nitride obtained from mechanically activated B_4Si.

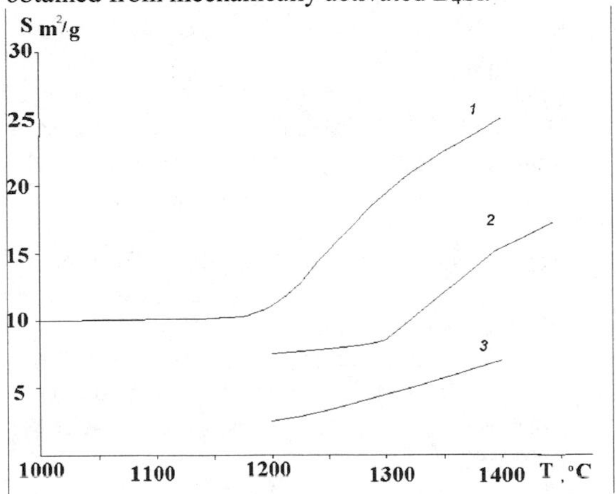

Figure 2 Specific surface area of boron silicide against temperature under nitrating in nitrogen:
1 – B_4Si (after mechanical activation); *2* – B_4Si;
3 – $B_{2,8}Si$

An analysis of measured specific surface area data of the obtained powders (Table 3) indicate that $S_{уд.}$ of the powders nitrided in ammonia are higher that those of powders nitrided in nitrogen. According to [9], this can be connected with the participation of reactions in a gaseous phase, that proceed during interaction of silicon and boron with products of incomplete dissociation of ammonia, in the processes of formation of nitride phases.

In the comparison of the specific surface area of all powders, much higher values for powders obtained by nitration of the mechanically activated boron silicide powders in ammonia attract attention. Their specific surface area 32—43 m^2/g, which corresponds to a particle size of 70 –50 nm of powders of the 60 BN—40 Si_3N_4 composite material.

Conclusions

1. It is shown that the nitration products of boron silicide powders in a nitrogen and ammonia atmospheres are composite powders of boron and silicon nitride. In this case, the temperatures of formation of BN и Si_3N_4 are substantially lower than those in the case of nitridation of elementary boron and silicon, which makes it possible to obtain nanosized powders of these nitrides.

2. In ammonia, the nitride formation process is more intensive than in nitrogen and allows to obtain thin powders with a specific surface area above 20 m^2/g.

3.It was established that the preliminary mechanical activation of boron silicide powder increased significantly it nitration rate, lowered the temperatures of formation of BN и Si_3N_4, and, as a consequence, made it possible to obtain this composition in the nanocrystalline state.

1. Zutz E. H., Swain M. V. Fracture toughness and thermal shock behavior of silicon nitride-boron nitride ceramics // J. Am. Ceram. Soc. — 1992.—**75**, No. 1.— P. 67-70.
2. Mazdiyasni K. S., Ruh R. High/low modulus Si_3N_4-BN composites for improved electrical and thermal shock behavior // J. Am. Ceram.Soc. — 1981. — **64**, No. 4. — P. 415—418.
3. Makarenko G.N. Materials of boron- silicon system // Refractory compounds. Production, structure and use. — Kyiv.: Naukova dumka, 1991. — C. 107—111.
4. Properties, production and use of refractory compounds. Hand-book. ed. Kosolapova T.Y. — M.: Metallurgy, 1986.
5. Gordienko S.P. Thermodynamic properties and high temperature behaviour of boron silicide // Borides and materials on their base. — Kyiv: IPM, 1994. — C. 52—57.
6. Thermodynamic properties of individual substances. Hand-book — M.: Science . — 1978.— 1, кн. 2. —326 с; 1979. — П, book.2. —339 с; 1981. — Ш, book 2. — 395 с.
7. Bolgar A.S., Blinder A.V. et al., Thermodynamic properties of boron silicide // Silicide and their use in technique. — Kyiv: IPM, 1990. — C. 78—87.
8. Bartnitskaya T.S., Vlasova M.V., Galchinskaya Yu. P. et al. Properties of silicon powders, prepared by milling in liquid nitrogen // Inorganical materials — 1993.— **29**, № 7. — C. 894—897.
9. Kosolapova T.Y., Andreeva T. V., Bartnitskaya T.S. et al. Non-metallic refractory compounds M.: Metallurgy, 1985.
10. Bartnitskaya T.S., Zelyavsky V.B., Kurdumov A.V. et al. Influence of real structure of silicon powders on the processes of nitrating // Powder metallurgy. — 1992.-№ 10. — C. 1—7.
11. Milinskaya I.N., Tomilin I.A Interaction of nitrogen with iron-silicon alloys // Izvestia of AN USSA Ser. physics — 1970.— XXX1V, № 2.— C. 255—261.

ACKNOWLEDGEMENTS

This work was done under National Project “Nanosystems, nanomaterials and nanotechnology”.

The authors thanks to Bartnitskaya T.S for discussion of the results

Simple and effective way to prepare $CuFe_2O_4/SiO_2$ nanocomposites by Sol-Gel method

I. Prakash*, P. Muralidharan*, N. Nallamuthu*, M. Venkteswarlu**, N. Satyanarayana*

* Department of Physics, Pondicherry University, Pondicherry- 605 014, India
Corresponding author: E-mail: nallanis2000@yahoo.com
** Present address: Department of Chemical Engineering, NTUST, Taipei, Taiwan

ABSTRACT

Nano-composite containing $CuFe_2O_4$ nano size crystals embedded in amorphous silica matrix was prepared by Sol-Gel method. The gel of composition 5%CuO-6%Fe_2O_3-89%SiO_2 was synthesized at ambient temperature and heat treated at different temperatures (303 K - 1073 K). X-ray diffraction (XRD), Fourier transform infrared (FTIR), Transmission electron microscopy (TEM) and Differential scanning calorimetry (DSC) were used to characterize the formation of nanocrystals during heat treatment. $CuFe_2O_4$ nanocrystals were formed from the heat treated gel around 1073 K. The particle size of uniformly distributed CuFe2O4 nanocrystals is estimated from TEM micrograph and is found to be 10 to 20 nm in diameter.

Keywords: Sol-Gel, $CuFe_2O_4/SiO_2$ Nano-composite, DSC, XRD, FTIR and TEM

1 Introduction

Copper ferrites with spinel structure are widely used in magneto-optic recording devices, colour imaging, bio processing, electrical switching, high frequency communication, magnetic refrigeration, etc. devices [1-3]. Recently, due to an advancement of technology, modern scientific world invites materials with better control over their properties, which can be achieved by hybridization of different phases. Composite material is one of those kinds and it combines among crystalline, amorphous and polymer phases to enrich and enhance the properties [4]. Nanomaterials show drastic change in physical and chemical properties over their respective bulk materials due to their quantum confinement [5-8]. Also, the control over the size and the polydispersity of the nanoparticles is important, since these can change the properties of nanomaterials drastically. Strong supporting matrices like glasses, ceramics and polymers are used to stabilize the nanocrystals, where, nanoparticles are embedded in the supporting matrix to form the nanocomposite structure. Thus, more efforts have been made to develop synthesis methods to prepare nanocrystalline and nanocomposite materials in various forms [9-13]. Sol-gel route has been used effectively to prepare nanocomposites, where, precursors can be mixed in solution form. Therefore, the reaction will be in ionic level and at low temperatures, which can provide good control over the Stoichiometric, homogeneity, structure, purity, etc. of the materials [14]. Hence, the present paper deals with the synthesis, by sol - gel route, and characterization, by XRD, FTIR, DSC and TEM, of uniformly distributed $CuFe_2O_4$ nanocrystals of 10 to 20 nm size in SiO_2 glassy matrix, forms nano-composite material.

2 Experimental

2.1 Sol – Gel synthesis

Xerogels of Copperferrosilicate were prepared by sol – gel process using precursors of analar grade tetraethylorthosilicate TEOS ($Si(OC_2H_5)_4$ (Acros organics), Ferric nitrate (s.d. fine-chem), Cupric nitrate (s.d. fine-chem) and ethanol. All precursors were mixed according to the composition 5% CuO – 6% Fe_2O_3 – 89% SiO_2, based on their molecular weight percentage. Fig. 1 shows the flow chart representation of various stages of Sol-Gel process involved in synthesis of $CuFe_2O_4/SiO_2$ nanocomposite. In fig.1, solution A contains TEOS, ethanol and water. The TEOS and

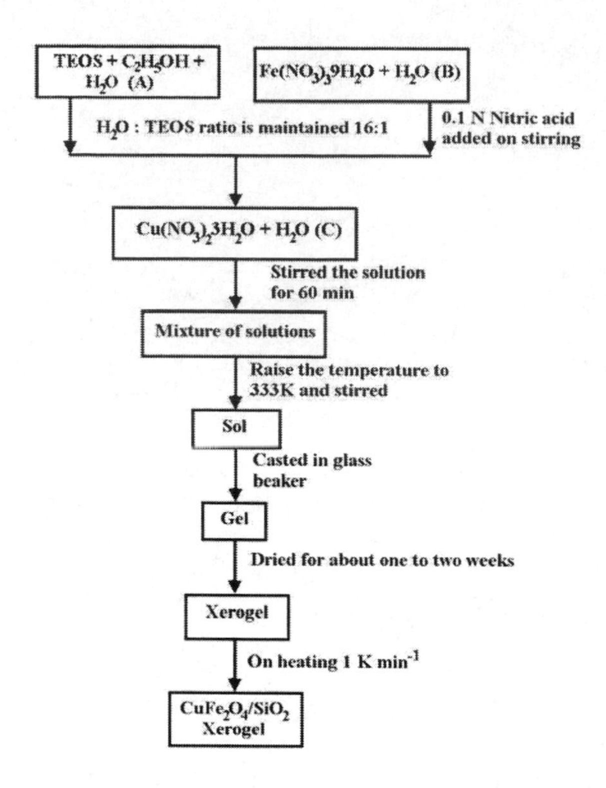

Fig.1. Flow chart for $CuFe_2O_4$ /SiO_2 nanocomposite sample preparation by sol-gel process

water ratio was maintained at 1:16 and 2.5 N nitric acid was added as catalyst. An equal volume ratio of ethanol as solvent was mixed with TEOS and water, which was stirred for an hour to get clear transparent solution. Solution B contains required amount of ferric nitrate dissolved in distilled water. Solution B was mixed with solution A and stirred for an hour. Solution C contains required amount of cupric nitrate dissolved in distilled water and was added to A and B solution stirred for two hours at 333 K. The final solution was poured in to a glass beaker covered with aluminum foil and allowed to form gel at ambient temperature. After gelation, holes were punched on the aluminium foil to allow ethanol and water evaporation, formed during condensation. The dried gel was heat treated at various temperatures at the rate of 1 K per minute and it is characterized by XRD, FTIR DSC and TEM to observe the formation of nanocrystals in amorphous SiO_2 matrix.

2.2: XRD, FTIR, DSC and TEM measurements

XRD patterns were recorded for the fine powdered dried gels, using Rigaku miniflex diffractometer with Cu K_α and λ=1.4158 A^o, as source wavelength. Dried gel sample and KBr powder were mixed in 1:20 ratio and grounded into fine powder. Using KBr press, thin transparent pellet samples, prepared at room temperature and higher temperatures, were made and were dried to reduce the moisture content. FTIR spectra were recorded on pellet samples using Schimadzu FTIR/8300/8700 spectrophotometer in the frequency range of 4000 – 400 cm^{-1} with 2 cm^{-1} resolution for 20 scans. The fine powdered dried gel sample of 3 mg was placed in the aluminum pan with lid and pressed to form a micro pellet. Pellet sample was heated at the rate of 10 K per minute from 303 K to 773 K under nitrogen atmosphere and recorded DSC curves using Mettler Toledo star (e) system module;821e/500/575/414183/528.

3 Results and discussions

Fig.2 shows the XRD patterns of the dried gel powders heat treated at 333 K, 473 K, 673 K and 1073 K. From fig.1, the dried gel powder heated up to 873 K showed broad peak centered at 22° is attributed to the characteristic diffraction of the amorphous phase of SiO_2. Around 1073 K, new crystalline peaks are observed at 36, 30, 43, 57 and 63 over the broad peak.

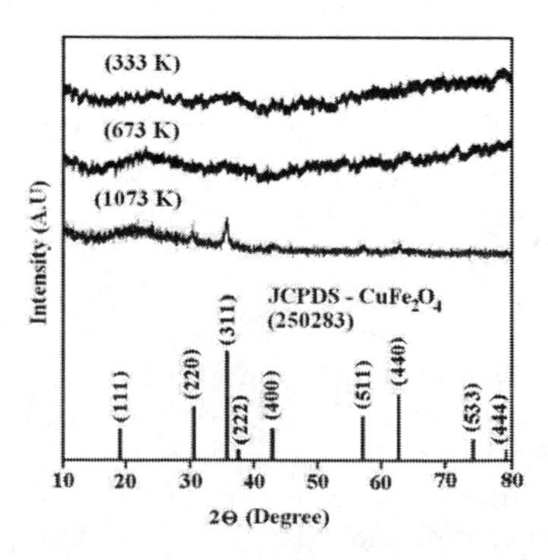

Fig.2. XRD spectra for $CuFe_2O_4$ /SiO_2 nanocomposite sample heat treated at various temperatures

The observed 2θ values of the peaks were compared with JCPDS data and confirmed the formation crystalline phase of $CuFe_2O_4$ cubic structure. The amorphous phase of SiO_2 retained till the sample heated up to 1073 K. Thus, the formation of $CuFe_2O_4$ crystals in the SiO_2 amorphous matrix forms a composite phase, having both crystalline and amorphous phases in a single material.

Fig. 3 shows the FTIR spectra of the dried gel powders heat treated at 333 K, 673 K and 1073 K and exhibit bands at 3424, 1634, 1383, 1080, 964,795 and 453 cm $^{-1}$. The IR bands observed at 3424 and 1634 cm^{-1} are respectively attributed to the OH stretching of water molecules and isolated silanol stretching. The band at 453 cm $^{-1}$ corresponds to the deformation mode of Si-O-Si. The peak at 1383 cm $^{-1}$ is due to the presence of nitrate groups in the sample. The FTIR spectra for higher temperature show the decrease of intensity for the bands at 3424 cm $^{-1}$, 1634 cm $^{-1}$ and 1383 cm $^{-1}$, which are due to the removal of water molecules and nitrate groups from the sample [15]. At 1073 K, all the above observed peaks were removed and two new peaks were observed, which indicate that all the impurities like nitrate and other organic groups were completely decomposed from the sample. The observed two new peaks 603 cm^{-1} and 467 cm^{-1} were attributed to the Fe_2O_3 & CuO vibrational modes, indicates the formation of $CuFe_2O_4$ structure.

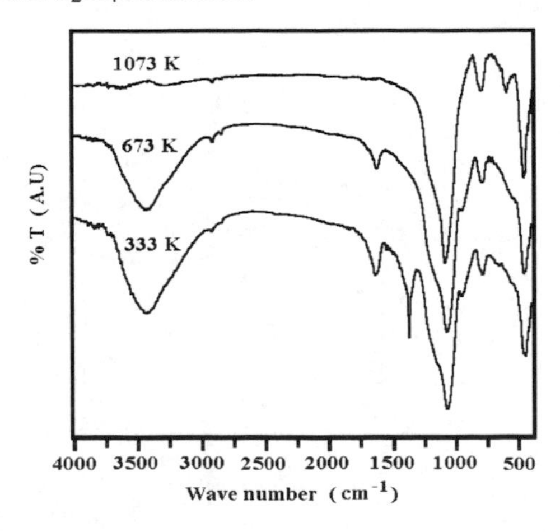

Fig.3. FTIR spectra for $CuFe_2O_4$ /SiO_2 nanocomposite sample heat treated at various temperatures

Hence, the XRD and FTIR results confirm the formation of $CuFe_2O_4$ crystals in the SiO_2 amorphous matrix and forms a $CuFe_2O_4$ /SiO_2 composite phase.

The observed wide endothermic peak, shown in fig.4, between 313 K and 373 K for the dried gel is attributed to the evaporation of ethyl alcohol, water molecules and other organic residues existing in the sample and it is also confirmed by FTIR. There is no exothermic crystallization peak was observed with in the measured temperature range.

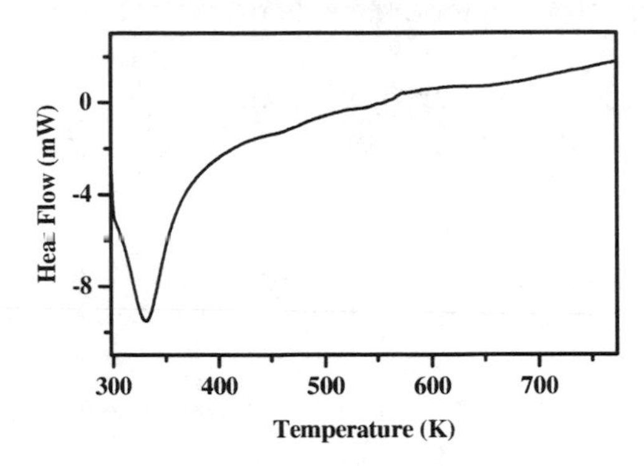

Fig. 4. DSC thermo gram of $CuFe_2O_4$/SiO_2 nanocomposite

3.1 Transmission Electron Microscopy Results

Fig.5 a & b show the TEM images of the $CuFe_2O_4$/SiO_2 composite taken at various spots. From Fig.5, the TEM images show the dark particles, $CuFe_2O_4$ crystals, dispersed uniformly in the white background amorphous SiO_2 phase. From the TEM images, the size of the $CuFe_2O_4$ crystallites is found to be of order of 10 to 20 nm. Hence, the prepared $CuFe_2O_4$ /SiO_2 compound by sol -gel route is found to be nanocomposite phase.

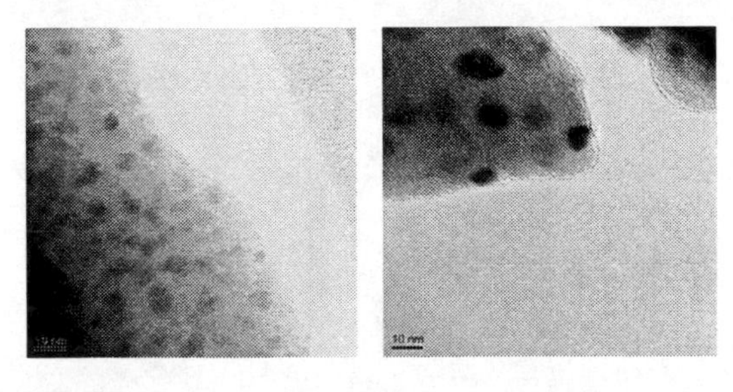

Fig. 5 a&b show TEM images of the $CuFe_2O_4$/SiO_2 composite taken at various spots

4. Conclusion

Nano-composite containing nano size $CuFe_2O_4$ crystals dispersed in SiO_2 glass matrix was prepared by Sol-Gel technique. The prepared dried gel heat treated up to 873 K showed only the amorphous phase of SiO_2. XRD and FTIR results of dried gel heat treated up to 1073 K confirm the formation of $CuFe_2O_4$ crystals in the SiO_2 amorphous matrix and forms a $CuFe_2O_4$ /SiO_2 composite phase. From the TEM images, the size of the $CuFe_2O_4$ crystallites is found to be of order of 10 to 20 nm. Hence, the prepared $CuFe_2O_4$ /SiO_2 compound by sol -gel route is found to be nanocomposite phase.

Acknowledgements
Authors are gratefully acknowledged DRDO, DST and CSIR for utilizing the research facilities available from the major research projects.

References

[1] J.L. Dorman (Ed), Magnetic Properties of Fine Particles, Fiorani, North Holland, Amsterdam, 1994.
[2] F. Bodker, S. Morup, S. Linteroth, Phys. Rev. Lett. 72 (1994) 282.
[3] S. Chikazumi, Physics of Ferromagnetism, Clarendon Press, Oxford, 1997, p. 201.
[4] Xiluan Duan, Duorong Yuan, Zhihong Sun, Haiqing Sun, Dong Xu, Mengkai Lv, J. Crystal Growth 252 (2003) 4.
[5] H.M. Gibbs, G. Khitrova, in; H.M. Gibbs, G. Khitrova, N. Peyghambarian (Eds.), Nonlinear Photonics, Springer, Berlin, 1990 (Chapter 1).
[6] A.I. Ekimov, A.L. Efros, A.A. Orushchenko, Solid State Commun. 56 (1985) 921.
[7] R.N. Bhargava, D. Gallenger, X. Hong, A. Nurmikko, Phys. Rev. Lett. 72 (1994) 416.
[8] A.P. Alivisatos, J. Phys. Chem. 100 (1996) 13226
[9] F.H. Dulin, D.E. Rase, J. Am. Ceram. Soc. 44 (1961) 493.
[10] Y.S. Chang, Y.H. Chang, I.G. Chen, G.J. Chen, Y.L. Chai, J. Crystal Growth 243 2002)319.
[11] N.N. Parvathy, G.M. Pajonk, A.V. Rao, J. Crystal Growth 179 (1997) 249.
[12] M. Moller, J.P. Spatz, Curr. Colloid. Interface Sci. 2 (1997) 177.
[13] N.F. Borrelli, D.W. Smith, J. Non –Cryst. Solids 180 (1994) 25.
[14] C.J. Brinker, D.E. Clark and D.R. Ulrich, Edited Better Ceramics through Chemistry,
[15] George Socrates, Infrared and Raman Characteristic Group Frequencies Tables and Charts, Third Edition, John Wiley & Sons, 2001.

Development of a Biologically Enhanced Carbon Nanotube Based Composites

R. Parker*, **R. O'Neal****, D. Fisher**, Kirk Davey*, R. Liang* and J. Edwards***

* Department of Industrial and Manufacturing Engineering, FAMU-FSU College of Engineering
Room 317B, 2525 Pottsdamer Street, Tallahassee, FL, rparker@eng.fsu.edu
**Department of Physics, FAMU College of Arts and Science,
205 Jones Hall, Tallahassee, FL, ray.oneal@famu.edu
***Department of Chemistry, FAMU College of Arts and Science,
219 Jones Hall, Tallahassee, FL, jesse.edwards@famu.edu

ABSTRACT

Composite materials continue to be important to the evolution of human space flight activity. The low weight/high strength ratio and thermal properties of carbon based composite materials in particular that make them well suited for use in extraterrestrial environments is widely known and understood. Much of the recent activity in carbon composite investigations relevant to space flight has focused on the potential of materials incorporating carbon nanotubes (CNTs) for enhancing mechanical, electronic and radiation attenuating properties of aerospace materials. We describe a CNT based, biologically enhanced composite that provides simultaneous functionality across all three of the aforementioned properties. Initial studies show that a mixture of biological polymers and CNTs will increase shielding radiation capability and provide a host of other functional properties. Additionally, it has been shown that the biological polymer can be modified and tuned to respond to different types of radiation.

Keywords: organic polymers, biological polymers, carbon nanotubes, multifunctional materials, composite materials

1 INTRODUCTION

High performance composites are currently being used in the marine, automotive, aerospace and defense industries. These industries demand materials with properties that are similar or better than conventional metals at a fraction of the weight. Advanced composite materials can meet these demands. Advanced composite materials are carbon or aramid fiber-based composites, metal matrix composites (MMC), ceramic matrix composites (CMC) and carbon-carbon (C-C) composites, which can be used to replace metal materials for structural applications [1]. However, there are still some weakness with advanced composites, such as the transverse properties, low toughness and high environmental sensitivity, which are mainly determined by the properties of the resin matrix. This proposed research will focus on using advanced composites, mainly biological polymers, along with nanoparticles to produce multifunctional composites, with enhanced material properties, for the development of value added material applications.

The development of biologically reinforced nano-composites is presently one of the most promising areas in materials science and engineering. The exceptional properties of nanoparticles have made them a focus of widespread research. Carbon nanotubes and other nanoparticles have the potential to greatly enhance the properties of composites when combined with biological polymers which were previously unobtainable properties can be developed [2]. Moreover, the biological polymer can be tuned through chemical modification. Using nanotubes in biologically enhanced composites provides the potential for improving resin-dominated properties, such as interlaminate strength, toughness, thermal and environmental durability. As an example, biological polymers are capable of convert radiation into electrical energy which can be later used via trickle charging. Consequently, whereas most resins would degrade via radiation over time, the biological substance would mediate such phenomena.

By combining nanoparticles with biological polymers for reinforcement, enhanced functionality composites can be produced, with superior properties to that of regular composites. In order to successfully produce multifunctional composites several problems must first be understood. (1) Dispersion of nanoparticles is a major concern due to the challenges that exist with the formation of agglomerates during the mixing of nanoparticles with the resin matrix. (2) With the addition of nanoparticles to the resin matrix there is a dramatic increase in the viscosity of the resin. This poses a problem with the processing of nano-composites. (3) The effects of nanoparticles on the properties of the resin matrix system need to be fully evaluated and understood.

2 PREPARATION OF SAMPLES

Two samples were used to evaluate the performance and properties. One is a modified form of buckypaper. Biological polymer is added to the initial CNT, surfactant and water mixture. The resulting mixture is sonicated and the paper is formed through filtration. The other sample preparation technique was to produce the aforementioned mixture with basic commercial grade gelatin (no sugar, no coloring). The resulting mixture was sonicated and cooled to gel.

A factorial design was established which included 1) magnetic alignment (0 1, 3, & 5 Tesla in strength and 2) concentration of the biopolymer (none, low, high).

3 OBSERVATIONS & DISCUSSION

The CNT based gel had a fairly homogenous distribution of CNT which was accomplished primarily through locking the CNT into place after sonication via the rapid set of gelation. That is, the onset of gelation was quickened as a consequence of the presence of the CNT. The CNT based gel coagulated within 90 seconds whereas the non-CNT system took several hours to coagulate in a 9°C refrigerator. The system's gel time with and without CNT did not change from with the addition of the biological polymer. That is, with CNT it took less than 90 seconds to gel and hours otherwise. Another important noteworthy phenomena was the resulting tenacity of the system upon gelation. Figure 1 shows a beaker of the CNT based gel which was held upside down for over 1.5 hours.

Figure 1: CNT based gel maintains its temperature and surface tension after 1.5 hours upside down

In addition, the in-situ CNT based gel system has attractive mechanical properties, magnetic alignment and biological enhancement seem to have increased the CNT based system. Table 1 shows that magnetism increases, the electrical resistivity of a system.

In Table 1, the y axis is computed by taking the ratio of the resistance perpendicular to the direction of the applied magnetic field to the resistance parallel to the direction of the applied magnetic field.

The analysis performed in Figure 2 was for a suspension using 40mg of CNT for a 1 liter solution. This solution was modified by reducing the volume of CNT by 80% and replacing it with biological polymer under 3 Tesla. The resulting electrical resistivity anisotropy ratio was calculated to be slight greater than 4. It is believed that the biological polymer allows for greater communication between CNT molecules thus showing lower resistivity in the aligned direction.

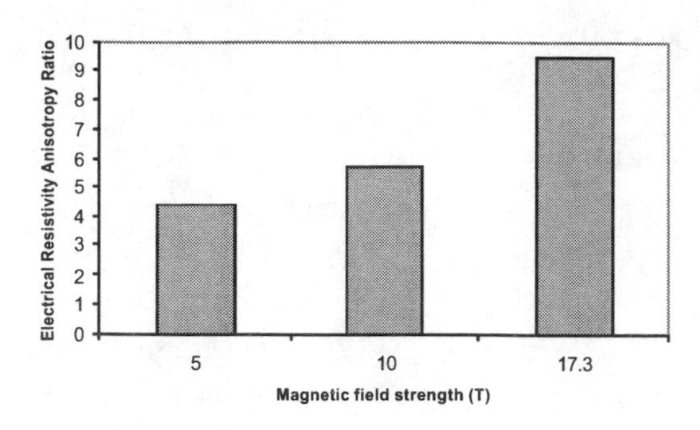

Figure 2: Impact of Magnetic field strength.

From the electrical resistivity and from Fig 3, it is clear that the molecular architecture of magnetically aligned CNTs is demonstrably different than randomly CNT in a buckypaper composite.

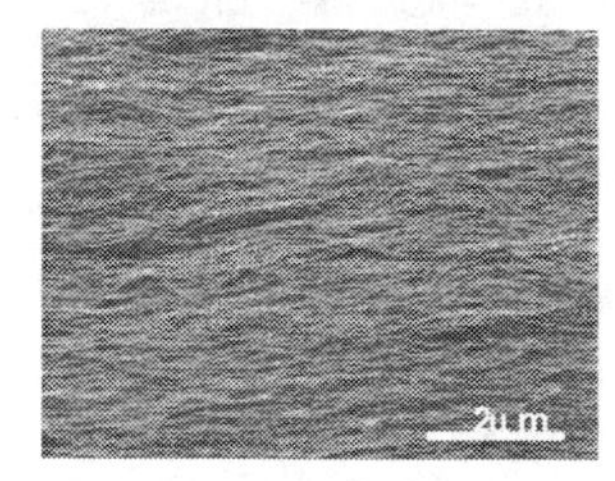

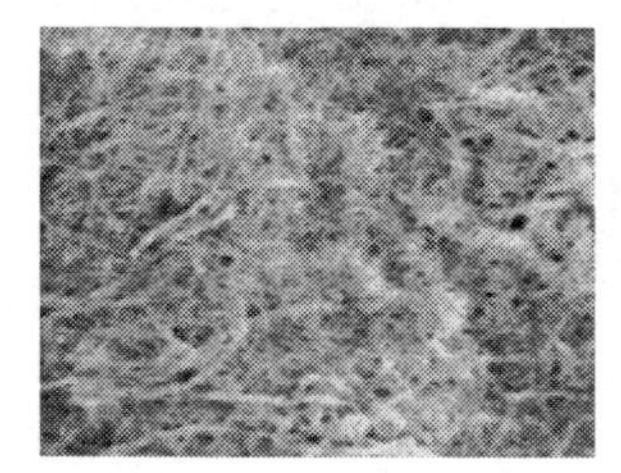

Figure 3: Visual Affect of 5 Tesla on buckypaper (left side is of magnetically aligned system where the right side is not aligned)

The biological polymer chosen is currently FAMU proprietary; however it's structure is chemical similar to Figure 4. The polymer has a serious of conjugated bonds which allow for electron migration, thus conductivity. The polymer is a semiconductor. Once doped, the biological polymer conductivity was measured at $10^{-2} - 10^{-3}$ Sm/ cm.

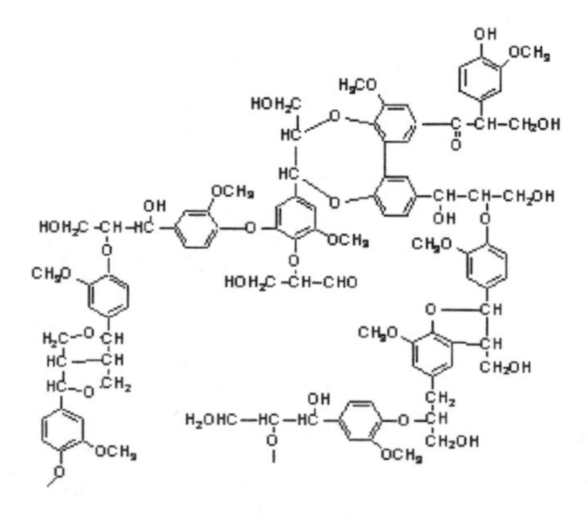

Figure 4: Proposed Conductive Biological Polymer

Other properties were measured from this biological polymer. These are that a 2 cm^2 panel of the system was capable of an observable photoconductivity (~0.656 V).

The sizeable photoconductivity observed is hypothesized to born from its color as seen in Figure 5.

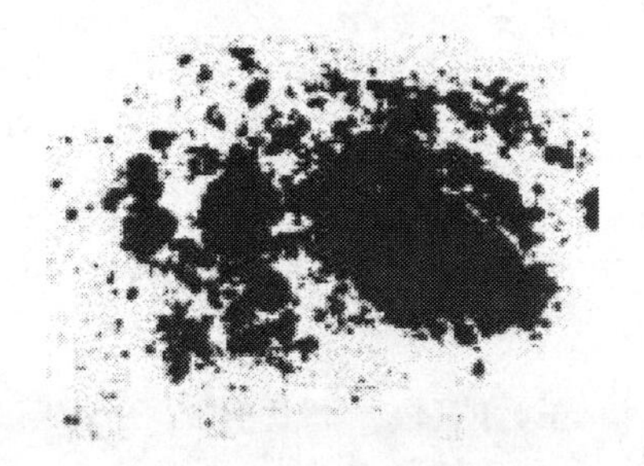

Figure 5: Brownish-Black colored biopolymer in powder form

The system seemed to persist at the same voltage which demonstrated that the system had capacitance. Specifically, the system was able to maintain a voltage of > 0.6 V over 30-40 minutes. Figures 6 & 7 show the semi-conducting nature of the systems which is thought to have aided the resistivity of the aligned CNT/biological polymer system.

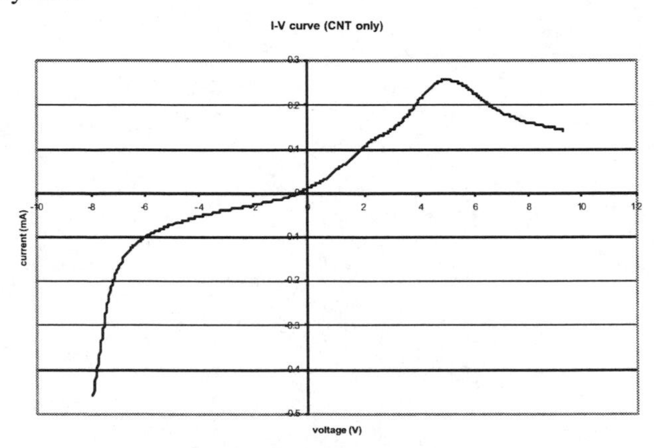

Figure 6: I-V curve of CNT only based gel

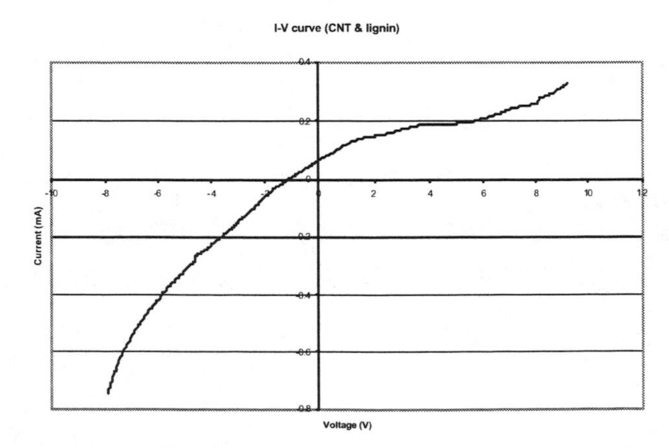

Figure 7: I-V curve of CNT with lignin based gel

A 10 micron sample was able to block 20% of beta radiation. Consequently, a radiological study was performed which calculated the radiation stopping power for the proposed materials. This calculation estimated the energy transferred to the material which may result in ionization of atoms, displacement of atoms and/or generation of secondary radiation (pair – production, etc...). Equation 1 is the Bethe-Bloch equation that allows us to determine the radiation energy transfer per unit length of an element. Equation 2, the Bragg Additivity formula, allows us to calculate the radiation energy deposition per unit length of the material. This is done by using the stoichiometry of the molecule being modeled. Polyethylene (or polyethene) has been defined as a benchmark for radiation blocking due to its high concentration of protons. For polyethene, the relative stoichiometry is C (1) H (2). That is on average, there are two hydrogen atoms for every carbon atom. From these equations, figures 8-10 were created. Figure 8 shows the stopping power of both the targeted biopolymer and CNT in the presence of alpha particle radiation. Figure 9 is for beta particle radiation, and figure 10 is for proton radiation.

$$\frac{dE}{dx} = Kz^2 \frac{Z}{A} \frac{1}{\beta^2} \left[\frac{1}{2} \ln \frac{2m_e c^2 \beta^2 \gamma^2 T_{max}}{I^2} - \beta^2 - \frac{\delta}{2} \right]$$ Equation 1

$$\frac{dE}{dx} = \sum w_j \left. \frac{dE}{dx} \right|_j$$ Equation 2

From this level of analysis, it seems as though the biopolymer composite has radiation stopping power comparable to that of polyethylene. Additionally, the biopolymer differences in the atomic composition provide some differences. One of these might be the level of saturation in the bonds and benzyl structures that exist in the proposed biological polymers. These structures might allow for increased stability, especially if the molecule is capable of distributing some of the radiation energy when bonded to CNTs.

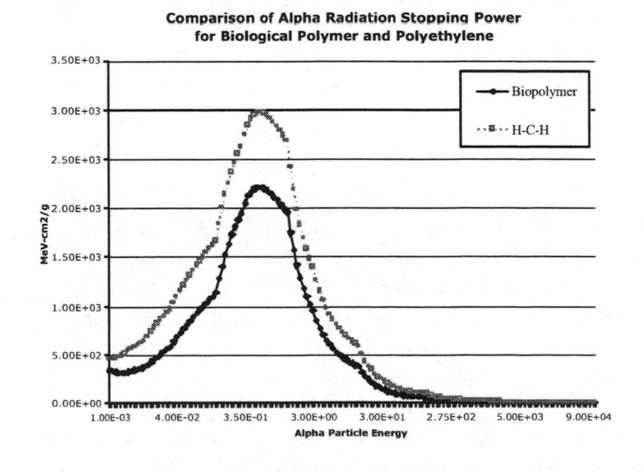

Figure 8: Comparative study of Alpha Radiation Stopping Power of the Proposed Biological Polymer and Polyethylene

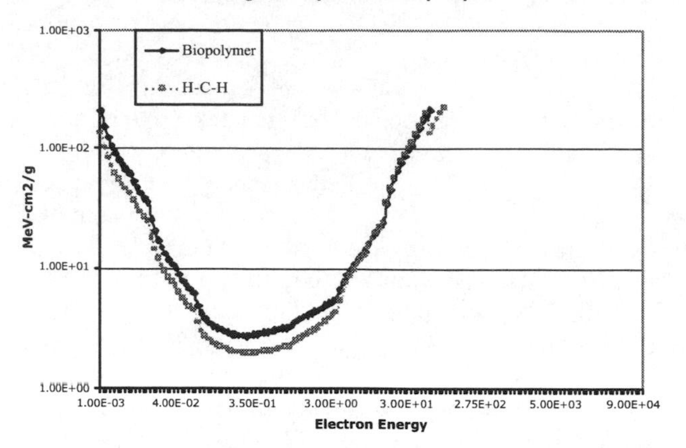

Figure 9: Comparative study of Beta Radiation Stopping Power of the Proposed Biological Polymer and Polyethylene

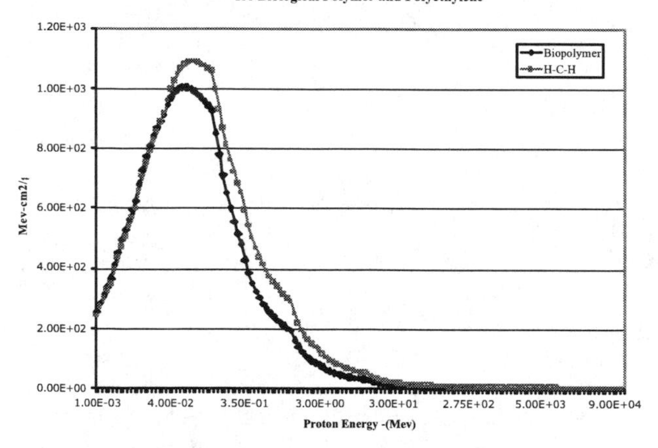

Figure 10: Comparative study of Proton Radiation Stopping Power of the Proposed Biological Polymer and Polyethylene

4 CONCLUSIONS

Biological polymer can greatly enhance the mechanical, electrical, optical and radiological properties of CNT based nano-composites. In the present study, it is shown that smaller concentrations of CNT can be used in the presence of biological polymers while simultaneously achieving similar properties using less energy and costs. Additionally, it is known that CNTs have an enormous unlocked mechanical strength. From these initial studies, the biological polymers seem to have the ability to help achieve some of these through polymer-tube entanglement. Lastly, the radiological properties for this system seem to be most interesting and should have great merit for interstellar travel.

5 FUTURE STEPS

Much work is required to fully develop, fabricate and characterize these biologically enhanced CNT nanocomposites. A technology platform to develop will be required. From these platforms, devices and applications can be built and tested. The promise of new electronic materials, energy systems and radiological materials can be achieved with further study.

REFERENCES

1. Liang, R. (2004). Class notes EMA 5182, Composite Material Engineering, FAMU-FSU College of Engineering
2. Gojny, F., Wichmann M., Kopke, U., Fiedler, B., & Schulte, K. (2004). Carbon Nanotube-Reinforced Epoxy Composites: Enhanced Stiffness and Fracture Toughness at Low Nanotube Content. *Composites Science and Technology*, 64, 2363-2371.

Structure and Rheology of Supercritical Carbon Dioxide Exfoliated Polymer Nanocomposites

G.K. Serhatkulu, S.E. Horsch, R.Kannan and E.Gulari

Wayne State University, Detroit, MI, USA, egulari@eng.wayne.edu

ABSTRACT

A major technological barrier to achieving superior properties of nano-structured composite materials is the difficulty of dispersing fillers uniformly in a host matrix. The existing nanocomposite manufacturing processes achieve property improvements through chemical modifications of filler structures and utilization of chemical interactions between the chemically modified fillers and matrix polymers. We have developed a novel and efficient supercritical fluid processing method for extensively exfoliating and coating nano-scale fillers for use in polymeric nanocomposites. The filler material, consisting of aggregated or layered particles, is contacted with a supercritical fluid containing a soluble organic material for a certain time, the soak step, followed by a catastrophic depressurization step. During the soak step, under the selected processing conditions, the mixture of the supercritical fluid and the organic material diffuses between the layers or bundles. Depending on the selected conditions and the solubility of the organic material in the supercritical fluid, the mixture can be a gas-expanded liquid phase or a dense-gas solution. The high diffusivity and low viscosity of the mixture enable efficient kinetics of this step. During depressurization, expansion of the gas between the layers or bundles pushes them apart causing extensive exfoliation while the organic material remains between the layers and bundles, coating the surfaces of the layers or bundles, thus preventing the reformation of the layered structure. We investigated the structure and rheology of several nanocomposites prepared by the supercritical carbon dioxide exfoliation and coating process: Thermoplastic polyurethane (TPU)/graphite nanocomposite Polydimethlsiloxane (PDMS)/natural clay nanocomposite Polyvinylmethylether (PVME)/I30P (a chemically modified clay). The morphology of the samples was studied by X-ray diffraction (XRD) and transmission electron microscopy (TEM) and the steady-shear and oscillatory-shear measurements were performed on the Rheometrics ARES and the Rheometrics RSA II rheometers, respectively.

Keywords: supercritical fluids, exfoliation, coating, clays, graphite, PDMS, PVME, TPU

1 INTRODUCTION

The field of polymer/silicate nanocomposites has recently gained substantial momentum. Polymer/clay nanocomposites with attractive property profile90s such as improved stiffness, barrier resistance, thermal stability, and flame retardancy have been developed with minimal increase in density due to the low inorganic loading requirements [1,2,3,4,5]. The dramatic enhancements in properties can be attributed to a decrease in particle size (particles ~ 100-1000 nm in diameter and 1 nm thick) and an increase in aspect ratio. If the layered silicate can be efficiently dispersed throughout the polymer matrix, a small fraction can easily generate significant interfacial area (~760m^2/g). Two recent research efforts have made significant contributions to the revival of these materials: First, the report of a Nylon-6 montmorillonite material from Toyota research where very moderate inorganic loadings resulted in concurrent and remarkable enhancements of thermal and mechanical properties [6,7]. And second, Giannelis et. al. found that polymers and clays could be melt-mixed, without organic solvents, resulting in intercalated nanocomposites [8].

Layered silicates are a family of smectic clays with a 2:1 layered structure and typical aspect ratios ranging from 100 to 1000, which is an order of magnitude greater than conventional fillers such as mica and talc. Polymer/silicate hybrids are typically classified into three categories according to the degree of silicate dispersion: immiscible, intercalated, and exfoliated [9]. *Immiscible* corresponds to a composite system where the layered fillers exist in their original state of aggregated layers with no penetration of the polymer matrix into the inter-gallery spacing (interaction is on the micro-scale level like traditional composites). Systems where the extended polymer chains penetrate between host layers, resulting in increased repeat stacking distances of a few nanometers, are *intercalated.* And nanocomposites where the individual filler layers are dispersed throughout a continuous polymer matrix are referred to as *exfoliated* [10]. Typically, a nanocomposite is considered to be exfoliated when the polymer increases the initial separation in the clay platelets an additional 80 Å [11]. Further characterizations include more complex morphologies like disordered intercalated phases, house of card structures, partially exfoliated phases, and structures resembling nematic phases [12]. In order to correctly characterize polymer/clay nanocomposites a variety of devices must be used. Of these devices wide-angle X-ray diffraction (WAXD) and transmission electron microscopy (TEM) are the most common. WAXD is best suited for determining the d-spacing in intercalated and immiscible nanocomposite systems.[13] And TEM is useful for providing a qualitative spatial relationship between clay platelets in the polymer matrix.

In addition to WAXD and TEM, rheology can be employed as another tool to characterize polymer/clay nanocomposites. In order to effectively process nanocomposites, the detailed melt rheological behavior of these materials must be understood. This is important not only for gaining a fundamental knowledge of the processability, but also for elucidating the structure–property relationships [13]

The use of supercritical fluid treatment appears to be another promising technique to produce intercalated/exfoliated polymer clay nanocomposites. Supercritical fluids (SCF) have been proposed as environmentally friendly solvents for a range of materials synthesis and processing applications including: polymerization processes, polymer purification and fractionation, coating applications, and powder formation [14]. In particular, supercritical carbon dioxide (scCO2) has emerged as a powerful alternative to traditional solvents due to its many desirable attributes such as low cost, abundance, and low toxicity. In general supercritical fluids offer mass transfer advantages over conventional organic solvents because of their gas-like diffusivity, low viscosity, and surface tension [15]. When CO_2 is raised above its critical point (T_c =31.1 °C; P_c= 73.8 bar; ρ = 0.472 g/mL), its physicochemical properties (i.e., density, dielectric constant) can be continuously tuned between vapor-like and liquid-like limits by varying the system pressure and/or temperature.

We have a novel method for developing new classes of supercritical fluid exfoliated nanocomposites. The supercritical fluid based process (**Figure 1**) involves contacting the aggregated or layered particles (A) with a supercritical fluid containing a soluble organic material for a certain time (B) followed by a catastrophic depressurization step (C). The particles can be aggregated or layered nanoplatelets, nanofibers, or nanotubes. The organic material can be a pre-polymer, polymer, or monomer soluble in the supercritical fluid under the processing conditions. At the elevated pressures, the supercritical fluid with the soluble organics diffuses between the layers or bundles. During depressurization, expansion of the gas (depressurized fluid) between the layers or bundles pushes them apart causing extensive exfoliation while the solubility of the organic material drastically drops causing the organic material to precipitate and deposit on the surfaces of the layers or bundles, thus preventing the reformation of the weak bonds. The supercritical fluid based process creates ready-to-use exfoliated and coated nanoplatelets, nanofibers, or nanotubes. The ready-to-use nanoparticles can further be introduced into any polymer matrix and in any specified quantity by standard methods of polymerization, melt mixing, or extrusion.

2 RESULTS AND DISCUSSIONS

We have conducted preliminary studies on the morphology and rheology of several nanocomposites manufactured by employing the $scCO_2$ exfoliation and coating process. These include: Polydimethlsiloxane (PDMS)/natural clay nanocomposite, Polyvinylmethylether (PVME)/I30P (a chemically modified clay) nanocomposite, Thermoplastic polyurethane (TPU)/ graphite nanocomposite.

The morphology of the samples was studied by X-ray diffraction (XRD) and transmission electron microscopy (TEM). Steady-shear and oscillatory-shear measurements were performed on the TA/Rheometrics ARES and the TA/Rheometrics RSA II rheometers, respectively.

2.1 Polydimethlsiloxane (PDMS)/natural clay nanocomposite (Polymer/clay dispersion in $scCO_2$):

This nanocomposite contains Cloisite Na^+ (from Southern Clay Products), which is a naturally occuring montmorillonite with a cation exchange capacity of 93meq/100g clay, in a PDMS matrix. The clay was first blended with a low molecular weight PDMS (MW=38,900 g/mole, the coating polymer) at a ratio of 50 wt% clay and 50 wt% polymer. The blended clay and coating polymer was introduced into the matrix PDMS (MW=400,000 g/mole) and processed with $scCO_2$ at 14 MPa and 75^0C (d_{CO2} =0.40g/cm^3) for 24 hours. Then the system was depressurized at a rate of 5.4 cm^3/s.

The structural features of Cloisite Na^+ and nanocomposites processed with $scCO_2$ are displayed in **Figure 2**. The XRD scan of the as-received Cloisite Na^+ has a characteristic peak at 2θ =6.8° corresponding to a basal spacing d_{001}=1.29 nm. Pure PDMS shows a broad peak at 2θ=12° (d=0.774 nm) attributed to its amorphous halo (Figure 2A). The XRD scan of the nanocomposite with 5 wt% clay content shows no basal spacing peak of clay, suggesting significant degree of exfoliation and dispersion. The XRD scan of the nanocomposite with 15 wt% clay content has a very low intensity broad peak at 2θ= 6.1° corresponding to d_{001}=1.44 nm. This peak emanates from the clay dispersion and indicates a small degree of intercalated structure. **Figure 3** shows the TEM of the nanocomposite. The single platelets and more complex structure are very clear from these images.

The linear viscoelastic properties of the PDMS/ Cloisite Na^+ nanocomposite with 15 wt% clay showed marked enhancement at all frequencies and loadings. An improvement in storage modulus (G') of about 100% over the matrix polymer was observed over the entire frequency range (**Figure 4**). This improvement is a factor of 2 higher than those observed for nanocomposites prepared by dispersing chemically modified clays in PDMS by solution mixing techniques.

2.2 Polyvinylmethylether(PVME)/I.30P nanocomposite:

This nanocomposite contains a chemically intercalated clay, I.30P, dispersed in a PVME matrix. I.30P has 25-30% of the natural clay cations replaced by trimethyl-octadecylamine onium ions $[CH_3(CH_2)_{17}N^+(CH_3)_3]$ resulting in doubling of the distance between natural clay layers. The clay was blended with PVME (MW=90,000 g/mole) and was processed with $scCO_2$ at 14 MPa and 75^0C for 24 hours followed by depressurization at 5.4 cm^3/s. The motivation for using PVME is its compatibility with polystyrene, an extensively studied commercial polymer.

The rheological characterization and the structure of the PVME/I.30P nanocomposite indicate a very significant improvement in the storage modulus and the presence of intercalated structures in the processed nanocomposite (**Figure 5**). The XRD scan of I.30P has a characteristic peak at $2\theta=3.8^o$ corresponding to d_{001}=2.7 nm. The XRD scan of the $scCO_2$ processed nanocomposite with 15 wt% clay content displays a highly intercalated structure, evident by the shift of the clay's characteristic sharp peak to lower 2θ value corresponding to d_{001}=3.6 nm. TEM images also confirms the intercalated structure (**Figure 6**). *A ten-fold increase* in the storage modulus, G', of the nanocomposite relative to pure PVME is measured over a wide frequency range. Some of the increase of the modulus is attributable to favorable interactions via H-bonding between the onium ions of the clay and the ether functions of PVME. Even in the absence of H-bonding, for $scCO_2$ processed PVME/natural clay nanocomposite, with 15 wt% clay content, greater than 100% improvement in modulus was observed.

2.3 *Thermoplastic polyurethane (TPU)/ Graphite nanocomposite*:

Thermoplastic polyurethanes are semirigid linear block copolymers consisting of soft segments, such as polyethers or polyesters and hard segments. The preparation of the TPU graphite nanocomposite combined the SCFP of graphite with prepolymer followed by the polymerization step. The graphite and the prepolymer, an ether-based polyol, were first processed with $scCO_2$ at 14 MPa and 75^0C for 24 hours and depressurized at 5.4 cm^3/s. The resultant prepolymer coated graphite was reacted with diisocyanates (hard segments) and chain extenders to form the TPU/graphite nanocomposite. The control experiment involved mechanically blending the prepolymer with graphite followed by the polymerization step.

In the XRD scan of the nanocomposite prepared by the conventional method, the characteristic peak of graphite at $2\theta=26.36^o$ (d_{002}=0.34nm) is clearly observed while the corresponding scan of the nanocomposite prepared by the supercritical carbon dioxide processing method shows a very small peak, indicating substantial reduction of the layered structure and better dispersion of graphite (**Figure 7B**). The rheological characterization of pure TPU, TPU/graphite nanocomposite prepared by conventional method and TPU/graphite nanocomposite prepared by the SCFP method were done at 190°C (**Figure 7A**). The storage modulus of the $scCO_2$ processed nanocomposite was improved by a factor of 5 over a wide frequency range when compared to the modulus of the nanocomposite prepared by the conventional method.

FIGURES:

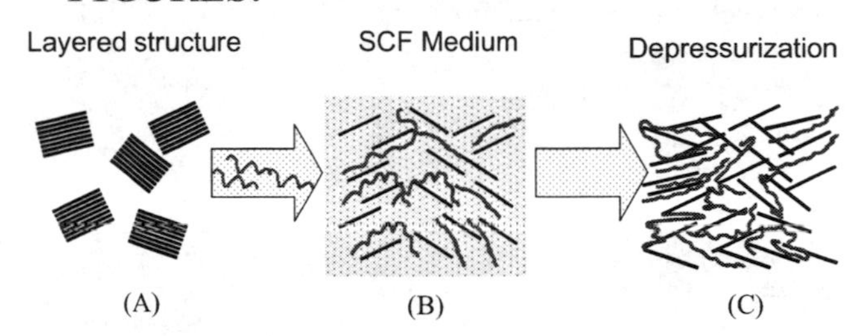

Figure 1: The supercritical fluid based process

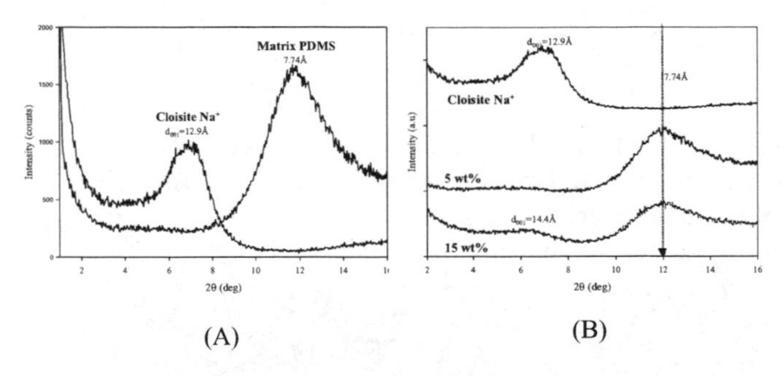

Figure 2: PDMS/Cloisite Na^+ system: A) XRD-scan of Cloisite Na^+ and pure PDMS matrix polymer. B) XRD scans of Cloisite Na^+ (as received), $scCO_2$ processed PDMS/ Cloisite Na^+ nanocomposite (5 wt%) and PDMS/ Cloisite Na^+ nanocomposite (15 wt%).

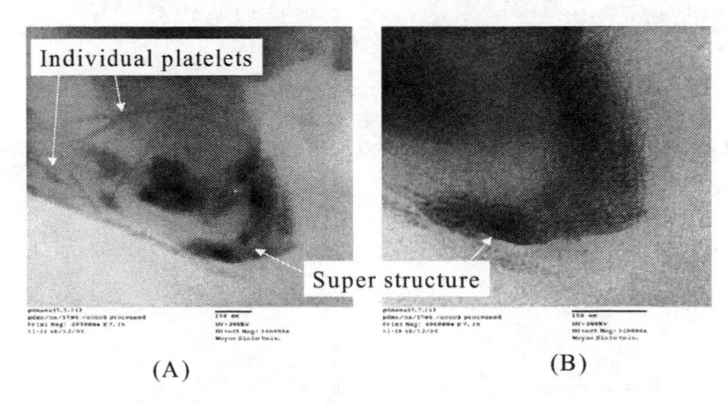

Figure 3: TEM images of PDMS/ Cloisite Na^+ nanocomposite

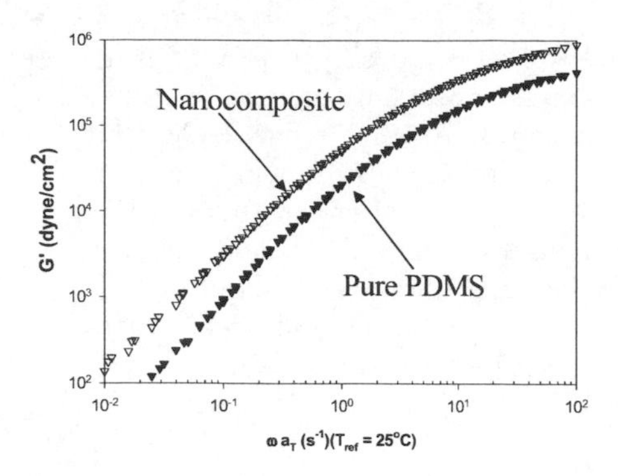

Figure 4: PDMS/Cloisite Na^+ nanocomposite (15 wt% clay): Storage modulus of pure PDMS (dark) and PDMS/Cloisite Na^+ nanocomposite (open) at 25°C. G' of nanocomposite was improved by approximately 100% compared to pure PDMS.

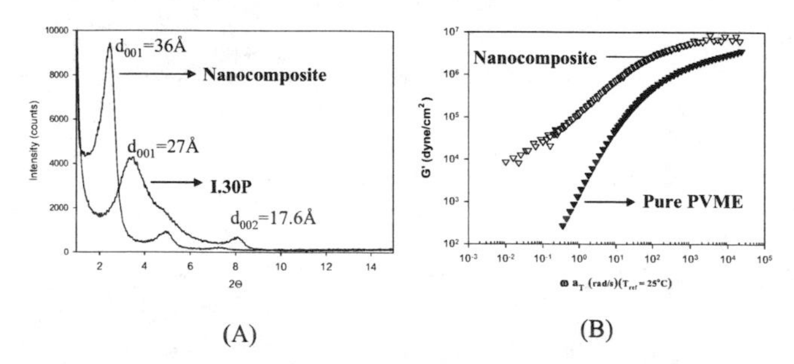

Figure 5: PVME/I.30P nanocomposite (15 wt% clay): A) XRD-scan of I.30P as received and PVME/I.30P nanocomposite. B) Storage modulus of pure PVME (dark) and PVME/I.30P nanocomposite (open) at 25°C.

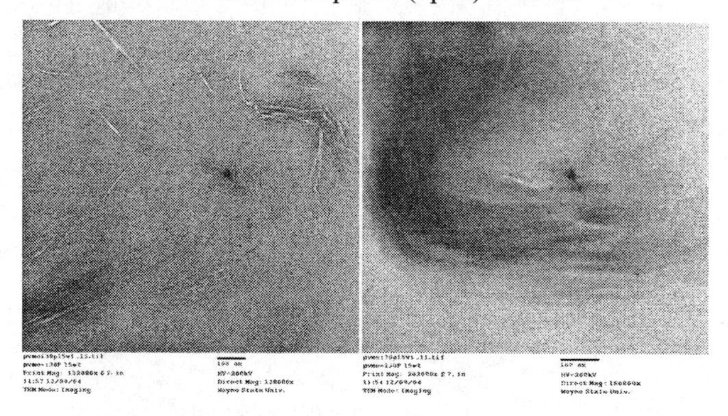

Figure 6: TEM images of PDMS/ Cloisite Na^+ nanocomposite

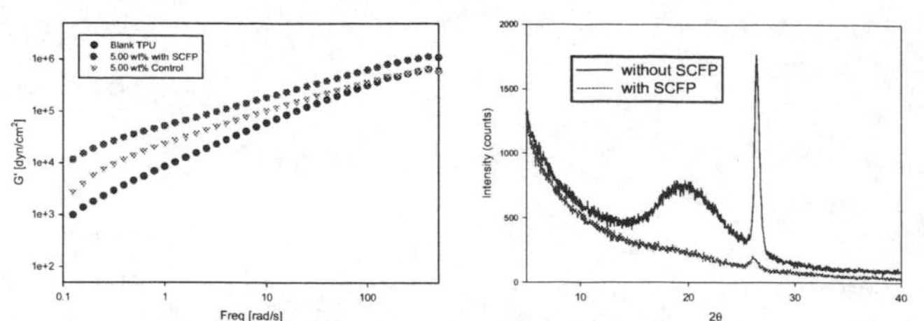

Figure 7: TPU/Graphite Nanocomposite (5 wt% graphite): A) Storage modulus of pure TPU (black), and control TPU/Graphite nanocomposite (green) and supercritical fluid processed (SCFP) TPU/graphite nanocomposite (red) at 190°C. B) XRD-scan of the control (black) and the SCFP (red) nanocomposite. d_{002}=0.34 nm peak of graphite is diminished for the SCFP nanocomposite indicating an exfoliated graphite structure.

REFERENCES

[1] P.L. LeBaron, Z. Wang, T.J. Pinnavaia, Applied Clay Science, 15, 11-29 (1999).

[2] K.A. Carrado, Applied Clay Science, 17, 1-23 (2000).

[3] G. Lagaly, "Smectic Clays as Ionic Macromolecules" in: Developments in Ionic Polymers, A.D. Wilson, H.T. Posser, Eds., Applied Science Publishers, London (1986).

[4] V. Castelletto, I.A. Ansari, I W Hamley, Macromolecules 36, 1694-1700 (2003).

[5] J. Ren, A.S. Silva, R. Krishnamoorti, Macromolecules, 33, 3739-3746 (2000).

[6] Y. Kojima, A. Usuki, M. Kawasumi, A.Okada, Y. Fukushima, T.T. Kurauchi, O.Kamigaito. J. Mater. Res., 8, 1179-1185 (1993).

[7] Y. Kojima, A.Usuki, M.Kawasumi, A. Okada, T.T. Kurauchi, O. Kamigaito, J. Polym. Sci. Part A: Polym. Chem., 31, 983 (1993).

[8] R.A Vaia, H Ishii, E.P Giannelis, Chem. Mater. 5,1694-1696 (1993).

[9] J.T. Yoon, W.H. Jo, M.S. Lee, M.B. Ko . Polymer, 42, 329–336 (2001).

[10] E.P. Giannelis, R.Krishnamorti, E. Manias, Advances in Polymer Science, 138,(1999).

[11] H.R. Dennis, D.L. Hunter, D. Chang, S. Kim, J.L. White, J.W. Cho, D.R. Paul,Polymer, 42, 9513-9522 (2001).

[12] R.A.Vaia, Polymer-Clay Nanocomposites. Edited by T.J.Pinnavaia and G.W.Beall, (2001) Chapter 12, pp 229-266

[13]A.B. Morgan, J.W. Gilman, J. of Appl. Polym. Sci, 87, 1329, 2003

[14] F.C. Kirby, M.A. McHugh, Chem. Rev, 99, 565-602, (1999).

[15] J.L.Panza and E.J.Beckman, Chapter 6 "Supercritical Fluid Technology in Materials science and Engineering" Edited by Ya-Ping Sun, Chapter 6, 255-284, (2002).

Organic, Cluster Assembled and Nano-Hybrid Materials produced by Supersonic Beams: Growth and Applications to Prototype Device Development

Salvatore Iannotta, Lucrezia Aversa, Andrea Boschetti, Silvia Chiarani, Nicola Coppedè, Marco Nardi, Alessia Pallaoro, Fabrizio Siviero, Tullio Toccoli, Roberto Verucchi

Istituto di Fotonica e Nanotecnologie – Trento Division
Via Sommarive, 18, 38050 Povo di Trento, iannotta@itc.it

ABSTRACT

An approach to the growth of films of π-conjugated organic materials, cluster assembled and nanohybrid materials combining supersonic free jets with a UHV deposition apparatus including surface characterization methods will be discussed. The unique control achievable with supersonic beams on initial kinetic energy, momentum and state of aggregation enables the growth of materials with controlled properties at different length scales. Results obtained with organic semiconductors and oligomers point out the crucial role of kinetic energy in growing organic crystalline films with well controlled morphologies and structures. By means of supersonic beams of clusters, nanocrystalline metal oxide films can be grown without annealing, so that grain size and morphology can be better controlled. In a co-deposition scheme these interesting features are combined in order to obtain a new class of hybrid functional materials with appealing properties for electronics, gas sensing and photovoltaic applications.

Keywords: growth, organic semiconductors, nanophase metal oxides, nanohybrid materials, gas sensors.

1 INTRODUCTION

The ability to synthesize nanostructured thin films with controlled structure and to tailor the needed interfaces is a key to develop new classes of devices. Indeed the properties of organic semiconductors as well as those of nanostructured metal oxides make them appealing for application in many fields, as electronics (Thin Film Transisitors, Organic Light Emitting Diodes), gas sensing (both air and Volatile Organic Compounds analysis) and solar energy conversion, nevertheless control on morphology and properties of thin films with thickness suitable for the use in real prototype devices is still hard to achieve.

As to organic molecules, it has been proved that electronic transport and optical properties depend strongly on molecular orientation and packing. A supersonic molecular beam growth (SuMBE) technique has been developed that ensures a substantial improvement of quality and control of the properties of thin films. Very interesting results have been obtained with molecules such as thiophene-based oligomers, which can be considered the prototypes of π-conjugated systems for studying optical and electrical properties [1], and with pentacene [2]. With regard to metal oxides, the deposition from Supersonic Cluster Beams (SCBD) has proven to be a viable bottom-up approach to the synthesis of films with controlled structure at the nano-level [3]. It will be shown how the appropriate combination of these molecular beam methods opens new perspectives in the intriguing field of hybrid materials in which inorganic structures (metal oxides) are functionalized by means of organic species. The combination of nanophase TiO_2 and metal phthalocyanines will be used as a test case.

2 EXPERIMENTAL

2.1 Supersonic beams of organic molecules and clusters

Supersonic free jets have been in the past largely exploited to prepare molecules in a well defined thermodynamic state for studies with time of flight methods [4]. Indeed molecules or clusters highly diluted in a supersonically expanding carrier gas exhibit a narrow velocity distribution, low divergence and, especially in the case of small molecules, alignment and a substantial relaxation of internal degrees of freedom. Therefore, when depositing species, control on the expansion's parameters gives unprecedented control on the initial state of the precursors.

The production of continuous supersonic beams of organic molecules is performed by means of a source consisting of a quartz tube in which a carrier gas (He, H_2, Ar) is seeded with species sublimated by Joule heating (see Figure1). The mixture then expands into vacuum through a nozzle. Kinetic energy as well as the degree of clustering can be tuned by changing the carrier gas, the nozzle diameter and the seeding parameters (source temperature, gas inlet pressure).

The deposition of clusters is performed via a Pulsed Microplasma Cluster Source (PMCS) [5], which has been developed in collaboration with the group directed by Prof. Milani at the University of Milan (Figure 1). Clusters are produced by quenching of the plasma in a buffer gas after a discharge between two electrodes hosted in a ceramic cavity. Virtually any conducting material can be vaporized, and the contamination of the gas with chemical species can be exploited to modify the nature of the aggregates (for example oxygen is introduced in order to obtain metal oxide clusters). The kinetic energy is in the eV/atom range, thus very interesting for studying cluster-surface interactions and for assembling nanostructured materials. Indeed cluster fragmentation is negligible, and a material preserving memory of its precursors is obtained. Control on the cluster size is attainable acting on the source operating parameters and by means of inertial aerodynamic separation effects: this is very important in order to control the structure and properties of the film, since many properties of these precursors are size-dependent.

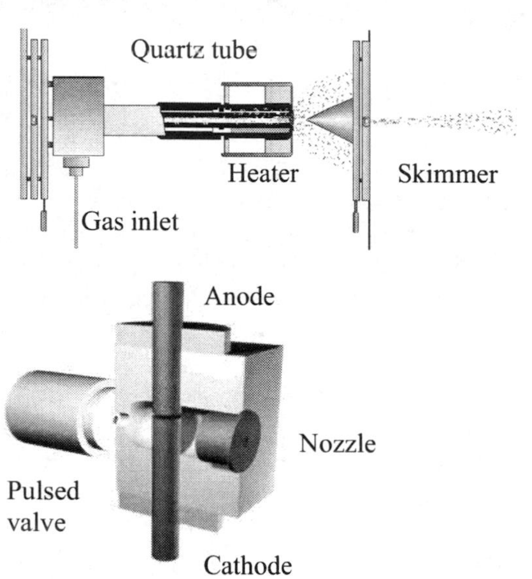

Figure 1: Schemes of the continuous supersonic molecular beam source (top) and of the PMCS (bottom).

2.2 "In situ" characterization and codeposition scheme

To better understand growth and properties of the nanostructures and interfaces we have developed a system with suitable "in situ" characterization tools. In particular this is necessary to link the properties of films and interfaces to the deposition parameters and to the initial state of the precursors, so that we can learn to grow "tailored" materials. In order to produce hybrid nanostructures we have made the apparatus to be operated with up to three different sources running simultaneously and aiming at the same focus point where the substrate is located. Layer by layer growth, blending, doping and direct synthesis of nanostructures of organic, inorganic or hybrid materials can be tailored by the beam properties and parameters. Different devices can be fully assembled and characterized "in situ".

The system can be operated in UHV and is equipped with a fast load lock entry. The deposition chamber is at present equipped with two supersonic sources and an electron beam evaporator, while a Knudsen cell and a third source can be mounted, all facing the sample. Due to the collimation of the beams, precise spots (nearly 1 cm in diameter) are produced on the sample, thus avoiding an extensive use of masks and shields. The sample is held by an x-y-z-θ–ϕ manipulator. The temperature of the substrate can be tuned in the range 120 – 800 K. The deposited films can be characterized in situ at several stages of growth by means of a Jobin-Yvon ellipsometer. A time of flight mass spectrometer (TOF-MS), developed on purpose to characterize both supersonic beams of clusters and organic molecules is available for beam characterization (cluster mass distribution, kinetic energy).

The study of electronic and chemical properties of surfaces and interfaces is performed by means of Auger, X-ray and UV photoemission spectroscopy. Low resolution SEM and LEED investigation of surface structure and morphology is also possible.

3 RESULTS

3.1 Organic thin film growth and prototype devices

With the SuMBE approach we have achieved very promising results in the synthesis of thin films of organic molecules such as pentacene [6] and thiophene-based oligomers [7,8]. As to the latter, both AFM and photoluminescence show a highly ordered structure and good optical response. The key to understand this outcome is in the features of the deposition technique: the kinetic energy (tens of eV) and the alignment that the molecules gain in the beam induce ordering on the surface. A clear evidence of this has been given by experiments on pentacene [9], demonstrating good electronic properties comparable to those of amorphous silicon. AFM characterization of films grown at increasing kinetic energies has shown that larger and larger micrometric crystalline terraces can be produced up to uniform crystalline films, where charges move without meeting grain borders. Such morphology has allowed obtaining OTFT with state of the art field effect mobility.

The deposition of metal phthalocyanines (MPc) is currently under study. In particular TiOPc SuMBE deposition has shown to procure access to different crystalline phases, including the phase II with improved absorption optical properties [10], and CuPc is being studied both for gas sensing (see § 3.4) and photovoltaic application.

3.2 Kinetic activated growth of nanostructures

Features of the SuMBE method can also be exploited to approach the growth of nanostructures difficult to obtain with traditional methods. Silicon carbide (SiC) synthesis on silicon substrates is a clear example of the potential of the technique. This is an attractive field, due to the possible integrations between the two materials, for example for a

NSTI-Nanotech 2005, www.nsti.org, ISBN 0-9767985-1-4 Vol. 2, 2005

brand new class of sensors, as well as MEMS based tools for harsh or bio-compatible environments. Some problems arise when SiC is grown heteroepitaxially on a Si substrate, because of the mismatches in lattice and thermal coefficients.

Using a highly collimated supersonic flux of C_{60} the kinetic energy of the fullerene cage can be tuned from 0.5eV up to 70eV, therefore well above the ~0.07eV of the conventional thermal evaporation techniques. We have grown SiC films on Si(111)-7×7 in UHV as a function of the C_{60} kinetic energy, at 800°C and 750°C substrate temperatures, the latter being a value at which carbide formation is not achievable by standard methods using such a precursor [11]. The carbide synthesis can be obtained by the kinetic activation of the process, while the electronic and structural properties of the film can be controlled by monitoring the beam parameters (flux and particles energy). Substrate temperatures ranging from 500°C down to Room Temperature were also explored, performing in situ surface electron spectroscopies (AES, XPS, UPS, LEED). The possibility to grow SiC at room temperature, where no thermal formation of the carbide is possible, is currently under investigation.

3.3 Cluster assembled titania and gas sensing applications

The development of the PMCS has opened new perspectives in the synthesis of cluster-assembled materials by means of the SCBD (supersonic cluster beam deposition) technique. This source is capable of delivering stable and intense cluster beams, so that the production of films several hundreds of nm thick is possible in a few hours. Very interesting results on the synthesis of nanostructured titania have been obtained. Particularly, the XRD, Raman and AFM characterization of as-deposited films show a highly porous structure with the presence of anatase, brookite and rutile crystals with size ≤20 nm (figure 2), and there are indications of an existing correlation between cluster size and crystalline phase [12]. Thus this deposition technique allows one to obtain a nanocrystalline porous film without any thermal annealing procedure, which would produce undesired grain growth and coalescence.

Gas sensing devices that include nanostructured TiO_2 as active medium exhibit performances at the state of the art as to sensitivity to VOC (ethylene and methanol) [12]. Thanks to the high effective area and to effects related to the nanometric grain size, such results are obtained at a temperature well below 300°C. The lowering of operation temperature with respect to sensors produced with more standard techniques yields advantages both in terms of stability and power consumption.

Figure 2 : AFM 500 x 500 nm^2 image of a nanostructured TiO_2 film grown by supersonic cluster beam deposition.

3.4 Hybrid materials: the organic-metal oxide interface and preliminary results in gas sensing

The concept of kinetic activation described in § 3.2 founds a very promising application in the field of the synthesis of hybrid nanostructures. Indeed chemical processes at the organic – inorganic interface can be activated by means of the supersonic beam deposition, as pointed out by photoemission experiments performed on the CuPc/TiO_2 system, whose results are summarized in figure 3. In the bottom XPS spectra the C1s and N1s core level excitations of a CuPc thin film are reported. Depositing nanostructured TiO_2 on the organic film does not produce significant changes, while in the case of the deposition of CuPc on the oxide a shift and a change in the shape of the levels is found, indicating that a chemical interaction takes place. This result confirms the crucial role of kinetic energy in view of the synthesis of a novel hybrid material in which the interaction between the organic and inorganic parts improves the properties of the film. In the codeposition scheme this interaction can be maximized while keeping control on the deposition parameters of both species.

Preliminary results on CuPc/TiO_2 hybrid gas sensors confirm that with the SuMBE/SCBD approach metal oxides can be sensitized by means of organic molecules. Figure 4 shows a SiO_2/Si substrate with Au interdigitated contacts (and Pt heaters on the back) on which a sensing layer has been deposited through a stencil mask. These hybrid sensors under methanol exposure have shown performances improving TiO_2-based devices from the point of view of sensitivity and operation temperature, and CuPc sensors in terms of baseline stability. Such results show that this is really a novel material in which qualities of the two precursor species are combined. The work on hybrid gas sensors is carried out in close collaboration with the group of prof. Siciliano at the Institute for Microelectronics and Microsystems (IMM-CNR) in Lecce.

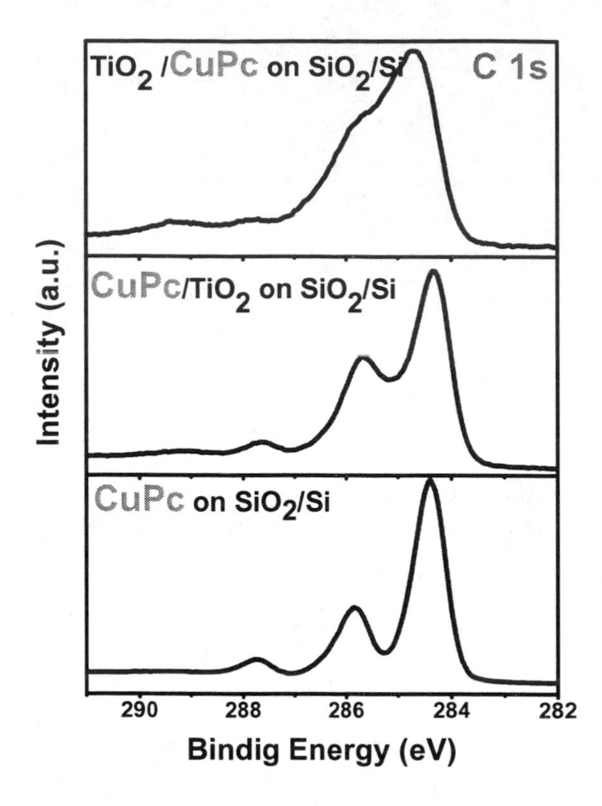

Figure 3 : XPS investigation of the C1s core level in the CuPc/TiO_2 hybrid system.

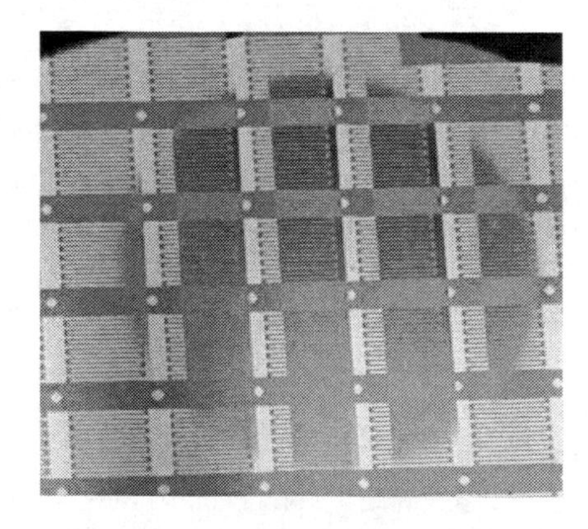

Figure 4 : Silicon substrate with patterns of Au contacts on which a sensing layer spot is clearly visible. A stencil mask has been used to select the sensor area.

4 CONCLUSIONS

The application of supersonic beam deposition methods of organic, inorganic and hybrid nanostructures is very promising. Thanks to the kinetic activation of ordering and chemical processes achievable with these techniques, unprecedented control on film structure and morphology is obtained. Moreover new perspectives in the synthesis of functional nanomaterials are opened by the codeposition scheme. The development of a system enabling in situ characterization of precursors, films and interfaces with several techniques (TOF-MS, ellispometry, electron spectroscopy, LEED) allows us to perform comprehensive studies of these materials. Applications of the deposition technique cover a wide range of applications: interesting results in the preparation of OTFT and gas sensors have been obtained and very promising preliminary results on hybrid sensors have been mentioned. The study of such hybrid systems in which inorganic nanostructures are functionalized by small organic molecules is the main objective of the future activity, together with the production of prototype devices based on these concepts, mainly gas sensors and PV cells.

ACKNOLEDGEMENTS

Authors are very grateful to C. Corradi and M. Mazzola for their precious technical support. The work is financially supported by: PAT – projects RASO, TASCA and NANOCOSHY; MIUR – projects FIRB SQUARE (RBNE01Y8C3_007) and MICROPOLIS (RBNE017JZF).

REFERENCES

[1] S. Iannotta and T. Toccoli, Journal of Polymer Science B 41, 2501, 2003.
[2] P. Milani and S. Iannotta, "Cluster Beam Synthesis of Nanostructured Materials", Spinger Verlag, 1999.
[3] G. Scoles edt., "Atomic and Molecular Beam Methods", Oxford University Press, Oxford, 1988.
[4] R.Ruiz, D.Choudhary, B. Nickel, T. Toccoli, K.C. Chang, A.C. Mayer, P. Clancy, J. M. Blakely, R. L. Headrick, S. Iannotta, and G.G. Malliaras, Chem. Mater., 16, 4497, 2004
[5] E. Barborini, P. Piseri, P. Milani, J. Physics D: Appl. Phys. 32, L105, 1999.
[6] L. Casalis, M. F. Danisman, B. Nickel, G. Bracco, T. Toccoli, S. Iannotta and G. Scoles, Physical Review Letters 90, 206101, 2003.
[7] A. Podestà, T. Toccoli, P. Milani, A. Boschetti, S. Iannotta, Surf. Sci. 464, L673, 2000.
[8] S. Iannotta, T. Toccoli, F. Biasioli, A. Boschetti, M. Ferrari, Appl. Phys. Lett. 76, 1845, 2000.
[9] F. De Angelis, T. Toccoli, A. Pallaoro, N. Coppedè, L. Mariucci, G. Fortunato and S. Iannotta, Synthetic Metals (in press 2004).
[10] K. Walzer, T. Toccoli, A. Pallaoro, R.Verucchi, T. Fritz, K. Leo, A. Boschetti and S. Iannotta, Surf. Sci. 573, 346, 2004.
[11] R. Verucchi, L. Aversa, G. Ciullo, A. Podestà, P. Milani, S. Iannotta, Eur. Phys. J. B 26, 509, 2002.
[12] T. Toccoli, A. Boschetti, L. Guerini, S. Iannotta, S. Capone, P. Siciliano, A. Taurino., IEEE Sensors Journal 3, 199, 2003.

Platinum Nanowire Actuator: Metallic Artificial Muscles

Shaoxin Lu, Kousik Sivakumar and Balaji Panchapakesan*

Delaware MEMS and Nanotechnology Laboratory, Department of Electrical Engineering, University of Delaware, Newark, Delaware, 19716, USA, * Email: baloo@eecis.udel.edu

ABSTRACT

In this paper, we report the fabrication of platinum nanowires using single wall carbon nanotubes as templates and their application as an electro-chemical actuator. Two regimes of actuation were seen in this actuator. In the "low" charge injection regime, quantum mechanical effects introduced by electrochemical double layer charge manipulation were the dominant actuation mechanism. Strain of about 0.04% without significant hysteresis was readily achieved. In the "medium" charge injection regime, electrostatic effect is believed to be the dominant mechanism producing strain of 0.22% which is much larger than those of the commercial piezoelectric materials.

Keywords: platinum nanowire, actuator, carbon nanotube

1 INTRODUCTION

The direct conversion of electrical energy to mechanical energy is of importance in many applications such as robotics, artificial muscles, optical displays and micro-mechanical devices and many material systems have been introduced as actuators to accomplish the energy conversion. Piezoelectric ceramics, shape memory alloys, magnetostrictive materials are well known conventional actuation materials and in recent years polymer actuators [1-3] have been proposed to be attractive alternatives. More recently, as new emerging materials, nano material systems such as carbon nanotubes [4-6] and metallic nano-particles [7, 8] have also been proposed as promising candidates for actuation technologies. Both these actuators from nano materials pose the form of electro-chemical actuator and actuate by means of an electrochemical double layer charging processes at the SWNTs (platinum nanoparticles) /electrolyte interface. It is well known that nano materials such as SWNTs and nanoparticles have extremely large surface area to volume ratio, which makes a large volume fraction of materials to be surface or interface atoms. In electrochemical actuators, the large electrochemically accessible surface area of these nano materials and the nanometer scale separation of charges between nano materials and electrolyte render the actuator a giant capacitance. The charge injection, when applying a voltage, will in turn modify the surface charge density and related surface properties causing the actuation. Strain values larger than 0.2% in SWNT actuator and 0.15% in platinum nano particle actuator (due to quantum mechanical effects induced by electrochemical double layer charging process) could produce dimensional changes large enough to do mechanical work [8]. This implies that nano material systems may open the way for better actuation technologies especially in the nano scale [9].

In this paper we demonstrate the electrochemical actuation of a new nanowire material—platinum nanowire. Single wall carbon nanotubes were used as templates for the fabrication of platinum nanowires. Then thin sheets of platinum nanowires were made to demonstrate the actuation. This actuator has the advantages of metallic nanowires such as high temperature endurance, chemical stability and low operating voltage of only a few volts. Due to the excellent mechanical properties of the inner SWNT templates, the actuator could have long-time stability against mechanical fatigue and defects, which are desired for many actuation applications such as artificial muscles. These platinum nanowires are also suitable candidates for nano scale actuators such as nano grippers due to their one-dimensional structure.

2 EXPERIMENTAL AND DISCUSSION

Commercially obtained carbon nanotubes made by pulsed laser ablation process were used in this work as templates for the fabrication of nanowires. Dihydrogen Hexachloroplatinate ($H_2PtCl_6 \cdot 6H_2O$), purchased from Alfa Aesar, was used as the platinum source. The platinum nanowires were synthesized by procedures described in reference [10], following which a nanowire sheet was made using vacuum filtration of the resulting platinum nanowire solution [4]. After subsequent rinsing with iso-propyl alcohol and DI water, the nanowire sheet was dried at 80ºC for 20 minutes to remove the remaining solution and organics in the sheet and further annealed in argon ambient at 750ºC for 30 minutes to enhance the mechanical strength of the sheets. The resulting sheet had an average thickness ranging from 15μm to 75μm, depending on the amount of nanowires used, and a density of ~ 5.1g/cm^3. Figure 1(a) is the SEM image of the platinum nanowires after annealing. The nanowires have diameters ranging from 60nm to 100nm and form highly entangled nanowire bundles. These sheets were used to examine the actuation properties without further optimization. The TEM image in Figure 1 (b) gives a better view of an individual platinum nanowire, showing that the nanotube templates were coated with layers of platinum nanoparticles. The inset shows a high-resolution image of the platinum nanoparticles with an average diameter of 8nm. This would result in a surface to volume ratio even higher than nanowires with a smooth coating of platinum, which is essential for actuation applications.

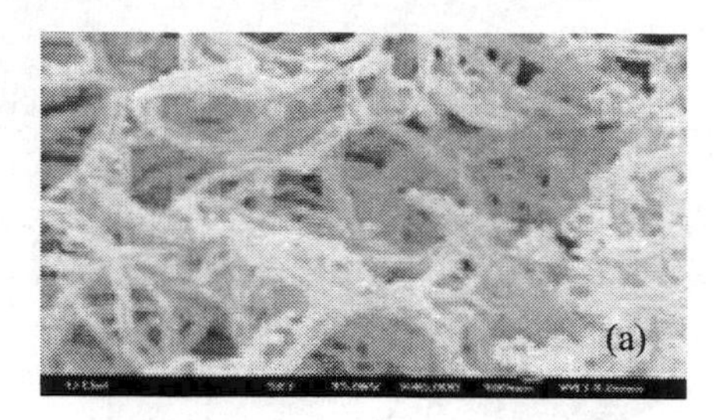

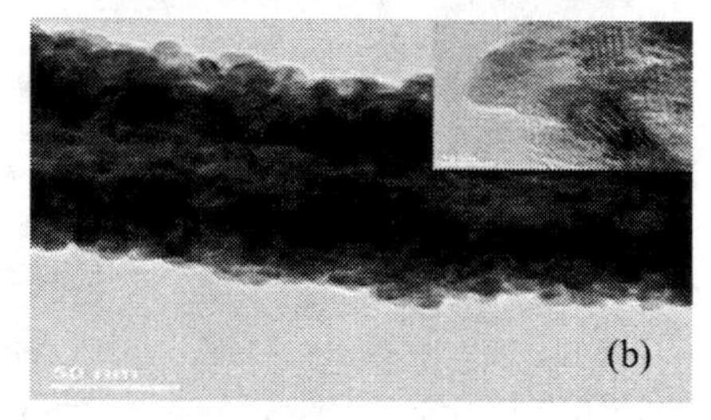

Figure 1: (a) SEM image and (b) TEM image of platinum nanowires. The inset in (b) shows platinum nanoparticles of size ranging from 5nm to 10nm.

A set up shown in Figure 2 was used to characterize the strain of the nanowires. A platinum nanowire sheet with dimensions of 20mm×2mm×50μm was attached to a strip of PVC of dimensions 50mm×3mm×100μm, which is vertically anchored to the bottom of the beaker. Another platinum nanowire sheet of much larger size than the actuator was used as the counter electrode. The bending of PVC strip was recorded using a digital camera mounted on a microscope and this displacement data was further used to characterize the strain of platinum nanowires. A third Ag/AgCl reference electrode was also inserted in the solution and all the voltages were measured versus the reference potential. 1M KOH solution was used as the testing electrolyte.

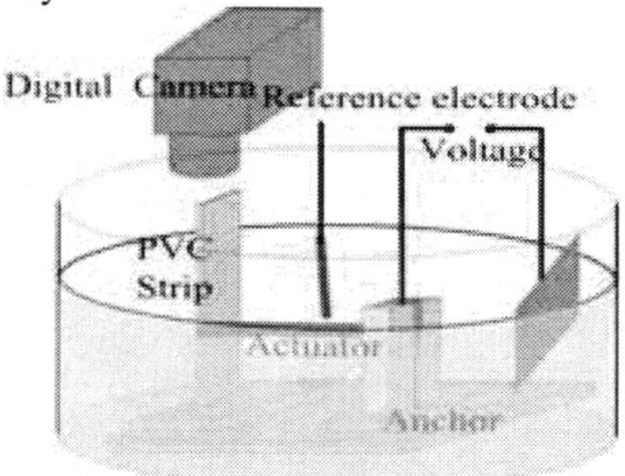

Figure 2: Experimental setup used to characterize the strain of platinum nanowires.

When platinum nanowires are biased, a potential difference forms at the interfaces between the nanowires and the solution. Thus an electrochemical double layer forms at the interface due to the capacitive charging processes, which is similar to those in SWNT [4] and platinum nano particle electro-chemical actuators [7]. It was suggested by Baughman [4, 8] and Weissmuller [7] that at "low" charge injection levels, quantum chemical effects lead to a strain and hence actuation in their actuators. The large surface area to volume ratio and the small dimensions of both the SWNTs and platinum nano particle networks that make the electrolyte-accessible area very large, together with the nanometer scale separation of charges at the interface between nano materials and electrolytes, lead to super capacitance in the actuator, which is the key factor for achieving high actuator strains at low voltages [8]. Similarly, super capacitance could also result in our platinum nanowire actuator from the electrochemical double layer due to the high surface area of these one-dimensional nanowires.

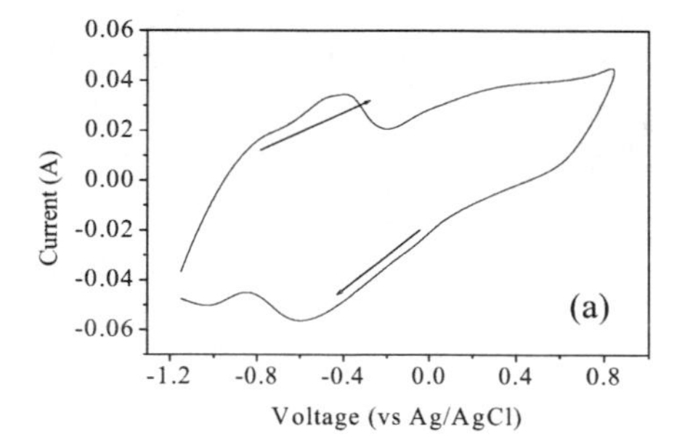

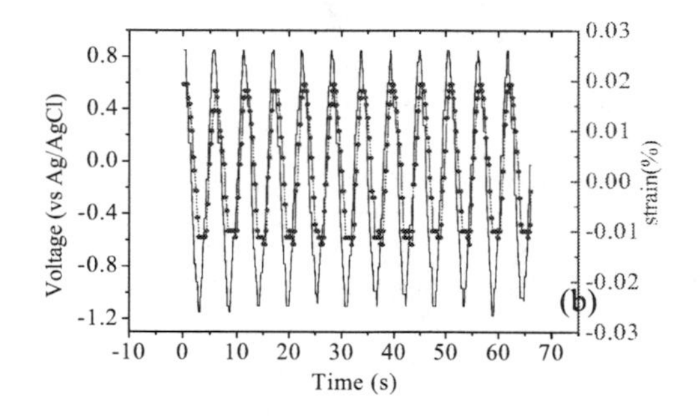

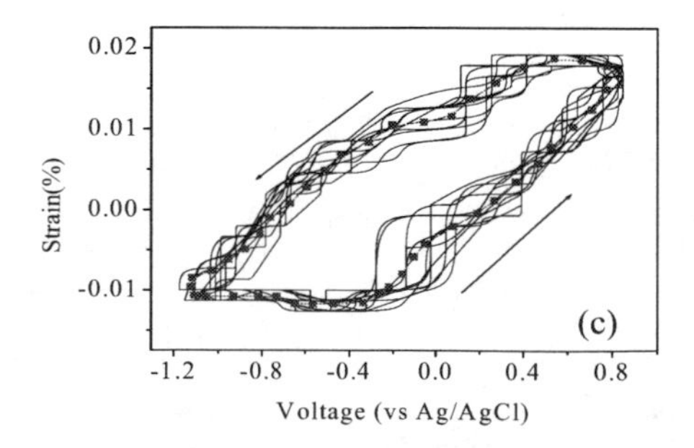

Figure 3: (a) In situ cyclic voltammograms of current measured during actuation. (b) Actuation responses versus time and applied voltage (c) Strain response of actuator versus the applied potential.

Figure 3(a) shows the recorded in situ cyclic voltammograms of current during the actuation when the applied voltage was swept from -1.15V to 0.85V at a scan rate of 710mV/s (0.18Hz), which is a typical curve for the actuator in different operation conditions. The strain response of the actuator during actuation is shown in Figure 3 (b) where the applied voltage and corresponding strain are plotted as a function of time. The cycles are quite repeatable with nearly the same amplitude and durations. When the strain response is plotted versus the applied voltage, a hysteric curve results as shown in Figure 3(c).

Comparing with the actuation response of platinum nanoparticles [7], the strain in platinum nanowires follows nearly the same pattern, indicating that the strain obtained from the platinum surface dominates over the inner carbon nanotube template. With an arbitrary zero strain point, a total strain of about 0.03% was acquired when the applied voltage is swept from -1.15V to 0.85V at scan rate of 710mV/s (0.18Hz). This obtained strain value is smaller than that of platinum nano particle actuator [7], however, the scan rate is more than 700 times faster (compared to less than 1mV/s rate in [7]), which is crucial in deciding the strain value as will be addressed later in this paper.

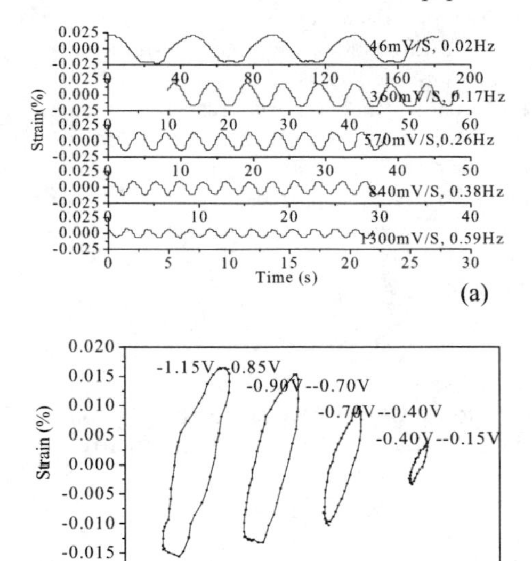

Figure 4: The dependence of strain on (a) potential scan rate (frequency response) and on (b) potential sweeping range.

To investigate the frequency dependence of the actuator, the voltage sweeping range was set to be constant from -0.7V to 0.4V (vs Ag/AgCl), and the strain responses measured at different voltage scan rates or different frequencies. As shown in Figure 4 (a), when the voltage scan rates varied from 1300 mV/s (0.59Hz) to 840 mV/s (0.38Hz), 570 mV/s (0.26Hz), 360 mV/s (0.17Hz) and 46 mV/s (0.02Hz), the total strain increased from 0.014% to about 0.042%, three times lager than the high frequency values. This negative frequency dependence of strain was also witnessed at other voltage sweeping ranges when operated at different frequencies, which means that the strain of platinum nanowires is not saturated under faster scan rate or high frequencies. If the quantum mechanical effect induced by modulation of surface charge density is responsible for the strain, the more the modulation by means of electrochemical double layer charging, which needs more time to accomplish under same driving voltage, the more will be the resulting strain. Consequently, larger strain will be acquired under longer charging time as in the case of low frequency actuations.

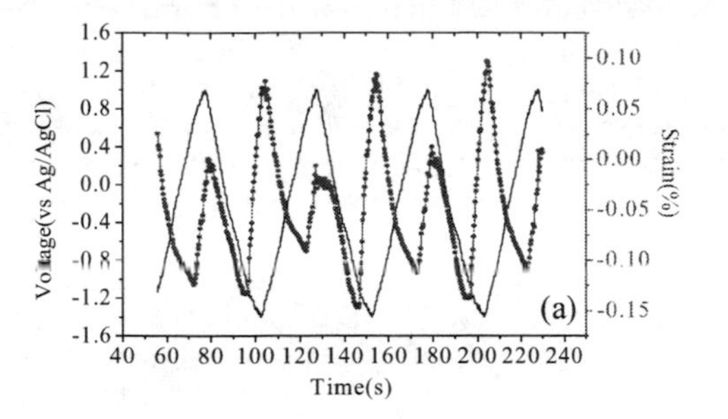

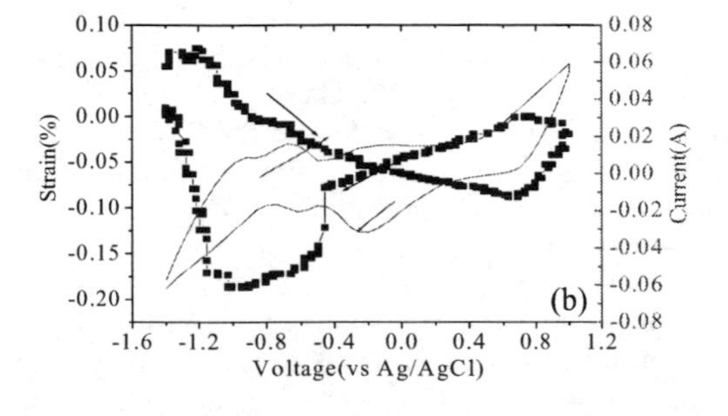

Figure 5: Actuation response at "medium" charge injection level. (a) Actuation responses versus time and applied voltage. (b) The strain is plotted as a function of applied potential (left) and current measured during actuation (right). The arrows indicate the direction of sweeping.

To investigate the voltage dependence of strain, the scan rate was set constant at 900mV/s and the strain of the actuator measured under the following voltage sweeping ranges: -0.4V to 0.15V, -0.7V to 0.4V, -0.9V to 0.7V and -1.15V to 0.85V. Figure 4(b) shows the strain response of different voltage sweeping range under the constant scan rate of 900mV/s. It can be seen when the voltage sweeping range was increased from -0.4V-0.15V to -1.15V-0.85V, the total strain (arbitrary zero point for strain value) increased from 0.008% to 0.033%, more than 4 times its value at smaller voltage sweeping range, which also led to the occurrence of hysteresis at the same time. As the potential sweeping range increases, the interfacial electrical field between platinum nanowires and electrolyte becomes stronger and causes increased charge injection and surface charge modulation. This increased charge injection consequently leads to a higher strain in the nanowire sheet. Further, under constant voltage scan rate, larger voltage sweeping range means longer time of sweeping, which would also cause an increase in charges injected and hence resulting in greater strain values.

From these results, at small voltage sweeping range (normally smaller than -1V to 0.8V) or fast scan rate (normally larger than 150mV/s), the charge injection level in electro-chemical double layer is relatively low due to the insufficient driving voltage or insufficient time for charging. The strain due to the quantum mechanical effect is proportional to the charge injection and is a dominant mechanism for the actuation [4, 7, 8]. In order to probe the actuation response at higher charge injection levels, which can be realized by increasing the voltage sweeping range or by decreasing the voltage scan rate, we measured the strain response of the actuator at an increased voltage sweeping range of -1.4V to 1V and a small voltage scan rate of 96mV/s. Figure 5 (a) shows the actuation responses (right) and the applied voltage (left) versus time. An important difference between this curve and that in "low" charge injection level is the double peaks in every cycle, which is a typical feature in actuation in this charge injection regime. It is known that strain due to the quantum mechanical effects will change sign when the charge injection changes sign [4, 8, 11, 12]. Due to this only one peak is obtained in every cycle. Instead, if the electrostatic effect dominates the actuation, then the strain value would go close to zero with minimum dimension change of actuator when it is at the potential of zero charge (PZC), and either positive or negative charge injection would cause expansion of the actuator resulting in strain peaks at both ends of the voltage sweeping range [4, 7, 8, 13]. As a result double strain peaks form during every cycle. Based on our experimental result, we believe that strain from electrostatic effect is dominant beyond a certain charge injection level and quantum mechanical effects are suppressed. Three cycles of actuation is shown in Figure 5(a) to be quite repeatable with nearly the same strain amplitude of ~0.22%. This strain value is much larger than the maximum strain of 0.15% obtained from platinum nanoparticles actuator [7]. Figure 5 (b) plots the strain response as a function of applied potential. The cyclic voltammograms of current is also shown in the graph for reference. The two strain peaks in a cycle are not symmetrical with a much larger peak amplitude at negative potential end and smaller peak amplitude at positive potential end. The reason for this asymmetry is not clear. One possible reason could be due to the unique structure of the nanowires made of platinum nanoparticles on SWNTs templates. The platinum nanowires could have an inherent Schottky barrier at the nanotube – platinum interface. Under moderate negative bias, more electrons may gain sufficient energy and reach the nanotubes through tunneling or thermal emission across the Schottky barrier. This injection of negative charges could lead to a quantum mechanical expansion of the nanotubes and add to the electrostatic strain experienced by the platinum coating to result in a bigger strain. Whereas, under positive bias, charge injection leads to shrinkage of the nanotubes and the overall effective strain of the nanowire (the platinum surface would still expand under positive bias due to electrostatic effect) is reduced resulting in a smaller peak under positive bias.

To the best of our knowledge, the platinum nanowire actuator reported here is the first nanowire actuator that exhibits strain comparable to commercial piezoelectric materials while requiring only a few volts to operate. The work explained here is only a preliminary demonstration of the actuation capabilities of the nanowire actuator. There is a lot of scope for improvement to obtain better actuation. Platinum nanowires of smaller diameters would have higher surface to volume ratio leading to higher surface charge modulation and hence larger strain. A better control of nanowire thickness can be achieved by more efficient separation of carbon nanotube templates in the solution [14]. We believe that with proper control over synthesis, nanowires as small as 10nm can be repeatably fabricated [15]. Moreover, in order to obtain better strain values, it is desirable to use directional, oriented platinum nanowires instead of the highly entangled ones for the construction of nanowire sheets, which could suppress the overall strain response. SWNT sheets grown with good directionality using CVD processes [16] can be used for the fabrication of platinum nanowire sheets. The use of proper electrolytes with wider stability window such as organic solutions is also an important factor in improving the performance of the actuator [5].

3 ACKNOWLEDGEMENTS

We acknowledge the partial funding provided by National Science Foundation Grant: CCR: 0304218.

REFERENCES

[1] A. S. Hutchison et al., Synth Met. 113, 121, 2000.
[2] H. B. Schreyer et al., Biomacromolecules 1, 642, 2000.
[3] D. L. Thomsen et al., Macromolecules 34, 5868, 2001.
[4] R. H. Baughman et al., Science 284, 1340, 1999.
[5] G. M. Spinks et al., Adv. Mater. 14, 1728, 2002.
[6] J. N. Barisci et al., Smart Mater. Struct. 12, 549, 2003.
[7] J. Weissmuller et al., Science 300, 312, 2003.
[8] R. H. Baughman, Science 300, 268, 2003.
[9] H. Gleiter et al., Acta Materialia 49, 737, 2001.
[10] K. Sivakumar et al., J. Nanosci. Nanotech. 5, 1, 2005.
[11] C. T. Chan et al., Phys. Rev. Lett. 58, 1528, 1987.
[12] S. Flandrois et al., Synth Met. 34, 399, 1990.
[13] R. H. Baughman et al., Phys. Rev. B 46, 10515, 1992.
[14] M. Zheng et al., Nature Mater. 2, 338, 2003.
[15] W.-Q. Han et al., Nano Letters 3, 681, 2003.
[16] K. Hata et al., Science 306, 1362, 2004.

Evaluation of the Degree of Exfoliation in Poly(ε-caprolactone) Nanocomposites Using by the Dynamic Mechanical Analysis

Jae Woo Chung*, Jae–Deok Jeon** and Seung–Yeop Kwak***

School of Materials Science and Engineering, Seoul National University, San 56-1, Sillim-dong, Gwanak-gu, Seoul 151-744, Korea
*cwfrank5@snu.ac.kr, **jdjun74@snu.ac.kr, ***sykwak@snu.ac.kr

ABSTRACT

Poly(ε-caprolactone) (PCL) nanocomposites, were prepared using layered silicates with the nonpolar and polar organic modifiers by melt mixing with an internal mixer, PCLOC25A and PCLOC30B, respectively. Additional heat treatment was imposed on the PCL nanocomposites for the complete exfoliation. WXRD patterns showed that d_{001} peaks of both nanocomposites were disappeared after the additional heat treatment. This means that the nanocomposites are exfoliated. However it was impossible to evaluate the degree of the exfoliation of both nanocompoistes by WXRD because d_{001} peaks are undetectable. Thus, linear viscoelastic and relaxation behaviors were measured by DMA. DMA results showed that PCL nanocomposites were completely exfoliated and PCLOC30B is more exfoliated than PCLOC25A because of the specific interaction between PCL and polar organic modifier. Finally, DMA was confirmed as a reliable mean to understand the degree of exfoliation of nanocomposites.

Keywords: nanocomposite, poly(ε-caprolactone), degree of exfoliation, dynamic mechanical measurement, polar interaction

1 INTRODUCTION

In the past decades, layered silicate-based polymer nanocomposites have attracted considerable attention because of the dramatic enhancement in strength, modulus, thermal resistance, and gas permeability barrier properties with far less amount of silicate content than that used in conventional filled polymer composites [1]. Layered silicates have layer thickness on the order of 1 nm and very high aspect ratios, e.g., 10-1000, and the interlayer spacing between stacked layers is also about 1 nm [2]. As such features of layered silicates, structures of nanocomposites of two different types such as the intercalation or exfoliation are achievable. The intercalated nanocomposites show regularly alternating layered silicates and polymer with a repeat distance of a few nanometers while the individual layers in exfoliated nanocomposites are irregularly delaminated and dispersed in a continuous polymer matrix [3]. This structural difference in nanocomposites plays a key role in the enhancement of properties. For the exfoliated nanocomposites, higher mechanical properties are expected because of the larger surface area between reinforcement phase and polymer matrix relative to intercalation nanocomposites [4]. Giannelis et al. reported that thermodynamically stable equilibrium states of the nanocomposites such as intercalated and exfoliated systems depend on various entropic and enthalpic factors [5]. However, since the total entropy change in the system is small, the change of enthalpic factors such as intermolecular interaction determines the structure of nanocomposite [5]. Thus, layered silicate structure depends on the establishment of very favorable polymer-surface interactions to overcome the penalty of confinement condition [6], and the specific interactions driven by polar interaction or hydrogen bonding play a significant role for enhancing dispersion of layered silicate [7]. The structure of nanocomposites has been usually investigated by means of wide angle X-ray diffraction (WXRD) and transmission electron microscopy (TEM). WXRD and TEM are the powerful tools to prove the direct information on the structure of nanocomposite, but do not provide information about properties caused by the silicate structure [8]. Besides, it is impossible to analyze the overall three-dimensional structure of the silicate layers in nanocomposites. Recently, the interest about the dynamic mechanical analysis for property evaluation of nanocomposite are increasing because it provides information about the particle size, the shape, and the surface characteristics of the dispersed phase as well as enabling the facile characterization of the silicate structure [2,9]. Therefore, the aim of this study is to investigate dynamic mechanical properties such as linear viscoelastic and relaxation behavior occurred by silicate structures and to evaluate the degree of the exfoliation of nanocomposites using by dynamic mechanical analysis (DMA).

2 EXPERIMENTAL

2.1 Materials

Poly(ε-caprolactone) (PCL), with a number average molecular weight, M_n, of 80,000 g/mol was purchased from Aldrich. Organically modified montmorillonites (MMT) were supplied from Southern Clay Products, Inc., USA, under the trade name of Cloisite®25A (OC25A) and

Cloisite®30B (OC30B). Organic modifiers in OC25A and OC30B were dimethyl hydrogenated tallow 2-ethylhexyl, 2MHTL8, and methyl tallow bis(2-hydroxyethyl), MT2EtOH, quaternary ammonium cations.

2.2 Preparation of nanocomposites

The PCL and organoclay were mechanically mixed with an internal mixer with roller rotors at a rate of 100 rpm at 180 °C for 450 s. The compositions of the PCL/organoclay nanocomposites are given in Table 1. After mechanical mixing, additional heat was imposed to PCL/organoclay nanocomposites until the acquirement of the subsequently exfoliated nanocomposite. The PCL nanocomposite with OC25A is named PCLOC25A, whereas that with OC30B is denoted PCLOC30B.

Samples	Composition (wt%)		
	Poly(ε-caprolactone)	Organoclay	
		Modifier	MMT
PCLOC25A	92.4	2.6	5
PCLOC30B	92.6	2.4	5

Table 1: Composition of PCL/Organoclay Nanocomposites

2.3 Measurements

X-ray diffraction measurements were performed using a MAC Science MXP 18A-HF X-ray diffractometer with CuK_α radiation, $\lambda = 1.5405$ Å, generated at 40 kV and 100 mA in order to verify the gallery height and the structure of layered silicate in nanocomposites. Diffraction spectra were obtained in a 2θ range of 1.5–10°, and the diffraction angle was scanned at a rate of 3°/min. Dynamic mechanical measurements were performed on a rheometrics mechanical spectrometer model 800 (RMS 800). Dynamic isothermal frequency sweeps were performed using a parallel geometry plates 25 mm in diameter at different temperatures ranging between 60 °C to 180 °C, with nitrogen gas purging and an angular frequency ranging from 10^{-1} to 10^{2} rad/s. The strain regime which can be regarded as the linear viscoelastic data was determined from strain sweep experiments at strain amplitudes at selected temperatures.

3 RESULTS AND DISCUSSION

3.1 Wide angle X-ray diffraction (WXRD)

The exfoliated structure of PCLOC25A and PCLOC30B caused by the additional heat imposition is provided by WXRD analysis, as shown in figure. 1 and 2. The d_{001} silicate peaks in both nanocomposites without additional heat treatments after compounding are still detectable, which suggests that these structures remain intercalated.

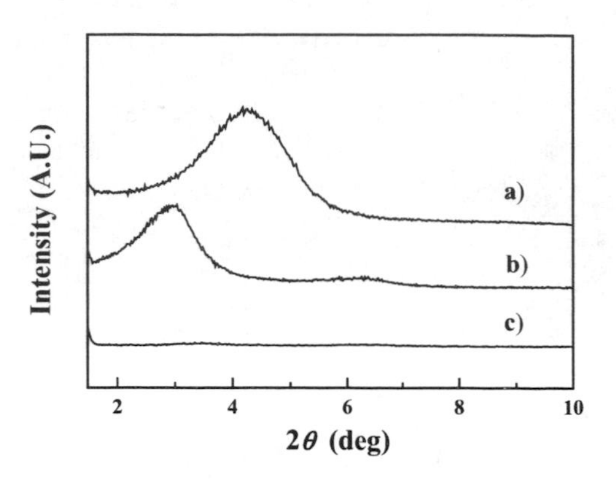

Figure 1: WXRD intensity profiles: a) OC25A, b) PCLOC25A before heat treatment, and c) PCLOC25A after heat treatment.

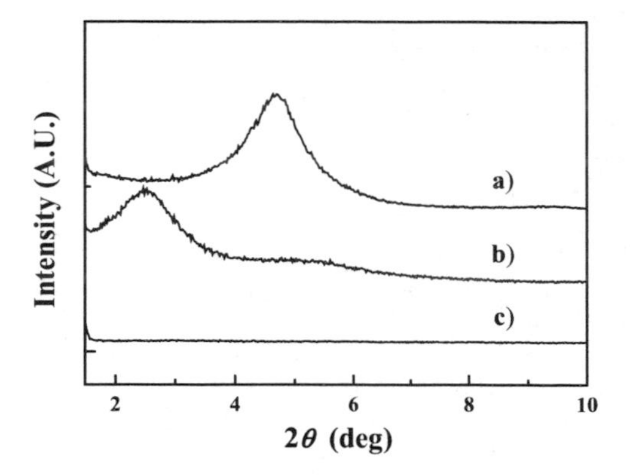

Figure 2: WXRD intensity profiles: a) OC30B, b) PCLOC30B before heat treatment, and c) PCLOC30B after heat treatment.

However, the d_{001} peaks of both nanocomposites with the additional heat imposition perfectly are disappeared. This indicates that PCL chains have penetrated further into the silicate interlayers and the both nanocomposites have completely the exfoliated structure due to the additional heat treatment.

3.2 Dynamic mechanical analysis (DMA)

From WXRD results, it was proved that both nanocomposites had fully exfoliated structure. However, the degree of exfoliation between the nanocomposites was not compared by WXRD data because the d_{001} silicate peaks in both nanocomposites with additional heat treatment were disappeared. Thus, dynamic mechanical measurement was performed. Figure 3 shows the dynamic mechanical behavior of the PCL and the PCL/organoclay nanocomposites. We observed that log $G''(\omega)$ versus log

$G'(\omega)$ plots for PCL and both nanocomposites were independent of temperature. This implies that PCLOC25A and PCLOC30B have no structural change from intercalation to exfoliation over the temperature range because both nanocomposites already are completely exfoliated by the additional heat treatment. In addition, figure 3 shows that PCLOC30B has a higher elastic property than PCLOC25A. Since the content of the organic modifiers contained in the nanocomposites is small, the improvement of the mechanical properties caused by organic modifiers could be ignored.

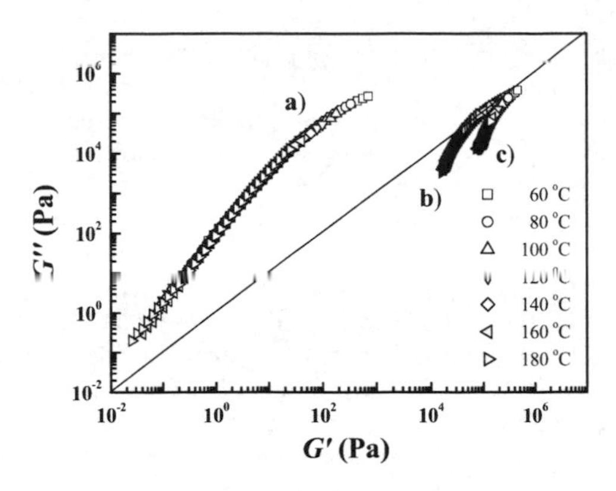

Figure 3: Plots of log $G''(\omega)$ versus log $G'(\omega)$ in the temperatures range of 60 to180℃: a) pure PCL, b) PCLOC25A, and c) PCLOC30B.

This implied that the increase of mechanical properties depends on the degree of exfoliation and more exfoliated silicate structure provides higher elastic properties. Actually, layered silicate structure depends on the establishment of very favorable polymer-surface interactions [6]. PCLOC30B exists the polar interaction between the carbonyl groups in PCL and the hydroxyl groups in organic modifier at the surface of the OC30B. Polar interaction may facilitate the penetration of PCL into the interlayer. Therefore, PCLOC30B is more exfoliated than PCLOC25A. These results show that the degree of exfoliation depends on specific interactions between PCL and organic modifiers. The dynamic viscoelastic master curves were formed by the application of time-temperature superposition (TTS). We observed that the pure PCL exhibited the terminal behavior. On the other hand, for the PCLOC25A and PCLOC30B displayed the non-terminal behavior at low frequency. Then, these low-frequency responses might be indicative of a solid-like behavior [3]. The dynamic complex viscosity master curves, η^*, for the pure PCL and PCL/organoclay nanocomposites are presented in figure 4. Krishnamoorti et al. reported that the rheological properties were unaffected by the choice of the chemical nature of the polymer in nanocomposites [10]. However, relationship between chemical nature of the polymer and organoclay influences on the degree of exfoliation. Thus, the high viscosity of PCLOC30B is due to the higher degree of exfoliation in PCLOC30B than PCLOC25A

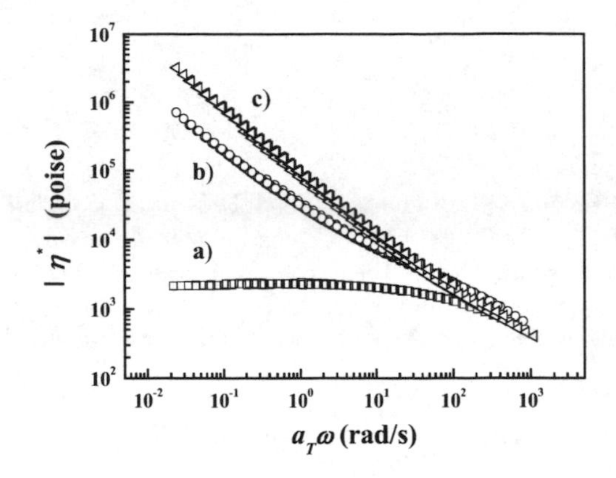

Figure 4: Reduced complex shear viscosities: a) pure PCL, b) PCLOC25A, and c) PCLOC30B.

Krishnamoorti et al. found that a_T values were unaffected by the presence of the silicate [6,10]. However, we found that PCLOC30B shows the different a_T values with pure PCL while PCLOC25A has the similar a_T values with pure PCL. This result is different from so far previously reported study by Krishnamoorti and co-works. The behavior may be due to formation of highly exfoliated structure in PCLOC30B occurred by the strong interaction between PCL and OC30B. Furthermore, we suggest that the strong interaction between PCL molecules and the modifier in PCLOC30B may form the pseudo-network-like interaction as shown in figure 5.

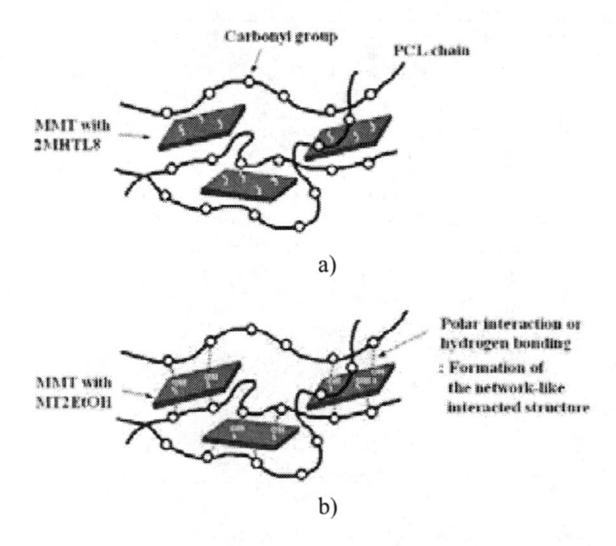

Figure 5: Comparison of the relationship between organoclay and PCL molecules depending on the interaction present in each nanocomposite: a) PCLOC25A and b) PCLOC30B.

Relaxation behaviors of both nanocomposites were evaluated by the activation energy, E_a. C_1 and C_2 were calculated by the nonlinear curve fitting of the WLF equation as follows:[11]

$$\log a_T = \frac{-C_1(T - T_0)}{C_2 + T - T_0} \quad (1)$$

Both C_1 and C_2 are constants, and T_0 is the reference temperature, in this case 120 ℃. Subsequently, E_a of flow can be determined using Equation (2), where the WLF equation is substituted into the horizontal shift factor in the Arrhenius equation, which links the activation energy with the horizontal shift factor:

$$\Delta E_a = R\left(\frac{d \ln a_T}{d(1/T)}\right) = 2.303R\left[\frac{C_1 C_2 T^2}{(C_2 + T - T_0)^2}\right] \quad (2)$$

The activation energies become thus dependent on the temperature. Their values are shown in figure 6.

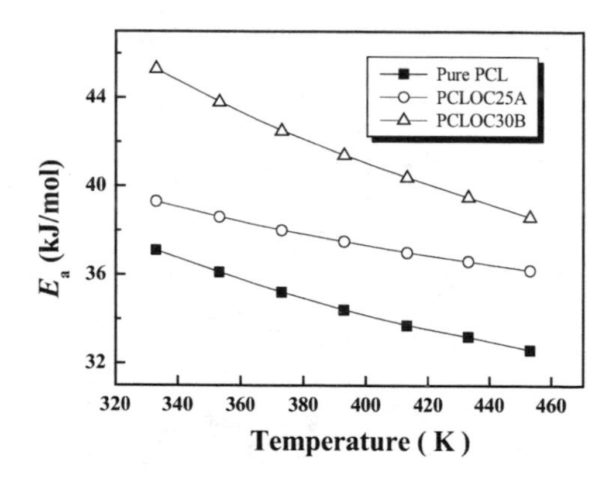

Figure 6. Activation energies for pure PCL, PCLOC25A, and PCLOC30B

The activation energy is a measure of molecular mobility and the energy barrier that must be overcome. It was confirmed from the plots of the activation energies as a function of the temperature that PCLOC30B had remarkably higher activation energies than PCLOC25A and pure PCL. It is attributed to the formation of the *pseudo*-network-like interacted structure caused by the polar interaction in PCLOC30B.

4 CONCLUSIONS

Poly(ε-caprolactone), PCL, compounds containing two different types of organically modified silicates, OC25A with nonpolar and OC30B with polar, namely PCLOC25A and PCLOC30B, were prepared by the additional heat imposition as well as the melt mixing using an internal mixer at 180 °C for 450 seconds. It was verified by WXRD that PCL/organoclay nanocomposites was completely exfoliated through the heat treatment. The log $G''(\omega)$ versus log $G'(\omega)$ plots acquired from dynamic mechanical measurement showed that both nanocomposites were independent of the temperature. These mean that both nanocomposites were exfoliated. Then, it showed that PCLOC30B was the more elastic than PCLOC25A. From master curves of the storage moduli and the complex shear viscosity, it was also found that the PCLOC30B with the polar interaction showed a better elasticity than PCLOC25A. These are considered as the fact that PCLOC30B has more exfoliated structure than PCLOC25A due to the polar interaction. In the plots of the temperature versus a_T values, PCLOC30B showed the different a_T values with pure PCL. Furthermore, the activation energy was revealed that the PCLOC30B had remarkably higher activation energies than PCLOC25A and pure PCL. It is attributed to the formation of the *pseudo*-network-like interacted structure by the polar interaction. Consequently, these results suggest that the polar interaction between the polymer and the organic modifier is important factor depending on the silicate structure. Moreover, dynamic mechanical analysis is confirmed as a reliable mean to understand not only the silicate structure but also the relationship between the structure and properties of the nanocomposites.

REFERENCES

[1] C. Wan, X. Quiao, Y. Zhang and Y. Zhang, Polym. Test 22, 453, 2003.
[2] S. S. Ray, K. Okamoto and M. Okamoto, Macromolecules 36, 2355, 2003.
[3] R. Krishmamoorti and E. P. Gianelis, Macromolecules 30, 4097, 1997.
[4] V. Krikorian and D. J. Pochan, Chem. Mater. 15, 4317, 2003.
[5] R. A. Vaia and E. P. Gianelis, Macromolecules 30, 7990, 1997.
[6] E. P. Giannelis, R. Krishnamoorti and E. Manias, Adv. Polym. Sci. 138, 107, 1999.
[7] S. W. Lim, J. W. Kim, I. Chin, Y. K. Kwon and H. J. Choi, Chem. Mater. 14, 1989, 2002.
[8] T. J. Pinnavaia and G. W. Beall "Polymer-Clay Nanocomposites," John Wiley & Sons Ltd, 238-249, 2000.
[9] M. J. Solomon, A. S. Almusallam, K. F. Seefeldt, A. Somwangthanaroj and P. Varadan, Macromolecules 34, 1864, 2001.
[10] J. Ren, A. S. Silva and R. Krishmamoorti, Macromolecules 33, 3739, 2000.
[11] J. D. Ferry, "Viscoelastic Properties of Polymers," John Wiley & Sons Ltd, 274-280, 1980.

The Role of Different Parameters in the Properties of Carbon Nanotube Polymer Nanocomposites.

C. Velasco-Santos[*,δ,&,+], A.L. Martínez-Hernández[*,δ,&,++] and V. M. Castaño [*,+++]

[*]Centro de Fisica Aplicada y Tecnologia Avanzada A.P.1-1010
Querétaro, Querétaro, 76000 Mexico, [+]carlosv@fata.unam.mx,
[++]analaura@fata.unam.mx, [+++]castano@fata.unam.mx
[δ]Departamento de Materiales y Mecatrónica
Instituto Tecnológico de Queretaro, Av. Tecnológico, Esquina Mariano Escobedo,
Col. Centro, Querétaro, Querétaro, 76000 México,
[&]Department of Material Science, University of North Texas,
Denton, TX 76203-5310, USA;

ABSTRACT

Carbon nanotubes are considered like one of the most interesting materials in the last fifty years. Inasmuch as since their discovery, have opened an interesting gate in nanoscience and in different fields of materials science. Their outstanding mechanical, thermal and electrical properties, together with their size and cristallinity, provide great surface area structures which represent excellent candidates in order to develop different applications. However the compatibility of these structures with other materials is required so as to take advantage of the amazing properties in other scale and reach in this way, multifunctional nanocomposite materials. This work presents diverse results which depend on different parameters that play an important role in the distribution, interaction and synergetic effect in composites developed with the incorporation of multiwalled nanotubes in different polymeric matrix.

Keywords: carbon nanotubes, nanocomposites, polymer matrix, thermal and mechanical properties, interaction, chemical functionalization

1 INTRODUCTION.

Research focus to create nanocomposites with different properties has increased in the last ten years with the discovery of impressive materials at nanometric level such as carbon nanotubes. To employ these nanoforms as reinforcement of polymers has opened the possibility to develop new ultra-strong and conductive nanocomposites never before seen. With this perspective, different studies have been developed in order to take advantage of the amazing mechanical and electronic properties of carbon nanotubes. Nevertheless the challenge to make a new age of multifunctional composite materials with these nanometric forms is beginning and it is needed to understand the behavior on the interface and different parameters that play an important role in the final properties in nanocomposites materials. This work presents diverse results which depend on important parameters that take part in the distribution, interaction and synergetic effect in composites developed with the incorporation of multiwalled nanotubes in different polymeric matrix. The nanocomposites evaluated were made by different processing approaches with the assistance of additives and ultrasound [1-2]. In addition is presented a chemical modification of these carbon structures in the surface and tips, showing that it is possible take advantage of the physical-chemical changes produced in the nanotubes when new moieties are attached in order to develop new polymeric nanocomposites with outstanding mechanical and thermal properties [3]. The interactions caused by functionalized carbon nanotubes with polymer chains are presented in figure 1. This interaction has been evaluated in our work, improving the thermal and mechanical properties, these results are compared with other obtained in different papers where chemical functionalization was developed [4-5] with diverse results. Also here are discussed the behavior and the results obtained in these materials when relatively high concentrations are incorporated in nanocomposites based carbon nanotubes, showing an interesting effect found in this and in other works.

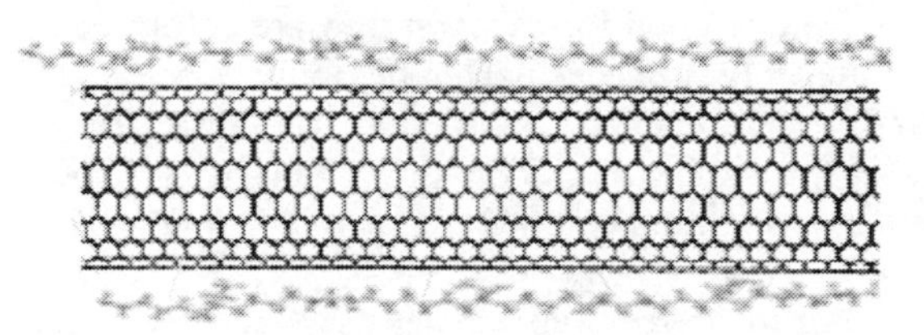

Figure 1. Schematic representation of functionalized carbon nanotube and polymer chains

2 CARBON NANOTUBES TYPES.

Carbon nanotubes have many characteristics that put themselves as a nanofiber with amazing properties that no other material has. The small size, structure, novel arrangements and excellent mechanical, thermal and electronic properties give as result a unique reinforcement that could produce interesting multifunctional composite materials.

In fact the fibers used in composites don't have the outstanding modulus and the conductive properties that CNs possess. For instance, carbon fibers in composites industrial production have functioned as excellent reinforcement for polymers, and produce materials with high quality, excellent properties, low density and high specific strength and specific modulus. Nevertheless CNs have superior properties than any carbon fiber. In spite of this fact it is important to consider that CNs also have a wide range of properties depending on the production approach, which is related with the graphitization degree and therefore with the structure and the properties of these materials. In this context there exist different studies that have shown diverse mechanical properties depending on the carbon nanotube types including the presented number of walls in the material [6-8].

In addition, there exist other important features that place CNs upon the reinforcement fibers used currently in polymer composites. In nanotubes, the diameter at nanometric scale migth be an important point in order to enhance the interface, due to the small size allows to have a closed interaction at the molecular scale, which could be very useful to diminish the zones without contact in the interface, this would enhance the interactions, providing compatibility between two materials: CNs and polymer matrix. CNs diameter size not only plays an important role to provide a scale near to the molecular interactions, also the small diameter together with long extension give high aspect ratio in nanotubes, sometimes more than 1000, this is an important requirement to reinforce composites.

Other important point is that CNs present the property of recovery their original forms when they are deformed by stress and after this is released, This flexibility places these materials in other context as possible reinforcement in composites, since neither carbon fibers nor other known reinforcement present this property [7].

All mentioned above sign on that CNs represent new possibilities to develop polymer composite materials, however the development of these composites and the goal of reach important links in the interface have found different barriers to overcome. Thus the processing of really new nanocomposites with relevant properties is beginning and it is needed in first place to develop different researches to understand the phenomenon that are produced in the interface between CNs and polymer. The comprehension of the parameters that are relevant in this zone will provide the knowledge to manage the nanocomposites properties and therefore it will allow to develop nanocomposites with diverse properties taking advantage in the best manner of CNs properties.

3 CARBON NANOTUBE POLYMER NANOCOMPOSITES.

In agree with the mentioned in the later section, carbon nanotubes type is a important parameter that must be considered. However there are other relevant points such as polymer kind, additives and processing methods, which play an important role in the interface and therefore in final properties of CNs nanocomposites. Thus, different approaches have been used to produce carbon nanotube polymer nanocomposites (CNPN), some of the processes used are: melt blend [9], dissolve and casting [1], "*in situ*" polymerization [3] and extrusion [10] among others. Next are discussed and analized the mechanical and thermal properties of two different types of nanocomposites, in both were utilized Multiwalled nanotubes (MWNTs), however in one case were used MWNTs without purification and methyl-ethyl methacrylate copolymer (MEMA) dissolved with acetone using additives and ultrasound in order to improve the distribution of CNs in the polymer. The second type of CN composites were developed using functionalized Multiwalled nanotubes (f-MWNTs) and "*in situ*" polymerization. This approach allows distribute nanotubes when monomer is in solution and the polymer is in grown. More experimental details can be found in references [1,3].

Figure 2 shows the storage modulus (E') obtained by DMA (Dynamical Mechanical Analysis) for the samples produced by dissolution of MEMA copolymer . The most outstanding modulus is obtained when only 1 wt % CN without additives (sample 1) were used, increasing the modulus by more than 200% at 40°C, which represents one of the most relevant increment reached until now at room temperature with the incorporation of very low quantity of CNs.

Figure 3 shows the results of storage modulus (E') obtained in DMA for the samples where an surfactant was used (0S, 1S, 5S, 7S, 10S), E' is increased in the samples with 1 wt % and 5 wt % of CN with respect to the sample with only 1 wt% of surfactant, nevertheless in the samples with 7 wt % (7S) and 10 wt % (10S), E' diminish with respect to the sample 5S and 1S. This behavior have been presented in other reports where more than 5 wt % of nanotubes were used, inasmuch as CNs form cluster zones which difficult the interaction at interface, avoiding that several nanotubes maintain contact with polymer matrix.

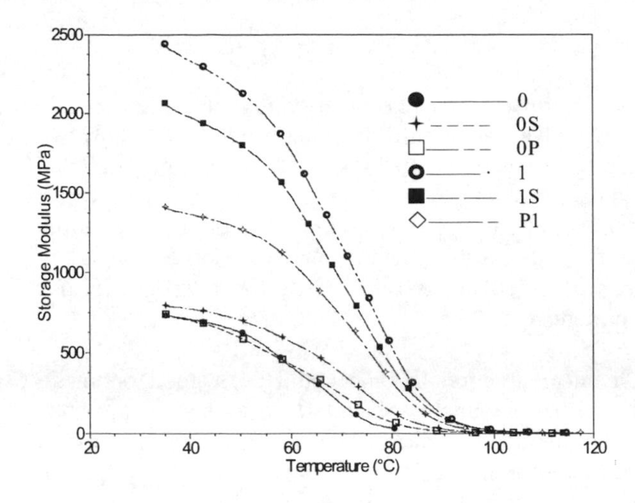

Figure 2. Storage modulus of carbon nanotube composites. samples: 0 (MEMA), 0S (MEMA - 1wt% surfactant), 0P (MEMA - 1wt% plasticizer), 1(MEMA - 1wt% CN), 1S (MEMA - 1wt% surfactant- 1wt% CN), 1P (MEMA – 1wt % plasticizer- 1wt% CN). [1] Copyright IOP

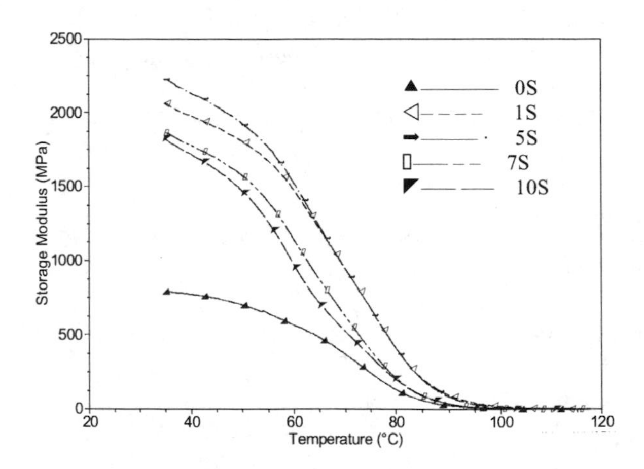

Figure 3. Storage modulus of carbon nanotube composites, samples: 0S (MEMA - 1wt% surfactant), 1S (MEMA - 1wt% surfactant- 1wt% CN), 5S (MEMA -5wt% surfactant- 5 wt% CN), 7S (MEMA -7wt% surfactant- 7wt% CN), 10S (MEMA -10wt% surfactant- 10wt% CN). [1] Copyright IOP

The thermal behavior for these samples were obtained by DMA, in this case glass transition temperature (Tg) results were calculated with the tangente delta maximum for all samples. Results shows that in spite of there is not a clear tendency in the behavior of Tg when nanotubes are incorporated, the samples that contain CNs have higher Tg than the polymer samples.

Figure 4 shows DMA results for samples where f-MWNTs were incorporated to poly (methyl methacrylate) (PMMA) polymer matrix. In this characterization, samples with 1 wt% of MWNTs and 1 wt% and 1.5 wt% of f-MWNTs were analyzed and compared with polymer matrix. It is evident that the sample which contains 1 wt% of the f-MWNT has a better behavior, in terms of storage modulus (E'), than the sample which contains 1 wt % of MWNT without chemical treatment. Sample with 1.5 wt% of f-MWNTs shows the same behavior than the sample with 1 wt% of f-MWNTs. However the most significant increment is reached at relatively high temperature. The curves show that the Storage modulus at 90°C for the samples which contain f-MWNTs raise in more than 1000 % with respect to the polymer sample, and in comparison with only 250% of the sample with MWNTs. This effect is caused by the interaction at molecular level between the chemical moieties on CNs surface with polymeric chains. Details of diverse chemical functionalizations on carbon nanotubes can be found in reference [11,12]. Spectroscopy characterization about the interaction between nanotubes and polymer is presented in detail in reference [3].

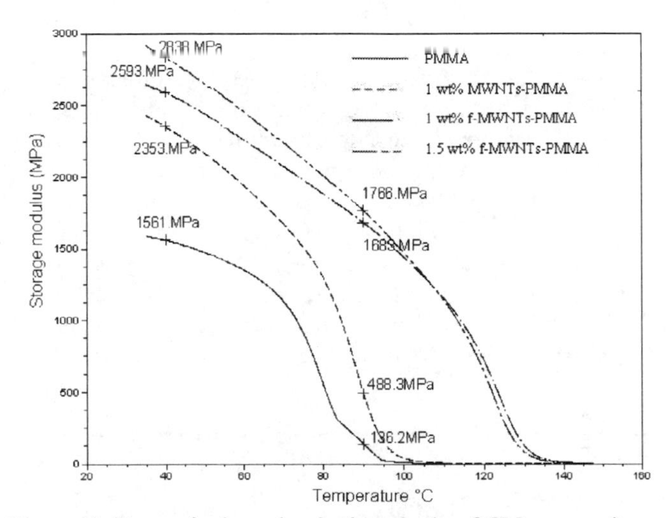

Figure 4. Dynamical mechanical analysis of CN composite. Storage modulus. [3] Copyright ACS

In addition, thermal results of nanocomposites than contain f-MWNTs shown a relevant increment in the Tg (obtained with the tangent delta maximum) around 40°C with respect to polymer samples. Nanocomposites which include MWNTs only raise Tg in 3°C with respect to polymer matrix.

Figure 5 shows an image obtained by Scanning Electron Microscopy (SEM) on fractured zone of a nanocomposite. This composite was developed with functionalized nanotubes. Here is notable that nanotubes interact with polymeric matrix at the interface level and take part in the stress when the material is fractured, proving that chemical interaction improve the transfer of mechanical load, which is reflected in mechanical and thermal properties of these f-MWNT nanocomposites.

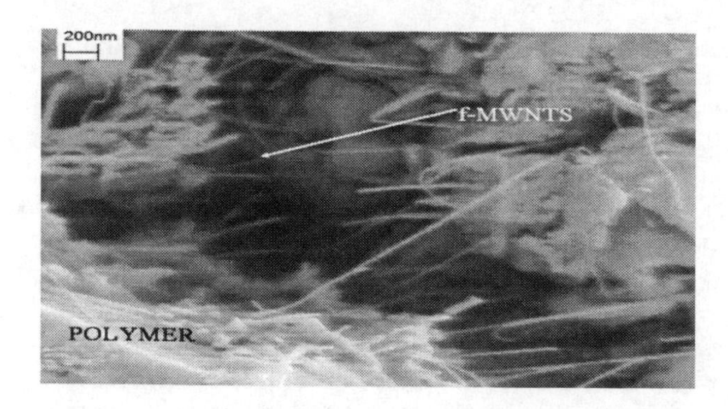

Figure 5. SEM image of a fissure after tensile test, in f-MWNT composite fracture zone, where it is possible to observe different fractured nanotubes. [11] Copyright Nova Science Publishers.

Chemical functionalization in this case shows that take advantage of moieties on CNs surface in order to interact with polymer matrix could be an important approach in order to form interesting multifunctional nanocomposites. Table 1 shows different results where functionalized nanotubes have been employed. Functionalization in these studies consists of: oxidized COOH terminated MWNTS (f-MWNTs) and PMMA as matrix in reference [3], fluorinated single walled nanotubes (fl-SWNT) and poly(ethylene oxide) as matrix in reference [5] and in situ functionalized imidazole moieties- MWNTs (i-MWNTs) and phenoxy resin as matrix. These researches present interesting improve of thermal and mechanical properties, however different results have been found in other work [4].

Parameter	f-MWNTs [3]	fl-SWNT [5]	i-MWNTs [13]
E' blank (MPa)	(at 40°C) 1561	(at 20°C) 400	(at 30°C) 2025
E' 1 wt % f-MWNT	(at 40°C) 2593	(at 20°C) 600	1.5% wt% (at 30°C) 2150
Increment in E' (1 wt % f-MWNT)	66%	50%	(1.5 wt %) 6%
Tan Delta Max. (Tg °C) blank	97.1	----	95
Tan Delta Max (Tg °C) 1 wt % f-MWNT	136.4	----	98
Increment in Tg °C	39.3°C	---	3°C

Table 1: Thermal and mechanical properties of functionalized nanotube composites.

4 CONCLUSIONS

CNs promise be an important nanomaterial in the field of new advanced multifunctional materials, however diverse results have been found in the research developed nowadays, showing that different parameters such as: polymer kind, nanotubes type, chemical functionalization, additives, processing method between other important points must be considered in order to form and control this kind of nanocomposites.

Chemical functionalization could be a good option to reach the best interaction in interface in nanocomposites, however chemical conditions and functional groups are in currently research around the work with the target of manage and control this parameter in the best way.

ACKNOWLEDGMENTS

To Mrs. Carmen Vasquez for her assistance in thermal analysis.

REFERENCES

[1] C. Velasco-Santos, A.L. Martínez-Hernández, F. Fisher, R.S. Rouff and V.M. Castaño, *J. Phys. D. Appl. Phys.* 36, 1423, 2003.

[2] C. Velasco-Santos, A.L. Martínez-Hernández and V.M. Castaño, *Composites Interf.* In press 2005.

[3] C. Velasco-Santos, A.L. Martínez-Hernández, Fisher, F.; Ruoff, R.S. and V.M. Castaño, *Chem. Mat.* 15, 4470, 2003.

[4] G. M. Odegard; S. J. V. Frankland and T. S. Gates, 44th AIAA/ASME/ASCE/AHS Structures, Structural Dynamics, and Materials Conference, Norfolk, Virginia, AIAA 2003, 1701.

[5] H.Z. Geng, R. Rosen, B. Zheng, H. Shimoda, L. Fleming, J. Liu and O. Zhou, *Adv. Mater.* 14, 1387, 2002.

[6] 21. M-F. Yu, O. Lourie , M.J. Dyer , K. Moloni , T.F. Kelly and R.S.Ruoff , *Science* 287, 637, 2000.

[7] T.W. Ebbesen H.J. Lezec, H. Hiura, J.W. Bennett, H.F. Ghaemi and T. Thio. *Nature,* 382, 54, 1996.

[8] J. Che, T. Cagin and W.A. Gooddard III, *Nanotechnology,* **11**, 65,2002.

[9] Z. Jin, K.P. Pramoda, G. Xu and S. Hong Goh, *Chem. Phys. Lett.* **337**, 43, 2001.

[10] M.L. Shofner, F.J. Rodriguez-Macias, R. Vaidyanathan and E.V Barrera, *Comp. Part. A.* 34, 120, 2003.

[11] C. Velasco-Santos, A.L. Martínez-Hernández and V.M. Castaño, *Chemical functionalization on carbon nanotubes: principles and applications* in "Trends in nanotechnology Research" (ed.) Eugene Dirote, Nova Science Publishers, New York 2004.

[12] C. Velasco-Santos, A.L. Martínez-Hernández, M. Lozada-Cassou, A. Alvarez-Castillo and V.M. Castaño, *Nanotechnology,* 13, 495, 2002.

[13] H. W. Goh; S. H. Goh; K.P. Pramoda and W.D. Zhang *Chem. Phys. Lett.*, 373, 277, 2003.

Improvements in High Specific Strength Epoxy-Based Composites using High Magnetic Fields

H. Mahfuz, S. Zainuddin, Vijaya K. Rangari, S. Jeelani
Tuskegee University's Center for Advanced materials (T-CAM), Tuskegee, AL 36088
Florida Atlantic University, Boca Raton, FL 33431, hmahfuz@oe.fau.edu
And
M. R. Parker, T. Al-Saadi
Department of Electrical and Computer Engineering, University of South Alabama, Mobile, AL 36688.

ABSTRACT

Changes occur in the bulk physical properties of polymers, when cured in the presence of high magnetic fields, has been known for some considerable time. It has also long been realized that polymer-based composites containing low-percentage nanoparticle dispersions have attractive mechanical characteristics such as high compressive/tensile strength and stiffness. This paper describes, for the first time, significant improvements in those mechanical properties when epoxy-based composites are cured in high uniform magnetic fields. An interesting additional finding, also described here, concerns concomitant improvements in the thermal stability of these systems.

Keywords: High magnetic fields, magnetic flocculation, nanoparticles, epoxy.

1 INTRODUCTION

Numerous studies have been carried out, over a period of several years, on the detection of changes in physical characteristics of polymers subjected to high magnetic fields during the curing stages. Characteristics changes investigated include viscoelasticity[1], stress-strain behavior[2], fracture toughness[3], and electrical polarization[4]. It has been widely demonstrated that liquid crystal thermosets (LCTs) reveal fundamental changes in structure when cured in high magnetic fields. These include the inducement of anisotropic thermal expansion properties[5]. More recent studies of LCTs by Douglas and various co-workers[6,7] have shown extremely large increases in the tensile elastic modulus, with values of up to 8.1 GPa for samples cured in fields of 12 T, compared to 3.1GPa for samples cured in zero field.

Additionally, a wide body of literature[8-13] provides convincing evidence that an infusion of nanoparticles (NPs) can bring about enhancement of various mechanical properties of polymer composites. Some of the present authors have recently successfully demonstrated enhancements in a wide range of mechanical and thermal characteristics (also discussed in detail below in this paper) of carbon NP- infused bulk composites[11], sandwich composites[12] and polyurethane foams[13].

The present paper describes an attempt to combine these two separate bodies of work and to use high magnetic fields as a means of obtaining superior distributions of NPs in epoxy-based composites. In the following are described the results of preliminary experiments designed to improve upon the bulk mechanical properties of such composites by subjecting the fluid mix, during the curing process, to high uniform magnetic fields. The aim here is to initiate the process of magnetic flocculation[14] of these NPs during the curing process and, in particular, during the pre-gel stage at which time the viscosity of the matrix is still relatively modest.

Magnetic flocculation is a term commonly used to describe the agglomeration of colloidal microscopic particles under the influence of a uniform applied magnetic field[5]. One of the present authors[14,15] has previously demonstrated that, provided sufficiently large magnetic fields are applied, both paramagnetic and diamagnetic particles of microscopic dimensions can be flocculated. The process is characterized by the formation of pairs of these microscopic particles joined end to end within the colloid and parallel to the field axis. Such a phenomenon is termed 'binary' flocculation. Under more favorable conditions, namely high magnetic volume susceptibility particles, coupled with high magnetic fields, low matrix viscosity and higher particle number densities, higher order forms of flocculation (i.e. short chains) become feasible. Monte Carlo simulations have also indicated a tendency of the chains to organize themselves in regularly spaced fashion in cross-section[16,17]. This latter point is consistent with optical transmission studies[17,18] on colloids in high magnetic fields. The task of promoting binary or higher order forms of magnetic flocculation in fluid nanocomposites is a daunting one. The small particle size, coupled with the low magnetic volume susceptibility of the majority of commercially available types, reduces the likelihood of uniaxial inter-particle dipolar interactions having the capability to dominate isotropic clustering induced by Van der Waals interactions. The latter effect compounds the problem further by effectively imposing an upper limit on composite number density (of order 1% by particle volume) due to the tendency of NPs in fluid composites to form the aforementioned clusters in the preliminary sonification and mechanical mixing stages.

In the experiments described here, the fluid matrix chosen was an epoxy resin system, SC-15, of the sort used in high specific mechanical strength systems[11-13]. This epoxy was used in conjunction with a standard (cycloaliphaticamine+ Polyoxyalkyleneamine) hardening agent, comprising approximately 30% of the total composite volume. With this mix, the composite changes, at room temperature, from a relatively low viscosity system, (SC-15, for example, has about 2.5 times the viscosity of water), to a gel in about one half hour and , eventually, to a solid matrix in 2 – 4 hours, although the complete curing may last up to 24 hours.

2 EXPERIMENTAL PROCEDURE

The preparation of the nanocomposites was undertaken in two separate stages. First, spherical NPs were added at a 1% loading, by weight, to (part A) SC-15 epoxy resin (60-70% diglycidylether of bisphenol A, 10-20% aliphatic diglyciylether, 10-20% epoxy toughener) and dispersed by acoustic cavitation. Two principal types of NPs were used: SiC and TiO_2, with respective diameters 29nm and 30 nm. This dispersion was carried out in a Sonic Vibra Cell Ultrasound liquid processor for about 30 minutes at room temperature. The vessel containing the composite was externally cooled during this process to prevent undesired temperature rises. Following infusion of the nanoparticles, this modified epoxy part A was mixed with part B hardener, comprising 70-90% cycloaliphatic amine and 10-30% polyoxylalkylamine. The hardener was added to part A at a volume ratio of 3/10. Mixing was promoted with a high-speed mechanical stirrer for about 5 minutes. The mixture was then vacuum-degassed for 10-15 minutes. After completion of this sample preparation step, the admixture was added to cylindrical PVC vessels of a length designed to match the most uniform portion of a high-field Bitter magnet. The curing process of the various samples took place within the magnet at various high fields and for a variety of times of exposure.

3 RESULTS AND DISCUSSION

Static compression tests were performed on test coupons of dimension 12.7 mm x 12.7 mm x 25 mm, cut from the cylindrical samples. Four replicate test coupons of each sample were examined on a hydraulically-controlled MTS machine to ASTM standards. In addition, dynamical mechanical analysis (DMA) tests were undertaken to determine the thermal transition temperature and the elastic response of polymers as a function of temperature. Thermo-gravimetric analyses were also carried out on all samples as a means of detecting any magnetic-field induced changes in the thermal stability of the composites.

The stress-strain characteristics of magnetically cured composites under quasi-static compression loading were investigated and the resulting mechanical strength and stiffness data (for 1% SiC NP loading) are shown, respectively, as a function of exposure time in a magnetic field of 16T in Figs. 1 and 2.

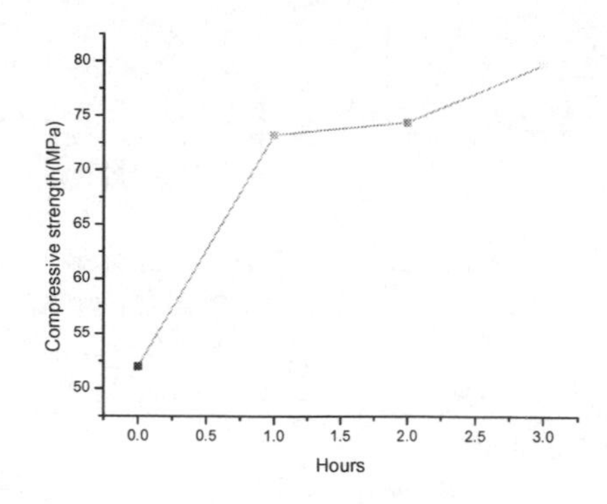

Fig.1. Shows increase in Strength with time at a Magnetic strength of 16 Tesla.

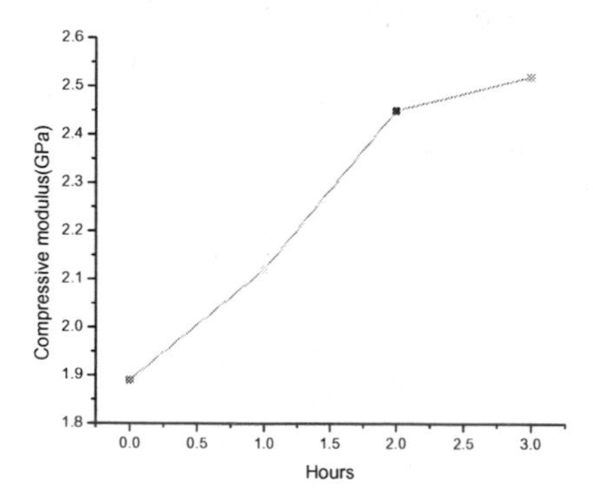

Fig.2. Shows increase in Stiffness with time at a Magnetic strength of 16 Tesla.

Here, it was seen that strength and stiffness of 1% SiC-epoxy composite prepared at 16T for 3 hours showed increase, respectively, by 35% and 72% compared to neat control sample values and 49% and 33% compared to control sample values of same % NP loading. Clearly, in both instances, experimental data give no hint of saturation of the magnetic field effect. Similar experiments, performed in a field of 21T for 3 hours, show even bigger improvements in these parameters with 1% TiO_2 NP loading (98% and 114%, respectively) over neat control sample values as shown in table 1.

Table 1: Compressive response of polymers cured in zero field and in magnetic field at magnetic strength, 21 Tesla.

Material	Magnetic curing (hrs)	Compressive Strength (MPa)	Gain/loss in %	Compressive Modulus (GPa)	Gain/loss in %
Neat SC-15	0	58.47	-	1.46	-
Neat SC-15	3	70.46	+20	2.17	+48.6
1%SiC	0	53.46	- 9	1.92	+34.2
1%SiC	3	104.85	+79	2.71	+85.6
1%TiO_2	0	65.96	+13	2.35	+60.9
1%TiO_2	3	116.1	+98	3.12	+114.7

It was observed that the neat epoxy polymer cured in an uniform magnetic field have significant increase over neat controlled polymers. With the increase in magnetic strength and cure time, properties enhanced significantly in almost linear fashion. In making a comparison between the flocculated and non-flocculated systems, at least half of the enhancement in properties is contributed by magnetic flocculation. It gives a clear indication that properties of polymers can be further enhanced by choosing better susceptible particles, higher magnetic strength and/or longer magnetic field curing.

A *Dynamic mechanical analysis* was also used to determine the elastic response of these nanocomposites by subjecting the test coupons in flexural mode. Fig. 2(a) and (b) shows, respectively, storage moduli for 1% SiC and 1% TiO_2 composites prepared under various field conditions. In Fig. 2(a) an increase in the storage modulus of around 226MPa was observed for SiC composite cured for 3 hrs in a 21T field. Storage modulus of polymers decreases as they approach their glass transition temperature. Significant increase in glass transition temperature was achieved that can be inferred From Fig. 2(b), Tan delta vs. temperature curves, 1% SiC composite cured for 3 hrs in a 21T field, showed increase of about 14^{0}C. Similar results were seen with TiO_2 composites similarly treated with a significant enhancement of mechanical stiffness (approximately 315Mpa).

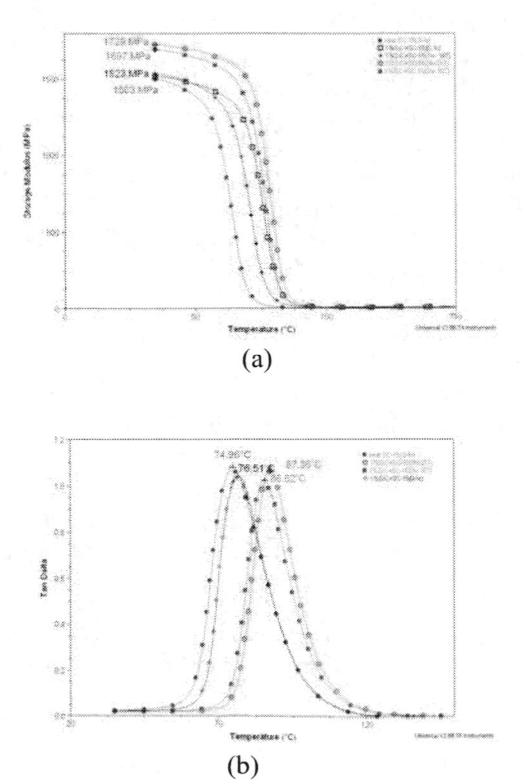

(a)

(b)

Fig.2 (a,b)Elastic response and glass transition temperatures of SiC polymer composites cured in an uniform magnetic field for different length of time.

It was observed that with increase in magnetic field curing, significant increase in storage moduli and glass transition temperature were achieved. This results clearly suggests that with the combination of field magnitude, time of exposure, effective flocculation and susceptibility of particles, the properties can be further enhanced.

Thermo-gravimetric analysis (TGA) was carried out for these systems to understand the effect of magnetic field on the thermal stability of the nanocomposites. Figure 3 (a,b)show, TGA outputs for neat epoxy and TiO_2–epoxy nanocomposites prepared under zero field. It was found that 50% by weight of the neat sample was decomposed at ~ 375°C temperature where as the 1% TiO_2 decomposed at ~ 378°C under similar conditions. This small difference in decomposition temperatures were due to the effect of nanoparticles presence in epoxy[10]. The Figure 3(c) represents the TGA output of 1% TiO_2 nanocomposite cured at 21T in a magnetic filed. In this graph 50% by weight of the sample was decomposed at ~ 397°C showing significant increase in decomposition temperature. This increase of about ~ 20°C clearly shows the effect of magnetic field curing and this increase in temperature may be due the alignment of particle as a result of magnetic flocculation.

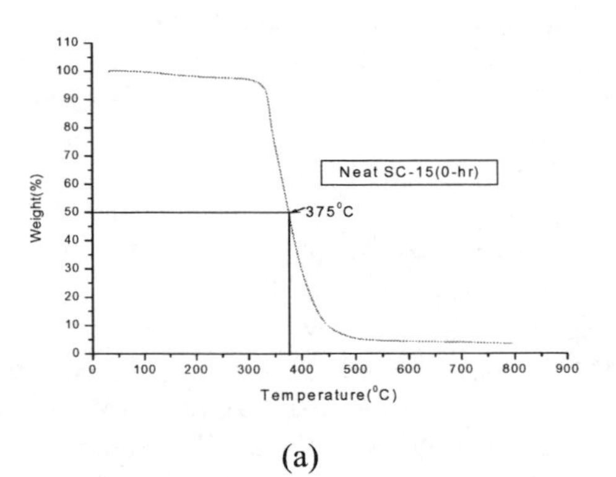

(a)

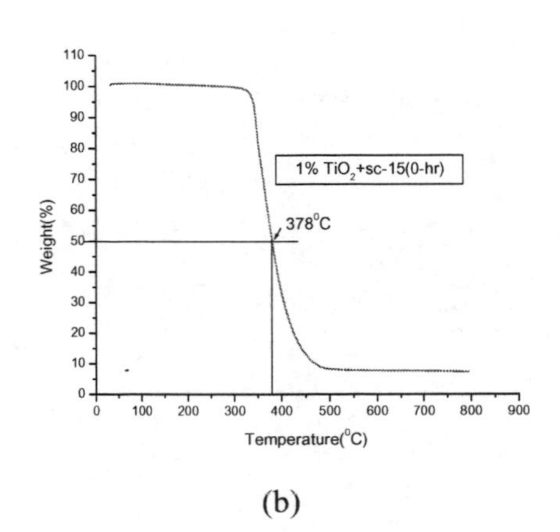

(b)

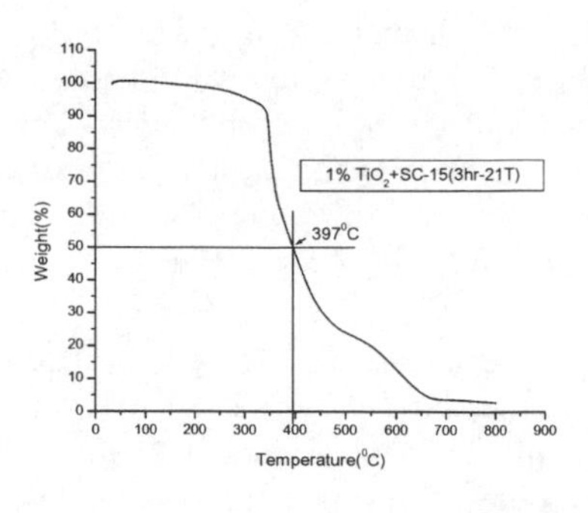

(c)

Fig.3 (a,b,c). TGA of zero field cured epoxy polymer and magnetically cured epoxy polymers at 21T.

4 CONCLUSION

The ability of high magnetic fields to affect, in a positive manner, the mechanical strength of an epoxy resin nanocomposite has been clearly demonstrated. The NP systems chosen here - diamagnetic SiC and paramagnetic TiO_2- have extremely weak magnetic susceptibilities (-6.3 x 10^{-6} and 3.54 x 10^{-6} cgs units, respectively). The fact that the results for TiO_2 are better than for SiC, even though the numerical susceptibility is smaller can be explained by the fact that the magnetic dipolar interaction depends on the *effective* volume susceptibility, χ_{eff} (i.e. on the difference between the susceptibility of the particle and that of the background fluid). The SC-15 epoxy is, clearly, diamagnetic, meaning that its volume susceptibility effectively adds to that of the TiO_2 and subtracts from that of the SiC. What has not yet been shown in this study is an optimized nanocomposite, given that higher fields applied during curing continue to show additional improvement. Clearly, the use of higher magnetic susceptibility nanoparticles would offer an easier opportunity for system optimization. Further studies of this kind are being planned. Furthermore, in the case of strongly paramagnetic substances, the likelihood of securing viable higher nanoparticle loading values in the mix is much increased. The dipolar interaction due to the magnetic field increases as $(\chi_{eff})^2$, meaning that initial spherical particle clusters, held together by Van der Waals forces, ought to be effectively distributed by the magnetic field into more favorable dispositions.

5 ACKNOWLEDGEMENTS

A portion of this work was performed at the National High Magnetic Field Laboratory, which is supported by NSF Cooperative Agreement No.DMR-0084173, by the State of Florida, and by the DOE. The sample holders and support system was designed by L. McCormick.

6 REFRENCES

1. S. Jaromi, W.A.G. Kuipers,B. Norder, W. J. Wijs, *Macromolecules*, 1995, 28, 2201.
2. A. Shiota, C. K. Ober, *J. Polymer Sci. Part B: Polym. Phys.*1998, 36, 31.
3. C. Carfagna, E. Amendola, M. Giamberini, *Liquid Crystalline Polymers:Proceedings of the international Workshop on Liquid Crystalline Polymers*, C. Carfagna Ed.; Pergamon Press: Oxford, U.K.1994, 69-85.
4. W-F. A. Su, *J. Polymer Sci., Part A: Polym. Chem.,* 1993, 31, 3251.
5. G. G. Barclay, C.K. Ober, K.I. Papathomas, D. W. Wang, *Macromolecules*,1992, 25, 2947.
6. B. C., Benicewicz, M. E., Smith, J. D. Earls, R. D. Priester, S. M Setz., R. S. Duran, E. P. Douglas. *Macromolecules* 1998, 31, 4730-473.
7. M.D.Lincoln, E.P.Douglas *Polymer Eng. and Sci.* 1999, 39, 1903-1912.
8. E. Reynaud, C. Gautier and J. Perez, "Nanophases in Polymers", Review Metallurgie, 96, pp169-176, 999.
9. L.W. Chun, Q. Z. Ming, Z. R. Min and F. Klaus, "Tensile Performance Improvement of Low Nanoparticle-Filled Polypropylene Composites", Composites Science and Technology, 62, pp1327-1340, 2002.
10. H. Mahfuz,, N. Chisholm, and S. Jeelani, "Mechanical Characterization of Nanophased Carbon Composites," 2002 ASME winter Annual Meeting, New Orleans, Louisiana.
11. H. Mahfuz, V. Rangari, A. Adnan, and S. Jeelani, "Carbon Nanoparticles/Whiskers Reinforced Composites and their Tensile Response," Composites Part A: Applied Science and Manufacturing, Vol. 35 (2004), pp519-527.
12. H. Mahfuz, M. Islam, V. Rangari, M. Saha, and S. Jeelani, "Response of Sandwich Composites with nanophased Cores under Flexural Loading,"Composites Part B: Engineering, in press, 2004.
13. H. Mahfuz, V. Rangari, M. Islam, and S. Jeelani, "Fabrication, Synthesis and Mechanical Characterization of Nanoparticles Infused Polyurethane Foams,"Composites Part A: Applied Science and Manufacturing, Vol. 35 (2004), pp453-460.
14. M. R. Parker, R. P. A. R. van Kleef, H.W. Myron, P. Wyder, "Particle Aggregation in Colloids in High Magnetic Fields", J. Mag. and Mag. Materials, Vol. 27, pp250-256, 1982.
15. R. P. A. R. van Kleef, H. W. Myron, P. Wyder, M. R. Parker, "Flocculation of Diamagnetic Particles in High Magnetic Fields", J. App. Phys., Vol. 54, pp 4223-4225, 1983.
16. J. J. Miles, R. W. Chantrell, M. R. Parker, "Model of Magnetic-Field Ordering in Dispersions of Fine paramagnetic Particles", J. App. Phys., Vol. 57(1), pp 4271-4273, 1985
17. A. Tasker, R. W. Chantrell, J. J. Miles, M. R. Parker, A. Bradbury, "Monte- Carlo Simulations of Light Transmission in Dispersions of Paramagnetic Particles", I.E.E.E. Trans. on Magn., Vol. 24, pp1671-1673, 1988.
18. J. J. M. Janssen, Ph.D. Dissertation, University of Nijmegen, 1989.

Cutting Fluorinated/Ozonated Single-Walled Carbon Nanotubes by Pyrolysis

Z. Chen*, Z. Gu**, K. Ziegler***, C. K. Chan****, J. Shaver*****,
R. H. Hauge****** and R. E. Smalley*******

Carbon Nanotechnology Laboratory, Department of Chemistry, Center for Nanoscale Science and Technology. Rice University, 6100 Main St. MS 100, Houston, TX, 77005.
*zychen@rice.edu, **zngu@rice.edu,***kziegler@rice.edu,****ckchan@rice.edu,
*****shaverj@rice.edu , ****** hauge@rice.edu,*******smalley@rice.edu

ABSTRACT

Pyrolysis and/or piranha treatment of fluorinated /ozonated single-walled carbon nanotubes (SWNTs) were found to have cut the nanotubes to short length (~100nm). Pyrolysis of partially fluorinated SWNTs (~C_5F) at 700°C cut the nanotubes without causing significant crosslinking among the SWNTs. When pyrolyzed at lower temperatures, the fluorinated SWNTs need room temperature piranha treatment to finish the cutting. SWNTs that was ozonated in perfluoropolyethers (PFPE) at room temperature has also been cut when pyrolyzed at 375°C. Functionalization of cut-SWNT with alkylhalide in Li/NH_3(l) was used to debundle the nanotubes into individuals in chloroform for the length measurement. The cut-SWNTs have been characterized by Raman and AFM. AFM shows cut-SWNTs with a length distribution mainly between 30nm and 100nm. Large scale cutting of SWNTs would lead to novel applications of nanotubes in microelectronics, biological imaging and sensing, drug delivery and composite material, etc.

Keywords: nanotubes, cutting, fluorination, ozonation, pyrolysis.

1 INTRODUCTION

Since their discovery in 1991[1], carbon tubes are one of the most promising constituents of carbon nanoscale composites because of their unique electrical, mechanical and optical properties. Because they are hollow and much smaller than the blood cells, some methods were developed to attach DNA and protein molecules to the inside and outside of the nanotubes, this allows us to target and destroy individual cancer or virus cells. [2]. However, the SWNTs with uniform length scale are required to accomplish these applications. Short SWNTs were obtained by chemical etching using H_2SO_4/HNO_3[3], Also Gu[4] got the bundled cut-SWNTs by fluorination and pyrolysis at 1000C. In this paper, we introduce the cutting of single-wall carbon nanotubes without crosslink to an average length of 100nm by pyrolysis. Recently the STM analysis of Kelly[5] shows that fluorine could be removed slowly by pyrolysis , So cutting of SWNTs can be viewed as a two step process----introduction of sidewall damages and exploiting the damage sites to create cut-SWNTs. [6] First of all, the purified single-wall carbon nanotubes need to be partially fluorinated or ozonated. The cut nanotubes(cut-SWNTs) can be produced directly by pyrolysis of fluorinated SWNTs(F-SWNTs) in an argon atmosphere at 700° C or by room temperature piranha treatment after lower temperature pyrolysis. Also ozonated SWNTs(O-SWNTs) has been demonstrated to be cut by pyrolysis at 375°C. Functionlization in Li/NH_3(l) by dodecyl groups[7] was used to debundle the SWNTs for characterization. Raman imaging and atomic force microscopy (AFM) has been successfully employed to study SWNTs' length distributions.

2 EXPERIMENTAL

2.1 Raw material and Purification

The SWNTs used in our experiment were produced by HiPco process [8]. Purification process [9] was used to remove the amorphous carbon and iron particles. The average length of the nanotubes is more than 200nm. All the AFM images in this paper are taken from SWNTs samples .that have been individualized with dodecylation in Li/NH_3(l)

2.2 Fluorination

The fluorination on the sidewalls of SWNTs was done at temperature between 50° C and 70° C in 10% Fluorine/90% Helium flow for 2~4 hrs. It was used to make F-SWNTs with a stoichiometry of C_5F. This process introduces functionlization sites on the sidewall of SWNTs.

2.3 Pyrolysis

The pyrolysis was done in a quartz boat in tube furnace. The sample was heated in an argon atmosphere at a rate of 20° C/min from room temperature to target temperature.

2.4 Piranha/Ammonium Persulfate Treatment

Piranha is 1:4 (vol/vol 30% H_2O_2/96% H_2SO_4 solution. Room temperature treatment for 1 hour was used. The solution was added to the noanotubes(1mL piranha: 1 mg nanotubes). Ammonium Persulfate solutions were prepared by dissolving 0.1g of the salt in 1mL of 96% H_2SO_4. The persulfate solution was then added to the nanotubes(1mL solution: 1mg of nanotubes). Both these two methods can cut the nanotubes.

2.5 Ozonation

Ozonation plus piranha treatment is another potential method of cutting SWNTs. Previous work has been done in methanol at -78° C [9, 10] though it is an unstable system due to the explosive nature of ozone/methanol when warmed. Here cutting was done with O_3 bubbling (3.7% O_3 in O_2) in nanotubes/PFPE solution with homogenizing at room temperature for three hours, clean out by quenched with Nanopure (Barnstead International, Dubuque, IA) water and ethanol(200 proof) for 10 times. After that, cut-SWNTs were extracted into ethanol and PFPE was separated from the two-phase system. The reason PFPE was used as solvent is O_3 has very high solubility in PFPE.

3 RESULTS AND DISCUSSION

AFM image of purified nanotubes---the precursor is shown in Fig. 1. Fig. 2 shows the length distribution of the precursor. Functionalization debundles the SWNTs to make it soluble in chloroform for AFM histogram measurement. The length distribution shows there are tubes whose lengths are between 50nm~1000nm. The tube lengths were measured by Nanotube Length Analysis package of SIMAGIS software. Typically more than 500 nanotubes need to be measured to get statistical results.

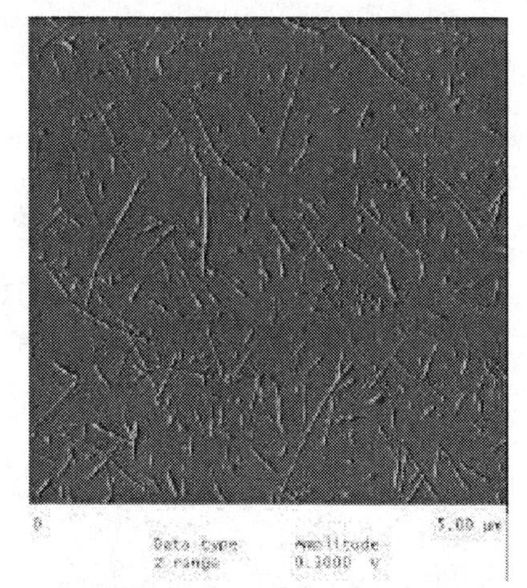

Fig. 1 AFM image of the purified SWNTs

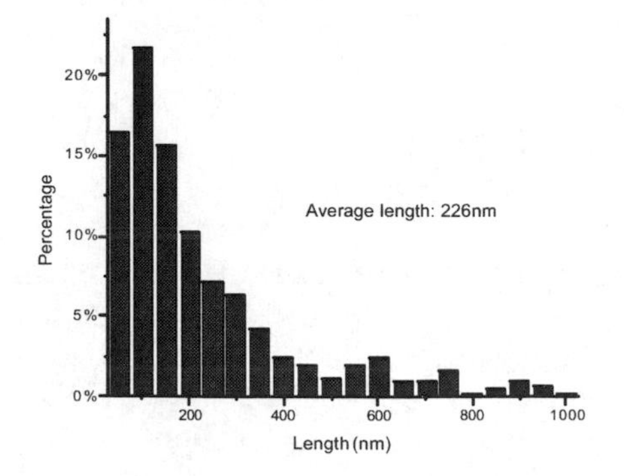

Fig. 2 Length distribution of purified SWNTs

To cut the nanotubes to shorter length, Pyrolysis of F-SWNTs in an argon atmosphere at 700°C can exploit the damage sites and produce the cut-SWNTs without crosslink but with shorter average length of 87nm. After pyrolysis, the SWNTs sample was functionalized with the same mechanism as mentioned above. Gu[4] pyrolyzed the F-SWNTs at 1000°C and cut the nanotubes to short bundled nanotubes with length of 50nm. We found that the pyrolysis beyond 700°C induce the crosslink. But the pyrolysis below 700°C is able to cut SWNTs without causing significant crosslinking between nanotubes.

Fig. 3 is the AFM image of cut-SWNTs by pyrolysis at 700°C. Fig. 4 shows the length distribution of the cut-SWNTs. Fig. 5 shows the Raman spectrum of purified SWNTs, fluorinated SWNTs, pyrolyzed SWNTs and functionalized SWNTs. The disorder peaks of SWNTs at different processing stages indicate that the sidewall functionalization by fluorination process, the exploiting the damage sites by pyrolysis process and covalent attachment of dodecyl groups by functionalization process respectively. The height data on Fig. 3 indicates that there is no apparent bundling of nanotubes, which suggests that there is no crosslinking during the pyrolysis.

Fig. 3 AFM image of cut-SWNTs(pyrolysis at 700°C of fluorinated SWNTs)

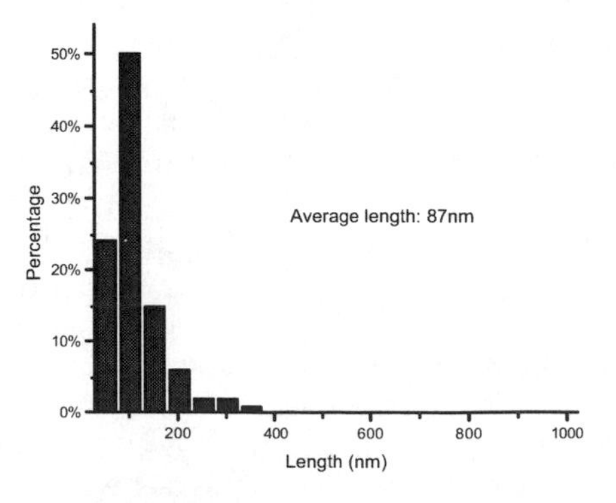

Fig. 4 Length distribution of cut-SWNTs(pyrolysis at 700°C of fluorinated SWNTs)

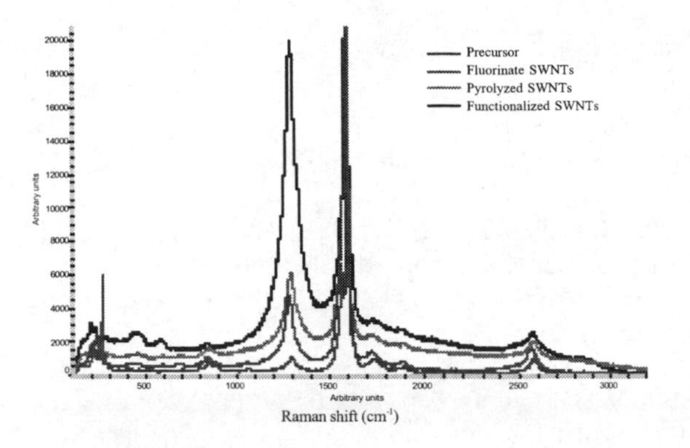

Fig. 5 Raman image of purified SWNTs, fluorinated SWNTs, pyrolyzed SWNTs and functionlized SWNTs.

Pyrolyzing SWNTs at lower temperature (~500°C) followed by room temperature piranha or ammonium persulfate treatment has also been tried, Fig. 6 shows the AFM image. It was found that with this process the SWNTs can be cut to shorter length than the precursor, but longer than the cut-SWNTs we got from 700° C pyrolysis of fluorinated SWNTs.

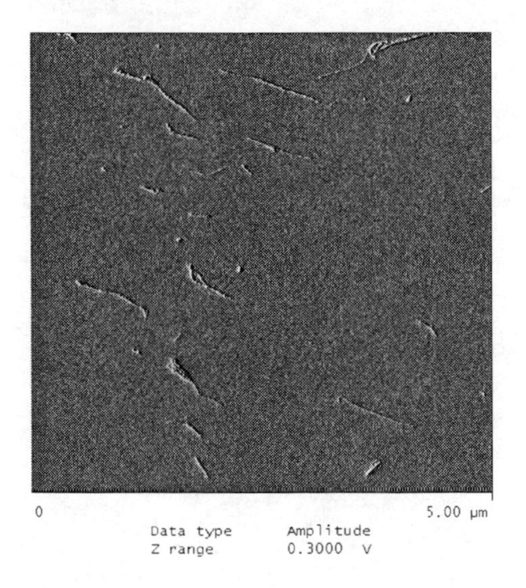

Fig. 6 AFM of Cut-SWNTs(pyrolysis at 500°C of fluorinated SWNTs followed by room temperature piranha treatment).

Besides the F-SWNTs, O-SWNTs can also be cut by pyrolysis. Same functonalization process has been done to debundle the cut-nanotubes for characterization (see AFM picture in Fig. 7). As shown in the Fig. 8, the average length of the cut-SWNTs is 103nm. So compare to precursor, the ozonated SWNTs were cut and we believe that cutting was occurred during the pyrolysis. So far we demonstrated the ozonation plus pyrolysis is another effective way to cut the SWNTs.

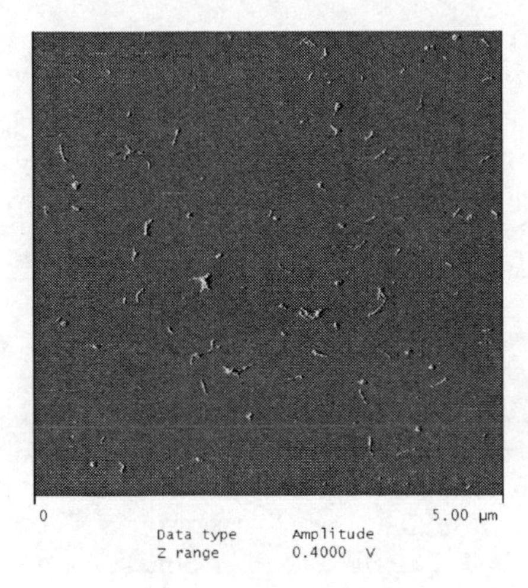

Fig. 7 AFM of Cut-SWNTs(ozonation followed by pyrolysis at 375°C)

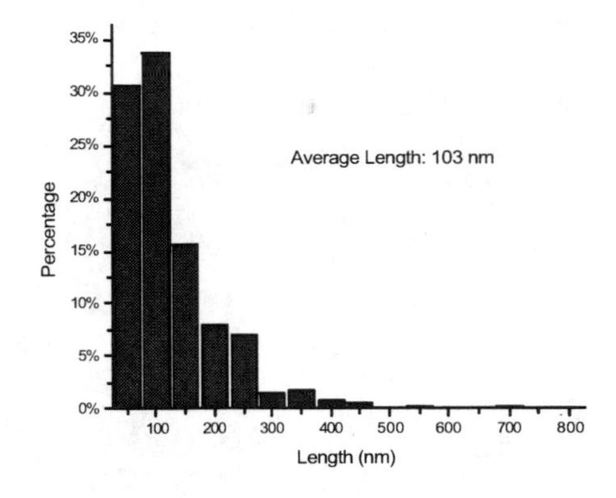

Fig. 8 Length distribution of cut-SWNTs(ozonation followed by pyrolysis at 375°C).

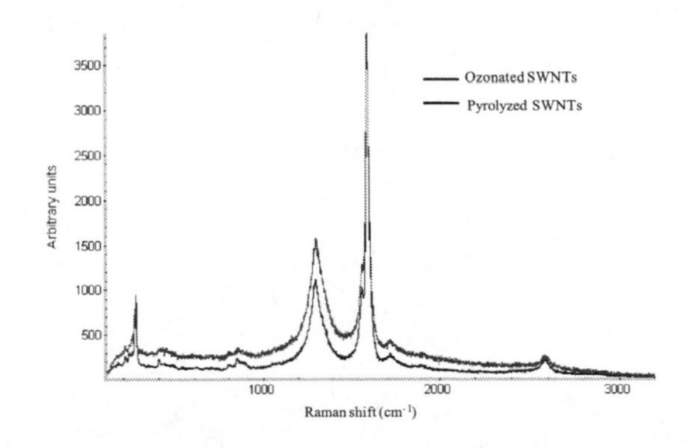

Fig. 9 Raman Spectrum of ozonated and pyrolyzed SWNTs

Raman spectrum in Fig. 9 shows the extensive ozonation of the nanotubes in PFPE solvents. The difference in the intensity of disorder peaks suggests pyrolysis cut the ozonated-SWNTs at damage sites.

4 CONCLUSION

Pyrolysis has been shown as an effective method of cutting. Fluorination and ozonation process can covalently attach fluorine and oxygen on the sidewalls of SWNTs. during the subsequent pyrolysis, C-C bonds are broken and cutting occurs at the damages sites. Pyrolysis at 700°C can cut the F-SWNTs without crosslink. At lower temperature, pyrolysis at 500°C apprarently create damage sites along SWNTs sidewalls, but those damage sites are not completely exploited to finalize cutting. One hour room temperature piranha treatment finalizes the cutting. Pyrolysis at 375°C can cut the ozonated SWNTs through the same mechanism. The overall carbon yield is 70-80% in all cutting methods described here. The final average length of the cut-nanotubes is ~100nm.

Acknowledgement. This work was supported by DARPA (Defense Advanced Research Projects Agency), Grant No.R14620 and AFOSR(Air Force Office of Scientific Research), Grant No. R14620, 489020.

REFERENCE:

[1] S. Iijima, Nature, 354, 56,1991.
[2] European Cells and Materials Vol. 3. Suppl. 2, 2002 (pages 84-87)
[3] J. Liu, A. G Rinzler, H. Dai, J. H. Hafner, R. K. Bradley, P. J. Boul, A. Lu, T. Iverson, K. Shelimov, C. B. Huffman, F. Rodriguez-Macias, D. T. Colbert, R. E. Smalley, Science 1998, 280,(5367),1253.
[4] Z. Gu, H. Peng, R. H. Hauge, R. E. Smalley and J. L. Margrave. *Nano Letters*, 2, 1009-1013 (2002).
[5] W.-T. Chiang, K. F. Kelly, C. Radloff, R. H. Hauge, E. T. Mickelson, J. L. Margrave, X. Wang, G. E. Scuseria, and J. Halas, "STM Images of Sidewall Fluorinated SWNTs," *Chemical Physics Letters,* (in press.).
[6] K.J. Ziegler, Z. Gu, J. Shaver, Z. Chen, E. L Flor, J. M. Tour, R.H. Hauge, and R. E. Smalley. Submitted to Nanotechnology, 2004
[7] F. Liang, A. K. Sadana, A. Peera, J. Chattopadhyay, Z. Gu, R. H. Hauge, and W. E. Billups, *Nano Letters*, 4 (2004): 1257-1260.
[8] P.Nikolaev, M.J. Bronikowski, R. K. Bradley, F. Rohmund, D. T. Colbert, K. A. Smith, R. E. Smalley, Chem. Phys. Lett., 313,91,1999
[9] I. W. Chiang, B. E. Brinson, A. Y. Huang, P. A. Willis, M. J. Brownikowski, J. L. Margrave, R. E. Smalley and R. H. Hauge. *Journal of Physical Chemistry B*, 105, 8297-8301 (2001).
[10] Banerjee, S.;Wong, S. S. , J. Phys. Chem. B 2002, 106,(47), 12144.
[11] Banerjee, S. ;Wong, S. S , Nano Lett. 2004, 4, 1445.

Fabrication of nanometer-sized structures by C-NEMS technology

Kuo-Sheng Ma*ζ, Guangyao Jia**ζ, Qingzhou Xu***, Hong Zhou**, Chunlei Wang**, Jim Zoval**, Marc Madou**[1]

*Department of Electrical Engineering & Computer Science
**Department of Mechanical & Aerospace Engineering
***Integrated Nanosystems Research Facility
Henry Samueli School of Engineering
University of California, Irvine

ABSTRACT

Herein, we report a C-NEMS (Carbon Nano Electrical Mechanical Systems) technique for the synthesis carbon structures by pyrolysis of thin film SU-8 photoresist patterned by electron beam lithography (EBL). We have demonstrated the technique to fabricate carbon nano wires of controllable dimension and location using a top-down methodology. The width and thickness of the wires are in the tens of nanometer range. Additionally, the technique is able to produce conducting aligned carbon nano wires with addressable positioning, controllable length, and high aspect ratio (length vs width). These carbon nano wires can be exploited in large-scale assembly to make highly integrated nano wire sensor arrays or electronic devices.

Keywords: nano wires, carbon NEMS, electron beam lithography (EBL), SU-8

1 INTRODUCTION

Carbon as material is broadly used in different applications. There are a wide variety of carbons available, ranging from bulk commodity carbons i.e. coke, graphite, etc. to the specialty formats such as carbon fibers or carbon nanotubes. With the increasing interest in various aspects of nano devices, carbon nanotubes or carbon nanofibers become highly attractive carbon formats in modern science and engineering. They have great potential to be used as nanoscale devices such as field effect transistors [1-3], field emitters [4,5] and chemical or biological sensors [6]. Within those different applications, it is becoming apparent that synthesis method is the key to motivate underlying research. In spite of the success of the laser ablation [7] or chemical vapor deposition (CVD) [8] methods in developing those carbon nano structures, those nanoscale building blocks still can not be synthesized with precisely controlled and tunable structure, morphology, and size. The ability to control the size, shape, and/or orientation of structural materials on the nanometer scale is very important since it determines nano device properties and applications. To meet this goal, it is necessary to develop methods that enable rational design and predictable synthesis of those building blocks. Besides, it is essential to explore the limits of functional devices built on those elements. In the future, manufacturing nano-systems will require the development of efficient and scalable strategies for assembly of building blocks into the complex architectures that enable high-density integration with desired functions.

Over the past several years, well-defined carbon structures have been obtained by pyrolyzing positive photoresists that are commonly used in optical lithography [9]. This "diamond-like" carbon has many advantages in different applications. First, it is atomically smooth (i.e. surface roughness of about 30Å) [10] as protective and/or tribological coatings. Second, it has wider stability window as electrodes in aqueous solution [9]. Higher potentials can be applied to electrodes of carbon without electrolysis of water. Third, the process of making these carbon structures is totally compatible with MEMS processes. There exists a wealth of knowledge on making "microstructures" by photolithography can be the starting point to make the carbon structures. Thin layer of photoresist was spin coated on silicon wafers and subsequent photolithographic patterning using well-developed technologies in MEMS or in semiconductor industry. The micro structures that are routinely produced can directly transfer to fabricate microscopic structures or devices in carbon that may be useful in many different applications. This method provides the controllability of dimension and location of the carbon devices. Besides, the capability to pattern the structures can be exploited in large-scale assembly to make highly integrated arrays or electronic devices.

Although the efforts in fabricating and investigating carbon structures that are obtained by pyrolysis of photoresists have been made, the pyrolyzing photoresist precursors of carbon, so far, are positive tone such as AZ 4330. The dearth of using negative photoresist will restrict many applications using pyrolysis carbon as devices material. For example, high aspect ratio structures made by negative photoresist SU-8 can not be formed by positive photoresist and then pyrolyzed to carbon. Recently, Madou et al. have reported successfully made high aspect ratio (aspect ratio > 10) carbon posts by pyrolyzing negative photoresist SU-8 for micro-batteries applications [11]. More material characterizations have been done on the carbon made by pyrolyzing commercially available negative photoresist SU-8. Besides, investigation of different process techniques on SU-8 making various carbon structures have been carried out.

The chemically amplified resist epoxy novolak SU-8 was first introduced by researchers at IBM [12]. It is a negative acting, solvent developed epoxy based resist with excellent UV sensitivity and high aspect ratios characteristics for MEMS processes. It has been used in many applications to fabricate thick or ultra thick structures by optical UV lithography [13-15]. However, there is much less effort aimed at developing this resist in thin film (thickness in sub micron range) process by using electron beam lithography (EBL). Pun and Wong [16] recently published the fabrication optical waveguides by electron beam lithography of a formulation of thinner layer SU-8. The minimum feature size of lines as narrow as 100 nm using the same patterning technique has also been published [17]. SU-8 has several advantages compared to the widely used e-beam resist, polymethyl methacrylate (PMMA). First, it has much higher electron sensitivity than PMMA. This reduced the necessary time to expose a given pattern which made it ideal for patterning over large writing fields. Second, it has a much lower contrast than pattering on PMMA. This leads to smooth surface roughness making multilevel exposures possible.

In this work we report the fabrication of 30-nm-wide lines with diluted SU-8 2 by e-beam lithography. Thin film SU-8 with thickness less than 20 nm has been achieved. The patterned SU-8 structures have been carbonized by thermal pyrolysis and different dimensions of carbon nano wires are reported. Our results demonstrated that, although SU-8 is design for thick film processes by optical lithography, it may be used for thin film e-beam lithography to make surface architectures with sub micron feature size. The technique is capable of producing conducting carbon nano wires with addressable positioning, controllable length, high aspect ratio (length vs width).

2 EXPERIMENT

2.1 Materials

For all experiments we used the commercially available NANO™ SU-8 from MicroChem Corp. (Newton, MA). Gamma-btytolactone (GBL) and SU-8 developer were purchased from MicroChem.

2.2 Electron beam lithography

All the pattern exposures were carried out with a FEI SIRION nanowriter system at 30 KeV. Beam current is 25 pA. Patterns are generated by JC Nobity Lithography system.

2.3 Procedures

The fabrication process sequence is schematically showed in Figure 1. Single-crystal Si (001) wafers were used as the substrate material throughout this work. Before the resist was dispensed, the substrates were cleaned using a RCA clean (5 parts deionized water, 1 part NH_4OH, 1 part H_2O_2) and dehydrated in a 100 °C oven for overnight. For all the experiments we used the commercially available NANO™ SU-8 formulation 2 from MicroChem Corp. (Newton, MA) as the original coating resist. It is composed of three components. The first component is an EPON epoxy resin. The chemical formula of EPON resin SU-8 is a multifunctional glycidyl ether derivative of bisphenol A Novolac epoxy oligomer. The second component is the photoinitiator triarylium-sulfonium salts. Epoxy resins are cationically polymerized by utilizing a photoinitiator which generates strong acid upon irradiation to the power source such as ultraviolet light or electron beam and the acid facilitates polymeric cross-linking during post-exposure bake. The third component is the gamma-butyrolactone (GBL), an organic solvent. The viscosity of resist is determined by the quantity of the solvent. The viscosity can also determine final thickness of the spin-coated film. The resists were spun onto the wafer by a 1275 rpm/s ramped to final 6000 rpm for 40 seconds. Before exposure, samples were pre-baked at 65 °C for 2 min, then at 95 °C for 2 min in the ovens. Post exposure bake (PEB) was carried out using ovens of 65 °C for 1 min and 95 °C for 1 min. Samples were developed for 30 second in MicroChem SU-8 developer and then blew dry under nitrogen flow. Before the pyrolysis process, samples were hard baked at 120 °C for 30 min.

In process, if pyrolysis occurs too fast, the carbon structures will peel off from substrate. Here, we used a two step heating profile to pyrolyze the photoresist. First, samples were placed in a Lindberg furnace, purged with forming gas (N_2:H_2 = 9:1), heated to the 200 °C and kept at the same temperature for 30 min. The purpose for this step is to release thermal stress of substrate and photoresist owning to the thermal expansion coefficient difference between two materials. Next, the temperature was increased to 900 °C and held for 1 h to pyrolyze the photoresist. Samples were removed from the furnace at room temperature.

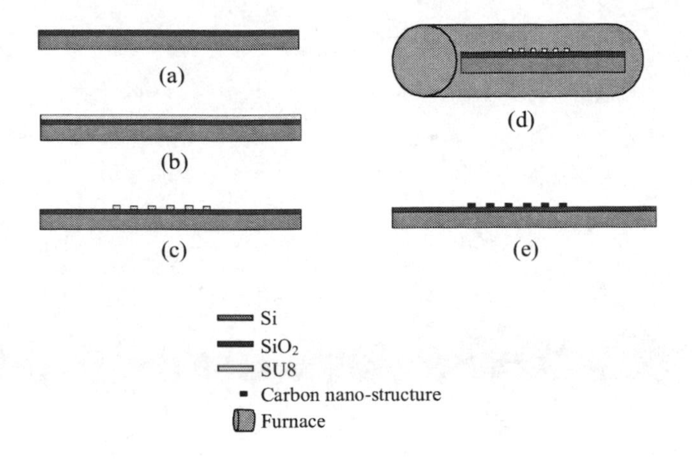

Figure 1: Schematic illustration of the process for fabrication of carbon nano-structures. (a) SU-8 photoresist spin coating. (b) SU-8 nanostructure patterning by Electron beam lithography (EBL). (c) Photoresist development. (d) Carbon nano-structures pyrolysis process. (e) Carbon nano-sturctures on substrate.

3 RESULTS AND DISCUSSIONS

In order to test the feasibility of the technique we proposed, first, the commercial available resist SU-8 2 has been coated and exposed by e-beam. Then, samples were pyrolyzed by the process mentioned above. Fig 2 (a) and (b) are SEM images of the resulting structures. (a). Three different carbon lines with width 300 nm, 400 nm, 500 nm (from left to right) and each width having three lines. (b). Arrays of aligned carbon nano wires. Each array has the same pattern as in (a).

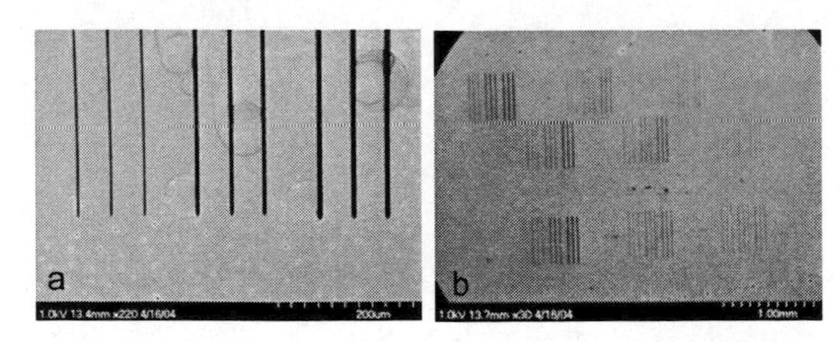

Fig 2

Upon inspection of the dimensions of the wires, we found that the thickness of the photoresist decreases dramatically during pyrolysis process, i.e., a decrease from 2 μm to about 500 nm. Assuming the width does not change, this corresponds to a weight loss of about 75% occurring in pyrolysis process. A weight loss of about 87% by pyrolyzing positive photoresist had been reported [18]. The patterned SU-8 lines have profiles more rectangular prior to heat-treatment. After pyrolysis, however, the lines have profiles wider at the base in contact with substrate and narrower at the top. The top regions of photoresist have more freedom to shrink than the photoresist at the structure/substrate interface. The volume change is easier on the top when the sample is slowly heated in the furnace.

E-beam lithography can be used with thin resists to generate very high resolution patterning. Although SU-8 is designed for thick film processes, its viscosity determined the thickness of coating films. In order to get thinner resist coating, we diluted the less viscose commercially available SU8 2 resist with GBL to a volume ratio (GBL: SU-8 2 (v/v)) 100. The thickness of carbon structures made by this photoresist formula have been investigated. AFM data (data not shown here) show the thickness to be around 10 nm. We believe this is the thinnest carbon film made by thermal pyrolysis and a top down approach. More work needs to be done to systematically predict and control the photoresist coating in order to get the desired thickness of carbon structures.

To assess the resolution limit of this resist, single pixel lines were used to expose the resist with varied doses. Fig 3 (a) and (b) display SEM images of the testing results. (a). paralleled carbon wires with different line widths. Clearly, the size of SU-8 patterns made by EBL depended on the dose modulation. The main purpose of this part of work was to determine the smallest possible feature size using our technique. (b) shows the smallest wire (30nm wide) made to date. Similar high resolution EBL results on SU-8 2000 were recently reported in the literature [17], which outline further interest in large arrays of nanoscale patterning.

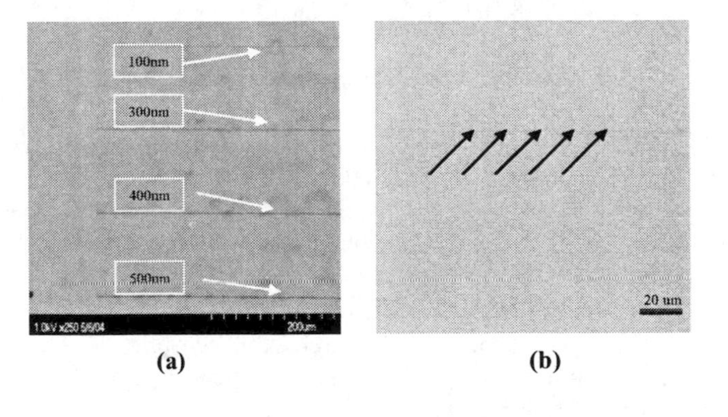

Fig 3

4 CONCLUSIONS

In summary, we have reported a technique for fabrication of carbon nanowires of controlled dimensions and location using a top-down methodology. The width and thickness of the wires are down to tens of nanometers range. Additionally, the technique is able to produce conducting carbon nanowires with addressable positioning, controllable length, and alignment. In addition, we have demonstrated the electron beam lithography patterning of thin film SU-8 with high aspect ratio lines and line width down to the tens of nanometers range. It should be possible to extend this technique to fabricate the nanometer structures in NEMS. Generally, the addressable and patternable carbon nanowires we developed can be exploited in large-scale assembly to make highly integrated nanowire sensor arrays or electronic devices.

REFERENCES

[1] S. J. Tans, A. R. M. Verschueren and C. Dekker, Nature, 292, 49 (1998)

[2] R. Martel, T. Schmidt, H. R. Shea, T. Hertel and P. Avouris, Appl. Phys. Lett., 73, 2447 (1998)

[3] C. Zhou, J. Kong, and H. J. Dai, Appl. Phs. Lett. 76, 1597 (2000)

[4] W. A. de Heer, A. Chatelain, and D. Ugarte, Science 270, 1179 (1995)

[5] A. G. Rinzler, J. H. Hafner, P. Nikolaev, L. Lou, S. G. Kim, D. Tomanek, P. Nordlander, D. T. Colbert, and R. E. Smalley, Science 269, 1550(1995)

[6] J. Kong, N. R. Franklin, C. Zhou, M. G. Chapline, S. Peng, K. Cho, and H. J. Dai, Science 287, 622 (2000)

[7] A. Thess, R. Lee, P. Nikolaev, H. J. Dai, P. Petit, J. Robert, C.H. Xu, Y. H. Lee, S. G. Kim, A. G. Rinzler, D. T. Colberrt, G. E. Scuseria, D. Tomanek, J. E. Fischer, and R. E. Smalley, Science 273, 483 (1996)

[8] J. Kong, H. T. Soh, A. M. Cassell, C. F. Quate and H. Dai, Nature 395, 878 (1998)

[9] J. Kim, X. Song, K. Kinoshita, M. Madou, and R. White, J of Electrochem. Soci. 145 (7), 2314 (1998)

[10] R. Kostecki, X. Song, and K. Kinoshita, Electrochemical and Solid-State Letters, 5 (6) E29 (2002)

[11] C. Wang, G. Y. Jia, L. Taherabadi, and M. Madou, IEEE Journal of Microelectromechanical Systems, in press

[12] K. Y. Lee, N. LaBianca, S. A. Rishton, S. Zoghamain, J.D. Gelorme, J. Shaw, and T. H.-P Chang, J. Vac. Sci. technol. B 13, 3012 (1995)

[13] G. M Kim, B. Kim, M. Liebau, J. Huskens, D. N. Reinhoudt, and J. Brugger, J. Microelectromech. Syst. 11, 175 (2002)

[14] J. Zhang, K. L. Tan, G. D. Hong, L. J. Yang, and H. Q. Gong, J. Micromech. Microeng. 11, 20 (2001)

[15] G. Genolet, M. Despont, P. Vettiger, D. Anselmetti, and N.F. de Rooij, J. Vac. Sci. Technol. B 18, 3431 (2000)

[16] W. H. Wong and E. Y. B. Pun, J. Vac. Sci. Technol. B 19, 732 (2001)

[17] M. Aktary, M. O. Jensen, K. L. Westra, M. J. Brett, M. R. Freeman, J. Vac. Sci. Technol. B 21 (4), L5 (2003)

[18] R. Kostecki, X. Song and K. Kinoshita, Electrochemical and Solid-StateLetters, 2 (9), 465 (1999)

Inelastic light scattering and light emission from single and double wall carbon nanotubes

Yan Yin[1], Stephen Cronin[2], Andrew Walsh[1], Alexander Stolyarov[2], Michael Tinkham[2], Anthony Vamivakas[3], R.R Bacsa[4], SelimÜnlü[3,1], Bennett Goldberg[1,3], Anna Swan[3] and Wolfgang Bacsa[3,5]

[1]Physics department, Boston University, Boston, MA 02215
[2]Physics department, Harvard University, Cambridge, MA 02138,
[3]Electrical and Computer Engineering Department, Boston University, Boston, MA 02215
[4]Nanolab Inc, Newton, MA 02458
[5]LPST-IRSAMC CNRS, Université, Paul Sabatier, 31062 Toulouse, France; bacsa@lpst.ups-tlse.fr

ABSTRACT

We use inelastic light scattering (Raman) to probe individual and isolated single wall carbon nanotubes suspended in air over trenches to explore the influence of the environment on the electronic transition energies. We find narrow and predominantly symmetric resonance profiles which are significantly downshifted as compared to isolated tubes in aqueous surfactant suspensions. A lower limit of the exciton binding energy in SWNTs is deduced by using the observed optical transition energies of SWNTs in SDS solutions. Photoluminescence is observed from narrow diameter double wall carbon nanotubes in aqueous surfactant suspensions.

Keywords: carbon nanotubes, optical transition energies, chirality, photoluminescence

1 INTRODUCTION

Carbon nanotubes are one dimensional model systems [1]. The large one dimesional singularites in the electronic density of states lead to optical transitions which are sensitive to the chirality of the carbon nanotube. With the help of maps of the photoluminescence signal as a function of excitation energy from a distribution of single wall carbon nanotubes, in a surfactant suspension, it was possible to assign the emission peaks to a particular chirality [2]. The assignment is based on comparing the distribution of emission peaks with the predicted distribution of values for all chiralities of semiconducting tubes using the tight binding model of graphene and applying the zone folding scheme. Although the theoretical values are significantly shifted to lower energies the correspondence of the distribution of photoluminescence peaks was sufficient to identify the tube chirality. It is believed that strong electron confinement in the tube leads to an increase of the electronic band gap and to the formation of excitons [3]. This is in part consistent with the higher experimental observed energies of the photoluminescence peaks. The photoluminescence excitations studies have been carried out on tubes surrounded by surfactant molecules to isolate them and prevent any agglomeration which prevents any radiative recombination. But the surfactant itself is believed to influence the optical transition energies. In this context we have carried out experiments of individual and isolated single wall carbon nanotubes suspended over trenches to measure the intrinsic optical properties of SWNTs. Furthermore we have used inelastic light scattering as a function of excitation energy to map the energetic positions of the optical transition energies (tunable resonance Raman spectroscopy). This has the advantage that singularities in the electronic density of states can also be detected for metallic tubes apart of the fact that resonance Raman profiles are narrower than the photoluminescence emission bands due to the fundamental differences of the underlining physical process. Double wall carbon nanotubes are of interest due to the fact that the internal tube is protected from the interaction with its environment. We present here first photoluminescence spectra from double wall carbon nanotubes (DWNTs).

2 EXPERIMENTAL

The samples are prepared by first etching 1-1.5 μm wide trenches with markers in quartz substrates by reactive ion etching (RIE) in CF_4 plasma. SWNTs are grown over the trenches by chemical vapor deposition in methane gas at 900°C using a 1nm thick film of Fe as the growth catalyst [4]. The tube concentration is lower than 1tube/ μm^2 as verified by scanning probe microscopy. A continuous wave tunable Ti-sapphire laser is used as the excitation laser source and we use a Renishaw RM1000B system with a motorized stage after custom-built modifications in order to achieve tunable Raman data collection for excitation wavelengths from 720 to 830*nm* (1.72 to 1.49*eV*). Filters are tuned by tilt and any beam offset is corrected by a matched filter tilt. The use of filters and a single grating offers a high through-put system and low signal to noise ratios. A 600 g/mm grating is used, yielding ~2-3 cm^{-1} /pixel. The laser beam is focused by a 100X objective on a single nanotube which is suspended over the trench in air. The FWHM of the Gaussian spot-profile is 0.42 μm at E_{laser}=875 nm. 1-2 *mW* constant excitation laser power was used during a mapping. Measurements of Stokes and anti-Stokes intensity ratios show that no heating of the nanotubes takes place under such power density. We observe that the Raman signal from tubes on the substrate, are strongly suppressed. The entire collected signal originates therefore from the suspended part of the tube. Typical Stokes radial breathing mode (RBM) count rates are 3-35 counts/second. We map out the resonant excitation profile (REP) by changing the excitation wavelength. Each REP is measured twice with staggered 2 *nm* separation in excitation wavelength to check the repeatability.

3 RAMAN MAPPING

Figure 1 shows a Raman map of the radial breathing mode of one individual and isolated carbon nanotube using a step size of 0.5 μm (comparable to lateral resolution). The RBM frequencies are used to assign specific chiral indices (n,m) to the observed optical resonance. We observe that the Raman signal of the radial breathing mode is localized to the section where the tube crosses the trench. The trench location is indicated with two horizontal lines in figure 1.

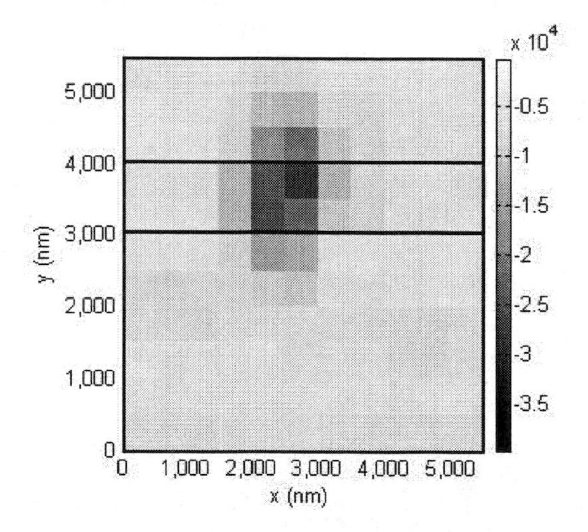

Figure 1 Spatial map of radial breathing mode intensity of one individual and isolated single wall carbon nanotube across a 1μm wide trench. Horizontal lines show the edge of the trench. Excitation wavelength: 798nm.

We attribute the disappearance of the Raman signal of the CNT to the interaction of the nanotube with the substrate and to local field variations. Incident and reflected waves are out of phase at the interface which leads to destructive interference and a reduction of the time averaged local field. The expected local field variation depends on the amplitude of the reflected wave.

Figure 2 shows a series of inelastic light spectra of the radial breathing mode for a fixed location on the nanotube by varying the excitation wavelength. The intensity of the radial breathing mode is enhanced as the excitation wavelength is in resonance with the one dimensional singularity of the tube. The radial breathing mode frequency is inversely proportional to the tube diameter and the observed resonance can then be related to the tube diameter. This can be repeated for individual and isolated tubes of different diameter to obtain a distribution of optical transition energies.

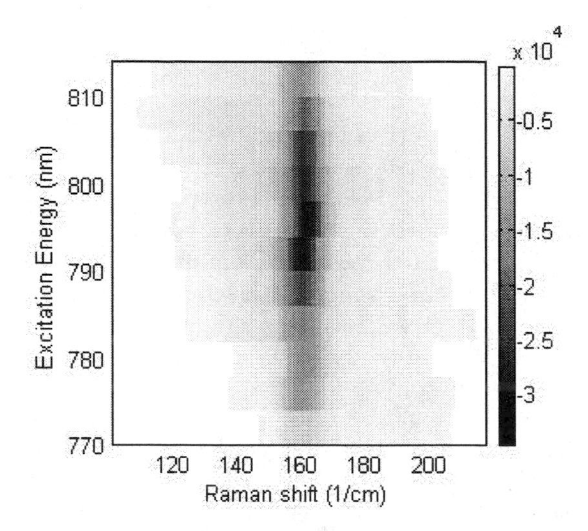

Figure 2 Spectra of radial breathing mode of individual and isolated SWNT as a function of laser excitation energy.

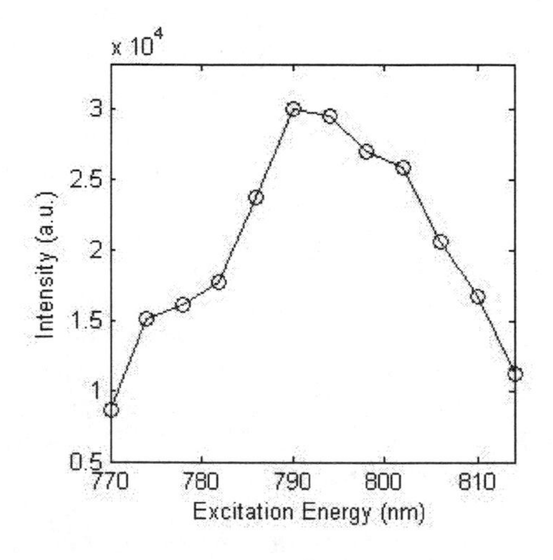

Figure 3 Resonance profile of radial breathing mode at 162cm^{-1} extracted from figure 2.

The resonance profile has contributions due to the resonant enhancement for the incoming and scattered light, separated by the addition of the phonon energy of the radial breathing mode. The resonance profile is then fitted using the time-dependent, third-order perturbation theory for the Raman process in a one dimensional (1D) systems [5,6]. We note that even though the van Hove singularities are asymmetric around the resonance, the resulting Raman intensity profile is symmetric [5].

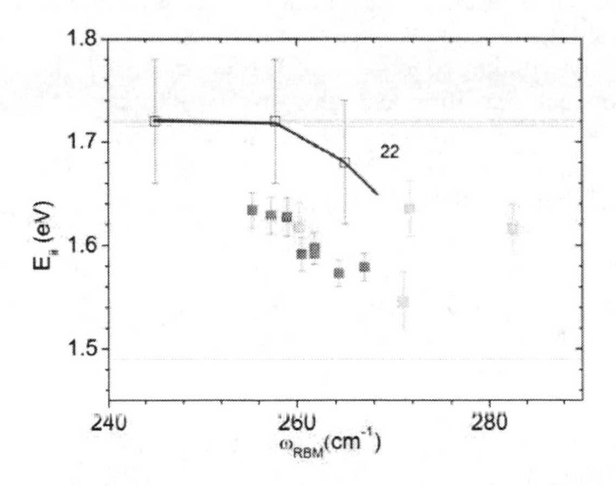

Figure 4 Measured E_{ii} vs. ω_{RBM} for the SWNTs measured by tunable RRS. The filled symbols are measured for SWNT suspended in air, squares denotes semiconducting tubes from RRS form tubes in SDS Ref [6, 7]. The numbers denote the 2n+m families. The energy ranges shown for each point are the spectral width.

The fitting determines the location of the E_{22} resonance energy and the spectral width η. The resonant profile has been recorded on a series of single and isolated SWNTs to determine the E_{22} and the corresponding width η. About half of the nanotubes have a η ranging from 8 to 18 meV and the remaining broadening parameters range up to 50 meV in decreasing numbers. The minimum η value we observed is 8.8 meV, similar to the only previously reported REP measurement of one single tube on a Si substrate [7]. The tubes with wider resonance profile are believed to be affected by defects. The narrow broadening (~10 meV) measured for ~1/3 of the tubes is either a measure of the intrinsic lifetime broadening for a single suspended tube, or, possibly broadened by the finite suspension length (~1 μm) across the trench. The overall width of the resonance for the RBM REP can be estimated by $\eta+E_{phonon}$, ~ 25-50 meV. PLE measurements from a single tube suspended in air show a symmetric E^S_{22} excitation profile with a width of 44 meV, about 5 times broader than our REP data (8). This difference can be explained by the difference in the underlining physical process. In PL, absorption, intra- and inter-band phonon relaxation and emission rates all involves real states, where transition rates can be calculated by Fermi's golden rule (9). On the other hand resonance Raman scattering is described by third order time dependent perturbation theory with a three-step phase coherent quantum mechanical process resulting in scattered Raman signal. In either the incoming or outgoing resonance situation, it is necessary for the excited electron to scatter from a real to a virtual intermediate electronic state (or vice-versa). This explains the narrower resonance Raman profile. Figure 4 shows a systematic downshift (~100 *meV*), compared with SDS data. We find that the average line broadening η is ~14 meV for the isolated tubes suspended in air which is significantly narrower than for HiPCO tubes wrapped in SDS where the resonant peaks are broadened by 65 meV [10] and 120 meV for bundles [11]. These two observations show clearly the influence of the environment on optical transition energies.

Moore *et al* demonstrated that changing the surfactant can change the resonance energy by up to 25 meV [12]. The resonance energies for bundles are found in between the reported SDS data and our data [11]. The dielectric constant ε in a medium influence the interaction between electrons and holes, and as the dielectric constant increases, the electron hole-pair binding energy from the Coulomb attraction is screened. Perebinos *et al.* calculated the scaling of exciton binding energy for nanotubes embedded in a dielectric medium [13]. They found a simple scaling relationship between the exciton binding energy and the dielectric constant, $E_b \propto R^{\alpha-2}\varepsilon^{-\alpha}$ for $4<\varepsilon<15$ where R is the tube radius and the scaling exponent, $\alpha=1.4$. Assuming an effective dielectric constant ε_{eff} for nanotubes in SDS, we can estimate the change in exciton binding energy with respect to a tube in air ($\varepsilon_{eff}=1$). Using $\alpha=1.4$ and $\varepsilon_{eff}^{SDS}\sim(1.5)^2$ one that E^{SDS}_b for a tube in SDS is only ~0.3 E^{air}_b, the binding energy for a tube in air. The observed shift of 0.1 eV, translates into a binding energy of ~0.15 eV provided the band edge energy position is not affected by the change in dielectric index which is much lower than expected. Spataru *et al* used an *ab initio* many-electron Green's function approach to calculate the exciton binding energy for an (8,0) tube, which was found to be $E_b^{8,0}\sim1$ eV. If we use this value and the scaling relationship for the radius [13] to estimate the binding energy for the (11, 0) zigzag tube in family 22, we obtain $E_b^{11,0} \sim E_b^{8,0}(R_{(8,0)}/R_{(11,0)})^{\alpha-2} \sim 0.8$ eV. The difference of ~ 0.6 eV between our observed shift and the calculated binding energy could be taken as a measure of the shift of the band edge with dielectric constant, not accounted for in the scaling calculation.

3 PHOTOLUMINESCENCE FROM DOUBLE WALL CARBON NANOTUBES

Double wall carbon nanotubes (DWNTs) are of interest due to the fact that the internal tube is protected from environmental effects and the intrinsic properties of internal carbon nanotubes are preserved. While photoluminescence have been observed in SWNTs in SDS, most DWNT samples do not disperse well in SDS and only very weak photoluminescence signals have been observed so far.

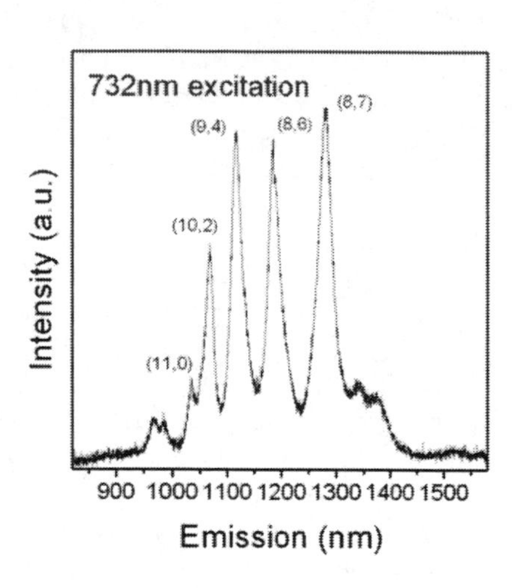

Figure 4 Photoluminescence spectrum of surfactant suspended double wall carbon nanotubes exited at 732nm.

However, DWNTs prepared by the CVD method disperse easily in other surfactants such as CTAB and DOC. With an improved suspension centrifuged to remove aggregates, an intense PL spectrum is obtained. Figure 4 shows the photoluminescence spectrum of a suspension of double wall carbon nanotubes (Nanolab Inc.) excited at 732nm. We observe several emission bands where three of them are clearly split in doublets. DWNT samples contain also single and triple wall carbon nanotubes (<30%). We can attribute the observed spectrum to tubes with more than one wall due to the fact that the surfactant is optimized for multiwall carbon nanotubes. The interaction of the two layers is expected to influence the emission spectrum. In general the chirality (helicity) of the two tubes is not commensurate and this opens the possibility that one or both of the two tubes are metallic or semiconducting. It is expected the tube-tube interaction depends on chirality. The interaction will be larger with metallic tubes and this could suppress any luminescence. This implies that the emission spectrum is broadened or red shifted when measuring a distribution of tubes with different diameters. We find that the experimental tube has narrower spectral lines which ar red shifted by 10-20meV. This indicates that the tubes which do emit are in an environment which is more uniform or DWNTs have a higher relative quality. The observed emission spectrum can be compared to the photoluminescence spectrum of single wall carbon nanotubes. The emission bands are labeled with chiral indices using the assignment made for SWNTs in SDS. The observed red shift is indicative of the tube interaction in DWNTs.

4 CONCLUSION

We have observed individual and isolated SWNTs using tunable resonance Raman scattering to determine the optical transition energies. We are able to give a lower limit of the exciton binding energy in SWNTs by using the observed optical transition energies of SWNTs in SDS solutions. We find that photoluminescence spectra of double wall carbon nanotubes are strikingly similar and red shifted as compared to SWNTs in SDS.

REFERENCES

[1] S. Reich, C. Thomsen, J. Maultzsch, *Carbon Nanotubes* (Wiley-Vch 2004)

[2] Sergei M. Bachilo, Michael S. Strano, Carter Kittrell, Robert H. Hauge, Richard E. Smalley, and R. Bruce Weisman, *Science* **298**, 2361 (2002).

[3] T. Ando, J. Phys. Soc. Jpn. 66, 1066 (1996)

[4] J. Lefebvre, J. M. Fraser, Y. Homma, and P. Finnie, App. Phys. A **78**, p. 1107 (2004)

[5] R. M. Martin and L. M. Falicov, *Topics in Applied physics Vol 8 - Light scattering in solids I*, chapter 3, page 79 to page 95

[6] M. Canonico, G. B. Adams, C. Poweleit, J. Menéndez, J. B. Page, G. Harris, H. P. van der Meulen, J. M. Calleja, and J. Rubio, *Phys. Rev. B.* **65**, 201402(2002)

[7] A. Jorio, R. Saito, J. H. Hafner, C. M. Lieber, M. Hunter, T. McClure, G. Dresselhaus, and M. S. Dresselhaus, *Phys. Rev. Lett.* **86**, 1118 (2001).

[8] J. Lefebvre, J. M. Fraser, Y. Homma, P. Finnie, *Phys. Rev. B* **69**, 075403, (2004)

[9] J. Jiang, R. Saito, A. Grüneis, S. G. Chou, Ge. G. Samsonidze, A. Jorio, G. Dresselhaus and M. S. Dresselhaus. *Phys. Rev. B* **71**, 045417 (2005)

[10] H. Telg, J. Maultzsch, S. Reich, F. Hennrich, and C. Thomsen, *Phys. Rev. Lett.* **93**, 177401(2004).

[11] C. Fantini, A. Jorio, M. Souza, M. S. Strano, M. S. Dresselhaus, and M. A. Pimenta, *Phys. Rev. Lett.* **93**, 147406 (2004)

[12] V. C. Moore, M. S. Strano, E. H. Haroz, R. H. Hauge, R.E. Smalley Nano Lett. 3, 1379, (2003)

[13] Perebinos, J. Tersoff and Ph Avouris. *Phys. Rev. Lett.* **93** (2004)

Fabrication of Suspended C-NEMS Structures by EB Writer and Pyrolysis Method

K. Malladi*, C. Wang** and M. Madou***
Department of Mechanical and Aerospace Engineering,
University Of California, Irvine, CA92697, USA
*kmalladi@uci.edu, **chunleiw@uci.edu, ***mmadou@uci.edu

ABSTRACT

Carbon micro- and nanostructures has received widespread interest recently due to their potential applications in biomedical devices, chemical sensors, and microelectronics. In this work, we successfully fabricated carbon-micro and nano electromechanical systems (C-MEMS/NEMS) by UV/EB lithography and pyrolysis method. Our starting material is a negative photoresist, SU-8. We tried to solve charging problem by forming a thin metal layer before EB writing using various methods, such as EB evaporation, sputtering system and thermal evaporation. By partly depositing a thin layer of a metal to prevent the repelling of negative charged electrons, we have successfully formed various suspended carbon structures, such as bridges and networks.

Keywords: carbon-microelectromechanical systems, suspended structure, pyrolysis, photoresist

1 INTRODUCTION

Carbon comes in many forms, such as: diamond,graphite, buckeyballs, nanotubes, nanofibers, glassy carbon, and diamond-like carbon (DLC). These materials are used in a variety of applications, based on their different crystalline structures and morphologies which enable very different physical, chemical, mechanical, thermal and electrical uses. Recently Carbon micro and nano structures have attracted a lot of research interest because of its potential applications in biological, chemical and electronic devices [1,2,3]. For example, in biosensing application these micro or nanostructures can alter the catalytic activity, affect specificity of biological systems and highly specific molecular recognition processes [4,5]. Microfabrication of carbon structures using current processing technology includes focused ion beam and reactive ion etching [6,7]. Recently a novel technology to microfabricate carbon micro structures based on UV lithography and pyrolysis process was reported [8-10]. However the idea of fabricating various complicated suspended three dimensional carbon structures is still a big challenge. Here suspended C-MEMS structures were fabricated by EB lithography and pyrolysis by accurately controlling the processing parameters.

2 FABRICATION & PROCESSING

The fabrication of the suspended C-MEMS structures involved mainly three steps: (1) UV lithography, (2) EB lithography, and (3) pyrolysis process.

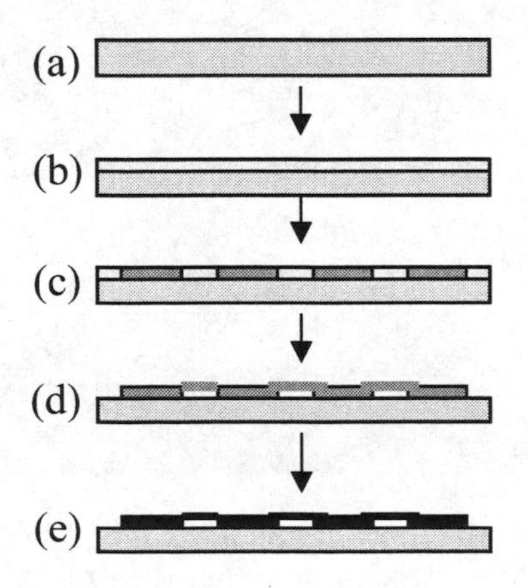

Figure 1:(a) Si substrate (100); (b) SU-8 film of 100μm thickness was spin-coated; (c) patterning the photoresist by UV exposure; (d) patterning suspended structures by EB writer, and then developing; (e) finally pyrolyzing the photoresist patterns to convert them to carbon structures.

The substrate we used were Si (100) surface, which was spin-coated with SU-8 negative chemically amplified resist. The SU-8 resist was then patterned to get SU-8 posts which have a diameter of 50μm and have a spacing of 50μm between them. The patterned photoresist posts (non-developed) were then transferred to E-beam lithography system to write the desired pattern on these posts. Eventually the unexposed photoresist was washed away in developing process to leave us the suspended SU-8 structures. The suspended carbon structures were obtained by pyrolysing the SU-8 structures in a diffusion furnace at about 900^0C for 1 hr in a forming gas environment. The process flow of the fabrication steps is shown in figure1.

3 RESULTS & DISCUSSION

Figure 2 shows the SU-8 suspended structures we obtained. Suspended thin SU-8 bridges of about 30 μm in width, 50 μm in length, and 10 μm in thickness were patterned between SU-8 post arrays. After pyrolysis, suspended carbon microstructures with about 15 μm in width, 50 μm in length, and 3-5 μm in thickness are shown in figure 3. It can be seen that after high temperature process the carbon posts and suspended bridges shrink isometrically and kept similar shapes as SU-8 structure. For some of the structures, because of the unbalanced drag force,

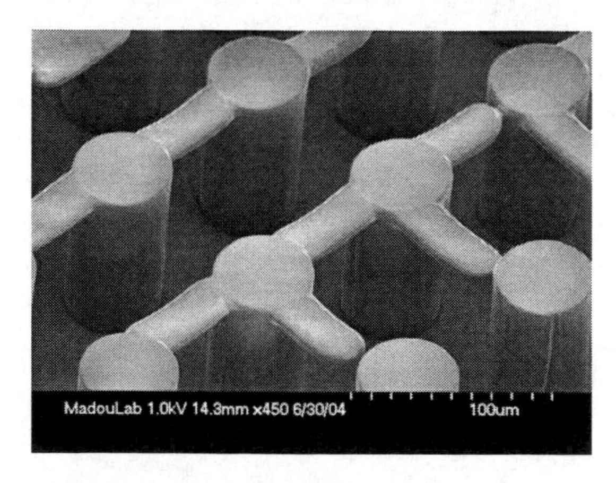

Figure 2: A typical SEM picture of SU-8 posts with suspended micro structures (before pyrolysis).

the posts were found to bend towards each other. A major problem here is charging up during EB writing on nonconductive SU-8 surface because of the accumulation of negative charged electrons. The repelling effect of the accumulated charge with the incoming electron results in further difficulty to focus and align the electron beam on surface.

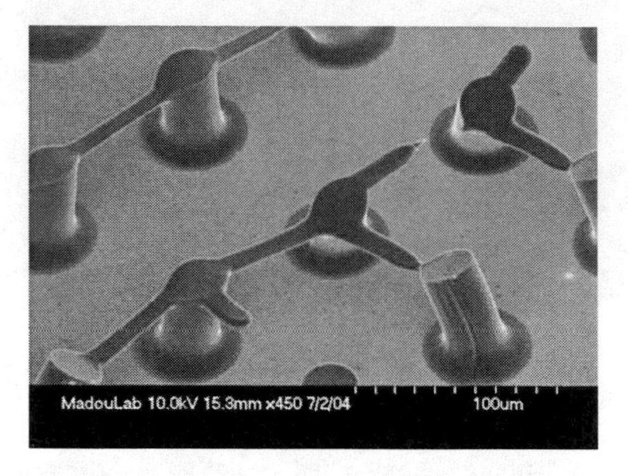

Figure3: A typical SEM picture of carbon posts with suspended micro structures (after pyrolysis).

In order to prevent the charging of the SU- 8, thin layers of gold (10 nm) were deposited by both sputtering method and e-beam evaporation method. After that EB writing was performed to pattern suspended structures without charge up effect. Unfortunately after removing Au layer by wet etching, it was observed that a thin SU-8 layer was formed which could not be removed by developer. Figures 4 and 5 show the thin SU-8 film thus formed. It was confirmed that underneath the thin sheet of SU-8 the posts were developed very well. Also, it was concluded that during both sputtering and EB evaporation processes, the SU-8 surface was attacked and exposed by the high energy ions and/or X-rays resulting in the surface cross linking and inability to pattern suspended structures.

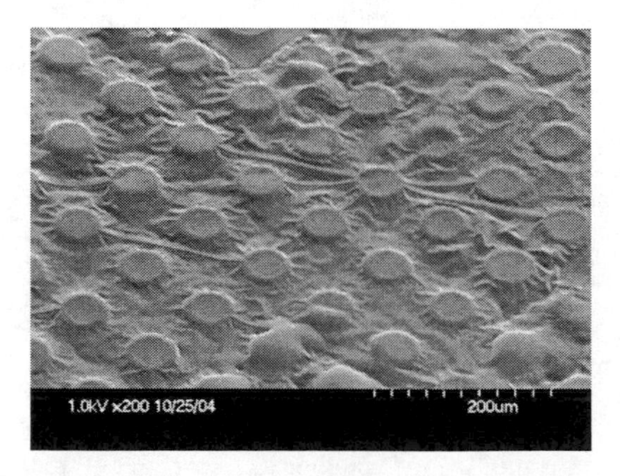

Figure 4: The typical SEM picture obtained from the SU-8 sample after e-beam writing, Au layer (formed by sputter deposition) removal, and developing.

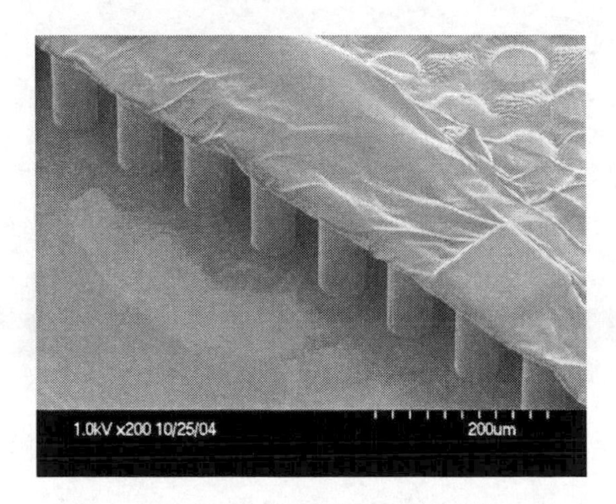

Figure 5: Typical SEM picture obtained from the SU-8 sample after e-beam writing, Au layer (formed by EB evaporation) removal, and developing.

In order to prevent both charge-up and complete cross linking of the surface, we partially mask the area of the SU-8 posts with a dummy wafer and then evaporate Au (10nm) on the unmasked part of the sample. Next, we do the routine aligning on metallized part and EB writing on unmetallized parts to obtain the desired suspended pattern. The EB written area was successfully developed. The SU-8 posts are then pyrolysed in an forming gas environment at 900^{0}C for one hour to convert it to carbon [9,10]. Figure 6 shows the perfectly horizontal suspended carbon microstructures with no bending of the carbon posts. It can be observed that the suspended carbon microstructures were about 10μm in diameter and about 50μm in length. In this case since there was equal drag force in both the directions there was no bending of the carbon posts. Figure 7 shows a ring type C-MEMS structure. Here there is inward bending of the carbon posts due to the radial nature of the drag force components. It can also be observed from figure 8 that the in case of the suspended structures that were patterened in a straight line. The bending of the posts occurred only at the edge posts because of unidirectional drag forces on these posts. Different suspended structures as desired can be generated by designing the pattern we want on the EBL.

4 CONCLUSIONS

In conclusion, suspended C-MEMS structures were formed by UV/EB lithography and pyrolysis method. We tried to solve charging problem by forming a thin metal layer before EB writing using various methods, such as EB evaporation, sputtering system. By partly depositing a thin layer of a metal to prevent the repelling of negative charged electrons, we have successfully fabricated various complicated suspended carbon structures.

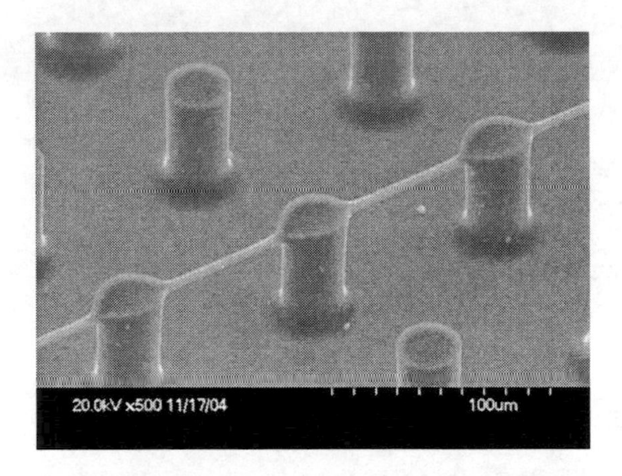

Figure6: Typical SEM picture horizontal carbon microstructures suspended between carbon posts obtained from the SU-8 sample after e-beam writing, Au layer (formed by EB evaporation) removal, and developing.

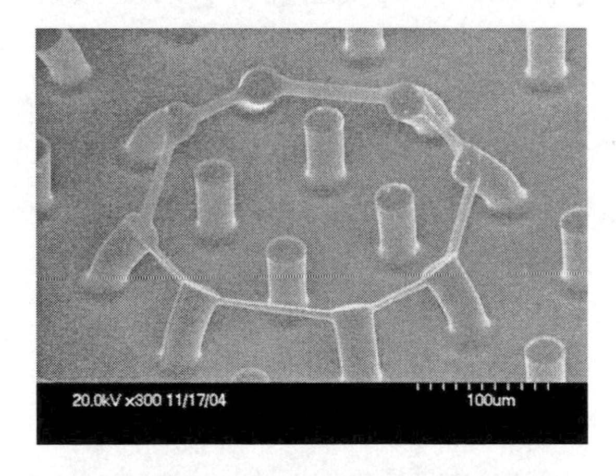

Figure 7: Typical SEM picture showing the Ring type C-MEMS structures fabricated by EBL and pyrolysis of SU-8 photoresist.

ACKNOWLEDGMENTS

This work is supported by the NSF grant, DMI-0428958. The authors would thank Dr.Quinzhou Xu, INRF, UCI, for his assistance in EBL operation and useful discussions.

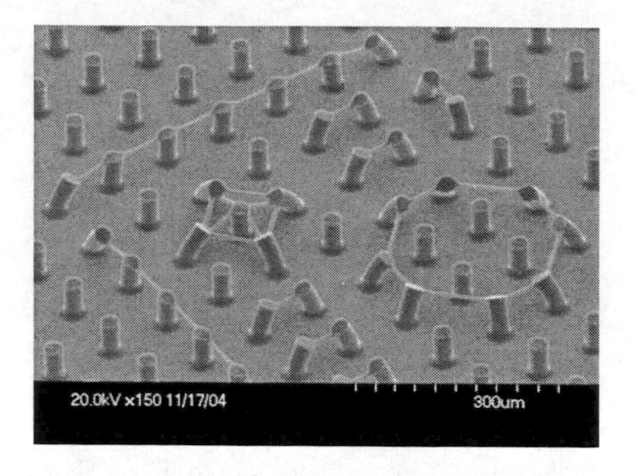

Figure 8: A SEM image of microstructures with various designs.

REFERENCES

[1] Bachtold, A.; Hadley, P.; Nakanishi, T.; Dekker, C. *Science,294*, 1317-1320, 2001.

[2] Duan, X. F.; Huang, Y.; Agarwal, R.; Lieber, C. M. *Nature,421*, 241-245, 2003.

[3] Rochefort,A; Di Ventra,M; and Avouris,Ph; Applied Physics Letters,Vol. 78, No. 17, p.2521,2001.

[4] Miri Yemini, Meital Reches, Judith Rishpon, and Ehud Gazit*, Nano Lett., Vol. 5, No. 1,183-186, 2005

[5] Reches, M.; Gazit, E. *Nano Lett.*, *4*, 581-585, 2004.

[6] Miura, N., Numaguchi, T., Yamada, Konagai, M., Shirakashi, J., 1998. Room temperature operation of amorphous carbon-based single-electron transistors fabricated by beam-induced deposition techniques. Japanese Journal of Applied Physics, Part 2: Letters, vol. 37, L423-L425.

[7] Tay,B,K,; Sheeja,D; Yu,L.J;, "On stress reduction of tetrahedral amorphous carbon films for moving mechanical assemblies", Diamond Relat.Mater., vol. 12, pp. 185-194, 2003.

[8] Kinoshita,K, X.Song, J.Kim, M.Inaba, Journal of Power Sources 81-82, 170 ,1999.

[9] Srikanth Ranganathan, Richard Mccreery, Sree Mouli Majji, and Marc Madou, Journal of the Electrochemical Society, 147(1), 277, 2000

[10] Chunlei Wang, Lili Taherabadi, Guangyao Jia, and Marc Madou, Electrochemical and Solid State Letters Electrochemical and Solid-State Letters, 7 (11), A435-A438, 2004

Mimicking Photosynthesis to Make Functional Nanostructures and Nanodevices

J. A. Shelnutt[*,**], Z. Wang[*,**], Y. Song[*], C. J. Medforth[*] and E. Pereira[***]

[*]Sandia National Laboratories, Albuquerque, NM, USA, jasheln@sandia.gov
[**]University of Georgia, Athens, GA, USA, jasheln@unm.edu
[***]Universidade do Porto, Porto, Portugal, efpereir@fc.up.pt

ABSTRACT

The processes and functional constituents of biological photosynthetic systems can be mimicked to produce a variety of functional nanostructures and nanodevices. The photosynthetic nanostructures produced are analogs of the naturally occurring photosynthetic systems and are composed of biomimetic compounds (*e.g.*, porphyrins). For example, photocatalytic nanotubes can be made by ionic self-assembly of two oppositely charged porphyrins tectons [1]. These nanotubes mimic the light-harvesting and photosynthetic functions of biological systems like the chlorosomal rods and reaction centers of green sulfur bacteria. In addition, metal-composite nanodevices can be made by using the photocatalytic activity of the nanotubes to reduce aqueous metal salts to metal atoms, which are subsequently deposited onto tube surfaces [2]. In another approach, spatial localization of photocatalytic porphyrins within templating surfactant assemblies leads to controlled growth of novel dendritic metal nanostructures [3].

Keywords: porphyrin, photocatalyst, nanotube, metal, nanodevice

1 PORPHYRIN NANOTUBES

Porphyrins and other tetrapyrroles such as chlorophylls play important functional roles in biological nanostructures and proteins of photosynthetic systems. The chlorophyll molecules are sometimes self-organized into nanoscale superstructures that perform essential light-harvesting and energy- and electron-transfer functions. An example of the biological nanostructures formed by chlorophyll is the light-harvesting chlorosomal rods of the photosynthetic pseudo-organelles of green-sulfur bacteria, called chlorosomes. The chlorosomal rods are composed entirely of aggregated bacteriochlorophyll molecules as illustrated

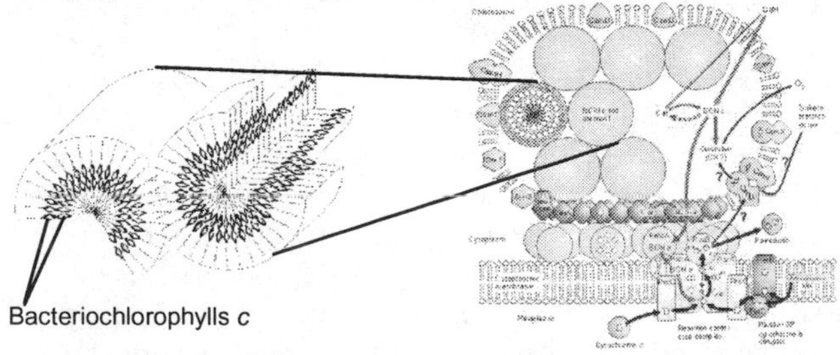

Figure 1. A chlorosome on the bacterial membrane contains the chlorosomal rods that harvest light.

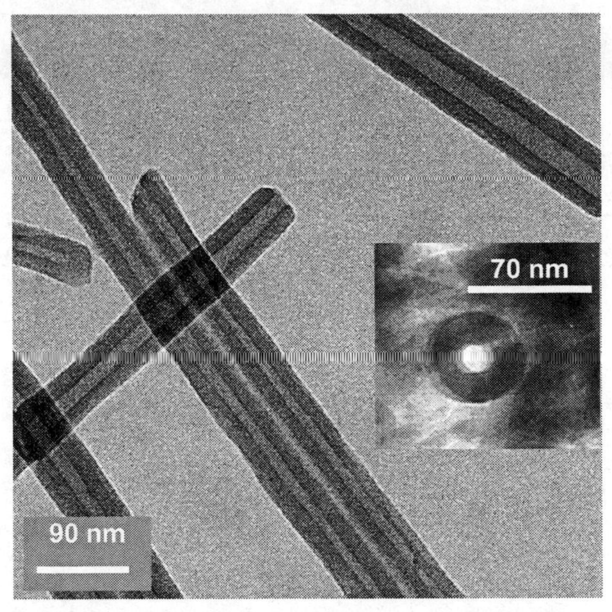

Figure 2. Transmission electron microscopy image of the porphyrin nanotubes. Inset: A tube trapped in a vertical orientation by a thick mat of tubes.

in Figure 1. Because of their desirable functional properties, porphyrins are attractive building blocks for a wide variety of functional synthetic nanostructures.

Recently, we discovered the first porphyrin nanostructures that possess well-defined morphologies suitable for technological applications [1,2]. An example is the porphyrin nanotubes shown in Figure 2 [1]. These sturdy nanostructures are suitable for incorporation into nanodevices for solar water splitting and other applications. The porphyrin nanostructures are biomimetic since the porphyrins are related to the chlorophyll molecules contained in systems found in nature that carry out light harvesting, charge separation, and energy conversion. Specifically, the porphyrin nanotubes shown in Figure 2 appear to mimic the light-harvesting and reaction-center functions of chlorosomes (Figure 1), and they possess other useful properties.

Our porphyrin nanotubes are prepared by ionic self-assembly of two oppositely charged porphyrins in aqueous solution and are composed entirely of porphyrins such as those shown in Figure 3. The nanotubes are just one member of a new class of porphyrin-based nanostructures that we are developing. We have found that the molecular building blocks (tectons) can be altered to produce a range

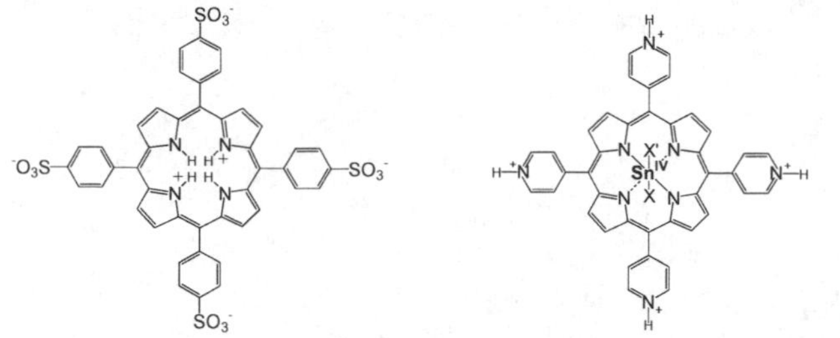

$[H_4TPPS_4]^{2-}$ $[Sn(X)(X')TPyP]^{4+/5+}$

Figure 3. Porphyrins used in making the porphyrin nanotubes shown in Figure 2. Sn(IV) will have ligands (X, X' = Cl, OH, H_2O) bound above and below the porphyrin plane. At pH 2 where the tubes are assembled, X = OH and X' = H_2O (net charge +5) or X = X' = OH (charge +4).

of nanostructures with different structural and functional properties. The porphyrin nanotubes in Figure 2 are hollow structures with uniform diameter and shape. The strong electrostatic forces between the porphyrin tectons (in addition to the van der Waals, hydrogen-bonding, axial coordination, and other weak intermolecular interactions that typically contribute to the stability of porphyrin aggregates) enhance the structural stability of these new nanostructures.

It has long been known that synthetic porphyrins under certain conditions form aggregates with interesting optical and electronic properties, and these aggregates have sometimes been found in the form of useful nanostructures including fibers, nanorods, or thin stripes on surfaces. However, the aggregates are more typically found as less useful forms such as nanoparticles, sheets, or fractal objects. The nanotubes, nanofibers, and structurally more complex nanostructures recently synthesized by our group satisfy the need for well-defined and robust porphyrin nanostructures that will be useful in many applications, including electronics, photonics, light-energy conversion, and catalysis. In particular, the porphyrin nanotubes and related porphyrin nanostructures possess shapes and functional properties that make them particularly useful for the construction of electronic and photonic nanodevices.

1.1 Synthesis of Porphyrin Nanotubes

The self-assembled porphyrin nanotubes are composed of two porphyrin molecules having opposite charges. The porphyrin nanotubes shown in Figure 2 are formed by mixing aqueous solutions of the two porphyrins shown in Figure 3. Typically, 9 mL of $H_4TPPS_4^{2-}$ solution (10.5 μM $H_4TPPS_4^{2-}$, 0.02 M HCl) was mixed with 9 mL of Sn(IV) tetrakis(4-pyridyl)porphyrin ($SnTPyP^{2+}$) dichloride in water (3.5 μM $SnTPyP^{2+}$), and the mixture was left undisturbed in the dark at room temperature for 72 h. Although the individual porphyrins by themselves exhibit negligible aggregation under these conditions (pH 2), the mixture of porphyrins immediately forms colloidal aggregates and, over time, provides a high yield (approximately 90%) of nanotubes. The remaining 10% of the porphyrin material is mostly rod-like with dimensions similar to those of the tubes, and thus appears to be collapsed tubes.

1.2 Structure of the Porphyrin Nanotubes

Transmission electron microscope (TEM) images of the porphyrin nanotubes are shown in Figure 2. They reveal that the nanotubes are micrometers in length and have diameters in the range of 50-70 nm with approximately 20-nm thick walls. Images of the nanotubes caught in vertical orientations (*e.g.*, see Inset of Figure 2) confirm a hollow tubular structure with open ends. Fringes with 1.7-1.8-nm spacing are seen both in end-on views and at the edges of the nanotubes in TEM images. The fringes probably originate from the heavy tin and sulfur atoms in the porphyrin stacks. The fringes seen in the TEM images, combined with the optical spectral results discussed below, are consistent with a structure composed of stacks of offset *J*-aggregated porphyrins (individual porphyrins are approximately 2 x 2 x 0.5 nm) likely in the form of cylindrical lamellar sheets. The lamellar structure could be similar to an architecture proposed for the stacking of bacteriochlorophyll molecules in the chlorosomal rods. X-ray diffraction studies (not shown) exhibit peaks in the low- and high-angle regions with line widths suggesting moderate crystallinity.

1.3 Composition of the Porphyrin Nanotubes

The composition of the nanotubes was determined by UV-visible absorption spectroscopy and energy dispersive X-ray (EDX) spectroscopy. The filtered nanotubes were dissolved at pH 12 and the ratio of the porphyrins was determined by spectral simulation using extinction coefficients for $H_2TPPS_4^{4-}$ (ε_{552} = 5500 $mol^{-1}dm^3cm^{-1}$) and $Sn(OH)_2TPyP$ (ε_{552} = 20200 $mol^{-1}dm^3cm^{-1}$), giving an approximate molar ratio of 2.4 $H_2TPPS_4^{4-}$ per $Sn(OH)_2TPyP$. EDX measurements of the S:Sn atomic ratio of the porphyrin tubes on the TEM grids also indicate a molar ratio of between 2.0 and 2.5. The observed ratio of the two porphyrins in the tubes (2.0-2.5) is related to the charges of the porphyrin species present at pH 2 (see Figure 3). As shown by acid-base titrations monitored by UV-visible spectroscopy, the porphyrin species present at pH 2 are $H_4TPPS_4^{2-}$ and a mixture of $Sn(OH)_2TPyP^{4+}$ and $Sn(OH)(H_2O)TPyP^5$, giving a ratio of charges of 2.0 or 2.5.

1.4 Properties of the Porphyrin Nanotubes

Porphyrin nanotubes have desirable electronic and optical properties making them suitable for incorporation into electronic and photonic devices. One of the interesting optical properties is the ability of the tubes to resonantly scatter light at the wavelength of the *J* band of the aggregated porphyrins forming the tube walls (particularly the 496-nm band; see Figure 4) [1]. The *J*-aggregate bands

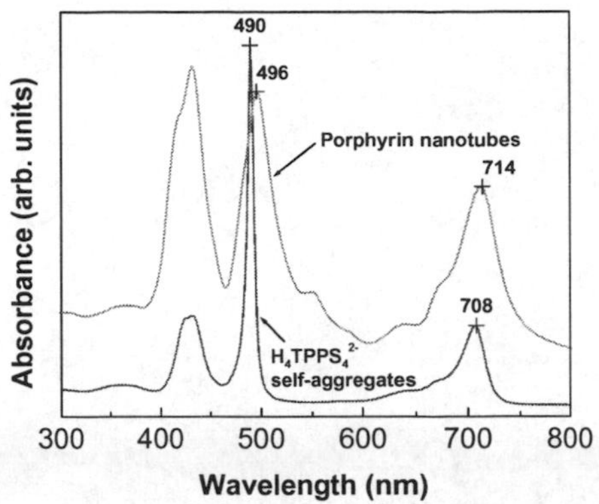

Figure 4. UV-visible absorption spectrum (red) of a colloidal suspension of the porphyrin nanotubes and a spectrum of a suspension of the $H_4TPPS_4^{2-}$ self-aggregates formed in 0.3 M HCl (blue) for comparison. The *J*-aggregate bands are near 496 and 714 nm.

Figure 5. Photographs showing a colloidal suspension of porphyrin nanotubes in transmitted white light (a) and the same suspension when viewed perpendicular to a beam of incandescent white light (b). The bright green colour observed in (b) is due to the intense resonance light scattering at the 496-nm *J*-aggregate absorption band.

result from the slipped or off-set stacking of the porphyrins, for which the strong coupling of transition dipoles of several adjacent molecules leads to intense light scattering from the interacting molecules.

The effect of the intense scattering is shown in the comparison of the photographs for weak transmitted light (Figure 5a) and for intense incandescent light directed at right angles to the observer (Figure 5b) for the suspension of the porphyrin nanotubes; the latter shows the intense green color of the resonance scattered light.

Another potentially useful property of the nanotubes is their ability to respond mechanically to light illumination. Even though they are stable for months when stored in the dark, irradiation of a suspension of the tubes for just five minutes using incandescent light from a projector lamp (800 nmol-cm^{-2}-s^{-1}) results in TEM images showing rod-like structures instead of tubes [1]. This response to light is reversible, as the tubes reform (self-heal) when left in the dark. The tubes have other interesting photochemical properties, including acting as photocatalysts for the reduction of metal complexes in aqueous solution or for the reduction of water to hydrogen (see below).

1.5 Control of Nanotube Structure

By altering the molecular structure of the porphyrin tectons, the dimensions of the nanotubes can be controlled. For example, nanotubes made with Sn tetra(3-pyridyl)-porphyrin instead of Sn tetra(4-pyridyl)porphyrin have significantly smaller average diameters (35 nm instead of 60 nm) [1]. Switching to the 3-pyridiyl porphyrin subtly repositions the charge centers and the associated H-bond donor atoms on the pyridinium rings, apparently changing the inter-porphyrin interactions sufficiently to alter the diameter while still allowing the tubes to form. Nanotubes are not produced by the 2-pyridyl porphyrin. Nanotubes are also obtained when Sn^{4+} is replaced with other six-coordinate metal ions (*e.g.*, Fe^{3+}, Co^{3+}, TiO^{2+}, VO^{2+}), but tubes have not been observed for four-coordinate metals (*e.g.*, Cu^{2+}) or the metal free porphyrin. A high degree of control over the structure of the tubes may be realized by modifying the tectons, including variation of the peripheral substituents of the porphyrin, the metal contained in the porphyrin core, and the nature of the axial ligands. Variation of the metal in particular will also alter the functional properties of the tubes.

2 PHOTOCATALYTIC NANOTUBES

Sn porphyrins are known to be good photocatalysts in homogeneous solutions, so we investigated whether the Sn porphyrins in the nanotubes would make the tubes themselves photocatalytic. To demonstrate photocatalytic activity of the tubes we examined the reduction of aqueous metal complexes of Au(I) and Pt(II) [2]. The photosynthetic properties of the nanotubes containing Sn porphyrins shown in Figure 2 have been used to synthesize functional nanotube-metal composite systems, which, in turn, may lead to solar nanodevices that mimic photosynthesis to produce hydrogen fuel by splitting water.

2.1 Gold Deposition on Porphyrin Nanotubes

When nanotubes are used to photoreduce the positively charged Au(I)-thiourea complex, the metal is deposited exclusively within the hollow interior of the nanotubes, forming a continuous polycrystalline gold nanowire that is of the same diameter as the tube core (Figure 6a). For those tubes that contain nanowires, only continuous gold wires are found, *i.e.*, multiple short segments are not observed in the same nanotube. In addition, the nanowires are typically terminated at one end of the nanotube with a gold ball of larger diameter than the tube. When the porphyrin nanotubes are dissolved by raising the pH, the gold wire and ball remain intact as shown in Figure 6b. In contrast with the thiourea complex, reduction of the *negatively* charged Au(I) thiosulfate complex produces gold particles

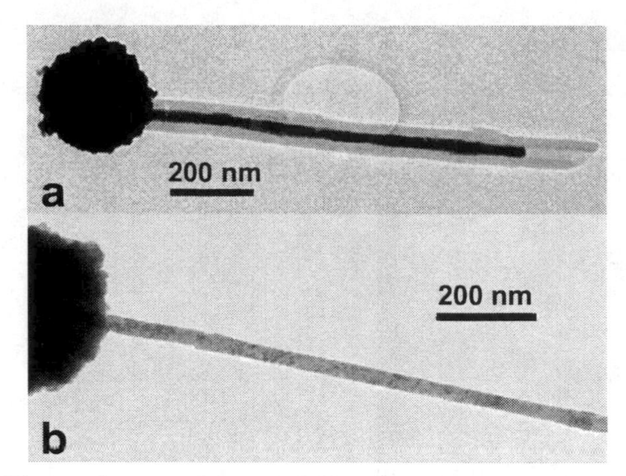

Figure 6. TEM images of (a) a gilded nanotube obtained using the Au(I) thiourea complex, and (b) a gold wire obtained after the porphyrin tube has been dissolved away at pH 10 by adding 0.05 M NaOH. The latter demonstrates the structural integrity of the free standing nanowires.

mainly on the outer surfaces of the tubes. Differences in electrostatic and other surface interactions with the complex may determine where metal is deposited. Directional electron/energy transport in the walls might also play a role.

2.2 Growth of Platinum Dendrites

Photocatalytic seeding and autocatalytic reduction of platinum and palladium salts leads to remarkable dendritic metal nanomaterials [3]. In micellar solutions, spherical metal nanodendrites are obtained. With liposomes as the template, dendritic platinum sheets in the form of thin circular disks or foam-like Pt nanomaterials (Figure 7) are obtained. Exquisite synthetic control over the morphology of these nanoscale dendrites, sheets, and nanostructured foams is realized by using photocatalytic tin porphyrins. Illumination conveniently and effectively produces a large initial population of catalytic growth centers in the surfactant assemblies. The initial concentration of these seed nanoparticles determines the average size and uniformity of the platinum dendrites, which grow by templating on the surfactant assemblies.

Figure 7. SEM image of platinum foam-like nanoballs templated on DSPC/cholesterol liposomes.

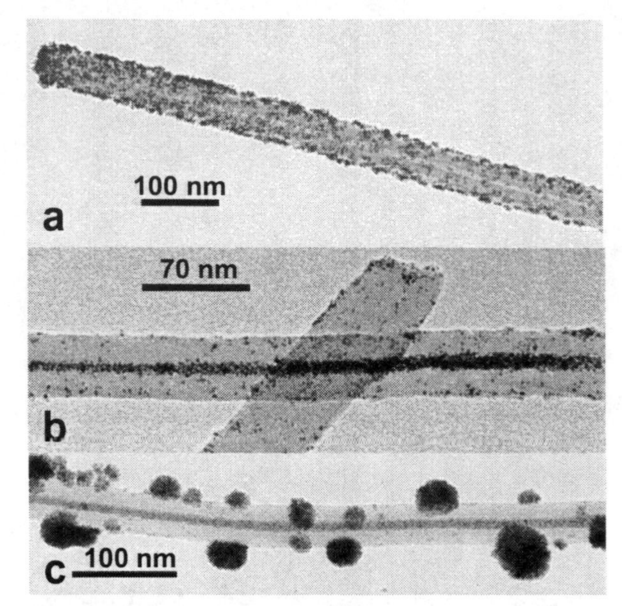

Figure 8. TEM images of (a) a platinized porphyrin nanotube with Pt nanoparticles distributed mainly on the outside surface, (b) a long Pt dendrite in the core of the tube obtained at a higher Pt concentration, and (c) a long dendrite in the core and Pt dendrites on the outer surface.

2.3 Platinum Deposition on Nanotubes

Figure 8 shows TEM images of some of the nanodevices that can be fabricated by growing Pt particles and dendrites on the porphyrin nanotubes. Figure 8b,c show tubes with dendritic Pt nanowires in the tube cores.

3 ACKNOWLEDGEMENTS

This work was partially supported by the Division of Materials Sciences and Engineering, Office of Basic Energy Sciences, U.S. Department of Energy, and by the Division of Chemical Sciences, Geosciences and Biosciences, Office of Basic Energy Sciences, U.S. Department of Energy (DE-FG02-02ER15369). Sandia is a multiprogram laboratory operated by Sandia Corporation, a Lockheed Martin Company, for the United States Department of Energy's National Nuclear Security Administration under Contract DE-AC04-94AL85000.

REFERENCES

[1] Wang, Z.; Medforth, C. J.; Shelnutt, J. A., *J. Am. Chem. Soc.* **2004**, *126*, 15954-15955.

[2] Wang, Z.; Medforth, C. J.; Shelnutt, J. A., *J. Am. Chem. Soc.* **2004**, *126*, 16720-16721.

[3] Song, Y.; Yang, Y.; Medforth, C. J.; Pereira, E.; Singh, A. K.; Xu, H.; Jiang, Y.; Brinker, C. J.; van Swol, F.; Shelnutt, J. A., *J. Am. Chem. Soc.* **2004**, *126*, 625-635

Nanomechanical Properties of Silica Coated Multiwall Carbon Nanotubes – Poly(methyl methacrylate) Composites

M. Olek*, K. Kempa**, S. Jurga***, M. Giersig*

*Center of Advanced European Studies and Research (CAESAR), Bonn, Germany, giersig@caesar.de
**Boston College, USA
*** Adam Mickiewicz University, Poznan, Poland

ABSTRACT

Nanomechanical properties of polymer composites, reinforced with silica *coated* multiwall carbon nanotubes (MWNT), have been studied using the nanoindentation technique. The nanohardness and the Young's modulus have been found to increase strongly with the increasing content of these nanotubes in the polymer matrix. Similar experiments conducted on thin films containing MWNT, but *without* silica shell, revealed that the presence of these nanotubes does not affect nanomechanical properties of the composites. While carbon nanotubes (CNT) have a very high tensile strength due to small nanotube stiffness, composites fabricated with CNT may exhibit inferior toughness. The silica shell on the surface of a nanotube enhances its stiffness and rigidity. Our composites, at 4wt% of the silica coated MWNT, display maximum hardness of 120 ± 20 Mpa, and the Young's modulus of 9 ± 1 GPa. These are respectively 2 and 3 times higher than those for the polymeric matrix.

Keywords: carbon nanotubes, nanoindentation, silica coating, nanohardness, elastic modulus, composites.

1 INTRODUCTION

The unique mechanical, electrical and optical properties [1-2] of MWNT make them very attractive for the fabrication of new advanced materials, particularly polymer composites with improved performance, or with new properties. Due to exceptionally high strength, and axial Young's modulus [1, 3], MWNT have been commonly considered as reinforcing fillers for high-strength materials. In this context, various polymers have been used as matrix materials, and different preparation techniques employed [4, 5]. In general, the tensile modulus and ultimate strengths of the CNT composites are reported to increase, although below the level of expectation.

The nanoindentation technique has been proven as a useful tool for determination of the mechanical properties of thin films, including polymers [6]. Pavoor et al. studied nanomechanical behavior of MWNT/polyelectrolytes composites produced by using the layer-by-layer (LBL) assembly deposition technique [7]. Utilization of the LBL technique has been proven to be an efficient method for incorporating CNT to a polymer matrix, with reduced phase segregation, high homogeneity, good dispersion and interpenetration of the nanocolloids and polymers, and high CNT concentration [5]. The nanoindentation studies of Ref. [7] showed that the hardness and Young's modulus of thin LBL MWNT/ poly(allylamin hydrochloride) (PAH) films are comparable to that of PAH. The individual MWNT can be displaced easily during indentation, leading to poor nanomechanical properties, close to those of the surrounding matrix.

In this study we perform a nanomechanical characterization of composites, reinforced with the silica coated MWNT ($MWNT@SiO_2$), within the poly(methyl methacrylate) (PMMA) matrix. Individual carbon nanotubes were coated with a uniform, thick layer of silica, and incorporated into PMMA. The $CNT\text{-}SiO_2$/PMMA composites were spin-coated on a silicon wafer substrate, and subsequently investigated by the nanoindentation technique. The hardness and elastic modulus were measured and compared with values obtained for the films made of uncoated MWNT in the PMMA matrix.

2 MECHANICAL CHARACTERISATION

Nanomechanical tests were carried out using the atomic force microscope (AFM) (NanoScope IV Digital Instruments) with conjugated TriboScope Nanomechanial Test Instrument from Hysitron Inc. The diamond conical and Berkovich tips were employed in this study as indenters. The calibration of indenters was carried out on soft material: poly(methyl methacrylate) was used as a standard material with an elastic modulus equal to 3.6Gpa.

A series of indentations were performed for each material throughout whole area, but in reasonable distance from the sample edges to avoid the edge influence on the mechanical properties of tested composites. In a typical experiment trapezoidal loading pattern was used. The peak loads were varied from 100μN to 1400μN, while the load/unload time was varied from 3s to 45s to maintain constant load/unload rate equal to 40uN/s. In general, indents with a contact depth ranging from 60-500nm were performed for each sample. To minimize the effect of the material creep, the hold time was incorporated at the maximum load. In all experiments, the 20s hold time was set. Prior to the indentation, the tip was used for a surface scanning, in order to find a reasonably smooth areas and to avoid large roughness influence on the mechanical properties. The in-situ imaging of the material's surface can also be used for visualization of the indent impressions into samples. To minimize any substrate contribution to the deformation response of tested materials, the thin films were indented within maximum displacement lower then 15% of the films thickness.

The data from the indents, performed under the same maximum load, were averaged to obtain the mean and the standard deviation for all samples. The fluctuations in the measured displacement, that occurred under the same maximum load are in the range of 1-15nm but not included in the graphs simply for reasons of clarity.

3 EXPERIMENTAL SECTION

MWNT (CVD method, purity > 95%, diameter 10-20 nm, length 5-20 um) were obtained from NanoLab (Newton, MA). In this study we produced two different composites, based on multiwall carbon nanotubes in PMMA, Mw=320000). The first one is based on functionalized MWNT that were blended with PMMA. In the second composite we employed silica coated MWNT, that were incorporated into a polymer (PMMA) matrix.

The first sample was prepared by an amide functionalization of MWNT in order to obtain soluble carbon nanotubes in organic solvents. To solubilize the MWNT, we used sample preparation according to the method of Ref. [8]. The MWNT dispersed in chloroform and the appropriate amount of poly(methyl methacrylate) (Mw=320000) were mixed in order to obtain the desired weight concentration of CNT with respect to PMMA. For nanomechanical tests we prepared five MWNT/PMMA samples with 1, 2, 3, 4 and 5wt% of MWNT.

In another sample silica-coated MWNT were used as a filler in the poly(methyl methacrylate) polymer. The coating process steps are as follows. The MWNT were functionalized with poly(allylamin hydrochloride) (PAH). CNT (50mg) were dispersed in a 0.5wt% PAH (Mw=70000) salt solution (0.5M NaCl, 500ml) and sonicated for 5h. Excess of polymer was removed by centrifugation (5 times) and washed with water. A residual black solid was re-dispersed in water, forming a stable, homogenous CNT suspension. The CNT water dispersion was transferred to silica sol (mixture of TEOS, H_2O, ethanol; mass ratio 2:1:4) in a 5:1 volume ratio. To prevent phase separation of TEOS and MWNT/water, the mixture was sonicated. After 2h the solution became homogenous and was set aside overnight at room temperature. After 12h, the mixture was centrifuged (4 times) to wash the carbon nanotubes with ethanol. The sediment was re-dispersed in a solution of ammonia in ethanol (4.2 vol. % ammonia (28 wt % in water) in ethanol). Immediately after this, TES solution (10 vol. % in ethanol) was added under stirring (5ml of TES in 500ml ethanol solution of CNT). The reaction mixture was stirred for another 8h and sonicated from time to time. Finally the CNT were washed with ethanol (4 times centrifuged) and again re-dispersed. The process described above leads to the formation of a uniform and thick layer of silica on every individual MWNT (Fig. 1). The modified MWNT were subsequently transferred to chloroform by functionalization with 3-aminopropyl trimethoxysilane (97%). An appropriate amount of PMMA was added to the silica-coated MWNT - chloroform solution to obtain the desired concentration of CNT. This mixture was further homogenized in an ultrasonic bath for 1h. For nanomechanical investigation, 5 different silica-coated MWNT/PMMA (MWNT-SiO_2/PMMA) composites were prepared with 1, 2, 3, 4 and 5 wt % of CNT in a polymer matrix.

The chloroform dispersions of both MWNT/PMMA and MWNT-SiO_2/PMMA composites were spin-coated on a silicon wafer substrate. In general, thin films with a thickness greater than 3μm were formed.

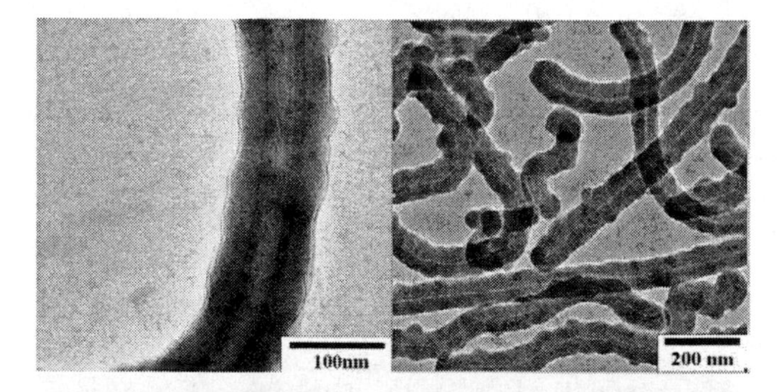

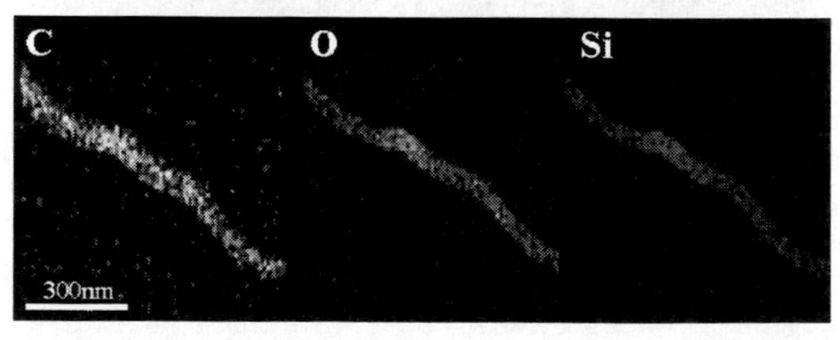

Figure 1 : TEM images and EDX mapping of silica coated multiwall carbon nanotubes.

4 RESULTS AND DISCUSSION

Fig. 2 shows the nanomechanical characterization of MWNT/PMMA composites. The homogeneity of our samples is confirmed by the relatively small standard deviation of data points. The hardness and Young's modulus are shown as a function of the contact depth for different MWNT content. The nanoindentation studies reveal that the mechanical properties of the MNWT composites are comparable to those of thin films of PMMA. Moreover, H and E_r as a function of contact depth exhibit exactly the same behavior as was shown for PMMA. There are no significant changes in H values with increasing concentration of carbon nanotubes in polymer. Young's modulus of thin films presents independent behavior on indentation displacement and E_r values are close to that obtained for PMMA.

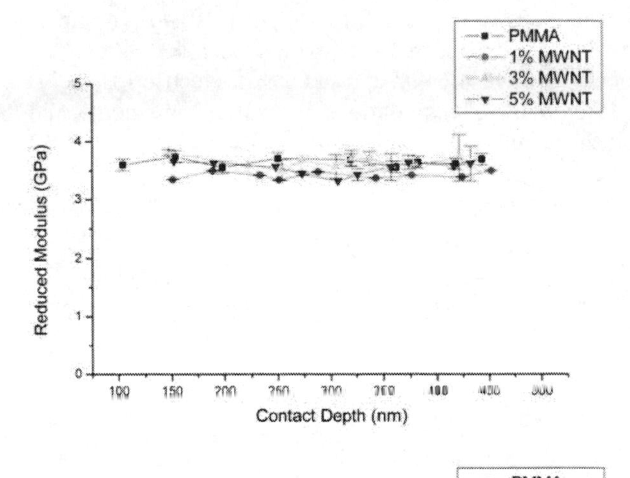

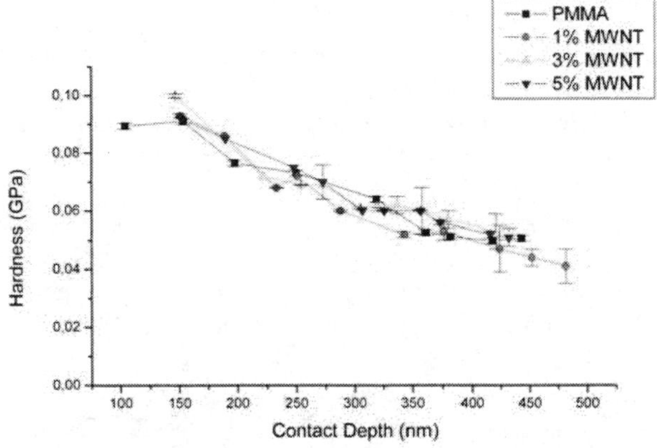

Figure 2: The elastic modulus and hardness for different CNT weight percentage content in MWNT/PMMA composites, as a function of the contact depth.

LBL technique has been proven an efficient method for incorporating carbon nanotubes into a polymer matrix, allowing for high composite homogeneity, good dispersion, interpenetration and high CNT concentration [5]. However, nanoindentation investigations of LBL - MWNT/PAH composites revealed, that the hardness and elastic modulus are close to those of the surrounding polymer matrix [7]. These results are consistent with our study, and confirm that high concentration and a homogenous distribution of CNT within a polymer matrix, as well as strong adhesion between the structural components do not ensure reinforcement of composites (in the nanomechanical sense). It was suggested, that flexibility of carbon nanotubes and their curvy morphology reduce the reinforcement action. The indenter can easily displace carbon nanotubes due to their flexibility and bending properties. As a result, the indenter "feels" essentially only resistance of the surrounding matrix, and the mechanical response of the composite is close to that of the polymer matrix. Wong and Sheehan determined the average bending strength for MWNT to be 14.2 $\pm$ 8.0GPa, i.e. several times smaller than for SiC nanorods [9]. Thus, nanomechanical improvement of a CNT/polymer composite examined by nanoindentation is limited by the relatively small bending strength of carbon nanotubes.

A completely different situation occurs for the silica coated CNT, as shown in Fig. 3. The MWNT@SiO_2 reinforced composites can exhibit much higher hardness and elastic modulus than PMMA. Both those quantities increase with increasing concentration of the coated MWNT in the polymer matrix. The results demonstrate the great influence of the silica reinforcing on the mechanical response of the CNT/polymer composite. Silica shell on a carbon nanotubes surface changes its mechanical properties. Such modified carbon nanotubes possess higher stiffness and are more rigid.

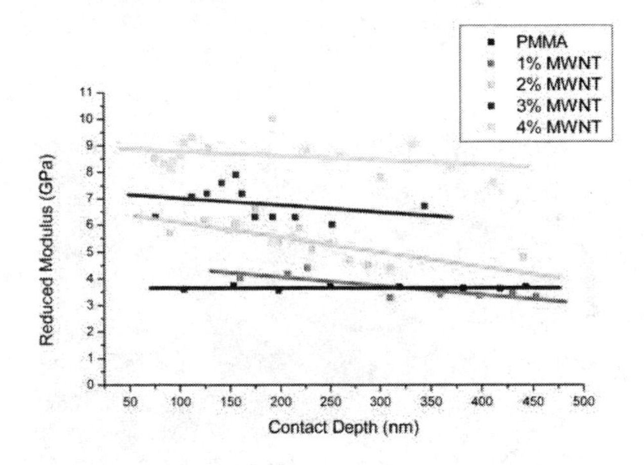

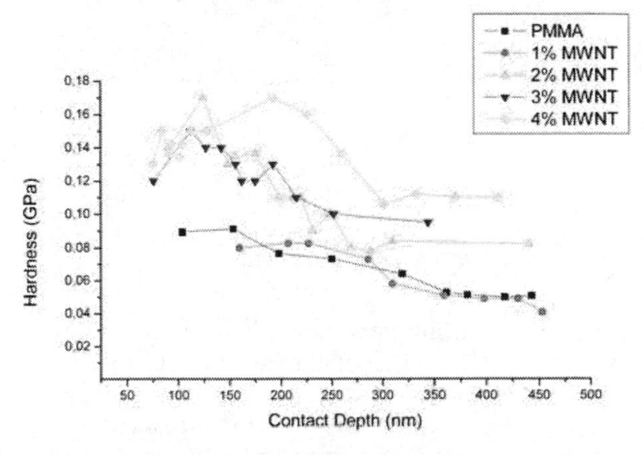

Figure 3: The elastic modulus and hardness for different CNT weight percentage content in silica coated MWNT/PMMA composites, as a function of contact depth.

Relatively large data scatter for the $MWNT@SiO_2$ films, as compared to MWNT/PMMA composites, indicate the presence of some inhomogeneities, and a nonuniform distribution of MWNT throughout samples (Fig. 4). The preparation method used in our study of $MWNT@SiO_2$/PMMA composite does not provide uniform distribution of CNT within the film, due to poor solubility of $CNT@SiO_2$ in chloroform. This leads to large errors in E_r and H. Nevertheless, the shown results demonstrate clearly a significant increase in the hardness and elastic modulus of $MWNT@SiO_2$/PMMA thin films, emphasizing the importance of the silica reinforcement of the carbon nanotubes. For example, Fig. 3 shows the Young's modulus for the 4wt% sample to be approximately 3 times as high as that for PMMA. For this CNT concentration, the hardness increases about two times in comparison to the polymer. We point out that this is not due to different preparation techniques used for the coated and uncoated CNT. Results of Ref. [7] clearly demonstrate, that even for a strong interconnectivity between components, and high homogeneity at very high wt% of CNT, the nanohardness and Young's modulus remain as low as that for a surrounding polymer. Presence of carbon nanotubes in polymeric materials does not improve nanomechanical properties due to high elasticity and small bending strength of CNT [9].

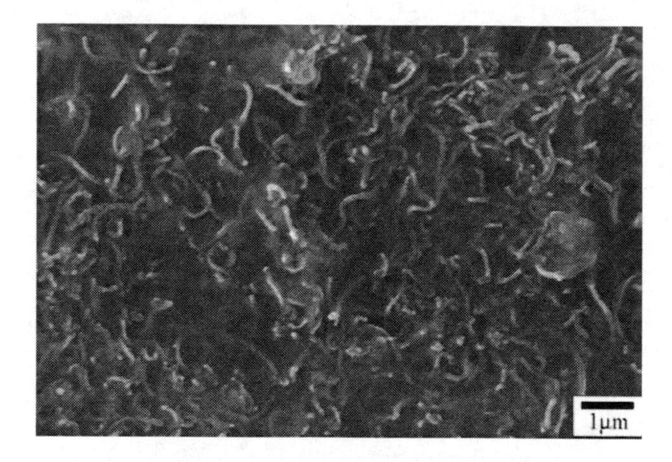

Figure 4: SEM image of the 3% silica coated MWNT composite.

5 CONCLUSIONS

The nanomechanical properties of the CNT/polymer composites (e.g. hardness) are not affected by the presence of CNT. We show, that silica coated MWCNT improve the nanomechanical properties of polymeric composites. Since the bending strength of CNT is improved by silica shell, the hardness and elastic modulus of MWNT/polymer composites increase with increasing content of MWNT in matrix. For example, a polymer composite at 4wt% of $MWNT@SiO_2$ displays an ultimate hardness of 120 ± 20 MPa and the Young's modulus of 9 ± 1 GPa. These are, respectively, 2 and 3 times higher as compared to those for the polymeric matrix. Silica coating of MWNT opens up possibilities for production of new advanced, reinforced materials for variety of applications. Since silica is an insulator, a coated CNT can be used as a coated nanowire for some nano-electrical applications. The electrical, as well as dielectric properties, of composites containing silica coated CNT are strongly affected by the insulator layer on each nanotube. It was recently shown [10], that a super-dielectric can be made this way, which has a very large, low frequency dielectric constant, and low dielectric loss. This may lead to novel applications in biology, medicine and nanolectronic devices.

REFERENCES

[1] Yu, M.-F.; Lourie, O.; Dyer, M. J.; Moloni, K.; Kelly, T. F.; Ruoff, R. S., Science, 287, 637-640, 2000.

[2] Ajayan, P. M. Chem. Rev., 99, 1787-1799, 1999.

[3] Salvetat, J.-P.; Kulik, A. J.; Bonard, J.-M.; Briggs, A. D.; Stöckli, T.; Metenier, K.; Bonnamy, S.; Beguin, F.; Burnham, N. A.; Forró, L. Adv. Mater. 11, 161-165, 1999.

[4] D. Li, X. Zhang, G. Sui, D. Wu, J. Liang, J. Mater. Sci. Let. 22, 791-793, 2003.

[5] M. Olek, J. Ostrander, S. Jurga, H. Möhwald, N. Kotov, K. Kempa, M. Giersig; Nano Letters, vol 4, No 10, 1889-1895, 2004.

[6] C. Klapperich, K. Komvopoulos, L. Pruit. J.Tribology, 123 624, 2001.

[7] P.V. Pavoor, B.P. Gearing, R.E. Gorga, A. Bellare, R.E. Cohen, J. App. Polym. Sci., 92, 439-448, 2004.

[8] Y. Qin, L. Liu, J. Shi, W. Wu, J. Zhang, Z.X. Guo, Y. Li, D. Zhu., Chem. Mater. 15, 3256-3260, 2003.

[9] E.W. Wong, P. E Sheehan. Science; Vol. 277, 1971.

[10] K. Kempa, M. Olek, M. Correa, M. Giersig, M. Cross, G. Benham, M. Sennett, D. Carnahan, T.Kempa, Z. Ren. " Dielectric Media Based on Isolated Metallic Nano-Particles", to be published.

Highly Dispersible Microporous Carbon Particles from Furfuryl Alcohol

Jin Liu[a], Jianfeng Yao[a], Huanting Wang[a,b]*, Kwon-Yu Chan[a]
[a]Department of Chemistry, The University of Hong Kong, China
[b]Australian Key Centre for Microscopy & Microanalysis, Electron Microscope Unit, The University of Sydney, Sydney, NSW 2006, Australia. Tel: +61 2 9351 7552 E-mail: huanting.wang@emu.usyd.edu.au

ABSTRACT

This paper reports the preparation of microporous carbon (MC) nanoparticles (25−90 nm) and spheres (260−320 nm) by carbonization of poly(furfuryl alcohol) (PFA) particles from furfuryl alcohol. Two synthetic routes were developed for forming "nonstick" PFA particles, which include (1) using silica as temporary barrier to isolate surface-reactive PFA nanoparticles and (2) a two-step polymerization of furfuryl alcohol (FA) involving slow polymerization (1st step) and sphere formation (2nd step). The high dispersibility of the MC spheres was realized by removing surface functional groups of the PFA spheres with the evaporation-induced concentrated sulfuric acid (2nd step). Both routes were effective in preventing irreversible aggregation of nanoparticles and spheres. Scanning electron microscopy, transmission electron microscopy, photon correlation spectroscopy and nitrogen adsorption were used to characterize the nanoparticles and spheres. The MC nanoparticles and spheres obtained showed good dispersibility in water and various organic solvents. They had a micropore volume of 0.13−0.14 cm^3/g, and a pore size of 0.49−0.56 nm. This synthetic routes presented here are very efficient for large-scale synthesis of MC nanoparticles and spheres.

KEYWORDS

Microporous carbon, Carbonization, Nanoparticle, Sphere, Furfuryl alcohol

1 INTRODUCTION

Microrporous carbons (MCs) have been long attractive for many important applications, e.g., as gas-selective adsorbents, membranes and catalyst supports and electrode materials for lithium ion batteries [1-6]. In particular, colloidal MC particles are of great interest because the diffusion of guest species through the MCs can be significantly manipulated by changing their particle sizes and shapes; the MC colloids would be useful as molecular sieves for developing the so-called mixed-matrix membranes that show high potential for industrial gas separations [6]. In addition, the MC particles with functionalized shells could provide with specific catalytic, magnetic, electronic, optical, or optoelectronic properties and thus broaden their uses. It is well known that MCs are usually prepared by carbonizing solid organic precursors, and only the rigid and crosslinked polymers and the pitch-like materials are widely chosen as the carbon source for desired microporous structures. Very recently, we have synthesized highly dispersible MC nanoparticles (25−90 nm) with irregular shapes by the pyrolysis of poly(furfuryl alcohol) (PFA) nanoparticles. Silica gel was used as the sacrificial barrier to prevent irreversible aggregation of highly reactive PFA nanoparticles throughout drying and carbonization [7]. We have also developed a novel two-step polymerization of furfuryl alcohol (FA) to the synthesis of colloidal MC spherical particles with larger size [8]. Here, we summarize the preparation and characterization of MC nanoparticles and spheres with high dispersibility from FA.

2 EXPERIMENTAL

Two synthetic routes to synthesize microporous carbon particles are described as follows.

The synthesis procedure of highly dispersible MC nanoparticles by using silica temporary barrier (Route 1) is illustrated in Figure. 1. Typically, 3 g of furfuryl alcohol, 3 g of P123 [(EO)20-(PO)70-(EO)20] (Aldrich), 1.4 g hydrochloric acid (36.0--38.0%), deionized water and ethanol were mixed under rigorous agitation at room temperature to form a homogeneous solution (Figure 1a). The cationic polymerization of FA was conducted at 30−40 °C for 1−2 days under continuous agitation, followed by heating at 60−70 °C overnight to complete polymerization under catalysis of HCl acid. The resultant PFA suspension obtained was cooled to room temperature (Figure 1b). To avoid the PFA particle aggregation through drying and carbonization, a temporary barrier technique was applied to the present system, and silica serves as a temporary barrier since it can be easily removed by using HF or NaOH solution. A solution of 3−6 g of tetraethyl orthosilicate (TEOS, Aldrich) in 10−20 g of ethanol was added into the PFA suspension and heated under stirring at 50−60 °C for 7−10 h allowing for hydrolysis of TEOS under acidic condition. The PFA nanoparticles covered with silica gel were retrieved by centrifugation. After the nanoparticles were completely dried at 100 °C overnight, a dark brown solid composite was obtained. The dried solid composite was carbonized by heating at a rate of 1°C/min, keeping at 200 °C for 2 h, and finally at 550 °C for 2 h under high-purity flowing Ar gas 2-4 to obtain the MC−silica nanocomposites. When the silica in the nanocomposite was completely dissolved in 10% NaOH solution, isolated MC nanoparticles without surface reactivity became dispersible, and they were retrieved by repeating a cycle of centrifugation, decanting, and ultrasonic redispersion in deionized water until the pH value reaches 7.0. The MC nanoparticles obtained were denoted as C-P-NP.

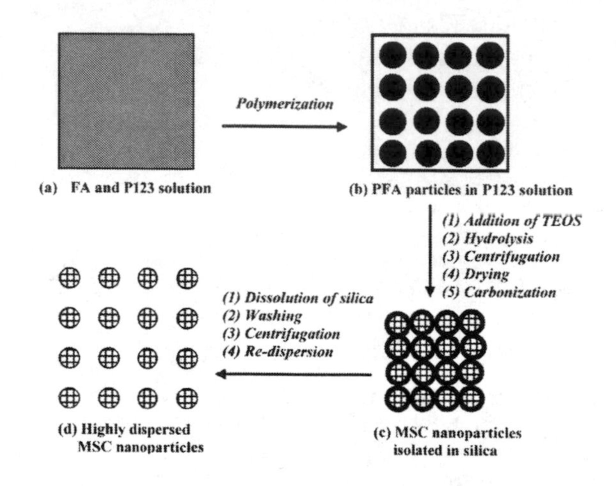

Figure 1 Synthesis procedure of highly dispersible MC nanoparticles by using silica as temporary barrier

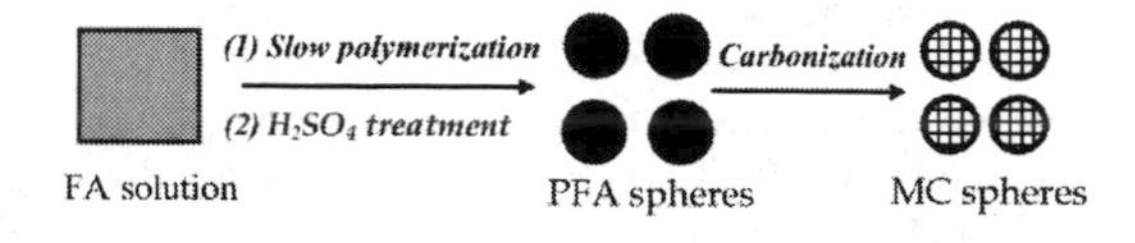

Figure 2 Synthesis of MC spheres by a two-step polymerization of FA combining with H_2SO_4 treatment

The MC spheres were synthesized by a two-step polymerization of FA, followed by retrieval of PFA spheres, drying and carbonization (Route 2, Figure 2). In the 1st step, the slow polymerization of FA was carried out in a FA-surfactant solution with a low concentration of HCl acid, and the FA pre-polymerized to form PFA oligomers. Amphiphilic triblock copolymers, Pluronic F127 (Sigma-Aldrich) and P123 were used as surfactants. Typically, 3 g of FA, 3 g of surfactant, 6 g of water, 1.4 g of HCl (36-38%, Merck) and 20 ml of ethanol were mixed in a 60 mL capped polypropylene bottle, and the obtained solution with a weight ratio of 1 FA: 1 surfactant: 6 ethanol: 2 H_2O: 0.17 HCl was continuously stirred at room temperature overnight. In the second step, H_2SO_4 solution with different concentrations at a volume ratio of 1 H_2SO_4 solution: 2 PFA solution was added into the above PFA solution under stirring. The mixture obtained was transferred into a glass beaker with a magnetic stirrer, and heated at 90 °C for 1 h allowing the solvents (water and ethanol) to evaporate. As a comparison, in the 2nd step, some syntheses were also carried out in the closed polypropylene bottles. The PFA spheres as-synthesized were collected by repeating three cycles of deionized water washing and centrifugation, followed by drying at 90−100 °C overnight. The MC spheres were finally prepared by carbonization of the PFA obtained with a heating program consisting of a heat rate of 1 °C/min, 200 °C for 2 h, and 650 °C for 5 h under high-purity flowing argon gas. The MC spheres by using P123 and F127 were denoted as C-P-RT and C-F-RT, respectively.

A Cambridge Stereoscan S440 scanning electron microscope (SEM) and a Philips Tecnai 20 transmission electron microscope (TEM) were used to examine the morphology and particle size of the samples. Nitrogen adsorption-adsorption experiments on the MCs were performed at 77 K on a Micromeritics ASAP 2020MC instrument to determine the Brunauer-Emmett-Teller (BET) surface area and micropore volume (t-plot method) and pore size distribution (Horvath-Kawazoe method based on slit-pore geometry). The samples were outgassed at 200 °C for 5 h before the measurement. Photon correlation spectroscopy (PCS) with a Zetasizer 3000 HSa equipment (Malvern Instruments Ltd. UK) at room temperature (25 °C) was used to determine the mean particle size and particle size distribution of MCs that were dispersed in ethanol. The wavelength of the internal laser was 633 nm and the selected measurement scattering angle was 90 degrees.

3 RESULTS AND DISCUSSION

3. 1 Electron microscopy and dispersibility

The size and dispersibility of MC nanoparticles were determined by extensive TEM observations. As can be seen from typical TEM images shown in Figure 3 (a, c), the MC nanoparticles have a mean size of 45 nm ranging from 25 to 90 nm and exhibit good dispersibility (Figure 3c). MC nanoparticles self assembled into an extremely uniform film with smooth surface throughout the whole carbon film surface after ethanol evaporation (Figure 3d). This again implies that MC nanoparticles were well dispersed in ethanol.

The dispersibility of MC nanoparticles in organic solvents such as ethanol, toluene and tetrahydrofuran by photon correlation spectroscopy is shown in Figure 3b. The MC nanoparticles are almost in the same size range of 20-100 nm, but their size distribution and peak size somewhat vary with the dispersion medium (solvent). It is well-known that colloidal particles are solvated leading to dispersion in the solvent. In addition to the presence of surface polarity, surface chemical heterogeneity, and surface flatness (pore), our carbon nanoparticles exhibit irregular shapes (e.g. nonsphere), this obviously gives rise to different "surface solvation effect" when the nanoparticles are dispersed in different solvents. Therefore, some deviations can be given from both calculation and slight difference in the dispersibility (reversible aggregation) when the scattering data obtained from different solvents are converted into the size and distribution. This particularly makes it difficult to precisely compare the size and distribution obtained from different solvents. Even so, it is still quite useful to measure the size and distribution of MC nanoparticles in various solvents. One can roughly see the dispersibility of the nanoparticles in different solvents. It is worth mentioning that the MC nanoparticles have rigid micropore frameworks through high-temperature carbonization, thus they are expected to be stable in organic solvents. This is evidenced by the fact that the pore volume and the surface area remain unchanged after the sample is dried from solvents. From PCS measurements, the mean particle sizes were calculated to be 35, 42 and 55 nm for ethanol, toluene, and tetrahydrofuran, respectively. Their particle size ranges are well consistent with TEM observations. This clearly indicates that there is not a significant amount of agglomerates observed, and the MC nanoparticles show good dispersibility in various organic solvents.

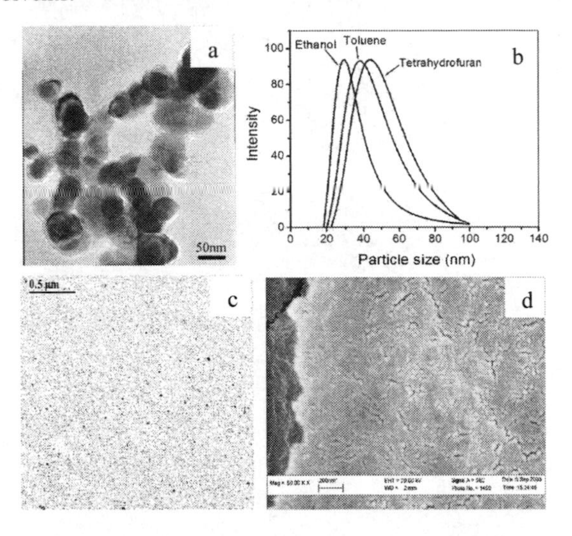

Figure 3 TEM images (a,c), FESEM image (d) and particle size distribution (b) of MC nanoparticles by Route 1.

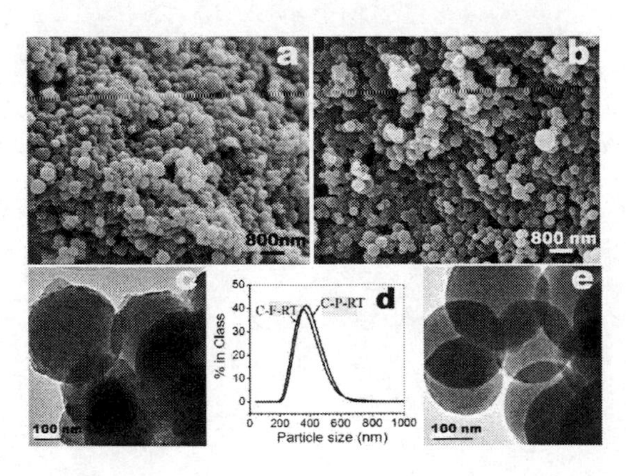

Figure 4. Electron microscopy images and particle size distribution (PSD) of the MC spheres by Route 2. (a, c) C-P-RT, FA polymerization with P123 at room temperature, (b,e) C-F-RT, FA polymerization with F127 at room temperature, C-R-RT (TEM), (d) PSD.

The MC spheres were observed with a scanning electron microscope and a transmission electron microscope. The SEM and TEM images (Figure 4) clearly show that both C-P-RT and C-F-RT spheres are fairly uniform. Their average sphere sizes are 320 nm, and 260 nm for C-P-RT and C-F-RT, respectively. The C-P-RT spheres have rough surface whereas C-F-RT spheres possess very smooth surface (Figure 4c, e). This is probably due to a lower cloud point of P123, which tends to phase separate at 90 °C (2nd step) and is less effective in functioning as the surfactant. The colloidal suspension of MC spheres in ethanol was studied by PCS method. Their particle size distribution is shown in Figure 4d. The sphere sizes are in the range of 270 550 nm, and 230—465 nm, and the average sizes are 370 nm and 340 nm for C-P-RT and C-F-RT, respectively. This confirms that both C-P-RT and C-F-RT exhibit good dispersibility.

3.2 Nitrogen adsorption

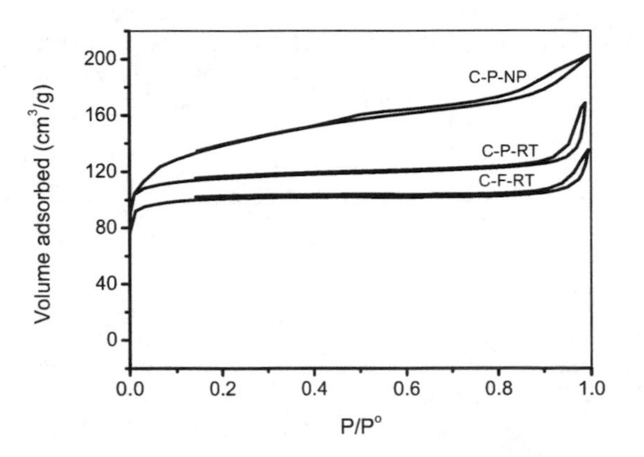

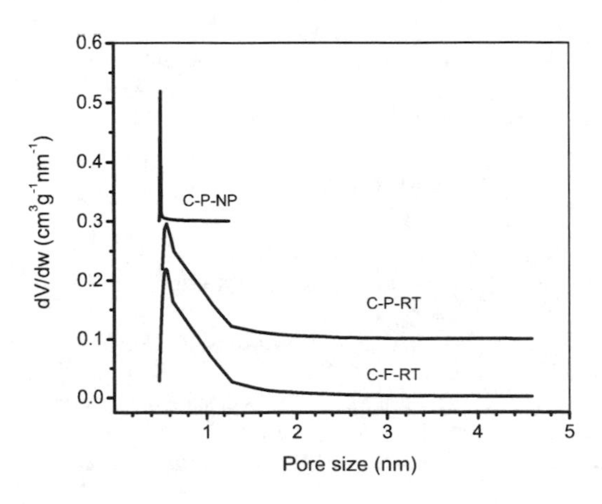

Figure 5. Nitrogen adsorption-desorption isotherms (upper) and pore size distribution (lower) of MC nanoparticles and spheres.

The washed MC nanoparticles were dried for N_2 adsorption-desorption measurement. The isotherm and the micropore size distribution of the sample (C-P-NP) are shown in Figure 5. The C-P-NP has a narrow pore size distribution, and its peak pore size is 5.0 Å corresponding to 4.8 Å for the PFA-derived molecular sieve carbons [1]. The micropore volume is calculated to be 0.13 cm^3/g. The BET surface area and the external surface area are determined to be 558 and 236 m^2/g, respectively. It can be concluded that well-defined micropores were formed in the nanoparticles, and the MC nanoparticles have very small particle sizes. On the other hand, the hysteresis loop at high relative pressures is a consequence of N_2 filling the large pores that may be associated with loose packing of highly dispersed MC nanoparticles. The nitrogen adsorption isotherms and pore size distribution of the MC spheres are also included in Figure 5. The BET surface area of the MC spheres is determined to 392 and 376 m^2/g for C-P-RT and C-F-RT, respectively. Obviously, the MC spheres possess smaller surface area as compared with C-P-NP because they have larger particle size. Both C-F-RT and C-P-RT have almost the same micropore volume (0.14 cm^3/g) and well-defined pore size distributions, and their pore sizes center at 0.56 nm.

4 CONCLUSIONS

We have successfully synthesized microporous carbon (MC) nanoparticles and spheres by polymerizing furfuryl alcohol in the P123 or F127 solution using silica as temporary barrier or H_2SO_4 treatment. The MC particles through carbonization exhibited a size range of 25−90 nm for the nanoparticles and 260−320 nm for the spheres, and they both had well-defined microporosity and high-dispersibility in water and organic solvents.

ACKNOWLEDGMENT

This work was supported in part by a seed grant from The University of Hong Kong. HW thanks the Australian Research Council for the QEII fellowship.

REFERENCES

1. H. C. Foley, *Micropor. Mater.* **1995**, *4*, 407-433.
2. M. B. Shiflett, H. C. Foley, *Science* **1999**, *285*, 1902-1905.
3. H. T. Wang, L. X. Zhang, G. R. Gavalas, *J. Membr. Sci.* **2000**, *177*, 25-31.
4. L. X. Zhang, K. E. Gilbert, R. M. Baldwin, J. D. Way, *Chem. Eng. Comm.* **2004**, *191*, 665-681.
5. D. Q. Vu, W. J. Koros, S. J. Miller, *J. Membr. Sci.* **2003**, *211*, 311-334.
6. Q. Wang, H. Li, L. Q. Chen, X. J. Huang, *Solid State Ionics*, **2002**, *152*, 43– 50.
7. J. Liu, H. T. Wang, L. X. Zhang, *Chem. Mater.* **2004**, *16*, 4205-4207.
8. J. F. Yao, H. T. Wang, J. Liu, K. Y. Chan, L. X. Zhang, and X. P. Xu, *Carbon* **2005**, in press.

Growth of individual vertically aligned nanotubes/ nanofibers with small diameter by PECVD on different metal underlayers

M.S. Kabir[1,*], R.E. Morjan[2], K. Svensson[4], O.A. Nerushev[2,3], P. Lundgren[1], S. Bengtsson[1], P. Enokson[1] and E. E. B. Campbell[2]

[1]Dept. of Microtechnology and Nanoscience (MC2), Chalmers University of Technology, SE-41296 Göteborg, Sweden

[2]Dept. of Physics, Göteborg University, SE-41296 Göteborg, Sweden

[3] Permanent address: Institute of Thermophysics, 630090, Novosibirsk, Russia

[4]Dept. of Applied Physics, Chalmers University of Technology

* Corresponding author: Email: mohammad.kabir@mc2.chalmers.se

ABSTRACT

The primary goal of this work is to achieve significant improvement of growth of carbon nanotubes (CNTs)/ carbon nanofibers (CNFs) on different metal underlayers that can provide a platform for electronic companies to adopt the technology into their existing manufacturing pipeline. In order to meet the requirements we have exploited a crucial component, inclusion of amorphous Si, as a buffer layer between the catalysts and the metal substrates to facilitate the growth of carbon nanotubes/fibers. Here we present some of the results on growth of carbon nanostructures on different metal underlayers for three different catalyst configurations: as film, as metal stripes and as dots of catalysts. Electrical characteristics of the grown structures as measured by STM are also addressed.

Keywords: cnf, cnt, vacnt, pecvd, ebl, stm.

1 INTRODUCTION

Carbon nanotubes have emerged as one of the most promising candidates for future nano-elements due to their one dimensional nature, and unique electrical, optical and mechanical properties. Hence, there is considerable interest on integrating carbon nanotubes into existing silicon based technology. Nanotubes have been used to demonstrate many applications including electronic applications: field emitters and/ or as sources for electron guns [1], NEMS [2], etc. For most of these and other applications, aligned /micro patterned carbon nanotubes are highly desirable [3]. Moreover, a prerequisite for exploring CNTs/ CNFs in an industrial process is to control mass production with high reproducibility. Controlled growth on patterned substrates is therefore an important issue and plasma enhanced chemical vapor deposition (PECVD) is perhaps the most promising technique to meet such requirements. To fully exploit this promising growth technique, it is necessary to investigate and optimize the growth of nanotubes on different types of metal and insulating substrates.

Cassell et al. have studied PECVD growth of CNFs on metal underlayers exploiting a combinatorial fabrication technique [4]. Recently we have performed a thorough investigation [5] on the effect of a thin intermediate a-Si layer for the PECVD-growth of nanotubes using Ni as catalyst on different metal underlayers. The Si/metal interaction occurring during growth plays a vital role in nanotube formation. One significant accomplishment achieved by a-Si insertion is the production of individual vertically aligned nanotubes/ nanofibers with diameters on the order of 10-20 nm, which is advantageous in many applications. Another important achievement with this technology is to be able to grow nanotubes on metal substrates which otherwise do not display CNF-growth with a Ni catalyst.

2 EXPERIMENTS

Oxidized silicon substrates, 1cm^2 in area, with an oxide thickness of 400 nm, were used. First the metal electrode layer was evaporated directly on the substrate by electron beam evaporation to a thickness of 50 nm. Sheet resistance measurements were carried out on the as-deposited films. An amorphous Si layer was evaporated before the Ni catalyst. For the case of stripe and dot patterns we used, electron beam lithography (EBL). A double layer resist, 10% co-polymer and 2% PMMA, was used as part of the lift off process.

A DC plasma-enhanced CVD chamber was used to grow the nanotubes. The experimental set-up and detailed growth procedure have been described previously [5]. The nanotube growth was carried out in a C_2H_2:NH_3 gaseous (1:5) mixture at 5 mbar chamber pressure at 700 ^{0}C for 10-20 minutes. After growth, the samples were cooled down to room temperature before air exposure. As-grown structures from pre-fabricated dots were then imaged with a JEOL JSM 6301F scanning electron microscope (SEM). All the experiments were performed repeatedly to verify the reproducibility. An STM probe was employed as the counter electrode to measure the electrical properties of the grown nanostructures.

3 RESULTS AND DISCUSSIONS

3.1 Growth from catalyst film

Figure 1 shows the results obtained after growth on different metal underlayers where amorphous Si was introduced between the metal layer and the catalyst. Significant improvement was achieved by introducing the Si layer. However, no growth was achieved for the case of Ti and Cr. The reason for lack of growth was attributed to Ti silicidation on the thick silicon oxide layer with a high release of oxygen that influences the Ni/Ti interface [5]. The rest of the metal underlayers seem to give growth of nanostrucures with reasonable density as depicted in Figure 2 (a). Pt gave very densely packed growth whereas Pd gave growth of vertical nanostructures together with many entangled nanotubes. These four metals reacted strongly with Ni and Si forming different kinds of silicides at the growth temperature. Therefore, silicidation processes occuring during the growth sequence are important for growing nanotubes on metal underslayers.

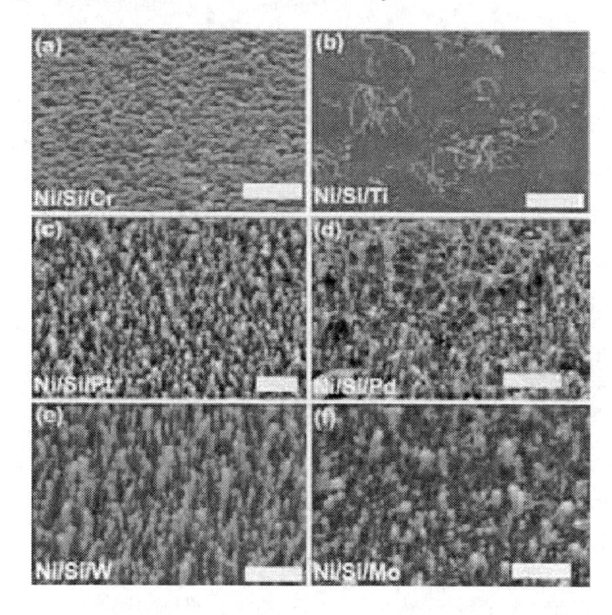

Figure 1: SEM micrograph of the samples after 15 minutes of CVD growth. The presence of Si facilitated the growth of nanotubes on some metal underlayers which was not possible in the other set of experiment (not shown here). All scale bars are 1μm.

The Density of the grown nanostructures is plotted for both the Si inclusion and Si exclusion cases. When the thin Si layer is included the density is much higher compared to no Si with a strong inclination towards forming structures with small diameters ≤ 20 nm. Only W and Mo metal underlayers show growth for the Si exclusion case with poor density and uniformity.

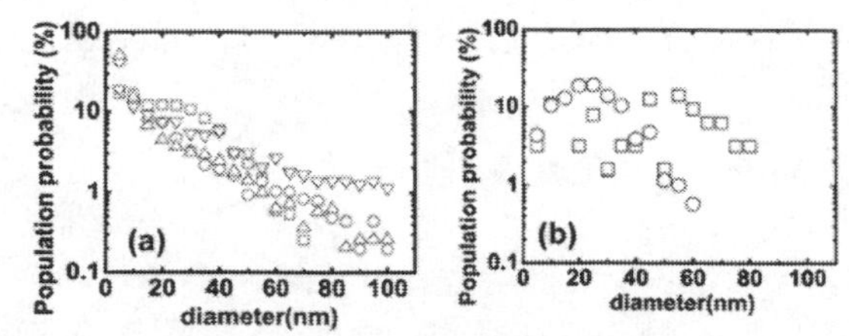

Figure 2: Size distribution of CNT: (a) metal underlayer with amorphous Si layer; square-Platinum-390 counts/μm^2; circle – Palladium – 226 counts/μm^2; up-triangle – Tungsten – 212 counts/μm^2; down triangle – Molybdenum – 89 counts/μm^2 and (b) metal underlayer without amorphous Si layer; square – Molybdenum – 5 counts/μm^2; circle – Tungsten – 73 counts/μm^2. No growth was achieved for other metal underlayers without the amorphous Si layer.

3.2 Growth on metal catalyst stripes

Growth from 1-10 μm wide catalyst stripes resembled the results obtained from films of catalyst as seen in Figure 3. CNTs/CNFs with small diameter and length are usually grown at the edges of the catalyst stripes. Though structures with large diameters are more visible in the figure, a large number of small diameter structures are present as can be seen in Figure 3 (b).

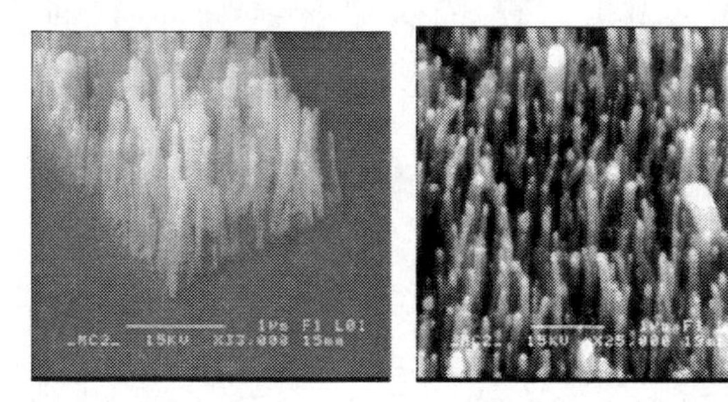

Figure 3: SEM micrograph of the samples after 15 minutes of CVD growth. Forest-like growth is observed. The length distribution is uneven (b) CNTs/CNFs grown on a 10 μm wide catalyst stripe.

The prevailing parameters that control the length distribution include the plasma power and gas ratio, the material below the plasma, species that are present in the plasma just above the material's surface due to secondary electron emission and different electric field distributions for different materials [6].

3.3 Growth on pre-fabricated dots

E-beam lithography (EBL) was employed for fabricating catalyst dots with different dimensions and pitches. Figure 4 portrays the results obtained from pre-fabricated catalyst dots of 100 nm and 50 nm diameter on a Mo metal underlayer. In each case there is a thin Si layer inserted between the metal underlayer and the catalyst. For both 100 nm and 50 nm sized catalyst dots, there are some instances of multiple CNFs growing from a single dot (100% and 2% respectively). CNFs grown from 50 nm dots are well aligned vertically with respect to the substrate. Most of the fabricated dots (more than 95%) acted as nucleation centers for growth. It is found from our measurements that the base diameter of isolated VACNFs in most cases was slightly larger than the tip diameter. 100 nm catalyst islands tended to split into several islands resulting in the growth of more than one nanotube/ nanofiber. The verticality of the grown structures is not as pronounced as for 50 nm dots. The Inset of figure 4 (a) shows one of such dots where multiple fibers have grown. Diameters of the grown structures are of the order of 30 nm and in some cases even smaller. Typically length variations from 400 nm to 800 nm were observed.

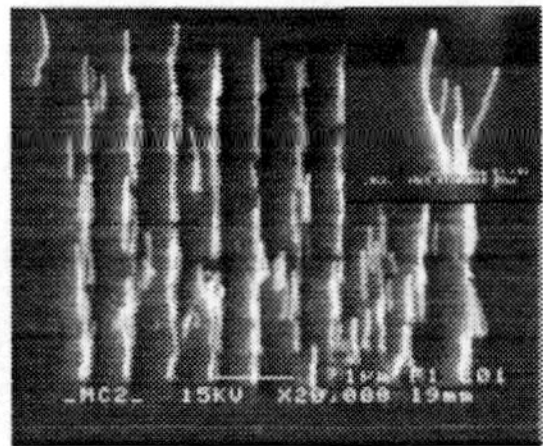

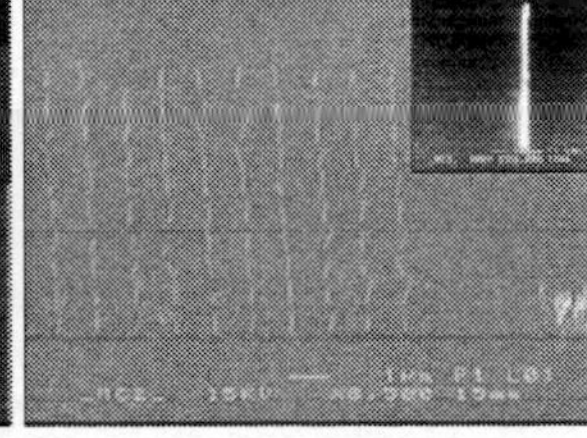

Figure 4: Multiwalled CNTs grown on pre-fabricated dots (a) 100 nm catalyst dots. Inset shows growth of multiple nanotubes. (b) 50 nm catalyst dots. Inset shows a high resolution image of an individual nanotube with ~50 nm diameter. In both cases a Mo metal substrate was used.

The inset of figure 4 (b) shows a high resolution image of an individual free standing CNF grown from a 50 nm catalyst dot. A tip growth mechanism prevails in all cases as the catalyst is riding at the top. The Mo underlayer provided a good base for the Ni and Si to interact and form catalyst particles which in most cases nucleated growth of free standing structures. However, size induced thermal expansion mismatch between the Ni/Si layer and the Mo layer let to break-up of the larger dots and subsequent growth of multiple structures As seen in the inset of figure 4 (b), the critical size for the nucleation of single CNTs was ~ 50 nm, which is much smaller than the previously published values [6].

3.4 Pitch limitations on growth

We performed a test on pitch induced limitations on growth and we find that it is possible to grow individual nanostructures down to 100 nm pitch for 50 nm dots, to a certain extent. Figure 5 portrays the obtained results. The grown structures are individual but tend to entangle with their nearest neighbors thereby putting a limit to the density attainable for growth of free standing individual carbon nanostructures.

Figure 5: Pitch limitations for growing individual CNFs. 50 nm dot size with 100 nm pitch was used in this experiment.

3.5 STM study on grown structures

Since the majority of the structures grown here are CNFs, a metallic behavior is anticipated. We have used an STM probe station to investigate the two-terminal electrical resistance at room temperature.

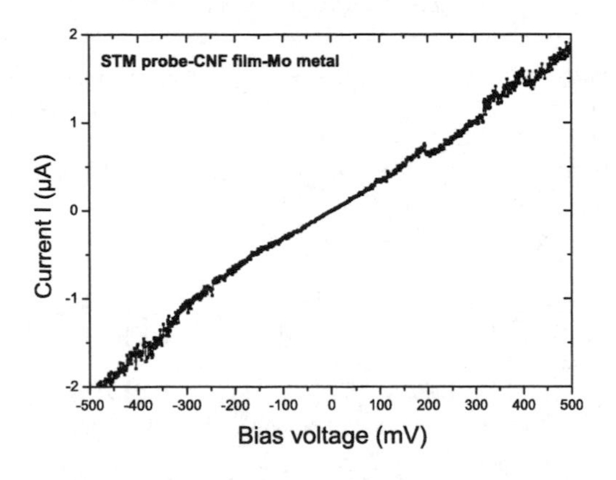

Figure 6: Typical IV characteristics of the measured CNF film after contacting the STM tip with the film

Figure 6 shows a typical current-voltage (*I-V*) measurement obtained from the CNF films. The measurements were carried out on grown structures from the 10 μm wide catalysts stripes. The resistance values were of the order of 200 kΩ which is slightly higher than the values reported by others for 4-probe measurements on individual structures removed from the substrates on which they were grown [4, 8].

Further measurements will be carried out to measure the conductivity of the free standing individual CNFs.

4 CONCLUSIONS

In conclusion, we have shown that it is possible to grow CNTs/ CNFs on different metal underlayers, a prerequisite for making electronic devices. The use of Si as an intermediate layer facilitated the growth of CNTs/ CNFs on four metal underlayers. Catalyst metal stripes showed similarity with the film growth case. Moreover, we show successful growth of free standing vertically aligned individual CNFs from pre-fabricated catalyst dots. Vertical growth and linear IV characteristics make these grown structures promising for example as possible replacements for Cu interconnects for multilevel circuit manufacturing technology.

ACKNOWLEDGEMENTS AND DISCLAIMER

Financial support from the Swedish Foundation for Strategic Research, VR: contract no. 621-2003-3539, ETEK and EC FP6 (contract no.FP6-2004-IST-003673, CANEL) is gratefully acknowledged. This publication reflects the views of the authors and not necessarily those of the EC. The community is not liable for any use that may be made of the information contained herein.

REFERENCES

[1] M.A. Guillorn, et al., J. Vac. Sci. Technol. B, 21, 35 (2004)
[2] Lee, S. W.; Lee, D. S.; Morjan, R. E.; Jhang, S. H.; Sveningsson, M.; Nerushev, O. A.; Park, Y. W.; Campbell, E. E. B. Nano Letters, 4 (10), (2004)
[3] Liming, Dai Pingang, He, Sinan, Li, Nanotechnology, 14, 10, (2003)
[4] Cassell, A. M, Qi, Ye, Cruden, B. A., Jun, Li, Sarrazin, P.C., Hou Tee, Ng, Jie, Han, Meyyappan, M., Nanotechnology, 15, 1, (2004)
[5] M.S. Kabir, R.E. Morjan, O.A. Nerushev, P. Lundgren, S. Bengtsson, P. Enokson and E.E.B. Campbell, Nanotechnology, 16, (2005)
[6] Merkulov, V. I. Lowndes, D. H. Wei, Y. Y. Eres, G. Voelkl, E. Applied Physics Letters, 76, 24, (2000)
[7] Merkulov, V. Michael J. GullioanI, Lowndes, D. H. Wei, Michael J. Simpson, Applied Physics Letters, 79, 8, (2001)
[8] Melechko, A.V., et al, J. Appl. Phys. 97, 041301 (2005)

Interactions between Carbon Nanotubes and Liquids: Imbibition and Wetting

A. Cuenat*, M. Whitby**, M. Cain*, D. Mendels* and N. Quirke**

*National Physical Laboratory, Division of Engineering and Process Control,
Hampton Road, Teddington, TW11 0LW UK, alexandre.cuenat@npl.co.uk,
**Department of Chemistry, Imperial College, London SW7 2AY, UK, m.whitby@imperial.ac.uk

ABSTRACT

We report on our progress to extend the current experimental measurements of interactions between fluid and materials to times and length scales characteristic of nanomaterials. The rapid imbibition of decane and non-polar fluids by carbon nanotubes is investigated to confirm recent theoretical results obtained using MD simulations. The wetting dynamics of carbon nanotubes and larger carbon fibres (~200 nm) are studied as a function of the liquid. We outline the main experimental challenges both in terms of time resolution and force sensitivity necessary for reliable experiments at these scales and present the first result of in-situ TEM observations of wetting of carbon nanotubes.

Keywords: carbon nanotube, wetting, imbibition, nano-fluidic

1 INTRODUCTION

Nano-tubular structures, and especially carbon nanotubes (CNT), have a large set of potential applications in nanofluidic devices. Most microfluidic devices have been fabricated in silicon or glass and are often limited to the micrometric range due to lithographic constrains. Recently, the capillary filling speed of water in nanochannels around 100 nm high has been measured [1] and the meniscus position qualitatively follows the Washburn model.

However, so far, no efficient design for 3D nanofluidic devices has been found and we may have to resort to manipulating "natural" nano-pipes. CNT provide a test platform that is smaller than capillaries used in fluid experiments and are thus the perfect tool to study confinement and fluid behavior at the nanoscale. In order to make use of these potential nanofluidic applications, it is necessary to understand the flow in a CNT channel and the influence of confinement.

Carbon nanotubes are already in use in composites where their combination of high aspect ratio, small size, high strength and stiffness, low density and high conductivity makes them perfect for improving the mechanical, thermal, and electrical properties of the materials. The three major problems in nanocomposite applications: dispersion, alignment and bonding of carbon nanotubes in the composite matrix, all require a better understanding of the wetting behavior of the nanotube surface.

In summary, CNT interactions with fluids, such as wetting, imbibition and solution rheology are critical in both nano-fluidic and composite applications. Furthermore, this interest extends to the many engineering applications where the wetting dynamics of homogeneous and structured surfaces is critical and to the physico-chemical properties of confined fluids. The present research aims at examining experimentally the interactions of carbon nanotubes with various types of fluids. Fluids filling the inside and surrounding the outside of nanotubes are being examined using a combination of scanning probe microscopy (SPM) and transmission electron microscopy (TEM).

2 FLUID-NANOTUBE INTERACTION

Capillary filling of nanotubes was first reported by Ajayan and Iijima [2] back in 1993. They observed, a very low filling efficiency. Ugarte et al [3] used TEM observation to show the filling of carbon nanotubes by a range of molten metal salts., but again for the smaller tubes (< 5 nm), the filling efficiency was poor.

Interaction of CNT with fluid was also probed by Barber et al [4] using carbon nanotubes attached to an AFM tip. They studied the dipping of the CNT into various organic liquids and measured the vertical forces as the probe was withdrawn from the surface. The diameter of the MWNT was 20 nm with a 2 μm length. They found that with polydimethyl-soloxane (PDMS) catastrophic flooding of the cantilever occurs with jump to contact; this indicated complete wetting of the nanotube. Using other liquids, similar flooding occurred where the nanotube tip was smaller than 0.5 μm. With a longer probe, an equilibrium meniscus was established below the attachment point on the cantilever and force measurements were successfully carried out.

2.1 Molecular Dynamics Simulations

In 2003, N. Quirke et al reported MD simulations of carbon nanotube imbibition [5]. Their key finding was a rapid filling of the pore with a velocity of about 445 ms-1. The filling did not follow the Washburn equation, but instead the filled length dependence was linear with time (rather than proportional to the square root).

In a second paper, N. Quirke et al [6] found that decane molecules were also attracted to the outside of the nanotube and wetted this surface. The velocity of the external wetting was much less than in the pore: 30 times slower in the case of the (13,13) nanotube.

3 EXPERIMENTS

The main objective of this research is to provide experimental evidence of rapid filling of CNT with inner diameter smaller than 5 nm in order to confirm the recent MD simulations [5,6]. The second objective is to provide quantitative data for the interactions between CNT and wetting fluids. The main experimental approach is based on the immersion into fluid of an open nanotube attached to the tip of an AFM. The resulting forces on the nanotube are measured as the probe is advanced, equilibrates and is withdrawn from the surface.

3.1 Sample preparation

The nanotubes are opened using standard chemical and physical treatment [3], so that their central channels are accessible for filling. The tubes are heated in air and oxygen at 700 °C to open the pores by oxidation of the tips. A further high temperature heating step at 2000 to 2100 °C in vacuum to graphitise amorphous carbon residues and to remove dangling bonds is done. This annealing step reduces the number of defects in the tubes, which may be important for unimpeded capillary flow. The opening of the tube and the quality of the inner wall are measured using TEM.

For SPM measurement, the open nanotubes are manipulated for attachment to the AFM tip by using a SEM based nano-manipulator (Zyvex Corp, Austin, TX, USA).

3.2 TEM observation

A first set of experiments is carried out to confirm imbibition. This involves immersing open carbon nanotubes into suitable liquids, such as decane and measuring the filling of the nanopore by direct TEM observation. These simple experiments determine that imbibition has taken place, but they do not allow for a precise determination of the timescale and of the force involved. TEM observations show liquid nonadecane droplets mixed with MWNT. The inner diameter of the nanotubes are around 5 nm with a 30 nm external diameter.

3.3 Experimental issues

A considerable number of questions and uncertainties surround these experiments. The first issue is the time resolution needed to measure the imbibition and capillary forces wetting the nanotube. It is clear that as soon as the CNT come into contact with the surface of the liquid, capillary forces cause spontaneous wetting of the entire nanotube. A way to solve this problem is to use a long enough carbon nanotube. The imbibition speed being an order of magnitude larger than the wetting speed, this should increase the time delay between the imbibition and the wetting signal. It may be necessary to functionalize the outside of the nanotube to reduce its wettability.

Regarding the comparison between the MD simulations and the experiments, the MD simulations are made for CNT without defects and with perfect openings at the end. Real nanotubes may have a variety of dangling bonds, irregularities, lattice defects and attached functional groups. Annealing of CNT is used to reduce these imperfections, but the possibility remains that they may hinder imbibition. It is interesting to note that only 1 in 30 of the open carbon nanotubes used by Ugarte et al were successfully filled [4].

Trapped air molecules inside the carbon nanotube may eventually impede the filling due to the fact that the top end of the nanopore is closed. In the existing MD simulations a vacuum is assumed inside the tube. Hopefully the pressure-build up effect will be small or only comes in to play after imbibition can be detected. It is however not clear whether the fluid dynamics associated with an open nanotube penetrating a thin layer of molecules allows imbibition to occurs spontaneously, or if a driving force is needed.

4 CONCLUSION

These results will be used to make the most of CNT properties and to explore their possible use as nano-test tubes filled with different fluids. We envisage the use of CNT nanotubes for fluid transport applications in the near future.

5 ACKNOWLEDGEMENT

The authors whish to acknowledge the contribution of Milo Schaffer and Mathhew Longhurst from Imperial College to this project.

REFERENCES

[1] N.R. Tas et al., "Capillary filling of water in nanochannels". Appl. Phys. Lett.,85, 3274, 2004.
[2] P.M. Ajayan et al., "Opening Carbon Nanotubes with Oxygen and Implications for Filling". Nature, 362, 522, 1993.
[3] Ugarte, D., et al., "Filling carbon nanotubes." Applied Physics a-Materials Science & Processing, 67, 101, 1998.
[4] A. H. Barber, S.R. Cohen and H. D. Wagner, "Static and DynamicWetting Measurements of Single Carbon Nanotubes," Phys. Rev. Lett, 92, 186103, 2004.
[5] S. Supple and N. Quirke, "Rapid Imbibition of Fluids in Carbon Nanotubes," Phys. Rev. Lett, 90, 214501, 2003.
[6] S. Supple and N. Quirke, "Molecular dynamics of transient oil flows in nanopores I: Imbibition speeds for single wall carbon nanotubes." J. Chem Phys., 121, 8571, 2004.

Large Scale Nanocarbon Simulations

S. Tejima*, M. Iizuka*, N. Park*, S. Berber**, H. Nakamura* and D. Tomanek**
Research Organization for Information Science & Technology

*Research Organization for Information Science &Technology, Tokyo, Japan
tejima@tokyo.rist.or.jp
**Physics and Astronomy Department, Michigan State University, MI, U.S.A
tomanek@pa.msu.edu

ABSTRACT

The mechanical response of carbon nanotubes to revere deformation has attracted much attention since their discovery in 1991. Carbon nanotubes have already demonstrated exceptional mechanical properties: their excellent flexibility during bending have been observed experimentally. Nanotubes combine high stiffness with elasticity and the ability to buckle and collapse in a reversible manner even largely axially compressed or twisted deformation. For these reasons, it has been suggested that carbon nanotubes could be promising candidates for a new generation of extremely light and super strong fiver. However, experiments probing the strength of nanotubes are very challenging, but to the difficulties in growing high quality, defect-free nanotubes of sufficient length and in measuring the strength of nanoscale objects. Theoretically, investigating the strength of carbon nanotubes requires modeling of inherently mesoscopic phenomena, such as plasticity and fracture on a microscopic compose of several thousands of atoms. The first principle methods based on the wave function of electrons are limited in the atomic structure of several hundreds atoms, but a large scale tight-binding simulation based on quantum orbit presents challenging up to tens of thousands of atoms. Our large scale simulation using Earth Simulator enables ourselves to reach this target. In simulations, elastic and buckling properties of nanotubes in difference radius and chirality are investigated on. When the change in length is small fracture under the compression, resisted load is proportional to compressed length. Above the first critical load, carbon nanotube occurs symmetrical buckling with keeping the elastic property. Above the second critical higher than first one, carbon nanotube dose asymmetrical buckling and into a fracture. The dependence of buckling point on length, radius and chirality of nanotubes will discuss in detail.

Keywords: large scale simulation, carbonnanotube, mechanical properties

1 INTRODUCTION

Carbon nanotubes and fullerenes have been expected to make a breakthrough in the new material development of nano-technology. Their remarkable properties such as great strength, light weight, special electronic structures, and high stability make carbon nanotubes the ideal material for a wide range of applications. A considerable number of potential applications have been reported about semiconductors-device, electronic field emitter, nano-diamond, battery, super strong threads and fibers, and so on.

Microscopic experiments are quite limited by hardness of measurements in nano space and by high experimental cost for the advanced microscope as scanning probe microscope (SPM). Material simulation based on quantum mechanics has turned out to be a very efficient methodology. The quantum model systems interests at microscopic level and are completely governed by a microscopic interaction between atoms. Therefore, the quantum simulation has the unique advantage in predicting nano scale properties, which are difficult to perform experiments.

The trials for so many simulations with any conditions give us valuable information on material properties and on design ways. However, to simulate the nano-scale phenomena on realistic time and space we need to deal with heavy computation. According to our estimation, in the case of the material formation from the bottom of atom scale to the sub-micron scale that could be applicable for sensor class, it would be required the Peta flops scale computing to deal with 10^8 atom system by the tigth-binding (TB) method. A recent higher performance computing provides a large-scale simulation of up to 10^4 atoms.

As for the high performance computational science, it is important to combine excellent physical ideas and optimizing program techniques on the simulation code. Thus, under the international collaboration in U.S.A. through Carbon Nanotube Simulation consortium, researchers, belong to different field launched the challenging large scale simulation for the nano-carbon materials using the Earth Simulator. In this project we developed and optimized a parallelized and vectorized CRTMD code for massively parallel and vector supercomputer, the Earth Simulator. Furthermore, by using the CRTMD code, we report some example in the applicability to the material science and design.

The CRTMD code, we have developed is shown to be suitable for the very large systems even though the lack of symmetrical arrangement. Our purpose is to give the clear explanation of properties and phenomena of nano-scale events and we deduce guiding principle to design new materials from nano-structures using super-computers.

Speedup by parallelization and optimization to more than 1000 processors on the Earth Simulator needs new algorisms to reduce the heavy computation time and cost. By optimization of code, we achieved performance of 7.1 Tera flops on a thermal conducting simulation.

This challenging collaboration has contributed for accelerating the investigation of the fundamental phenomena of nano-scale science and technology.

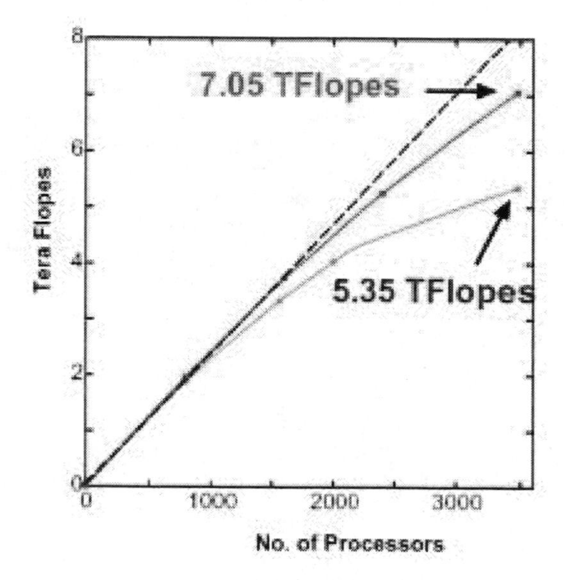

Figure.1 A sustained performance for the nanotube simulation with 48600 atoms

2 THEORY

A large scale molecular-dynamics simulation is suitable for studying micro-level physical properties at some conditions for temperature and pressures. Though the energy band structure can in principle be determined using ab *initio* techniques, we prefer a tigth-binding parametrization for its computational efficiency. Calculations based on the tight-binding formalism are much easier to perform than analogous ab *initio* calculations especially when describing large units cells and low symmetry situations. The tigth-binding Hamiltonian is described using a linear combination of atomic orbitals (LCAO) Hamiltonian,

$$H = \sum_{i,\alpha} \varepsilon_{i,\alpha} c_{i\alpha}^{\dagger} c_{i\alpha} + \sum_{i,j,\alpha,\beta} \varepsilon_{i\alpha,j\beta} c_{i\alpha}^{\dagger} c_{j\beta} + h.c. \tag{1}$$

We have used a simple two-center Slater-Koster parametrization for our four state (s, p_x, p_y, p_z) nearest-neighbor tight-binding Hamiltonian. The parameters have been obtained from a global fit to Local Density Approximation calculations for the electronic structure of C_2, a graphite layer and bulk diamond for different nearest-neighbor distances.

Our approach is based on the recursion method (RM) to calculate the electronic local density of states (DOS) and energies from selected elements of the one-electron Green's function . The RM eliminates the computational consuming load time for conventional diagonalization techniques to determine local DOS. They allow us to calculate the electron density with a workload which scales linearly with N not be proportional to N^3. This computational requirement is not acceptable when studying the dynamics of very large structure.

In matrix notation, the tridiagonalized Hamiltonian describing the local cluster centered at site i=0 is given by

$$U^{\dagger} H U = H_{TB} = \begin{pmatrix} a_0 & b_1 & 0 & \cdots \\ b_1 & a_1 & b_2 & \cdots \\ 0 & b_2 & a_2 & \cdots \\ \vdots & \vdots & \vdots & \vdots \end{pmatrix}. \tag{2}$$

The single Green's function at site i=0 is following:

$$G_{00}(E) = \left(\frac{1}{E - H_{TB}}\right)_{00} = \begin{pmatrix} E-a_0 & b_1 & 0 & \cdots \\ b_1 & E-a_1 & b_2 & \cdots \\ 0 & b_2 & E-a_2 & \cdots \\ \vdots & \vdots & \vdots & \vdots \end{pmatrix}_{00}^{-1} = \cfrac{1}{E-a_0-\cfrac{b_1^2}{E-a_1-\cfrac{b_2^2}{\ddots}}}, \tag{3}$$

Using Green's function, we derive the local density of states at site i=0 as

$$D_i(E) = -\lim_{\delta \to 0} \frac{1}{\pi} \operatorname{Im} G_{00}(E + i\delta). \tag{4}$$

Band energy is

$$E_{i,band} = \int_{-\infty}^{E_F} (E-E_{0,i}) D_i(E) dE \tag{5}$$

where Fermi energy is determined by

$$N_i = \int_{-\infty}^{E_F} D_i(E) dE \tag{6}$$

Here, N_i is the number of electrons. Finally, the force can be obtained by taking a negative gradient of the total energy with respect to each atomic position,

$$\vec{F}(r) = -\sum_i (\nabla E_{i,rep} + \nabla E_{i,band}). \quad (7)$$

The local repulsive energy functional $E_{i,rep}$ also depends on the local atomic density. The algorism to decide the forces on each atom is highly suitable for parallel computing, as the charge density can be calculated independently at each point of grid. This implementation makes the strong advantage of performing molecular dynamics calculations on massively parallel computers.

3. MECANICAL PROPERTIES OF CNT

The mechanical response of carbon nanotubes to extreme deformation and strain has attracted much attention since their discovery in 1991. It has been suggested that the tensile strength of carbon nanotubes might exceed that of other known fibers because of the strong interaction of the carbon-carbon bond. And under largely axially compression CNTs show structural buckle and collapse retaining sp_2 carbon network without a fracture. In addition, there is the excellent flexibility during bending. For these reasons, CNTs could be promising candidates for bundled composite materials in fibers and for atomic-force microscope (AFM) tips using isolated CNT.

The mechanical properties of nanotubes are predicted to be sensitive to details of their structure and to the presence of defects. It means that understanding on individual nanotubes is essential to examine these properties.

However, experiments probing the strength of a nanotube are very challenging, but to the difficulties in growing high quality, defect-free nanotubes of sufficient length and in measuring the strength of nanoscale objects. Theoretically, investigating the strength of carbon nanotubes requires modeling of inherently mesoscopic phenomena, such as plasticity and fracture on a microscopic compose of several thousands of atoms. The first principle methods based on the wave function of electrons are limited in the atomic structure of several hundreds atoms. In case of Young's modulus of CNTs, they are derived from a small strain in length using tens of atoms in unit cell. But for large deformations like an undulation under high pressures, more than thousands of atoms are required to confirm spatial extended deformation character along long CNTs. A large scale tight-binding simulation based on quantum orbit presents challenging up to tens of thousands of atoms. Our large scale simulation using Earth Simulator enables ourselves to reach this target.

A lot of theoretical investigations have focused on electric band structure of carbon nanotubes with metallic or semiconductor and on basic mechanical properties such as Young's modulus. The reason why these simulations are developed is that a small calculation treating with only unit cell atoms is good enough to solve it.

One of the most importance of the mechanical property is the critical buckling load and strain for stretched and compressed CNTs. We investigated the undulation process and broken process of long nanotubes under axial pressures and stretches, respectively, by help of a large scale tight binding simulation.

For example, the number of atoms for the (10,10) nanotube is 2200. This size is ten times as large as conventional calculations. Undulating compressed structure obtained from our simulations are very different from the previous one. Initial linear elastic deformations are observed up to the critical strain 92% beyond which nonlinear responses set in. In the nonlinear response regime at symmetric deformed structures such as buckles have been observed in simulation. The next buckling in (10,10)CNT of strain of 91.5% presents asymmetric structure.

The strech of CNTs eventually reach fracture at 240 % in length from original one. Surprisingly, from the energy dependenc, mixing state with sp_2 hexagon and sp chain between 150 to 240% kept the properties elastic.

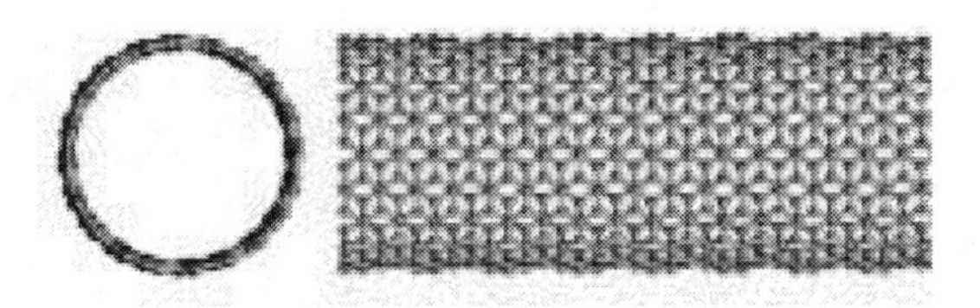

Figure.2 Linear elastic deformations at the 92 % compressed CNT.

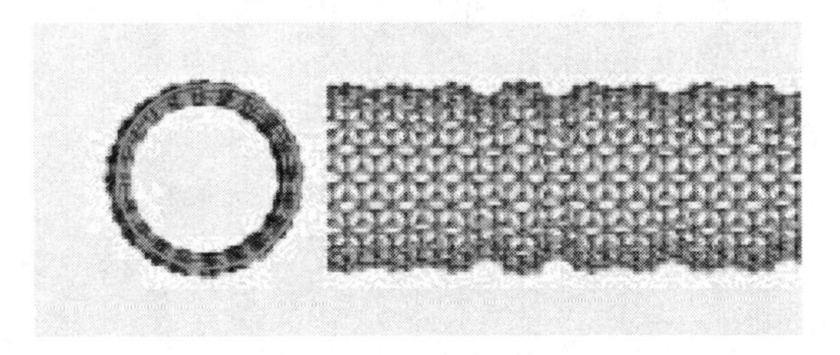

Figure.3 An asymmetric structure at the 91.5% compressed CNT.

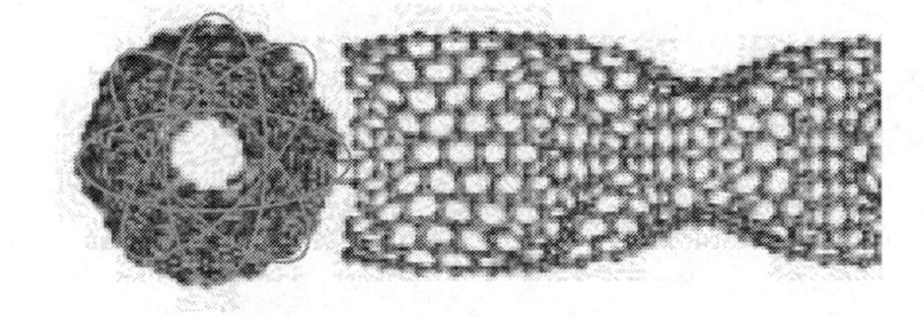

Figure.4 A fractured structure for the 90 % compressed CNT.

REFERENCES

[1] W.Zhong, D.Tomanek and G.Bertsch, "Total Energy Calculation for Extremly Large Cluster (The Recursive Approch)," Solid State Phys. **86**, 607, (1993)
[2] S.Tejima *et al.*, "Massively Parallel Simulation on Large-Scale Carbon Nanotubes", NANOTECH, San Francisco, CA., (2003, February, 2003)
[3] D. Srivastava, M. Menon, and K. Cho, "Nanoplasticity of Single-Wall Carbon Nanotubes under Uniaxial Compression," Phys. Rev. Lett. 83, 15, 2973, (1999)

Multi-Scale Modeling of Processing of Carbon Nanotubes

Hyuk Soon Choi, Kwang Hee Kim, Ki-Ha Hong, Jongseob Kim, Hyo Sug Lee and Jai Kwang Shin
Samsung Advanced Institute of Technology, San 14-1 Nongseo-Ri, Giheung-Eup, Yongin-Si
Gyeonggi-Do, Republic of Korea 449-772

A. V. Vasenkov, A. I. Fedoseyev, and Vladimir Kolobov
CFD Research Corporation, 215 Wynn Drive, Huntsville, AL 35805
Gyeonggi-Do, Republic of Korea 449-772

ABSTRACT

As technologies are continuously advancing, new semiconductor materials and novel devices are being developed which are indispensable for electronic applications. Progress in the computer aided design of fabrication of nanoscale materials lags behind experimental advances, since atomistic simulation methods are computationally impractical, while mesoscopic simulation methods are not capable of capturing nanoscale effects. Multi-Scale Computational Framework is introduced to conduct over 3 scales, continuum model of reactor-scale processes, kinetic Monte Carlo for the growth of nanostructures, and molecular dynamics for the prediction of reaction rates. "Gap-tooth" algorithm and "coarse time-stepper" method are employed to couple the scales. The results of multiscale simulation for growth of carbon nanotubes by plasma enhanced chemical vapor deposition are presented.

Keywords: Multiscale Simulation, Coarse Time-Stepper, Gap-Tooth, Plasma Enhanced Chemical Vapor Deposition (PECVD), Carbon Nanotube (CNT).

1 DESCRIPTION OF THE FRAMEWORK

The rapid development of nanotechnology has created significant interest to predicting the behavior of materials from the atomic to the engineering scales. For simulation of processes over the length and time scales that are a million times disparate, Multi-Scale Computational Framework (MSCF) has been developed by coupling computational tools of different types. The computational fluid dynamics (CFD) simulation was employed in reactor-scale, while atomistic kinetic Monte Carlo (kMC) and molecular dynamic (MD) simulations were performed in microscopic region to describe behaviors of atoms and molecules. The feasibility of the developed framework was shown using reactor-scale CFD-ACE software with plasma simulation capabilities [1], NAMD code designed for molecular systems [2], and kMC-FILM software developed for the deposition of nanostructured films and growth of carbon nanotubes.

The interactions between different kMC modules can be outlined as follows. The Interface control module receives initial data, transfers these data to other modules and also calls other modules to perform simulations of different processes. The Transport module calculates the transport of gas species from the source plane to the surface. The surface kinetic module calculates surface reactions between gas and surface species and computes surface fluxes and fluxes of carbon absorbed on the surface of catalyst. The surface fluxes are used by the FILM/CNT growth module and the carbon fluxes are sent to the Catalytic Interface module. The Catalytic Interface module calls continuum solver to simulate the diffusion of absorbed carbon through the catalyst and compute fluxes of carbon atoms to be incorporated into CNTs. The CNT/FILM module simulates the growth of CNT on the surface of catalyst and the formation of crystal film on other surfaces. The Output module writes the growth and visualization data to output files.

The Gap-tooth module was used to couple kMC simulations in teeth with CFD simulations in large gaps particularly, incoming microscopic surface fluxes and number of particles to a tooth at each tooth boundary were updated using the microscopic surface fluxes and number of particles outgoing from the nearest.

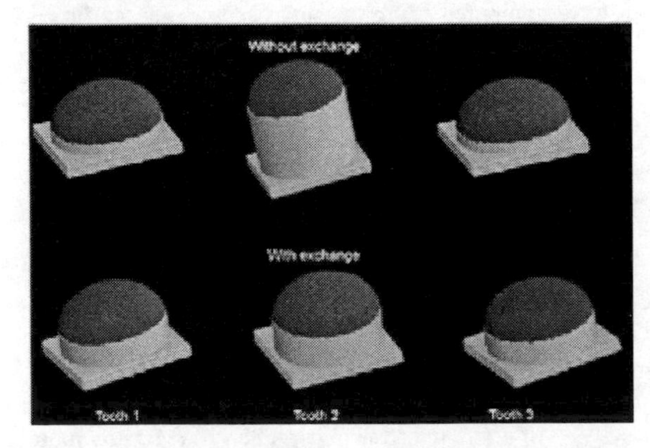

Figure 1: Growth of carbon nanotubes with (top) and without (bottom) gap-tooth method for teeth of low and high reactant fluxes from reactor.

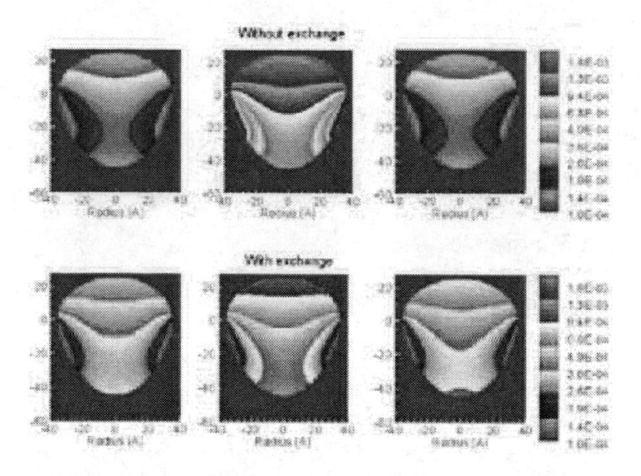

Figure 2: Carbon concentration profiles in the catalyst of teeth of low and high reactant fluxes from reactor with (top) and without (bottom) gap-tooth method .

To verify effect of gap-tooth method, the growth and carbon concentration profile were calculated for the carbon nanotubes with low and high reactant fluxes from reactor scale. Tooth 1 and 3 have low C_2H_2 flux of $1.0*10^{16}$ /cm^3 and tooth 2 has high C_2H_2 flux of $2.0*10^{17}$/cm^3, 20 times larger than tooth 1 and 3, while other reactant species fluxes are kept identical. Fig. 1and Fig. 2 show the effect of gap-tooth method to distribute the reactant hydrocarbon flux to nearest teeth, i.e. carbon nanotubes, by allowing the transport the reactants between teeth. Resultantly, the growth rate of carbon nanotube with high flux decreases and those of carbon nanotube with low flux increases by getting reactant species from high flux neighbor.

Two choices of flux interpolation were implemented in the Gap-tooth module: linear flux interpolation and quadratic flux interpolation. in both cases, incoming microscopic surface fluxes and number of particles to a tooth were computed using the microscopic surface fluxes and number of particles outgoing from the nearest teeth as described in earlier report [3].

The coarse timestepper module was developed to bridge the differences in the time scales between kMC and MD. The major steps of the Coarse Timestepper module are:

- Set initial configurations for NAMD simulations by cloning MD spatial coordinates from those of corresponding kMC simulations.
- Perform the evolutions of molecular systems. The evolutions were performed for the time T_{MD} long enough compared to MD time step τ_{MD} of 10 femto seconds and short enough compared to kMC time step τ_{KMC} of a few microseconds. Each NAMD ensemble was at fixed volume, temperature, and the number of particles.
- Calculate rates, and their derivatives in respect to time. Accuracy required for the prediction of reaction rates was achieved due to the use of reactive MD approach. This approach employs pre-defined criteria in bond-breaking/bond-making routine and it has been successfully used for computing rates of thermal decomposition reactions in polymers [4].
- Project reaction rates to the time $t+\tau_p$ such that $T_{MD} \le \tau_p \le \tau_{KMC}$. The projection is achieved using the Newton-Raphson method.

The interactions between different modules of MSCF over a mesoscopic time step τ are conducted as follows. MD was performed in each tiny tooth to compute the rates of surface processes and their derivatives in time for a nanoscopic time. The MD results were transferred to the Coarse timestepper module which calculated the time-dependent rates using the Newton-Raphson method. The rates were used by kMC to model system evolution during a microscopic time step. Subsequent to each kMC iteration, the incoming fluxes for each tooth were updated by the Gap-tooth module from the fractions of outgoing kMC fluxes obtained in the nearest teeth and MD was called by the Coarse timestepper module for updating time-dependent rates. Once the simulation time in kMC reaches $t+\tau$, the CFD solver was called by the Gap-tooth module to simulate reactor-scale processes and compute the fluxes of absorbed species for kMC.

2 GROWTH SIMULATIONS OF CARBON NAOTUBES BY PLASMA ENHANCED CHEMICAL VAPOR DEPOSITION (PECVD)

MSCF was applied to simulate the plasma assisted growth of aligned CNTs in an Inductively Coupled Plasma (ICP) reactor in a CH_4/H_2 gas mixture in Fig. 3. In this reactors, plasma is sustained by radio frequency (RF) electro-magnetic fields created by an RF current in a coil [2].

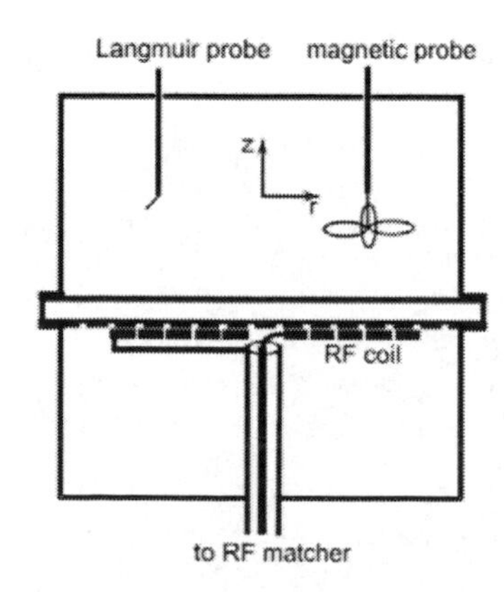

Figure 3: The schematic of ICP reactor.[5]

The calculated electron concentration, gas temperature and concentrations of C_2H_2 and CH_3 species are shown in Fig. 4 for a gas pressure of 100 mTorr, 100 W power adsorbed in plasma, and a driving frequency of 6 MHz. The substrate temperature was kept at 1000 K. The electron density peaks in the center of the reactor, the gas temperature has two peaks: one near the substrate and the other near the coils. The C_2H_2 and CH_3 densities peak near the sidewall where the gas temperature is low.

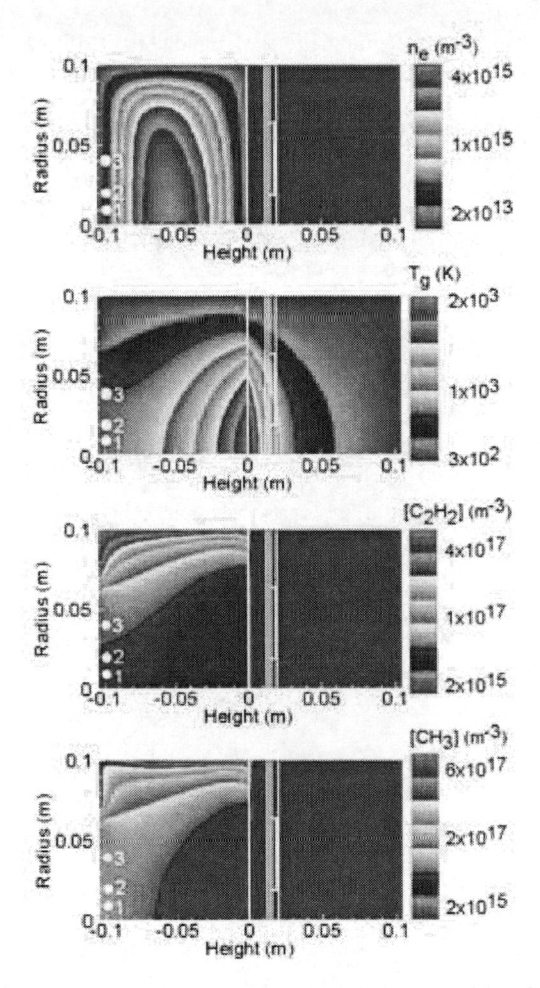

Figure 4: Plasma and gas properties computed by CFD-ACE for ICP plasma sustained in CH_4/H_2. White spots represent locations of monitor points where fluxes of radicals were transferred from CFD to kMC solvers.

The fluxes of different radicals computed by CFD-ACE at selected monitor points were used to update on-the-fly kMC-FILM fluxes. The growths of CNTs of different diameters were calculated using these fluxes as shown in Fig. 5. For example, CFD-ACE fluxes at the 1st monitor point were used for the growth of two CNT of small diameter (35 Å). At the same time, CFD-ACE fluxes at the 2nd monitor point were used for the growth of two CNT of small diameter (35 Å) and one CNT of large diameter (58 Å). Finally, CFD-ACE fluxes at the 3d monitor point were used for the growth of two CNT of large diameter (58 Å).

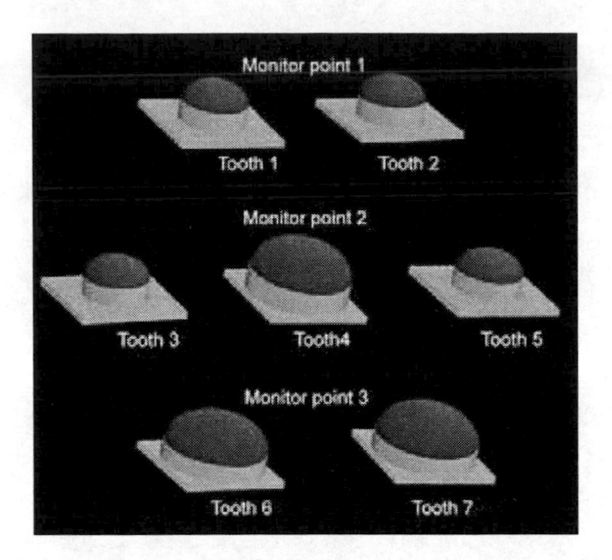

Figure 5: The growth of CNTs on the catalyst of different size at chosen monitor points.

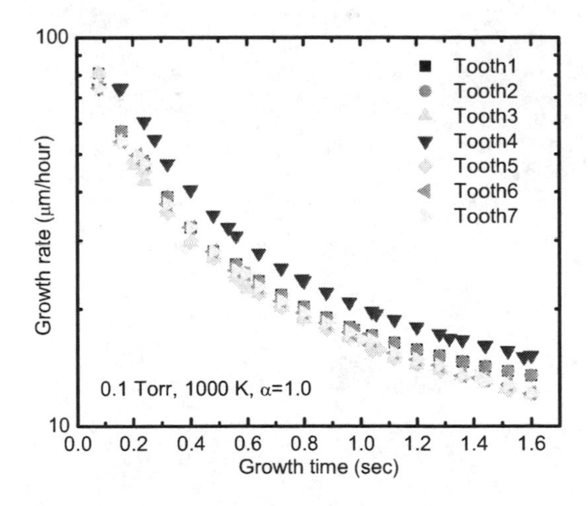

Figure 6: The growth rates. The largest growth rate is in the 4th tooth due to data exchange between 3rd, 4th and 5th teeth. The growth rates in 1st and 2nd teeth are slightly higher than those in the 6th and 7th teeth.

The growth rates of CNTs in seven teeth are shown in Fig. 6. It was observed that the growth rates of CNTs in the 1st and 2nd teeth with small catalyst are larger than those in the 6th and 7th teeth with large catalyst. In contrast, the growth rate in the 4th tooth with large catalyst is larger than that in the 3rd and 5th teeth with small catalyst. The reason for the described trends is as follows. The light hydrocarbon specie can react on both catalyst surface and substrate surface, while heavy hydrocarbon radicals dominantly react on the surface of catalyst. Consequently, in the teeth with small catalyst, the ratio of heavy species to light species increases with time since heavy species react only on a portion of the tooth's surface. As the fraction of heavy species increases, the growth rates of CNT in 1st and 2nd teeth also increase compared to those in the 6th and 7th teeth. This is due to the thermal decomposition of heavy hydrocarbon radicals on the surface of catalyst. At the same time, data exchange between 3rd, 4th, and 5th teeth increases the ratio of heavy species to light species in the central tooth and decreases this ratio in the side teeth.

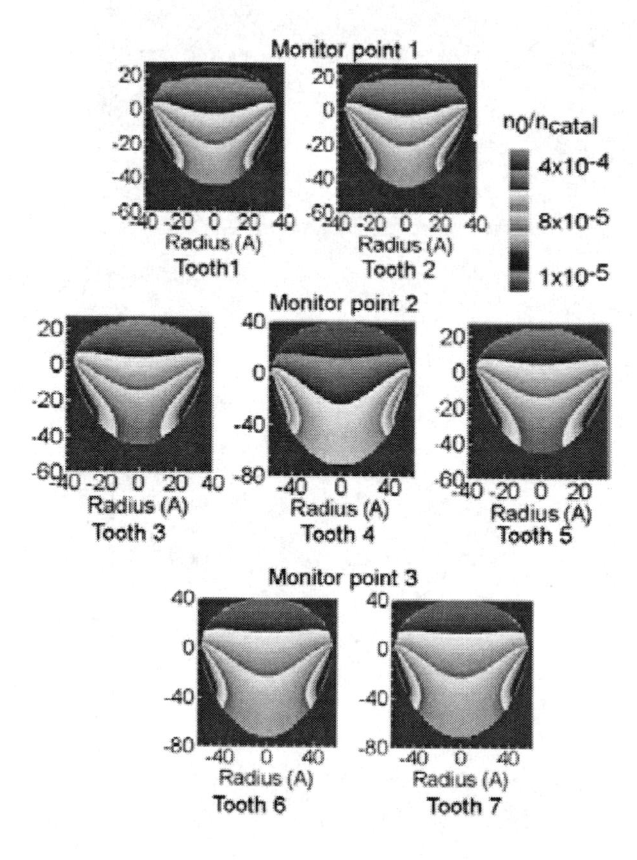

Figure 6: The concentrations of carbon inside of catalysts of different size.

The concentration of carbon inside catalytic particle is shown in Fig. 6. Similar to the growth rate, the peak concentration is achieved inside the catalytic particle located in the 4th tooth. The peak concentration in this tooth is about two times larger than that in the 3rd and 5th teeth. This is due to the increase of the ratio of heavy hydrocarbon to light hydrocarbons in the 4th tooth and decrease of this ratio in the 3rd and 5th teeth as a result of boundary data exchange. The concentrations of carbon in the 1st and 2nd teeth are 30% larger than those in the 6th and 7th teeth as a result of differences in both catalyst size and input fluxes computed by CFD-ACE.

3 CONCLUSIONS

The multi-scale computational framework (MSCF) which couples continuum model, kMC simulation, and molecular dynamics method has been developed for nanostructured materials' fabrication. The validation was performed for the plasma-assisted growth of vertically aligned CNTs in a realistic CH_4/H_2 ICP system. It was found that boundary data exchange between regions where atoms self-assemble into CNTs of different size substantially affected the simulation results.

REFERENCES

[1] V. I. Kolobov, Comput. Materials Sci. **28,** 302, 2003
[2] L. Kale, R. Skeel, N. Bhandarkar, R. Brunner, A. Gursoy, N. Krawetz, J. Phillips, J. Shinozaki, K. Varadarajan, and K. Schulten, J. Comput. Physics **151,** 283, 1999.
[3] C. W. Gear, J. Li, and I. G. Kevrekidis, Phys. Lett. A **316,** 190, 2003.
[4] S. I. Stoliarov, P. R. Westmoreland, M. R. Nyden, and G. P. Forneyb, Polymer **44,** 883, 2003.
[5] Godyak, V. A., R. B. Piejak, Plasma Sources Science and Technology **11**(4): 525-543, 2002.

Role of Nanoscale Topography on the Hydrophobicity:

A Study of Fluoro-Based Polymer Film on Silicon and Carbon Nanotubes

A.R. Phani*, L. Lozzi, M. Passacantando and S. Santucci

Department of Physics-University of L'Aquila and INFM-CASTI Regional Laboratory
Via Vetoio 10 Coppito 67010 L'Aquila ITALY, arp@net2000.ch

ABSTRACT

The unique electronic, mechanical and chemical properties of carbon nanotubes making them most promising candidate for the building blocks of molecular-scale machines, and nanoelectronic devices. On the other hand, highly hydrophobic films are being actively considered in silicon based micro-electromechanical systems, nanotechnology based devices, optoelctronic devices, or biomedical devices to reduce adhesion that may be encountered during wet processing. In order to fill the gap, and fulfil the requirements, it could be proved that morphological changes in the nanometer range influences the water contact angles and their hysteresis of low-surface energy materials. Thin films of fluorine based block co-polymer itself forms nano-hemispheres (similar to lotus leaf) at and above 100°C favoring an increase in the water contact angle from 122° (25°C) to 138° (400°C). The structural, optical, mechanical and hydrophobic properties of fluorine based block co-polymer are also discussed. By applying nanolayered (5 nm) fluorine-based block co-polymer film on a aligned carbon nanotubes (CNT) morphology with a certain roughness, the advancing contact angle for water on fluoro-based polymer film on a nearly atomically flat Si wafer increased from 122 ° to 138° (close to super hydrophobicity) and 150° on the rough asparagus-like structure of CNT has been observed.

Keywords: carbon nanotubes, block-co-polymer, hydrophobicity, contact angle, nano-hemispheres

1 INTRODUCTION

It is well known, that the hydrophobicity of surface is strongly affected by the chemical composition and topographical appearance. Considering only the chemical factor, contact angles of around 120° can be obtained for materials with lowest surface energy (6.7 mJ/m^2 for a surface with regularly aligned closer-hexagonal packed – CF_3 groups) For achieving higher contact angles (super hydrophobicity), a certain topography is required. For the topographical modification of low-surface energy materials, in the last years two main approaches have been investigated. The first approach considers the structuring or texturing of the material itself during or after depositing on a smooth substrate through, e.g. varying the deposition conditions [1], post treatment with ion or electron beam, and plasma etching [2]. In other case, the low energy materials were deposited on substrate with certain structure or texture, which were obtained by methods like embossing or laser ablation [3] photolithography [4] or deposition of rough interlayer by thin film technology. It was proved that, by combining the chemical structure (by using sol-gel technique) with nano-topography (by using sputtering), one could achieve super hydrophobicity [5].

Many deposition techniques including physical vapor deposition (PVD), chemical vapor deposition (CVD) or plasma polymerization, have been used to produce hydrophobic coatings on glass, silicon and other metallic substrates. Unlike PVD, CVD, or plasma polymerization techniques, with the sol-gel technique it is possible to modify the surface of any material (e.g. glass, metal, polymers) with organic polymers. This process is cost effective with easy operation at low or even at ambient temperature. One main advantage of the process is, the possibility to coat three-dimensional substrates (or irregular geometries) with high homogeneity in film thickness and good adhesion.

In the present work, a clean, and economic process is introduced to prepare temperature-resistant, hydrophobic thin films on Si (100) and CNT previous deposited on SiO_2/Si substrates. Keeping in view the necessity of the hydrophobicity on nanoelectronics made from carbon nanotubes, an attempt has been given to spin coat a nanolayer of (5 nm) of hydrophobic fluorine containing block-co-polymer poly [4,5- difluoro 2,2-bis (trifluoromethyl)-(1,3 dioxole)-co-tetrafluoroethylene] here after named as (TFD-co-TFE) on to bare Si (100) substrates as well as previously grown carbon nanotubes on SiO_2/Si substrates. Films deposited on quartz, glass surfaces [6] and polished AISI 440C steel surfaces [7] and their structural, chemical characterization, wettability properties have been reported. The basis of this work is the use of fluorine containing block-co-polymer namely poly [4,5- difluoro 2,2-bis (trifluoromethyl)-(1,3 dioxole)-co-tetrafluoroethylene] (TFD-co-TFE). Compared to perfluoroalkylsilane or hexamethyldisilazane, TFD-co-TFE exhibits higher mechanical strength, chemical durability, and temperature stability. The importance of the compound is the stability to high temperatures up to 400 °C, which could be due to carbon-oxygen link of the fluorine containing block-co-polymer structure as depicted in the figure 1.

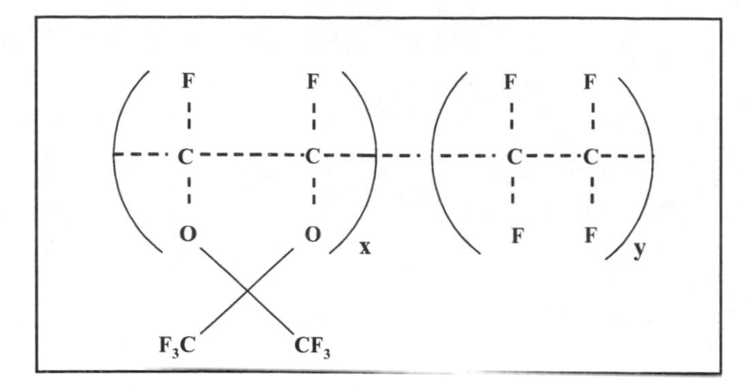

Figure 1: Structure of the fluorine-based block-co-polymer used as a precursor for the preparation of TFD-co-TFE films by sol-gel spin coating process

Sol-gel thin films based on this compound have been prepared by a spin coating technique followed by annealing at low temperatures. It was observed that as the annealing temperature increased from 100 °C to 400 °C, formation of nano-hemisphere-like structures were formed on the film surfaces of TFD-co-TFE coated on Si (100).

3 EXPERIMENTAL PROCEDURE

Calculated quantities (0.1 M) of poly [4,5 difluoro 2,2-bis (trifluoromethyl) - (1,3 dioxole) – co -tetrafluoroethylene] (amorphous copolymer: 65 mole % dioxole, T_g: 160 °C) were dissolved in special fluorinated solvent (Fluorinert FC-70) in a 100 ml round bottom flask. Both reagents were obtained from FLUKA Chemicals and used without further purification. To the above solution few drops of stabilizing agent was added and stirred at room temperature for about 6 hours in nitrogen atmosphere to prevent particle contamination. The contents were refluxed at 40°C for 4 h. The final solution was filtered in order to remove any particulates formed during the process. The obtained filtered stock solution was used for spin coating the substrates. A schematic flow chart diagram for the preparation of the hydrophobic thin films by sol-gel spin coating technique was shown in figure 2.

Polished Si (100) substrates and CNT grown on SiO_2/Si substrates with surface roughness R_a = 1.5 nm 36 nm, respectively and have been used in the present study. The deposition parameters for the growth of CNT on SiO_2 /Si substrates by radio frequency plasma chemical vapor deposition technique were described elsewhere [8]. Prior to spin coating, Si (100) substrates were rinsed with isopropanol and then placed in a trough containing isopropanol and subjected to ultrasonic cleaning for 15 min in order to remove any organic matter and dust particles present on the surface. Afterwards the substrates were again rinsed with isopropanol and then blown dry with high purity nitrogen gas (99.95%).

The clean substrates were immediately placed in the spin coating chamber. Using 50 microliters pipette 10-15 microliters of the stock solution was placed on the surface of the substrate and spun with 1000 rpm speed for 30 sec (cylce-1) to spread the solution on the entire surface of the substrate. This step was followed by 3000 rpm for 30 sec (cycle-2) in order to evaporate the solvent. The steps consisting of cycle-1 and cylcle-2 were repeated 10 times for the Si (100) substrates and one time for CNT / SiO_2 /Si substrates in order to get 50 nm thickness and 5-8 nm nanolayer thickness. At the end, the obtained xerogel films deposited on Si (100) were subjected to annealing for 1 hour at different temperatures ranging from 100 °C to 400 °C in argon atmosphere.

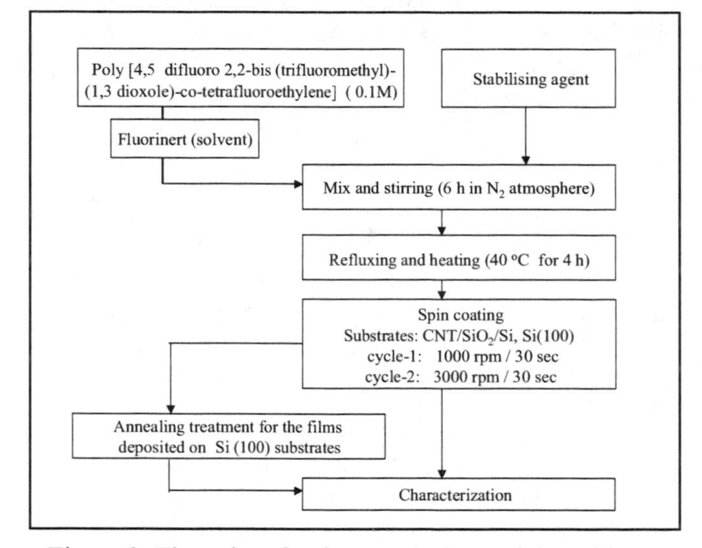

Figure 2: Flow chart for the preparation and deposition of TFD-co-TFE thin films by sol-gel spin coating process

The microstructure, elemental composition, and surface topography of the TFD-co-TFE xerogel films as well as annealed thin films were characterized by employing scanning electron microscopy (SEM) attached with energy dispersive X-ray (EDX) analysis, and atomic force microscopy (AFM) techniques, respectively.

4 RESULTS AND DISCUSSIONS

4.1 Scanning Electron Microscopy

SEM images and the corresponding energy dispersive X-ray (EDX) spectra of the TFD-co-TFE film deposited on Si (100) at room temperature and annealed at 400 °C for 1 h in argon atmosphere have been shown in figure 3 and 4, respectively. The xerogel (as deposited) films were uniform with flat and smooth structure.

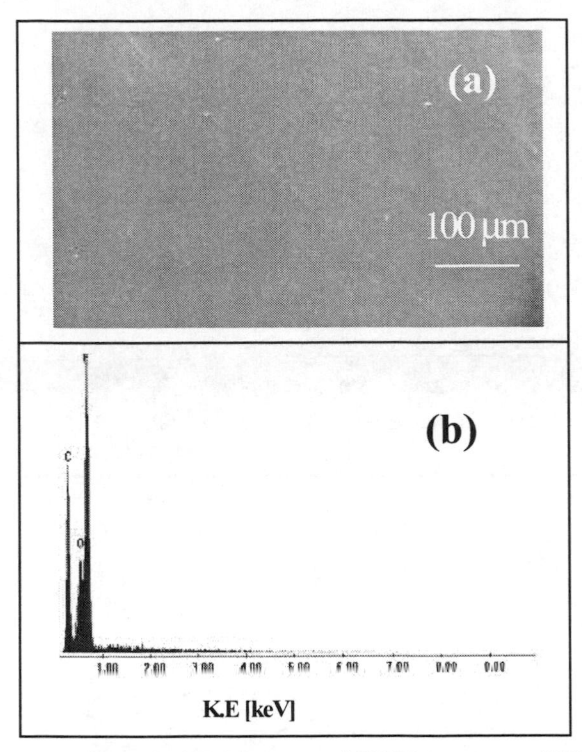

Figure 3: SEM image and EDX spectrum of TFD-co-TFE film deposited on Si (100) at room temperature by sol-gel spin coating process

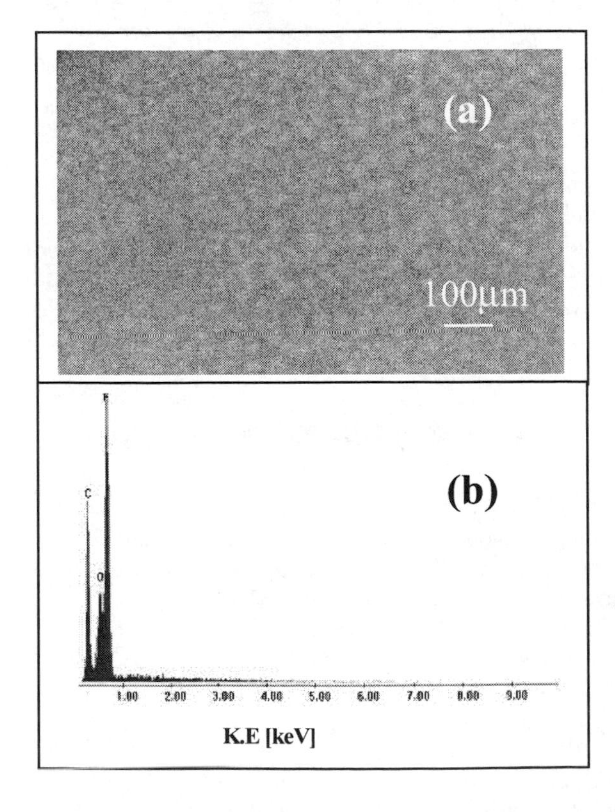

Figure 4: SEM image and EDX spectrum of TFD-co-TFE film deposited on Si (100) by sol-gel spin coating process and annealed at 400 °C for 1 h

It became difficult to take the SEM images because of the interaction of the high electron beam on the sample, which caused the film to leave with a small hole. No morphological changes in the surface were observed for the annealed samples at other temperatures except in the growth of the nano-hemisphere-like structures. Figure 3b and 4b represents SEM-EDX spectra of the TFD-co-TFE film on Si (100) for both xerogel film and xerogel film annealed for 1 h at 400 °C in an argon atmosphere. The X-ray pattern confirms the presence of carbon and fluorine along with oxygen content in the films. The weight percent of the C, F and O in the film were 40.2 wt.%, 42.3 wt.% and 17.5 wt.%, respectively. Similar behavior was also observed for the annealed samples as shown in figure 4b.

The as deposited carbon nanotube structure on SiO_2 /Si substrate grown at substrate temperature 650 °C has been shown in the figure 5. It is evident from the figure that the growth of carbon nanotubes is rough and asparagus-like structure.

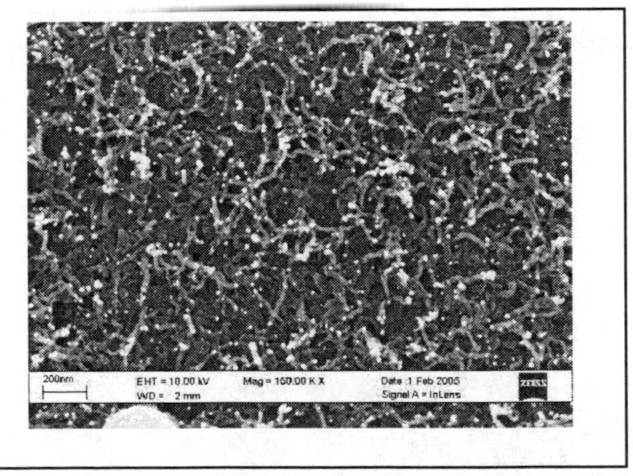

Figure 5: SEM image of roped CNT on SiO_2/Si substrate grown at 750°C by radio frequency pulsed plasma chemical vapor deposition technique

4.2 Atomic Force Microscopy

Figure 5a and 5b represents the topographical AFM images of the TFD-co-TFE films deposited on Si (100) substrates for both xerogel film and xerogel film annealed for 1 h at 400 °C in an argon atmosphere. The xerogel films have exhibited a flat like structure with little formation of nano-hemisphere like structures as indicated by the white circles (figure 5a). As the annealing temperature was increased from 100 °C to 400 °C, the growth of nano-hemisphere-like structures was increased as shown in the figure 5b, indicated by white circles.

The roughness of the films has been measured by AFM technique. The roughness of the films (R_a) deposited on Si (100) substrates has changed with annealing temperature from 8 ± 1 nm for xerogel film to 32 ± 1 nm for the film annealed at 400 °C as shown in the figure 6. The growth size of the nano-hemispheres increased from 6 ± 1 nm for xerogel film to 26 ± 1 nm for film annealed at 400 °C.

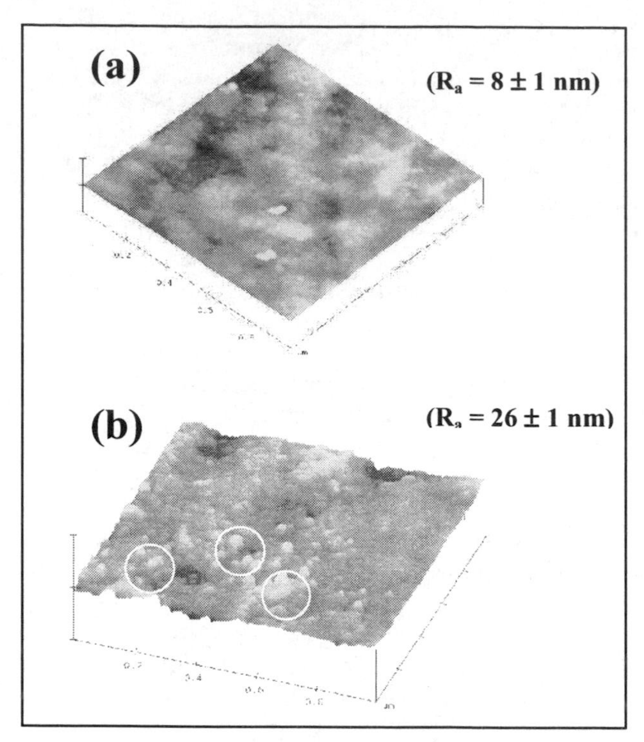

Figure 5: AFM topographical images of TFD-co-TFE thin films on Si (100) substrate by sol-gel spin coating process (a) As deposited (xerogel) film and (b) Xerogel film annealed at 400 °C / 1 h
(scan area 1 μm × 1 μm, Z axis = 100 nm)

4.3 Hydrophobicity

Contact angle of the films deposited at room temperature and annealed at different temperatures as well as substrate heated at various temperatures have been measured by spheroidal segment method using a contact angle measurement system (Circular Curve Fit Option). The contact angle, θ, was derived from advancing contact angle, θ_a, and reducing contact angle θ_b, by the following equation: $\text{Cos}\,\theta = \theta_a + \theta_b / 2$ or $[(\gamma_{SV} - \gamma_{LV}) / \gamma_{LS}]$, where γ_{SV}, γ_{LV}, and γ_{LS} are free energies of solid-vapor, liquid-vapor and liquid-solid interfaces, respectively. Water contact angle versus annealing temperature for the TFD-co-TFE films on Si (100) substrates have been presented in figure 6a. From figure 6a, it was evident that TFD-co-TFE films on Si (100) as deposited have shown water contact angle 122° and as the annealing temperature increased from 100 °C to 400 °C the contact angle was increased from 122° to 138°. This could be due to the growth of nano-hemisphere like structures at 100 °C and above. The room temperature water contact angles of TFD-co TFE film deposited on Si (100) and CNT /SiO_2/Si were 123 and 150° as shown in figure 6b and 6c, respectively.

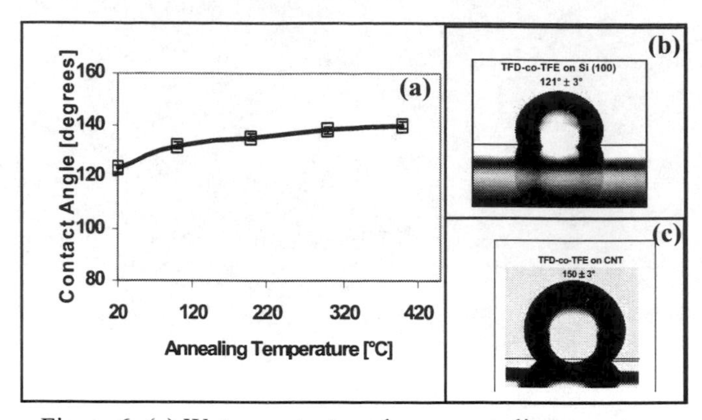

Figure 6: (a) Water contact angles *vs* annealing temperatures of TFD-co-TFE films deposited on Si (100). Water droplet pictures taken during measurement at room temperature (b) TFD-co-TFE film coated on Si (100) and (c) TFD-co TFE film coated on CNT/SiO_2/Si

5 CONCLUSIONS

Poly [4,5-difluoro 2,2-bis (trifluoromethyl)-(1,3 dioxole)-co-tetrafluoroethylene] (TFD-co-TFE) films have been deposited by sol-gel spin coating technique on Si (100) and CNTprevious grown on SiO_2/Si substrates. The xerogel (as deposited) films on Si (100) have been subjected to annealing for 1 h at different temperatures ranging from 100 °C to 400 °C in an argon atmosphere. SEM-EDX spectra confirmed the presence of carbon (~41 wt%), fluorine (~41 wt%) and oxygen (~18 wt%) in the deposited and annealed films. AFM topographical images revealed the formation of nano-hemisphere-like structures with increasing annealing temperatures. The size of the nano-hemispheres increased from 8 nm for xerogel film to 28 nm for film annealed at 400 °C on Si (100). The room temperature water contact angles of TFD-co TFE film deposited on Si (100) and CNT /SiO_2/Si were 123 and 150°, respectively.

REFERENCES

[1] G. Cicala, A. Milella, F. Palumbo, P. Favia, R. d'Agostino, Diamond and related Materials 12, 2020, 2003.
[2] Y. Inoue, Y. Aoshimura, Y. Ikeda, A. Kohno, colloids and Surfaces B: Biointerfaces 19, 257, 2000.
[3] D. Oner, T. J. McCarthy, Langmuir 16, 777, 2000.
[4] T. Uelzen, J. Muller, Thin Solid films, 434, 311, 2003.
[5] Y. Gerbig, A.R. Phani, H. Haefke, Applied Surface Science, 242, 251, 2004.
[6] A. R. Phani, M. Passacantando, S. Santucci, Materials Chemistry and Physics, 2005 (unpublished).
[7] A.R. Phani and H. Haefke, Materials Letters, 58, 3555, 2004.
[8] L. Valentini, J.M.Kenny, L. Lozzi, S.Santucci, J. Appl. Phys. 92, 6188, 2002

Amorphous Diamond Electron Emission for Thermal Generation of Electricity

James Chien-Min Sung[*,1,2,3], Kevin Kan[1], Michael Sung[5], Jow-Lay Huang[4],
Emily Sung[6], Chi-Pong Chen[1], Kai-Hong Hsu[3] and Ming-Fong Tai[7]

Address: KINIK Company, 64, Chung-San Rd., Ying-Kuo, Taipei Hsien 239, Taiwan, R.O.C.
Tel: 886-2-2677-5490 ext.1150
Fax: 886-2-8677-2171
e-mail: sung@kinik.com.tw

[1] KINIK Company, 64, Chung-San Rd., Ying-Kuo, Taipei Hsien 239, Taiwan, R.O.C.
[2] National Taiwan University, Taipei 106, Taiwan, R.O.C.
[3] National Taipei University of Technology, Taipei 106, Taiwan, R.O.C.
[4] National Cheng-Kung University, Tainan, Taiwan, R.O.C
[5] Massachusetts Institute of Technology, Cambridge, Massachusetts, U.S.A.
[6] Johnson and Johnson, Freemont, California, U.S.A.
[7] Department of Electronic Engineering, Wufeng Institute of Technology, Chia-yi 621, Taiwan, R.O.C.

ABSTRACT

Amorphous diamond represents a class of material of its own. It may be viewed to be a composite of metal (graphite) and insulator (diamond), but together they form a unique passage for electrons to flow through and to emit in vacuum. Amorphous diamond contains much defect so its electrical resistance is intermediate between metal and semiconductor. However, its ability to emit electrons in vacuum as cold cathode outstrips almost any class of materials. The easiness for electrons to flow through amorphous diamond and fly toward an anode across vacuum makes it an ideal material for electrical generator. In fact, the electricity generation can be so easy that amorphous diamond may become the most efficient solar cell ever invented. Moreover, by reversing the role of electricity generation, an amorphous diamond film may become an electron radiator. Such a radiator may dissipate heat much faster than the most advanced heat spreader (e.g., diamond substrate or heat pipe) currently being investigated.

Recent experimental results has confirmed that the current of electron emission in vacuum has increased two orders of magnitude when amorphous diamond is heated to 300 °C. Such a dramatic increase of current indicates that thermal energy can indeed shake off electrons in carbon atoms and accelerate them toward an anode across a vacuum. This phenomenon is consistent with the proposal that amorphous diamond can indeed be made of solar cells and/or heat spreaders.

Keywords: amorphous diamond, solar cell, diamond electrode, heat spreader

1 THE ELECTRON LADDER

Electronic energy states of atoms in a solid often form continuous bands. In the case of a metal (e.g., Cu), the conduction band overlaps with the valence band, so electrons can move up and down with no hindrance. In this case, although an electron can move to a higher orbital by gaining energy (e.g., by heating), it will drop back to the Fermi level instantly when the energy is lost. On the other hand, for an insulator (e.g., diamond), the conduction band is separated well above the valence band. Because of the presence of a large energy barrier in between, no electron will move to the conduction band unless it gains a quantum energy equal or higher than the band gap (5.45 eV for diamond). The band gap can prevent electrons from moving in either direction, so an electron once moves to the conduction band may stay there at least momentarily before falling down. If an insulator is doped with atoms of a different element (e.g., Li in diamond), a semiconductor is formed. In this case, an electron may assume an energy state that resides within the forbidden range of the host material. This electron may move up to the conduction band by acquiring a moderate amount of energy (e.g., 0.1 eV for Li in diamond) and stay there momentarily.

A diamond semiconductor contains a chemical dopant that can provide only one energy rung to the wide open band gap. Hence, the likelihood for an electron to climb up the energy barrier and reach the conduction band is slim. However, if a multitude of energy rungs can be inserted in the band gap, an electron may climb up the energy ladder with much less effort. In this case, more electrons may reach the conduction band at ease.

In order to insert this energy ladder, chemical

dopants with different energies must be adapted, but known dopants for diamond are very limited (e.g., B, N, P) so they are not sufficient to create such an energy ladder. On the other hand, physical dopants with varying electron energies may be assembled to form such an energy ladder. The electron energy in a carbon atom can be varied by distorting the tetrahedral coordination of the diamond lattice in different degree. Hence, in an amorphous diamond that contains carbon atoms in distorted tetrahedral coordination, the band gap is full with energy rungs for electrons to occupy (Figure 1). Such energy rungs are known as defect band. Amorphous diamond possessing such an energy ladder is not an electrical conductor like graphite, nor is it an insulator like diamond, but a semi-metal that can transmit electricity intermittently.

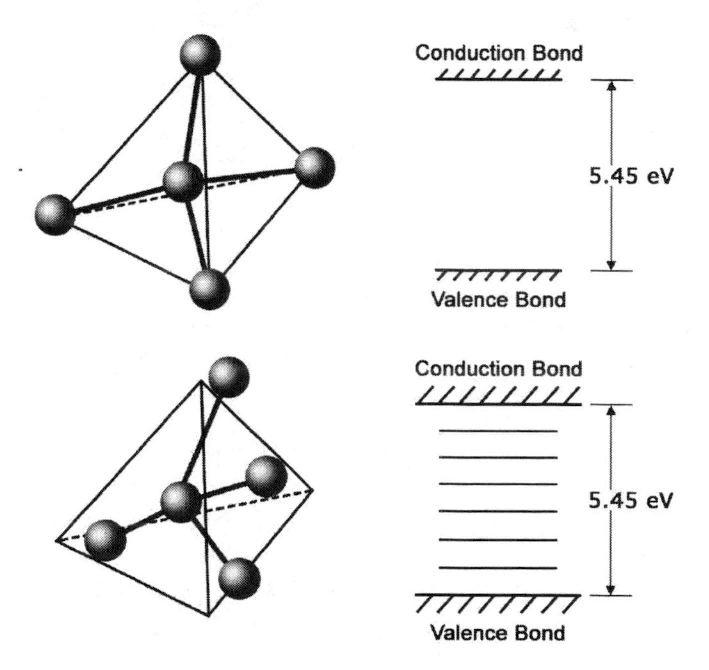

Figure 1: The symmetrical tetrahedral coordination of carbon atoms in diamond can form a wide band gap (top diagram), but the distorted tetrahedral coordination of carbon atoms in amorphous diamond will incorporate an energy ladder (defect band) in the band gap so electrons can climb up to the conduction band by leverage (bottom diagram).

In order to generate electricity in a solid, their electron must be excited to a higher energy. One way to do so is to emit electrons in vacuum and then collect these electrons by an anode. The electron energy at the Fermi level of an atom is much lower than that in vacuum. The difference between the two is known as work function that is the energy barrier to be overcome for an electron to enter vacuum. However, if an electron can climb up the energy ladder to reach the conduction band, its energy is now close to the vacuum level and hence with a little help from an external field, it could fly readily toward an anode across vacuum. Thus, for electrons to emit in vacuum from a metal, a large energy must be acquired (e.g., by heating to a high temperature as in the case of cathode ray) to overcome the work function, but very little energy may be needed to shoot electron out in vacuum from an amorphous diamond (e.g., by heating to a low temperature). In the case of diamond, unless there is a passage for electron to move through, no electron will be emitted.

When a solid is irradiated by a spectrum of electromagnetic radiation (e.g., by bathing in sun shine), the photon energy is absorbed by both electrons and atoms. If only electrons acquire the energy, the electrons will move with no resistance so the solid is a superconductor of electricity. On the other hand, if only atoms acquire the energy, the phonons will transmit with no hindrance so the solid is a superconductor of heat. In almost all cases, the light energy is distributed between excited electrons and vibrating atoms so the solid is not a superconductor of either electricity or heat.

The material will respond differently upon the impingement of electromagnetic radiation. For an opaque metal, photons can agitate atoms and force some electrons to move, for transparent diamond, most photons will pass through without being absorbed, and for doped silicon, some photons will be absorbed to form free electrons. However, an amorphous diamond will convert a significant amount of heat into the kinetic energy of electrons. In essence, a thermal electric effect is in action. The conversion of low thermal energy to high electricity energy seem to defy the second law of thermodynamics, but this apparent violation is explained as higher energy phonons are dissociated into lower energy phonons so the entropy in the form of heat is produced at both cathode and anode. Different materials may generate electricity from electromagnetic radiation based on different mechanisms.

2 AMORPHOUS DIAMOND THERMAL GENERATOR

The idea that diamond micro tips may be used as a solar cell was first proposed by Fisher. According to this model, a micro tip array made of boron doped diamond may be used as an efficient field emission device. The current can be induced by incorporating a gate anode for the application of an external field. The diamond microtips are mounted on a cathode. Fisher proposed to heat these diamond micro tips to about 1000 ^{o}C by focusing the sun light on a heat absorber that is coated on the cathode. He estimated that such a solar cell may achieve an efficiency as high as 50% that is at least double the energy conversion efficiency of the most advanced semiconductor solar cells currently available (e.g., made by GaAs).

The nano-tipped amorphous diamond has a diamond tip density about ten thousands times more than above mentioned micro-tipped doped diamond. Moreover, each tip is also much sharper and hence with a higher concentration of electrical field intensity. Hence, nano-tipped amorphous diamond can generate much more electricity than micro-tipped doped diamond.

Furthermore, the electrical resistance in amorphous diamond is much lower than doped diamond due to its high defects concentration. In addition, amorphous diamond can be deposited like spray painting, hence it is much cheaper to produce. In contrast, micro-tipped diamond relies on sophisticated lithography technology that is not only slow but also costly. Hence, amorphous diamond solar cells would be much more practical to be mass produced. The schematic of the design of an amorphous diamond solar cell.

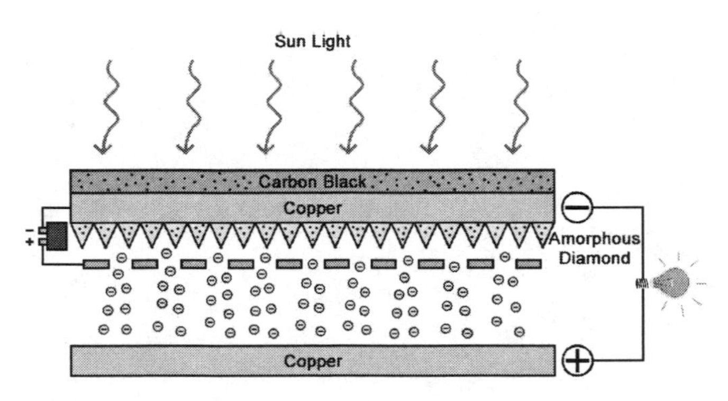

Figure 2: The schematic design of an amorphous diamond solar cell that may be operated at a temperature as low as 300 ^{o}C. The energy conversion efficiency may be up to 50%. Although this drawing portrayed a triode design, the two anodes may be combined in such a way that electrical current is drawn mostly from the heat rather than from the applied field.

The above solar cell is based on sharp nano-tips of carbon for efficient electron emission. Similar concepts may also apply by using carbon nanotube (CNT) bundles for shooting out electrons in vacuum. However, the emission of CNT bundles has several drawbacks. Firstly, the electrical resistance various among different CNTs, so the distribution of electrical current is intrinsically uneven. Secondly, CNTs are hallow tubes so electrons migration tends to concentrate on the thin wall. As a consequence, the tubes can be overloaded and burn out easily. Thirdly, the electrostatic repulsion can force CNT to erect periodically. Such movements make CNTs susceptible to fatigue deterioration. Hence, the aging effect of electron emission is inevitable.

In contrast, the electron emission performance using amorphous diamond nano-tips is much more reproducible. Moreover, amorphous diamond can be readily coated in large area at very low temperature ($<150\ ^{o}C$) that is significantly cooler than the deposition temperatures ($>600\ ^{o}C$) of CNTs. The lower cost combined with lower process temperature make the amorphous diamond a practical choice for industrial applications (e.g., to coat on the transparent conductor of ITO or on LCD glass panels).

In summary, amorphous diamond and CNT nano-tips may be used as solar cells. They may also be coated on panels that line up a heat producing system (e.g., an incinerator or even a nuclear power reactor) for electricity generation (Figure 3). Such a power generator does not contain moving parts that are typical for most power generators (e.g., turbine machine). The quiet power generator is not only maintenance easy, but also environmentally friendly.

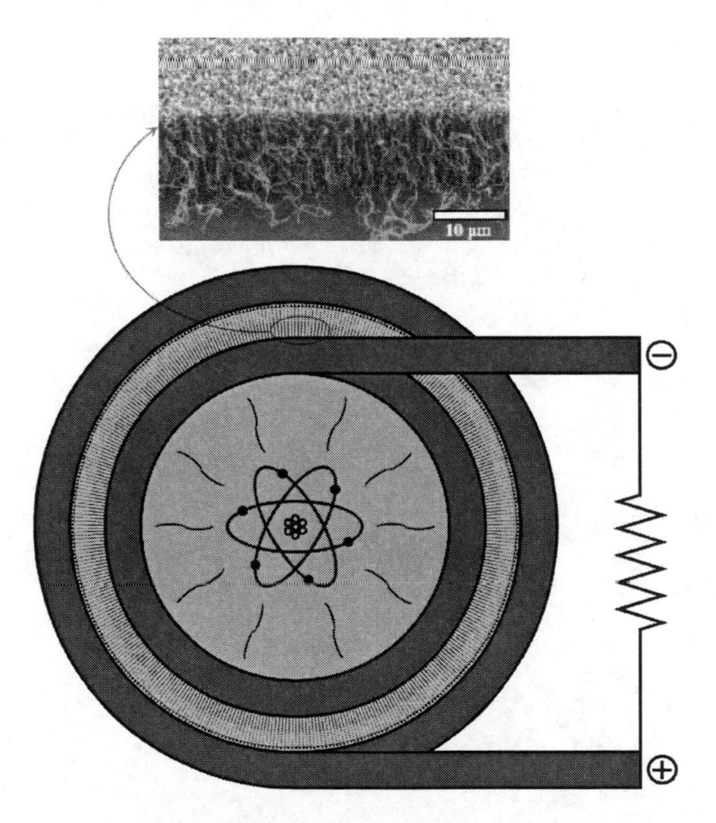

Figure 3: The idea of making a nuclear power plant safe by using amorphous diamond or CNT nano-tips to convert heat to electricity.

3 CONCLUSION

Amorphous diamond nano-tips (or CNTs) can be efficient electron emitters in vacuum. The easy for electrons to flow through the circuitry can make them the ideal thermal generator. This energy lever can allow low energy phonons to accumulate their energy and move electrons to fly across vacuum. The accumulation of phonon energy may appear to circumvent the second law of thermodynamics but in reality more lower energy heat are generated at both cathode and anode so the entropy of the entire system still rises. The nano-tip electron emitters envisaged can generate electricity from a variety of heat sources (sun light, incinery, hot spring, burning of coal, nuclear reactor…etc.). Moreover, they can be used as efficient chilling surfaces. Such surfaces are indispensable for future generations of computer CPU, high power electronics, laser diode, and MEM devices, to name just a few examples of exotic applications.

REFERENCES

[1] Ming-Chi Kan, Jow-Lay Huang, James C. Sung, Ding-Fwu Lii, Kue-Hsien Chen: International Conference of New Diamond Science & Technology, Melbourne, Australia (July, 2002).

[2] T. S. Fisher: Applied Physics Letters 79, 22 (2001) 3699-3701.

[3] James C. Sung: Industrial Materials (Chinese), September (1998) 114-123.

[4] James C. Sung: Industrial Materials (Chinese), October (1998) 152-164.

[5] V. A. Solntsev and A. N. Rodionov: Solid State Electronics 45 (2001) 853-856.

Near-Field Electrodynamics of Atomically Doped Carbon Nanotubes

I.V Bondarev* and Ph. Lambin**

* Institute for Nuclear Problems, Belarusian State University
Bobruiskaya Str. 11, 220050 Minsk, Belarus, bondarev@tut.by
** Facultés Universitaires Notre-Dame da la Paix
61 rue de Bruxelles, 5000 Namur, Belgium, philippe.lambin@fundp.ac.be

ABSTRACT

We develop a quantum theory of near-field electrodynamical properties of carbon nanotubes. The theory is used to investigate spontaneous decay dynamics and van der Waals (vdW) attraction of a two-level atomic system located close to a nanotube surface. We demonstrate a strictly non-exponential spontaneous decay dynamics of the upper state and nonlinear vdW coupling of the lower (ground) state of the coupled "atom–nanotube" system.

Keywords: carbon nanotubes, strong atom-field coupling, atomic spontaneous decay, van der Waals interactions

1 INTRODUCTION

Carbon nanotubes (CNs) are graphene sheets rolled-up into cylinders of approximately one nanometer in diameter. Extensive work carried out worldwide in recent years has revealed the intriguing physical properties of these novel molecular scale wires [1]. Nanotubes have been shown to be useful for miniaturized electronic, mechanical, electromechanical, chemical and scanning probe devices and materials for macroscopic composites [2]. Important is that their intrinsic properties may be substantially modified in a controllable way by doping with extrinsic impurity atoms, molecules and compounds [3]. Recent successful experiments on encapsulation of single atoms into single-wall CNs [4] and their intercalation into single-wall CN bundles [3], [5] stimulate an in-depth analysis of atom-vacuum-field interactions in such systems.

The relative density of photonic states (DOS) near a CN effectively increases due to the presence of additional surface photonic states coupled with CN electronic quasiparticle excitations [6], [7]. This causes an atom-vacuum-field coupling constant (which is proportional to the photonic DOS – see later) to be very sensitive to an atom-CN-surface distance. At small enough distances, the system may exhibit a strong atom-field coupling regime giving rise to rearrangement ("dressing") of atomic levels by vacuum-field interaction, so that standard weak-coupling-QED-based atom-field interaction models (see [8] for a review) are, in general, inapplicable for an atom in a close vicinity of a carbon nanotube. To give this issue a proper theoretical consideration, we develop a universal quantum theory of atom-field interactions in the presence of an infinitely long single-wall CN, which is valid for both strong and weak atom-field coupling. By applying this theory to two particular problems, the spontaneous decay of the excited state and the vdW attraction of the ground state of the two-level atom close to the nanotube surface, we demonstrate the strictly non-exponential spontaneous decay dynamics and nonlinear vdW coupling of the coupled "atom–nanotube" system.

2 BRIEF SKETCH OF THE MODEL

Consider a neutral atomic system with its center of mass positioned at the point $\mathbf{r}_A$ near an infinitely long single-wall CN. Assign the orthonormal cylindric basis $\{\mathbf{e}_r, \mathbf{e}_\varphi, \mathbf{e}_z\}$ with $\mathbf{e}_z$ directed along the CN axis. The total nonrelativistic Hamiltonian of the whole system can then be represented in the form (electric-dipole approximation, Coulomb gauge, CGS units) [8], [9]

$$\hat{H} = \hat{H}_F + \hat{H}_A + \hat{H}_{AF}^{(1)} + \hat{H}_{AF}^{(2)}, \tag{1}$$

where

$$\hat{H}_F = \int_0^\infty d\omega\hbar\omega \int d\mathbf{R}\, \hat{f}^\dagger(\mathbf{R},\omega)\hat{f}(\mathbf{R},\omega), \tag{2}$$

$$\hat{H}_A = \sum_i \frac{\hat{\mathbf{p}}^2}{2m_i} + \sum_{i<j} \frac{q_i q_j}{|\mathbf{r}_i - \mathbf{r}_j|}, \tag{3}$$

$$\hat{H}_{AF}^{(1)} = -\sum_i \frac{q_i}{m_i c}\, \hat{\mathbf{p}}_i \cdot \hat{\mathbf{A}}(\mathbf{r}_A) + \hat{\mathbf{d}} \cdot \nabla\hat{\varphi}(\mathbf{r}_A), \tag{4}$$

$$\hat{H}_{AF}^{(2)} = \sum_i \frac{q_i^2}{2m_i c^2}\, \hat{\mathbf{A}}^2(\mathbf{r}_A) \tag{5}$$

are, respectively, the Hamiltonian of the vacuum electromagnetic field modified by the presence of the CN, the Hamiltonian of the atomic subsystem, and the Hamiltonian of their interaction (separated into two contributions according to their role in the atom-vacuum-field interaction – see later). The operators $\hat{f}^\dagger$ and $\hat{f}$ in Eq. (2) are those creating and annihilating single-quantum electromagnetic excitations of bosonic type in the CN and the inner integral is taken over the CN surface assigned by the vector $\mathbf{R} = \{R_{cn}, \phi, Z\}$ with R_{cn} being the CN radius. In Eqs. (3)-(5), m_i, q_i, $\hat{\mathbf{r}}_i$ and $\hat{\mathbf{p}}_i$ are, respectively, the masses, charges, coordinates (relative to $\mathbf{r}_A$) and momenta of the particles constituting the atomic subsystem, $\hat{\mathbf{d}} = \sum_i q_i \hat{\mathbf{r}}_i$ is its electric dipole moment operator, $\hat{\mathbf{A}}$ and $\hat{\varphi}$ are the vector potential and the scalar potential of the CN-modified electromagnetic field.

We simplify the Hamiltonian (1)-(5) by applying the two-level approximation to obtain [9]

$$\hat{H} = \int_0^\infty d\omega\hbar\omega \int d\mathbf{R}\, \hat{f}^\dagger(\mathbf{R},\omega)\hat{f}(\mathbf{R},\omega) + \frac{\hbar\tilde{\omega}_A}{2}\hat{\sigma}_z \qquad (6)$$
$$+ \int_0^\infty d\omega \int d\mathbf{R} \left[\mathrm{g}^{(+)}(\mathbf{r}_A,\mathbf{R},\omega)\,\hat{\sigma}^\dagger - \mathrm{g}^{(-)}(\mathbf{r}_A,\mathbf{R},\omega)\,\hat{\sigma} \right] \hat{f}(\mathbf{R},\omega) + \mathrm{h.c.}.$$

Here, the Pauli operators $\hat{\sigma}_z = |u\rangle\langle u| - |l\rangle\langle l|$, $\hat{\sigma} = |l\rangle\langle u|$, $\hat{\sigma}^\dagger = |u\rangle\langle l|$ describe electric dipole transitions between the two atomic states, upper $|u\rangle$ and lower $|l\rangle$, separated by the transition frequency ω_A. This ('bare') frequency is modified by the interaction (5) which, being independent of the atomic dipole moment, does not contribute to mixing the $|u\rangle$ and $|l\rangle$ states, giving rise, however, to the new *renormalized* transition frequency $\tilde{\omega}_A = \omega_A[1 - 2/(\hbar\omega_A)^2 \int_0^\infty d\omega \int d\mathbf{R}\, |\mathrm{g}^\perp(\mathbf{r}_A,\mathbf{R},\omega)|^2]$ in the second term of Eq. (6). On the contrary, the interaction (4), being dipole moment dependent, mixes the $|u\rangle$ and $|l\rangle$ states, yielding the third term of the Hamiltonian (6) with the interaction matrix elements $\mathrm{g}^{(\pm)} = \mathrm{g}^\perp \pm (\omega/\omega_A)\mathrm{g}^\parallel$, where $\mathrm{g}^{\perp(\parallel)} = -i4\omega_A d_z \sqrt{\pi\hbar\omega \mathrm{Re}\sigma_{zz}(\omega)}\; G_{zz}^{\perp(\parallel)}/c^2$ with $G_{zz}^{\perp(\parallel)}(\mathbf{r}_A,\mathbf{R},\omega)$ being the zz-component of the transverse (longitudinal) Green tensor of the electromagnetic subsystem and $\sigma_{zz}(\omega)$ standing for the CN surface axial conductivity (we neglect the azimuthal current and radial polarizability of the CN). The functions $\mathrm{g}^{\perp(\parallel)}$ (and $\mathrm{g}^{(\pm)}$, respectively) have the following general property $\int d\mathbf{R} |\mathrm{g}^{\perp(\parallel)}(\mathbf{r}_A,\mathbf{R},\omega)|^2 = (\hbar^2/2\pi)(\omega_A/\omega)^2 \Gamma_0(\omega)\xi^{\perp(\parallel)}(\mathbf{r}_A,\omega)$ with $\xi^{\perp(\parallel)} = \mathrm{Im}\, G_{zz}^{\perp(\parallel)}/\mathrm{Im}\, G_{zz}^0$ representing the transverse (longitudinal) position-dependent (local) photonic DOS and $\Gamma_0(\omega) = 8\pi\omega^2 d_z^2 \mathrm{Im}\, G_{zz}^0(\omega)/3\hbar c^2$ being the atomic spontaneous decay rate in vacuum [7]. Thus, the total energy of the system with the Hamiltonian (6), in general, and the frequency $\tilde{\omega}_A$, in particular, are expressed in terms of only one *intrinsic* physical characteristic of the electromagnetic subsystem – the local photonic DOS.

3 PARTICULAR PROBLEMS

Based on the Hamiltonian (6) and taking a possible degeneracy into account of the upper level and lower level of the coupled "atom–nanotube" system at small $\tilde{\omega}_A$ (this takes place at small atom-CN-surface distances where the local photonic DOS strongly increases as is seen from the definition of $\tilde{\omega}_A$), we derive equations for the time evolution of the population probability $C_u(t)$ of the excited state and for the vdW energy $E_{vw}(\mathbf{r}_A)$ of the ground state of the two-level atom coupled with the CN modified vacuum electromagnetic field. The former is the result of the solution of the time-dependent Srödinger equation of the form [7], [10]

$$C_u(t) = 1 + \int_0^t K(t-t')\, C_u(t')\, dt', \qquad (7)$$
$$K(t-t') = \frac{1}{\hbar^2}\int_0^\infty d\omega\, \frac{e^{-i(\omega-\tilde{\omega}_A)(t-t')} - 1}{i\,(\omega-\tilde{\omega}_A)} \times \int d\mathbf{R}\, |\mathrm{g}^{(+)}(\mathbf{r}_A,\mathbf{R},\omega)|^2.$$

The latter comes from the solution of the secular equation yielding the total ground-state energy E of the coupled "atom–nanotube" system in the form [9]

$$E = -\frac{\hbar\tilde{\omega}_A}{2} - \int_0^\infty d\omega \int d\mathbf{R}\, \frac{|\mathrm{g}^{(-)}(\mathbf{r}_A,\mathbf{R},\omega)|^2}{\hbar\omega + \dfrac{\hbar\tilde{\omega}_A}{2} - E}, \qquad (8)$$

which the atom–nanotube vdW energy $E_{vw}(\mathbf{r}_A)$ is determined from by the relation $E = -\hbar\omega_A/2 + E_{vw}(\mathbf{r}_A)$ with the first term being the unperturbed ground-state energy of the two-level atom.

In view of the general integral property of the matrix elements $\mathrm{g}^{(\pm)}$ given in the end of Section 2, both Eq. (7) and Eq. (8) are represented in terms of the local (position-dependent) photonic DOS and, thus, are valid for both strong ($\xi^{\perp,\parallel} \gg 1$) and weak ($\xi^{\perp,\parallel} \sim 1$) atom-field coupling.

Using Eqs. (7) and (8), we have simulated the upper state spontaneous decay dynamics and the ground state vdW energy of an atom situated outside the (9,0) CN. The local photonic DOS functions $\xi^{\perp,\parallel}$ were computed in the same manner as it was done in Refs. [6],[7]. The free-space spontaneous decay rate was approximated by the expression $\Gamma_0(\omega) \approx \alpha^3\omega$ ($\alpha = 1/137$ is the fine-structure constant) valid for hydrogen-like atoms [11]. Figure 1(a) shows the transverse local photonic DOS for the atom at several distances from the CN surface. The DOS function is seen to increase with decreasing the atom-CN-surface distance, representing the increase of the atom-field coupling strength as the atom approaches the CN surface. The vertical dashed line indicates the dimensionless bare atomic transition frequency $x_A = \hbar\omega_A/2\gamma_0$ ($\gamma_0 = 2.7$ eV is the carbon nearest neighbor hopping integral) for which the upper state population probabilities and the lower state vdW energies shown in Fig. 1(b),(c) were calculated. The frequency $x_A = 0.33$ is the peak position (at least for the shortest atom-surface distance) of the local photonic DOS. Note that $x_A \sim 0.5$ are typical for heavy hydrogen-like atoms such as Cs (supposed to be non-ionized near a CN). Figure 1(b) shows the upper state population probabilities $|C_u|^2$ as functions of the dimensionless time $\tau = 2\gamma_0 t/\hbar$ and the

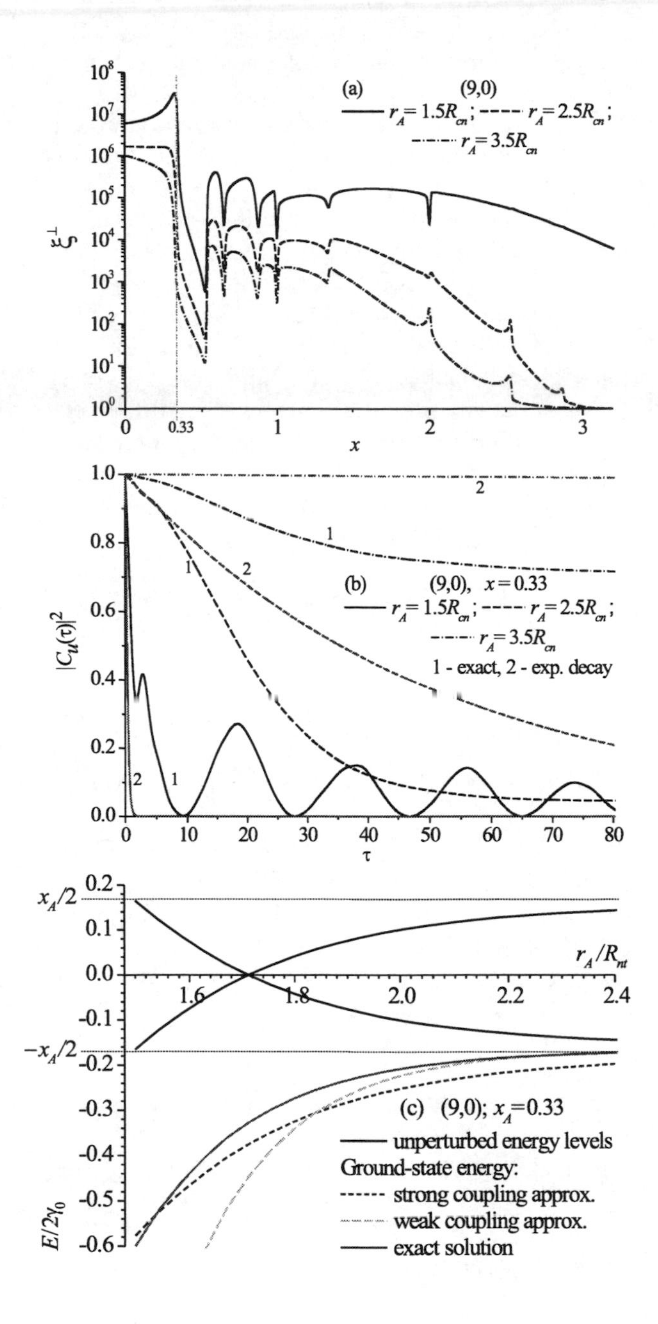

Figure 1: (a),(b) Transverse local photonic DOS's and upper-level spontaneous decay dynamics, respectively, for the two-level atom situated at three different distances outside the (9,0) CN. (c) Dimensionless "unperturbed" energy levels [the eigenvalues of the Hamiltonian in the first line of Eq. (6)] and total-ground state energy of the coupled "atom–nanotube" system as functions of the atomic position outside the (9,0) CN. The ('bare') atomic transition frequency is indicated by the vertical dashed line in Fig. 1(a); $x = \hbar\omega/2\gamma_0$ and $\tau = 2\gamma_0 t/\hbar$ are the dimensionless frequency and time, respectively, with $\gamma_0 = 2.7$ eV being the carbon nearest neighbor hopping integral.

atom-CN-surface distance in comparison with those obtained in the Markovian approximation which is known to yield the exponential decay dynamics (see, e.g., [12]). The actual spontaneous decay dynamics is clearly seen to be non-exponential. Very clear underdamped Rabi oscillations are observed for the shortest atom-surface distance, indicating *strong* atom-field coupling with strong non-Markovity. Similar to what takes place in photonic crystals [13], this is because of the fact that, due to the rapid variation of the local photonic DOS in the neighborhood of this frequency, the correlation time of the electromagnetic vacuum is not negligible on the time scale of the evolution of the atomic system, so that atomic motion memory effects are important and the Markovian approximation is inapplicable. Figure 1(c) shows the dimensionless total ground-state energy of the coupled "atom–nanotube" system and two "unperturbed" energy levels [the eigenvalues of the Hamiltonian in the first line of Eq. (6)] as functions of the atomic position outside the (9,0) CN. As the atom approaches the CN surface, its "unperturbed" levels come together, then get degenerated and even inverted at a very small atom-surface distance. As this takes place, the weak coupling approximation for the ground-state energy is known to diverge, whereas the strong coupling approximation yields a finite result (see, e.g., [11]). The *exact* solution given by Eq. (8) reproduces the weak coupling approximation at large and the strong coupling approximation at short atom-surface distances, respectively.

4 CONCLUSION

We have developed the quantum theory of near-field electrodynamical properties of carbon nanotubes and investigated spontaneous decay dynamics of excited states and vdW interaction of the ground state of a two-level atomic system (an atom or a molecule) close to a single-wall carbon nanotube. We have demonstrated a strictly non-exponential spontaneous decay dynamics and the inapplicability of weak-coupling-based vdW interaction models in a close vicinity of the CN surface where the local photonic DOS effectively increases, giving rise to an atom-field coupling enhancement. In certain cases, namely when the atom is close enough to the nanotube surface and the atomic transition frequency is in the vicinity of the resonance of the local photonic DOS, the system exhibits vacuum-field Rabi oscillations – a principal signature of *strong* atom-vacuum-field coupling. The non-exponential decay dynamics gives place to the exponential one if the atom moves away from the CN surface. Thus, the atom-vacuum-field coupling strength and the character of the spontaneous decay dynamics, respectively, may be controlled by changing the distance between the atom and CN surface by means of a proper preparation of atomically doped CN systems.

We would like to emphasize a general character of the

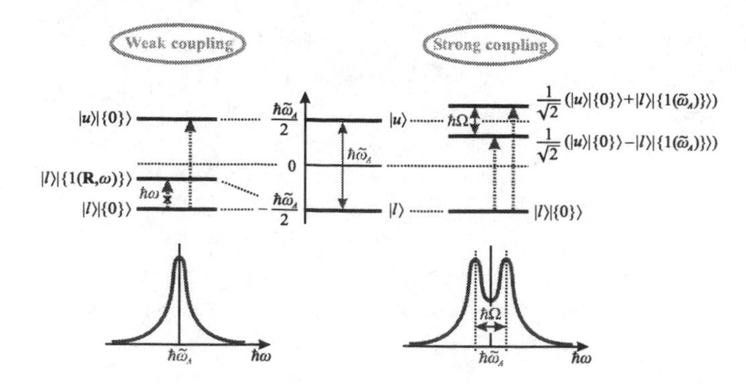

Figure 2: Schematic of the energy levels and the absorbtion line-shapes expected in the optical absorbance experiment with atomically doped CNs. In the center are the "unperturbed" atomic levels [the eigenvalues of the Hamiltonian in the first line of Eq. (6)]. On the left and on the right are the levels of the coupled "atom–nanotube" system in the weak and strong atom-vacuum-field coupling regime, respectively. The system is excited by an external optical radiation. The allowed and forbidden (crossed) optical transitions are shown by vertical dashed arrows.

conclusion above and its fundamental and applied significance. We have shown that similar to semiconductor microcavities [14], [15] and photonic band-gap materials [13], [16], carbon nanotubes may qualitatively change the character of the atom-electromagnetic-field interaction, yielding strong atom-field coupling. The study of such phenomena was started awhile ago in atomic physics [17] and still attracts a lot of interest in connection with various quantum optics and nanophotonics applications [18], [19] as well as quantum computation and quantum information processing [20], [21]. The fact that the carbon nanotube may control atom-electromagnetic-field coupling opens routes for new challenging applications of atomically doped CN systems as various sources of coherent light emitted by dopant atoms.

Strong atom-vacuum-field coupling we predict will yield an additional structure in optical absorbance spectra (see, e.g., [22]) of atomically doped CNs in the vicinity of an atomic transition frequency. The illustration is shown in Fig. 2. Weak non-Markovity of the atom-vacuum-field interactions (that yielding non-exponential spontaneous decay dynamics with no Rabi oscillations) will cause an asymmetry of an optical spectral line-shape (not shown in Fig. 2) similar to that taking place for the exciton optical absorbtion line-shape in quantum dots [23]. Strong non-Markovity of the atom-vacuum-field interactions (yielding non-exponential spontaneous decay dynamics with fast Rabi oscillations) originates from strong atom-vacuum-field coupling with the upper state of the system splitted into two "dressed" states (Fig. 2, on the right). This will yield a two-component structure of optical absorbance/reflectance spectra similar to that observed for excitonic and intersubband electronic transitions in semiconductor quantum microcavities [14], [15].

Similar manifestations of strong atom-field coupling may occur in many other atom-electromagnetic-field interaction processes in the presence of CNs, such as, e.g., atomic states entanglement, interatomic dipole-dipole interactions, cascade spontaneous transitions in three-level atomic systems, etc. A further intriguing extension of the present work could also be the study of the vdW interactions of excited atomic states where, even in the weak atom-field-coupling regime and in the simplest case of an atom near a planar semi-infinite medium, very interesting peculiarities (e.g., an oscillatory behavior) were recently shown to exist [24].

REFERENCES

[1] H. Dai, Surf. Sci. 500, 218, (2002).
[2] R.H. Baughman *et al.*, Science 297, 787, 2002.
[3] L. Duclaux, Carbon 40, 1751, 2002.
[4] G.-H. Jeong *et al.*, Phys. Rev. B 68, 075410, 2003.
[5] H. Shimoda *et al.*, Phys. Rev. Lett. 88, 015502, 2002.
[6] I.V. Bondarev, G.Ya. Slepyan, and S.A. Maksimenko, Phys. Rev. Lett. 89, 115504, 2002.
[7] I.V. Bondarev and Ph. Lambin, Phys. Rev. B 70, 035407, 2004.
[8] S.Y. Buhmann, H.T. Dung, and D.-G. Welsch, J. Opt. B: Quantum Semiclass. Opt. 6, S127, 2004.
[9] I.V. Bondarev and Ph. Lambin, Solid State Commun. 132, 203, 2004.
[10] I.V. Bondarev and Ph. Lambin, Phys. Lett. A 328, 235, 2004.
[11] A.S. Davydov, Quantum Mechanics (Pergamon, 1967).
[12] L. Knöll, S. Scheel, and D.-G. Welsch, in: Coherence and Statistics of Photons and Atoms, edited by J. Peřina, Wiley, 2001.
[13] M. Florescu and S. John, Phys. Rev. A 64, 033801, 2001.
[14] J.P. Reithmaier *et al.*, Nature 432, 197, 2004.
[15] D. Dini *et al.*, Phys. Rev. Lett. 90, 116401, 2003.
[16] T. Yoshie *et al.*, Nature 432, 200, 2004.
[17] S. Haroche and D. Kleppner, Phys. Today 42, No. 1, 24, 1989.
[18] T. Asano and S. Noda, Nature 429, 6988, 2004.
[19] J. Vučković and Y. Yamamoto, Appl. Phys. Lett. 82, 2374, 2003.
[20] X. Li *et al.*, *Science* 301, 809, 2003.
[21] J.M. Raimond, M. Brune, and S. Haroche, Rev. Mod. Phys. 73, 565, 2001.
[22] Z.M. Li *et al.*, Phys. Rev. Lett. 87, 127401, 2001.
[23] I.V. Bondarev *et al.*, Phys. Rev. B 68, 073310, 2003.
[24] S.Y.Buhmann *et al.*, Phys. Rev. A 70, 052117, 2004.

Continuous Mass Production of Fullerenes and Fullerenic Nanoparticles by 3-Phase AC Plasma Processing

F. Fabry*, T. M. Gruenberger*, J. Gonzalez Aguilar**, H. Okuno****, E. Grivei*, N. Probst*, L. Fulcheri**, G. Flamant*** and J.-C. Charlier****

*Timcal Belgium N.V. – Appeldonkstraat, 173 – B-2830 Willebroek, Belgium, Tel: +32 3 886 71 81 Fax: +32 3 860 16 30, f.fabry@be.timcal.com
**Ecole des Mines de Paris, ENSMP, rue Claude Daunesse B.P. 207, F-06904 Sophia Antipolis, France,
***PROMES-CNRS, B.P. 5 Odeillo, F-66125 Font Romeu, France,
****PCPM, Université Catholique de Louvain, B-1348 Louvain-La Neuve, Belgium,

ABSTRACT

A patented plasma technology has been developed over the last ten years for the continuous mass production of fullerenes and fullerene soot by a French-Belgian consortium. The process combines the high temperature arc method with continuous gas phase synthesis by injecting solid carbon precursors into a thermal arc plasma. This 3-phase AC plasma process can be considered as a highly flexible process with an enormous potential for further up-scaling to an industrial size at commercially viable cost.

In this paper, the plasma process and results on typical process conditions prevailing are detailed. The main results presented concern the study of the C/He ratio during the fullerene synthesis and its influence on the fullerene yield and on the C_{70}/C_{60} ratio. Price and production short term perspectives for both products are presented.

Keywords: mass production, thermal plasma, fullerene, fullerene soot.

1 INTRODUCTION

For a long time, laser, solar and electric arc processes have been used for fullerene synthesis. Although the fullerene concentration in the soot may be quite high, fullerene production rates remain low, as the material flow is limited to the ablation of a graphite target. For this reason, these processes can hardly be scaleable for industrial exploitation.

In the meantime, the combustion method developed at the Massachusetts Institute of Technology (MIT) by Jack Howard [1] has overcome these problems using an incomplete combustion process. However, the question of greenhouse gas emissions, mainly CO_2, associated with this process may be problematic in the frame of an industrial exploitation. Indeed, only a very small fraction of the hydrocarbon fuel is converted into valuable fullerenes, the remaining fraction being burned and released into the atmosphere [2].

A fundamentally different approach based on a patented plasma technology has been developed over the last ten years by the core partners of a consortium (TIMCAL BELGIUM N.V., EMP-ARMINES and CNRS) [3]. The process combines the high temperature arc method with continuous gas-phase synthesis by injecting solid carbon precursors into a thermal arc plasma (Figure 1).

Figure 1

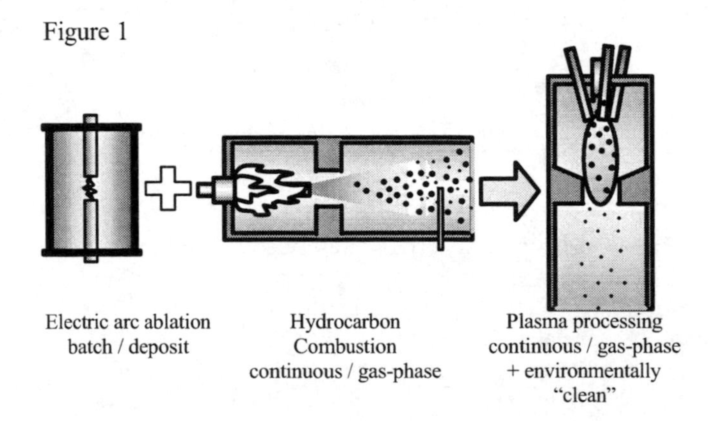

The plasma technology is totally clean and environmental friendly since zero greenhouse gas emission is produced on site. The success of this approach can be found in the use of a 3-phase AC plasma technology particularly suited for large capacity processing (typically employed in metallurgy).

The possibility to produce large quantities of fullerenes at commercially viable cost will bring their applications to emerge. Especially concerning C_{60}, enormous progresses have been done over the last years in pharmaceutics [4], cosmetics, electronics, etc and industrial users of these molecules are waiting for their abundant availability.

In addition fullerene soot, co-product of the plasma process, is a carbon black-like material that has shown interesting reinforcing properties in rubber and could consequently be of interest for tire and composite industry.

2 PLASMA PROCESS

This 3-phase AC plasma system, initially developed and optimized for the synthesis of novel grades of carbon black [5, 6], has been modified and adapted for the continuous synthesis of fullerenes and fullerene soot [7]. These adaptations lead to the process scheme shown in Figure 2.

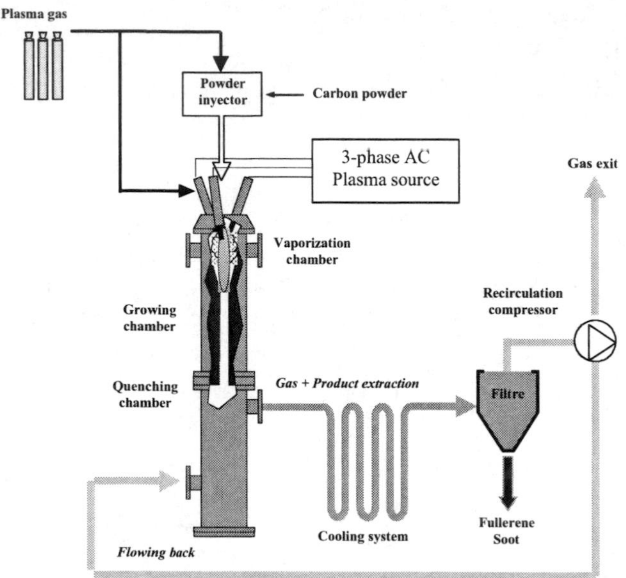

Figure 2: Scheme of the plasma facility configuration for fullerene production

The process can be briefly described as follows: A thermal plasma is generated by an arc discharge between three graphite electrodes placed in the upper part of the reactor. A special powder injection system is employed to mix a solid carbonaceous precursor (powder) with a suspending gas to transport the mixture inside the reactor. This aerosol flows across the arc reaching the high temperature region in the reactor. Due to the high enthalpy density obtained, the solid carbon is vaporized completely while passing through the graphite nozzle. The internal shape of the reactor has been designed to improve the conditions for solid carbon vaporization by a strong confinement of the gas flow. A quenching/sampling system collects the high temperature gas at a predetermined position, cools it down rapidly to the conditions chosen for the production of fullerenes (quenching) and extracts the resulting product from the reactor. The gas is filtered and a part of it is re-injected into the reactor. The fraction that corresponds to the flow rate initially entering the reactor as plasma gas is exhausted. This set-up (with recirculation) allows the extraction of a gas volume superior to the initial flow of plasma gas and therefore disconnects the dependency of these two parameters, which leads to an additional degree of freedom in relation to process operation. Absolute air-tightness of the system is monitored by continuous measurement of the oxygen concentrations at several points of the system.

The plasma process addresses two types of products:
- Fullerenes, mainly C_{60} and C_{70},
- Fullerenic nanostructures based on carbon nanoparticles (carbon-black-like) with particular fullerenic surface structure.

3 PROCESS OPERATING CONDITIONS

A large number of process parameters was investigated. A brief overview is given in Table 1.

Table 1: Main process parameters investigated for fullerene production.

Nature of the carbon precursor	2 carbon black grades 2 acetylene black grades Graphite powder
Nature of plasma gas	Helium Argon Nitrogen
Electric parameters	Arc current, arc voltage
Flow rates	Plasma gas Carbon precursor
Injection conditions	Velocity Location
Quenching and cooling conditions	Velocity (Flow rate) Location of product extraction

The main results presented in this paper concern the study of the C/He ratio during the fullerene synthesis and its influence on the fullerene yield and on the C_{70}/C_{60} ratio. Therefore, the plasma zone has been characterized also by optical emission spectrometry.

4 PROCESS CHARACTERIZATION

- Temperature Measurement by Optical Emission Spectrometry.

As part of the plasma zone characterization during process operation, optical emission spectroscopy was employed. The experiments were performed with helium as plasma gas. During fullerene synthesis, C_2 is produced from carbon particle vaporization as well as from electrode ablation. This chemical species is commonly used for temperature measurement. The details about the measurements and the analyses methods are presented in ref. [8].

Nozzle entrance zone

- Without carbon injection, the C_2 rotational temperature depends on the arc intensity, but for a given intensity the rotational temperature is not influenced by the gas flow variation.
- With carbon injection, the C_2 rotational temperature seems to decrease with the carbon mass flow rate. At constant gas flow rate, the increasing of the carbon gas flow

rate raises the C/He ratio and then C_2 clusters are in the presence of an important density of solid carbon particles which inhibits their formation. This phenomenon is accentuated with a current of 350 A for which there is a high current density at the three electrode tips. At the high values of current, a strong erosion of the electrodes is observed as well as a scraping of small solid carbon particles, which are also an additional source of carbon increasing the C/He ratio and again inhibiting the formation of C_2.

For the low gas flow rates, although the C/He ratio increases, the process has a different behavior. In this case, by decreasing the gas flow rate, the enthalpy density is increased, involving higher plasma temperatures for the same electrical power input. In the same time, the residence time of the solid particles in the high temperature plasma zone increases. Thus, the carbon vaporization is improved and C_2 concentration as well as C_2 rotational temperature increase. However, due to the increase of the temperature gradient towards the graphite walls and the slow flow of the vaporized carbon towards the quenching zone, the carbon vapor condenses favorably on the graphite wall of the reactor and reduces the process effectiveness for the fullerene synthesis in gas-phase.

Thermodynamic calculations were performed to determine stable carbon species concentrations versus temperature. Solid carbon (graphite) is the stable carbon phase at temperatures lower than 4100 K. Above this temperature, carbon species are gaseous. C_2 temperature measurements show that the temperature in the nozzle entrance zone is greater than the condensation temperature of graphite whatever the operating conditions. Consequently, the conditions are favorable for carbon vaporization in the whole upper zone between the arc and the nozzle entrance.

5 PREPARATION OF THE FULLERENE SAMPLES

In the plasma process, insoluble carbon soot is generated together with soluble fullerenes. The fullerenes have been extracted from the fullerene soot with organic solvents (toluene) by applying Soxhlet-extraction. The C_{60} and C_{70} fullerene molecules present in the separated fractions have been qualitatively and quantitatively identified and measured by UV/VIS spectroscopy.

6 RESULTS AND DISCUSSION

The yield dependence on the most influential process parameters is presented in ref. [9]. To summarize, the nature of the plasma gas appears to be the most critical one. Yields have been found to be about one order of magnitude higher when using helium as plasma gas rather than argon or nitrogen, a fact generally known for the arc or laser process. The nature of the carbon precursor, showed no drastic influence on the process performance.

The novel results presented in this paper concern the study of the C/He ratio during the fullerene synthesis and the influence of this parameter on the fullerene yield and on the C_{70}/C_{60} ratio.

- Influence of the C/He ratio

The influence of the C/He ratio on the fullerene content and the C_{70}/C_{60} ratio are presented in Figures 3 and 4.

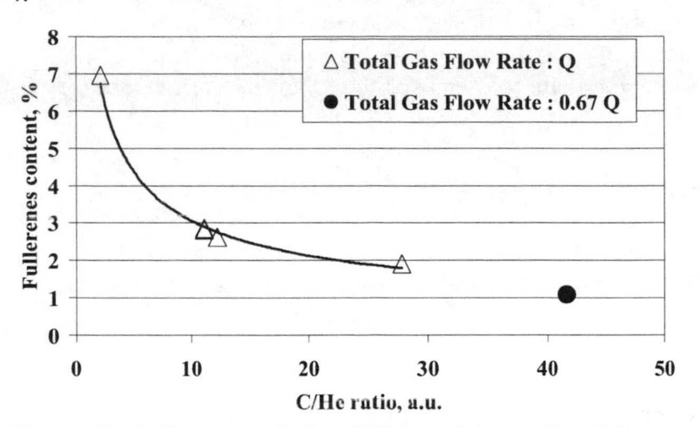

Figure 3: Influence of the C/He ratio on the fullerene content

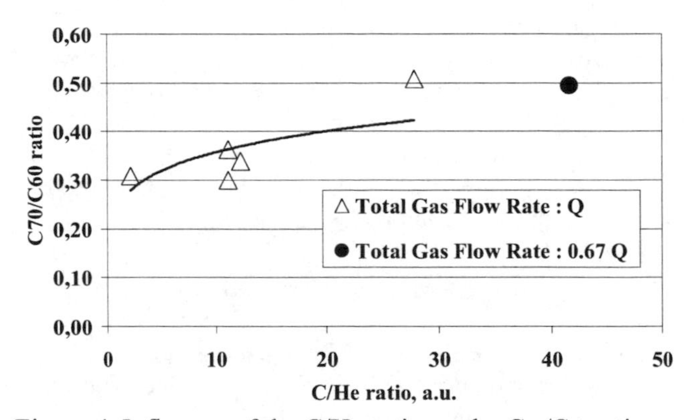

Figure 4: Influence of the C/He ratio on the C_{70}/C_{60} ratio

The fullerene content represents the weight fraction of all the fullerenes extracted by Soxhlet technique. To verify the reliability of the plasma process, three experimental runs with a C/He ratio of around 11 (arbitrary unit) have been performed. The two graphs show that a good reproducibility of the results is obtained.

The triangles symbols on the curves represent some samples obtained for five different experiments for which the operating conditions are the same excepted for the carbon mass flow rate. The filled points represent an experiment performed with the same operating conditions than the triangles at C/He = 28 but with a diminution of the plasma gas flow rate, Q by a factor of 0.67.

These figures show that when the C/He ratio increases, the fullerene content drops whereas the C_{70}/C_{60} ratio increases. A significant influence of the carbon mass flow rate on the fullerene yield is observed.

On the one hand, when increasing the carbon mass flow rate at a given gas flow rate, yield is reduced. But even if the yield appears lower, absolute production rate of fullerenes still increases when increasing carbon flow, if sufficient energy is available. On the other hand, when increasing the helium flow at a given carbon mass flow rate, yield is increased. In this case also, the process limitation is the energy provided to the gas by the plasma source. In conclusion of these results, to improve the performance of the plasma process for the production of fullerenes, it is important to reduce the C/He ratio for the fullerene yield and to increase the carbon mass flow rate for the productivity. The solution is to increase considerably the plasma gas flow rate, the limit being the energy provided by the plasma source (at a maximum current between 250 A and 300 A). In a next stage, the entire production system including periphery components will be further optimized to allow undercutting the present price of fullerenes and fullerene soot by a factor of ten in the medium term.

Concerning the influence of the C/He ratio on the C_{70}/C_{60} ratio, we can say that we have a behavior similar to the influence of the C/O ratio on the C_{70}/C_{60} ratio observed with the combustion process [2]. The higher the C/He ratio is, i.e. the higher the carbon concentration in the reactor, the higher the C_{70}/C_{60} ratio and the lower the C_{60} concentration. As concluded previously with the C_2 measurement, the influence of the C/He ratio on the C_2 concentration is the same. The higher the C/He ratio is, the lower the C_2 concentration. We can assume that the presence of C_2 in the high temperature zone of the reactor could contribute preferentially to the formation of C_{60} fullerenes.

7 FULLERENE SOOT

Fullerene soot (carbon-black-like nanoparticles with fullerenic surface structure) constitutes a completely new product family requiring intensive characterization and application testing. Plasma produced nanoparticles present some good reinforcement performances and are particularly well suited to be used as additives in various material applications like polymer composites and electrochemical components. Their commercial impact will of course be magnified if they can be the starting point of a new filler family. Performance fillers represent very important application volumes, where these new materials could progressively participate. Fullerene soot is of high interest for rubber reinforcing and as such for the tire and composite industry.

8 CONCLUSIONS AND PERSPECTIVES

The plasma process is addressing two types of products:

- Fullerenes, mainly C_{60} and C_{70},
- Fullerene soot based on carbon nanoparticles (carbon-black-like) with particular fullerenic surface structure.

So far the process of continuous fullerene synthesis is not fully optimized, but current yields typically of the order of 5% (toluene extractable) are obtained. At carbon flow rates of several hundred grams per hour, in extreme cases up to 1 kg/h, fullerene production rates of the order of 10 g/h can be obtained. No other technology has so far been able to reach such high fullerene production rates on a continuous basis and at atmospheric pressure.

The 3-phase AC plasma process can be considered as an improved highly flexible process with an enormous potential for further up-scaling to an industrial size and fullerene production at commercially viable cost. Major breakthroughs are expected through the implementation of a new operation mode at high gas flow rate, which will allow the treatment of a ten-fold product quantity at the present scale. Based on our knowledge of the pilot plasma process, an industrial unit based on the plasma technology will allow at a medium term the processing of 1 ton of fullerenes and 100 tons of fullerene soot per year, respectively and a reduction of the present price by a factor of ten is expected.

ACKNOWLEDGEMENT

This specific research and technological development programme has been supported by the European Commission under the Competitive and Sustainable Growth Programme, contract PLASMACARB, G5RD-CT-1999-00173

REFERENCES

[1] US Patent 5,275,729.

[2] H. Takehara, M. Fujiwara, M. Arikawa, M. D. Diener, J. M. Alford, Carbon 43, 311-319, 2005.

[3] Fulcheri, L., Schwob, Y., Fabry, F., Flamant, G., Chibante, L. F. P., and Laplaze, D., Carbon 38, 797 2000.

[4] e.g. US patent 6,613,771 and S. Bosi, T. Da Ros, G. Spalluto, M. Prato, Europ. J. Med. Chem., 38, 913 (2003).

[5] PCT/EP94/00321, AC Plasma Technology for Carbon Black and Carbon Nanoparticles, 1993.

[6] L. Fulcheri, N. Probst, G. Flamant, F. Fabry, E. Grivei, and X. Bourrat, Carbon 40, 169, 2002.

[7] PCT/EP98/03399, Fullerene Production in AC Plasma, 1997.

[8] S. Abanades, J. M. Badie, G. Flamant, L. Fulcheri, J. Gonzalez-Aguilar, T. M. Gruenberger, F. Fabry, High Temp. Material Processes, 7, 1, 43-49, 2003.

[9] T. M. Gruenberger, J. Gonzalez-Aguilar, L. Fulcheri, F. Fabry, E. Grivei, N. Probst, G. Flamant and J.-C. Charlier, edited by Kuzmany et al., AIP Conference Proceedings (2002).

Experimental Observation of Carbon and Carbon-Metal Nanotoroids and New Carbon-Metal Superstructures at Nanoscale

V. Kislov, B. Medvedev, I. Taranov

Institute of Radioengineering & Electronics, RAS,
125009, 11 Mokhovaya bld. 7, Moscow, Russia, kislov@cplire.ru

ABSTRACT

We report on experimental observation (by AFM, STM and HRTEM methods) of nanotoroids for both carbon and carbon-metal superstructures produced by methods of arc discharge and laser ablation. Size of superstructures is ~ 10 nm for carbon and ~ 30 nm for carbon-metal (outer diameter), with inner diameter ~ 1/3 of the total. We also discovered carbon-metal nanocapsules with shapes like nanodisks, reminiscent of human erythrocytes, and sizes ~ 30 nm. The influence of pressure, humidity and temperature was investigated. Effect of reversible transition of topology (sphere - toriod) on the same sample were oserved depending on external parameters. New types of topologically closed carbon and carbon-metal nanostructures are discovered. We also discuss some technological aspects of the growth of such toroidal nanostructures.

Keywords: carbon, metal, nanoclusters, nanotoroid, superstructures.

We report experimental observation (by AFM, STM and HRTEM methods) of nanotoroids for both carbon and carbon-metal superstructures produced by methods of arc discharge and laser ablation. Size of superstructures is ~ 10 nm for carbon and ~ 30 nm for carbon-metal (outer diameter), with inner diameter ~ 1/3 of the total.

Initial observation of toroids were made for pure carbon, but gradual increase of metal to some optimal concentration dramatically increased the yield of toroidal structures. The influence of pressure, humidity and temperature was also investigated. Also effects of reversible transition of topology (sphere - toroid) (Fig. 1. - 5.) on the same sample were observed depending on external parameters.

The possibility of such structures was discussed earlier [1, 2], but we also discovered carbon-metal nanocapsules among toroids, with shapes like nanodiscs, reminiscent of human erithrocites. Changing concentrations and metals (Mo, Cr, etc.) leads to differences in shapes and sizes of superstructures.

Therefore, new types of topologically closed carbon and carbon-metal nanostructures are discovered, with the theoretical model presented elsewhere [3, 4]. We also discuss some technological aspects of the growth of such toroidal nanostructures.

A continual model of topologically closed carbon and carbon-metal nanostructures is presented. This model considers mechanical properties of chemical bonds and external pressure of gas phase. A sequence of transitions for various topological forms of possible superstructures is described for continuous growth of external pressure (from

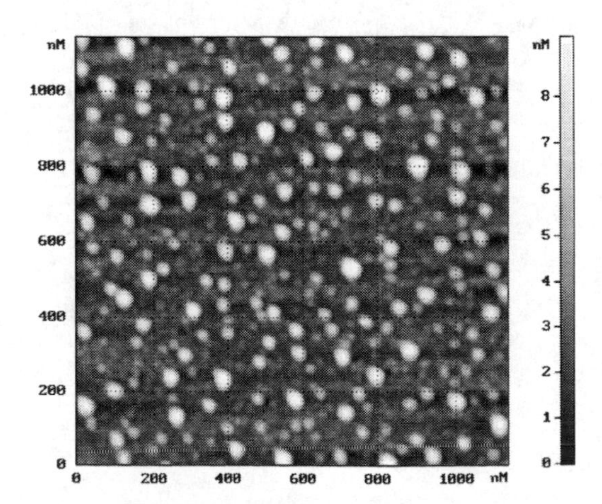

Fig. 1. AFM Image spheroidal of metal-carbon superstructures

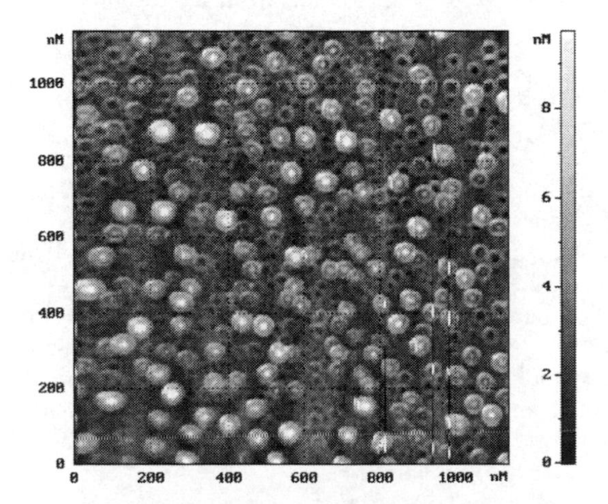

Fig. 5. AFM Image toroidal of metal-carbon superstructures

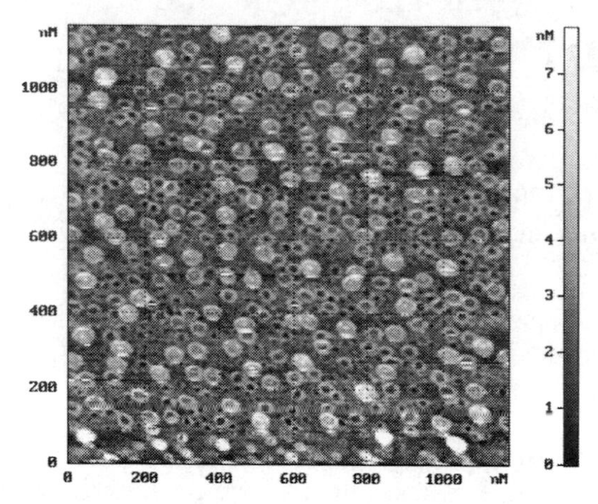

Fig. 2. AFM Image toroidal of metal-carbon superstructures

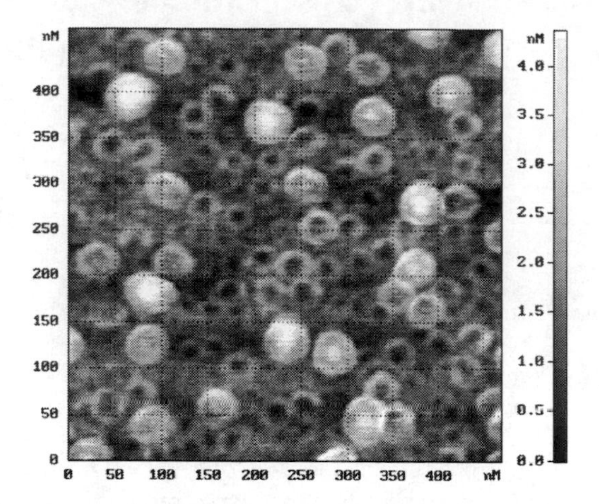

Fig. 3. AFM Image toroidal of metal-carbon superstructures

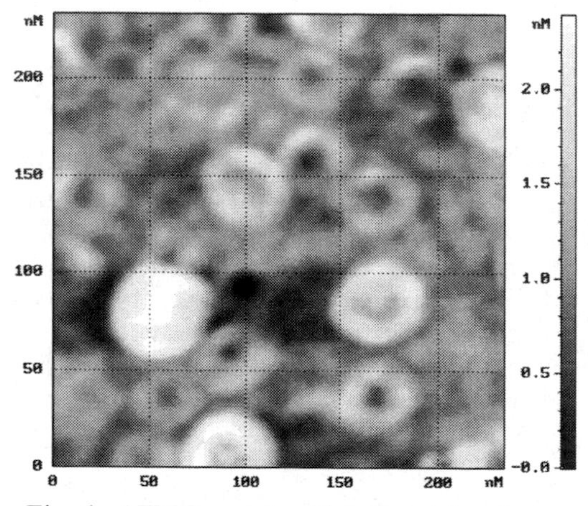

Fig. 4. AFM Image toroidal of metal-carbon superstructures

vacuum to atmospheric): from sphere to slightly pressed shape, then via growing negative curvature of "anti-caps" to toroidal structures and then via flattening and thinning of toroids and growth of their diameter to nanorings ("crop-circle" structures [5]). We also present some technological details of the growth of such toroidal nanostructures and discuss a number of possible applications. We believe that the very facts of existence of optimal carbon-metal concentrations and specific external parameters for production of nanotiroids and other superstructures should have important implication for theoretical understanding of superstructure formation.

REFERENCES

[1]. M. Terrones, W.K. Hse, J.P. Hare, H.W. Kroto, H. Terrones, D.R.M. Walton, Graphitic structures: from planar to spheres, toroids and helices, Phil. Trans. R. Soc. Lond. A, 354, 2025 (1996).

[2]. V. Kislov, Rings of the strings – topologically closed superstructures of nanoclusters, Nano-8, 8-th International Conference on Nanometer-Scale Science and Technology, June 28 - July 2, 2004, Venice, Italy

[3]. Kislov, I. Taranov, Topological Forms of Carbon and Carbon-Metal Superstructures: from Nanotubes to Toroids and Nanodiscs, 2005 MRS Spring Meeting, March 28 - April 1, San Francisco, CA, USA

[4]. V. Kislov, B. Medvedev, Experimental observation of carbon and carbon-metal nanotoroids, 2005 MRS Spring Meeting, March 28- April 1, San Francisco, CA, USA.

[5]. J. Liu, H. Dai, J. H. Hafner, D. T. Colbert, R. E. Smalley, S. J. Tans, C. Dekker, Fullerene "crop circles", Nature, 1997, **385**, 780.

Effects of the process parameters on the carbon nanotubes growth by thermal CVD

C. Verissimo[*], S. A. Moshkalyov[*], A. C. S. Ramos[**], J. L. Gonçalves[*], O. L. Alves[***] and J. W. Swart[*]

[*]Centro de Componentes Semicondutores – CCS, UNICAMP
C.P. 6061, CEP 13083-870, Campinas, SP, Brasil, carla@iqm.unicamp.br, stanisla@led.unicamp.br
[**]Instituto de Física "Gleb Wataghin" – IFGW, UNICAMP, CEP 13083-970, Campinas, SP, Brasil.
[***]Laboratório de Química do Estado Sólido – LQES , Instituto de Química – IQ, CP 6154,
CEP 13083-970, UNICAMP, Campinas, SP, Brasil.

ABSTRACT

Multiwall carbon nanotubes were grown by thermal chemical vapor deposition with a metal catalyst. This approach employs a metal film deposition onto a SiO_2/Si substrate followed by a thermal annealing to generate metal particles able to convert carbon containing gas (methane or acetylene) into carbon nanotubes. Nickel was used as a metal catalyst, and the influence of different growth parameters on the CVD process was investigated. The studies have shown that the effect of the methane concentration depends essentially on the process temperature. We have also observed that the catalytic capacity of the nickel depends on the carbon gas precursor reactivity. Moreover, it is generally assumed that the metal particle size determines the carbon nanotube diameter but, according to our research, this seems to be valid only for relatively small metal particles when the tip growth mechanism predominates. For relatively large metal particles, a hybrid growth mechanism (probably involving both tip and base growth models) can be considered, and the size of the particle does not determine the diameter of the resulting nanotube.

Keywords: carbon nanotubes, CVD, metal catalysis, metal thin films, growth mechanism

1 INTRODUCTION

Carbon nanotubes (CNTs) have received much attention due to their extraordinary properties and potential technological applications [1]. The development of technologies involving carbon nanotubes can result in nano-devices characterized by high performance, low mass and low consumption of energy. Thermal chemical vapor deposition (CVD) has been shown to be an efficient and versatile technique for the carbon nanotubes synthesis [2,3,4,5]. However, many challenges (in particular, the lack of precise control of the nanotube growth during the CVD process) still have to be addressed.

The identification and control of numerous growth parameters is crucial for the development and optimization of the synthesis process, study of growth mechanisms and nano-devices fabrication. Therefore, the detailed study of the effect of experimental conditions such as time, temperature, metal catalyst, carbon gas precursor, among others, is of great relevance and now it is the subject of many experimental researches.

In this work, the influence of different experimental parameters on the carbon nanotube growth by thermal CVD using nickel metal catalyst was investigated.

2 EXPERIMENTAL

Electron beam evaporation was used for thin (6 nm) nickel film deposition onto silicon wafers covered by a 50 nm thick thermally grown silicon oxide layer. Then the wafers were cut in small (~5x5 mm) samples, followed by their annealing in a quartz tube furnace at 700 °C in a H_2 flow of 400 sccm during 30 minutes. Two different CVD processes were used. In the growth method 1, the hydrogen was replaced by NH_3 flow of 400 sccm during 5 minutes. Then, the temperature was raised to 800-950 °C, and the synthesis was started by addition of 25-100 sccm of CH_4 for 30 minutes. The H_2/CH_4 gas ratios were 4:1, 8:1 and 16:1. After the first 5 minutes of the synthesis phase, NH_3 was replaced by the same flow of H_2. When C_2H_2 was used as the carbon gas precursor (growth method 2), the flow of H_2 was increased to 600 sccm after the annealing, and a flow of 6 sccm of C_2H_2 was introduced for 15 minutes keeping the temperature at 700 °C. A high resolution scanning electron microscope FEG-SEM JSM 6330F was used for structural analysis of samples. Raman spectroscopy was performed employing a Renishaw System 3000 Raman Imaging Microscope, using the 632.8 nm line of a He-Ne laser.

3 RESULTS & DISCUSSION

During the annealing process, nickel thin films supported on SiO_2/Si substrates form isolated metal particles. The characteristic of this thermal treatment is that the variation of particle diameters is relatively wide, and thicker films lead to a wider distribution of particle sizes [3,4,5,6]. Scanning electron micrographs obtained for Ni film with a thickness of 6 nm annealed at 700 °C show round (disk-like) particles with diameters ranging from 10 to 100 nm (Figure 1). Different size distributions were found for the catalyst particles located at the edge and in the center of the sample, with the distribution in the center being narrower (see Figure 1 A and B). The reasons for this are not clear at the moment. According to our observations (to be published elsewhere [7]), the metal film tends to break more readily (faster and at lower temperatures) at the edge of the substrate as compared with its center. This phenomenon could explain the different metal particle size distribution over the substrate.

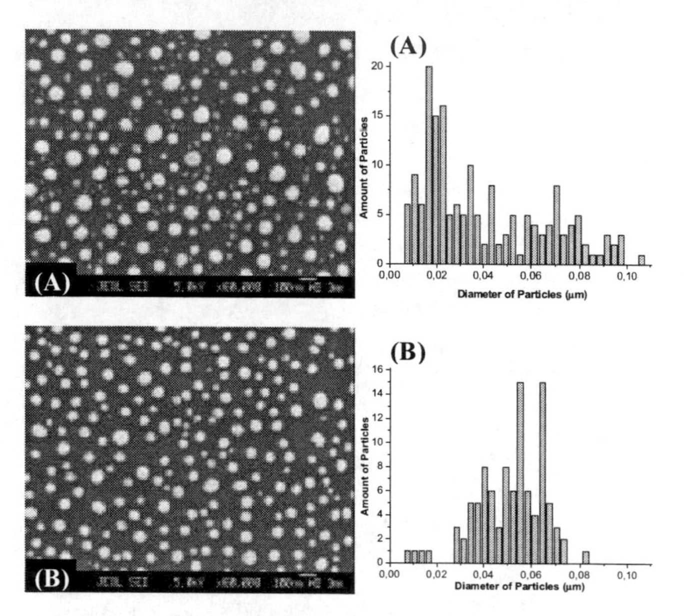

Figure 1. SEM images of Ni film after annealing process: (A) edge and (B) center of the substrate. In the central region of the substrate (histogram B) 96 % of particles present diameters ranging from 27 to 74 nm against 43 % of particles at the edge (histogram A).

After the annealing process, H_2 flow was replaced by NH_3 flow for cleaning of the metal particles before the CNT growth according to the method 1 described earlier. Then, the growth temperature was set and methane was introduced inside the furnace as the carbon precursor gas for the CNT growth. In the beginning of the growth process, NH_3 gas was employed to prevent a possible poisoning of the metal catalyst, and after that the NH_3 was replaced by H_2 again.

Scanning electron micrographs showed multiwall carbon nanotubes (MWNTs) with diameters larger than 20 nm for all samples. The characteristic profile of Raman spectra corroborates the formation of MWNTs (Figure 2). As expected, the Raman spectrum of MWNTs is similar to that of the polycrystalline graphite, since carbon nanotubes are constituted by graphene cylindrical walls. Therefore, spectra shown in Figure 2 consist of two bands, a G-line at 1579 cm^{-1} associated to graphitic structure and a D-line near to 1330 cm^{-1} accompanied by a second order D' band at 1612 cm^{-1} related to disordered graphitic structure [8]. The ratio of the D- and G-lines intensities (I_D/I_G) is used to

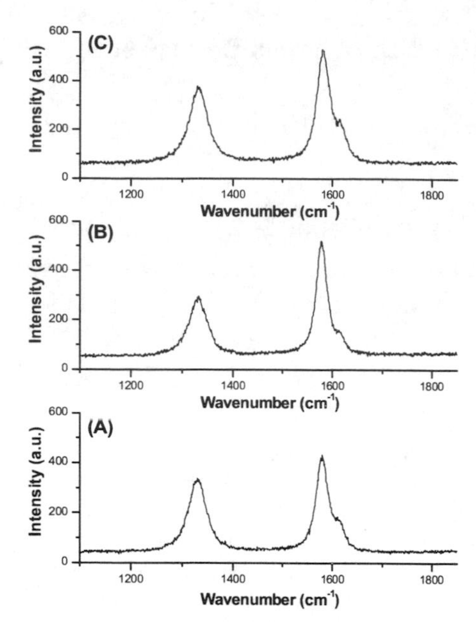

Figure 2- Raman spectra of CNTs grown using a gas ratio of $H_2/CH_4 = 4:1$ at different temperatures: (A) 850 °C, (B) 900 °C, and (C) 950 °C.

characterize the defects in carbon nanotubes. Higher I_D/I_G ratio implies in more structural defects in CNTs. As expected, sample prepared at lower temperature (850 °C – Figure 2 A) presented the highest I_D/I_G ratio (0.75), since lower temperatures do not lead to a high graphitization of the carbon walls structure. However, the smallest I_D/I_G ratio (0.52) was observed for the sample prepared at 900 °C whereas the I_D/I_G ratio for the sample prepared at 950 °C was slightly higher (0.65). Further investigations involving purification of the CNTs and TEM analysis, to provide additional information on the possible changes in CNTs structural properties, are in progress. Due to low thermal stability of methane, experiments at temperatures as high as 950 °C could result in decomposition of the gas and deposition of different amorphous carbon materials over the carbon nanotubes.

Regarding the influence of the main growth parameters on the CNT formation, the effect of the methane concentration depends essentially on the process temperature. In experiments at temperatures of 850, 900 and 950 °C, higher concentration of methane during the growth process resulted in a higher CNT density (Figure 3) indicating that the growth process is limited by carbon supply. At 850 °C, CNTs were not formed for the most diluted H_2/CH_4 mixture. Furthermore, the formation of CNTs was not observed at the lower temperature (800 °C), independently on the methane concentration. These results indicate the decrease of the catalytic capability of the metal with the growth temperature reduction when methane is used as a carbon feedstock gas.

Since the basic interest here is the development of procedures able to lead to a controlled synthesis of CNTs, it is important to describe a curious phenomenon observed for

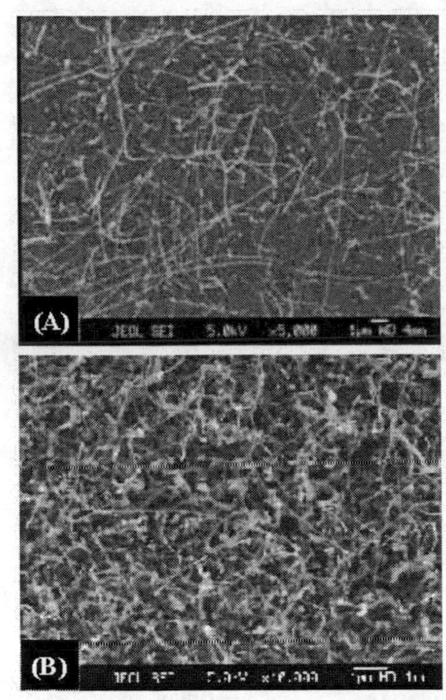

Figure 3. SEM images of CNT grown at 950 °C and using different H_2/CH_4 gas ratios: (A) 16:1 and (B) 4:1.

most of the samples while using methane and nickel as a catalyst (note that the effect is not observed for iron catalyst, see elsewhere [7]). The growth of CNT does not occur evenly over the entire substrate, but mainly in a region close to the edges. In fact, there is a gradient of the CNT nucleation density which decreases from the edge to the central region of the substrate, as shown in Figure 4. Investigations are in course to explain and overcome this phenomenon. A possible cause may be correlated with the presence of residual oxygen inside the furnace or the release of oxygen due to possible decomposition of SiO_2 layer catalyzed by the metal [9]. In this case, the metallic nickel could catalyze the following conversion: $SiO_2 \rightarrow \frac{1}{2} O_2 + SiO$, and the concentration of oxygen released from the silicon oxide could be higher at the center then at the edges. In turn, this would result in formation of the nickel oxide (which is known to be much less efficient as a catalyst as compared with the metallic nickel [10]), the process being stronger in the central part of the sample. Literature data also suggest that this non-homogeneity in the CNT growth may be caused by nickel oxidation. Yen *et al* [11] observed worm-like carbon fibers, similar to carbon structures shown in Figure 4 B, when nickel particles were partially oxidized.

In striking contrast to the case of methane as a carbon feedstock gas, experiments performed with acetylene (the growth method 2), led to a high density of CNTs grown evenly over the entire substrate. This result indicates the influence of the carbon precursor reactivity. Due to higher reactivity of acetylene compared to methane [12], essentially softened experimental conditions – such as

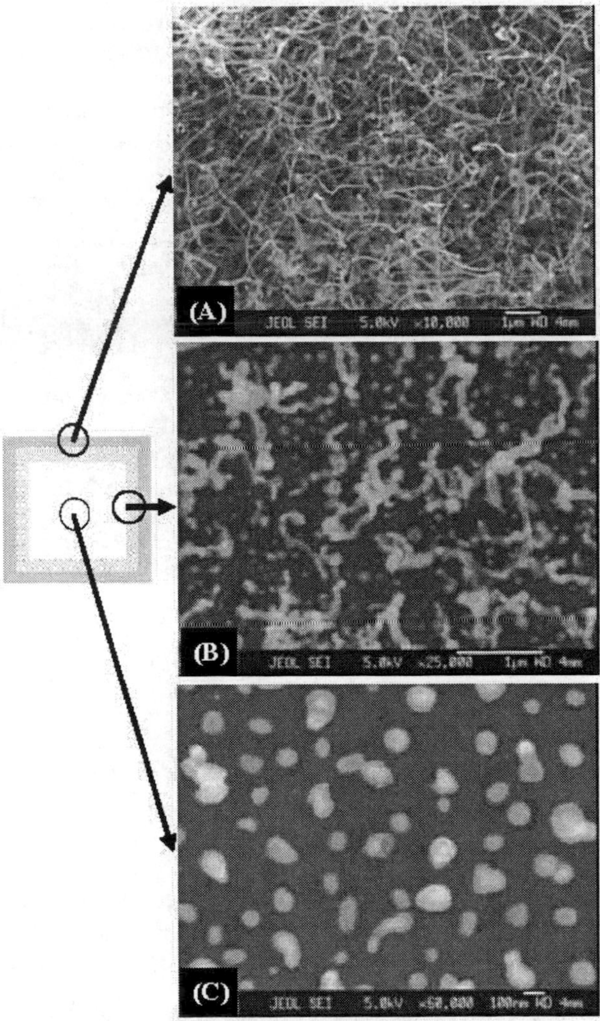

Figure 4. SEM images of CNT grown at 900 °C using H_2/CH_4 = 16:1. Micrograph (A) was obtained at the edge whereas images (B) and (C) were obtained a few microns away of the edge and in the central region of the substrate, respectively. The off-set drawing represents a substrate and the gradient of the CNT density (out of scale).

lower growth temperature and concentration of carbon (H_2/C_2H_2 = 100:1) – also resulted in the CNT growth. Probably, the enhanced acetylene reactivity makes the whole growth process less sensitive to the metal catalytic capacity, thus resulting in more uniform CNT growth in all regions of the substrate.

The studies performed with methane also showed other interesting result. Under our experimental conditions, small Ni catalyst particles were usually observed at the tip of the CNTs suggesting the tip growth mechanism. However, in some cases, CNTs with diameters between 20-80 nm grown from metal islands of larger dimensions (up to ~300 nm) were observed with another metal particle at the bottom of the tube (Figure 5). This is in contrast to most observations where the catalyst particle is observed at either the tip or bottom of the nanotubes and thus suggests a complex growth mechanism. The growth could involve both base and tip growth (hybrid growth mechanism) taking into account the liquid-particle model [13,14]. According to Nerushev *et al* [13], who observed a similar phenomenon, the formation of small layers of graphene over a relatively big and flat metal particle, followed by the motion and coalescence of some of these graphene layers, can be the initial stage of this event.

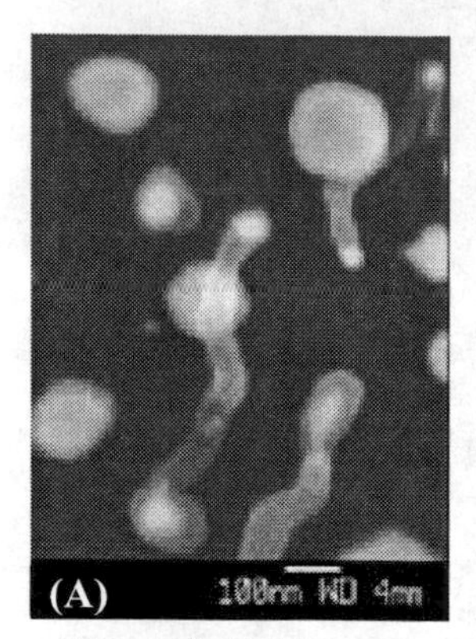

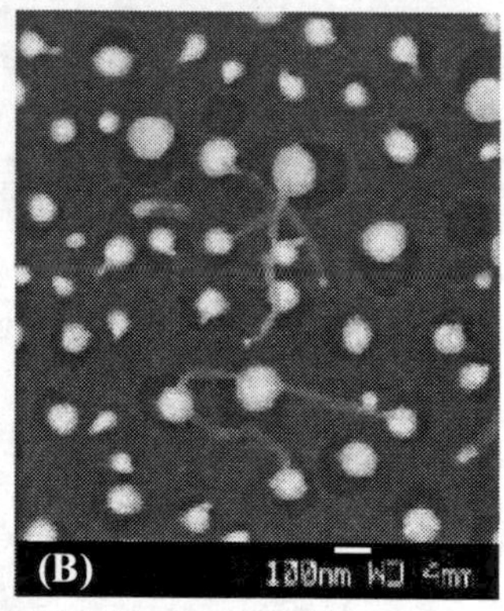

Figure 5. SEM images of CNT grown at 900 °C using different H_2/CH_4 ratios: (A) 4:1 and (B) 16:1.

The size and shape of catalyst particles are usually considered to be important growth parameters [13,14,15]. More specifically, in the literature it is generally assumed that the metal particle size determines directly the resulting CNT diameter, independently on the growth mechanism (either tip or base growth modes are considered). However, due to our data this seems to be valid only for relatively small particles. For large metal particles (> 100 nm), the CNT diameter appears to be no longer dependent on the particle size, and more complex mechanism apparently determines the CNT growth. Therefore, the size of resultant catalyst particles can determines the growth mechanism and characteristics of the carbon nanotubes.

4 CONCLUSIONS

In the thermal chemical vapor deposition method, the carbon nanotube growth is very sensitive to the experimental conditions. A specific combination of different growth parameters has to be reached to obtain the desired final material. Moreover, it was shown that catalytic metal islands with diameters bigger than 100 nm tend to form CNTs of considerably smaller diameters with metal present at both ends of the nanotubes suggesting a complex growth mechanism. In other words, the initial size of the metal particles can influence not only the diameter but also the growth mechanism of carbon nanotubes.

5 ACKNOWLEDGEMENTS

The work was financially supported by CNPq and Instituto do Milênio SCMN. The authors would like to thank LME/LNLS, Campinas, Brazil, by the use of scanning electron microscope, and LEM/USP, São Paulo, Brazil, by the use of Raman spectrometer.

6 REFERENCES

[1] (A) S. B. Sinnott and R. Andrews, Critical Reviews in Solid State and Materials Science, 26, 145, 2001; (B) M Terrones, International Materials Reviews, 49, 325, 2004; (C) M. S. Dresselhaus, G. Dresselhaus, J. C. Charlier and E. Hernandez, Philosophical Transactions of Royal Society of London A, 362, 2065, 2004.

[2] Y. Huh, J. Y. Lee, J. Cheon, Y. K. Hong, J. Y. Koo, T. J. Lee and C. J. Lee, Journal of Materials Chemistry, 13, 2297, 2003.

[3] M. Chhowalla, K. B. K. Teo, C. Ducati, N. L. Rupesinghe, G. A. J. Amaratunga, C. Ferrari, D. Roy, J. Robertson and W. I. Milne, Journal of Applied Physics, 90, 5308, 2001.

[4] Th. D. Makris, R. Giorgi, N. Lisi, L. Pilloni, E. Salernitano, F. Sarto and M. Alvisi, Diamond and Related Materials, 13, 305, 2004.

[5] S. A. Moshkalyov, A. L . D. Moreau, H. R. Guttiérrez, M. A. Cotta and J. W. Swart, Materials Science and Engineering B, 112, 147, 2004.

[6] J. D. Carey, L. L. Ong and S. R. P. Silva, Nanotechnology, 14, 1223, 2003.

[7] C. Verissimo *et al*, to be published.

[8] (A) V. P. Dymont, M. P. Samtsov and E. M. Nekrashevich, Technical Physics, 45, 905, 2000; (B) T. de los Arcos, M. G. Garnier, P. Oelhafen, D. Mathys, J. W. Seo, C. Domingo, J. V. García-Ramos, S. Sánchez-Cortés, Carbon, 42, 187, 2004.

[9] Y. Yao, L. K. L. Falk, R. E. Morjan, O. A. Nerushev and E. E. B. Campbell, Journal of Material Science: Materials in Electronics, 15, 533, 2004.

[10]G. Bertoni, C. Cepek, F. Romanato, C. S. Casari, A. Li Bassi, C. E. Bottani and M. Sancrotti, Carbon, 42, 423 2004.

[11] J. H. Yen, I. C. Leu, C. C. Lin and M. H. Hon, Diamond and Related Materials, 13, 1237, 2004.

[12] M. P. Siegal, D. L. Overmyer and F. H. Kaatz, Applied Physics Letters, 84, 5156, 2004.

[13] O. A. Nerushev, S. Dittmar, R.-E. Morjan, F. Rohmund and E. E. B. Campbell, Journal of Applied Physics, 93, 4185, 2003.

[14] E. F. Kukovitsky, S. G. L'vov, N. A. Sainov, V. A. Shustov and L. A. Chernozatonskii, Chemical Physics Letters, 355, 497, 2002.

[15] (A) S. Helveg, C. López-Cartes, J. Sehested, P. L. Hansen, B. S. Clausen,cJ. R. Rostrup-Nielsen, F. Ablid-Pedersen and J. K. Norskov, Nature, 427, 426, 2004; (B) C. L. Cheung, A. Kurtz, H. Park and C. M. Lieber, J. Phys. Chem. B, 106, 2429, 2002.

Structural Study of Amorphous Carbon using Adaptive Interatomic Reactive Empirical Bond-order Potential Model

M. Todd Knippenberg, Oyeon Kum, and Steven J. Stuart

Department of Chemistry, Clemson University, Clemson SC 29634 USA

ABSTRACT

The structural properties of amorphous carbon (a-C) systems with densities of 2.0, 2.6, and 3.0 g/cm^3 were studied using classical molecular dynamics with the adaptive interatomic reactive empirical bond-order (AIREBO) potential model. The AIREBO model enhances the locally reactive bond-order REBO potential for covalent materials to include long range van der Waals and torsional interactions that are quite important in an a-C system. The shape and position of the first peak of the pair correlation function were found to be quite sensitive to the density, with the peak position shifting towards shorter distances as the density increases. However, the simulated structure showed predominantly threefold coordination at the highest density of 3.0 g/cm^3, with higher sp^2 fractions than observed in ab initio models. The distribution of sp^2 and sp^3 carbons was visible in the pair correlation functions at 3.0 g/cm^3. Ring statistics showed relatively little density dependence up to five-membered rings and a significant number of larger rings. We observed that the AIREBO model demonstrates structural properties for a-C that differ from those obtained from ab initio and tight-binding models.

Keywords: structural study, amorphous carbon, AIREBO potential, coordination numbers, ring statistics

1 INTRODUCTION

Carbon has the ability to exist in a remarkable variety of nonequilbrium crystalline and noncrystalline structures at a wide range of temperatures and pressures. From both the scientific and application points of view, amorphous carbon (a-C) is one of the more important of these, and the microscopic structural properties of this noncrystalline form of solid carbon are quite intriguing and useful. For example, diamondlike a-C films are found to be hard, optically transparent, and chemically inert, which makes them important for applications in coating technology.

In the last couple of decades, extensive experimental investigations [1]–[3] and theoretical studies using computer simulations [4]–[9] have been devoted to elucidating the microstructure and bonding nature of this disordered form of carbon. While these studies have provided useful data about the structure of a-C, the microscopic bonding details are still not well understood. Many properties such as the relative concentration of atoms with various coordination nubmers in a-C are still controversial in the literature [6]–[8].

The a-C structure is stable in a wide range of densities varying from lower than 2.0 g/cm^3 up to higher than 3.0 g/cm^3. The physical properties are quite different for different densities. For example, at a low density of approximately 2.0 g/cm^3, it is an electrically conductive, optically absorbing material with a majority of atoms bonded in a graphitelike (sp^2) configuration. On the other hand, at densities of approximately 3.0 g/cm^3, the material becomes electrically insulating with the properties of a wide-band-gap semiconductor, including an optical band gap giving significant visible light transmission. The hardness at this density is very high, with a majority of atoms in the diamondlike (sp^3) configuration.

Thus the a-C system is a challenging system for simulation study because the carbon atoms can show bonding states ranging from sp up to sp^3. For an accurate computational model, it is therefore necessary to have a framework that can describe all bonding possibilities simultaneously. The system size may affect the results because of the wide heterogeneity in coordination number and chemical environments. In this study, we investigated structural properties of a-C using the adaptive interatomic reactive empirical bond-order (AIREBO) potential model [10], which treats both covalent bonding as well as long-range non-bonded interactions. We chose three systems with densities of 2.0, 2.6, and 3.0 g/cm^3 spanning a wide range of properties. This work addresses the structural trends in different a-C samples by comparing radial distribution functions, coordination numbers, and ring statistics [11] against existing experiments and other simulation models.

2 THE MODELS

The accuracy of atomistic molecular dynamics simulations depends directly on the use of appropriate interatomic energies and forces. These interactions are generally described using either analytical potential energy expressions or semi-empirical electronic structure methods, or obtained from a total-energy first principles theory. Though the last approach is not (as) subject to errors that can arise from assumed functional forms and parameter fitting usually required in the first two methods, there are still clear advantages to classical potentials for large systems and long simulation times.

The AIREBO potential [10] is an enhanced classical potential model for reactive covalent systems that includes both long range interaction forces and covalent bond breaking and forming. Chemical bonds vary in strength depending on the local bonding environment, via a bond-order potential. It has appropriately fitted parameters to describe the attenuation of the Lennard-Jones repulsive barrier at short distances, which would otherwise cause problems when modeling van der Waals interactions in condensed phases such as a-C. However, to date there have been no simulations of a-C using the AIREBO potential.

The simulation systems contained 2048 atoms for each density, in a simulation cell of constant volume. These systems were heated to 5000 K with a Langevin thermostat, under which conditions they were liquid and diffusive. Once molten, the samples are equilibrated at 5000 K for 100 ps. The liquid samples were then cooled down to 300 K over 10 ps for a cooling rate of 5×10^{14} K/s, comparable to previous stidues [6]. Following this, the systems were equilibrated for a further 50 ps to anneal away any transient structures. Finally, we calculated the temporal averages over another 50 ps. The final structures are shown in Fig. 1 for each density with color coding indicating the different coordination numbers.

3 RESULTS AND DISCUSSION

Figure 2 shows the radial distribution functions for the three different densities of 2.0, 2.6, and 3.0 g/cm^3 at a temperature of 300 K for the 2048-atom a-C systems. Each distribution was averaged over 50 ps after the systems were equilibrated at ambient temperature. The most interesting feature in this plot is that the positions of the first peaks of the radial distributions shift systematically towards smaller distances as the density of the system is increased. This is the reverse of the result observed in Wang and Ho's tight-binding model [5]. This result occurs because the sp^3 fraction increases to only 17% at 3.0 g/cm^3 in these studies, in contrast to Wang and Ho's results shown in Table 1. As seen in Fig. 1, the majority of the bonding topology observed here is sp^2 (threefold) at all densities. Thus, the position of the first peak of the radial distribution function results primarily from the increase in density. The small shoulder in the radial distribution function at distances slightly beyond 1.5 Å indicates the growing sp^3 fraction with increasing density (cf. Table 1). Table 1 com-

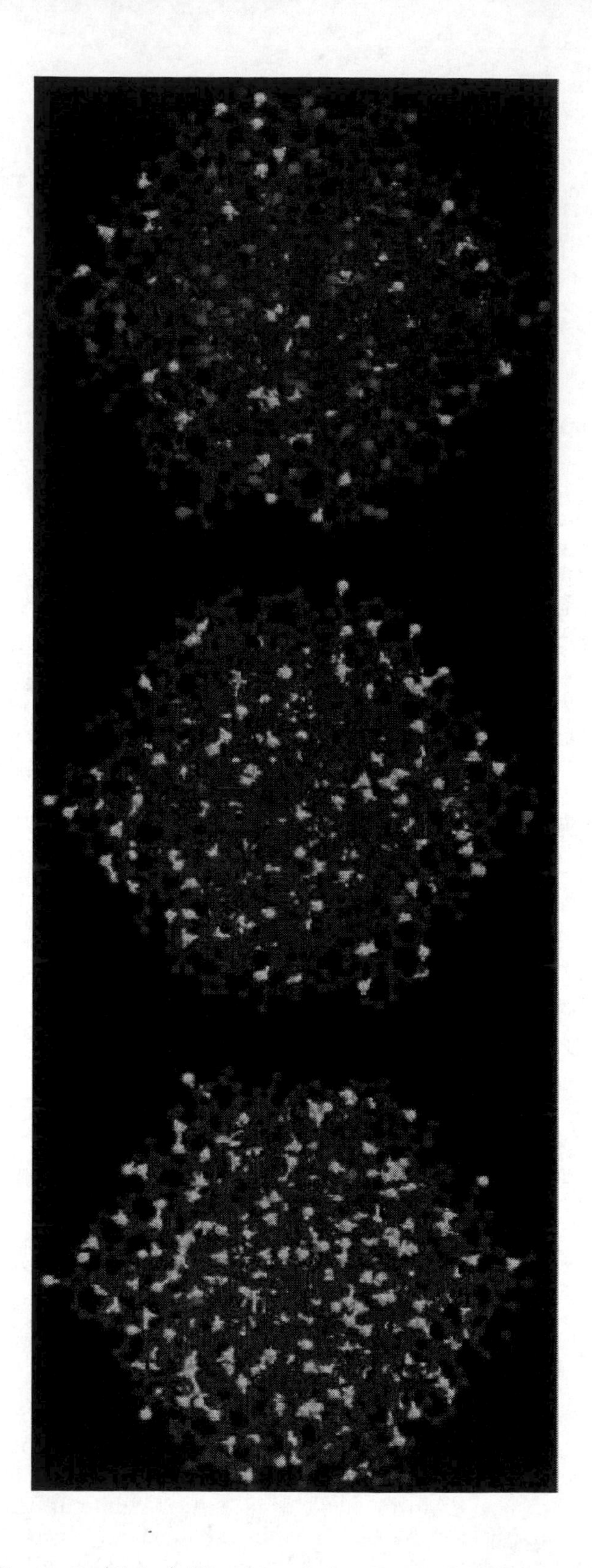

Figure 1: (Color) The annealed a-C samples. From top to bottom, the densities are 2.0, 2.6, and 3.0 g/cm^3. Colors show the coordination numbers: red for two-fold coordinated sp atoms, blue for three-fold coordinated sp^2, and green for four-fold sp^3 bonding. While the densities vary in a wide range from 2.0 to 3.0 g/cm^3, the blue sp^2 coordination dominates at all densities.

NSTI-Nanotech 2005, www.nsti.org, ISBN 0-9767985-1-4 Vol. 2, 2005

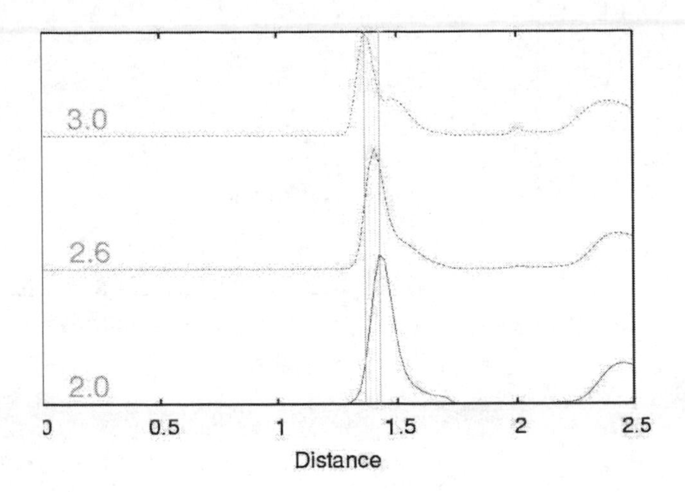

Figure 2: The radial distribution functions for densities of 2.0, 2.6, and 3.0 g/cm^3. The first peak (sp^2-bonded carbons) moves to progressively shorter distances with increasing density, as a shoulder grows in representing sp^3-bonded carbons.

pares the concentration of various coordinated atoms in a-C network with those from ab initio, tight-binding, and experiments. The chemical bonding between carbon atoms is an inherently quantum mechanical process. While the AIREBO potential includes parameters that describe the variation in covalent bonding in an average way, it does not explicitly treat electronic effects. Thus, the discrepancy between the AIREBO and other results may reflect an inability of bond-order models to capture subtle electronic effects. Simulations with the reactive empirical bond-order (REBO) model (not presented here) also show similar results. Also, the a-C is a metastable, nonequilibrium structure that is formed kinetically. Although other simulation studies have followed similar quenching procedures to generate a-C structures, these differ markedly from experimental deposition methods; the variations could be due to kinetic effects in how the structures were produced.

Ring statistics are a useful tool to investigate physical properties of amorphous solids and have become the generally accepted measure of medium-range order. However, there is unfortunately no general theory that explains which rings are relevant to the physical properties of the solid represented. In this study, we calculated the shortest-path rings, which exclude all rings with bridges that reduce their length, while preserving at least half of the original ring as defined in Ref. 11. Figure 3 shows two of the larger rings observed.

Figure 4 shows the distribution of the ring sizes for each of the three densities. For hydrocarbon systems, the ring size that exhibits the smallest strain energy for both sp^2 and sp^3 carbon is a six-membered ring. However, in this amorphous environment, there is a substantial population of rings at a wide range of sizes.

Table 1: Percentage of atoms with various coordination in the amorphous carbon networks. The density is in units of g/cm^3. KKS is this work.

ρ	Coords.	KKS	MM[1]	WH[2]	HA[3]	FE[4]
2.0	sp	6.69				
	sp^2	88.96			78.0	
	sp^3	4.3			22.0	
2.6	sp	0.44				
	sp^2	89.35			31.0	
	sp^3	10.2			67.0	
3.0	sp	0.34		2.3		
	sp^2	82.42		64.7	6.0	
	sp^3	17.23	65.0	33.0	94.0	81.0

1. Ref. 6; 2. Ref. 5; 3. Ref. 3; 4. Ref. 2.

Small 3-, 4- and 5-membered rings exist in low numbers, due to their higher strain energy. There is little dependence on density for these small rings. A higher-than-expected population of 3-membered rings indicates that the AIREBO potential underestimates the strain energy for cyclopropyl systems. The rings that occur in the highest numbers are 7-membered rings. This differs slightly from previous studies [5]–[9], [11], which found primarily 5- and 6-membered rings. We find a non-negligible number of larger rings, especially at the lowest density. In the low-density system, these represent traversals around small pockets or voids in the a-C. There are fewer voids, and thus fewer large rings, in the denser systems. This effect has also been observed in previous simulations at low densities. There is some even-odd alternation, in which rings with an odd number of members are more prevalent than even rings of about the same size. This phenomenon persists for all ring sizes and densities, and is a side effect of the method by which the shortest-path rings are defined [11]. The requirement that a shorter ring contain at least half of the atoms in a longer ring in order to disqualify it has a different effect on even and odd rings: rings of length $2N$ and $2N-1$ both need a shorter ring with at least N atoms in common in order to be disqualified. Thus a higher proportion of the even rings are disqualified, leaving a population slightly enriched in odd rings.

4 CONCLUSIONS

We have studied structural properties of a-C systems as modeled with a classical bond-order potential at three densities of 2.0, 2.6, and 3.0 g/cm^3. We used the AIREBO potential model including long range van der Waals interactions, which has computational advantages for the large 2048-atom system size, but relies on a fitted functional form.

We calculated pair distribution functions, coordination number ratios, and ring statistics. The AIREBO model showed quantitatively different results from those

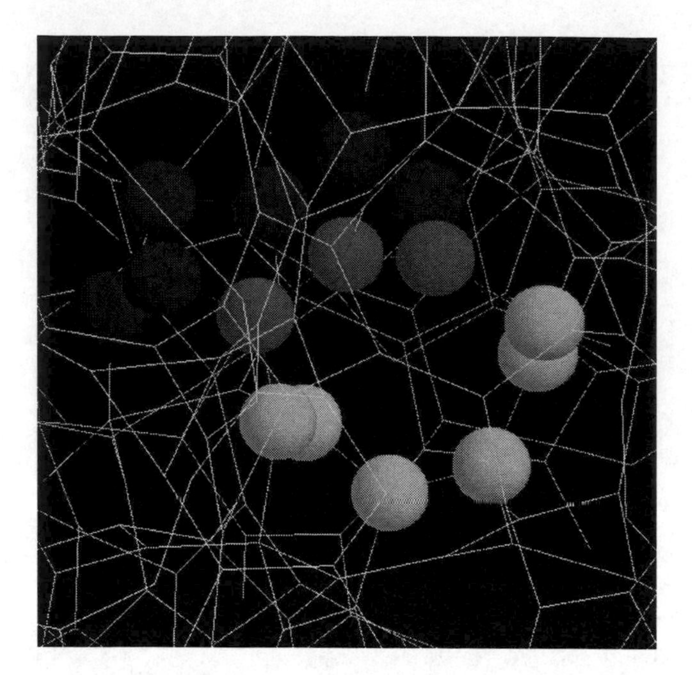

Figure 3: (Color) An example of two nine-membered (shortest-path) rings at a density of 2.0. Three atoms (red color) are common to both rings. Both rings satisfy the shortest-path condition of Ref. 11.

of experiments and other simulations such as ab initio and tight-binding models. The first peak of the radial distribution moved to shorter distances as the system density increases, in contrast to previous studies. This phenomenon is explained by the small fraction of sp^3 coordination generated by the AIREBO model. With this model, even at the highest density of 3.0 g/cm^3, the fraction of sp^3 coordination was below 20%, while values from other studies ranged from 33% to 94%.

The ring statistics showed a peak for moderate ring sizes of 7 atoms, slightly different from previous results which favored 5- and 6-membered rings. This likely indicates a different quantitative treatment of strain energy in the classical, tight-binding, and ab initio models. It is also worth noting that in previous studies, larger rings are found more commonly in studies that utilized larger numbers of atoms, so there is possibly some system size effect that has not been fully acknowledged. Small rings were not favored due to their higher strain energy, and larger rings were not favored due to the high cost of introducing voids into the a-C. The larger rings were less common in the denser systems.

It is not clear whether the differences between this study and earlier studies arose from the inability of the classical model to account for electronic effects treated in the ab initio and semi-empirical models, or whether there are differences in the kinetic pathway by which the systems were formed, or whether there are system size effects that affected earlier studies, most of which used

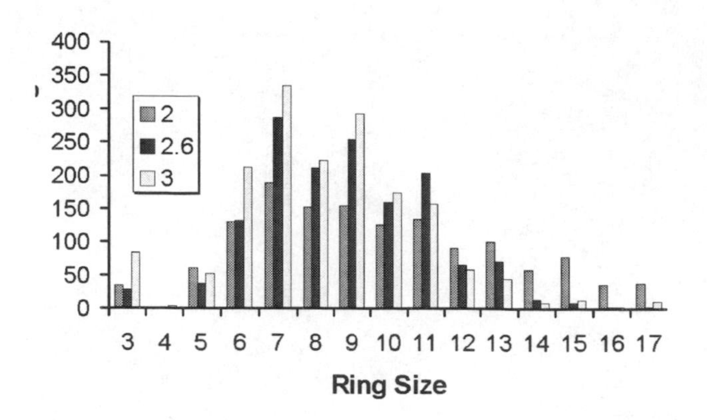

Figure 4: (Color) : Distribution of shortest-path ring sizes for each of the three densities in the AIREBO a-C systems.

216 or fewer atoms. The ability to use a classical model to study this important system would represent a significant advance, so a detailed understanding of these issues is crucial. Further study to elucidate the dependence on potential model, system size, and kinetic pathway are currently underway.

REFERENCES

[1] K. W. R. Gilkes, P. H. Gaskell, and J. Robertson, Phys. Rev B. **51**, 12303 , 1995.

[2] A. C, Ferrari, Libassi, A, Tanner, B. K, Stolojan, V, Yuan, J, Brown, L. M, Rodil, S. E, Kleinsorge, and J. Robertson, Phys. Rev. B. **62**, 11089, 2000.

[3] R. Haerle, E. Riedo, A. Pasquarello, and A. Baldereschi,Phys. Rev. B, **65**, 045101, 2001.

[4] J. Tersoff, Phys Rev. B. **37**, 6991, 1988.

[5] C. Z. Wang and K. M. Ho, Phys. Rev. B **50**, 12429, 1994.

[6] N. A. Marks, D. R. McKenzie, B. A. Pailthorpe, M. Bernasconi, and M. Parrinello, Phys. Rev. Lett. **76**, 768, 1996.

[7] D. G. McCulloch, D. R. McKenzie, and C. M. Goringe, Phys. Rev. B, **67**, 2349, 2000.

[8] G. Opletal, T. Petersen, B. O'Malley, I. Snook, D. G. McCulloch, N. A. Marks, and I. Yarovsky, Molecular Simulation, **28**, 927, 2002.

[9] C. Mathioudakis, G. Kopidakis, P. C. Kelires, C. Z. Wang, and K. M. Ho, Phys. Rev. B **70**, 125202, 2004.

[10] S. J. Stuart, A. B. Tutain, and J. A. Harrison, J. Chem. Phys. **112**, 6472, 2000.

[11] D. S. Franzblau, Phys. Rev. **44**, 4925, 1991.

Hybrid Nanomaterials for Active Electronics and Bio-Nanotechnology

Xu Wang[*], Rajeev R. Pandey[**], Roger Lake[**], Cengiz S. Ozkan[***]
[*]Department of Chemical and Environmental Engineering, xwang0@engr.ucr.edu
[**]Department of Electrical Engineering, raj@ee.ucr.edu, rlake@ee.ucr.edu
[***]Department of Mechanical Engineering,cozkan@engr.ucr.edu
University of California Riverside, Riverside, CA 92521, USA

ABSTRACT

In this paper, the metallization of single-walled carbon nanotube/single stranded DNA has been reported. At first, SWCNT-ssDNA conjugation is produced, using a well-known chemical pathway which is based on attaching amino compounds to carboxyl-terminal carbon nanotubes. Then the conjugation is metallized with platinum particles. We prove that although the addition of platinum particles has an effect on DNA by Raman spectrum, we still get perfect metallized SWCNT-ssDNA hybrid, which is characterized by SEM, TEM and FTIR. In addition, we have analyzed the metallization mechanism of SWCNT-ssDNA assemblies with HOMO-LUMO calculations.

Keywords: DNA, hybrid, metallization, platinum

1 INTRODUCTION

Single-walled carbon nanotubes (SWCNTs) exhibit special structural features and unique electronic properties that have focused interest in their application as active components in future solid state nanoelectronics[1] and optoelectronics[2]. However, the difficulty in precise localization and interconnection of nanotubes impedes further progress toward larger-scale integrated circuits. Self-assembly based on molecular recognition provides a promising approach for constructing complex architectures from molecular building blocks, such as SWNTs, bypassing the need for precise nanofabrication and mechanical manipulations[3]. DNA is a naturally occurring polymer that plays a central role in biology. Many unique properties of DNA have inspired a search for non-biological applications for it. Molecular recognition between complementary strands of a double-stranded DNA has been used to construct various geometric objects at the nanometer scale[4].The stacking interaction between bases in DNA has prompted the exploration of its electronic properties for possible use in molecular electronics[5].So far, building nanoscale structures with DNA is getting more and more attention. It is stimulated by the enormous progress in utilizing DNA either for the controlled self-assembly of molecularly designed nanostructures[6], or as template for thoroughly engineered inorganic structures[7]. These studies illuminate our work focusing on developing a controllable assembly system about metallization of SWCNT-ssDNA conjugation. In our work, we report the first time for metallizing SWCNT-ssDNA assembly with platinum. Considering there will be some effect on DNA structure after being metallized, we first synthesized the SWCNT-ssDNA assemblies and then metallized them. Metallized SWCNT-ssDNA complex will get more electrical properties. This will be largely stimulated by many proposed applications of SWCNT-ssDNA assemblies in future nanoscale electronic devices based on their unique electrical properties, biological properties and nanometer sizes.

2 EXPERIMENT AND DISCUSSION

Single wall carbon nanotubes are purchased from Nanolab, INC. One strand DNA is from Sigma Genosys with nine bases 5' (NH_2)GCATCTACG. Single wall carbon nanotubes are oxidized in HNO_3. Oxidization will produce enough carboxyl groups at the end of CNTs, which changes the ends of the CNTs from hydrophobic to hydrophilic. DNA will react with carboxylic acid groups on SWNTs by the amine group at the end. Scheme 1 illustrates the basic chemical pathway which we have used in this work. For the SWCNT-ssDNA conjugation, we use 1-ethyl-3-(3-dimethylaminopropyl) carbodiimide (EDC, Pierce Chemicals, Inc.) in the presence of N-hydroxysuccinimide (sulfo-NHS, Pierce Chemicals.Inc.) in DI water. SWCNT-ssDNA conjugates are characterized by scanning electron microscopy and transmission electron microscopy. Figure 1A, B shows SEM images of SWCNT-ssDNA conjugates. The ends of the oxidized SWCNTs produce multiple carboxylic groups, which results in the conjugation of DNA at the ends. SWCNT-ssDNA conjugation metallization is accomplished in a two-step chemical reduction and deposition of metallic colloids[8]. Metallic clusters are first formed on the DNA, and then used as nucleation sites for selective metal deposition. Second, metal ions are reducted into metallic particles. The metallization process is uniform over the entire DNA. Fig1C, D shows SEM images of SWCNT-ssDNA conjugates metallized by platinum. We notice that a lot of small particles distribute along the sidewall of CNTs, which might be Pt particles absorbed on or inserted in CNTs. In order to prove this, we take TEM images and EDS analysis. From Fig2A, DNAPt cluster is located at the end of CNT. Except that, there are many black particles at the surface of CNTs. From the high resolution image, Fig2B, it is clear to see ssCNT is surrounded by black grains. And its Energy Dispersive Spectroscopy (EDS) tells us those black grains are Pt particles. Pt particles on ssDNA and ssCNT help to increase the electrical conductivity of SWCNT-ssDNA conjugates. Our future research will focus on the electrical characterization of metallized SWCNT-ssDNA.

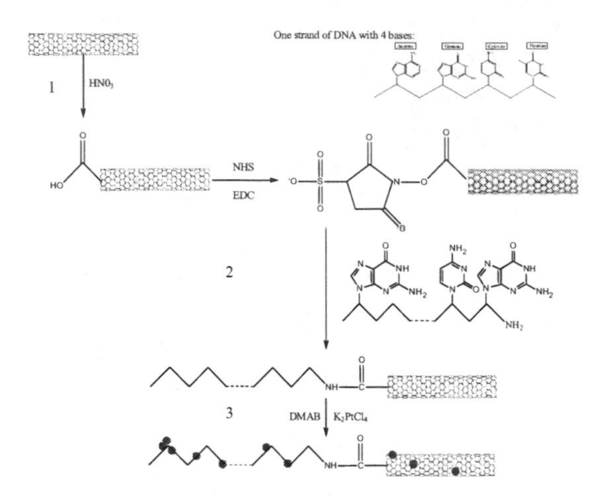

Scheme 1: Chemical synthesis of metallized SWCNT-ssDNA hybrid. Step 1 is the selective oxidation of CNT at ends to produce carboxyls. In step 2, DNA is chemically connected with CNTs at the end by amide group. Step 3 shows that in the presence of the reductant DMAB, Pt nanoparticles are bound to both DNA and CNT.

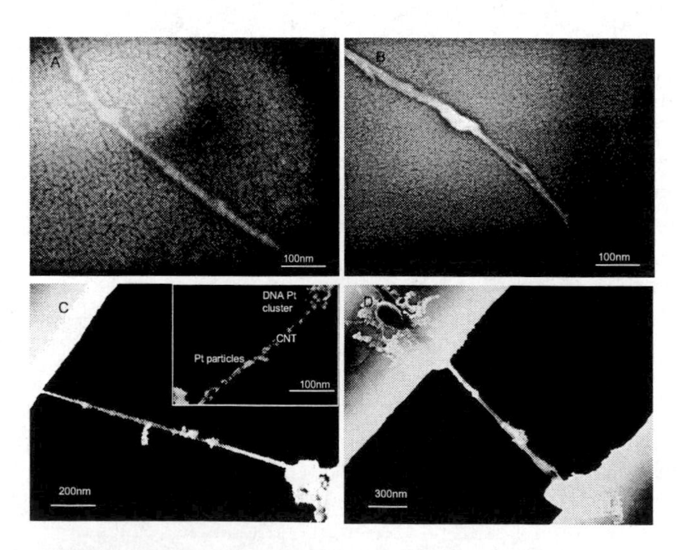

Figure 1: SEM images of SWCNT-ssDNA conjugation (A, B) and metallized SWCNT-ssDNA hybrid (C, D). (A,B) Two CNTs are connected by DNA, and only CNT's end functionalization occurs. (C,D) Metallized SWCNT-ssDNA hybrid traverses across interconnect lines on a Si/SiO_2 substrate, and its high-resolution SEM image.

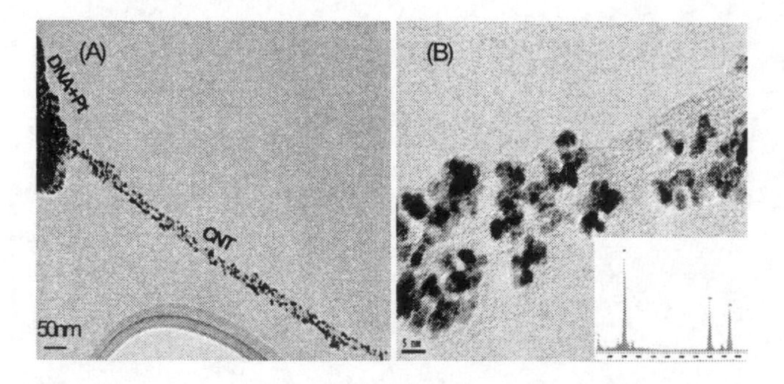

Figure 2: (A) TEM image of metallized SWCNT-ssDNA hybrid. The end of CNT is connected by DNAPt cluster and Pt particles locate on SWCNT-ssDNA surface. (B) High resolution of metallized SWCNT-ssDNA. Its EDS shows that except of carbon and copper, there are two strong platinum peaks. So we can tell the sample is made up of Pt and C.

In order to explore the effect that Pt has on DNA and the metallization mechanism, we use Raman Spectrum, Infrared Spectroscopy, and HOMO-LUMO computation to analyze. Raman spectroscopy provides a powerful tool to identify specific structure in samples. In Fig3, the red line is SWCNT-ssDNA spectrum, while the black one is metallized SWCNT-ssDNA spectrum. There are four peaks in SWCNT-ssDNA conjugation, 1594 cm^{-1} is the G-band of CNT, 1508 cm^{-1} is base electronic structure in DNA (stacking,ligtion,ect), 1361 cm^{-1} is the D-band of CNT, and 1153 cm^{-1} is PO_2^{-1} interaction in DNA backbone[9,10]. There are only two peaks in metallized SWCNT-ssDNA hybrid. The peaks of D-band and G-band of CNT still exist. The peaks of base electronic structure in DNA and $PO2^{-1}$ disappear. It shows that there are some changes for DNA's structure due to the existence of Pt nanoparticles. With infrared spectroscopy (FTIR), we find pronounced amide peak at 1646.9 cm^{-1}, which means covalent attachment between carboxyls on oxidized ends of CNTs and amine groups on DNA's end. In addition, we also find that 2971.7 cm^{-1} is the peak of C-H in aromatic ring, 1714.4 cm^{-1} is the peak of C=O in –COOH[11], shown in Fig4. So the addition of Pt does not affect SWCNT-ssDNA assembly.

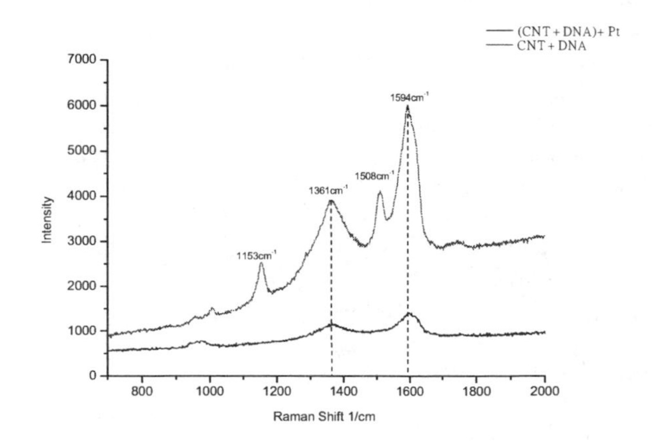

Figure 3: Raman spectrum of (SWCNT-ssDNA)-Pt/ SWCNT-ssDNA assembly.

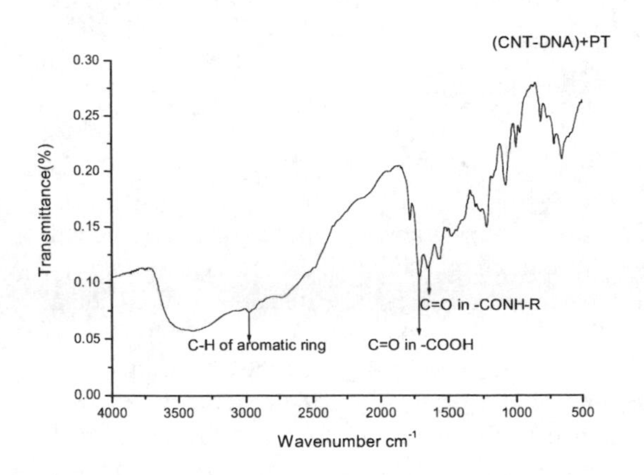

Figure 4: Fourier transform infrared spectra metallized SWCNT-ssDNA hygrids.

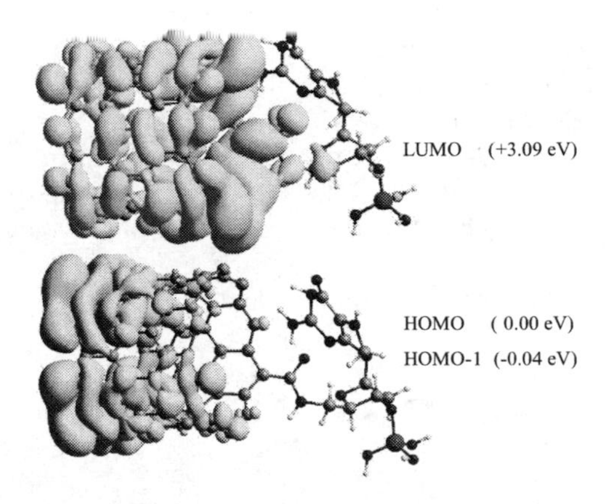

Figure 5: HOMO-LUMO calculation of CNT linked by an amide linkage to a deoxyribose nucleoside with Guanine base

3 CONCLUSION

We have shown that SWCNT/ssDNA is metallized with Platinum successfully. Because of the mild and well-controlled oxidation of the SWCNT, conjugation occurs at the ends of SWCNTs, which is very important in keeping the electronic properties of the CNTs. Although metallization has an effect on ssDNA's structure, it will not destroy SWCNT-ssDNA assemble's structure. Platinum has affinities for both DNA and CNT surfaces. Therefore, metallized SWCNT-ssDNA hybrid can improve the conductivity of SWCNT-ssDNA conjugation and can be used as building blocks for various nanoscale electronic devices.

REFERENCES

[1]. K.Tsukagoshi, N. Yoneya,S. Uryu,Y.Aoyagi,A.Kanda, Y.Ootuka,B.W. Alphenaar, Physica B: Condensed Matter 2002 ,323 ,107-114.

[2]. L.Dai,P.He,S.Li,Nanotechnology 2003 ,14 ,1081-1097.

[3]. J. R. Heath, M. A. Ratner, *Phys. Today* 2003, 43 (May 2003).

[4]. Seeman, N. C.DNA engineering and its application to nanotechnology. *Trends Biotech.* 17, 437–443 (1999).

[5]. Arkin, M. R. *et al.* Rates of DNA-mediated electron transfer between metallointercalators. *Science* 273, 475–480 (1996).

[6]. N. C. Seeman, Annu. Rev. Biophys. Biomol. Struct. 27, 225 ~1998.

[7]. J. L. Coffer, S. R. Bigham, X. Li, R. F. Pinizzotto, Y. Rho, G. Young, R. M. Pirtle, and I. L. Pirtle, Appl. Phys. Lett. 69, 3851 ~1996!.

[8]. W. Pompe, M. Mertig, R. Kirsch, R. Wahl, L. C. Ciacchi, J. Richter, R.Seidel, and H. Vinzelberg, Z. Metallkd. 90, 1085 (1999).

[9]. Raman spectroscopy in biology : principles and applications / Anthony T. Tu. Publisher New York : Wiley, c1982.

[10]. Jorio, A; Souza Filho, AG; Dresselhaus,G; Dresselhaus,MS; Swan,AK; Goldberg,BB; Pimenta,MA; Hafner,JH; Lieber,CM. Phys.Rev.B, 2002,65(15),155412

[11]. Infrared spectroscopy: applications in organic chemistry [by] Margareta Avram [and] Gh. D. Mateescu. Translated by Ludmila Bîrlădeanu.

[12]. Wadt, W. R.; Hay, P. J. Ab initio effective core potentials for molecular calculations. Potentials for main group elements sodium to bismuth. *J. Chem. Phys.* 1985, 82, 284-298.

[13]. Frisch, M. J. *et al*, Gaussian Inc., Pittsburgh PA 2003

The Structure and Properties of ta-C Film with Dispersion of Incident Beam Energy

S.-H. Lee[*,**], T.-Y. Kim[*], S.-C. Lee[*], Y.-C. Chung[**], D.W. Brenner[***] and K.-R. Lee[*]
[*] Future Technology Research Division, Korea Institute of Science and Technology,
P.O.Box 131, Cheongryang, Seoul, 130-650, Korea, krlee@kist.re.kr
[**] Hanyang University, Seoul, Korea
[***] North Carolina State University, Raleigh, NC, USA

ABSTRACT

Structures and properties of tetrahedral amorphous carbon (ta-C) films are investigated as a function of Gaussian distribution of incident carbon beam energy. The ta-C films are synthesized by controlling the standard deviation (σ) of Gaussian distribution from 0 to 10. The Brenner type interatomic potential was used for carbon-carbon interaction. Densities (3.14 ± 0.03 g/cm^3) and sp^3 bond fractions (53.7 ±1.7 %) were not significantly changed with varying standard deviations (σ). On the contrary, the compressive residual stresses of ta-C film were changed remarkably with changing the standard deviation. The residual stress was reduced from 6.0 to 4.2 GPa with the standard deviations (σ). The decrease of residual stress corresponds to the disappearance of a satellite peak of the second nearest neighbor of the radial distribution function. The relationship between the structure of ta-C film and dispersion of incident beam energy was investigated.

Keywords: Tetrahedral amorphous carbon, Molecular dynamics simulation, Dispersed incident energy, Compressive residual stress

1 INTRODUCTIONS

Tetrahedral amorphous carbon (ta-C) films made by filtered vacuum arc of graphite have been used for the protective coatings of various tools and components owing to their superior mechanical and optical properties[1-4]. However, the residual compressive stress should be reduced for the applications since the high level of the residual stress induces spontaneous delamination of the film from a substrate. Therefore, many studies focused on the structural factor to be significant to control the residual stress.

Molecular dynamics studies are very useful methods to investigate the structural properties and stress behavior of ta-C film in the atomic scale. Growth behavior of ta-C film was successfully simulated by using energetic carbon atoms which bombard the diamond or silicon substrate[5,6]. However, most previous simulation studies have dealt with monochromatic kinetic energy of the deposited carbon atoms. Monochromatic incident energy is suitable for modeling calculation of deposition system. However, the incident energy is not monochromatic in real situation. In the case of filtered cathodic vacuum arc (FCVA) system, the incident carbon ion has the energy range of incident beam. It is the difference between the most experimental conditions and modeling calculations. It is required to investigate the effect of the energy dispersion of incident beam.

In the present work, we synthesized the ta-C films with the varying the standard deviation of the incident energy, which is assumed to follow the Gaussian distribution with standard deviation (σ). In this case, we investigated the structures and properties such as density, sp^3 bond fraction, radial distribution function (RDF) and residual stress of ta-C film and compare with that of ta-C film using by monochromatic incident energy. The structure, density and sp^3 bond fraction did not show significant behavior. However, the residual stress of ta-C film changed with standard deviation (σ) of Gaussian distribution. We observed the satellite peak of the second nearest peak from radial distribution function showed the same tendency with residual stress of ta-C film. We investigated the physical meaning of these satellite peaks and the relationship between these satellite peaks and the residual stress of ta-C film.

2 CALCULATIONS

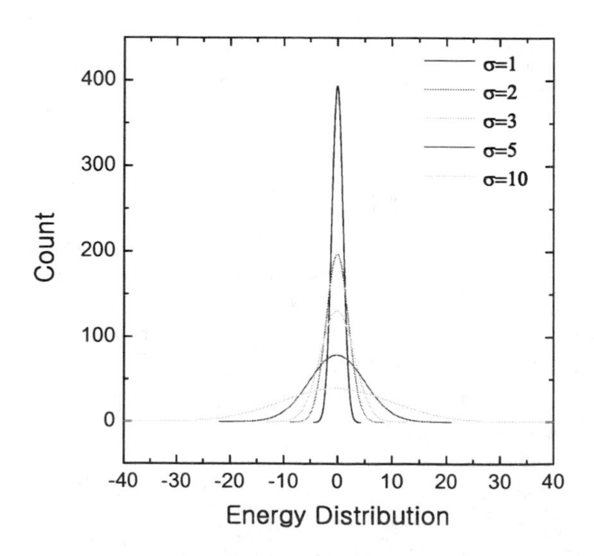

Figure 1. Gaussian distribution with standard deviation (σ)

The initial substrate is of a perfect $6a_0 \times 5\, a_0 \times 6\, a_0$ (Here, a_0 is equilibrium lattice constant of bulk diamond 3.567 Å) diamond lattice consisting of 1512 carbon atoms with 72 atoms per layer. Before the deposition, the substrate is sufficiently relaxed to its minimum potential configuration and almost equilibrated at 300 K for 1 ps. The ta-C films were produced on a diamond (100) surface by bombardment of 3000 carbon neutrals with the average incident energy (75eV). Energetic carbon neutrals are deposited onto the substrate with normal incidence and randomly chosen location. The Brenner interatomic potential was used for the carbon atoms[7,8].

In order to investigate the structural properties of ta-C film as a function of the dispersion in the incident kinetic energy, the ta-C film was growth using by the Gaussian distribution function for incident carbon atom energy. The highly dense and residual stressed ta-C films were synthesized, when the incident energy is between 50 and 100 eV. Therefore, the average valuc of incident energy was chosen 75 eV. The standard deviation (σ) of the Gaussian distribution was controlled from 0 to 10 as shown in Fig. 1. The incident energy of the each atoms is dispersed with the standard deviation (σ) of the Gaussian distribution function.

3 RESULTS AND DISCUSSIONS

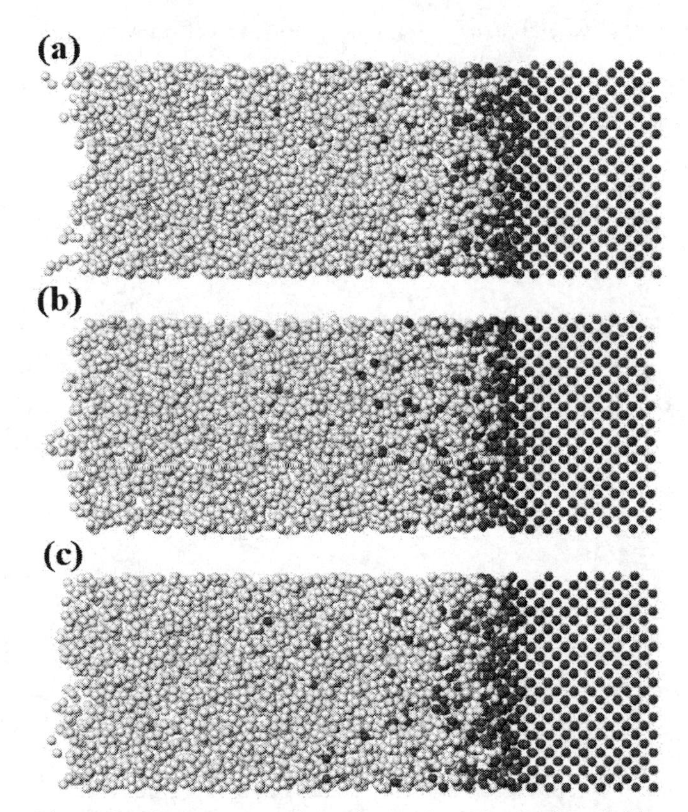

Figure 2. The atom configurations of synthesized ta-C films at each standard deviation (σ) 0 (a) 3 (b) 10 (c)

Figure 2 shows the atoms configuration of synthesized ta-C films at each standard deviation (σ) after 3000 carbon atoms deposited to substrate. White and grey balls of the configuration, which are shown in Fig. 2, represent incident carbon atoms and substrate carbon atoms, respectively. Due to the average incident energy (75 eV) is much higher than the cohesive energy of diamond (7.4-7.7 eV/atom)[9,10], intermixing occurs between film and substrate. The incident atoms bombarded to the substrate atom and the substrate surface was amorphized due to the significant agitation of the substrate atoms. Since the amorphous structure generated by the bombardment cannot return to their equilibrium phase at this temperature, a highly stressed and dense surface layer was formed on the substrate surface. However, the atom configurations with the standard deviation did changed largely that is shown in Fig. 2.

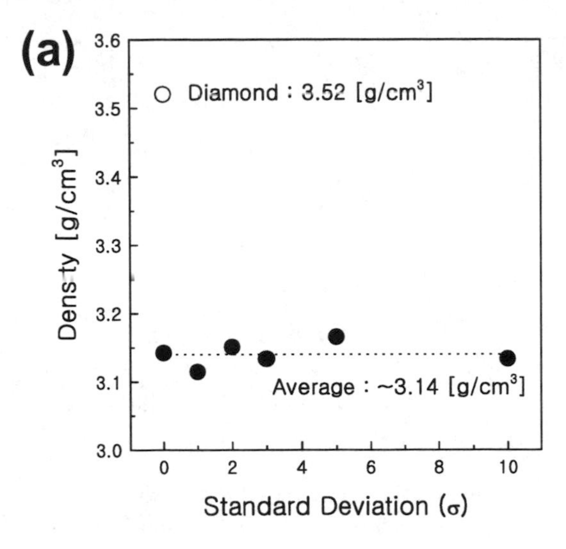

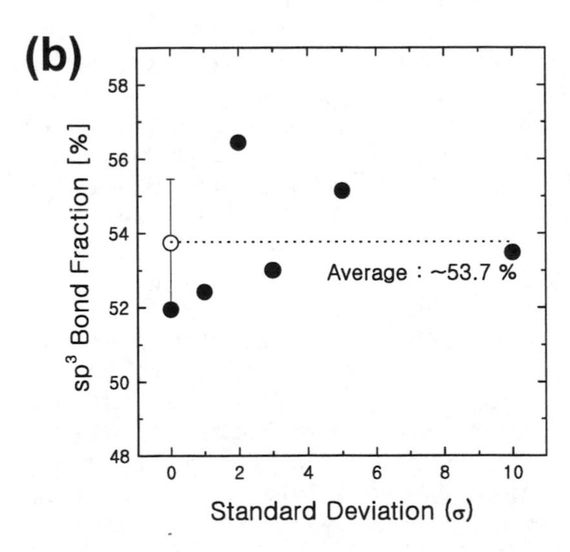

Figure 3. Density (a) and sp^3 bond fraction (b) of ta-C film with standard deviation (σ)

Figure 3 shows the behavior of density and the sp^3 bond fraction with standard deviation (σ). The average density and the sp^3 bond fraction of ta-C films deposited are 3.14 g/cm^3 and 54 %, respectively. They did not show any significant changes as the standard deviation varied.

However, the residual stress decreased with the standard deviation of incident beam energy and exhibited a minimum when the standard deviation (σ) was 3 as shown in Fig. 4.

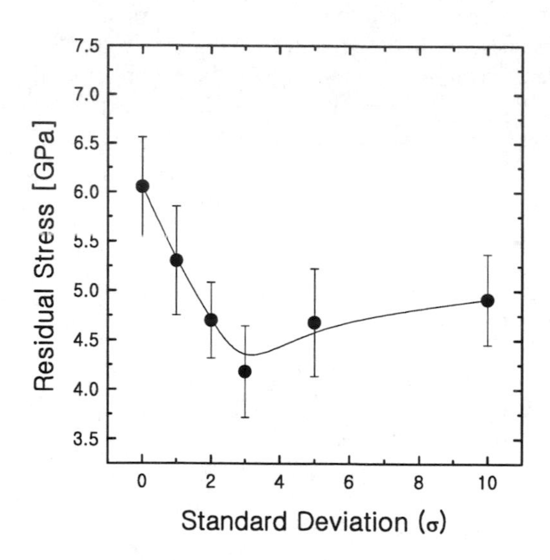

Figure 4. Residual stress behavior of ta-C film with standard deviation (σ)

It has been generally believed that the residual stress has a close relationship with the density and sp^3 bond fraction of film. It is well known that the high residual stress of ta-C film is due to high density and sp^3 bond fraction of ta-C film. However, in spite of the reduction of residual stress of ta-C from 6.0 to 4.2 GPa, the density and sp^3 bond fraction did not significantly change. Therefore, the reduction of the residual stress cannot be understood from the density and sp^3 bond fraction arguments.

In order to reveal the reason for the stress reduction, the radial distribution function (RDF) of these films were investigated. Figure 5 showed the radial distribution functions of ta-C film with standard deviation (σ) except the substrate atoms and surface atoms of ta-C film. The plots clear show the typical behavior of the amorphous structures of ta-C film. The satellite peak of second nearest peak (as indicated as arrow in Fig. 5) was observed. These peaks can be also observed from the previous molecular dynamics studies[5,9,12,13]. In our previous work suggested that the meta-stable state is generated by the localized thermal spike due to the collision of the high-energy carbon atoms, which induced the distortion of bond angle and bond distance between carbon atoms[14]. The satellite peak of second nearest neighbor represented the number of the strained bond defects by distorted the bond length and angle, when the ta-C films were synthesized. Therefore, the intensity of strained bond defects has the relationship between the residual stress of ta-C film. The intensity of the satellite peak at 2.1 Å of the second nearest neighbors showed the same tendency with stress behavior of ta-C film as shown in Fig. 6.

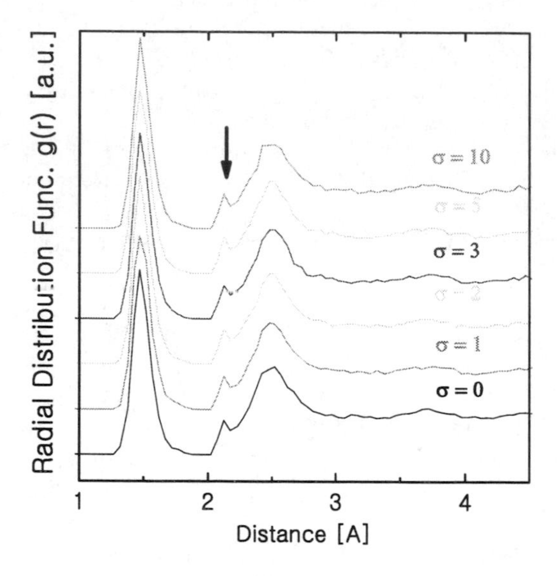

Figure 5. Radial distribution functions of ta-C film with standard deviation (σ)

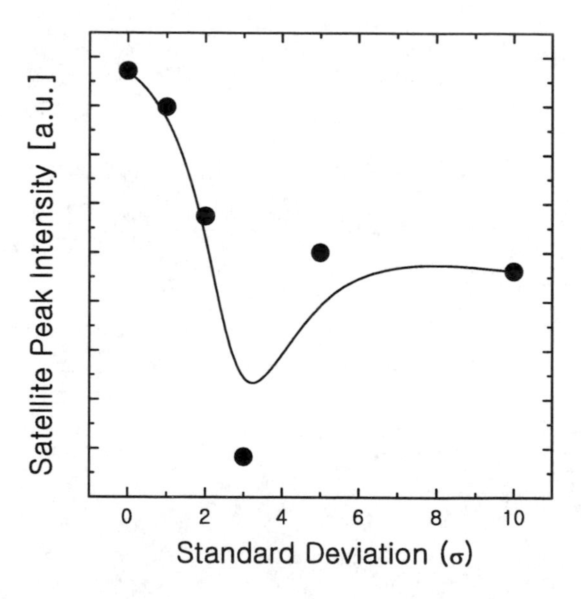

Figure 6. Satellite peak intensity behavior of ta-C film with standard deviation (σ)

4 CONCLUSIONS

We investigated the structure and properties of ta-C films using dispersed incident energy instead of monochromatic incident energy. The density and sp^3 bond fraction of synthesized ta-C film did not showed significant behaviors with standard deviation (σ). However, the residual stress of ta-C films reduced and had the minimum value when the standard deviation (σ) was 3. The intensity of satellite peak of second nearest peak, which means that of strained bond defects, shows the same tendency with residual stress. The strained bond defects are also the one of the reason of generation residual stress of ta-C film. The present simulation result shows that the residual stress can be reduced by controlling the number of strained bond defects. It is provided by the energy dispersion of the incident carbon atom

.

REFERENCES

[1] R. Lossy, D.L. Pappas, R.A. Roy, J.J. Cuomo and J. Bruley, J. Appl. Phys., 77, 4750, 1995.

[2] R.L. Boxman, V. Zhitomirsky, B. Alterkop, E. Gidalevich, I. Beilis, M. Keidar and S. Goldsmith, Surf. Coat. Technol., 86-87,243, 1996.

[3] N. Mustapha and R.P. Howson, Surf. Coat. Technol., 92, 29, 1997.

[4] H.C. Miller, J. Appl. Phys., 52, 4523, 1981,

[5] H.P. Kaukonen and R.M. Nieminen, Phys. Rev. Lett., 68, 620, 1992.

[6] M. Kaukonen and R.M. Nieminen, Phys. Rev. B, 61, 2806, 2000.

[7] D.W. Brenner, Phys. Rev. B, 42, 9458, 1990.

[8] D.W. Brenner, Phys. Status. Solidi., 217, 23, 2000.

[9] H.U. Jager and K. Albe, J. Appl. Phys., 88, 1129, 2000.

[10] P.C. Kelires, Phys. Rev. B, 47, 1829, 1993.

[11] M.T. Yin and M.L. Cohen, Phys. Rev. B 29, 6996, 1984.

[12] N.C. Cooper, M.S. Fagan, C.M. Goringe, N.A. Marks and D.R. Mckenzie, J. Phys. Condens. Matter, 14, 723, 2002.

[13] N.A. Marks, J. Phys. Condens. Matter, 14, 2901, 2002.

[14] S.H. Lee, C.S. Lee, S.C. Lee, K.H. Lee and K.R. Lee, Surf. Coat. Technol., 177-178, 812, 2004.

Computing Fullerene Encapsulation of Non-Metallic Molecules: $N_2@C_{60}$

Zdeněk Slanina*,** and Shigeru Nagase*

*Department of Theoretical Molecular Science
Institute for Molecular Science
Myodaiji, Okazaki 444-8585, Aichi, Japan
**Institute of Chemistry, Academia Sinica
128 Yen-Chiu-Yuan Rd., Sec. 2
Nankang, Taipei 11529, Taiwan - ROC

ABSTRACT

As $N_2@C_{60}$ has been observed, there is also a clear interest in computational description. Owing to a relationship to $N@C_{60}$, they should have some potential for nanoscience studies. The computations deal with a search for stationary points by geometry optimizations using density-functional theory (DFT) methods. In the optimized stationary points, the analytical harmonic vibrational analysis was carried out in order to check their physical nature. The lowest-energy structure has the N_2 unit oriented towards a pair of parallel pentagons so that the complex exhibits D_{5d} symmetry. The highest computational level estimates the stabilization upon encapsulation by an energy term of -9.3 kcal/mol. Moreover, the entropy term for the encapsulation is also evaluated which leads to the standard Gibbs-energy change upon encapsulation at room temperature of -3.3 kcal/mol. The computed structural and vibrational characteristics are reported, too, and possibilities for encapsulation of other small molecules are discussed.

Keywords: Carbon-based nanotechnology; molecular electronics; non-metal fullerene encapsulations; optimized syntheses; Gibbs-energy evaluations.

1 INTRODUCTION

$N_2@C_{60}$ and $N_2@C_{70}$ were first prepared by Peres *et al.* [1] using heating under high pressure. Out of two thousand C_{60} molecules about one was observed to incorporate N_2. The nitrogen molecule containing endohedrals were present even after several-hours heating at 500 K. $N_2@C_{60}$ was also reported [2,3] in the chromatographic separation after the nitrogen-ion implantation into C_{60}. This ion bombardment is primarily used for the $N@C_{60}$ production [4,5] though with very low yields ($Li@C_{60}$ was also prepared by the ion bombardment method [6]). Still, $N@C_{60}$ and its derivatives have been studied vigorously [7-11]. Other non-metallic endohedrals are represented by complexes of fullerenes with rare gas atoms, in particular with He [12-16]. Very recently, molecular hydrogen [17] and even a water molecule [18] were placed inside open-cage fullerenes. This paper presents computations on $N_2@C_{60}$ in order to judge its stability and compare with other available endohedral data [19-23], especially for $N@C_{60}$.

** *The corresponding author e-mail: zdenek@ims.ac.jp*

2 COMPUTATIONS

The computations started with a search for stationary points by geometry optimizations using density-functional theory (DFT) methods. The standard 3-21G basis set was used at this stage and applied in two different DFT approaches. One of them employs Becke's [24] three parameter exchange functional with the non-local Lee-Yang-Parr [25] correlation functional (B3LYP) with the above basis set, i.e., the B3LYP/3-21G treatment. The other approach uses Perdew and Wang's [26] exchange and correlation functionals known as PW91. The geometry optimizations were carried out using the analytical energy gradient. In the optimized stationary points, the analytical harmonic vibrational analysis was carried out in order to check their physical nature by the number of imaginary vibrational frequencies. The ultra-fine grid in numerical integrations of the DFT functional and the tight SCF convergency criterion were used. The reported computations have been carried out with the Gaussian program package [27].

In the optimized B3LYP/3-21G and PW91/3-21G geometries, additional single point energy calculations B3LYP/6-31G* and PW91/6-31G*, respectively, were carried out using the standard 6-31G* basis set. Moreover, the second order Møller-Plesset (MP2) perturbation treatment [28,29] with frozen core (FC) option, in which inner-shells are excluded from the correlation calculation, was also performed with the same basis set (MP2=FC/6-31G*). The basis set superposition error (BSSE) was estimated by the Boys-Bernardi counterpoise method [30].

In addition to the interaction (dimerization) energy, the standard entropy change upon the encapsulation was also evaluated (the 1 atm standard-state choice). The conventional rigid-rotator and harmonic-oscillator partition functions [31] were used and no frequency scaling was applied.

3 RESULTS AND DISCUSSION

The B3LYP treatment is considered as the most reliable choice for the thermochemical computations [32] and the B3LYP functional has been the most common DFT option in the computations [33,34] of fullerenes and endohedrals. However, it was pointed out recently [35,36] that the Becke exchange functional due to its erroneous asymptotic behavior does not produce physically acceptable description of the attraction between weakly bonded systems. It was shown [37] that the

BLYP or B3LYP treatments do not show weak energy minima in the long-range region otherwise well established by the MP2 or coupled cluster calculations. It was suggested [35-37] to use Perdew and Wang's [26] exchange and correlation functionals PW91 instead. The PW91 functional at least partly recovers the attraction [37]. Its error in the computed binding energies [37] for the hydrogen-bonded complexes was 20% in the worst case though somewhat bigger in some still weaker complexes. With this picture, the DFT computations here are carried out side by side with the B3LYP and PW91 functionals.

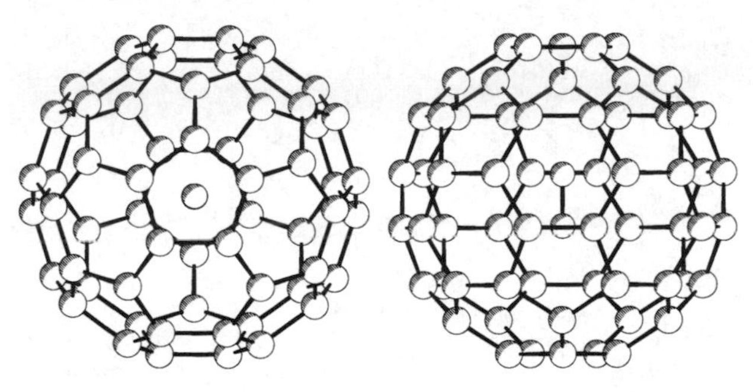

Fig. 1. Two views of the PW91/3-21G optimized structure of $N_2@C_{60}$

In both the B3LYP/3-21G and PW91/3-21G treatments, the lowest-energy structure $N_2@C_{60}$ found has the N_2 unit oriented towards a pair of parallel pentagons so that the complex exhibits D_{5d} symmetry (5:5 structure, see Fig. 1). The stationary point was reached without symmetry constraints so that some numerical inaccuracy could influence the point-group determination. In fact, the G98 symmetry finder extracts only C_i symmetry. However, application of a procedure [38] allowing for variable precision of the optimized coordinates concludes the D_{5d} symmetry. Moreover, N_2 is to exercise motions around the equilibrium position with an averaging effect. It has been known [39] that more symmetric structures for endohedrals can in fact be saddle points. Then, it is sometimes actually more effective to optimize their structures without symmetry constraints. In any case, the vibrational analysis is an important check for the optimized endohedral structures. The harmonic vibrational analysis confirms that the 5:5 structure is indeed a local energy minimum though the interaction hypersurface is very shallow.

Another optimized $N_2@C_{60}$ stationary point has the N_2 unit oriented towards a pair of parallel hexagons - a 6:6 structure. The 6:6 structure is however not a local energy minimum but a saddle point. Still, it is very close in energy to the 5:5 structure. In both the B3LYP/3-21G and PW91/3-21G approaches the 6:6 species is located less than 0.1 kcal/mol above the 5:5 structure. This feature documents that the interaction hypersurface is very shallow and at elevated temperatures the N_2 unit should exhibit large-amplitude motions. In the 5:5 equilibrium position the closest computed N-C distance is 3.046 and 3.049 Å in the B3LYP/3-21G and PW91/3-21G method, respectively. The Mulliken partial charge on the N atoms is positive but very small, about 0.01, so that the charge transfer to the cage is negligible.

The computed interaction or stabilization energies have thermochemically the meaning of the potential-energy change along the formal encapsulation process:

$$N_2 + C_{60} = N_2@C_{60}. \qquad (1)$$

In the B3LYP/3-21G geometry, the interaction energies are positive before and after the BSSE correction, this being true for both the B3LYP/3-21G and B3LYP/6-31G* approaches. Hence, this computational level would not speak for a $N_2@C_{60}$ formation. In the PW91/3-21G geometry, there is a stabilization effect at the PW91/3-21G level before the BSSE correction, but vanishes after the Boys-Bernardi counterpoise method is applied. The PW91/6-31G* approach does not show a stabilization even before the BSSE correction. There is however a substantial change at the MP2=FC/6-31G* level with a stabilization of -17.5 kcal/mol before the BSSE correction. This value corresponds to the -7.3 kcal/mol computed [22] at the same level for $N@C_{60}$. After addition of the BSSE correction, the MP2=FC/6-31G* stabilization for $N_2@C_{60}$ is reduced to still significant -9.3 kcal/mol. It is however likely [21] that the value will still be modified when larger basis-set computations will be possible.

There are 180 vibrational modes in $N_2@C_{60}$. One of them is the N-N bond stretching mode, five are vibrations of the N_2 unit against the cage, and the remaining 174 modes are skeletal vibrations of the C_{60} cage. The N-N bond stretching frequency is only slightly affected by the encapsulation. Interestingly enough the N-N frequency increases in the B3LYP/3-21G approach but decreases in the PW91/3-21G treatment. The shift is large enough to be significant in observation. The changes in the N-N frequency upon encapsulation are parallel with the changes in the N-N bond distance as in the B3LYP/3-21G computations the bond is slightly compressed while in the PW91/3-21G case the bond is expanded by about 0.0003 Å. The five frequencies for the N_2-cage vibrations are low but they could be observed as it was recently possible [41,42] for $La@C_{82}$. The inter-system frequencies at the B3LYP/3-21G level are 25.4, 38.9, 83.6, 84.6, and 126.1 cm^{-1} while for the PW91/3-21G approach they read 45.9, 49.1, 55.9, 65.3, and 106.7 cm^{-1}.

The high symmetry of C_{60} considerably simplifies [43] its vibrational spectrum, for example, only four T_{1u} three-fold degenerate modes are actually active in its IR spectrum. Once the high symmetry is reduced upon introduction of N_2, the symmetry selection rules will not be, strictly speaking, the same. However, in fact out of the 174 cage modes in $N_2@C_{60}$ just twelve show significant IR intensities. Those IR active modes are actually rooted in the original T_{1u} three-fold degenerate modes of C_{60}. For example, at the B3LYP/3-21G level we deal with the following four frequency triads: (508.4; 508.8; 509.8), (581.6; 581.6; 581.7), (1175.6; 1175.8; 1177.0), and (1451.4; 1451.5; 1451.7) cm^{-1}. The splitting could be distinguished in observed vibrational spectra.

The stabilization energy from Eq. (1) is the potential-energy change along the reaction. In order to get the related enthalpy change ΔH^o_T at a temperature T, the zero-point vibrational energy and heat-content functions are to be added though the change will be small. Once the corresponding entropy change ΔS^o_T is evaluated, one can deal with the thermodynamics controlling Gibbs-energy term ΔG^o_T. Using the partition functions from the PW91/3-21G calculations, the

$T\Delta S^o_T$ term comes as -5.9 kcal/mol for room temperature. Let us mention that the entropy evaluations require a due care paid to the symmetry numbers as quantum chemistry programs rarely produce the correct symmetry number [45,46] of 60 for C_{60}. If this entropy term is combined with the BSSE corrected MP2=FC/6-31G* stabilization energy, the ΔG^o_T term for Eq. (1) at room temperature reads -3.3 kcal/mol which is the driving thermodynamic force for the $N_2@C_{60}$ formation according to the present computational data. The value should be further checked, especially with respect [47] to the BSSE correction and basis-set finiteness in general, when computationally feasible. There are other interesting non-metal fullerene encapsulates to be studied like $H_2O@C_{60}$ or $NH_3@C_{60}$ (Fig. 2).

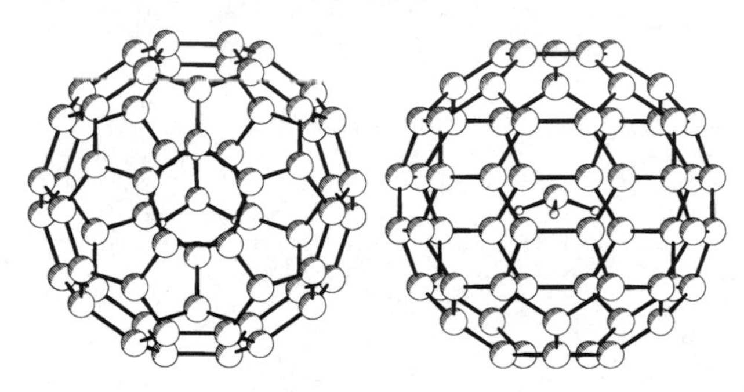

Fig. 2. Two views of the B3LYP/3-21G optimized structure of $NH_3@C_{60}$.

Interestingly enough, the entropy change for the $N_2@C_{60}$ formation is quite similar to the dimerization entropy for water in gas phase [48,49] while the stabilization energy for the water dimer is somewhat smaller. There are no entropy data available yet for formation of other fullerene endohedrals, however, it is clear that the stabilization ΔG^o_T terms would represent a unified thermodynamic stability measure. For a complete picture, the ΔG^o_T terms or the encapsulation equilibrium constants should be combined [50] with the typical partial pressures accessible to the fullerene cage and encapsulate under particular experimental conditions. There are however essential kinetic [51,52] and in particular catalytic [16,53-55] aspects obviously involved, too. All the steps represent important factors to be mastered before more practical applications of computational nanoscience and nanotechnology can be launched [56-60].

ACKNOWLEDGMENTS

The reported research has been supported by a Grant-in-aid for NAREGI Nanoscience Project, and for Scientific Research on Priority Area (A) from the Ministry of Education, Culture, Sports, Science and Technology of Japan.

REFERENCES

[1] T. Peres, B. P. Cao, W. D. Cui, A. Khong, R. J. Cross, M. Saunders and C. Lifshitz, Int. J. Mass Spectr. 210, 241, 2001.

[2] T. Suetsuna, N. Dragoe, W. Harneit, A. Weidinger, H. Shimotani, S. Ito, H. Takagi and K. Kitazawa, Chem. Eur. J. 8, 5079, 2002.

[3] T. Suetsuna, N. Dragoe, W. Harneit, A. Weidinger, H. Shimotani, S. Ito, H. Takagi, and K. Kitazawa, Chem. Eur. J. 9, 598, 2003.

[4] T. A. Murphy, Th. Pawlik, A. Weidinger, M. Höhne, R. Alcala and J.-M. Spaeth, Phys. Rev. Lett. 77, 1075, 1996.

[5] C. Knapp, K.-P. Dinse, B. Pietzak, M. Waiblinger and A. Weidinger, Chem. Phys. Lett. 272, 433, 1997.

[6] A. Gromov, N. Krawez, A. Lassesson, D. I. Ostrovskii and E. E. B. Campbell, Curr. App. Phys. 2, 51, 2002.

[7] B. Pietzak, M. Waiblinger, T. A. Murphy, A. Weidinger, M. Höhne, E. Dietel, and A. Hirsch, Chem. Phys. Lett. 279, 259, 1997.

[8] E. Dietel, A. Hirsch, B. Pietzak, M. Waiblinger, K. Lips, A. Weidinger, A. Gruss and K.-P. Dinse, J. Am. Chem. Soc. 121, 2432, 1999.

[9] B. P. Cao, T. Peres, R. J. Cross, M. Saunders and C. Lifshitz, J. Phys. Chem. A 105, 2142, 2001.

[10] K. Kobayashi, S. Nagase and K.-P. Dinse, Chem. Phys. Lett. 377, 93, 2003.

[11] T. Wakahara, Y. Matsunaga, A. Katayama, Y. Maeda, M. Kako, T. Akasaka, M. Okamura, T. Kato, Y. K. Choe, K. Kobayashi, S. Nagase, H. J. Huang, and M. Atae, Chem. Commun. 2940, 2003,

[12] M. Saunders, H. A. Jiménez-Vázquez, R. J. Cross and R. J. Poreda, Science 259, 1428, 1993.

[13] M. Saunders, H. A. Jiménez-Vázquez, R. J. Cross, S. Mroczkowski, D. I. Freedberg and F. A. L. Anet, Nature 367, 256, 1994.

[14] M. Saunders, R. J. Cross, H. A. Jiménez-Vázquez, R. Shimshi and A. Khong, Science 271, 1693, 1996.

[15] R. J. Cross, M. Saunders and H. Prinzbach, Org. Lett. 1, 1479, 1999.

[16] Y. Rubin, T. Jarrosson, G.-W. Wang, M. D. Bartberger, K. N. Houk, G. Schick, M. Saunders and R. J. Cross, Angew. Chem. Int. Ed. 40, 1543, 2001.

[17] M. Carravetta, Y. Murata, M. Murata, I. Heinmaa, R. Stern, A. Tontcheva, A. Samoson, Y. Rubin, K. Komatsu, and M. H. Levitt, J. Am. Chem. Soc. 126, 4092, 2004.

[18] S.-I. Iwamatsu, T. Uozaki, K. Kobayashi, S. Re, S. Nagase and S. Murata, J. Am. Chem. Soc. 126, 2668, 2004.

[19] C. I. Williams, M. A. Whitehead, and L. Pang, J. Phys. Chem. 97, 11652, 1993.

[20] J. C. Greer, Chem. Phys. Lett. 326, 567, 2000.

[21] J. A. Larsson, J. C. Greer, W. Harneit and A. Weidinger, J. Chem. Phys. 116, 7849, 2002.

[22] J. M. Park, P. Tarakeshwar, K. S. Kim and T. Clark, J. Chem. Phys. 116, 10684, 2002.

[23] B. Pietzak, A. Weidinger, K.-P. Dinse and A. Hirsch, in *Endofullerenes - A New Family of Carbon Clusters*, Eds. T. Akasaka and S. Nagase, Kluwer Academic Publishers, Dordrecht, 2002, p. 13.

[24] A. D. Becke, J. Chem. Phys. 98, 5648, 1993.

[25] C. Lee, W. Yang and R. G. Parr, Phys. Rev. B 37, 785, 1988.

[26] J. P. Perdew and Y. Wang, Phys. Rev. B 45, 13244, 1992.

[27] M. J. Frisch, G. W. Trucks, H. B. Schlegel, G. E. Scuseria, M. A. Robb, J. R. Cheeseman, J. A. Montgomery, Jr., T. Vreven, K. N. Kudin, J. C.

Burant, J. M. Millam, S. S. Iyengar, J. Tomasi, V. Barone, B. Mennucci, M. Cossi, G. Scalmani, N. Rega, G. A. Petersson, H. Nakatsuji, M. Hada, M. Ehara, K. Toyota, R. Fukuda, J. Hasegawa, M. Ishida, T. Nakajima, Y. Honda, O. Kitao, H. Nakai, M. Klene, X. Li, J. E. Knox, H. P. Hratchian, J. B. Cross, C. Adamo, J. Jaramillo, R. Gomperts, R. E. Stratmann, O. Yazyev, A. J. Austin, R. Cammi, C. Pomelli, J. W. Ochterski, P. Y. Ayala, K. Morokuma, G. A. Voth, P. Salvador, J. J. Dannenberg, V. G. Zakrzewski, S. Dapprich, A. D. Daniels, M. C. Strain, O. Farkas, D. K. Malick, A. D. Rabuck, K. Raghavachari, J. B. Foresman, J. V. Ortiz, Q. Cui, A. G. Baboul, S. Clifford, J. Cioslowski, B. B. Stefanov, G. Liu, A. Liashenko, P. Piskorz, I. Komaromi, R. L. Martin, D. J. Fox, T. Keith, M. A. Al-Laham, C. Y. Peng, A. Nanayakkara, M. Challacombe, P. M. W. Gill, B. Johnson, W. Chen, M. W. Wong, C. Gonzalez and J. A. Pople, Gaussian 03, Revision C.01, Gaussian, Inc., Wallingford, CT, 2004.

[28] C. Møller and M. S. Plesset, Phys. Rev. 46, 618, 1934.

[29] K. Raghavachari and J. A. Pople, Int. J. Quant. Chem. 14, 91, 1978.

[30] S. F. Boys and F. Bernardi, Mol. Phys. 19, 553, 1970.

[31] Z. Slanina, Int. Rev. Phys. Chem. 6, 251, 1987.

[32] L. A. Curtiss, K. Raghavachari, P. C. Redfern and J. A. Pople, J. Chem. Phys. 106, 1063, 1997.

[33] J. Cioslowski, *Electronic Structure Calculations on Fullerenes and Their Derivatives*, Oxford University Press, Oxford, 1995.

[34] G. E. Scuseria, *Modern Electronic Structure Theory, Part I*, Ed. D. R. Yarkony, World Scientific, Singapore, 1995, p. 279.

[35] Y. K. Zhang, W. Pan and W. T. Yang, J. Chem. Phys. 107, 7921, 1997.

[36] T. A. Wesolowski, O. Parisel, Y. Ellinger and J. Weber, J. Phys. Chem. A 101, 7818, 1997.

[37] S. Tsuzuki and H. P. Lüthi, J. Chem. Phys. 114, 3949, 2001.

[38] M.-L. Sun, Z. Slanina, S.-L. Lee, F. Uhlík and L. Adamowicz, Chem. Phys. Lett. 246, 66, 1995.

[39] Z. Slanina, K. Kobayashi and S. Nagase, J. Chem. Phys. 120, 3397, 2004.

[40] K. P. Huber and G. Herzberg, *Molecular Spectra and Molecular Structure, IV. Constants of Diatomic Molecules*, Van Nostrand Reinhold Company, New York, 1979, p. 716.

[41] S. Lebedkin, B. Renker, R. Heid, H. Schober and H. Rietschel, Appl. Phys. A 66, 273, 1998.

[42] K. Kobayashi and S. Nagase, Mol. Phys. 101, 249, 2003.

[43] Z. Slanina, J. M. Rudziński, M. Togasi and E. Ōsawa, J. Mol. Struct. (Theochem) 202, 169, 1989.

[44] C. I. Frum, R. Engleman, Jr., H. G. Hedderich, P. F. Bernath, L. D. Lamb and D. R. Huffman, Chem. Phys. Lett. 176, 504, 1991.

[45] Z. Slanina, J. Mol. Struct. (Theochem) 185, 217, 1989.

[46] Z. Slanina, F. Uhlík, J.-P. François and E. Ōsawa, Croat. Chem. Acta 73, 1047, 2000.

[47] F. B. van Duijneveldt, J. G. C. M. van Duijneveldt-van de Rijdt and J. H. van Lenthe, Chem. Rev. 94, 1873, 1994.

[48] Z. Slanina, Chem. Phys. Lett. 127, 67, 1986.

[49] Z. Slanina, Chem. Phys. 150, 321, 1991.

[50] Z. Slanina, J. Chem. Educ. 68, 474, 1991.

[51] T. R. Walsh and D. J. Wales, J. Chem. Phys. 109, 6691, 1998.

[52] K. Kobayashi and S. Nagase, in *Endofullerenes - A New Family of Carbon Clusters*, Eds. T. Akasaka and S. Nagase, Kluwer Academic Publishers, Dordrecht, 2002, p. 99.

[53] B. R. Eggen, M. I. Heggie, G. Jungnickel, C. D. Latham, R. Jones and P. R. Briddon, Science 272, 87, 1996.

[54] S. Patchkovskii and W. Thiel, J. Am. Chem. Soc. 118, 7164, 1996.

[55] S. Patchkovskii and W. Thiel, Helv. Chim. Acta 80, 495, 1997.

[56] J. K. Gimzewski, in *The Chemical Physics of Fullerenes 10 (and 5) Years Later*, Ed. W. Andreoni, Kluwer, Dordrecht, 1996, p. 117.

[57] W. Harneit, M. Waiblinger, C. Meyer, K. Lips and A. Weidinger, in *Recent Advances in the Chemistry and Physics of Fullerenes and Related Materials*, Vol. 11, *Fullerenes for the New Millennium*, Eds. K. M. Kadish, P. V. Kamat and D. Guldi, Electrochemical Society, Pennington, 2001, p. 358.

[58] A. Ardavan, M. Austwick, S. C. Benjamin, G. A. D. Briggs, T. J. S. Dennis, A. Ferguson, D. G. Hasko, M. Kanai, A. N. Khlobystov, B. W. Lovett, G. W. Morley, R. A. Oliver, D. G. Pettifor, K. Porfyrakis, J. H. Reina, J. H. Rice, J. D. Smith, R. A. Taylor, D. A. Williams, C. Adelmann, H. Mariette and R. J. Hamers, Phil. Trans. Roy. Soc. London A 361, 1473, (2003).

[59] K. E. Drexler and R. E. Smalley, Chem. Engn. News 81, No. 48, 37, 2003.

[60] Z. Slanina, K. Kobayashi and S. Nagase, in *Handbook of Theoretical and Computational Nanotechnology*, Eds. M. Rieth and W. Schommers, American Scientific Publishers, Stevenson Ranch, CA, 2005.

An Innovative Ignition Method using SWCNTs and a Camera Flash

Stephen A. Danczyk and Bruce Chehroudi

Air Force Research Laboratory
10 E Saturn Blvd
Edwards AFB Ca 93526
Stephen.Danczyk@Edwards.af.mil
Bruce.Chehroudi@Edwards.af.mil

ABSTRACT

This paper describes an ignition method that uses a simple camera flash and single-walled carbon nanotubes (SWCNTs) to ignite various fuels. This method has been used to ignite both solid and liquid fuels. The effects of iron (Fe) nanoparticles (embedded in the SWCNTs) concentration on the ignition process have been studied. The application of this nano-technology based ignition method may also be extended to achieve distributed ignition that would allow ignition in numerous locations simultaneously.

Keywords: nanotube, ignition, combustion

INTRODUCTION

Many industrial processes that utilize chemical reactions in their applications often require an initiation stimulus to start the conversion of the chemicals to desired products. A process which initiates the combustion of fuels is commonly referred to as ignition. A device which achieves this goal is a critical system component for most combustion processes especially in mobile or stationary power producing machines. For example, improper ignition during the firing of a rocket engine for lift-off can lead to combustion instability causing a catastrophic engine failure and possible loss of the spacecraft and human life, see Harrie and Reardon [1]. Also, the ignition characteristics in a gasoline-fueled automotive engine can strongly affect the fuel's burn rate, the chamber's combustion efficiency, and the exhaust emission profile.

Although many ignition methods exist, by far the most popular one is the electric spark igniter. It requires high-energy input supplied by a high-voltage circuitry often consisting of heavy components, and by its nature is a single-point stimulus method. Other ignition methods such as plasma jet injection or flame jet initiation, and high-power laser ignition are all bulky, heavy and expensive to operate. Other approaches to ignition can be through the usage of pyrophoric charges or via the mixing of hypergolic chemical components, or through activation over a catalytic bed, all of which either make use of hazardous chemicals and/or of highly specialized materials or sophisticated mixing machinery. Again these are either single-point initiation methods, as in the case of focused-beam laser ignition, or limited to a narrowly-defined region in the combustion chamber, uncontrollable by the operator.

Another disadvantage of these ignition systems, with the exception of perhaps lasers, is that once they are installed on an engine, the ignition location remains fixed with respect to the combustion chamber. It is often preferable to have a plurality of ignition points to initiate a uniform or well-distributed combustion initiation. However, using the above-described fixed-point ignition techniques, multi-point ignition within a chamber can only be achieved by repeated implementation of the same ignition hardware, which spatially can be very restrictive to attain in addition to increasing the engine size and mass. Additionally, the ability to select and continuously vary the ignition locations or regions in an engine as a parameter, or "distributed ignition," is a critical and useful engineering design strategy for developing highly-efficient and possibly more stable combustion chambers. Current ignition methods are known to possess one or more other disadvantages such as causing combustion instability, start-up transients which not only can bring severe damage but also degradation in engine efficiency and emission of pollutants, see Harrje and Reardon [1]. Thus, an ignition method is needed which is effective, multi-point or distributed in nature, while allowing design versatility in decreasing engine size and overall mass. At the same time, such a method should exhibit increased ignition efficiency and contribute positively towards reduction of harmful pollutants.

USING SWCNTS AND A CAMERA FLASH TO IGNITE FUELS

Ajayan, et al. [2] were first to report that as-produced SWCNTs ignite when subjected to a common photographic flash at a close range. This effect was observed for dry, "fluffy" SWCNTs and was diminished somewhat for compacted material. Ignition did not occur for similar materials such as multi-walled nanotubes (MWCNTs), graphite powder, fluffy carbon soot, and C_{60}. The observed structural changes in the SWCNTs during an ignition process indicate a temperature in excess of 1500°C.

Ajayan et al. [2] explained this process through generation of a heat pulse, created through light absorption by the nanotubes, combined with inefficient heat dissipation due to the structure of the nanotubes. Braidy et al. [3] confirmed the flash ignition effect on SWCNTs but also reported the presence of iron oxide particles in the combustion byproducts. These were reported to be predominantly Fe_2O_3 with a small amount of Fe_3O_4. The diameters of these oxide particles varied depending on their proximity to the nanotube bundles. Particles within the bundles ranged from 15 to 20 nm in diameter, while free-standing fused grains, which were not enclosed by nanotube bundles, ranged from 15 to 50 nm in diameter. The electron energy loss spectroscopy (EELS) was used to show that both types were indeed Fe_2O_3. Both morphology types were considerably larger than the nanoparticles of Fe in the pristine SWCNTs, which were 1-5 nm in diameter. Smits et al. [4] conducted experiments to determine the cause for ignition of the SWCNTs. They tested three different samples: (a) as-produced SWCNTs synthesized by the HiPco process (from Carbon Nanotechnologies Inc.), (b) the same material obtained but after a purification process, and (c) 99.5% pure, 6 -10 μm diameter Fe powder. All three samples were subjected to identical flash lighting and subsequently micro analyzed. When flashed, both the as-produced SWCNTs and the Fe powder ignited, while the purified SWCNTs showed no reaction. Burning of the unpurified SWCNTs and Fe powder occurred slowly in small, localized locations, giving off a faint red-orange glow. The burning spread slowly, typically lasting 1-3 s before dying out. Burning of the Fe powder could be extended to several seconds by physical agitation. Examination of the nanotube samples by unaided eyes revealed relatively large clusters of hardened, orange material in the burned areas. TEM analysis of the as-produced nanotube samples before burning revealed large quantities of Fe nanoparticles 3-8 nm in diameter. The Fe particles appear to be contained within the nanotube bundles and along their exterior surfaces. After flashing the as-produced SWCNTs, the Fe nanoparticles changed from evenly-distributed throughout the sample to large clusters of particles with substantially increased diameters. The particle sizes after burning have increased from 3 to 8 nm in diameter to well over 100 nm. The structure of SWCNTs had been altered by exposure to high temperature and their diameters were no longer uniform. The large increase in particle size suggests that the Fe particles melted and coalesced which imply temperatures in excess of the melting point of Fe, or 1538°C. It seems that radiant energy transfer from the flash to the SWCNTs and the Fe nanoparticles causes different responses, and it is likely that the heat dissipates in the highly conductive and interconnected CNT bundles whereas it is almost entrapped locally in the Fe nanoparticles. The nanoparticles are also encapsulated in carbon atom layers which may serve to insulate the Fe. Catalytic Fe nanoparticles, created during the SWCNT synthesis process, with sizes on the order of those found in the as-produced SWNCTs are highly pyrophoric with a propensity to spontaneously ignite in the presence of oxygen, resulting in oxides of iron (Fe_2O_3 and Fe_2O_4). Bare Fe nanoparticles observed on the exterior surfaces of the SWNCT bundles are most susceptible to oxidation. Sub-surface Fe nanoparticles are likely to be involved in the oxidation process as well. Chiang et al. [5] describe that the carbon shells are known to be permeable to oxygen and allow oxidation of Fe nanoparticles. The first step in the purification of SWCNTs involves the oxidation of encapsulated Fe nanoparticles. The remaining Fe nanoparticles left after the purification process are less likely to oxidize during flash exposure. Smits et al. [4] found that the exothermic oxidation of Fe released sufficient heat to cause melting, coalescence and oxidation of nanoparticles, which caused aggregation into clusters of large particles. The heat released during the oxidation is sufficient to promote transformations within the adjacent SWCNTs, including fusing some SWCNTs into larger tubes. This indicates temperatures in excess of 1500 °C. Iron melts at 1538 °C and the fully-oxidized Fe_2O_3 compound decomposes at 1565 °C, which is likely the temperature range of the reaction. In addition to structural modifications of the SWCNTs during the ignition and combustion processes, some material is reduced to an amorphous state. Amorphous carbon coatings are found on most of the Fe_2O_3 particles. These may have condensed on the surface of the Fe_2O_3 particles during cooling of the oxide particles. Smits et al. [4] believe that the flash ignition of SWCNTs should be attributed to the pyrophoric nature of fine Fe particles within the nanotube bundles, rather than to any property of the SWCNTs themselves. However, the SWCNTs are providing a medium to stabilize these nanoparticles not to spontaneously initiate ignition until they are exposed to an appropriate stimulus such as energy from an ordinary camera flash.

We have used this observation to develop an ignition method that is capable of igniting liquid hydrocarbons, such as RP-1 and methanol, and solids, such as Potassium-Chlorate ($KClO_3$) and wax paper with SWCNTs and a camera flash. A U.S. patent has already been applied by Chehroudi et al. [6] and is pending. A small amount (a fraction of a milligram) of SWCNTs along with a few drops of fuel could be ignited with a camera flash placed a few millimeters above the SWCNTs. The camera flash would ignite the SWCNTs in multiple locations and the heat released by the oxidizing Fe nanoparticles would ignite the fuel. As the fuel burned off, more of the SWCNTs would ignite until the entire amount was consumed. The fuel would usually burn off much faster than the SWCNTs. As the SWCNTs burned, there appeared a faint glow in the areas where it ignited first and the mixture would turn from black to an orange in color. What remained after the end of the combustion process was a much denser and harder mixture. To have a successful ignition, the SWCNTs had to be arranged in a

1) Main fuel feed system
2) Oxidizer feed system
3) Reservoir for the fluidized mixture of fuel and SWCNTs
4) Fuel injector assembly
5) A flash-light unit
6) The spatial coverage angle of the flash-light unit (adjustable)
7) Engine assembly
8) Fuel droplets containing SWCNTs inside (sizes of the SWCNTs are highly exaggerated)
9) SWCNTs inside a drop and in its surrounding as drops vaporize (sizes of the SWCNTs are highly exaggerated)
10) Exhaust burned gases
11) Combustion chamber

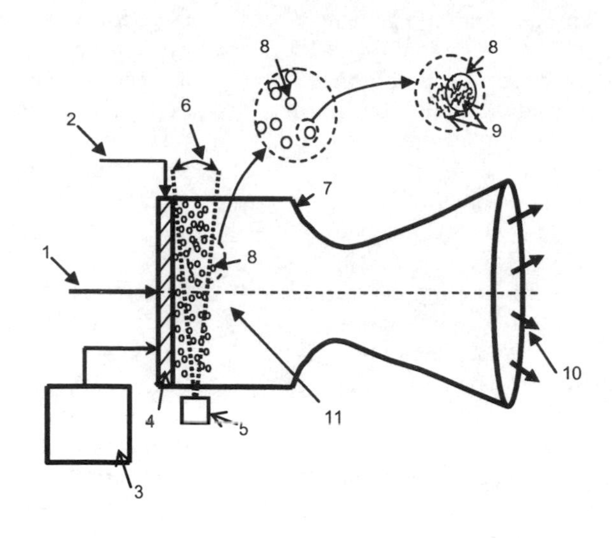

Figure 1. Ignition of a rocket engine using SWCNTs and a laser.

"fluffy" pile and as the ignition sites moved from one location to another, one could observe the pile collapsing on itself. Ignition did not occur if the pile of SWCNTs were compacted before being exposed to a flash light. There is no reason to believe that the metal oxides which remained in our tests are any different than what Brady et al. [3] and Smits et al. [4] described. The orange-colored material is most likely iron oxides Fe_2O_3 and Fe_2O_4. The iron oxide particles that remained after our combustion were much larger than any particles that were in the initial mixture. This reinforces the observation by Smits et al. [4] that the Fe particles melted and coalesced as the oxidation continued. Since the nanosized Fe particles are pyrophoric, it is believed that the SWCNTs must stabilize the Fe particles as mentioned before. What has not yet been determined is the mechanism by which the flash either damages the SWCNTs, exposing the Fe particles to the oxygen, or somehow makes the Fe particles more reactive.

We have found that the SWCNTs were only able to ignite the fuels when the SWCNTs were in contact with the air or immersed in an oxygen-rich environment. The SWCNTs did not ignite the fuels if they were completely submerged in liquid fuels. If they were submerged completely, then only when the liquid fuel was allowed to vaporize and consequently expose the SWCNTs to sufficient oxygen, was the mixture able to be ignited with a flash.

The effects of the concentration of the Fe nanoparticles on the ignition process were also studied. As-produced SWCNTs synthesized by the HiPco process (from Carbon Nanotechnologies Inc.) has Fe concentrations of approximately 30% by weight. These SWCNTs are very "fluffy" and reliably ignite in atmospheric air using a camera flash. Carbon Nanotechnologies Inc., also purified SWCNTs to an Fe concentration of approximately 3% by weight. Samples of these SWCNTs were much less "fluffy" than the as-produced SWCNTs. The as-produced SWCNTs occupied approximately five times the volume of the purified SWCNTs. To reliably ignite the purified SWCNTs, they had to be exposed to an oxygen-rich environment. In this case, a small amount of oxygen was flown over the sample while it was flashed. However, if the SWCNTs were purified once more to an Fe concentration of 1.5% by weight, the SWCNTs would not ignite even in an oxygen-rich environment.

The smallest amount of SWCNTs needed to reliably ignite the fuels tested here has not been established yet. In this preliminary study, we have focused on using amounts that could reasonably be assumed to exist in a single large droplet (approximately 100 μm). It is believed that as the Fe concentration in the sample is increased, the amount of SWCNTs can be decreased. It has been observed that when flashing SWCNT samples without the fuel, very small amounts of SWCNTs will ignite. These amounts are hardly visible to the naked eye and it is not unreasonable to assume that the SWCNTs could be agglomerated in such a way to minimize the amount of SWCNTs needed to reliably initiate ignition.

Figure 1 shows a concept ignition system for a rocket engine using SWCNTs and a short-duration light source. The SWCNTs would be mixed with the fuel and as the mixture atomized forming droplets and vaporized the SWCNTs would be exposed to the oxidizer which can then be ignited with the light source. This technique could offer the ability to produce distributed ignition with multiple droplets being ignited throughout the light-exposed volume (or plane if a laser beam is used). The expanded views in the inset show the individual droplets containing SWCNTs vaporized to a sufficient level where they leave SWCNTs free to have full contact with oxidizer molecules and be ignited upon light flashes. The SWCNTs could also be injected into or fluidized with the oxidizer which would eliminate the need for droplet vaporization necessary in the previous method to expose the SWCNTs to oxidizer molecules.

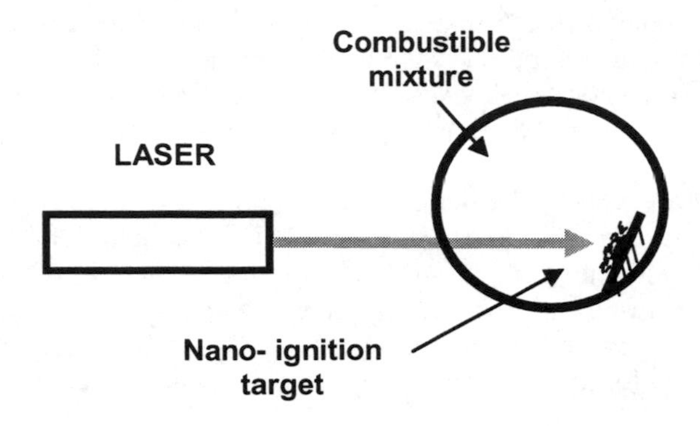

Figure 2. Laser ignition system using a target coated with SWCNTs

Figure 2 shows another low-power laser ignition conceptual design. The current laser ignition systems use a high-power laser that is focused on a metal target. The laser must have enough energy to produce a plasma cloud when the laser pulse impinges the target. This plasma then ignites the combustible mixture in the vicinity of the target causing combustion in the chamber. By using a plate coated with SWCNTs as the laser target material, one could produce a low-energy ignition system since it was shown that SWCNTs can be ignited with a laser power as low as 100 mW/cm^2.

CONCLUSION

It has been demonstrated that SWCNTs and an ordinary camera flash can be used as an ignition source for a variety of fuels. It is believed that the Fe nanoparticles in the SWCNTs are the entities actually ignited by the flash and the burning of these Fe particles lead into the ignition of the fuels. As-produced SWCNTs with Fe concentrations of 30% by weight are reliably ignited in atmospheric air. As the Fe concentration is reduced to approximately 3% by weight, the SWCNTs must be in an oxygen-rich environment to reliably ignite. If the Fe concentration is reduced to 1.5% by weight, the SWCNTs can no longer be ignited. The exact mechanism which causes the camera flash to ignite the Fe particles has not yet been determined. Once this mechanism is clarified, it is believed that we will be in better position to customize the SWCNTs to produce the desired ignition. Two industrial applications were briefly explained which show the exciting potential of this nano-technology based novel ignition technology.

REFERENCES

[1] Harrje, D. T. and Reardon, F. H. Liquid propellant rocket combustion instability, NASA SP-194, Scientific and Technical Information Office, Washington, D. C., 1972.

[2] Ajayan, P.M., Terrones, M., de la Guardia, A., Huc, V., Grobert, N., Wei, B.Q., Lezec, H., Ramanath, G., Ebbesen, T.W., Science 296 (2002) 705.

[3] Braidy, N, Botton, G. A., and Adronov, A., Nanoletters 2 (11) (2002) 1277-1280.

[4] Smits, J., Wincheski, B., Namkung, M., Crooks, R., Louise, R., Material Science and Engineering A358 (2003) 384 -389.

[5] Chiang, I.W., Brinson, B.E., Huang, A.Y., Willis, P.A., Bronikowski, M.J., Margrave, J.L., Smalley, R.E., Hauge, R.H., J. Phys. Chem. B 105 (2001) 8297 – 8301.

[6] Chehroudi, B., Vaghjiani, G. L., and Ketsdever, A. Method for distributed ignition of fuels by low-energy light sources, United Stated Patent Office, Patent Pending, 2004.

Counting Point Defects in Carbon Nanotube Electronic Circuits

Yuwei Fan, Nathan Emmott, and Philip G. Collins

Department of Physics and Astronomy, University of California, Irvine

ABSTRACT

Carbon nanotubes are often imagined to be pristine, defect-free objects, but different types of synthesis and processing are known to result in materials of different qualities. We have developed a method for the quantitative characterization of nanotube defect densities which can readily be used to compare nanotubes from different batches or processes. The method relies on the enhanced chemical reactivity of defect sites as compared to the graphene lattice. By tailoring the potentials used in electrochemical deposition, we selectively seed the growth of metal particles at reactive defect sites without decorating the bulk of a carbon nanotube. Using this technique, nominally identical synthesis runs have produced defect densities varying from 1 defect per 100 nm down to 1 defect per 10,000 nm. Because the particles can be grown arbitrarily large, simple optical imaging can be used to generate wafer-scale statistics.

Keywords: Single-walled carbon nanotubes, defects, electrochemical deposition

1 INTRODUCTION AND MOTIVATION

Single-walled carbon nanotubes (CNTs) are molecular wires with highly robust chemical structure, and their quasi-one-dimensional electronic properties make them promising candidates for nanoelectronics applications [1]. However, the presence or absence of defects in CNTs can significantly change this promise, either for the better or the worse. For example, defects enable chemical functionalization schemes which further tailor electronic properties [2,3]. Efforts to understand CNT-based chemical sensors [4], hydrogen storage [5], and electronic transistors [6,7] have all implicated defects as components of these complex systems.

Despite the importance of defects, there has been little progress characterizing their density or distribution in CNT materials. On an individual basis, single CNTs have been investigated using high resolution scanning tunneling microscopy (STM) and spectroscopy (STS) [8-12], and indirectly probed by scanning gate microcopy (SGM) [6,13]. However, little is known about the density, distribution, and chemical nature of defects in CNTs. This situation is mainly due to the lack of fast and reliable methods to locate defect sites, especially on a useful scale for comparing one batch with another.

Electrochemical deposition is a flexible and powerful solution to this problem. The enhanced chemical reactivity of CNT point defects allows various electrochemical processes to proceed at these sites without modifying the remainder of the CNT. Using the deposition of metals, for example, point defects can be made easily visible to optical or electron microscopes. Selective deposition has been previously demonstrated using the point defects and step edges of a highly oriented pyrolytic graphite (HOPG) crystal [14,15], which can be considered a model test system for CNTs. Furthermore, the metal deposition is completely reversible using proper oxidation potentials, so the technique is a nondestructive one as well. These properties make electrodeposition a promising technique for wafer-scale quantitative analysis of CNT electronic devices and for further research into the roles which defects play.

2 EXPERIMENTAL SETUP

Single-walled carbon nanotubes were produced by chemical vapor deposition (CVD) on a thermally oxidized silicon substrate, with a SiO_2 thickness of 200nm. Ti-Au electrodes defined by optical lithography connected the CNTs. The devices were tested and modified in a custom electrochemical cell, as depicted in Figure 1. The cell volume of about $3cm^3$ is sufficient to allow manipulation of probing needles to contact various parts of the CNT circuit. The needles themselves are polymer-coated to minimize leakage currents. The CNT circuit was configured as a working electrode, and platinum wires were used as counter and reference electrodes. The electrochemical deposition potentials were controlled by a customized potentiostat system.

In order to deposit nanoparticles selectively on only the CNT defects, a tri-potential pulse sequence was used [14,15]. First, an oxidizing pulse (A) prepares the CNT surface. A second, reducing pulse (B) strips oxygen atoms from carbonyls, ethers, hydroxyls, or any other oxygen-containing functionalities at defect sites. The height of pulse B can be adjusted so that metal deposition is seeded at these reactive sites while the CNT sp^2 lattice remains unaffected. Finally, a third potential (C) is used for further metal deposition at the seeded sites.

 NSTI-Nanotech 2005, www.nsti.org, ISBN 0-9767985-1-4 Vol. 2, 2005

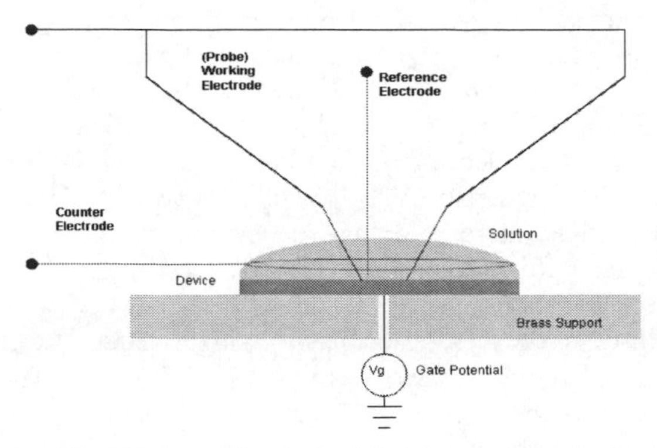

Fig.1 Schematic view of the electrochemical cell

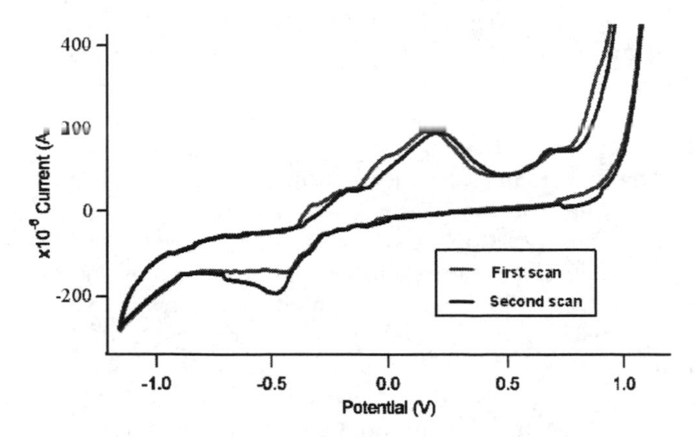

Fig.2 Cyclic votammetry of HOPG surface

As one example of this technique, nickel dots were grown on CNT defect sites. For Ni deposition, the electrochemical solution was composed of 1mM $NiSO_4$ and 0.1M Na_2SO_4 at a pH of 5.0. To assess the appropriate electrochemical potentials, cyclic voltammograms were first obtained on clean graphite (HOPG) surfaces. As shown in Figure 2, the data exhibit a peak at -0.4 V corresponding to the deposition of Ni and oxidation peaks at +0.3 V and +0.7 V. With some iterative testing, an optimized pulse sequence was found to be A = (+0.8V, 5s), B = (-1.1V, 10ms), C = (-0.9V, 30-300s). The duration of the C pulse determines the size of the final nanoparticles, and can be tailored to make the particles 5nm, 50nm, or 500nm in diameter.

Successful deposition requires care in a number of experimental techniques. The oxidation pulse is essential for Ni attachment, probably because it prepares the defect sites into a reproducible chemical state with a low nucleation overpotential. Furthermore, the solutions must be purged with N_2 before each use in order to minimize dissolved O_2 in the cell. Finally, the sample must be thoroughly rinsed with ultrapure water following each deposition, or else salt precipitates foul the surface and complicate comparative imaging.

3 RESULTS AND DISCUSSION

Figure 3 consists of atomic force microscopy (AFM) images of a single CNT device before and after various electrochemical modifications. The catalytic particle used to initiate the growth of the CNT is visible as a bright spot in the lower left-hand side of each image. The vertical stripe on the right-hand side of each image is the first of multiple electrodes connecting the CNT into a measurement circuit. The complete length of the CNT exceeds 20 μm, though only the initial 2.5 μm is visible in this image.

Following the tri-potential sequence described above, the chemically-reactive sites of this particular CNT can be readily identified by four new Ni nanoparticles (Fig. 3b) attached to the CNT wall. The average height of the four particles is 12.8 nm after a deposition time of 30 s. The average distance between the four particles is measured to be 310 nm.

In order to test the reliability and robustness of the electrochemical identification process, various experiments have been completed. Repeating the deposition process on the same CNT allows the particles to grow larger (Fig. 3c), but does not increase the number of particles attached. This observation is consistent with the premise that the deposition is selective, and that the chosen pulse sequence decorates point defects with a very high probability. After imaging, the removal of the deposited Ni can be accomplished using either a 90 s electrochemical oxidation or an acid wash (18% HCl for 10 minutes). In either case, the CNT is returned to its pristine state (Fig. 3d). Finally, a new deposition sequence can be initiated to grow Ni nanoparticles on the cleaned CNT. The new Ni particles deposit at identical positions and at the same rate as before. The one-to-one correspondence is excellent in all of the samples tested, through multiple cycles of deposition and removal. However, it is noteworthy that the acid treatment appears to also produce new reactive sites with a relatively low probability. Fig 3c demonstrates this effect, with an additional fifth Ni nanoparticle formed after repeated HCl washes.

The selectivity of the metal deposition is found to be very sensitive to the pulse profile. As in earlier studies with HOPG [14,15], the CNTs can be completely coated if the overpotential is too high and not decorated at all if the potential is too low. The oxidation of the carbon nanotube surface at +0.8 V is also a critical step. Presumably, this oxidation produces carbonyls, ethers, hydroxyls, or other oxygen-containing functionalities at the defects position, thereby increasing the affinity of metal adatoms at these sites. Experiments showed that all three pulses (A), (B), and (C) were necessary for the reproducible decoration of individual sites, with pulse (C) the most important for tuning the selectivity of the deposition.

To further confirm the hypothesis that the tri-potential sequence is selective to defect sites, we compared the process with an alternate approach previously studied. Using a gas phase reaction, selenium nanoparticles were

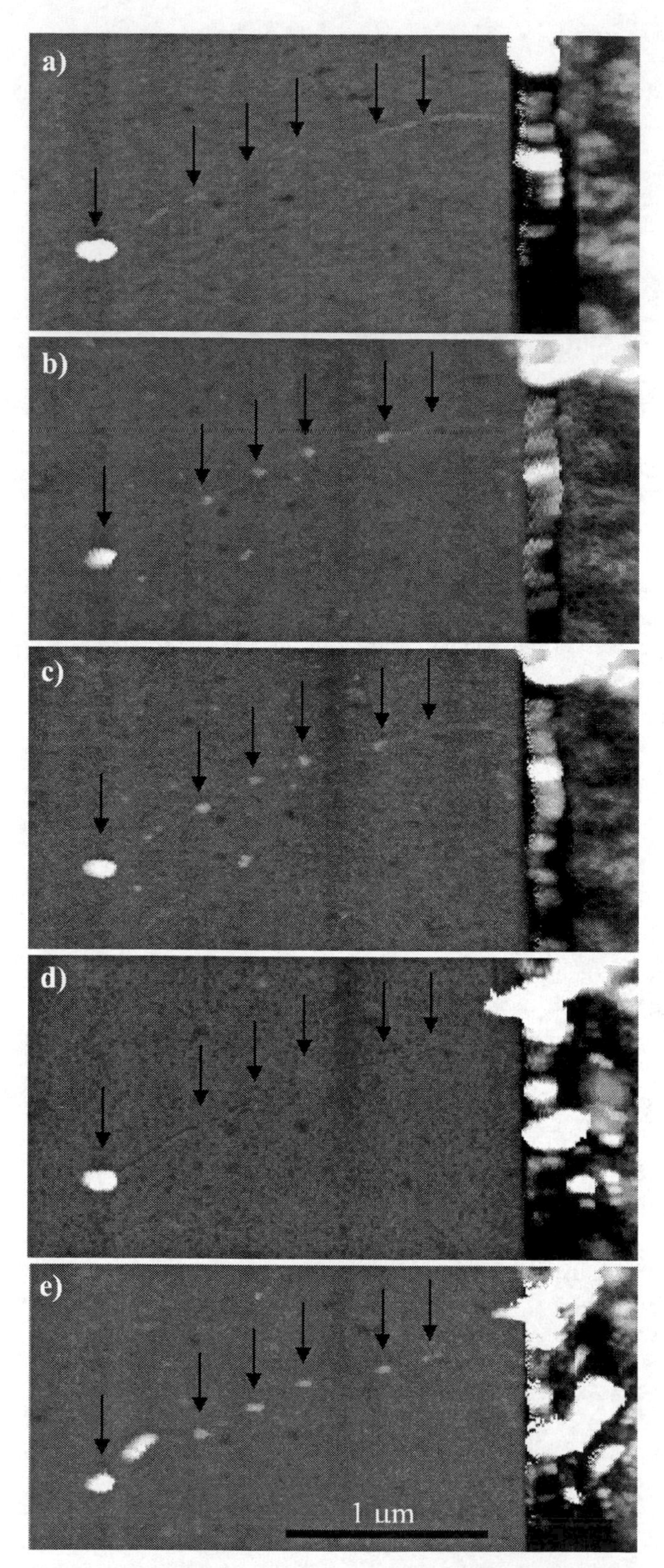

Fig.3 A carbon nanotube working electrode (a) before and (b, c) after selective deposition cycles. (d,e) Reversibility and reproducibility of the deposition. Vertical arrows guide the eye to identical positions in each image.

selectively precipitated onto CNTs from H_2Se [16]. We have reproduced this work and found a one-to-one correspondence on a given CNT between Se and Ni particle attachment sites. The electrochemical technique has the advantage that the deposition potential can be tailored to a wide variety of metals including Au, Ag, and Cu. Furthermore, the electrochemical technique can be re-verified in cases where a CNT appears to have zero defects because a slight increase in the deposition overpotential results in the uniform coating of the entire CNT.

The sample shown is representative of dozens of single- and multi-CNT devices tested from a single wafer. Over one hundred defect sites were readily found. The separation distance between consecutive defects was measured to have a mean value of 360 nm and a standard deviation of 330 nm. This mean is consistent with the reported phase coherence length for CNTs, which is estimated to be in the range of 150-750 nm [17-19]. The high value of the standard deviation of our measurements indicates the wide variability in defect spacing. In particular, curved CNT segments were observed to have higher defect densities than straight segments, possibly indicating that small fluctuations in the CVD environment can directly introduce these sites during growth.

Despite this variation, however, the simple mean is still likely to be a useful parameter for quantitative comparisons. In experiments which are still preliminary, we have reproduced the measurements on two additional wafers containing CNTs synthesized under different conditions. The mean density of defect sites varies considerably between the samples, ranging from a low value of 100 nm^{-1} to values exceeding 1000 nm^{-1}. Therefore, this method described may be quite useful for the quantitative comparison of CVD, arc discharge, and laser ablation synthesis methods. It may also identify differences in nominally identical CVD runs, as would be necessary for process and quality control of nanotube electronics.

A complexity and potential limitation of this electrochemical modification technique is the sensitivity of the CNTs to gating by the solution. Various researchers have reported CNT gating using electrolyte solutions [20], including a case combining electron transport measurements and resonant Raman spectroscopy [21]. Significant changes occur to the CNT's electronic transport properties and phonon spectra, and these changes are generally attributed to dopant shifts to the CNT Fermi level. In our experiment, the (A) oxidation pulse acts as a negative gate, generally increasing the conductivity of the CNT. The (C) deposition pulse, on the other hand, can potentially be rendered ineffective if it completely depletes the CNT of carriers. Experimentally, the most effective pulse heights and widths did not appear to be sensitive to this liquid gating effect. The problem may have been alleviated by the use of a global backside gate (the heavily doped Silicon substrate), which was maintained at ground potential throughout the deposition. Nevertheless, the possible interference by liquid gating remains an issue which deserves further investigation.

4 SUMMARY

In conclusion, a simple and direct characterization technique is demonstrated for counting defects in large numbers of CNT electronic devices. A tri-potential deposition pulse sequence reversibly and reproducibly decorates defect sites with metal nanoparticles, as demonstrated here using Ni. The particles can be grown as large as necessary for a particular imaging technique, and can then be easily removed. The average spacing between defect sites was measured to be 360 nm, but it is likely that this number is highly sensitive to the particularities of the synthesis. Indeed, the method may be a very sensitive way to compare CNTs from one batch to another.

REFERENCES

[1] M. Dresselhaus, G. Dresselhaus, and Ph. Avouris, "Carbon Nanotubes: Synthesis, Structure Properties and Applications, Eds," Springer-Verlag, Berlin, 2001.

[2] A. Hirsch, Angew. Chem. Int. Ed., 41, 1853, 2002.

[3] S. Banerjee, T. Hemraj-Benny, and S. S. Wong, Adv. Mater., 17, 17, 2005.

[4] R. J.Chen, S. Bangsaruntip, K. A. Drouvalakis, N. W. S. Kam, M. Shim, Y. Li, W. Kim, P. J. Utz, H. Dai, Proc. Nat.Acad. Sci. USA, 100, 4984, 2003.

[5] K. Atkinson, S. Roth, M. Hirscher and W. Gruenwald, Fuel Cells Bulletin, 38, 9, 2001.

[6] M. Bockrath, W. Liang, D. Bozovic, J. H, Hafner, C. M. Lieber, M. Tinkham, H. Park, science, 291, 283, 2001.

[7] D. Bozovic, M. Bockrath, J. H. Hafncr, C. M. Lieber, H. Park, M. Tinkham, Appl. Phys. Lett., 78, 3693, 2001.

[8] A. Rubio. Appl. Phys. A, 68, 275, 1999.

[9] M. Ishigami, H. J. Choi, S. Aloni. S. G. Louie, M. L. Cohen, and A. Zettl, Phys. Rev. Lett., 93, 196803_1, 2004.

[10] D. Orlikowski, M. B. Nardelli, J. Bernholc, C. Roland, Phys. Rev. B, 61, 14194, 2000.

[11] H. Kim, J. Lee, S. J. Kahng, Y. W. Son, S. B. Lee, C. K. Lee, J. Ihm, and Y. Kuk, Phys. Rev. Lett., 90, 216107_1, 2003.

[12] F. Wei, J. Z. Zhu, H. M. Chen, J. Phys.: Condens. Matter, 12, 8617, 2000.

[13] M. Freitag, A. T. Johnson, S. V. Kalinin, and D. A. Bonnell, Phys. Rev. Lett., 89, 216801_1, 2002.

[14] E. C. Walter, B. J. Murray, F. Favier, G. Kaltenpoth, M. Grunze, and R. M. Penner, J. Phys. Chem. B, 106, 11407, 2002.

[15] E. C. Walter, M. P. Zach, F. Favier, B. J. Murray, K. Inazu, J. C. Hemminger, and R.M.Penner, Chemphyschem, 4, 131, 2003.

[16] Y. Fan, M. Burghard, K. Kern. Adv.Mater., 14, 2, 130-133, 2002.

[17] H. Stahl, J. Appenzeller, R. Martel, P. Avouris, B. Lengeler, Phys. Rev. Lett., 85, 5186 2000.

[18] H. R. Shea, R. Martel, P. Avouris, Phys. Rev. Lett., 84, 4441, 2000.

[19] W. Liang, M. Bockrath, D. Bozovic, J. H. Hafner, M. Tinkham, H. Park, Nature, 411, 665, 2001

[20] M. Kruger, M. R. Buitelaar, T. Nussbaumer, C. Schonenberger, L. Forro, Appl. Phys. Lett. 78, 1291, 2001.

[21] S. B. Cronin, R. Barnett, M. Tinkham, S. G. Chou, O. Rabin, M. S. Dresselhaus, A. K. Swan, M. S. Uenlue, and B. B. Goldberg, Appl. Phys. Lett., 84, 2052, 2004.

This work has been supported by NSF-DMR and an ACS-PRF Grant.

Molecular Dynamics Study of Electromechanical Nanotube Random Access Memory

Oh-Keun Kwon[*], Jeong Won Kang[**], Ki Ryang Byun[**], Jun Ha Lee[***], Ho Jung Hwang[**]

[*] Department of Ecommerce, Semyung University, Jecheon 390-711, Korea, kok1027@semyung.ac.kr
[**] Department of Computer System Engineering, Sangmyung University, ChungNam 330-720, Korea
[***] School of Electrical and Electronic Engineering, Chung-Ang University
221-1 HukSuk-Dong, DongJak-Ku, Seoul 156-756, Korea, gardenriver@korea.com

ABSTRACT

A nanoelectromechanical (NEM) switching device based on carbon nanotube (CNT) was investigated using atomistic simulations. The model schematics for a CNT-based three-terminal NEM switching device fabrication were presented. For the CNT-based three-terminal NEM switch, the interactions between the CNT-lever and the drain electrode or the substrate were very important. When the electrostatic force applied to the CNT-lever was the critical point, the CNT-lever was rapidly bent because of the attractive force between the CNT-lever and the drain; then, the total potential energy of the CNT-lever was rapidly increased and the interatomic potential energy of the CNT-copper was rapidly decreased. The energy curves for the pull-in and the pull-out processes showed the hysteresis loop that was induced by the adhesion of the CNT on the copper, which was the interatomic interaction between the CNT and the copper.

Keywords: nanoelectromechanical switch, carbon nanotube, nanorelay, three-terminal switch, molecular dynamics simulation

1 INTRODUCTION

Microelectromechanial systems (MENS) have already had a significant impact on medical, automobile, aerospace, and information technology areas [1]. Nanoelectromechnical systems (NEMS) are about a thousand times smaller than MEMS and have the potential to enable revolutionary technology for various areas. Hierold [2] discussed that microsystems will go nanosystems provided that self-assembly of nanostructure becomes a will-controlled fabrication technology. To date, NEMS are rapidly growing area of research with considerable potential for future applications [3]. The basic idea underlying NEMS is the strong electromechanical coupling in devices on the nanometer scale in which the Coulomb forces associated with device operation are comparable with the chemical binding forces.

Carbon nanotubes (CNT) [4] are excellent candidates for NEMS devices not only because of their excellent electronic and mechanical properties, but also because of the significant progress that have been made in the last few years in fabrication of carbon nanostructures [5]. This is a consequence of their well-characterized chemical and physical structures, low mass and dimensions, exceptional directional stiffness, and range of electronic properties [6]. Several possible devices based on CNTs have been investigated: nano-bearings [7-9], nano-gears [10,11], constant-force nano-springs [12], mechanical nano-switch [13], electrical nano-switch and nano-drill [14], gigahertz oscillators [15-19], data storage nano-devices [20-22], etc.

Some prototype CNT-based NEMS have already been demonstrated, such as nanotweezers [23], a random access memory [24], and sensors [25]. Recently the two-terminal CNT switch [26] and the three-terminal CNT nanorelay [27-29] have been studied theoretically. The nanorelay was shown theoretically to act as a switch in the GHz regime and to be potentially suitable for applications such as logic devices, memory elements, pulse generators, and current or voltage amplifiers [28].

However, the analysis of the impact of short-range forces on device design and performance has never been discussed in the previous works. Another effect including the possibility of field emission has been neglected. In this paper, we show model schematics for a three-terminal nanoelectromechancial (NEM) switch and then also perform classical molecular dynamics simulations for the NEM switch.

2 MODEL SCHEMATICS

Figure 1 shows the model schematics for a CNT-based NEM switching device fabrication. After the oxide film growth on the substrate (Fig. 1(a)), the gate electrode is formed as shown in Fig. 1(b). The drain electrode is also formed on the substrate as shown in Fig. 1(c). Another oxide film is also grown (Fig. 1(d)), and a CNT is deposited on the oxide film as shown in Fig. 1(e). The source electrode is deposited to cover one-side of the CNT (Fig. 1(f)). Finally, when the second oxide film is removed by the oxide etching processes in the condition that the etching gases cannot affect the CNT, the NEM switching device called the nanorealy can be fabricated as shown in Fig. 1(g). A conducting CNT is connected to a source electrode and suspended above the surface of a substrate, above gate and drain electrodes. The key components are a movable CNT as switching bar, a gate electrode for position control of the movable CNT, and a drain electrode. Electrical charge is induced in the suspended CNT by a voltage applied to the gate electrode. The resulting capacitive force between the CNT and the gate bends the CNT and brings the CNT end into electrical contact with the drain electrode.

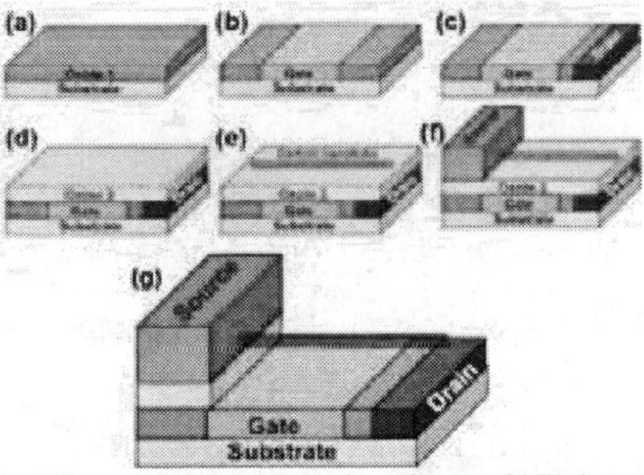

Figure 1. The model schematics for a CNT-based NEM switching device fabrication. (a) Oxide film growth on the substrate, (b) gate electrode formation, (c) drain electrode formation, (d) another oxide film growth, (e) CNT deposition on the oxide film, (f) source electrode formation to cover one-side of the CNT, and (g) finally, the second oxide film removal.

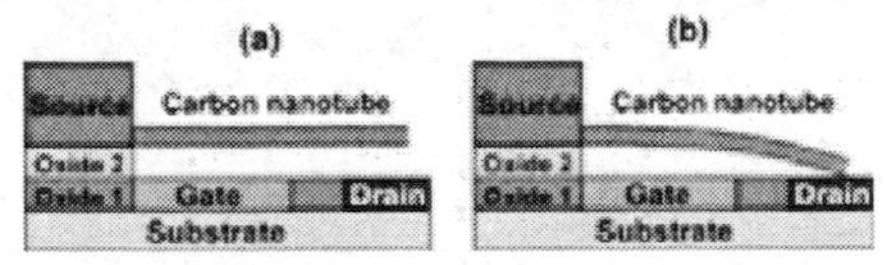

Figure 2. The physical operation of a CNT-based NEM switch. (a) Turn-on and (b) turon-off states.

Figure 2 shows the physical operation of a CNT-based NEM switch. When the potential different is achieved between the CNT and the gate electrode and between the CNT and the drain electrode, electrostatic charges are induced on both the CNT and the electrodes. The electrostatic charges give rise to electrostatic force, which deflect the CNT. In addition to electrostatic forces, depending on the gap between the CNT and the electrodes, the interatomic interactions also act on the CNT and deflect the CNT. The electrostatic and the interatomic forces make the CNT to bend toward the drain. Counteracting the electrostatic and interatomic forces are elastic forces, which try to restore the CNT to its original straight position. For an applied voltage, an equilibrium position of the CNT is defined by the balance of the elastic, electrostatic, and interatomic forces.

When the applied potential difference between the CNT and the gate electrode exceeds a certain potential, the CNT becomes unstable and collapses onto the substrate. The potential difference, which causes the CNT to contact onto the drain, is defined as the pull-in voltage. When the pull-in voltage is applied, the CNT comes in contact with the drain electrode, and the device is said to be in the ON state (shown in Fig. 2(b)). When the potential is released and the CNT and the drain electrode are separated, the device is said to be in the OFF state (shown in Fig. 2(a)).

Due to the exponential dependence of the tunneling resistance on tube deflection, there is a sharp transition from a non-conducting (OFF) to a conducting (ON) state when the gate voltage is varied at fixed source-drain voltage. The sharp switching curve allows for amplification of weak signals superimposed on the gate voltage.

3 MODEL STRUCTURE AND INTERATOMIC POTENTIALS

For C-C interactions, we used the Tersoff-Brenner potential function that has been widely applied to C systems [31-33]. The long-range interactions were characterized with the Lennard-Jones 12-6 (LJ12-6) potential that was continually connected by the cubic spline functions with the Tersoff-Brenner potential such as methods by Mao *et al* [39] when the interatomic distance (r) is between 2.0 and 2.7 Å. The cutoff distance of the LJ12-6 with parameters $\varepsilon_C = 0.0024$ eV and $\sigma_C = 3.37$ Å was 10 Å. We assumed the Cu nanowires as the conductor material. For Cu–Cu and Cu–C, we used the Mores-type potential, a pair interatomic potential function, which have been widely used in many atomistic studies for nanoindentations and nanomechanics [34,35].

The MD simulations used the same MD methods as were used in our previous works [36,37]. The MD code used the velocity Verlet algorithm, and neighbor lists to improve computing performance. MD time step was 5×10^{-4} ps. A Gunsteren-Berendsen thermostat was used to control temperature for all atoms except for fullerenes. The structure was initially relaxed by the steepest descent method; then the atoms of both edges were fixed during the MD simulations and on the other atoms, MD methods were applied.

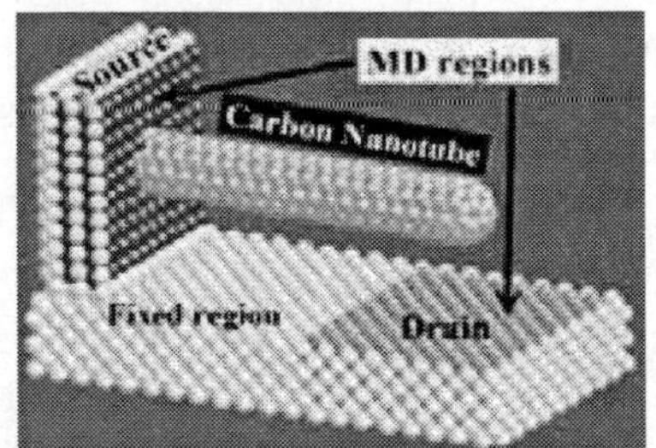

Figure 3. The atomic structure for the three-terminal NEM switching device. This structure is the same as the structure of Fig. 3 except for the substrate including drain electrode. The substrate was composed of 2400 copper atoms in $26\times70\times12$ Å^3. The drain region was 50 to 70 Å of the substrate.

Figure 3 shows the atomic structure for the three-terminal NEM switching device. The length of the (5,5) CNT was approximately 60 Å. The (100) copper surface acting as the source electrode for the three-terminal switch device was composed of 663 atoms, 26 Å$\times$26 Å$\times$9 Å. The

CNT is inserted at the center of the copper surface; then, the 9 Å of the CNT overlapped the copper surface. The substrate was composed of 2400 copper atoms, 26 Å × 70 Å × 12 Å. The boundary layers of the copper were fixed in the MD simulations and the other copper atoms were under the constraint dynamics to a constant temperature in the MD simulations. Ten carbon atoms of the bottom were fixed during the MD simulations but the other carbon atoms were applied to free MD simulations without any constraint dynamics. The drain region was 50 to 70 Å of the substrate. Copper atoms composed of the substrate except for the drain region were fixed in the MD simulations. The height of the CNT was 18 Å higher than the substrate.

4 RESULTS AND DISCUSSION

We investigated a three-terminal switch of the CNT-lever shown in Fig. 3 using the SD and the MD simulations. For the three-terminal switching devices, which have investigated by theoretical methods and by experimental methods, the interaction between the CNT-lever and the drain electrode or the substrate is very important. Dequesnes et al. [26] investigated the significance of the vdW interactions in the design of the two-terminal NEM switches. Kinaret et al. [27] investigated the three-terminal NEM switch using a simple model and discussed the characteristics of the device. Ke and Espinosa [28] developed the model for the switchable CNT-lever device including the concentration of electrical charge and the vdW force. Jonsson et al. [29] showed that the short range and the vdW forces have a significant impact on the characteristic of the three-terminal NEM switch. Therefore, for the three-terminal NEM switch as shown in Fig. 4, the significance of the interaction between the CNT-lever and the drain electrode was investigated using the SD and the MD simulations.

Figure 4 shows the pull-in of the three-terminal NEM switch using the SD simulation. We calculated the total potential energy of the CNT-copper interactions (Fig. 4(a)) and the total potential energy of the CNT (Fig. 4(b)) as a function of the external force per atom. The three snapshots in Fig. 4 show the atomic structures corresponding to the indicated external forces. The interatomic potential energy between the CNT and the source is –21.95 eV. The interatomic potential energy between the CNT and the drain is –16.11 eV when the CNT-lever is fully contacted with the drain. When the external force per atom is 0.0014 eV/Å, the CNT-lever is rapidly bent because of the attractive force between the CNT-lever and the drain; then, the total potential energy of the CNT-lever is rapidly increased and the total potential energy of the CNT-copper is rapidly decreased. Therefore, the long range or the short range interactions between the CNT-lever and the metal electrode are very important for the three-terminal NEM switch to operate such as the investigations of the previous papers.

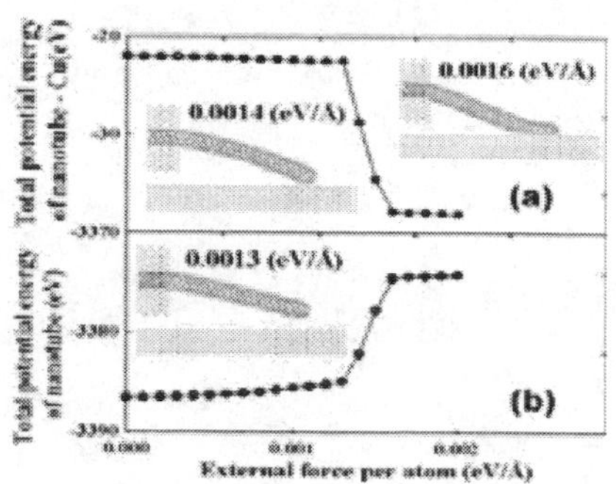

Fig. 4. The pull-in of the three-terminal NEM switch using the SD simulation. (a) The total potential energy of the CNT-copper interactions and (b) the total potential energy of the CNT as a function of the external force per atom. The three snapshots show the atomic structures corresponding to the indicated external forces.

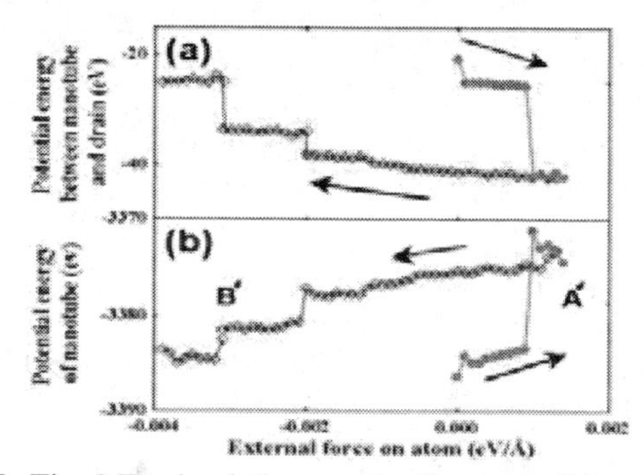

Fig. 5. The MD simulation results for the pull-in and the pull-out of the CNT-lever. (a) and (b) show the potential energies of the CNT-drain interaction and the CNT, respectively. The pull-in force increased from 0 to 0.0014 eV/Å and the pull-out force decreased to –0.0039 eV/Å

We performed the MD simulation for the pull-in and the pull-out of the CNT-lever as shown in Fig. 5. Figures 5(a) and 5(b) show the potential energies of the CNT-drain and the CNT, respectively. The pull-in force increased from 0 to 0.0014 eV/Å and the pull-out force decreased to –0.0039 eV/Å. When the external force per atom was 0.001 eV/Å, the CNT-lever contacted with the drain. The pull-in force for the switch turn-on was above 0.001 eV/Å per atom. In the MD simulation for the pull-out of the CNT-lever, three stages are found such as full-contacting, edge-contacting, and non-contacting. Figure 6(a) and 6(b) indicate the structures corresponding to the labels A' and B' in Fig. 5 (b). Figures 6(a) and 6(b) show the full contacting and the edge-contacting modes. In the pull-out simulations, the full contacting mode is until –0.0019 eV/Å, the edge-contacting mode is –0.002 to –0.0031 eV/Å, and the non-contacting mode is below –0.0032 eV/Å. Therefore, the pull-out force for the switch turn-off is below –0.0032 eV/Å. The energy curves for the pull-in and the pull-out processes show the hysteresis loop shown in Fig. 5. The difference between the turn-on and the turn-off forces, called the hysteresis loop, is induced by the adhesion of the CNT on the copper, which is the interatomic interaction between the CNT and the copper. The hysteresis loop has been found in the previous theoretical works [27-29] and the previous experimental work [30]. This difference in the hysteresis loop makes the three-terminal NEM switch device to utilize a memory device [28].

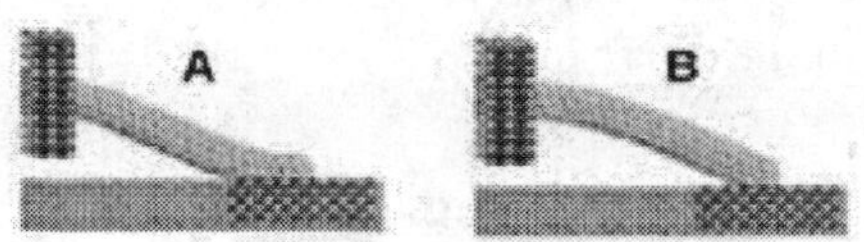

Fig. 6. (a) and (b) indicate the structures corresponding to the labels A' and B' in Fig. 12 (b).

5 SUMMARY

We investigated a nanoelectromechanical (NEM) switching device based on carbon nanotube (CNT) using atomistic simulations. We presented the model schematics for a CNT-based three-terminal NEM switching device fabrication. The CNT-based NEM switch should be operated when the electrostatic force acting on the CNT-lever is below the critical point. The electrical-induced potential energy was changed to the mechanical energy. For the CNT-based three-terminal NEM switch, the interaction between the CNT-lever and the drain electrode or the substrate was very important. When the external force per atom was the critical point, the CNT-lever was rapidly bent because of the attractive force between the CNT-lever and the drain; then, the total potential energy of the CNT-lever was rapidly increased and the total potential energy of the CNT-copper was rapidly decreased. The energy curves for the pull-in and the pull-out processes showed the hysteresis loop that was induced by the adhesion of the CNT on the copper, which was the interatomic interaction between the CNT and the copper. The three-terminal NEM switch device could be applied to a memory device because of the difference in the hysteresis loop. For various materials for the drain, the operating properties of the NEM switch should be investigated in further works.

REFERENCES

[1] G.T.A. Kovacs, "Micromachined Tranducers Sourcebook", New York, McGraw Hill, 1998.
[2] C. Hierold, J. Micromech. Microeng. 14 (2004) S1.
[3] A.N. Cleland, "Foundations of Nanomechanics", Springer-Verlag, Berlin, 2003.
[4] S. Iijima, Nature 354 (1991) 56.
[5] "Handbook of Nanoscience, Engineering, and Technology" edited by W.A. Goddard, D.W. Brenner, S.E. Lyshevski and G. J. Iagrate, New York, CRC Press, 2003.
[6] D. Dian, G.J. Wagner, W.K. Liu, M.-Y. Yu and R.S. Ruoff, Appl. Mech. Rev. 55 (2002) 495.
[7] R.E. Tuzun, D.W. Noid and B.G. Sumpter, Nanotechnology 6 (1995) 64.
[8] R.E. Tuzun, K. Sohlberg, D. W. Noid and B.G. Sumpter, Nanotechnology 9 (1998) 37.
[9] J.W. Kang and H.J. Hwang, Nanotechnology 15 (2004) 614.
[10] D. Srivastava, Nanotechnology 8 (1997) 186.
[11] T. Cagin, A. Jaramillo-Botero, G. Gao and W.A. Gaddard, Nanotechnology 9 (1998) 143.
[12] J. Cumings and A. Zettl, Science 289 (2000) 602.
[13] L. Forro, Science 289 (2000) 5479.
[14] Yu.E. Lozovik, A.V. Minogin and A.N. Popov, Phys. Lett. A 313 (2003) 112.
[15] Q. Zheng and Q. Jiang, Phys. Rev. Lett. 88 (2002) 045503.
[16] Q. Zheng, J.S. Liu and Q. Jiang, Phys. Rev. B 65 (2002) 245409.
[17] Y. Zhao, C.-C. Ma, G.H. Chen and Q. Jiang, Phys. Rev. Lett. 91 (2003) 175504.
[18] S.B. Legoas, V.R. Coluci, S.F. Braga, P.Z. Coura, S.O. Dantas and D.S. Galvao, Phys. Rev. Lett. 90 (2003) 055504.
[19] S.B. Legoas, V.R. Coluci, S.F. Braga, P.Z. Coura, S.O. Dantas and D.S. Galvao, Nanotechnology 15 (2004) S184.
[20] W.Y. Choi, J.W. Kang and H.J. Hwang, Physica E 23 (2004) 125.
[21] H.J. Hwang, K.R. Byun and J.W. Kang and, Physica E 23 (2004) 208.
[22] J.W. Kang and H.J. Hwang, Physica E 23 (2004) 36.
[23] P. Kim and C.M. Lieber, Science 286 (1999) 2148.
[24] T. Rueckes, K. Kim, E. Joselevich, G.Y. Tseng, C.-L. Cheung, and C.M. Liever, Science 289 (2000) 94.
[25] P.G. Collins, K.B. Bradley, M. Ishigamo, and A. Zettl, Science 287 (2000) 1801.
[26] M. Dequesnes, S.V. Rotkin and N.R. Aluru, Nanotechnology 13 (2002) 120.
[27] J.M. Kinaret, T. Nord, and S. Viefers, Appl. Phys. Lett. 82 (2003) 1287.
[28] C. Ke and H.D. Espinosa, Appl. Phys. Lett. 85 (2004) 681.
[29] L.M. Jonsson, T. Nord, J.M. Kinaret, and S. Viefers, J. Appl. Phys. 96 (2004) 629.
[30] S.W. Lee, D.S. Lee, R.E. Morjan, S.H. Jhang, M. Sveningsson, O.A. Nerushev, Y.W. Park, and E.E.B. Campbell, Nano Lett. 4 (2004) 2027.
[31] J. Tersoff, Phys. Rev. B 38 (1988) 9902.
[32] J. Tersoff, Phys. Rev. B 39 (1989) 5566.
[33] D.W. Brenner, Phys. Rev. B 42 (1990) 9458.
[34] S. Dorfman, K.C. Mundim, D. Fuks, A. Berner, D.E. Ellis, and J. Van Humbeeck, Mater. Sci. Eng. C 15 (2001) 191.
[35] L.-L Wang and H.-P. Cheng, Phys. Rev. B 69 (2004) 045404.
[36] J.W. Kang and H.J. Hwang, Nanotechnology 14 (2003) 402.
[37] J.W. Kang and H.J. Hwang, Nanotechnology 15 (2004) 115.

Application of Encapsulated PECVD-grown Carbon Nano-Structure Field-Emission Devices in Nanolithography

J. Koohsorkhi*, Y. Adbi*, S. Mohajerzadeh*, J. Derakhshandeh*, H. Hoseinzadegan*, A. Khakifirooz* and M.D. Robertson**

*Thin Film Laboratory,
Department of Electrical and Computer Engineering,
University of Tehran, Tehran, Iran, smohajer@tfl.ir
**Department of Physics,
Acadia University, Nova-Scotia, Canada, michael.robertson@acadiau.ca

ABSTRACT

A novel technique for submicron and nanolithography using vertically grown nickel seeded carbon nano-structures (CNS) on silicon substrates is reported. The field emission characteristic of carbon tips encapsulated in a titanium-dioxide insulator is utilized to create the nano-scale features. The electrical behavior of nano-structures indicates a precise control of the emission current with the surrounding gate electrode. Also by applying 100V between the cathode and resist-coated anode, lines with widths between 100 and 200nm have been drawn.

Keywords: carbon nano-tubes, nano-lithography, PECVD, vertical growth, encapsulation

1 INTRODUCTION

Carbon nanotubes and carbon nano-structures have drawn significant attention due to their various potential applications as high performance electronic devices, field emission displays, sensors and memory storage[1-2]. In particular vertically grown nano-fibers are of great importance in emission-based devices where nano-size tips are utilized to generate and control the emission current from tip towards the anode[3]. Several methods are being used to realize carbon nanotubes and nanofibers among which plasma-enhanced chemical vapor deposition is a well-established technique [4]. In this paper we report a novel self-defined PECVD-growth technique which employs the encapsulation of nano-tubes and allows formation of an electron-beam. The emission of electrons from the encapsulated tips is then exploited to perform low energy electron-based lithography. The structure proposed in this paper allows fabrication of field emission displays with a proper control on the emission level with the aids of an integrated gate electrode. Since a self-defined structure is used there is no need to lithography for the alignment of the gate with the central nano-tube. Using this approach, lines as thin as 100nm have been obtained although smaller features are expected. The electrical as well as physical characteristics of the grown structures have been investigated.

2 EXPERIMENTAL

Fig.1 shows the operation of the device where the electron beam is generated by the encapsulated carbon tips with a self-defined integrated electrostatic lens. The grown carbon-based nanostructures are responsible for electron emission and the surrounding metal, separated from the inner CNS by an insulating dielectric layer, acts as a self-defined gate to control the level of electron emission as well as a tool for beam shaping and focusing.

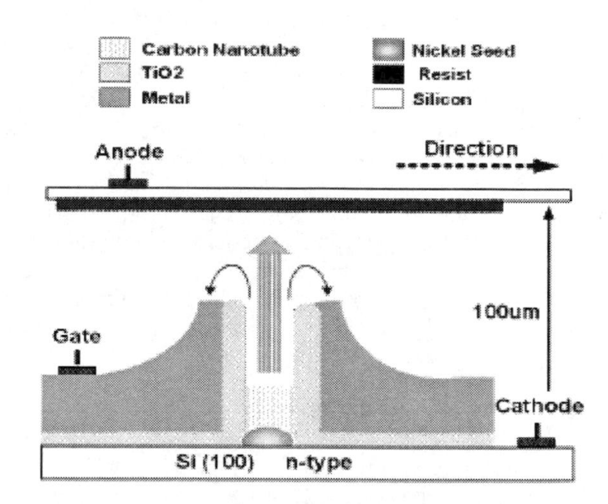

Figure 1: The schematic setup for the nano and submicron lithography using carbon nano-tubes.

The first step in the fabrication process is e-beam evaporation of a 5nm thick layer of nickel on (100) silicon substrates. The Ni-coated samples are then patterned using standard photo-lithography and are placed in a DC-PECVD chamber to perform the CNT growth. The pre-growth treatment is done at a pressure of 1.6torr in the presence of H_2 and at 650°C. After 15mins of H_2 blow, plasma turns on with a current of 30mA to form nano-islands with a typical size of 10 to 50nm. Immediately after, acetylene (C_2H_2) is introduced into the chamber to initiate the growth at a pressure of 1.8torr. Fig.2 shows an SEM image of vertical CNTs grown with a high plasma density, indicating a near-vertical growth geometry.

Figure 2: A close to vertical growth of CNT's on silicon substrates using the PECVD apparatus.

Fig.3 depicts the growth of conical nano-structures, achieved by varying the substrate temperature during the growth. The inset in this figure shows the sparsely distributed island growth, resulted from a reduced nickel thickness providing a structure suitable for nano-lithography applications.

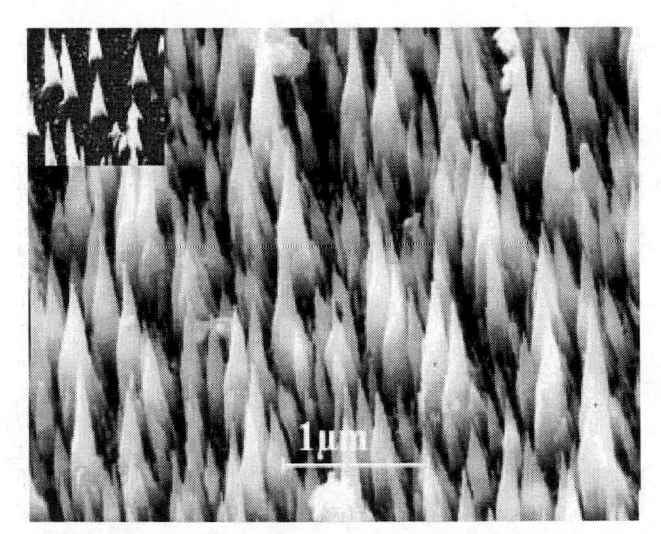

Figure 3: Growth of conical CNSs on Si substrate. Inset displays the sparse distribution of CNSs.

After unloading the samples, they are coated with 200nm of titanium-oxide using an atmospheric pressure CVD reactor at 220°C. The gate electrode is composed of 0.1μm-thick layer of Cr (or Ag) deposited by e-beam evaporation at 350°C. Final fabrication of the device requires one step of chemical mechanical polishing followed by plasma ashing to open up the nano-tips and to form partially hollow nano-pipes. This step is crucial in obtaining electron emission with a reproducible beam-shape and it is typically achieved by oxygen plasma ashing. Fig.4 shows an SEM image of a completed cluster of CNS's where all the fabrication steps were accomplished. The inset in this figure shows the magnified view of the completed nano-tips.

Figure 4: Final nano-tubes with metal and dielectric coatings. Photolithography of nickel seed layer can be used to form individual tips in desired locations.

Fig.5 depicts the electron emission current, measured at anode side against the gate voltage. As can be seen in this figure, increasing the gate voltage causes the emission current to drop significantly. The inset in this figure depicts the emission current in a semi-log scale. At a gate voltage of 10V, the emitted current begins to drop by more than three orders of magnitude.

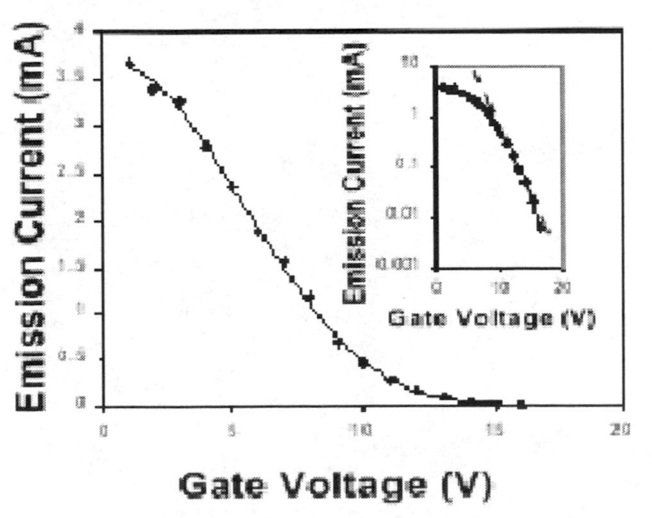

Figure 5: Emission current versus gate voltage. Inset shows the same results in a semi-log scale indicating the proper control of the gate on the emission current.

The effect of anode voltage on the current is also plotted in Fig.6 evidencing an almost linear variation of the current with respect to the anode-cathode voltage. As observed from Fig.5 and 6, the level of emission current is mainly controlled by the surrounding gate with less dependence on the anode voltage. At anode voltages below 50V, the emission current is almost constant and it is controlled mainly with the gate voltage.

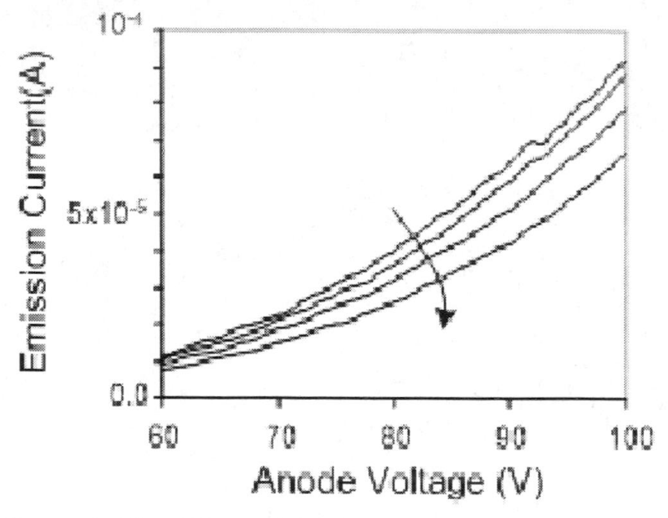

Figure 6: The Effect of anode voltage on the emission current. The arrow in this figure shows the effect of gate voltage.

By applying a proper voltage between carbon-tip and resist-coated substrate, electrons emit from the negative side onto the positive one and nano-size features are emerged. In this experiment we have used standard photo-resist.

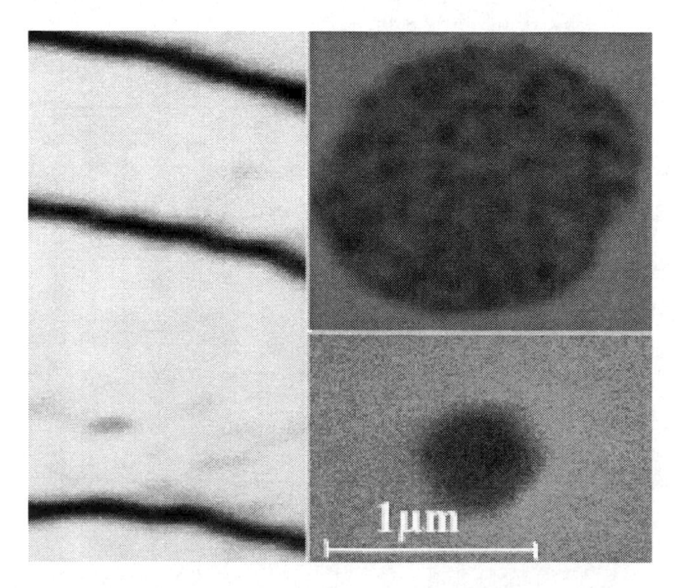

Figure 7: Evolution of nano-metric and submicron features by scanning the carbon emitter onto the surface of the top substrate using mechanical manipulation.

Using this technique, we have drawn lines with a width of 120nm and a length of 5 to 10μm. Also round dots smaller than 0.4μm were realized as shown in Fig. 7. The images at the right side of Fig.7 correspond to spots generated by holding the resist-coated substrate opposite to the electron emitting surface. The single spot at the bottom has a diameter of 400nm, whereas the top image belongs to the case where a cluster of nano-tubes have bombarded the resist-coated substrate.

Figure 8: Submicron lithography with scanning the surface of the sample.

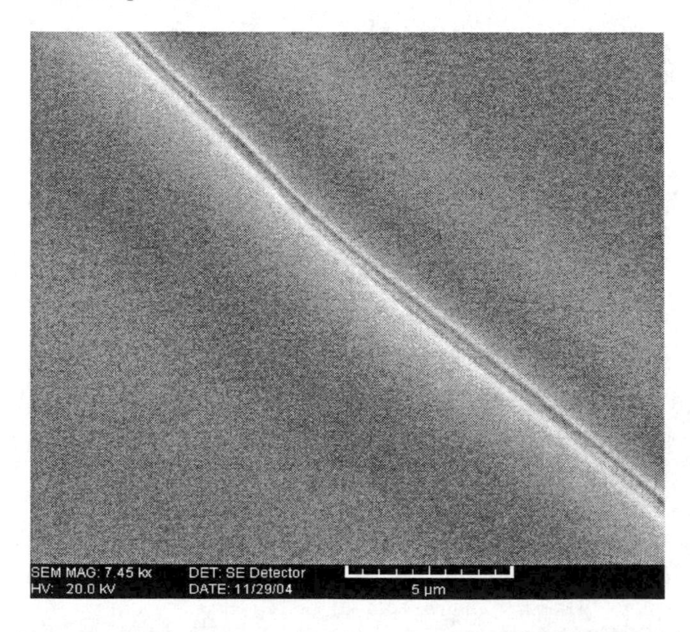

Figure 9: Straight lines with a length of 500um have been drawn using mechanical manipulation of nano-tubes.

Straight lines with thicknesses around 200nm have also been achieved (Fig.8 and 9). Also as seen from Fig.9, long lines can be realized using this approach which could be used for nano-wires. The width of the features depends mostly on the diameter and shape of the grown CNS's. By extending this approach, we have been able to draw lines with a thickness of 48nm using conventional photoresists. Further reduction of the size of the lines and better mechanical manipulation of the tips are being pursued. We are also trying to realize deep sub-micron MOSFET devices using this nano-lithography technique.

3 ACKNOWLEDGEMENT

Authors are indebted to Professor E. Asl. Soleimani for his support and encouragement. This work has been supported by a grant from Center of High Technology of the Ministry of Industry, and partial support from Research Council of the University of Tehran, Iran.

REFERENCES

[1] S. Hofmann, C. Ducati, B. Kleinsorge, and J. Robertson; Appl. Phys. Lett., Vol **83**, 4661, 2003.

[2] R. N. Franklin, Q. Wang, T. Tombler, A. Javey, M. Shim, H.Dai, Appl. Phys. Lett, Vol. **81**, 913, 2002.

[3] M.A. Guillorn, X. Yang, A.V. Melechko, D.K. Hensley, M.D. Hale, V.I. Merkulov, M.L. Simpson, L.R. Baylor, W.L. Gardner and D.H. Lowndes, Journal of Vacuum Science & Technology B Vol. **22**, 35-9, 2004.

[4] H. Hesamzadeh, B. Ganjipour, S. Mohajerzadeh, A. Khodadadi, Y. Mortazavi and S. Kiani, Carbon Vol. **42**, 1043–1047, 2004.

Single Wall Carbon Nanotube Optical Actuators: Towards Optical Artificial Muscles

Shaoxin Lu and Balaji Panchapakesan*

Delaware MEMS and Nanotechnology Laboratory, Department of Electrical Engineering, University of Delaware, Newark, Delaware, 19716, USA, * Email: baloo@eecis.udel.edu

ABSTRACT

Optically driven nanotube actuators have been fabricated from single wall carbon nanotube -acrylic elastomer bimorph sheets. They were shown to generate higher stress than natural muscles and higher strains than high-modulus ferro-electric materials. Strain measurements revealed about 0.3% elastic strain under visible light intensities of $120mW/cm^2$. A nanotube gripper based on this optical actuation principle is demonstrated to show manipulation of small objects. The principle of actuation is based on a combination of optical, thermal, electrostatic and elastic effects in the nanotube–elastomer sheets. Optically driven actuators using nanomaterials may eventually provide higher strains and lower power requirements than currently known actuation technologies.

Keywords: carbon nanotubes, acrylic elastomer, optical actuator

1 INTRODUCTION

The direct conversion of different types of energy to mechanical energy is of importance in many applications such as robotics, artificial muscles, optical displays, prosthetic devices, optical communication, micro-mechanical and microfluidic devices. Although piezoelectric ceramics, shape memory alloys, magnetostrictive materials are well known and their applications commercialized, they are handicapped by the maximum allowable operational temperatures, requirement of high voltages, and limitations on the work densities per cycle. In recent years, carbon nanotubes [1], metallic nano-particles [2] and polymer actuators [3, 4] have been proposed to be attractive alternatives to overcome these problems. Of these materials systems, polymer materials, which have stroke, force, and efficiency similar to that of human muscles, are very promising because of their low cost and wide choice of polymer materials.

Since its discovery, single wall carbon nanotubes (SWNTs) have attracted much attention owing to their excellent electrical, mechanical, optical and thermal properties [5-8]. A combination of the excellent optical and thermal properties of SWNTs could form the basis for future applications in the field of opto-thermal transduction at the micro- and nano-scale.

Many attempts have been made in the past to combine polymer material systems with SWNTs aiming to improve the properties of either system. They have been combined to improve electrical properties [9], mechanical properties of the composite [10], strain response of polymer actuator [11] and used as the functional layer in organic light emitting diodes [12,13]. However, the entire spectrum of polymer nanotube composite actuators has not been fully explored.

The new actuators we describe here use carbon nanotube-acrylic elastomer bimorph structure to realize the conversion from optical to mechanical energy. It is different from nanotube reinforced polymer actuators described recently [14] as the nanotubes are not mixed with the polymer materials to form composites but mechanically bonded to the acrylic elastomer to form a bimorph structure, which doesn't change the properties of polymer material. The primary mechanism of actuation is a combination of thermal and electrostatic mechanisms that are evidenced in nanotubes upon light absorption. The actuators overcome some of the fundamental limitations over piezoelectric, ferroelectric actuator materials and carbon nanotube based electro-chemical actuation technologies. The actuators respond to any remote source of light causing movements, eliminating the complicated electrical connections needed for electrically driven actuators.

2 EXPERIMENTAL AND DISCUSSION

The optically driven nanotube actuators as described here were surprisingly simple to fabricate. 16mg of SWNTs commercially obtained from Carbon Nanotechnologies was dispersed in 100ml iso-propyl alcohol and agitated for 20 hours to disperse the nanotubes uniformly in the solution to a final concentration of 0.16mg/ml. The SWNT suspension was then vacuum filtered through a poly (tetrafluoroethylene) filter (47 mm in diameter) [1]. The resulting SWNT sheet on the filter was rinsed twice with iso-propyl alcohol and DI water and then dried at 80°C for 2 hours to remove any remaining organic residues in the sheet. After drying, the SWNT sheet was peeled off the filter with a final thickness ranging from 30μm to 40 μm and a bulk density of about $0.3g/cm^3$. Figure 1(a) shows the optical image of a free standing SWNT sheet or "Bucky Wafer" made by vacuum filtration. Figure 1 (b) is the scanning electron microscope (SEM) image of the SWNT sheet composed of highly entangled SWNT bundles. This SWNT sheet was used in making the bimorph actuator without further optimization. The acrylic elastomer was purchased from 3M, sold as137DM-2. The material is available as a precast adhesive tape with 12.5mm in width and about 80 μm in thickness. A thin acrylic elastomer film derived from the adhesive tape, 30μm thick and 30mm×2mm in dimensions, was attached to a piece of SWNT sheet of similar dimensions by direct physical contact. The resulting bimorph structure was used to study the light induced actuation.

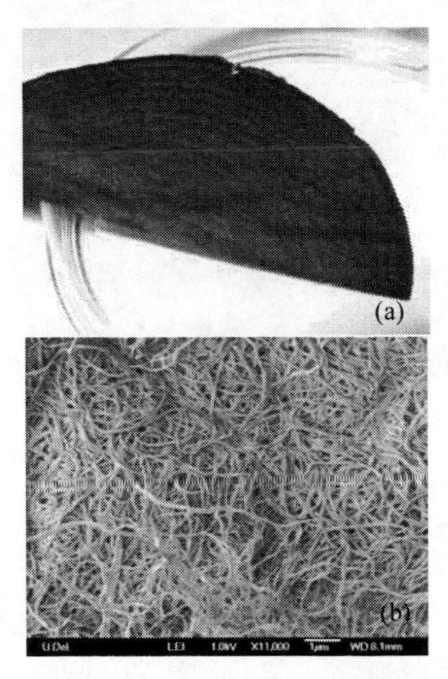

Figure 1: (a) SWNT sheet made by vacuum filtration. (b) SEM image of SWNT sheet composed of highly entangled SWNTs bundles.

As a first demonstration of actuation, a cantilever structure was made by attaching the bimorph actuator to a 100 μm thick PVC film. Figure 2(a) shows the schematic arrangement of the entire setup with the cantilever structure anchored on a base that bends in a direction normal to the cantilever surface. The structure of the bimorph actuator is shown in the inset. A halogen lamp of tunable intensity was used as the light source and was incident normal to the surface of the cantilever. The light intensity was recorded using Newport 1815-C intensity meter. A digital camera was used to characterize the displacement of the bimorph structure.

The displacement of the cantilever under optical illumination is shown in Figure 2(b). An intensity of 60mW/cm^2 was used to actuate the cantilever for four cycles. During the period of light exposure, the structure bends to the PVC side, indicating an increase in the length of the bimorph structure. Once the light source was turned off, the bimorph structure contracted back to its original length and the cantilever to its initial position, suggesting elastic deformations of the actuator upon illumination. The actuation was quite repeatable from cycle to cycle with nearly the same displacement amplitude. A maximum displacement of 4.3mm was achieved for a cantilever length of 30mm.

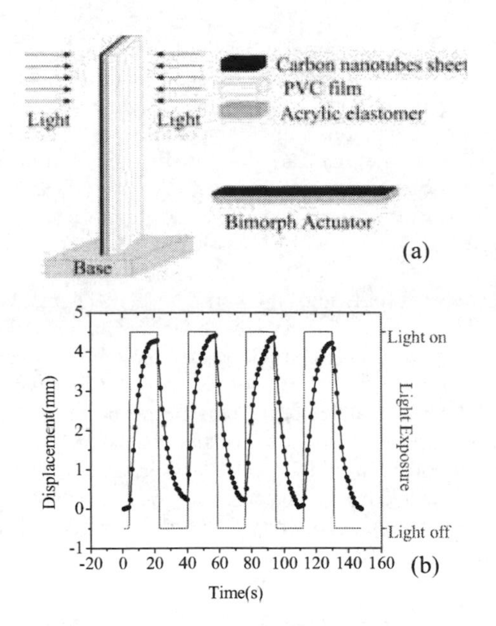

Figure 2: (a) Shows a cantilever vertically anchored on a base. The cantilever composed of a bimorph actuator (shown in the right lower part) and a 100μm thick PVC film. (b) Shows the displacement of the cantilever measured when light was turned "on" and "off".

In order to characterize the strain instead of the displacement suffered by the actuator under optical illumination, another experiment was set up as shown in Figure 3(a), in which the bimorph actuator was doubly clamped between a vertical anchor and the PVC film. The PVC film was 100μm thick and was fixed vertically to the base. The stress on the bimorph actuator (30mm×2mm), due to light incident normal to its surface, bent the PVC film. The displacement of the PVC film was recorded by a digital camera and was used to characterize the change in length of the bimorph actuator. Figure 3 (b) shows six cycles of strain response of the actuator under different light intensities. The strain cycles were quite repeatable with nearly the same strain amplitude for any given intensity. It can be seen that the strain values are positive suggesting that the SWCNT – acrylic elastomer bimorph actuator expands during optical exposure and returns back to its original strain free position when the source is turned off.

Acrylic elastomers have earlier been used as dielectric electroactive polymers as they produce higher strain and possess higher elastic energy density than any other dielectric elastomer [4, 15-17]. In the case of optical actuation, bonding the nanotube sheets with acrylic elastomer results in a new type of actuator that utilizes the opto-thermal transitions and electrostatic effect in carbon nanotubes. SWNTs are prefect black bodies and exhibit excellent optical and thermal absorption properties. Ajayan *et. al.* showed that fluffy SWNTs can burn when exposed to the flash light of a photographic flash camera [6]. This suggests that SWNTs can absorb significant amount of photon energy and convert it to thermal energy. When SWNTs absorb photons, the temperature of nanotube bundles increases dramatically owing to their extremely high thermal conductivity along the tube axis [18, 19]. According to Savas *et.al.*, the room temperature thermal conductivity of isolated SWNTs is 6600W/mK, much larger than that of pure diamond [18]. Although the thermal conductance is significantly lowered in SWNT bundles because of the anisotropic nature of thermal conductivity and thermal barriers between nanotubes, they are still very good thermal conductors [20]. In the presence of oxygen, nanotubes burn out at temperatures around 600°C. If the nanotube bundles are 'fluffy', then the dissipation of heat between nanotubes would be poor and lead to a drastic increase in localized heating achieving high temperatures. And in the presence of oxygen, the enormous heat confinement can burn the nanotubes. However, if the SWNTs were packed tightly together and were in close contact with each other, or if there was a substrate which could absorb the heat and act as a sink, then the thermal energy generated by the SWNTs can be rapidly dissipated to the other nanotubes or the substrate and make the bundles less prone to burning. Under optical illumination, nanotubes absorb the photons thereby causing the localized heating of nanotubes. But the temperature of the nanotubes never really increases drastically as the heat is readily transferred to the acrylic elastomer which is in direct contact with the nanotube sheet. This leads to a thermal expansion of the elastomer causing strain and thus actuating the bimorph structure. An SEM of the SWNT sheet was performed after two hundred cycles of exposure to light to see if the nanotubes had burned. The images obtained were similar to those obtained before light exposure showing no signs of any physical damage to the nanotubes. Figure 3(b) shows the strain in the actuator under light intensities of 70mW/cm^2, 40mW/cm^2 and 20mW/cm^2. It is apparent that the higher the intensity of light incident on the sample, the greater is the strain amplitude measured. Figure 3(c) shows the curve of strain versus incident light intensity ranging from 0 to 120mW/cm^2. It can be construed that for smaller intensities the rate of strain induced is higher than at higher intensities. At higher intensities, optical scattering and reflection and heat dissipation due to the temperature rise become more serious and in turn slow down the rate of increase of strain. A maximum strain of ~0.3% is acquired under visible light intensities of 120mW/cm^2, which is comparably larger than that from piezo-electric materials.

Having shown the origins of thermal effects, the stretching of carbon nanotubes sheets as well as acrylic elastomer due to electrostatic effects was also investigated. The reason for such investigation stems from Zhang et al [7] found that SWNTs could stretch when shining light on it, which is quite similar to the stretching under static electrical field. They concluded that the stretching was due to the locally imbalance in the charge density due to the intertube interaction, the charge transfer from metallic tubes to the semiconductor ones, interbundle or intertube barriers and poor intertube and interbundle contacts, which results in local electrical field inside the SWNTs bundles.

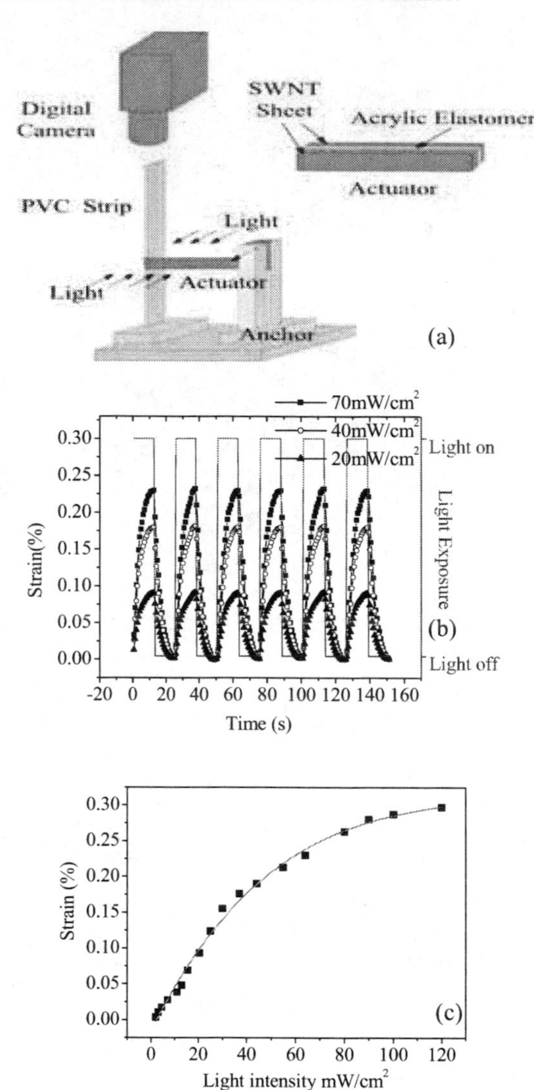

Figure 3: (a) Experimental setup for strain characterization. (b) Strain in the actuator measured under light intensities of 70mW/cm^2 (square), 40mW/cm^2 (circle), 20mW/cm^2 (triangle). (c) The strain as a function of light intensity.

Same electro-static effect could also happen in our actuator. The excess charge induced by light exposure on the carbon nanotubes distribute unevenly in the bundles and could induce local electrostatic fields, which stretch the carbon nanotubes sheet and increase its total length. Since the nanotubes are randomly aligned, the contribution of electrostatic strain in our actuator may be small compared to thermal actuation effects. However, for single nanotube actuators or small bundles of nanotubes, the electrostatic effects may dominate more than thermal effects in the actuator [7]. Further, the local electrical field at the surface between the SWNTs sheet and acrylic elastomer film could also exert Maxwell stress on the acrylic elastomer and result in the stretching of acrylic elastomer, which add up to the total strain in our samples.

The nanotube sheets that we used for our experiments were produced by the same methods as described in Reference [1]. Assuming the same value of Young's modulus of 1 Gpa for the nanotube sheets, and 0.5 Mpa for the Young's modulus of the acrylic elastomer, the Young's modulus of the bimorph structure can be calculated to be ~ 270 Mpa. The maximum stress generated in the actuator for a maximum strain of 0.3% was therefore 0.8 Mpa which is significantly higher than the peak capacity of human skeletal muscles (0.3 Mpa) and similar to the first electro-chemically driven nanotube actuators. For more optimized nanotube sheets and using ceramic materials instead of elastic polymers, one can improve the Young's modulus of the bimorph that can result in higher stress than for the non-optimized nanotube bundles as reported here. For making higher strength nanotube sheets, new processing methods are needed compared to the vacuum filtration technique which has its limitations.

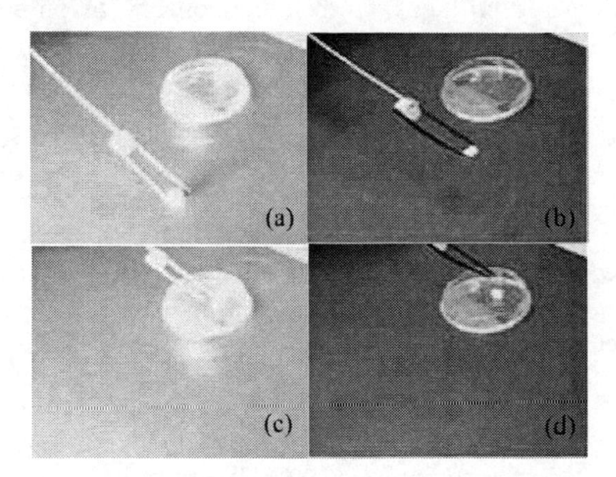

Figure 4: Showing a simple demonstration of a gripper manipulating small objects.

As a simple demonstration of the potential of this technology in different fields, a millimeter scale gripper was fabricated using the bimorph actuator which was subsequently used to manipulate small objects. The sequence of gripper manipulation is shown in Figure 4. The dimensions and structure of the bimorphs in the gripper are the same as discussed earlier and shown in Figure 2(a). The bimorph structures are attached to either side of a probe station arm such that the gripper is closed when the light source is switched off and open when the light is turned on. This setup was used to grip a piece of aluminum oxide, 4mm x 2mm in dimension and 0.3gms in weight and place it on a petri dish as shown in the Figure 4. When light is shining on the gripper, it is opened to grip the particle. After the particle is moved to the position above the petri dish, the gripper was again opened by light exposure to release the particle from it.

The demonstrations of these millimeter-scale grippers are the first prototypes showing the ability to manipulate small objects very easily using this technique. Much more work is required before the potential advantages of this technology can be experimentally assessed in practical devices. With just modest improvements, one can use this technology for cell manipulation and optically driven micro-cantilevers for medical catheter applications. The actuators are easy to fabricate and versatile compared to many other actuation technologies. It can be integrated with optical sources such as semiconductor lasers and LEDs directly to form devices on a single chip as the fabrication techniques of these are quite mature. Robotic structures from nanotubes could also be possible and could be driven optically that can be used for interplanetary space explorations.

3 ACKNOWLEDGEMENTS

Funding for this research was partially supported by the National Science Foundation Grant CCR: 0304218

REFERENCES

[1] R.H. Baughman et al., Science 284, 1340, 1999.
[2] J. Weismuller et al., Science 300, 312, 2003.
[3] D.L. Thomsen et al., Macromolecules 34, 5868, 2001.
[4] R. Pelrine, R. Kornbluh et al., Science 287, 836, 2000.
[5] S.Iijima, Nature 354, 56, 1991.
[6] P.M Ajayan et al., Science 296, 705, 2002.
[7] Y.Zhang, S.Iijima, Phys. Rev. Lett. 82, 3472, 1999.
[8] Kataura H et al., Synth. Met. 103, 2555, 1999.
[9] H. Zengin et al., Adv. Mater. 14, 1480, 2002.
[10] Z. Jia et al., Mater. Sci. Eng. A 271, 395, 1999.
[11] M. Tahhan et al., Smart Mater. Struct. 12, 626, 2003.
[12] H.S. Woo et al., Synth Met. 116, 369, 2001.
[13] P. Fournet et al., J. Appl. Phys. 90, 969, 2001.
[14] H. Koerner, G. Price et al., Nature Mater. 3, 115, 2004.
[15] B.C. Yoseph, Proc. 42nd AIAA Structures: Struct. Dyn. Mater. Conf. 1492, 2001.
[16] W. Ma, L.E. Cross, Appl. Phys. A 78, 1201, 2004.
[17] K. Guggi, S.L. Peter, K. Roy et al., J. Intell. Mater. Sys. Struct. 14, 787, 2003.
[18] S. Berber et al., Phys. Rev. Lett. 84, 4613, 2000.
[19] M. A. Osman et al., Nanotechnology 12, 21, 2001.
[20] J Hone et al., Phys. Rev. B. 59, R2514, 1999.

Directed Self-Assembly of Virus-Based Hybrid Nanostructures

Chunglin Tsai* and Cengiz S. Ozkan**

*Department of Electrical Engineering,
**Department of Mechanical Engineering,
University of California at Riverside, Riverside, CA, USA,
*ctsai@ee.ucr.edu, **cozkan@engr.ucr.edu

ABSTRACT

Viruses of various geometrical shapes have been exploited as higher hierarchical biomolecules in self-assembled nanoelectronic structures. We have demonstrated several organic virus particle and inorganic nanoparticle peptide-directed conjugations, including cylindrical tobacco mosaic virus (TMV) with quantum dots (QD) and single-walled carbon nanotubes (SWCNT) and icosahedral poliovirus (PV) with SWCNTs using ethylene carbodiimide coupling (EDC) procedure. In order to exploit these nanostructures as interconnects in nanoelectronics, metallization was also performed by reducing platinum particles onto these conjugations to make them conductive. Characterizations such as scanning and tunneling electron microscopy, fluorescent imaging and Fourier transform infrared (FTIR) spectroscopy were shown to prove the organic-inorganic connected heterostructures

Keywords: biological self assembly, functional carbon nanotubes, tobacco mosaic virus, poliovirus

1 INTRODUCTION

Virus-based self-assembly has aroused numerous interests in the field of nanoelectronics in recent years. Viruses are composed of a nucleic acid core surrounded by capsid which is made up of encoded proteins. Different viruses have different ways to encode the protein structure of the capsid in that they have various shapes. Viruses of different geometrical shapes have been exploited as higher hierarchical biomolecules in self-assembled nanostructures due to their manifold configurations, well characterized surface protein properties, and nanoscale dimension. Along with the special antibody-antigen recognition ability and site-directed conjugation, viruses exhibit the uniqueness as self-assembled materials. Many self-assembled or directed self-assembled viral nanostructures have been reported, such as viral self-assembled semiconductive and magnetic nanowires [1-3] and oriented quantum dot nanowires [4, 5]. Viral nanosensors based on the principles of viral induced self-assembly were also demonstrated [6]. These results have proven that virus-based biotemplate is a promising technique in directed nanostructure synthesis.

In this paper, we present several peptide-directed organic-inorganic heterto-nano-structures, including cylindrical tobacco mosaic virus (TMV) with quantum dots (QD) and single-walled carbon nanotubes (SWCNT) and icosahedral poliovirus (PV) with SWCNTs using ethylene carbodiimide coupling (EDC) chemistry [7]. Meanwhile, electrodeless chemical deposition [8] of platinum metallized poliovirus and tobacco mosaic virus particles is also reported.

2 MATERIALS AND METHODS

2.1 Coupling of viruses and nanoparticles

For exploring possible combinations of virus particles with inorganic nanoparticles, we chose wild type icosahedral poliovirus [9] and wild type rod-shaped tobacco mosaic virus [10] due to their dissimilarity in physical shape. Among three poliovirus serotypes, type one was selected owing to it generality. Type one poliovirus was replicated in monkey kidney cells and modified for vaccine use so that they are less toxic than normal wild type polioviruses and not harmful for human beings with vaccine shot. The original poliovirus samples that we acquired had the concentration of 10^5 particles in one μL. Tobacco mosaic viruses purchased from American Type Culture Collection (ATCC) were common strain with concentration of 26 μg/μL.

Single-walled carbon nanotubes that we purchased were refluxed in 2N HNO_3 for 5 hours followed by 30 minutes sonication so that we had proper oxidizes sites on SWCNTs. After sonication, SWCNTs were filtered using a vacuum filter unit with one micron Millipore filter. During the filtration, excess amount of deionized water was added at a slow rate to ensure completely removal of the acid from nanotubes. Filtered SWCNT cake was then vacuum dried overnight and added to deionized water for making nanotube suspension.

In order to functionalize quantum dots with carboxyl terminal groups on the surface, we mixed 50 μL ZnS capped CdSe quantum dots (Core Shell Topo Evidots, Evident Technologies) with 5 mg/mL concentration in toluene solvent with 11 μL mercaptoacetic acid (MAA) to activate the surface of QD for four hours. Then, the supernatant was removed from the sample after it was centrifuged for three minutes at 13,000 rpm. The pellet left after centrifugation was re-suspended with 50 μL MeOH and repeated the centrifugation-resuspension process several time in order to completely remove the residual MAA. After MeOH wash, QD pellets were re-suspended in deionized water and transferred to Millipore tube (Millipore UFV5 BQK 25) for centrifugal filtration. The QD solution fluorescence was examined by ultraviolet excitation under optical microscope.

After the carboxyl functionalization of SWCNT and QD, they were added to deionized water followed by EDC (0.5M) addition to activate the carboxyl groups. Sulfo-NHS was added later on to initiate the covalent conjugation between carboxyl and amino groups. Virus was added to the coupling solution and incubated for 4 hours.

The mixture was desalted by centrifugal filtration for scanning electron microscopy (Zeiss Supra 55) and tunneling electron microscopy (Philips CM300) sample. Solution was dropped on a Si chip and dried in a desiccator for SEM analysis. Figure 1 summarized the process for EDC coupling between virus particles and inorganic nanoparticles. Fourier transform infrared spectra using AgCl window were acquired by dropping each sample onto the window analyzed by Bruker Equinox 55 FTIR.

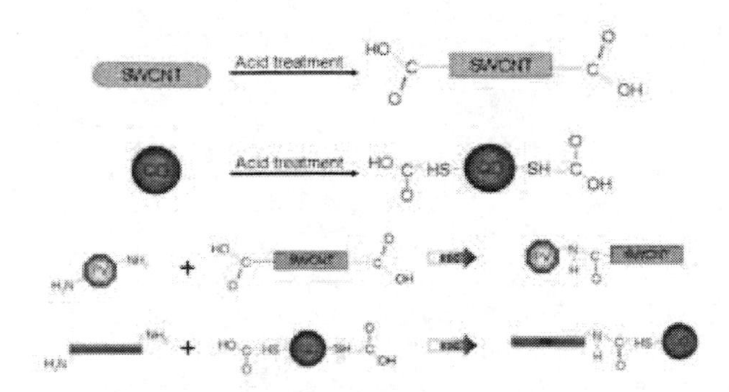

Fig. 1 (a) single-walled carbon nanotube oxidized by HNO_3 in order to open the cap at both end and introduce carboxyl functional group on nanotube (b) quantum dot functionalized with amino groups so that it can be used in combination formed by EDC chemistry (c) EDC coupling of poliovirus with carboxylated SWCNT (d) EDC coupling of tobacco mosaic virus with carboxylated QD

2.2 Platinum metallization of viruses

Virus metallization was originally accomplished by adding 65 μL of 1 mM Potassium Tetrachloroplatinate (K_2PtCl_4) to 1μL of virus solution. The mixture was kept at room temperature for aprroximately 4 hours to let virus capsid be activated by Pt ion. After that, 1 μL 10 mM Dimethylaminobenzaldehyde (DMAB) was added to the sample to provide a reduction bath for Pt ions to form metal clusters on the surface of virus particles. The sample was kept 27 degrees centigrade for 2 to 3 hours. The schematic of the Pt metallization is shown in Figure 2. We also manipulated several parameters such as temperature of the reduction bath, concentration of every chemical and virus sample, and the time of incubation of sample mixtures. The samples were characterized by scanning and tunneling electron microscopies.

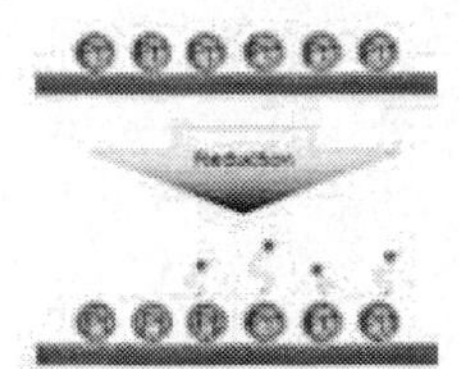

Fig. 2 Platinum ions first activated the surface proteins of viruses before reduction bath applied. After activation, DMAB was added to reduce Pt ions which attached on the capsids of viruses to Pt atom.

3 RESULTS AND DISCUSSION

For the purpose of self-assembled conjugation, we have applied several chemical reagents for different forms of conjugations, including EDC coupling with the aid of sulfo-NHS to stabilize the O-acylisourea generated by the reaction of EDC and carboxylate group and the reduction of platinum ion solution to deposit platinum clusters on the protein surface of virus. The results have been characterized by optical microscopy, scanning and tunneling electron microscopies, and Fourier transform infrared spectroscopy.

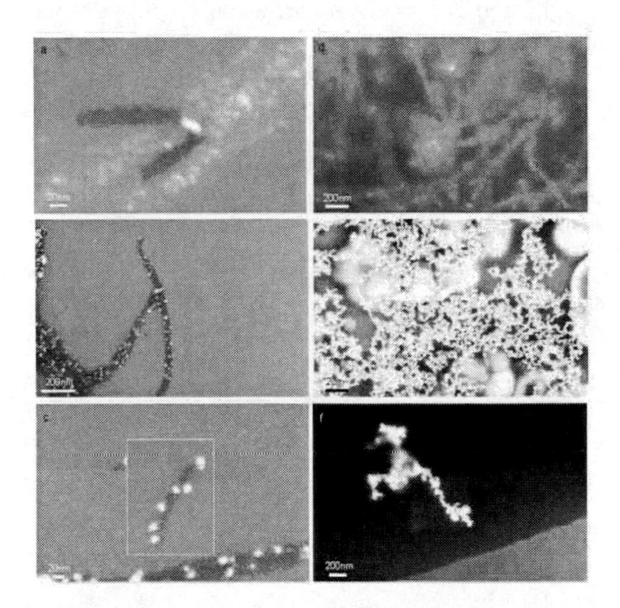

Fig. 3 SEM images of (a) TMV-QD-TMV structure (b) TMV with platinum clusters on capsid. (c) Close up image of a single TMV particle with six tiny platinum clusters on the surface. For self-assembled interconnect purposes, we can obtain conductive rod-shaped virus particle without sacrificing the targeting ability of virus by minimum metallization. (d) PV particles attached at the end and sidewall of carbon nanotubes. (e) Lower magnification of metallized PV particles is illustrated in this image. (f) A standout metallized PV aggregation. The size of the metal clusters depends upon how long the metal ion solution activates the virus sample. The longer the time, the larger the metal particle will be after reduction bath is introduced.

The scanning electron microscopy study of several conjugation formats of viruses with inorganic nanoparticles is illustrated in figure 3. In figure 3(a), a TMV-QD-TMV formation was formed by a QD cluster linking two small bundles of two to three TMV particles. Poliovirus particles attached on the end and the sidewall of SWCNT by EDC coupling is shown in figure 3(d). As we can tell, most of the PV particles aggregated at the end of carbon nanotubes. But there are also some virus particles affix to the sidewall of nanotubes due to the sidewall oxidization. We can control the oxidized sites by altering the oxidizing time and the acid normality. Therefore, depending upon the application, sidewall attachment on carbon nanotube is an alternative.

Figure 3(b) and 3(c) demonstrate the Pt metallization on TMV particles with Pt particle size of approximately 20 nm in diameter. The size of metal cluster is determined by the time that metal ions activate the protein surface and also the time of reduction bath. The SEM images shown in the paper imply the possibility of minimum metallization of virus particles so that the virus is conductive for the use of nanoelectronic interconnects while the recognition capability is maintained for selectively binding with its antibody counterpart to build a more complicated two or three dimension circuit network. Poliovirus particles metallized by Pt is shown in figure 3(e) and 3(f) with different magnification. Since PV particle has diameter of 30 nm, metal particles with similar size covered all around the virus capsid. With the freedom to control the degree of metallization and to choose the versatile geometry of viruses, such as rod-shape TMV and icosahedral PV, the configuration of bio-nanoelectronics is beyond one's imagination.

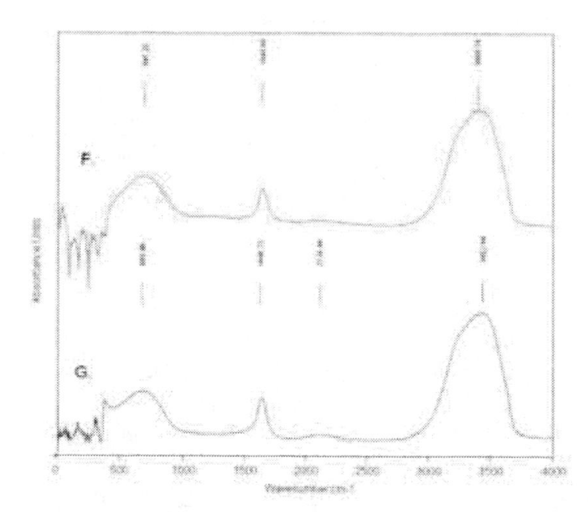

Fig. 4 Fourier transform infrared spectra of TMV-QD using AgCl window for background. (a) after EDC coupling reaction applied. (b) before applied EDC coupling. The alkyne bond in TMV was suppressed after conjugation.

FTIR spectra of TMV-QD covalent coupling using AgCl window shown in figure 4 indicates the amide (1645 cm-1) of TMV and R-CONH-R'(3405 cm-1) bonding, compared to FTIR spectrum of TMV in figure 4(b). Alkynes bond in TMV is suppressed after conjugation. The hybridization of Pt and QD cluster with TMV is also examined under tunneling electron microscopy as shown in figure 5. Figure 5(a) indicates QD cluster surrounding TMV particle bundles forming a large QD-TMV network. The metallization of TMV by Pt ion is shown in figure 5(b). The dark spots in the TEM image are the platinum metal clusters lying quite uniformly on the scaffold constructed by TMV particles. Furthermore, to confirm the existence of QD, we identified the QD in our covalent coupling virus samples by fluorescence imaging shown in figure 6. Although the visualization of virus particles is not feasible under optical microscope, we still can see, in figure 6(a), the fluorescence of QD forming a line in QD-TMV sample comparing to QD-PV sample which just randomly dispersive on the substrate.

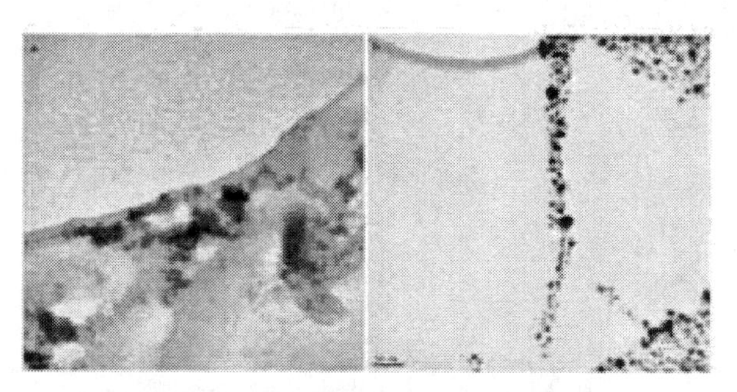

Fig. 5 Tunneling electron microscopy images of (a) TMV-QD (b) TMV-Pt under 100kV. Because of the EDC coupling chemistry, QD and TMV tends to bundle together forming a network. Instead in Pt reduction on TMV particle, Pt clusters distribute more uniformly along the axis of TMV particle.

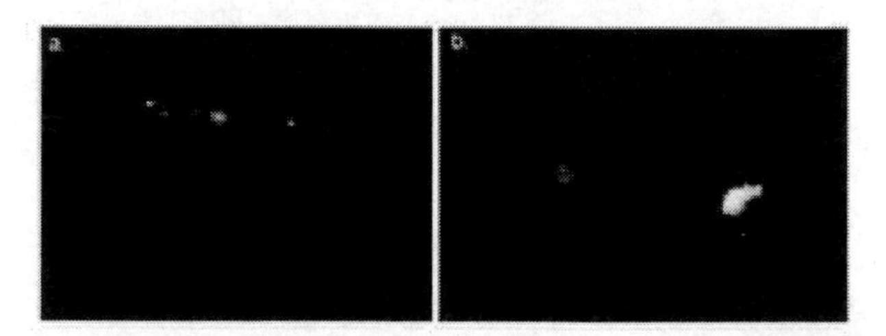

Fig. 6 Optical microscopy images of (a) TMV-QD (b) PV-QD. In (a), there are several small bright green spot forming a belt which could be the TMV underneath. We did not observe this belt-like bright cluster arrangement in PV-QD sample.

We have reported not only the covalent conjugation of different viruses with SWCNTs and QDs using EDC with sulfo-NHS but also the metallization of viruses by Pt salt reduction. Besides various approaches of hybridization, we have also provided several degrees of freedom to manipulate the structure of complexes, such as the choice of sidewall attachment of carbon nanotubes and

NSTI-Nanotech 2005, www.nsti.org, ISBN 0-9767985-1-4 Vol. 2, 2005

the targeting ability of minimum metallized virus particles. Future work to utilize antibody for the selective metallization and bioconjugation is underway.

4 ACKNOWLEDGEMENT

We acknowledge the financial support of this work by Functional Engineered Nano Architectonics (FENA) in Microelectronics Advanced Research Corporation (MARCO).

REFERENCES

[1] Mao, C., et al., *Virus-based toolkit for the directed synthesis of magnetic and semiconducting nanowires.* SCIENCE, 2004. **303**(5655): p. 213-217.

[2] Knez, M., et al., *Biotemplate synthesis of 3-nm nickel and cobalt nanowires.* NANO LETTERS, 2003. **3**(8): p. 1079-1082.

[3] Flynn, C., et al., *Viruses as vehicles for growth, organization and assembly of materials.* ACTA MATERIALIA, 2003. **51**(19): p. 5867-5880.

[4] Mao, C., et al., *Viral assembly of oriented quantum dot nanowires.* PROCEEDINGS OF THE NATIONAL ACADEMY OF SCIENCES OF THE UNITED STATES OF AMERICA, 2003. **100**(12): p. 6946-6951.

[5] Lee, S., et al., *Ordering of quantum dots using genetically engineered viruses.* SCIENCE, 2002. **296**(5569): p. 892-895.

[6] Perez, J., et al., *Viral-induced self-assembly of magnetic nanoparticles allows the detection of viral particles in biological media.* JOURNAL OF THE AMERICAN CHEMICAL SOCIETY, 2003. **125**(34): p. 10192-10193.

[7] Ravindran, S., et al., *Covalent coupling of quantum dots to multiwalled carbon nanotubes for electronic device applications.* NANO LETTERS, 2003. **3**(4): p. 447-453.

[8] Richter, J., et al., *Nanoscale palladium metallization of DNA.* ADVANCED MATERIALS, 2000. **12**(7): p. 507-+.

[9] Belnap, D., et al., *Three-dimensional structure of poliovirus receptor bound to poliovirus.* PROCEEDINGS OF THE NATIONAL ACADEMY OF SCIENCES OF THE UNITED STATES OF AMERICA, 2000. **97**(1): p. 73-78.

[10] Knapp, E. and D.J. Lewandowski, *Pathogen profile Tobacco mosaic virus, not just a single component virus anymore.* MOLECULAR PLANT PATHOLOGY, 2001. **2**: p. 117-123.

Carbon Nanotubes as Machine Elements – A Critical Assessment

P.V.M. Rao, Manvinder Singh, Vijay Jain

Department of Mechanical Engineering,
Indian Institute of Technology, Delhi
New Delhi -110 016, India
pvmrao@mech.iitd.ernet.in

ABSTRACT

Gears, Bearings, Springs, Fasteners etc. are some of the typical machine elements used to build machines and mechanical systems. The function of these machine elements is to transmit motion, to support moving members, to store energy, to join two components etc. Carbon nanotubes can be used as these machine elements when building nano-machines or nano-mechanical systems. In this paper we review and do a critical assessment of past work done on use of carbon nanotube as machine elements. The suitability, advantages and limitation of carbon nanotube for each category of machine elements is discussed. The paper also discusses future directions of research in building nano-machines and nano-mechanical systems using carbon nanotube based machine elements.

Keywords: nanotubes, nano-mechanical systems, nano-machines, carbon nanotube.

1 INTRODUCTION

Machines built at the molecular level have been contemplated extensively in the past few years [1]. Carbon nanotubes can be used as building blocks for realizing machines and mechanical systems at a nano level. Ever since the preparation of carbon nanotubes (CNTs), also known as buckytubes, first reported by Iijima [2], they have attracted unprecedented attention for there unique attributes namely mechanical, electrical and physical attributes [3,4]. CNTs are needle like tubes and can be thought of as sheets of graphite rolled into cylinders and capped at both ends. CNTs can be single wall (SWNTs), double walled (DWNTs) or multiple walled (MWNTs). A wide variety of applications have been suggested for CNTs, however the major emphasis has been on applications like composites, hydrogen storage, electrochemical uses etc. [5]. CNTs as building blocks of movable machine parts is another major field [6] which needs to be explored.

The major machine elements which have been conceptualized at the nano-levels and are built using carbon nanotubes are springs [7-10], bearings [11-15], gears [16,17] and threaded elements [6,18]. Apart from these other related mechanical components have also been fabricated or proposed using carbon nanotubes. A summary of common mechanical elements built using carbon nanotubes is presented in Table 1.

Carbon nanotubes (Figure 1) have an interesting combination of properties that make them viable for machine elements [5]. CNTs have a lows inter shell co-efficient of friction [19,20] which make them suitable for rotational and sliding machine elements, primarily linear and rotational bearings [11,13-15]. Efforts have been made to accurately determine and study its dependence on factors like tube configuration and temperature [21] as well as pressure [22]. CNTs also have a high aspect ratio, high strength, high stiffness, low density which makes them suitable for a class of machine elements.

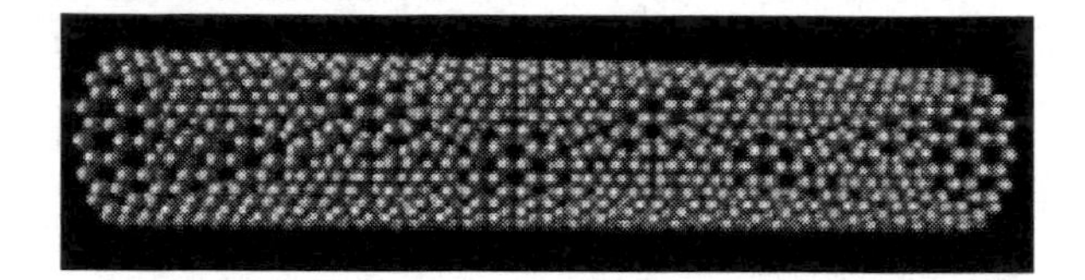

Figure 1: Single Wall Carbon Nanotube

Nanotubes have a very broad range of electronic, thermal, and structural properties that change depending on the different types of nanotube (defined by its diameter, length, and chirality, or twist). Besides having a single cylindrical wall, with nanotubes (SWNTs) one can have multiple wall nanotubes (MWNTs) -- cylinders inside other cylinders. As strength is one of the critical properties required by many mechanical elements, the following are the accepted values for the elastic properties of CNTs.

- Young's Modulus (SWNT) 1 TPa [35-39]
- Young's Modulus (MWNT) 1.28 TPa [37]
- Maximum Tensile Strength 30 GPa [40]

In this paper we look at possible applications of carbon nanotubes as machine elements. For each type of machine elements, we have discussed the suitability of carbon nanotube in terms if its strengths and weaknesses. Despite the progress, a lot of research is still needed to explore their possibilities for existing and newer applications.

No	Element	Type	Properties	Major issues	References
1	Linear Bearings	MWNT, 'Bamboo Configuration' [12]	• Controlled and reversible telescopic extension. • No wear or Fatigue.	Capped Ends as major hindrance	[11], [7], [13]
2	Linear Springs	MWNT SWNT	• High Mechanical Energy storage. • Controlled & reversible telescopic extension. • Large & reversible volume reduction due to crush & flattening of cross-section.		[7], [8]
3	Torsional Springs	MWNTs	• High Aspect Ratio Beams. • Chemically inert and no surface roughness.	Disordered form of coupling in MWNTs. Similar devices may be produced using SWNTs	[9], [23], [24]
4	Rotational Bearings	MWNTs, DWNTs	• Low Inter shell friction. No wear	Critical threshold exists for the maximum strain energy.	[14], [15]
5	Gears	MWNTs.	• High Young's Modulus • Electronic properties vary as a function of diameter and chirality	Work concentrated at simulation level.	[16], [17]
6	Conduits	SWNTs	• Structure of nanotubes.		[25]
7	Bolt/Nut Pair	DWNTs	• Potential relief analogous to a thread of a bolt		[6], [18]
8	Drive Shaft	MWNTs	• High Stiffness • High aspect ratio		[26]
9	Ropes	SWNTs	• Deformability of Cross-section. • High strength and Stiffness.	Larger diameter nanotubes are required.	[27], [28]
10	Tweezers	Two CNTs attached to electrode.	• Mechanical toughness • Electrical conductivity.	Used for mechanical manipulation and for electrical applications	[29]

Table 1: Carbon Nanotubes as Machine Elements

2 CARBON NANOTUBES AS MACHINE ELEMENTS

2.1 Springs and Bearings

Bearings and springs are by far the two most well understood carbon nanotube machine elements. It is now possible to make the layers of a MWNT slide over one another in a controlled fashion, thus leading to the fabrication of linear bearings and springs which are completely wear-free due to the extremely small friction force [7,11,13]. The springs formed from MWNTs are constant force springs. In a very similar fashion this low inter-shell friction can be utilized to make rotational bearings from DWNTs or MWNTs [14]. Thermal effects on these bearings have also been studied through molecular dynamics simulations at finite temperature.[15]

Applying pressure on the CNT material can also store mechanical energy. The density of CNT tends towards that of graphite on application of pressure because of flattening of tube cross-section. The reversible work done was found to be 0.18 eV/C atom and pressure at maximum compression was 29 Kbar [8].

Torsional springs have also been successfully made from carbon nanotubes [9,23,24]. The estimation torsional spring constant as well as shear moduli through controlled experiments has also been reported[23]. The inter-shell coupling force however is found to vary in MWNTs due to disordered bonds present in between layers. Thus there is a need to develop these devices by using SWNTs which are free from these defects [9].

2.2 Gears

If we bond rigid molecules like benzene to a carbon nanotube shaft by techniques such as Scanning Tunnel Microscopy (STM), it may be possible create nano-gears. [16] Currently research on this topic is limited to molecular dynamics simulations [17] but potential applications if successfully implemented are numerous. Ref. 16 also studies the chemical feasibility of bonding atoms to the shaft. Also presented are some processes to manufacture such a system. To run such a system we can place free positive and negative charges along the "gear teeth". These charges can easily be introduced through simple functional group substitution. Srivastava [17] has shown molecular dynamics simulations for such a system where a laser electric field is applied to generate motion.

2.3 Nut-Bolt Pair

An interesting observation in DWNTs is that they have a potential relief analogous to a thread of a bolt.[18] It has been proposed to use this property to generated nanotube devices which have linear as well as rotational motion between shells. Thus acting as a Nut and Bolt pair where the relative motion of the nanotube walls takes place along these "thread" lines.[6] The dynamics of such a motion have also been studied and some applications like nano-resistor and nano-drill have been proposed.[6]

2.4 Other Machine Elements

Besides the components discussed above, CNTs have also been proposed for use as other machine elements in nano mechanical systems not necessarily involving motions. These are conduits, drive shaft, ropes, tweezers etc. Nanotweezers developed are Nano Electro-Mechanical Systems (NEMS) for manipulation and interrogation of nanostructures. These nanotweezers are made of two carbon nanotube probes forming the arms of the tweezers [29]. Sub-micron clusters and nanowires were successfully grabbed and manipulated by these nanotweezers. Nanotubes are a good choice for construction of nanotweezers because they exhibit remarkable mechanical toughness and also electrical conductivity.

It has also been found that carbon Nanotubes can be used for transporting nano-materials, thus acting as conduits, by application of a voltage gradient [25]. Nano-ropes can be built from nanotubes and used for the purpose of load transfer.[28] This however is not a true mechanical machine element and should be dealt with in the field of composites.

3 SCOPE FOR FUTURE WORK

Some of the important barriers for realization of the true potential of carbon nanotubes are cost, polydispersity in nanotube type and limitations in processing and assembly. [5]. Attempts have been made to develop methods for continuous production [30]. Some element specific hindrances need to be worked upon.

There is an urgent need to model and simulate performance of carbon nanotube based machine elements considering mechanical, electros-static and electromechanical forces. The results of these simulations will pave the way for building nanomechanical systems purely based on carbon nanotubes.

Assembly and fabrication of nanomachines based on carbon nanotubes is another important issue which has not been dealt in the literature. Maintenance and environmental factors are crucial for successful realization of such machines which need to be addressed.

We plan to conceptualize, simulate and build complete mechanical systems using nanotubes. The work of modeling and simulation of carbon nanotube based machine elements is underway and results of which will be communicated in future.

REFERENCES

[1] Drexler K. E., "Nanosystems: Molecular Machinery, Manufacturing and Computation", New York: Wiley, 1992.

[2] S. Iijima, Nature 354, 56, 1991.

[3] DresselhausM S, Dresselhaus G and Avouris Ph, "Carbon Nanotube: Synthesis, Structure, Properties and Applications", Berlin: Springer, 2001.

[4] Dong Qian, Gregory J Wagner, and Wing Kam Liu, Min-Feng Yu, Rodney S Ruoff, "Mechanics of carbon nanotubes", Appl. Mech. Rev. 55, 495 (2002)

[5] Baughman et al., "Carbon Nanotubes--the Route Toward Applications," 297, 787-792, Science, 2002.

[6] Yu.E. Lozovik, AV.Minogin, AM.Popov, "Nanomachines Based on Carbon Nanotubes," Phys. Lett. A 313, No1-2, 112 – 121, 2003.

[7] A. Zettl and John Cumings, AIP Conference Proceedings Vol 544(1), 526-532. 2000.

[8] Chesnokov SA, Nalimova VA, Rinzler AG, Smalley RE, and Fischer JE, "Mechanical Energy storage in Carbon Nanotube Springs ," Phys. Rev. Lett., 82, 343–346, 1999.

[9] S. J. Papadakis, A. R. Hall, P.A.Williams, L. Vicci, M.R. Falvo, R. Superfine and S.Washburn, "Resonant Oscillators with Carbon-Nanotube Torsion Springs," Phys. Rev. Lett., 93(14), 146101-1, 2004.

[10] Baughman et al., "Carbon Nanotube Actuators,", Science, 284, 1340-1344, 1999.

[11] Cummings, J and Zettl, A., "Low friction nanoscale linear bearing realized from multiwall carbon nanotubes," Science, 289, pp 602-204, 2000.

[12] Iijima S, Ajayan PM, Ichihashi T., "Growth model for carbon nanotubes," Phys Rev Lett., 69(21), 3100-3103, 1992.

[13] A. Kolmogorov and V. H. Crespi, "The smoothest bearings: interlayer sliding in multiwalled carbon nanotubes," Phys. Rev. Lett., 85, 4727–4730, 2000.

[14] Bourlon B, Glattli DC, Miko C, et al., "Carbon nanotube based bearing for rotational motions", Nano Letters, 4, 709-712, 2004.

[15] Sulin Zhang, Wing Kim Liu, and Rodney S. Ruoff "Atomistic Simulations of Double-Walled Carbon Nanotubes (DWCNTs) as Rotational Bearings," Nano Letters 4, 293-297, 2004.

[16] Han, J., Globus, A., Jaffe, R. and Deardorff, G., "Molecular dynamics simulations of carbon nanotube-based gears," Nanotechnology, 8, 95-102, 1997.

[17] Srivastava, D. "A Phenomenological Model of the Rotation Dynamics of Carbon Nanotube Gears with Laser Electric Fields", Nanotechnology, 8, 186, 1997.

[18] R.Saito, R. Matsuo, T. Kimura, G. Dresselhaus, M.S. Dresselhaus, Chem. Phys. Lett., 348, 187, 2001.

[19] M.F. Yu, O. Lourie, M.J. Dyer, K. Moloni, R.S. Ruoff_, Science, 287, 5453, 637, 2000.

[20] M.F. Yu, B.I. Yakobson, R.S. Ruoff_, J. Phys. Chem. B 104, 37, 8764, 2000.

[21] J. Servantie, P. Gaspard, "Methods of Calculation of a Friction Coefficient: Application to Nanotubes," Phys. Rev. Lett., 91(18), 185503-4, 2003.

[22] Xinling Ma, Hongtao Wang, and Wei Yang, "Tribological Behavior of Aligned Single-Walled Carbon Nanotubes," Journal of Engineering Materials and Technology, 126, 258-264, 2004

[23] Williams PA, Papadakis S.J., Patel A.M, Falvo M.R., Washburn S, Superfine R., "Torsional response and stiffening of individual multi-walled carbon nanotubes," Phys Rev Lett, 89(25), 255502, 2002.

[24] Williams, P.A., Papadakis, S.J. Patel, A.M.; Falvo, M.R. Washburn, S. Superfine, "Fabrication of nanometer-scale mechanical devices incorporating individual multiwalled carbon nanotubes as torsional springs," R. Applied Physics Letters, 82, 805-807, 2003.

[25] B.C. Regan, S. Aloni, R.O. Ritchie, U. Dahmen, and A. Zettl, "Carbon nanotubes as nanoscale mass conveyors," Nature, 428, 924-927, 2004.

[26] Jian-Min Li, "Fabrication of a carbon nanotube drive shaft component," Nanotechnology, 15, 551-554, 2004.

[27] Jian Ping Lu, "Elastic Properties of Carbon Nanotubes and Nanoropes", Phys. Rev. Lett. 79, 1297–1300, 1997.

[28] Dong Qian et. al., Composites Science and Technology, 63, 1561–1569, 2003.

[29] P. Kim, C. M. Lieber, "Nanotube Nanotweezers," Science, 286, 2148, 1999.

[30] Ishigami et. al. Chemical Physics Letters 319, 457–459. 2000.

[31] Su C.J., Hwang D.W., Lin S.H., Jin B.Y. and Hwang L.P., "Self-organization of triple-stranded carbon nanoropes"

[33] Peter J. F. Harris. "Carbon Nanotubes and Related Structures: New Materials for the Twenty-First Century"

[34] M. Ishigami, John Cumings, A. Zettl, S. Chen "A simple method for the continuous production of carbon nanotubes"

[35] "Energetics, Structure, Mechanical and Vibrational Properties of Single Walled Carbon Nanotubes (SWNT)", Guanghua Gao, Tahir Cagin, William A. Goddard III

[36]"Young's Modulus of Single-Walled Nanotubes", E. Dujardin, T. W. Ebbesen, and A. Krishnan, P. N. Yianilos, M. M. J. Treacy,

[37] "Nanotubes: Mechanical and Spectroscopic Properties". E. Hernández and Angel Rubio

[38]"Physics News Update, The American Institute of Physics Bulletin of Physics News, Number 279 (Story #2)", July 15, 1996, Phillip F. Schewe and Ben Stein.

[39] Lecture given at Michigan State University by Phaedon Avouris, a nanotube researcher at the IBM labs. [2000]

[40] Min-Feng Yu, Bradley S. Files, Sivaram Arepalli, Rodney S. Ruoff, Phys. Rev. Lett. 84, 5552 (2000).

Frequency Stability and Noise Characteristics of Ultra-High Frequency (UHF) Nanoelectromechanical Resonators

X.L. Feng* and M.L. Roukes**

*Electrical Engineering, **Applied Physics, Physics & Bioengineering
California Institute of Technology, MC 114-36
Pasadena, CA 91125, USA, *Email: xfeng@caltech.edu

ABSTRACT

Nanoelectromechanical silicon carbide doubly-clamped beam vibrating resonators operating in the ultra-high frequency (UHF) band, with resonance frequencies of 395MHz, 411MHz, 428MHz, and 482MH, and quality factors (Q's) of 2000~3000, have been demonstrated. The readout of electromechanical resonances from the nanodevices is integrated with a low-noise phase-locked loop to implement *real-time* resonance frequency locking and tracking, with which the frequency stability and phase noise of the resonators are measured. The measured frequency stability shows typical crystal resonator behavior, and represents *unprecedented* mass sensitivity (in 10^{-20}g ~ 10^{-21}g scale) achieved by nanoelectromechanical resonant mass sensors. The achievable lowest phase noise of the system is ultimately determined by the thermomechanical noise of the resonator device, but presently limited by the thermal noise and other noise processes of the measurement electronic system.

Keywords: nanoelectromechanical systems, resonator, mass sensor, frequency stability, phase noise

1 INTRODUCTION

Nanoelectromechanical systems (NEMS) are emerging as great candidates for a variety of technological applications ranging from sensors and actuators to signal processing and communications [1]. Within the past decade, it has already been firmly established that advances in the mainstream microelectromechanical systems (MEMS), as enabling technologies, are demonstrating immense potentials for signal processing and communications based on micromachined devices and systems [2]. Facilitated by the advanced nanofabrication technologies, as the MEMS devices are scaled down into their NEMS counterparts, smaller device size and mass, lower power consumption, higher operating frequencies, higher responsivity and sensitivity, and other superb properties are attained [1]. In particular, nanofabricated electromechanical resonators (i.e., NEMS resonators) routinely operate at higher frequencies (often in the VHF and UHF bands) than their MEMS counterparts do. Therefore, radio-frequency (RF) NEMS resonators are promising to be the future on-chip high-Q resonators for frequency generation and conversion in RF electronics and wireless communication systems. Moreover, NEMS resonators' high operating frequencies, small mass, and high-Q's, also make them natural choices for resonant mass and force sensors with ultrahigh sensitivities [3,4], and for approaching quantum limits in fundamental physics measurements [5,6]. In almost all these cases, high-frequency, high-Q and low-noise operation are desirable, while on the other hand the implementations with nanoscale devices impose tremendous challenges. Thus, new and elaborate engineering is *crucial* to realize the above projected applications with NEMS resonators. Particularly, as the ultimate performance of NEMS is limited by various fundamental noise processes [7], it is of significance to get a comprehensive understanding of the frequency stability and noise processes of NEMS resonators. In this work, we present the initial experimental study of frequency stability and phase noise of UHF NEMS resonators.

2 UHF NEMS RESONATOR DEVICES

In order to make high-frequency mechanical resonators, it is favorable to choose materials with high modulus-to-density ratios, as resonance frequencies of most flexural modes follow $f \propto \sqrt{E/\rho}$, with E the elastic modulus and ρ the density. Hence, new promising materials such as SiC, diamond, and carbon nanotubes are better choices than conventional Si and GaAs in achieving higher-operating frequencies. In the present study, monocrystalline 3C silicon carbide (3C-SiC) thin layer epitaxially grown on silicon is used as the structural material of the vibrating UHF NEMS resonator devices, as it has been proven to be remarkably suitable for making UHF NEMS resonators [8]. The device nanofabrication process, as described in detail in [9], consists of definition of micron-scale by photolithography, the pattern transfer of the nanoscale device via electron-beam lithography and metallization, and finally a two-step electron-cyclotron resonance (ECR) plasma etching to suspend the NEMS device. To minimize the mass loading effect upon the SiC structural layer and to facilitate UHF operation, 10nm Ti atop of 30nm Al is used as metallization layer, instead of the widely used but much heavier Au. Ti layer is used as a passivation to prevent the oxidation of Al.

The NEMS resonator devices are designed to be doubly-clamped beams as this simple design allows for better geometry and dimension control, and thus relatively accurate resonance frequency control in fabricating a family of UHF devices for a collective and comparative study. The specs of these devices are summarized in Table 1 (a VHF device, at 295MHz, is also included).

Table 1: UHF NEMS Resonator Devices Specs.

Resonance Frequency (MHz)	Length (μm)	Width (nm)	Thickness (nm)	Device Mass (fg, 10^{-15}g)
295	2.65	180	80	158.3
395	1.75	120	80	74.4
411	1.7	120	80	72.3
420	1.8	150	100	103.6
428	1.65	120	80	75.5
482	1.55	120	80	70.9

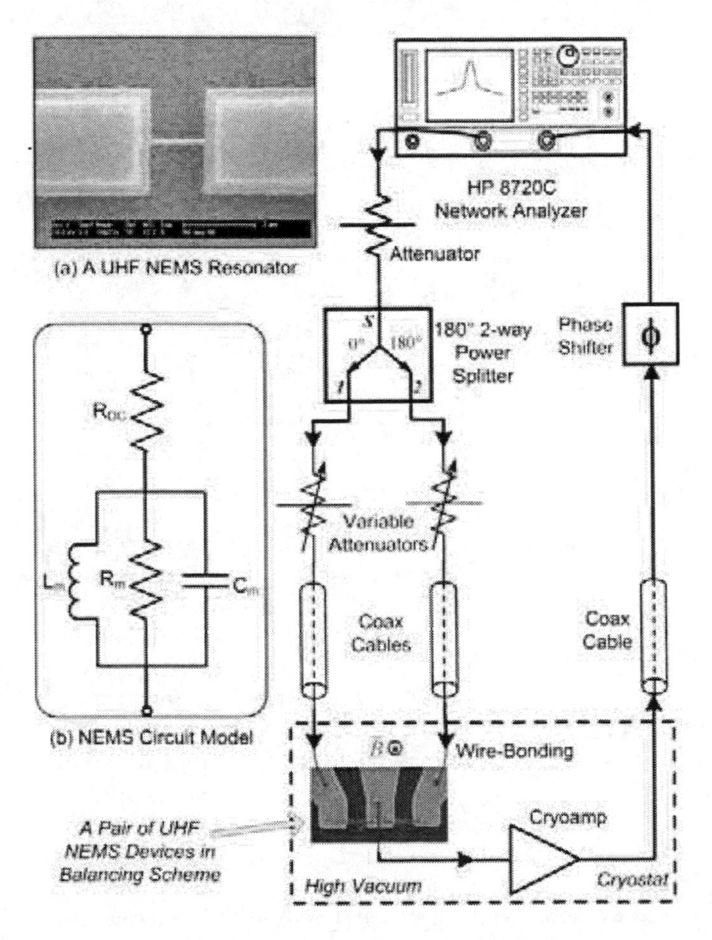

Figure 1: Schematic of the electromechanical resonances readout scheme, with a balanced electronic detection circuit specifically modified and optimized for UHF NEMS. *Inset (a)*: SEM image of a typical UHF SiC NEMS (top view, the etched undercut indicating that device is freely suspended). *Inset (b)*: the parallel LRC tank circuit model for a magnetomotively-transduced NEMS. The total impedance of the device includes the DC impedance R_{DC} and the electromechanical impedance $Z_m(=R_m//(1/j\omega C_m)//j\omega L_m)$.

The NEMS resonators are actuated and the resonance signals are detected with the magnetomotive transduction scheme [10], in which the NEMS devices are preserved in high vacuum ($<10^{-7}$ Torr) and cryogenically cooled down in liquid helium where a superconducting magnet provides the strong magnetic field for the magnetomotive transduction. In this work, all samples are positioned so that the magnetic field is perpendicular to the sample plane, and in-plane fundamental flexural mode is excited and picked up for all NEMS devices. To date, it is still only magnetomotive transduction that has successfully allowed nanomechanical resonators to operate in the UHF band.

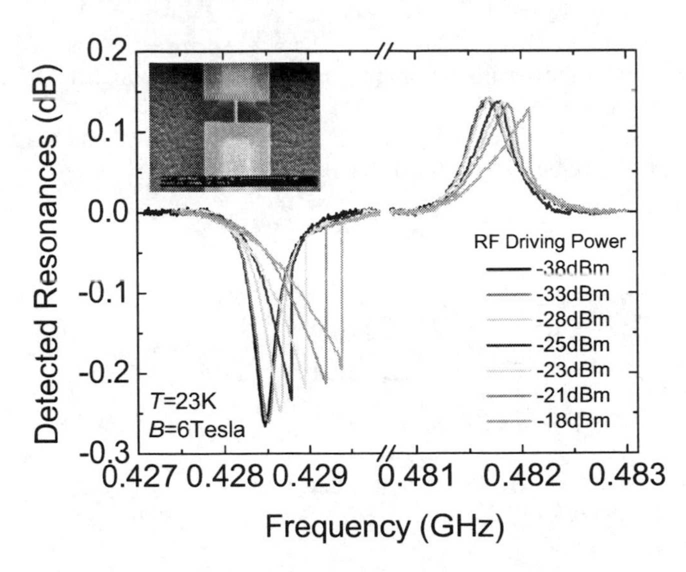

Figure 2: Detected electromechanical resonances of a pair of NEMS resonators, at 428MHz (Q~2500) and 482MHz (Q~2000) respectively. Shown are the resonance curves as the driving RF power is increased. *Inset*: SEM image (tilted view) showing the suspension of a typical UHF NEMS.

To read out the vanishingly-small resonances signals from these tiny UHF NEMS devices (the signal amplitude decreases as the device size decreases), the basic idea of a prototype balanced-bridge detection circuit [11] (initially demonstrated for HF and VHF NEMS) is implemented, modified and engineered specifically for the readout of UHF electromechanical resonances. As shown in Figure 1, the source RF power from the network analyzer (HP 8720C) is split into out-of-phase two branches to drive the two devices which have been deliberately designed to have different but close dimensions to attain different resonance frequencies as well as close DC impedance, for better impedance balancing between the two branches off the resonances. Variable attenuators are applied in each branch to attain better balancing, as the differences in the coax cable and other components between both branches always induce amplitude and phase change. With the background signal (off resonances signal, arising from the embedding impedances of the measurement system) minimized by this balancing scheme (ideally, off resonances background should be at *virtual ground*, i.e., $-\infty$ dB in transmission S_{21}), the vanishingly-small resonances signals become visible (better than those from the direct measurements of reflection or transmission with only one single device) and it is possible to extract the resonances and process them to realize more advanced functions.

Figure 2 shows the detected resonances signals from the pair of 428MHz and 482MHz devices. The off-resonances background signal has been subtracted and the plotted resonances curves reflect the fact that the devices are driven out-of-phase. It is also shown that as the RF power sent to the device is increased, the device is driven into nonlinear regime. From the experimental data, the onset of nonlinearity for the 428MHz device is about -28dBm, while that for the 482MHz device is about -18dBm. This is because the shorter the beam, the stiffer it is and the higher power it can handle. Moreover, the smaller resonance peak of the 482MHz is also because the device is shorter and stiffer as compared with the 428MHz device.

3 FREQUENCY STABILITY

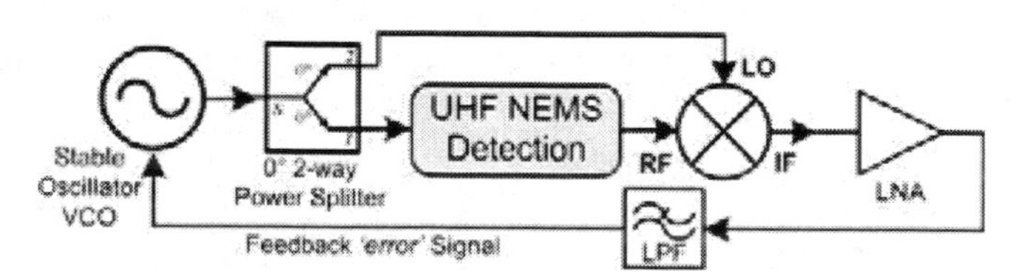

Figure 3: UHF NEMS detection integrated with low-noise phase-locked loop for precise NEMS resonance frequency locking and tracking in *real-time*.

Among many engineering challenges to boost NEMS for more practical applications, it is of great importance to develop the generic protocol of integrating NEMS resonators with feedback and control systems. Low-noise, precise and reliable frequency locking and tracking are the keys for *real-time resonant sensing* applications. With the extraction of UHF resonances signals addressed above, it becomes possible to apply feedback and control upon the NEMS resonances signals. As shown in Figure 3, we have successfully demonstrated a low-noise phase-locked loop (PLL) integrated with UHF NEMS, which is an upgraded version of the VHF NEMS-PLL system [12].

Frequency stability of the NEMS resonators is measured with the NEMS-PLL system at stabilized temperatures. A universal counter (Agilent 53132A) is used to carry out this time-domain measurement, and the frequency stability is evaluated by the statistics of the measurement ensemble. A widely-used criterion for frequency stability (or, instability) is the Allan deviation [7,13]. For a finite measurement ensemble with N samples, the Allan deviation is ($1/\sqrt{2}$ times the standard deviation of the fractional frequency variation)

$$\sigma(\tau) \cong \frac{1}{\sqrt{2}}\left[\frac{1}{N}\sum_{i=1}^{N}\left(\frac{\bar{f}_{i+1}-\bar{f}_i}{f_0}\right)^2\right]^{\frac{1}{2}}, \qquad (1)$$

where f_0 is the resonance frequency, and $\bar{f}_i$ is the measured (averaged) frequency in the ith time interval.

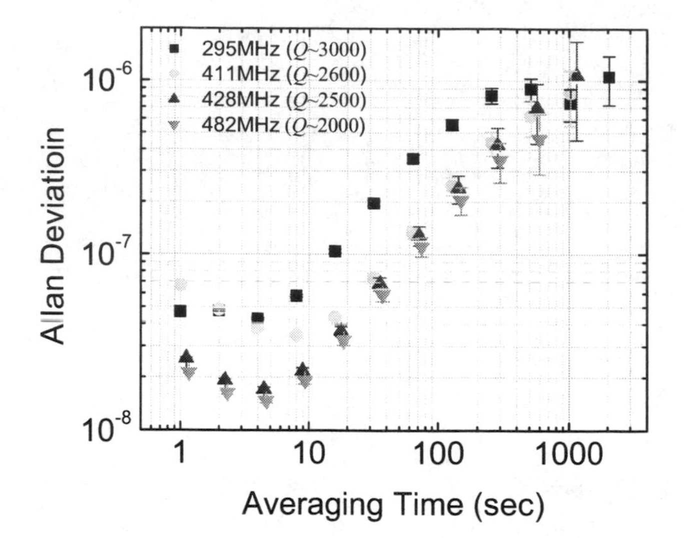

Figure 4: Measured Allan deviation data, showing the frequency stability (instability) of a family of UHF NEMS.

Table 2: Performance of UHF NEMS Resonators & PLL.

f_0 (MHz)	Q	σ (τ=1sec)	DR (dB)	Mass Sensitivity (1zg=10^{-21}g)
295	~3000	4.7×10^{-8}	80	~15 zg
420	~1200	3.1×10^{-7}	90	~67 zg
411	~2600	6.6×10^{-8}	85	~10 zg
428	~2500	2.5×10^{-8}	90	~4 zg
482	~2000	2.1×10^{-8}	98	~3 zg

As plotted in Figure 4, the measured Allan deviation as a function of averaging time, for several UHF NEMS resonators, shows the characteristics of a typical crystal resonator, with similar tendency as that of quartz crystal resonator. The data show that the short term frequency stability of the NEMS resonators is optimized to be in the 10^{-8} to 10^{-7} range, with minimum Allan deviation values at about τ=5sec averaging time for all these resonators. It is probably the most intriguing promise of these UHF NEMS-PLL systems, that the measured frequency stability is translated into *unprecedented* mass sensitivity if the devices are used as inertial mass sensors, based on the analyses in [3]. The measured Q's, Allan deviation at 1sec averaging time, achieved dynamic range for the UHF NEMS-PLL systems, and the corresponding mass sensitivity values calculated from these measured quantities, are collected in Table 2. These mass sensitivity values go deep into the *zepto-gram* (10^{-21}g) scale and again manifest that single-molecule/atom mass diction with UHF NEMS are possible and applicable. Also it is clearly verified that frequency stability and thus the overall mass sensing performance rely on a combination of high-frequency and high-Q. Therefore, scaling up operating frequency and simultaneously retaining high-Q remains a great challenge for NEMS mass sensor engineering. Besides, to fully understand the origin and mechanism of the deteriorating long-term stability shown in Figure 4, and to develop protocols for optimizing both short-term and long-term stability, study upon various possible drifting and aging effects in the system is needed.

4 PHASE NOISE OF UHF NEMS

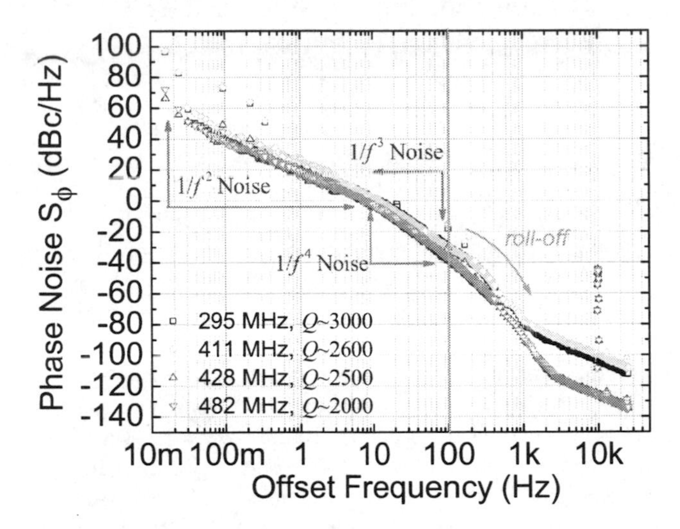

Figure 5: Phase noise performance of generations of UHF NEMS resonators. Shown are the measured phase noise spectrum curves versus offset frequency from the carrier.

The phase noise spectrum is also measured using the NEMS-PLL scheme in Figure 3. The noise spectrum of the control voltage of the VCO, $S_V(\omega)$ is directly measured by a dynamic signal analyzer (HP 35665A), at the port where the error signal is fed back to the VCO as control voltage. Hence the frequency noise spectrum of the VCO output induced by the VCO input control voltage noise is

$$S_f(\omega) = K_V^{\ 2} S_V(\omega), \tag{2}$$

where K_V [Hz/volt] is the gain of the VCO in the frequency modulation mode. Thus the equivalent phase noise spectrum is [13]

$$S_\phi(\omega) = \frac{1}{\omega^2} S_f(\omega), \tag{3}$$

where $\omega = 2\pi f$ is the offset frequency in radius (with f in Hz). The measured phase noise as a function of offset frequency from carrier is collected in Figure 5 for several NEMS resonators. It is observed that in the range of 0.1Hz to 10Hz, the phase noise has $1/f^2$ behavior; while in the range of 10Hz to 100Hz, it approximately follows $1/f^3$ (for the 295MHz and 411MHz NEMS with PLL) or $1/f^4$ (for the 428MHz and 482MHz NEMS with PLL). The far-from-carrier roll-off is attributed to the measurement system time constant. Analyses show that the ultimate phase noise performance is limited by thermomechanical noise of the device itself; while here in the practical system, as there is a mismatch between the noise floor of the preamplifier and that of the NEMS device, the real phase noise is currently limited by the thermal noise of the preamp plus other noise processes in the measurement electronic system.

5 CONCLUSIONS

We have demonstrated generations of UHF silicon carbide NEMS resonators. The successful integration of a low-noise phase-locked loop with these resonators has been employed to directly characterize the frequency stability and phase noise performance of the NEMS. Roadmaps of Allan deviation and phase noise for 300~500MHz UHF NEMS are built. The measured frequency stability is translated into unprecedented mass sensitivity and the data show that ultrasensitive mass detection based on UHF NEMS resonators is very intriguing and is promising for approaching single-molecule/atom sensitivity. The unique NEMS-PLL integration allows for real-time, low noise detection of miniscule mass loading and fluctuation upon NEMS devices. The study shows that high-frequency and high-Q engineering is crucial for both sensitive detection and low phase noise, stable oscillator applications.

Acknowledgement: This work has been generously supported by DARPA/MTO. We thank S. Stryker for his help with the experimental apparatus.

REFERENCES

[1] M.L. Roukes, "Nanoelectromechanical Systems", *Tech. Digest of the 2000 Solid-State Sensor and Actuator Workshop* (*Hilton Head'00*) 1, 2000.

[2] C. T.-C. Nguyen, *Tech. Proceedings of the 2003 Nanotechnology Conference and Trade Show* (*Nanotech2003*) 452, 2003.

[3] K.L. Ekinci, Y.T. Yang, M.L. Roukes, *J. Appl. Phys.* **95**, 2682, 2004.

[4] K.L. Ekinci, X.M.H. Huang, M.L. Roukes, *Appl. Phys. Lett.* **84**, 4469, 2004.

[5] M.D. LaHaye, O. Buu, B. Camarota, K.C. Schwab, *Science* **304**, 74, 2004.

[6] K. Schwab, E.A. Henriksen, J.M. Worlock, M.L. Roukes, *Nature* **404**, 974, 2000.

[7] A.N. Cleland, M.L. Roukes, *J. Appl. Phys.* **92**, 2758, 2002.

[8] X.M.H. Huang, C.A. Zorman, M. Mehregany, M.L. Roukes, *Nature* **421**, 496, 2003.

[9] X.M.H. Huang, C.A. Zorman, M. Mehregany, M.L. Roukes, *Transducers'03* **1**, 722, 2003.

[10] A.N. Cleland, M.L. Roukes, *Sensors & Actuators A* **72,** 256, 1999.

[11] K.L. Ekinci, Y.T. Yang, X.M.H. Huang, M.L. Roukes, *Appl. Phys. Lett.* **81**, 2253, 2002.

[12] C. Callegari, Y.T. Yang, X.L. Feng, M.L. Roukes, *to be published*, 2005.

[13] J.A. Barnes, A.R. Chi, L.S. Cutler, *et al.*, *IEEE Trans. Instr. & Meas.* **IM-20,** 105, 1971.

Flame Made Nanoparticles for Gas Sensors

L. Mädler[*†], S.E. Pratsinis[*], T. Sahm[**], A. Gurlo[**], N. Barsan[**], U. Weimar[**]

[*] Particle Technology Laboratory, Swiss Federal Institute of Technology (ETH) Zurich, Sonneggstrasse 3, CH-8092 Zürich, Switzerland, maedler@ptl.mavt.ethz.ch, pratsinis@ptl.mavt.ethz.ch

[†] currently at Department of Chemical Engineering, University of California, Los Angeles, 405 Hilgard Avenue, Los Angeles, CA 90095 USA, lutz@seas.ucla.edu

[**] Institute of Physical and Theoretical Chemistry, University of Tübingen, Auf der Morgenstelle 8, D-72076 Tübingen, Germany, thorsten.sahm@ipc.uni-tuebingen.de, alexander.gurlo@ipc.uni-tuebingen.de, nb@ipc.uni-tuebingen.de, upw@ipc.uni-tuebingen.de

ABSTRACT

Tin oxide nanoparticles for gas sensing application have been synthesized with an aerosol method. The particles were manufactured with the versatile Flame Spray Pyrolysis (FSP) method producing highly crystalline powders with closely controlled primary particle and crystallite size of 10 nm and 17 nm. The single crystalline particles were only slightly aggregated and directly used for thick film sensor deposition by screen printing. The flame made SnO_2 nanoparticles showed high and rapid response to reducing gases such as CO. Furthermore, the aerosol generated by the dry FSP method was directly, *in-situ* thermophoretically deposited onto interdigitated Pt-electrodes to form a porous, thick film of controlled thickness within the active sensor area. Tin oxide grain size (10 nm) and a high film porosity (98 %) was obtained These sensors exhibited high carbon monoxide (CO) sensor signals (8 for 50 ppm CO in dry air at 350°C), good reproducibility, high analytical sensitivity and a remarkably low detection limit (1 ppm CO in dry air at 350°C).

Keywords: tin dioxide, platinum, gas sensors, flame spray pyrolysis, in-situ deposition

1 INTRODUCTION

Flame aerosol technology is one of the most widely used synthesis routes in manufacturing of commercial quantities of nanoparticles [1]. The application of flame spray pyrolysis further broadens the spectrum of flame made powders and their use in various applications as there are more liquid than gaseous precursors available [2]. Even particles with pre-defined stoichiometric composition can be produced [3-5].

Metal oxides in general and SnO_2, in particular, have attracted the attention of many users and scientists interested in gas sensing under atmospheric conditions. SnO_2 sensors are the best-understood prototypes of oxide-based gas sensors [6]. The performance of metal oxide gas sensors is strongly related to the properties of the ceramic such as grain size, morphology, surface groups, etc. It has been shown that flame spray pyrolyis enables the controlled synthesis of such materials and bears the advantage of complete manufacture of nanopowder in a single high temperature process step leading to controlled microstructure and noble metal loadings if desired [7, 8]. Here, single crystalline tin oxide particles with specific averages particle sizes of 10 and 20 nm were produced using the versatile FSP technique. The particles were only slightly aggregated and were directly used for thick film sensor deposition by screen printing. The size effect has been explored for CO detection [9].

Furthermore, state-of-the-art sensors have important technical limitations that are generally related to the way in which the sensitive materials are processed. For example, the wet chemistry methods employed for both preparation and functionalization of base materials are difficult to control and as a result both the size distribution in the base material, and the amount and distribution of the noble metals additives, are rather broad. This results in significant variation of gas-sensing properties from batch to batch (30 % variation is common in the industry). The fabrication of the sensitive materials is labor and time intensive, with typical batch production times on the order of days with small batch volumes in the range of 100 g [10]. The deposition of sensing films, either by classical screen-printing or more sophisticated drop deposition techniques, is performed after the additional step of combining the sensitive material with organic carriers. This increases processing time and costs related to deposition equipment and handling. Recent advances have been made using electrospray deposition for sensor applications but limitations include processing time and the required post-processing to obtain nano-crystalline material [11]. Additionally, variations in the deposition parameters such as a new film or varying film thickness are difficult to implement and require repetition of the full process.

2 EXPERIMENTAL

2.1 Particle Production

Tin(II) 2-ethylhexanoic acid (Aldrich) was diluted in toluene to obtain a 0.5 M precursor solution. The precursor was fed into a flame spray pyrolysis (FSP) reactor by a syringe pump with a rate of x=5 or 8 ml/min and was dispersed by y=5 or 3 l/min of oxygen, respectively, into fine droplets by a gas-assist nozzle. The conditions will later be referred later as (x/y). Cooling of the reactor avoided any evaporation of the precursor within the liquid feed lines or overheating of the nozzle. The spray flame was maintained by a concentric supporting flamelet ring of premixed methane / oxygen (CH_4 = 1.5 l/min, O_2 = 3.2 l/min). In order to assure enough oxidant for complete conversion of the reactants, an additional outer oxygen flow (5 l/min) was supplied. For Pt/SnO_2 synthesis, appropriate amounts of platinum acetylacetonate ($Pt(acac)_2$, Strem, purity > 98 %) were added to the solution.

The powder was collected with the aid of a vacuum pump on a glass fiber filter (GF/D Whatman, 257 mm in diameter). During the experiment the filter was placed in a water cooled holder 400 mm above the nozzle where the off-gas temperature was maintained below 200 °C [12].

2.2 Dry depsoition

Dry aerosol synthesis applying the Flame Spray Prolysis (FSP) technique [13] has been used for direct (in-situ) deposition of pure and functionalized (doped) sensing materials which eliminates theses difficulties and functionalization of the sensing films can be realized during one proceeding step on ceramic (planar) and micro-machined substrates. The method is in principle applicable to all materials that are able to be synthesized by FSP [2]. Each sensor substrate consists of interdigitated Pt-electrodes on the front side and heater on the back side and an active sensing area of 7.0 x 3.5 mm^2. A mask was used to deposit the particles within the desired sensor area. The substrate was mounted on a water-cooled copper block equipped with a K-type thermocouple to control the substrate temperature during the deposition process. In this study, the substrate temperature was maintained at T_{sub} = 120 °C in order to avoid water condensation on the substrate. The deposition substrate was centered 200 mm above the nozzle facing it. At this position the gas temperature in front of the substrate was T_{gas} = 500 °C using the reactor settings described above. Both temperatures (T_{sub} and T_{gas}) were maintained throughout the deposition process.

2.3 Particle Characterization

X-ray diffraction patterns were recorded with a Bruker AXS D8 Advance (40 kV, 40 mA) and used to obtain the crystallite size (d_{XRD}) based on the fundamental parameter approach and the Rietveld method with the structural parameters of casserite (ICSD Coll. Code: 084576) [14, 15]. The BET powder specific surface area (SSA), was measured by nitrogen adsorption at 77 K after degassing the sample, for at least 1 hour at 150 °C in nitrogen. Assuming monodisperse spherical primary particles within an aggregate, the equivalent average primary particle diameter d_{BET} is calculated by $d_{BET} = 6 / (SSA \cdot \rho_p)$, where ρ_p is the density of SnO_2 (6.85 g/cm^3). The product powder was further analyzed by transmission electron microscopy.

2.4 Sensor Characterization

DC electrical measurements (sensor tests) were performed to monitor the sensor response to CO in dry synthetic air and in synthetic air with 50% relative humidity at 20°C. The sensing layers were fabricated by screen printing (in the case of CO sensors) on alumina substrates with interdigitated Pt-electrodes on the front side and heater on the back side [6]. After deposition, the sensors were annealed for 10 min in a belt oven at 500°C in air. For the comparison tests, commercially available SnO_2 powder (Aldrich, mesh 325, d = 330 nm) was also used for the sensor deposition [12].

The measurements were performed with a set of two identical sensors placed symmetrically in a teflon-made test chamber and operated under the same conditions. The operating temperature of the sensors was adjusted between 200 and 400°C. The sensor signal is given in the following as the resistance ratio R_{air}/R_{gas} for CO where R_{gas} and R_{air} denote sensor resistance in the presence and in the absence of CO. Gas mixing achieved using a combination of computer controlled mass flow controllers and computer controlled valves. The sensors were exposed to CO (500 and 1000 ppm) in dry and humid (50% r.h.) synthetic air. The humidity was adjusted by bubbling synthetic air through a column of water and subsequently mixing it with dry air. Defined concentration CO and were obtained in the PC controlled gas mixing bench by mixing certified CO test gas.

3 RESULTS AND DISCUSSION

Figure 1 shows a TEM image of the as prepared powders (5/5). The aggregated powder consists of highly crystalline polyhedral primary particles with size of d_{BET} = 9.9 nm and d_{XRD} = 10.7 nm (5/5) and powders which had a longer residence in the flame resulting in d_{BET} = 19.7 nm and d_{XRD} = 19.8 (8/3) (Fig. 1). The close correlation of the average primary particle and the average crystallite sizes give strong evidence that the primary particles are single crystals with a low degree of aggregation.

Similar results were obtained by Hall et al. 2002 [16] in a low pressure flat flame burner using also the tri-methyl-tin but at much lower concentrations. The authors showed good control of the primary particle size from about 6 to 15 nm by varying the precursor concentration in the flame.

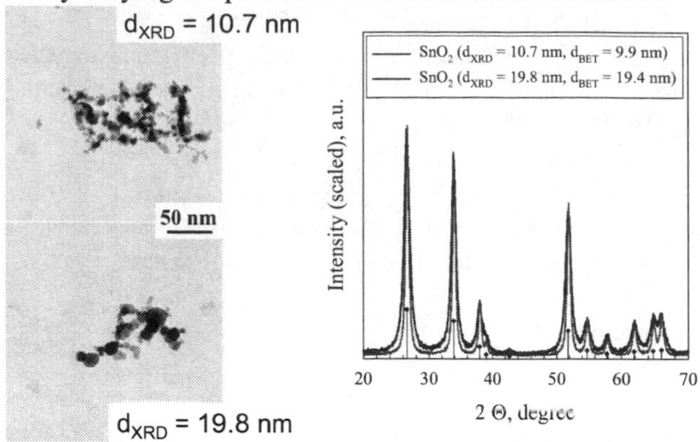

Figure 1. Increasing the precursor (solvent=toluene) flow rate from 5 to 8 ml/min increased the average primary particle size determined by BET from about 10 to 20 nm. This size increase is clearly visible in the TEM analysis. The average crystal size obtained by X-ray analysis (fundamental parameter approach and Rietveld method) resulted in nearly identical average sizes of 10 and 20 nm, respectively, indicating single crystalline particles [9].

Therefore, the tin oxide particles produced with the flame spray process are similar to the particles synthesized in the vapor flame reactor corroborating the fact that the precursor reaction and particle formation takes place within the gas phase and that all liquid precursor left the droplet environment before the formation of the tin oxide. However, the vapor flame made tin oxide powders were not tested for their sensor performance and therefore the versatile FSP technique of SnO_2 was explored.

Exposing the sensor with reducing gases (CO) decreases the resistance of the sensors which is a typical behavior for the tin dioxide as an n-type semiconductor. The measured sensor resistance in the presence of CO increases with decreasing particle size (Fig. 2). Commercial powder from Sigma Aldrich with an average size of 300 nm has two orders of magnitude lower total resistance which is an advantage in signal processing. However, the sensor signal is drastically increased for smaller (nanoscale) particles. This increase results from the higher specific surface area within the sensing layer and also from the relatively larger depletion layer size in comparison with the particle size itself.

The direct deposition from the aerosol phase resulted in fully formed functionalized sensing films. The *in-situ* prepared sensors of Pt doped SnO_2 exhibited high carbon monoxide (CO) sensor signals (8 for 50 ppm CO in dry air at 350°C), good reproducibility, high analytical sensitivity and a remarkably low detection limit (1 ppm CO in dry air at 350°C). The *in-situ* platinum doping enhanced the overall sensor performance (figure 3) [9].

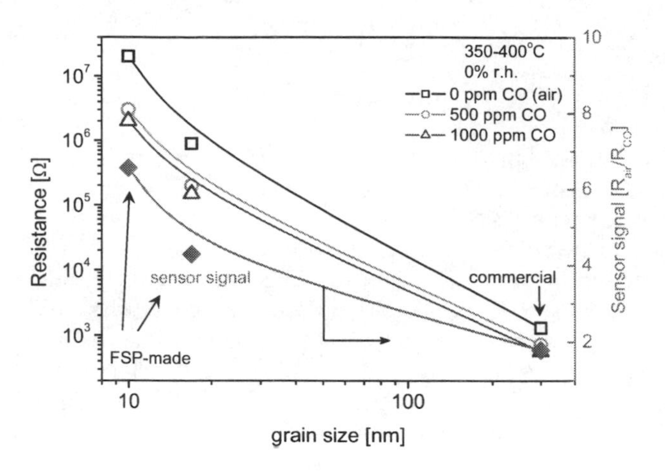

Figure 2. SnO_2 particles of different sizes (about 10 nm for 5/5, about 20 nm 8/3, and about 300 nm Sigma Aldrich, respectively) are screen printed on interdigital electrodes. The measured sensor resistance increases with decreasing particle size. The commercial powder with an average size of 300 nm has two orders of magnitude lower total resistance which is an advantage in signal processing. The sensor signal e.g. for 500 ppm (plotted on the right hand site) is drastically increased for smaller (nanoscale) particles [9].

4 CONCLUSION

In summary, we demonstrated that flame spray pyrolysis can be successfully used for the preparation of SnO_2 nanoparticles for gas-sensing applications. Single crystalline tin oxide particles with size of 10 to 20 nm were produced using the versatile FSP technique. The fabricated sensors show high sensitivity and fast response CO. Direct control of particles sizes with the FSP parameters has a promoting effect.

Nano-crystalline tin-oxide can be directly *in-situ* deposited as porous films onto alumina sensor substrates by thermophoresis. The as-obtained sensors exhibit extremely good homogeneity of the sensing film and good sensor performance. The *in-situ* prepared sensors of Pt doped SnO_2 are reproducible and have a very low detection limit for CO (down to 1 ppm) with high sensor response (up to 8).

A simple model for the sensor film growth rate by particle deposition was developed based on diffusion and thermophoresis. Its predictions were in excellent agreement with microscopic measurements of the film thickness. Control of the film thickness during the deposition process is an effective tool for tuning sensor performance besides its chemical composition. Furthermore, in principle, it is possible and simple to deposit a combination of various films having different functions (filtering, sensing) by the same deposition process, enabling direct construction of fully functional sensors in very short times using a simple and clean fabrication process.

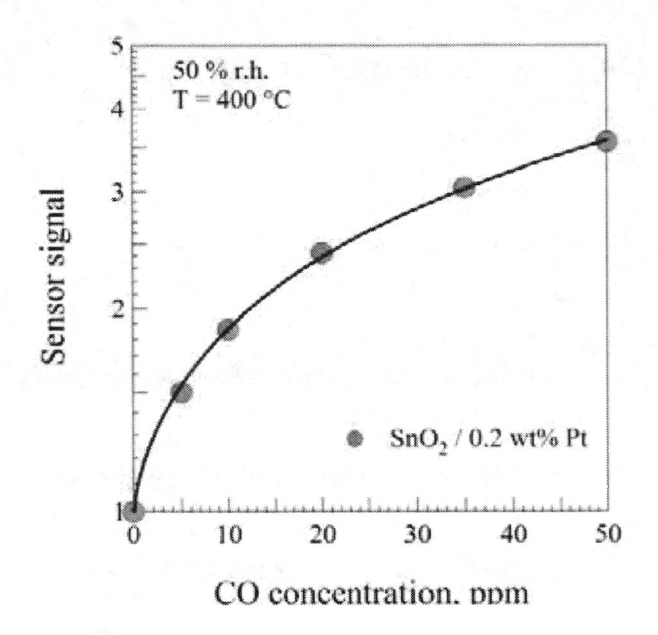

Figure 3. Sensor signal to CO for in-situ (directly) simultaneously deposited Pt-doped SnO_2 (0.2 wt% Pt) operated at 400°C in 50 % r.h. (at 20°C) air. The thickness of the sensing films is 30±3 µm (deposition time: 180 s).

ACKNOWLEDGEMENTS

We gratefully acknowledge the technical support of Dr. M. Müller and Dr. F. Krumeich (ETH) operating the TEM facilities and stimulating discussions with Dr. A. Rössler and Dr. M.J. Height (ETH).

REFERENCES

[1] S. E. Pratsinis, "Flame Aerosol Synthesis of Ceramic Powders," *Prog. Energy Combust. Sci.*, vol. 24, pp. 197-219, 1998.

[2] L. Mädler, "Liquid-fed aerosol reactors for one-step synthesis of nanostructured particles," *KONA*, vol. 22, pp. 107-120, 2004.

[3] R. Strobel, W. J. Stark, L. Mädler, S. E. Pratsinis, and A. Baiker, "Flame-made platinum/alumina: structural properties and catalytic behaviour in enantioselective hydrogenation," *Journal of Catalysis*, vol. 213, pp. 296-304, 2003.

[4] L. Mädler, W. J. Stark, and S. E. Pratsinis, "Rapid Synthesis of stable ZnO Quantum Dots," *Journal of Applied Physics*, vol. 92, pp. 6537-6540, 2002.

[5] W. J. Stark, L. Mädler, M. Maciejewski, S. E. Pratsinis, and A. Baiker, "Flame synthesis of nanocrystalline ceria-zirconia: effect of carrier liquid," *Chemical Communications*, pp. 588-589, 2003.

[6] N. Barsan and U. Weimar, "Understanding the fundamental principles of metal oxide based gas sensors; the example of CO sensing with SnO_2 sensors in the presence of humidity," *Journal of Physics-Condensed Matter*, vol. 15, pp. R813-R839, 2003.

[7] L. Mädler, W. J. Stark, and S. E. Pratsinis, "Flame-made ceria nanoparticles," *Journal of Materials Research*, vol. 17, pp. 1356-1362, 2002.

[8] L. Mädler, W. J. Stark, and S. E. Pratsinis, "Simultaneous deposition of Au nanoparticles during flame synthesis of TiO_2 and SiO_2," *Journal of Materials Research*, vol. 18, pp. 115-120, 2003.

[9] T. Sahm, L. Mädler, A. Gurlo, N. Barsan, S. E. Pratsinis, and U. Weimar, "Flame spray synthesis of tin oxide nanoparticles for gas sensing," *Mater. Res. Soc. Symp. Proc.*, vol. 828, pp. A1.3.1-A1.3.6., 2004.

[10] J. Kappler, *Characterization of high-performance SnO2 gas sensors for CO detection by in-situ techniques*. Aachen: Saker Verlag, 2001.

[11] Y. Matsushima, Y. Nemoto, T. Yamazaki, K. Maeda, and T. Suzuki, "Fabrication of SnO_2 particle-layer on the glass substrate using electrospray pyrolysis method and the gas sensitivity for H-2," *Sensors and Actuators B-Chemical*, vol. 96, pp. 133-138, 2003.

[12] T. Sahm, L. Mädler, A. Gurlo, N. Barsan, S. E. Pratsinis, and U. Weimar, "Flame spray synthesis of tin dioxide nanoparticles for gas sensing," *Sensors and Actuators B-Chemical*, vol. 98, pp. 148-153, 2004.

[13] L. Mädler, H. K. Kammler, R. Mueller, and S. E. Pratsinis, "Controlled synthesis of nanostructured particles by flame spray pyrolysis," *Journal of Aerosol Science*, vol. 33, pp. 369-389, 2002.

[14] A. A. Bolzan, C. Fong, B. J. Kennedy, and C. J. Howard, "Structural studies of rutile-type metal dioxides," *Acta Crystallographica Section B-Structural Science*, vol. 53, pp. 373-380, 1997.

[15] R. W. Cheary and A. Coelho, "A Fundamental Parameters Approach to X-Ray Line-Profile Fitting," *Journal of Applied Crystallography*, vol. 25, pp. 109-121, 1992.

[16] D. L. Hall, P. V. Torek, C. R. Schrock, T. R. Palmer, and M. S. Wooldridge, "Gas-phase combustion synthesis of tin oxide nanoparticles," *Metastable, Mechanically Alloyed and Nanocrystalline Materials*, vol. 386-3, pp. 347-352, 2002.

Nanomechanical Sensor Platform Based On Piezo-Resistive Cantilevers

J. Thaysen, M. Havsteen Jacobsen and L. Kildemark Nielsen

Cantion A/S
Ørsteds Plads bldg. 347, 2800 Lyngby, Denmark, jt@cantion.com

ABSTRACT

This paper presents Cantion's cantilever-based bio/chemical sensor. The cantilevers have integrated readout and can be used both in liquids and gases. The sensing principle is based on a cantilever bending due to a change in surface stress. Such surface-stress change is obtained during molecular interaction which has made it possible to measure DNA hybridization, antigen-antibody interactions etc. This paper presents Cantion's technology and instruments and furthermore discusses different types of applications realized on the platform. Finally, the perspective of using Cantion's cantilever sensor technology for the next generation handheld application is discussed.

Keywords: cantilever, sensor, nanomechanical, label-free, bio/chemical

1 INTRODUCTION

The cantilever-based bio/chemical sensing principle was discovered by T. Thundat *et al.* [1] in 1994 where they found that AFM cantilevers coated with gold would deflect when exposed to for example mercury due to a change in surface stress. They suggested that this could be used as a novel type of sensitive chemical sensor. Since then, a lot of different applications have been demonstrated on the cantilever platform, such as: DNA hybridization [2], antibody-antigen reaction [3] and bacteria detection [4]. Because the sensor technology is based on AFM most of the published work has been based on an optical readout scheme.

Cantion's technology is based on nanomechanical cantilevers as bio/chemical sensors with integrated readout. Which, in contrast to the optical readout scheme known from AFM, makes it possible to make a complete sensor solution with a very small footprint.

Cantion's cantilevers are 100 μm long, 50 μm wide and 0.5 μm thick. The upper side of the cantilever can be coated with a detector layer. This layer has the ability to recognize and interact with molecules of interest. When molecules bind to the detector layer, a change in the surface stress is induced, see figure 1. This effect is usually not observed when the detector layer is placed on a solid surface. However, on the very small and flexible cantilevers this induces a deflection of the cantilever in the order of a few nanometers. The cantilever deflection is picked-up in the integrated piezo-resistor, which then changes its electrical resistance. Thereby the molecular interaction is transduced through a nanomechanical sensor into an electrical signal.

Cantion's sensor can be operated both in liquids and gases.

The generated surface stress on the cantilever is not due to a change in mass but due to the molecular interactions. For example, if the molecules on the surface are electrically charged they will repel each other and thereby bend the cantilever downwards. Other general surface stress generating mechanism are steric hindrance, entropical changes etc. This actually means that the cantilever sensor is able to use many different types of detector layers such as polymers, self assembled monolayers, DNA, proteins etc.

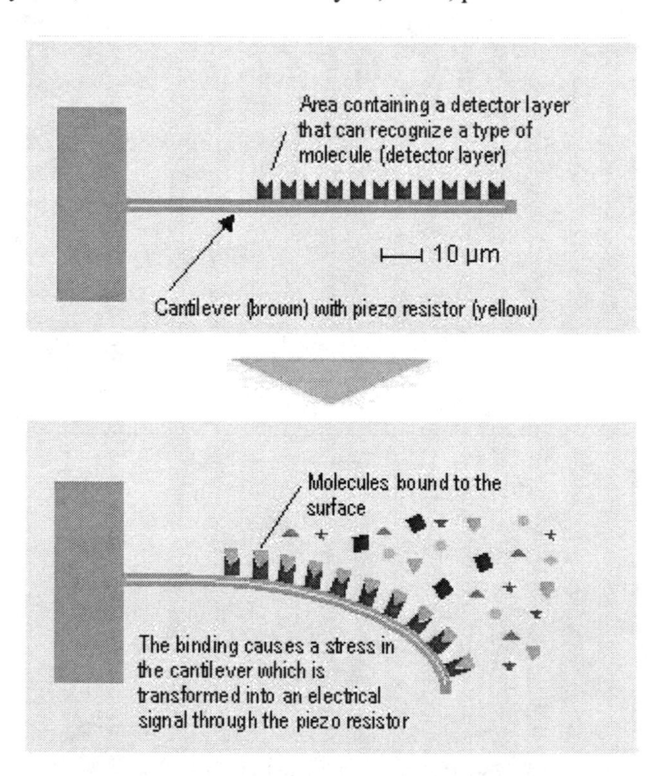

Figure 1: The basic principle for the surface stress sensitive piezoresistive cantilever

The nanomechanical cantilever sensor is by nature sensitive to many different parameters, such as temperature, mechanical vibrations, electrical fields etc. Using a differential readout method these effects can be significantly reduced. The differential configuration consists of two cantilevers, one as a measurement cantilever and one as the reference cantilever.

The signal from the two cantilevers can be subtracted in a Wheatstone bridge configuration such that only the differential signal is measured. In order to further minimize the parameters from the environment, Cantion has integrated arrays of cantilevers into a micro fluid handling system. Thermal gradients are thereby significantly reduced compared to larger systems. Furthermore, the laminar flow obtained in micro fluid systems eliminates noise sources from turbulences, which would be a significant noise factor for the cantilevers.

2 INSTRUMENTS

In order to perform reliable and reproducible experiments, Cantion offer an product portfolio consisting of 3 main products:

2.1 Canti™Chip 4

Cantion's Canti™Chip 4 is designed for sensitive detection of surface stress caused by molecular interactions on the cantilever surface. The chip has 4 cantilevers placed in a micro channel. The SEM micrograph in figure 2 shows the chip where 3 of the cantilevers are visible. The chip is packaged unto a printed circuit board for easy handling. The Canti™Chip 4 is optimized for measurement in both gases and liquids and is designed as an open platform, which allows the user to functionalize each cantilever individually.

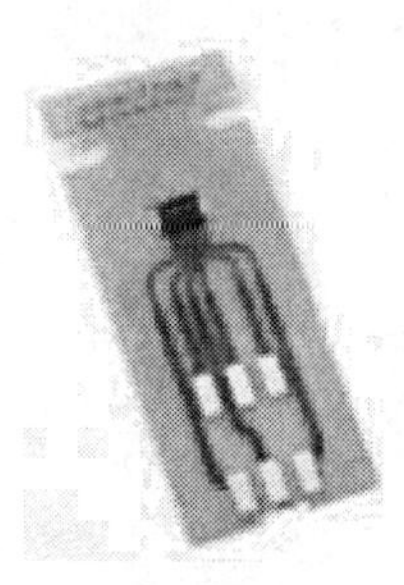

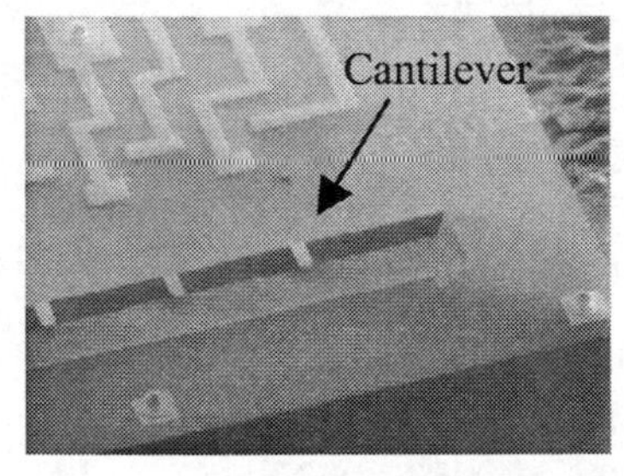

Figure 2: Optical image of chip placed on a PCB board (Canti™Chip 4) and SEM micrograph of the cantilevers

2.2 Canti™Spot

The cantilevers are functionalized using an inkjet printing principle. This is done by aligning each cantilever under a dispenser head whereafter about 100 pl of a given reagent is dispensed on the cantilever. The dispensed volume covers the topside of the cantilever only. The cantilever surface, for example coated with gold, reacts with the reagent and thereby functionalizes the cantilever.

Cantion has developed a research platform for the reagent dispensing, called Canti™Spot. The instrument is shown in figure 3 and consists of a table where the Canti™Chip 4 is placed. By using the xyz microposition system it is possible to align each cantilever underneath the dispenser head.

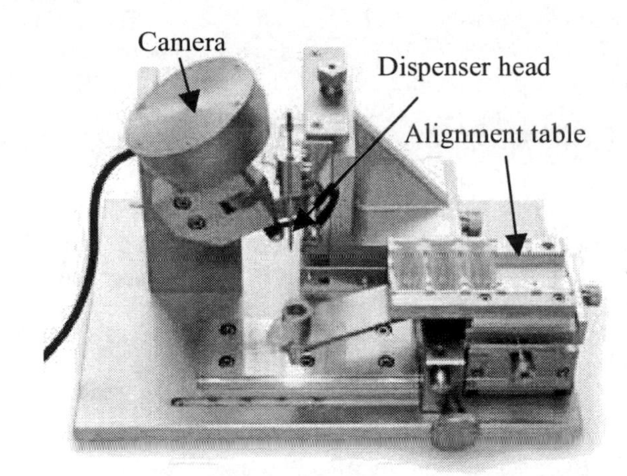

Figure 3: Canti™Spot instrument for functionalizing each cantilever individually. The dispenser head delivers 100 pL of reagent on each cantilever

2.3 Canti™Lab 4

After the Canti™Chip 4 has been functionalized it is ready for performing the experiment. For that purpose Cantion has developed a highly flexible laboratory platform called Canti™Lab 4, see figure 4. The instrument both included the fluid handling system and the electrical readout from the Canti™Chip 4.

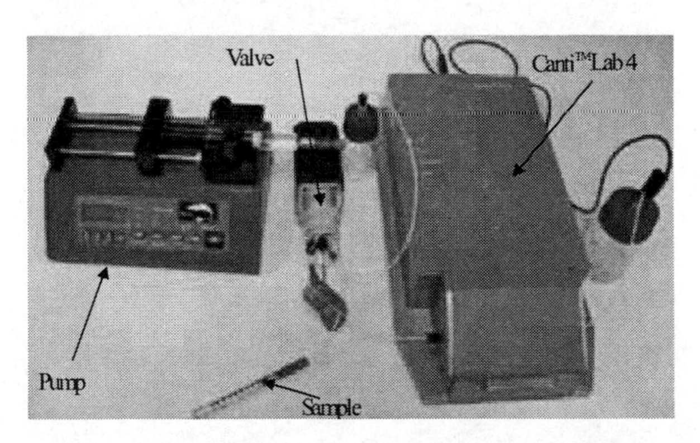

Figure 4: Canti™Lab 4 is a highly flexible laboratory equipment optimized for easy and reliable assay development on the Canti™Chip 4

The sealed interaction cell obtained when the Canti™Chip 4 is inserted in the Canti™Lab 4 has a volume of about 500 nL. This small volume decreases the required sample, which for example is important if the sample of interest is expensive. The small interaction cell furthermore allows laminar flow. The Canti™Lab 4 also features an external liquid system that consists of a programmable pump and valve. Finally, the instrument allows both differential and absolute signals from the cantilever to be measured simultaneously.

3 PROVEN TECHNOLOGY

Different types of assays have been tested on Cantion's sensor platform and a couple of these are discussed in order to demonstrate the potential of the technology.

3.1 Antibody-antigen assay

The first example is an antibody-antigen assay. Usually, such an assay is made by ELISA or similar where a labeled secondary antibody is required. We demonstrate that label-free detection is possible on Cantion's sensor platform.

C-reactive protein (CRP) is an acute phase protein, which is found in the human blood at very low concentrations (<0.1 mg/l). However, if a patient suffers from a bacterial infection the CRP level will increase dramatically in concentration up to 300 mg/l.

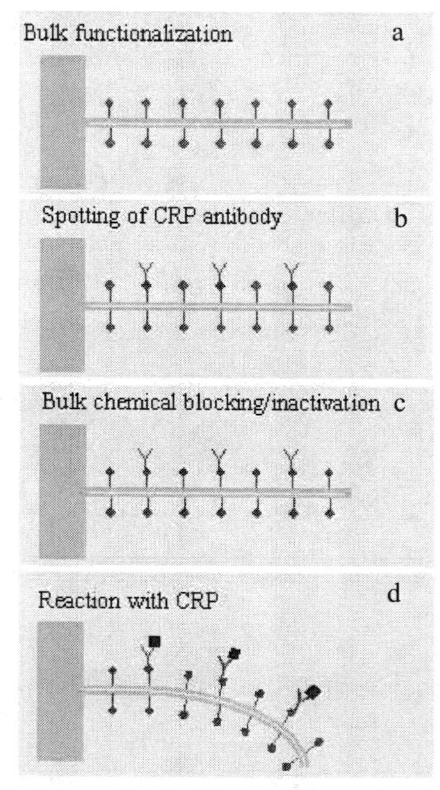

Figure 5: Schematic of CRP experiment. a) activation of silicon nitride, b) functionalization of cantilever by inkjet dispensing, c) blocking of active sites, d) CRP experiment in Canti^{TM}Lab 4

In order to obtain covalently bound CRP-antibodies to the cantilever surface an activation step of the inert silicon nitride surface is required. Eletrophilic groups were obtained on the cantilever surface by an activation procedure developed by Cantion. Hereafter, the Canti^{TM}Spot was used for functionalizaton of the cantilever with CRP antibodies. The backside of the measurement cantilever and the complete reference cantilever was then inactivated by a blocking agent.

The functionalized Canti^{TM}Chip 4 was inserted in the Canti^{TM}Lab 4. First, a stable base line was obtained by flowing buffer through the system. After 500 sec the valve was opened and the buffer + CRP is introduced.

Different concentration of CRP was measured by regenerating the Canti^{TM}Chip 4 after each experiment. In figure 6 three different concentration curves are shown: 0, 5, 100 mg/l. It can be seen that the amplitude of the cantilever signal increases as a function of concentration. The selectivity of the assay was investigated by introducing buffer + 100 mg/l BSA. This did not give rise to any signal (curve was similar to 0 mg/l).

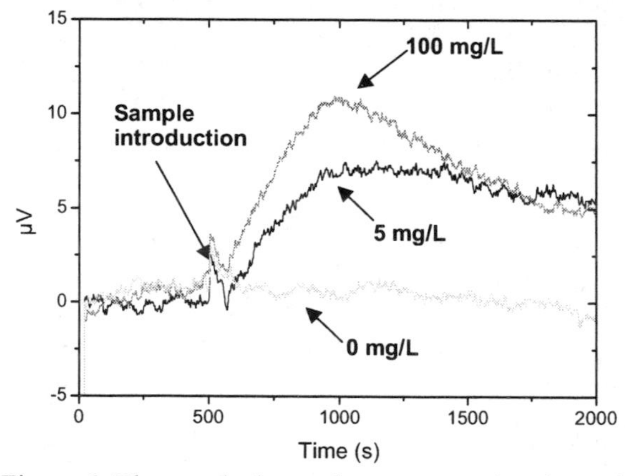

Figure 6: The graph shows the concentration dependent signal. The signal monitored is the change in resistance of the piezo-resistor which is directly proportional to the cantilever bending.

3.2 DNA assay

Another interesting assay developed on Cantion's platform is DNA hybridization. In this case, the cantilever surfaces were coated by gold to facilitate immobilization of thiol-modified oligonucleotides on the surface. In this experiment we used one type of 12-mer oligo on the measurement cantilever and another 12-mer oligo on the reference cantilever. The oligoes were dispensed on to the cantilevers by the use of the Canti^{TM}Spot.

After a stable baseline in a buffer solution was obtained, the valve was opened and buffer + 1 µM of DNA complementary to that on the measurement cantilever was introduced. In figure 7 hybridization curves from 3 different chips is shown. As seen from the graph, the signals are reproducible in terms of amplitude and the required time in order to reach steady-state. A non-complementary DNA strand was also introduced to chip 3 and as seen from the figure this did not give rise to any signals.

3.3 Other types of assays

L. Pinnaduwage *et al.* [5] from Oak Ridge National Laboratory (ORNL) has used Cantion's technology for development of a highly sensitive plastic explosive sensor in ambient air. They were able to obtain a sensitivity of sub-ppt, which is far better than achieved with competing technologies. The explosive detection project on Cantion's platform was so successful that it won an R&D 100 award in 2004.

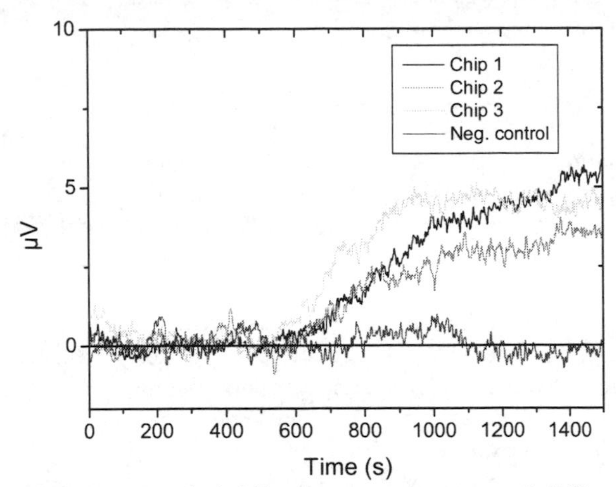

Figure 7: DNA hybridization experiments on 3 different chips. It can be seen that the signals are very reproducible. More signals from other chips showed the same trend.

R. Mukhopadhyay *et al.* [6] from Aarhus University developed a sensitive SNP in DNA assay using Cantion's technology.

Furthermore, Cantion's sensor platform has been used in various other applications, such as alcohol detection, enzyme-substrate detection and chemical warfare agent detection.

4 PERSPECTIVES

The electrical readout principle used in Cantion's sensors makes the technology very robust in different environments and the small sensor and readout footprint makes it ideal for handheld applications. Cantion has developed a prototype of such a handheld device for detection of various gases, see figure 8. The very versatile detection principle used as a handheld device addresses different applications such as detection of explosives, bio/chemical warfare agents, medical point-of-care and food quality control.

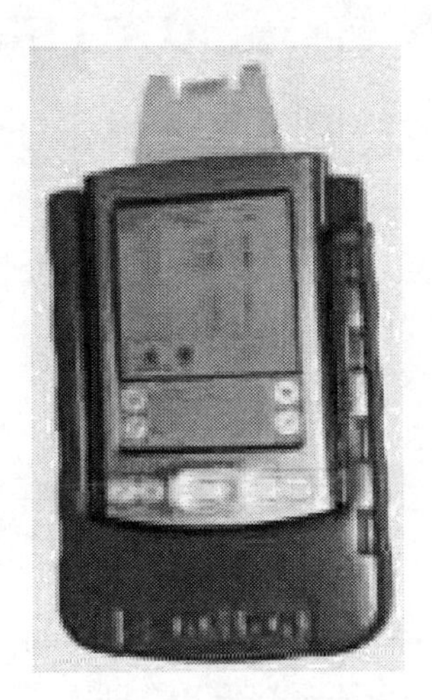

Figure 8: Prototype of handheld device based on Cantion's cantilever technology for gas sensing applications.

5 CONCLUSIONS

The paper have presented Cantion's cantilever based bio/chemical sensor with integrated readout. The cantilever sensor is a true label-free detection platform where different types of assaya have already been proven, such as DNA hybridization and antibody-antigen detection in liquid and explosives detection in gas. Due to the electrical readout it is further possible to minimize the sensor platform such that it becomes ideal for handheld applications.

REFERENCES

[1] T. Thundat et al., Applied Physic Letters, Vol 64 (21), 1994, pp. 2894-2896
[2] J. Fritz et al., Science, vol 288 (5464), 2000, pp. 316-318
[3] G. Wu et al., Nature Biotechnology, vol 19 (9), 2001, pp.856-860
[4] B. Weeks et al., Nanotechnology Conference proceeding (Nanotech 2003), Vol 1, 2003 pp. 123 – 125
[5] L. A. Pinnaduwage et al., Rev. of Scientific Instruments vol. 75 (11) Nov 2004, pp. 4554-4557
[6] R. Mukhopadhyay et al. 1st Workshop on Nanomechanical Proceedings, Nov. 2004, pp.11

NANOSTRUCTURED GAS MICROSENSOR PLATFORM

Dmitri Routkevitch, Oleg Polyakov, Debra Deininger, Clayton Kostelecky

Synkera Technologies Inc.
2021 Miller Drive, Suite B, Longmont, CO, USA, droutkevitch@synkera.com

ABSTRACT

Nanostructured materials, with their small grain size, large number of grain boundaries, and high specific surface area, hold promise to enable significant performance benefits for solid-state gas sensors. To fully realize this potential, precision nanoscale engineering of the morphology and composition of sensing materials is needed. Furthermore, multiple challenges are associated with integration of nanostructured materials into reliable and manufacturable microsensors. We will overview our efforts on addressing these challenges by developing a novel gas microsensor platform based on nanostructured alumina ceramic.

Keywords: nanoporous anodic alumina, solid state gas sensors, microsensors, ceramic MEMS

1 PLATFORM

Synkera's novel gas microsensor platform [1] is based on nanostructured anodic aluminum oxide (AAO). Due to its self-organized nanoscale morphology, formed by uniform and parallel nanopores (Figure 1), AAO is an attractive host for templated nanofabrication [2], and is well recognized and widely used in both fundamental research and application development. We use AAO as a host for the synthesis of sensing materials with composition and morphology tailored at the nanometer scale.

In addition, intrinsic anisotropy of morphology and chemistry of anodic alumina enables its micromachining via a flexible process [3] that enable fabrication of robust sensor substrates (Figure 2) that support high temperature low power operation.

Fabrication of gas microsensor from AAO includes the following steps [1]:

(1) synthesis of nanoporous anodic alumina with required thickness and pore diameter;
(2) micromachining of sensor substrates equipped with microheater and sensing electrodes;
(3) conformal deposition of high surface area (up to 100 m^2/g) nanostructured sensing materials onto the walls of the nanopores, and
(4) sensor packaging.

The resulting solid-state sensing element is formed by high density arrays of nanotubules intrinsically integrated into micromachined ceramic substrates. The effects of sensor design, specific surface area, deposition processes and other factors on sensor fabrication and performance were thoroughly evaluated and will be summarized in this presentation.

2 CASE STUDIES

Several types of gas microsensors are currently under development at Synkera Technologies using presented platform, including metal oxide conductimetric, catalytic combustion and electrochemical sensors and sensor arrays. Targeted applications include detection of humidity/moisture, air quality (formaldehyde, CO, volatile organic contaminants (VOC)), combustible gases (methane, hydrogen), hydrogen sulfide, and emission monitoring.

Some of the highlights are presented in Figure 3 - Figure 6. One example includes sensing of low level (0.1-100 ppm) of water vapor. By optimizing nanoscale morphology and surface chemistry of blank anodic alumina and using advanced operating modes, significant reduction in the detection limit were realized (Figure 3), opening opportunities for high altitude low temperature balloon-born meteorological measurements, process control and other demanding applications.

Another example is the development of air quality sensors (Figure 4, Figure 5). Using different methods for sensing materials deposition, optimizing sensing materials composition and doping, and employing precise temperature control, several routes for highly specific detection of gases of interest (formaldehyde, VOC's and CO) were discovered.

In yet another example, excellent sensitivity and reversibility for the detection of hydrogen sulfide was demonstrated with AAO/SnO_2/polymer sensors (Figure 6).

3 ADVANTAGES

Described microsensor architecture has numerous intrinsic advantages over conventional sensors. It enables low thermal mass and low power consumption, broadens the operating temperature range, and provides capability of regeneration by internal heating to high temperature. In many cases it also improves response time, sensitivity and selectivity.

This work was supported by DOE (DE-FG03-99ER82842, DE-FG03-99ER82839), NSF (DMI-9861546), NIH (1R43-ES10739) and NIST (SB1341-02-W-1073).

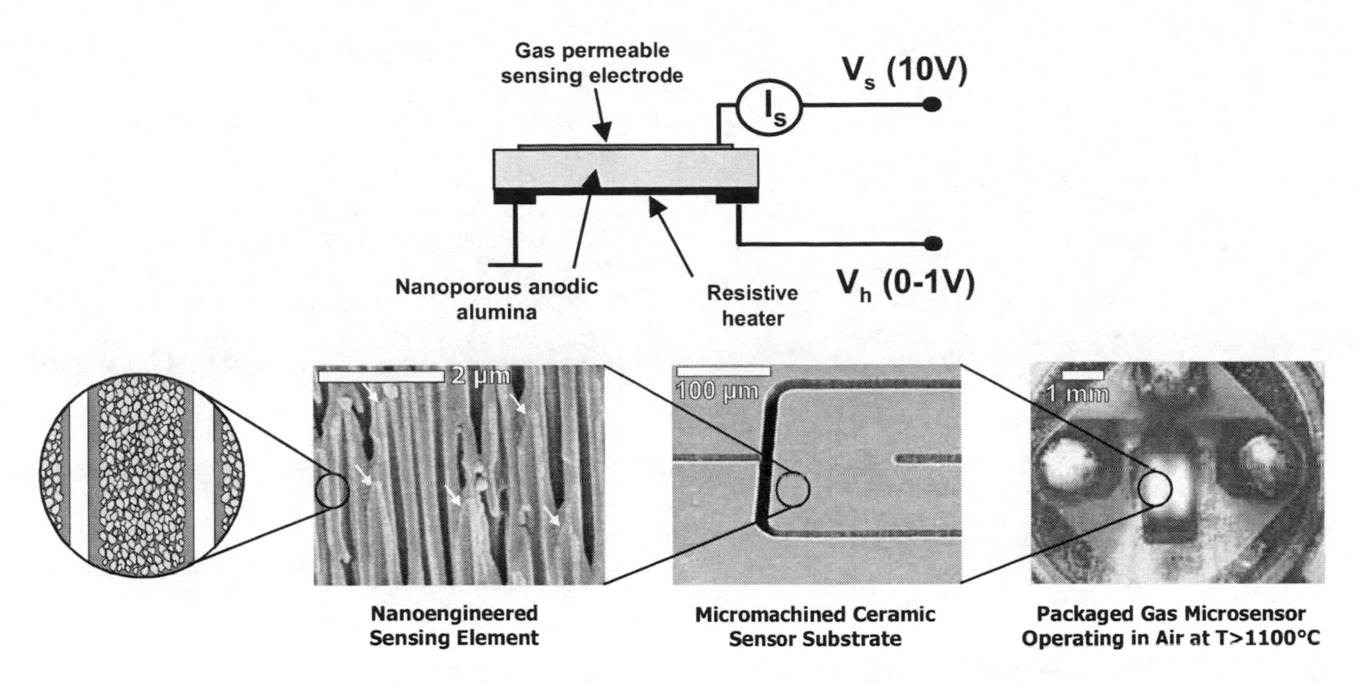

Figure 1: Outline of the Synkera's nanoengineered ceramic platform for gas microsensors.

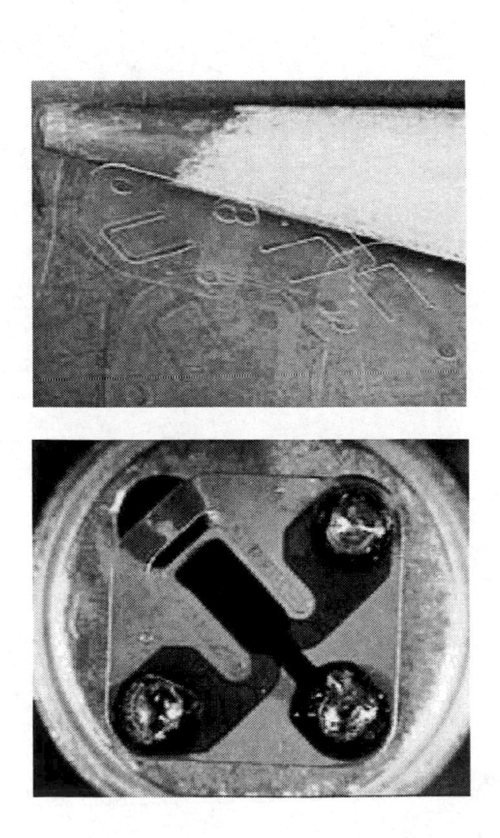

Figure 2: Synkera's micromachined AAO-based sensor substrates and packaged microsensor.

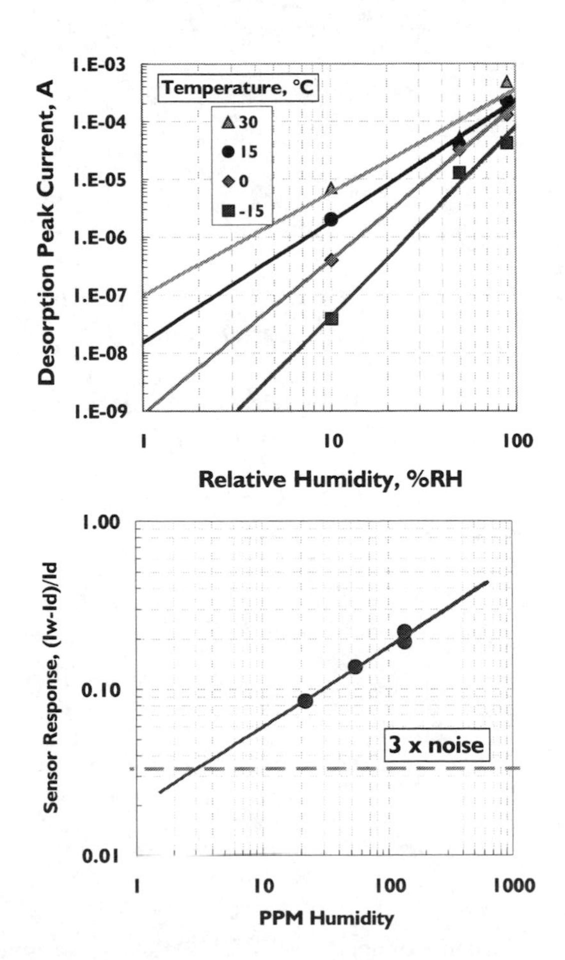

Figure 3: Response of moisture sensor at different temperatures and humidity levels.

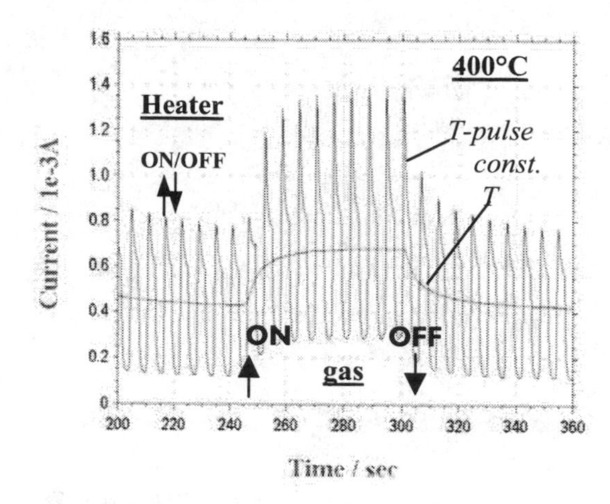

Figure 4: Response of AAO/SnO_2 – based sensor to 2.5 ppm HCHO at constant temperature and in pulsed temperature mode.

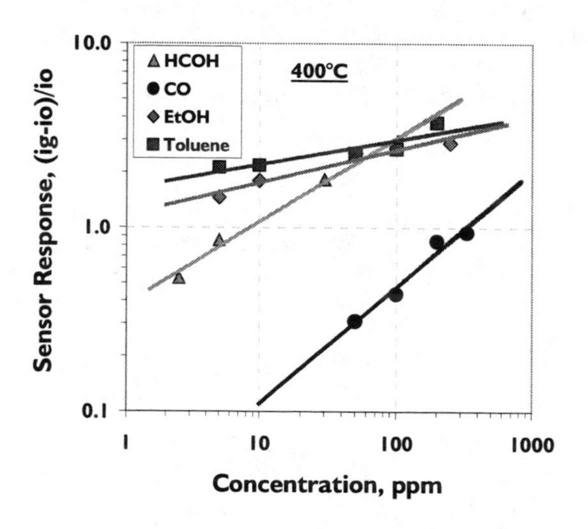

Figure 5: Sensitivity plots for detection of different gases by AAO/SnO_2 – based sensor.

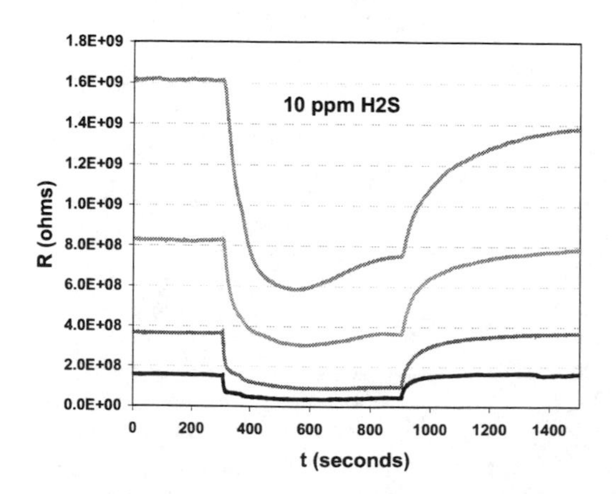

Figure 6: Response of doped AAO/SnO2/polymer – based sensor to 10 ppm of hydrogen sulfide at different temperature.

REFERENCES

[1] D. Routkevitch, P. Mardilovich, A. Govyadinov, S. Hooker, S. Williams, *US patent* No 6,705,152 B2, filed 05/07/1999; granted 03/16/04.

[2] H. Masuda, M. Satoh, *Jpn. J. Appl. Phys.* 1996, **35**, L126; D. Routkevitch, T. Bigioni, M. Moskovits, J. M. Xu, *J. Phys. Chem.* 1996, **100**, 14037.

[3] D. Routkevitch, A. Govyadinov, P. Mardilovich, High Aspect Ratio High Resolution Ceramic MEMS, Microelectromechanical Systems (MEMS) -- 2000, Proc. ASME Int. Mech. Eng. Congress, Nov. 5-10, 2000, Orlando, Florida, Vol. 2, ASME, New York, 39-44 (2000).

RuO_2 and Ru nanoparticles for MISiC-FET gas sensors

Anette Salomonsson*, Somenath Roy*, Christian Aulin**, Lars Ojamäe**, Per-Olov Käll**, Michael Strand***, Mehri Sanati*** and Anita Lloyd Spetz*

*S-SENCE and Division of Applied Physic, Linköping University, SE-581 83 Linköping, Sweden, ansea@ifm.liu.se, aszansea@ifm.liu.se

*** Physical and Inorganic Chemistry, Linköping University, SE-581 83 Linköping, Sweden, aulin@ifm.liu.se, lars@ifm.liu.se, pokje@ifm.liu.se

****Division of Chemistry, Växjö University, SE-351 95 Växjö, Sweden, michael.strand@vxu.se, mehri.sanati@vxu.se

ABSTRACT

Nanotechnology or nanoscience refers to the fact that novel and unique properties of materials are obtained which are due to the small scale. When the particle size reaches nanoscale dimensions, the particle itself could be considered as a surface in three dimensions. For instance, the grain size can affect the conductivity and the resistivity of the material [1].

In this work we have studied catalytically active nanoparticles of RuO_2 and pure Ru as gate material. The particles were synthesized by wet chemical procedure. These materials were chosen because they are conducting and catalytically active. A distinct shift in the depletion area of the capacitance versus voltage curve is observed when measured in oxidizing (synthetic air) and reducing gases (H_2, NH_3, CO and C_3H_6). For RuO_2 particles, the shift is typically ~200 mV for 1% H_2 in synthetic air and ~140 mV for Ru.

Keywords: nanoparticles, Ru, RuO_2, gas sensors, field effect devices, catalytic metal.

1. INTRODUCTION

Metal oxide nanoparticles, having very high surface area and versatile catalytic properties, are promising as gas sensing materials. The metal oxide nanoparticles show interesting behaviors when used as sensing materials for high temperatures (typically at 300 - 400 °C). When used as resistive sensing element, oxide nanoparticle layers exhibit high sensitivity (down to ppb level) and enhanced selectivity to the target gas [1].

Metal Insulator Silicon Carbide Field Effect transistor (MISiC-FET) devices are used as gas sensors for reducing gases in harsh environment like engine exhaust and flue gases. By using metal oxide nanoparticles as the gate material of SiC-FET sensors the high surface area and the specific properties of small particles are benefited [2]. A variety of sensing layers may be explored with an intention to enhance the selectivity and long term stability. The gate material in SiC-FETs needs to be conducting and this is even more important for capacitor devices. In case of porous catalytic metals as gate material needed e.g. for detection of ammonia the metal tend to sinter at high operation temperature, which causes drift [3]. By using support in the form of nanoparticles impregnated with catalytic metals, the sintering effect could be reduced.

The excellent properties of silicon carbide (SiC) e.g., wide bandgap, high melting point and chemical inertness make it especially suitable as sensor material for rough and corrosive environments. Metal-insulator-SiC field effect transistor (MISiC-FET) sensors, with buried source, drain and channel region have already been tested with success in several industrial applications [4].

Triple phase boundaries have been proven to be significant for generation of specific gas response e.g. for NH_3 [5]. Use of particles as the gate material increases the occurrence of triple phase boundaries where the catalytic gate material, the insulator and the test gas molecules are in contact. Nano-dimension of the particles has the potential to introduce novel features in sensing parameters like selectivity and speed of response as compared to those for the conventional thin film sensing layers [6].

A number of techniques may be adopted to produce a variety of nanoparticles with unique features for optimizing selectivity, sensitivity and thermal endurance of the gas sensors. Commercially available oxide nanoparticles (e.g. Al_2O_3, SiO_2, TiO_2) might be impregnated with catalytic metals [7]. Aerosol technology [8] can also be used to produce catalytically active particles [9]. Other techniques like flame spray synthesis [10], spray pyrolysis [11] and sol-gel spin coating method [12] have also been used to produce nano-dimension particles with specific gas response properties.

In this work we have tested wet chemical synthesis of catalytically active and electrically conducting particles of RuO_2 and Ru as gate materials in MISiC capacitor devices.

2. EXPERIMENTAL DETAILS

2.1 Gas sensing principle

The sensor mechanism origins from gas molecules, which adsorb and dissociate on the catalytically active surface. The operating temperature together with the material properties controls the adsorption rate as well as the degree of dissociation. The dissociated hydrogen atoms will diffuse to the metal-insulator interface. There they form a dipole layer, of e.g. OH groups [13] located on the insulator surface [14], which shows up as a voltage shift of the C(V) characteristics of the device, and thereby changes in the surrounding pressure of hydrogen containing gases can be continuously monitored. Some molecules, like ammonia, requires the catalytically active material, the insulator and the gas phase to be present simultaneously. These special sites (triple phase boundaries) can be created by using a porous material or particles. The detection mechanism of the field-effect gas sensors can then be explained in terms of the modulation of an electric field in the underlying insulating layer (SiO_2) due to the specific change in electric charges on the insulator surface. As a result, a shift in the sensor output voltage (sensor signal), is observed. This means that the sensor response characteristics are defined by the properties of both the conducting gate material and the insulator material of the gate. [15].

2.2 Methods for synthesizing RuO_2 and Ru nanoparticles

By using a synthesis method based on NH_3 (25%) precipitate agent, rutile RuO_2 can be produced. The formed amorphous precursor of hydrated ruthenium oxide, "$Ru(OH)_3$", continuously loses water when heated. The crystallization process starts when nearly all the water has been removed from the sample. At various temperatures above 300 °C, crystallization takes place [16]. The rutile RuO_2 nanoparticles were examined by TEM which showed that the particles size are about 20 nm.

In the synthesis of nanocrystalline metallic Ru powder tetraethylene glycol was used as solvent. In the solution $RuCl_3$ (0.20 g) and $C_8H_{18}O_5$, (20 ml) was dissolved and palmitic acid (0.49 g) was added as capping molecule. The suspension was heated to ~40 °C. H_2O_2 (30%) (2 ml) was added slowly and drop wise to the solution which was heated to ~310 °C under O_2-purge. During heating, the solution turned from dark red to dark green. Finally, a black precipitate was obtained, which was separated from the solvent by centrifuging, and later washed twice with deionised water. XRD pattern showed discrete metallic Ru peaks, without the formation of any other crystalline by-products. The average crystallite size was estimated to ~5.0 nm from Scherrer´s equation.

2.3 Fabrication of the sensor device

A schematic of a sensor is shown in Fig. 3, the sensors used in this study were capacitors based on 4H-SiC with a RuO_2 or Ru as the gate material.

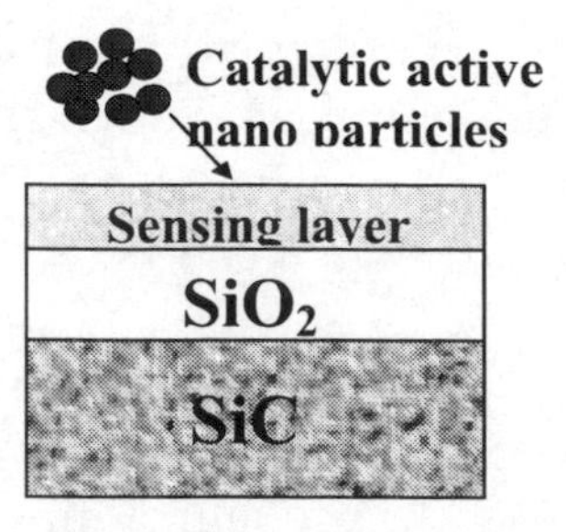

Fig. 3. Schematic picture of a MISiC capacitor sensor.

The total thickness of the insulator was 80 nm and consists of a sandwich structure of SiO_2 and Si_3N_4, this was chosen to improve the device performance. The process started by first growing a thermal oxide of SiO_2 onto the n-doped 4H-SiC surface, followed by a densified layer of Si_3N_4, which also gives a top layer of SiO_x upon oxidation. The SiC is electrically connected via an ohmic backside contact, consisted of alloyed Ni with 50 nm $TaSi_x$ and 400 nm Pt deposited on top (as corrosion protective layer). As bonding pad on top of the sensor a Ti/Pt layer was formed. In this study the active gate materials that are used are RuO_2 and Ru nanoparticles. The active gate region was formed by drop deposition of particles in a suspension of methanol and deionised water using a micropipette. A constant volume (3 μl) of the suspension was taken to define a gate area of ~1 mm diameter for each capacitor. Post-deposition annealing (400 °C for 30 min) was performed to enhance the stability of the particles on the SiO_2 surface. This thermal treatment was found also effective for consistent electrical behaviour of the gate material.

The holders were mounted in leak tight aluminium capsules, which are connected to a gas flow line. The gases were primed across the sensor surfaces using a computer-controlled gas mixing system.

The electrical properties were evaluated using a Boonton 7200 capacitor meter and a Hewlett Packard 41924 LF impedance analyzer. The capacitance-voltage (C(V)) characteristics were obtained at 1 MHz. Microstructure of the gate material was characterized by X-ray diffraction (XRD) studies using Philips PW 1729 X-ray generator (40 kV, 40 mA), by transmission electron microscopy (TEM) using Philips EM 400 T, operated at 120 kV using LaB_6 filament and scanning electron microscopy (SEM) using a LEO 1550 FEG microscope.

2.4 Mounting and measurements

A Pt-100 element together with the sensor chip was glued onto a ceramic micro-heater and placed on a 16-pin holder, Fig. 4. The Pt-100 element was used as temperature detector [4], and to withstand temperatures up to 400 °C there is an air gap between the holder and the heater. To create electric connections gold bonding are used.

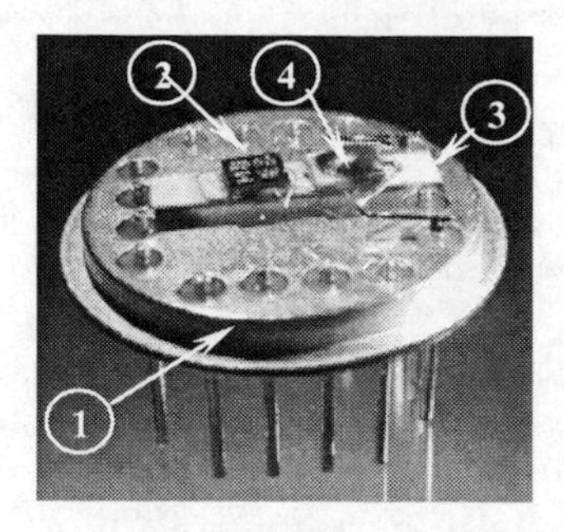

Fig. 4. Mounting of the MISiC capacitor sensors, (1) 16-pin holder, (2) sensor chip, (3) ceramic heater, (4) Pt-100 temperature detector

3. RESULTS AND DISCUSSION

Two kinds of MISiC capacitors, with RuO_2 and Ru as gate materials, were exposed to different gases. The sensors show the same sensitivity pattern but with some small material related differences.

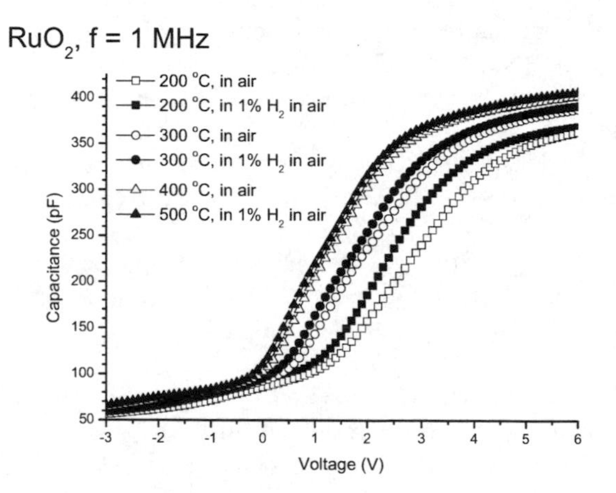

Fig. 5. C(V) curves for RuO_2 capacitor at three temperatures.

The C(V)-curves of the different sensors in different ambients and different temperatures are shown in Figs. 5 and 6. The working point (sensor signal) of the two materials differs in capacitance, for the RuO_2-device ~250 pF is chosen (Fig. 5) and for Ru ~150 pF (Fig. 6). The gate areas in these devices are not well defined due to the deposition method, and this explains the result.

RuO_2 shows the highest response (the voltage shift in the depletion region at a constant capacitance) at 200°C and Ru at 300°C. It seems like RuO_2 nanoparticles has a higher catalytic activity than Ru. The level of the response is however significantly lower for the RuO_2 and Ru materials as compared to porous Pt. This and the lower optimum temperature for the response for RuO_2 as compared to Ru indicate that oxidation and reduction of the material may be involved in the detection mechanism. Comparing the flat band voltages of the C(V)-curves of the two materials it seems like RuO_2 has a somewhat higher work function as compared to Ru. This will be further studied by XPS.

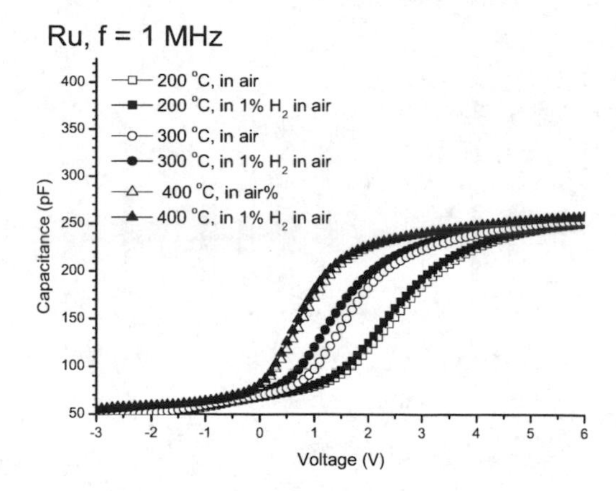

Fig. 6. C(V) curves for Ru capacitor at three temperatures.

The response to typical exhausts and flue gases [17] of the RuO_2 and Ru sensors are compared in Figs. 7 and 8. None of the sensors show any response to CO, while both of them significantly respond to H2, propene and NH3.

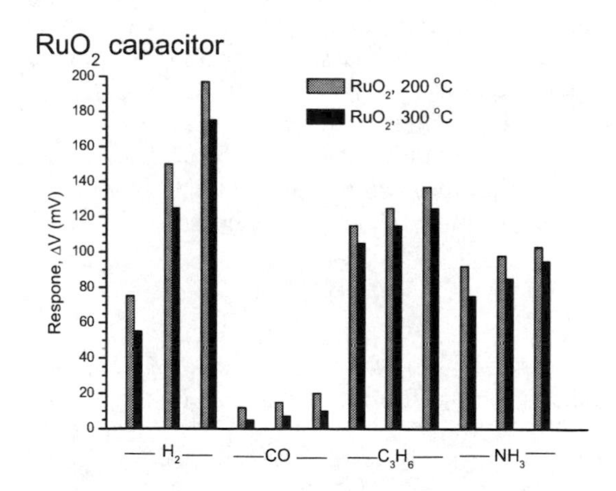

Fig. 7. The gas response for a Ru capacitor (at a constant capacitance) to: 1000, 5000 and 1% H_2, 50, 100, 250 ppm CO, 250, 500 and 1000 ppm C_3H_6 and NH_3 all gases are mixed with synthetic air.

The trend already seen in the C(V)-curves is also shown here, while RuO_2 show the highest response at 200°C, Ru has a higher response at 300°C. This may be used in sensor arrays to get sensors slightly different in selectivity. However, these materials show rather similar behaviour to e.g. sensors with porous Pt gates, which is used as standard sensors today [4]. Non-conducting materials impregnated with catalytic metals, e.g. synthesized by aerosol technology, are expected to give more selective sensors.

Ru capacitor

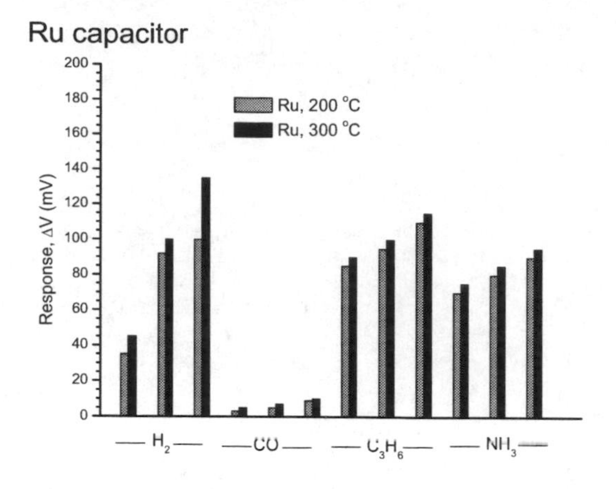

Fig. 8. The gas response for a RuO_2 capacitor (at a constant capacitance), gases as in fig. 7.

4. CONCLUSION

MISiC capacitors with catalytic active nanoparticles as gate material exhibit sensitivity to hydrogen containing gases. The sensitivity pattern depends on the sensing material and the operating temperature of the device. The maximum operating temperature is 300°C for the Ru capacitor and 200°C for RuO_2 capacitor. The sensitivity pattern to reducing gases is similar to Pt-gate field effect sensors, which has the maximum operating temperature at 200°C. There are indications that oxidation and reduction of the Ru / RuO_2 material may be involved in the detection mechanism. This will be further studied by XPS.

Since a wide variety of materials may be used, the metal oxide nanoparticles are potential candidates to be used as gate material for selective field-effect devices.

REFERENCES

[1] M. I. Baraton, L. Merhari, "Nanoparticles-based chemical gas sensors for outdoor air quality monitoring microstations", Materials Science and Engineering B, 112, 206-213, 2004.

[2] A. Salomonsson, S. Roy, C. Aulin, J. Cerdà, P.-O. Käll, L. Ojamäe, M. Strand, M. Sanati, A. Lloyd Spetz, "Nanoparticles for long term stable, more selective MISiCFET gas sensors", Sensors and Actuators, in press, 2004.

[3] H. Wingbrant and A. Lloyd Spetz, "Dependence of Pt Gate Restructuring on the Linearity of SiC Field Effect Transistor Lambda Sensors", Sensor Letters, vol. 1, 37-41, 2003.

[4] A. Lloyd Spetz and S. Savage, "Advances in FET chemical gas sensors", in Recent Major Advances in SiC, Eds. W.J. Choyke, H. Matsunami and G. Pensl, Springer, Berlin, 879-906, 2003.

[5] M. Löfdahl, C. Utaiwasin, A. Carlsson, I. Lundström and M. Eriksson, "Gas response dependence on gate metal morphology of field-effect devices", Sensors and Actuators B 80, 183-192, 2001.

[6] S.K. Shukla, G.K. Parashar, A.P. Mishra, Puneet Misra, B. C. Yadav, R.K. Shukla, L.M. Bali, G.C. Dubey, "Nano-like magnesium oxide films and its significance in optical fiber humidity sensor", Sensors and Actuators B 98, 5-11, 2004.

[7] P. Papageorgiou, D. M. Price, A. Gavriilidis, A. Varma, Preparation of Pt/γ-Al2O3 Pellets with Internal Step-Distribution of Catalyst: Experiments and Theory, Journal of Catalysis 158, 439-451, 1996.

[8] M. Strand, J. Pagels, A. Szpila, A. Gudmundsson, E. Swietlicki, M. Bohgard, and M. Sanati, "Fly Ash Penetration through Electrostatic Precipitator and Flue Gas Condenser in a 6 MW Biomass Fired Boiler", Energy & Fuel 16, 1499-1506, 2002.

[9] W. J. Stark, K. Wegner, S. E. Pratsinis, and A. Baiker, "Flame Aerosol Synthesis of Vanadia-Titania Nanoparticles: Structural and Catalytic Properties in the Selective Catalytic Reduction of NO by NH3". Journal of Catalysis 197, 182-191, 2001.

[10] T. Sahm, L. Mädler, A. Gurlo, N. Barsan, S.E. Pratsinis, U. Weimar, "Flame spray synthesis of tin dioxide nanoparticles for gas sensing", Sensors and Actuators B 98, 148-153, 2004.

[11] G. Korotcenkov, V. Brinzari, A. Cerneavschi, M. Ivanov, A. Corent, J. Morante, A. Cabot, J. Arbiol, "In2O3 films deposited by spray pyrolysis: gas response to reducing (CO, H2) gases", Sensors and Actuators B 98, 122-129, 2004.

[12] Y.G. Choi, G. Sakai, K. Shimanoe, N. Yamazoe, "Wet process based fabrication of WO3 thin film for NO detection", Sensors and Actuators B 101, 107-111, 2004.

[13] M. Wallin, H. Grönbeck, A. Lloyd Spetz, M. Skoglundh, "Vibrational study of ammonia adsorption on Pt/SiO2", Applied Surface Science 235, 487-500, 2004

[14] J. Fogelberg, M. Eriksson, H. Dannetun, L.-G. Petersson, "Kinetic modeling of hydrogen adsorption/absorption in thin films on hydrogen-sensitive field-effect devices: Observation of large hydrogen-induced dipoles at the Pd-SiO2 interface", Journal of Applied Physics 78(2), 988-996, 1995

[15]A. Salomonsson, D. Briand, A. E. Åbom, I. Lundström, M. Eriksson, "The influence of the insulator surface properties on the hydrogen response of field-effect gas sensors", in manuscript.

[16] J. Málek, A. Watanabe, T. Mitsuhashi, "Crystallization kinetics of amorphous RuO2", Thermochimica Acta 282/283, 131-142, 1996.

Gold Removal and Recovery using Mesoporous Silica Adsorbents

K. F. Lam[1], K. L. Yeung[2] and G. McKay[3]

The Hong Kong University of Science and Technology (HKUST)
[1]Environmental Engineering Program, HKUST, Kowloon, HK, SAR-PR China, louislam@ust.hk
[2]Chemical Engineering, HKUST, Kowloon, HK, SAR-PR China, kekyeung@ust.hk
[3]Chemical Engineering, HKUST, Kowloon, HK, SAR-PR China, kemckayg@ust.hk

ABSTRACT

Two gold-selective adsorbents were prepared from MCM-41 by grafting amino and thiol groups. The mesoporous silica adsorbents possess higher adsorption capacity for gold compared to activated carbons. The OMS-NH_2 adsorbent was able to selectively remove and adsorb gold from solutions containing copper. Pure gold was recovered by a simple acid wash and the regenerated adsorbents could be reused. The OMS-SH also selectively adsorb gold, but always with a small quantity of copper (ca 1-5 at. %). In addition, the metals were irreversibly adsorbed on OMS-SH and can not be easily recovered.

Keywords: MCM-41, adsorption, separation, binary mixture, copper, gold

1 INTRODUCTION

The ordered mesoporous materials (e.g., M41S, FSM, HMS and SBA) belong to an important class of molecular sieve materials. Their large surface areas, ordered pore structures and nanometer-sized pores offer a unique environment for chemical separation and reaction [1,2]. The concept of 'supra-molecular templating' has enabled the design of mesoporous silica with adjustable pore sizes and structures [3]. Chemical modifications of the pore channels with metals, metal oxides and organic moieties [4-6] led to new materials with unique physical, chemical and catalytic properties.

The cylindrical pore structure and high degrees of pore symmetry found in MCM-41 make it an ideal system for testing new and existing adsorption and diffusion models. The simple pore geometry allows for easier mathematical description and the amorphous pore wall is a good approximation of an ideal Langmuir surface. The width of the pore channel restricts the size and shape of the molecules that can enter and leave the pores. This gives rise to molecular sieving effects that have many beneficial applications in separation. The chemical environment of the pore channels can be simply manipulated to affect the adsorption and diffusion of molecules by grafting surface chemical moieties. This provides new opportunities for solving difficult separation problems.

In the last few years there have been significant advances in the synthesis of ordered mesoporous silica (OMS). The two main obstacles for the widespread industrial application of OMS are its cost and stability. In the manufacture of the ordered mesoporous silica, the surfactant template accounts for more than 80 % of the cost. Nondestructive methods for the removal and recovery of the surfactants and the use of cheaper polymer substitutes have substantially cut the cost of OMS [7]. Better thermal and hydrothermal stability is obtained by post-synthesis treatment of the OMS with metal alkoxides, salt solution and trimethylsilylation [8,9]. Although further cost reduction is needed for the economical application of OMS technology to general environmental problems, there are many specific cases where OMS technology is urgently needed. Feng and coworkers [2] are among the first to demonstrate the use of OMS for the removal of mercury and other toxic heavy metals from water. It is found that the modified mesoporous silica has high adsorption capacity for mercury ions. Our group has shown that it is possible to tailor OMS adsorbent to remove specific dye from mixtures, enabling their recovery and reuse [10]. The increasing use of precious metals such as gold and platinum in microelectronics, sensors and catalysts pose both economic and technical challenges for their recovery, purification and reuse. This work reports the use of OMS adsorbents for the selective removal and recovery of gold from solutions containing a base metal (i.e., copper).

2 EXPERIMENT

The synthesis of MCM-41 and its modifications to produce selective adsorbents are described along with the details of the single and binary components adsorption experiments.

2.1 Synthesis

The MCM-41 powder was prepared from an alkaline solution containing tetraethyl orthosilicate (TEOS, 98%, Aldrich), cetyltrimethylammonium bromide (CTABr, 99.3%, Aldrich), ammonium hydroxide (NH_4OH, 28-30wt%, Fisher Scientific) and double distilled, deionized water. Plate-shaped crystals with an average diameter of 0.9 $\pm$ 0.1μm and a thickness of about 0.1μm were crystallized from a synthesis solution with a molar composition of 6.58 TEOS: 1 CTABr: 292 NH_4OH: 2773 H_2O. The synthesis solution was stirred for 24 hours at room conditions after mixing all the chemical components. The precipitate formed in the solution was filtered and washed with DDI water. After drying, the powder was calcined at 823 K for 24 h to remove the organic template molecules. Characterizations were conducted on the calcined powder. The recovered powder displayed the characteristic X-ray diffraction pattern of MCM-41 and the nitrogen physisorption measurement (Coulter SA 3100) gave a BET surface area of 1071 m^2/g and a pore volume of 0.3805 cm^3/g. An average pore size of 27.6 Å was calculated from the XRD (Philip 1080) and N_2 physisorption data. Routine chemical analysis of the sample by X-ray photoelectron spectroscopy (XPS, Physical Electronics PHI 5600) detected only silicon and oxygen atoms with carbons from adsorbed carbon dioxide and ambient hydrocarbons as the main surface impurities. The FTIR (Perkin-Elmer model GX 2000) spectrum of the powder displayed the characteristic bands for Si-OH groups at 3675 cm^{-1}, 950 cm^{-1} and 800 cm^{-1}.

2.2 Post-modification

Amino and thiol-containing mesoporous silicas were prepared using the following procedures. $OMS-NH_2$ was obtained when 2.5 g of MCM-41 was refluxed in 250 ml dry toluene solution containing 0.1 mole of 3-aminopropyltrimethoxysilane (APTS, 97%, Aldrich) at 383 K for 18 h. After cooling to room temperature, the powder was collected by series of centrifugation and washing steps. OMS-SH was prepared using the same procedure but replacing APTS with (3-mercaptopropyl) trimethoxysilane (MPTS, 95%, Aldrich). The resulting adsorbent powders were analyzed by FTIR and thermogravimetric method to identify and quantify the grafted surface moieties.

2.3 Adsorption

Single component adsorption isotherms were obtained for 0.1 g of adsorbent in 100 ml aqueous solutions containing 0.5 to 4 mM of metal ions. Gold and copper solutions were prepared from $AuCl_3$ (99% Aldrich) and $CuCl_2$ (98%, Aldrich) salts, respectively. The batch adsorption experiments were at room temperature (295 ± 2 K) for 14 days with constant agitation. Metal concentrations were determined using an inductively coupled plasma-atomic emission spectrometer (ICP-AES) using reference standard solutions of gold and copper.

The adsorption behavior and selectivity of $OMS-NH_2$ and OMS-SH adsorbents were determined for binary solutions of gold and copper salts. The effects of pH, total metal concentration and metal composition were investigated. In the experiment, 0.1 g of adsorbent was added into 100 ml of metal salt solution. The pH of the solution was adjusted using hydrochloric acid. The pH at the start and at the end of the experiment rarely deviated by more than ± 0.2. The adsorption was conducted in a shaker bath at 295 ± 2 K. Experiments were carried out to study the effects of metal composition on adsorption. Solutions with a total metal concentration of 2 mM, but different metal compositions were prepared. The adsorption was conducted at a fix pH (2.5) and temperature (295 K).

After the adsorption, the used adsorbents were collected by centrifugation and regenerated by washing with 1 M hydrochloric acid.

3 RESULTS AND DISCUSSION

The $OMS-NH_2$ displayed an I.R. absorbance peaks at 3360cm^{-1} together with 3288 cm^{-1} and 1600 cm^{-1} belonging to the -NH2 and C-N bond, respectively. A signal at 2580 cm^{-1} assigned to S-H stretching vibration was detected in OMS-SH. Table 1 summarizes the physical and chemical properties of the OMS adsorbents.

Sample	Surface Area	Pore Size[1]	Concentration of Functional Group		Isoelectric point
	(m^2/g)	(nm)	mmol/g	groups/nm^2	pH
MCM-41	1150	3.01	- -	- -	2.1
$OMS-NH_2$	774	2.57	2.23	1.7	3.6
OMS-SH	887	2.76	1.27	0.9	4.0

Table 1: Physical and chemical properties of the mesoporous silica adsorbents.

In a simple stability test in aqueous solution, it shows that the mesoporous silica adsorbents were stable over a broad range of pH values (2 to 10), however at pH > 10, the organic moieties detach from the MCM-41 as the siliceous wall dissolved at these high pH. Both $OMS-NH_2$ and OMS-SH were also stable in solvents (e.g., toluene, acetone and ethanol), but were vulnerable to oxidation by strong oxidizing agents (e.g., H_2O_2).

Sample	Adsorption capacity (mmol/g)/[mg/g]	
	Au^{3+}	Cu^{2+}
MCM-41	~	~
$OMS-NH_2$	1.69 [333]	0.82 [52]
OMS-SH	1.60 [315]	0.28 [18]

Table 2: Single component adsorption capacities for mesoporous silica adsorbents

From the adsorption results shown in Table 2, despite the enormous surface area, gold and copper adsorptions on

NSTI-Nanotech 2005, www.nsti.org, ISBN 0-9767985-1-4 Vol. 2, 2005

MCM-41 are negligible compared to the other two adsorbents. OMS-NH_2 and OMS-SH have comparable adsorption capacity for gold (Table 2), which is four times larger than the best value reported for activated carbon [11, 12]. OMS-SH adsorbs less copper suggesting that it can be the more selective gold adsorbent of the two. These results show the importance of the surface functionality for metal adsorption.

Figure 1 plots the adsorption data of OMS-NH_2 and OMS-SH for the binary metal solutions. Metal adsorption at different pH was conducted using solutions containing equimolar concentrations of gold and copper salts. It is clear from Fig. 1a that only gold ion is adsorbed by OMS-NH_2 in complete exclusion of copper. Also, analysis indicated that complete gold removal is achieved over the whole experimental pH range. OMS-SH also selectively removes and adsorbs gold from the solution (Fig. 1b), but at low pH (< 2.5) small quantities of copper (2.5 mg/g) is also adsorbed. Figures 1c and 1d plot the metal adsorbed as a function of metal composition in the solution. Gold-only removal without copper adsorption is obtained using OMS-NH_2 adsorbent. More than 90 % of the gold is adsorbed from the solution as shown Fig. 1e, and the removal efficiency reaches 100 % for solutions with dilute gold concentration making it an attractive adsorbent for gold recovery from dilute mining and recycling wastes. The OMS-SH adsorbent also selectively adsorbs gold from the solution (Fig. 1d), but the removal efficiency is lower (Fig. 1f) and small quantities of copper are always adsorbed.

More than 70 % of the adsorbed gold was recovered from OMS-NH_2 after a single wash with 1 M HCl solution. It was found that the gold concentration was increased tenfold after the adsorption and regeneration process. In contrast, less than 5 % of the adsorbed metals in OMS-SH were recovered even using more concentrated acid wash solutions (i.e., 5 M HCl). X-ray photoelectron spectroscopy suggested that gold and copper underwent a change in oxidation state during their adsorption on OMS-SH, which maybe responsible for the irreversible adsorptions of these metals. Therefore, although OMS-SH can selectively adsorb gold, it is not suitable for industrial use since the metal can not be easily recovered. The gold recovered from OMS-NH_2 is pure, requiring no further purification and can be reduced into metallic form using sodium borohydride. The regenerated adsorbent can be reused without loss in capacity by simply taking the precaution to pre-rinse with a small amount of ethanol.

4 CONCLUSIONS

This work clearly demonstrated the potential use of selective adsorption for the recovery of precious metals such as gold from the solution containing other metals. The OMS-NH_2 proved to have a large capacity and excellent affinity for gold, enabling its separation from mixtures and recovery as high purity gold.

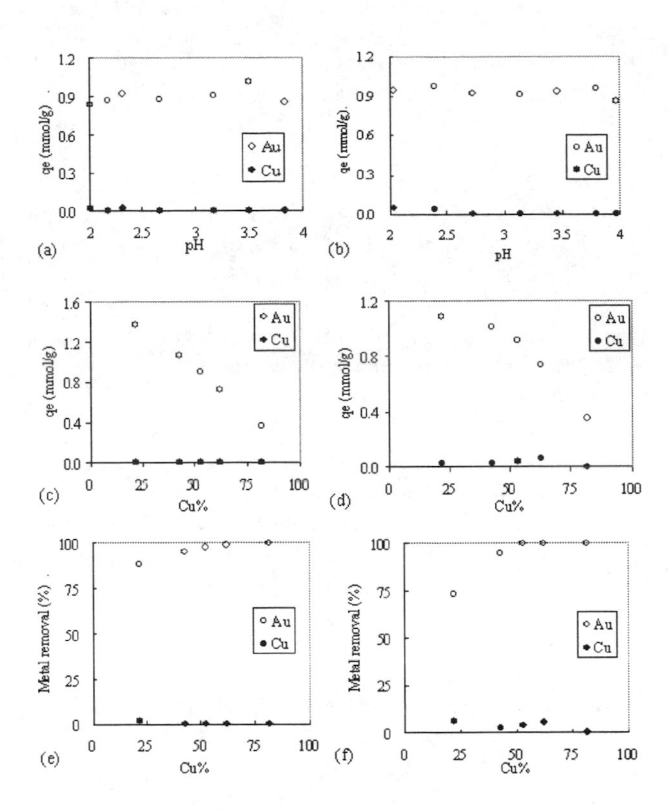

Figure 1: Binary metal adsorptions of copper and gold solutions at different pH for (a) OMS-NH_2 and (b) OMS-SH; Metal Adsorption vs At.% Cu for (c) OMS-NH_2 and (d) OMS-SH; Metal Removal vs At.% Cu for (e) OMS-NH2 and (f) OMS-SH. (q_e is the adsorption capacity of metal ion in mmol/g).

5 ACKNOWLEDGEMENTS

The authors gratefully acknowledge the funding from the Hong Kong Research Grant Councils (grant RGC-HKUST 6037/00P). Mr. K. F. Lam would also like to thank the financial support from the Environmental Engineering Program and the Institute of Nano Science and Technology. We also thank the Material Characterization and Preparation Facility at the HKUST for the use of XRD, TEM, SEM, EDXS, XPS and TGA/DTA equipment.

REFERENCES

[1] A. Corma, M. T. Navarro and J. P. Pariente, "Synthesis of an ultralarge pore titanium silicate isomorphous to mcm-41 and its application as a catalyst for selective oxidation of hydrocarbons" J. Chem. Soc., Chem. Commun., 2, 147, 1994.

[2] X. Feng, G. E. Fryxell, L. Q. Wang, A. Y. Kim, J. Liu and K. M. Kemmer, "Functionalized monolayers on ordered mesoporous supports" Science, 276, 923, 1997.

[3] J. C. Vartuli, K. D. Schmitt, C. T. Kresge, W. J. Roth, M. E. Leonowicz, S. B. McCullen, S. D. Hellring, J. S. Beck, J. L. Schlenker, D. H. Olson and E. W. Sheppard, "Effect of surfactant silica molar ratios on the formation of mesoporous molecular-sieves - inorganic mimicry of surfactant liquid-crystal phases and mechanistic implications" Chem. Mater., 6, 12, 2317, 1994.

[4] K. A. Koyano, T. Tatsumi, Y. Tanaka and S. Nakata, "Stabilization of mesoporous molecular sieves by trimethylsilylation" J. Phys. Chem. B, 101, 46, 9436, 1997.

[5] R. S. Mulukutla, K. Asakura, S. Namba and Y. Iwasawa, "Nanosized rhodium oxide particles in the MCM-41 mesoporous molecular sieve" Chem. Comm., 14, 1425, 1998.

[6] M. Okumura, S. Tsubota, M. Iwamoto and M. Haruta, "Chemical Vapor Deposition of Gold Nanoparticles on MCM-41 and Their Catalytic Activities for the Low-temperature Oxidation of CO and of H_2" Chemical Letter, 4, 315, 1998.

[7] S. Kawi and M. W. Lai, "Supercritical fluid extraction of surfactant template from MCM-41" Chem. Comm., 13, 1407, 1998.

[8] R. Ryoo and S. Jun, "Improvement of hydrothermal stability of mcm-41 using salt effects during the crystallization process" J. Phys. Chem. B, 101, 317, 1997.

[9] T. Tatsumi, K. A. Koyano, Y. Tanaka and S. Nakata, "Stabilization of M41S Molecular Sieves by Trimethylsilylation" Stud. In Surf. Sci. and Catal., 117, 143, 1998.

[10] K. Y. Ho, G. McKay and K. L. Yeung, "Selective adsorbents from ordered mesoporous silica" Langmuir, 19, 3019, 2003.

[11] D. W. Darnall, B. Greene, M. T. Henzl, J. M. Hosea, R. A. McPherson, J. Sneddon and M. D. Alexander, "Selective recovery of gold and other metal ions from an algal biomass" Environmental Science and Technology, 2, 206, 1986.

[12] W. Nakbanpote, P. Thiravetyan and C. Kalambaheti, "Comparison of gold adsorption by Chlorella vulgaris, rice husk and activated carbon" Minerals Engineering, 15, 549, 2002.

Preparation of Zeolite Microspheres and their Application in Biology

L.W.Wong [1] and K.L.Yeung[2]

The Hong Kong University of Science and Technology (HKUST)
[1]Bioengineering Graduate Program, HKUST, Kowloon, HK, SAR-PR China, lingwong@ust.hk
[2]Department of Chemical Engineering, HKUST, Kowloon, HK, SAR-PR China, kekyeung@ust.hk

ABSTRACT

Hollow zeolite microspheres were prepared by growing a thin zeolite membrane shell on seeded polystyrene beads. The removal of the organic growth directing agent and polystyrene bead was successfully performed without damaging the thin zeolite shell. The storage and delivery of sodium pervanadate, a potent tyrophosphatase inhibitor to MDCK cells was successfully carried out. Thus, demonstrating the potential use of these materials for biological studies of cell signaling.

Keywords: hollow zeolite microspheres, storage, delivery, water-soluble organic compounds, bioactive molecules.

1 INTROUDUCTION

Chemical storage and delivery system is an important technology not only for drug delivery, but also for biological research. It is a critical technology for the emerging "laboratory-on-a-chip" devices that include μ-TAS, DNA-Chips and Bio-MEMS. These microdevices are expected to duplicate and automate the procedures used in chemical analysis, biological assay and medical diagnosis. This means that chemical reagents needed for the separation, purification, isolation, conversion, amplification and detection must be stored in the microdevice and made available on demand during the operation. Chemical storage and delivery system that respond to changes in environmental conditions, such as temperature [1], pH [2], light [3], electric field [4], and certain chemicals [5] have been the focus of many researches. The selection of an appropriate method for preparing drug-delivery vehicles depends on the physicochemical properties of the polymer used for fabricating the vehicles and the drug itself. However, the methods available for incorporating the hydrophilic compounds are few and difficult to implement. One major challenge is to produce delivery vehicle that can carry water soluble drugs and does not tend to aggregate. Membrane-enclosed microspheres based on nanoporous zeolites and molecular sieves are potential candidate for chemical storage and release, either alone or onboard a lab-on-a-chip device. The transport through zeolites is governed by the structure and chemistry of the pore channels. The zeolite pore structure restricts the size and shape of the molecules that can access the zeolite pores giving rise to molecular sieving properties. The narrow channels also restrict the movement of bulky molecules and constrain their orientation within the pore. The chemical environment within the zeolite pore channel is governed by the structure of the zeolite framework, the nature of the framework substitution ions and the presence of counterions. The interactions between the diffusing molecules and these adsorption sites have resulted in many unique separation properties. Here, we report on the fabrication of hollow zeolite microspheres that posses ordered pores size and with long term chemical and biological stability. The growth directing agent and polymer core are removed by ozone treatment, which is carried out in solution under low temperature. The procedure avoids the problem of agglomeration and the need for redispersion that plagued the other methods.

2 METHODOLOGY

2.1 Construction of zeolite-coated hollow capsule

Assembly of polyelectrolytes was started by adding 1 ml of Poly(diallyldimethylammonium chloride)PDADMAC (1mg/ml in 0.5 M NaCl) to a 100 μl suspension of 4.5 μm polystyrene (PS) particles (1 wt%, ~ 10^{11} particles mL^{-1}). The suspension was stirred at room temperature for 15 minutes to allow adsorption, and then centrifuged at 6,000g for 5 minutes. Following the removal of supernatant, the coated particles were washed three times with 0.5 M NaCl to remove excess polyelectrolyte. A 1 ml aliquot of (polystyrene sulfonate) PSS (1 mg ml^{-1} in 0.5 M NaCl) was next added to form a second polyelectrolyte layer of negative charge on the particle surface. The centrifugation and washing steps and the sequential adsorption of oppositely charged polyelectrolytes were repeated to obtain PS particles with three layers of polyelectrolyte coating, herein referred to as PS/(PDADMAC/PSS/PDADMAC). The coated particles were re-suspended in 900 μl of double distilled, deionized (DDI) water and 100 μl of 100 nm TPA-Sil-1 seeds (2 wt %) was added. The suspension was stirred at room temperature for overnight. The consecutive assembly of polyelectrolyte multilayer onto PS particles was examined by Zeta Potential Analyzer (Brookhaven, Zetaplus), Scanning Electron Microscope (SEM, Philips XL-30) and Transmission Electron Microscope (TEM, Philips CM-20). Zeolite seed-coated PS particles were transferred to a synthesis solution containing molar ratio of 40 SiO_2: 5 TPA_2O: 20,000 H_2O. The synthesis was conducted at 403 K for 24 h. The zeolite microspheres were recovered and purified by a series of centrifugation and washing steps. The organic growth-directing molecules and the polymer core were removed by bubbling ozone to the microsphere suspension at 353 K for 20 h. The suspension was then centrifuge to recover the hollow zeolite microspheres. The entire preparation process for hollow zeolite microspheres is depicted in Fig. 1.

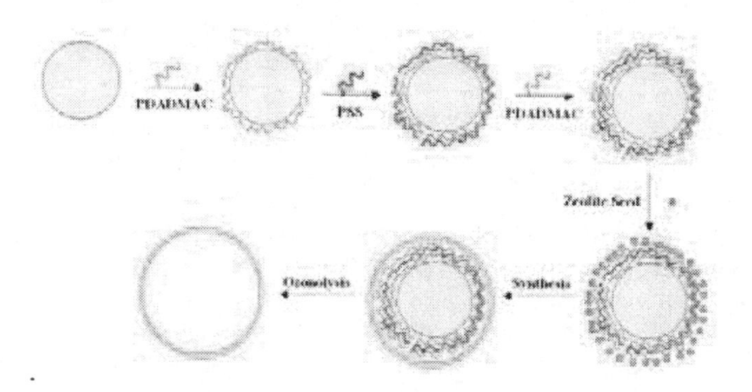

Figure 1. Illustration of the procedure for the preparation of hollow zeolite microspheres.

2.2 Use of zeolite microspheres to study cell signaling

A 1 M sodium orthovanadate was mixed with 1.7% H_2O_2 at a 30:1 ratio to produce sodium pervanadate, which known to suppress the phosphokinase activity in cells. Hollow zeolite microspheres were equilibrated with the solution for half an hour, spin down and washed. This procedure was repeated three times. The loaded microspheres were applied to Madin-Darby canine kidney (MDCK) cells and incubated for 1 h. The MDCK cells were then fixed with 96% ethanol at 253 K for 5 minutes and then stained with anti-phosphotyrosine antibodies ($mAb4G_{10}$, Upstate Biotechnology) and FITC-conjugated anti-mouse secondary antibodies.

3 RESULTS AND DISCUSSIONS

Polystyrene (PS) beads were selected as the template for preparing hollow zeolite microspheres. Polystyrene beads with narrow size distributions are readily available commercially. The sulfate-stabilized 4.5 μm PS beads have a measured surface zeta potential of -74.1 mV (Fig. 2). This means that the beads have the same negative charge as the zeolite seeds. Seeds were needed to provide the nucleation sites for zeolite growth. The beads were seeded with 100 nm-sized TPA-Sil-1 using the established layer-by-layer (LBL) method. The polymer beads were coated with PDADMAC to change surface from a negative to a positive charge (Fig. 2). However, to ensure a uniform charge distribution over the beads' surface, a second layer of PSS was added. This resulted in a negatively charged surface. A final layer of PDADMAC gives the beads a uniform positive charge needed for seed deposition. Figure 2 plots the changes in the surface Zeta potential of the beads after each coating steps. Precaution must be taken to wash away the excess polyelectrolyte species between coating steps. These make sure that a monolayer is formed with each coating [6,7].

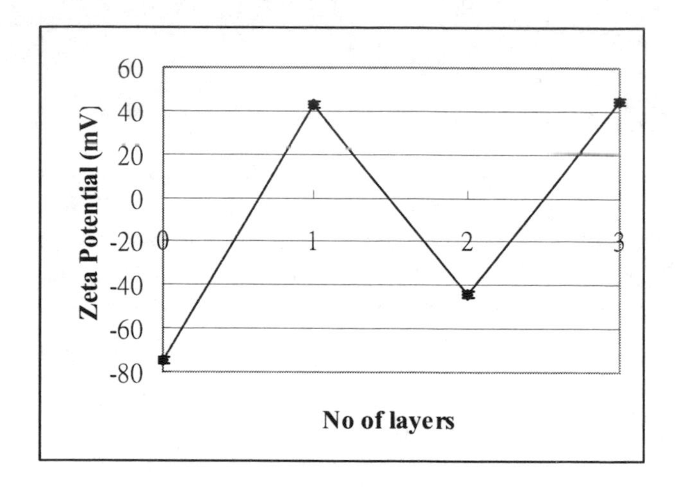

Figure 2. A plot of Zeta Potential for 0-uncoated PS beads, 1-PS/PDADMAC, 2-PS/(PDADMAC/PSS) and 4-PS/ (PDADMAC/PSS/PDADMAC)

Figure 3a displays the PS beads seeded with 100 nm zeolite seeds. The seeds form a uniform, densely packed layer on the polyelectrolyte-coated bead as shown in a higher magnification SEM picture in Fig. 3c. The zeolite shell was then grown from a synthesis solution containing both the silica source, tetraethyl orthosilicate (TEOS) and the growth directing agent, tetrapropylammonium hydroxide (TPAOH). The TPAOH also served as a source of hydroxide ions for maintaining the alkalinity of the solution. The morphology of the Sil-1 zeolite grain depends on the composition of the synthesis solution. Larger amount of TPAOH in the solution led to a larger grain size and square shape. Dilute synthesis solutions often resulted in elongated zeolite grains. A smooth and well-intergrown Sil-1 zeolite shell was obtained from synthesis composition of 40 SiO_2: 5 TPA_2O: 20,000 H_2O as shown by the zeolite microspheres in Fig. 3b & 3d.

The TPA^+ molecules trapped with the zeolite pores and the polystyrene core enclosed by the thin zeolite shell were removed using a novel method. Ozone gas was bubbled into the microsphere suspension at 353 K. The ozone oxidizes and removes the TPA^+ molecules freeing the zeolite pores. The ozone then diffuses and reacts with the polymer core, breaking the polymers into shorter and more soluble carboxylic acid chains that could easily dissolve in water and diffuse out of the zeolite shell. The whole process is conducted at low temperature, minimizing the damage to the microspheres. Unlike the current high temperature treatment process, less than 1 percent of the hollow zeolite microspheres were damaged compared to the usual 10-20 % values reported in the literature [8-10]. Also by keeping the microspheres in suspension throughout the preparation prevented their agglomeration and dispersed particles were obtained.

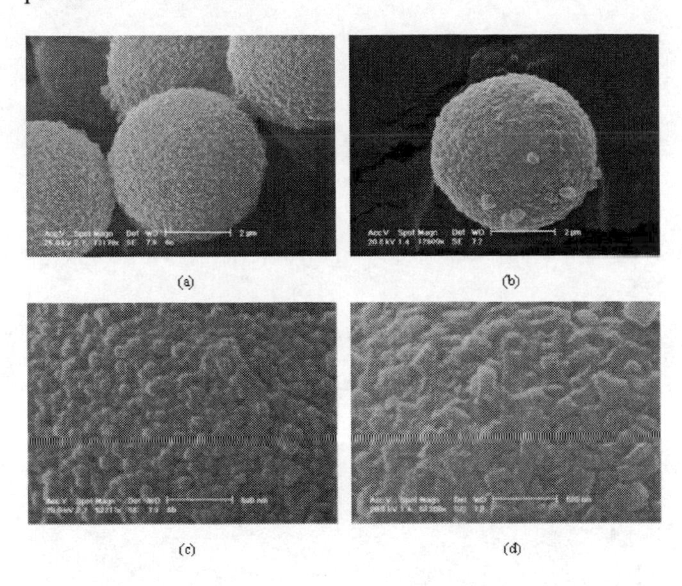

Figure 3. Scanning electron micrographs of (a, c) zeolite seeded microsphere and (b,d) zeolite microsphere.

Figure 4a shows a SEM picture of a hollow zeolite microsphere. The average sphere diameter is 4.9μm from the original 4.5μm PS beads. This means that the zeolite shell is roughly 200nm thick. Transmission electron microscopy was used to confirm that the polymer core was successfully removed from the hollow zeolite microspheres as shown in Fig. 4b.

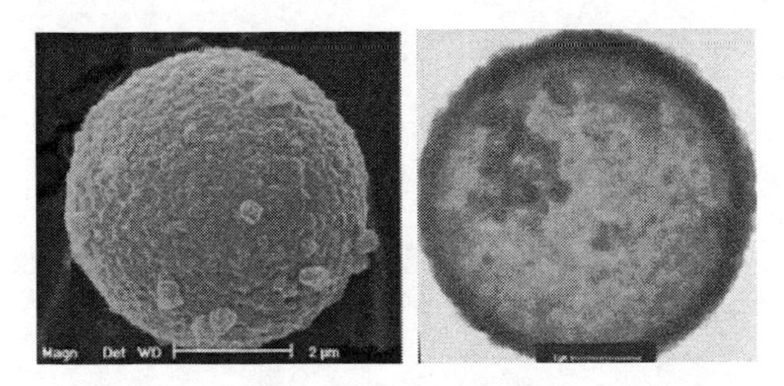

Figure 4. Hollow zeolite microspheres.

The potential use of hollow zeolite microspheres for biological studies was illustrated in the study of cell signaling. The hollow zeolite microspheres carry and locally deliver signalling molecules that mimicks the *in vivo* signalling process between chemicals and regulating enzymes. The hollow zeolite microspheres were loaded with sodium pervanadate (i.e., Na-orthovanadate + H_2O_2). Sodium pervanadate (PV) is known to inhibit the activity of tyrosine phosphatase that leads to the accumulation of tyrosine in phosphorylated form. The PV loaded zeolite microspheres were applied on Madin-Darby canine kidney (MDCK) cell as shown in Figure 5. The marker used to assess the action of the pervanadate PTPase inhibitor was anti-phosphotyrosine (PY) antibody. The increases in the PY staining at microsphere-cell contacts indicate a successful local delivery of PV.

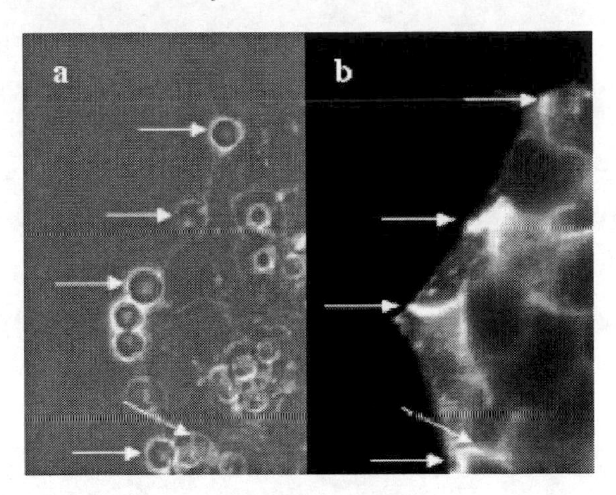

Figure 5 (a) Optical and (b) fluorescent microscope pictures of microspheres deposited on MDCK cells. It is clear that there is an increase in the phosphotyrosine (PY) labelling at the point of contacts in (b).

4 CONCLUDING REMARKS

Hollow zeolite microspheres can carry and deliver small, water-soluble molecules such as sodium pervanadate, a potent inhibitor of the cellular signaling proteins called tyrophosphatase. Local delivery of PV resulted in a clear increase in phosphotyrosine at the microsphere-cell contacts. These microspheres could prove to be an invaluable tool for investigating cell signaling such as the post-synaptic formation of neuromuscular junction (NMJ). An understanding of NMJ is critical for combating many serious diseases including NMJ disorder, muscular dystrophy and Myasthenia Gravis. It may also provide insights for curing botulism and snakebite that act on NMJ.

REFERENCES

[1] R Yashida, K Saka, T Okano, Y Sakurai, J Pharm Sci, 72, 12, 1981.
[2] DC Dong, AS Hoffman, Proc Int Symp Controlled released Bioact Mater, 17, 325, 1990.
[3] K Ishihara, N Hamada, S Kato, I Shinohara, J Polym Sci Polym Chem Ed, 22 , 881, 1984.
[4] A Suzuki, T Tanaka, Nature, 346, 345, 1990.
[5] IC Kwon, YH Bae, SW Kim, Nature 54, 291, 1991.
[6] G Decher, Science, 277, 1232, 1997.

[7] N G Hoogeveen,, M A C Stuart, G Fleer, M R Böhmer, Langmuir, 12, 3675, 1996.
[8] F Caruso, R A Caruso, H Mohwald, Science 282, 1111, 1998.
[9] V Valtchev, S Mintova, Microporous and Mesoporous Mater., 43, 41, 2001.
[10] Y Li, J Shi, Z Hua. H Che, M Ruan, D S Yan, Nano. Lett. 3, 9, 2003.

Polyaniline Nanofibers and Composite Materials for Chemical Detection

Bruce H. Weiller[‡*], Shabnam Virji[‡†], Christina Baker[†], Jiaxing Huang[†#], Dan Li[†], and Richard B. Kaner[†*]

[‡]Materials Processing and Evaluation Department, Space Materials Laboratory,
The Aerospace Corporation, P.O. Box 92957 / M2-248, Los Angeles, CA 90009-2957
[†]Department of Chemistry and Biochemistry, Exotic Materials Institute and
California Nanosystems Institute, University of California, Los Angeles, Los Angeles, CA 90095-1569

ABSTRACT

Recently we developed a simple, template-free chemical synthesis for polyaniline nanofibers that is selective for nanofibers, can be readily scaled to make large quantities and can be controlled to selectively produce nanofibers with narrow size distributions. Chemical sensors fabricated from polyaniline nanofibers have significantly better performance than conventional material in both sensitivity and time response. The high surface area, small diameter, and porous nature of the nanofiber films allow for facile diffusion of vapors, which is responsible for the enhanced performance. Most recently we have shown that composites with polyaniline are useful to detect analytes that do not give a significant response with unmodified polyaniline. These include fluoroalcohol additives for hydrazine detection and inorganics for hydrogen sulfide detection. Polyaniline nanofibers are superior materials that have excellent potential for many chemical detection applications.

Keywords: chemical sensors, polyaniline, nanofiber, conducting polymer.

INTRODUCTION

Conducting polymers hold much potential to accomplish the chemical detection tasks required for chemical agents, toxic industrial chemicals or explosives. Conducting polymers are unique because they show very large electrical property changes when they are chemically treated with oxidizing or reducing agents. After chemical treatment with protonating, deprotonating, or reducing agents, these polymers can change from an initially electrically insulating state to a conducting state [1] with a change in conductivity that can approach ten orders of magnitude. This transition can be used to perform very sensitive chemical sensing [2,3] or biosensing [4,5,6]. Other vapor interactions with the polymer also cause smaller conductivity changes; these include vapor induced swelling or conformational changes that affect the interchain conduction.

Of the conducting polymers, polyaniline is one of the most promising from an applications standpoint since it is very stable in air and undergoes doping and dedoping by simple acids and bases. Other conducting polymers require stronger oxidizing or reducing agents to be converted between conducting and insulating forms. However, conventional films of polyaniline have not found much use as chemical sensors due to relatively low sensitivity or time response.

Recently, we have successfully developed a chemical synthesis to make polyaniline nanofibers in bulk quantities [7,8]. Instead of using the traditional homogeneous aqueous solution of aniline, acid, and oxidant, the polymerization is performed in an immiscible organic/aqueous biphasic system. The products are polyaniline nanofibers with nearly uniform diameters between 30 and 50 nm with lengths varying from 500 nm to several micrometers. Gram-scale products can be synthesized that contain almost exclusively nanofibers. The nanofibers can be modified by a variety of dopants and dedoping doesn't appear to affect their morphology. This novel, yet simple synthetic method makes polyaniline nanofibers readily available for development of inexpensive chemical sensors. Because the nanofibers can be isolated, purified and chemically modified, they can be tailored to provide response to new classes of chemicals as needed for homeland security and other applications.

EXPERIMENTAL

The emeraldine base form of polyaniline was chemically synthesized from aniline by oxidative polymerization using ammonium peroxydisulfate in an acidic media. The polyaniline nanofibers were synthesized in an aqueous-organic two-phase system and purified by dialysis. Polyaniline solutions were made by either dissolving polyaniline in hexafluoroisopropanol (HFIP, 2 mg/mL) or by dissolving polyaniline in N-methylpyrrolidinone (NMP, 1 mg/mL). Filtered solutions were then used to cast films onto substrates by dropping the desired amount of solution from a disposable microliter pipette. The films were air dried (HFIP) or dried in a 60°C oven overnight (NMP). The nanofiber suspensions

* bruce.h.weiller@aero.org, kaner@chem.ucla.edu
[#] Current address: Department of Chemistry,
University of California, Berkeley, CA 94720

obtained after synthesis and dialysis were diluted with water and drop-cast onto the substrates. Film thicknesses were controlled by dilution of the original synthesized suspensions using a constant volume for casting.

Interdigitated electrode sensor substrates were fabricated using standard photolithographic methods at The Aerospace Corporation. The electrode geometry consists of 50 pairs of fingers, each finger having dimensions of 20 μm x 4970 μm x 0.18 μm and a 10 μm gap. Film thicknesses were measured with a profilometer (DekTak II). Electron microscope images were obtained using field-emission scanning electron microscope (JEOL JSM-6401F). Electrical resistances (DC) were measured with a programmable electrometer. Instruments were controlled and read by computer using a GPIB interface and Labview software.

Acid and base gas exposures used certified gas mixtures of HCl, NH_3 or H_2S in nitrogen. For hydrazine exposures, a permeation tube of hydrazine was used with a certified emission rate. Mass flow controllers were used to meter separate flows of nitrogen buffer gas and the gas mixture.

RESULTS AND DISCUSSION

Figure 1a shows a typical TEM image of the polyaniline nanofibers after dialysis. The lengths of the fibers range from 500 nm up to several microns. The nanofibers tend to agglomerate into interconnected nanofiber networks, rather than bundles. A closer look at the nanofibers reveals that many of them are twisted, as shown in the inset. The sample uniformity and narrow diameter distribution are also confirmed by field emission SEM images. Figure 1b shows a typical SEM secondary electron image of a nanofiber thin film cast on an electrode substrate from dialyzed suspension. Dedoping has no noticeable effect on the fiber morphology.

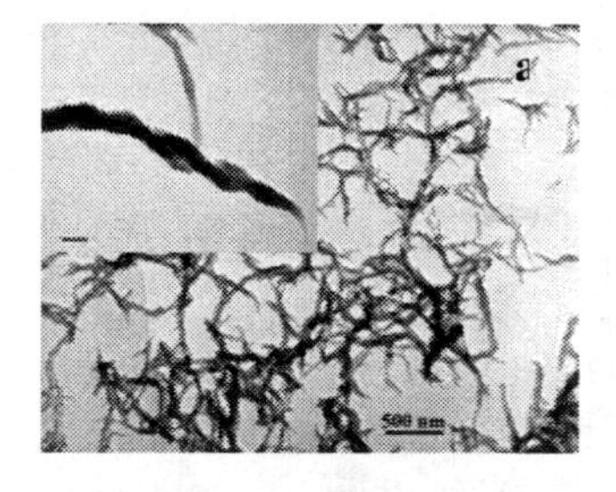

Figure 1. a) TEM images of polyaniline nanofibers cast from suspension. The inset shows a twisted fiber (scale bar = 50 nm), b) SEM secondary electron images of a thin film deposited on sensor substrates. The inset shows a cross-sectional view of the film on the glass substrate (scale bar = 200 nm).

Figure 2 shows the resistance changes of a dedoped film upon exposure to 100 ppm of HCl (left) and a fully HCl doped film exposed to 100 ppm of NH_3 (right). The nanofiber thin film responds much faster than a conventional film to both acid doping and base dedoping even though it is more than twice as thick. This is likely due to the small diameters of the nanofibers that gives rise to a high surface area within the film that can be accessed by the gas vapors. With the small diameter of the fibers, it takes the gas molecules a much shorter time to diffuse in and out of the fibers. This also leads to a much greater extent of doping or dedoping over short times for the nanofiber films. The nanofiber films have better performance in both sensitivity and time response compared to conventional films. In addition, the nanofiber films show no thickness dependence to their sensitivity or time response whereas the conventional films show a strong dependence on the thickness. We have not tested the lower detection limit for HCl but reasonable assumptions indicate detection levels in the ppt range are feasible. Many of the toxic industrial chemicals of concern are strong acids or hydrolyze to form strong acids. Therefore polyaniline nanofibers should be useful to detect many of them.

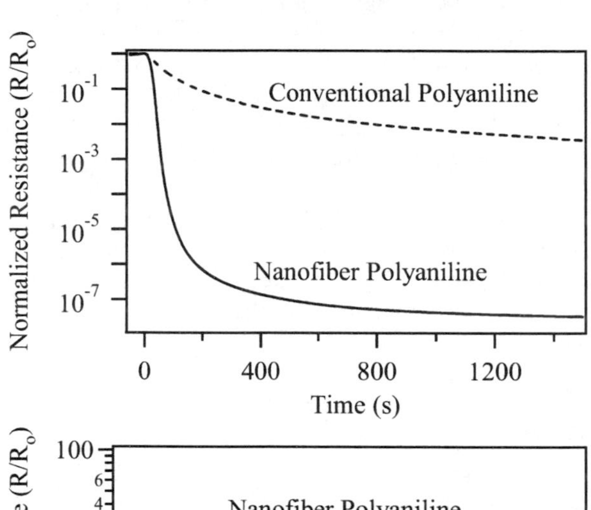

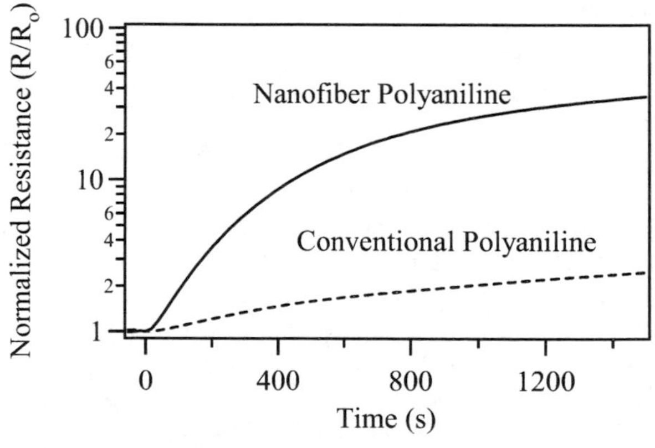

Figure 2. Resistance changes of a nanofiber emeraldine base thin film (solid line) and conventional (dotted line) emeraldine thin film upon exposure to (top) 100 ppm HCl vapor and (bottom) 100 ppm NH_3 vapor. R/R_0 is the resistance (R) normalized to the initial resistance (R_0) prior to gas exposure.

As shown above, when protonated polyaniline interacts with base, it becomes de-protonated, reverses the effect of acid and loses its conductivity. In contrast to the high concentrations of ammonia shown in Figure 2, low concentrations of n-butylamine (24 ppb) have also been detected with *unoptimized* nanofiber sensors. Once we optimize the nanofiber sensors, we believe it will be possible to detect very low levels of basic vapors such as many of the chemical agents of concern.

The response time of the films is a central issue to sensing applications. The acid doping response is expected to be very rapid since it involves the diffusion of protons into the film. The data shown in Figure 2 was taken at relatively low flow rates in a large volume cell hence the fastest data shown is flow rate limited. The data is also plotted over many orders of magnitude on a log scale which makes the time response data appear slow. Figure 3 shows that when the change in resistance of both the nanofiber and polyaniline films are plotted on linear scales, we find that the response time (90%) is ~2 sec for the nanofiber film and ~30 sec for the conventional film. However, the response time is still instrumentally limited. In addition, when very high flow rates are used, resistance changes of 10^7 occur in less than 4 seconds for the nanofiber films. This is consistent with the other data supporting that the nanofibers should and do respond very rapidly. The true response time of the nanofibers should be very fast; work in progress is geared toward an intrinsic time response measurement.

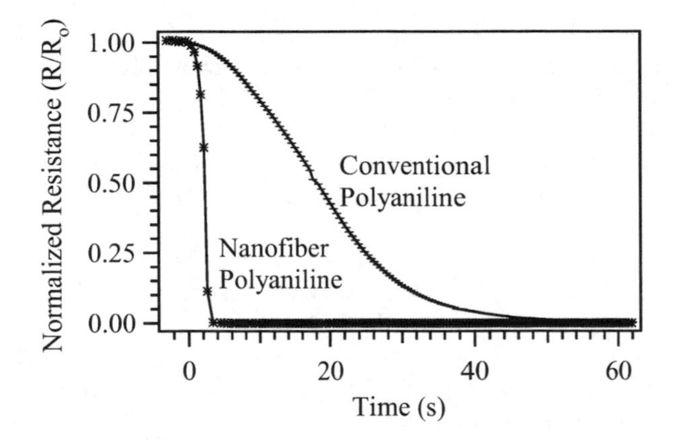

Figure 3. Response time of a nanofiber and conventional polyaniline sensor to 100 ppm HCl.

Hydrazine is a toxic industrial chemical that is used as a rocket fuel and is explosive. Figure 4 shows that polyaniline nanofibers respond to 3 ppm of hydrazine with about a 25-fold increase in resistance. Surprisingly the response of a conventional polyaniline film cast from N-methyl pyrrolidinone (NMP) is negligible ($R/R_0 \sim 1.25$) as shown [9]. We expect the minimum detection limit in the low ppb or even ppt level for hydrazine. Polyaniline is known to undergo a redox reaction with hydrazine. We believe that the nanometer morphology and large surface area of our sensors allows this reaction to occur very rapidly giving a large response in a relatively short time. We expect this type of mechanism to be useful for the detection of other toxic industrial chemicals that can undergo redox reactions with the polyaniline nanofibers.

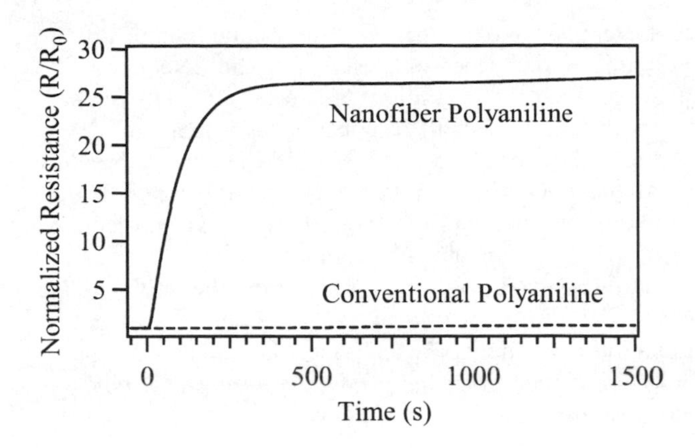

Figure 4. Response of polyaniline nanofibers and a conventional film to 3 ppm of hydrazine.

Figure 5 shows the effect of using a fluorinated alcohol additive on the response of a conventional polyaniline film. When hexafluoroisopropanol (HFIP) is used as a solvent to cast films of polyaniline, we observed a very different response to hydrazine. Instead of reduction of the polyaniline to a nonconducting form, we find a response consistent with doping of the polyaniline. When most of the HFIP is removed from the film by oven drying, we see a reduced response of the sensor. Furthermore, studies using HFIP diluted in NMP shown a similar effect. Similar results were obtained using a chemical analog of HFIP, hexafluoro-2-phenylpropanol (HFPP) [10].

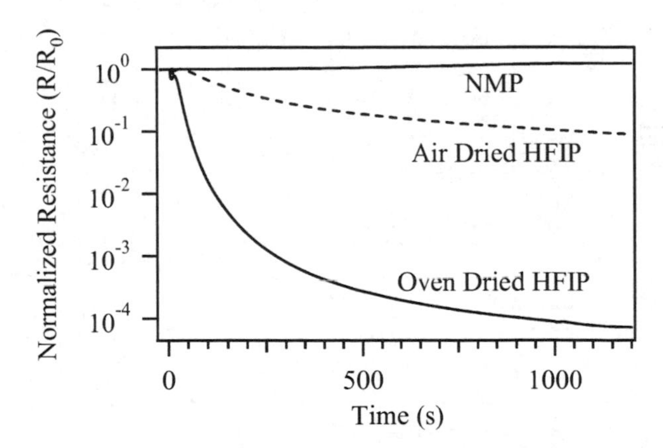

Figure 5. Response of conventional dedoped polyaniline films with added HFIP and NMP upon exposure to hydrazine.

All of these results strongly suggest that the hydrazine is reacting with the HFIP or HFPP to create a strong acid that dopes the polyaniline. The most likely product of this reaction is HF however, literature searches found no reports of such a reaction. Separate experiments show that aqueous hydrazine reacts strongly with HFIP with a drop in the pH of the solution consistent with this proposal. Furthermore, anhydrous hydrazine can be used to defluorinate fluorinated carbon nanotubes with proposed formation of HF [11].

We have also shown that new inorganic composites with polyaniline nanofibers are useful for the detection of hydrogen sulfide [12]. Figure 6 shows that untreated nanofibers give only a small response to hydrogen sulfide. This is due to the weak acidity of H_2S and the lack of direct doping of the polyanilne. When a composite material is formed between polyaniline nanofibers and inorganic complexes (ML_n), this new material shows a dramatic response to H_2S, $R/R_0 < 10^{-4}$. As before the inorganic material reacts with H_2S to create a strong acid that dopes the polyaniline. The approach takes advantage of the tremendous conductivity change of polyaniline and holds great promise for the detection of many different chemicals.

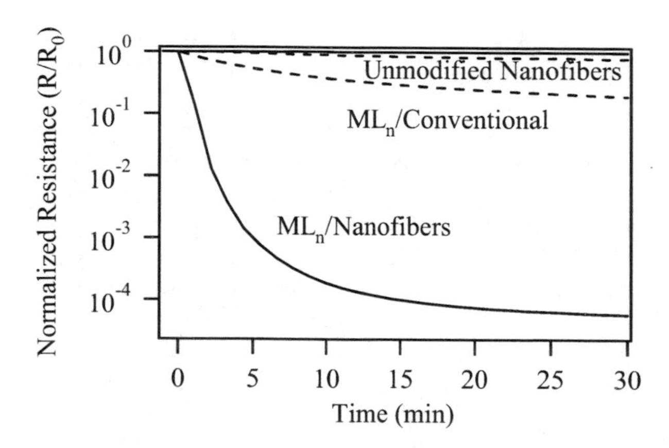

Figure 6. Resistance changes of the polyaniline films upon exposure to hydrogen sulfide; doped polyaniline nanofibers (—), dedoped polyaniline nanofibers (---), and conventional polyaniline containing metal ligand complex (ML_n) (---) and polyaniline nanofibers containing ML_n (—). The H_2S concentration was 10 ppm with 45% relative humidity, all film thicknesses were 0.25 μm.

CONCLUSIONS

In conclusion, when the nanofibers are compared to conventional polyaniline films, they give significantly better performance in both sensitivity and time response for all sensor detection applications examined. We believe this is due to the high surface area, small nanofiber diameter, and porous nature of the nanofiber films. Recent results indicate that composites with polyaniline are useful to detect analytes that do not give a significant response with unmodified polyaniline. These include fluoroalcohol additives for hydrazine detection and inorganics for hydrogen sulfide detection. Therefore polyaniline nanofibers appear to be a superior chemical sensor material for many analytes and have excellent potential for many chemical detection applications including homeland security.

ACKNOWLEDGEMENTS

We thank the NSF for funding through an IGERT fellowship (S.V.), the Microelectronics Advanced Research Corp. (R.B.K.) and The Aerospace Corporation's Independent Research and Development Program (B.H.W.).

REFERENCES

[1] Kohlman, R. S.; Epstein, A. J. Handbook of Conducting Polymers; 2nd ed.; Marcel Dekker: New York, 1998.

[2] Sukeerthi, S.; Contractor, A. Q. *Anal. Chem.* **1999**, *71*,2231-2236.

[3] Bossi, A.; Piletsky, S. A.; Piletska, E. V.; Righetti, P. G.; Turner, A. P. F. *Anal. Chem.* **2000**, *72*, 4296-4300.

[4] Karyakin, A. A.; Bobrova, O. A.; Lukachova, L. V.; Karyakina, E. E. *Sens. Act. B* **1996**, *33*, 34-38.

[5] Karyakin, A. A.; Vuki, M.; Lukachova, L. V.; Karyakina, E. E.; Orlov, A. V.; Karpachova, G. P.; Wang, J. *Anal. Chem.* **1999**, *71*, 2534-2540.

[6] Sergeyeva, T. A.; Lavrik, N. V.; Piletsky, S. A.; Rachov, A. E.; El'skaya, A. V. *Sens. Act. B* **1996**, *34*, 283-288.

[7] Huang, J.; Virji, S.; Weiller, B. H.; Kaner, R. B. *J. Am. Chem. Soc.* **2003**, *125*, 314-315.

[8] Huang, J.; Virji, S.; Weiller, B. H.; Kaner, R. B. *Chem. Eur. J.* **2004**, *10*, 1314-1319

[9] Virji, S.; Huang, J.; Kaner, R. B.; Weiller, B. H. *Nano Lett.* **2004**, *4*, 491-496

[10] Virji, S.; Kaner, R. B., Weiller, B. H. *Chem. Mater.* in press.

[11] Mickelson, E. T.; Huffman, C. B.; Rinzler, A. G.; Smalley, R. E.; Hauge, R. H.; Margrave, J. L. *Chem. Phys. Let.* **1998**, *296*, 188-194.

[12] Virji, S.; Huang, J.; Kaner, R. B., Weiller, B. H. *Small,* submitted for publication.

Calculating Deflection of Micro-Cantilever with Self-Assembled Monolayer Molecules Using Molecular Dynamics Simulations

Heng-Chuan Kan[1], Yeng-Tseng Wang[2], Shang-Hsi Tsai[3]

National Center for High-performance Computing
No.21 Nan-ke 3rd Rd., Hsin-Shi, Tainan County 74147, Taiwan, R.O.C.
TEL:+886-6-5050940 ext. 742 FAX:+886-6-5050945

[1]n00hck00@nchc.org.tw, [2] c00jsw00@nchc.org.tw, [3] shangsi@nchc.org.tw

Abstract

By evaluating the change of Gibbs free energy of the Self-Assembled Monolayer (SAM) molecules on the surface of the micro-cantilever, the deflection of the micro-cantilever can be calculated according to the Stoney's equation. All computations of the free energy changes were performed isothermally with periodic boundary conditions (PBC) by the commercial Molecular Dynamics (MD) simulation software CHARMM in associated with the specific molecular topology and force fields for alkanethiolic molecules $(HS)(CH_2)_7CH_3$. The alkanethiolic molecules are respectively arranged in squared 4x4 and 8x8 array formats adsorbed on the Au atoms coated on the surface of the micro-cantilever. The effects of the seeding density of the adsorbed molecules are taken into consideration by different separations between Au atoms.

Keywords: Molecular Dynamics (MD) Simulation, Self-Assembled Monolayer (SAM), CHARMM, Stoney's Equation, Dupre Relation.

1. Introduction

Self-Assembled Monolayer (SAM) by spontaneous adsorption of molecules on microscale substrates has attracted a huge interest due to their scientific importance and potential use in technological applications and health care industry [1-4]. Diverse selections of SAM in accompanying with various bio-molecules to form biological probes on the substrate surface have made the micro-cantilever an excellent device for studies of interactions between solid surfaces and biological systems, such as molecular recognition, selective binding of enzymes to surfaces, chemical force microscopy, corrosion protection, and pH-sensing. It has been shown [5] that intermolecular interactions resulting from absorption of SAM molecules and probe molecules on the surface of a micro-cantilever induce the changes in surface stresses, which directly contribute to the deflection of the micro-cantilever. Therefore, the deflection of micro-cantilever can, in principle, be expressed in terms of the properties mentioned above like:

δ = δ (topology, number density, configuration, conformation, thickness and patterns...)

In present study the deflection of the micro-cantilever is analyzed, which is coated on the upper surface an Au (111) layer adsorbed with $(HS)(CH_2)_7CH_3$ as SAM molecules. The deflection is calculated by the Stoney's equation [6, 7], which is related to the change of Gibbs free energy according to the Dupré relation [7] by the thermodynamics laws [5]. The change of Gibbs free energy is obtained by the MD simulation by the commercial software CHARMM [8, 9].

2. Formulations

2.1 Thermodynamics relation and mechanical analysis

According to the thermodynam ics laws, the change of Gibbs free energy can be expressed as [5]

$$dG = -SdT + dW + \mu dN \qquad (1)$$

where S is the entropy, T the temperature, W the work done on the system, μ the chemical potential, N the number of particles, respectively. Suppose that the process is isothermal and no net particle flux through the boundary, Eq.(1) can be simplified by setting $dT = dN = 0$:

$$dG = dW \qquad (2)$$

On the other hand, from experiments one can assume that the work dW on the molecular system is done by the micro-cantilever from the adsorption of the molecules on the micro-cantilever surface. Thus the work can be expressed as the scalar product of the tangential stress $\sigma_\parallel$ along the surface and the surface deformation $dA_\parallel$ (extension or contraction), which is caused by the micro-cantilever deflection resulted from adsorption:

$$dW = \sigma_{\|} \cdot dA_{\|} \qquad (3)$$

Substituting (3) into (2) and follows the Dupré relation [7]:

$$N_a \Delta \bar{G} = \Delta\sigma \qquad (4)$$

where Δσ stands for the surface tension and N_a and $\Delta\overline{G}$ are the number of adsorbed molecules and the change of Gibbs free energy per unit surface area, respectively.

2.2 The Stoney's equation

The geometry of the micro-cantilever is shown in Figure 1. The deflection δ of the micro-cantilever can be calculated by the Stoney's equation [6,7]:

$$\delta = \frac{3l^2}{t_m^2}\frac{1-\upsilon}{E}\Delta\sigma \quad (5)$$

where E and υ are the Young's modulus and Poisson's ratio, respectively; and their values are given in Table 1. Combining (4) and (5) gives the relation for the deflection of the micro-cantilever and the change of Gibbs free energy:

$$\delta = \frac{3l^2}{t_m^2}\frac{1-\upsilon}{E}N_a\Delta\bar{G} \quad (6)$$

2.3 Molecular Dynamics simulation

All computations were performed by the commercial software CHARMM [9]. The MD simulations follow the trajectories of the alkanethiolic molecules $(HS)(CH_2)_7CH_3$ adsorbed on the Au(111) surface and with applied Periodic Boundary Conditions (PBC) at a temperature of 300K with a time step of 1fs (femto second) = 0.001ps (pico second) are specified in present work. Starting from the 4x4 and 8x8 arrays of alkanethiolic structures with the separation distances between sulfur atoms of 5, 6, 7, 8, and 9Å. The sulfur atoms of each alkanethiol molecular structure have been fixed with the bond constraints using the SHAKE algorithm [10]. A cut-off ratio of 14Å has been used for Lennard-Jones function. The setting of Periodic Boundary Condition was applied as the cubic box covering the entire molecular system with additional 2Å on each spatial direction. The minimization of potential energy for the alkanethiolic structures has been carried out with a combination of two algorithms: 20ps with steepest descent, 80ps with adopted basis Newton-Ralphason minimization. The heating procedure of the molecular system ranges from 240K to 300K with increments of 3K over 0.2ps. An equilibration process of system was performed with 50ps. The simulation process of the molecular system was performed with 4.5ns and 5.5ns for 4x4 and 8x8 arrays of alkanethiolic structures, respectively.

The bond lengths are constrained by the SHAKE algorithm [10] with the Au-S, S-H, S-C, C-C and C-H bonds held fixed at $d_{Au\text{-}S}$ = 2.531Å, $d_{S\text{-}H}$ = 1.325Å, $d_{S\text{-}C}$ = 1.836Å, $d_{C\text{-}C}$ = 1.531Å and $d_{C\text{-}H}$ = 1.111Å, respectively. The intramolecular potentials contain the bending $V_b(\theta)$ and dihedral $V_d(\phi)$; while the van der Waals effects are represented by the Lennard-Jones potential $V_{LJ}(r_{ij})$ with a cut-off ratio of 14 Å, as defined below [11,12,13]:

$$V_b(\theta) = \frac{1}{2}k_\theta(\theta-\theta_0)^2 \quad (7)$$

$$V_d(\phi) = k_d[1+\cos(n\phi-\phi_0)] \quad (8)$$

$$V_{LJ}(r_{ij}) = \varepsilon_{ij}\left[\left(\frac{r_{\min}}{r}\right)_{ij}^{12} - 2\left(\frac{r_{\min}}{r}\right)_{ij}^{6}\right] \quad (9)$$

where θ is, for example, the C-C-C or S-C-C angle, with corresponding force constant k_θ and equilibrium angle θ_0; while ϕ is the torsion angle in associated with the constants k_d, n and equilibrium angle ϕ_0; and $(r)_{ij}$ is the distance between the i-th and j-th atoms. Note that the ε_{ij} and $(r_{min})_{ij}$ in (9) are defined as

$$\varepsilon_{ij} = \sqrt{\varepsilon_i\varepsilon_j}$$
$$(r_{\min})_{ij} = \frac{1}{2}\left[(r_{\min})_i + (r_{\min})_j\right] \quad (10)$$

The parameters used for calculating $V_{LJ}(r_{ij})$ are listed in Table 2, where the atoms connected to the considered atoms are represented in the parentheses.

3. Results and Discussions

The free energy change is obtained by the MD simulations in the sampling rate of 0.1 ns, i.e., the results are averaged for every 10^5 time steps. The potential of the mean force (PMF) is calculated by the Weighted Histogram Analysis Method (WHAM) [14, 15] and the results are shown in Figure 2. The deflection δ of the micro-cantilever is calculated by (6) and the results are shown in Figure 3.

As shown in the results, the calculated Gibbs free energy change, and the deflection as well, for both cases decrease as the separations between the adsorbed molecules increase. This can be interpreted as that the micro-cantilever is deflected by the adsorbing molecules through the work done on the Au atoms coated on the micro0cantilever surface. The applying work is related to the Gibbs free energy change according to the thermodynamics laws under constant (N, T) condition. On the other hand, the contribution to the Gibbs free energy change of the entire molecular system is diminishing as the seeding density becomes lower. Therefore, the micro-cantilever will be flatten-out as the seeding density of the adsorbing molecules is getting dilute.

4. Conclusion

As shown by the results the deflection of the micro-cantilever can be calculated by the Gibbs free energy change obtained by the MD simulation, whose results will be improved as the magnification of the simulated region, i.e., increasing the seeding density, and the application of more accurate interaction models.

Acknowledgements

This work is supported by the Ministry of Economic Affairs, Department of Industrial Technology of Taiwan through the Grant 91-EC-17-A-05-S1-0017.

References

1. G.M. Whitesides, "Self-Assembling Materials", *Sci. Am.*, **273**, pp.146-149 (1995).
2. G.P. Lopez et al, "Convenient Methods for Patterning the Adhesion of Mammalian Cells to Surfaces Using Self-Asselbled Monolayers of Alkanethiolates on Gold", *J. Am. Chem. Soc.*, **115**, pp. 5877-5878 (1993).
3. Li Deng, M. Mrksich, G.M. Whitesides, "Self-Assembled Monolayers of Alkanethiolates Presenting Tri(propylene sulfoxide) Groups Resist the Adsorption of Protein", *J. Am. Chem. Soc.*, **118**, pp. 5136-5137 (1996).
4. A. Ulman, "Formation and Structure of Self-Assembled Monolayers", *Chem. Rev.*, **96**, 1533 -1554 (1996).
5. L.D. Landau et al, *Statistical physics*, Pergamon Press Ltd., 1969.
6. G.G. Stoney, "The Tension of Metallic Films Deposited by Electrolysis", *Proc. R. Soc. London*, **A 82**, pp. 172-177 (1909).
7. N.V. Lavrik, C.A. Tipple, M.J. Sepaniak and P.G. Datskos, "Gold Nano-Structures for Transduction of Biomolecular Interactions into Micrometer Scale Movements", *Biomedical Microdevices*, **3:1**, pp. 35-44 (2001).
8. http://www.pharmacy.umaryland.edu/faculty/amackere/
9. C.L. Brooks, R.E. Bruccoleri, B.D. Olafson, D.J. Statos , S. Swaminathan, M. Karplus, "CHARMM: a program for macromolecular energy, minimization, and dynamics calculations", *J. Comp. Chem.* ,**4** , pp. 187-217(1983).
10. J.-P. Ryckaert, G. Ciccotti, snd H.J.C. Berenden, *J. Comput. Phys.* 23, pp. 327 (1977).
11. J. Hautman and M.L. Klein, "Simulation of a Monolayer of Alkyl Thiol Chains", *J. Chem. Phys.*, **91**, pp. 4994-5001 (1989).
12. A.V. Shevade, J. Zhou, M.T. Zin and S. Jiang, "Phase behavior of Mixed Self-Assembled Monolayers of Alkanethiols on Au(111): A Configurational-Bias Monte Carlo Simulation Study", *Langmuir*, **17**, pp. 7566-7572 (2001).
13. A.R. Bizzarri, G. Constatini and S. Cannistraro, "MD Simulation of a Plastocyanin Mutant Adsorbed onto a Gold Surface', *Biophys. Chem.*, **106**, pp. 111-123 (2003).
14. S. Kumar, D. Bouzida, R.H. Swendsen, P.A. Kollman, J.M. Rosenberg, "The weighted histogram analysis method for free-energy calculations on biomolecules", *J. Comput. Chem.*, **13**, pp.1011-1021 (199).
15. S. Kumar, J.M. Rosenberg, D. Bouzida, R.H. Swendsen and P.A. Kollman, "Multidimensional free-energy calculations using the weighted histogram analysis method", *J. Comput. Chem.*, **16**, pp. 1339-1350 (1995).

Table 1. Parameters of micro-cantilever.

$E(Nt/\mu m^2)$	0.15
t_m (μm)	0.6
l (μm)	180
υ	0.26

Table 2. Parameters for the Lennard-Jones potential.

	ε_{ij} *(kcal/mole)*	$(r_{min})_i/2$(Å)
S	-0.47	2.20
C(S)	-0.11	2.20
H(S)	-0.10	0.45
H(C)	-0.022	1.32

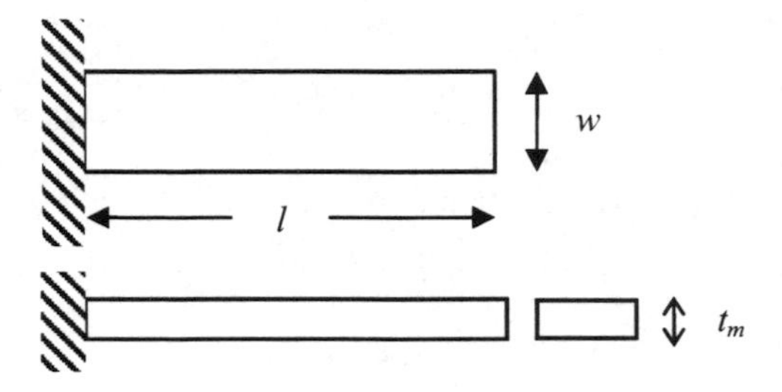

Figure1. Geometry of micro-cantilever.

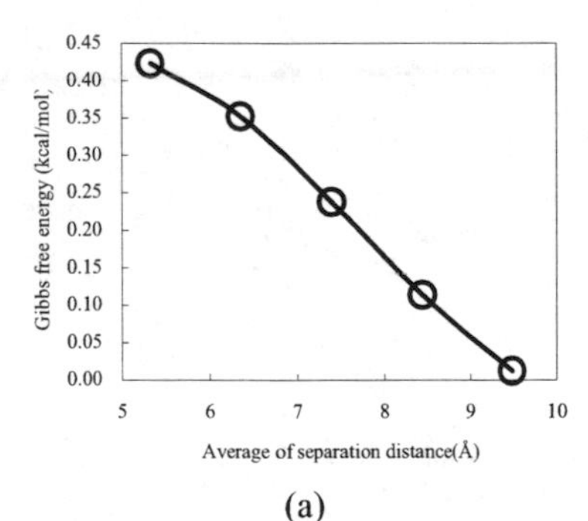

(a)

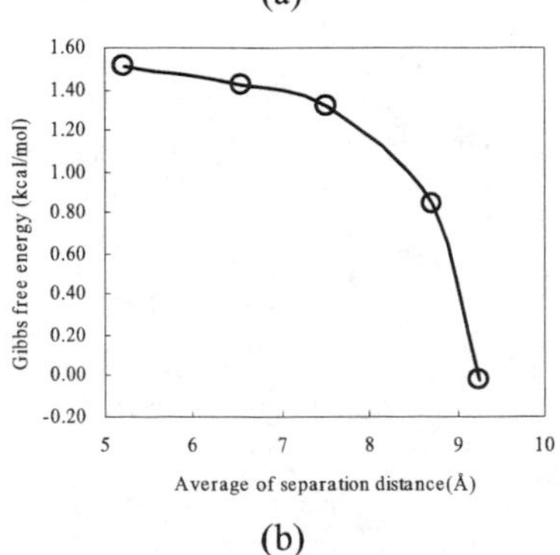

(b)

Figure 2. Potential of mean force (PMF): (a) 4x4 case; (b) 8x8 case.

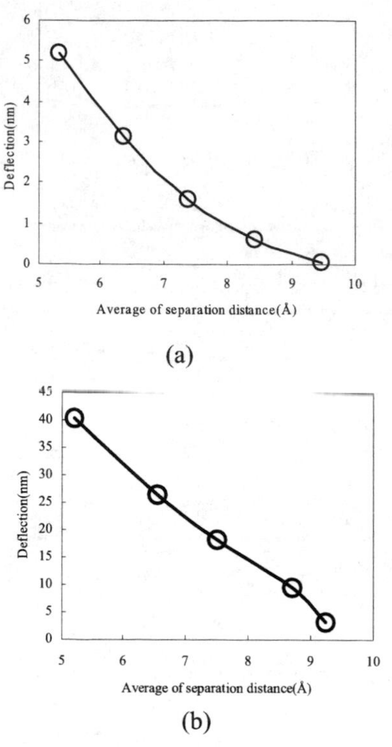

Figure 3. The deflection δ of micro-cantilever: (a) 4x4 case; (b) 8x8 case.

Preparation and property of novel CMC tactile sensors

Xiuqin Chen, Shaoming Yang, and Seiji Motojima,

Faculty of Engineering, Gifu University, Gifu 501-1193, Japan, xqchen@apchem.gifu-u.ac.jp

Abstract

In this study, carbon microcoils (CMCs) which could be extended and contracted were used to manufacture thin film CMC/polysilicone composite sensor elements that was found to have a very high tactile sensor performance. This is because the extension of CMC was very highly sensitive to an applied load. That is, inductance (L) capacitance (C), and resistance(R) of CMC/polysilicone composites extensively changed with the extension and contraction of the composite under applied loads.

CMC tactile sensors could detect very low applied loads on milligram orders. The sensitivity of CMC sensors is 1,000-10,000 times higher than that of commercially available tactile sensors. Accordingly, the CMC tactile sensors have potential applications to various medical instruments such as endoscopes, catheters, manipulation sheets, etc., or as artificial skin of humanoid robot, etc.

1 ITRODUCTION

Carbon microcoils/nanocoils (CMCs) have an interesting morphology with a 3D-herical/spiral structure similar to DNA or proteins. The super-elastic CMCs (SECMCs) have a double-helix conformation with a coil diameter of 10-20 μm as shown in Figs. 1-2. The SECMCs have a very high elasticity, and their electrical parameters; inductance (L), capacitance (C) and electrical resistivity (R), are changed by the extension and contraction. It is well known that the Meissner's corpuscles, which is the most important tactile receptor of human skin and have helical forms of micron sizes. It is considered that CMCs can be used to manufacture artificial tactile sensors due to their similar helical forms [1]. In this study, novel CMC tactile sensor elements were prepared by embedding 1-5 wt% CMCs in elastic polysilicone resin. The changes of LCR parameters of the CMC sensor elements under applying static loads or other stimulations, were measured using an impedance analyzer.

2 EXPERIMENTAL

CMCs can be synthesized by catalyzed CVD process using acetylene as a carbon source and sulfur compounds as an impurity at reaction temperature 700-800 ℃. By the standard reaction conditions, the obtained CMC were usually regular coiled with a constant coil diameter of 1-6 μm, a coil gap of nearly zero through a coil length, and a coil length of 1-5 mm in 2hrs reaction time. Fig. 1 shows the representative regular flat-fiber CMCs with a coil diameter of about 4 μm obtained using Ni powder catalyst at 750-800℃. The double-helix structure can be clearly seen from the cross-section image.

On the other hand, the CMCs with a large coil diameter of 10-20 μm could be obtained at lower reaction temperature of 700-730℃ and lower gas flow rates. Their representative images and the enlarged view are shown in Fig. 2. Because of the super elasticity, this kind of CMCs is referred to as super-elastic CMCs (SECMCs).

In this study, only SECMCs were used. Two kinds of sensor elements were prepared: (1) CMC composite sheets(sheet sensor elements): CMCs were embedded into polysilicone (Shin- Etsu, KE-103) by 1-5wt%, and obtained the CMC composite sheets of $10x10x2mm^3$. The CMCs were uniformly dispersed in the polysilicone matrix, and respective CMCs did not contact each other, and percolation structure was not observed. (2) CMC composite thin films sensor elements: including random-CMC sensor films and array-CMC sensor films (sensor elements). For the latter, because SECMCs are relatively easily arrayed among polysilicone resin by extension (Fig. 2b), thus, array-SECMCs/polysilicone composites were manufactured, and whose structure is similar to skin.

The measurement of sensor properties was carrying out as the following: loads were applied vertically on the CMC sensor elements by a manipulator, AC voltage of 5V (200 KHz) was applied on the sensors through two electrodes (separation: 2.5 mm), and the output was measured using impedance analyzers, Agilent 4294A or Agilent OSCILLOSCOPE 54621. When using the former, L, C, R can be produced respectively; when using the latter the respective parameters were transformed into a DC voltage output of (LC) and R respectively. The image of the array-CMCs among the matrix was observed by microscopic OLYMPUS U-LH100-3.

3. RESULTS AND DISCUSSION

3.1 Sheet sensor elements

A load of 5 g was applied on the surface of composite sheets of 10x10x2 mm^3 through a wood needle controlled by a manipulator and output change was measured by Agilent OSCILLOSCOPE 54621. From Fig. 3, it can be seen that a strong output (voltage, i.e.V) of the (LC) and R parameters significantly produced as soon as the loads were applied. It can be seen that very sharp output peaks can be obtained when the CMC sensor element was touched by a needle, and that the sharp output signal line quickly decreased and disappeared after releasing the applied load. The response time was on a millisecond order. A continuously applied load resulted in a continuous shift in the output lines. It is noticed that the intensity of R peaks are stronger than those of LC peaks.

Fig. 4 shows the changes of C parameter under applying various loads on the mgf～gf order. It can be seen that C parameter was dramatically changed under applying a load. When the load was as small as 10 mg, the output is 12 pF. The outputs increased with the load increase slightly in gf order.

3.2 Film sensor elements

Very thin CMC sensor elements with different array of the CMCs in the matrix were manufactured with a size of 2x2x0.1 mm^3. A load of 5 g was applied to the sensor element using a wood rod with a diameter of 1 mm. For the random-CMC elements, the output of the (LC) parameter was about 1 V, although the R parameter output was small (Figure is omitted). For array-CMC sensor, in which the slightly extended CMCs were embedded in the matrix (The image of extended SECMCs among polymer matrix is shown in Fig.5c.), the CMCs were vertically arrayed in the matrix (Set A) or parallel (Set B) to the direction of the electrodes as shown in Fig. 5 (arrayed-CMC sensor). When the CMCs were partly arrayed in the vertical direction to the electrodes, the LC parameter output was 5v, and the R parameter output was 0.5v. However, when the arraying direction of the CMCs was parallel to the electrodes, no output was observed. This results indicates that by arraying the CMCs vertical to the electrodes, very high tactile sensor properties can be obtained.

M. Konyo et al. proposed that for high CMC addition, resistance is more important than capacitance, while for low CMC addition catacitance is more important than resistance. In the case of random-CMC sensors, the CMC addition amount is larger than in the case of array-CMC sensors. Thus, in Fig.3, more attention should be paid to R change, while in Fig.5, more attention should be paid to LC change.

3.3 Comparison of different stimuli

Fig.6 shows the change of L parameter of the CMC sensor elements under approximating a hand and a heated solder tong, as well as under applying static load of 200mgf. Strong signal changes are observed when a hand or heated solder tong is approximated to the sensor elements. An IR ray is emitted from a hand. Furthermore, it was observed that L and R signal changes was observed when cellular phone or sound was approximated. That is, the CMC tactile sensors can be detected various stresses, temperature, IR, EM waves, etc. with very high detection sensitivity and high discrimination ability. Accordingly, the CMC sensor elements has high potential applications as tactile sensors for endoscopes, catheters, manipulation sheet, etc., or as artificial skin of a humanoid robot, detection sensors of humans buried in debris by earthquake, and various industrial sensors. It is considered that these properties may be affected by the formation of hybrid LCR oscillation circuit between CMC and dielectric elastic matrix.

REFERENCES

[1] Jonathan Engel, Jack Chen, Zhifang Fan and Chang Liu, Sensors and Actuators A: Physical, 117, 50-61.

[2] X. Chen, S. Motojima and H. Iwanaga, J. Cryst. Growth, 237-239, 1931(2002).

[3] X. Chen, S. Yang, M. Hasegawa, and K. Takeuchi, S. Motojima, Proceeding of International Conference on MEMS, NANO, and Smart Systems, August 25-27, 2004, Banff, Alberta-Canada, pp486-490, 2004 IEEE,

[4] S. Motojima, X. Chen, S. Yang and M. Hasegawa, 9th Int. Conf. on New Diamond Science and Technology, Diamond and Related Materials, 13, 1989(2004).

[5] K. Kawabe, C. Kuzuya and A. Ueda, Materials Integration, 17, 9(2004).

[6] M. Konyo, unpublished data.

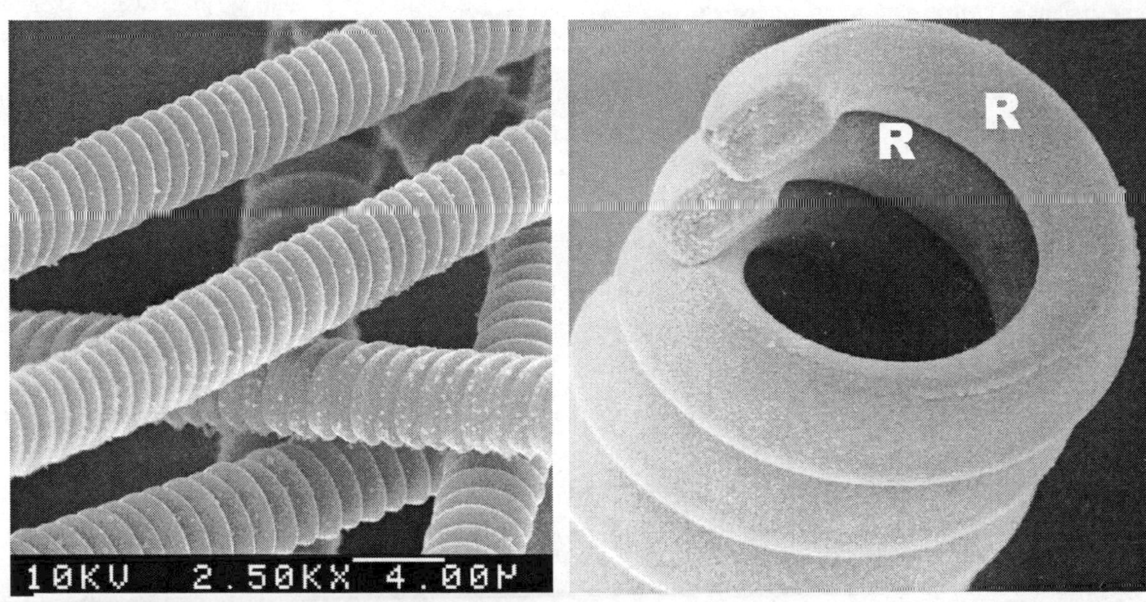

Fig.1 Regular double-helix carbon microcoils:coil diameter, 4μm

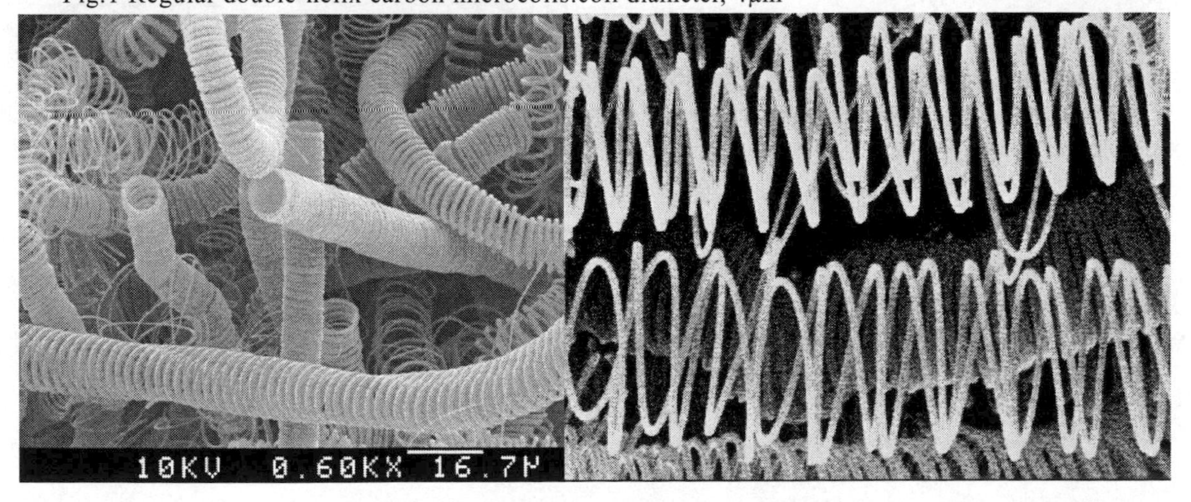

Fig.2 Irregular double-helix carbon microcoils with a super-elastic property (SECMC), coil diameter: 10-15μm.

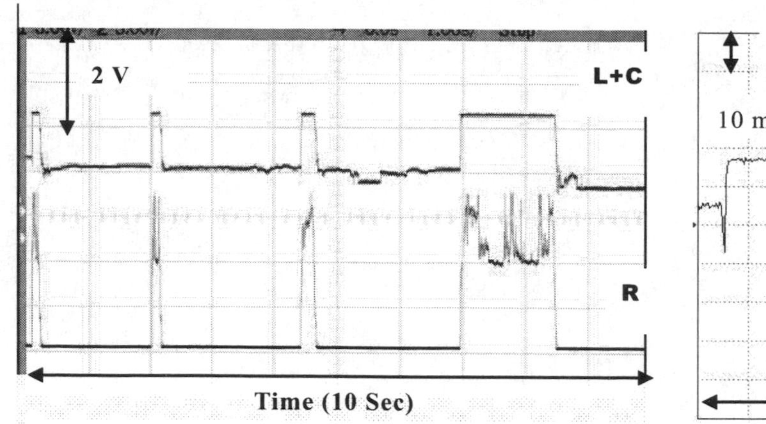

Fig.3 Change in LCR parameters of the SECMC (composite sheet) tactile sensors to a needle with a load of 5 g, the output of L+C and R respectively.

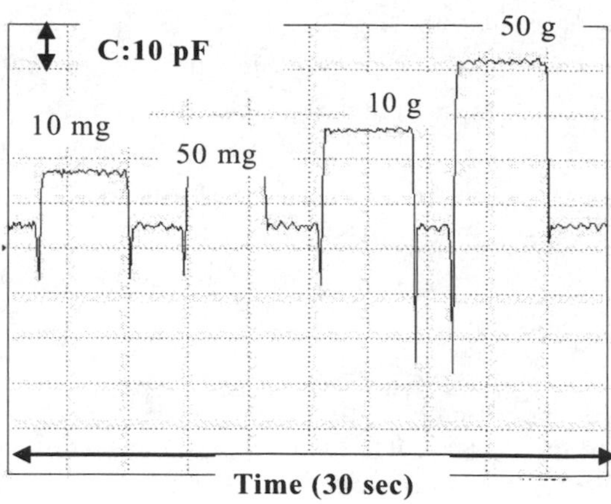

Fig.4 Capacitance change of SECMC sensor when applied different loads.

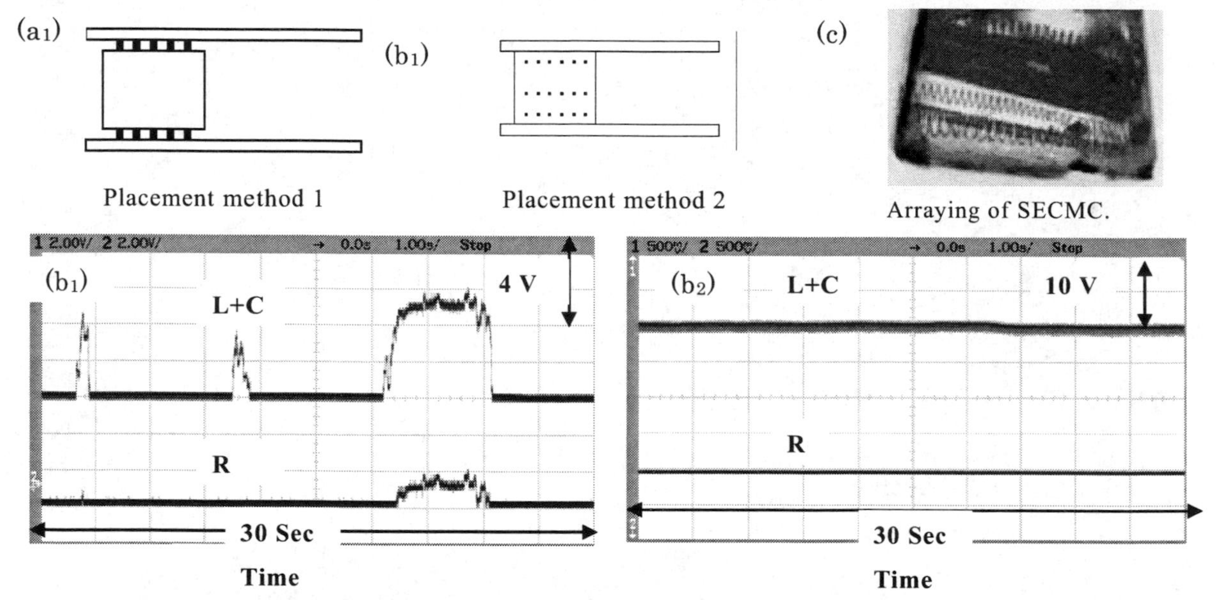

Fig.5 Change in LCR parameters of the SECMC (composite films) tactile sensor with CMC arraying. (a_2): when the setup of the sensor and electrodes was done as in a_1. (b_2): when the setup of the sensor and electrodes was done as in b_1. (c): arraying of SECMC in the film sensor element.

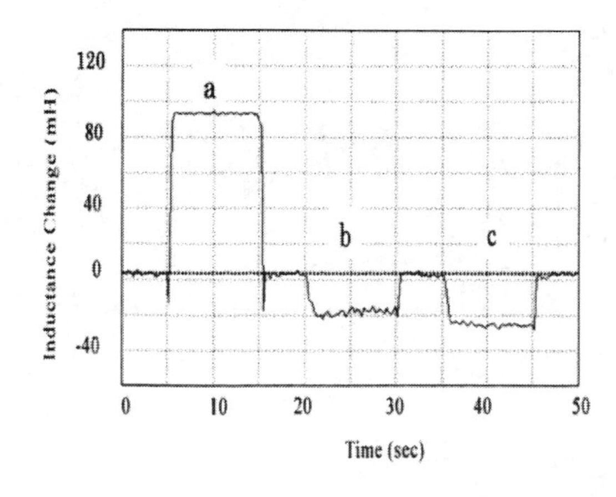

Fig. 6 Inductance changes under applying small loads and approximating of a hand and heated solder tong.
Dotted line indicates without applying load.
(a) applying of a load 200mgf, (b) approximating of a hand, (c) approximating of a heated solder tong.

A Nanoelectronic Device Based on Endofullerene Peapod: Model Schematics and Molecular Dynamics Study

Jeong Won Kang[*], Ki Ryang Byun[*], Jun Ha Lee[**], Hoong Joo Lee[**], Ho Jung Hwang[*]

[*]School of Electrical and Electronic Engineering, Chung-Ang University
221-1 HukSuk-Dong, DongJak-Ku, Seoul 156-756, Korea, gardenriver@korea.com
[**] Department of Computer System Engineering, Sangmyung University
ChungNam 330-720, Korea

ABSTRACT

We investigated a nanoelectronic device based on multi *endo*-fullerenes shuttle memory element based on nanopeapods using classical molecular dynamics simulations. We suggested the model schematics of *endo*-fullerene shuttle memory device fabrication. The *endo*-fullerene shuttle memory element could operate a nonvolatile nano memory device. The switching speed, the applied force field, and the active region should be considered to design the *endo*-fullerene shuttle memory element.

Keywords: nanopepod memory device, bucky shuttle memory device, nano nonvolatile memory, endo-fullerene, molecular dynamics simulation

1 INTRODUCTION

The nanospaces inside nanocapsules or nanotubes have opened various applications as storage materials with high capacity and stability. In particular, self-assembled hybrid structures called "carbon nanopeapods" have been reported [1-3]. Recently, "boron-nitride nanopeapods" were also synthesized [4]. The application of nanopeapods ranges from nanometer-sized containers of chemical reactant to data storage [5,6]. Kwon *et al* [5] investigated 'bucky shuttle' memory device, which acted as nanometer-sized memory element, using molecular dynamics (MD) simulations. A lot of *endo*-fullerenes and *endo*-fullerene-encapsulated carbon nanotubes (CNTs) have been synthesized and investigated by experimental and theoretical methods [7-11]. Several field-effect molecule-shuttle memory elements based on nanopeapods have been investigated using classical MD simulations in our previous works [12-17].

While aligned bucky shuttle structures are difficult to be in self-assembly, nanopeapods can be synthesized in the aligned structures using bundles of single-walled nanotubes. If some processes, such as *endo*-fullerenes intercalation control, nanolithography, nanotube etching or cutting, nanotube capping or metal fillings for electrodes, are treated appropriately to the aligned nanotubes, the aligned bucky shuttles can be synthesized. Hence molecule-shuttle memory devices based on nanopeapods have been expected to bee realized by nanoscience and nanotechnology. This paper, using classical MD simulations, shows model schematics and probability of the molecule-shuttle electronic devices based on *endo*-fullerene nanopeapods.

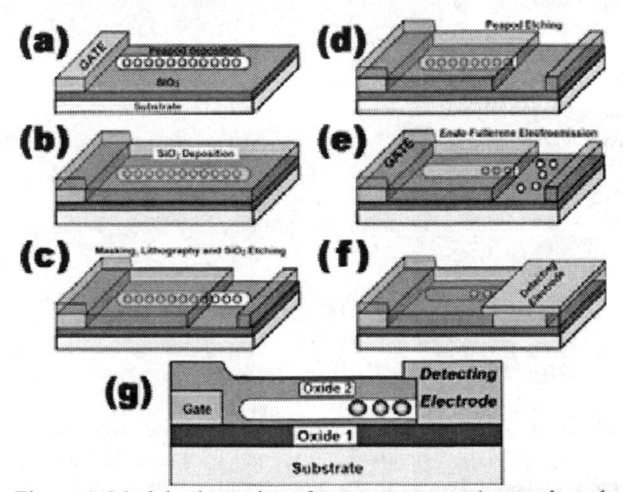

Figure 1. Model schematics of nano-memory-element based on nanopeapod. (a) The gate electrode is fabricated on the substrate covered with SiO_2 thin film; then nanopeapods are deposited on the SiO_2 thin film. (b) SiO_2 film growth on the surface. (c) A part of the nanopeapod is exposed by the SiO_2 removal process. (d) The exposed region of the nanopeapod is removed by the carbon etching processes. (e) Using the *endo*-fullerene electroemission process, some *endo*-fullerenes are emitted from the nanopeapod under the gate bias. (f) The electrode to detect the current impulse is deposited in the etched SiO_2 region. (g) The cut side view of the final structure.

2 MODEL SCHEMATICS

Figure 1 shows the model schematics of *endo*-fullerene shuttle memory device fabrication. The gate electrode is fabricated on the substrate covered with SiO_2 thin film; then nanopeapods are deposited on the SiO_2 thin film shown in Fig. 1(a). After the nanopeapod deposition, as Fig. 1(b) shows, SiO_2 film is grown on the structure of Fig. 1(a). On the opposite side of the nanopeapod for the gate electrode along the tube axis, the SiO_2 etching process is achieved after masking and lithograph processes; then a part of the nanopeapod is exposed by the SiO_2 removal process shown in Fig. 1(c). The exposed region of the nanopeapod is removed by the nanotube etching processes; then, the structure shown in Fig. 1(d) is achieved. At this process, the edge of the nanotube should be opened to achieve the *endo*-fullerene electroemission from the nanotube. When the nanotube is fully filled with the *endo*-fullerenes, the *endo*-fullerene electroemission from the nanotube should be achieved for *endo*-fullerenes to migrate in the nanotube. Using the *endo*-fullerene electroemission process, some *endo*-fullerenes are emitted from the nanopeapod under the gate bias shown in Fig. 1(e). After the *endo*-fullerene electroemission process, the electrode to detect the current impulse is deposited in the etched SiO_2 region shown in Fig. 1(f). Figure 1(g) shows the cut side view of the final structure of Fig. 1(f). Since fully ionized *endo*-fullerene can be accelerated under external force fields, the *endo*-fullerene can be shuttled by the alternative external force fields. When the bits of the molecular-shuttle memory elements are defined by the positions of the *endo*-fullerenes, the data writing/erasing can be achieved from the gate bias. The positions of the *endo*-fullerenes can be controlled by the gate bias whereas the bit as the stored data can be detected by the current pulse of the detecting electrode as suggested by Kwon *et al.* [5].

3 INTERATOMIC POTENTIALS

3.1 Charge of endo-fullerene in nanotube

A lot of *endo*-fullerene peapods have been investigated in experimental and theoretical studies. In this work, we assume that the charge of the *endo* metal encapsulated in fullerene is fully ionized such as F^-, Ne, Na^+, K^+, Mg^{2+}, and Al^{3+} [18]. In order to move the encapsulated *endo*-fullerenes, the *endo*-fullerenes should carry a net charge. As the number of carbon atoms composed of fullerenes increases, the electron affinity generally increases [19]. The electron affinities of the *endo*-fullerenes are generally higher than those of the empty fullerenes [20-23]. Some *ab initio* calculations for the *endo*-fullerene peapods have shown the charge transfer from nanotubes to *endo*-fullerenes or fullerenes [24-26]. Chen and Lue [27] showed that the electron affinities of the CNTs were 1.0 ~ 1.5 eV by the theoretical fitting based on field emission experiments of CNTs. The above results imply that the charge of the encapsulated *endo*-fullerene is negative due to the charge transfer from nanotubes to *endo*-fullerenes. However, contrary results for the charge transfer have been reported. Cioslowski *et al.* [28] showed that the electron affinities of nanotubes with a finite length were generally higher than those of empty fullerenes. Kazaoui *et al.* [29] estimated that the average electron affinity of semiconductor single-wall CNTs was 4.8 eV by *in situ* measurements of optical absorption spectra. Hirahara *et al.* [30] discussed the possibility of the charge transfer from Gd@C_{82} *endo*-fullerenes to single-wall CNTs. These results imply that the charge of the encapsulated *endo*-fullerenes can be positive. Therefore, we think that the charge transfer of *endo*-fullerene peapods should be investigated in further studies. Although there are contrary results for the charge transfer, the *endo*-fullerenes can have either a negative or a positive net charge. In this work, we follow the assumption by Kwon *et al.* [5] for the charge of the *endo*-fullerenes encapsulated in CNTs. In $K@C_{60}$ *endo*-fullerenes, which are known to form spontaneously under synthesis conditions in the presence of potassium, the valence electron of the encapsulated potassium atom is completely transferred to the C_{60} shell, and then $K^+@C_{60}^-$ is formed [31-34]. Kwon *et al.* assumed that the C_{60} shell is likely to transfer the extra electron to the graphitic outer capsule. The extra electron on the outer capsule will likely be further transferred to the structure that holds the bucky shuttle device. Therefore, Kwon *et al.* modeled the dynamics of the $K@C_{60}^+$ ion in the neutral carbon capsule by uniformly distributing a static charge of $+1e$ over the C_{60} shell. However, in our MD code, the *endo* metals were not included for computational efficiency, whereas the mass of C_{60} increased with the mass of the potassium atom and the charge of the C_{60} was assumed to be $+1e$ and was uniformly distributed on the C_{60}, such as in the previous study by Kwon *et al.* Therefore, the charge per carbon atom was assumed to be $+e/60$.

For C_{60} fullerene as a Faraday cage [35], when the applying electrostatic force field is 1 V/Å, the electrostatic force acting on K^+ is around 0.25 V/Å because of the field screening of the C_{60} fullerene. Therefore, the average electrostatic force field acting on carbon atom composed of the C_{60} fullerene will be 0.004 167 V/Å.

3.2 Interatomic potentials

For C-C interactions, we used the Tersoff-Brenner potential function that has been widely applied to C systems [36-38]. The long-range interactions were characterized with the Lennard-Jones 12-6 (LJ12-6) potential that was continually connected by the cubic spline functions with the Tersoff-Brenner potential such as methods by Mao *et al* [39] when the interatomic distance (r) is between 2.0 and 2.7 Å. The cutoff distance of the LJ12-6 with parameters $\varepsilon_C = 0.0024$ eV and $\sigma_C = 3.37$ Å was 10 Å. We assumed the Cu nanowires as the conductor material. For Cu–Cu and Cu–C, we used the Mores-type potential, a pair interatomic potential function, which have been widely used in many atomistic studies for nanoindentations and nanomechanics [40,41].

The MD simulations used the same MD methods as were used in our previous works [42,43]. The MD code used the velocity Verlet algorithm, and neighbor lists to improve computing performance. MD time step was 5×10^{-4} ps. A Gunsteren-Berendsen thermostat was used to control temperature for all atoms except for fullerenes. The structure was initially relaxed by the steepest descent method; then the atoms of both edges were fixed during the MD simulations and on the other atoms, MD methods were applied.

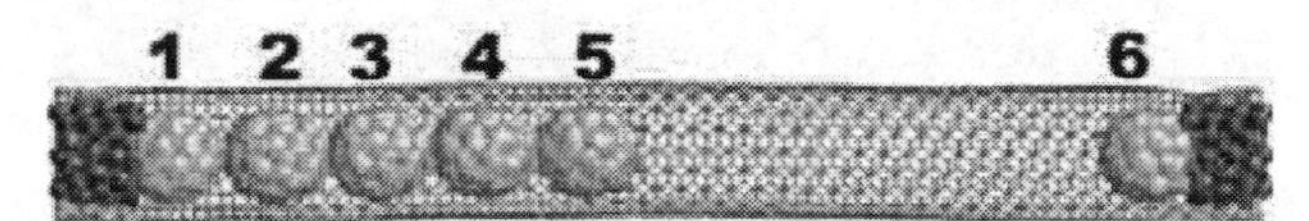

Figure 2. (a) The positions of the C_{60}^{+}s and (b) the total potential energy as a function of MD time when the external applying field is –0.2 V/Å.

4 RESULTS AND DISCUSSION

Figure 2 shows the initial structure for the MD simulations of the *endo*-fullerene shuttle in the nanotube. Six *endo*-fullerenes were encapsulated into a (10, 10) CNT with 117 Å length, and both Cu electrodes with 7 Å was obtained from our previous work [43] that shows the Cu nanowires encapsulated in the armchair CNT. The minimum potential energies were found near the both ends of the nanopeapod and the activation energy barrier was 0.399 eV [14]. Since the binding energy between a Cu electrode and a C_{60} was higher than that between C_{60}s, active elements became the structure as shown in Fig. 2 after alternative force fields were successively applied. Two *endo*-fullerenes labeled by 1 and 6 were always attached at the both end electrodes and the shuttle media were always the central four *endo*-fullerenes.

The position control of the *endo*-fullerenes by the electromigration processes is the basis of the bucky-shuttle memory device of Kwon *et al.* [5]. Figure 4 shows the total potential energy of the system (Fig. 3(a)) and the position variations of the shuttle fullerenes along the tube axis (Fig. 3(b)) as a function of MD time. The applied force field is shown in Fig. 3(c). External force field was initially 0.06 V/Å to 5 ps, increased to 0.12 V/Å by a exponential function to 6 ps, continually maintained 0.12 V/Å to 25 ps, and then decreased to 0.06 V/Å by an exponential function to 30 ps. The applied force fields were disappeared after 30 ps. Peaks in the total potential energies were found when the *endo*-fullerenes collided with the electrode or with each other.

Times required for entire bit flop were about 80 ps due to the rebound events of the *endo*-fullerenes, as shown in Fig. 3. Though the external force field vanished, the *endo*-fullerenes were safely settled at the expected bit flop. This means that the force fields for bit writing/erasing need not be applied during entire bit flop. Since the activation energy barrier of an *endo*-fullerene inside the (10, 10) CNT was 0.399 eV [14], the memory system based on *endo*-fullerene peapods can be applied to a nonvolatile nano memory element. When the magnitude and the duration of the external force fields are appropriately applied to this system, the rebound events during the bit flop can be reduced and the bit flop speed can be also upgraded. When the active region and the number of the *endo*-fullerenes are also properly selected, the rebounded events may be unimportant in the operation of this system. When the applied force fields were very low in the conditions that the *endo*-fullerenes can be switched, these rebounded events were not found. The switching speed was very low under the very low applied fields. When high external force fields were applied, the speeds of the *endo*-fullerenes were also high whereas the entire bit flops were achieved after a long time because of several rebounded events. Therefore, to design the system suggested in this work, the switching speed, the applied force field, and the active region should be considered.

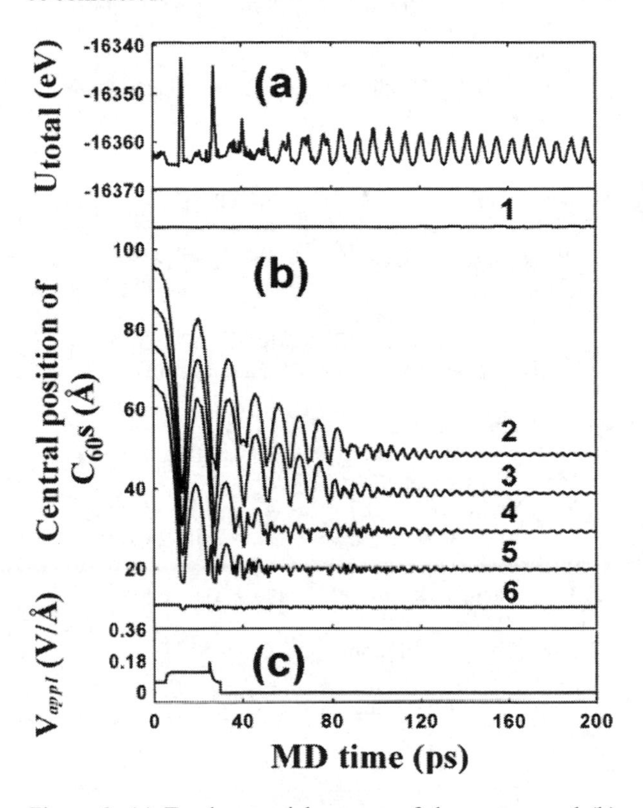

Figure 3. (a) Total potential energy of the system and (b) the position variations of the C_{60}^{+}s along the tube axis as functions of MD time and external force field.

5 SUMMARY

In summary, we studied the *endo*-fullerene shuttle memory element based on nanopeapods using classical molecular dynamics simulations. We suggested the model schematics of *endo*-fullerene shuttle memory device fabrication. The systems could operate nonvolatile nano memory devices when the positions of *endo*-fullerenes were controlled by gate bias. Our molecular dynamics simulations showed that the switching speed, the applied force field, and the active region should be considered to design the nano device suggested in this work.

ACKNOWLEDGEMENTS

This research was supported by grant No. R01-2004-000-10864-0 from the Korea Science and Engineering Foundation.

REFERENCES

[1] B.W. Smith, M. Monthioux, D.E. Luzzi, Nature 396 (1998) 323.
[2] B. Burteaux, A. Claye, B.W. Smith, M. Monthioux, D.E. Luzzi, J.E. Fischer, Chem. Phys. Lett. 310 (1999) 21.
[3] B.W. Smith, M. Monthioux, D.E. Luzzi, Chem. Phys. Lett. 315 (1999) 31.
[4] W. Mickelson, S. Aloni, W.Q. Han, J. Cumings, A. Zettl, Science 300 (2003) 467.
[5] Y.K. Kwon, D. Tománek, S. Iijima, Phys. Rev. Lett. 82 (1999) 1470.
[6] R.F. Service, Science 292 (2001) 45.
[7] J. Cioslowski, E.D. Fleischmann, J. Chem. Phys. 94 (1991) 3730.
[8] J. Sloan, S. Friedrichs, R.R. Meyer, A.I. Kirkland, J.L. Hutchison, M.L.H. Green, Inorganica Chimica Acta 330 (2002) 1.
[9] M. Monthioux, Carbon 40 (2002) 1809.
[10] D. Qian, G.J. Wagner, W.K. Liu, M-F. Yu, R.S. Ruoff, Appl. Mech. Rev. 55 (2002) 495.
[11] P.W. Chiu, G. Gu, G.T. Kim, G. Philipp, S. Roth, S.F. Yang, S. Yang, Appl. Phys. Lett. 23 (2001) 3845.
[12] J.W. Kang, H.J. Hwang, J. Phys. Soc. Japan. 73 (2004) 1077.
[13] J.W. Kang, H.J. Hwang, J. Korean Phys. Soc. 44 (2004) 879.
[14] W.Y. Choi, J.W. Kang, H.J. Hwang, Physica E 23 (2004) 135.
[15] J.W. Kang, H.J. Hwang, Carbon 14 (2004) 3018.
[16] J.W. Kang, H.J. Hwang, Jpn. J. Appl. Phys. 73 (2004) 4447.
[17] J.W. Kang, W.Y. Choi, H.J. Hwang, J. Comp. Theori. Nanosci. 1 (2004) 199.
[18] J. Cioslowski, E.D. Fleischmann, J. Chem. Phys. 94 (1991) 3730.
[19] O.V. Boltalina, E.V. Dashikova, L.N. Sidorov, Chem. Phys. Lett. 256 (1996) 253.
[20] M.-H. Du, H-P. Cheng, Phys. Rev. B 68 (2003) 113402.
[21] J. Lu, X. Zhang, Z. Zhao, Solid State Commu. 110 (1999) 565.
[22] I.N. Ioffe, A.S. Levlev, O.V. Boltalina, L.N. Sidorov, H.C. Dorn, S. Stevenson, G. Rice, Inter. J. Mass Spectrometry 213 (2002) 183.
[23] K. Kobayashi, Y. Sano, S. Nagase, J. Comp. Chem. 22 (2001) 1353.
[24] S. Okada, M. Otani, A. Oshiyama, Phys. Rev. B 67 (2003) 205411.
[25] S. Okada, S. Saito, A. Oshiyama, Phys. Rev. Lett. 86 (2001) 3835.
[26] Y. Cho, S. Han, G. Kim, H. Lee, J. Ihm, Phys. Rev. Lett. 90 (2003) 106402.
[27] S.-Y. Chen, J-T. Lue, Phys. Lett. A 309 (2003) 114.
[28] J. Cioslowski, N. Rao, D. Moncrieff, J. Am. Chem. Soc. 124 (2002) 8485.
[29] S. Kazaoui, N. Minami, N. Matsuda, H. Kataura, Y. Achiba, Appl. Phys. Lett. 78 (2001) 3433.
[30] K. Hirahara, K. Suenaga, S. Bandow, K. Kato, T. Okazaki, H. Shinohara, S. Iijima, Phys. Rev. Lett. 85 (2000) 5384.
[31] Y.S. Li, D. Tománek, Chem. Phys. Lett. 221 (1994) 453.
[32] G. Gao, T. Cagin, W.A. Goddard, Phys. Rev. Lett. 80 (1998) 5556.
[33] G. Chen, Y. Guo, N. Karasawa, W.A. Goddard, Phys. Rev. B 48 (1993) 13959.
[34] Y. Guo, N. Karsawa, W.A. Goddard, Nature 351 (1991) 464.
[35] P. Delaney, J.C. Greer, Appl. Phys. Lett. 84 (2004) 431.
[36] J. Tersoff, Phys. Rev. B 38 (1988) 9902.
[37] J. Tersoff, Phys. Rev. B 39 (1989) 5566.
[38] D.W. Brenner, Phys. Rev. B 42 (1990) 9458.
[39] Z. Mao, A. Garg, S.B. Sinnott, Nanotechnology 10 (1999) 273.
[40] T.H. Fang, C.I Weng, Nanotechnology 11 (2000) 148.
[41] L. Zhang, H. Tanaka, Wear 211 (1997) 44.
[42] J.W. Kang, H.J. Hwang, Nanotechnology 15 (2004) 614.
[43] W.Y. Choi, J.W. Kang, H.J. Hwang, Phys. Rev. B 68 (2003) 193405.

Enhanced Mass Sensing for Nanomechanical Resonators

I. De Vlaminck*, K. De Greve*, R. Naulaerts, V. Sivasubramaniam, L. Lagae, H.A.C. Tilmans, G. Borghs

IMEC, Kapeldreef 75, B-3001 Leuven, Belgium

*also at ESAT/INSYS, KULEUVEN, Belgium

ABSTRACT

A new method for improving the mass detection limit of nanomechanical resonators in crystalline silicon, compatible with scaling, is demonstrated. Clamped-clamped silicon beam resonators are fabricated with a non-rectangular, 'diabolo' cross-section starting from SOI wafers. The simple, single mask process for making these resonators is demonstrated. The non-conventional cross-section results in a change of almost all functional parameters, frequency, effective mass, and signal to noise ratio, thereby decreasing the mass detection limit of the resonator up to a factor of 2.7 as compared to similar devices without the additional step. A comparison between beams with a rectangular cross-section and beams with the novel diabolo cross-section by finite element modeling confirms the predictions on frequency gain made by the analytical model. First experimental results on resonators with a rectangular and diabolo cross-section confirm the predictions made by modeling.

Keywords: nems, mass sensing, diabolo cross-section, mass detection limit.

1 INTRODUCTION

The large sensitivity of the resonance frequency of mechanical resonators to applied forces and adsorbed masses is widely exploited in sensor applications. In this context, scaling of the resonant elements has proven to be a powerful method for increasing this sensitivity. Using the scaling scheme, researchers were able to report astonishing force [1] and mass detection limits (in the attogram range and below [2,3]). These tiny mass detection limits open the doorway to an intriguing world of new biomedical applications. Further scaling of the system is expected to increase the sensitivity even more, down to single molecule mass sensitivity [4].

However, scaling limits are popping up: firstly the actuation and detection of the resonant motion of small-scale resonators becomes increasingly difficult [5], secondly the quality factor tends to deteriorate when the surface to volume ratio increases upon scaling [6].

In this communication, a new method for improving the mass sensitivity of nanomechanical resonators in crystalline silicon, compatible with scaling, will be presented. State of the art clamped-clamped silicon beam resonators are fabricated with a non-rectangular cross section starting from SOI wafers. Although the process for making these resonators differs from the standard process for making freestanding structures starting from SOI wafers [6,7] by only a single, unmasked wet etch step, almost all functional parameters are changed, thereby significantly increasing the mass sensitivity of the resonator as compared to similar devices without the additional step.

An analytical model is presented in which the parameters of importance are expressed as function of the dimensions and shape of the resonators. A detailed finite element model (FEM) study is discussed, confirming the predictions made by the analytical model on frequency gain and providing additional insight in short length effects.

The resonant behaviour of rectangular and non-rectangular resonators was studied by means of the magnetomotive method for actuation and detection.

2 FABRICATION

We use (100) silicon on insulator (SOI) wafers for fabricating all resonators. The buried oxide has a thickness of 400 nm, the silicon layer a thickness of 210 nm. The fabrication of the resonators with a non-rectangular cross-section is outlined schematically in Fig. 1A-D. Electron beam lithography is used to define the desired pattern in a polymethyl (methacrylate) resist layer. Evaporation and lift off of a Cr (10 nm)/ SiO_2 (70 nm) etch mask is performed. The pattern is transferred into the silicon by a reactive ion etch in a SF_6-O_2-Ar plasma. In regular process schemes an underetch of the sacrificial layer would conclude the resonator fabrication. In this scheme a wet, anisotropic KOH etch of the silicon is performed first. This wet etch, retarding once the (111) crystal planes are reached, defines the non-rectangular cross-section of the resonators which are aligned along the <110> direction. The Cr layer at the top and the buried SiO_2 at the bottom of the silicon provide excellent interfaces, prohibiting the silicon at the top and bottom from being attacked by the KOH solution. After the subsequent, HF-based underetching of the devices and rinsing with H_2O, acetone and isopropyl alcohol, the hard mask is removed and a layer of Al (30 nm) is deposited.

As the sidewalls of the resonator are enclosed by (111) crystal planes, a restriction is imposed on the width to thickness ratio, w/t; this ratio should be larger than $\cot(\alpha)$ for the cross-section to be uninterrupted, 'diabolo' like (the angle between the (111) planes and the (100) planes being α = 54°7', Fig. 1C and 1D). In Fig 1E. an electron micrograph of a diabolo cross-sectioned device is shown.

We fabricated reference resonators with a rectangular cross-section using an identical process except for the wet KOH etch which is left out.

3 MODELING

The ultimate mass detection limit imposed on a nanomechanical resonator depends strongly on the dominant noise regime. Ultimate detection limits have been theoretically described for various noise sources [4,2], with the thermomechanical noise as the fundamental, measurement setup independent limit.

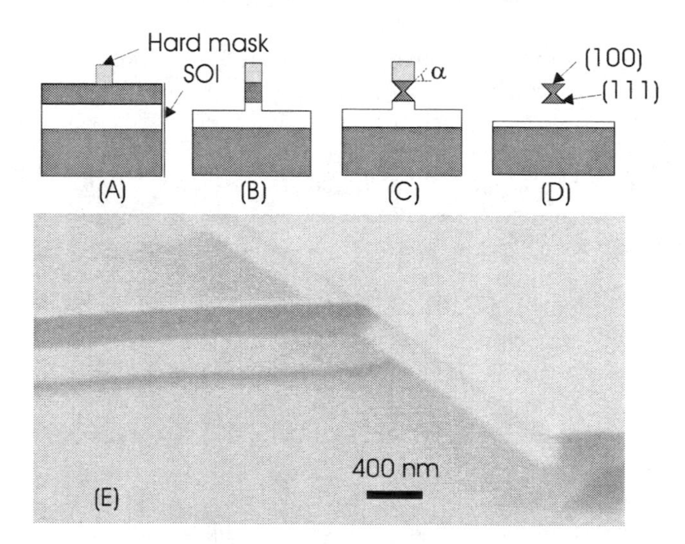

Fig. 1: Fabrication sequence of the nanomechanical resonators with the diabolo crosses section. (A) Lithographic definition of the resonator, (B) dry etch of the silicon, (C) anisotropic KOH etch, (D) removal of the hard mask and HF etch of the sacrificial silicon dioxide, (E) scanning electron micrograph of a released clamped-clamped beam resonator (0.21 X 0.21 X 12.5 μm).

The limit of detection for this thermomechanical noise limited regime can be described as [4]:

$$\delta M = 2M_{eff}\left(\frac{E_{th}}{E_c}\right)^{\frac{1}{2}}\left(\frac{\Delta f}{Q\omega_0}\right)^{\frac{1}{2}} \tag{1}$$

where M_{eff}, E_{th}, E_c, Δf, Q, ω_0 are the effective mass, the thermal energy ($E_{th}=k_bT$), the maximum drive energy, the measurement bandwidth, quality factor, and the resonance circular frequency respectively.

The effective mass, maximum drive energy and resonance frequency can all be expressed as a function of the cross-sectional area and inertial bending moment related to the shape of the resonator. We show that the diabolo cross-section of the resonator changes all these functional parameters determining the detectable mass limit in a beneficial way compared to a resonator with the same outer dimensions and a conventional, rectangular cross-section. The cross-sectional areas (A_{rect},A_{diab}) and inertial bending moments (I_{rect},I_{diab}) corresponding to the shape and dimensions of the diabolo and rectangular beam respectively. When out of plane bending is considered these can be expressed as:

$$A_{diab} = wt - \cot(\alpha)\frac{t^2}{2}, I_{diab} = \frac{wt^3}{12} - \cot(\alpha)(\frac{t^4}{48}) \tag{2}$$

$$A_{rect} = wt, I_{rect} = \frac{wt^3}{12}. \tag{3}$$

If w has the minimal allowed value of $t\cdot\cot(\alpha)$ the effective area and inertial bending moment are minimal.

The effective mass M_{eff} is proportional to A, for $w= t\cdot\cot(\alpha)$ it is a factor of two smaller than for a rectangular beam.

The resonance frequency ω_0 is proportional to $\sqrt{(I/A)}$, (in the Euler-Bernoulli beam theory) [8]. Therefore the resonance frequency of the diabolo beam can be up to a factor of 1.22 higher.

The maximum signal to noise ratio, E_c/E_{th}, ($E_c= M_{eff}\omega_0^2\langle x^2\rangle$, with x the vibration amplitude) corresponds to the maximum vibration amplitude $\langle x\rangle=\langle x_{max}\rangle$ leading to a predominantly linear response. $\langle x_{max}\rangle$ is known to be proportional to $\sqrt{(I/A)}$ for a double clamped beam resonator [9], therefore E_c/E_{th} is proportional to I^2/A, for $w=t\cdot\cot(\alpha)$ it is a factor of 1.1 larger.

The quality factor is determined by the dominant energy loss mechanisms (for an overview of the various loss mechanisms we refer to ref. 10). In our case, the energy loss is assumed to be surface loss dominated, as is the case for similar sized resonators reported on in literature, which are measured under the same conditions (room temperature and low pressure). No quantitative models exist for this mechanism; therefore no shape dependencies are expressed. Looking at all proportionalities we write:

$$\frac{\delta M_{diab}}{\delta M_{rect}} \sim \left(\frac{A_{diab}}{A_{rect}}\right)^{\frac{7}{4}}\left(\frac{I_{rect}}{I_{diab}}\right)^{\frac{5}{4}}\left(\frac{Q_{rect}}{Q_{diab}}\right)^{\frac{1}{2}} \sim \kappa\left(\frac{Q_{rect}}{Q_{diab}}\right)^{\frac{1}{2}} \tag{4}$$

$$\kappa = \left(\frac{A_{diab}}{A_{rect}}\right)^{\frac{7}{4}}\left(\frac{I_{rect}}{I_{diab}}\right)^{\frac{5}{4}} \tag{5}$$

using (1). In Fig. 2 the quality-factor-independent ratio of mass detection limits, κ, is plotted versus w. The mass detection limit of the diabolo cross-sectioned resonator is up to a factor of 2.7 lower than for the rectangular resonator, ignoring any differences in quality factor.

3.1 Finite Element Modeling (FEM)

The analytical model for the resonance frequency used earlier was based on Euler-Bernoulli beam theory. It does not take into account rotary inertia effects and shearing deformations, nor the effect of finite stiffness of the clamping region (these effects all become increasingly important for shorter beam lengths) [8].

To investigate these effects closer, Finite Element Models were built and studied using the Msc. Marc software. A modal analysis was performed to investigate the resonance frequency of the desired resonant vibration modes. A convergence analysis was performed as to ensure relative errors were no larger than 10^{-3}.

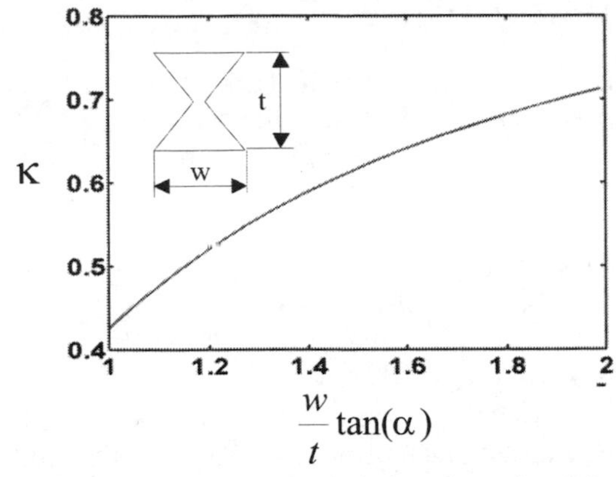

Fig. 2: κ vs. (w/t)·tan(α), (κ being the ratio of the mass detection limit of the diabolo cross-sectioned resonator and the mass detection limit of a rectangular resonator, independent of quality factor as defined in (5)).

Two different kinds of boundary conditions were imposed on both the models for the rectangular resonators and diabolo resonators. In the first the end faces of the resonator were fixed, mimicking an infinitely stiff or perfect clamping, in the second the bottom faces of the clamping region, this time included in the model, were partly fixed as to mimic a certain underetch and finite stiffness of the clamping. In Fig. 3, the models for the diabolo resonator and rectangular resonator and the different ways of modeling clamping are illustrated. The material parameters for silicon used in the calculations (taking into account the anisotropy of the material) are ρ=2330 kg/m^3, c_{11}=165 GPa, c_{12}=64 GPa, c_{44}=80 GPa.

The thickness of the resonators was chosen to be 210 nm, corresponding to the device thickness in the experiments. For resonators with a width of 150 nm (= t·cot(α)), the length was varied. In Fig. 4, the results of the FEM study are summarized and compared to the results obtained in the analytical model.

For the case of clamping with infinite stiffness, the ratio of the resonant frequencies $\omega_{diab}/\omega_{rect}$ deviates strongly from the result obtained by the analytical model ($\omega_{diab}/\omega_{rect}$=1.22) for short beam lengths. This is due to rotary inertia effects and shearing deformations, which are not taken into account in the analytical model.

The effect of rotary inertia on the resonant frequency is more outspoken for a diabolo beam than for a rectangular resonator because the rotary inertia effect is of a larger relative importance for the diabolo (again the ratio I/A), therefore the ratio of the resonant frequencies drops.

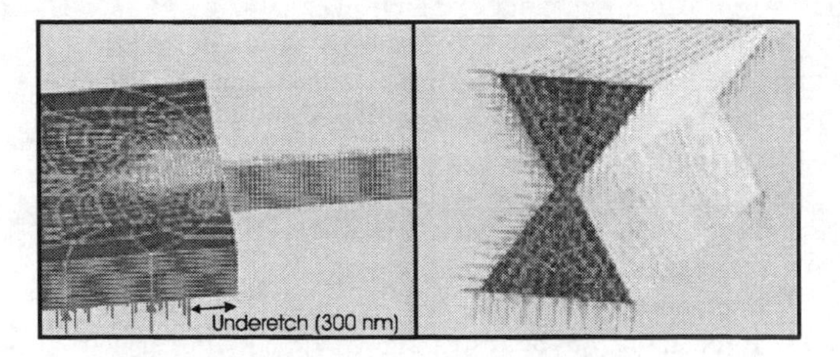

Fig. 3: (A) Model for a rectangular resonator. In this case the clamping region is included in the model. Zero-displacement boundary conditions for all directions (indicated by arrows at the nodes) are imposed at the bottom of the clamping region. Part of bottom face of the clamping region was not fixed, an underetch and finite stiffness of the clamping are mimicked as such. (B) Model for a diabolo beam resonator with the end faces fixed mimicking an infinitely stiff clamping.

In the case where the effect of finite stiffness of clamping was included in the model (underetch 300 nm), the deviation is smaller and only of importance for small and impractical beam lengths. The finite stiffness of the clamping results in a lower frequency for both rectangular and diabolo beams since part of the clamping will deform making the effective beam length larger. However this effect is less pronounced for the diabolo beam because it has a lower stiffness and will cause a smaller relative deformation of the clamping.

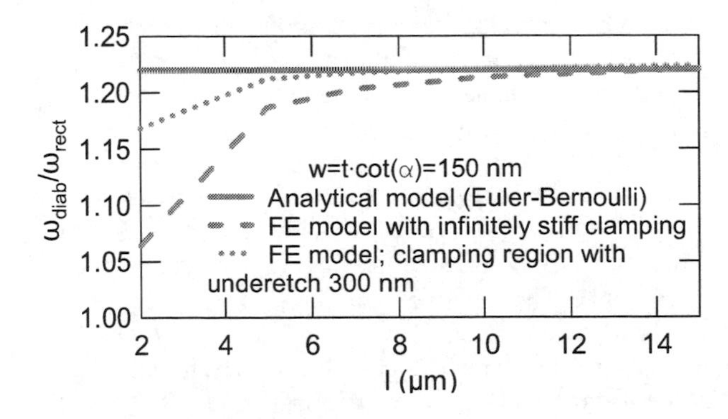

Fig.4: Comparison of the ratio $\omega_{diab}/\omega_{rect}$ calculated by the analytical and FEM models as function of beam length, l. The full line shows the result for the analytical model the dashed line shows the results for the FE models with infinite stiff clamping (indicating the effect of rotary inertia and shearing deformations), the dotted line shows the results for the FE models also including the finite stiffness of clamping (underetch 300 nm)

4 EXPERIMENTAL

Resonators with a diabolo cross-section and rectangular cross-section were fabricated using the aforementioned process schemes. The structures were made in the same process run, using the same dry etch plasma and wet etch steps except for the KOH etch which is only applied for making the diabolo cross-sectioned resonators. Before measuring the devices, we annealed them at 165°C for 12 hours. Annealing partly removes the as-deposited stress in the aluminum metal layer. An anneal-temperature dependent residual stress remains [11].

All experiments were performed using the well-described magnetomotive detection method [12]. An ac-Lorentz force acting on the resonators when applying an ac-current in a magnetic field drives the resonators. A change in resistance of the device due to oscillation in the magnetic field as function of the amplitude of vibration is sensed in a reflection based approach and provides a means of measuring the frequency spectrum. Measurements were performed at room temperature and the pressure inside the chamber was kept below 10^{-4} mbar. The applied field was 1.4 Tesla, devices were actuated out of plane in the linear operation regime.

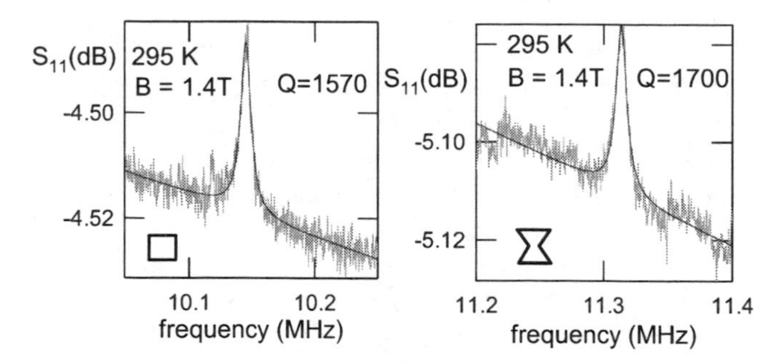

Fig. 5: frequency spectrum for a rectangular and a diabolo beam measured at room temperature and Lorentzian fit curves. Both devices have the same outer dimensions (0.23 X 0.21 X 12.5 μm). The applied field is 1.4 Tesla.

In Fig. 5 the measured frequency spectrum of a rectangular and a diabolo beam are shown. Both devices have the same outer dimensions (width 230 nm, thickness 210 nm and length 12.5 μm). The relative gain in frequency for the diabolo over the rectangular beam measured, factor 1.11, is in accordance to the expected gain for these dimensions. The surface-to-volume ratios are 0.018 nm^{-1} and 0.036 nm^{-1} for the rectangular and diabolo beam respectively. Although the surface to volume ratio is larger for the diabolo, the quality factor is slightly higher for the diabolo resonator. A thorough study towards the shape, dimension and temperature dependency of the quality factor is ongoing.

5 SUMMARY AND CONCLUSION

A novel type of nanomechanical resonator with a non-rectangular cross-section is demonstrated leading to a significantly improved mass detection limit. The fabrication process for these resonators is simple, involving a single additional, unmasked, wet etch step, and is compatible with scaling. Analytical as well as finite element modeling established that the mass detection limit can be lowered by as much as a factor of 2.7, when ignoring any changes in quality factor. First experimental results confirm the predicted gain in frequency and indicate that the quality factor is slightly higher for a diabolo shaped resonator. An elaborate study towards the shape, size and temperature dependence of the quality factor is ongoing and may provide more insight in the mechanisms governing dissipation in nanoresonators.

ACKNOWLEDGEMENT

We thank Jos Moonens and Hans Costermans for their e-beam service, Bart Vandevelde, Mario Gonzalez and Dominiek Degryse for assistance in the finite element modeling, and Johan Feyaerts and Erwin Vandenplas for technical support. IDV acknowledges financial support from the I.W.T. (Belgium), KDG acknowledges financial support from the F.W.O. (Belgium).

REFERENCES

[1] A.N. Cleland and M.L. Roukes, Nature **392**, 160 (1998).

[2] K.L. Ekinci, X.M.H. Huang and M.L. Roukes, Appl. Phys. Lett. **84**, 4469 (2004).

[3] B. Ilic, H. G. Craighead, S. Krylov, W. Senaratne, C. Ober and P. Neuzil, J. of Appl. Phys. **95**, 3694 (2004).

[4] K.L. Ekinci, Y.T. Yang, and M.L. Roukes, J. Appl. Phys **95**, 2682 (2004).

[5] M.L. Roukes, Technical digest of the 2000 Solid-State and Actuator workshop (2000), Hilton Head Isl, and SC, 4-8 June 2000

[6] D.W. Carr, S. Evoy, L. Sekaric, H.G Craighead, and J.M. Parpia, Appl. Phys. Lett. **75**, 920 (1999).

[7] S. Evoy, D.W. Carr, L. Sekaric, A. Olkhovetz, J. M. Parpia and H.G. Craighead, J. of Appl. Phys. **86**, 6072 (1999).

[8] W.J. Weaver, S.P. Timoshenko and D.H. Young, *Vibration problems in engineering* 5th Edition, John Wiley & sons, New York (1990).

[9] H.A.C. Tilmans, M. Elwenspoek and J.H.J. Fluitman, Sens. actuators A **30**, 35 (1992).

[10] P. Mohanty, D. A. Harrington, K. L. Ekinci, Y. T. Yang, M. J. Murphy and M. L. Roukes, Phys. Rev. B **66**, 085416 (2002).

[11] S. Hyun, W. L. Brown and R. P. Vinci, Applied Physics Letters **83**, 4411 (2003).

[12] A. N. Cleland and M. L. Roukes, Appl. Phys. Lett. **69**, 2653 (1996).

Micromachined Nanocrystalline Silver Doped Tin Oxide Hydrogen Sulfide Sensor

Jianwei Gong[1], Quanfang Chen[1]*, Ming-Ren Lian [2], Nen-Chin Liu[2]
[1] Department of Mechanical, Materials & Aerospace Engineering, University of Central Florida, Orlando, FL 32816-2450, USA
[2] TYCO SAFETY PRODUCTS-Sensormatic, 6600 Congress Avenue, Boca Raton, FL, 33487 U.S.A

ABSTRACT

We report here an ultra high sensitive hydrogen sulfide micromachined gas sensor based on SnO_2-Ag gas sensing material was fabricated on ceramic wafer and quartz wafer. A polymeric sol-gel process has been successfully developed in fabricating thin film SnO_2-Ag nanocomposite, which shows excellent sensing characteristics upon exposure to as low as 1 ppm concentrations of H_2S at low working temperature of 70C. This sensor shows ultra high sensitivity to H_2S even at extreme (both dry and wet) humidity conditions compared to published results and testing commercial H_2S sensors. SEM was used to investigate surface morphology and crystalline of the film. Selectivity of this gas sensor was studied with comparison with several commercial H_2S sensors; the result shows that this sensor is less sensitive to common interference gases like Cl_2, HCl, SO_2, C_6H_{14}, CH_4, CO, C_3H_8 etc. which may cause false alarm in real applications.

Keywords: MEMS, SMO, Nanocrystalline SnO_2, SWNT Sol Gel

1. INTRODUCTION

Hydrogen Sulfide is a colorless, flammable toxic gas, occurring naturally in crude petroleum, natural gas, volcanic gases, and hot springs with smells like rotten eggs. It can result from bacterial breakdown of organic matter or produced by human and animal wastes. Other sources are industrial activities, such as food processing, coke ovens, craft paper mills, tanneries, and petroleum refineries. H_2S monitoring is very important in industrial areas such as oil and natural gases exploitation plants and coal manufacturing. Electro chemical cell H_2S sensors are mostly widely used commercial sensors, however, it suffers from problems working in extreme environmental conditions (like dry dessert) and short life time etc.

Semiconductor metal oxide (SMO) chemical sensors have shown advantages in commercialization prospect and market potential. These advantages include long lifetime, fast response and recovery time, low cost, simple electronic structure and low maintenance [1]. Most research and development of SMO H_2S gas sensors are focused on SnO_2 based and WO_3 based; with few other types reported such s $FeNbO_4$, $BaTiO_3$, and TiO_2. Reported SnO_2 based H_2S sensor can be classified as sintered pastes [2], thick [3] and thin film structures using mixed SnO_2-CuO powders, Cu- SnO_2 bi-layers [4], CuO-SnO_2 hetero-contacts [5], chemically fixed CuO on SnO_2 [6-8, 11, 22], Ag_2O- SnO_2, Ag- SnO_2 and SnO_2 -Ag- SnO_2 composite film gate MIS diodes [9, 10, 12]. Other doping material like ZrO to SnO_2 was also reported [13]. WO3 based SMO gas sensor has also been extensively studied as H_2S gas sensor [14-19], Pt and Au are reported to be used as dopant in WO_3 thin film or thick film gas sensor which can detect 1ppm accuracy of H_2S.

The major problems of existing H_2S SMO gas sensor includes cross sensitivity, modest sensitivity, moisture influence, power consumption. In this paper, we report a mcirofabricated SnO_2-Ag H_2S sensor with advantages of ultrahigh-sensitivity (even in dry conditions), less cross sensitive to interfering gases, low working temperature and power consumption. The sensing material is made of the proper combination of silver as a doping in the nanocrystalline SnO_2, using the sol-gel method. The extreme sensitivity can be contributed to appropriate dispersion of Ag on nanocrystalline SnO_2 grains and uniformity of the nanoporous structure of the films. The

*Author to contact: qchen@mail.ucf.edu, Phone: 407-823-2152, Fax: 407-823-0208

response of fabricated sensors to staircase concentration of H_2S, relative humidity change and other interference gases, is discussed in this paper.

2. EXPERIMENTAL

2.1 Gas Sensor Design and Fabrication

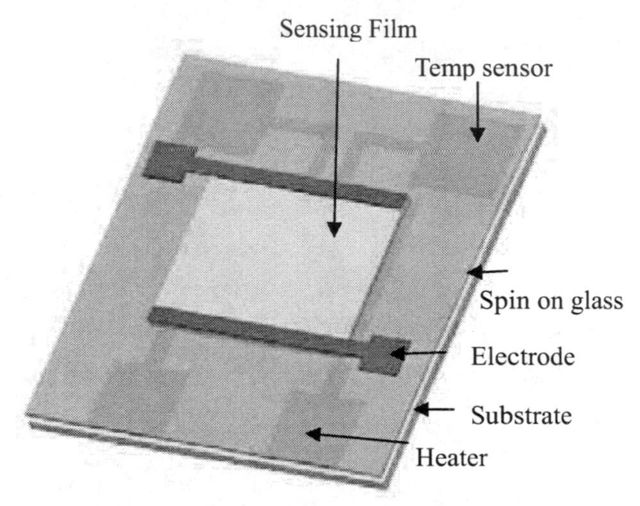

Fig. 1 Micromachined gas sensor developed

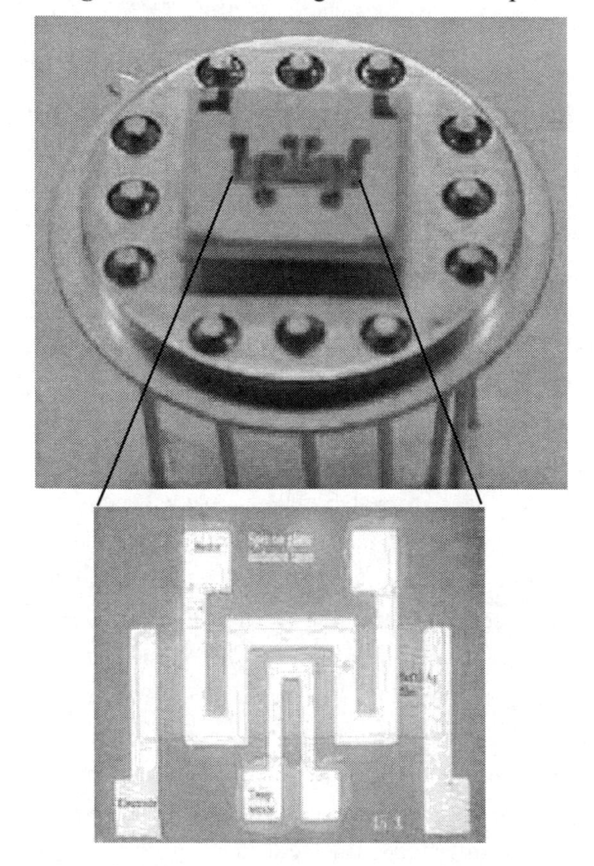

Fig. 2 Fabricated gas sensor

Figure 1 shows a schematic diagram of a micro fabricated SMO ceramic substrate sensor, which includes five major parts, namely a substrate, a SMO sensing film, sensor electrodes, a Pt heater and a Pt temperature sensor. The detail of the fabrication process has been reported in [22]. Figure 2 shows the fabricated device fixed on 12 legs package, 1 mil diameter aluminum wires are bonded using an ultrasonic wire bonder.

2.2 SnO_2-Ag thin film fabrication by Sol Gel process

Tin isopropoxide ($Sn(O^iPr)_4$) was dissolved in anhydrous ethanol. A complexing agent, acetylacetone (AcAc), was added to stabilize the hydrolysis of tin isopropoxide. After complete the mixing using a magnetic stirring, hydrolysis was performed by adding distilled water with an appropriate ratio. A clear and stable sol was formed after continuous stirring. The viscosity of the sol was adjusted with addition of PVA (poly vinyl alcohol). PEG (poly ethyl glycol) was added to improve the plasticity of the coating to prevent cracking during the firing process. $AgNO_3$ was added into the SnO_2 sol and followed by a magnetic stirring (for 1 day), while HNO_3 was used to stabilize the SnO_2 sol. Figure 1 shows SEM picture of SnO_2 film doped with Ag.

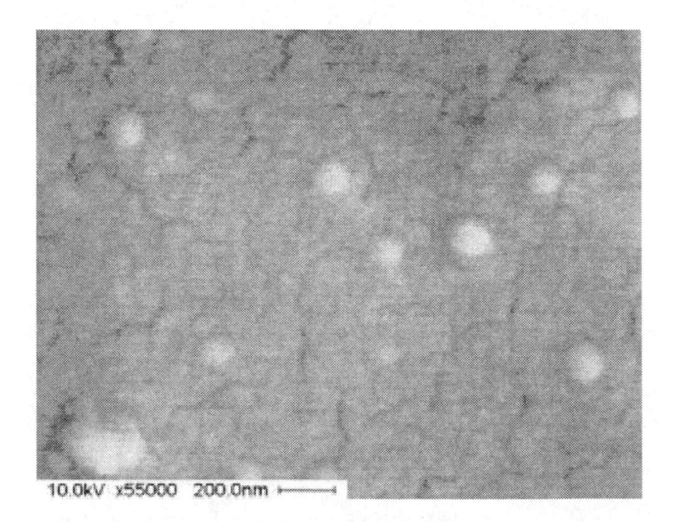

Fig.3 SEM picture of Ag doped SnO_2 film

Nanostructured SnO_2 thin film was fabricated by a spin coating and a subsequent sintering process. A ramp spin coating method was used to increase spinning speed gradually up to 1,000 rpm and keep spinning for 30 seconds in total. After air-dry the film for 1 min, the coated film was dried in an oven at 100 degree °C for 30 min. The film sintering was conducted in a furnace (Thermolyne) with a heating rate of 2°C/min incensement to 500°C and kept it at this temperature for 2 hours.

2.3 Gas Sensor Test System

The testing system includes hardware setup and software program. Hardware setup comprises of five major components, namely, gas mixing and environmental condition control, gas testing chamber, data acquisition, signal output (control), temperature feedback control circuits [21], and a computer, as are shown Figure 4. The control program running in PC was written using LABVIEW. The main function includes data acquisition, data collection and controlling parameters of the testing system like Mass Flow Controller setting, feedback control circuit working temperature setting, etc.

The gas sensitivity was determined by:

$$\text{Sensitivity} = (R_1-R_0)/R_0 \times 100$$

Where R_1 is resistance values of gas sensing thin film in air, and R_0 represents resistance values in gas environment.

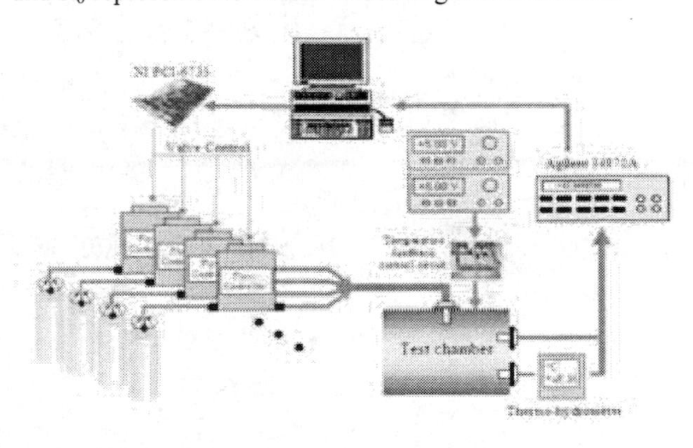

Fig. 4 Testing System Setup

3. RESULT AND DISCUSSION

Two fabricated SnO_2 sensors doped with Ag are tested in Hydrogen Sulfide. UCF1 is a sensor made with a quartz substrate, and UCF2 is small sensor made with a ceramic substrate. For better understanding our sensor's performance, we have included some commercial sensors include Figaro (Tin oxide) H_2S sensor TGS825, and Bacharach (SMO) H_2S sensor for comparison.

A staircase concentration test was conducted for H_2S testing. The procedure was to increase the concentration from lowest level to highest level. At each level, keep exposing sensors to H_2S gas for half an hour before increasing to the next concentration level. UCF1 sensor is working at 64C and UCF2 sensor is work at 79C. Figure 5 shows the four sensors response in a staircase concentration experiment.

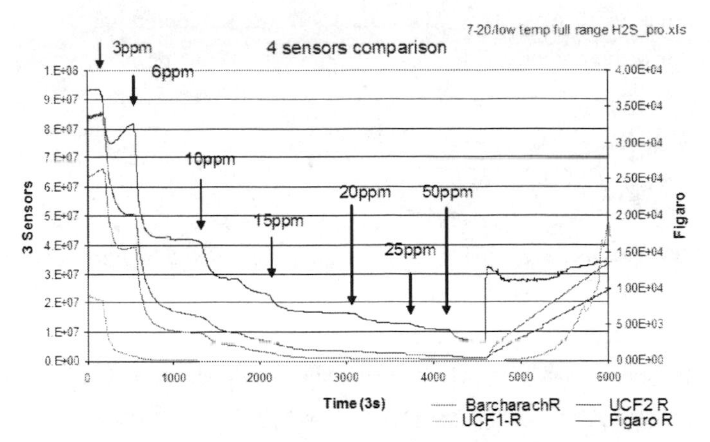

Fig. 5 Step Concentration H_2S Test Result

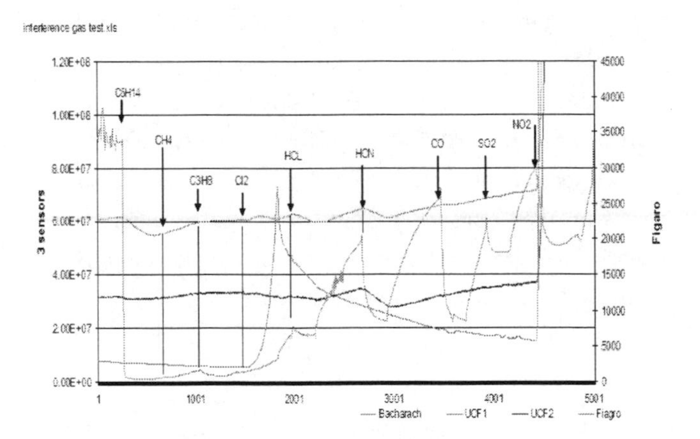

Fig. 6 Selectivity Test Result

Figure 5 and Table 1 summarized sensitivities of four gas sensors at different H_2S concentrations. It is clear that our fabricated sensor, the UCF1 sensor, has the best sensitivities over other three sensors. UCF2 and Bacharach sensors have comparable sensitivities.

Gases like Cl_2, HCl, SO_2, C_6H_{14}, CH_4, CO, C_3H_8, NO_2 etc. are common interference gases, which may cause false alarm in real applications. In this study, we conducted test for the following concentrations of interference gases: Cl_2 (10ppm), HCl (10ppm), NO_2 (5ppm), HCN (10ppm), SO_2 (9ppm), C_6H_{14} (0.3%), CH_4 (2.2%), CO (50ppm), C_3H_8 (0.44%). Figure 6 shows the interference gas test result, it is obvious that our fabricated sensors are less sensitive to common interference gases compared to Figaro and Bacharach sensors. For NO_2 case, all sensors have large response. However, since the NO_2 is oxidizing gas, exposure to them will make the SMO sensor's resistance increasing instead of decreasing. Therefore, the risk of causing false alarm is small.

4. CONCLUSION

We have developed microfabricated Ag-SnO_2 thin film gas sensor with ultra high sensitivity and selectivity to H_2S detection. Ag doped nanocrystalline SnO_2 gas sensing material was successfully prepared with sol gel processes. The Ag-SnO_2 films showed better sensitivity compared to pure SnO_2 due to proper dispersion of Ag_2O particles between nanocrystalline SnO_2 grains and formation of p-n hetero junctions. The H_2S measurements were carried out for staircase concentration at an operating temperature as low as 64 C. A high sensitivity was obtained for 1ppm H_2S. The selectivity of the sensors was studied for various inference gases; the result shows that our fabricated sensors are less sensitive to common interference gases compared to some commercial H_2S sensors like Figaro and Bacharach sensors.

ACKNOWLEDGEMENTS

This research was made possible through the generous support in part of University of Central Florida in-house research and in part of TYCO SAFETY PRODUCTS-SENSORMATIC. The authors would also like to thank Mr. Ravi Todi, Mr. Kevin R. C., Mr. Quinones Raul and Mr. Stephen Corry, UCF-Cirent Materials Characterization Facility and TYCO SAFETY PRODUCTS-SENSORMATIC for their instrument supports.

REFERENCES

1. J. E., Haugen and K, Kvaal, Meat. Science Vol. 49, No. Suppl. 1, pp. 273-286 (1998)
2. S. Kanefusa, et al., IEEE Trans. Elect. Devices 35, 65 (1988)
3. V. Lantto, et al., Sens. Act. B 4, 451 (1991)
4. J. P. Li., et al., Sens. Act. B 65, 111 (2000)
5. R. B. Vasiliev, et al., Mater. Sci. Engg. B 57, 241 (1999)
6. T. Maekawa, et al., J. Mater. Chem. 4, 1259 (1994)
7. D. J. Yoo, et al., Jpn. J. Appl. Phys. 34, L455 (1995)
8. A. Chowdhuri, et al., IEEE sensors 2002, June Orlando, FL, USA
9. X.L. He, et al., IEEE sensors 2002, June Orlando, FL, USA
10. C.H. Liu, et al., Thin Solid Films, 304, 1997, 13-15
11. X. H. Zhou, et al., Materials Science and Engineering B99 (2003) 44 -/47
12. http://people.cecs.ucf.edu/sseal/nano.htm#silver

	3(ppm)		**6**(ppm)		**10**(ppm)		**15**(ppm)	
	S_1 (%)	S_2	S_1 (%)	S_2	S_1 (%)	S_2	S_1 (%)	S_2
Figaro	19%	0.14	54%	1.28	75%	3.09	82%	4.7
Bacharach	40%	0.86	85%	5.91	95%	23.2	98%	83
UCF-1	90%	9.5	99%	117.4	99%	242.7	99%	372
UCF-2	40%	0.86	82%	4.67	92%	12.7	96%	24.6

	20(ppm)		**25**(ppm)		**50**(ppm)	
	S_1 (%)	S_2	S_1 (%)	S_2	S_1 (%)	S_2
Figaro	82%	6.22	82%	7.78	82%	13.3
Bacharach	99%	108.3	99%	137.6	99%	204.6
UCF-1	99%	437.6	99%	508.4	99%	640.9
UCF-2	97%	35.8	98%	52.1	99%	91.2

Table 1: Sensitivity Summary of Step Concentration Test

13. S. Kanefusa, et al., IEEE Trans. Electron Devices, vol. ED-35, no. 1, (1988)
14. W. H. Tao and C. H. Tsai, Sens. Act. B 81, 237-247 (2002)
15. M. Penza, et al., Sens. Act. B 81, Issue 1, 15 115-121 (2001)
16. J. L. Solis, et al., Sens. Act. B 77, Issues 1-2, 15 316-321(2001)
17. M. D. Antonik, et al., Solid Films, 256, Issues 1-2, 1 247-252 (1995)
18. B. Frühberger, et al., Sens. Act. B 31, Issue 3, 167-174 (1996)
19. I. Jiménez, et al., Sens. Act. B 93, Issues 1-3, 475-485 (2003)
20. B.K.Miremadi, K.Colbow: Sensors and Actuators B46 (1998) 30–3
21. S. Seal, S. Shukla, Nanocrystalline SnO gas sensors in view of surface reactions and modifications, JOM 54 (9) (2002) 35–38, 60.
22. J. W. Gong, et al., Sens. Act. B 102, Issue 1, 117-125 (2004).

Multiscale Modelling of Self-Organized Mono-layer Surface Atomic Clusters

Qiyang Hu[†], Nasr M. Ghoniem[†] and Daniel Walgraef[‡]

[†]Mechanical and Aerospace Engineering Department
University of California, Los Angeles, CA 90095-1597, USA
[‡] Center of Nonlinear Phenomena and Complex Systems
Free University of Brussels, Brussels, Belgium.

ABSTRACT

We present here a novel multiscale modelling approach to investigate the conditions for atomic cluster self-organization on atomically flat substrates during epitaxial deposition processes. A phase field model is developed for the free energy of the system, which includes short-range as well as long-range elastic interactions between deposited atoms and atom clusters. The effects of externally applied periodic strain fields through the substrate are also investigated. At very low atomic coverage, a Kinetic Monte Carlo method is used to determine the nucleation conditions for atomic clusters forming on surfaces with periodic strain fields.

Keywords: multiscale modelling, atomic clusters, kinetic Monte Carlo, adatom interactions, self-organization

1 Introduction

Self organization of surface atomic clusters, which is an important aspect of future photonic and electronic devices, is driven by natural instabilities in reaction-diffusion kinetics, and by periodic external force fields. Examples are periodic fields generated by an interfacial dislocation array[1] or by a surface laser field[2]. In the present approach, we present a statistical mechanics model to describe the dynamics of deposited adatoms clustering and the formation of spontaneously ordered structures. The model builds on Suo's approach [3], and is an extension of Walgraef's model [4]. Specifically, we include the effects of external elastic fields and long-range interactions between adatom clusters. In addition to the continuum approach, we investigate the mechanism of clustering at the atomic scale using the Kinetic Monte Carlo method. Comparison and connections between these two methods are discussed in the last section.

2 Phase Field Modelling

2.1 Theoretical Model

We divide a macroscopic surface area into a set of mesoscopic cells and write the partition function $Z^{(i)}$ using the number of substrate sites ($N_S^{(i)}$) and the number of adsorbates within the cell ($N_A^{(i)}$) in each ith mesoscopic cell. By applying the mean field approximation and Sterling's formula, we obtain the total free energy of a system in a form similar to reference [4]. In the present case, the interaction between individual adatoms and the substrate is ignored and the atomic cluster is considered as the parts of the substrate surface to store the elastic energy generated by the external strain field. Thus the energy per i^{th} atom (W_S^i) is approximated as: $W_S^i = \tau \cdot \varepsilon^{(ext)} a^2$, where $\varepsilon_{ext} = \varepsilon - \varepsilon^{(0)}$ is the external strain field applied in the substrate and a is the lattice constant [5]. The dynamic equation for surface atoms based on the conservation of the adsorption (α) and desorption rates (β) and atomic mass current ($\vec{J}$) can be written as:

$$\begin{aligned}\partial_t c &= \frac{1}{\alpha+\beta}(c_0 - c) \\ &\quad + \vec{\nabla} \cdot \left\{ \frac{D}{k_B T} \vec{\nabla} \left[\varepsilon^{ext}\tau + k_B T \ln\left(\frac{c}{1-c}\right) \right.\right. \\ &\quad \left.\left. - \varepsilon_0 c - \xi_0^2 \nabla^2 c \right] \right\} \qquad (1)\end{aligned}$$

where $\epsilon_0 = -\sum_j \epsilon_{ij}$ and $\xi_0^2 = \gamma\epsilon a^2$, in which γ is the lattice coordination number, ϵ is the pair interaction energy, a is the lattice constant and positive sign of energy is for attractive interactions.

2.2 Numerical Solution

A Fourier-spectral method is adopted to numerically solve equation (1) with periodic boundary conditions [6]. The 1st order ordinary differential equation (ODE) in Fourier space is solved by the DVODE package with fixed-leading-coefficient implementation [7]. Figure (1) shows a numerical solution for the case of Si layers deposited on a periodically strained surface subjected to a 1-D sinusoidal strain field over the substrate ($\varepsilon_{\max} = 0.05$), in which we choose $a_{Si} = 5.4\,\mathring{A}$, $\tau \approx 0.02\,\mathrm{s}$, $\epsilon_0 \approx 0.22\,\mathrm{eV}$, at temperature $T = 400\,\mathrm{K}$ and the mean coverage of 0.2. Figure (2) shows the same situation but in a 2-D case. In Figure (3), we calculated a surface quantum dot pattern on the 80 nm SiGe buffer layer with two perpendicular interface dislocations underneath. The uniform coverage is 0.15. The alignments of Ge dots and the existence of corresponding denuded zones are consistent with the experiments of Kim *et al.* [1] which validate the present approach.

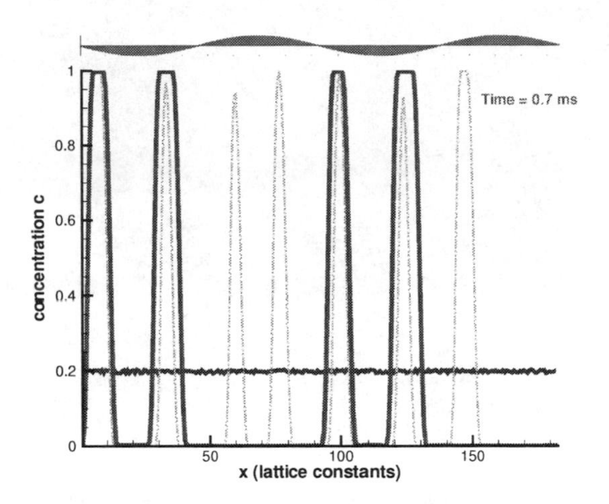

Figure 1: Numerical solution for the dynamic model with a 4-period sinusoidal external field. The initial conditions are small perturbations around the mean concentration. Concentration profiles are shown for the case with (red solid lines) and without (blue dashed lines) a periodic external field.

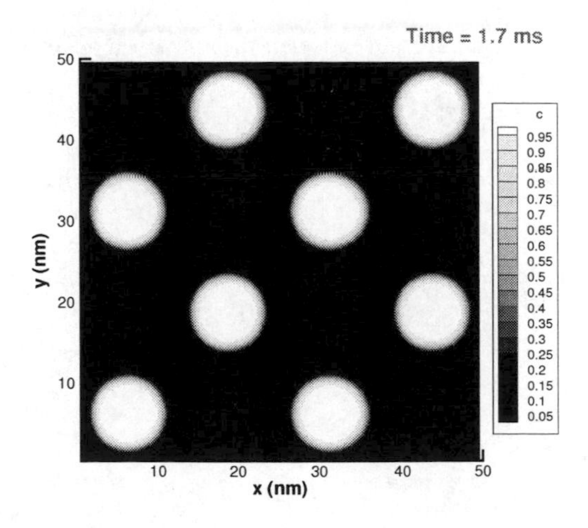

Figure 2: 2-D numerical solution for the same conditions as Fig.(1).

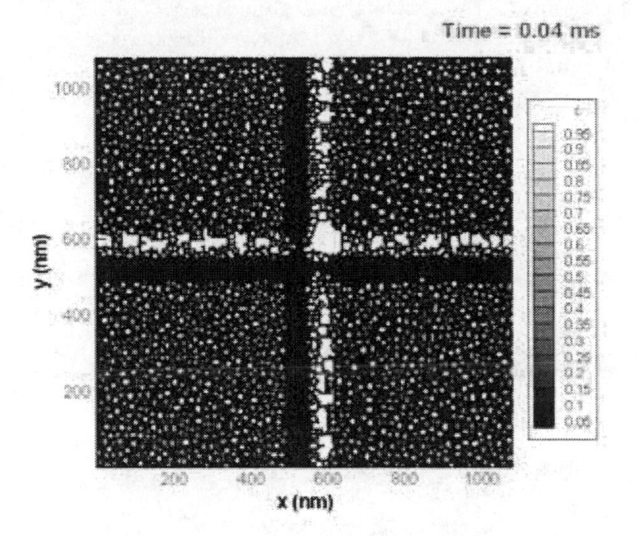

Figure 3: Numerical results for the model with two perpendicular interfacial dislocations underneath a 80 nm SiGe buffer layer. A grid of 128 × 128 is used.

3 Atomistic Modelling

3.1 Island Diffusion Kinetics

Our investigation shows that atomic island diffusion is mostly affected by the elastic field, causing spontaneous self-organization. In order to consider the strain effect on biased cluster diffusion, a simple model similar to the one proposed by Mattson et al. [8] is adopted. We consider an island fixed on the surface and calculate the binding energy for different island sizes. Here different from the metal-on-metal epitaxy in Mattson's work, the Modified Embedded Atom Methods (MEAM) for a Si surface is used[9]. By fitting the MEAM potential, we obtain the change on binding energy ($E_{bc}^{(i)}$) of a static island with size N and strain fields $\varepsilon_x(x,z)$, $\varepsilon_z(x,z)$ in a 3rd order polynomial function:

$$E_{bc}^{(i)}(N,\varepsilon) \approx (0.02N^3 - 0.68N^2 + 29.55N - 72.56) \cdot [\varepsilon_x(x,z) + \varepsilon_z(x,z)] \quad (2)$$

where the energy is in the unit of eV. The energy barrier of an island on a strained surface can be written as:

$$E_a^{(i)} = E_{a0}^{(i)}(N) + E_{bc}^{(i)}(N,\varepsilon(\overline{x},\overline{y})) - E_{bc}^{(i)}(N,\varepsilon(x,y)) \quad (3)$$

where $E_{a0}^{(i)}$ is the activation energy of the island without a strain field[10]. Here, we assume that the semiconductor surface has a fixed value: $E_{b0}^{(i)} \approx 0.79\,\text{eV}$.

To obtain the probability of island diffusion "events", an opposite thinking as Shöll did in his paper is applied.

First, From "random cluster scaling theory" (RCST), the following relationship is used for the diffusion coefficient of island diffusion $D^{(i)}$[11]:

$$D^{(i)} \sim N^{-3/2} e^{-E_a^{(i)}/k_B T} \tag{4}$$

If the random walk is uncorrelated, we can also have:

$$D^{(i)} \propto \langle \nu_h \rangle \langle \delta d_{c.m.}^2 \rangle \tag{5}$$

where $\langle \nu_h \rangle$ is the jump rate for island diffusion; $\langle \delta d_{c.m.}^2 \rangle$ is the mean-square displacement of the island mass center per events. In our model, every jump distance is assumed to be fixed that:

$$\langle \delta d_{c.m.}^2 \rangle = \text{const.} \tag{6}$$

Obviously, we have:

$$\langle \nu_h \rangle = \nu_0 N^{-3/2} e^{-E_a^{(i)}/k_B T} \tag{7}$$

We choose the rate constant as: $\nu_0 = 10^{13}\,\text{sec}^{-1}$. The evaporation process in an island is simulated by the chemical kinetics analysis see [12], [13]. Based on the fact that the evaporation is a first-order rate process, we adopt the following relation as the following:

$$p(t)\,dt = k_e dt \exp[-k_e t] \tag{8}$$

Here, $p(t)dt$ is the probability that an island with size N will emit one atom at a time between t and $t + dt$. k_e is the evaporation rate constant and is dependent on the size N and temperature T with the form of:

$$k_e = A \exp[-E_e/k_B T]\, N^{1/2} \exp\left[B \big/ N^{1/2}\right] \tag{9}$$

where A, B and E_e are constants and set to be: when $T < 650\,\text{K}$, $A = 0.063$, $B = 4.07$; when $650\,\text{K} < T < 950\,\text{K}$, $A = 0.051$, $B = 4.87$; when $T > 950\,\text{K}$, $A = 0.086$, $B = 4.55$.

3.2 KMC Simulation Results

Our simulation follows a standard Kinetic Monte Carlo method on $350 \times 350\,\text{nm}^2$ surface area. Two infinitely long straight dislocation lines are buried 80nm underneath the surface at $x = 250\,\text{nm}$ and $z = 250\,\text{nm}$, respectively. Figure 4 shows the diffusion process of 500 atoms on the top of the surface at a temperature of 650°C. The background contours are the strain field imposed by the interfacial dislocation network. The white dot denote the atoms. The clusters of adatoms which contain more than 6 atoms are declared by the block arrows pointing to the nearest spot.

It is clear that by introducing island diffusion, the Ge atom clusters tend to migrate toward the maximum compression area on the Si/SiGe/Ge surface, even though the external strain field is weak at the surface. It is

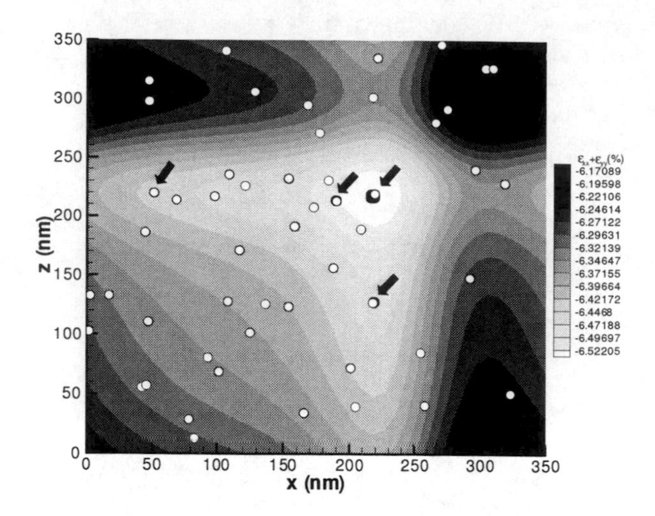

Figure 4: KMC simulation for 500 atoms on surface at 650°C after 0.033 sec

consistent with the expectation made in the above subsection. Due to the emission effect, there always exists single atoms at the same time.

Figure 5 compares the average mean square displacement between an atom diffusion only model and atom-cluster diffusion model. The atom-cluster coupling diffusion model shows a higher mobility and diffusion length and much lower island density. This indicates the self organization effects during the diffusion are taking place. This can also explain why experimentally quantum dots nucleate at places which have a larger distance than the average diffusion length of single atoms [14], [1]. It has been seen that the key role in the simulation is the island diffusion process. Our MEAM calculation shows that in semiconductor system, the self-organization effect tends to be stronger than in metallic systems.

We also compared the diffusion processes at three different temperatures. The results are shown in Figure 6, 7. It can be concluded that higher the temperature can equilibrate the atoms much quicker. Also it reveals that the atom emission process from islands just provides a fluctuation effect and has no remarkable influence on the whole process, which means ignoring the evaporation effect is a reasonable approximation as Bogicevic *et al.* did[15].

4 Conclusions

With the multiscale modeling of self-organized surface atomic clusters presented above, we conclude that a weak external period strain field have a significant effect on self-organized surface adatom clusters, especially in the low coverage case. The clusters are formed in the compressive stress region. With the comparison of

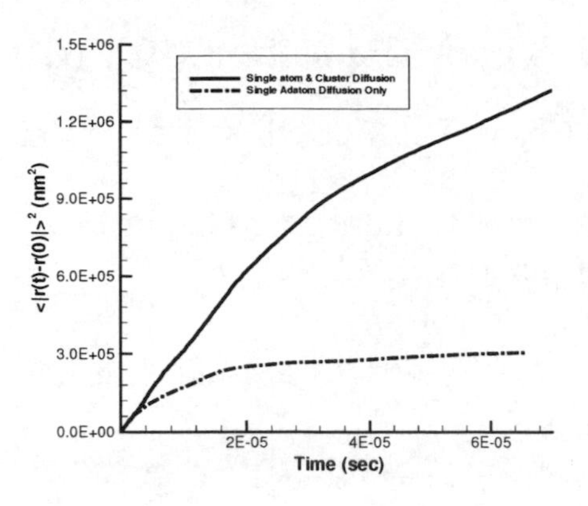

Figure 5: The comparison of average mean square displacement of atom-diffusion-only model and atom/cluster-diffusion model

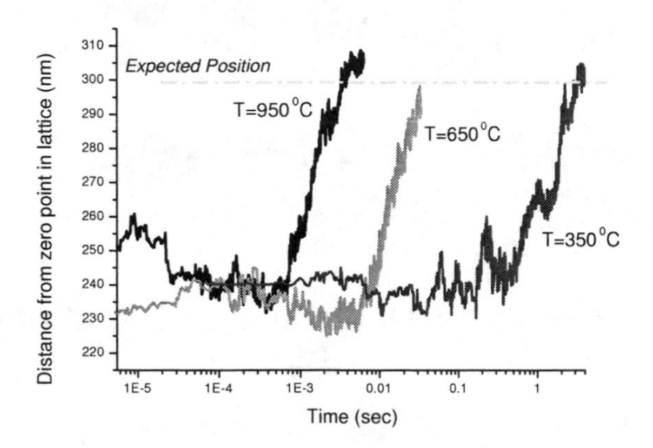

Figure 6: Comparison of distance from the zero point in each cell at three different temperatures

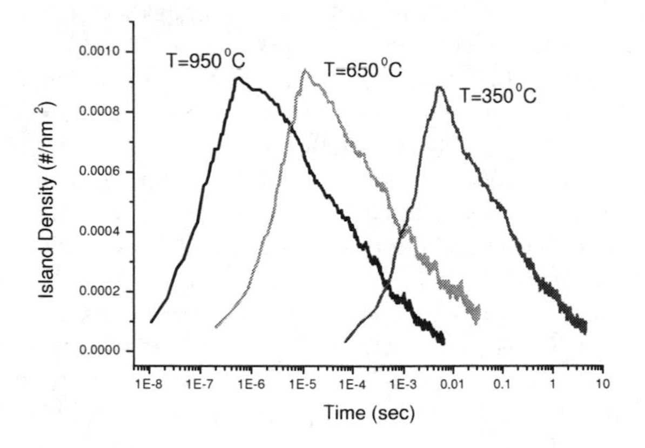

Figure 7: The comparison of Ge island density at three different temperatures

the experimental data, it confirms that the macroscopic elastic theory is valid in evaluating the interaction energy between microscopic adatoms and substrates. We also find that island diffusion plays a dominant role in the early stage of the quantum dot nucleation.

Acknowledgments: This work was supported by the National Science Foundation (NSF) through grant DMR-0113555.

REFERENCES

[1] H.J. Kim, Z.M. Zhao, and Y.H. Xie. *Physical Review B*, To be published.

[2] J.J. McClelland, R.E. Scholten, E.C. Palm, and R.J. Celotta. *Science*, 262:877–80, 1993.

[3] Z. Suo and W. Lu. *Journal of Nanoparticle Research*, 2:333–44, 2000.

[4] D. Walgraef. *Philosophical Magazine*, 83:3829–46, November 2003.

[5] V. I. Marchenko and A. Ya. Parshin. *Sov. Phys. JETP*, 52(1):129–31, July 1980.

[6] L. Q. Chen and J. Shen. *Computer Physics Communications*, 108:147–58, 1998.

[7] P.N. Brown, G.D. Byrne, and A.C. Hindmarsh. *SIAM J. Sci. Stat. Comput.*, 10:1038–51, 1989.

[8] T.R. Mattsson and H. Metia. *Journal of Chemical Physics*, 113(22):10323–32, 2000.

[9] M. I. Baskes. *Physical Review B*, 46(5):462727–42, August 1992.

[10] G. Mills, T.R. Mattsson, L. Mollnitz, and H. Metiu. *Journal of Chemical Physics*, 111:8639–50, 1999.

[11] D.S. Sholl and R.T. Skodje. *Physical Review Letters*, 75(17):3158–61, October 1995.

[12] H. Shao, P.C. Weakliem, and H. Metiu. *Physical Review B*, 53(23):16041–9, June 1996.

[13] T.R. Mattsson, G. Millis, and H. Metiu. *Journal of Chemical Physics*, 110(24):12151–60, June 1999.

[14] H.J. Kim, J.Y. Chang, and Y.H. Xie. *Journal of Crystal Growth*, 247:251–4, 2003.

[15] A. Bogicevic, S. Liu, J. Jacobsen, B. Lundqvist, and H. Metiu. *Physical Review B*, 57(16):R9459–62, 1998.

Polymer Nanocoatings by Initiated Chemical Vapor Deposition (iCVD)

Prof. Karen K. Gleason*, Dr. Hilton G. Pryce Lewis**, Kelvin Chan*, Dr. Kenneth K.S. Lau*, Yu Mao*

*Massachusetts Institute of Technology, 77 Massachusetts Avenue, Cambridge, MA, kkg@mit.edu
**GVD Corporation, 19 Blackstone Street, Cambridge, MA 02139

ABSTRACT

Initiated chemical vapor deposition (iCVD) is a novel process capable of producing a range of polymeric and multifunctional nanocoatings. The process utilizes hot filaments to drive gas phase chemistry which enables the deposition of true linear polymers rather than the highly crosslinked organic networks often associated with plasma enhanced CVD. Importantly, the object to be coated remains at room temperature, which means that nanothin coatings can be prepared on materials ranging from plastics to metals. The process is also conformal, which means it provides uniform coverage on objects which have small, complex, 3D geometries.

Keywords: chemical vapor deposition, polymer, surface modification, polytetrafluoroethylene

1 INTRODUCTION

For iCVD of polymeric films (Fig. 1), the resistively heated filament serves to drive the decomposition of

- a precursor gas, forming a polymerizing monomeric species, and/or
- an initiator, forming a reactive radical promoting polymerization.

The resulting pyrolysis products adsorb onto a substrate which is generally maintained at room temperature by backside water cooling and react to form a film [1-6]. Since the growth of many polymeric films are subject to absorption limitations, cool substrates are essential to rapid film growth. The filament to substrate stand-off is typically between 1 and 4 cm.

The iCVD method is a nanocoating process in the sense that it can be used (1) to encapsulate objects containing nano-sized features, and (2) produce surfaces with thicknesses in the nanometer range. It is particularly valuable for its ability to create ultrathin layers of insoluble polymers. The combination of chemical flexibility and generic substrate requirements opens up new markets for polymer nanocoating technology.

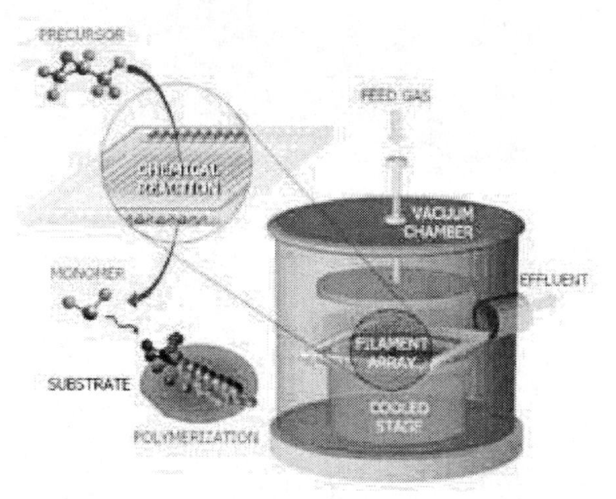

Figure 1. Process schematic for Initiated Chemical Vapor Deposition (iCVD).

2 POLY(TETRAFLUOROETHYLENE) (PTFE)

PTFE is one of the most useful and ubiquitous polymers. It is renowned for its excellent chemical and thermal resistance, biostability, low friction characteristics, and dielectric performance. However, PTFE's stability means that conventional coatings must be prepared from aqueous powder dispersion and cured at high temperatures (300°C+). This limits the nature of substrates which can be coated to those able to withstand curing, and limits the minimum feature sizes which can be used. In contrast, iCVD can be utilized to grow PTFE nanoncoatings directly on the surface of devices. Both porous and dense coatings can be prepared, and these coatings share all of the properties of conventional PTFE surfaces. The process is economical and capable of coating materials at high rates and efficiencies.

The iCVD of PTFE [7,8] has recently been used to create ultrathin (<100 nm) coatings which conformally coat carbon nanotubes (Fig. 2) [9]. In contrast, conventional coating methods result in PTFE coating thicknesses of > 10 μm. Hydrophobic nanocoatings can be deposited on nearly any object, as the substrate remains at ambient temperature during the PTFE iCVD process.

For the iCVD of PTFE, an array of stainless steel filaments, resistively heated to 500 °C, thermally decomposes hexafluoropropylene oxide (HFPO). The filament segments form a parallel array suspended above the deposition surface. Difluorocarbene (CF_2) forms and polymerizes into PTFE on a substrate that is held at room temperature using backside water cooling. The filament to substrate distance was 1.5 cm.

An initiator, perfluorobutane-1-sulfonyl fluoride, promotes the polymerization. The HFPO and the initiator have flow rates of 23 and 6 sccm, respectively, into a chamber held at 0.5 Torr. Fourier transform infrared (FTIR) spectroscopy of the iCVD PTFE coating on the nanotubes shows strong symmetric and asymmetric CF_2 stretches in the 1250-1150 cm^{-1} region, the same signature as for bulk PTFE [10].

Because of its extremely low surface energy (18 mN/m), smooth PTFE surfaces exhibit a high contact angle with water, 108°[11]. The advancing and receding contact angles of the PTFE iCVD treated carbon nanotube forest are even greater: 170° and 160°, respectively, resulting in nearly spherical water droplets on a macroscopic level when water is deposited on the surface, as shown in Figure 3. This superhydrophobicity is most likely a consequence of secondary roughness produced by the variation in heights between nanotubes [9].

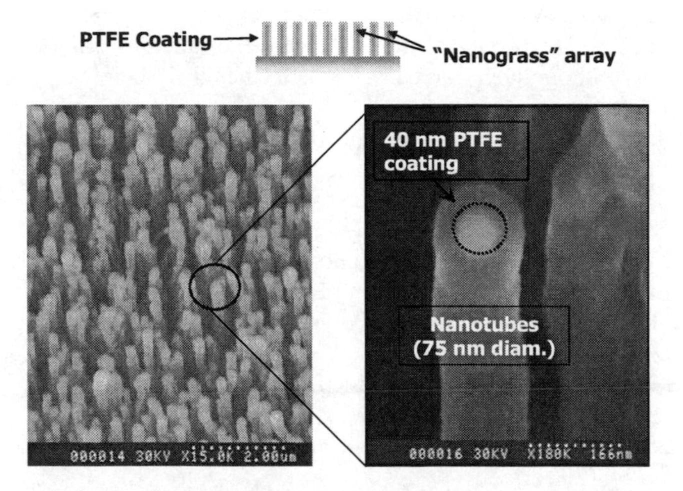

Figure 2. Conformal PTFE coating on an array of carbon nanotubes. Nanotubes have been individually "shrink-wrapped" to provide chemical resistance and hydrophobicity.

Figure. 3 A nearly spherical water droplet resting on a forest of iCVD coated carbon nanotubes.

3 POLY(GLYCIDYL METHACRYLATE) (PGMA)

PGMA contains a pendant epoxy group which can be converted into different kinds of functionalities through ring-opening reactions. In particular, the crosslinking reaction of the epoxy group in PGMA under electron-beam exposure creates the potential for a high-sensitivity negative tone electron-beam resist [12].

iCVD PGMA thin films is a dry processing alternative to the conventional spin casting of resists from solution, representing a potential reduction in volatile organic compounds (VOC) emissions. Using an initiator resulted a low energy vapor deposition process which selectively drives the polymerization of GMA while retaining both its irradiation-sensitive pendent epoxy groups and the linear, uncrosslinked, polymeric structure [4].

PGMA films were deposited on silicon wafers resting on a stage maintained at 25°C using a parallel array of Nichrome filaments (80% Ni/20% Cr) at temperatures between 180°C and 250°C. The power required was less than 3 W. The distance between the filament and the stage was 2.2 cm. Pressure in the vacuum chamber was maintained at 0.5 torr using a butterfly valve. The flow rate of glycidyl methacrylate (GMA) precursor was 2 sccm, while the flow rate of tert-butyl peroxide (TBP) was varied between 0 and 1.7 sccm. The presence of the TBP initiator was found to be essentially for rapid growth at low filament temperatures. Decomposition of the TBP initiator begins at temperatures as low as 150 °C. The formation of tert-butyl radicals initializes the polymerization of GMA, which greatly enhances the film growth rate.

Solution state ^{1}H NMR confirmed the retention of pendent epoxide groups and the linear polymeric structure in the iCVD PGMA films. The ability to completely dissolve the iCVD film in $CDCl_3$ provides strong evidence of the lack of crosslinking. The chemical shifts and the relative areas of these peaks are almost identical to conventionally synthesized PGMA chemical structure [13]. The intensity differences at around 2 ppm are due to differences in tacticity.

Additionally, the number-average molecular weight (M_n) of PGMA films can be systematically varied from 3,000 g/mol to over 33,000 g/mol by adjusting the filament temperature, the flow ratio of initiator to precursor, and the configuration of the reactor. The ability to control film molecular weight in-situ allows optimization of the tradeoff between the sensitivity improvement and the resolution loss due to the swelling problem in negative-tone resists [12]. The electron-beam sensitivity of iCVD PGMA film (27 $\mu C/cm^2$) was similar to that of PGMA (30 $\mu C/cm^2$) prepared from solution polymerization with approximately the same molecular weight (M_n~11,000 g/mol) [14]. As film molecular weight decreases, the electron-beam sensitivity decreases, but the resolution is greatly improved. The smallest feature obtained for PGMA with M_n 11,000 is

500 nm, while the PGMA film with M_n 4,700 g/mol resolved 80 nm features.

4 ADDITIONAL POLYMERS

Deposition of organosilicon iCVD films onto room temperature substrates from the precursors hexamethylcyclotrisiloxane (D3) and octamethylcyclotetrasiloxane (D4) occurs at high rates (>1 micron/min) and permits systematic control over the incorporation of cyclic and linear siloxane structures [15]. Filament temperatures ranged from 800 to 1200 °C.

iCVD has also successfully produced copolymer thin films consisting of fluorocarbon and organosilicon groups, where the presence of covalent bonds between the fluorocarbon and organosilicon moieties in the thin films has been confirmed by FTIR, X-ray photoelectron spectroscopy (XPS) and solid-state NMR [16]. Employing an initiator allowed deposition at relatively low filament temperatures (370 °C), and under these conditions, chemical characterization showed that polymerization occurs across the vinyl bonds of V_3D_3, a D3 analog with a vinyl group replacing a methyl group on each Si The resulting films consisted of polymer chains with carbon backbones and siloxane rings as pendant moieties [3]. Additionally, fluoroorganosilicon iCVD copolymer films can be modified to facilitate the surface attachment of peptides such as poly-L-lysine and the arginine-glycine-aspartic acid (RGD) tripeptide [17].

iCVD has also been demonstrated as a novel method for making thin sacrificial layers of polyoxymethylene (POM) [18]. Trioxane, a cyclic trimer of formaldehyde, decomposes cleanly into three molecules of formaldehyde with no side reactions and the kinetics of this reaction is well-known [19]. The decomposition of trioxane is driven by the hot filaments and polymerization onto the cool substrate can be enhanced by a variety of initiators.

4 ACKNOWLEDGEMENTS

We gratefully acknowledge the support of the NSF/SRC Engineering Research Center for Environmentally Benign Semiconductor Manufacturing, the National Institutes of Health under contract NO1-NS-9-2323, and the Cambridge-MIT Institute (CMI) Project of Carbon Nanotube Enabled Materials. This work made use of MRSEC Shared Facilities supported by the National Science Foundation under Award Number DMR-9400334, MIT's shared scanning-electron-beam-lithography facility in the Research Laboratory of Electronics, and the NMR facilities of Chemistry Department at MIT. We also thank Professor Chris K. Ober's group at Cornell University, Professor Gareth H. McKinley's group at MIT and Professor William I Milne's group at Cambridge University for their collaborative efforts.

REFERENCES

1. K.K.S. Lau, H.G. Pryce Lewis, S.J. Limb, M.C. Kwan, and K.K. Gleason, Thin Solid Films 395 (2001) 288.
2. H.G.P. Lewis, J.A. Caulfield, and K.K. Gleason, Langmuir 17 (2001) 7652.
3. S.K. Murthy, B.D. Olsen, and K.K. Gleason, Langmuir 18 (2002) 6424.
4. Y. Mao and K.K. Gleason, Langmuir 20 (2004) 2484.
5. S.J. Limb, K.K.S. Lau, D.J. Edell, E.F. Gleason, and K.K. Gleason, Plasmas Polym. 4 (1999) 21.
6. H.G. Pryce Lewis, G.L. Weibel, C.K. Ober, K.K. Gleason, Chem. Vap. Dep. 7 (2001) 195.
7. S J. Limb, C.B. Labelle, K.K. Gleason, D.J. Edell, E.F. Gleason, Appl. Phys. Lett. 68 (1996) 2810.
8. K.K.S. Lau, J.A. Caulfield, K.K. Gleason, Chem. Mater. 12 (2000) 3032.
9. K.K.S. Lau, J. Bico, K.B.K. Teo, M. Chhowalla, G.A.J. Amaratunga, W.I. Milne, G.H. McKinley, and K.K. Gleason, Nano Lett. 3 (2003)1701.
10. C.Y. Liang, S. Krimm, J. Chem. Phys. 25 (1956) 563.
11. H.W. Fox, W. A. Zisman, J. Colloid Sci. 5 (1950) 514.
12. L.F. Thompson, C.G. Willson, Introduction to Microlithography, 2nd ed. American Chemical Society, 1994.
13. M. Gromada, L. le Pichon, A. Mortreux, F. Leising, and J.F. Carpentier, J. Organomet. Chem. 44 (2003) 683.
14. Y. Mao, N. Felix, P. Nguyen, C. K. Ober, K.K. Gleason, submitted to J. Vac. Sci. Technol. B, 2004.
15. H.G.P. Lewis, T.B. Casserly, and K.K. Gleason, J. Electrochem. Soc. 148 (2001) F212.
16. S.K. Murthy, K.K. Gleason, Polym. Mater. Sci. Eng. 87 (2002) 151.
17. S.K. Murthy, B.D. Olsen, and K.K. Gleason, Langmuir 20 (2004) 4774.
18. L.S. Loo and K.K. Gleason, Electrochem. Solid State Lett. 4 (2001) G81.
19. H.K. Aldridge, X. Liu, M. C. Lin, and C.F. Melius, Int. J. Chem. Kin. 23 (1991) 947.

Coating of Glass Surfaces with Nanoparticles of different Materials Synthesized in Microemulsions

A. López-Trosell* and R. Schomäcker*

* Department of Chemistry, TU-Berlin,TC 8
Strasse des 17. Juni 124, 10623-Berlin, Germany
Lopez.Alejandra@chem.tu-berlin.de and Schomaecker@tu-berlin.de

ABSTRACT

In this paper, a new method is presented to coat glass substrates with nanoparticles of different materials such as manganese perovskite ($Sr_{0.5}Ca_{0.5}MnO_3$), zirconium oxide (ZrO_2), yttrium iron garnet ($Y_3Fe_5O_{12}$) and palladium (Pd). The particles were synthesized using w/o-microemulsions, which were also employed to coat the surfaces. The laboratory-scale method involved five basic steps (surface cleaning, synthesis of the nanoparticles, coating, drying and calcination). Finally, the coated substrates were characterized by X-ray diffraction (XRD) and scanning electron microscopy (SEM).

Keywords: coating, microemulsions, nanoparticles

1 INTRODUCTION

Nanostructured coating on solid substrates has a significant importance due to the possibility to synthesize materials with specific physical-chemical properties such as magnetic, catalytic, electronic, mechanical, etc. These properties are attractive for several industrial applications. However, they depend strongly on the size and morphology of the particles. Therefore, the control of the parameters of the particles provides a way to enhance the features of the coated surfaces.

The following section delineates the procedure for coating glass substrates with nanoparticles of different materials. The laboratory-scale method involved five basic steps (surface cleaning, synthesis of the nanoparticles, coating, drying and calcination).

Reverse micelles have been successfully used to synthesize nanoparticles with a good control of size and morphology. Microemulsions form spontaneously and present behaviors that change systematically with temperature. With these properties, microemulsions represent a suitable medium for the deposition of particles on cleaned surfaces to yield thin films. The advantage of this method over other processes, which involve solution deposition (e.g. sol-gel [1]), is the wide variety of compounds that are possible to synthesize employing microemulsions, as well as the possibility to change the composition of the microemulsions in order to favor the deposition. Additionally, the use of reverse micelles for coating the surfaces is inexpensive and does not require special equipment.

2 EXPERIMENTAL PART

The method to coat the surfaces with nanoparticles is a five-steps procedure; cleaning the glass surfaces, synthesis of the nanoparticles, coating, drying and calcination. The chemicals employed to carried out the experimental part were supplied by Fluka and were used without further purifications.

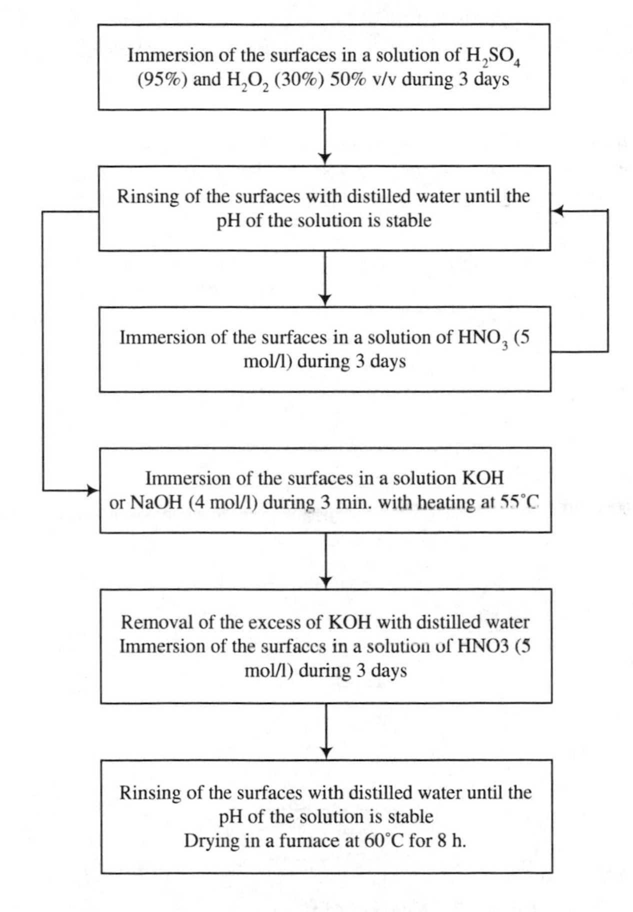

Figure 1: Diagram of the surface cleaning process.

2.1 Surfaces Cleaning

Glass surfaces were supplied by Mezel-Glaser. The surfaces were cleaned employing the process described in [2].

This process is based on cleaning the surface with different acid-basic solutions, followed by rinsing the surfaces with sufficient amount of distilled water untill the pH of the solution is stable. Afterwards, a dry process in a furnace for eight hours takes place. An Scheme of the cleaning process is giving in figure 1. Well cleaned surfaces exhibit contact angles of 0° between the surface and distilled water.

2.2 Particle Synthesis

Particles of palladium, zirconium oxide, yttrium iron garned and perovskite $Ca_{0.5}Sr_{0.5}MnO_3$ were synthesized in microemulsions according to synthesis proposed in [3], [4], [5] and [6], respectively.

The syntheses of the different materials were carried out in a semi-batch reactor at constant stirring rate, feed rate, and temperature. In general, the synthesis in reverse micelles is achieved by preparing two microemulsions with identical composition (W_o, α and γ) but with different aqueous phases. Microemulsion "A" contains the solutions with the salts while microemulsion "B" contains the appropriated precipiting agent. For example, if the reaction occurs via oxidation the aqueous phase is a solution with the oxidizing agent. These microemulsions are mixed in the reactor. A four pitched blade turbine impeller was used as agitator. The feed input was located near the agitator and above the liquid level. The stirring rate, feed rate and temperature were constant. These conditions were chosen according to the microemulsion feature for each synthesis and following the procedure described in [7].

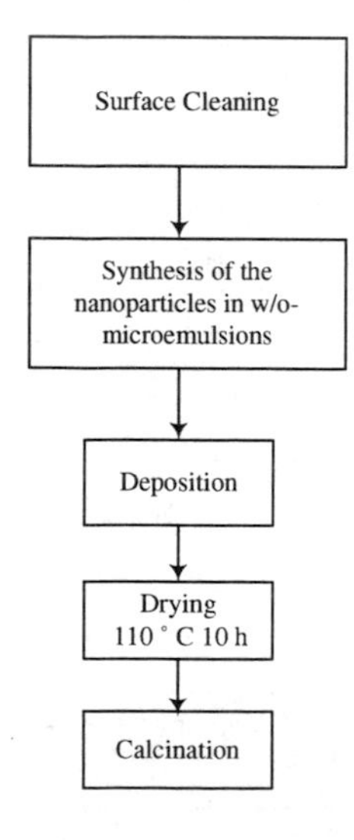

Figure 2: Diagram of the coating process.

2.3 Coating and Drying Process

The microemulsion containing the particles was directly dripped onto the cleaned glass surfaces. Two processes further followed. In the first one, surfaces were sloped at 45° and the microemulsion was spread from the top to the bottom of the surfaces. In the second process, the glass surfaces were coated with a defined volume of the microemulsion without sloping the surfaces. Afterwards, the surfaces were dried in a furnace at 110^oC for 10 hours. Finally, some surfaces coated in the last way were briefly immersed in a mixture of water-methanol $(3:1)$ to remove the excess of particles.

2.4 Calcination Process and Characterization of the Samples

Dried surfaces were calcinated to pyrolyze the surfactants as well as to obtain the desired crystall structure of the materials. The heat treatment did not exceed 600^oC, except for the yttrium iron garnet. YIG required a higher temperature for crystallization to the desired phase. The samples were characterized by XRD using an X-ray diffractometer Siemens D500 with Cu anode (radiation $K_{\alpha 1}$ of $\lambda = 0.154\ nm$) and by scanning electron microscopy (SEM) using a Hitachi S-520. Figure 2 resumes the steps of the employed method.

3 RESULTS AND DISCUSSIONS

Figure 3 shows the X-ray patterns for the different coated surface. The results were compared with the patterns in "EVA" program to identify the correct structure. The average crystall size D_{hkl} was calculated using the Debye-Scherrer's equation 1 with the full width at the half maximum β (in rad.) of the x-ray diffraction peak.

$$D_{hkl} = \frac{K\,\lambda}{\beta\ Cos\theta} \quad (1)$$

Where K is a constant (often taken as 1), λ represents the X-ray wavelength, β and θ are the full width at the half maximum (in rad.) and the Bragg angle, respectively.

The results calculated using Debye-Scherrer's equation are in good agreement with those obtained by SEM. Table 1 summarized the particle size obtained by XRD and by SEM as well as the thickness of the films for the surfaces coated without slope (second procedure).

Table 1: Particle diameters obtained by $^*XRD\ (D_{hkl})$, $^\ddagger SEM\ (D_p)$ and †thickness of the film (D_F) in nanometers.

Material	$^*D_{hkl}$	$^\ddagger D_p$	$^\dagger D_F$
Pd	30	32	112
ZrO_2	27	30	300
$(Y_3Fe_5O_{12})$	20	22	450
$(Sr_{0.5}Ca_{0.5}MnO_3)$	44	46	470

The SEM analysis shown that surfaces coated by spreading the microemulsion on the sloped surfaces (first procedure) were not homogeneous and the films presented several fissures. In contrast to this behavior, surfaces coated in the second procedure (without slope) were homogeneous films.

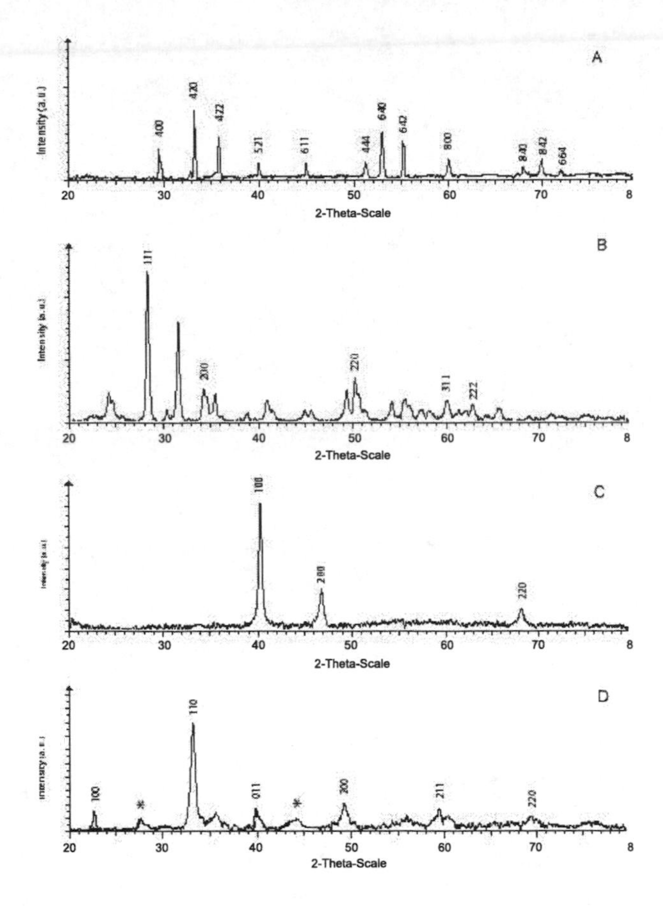

Figure 3: XRD patterns of (A) YIG, (B) $ZrO2$, (C) Pd and (D) Perovskite layers at glass surfaces.

Figure 4 shows the surface topography of the glass coated with yttrium iron garnet and the thickness of the film. Additionally, a relationship between the employed volume of the microemulsion and the final thickness of the film was found. Table 2 compares the employed microemulsion volume and the thickness of the film for the surfaces coated with perovskite.

Table 2: Volume of employed microemulsion and the thickness of the perovskite film (D_F).

Volume (ml)	D_F (nm)
0.2	140
0.4	246
0.6	470

Coated surfaces obtained by the second procedure (without slope) and treated with methanol-water yield homogeneous dispersions of nanoparticles. Figure 5 presents a glass surface coated with palladium nanoparticles and treated in this way.

The particle sizes were significantly smaller than those found normally when the microemulsion is removed and the precipitated particles calcined, probably because the particles

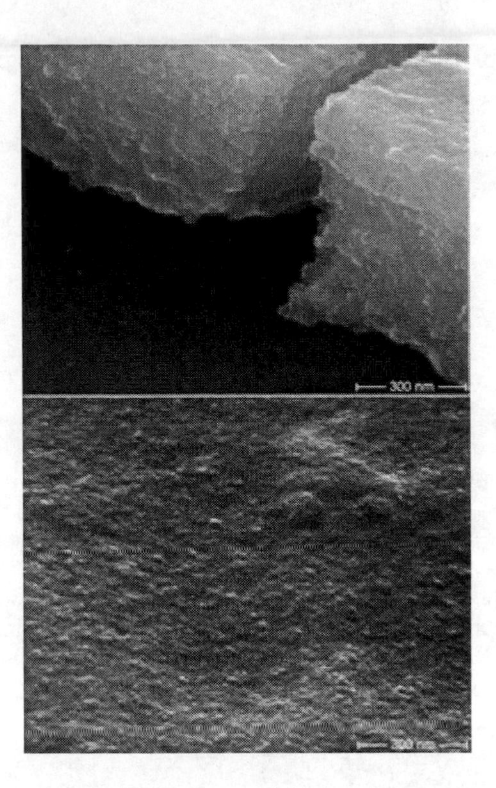

Figure 4: SEM of YIG nanoparticles attached to a glass surface.

are fixed to the substrate before removing the surfactant. Figure 6 shows the possible release of the particles on the substrate by using microemulsions for precipitation and coating.

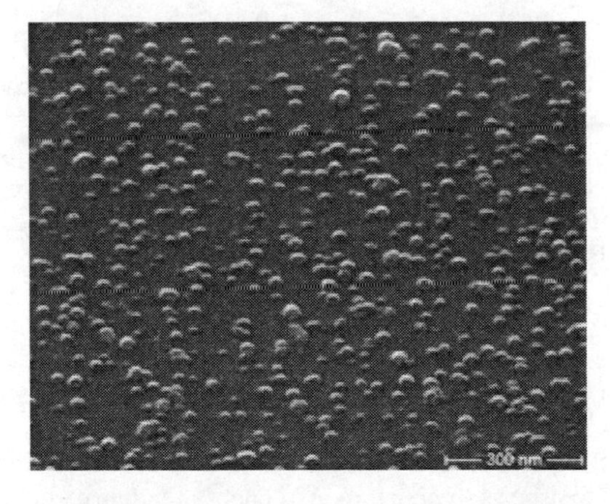

Figure 5: SEM of Pd nanoparticles attached to a glass surface.

The advantage of this method over other processes, which involve solution deposition (e.g. sol-gel [1]), is the wide variety of compounds that are possible to be synthesized, as well as the possibility to change the composition of the microemulsions in order to favor the deposition. This method is inexpensive and does not require special equipment.

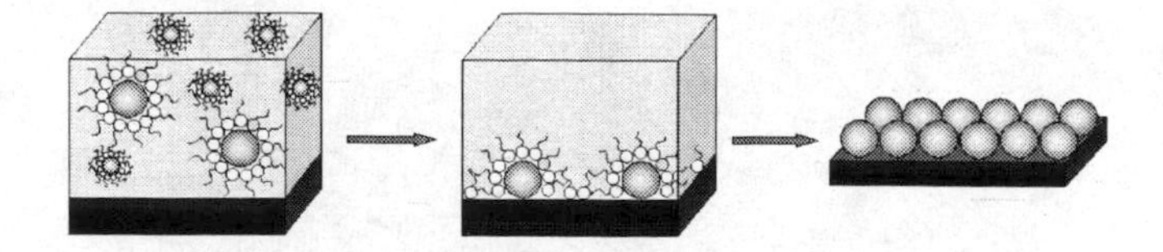

Figure 6: Diagram showing the possible release of the particles on the substrate.

4 CONCLUSIONS

A five-step method for coating of glass surfaces with nanoparticles of different materials was presented. Microemulsions containing the particles were a suitable medium for the deposition. The films were homogeneous and the thickness depended on the volume of employed microemulsion. The method presents some advantages. For example, the wide variety of compounds that can be synthesized in reverse micelles as well as the possibility to change the composition of the microemulsion in order to control the deposition. In addition to these advantage, the method is inexpensive and does not require special equipment.

REFERENCES

[1] Schwartz, R.W.: Chemical solution deposition perovskite thin films. Chem. Mater. **9** (1997) 2325–2340

[2] Lopez-Trosell, A., Araujo, Y.C.: Metodologia experimental para la determinacion de coeficientes de esparcido en sistemas liquido/liquido/solido. Technical Report INT-6226, PDVSA-Intevep (1999)

[3] Lade, M., Mays, H., Schmidt, J., Willumeit, R., Schomaecker, R.: On the nanoparticle synthesis in microemulsions: detailed characterization of an applied reaction mixture. Colloids and Surfaces A: Physicochem. Eng. Aspects **163** (2000) 3–15

[4] Guo, R., Qi, H., Chen, Y., Yang, Z.: Reverse microemulsion region and composition optimization of the aeo_9/alcohol/alcane/water system. Material Research Bulleting (2003)

[5] Vaqueiro, P., Lopez, M.A., Rivas, J.: Synthesis of yttrium iron garnet nanoparticles via coprecipitation in microemulsions. J. Mater. Chem. **7** (1997) 501–504

[6] Lopez-Trosell, A., Schomaecker, R.: Nsti nanotech (2005) Under Submission.

[7] Schmidt, J., Guesdon, C., Schomaecker, R.: Engineering aspects of preparation of nanocrystalline particles in microemulsion. J. Nanoparticle Reserch **1** (1999) 267–276

Ferroelectric properties of FIB-prepared single crystal $BaTiO_3$ nanocapacitors

A. Schilling[1?], M. M. Saad*, P. Baxter*, T. B. Adams*, R. M. Bowman*, J. M. Gregg*, F. D. Morisson** and J. F. Scott**

* Nanotec Northern Ireland, Queen's University, Belfast, UK
** Symetrix Centre for Ferroics, University of Cambridge, UK

ABSTRACT

Parallel-plate capacitors with dielectric thickness varying between ~500 nm and <100 nm have been fabricated from $BaTiO_3$ bulk single crystals using focused ion beam (FIB) milling. Electrical measurements suggest that the dielectric response is the same as found in bulk. This contrasts with most experimental observations made in the last 40 years, performed on conventionally deposited thin film capacitor heterostructures of similar dimensions. The implication is that all the functional degradation seen to date, on reducing dielectric thickness, is due to extrinsic sources, and not to either intrinsic size effects or fundamental physics associated with the ferroelectric-electrode boundary. This paper describes two specific approaches to FIB fabrication of the single crystal *thin film* capacitor structures, and presents details of their structural and functional characterization.

***Keywords*:** ferroelectric, nanocapacitor, $BaTiO_3$, focused ion beam, dielectric constant, loss tangent.

1 INTRODUCTION

Unfortunately for ferroelectric memories (FRAMs), the continual effort to reduce the size of electronic devices has major ramifications, as size reduction appears to be associated with severe degradation in dynamic functional properties in the ferroelectric material itself [1,2]. In particular, the dielectric constant suffers a dramatic collapse by orders of magnitude in moving from bulk to thin film geometries [3,4] and the coercive field increases considerably [5]. However, to date it is unclear to what extent property degradation is due to intrinsic issues [6,7] of reduced size (i.e., critical volumes, interface physics etc) and to what extent it is due to extrinsic issues [7-10] associated with fabrication methodology (i.e., substrate strain, vacancy defects, grain boundary etc).

In this study, an attempt is being made to examine the functional properties of extremely thin $BaTiO_3$ lamellae cut from bulk single crystals using FIB milling. The experiment has been undertaken in order to observe functional behaviour in the absence of many of the factors which are not well controlled in conventional fabrication of thin film capacitor heterostructures (or are unavoidable), but which have been suspected as contributing to functional degradation. For example: (i) microstructure-related regions of reduced permittivity [11]; (ii) strain clamping of ferroelectric to substrate [12]; (iii) gradient terms in the ferroelectric induced through defect concentration gradients [10] or interfacial strain relaxation [13].

The idea in examining single crystal *thin film* capacitors is to observe scaling effects on ferroelectric dynamic functional behaviour in the cleanest possible test-system, providing a paradigm for the best possible film properties obtainable which will be of use to those involved in the thin film growth community and to those seeking to use ferroelectric thin films in microelectronic applications.

2 EXPERIMENTAL

Commercial $BaTiO_3$ single crystals were used as a starting material. In order to fabricate $BaTiO_3$ single crystal nanocapacitors by an FEI200TEM FIB microscope, two methods were employed. Firstly, the $BaTiO_3$ single crystal lamellae were milled at the edges of the $BaTiO_3$ bulk single crystal. After a thermal anneal, gold was evaporated onto the crystal from two directions, such that both sides of the dielectric lamellae were fully coated. The coated crystal was again placed in the FIB, but with the polished crystal surface perpendicular to the Ga-beam. Rectangular electrodes (5 x 7 μm^2) on the top surface of the lamellae were isolated by milling through to the dielectric. Further milling defined gold strips that connected the lamellae electrodes to contact pads. A micromanipulator was then used to make electrical contact. Details about the fabrication of the edge-prepared lamellae and specific issues associated with Ga-removal were presented elsewhere [14]. Figure 1 shows an example of an edge-prepared $BaTiO_3$ capacitor. In the second fabrication process, the single crystal lamellae were prepared as previously, given a similar thermal anneal, but were then covered with Pt sputtered from an angle of 20° to the lamellar milled surfaces. Then the lamellae were cut free from the parent single crystal (again using FIB) and lifted out, using a micromanipulator and fine glass rod, onto a

[1] E-mail: a.schilling@qub.ac.uk

commercial Pt/Ti/SiO_2/Si wafer and coated with Au by thermal evaporation. The isolation of the capacitor top electrode was done by FIB. An example is shown in figure 2. Electrical measurements performed on such free-standing capacitors were performed by either micromanipulator contact (as before, but without separate contact pads), or through use of an atomic force microscope with conducting tips.

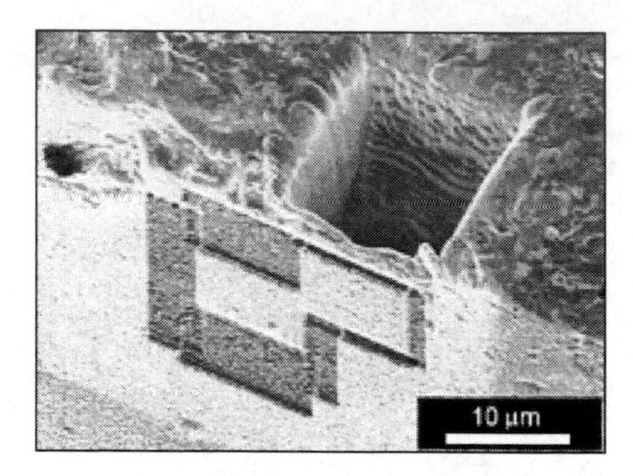

Figure 1: FIB image of an edge-prepared $BaTiO_3$ capacitor.

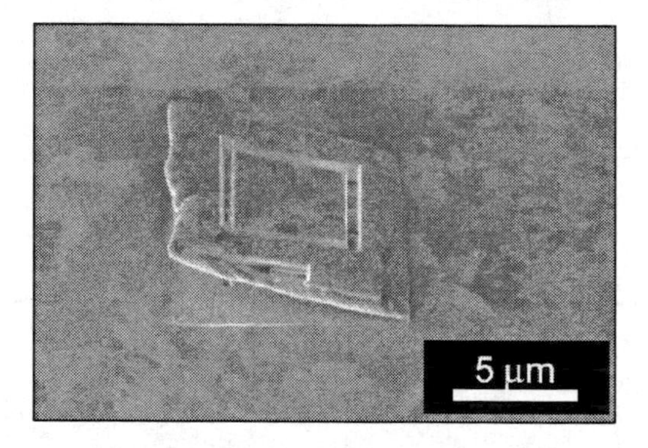

Figure 2: FIB image of a lifted out $BaTiO_3$ capacitor standing free on a substrate.

The capacitance and loss tangent were measured using an HP4284A LCR meter and an HP4192A impedance analyzer with applied voltages of 100 mV.

3 RESULTS AND DISCUSSION

Figure 3 illustrates the temperature dependence of the dielectric constant found in lamellar single crystal capacitors prepared using the first of the methodologies described above. As can be seen, the response is similar to bulk single crystals, rather than conventional thin films. An important feature is that no broadening or temperature shift of the dielectric peak or loss tangent is present. Peak dielectric constant is about 25,000. The dielectric constant goes through a very sharp peak at the Curie temperature, mirroring the response that is typically seen in bulk single crystals. The temperature of the Curie anomaly was found to be about 395 ± 5 K for all the different capacitor thicknesses ranging from ~450 nm down to ~75 nm.

Figure 4 illustrates the first order transition behaviour in the Curie-Weiss analysis with $T_0 < T_c$ measured on a thin edge-prepared $BaTiO_3$ nanocapacitor. More details on the electrical measurements performed on edge-prepared $BaTiO_3$ single crystal nanocapacitors and a summary of the current literature pertinent to thin film effects on the nature of the permittivity anomaly observed around the Curie temperature can be found in ref [15].

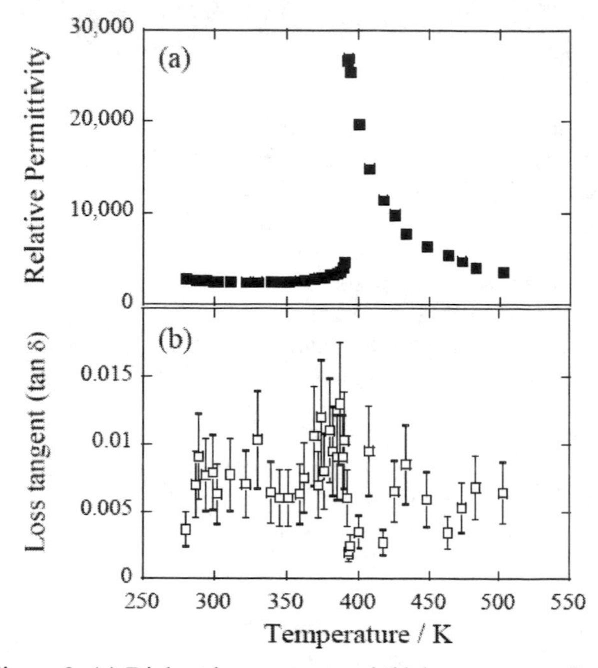

Figure 3: (a) Dielectric constant and (b) loss tangent of an edge-prepared $BaTiO_3$ capacitor with thickness of about 75 nm plotted at 10 kHz.

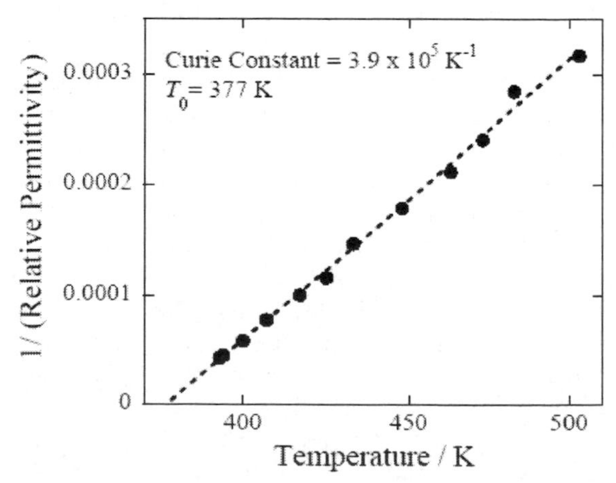

Figure 4: Curie-Weiss plot for the dielectric response.

In addition, I-V measurements were performed on these capacitors, an example of which is shown in figure 5. It is important to note that the data appears to separate into three regimes: a space charge limited current regime above

200 MV/m, an intermediate linear regime, and a non-ohmic regime with trap discontinuities at very low voltages. These features differ from thin films. Details about this work can be found in Ref. [16].

Overall, comparing the functional response in conventionally fabricated thin film heteroepitaxial capacitor systems with those made from single crystals presented in this paper, we observe that the dielectric broadening, apparent second order phase transition behaviour and collapse in dielectric constant present in conventional thin films are likely to be associated with extrinsic factors. They appear not be the result of size effects, nor due to fundamental factors associated with electrode-ferroelectric interface. Rather, they may be due to homogeneous strain associated to coupling to a substrate, grain boundaries and microstructural effects, or due to gradient terms related to chemical defects, oxygen vacancies or strain gradient flexoelectricity present in epitaxial thin films [10-13].

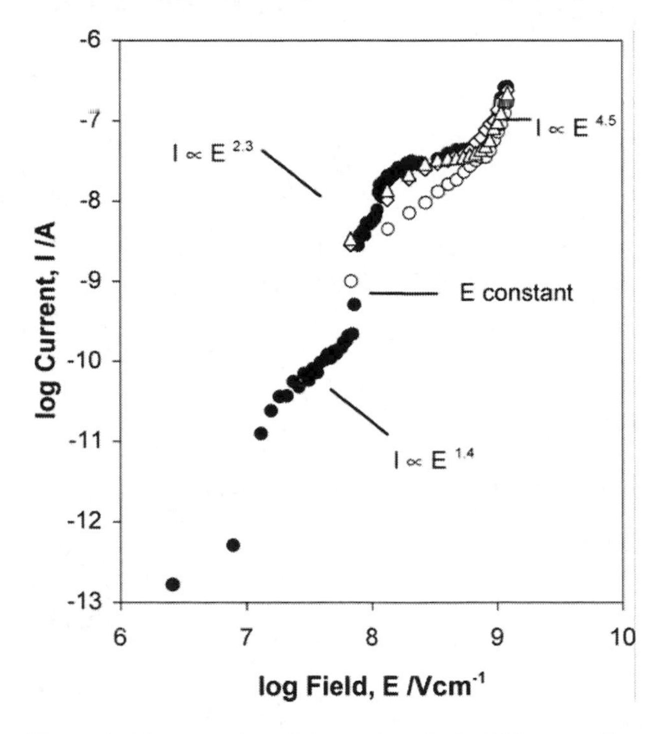

Figure 5: Three cycles I(V) graphs of a $BaTiO_3$ capacitor with a thickness of about 75 nm (Notation: circles – 1st cycle, triangles – 2nd cycle, diamante – 3rd cycle, and filled circles – lower fields).

Some concern was felt that data obtained from FIBed lamellae, that were in close proximity to bulk single crystal material, could include contributions from stray capacitance that had not been accounted for in our interpretation of data presented above. Hence the drive to investigate $BaTiO_3$ nanocapacitors that had been lifted away from the parent single crystal and placed on a different substrate.

Before detailed functional analysis of these stand-alone capacitors has been performed, ongoing efforts have concentrated on studying Ga-removal from the two sides milled $BaTiO_3$ lamellae before covering them with electrode and lifting them out. There is a notable difference in the amount of Ga ions impregnating the surface layers of the $BaTiO_3$ in milling not one, but two sides of the lamellae (in comparison with the first fabrication route where only one side is milled away). The considerable amount of Ga - induced wall damage and Ga implantation was overcome initially by annealing at 700°C for 1h. However, our recent work has shown that although Ga may be effectively expelled from the $BaTiO_3$ crystals, it appears to form oxide particles attached onto the $BaTiO_3$ lamellae. This had not been previously noted.

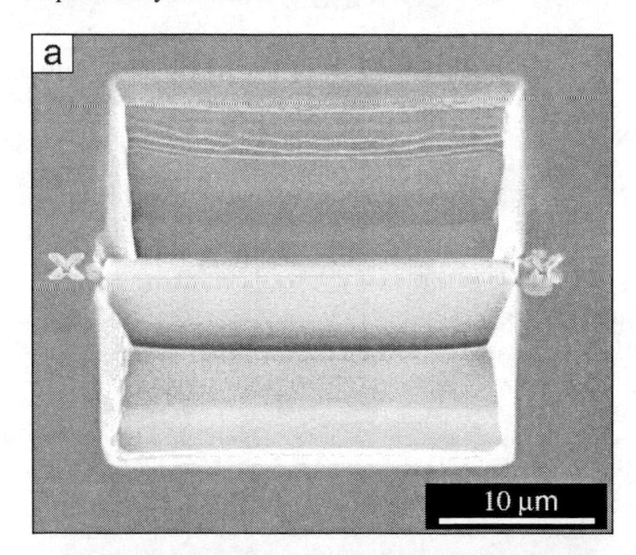

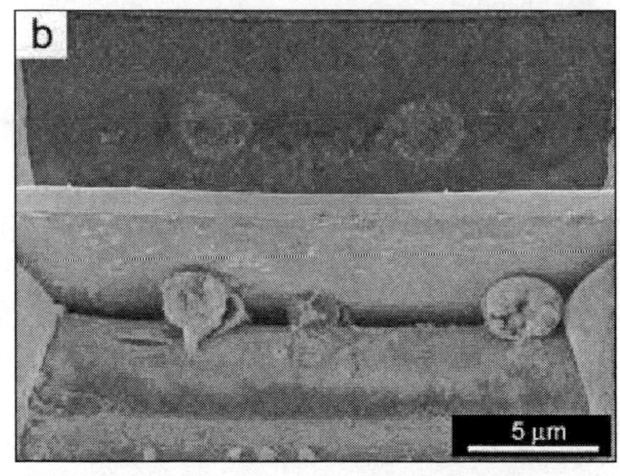

Figure 6: (a) FIB image of a $BaTiO_3$ lamella before annealing, and (b) SEM image of the same lamella after annealing at 700 °C for 1 h.

Figure 6 shows a FIB-milled $BaTiO_3$ lamella before and after annealing at 700 °C for 1h. The lamella is tilted 30 degrees for a better view. An important feature is the formation of large particles on the sides of lamella. Energy Dispersive X-ray (EDX) microanalysis identified the presence of Ga in large proportion up to 53 % in the large particles and about 10 % along the lamella wall. Moreover, higher temperature (up to 800 °C) or longer annealing time (up to 3 hours) did not show any differences in the $BaTiO_3$ lamellae morphology. In conclusion, annealing treatment performed at higher temperatures or for longer time did not result in successful Ga-removal. Presently, chemical routes are in progress in order to remove the Ga contamination, followed by annealing treatment to recrystallize the possible amorphous layers on the lamella walls. This chemical method includes using different acids (i.e., HCl, HNO_3, KOH, or ultimately HF). Details will be presented in a future article.

4 CONCLUSION

$BaTiO_3$ single crystal nanocapacitors with thickness ranging from 500-450 nm to 70-50 nm have been successfully prepared by FIB milling. Two routes have been considered. One was fabricating edge-prepared $BaTiO_3$ nanocapacitors, the other was lifting out $BaTiO_3$ nanocapacitors. Dielectric measurements and I(V) measurements performed on edge-prepared $BaTiO_3$ nanocapacitors with different thickness suggest that the reduction in dimensionality did not affect the dielectric properties, suggesting that so-called size effects on dynamic functional behaviour seen in conventionally deposited thin film systems are not the result of intrinsic size-related behaviour, nor due to unavoidable ferroelectric-electrode boundary physics. Checking of such a conclusion is underway through the investigation of genuinely *stand-alone* single crystal nanocapacitors size effects as results of intrinsic contributions.

ACKNOWLEDGEMENTS

This work has been supported by the Engineering and Physical Sciences Research Council. The authors are grateful to the staff of the International Research Centre for Experimental Physics for their technical assistance.

REFERENCES

[1] J. F. Scott, *Ferroelectric Memories*, Springer – Verlag, Berlin, 2000.

[2] A. I. Kingon, J. –P. Maria, and S. K. Streiffer, *Nature* **406**, 1032, 2000.

[3] C. B. Parker, J.-P. Maria, and A. I. Kingon, Appl. Phys. Lett. **81**, 340, 2002.

[4] C. A. Mead, Phys. Rev. Lett. **6**, 545, 1961.

[5] A. K. Tagantsev and I. A. Stolichnov, Appl. Phys. Lett. **74**, 1326, 1999.

[6] M. Ainzenman and J. Wehr, *Phys. Rev. Lett.* **62**, 2503, 1989.

[7] J. Cardy and J. L. Jacobsen, *Phys. Rev. Lett.* **79**, 4063, 1997.

[8] K. J. Choi et al, *Science* **306**, 1005, 2004

[9] N. A. Pertsev, A. G. Zembilgotov and A. K. Tagantsev, *Phys. Rev. Lett.* **80**, 1988, 1998.

[10] A. M. Bratkovsky and A. P. Levanyuk, *Preprint* cond-mat/0402100, 2004.

[11] C. Basceri, S. K. Streiffer, A. I. Kingon and R. Waser, *J. Appl. Phys.* **82**, 2497, 1997.

[12] V. Nagarajan, A. Royburd, A. Stanichesky, S. Prasetchoung, T. Zhao, L. Chen, J. Melngailis, O. Auciello and R. Ramesh, *Nature Materials* **2**, 43, 2003.

[13] S. Y. Cha, B.-T. Jang and H. C. Lee, Jpn. J. Appl. Phys. 38, L49, 1999.

[14] M. M. Saad, R. M. Bowman and J. M. Gregg, *Appl. Phys. Lett.* **84**, 1159, 2004.

[15] M.Saad, R. M. Bowman, J. M. Gregg, F. D. Morrison and J. F. Scott, *J. Phys. Cond. Mat.* **16**, L451, 2004.

[16] F. D. Morrison, P. Zukko, J. Jung, J. F. Scott, P. Baxter, M. M. Saad, J. M. Gregg, and R. M. Bowman, *in press in Appl. Phys. Lett.*, 2005.

Formation of Biotinylated Alkylthiolate Self-Assembled Monolayers on Gold

C. A. Canaria[*], J. R. Maloney[**], C. J. Yu[***], J. O. Smith[***], S. E. Fraser[***], and R. Lansford[***]

[*]Department of Chemistry, [**]Department of Physics, and [***]Department of Biology
California Institute of Technology
1200 E. California Blvd., Pasadena, CA 90016

ABSTRACT

The formation of mixed self-assembled monolayers (SAMs) from novel biotinylated alkylthiols (BAT) and diluent triethyleneglycol (TEG) alkylthiols on gold has been accomplished using aqueous solvents. We compare the effects of aqueous and ethanolic solvents on the binding performance of mixed SAMs and the integrity of monolayers formed. We use cyclic voltammetry (CV) and fluorescence microscopy to characterize the films. Comparison of fluorescence data between solution mixtures of alkylthiols in various concentrations of ethanol and water solvent indicates that functional SAMs may be formed at ethanolic solvent compositions as low as 2% in water. In addition, CV data indicate that SAMs adsorbed from aqueous solution form faster and have fewer defects than those formed using ethanol.

Keywords: SAM, monolayer, streptavidin, biotin, gold

1 INTRODUCTION

Understanding of the interactions of proteins and other biological molecules with surfaces is vital for the development of detection systems and assay platforms. These relationships are frequently quite complex, involving hydrophobic interactions, electrostatic interactions, van der Walls forces and covalent chemical bonds [1]. We can utilize these interactions in a surface capture device by modifying the surface substrate with thin films and monolayers. Employing self-assembled monolayers (SAMs), in particular, enables substrates to exhibit a variety of chemical properties and reactivites. In a SAM, alkyl thiols and disulfides form stable film structures on Au via a Au-S interaction. In literature, SAMs employing biotinylated thiolate molecules in gold and streptavidin sandwiches have been studied and utilized for the attachment of DNA and other biological molecules [2]. The development of such surfaces is a contribution towards bioassay technologies such as DNA chips, protein chips and small molecule biosensors.

We are building a new family of biofunctionalized nanoelectromechanical systems (BioNEMS) that will carry the analysis of biological processes to the stochastic limit. To carry out these experiments, we are building a device with a large array of NEMS cantilevers in a small liquid volume and have designed and synthesized alkylthiol molecules that are specifically attached in the form of self-assembled monolayers to Au pads located at the cantilever tips. The nature of our BioNEMS system requires more stringent conditions than previously demonstrated in literature. As biological assay platforms evolve into integrated microfluidics-packaged chips, the in situ formation of SAMs in devices needs to be addressed. Traditionally, SAMs are adsorbed from ethanol, DMSO, or other organic solvent systems. Many of these solvents will swell polydimethylsiloxane (PDMS) polymers or evaporate from microfluidic channels before well assembled monolayers can be formed [3]. Aqueous media, however, is compatible in PDMS and does not evaporate out of the polymer as quickly. Precedence in literature demonstrates the use of surfactants in water to aid in solvation of longer chain alkane thiols and their adsorption onto gold surfaces [4]. Additional studies by Yang, et al. indicate that SAMs are more stable when stored in water, as opposed to organic solvent [5]. Submersing SAMs on gold in water after a short thiolate adsorption time has also been shown to promote more crystalline packing of alkanethiolate chains. The hydrophobic interactions between the long alkyl chain and water are believed to promote organization of the monolayer on the surface.

2 MATERIALS AND METHODS

Silicon wafers were purchased from Wafer World. Chromium was purchased from R. D. Mathis Company and gold shots from Refining Systems, Inc. Phosphate buffered saline (PBS) was prepared as 0.139 M NaCl, 2.68 mM KCl, 8.1 mM Na2HPO4 and 1.1 mM K2HPO4 (Malinckrofdt) in Nanopure water. Potassium ferrocyanide and potassium ferricyanide were purchased from Aldrich. Absolute ethanol was purchased from Aaper Alcohol and Chemical Company (Kentucky.) Cy3 labeled streptavidin was purchased from Zymed, Inc.

2.1 Alkylthiol Reagents

Reagents BAT and TEG were synthesized in house with reagents purchased from Sigma and Aldrich (Figure 1) [6].

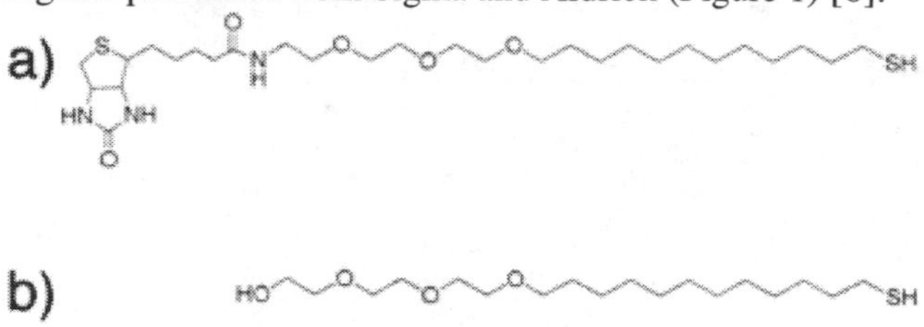

Figure 1. a) biotinylated tri(ethylene glycol) dodecylthiol (BAT). b) tri(ethylene glycol) dodecylthiol (TEG).

2.2 Preparation of Substrates and Monolayers

Silicon substrates were photo-patterned with ~7 nm Cr and ~100 nm Au by thermal evaporation. Au patterned substrates underwent plasma treatment at an oxygen flow rate of 0.8 L/minute in a UV ozone cleaner (SAMCO UV & Ozone Dry Stripper, Model UV-1) at 100° C for 30 minutes followed by a 2 minute nitrogen purge. Au substrates were then either submerged in thiol solution or underwent oxidative CV scans out to 1.2 V (vs Ag/AgCl, sat. KCl) in a solution 30 mM ferrocyanide/PBS of to remove any remaining surface contaminants. These electro-cleaned samples were then rinsed in copious amounts of water and ethanol, dried under a stream of Argon and immediately placed in thiol solution. Thiol solution compositions (BAT and TEG) were varied for a total disulfide concentration of 0.2 mM. Varied solvent compositions were attained by diluting an ethanolic stock solution of thiols in Nanopure water. After adsorption, samples were rinsed in ethanol and dried with Argon.

2.3 Characterization with Electrochemical Methods

The conductive properties of the Au substrate allow us to exploit electrochemical measurements in order to probe the properties of alkylthiolate assemblies on the surface. The structural integrity of the adsorbed monolayer may be characterized using CV methods [7]. During a CV scan, a tightly assembled monolayer will insulate the Au surface against electron transfer with a redox-active molecule in solution. Any defects in the thin film will be detected by CV and characterized by current flow. Electrochemistry measurements were carried out with a CH Instruments Model 600B potentiostat (Austin, TX.) A conventional three-electrode electrochemical cell was constructed with a platinum wire/mesh counter electrode and Ag/AgCl reference electrode in saturated KCl. The Au substrate served as the working electrode. CV measurements were taken in an electrolyte solution of 30 mM potassium ferrocyanide at a scan rate of 100 mV/s from –0.2V to 1.2V.

2.4 Protein-Binding Assay

Au substrates hosting BAT/TEG SAMs were submersed in a 200 nM solution of Cy3-labeled streptavidin in PBS for 30 minutes at room temperature. The samples were removed and washed five times with 1 ml of PBS, then stored in PBS for immediate analysis by fluorescence microscopy.

2.5 Fluorescence Microscopy

Images were acquired on an upright Zeiss Axioplan2 infinity corrected microscope (*Zeiss, Germany*) and acquired with a monochrome CCD Zeiss Axiocam HRm camera. Zeiss Plan-Neofluar objectives 10x/NA 0.3 and 20x/NA 0.5 were used in conjunction with a Chroma (*Rockingham, VT*) Cy3 filter set. A mercury arc lamp served as the excitation source. Images were acquired in 8-bit monochrome resolution and 1030 x 1300 pixel resolution.

3 RESULTS

3.1 Effect of Adsorption Solvent on SAM Formation Time

The adsorption of self-assembled monolayers is a versatile method for altering the chemical and physical properties of a surface. For Au substrates, SAMs are typically adsorbed from organic solutions. However, commonly used organics such as DMF and DMSO have several deleterious effects on our BioNEMS devices including swelling and solubilizing the PDMS microfluidics layers [3]. We have found that SAMs formed from BAT or TEG reagents may be adsorbed from ethanolic and aqueous solutions. Thus, we tested several different concentrations of aqueous ethanol solvents for thiol solubility and the ability of BAT and TEG to successfully form SAMs on Au substrates.

Alkylthiolates are believed to adsorb immediately onto the Au surface in a disordered fashion [8-11]. During the first few minutes, chemisorbtion of alkane thiolates results in 80-90% coverage of the substrate. In a longer subsequent adsorption stage, the alkane thiolates self-assemble into a more organized and insulating film [12]. Variables such as temperature, thiol concentration and solvent composition can affect the rate of monolayer formation [13].

CV techniques were used to evaluate surface coverage of Au substrates after adsorption of thiolates. Adsorption of BAT and TEG reagents after 12 hours at room temperature and in 0.2 mM thiol results in comparable insulation of Au substrates when using either ethanolic or aqueous solvents. In order to compare the effects of solvent composition on SAM quality, we terminate monolayer formation after 1-2 hours and observe insulating properties of the monolayer at this intermediate timepoint. Comparison of CV traces for SAMs formed in water and ethanol reveals that the samples treated in aqueous solution result have more diminished cathodic and anodic peaks, indicating greater monolayer coverage on the Au substrate. In particular, TEG SAMs form at a faster rate than mixed BAT/TEG SAMs or pure BAT SAMs. (Figure 2).

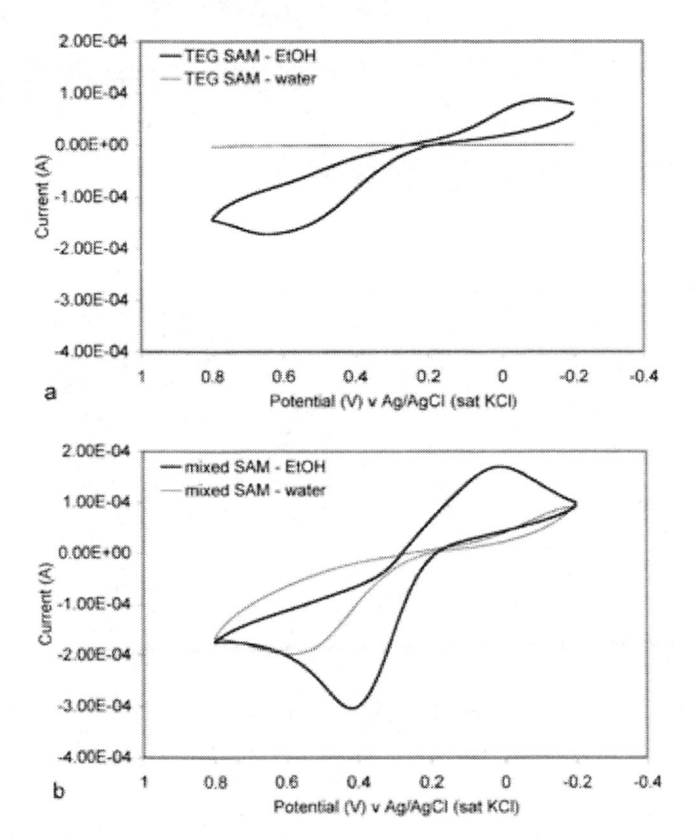

Figure 2. CV traces for Au electrodes covered with TEG and mixed biotin-TEG monolayers in Fe2+/3+ and PBS. (a) TEG monolayers adsorbed from ethanol and water solutions. (b) Mixed biotin and TEG monolayers adsorbed from ethanol and water solutions.

3.2 Binding of Fluorescent Proteins to SAMs

In order to assay the functionality of the BAT SAMs prepared, we use fluorescence microscopy to qualitatively compare the relative fluorescence intensities of Cy3-labled streptavidin adsorbed on SAM-covered Au substrates. Based on CV traces, monolayers adsorbed from water formed more complete monolayers than from ethanol. Au chip substrates were immersed in mixed thiol solutions ([BAT] + [TEG] = 0.1 mM) of varying solvent composition (ethanol, aqueous ethanol, water) for 12 hours and subsequently incubated with Cy3-labeled streptavidin. Based on the fluorescent protein-binding assay, the monolayers formed from mixed water/ethanol solutions yielded consistently lower fluorescence intensities than those in water or even ethanol solutions (Figure 3). SAMs prepared at BAT/TEG ratios between 0.0625 to 0.25 behaved with similar trends. The fluorescence levels from adsorbed Cy3-streptavidin on the monolayers was 50-70% lower for SAMs adsorbed from aqueous ethanol than from ethanol or water solvents. At BAT/TEG ratios of 0.5 or 1, the trend breaks down, presumably due to the more disordered nature of these SAMs [14]. SAMs made from BAT reagent alone have been shown to be disordered, with biotin units buried in the monolayer, resulting in a loss of specific protein binding and an increase in non-specific binding.

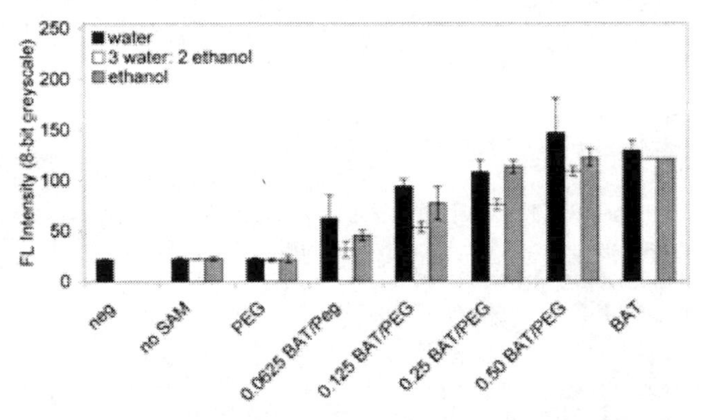

Figure 3. Fluorescence intensities plotted in bar graph displaying signal for each thiol adsorption condition.

However, these monolayers are still functional and are shown to specifically bind streptavidin. It is possible that the BAT and PEG reagents are most soluble in aqueous ethanol, making the solvation of the thiols more energetically favorable than adsorbtion onto the surface.

4 CONCLUSIONS

We demonstrate here that functional biotinylated monolayers may be adsorbed onto Au from aqueous solvents using methods that are compatible with a packaged BioNEMS device. These biotinylated monolayers are capable of binding fluorescently labeled streptavidin, as shown from fluorescence microscopy data. Electrochemical techniques confirm that monolayer coverage in different solvents is comparable, if not better in aqueous solvents.

REFERENCES

[1] Nuzzo, R. G.; Allara, D. L., Adsorption of bifunctional organic disulfides on gold surfaces. *J. Am. Chem. Soc.* **1983,** 105, (13), 4481-4483.
[2] Schumaker-Parry, J.; Zareie, M.; Aebersold, R.; Campbell, C., Microspotting Streptavidin and Double-Stranded DNA Arrays on Gold for High-Throughput Studies of Protein-DNA Interactions by Surface Plasmon Resonance Microscopy. *Anal. Chem.* **2004,** 76, 918-929.
[3] Lee, J. N.; Park, C.; Whitesides, G. M., Solvent Compatibility of Poly(dimethylsiloxane)-Based Microfluidic Devices. *Anal. Chem.* **2003,** 75, 6544-6554.
[4] Yan, D.; Jordan, J.; Burapatana, V.; Jennings, G., Formation of n-Alkanethiolate Self-Assembled Monolayers onto Gold in Aqueous Micellar Solutions of n-Alkyltrimethylammonium Bromides. *Langmuir* **2003,** 19, 3357-3364.
[5] Yang, G.; Amro, N.; Starkewolfe, Z.; Lui, G.-Y., Molecular-Level Approach to Inhibit Degradations of Alkanethiol Self-Assembled Monolayers. *Langmuir* **2004,** 20, 3995-4003.
[6] Canaria, C.; Smith, J.; Yu, C.; Fraser, S.; Lansford, R., New syntheses for 11-(mercaptoundecyl)triethylene glycol and mercaptododecyltriethyleneoxy biotin amide. *Tetrahedron Lett.* **submitted 2005**.
[7] Porter, M. D.; Bright, T. B.; Allara, D. L.; Chidsey, C. E. D., Spontaneously organized molecular assemblies. 4. Structural characterization of n-alkyl thiol monolayers on gold by optical ellipsometry, infrared spectroscopy, and electrochemistry. *J. Am. Chem. Soc.* **1987,** 109, 3559-3568.
[8] Dannenberger, O.; Buck, M.; Grunze, M., *J. Phys. Chem. B* **1999,** 103, 2202.
[9] Peterlinz, K. A.; Georgiadis, R., In Situ Kinetics of Self-Assembly by Surface Plasmon Resonance Spectroscopy. *Langmuir* **1996,** 12, 4731-4740.
[10] Bain, C. T., EB; Tao, Y; Evall, J; Whitesides, GM; Nuzzo, RG, Formation of monolayer films by the spontaneous assembly of organic thiols from solution onto gold. *J. Am. Chem. Soc.* **1989,** 111, (1), 321-335.
[11] DeBono, R. F.; Loucks, G. D.; Manna, D. D.; Krull, U. J., *Can. J. Chem.* **1996,** 74, 677.
[12] Hahner, G.; Woll, C.; Buck, M.; Grunze, M., *Langmuir* **1993,** 9, 1955.
[13] Bain, C. D.; Evall, J.; Whitesides, G. M., Formation of monolayers by the coadsorption of thiols on gold: variation in the head group, tail group, and solvent. *J. Am. Chem. Soc.* **1989,** 111, 7155-7164.
[14] Nelson, K. E.; Gamble, L.; Jung, L. S.; Boeckl, M. S.; Naeemi, E.; Golledge, S. L.; Sasaki, T.; Castner, D. G.; Campbell, C. T.; Stayton, P. S., Surface characterization of mixed self-assembled monolayers designed for streptavidin immobilization. *Langmuir* **2001,** 17, 2807-2816.

Functional nano-structured surfaces for protein based sensors.

A.Valsesia, , T. Meziani, P.Colpo, M.Lejeune, G. Ceccone, F.Rossi

European Commission Joint research Centre,
Institute for Health and Consumer Protection,
21020 Ispra (Varese), Italy

ABSTRACT

The reduction of the typical length scale in the creation of patterned surfaces is of high interest in the field of biosensor and bio-interacting materials. During the last few years several researchers explored the influence of the nano-structured materials on the protein adsorption or on the cell adhesion assays. The technological difficulty consists in the creation of nano-structures with controlled physico-chemical and geometrical properties, by exploiting the current nano-fabrication techniques, which are potentially compatible with industrial scaling. Several techniques of generation of nanostructured surfaces with chemical contrast for biosensor applications have been developed. In this work is described the fabrication process and characterization of Poly Acrylic Acid nano-domes and gold nano-wells using colloidal lithography

Keywords: colloidal lithography, surface functionalization

1 INTRODUCTION

Patterning of surface with active and non-active spots at sub-micron level is one of the main issues for the development of protein and cell based sensors for drug screening application. The vision is that nano-patterned surfaces with high chemical contrast may allow the triggering of specific interactions exclusively and therefore improve drastically the signal to noise ratio of the bio-analytical devices. Many studies have been undertaken to develop reliable method of chemical nanopatterning and to study the effect of the nano-structured materials on the protein adsorption or on the cell adhesion. Several approaches have been used successfully but many issues still to be addressed. On one hand, the excellent results obtained by using a bottom-up approach (molecular assembly, auto-nano-fabrication) are now far to be scaled to high-throughput systems. On the other hand, top-down approach is often too expensive and time consuming (e.g. electron beam lithography) otherwise is giving very promising results.
In this work, we present a novel method combining well-established techniques for material processing with low-cost and fast fabrication steps such as plasma deposition and etching techniques, using Nanosphere masking of polymeric materials. Several type of nanostructures (ranging from 70 to 250 nm) such as nano-domes and nano-wells consisting of materials with chemical and/or biological functionalities suitable for biosensor applications are produced in our laboratory. In this work some examples of this research activity will be illustrated. In particular the fabrication process and characterization of Poly Acrylic Acid nanodomes and of gold nano-wells will be described and discussed. The selective biological response the nano-patterned surface is demonstrated with protein assays i.e. BSA (Bovine Serum Albumin) is selectively bound to the functionalized nanostructures, whereas no protein adhesion is detected in the surrounding anti fouling matrix.

2 MATERIALS AND METHODS

PAA films have been deposited by PE-CVD in a capacitively coupled plasma reactor, previously described in ref. [1]. Gold film was deposited by classical magnetron sputtering. SiOx films were deposited by PECVD from HMDSO precursor (Sigma Aldrich) and oxygen etching was carried out in high density plasma source previously described in ref. [2].
PS (polystyrene) colloidal particles monodispersed in salt solution (Aldrich, Average Diameter: 500 ± 50 nm Concentration: 2% of solid content) were deposited onto the polymeric substrates by spin coating. A small drop (1 μl) of nano-particles solution (thermalized at the room temperature) was casted on the substrate with the spin-coater off and then accelerated (Average acceleration was varied between 1000 and 2000 $rp(m)^2$) to the final velocity varying between 500 and 5000 rpm). The spin-coater was stopped after the solvent evaporation.
BSA, as received from the supplier (Aldrich) was dissolved in PBS buffer solution (Fluka, pH = 7.4) with a concentration of 40 μg/ml. The samples were incubated in BSA solution for 30 min, washed in milli-Q water and then dried under a pure N_2 gas flow.

3. RESULTS AND DISCUSSION

3.1 Formation of colloidal mask

Nanosphere lithography is a reliable method to produce nano-topography over large area surfaces [3,4]. It uses the ability of the nano-sized particles to organize themselves on a surface to form some nano-masks. After plasma etching or deposition operation, nano-features can be transferred on the substrate. The nano-spheres adsorption on the surface

can be done by spin coating on surface treated by surfactant [3] or electrostatic adsorption using the electric charge contrast between particles and the substrate [4]. In both methods, the particles adsorption is controlled by the physico- chemical properties of the surfaces.
These two methods present some drawbacks for our application since both use chemical products, which can modify the chemical properties of the functionalized surface. In this paper, we use the PAA films hydrophilic character for the deposition of the nanospheres on the surface to avoid the use of a surfactant.

3.2 Polymeric nano-domes

The aim of this paragraph is to describe a method to create functional nano-domes over an anti-fouling matrix. The process is described in figure 1. The colloidal mask was created on the PAA layer deposited on a PEG layer. The oxygen plasma etches the PS and the unmasked PAA between the particles. An accurate control of the plasma etching time was necessary to avoid the over etching of the PAA. Indeed, the PAA over etching is harmful since the oxygen plasma strongly reduces the concentration of the functional PAA group on the surface. The etching time to avoid acrylic acid etching was 90 seconds. Some residual PS beads remained on the top of the sculptured PAA nano-domes. These residual beads are rinsed away by an ultrasonic bath in ultra-pure water. A SEM picture of the resulting nano-structured surface is shown in figure 2.

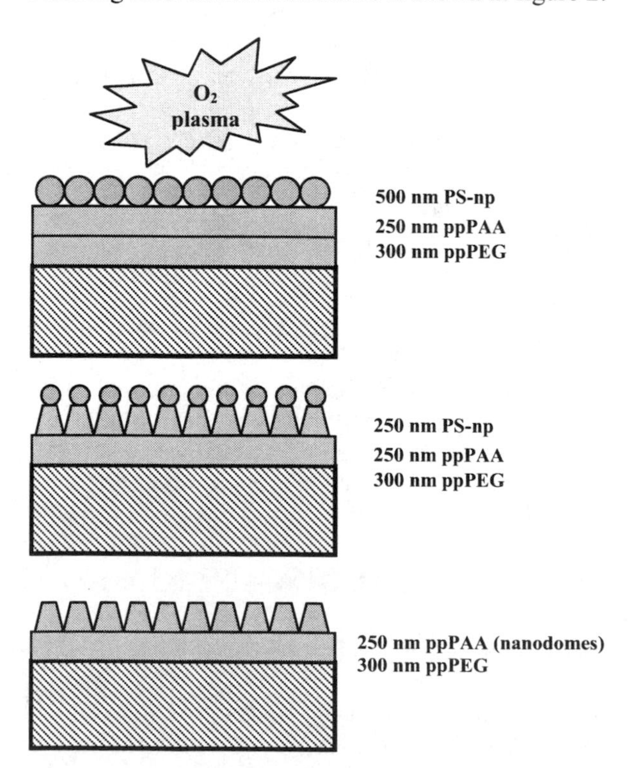

Figure 1 : Method of Polymeric nano-domes creation

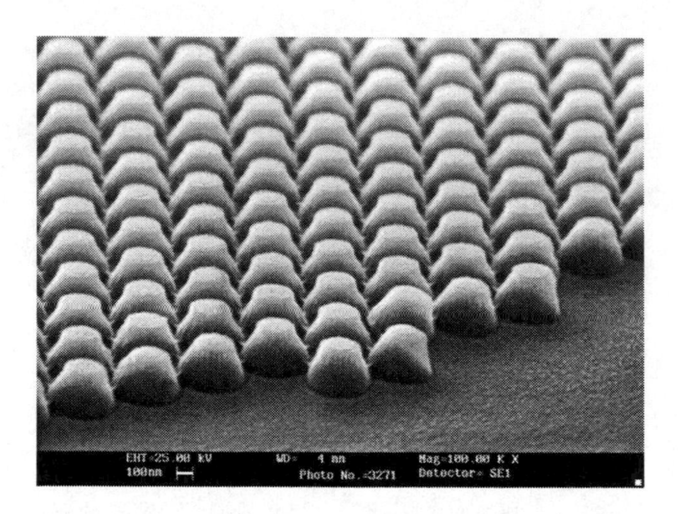

Figure 2: PAA nanodomes over antifouling matrix

The resulting surface presents 2-D crystalline PAA nano-domes (diameter: 250 nm) on a anti fouling surface..

3.3 Metallic nano-wells

In this case the aim is to create a contrast between a metal (e.g. Gold) and a silica surface in order to induce the selective absorption of Self Assembled Monolayers (SAM) with contrasted functionalities.
The process is described in figure 3. A thin (30 nm) PAA buffer layer is deposited on the gold surface in order to induce the formation of the PS nanomasks (Figure 3a). The PS beads are then etched by oxygen plasma as well as the unmasked PAA (Figure 3b). The plasma process is stopped before the complete removal of the particles. By this method it is also possible to control the well size and average distance. After the mask modification by oxygen plasma, the SiOx layer is deposited by PECVD (Figure 3c). The residual PS beads and PAA are then removed by an oxygen plasma cleaning process which does not affect the SiOx and gold surfaces (Figure 3d). The result (figure 4) is a nano-patterned surface with gold nano-wells over a SiOx matrix. The dimensions of the patterned can be modulated by changing the PS size and the plasma process parameters (etching and deposition times). These gold nano-wells are promising candidate for nano-electrodes to be used in electrochemical based sensors.

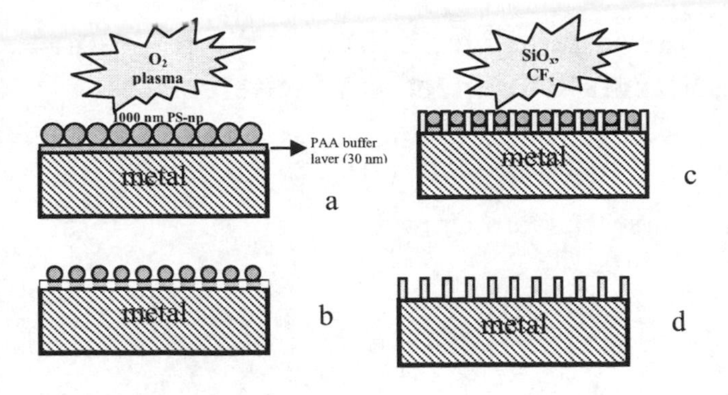

Figure 3: Method of nano-well creation.

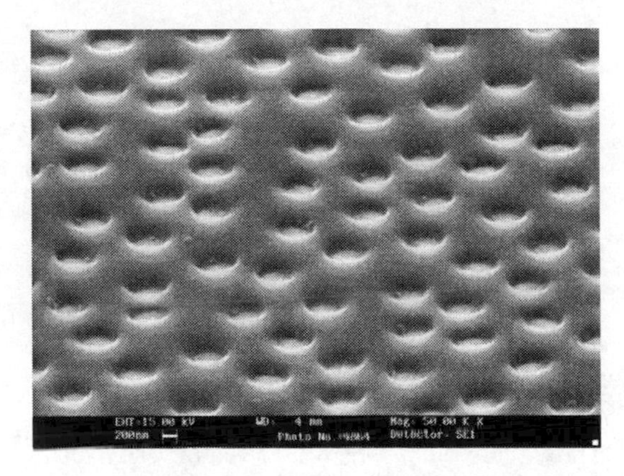

Figure 4 : Gold nano- well on SiOx matrix (diameter 250 nm)

3.4 Protein interaction assays

The first assays were done with proteins (BSA). Figure 5 shows that spherical BSA clusters are preferentially bounded to the functional plateau... The results outlined in this work are very promising for the biochip application; a great opportunity for these applications is the possibility to control the surface distribution of the nano-structures in a macroscopic area of the devices

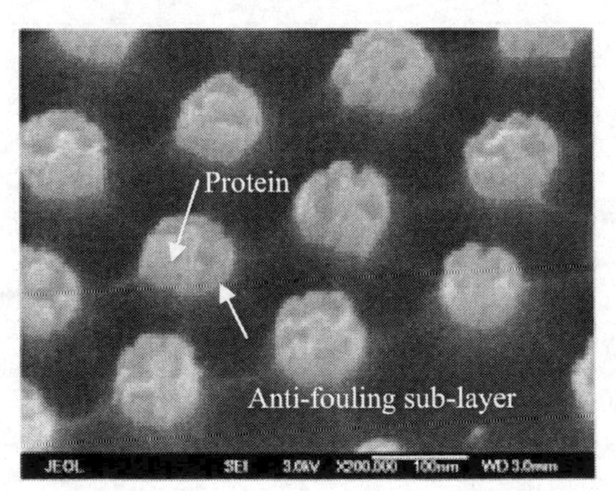

Figure 5: BSA protein selectively bounded on domes top surface.

REFERENCES

[1] Rossini N.; Colpo P.; Ceccone G.; Kandt K.D.; Rossi F., *Mater. Sci. and Eng. C* **2003,** *23*, 353-358

[2] Meziani T. ; Colpo P. ; Rossi F., *Plasma Sources Sci. Technol.* **2001,** *10*, 276–283

[3]. Hulteen J. C. ; Van Duyne R P., *J. Vac. Sci. Technol. A* **1995,** *13(3)*, 1553-1558

[4] Hanarp P.; Sutherland D. S.; Gold J.; Kasemo B., *Colloids and Surfaces A: Physicochem. Eng. Aspects* **2003,** *214,* 23-36

Low-Temperature Deposition of High-Quality, Nanometer-Thick Silicon Nitride Film in Electron Cyclotron Resonance (ECR) Plasma-Enhanced CVD System

Ling Xie, Jiangdong Deng, Steve Shepard, John Tsakirgis, and Erli Chen

Center for Imaging and Mesoscale Structures, Harvard University
9 Oxford St, Cambridge, MA 02138
Tel: (617)496-9069, Fax: (617)496-4654, e-mail: lxie@cims.harvard.edu

ABSTRACT

Ultra-thin silicon nitride films were deposited in an electron cyclotron resonance (ECR) plasma-enhanced chemical vapor deposition (PECVD) system. High quality films with low leakage current, high dielectric strength, and nanometer-scale thickness were achieved at deposition temperatures below 150°C. The study showed that, among all deposition parameters, ECR power played a dominant role in determining film quality.

Keywords: nanometer dielectric films, electron cyclotron resonance, low temperature PECVD, silicon nitride films

INTRODUCTION

One of the challenges facing nanofabrication is to deposit dielectric films with thicknesses compared to the critical dimension of devices under fabrication, typically in a range of a few nanometers, for purposes such as gate dielectrics, surface passivation layers, diffusion barriers, charge isolation gaps, etc. In addition, many applications use a variety of substrates that require deposition temperatures below 200ºC. Although low-temperature physical vapor deposition, such as plasma sputtering, has been studied [1, 2], plasma enhanced chemical vapor deposition (PECVD) is more attractive due to its ability of achieving high quality films [3-5]. However, to our knowledge, these studies were carried out at a temperature range of 300 – 400ºC.

On the other hand, it has been demonstrated that electron cyclotron resonance (ECR) can produce high-density plasmas with low ion energy at low chamber pressure, allowing the deposition of high-quality films at relatively low temperatures [6]. In this work, we demonstrate the feasibility of using ECR-PECVD method to deposit ultra-thin (nanometer scale) silicon nitride films at temperatures below 150ºC.

EXPERIMENT

Silicon nitride films were deposited in an ECR-PECVD system equipped with a helium-cooled chuck whose temperature was controlled between -15 to 80°C. Substrate surface temperature was measured using OMEGA Irreversible Temperature Indicators, which are placed on the top surface of a sample. The system is integrated with a loadlock that allows fast system pumping-down and low base pressure (~ 10^{-7} Torr).

3% SiH_4 and N_2 were used as the active gases in this study. The typical deposition conditions are listed in Table 1. SiH_4 flow rate, microwave and rf power, and chamber pressure were eventually optimized to improve film properties, such as film dielectric breakdown voltage. The temperatures on substrate surfaces were controlled below 150°C throughout the entire study.

Table 1 Typical deposition conditions

Parameter	Value
Microwave power (W)	265
RF bias (W)	0
Pressure (mTorr)	10
3% SiH4 flow rate (sccm)	55
N2 flow rate (sccm)	5.8
Ar flow rate (sccm)	20
He backside cooling pressure (Torr)	10

Si was chosen for this study simply because it is the most popular substrate used today. In particular, n-type, low-resistivity substrates were used. The low resistance substrate also served as one of the electrodes used for leakage current and breakdown voltage measurements. However, there is a drawback in using Si substrate, caused by its native oxide. Since the thickness of the films under study is comparable to that of the native oxide, its impact on film properties and, most importantly, the accuracy of estimating the film dielectric strength have to be addressed carefully.

WVASE32 Variable-Angle Spectroscopic Ellipsometer was used to measure film thickness. X-ray photoelectron spectroscopy (XPS) is an ideal tool for analyzing chemical bonding structures in nanometer-scale thin films. In this study, Surface Science SSX-100 XPS system with Al Kα x-ray source was used.

After deposition, 0.7μm-thick aluminum pads with different areas were fabricated on the top of the just-deposited films. The Al metal film, silicon nitride film, and low-resistive Si substrate form Metal-Insulator-Si (MIS) capacitors. Leakage current density and breakdown voltage were measured using these capacitors with a Keithley 2400 source meter.

RESULTS AND DISCUSSION

In order to determine the impact of the native oxide, two types of substrates were prepared. One was directly from as-purchased Si wafers, but cleaned in acetone then methanol solutions. The other type was additionally dipped in buffed HF solution for 20 seconds before immediately loading into the deposition chamber.

Fig. 1 shows the Si 2p XPS spectrum of a silicon nitride film deposited on the acetone/methanol cleaned substrate with deposition conditions shown in Table 1. A strong chemically shifted Si 2p peak at the binding energy of 102.0 eV, resulting from silicon nitridation, was observed. Since the thickness of the silicon nitride film is thinner than the escape depth of photoelectrons from the native oxide and Si substrate, a Si-O peak at 103.0 eV contributed by the native oxide and a Si-Si peak at 98.9 eV contributed by the Si substrate were also observed. The N 1s spectrum of the sample is shown in Fig. 2. A small amount of N-H bonds exist in the film, as indicated by the peak at 400.7 eV, caused by H^+ from the SiH_4 gas.

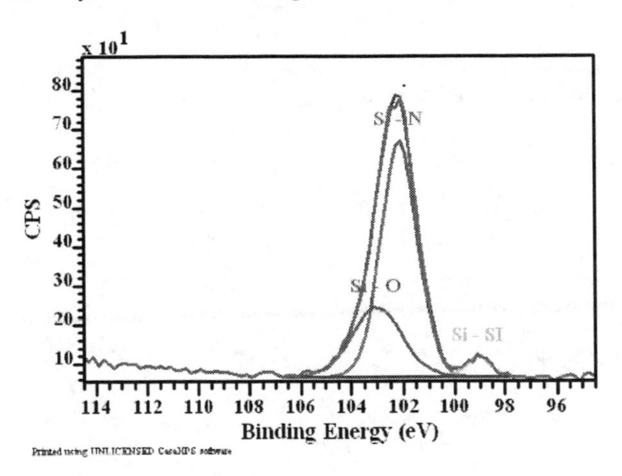

Fig. 1 Si 2p XPS spectrum from a sample deposited on Si substrate cleaned with acetone/methanol

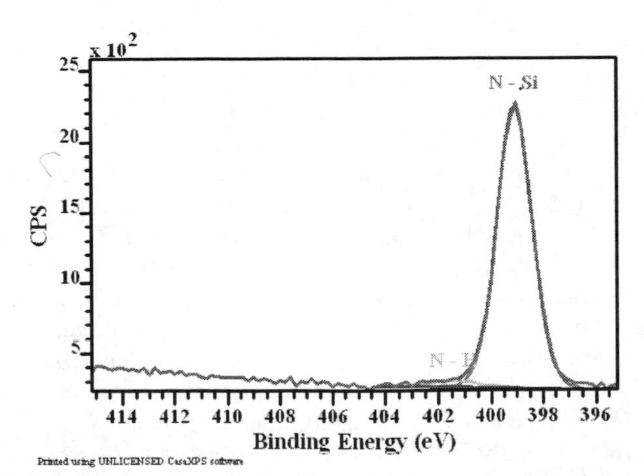

Fig. 2 N 1s XPS spectrum of the silicon nitride film shown in Fig. 1

The Si 2p XPS spectrum of the nitride film deposited on a Si substrate with HF dip is shown in Fig. 3. The deposition conditions are the same as the one shown in Fig 1. A Si-O peak still exists but its intensity is much smaller than the one in Fig. 1, indicating thin oxide still exists on the Si substrate. It is well known that, in the atmosphere, native oxide grows immediately on a Si substrate after dipped in HF. It is a challenge to accurately measure the thickness of such a thin layer using Ellipsometry. However, its thickness can be estimated using the core level ratio of the Si-O peaks shown in Figs. 1 and 3. Later, we will show that the native oxide thickness on a substrate without HF dip is 3.0 – 3.5 nm. Using this number, we estimated the native oxide thickness on a Si substrate with HF dip is between 0.6 to 0.7 nm.

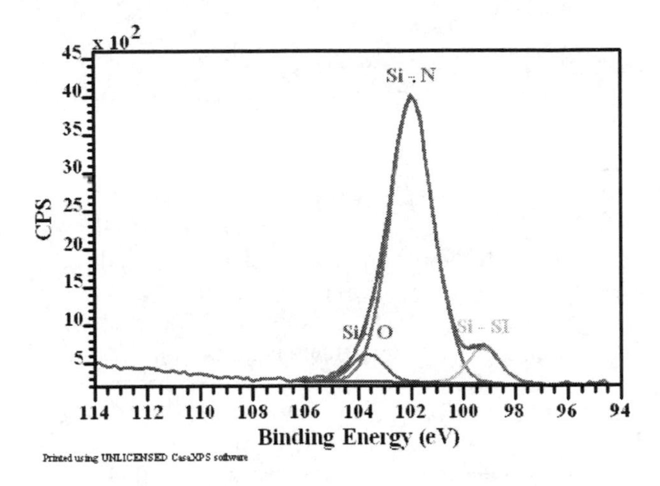

Fig. 3 Si 2p XPS spectrum of silicon nitride film on Si substrate cleaned with HF dip

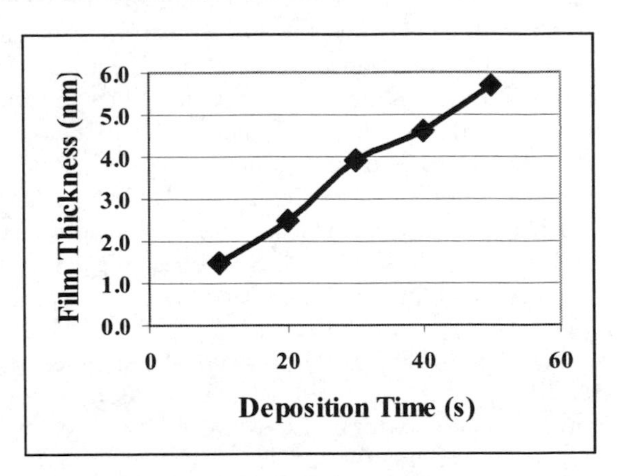

Fig. 4 Silicon nitride film thickness as a function of deposition time determined using ellipsometry

Film thicknesses were controlled by deposition time. The spectroscopic ellipsometer was then used to determine the thicknesses of both silicon nitride films and the native oxide as well. The dependence of silicon nitride thickness on deposition time is shown in Fig. 4. Film thickness increases linearly with deposition time at a deposition rate of 0.11 nm/s. On the other hand, as shown in Fig. 5, the native oxide on substrates without HF dip are 3.0 – 3.5 nm thick, independent from the deposition time. It is understandable since no oxygen existed in the deposition chamber during the deposition.

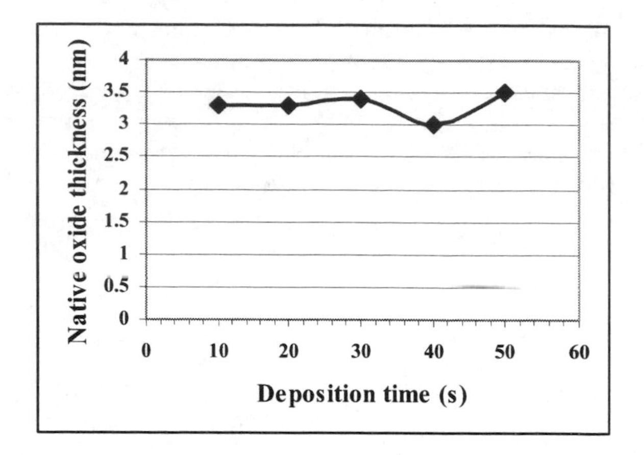

Fig. 5 Native oxide thickness vs. nitride deposition time

Fig. 6 shows the curves of current density vs. voltage (J-V curves) for the nitride films deposited with different ECR powers. The deposition time of these films was 30 s, corresponding to a nitride thickness of 3.8 nm, as shown in Fig. 4. In this case, Si substrates were not dipped in HF. As expected, ECR power played a significant role in determining the film quality. When the ECR power was greater than 265 W, the films with high breakdown voltage and low leakage current were obtained. With lower ECR power, a high percentage of MIS capacitors failed at very low bias voltage. It also shows that when the quality of the silicon nitride films is poor, the quality of the entire film stack, native oxide plus the nitride film, is also poor. It confirmed the well known fact that the electric strength of the native oxide is poor. In addition, since the dielectric constant of silicon nitride is almost twice of that of silicon oxide, the electric field in the native oxide is much higher than in the nitride film during the test. Therefore, the native oxide is expected to breakdown at much lower voltage than the nitride film does. Thus, the excellent dielectric strength at high ECR power is mainly a contribution of the silicon nitride layer.

Fig. 7 shows the J - V curves of the samples deposited with different SiH_4 flow rates. The deposition time for these samples was 40 s with an ECR power of 265 W. When the flow rate increased from 55 to 65 sccm, slight increase in leakage current and decrease in breakdown field were observed. A flow rate of 55 sccm resulted in a breakdown field of 8.5 MV/cm, comparable to the published results on the nitride films deposited at much higher temperature [3]. Similarly, the leakage current density is also low, about 1.0×10^{-5} A/cm^2 at an electric field of 5 MV/cm, within the same magnitude of silicon nitride film formed at much higher temperatures [3].

The breakdown voltage as a function of the capacitor area was measured and is shown in Fig. 8. The breakdown voltage decreases from 6.2 to 4.3 V when the capacitor area increases from 7.5×10^{-3} to 1.0 mm^2, indicating weak spots exist in the deposited films, which are the places where dielectric breakdown starts. As the capacitor area increases, the number of weak spots in the capacitor also increases, causing the average breakdown voltage to drop. However, even for the largest capacitor, the breakdown field still remained at ~ 6.0 MV/cm, indicating the ultra-thin films are continuous and pinholes absent.

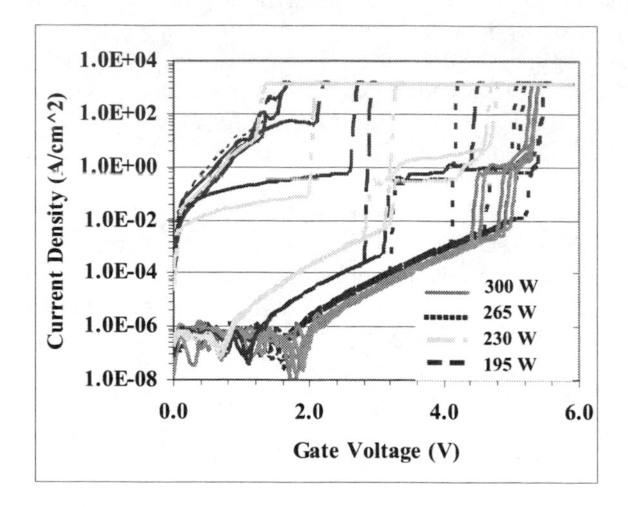

Fig. 6 Current density vs. voltage for the samples deposited at different ECR powers

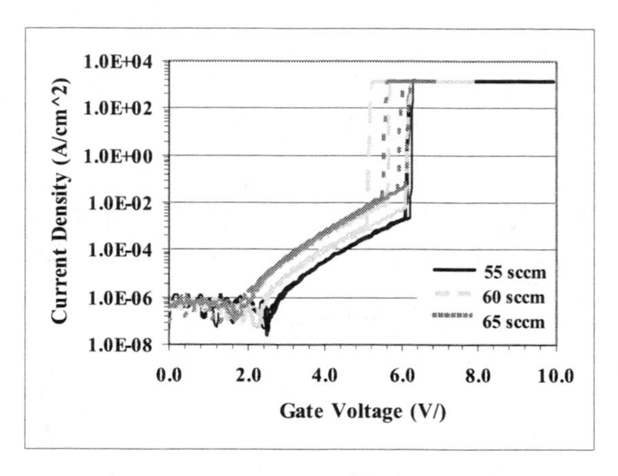

Fig. 7 J – V curves of silicon nitride films deposited at 265W ECR power, but different SiH_4 flow rates. The deposition time was 40 s for all the films.

The breakdown voltages of two groups of samples, prepared with and without HF substrate dip, are compared in Fig. 9. Nitride films on the substrates dipped in HF did show lower breakdown voltages compared with that on the substrates without dipping in HF. Two possible reasons are attributed to the observations. First, the additional native oxide reduces the electric field in the nitride film significantly because of the large difference in their dielectric constant. As a result, a larger voltage is needed in order to break down the nitride film when a thicker native oxide exists. Second, the quality of films on substrates with and without HF dip may be different. In fact, calculations showed that the dielectric strength of the films deposited on HF dipped substrates is actually slightly higher than that deposited on the acetone/methanol cleaned substrates.

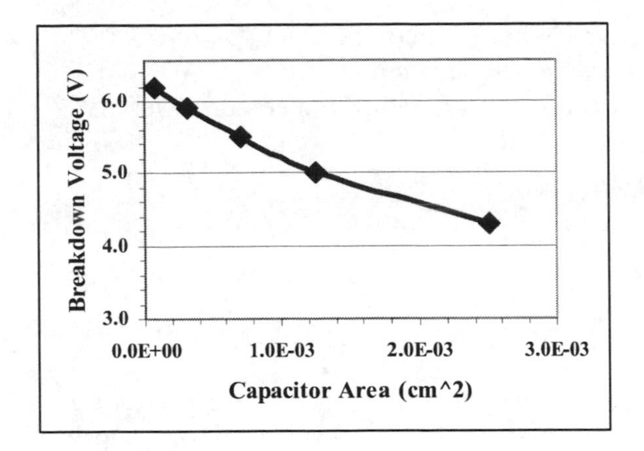

Fig. 8 Breakdown voltage as a function of capacitor area

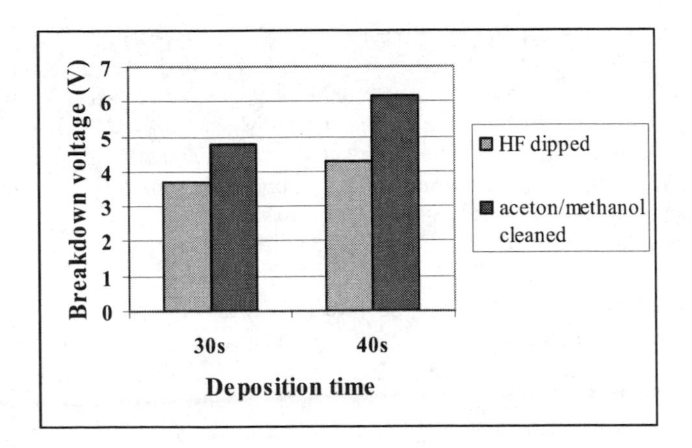

Fig. 9 Breakdown voltage comparison between silicon nitride films deposited on silicon substrates with and without HF dip

Electric strength of the films with different thicknesses but same deposition conditions was also compared. Fig. 10 shows the breakdown fields of ultra-thin (4.6 nm thick) and thin (50 nm) films deposited at optimized conditions. Breakdown fields decreased from 10 to 8.5 MV/cm when the film thicknesses decreased from 50 to 4.6 nm. The two reasons for this are: first, at the initial phase of deposition, equilibrium plasma conditions are not yet established. As a result, the first several atomic layers of the film are not stoichiometrically optimized, which degrades the integrity of the film, particularly when the film is only several nanometer thick. Second, as discussed previously, the impact of the native oxide is more profound in thinner film than in thicker film. Nevertheless, the data shows that the quality of the ultra-thin nitride films is comparable to that of the much thicker ones.

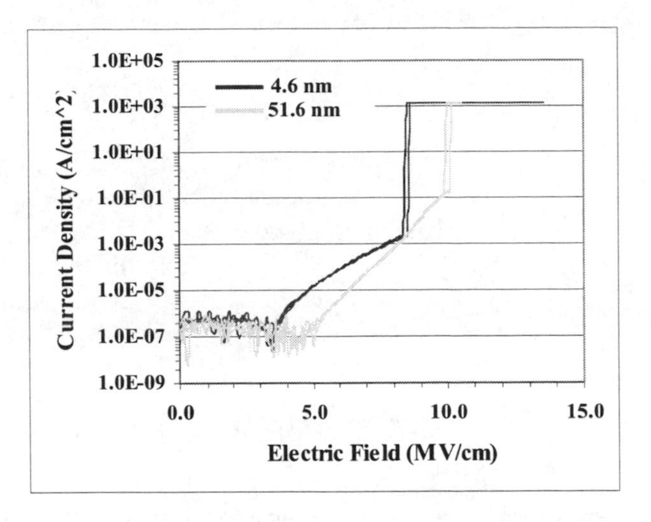

Fig. 10 Breakdown field comparison between ultra-thin and thin nitride films deposited with the same condition

SUMMARY

Ultra-thin silicon nitride films were deposited at temperatures below 150°C using an ECR-PECVD system with $SiH_4/N_2/Ar$ gases. Through deposition condition optimization, high quality films with thickness only several nanometers were achieved. Electrical characterization indicated that the leakage current and dielectric strength of the films are comparable to that of those films deposited at much higher temperatures. Analysis also indicated that no pinholes exist in such thin films.

REFERENCES

[1] R.D. Gould, S.A. Awan, "Dielectric properties of RF-sputtered silicon nitride thin film with gold electrodes", Thin Solid Films 433 (2003) 309 – 314

[2] M. Vila, C. Prieto, R. Ramirez, "Electrical behavior of silicon nitride sputtered thin films", Thin Solid Films 459 (2004) 195 – 199.

[3] K. Sekine, Y. Saito, M. Hirayama, and T. Ohmi, "Highly robust ultrathin silicon nitride films grown at low-temperature by microwave-excitation high-density plasma for giga scale integration', IEEE Transaction on Electron Devices, Vol, 47, No. 7, July 2000.

[4] I. Hallakoun, I. Toledo, J. Kaplun, G. Bunin, M. Leibovitch, Y. Shapira, "Critical dimension improvement of plasma enhanced chemical vapor deposition silicon nitride thin films in GaAs devices", Materials Science & Engineering B102 (2003) 352 – 257.

[5] D. Criado, I. Pereyra, and M.I. Alayo, "Study of nitrogen-rich silicon oxynitride films obtained by PECVD", Materials Characterization 50 (2003) 167 – 171.

[6] S. Sitbon, M,C. Hugon, and B. Agius, "Low temperature deposition of silicon nitride films by distributed electron cyclotron resonance plasma-enhanced chemical vapor deposition," J. Vac. Sci. Tech., A13(6) (1995), 2900.

Development of a new treatment to induce anatase growth on Ti

M. Pedeferri*, C. E. Bottani**, D. Cattaneo**, A. Li Bassi**, B. Del Curto*, M. F. Brunella*

*Dept. of Chimica, Materiali e Ingegneria Chimica, Politecnico di Milano,
Via Mancinelli 7, 20131 Milan – Italy, mariapia.pedeferri@polimi.it
** Dept. of Ingegneria Nucleare and NEMAS – Center of Excellence for NanoEngineered MAterials and Surfaces, Politecnico di Milano
Via Ponzio 34/3, 20133 Milan – Italy, carlo.bottani@polimi.it

ABSTRACT

The work deals with the evaluation of the application of a novel treatment of the titanium surface in order to increase anatase growth. Two different anodizing methodologies, previously developed [6] were considered, followed by a thermal post treatment able to convert the amorphous oxide film, obtained by anodizing, in a crystalline anatase structure. X ray thin film diffractometry and Raman spectroscopy have been used to identify anatase on the treated titanium surface. Combining the information obtained by the two analysis techniques allowed to verify that the best anatase crystallinity could be obtained by the treatment 1A followed by heat treatment.

Keywords: titanium, anatase, anodizing, spectroscopy, diffraction

1 INTRODUCTION

Titanium is a corrosion resistant and biocompatible material due to the titanium oxide film (about 2 nm thick) formation. The oxide film thickness can be increased by anodic oxidation techniques (anodizing) to hundreds of nanometers. Anodized titanium can assume different brilliant colors due to light interference phenomena. Colored titanium is appreciated for its aesthetical properties and has found many applications in design.

The anodizing treatment is already well known in the scientific background and is going to be transferred as industrial process for the treatment of bi- and tri-dimensional pieces as well as of tapes.

Titanium oxide produced during anodizing is essentially an amorphous oxide, but it can be transformed into crystalline TiO_2. Among the crystalline types the allotropic structure anatase is the most interesting one because of its photocatalytic and antibacterial properties [1,2]. Many works in the latest years have dealt with the studies of different treatments to utilize titanium dioxide as an agent to reduce environmental pollution [3], obtain sterilized surface [4] and inducing osteointegration of orthopedic and dental devices [5]. These not completely studied properties have enhanced titanium dioxide employment in the fields of environmental health and bioarchitecture. Aim of the work has been the evaluation of the application of a novel treatment of the titanium surface in order to increase anatase growth.

2 MATERIALS AND METHODS

After preliminary tests, two different anodizing methodologies, previously developed [6] were considered, followed by a suitable post treatment able to convert the amorphous oxide film, obtained by anodizing, in a crystalline anatase structure. The adopted methodology consist in two different types of titanium surface pretreatment named 1A and 2A followed by anodizing in a phosphoric acid solution at a potential of 130 V and a post heat treatment at 400 °C for two hours. X ray thin film diffractometry (PHILIPS PW 3020 Cu K_α radiation, parallel beam configuration) and Raman spectroscopy were used to characterize six sample conditions:

- NA-NT not anodized not heat treated (Ti as received);
- NA-YT not anodized heat treated;
- 1A-NT type 1 anodizing not heat treated;
- 1A-YT type 1 anodizing heat treated;
- 2A-NT type 2 anodizing not heat treated;
- 2A-YT:type 2 anodizing heat treated.

3 RESULTS AND DISCUSSION

As already mentioned the anodizing treatments produce a colored surface and the post treatment only slightly modifies the colors. This is an important result because of the possible aesthetic application of titanium products treated by the developed methodology.

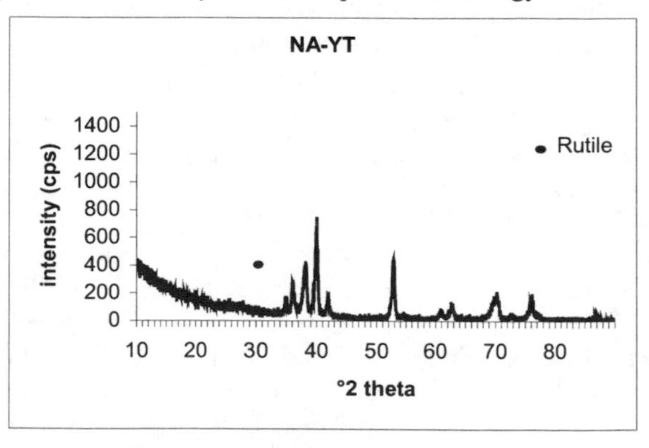

Figure 1: X ray thin film diffraction pattern.

The growth of titanium dioxide in the anatase allotropic form has been evaluated by X ray thin film diffraction and Raman spectroscopy, the results of the analysis are shown in the following sections.

3.1 X ray Thin Film Diffraction

In the figures from 1 to 3 the thin film X ray difraction patterns are showed. Diffraction pattern shows that no peaks other than titanium ones are present on the sample not treated, that is on the titanium strip. A small peak, corresponding to the most intense peak of rutile, appears in the diffraction pattern of the sample not anodized, but thermal treated (fig. 1).

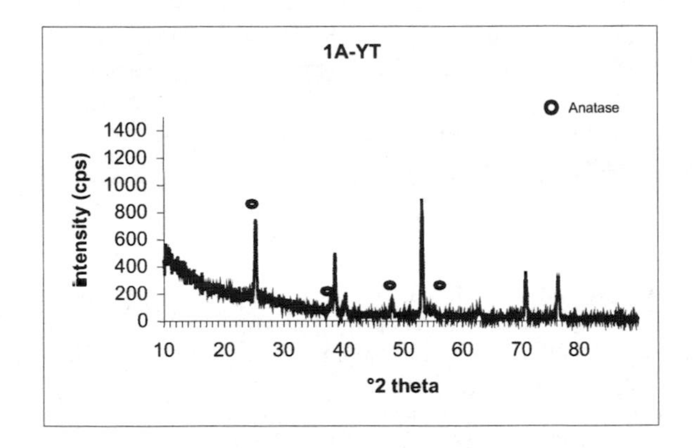

Figure 2: X ray thin film diffraction pattern.

The diffraction patterns, obtained from the samples anodized but not thermally treated (1A-YT, 2A-YT), show a small and broad peak in the angular range 25-30 °2Theta, suggesting the presence of an amorphous compound. The thermal treatment produces anatase growth on the samples anodized (1A-YT e 2A-YT) as can be seen from the diffraction patterns of figures 2 and 3. The main anatase peak is narrower and more intense in the case of treatment 1A than 2A, suggesting the presence of crystalline anatase.

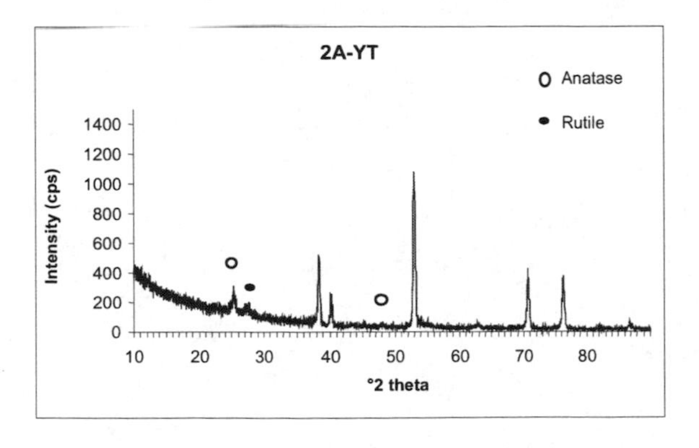

Figure 3: X ray thin film diffraction pattern.

3.2 Raman Spectroscopy

In figure 4 the Raman spectra of not anodized samples, before and after post treatment (NA-NT and NA-YT), are compared. As can be seen NA-NT is an essentially metallic sample. In the spectrum of sample NA-YT rutile peaks (442 and 611 cm^{-1}) are evident. The broad band at 200-300 cm^{-1} is typical of defective and/or substoichiometrics structures.

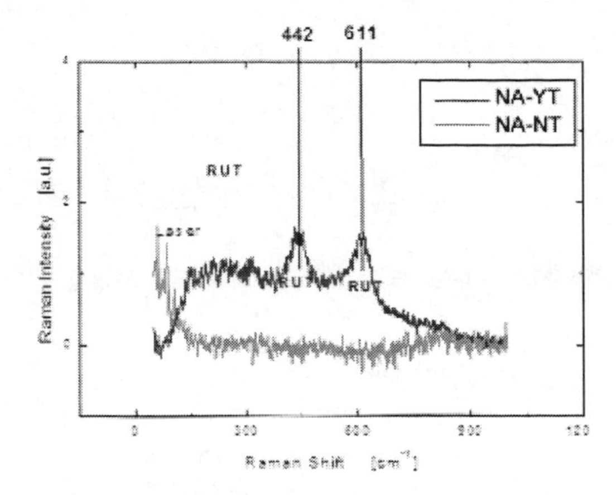

Figure 4: comparison of Raman spectra.

From comparison of Raman spectra of 1A anodized samples, before and after heat treatment (fig. 5), it can be noted that the spectrum of the not post treated sample is typical of an amorphous material, as demonstrated by the broad bands, while the five peaks present in the spectrum of the heat treated sample are proper of anatase. The peaks position is in good agreement with literature.

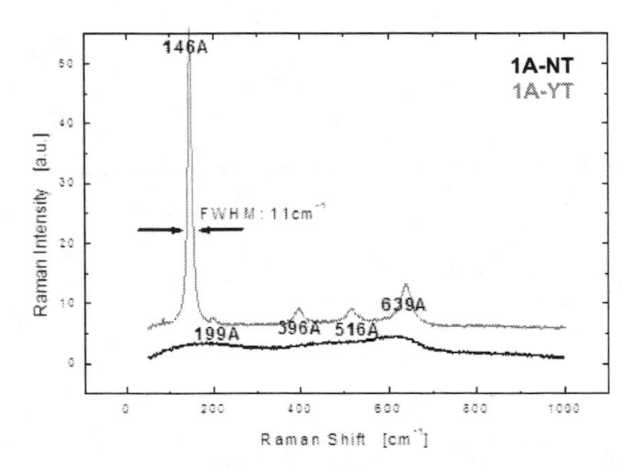

Figure 5: comparison of Raman spectra.

Figure 6 shows Raman spectra of 2A anodized samples before and after heat treatment. The spectrum of the not post treated sample consists of broad bands and a low intensity peak at 148 cm^{-1} corresponding to anatase. This suggest that the material is mainly amorphous with inclusions of anatase mono crystals. In the spectrum of the post treated sample four anatase peaks are present. The most intense peak is shifted toward the highest frequencies and broadened, this can be related to the phononic entrapping typical of a crystalline phase constituted of crystals whose dimension are of about tens nanometers. As a consequence it can be assumed that the analyzed material is composed by anatase nanocrystals.

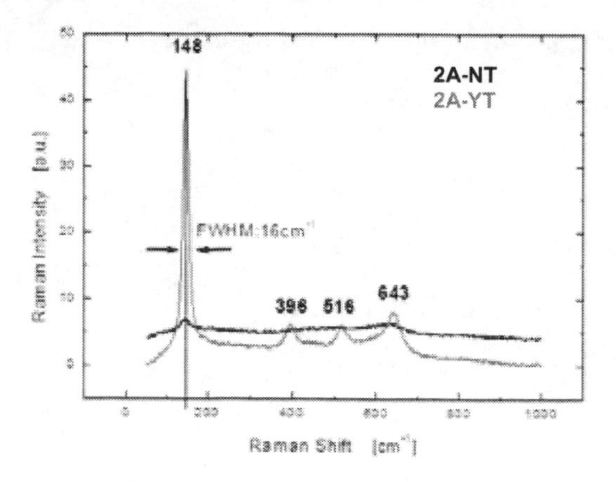

Figure 6: comparison of Raman spectra.

From the results obtained by Raman analysis of the not heat treated samples it can be summarized that:

- in the spectrum of the NA-NT sample no broad bands nor peaks are present as can be expected from metallic material, on the contrary both 1A-NT and 2A-NT spectra show broad bands related to amorphous phases (fig. 7);
- the 2A-NT spectrum shows one peak related to anatase: it could be an amorphous phase with nanocrystals of anatase. This result is partially in agreement with X ray diffraction analysis.

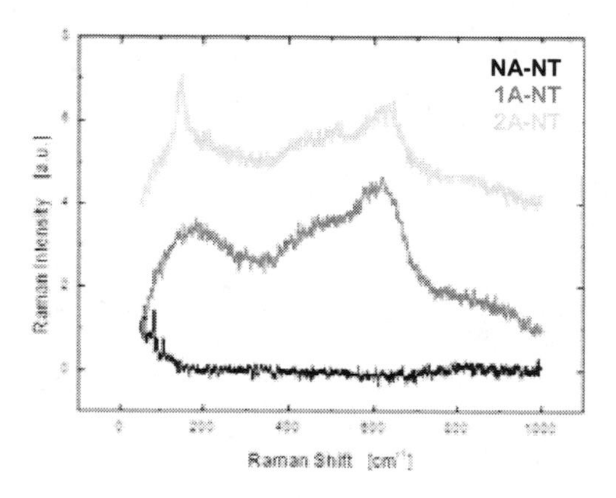

Figure 7: comparison of Raman spectra.

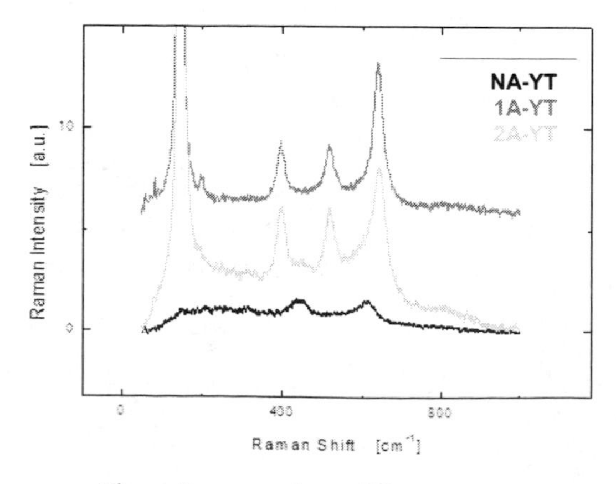

Figure 8: comparison of Raman spectra.

From the results obtained by Raman analysis of the heat treated samples it can be summarized that the difference in the position and width of the more intense anatase peak between the 1A and 2A anodized samples is due to different size of the crystalline domains in the anatase phase. Raman spectroscopy indicates that anatase crystalline domains of sample 2A-YT have a smaller dimension than those of sample 1A-YT.

4 CONCLUSIONS

Raman spectroscopy could give complementary informations to X ray diffraction analysis. Both the analysis techniques indicate that the anodizing 1A treatment followed by an appropriate heat treatment give rise to the formation of crystalline anatase. As a matter of fact the two analysis techniques equally indicate that the treatment 1A followed by heat treatment could be applied to improve anatase growth on titanium surface.

REFERENCES

[1] Ichiura H, Kitaoka T, Tanaka H Chemosphere; **51**, 855-60 (2003).

[2] Wang W, Chiang LW, Ku Y, J Hazard Mater., **10**, 133-46 (2003).

[3] Iguchi Y, Ichiura H, Kitaoka T, Tanaka H, Chemosphere, Dec;53, 1193-9 (2003).

[4] Kuhn KP, Chaberny IF, Massholder K, Stickler M, Benz VW, Sonntag HG, Erdinger L Chemosphere, Oct; **53**, 71-7 (2003).

[5] Yang B, Uchida M, Kim HM, Zhang X, Kokubo T, Biomaterials, **25,** 1003-10 (2004).

[6] EP 1 199 385 A2 - 19.10.2001 "Method of coloring titanium and its alloys through anodic oxidation".

[7] T. Oshaka, et al., J. Raman Spectrosc. **7**, 321 (1978)

[8] D. Bersani, P.P.Lottici, Appl. Phys. Lett. **72**, 73-75 (1998)

Silicon Surface Modification with Silane-Functionalized Polymers

Kalpana Viswanathan, Timothy E. Long and Thomas C. Ward

Department of Chemistry
Macromolecules and Interfaces Institute (MII)
Virginia Tech Blacksburg, VA, 24061 USA telong@vt.edu

ABSTRACT

The synthesis of trimethoxysilane end-capped linear polystyrene (PS), star-branched PS, and subsequent silicon (Si) surface modification with the linear and star polymers are described. Trimethoxysilane terminated PS was synthesized using *sec*-butyl lithium initiated anionic polymerization of styrene and subsequent end-capping with *p*-chloromethylphenyl trimethoxysilane (CMPTMS). Acid catalyzed hydrolysis and condensation of the trimethoxysilane end-groups resulted in star-branched PS. This is the first report of core-functionalized star-shaped polymers as surface modifiers and the first comparative study showing differences in surface topography between star and linear polymer modified surfaces. Comparison of polymer film properties to polymer dimensions in dilute solution revealed that the linear PS chains were in the intermediate brush regime and the star-branched PS produced a surface with covalently attached chains in the mushroom regime.

Keywords: anionic polymerization, hydrolysis and condensation, star polymers, sol-gel chemistry, surface modification, polymer mushrooms and brushes

1 INTRODUCTION

Solid surfaces are often chemically modified to improve properties such as adhesion, lubricity, wettability, biocompatibility, and environmental resistance.[1] Self-assembled polymeric films offer significant advantages compared to other conventional modification techniques and thus are widely used as surface modifiers. While a polymeric surface modifier dictates the chemical composition of a modified surface, the chain length of the polymer also determines the film thickness. In addition, the density of polymer chains and concentration of functional groups on a modified surface may be further controlled using branched and dendritic polymer films.[2-4] There are many approaches for modifying solid surfaces with polymers, which include physisorption, covalent attachment, and electrostatic adsorption.[5] Physisorption is relatively easy, but suffers from thermal and solvolytic instabilities due to the absence of stable covalent bonds with the surface.[6] Covalent grafting of polymer chains is therefore preferred. Two commonly employed approaches to covalently graft polymer chains are the "grafting to" and the "grafting from" techniques.

This paper reports the use of the "grafting to" approach to anchor core-functionalized star-shaped PS to Si. In recent years, many studies involved the use of branched polymers as surface modifiers. This is partly attributed to the high density of functional groups in branched polymers, making them attractive candidates for chemical sensors, drug delivery agents, nanoscale catalysts, and smart adhesives.[7] Most earlier surface modification investigations with branched polymers have employed dendrimers and hyperbranched polymers. Studies on surface attached star polymers has mainly concentrated on the use of PEG-based star polymers to impart enhanced biocompatibility to silicon surfaces compared to linear polymers.[8,9] This paper is the first report on the modification of Si surfaces with star-shaped PS, which contained surface reactive silanol functionalities in the core. Synthesis of the star-branched PS was achieved using the sol-gel process resulting from acid catalyzed condensation of trialkoxysilane-functionalized linear PS obtained using living anionic polymerization. The sol-gel condensation reported herein uniquely results in a core-functionalized star polymer. Despite the steric hindrance associated with the accessibility of the core silanols, core-functionalized polymers were attached to Si wafers and presented a significantly different surface topography compared to linear PS analogues. The differences in the surface characteristics of linear and star PS modified surfaces are discussed below.

2 EXPERIMENTAL

2.1 Materials

Styrene (99%, Aldrich) was stirred over CaH_2 for 3-4 days, distilled under reduced pressure (0.1 mm Hg, 10 °C) after repeated degassing and freeze-thaw cycles, and stored at – 25 °C until further use. Styrene was distilled from dibutylmagnesium (DBM) under similar conditions immediately prior to polymerization. DBM (FMC Corporation Lithium Division, 25% solution in heptane) and *sec*-butyllithium (Aldrich, 1.4 M in cyclohexane) were used without further purification. *p*-chloromethylphenyl trimethoxysilane [CMPTMS] (95%, Gelest, Inc.) was stirred over CaH_2 for a day and vacuum-distilled (0.1 mm Hg, 50 °C) immediately prior to use. THF (EMD Chemicals), which was the polymerization solvent, was distilled from sodium in the presence of benzophenone immediately prior to use. Methanol (EMD Chemicals) was stored over 5 Å molecular sieves.

Silicon wafers used as substrates were graciously provided by Hewlett Packard Company. Conc. H_2SO_4, NH_4OH (VWR International), 30% H_2O_2, methanol, and dichloromethane (EMD Chemicals) were used as received. Milli-Q water was used for surface cleaning.

2.2 Synthesis of Trimethoxysilane End-capped Polystyrene

The synthesis of trimethoxysilane end-capped PS was achieved using anionic polymerization of styrene, which was initiated with *sec*-BuLi in THF at -78 °C and subsequently end-capped with a molar excess of CMPTMS as described by Long *et al.*[10] The functionalized polymers were isolated by precipitation into dry methanol. The polymers were then dried at 100 °C overnight.

2.3 Hydrolysis and Condensation of Trimethoxysilane-Functionalized Polystyrene

Hydrolysis was performed in THF at 20 wt% solids and 1 N HCl solution was added as the catalyst. The reaction mixture was stirred for 24 to 36 h at 25 °C and precipitated into methanol. The polymer was then dried at 100 °C for 12 h. The synthetic scheme for the polymerization, hydrolysis, and condensation is summarized in Scheme 1.

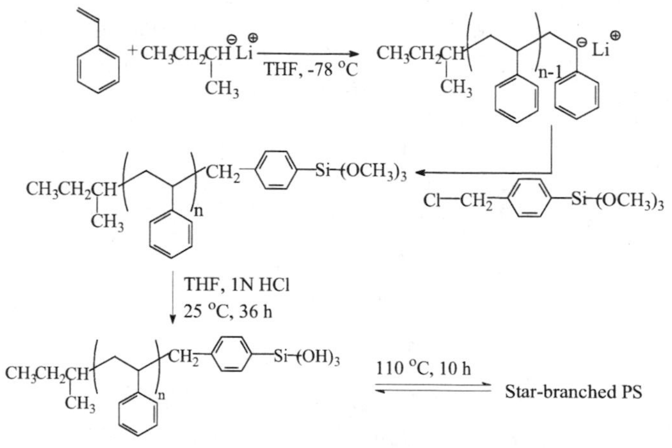

Scheme 1: Synthesis of phenyltrimethoxysilane end-capped PS via living anionic polymerization and subsequent acid catalyzed hydrolysis and condensation to branched PS

2.4 Polymer Characterization

Triple detection size exclusion chromatography (SEC) was used to determine polymer molecular weights and molecular weight distributions using a Waters 171 SEC system with a Waters 2410 refractive index detector, a Wyatt Technology Minidawn MALLS detector, and a Viscotek 270 viscosity detector. The data from the viscosity detector was also used to calculate the branching coefficient (g′) for the star-branched polymers. SEC measurements were performed at 40 °C in THF at a flow rate of 1 mL/min. ^1H NMR and ^{29}Si NMR spectroscopy were obtained on a 400 MHz Varian UNITY using $CDCl_3$ as the solvent.

2.5 Substrate Treatment

Silicon wafers were cut into $1cm^2$ pieces and sequentially sonicated for 5 min in dichloromethane and methanol. After blowing dry with nitrogen, the wafers were cleaned with NH_3:H_2O_2:H_2O (v/v/v: 1/1/5) at 60 °C for 15 minutes. After copiously rinsing with milli-Q water, the wafers were cleaned in a freshly prepared Piranha solution (conc.H_2SO_4:H_2O_2; v/v: 70/30) at 90 °C for 30 minutes. The wafers were then rinsed with milli-Q water several times, blown dry with nitrogen, and immediately spun cast with the polymer solutions. (*Caution: Piranha solution reacts violently with many organic materials and should be handled with care.)*

2.6 Polymer Brush Preparation

The trimethoxysilane terminated PS and the star PS were dissolved in toluene (1 wt% solution). These polymer solutions were filtered through a 0.2 μm PTFE syringe filter and spun cast at 2500 rpm onto silicon wafers that were cleaned as described above. The wafers were then heated at 145 °C for 24 h to enable the polymer end-groups to react with surface hydroxyls. The wafers were then exhaustively sonicated in toluene to remove any physically adsorbed polymer prior to characterization.

2.7 Surface Characterization

Topographic information on the polymer-modified silicon wafers was obtained using a Digital Instruments Nanoscope III atomic force microscope and software. TAP 300 Si_3N_4 tips with a spring constant of 40 N/m were used and set point to free amplitude ratios of 0.5 to 0.6 were used in the experiments. Stati water contact angles on the surfaces were measured in the sessile drop mode on a FTA-125 contact angle goniometer with a syringe-driven droplet. Ellipsometric data were obtained on a Beaglehole ellipsometer with a He-Ne laser (λ=632.8 nm) and an angle of incidence of 70°. Refractive indices of SiO_2 and PS were determined to be approximately 1.50 in the calculation of polymer layer thickness.

3 RESULTS AND DISCUSSIONS

Living anionic polymerization leads to well-defined polymers and enables quantitative functionalization of the living chain ends with a variety of reactive groups.[11] Thus, trimethoxysilane functionality was introduced at the polymer chain ends for subsequent silicon surface modification. ^1H NMR and ^{29}Si NMR spectroscopic analyses were used to confirm the yield of the end-capping procedure. After confirming the success of this functionalization approach, a series of molecular weights were targeted as shown in Table 1. It was necessary to add an excess of CMPTMS in order to quantitatively end-cap the living chain ends.

Table 1: Trimethoxysilane end-capped PS of different molecular weights

[CMPTMS]: [*sec*-BuLi]	M_n (NMR)	M_n (SEC)	M_w/M_n	% end-capping
1.5:1	3,200	3,590	1.22	99
1.5:1	3,900	3,100	1.27	97
1.5:1	2,900	3,600	1.20	98
10:1	11,000	11,000	1.05	95
10:1	9,200	9,200	1.10	100

The hydrolysis and condensation of linear PS-$Si(OMe)_3$ resulted in star-shaped polymers. The number of arms in the star-branched polymers was determined from the MALLS SEC M_w of the precursor oligomers and the star polymers. In order to confirm a branched topology, the branching coefficient g′ was calculated for these polymers from the intrinsic viscosity values obtained from the SEC viscosity detector. The data that were obtained for the linear and the star-branched PS are summarized in Table 2. The low g′ values confirmed the presence of branching in these polymers, which agrees well with both theory and earlier experimental data for star-shaped polymers.[12,13] The polymers that are described in Table 2 were synthesized from the same precursor PS (M_w=3,800) and all the polymers obtained had an average of 5-6 arms.

Table 2: Molecular weight data, degree of branching and the calculated number of arms for star branched PS synthesized by hydrolysis and condensation of PS-$Si(OMe)_3$ of M_w 3,800

Sample name	M_w (SEC)	$g'=([\eta]_{br}/[\eta]_l)_M$	Number of arms
PS-3k-linear	3,800	n/a	n/a
PS-24k-star	24,100	0.65	4.2
PS-21k-star	21,700	0.73	3.4
PS-19k-star	18,500	0.68	3.5
PS-18k-star	17,600	0.67	3.5

Tapping mode AFM, contact angle goniometry, and ellipsometry were used to examine surfaces modified with the linear and the star-branched PS. Many earlier studies have described the attachment of hyperbranched polymers, and dendrimers. Star PEO modified surfaces were also recently studied for their protein resistance. Many surface-sensitive techniques such as ellipsometry, contact angle goniometry, and fluorescence microscopy were used to characterize such surfaces; however, information on the surface topography after modification with the star polymers was not reported.[8, 9]

In this study, the "grafting to" approach was used to attach hydrophobic star polymers to Si wafers. The hydrolysis and condensation of trialkoxysilane-functionalized polymers leads to star-shaped polymers with excess unreacted silanol groups in the core that are capable of reacting with silica surfaces. The synthesis of these star polymers is relatively facile and does not require additional functionalization steps to introduce surface reactive functionality. Thus it was interesting to study these silanol-containing star polymers as surface modifiers and observe differences in the star and linear PS modified surfaces in terms of the surface properties. Water contact angle on polymer modified Si wafers were greater than 90° irrespective of the PS topology, indicating the successful modification of surfaces with the polymers. However, contact angle measurements did not differentiate the linear and star polymer modified surfaces. Thus, AFM analysis was used to examine the surface topography.

AFM images of the Si surface modified with a linear PS (M_w =18,000) and a star PS (M_w =18,500) are shown in Figure 1. Although polymer topology did not influence surface roughness, star polymer modified surfaces exhibited distinct and deeper features than linear polymer modified surfaces. The well-spaced crevices in the topographic image of the star polymer modified surfaces were also observed as distinct mounds in the phase images. The tall features that were observed in the present study were attributed to inadvertent dust particles trapped in the film. The mounds observed in this case resemble mushrooms and were attributed to topology-induced reduction in chain grafting density.

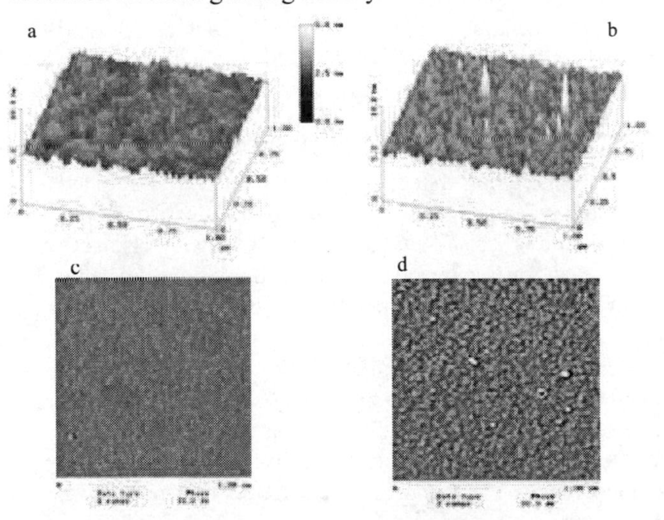

Figure 1: Tapping mode AFM topographic (a&b) and phase (c&d) images on silicon wafers modified with linear [a&c] and star [b&d] PS of M_w= 18,000. Both surfaces show an RMS roughness value of 0.28 nm

In order to prove that the observed topographic features on the star PS modified surfaces corresponded to the mushroom regime, the thickness of the linear and star PS films on Si wafers were compared to their respective radii of gyration (R_g), which was obtained in a good solvent.

Table 3: Comparison of polymer film thickness, and R_g values for linear and star PS

Sample name	R_g (nm)	Ellipsometric thickness (nm)
PS-3k-linear	2.3	4.1 ± 0.1
PS-10k-linear	3.2	7.5 ± 0.1
PS-17k-linear	5.5	10.8 ± 0.1
PS-22k-star	4.2	3.8 ± 0.1
PS-18k-star	3.8	3.2 ± 0.1

As seen in Table 3, the thicknesses of the linear polymer films were nearly double the respective polymer R_g values, which is characteristic of a polymer chain in a moderately stretched brush conformation.[14] However, the thicknesses of the star PS films were lower than their respective R_g values. In the mushroom regime, the polymer chains have dimensions similar to their unperturbed radii of gyration. Taking into account the swollen dimensions of the polymers in THF, which is a good solvent for PS, the observed similarity in the R_g and thickness values indicated that the star polymers are in the mushroom regime. Although the comparison of polymer chain dimensions on the surface is made to the chain dimensions in a good solvent, the arguments that hold for swollen chains will also be true for unperturbed chains, as in the latter case the chain dimensions are smaller than those of the swollen chains. Any R_g value lower than those shown in Table 3 further asserts the conclusion that the linear chains were stretched and the star polymers deposited as mushrooms. It should also be noted that R_g values from intrinsic viscosity measurements for linear polymers are reliable, but the same does not hold for star polymers.[15] In order to obtain an estimate of the R_g values for star polymers, a "g" value, where "g" is described as $[\langle R_g^2\rangle_{br}/\langle R_g^2\rangle_{lin}]_M$; of 0.5 was assumed, as suggested by Douglas *et al.* and R_g for the star polymers was calculated.[14] The calculated R_g values were similar to those obtained experimentally. Therefore, it was concluded that the star PS chains were in the mushroom regime.

It is believed that this work represents the first study that shows a comparison of the surface topography of Si surfaces modified with both core-functionalized star polymers and linear polymers of nearly equivalent molar mass. From the present investigation, we conclude that use of core-functionalized star polymers as surface modifiers will lead to surfaces that are modified with polymer chains in a mushroom conformation. Such star polymers deposited as mushrooms can function as adhesion promoters in conjunction with added free chains. Also, it is possible to obtain surfaces with a large number of exposed and accessible functional groups for further functionalization; this is achieved by controlling the number of arms in the star polymers and the use of functionalized initiators to synthesize the star polymers. Many groups have demonstrated the capability of dendrimers and hyperbranched polymers to function as sensors, nanoscale catalysts, and smart adhesives due to the presence of large number of accessible functional groups. Traditional dendrimer synthesis is relatively more tedious and hyperbranched polymer properties are less well defined. On the other hand, the synthesis of well-defined star polymers is relatively simple and can afford surfaces that have a large number of exposed endgroups.

ACKNOWLEDGEMENTS

This material is based upon work supported by, the US Army Research Laboratory and the US Army Research Office under grant number DAAD19-02-1-0275 Macromolecular Architecture for Performance (MAP) MURI. The authors also acknowledge FMC Lithium Division for the donation of alkyllithium and dibutylmagnesium reagents and Tom Glass for ^{29}Si NMR analysis.

REFERENCES

[1] Claes, M., *et al.*, Macromolecules, 36, 5926, 2003.
[2] Zhou, Y., *et al.*, J. Am. Chem. Soc., 118, 3773, 1996.
[3] Peez, R.F., *et al.*, Langmuir, 14, 4232, 1998.
[4] Tully, D.C., *et al.*, Adv. Mater., 11, 314, 1999.
[5] Tully, D.C., *et al.*, 11, 2892, 1999.
[6] Zhao, B. and Brittain, W. J. Macromolecules, 33, 342, 2000.
[7] Li, J., *et al.*, Langmuir, 16, 5613-5616, 2000.
[8] Groll, J., *et al.*, J. Amer. Chem. Soc., 126, 4234, 2004.
[9] Sofia, S.J., Premnath, V., Merrill, E. W. Macromolecules, 31, 5059, 1998.
[10] Long, T.E., Kelts, L. W., Turner, R. S., Wesson, J. A., Mourey, T. A. Macromolecules, 24, 1431, 1991.
[11] Quirk, R.P. and Lee, Y. Macromol. Symp., 157, 161, 2000.
[12] Roovers, J.E.L. and Bywater, S. Macromolecules, 5, 384, 1972.
[13] Douglas, J.F., Roovers, J. E. L. and Freed, S. T. Macromolecules, 23, 4168, 1990.
[14] Milner, S.T. Science, 251, 905, 1991.
[15] Wang, W.-J., et al. Polymer, in press, 2004.

Molecular Fan: A Heat Sink for Nanoelectronic Devices

Charles A. Sizemore, Michael L. Recchia, Taesam Kim, and Chhiu-Tsu Lin*

Department of Chemistry and Biochemistry
Northern Illinois University
DeKalb, Illinois 60115-2862, USA, ctlin@niu.edu

ABSTRACT

An innovative heat-dissipation surface containing active functional groups (CH, methyl, phenyl, and cyclohexyl) is designed to act as a molecular cooling fan, termed "molecular fan". Two different types of coatings are used to prepare molecular fan; the first is a thin (~ 1 μm) optically transparent sol-gel based coating; and the second is a 5-10 μm thick polymer emulsion. Use of this "molecular fan" coating on one-side of test panel lowers the equilibrium temperature of heat sinks by 5-12 °C. The efficiency of "molecular fan" coating is shown to depend on the active vibrational modes, heat sink substrates, coating adhesion and film thickness, and conductive property of coatings.

Keywords: molecular fan, nanocoating, molecular vibrations, heat sink, radiative cooling

1. INTRODUCTION

Currently, electrical components are cooled by conduction of heat to a heat sink, which is cooled by convection in air. The warm air is removed from the system using a mechanical fan. For future nanoelectronic (or optoelectronic) devices, simple convection and mechanical fan will not be able to keep up with the increased heat density. In molecular fan, the densely oriented molecular functional groups are prepared on the coated heat sink surface that absorbs the energy from heat source and leads to fast vibrations. The vibrating part of the molecule then effectively releases the heat via radiative/nonradiative processes, greatly reducing or eliminating the need for cooling by convection and mechanical fans.

All machines, including computers, produce heat that usually must be dissipated in order to prevent overheating. This heat usually comes from friction (in mechanical systems) or ohmic heating (in electronic systems). In electronic applications, excess heat reduces efficiency and can eventually cause total failure of the device [1]. As technology progresses and silicon chips have more components in a smaller area, the heat produced by the chips per volume will increase [2]. An innovative device and/or mechanism is critical for dissipating the heat and effectively lowering the temperature of the system.

There are three common mechanisms by which an object can release heat energy: conduction, convection, and radiation. The best way to draw heat away from a component is conduction, as is done in the regular heat-sink-and-fan arrangement. In convection, a hot region of a gas or liquid moves away from the source of heat. The hot fluid is replaced by cool fluid, which is then heated. In electronic instrument, such as computer, mechanical fans are positioned near these heat sinks to draw the warm air away from the heat sink by convection. This cyclical method of cooling is the common way for a modern machine to disperse heat [3-6].

There are several problems with the current heat release arrangement. One is that of the space required to add more fans. As the components become more powerful, they tend to release more heat and require more circulation to cool. Eventually the number of fans in the computer becomes cumbersome. Because of this, when the heat is to be transferred to the surroundings and away from the computer, convection is not the best solution. Another problem is present in miniaturization. As components, and therefore the computers and machines themselves, get smaller and smaller, they will have less room for fans while concentrating heat production into a smaller area [7-8]. This will result in a small center for a great deal of heat, and will require more efficient cooling than modern machines. Eventually machines would have fans as their largest components. Finally, mechanical fans create their own heat when they convert magnetic potential into kinetic energy. This amount of heat is small, and is almost immediately dispersed by the action of the fan itself. With many fans in a small area, however, this heat can actually cause an increase in temperature over time, reducing the rate of cooling of heat sinks and components.

Radiation is the best and most efficient way to transfer heat energy [9]. Conduction of heat away from the component to the surface of the coating is still necessary, whether by a heat sink or by simply coating the component itself. Once a coating molecule has absorbed the energy, it becomes excited and vibrates rapidly. This excess energy can be released by the emission of a photon of infrared light by the surface molecule. This process is more or less independent of many of the variables that limit the ability of mechanical fans to disperse heat, such as ambient temperature, air pressure and circulation, and humidity; however, good conditions will still benefit the molecular fan, since a portion of thermal energy will still be dispersed by convection.

In this paper, we describe the design of an innovative heat-dissipation surface that acts as a molecular cooling fan. The coated heat-sink surface is selected to display active CH, methyl, phenyl, and cyclohexyl functional groups which are monitored by Raman spectroscopy. The cooling efficiency of molecular fan is investigated in relation to the active vibrational modes, heat sink substrates, coating adhesion and film thickness, and conductive property of coatings. The colored molecular fan is also fabricated and its cooling efficiency is also evaluated.

2. EXPERIMENTAL DESIGN

Two major types of coatings were investigated. One is a transparent, durable coating formed using sol-gel techniques. The other is an organic/inorganic hybrid coating formed using one of two organic aqueous emulsions, an acrylate and a urethane. The sol-gel coatings were formed using silanes, alcohols, water, and potentially, acid catalysts, rheological agents, and/or wetting agents. Formulation was similar to other sol-gel coatings that show good adhesion (surface bonding), hardness (crosslinking), and transparency. Sol-gels were made of 20 wt% silane (Aldrich, Gelest), 60 wt% ethanol, and 20 wt% water, with small amounts (1 wt% or less) of any other additives. Four different combinations of silane precursors were tested, all including 3-glycidoxypropyltrimethoxy silane at 15 wt% of the total formulation. Each of the four sol-gels also contained 5 wt% of one of the following: tetraethylorthosilicate ("TEOS", Aldrich), methyltrimethoxysilane ("MTMOS", Aldrich), phenyltriethoxysilane (Gelest), and cyclohexyltrimethoxysilane (Gelest). The dry film thickness of sol-gel coating is less than 1.0 m.

The organic/inorganic hybrid coatings were made using an organic aqueous emulsion in conjunction with water and a small amount of co-solvent to enhance drying/curing. The coating solution consisted of 22 wt% organic oligomers (acrylic/styrene or urethane, Alberdingk, NeoCryl), ~63 wt% water, and ~15 wt% propylene glycol butyl ether ("PnB", Aldrich). A wetting agent was added (less than 1 wt%) to improve the appearance of the coating by removing fisheyes and other defects. In some cases, other conductive particles such as carbon black (Akzo Nobel) or titanium oxide (Ishihara Sangyo Kaisha, Ltd.) were used in an attempt to increase the surface conductivity of the coating. Some color molecular fans of the urethane coating were also made by adding a very small amount (0.05 wt%) of pH indicators or fluorescent dye. Two dry film thicknesses of the organic/inorganic coatings were applied on aluminum, copper, and cold-rolled steel heat sink coupons: one is about 4.5-5.0 mm and the other is about 9.5-10.5 mm.

The main type of testing was done by measuring temperature as a function of time. An aluminum (or other metal alloy) block was cut to have the same cross-section area as our test substrates (panels). The substrate was placed on the aluminum block with good thermal contact between the block and the panel. The outer part (one-side only) of the panel had been previously coated with the coating to be tested. A very small part (about 4 cm^2) of the coated side was left uncoated so that a temperature

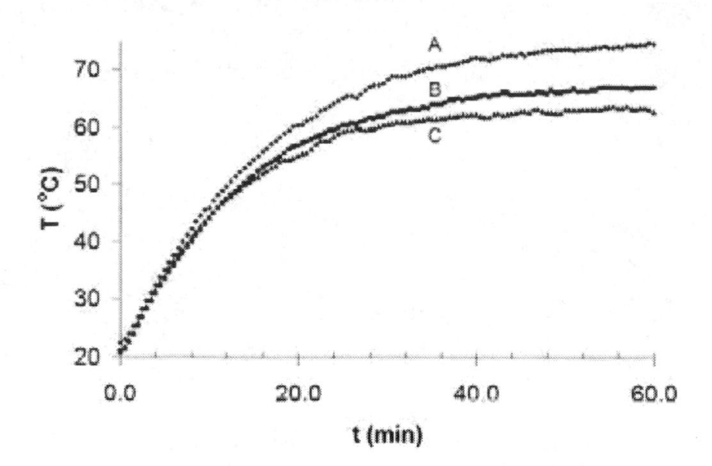

Figure 1. Cooling temperature vs, time of molecular fan made of AE coating with and without conductive carbon black

transducer could be placed, in good thermal contact, on the panel. The potential of the transducer was measured with a multi-meter that was interfaced with a PC to record data. To account for potential problems with ambient temperature fluctuation, an uncoated control panel was run every day that coated panels were measured. The surface vibrations of molecular fan were characterized by a microfocus Renishaw laser Raman imaging microscope.

3. RESULTS AND DISCUSSION

Most of the molecular fan coatings showed the ability to cool heat sink (aluminum, copper, steel, or plastic) substrates, with a cooling temperature ΔT = 7-12 °C for a

Table 1. Molecular fan based on acrylic emulsion (AE),urethane emulsion (UE), and sol-gel (SL) coatings.

	ΔT in °C			
	AE		AE with conductive carbon black	
	4.5 – 5.0 µm	9.5-10.0 µm	4.5 µm	10.0-10.5µm
Al coupon	6.6	7.3	7.5	11.1
Cu coupon	-	-	9.6	-
Steel coupon	-	-	11.8	11.8
UE on Al coupon (dry film thickness, DFT = 4.5 – 5.0 µm)				
UE transparent (clear fan)				7.7
UE with conductive TiO_2 (silver-white fan)				9.4
UE with aniline green (blue-green fan)				8.3
UE with fluorescein (light green fan)				8.1
UE with methyl red (orange red fan)				7.1
UE with bromomethyl blue (yellow fan)				7.8
UE with rhodamine B (pink fan)				6.7
SL on Al coupon (transparent, DFT less than 1.0 µm)				
Tetraethylorthosilicate				7.5
Methyltrimethoxysilane				4.8
Phenyltriethoxysilane				4.0
Cyclohexyltrimethoxysilane				4.0

single-side coating. Figure 1 shows the results of two of these coatings, acrylic emulsion formulations with and without conductive carbon black nanoparticles. Curve 1A shows the temperature as time progresses of a bare aluminum panel. It heats up from room temperature rapidly for the first twenty minutes and then begins to stabilize as the panel equilibrates with its surroundings, the equilibrated temperature is 74.5 °C. Curve 1B shows the same experiment with a heat-sink panel that was coated with a 9.5-10.0 mm thin layer of acrylate-based molecular fan with no conductive carbon black. It follows the same general pattern, but equilibrates at a temperature of 67.2 °C that is 7.3 °C cooler than that of the uncoated panel. Curve 1C shows the results of one of the best molecular fan coatings, an acrylate-based film of 10.0-10.5 mm thick with a small amount of added conductive carbon black. The coating equilibrates at a temperature of 63.4 °C which is 11.1 °C cooler than the uncoated aluminum control (curve 1A). It is suggested that a conductive molecular fan gives higher cooling efficiency of heat sink.

The results of the cooling experiments for molecular fan based on acrylic emulsion (AE), urethane emulsion (UE), and sol-gel (SL) coatings are summarized in Table 1; the numbers (ΔT in °C) are the differences between the "average equilibrium temperature" (average of last ten temperatures measured) of the molecular fan sample and the bare aluminum control. The top portion of Table 1 is the results for AE molecular fan, the cooling efficiency is enhanced for coating with conductive carbon black as also show in Figure 1. There is a slight enhancement in cooling efficiency for AE coating on steel and copper coupons as compared to aluminum coupon. The middle portion of Table 1 shows the results for UE molecular fan. Similar to AE molecular fan, UE coating with conductive TiO_2 particles gives a higher cooling efficiency than that of UE transparent clear fan. The molecular fan made of UE coating and organic dyes (aniline green, fluorescein, methyl red, bromomethyl blue, and rhodamine B) gives a bright color coating and good cooling efficiency. The bottom portion of Table 1 gives the results of sol-gel molecular fan. The coating is less than 1 mm thick and optically transparent. The cooling efficiency is sensitive to the molecular functional groups, following the order CH > CH_3 > phenyl ≈ cyclohexyl. The lighter functional group can vibrate faster upon heating, thus provides a higher efficiency radiative cooling.

In almost all cases, the thinner coatings showed less cooling than their thicker counterparts, as can be seen in Figure 2. Curve 2A shows the temperature as time progresses of a bare aluminum panel, with an equilibrated temperature at 74.5 °C. Curves 2B and 2C are molecular fan made of two film thickness of AE coating with conductive carbon black particles, where curve 2B is 4.5 mm and curve 2C is 10.0-10.5 mm thick. We believe that the thicker coatings offer better coverage of the substrate, filling the microscopic pores that a thinner film might leave and increasing the amount of fans available to dissipate heat.

The intention of using organofunctionalized silanes was to introduce groups with different vibrational frequencies into the coating to increase infrared emission, thereby releasing more energy over time and increasing cooling power. The radiative cooling (ΔT) is 7.5 °C for tetraethylorthosilicate coating, 4.8 °C for methyltrimethoxylsilane, and 4.0 °C for both phenyltriethoxysilane and cyclohexyltrimethoxysilane coatings. The Raman spectra of four sol-gel coatings are displayed in the bottom portion of Figure 3. The most intense (Raman active) vibrations are the carbon-hydrogen

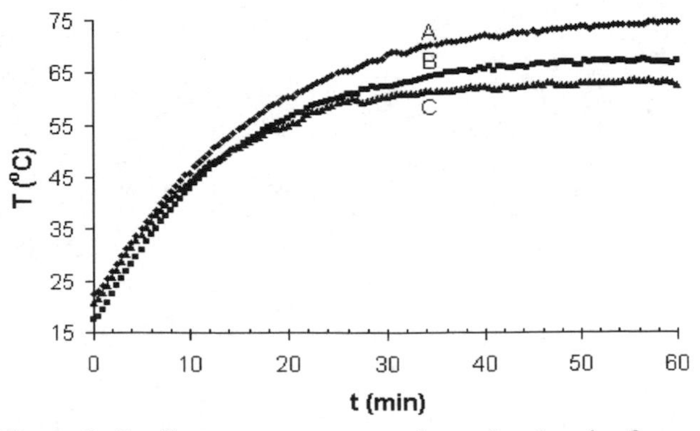

Figure 2. Cooling temperature vs. time of molecular fan for two thickness of AE coating with and without conductive carbon black.

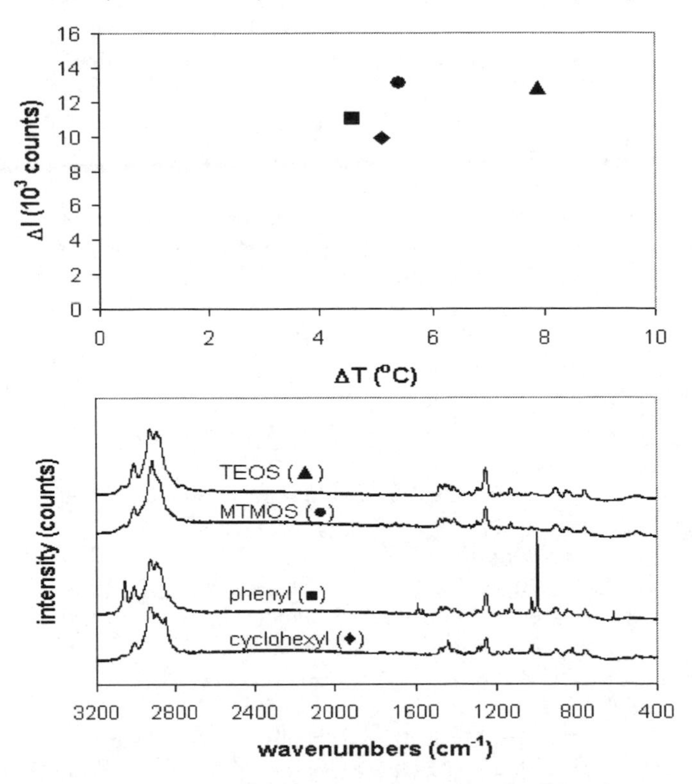

Figure 3. Four sol-gel fans with different molecular functional groups: Top – Raman intensity of CH vibrational mode vs. cooling temperature; Bottom: Raman spectra

stretches at 2930, and 2890 cm^{-1}. The top portion of Figure 3 is a plot of the integrated Raman intensity in the spectral range of 2930-2890 cm^{-1} vs. cooling efficiency (ΔT in °C) for four sol-gel molecular fans. The correlation is good that the lighter functional group has a higher efficiency radiative cooling

4. CONCLUSIONS

A set of molecular fan has been fabricated by using acrylic emulsion, urethane emulsion, and sol-gel coatings. The molecular fan has a coating layer of less than 1 mm or thicker layer 10 mm, and can be an optical transparent layer, a conductive black layer, a conductive silver-white layer or a bright color layer. With only single-side coating, the cooling efficiency of molecular fan is excellent at ΔT = 10-12 °C. The molecular fan can be fabricated on metal (such as aluminum, copper, steel, etc) and plastic (such as PC, PMMA, PET, etc.). The molecular fan with a lighter functional group gives a higher efficiency radiative cooling, due to a faster vibration upon heating.

5. REFERENCES

[1] Roig, J., Flores, D., Urresti, J., Hidalgo, S., and Rebollo, J., "Modeling of non-uniform heat generation in LDMOS transistors," *Solid-State Electronics*, 2005, *49*, 77-84.

[2] Chen, Y.C., Zwolak, M., and Di Ventra, M., "Local Heating in Nanoscale Conductors," *Nano Letters*, 2003, *3*, 1691-1694.

[3] Churchill, S.W., "The Conceptual Analysis of Turbulent Flow and Convection," *Chem and Eng Processing*, 1999, *38*, 427-439.

[4] Bessaih, R., and Kadja, M., "Turbulent natural convection cooling of electronic components mounted on a vertical channel," *Applied Thermal Engineering*, 2000, *20*, 141-154.

[5] Sathe, S.B., and Tong, T.W., "Comparison of Four Insulation Schemes for Reduction of Natural Convective Heat Transfer in Rectangular Enclosures," *International Communications in Heat and Mass Transfer*, 1989, *16*, 795-802.

[6] Tan, F.L., and Tso, C.P., "Cooling of mobile electronic devices using phase change materials," *Applied Thermal Engineering*, 2004, *24*, 159-169.

[7] Fan, J.R., Hu, G.L., Yao, J., and Cen, K.F., "A Two-Dimensional Mathematical Model of Liquid-Feed Direct Methanol Fuel Cells," *Energy & Fuels*, 2002, *16*, 1591-1598.

[8] Schill, A.W., and El-Sayed, M.A., "Wavelength-Dependent Hot Electron Relaxation in PVP Capped CdS/HgS/CdS Quantum Dot Quantum Well Nanocrystals," *J. Phys. Chem. B*, 2004, *108*, 13619-13625.

[9] Orel, B., Gunde, M. K., and Krainer, A., "Radiative Cooling Efficiency of White Pigmented Paints," *Solar Energy,* **1993,** 50, 477-482.

Microwave Irradiation an Alternative Source for Conventional Annealing: A Study of Aluminum Oxide Thin Films by a Sol-gel Process

A. R. Phani*, and S. Santucci

Department of Physics-University of L'Aquila and INFM-CASTI Regional Laboratory
Via Vetoio 10 Coppito 67010 L'Aquila ITALY, arp@net2000.ch

ABSTRACT

The lower temperature and shorter time with microwave irradiation might be ascribed to the activating and facilitating effect of microwave on solid phase diffusion. Using microwave-heating process, it is possible to achieve enhanced mechanical properties such as higher hardness, improved scratch resistance and structure texturing. In the present investigation, thin films of Al_2O_3 have been prepared by simple and cost effective sol-gel process on quartz substrates, and as deposited films are subjected to annealing in microwave irradiation at different powers. It is evident that there is a dramatic change in the mechanical properties of the films irradiated in microwave compared to the conventional annealing temperature.

Key words: microwave irradiation, annealing, sol-gel process, hardness, surface texturing

1 INTRODUCTION

Microwave irradiation is becoming an increasingly popular method of heating in the laboratory to synthesize biological [1], organic [2], and inorganic materials [3]. Coatings with surface laser treatments [4] or microwave exposure [5] leads to enhance mechanical and tribological properties. There has been less attention in the literature that sol-gel films treatment with microwave irradiation. In the recent years, many investigators have utilized microwave technique to synthesize powder perovskite solid compounds such as $LaCuO_4$ [6], and high temperature superconducting materials such as $YBa_2Cu_3O_{7-x}$ [7]. Present authors have also applied this technique on multilayers of Al_2O_3 / ZrO_2 deposited by a sol-gel process and found enhanced mechanical properties [8].

Unlike conventional annealing process, microwave technique offers a clean, cost effective, energy efficient, quite faster, and convenient method of heating, which results in higher yields and shorter time reactions. This will have strong impact on coatings applied in nanotechnology, microtechnology, and biotechnology industries. This could be a great advantage for the hard, protective, corrosion, abrasion, wear resistance, thermal barriers, optical, optoelectronic, microelectronic, polymer thin films that are applied on plastic, quartz, glass substrates where conventional annealing process is a limiting factor. Keeping in view the necessity of the coatings that are applied on plastic substrates, microwave irradiation has been chosen as an alternative method for annealing process. In the present investigation, thin films of Al_2O_3 have been deposited on quartz substrate at room temperature by sol-gel dip coating process. Two sets of films are treated as follows: one set by annealing at different temperatures ranging from 200 °C to 800 °C for 5 h in nitrogen atmosphere, and second set of films are exposed to (2.45 MHz) at different powers ranging from 140 W to 800 W for 10 min. The conventional annealed films as well as microwave-irradiated films have subjected to characterization for structural and mechanical properties.

3 EXPERIMENTAL PROCEDURE

The preparation and deposition of Al_2O_3 thin films by sol-gel dip coating process have been shown in the figure 1.

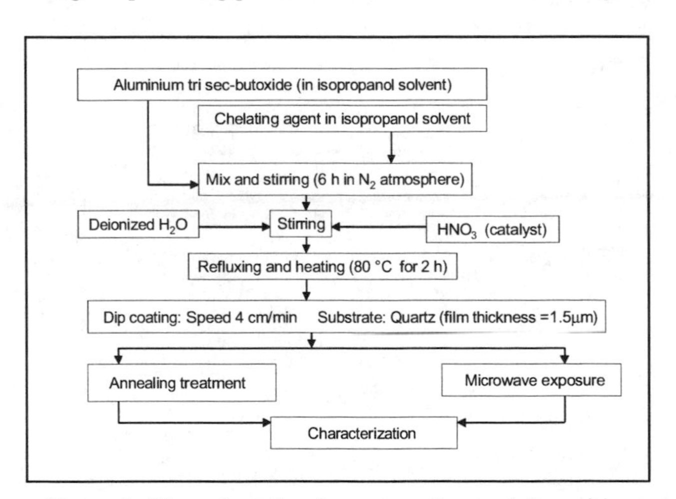

Figure 1: Flow chart for the preparation and deposition of Al_2O_3 thin films by sol-gel dip coating process

Calculated quantity of aluminum sec-butoxide (Chemat, 99.98 % purity) has been dissolved in isopropanol (99.99%) solvent along with acetylacetone as a complexing / chelating agent. The molar ratio of aluminum sec-butoxide and acetylacetone is 20:0.5. The contents are stirred at room temperature for about 6 h. To the above contents calculated quantity of HNO_3 is added drop wise as a catalyst and followed by deionized water using dropping funnel for stabilization and hydrolysis of the sol, respectively. Further the contents are refluxed at 80 °C for 2h to complete the reaction and later cooled to room temperature. The contents are filtered using Wattman filter paper in order to remove any particulates formed during the reaction. The obtained stock solution is used for the deposition of quartz substrates by dip coating process.

Cleaned quartz substrates (Ra = 1.4 nm) are clamped to pin and dipped in to the sol-gel stock solution at the rate of 4 cm/min. The process is repeated in order to get thickness around 1.5 microns. The deposited samples are subjected to two sets treatments as follows: one set by annealing at different temperatures ranging from 200 °C to 800 °C for 5 h in nitrogen atmosphere, and second set of films are exposed to (2.45 MHz) at different powers ranging from 140 W to 800 W for 10 min. The conventional annealed films as well as microwave-irradiated films have subjected to characterization for structural and mechanical properties.

4 RESULTS AND DISCUSSIONS

4.1 X-Ray Diffraction

XRD pattern for the Al_2O_3 thin films for both annealed and microwave exposed are shown in the figure 2a and 2b, respectively. The X-ray source is a sealed X-ray 1.5 kW Cu radiation (λ=1.5406Å).

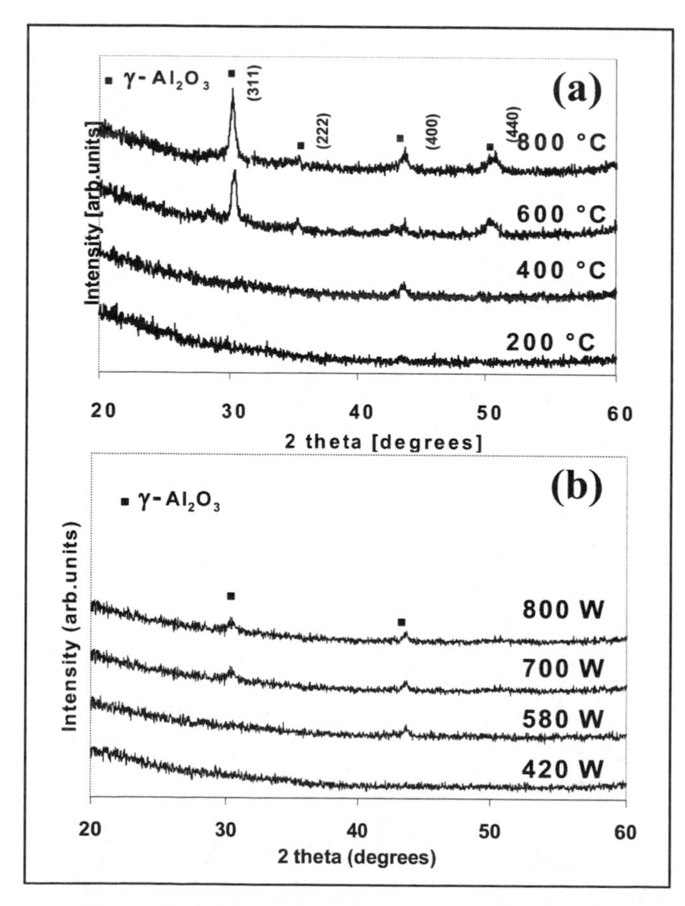

Figure 2: XRD pattern of Al_2O_3 thin films on quartz substrate by sol-gel dip coating process

(a) Annealed at different substrate temperatures

(b) Exposed to different microwave powers

Sharp and low intensity peaks belonging to γ-Al_2O_3 are observed in the annealed samples. Films annealed at 200°C are amorphous and as the substrate temperature increased from 400°C to 800 °C, γ-Al_2O_3 has grown preferentially in (311) orientation. On the other hand films exposed to 420 W and below have shown amorphous nature of γ-Al_2O_3 and as the power increased from 580 W and above there is slight formation of γ-Al_2O_3 in (311) and (400) planes. All peaks belonging to γ-Al_2O_3 phase are well matched with database in JCPDS (card #41-1432).

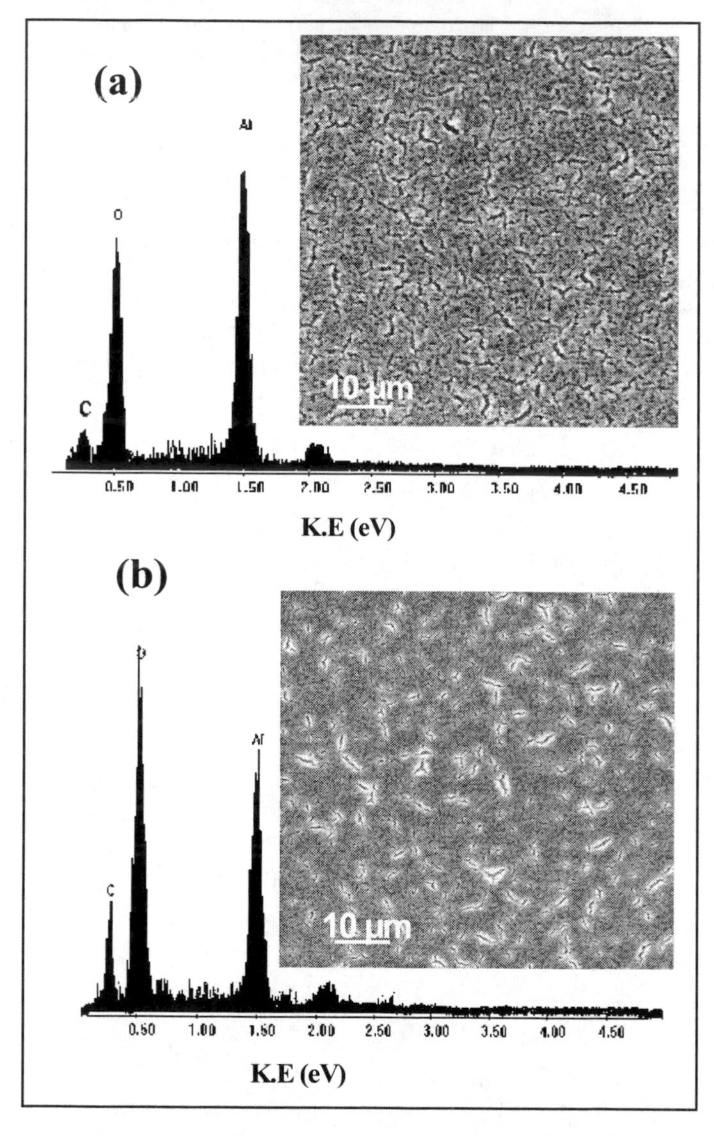

Figure 3: SEM-EDX pattern of Al_2O_3 thin films on quartz substrate by sol-gel dip coating process

(a) Annealed at 800 °C substrate temperatures

(b) Exposed to 800 W microwave power

4.2 Scanning Electron Microscopy

Scanning electron microscopy images and corresponding energy dispersive X-ray analysis of the Al_2O_3 thin films for the annealed and exposed to microwave are shown in the figure 3a and 3b, respectively. The presence of Al and O is evident in all the spectra with small amount of unburnt carbon in the films. Films annealed at different temperatures have shown dense structures with some cracks on the surface. The crack width ranges from 1 to 1.2 microns. On the other hand films exposed to different microwave powers have shown rupturing of the film in several areas. This may be due to abrupt heating of the film. The rupture width ranges from 0.5 to 1 micro in size.

4.3 Atomic Force Microscopy

Roughness (R_a) and grain size of Al_2O_3 thin films are measured by atomic force microscopy technique. Films annealed at different substrate temperatures have exhibited fine spherical grain structure with size ranging from 12 ± 2 nm to 36 ± 2 nm with increasing annealing temperature. On the other hand films exposed to lower microwave powers (up to 420 W) have shown similar behavior to that of annealed films. With further increasing in microwave

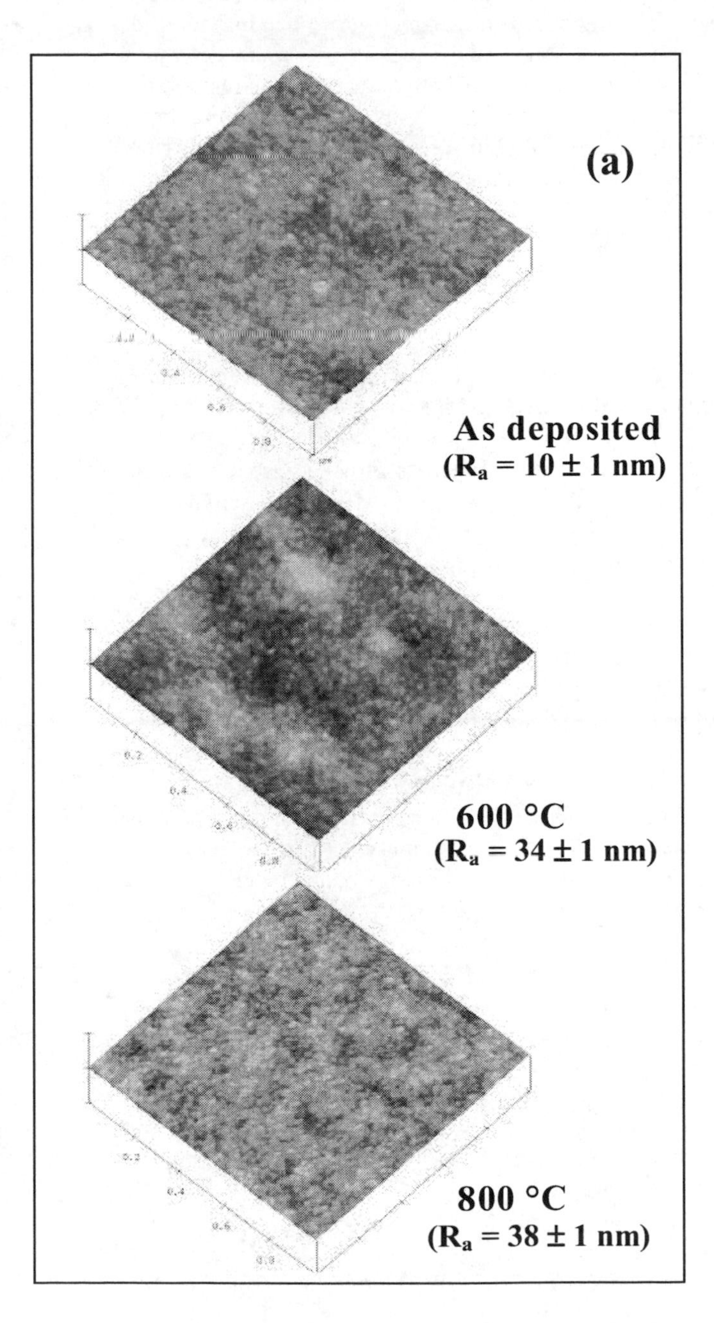

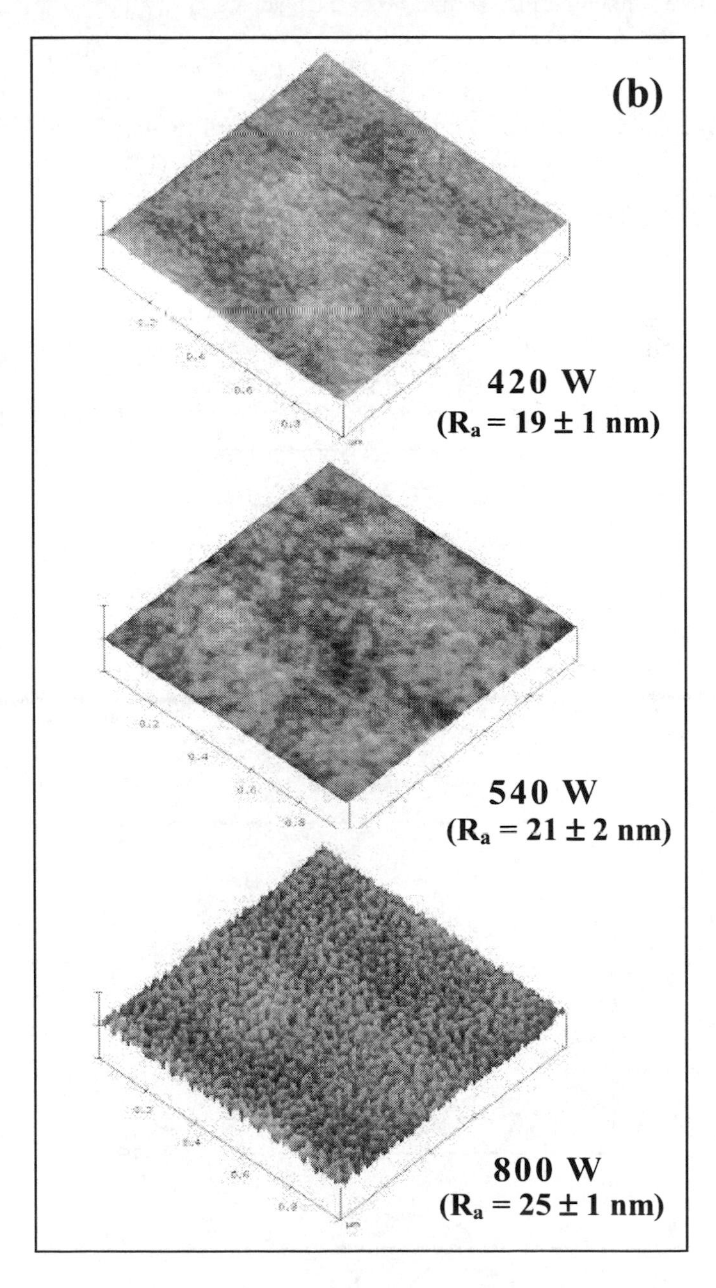

Figure 4: AFM topographical images of Al_2O_3 thin films on quartz substrate by sol-gel dip coating process
(a) Annealed at different substrate temperatures
(b) Exposed to different microwave powers

power (540 W) led to rupturing of the spherical structure in to open shell-like structures. At 800 W microwave power the open shell-like structures further opened to form needle -like structures. This suggests that one can change the surface structure or surface texturing by using microwave power.

4.4 Hardness

Nanoindentation measurements are performed with Berkovich indenter. The films are tested using a multiple loading sequence program with loading 5 mN. The maximum indentation depth is 260 nm, generally less than 10% of the typical film thickness. For each film 8 multiple loading indentations are measured each separated by 1000 microns. Al_2O_3 films have shown increase in hardness 2.3 to 4.4 GPa with increasing annealing temperature. On the other hand films exposed to microwave have shown hardness values from 4.2 GPa to 8.3 GPa with increasing power. This indicates that microwave irradiation is alone sufficient in order to get higher hardness compared to conventional annealing process. To understand whether this is surface effect or bulk, higher loads are applied up to 10 mN, where similar hardness (8.3 GPa) values are obtained with penetration depth of 340 nm for the film exposed to 800 W power, indicating the change in hardness is in bulk.

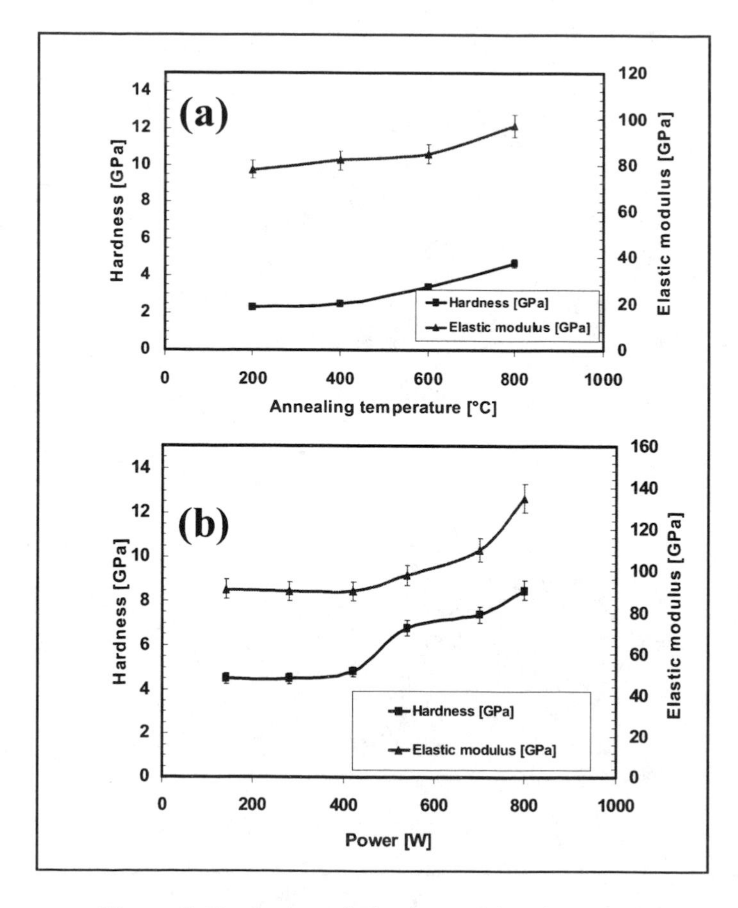

Figure 5: Hardness and Elastic modulus data of Al_2O_3 thin films on quartz substrate by sol-gel dip coating process (a) Annealed at different substrate temperatures (b) Exposed to different microwave powers

5 DISCUSSION

The ability of a material to absorb microwave energy is being expressed by its dielectric loss factor, which is always combined with the dielectric constant. In the present case, the plausible mechanism could be the conductivity of alumina increases with temperature as electrons are promoted into the conduction band from the O 2p valence band leading to increase in the dielectric constant. The increase in the dielectric properties with temperature is especially important in the microwave heating of solids, as it introduces the phenomena of thermal runaway. Microwave heating of alumina increases the temperature so too does the dielectric loss factor and heating becomes more effective which in turn has effect on the formation of dense structure causing increase in hardness of the alumina.

6 CONCLUSIONS

In conclusion, thin films of Al_2O_3 are deposited on quartz substrates at room temperature by sol-gel dip coating process and subjected to annealing at different temperatures and exposed to different microwave powers. The annealed films have shown γ-Al_2O_3 phase at and above 400 °C substrate temperature, whereas films exposed to microwave have exhibited low intensity γ-Al_2O_3 phase above 420 W microwave power. The films annealed at different temperatures are having grain size ranging from 12 nm to 32 nm with increasing substrate temperature. Films exposed to different microwave powers are having grain size from 12 nm to 20 nm. There is two fold increase in hardness, 8.2 GPa for the films exposed to microwave at 800 W compared to 4 GPa for film annealed at 800 °C.

Acknowledgements

The present work has been carried out during the one of the authors (A.R. Phani) employment at Center for Swiss Electronics and Microtechnology, Neuchatel, Switzerland, is gratefully acknowledged.

REFERENCES

[1] B. Wan, H. Lai, Bioelectromagnetics 21, 52, 2000.

[2] P. Lindtrom, J. Tierney, B. Wathey, J. Westman, Tetrahedron, 57, 9225, 2001.

[3] W.C. Lee, K.S. Lui, I.N. Lin, Jap. J. Appl. Phys, 38, 5500, 1999.

[4] K.L. Wang, Q.B. Zhang, Y. M. Zhu, Appl. Surf. Sci, 174, 191, 2001.

[5] S. Shigeng, Phys. Stat. Sol.A Appl. Res, 164, 779, 1997.

[6] Y.F. Liu, X.Q. Liu, G.Y. Meeng, Materials Letters, 48, 176, 2001.

[7] T. Kimura, H. Takizawa, K. Uheda, M. Shimada, J.Am.Ceram.Soc, 81, 2961, 1998.

[8] A.R. Phani and H. Haefke, MRS Fall 795, U12.11.1, 2003.

Durable Anti-Stiction Coatings by Molecular Vapor Deposition (MVD)

Boris Kobrin[a], Jeff Chinn[a], and Robert W. Ashurst[b]

[a] Applied Microstructures, Inc., 4425 Fortran Drive, San Jose, CA 95134, info@appliedmst.com
[b] Department of Chemical Engineering, Auburn University, Auburn, Al 36849, ashurwr@eng.auburn.edu

ABSTRACT

This paper reports on the results of an improved surface modification method called Molecular Vapor Deposition (MVD)[1]. MVD allows for the creation of molecular organic coatings which are denser and more durable than those obtained by current liquid or vapor-phase methods. This improvement has been achieved using a "sequential" or "layered" vapor deposition scheme of two different molecular films. The first molecular coating is a "seed" or adhesion promoter layer which is used to increase the binding sites for the subsequent functional molecular layer. The resulting surface coatings were observed to have improved stability to immersion applications, higher temperature stability and overall improved durability as a result of the increased surface coverage when compared to standard self-assembled monolayers (SAMs). These new film capabilities will have significant importance in improving the functionality and reliability of many micro- and nano-scale devices. The sequential approach with the seed layer has also been used to deposit molecular coatings on a variety of substrate materials (such as polymers, plastics and metals) which normally do not allow high quality surface coatings.

Keywords: self-assembled monolayer, anti-stiction coating, Molecular Vapor Deposition, MVD, MEMS

1. INTRODUCTION

Self-assembled monolayer (SAM) coatings have been a common technique employed in MEMS, optics, and life science applications for the modification of surfaces. However, there have been several commercialization hurdles which have stymied the widespread adoption of this technology. One major hurdle is the development of a production worthy, repeatable application technique. The application of these films in manufacturing is extremely difficult due to particulation problems caused by the high sensitivity of the reaction to environmental humidity[2]. For MEMS devices, reliability is often improved by the application of an anti-stiction layer. However, the use of solution-based films has been limited by the quality, scalability, and reproducibility of the films that are produced by elaborate wet processing. More recently, a variety of monolayer systems including self-assembled monolayers (SAMs) have been deposited in the vapor phase in an effort to further improve the performance of the MEMS devices and to eliminate in-use stiction[3,4]. The vapor-phase deposition processes of these anti-stiction films have been shown to produce higher quality films and appear to be an attractive in MEMS manufacturing.

Another technical obstacle to a wider acceptance of SAMs has been the lack of chemical stability of the coatings in certain critical environments. For example, SAM coatings usually cannot withstand liquid immersion applications for long periods of time. The coatings have been observed to detach[5] from the target surface. This phenomenon has been attributed to polar molecules penetrating through the film and breaking the surface attachment bond. Additionally, some SAM coatings do not survive thermal environments necessary in backend manufacturing processes (e.g. die attach, wire bonding, soldering, etc.) where temperatures may exceed 250°C, even though the elevated temperature may last for very short periods of time.

In this paper, the development of new SAM coatings using a "sequential" vapor technique which can be deposited on almost any substrate material and withstand harsh environments is reported. This improvement has been achieved using a sequential or "layered" vapor deposition scheme[6] of two different molecular films. In this multi-step process, an initial "seed" or adhesion promoter layer is used to increase the binding sites for the subsequent functional SAM coating. Subsequently, the deposited SAM layer that forms is denser and is believed to be better attached to the surface depending on the substrate material. The resulting molecular coatings are observed to have improved stability to immersion applications, overall high temperature stability and improved durability as a result of the improved surface coverage when compared to standard self-assembled monolayers (SAMs).

2. EXPERIMENTAL

A schematic diagram of the MVD coating apparatus is shown on Fig. 1. The vapor deposition chamber is evacuated by a mechanical dry pump and can be purged with nitrogen. SAM precursors and catalytic agents are delivered using a temperature controlled vapor delivery system which assures the accurate delivery of material during the reaction. A remote O_2 plasma treatment is used for substrate surface conditioning prior to the organic layer deposition. The process is programmable and executable by the controller. The chamber walls and the delivery lines are heated above room temperature to eliminate condensation of the reactants and residual vapor contamination. The reaction pressure can be controlled within a relatively wide range depending on the initial vapor pressure of the precursor used in the reaction but is typically maintained in the range of 0.1 to 50 Torr.

The following organo-chloro-silane chemistries were used for deposition of the anti-stiction coatings: tridecafluoro-1,1,2,2-tetrahydrooctyltrichlorosilane (FOTS) and heptadecafluoro-1,1,2,2-tetrahydrodecyltrichlorosilane (FDTS). The reaction was carried out using 18 MΩ-cm de-ionized water vapor as a catalytic agent. Both the precursors and catalytic agent, were degassed in vacuum prior to their use. The resulting FDTS and FOTS films were annealed after deposition at 120°C for about 30 minutes. Here, only the use of trichloro-silane based precursors is reported, but methoxy-silanes, ethyloxy-silanes, thiols and other common SAM precursors have been successfully used with this technique.

Water, hexadecane and diiodomethane contact angles were measured on Si(100) wafers using a Rame-Hart-100 goniometer equipped with DROP-image software. The film composition was obtained by X-ray photoelectron spectroscopy (XPS) performed on an Omicron EAC-125 hemispherical energy analyzer with a DAR-400 X-Ray source. The excitation source was Mg (Kα) at 1253.6 eV and the XPS system has a base pressure of less than 3 x 10^{-9} Torr.

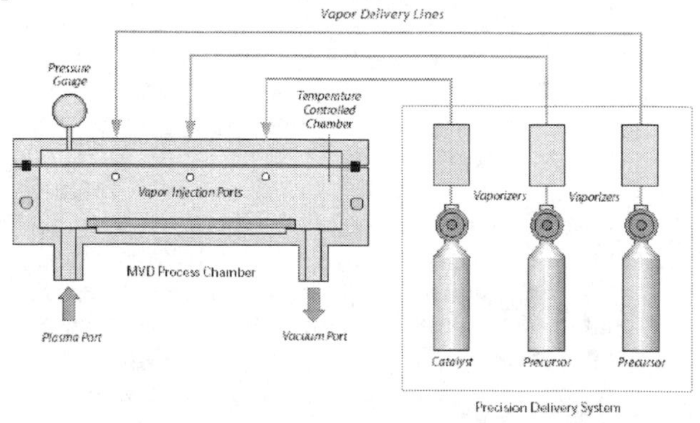

Fig. 1: Schematic of the MVD vapor deposition system

3. RESULTS AND DISCUSSION

In the case of a smooth, non-porous silica surface, it is generally accepted that there are 4-5 hydroxylated silanol sites present per nm^2 as compared to the theoretical limit[7] of 7.8 sites/nm^2. Thus, under standard deposition conditions, the surface coverage is limited by the surface binding sites, which are often less than ideal. In materials such as metals or polymers, there are often very few reactive sites and thus the surface coverage by a SAM coating is generally poor. Using a sequential vapor exposure method, an intermediate adhesion layer was created in order to increase the density of the reactive sites prior to the surface coating. As evidence of this increased surface coverage or film density, XPS analysis[8] shows that the ratio of total fluorine to the total silicon peaks for a FOTS SAM coating on a seed layer is approximately 15% higher than that of a liquid-phase or vapor-phase deposited film on a glass surface. This indicates an increase in density of the SAM coating in terms of molecular coverage. In Table-1, the improved contact angle of the sequential coating, using a diiodomethane test, is another independent verification of this increased surface coverage.

Coating Technique	VSAM	Sequential	Comments
Water Contact Angle	110	110	Polar + Hydrogen Bonding
Diiodomethane	84	94	Polar / No Hydrogen Bonding
Hexadecane	75	75	Non Polar / No Hydrogen Bonding

Table 1. Sessile drop contact angle analysis of an FOTS layer with and without seed layer as measured by different liquids.

Many different precursors can be used as a seed layer during a sequential deposition process. A typical seed layer is a in-situ vapor deposited silane containing layer prior to deposition of the SAM layer in the same chamber. In this work, an organo-silicon chloride was used as the seed layer. In our studies, the thickness of seed layer is important to increase in the starting hydroxyl (-OH) density. The –OH surface coverage can be measured by the water contact angle of the surface.

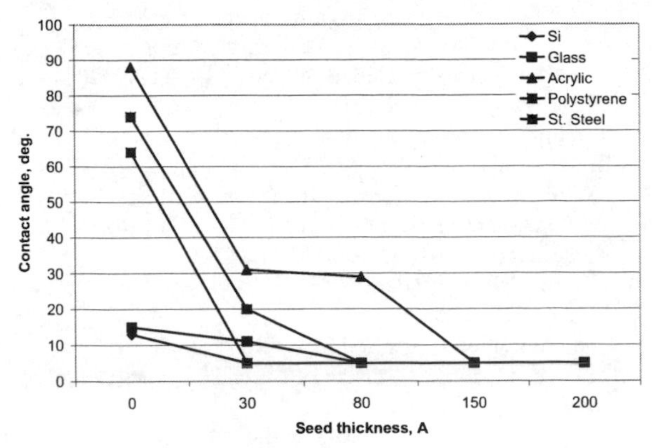

Fig. 2 Seed layer hydrophilicity (measured by water contact angle) on different materials as a function of thickness.

Completely hydroxylated surfaces are hydrophilic (Fig. 2) in nature. For example, a hydroxylated Si wafer is very hydrophilic (<5 deg water contact angle). This can be achieved by exposing a clean silicon surface to an oxygen plasma. By vapor depositing an adhesion promoter layer, the silanol density can be increased. In the case of a stainless steel material, a seed layer of approximately 30Å is used to obtain to a completely hydrophilic surface as measured by the water contact angle. Glass and polystyrene materials were observed to be very hydrophilic with an ~80Å seed layer, and an acrylic required ~150Å of seed layer to have hydrophilic properties. These experiments also show that different materials require specific thickness of seed layer for complete coverage of organic layers. After having increased the silanol density with a seed deposition layer, the SAM precursor can be deposited, exploiting the maximum number of available surface binding sites.

Fig. 3 demonstrates the effectiveness of the seed layer in a different way. Using a baseline process of FDTS, and varying the thickness of the seed layer, it is observed that any chosen material can be made to have water contact angles comparable to FDTS deposition performed on a silicon substrate, approximately 110°. For instance, an 80Å seed layer results a complete coverage contact angle of FDTS on stainless steel. Alternately, the contact angle associated with complete FDTS coverage is not achieved until 150Å of the seed layer is deposited on glass and polystyrene while acrylic required nearly 400Å.

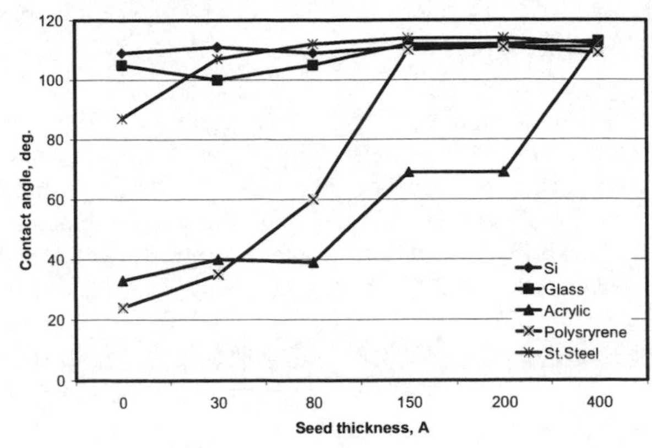

Fig. 3. Dependence of FDTS layer coverage (contact angle) on thickness of seed layer for different materials

Sequential or "layered" coatings can also improve the SAM's stability to resist harsh environments, such as exposure to moist environments or in aqueous immersion applications. As an example, a standard vapor deposited SAM coating of FOTS on TiN typically exhibits a water contact angle of ≥120°, which is indicative of full surface coverage. The stability of this coating in ambient air is excellent. However, upon immersion in water or other polar solvents, the film rapidly degrades over time as shown in Fig. 4. It is speculated that small polar molecules (e.g. water) can diffuse through the interstices of long chain surface coating and attack the covalent bonds which anchor the molecule to the surface. This phenomenon can be greatly reduced or even eliminated by the MVD method, using an intermediate sequential seed layer to increase the packing density. The stability and durability of the MVD coating is improved even after a prolonged (2 weeks) immersion in DI-water.

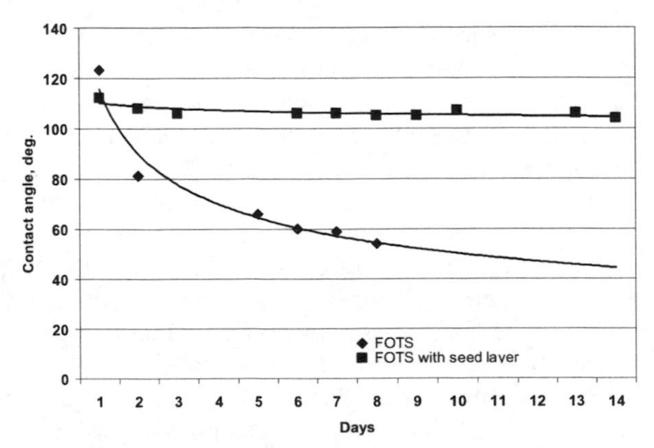

Fig. 4 FOTS coating stability in DI water (TiN substrate with and without seed layer)

The stability of MVD coatings in DI water has been investigated for the same set of materials: Si, glass, acrylic, polystyrene and stainless steel. Since polymers can not be annealed after the deposition, for consistency of this study deposited films on Si, glass and stainless steel have not annealed.

The sequential coating also has greater thermal stability on various materials other than silicon, which is important in many MEMS/MOEMS devices. In Fig.5, an FOTS sequential coating on aluminum demonstrates its ability to retain contact angle under elevated temperatures (250°C).

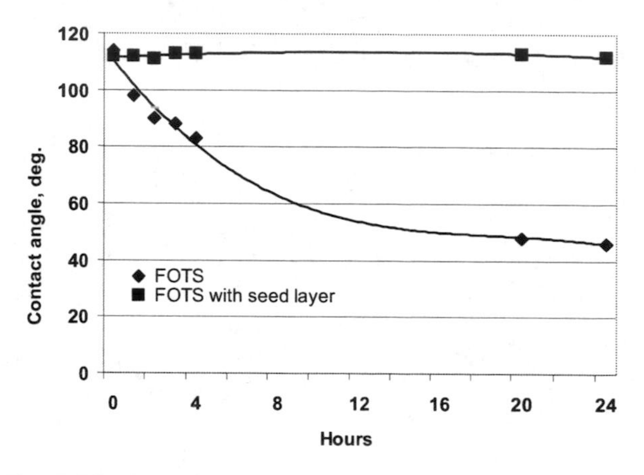

Fig. 5 FOTS coating thermal stability (Al substrate with and without seed layer)

MVD coating has been proven to be more abrasion resistant. Nano scratch tester experiments performed on a regular FDTS and MVD multi(5)-layer FDTS coating indicate that critical force for inducing coating damage is 2.34 times higher for MVD coating in comparison to the regular FDTS.

SUMMARY

The Molecular Vapor Deposition (MVD) technique using a sequential deposition approach allows for the synthesis of self-assembled coatings on a variety of new materials. Depending on the material to be coated, the native reactive surface site density may be low, which may be augmented by the deposition of a seed layer. This can be achieved using an organo-silane layer deposition which effectively increases the surface silanol density. Thus, a full coverage SAM coating can be realized on many different materials using a manufacturable and repeatable process. Additionally, the use of a seed layer allows for a more robust molecular film which is manifested by improved thermal and immersion stability compared to standard liquid based synthesis approaches.

References

[1] Molecular Vapor Deposition (MVD) is a trademark of Applied MicroStructures, Inc.

[2] W. R. Ashurst, C. Yau, C. Carraro, R. Maboudian, M. T. Dugger, "Dichlorodimethylsilane as an anti-stiction monolayer for MEMS: a comparison to the octadecyltrichlorosilane self-assembled monolayer", J. Microelectromechanical Systems, 10, 41-49, 2001.

[3] T. M. Mayer, M.P. de Boer, N.D. Shinn, P.J. Clews, "Chemical vapor deposition of fluoroalkylsilane monolayer films", J. Vac. Sci. Technol. (B), 18, 2433-2440, 2000.

[4] W. R. Ashurst, C. Carraro, R. Maboudian, W. Frey, "Wafer level anti-stiction coatings for MEMS", Sensors and Actuators (A), 104, 213-221, 2003.

[5] Thesis, Elizabeth Parker, UC-Berkeley, 2004. (eparker@kcp.com)

[6] Patent Pending, Applied MicroStructures, Inc.

[7] Lier, K.P., The Chemistry of Silica, John Wiley and Sons, New York, 1979

[8] Unpublished work: W.R. Ashurst – UC Berkeley (Currently at Auburn University)

In-situ Quantitative Integrated Tribo-SPM Nano-Micro-Metrology

N. Gitis, A. Daugela and M. Vinogradov

Center for Tribology, Inc.
1715 Dell Avenue, Campbell, CA 95008, USA, info@cetr.com

ABSTRACT

A novel quantitative nano+micro-tribometer with integrated SPM and optical microscope imaging has been developed to characterize numerous physical and mechanical properties of liquid and solid thin films and coatings, with in-situ monitoring their changes during micro and nano indentation, scratching, reciprocating, rotating and other tribology tests. Both the materials properties and surface topography can be assessed periodically during the tests. The integrated multi-sensing tribo-metrology is illustrated with two examples, of nano-indentation characterization of silicon wafer based coatings and micro-scratch characterization of diamond-like carbon coatings on magnetic media on the same instrument

Keywords: nano-tribology, tribometer, nano-indentation

1 INTRODUCTION

Quantitative nanometer resolution metrology tools have become a standard in semiconductor, data storage and other hi-tech industries, where products have to be tested for thin-film properties. Though it is critical to characterize advanced thin films and coatings, today's off-line nano-scale metrology tools can capture only a limited number of parameters. There is an immediate need for process control instruments capable of in-situ nanometer scale quantitative characterization at different stages of manufacturing process. Latest advances in nano and micro tribology forced the development of integrated instrumentation. A new generation of innovative optical/SPM instruments for chemical and mechanical characterization of surfaces is being developed for tribological testing applications. A combinatorial approach in micro-scale tribology tests [1] indicated the need for in-situ characterization instruments integrated into a tribometer. A recent nano-scale tribometer combination with AFM instrument has been reported [2]. Therefore, the number of novel tribometer applications is growing fast and covering fast changing industries, such as biomedical [3].

Performance of a quantitative nano+micro-tribometer [4], integrated with AFM and high resolution optical microscope, is demonstrated in two examples, on silicon wafer coatings, where quantitative materials properties were derived at several intermediate characterization steps with nano-indentation, and on diamond-like carbon coatings on magnetic disks, where disk durability was characterized by micro-scratching with a patented variable-angle blade micro-scratcher

2 NANOINDENTATION

2.1 Instrumentation

Photo of the newly developed instrument is shown in Fig. 1. The Universal Nano+Micro Materials Tester (UNMT) consists of a fully automated precision electro-mechanical tester [4], capable of performing numerous multi-axial linear and rotary tribological tests, 3-μm resolution optical microscope, closed-loop interchangeable SPM scanner and a nano-indentation instrument Nanohead.

Figure 1: Photo of UNMT with integrated AFM and optical microscope.

The UNMT has easily interchangeable rotary and linear, including fast oscillations, lower and upper drives, that provide a speed range from 0.001 (0.1 μ/s) to 10,000 rpm (50 m/s). It can apply a servo-controlled load that can be programmed to be constant or variable, in several ranges from 0.1 μN to 0.1 kN. The AFM and Nanohead can be installed onto the motorized Z-stage using a quick-release connector. All the UNMT motions and signals are controlled with a dedicated PC and sophisticated control software package. Both the Nanohead and the AFM have sub-nanometer resolution in terms of displacement noise floor. The Nanohead has a sub-micro-Newton force noise-floor, maximum ranges of Z–displacement and force of 300 μm and 500 mN, respectively. Instrument calibration and mechanical properties extraction for the Nanohead are performed automatically according to the ISO 14577 standard for instrumented nano-indentation [5].

2.2 Methodology

The Nanohead–1 shown in Fig. 2 is a sub-nanometer resolution nano-mechanical instrument integrated into the Universal Nano+Micro Tester and capable of performing nano-indentation tests, where the applied load F_z and penetration depth h are continuously monitored. The load versus depth plots are generated and processed from the collected data. The sample hardness H and the reduced elastic modulus E_r are calculated from the unloading segment of the curve as below. The reduced modulus is defined as follows:

$$E_r = S\frac{\sqrt{\pi}}{2\sqrt{A}} , \qquad (1)$$

where S is the unloading stiffness and A is the projected contact area. The stiffness S is calculated by fitting the unloading curve to the power law curve.

The nano-indentation hardness is defined by the ratio of the maximum load to the projected contact area:

$$H = \frac{F_{Z\max}}{A} . \qquad (2)$$

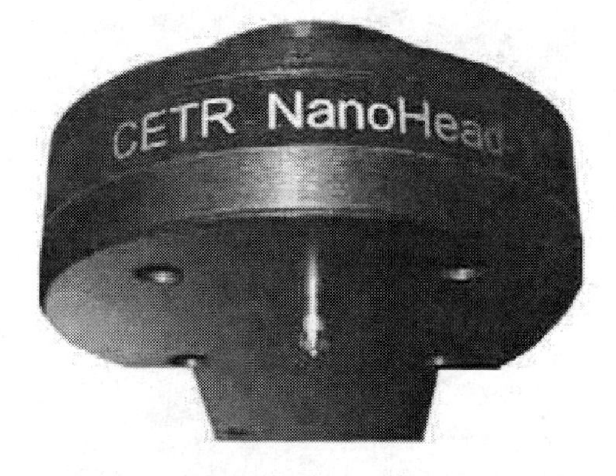

Figure 2: Photo of the UNMT nano-indentation module.

2.3 Experimental

Eight silica nitride and titanium nitride coated samples were blindly tested and characterized using the methodology and the instrument described above. Over 120 indentation tests were performed on all the samples. The samples were individually mounted on AFM specimen stubs and tested using a Berkovich tip, a 3-sided tip with an included angle of 142°, with a tip radius of approximately 80 nm.

The samples were tested using a trapezoidal loading profile that loaded in 5 s, held the maximum load for 5 s to allow for creep, and unloaded in 5 s. Several tests were conducted at various loading rates to ensure that the sample was not strain-rate dependant and so was not causing hardening at higher loading rates. The majority of the tests were performed at 3 N, but additional tests were performed at 1.5 and 4.5 N loads to ensure that resulting indentation depth was in the range of 150 nm, that is less than 5% of the thickness of coatings (4 mm). For the silica based coatings, substrate effects are observed on the data when indentation depth is approaching 30 – 40 % of the total coating thickness. Overall, different types of samples produced different reduced elastic modulus and nano-hardness values, indicating that both the nano-indentation technique and the instrument can distinguish different samples.

The results are summarized in the Table 1, yielding the following observations:

1. Similarly patterned TiN coated samples #5 and #6 have Er and H differences of 10%. Indentation into the silicon substrate of the sample #5 gave Er=72 GPa and H=10.5 GPa values, that are in the range of the typical values for silicon.
2. Mechanical properties of the pair of similarly patterned SiN coated samples #7 and #8 were found to be almost the same, suggesting that the same type of coating process was used.
3. TiN coated sample #1 had the highest hardness, though its Er was in the same range as for #7 and #8.
4. The softest was Cu coating #2, which had Er and H values of the order of magnitude lower than the rest of the samples.
5. Sample #4 had Er and H similar to the bare silicon.

A typical loading-unloading curve for the SiN coated wafer indent is shown in Fig 3.

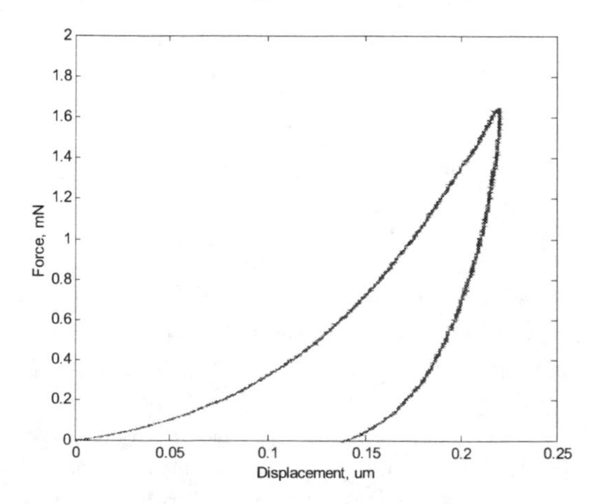

Figure 3: Loading-unloading curve on Si wafer.

Samples	#1 (SiN)	#2 (Cu)	#3 (SiN)	#4 (Si)	#5 (TiN)	#6 (TiN)	#7 (SiN)	#8 (SiN)
REM, Er, [GPa]	105.4±5.1	14.3±0.7	89.1±2.28	74.2±1.78	110.26±4.8	128.9±6.9	95.58±5.1	94.59±2.4
Hardness, H, [GPa]	16.64±0.43	0.87±0.04	13.71±0.36	12.5±0.55	15.21±0.78	16.01±0.4	14.33±0.72	13.83±0.3

Table 1: Mean values of elastic modulus and hardness with corresponding data scattering

3 MICROSCRATCH

3.1 Instrumentation

Micro and nano scratch testing is frequently used to investigate the behavior of thin films under various loading conditions. In particular, scratch testing can be used to investigate coating adhesion and failure modes. Many types of scratch probes are utilized for different types of coatings such as the DLC (diamond-like carbon) coating on magnetic disks. A tungsten carbide micro-blade with adjustable-angle holder has been used to evaluate durability of both freshly deposited bare overcoats and finished (lubricated and burnished) disk surfaces. The 0.8-mm micro-blade was chosen based on the contact stress analysis. For spherical or cylindrical contact geometry, the contact stresses are distributed deeper than a few nanometers of film thickness. For evaluation of thin films, however, the contact stresses should be concentrated within or near the films, which is achieved with the special micro-blade geometry. The schematic of the UNMT used in this study is shown in Fig. 4. The test procedure involved servo-controlled loading, multiple sensors, and precision motion.

3.2 Experimental

The fast and quantitative micro-scratch method was applied for ultra-thin coatings of 4.5 nm, covered with a mono-layer 1.5-nm lubricant. Three types of lubricated disks, two samples of each, were tested, named as Bias Carbon at 300V (samples 300_1 and 300_2), Bias Carbon at 150V (samples 150_1 and 150_2), and Bias Carbon at 50V (samples 50_1 and 50_2). Here, bias means film deposition voltage. The tests were performed at both A and B sides of the disk, three tests on each side at three different locations, OD (disk radius 47.5 mm), MD (disk radius 32 mm), and ID (disk radius 16 mm). During the test, while the micro-blade moved slowly against the film coating, progressive materials removal occurred. As shown in Figure 5, the measured coefficient of friction COF, normal force Fz, and electric contact resistance ECR were monitored with time. The normal force was gradually increased from 20 mN to 1000 mN. The coating film was cut through in 36 s, which corresponded to a critical load of about 300 mN. At that critical load, COF shifted to a higher value with a different slope, while ECR dropped to practically zero, because the micro-blade made contact with the conductive magnetic film after cutting through the coating.

The fact of the ultra-thin coatings break-through at the observed critical loads was confirmed by the integrated AFM images.

The UNMT allows for mapping of the tested surfaces, with 3-dimensional maps of coating durability, with a statistical analysis data for both the radial and the circumferential directions.

Wear tracks were examined after the test, using the Optical Surface Analyzer (made by Candela Instruments). They confirmed that wear tracks were very repeatable in both track pattern and track depth.

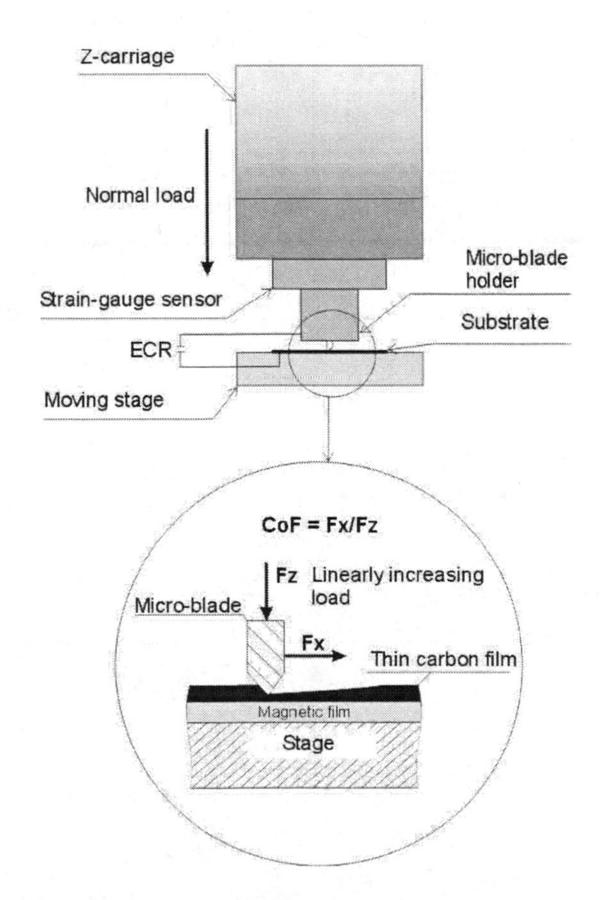

Figure 4: Schematics of UNMT micro-scratch module

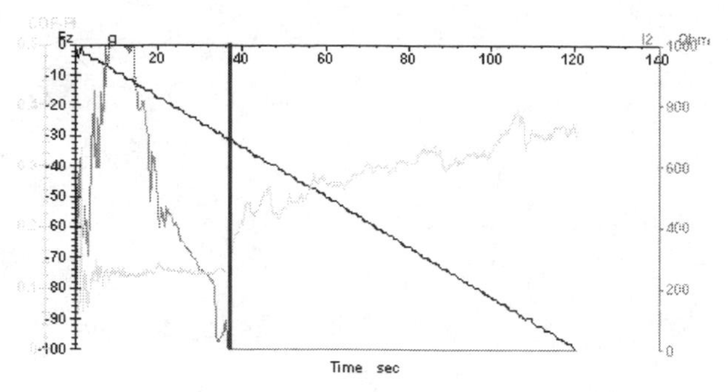

Figure 5: Coefficient of friction, contact electrical resistance and ramping vertical force versus time.

Table 2 summarizes the critical loads for the disk overcoats at different disk radii. The scratch resistance was the highest for the disks with the 300V-bias carbon, slightly lower for the disks with 150V-bias, and the lowest for the disks with 50V bias.

The Table 2 shows some radial non-uniformity of disk durability, with the scratch-resistance being the highest at the MD of the disks, lower at the ID and the lowest at the OD; this was then related to a sputtering chamber fixture, which was re-designed based on this durability data.

4 CONCLUSION

The Universal Nano+Micro Tester with integrated nano-indenter, micro-scratcher, optical microscope, AFM and many other tribology modules is a powerful tool for nano and micro mechanical characterization of thin films and other nano-technology specimens.

In the example with nano-indentation, both hardness and reduced elastic modulus indicated differences in mechanical properties between different SiN and TiN coated blindly selected samples. In terms of the data scattering, the reduced elastic modulus measurement results were more pronounced than the nano-hardness ones.

Micro-scratch experiments resulted in finding critical loads to breakthrough the ultra-thin films. Simultaneous coefficient of friction and contact electrical resistance measurements helped to "fingerprint" the samples.

5 REFERENCES

[1] Eglin, M., Rossi, A., Spenser, N.D., 2003. A Combinatorial Approach to Elicidating Tribological Mechanisms, Tribology Letters, 15, 3, pp. 193-198.

[2] Liu, X., Gao, F., 2004, A Novel Multi-function Tribological Probe Microscope for Mapping Surface Properties, Meas. Sci. Technol., 15, pp. 91-102.

[3] Gitis, N., 2004, Tribometrological Studies in Bioengineering, Proceed. SEM X Congress on Experimental. Mechanics, Costa Mesa, June 7–10, pp. 1 – 11.

[4] Gitis N., Daugela A., Vinogradov M., Meyman A., 2004. In-situ Quantitative Integrated Tribo-SPM Nano/Micro Metrology. Proceed. ASME/STLE Tribology Conf., Long Beach, October 24-27.

[5] ISO Standard 14577, 2002. Metallic Materials - Instrumented Nano-indentation for Hardness and Materials Parameters. Parts: 1, 2, 3.

Sample	Critical Load (g), at Disk Radius						
	16 mm		32 mm		47.5 mm		Average
	Side A	Side B	Side A	Side B	Side A	Side B	
300_1	32	35	44	45	28	31	[illegible]
300_2	35	36	46	40	34	41	
150_1	37	39	41	49	25	35	[illegible]
150_2	28	29	36	44	25	30	
50_1	30	20	25	30	18	20	
50_2	19	24	25	18	26	30	
Average	30.3		36.9		28.6		

Table 2. Micro-Scratch Test Results For Ultra-Thin DLC Coatings

Field emission of nanostructured AlQ_3 amorphous film and the heat treatment effect

C.-P. Cho and T.-P. Perng*

Department of Materials Science and Engineering
National Tsing Hua University, Hsinchu, Taiwan, *tpperng@mx.nthu.edu.tw

Abstract

Amorphous AlQ_3 film with nanostructured protrusions on the surface was fabricated by vapor condensation. It exhibits better field emission property than those of AlQ_3 nanowires and nanoscaled crystalline film. The film becomes crystalline and the field emission becomes worse after heat treatment. The field emission property is rather related to the crytallinity, and the probable reasons are discussed.

Keywords : AlQ_3, thin film, amorphous, field emission.

1. Introduction

Since high efficiency and good performance of a bilayer organic light-emitting diode (OLED) was reported by Tang and VanSlyke in 1987 [1], there has been a particular interest in tris-(8-hydroxyquinoline) aluminum (AlQ_3)-based devices for development of large-area displays [2]. AlQ_3 is one of the most important electron transport and emitting materials for OLEDs. Lots of attention has been paid to this organometallic semiconductor due to the unique properties such as excellent flexibility, high photoconductivity, and nonlinear optical behavior. It is then of great interest to prepare AlQ_3 nanostructures and study their optoelectronic properties for potential applications in nano-optoelectronic devices [3, 4].

Previously it has been revealed that crystalline AlQ_3 film could be prepared by vapor condensation [5, 6], and some nanostructures of AlQ_3 could be synthesized on a liquid nitrogen-cooled substrate [7, 8]. The field emission properties of nanostructured film and nanowires of AlQ_3 have also been reported [5, 8]. It was subsequently found that the field emission performance of amorphous AlQ_3 thin film was superior to that of crystalline film. However, the field emission of the amorphous AlQ_3 film becomes worse after heat treatment. It has been frequently reported that the optoelectronic properties of materials are highly related to their crystallinity. To investigate the relationship between the crystallinity and optoelectronic properties of AlQ_3 is necessary before further practical applications. In this study, the field emission of AlQ_3 materials are examined and found to be related to their crystallinity.

2. Experimental

The amorphous AlQ_3 film with nanostructured protrusions on the surface was fabricated by vapor condensation in a vacuum of 3×10^{-4} Pa. Commercial AlQ_3 powder (TCI Ltd., T1527) was placed in a graphite boat, and the silicon or ITO-coated glass substrate was placed under the stainless steel cold trap which was 10 cm above the graphite boat. AlQ_3 powder was sublimed when the temperature of the graphite boat reached 400°C. The temperature of the boat could be regulated by a power supply together with a K-type thermocouple. The detailed setup has been presented previously [7]. Field emission scanning electron microscopy (FESEM) was employed to examine the surface morphology, and atomic force microscopy (AFM) was utilized to examine the surface topography of the AlQ_3 film. A low angle X-ray diffraction (LAXRD) spectrometer equipped with a Cu target and an X-ray generator of 18 kW (400mA) was utilized to examine the crystallinity. Heat treatment at 190°C for 150 mins was executed in Ar under 1.33×10^4 Pa. The field emission properties before and after heat treatment were measured with a 50 μm gap between the anode and the cathode under a base pressure of 1.0×10^{-5} Pa. The anode-cathode configuration had a sphere-to-plate geometry. The high voltage was supplied by a power source, Keithley 237, and the current under the increasing applied voltage was recorded with a good accuracy to 10^{-13} A.

3. Results and Discussion

The amorphous thin film has a smooth surface with uniform morphology (Fig. 1a). The AFM image shows that there are many small nanoprotrusions on the surface, with the diameter of each protrusion being approximately 60 nm (Fig. 1b). The amorphism of the AlQ_3 thin film was demonstrated by the XRD pattern, as shown in curve A of Fig. 2. After heat treatment in Ar of 1.33×10^4 Pa at 190°C for 150 min, highly crystalline AlQ_3 can be obtained, as shown in curve B of Fig. 2.

A turn-on field of 6.5 $V\mu m^{-1}$ for field emission test is observed on the amorphous thin film as the current density reaches 10 $\mu A/cm^2$, as displayed in Figure 3. The inset shows the Fowler-Nordheim plot which yields a straight line at high applied fields displaying a typical field emission behavior. The amorphous AlQ_3 thin film exhibits a better field emission property than that of nanoscaled crystalline film with a turn-on field of 10 $V\mu m^{-1}$ as reported previously [5]. The highly crystalline AlQ_3 obtained from heat treatment also revealed a field emission characteristic, but much worse than those of nanostructured amorphous AlQ_3 and nanoscaled crystalline thin film. It exhibits a current density of 1 $\mu A/cm^2$ at a field of 18 $V\mu m^{-1}$. Apparently, the field emission property is quite related to the crystallinity, and higher crystallinity is detrimental to field emission. The domain size might have become larger and the domain boundaries decrease after heat treatment, so the number of channels for transporting electrons is reduced. Moreover, the domain boundary exhibits a larger geometric enhancement factor than that of the domain center. Therefore, it is deduced that smaller protrusions and more

(a)

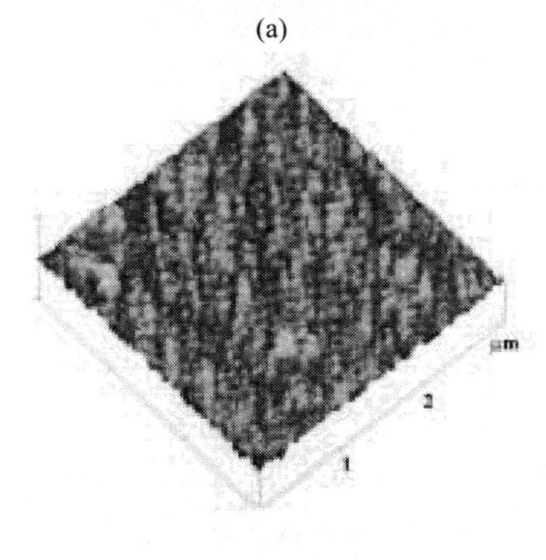

(b)

Fig. 1. (a) FESEM image shows a smooth surface with uniform morphology of the amorphous AlQ_3 thin film.
(b) AFM image of the amorphous AlQ_3 thin film. The diameter of each protrusion is approximately 60 nm.

domain boundaries lead to better field emission of the as-prepared amorphous thin film.

4. Conclusion

In this study, nanostructured AlQ_3 amorphous thin film has been demonstrated to exhibit better field emission than those of as-prepared crystalline AlQ_3 film and the crystalline one obtained from heat treatment. More domain boundaries, larger geometric enhancement and the nanoprotrusions would contribute to better field emission of the amorphous thin film. The follow-up investigations have been carried out and will be reported elsewhere.

5. Acknowledgement

This work was supported by the National Science Council of Taiwan under Contract No. NSC 93-2216-E-007-034 and Ministry of Education of Taiwan under Contract No. A-91-E-FA04-1-4.

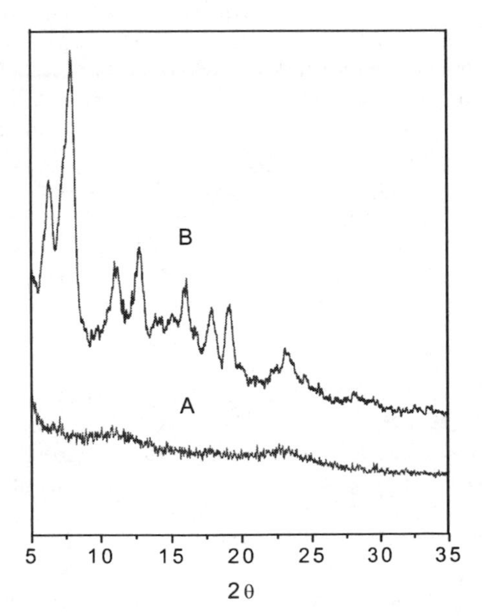

Fig. 2. LAXRD patterns of AlQ_3 thin film. Curves A and B are before and after heat treatment, respectively.

References

[1] C. W. Tang and S.A. VanSlyke, Appl. Phys. Lett. **51**, 913 (1987).

[2] J. R. Sheats, H. Antoniadis, M. Hueschen, W. Leonard, J. Miller, R. Moon, D. Roitman, and A. Stocking, Science **273**, 884 (1996).

[3] Y. Cao, I. D. Parker, G. Yu, C. Zhang, and A. J. Heeger, Nature **397**, 414 (1999).

[4] A. Noy, A. E. Miller, J. E. Klare, B. L. Weeks, B. W. Woods, and J. J. DeYoreo, Nano Lett. **2**, 109 (2002).

[5] J. J. Chiu, W. S. Wang, C. C. Kei, C. C. Cho, T. P. Perng, P. K. Wei, and S. Y. Chiu, Appl. Phys. Lett. **83**, 4607 (2003).

[6] J. F. Moulin, M. Brinkmann, A. Thierry, and J. C. Wittmann, Adv. Mater. **14**, 436 (2002).

[7] J. J. Chiu, W. S. Wang, C. C. Kei, and T. P. Perng, Appl. Phys. Lett. **83**, 347 (2003).

[8] J. J. Chiu, C. C. Kei, T. P. Perng, and W. S. Wang, Adv. Mater. **15**, 1361 (2003).

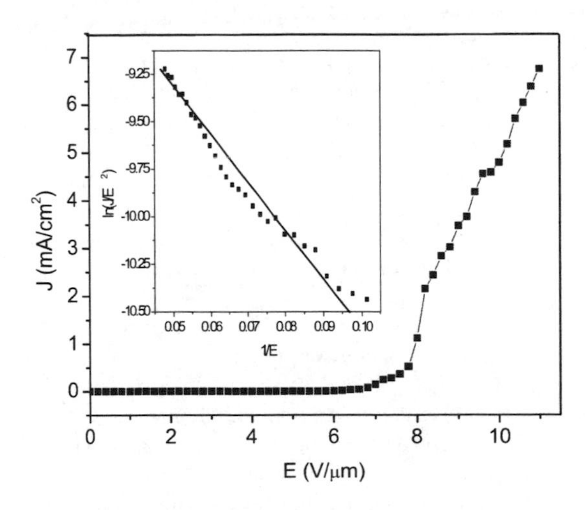

Fig. 3. Field emission J-E curve. The inset shows the Fowler-Nordheim plot which yields a straight line at high applied fields displaying a typical field emission behavior.

Surface Roughness as a Function of the DLC Thickness Coating

M.C. Salvadori*, D.R. Martins**, M. Cattani*

*Institute of Physics and **Escola Politécnica, University of São Paulo, Brazil
mcsalvadori@if.usp.br

ABSTRACT

The evolution of surfaces roughness coated with DLC as function of the film thickness is analyzed. We have used substrates with three different roughness, between hundreds of nanometers and just few nanometers. The surface roughness shifts as a function of the DLC thickness coating have been obtained. For the substrates with original roughness 393 and 278 nm, the roughness shift increased with the DLC thickness up to a maximum value and then decreased. For the substrate with original roughness around 4 nm, the roughness shifts oscillates but have no systematic tendency to decrease or increase.

Keywords: DLC Coating; Thin Films; Roughness.

INTRODUCTION

Diamond-like carbon (DLC) is a superhard material containing both graphitically bonded carbon (sp^2) and diamond-bonded carbon (sp^3). The higher the sp^3:sp^2 ratio, the more diamond-like the material performs, quite generally for all its properties. The material may be hydrogenated, however the highest sp^3:sp^2 materials are hydrogen-free. Properties of DLC and methods for forming this important thin film material have been widely discussed in the literature [1].

DLC is a very interesting material to be used as a protective coating since it is hard, it has low thermal expansion, high thermal conductivity, low friction and chemical inertness. In the case of MEMS, DLC coating is especially important due to the possibility to deposit very thin coatings, avoiding changing significantly the surface morphology. But a new aspect, not yet explored, is the surfaces roughness (Δ) modification for a large range of original substrate roughness (Δ_O).

In this work we have analyzed the evolution of surfaces roughness coated (Δ_C) with DLC as function of the film thickness. This study explores substrates with initial roughness (rms) Δ_O between hundreds of nanometers and just few nanometers.

The results are discussed taking to account the thin films growth dynamics [2].

MATERIALS AND METHODS

In this work we have used three substrates, two of policrystalline diamond films and one of silicon. Our intention is to have substrates with three different roughnesses Δ_O.

The diamond films were deposited by microwave plasma assisted chemical vapor deposition (CVD). The equipment used is described in reference [3]. A silicon substrate 17 x 17 mm^2 was polished by diamond powder, then it was cut in pieces and washed in acetone using ultrasonic bath. The silicon pieces were placed on the CVD substrate holder and a diamond film was grown using following parameters: 300 sccm hydrogen flow rate, 1.5 sccm methane flow rate (0.5-vol% methane in hydrogen), 1.0×10^4 Pa chamber pressure, 1128 K substrate temperature, with a nominal 700 W microwave power. The substrates, obtained in this way, have Δ_O about hundreds of nanometers.

The silicon substrates were wet etched, using a room temperature mixture of hydrofluoric acid, nitric acid and acetic acid. The silicon pieces were immerged in the acid solution during about 5 minutes. The silicon samples in this way have Δ_O about few nanometers.

Each sample was characterized by atomic force microscopy (AFM) in contact mode. The microscope used was a Scanning Probe Microscope, NanoScope IIIA, Digital Instruments. A silicon nitride tip was used, with a highest measurable angle of 65°. In our AFM images the measured angles are much smaller than 65° and the smallest pixel size used was about 30 nm, so it was not necessary to take into account the convolution effect of the tip shape and surface profile. For each sample, three images were taken and an average roughness was obtained.

NSTI-Nanotech 2005, www.nsti.org, ISBN 0-9767985-1-4 Vol. 2, 2005

For each piece of the three substrates, different DLC thicknesses were deposited. The equipment used for the DLC deposition was a metal plasma immersion ion implantation and deposition (MePIIID) system [4]. This technique is highly effective for producing high quality DLC films, and has been described in detail elsewhere [1]. In this approach, a carbon plasma is formed from a vacuum arc plasma gun [5]. Hydrogen-free DLC films have been made that are ion stitched to the substrate. In this work, the parameters used for the DLC deposition were: 200 A for the arc current, with 5 ms for arc duration and the frequency of the pulses was 1 Hz.

Finally the DLC films on the samples were characterized by AFM. For each sample, three images were taken and an average roughness was obtained.

RESULTS

The purpose in using the substrates describe above was to have substrates with Δ varying from hundreds of nanometers up to few nanometers.

Figures 1, 2 and 3 present the surface roughness shifts (Δ_C-Δ_O) as a function of the DLC thickness coating. As one can see, for the substrates with Δ_O = 393 nm and Δ_O = 278 nm, the roughness shift (Δ_C-Δ_O) increased with the DLC thickness up to a maximum value and then decreased. For the substrate with $\Delta_O \approx 4$ nm, the roughness shift fluctuated without any systematic tendency to decrease or increase.

These results can be interpreted using some of our previous papers [6, 7]. In one of theses paper [6] we have coated silicon AFM tips with DLC, with different film thicknesses. We verified that in the tip extremity there was a higher DLC deposition rate than on the lateral surfaces of the tip, as illustrated in the scheme presented in figure 4. In this way, for pointed structures, as we have for the diamond films, the DLC films introduce broadening of the sharp edges of the faceted crystals, increasing the roughness. This explains why the Δ_C increases for small DLC film thickness. On the other side, for thicker films, the Δ_C begins to decrease (see figures 1 and 2).

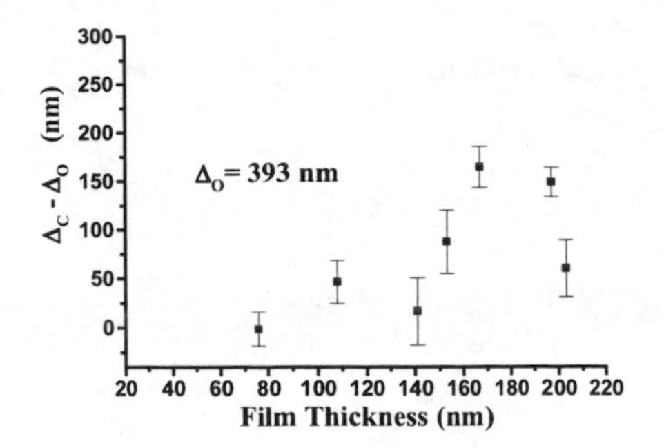

Figure 1: Plot of the roughness shifts (Δ_C-Δ_O) as a function of the DLC thickness coating, using as substrate CVD diamond, with Δ_O = 393 nm.

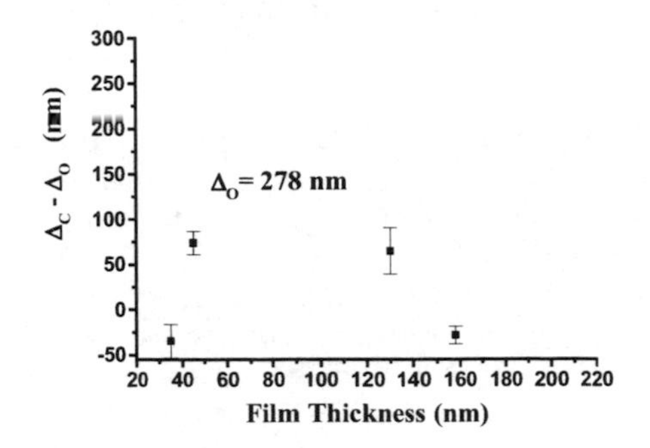

Figure 2: Plot of Δ_C-Δ_O as a function of the DLC thickness coating, using as substrate CVD diamond, with Δ_O = 278 nm.

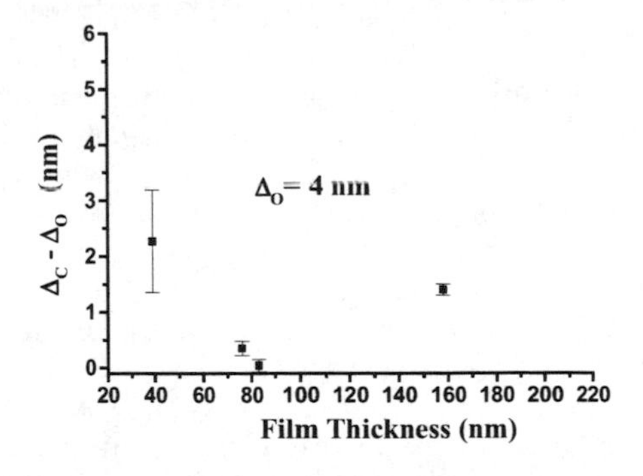

Figure 3: Plot of Δ_C-Δ_O as a function of the DLC thickness coating, using as substrate silicon surface etched with acid, with $\Delta_O \approx 4$ nm.

In the other previous paper [7] flat silicon (monocrystal) have been coated with DLC, with thickness between 30 and 200 nm. In this case we verified that the DLC roughness was about 0.5 Å, corresponding to the original substrate roughness. This fact is coherent with the result, seen in figure 3, obtained here for the silicon substrate wet etched. In this case Δ_C-Δ_O, due to the DLC coating, fluctuates without systematic tendencies to decrease or increase.

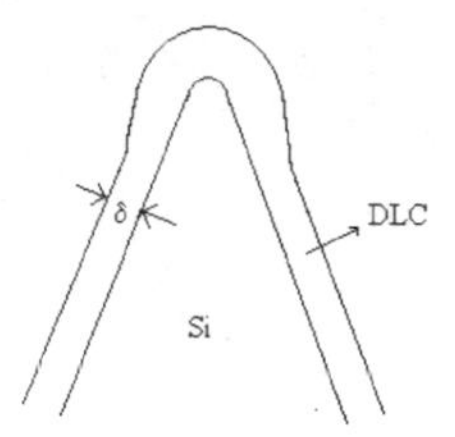

Figure 4: Scheme illustrating that in the tip extremity the DLC deposition is thicker than on the lateral surfaces of the tip (δ). This was observed in a previous paper [6].

DISCUSSION AND CONCLUSION

The diamond films with different amorphous carbon concentrations have been studied using growth dynamics [8, 9]. It was verified that the ballistic deposition model (BD) is obeyed, belonging to the same universality class of the stochastic Kardar-Parisi-Zhang (KPZ) equation. The DLC films are basically amorphous carbon with diamond carbon bonds, so we expect that they also obey the ballistic deposition model. As it is extensively described in the literature [2], in the ballistic deposition the films have a vertical and a lateral growing. This last effect is generated by the capture of the arriving particles by the top of the structures on the surface. In the KPZ equation it is described by a non-linear term. This lateral growing is responsible for the broadening of the extremities of the sharp edges on the diamond crystallites, as described above. So, in the very beginning of the DLC film deposition on a rough surface like diamond with sharp edges, we have a rapid increase in the surface roughness Δ_C. As the DLC film thickness increases, due to the lateral spreading of the deposited particles and the formation of holes in the film structure [2], Δ_C stops to increase and tends to decrease. This is seen in figures 1 and 2. On the other hand, when DLC is deposited on a smooth surface, like the silicon with low Δ_O, with no sharp edges, Δ_C is practically constant with the film thickness. It does not increases and decreases, as observed in the cases of rough diamond substrates.

It is important to note that these roughness features are valid for small DLC film thickness, as presented in this paper. In these cases the films growing have a memory of the initial substrate roughness Δ_O [10, 11]. For thicker films this initial memory is lost and the DLC film roughness Δ_C will tend to increase as the film thickness increases [2, 10, 11].

ACKNOWLEDGMENTS

The authors thank the FAPESP and the CNPq for the financial support.

REFERENCES

[4] A. Anders, Surf. Coat. Technol. 93, 158-167 (1997).
[1] See, for instance, Proceedings of the 10th European Conference on Diamond, Diamond-like Materials, Carbon Nanotubes, Nitrides and Silicon Carbide, in Diamond Relat. Mater. 9, 231-1306 (2000) [Vol. 9, Nos.3-6, April/May, 2000].
[2] A.-L. Barabási and H. E. Stanley - "Fractal Concepts in Surface Growth". Cambridge University Press (1995).
[3] M. C. Salvadori, V. P. Mammana, O. G. Martins and F. T. Degasperi, Plasma Sources Science and Technology 4, pp 489-493 (1995).
[5] R.A. MacGill, M.R. Dickinson, A. Anders, O.R. Monteiro and I.G. Brown, Rev. Sci. Instrum. 69, 801-803 (1998).
[6] M.C. Salvadori, M.C. Fritz, C. Carraro, R. Maboudian, O.R. Monteiro and I.G. Brown, Diamond and Related Materials 10, 2190-2194 (2001)
[7] M. C. Salvadori, R. M. Galvão, O. R. Monteiro and I. G. Brown, Thin Solid Films 325, 19-23 (1998).
[8] M. C. Salvadori, M. G. Silveira and M. Cattani, Physical Review E 58, 6814-6816 (1998).
[9] M.C. Salvadori, L.L. Melo, D.R. Martins, A.R. Vaz and M. Cattani, Surface Review and Letters 9, 1409-1412 (2002).
[10] M.C. Salvadori, A.R. Vaz, R.J.C. Farias, M. Cattani, Surface Review and Letters 11, 223 – 227 (2004).
[11] T.J. da Silva and J.G. Moreira, Physical Review E 56, 4880-4883 (1997).

A Novel Process for Flexible Nano Electronic Web Production

G. Picard, J. Schneider and M.F. Seye
Nanometrix Inc., 329 West, rue de la Commune, suite 200, Montréal (Québec), Canada H2Y 2E1, gillespicard@nanometrix.com

ABSTRACT

A new concept to mass produce smooth ultra-thin polymeric films for nano electronics is presented. The new method versatility is such that any kind of surfaces can be coated with an equal linear speed. The polymeric film thickness can be predetermined with nanometer precision, from 100 nm down to only 1 nm. Three polymeric solutions were thinned down to nanometric scale, namely: PVDF, PMMA and PVPh. The substrates were chosen according to their near future impact, namely rigid 300 mm wafers for EUV lithography and rolls of flexible Mylar for flexible electronics. The results are very impressive, as the substrates were all coated with uniform ultra-thin films at a rate of one square meter per minute, with atomic smoothness. The defects density is quite low, because the method works in high compression. This technological breakthrough, due to a systematic structural approach, enables the most far reaching road maps in nano electronics for the next decades.

Keywords: nanotechnology, electronics, web, monolayer, ultra-thin film

1 INTRODUCTION

A novel method for high speed monolayer or ultra-thin film preparation was designed ex novo to match with today's industrial production line standards. This effort responds to an increasing demand for efficient fuel cells, high performance catalysis, optical devices, flat displays and micro- and nano-electronics. Indeed thinness, uniformity, orientation and organization are all synonymous of performance and economics. Since the trend is carrying us toward the ultimate limit, the atomic level, it becomes much wiser to built films from bottom up using elements in a similar fashion than bricks and mortar.

Nanometrix' process is very versatile: electronics, optics, photonics, fuel cell, and biotech applications are being developed in our laboratories. The materials being processed are any solutions, colloidal suspensions or fine powders. Lipids, nano diamonds, quantum dots, platinum dots, SiO_2, TiO_2, latex, PMMA, PVDF, poly (4-vinylphenol), fluorescent, thermo- and photo chromic beads were deposited onto meter long flexible substrates or wafers of Ø 4", 6" and 8". In total, this represents a size range from one nanometer to 300 microns.

More specifically, the advantages of ultra-thin films for micro electronics, in particular micro lithography, have been mentioned for decades [1-2]. Already in 1983, monolayers were proof tested and it became obvious from the point of view of specialists in monolayers that ultra-thin films ($\leq$ 10 nm) could be tailored to any specific requirement down to one single monolayer ($\approx$ 1 nm).

A decade later, in 1994, a scientific publication entitled "Superiority of Langmuir-Blodgett resist films in electron beam lithography as demonstrated by the backscattering yield" was showing that for making ultra-thin films the bottom up approach still remains the best choice for the future.

In today's Nanometrix labs, ultra-thin films are mass produced at linear speed of a meter per minute, with thinness reaching one nanometer. This document presents three bench tested polymers: PMMA, PVDF and poly (4-vinylphenol) (PVPh). The first polymer, PMMA, was successfully thinned down to 1.3 nm with atomic flatness. The film quality across several square meters was constant.

In the two latter cases, we bench tested the film thickness against one variable, polymer initial concentration. PVDF films with a thickness range from 2 nm to 80 nm are mass produced, and poly (4-vinylphenol) film thickness ranges from 2 to 20 nm in very same conditions. All films were deposited onto wafers to demonstrate their potential use in micro- nano-electronic business, or onto quartz slide for spectroscopic investigation. It is noteworthy that only one single coating passage is enough for obtaining these results.

2 FUNDAMENTAL PRINCIPLES FOR HIGH QUALITY ULTRA-THIN FILM

Nanometrix presents in this document a breakthrough technology, a new thin film generator concept to make monolayers and ultra-thin films as well, suitable for nano electronics web processing. In order to avoid confusion with the terminology, a monolayer is defined as a one layer thick assembly of elements, while an ultra-thin film is qualified only by its thickness. In this work, an ultra-thin film is always made of long polymeric chains, on purpose intermingled or not to obtain any desired thicknesses.

Therefore, we first explain the basic ideas behind the Nanometrix method. This is followed by examples of thin films having an immediate impact on nano electronics. A discussion concludes this document with the concept of industrial production line of tailored nano coats onto wafers.

2.1 Nanometrix' first fundamental principle: "Build monolayers one element at a time"

To mass produce continuous, well packed, uniform monolayers, stacking orderly the material side-by-side is the only way. In Nanometrix' proprietary process, the gas-liquid interface natural properties like flatness, mobility and tension, together with other natural forces like gravity are elegantly used. The material attached at the interface is hydro dynamically driven one after the other toward and onto a film formation line. The pressure applied onto the film long axis is kept constant while a conveyer transfers the film from the liquid surface toward a solid substrate.

2.2 Nanometrix' second fundamental principle: "High speed beats self-organization"

As a matter of fact, the process was invented to cure a long lasting basic problem encountered with quasi-static monolayer instrumentations: self-organization [3]. High speed hydrodynamics for mass production also cures defects in monolayers. Because of the speed, molecules or other material like polymer chains are driven at the interface in only one direction, preventing self-organization by random motion. Therefore, mass production and quality match perfectly.

2.3 Nanometrix' third fundamental principle: "High compression means solid state"

The particle assembly is afterward put under an ever increasing pressure. In the case of an ultra-thin film, it becomes solid on the liquid phase and pressurized. The pressure inside a monolayer is about 300 atmospheres for a 1 nm layer. Still under this high pressure, the film is transferred onto a solid surface.

Such a compression level means that no lateral forces on the solid surface are likely to rupture the thin film. In fact, thinness as small as one nanometer with PMMA were obtained, with no pin holes or else.

It is noteworthy that spin coating is a process that works in expansion. Therefore, strong disjoining (or rupturing) pressures and shear stresses between the thin liquid film interfaces (gas-liquid and liquid-solid) create hills and valleys, or worse, pin holes. Also the presence of irregularities like dusts or scratches at nanometer scale acts like needle on a balloon: the membrane blows apart.

3 POLYMER FILMS ONTO RIGIDS

Three polymers were bench tested to see the potential impact of Nanometrix method in nano- to micro-electronics. The polymers are namely PVDF, PVPh and PMMA. The polymers were transferred either on flexibles and rigids, e.g. silicon wafer and glass slide.

In the case of PVDF, the polymer is well known for its dielectrics properties. It is used to isolate conducting materials one to the other and can therefore be used for solid state mass storage memory. By using DMF as a solvent, we spread it at the air-water interface and made a continuous production at a rate of one meter square per minute. The area was sampled for TEM and AFM examination. The results are that the thickness was uniform at 42.3 nm with less than one nanometer roughness. The thickness could be adjusted with the choice of the initial solution concentration.

With PMMA, the target was to reach the lowest limit possible. Indeed, this is a classical polymer for spin coating industry. Large surface like 30 cm wide wafers are difficult to coat uniformly. Using PMMA/chloroform solution, we were able to produce a layer of 1.3 nm with a surface roughness of only 0.1 nm, i.e. an atom thick. The production rate of such film was two square meters per minutes, which is the upper speed limit of our machinery (electric motor). It is quite possible to go much faster if desired.

A polymer has been suggested by IBM for Extreme Ultra-Violet (EUV) lithography [4]. This is PVPh. The polymer has a strong absorption band around 198 nm. We used this specific absorption band to quality control by spectro photometric measurements the thickness uniformity on clean quartz slide. Furthermore, TEM and AFM microscopes were also used to evaluate matter organization at molecular level.

The film thickness was measured against the initial polymer concentration, and the results are reported at Figure 1. It can be seen that the relationship is linear till saturation, where the solution solidifies too quickly at the air-water interface. At the lowest possible concentration, the intercept value represents the thickness of the thinnest layer possible.

This method allows the preparation of multilayers by piling up uniform films, one layer on top of another one. This was demonstrated using a 20 mg/ml solution of PVPh in DMF. The thickness was monitored with photometric measurements of the transmitted light at 198 nm to quantify the amount of matter deposited onto the quartz slide. The multi-film number and thickness are illustrated at the Figure 2.

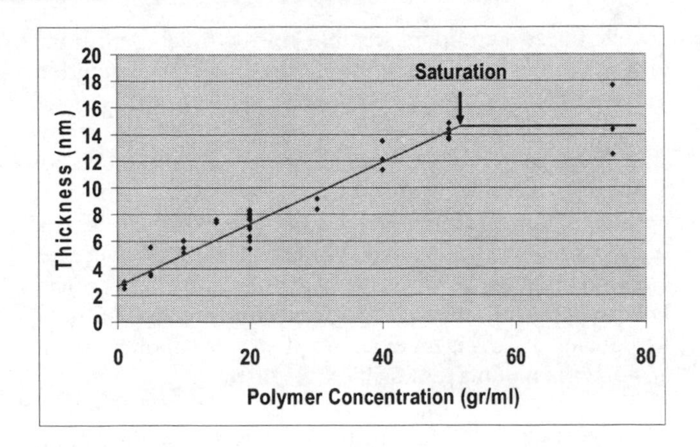

Figure 1: Film thickness vs. concentration. The film thickens linearly till saturation. In the linear part, the film remains smooth. The upper limit can be much higher if desired by choosing another solvent.

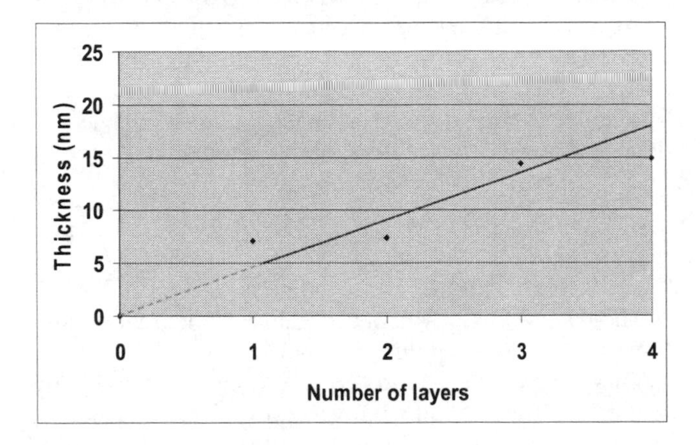

Figure 2: Multi-film thickness against the number of layers deposited onto the quartz slide. The intercept crosses the origin, meaning the superposition is good.

It is noteworthy that different polymers could have been used for making the stratification. Therefore, they could have been piled up alternately. Doing so, a whole set of new devices could be fabricated.

Concomitantly to processing material, nano scale planerizing surfaces of rough substrates (plastic films like PET, PC, etc.) by liquid substrate entrapment can also be performed. This will be better explained in the next section. Since thinness means also flexibility for rigid material, Nanometrix offers to the nano electronic business a versatile tool for the present and near future in nano electronics.

4 POLYMER FILMS ONTO FLEXIBLES

Nanometrix' process for making monolayers has several intrinsic advantages when the use of plastic rolls becomes an issue. Today's scientific publications are describing devices that a detail description of each of them would require several books. However, all of them are based on the use of electronics elements that are themselves based on the superposition of ultra-thin films. Therefore, the basic unit of construction of such devices is the ability to make thin films. In several cases, there are big issues, as thinness, surface roughness, uniformity and quality control are problems that must all be addressed at the same time. Nanometrix offers a simultaneous solution for all these concerns:

4.1 It works at room temperature

The process being based on driving particles onto a liquid surface, water or an aqueous solution is the preferred choice. This means that all organic material can be used to make devices based on ultra-thin film. For instance, PMMA ultra-thin films of $\approx$ 1 nm were created and transferred onto a flexible substrate at a rate of one square meter a minute.

4.2 This is a roll-to-roll process

Typically, at Nanometrix's facility, when a flexible substrate is used, we install plastic rolls on one end, and rolled up the ultra-thin film coated onto plastic on the other end. The process being computer controlled this works on automatic mode. It is interesting to note that such a process is easily upgraded to one linear meter per second, which is fully integrated in present industrial production line. Therefore, standard pre- and post-processing devices are fully exploited in the treatment of web surfaces for adhesion and other properties.

4.3 The deposition can be conformal

The monolayer is deposited onto the plastic surface with a few nanometer liquid thin films coming from the subphase. Evaporation through the ultra-thin film creates a suction that tends to make the monolayer to follow each surface features. This is important when micro-electronics are based on patterning.

4.4 Smoothing out plastic surfaces

If desired, defects like spikes, pin holes, cracks and so can be flooded while making the monolayer of ultra-thin film. This can be done only because between the monolayer and the flexible plastic is a thin layer of liquid subphase that is trapped during the transfer. By adjusting the production parameters, this entrapped layer is mobile and floods all the undesired features. The pressure of the monolayer keeps the whole film together without being ripped off by molecular forces, which makes a perfectly smoothed out surface. Therefore, a perfectly flat monolayer is deposited on an initially rough plastic surface.

4.5 Bridging over valleys

By making the right choice of molecules to be assembled, the monolayer can play the role of a canopy that span above grooves. Tests with a glass surface have demonstrated that the level of diffusion is much less, meaning the gaps were covered successfully.

4.6 Alternating organics and inorganics

With Nanometrix' process, both types of material can be used at the same time on the same surface. This means that semi-conductors for OLED can be mass produced with multilayers made of different materials at rates well above one meter per minute (at the present state).

4.7 Moisture or oxygen barrier

In the same way, nanometric silicate spherules were also deposited onto plastic surfaces. Again, these spherules can play an important role in electric isolator, as well as moisture or oxygen barriers to protect luminescent elements on plastic foil. This is a well know problem that can be solved very easily with Nanometrix technology.

5 OLED: AN APPLICATION EXAMPLE

Organic light-emitting display (OLED) has been intensively studied in the past 20 years. OLED offers a number of advantages and potential breakthroughs in the display technologies. OLED is a self emissive display where no backlighting is needed as in LCD, lower (about 1/2) power consumption, wide viewing angle, and most importantly, OLED offers a potential for thin, lightweight and flexible displays.

However, when flexible substrates are used, there are several problems to be solved. One is the surface regularity. Extruded plastic films contain a surface roughness that is in some cases too large. The second is that a moisture barrier must be put onto the plastic sheet. Here, the best barrier is glass. However, it is well know that glass does not tolerate bending and twisting. Therefore, the substrate must have a much restricted flexibility.

Nanometrix has a solution for both problems. Nanometrix has demonstrated its ability to make ultra-thin film of any kind of polymers onto any kind of substrates. The ultra-thin films are in the range of tens of nanometer in thickness, with a roughness never exceeding one nanometer (basically the molecular size of the monomer).

Nanometrix can make monolayers of glass with thicknesses in the nanometer range. It is admitted that glass is a good barrier for moisture, but is brittle and can only sustain small strains. In an OLED that uses glass as a substrate it is the glass that limits the flexibility of the device. Plastics are more flexible but too much permeable to water. Here again, Nanometrix brings his own solution to the problem.

Tests were made with 100 nm diameter silicate stacked in a two dimension assembly (a monolayer) and deposited onto a plastic foil. The openings in between spherules are filled with another silicate precipitate. The monolayer was therefore analyzed with an AFM and an optical microscope. The result is that uniformity and flexibility was obtained. This is an example of specific application of Nanometrix monolayers for flexible display.

Another issue is the strain the layer can sustain. Again, Nanometrix has a solution based on the extreme thinnest the layer can be. Indeed, thin glass available in the market is about 50 micron. Nanometrix can make glass sheets of only 50 nm. Therefore, because the strain on the glass substrate of thickness h for a given radius of curvature R is: $\varepsilon = h/R$, this means that at 50 nm thick, the strain problem is not an issue anymore, as the radius of curvature the thin layer can sustain is one thousand time smaller. Therefore, if the glass sheet is thin enough, say in the tens of nanometers, full flexibility is obtained.

6 CONCLUSION

The Schneider-Picard method is a breakthrough technology. Atomic smooth organic or inorganic thin films with nanometer thicknesses can be mass prepared. Coatings onto any rigids or flexibles are routinely performed. The method great versatility as well as ease of use makes the near future promising for significant advancements. Therefore, all road maps by micro electronic industry as well as flexible electronic devices industries are from now within reach.

REFERENCES

[1] A. Barraud, Polymerization in Langmuir-Blodgett films and resist applications, Thin Solid Films, Volume 99, Issues 1-3, 14 January 1983, pages 317-321.

[2] W. Lu, H. Y. Shen, N. Gu, C. W. Yuan, Z. H. Lu and Y. Wei, Superiority of Langmuir-Blodgett resist films in electron beam lithography as demonstrated by the backscattering yield, Thin Solid Films, Volume 243, Issues 1-2, 1 May 1994, Pages 501-504.

[3] H. E. Ries, Jr. , M. Matsumoto, N. Uyeda and E. Suito, Electron micrographs of cholesterol monolayers, Journal of Colloid and Interface Science, Volume 57, Issue 2, November 1976, pages 396-398.

[4] R.D. Allen, Wallraff, G.M., Hofer, D.C. and Kunz, R.R., IBM Journal of Research and Development, Vol.41, N.1/2, 1997, Optical Lithography.

Transition Mechanisms and Phases of Hexane Physisorbed on Graphite

C.L. Pint[*,†], M.W. Roth[*] and C. Wexler[**]

[*]University of Northern Iowa, Cedar Falls, IA, 50614
[**]University of Missouri-Columbia, Columbia, MO, 65211
[†]Electronic mail: cpint@uni.edu

ABSTRACT

We present the results of molecular dynamics (MD) simulations of hexane adsorbed onto graphite at submonolayer ($\rho<1$), monolayer ($\rho=1$), and supermonolayer ($\rho>1$) coverages. We find the presence of three phases, the low temperature solid herringbone (HB) phase, the intermediate-nematic (IN) phase, and the isotropic fluid phase. For submonolayer coverages, we find that the IN phase disappears except for coverages very close to completion, however, for $\rho>1$ we find an augmented IN phase with most molecules rolled normal to the substrate. We also observe for $\rho<1$ the presence of fluid-filled domain walls when taking the tetrahedral symmetry of the endgroups into account. Through variations we study the formation of gauche molecules and conclude that both phase transitions observed exhibit much out-of-plane tilting, with only a *small* presence of conformation changes near the melting transition. We conclude that in-plane space plays an important role in the phases and transitions in this system.

Keywords: molecular dynamics, adsorption, *n*-alkanes, hexane, graphite

1 INTRODUCTION

Hexane, (C_6H_{14}) is a member of the family of straight-chained *n*-alkanes whose members differ mainly in their length and whose study is appealing due to the vast industrial applications (e.g. adhesion, lubrication, wetting, etc.) that these molecules exhibit as adsorbates. Hexane is a short-chained even numbered *n*-alkane whose properties are general to a subset of similar even *n*-alkanes [1] with $6<n<10$ in that the phases exhibited by each involve an isomorphic herringbone (HB) phase at low temperatures, followed by a transition into an intermediate phase, where experiment observes a solid and liquid in coexistance.

Experimentally, the behavior of monolayer and submonolayer hexane on graphite is well studied [1-3]. These studies reveal that for monolayer hexane, there are three distinct phases observed. The first is the solid HB phase that exists at low temperatures and is commensurate with the graphite substrate. With increasing temperature, and by T=150K, they [2-3] observe a transition into a rectangular phase that seems to coexist with a liquid. This phase persists until T=176K where all order is lost in the system and melting into an isotropic fluid ensues.

Similarly for submonolayer hexane, experiment [1-3] suggests the molecules arrange in a uniaxially incommensurate (UI) structure on the substrate at low temperatures, evolving continuously into a fully commensurate structure at completion. Further study [3] of this structure gives indication of low-density light domain walls existing below coverages of ρ=0.92 that seem to be filled by a fluid. These domain walls are observed to be commensurate, except near the domain walls, where the molecules relax from commensurate positions, and melting is observed to occur from such structures.

Previous simulations [4-8] have been carried out studying monolayer hexane on graphite, which confirm the presence of the three phases and present an elegant theory by which melting in such systems takes place by means of a "footprint reduction," [4] rather than exhibiting properties of the KTHNY theory of melting in 2D. These simulations initially observe that conformation changes are responsible for the melting transition in hexane. However, more recent simulations [5-7] show that through the use of parameters more accurately representing the surface interaction [5], the melting transition occurs at a lower temperature with a large presence of molecules tilted out of the surface plane. There has been no known theoretical work completed regarding the anomalous behavior of submonolayer or supermonolayer hexane on graphite.

2 MODEL AND METHODS

The model that is used in all simulations has been described in detail elsewhere [8] and so will only be briefly described here. In all cases, we use a molecular dynamics method to simulate 112 hexane molecules at coverages of ρ =0.87 to ρ =1.05 on the graphite surface. The UA model [9] is used in most cases, which combines each methyl and methylene group into a single psuedoatom, with the two different groups distinguished only by mass and a different Van Der Waals radius. For submonolayer hexane, the anisotropic united-atom (AUA4) model [10] is also used which is anisotropic in that it treats the tetrahedral symmetry in the methyl groups and extends the psuedoatom position closer to the hydrogen atoms in the methylene groups. To modify the coverage, the computational cell is expanded along the *b*-direction of the oblique unit cell to

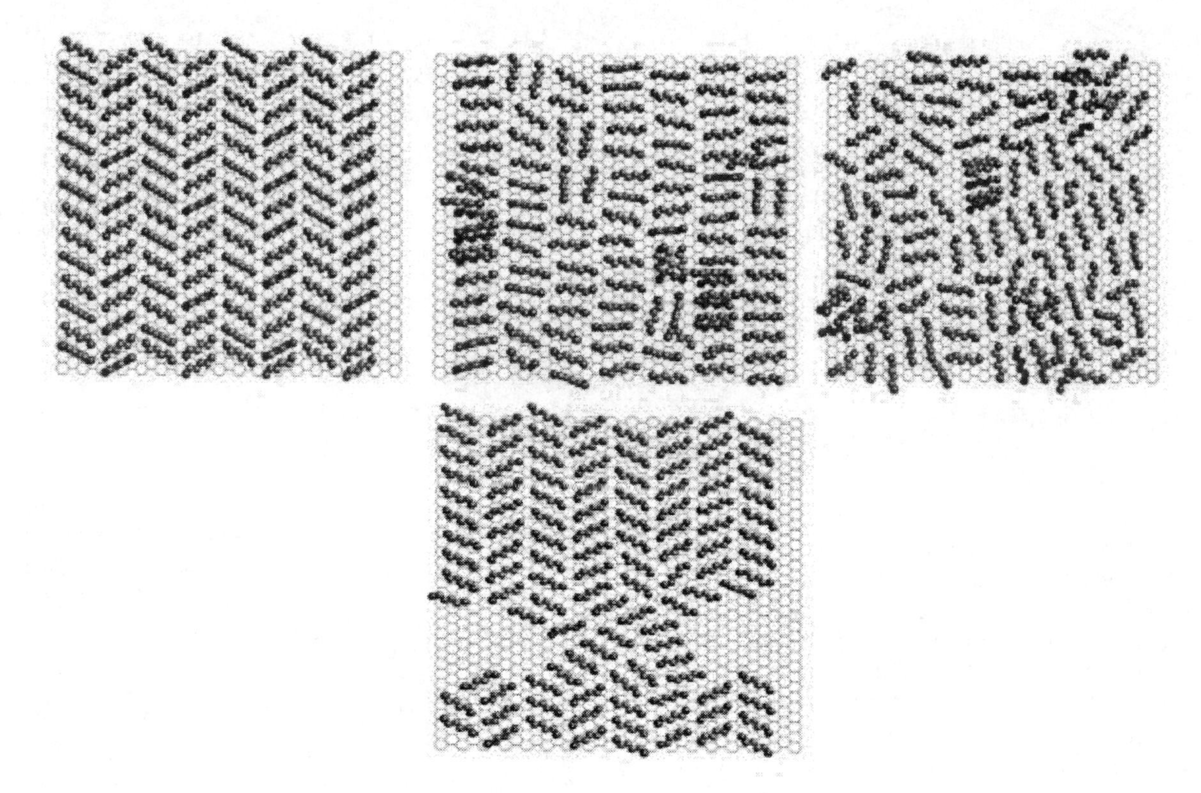

Figure 1. (Top) Snapshots of the three phases observed in hexane on graphite. These are the low temperature herringbone phase (left), the intermediate nematic phase (middle), and the isotropic fluid phase (right). Also, (bottom) the low temperature phase for ρ=0.875 is shown that is comprised of commensurate herringbone solid structures with ρ=1 on either side of a low-density fluid filled domain wall.

preserve the commensurability in the a-direction (which is most important). Periodic boundary conditions are used in the plane parallel to the substrate, and each simulation is begun from a low-temperature HB phase consistent with experimental observations [3]. To maintain temperature stability, the velocities are rescaled to satisfy equipartition for the center-of-mass, translational, and rotational velocities. A velocity Verlet RATTLE [11] algorithm is used to integrate the equations of motion as well as to constrain the fast C-C stretching modes to their equilibrium value. All simulations are carried out for a total of 700 ps with a period of 500 ps over which averages are taken.

The intermolecular interactions are modeled by a 12-6 Lennard Jones pair potential with Lorentz-Berthelot rules describing mixed interactions. The surface interaction is modeled by a Fourier expansion proposed by W.A. Steele [12] that assumes an isotropic surface of infinite extent in the xy plane, and semi-infinite in the $-z$ direction. Since the bond-lengths are constrained, the intramolecular potential is composed of two terms representing the C-C-C bond angle bending [9] and the torsion of the molecules [6,13]. Again, details regarding these interactions and the para-meters involved in their potentials are given elsewhere [8].

3 RESULTS AND DISCUSSION

A series of snapshots are presented in figure 1 that represent (top) the three phases that are observed in the study of hexane on graphite and (bottom) the domain walls observed when the AUA4 model is used at the submonolayer coverage of ρ=0.875. These give a good representation of the structure of the system in each observed phase. The behavior of submonolayer hexane with the UA model is similar to that of monolayer hexane at low temperatures with the exception that HB phase is "stretched" in the y-direction (uniaxially incommensurate).

In figure 2, the bond-roll angle, $P(\Psi)$, distribution is presented where Ψ is defined as:

$$\psi = \cos^{-1}\left[\frac{\{(\vec{r}_{j+1}-\vec{r}_j)\times(\vec{r}_{j-1}-\vec{r}_j)\}\cdot\hat{z}}{\left|(\vec{r}_{j+1}-\vec{r}_j)\times(\vec{r}_{j-1}-\vec{r}_j)\right|}\right]. \quad (1)$$

In equation 1, the values of $\vec{r}$ represent the psuedoatom positions of three consecutive psuedoatoms in a molecule. The cross product therefore involves two vectors that extend from an innermost psuedoatom to two bonded neighbors. The dot product in the numerator is zero when the backbone of the molecule is rigid with the surface (when the molecule is rolled normal to the substrate), and Ψ=90°. If the reverse is true, where the molecule lies flat in the surface plane, the quantity inside the brackets in

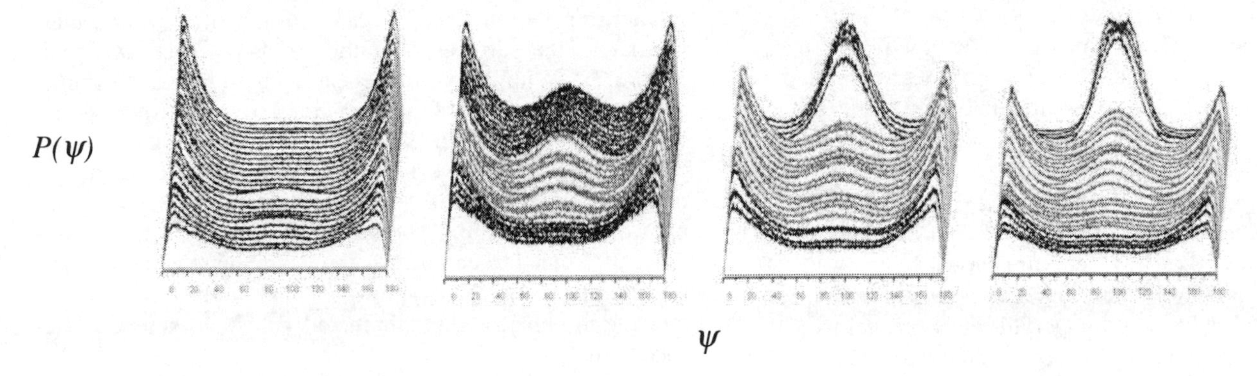

Figure 2. Bond roll angle distributions $P(\Psi)$ for submonolayer (two on left), and supermonolayer (two on right) densities. The value of Ψ is 90° when the molecule is perpendicular to the plane of the substrate, and is 0° and 180° when the molecule is parallel to the plane of the substrate. The blue lines correspond to temperature points in the herringbone, the green lines correspond to points in the nematic, and the black lines correspond to the isotropic fluid. The second plot corresponds to the submonolayer density exhibiting the weak nematic phase. The average density increases from left to right.

equation (1) assumes a value of 1, which gives Ψ=0° or 180°. Therefore, a large peak at Ψ=90° in figure 2 for supermonolayer coverages indicates that the majority of the molecules are rolled normal to the substrate, whereas for submonolayer coverages, $P(\Psi)$ indicates that there is very little to no in-plane rolling of the molecules. Although the presence of the intermediate phase is clearly augmented from an increasing coverage by inspection of figure 2, table 1 gives a quantitative representation of the simulated transition temperatures that are observed for each coverage studied with the UA model. These transition temperatures are composed through study of averaged order parameters, energies, and correlation functions that are monitored through each simulation. The results seem to suggest that the intermediate phase is dramatically affected by the increase/decrease of coverage, however, the melting transition temperature stays fairly consistent until the lowest coverage, which we find is where the UA model breaks down and fails to represent this system in terms of both structure and transition temperatures.

3.1 Observed Phases

In most cases, the three phases that we observe are those that have been both observed experimentally as well as in previous theoretical work. The intermediate phase, however, we find exhibits order that is similar to a liquid crystal, and although systems of 112 molecules are not large enough to fully indicate that there is fluid coexisting with a solid, simulations of 336+ hexane molecules shows that between T=160-175K, there is a significant amount of disordered molecules that exist around a cell of orientationally ordered molecules. This indicates that the nematic phase shown in figure 1 is representative of the solid-liquid coexistance region that has been observed experimentally [1-3] to exist between the solid and liquid

ρ	T_1 (K)	T_2 (K)
0.875	N/O	155±3
0.9	N/O	172±3
0.93	N/O	174±3
0.965	155±5	175±3
1.0	138±2	176±3
1.02	122±3	172±3
1.035	98±3	175±3
1.05	85±3	175±3

Table 1. Intermediate and melting transition temperatures, T_1 and T_2, for each coverage studied. The use of "N/O" refers to a transition that is not observed. All temperatures are in units of Kelvin and all uncertainties are noted.

phase. The observation that this coexistance region continuously loses solid properties until the melting transition is in agreement with the fact that we only observe coexistance over the latter portion of the nematic phase.

There are some distinct differences that are present structurally in the intermediate phase as the coverage is increased. First of all, we find the intermediate phase is preempted by layer promotion that is a result of molecular interactions that cause molecules to tilt out of the surface plane and promote to the bilayer. This action causes vacancies in the monolayer and if the molecules are not tightly packed (i.e. coverages greater then completion), then in some cases, the molecules will be free to orient normal to the director angle. This is evident in figure 1 where five groups of 2 hexanes orient normal to the monolayer system orientation. However, for higher coverages, the molecules *all* tend to a single orientation because there is less in-plane space for them to undergo such an orientational change.

This gives rise to an increased value for the nematic order parameter as well (not shown).

The other phase that is of interest is the low-temperature HB phase. For submonolayer hexane, through use of the UA model, the HB phase is uniaxially incommensurate under all circumstances until the model breaks down at a coverage of ρ=0.875. Experiment suggests that a UI HB structure is observed for coverages greater than ρ=0.92, but below this coverage, the formation of light domain walls with a proposed fluid occupying the domain wall is observed. Simulations that use the AUA4 model for ρ=0.875 show that the low-temperature HB phase involves the presence of domain walls with domain wall spacing of about 8-12a_g, which seems to be consistent with experiment for the most part. Also, the molecular relaxation near the domain wall is also observed, as shown in figure 1. However, the melting temperature of the system is very low with AUA4 model, but this could arise from a number of effects including finite size effects and periodic boundary conditions as well as the initial UI configurations that all simulations are started from.

3.2 Observed Transitions

The transition temperatures observed are presented in table 1 with their respective uncertainties. The melting transition in all cases seems to be consistent at about T=175K for all coverages that are studied. However, the intermediate transition temperatures are far more dependent upon coverage than the melting temperatures.

Previous work on this system has found that it obeys the space reduction theory previously proposed [4-7], but it has been unclear how this space reduction takes place. We find that the dihedral distribution (not shown) gives indication that there are gauche molecules present near the melting transition, but this gives no quantitative explanation for how these affect the transition. To elucidate this melting behavior, we simulate the monolayer with constants in the torsional potential increased by 10x (to constrain the molecules from undergoing conformation changes). We find that the melting transition is increased by ca. 20K, which indicates that gauche defects do contribute to the melting transition, but the contribution is not very significant due to the effects of out-of-plane tilting. This shows the footprint reduction in this system is cooperative between these two space reduction mechanisms with out-of-plane tilting playing the lead role.

For the intermediate phase, we find that the footprint reduction is *also* satisfied through layer promotion. The promotion of molecules to the 2nd layer creates a number of in-plane vacancies that preempt a phase transition into the intermediate phase. We notice that the intermediate phase transition occurs at lower temperatures when the in-plane space is reduced, which means that the molecules are undergoing more intermolecular "collisions", which provide them with more kinetic energy to tilt out of the surface plane at lower temperatures and for some molecules to simultaneously promote to the 2nd layer. We also find a relationship between the in-plane rolling of the molecules normal to the substrate and the in-plane space available (figure 2). With less in-plane space, the molecules assume a less energetically favorable position (rolled normal to the substrate) which indicates that the intermolecular kinetic energy is higher in the system, however, upon the transition into the nematic-ordered phase, the molecules relax and roll back on their sides until the melting transition. This is also evident from inspection of the average Lennard-Jones interaction energy (not shown), which indicates that the interaction energy is minimized at this intermediate transition.

Therefore, we find that the transitions in this study and the intermediate phase are a result of the effects that come about due to space reduction in the system. We conclude that understanding the effects of in-plane space in this and similar systems is a fundamental part of understanding the behavior of the transitions and the phases.

The authors thank H. Taub, F.Y. Hansen, and G. Peters for enriching discussions. Acknowledgement is made to the Petroleum Research Fund, administered by ACS, for support of this work. CW acknowledges support from U. of Missouri Research Board and Research Council. CP and MR acknowledge UNI for support through Summer Research Fellowships.

REFERENCES

[1] T. Arnold, R.K. Thomas, M.A. Castro, S.M. Clarke, L. Messe, and A. Inaba, Phys. Chem. Chem. Phys. 4, 345-51, 2002.

[2] J. Krim, J. Suzanne, H. Shechter, R. Wang, and H. Taub, Surf. Sci. 162, 446, 1985.

[3] J.C. Newton, Ph.D. dissertation, University of Missouri-Columbia, unpublished, 1989.

[4] F.Y. Hansen and H. Taub, Phys. Rev. Lett. 69, 652, 1992.

[5] E. Velasco and G.H. Peters, J. Chem. Phys. 102, 1098, 1995.

[6] G.H. Peters and D.J. Tildesley, Langmuir 12, 1557 1996.

[7] G.H. Peters, Surf. Sci. 347, 169, 1996.

[8] M.W. Roth, C.L. Pint, C. Wexler, Phys. Rev. B (in press)

[9] M.G. Martin and J.I. Siepmann, J. Phys. Chem. 102, 2569, 1998.

[10] P. Ungerer, C. Beauvais, J. Delhommelle, A. Boutin, B. Rousseau, and A.H. Fuchs, J. Chem. Phys. 112, 5499, 2000.

[11] M. P. Allen and D.J. Tildesley, "Computer Simulation of Liquids," Clarendon Press, New York, 1988.

[12] W.A. Steele, Surf. Sci. 36, 317, 1973.

[13] P. Padilla and S. Toxvaerd, J. Chem. Phys. 94, 5650, 1991.

Fabrication of Hydrophobic Film Using Plant Leaves from Nature

Seung Mo Lee, Hyun Sup Lee and Tai Hun Kwon

Dept. of Mechanical Engineering, Pohang University of Science and Technology (POSTECH),
San 31 Hyoja-dong, Namgu, Pohang, Kyungbuk, 790-784, Korea, thkwon@postech.ac.krT

ABSTRACT

In this study, we have investigated the feasibility of reproducing the hydrophobic characteristics of plant leaves from nature by replicating the micro/nano features of a leaf surface in UV embossed film. Films were fabricated by means of UV nanoimprint lithography (UV-NIL) technology. By varying UV-NIL processing conditions, we were able to produce hydrophobic films which replicated the micro/nano structure of leaves. The surface topology of those replicated films was observed by scanning electron microscope (SEM) and atomic force microscope (AFM). Also for the purpose of evaluating the hydrophobicity of the replicated film surface, contact angle (CA) was measured. We propose a new measure of replication quality to evaluate how well the film can reproduce the hydrophobic characteristics of natural plant leaves. In terms of these measurements of CA and replication measure (r), we were able to confirm that UV-NIL could be a possible tool to produce a hydrophobic film by replicating surface of plant leaves.

Keywords: UV-nanoimprint lithography, Plant leaf, Hydrophobicity, Contact angle, Replication measure

1 INTRODUCTION

For the past decade, the secret of biological surfaces has been elucidated with the help of scanning electron microscope (SEM). So recently, botanists have investigated hydrophobicity of plant leaves with high CA [1,2]. And hydrophobic films with high contact angle (CA > 150^0) have attracted many researchers' attention for various practical applications [3]. Some researchers developed hydrophobic films by utilizing chemical treatment. And various methods have been reported in the literature to construct hydrophobic surface which have micro and nano structures similar to those in the plant leaves. Conventionally, hydrophobic surfaces are fabricated with the help of chemical treatment by changing surface energy of materials or by modifying the surface roughness, for instance, etching of polypropylene (PP) [4], plasma-enhanced chemical vapor deposition [5], and carbon nano tube aligning [6] etc. Some of these methods produce hydrophobic surfaces by controlling the surface topography through complex chemical processes. These methods mentioned above are generally time-consuming or costly. In contrast to these methods, this article suggests a new and simple method to produce a hydrophobic rough surface based on hydrophobic plant leaves by means of UV-NIL.

2 EXPERIMENTAL

The hydrophobicity of a material is commonly measured by the CA that a water droplet makes with the surface. Before performing imprint experiments, we measured the water contact angles of many plant leaves and selected just four plant leaves for further investigation due to their excellent water repellency in this study: *Bamboo, silver maple tree, love grass, tulip tree*. CA values were measured both on upper-surface and on under-surface in the natural state. The CA measurement equipment used in this study was a Krüss Drop shape Analysis system (Krüss GmbH Germany, model: DSA 10-Mk2). Unfortunately, this contact angle meter was suitable only for measuring equilibrium contact angles, not advancing or receding contact angles. CA values were measured at ten distinct positions using water 5 μl in volume and averaged for each case. Through this measurement, we were able to find that CA of under-surfaces is higher than that of upper-surfaces in general (table 1). Based on this result, we selected the under-surface of leaves rather than the upper-surface in our subsequent imprint experiments.

2.2 SEM Images of Plant Leaves

Figures 1 – 4 show SEM images of green plant leaves at various magnifications. Green plant leaves have a double-structured surface which is the combination of micro- and nanostructures (epicuticular wax crystals [2]). These four kinds of plant leaves had their own surface structure based on combination of micro/nano sized structures: *tulip tree,* 5 μm ~ 15 μm sized hemisphere structure which has nano scale particles on its surface; *silver maple*, 10 μm ~ 20 μm sized small dots and about 100 μm sized veins of leaf which have many nano scale wrinkles; *bamboo,* long veins of leaf and about 5 μm sized protruding pattern on which rose quartz like nanostructure exists; *lovegrass*, caterpillar looking long veins of leaf which has 5 μm ~ 10 μm sized protrusions over which banana looking nano structures are sparsely spread.

2.3 Replication of Plant Surfaces via UV-NIL Technology

The primary aim of this study is firstly to fabricate a rough surface patterned after hydrophobic plant leaves' under-surface via the UV-NIL technique and secondly to check the hydrophobicity of the replicated surface. So, for this study, our group designed and constructed the UV-NIL equipment as schematically shown in figure 5. This UV-NIL process was carried out in the following sub-processes: setting up the processing sequence via a controller, attaching a plant leaf on a jig, fixing the jig on the vacuum chuck, dispensing the UV curable photopolymer on the slide glass, removing air from a vacuum chamber, applying pressure to the pneumatic cylinder, exposing the photopolymer to UV light and finally detaching the replicated film from the jig. In this UV-NIL experiment, we used *RenShape SL 5180* UV curable photopolymer (Vantico Inc., density: 1.15×10^3 kgm^{-3}, viscosity: 240 cps, critical Exposure: 13.3 $mJcm^{-2}$). One should be able to determine the appropriate processing conditions for a successful replication of leaf surface via UV-NIL technology. The present study with the equipment we developed found the most appropriate processing conditions as follows: the UV light exposure time of 600 seconds and the applied pressure of 150 kPa. It may be mentioned that all processes were performed at room temperature. All thin films mentioned in this article were deposited on flat slide glass. After we completed the fabrication process, we kept the replicated films in vacuum desiccator to prevent the film from being exposed to air and dust.

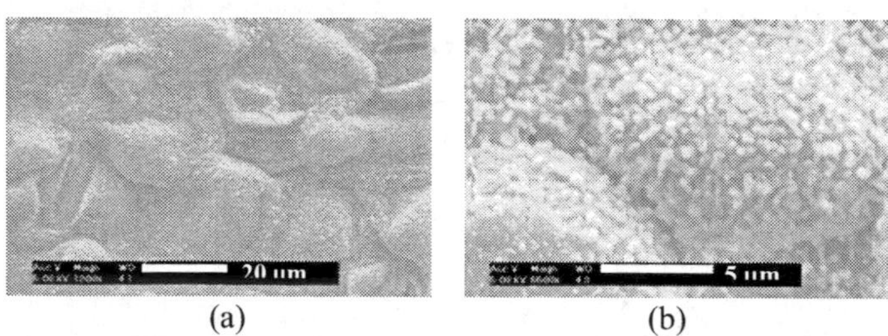

(a) (b)

Figure 1. SEM images of the *tulip tree* leaf under-surface at various magnifications: (a) 1200×, (b) 6500×.

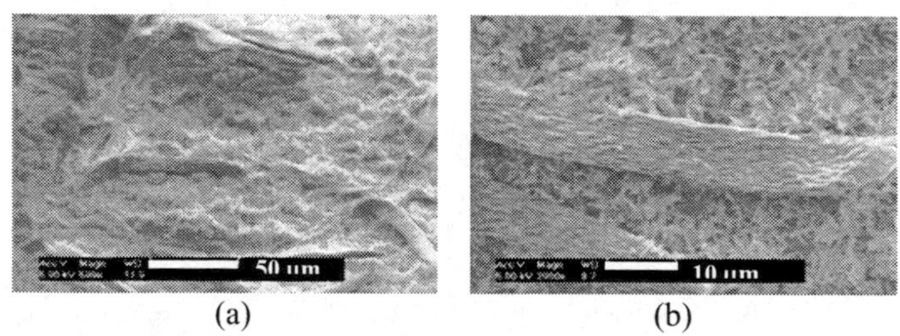

(a) (b)

Figure 2. SEM images of the *silver maple tree* leaf under-surface at various magnifications: (a) 500×, (b) 2000×.

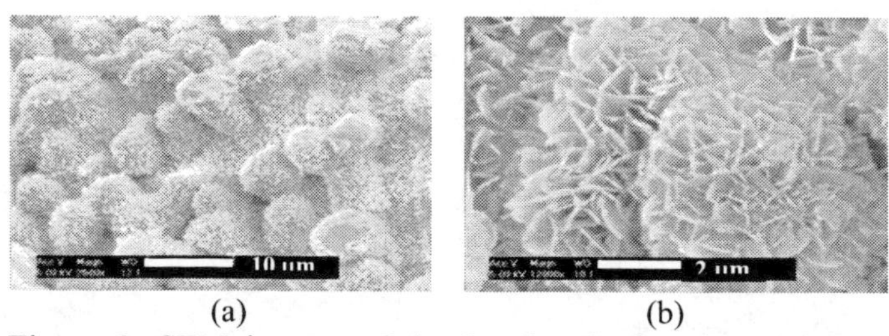

(a) (b)

Figure 3. SEM images of the *bamboo* leaf under-surface at various magnifications: (a) 2500×, (b) 12000×.

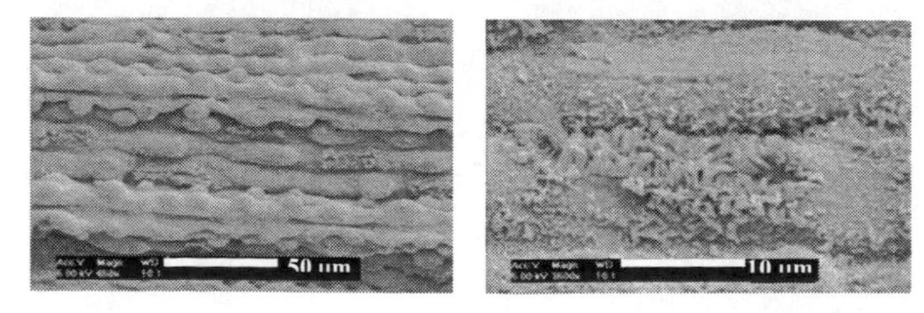

(a) (b)

Figure 4. SEM images of the *lovegrass* leaf under-surface at various magnifications: (a) 650×, (b) 3500×.

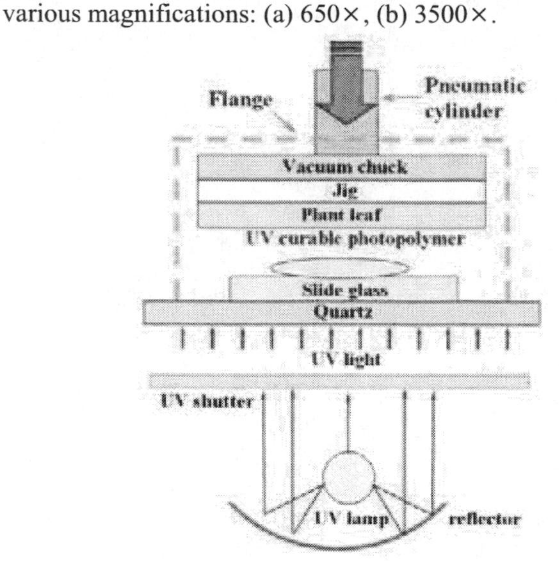

Figure 5. Schematic of UV-NIL equipment.

2.4 SEM Images of Replicated Films

Figures 6 - 9 show SEM images of replicated films. Needless to say, since plant leaves had protruding patterns, replicated films naturally had porous and rough surface structures. However, while the microstructure was generally well replicated, those nanostructures were not so well reproduced on the cured polymer film surface. This seemed to have happened because the nanostructure of plant leaf was not rigid enough to endure the pressure applied during the UV-NIL process and thereby might have collapsed or broken, resulting in poorer replication of nanostructures. Another possible reason might be that UV cured photopolymer once infiltrated inside nanoscale holes or indented surface in a liquid state remains in the leaf surface when it is cured and the film is detached from the plant leaf.

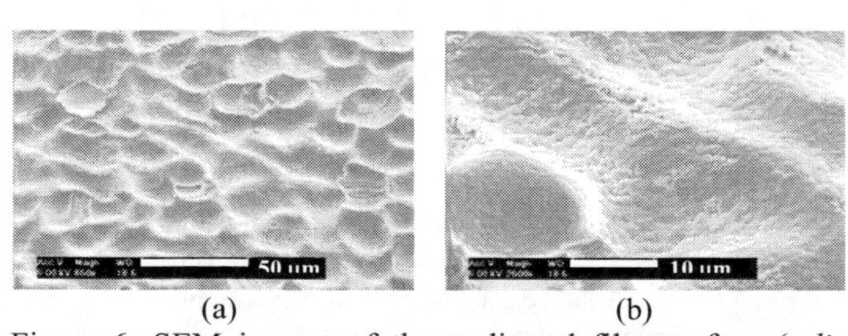

(a) (b)

Figure 6. SEM images of the replicated film surface (*tulip tree*) at various magnifications: (a) 650×, (b) 2500×.

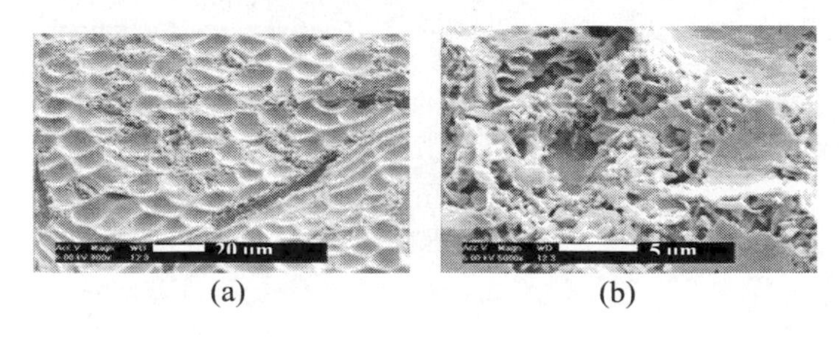

(a) (b)

Figure 7. SEM images of the replicated film surface (*silver maple tree*) at various magnifications: (a) 800×, (b) 5000×.

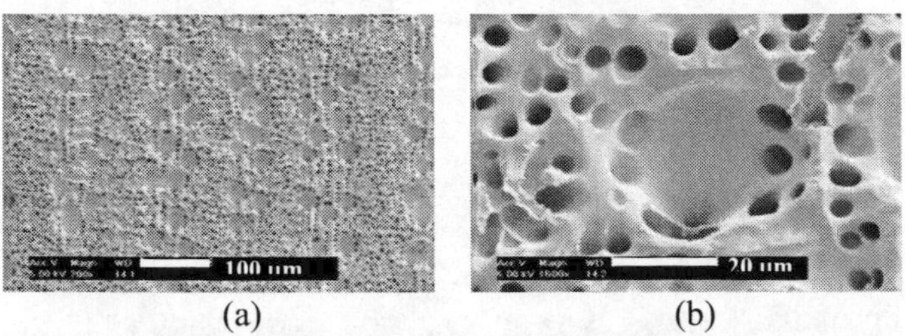

Figure 8. SEM images of the replicated film surface (*bamboo*) at various magnifications: (a) 200×, (b) 1500×.

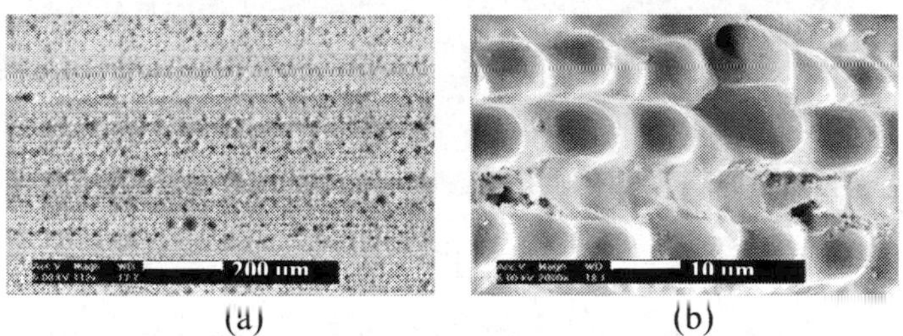

Figure 9. SEM images of the replicated film surface (*lovegrass*) at various magnifications: (a) 112×, (b) 2000×.

3 ANALYSIS OF REPLICATED FILMS

3.1 Measure of Replication Quality

It might be of interest to find how well the nano/micro combined structure is reproduced in the replicated films. For the purpose of quantifying the replication quality of replicated films, we proposed a new replication measure, *r*, defined by the surface area ratio, i.e.

$$r = \frac{\text{surface roughness of a replicated film}}{\text{surface roughness of a plant leaf}} = \frac{\text{scanning surface area of a replicated film / projected scanning surface area}}{\text{scanning surface area of a plant leaf / projected scanning surface area}}$$

As schematically depicted in figure 10, a replication measure *r* represents a kind of degree of replication. We propose to measure the surface area of both the plant leaves and their replicated films with the help of a function of AFM. Scanned AFM images in an observed area topographically represent plant surface, indicating its dimensional feature qualitatively similar to SEM. But the quality of the AFM image may become poorer than SEM due to the limitation of AFM in dealing with very steep surface and the frequent failure of the AFM probe tip during the scanning. However, SEM image does not provide a quantitative measure of three dimensional topography of the surface of interest. In this regard, we propose to use an AFM function to measure the surface area. It may be mentioned that *r* is larger than zero and smaller than unity. If *r* approaches to unity (when area of the mold surface is equal to that of the replicated polymer, i.e. $r \approx 1$), it means complete replication quality. And as *r* decreases, replication quality gets poorer. In the measurement of surface area, we scanned an area of 100 µm × 100 µm in non-contact mode of AFM. We obtained surface areas at 10 different sampling locations for each sample and calculated average and standard deviation of the measured areas. Figure 11 shows typical AFM images of *bamboo* leaf and its replicated film surface. Table 2 summarizes measurement results in terms of the average surface area with standard deviation and replication measure *r*. Due to the failure of probe tips, unfortunately, we could not obtain AFM images of *silver maple tree* leaf of which the data is thus not included in table 2. *r* values higher than *0.96* for all cases seem to indicate a satisfactory replication via UV-NIL. Undoubtedly, value of *r* cannot be regarded as an absolute estimation of replication quality. However, considering the difficulty of quantifying complicated topography of the natural plant leaf surface and their replicated films, the replication measure seems to be a relatively meaningful method to evaluate the replication quality to some extent. From this measurement, we were able to confirm the feasibility of producing replicated films by imprinting plant leaves whose surface is composed of micro and nano combined structures.

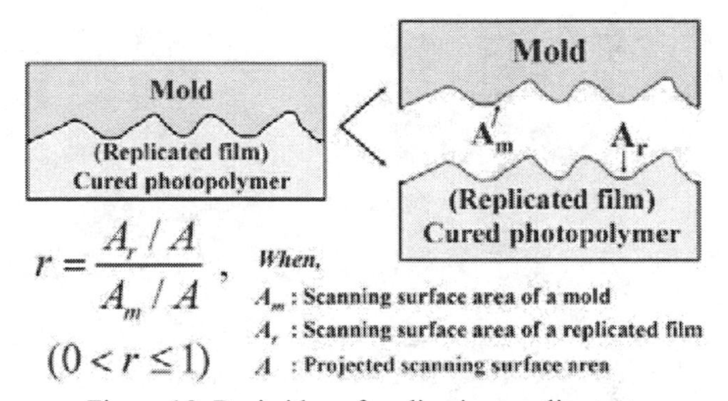

Figure 10. Basic idea of replication quality test.

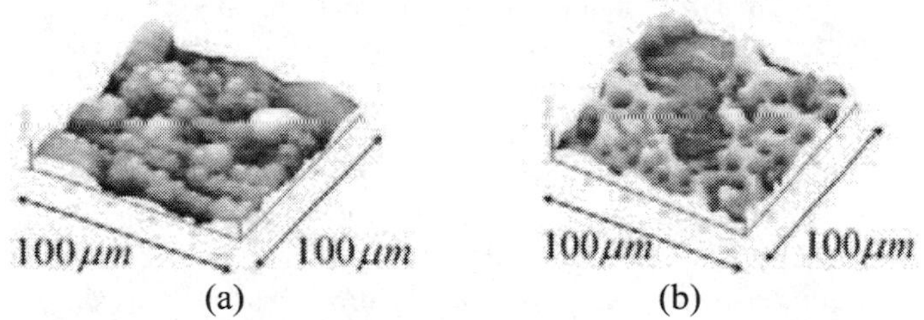

Figure 11. AFM images of two samples: (a) under-surface of *bamboo* leaf, (b) its replicated film surface.

3.2 Hydrophobic Characteristics

CA could be a measure of hydrophobicity of a replicated film surface. For a reference value to be compared with replicated films, we measured a CA of a flat and smooth surface of UV cured photopolymer materials as follows. For this purpose we fabricated a cured flat polymer surface over a self assembled monolayer (SAM) laid silicon wafer by means of UV-NIL equipment. A SAM was laid on the silicon wafer surface through the vapor evaporation method with 1H,1H,2H,2H-Perfluorodecyltrichlorosilane, $C_{10}H_4C_{13}F_{17}Si$ to help release of the cured photopolymer from the silicon wafer.

	Tulip tree	Silver maple tree	Lovegrass	Bamboo
CA of upper-surface of leaves	123.41 ± 4.50^0	115.07 ± 3.26^0	116.80 ± 3.66^0	102.49 ± 3.39^0
CA of under-surface of leaves	134.74 ± 1.71^0	136.42 ± 2.39^0	142.26 ± 5.77^0	136.25 ± 2.97^0
CA of replicated films	117.28 ± 3.98^0	121.95 ± 3.98^0	131.38 ± 3.79^0	126.71 ± 3.43^0

Table 1. Comparison of CA between green leaves and the replicated films, when a water drop is about 3 mm in diameter and 5 μl in volume

	Tulip tree	Lovegrass	Bamboo
Average surface roughness of a plant leaf	1.11199	1.21503	1.44064
Standard Deviation	0.04569	0.05743	0.10414
Average surface roughness of a replicated film	1.07953	1.17339	1.39929
Standard Deviation	0.01658	0.05728	0.08904
Replication measure	0.97081	0.96551	0.97126

Table 2. Replication measure r of each replicated film when projected scanning area is $10^{10}nm^2$.

(a) (b)

Figure 12. Photos of a water drop: (a) on under-surface of lovegrass leaf, (b) on its replicated film surface.

With the help of SAM, the surface of the UV cured photopolymer maintained the smoothness of the silicon wafer surface. CA value of a flat and smooth surface of the UV cured photopolymer was found to be $\theta_Y = 80.46 \pm 0.92^o$.

We have measured CA values for all the replicated films. As a typical illustration, figure 12 shows photos of a water drop on under-surface of lovegrass and on its replicated film. Table 1 summarizes the measured CA values for comparison between leaves and replicated films. And CA value is represented by $\theta = X \pm Y$, with X and Y indicating the average value and standard deviation, respectively. Due to the large water drop size (about 3mm in diameter and 5μl in volume) provided by the contact angle meter used in this study, CA values of plant leaves range between 135° and 142° (less than 150° which is the data reported in literature [1]). CA differences between leaves and replicated films were relatively small (about 13° in average). In table 1, *lovegrass* leaf and its replicated film had the highest CA. On the other hand, *tulip tree* leaf and its replicated film had the lowest one.

4 CONCLUDING REMARKS

In this study, we investigated the feasibility of making hydrophobic rough films by replicating plant leaves from nature via UV-NIL technology without resorting to any chemical treatments. Appropriate processing condition of UV-NIL enables us to successfully replicate hydrophobic films from four typical plant leaves. Furthermore, to quantify the quality of replicated films we proposed a new replication measure, r. r values obtained in this study were found to be around 0.97, clearly indicating good replication. The measured CA also shows the hydrophobicity of the replicated films. Good quality of replicated films in this study seems to confirm that UV-NIL is a new and simple fabrication method to produce hydrophobic films by simply mimicking nature.

ACKNOWLEDGEMENTS

This research was supported partially by a grant (2000-N-NL-01-C-148) via the National Research Laboratory Program and a grant (M102KN010003-04K1401-00312) from Center for Nanoscale Mechatronics & Manufacturing, one of the 21st Century Frontier Programs, which are supported by Ministry of Science and Technology, Korea.

REFERENCES

[1] Neinhuis C and Barthlott W, *Ann. Bot.*, **79,** 667, 1997
[2] Barthlott W and Neinhuis C, *Planta*, **202**, 1, 1997
[3] Onda T, Shibuichi S, Satoh N and Tsujii K, *Langmuir*, **12**, No. 9, 2125,, 1996
[4] Youngblood J P and McCarthy T J, *Macromolecules*, **32,** 6800, 1999
[5] Hozumi A and Takai O, *Thin Solid Films*, 303**,** 222, 1977
[6] Li S, Li H, Wang X, Song Y, Liu Y, Jiang L and Zhu D, *J. Phys. Chem. B*, **106**, 9274, 2002

Cermet Ceramic Coating on Diamond Dresser for In-Situ Dressing of Chemical Mechanical Planarization

James Chien-Min Sung[*,1,2,3], Kevin Kan[1]

Address: KINIK Company, 64, Chung-San Rd., Ying-Kuo, Taipei Hsien 239, Taiwan R.O.C.
Tel: 886-2-2677-5490 ext.1150
Fax: 886-2-8677-2171
e-mail: sung@kinik.com.tw

[1] KINIK Company, 64, Chung-San Rd., Ying-Kuo, Taipei Hsien 239, Taiwan R.O.C.
[2] National Taiwan University, Taipei 106, Taiwan, R.O.C.
[3] National Taipei University of Technology, Taipei 106, Taiwan, R.O.C.

ABSTRACT

In recent years, copper circuitry has been replacing aluminum wires and tungsten via as the dominant interconnections. The polishing of copper circuitry requires using highly acidic solution (e.g. pH = 4) and with strong oxidizing agent (e.g. hydrogen peroxide). In order to withstand the corrosive reactions, the bonding metal of the CMP pad conditioners must be corrosion resistant. Otherwise, it will be dissolved in the acidic slurry that may contaminate the delicate circuitry. Ideally, the diamond pad conditioner is best shielded from contacting with the acidic slurry. The chemical shield must be made of a hard substance so it can withstand the polishing action of the slurry. In addition, this hard substance ought to be highly inert so it will not react with the chemicals in the slurry.

The best chemical shield is diamond-like carbon (DLC) and the second best is ceramic coating. Kinik Company pioneered diamond pad conditioners protected by DLC barrier (DiaShield® Coating) back in 1999 (Sung & Lin, US Patent 6,368,198) and there has been no follower since then. Kinik's offered two varieties of DiaShield® Coatings: ultrahard amorphous diamond and superhard hydrogenated DLC. Recently, Kink also evaluated Cermet Composite Coating (CCC or C^3, patent pending). C^3 is unique that the coating composition grades from a metallic (e.g. stainless steel) under layer to a ceramic (e.g. Al_2O_3 or SiC) exterior. The metallic under layer can form metallurgical bond with metallic matrix on the diamond pad conditioner. The ceramic exterior is both wear and corrosion resistant. The gradational design of C^3 coating will assure its strong adherence to the substrate because there is no weak boundary between coating and substrate.

By dipping diamond pad conditioners of various designs in acidic solution (e.g. copper cleaning solution) for extended period of time (e.g. 50 hours) the chemical inertness of various matrix materials are determined with the decreasing ranking as: hydrogenated DLC > C^3 coating > amorphous diamond > sintered nichrome > brazed alloy > electroplated nickel.

Keywords: DLC, cermet, CMP, coating

1 The IN-SITU DRESSING OF CMP

Chemical mechanical planarization (CMP) is the must step of flattening wafer and smoothening its surface for the manufacture of sophisticated integrate circuitry (e.g. ULSI). During the process a coated wafer is pressed against a rotating polyurethane pad (e.g. IC1010 manufactured by Rohm Haas Electronic Materials) and it is gradually polished. The pad top is permeated with a slurry that contains both dissolved chemicals (e.g. hydrogen peroxide) and suspended abrasives (e.g. fumed silica). The chemical is used to react with the wafer coating; and the abrasive is employed to remove the reactant, as well as other protruded part.

The CMP process cannot be sustained because the pad surface will be deposited with polishing debris. Moreover, the pad top itself will be polished by the wafer so it becomes smooth. The dirt loaded smooth pad cannot polish the wafer effectively. A diamond dresser, also known as pad conditioner, is used to scrape off the debris from the pad top. At the same time, the diamond grits on the dresser will cut into the pad top (e.g. 20 microns) and carve out the texture required for holding the slurry and for abrading the wafer surface.

The diamond dresser is made by attaching diamond grits (e.g. 200 microns) on a substrate (e.g. stainless steel). In general, the diamond grits are embedded in a metal matrix (e.g. nickel alloy) that either bonds to diamond or locks up the grit in place. The metal matrix can be applied by either brazing or electroplating. The former process is more desirable as the matrix is once in a molten state that can wet diamond and form strong chemical bond at the interface. In this case, diamond grit will not be pulled out during the polishing process. Moreover, the molten matrix will form a wetting profile that rises toward diamond but recesses away from diamond. The rise will reinforce the support of diamond; and the recess will facilitate the flow of slurry, as well as the removal of the debris.

It is preferably to dress the pad in-situ, i.e. while the wafer is polished. In this case, the pad top can be cleaned in real time and its surface texture can be maintained continually. However, for polishing metallic circuitry (e.g. copper), the slurry used is acidic (e.g. pH = 3) and the chemical is corrosive. As a result, the metal matrix of the dresser may be dissolved. Without the adequate support of the metal matrix, the diamond grit will soon fall out. The detached diamond grit may be embedded in the pad and cut deep in the wafer that cannot be tolerated by CMP operation.

In order to prevent the dissolution of the metal matrix, the diamond dresser is removed from the pad when the slurry is used in polishing wafer. After the wafer reaches the end point of polishing, it is taken away and the slurry feeding is temporarily suspended. Only at this stage, the diamond dresser is used to dress the pad. This ex-situ dressing operation can result in significant loss (e.g. 1/4) of CMP throughput. Moreover, because the pad texture is not optimized for polishing wafer, the polishing rate will decline substantially and the wafer thickness may not uniform either.

In order to protect the metal matrix from the attack of the slurry, a diamond-like carbon (DLC) coating is applied to the surface of diamond dresser. DLC is superhard so it can resist the abrasion of the slurry. In addition, DLC is highly inert and it is non-reactive to almost all chemicals and acids. Consequently, a DLC coated diamond dresser can be used to dress the polishing pad in-situ. As the result, the CMP productivity can be increased, so is the wafer quality.

2 CERMET CERAMIC COATING (C^3)

Although DLC is an ideal chemical barrier that may protect the metal matrix from corrosion, it suffers several drawbacks that may offset its advantage for coating diamond dressers. Firstly, DLC, in addition to deposit on the metal matrix, it will also adhere onto diamond grits. The DLC will wrap around diamond grits so their sharp cutting tips become blunt. The dull diamond tip can no longer penetrate the pad effectively, as a result, polishing rate of the wafer will decline.

Secondly, the adhered DLC on diamond grits may flake off during the dressing action. The separated DLC shreds may mix with the polishing debris that can scratch the expensive wafer.

Thirdly, because DLC is very different from metal in physical characteristics (e.g. thermal expansion) and chemical properties (e.g. reaction compatibility), their interface is weak. The poorly adhered DLC may also flake off the metal matrix. The detached DLC may scratch the wafer. Moreover, the exposed metal matrix is likely to be etched by the corrosive slurry.

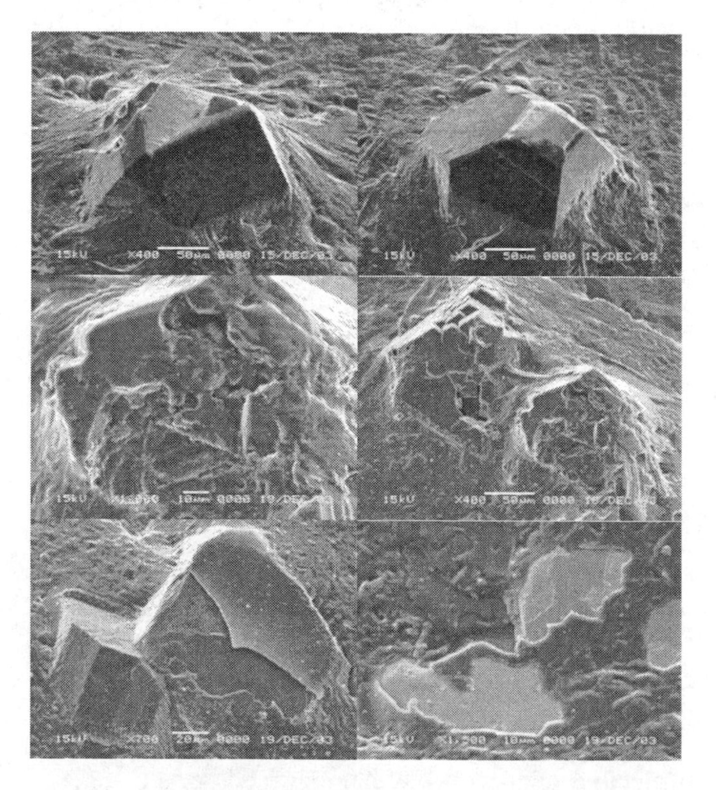

Figure 1: The coating of DLC on diamond grits and the metal matrix and its detachment after CMP polishing. The top diagram shows the flaking of DLC coating from the surface of euhedral diamond crystals; and the middle diagram, the flaking from grits of irregular shapes. The bottom diagram shows the peeling of

DLC from the metal matrix.

In order to improve the adherence of the coating material on the metal matrix and prevent it from coating onto diamond grits, a new coating technology is developed. This time, a metal film (e.g. stainless steel) is deposited first so it can form metallurgical (diffusion) bonding with the metal matrix. The metal film cannot adhere to diamond grits, however. While the metal atoms are showering onto the diamond dresser, a plasma of ceramic material (e.g. Al_2O_3) is also introduced in the atmosphere. Moreover, the concentrations of metal and ceramic are adjusted in such a way that metal underlayer can grade into ceramic top surface.

The above described cermet (ceramic-metal) composite coating (CCC or C^3) has several advantages. Firstly, its base is metal so it can adhere firm to the metal matrix of the diamond dresser. Secondly, because the metal grades into ceramic, there is no boundary in the coating. In other words, the vulnerable weak interface is eliminated. Thirdly, the ceramic is acid proof and it is corrosion resistant. Moreover, the composition of the ceramic as well as metal can be changed to suit specialty requirements (e.g. silicon carbide may be used instead of alumina as the top surface).

The CCC has been applied to diamond dressers and the effect is apparent. The metal underlayer does not adhere to diamond grits so the risk due to the loss of CCC there is minimized. Moreover, because the sharp edges of the diamond are exposed, their dressing ability is maximized. Hence, the pad dressing can be effective, and the CMP polishing may be efficient.

Figure 2: The diamond grit on the dresser that is coated with CCC is free from over coating after being rubbed against a pad. The exposure of the fresh diamond can also penetrate the pad top easily and hence it can create the optimal texture for efficient polishing of wafer.

The CCC can adhere firm on the metal matrix so the used diamond dresser shows no sign of coating loss. Because the metal matrix is covered during entire period of CMP operation, it is not etched.

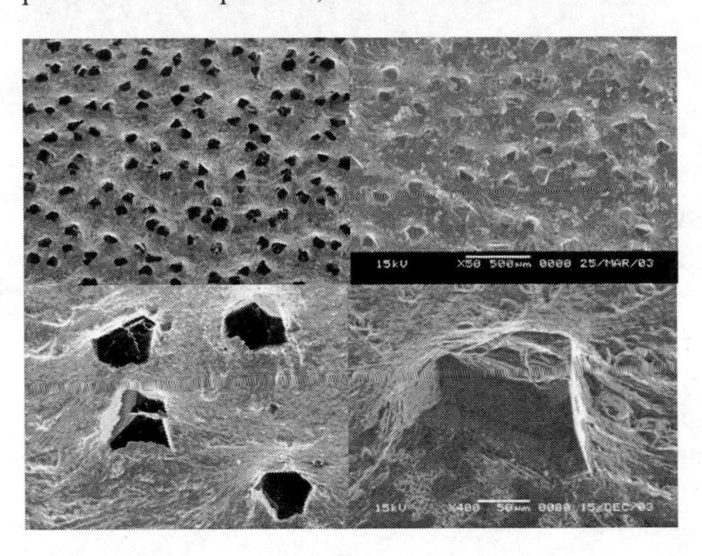

Figure 3: The intact appearance of CCC after a diamond dresser went through a CMP process (left diagram). In contrast, the peeling is serious with a DLC coated diamond dresser (right diagram).

In summary, the CCC is more preferred over DLC coating for protecting diamond disks. The advantage of CCC is in its selective ability to adhere metal matrix and not diamond grits. Moreover, CCC is versatile as its geometry (e.g. thickness) and composition can be engineered. It can be specifically designed to meet the in-situ dressing in caustic environment (e.g. copper CMP).

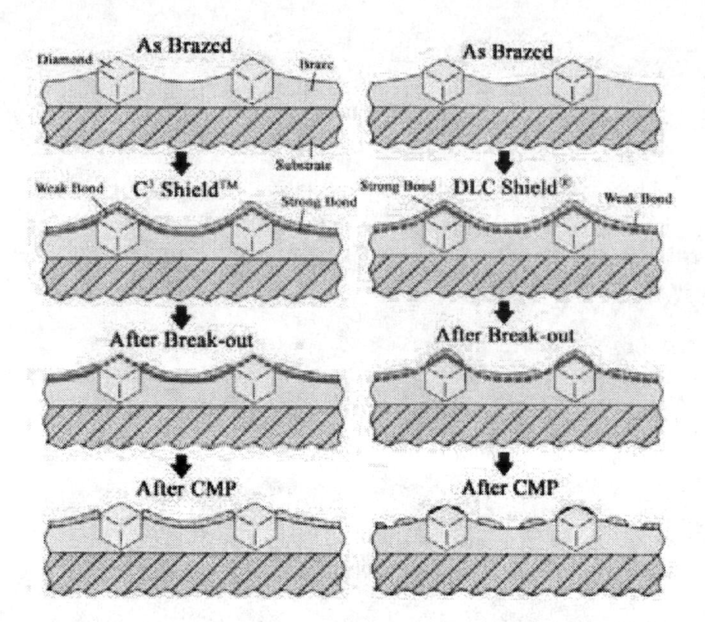

Figure 4: The contrast of CCC and DLC coating on diamond dresser. The CCC can adhere firm on metal matrix but it can be removed from diamond grits (left diagram). DLC coating is just the opposite, it will flake off from both metal matrix and diamond grits (right diagram).

REFERENCES

[1] Sung, C. M., U. S. Patents 6,039,641, 6,286,498 and 6,679,243.

[2] Sung, C. M. and Frank S. Lin, U. S. Patent 6,368,198.

Wurtzitic Boron Nitride on Diamond: The Ultimate Epitaxial Wafer for "Semiconductor on Insulator"

James Chien-Min Sung

Address: KINIK Company, 64, Chung-San Rd., Ying-Kuo, Taipei Hsien 239, Taiwan, R.O.C.
Tel: 886-2-2677-5490 ext.1150
Fax: 886-2-8677-2171
e-mail: sung@kinik.com.tw

[1] KINIK Company, 64, Chung-San Rd., Ying-Kuo, Taipei Hsien 239, Taiwan, R.O.C.
[2] National Taiwan University, Taipei 106, Taiwan, R.O.C.
[3] National Taipei University of Technology, Taipei 106, Taiwan, R.O.C.

ABSTRACT

In last four decades, the number of transistors of integrated circuits (IC) has been following Moore's Law to double in about every 18 months. This relentless densification of circuitry would require technological breakthroughs of many fronts such as nanom chemical mechanical planarization (CMP) for achieving a super smooth surface on an ultra flat wafer, and UV photolithography for imprinting nanom copper wires on it. However, with the circuitry going beyond ultra large-scale integration (ULSI), the noise level in the chip will increase that may challenge the material limits of semiconductor designs. The Second Law of Thermodynamics requires entropy to increase dramatically in a chip of super ULSI. Entropy may manifest as waste heat or scattered phonons, i.e., random vibration of atoms; and chaotic charges or intrinsic carriers, i.e. quantum fluctuation of defects. Both of them will increase the background noise that may overshadow the electrical signals that are processed in the future chip.

The ULSI chips for computer CPU have been using copper heat spreader to avoid overheating. However, copper with its thermal conductivity of 400 W/mK is proven incapable to chill the next generation Pentium IV with a speed of 4 GHz and a power of 120 W. Hence, Intel has decided in 2004, for the first time since Moore's Law took effect forty years ago, to use dual chips of slower speed and lower power for the next generation CPU.

In addition to be plagued by heat, the semiconductors are also haunted by the surge of intrinsic charge carriers that tend to concentrate with temperature. As a result, the current Pentium IV CPU is leaking 1/4 of its electricity into the background noise. The very material that constitutes the semiconductor is too conductive for packing transistors required by post ULSI chips.

Both problems of phonon scattering and charge fluctuations can be reduced by using more stable crystal lattice with strong atomic bonds. Diamond has the most stable lattice with the strongest atomic bonds. Hence, diamond is not only super hard, but also super resistant in electron movements. Moreover, diamond is super fast in transmitting sound and in ridding heats. In addition, diamond crystal lattice remains intact at temperatures that will burn out all other semiconductors. All these extreme attributes will make diamond the best friend of semiconductors. In fact, diamond itself is the ultimate semiconductor with its future performance that will make current semiconductors primitive like stone-age tools.

The highest thermal conductivity makes diamond the ideal heat spreader. The lowest charge fluctuation makes diamond the dream barrier substrate. Hence, semiconductor on diamond (SOD) can be the best semiconductor on insulator (SOI) that is envisaged for making future ULSI, laser diodes, LED, microwave (MW) generators, and other signal processing or optic-electronic devices.

Recently breakthroughs of CVD diamond film deposition and polishing technologies have made diamond substrates engineering possibility rather than research curiosity. For example, Kinik Company is now offering cost effective diamond wafers up to four inches that can substitute current substrate materials (e.g. sapphire, silicon carbide). Moreover, the super smooth (e.g. Ra about 2 nm) polycrystalline diamond film can be coated with wurtzitic boron nitride wBN) to become preferably oriented with hexagonal planes. The isostructural hexagonal AlN can be further over coated on wBN to make the surface fully compatible with nitride semiconductors that are used for fabricating high power semiconductors. The gradation from wBN to AlN with random to preferred crystal orientation is ideal for epitaxial deposition of Si, SiC, GaN, InN, GaAs or other semiconductors. Consequently, Kinik' DiAlN® wafers are suitable to make next generation devices of ULSI, LD, LED, MW that are capable to operate at high frequency and large power. For example, DiAlN® wafers may be used to fabricate UV or even X-ray LED by widening the band gap of nitride semiconductors such as by doping GaN with smaller Al or B atoms.

The DiAlN wafer is also the dream substrate for making surface acoustic wave (SAW) filters with ultra high frequencies (e.g. 20 GHz) and larger power (e.g. for satellite communications). wBN is the strongest piezoelectric material; and, AlN, the second strongest. They can convert electromagnetic signals into sound waves, and vice versa. Current ceramic SAW filters (e.g. $LiTiO_3$, $LiNbO_3$) are thermal insulators that tend to heat up during the operation. They are also ionic conductors that will leak out electricity. In contrast, DiAlN is the most efficient thermal conductor and electrical resistor so it can be the dream SAW filters for future telecommunications.

Keywords: diamond film, wurtzite BN, AlN, ULSI, LED, SAW filter

1 THE SUPER SMOOTH CVD DIAMOND FILM

Diamond is the best insulator of all materials so it can be the ideal substrate for SOI. Diamond has a thermal conductivity that may approach 2000 W/mK (e.g. for perfect diamond) that is at least 10 times higher than common oxides. Consequently, SOD cannot only be operated with higher speed, but also with lower temperature.

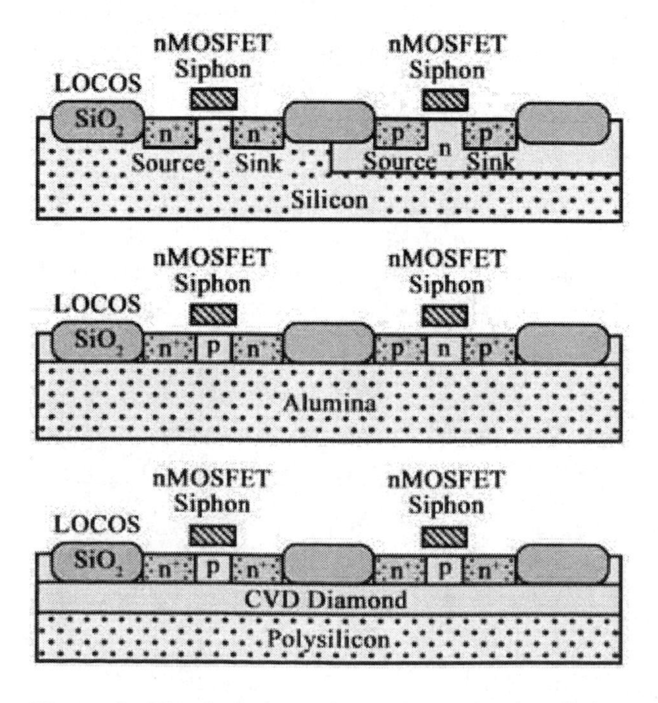

Figure 1: The isolation of transistors by insulating with oxide (top diagram) and the use of oxide substrate for further insulation (middle diagram). Semiconductor on diamond(SOD) is the most effective silicon on insulator (SOI). It also improves the heat dissipation.

Diamond film for SOD can be deposited on a suitable substrate (e.g. silicon wafer) by chemical vapor deposition (CVD). The CVD diamond film is then polished to achieve a mirror finish (e.g. Ra < 5 nm). Subsequently, semiconductor (e.g. silicon) can be deposited onto the smooth diamond surface.

2 EPITOMICAL WAFERS FOR LED

Current LED (or LD) wafers use sapphire or silicon carbide (SiC). AlN/wBN coated SOD are much superior than these commercial substrates. SOD is more insulating and it can dissipate heat fast, hence LED with SOD can operate at a much higher power. In addition, because the lattice mismatch is better, LED with SOD is much more energy saving and it can last much longer time. Furthermore, with the thin layers of AlN/wBN/diamond film used, the cost of SOD is competitive to thick single crystals of either sapphire or SiC.

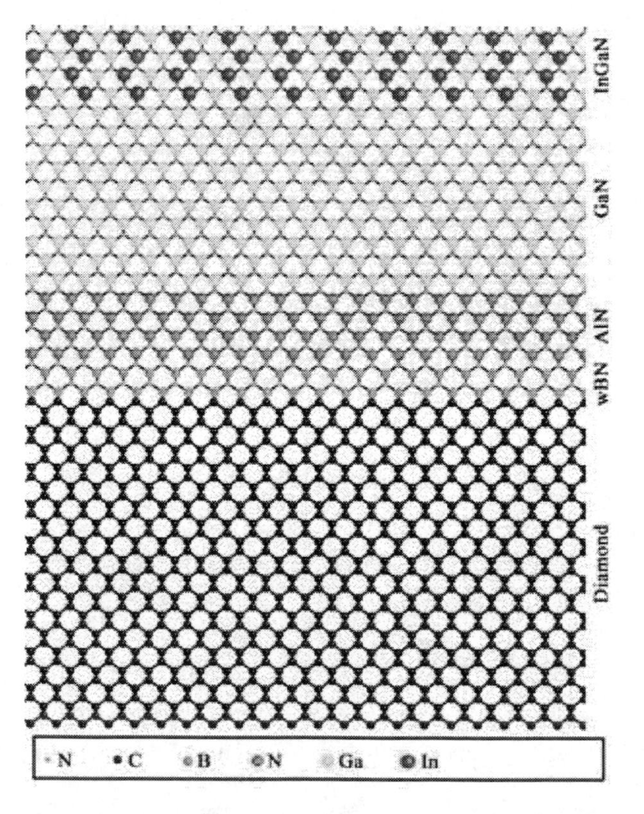

Figure 2: This atomic super lattice illustrates the gradation of the buffer layer to convert diamond lattice to wBN and then to AlN. Subsequently, GaN and InGaN may be deposited as LED or LD. The

relative atomic size is also shown as legends. The gradation of atomic size (as shown by colored circle) will minimize the boundary stress as well as the dislocation density. The lower the dislocation density, the more reliable is the device.

Based on the above epitaxial gradation, one example of making laser diode with SOD is illustrated below. Current LED are monochromatic so the white light must be produced by mixing colors (e.g. Red, Green, Blue). It is possible to dope nitride semiconductors (e.g. GaN) with smaller Al or even B atoms. The result is to bring atoms closer with the consequence of increasing the band gap. Hence, the wavelength of the light produced can be effectively shortened. With the doping of Al atoms, a UV LED can be fabricated. The UV LED can be used to excite phosphor with white fluorescence light. This LED fluorescence light is not only brighter than conventional fluorescent lights, but also it avoids the use poisonous mercury as the source of UV. Moreover, if B doping is also added, an LED may also become a convenient X-ray source without involving cumbersome vacuum chamber. The X-ray LED can be used in X-ray lithography for fabricating true nanom structures of MEMs, and the ultimate semiconductor chips that contain single electron transistors.

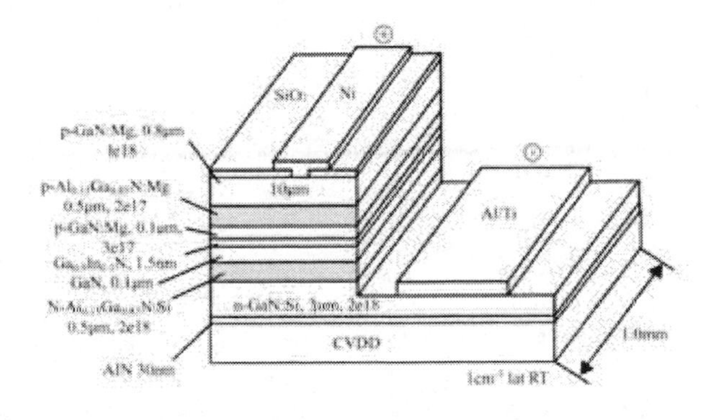

Figure 3: One example of making laser diode with SOD.

Although SOD is intended to be an insulator, for certain applications, the substrate may require electrical conduction as well (e.g. as an electrode). In this case, a metal conductor (e.g. copper) can be inserted in the diamond film to allow the passage of electrical current. The conductor can be isolated by insulating diamond or they can form alternative partitions. One example to fabricate such a conducting SOD is illustrated below.

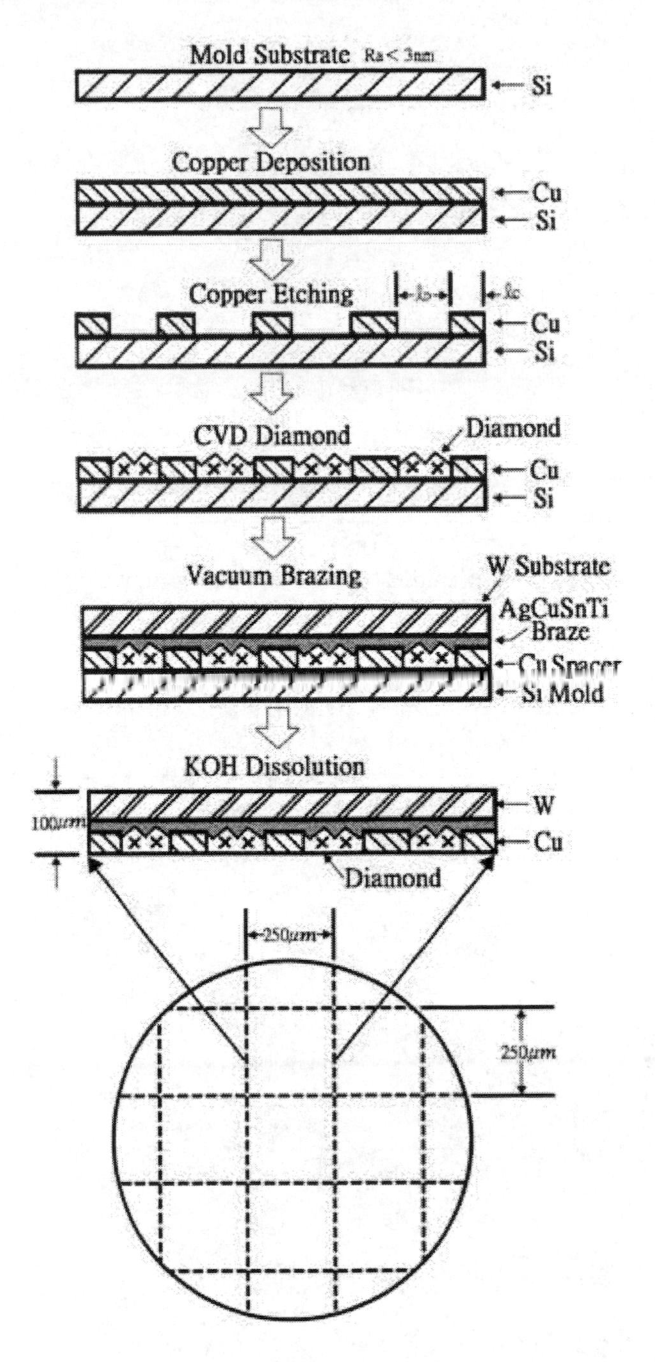

Figure 4: One example of fabricating SOD with an electrode for each square of diamond. Electrically conducting SOD is a useful design for LED or LD.

3 EPITOMICAL WAFERS OF SOD

The SOD with or without AlN/wBN coating can be used for a variety of applications. These applications can be much more powerful and reliable than current materials. For example, Intel's failure to make 4 GHz CPU in 2004 as described above can be corrected by making the 90 nm ULSI on SOD. Some examples of epitaxial wafer applications are illustrated as follows.

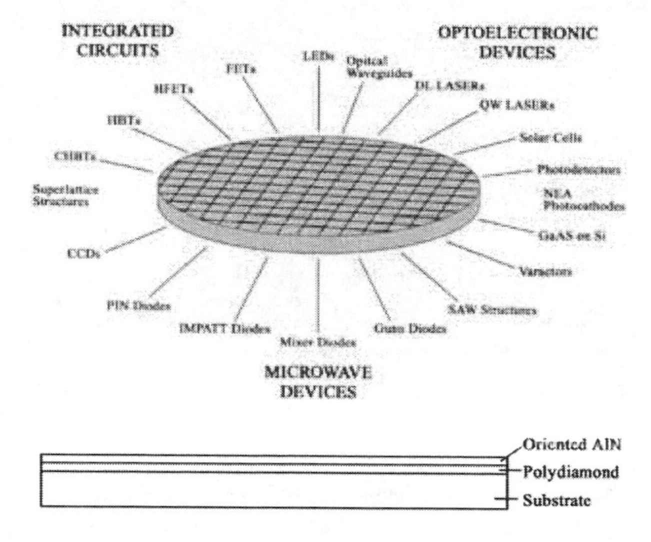

Figure 5: The examples of using SOD for making future semiconductor devices for ULSI, LED, MW, and much more.

Diamond epitaxial wafers up to six inches in size have been manufactured by Kinik Company with superb surface finish (e.g. Ra = 3 nm). They are ideal substrates of various applications.

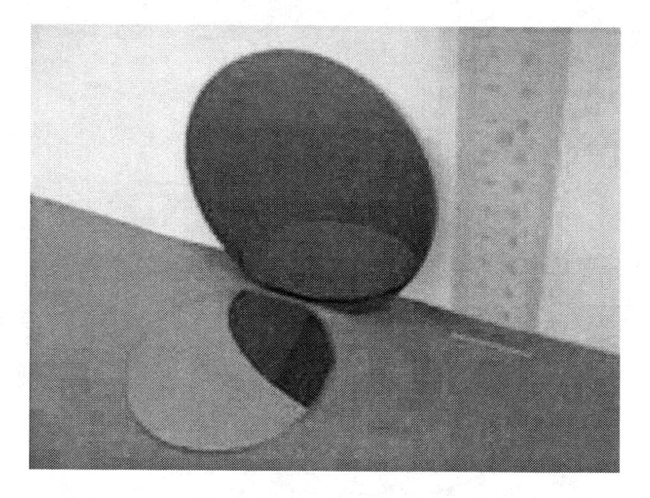

Figure 6: Diamond epitaxial wafers manufactured by Kinik Company in Taiwan.

Nano-crystalline Ta_2O_5 films deposited by DC pulsed magnetron sputtering

V. Ligatchev[*], Rusli[*], Lok Boon Keng[**], Lai Foo Khuen[**], Yar Rumin[*], Jihua Zhao[*]

[*] Nanyang Technological University,
School of Electrical and Electronic Engineering, 639798 Singapore, evaleri@ntu.edu.sg
[**] Singapore Institute of Manufacturing Technology
71 Nanyang Drive, 638075 Singapore, bklok@SIMTech.a-star.edu.sg

ABSTRACT

Tantalum pentaoxide (Ta_2O_5) has attracted significant research interest due to its excellent dielectric properties that are suitable for new generation of gigabit dynamic random access memories (DRAMs). In spite of its actual implementation in DRAMs, the relations between its atomic structure, morphology, and macroscopic properties are still not clearly understood. In this work, we have investigated the relationship between the nano-morphology and optical properties of thin Ta_2O_5 layers (95 - 110 nm), deposited at room temperature using pulsed DC magnetron sputtering (PDCMS) in an argon-oxygen (Ar-O_2) gas mixture. As-deposited Ta_2O_5 samples prepared at different Ar/O_2 flow ratios (0.55 – 1.00) were studied, as well as samples post-annealed in an oxygen ambient at 350, 600, and 900 ^{0}C. An increase in the average size of the nano-crystalline grains from 1-2 nm in the as-deposited layers to 10-15 nm in those annealed at 900 ^{0}C is noted, and has resulted in a considerable decrease in their Tauc's bandgap. The results are interpreted within the framework of the effective mass approximation.

Keywords: tantalum pentaoxide, nano-morphology, Tauc gap, electron confinement,

1 EXPERIMENTAL

Three sets of the Ta_2O_5 samples have been prepared using the PDCMS system Unaxis LLS EVO by sputtering Ta target in an Ar-O_2 gas mixture. The Ar flow rate was fixed at 100 sccm, while the flow rate of the oxygen was varied from 55 to 100 sccm for the different sample sets (see Table 1). Four-inch single-side-polished silicon wafers of p-type (7-20 Ohm*cm, <100>-oriented) were used as the substrates. The temperature of the substrate holder was not controlled, and was nominally close to the room temperature. The deposition rates for our Ta_2O_5 films are in the range of 25 – 35 nm/min. After deposition, one sample from each of the four sets was kept untreated (reference), whereas the rest were annealed in a clean furnace with an oxygen flow rate of 100 sccm at different temperatures of 350, 600, and 900 ^{0}C (see Table 1). The heating and the cooling rates of the annealing process were set at 5^{0}C/min and 2 ^{0}C/min respectively. The annealing time was fixed at 30 minutes.

Table 1

Oxygen flow rate during deposition	55 sccm	80 sccm	100 sccm
Annealing temperature, ^{0}C	As deposited	As deposited	As deposited
	350	350	350
	600	600	600
	900	900	900

2 RESULTS AND DISCUSSION

In contrast to previous studies [1] which reported amorphous structure of Ta_2O_5 at temperatures below 500^{0}C, our XRD data clearly demonstrates the presence of a nanocrystalline phase in all the samples, including those that were as-deposited, as can be seen in Figs. 1 (a) and (b). Unfortunately, we are not able to identify reliably peaks in Fig. 1 (a), though similar XRD peaks have been recently reported for Ta_2O_5 in ref. [2]. On the other hand, the XRD peaks in Fig. 2 (b) are 'standard' for studied material and originate from different atomic planes of orthorhombic β-Ta_2O_5 structure [2]. The average size of the nano-crystals, L, was estimated using the Scherrer formula [3] to range from 1.1 to 2.3 nm for the as-deposited samples and those annealed at 350 and 600 ^{0}C. For films annealed at 900 ^{0}C, the average crystalline size increases and varies from 10 to 16 nm, as illustrated in Figs. 2 (a) and (b).

Nano-scaled morphology has been observed by AFM technique on the surfaces of all the Ta_2O_5 films studied using the 'deflection' mode of the SHIMADZU SPM 9500J2 equipment. Typical images are given in Figs. 3 (a) and (b). As can be seen from these images, the average size of the nano-morphology depends on both the Ar/O_2 flow ratio and the annealing temperature. In particular, the samples annealed at 900 ^{0}C reveal nano-crystals with sizes of 5-15 nm, consistent with the results deduced from the XRD data.

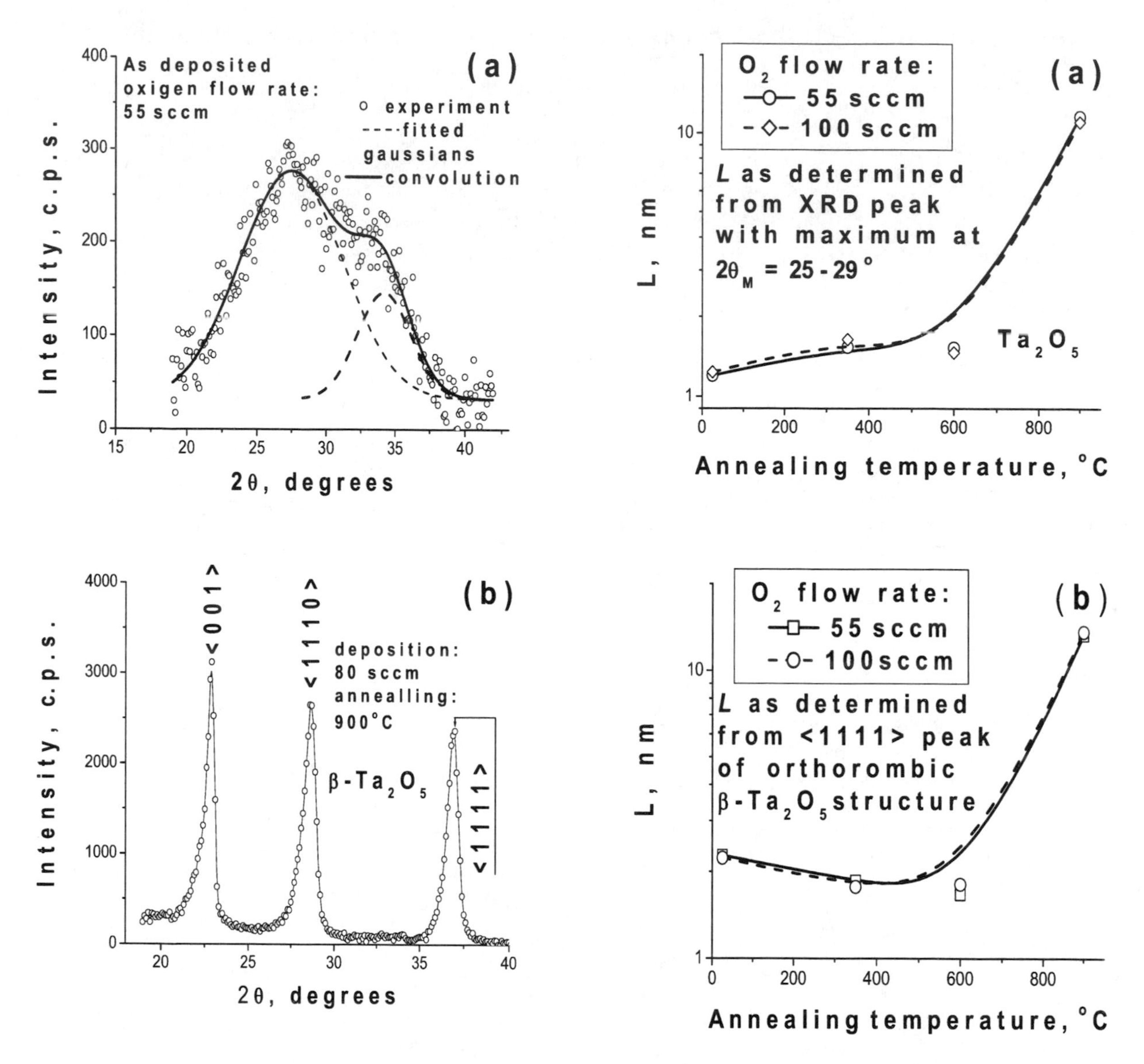

Figures 1 a, b: Typical XRD curves obtained from (a) as deposited and (b) annealed samples. More details on the deposition and annealing regimes of the samples are given directly in the figures. In figure (a) experimental XRD spectrum is deconvoluted into two Gaussian peaks. Full width at halfmaximum (FWHM) of the peak(s) is employed for estimations of average size *L* of nano-crystals in the studied materials in accordance to the Scherrer formula, ref. [3]. The obtained *L* values are plotted in two next figures.

Figures 2 a, b: Influence of the annealing temperature on the average size *L* of nano-crystals in studied films. The *L* values are determined from FWHM of different XRD peaks in the samples, deposited at oxygen flow rates of 55 and 100 sccm.

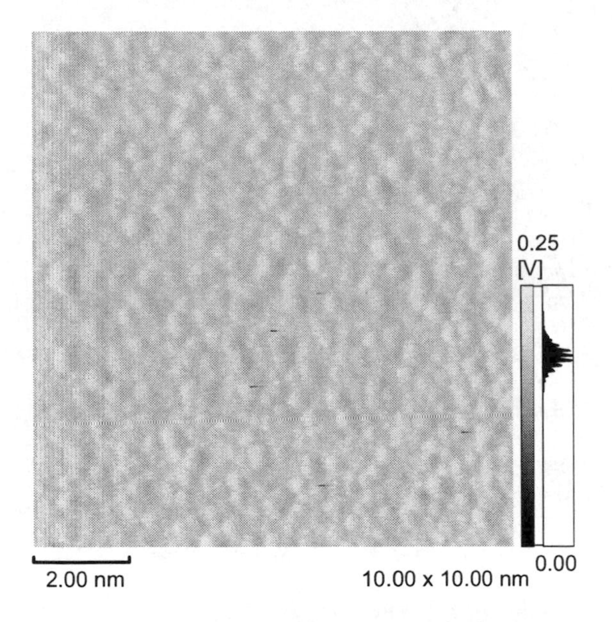

Figure 3 a: nano-morphology of the 'reference' Ta_2O_5 sample, deposited at the oxygen flow rate of 80 sccm.

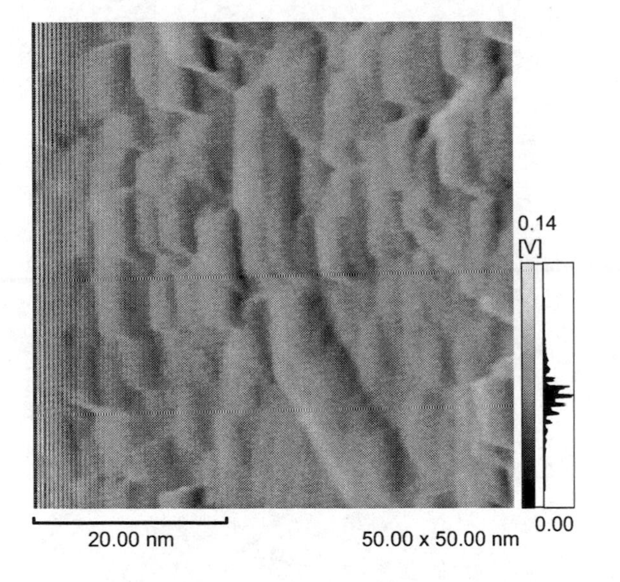

Figure 3 b: nano-morphology of Ta_2O_5 sample, deposited at the oxygen flow rate of 80 sccm and thereafter annealed at 900 °C.

Optical characterization of the Ta_2O_5 films have been carried out in the wavelength range 250 nm < λ < 1100 nm using VASE spectroscopic ellipsometer. The complex refractive indices ($n - ik$) of the samples were determined using the WVASE32 software as a function of wavelength from the experimental ellipsometry data. We found that the Cauchy dispersion relations, as given below, can be used to satisfactorily fit the data.

$$n(\lambda) = A_1 + \frac{B_1}{\lambda^2} + \frac{C_1}{\lambda^4} \quad (1a)$$

$$k(\lambda) = A_2 + \frac{B_2}{\lambda^2} + \frac{C_2}{\lambda^4} \quad (1b)$$

A_1, B_1, C_1, A_2, B_2, C_2 are unknown parameters to be determined through the fitting process. Typical n and k spectra are plotted in Fig. 4 (a) as a function of photon energy $h\nu$. The absorption coefficients of the films were also calculated using the relation, $\alpha = 4\pi\ k\ /\ \lambda$ [4]. Application of a standard Tauc's procedure [5] allows one to determine the Tauc gap E_G of the Ta_2O_5 films. The variation of E_G versus annealing temperature for the different sets of samples is given in Fig. 4 (b).

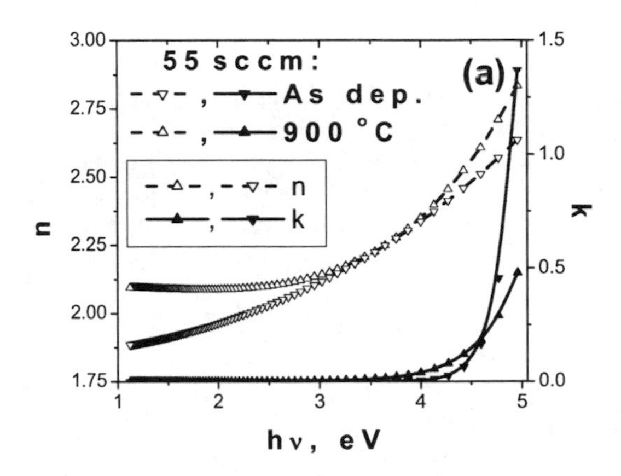

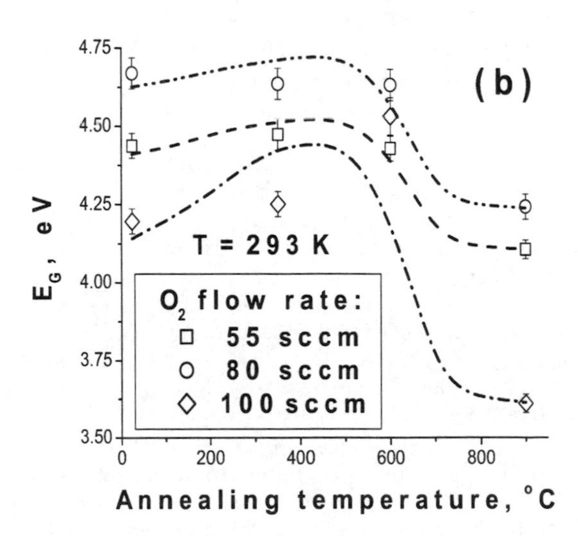

Figures 4 a, b: (a) typical $n(h\nu)$ and $k(h\nu)$ dependencies for studied samples (b) influences of deposition and annealing conditions on Tauc's bandgap width of the Ta_2O_5 layers.

As seen from Fig. 4(b), E_G varies over a small range (4.17 – 4.69 eV) for the as-deposited and annealed samples at temperatures less than 600 ^{0}C. However, it drops significantly (up to 3.6 eV) for films annealed at 900 ^{0}C. Such variation in E_G can be attributed to the quantum confinement of electron wavefunctions within the nano-scaled particles. Quantitatively, this effect can be described within the framework of effective mass approximation [6]:

$$E_G(L) = E_G(\infty) + \frac{h^2}{2m^*}\left(\frac{1}{L}\right)^2 \qquad (2)$$

where $E_G(\infty)$ is a reference bandgap, which formally corresponds to an infinite size of crystals in the material, m^* and L are the electron effective mass and the average size of the nanocrystals respectively and h is the Planck's constant.

In our simulations the L values deduced from the width of the <1111> peak are employed (see Fig. 2 (b)). The simulated results are shown in Fig. 4(b), together with the experimental data. Reasonable agreement between the simulated and experimental E_G can be seen for the as-deposited and annealed Ta_2O_5 samples. Value of the simulation parameters are listed in the second Table:

Table 2

Simulation parameters:	Oxygen flow rate, sccm:		
	55	80	100
$E_G(\infty)$, eV	4.10	4.23	3.60
m^* / m_0	0.70	0.35	0.60

m_0 in the Table 2 is a mass of a free electron. The $E_G(\infty)$ value of 4.23 eV is in a pretty good agreement with optical gap of crystalline Ta_2O_5 material, reported in ref. [7]. Alterations in the simulation parameters for two other sets are caused by significant changes in physical properties of Ta_2O_5 films, deposited at different flow rate of the oxygen. In particular, lack or excess of oxygen in the atomic structure of the films deposited at the flow rates of 55 and 100 sccm could originate considerable alterations in so-called 'ionic' (asymmetric) components of valence bonds (interatomic potential) in the material. In accordance to the 'dielectric' theory of semiconductors, described in ref. [8], variations in the 'ionic' component should cause significant changes in the optical gap width [quantitatively represented here with the $E_G(\infty)$ parameter]. As we can see from the Table 2, such changes really happen in the sample sets, deposited at different flow rates of the oxygen.

The results of this work suggest that it is possible to have nanocrystalline phase in Ta_2O_5 films deposited using the PDCMS. The size of the nanocrystals and the optical properties of the films are affected by high temperature annealing. Therefore, in principle it is possible to control their nano-morphology, and consequently tune and optimize their macroscopic properties for their practical applications in DRAMs.

REFERENCES

[1] T. Dimitrova, K. Arshask, E. Atanasova, Thin Solid Films, 381, 31 - 38, 2001.

[2] M. Kerlau, O. Merdrignac-Conanec, M. Guilloux-Viry, A. Perrin, Solid State Sciences, 6, 101-??, 2004.

[3] H.P. Klug, L.E. Alexander, *X-ray Diffraction Procedures for Polycrystalline and Amorphous Materials*, John Wiley & Sons, New York (1974), 966 p.

[4] K.W. Boer, Survey of Semiconductor Physics, Van Nostrand Reinhold (1990) 1423 p.

[5] J. Tauc, in *Amorphous and Liquid Semiconductors*, eds. J. Tauc, Chap. 4, (Plenum Press, New-York) 1984, 412 p.

[6] P.E. Batson, J.R. Heath, Phys. Rev. Lett. 71, 911-14 1993.

[7] F.E. Ghodsi et al. Thin Solid Films, 295, 11-15, 1997.

[8] J.C. Phillips, Rev. Mod. Phys. 42, 317 – 56, 1970.

Anodisation of Thin Film Aluminum.

C Douglas†, P Evans*† and R.J. Pollard*†

* Nanotec Northern Ireland, Belfast, UK
† Queen's University of Belfast, Belfast BT7 1NN, r.pollard@qub.ac.uk

ABSTRACT

We have investigated the single step anodisation of aluminum films in sulfuric, oxalic and phosphoric acid. In general increasing anodisation voltage is found to lead to increasing volume expansion, for a given acid. It is also shown that it is important to match the anodisation conditions to the electrolyte to maximize the ordering for a given pore size. Preliminary work on the electrodeposition of magnetic materials into the pores has also been carried out.

Keywords: porous alumina, nanopores, anodisation, nanowires

1 INTRODUCTION

It has been known since the fifties that anodisation of aluminum leads to a porous aluminum oxide structure [1]. More recently it has been shown that a two step anodisation process leads to greater pore ordering [2]. As conventional lithographic techniques reach their lower limit in terms of feature size, there has been increased interest in anodic alumina as a quick and economic method for producing nanostructured material over a large area. Here we have examined the single step anodisation of Al films grown by sputter deposition.

2 EXPERIMENTAL

Aluminium films were grown on glass substrates by DC magnetron sputter deposition. Anodisation was carried out at constant voltage with a platinum counter electrode. The electrolyte was kept cooled and the temperature monitored throughout the anodisation process. Cross-sections were prepared for transmission electron microscopy (TEM) using a Philips focused ion beam system.

3 RESULTS

3.1 Growth of porous alumina

Figure 1 shows current-time curves for the anodisation of a 300nm Al film on glass in (a) 0.3M Sulfuric (b) 0.3M Oxalic acid and (c) 10% phosphoric acid under typical conditions. It can be seen that the curves in sulphuric and oxalic acid show the same features although the time scale is different. **Stage I** is a decrease in the current until a minimum is reached at which point it increases (**stage II**) to a plateau (**stage III**). Finally the current decreases, (**stage IV**) to the final small residual value. The current in phosphoric acid shows only a decrease to a plateau and then the final decrease.

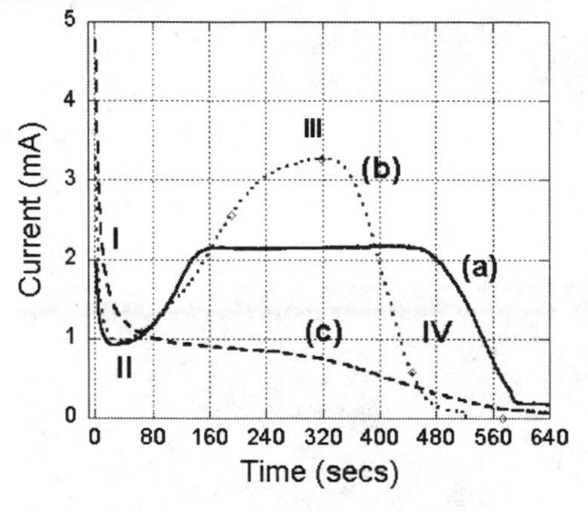

Figure 1: Current time curves for anodisation for (a) 0.3m Sulfuric acid 20V (b) 0.3m Oxalic acid 40V (c) 10% phosphoric acid 100V

TEM cross-sections were prepared at the various stages of growth to help explain these I-t curves. The initial large current is due the initial formation of an aluminium oxide layer, the current decreases as this layer is more resistive than aluminium. The pores start to form and the current increases due to dissolution of the alumina (stage II). The thickness of this initial oxide layer is ~88nm for 10% phosphoric acid (60V), 74nm for oxalic (60V) and 21nm for sulfuric(15V). A plateau is reached when the dissolution of the alumina to form the pores is in equilibrium with the growth of the aluminium oxide in front of the advancing pores. Finally, when all the aluminium has been used-up the current drops, leaving a 'barrier layer' of oxide at the bottom of the pores. This final barrier layer is 16nm thick for film anodised in sulfuric acid (15V), 42nm for oxalic acid (60V) and 89nm for phosphoric (60V), following the same trend as the initial oxide layer prior to pore growth. Figure 2 shows a TEM cross-section of a film anodised in 0.3M oxalic acid (60V 4°C) where the barrier layer can be clearly seen.. Figure 3 shows the surface of a film anodised in 0.3M sulfuric acid (20V, 3°C) after milling.

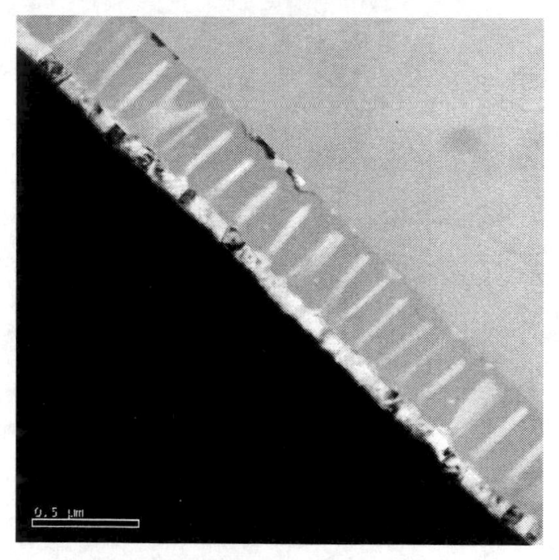

Figure 2: TEM cross-section of porous alumina anodised in 0.3M Oxalic acid at 60V (40 nm pores). The barrier layer at the bottom of each pore can be clearly seen.

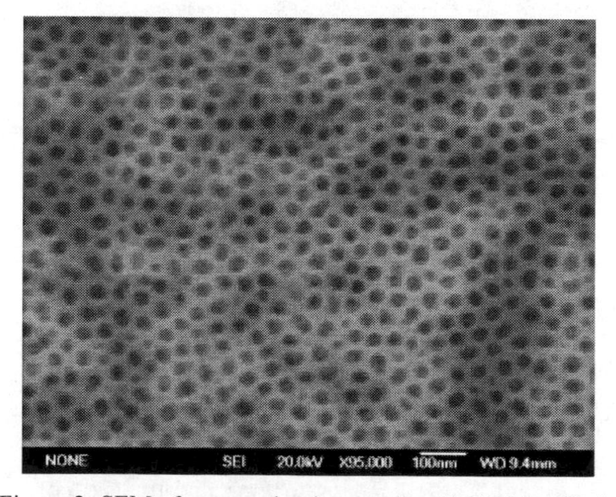

Figure 3: SEM of porous alumina anodised in 0.3M sulfuric acid at 20V (15 nm pores).

The growth of the pores in the phosphoric acid is found to proceed in the same way as the other two acids, but the initial minimum in the current-time curve is not seen, possibly due to the greater oxide thickness.

A volume expansion is expected when Al forms Al_2O_3 and is indeed observed in the cross-sections. The volume expansion is found to show a linear response with applied voltage. The degree of volume expansion is found to be similar for oxalic and phosphoric, ~1.2 for an anodisation voltage of 60V and 1.6 for phosphoric at 100V with a similar expansion being predicted for oxalic acid by linear extrapolation from the relationship at lower voltages. The voltage expansion for sulfuric is seen to increase much more rapidly with voltage, 1.2 at 15V and 1.8 at 25V. The mechanical stress from the expansion of the oxide layer is believed to be the result of repulsive forces between neighboring pores during growth that leads to self-organized formation [3]. A moderate expansion of 1.2 is associated with highly ordered growth in oxalic and sulfuric [3]. An expansion of 1.6 at 100 volts phosphoric was obtained for the best-ordered films and this considerable expansion may be allowed due to the slower conversion rate during anodisation. The anodisation rate in phosphoric is three times slower at 100 volts than oxalic at 60 volts and sulfuric at 15 volts, which have lower volume expansions. So it must be the rate rather than the amount of volume expansion when anodisating in sulfuric and oxalic that may cause limitations at higher voltages due to the adhesion to the substrate.

3.2 Electrodeposition into templates

Electrodeposition is an inexpensive method of producing high quality magnetic material [4]. It differs from vacuum deposition techniques in that the electrical properties of the substrate play an important part in the growth. In order to electrodeposit into the pores a conductive backing layer, such as gold, is needed. Ar^+ ion milling at normal incidence removes the oxide barrier layers at the bottom of the pores and exposes the conductive Au layer. The template can then be electrodeposited onto with the material growing on the conducting gold but not the insulating aluminium oxide. This allows wires with large aspect ratios to be formed.

Figure 4 shows the current time curve for Ni wires grown at –1V (vs. SCE) from a 0.1M $NiSO_4$ solution. The alumina template had 40nm pores. There is a large initial current, which decreases as the concentration of ions near the surface of the growing wires decreases until it reaches a steady value governed by diffusion (point 1). When the wires reach the surface of the template there is an increase in current due to the increase in surface area (point 2). Figure 5 shows a cross section of a nanowire stopped shortly after the current dropped to a steady value whilst figure 6 shows what happens if the wires are allowed to grow until they reach the surface.

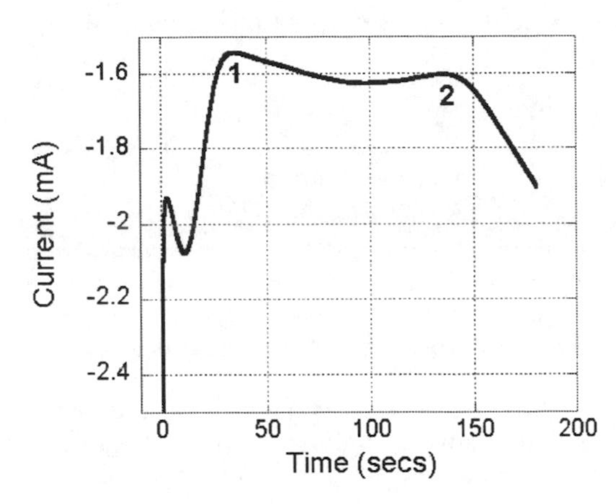

Figure 4: Current vs time for Nickel nanowires grown in 40nm pores (oxalic 60V). Point 2 corresponds to the point the wires reach the surface

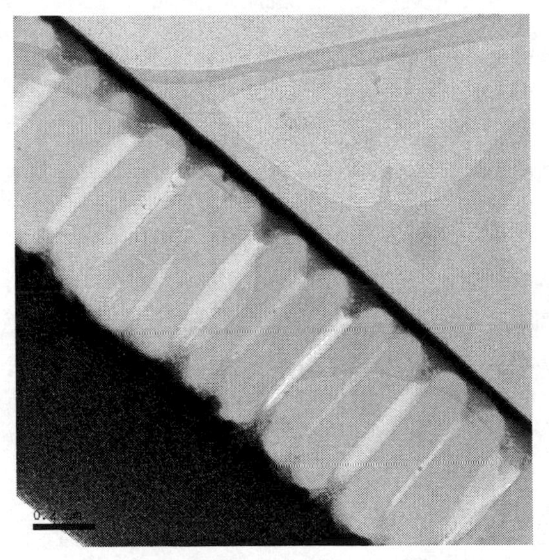

Figure 5 Ni wires electrodeposited in 40nm pores (oxalic 60V). The growth was stopped shortly after the wires started growing (point 1 in fig 4)

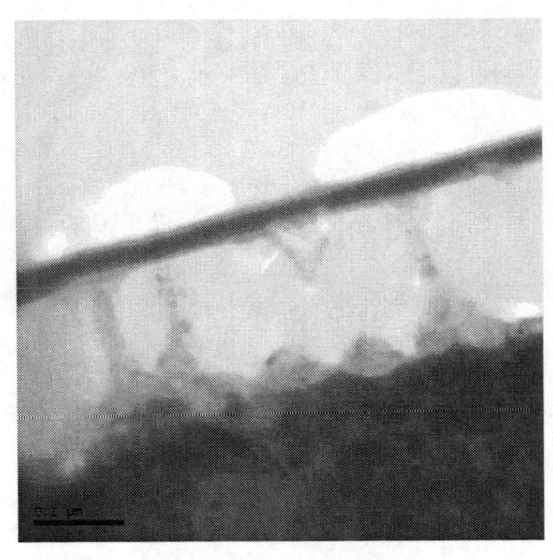

Figure 6 Ni wires electrodeposited in 40nm pores (oxalic 60V). The wires were allowed to reach the surface (point 2 in fig 4) before the growth was stopped

REFERENCES

[1] F. Keller, M. Hunter, and D. L. Robinson, *J.Electrochem. Soc.*, **100**, 411 (1953).
[2] H Masuda and K Fukada, Science, **268** 1466 (1995)
[3] O. Jessensky, F. Muller, & U. Gössele, Appl. Phys. Lett, 72 1173 (1998)
[4] Schwarzacher W, Kasyutich O.I., Evans P.R., Darbyshire M.G., Yi G., Fedosyuk V.M., Rousseaux F., Cambril E., Decanini D. J. Magn. Magn. Mater **198-199** 185 (1999)

Preparation of PZT nanodots on Nano-Templated Substrates

M.J.A. McMillen†, A. Lookman†, C. Douglas†, P. Evans*†, R.M.Bowman*†, J.M.Gregg*† and R.J. Pollard*†

* Nanotec Northern Ireland, Belfast, UK
† Queen's University of Belfast, Belfast BT7 1NN, r.m.bowman@qub.ac.uk

ABSTRACT

There is considerable interest in creation of nanoscale capacitor arrays. For example, in FeRAM technology, which has great potential, significant capacitor cell size reduction and transition to 3D structures, can be achieved within existing CMOS procedures. However, cheap and simple methods to create nanoscale capacitors would be attractive. Lead zirconate titanate (PZT) films were deposited on alumina substrates using a solution deposition method involving dip-coating. The method was based on the pre-calcination of oxides, which were then dissolved in nitric acid. XRD θ–2θ scans showed that deposition improved with higher concentrations of PZT solution. AFM roughness analysis of the as-deposited films gave an r.m.s. roughness of approximately 39 nm, which was found to improve to approximately 6 nm on post-annealing. TEM imaging and EDX elemental analysis confirmed the presence of PZT on the deposited samples. Having found the optimum conditions for solution deposition, nano-dots were prepared by deposition on porous anodic alumina substrate templates. Further TEM imaging and EDX elemental analysis showed the presence of the PZT nano-dots aligned with the pores.

Keywords: ferroelectric, nanodot, porous-alumina, substrates, template.

1 INTRODUCTION

The smooth operation and growth of modern economies is increasingly dependant on memory and data storage. Consequently, technological advancement in this area is of great significance [1], and a variety of possible future memory technologies are now being driven forward [2]. Amongst these are systems based on ferroelectric thin films (FeRAM), where the difference between two polarisation states allows binary data storage [3]. FeRAM is attractive: read and write times are comparable to conventional DRAM and much faster than Flash; significantly, storage is non-volatile and write-cycle endurance (fatigue resistance) has been improved to the extent that now it is orders of magnitude better than Flash [2]. However, FeRAM is a still-maturing technology. Despite 2002 being seen as "a coming of age" [4], it will need further development if it is to supplant the mainstream established Si-based systems. In particular the following issues remain: materials integration (perovskites integrated with CMOS); storage densities (capacitor cell/footprint sizes must be reduced and moved into 3D configurations, by 2007); and, in common with all microelectronics, reduction of production costs is a major motivation [2].

To address this latter challenge it may well be necessary to explore alternative process and fabrication methodologies. There has already been preliminary work in this area [5] by examining the self-assembly of top electrode material on a continuous film of ferroelectric (see also [6]), but other approaches included creation of islands of ferroelectric material on a continuous lower layer [7]. The field is, however, only in infancy.

A particularly promising route that may be explored is that due to the work of Masuda et al. [8] who developed a two-stage anodisation process that produces an alumina film with a regular distribution of pores on an aluminum substrate. Due to their spatially periodic index of refraction, these porous alumina structures display a stop band in their electromagnetic transmission properties and therefore have been investigated as photonic crystals [9]. However, they also offer a nanoscale template that could be used with a range of nanotechnology applications such as magnetic storage, solar cells, carbon nanotubes, catalysts and metal nanowires.

We have already successfully prepared porous alumina substrate templates [10]. We have also extended the procedure to create arrays of nano-magnetic Co elements [11].

A particular challenge is to find a process that can prepare ferroelectric media, for example lead-zirconate-titanate (PZT), which can be incorporated easily into a self-assembly methodology. Conventional vapour deposition techniques such as sputtering or pulsed laser [12] are perhaps not so promising. A solution deposition technique may be more appropriate.

The most frequently used solution preparation methods for ferroelectric thin films are; Sol-gel processes, using alkoxide compounds as starting precursors, hybrid processes, which use chelating agents such as acetic acid and metalorganic decomposition (MOD), which uses carboxylate compounds [13]. An alternative solution-deposition method has been developed recently, based on the pre-calcination of oxides or carbonates to be used as the precursor material for the solution. This method is based on the fact that not all precursor oxides (e.g. PbO, TiO_2, ZrO_2, Bi_2O_3) are soluble in acid media, but a reacted oxide may be soluble [14]. The method has been used to prepare

PZT and bismuth titanate (BIT) thin films with good quality, homogeneity and the desired stoichiometry [14].

In this paper we describe preliminary work in the creation of a basic element of a self-assembled capacitor structure by combining a porous alumina substrate template process with solution-deposition. The porous alumina films were coated with the ferroelectric PZT using a solution deposition method. X-ray diffraction was used to determine the crystallographic orientation of the PZT films. Transmission electron microscopy was used to investigate the formation of ferroelectric films over the alumina pores.

Potential advantages of this technique are that the resulting capacitor structures could be electrically isolated from each other, and that their architectures may well made to be reminiscent of those familiar to industry.

2 EXPERIMENTAL METHODS

2.1 Solution deposition of lead zirconate titanate

In the present work, PZT powder was prepared from oxides and dissolved in nitric acid (HNO_3) to produce a PZT solution for dip-coating substrates.

The preparation of the PZT solution for deposition was carried out as follows. Lead oxide, titanium oxide and zirconium oxide were mixed at a ratio of Zr/Ti = 53/47. This mixed powder was then calcined in an electric furnace at 850°C for 3.5 hours in air ambient. The result was approximately 20 g of $PbZr_{0.53}Ti_{0.47}O_x$ powder.

Then 2g of this powder was dissolved in 60mL of diluted acid solution (10% HNO_3, 90% distilled water) for 1 hour at 80°C. The solution was allowed to cool to room temperature and then further diluted with distilled water to complete a PZT stock solution with concentration $12.5gL^{-1}$.

This stock solution was then diluted further to produce solutions with concentrations of $6.25gL^{-1}$ and $3.125gL^{-1}$; this would allow investigation of how the film properties varied with the concentration of solution.

The films were initially deposited on r-plane Al_2O_3 substrates to determine the optimum conditions for deposition and anneal. The substrate was heated on a hotplate in air to approximately 200°C for 15 minutes. It was then immediately dipped in the solution and removed.

The films were also deposited onto porous alumina templates prepared by anodisation of thin film aluminium as described elsewhere [15]

2.2 Characterization

The films were characterized by a range of techniques; including X-ray diffraction (XRD), atomic force microscopy (AFM), transmission electron microscopy (TEM), and energy dispersive X-ray analysis (EDX). XRD θ-2θ scans were performed on a Bruker-AXS D8, the surface roughness and topography was determined using a Digital Instruments D3000 AFM in tapping mode. The microstructure was revealed by TEM on a FEI Tecnai F20, scanning mode (STEM) was also used. The TEM cross-sectional samples were prepared on a FEI FIB200TEM focused ion beam microscope.

3 RESULTS

3.1 Dip coating of PZT on alumina substrates

Various anneal protocols were investigated; including immediate post-deposition high temperature annealing or being allowed to dry in air for approximately 3 hours before the annealing. An as-deposited film showed presence randomly oriented PZT in (100), (110) and (111) orientations (Fig. 1). The (110) and (111) peaks are not as clearly visible in scans of films where the solution was cold during dip-coating.

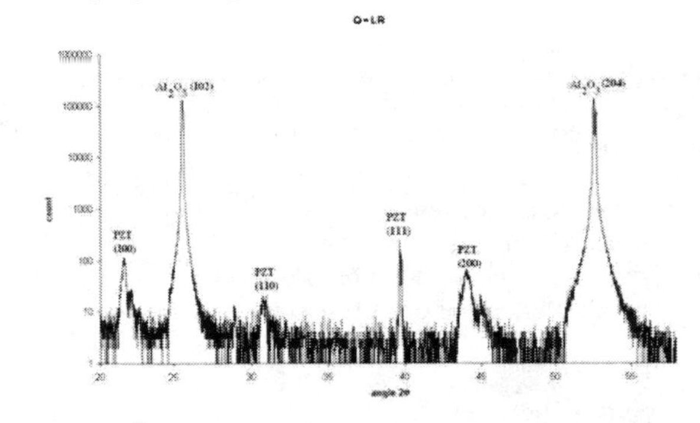

Figure 1: A θ-2θ scan of an as-deposited PZT film.

The various PZT XRD reflections were lost if the substrate and the as-dipped film were immediately transferred on to a hotplate until the coating dried. This suggesting that the immediate anneal on the hotplate after dip-coating does not encourage the growth of PZT. Also it was observed that as the concentration of the PZT solution decreased, the various PZT XRD peak intensities decrease, until a film made with a concentration $3.125gL^{-1}$ showed no PZT fingerprints..

A series of high temperature anneals following the room temperature drying indicated that optimum results were obtained at around 500°C for 2.5 hours in air. A XRD trace of a film prepared with a $12.5gL^{-1}$ solution is illustrated in Fig. 2.

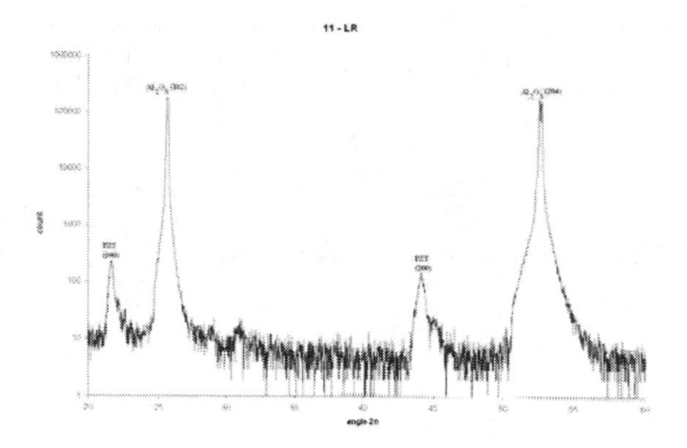

Figure 2: A θ-2θ scan of a 12.45gL^{-1} solution PZT film after the 500°C anneal.

AFM roughness analysis on films prior to annealing showed a very coarse surface topography, with r.m.s. roughness of order 39nm. This was improved by introduction of the high temperature anneal. In Fig. 3 an image and data are presented illustrating an r.m.s. roughness of 6nm. This film was prepared from a 12.5gL^{-1} solution and subject to the 500°C anneal described above. The AFM images show the presence of large crystallites on the film surface. It is not clear if the thin PZT film extends, uniformly, over the whole of the substrate surface, or if the peaks on the XRD scans are due only to isolated, oriented crystals.

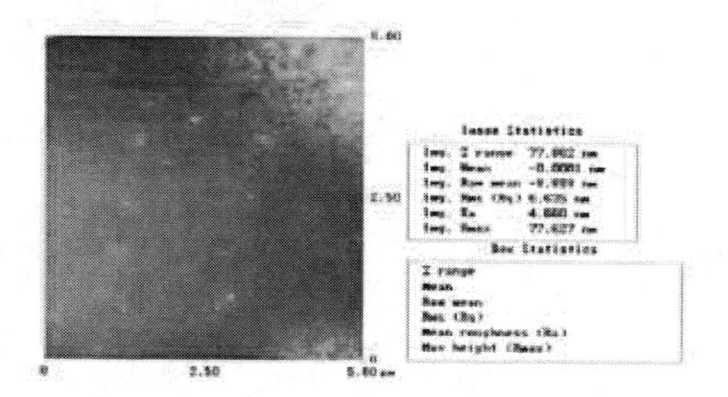

Figure 3: AFM topography of the 12.45gL^{-1} solution PZT film after the 500°C anneal.

A film annealed for 30 minutes on a hotplate was imaged by TEM and no PZT coating was observed at any point on the sample. A STEM image is shown in Fig 4. EDX analysis at a point between the substrate and the protective gold layer also failed to produce evidence of a PZT film. This again would suggest that such film is certainly not uniform, and may exist only as isolated crystals on the substrate surface as suggested by AFM.

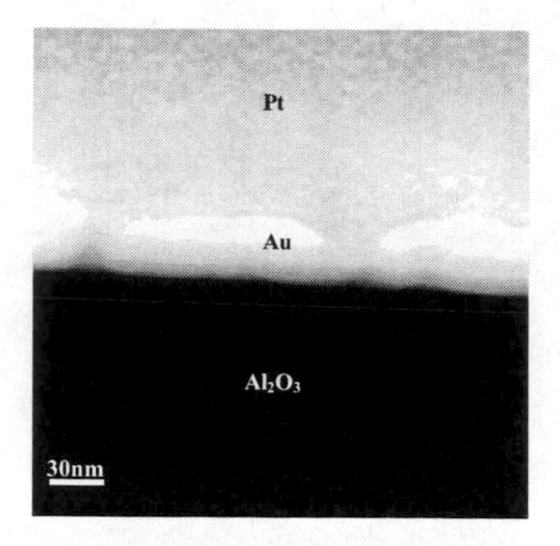

Figure 4: STEM image showing absence of PZT at interface with Al_2O_3 substrate. The Au and Pt layers are to protect the sample and provide contrast.

3.2 Dip-coating on porous alumina

Having established the behavior of dip-coating PZT on Al_2O_3 substrates the process was then transferred to the coating of the porous anodic alumina (Section 2.2). Conditions used included the 12.5gL^{-1} PZT solution in combination with high temperature annealing around 500°C.

In Fig. 5 the XRD θ-2θ scan of a PZT film on a porous anodic alumina substrate is shown. It confirms the presence of PZT, with (100), (110) and (111) orientations visible.

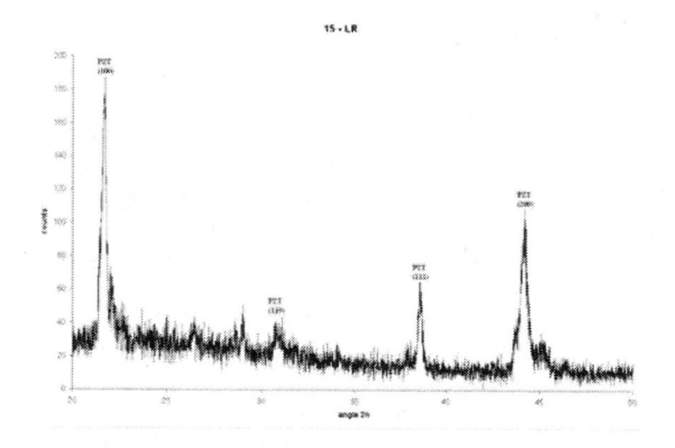

Figure 5: A θ-2θ scan of the 12.5gL^{-1} solution PZT film on porous anodic alumina substrate following anneal at 500°C.

Cross-sectional TEM imaging of the sample reveals some interesting features. In Fig 6. the pores in the alumina may be clearly seen, with PZT nano-dots visible just below the darker gold layer. At lower magnification we see a similar distribution of PZT adjacent to the nano-pores. An EDX spectrum was obtained at a point in one of the pores and small peaks are visible for Zr and Ti, indicating the presence of PZT nano-dots.

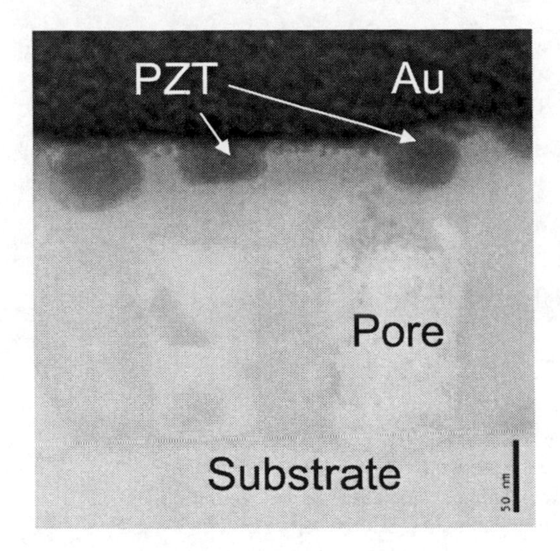

Figure 6: A cross-sectional TEM image showing the presence of PZT nano-dots aligning with the pores on the porous anodic alumina substrate.

4 DISCUSSION

A solution deposition method was implemented for the preparation of $PbZr_{0.3}Ti_{0.7}O_3$ thin films. This approach was based on the pre-calcination of oxides which were then dissolved in nitric acid. Using XRD, it was observed that higher concentrations of solution resulted in the best deposited films and that optimum anneal temperature was 500°C. AFM roughness analysis showed that the surface roughness was greatly improved by annealing the films. TEM imaging and EDX analysis confirmed the presence of PZT on the deposited samples.

The PZT was solution deposited on anodic porous alumina substrates. TEM imaging and EDX analysis were used to examine the formation of the film over the porous structure, and PZT nano-dots were confirmed to be present.

Future, and ongoing, work involves deposition of metals for plugs, and both lower and upper electrodes to provide electrical contacts. The hexagonal symmetry of the anodic pores makes registry an interesting avenue to explore.

REFERENCES

[1] Anon, "File that; Data Storage", The Economist 370(8365), 72, 2003.

[2] Anon, "International technology roadmap for semiconductors" at http://public.itrs.net

[3] J.F.Scott, "Ferroelectric memories", Springer-Verlag, Berlin.

[4] Anon, "MRAM, FeRAM are at the finish line", at http://www.eet.com/story/OEG2003032S0047

[5] M. Alexe, J. F. Scott, C. Curran, N. D. Zakharov, D. Hesse & A. Pignolet, Appl. Phys. Lett., 73, 1592, 1998; M. Alexe, A. Gruverman, C. Harnagea, N. D. Zahkarov, A. Pignolet, D. Hesse & J. F. Scott, Appl. Phys. Lett. 75, 1158, 1999; M. Dawber, I. Szafraniak, M. Alexe & J. F. Scott, J. Phys.: Condens. Matter, 15, L667, 2003.

[6] E. Vasco, R. Dittman, S. Karthauser & R. Waser, Appl. Phys. Lett., 82, 2497, 2003.

[7] A. Seifert, A. Vojta, J. S. Speck & F. F. Lange, J. Mat. Res., 11, 1470, 1996.

[8] H. Masuda and K. Fukuda, Science 268, 1466, 1995.

[9] H. Masuda, M. Ohya, H. Asoh, M. Nakao, M. Nohtomi and T. Tamamura, Jpn. J. Appl. Phys. 38, L1403, 1999.

[10] C. Douglas, R. Evans, and R.J.Pollard, "Anodisation of aluminium", to be presented at Nanotech 2005.

[11] R.J.Pollard, "Method of manufacturing nanostructure arrays", patents pending.

[12] S.S. Roy, C. Morros, R.M. Bowman and J.M. Gregg, Appl. Phys. A, DOI: 10.1007/s00339-004-2900-y, 6 July 2004.

[13] E. B. Araújo and J. A. Eiras, Mater. Res. 2, 17, 1999.

[14] E. B. Araújo and J. A. Eiras, J. Eur. Ceram. Soc. 19, 1453, 1999.

[15] C. Douglas, PhD thesis, Queen's University of Belfast 2004.

SPM Investigation of the Electron Properties YSZ Nanostructured Films

D.A.Antonov, O.N. Gorshkov, A.P. Kasatkin, G.A.Maximov, D.A.Saveliev, D.O.Filatov

Research and Educational Center for Physics of solid state nanostructures,

University of Nizhny Novgorod, 23 Gagarin Ave., Nizhny Novgorod 603 950 Russia,

antonov@phys.unn.ru.

ABSTRACT

The electronic properties of the yttria stabilized zirconia (YSZ) films with Zr nanoclusters formed by implantation of Zr ions have been studied by combined Atomic Force Microscopy / Scanning Tunneling Microscopy (AFM\STM). The feedback was maintained by AFM in Contact Mode using a conductive cantilever. Simultaneously the I-V curves of the junction between the probe and the conductive substrate through the nanostructured YSZ film were measured. It is the evidence that we observed the localized channels of the tunneling current attributed to the successive tunneling through Zr clusters. The lateral sizes of the channels where 1-10 nm in thick (24 nm) YSZ films. The effects of resonant tunneling and Coulomb blockade of tunneling through the Zr clusters have been observed.

Keywords: nanoclusters, YZS films, STM, AFM, coulomb blockade, resonance tunneling.

1 INTRODUCTION

Formation of the metal and semiconductor nanoclusters in the dielectric oxide matrices is a promising scientific direction developed in the last years with the purpose of making new materials for opto- and microelectronics [1]. One of the ways of making such systems is implantation of ions of the metal or semiconductor in the oxide films [2-5].

Scanning Tunneling Microscopy (STM) is used widely for investigation of the electronic properties of the clusters deposited on the conductive substrates [6] as well as embedded in the dielectric films [7-8]. Such effects as Coulomb blockade, resonant tunneling, and increasing effective gap with decreasing cluster sizes have been observed in these systems [9]. However, investigations of the clusters embedded in the dielectric films by STM encounter a number of problems. One of them is difficulty to maintain a stable tunneling current feedback through the thick dielectric films. In the present work we applied combined Scanning Tunneling/Atomic Force Microscopy (STM/AFM) to investigate the electronic properties of the Zr nanoclusters formed in yttria stabilized zirconia (YSZ) films by implantation of Zr ions. We used a standard beam deflection AFM with a conductive cantilever. The feedback was maintained by AFM technique in Contact mode. Simultaneously the I-V curves of the tunneling contact between the cantilever and p^+-Si substrate through the nanostructured YSZ film were recorded in every point of the scan. This approach has allowed us to decouple maintaining the feedback and measuring the tunneling current and to observe tunneling through the Zr

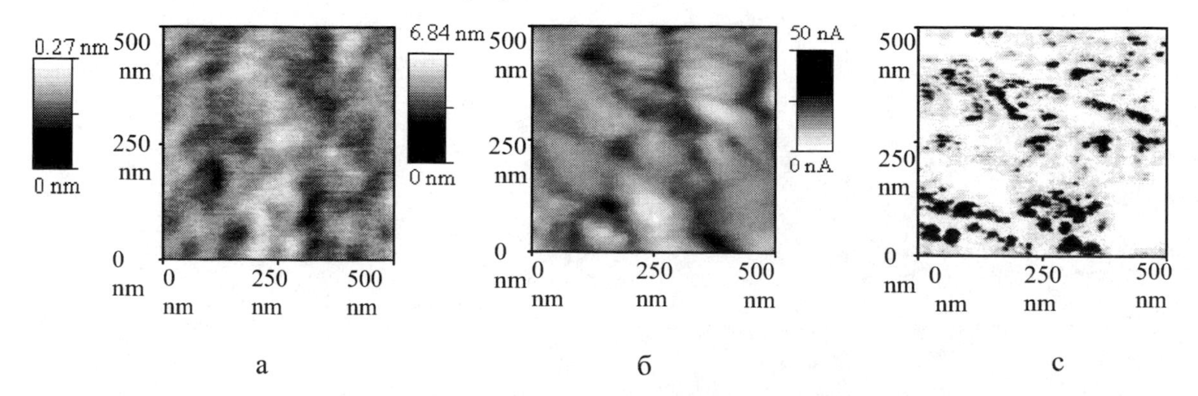

Fig. 1. The surface topography (*a, b*) and the current image (*c*) of YSZ film before (*a*) and after (*b,c*) implantation of Zr. The gap voltage *U*=+4V.

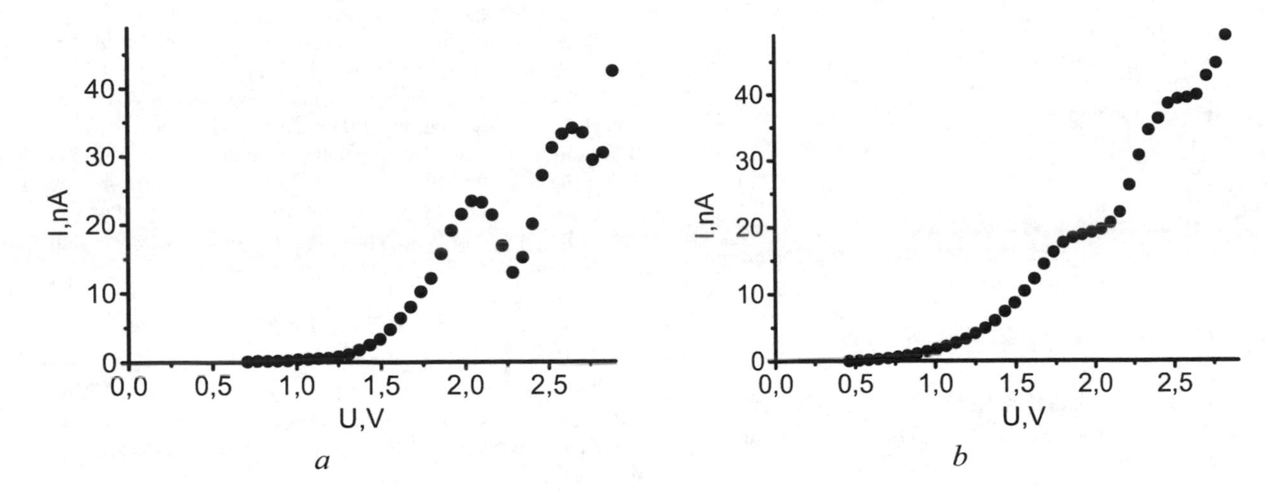

Fig. 2. Typical I-V curves recorded on the separated channels of the tunneling current; in the spots with sizes < 2 nm (a), in the spots with size 2-4 nm (b).

nanoclusters including Coulomb blockade and resonant tunneling effects.

2 EXPERIMENT

The YSZ films were deposited on p^+-Si (001) substrates by magnetron sputtering of a target from 90 mol. % ZrO_2 + 10 mol. % Y_2O_3. The thickness of the films was 24 nm. The implantation was performed on the pulsed ion beam accelerator. The averaged ion energy was 190 keV, the ion dose was ~ 10^{17} cm^{-2}. The optical absorption spectra of the implanted YSZ demonstrated an absorption band in the wavelength interval of 400 ÷ 650 nm conditioned by Zr clusters with average radius ~ 1nm [5].

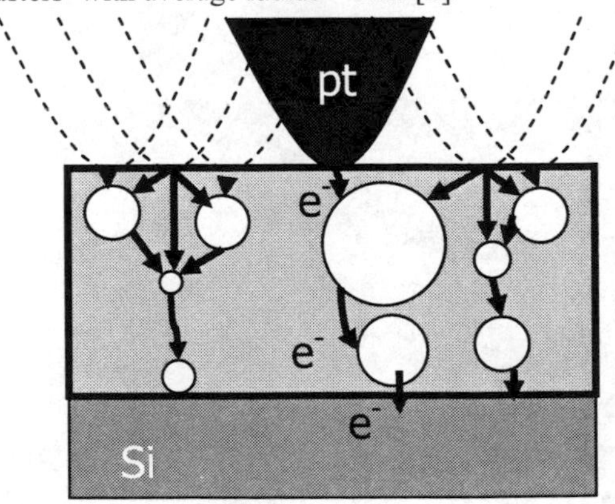

Fig. 3. Schematic of the mechanism of the electron tunneling through the dielectric film with the metal nanoclusters.

The morphology and electronic properties of the implanted films were studied by Omicron® Ultra High Vacuum (UHV) AFM/STM LF1 installed into Omicron® Multiprobe S™ UHV system. Soft p^+-Si I-type cantilevers coated with Pt were employed. The accuracy of the tunneling current measurements was ~10 pA.

3 RESULTS AND DISCUSSION

In Fig.1 the morphology of YSZ film before (*a*) and after (*b*) implantation of Zr ions is presented. Implantation results in increased surface roughness and formation of the areas of conductivity with lateral sizes ranging from 1 to 20 nm represented as the dark spots in the current image (Fig.1*c*). It should be noted that no correlation between morphology and the current image was observed. No current was observed before implantation even at the highest possible values of the gap voltage U (±10 V).

In Fig.2 typical I-V curves taken in the spots with high conductance (the tunneling current channels) are shown. In the I-V curves taken in the spots with small sizes (< 2 nm) the peaks typical for resonant tunneling through the nanometer sized clusters with the discrete energy spectrum [9] have been observed (Fig.2*a*). In the I-V curves taken in the spots with larger size (2-4 nm) the steps typical for the Coulomb blockade of the tunneling have been observed (Fig.2*b*). To estimate the cluster size the asymmetric contact model [9-10] was used for coulomb blockade of the tunneling through the junction with a largest resistance (this junction behaves like a bottleneck of the tunneling current [7]). Assuming the cluster to be a metal balls with the radius r, its capacitance C can be expressed as follows:

$$C = 4\pi\varepsilon_0\varepsilon r \quad (1)$$

Here ε_0 is the electric constant and $\varepsilon = 4$ is the dielectric constant of the material (YSZ) surrounding the metal clusters. On the other hand, C is related to the period of the steps in the I-V curve V (taking into account the temperature broadening of the steps ΔV) by expression

$$C = \frac{e^2 \Delta V}{2VkT}, \qquad (2)$$

where e is the elementary charge, k is the Boltzman's constant, and T = 300K is the measurement temperature. Finally,

$$r = \frac{e^2 \Delta V}{8\pi\varepsilon_0 \varepsilon VkT} \qquad (3)$$

For $V \approx 0.66$V and $\Delta V \approx 0.28V$ (from Fig.2*b*) the estimate of the cluster size gives $r \approx 2.9$ nm which is in a reasonable agreement with the STM/AFM and optical spectroscopy data.

In most cases the lengths of the steps on the I-V curves were not equal as it should be when tunneling through a single cluster took place. To explain this effect it was suggested that the size and number of clusters forming a tunneling current channel can change with changing gap voltage. A scheme of the tunneling in a film with random spatial distribution of the clusters is shown in Fig.3. Since the total conductivity of the cluster chain is determined by a "bottleneck" cluster having the highest tunneling resistance [7], switching of the current channels would make different clusters to determine the step parameters.

In Fig.4 the current images of an area of the increased conductivity with lateral sizes ~10 nm at different gap voltages U are presented. At low U the channels with the lowest tunneling resistance appeared first (Fig.4*a*). The lateral sizes of the current channels at low U were within 1 to 3 nm.

Increasing U resulted in switching on new current channels (Fig. 4*b*). At further increasing of U the channels began to overlap (Fig.4*c*).

Annealing the Zr implanted films at the temperatures T=300-540 C resulted in aggregation of the small clusters in the larger ones with the sizes ranging from 10 to 40 nm. No peaks or steps were observed in the I-V curves measured on these aggregated clusters. This effect confirms that formation of the current channels is caused by tunneling of the electrons through Zr clusters in the dielectric film but not through the local electronic states of the radiation defects created in the YSZ matrix by ion bombardment.

4 CONCLUSIONS

The results of this work have demonstrated that combined AFM/STM technique allows to study the electronic properties of the nanometer sized clusters in the dielectric films. Using this technique allowed us to study the electron tunneling through Zr nanoclusters formed by ion implantation in YSZ films and to observe the effects of Coulomb blockade and resonant tunneling through the metal clusters.

Authors are very grateful to Dr. V.A. Kamin who formed the YSZ films and to Dr Yu. A. Dudin who performed the ion irradiation of the films.

5 ACKNOWLEDGEMENTS

Authors gratefully acknowledge the financial support by joint Russian-American Program "Basic Research and Higher Education" (BRHE) sponsored by Russian Ministry of Education and by US Civilian Research and Development Foundation (CRDF), Award No. REC-NN-001.

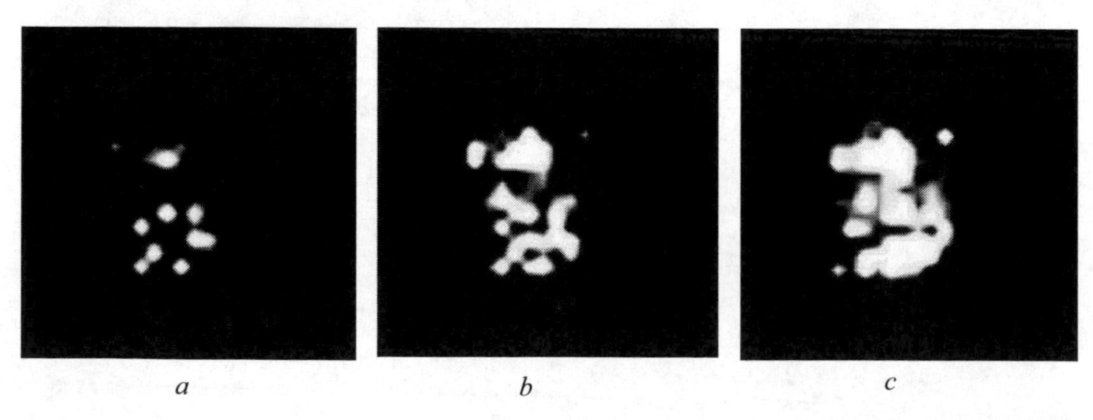

Fig. 4. Current image of a YSZ after Zr implantation at different gap voltages. Gap voltage U (V): *a* - 0.88, *b* - 1.90, *c* - 3.72. Frame size 35 × 35 nm^2

REFERENCES

[1] V.P.Dragunov, I.G.Neizvestny, B.A.Gridchik, Basics of Nanoelectronics,. Novosibirsk 2000.
[2] G.W.Arnold, J.Appl.Phys., 46, 4466, 1975.
[3] P.Mazzoldi, A.Boscolo-Boscoletto, G.Battaglin et al., Advanced Materials in Optics, Electro-Optics and Communication Technologies, Vincenzini P. Techna Srl, 149, 1995.
[4] D.I.Tetelbaum, O.N.Gorshkov, S.A.Trushin et al., Chemistry and Application of Nanostructures, 239, 2003.
[5] O.N.Gorshkov, V.A.Novikov, A.P.Kasatkin, Inorganic Materials, 35, 502, 1999.
[6] C.Binns, Surf. Sci. Reports, 44, 1, 2001.
[7] H.Imamura, Phys. Rev.B, 61, 46, 2000.
[8] E.Bar-Sadeh, Y.Goldstein, Phys. Rev. B, 50, 8961, 1994.
[9] V.Ya. Demihovsky, G.A. Vugalter «Physics of quantum low-dimensional structures», M. «LOGOS» 211, 2000.
[10] I.O.Kulik, R.I..Shcehter, JETP, 62, 623, 2000.

Optical Characterization of the Au Nanoparticle Monolayer on Silicon Wafer

Da-Shin Wang[1], Light Chuang[2] and Chii-Wann Lin[3]

[1]National Taiwan University, Center of Nanoscience and Technology
amydsw@ibme.mc.ntu.edu.tw
[2] National Taiwan University, Inst. of Biomedical Engineering
light@ibme.mc.ntu.edu.tw
[3] National Taiwan University, Dept. of Electrical Engineering, Taiwan.R.O.C.
cwlinx@ntu.edu.tw

ABSTRACT

This paper mainly studies the optical property of gold nanoparticle films and its variations under different surface coverages. Au nanoparticle monolayers with different surface densities are prepared and its optical constants are decided by variable-angle ellipsometry at 530nm and 643nm, respectively. The extinction coefficient shows a linear correlation with the surface coverage, nonetheless the refractive index n presents a more complex behavior. This measured optical constants are compared to the theoretical estimations of effective medium theory(EMT); both Maxwell-Garnett and Bruggeman models are employed. Neither of these two theoretical predictions fit in with the experimental results. When the surface coverage is nearly full, the optical constant does not approaches to the bulk values as predicted by EMT, which implies the Au nanoparticle film has distinct optical property from the bulk gold and calls for a deeper understanding in its electronic structure.

Keywords: effective medium, nanoparticle, film, refractive index

1 INTRODUCTION

In recent years, self-assembly technique has been widely used in fabricating mono- or multi-layers of Au nanoparticle layer for various applications. The nanoparticle layer is covalently linked with a linker layer on the substrate, so its structure is more stable and the spacing can be controlled by the length of linkers. It has been shown that Au nanoparticle layer made this way has promising applications in biosensor, since the optical absorption of this layer is very sensitive to the optical properties of ambient medium . The optical characterization of this layer and understanding how it changes with ambient environment are crucial in elucidating the sensing mechanism and developing better devices. However, up to now there have been very few reports on the optical constant, N=(n+ik), of the covalently -linked nanoparticle monolayer[1]. .

Before performing the measurement of optical constant for this nanoparticle film, there is another factor needed to be considered. Different from typical gold film, Au nanoparticle film is discontinuous so has a lot of boundaries, which are in contact with the ambient medium. What we measure as the optical property is actually due to the nanoscale mixing effects of Au nanoparticles and the medium. That may be the reason why the optical absorption is so sensitive to the medium. Different surface mixing conditions give rise to different values for optical constant. For the optical constant to be unambiguously referred, the mixing factors, such as surface density, the shape and size of the grain, need to be specified. So here surface coverage (or density) is chosen as a mixing factor, the optical constant is studied in terms of the surface coverage.

In this study, the surface coverage is decided by counting the covering area in AFM images. When the coverage is nearly full, the optical constant is not approaching that of bulk gold, as we supposed beforehand. The measured values are also compared with the results estimated by two effective medium theories, Maxwell-Garnett and Bruggeman models. Both fail to predict the experimental data. It suggests the nanoparticle film has distinct optical properties which cannot simply approximated by linearly adding and averaging two bulk materials, as it is done in classical effective medium models.

2 METHOD

In this section, the experimental methods will be described. It includes three parts: i) the chemical preparation of Au nanoparticle film with different surface coverages, ii) the ellipsometric measurement to obtain the optical constant n and k, iii) the AFM measurement of the the height and the surface coverage.

2.1 Preparation of Au nanoparticle film

Derivatization of silicon wafer Polished silicon wafer is used as the substrate for its flatness and suitability for ellipsometry. It is first cleaned in piranha solution at 60℃ (H_2SO_4/H_2O_2) for 30min, and is then derivatized by immobilizing APTMs (3-Amino-propyl-trimethoxysilane)

on it. APTMs is purchased from Sigma and then further diluted to 1.1mM in ethanol. The derivation of silicon wafer is made by immersing the wafer in 1.1mM APTMs for 3hrs. After it, the substrate is thoroughly washed by ethanol and H2O. It is stored in H2O before use.

Immobilization of Au nanoparticles Colloid Au (20nm) is purchased from Sigma. The undiluted solution has a absorbance value 1.67 at 531nm. The different coverages of nanoparticle film are made possible by immersing silicon substrate in different concentrations of colloidal Au. The Au colloidal solution is first diluted as 1/2,1/3,1/4,1/5 of original concentration. Five silicon substrate is prepared. The substrates are first dried by nitrogen and then are immersed 1,1/2,1/3,1/4,1/5 diluted colloidal Au solution and sit for six hrs. The samples are then washed by H2O and dried by nitrogen for further investigation.

2.2 Ellipsomeric measurement

A variable-angle ellipsometer made by Nanofilm, Germany is used to study the optical properties of the nanoparticle film. The light source is 530nm and 643nm laser. For each sample , ellipsometric measurement is performed at various angle of incidence from 42° to 62°, at 5° intervals. Taking the measured height 17.5nm in AFM as the thickness, n(refractive index) and k(extinction coefficient) are obtained by employing the 3-layer Fresnel equation to fit the ellipsometric parameters Delta and Psi. Very good fit (MSE~10e-1) are obtained, which indicates the validity of the fitted n and k. For each sample, the measuring area is around 1*1mm; three positions are chosen for measurement and the average n,k and standard deviation of are calculated.

2.3 Surface coverage measurement

DME(Danish Micro Engineering) atomic force microscope is used to measure the surface topology of the nanoparticles film. Tapping-mode is adopted in whole scanning. The images are taken in 2*2um and 1*1um regions. The height is obtained by averaging the difference of 5 highs and lows in each profile. The 17.5±0.6nm is the average of 20 profiles taken from 5 images. The surface coverage of each image is decided by counting the pixel which value is above set threshold; the number of these pixels divided by total pixels is the surface coverage.

3 RESULTS AND DISCUSSION

Different coverages are attained by immersing substrate in different concentrations of colloidal gold as expected. The different coverages are observed by AFM; the images (2*2um) are shown as in the figure 1. (a) 97% (b) 72.2% (c) 58.1% (d) 49.8% (e) 9% (g) line profile of (b) figure

Fig 1.(a) Immersion in undiluted colloidal gold makes the 97% surface coverage

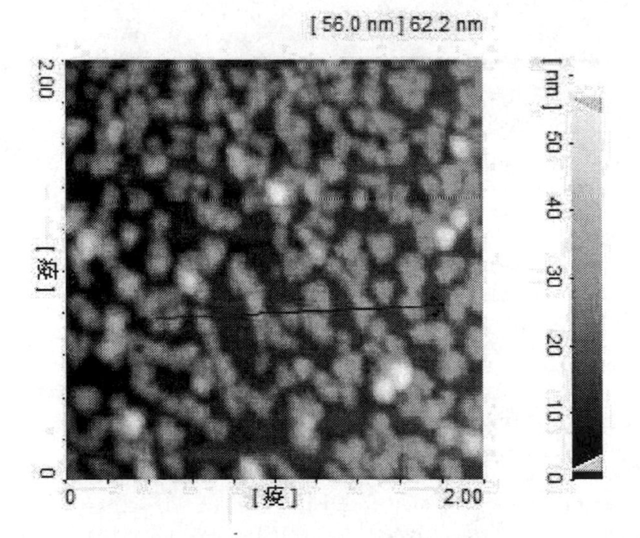

Fig.1.(b) Immersion in 1/2 diluted colloidal gold makes 72.2% coverage. A line is drawn to show the profile. See fig.1(g)

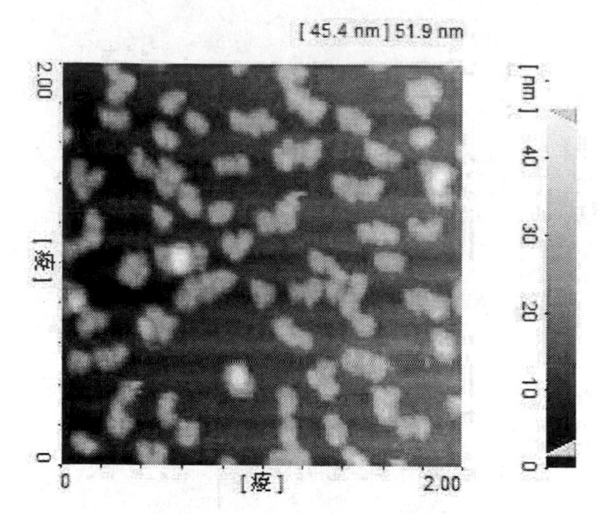

Fig.1(c) Immersion in 1/3 diluted colloidal gold makes 58.1% coverage

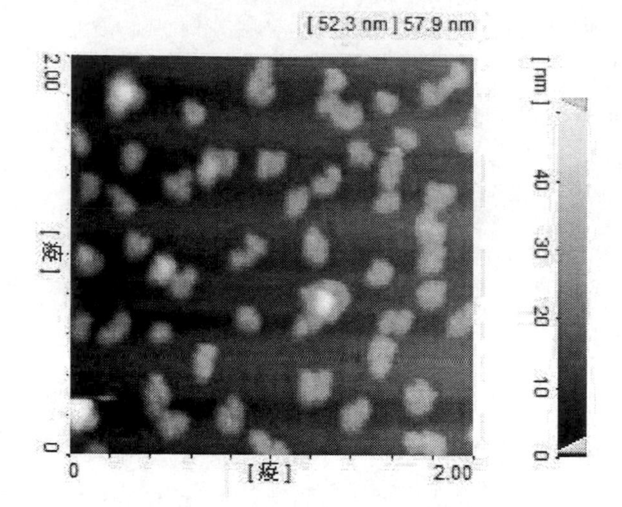

Fig.1(d) Immersion in 1/4 diluted colloidal gold makes 49.8% coverage

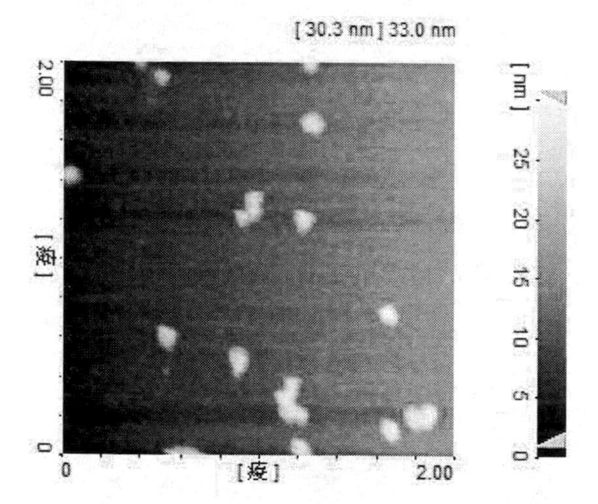

Fig.1(e) Immersion in 1/5 diluted colloidal gold makes 9%coverage

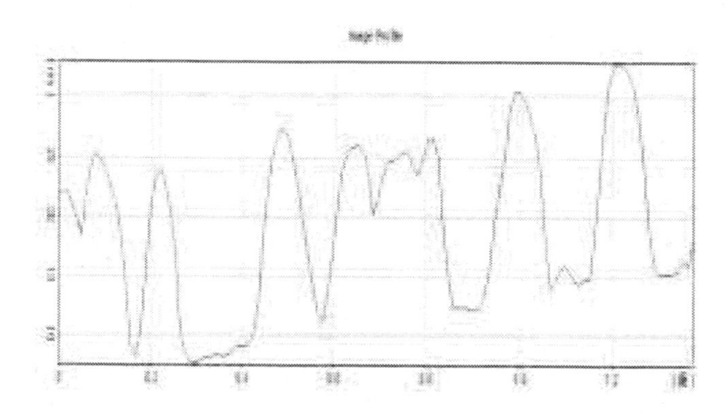

Fig.1(f) Line profile of Fig.1(b)

The measured n and k varies with surface coverage as expected. The extinction coefficient k shows an approximately linear relation with the surface coverage. The relation of k value vs. surface coverage at 530nm and 643nm is shown below in Fig.2. The correlation coefficient r is 0.96, indicating the high correlation between these two variables: k and surface coverage. Though large numbers of data are required to test to see if the relation is truly linear, and statistical work is needed for putting such an application into practice, this simple result at least sheds a little light on the future direction.

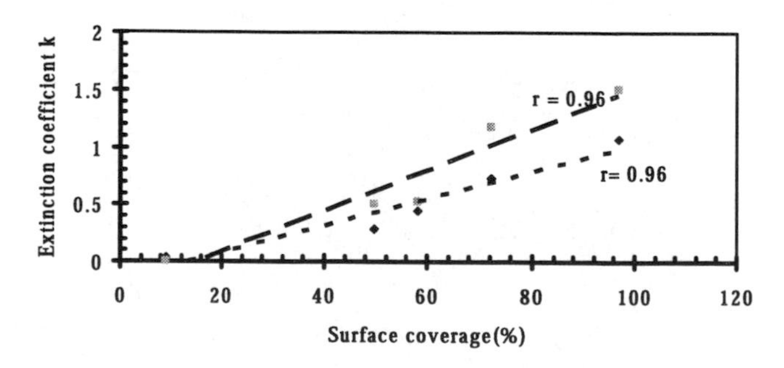

Fig.2 **Square**: k values at 643nm **Diamond:** k values at 530nm

However, the refractive index does not exhibit a simple monotonously increasing or decreasing relation(See Fig.3).

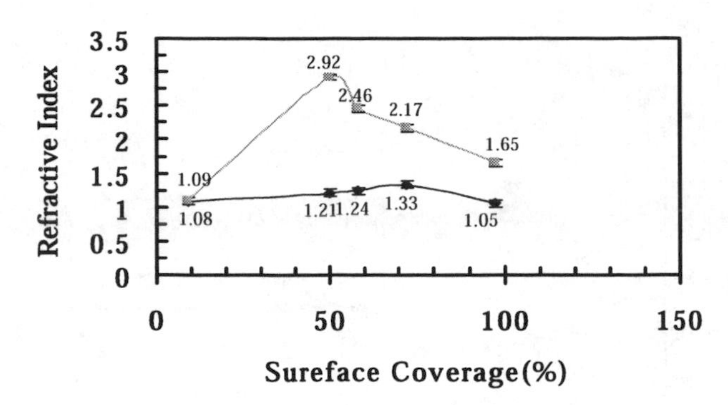

Fig.3 **Square**: n values at 643nm. **Triangle**: n values at 530nm

The error bar indicates the variation between different points on the same sample; for each sample, 3 different points are measured. Only slight changes of Delta(< ±0.5) and Psi(< ±0.1) are observed, so the variation of millimeter –average n,k is small as ±0.05.Effective medium models are employed in hopes of explaining the complex behavior. The value of surface coverage is substituted for the volume fraction f in Maxwell-Garnett and Bruggeman models to decide the effective n and k of this discontinuous film. Equation 1 is the expression of Maxwell-Garnett formula in this example(choosing Au as host material and air as the ambient).

$$\frac{\varepsilon_{eff}-\varepsilon_{au}}{\varepsilon_{eff}+2\varepsilon_{au}}=(1-f)\frac{\varepsilon_{air}-\varepsilon_{au}}{\varepsilon_{air}+2\varepsilon_{au}} \quad (1)$$

Equation 2 is Bruggeman formula;

$$f\frac{\varepsilon_{au}-\varepsilon_{eff}}{\varepsilon_{au}+2\varepsilon_{eff}}+(1-f)\frac{\varepsilon_{air}-\varepsilon_{eff}}{\varepsilon_{air}+2\varepsilon_{eff}}=0 \quad (2)$$

ε_{eff} is decided accordingly and the corresponding n and k ($\varepsilon=(n+ik)^2$) is then calculated. The results are shown in Fig.4

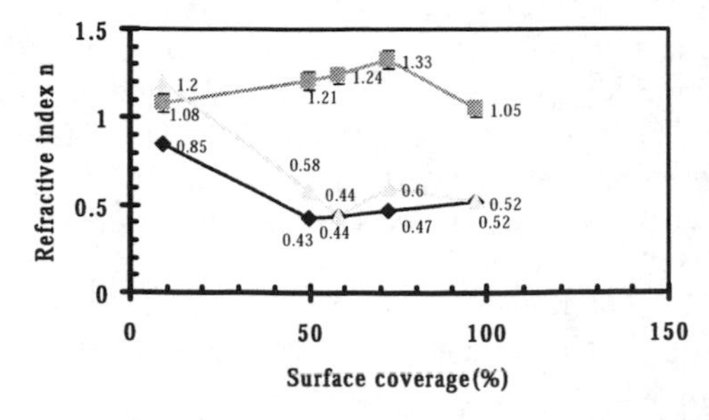

Fig.4 **Square:** n values measured at 530nm **Triangle:** Bruggeman's prediction **Diamond:** MG's prediction

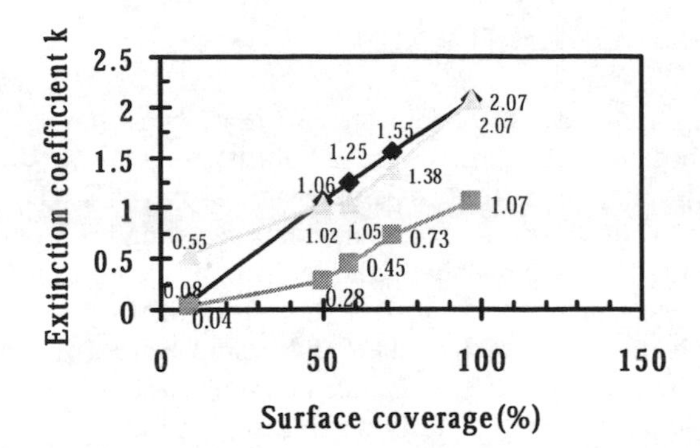

Fig.5 **Square**: k values measured at 530nm **Diamond:** MG's prediction **Triangle:** Bruggeman's prediction

Both MG and Bruggeman models predict an increasing k with the surface coverage(See Fig.5). But for the part of n, MG and Bruggeman models neither successfully describe its relation with surface coverage nor give prediction close to the measured values. As to the optical constant when the coverage is full, both MG and Bruggeman models give the bulk result; ie, n and k are 0.52 and 2.13; this disagrees with the measured n and k values ,1.05 and 1.07. it is very different from that of bulk gold, which has the value as 0.52 and 2.13. It is supposed that with the presence of boundaries at nanoscale, the electrons in the nanopaticle film are no longer in an infinite crystal potential field, as they are in bulk gold. The electronic band structure has become different with the existence of a different potential field. This difference then gives rise to the distinct optical properties as observed in experiments. This thinking also implies that the mixing factors, including grain size and shape ,density, and topology, etc, may play important roles in the optical properties.

4. CONCLUSION

From the data above, the extinction coefficient appears to be a candidate for quantifying the surface density, which surely is a measure of surface amount as long as the distribution is random. As to the n, k values of the Au nanoparticle film, the measurement does not agree with the theoretical approximation, which prediction approaches to the bulk gold when the surface coverage gets saturated. But experiment shows that Au nanoparticle film has n,k very different from that of bulk gold. It seems to imply that this Au nanoparticle film, composed of grains at nanoscale, has different electronic structure from bulk gold, and its optical effects cannot be reduced to the linear addition of the effects of two bulk material, gold and air.

REFERENCES

(1) [1] Zhang, H.L.; Evans,S.D.; Henderson, J.R. Advanced Materials. 2003,15,No.6,p.531-534

Nanocharacterization Using Secondary Ion Mass Spectrometry (SIMS)

Steven W. Novak and Charles W. Magee
Evans East, 104 Windsor Center Drive, East Windsor, NJ snovak@evanseast.com

ABSTRACT

Secondary Ion Mass Spectrometry (SIMS) using ultra-low energy beams (<1keV) gives the capability to measure structures within films as thin as 1nm. By measuring SiO_2 calibration films of known thickness and extrapolating back to zero thickness, we can calculate the ion mixing depth of the SIMS measurements. These data show that SIMS can achieve depth resolution of less than 1nm using beams of 700-300eV. By using an inorganic elemental tag, an isotopic tag or staining, low energy depth profiling can also discern the structure of planar organic films. We give examples of Zn-doped layers of tetraphenylporphyrin (TPP) and a self-ordered polystyrene/polybutadiene copolymer film stained using OsO_4. Standard SIMS measurements have also been carried out on doped Si nanowires to measure boron doping levels.

Keywords: sims, gate oxide, copolymer, nanowires

1 INTRODUCTION

The technique of secondary ion mass spectrometry (SIMS) is one of the few tools having the requisite resolution to characterize nanomaterials. Because nanometer scale resolution is available only in the depth dimension, SIMS has not been widely applied to analysis of nanomaterials. It is the purpose of this article to show that experiments can be devised in ways that take advantage of SIMS depth resolution to allow characterization of several types of nanostructured materials.

The SIMS technique consists of bombarding a solid with an energetic ion beam and detecting the ions emitted from the surface using a mass spectrometer [1]. It has been applied most widely for trace element analysis of semiconductors and other materials used in the manufacture of microelectronics, however it can be used to measure any solid material. SIMS is most often used to acquire depth resolved information, presented as a plot of ion intensity or concentration versus depth, but can also be used to measure the bulk concentration of trace impurities in a solid. It is the combination of sensitivity and high depth resolution that makes SIMS valuable for characterizing nanostructures.

2 SIMS DEPTH RESOLUTION

We first need to demonstrate that SIMS has nanometer scale resolution. In recent years nanometer scale SiO_2 layers are being used as gate dielectrics in MOS capacitors [2]. These oxides are now on the order of 2nm thick and

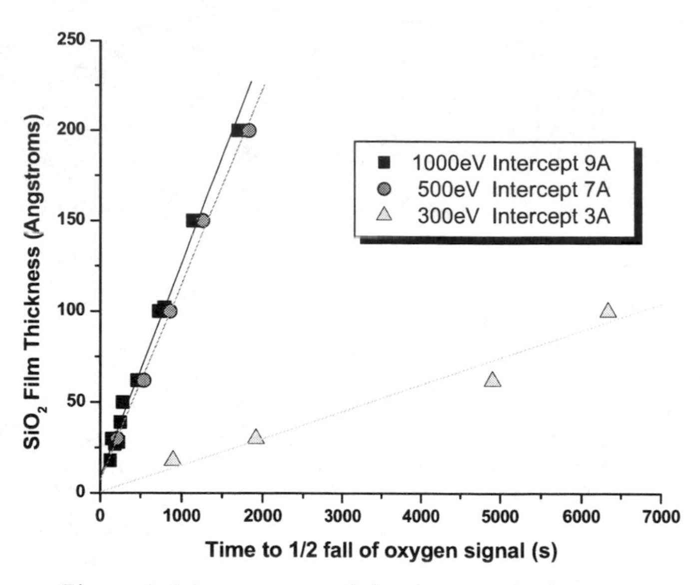

Figure 1. Measurement of the time required to sputter through oxide films versus film thickness. All measurements done at an incidence angle of 75 degrees from normal.

must be tightly controlled in uniformity. Because they are formed on a highly polished Si wafers they provide ideal calibration samples to measure the resolution of the SIMS technique. By plotting the time required to sputter through the films against their known thickness, we can extrapolate back to zero time to calculate the thickness of the ion mixed layer induced by ion bombardment. [3]. Figure 1 shows a series of measurements on oxide films of various thickness acquired using Cs ion bombardment at three impact energies. Extrapolating these trends back to zero time indicates ion mixing depths of 0.9nm for 1keV, 0.7nm for 500eV and 0.3nm for 300eV bombardment energies. The mixing depth is usually considered equivalent to the depth resolution of the technique.

Thus we can demonstrate that SIMS has sub-nanometer depth resolution for oxide films on well-controlled polished Si wafers. This level of depth resolution is available to measure any solid layered structure, providing the layering is planar over the area of analysis and the substrate is flat. An example of a high-resolution measurement is given in Figure 2. This depth profile shows a 3nm thick SiO_2 film that has been treated to introduce nitrogen into the film. The high resolution of the measurement allows us to see that nitrogen is accumulated at both the surface of the film and the substrate interface.

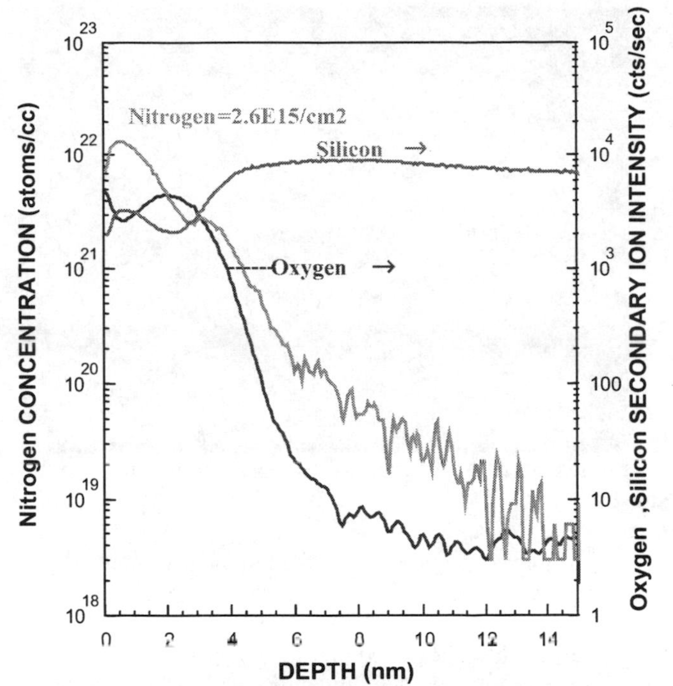

Figure 2. SIMS depth profile of a nitrided SiO_2 film.

3 DEPTH PROFILING LAYERED POLYMER STRUCTURES

Nanostructured materials are being formed from layered polymer films and copolymer blends because these materials can be tailored to form various structures depending on the molecular weight of the polymers and percentage of each in a blend [4]. In order to use SIMS to characterize such materials we must have way to distinguish the different compounds and we must have a planar structure. There are three options to tag the components of a polymer structure such that SIMS can distinguish them: inorganic tagging, staining and isotopic tagging. Only the first two techniques are discussed here as many isotopic tagging experiments have been used to show details of copolymer blends [5].

3.1 Depth Profiling Using an Inorganic Tag

One method by which SIMS can discern two organic materials in a layered structure is to tag one with an inorganic element. Figure 3 gives an example of a sample in which alternating layers of 5,10,15,20 tetraphenylporphyrin have been deposited on a Si wafer substrate [6]. Layers doped with Zn have been deposited between undoped layers by the technique of atomic layer epitaxy.

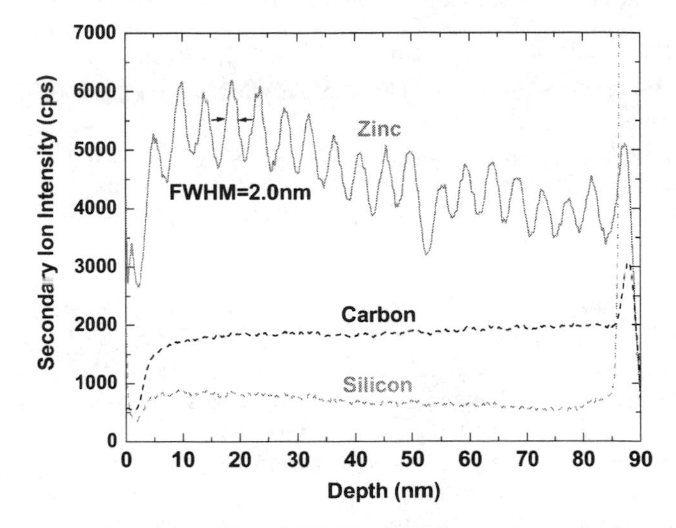

Figure 3. Depth profile of TPP:Zn/TPP layered sample by SIMS.

This profile allows us to demonstrate that layers of about 2nm thickness were deposited, count the number of layers and state that the layer thickness is quite uniform through the deposition. Many other elemental tags could probably be used in this application successfully, such as F or S, depending on the interaction with the polymers used in the structure. A flat highly polished substrate is essential.

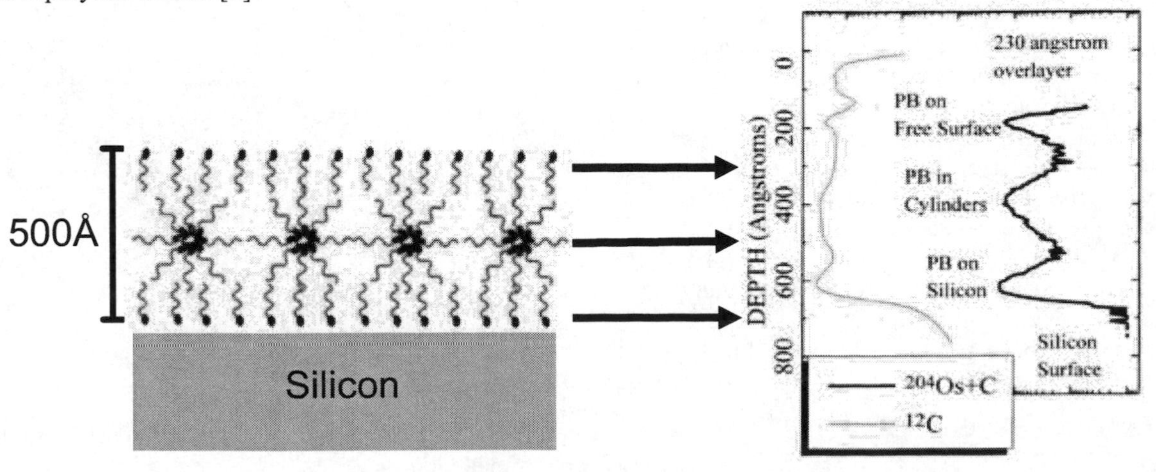

Figure 4. Cartoon structure of a PS/PB copolymer layer and corresponding SIMS depth profile of a stained film.

3.2 Delineation of Structure using Staining

Copolymer blends are being investigated as a way to form regularly patterned templates for such uses as nanolithography [7]. Copolymers consist of a mixture of two dissimilar polymers that spontaneously order into nanometer scale structures at equilibrium. Changing the relative proportion of the polymers yields different exsolution structures, ranging from lamellae to ordered spheres to a complex three-dimensional lattice [8]. SIMS can discern the structure of the exsolved films because of its high depth resolution. Many polymers are compositionally quite similar, however, so it is difficult or impossible to detect the individual polymers on the basis of elemental signals. It is possible to selectively tag doubly bonded polymers using standard osmium tetraoxide staining that is widely used for staining biological structures. By monitoring the Os signal, a SIMS depth profile will show the location of the doubly bonded polymer.

Figure 4 gives an example of a copolymer film that consists of polystyrene (PS) and polybutadiene (PB) deposited on a Si wafer. After annealing the film consists of cylindrical polybutadiene agglomerations surrounded by polystyrene. The overall structure of the film was confirmed by plan view TEM, however the existence of the top and bottom PS/PB layers is not evident from TEM examination. A depth profile of an OsO_4 stained film is shown on the right side of Figure 3. Because PB contains doubly bonded C and PS does not, the OsO_4 should selectively stain the PB. The Os+C signal clearly shows concentrations of Os at the top and bottom of the film, corresponding to the additional layers, as well as the broader peak in the center of the film that corresponds to the PB cylinders. Thus SIMS was able to confirm details of a three layer film that were not available in TEM.

4 MEASURING DOPING IN Si NANOWIRES

The formation of Si nanowires is being investigated as a route to further miniaturizing semiconductor devices and to examine Si properties in confined systems [9]. One method of forming nanowires uses an alumina template with a gas-phase catalyzed reaction [10]. Regularly spaced pores in the alumina film are the template for growth of the nanowires, which extend upward from the surface of the film. The nanowires form a dense partially oriented aggregate on the surface of the template (Figure 5). This growth method produces wires that are approximately 80nm in diameter and up to 30 micrometers long. Both B and P doped wires have been produced but only the B-doped wires are discussed here.

Figure 5. SEM image of Si nanowires on alumina substrate.

SIMS measurements are the standard method to determine the concentration of dopants (boron in this case) in silicon; however it was not clear if standard SIMS techniques could be used to analyze the nanowires. Initially we simply analyzed the aggregate of Si wires as in a standard depth profile and quantified the boron signal as in a typical profile. The results of SIMS measurements of 5 different nanowire arrays grown with different dopant gas flows are shown in Figure 6. The good correlation between gas composition and doping measured by SIMS indicates that changes in the gas composition change doping within the wires and that we appear to be measuring doping of the wires well. Four point probe measurements of individual wires give resistivities that correlate well with the SIMS measurements, although they do not provide absolute resistance measurements to allow calculation of B doping. The minimum B level measured here, about $1x10^{18}$ atoms/cm^3 is the minimum measurement limit due to a background signal from B on the surface of the wires and the alumina substrate. Similar measurements have been performed on phosphorus-doped Si nanowires.

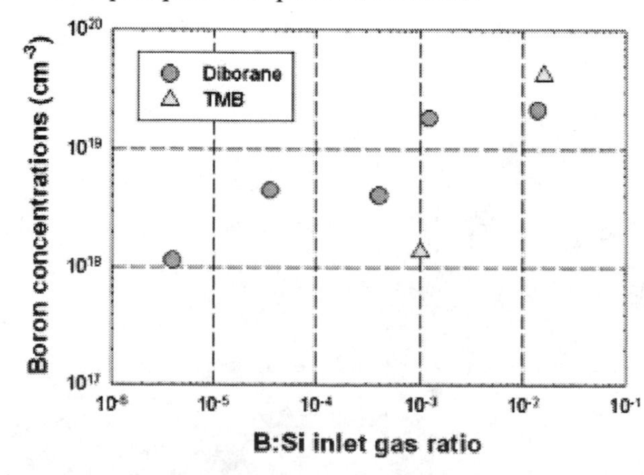

Figure 6. SIMS measurements of B doping in Si nanowire aggregates grown with two different gas sources. TMB=trimethylboron.

5 CONCLUSIONS

The SIMS technique has demonstrated depth resolution of better than 1nm when using beam energies less than 1keV and a high incidence angle. These conditions have allowed us to analyze 2nm thick oxynitride films, films containing 2nm doped organic films and a self patterned polystyrene/polybutadiene block copolymer film 50nm thick. In addition standard SIMS measurements have been used to detect B and P doping variations in Si nanowires. Many other types of nanostructured materials can be analyzed by SIMS, depending on the inventiveness of the investigator. To obtain the best resolution for depth profiling layered films it is imperative that flat highly polished substrates are used.

REFERENCES

[1] R.G. Wilson, F.A. Stevie, C.W. Magee, Secondary Ion Mass Spectrometry: A Practical Handbook for Depth Profiling and Bulk Impurity Analysis, John Wiley, 1989

[2] M.P. Seah et al., Surf. Interface Anal; **36**: 1269–1303. (2004)

[3] I.S.T. Tsong, J.R. Monkowsk and D.E. Hoffman, Nucl. Inst. Meth. **182/183** 237 (1981)

[4] M.J. Fasolka and A.M. Mayes, Ann. Rev. Mater. Res. 2001, **31**:323-355

[5] T.P. Russell, V.R. Deline, V.S. Wakharkar, G. Coulon, MRS Bull., Oct. 1989; Schwarz et al., Molec. Phys. **76**, 937-950 (1992); H. Yokoyama, T.E. Mates, E.J. Kramer, Macromolecules, **33**, 1888-1898 (2000)

[6] T. Nonaka et al., J. Appl. Phys. **73**, 2826-2830 (1993)

[7] M. Park, C. Harrison, P.M. Chaikin, R. A. Register, D.H. Adamson, Science, **276**, 1401-1404.

[8] G.H Fredrickson, S.F. Bates, Ann. Rev. Mater. Sci. 1996. 26:501-550.

[9] Y. Cui and C.M. Lieber, Science **291** i5505 p851 (2001)

[10] K.K. Lew, L. Pan, T.E.Bogart, S.M. Dilts, E. C. Dickey, J. Redwing, Y. Wan, M. Cabassi, T.S. Mayer, S.W. Novak, Appl. Phys. Lett. **85**, 3101 (2004)

Surface Reactions of Linear Atomic Metal Wire Complexes

Chien-Hua Lung, Che-Chen Chang*

Department of Chemistry, National Taiwan University, Taipei, Taiwan 10617
ptrci@yahoo.com.tw, cchang@ntu.edu.tw

ABSTRACT

The communication among nanostructures and between these structures and the outside world becomes very critical as the key features in electronics devices continue to shrink in the double-digit nanometer regime. To explore the possible formation of molecular metal wires by CVD of organometallic complexes, the present work studies the adsorption and the thermal reaction, on the gallium arsenide surface, of the complex which contains the linear atomic metal chain wrapped helically by four ligands. Studies using synchrotron radiation-induced XPS, SIMS, and TPD showed that the pyridylamino trinuclear chromium chelate may adsorb at 105 K on the sample surface with the chelate structure remaining intact. The terminal bond of Cl to the metal chain was cleaved upon adsorption and the Cl surface population increased with the exposure. Molecular adsorption of intact chelates took place at exposures of more than saturation coverage. Increasing the substrate temperature to ~340 K caused the detachment of one of the four ligands from the chelate. Disruption of the chromium string occurred at 540 K, which led to desorption of chromium-containing species. The chemical bonding and the reaction chemistry involved in the formation of the linear metal atom wire are discussed.

Keywords: nanowire, linear atomic metal chain, chemical vapor deposition, surface reaction, metal-semiconductor chemistry

1 INTRODUCTION

In the last two decades, the continued miniaturization of electronic and optoelectronic devices has motivated a worldwide extensive research effort on low-dimensional materials. Along with the development of novel methods to prepare these materials, there is also a great need in understanding the chemistry and physics involved in the preparation. In particular, understanding the chemical reaction and bonding involved in the synthesis of the one-dimensional nanostructures, such as nanowires, nanorods, and linear atom chains, poses exciting challenges from both fundamental and application points of view.

In order to prepare conducting wires in the nanometer scale, various experimental approaches have been developed. Among them, the laser-assisted catalytic growth [1] method is a straightforward approach. It allows a composite target, consisting of a catalyst and the material of interest, to be laser-ablated in a heated flow tube. Clusters generated by laser then grow to form nanowires when they become supersaturated.

Scanning the voltage-applied tip over the surface can also be utilized to create lines of dangling bonds [2] and thus nanowires on the surface. For example, silicon conductive wires were produced by selectively desorbing H atoms from a fully hydrogenated Si(100) surface. It was demonstrated [2] that an OR gate could be formed using these dangling-bond wires.

The superlattice nanowire pattern transfer process [3] was also developed to synthesize high density nanowire arrays. In this process, the molecular beam epitaxy technique was employed to grow alternating layers of the gallium arsenide/aluminum gallium arsenide superlattice on the gallium arsenide substrate. Hydrofluoric acid was then used to selectively etch the aluminum gallium arsenide layers to a certain depth. Subsequently, metal wires were evaporated onto the top of the gallium arsenide layers while the superlattice was oriented at a certain angle with respect to the metal vapor flux. The resulting superlattice was then attached to an adhesive layer on silicon and etched to release metal wires. The nanowires fabricated using this process were well aligned [3].

Other nanowire synthesis strategy includes the use of a selective electrodeposition method [4]. The method allows metals to be deposited onto the cleaved edge of a semiconductor wafer onto which extremely thin layers of various dopant levels have been grown by molecular beam epitaxy. During electrodeposition, electrons drop from the heavily doped layer into an undoped layer because of the lineup of the particular conduction band and valence band in this modulation-doped structure. Thus, the atomic precision of the grown semiconductor layers is converted in this method to the fabrication of metallic nanowires, and ultrathin metal wires may be produced.

Since the electronic and optoelectronic devices produced nowadays rely frequently on the process of chemical vapor deposition, it is of interest to explore if conducting nanowires, especially in the form of atomic wires, may also be synthesized via the chemical vapor deposition approach on the semiconductor surface. Typically, chemical vapor deposition allows a metallic thin layer to be formed on the substrate by exposing the intended substrate surface to the thermal metal source or by depositing organometallic compounds on the surface [5]. The problem associated with such a thin-layer synthesis strategy for atomic metal wire formation is that the metal particles in contact with the surface are mostly either in the form of individual atoms or clusters. It is expected that, instead of forming a linear nanostructure, these particles will be distributed randomly on the isotropic surface.

To circumvent the problem, oligonuclear metal-chain complexes are employed in this study as nanowire precursors. A variety of metal string chelates, which have an extended metal-metal bond supported by oligo-α-pyridylamino ligands, have been successfully synthesized by S. M. Peng, et al. [6]. The success makes our strategy of forming on the solid surface linear metal wires of the single-digit angstrom dimension a viable one, since in these chelates, the ligands bind metal ions in a linear fashion. Exploring different aspects of the interaction and bonding involved in the reaction of the oligonuclear metal string complexes on the semiconductor surface holds the promise for understanding the fundamental processes taking place at the interface between metals and semiconductors. It may open up the possibility of experimentally examining the interaction of one-dimensional metal bonds with the solid surface.

2 EXPERIMENTAL

The specific type of chelates used in this study was a trinuclear chromium atom chain complex, tetrakis(2,2'-dipyridylamino)chromium(VI) chloride. The complex had a central linear chain comprised of three Cr atoms and terminated at both ends by Cl atoms. The central metal chain was encaged by four dipyridylamino ligands, which wrapped around the chain helically in a syn-syn configuration. Each ligand contained two terminal pyridyl groups, bridged together by one N atom, and had the hydrocarbon portion of the group pointing away from the central metal chain. The two pyridyl groups bonded separately to the two terminal Cr atoms of the chain and the bridging N atom bonded to the middle Cr atom.

The experiments were performed in two different ultrahigh vacuum analysis systems. One of them was pumped by a turbomolecular pump, with a base pressure of 1×10^{-10} torr achieved after baking out. The details of the system design were described elsewhere [7]. The vacuum of the other system was maintained by an ion pump and a titanium sublimation pump. The system was used for XPS studies.

A Ga-terminated, un-doped GaN(0001), acquired from Cree, Inc. was used as the substrate. Following a wet-chemical cleaning process, the substrate surface was sputtered by nitrogen ions after it was placed under ultrahigh vacuum to ensure that as less metallic gallium was formed [8]. Repeated cycles of nitrogen ion sputtering (500 eV, 0.7 μA beam current, 20 min) and annealing were carried out, with the annealing temperature of below 1050 K so as not to cause GaN decomposition in the high temperature region [9]. A hexagonal 1x1 LEED pattern with satellite spots was obtained after sample cleaning, indicating the presence of a well-ordered but facetted surface. No oxygen and carbon contamination could be detected by AES.

Metal string chelates were admitted via a Knudsen cell deposition source onto the sample surface. Molecular evaporation of the trichromium complex was verified by depositing in vacuum a thick film of the complex on a stainless steel plate. Subsequent TGA and FAB measurements showed identical spectra to those taken before evaporation. The exposure pressure was measured with an ionization gauge and was kept below $\sim 2 \times 10^{-9}$ torr. The dose reported in this study is the corresponding background exposure at the sample surface and expressed on the unit of langmuir ($1 \text{ L} = 1 \times 10^{-6}$ torr-s).

SIMS measurements were performed with the sample in line-of-sight of the mass spectrometer. A primary beam of Ar ions of 2 keV energy was used to bombard the sample, with the impact angle measured from the surface fixed at 45 degrees. During the measurement, the ion flux was held within the $0.5\text{-}1.5 \times 10^{-9}$ Amp/cm^2 range. TPD experiments were conducted with the sample placed ~0.1 in. away from the skimmer mounted on the mass spectrometer. A heating rate of ~2.5 K/s was used in all TPD experiments. XPS studies were performed at 6m LSGM beamline of National Synchrotron Radiation Research Center, Taiwan, with the photon energy of 250 eV. The incident angle of photons to the surface normal was 45°, and photoelectrons were collected with the analyzer normal to the sample surface. The collected spectra were numerically fitted with the Gaussian- Lorentzian function, after Shirley background subtraction. The binding energies in the spectra are referred to metallic Ga $3d_{5/2}$ at 18.9 eV [10].

3 RESULTS AND DISCUSSION

The chemical process involved in the reaction of the trichromium chelate on the GaN(0001) surface was investigated using SIMS. Shown in Fig. 1 are the SIMS spectra taken from a GaN(0001) surface exposed at 105 K to several different doses of the trichromium chelate and its ligand, dipyridylamine, respectively. For the surface exposed to the trichromium chelate at the low dose of 0.45 L (Fig. 1a), the positive SIMS spectrum obtained was dominated the sputter desorption of Ga^+ isotopes, measured at m/e 69 and 71, from the substrate. A number of small peaks at m/e 52, 78, 130, 172, and 222 were also identified in the spectrum. As mentioned before, there are four dipyridylamino (m/e 170) ligands in the trichromium chelate wrapping helically around the trinuclear Cr metal atom chain. Considering various mass combinations for the possible sputtered species made up from segments of the chelate under ion bombardment showed that the m/e 172 peak observed in Fig. 1a was due to the contribution from desorption of the ligand which obtained two hydrogen atoms during sputtering. The mass signal observed at m/e 78 may be attributed to the pyridyl unit breaking away from the dipyridylamino ligand. The m/e 52 peak was ascribed to the emission of Cr^+ ions from the complex-exposed surface. Small signals peaked around m/e 78, 130, 172, and 222 were also observed in the spectrum.

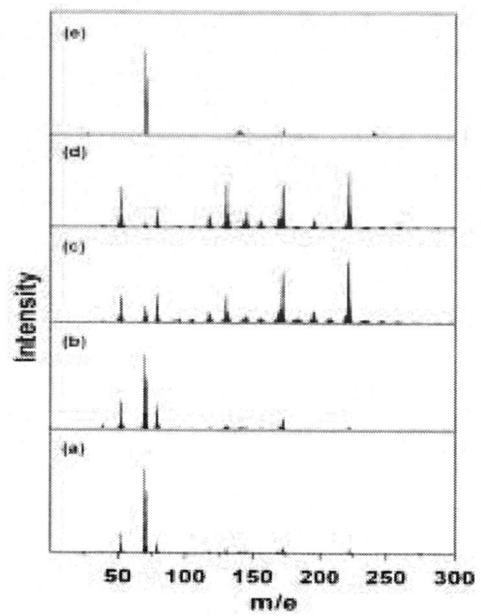

Figure 1: SIMS spectra of a GaN(0001) surface exposed to (a) 0.46L, (b) 1.4L, (c) 3.6L, (d) 5.4L of pyridylamino trichromium chelate and (e) 0.025L of dipyridylamine, respectively. All spectra are recorded at 105 K.

As shown in Fig. 1, all the SIMS peak intensities, except the Ga^+ peak intensity, increased as the chelate exposure was increased to 3.0L. It indicated that the chelate may adsorb at increasing exposures on the sample surface at 105 K. In order to deduce the reaction process taken place on the surface upon chelate adsorption, SIMS spectra of the model compound of dipyridylamine, which is the ligand molecule of the chelate, adsorbed on the sample surface were taken for comparison with those obtained from the surface exposed to the chelate. As shown in Fig. 1e, which was normalized to the m/e 69 signal intensity of Fig. 1a, the major peaks observed in the SIMS spectrum obtained from the exposure of ligand molecules on the surface included those due to the sputtering of Ga_2 (m/e 138, 140, 142), the protonated ligand (m/e 172), and the Ga-ligand adduct (m/e 240, 242). The spectrum contains no signals from the fragments of the ligand molecule, indicating that the ligand exposure to the sample surface resulted in molecular adsorption. The observation of the m/e 240 and 242 peaks, with the intensity ratio close to that for the peaks measured at m/e 69 and 71, revealed that the ligand molecule may adsorb on the Ga site of the surface. As shown in Figs. 1a and 1b, there were no signals detected at these two m/e values from the sample surface exposed to the chelate, however. The presence of the protonated ligand and the absence of Ga-ligand adduct signal in the SIMS spectrum taken from the chelate-exposed surface indicated that the chelate structure may remain intact upon adsorption at 105 K on the sample surface. The ligands of the adsorbed chelate, however, may be sputter desorbed from the adsorbed molecules during SIMS measurement to give the observed peak at m/e 172. The attribution of the major SIMS peaks obtained in Figs. 1a and 1b to the bombardment-induced bond rupture, due to the dissipation of the probe ion energy in the large admolecule, of the stable chelates adsorbed on the sample surface also explained the presence of the pyridyl-Cr (m/e 130) and ligand-Cr (m/e 222) adduct peaks, instead of the pyridyl-Ga and ligand-Ga peaks, in the SIMS spectrum shown in Figs. 1a and 1b. It should also be pointed out that the observation of the Cr and Cr-adduct signals in the SIMS spectrum does not imply that most chelates adsorbed on the surface were completely dissociated to expose their central metal atoms to the probe ion. The lower ionization energy of the metal atom like Cr, as compared to the ionization energy of organic species, may result in the relatively large peak intensity observed in the SIMS spectrum at m/e values related to Cr containing species even at the surface electronic condition when the bombardment-induced Cr atom desorption is smaller than that of the organic neutrals.

As the chelate exposure was increased to more than 3.0L, there was a change in the SIMS intensity distribution. The relative intensities of the Cr, pyridyl-Cr, and ligand-Cr peaks increased substantially, as compared to the protonated ligand peak intensity, whereas the one of the pyridyl peak decreased. It indicated that there may be a change in the chelate adsorption behavior on the sample surface at the exposure higher than 3.0L.

The adsorption chemistry of the trichromium complex on the sample surface was further investigated using XPS. As shown in Fig. 2, the Cl $2p_{3/2}$ XPS peak obtained at low

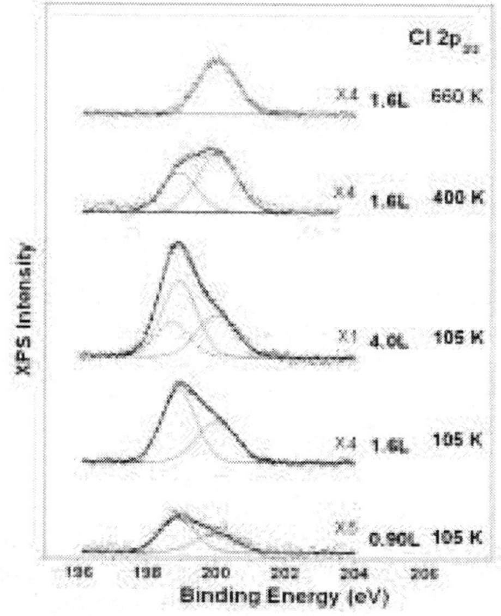

Figure 2: XPS spectra of Cl $2p_{3/2}$ obtained from the GaN(0001) surface exposed at indicated substrate temperatures to the indicated dosages of the trichromium chelate.

chelate exposure of 0.94 L at 105 K can be fitted with two component peaks at binding energies of 198.9 eV and 200.0 eV, respectively. Both peaks grew in intensity with the chelate exposure up to 2.9 L, with the ratio of their peak areas remained roughly constant. Because the two Cl atoms in the chelate molecule have the identical bonding configuration and, as discussed above, the chelate structure of the molecule remains intact upon adsorption at 105 K, the observation of two features in the Cl $2p_{3/2}$ spectrum indicates that in the low exposure region, some of the terminal Cr-Cl bonds of the chelate adsorbed on the surface may be disrupted during the reaction to yield two different Cl chemical states. The electronegativity of Cl may cause the disrupted atom to bond on the positive site, the Ga atom, of the polar GaN surface and result in the observation of the Cl $2p_{3/2}$ subpeak measured at 200.0 eV. Because of the lower electronegativity of Cr than Ga, the subpeak observed at the lower binding energy of 198.9 eV in Fig. 2 may be associated with the undisrupted Cl atom remaining bonded to the end of the Cr atom chain in the trichromium chelate.

As shown in Fig. 2, additional exposures of the chelate higher than 3.0 L did not contribute to the XPS intensity of Cl $2p_{3/2}$ at 198.9 eV and 200.0 eV. Instead, it caused the intensity at the binding energy (198.7 eV) near to the one assigned to the disrupted Cr-Cl bond to increase continuously. Since the small difference in binding energy for the XPS intensity measured at 198.9 eV and 198.7 eV is insignificant, molecular adsorption of chelates may start to take place on top of the first adsorbate layer at high exposures. The molecular adsorption of chelates at the exposures of larger than 3.0L also caused the large change in the SIMS intensity distribution observed in Fig. 1.

The chemical process involved in the thermal reaction of the trichromium complex on the GaN(0001) surface was investigated using TPD. Fig. 3 represents the TPD spectra taken at m/e 52 and m/e 170, respectively, from the sample

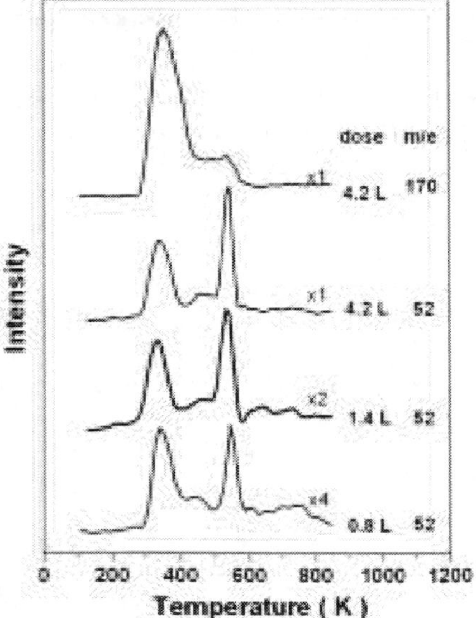

Figure 3: TPD spectra taken at m/e 52 and m/e 170, respectively, from the GaN(0001) surface exposed at 105 K to the indicated doses of pyridylamino trinuclear chromium chelate.

surface exposed at 105 K to trichromium complex of 0.8 L to 3.9 L. At low exposure of 0.8 L, the desorption feature in the m/e 52 TPD curve showed the peak temperatures of desorption occurring at ~340 K and ~540 K. The intensity of both peaks increased as the exposure was increased. Comparing the m/e 52 and m/e 170 TPD spectra revealed similar desorption profile at ~340 K. Since m/e 52 was one of the fragments that may be produced from the ligand molecule in the electron-impact ionizer installed in the TPD spectrometer, the observed similarity indicated that the m/e 52 desorption peak observed at ~340 K may be attributed to the fragmentation of the ligand (m/e 170). Since the TPD spectra measured at different m/e values all showed negligible signals at substrate temperatures of less than ~340 K, thermal decomposition of the trichromium chelate thus started on the sample surface at or below ~340 K via the detachment of the ligand.

Figs. 2 and 3 also showed that at substrate temperatures between ~340 K and ~540 K, further decomposition of the chelate took place, which led to the increase in population of the disrupted Cl on the surface (Fig. 2) as well as the desorption of species containing m/e 52, possibly Cr (Fig. 3). Fig. 2 showed that all the ligands were detached from the metal chain at 660 K. The larger increase at higher exposure in intensity shown in Fig. 3 of the high-temperature peak than that of the low-temperature one indicated that more than one m/e 52 particle may be present in the species desorbed at ~540 K. The detachment thus occurred along with the cleavage of the Cr-Cr bond.

This work was supported by ROC National Science Council (NSC93-2113-M002-003, NSC93-2120-M002-011). The authors gratefully thank Dr. Wei-Hsiu Hung and Dr. Yaw-Wen Yang of NSRRC for allowing us to use their instruments. The synchrotron radiation beam time allocated to this work by NSRRC is also acknowledged.

REFERENCES

[1] S. T. Lee, Science 299, 1874, 2003.

[2] L. Soukiassian, et. al. Surf. Sci. 528, 121, 2003.

[3] J. Heath, et. al. Science 300, 112, 2003.

[4] G. Fasol, Science 280, 545, 1998.

[5] P. Grunberg, R. Schreiber, Y. Pang, M.B. Brodsky, and H. Sow-ers, Phys. Rev. Lett. 57, 2442, 1986.

[6] S.-M. Peng, C.-C. Wang, Y.-L. Jang, Y.-H. Chen, F.-Y. Li, C.-Y. Mou, M.-K. Leung, J. Magnet. Magnet. Mater. 209, 80, 2000.

[7] C.-C. Chang, I.-J. Huang, C.-H. Lung, H.-Y. Hwang, L.-Y. Teng, J. Phy. Chem. 105, 994, 2001.

[8] Y. Ould-Metidji, L. Bideux, D. Baca, B. Gruzza, V. Matolin, Appl. Surf. Sci. 212/213, 614, 2003.

[9] H. W. Choi, M. A. Rana, S. J. Chua, T. Osipowicz, J. S. Pan, Semicond. Sci. Technol. 17, 1223, 2002.

[10] Y. H. Lai, C. T. Yeh, J. M. Hwang, H. L. Hwang, C. T. Chen, W. H. Hung, J. Phys. Chem. B 105, 10029, 2001.

In-situ Study of SAMs Growth Process by Cross Analysis of AFM Height and Lateral Deflection

[1]Chun-Lung Wu, [1,2] *Fan-Gang Tseng, and [2]Ching-Chang Chieng

[1]Institute of Microelectronmechanical System,
[2] Engineering and System Science Department
National Tsing Hua University, Taiwan, ROC

ABSTRACT

This paper introduces a novel way to investigate SAMs growth process by cross analysis of the information of AFM height and lateral deflection scanning results. The traditional ways by analysis of AFM height and histogram data can not differentiate the molecular growth detail behavior of alkylsinae SAMs because the images captured by AFM are static and lack molecular level information. Thus, this paper proposes to employ the standard deviation of AFM data to analyze the in-situ growth behaviors of alkylsinae SAMs in comparison with the local height and lateral deflection data. The analysis results demonstrated close correlation of the SAMs growth process among the aforementioned data analysis methods.

Keywords: *In-Situ, AFM, growth, self assembly monolayer, lateral deflection*

1. INTRODUCTION

Self assembly monolayers (SAMs) has been studied since 1945 [1], and play an important role in the modification of nano-properties for micro devices, owning to their easy preparation, capability of the design of functional groups on the molecule head and tail, chain length and specie selection, and simple fixing method on substrate, thus extensively applied to biotechnology, nanotechnogy, nanopattern [2], material science and Microelectronmechanical System (MEMS)[3].

Among different SAMs' systems, the growth mechanism of alkanethiol SAMs on gold substrate were most studied [4,5], partially due to the more regular gold crystalline surface for AFM and conductive substrate ready for STM studies. However in alkylsinae SAMs, they are usually applied on surfaces with oxides, not a surface with regular crystal structures and conductive. As a result, the surface roughness of oxides, usually close to the height of SAMs, will greatly affect AFM measurement. Yet alkylsinae SAMs are commonly applied to silicon or glass substrate based micro devices. Therefore, this study focus on the investigation of the in-situ self-assembly processes of alkylsinae SAMs, including nucleation, growth and integration processes by AFM techniques and new data analysis methods.

There have been many progresses in characterization of SAMs growth process in the past 15 years [6] owing to the invention of new tools, such as Atomic Force Microscope (AFM), XPS, ellipometry, etc. However most of those methods measuring the average properties such as morphology or height of sample in at least μm^2 area, can not bring out detail information about nucleation, clustering, growth progress, and their molecular behaviors in nano scale. Although Brewer had investigated SAM phase separation by chemical force microscopy at 2004 [7], the histograms of pull-off forces still can't explain molecular dynamic behaviors. In this research, we propose a novel method that can carry out more nano scale information on SAM growth mechanism by cross analysis of AFM height and lateral deflection in raw data and standard deviation at same time for different SAMs growth stages.

2. MATERIALS AND METHODS

2.1 Sample Preparation

Commercially available Si(100) wafers were cut into pieces of $1.5x1.5cm^2$ and subsequently cleaned and oxidized in a freshly prepared 4:1 mixture of 96% sulfuric acid and 30% hydrogen peroxide (piranha solution) at 100°C 10min. After the substrates have been allowed to cool to room temperature, the samples were rinsed with DI water and dried in N_2 gas. After drying, the oxidized silicon samples were immediately immersed into a freshly prepared 10^{-3}(v/v) solution of OTMS (Octadecyltrimethoxysilane, Fluka 74763) in dried ethanol at room temperature (25°C).

2.2 Measurement

In situ AFM measurement were performed using JPK NanoWizard AFM (Germany). The AFM operated in contact mode by ultrashap silicon cantilever (CSC38/AIBS of Micromash, Russia). In-situ image were continuously captured in a 256-second period, and the whole experiment acquired 67 images within 17152 seconds. Scan size was 5um x 5um with a x-y resolution of 256x 256 pixels. All the information measured by height mode and lateral deflection mode were obtained at the same time. (Fig.1)

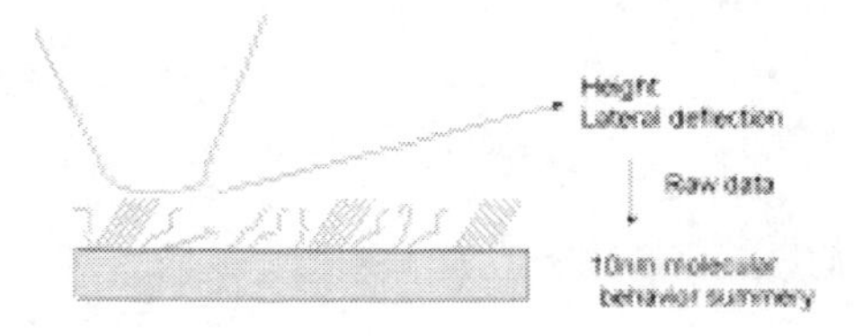

Fig.1 Schematic of multiple-channel AFM image analysis: height and lateral deflection analyzed at one time.

2.3 Data Process and Analysis

All images have been adjusted by linear leveling, and transferd into to 256x 256 raw data. Because the height and lateral deflection are not under the same reference, the analysis of standard deviation (STD) of the whole raw data provides the distribution of SAMs status on the measured area and information of integral molecular behavior in stead. On the other hand, the analysis of the relative height and lateral defection variation in each pixel can indicate molecular status at the local region for comparasion. The analysis flowchart is shown in Fig.2.

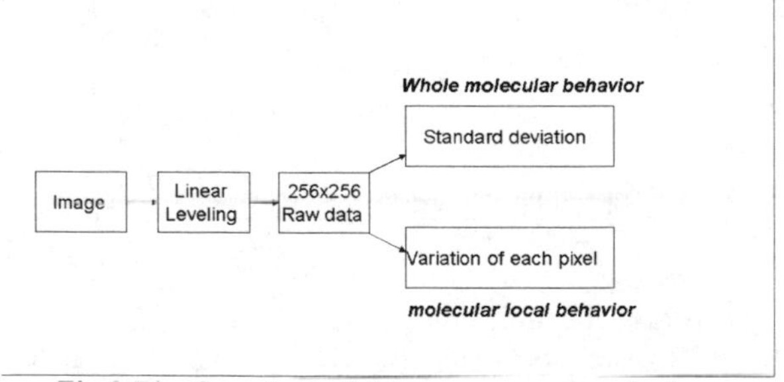

Fig.2 The flowchart of data process and analysis.

3 RESULT AND DISCUSSION

By analyzing the raw data of the whole image, the standard deviation (STD) increases very fast then decreases to the initial one in both height mode and deflection mode scanning results, as shown in Fig. 3 and 4. The higher standard deviation mean the larger variation resulted from molecular random distribution on substrates. On the contrary, when the SAMs grow into a complete nano-film and fill up the substrates, the STD decreases into that of the initial stage, meaning the surface properties, including roughness (from the height data) and friction forces (from the lateral deflection data), become very uniform as those on the bare substrate surface. Therefore, the growth process in Fig. 3 can be classified into 4 stages based on the cross correlation between the STD of height and lateral deflection data. In the growth process, the STD of height appears rapid increasing stage (A-B), decreasing stage (B-C), standing stage (C-D) and decreasing stage (D-F), in Fig 3, meaning that group SAMs molecular nucleated first (A-B), partially leaving (B-C), forming cluster (C-D) and filling up substrate (D-F). The lateral deflection in Fig. 4 means fluctuated friction forces on surfaces with high roughness and low stability at the stage of A-C, the friction forces are then rapidly reduced owning to more stable cluster formation on the surface at the stages of B-D, and then the friction further reduce into lower level than that of the bare silicon dioxide surface owning to complete and flat film formation as well as lower friction on OTMS film than that of silicon surface because of more hydrophobic functional group on the tip of OTMS. The process is schematically shown in Fig. 5.

To better verify this point of view, in addition to the average STD throughout the whole chip, the relative heights and lateral deflections of individual pixels in nano scale are also analyzed and the results are shown from Fig. 6 to Fig. 10. The local height information in Fig. 6 shows that molecules stack firstly in group, leave, then gradually stack again, and eventually form complete films on the substrate. This process is schematically shown in Fig. 7, and demonstrates a similarity to that in Fig. 5.

The height of SAMs gradually increasing in Fig. 6 and 8 shows molecule standing-up from lying and scattering status, and gradually form complete film when time goes by. The existing of Multiple layers can also be verified form the height variation couple times of molecular length. All those individual behavior have been averaged out in the overall behavior in Fig. 3 and 4.

Not like the height one, local lateral deflection information does not reveal large variation versus time in Fig. 9 and 10, yet the local fiction forces are lower than that of bare silicon dioxide surface (decreases from 0.04 to -0.08), consistent with the surface properties for silicon dioxide and OTMS and the data analysis in Fig. 4.

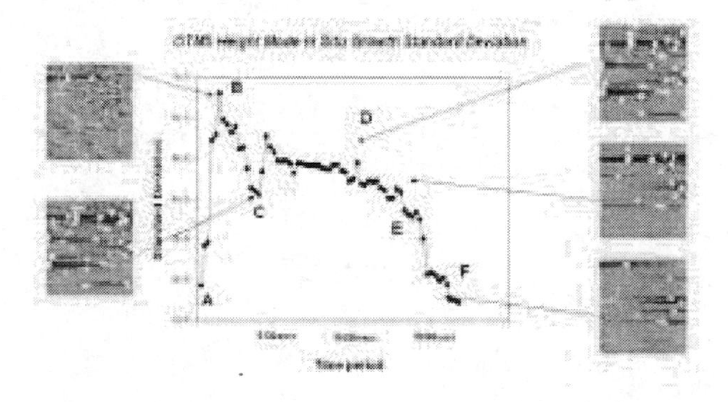

Fig.3 The standard deviation of height mode AFM scanning data at different time stages of SAMs growth.

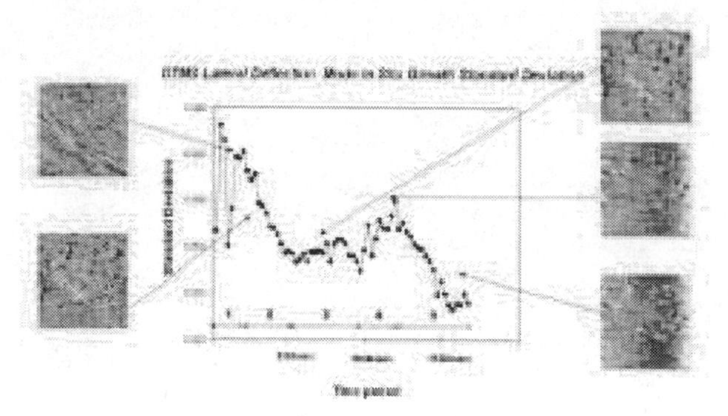

Fig.4 The standard deviation of lateral deflection mode AFM scanning data at different time stages of SMAs growth.

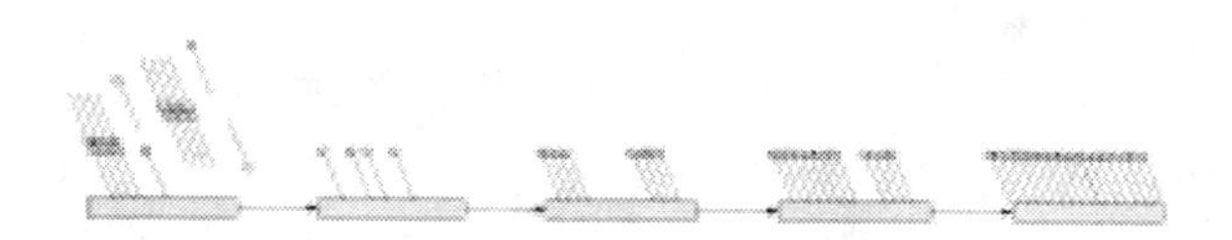

Fig.5 Schematic Diagram of SAMs growth process.

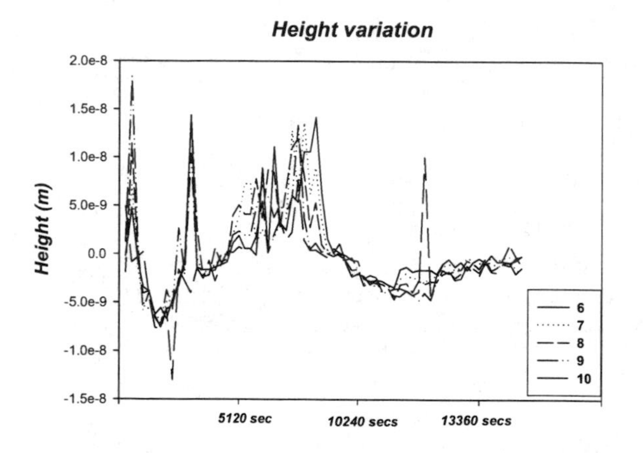

Fig.6 The height variation data of SAMs growth in an area containing 6th-10th pixel that were captured randomly from 65536 pixels at different time.

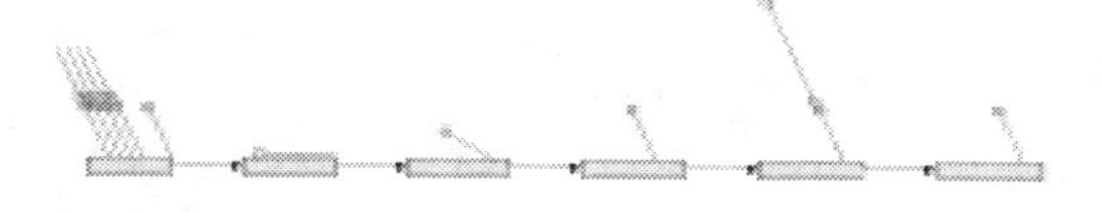

Fig.7 Schematic diagram shows the local behaviors of molecules.

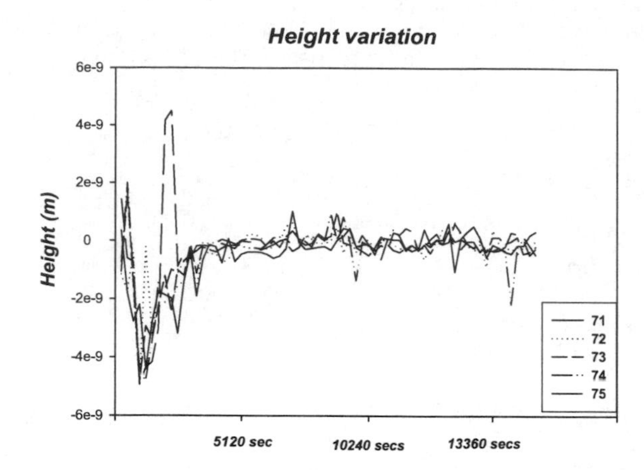

Fig.8 The height variation data of SAMs growth in an area containing 71th-75th pixel that been captured randomly from 65536 pixels at different time

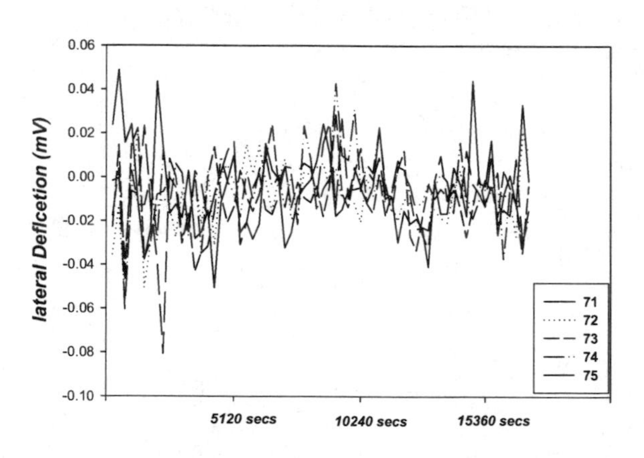

Fig.9 The lateral deflection variation data of SAMs growth in an area containing 6th-10th pixel that been captured randomly from 65536 pixels at different time.

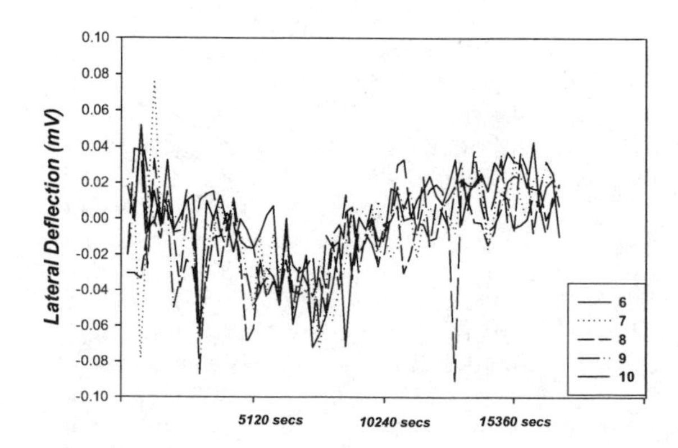

Fig.10 The lateral deflection variation data of SAMs growth in an area containing 71th-75th pixel that been captured randomly from 65536 pixels at different time.

4 CONCLUSION

This paper introduces a novel way to investigate SAMs growth process by cross analysis the information of AFM height and lateral deflection scanning results. The analysis results show that the growth of alkylsinae SAMs have 4 different stages including group SAMs molecular nucleated, partially leaving, forming cluster and filling up substrate. Those results are also verified by the local height and lateral deflection data. This method provide an alternative to the traditional AFM analysis ways on only morphology and histogram of scanned image, carrying out more meaningful information on the explanation of SAMs growth process.

REFERENCES

[1] Bigelow, W. C.; Pickett, D. L.; Zisman, W. A. *J. Colloid Interface Sci.*, *1,* 513, **1946**.

[2] **Kumar, A.; Biebuyck, H. A.; Whitesides, G. M.** *Langmuir* **1994**, *10,* 1498.

[3] Huiwen Liu, Bharat Bhushan, *Ultramicroscopy* 91, 185-202, **2002**

[4] Ryo Yamada and Kohei Uosaki, *Langmuir* **2001,** *17,* 4148-4150, **2001**

[5] C. O'Dwyer,* G. Gay, B. Viaris de Lesegno, and J. Weiner, *Langmuir, 20,* 8172-8182, 2004

[6] Ulman, A. *An Introduction to Ultrathin Organic Films*; Academic Press: Boston, **1991**.

[7] Nicholas J.Brewer, Graham J.Leggett, *Langmuir* **2004,** *20,* 4109-4115

Reflow Microfabrication for Refractive Optical Components: Processes and Materials

R.W. Johnstone and M. Parameswaran

Institute for Micromachine and Microfabrication Research, School of Engineering Science
Simon Fraser University, 8888 University Dr., Burnaby, BC, Canada, V5A 1S6

ABSTRACT

With standard surface-micromachining, free-space micro optical benches can be constructed that contain mirrors and diffractive optics. This paper outlines a novel microfabrication process for integrating refractive optical elements, such as lenses, into surface-micromachining processes. The process involves melting the optical material after the raised structures have been assembled. The resulting lenses can have a wider range of focal lengths, and can be made from a wider variety of materials than previous work. This technique will allow integration and customization of micro-optical and micro-photonic systems for various application.

Keywords: reflow fabrication, micro-optics, refractive lenses, surface-micromachining

1 INTRODUCTION

The seminal paper on the topic of free-space micro-optical systems was written by Pister *et al* [1], which demonstrated several optical components that were fabricated using surface-micromachining [2], [3]. Surface-micromachining is a planar fabrication process, and excels at constructing integrated mechanical systems in the two dimensions parallel to the wafer surface. However, surface-micromachining does suffer an important disadvantage in that far less control is available in the third dimension. Further, the total available height is usually limited to less than 10μm.

The paper by Pister *et al* introduced surface-micromachined hinges, allowing the construction of fully 3D structures. With hinges, components are fabricated flat and then raised to a standing position after release [1]. Geometric control during processing is still only in the two dimensions parallel to the wafer surface, and surface-micromachining is still only capable of constructing flat structures. However, these flat structures can be oriented in new positions during the assembly process.

The literature not only discusses various types of surface-micromachined hinges, but also demonstrates several novel surface-micromachined structures, including several optical components. In particular, the method has been used to construct surface-micromachined mirrors and diffractive optics [4]–[9].

The main difficulty with the optical components available to date is the lack of refractive optical components, particularly lenses. While diffractive elements, such as Fresnel-zone plates, can approximate refractive lenses, they only work at a single wavelength and suffer high optical losses [5], [10]. Even with novel materials, standard surface-micromachining techniques cannot be used to construct refractive components because they contain no provisions for curved surfaces. Therefore, we are investigating a new method of microfabrication that will allow us to fabricate miniature refractive components.

Integrated refractive components have been reported. King *et al* [10] demonstrated a refractive lens fabricated using photoresist, but significant improvements can be made. The process we are developing, outlined below, will be capable of a greater range of F#'s and will use better optical materials.

2 FABRICATION PROCESS

Reflow microfabrication is a modification of standard surface-micromachining techniques [2], [3], and does not limit the construction of structures available using conventional surface-micromachining. In our proposed fabrication process, the wafers undergo a complete surface-micromachining process with at least two moveable structural layers, such as PolyMUMPs^{TM}. An additional structural layer, called the 'reflow' layer, is then deposited upon the wafer and patterned, before the release step. Because this layer is composed of the material that will be used to construct the refractive components, and so it should be a suitable optical material. Additionally, the material for the 'reflow' layer is chosen to have a melting point below that of the other structural materials in the surface-micromachining process. Processing then continues with the removal of the sacrificial material [2], [3] and the assembly of the 3D structures [1].

The entire system, containing the assembled structures, is then heated until the 'reflow' material melts. Surface tension pulls the 'reflow' material into droplets (figure 2). For miniature devices, gravity plays a negli-

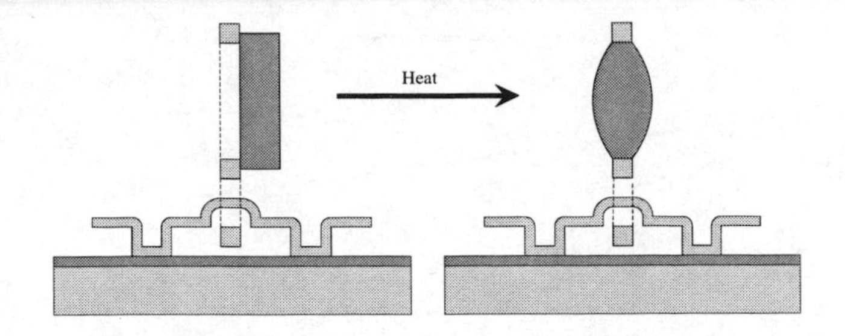

Figure 1: Illustration reflow fabrication process for forming refractive lenses.

gible role in determining the final shape of the droplets. The droplets' shapes are thus determined by surface tension, and so the liquid-air interfaces should be spherical. Therefore, carefully constructed droplets can be used to construct refractive components.

The key difference between this process and the process outlined by King *et al* [10] is the timing of the reflow. By performing the reflow after assembly, lenses can be formed with two spherical surfaces instead of one. The additional spherical surface allows for lenses with much smaller focal lengths.

3 THEORY

3.1 Focal length and droplet volume

The focal length of a thin lens can be determined from the radii of curvature of the lens' two surfaces [11].

$$\frac{1}{f} = \frac{n_2 - n_1}{n_1}\left(\frac{1}{R_1} + \frac{1}{R_2}\right) \tag{1}$$

In equation 1, f is the focal length, n_1 is the index refraction of the ambient material (air), n_2 is the index of refraction of the lens material, and R_1 and R_2 are the radii of curvature for the lens's two surfaces.

However, since the pressure difference between the atmosphere and the droplet will be constant at equilibrium, it follows from the Laplace-Young equation [12] that the two radii of curvature must also be equal. Equation 1 can then be rearranged into the following equation:

$$R = 2\frac{n_2 - n_1}{n_1} f \tag{2}$$

One would now like to determine the droplet's total volume. However, a more careful look at the droplet's contact angle is necessary before one can calculate the droplet's volume. The contact angle is a material property, and is fixed independent of fluid volume or radius of curvature. On a flat surface, the radius of curvature and the contact angle would fix the diameter of the lens. However, the aperture diameter and the focal length are central to lens design, so a way to vary the contact angle is necessary. One can vary the effective angle by changing the local slope of the surface. This is most easily accomplished by using a raised platform (figure 3.1). Although drawn as a sharp corner, a fillet, although microscopic, exists at the corner. This allows the effective radius of curvature to vary, even if the real contact angle is fixed.

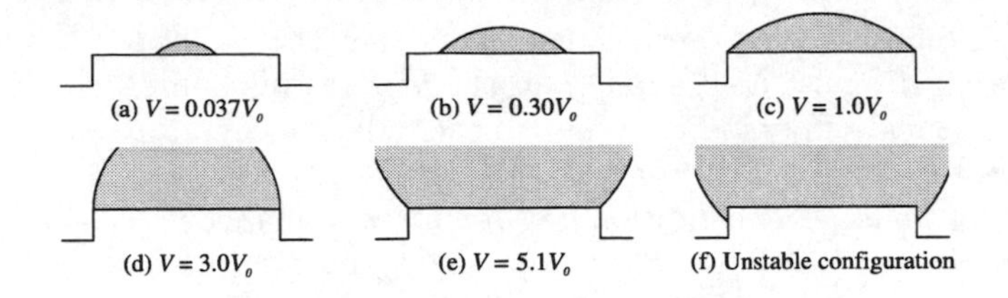

Figure 2: Illustration of heat treatment in reflow fabrication process

$$\theta^* \in [\theta, \theta + \pi] \tag{3}$$

Above, θ^* is the effective contact angle, and θ is the real contact angle. This concept can be extended to reflow lenses, which are supported by a circular annulus (figure 2). Note that in this case, there are two corners, an inside and an outside, at which the effective contact angle can vary. With both inside and outside corners, the effective contact angle can vary between $\theta - \pi$ and $\theta + \pi$.

Assuming that the droplet is attached to the inside corner of the supporting annulus, the total droplet volume can be calculated using the following formula. In the opposite case, where the droplet is attached to the outside corner, the following formula is still used, but the volume of the annulus is subtracted from the result.

$$V = t\pi r^2 + 2\int_{R_0}^{R} \pi\left(R^2 - r^2\right) \mathrm{d}r \tag{4}$$

$$V = t\pi r^2 + \frac{4}{3}\pi R^3 + \frac{2}{3}\pi R_0^3 - 2\pi R^2 R_0 \tag{5}$$

In equation 4, the first term represents the cylindrical volume defined by the interior of the annulus, and depends on the thickness of the annulus, t, and the inner radius. The second term accounts for the two spherical sections that make up the droplet's volume outside the annulus' inner space, and depends on the radius of curvature, R, and the distance from the centre of curvature to the spherical section, R_0. This distance can be calculated by noting that the radius of curvature, the lens radius, and R_0 form a right-angle triangle. This results in the following equation for R_0:

$$R_0 = \sqrt{R_2 - r_2} \tag{6}$$

Equation 6 imposes the condition that the radius of curvature must be greater than the lens' radius. Physically, this simply indicates that if the radius of curvature was too small, the droplet would not reach the supporting annulus. The condition $r < R$ can be combined with equations (1) and (2) to develop a lower limit on the F# of lenses, where is F# is the ratio of the focal length to lens diameter.

$$F\# > \frac{n_1}{4\,(n_2 - n_1)} \tag{7}$$

Equation (7) indicates that materials with high indices of refraction make better lens materials since they allow for lenses with shorter focal lengths. However, this advantage must be traded against the higher reflective loses that will occur [11]. The lens material thus represents a trade-off.

As previously mentioned, the limits expressed by equation (7) are the result of manufacturing limits. These limits are also represented in the range of possible droplet volumes. To achieve the minimum possible focal length, one must have the largest allowed droplet size, which occurs when $R = r$, as discussed when deriving equation (7). Similarly, to achieve the largest possible focal length, one must have the smallest allowed droplet size, which occurs in the limit as $R \rightarrow \inf$.

$$V_{\min} = t\pi r^2 \tag{8}$$

$$V_{\max} = t\pi r^2 + \frac{4}{3}\pi r^3 \tag{9}$$

The above two equations are important because they indicate the range of volumes a fabrication process must support in order to achieve the full range of focal lengths possible. The 'reflow' material will be patterned, so volume can be easily removed from the starting material, and hence from the resulting droplet. Assuming that all the material for the droplet will be deposited over the aperture of the lens, then equation (9) can thus be used to determine the necessary thin-film thickness necessary for the 'reflow' material.

$$h = \frac{V_{\max}}{\pi r^2} = t + \frac{4}{3}r \tag{10}$$

For equation 10, the thickness of the 'reflow' material, h, is a function of two terms. The first term will constant for all lenses on a single chip, and so does not pose a problem. However, the second term depends on the lens' aperture, and this parameter can change from lens to lens. A particular process will therefore be optimal for a particular lens diameter; lenses with smaller apertures can be fabricated, but not larger apertures.

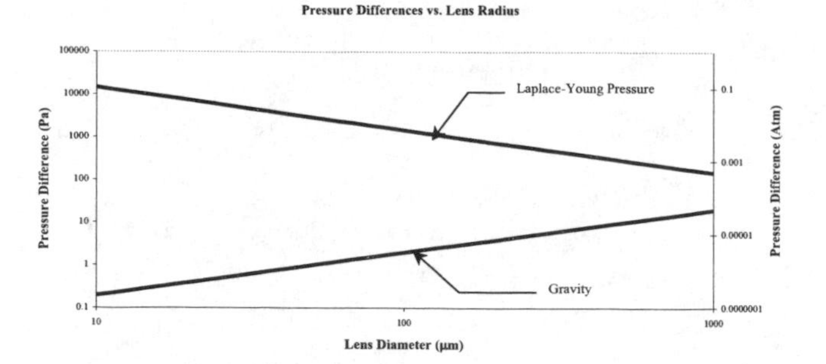

Figure 3: Plot of pressure differences caused by (a) curvature of the droplet surface, known as the Laplace-Young pressure, and (b) pressure gradient induced in the droplet by gravity. The droplets were of water.

3.2 Lens shaping by surface tension and gravity

In the previous section, the effect of gravity on the lenses' shapes was neglected. For surface-tension to dominate the shape of the lens, pressure changes in the body of the droplet due to gravity must be negligible. Both of these pressures are easily calculated for droplets of various sizes (figure 3.2).

By comparing the pressure differences due to the Laplace-Young equation [12] and gravity, a limit on droplet sizes can be obtained. We introduce a dimensionless number that relates the importance of surface tention and gravity in determining a droplet's shape:

$$\eta \equiv \frac{\rho g r R}{\sigma} \tag{11}$$

Above, η is the ratio of the Laplace-Young and gravity pressure differences, which depends Of key note is r, the lens' aperture radius, and R, the lens' radius of curvature. Equation (14) represents a limit on the product of these two design variables.

The results in figure 3.2 were verified by performing a series of volume-of-fluid (VOF) analyses using ANSYS™. In these simulations, the gravitational acceleration was varied between its true value of 9.81 ms^{-2} and 9.81×10^5 ms^{-2} for a lens with a radius of 10 μm. Distortions of the lens' shape were not visibly noticeable until gravity had been increased by a factory of 10^5, not the 10^4 as predicted by figure 3.2, although optically significant distortions did appear earlier.

4 MATERIALS

We are currently investigating both inorganic and organic materials for use as the reflow material in this process. We are currently focusing on poly(methyl methacrylate), or PMMA, as our organic material. A particular inorganic material has not yet been choosen, but we are planning to select a halogenated salt.

Most of our work to date has focused on PMMA. This thermoplastic is a common optical material, and is also know as Plexiglass. PMMA was chosen because it is a thermoplastic, and so can be melted and reformed. Further, PMMA exhibits a true melting point in addition to a glass transition temperature. This makes it a good candidate for reflow microfabrication.

While the PMMA does reflow, our efforts at fabrication have been hampered by the extremely high viscosity of PMMA. For example, the zero-shear viscosity of PMMA with a molecular weight of 10^5 is approximately 10^6 Pas at 190°C [14]. For comparison, the viscosity of water at room temperature is near 10^{-3} Pas. The reflow process thus takes a long time. The viscosity can be lowered by increasing the temperature, but oxidation and decomposition are a concern. Further work at higher temperatures is planned.

For inorganic reflow materials, we also plan to develop processes using halogenated salts. Many of these materials are commonly used as optical materials. We wish to keep the process as generic as possible, so that all of these materials may be used. This would offer a range of transmission windows and refractive indices.

Unfortunately, many of the halogenated salts are soluble in water. This complicates manufacturing, as the surface-micromachined chips will still need to be released after deposition and pattering of the reflow layer, and the solvent of common oxide etchants is water. However, the manufacturing process can compensate. Currently, we are considering two different methods of preventing damange if the halogenated salt is water soluble:

- It may be possible to pre-load the solutions with the appropriate ions. This would lead to a solution already at the solubility limit, and prevent etching of the reflow material. However, cross-solubility products with the desired etchants must be considered.

- The reflow material can be encapsulated during release. The simplest approach would be to use photoresist. While this limits solution interactions, it would require an additional mask step.

5 CONCLUSION

At micro-scale lengths, the shape of liquid droplets is influenced far more by surface tension than gravity. This fact can be used to construct refractive optics for miniature free-space optical systems. Lenses can be constructed using standard planar processing techniques and a simple post-process heat treatment.

The focal length of the fabricated lenses can be determined using the geometry of the supporting annulus and the droplet's volume. Designers could thus control both the aperture and focal length of the fabricated lenses.

REFERENCES

[1] K.S.J. Pister, M.W. Judy, S.R. Burgett, and R.S. Fearing. "Microfabricated Hinges," *Sensors and Actuators A.* vol. 33, no. 3, pp. 249-256 (1992).

[2] R.T. Howe. "Surface micromachining for microsensors and microactuators," *Journal of Vacuum Science and Technology B.* vol. 6, no. 6, pp. 1809-1813 (1988).

[3] J.M. Bustillo, R.T. Howe, and R.S. Muller. "Surface micromachining for microelectromechanical systems," *Proceedings of the IEEE.* vol. 86, no. 8, pp. 1552-1574 (1998).

[4] K.Y. Lau. "MEM's the word for optical beam manipulation," *IEEE Circuits and Devices.* vol. 13, no. 4, pp. 11-18 (1997).

[5] M.C. Wu. "Micromachining for optical and optoelectronic systems," *Proceedings of the IEEE.* vol. 85, no. 11, pp. 1833-1856 (1997).

[6] R.S. Muller and K.Y. Lau. "Surface-micromachined microoptical elements and systems," *Proceedings of the IEEE.* vol. 86, no. 8, pp. 1705-1720 (1998).

[7] A. Friedberger and R.S. Muller. "Improved Surface-Micromachined Hinges for Fold-Out Structures," *Journal of Microelectromechanical Systems.* vol. 7, no. 3, pp. 315-319 (1998).

[8] Y.W. Yi and C. Liu. "Assembly of micro-optical devices using magnetic actuation," *Sensors And Actuators A.* vol. 78, no. 2-3, pp. 205-211 (1999).

[9] L. Zhou and J.M. Kahn. "Corner-Cube Retroreflectors Based on Structure-Assisted Assembly for Free-Space Optical Communication," *Journal of Microelectromechanical Systems.* vol. 12, no. 3, pp. 233-242 (2003).

[10] C.R. King, L.Y. Lin, and M.C. Wu. "Out-of-plane refractive microlens fabricated by surface micromachining." *IEEE Photonics Technology Letters.* vol. 12, no 10, pp. 1349-1351 (1996).

[11] F.L. Pedrotti and L.S. Pedrotti. *Introduction to Optics: Second Edition.* Prentice Hall, Englewood Cliffs, 1987.

[12] C. Isenberg. *The Science of Soap Films and Soap Bubbles.* Dover Publications, New York, 1992.

[13] G. Elert, Ed. *Density of Air.* [Online] Accessed: 2004/10/29. Available: http://hypertextbook.com/facts/2000/RachelChu.shtml

[14] K. Fuchs, C. Friedrich, and J. Weese. "Viscoelastic Properties of Narrow-Distribution Poly(methyl methacrylates)," *Macromolecules.* vol. 29, no. 18, pp. 5893-5901 (1996).

NEMS Mass Sensor by Focused Ion Beam Fabrication

B. Boonliang*, P. D. Prewett, M. C. L. Ward and P. T. Docker

The University of Birmingham, Research Centre for Micro-Engineering and Nanotechnology, Birmingham, B15 2TT, UK, *BXB915@bham.ac.uk

ABSTRACT

The possibility of using paddle-type resonators for mass/chemical sensor applications is explored. An analytical model of a magneto-motive-driven paddle resonator is derived to determine intrinsic behaviour, response and sensitivity to mass adsorption. Confirmation of the model was carried out using the ABAQUS FEA package. Preliminary devices have been manufactured using focused ion beam fabrication techniques.

Keywords: FIB fabrication, mass sensor, nano-resonator

1 INTRODUCTION

At present, chemical detection technologies are principally based on adaptation of laboratory techniques. A number of spectroscopic methods that utilise optical absorption, light scattering, luminescence, atomic fluorescence or refractive index changes have been explored [1]. However, these methods are still based on laboratory analysis on extracted samples. Hence they do not provide real-time data and are generally complicated, time-consuming and expensive.

Recently, there has been a rising demand for real-time in situ chemical detection technologies, as monitoring of specific substances is vital in many industrial and research fields, ranging from clinical analysis, environmental control to industrial processes. The demands also extend to safety and military services, especially for hazardous materials, contraband and explosive chemicals.

Interest in sensors and actuators and the rapid growth of nanotechnology have led to a new horizon for the development of sensor devices. The availability of new fabrication technologies is promoting the growth of micro- and nano-electromechanical systems (MEMS and NEMS), oscillators and resonant systems.

Many researchers have shown that microcantilevers (MC) in dynamic mode are a major candidate for such a task [2, 3], providing exceptionally high sensitivity to additional mass [4].

This paper explores another type of resonator, namely a paddle resonator, which operates in dynamic mode through torsional vibration of its shafts. It offers better response to changes in additional mass with structures of similar size to MC. Focused Ion Beam (FIB) is the selected manufacturing technique. It is an extremely versatile fabrication tool which has the capabilities for milling, deposition and inspection in nanometer-scale.

2 DESIGN PRINCIPLES

2.1 Structural Design

Essentially, a paddle resonator is a double-clamped beam with large plate at the mid-point. It is designed to resonate at fundamental frequency in torsion through the beams. The plate is to be coated by chemically selective polymer compounds for detection. The schematic drawing of the structure is shown in Fig 1. This structure exhibits many possible advantages over MC, such as;

- Larger area of detection with similar sized structure
- Linear response to mass addition
- Less intrinsic bending in the beams due to the weight of the detecting plate
- No stress induced effect, upon detection, as beams do not have chemical layer coating
- Utilise bending moment to drive, therefore less power required for significant movement

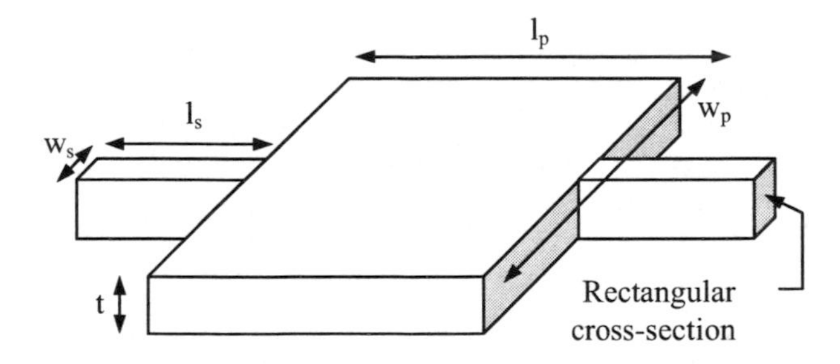

Fig. 1. Schematic drawing of the paddle resonator

From Fig. 1, l_p and w_p are length and width of the plate, l_s and w_s are length and width of the shaft and t is thickness of membrane.

The movement in the beams can be described by the torque-deflection equation:

$$T = GJ_t \frac{\partial \theta}{\partial x} \tag{1}$$

Where T is torque, G is shear modulus, θ is angle of twist and x is distance along the beam. J_t is the *"Torsional Parameter"* of a non-circular bar which is defined below [5].

$$J_t = ab^3\left[\frac{16}{3} - 3.36\frac{b}{a}\left(1 - \frac{b^4}{12a^4}\right)\right] \quad (2)$$

(for rectangular cross-sectional beam of a ≥ b)

2a

2b

By considering a uniform shaft segment of length l_s, with overall angular deformation θ, the *torsional stiffness*, K_t may be written as:

$$K_t = \frac{T}{\theta} = \frac{GJ_t}{l_s} \quad (3)$$

The mass of the plate m_p generates rotational resistance or *"Polar Mass Moment Of Inertia"* J_P, given by:

$$J_p = \frac{m_p}{12}\left(w_p^2 + t^2\right) \quad (4)$$

During vibration, it is assumed that the plate is a rigid structure rotating about the weightless shafts. The natural frequency f_n of the device is defined by:

$$f_n = \frac{1}{2\pi}\sqrt{\frac{2K_t}{J_P}} \quad (5)$$

2.2 Driving Force

The structure is to be driven at its natural frequency by the Lorentz force. The proposed layout of the drive-system is shown in Fig. 2. The tracks are to be FIB-deposited Platinum (Pt). The top track carries an AC current in the presence of perpendicular external magnetic field producing up-/downwards oscillating forces.

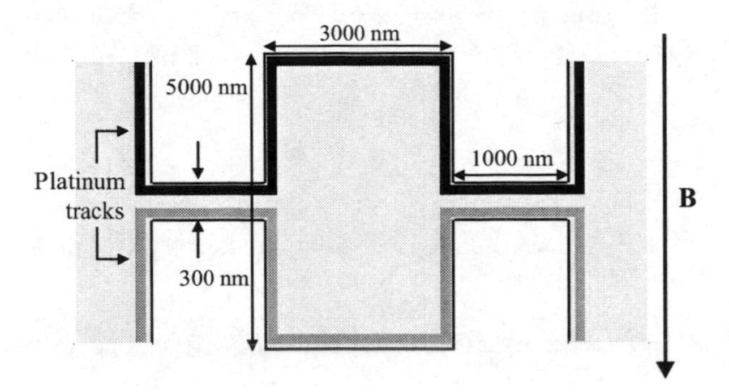

Fig. 2. Schematic drawing of paddle resonator and Pt tracks with dimensions

The top track of length l_{Pt}, carrying current I in magnetic field B, generates torque T producing a twisting angle θ, which can be described by the equation below.

$$T = \frac{GJ_t\theta}{l_s} = Fd \quad (6)$$

Where d is half the width of the plate and F is Lorentz force which is;

$$F = BIl_{Pt} \quad (7)$$

By rearranging the Equation (7) and assuming a Q factor, the angle of twist (maximum amplitude) can be written, in terms of all factors, as:

$$\theta = \frac{QBIl_{Pt}l_s d}{2GJ_t} \quad (8)$$

2.3 Pick-Up System

Monitoring change in natural frequency can be done through exploitation of electromagnetic induction.

The Faraday-Lenz law of Electromagnetic Induction states that:

$$\varepsilon = -\frac{d\Phi_b}{dt} \quad (9)$$

Where ε is induced emf and Φ_B is magnetic flux through a finite area. During resonance, the non-current-carrying bottom track in Fig. 2.2 cut magnetic field line perpendicularly, generating an emf of:

$$\varepsilon = -B\, l_{Pt} r\frac{d\theta}{dt} \quad (10)$$

By assuming the excitation to be simple harmonic with $\theta = \theta_{\max} \sin pt$ and $p = 2\pi f$. Equation (10) can be rewritten to get the final form of ε;

$$\varepsilon = -B\, l\, r\, \theta_{\max}\, p \cos pt \qquad (11)$$

The induced emf is to be monitored through a controlled feedback loop, which will readjust the driving frequency automatically upon the detection of a shift in natural frequency. Thus the shift can be recorded then converted to additional mass through calculation.

3 FABRICATION

The resonant device is made from 200 nm-thick Si_3N_4 membrane window supplied by Silson Ltd. All fabrication is done using the FEI Strata™ DB 235. The process has 4 steps involving both Pt deposition and Si_3N_4 millings.

The SEM images of the device, Fig. 3, show a FIB-fabricated paddle resonator. The Pt tracks are the smallest feature, roughly 75 nm in width and 75 nm in height. The process also exhibits good reproducibility and multiple-device-manufacturing capability.

A smaller resonator, fabricated from 100 nm-thick Si_3N_4 membrane window to further show the capability of the FIB, is shown in Fig. 4, with a plate dimension of 1000 nm x 1000 nm and 400 nm-long beams.

4 ANALYSIS AND RESULTS

By analytically modelling the theoretical behaviour of the device with the dimensions (as in Fig. 2), the natural frequency of the Si_3N_4 paddle (i.e. without the Pt tracks) is predicted to be approximately 10.6 MHz. The ABAQUS Finite Element Analysis package confirms that the device will be in torsional vibration at fundamental frequency of 12.5 MHz. This indicates a reasonable correlation between the analytical model and the FEA methods. The difference of 2 MHz is believed to be explained by stress in the structure upon rotation in the shafts, which the FEA package took into account. With the addition of Pt tracks, the analytical model predicts that the natural frequency would drop down to about 8.8 MHz, compared to the FEA prediction of 11.5 MHz.

Hence, the analytical model is believed to be sufficient enough to preliminarily indicate more complex behaviours of the device i.e. with proposed drive and pick up systems.

The model prediction of the device's response to mass adsorption is shown in Fig. 5. Sensitivity is approximately 300 Hz per femtogram. This is based on the assumptions of B of 1 T, input I of 100 μA and Q of 10000. This produce roughly 10° angle of twist which results an induced emf of roughly 70 μV peak-to-peak.

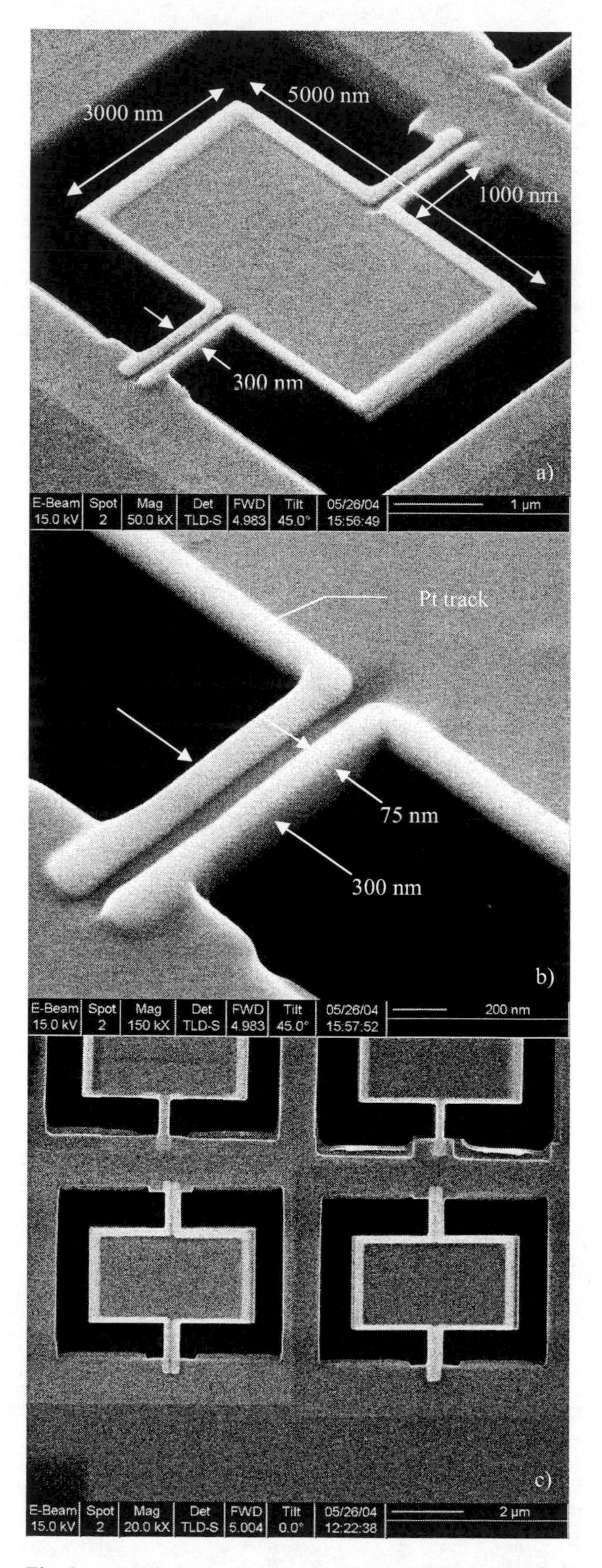

Fig. 3. a) Whole device, b) Close-up of shaft and tracks, c) 2 resonators made in one process

By applying the same analytical model to the device in Fig. 4, the smaller paddle resonator would have natural frequency of 79.9 MHz, with mass sensitivity of 133 Hz per attogram.

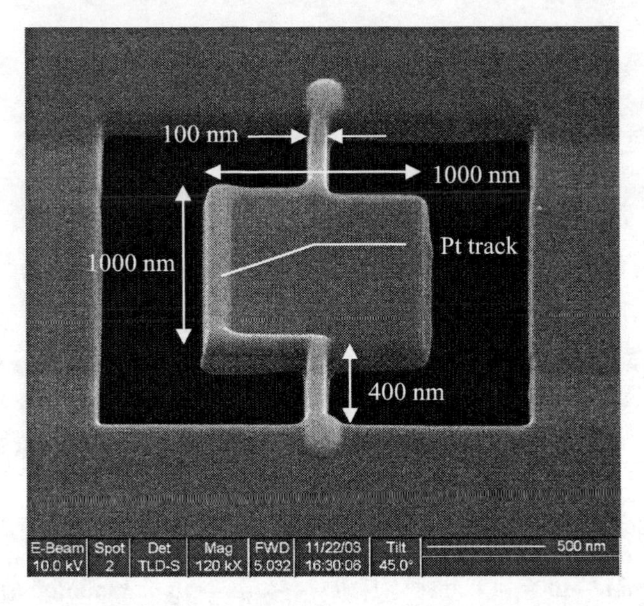

Fig. 4. Submicron FIB fabricated paddle with Pt track

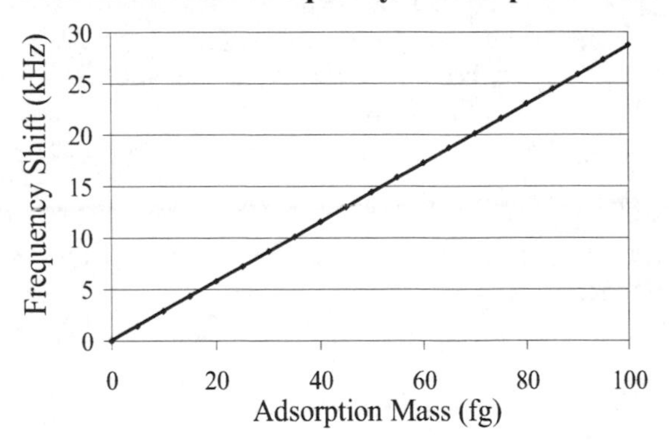

Fig. 5. Plots of frequency shift upon mass adsorption showing a linear response of the resonator to mass addition

CONCLUSION

Analysis of a simple paddle resonator using analytical and FEA modeling show acceptable agreement. The device, driven to resonance by Lorentz forces and with manageable electromagnetic induction readout, shows high sensitivity to mass adsorption with linear response. FIB milling and deposition have been used successfully to fabricate initial experimental devices in Si_3N_4 membranes, with good integrity and reproducibility. The capability of FIB for fabrication of submicron (NEMS) prototype has been demonstrated.

REFERENCES

[1] P. G. Datskos, M. J. Sepaniak, C. A. Tipple and N. Lavrik, Sensors and Actuators B 76, 393-402, 2001

[2] J. Mertens, E. Finot, T. Thundat, A. Fabre, M. H. Nadal, V. Eyraud and E. Bourillot, Ultramicroscopy 97 (1-4), 119-126, 2003

[3] A. Passian, G. Muralidharan, A. Mehta, H. Simpson, T. L. Ferrell and T. Thundat, Ultramicroscopy 97 (1-4) 391-399, 2003

[4] B. Ilic, H. G. Craighead, S. Krylov, W. Senaratne, C. Ober and P. Neuzil, Journal of Applied Physics, Vol. 95, 3694-3703, 2004

[5] R. J. Roark, "Roark's formulas for stress and strain". 6th Edition, McGraw-Hill, 1996

Bio-assembly of Nanoparticles for Device Applications

Krishna V. Singh[1], Khan Alim[2], Xu Wang[1], Alexander Balandin[2], Cengiz S. Ozkan[3] and Mihrimah Ozkan[2, 3*]

[1] Department of Chemical and Environmental Engineering
2 Department of Electrical Engineering
3 Department of Mechanical Engineering
University of California Riverside, Riverside, CA 92521

ABSTRACT

Conjugation of carbon nanotubes (CNTs) with biomolecules results in highly functionalized CNTs which serve as the templates for self assembly of nanostructured materials and as elements of nanoelectronic circuits. Here, we report the synthesis of novel nanocomponents by conjugating single walled carbon nanotubes (SWNTs) with peptide nucleic acid (PNA), an artificial DNA analogue by using carbodiimide coupling. Raman spectroscopy was employed as initial characterization technique for these SWNT bio-conjugates. Finally, their formation was confirmed by using transmission electron microscopy (TEM).

Keywords: self assembly, carbon nanotubes, peptide nucleic acid, electron microscopy, nanodevices.

1. INTRODUCTION

Due to their excellent chemical and electrical properties, carbon nanotubes (CNTs) have been used to realize a number of nanodevices [1, 2], but there is a need for integrating these devices into large structures. Self assembly by molecular recognition provides us a robust yet cost effective alternative for this purpose [3]. Nucleic acids especially DNA, because of their inherent molecular recognition, nanoscale dimensions and easy chemical manipulation are the best candidates for imparting molecular recognition to CNTs [4, 5]. But use of DNA is limited due to its charged backbone, low thermal and chemical stability. To overcome these limitations of DNA, this work aims to utilize the superior chemical and structural properties of its artificial analogue, peptide nucleic acid (PNA). PNA has an uncharged backbone, which is made from repeating N-(2-aminoethyl)-glycine units linked by peptide bonds [6]. Few of the advantages of PNA are shorter probe length, stronger hybridization, higher thermal stability, greater resistance to acidic conditions and enzyme degradation, and higher shelf life [7]. In this work we synthesized and characterized SWNT-PNA nanocomponents.

2. MATERIALS AND METHODS

There is very little work done to realize the potential of PNA outside the scope of conventional biology [8]. This work aims to exploit PNA's ease of functionalization by conjugation to develop SWNT-PNA bioconjugates. As received SWNTs (Sigma Aldrich) were oxidized by refluxing with 1 M HNO_3 for 12 hrs. After sonication in acidic mixture and filtration, the CNT cake was suspended in dimethylformamide (DMF, 99.5%) and incubated for 30 min in 2 mM 1-ethyl-3-(3-dimethylaminopropyl) carbodiimide hydrochloride and 5 mM *N*-hydroxysuccinimide (NHS) to form SWNT bearing NHS esters [8].
The PNA (sequence: NH_2-Glu–GTGCTCATGGTG-Glu-NH_2) probe purchased from Applied Biosystems Inc. was functionalized on both the ends by glutamate amino-acid residue (Glu) as amine present on PNA backbone is not sufficiently reactive for the conjugation purposes. The SWNT-PNA bioconjugates were formed by reacting SWNT-NHS esters with PNA in DMF for 1 hour. For further characterization these nanocomponents were transferred to water.

3. RESULTS AND DISCUSSION

In this work we attempt to find characterization techniques which are simple, less costly but effective as well. We decided to use Raman spectroscopy as our primary characterization technique as it is a powerful, quick, nondestructive yet simple tool for identifying specific materials in complex structures. A high-resolution micro-Raman spectrometer Renishaw 2000 was employed for this study. The spectral resolution of the instrument was about 0.01 cm^{-1}. The spectra were taken under the visible ($\lambda = 488$ nm) laser excitation and 1800 gratings at room temperature. All spectra were taken using the backscattering geometry setup.
The obtained Raman spectra of SWNTs, SWNT-PNA are shown in Fig. 1. One can identify in the SWNT spectrum

* Corresponding author: Prof. Mihrimah Ozkan, mihri@ee.ucr.edu

the radial breathing mode (RBM), disorder induced D-band and G-band at 157.78 cm^{-1}, 1346 cm^{-1} and 1582.4 cm^{-1}, respectively. In SWNT-PNA conjugation spectrum the D-band and G-band of SWNT are observed at 1354.6 cm^{-1} and 1592.4 cm^{-1}, respectively. A possible reason for the disorder peak shift is the conjugation of SWNTs with PNA. The observed peak at 1569.8 cm^{-1} characterizes the basic electronic structure of nucleic acid [9], in our case PNA. The presence of PNA is further indicated by the absence of peak(s) for phosphate backbone, 1050-1150 cm^{-1}[9], which differentiates it from DNA. A very weak RBM band of SWNT is observed in this case, which is reasonable since the low-frequency radial breathing mode is expected to be sensitive to the SWNT conjugation with other nanomaterials. Since the SWNT-PNA sample was prepared on Si substrate, a strong Si peak was also observed.

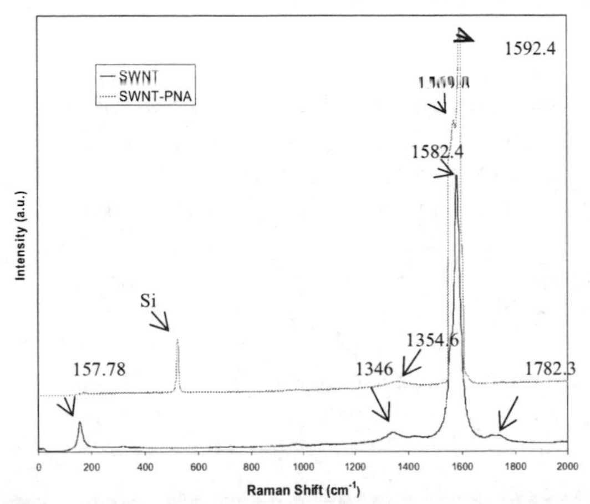

Figure 1. Raman spectra of SWNT (blue) and SWNT-PNA bioconjugates (red). The observance of a new peak (1569.8 cm^{-1}) and shifting of D- band and G- band of SWNTs indicate the conjugation of SWNTs with PNA molecules.

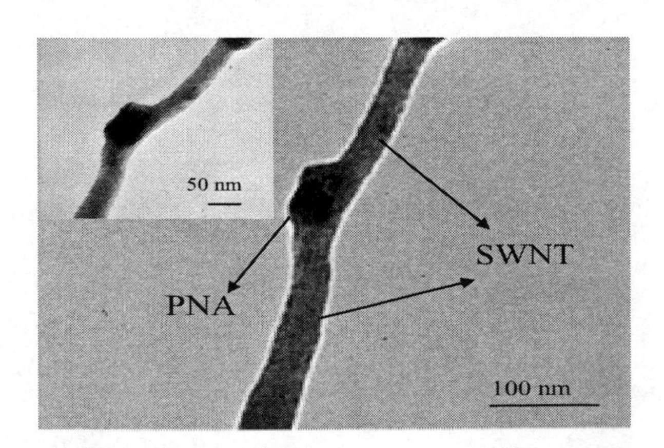

Figure 2. TEM micrograph of a SWNT-PNA bioconjugate. SWNTs are joined by PNA molecules (dark cluster). The high resolution image (inset) of the same component clearly shows that the dark cluster of PNA molecule is more in contrast than SWNTs and joins them.

To support the observations of Raman spectroscopy and confirm the conjugation of PNA molecules with SWNTs, these nanocomponents were visualized under TEM. TEM image (Fig. 1b) clearly shows the joining of two SWNTs ropes with a PNA cluster, which appears as dark spot in the image. PNA molecules form clusters as they aggregate because of their neutral backbone and functionalization with glutamate on both the ends.

We acknowledge the financial support for this research by Microelectronics Advanced Research Corporation (MARCO) and its focus center on Functional Engineered Nano Architectonics (FENA).

REFERENCES

[1] H. W. Ch. Postma, T. Teepen, Z. Yao, M. Grifoni, and C. Dekker, Science **293**, 76 (2001).

[2] S. J. Tans, A. R. M. Verschueren, and C. Dekker, Nature **393**, 93 (1998).

[3] G.M. Whitesides, J.P. Mathias and C.T. Seto, Science **254**, 1312 (1991).

[4] K. Keren, R. S. Berman, E. Buchstab, U. Sivan, and E. Braun, Science **302**, 1380 (2003).

[5] J. D. Le,Y. Pinto,N. C. Seeman, K. Musier-Forsyth, T. A. Taton, and R. A. Kiehl, Nano Lett. **4**, 2343 (2004).

[6] P.E. Nielsen, Peptide Nucleic Acids Methods and Protocols (Human Press, New Jersey, 2002).

[7] A. Ray and B. Norde´N, Faseb J. **14**, 1041 (2000).

[8] K. A. Williams, P. T. M. Veenhuizen, B. G. de la Torre, R. Eritja, and C. Dekker, Nature **420**, 761 (2002).

[9] J. Duguid, V.A. Bloomfield, J. Benevides and G.J. Thomas Jr., Biophys. J. **65**, 1916 (1993).

[10] J.J.P. Stewart, J. Comp. Chem. **10**, 209 (1989).

[11] J.J.P. Stewart, J. Comp. Chem. **10**, 221 (1989).

[12] M.J. Frisch et al., Gaussian 03, Revision B.03 (Gaussian Inc., Pittsburgh, 2003).

Layer-by-Layer Nano-assembled Polypyrrole Humidity Sensor

R.Nohria, R.K.Khillan, Y. Su*, Y. Lvov, K. Varahramyan
Institute for Micromanufacturing, Louisiana Tech University, Ruston, LA 71272
*yisu@latech.edu

Abstract

In this paper we demonstrate highly sensitive and fast response humidity sensors using layer-by-layer (LbL) nano-assembled films of Polypyrrole (PPY). Inkjet (IJ) printed and spin coated PPY films were used for sensitivity and response time comparisons. The change in electrical sheet resistance of the sensing films was monitored as the device was exposed to humidity. The LbL nano-assembled films of PPY showed the better response in terms of response time, linearity and sensitivity. An intended application for these LbL nano-assembled devices is in disposable handheld instruments to monitor the presence of humidity in humidity sensitive environments.

Key Words: - Polypyrrole, humidity sensor, Ink jet printing layer-by-layer nano-assembly and spin coating.

Introduction

The importance of humidity sensing has been well understood and much research has been focused on the development of humidity sensitive materials. In recent years the influence of humidity has been given top priority in moisture sensitive areas such as high voltage engineering systems, storage areas, computer rooms etc. [1,2]. There has been a considerable interest in exploiting organic substances and doped conducting polymers [3-5] for humidity and gas sensing. Advantages with polymers as sensing materials are light weight, flexible, low cost and simple process for fabrication of sensor arrays [6]. A number of different techniques such as electrochemical polymerization, chemical and electrochemical deposition and spin coating have been explored for the fabrication of polymer sensors [7-8].

The main issue of polymer sensors is senstivity of the deposited polymer, which is determined by the thickness of the sensing layer, sensing area and surface roughness. LbL is a unique method for the deposition of composite and polymeric films with controlled thickness at a nanometer range. Ink jet printing is another way of depositing polymers in nanometer range with high degree of surface roughness.

In this paper, LbL nano-assembly is used for deposition of ultrathin Polypyrrole (PPY) film for humidity sensing application. Response time and sensitivity of LbL nano-assembled films was compared to IJ printed and spin coated PPY films. Sensitivity, response time and degradation of the polymer sensors have also been investigated.

Experimental Details

PPY was purchased from Sigma-Aldrich. Before the nano-assembly of the PPY on the glass substrate, the substrate was cleaned with acetone for 2 minutes and then rinsed in de-ionized water for 5 minutes. Five alternate layers of poly (allyl amine) hydrochloride (PAH) and poly(styrenesulfonate) (PSS) $(PSS/PAH)_5$ were deposited as the precursor layers for nano-assembly of PPY. After the deposition of the precursor layers the PAH was replaced by the PPY. Ten bilayers of PPY/PSS were deposited on the precursor layer. Fig.1 shows a schematic cross-section of polymer sensors fabricated by the LbL assembly.

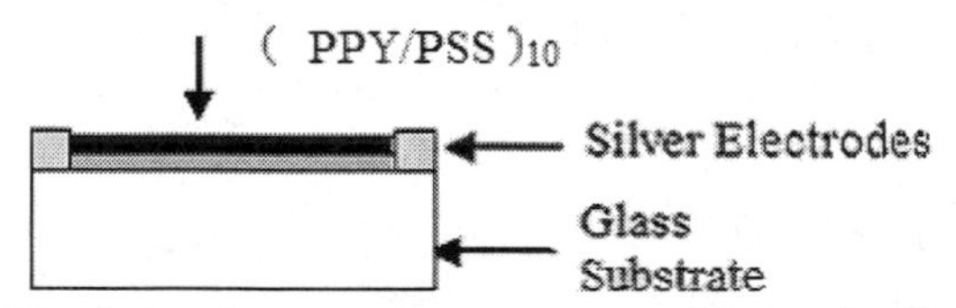

Fig.1 A schematic cross section of Layer-by-Layer assembled polymer sensor

PPY was spin coated on the glass substrate at 3000 rpm for 25s. The thickness of the film was measured to be 120nm. Water soluble conductive polymer PPY was used as an ink to be inkjet printed on glass substrate. Two layers of PPY as shown in Fig. 2 were printed at the substrate temperature of 55^0C. The thickness of IJ printed sensor layer is about 150 nm as measured by the tencor profilometer.

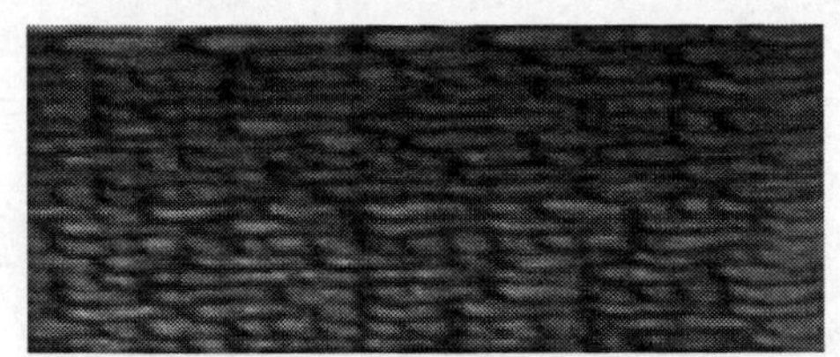

Fig. 2 Inkjet printed PPY layers

Silver conductive adhesive purchased from Sigma Aldrich was used to make contact electrodes. The deposited polymer material was cured at 100°C for 2 minutes in air ambient and then was stored in vacuum for 24 hrs for removing residual solvent in the polymer films. The Attenuated Total Reflection-Infrared (ATR-IR) spectroscopy was used to study the sensing mechanism in the polymer after the absorption of humidity using a Thermo Nicolet Nexus 470 Fourier Transform Infrared Radiometer (FTIR) equipped with ZnSe ATR crystal.

The electrical characteristics of the devices were tested as a function of relative humidity (RH%) in a home built testing chamber used for generation of controlled humidity. Relative humidity inside the chamber can be adjusted in a range from 45% to 90%. Relative humidity inside the chamber was monitored by a standard precalibrated humidity meter. The baseline resistance of the polymer sensors was established at room temperature in air ambience. After the testing of the device the sensor was stored in dry environment.

Results and Discussion

Change in sheet resistance, response time, sensitivity and degradation of the PPY polymer sensor were monitored. The structure of PPY is shown in Fig.3. The change in the sheet resistance was measured for each of these devices as the relative humidity was varied. The experimental results show that the sheet resistance of polymer sensors reduced with increase of relative humidity.

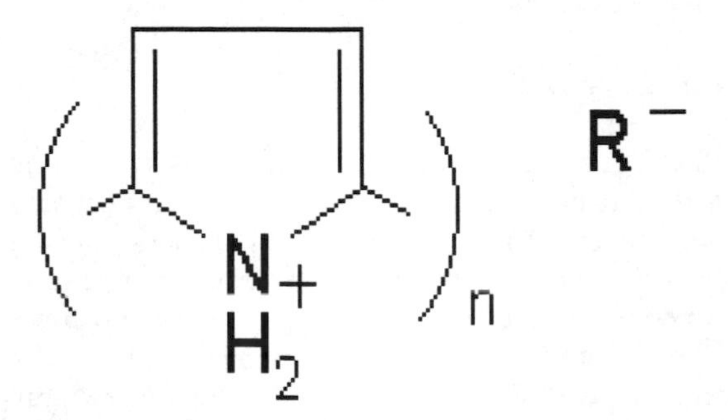

Fig. 3 Structure of PPY

Fig.4 shows output signal from the polymer sensors in term of the change in sheet resistance versus relative humidity. The response time is measured to show the change in sheet resistance after 5% change in relative humidity. The sensitivity of the sensor was defined to be a ratio of the relative change in sheet resistance and 5% humidity change. The degradation is monitored as change in base resistance of the sensing polymer versus time. The recovery time is the time for the sensor's output returning to its original valve after the humidity source is removed.

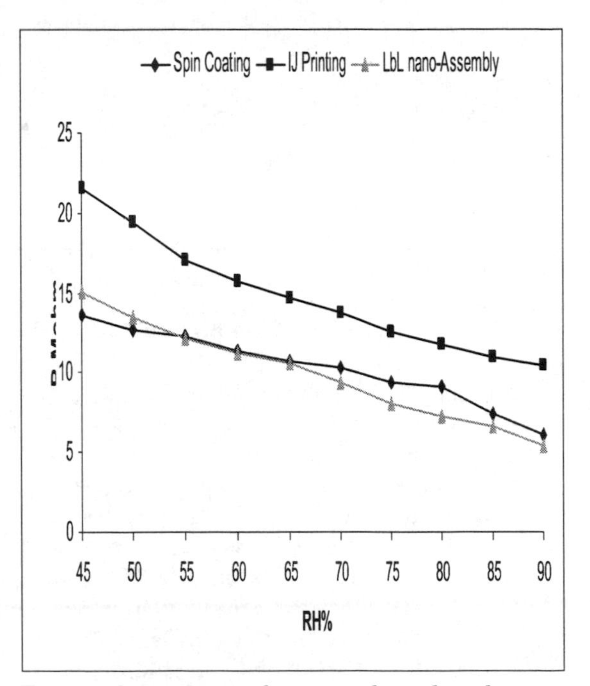

Fig 4. Output signal versus humidity for spin coating, Inkjet printing and LbL assembled sensors.

The LbL nano-assembled films of PPY showed the better response in terms of response time, linearity and sensitivity. The base resistance drifting is also less for LbL assembled humidity sensors as compared to that of the inkjet printed and spin coated sensors. The sensing response time was measured to be 25s, 35s and 57s for LbL assembled, Inkjet printed and spin coated sensors, respectively. Overall recovery time is within 60 seconds, and LbL assembled sensors show shorter recovery time. The water molecules contributed to the initial decrease in resistance [10]. The decrease in resistance shown in Fig. 5 is likely attributed to polarization of electronic charge of the adsorbed water molecules resulting in additional free hole charge carriers which cause an increase in conductivity and thus a decrease in resistivity.

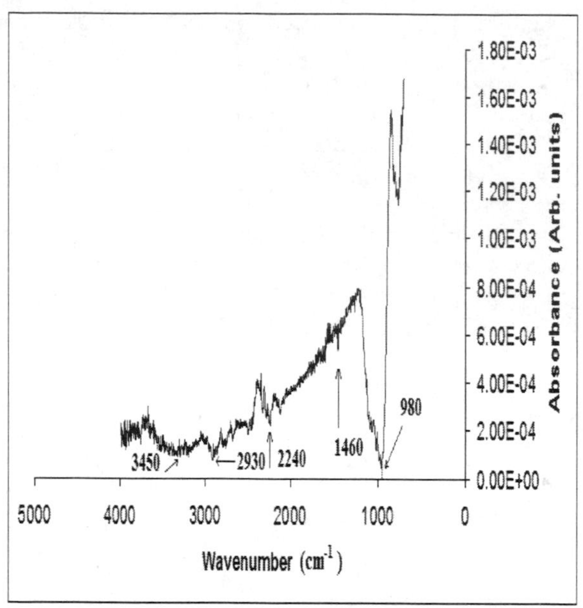

Fig 5. ATR-IR spectrum of PPY

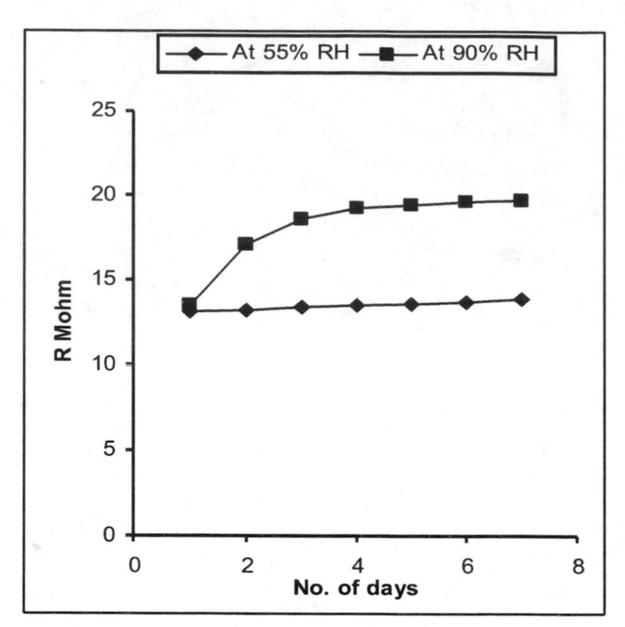

Fig. 6 Degradation behavior of PPY for 7 days.

The ATR-IR experiments were performed after the absorption of humidity [9]. Fig. 4 shows the ATR-IR spectrum of the polymer before and after the absorption of the humidity for 20 s.

The significant modes of various peaks are summarized in Table-1. Optical absorbance is determined by the molecular structure of polymers. The absorbance peaks at 980 cm^{-1}, 2230 cm^{-1} and 1460 cm^{-1} N-H and C-H deformation vibrations respectively. The peaks at 2930 cm^{-1} and 3450 cm^{-1} are attributed to C-H vibrations and free O-H vibrations respectively.

Wave numbers (cm^{-1})	Significant Modes
980, 2230	N-H vibrations
1460	C-H deformation vibrations
2930	C-H bond stretching
3450	O-H bond stretching

Table 1 Significant modes of various peaks at different wave numbers

The humidity sensitivity was measured to be 11%, 10% and 8% for Inkjet printed, LbL assembled and spin coated humidity sensors, respectively. The LbL assembled sensors were found to be working for 10 days whereas inkjet printed and spin coated sensors were found to be working for 7 days. Fig. 6 shows the degradation behavior of PPY by exposing it in air for 7 days by exposing a sample at constant 50% humidity.

The total change in resistance for 7 days was 5.7% with about 1% change in resistance per day. The change in resistance was due to the moisture present in the air. The degradation behavior studied by testing the sample to 90% RH and then exposing it to an open environment at a constant 50% humidity is also shown in Fig.5. The change in resistance was 47.3%. This shows that the moisture absorbed by the sample during testing caused the oxidation of the polymer material. The degradation effect was critical for first three days and after that change in resistance per day was less than 1%. The Inkjet printed and LbL assembled sensors had the consistent response while the spin coated sensors showed degradation at higher humidity range as shown in Fig. 4. The overall experiment results show that the LbL assembled sensors have better humidity sensing performance than those of spin coated sensors and spin coated sensors due to shorter response time and recovery time.

Conclusion

PPY has been deposited using LbL, Inkjet printing and spin coating techniques for humidity sensing application. The LbL assembled sensors showed better sensing performance in terms of response time and recover time due to thin film deposition. The sensitivity of the sensor is improved by thin film deposition using LbL as compared to spin coating.The inkjet printed sensors showed a little better sensitivity than LbL due to the surface roughness. ATR-IR was used to investigate polymer molecular structure change after humidity sensing. The decrease in resistance is attributed to polarization of electronic charge of the adsorbed water molecule. Inkjet printing, LbL assembly and spin coating open an alternative way for fabrication of polymer sensors. These polymer sensors can be used as disposable handheld instruments due to low cost and light weight.

Acknowledgement

The authors thank the Institute for Micro-manufacturing for providing financial and technical support to this project. We are grateful to Dr. Michael McShane, Rohit Srivastav and Rajendra Aithal for helping with ATR-IR experiments.

References

[1]. Trupti Maddanimath, I. S. Mulla, S. R. Sainkar, K. Vijayamohanan, K. I. Shaikh, A. S. Patil and S. P. Vernekar, Humidity sensing properties of surface functionalised polyethylene and polypropylene films, Sensors and Actuators B: Chemical, 81 (2002) 141-151.

[2]. Shilpa Jain, Sanjay Chakane, A. B. Samui, V. N. Krishnamurthy and S. V. Bhoraskar, Humidity sensing with weak acid-doped polyaniline and its composites, Sensors and Actuators B: Chemical, 96 (2003) 124-129.

[3]. D. Hodgins, The electronic nose using conducting polymer sensors, Sensor review 14/4 (1994) 28-31.

[4]. Basudam Adhikari and Sarmishtha Majumdar, Polymers in sensor applications, Progress in Polymer Science, 29, (2004) 699-766.

[5]. W. Bourgeois, P. Hogben, A. Pike and R. M. Stuetz, Development of a sensor array based measurement system for continuous monitoring of water and wastewater , Sensors and Actuators B: Chemical, 88 (2003) 312-319.

[6]. Seung-Yeol Son and Myoung-Seon Gong, Polymeric humidity sensor using phosphonium salt-containing polymers, Sensors and Actuators B: Chemical, 86 (2002) 168-173.

[7]. Gabor Harsanyi, Polymer films in sensor application: A review of present uses and future possibilities Sensor review, 20/2 (2000) 98-105.

[8]. J. P. Lukaszewicz, Controlling of surface and humidity detecting properties of carbon films — selection of a precursor for carbonization, Thin Solid Films 391 (2001) 270-274.

[9]. John Coates Encyclopedia of analytical chemistry © John Wiley and Sons Ltd, Chicester, 2000.

[10]. K.Suri, S.Annapoorni, A.K.Sarkar, R.P. Tandon, Gas and humidity sensors based on iron oxide- polypyrrole nanocomposites, Senors and Actuators B: Chemical, 81 (2002) 277-282.

A micromanipulation method based on the capillary force by phase transition

O. Katsuda [*], S. Saito [**] and K. Takahashi [***]

[*] Tokyo Institute of Technology, Japan, katsudao@ide.titech.ac.jp
[**] Tokyo Institute of Technology, Japan, saitos@mep.titech.ac.jp
[***] Tokyo Institute of Technology, Japan, takahak@ide.titech.ac.jp

ABSTRACT

In this study, our group manipulates micro-objects with capillary force by condensing water from the atmosphere. This condensed water forms a water bridge. The water bridge can be controlled by evaporation and condensation. A Peltier device is used for temperature control in order to achieve water phase transition. By using this method, we have achieved repeatable micromanipulation with simple equipments.

Keywords: Micromanipulation, Liquid bridge, Phase transition, condensation, Micro gravity

1 INTRODUCTION

The size of an object has decreased in the electrical and mechanical engineering field for fabricating highly functional microelectro-mechanical systems and photonic crystals[1]. In micromanipulation, even if we can pickup micro-objects, it is very difficult to detach the micro-objects because adhesional force is dominant[2]. Tanikawa et al. have detached micro-objects by using a micro-drop[3]. This indicates that capillary force is effective in micromanipulation. Obata et al. have shown that capillary force can be controlled by regulation of liquid bridge volume[4]. J. Liu et al., have fabricated a manipulation system named freeze tweezer based on freezing force to manipulate a wide variety of objects[5]. Freeze tweezer has been controlled temperature by means of the Joule-Thompson throttling effect. However, they have not solved the detaching problem of adhesion phenomenon completely.

2 MANIPULATION MODELS

2.1 Pickup Operation

(I) The probe is positioned above the target. (II) Cooling the probe. Water condenses on the surface of the probe. A waterdrop forms. (III) The probe comes down. Liquid bridge forms. (IV) The object is picked up by means of the capillary force.

2.2 Place Operation, Case A

PLACE operation: The placing operation has been divided into two scenarios. Case A: (Va) Heating of the probe. The waterdrop evaporates. The object remains attached to the probe due to adhesional force. (VIa) The probe comes down. The object contacts with the substrate. (VIIa) Raising the probe, but the object remains attached to the substrate since the object-substrate adhesional force is larger than the probe-object adhesional force because it depends on curvature.

2.3 Place Operation, Case B

Case B: in this case the probe-object adhesional force is larger than that in Case A. (Vb) Cooling of the probe. Fair quantity water condenses. (VIb) The probe comes down. The object contacts with the substrate. An additional water bridge forms between object and substrate. (VIIb) Raising the probe. The probe-object liquid bridge collapses, consequently the object remains on the substrate.

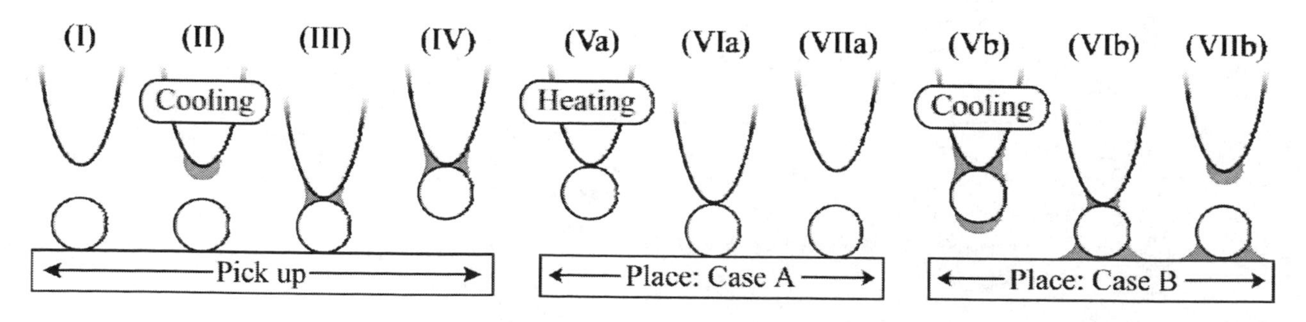

Figure 1: Schematic illustration of pick/place procedure on the experiment.

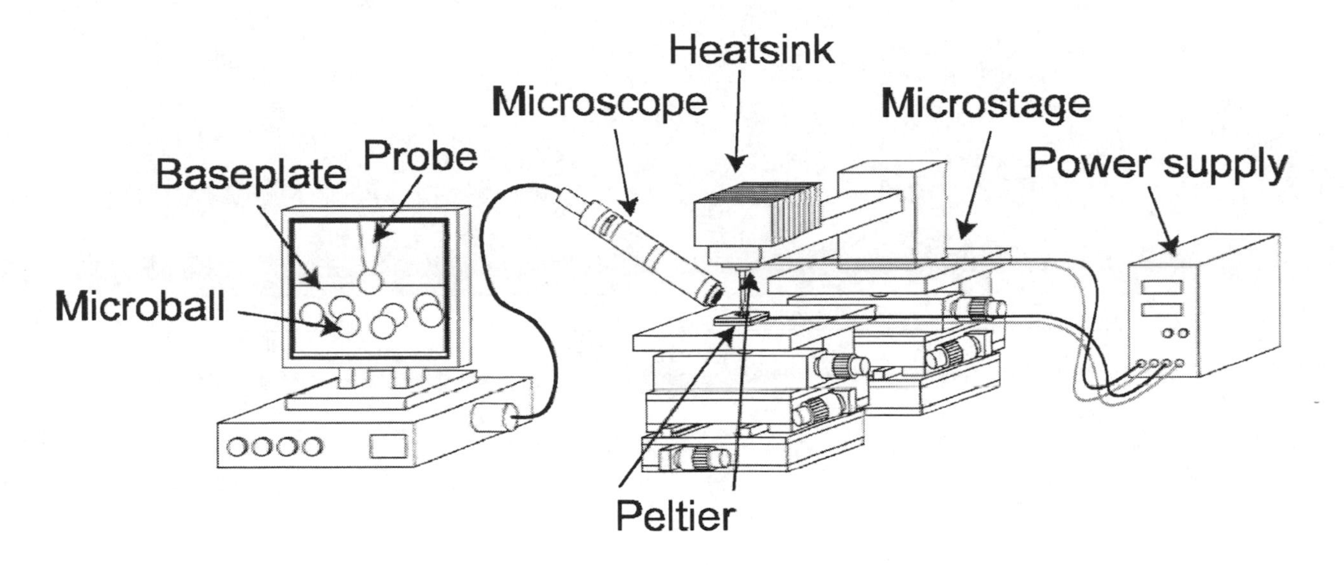

Figure 2: Experimental devices

3 EXPERIMENTS

3.1 Experimental Devices

Figure 2 illustrates experimental system we used. A stainless-steel needle or a tungsten needle was used as the manipulation probe. Solder balls (Sn:96.5%, Ag:3.5%) with a 30 μm diameter were used as micro objects. A mirror-finished stainless-steel plate or a silicon wafer were used as the substrate. An optical video microscope was used to observe. Two peltier devices and heatsink were used to cool and heat the probe and the substrate.

Figure 3 is overview of the experimental devices. Figure 4 is a close-up side view of probe and substrate.

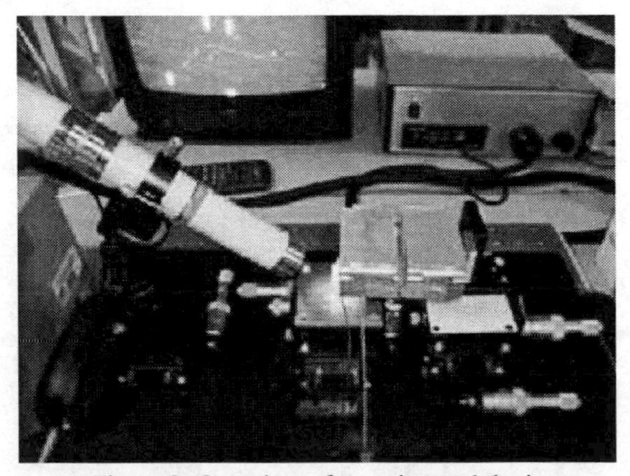

Figure 3: Overview of experimental devices

Figure 4: Close-up side view of probe and substrate

3.2 Experimental Procedure

We experiment two procedures. Figure 5 illustrates experimental simple procedure we did. On Experiment 1, Tungsten we used as the probe and Stainless-steel we used as the substrate. On Experiment 2, Stainless-steel we used as the probe and Si-wafer we used as the substrate.

3.3 Experimental Results

Figure 6 shows various stages of Experiment 1. We explain each stage by following texts.

(1) The probe is positioned above the target.
(2) Cooling the probe. Water condenses on the surface of the probe.
(3) The probe comes down. Liquid bridge forms. The object is picked up by means of the capillary force.

Exp. 1 Pick & Place : 1

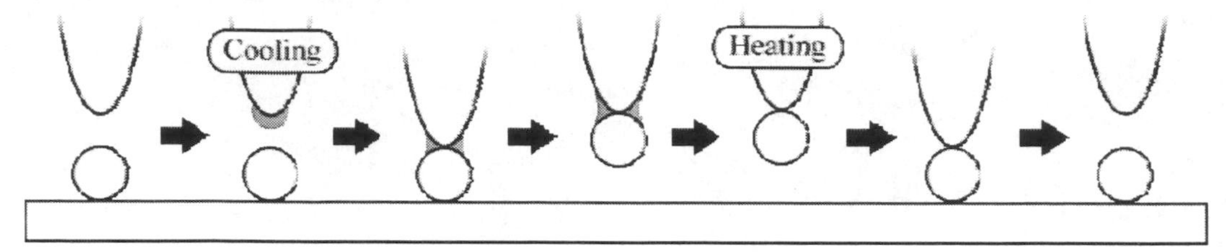

(Probe : Tungsten, Substrate : Stainless-steel)

Exp. 2 Pick & Place : 2

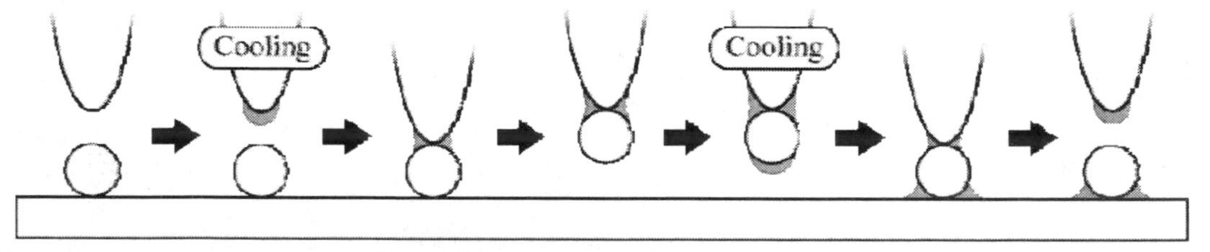

(Probe : Stainless-steel, Substrate : Si-wafer)

Figure 5: Experimental procedures

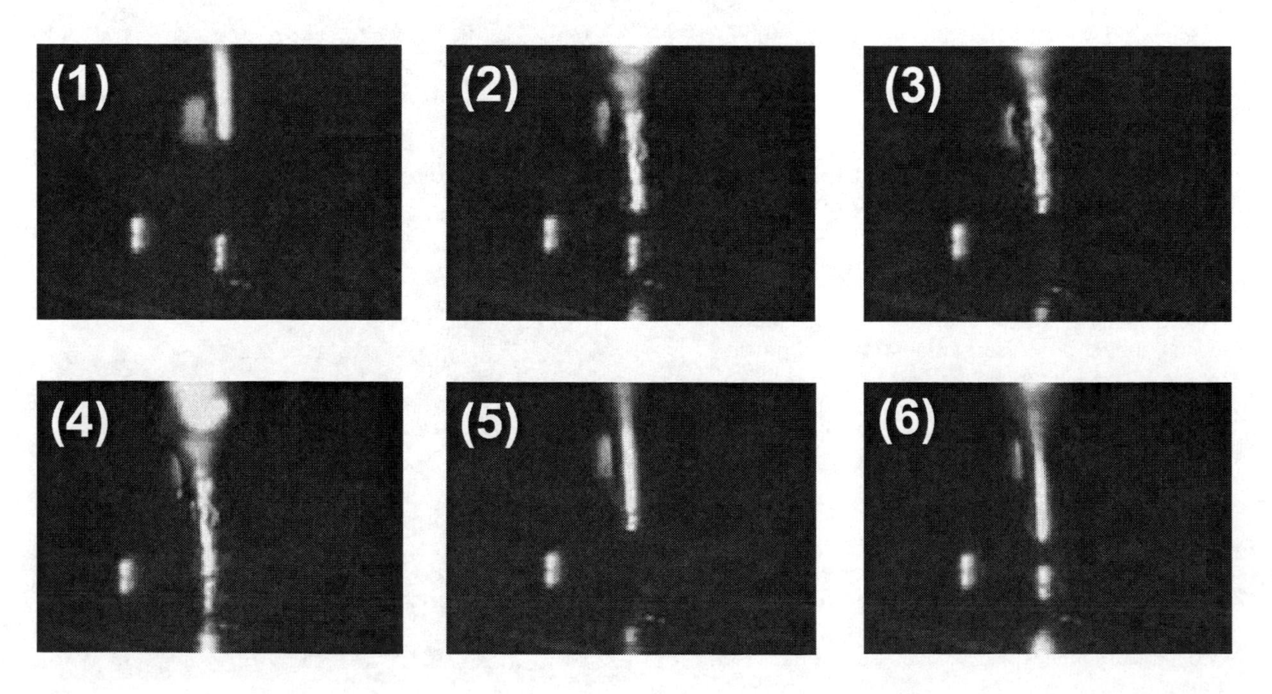

Figure 6: Various stages of Experiment 1

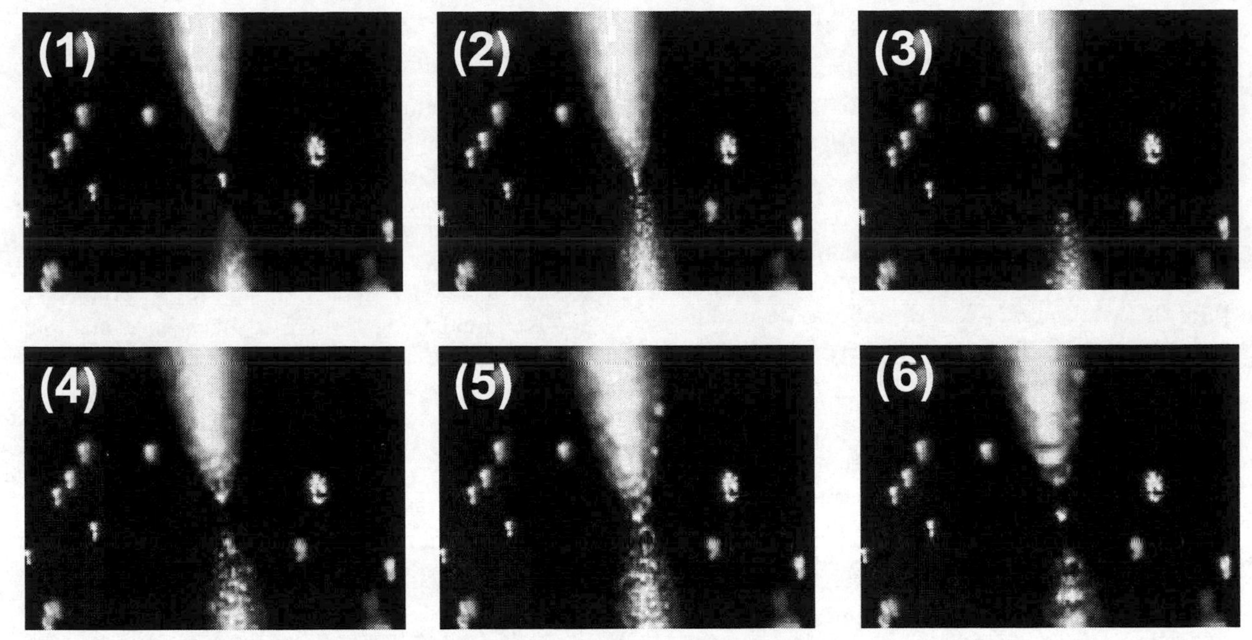

Figure 7: Various stages of Experiment 2

(4) The probe is positioned above placing point.
(5) Heating of the probe. The Water evaporates.
(6) The probe comes down. The object contacts with the substrate. Raising the probe, but the object remains attached to the substrate.

Figure 7 shows various stages of Experiment 2. We explain each stage by following texts.
(1) The probe is positioned above the target.
(2) Cooling the probe. Water condenses on the surface of the probe.
(3) The probe comes down. Liquid bridge forms. The object is picked up by means of the capillary force.
(4) The probe is positioned above placing point. Cooling of the probe. Fair quantity water condenses.
(5) The probe comes down. The object contacts with the substrate. An additional water bridge forms between object and substrate.
(6) Raising the probe. The probe-object liquid bridge collapses, consequently the object remains on the substrate.

4 CONCLUSION

Excellent repeatability was achieved, thus we insist this wide applicable method can be effective in micro-manipulation automation.

REFERENCES

[1] K. Aoki, et al., Nature Materials, vol.2, no.2, pp.117-121, 2003.
[2] R.S. Fearing, Proc. IEEE Micro Electro Mechanical Systems, pp.212-217, 1995.
[3] T. Tanikawa, Y. Hashimoto, and T. Arai, Proc. IEEE Intl Conf. on Intelligent Robots and Systems, pp.776-781, 1998.
[4] K. J. Obata, et al., Journal of Fluid Mechanics, vol. 498 pp.113-121, 2004
[5] J. Liu, Y. Zhou, and T. Yu, Journal of Micromechanics and Microengineering, vol.14, no.2, pp.269-276, 2004.

Evaluation of manipulation probes for expanding the range of capillary force

K. J. Obata*, S. Saito** and K. Takahashi***

* Tokyo Institute of Technology, Japan, obatak@ide.titech.ac.jp
** Tokyo Institute of Technology, Japan, saitos@mep.titech.ac.jp
*** Tokyo Institute of Technology, Japan, takahak@ide.titech.ac.jp

ABSTRACT

In order to perform reliable micromanipulation, we need to use a force that is both controllable and greater than the adhesional force. We have proposed a manipulation method based on capillary force. For reliable micro-manipulation, a wide range of capillary force is required to pick/place a micro-object. This paper evaluates the capillary force generated by a liquid bridge between a spherical object and a manipulation probe with a concave tip. As the radius of the concavity being close to the object, the capillary force can be increased drastically, and it can also be controlled through a wide range by means of the liquid volume. There is a drastic decrease of the capillary force with the liquid volume when the hemline of the meniscus is on the brim of the concave, and it can be shifted by means of the depth control of the concavity. The mechanism of this increase of the capillary force, the capillary force curve, and the profile of the liquid bridge are also presented.

Keywords: micromanipulation, capillary forces, adhesion, numerical methods

1 Introduction

It is difficult to manipulate a micro-sized object precisely with conventional tools, *i.e.* tweezers can pick up the micro object but the object still adheres to any of the fingers even if the tweezers opens for releasing it(1). In the micro world, the gravity effect becomes extremely small compared to the adhesional one. Therefore, in order to perform reliable micro-manipulation, we need to use a force that is both controllable and potentially greater than the adhesional force. Saito *et. al* have analyzed the mechanical force required to slip and roll an object by considering adhesional effect, under a scanning electron microscope, and proposed a method of manipulation using a needle shaped tool. If the required force is compressive and larger than the strength of the object, their method might break a brittle object(2; 3). Takahashi *et. al* have evaluated the force generated by Coulomb interaction and estimated the voltage required to detach an adhered particle(4). If the required voltage is too large, Takahashi's method might cause electric discharge and melting of the object(5).

Our group has analyzed the capillary force generated by a liquid bridge between a spherical object and a plate, and has proposed the micro-manipulation method based on the regulation of two liquid bridge volumes(6). The first liquid bridge is formed between a manipulation probe and the object to pick up it, and the second bridge between a substrate and the object is provided in order to collapse the first one during the place manipulation. We have pointed out these critical volumes required to pick/place the object in the previous paper. To remove the spherical object from the substrate, the applied force must be greater than the adhesional force. The adhesional force can be obtained from JKR theory(7);

$$F_{\text{adh.}} = \frac{3}{2}\pi R \Delta\gamma, \tag{1}$$

where R is the radius of a spherical object and $\Delta\gamma$ is the work of adhesion. Experimentaly, the value of $\Delta\gamma$ is highly depending on the surface condition. Therefore, for reliable manipulation, the capillary force should be greater than adhesional force even if $\Delta\gamma$ is on the order of 1 Nm^{-1} and also should be regulated to be smaller than the second capillary force in order to collapse the first bridge (see figure 1(V) and (VI)).

In this paper, a manipulation probe with a concave tip, as shown in figure 1, is introduced in order to expand the capillary force range. The concavity is expected to generate a greater capillary force. The capillary force would remain small when the liquid overflow into the flat surface. A numerical analysis of sphere-concavity-plate model is presented to evaluate the capillary force. Note that we are not concerned here with any mechanical ways or problems of the liquid flow to the surface of the manipulation probe or substrate.

2 Analysis of the liquid bridge

Figure 2 shows an axisymmetric model for the analysis of a liquid bridge between a spherical object and a concave shaped probe, where X and Z are cylindrical coordinates of the profile of the meniscus, which have the origin at the bottom of the concavity, R is the radius of the object, R_{p} is the radius of the concavity, D is the distance from the probe to the object, φ is the filling angle of the object, F is the attractive force acting on

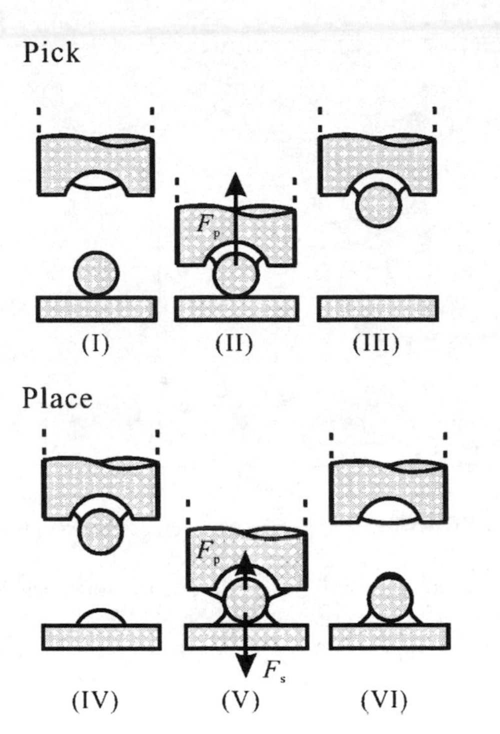

Figure 1: Schematic illustration of pick/placing procedure: (I)positioning (II)lowering (III)picking up (IV)positioning (V)lowering (VI)placing.

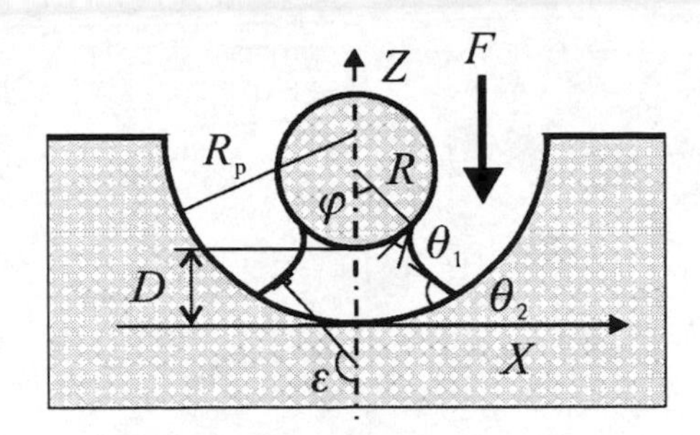

Figure 2: Liquid bridge between a spherical object and a concave shaped probe

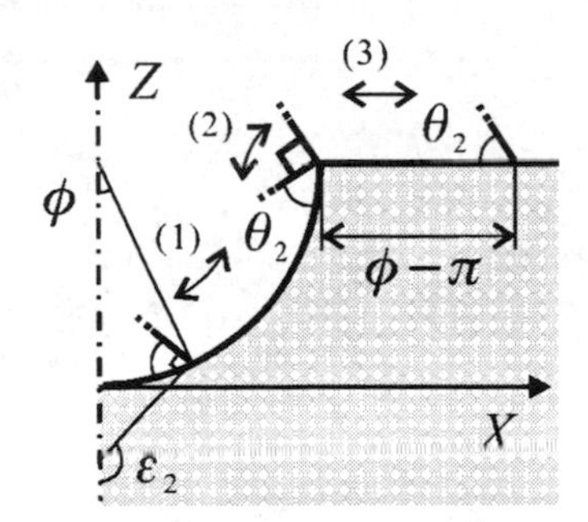

Figure 3: Boundary condition on the concaved probe surface

the object, and V is the volume of the liquid bridge between two solids. The meniscus forms contact angles θ_1 at the object and θ_2 at the probe. We make the following assumptions as simillar as our previous paper (6); 1. The influence of gravity is negligible. 2. The dynamic flow of the liquid is negligible. 3. The volume of the liquid is conservative. 4. The contact angles are determined by Young-Dupré equation (8). 5. The object and the probe are rigid.

Capillary force F can be expressed as the sum of the pressure difference force and the surface tension force:

$$F = -\Delta P \pi {X_1}^2 + 2\pi\sigma X_1 \sin\varepsilon_1, \tag{2}$$

where ΔP is the pressure difference between inside and outside the liquid, X_1 is the radius of the contact circle on the spherical object, σ is the surface tension, and ε_1 is the inclination of the meniscus at the contact circle on the object. According to Young-Laplace equation, the value of ΔP can be obtained from the stability profile of the liquid bridge:

$$\Delta P = 2H\sigma, \tag{3}$$

where H is the local mean curvature. Since ΔP is a constant, the surface of the meniscus has the same mean curvature at any local point. As shown by Orr(9), the value of H in (3) can be expressed by geometrical parameters as

$$2H = \frac{\mathrm{d}}{\mathrm{d}X}(\sin\varepsilon) + \frac{\sin\varepsilon}{X}, \tag{4}$$

where ε is the angle between the normal to the meniscus and the vertical axis. Since the left-hand side of this equation is constant, it can be solved as a two-point-boundary value problem, for which the boundary conditions are the inclinations ε and X coordinates of the menisci on the solid surfaces. These inclinations are determined by the slopes of the solid surfaces and the respective contact angles θ_1 and θ_2 (see figure 2).

Figure 3 shows three boundary states on the probe surface, which appeares (1)on the concavity, (2)at the brim, and (3)on the flat surface. If the hemline of the meniscus is on the concavity (see (1) in figure 3), the boundary conditions can be written as

$$\left.\begin{array}{ll} \varepsilon_1 = \theta_1 + \varphi, & X_1 = R\sin\varphi, \\ \varepsilon_2 = \pi - \theta_2 + \phi, & X_2 = R_\mathrm{p}\sin\phi, \end{array}\right\} \tag{5}$$

where ϕ is the filling angle of the concavity. When the meniscus reaches the brim of the concavity as (2) in figure 3, the boundary condition on the probe surface is

$$\varepsilon_2 = \pi - \theta_2 + (\pi - \phi), \quad X_2 = R_\mathrm{p}, \tag{6}$$

where ϕ shows the increasing angle at the brim of concavity. In the case of (3), the boundary can be shown

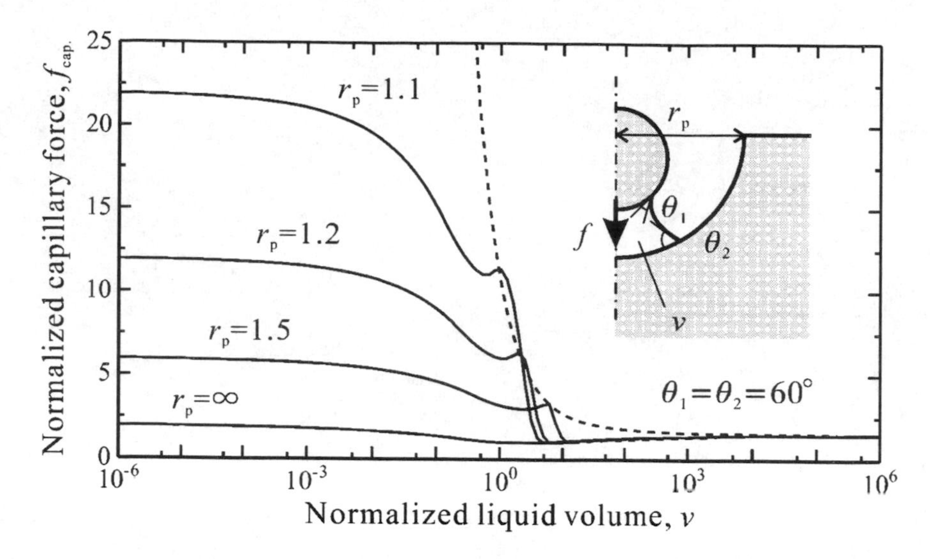

Figure 4: Relation between the maximum capillary force and the liquid volume for each radius of the concavity and for $\theta_1 = \theta_2 = 60°$ ($r_{\text{p}} = \frac{R_{\text{p}}}{R}, f_{\text{cap.}} = \frac{F_{\text{cap.}}}{\pi R\sigma}, v = \frac{V}{R^3}$). The two broken lines represent the thresholds of the boudary changes on the probe surface. The second boundary change occurs at the bottom of the each solid line (the second broken line is behind the solid lines).

as

$$\varepsilon_2 = \pi - \theta_2, \quad X_2 = R_{\text{p}} + (\phi - \pi), \tag{7}$$

where ϕ becomes the parameter describing X coordinate.

The boundary-value problem has the solution(9). The meniscus profile (X, Z), the distance D, the liquid volume V, and also the capillary force F can be calculated from given four parameters; contact angles θ_1 and θ_2, the filling angle φ, and the parameter ϕ. If the volume V is given in advance instead of the parameter ϕ, ϕ must be determined so that V could be equal to the given value. Then, the relation between D and F, which has the conservative liquid volume and given contact angles, can be plotted as a function of the filling angle φ.

To generalize the following discussion, all the parameters are normalized as

$$z = \frac{Z}{R}, x = \frac{X}{R}, d = \frac{D}{R}, f = \frac{F}{\pi R\sigma}, v = \frac{V}{R^3}, r_{\text{p}} = \frac{R_{\text{p}}}{R}. \tag{8}$$

3 The shape effect for capillary force

The liquid bridge is treated as static. Thus the capillary force curves are just plotted in its stable equilibrium state. The curves show the value of the external force that would equilibrate the capillary force. If the external force is larger than the maximum value of the curve, the liquid bridge is extended until eventually the bridge collapses. Therefore the maximum capillary force is essential to predict which bridge will collapse. We set the maximum capillary force as $F_{\text{cap.}}$, instead of $F_{\max}$ used in Ref. (6).

The solid lines in figure 4 shows the relationships between the normalized maximum capillary force $f_{\text{cap.}}(\equiv \frac{F_{\text{cap.}}}{\pi R\sigma})$ and the liquid volume v assuming $\theta_1 = \theta_2 = 60°$ for example. Each line has a different concave radius r_{p}. The indication $r_{\text{p}} = \infty$ represents the sphere-plate model used in Ref. (6). As the radius of concavity being close to the sphere radius, the maximum capillary force grows more and more when the liquid volume is relatively small compared to the object. When the volume is close to zero, the capillary force can be written as

$$f_{\text{cap.}} = \frac{2r_{\text{p}}}{r_{\text{p}} - 1}(\cos\theta_1 + \cos\theta_2). \tag{9}$$

The increasing ratio compared to sphere-plate model is $\frac{r_{\text{p}}}{r_{\text{p}}-1}$. This represent that we can achieve high reliability of the picking manipulation by means of the design of the probe shape.

If the volume is sufficiently large, the capillary force is almost independent of the concave radius, since the liquid must overflow into the flat surface. Thus, with increasing v, the capillary force reaches equivalent to that of the sphere-plate model as

$$f_{\text{cap.}} = 1 + \cos\theta_1. \tag{10}$$

Therefore, we can control the capillary force within the range from (3.2) to (3.1) at least by means of the regulation of the liquid volume.

The broken lines represents the thresholds of the boundary changes on the probe surface as shown in fig-

NSTI-Nanotech 2005, www.nsti.org, ISBN 0-9767985-1-4 Vol. 2, 2005

ure 3. The second threshold is on the bottom of the solid lines. If the hemline of the meniscus is on the concavity, the capillary force remains large level. As increasing v, the boundary condition changes from the concavity state to the brim state. In the brim state, the capillary force decreases drastically with v. Finally, in the plate state, the capillary force is equivalent to the sphere-plate model. Therefore, for pick/place manipulation, we can evaluate the required liquid supply by comparing these thresholds. Furthermore, the drastic decrease can be shifted to smaller liquid volume by making the concavity shallower. We can design the probe shape so that a liquid supplying device could make the drastic decrease of the capillary force. In addition, when the concavity is shallower than hemisphere, the local peak of the capillary force can not be observed. It might be one of the characteristics of the hemispherical concavity shape. In the next section, we clarify the mechanism of the increase of the capillary force with showing the capillary force curve.

4 Conclusion

In this study, the probe shape for micro-manipulation was discussed for increasing the range of the capillary force. The capillary force generated by the liquid bridge between the spherical object and the probe with a concave tip was analized to find a proper shape of the probe. As the radius of concavity being close to the sphere radius, the maximum capillary force grew more and more when the liquid volume was relatively small compared to the object. The increasing ratio compared to sphere-plate model was $\frac{r_p}{r_p - 1}$, where r_p is the concave radius. This represent that we can achieve high reliability of the picking manipulation by means of the design of the probe shape. If the volume was sufficiently large, the capillary force was approximately equivalent as the sphere-plate model, since the liquid must overflow into the flat surface. Therefore, we can control the capillary force with the wide range by means of the regulation of the liquid volume. The capillary force drastically decreased with the liquid volume, when the hemline of the meniscus was on the brim of the concavity. This decrease can be shifted to the other volume area by means of the depth control of the concavity. The object can float even before this drastic change of the capillary force. This would be the great advantage of our method for both supplying liquid and less mechanical damage.

This study was supported by a Grant-in-Aid for Scientific Research from the Ministry of Education, Culture, Sports, Science, and Technology.

References

[1] R.S.Fearing, "Survey of Sticking Effects for Micro Parts Handling," *Proceedings of IEEE/RSJ International Conference on Intelligent Robots and Systems*, 212-217, 1995

[2] Shigeki Saito, Hideki T. Miyazaki, and Tomomasa Sato, "Micro-object Pick and Place Operation under SEM based on Micro-physics," Journal of Robotics and Mechatronics, 14, 227-237, 2002

[3] Shigeki Saito, Hideki T.Miyazaki, Tomomasa Sato, and Kunio Takahashi, "Kinematics of mechanical and adhesional micromanipulation under a scanning electron microscope," Journal of Applied Physics, 92(9), 5140-5149, 2002

[4] Kunio Takahashi, Hideaki Kajihara, Masataka Urago, Shigeki Saito, "Voltage required to detach an adhered particle by Coulomb interaction for micromanipulation," Journal of Applied Physics, 90(1), 432-437, 2001

[5] Shigeki Saito, Hideo Himeno, and Kunio Takahashi, "Electrostatic detachment of an adhering particle from a micromanipulated probe," Journal of Applied Physics, 93(4), 2219-2224, 2003

[6] Kenichi J. Obata, Tomoyuki Motokado, Shigeki Saito and Kunio Takahashi, "A Scheme for micromanipulation based on capillary force," Journal of Fluid Mechanics, 498, 113-121, 2004

[7] K.L.Johnson, K.Kendall, and A.D.Roberts, "Surface energy and the contact of elastic solids," Proc. R. Soc. Lond. A., 324, 301-313, 1971

[8] Jacob N.Israelachvili, Intermolecular and Surface Forces, 301-322, 1985

[9] F.M.Orr, L.E.Scriven, and A.P.Rivas, "Pendular rings between solids: meniscus properties and capillary force," Journal of Fluid Mechanics, 67(4), 723-742, 1975

Immobilization of Polydiacetylene Sensor on Solid Substrate

Young Bok Lee*, Ju Mi Kim*, Kun Ha Park*, Yong Goo Son*,
Dong June Ahn**, and Jong-Man Kim*

*Department of Chemical Engineering Hanyang University, Seoul 133-791, Korea,
lybok0526@ihanyang.ac.kr, jmk@hanyang.ac.kr
**Department of Chemical and Biological Engineering, Korea University, Seoul 136-701,
dja@korea.ac.kr

ABSTRACT

Immobilized polydiacetylene vesicles on solid substrates were successfully prepared by utilizing imine formation or Diels-Alder reaction. Incubation of an aldehyde-modified glass substrate in an aqueous solution containing self-assembled diacetylene vesicles allowed efficient formation of immobilized polymer vesicles. In addition, preparation of immobilized polydiacetylene vesicles on solid substrate was achieved, for the first time, using the Diels-Alder reaction with diacetylene monomers having maleimido headgroups and a furan-modified glass substrate. Patterned fluorescent images were obtained with immobilized vesicles by selsctive irradiation through a photomask followed by heating the glass substrate at 100 °C.

Keywords: immobilization, polydiacetylene, vesicles, fluorescent images

1 INTRODUCTION

Recently, the development of efficient sensors utilizing conjugated polymers as sensing matrices has gained much attention among many researchers[1]. Especially, polydiacetylene (PDA)-based sensors for the detection of biologically important species have been intensively investigated due to the unique color changing properties upon stimulation[2]. The advantage of the nanostructured polydiacetylenes as biosensors comes from the fact that visible color change from blue to red occur in response to a variety of environmental perturbations, such as temperature, pH, and ligand-receptor interactions.

The vast majority of polydiacetylene-based sensors reported to date have been prepared in the form of liposomes in aqueous solutions or thin films on solid supports using Langmuir-Blodgett or Langmuir-Schaefer methods. Recently, immobilization of polydiacetylene vesicles onto gold film was reported[3]. Very recently, we reported immobilization of polydiacetylene vesicles on aldehyde substrate via imine formation[4]. Here, we present immobilization and patterned fluorescence images of polydiacetylenes on solid substrate using two different strategies.

2 EXPERIMENTAL

2.1 Preparation of diacetylene monomers

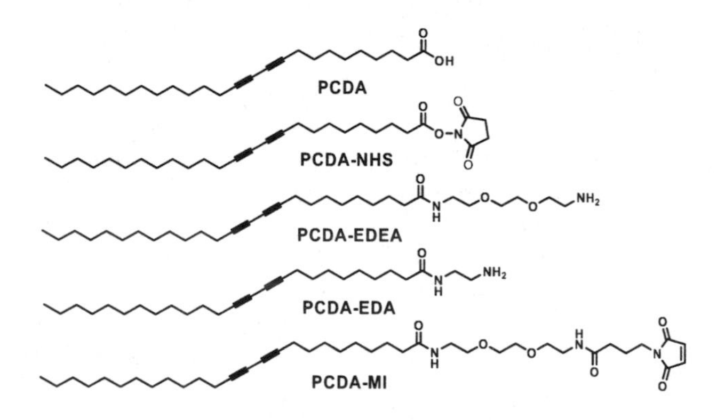

Figure 1: Diacetylene monomers used in this study

PCDA-NHS

To a solution of PCDA (4.00 g, 10.68 mmol, GFS Chemicals) in CH_2Cl_2 (40 mL) was added N-hydroxysuccinimide (NHS) (1.35 g, 11.75 mmol) and 1-(3-dimethylaminopropyl)-3-ethyl carbodiimide hydrochloride (EDC) (2.46 g, 12.82 mmol). The solution was allowed to stir at ambient temperature for 3 h followed by removing the solvent *in vacuo*. The residue was extracted with ethyl acetate and water. The organic layer was dried with $MgSO_4$ and the solvent was removed *in vacuo* to give a white powder (4.83 g, Y=96%)

PCDA-EDEA

To a solution of PCDA-NHS (5.03 g, 10.68 mmol) in CH_2Cl_2 (200 mL) was added a solution of EDEA (15.83 g, 106.80 mmol) in CH_2Cl_2 (100 mL). The mixture was allowed to stir at ambient temperature for 3 h followed by removing the solvent *in vacuo* much as possible and poured into ethyl acetate. After the precipitate was filtered off, the filtrate was concentrated *in vacuo* to give a white powder (4.09 g, Y=76%).

PCDA-EDA

To a solution of PCDA-NHS (2 g, 4.24 mmol) in CH_2Cl_2 (80 mL) was added a solution of EDA (5.10 g, 84.80 mmol) in CH_2Cl_2 (100 mL). The mixture was allowed to stir at ambient temperature for 6 h followed by removing the solvent *in vacuo* much as possible and poured into ethyl acetate. After the precipitate was filtered off, the filtrate was concentrated *in vacuo* to give a white powder (1.14 g, Y= 65 %).

PCDA- MI

Coupling of maleic anhydride with aminobutyric acid yielded 4-maleimidobutanoic acid (MI-BA) in two steps. The acid MI-BA was converted to the activated ester MI-BA-SI in the presence of N-hydroxysuccinimide. The desired diacetylene monomer PCDA-MI was obtained by coupling the activated ester MI-BI-SI with the amine-terminated diacetylene lipid PCDA-EDEA.

2.2 Preparation of liposome

Diacetylene monomer was dissolved in chloroform in a test tube. The solvent was evaporated by a stream of a N_2 gas and buffer solution was added to the test tube to give desired concentration of a lipid (1 mM). The resultant suspension was sonicated for 15 min at a temperature of around 80 ℃. Following sonication, the solution was filtered to remove dispersed lipid aggregates by using a 0.8 ㎛ filter and cooled at 4 ℃ for overnight. Polymerized diacetylene liposomes are prepared by UV irradiation (1 mW/cm^2) with 254 nm.

2.3 Preparation of glass substrate

Amene/aldehyde-modified glass was purchased from CEL Associate. The furan-modified glass was obtained by incubation of amine-modified glass in a solution containing FA-SA-NHS (5 mM in CH_2Cl_2) for 1day.

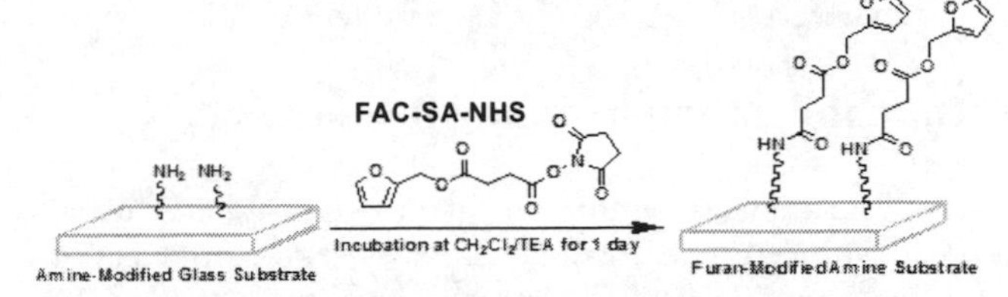

Figure 2: Furan-modification of an amine glass substrate

2.4 Immobilization of polydiacetylene liposome on glass substrate

Imine formation : an aldehyde-modified glass was placed in the diacetylene liposome (PCDA-EDEA : PCDA-EDA = 50 : 50 mol%) for 1day. The glass was sonicated to remove unreacted liposome for 1min and further washed with excess deionized water and dried under a stream of nitrogen.

Diels-Alder reacation : a furan-modified glass was placed in the diacetylene liposome (PCDA-MI : PCDA-EDEA = 50 : 50 mol%) for 1day. The glass was sonicated to remove unreacted liposome for 1min and further washed with excess deionized water and dried under a stream of nitrogen.

2.5 Fabrication of fine fluorescent patterned images

A photolithographic method shown in Figure 3 was employed for patterned fluorescent images.

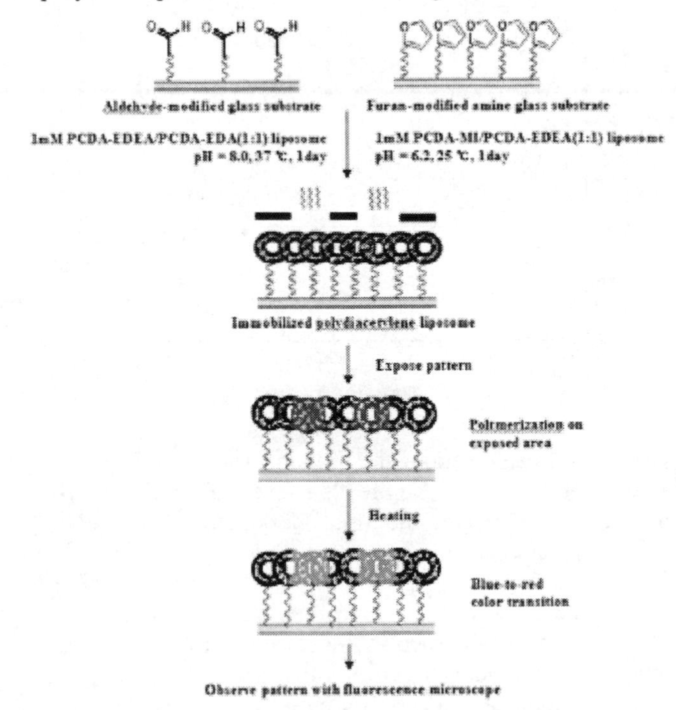

Figure 3: Procedures for patterned fluorescence images.

3 RESULTS AND DISCUSSION

3.1 Immobilization by imine formation

In order to generate immobilized polydiacetylenes on the solid substrate, amine-terminated diacetylene monomers PCDA-EDEA and PCDA-EDA were prepared. A blue-colored and stable polydiacetylene vesicle solution was obtained with PCDA-EDEA when the monomer was subjected to the routine procedures for polydiacetylene vesicle formation in aqueous solution. On the contrary to PCDA-EDEA, the diacetylene monomer PCDA-EDA was converted to only unstable self-assembled diacetylenes which eventually led to solid aggregates. Immobilization on the aldehyde substarte was attempted with self-assembled diacetylene liposomes prepared from various ratios of monomers between PCDA-EDEA and PCDA-EDA. We found best results were obtained with liposomes prepared with a 1:1 mixture of the two monomers.

Figure 4 shows visible spectra after irradiation with UV light for 10 min of the glass substrates immobilized with diacetylene vesicles. It clearly demonstrates that immobilization on the aldehyde substrate (Figure 4A) is more effective than that on a unmodified clean glass substrate (Figure 4B). The control experiment shows that non-specific physical adsorption of the liposomes is negligible.

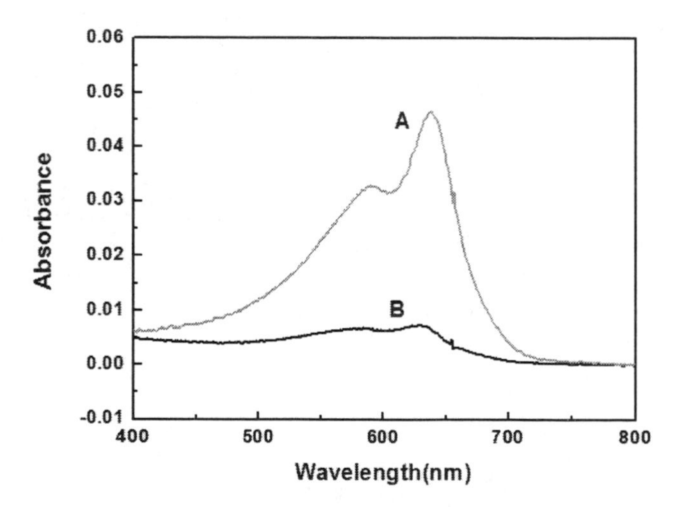

Figure 4: Visible spectra of the immobilized glass substrate after irradiation with UV-light (254 nm, $1 mWcm^{-2}$) for 10 min. (A: with aldehyde-modified glass substrate, B: with unmodified glass substrate)

Next phase of current investigation focused on the generation of patterned fluorescent images with immobilized polydiacetylenes. For this purpose, a photolithographic method was employed. Accordingly, an aldehyde-modified glass substrate immobilized with diacetylene liposomes derived from PCDA-EDEA and PCDA-EDA (1:1 mixture) was irradiated with UV light for 2 min through a photomask. This process would induce photopolymerization of the immobilized diacetylene vesicles in the exposed areas. The polydiacetylene vesicle-immobilized glass substrate was then heated at 100 °C for 10 sec to induce the blue-to-red color shift of the polydiacetylene molecules. Since polydiacetylene in the blue phase is nonfluorescent while it is strongly fluorescent in the red phase, it should be possible to observe the patterned fluorescence images. Figure 5 shows patterned fluorescence images observed under a fluorescent microscope. The red (bright) areas are exposed areas with UV light.

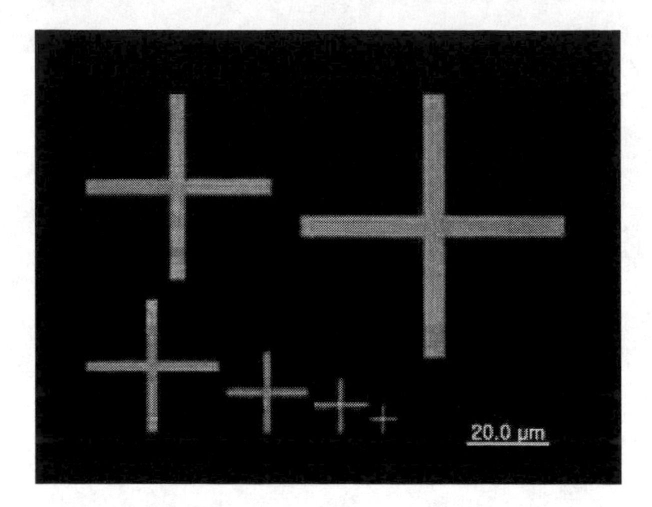

Figure 5: Patterned fluorescent images obtained with PCDA-EDEA/PCDA-EDA (1:1 w/w) as described in the text.

3.2 Immobilization by Diels-Alder reaction

A Diels-Alder reaction is a cycloaddition reaction between a diene and a dienophile. The advantages of bioconjugation utilizing the Diels-Alder reaction are 1) showing little effect of the pH of the reaction medium on the reaction rate, 2) producing no by-products, 3) having rate accelerating effect by hydrophobic interaction between diene and dienophile in aqueous solution. Accordingly, efficient bioconjugation of saccharides to protein and peptides to solid substrates have been reported[5].

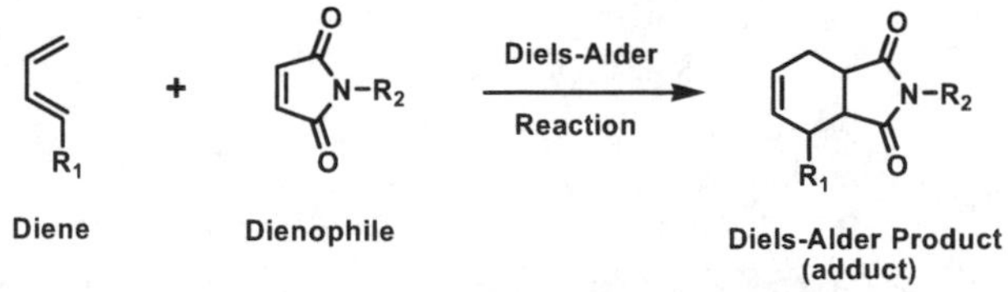

Diene + Dienophile —Diels-Alder Reaction→ Diels-Alder Product (adduct)

Figure 6: The Diels-Alder reaction between a diene and a dienophile.

Since the bioconjugation using the Diels-Alder reaction has attractive nature, we felt it would be useful if we could immobilize polydiacetylene vesicles using the same approach. For this purpose, we have the modified glass substrate with furan moieties which can form Diels-Alder adducts with polydiacetylene vesicles. The diacetylene monomer PCDA-MI was designed so that the terminal maleimido group can form adduct with the furan group.

In order to obtain immobilized polydiacetylene vesicles via the Diels-Alder reaction, the furan-modified solid substrate was incubated in the liposome solution prepared with a mixture of diacetylene monomers PCDA-MI and PCDA-EDEA (1:1 w/w). The comonomer PCDA-EDEA was introduced to avoid dense packing of the maleimido moieties on the surface of the vesicles which might have bad effect on the immobilization of the vesicles.

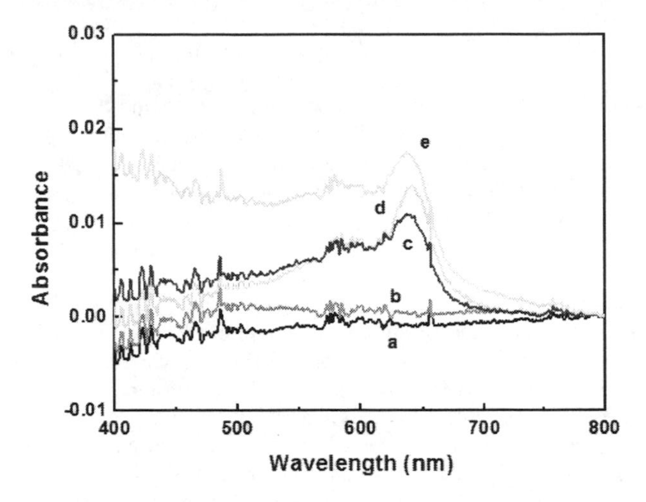

Figure 7: Visible spectra of furan-modified amine glass substrates (a, b and c) and of amine-modified glass substrates (d and e) after immobilizing vesicles prepared with a) PCDA-MI , b) PCDA-EDEA, c) 1:1 mixture of PCDA-MI and PCDA-EDEA, d) PCDA-MI, and e) 1:1 mixture of PCDA-MI and PCDA-EDEA.

As can be seen in Figure 7, immobilization of polydiacetylene vesicles prepared with PCDA-MI alone on the furan-modified solid substrate was failed (Figure 7, a). This is presumably due to the lack of the space needed to form adducts by the Diels-Alder reaction. Immobilization of the vesicles prepared with 100% PCDA-EDEA was not observed (Figure 9, b), demonstrating amine moieties on the vesicles do not interact with furan groups on the solid substrates. Interestingly, the polydiacetylene vesicles prepared with PCDA-MI was successfully immobilized on the amine-modified glass substrate. We believe this could be due to covalent bond formation between the double bond of the amido group and the free amine on the solid support.

In order to confirm the immobilization via Diels-Alder reaction is effective, generation of patterned fluorescent images on solid substrate was attempted. Thus, the furan modified glass substrate was incubated in a diacetylene liposome solution. After immobilization, the solid substrate was irradiated with UV light followed by heating at 100 °C. As can be seen in Figure 8, finely resolved patterned fluorescent images were obtained.

4 CONCLUSION

We have prepared diacetylene monomers and immobilized the resulting self-assembled diacetylene liposomes onto the glass substrates via imine formation or by Diels-Alder reaction. Both methods were found to be effective for the immobilized polydiacetylene vesicles. Fine fluorescent images were obtained when the immobilized liposomes were irradiated with UV light through a photomask followed by heating the glass substrate at 100℃. The results described in this study should be potentially useful for fabricating stable liposome array sensor systems.

Figure 8: Patterned fluorescent images obtained by sequential immobilization by DIels-Alder reaction, selective irradiation and heating of the glass substrate as described in the text.

REFERENCES

[1] D. tyler McQuade, Anthony E. Pullen and Timothy M. Swager, Chem. Rev. 100, 2537, 2000.

[2] Jelinek Raz, Kolusheva Sofiya, Biotechnology Advances, 19, 109, 2001

[3] Stanish I., Santos J. P., Sigh A., J. Am. Chem. Soc., 123, 1008, 2001

[4] Kim Jong-Man, Ji Eun-Kyung, Woo Sung-Min, Lee Haiwon and Ahn Dong Jun, Adv. Mater. 15, 1118, 2003

[5] Pozsgay V., Vieira N. E., Yergey A., Organic. Letters, 4, 3191, 2002

Polymer-Controlled Growth of CdSe Nanoparticles into Micro- and Nanowires

A. Fahmi*, U. Oertel**, V. Steinert**, P. Moriarty* and M. Stamm**
*Department of Physics, University of Nottingham, Nottingham NG7 2RD, United Kingdom. Amir.Fahmi@nottingham.ac.uk
**Leibniz-Institut für Polymerforschung Dresden e.V. (IPF), Hohe Straße 6, 01069 Dresden, Germany
Stamm@ipfdd.de

ABSTRACT

This work reports an easy and effective method to prepare CdSe semiconductor nanoparticles stabilized by an amphiphilic diblock copolymer polystyrene-block-polyvinylpyridine [PS-b-P4VP] in a nonaqueous medium [1]. The CdSe nanoparticles are held in the polymeric matrix via the lone pairs of electrons in the pyridine rings of the block copolymer. We find that the structure of the polymeric CdSe nanoparticles depends strongly on solvent polarity. For instance, spin coating of the stock CdSe(PS-b-P4VP) in DMF onto native oxide terminated Si-wafers lead to a core-shell structure (CdSe nanoparticles in the core, encased in a shell of the block copolymer).However, a non-polar solvent can be used to switch the structure from core-shell to flat ring structure decorated with CdSe nanoclusters on the edge. After a specific thermal treatment of CdSe/PS-b-P4VP thin film CdSe nano- and micro wires were formed. The aggregation of CdSe nanoparticles during the thermal elimination of the polymer phase causes the growth of wires [2,3]. The wires have a diameter between 0.4- 2 µm (depending on the film thickness) and a length of 250 µm. SEM was used to observe the mechanism of the wire formation on the Si-wafer substrate. It is believed that the thermal treatment of the polymer CdSe complex leads to dewetting of the polymer phase, which causes the CdSe nanoparticles to aggregate into wires.

Keywords: block copolymer, CdSe nanoparticles, Nanoring, core-shell structure, CdSe wires

Introduction

Fabrication of semiconductor nanostructures has attracted considerable attention owing to their potential applications in electronics, optics, catalysis, ceramics and magnetic storage [4,5]. Nanostructured materials often display optical, electronic, and structural properties different from those of the bulk [6,7,8,9]. For example, nanoparticles may adopt various shapes, and the form adopted plays a critical role in determining these properties. The unusual properties of nanoparticles can be attributed to two main factors: high surface area to volume ratio and the quantum confinement of electronic states.

During the past decade, many synthetic methods based on polymer materials have been developed, among which the synthesis of nanoparticles in a polymer matrix is prominent [10,11,12,13]. The use of a polymeric matrix as a medium for metal nanoparticle formation does not only provide improved stabilisation and fine control of the growth of the nanoparticles, it also imparts new properties to the polymeric material [14,15]. Block copolymer micelles, especially those formed by well-defined block copolymers, with well-chosen block length and chemical composition in selective solvents turned out to be an excellent model system. In particular, it is easy to delineate their advantages in colloid synthesis, compared to classical stabilisation systems by surfactants or in microemulsion. In principle, microphase-separation of the block copolymer could be harnessed to promote the ordering of the particles and thereby selectively create high degree of organisation in one block of the diblock copolymer (Fig.1). In such a system, the unpolar blocks form the corona, which provides solubility and stabilisation, while the polar block forms the core, which is able to dissolve metal compounds due to coordination.

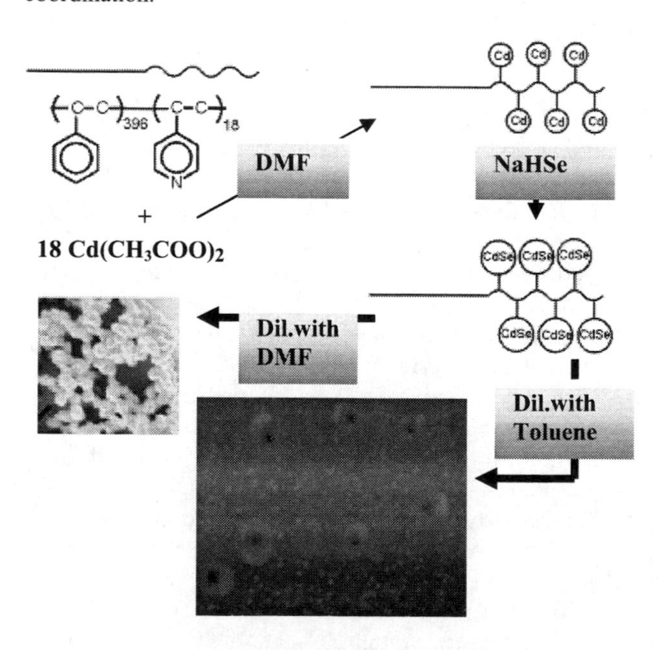

Fig . 1: demonstrating the switching process by dilution with different solvents

Such a micelle core can be considered as a quasi nanoreactor, where nucleation and growth of metal particles upon reduction are restricted to the mesoscale level.

NSTI-Nanotech 2005, www.nsti.org, ISBN 0-9767985-1-4 Vol. 2, 2005

This study describes a method to prepare CdSe nanoparticles with controllable size and stability as shows in (Fig 1). We used Poly(styrene-block-4-vinylpyridine) (PS-b-P4VP) block copolymer to form compound micelles upon a complex of 4VP with a stoichiometric molar amount of cadmium ions (Cd^{2+}) in dimethylformamide (DMF).
The formation of nanoparticles in PS-b-P4VP was studied very thoroughly and enabled the identification of the principles underlying the control of the particles size. Dissolution of metal salts for example in the solution of PS-b-P4VP micelles (that are otherwise insoluble in the solvents) means that these are salts incorporated only in the micelle core, due to the coordination with corresponding groups of polymer[11,13]. For instance, CdSe nanoparticles grow in the core of the micelles after inserting sodium hydrogen selenide (NaHSe) into the solution of PS-b-P4VP(Cd) to produce a yellow colour as evidence for CdSe formation.

DETERMINATION OF THE CORE-SHELL STRUCTURE OF CdSe

Fig. 2 shows a SEM image of CdSe/P4VP-b-PS film spin cast onto native oxide terminated Si-wafers. The image demonstrated a spherical structure with a reasonable diameter around 100 nm.

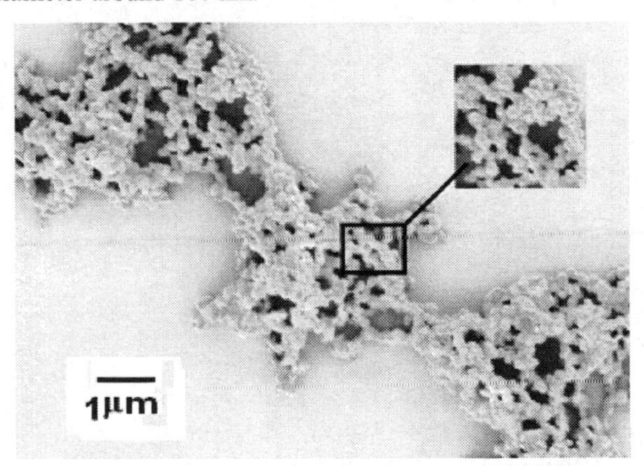

Fig. 2: SEM micrograph of core-shell structure CdSe in core and the shell occupied from block copolymer

FLAT RING CdSe STRUCTURE

Additional insight into the change in structure can be gained after dilution with a non-polar solvent (toluene). Fig. 3 shows an AFM image obtained from cast film on Si-wafer. The film shows that ionomer diblock copolyer micelles in toluene can exhibit extraordinary kinetic stability. The film allows the forming of flat uniform ring structures of the block copolymer decorated with CdSe nanoparticles on the edge. A cross-section demonstrates that the flat ring structures have a diameter of 250 nm and ≈2 nm height. However, the CdSe-nanoparticles occupied a sector onto the ring structures with ≈ 4 nm height. Moreover, XPS results have confirmed the structure switching on the substrate (It does not appear in this manuscript).

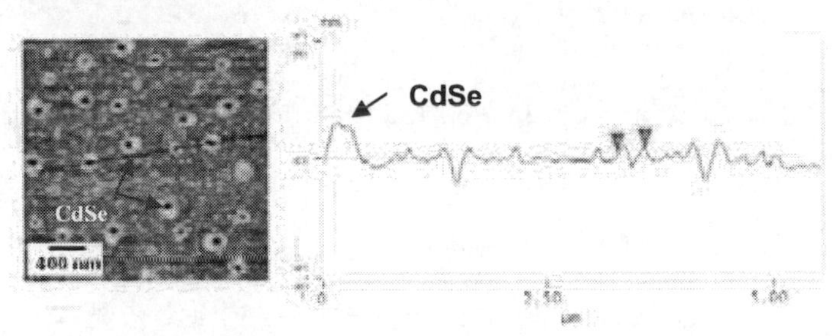

Fig. 3: AFM image of CdSe/P4VP-b-PS ring structure in ultrathin film was spin- casting onto native oxide Si-wafer

THERMAL DEGRADATION AND PARTICLES AGGREGATION

Thermal degradation of CdSe/PS-b-P4VP thin film (spin-cast) on native oxide Si-wafers has produced CdSe wires in micro- and nanometer scales. The aggregation of CdSe nanoparticles during the thermal elimination of the polymer phase caused the growth of wires [3]. The wires have a diameter between 0.4- 2 µm (depending on the film thickness) and length of 250 µm. SEM was used to observe the CdSe wires on Si-wafer substrates (Fig 4).

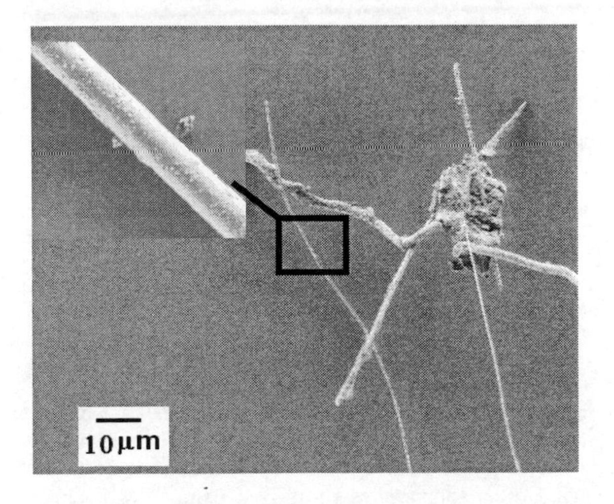

Fig. 4: SEM micrograph of CdSe microwires on Si-wafer developed after thermal treating of thin film of CdSe/P4VP-b-PS.

CdSe complex leads to a dewetting of the polymer phase, which causes the CdSe nanoparticles to aggregate into wires. This strategy could open a new avenue to produce nanoparticle assemblies derived from core-shell and hollow colloids into semiconductors wires, which may provide new application in areas such as photonics, photoelectronics and microelectronics.

We have performed a preliminary electrical transport measurement to confirm the conductivity of the CdSe wirees. The insert in Fig. 5 shows an optical image detailing the microwire alignment between two parallel AuTi electrodes supported on a silicon oxide substrate (the thickness of the oxide was more than 200 nm). The current changes were measured with applied potential up to 10 V across the CdSe-wire at 4.2 K. Fig. 5 shows a measurement of the current flow through the wire as a function of the voltage applied between the AuTi electrodes. Although the asymmetry in the I(V) characteristic arises from variations in the quality of the contacts, it is clear that the wire conduct. A probe measurement are planned (so as to alleviate the contact problem).

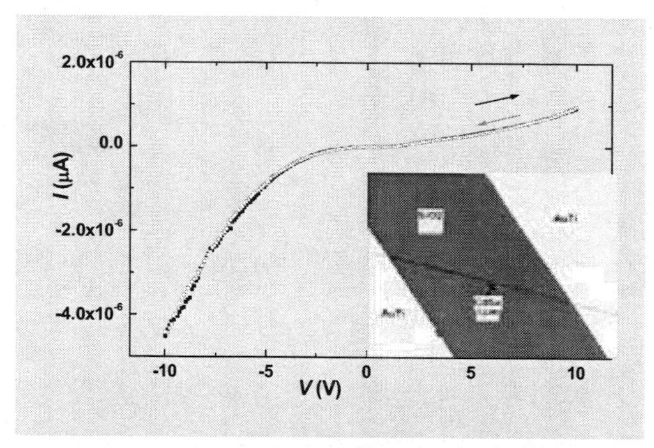

Fig.5: Electrical conductivity measurement on an individual CdSe wires that 1μm in diameter. An optical micrograph (insert) detailing the alignment of the CdSe wire with respect to the two parallel AuTi electrodes.

CONCLUSION

An amphiphilic block copolymer (PS-b-P4VP) has been used to form ''nanoreactors'' for the synthesis of CdSe nanoparticles in a non-aqueous medium. Switching the structures from core-shell into flat ring structures of the polymeric CdSe nanoparticles strongly depends on solvent polarity. A specific thermal treatment of PS-b-P4VP/CdSe thin film produces nano- and micro CdSe wires. We found that the wires diameter depended on the film thickness. It is believed that the thermal treatment of the polymeric CdSe complex leads to dewetting of the polymer phase, which causes the CdSe nanoparticles to aggregate into wires.

An initial experiment to measure the conductivity of the CdSe wire has shown current flow (across 1μm diameter CdSe wire) through the two AuTi electrodes. We believe this simple approach for producing CdSe wires, which is a functional material, is needed in future for many applications such as photonics, photoelectronics and microelectronics.

ACKNOWLEDGMENT

The authors thank the European Commission for funding the SUSANA Research and Training Network (Supramolecular Self-Assembly of Interfacial Nanostructures) Project No. HPRN-CT-2002-00185. we would like to thank Dr. F.Pulizzi and D. Taylor for the conductivity measurement.

REFERENCES

[1] A. Fahmi, U. Oertel, V. Steinert, C. Froeck, M. Stamm Macromol. Rapid Commun. 24, 625–629, 2003.
[2] Amir Fahmi "Nanotemplates via self-assembly of diblock copolymers and their metallization" Dissertation Technische Universitaet Dresden Germany 2003
[3] A. Fahmi, U. Oertel, V. Steinert, Philip moriarty, M. Stamm under preparation
[4] A. Henglein; Chem. Rev. 89, 1861, 1989.
[5] Y. Sun, J. E. Riggs; Int. Rev. Phys. Chem. 18, 43,1999.
[6] L. N. Lewis; Chem. Rev. 93, 2693, 1993.
[7] A. P. Alivisators; Science 271, 933,1996.
[8] L. E. Brus; J. Chem, Phys. 80, 4403, 1984.
[9] Y. Wang, N. Herron; J. Phys. Chem. 95, 525, 1991.
[10] W. Casert; Macromol; Rapid Commun. 21, 705, 2000.
[11] R. S. Kane, R. E. Cohn, R. Silbey; Chem. Mater. 8. 1919, 1996.
[12] J. Yue, V. Sankaran, R. E. Cohn, R. R. Schrock; J. Am.Chem. Soc. 115, 4409, 1993.
[13] M. Moffitt, A. Eisenberg; Chem. Mater. 7, 1178, 1995.
[14] M. Breulmann, H. Cölfen, H. –P. Hentez, M. Antonietti; Adv. Mater. 10, 237, 1998.
[15] M. Michaelis, A. Henglein; J. Phys. Chem. 96, 4719 1992.

Adhesive and Conductive – Inkjettable nano-filled inks for use in microelectronics and microsystems technology

Edouard Marc Meyer**, Andreas Arp***, Francesco Calderone**, Jana Kolbe*, Wilhelm Meyer***, Helmut Schaefer*, Manuela Stuve*

*IFAM - Fraunhofer Institut fuer Fertigungstechnik und Angewandte Materialforschung, Klebtechnik und Oberflaechen, Wiener Strasse 12, D-28359 Bremen, Germany

**Metalor Technologies SA, Avenue du Vignoble, CH-2009 Neuchâtel, Switzerland

***Microdrop Gesellschaft für Mikrodosiersysteme GmbH, Muehlenweg 143, D-22844 Norderstedt, Germany

ABSTRACT

Current technology, Inkjet is an accepted technology for dispensing small volumes of material (50 – 500 picolitres). Currently traditional metal-filled conductive adhesives cannot be processed by inkjetting (owing to their relatively high viscosity and the size of filler material particles). Smallest droplet size achievable by traditional dispensing techniques is in the range of 150 µm, yielding proportionally larger adhesive dots on the substrate. Electrically conductive inks are available on the market with metal particles (gold or silver) <20 nm suspended in a solvent at 30-50%. After deposition, the solvent is eliminated and electrical conductivity is enabled by a high metal ratio in the residue. Some applications include a sintering step. These nano-filled inks do not offer an adhesive function. Work reported here presents materials with both functions, adhesive and conductive. This newly developed silver filled adhesive has been applied successfully by piezo-inkjet and opens a new dimension in electrically conductive adhesives technology.

The present work demonstrates feasibility of an inkjettable, isotropically conductive adhesive (ICA) in the form of a silver loaded resin with a 2-step curing mechanism: In the first step, the adhesive is dispensed (jetted) and pre-cured leaving a "dry" surface. The second step consists of assembly (wetting of the 2nd part) and final curing.

Keywords: conductive, nano-filled, silver, inkjet, microelectronics

1 DESCRIPTION

Generally speaking, jetting of liquids by piezo-actuators can generate droplets of as low as 30 pl whereas dispensing with the most recent valve equipped instruments produces droplet sizes of 10 nl typically. Inkjet nozzles have diameters ranging from 100 µm to 30 µm and smaller. It is widely accepted that the maximum silver filler particle size should not exceed 4 µm (with spherical morphology) for inkjetting. Also, viscosity of the loaded adhesive should be low, typically <100 mPas in order to be ink-jettable. Thus the unloaded resin material needs to have a very low base viscosity. The results reported here have been obtained with a 75 µm nozzle. Further work with a 50 µm inkjet nozzle is in progress. Preliminary results show that adhesive droplet sizes can be further reduced.

2 SILVER POWDER PROPERTIES AND REQUIREMENTS

Under these conditions, special care has to be taken to prevent agglomeration and sedimentation of the filler particles in the resin matrix: agglomeration results in large particles blocking the nozzle, sedimentation modifies the loading ratio of silver locally and may also obstruct the ink-jet nozzle. High purity silver particles of small size possess a chemically and metallurgically very active surface. Unprotected it may cold-weld and sinter already at room temperature under high shear stress conditions, as they occur during jetting inside the nozzle. These factors require an effective protection coating on the silver surface which at the same time needs to enhance dispersion properties of the powder particles in the resin matrix. However, a balance has to be found between these "positive" properties and the detrimental electrical insulation typically provided by an effective surface coating. Also, the coating chemicals have to be neutral in terms of curing reactions and viscosity stability. An additional feature of the filler particles is reflectivity. Especially in the case of photo-initiated curing mechanisms, this property is key to obtain curing in depth of the adhesive layer. The success of the powder is determined by the final properties of the material, i.e. electrical conductivity and mechanical bond strength. The silver powder that fulfils all the above mentioned requirements is of >99.997% purity and has been developed specifically for this project.

3 2-STEP CURING MECHANISM – A COMBINATION OF UV AND THERMAL CURE

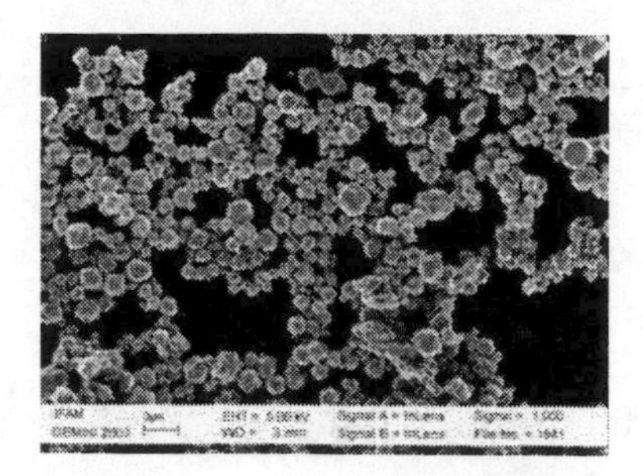

Figure 1: SEM of Silver powder. Main characteristics of the silver powder. Specific surface area: 0.55 m^2/g (single point BET); Tap density: 7.0 g/cm^3 (Tap pack); PSD D50: 2.4 μm (light scattering); Ag purity: >99.997%.

The technical approach retained to match the specificities of ink-jetting is based on a 2-step curing system: Dispensing into a final form, pre-cure to a material that can be easily handled (the jettable material would not allow easy processing outside of the inkjet itself), followed by the component assembly step and final curing of the conductive adhesive.

The resin system was designed with a first UV-cure with radical mechanism. After inkjet deposition on the substrate, the adhesive is partially cured by UV (Prepreg) and can be processed immediately or stored at <4°C for processing at a later stage. Because of its radical nature, the pre-cure step has to be performed under inert atmosphere, e.g., nitrogen, as oxygen could interfere in the curing reaction. At this stage the adhesive surface is "dry", similar to a thermoplastic material containing unreacted epoxy functional groups.

The second (and final curing) step involves heating of the adhesive (i.e., melting), followed by wetting of the 2nd part (component assembly) and cross-linking of the epoxy groups. Bonding of the 2nd part (component) has to be performed under pressure as thermoplastic materials exhibit very high viscosities. Final curing is achieved in a classical furnace/oven. The cross-linking ratio determines the mechanical properties and can be fine tuned through the number of epoxy groups in the initial resin material. At the end of the process a 3-dimensional network of cross-linked polymer chains is formed.

The 2-step cure is based on a specifically developed Acrylate-Epoxy-Resin matrix with Newtonian properties. It has a very low base viscosity, typically in the range of 3 mPas. To ensure high electrical conductivity in the final product the silver filler particle concentration must overcome the percolation threshold. Therefore, a loading of 70% by weight was chosen as a minimum silver filling rate of the adhesive (approx 20% by volume).

A further constraint is the fact that the metallic silver particles are not transparent to UV. Therefore UV light cannot penetrate deeply into the formulation to initiate the first radical curing reaction. Again, not every silver powder can be used. Surface morphology plays an important role and some degree of UV reflectivity could be achieved with the selected Ag material. Special attention had to be paid to include various UV initiators. They are essential for propagation of the reaction into the resin matrix shadowed by the filler particles. With the present adhesive, curing of layers up to 30 μm thick was achieved successfully. (This fact is noteworthy, as to the best of our knowledge, there are no commercially available conductive adhesives based on radical UV curing.)

The second, thermal curing step is accelerated by imidazoles. A specific electrical resistance of 10^{-4} Ωcm and bond strength of 10-15 N were achieved with SMD-resistors (case size 1206) on copper.

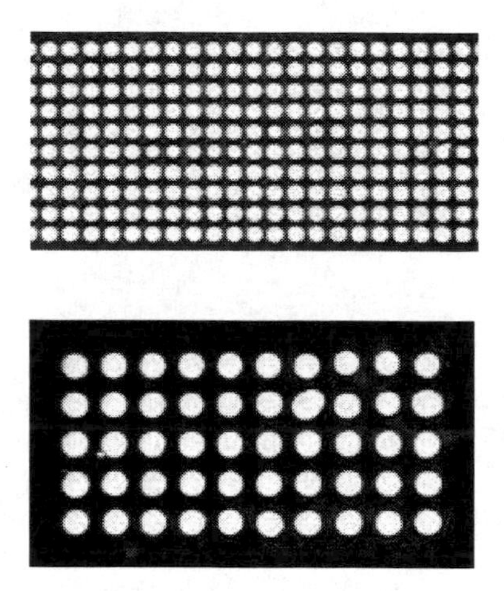

Figure 2 & 3: Array of ICA dots deposited by inkjet, pitch size 200 μm.

4 JETTING OF THE CONDUCTIVE ADHESIVE

Inkjet dispensing was performed using an innovative design based on a glass capillary nozzle, surrounded by a piezo-actuator.

5 POSITIONING SYSTEM

The experiments were conducted with the AutoDrop positioning system MD-P-801. This positioning system is designed for high accuracy with 1 μm repetition accuracy. The necessary stiffness is achieved by using granite components for the base plate as well as for the bridge which carries the z-axis. The substrate is fixed on

Figure 4 : AutoDrop positioning system

the x-y-table and is moved relative to the dispenser heads, while the z-axis which moves the dispenser heads up and down is fixed. This design minimizes the moved mass and improves positioning accuracy considerably. It is possible to move and control up to 8 dispenser heads or pipettes in parallel. Dosing can be done in "start-stop"-mode or in-flight, i.e. while the head is moving.

6 WORKING PRINCIPLE OF THE MICRODROP SYSTEM

The microdrop technology (Dispenser Patent n° DE 10153708) uses the same principle as that of the ink jet printing. However, the dispensing system developed by microdrop GmbH differs fundamentally. The core of the microdrop dispensing head consists of a glass capillary which is surrounded by a tubular piezo actuator. At one end, the capillary is formed to a nozzle (diameter 30 to 100 µm). Applying a rectangular voltage pulse, the piezo actuator contracts and creates a pressure wave which propagates through the glass into the liquid. At the nozzle the pressure wave is transformed into a motion of the liquid which is accelerated with up to 100 000 g. A small liquid column leaves the nozzle, breaks off and forms a droplet which flies freely through the air. Droplet size is mainly determined by the nozzle diameter. Volumes of 30 up to 500 picoliters can be dispensed. Ejection frequency can be varied between 0 and 2 kHz. Volume variation is approx. 1 %. Nozzle diameter can be produced with high accuracy (±1 µm) depending on the desired drop volume.

A stroboscopic video camera monitors the process of droplet formation. Varying the time delay between the signals to the piezo actuator and the strobe allow the operator to monitor the entire process of droplet formation.

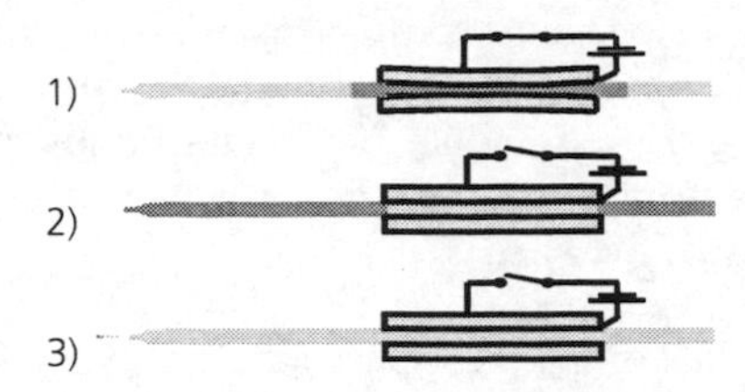

Figure 5 : 1) Voltage is applied to the piezo actuator; contraction of the actuator induces a pressure wave.

2) Voltage is switched off and the piezo actuator relaxes to normal state, the pressure wave propagates through the capillary to the nozzle.

3) The accelerated fluid jet at the nozzle outlet forms a drop which freely flies away from the nozzle.

7 INKJETTING TEST RESULTS

The samples of the dot array (ref. Fig. 2 and 3) were produced on the positioning system. It is the platform for dispensing the ICA on different microelectronic devices.

A special nozzle with a reduced flow resistance for fluids with a viscosity up to 200 mPas at room temperature was used. This innovative design prevents cavitation and agglomeration of particles which occurred with standard nozzles. Cavitation and agglomeration of particles destroy the jet and create a multidirectional spray.

Figure 6: Jetting of a transparent fluid (ethylene glycol)

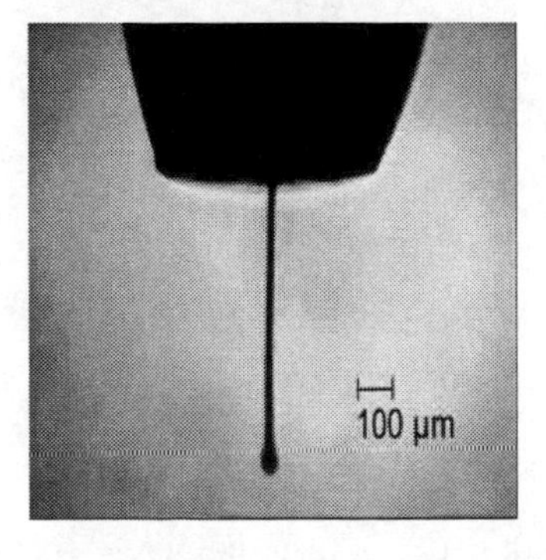

Figure 7: Jetting of ICA with a 75 µm nozzle

Successful dispensing of ICA is shown in Fig. 7. Droplet size is nearly 500 pl with the 75 μm diameter nozzle creating 130 μm dots on a glass substrate. The volume results from the size of the jet, which is dependent on the viscosity of the ICA.

The viscosities of the presented adhesives vary from (up to) 1 Pas at very low shear rate to 20 – 40 mPas at high shear rate. The shear rate occurring in the nozzle during jetting is estimated to reach 2500 s-1. Under these drastic conditions, silver particles tend to agglomerate. As mentioned above - silver being a ductile metal with high tendency of cold welding - its surface has to be coated for mechanical protection. However this protection does not remain functional in the cured material, otherwise it would prevent electrical conductivity. In the jetting conditions, care has to be taken to avoid separation (sedimentation) of the silver particles from the resin matrix. This is not an easy task as specific densities of the ink components are quite different: just above 1 g/cm^3 for the organic matrix and >10 g/cm^3 for silver!

8 OUTLOOK

Future work will be done with a 50 μm nozzle. This will reduce the dot diameter. The current objective is to generate 80 μm dots.

Additionally a new dispenser system (Dispenser Patent n° DE 10153708) is under development. It is characterized by a reduced dead volume and integrates a fluid circulation device. With this system the sedimentation of particles in the fluid can be prevented. First tests with ethylene glycol and fluids with low particle content have been performed already. They show a similar performance as the system described above. The optimization is still ongoing.

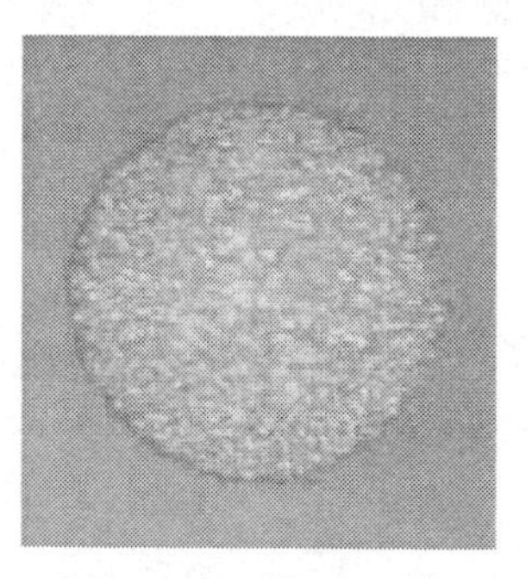

Figure 8: Single ICA dot deposited by inkjet

9 CONCLUSIONS

Arrays of ICA dots of 130 μm diameter with a pitch of 200 μm have been produced reproducibly. Specific electrical resistance of 10^{-4} Ωcm and bond strengths of 10-15 N were achieved with SMD-resistors (case size 1206) on copper substrate.

This joint development of resin system, silver particles and inkjet device has opened the door to apply the technology for electrically conductive joining in the microsystem and microelectronic fields.

10 ACKNOWLEDGEMENT

The present work was supported by the European Commission (Growth-Program) under G1RD-CT-2002-00656 and by the Swiss Confederation under OFES 01.0575. Thanks also to Dr. C. Dullaghan and K. McNeilly (both Metalor) for their valuable contributions.

Development of a Scaleable Interconnection Technology for Nano Packaging

K.-F. Becker, T. Löher, B. Pahl, O. Wittler, R. Jordan, J. Bauer, R. Aschenbrenner, H. Reichl

BECAP – Berlin Center for Advanced Packaging

Fraunhofer Institute for Reliability and Microintegration – TU Berlin Microperipherics Center

Gustav-Meyer-Allee 25, D-13355 Berlin, Germany

phone: +49-30/464 03 242 fax.: +49-30/464 03 254 e-mail: becker@izm.fhg.de

ABSTRACT:

Microelectronics miniaturization is following Moore's law since the mid sixties and over the years it has always been possible to follow it without meeting fundamental technological limits. This might be in question for future applications, where SIA roadmap shows a red brick wall for the further development of microelectronics without fundamentally new approaches. These new approaches as single atom and CNT based transistors do all target on maximum integration on chip level, leading to increased interconnect density and thus to a miniaturization of the individual contact. Parallel to the miniaturization of the interconnects the development of scaleable interconnect technology is necessary, providing reliable infrastructure for future packaging needs.

At Fraunhofer IZM various approaches towards a scaleable interconnect technology are researched. This paper will describe the development of reactive interconnects, i.e. contacts that need no external energy source, but release the energy for solder interconnect formation by exothermic reaction of a nano-enhanced encapsulant. Potential areas of application are interconnection of thermally sensitive devices as bio sensors or interconnection on low cost substrates, e.g. for the smart card applications or the expanding RF ID tag market. The current status of technological developments for the realization of reactive interconnects is described, including interconnect bumping, reactant application and interconnect formation. These technological processes are backed by thermal simulation. Summarized, this paper shows the potential of reactive interconnects being a "drop in" solution that not only allows the cost effective packaging of today's µscale ICs but also tomorrows nano-scale devices.

1. Introduction

The consequences of downsizing in assembly and interconnection as well as in packaging technologies are twofold: on one side the I/O density on chips is increased, causing a lateral x-y reduction in pitch and pad size of the interconnections, the driving force here is the increase of functionality per area due to progress in chip fabrication. Size reduction in the third dimension, z, is a consequence of reduced overall system thickness after assembly: with chips 10-20 μm thick and substrates in the same range the gap between substrate and chip should be kept as small as possible, i.e. in a range well below 10 μm, reaching below the limits of today's μPackaging into the nano range.

The shortest and thinnest interconnections between chips can be realized using flip chip technology. Here a chip is directly connected to the substrate by a solder bump. Depending on the technology pitches down to 40 μm (electroplating) and solder bumps with a height of 10 – 40 μm can be fabricated. For cost driven applications pitches of 80 μm are state of the art (stencil printing), bump height are around 30 to 40 μm [1]. Such solder bump heights result in a large gap between the chip and the substrate.

Considering the size reduction as well as the most cost effective process path the under bump metallization using the electroless deposition of Ni. Formation of solder caps by immersion solder bumping and subsequent interconnection by thermode bonding is a highly promising process chain for large scale low cost ultra-thin electronic systems leading the way towards pitches below 20 μm.

If this bumping process is combined with a low temperature melting solder, similar to woods metal, with a resulting melting point well below 100 °C, low temperature joining becomes possible allowing the use of low cost flex substrates with glass temperatures just above 100 °C. Typical joining process would be reflow soldering, that allows the use of existing equipment and of process know how.

To drive low temperature joining one step further it is possible to use reactive interconnects or reactive materials, resp., that provide a sufficient amount of energy at the solder bumps and allow solder melting and interconnect formation. This might not only be useful for the assembly on low cost substrate but can also be an interconnection technology for temperature sensitive ICs, e.g. biosensors or the like.

The experiments described are part of the feasibility investigations of low temperature joining and reactive interconnect processes.

IC preparation

For experiments 5x5 mm² dies with 45 μm diameter and 100 μm pitch with Al bond pads (Al0.7Si0.5Cu) and glass passivation over the active chip area.

Electroless Ni Deposition

After an initial wafer cleaning the electroless Ni deposition sequence starts with a zincation process of the Al. The surface is roughened (Al loss 200 -300 nm) and covered with Zn. Subsequently in the Ni bath zinc is exchanged by nickel. After the initial exchange Ni deposits through an autocatalytic process. Due to Ni-bath chemistry phosphorous is co-deposited along with the Ni. In the present case the phosphorous content is around 10 % by weight. The nickel thickness can be determined with an accuracy of 0.3 μm by adjusting the dwell time in the Ni bath. The Ni-surface is rough and exhibits spherical protrusions with heights in the range of 0.1 μm. The protrusions exceed this value if bath parameters are not appropriately adjusted. As a final step the Ni-surface is chemically coated with a 80 nm Au layer in order to prevent Ni oxidation.

Immersion Solder Bumping

Immersion solder bumping has turned out to give the highest yield solder temperatures at about 60 to 80 °C above the melting point of the respective solder [2, 3]. The liquid solder is capped with a thick layer of glycerol. A schematic graph of the soldering apparatus is depicted in Figure 1. Beside serving as a protection against contamination of the liquid solder it is used a heat reservoir to adjust the wafer temperature to the solder temperature. Glycerol also serves as a soft fluxing agent. Immersion speed into the solder is in the range of 3 to 10 mm/s. Hold time in the solder is 5 to 20 seconds. The pulling speed out of the solder is 10 to 20 mm/s.

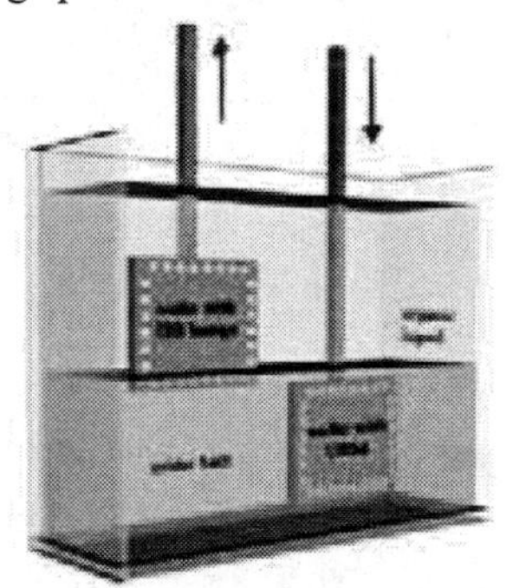

Figure 1: Schematic drawing of the immersion soldering.

The solder attaches to the wettable Ni (Au) surfaces on the and forms a cap shaped solder deposit. The outcome of the soldering process is investigated mainly by profilometer scanning, SEM cross sectioning. The morphology of the solder cap depends on the size and shape of the contact pad and also on the intermetallic phases that may be formed during the wetting process.

For chip bumping an InBiSn solder with a composition of In 51,0 - Bi 32,5 - Sn 16,5 wt%. This low temperature solder has a melting point T_m of 60,5 °C, a density of 7,5 g/cm^3 and a thermal conductivity of λ (@ 53 °C) of 17,8 W/m K. An example for such a solder bump is depicted in Figure 2, EDX-scans illustrate the composition of the major inter metallic compounds, the InBi-rich phase and the SnIn-rich phase

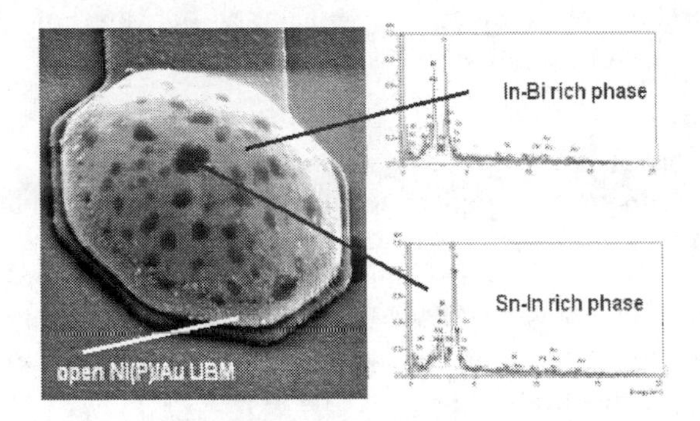

Figure 2: SEM micrograph of an immersion bump of InBiSn on a 40 μm Al pad

From related experiments, it was found that for solder caps in a profilometer scan line over 30 μm pitch solder bumps the variation in cap heights among caps is rather large, i.e. in the order of ± 30 %. Moreover, outliers with solder caps much higher than average are found. The large spread in cap heights seems to be detrimental on first sight. However, in the subsequent thermode bonding process most of the solder will be squeezed from the middle to the edges of the contact pad until a homogeneous solder thickness between the contact pad is reached.

Thermode Bonding

For first experiments concerning the joining behavior a thermode bonding process joining chip on chip was used. Thermode bonding in itself is not a cheap process, however, high accuracy placement and the option to deliberately set force and temperature gradients during the bonding process is not possible with other techniques. In large scale production the cost can finally be reduced by appropriate process optimization.

The thermode bonding technology is based on very fast reflow soldering by pulse heating. We used the flip chip bonder FC150 from SUSS MicroTec. The fast process allows the use of low cost materials with low temperature resistance for flip chip soldering at high temperatures without damage of a flex or other sensitive materials. In our processes no underfiller was used, as focus was the determination of contact . A low temperature flux was used.

Examples and Reliability Results of Thermode Bonded Systems

During the immersion soldering process the Ni-UBM is exposed to the liquid solder for 30 to 60 seconds. For some investigated solder alloys a considerable formation of intermetallics has been observed. Solders as SnPb, SnBi, InBiSn display only moderate intermetallics formation.

A typical cross section of an assembly at 100 μm pitch is shown in Figure 3. The left image shows a joint with optimum contact, while on the right side some oxide entrapment was visible. Reflow had taken place at 120 °C, under inert atmosphere. Joint thickness of these first samples is in the range of 20 μm including the Ni bumps. Figure 4 shows a row of solder interconnections generated by reflow with and additional scrubbing, i.e. a relative movement of the top IC versus the bottom one. This was introduced to minimize bond thickness and to get rid of the oxide entrapments.

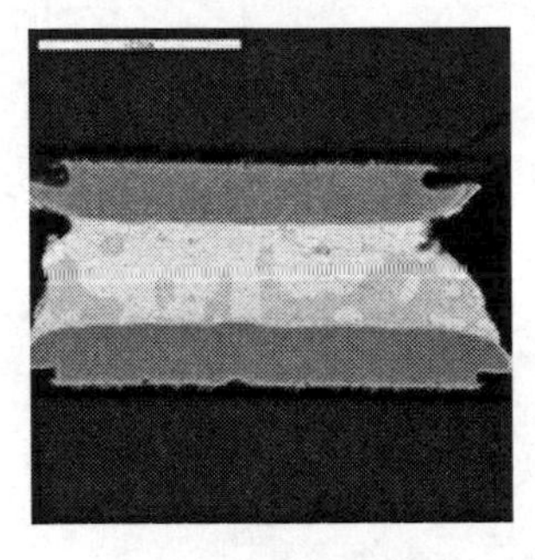

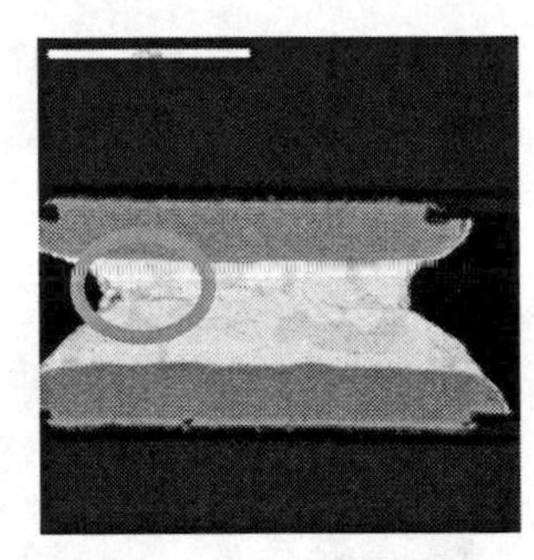

Figure 3: InBiSn (Tm=62 °C) placing and reflow at 120 °C. Close contact (l.) and contact with a thin oxide layer (r)

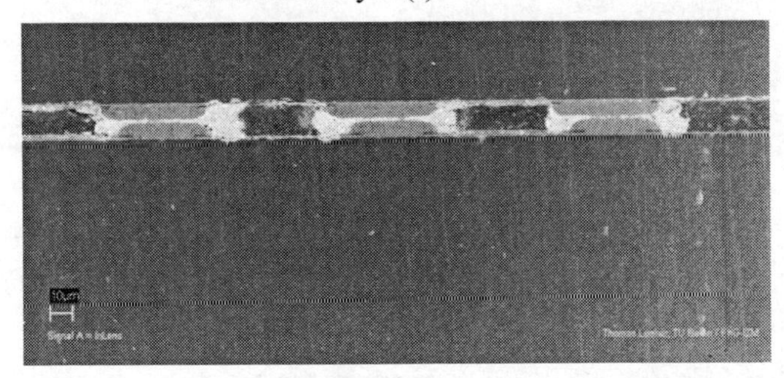

Figure 4: InBiSn (T_m=62 °C) interconnection with Ni(Au)-UBM pad size 45 μm, pitch 100 μm generated in scrubbing mode

These experiments showed that a successful formation of interconnects is possible with the low T_m solder. Further work was done to evaluate the potential of reactive systems to introduce the energy necessary for melting the solder.

Reactive System

To identify reactive material with the technical requirements:

- high thermal energy production in small space
- specific amount of heat
- controlled reaction rate
- spontaneous thus controlled ignition

• no or easy to clean reaction products

A potential system would most likely be a two component system, that reacts exothermically with low gas production. The system should have a low activation energy and should be moderate kinetically inhibited, to prevent overheating.

A first thermal simulation using a 2D model was performed to yield proof of concept. Boundary conditions were:

- No heat transfer to environment
- No melting energy
- Initial temperature is 25 °C (298 K)
- All Interconnects show ideal thermal conductivity

Aim was to very bump and chip temperature for different amounts of energy. With the boundary conditions depicted in Figure 5, it was determined, that a short heat pulse of 100 ms with an energy of 5 mJ should be sufficient for solder melting and interconnect formation.

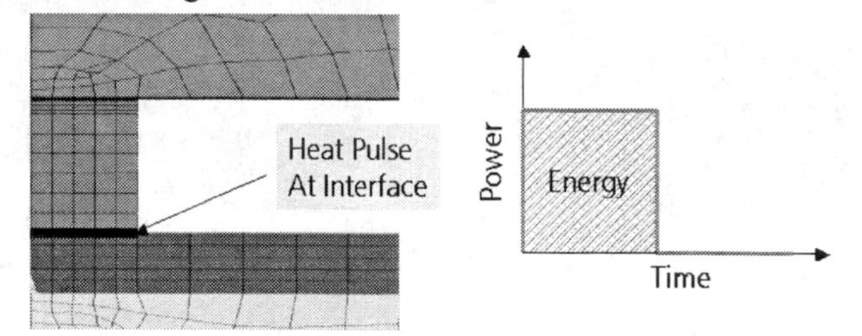

Figure 5: Boundary conditions of 2D thermal modeling

With the results of thermal simulation a market research was done and a system was identified, that matched the demands and was available for evaluation. This system consisted of commercially available substances, cyanate and amine, that allowed easy handling of liquid monomers and that showed quick polymerization after mixing. A highly exothermal reaction enthalpy was found with the system.

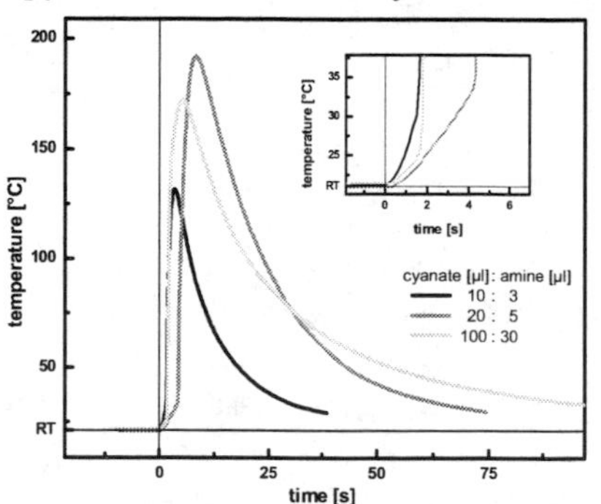

Figure 6: Temperature pattern during polymerization

First soldering evaluation was performed with this system and it was found that solder bumps could be remolten using this reactive system. To this purpose a mixture of component A and component B was created by jetting the components on a solder depot. After an autocatalytic phase of 1 to 5 s within 1 s peak temperatures between 120 °C and 190 °C are reached (Figure 6), where it is assumed that T-variations are due to the complex reaction mechanism. Nevertheless, solder T_m is reached in any case, the temperatures are reached for mixing ratios of 2:1 to 6:1 (cyanate:amine) and a total volume of 12,5 µl to 130 µl.Future work will deal with the transfer of this first laboratory results to actual test assemblies to evaluate the reliability potential of the process and technology.

Conclusions

We have shown that a combination of electroless Ni(P) UBM, immersion solder bumping and thermode bonding using a low temperature solder is a potential alternative for the assembly on temperature sensitive substrates or on temperature sensitive devices as biosensors. Additionally a feasibility evaluation of reactive interconnect technology has been performed that yielded promising results. We see this technology as one building block of packaging technology for miniaturized systems developing from a micro to nano scale. A large contribution of micro and nano technology is seen for the further evolution of packaging technology; actually, most of the obstacles predicted by state-of-the-art roadmaps for packaging technology can only be solved using nano technological means. Amongst all semiconductor-based research, the important contribution of packaging to a successful system in package realization should not be underestimated.

References

1. D. Manessis, R.P., A. Ostmann, R. Aschenbrenner, and H. Reichl. *Stencil Printing for 100 µm Flip Chip Bumping*. Proceedings of *IMAPS International*. 2003. Boston: IMAPS.
2. M. J. Brady, A.D., *Flip Chip bonding with solder dipping*. Rev. Sci. Instrum. 56, 1985. **7**: p. 1459-1460.
3. K. L. Lin, J.W.C., *Wave Soldering Bumping Process Incorporating Electroless Nickel UBM.* IEEE Transactions on Components and Packaging Technology, 2000. **23**(1): p. 143.
4. K. Zeng, K.N.T., *Six cases of reliability study of Pb-free solder joints in electronic packaging technology.* Mat. Sci. and Eng., 2002. **R 38**: p. 55 - 105.

Investigation of Patterned Media Concept for Very High Magnetic Storage Density

B. D. Horton[*] and J. L. Streator[**]

[*]Caterpillar, Inc., Cary, NC 27511, horton_brian@cat.com
[**]G. W. Woodruff School of Mechanical Engineering, Georgia Institute of Technology
Atlanta, GA 30332-0405, jeffrey.streator@me.gatech.edu

ABSTRACT

A patterned media concept for very high density magnetic recording is investigated. Extremely dense patterns, corresponding to 700 Gbit/in^2, are created via nanometer scale self assembled thin film lithography with wet chemical etching. Lower density patterns are created with focused ion beam etching. Strain gage, capacitance and acoustic emission sensors are used to characterize normal load, tangential load, and interface contact. It is demonstrated that loss of hydrodynamic lubrication is small for small pattern regions with high conserved surface area ratio. Flyability results suggest that patterned media look promising for magnetic recording applications.

INTRODUCTION

Since the first hard disk device, the IBM RAMAC, the density of magnetic storage has grown 25% or more per year. Recently, that growth rate has increased to 100% per year [1]. At this rate, the physical limit of areal density in longitudinal media, known as the superparamagnetic limit, will be reached in the near future. Superparamagnetism becomes important when the stored energy per magnetic grain starts competing with thermal energy [2]. In this condition, magnetic domains representing bits of data can spontaneously change polarity causing loss of stored information. An upper limit to densities achievable in longitudinal media has been estimated at 36 Gbit/in^2 [3].

Patterned magnetic storage media separates the bit domains so that exchange coupling can be limited and improve thermal stability. Yet, it is not known if current technology slider heads can generate sufficient hydrodynamic pressure fields to avoid head to disk contact, or "fly" over the media. Therefore, the focus of this work is to experimentally characterize the flyability of current production slider heads over patterned media.

Keywords: magnetic storage, patterned media, thin film self assembly

1 TEST APPARATUS

A test stand was designed and built to test the flyability of current generation sliders over patterned media. The test stand measures friction, acoustic emission, and head to disk interface capacitance. In doing so it has three concurrent measurements to determine if the slider head is flying over the disk, sliding in contact with the disk, or some combination of the two. In addition the apparatus measures the normal load on the slider head and rotation rate of the disk.

Eight individual strain gages, (Vishay Micro Measurements, Raleigh, NC) were attached to a dual cantilever beam and wired into two sets of four active arm Wheatstone bridge configurations. The dual cantilever beam allows measurements of both friction and normal load acting on the slider head. Direct physical contact between the slider and the disk is associated with high friction as compared to when the slider is flying.

A second type of flyability sensor involved the measurement of electrical capacitance. For this purpose the slider-disk interface was made to be a part of an electrical circuit, the slider and disk each serving as capacitor plates. Since, with parallel plates, the capacitance is inversely proportional to the spacing, interface capacitance is indicative of fly height. A circuit model outlined by Streator et al. [4] was used to find the capacitance from the degree of voltage attenuation in voltage divider circuit.

In conjunction with friction and capacitance transducers, acoustic emission sensors (Physical Acoustics, Princeton Jct., NJ) were also employed. These sensors were used to qualitatively compare the intensity of vibrations at the head to disk interface for different testing conditions. The chosen sensor exhibited good response at frequencies from 100 kHz to 800 kHz.

An optical reflectivity sensor focused on a reference mark on the disk surface provided a means to detect the period of disk rotation as well as determine the angular position of the disk corresponding to a particular feature in a recorded signal (i.e., friction, capacitance, or acoustic emission). The ability to identify the disk location in question was particularly helpful for experiments involving focus ion beam etch patterns that covered only a small portion of the disk.

2 EXPERIMENTAL METHODOLOGY

Experimental specimens were prepared using two methods. Focused ion beam (FIB) milling was used on 1" diameter commercially available magnetic disks, while thin film self-assembly was used to pattern on silicon wafer substrates.

2.1 Focused Ion Beam Patterns

Two different FIB patterns were designed. A low density, relatively large pattern area was made such that the entire slider would be over the patterned area at one time. The low density pattern measured 1.23 mm by 1.82 mm. Comparatively, the slider had a footprint of 1.0 mm by 1.3 mm. A section of a representative low-density pattern is illustrated in Figure 1. In this figure, one observes squares delineated by etched grooves. Each square represents a single recording domain or bit. As observed there are approximately 16 bits in the 100 μm^2 area, which corresponds to a 100 $Mbit/in^2$. A second pattern was made at the highest density capability of the FIB technique. The high density pattern corresponds to a density of 10 $Gbit/in^2$, and covers a disk region that is 1.1 mm by 0.064 mm.

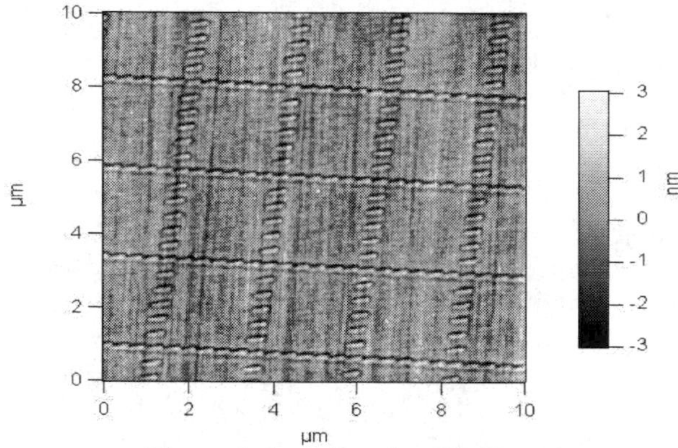

Figure 1: Low Density FIB Sample

2.2 Thin Film Self-Assembly Patterns

The thin film self-assembly process, which is based largely on previous work by others [5, 6], is described in detail in Figure 2. The RSMMA listed in the figure denotes a random copolymer while the SMMA denotes a diblock copolymer (Polymer Source, Inc., Montreal, Canada). After the last step in Figure 2 (Ti sputter), the pattern is ready to be used as a test specimen.

An example of the finished product is depicted in Figure 3. In addition to the smaller, dark circular domains, which comprise the nominal high density pattern, there are a number of defects, where the holes have merges to form serpentine grooves. The nominal density is 700 $Gbits/in^2$

2.3 Flyability Measurements

Post processing was performed on data from the three sensors to determine flyability. Friction measurements were averaged over the sample period. A technique outlined by Streator et al. [7] was used to mathematically remove cantilever beam vibration effects from the friction data.

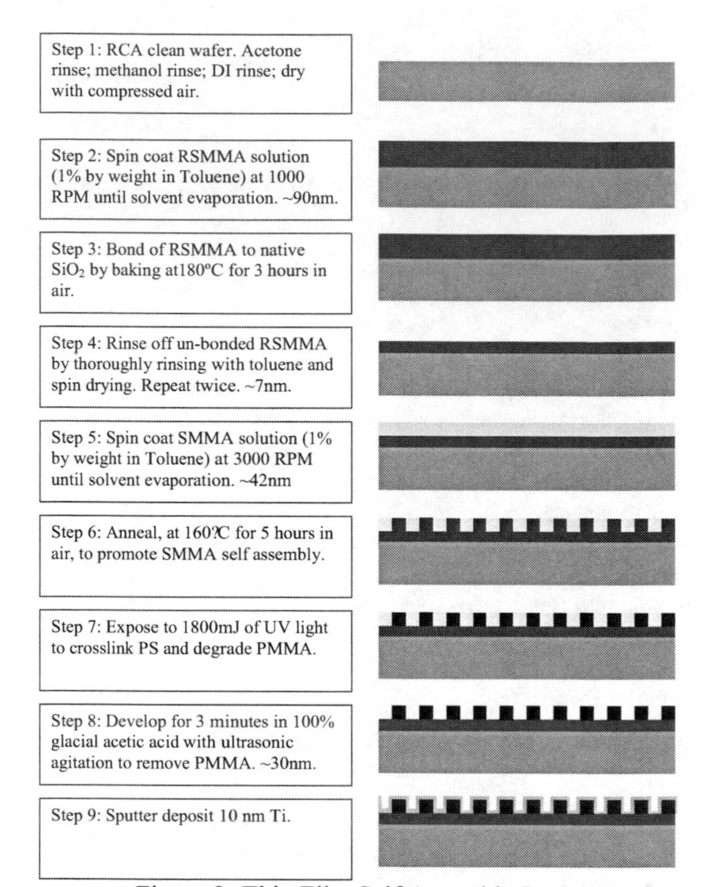

Step 1: RCA clean wafer. Acetone rinse; methanol rinse; DI rinse; dry with compressed air.
Step 2: Spin coat RSMMA solution (1% by weight in Toluene) at 1000 RPM until solvent evaporation. ~90nm.
Step 3: Bond of RSMMA to native SiO2 by baking at180°C for 3 hours in air.
Step 4: Rinse off un-bonded RSMMA by thoroughly rinsing with toluene and spin drying. Repeat twice. ~7nm.
Step 5: Spin coat SMMA solution (1% by weight in Toluene) at 3000 RPM until solvent evaporation. ~42nm
Step 6: Anneal, at 160?C for 5 hours in air, to promote SMMA self assembly.
Step 7: Expose to 1800mJ of UV light to crosslink PS and degrade PMMA.
Step 8: Develop for 3 minutes in 100% glacial acetic acid with ultrasonic agitation to remove PMMA. ~30nm.
Step 9: Sputter deposit 10 nm Ti.

Figure 2: Thin Film Self Assembly Process

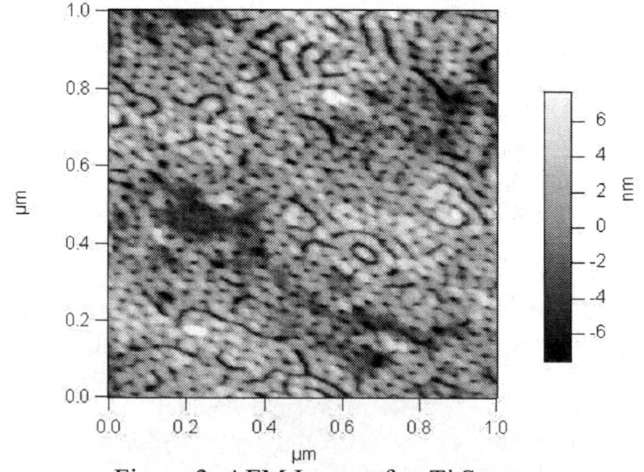

Figure 3: AFM Image after Ti Sputter

Capacitance calculations start with FFT of 10 data point windows calculated for both input and output data. From the FFTs, the maximum amplitude signal is chosen. Next, the input voltage amplitude is divided by its corresponding output voltage amplitude. This voltage ratio is used to calculate slider head to disk interface capacitance that is summarized, like friction, by taking the average.

The output of the acoustic emission sensor is left as voltage and is evaluated only relative to baseline data. Root mean square (RMS) AE voltage signal is plotted.

RESULTS AND DISCUSSION

Baseline data, Figures 4, 5 and 6, demonstrate slider lift off as speed increases. Figure 4 shows friction as a function of sliding speed for two data series. The first is no-load data taken with the slider head out of contact with the disk. Compared to that result is loaded data where the slider has been loaded to full design load, 2.5 g. The friction at speeds lower than ~7 m/s is significantly higher that above 7 m/s. This difference in friction measurements is interpreted as the difference between not flying and flying. The friction is constant from 7-11 m/s then increases slightly but steadily until 31 m/s, which is like due to viscous effects from the air film. This result is consistent with information on take off speed provided by the head manufacturer (Western Digital, Lake Forest, CA).

Figure 5 shows capacitance as a function of sliding speed. Since capacitance can be considered as inversely related to fly height, one arrives at the conclusion that the loaded capacitance data series in Figure 5 shows take off speed at the same point as indicated by friction, ~7 m/s. Past 7 m/s capacitance decreases slightly to a minimum at 31 m/s.

Figure 6 shows acoustic emission as a function of sliding speed for no load and loaded conditions. RMS acoustic emission is considered an indicator of the intensity and frequency of head to disk intermittent contact. The plateau observed at around 7 m/s is consistent with the apparent take off speed suggested by the friction and capacitance records.

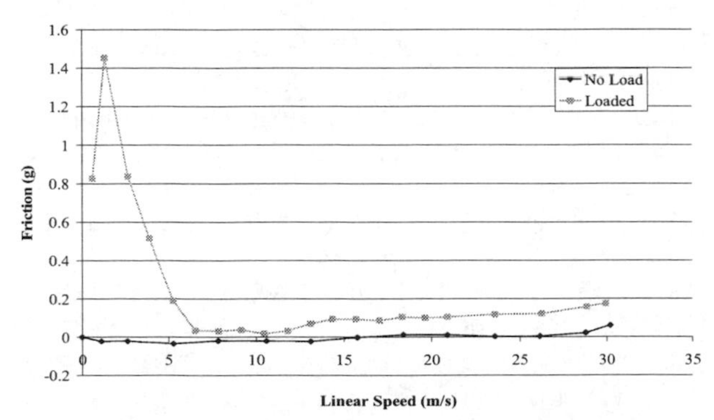

Figure 4: 4" Disk Baseline Data – Friction

Figures 7, 8 and 9 characterize the high density FIB results. Results are shown for both patterned and non-patterned samples. The data correspond to localized regions of 1 mm length containing the pattern. In Figure 7, the patterned curve generally looks similar to but less smooth than the non-patterned line. Additionally, the point where friction appears to show lift off is slightly delayed.

High density FIB capacitance results, presented in Figure 8, show a dip at 5 m/s which corresponds to friction and AE decreases and could be considered lift off.

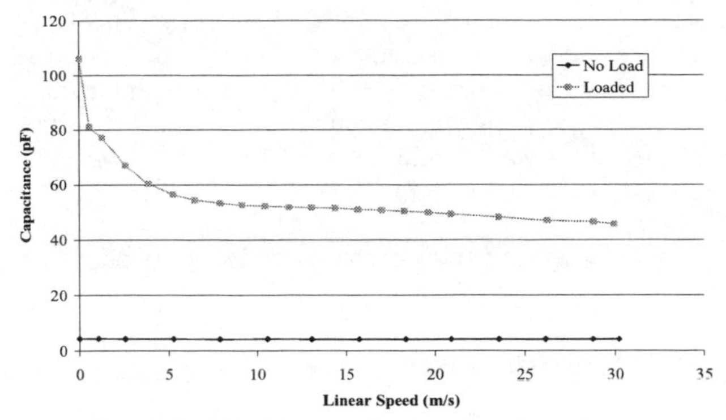

Figure 5: 4" Disk Baseline Data – Capacitance

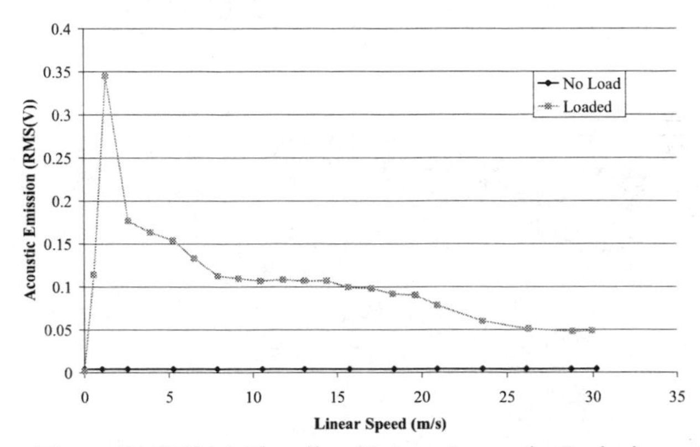

Figure 6: 4" Disk Baseline Data – Acoustic Emission

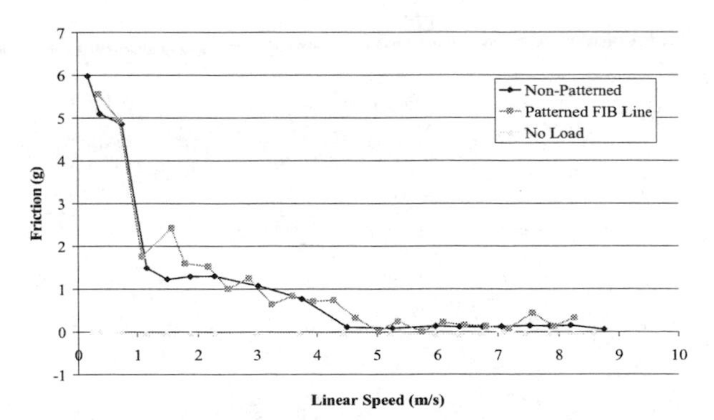

Figure 7: FIB High Density Results – Friction

In Figure 9, high density FIB regional acoustic emission results, deviation from the non-patterned AE data starts at 1.8 m/s and continues until apparent lift off at 4.6 m/s, which is delayed slightly compared to the non-patterned media.

Figure 10 and Figure 11 present thin film pattern data. Half-load patterned data is plotted against full-load non-patterned data and no-load data. Full-load data does not indicate lift off. Yet friction, in Figure 10, for the slider at half design load appears to have a lift-off profile. The significant drop in friction is complete by 16 m/s. The corresponding acoustic emission data, however, does not show the same trend.

One key difference between the FIB samples, that support stable flyability, and the thin film patterned sample, that does not clearly support flyability, is the difference in the fraction of non-etched media area. This can be termed the conserved surface area ratio (CSAR). If it is assumed that areas that are etched support little hydrodynamic pressure then the CSAR could be predictor of how well a particular patterned sample will allow a slider head to fly.

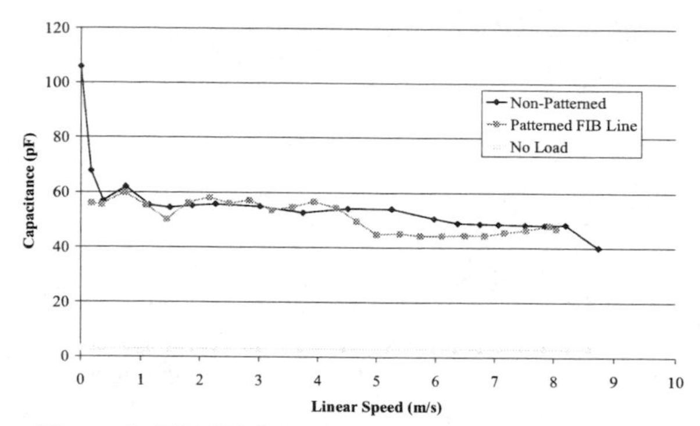

Figure 8: FIB High Density Results – Capacitance

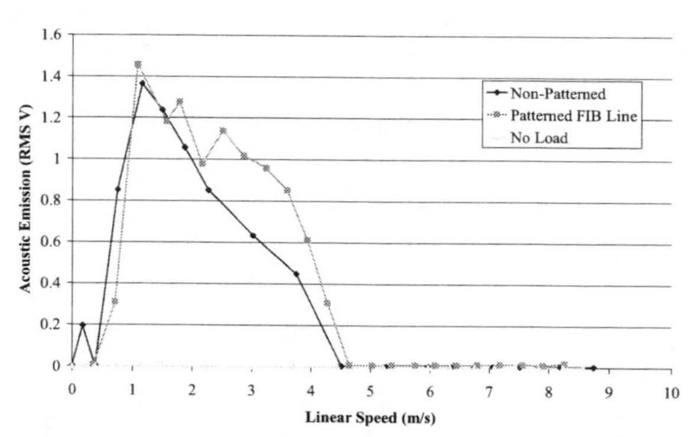

Figure 9: FIB High Density Results – Acoustic Emission

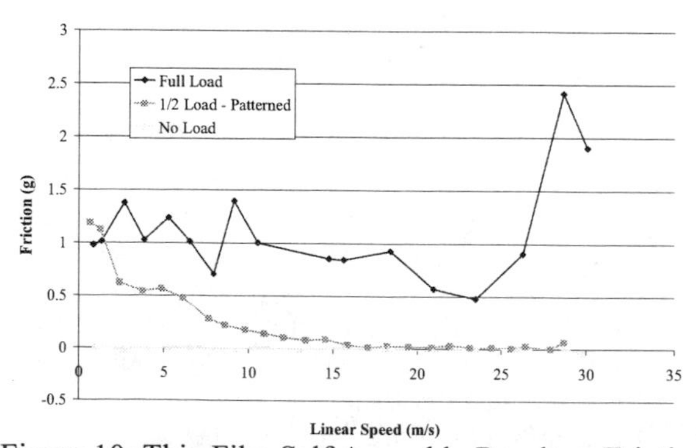

Figure 10: Thin Film Self Assembly Results – Friction

Considering CSARs for the three types of samples, it is more easily understood why the FIB samples provide better support for hydrodynamic lubrication. The conserved area ratios for the three types of patterned samples are 0.69, 0.5 and 0.15 for the FIB low density, FIB high density, and thin film pattern media, respectively.

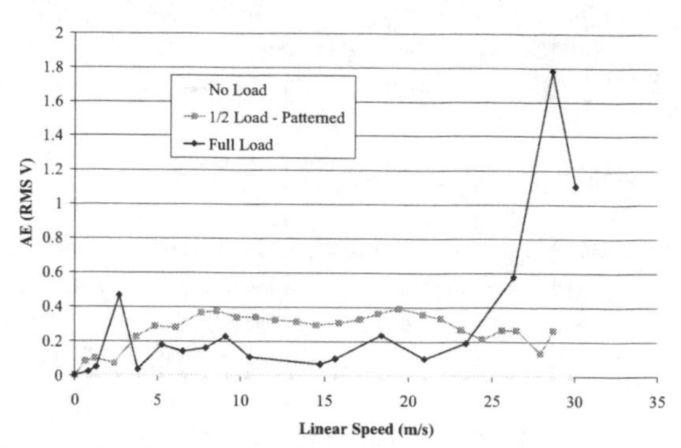

Figure 11: Thin Film Self Assembly Results – Acoustic Emission

SUMMARY

The ability to achieve lift off was not affected by the patterned area on either FIB sample. However, the thin film self assembled sample was not able to achieve lift off with full slider design load. However, at half design load the friction sensor appears to lift off by 16 m/s, which is higher than what was found with the baseline data a full load. A general predictor of a patterned media's ability to support hydrodynamic pressure is introduced as conserved surface area ratio (CSAR). High density patterned media look promising but may require a lower slider load.

REFERENCES

[1] A. Moser, K. Takano, D.T. Margulies, M. Albrecht, Y. Sonobe, Y. Ikeda, S. Sun, E.E. Fullerton, Journal of Physics D: Applied Physics, v. 35, pp. R157-R167, 2002.

[2] D. Weller, A. Moser, IEEE Transactions on Magnetics, v. 35 n. 6, pp. 4423-4439, 1999.

[3] S.H. Charap, P.-L. Lu, Y. He, IEEE Transactions on Magnetics, v. 33 n. 1, pp. 978-982, 1997.

[4] J.L. Streator, J. Huang, J. Zheng, Thinning Films and Tribological Interfaces, D. Dowson et al. (eds.) pp. 285-291, 2000.

[5] K.W. Guarini, C.T. Black, S.H.I. Yeung, Advanced Materials, v.14 n. 18, pp.1290-1294, 2002.

[6] K.W. Guarini, C.T. Black, K.R. Millkove, R.L. Sandstrom, Journal of Vacuum Science Technology, v. 19 n. 6, pp. 2784-2788, 2001.

[7] J.L. Streator, Journal of Tribology, v. 114 n. 2, pp. 360-369, 1992.

ACKNOWLEDGEMENT

The authors would like to thank student Christian Freyman at Northwestern University for assistance with the FIB sample, Prof. Bill King and student Harry Rowland (at Georgia Tech) for help with the self-assembly technique and AFM measurements. This work was supported by NSF Grant # 0205869.

Wafer Scale Fabrication of Nano Probes for Atomic Force Microscopy

Qi Laura Ye*[,1,2], Hongbing Liu[2], Alan M. Cassell[1,2], Kuo-Jen Chao[3], and Jie Han[1,2]

[1] Center For Nanotechnology, NASA Ames Research Center, Moffett Field, California 94035, USA
[2] Integrated Nanosystems, Inc., NASA Research Park, Moffett Field, California 94035, USA
[3] Charles Evans & Associates, 810 Kifer Road, Sunnyvale, CA 94086, USA
* Corresponding author: Phone: (650) 604-0497, Fax: (650) 604-0987, E-mail: qye@mail.arc.nasa.gov

ABSTRACT

We have developed an innovative wafer scale fabrication method for making carbon nanotube (CNT) probes for atomic force microscopy imaging. Our method combines nanopatterning and nanomaterials synthesis with traditional silicon micromachining technologies. Our fabrication process has produced 244 CNT nano probes per 4-inch wafer with control over the CNT location, diameter, length, orientation, and crystalline morphology. CNT nano probes with diameters of 40-80 nm and lengths of 2-6 μm are found to be functional nano probes with no need for shortening. This reliable and true bottom-up wafer scale integration and fabrication process provides a new class of high performance nanoprobes. AFM imaging results show that the CNT probes are strong, wear-resistant, and capable of high resolution AFM imaging.

Keywords: wafer scale fabration, carbon nanotube AFM probes, nanopatterning, nanomaterials synthesis

1 INTRODUCTION

Carbon nanotubes (CNTs) possess remarkable electrical, mechanical, and thermal properties [1]. The idea of using carbon nanotubes as nano probes in scanning probe microscopy was first introduced by Dai et al. in 1996 [2]. The intrinsic nanometer scale diameter, high aspect ratio, and strong mechanical robustness of CNTs make them ideal for high lateral resolution imaging [3,4] and deep trench/via critical dimension imaging [5] in semiconductor in-line processing applications. CNT probes are also highly desired in biological and chemical applications [6-8] where gentle probe-sample interactions are required. When CNT probes approach the sample surface during AFM tapping mode imaging, the CNT buckles elastically which restricts the maximum force that can be applied to soft samples. CNT probes can also be functionalized at the tube open ends [9,10]. They can be made into multi-purpose nano probes by imaging and sensing at the same time, and probing and manipulating materials at the same time. Therefore, CNT probes may be the ultimate probing tips for AFM.

The unique advantages of CNT probes have attracted many efforts to fabricate CNT based imaging probes. Most reported work so far has relied on a "pick and stick" approach to manually attach multi-wall CNTs on silicon pyramid tips under an optical microscope [2-8]. To improve this approach, an electric field has been applied to move multi-wall CNTs on a conducting film to a silicon pyramid tip [11]. However, these approaches are tedious hit-or-miss manual assembly processes, and can only fabricate CNT probes one at a time. Advances have been made by direct catalytic growth of CNTs on silicon pyramid tips using thermal chemical vapor deposition (CVD) methods, either through coating catalysts all over the silicon pyramids or through creating nanopores on top of the flattened silicon pyramid tips [12-14]. Wafer scale production of CNT probes have also been attempted [15,16]. These approaches have scaled up the CNT probe fabrication processes and increased the fabrication throughput. However, thermal CVD growth has little control over the CNT location, density, length, and orientation. It is extremely difficult to obtain individual free standing and well-oriented multi-wall CNTs using thermal CVD. Readily usable probe yields are very low. In addition, these processes all rely on commercially available silicon probes or prefabricated commercial silicon probe wafers. At the end of the fabrication, these processes still require one at a time manipulation approach to remove extra CNTs and/or to shorten the remaining CNTs for probe use. In order to solve the problems that exist with current CNT probe technologies, we have developed an innovative bottom-up wafer scale fabrication method for the reliable mass fabrication of CNT probes through integration of nanopatterning and nanomaterials synthesis with traditional silicon cantilever microfabrication technology.

2 EXPERIMENTS

Our method consists of six major steps: (1) wafer-scale nanopatterning and registration; (2) catalyst deposition and protection; (3) silicon microfabrication of cantilevers; (4) protection scheme release; (5) directional growth of CNTs from silicon cantilevers using PECVD method; and (6) CNT probe characterization, performance evaluation, and testing.

E-beam lithography is used to define the nano-sized catalyst spots on whole wafers. We used a high-speed electron beam writer Hitachi HL-700F e-beam system to pattern catalyst features on a 4-inch Silicon-On-Insulator (SOI) wafer (10 µm Si device layer, 1 µm oxide layer, and 380 µm Si handle layer). To investigate the optimal catalyst feature sizes for achieving single isolated CNT growth per catalyst site, we patterned catalyst dot sizes ranging from 50 nm to 300 nm in diameter with an interval of 50 nm. On a 4-inch SOI wafer, 244 dots were e-beam patterned for 244 cantilever tip locations at the end of the cantilever diving boards. As a control, we also patterned eight 1 µm catalyst sites at certain spots on the wafer. During the same e-beam lithography step, micron scale global and local alignment marks were also patterned over the whole wafer to register these nanometer sized dot arrays. A highly sensitive resist layer composed of poly(methylmethacrylate) (PMMA) (100 nm PMMA layer thickness) was applied to the 4-inch SOI wafer and then exposed at an e-beam dosage of 900 μC/cm^2. The nanometer dot arrays along with the global and local alignment marks could be written in less than 20 minutes. The stitching error across a 4-inch wafer was found to be only 0.5 μm. By writing the alignment marks in the same e-beam lithography step that defined the nano-catalyst spots, we successfully integrated the nano and microfabrication processes.

After nanopatterning and e-beam resist layer developing (1:3 methyl isobutyl ketone : isopropyl alcohol) , a short O_2 plasma descum was performed (150W, 250 mTorr, 10 sec) to clean the wafer prior to metallization. 20 nm Cr or Ti was evaporated on the wafer as barrier layer. 20 nm Ni was deposited on top as CNT growth catalysts by electron beam evaporation. The e-beam resist layer was lifted-off in acetone, leaving the catalyst dots deposited at the defined spots on the wafer. The whole wafer was then thoroughly cleaned, rinsed, and dried for microfabrication.

In order for the catalyst seeds that define the CNT locations to survive harsh dry and wet etch chemicals used in microfabrication, we developed catalyst protection schemes before commencing the conventional microfabrication of silicon cantilevers. PECVD-grown Si_3N_4 protection layer with a thickness of 200 nm was deposited on the front-side for protecting deep reactive ion etching (DRIE). ProTEK chemical series (ProTEK EXP02103-18 etch protectant) from Brewer Science, Inc., was coated on the backside for protecting backside KOH wet etching. 500 Å ProTEK protectant survived 33% KOH etching at 80°C for more than 6 hours.

The cantilevers were fabricated from SOI wafers through a simple two-mask fabrication process. The front-side mask defined the outline of the cantilevers (cantilever body, cantilever beam, and the tabs that held the cantilevers to the silicon wafer frame). The backside mask shaped the body of the cantilevers. Each fabrication step was aligned to the global alignment marks that were initially patterned by e-beam lithography. This insured that the catalyst sites and the final CNTs were located at the exact locations on the cantilever diving boards where they were designed. The process overall layer-by-layer alignment accuracies were found to be within ±1 μm.

We used a home-made 4-inch hot filament direct current PECVD reactor for CNT growth which permitted vast flexibility in process parameter control and monitoring. In addition to normal process variables such as pressure, plasma power, and gas mixture flow and composition, it was possible to also vary the electrode gap, electrode geometry (for angled CNT growth), substrate temperature, and hot filament heating. We conducted detailed CNT growth studies using high throughput combinatorial methodologies [17,18] to achieve individual isolated CNT growth at pre-defined locations on silicon cantilevers. The CNT probes possessed the desired properties for diameter, length, orientation, and crystallinity.

3 RESULTS

Figure 1 shows SEM image of an individual CNT grown from a 200 nm catalyst site on a cantilever beam. Only one isolated individual CNT has been grown from the catalyst site. The individual CNT is found to be 60 nm in probe diameter and 5 μm in length. The tube axis is tilted 13° towards the front of the diving board end with respect to the cantilever beam surface normal. This 13° angle is required because most commercial AFMs are equipped with tip mounting holders that tilt the cantilevers at 13° relative to the image surface and AFM scanning head. This controlled angle growth is only possible in PECVD where an electric field is present in the plasma discharge to direct the nanotubes to grow and align parallel to the electric field.

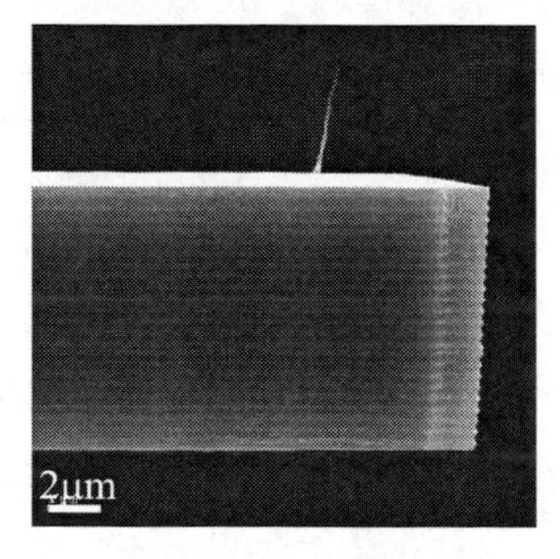

Figure 1. SEM image of CNT AFM cantilever probe with an individual CNT grown from a 200 nm catalyst site on a cantilever beam. Side view with ×3500 magnification. One single isolated CNT of 60 nm in probe diameter and 5 μm in length is grown per site from 200 nm catalyst spot size. The CNT is oriented at a 13° angle with respect to the cantilever beam surface normal.

Among the varied catalyst dot sizes, individual CNTs were obtained only from 50 nm to 200 nm catalyst sizes. 250 nm and 300 nm catalyst sizes gave multiple CNTs. Using the optimized catalyst formulation (20 nm Cr/Ti and 20 nm Ni) and growth conditions, the best CNT growth occurred at a gas mixture of 80 sccm NH_3 and 22.5 sccm C_2H_2 at a chamber pressure of 4 Torr, following a hot filament pretreatment (without plasma) with 80 sccm NH_3 for 10 min. Growth of well-aligned individual CNTs occurred at a bias voltage of -550 V (360 W, 690 mA, 800 Ω). In order to achieve a 3 to 5 μm length of CNT which we found best for AFM imaging, 10 minutes growth time was needed using the above process conditions. CNT growth rate was found to be ~550 nm/min, after 1 min induction/nucleation time.

To assess the crystalline quality of the individual CNTs on cantilevers after growth, high-resolution transmission electron microscopy (TEM) analysis was performed. Figure 2 shows the TEM images of individual CNT with a well aligned and non-entangled multi-wall carbon nanofiber bamboo morphology. The individual CNT probe diameter of 60 nm is confirmed in these TEM images. We can clearly see the multi-wall CNT wall structures and the defected crossover parts. This bamboo morphology is normally achieved in tip growth mechanism at high bias in PECVD. Because of this unique bamboo morphology, our CNT probes are typically much stiffer than pure multi-wall CNTs of the same diameter and length. The clear TEM images in Figure 2 indicate that there was no electron beam induced thermal vibration observed with this 5 μm long CNT. Ni catalyst is wrapped with thin graphite layers at the very end. There is no need to etch away this Ni particle before the CNT probe can be used for AFM imaging.

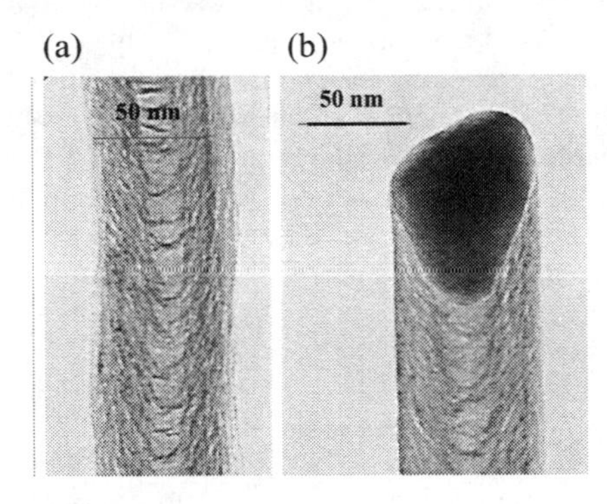

Figure 2. Cross-sectional TEM images of an individual CNT with bamboo morphology. (a) CNT probe body part with ×100K magnification. Multi-wall CNT wall structures with defected crossover parts clearly visible. (b) CNT tip end with ×100K magnification. Ni catalyst is wrapped with thin graphite layers at the very end. The clear TEM images here indicate that there is no electron beam induced thermal vibration observed with this 5 μm long CNT.

The CNT probe evaluation and testing were conducted on commercial AFMs (Digital Instruments Multimode AFM with a Nanoscope IIIA controller and Digital Instruments Dimension 5000 system). Figure 3 shows AFM images of a photoresist patterned 1 μm trench using CNT AFM probe made from this work. Figure 3(a) shows the interaction curve derived from tapping mode imaging (amplitude and deflection signals). It exhibits well-defined amplitude-dampening event resulted from the CNT AFM probe tip pushed towards and extracted from the sample surface. The fluctuation in the amplitude vs. distance signal indicates the presence of a CNT [2]. Figure 3(b) and 3 (c) show 2D and 3D AFM images of this photoresist patterned 1 μm trench using a CNT probe that is 60 nm in tip diameter and 5 μm in length. As seen in Figure 3 (b) and 3(c), the CNT probe is able to fully resolve the sidewalls and the bottom of this 1 μm trench that is 450 nm deep. The bottom of the trench is clearly visible. The trench slope angles were measured to be 80 and 69 degrees through the line cut section analysis, very well matched with the resist etching profile characterized by cross-sectional SEM.

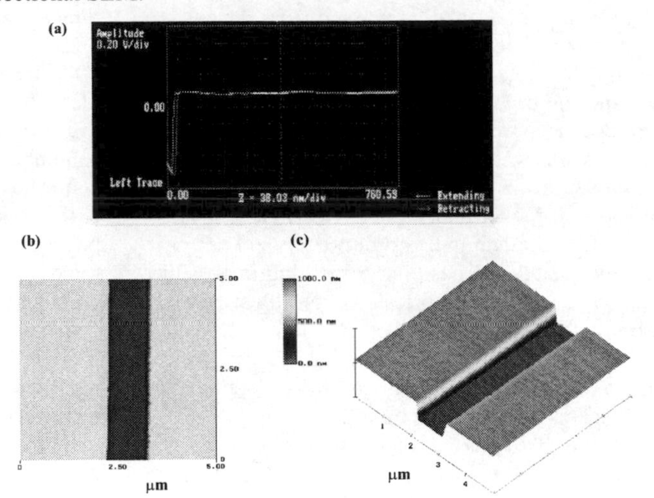

Figure 3. AFM images of a 1 μm trench using CNT AFM probe made from this work. (a) The force vs. distance curve. (b) 2D AFM image of 1 μm trench. (c) 3D image of 1 μm trench. The CNT probe is 60 nm in tip diameter and 5 μm in length. It is able to fully resolve the sidewalls and the bottom of this 1 μm trench that is 450 nm deep. The bottom of this 1 μm trench is clearly visible.

4 DISCUSSION

Our wafer scale growth optimization centered on controlling the CNT parameters from the nanometer sized catalyst spots over 4-inch wafer scale. Based on SEM and TEM analysis, an assessment of the wafer scale CNT probe yield and variations in diameter, length, orientation, and crystallinity has been recorded. Using the optimized catalyst formulation and growth conditions, for released CNT AFM cantilever probes, we have achieved an individual CNT growth yield of 85-90% from 100 to 200 nm catalyst sites. In a typical growth run, the base diameter of an individual CNT probe varies from 60 nm to 80 nm, while the tip diameter varies from 40 to 60 nm. The CNT lengths show a distribution of ±20%. The orientation of individual CNTs on cantilevers varies from 10° to 20° with respect to the cantilever beam surface normal. Our PECVD growth provides mostly bamboo structure CNTs with their crystalline morphologies very similar to the ones demonstrated in Figure 2(a) and 2(b). Across the wafer, CNT probes maintain reasonably consistent diameter, length, orientation, and bamboo morphology. Careful inspection of cantilevers from different areas of the 4-inch wafer has allowed us to analyze statistics for quality control and assurance. We are currently working on further improvement of our process control and yield.

Before we finalized our fabrication process, we investigated various other approaches for obtaining the desired CNT probes on cantilevers. We found that growing CNTs prior to cantilever microfabrication posed many processing challenges due to the inherent mechanical instability of individual CNTs exposed to liquid based processing agents or to CNT protection and stripping chemicals. Preliminary tests showed that upon exposing the as-grown CNTs to photoresist coating and removal or to nitride layer protection and stripping, the CNTs would be heavily damaged or totally removed from the silicon substrate. Since vertical CNT orientation was a prerequisite for obtaining usable probes, growing the CNTs at the end of the fabrication process proved to be a better approach with a reasonable yield of usable CNT probes per wafer. We also found that our wafer scale catalyst deposition and liftoff steps could not follow cantilever microfabrication. Resist layers must be extremely consistent in this process to yield uniform spot sizes for catalyst deposition, something impossible to achieve on a post-fabricated cantilever wafer. Therefore, we had opted to deposit and protect the catalyst ahead of cantilever fabrication, in predefined diameter and placement on final cantilever beam. This approach seems to be the best option.

It is known that a carbon nanotube probe is able to fully trace the bottom of the trenches without producing image artifacts [2,3,5]. CNT probes can also dramatically improve the imaging resolution and lifetime [2-5]. The image obtained by the conventional silicon probe is limited by its geometry. The micromachined silicon pyramid is physically unable to accurately trace the topographic variations of the surface, especially in critical dimension measurements. Probes fabricated from silicon tend to have limited lifetime due to the inherent chemical/mechanical instability of silicon. The silicon probe is brittle, and as such, even when an ultra sharp silicon probe is used for high aspect ratio feature imaging, the probe could easily become worn out or rendered unusable due to mechanical breaking or tip wear. By contrast, the remarkable flexibility of the hexagonal graphitic network allows the CNTs to sustain large distortions via axial compression, such as those encountered with tapping mode imaging [19,20]. Published experimental work has demonstrated that axial deformations are elastic with no atomic defects forming in the hexagonal lattice [21,22].

While most pure multi-wall CNTs with lengths greater than 3 μm require shortening for AFM imaging [2-5], we are able to collect satisfactory AFM images using our fabricated CNT probes with lengths as long as 6 μm. Therefore, we can eliminate the shortening step in our process. Furthermore, our as-grown CNT probes with length between 2 to 6 μm are found to be imageable. This reduces the CNT length uniformity requirement for wafer scale individual CNT growth. Our process yield is therefore greatly improved.

5 CONCLUSION

We demonstrate here an example of integrating CNTs with silicon cantilevers for reliable fabrication of CNT based probes at wafer scale for AFM imaging applications. In our case, the nano-micro integration is achieved through catalyst nanopatterning and registration at wafer scale and through effective nano-catalyst protection and release before and after microfabrication. Our wafer scale fabrication method provides CNT probes that are directly grown from the silicon cantilevers at wafer scale. CNT probe locations and diameters are defined by e-beam lithography. CNT length, orientation, and crystalline quality are controlled by plasma enhanced chemical vapor deposition (PECVD) method. PECVD method enables us to grow well-aligned single carbon nanotubes on individual catalyst sites on silicon cantilevers with control over CNT location, diameter, length, and crystallinity. Due to the crystalline morphology of our PECVD-grown CNTs, there is no need in our process to conduct post fabrication treatment to remove and/or to shorten the CNT probes. With effective catalyst protection schemes, this fabrication process is very similar to conventional approach for fabricating wafer scale silicon AFM probes. Process control is therefore feasible and the overall yield is greatly improved. The probes made from this method display good image acquisition characteristics. No shortening or post-fabrication treatment is necessary. AFM scanning tests show imaging capabilities of these probes in critical dimension measurements. This is a truly bottom up wafer scale CNT AFM probe fabrication method.

Acknowledgment. This work was supported by National Science Foundation Grant No. 0320512 to Integrated Nanosystems, Inc., and by NASA contracts No. NAS2-99092 and No. NAS2-03144 to Eloret Corporation and to the University of California, Santa Cruz.

REFERENCES

[1] Dresselhaus, M. S.; Dresselhaus, G.; Avouris, Ph. (Eds.). Carbon Nanotubes, Topics Appl. Phys. 2001, 80, Springer.
[2] Dai, H.; Hafner, J. H.; Rinzler, A. G.; Colbert, D. T.; and Smalley, R. E. Nature 1996, 384, 147.
[3] Nguyen, C. V.; Chao, K.; Stevens, R. M.; Delzeit, L.; Cassell, A.; Han, J.; and Meyyappan, M. Nanotech. 2001, 12, 363.
[4] Nguyen, C. V.; So, C.; Stevens, R. M.; Li, Y.; Delzeit, L.; Sarrazin, P.; and Meyyappan, M. J. Phys. Chem. B 2004, 108, 2816.
[5] Nguyen, C. V.; Stevens, R. M. D.; Barber, J.; Han, J.; and Meyyappan, M. App. Phys. Lett. 2002, 81, 901.
[6] Wong, S. S.; Harper, J. D.; Lansbury, P. T. Jr.; and Lieber, C. M. J. Am. Chem. Soc. 1998, 120, 603.
[7] Nishijima, H.; Kamo, S.; Akita, S.; and Nakayama, Y. App. Phys. Lett. 1999, 74, 4061.
[8] Stevens, R. M.; Nguyen, C. V.; and Meyyappan, M. IEEE Trans. on Nanobioscience 2004, 3, 56.
[9] Wong, S. S.; Joselevich, E.; Woolley, A. T.; Cheung, C. L.; and Lieber, C. M. Nature 1998, 394, 52.
[10] Wong, S. S.; Woolley, A. T.; Joselevich, E.; Cheung, C. L.; and Lieber, C. M. J. Am. Chem. Soc. 1998, 120, 8557.
[11] Stevens, R.; Nguyen, C.; Cassell, A.; Delzeit, L.; Meyyappan, M.; and Han, J. App. Phys. Lett. 2000, 77, 3453.
[12] Hafner, J. H.; Cheung, C. L.; and Lieber, C. M. Nature 1999, 398, 761.
[13] Hafner, J. H.; Cheung, C. L.; and Lieber, C. M. J. Am. Chem. Soc. 1999, 121, 9750.
[14] Cheung, C. L.; Hafner, J. H.; and Lieber, C. M. PNAS 2000, 97, 3809.
[15] Franklin, N. R.; Li, Y.; Chen, R. J.; Javey, A.; and Dai, H. App. Phys. Lett. 2001, 79, 4571.
[16] Yenilmez, E.; Wang, Q.; Chen, R. J.; Wang, D.; and Dai, H. App. Phys. Lett. 2002, 80, 2225.
[17] Cassell, A. M.; Ye, Q.; Cruden, B. A.; Li, J.; Sarrazin, P. C.; Ng, H. T.; Han, J.; and Meyyappan, M. Nanotech. 2004, 15, 9.
[18] Cassell, A. M.; Ng, H. T.; Delzeit, L.; Ye, Q.; Li, J.; Han, J.; and Meyyappan, M. Appl. Cat. A: General 2003, 254, 85.
[19] Lee, S. I.; Howell, S. W.; Raman, A.; Reifenberger, R.; Nguyen, C. V.; and Meyyappan, M. Nanotech. 2004, 15, 416.
[20] Yu, M. F.; Kowalewski, T.; and Ruoff, R. S. Phys. Rev. Lett. 2000, 85, 1456.
[21] Hertel, T.; Martel, R.; and Avouris, P. J. Phys. Chem. B 1998, 102, 910.
[22] Ruoff, R. S.; Qian, D.; and Liu, W. K. C. R. Physique, 2003, 4, 993.

Effect of Etchant Composition and Equipment Parameters on Silicon Etch Rate

D.Yellowaga[*], J. Starzynski[**] , B. Palmer[*], J. McFarland[*] and S. Drews[***]

[*]Honeywell Electronic Chemicals
6760 W. Chicago St, Chandler, AZ, USA 85226
[**]Honeywell Electronic Chemicals
12001 State Highway 55, Plymouth, MN, USA
[***]SEZ America
4829 S. 38th St. Phoenix, AZ, 85040

ABSTRACT

The drive for faster and smaller electronics has created an increased need for denser die stacking and smaller die packaging. Thinner substrates are in paradox with the other trend in advanced IC technologies: increased wafer diameter. In order to meet the requirements for packaging, stacking, mechanical reliability and faster switching speeds, the thickness of the wafer must be reduced before the device is diced for packaging. Currently, approximately 25% of all finished devices require thickness reduction, and that amount is expected to approach 80% within 5 years. Physical grinding of the wafers introduces damage in the back surface of the silicon that leads to early failures once the device has been packaged. It has been shown that by removing as little as 8μ of damaged silicon leads to an 80% reduction in the amount of stress on the wafer [1]. In this paper, factors that affect the etch rate and uniformity of wafer thinning etchants as a function of equipment parameters and acid composition were determined.

Keywords: wafer thinning, backside etch, grinding, stress relief

1 INTRODUCTION

Wet etch methods for reducing the thickness of a silicon wafer have been gaining popularity over traditional thinning methods of physical grinding or plasma treatment. Wet etch thinning of wafers is quick, offers yield improvements over grinding methods, and allows more expensive plasma equipment to be freed up for other processes. Wet etch wafer thinning products on the market today are typically an acidic mixture of 3 or 4 components, comprising chelating acids, oxidizing acids and diluent acids. Diluent acids are chosen for water absorption or viscosity properties, and their ability to control dissociation of other acids in the mixture [2,3]. Adding acids of increased viscosity also reduces the surface roughness resulting from the etch [2]. Etchants are usually applied to the back side of a wafer in a single wafer spray tool before the wafer is diced for packaging. Etchant compositions can be modified to achieve different etch rates and surface textures for silicon. In the case of a bulk silicon etch, acid compositions are chosen to have a high etch rate, but minimal dishing and wafer bow effects when applied in the toolset. Silicon polish etchants have compositions chosen to increase the overall viscosity of solution, which acts to delineate cracks formed during physical grinding of the silicon surface. Texture etchants are designed to maximize surface roughness of the wafer surface, to increase adhesion for metal layers applied to the back of the wafer.

Equipment technology available for the application of wet chemical etchants for wafer thinning is mature and robust, but one of the obstacles to wet etch wafer thinning becoming more widely accepted is the difficulty in controlling the manufacturing process for the etchants. Blending of the acids typically contained in these mixtures can be highly exothermic, and more volatile components in these mixtures can be lost. This creates very large lot to lot inconsistencies of product, which leads to changes in etch rate and uniformity when applied to the wafer [4].

2 EXPERIMENTAL

Variations in composition of Honeywell Wafer Thinning Products Bulk Silicon Etchant, Silicon Polish Etchant 1, Silicon Polish Etchant 2, and Silicon Texture Etch were tested on <100> silicon. Etch rates for <100> silicon were measured according to the standard method DIN 50453, which is a gravimetric method whereby silicon coupons of known dimensions are weighed pre and post etchant application. Variations in etchant compositions to be tested were determined using an extreme vertices mixture design of experiments using Minitab version 14.12. Etch rate results were analyzed using stepwise regression. For etch rate and uniformity testing for Honeywell Wafer Thinning Products, etchants were applied to 8" <100> silicon wafers in a SEZ 203, which is a single wafer spin processor for advanced frontside and backside etching, stripping, cleaning, wafer reclaim and film removal applications. Equipment parameters were varied as part of a Box-Behnken surface response design, and resulting etch rates and etch uniformities were determined using a FSM 413C optical wafer thickness mapper. Surface roughness after etch was measured using a KLA Tencor P2 long scan profiler.

3 RESULTS

For the Bulk Silicon Etchant, the amount of the 4 different components were varied by approximately 6 to 8 percent by volume, which gave percent by weight variations of 3 to 7 %, depending on the acid. The resulting formulations had etch rates ranging from just under 100 µm/min to over 140 µm/min (see figures 1 to 6). Etch rates obtained in a beaker are much higher than would be obtained in a full wafer tool, as the toolsets apply active cooling to the opposite side of the wafer during the etch process, and the etch is very exothermic. It was found that the etch rates were heavily dependent on all of the acids present in the mixture. There was also found to be a positive correlation between the etch rate and water content by weight of the bulk silicon etchant. Increasing water content will increase the degree of dissolution of the acids in the formulation, which accounts for the increase in etch rate [3]. When the Bulk Silicon Etchant was applied in the SEZ 203 spin processor, it was possible to obtain etch rates between 26 and 40 µm/min on an 8" wafer by simply adjusting the equipment parameters. Etch uniformity was excellent for almost all equipment parameters tested, showing that there was a broad process window for the Bulk Silicon Etchant. The etch rate of silicon in the Bulk Silicon Etchant is first order as is shows Arrhenius behavior as illustrated by a plot of the natural log of the etch rate versus the inverse of the temperature in Kelvins.

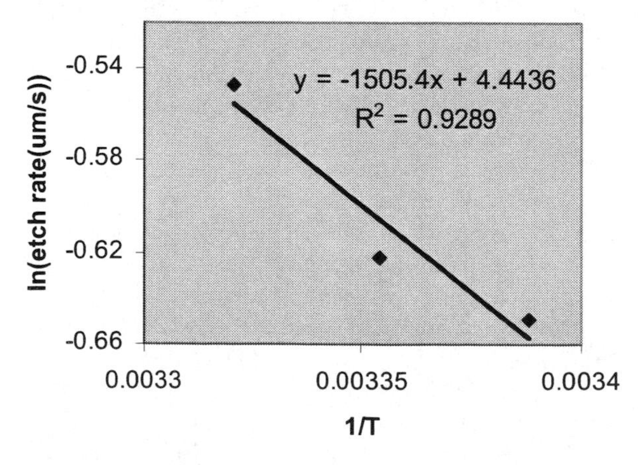

Figure 1: Arrhenius behavior of Bulk Silicon Etchant.

For Silicon Polish Etchant 1, the amount of the 4 different components were varied by approximately 5 to 8 percent by volume, which gave percent by weight variations of 2 to 5 %, depending on the acid. Etch rates in the DIN 50453 testing ranged from 3 to 6 µm/min. When the Silicon Polish Etchant 1 was applied in the SEZ 203 spin processor, it was possible to obtain etch rates between 12 and 17 µm/min. Once again, etch uniformity was excellent for a wide range of equipment parameters tested. The etch rate of silicon in the Silicon Polish Etchant 1 also showed Arrhenius behavior.

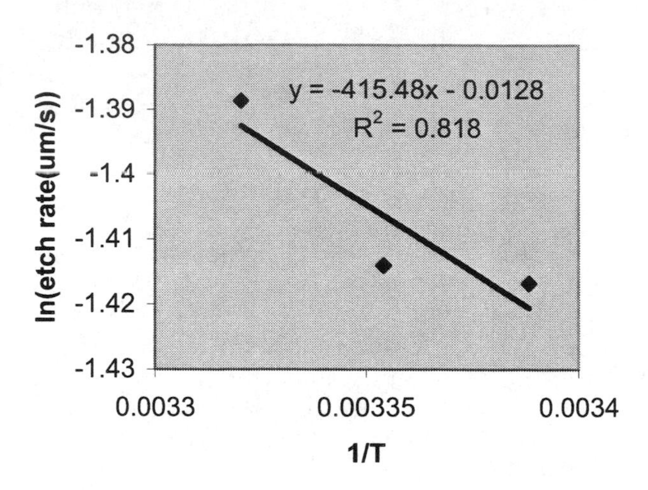

Figure 2: Arrhenius behavior of Silicon Polish Etchant 1

In the case of Silicon Polish Etchant 2, the amount of the 4 different components were varied by approximately 5 to 10 percent by volume, which gave percent by weight variations of 2 to 6 %, depending on the acid. Resulting etch rates are in the range of 0.5 to 3 µm/min in the DIN test. When the Silicon Polish Etchant 2 was applied in the SEZ 203 spin processor, it was possible to obtain etch rates between 5 and 8 µm/min. Etch uniformity was very good for a wide range of equipment parameters tested. The etch rate of silicon in the Silicon Polish Etchant 2 did not have as good of fit in the Arrhenius plot, suggesting that second order effects on etch rate may be occurring.

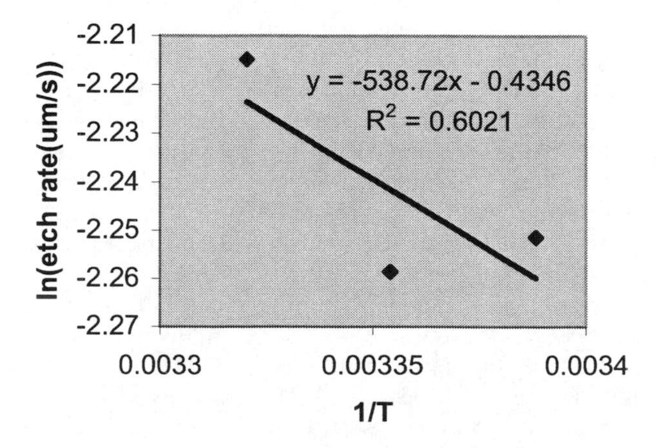

Figure 3: Arrhenius behavior of Silicon Polish Etchant 2

For Silicon Texture Etchant, measurements for composition variations and etch rate and surface roughness were all measured on 8" <100> wafers processed in the SEZ 203. Etch rates between 4 and 8 µm/min were achieved through equipment parameter manipulation. Surface roughnesses of up to 2400 R_a were also possible through variations in equipment settings. The etch rate of silicon in the Silicon Texture Etchant showed excellent Arrhenius behavior.

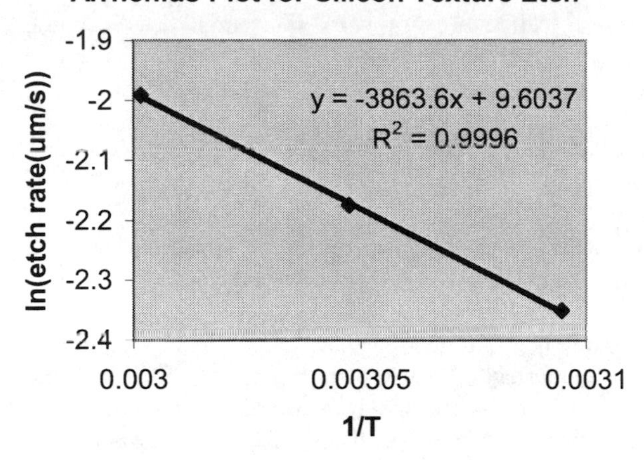

Figure 4: Arrhenius behavior of Silicon Texture Etchant

4 SUMMARY

As shown from the experimental results, changes of a few percent in acid composition can lead to changes in the etch rates that could create issues in a manufacturing environment. Honeywell Wafer Thinning Products are produced under strict manufacturing controls, and show excellent batch to batch uniformity amongst constituents. All of the etchants except for the Silicon Polish Etchant 2 showed Arrhenius behavior, which means the etch rate mechanism is first order. This is desirable as first order mechanisms are more easily predicted and are easier to control. It was also shown that etch rates and surface finish could be easily controlled through equipment parameter settings. Overall, Honeywell Wafer Thinning Products showed excellent uniformity over a wide range of equipment parameters, proving that there are wide process windows for these products.

5 ACKNOWLEDGEMENTS

The authors would like to thank Aaron Bicknell, Applications Engineer, and Gale Hansen, Metrology Engineer from the SEZ America Research Lab in Phoenix, Arizona for their invaluable assistance in this project.

REFERENCES

[1] C. McHatton and C. M. Gumbert, *Solid State Technology,* **41**, 11, (1998).

[2] M. S. Kulkarni, H. F. Erk, *J. Electrochem. Soc.* **147**, 176 (2000).

[3] H. Robbins, B. Schwartz, *J. Electrochem. Soc.* **107**, 108 (1960).

[4] W. Seivert, K. Zimmermann, J. A. McFarland, M. A. Dodd, United States Patent Application Publication US20030230548.

Fabrication of Complex Diffractive Structures in an Organic-Inorganic Hybrid and Incorporation of Silver Nanoparticles

F. H. Scholes*, F. L. Smith* and S. A. Furman*

*CSIRO Manufacturing and Infrastructure Technology, Private Bag 33, Clayton South MDC, Victoria 3169, Australia, fiona.scholes@csiro.au

ABSTRACT

We report on the fabrication of photo-patterned structures with sub-micron feature sizes in a sol-gel derived organic-inorganic hybrid material. Patterns with a high level of design complexity are presented, in the form of a diffractive optically variable device. We also demonstrate how the material can serve as a matrix for nano-scale components, via the incorporation of Ag nanoparticles. It is shown that Ag nanoparticles can be used to modify the optical properties of this photo-patternable material.

Keywords: sol-gel, organic-inorganic hybrid, silver nanoparticles, photo-patterning

1 INTRODUCTION

Research activities have been expanding rapidly in recent years at the interface between micro- and nano-technologies. In particular, organic-inorganic hybrids have attracted considerable interest in this area. These materials, which are synthesized by the highly versatile sol-gel process, consist of interpenetrating organic and inorganic networks at the nano- to molecular size scale. By incorporating photo-polymerizable groups into the organic network, the material can be patterned with light in a similar way to a photo-resist [1]. Thus, micro-scale relief structures can be generated in this glass-like material. Importantly, this can be achieved without needing harsh chemical etchants such as HF.

Photo-patternable organic-inorganic hybrids have been used to fabricate simple micro-optical elements such as microlenses [2], diffraction gratings [3-5] and stacked optical waveguides [6]. Photo-patterning has been achieved by conventional amplitude masking [2-4, 6], phase masking [7], laser direct writing [8] and holographic interference methods [9, 10]. In addition, electron beam lithography has been demonstrated as an alternative patterning method [5]. Only very simple relief structures have been produced to date, with the highest resolution demonstrated in basic diffraction gratings comprised of lines ~500 nm in width [7]. In this paper, we demonstrate the fabrication of more complex diffractive structures, in the form of a diffractive optically variable device (OVD).

Another active area of sol-gel research has been the fabrication of nanocomposite materials consisting of noble metal nanoparticles embedded in the sol-gel matrix. In particular, there has been considerable interest in sol-gel glasses doped with Ag nanoparticles, due to their nonlinear optical properties [11] and surface-enhanced Raman activity [12, 13]. Such properties make these materials important in a wide range of emerging technologies, such as optoelectronics (e.g. optical switches, nonlinear waveguiding devices) and chemical sensors. Here, we demonstrate how Ag nanoparticles can be incorporated into a photo-patternable, sol-gel derived, organic-inorganic hybrid material. While there are numerous reports in the literature on Ag-doped glass films, this is the first time that photo-patterning has been demonstrated in such a material.

2 ORGANIC-INORGANIC HYBRID FORMULATION

The organic-inorganic hybrid sol was prepared by dissolving 5 mL 3-methacryloxypropyltrimethoxysilane (MAPTMS) in 25 mL isopropanol and hydrolyzing with 1.1 mL 0.2 M HCl (aq). A second solution was prepared consisting of 2.0 mL titanium(IV) isopropoxide ($Ti(OCH_2CH_2CH_3)_4$) mixed with 4.3 mL acetylacetone. Both solutions were stirred for 1 hr. The two solutions were then mixed together, 0.75 g of the photo-initiator 1-hydroxy-cyclohexyl-phenyl-ketone was added and the resulting solution stirred for 1 min. All solution preparation was carried out under dim lighting to avoid photo-polymerization. The sols were then allowed to age at room temperature for 16 hr.

3 PHOTO-PATTERNING COMPLEX DIFFRACTIVE STRUCTURES

In order to photo-pattern complex diffractive structures, aged sols were filtered through 0.25 μm membrane filters and spin-coated for 1 min at 3500 rpm onto glass microscope slides or Cu sheet. The spin-coated substrates were then baked at 80°C for 30 min. This was done to drive away solvent, and to promote condensation polymerization reactions leading to the formation of an inorganic (≡Si–O–Si/Ti≡) network structure.

After cooling to room temperature, the spin-coated substrates were exposed through an amplitude mask in a highly collimated UV exposure unit (Hg lamp, 2.2 W/cm^2, 5 min exposure time). Exposure to UV light caused cross-linking of methacrylate groups via addition polymerization, resulting in the formation of an organic network structure.

The mask used was prepared from a standard photo-mask blank and was patterned by electron beam lithography. The design of the mask incorporated complex patterns of straight grooves of various lengths and spacing. The orientation of the lines was also varied to produce different images at different viewing angles. This type of mask design has been used in anti-counterfeiting applications and is described in detail elsewhere [14, 15]. The total image area was 2.5 × 3 cm.

Films were then developed in ethanol in order to remove unexposed material (i.e. areas devoid of organic network structure). Following development, the films were baked at 80°C for 1 hr. This was done to drive condensation polymerization reactions to completion, thereby strengthening the organic-inorganic hybrid network. The resulting photo-patterned features had a height of ~600 nm, as determined by AFM measurements.

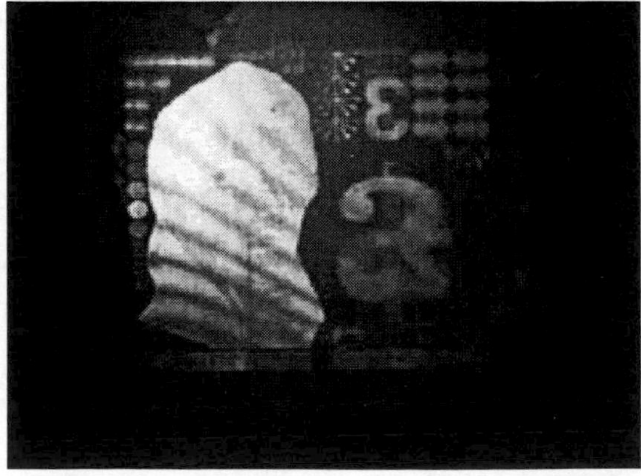

Figure 1: Diffractive OVD photo-patterned into organic-inorganic hybrid material on Cu substrate, as viewed with two different angles of white light illumination.

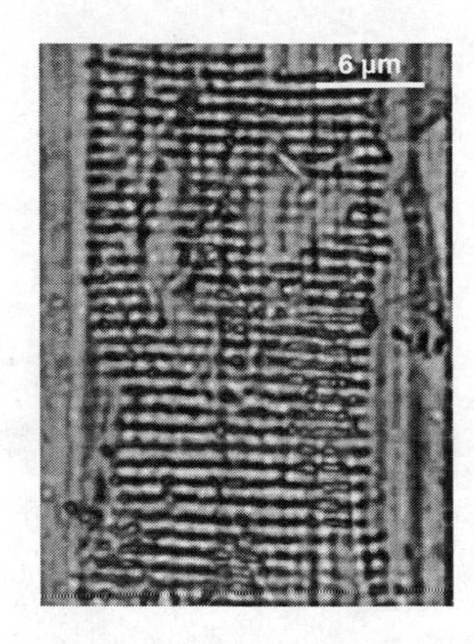

Figure 2: Sub-micron photo-patterned features in the organic-inorganic hybrid material on a Cu substrate.

Figure 1 shows an image photo-patterned into the organic-inorganic hybrid material on a Cu substrate, as viewed with two different angles of white light illumination. A portrait with gray-scale toning can be seen in the upper image, while diffractive effects are clearly evident in the image below. These optical effects result from the different lengths, spacing and orientation of sub-micron grooves in the photo-patterned organic-inorganic hybrid material.

A portion of the image is shown at higher magnification in Figure 2. Although some regions have delaminated due to the surface roughness of the Cu substrate, sub-micron line spacing can clearly be observed. Thus, we have achieved photo-patterned features in an organic-inorganic hybrid material with feature sizes comparable to previous work [7], but in a design of much higher complexity, and across a large image area.

4 INCORPORATING SILVER NANOPARTICLES

We have also been able to incorporate Ag nanoparticles into this photo-patternable organic-inorganic hybrid material. This is achieved by dissolution of $AgNO_3$ in the precursor sol. To the basic formulation described above, 0.15 g $AgNO_3$ (s) is added immediately after addition of the photo-initiator. The solution is then stirred rapidly for 15 min before ageing in darkness at room temperature for 16 hr.

Importantly, in the process of photo-patterning the host material, Ag^+ ions incorporated into the organic-inorganic network structure are photo-reduced, thereby producing metallic Ag nanoparticles. Figure 3 shows a transmission electron micrograph of the photo-patterned Ag-doped hybrid material. Ag nanoparticles over a range of sizes are clearly visible, with some of the larger particles displaying crystalline diffraction contrast.

Figure 4 shows the UV-visible absorption spectra of the undoped and Ag-doped organic-inorganic hybrid materials. The material incorporating Ag nanoparticles shows a weak absorption band with a peak maximum at 420 nm, which can be attributed to the surface plasmon resonance characteristic of Ag nanoparticles.

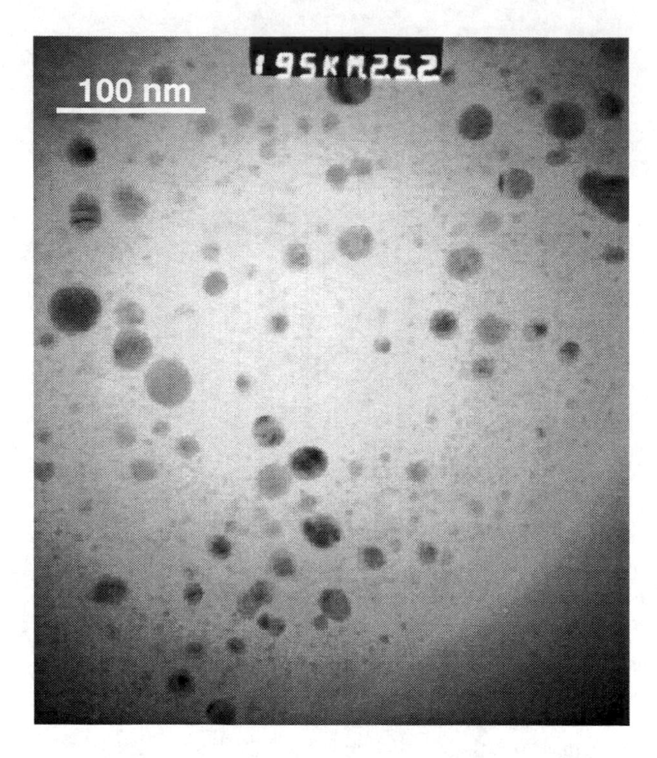

Figure 3: Transmission electron micrograph of photo-patterned organic-inorganic hybrid material doped with Ag nanoparticles.

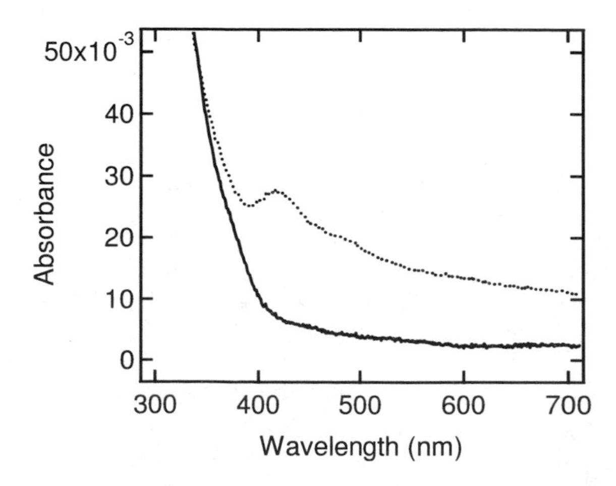

Figure 4: UV-visible absorption spectra of undoped hybrid material (solid line) and Ag-doped hybrid material (dotted line).

5 CONCLUSIONS

We have demonstrated that complex diffractive structures can be photo-patterned into an organic-inorganic hybrid material. The ability to combine both sub-micron feature sizes and a high level of design complexity in organic-inorganic hybrids may prove useful for the fabrication of a variety of devices, such as phase masks, X-ray optics and security elements.

We have also demonstrated how Ag nanoparticles can be incorporated into a photo-patternable organic-inorganic hybrid. By utilizing the unique properties of Ag nanoparticles, this novel material may find use in a range of emerging technologies such as optoelectronics (e.g. optical switches, nonlinear waveguiding devices) and chemical sensing (e.g. as a lab-on-a-chip substrate for detecting chemical species by surface-enhanced Raman scattering).

REFERENCES

[1] K.H. Haas, H. Wolter, Curr. Opin. Solid State Mater. Sci. 4 (1999) 571.

[2] H. Jiang, X. Yuan, Z. Yun, Y.-C. Chan, Y.-L. Lam, Mater. Sci. Eng. C 16 (2001) 99.

[3] P. Äyräs, J.T. Rantala, S. Honkanen, S.B. Mendes, N. Peyghambarian, Opt. Commun. 162 (1999) 215.

[4] O. Soppera, C. Croutxe-Barghorn, C. Carre, D. Blanc, Appl. Surf. Sci. 186 (2002) 91.

[5] W.X. Yu, X.-C. Yuan, Opt. Express 11 (2003) 899.

[6] U. Streppel, P. Dannberg, C. Wachter, A. Brauer, L. Frohlich, R. Houbertz, M. Popall, Opt. Mater. 21 (2003) 475.

[7] D.J. Kang, J.U. Park, B.S. Bae, J. Nishii, K. Kintaka, Opt. Express 11 (2003) 1144.

[8] W.X. Yu, X.C. Yuan, N.Q. Ngo, W.X. Qui, W.C. Cheong, V. Koudriachov, Opt. Express 10 (2002) 443.

[9] P. Cheben, M.L. Calvo, Appl. Phys. Lett. 78 (2001) 1490

[10] D.J. Kang, J.K. Kim, B.S. Bae, Opt. Express 12 (2004) 3947.

[11] P. Prem Kiran, G. De, D. Narayana Rao, IEE Proceedings: Circuits, Devices and Systems 150 (2003) 559.

[12] S. Lucht, T. Murphy, H. Schmidt, H. Kronfeldt, J. Raman Spectrosc. 31 (2000) 1017.

[13] K. Maruszewski, M. Jasiorski, D. Hreniak, W. Strek, K. Hermanowicz, K. Heiman, J. Sol-Gel Sci. Technol. 26 (2003) 83.

[14] P.W. Leech, H. Zeidler, Microelectr. Eng. 65 (2003) 439.

[15] R.A. Lee, Microelectr. Eng. 53 (2000) 513.

Fabrication of Nanochannels with Microfluidic Interface Using PDMS Casting on Ti/Si Nanomold

M.J. Rust, S. Subramaniam* and C.H. Ahn

Department of Electrical & Computer Engineering and Computer Science
*Department of Chemical and Materials Engineering
University of Cincinnati, 814 Rhodes Hall, Cincinnati, OH, USA, rustmj@ececs.uc.edu

ABSTRACT

In this work, the fabrication of nanochannels with microfluidic interface using poly(dimethylsiloxane) (PDMS) casting on Ti/Si nanomold is presented. This new method combines e-beam lithography, Ti metal deposition, photolithography, and casting in PDMS. The combination of nanofabrication with PDMS casting allows the patterning of high-resolution nanostructures while also maintaining high throughput. Additionally, the microfluidic interface simplifies integration with current testing and characterization procedures. This allows the rapid fabrication of nanochannels for basic nanofluidic transport studies, integration with current microfluidic devices, and applications in drug delivery.

Keywords: nanochannels, nanofluidics

1 INTRODUCTION

Biochemical analysis systems have seen remarkable changes in recent years due to the development of microfluidic technology. Microfluidics has opened several new research areas including micro total analysis systems (uTAS) or lab-on-a-chips [1]. These lab-on-a-chip devices have enabled the entire analysis of chemical and biological samples to be performed on a single lab chip, thus significantly reducing analysis times, reagent use, and analysis error [2]. The recent emergence of nanotechnology has enhanced uTAS fabrication possibilities to include nanofluidic devices and systems [3]. These nanofluidic devices offer a wide range of new opportunities for uTAS since many biological processes involve structures in the sub-micron to nanometer scale [4]. The ability to handle and probe samples at their fundamental length scales can thus enable the study of single molecules and cells [5].

While interest in nanofluidic devices has been growing, there has also developed a need for nanofluidic devices to enable studies of nanotransport phenomena [5]. There is large demand that these devices be fabricated quickly, at low cost, and easily integrated with current microfluidic technology, so as to simplify testing and characterization procedures. Poly(dimethylsiloxane) (PDMS) has been considered as a fabrication material for these prototype devices because of its attractive properties, such as low cost, high throughput fabrication, optical transparency, and flexible surface chemistry [6]. Examples of nanochannels fabricated in PDMS have emerged with channel widths below 100 nm [7]. However, it remains a challenge to interface these channels with micro- and macrofluidics and interconnects to create an integrated device.

In this work, a new method for fabricating nanochannels with microfluidic interface is presented. The new method combines e-beam nanolithography and photolithography to fabricate a Ti/Si nanomold for casting in PDMS. This combination allows the patterning of high-resolution nanostructures while also integrating nano- and microfluidics in a high throughput approach. The new Ti/Si nanostructure shows excellent robustness as a nanomold for mass nanoreplica. Additionally, the use of microphotolithography allows the fabrication of many different nanochannel designs while keeping the microfluidic interface constant. This allows the rapid fabrication of nanochannels for basic nanofluidic transport studies, integration with current microfluidic devices and uTAS, and applications in drug delivery.

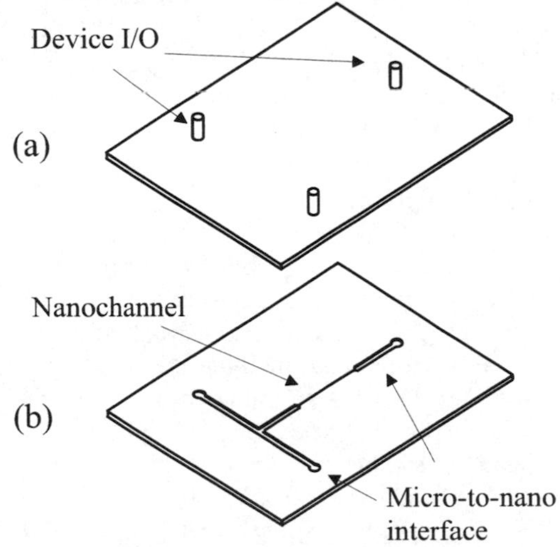

Figure 1: Conceptual schematic of nanofluidic device with microfluidic interface: (a) PDMS cover slip with PEEK tubing used as interconnects for device I/O and (b) PDMS fluidic device with nanochannel and micro-to-nano interface. Layers (a) and (b) are bonded together for a two-layer sealed device.

2 METHOD AND FABRICATION

The fabrication method is a three-stage process involving nanofabrication, microfabrication, and PDMS casting (see Figures 1 and 2). The process begins with the fabrication of a Ti/Si mold that contains both nanofeatures and microfeatures. First, nanofabrication is performed using e-beam lithography, Ti metal deposition, and liftoff to create nanopatterned Ti on Si substrate. These features will eventually be replicated in PDMS to form nanochannels. Microfabrication techniques including photolithography, Ti metal deposition, and liftoff are then used to create micropatterned Ti on the Si substrate. These microfeatures are aligned with the nanofeatues and will ultimately be replicated in PDMS, forming the micro-to-nano interface that couples sample inputs and outputs with the nanochannels. Once the mold fabrication is complete, the entire fluidic system is replicated in PDMS by a casting technique. The fluidic system is then sealed with a PDMS cover piece.

2.1 Nanofabrication

The fabrication process begins with the fabrication of nanofeatured Ti on Si mold (see Figure 2). A 2-inch Si wafer is selected and cleaned using DI H_2O:H_2O:NH_4OH = 5:1:1 by volume at 75 °C for 15 minutes. Positive electron sensitive resist poly(methyl methacrylate) (PMMA) is spin-coated at 4,000 rpm and then baked at 180 °C for 2 minutes. Electron beam lithography is performed using the Raith-150 EBL system, exposing the PMMA resist to an electron beam with a spot size of 10 nm. After exposure, the pattern is developed in a solution of methyl isobutyl ketone (MIBK):isopropyl alcohol (IPA) = 1:3 by volume for 30 seconds, followed by IPA solution for 15 seconds and blow drying with N_2. A Ti metal layer (1,000 Å) is deposited on the patterned wafer using an e-beam evaporator. The sample is then placed in acetone for lift-off.

2.2 Microfabrication

After the nanopatterns have been created, microfabrication techniques are used to produce the micro-to-nano interface (see Figure 2). The nanopatterned sample is cleaned with acetone and methanol. Then positive photoresist (Shipley 1818) is spin-coated on the surface at 3,000 rpm, followed by soft bake at 60 °C in an oven for 30 minutes. Next, the sample is immersed in chlorobenzene for 45 seconds and then dried in an oven at 120 °C for 1 minute. The sample is then exposed to UV light (300-460 nm wavelength) for 10 seconds and developed in DI H_2O:Microposit 351 = 3:1 by volume solution for 1 minute. After drying with N_2, a Ti metal layer (1,000 Å) is deposited using an e-beam evaporator. The sample is then baked at 120 °C for 2 hours, followed by dipping in acetone for liftoff.

2.3 PDMS Casting

After nano and micro patterning of the mold is completed, it is replicated in PDMS using a casting technique [6]. First, a PDMS mixture is made with Sylgard 184:curing agent = 10:1 by weight. The mold is taped to the bottom of a plastic Petri dish and coated with the PDMS mixture. The sample is left for 12 hours to let the air bubbles rise out and then cured at 65 °C for at least 1 hour. The PDMS is then peeled off from the mold and sealed with another slab of PDMS.

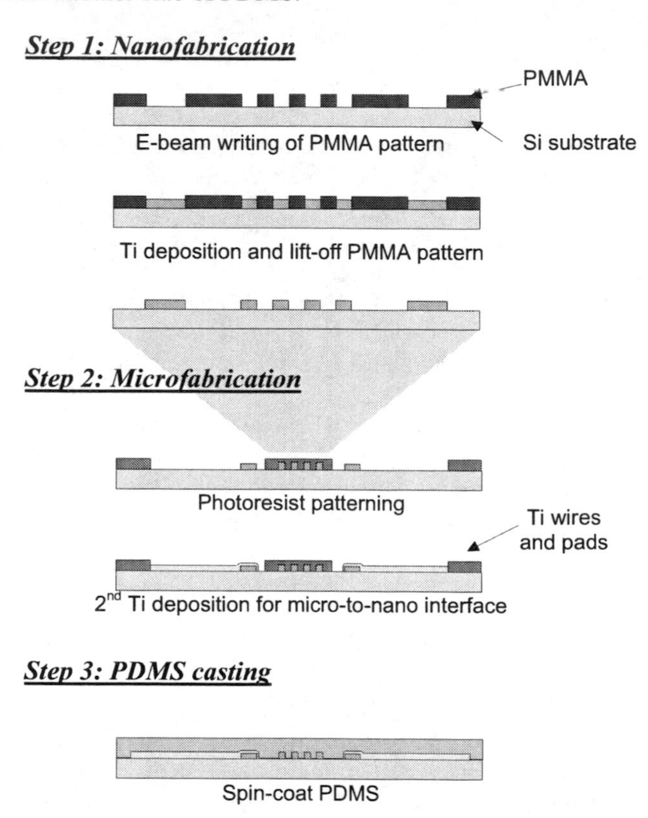

Figure 2: Fabrication process for PDMS nanochannels with microfluidic interface.

3 EXPERIMENTAL RESULTS

Several methods were used to characterize the Ti/Si nanomold and PDMS nanochannel dimensions. First, scanning electron microscopy (SEM) was employed to investigate the nanofeature dimensions of the Ti/Si mold. Arrays of Ti nanopatterns were fabricated with varying structure width from 100 – 500 nm. The resulting SEM measurements can be found in Table 1. The mold nanopatterns were plotted against the original design pattern for the e-beam lithography software as shown in Figure 3.

Comparing the original nanopattern design for e-beam lithography and resulting dimensions of the mold nanostructures after nanofabrication (see Table 1) shows the expected linear correlation, but the mold dimensions are larger than the originally designed pattern. This pattern expansion is most likely due to the proximity effect in e-beam lithography of very thin resists, in which backscattered electrons cause additional resist exposure [8]. To account for this pattern expansion, the chart shown in Figure 3 enables the design of patterns for e-beam lithography that will result in the desired nanofeatures for the mold.

Design size	Mold Size
100	153
200	244
300	347
400	459
500	568

Table 1: Results showing pattern transfer from original design to Ti/Si mold. The dimension for comparison is feature width (nm).

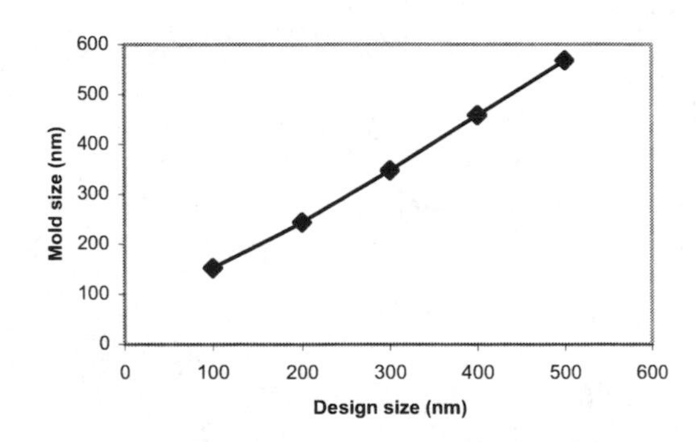

Figure 3: Comparison of original pattern design and resulting mold structures after nanofabrication.

Nanostructures with 459 nm feature size (see Figure 4) were selected for mold replication in PDMS, and atomic force microscopy (AFM) was used to measure the PDMS nanochannel dimensions before the device was sealed. A Digital Instruments Dimension 3100 AFM was used to perform the imaging in tapping mode to obtain surface and cross-sectional images of the PDMS nanochannels (see Figures 5 and 6). Table 2 shows the results of pattern transfer from the Ti/Si mold to PDMS nanochannels. The 459 nm wide mold structure resulted in a 445 nm wide channel width at the bottom of the PDMS nanochannel. The channel depth was measured at 94 nm, which correlates well with the Ti mold structure height of 100 nm. The channel widened to 908 nm at the top of the channel (see Figure 6), resulting in a trapezoidal channel profile. This is likely due to a trapezoidal mold structure profile that results after liftoff.

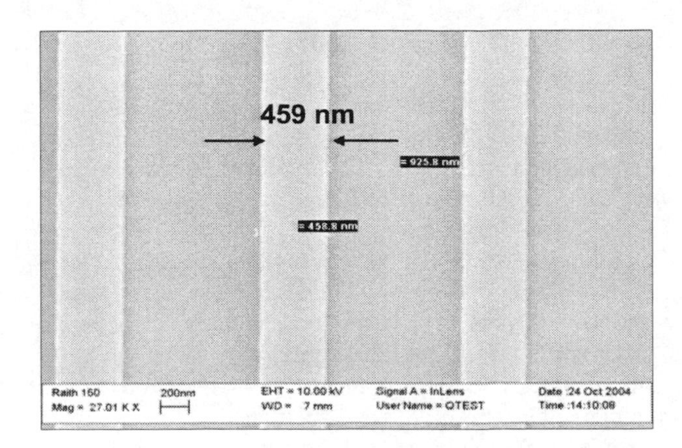

Figure 4: SEM image of Ti/Si mold showing nanostructures array of 459 nm width fabricated by e-beam lithography, metal deposition, and liftoff.

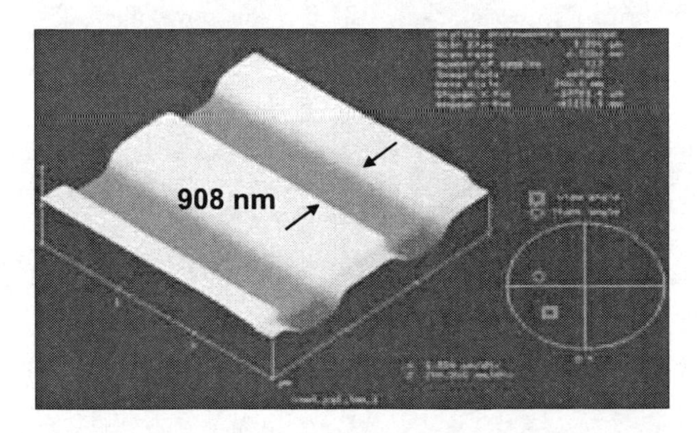

Figure 5: AFM tapping mode surface image of nanochannels in PDMS. Measurement indicated is width at the top of the channel.

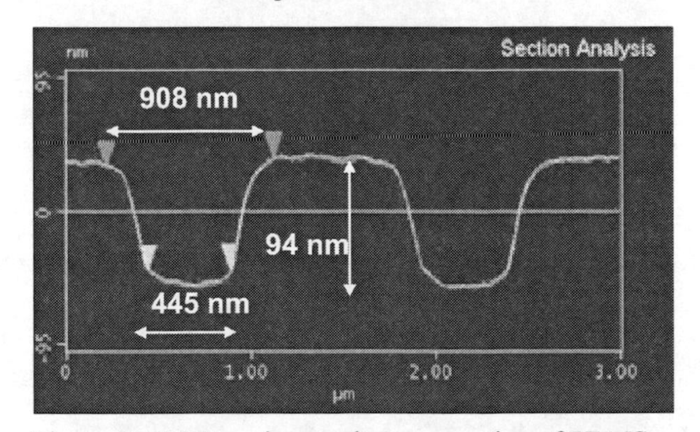

Figure 6: AFM tapping mode cross-section of PDMS nanochannels. Measurements indicated are width at the top and bottom of the channel and channel height.

Ti/Si Mold		PDMS Nanochannels	
Width	Height	Width	Height
459	100	445	94

Table 2: Results showing pattern transfer from Ti/Si mold to PDMS nanochannels. Dimensions are in nm.

4 DISCUSSION

The mold for PDMS casting consists of a nano-structured region and micro-to-nano interface. A vital part of the fabrication process is the alignment of the micro patterns with the underlying nanofeatures. If misalignment occurs during the photolithography process, the resulting fluidic circuit in PDMS will be incomplete, causing device failure. To ensure a successful alignment, multipoint alignment markers were incorporated in the nanopattern and micromask for photolithography. During the fabrication of test devices, almost 100% fabrication yield was maintained as a result of these methods.

One of the most important aspects of this work is the flexibility in design. Since the microfabrication and nanofabrication process steps are performed independently, one can design any nanostructure desired in the nanofabrication process and still use the same micro-to-nano interface scheme. This enables the designer to fabricate and test many different nanofluidic devices in parallel, thus reducing cycle times in prototype design. The structure shown in Figure 7 is being developed to investigate mixing properties of nanochannels as a function of channel dimensions such as length, width, and flow path.

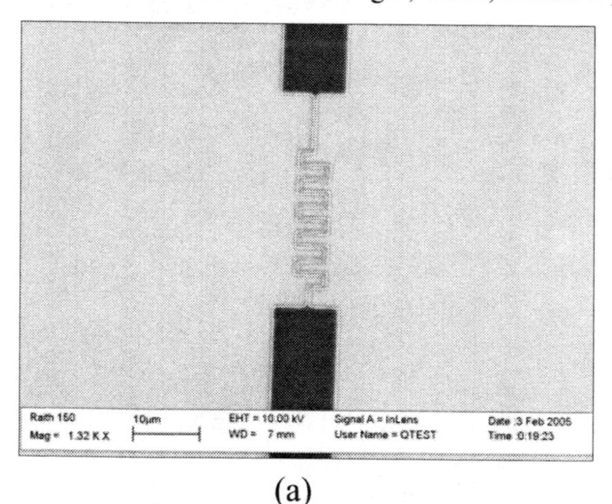

(a)

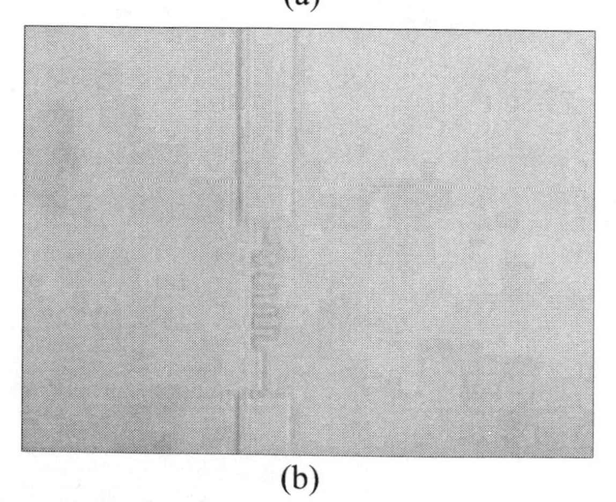

(b)

Figure 7: Device images: (a) SEM image of Ti/Si nanomold and (b) optical micrograph of PDMS channel showing serpentine design with 500 nm width and 100 nm height.

5 CONCLUSIONS

In conclusion, a new method for fabricating nanochannels with microfluidic interface has been presented and successfully demonstrated for the fabrication of a nano/microfluidic chip. The new method combines e-beam lithography and microphotolithography to fabricate a Ti/Si nanomold for casting in PDMS. This combination allows the patterning of high-resolution nanostructures while also integrating nano- and microfluidics in a high throughput manner. Additionally, this method allows the fabrication of many different nanochannel designs while keeping the interconnect scheme constant. This allows the rapid fabrication of nanochannels for basic nanofluidic transport studies, integration with current microfluidic devices and uTAS, and applications in drug delivery.

ACKNOWLEDGEMENTS

The authors gratefully acknowledge Mr. Xiaoshan Zhu, and Mr. Ron Flenniken for their technical assistance. M.R. thanks the National Science Foundation for support.

REFERENCES

[1] A. Manz, N. Graber, H.M. Widmer, Sens. Actuators B, Vol. 1, pp. 244-248, 1990.

[2] C. H. Ahn, J.-W. Choi, G. Beaucage, J.H. Nevin, J.-B. Lee, A. Puntambekar, J.Y. Lee, "Disposable smart lab on a chip for point-of-care clinical diagnostics," Proceedings of the IEEE, Vol. 92, pp. 154-173, 2004.

[3] J. C. Eijkel, J. Bomer, N.R. Tas, A. van den Berg, "1-D nanochannels fabricated in polyimide," Lab on a Chip, Vol. 4, pp. 161-163, 2004.

[4] J.O. Tegenfeldt, C. Prinz, H. Cao, R.L. Huang, R.H. Austin, S.Y. Chou, E.C. Cox, J.C. Sturm, "Micro- and nanofluidics for DNA analysis," Anal. Bioanl. Chem, 378, pp. 1678-1692, 2004.

[5] W. Li, J.O. Tegenfeldt, L. Chen, R.H. Austin, S.Y. Chou, P.A. Kohl, J. Krotine, J.C. Sturm, "Sacrificial polymers for nanofluidic channels in biological applications," Nanotechnology, Vol. 14, pp. 578-583, 2003.

[6] J.M. Ng, I. Gitlin, A. D. Stroock, G. M. Whitesides, "Components for integrated poly(dimethylsiloxane) microfluidic systems," Electrophoresis, Vol. 23, pp. 3461-3473, 2002.

[7] T. W. Odom, J.C. Love, D.B. Wolfe, K.E. Paul, G.M. Whitesides, "Improved pattern transfer in soft lithography using composite stamps," Langmuir, 18, pp. 5314-5320, 2002.

[8] M.A. McCord and M.J. Rooks, "Electron Beam Lithography," in Handbook of Microlithography, Micromachining, and Microfabrication, P. Rai-Choudhury, Ed., Vol. 1, Chap. 2, pp. 139-251, SPIE Optical Engineering, Bellingham, WA, 1997.

Direct Patterning of Functional Materials via Atmospheric-Pressure Ion Deposition

T.E. Hamedinger*, T. Steindl*, J. Albering**, S. Rentenberger*** and R. Saf*

* Institute for Chemistry and Technology of Organic Materials, Graz University of Technology, Stremayrgasse 16/1, 8010 Graz, Austria, thomas.e.hamedinger@tugraz.at
** Institute for Chemistry and Technology of Inorganic Materials, Graz University of Technology, Stremayrgasse 16/3, 8010 Graz, Austria
* Institute for Solid State Physics, Graz University of Technology, Petersgasse 16, 8010 Graz, Austria

ABSTRACT

Interest in thin films of functional materials has increased enormously in recent years because of the wide range of possible applications. Here we report an experimental setup we have termed Atmospheric-Pressure Ion Deposition (APID) as a novel technique that allows highly controllable and soft processing of various materials (e.g. polymers, bio molecules and inorganic materials) into thin and ultrathin structured films. The technique is based on an electrospray process. Microdroplets are initially formed and dried, generating ions that are extracted by electrostatic lenses. Thin structured films are then produced by the deposition of the resulting ion beam onto a moveable target. The technique offers several interesting features, including precise control of film thicknesses. Experiments investigating structured deposition of various kinds of functional materials are reported. This might provide a simple approach creating thin structured films and composites that are currently unattainable.

Keywords: vapor deposition process, thin structured films, devices

1 INTRODUCTION

The industrial demands for new advanced functional materials have driven the development of thin film technology. Today, thin films are used in a wide range of industrial applications, e.g. (bio)sensors, devices (organic light emitting devices (OLEDs), solar cells, transistors), optical applications and coatings [1]. A still challenging problem is the processing of these materials to thin films with variable and defined chemical composition in all three dimensions. However every single technique that is applied for thin film deposition (e.g. spincoating, chemical vapour deposition) shows limitations: the area that can be coated is limited, the use of solvents often makes the assembly of multilayered systems impossible, and molecular weight of the processed material is limited. Independent of the specific application, it is important to keep in mind that the way these films are produced is often the key to optimal exploitation of the intrinsic potential of the materials involved. To overcome such limitations we have introduced Atmospheric-Pressure Ion Deposition (APID) [2] as a new extremely soft technique that uses the same ionization process as ESI-MS and is capable of processing a large spectrum of materials such as polymers, biomolecules and inorganic materials.

2 EXPERIMENTAL SETUP

APID - an ion beam technique that operates under atmospheric pressure - is based on the electrospray-ionization (ESI) process well known from Mass Spectrometry (MS) [3]. A schematic diagram of the experimental setup is shown in Fig.1.

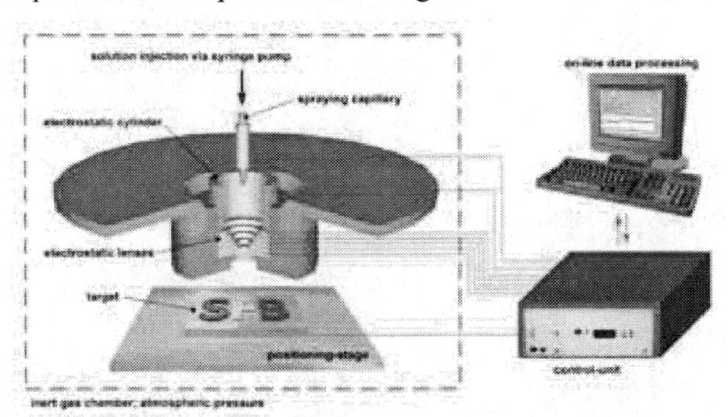

Figure 1: Experimental Setup of Atmospheric-Pressure Ion Deposition

A constant flow of a solution containing the material to be processed is injected into the system through an ESI-MS capillary. The solution leaving the needle is dispersed into an aerosol of charged micro-droplets. In contrast to comparable methods like spincoating these charged droplets are appropriately dried under atmospheric-pressure to produce singly and/or multiply charged ions. The electrostatic lenses are used to extract these ions out of the cylindrical electrode, and to focus the resulting ion-beam towards the target were the ions are deposited.

3 SIMULATION OF ION-TRAJECTORIES

In principle the technique can be divided into two main processes: the electrospray ionisation and the extraction/deposition of ions onto a target. Both can be controlled and monitored in real time. Simulations of ion trajectories with SIMION 7.0 were used to find a suitable experimental setup where focussing is adjustable. To simulate ion trajectories under atmospheric-pressure conditions it was necessary to implement an additional program that included the collisions of ions with the surrounding gas. The simulation predicts that the electrostatic field close to the first electrostatic lens is responsible for the extraction of the ions and can be tuned without significant effects on the electrospray process.

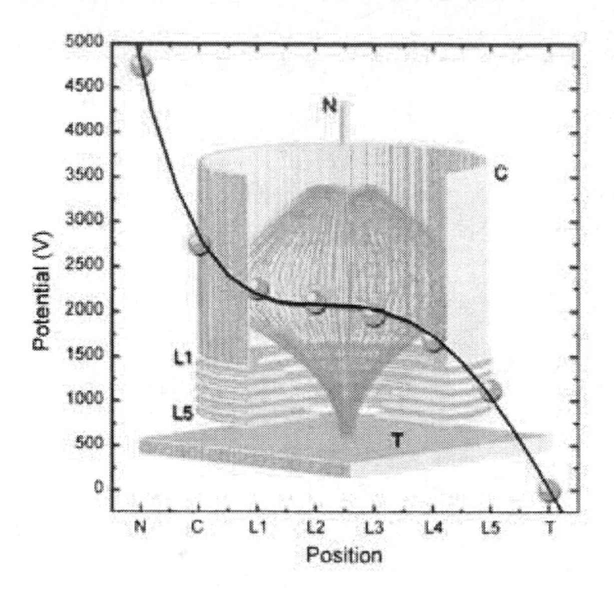

Figure 2: Ion-trajectories simulations with SIMION 7.0 (inset) and potential gradient from needle (N), cylinder (C), electrostatic lenses (L1-L5) to target (T)

Fig. 2 shows a potential gradient from needle to target that was applied during deposition and as an inset a three-dimensional representation of an electrode arrangement. By variation of the potentials applied to the electrostatic lenses and some other process parameters lateral resolutions up to $\approx$10,000 lines.cm^{-1} which is far beyond a standard laser printer (600 dpi) could be achieved (Fig. 3).

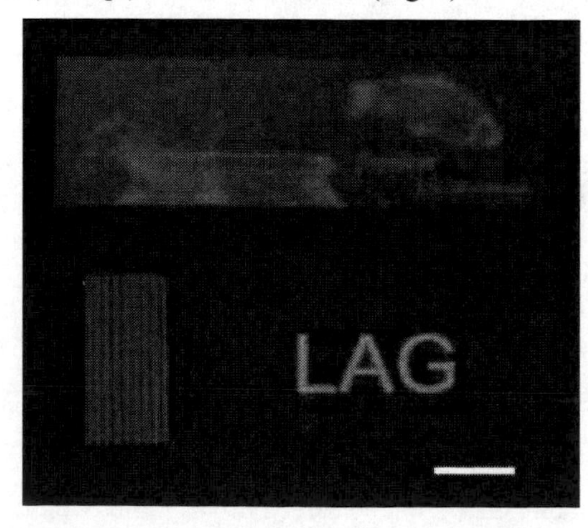

Figure 3: Ultra thin structured films of advanced functional materials with a resolution up to 1,000 lines.cm-1.

Emission photograph of the resulting polymer films (scale bar: 1000 μm).

4 THICKNESS CONTROL VIA MOTION CONTROL

Films having a constant thickness are practically impossible without movement of the target. For structured deposition and exact control of film thickness it was necessary to implement the possibility of target positioning according to given parameters. The algorithm used is shown schematically in Fig. 4.

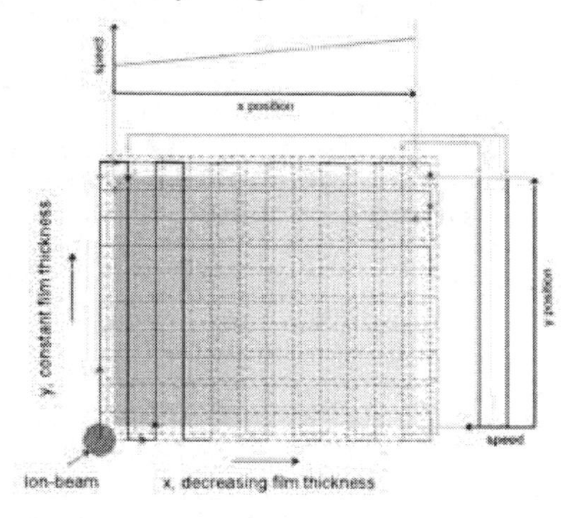

Figure 4: Target movement algorithm

"Line-scans" of the ion-beam, both in the *y*- and the *x*-direction, were performed over the whole area in order to deposit a first "layer". The same algorithm - appropriately shifted in *x* and *y*-direction - was applied to deposit additional layers, as indicated once with the grey grid. During the procedure the speed of movement determined how long the ion beam was positioned over a specific spot, and, consequently, how much material was deposited. This algorithm together with a controlled variation of the speed could be exploited to vary the film thickness in the range of ±1 nm. To facilitate the production of structured films, an additional procedure - that permitted the import of grey-scale pictures, assignment of desired film thickness to individual colors and appropriate translation into stage movements - was implemented into the software (Fig. 5).

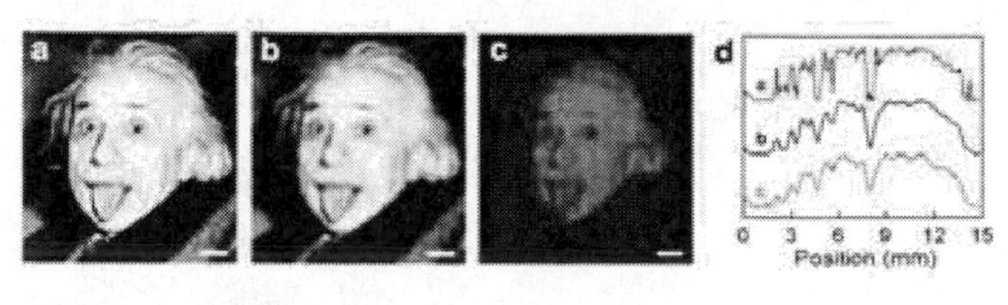

Figure 5: Production of a 3d structured thin film of electro active polymer [2]. (a) An image of Einstein (15x15 mm) was used as template. Grey scales were assigned to desired film thicknesses. (b) Calculation of the image that should be obtained under the assumption of a lateral resolution of 100 lines.cm-1. (c) Emission photograph of the resulted

thin structured film. (d) Comparison of thickness profiles and emission intensity along a horizontal line through the eyes in (a)-(c); scale bar: 2 mm

5 TAILORING FILM MORPHOLOGY

Each application of thin films requires distinct surface properties. With APID it is possible to tailor these properties by variation of different process-parameters (e.g. concentration, flow rate, etc.). To study the effects of these different settings during the APID-processing atomic force microscopy (AFM) measurements in the so-called "tapping-mode" were performed on a Digital Instruments Nanoscpope IIIa AFM under ambient conditions. AFM scanning was performed with a typical scan frequency of 0.5-1 Hz. The images were analyzed and processed using the standard software (Nanoscope V5.12r5) supplied with the control electronics.

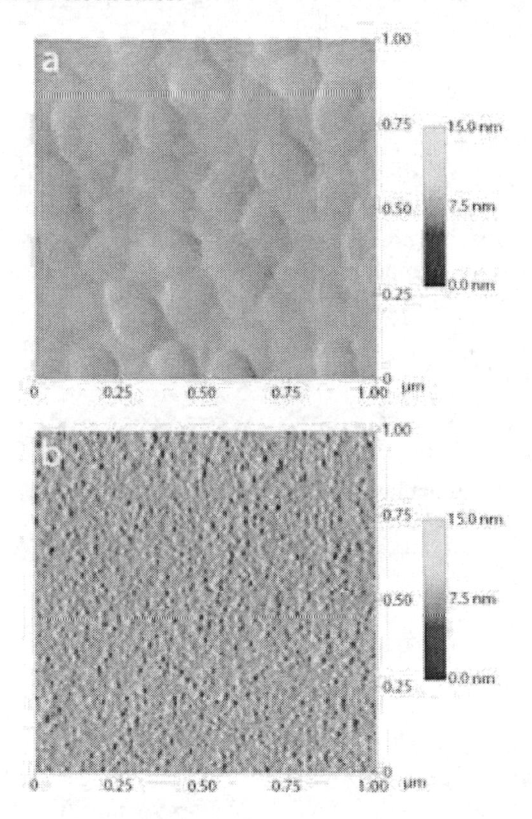

Figure 6: Tailoring surface morphology by variation of process parameters. (a) Thin film of a conductive polymer on ITO/glass with a roughness (r.m.s.) of ca. 0.4 nm (b) Thin film of the same material as in (a) on ITO/glass with a roughness (r.m.s.) of ca. 4 nm.

After interpretation of the images taken during AFM measurements two main trends could be deduced. Low concentrations of the material that is processed in combination with low flow rates give very smooth surfaces (Fig. 6a) with a roughness of ca. 0.4 nm (r.m.s.) whereas for high concentration in conjunction with higher flow rates the opposite is the case. During the deposition under these settings films with much rougher surfaces (Fig. 6b) e. g. ca. 4 nm (r.m.s.) are formed.

6 EXAMPLES OF APPLICATIONS

APID allows highly controllable and soft processing of various materials (e.g. polymers, bio molecules and inorganic materials) into thin and ultrathin structured films. Even multi-component systems of materials with comparable solubility can be prepared.

6.1 Multilayered Systems

A polymer multilayer with application in OLEDs was realized by combining spincoating and APID. The used materials are shown in scheme 1.

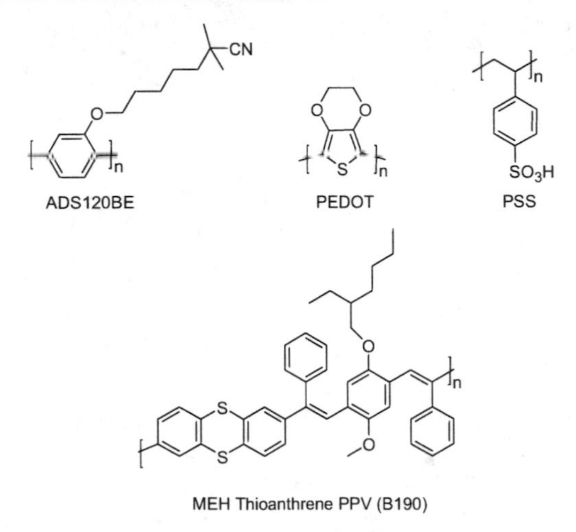

Scheme 1: Materials used for the multilayered system

A layer with a thickness of around 40 nm of PEDOT/PSS was spincoated out of aqueous solution on an Indium Tin Oxide (ITO) coated polycarbonate (PC) slide. The layer was dried in an evacuated oven for 4 hours at 120 °C. On this intermediate layer a second layer with a thickness of 40 nm of MEH Thiaoanthrene PPV (B190) was brought up by spincoating. The finishing layer of Poly[2-(6-cyano-6-methylheptoxy)-1,4-phenylene)] (ADS120BE) with a thickness of 40 nm was deposited by APID. The polymer was dissolved in a mixture of acetone and distilled water 99:1 (v/v). The whole setup is shown in Fig 7.

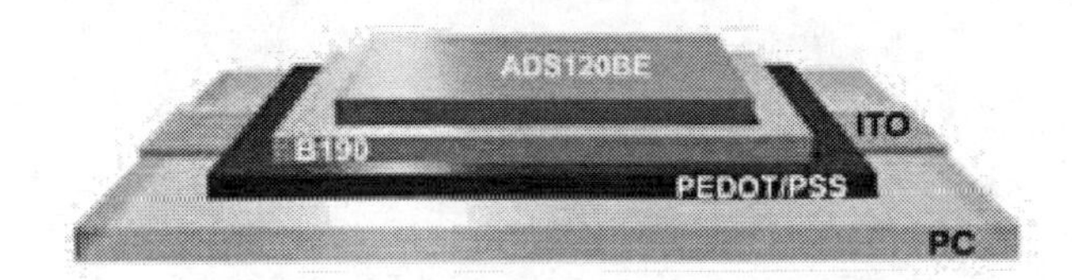

Figure 7: Setup of a multilayer system used for OLEDs

The multilayer was characterized before and after deposition of the APID layer. The measurements were performed on a Shimadzu RF-5310 IPC Spectro-fluorometer. The spectra are shown in Fig 8. The emission spectra (full red line) taken after the deposition shows a second peak at 3.1 eV what is characteristic for the material ADS 120BE. The peak of the MEH Thioantrene PPV is slightly shifted to lower wavelengths (2.4 eV) but in form and height identical with the spectra taken before deposition of the APID layer (full black line). The excitation spectra after the deposition (dashed red line) also showed a second peak caused by ADS120BE at 4.3 eV.

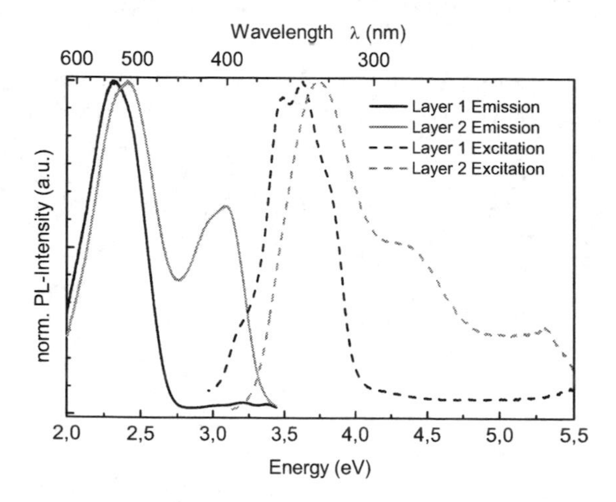

Figure 8: Emission- and excitation spectra of a multilayer system prepared by APID. The red lines show the measurements taken after the APID layer was deposited.

It can be seen, that the deposition of an APID layer does not change the optical properties of the former built up polymer layer. It seems that these layers are neither damaged nor dissolved, what can be seen as an evidence of an existing multilayer structure.

Another example is shown in Figure 9. An antidromic double-wedge consisting of polymers with equal solubility was produced by depositing a film of the blue polymer whose thickness continuously increased into the x-direction followed by a film of the red polymer whose thickness increased into the opposite direction.

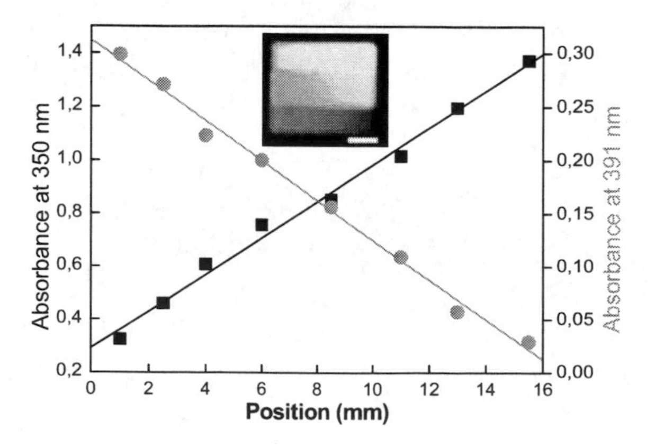

Figure 9: UV/VIS characterization of a double wedge of materials with equal solubility (inset: emission photograph, scale bar: 2 mm)

The graph shows the absorption measured at two wavelengths that correspond to λ_{max} of the two polymers.

Possible applications for structures like this might be found in organic light emitting devices (OLEDs) or in other organic electronic devices like organic field effect transistors (OFETs).

7 CONCLUSION

Direct patterning of advanced functional materials and the production of thin structured films with variable and defined chemical composition in all three dimensions is still a challenging problem. Therefore APID seems promising since being able to use multi-component chemical systems rather simply. We have demonstrated that presently the spectrum of materials that can be processed by this new approach is quite large and should be extended further. Adapted from the electrospray-ionization process APID is a soft technique where usually no chemical modifications will occur. Bearing in mind that APID includes a drying step prior to deposition, even the production of composites impossible today appears to be realistic. Moreover, the technique functions under atmospheric pressure, an important advantage over methods that need expensive vacuum equipment.

8 ACKNOWLEDGEMENT

The financial support by the Austrian Science Fund (FWF, Vienna) (SFB Electroactive Materials, project F922) is gratefully acknowledged.

REFERENCES

[1] Fraxedas, J. “Perspectives on thin molecular organic films” Adv. Mater., 14, 1603-1614, 2002.

[2] Saf, R., Goriup, M., Steindl, T., Hamedinger, T. E., Sandholzer, D., Hayn, G. "Thin Organic Films by Atmospheric-Pressure Ion Deposition." Nature Mat. , 3, 323-329, 2004.

[3] Saf, R., Mirtl, Ch., Hummel, K. “Electrospray ionization mass spectrometry as an analytical tool for non-biological monomers, oligomers and polymers” Acta Polym., 48, 513-526, 1997.

A Gate Layout Technique for Area Reduction in Nano-Wire Circuit Design

Hamidreza Hashempour* and Fabrizio Lombardi**

* LTX Corp., San Jose, CA 95134, USA, hhashemp@ece.neu.edu
** Northeastern University, Boston, MA 02115, USA, lombardi@ece.neu.edu

Abstract

This paper presents an homogeneous (array-based) approach for designing and manufacturing digital circuits using nano-tubes/nano-wires. As "a strategy for developing integrated devices with many individual elements has yet to be formulated" [1], it is evident that such an environment is a necessity for designing circuits using nano-wires. At logic level a novel formulation for area reduction is proposed and solved in polynomial time using a heuristic technique. The objective is to provide a further insight on the applicability of Moore's law to nanotechnology by evaluating the effects of area on logic design.

Keywords: Nano-wire, Nanotube, Circuit Design, Layout Optimization, Physical Area

1 Introduction

The feature size of basic devices (such as transistors) has constantly been decreasing over the past years. Today, transistors with gate lengths below 50nm can be fabricated and exhibit excellent electrical characteristics [2]. This trend has resulted in an almost exponential growth in integration level of electronic chips and integrated circuits, often referred to as Moore's law [3]. The architectural, and fundamental limits to this growth have been revised several times to account for new technologies; novel technological concepts (based on nano-devices and nano-electronics) are projected to be of primary importance [4] [5] [6] for future systems. Moreover, it is expected that so-called emerging technologies will not be limited by the fundamental barriers which are encountered today in for VLSI.

While fabrication of nano-devices presents considerable challenges, the high-level architectural organization at both device and circuit levels must be addressed. The unprecedented density and integration of these circuits necessitate novel arrangements for connections among nano-devices as well as input/output (I/O) of the chip; new techniques for handling a large number of connections (inclusive of electrical wires) are required to avoid undue heating, interference or "cross-talk" among them and exhibit proper control [7].

This paper presents novel techniques; a homogeneous array based methodology for nano-circuit implementation using nano-wires (e.g. carbon nanotubes) is presented. Initially, a circuit is represented using a set of standard Sum of Products (SoP). A minimal set of the products is constructed to cover all SoPs using a 2-level logic optimizer. The homogeneous array is then constructed by mapping each product term to a nano-wire. A nano-wire implements and connects multiple nano-devices to reduce the number of required contacts. A placement of the contacts for driving the inputs to the array and nano-devices is then formulated as a combinatorial Traveling-Salesman Problem (TSP). This instance is solved using the Lin-Kernighan heuristic to establish the placement of the nano-devices and their gate contacts, thus achieving a substantial saving in the physical layout for masking. This is required due to the difference in size between a nano-device metal gate and the nano-wire in its implementation.

2 SoP Based Array Design

Using nanotechnology, the active elements of a circuit (e.g. its transistors) can be made very small; however, the information transfer among these elements and the extraction of output/input signals from the circuit present difficult challenges due to the physical limitation of the contacts, i.e. the size of a contact is relatively large compared to a nano-device, thus often degrading the area benefits associated with a nano-scaled layout. An intra-molecular circuit implementation provides a promising design alternative; simple INV and NOR2 circuits have been proposed in [8] [9].

These techniques can be extended to circuits by utilizing basic boolean transformations, such as Sum of Products (SoP). For a SOP, each product term is mapped to a single nano-wire and implemented in a fashion that is reminiscent of Pass-Transistor logic design, as commonly encountered in CMOS. In this arrangement, one end of the nano-wire is constantly driven to logic 1 (high), while the other end represents the logic value of the product term. Nano-devices (as transistors) are placed along the nano-wire, Figure 1 shows a product term of 3 nano-devices.

Multiple nano-wires (each implementing a product

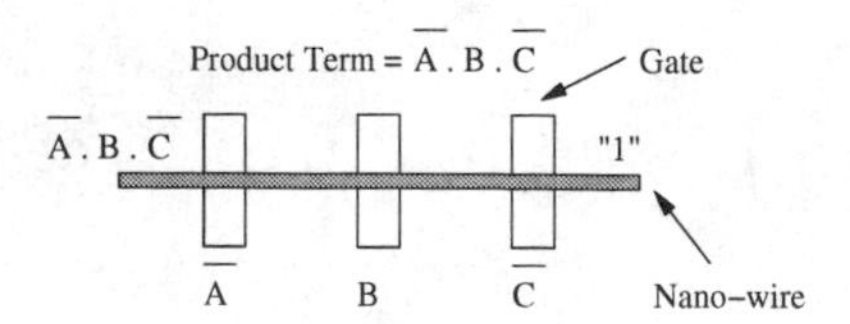

Figure 1: Nano-wire for constructing a product term

term) can be assembled into a nano-circuit; the gate of a nano-device occupies a rectangular area of metal placed over a nano-wire to form a nano-device, similar to traditional CMOS technology, i.e. a polysilicon crossing over an active diffusion area makes a MOS device. The metal gates are arranged in an homogeneous pattern to build the product terms and implement the nano-circuit. The important benefit of this approach is that *the nano-circuit design is effectively accomplished by a conventional process*, because patterning and etching of metal gates already use this technology. In this case the requirements are as follows: 1) the nano-wire must act in a fashion similar to a depletion mode MOS device, i.e. conducting with no field effect and semi-conducting in the presence of the field effect; 2) the on-resistance must be low, so many nano-devices can be connected in series with little effect on logic levels; 3) the threshold voltage must be low, thus allowing non-complementary operation in the nano-devices. This approach is different from [8] [9], because in the proposed design, the behavior is not complementary and complex (top) gates can be generated at relative ease (as multiple nano-devices can be placed along the same nano-wire).

The proposed approach offers substantial benefits because the source-drain contacts of the device are effectively removed (because the nano-wire acts as contacts between the source and drain of the nano-devices) and there is little added complexity in pattern-based processing steps for the nano-wires at manufacturing (in particular there is no need for complex patterning as in the active diffusion areas of conventional CMOS processes). Moreover, the *Schottky* barrier effect ([10]) is considerably reduced (i.e. it is distributed among multiple nano-devices).

2.1 Sizing

A problem which is commonly encountered in the layout for physical design, is size matching because even *the smallest feature sizes available today are substantially larger than a nano-wire diameter* (e.g. 130nm feature size and a 1.5nm wide nano-wire), thus degrading the possible benefit in area (and ultimately density) due to the small size of the nano-wires. Figure 2 shows two horizontally aligned nano-wires, the vertical adjacency between gates causes them to be spaced further away, thus adding to the unutilized (wasted) area. However this arrangement can be also viewed in a diagonal direction. In this case, the diagonal adjacency allows more closeness between nano-wires and gates, thus enhancing area utilization. This example highlights the importance of gate placement for efficient layout design.

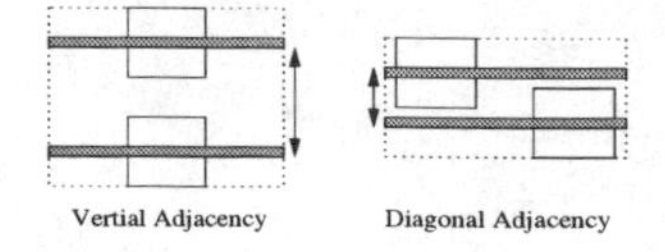

Figure 2: Effects of adjacency in gate design

Therefore, placement is a tight requirement for accomplishing closeness (diagonal adjacency) among gates. Figure 3 (a) shows a simple circuit made of three product terms, while Figure 3 (b) shows a possible modification to increase the closeness among nano-wires, thus effectively compressing the layout.

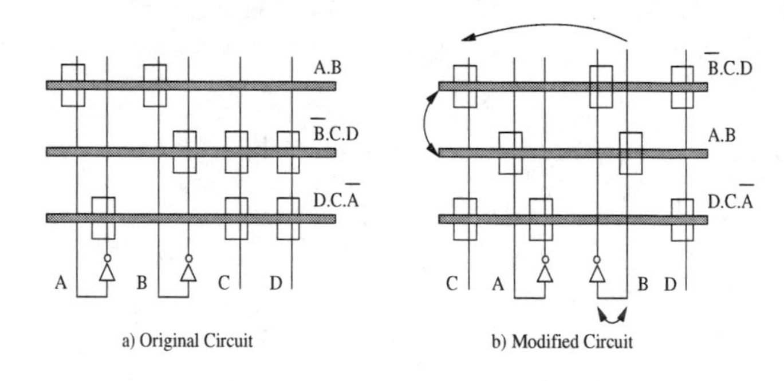

Figure 3: Increasing closeness among gates

3 Proposed Combinatorial Approach

As shown previously, modifications to the layout are required to overcome sizing differences; this process involves moving rows (product term nano-wires) and columns (gate lines). This can be solved using a graph approach in which each row/column is mapped to a graph node; an edge is placed between every pair of nodes (complete graph) with a weight given by the number of adjacent gates in the corresponding rows/columns. These graphs are referred to as adjacency graphs. Figure 4 shows the graph representations of the nano-circuit of Figure 3.

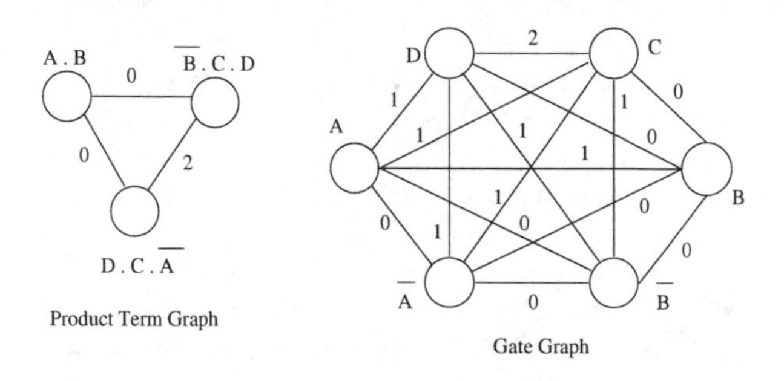

Figure 4: Adjacency graphs

NSTI-Nanotech 2005, www.nsti.org, ISBN 0-9767985-1-4 Vol. 2, 2005

The Minimum *Hamiltonian* cycle of these graphs corresponds to the optimal placement of the product terms (nano-wires or rows) and gates (columns). However, an optimum solution to this problem requires an exponential complexity (i.e. it is NP complete). The so-called Lin-Kernighan heuristic algorithm [11] is utilized in this paper; this algorithm has polynomial time with near optimal solution in most cases.

3.1 Example: the c17 benchmark

Consider a simple combinational circuit from the ISCAS85 benchmark set as shown in Figure 5. Each of the two outputs (g_{22} and g_{23}) can be expressed in SoP form of five inputs as,

$$g_{22} = g_1 \cdot g_3 + g_2 \cdot \overline{g_3} + g_2 \cdot \overline{g_6} \quad (1)$$

$$g_{23} = g_2 \cdot \overline{g_3} + g_2 \cdot \overline{g_6} + g_7 \cdot \overline{g_3} + g_7 \cdot \overline{g_6} \quad (2)$$

There are seven product terms, but two of them are redundant, i.e. they are used in both SoPs ($g_2 \cdot \overline{g_3}$ and $g_2 \cdot \overline{g_6}$). So, only five product terms must be implemented. In general, SoPs can be minimized using a 2-level logic optimization tool, such as Espresso [12]. The corresponding adjacency graphs can be constructed as in Figure 6. The arrow lines represent the solution to the combinatorial problems for an optimal gate placement. Figure 7 shows the original and the optimal implementations of the product terms for the c17 benchmark. The original circuit had 8 adjacencies, four of them are along the diagonals. The optimal circuit has only four adjacencies, all of them are in the diagonal directions.

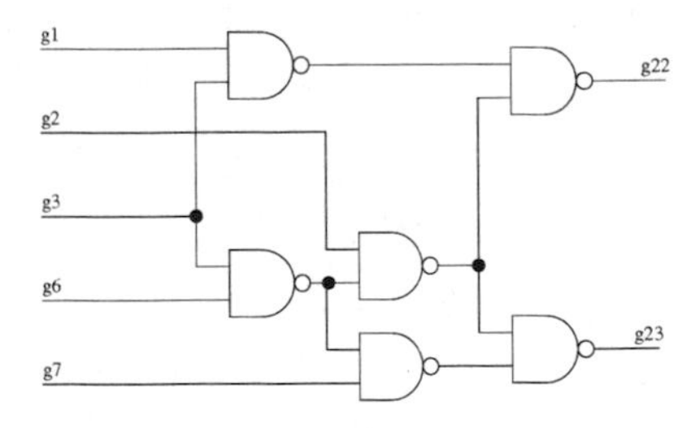

Figure 5: c17 benchmark netlist

4 Conclusion

This paper has presented an homogeneous (array-based) approach for designing and manufacturing digital circuits using nano-tubes/nano-wires. A methodology has been proposed for placement and reduced patterning effort for nano-wires to implement combinational circuits in Sum-of-Products (SoP) form. A combinatorial approach has been proposed for its solution; as an optimal technique has exponential complexity (due to NP

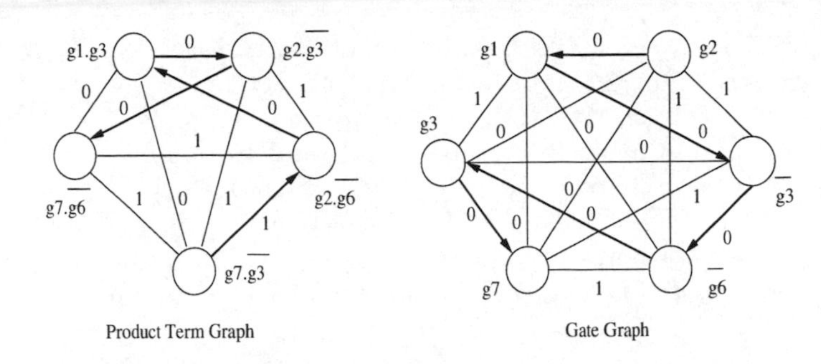

Figure 6: c17 benchmark adjacency graphs

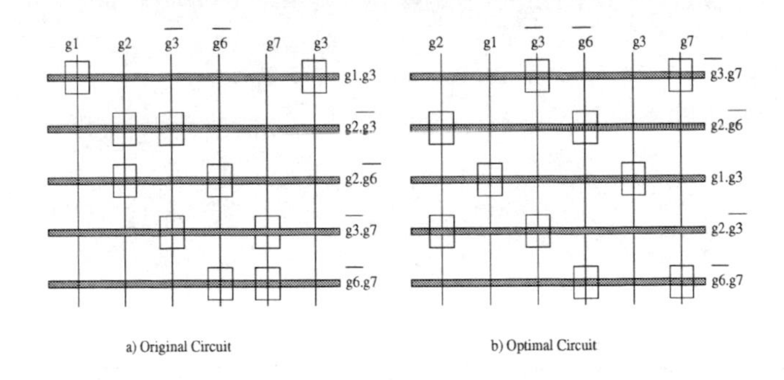

Figure 7: Optimal gate placement for c17

completeness), an heuristic based technique with polynomial complexity has been presented for gate placement to reduce the area layout of the circuit.

REFERENCES

[1] C. Dekker "Carbon Nanotubes As Molecular Quantum Wires," *Physics Today,* pp. 22-28, May 1999.

[2] R. Chau, J. Kavalieros, B. Roverts, R. Schenker, D. Lionberger, D. Barlag, B. Doyle, R. Arghavani, A. Murthy, and G. Dewey, "30-nm physical gate length CMOS transistor with 1.0-ps nMOS and 1.7-ps pMOS gate delays," *IEDM Tech. Dig.,* pp. 45-48, 2000.

[3] G. E. Moore, "Lithography and the Future of Moore's Law," *Proc. of SPIE, Advances in Resist Technology and Processing,* Vol. 2438, pp. 2-17, 1995.

[4] L. Risch, "How small can MOSFETs Get?," *Proc. of SPIE, Advances in Microelectronic Device Technology,* Vol. 4600, pp. 1-9, 2001.

[5] D. J. Frank, R. H. Dennard, E. Norwak, P. M. Solomon, Y. Taur, and H. S. P. Wong, "Device Scaling Limits of Si MOSFETs and their Application Dependencies," *Proc. of IEEE,* Vol. 89, pp. 259-288, March 2001.

[6] D. Sylvester and K. Keutzer, "Getting to the Bottom of Deep Submicron," *Proc. of Intl. Conf. on Computer Aided Design,* pp. 203-211, 1998.

[7] A. Dehon, "Array-Based Architecture for FET-Based, Nanoscale Electronics," *IEEE Trans. on Nanotechnology,* Vol. 2, No. 1, pp. 23-32, March 2003.

[8] A. Bachtold, P. Hadley, T. Nakanishi, and C. Dekker

"Logic Circuits with Carbon Nanotube Transistors," *Science,* Vol. 294, pp. 1317-1319, Nov. 2001.

[9] V. Derycke, R. Martle, J. Appenzeller, and P. Avouris "Carbon Nanotubes Inter- and Intramolecular Logic Gates," *Nano Letters,* Aug. 2001.

[10] P. L. McEuen, M. S. Fuhrer, and H. Park "Single-Walled Carbon Nanotube Electronics," *IEEE Trans. on Nanotechnology,* Vol. 1, No. 1, pp. 78-85, March 2002.

[11] *http://www.akira.ruc.dk/ keld/research/LKH/*

[12] P. McGeer, J. Sanghavi, R. Brayton and A. S. Vincentelli, "ESPRESSO-SIGNATURE: A New Exact Minimizer for Logic Functions," *IEEE Trans. on Very Large Scale Integration,* Vol. 1, No. 4, pp. 432-440, Dec. 1993.

Pressure sensor elements integrated with CMOS

J. Kiihamäki*, T. Vehmas*, T. Suni*, A. Häärä*, M. Ylimaula* and J. Ruohio**

VTT Information Technology, P.O. Box 1208, FIN-02044 VTT, FINLAND, jyrki.kiihamaki@vtt.fi
*VTI Technologies Oy, P.O. Box 27, FIN-01621 Vantaa, FINLAND

ABSTRACT

We report the measurement and fabrication results of monolithically integrated capacitive pressure sensor elements. The device fabrication process is based on novel plug-up process [1], which enables monolithic integration of sensors and CMOS in a modular fashion.

The electrical measurement results including pressure dependence, temperature dependencies of the various sensor geometries, and the effects of pre-bond tailoring of SOI wafers are presented in this paper.

Keywords: MEMS, SOI, monolithic integration

1 INTRODUCTION

Monolithic integration of MEMS and electronics is sought for miniaturization of microsystems. Single chip realization of a pressure sensor and its interface circuitry could be suitable for applications such as handheld devices, where small size is of great importance.

A simple monolithically integrated pressure sensor circuit fabricated on SOI was presented in previous work [2]. The capacitive sensors are made from the vacuum cavities embedded in the SOI structure, a schematic cross-section of a cavity is shown in Fig.1.

2 FABRICATION

The integration approach used in this work is ′MEMS first′, where the cavities for sensor elements are fabricated before CMOS. SOI wafers are used as starting material. The substrate and structure layer doping level were selected to simultaneously obtain good sensor performance and to facilitate CMOS fabrication. A non-patterned SOI substrate was used. For improving the performance of capacitive sensors a pre-bond boron implant was performed on some of the wafers during the SOI manufacturing process. The enhanced surface doping of MEMS devices is desired to void formation of voltage and temperature sensitive depletion regions and other phenomena associated with low doping.

2.1 Cavity formation

The key issue in this process is the controlled formation of hermetically sealed cavities in the SOI structure and the resulting smooth single crystal surface after etchback steps, which enables versatile backend processing.

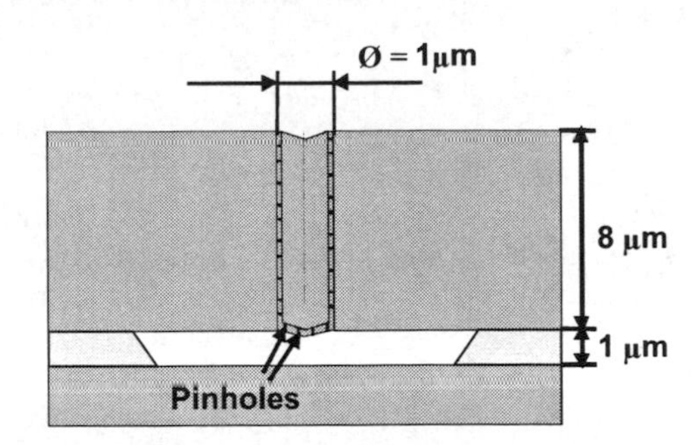

Figure 1: Cross-section and dimensions of a polysilicon plug and SOI structure used in this work.

The cavities are wet etched through a thin permeable polysilicon film [3], which contains small holes. The polysilicon deposition conditions are controlled in such a way that deposition occurs between amorphous and polycrystalline deposition regimes. The obtained film contains enough pinholes when the film thickness does not exceed the grain size [4]. A SEM micrograph of the thin polysilicon film grain structure is shown in Fig. 2.

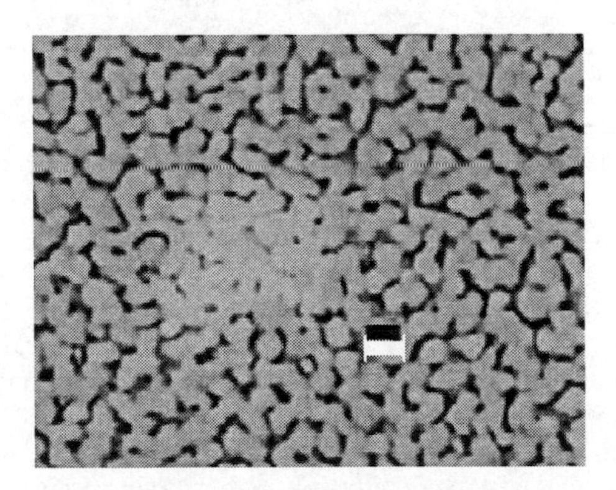

Figure 2: Grain structure of a permeable polysilicon film. The white marker shown is 200 nm.

The cavities are then closed by depositing a thick conformal polysilicon film, which blocks the pinholes as well as the one micron sized dry etched holes forming polysilicon plugs in the SOI structure layer.

2.2 CMOS

The CMOS used in this work is 1.2 μm gate length analog oriented molybdenum gate bulk CMOS. Besides CMOS this process contains bipolar transistors, thin film resistors, capacitors and optional EEPROM, and high voltage NMOS [5].

3 DEVICE DESIGN

The MEMS part of the device requires four extra mask layers to conventional CMOS: 1) the cavity defining the array of polysilicon plugs, 2) substrate contact, 3) trench for dielectric isolation, and 4) removal of extra layers from micromechanical device areas after CMOS.

A top view of a circular sensor and reference capacitor structures are shown in Figs. 3a and 3b. The ambient pressure deflects the structure layer of the vacuum cavity in the sensor, which appears as a darker area in the central parts of the sensor, while the vented reference structure is not deflected.

3.1 Pressure sensors

The pressure sensors are circular shaped cavities. The diameter of the cavity is about 300 μm, the size at which the used structure touches the substrate when an external pressure of 6 bar is applied. The cavities are isolated from the environment with dielectric isolation, a trench etched through the structure layer and refilled with silicon dioxide. The sensor element is surrounded by a hexagonal guard ring formed between two isolation trenches. The CMOS metallizations and substrate contacts are used for interconnecting the pressure sensors.

3.2 Reference elements

The integrated reference for temperature compensation of the capacitive element can be an oxide capacitor or any structure that does not change with the applied pressure, but where capacitance changes similarly to a pressure sensor with temperature. We made simple reference devices by opening the pressure sensor cavity using a lithographically defined and etched vent hole. A reference with an open cavity may not be optimal in real device that can be exposed to dirt and moisture, but measurements of the opened device provide interesting information about the structure when comparing the measurements with the closed structure.

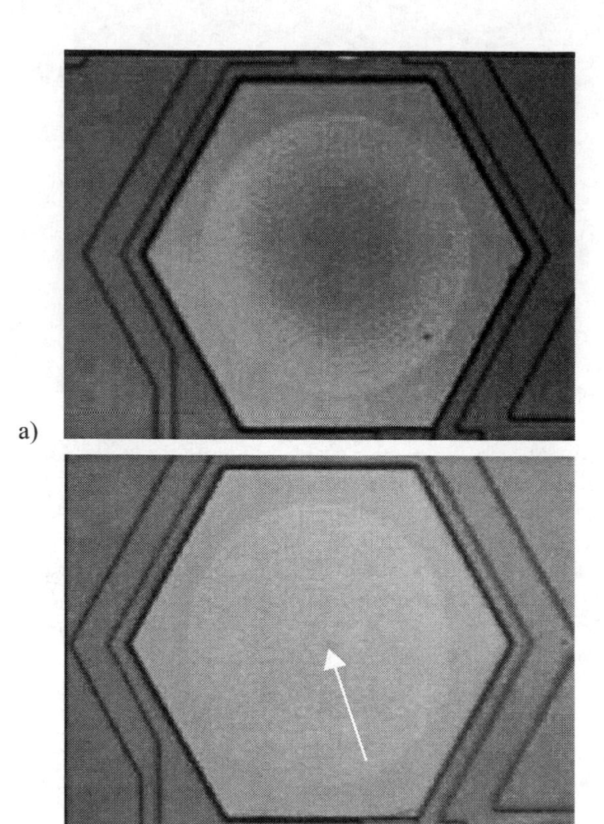

Figure 3: IR photographs of a pressure a) sensor cavity and b) reference element with vent hole in the center (indicated by the arrow).

4 MEASUREMENT RESULTS

4.1 CMOS

The properties of CMOS were not affected by the fabrication of MEMS elements. The leakage currents remained at low level. The relatively thick SOI layer did not change the process results significantly. Only the breakdown voltages of some inherent diode structures were lower in the SOI realization when compared to bulk wafers.

4.2 Pre-bond implant

The non-implanted wafers showed higher sensitivity to temperature and stronger asymmetric bias dependence than the devices with pre-bond implantation. The C-V curves of buried oxide capacitances are shown in Figs. 4a and 4b.

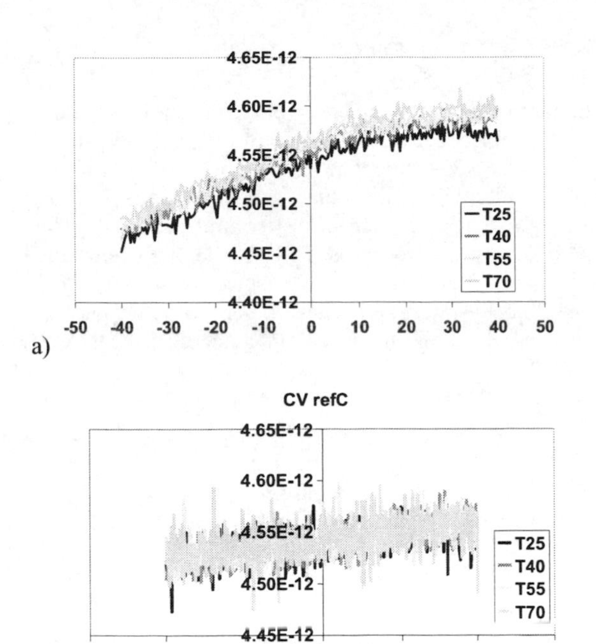

Figure 4: a) C-V characteristics of an oxide capacitor made made a) on standard SOI, b) on pre-bond implanted SOI. Temperature (°C) is used as a parameter.

4.3 Pressure sensors and references

The diced and encapsulated pressure sensor devices were measured in a weather chamber at VTI Technologies. The external pressure and temperature were controlled and the sensor capacitance was measured with an LCR bridge.

A typical measured pressure dependence of a sensor element is shown in Fig. 5a and the temperature stability of the reference element in Fig. 5b. The achieved temperature sensitivity of the sensor elements was excellent, limited only by material the properties of silicon and silicon dioxide.

The pressure dependence of the capacitance can be satisfactorily modeled with a modified parallel plate capacitor model using the following equation:

$$C = C_{00} + \frac{C_0}{1 - C_0 p / K} + \frac{aC_0}{1 - C_0 p / bK}, \quad (1)$$

where C_{00} is the pressure independent stray capacitance, C_0 the pressure dependent capacitance, p pressure, K reduced sensitivity, and a and b are sensor geometry dependent constants (the fitted parameters for the device of Fig. 5a. are as follows C_{00} = 2,453 pF, C_0 = 0.112 pF, K – 83.2 kPa·pF, a = 3.15 and b = 2.2).

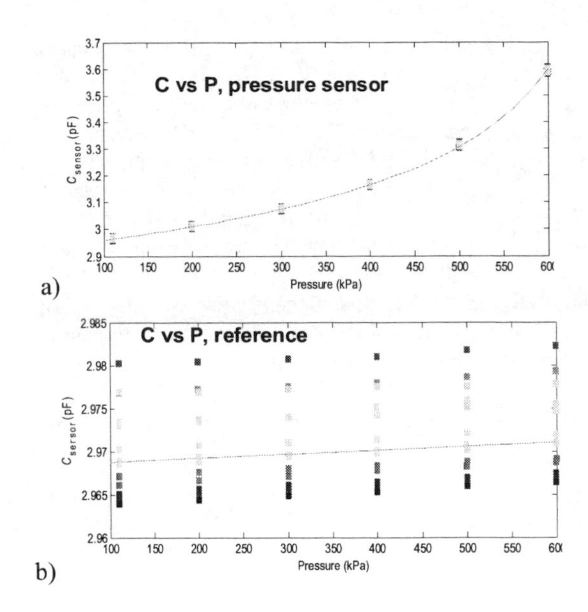

Figure 5: a) Pressure dependence of sensor element. a) pressure dependence of reference element. Temperature as a parameter, from -40°C to +125°C.

The temperature dependence of the sensor capacitance at 5 bar pressure is shown more closely in Fig 6. The capacitance sensitivity is about 43 ppm/K at 5 bar pressure and about 30 ppm/K at atmospheric pressure. The temperature sensitivity of thevented structure is about 30 ppm/K throughout the measurement range.

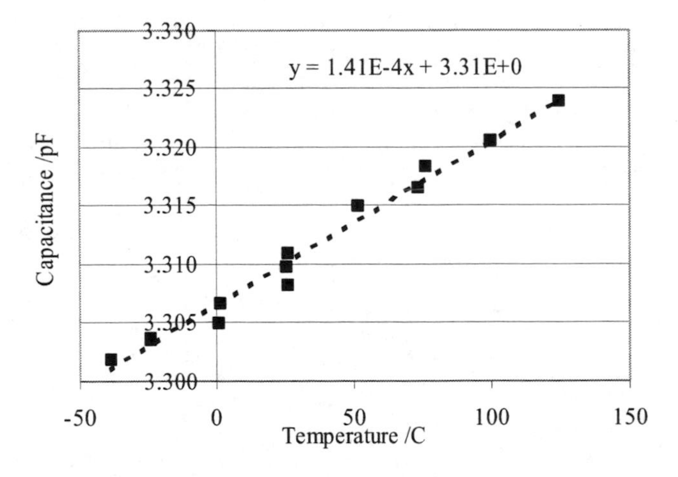

Figure 6: Temperature dependence of sensor capacitance at 5 bar pressure.

In the future temperature effects will be minimized with monolithic integration of readout electronics and temperature compensation circuitry.

The resonance frequency of another pressure sensor as a function of external pressure was measured at room temperature using an impedance analyzer. The measurement result is shown in Fig. 7. The resonance frequency of the closed cavity decreases as the pressure is increased, because of the electrostatic spring softening effect. The applied constant bias produces a higher electric field strength between the electrodes at higher pressure as the membrane is deflected closer to the bottom electrode. With the open reference structure the gap remains unaltered by the pressure. Instead, the effective spring constant increases with pressure, because the air in the gap acts like a spring. The change of resonance frequency or damping factor can be used for pressure sensing in applications where capacitive sensing is not applicable.

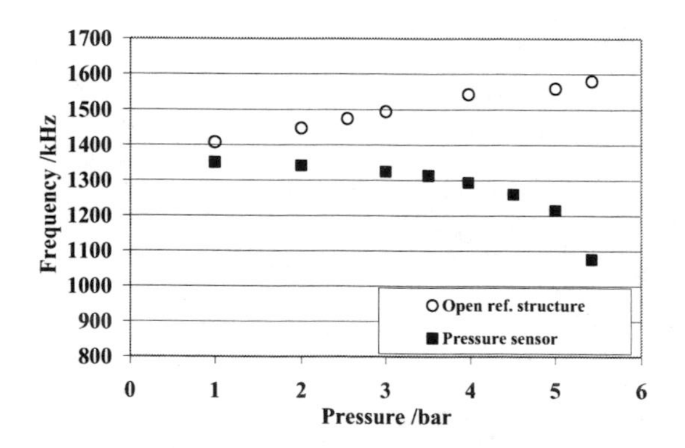

Figure 7: Resonance frequency of sensor and vented reference structures as a function of applied pressure, DC-bias 50 V.

Besides the circular sensor, oval shaped sensors were also designed. The pressure response of the oval shaped resonator is plotted in Fig. 8. The large structure touches the substrate at a low pressure of about 3.5 bar. The pressure dependence of the long sensor has improved linearity and dynamic range compared to a circular sensor.

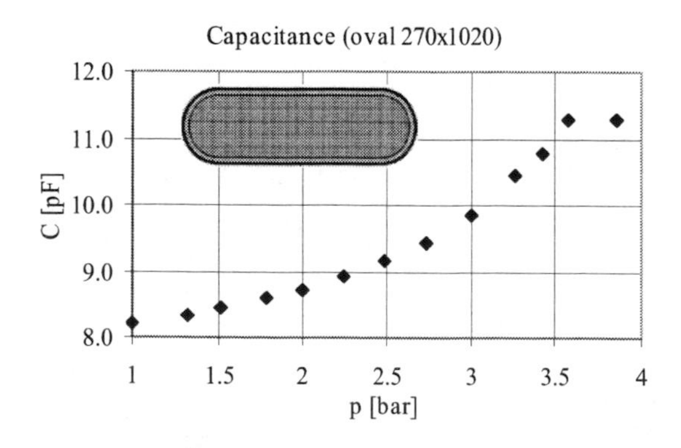

Figure 8: Pressure dependence of an oval shaped cavity (length 1020 and width 270 micrometers).

5 CONCLUSIONS

The fabricated pressure sensor elements show excellent electrical performance. The temperature sensitivity of the capacitance is below 50 ppm/K, which is very near the theoretical minimum value that can be obtained using silicon and silicon dioxide as structural material. The operation of the co-processed CMOS did not show any signs of degradation due to sensor fabrication. Fabrication of monolithically integrated pressure sensors is technically feasible using the plug-up integration concept.

ACKNOWLEDGEMENTS

This work was partly funded by the National Technology Agency of Finland, VTI Technologies, Okmetic, and Micro Analog Systems.

REFERENCES

[1] Kiihamäki, J., Dekker, J., Pekko, P., Kattelus, H., Sillanpää, T., Mattila, T. " 'Plug-Up' - A new concept for fabricating SOI MEMS devices", Microsystem Technologies Vol. 10 (2004) No. 5, pp. 346-350.

[2] Ylimaula, M., Åberg, M., Kiihamäki, J., Ronkainen, H., "Monolithic SOI-MEMS capacitive pressure sensor with standard bulk CMOS readout circuit", Proc. 29th European Solid-State Circuits Conference, ESSCIRC 2003, pp. 611-614.

[3] Lebouitz, K.S. Howe, R. T., Pisano, A. P., "Permeable Polysilicon Etch-Access Windows for Microshell Fabrication," Technical Digest, Transducers '95, Stockholm, Sweden, June 25-29, (1995), pp. 224-227.

[4] Dougherty, G. M., Sands, T. D., Pisano, A. P., "Microfabrication Using One-Step LPCVD Porous Polysilicon Films", J. Microelectromechanical Systems, Vol. 12, (2003), pp. 418-424.

[5] Kiihamäki, J., Ronkainen, H., Pekko, P., Kattelus, H., Theqvist, K., "Modular integration of CMOS and SOI-MEMS Using 'Plug-Up' concept", Digest of Technical Papers The 12th International Conference on Solid-State Sensors, Actuators and Microsystems, Transducers´03, (2003), pp. 1647 - 1650.

Development of Self-Assembled Robust Microvalves with Electroform Fabricated Nano-Structured Nickel

Bo Li and Quanfang Chen

Department of Mechanical, Materials & Aerospace Engineering
University of Central Florida (UCF)
PO Box 162450, Orlando, FL 32816-2450, USA, qchen@mail.ucf.edu

ABSTRACT

Self-assembled robust micro check valves with large flow rates (>10 cc/second, displacement related), high-pressure support ability (>10 MPa) and high operational frequencies (>10 kHz) made of nano-structured nickel were presented in this paper. The microvalve consists of an array of 80 single micro valves to achieve the required flow rates. A novel in situ UV-LIGA process was developed for the fabrication. Self-assembling was realized by guiding the electroforming process during the fabrication process. Test results show that the forward flow rate is about 19 cc/second under pressure of 90Psi. The backward flow rate is negligible. The reliability of the valve was tested by a specific loading/unloading sequence. Results show that the flow rates were repeated very well over a large range of tested pressure differences.

Keywords: self-assemble, microvalve, nano-Structured nickel, high frequency, high pressure support

1 INTRODUCTION

Hydraulic actuators, which convert hydraulic pressure into linear or rotary motions, have been used in a variety of applications. They are capable of producing large pressures/forces with compact size and high reliability. The compact robust hydraulic actuators are very important for space related applications due to their abilities in producing much larger forces/pressures per volume/mass than other existing technologies. They also play an important role in reducing the launching cost of spacecrafts, as any reduction in mass and/or power consumption for space instruments or subsystems will result in an exponential savings of the launch cost.

A compact hydraulic actuator mainly consists of a pump and inlet/outlet microvalves (Figure 1). The requirements for the compact pumping components include large pressure/force outputs, large flow rates and high working frequencies. The compact hydraulic actuators consisting of piezoelectric stacks and robust microvalves can produce power per unit volume ratio 100 to 1000 times greater than their electrostatic counterparts. The pumps (pushers) are fabricated with smart (active) materials, such as PZT, due to their simplicities in design and high power densities. On the other hand, microvalves are the key components for fluid management within the hydraulic actuator in order to meet the rigorous requirements in various space applications. The function of the microvalves is to manipulate fluid flow and to switch the flow directions (forward or backward). Obviously, the valves also have to bear the load (pressure) when they are in the closed state. Therefore, requirements for compact valves are stricter than those for the pumping components (pusher). These requirements include high flow rates (>10 cc/sec), high-pressure support ability (>10 MPa) and high operational frequencies (>10 kHz). As stated [1], active micro valves cannot fulfill all these requirements. Therefore, passive valves are promising for this application. The emerging MEMS techniques provide the opportunity to design and fabricate highly functional passive valves. Scaling laws have proven that mechanical systems could be improved greatly when the physical dimensions shrink [2], and they should apply equally to robust micro valves.

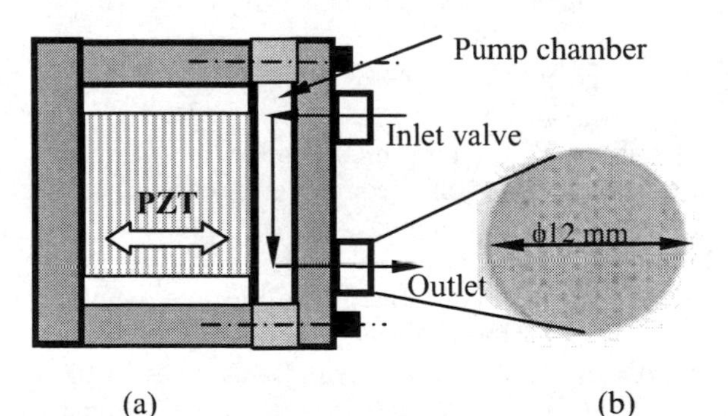

(a) (b)

Figure 1, Robust pump and microvalve.
(a) pump, (b) microvalve

Most currently developed microvalves use silicon and/or other non-metallic materials such as polymers for the structural and/or functional materials, due to the ease of fabrication. Such microvalves cannot survive without failure under severe dynamic loading conditions encountered in a robust pump (Figure 1a), due to the inherent materials' mechanical properties. Therefore, robust metallic microvalves are greatly needed for such applications since both the ultimate strength and the fracture toughness are important in these cases. Nickel is extremely suitable for the fabrication of these valves because of its high mechanical strength and ease of fabrication. The well-developed LIGA and UV-LIGA process can produce high-precision micro-scale components with vertical sidewalls that can be assembled or self-assembled to complex micro devices and systems. If properly processed, nano-structured nickel (Figure 2, right-upper corner) can be formed by the electroplating/electroforming process, which can in turn improve its mechanical performance. The authors have developed a novel mechanical tensile testing method to test the mechanical properties of the electroformed nano-structured nickel. The results are shown in Figure 2. The ultimate strength of this material can be extracted from the stress-strain curve, which is larger than 1Gpa. The Yield strength is more than 200 MPa. Using the nano-structured nickel as the building material, the authors have developed a pure nickel microvalve array, which is reported in this article.

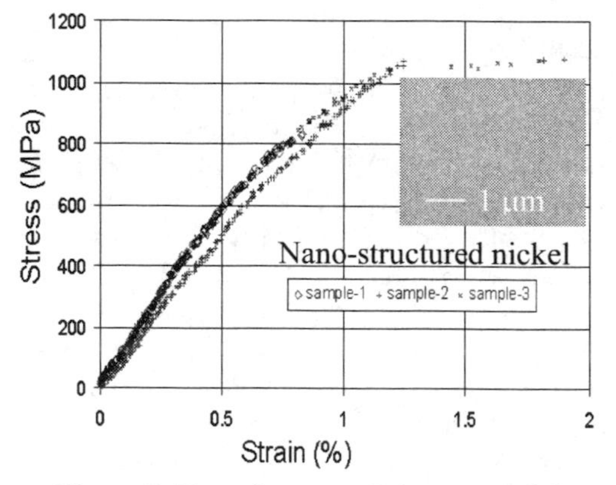

Figure 2, Tested stress-strain curve of the nano-structured nickel

2 MICROVALVE DESIGN

The design rules for the pure nickel micro mechanical passive valves are to meet the requirements of compact hydraulic actuators being developed for space applications. The concerns in mechanical design were to reduce the maximum stress of key components during the whole operation cycle. In order to match the operation of the PZT pump, the microvalve has to work at high frequencies (> 10 kHz). Therefore the natural frequencies should be much higher than 10 kHz to avoid resonance. This is insured by using the minimization. An 80-single-valve array is adopted to fulfill the flow rate requirement.

As shown in Figure 3, the designed microvalve is a normally closed micro check valve, which consists of an inlet channel, a micro nickel valve flap, and a valve stopper which houses the outlet channels. The valve flap is lifted by the pressure difference produced by the PZT pusher. Fluid is directed through the inlet channel and passes through the valve flap and leaves the valve through the outlet channels. A gap of 10 µm was designated between the valve flap and the stopper, which is used to prevent the potential tear off of the valve flap in the case of severe situation. The valve flap (Figure 3) is linked with four identical cantilevered micro beams (springs), which holds it to the valve substrate elastically. The valve returns closed by the spring force developed in the four beams, in addition to the reverse pressure difference created by the PZT's contraction (Figure 1).

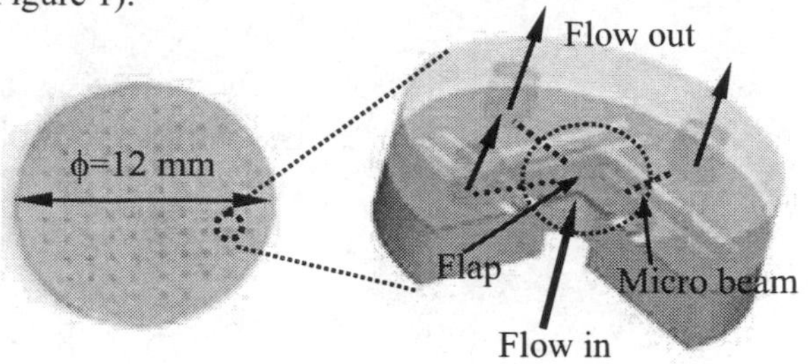

Figure 3, Concept sketch of the valve

The size of the valve-flap (square) is 300 µm by 300 µm and is held by four identical micro beams (600 µm x 50 µm x 10 µm each). The valve flap is reinforced with a cross pattern which is a15 µm thick sitting on a 10 µm membrane. The function of the cross pattern is to improve its pressure support ability in the closed state, as well as to overcome the stiction problem when the flap touches the stopper in the open state. Thick nickel layers (500 µm thick) were designed both as the supporting substrate and the stopper. All these structures are made of electroformed nickel with a novel in situ UV-LIGA process.

3 FEA SIMULATION

Static finite element simulation was performed to find the resonant frequency and to predict the stress distribution over the valve flap and the nickel substrate, both under open and closed conditions. The nickel properties used in the simulation are from the mechanical testing (Figure 2).

3.1 Resonant Frequency

The resonant frequency of the micro-beams-flap-system was predicted to be much higher than its macro counterparts due to scaling laws. Finite element simulation results showed that the system has a 1st mode resonant frequency of 18 kHz, which is much higher than the required working frequency (10 kHz). Therefore, no resonance will be induced when the valve works at 10 kHz.

3.2 Stress in the Valve Flap When Fully Opened

The stress distribution is another important concern in the design of the valve flap. Large stresses and stress concentrations need to be avoided. In the simulation (Figure 4), the ends of the micro beams were fixed, and pressure was applied over the valve flap. The stress distribution is shown in Figure 4, where the displacement of the valve flap is 10 µm, defined by the valve stopper. The maximum von misses stress is 128 MPa, located at the ends of the beams. This value is much smaller than tested strength values of electroformed nickel.

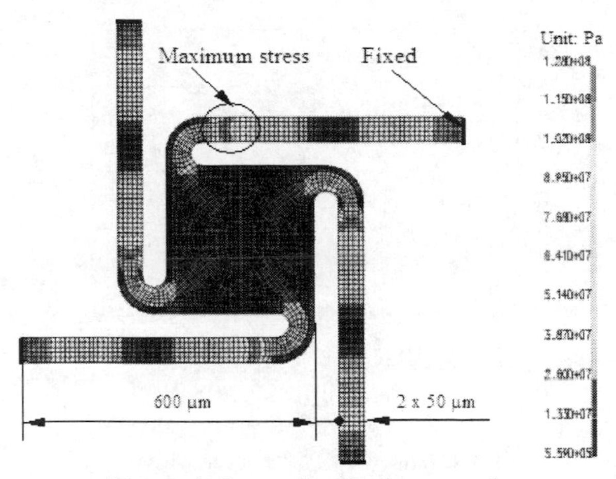

Figure 4, Stress distribution over the flap while it is fully opened

3.3 Stress in the Valve Flap When Closed

The valve flap was predicted to support high pressures (> 10 MPa) when it is closed. Therefore, finite element simulation was conducted to analyze the stress generated in this state. In the simulation (Figure 5), the ends of the micro beams were fixed and pressure was applied over the valve flap. The stress distribution is shown in Figure 5, where the applied pressure was 10 MPa. The maximum von misses stress was found to be 132 MPa, located in the center of the flap. This value is small in comparing to the strength of electroformed nickel.

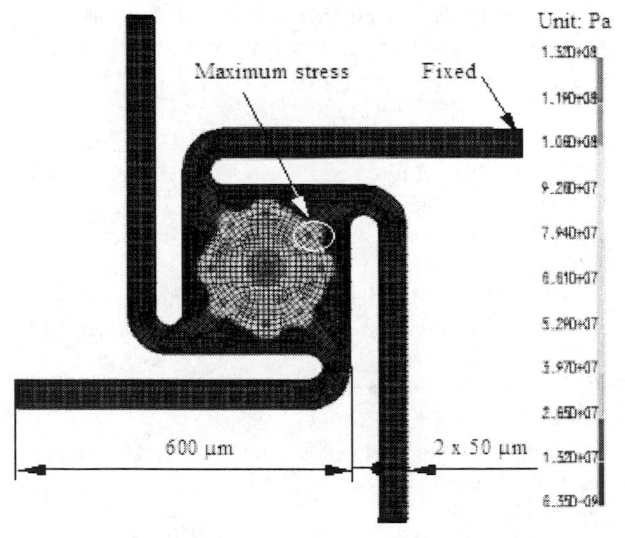

Figure 5: Stress distribution over the flap while it is closed

3.4 Stress in the Nickel Supporting Substrate

The nickel substrate (500 µm thick) needs to support large pressure, especially in the closed state. Stress analysis was performed to find the maximum stress and the distribution. The valve substrate is shown in Figure 6 (left). Thanks to geometric symmetry, only one quarter of the stress across the valve needed to be analyzed. A pressure difference of 10 MPa was applied to one side of the nickel piece. The results are shown in Figure 6. The maximum von misses stress was identified as 119 MPa. This value is in the safe range of the nickel material used. The use of nickel as the supporting substrate overcame the brittleness of the silicon substrate in the silicon-nickel valve [1]. This valve is much easier to handle, both in the process of fabrication and in the process of testing.

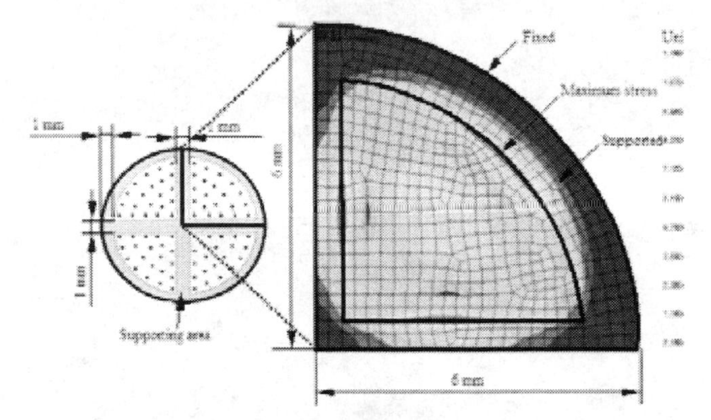

Figure 6, Stress distribution over nickel substrate under 10 MPa pressure while closed

4 FABRICATION PROCESS

The compact microvalve array had been fabricated by a newly developed innovative in situ UV-LIGA process. The valve inlet and outlet channels were defined by SU-8 molds [4] and the valve flap was defined by Photoresist molds. Chemical mechanical polishing (CMP) was applied during the fabrication. After electroforming, all SU-8 and photo resist molds were removed by using SU-8 remover and positive photoresist stripper. Finally, separated valve arrays were received. Therefore, this is a bottom–up self-assembled fabrication process. Neither etching nor bonding (or packaging) is involved during the fabrication. Figure 7 shows the fabricated valve array (80). Figure 8 shows the inlet/outlet view of the finished micro valve.

5 TESTING

To characterize the fabricated microvalve, a static fluidic test was performed by raising the pressure from low to high for both forward and backward flow. A specially designed valve holder was used to hold the valve while testing. Test results of flow rate versus applied pressure difference are provided in Figure 9. From the results, the valve's crack pressure is about 5 psi. The forward flow rate tested is approximately 19 cc/second under a pressure difference of 90 Psi, which is the largest pressure provided by the commercial air compressor used during testing. The backward flow rate is about 0.023 cc/s, which is negligible (<0.13%). It also can be seen that the measured flow rate is roughly proportional to the pressure difference applied, as predicted by Poiseuille's law. Based on Poiseuille's equation, it is reasonable to expect much higher flow rates if higher pressures are applied. On the other hand, the valve sealing is very good, as the backflow rate is negligible.

A specific loading/unloading sequence was designed to test the reliability of the valve (Figure 10). Results show that the flow rates are very well repeated over a large range of tested pressure differences. This implies that no damages or permanent deformations occurred during the test.

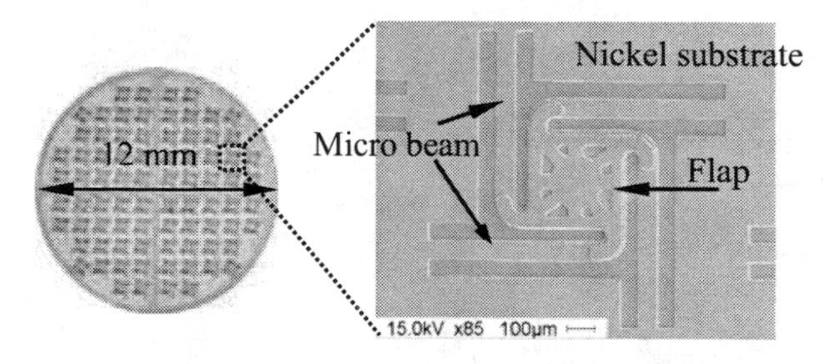

Figure 7, Fabricated micro valve array (80)

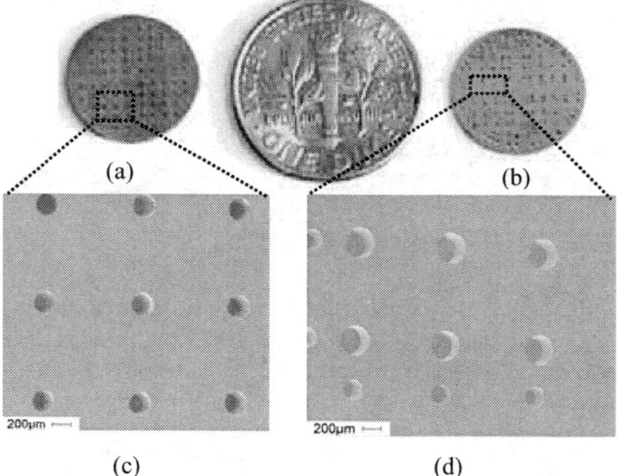

Figure 8, Fabricated microvalve. (a) Inlet view (b) Outlet view. (c) Inlet channels (d) Outlet channels.

6 SUMMARY

A self-assembled robust passive large flow rate, high-frequency and high-pressure solid nickel micro check valve has been developed for piezoelectrically actuated pumps or hydraulic actuators. The valve's reliability is assured by using a mechanically robust nickel valve stopper. Test results show that the forward flow rate is roughly proportional to the pressure applied. It is about 19 cc/s at a pressure difference of 100 Psi applied. The backward flow rate is negligible. The fabrication of the valve employed a novel in situ UV-LIGA process, where the valve stopper was self-assembled to the valve array without any additional bonding process. Therefore, this is a bottom–up self-assembled fabrication. No additional bonding (or packaging) process was involved during the fabrication.

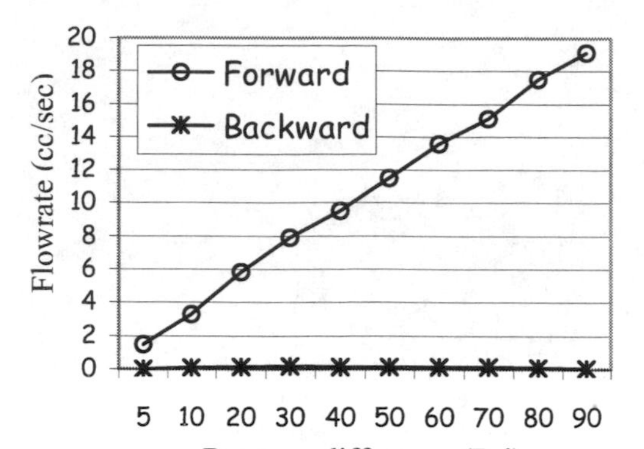

Figure 9, Tested flow rate versus pressure applied of the microvalve

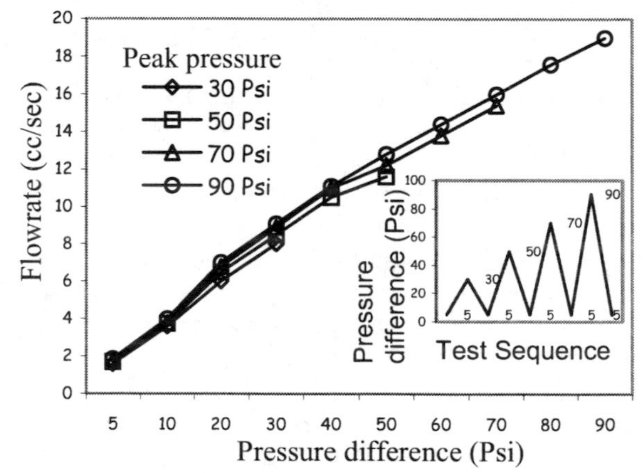

Figure 10, Repeated Test of forward flow rate under loading/unloading conditions

REFERENCES

[1] Bo Li and Quanfang Chen, Sensors and Actuators A: Physical, Volume 117, Pages 325-330, 2005.

[2] Marc Madou, "Fundamentals of Microfabrication", 2002.

[3] Quanfang Chen, Da-Jeng Yao, C-J Kim, and Greg P. Carman, Sensors and Actuators, Volume 73, Pages 30-36, 1999.

[4] Bo Li and Quanfang Chen, Proceedings of SPIE, Volume 5344, Pages 147-154, 2004.

[5] William L. Benard, Harold Kahn, Arthur H. Heuer, and Michael A. Huff, JMEMS, Volume 7, Pages 245-251, 1998.

[6] Donald E. Wroblewski, ASME International Mechanical Engineering Congress and Exposition, Volume 66, Pages 145-151, 1998.

DIODE LASER BONDING OF PLANAR MICROFLUIDIC DEVICES, BIOMEMS, DIAGNOSTIC CHIPS & MICROARRAYS

Dr. Jie-Wei Chen, Leister Process Technologies, Riedstr. CH-6060, Switzerland
Jerry Zybko, Leister Technologies, LLC, Glen Ellyn, IL 60137, U.S.A.
James Clements, NanoSciences, Inc. Aliso Viejo, CA 92656, U.S.A.

Abstract

The assembly of plastic microfluidic devices, MOEMS and microarrays requiring high positioning and welding accuracy in the micrometer range, has been successfully achieved using a new technology based on laser transmission welding combined with a photolithographic mask technique. This paper reviews a laser assembly platform for the joining of microfluidic plastic parts with its main related process characteristics and its potential for low-cost and high volume manufacturing. The system consists of a of diode laser with a mask and an automated alignment function to generate micro welding seams with freely definable geometries. A fully automated mask alignment system with a resolution of $< 2\ \mu m$ and a precise, non-contact energy input allows a fast welding of micro structured plastic parts with high reproducibility and excellent welding quality.

Background

The research and development of microfluidic device, so-called lab-on-a-chip technology, is one of the fastest growing areas of medical and biological diagnostics for a variety of applications including DNA analysis, drug discovery and clinical diagnostics [1,2]. Up to now, the preferred materials have been silicon, glass or quartz, mainly because micro fabrication methods for these materials have been extensively developed in the microelectronics industry. However, for many applications, these materials, and associated fabrication processes, are not cost-effective for commercial production and typically for single-use devices.

Plastics are playing an important and ever-increasing role in microtechnology - especially in low-cost, mass-produced applications. It is relatively easy to produce micro structure on the plastic substrates with complex patterns of 50-100μm-sized channels using state-of-the-art replication techniques such as injection molding, hot embossing [3] and UV molding. These substrates are typically used to produce devices on which reactions and high-efficiency electrophoretic separations of biomolecules have been achieved in timescales of seconds to minutes.

Plastics like polymethylmethacrylate (PMMA), polycarbonate (PC) and Cyclopolyolefinpolymere (COC) have been intensively investigated because of their resistance to certain chemicals and biocompatibility. The complete fabrication approach of such a microfluidic device involves two primary steps: (1) formation of micro channels in a plastic base wafer or layer, and (2) bonding of the base layer with a cover sheet to form closed channels. There are several joining procedures for plastic parts including glue cured by heat or UV light, ultrasonic welding and hot gas welding. However, most of these methods cannot, or can only with great difficulty, be adapted to micro structured plastic parts due to the dispensing problems, the use of additional material with different chemical and surface properties, and the lower precision of energy deposition.

The goal of this research was to develop an assembly technology for micro structured parts made from plastics with the goal of producing single-use fluidic devices at an acceptable cost level. The assembly process and the related equipment presented are based on the laser through transmission welding (TTIr) principle. This report describes a device fabrication process and packaging results.

Mask assisted laser transmission welding

Laser welding of plastics was developed in the 1960's, but not until recently has it become a cost effective technology which offers a wide range of advantages because of the non-contact and localized energy input by laser beam [4]. Laser represents a source of clean, controllable, and concentrated thermal and photochemical energy. Due to the recent progress in semiconductor laser technology, compact, economical and reliable laser sources in the NIR range (700 to 1500 nm) are now available which promote the direct use for the material processing with desired power and beam quality.

The principle of the TTIr welding process is indicated in Figure 1. The two plastic parts to be joined must have different optical transmission characteristics at the laser wavelength, one must be transparent or translucent and the other absorbent. The energy of the laser light is transmitted through the transparent part with minimal loss and converted into heat in the absorbing part. By applying a clamping pressure, physical contact between the two parts is ensured and the transparent part is heated by thermal conduction. Surface melting under the illumination of laser beam results in the excitation of convective fluxes within the liquid layer. These liquid layers from both welding parts permit the physical mixture within a cavity at the contact interface and the welding is initiated.

When very small and high accuracy welding seams in the form of lines or areas are required, exact local discrimination of the deposited laser energy must be ensured. With the mask technique the shape of welding seams can be controlled with a high resolution [5]. The principle is indicated in Figure 2. A reflective or absorbing mask is placed between the welding part and the laser source, which generates the lateral energy distribution at the welding surface. The mask is illuminated by a diode bar focused to a line which scans over the mask. The precision of the welding process depends both on the quality of the mask and beam quality of the laser. The mask can be exchanged quickly, allowing for great flexibility in production.

Process equipment

All assembly procedures of microfluidic devices described in this paper are performed by the mask welding system illustrated in Figure 3. The basic process equipment consists of a holder for welding parts, a laser source with optics, a mask, and a vision system for mask alignment. High power diode laser with an emission wavelength in the NIR range is used, typically between 808 nm and 980 nm. For a rapid heating and cooling process of plastic, a power in the range of 80-120 W is required.

In the mask welding process, the welding parts are placed underneath a transparent glass sheet with clamping pressure to assure contact of the mating parts. The mask is fixed on a 3-axis stage (X-Y-ϕ) allowing for precise alignment of the mask relative to the parts. The vision system moving between the mask and holder of welding parts, measures the locations of mask and welding part by using a two point measuring principle and well defined position marks selected directly from the micro structures on the welding part. The locating of defined position marks is carried out by using a pattern matching procedure correlating information about the location and orientation of a known object. According to the measured deviation of the positions the mask will be adjusted with an accuracy of 2 μm to the micro structured welding parts. After the adjustment process the distance between mask and welding part will be minimized. The laser welding process is carried out by scanning of laser source over the mask. This work investigated various levels of system integration that may improve the performance of the system for the mass-fabrication of microfluidic device.

Plastic microfluidic device

There are number of diagnostic devices that require micro channels. In determining the manufacturing method of the channel structure and the packaging of the device the dimensional tolerances, material specifications, and costs should be taken into consideration. For example, a device for the injection and pumping of a precise volume of liquid is achieved through a combination of hydrophobic and hydrophilic regions defined inside the micro-channels. Solutions placed at the inlet port are drawn in by capillary action. Once the liquid is positioned at the desired point, definite volume drops of approximately 10~20 nl can be injected using air pressure from the side-channel. A device for such a task usually consists of a solid plastic base on top of which a thin plastic film is bonded. In the top surface of the plastic base are numerous channels through which the fluids will flow during the handling and analysis process. The plastic film forms the top boundary of these channels. Figure 4 shows some typical macro and micro channel structures of polycarbonate (PC) microfluidic parts which are 14 mm × 10 mm in size and 2.5 mm in thickness (Fig. 4a). This sample has an injection channel and reservoir at the ends of the channel. It consists of a base plate and two tow cover sheets on both sides of the base plate. The cover sheets are 1 mm in thickness. The base plate was an injection molded part. The fine structure of the channel has a dimension of only 100 μm.

Common requirements for assembly of the above mentioned micro fluidic device are joining area, gas and liquid tight joints, and resistance to a few bars. Welding the film to the base must not block any channels, and there can be no delaminations between the film and base, which would permit leakage between adjacent channels. The overall cross-sections of the channels must not be changed during the welding process because of their influence on the effect of capillary action and the transport behavior of the fluids. An exact generation of the welding seams is needed.

Results and discussion

Controlling of melt spreading

The concept of mask welding in micrometer range has implications relating to the rapid heating and cooling of plastic during the welding process. The key factors in establishing the desired welding accuracy are not only the quality of mask and beam shape of the laser, which mainly determine the precise heat transfer, but also the speed of heat treatment and response of material due to the melt flow.

In order to achieve an optimum weld quality for small structures, the melt flow was investigated using both of non-structured (flat) and structured (with micro channels) plastic parts. Figure 5 shows selected test welding seams. Welding structures with a minimum size of 100 μm have been achieved. The behavior of melt flow depends strongly on the welding velocity and viscosity of plastic materials with a characteristic speed of extension that depends additionally on the laser-light intensity and illumination time. The control of this flow was achieved via process optimizing (Fig. 6).

The non-structured plastic welding parts were scanned by a laser source with a constant power of 80 W, equipped with a 150 μm slit mask. The width of the non-welded line was measured as a function of scan velocity. A nearly linear behavior is shown within the narrow range. With scan velocities less than 30 mm/s no well definable structure could be measured due to excessive melt beyond the mask area. For scan velocities greater than 50 mm/s no melt fluidic phase was produced by the laser illumination. It was observed that the speed of melt flow is about 2~3 mm/s. In order to achieve an accuracy of welding seams within 5~10 μm, a maximum illumination time of 3.5 ms is required. A scanning procedure with a line focused laser beam and a high energy intensity allows for a short illumination time.

The use of a laser beam to excite the melting fluxes within the liquid layer plays an important and decisive role for the heat transport, which ultimately determines the quality of the weld. Most plastic materials absorb the laser beam in NIR range only on the top surface with a depth of about a few micrometers. The deep penetration of laser energy can be reached by thermal interaction through liquid plastic fluidics. The quick heating and cooling process is of benefit to the restriction of melt spreading but it may result in a very thin effective convection zone which reduces the welding strength. The compromise must be taken by the excessive coverage of mask to reserve enough places for the expansion of plastic melt fluxes because the melt spreading cannot be completely ruled out by process optimization. The typical dimension of the over coverage is about of 10 to 20 μm, which is dependent on the material to be welded and should be integrated during the mask design phase. In addition, high viscosity of liquid plastic is often of benefit to the restriction of melt spreading due to the slower melt flow.

Figure. 7 shows a microstructure assisted mechanism, which makes it possible to control or stop the spreading of the liquid melt. This structure forms a thin air slit next to the illumination area. As soon as the joining area is illuminated with the laser beam, the liquid melt is guided in a controlled manner into this air slit space. The form of this assisted structuring determines the shape of the edge of melt spreading.

Surface deformation and material compatibility

Because the mask welding process is based on a discriminative heating procedure, the difference of temperature on the welding parts will result in residual stresses[6]. Figure 8 shows the welding seam achieved with a slit mask. A waveform structure at the boundary of welding zone is visible. The reason for that is the temperature gradient near the boundary zone, where the plastic is not efficiently heated to melting point due to the cool surrounding. A growth at the absorbing part with height of 250 nm and a sinking of the surface at the same place on the transparent part can be clearly observed. During the laser illumination the material is melted with overpressure within the interface region. Due to this overpressure the melt is expanded horizontally in all possible directions on the contact surface of welding parts and solidifies with this waveform at the boundary zone. Therefore the boundary of the welding zone is very well defined by this melt wave structure and some fine cornered structures are rounded off in the range of 5~10 μm.

In order to minimize the residual stresses in the microfluidic device, the welding parts with similar melting point is desired. In addition, the material properties, such as thermal expansion and surface tension, also play a very important role and should be considered. One of the successful treatments for reduction of such a stress is the temper of welded devices at moderate low temperature, which has been found very suitable for the mass-production of microfluidic devices.

Process and welding quality controlling

Until recently, there is no active method of process control with the use of the mask technique. However, the optical appearance after the welding process via a vision system offers a capability of quality control. Using suitable illumination techniques, some welding defects can be optically viewed because the air-filled gaps such as delaminations, voids and material decomposition due to the overheating have different reflective conditions in comparison with that of good welded area. The degree of reflection from an interface depends on the difference in welding quality between the two materials involved. Even minute miss-matching of the mask can be clearly detected. It should be mentioned that good looked welding sample may also be poorly welded. To avoid such a problem the welding parameter regarding the scan velocity and the laser power should be finely determined according to the tractive-force-test.

Due to the two points measure principle, the vision system for the mask alignment provides additional controlling mechanisms for the deviation of micro structured plastic parts in comparison to the microstructures on the mask. The measured tolerance during the alignment procedure can then generate error information and show influence on the accuracy of mask alignment process.

Conclusions

A microfluidic device with a channel width of 100 μm has been successfully welded and packaged with desired accuracy using TTIr with a masking based system. The specially designed arrangement with automated alignment and system calibration of welding procedures allows for the use of moderate laser power for a very fast heating and cooling process. For a device with an area of 10X10 mm^2 12 sec. is required, though the real welding time takes only 220 ms.

References

[1] J.M. Köhler, U. Dillner, A. Mokansky, S. Poser and T. Schulz, "Micro channel reactors for fast thermocycling", 2nd International Conference on Microreaction Technology", pp.241-247 (1998).

[2] J. Voldman, M.L. Gray and M.A. Schmidt, "Liquid mixing studies with an integrated mixer/valve", Proceedings of the μTAS '98 Workshop, pp.181-184 (1998).

[3] H. Becker, W. Dietz and P. Dannberg, "Microfluidic manifolds by polymer hot embossing for μ-TAS applications", Proceedings of the μTAS '98 Workshop, pp.253-256 (1998).

[4] H. Pütz, D. Hänsch, H.G. Treusch and S. Pflueger, "Laser welding offers array of assembly advantages", Modern Plastics, pp.121-124 (1997).

[5] J.-W. Chen and O. Hinz, "Feinste Fügung", Technologie Bilanz Switzerland, p.33 (2000).

[6] D. Grewell, *Applications with Infrared Welding of Thermoplastics*, 57th Annual Technical Conference for the Society of Plastic Engineers, 1999.

Key words

laser welding, mask technique, microfluidic, polymer

Fig. 1. Principle of laser transmission welding.

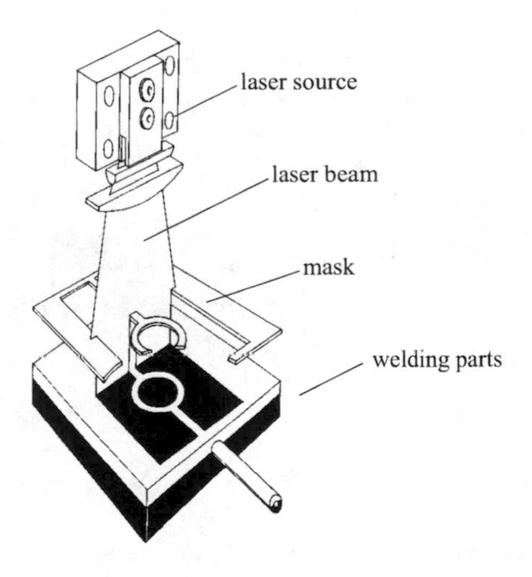

Fig.2. Mask welding. A line focused laser beam is scanned over a mask, which discriminates the energy to the desired weld area.

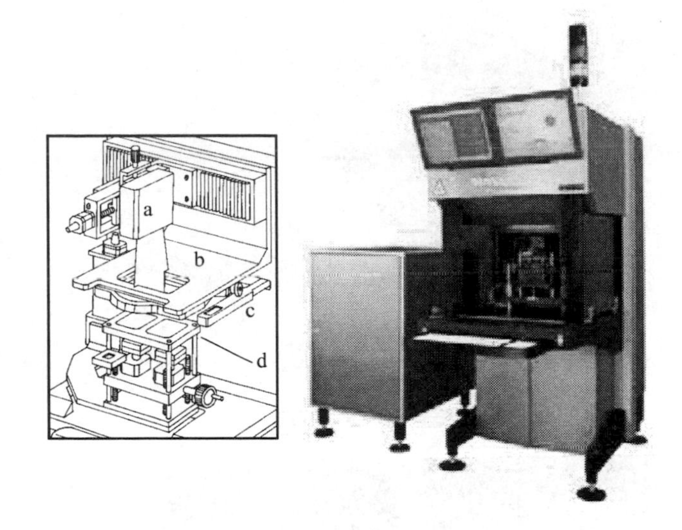

Fig.3. Mask welding system for assembly of plastic microfluidic device. a) laser source, b) micro alignment, c) vision system and d) clamping pressure system.

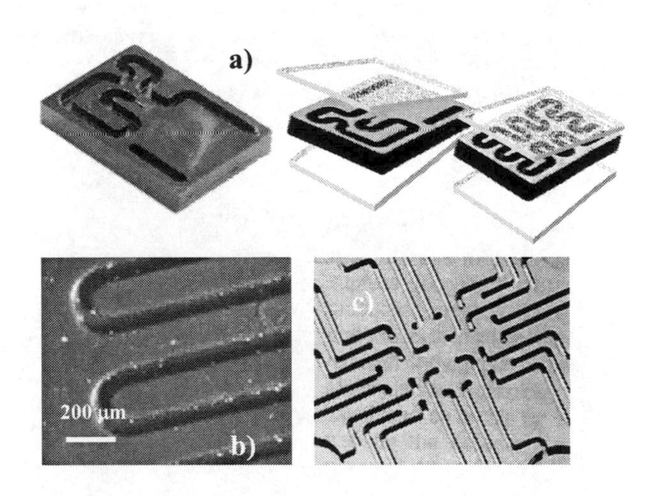

Fig.4. The key component for a microfluidic handling system with macro and micro channels.

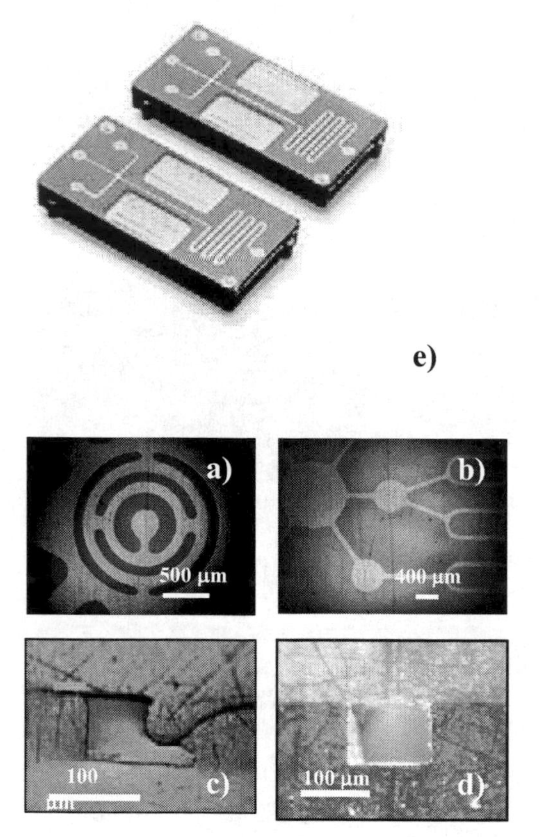

Fig.5. Test welding seams containing various micro structures with different dimensions. a) and b): Welding on non-structured plastic parts, The black areas are welded. c): Micro channels with melt spreading due to the miss-matching of mask. d): Micro channel welded using the excessive coverage of mask. e): Complete welded sample.

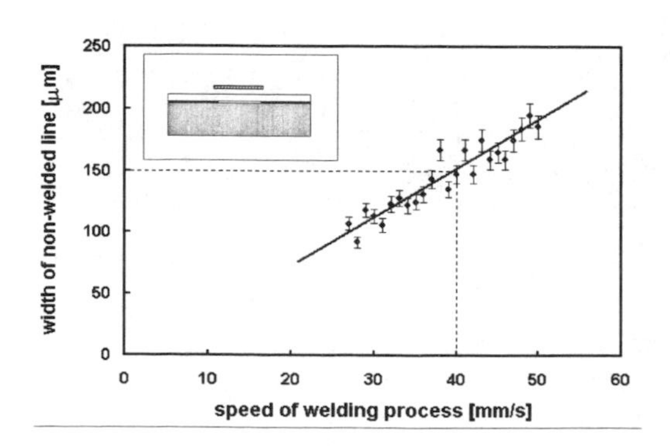

Fig.6. Melt spreading vs. process time

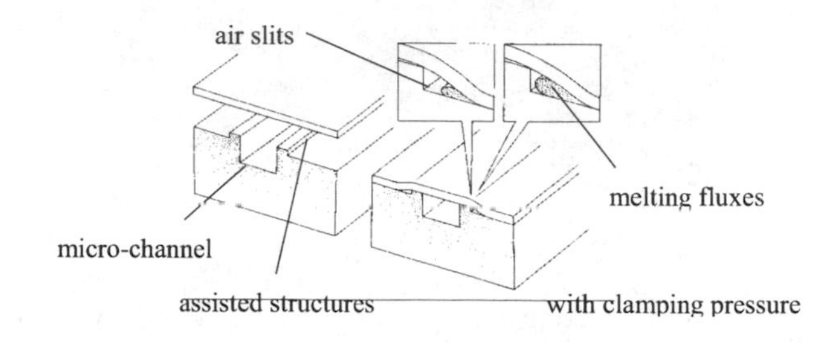

Fig.7. Controlling of melt spreading using assisted micro structures.

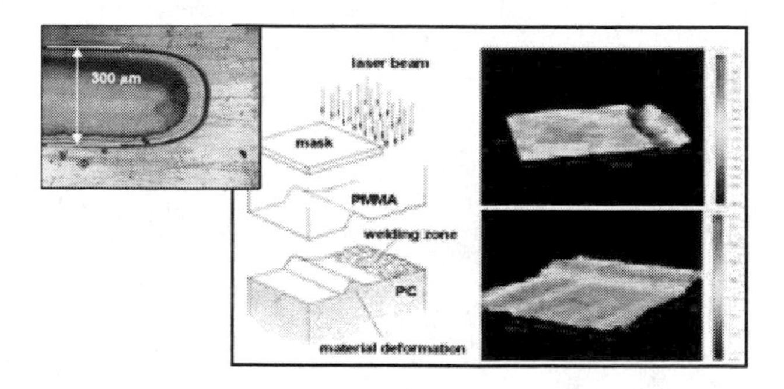

Fig.8. Material deformation on the edge of the welding zone due to the temperature gradient.

A Unified Compact Model for Electrostatic Discharge Protection Device Simulation

Hung-Mu Chou[1], Yen-Yu Cho[2], Jam-Wen Lee[3], and Yiming Li[3,4]

[1]Department of Electrophysics, National Chaio Tung University
[2]Department of Computer and Information Science, National Chaio Tung University
[3]Microelectronics and Information Systems Research Center, National Chaio Tung University
[4]Department of Computational Nanoelectronics, National Nano Device Laboratories
P.O. BOX 25-178, Hsinchu City, Hsinchu 300, Taiwan; Email: ymli@faculty.nctu.edu.tw

ABSTRACT

Snapback phenomenon plays an important role for electrostatic discharge (ESD) protection design, in particular for very large scale integration (VLSI) circuits. In this paper, we proposed a unified ESD model for metal-oxide-silicon field effect transistor (MOSFET) and silicon current rectify (SCR) devices. This new model characterizes the snapback characteristics and can be directly incorporated into ESD circuit simulation for whole chip ESD protection circuit design.

Keywords: compact model, ESD protection, circuit simulation, MOSFET, SCR, snapback, whole chip design

1 INTRODUCTION

In modern micro- and nano-electronics manufacturing, whole chip ESD protection circuit design is necessary for obtaining robust electrical performance [1-5]. In designing the whole chip ESD protection circuit, owing to the circuit complexity, an accurate and efficient computer-aided design (CAD) tool is not only helpful but also essential [5-10]. Several ESD models for MOSFET or SCR devices have been proposed [11-15]; unfortunately, they are constructed upon simplified bipolar junction transistor (BJT) models. Most of these BJT-based ESD models results from the fact that BJT models can only describe device behavior under the normally operation region. Therefore, they lack physical meaning in the ESD region and may trouble circuit design applications.

We propose here a unified ESD model for MOSFET and SCR devices, shown in Figs. 1-3, with a set of algebraic equations and limited parameters. Moreover, application of our recently developed automatic and intelligent parameter optimization technique, a series of comprehensive simulation is performed, which confirms the proposed model can be successfully applied on MOSFET and SCR devices for preliminary ESD simulation. This paper is organized as follows. In Sec. 2, we state the proposed ESD model. In Sec. 3, we show the results and compare with the measured data for MOSFET and SCR devices. Finally, we draw conclusions.

2 THE PROPOSED ESD MODEL

The developed ESD model is basically developed on the device physics that parasitic BJT will breakdown under the ESD stress. Figure 4 shows the characteristics of ESD considered in our model. According to the mechanism, we can simply formulate the snapback current - voltage (IV) characteristics by using a current controlled voltage source (the breakdown voltage of parasitic BJT) V_{BCE}, the series resistance caused by R_d and the external resistance R_x resulted from measurement instruments. The IV follows the equation (1). The current controlled voltage source V_{BCE} could be simply expressed by the following equation (2) where B can be obtained from Eq. (3), in which R and C can be simply solved from Eqs. (4) and (5).

$$I = \frac{V - V_{BCE}}{(R_x + R_d)} \quad (1)$$

$$V_{BCE} = V_{BCE0}(1 - B^{-1})^n \quad (2)$$

$$R(\frac{I}{B}) + \ln(\frac{I}{B}) + C = 0 \quad (3)$$

$$R = \frac{Z_s + r_e}{V_t * \alpha_b} \quad (4)$$

$$C = Z_s * \frac{I}{V_t} + \ln(I_s * \alpha_b) \quad (5)$$

There are six parameters have to be optimized in the equations above [16]; they are R_d, n, Z_s, r_e, I_s and α_b. In the model, R_x reflects the impedance of transmission line which equal to 50ohm, and V_t is the thermal voltage. The other parameters all have their physical meaning: n is the idea factor of pn diode, Z_s is the substrate resistance, r_e is emitter resistance, I_s is the revise saturation current, and the α_b is the common base current gain. Our model could be simply performed by considering the optimization of 6 parameters through the comparing between solved IV data to measured data. This approach enables the simulation of the whole chip ESD robustness.

3 RESULT AND DISCUSSIONS

We compare the modeled and the measured snapback characteristics of MOSFET device in Fig. 5 and SCR devices from the Figs. 6 to Fig. 10. It is clearly found that our model precisely describes the snapback behavior of both the MOSFET and SCR devices under ESD events.

Figure 5 compares the modeled and the measured snapback characteristics of the MOSFET device; it demonstrates that our model successfully achieves four major features. Those are trigger-on voltage, snapback slope, holding voltage and turn-on resistance. With correctly modeling those four characteristics, the designers could easily find out if the design margin is enough for avoiding both the ESD damage and latchup; moreover, the efficiency of both protection-activating and layout-drawing can be also optimized through the simulation.

Figures 6 to 10 presents the snap back IV characteristics of the SCR devices with different emitter to base spacing. It could be easily found that all the SCR devices have similar turn on voltage (~12V), but the holding voltages are different. The holding voltage will significantly increase with the decreasing of the emitter to base spacing. With an estimating, about 3 volt of holding voltage difference could be found. Modeling reflects the measurement results. It could be also found that a decreasing of emitter to base spacing will cause a higher holding voltage. This result is caused from the fact that the built-in potential between p-well and n-well is dominated by minority carriers which also influence on our model through the parameter n and αb. Make it more clearly that the higher minority carrier existing in the well region will cause a lower built-in potential; consequently, result in a lower holding voltage. Owing to a higher efficiency in sinking out the minority carrier, the SCR with the shorter emitter to base distance will certainly has a higher holding voltage. Finally, it is also noticeable that the jiggled IV curve is not the nature characteristics of device but caused from the fluctuation of data sampling.

Tables 1 and 2 illustrate the optimized parameters of all devices [16]. Among these results, we also evaluate the convergence property of the entire parameter range; after a comprehensive calculation, no singular point can be found. Moreover, it could be also drawn that every parameter set will converge at few inter-loops. Here, we can conclude that our model is a SPICE circuit simulation compatible result with no convergence difficulties.

4 CONCLUSIONS

In this paper, we have briefly presented an attractive unified ESD model for both MOSFET and SCR protection circuit design. The results have demonstrated good accuracy and high capability for both devices. We conclude that the developed ESD model could be performed in SPICE circuit simulator for desirable agreement with the measured ESD characteristics. Compared with conventional models, this model provides a novel way to ESD simulation and shows good computational efficiency.

5 ACKNOWLEDGMENTS

This work is supported in part by the TAIWAN NSC grants NSC-93-2215-E-429-008 and NSC 93-2752-E-009-002-PAE. It is also supported in part by the grant of the Ministry of Economic Affairs, TAIWAN under contracts No. 92-EC-17-A-07-S1-0011 and No. 93-EC-17-A-07-S1-0011.

REFERENCES

[1] N. M. Iyer and M. K. Radhakrishnan, Proc the 16th Int. Conf. VLSI Design, 20, 2003.

[2] S. Bonisch and W. Kalkner, Proc the 2003 IEEE Int. Symp. Electromagnetic Compatibility, 1, 37, 2003.

[3] S. Bonisch, W. Kalkner, and D. Pommerenke, IEEE Trans. Plasma Science, 31, 736, 2003.

[4] H. Feng, G. Chen, R. Zhan, Q. Wu, X. Guan, H. Xie, A. Z. H. Wang, and R. Gafiteanu, IEEE J. Solid-State Circuits, 38, 995, 2003.

[5] K. Wang, D. Pommerenke, R. Chundru, T. Van Doren, J. L. Drewniak, and A. Shashindranath, IEEE Trans. Electromagnetic Compatibility, 45, 258, 2003.

[6] J. Lee, K.-W. Kim, Y. Huh, P. Bendix, and S.-M. Kang, IEEE Trans. Computer-Aided Design of Integrated Circuits and Systems, 22, 67, 2003.

[7] M.-D. Ker and K.-C. Hsu, IEEE Trans. on Electron Devices, 50, 397, 2003.

[8] Y. Li, J.-W. Lee, and S. M. Sze, Jpn. J. Appl. Phys., 42, 2152, 2003.

[9] J.-W. Lee and Y. Li, Proc. the 2003 Third IEEE Conf. Nanotech., 639, 2003.

[10] J.-W. Lee, Y. Li, A. Chao, and H. Tang, Jpn. J. Appl. Phys., 43, 2302, 2004.

[11] K. Wang, D. Pommerenke, and R. Chundru, Proc. the 2002 IEEE Int. Symp. Electromagnetic Compatibility, 1, 93, 2002.

[12] M. Mergens, W. Wilkening, S. Mettler, H. Wolf, A. Stricker, W. Fichtner, Proc. Electrical Overstress / Electrostatic Discharge Symp., 1, 1999.

[13] G. Bertrand, C. Delage, M. Bafleur, N. Nolhier, J. Dorkel, Q. Nguyen, N. Mauran, D.Tremouilles, P. Perdu, IEEE J. Solid-State Circuits, 36, 1373, 2001.

[14] J. Li, S. Joshi, and E. Rosenbaum, Proc. the IEEE Custom Integrated Circuits Conf., 253, 2003.

[15] H. Wolf, H. Gieser, and W. Stadler, Proc. Electrical Overstress / Electrostatic Discharge Symp., 271, 1998.

[16] Y. Li and Y.-Y. Cho, Jpn. J. Appl. Phys., 43, 1717, 2004.

[17] Y. Li, Y.-Y. Cho, C.-S. Wang, and K.-Y. Huang, Cho, Jpn. J. Appl. Phys., 42, 2003, 2371.

[18] Y. Li, C.-T. Sun, and C.-K. Chen, in "Advances in Soft Computing-Neural Networks and Soft Computing," Edited by L. Rutkowski and J. Kacprzyk, Physica-Verlag, 364, 2003.

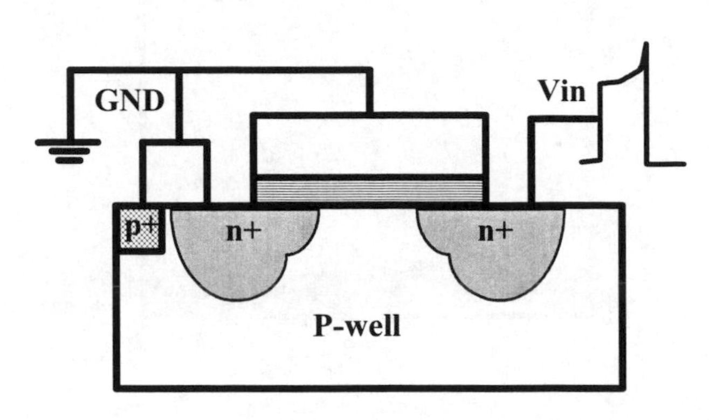

Figure 1: The device structure of the experimental MOSFET device.

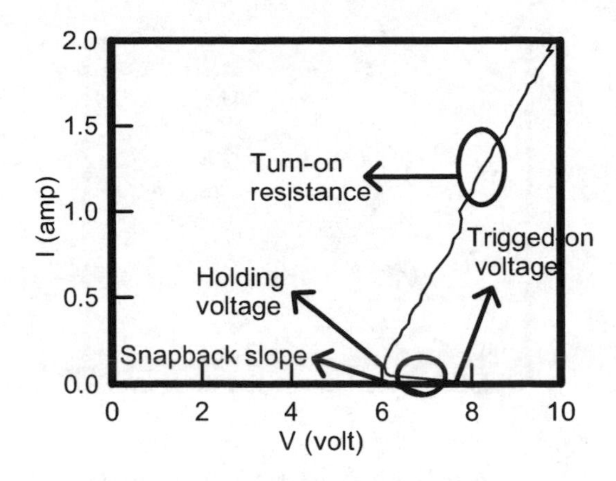

Figure 4: The characteristics of ESD considered in our model.

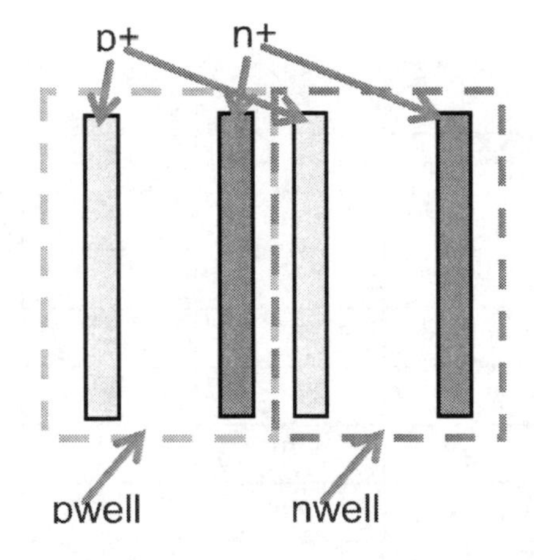

Figure 2: The top layout view of the investigated SCR device.

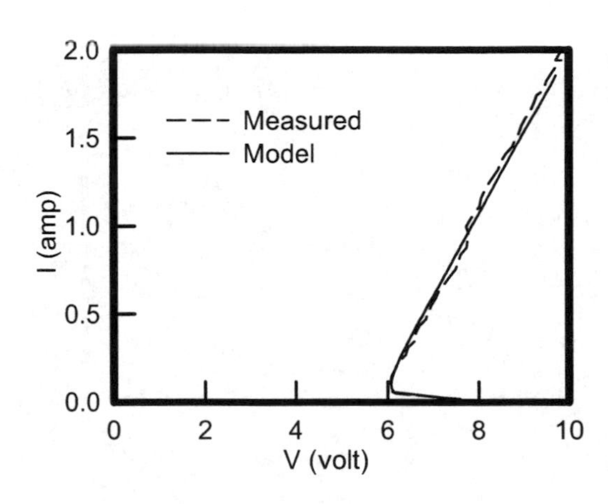

Figure 5: Comparison between the measurement and simulation of the MOSFET device.

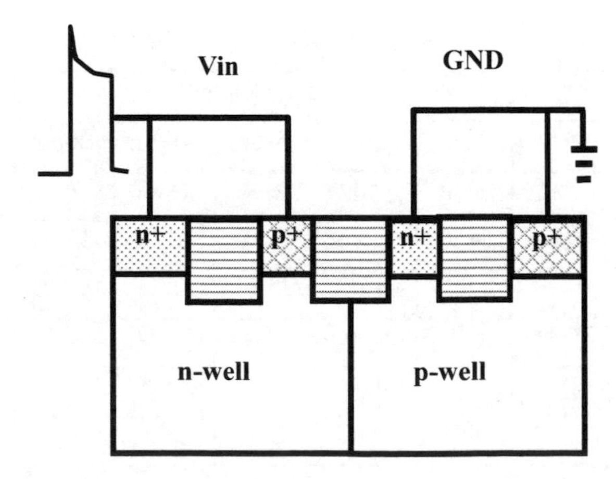

Figure 3: The device structure of the experimental SCR device.

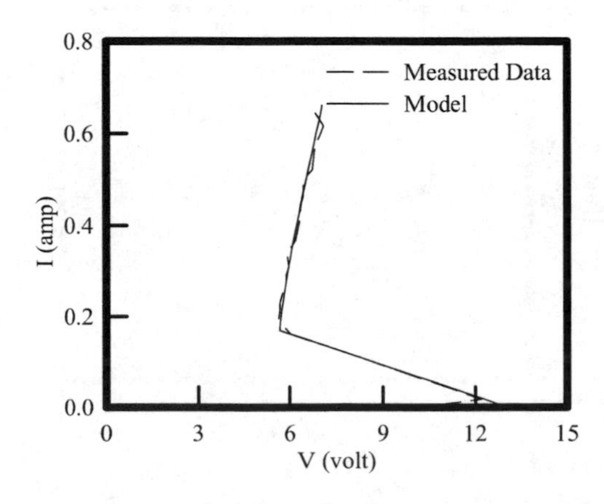

Figure 6: Comparison between the measurement and simulation of the 1μm SCR device.

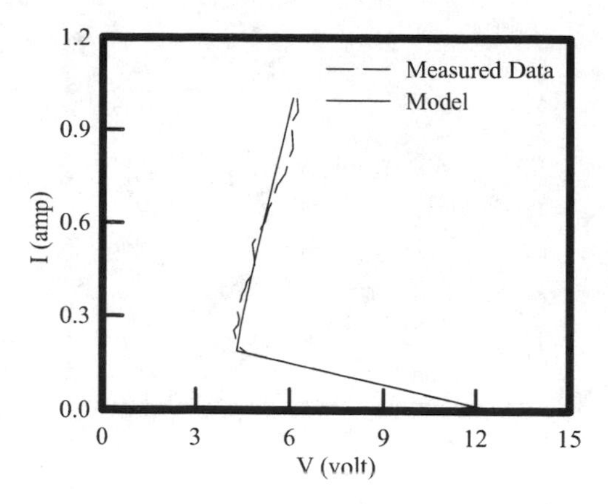

Figure 7: Comparison between the measurement and simulation of the 2μm SCR device.

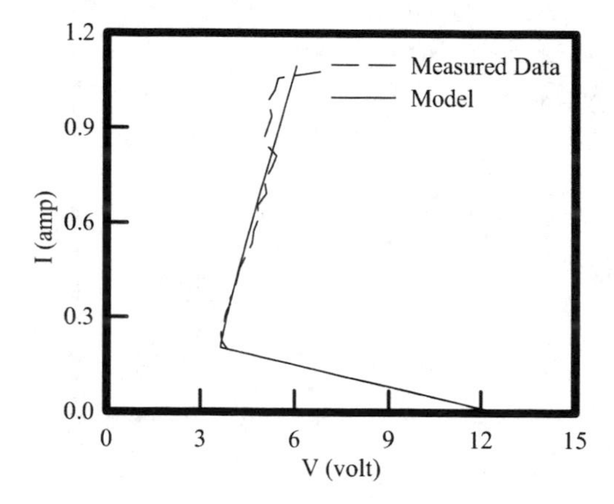

Figure 8: Comparison between the measurement and simulation of the 5μm SCR device.

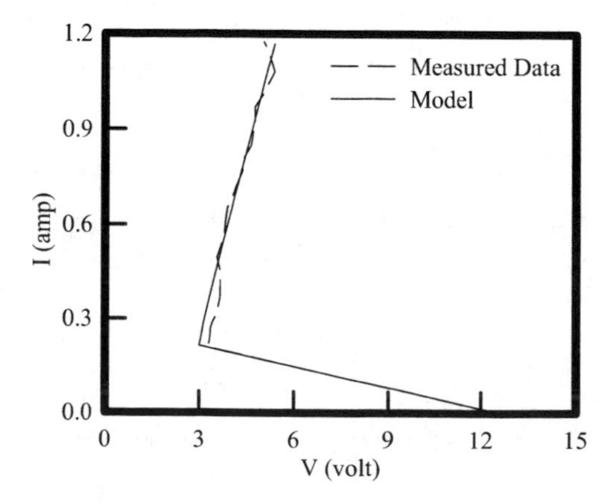

Figure 9: Comparison between the measurement and simulation of the 10μm SCR device.

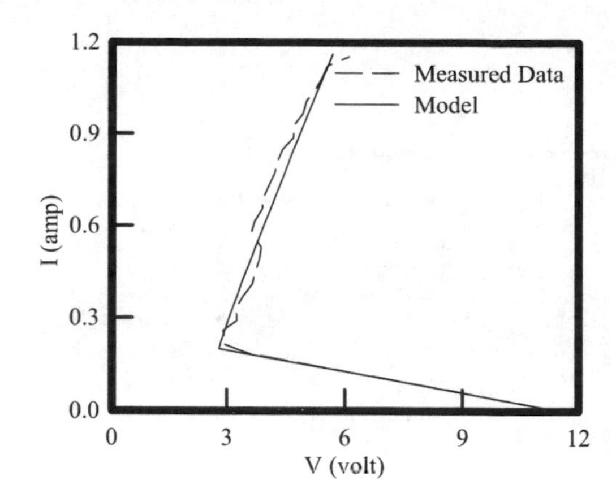

Figure 10: Comparison between the measurement and simulation of the 20μm SCR device.

Paramete	Unit	Value
R_d	ohm	2.5
V_{BCEO}	V	7.5
n	none	1
Z_s	ohm	100
R_e	ohm	5
α_b	none	0.9

Table 1. A set of the optimized parameters for MOSFET.

	Device Geometry				
Parameter	1μm	2μm	5μm	10μm	20μm
R_d	3.30	2.42	2.97	2.68	3.13
V_{BCEO}	13.08	12.59	13.08	13.08	11.49
n	2	1.65	2.09	2.23	2.05
Z_s	50	82.60	54	63.9	96.39
R_e	27	21.47	18.77	24.63	23.26
α_b	0.61	0.67	0.71	0.76	0.71

Table 2: A set of the optimized parameters for SCR with different dimensions.

Direct Wafer Polishing with 5 nm Diamond

James Chien-Min Sung[*,1,2,3], Ming-Fong Tai[4]

Address: KINIK Company, 64, Chung-San Rd., Ying-Kuo, Taipei Hsien 239, Taiwan, R.O.C.
Tel: 886-2-2677-5490 ext.1150
Fax: 886-2-8677-2171
e-mail: sung@kinik.com.tw

[1] KINIK Company, 64, Chung-San Rd., Ying-Kuo, Taipei Hsien 239, Taiwan, R.O.C.
[2] National Taiwan University, Taipei 106, Taiwan, R.O.C.
[3] National Taipei University of Technology, Taipei 106, Taiwan, R.O.C.
[4] Department of Electronic Engineering, Wufeng Institute of Technology, Chia-yi 621, Taiwan, R.O.C.

ABSTRACT

CMP for making future semiconductor chips with nanom (nano meter) feature sizes can be accomplished by using nanom diamond particles embedded in an organic matrix (e.g. epoxy). Such nanom diamond particles are derived from the detonation of dynamite (e.g. TNR and RDX) in oxygen deficiency atmosphere. The nanom diamond particles are formed instantaneous from the residue carbon during the transient ultrahigh pressure and temperature. These nanom diamond particles are defect ridden and they are coated with a softer carbon coating (e.g. bucky balls and nano tubes). The softer carbon coating can lubricate the cutting edge in-situ during the action of nanom polishing. The nanom diamond has an intrinsic tight size distribution (4-10 nanoms) so the scratch of delicate semiconductor chip (e.g. IC with copper circuitry) is avoided. Moreover, the nanom diamond itself contains built-in defects that will allow nanom chipping so the abrasive can be self-sharpened for continual polishing with high efficiency. In addition, the nano radius of the nanom diamond can polish the wafer in the ductile domain so chipping of the polished surface is avoided. The result would be a clean and smooth surface with minimal mechanical degradation or thermal damage. The resinoid matrix that holds nanom diamond is impregnated with nanom metal particles (e.g. Ni) that can be dissolved by acidic slurry. Alternatively, the epoxy matrix may also incorporate nanom salt particles (e.g. NaCl) that can be dissolved in water. The dissolution of non-carbon nanom additives will expose new nanom diamond particles continually so the efficient polishing can be sustained.

Keywords: nanom diamond, CMP, ULSI, wafer polishing, moore's law

1 NANOM DIAMOND POLISHING

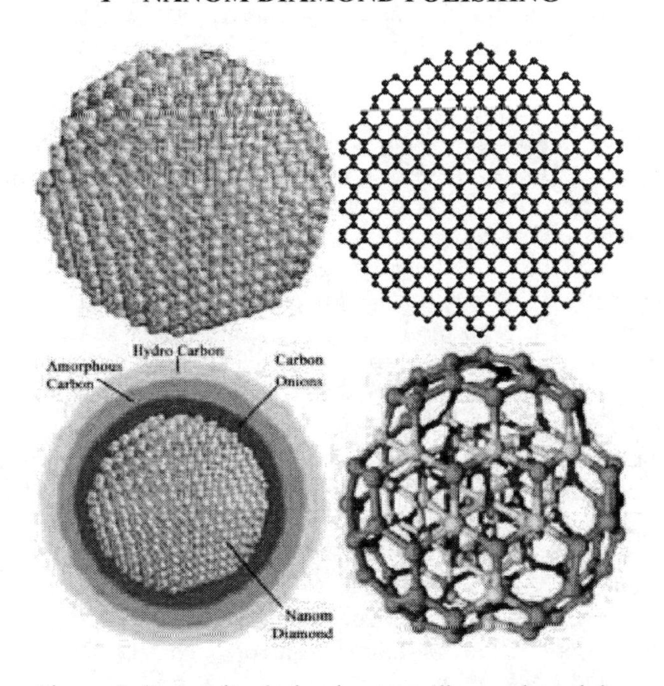

Figure 1: Dynamite derived nanom diamond particles contains high concentration of defects (top left) that can allow gradually nanom chipping during polishing. The particle is about 5 nm across in size with about 30,000 carbon atoms (top right). The nanom diamond particle is coated with non-diamond carbon soot that may serve as a damper for reducing the impact force (bottom left). The carbon soot may also form bucky ball that encloses the nanom diamond particle (bottom right).

Nanom diamond is small and spherical with tight size range, so it is ideal for achieving super smooth polishing. Because it is superhard, it can penetrate effectively into the wafer surface and chip off the protruded points. Hence, the polishing rate would be an order of magnitude higher.

Moreover, as diamond is much harder than the

wafer material, it can penetrate into the work material with the least energy so the surface damage due to plastic deformation or thermal degradation would be minimal. In other words, the polished surface would be clean and smooth with little or no dragging or smearing.

In is well known that during the cutting process, the damage of work material is inversely related to the sharpness of the cutting edge. Hence, cutters with smaller radii can make cleaner cutting surface. Consequently, nanom diamond particles will not drag on the wafer surface during polishing so the wafer surface would be smoother.

In addition, the nanom diamond particles are so small that only a few layers of atoms will be removed from the wafer each time. Such a nanomatic cutting operation will render the wafer ductile. In other words, the wafer surface is shredded off rather than chipped off. The nanom peeling of wafer material can generate the least heat so it is also a cold polishing process. In contrast, the larger particles will cause much mechanical damages to the wafer surface due to plastic deformation and brittle chipping. If the hardness of the abrasive is not superhard, as in the case of using ceramic fines, both the wafer surface and the abrasives will smear and fracture. As a result, significant heat is produced at the contact area that may rise the temperature momentarily. The transient hot spot may not only leave behind a thermal damage zone on the wafer surface, but also could soften or even melt the polyurethane matrix. Such a thermal shock is not acceptable when the wire width is narrowed down to 90 nm or less.

In contrast, the cold polishing of nanom diamond produce no trace of thermal damage. Moreover, the ductile polishing can make the surface super smooth like a shinning mirror. This is the ultimate CMP process for making all transistors identical with perfect flat circuitry, i.e., absolutely no dishing. Hence, nanom diamond CMP process can be the enabling technology for advancing Moore's Law to make transistors with feature sizes smaller than 65 nm.

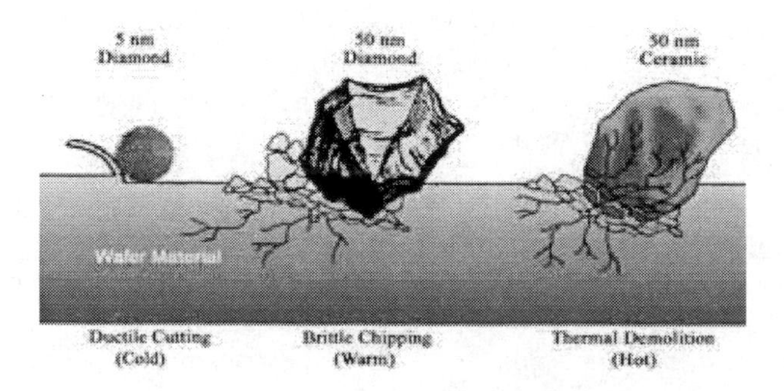

Figure 2: Nanom diamond particles are superhard and super sharp so they can remove only a few atomic layers of the wafer materials. As a result of this restricted penetration, the polishing is achieved by ductile peeling of the normally brittle material. Such a cutting mode will damage the least the fragile low K dielectric on the wafer. Consequently, the finished surface is both ultra precise and supersmooth. In contrast, the irregular shaped and over sized diamond superabrasive or ceramic abrasives cannot polish the wafer in the ductile domain. The forced penetration of these large particles will cause brittle chipping of the low K dielectric, in the case of superhard diamond, or thermal degradation, in the case of ceramic abrasives.

2 DIRECT POLISHING OF WAFER BY NANOM DIAMOND

As nanom diamond is much harder than conventional ceramic abrasives (e.g. silica, ceria, alumina, silicon carbide) used for CMP, it can penetrate deeper into the work material so the polishing rate would be higher. Moreover, as the plastic deformation or thermal damage is less on the polished surface, the smoothness of the wafer surface is greatly improved.

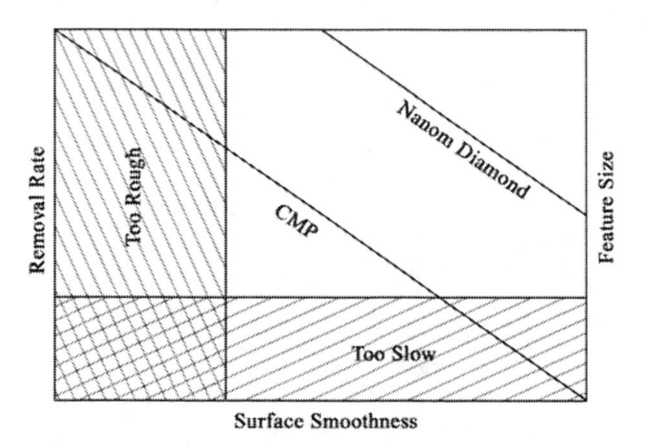

Figure 3: The polishing rate of conventional ceramic abrasive for CMP is compromised by the super smooth finish required for making future chips. The polishing rate of superhard nanom diamond is significantly faster and the polishing action is much less damaging, so nanom diamond polishing can produce super smooth wafer surface without sacrificing much of the productivity.

Although nanom diamond may be ideal for polishing wafers, it is too expensive to be mixed with slurry. The abrasive particles in slurry can move around freely so most of them are not used in polishing, but they are lost with the flow of the slurry. In order to minimize the loss of expensive nanom diamond, the particles must be impregnated in a solid carrying medium (bonding matrix), such as resin or polyurethane.

Because nanom diamond particles are exceedingly small, their exposure above the bonding matrix for effective cutting is minimal. Consequently, dragging of wafer against the bonding matrix is inevitable. Moreover, the polished debris cannot be removed around nanom diamond particles. The over coated polishing matrix must be removed by an effective dressing action. It is preferable that the amount of dressing is controllable as nanom diamond particles may be lost by excessive exposure.

3 NANOM DIAMOND POLISHING PAD

Because nanom diamond is relatively expensive (e.g. $1/carat or $5/g) and it is about 100 times more costly than conventional nanom abrasives, it is not suitable to be used as free slurry. Consequently, the nanom particles must be embedded in a fixed matrix. The use of fixed matrix will not only preserve the nanom diamond, but also allow it to cut more aggressively into the wafer surface. Consequently, the polishing rate would be further increased.

Nanom diamond particles have a tremendous amount of surface area (e.g. 300 m^2/g), so it can be attached firm in a suitable resin matrix (e.g. epoxy). This is further enhanced by the non-diamond carbon coating that is wettable by most organic materials such as resin or epoxy. The wetting will increase the adherence strength significantly due to the formation of chemical bonds between carbon and matrix atoms.

The nanom diamond impregnated matrix can be selectively coated on a pad surface such as that by forming "poles" that protrude above the surface. These poles can exert pressure on the wafer to be polished. A convenient method to stick poles on the pad is by screen-printing. In this case, nanom diamond particles are first dispersed (e.g. in an emulsifier in a liquid (e.g. acetone) and mixed in with a binder mixture (e.g. amino resins, phenolic resin, epoxy resin, arcrylated polyester resin, or others) to form a suitable paste with the right viscosity. The paste is then printed on a proper pad material (e.g. myler, PET) through a screen (e.g. reinforced laminate board) that contains patterned holes with the size and distribution that are intended for making poles. After the printing, the precursor materials of poles can be cured by evaporation of solvent under an infrared lamp, by ultraviolet polymeration or cross-linking of monomers or oligomers.

The poles do not have to be formed on a polishing pad; it may also be coated on a flat wheel or a cup wheel. In this case, the nanom CMP is accomplished by fast rotation of the wheels against the wafer. The operation is similar to grinding, but the polishing action is orders slower in material removal, and significantly lower in contact pressure.

The poles carrying pad must be properly supported by subpad materials in order to achieve local planarization and global uniformity. Certain designs have been used to achieve the optimal compromise of the two planarization scales. Thus, U. S. Patent 6,632,129 described the use of an intermediate layer of segmented tiles between fixed abrasive pad and a resilient subpad to allow local adjustment of planarization over the global scale. In this case, the rigid tiles (e.g. Young's modulus > 200 MPa) can tilt locally as a whole over an elastic support material (e.g. Young's modulus <100 MPa). All layers can be conveniently glued by pressure sensitive adhesive such as latex crepe, rosin, acrylic polymers and copolymers, alkyd adhesives, rubber adhesives.

Suitable rigid material may be made of polycarbonates, polyesters, polyurethanes, polyolefins, polyperfluoroolefins, polyvinyl chlorides, and other thermo-plastic materials. Alternatively, thermosetting copolymers may be used, such as epoxies, polyimides, polyesters, and copolymers with at least two different monomers (e.g. terpolymers, tetrapolymers).

In contrast to the above patent teaching, it is desirable to support the poled pad with either more than one layer of segmented tiles or more than one size of segmented tiles in order to smooth further out the stress that may be built on the interface between poled pad and tiles, and between tiles and subpad. In this case, the local planarization can be uniformly cushioned by the resilient subpad so the polished wafer will not see abrupt changes of local planarization markings.

4 SELF-SHARPENING NANOM DIAMOND PAD

In order to expose the nanom diamond particles that are embedded in the matrix, the matrix may be selectively dissolved in an organic solvent. The solubility of the matrix in the solvent is higher when the pressure is increased. As a result, the matrix around each nanom diamond particle that is pressing against the wafer will be etched away preferentially. The exposed nanom diamond particle can then cut into the wafer more effectively.

Alternatively, the matrix that holds nanom diamond particles may also be added with another ingredient that can be dissolved uniquely by another solvent. For example, the matrix may contain nanom nickel particles that may be dissolved by nitric acid diluted in water. The preferential dissolution of nickel particles around nanom diamond particles may expose the latter for fast polishing action.

In summary, by incorporating nanom diamond in fixed matrix can allow fast polishing of future wafers without causing scratching of super smooth surface. Because nanom diamond is superhard with much higher wear life, and the fact that fixed abrasive is significantly less waste that free moving slurry, the work life of nanom diamond may be a thousand times more than that of conventional abrasive. This long life will make the nanom diamond more economical because it is only about 100 times more expensive in price. In addition to cost less, nanom diamond cannot scratch the wafer so the production yield of the polished wafer is higher. Moreover, nanom diamond can polish with minimal energy so the surface damage of the wafer is avoided. Furthermore, nanom diamond can polish in the ductile domain. As a result, the wafer surface will be super smooth. Thus, the direct polishing of wafers by nanom diamond has the potential to make future chips required by Moore's Law in cost, quality, and throughput.

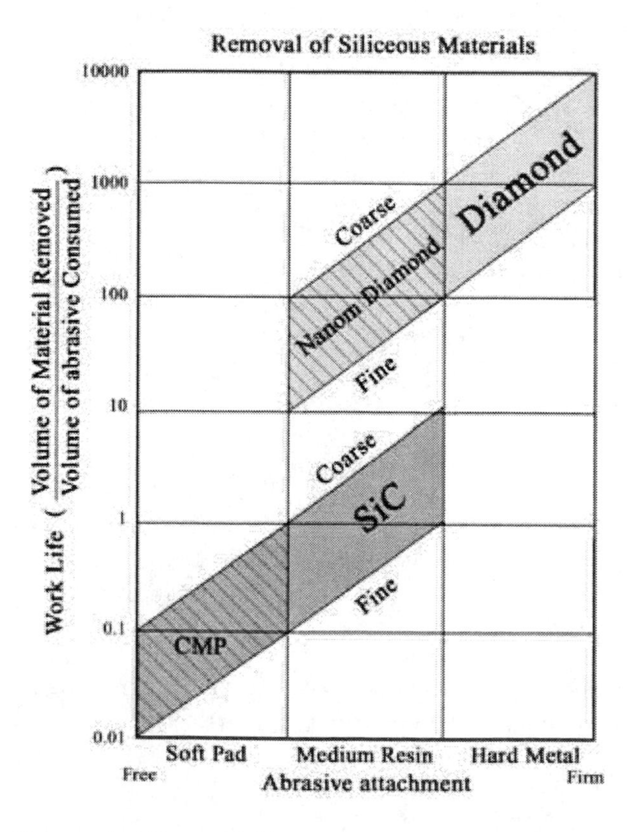

Figure 4: The polishing life of nanom diamond versus conventional abrasive labeled as CMP. The figure shows that the firm attachment of superhard nanom diamond can last about 1000 times longer than free moving ceramic abrasive particles.

Fabrication of Well-aligned and Mono-modal Germanium Dots on the Silicon Substrate with Trench-ridge Nano-structures

Yan-Ru Chen, Y. H. Peng, and C. H. Kuan
Graduate Institute of Electronics Engineering and Department of Electrical Engineering, National Taiwan University, Taipei, Taiwan, Republic of China
P. S. Chen
Electronics Research & Service Organization, Industrial Technology Research Institute, Hsinchu, Taiwan, Republic of China

We perform the ability to fabricate germanium (Ge) dots on the silicon substrate with trench-ridge nano-structures by integrating electron beam lithography system (E-beam), reactive ion etching (RIE), and ultra-high-vacuum chemical vapor deposition (UHV-CVD). Two experiments are shown, and a model is hypothetically set up based on our analysis according to the statistics of dots. Most importantly, well-aligned and mono-modal dots are realized successfully by manufacturing about 30nm-width trenches, and we have the suggestion strongly that it is achieved by exploiting the high surface mechanical energy of ridge edges.

Keywords: Ge dots, aligning, nano-structures, E-beam

1 INTRODUCTION

Semiconductor dots are interesting systems because of their potential applications in novel devices. Among them, germanium (Ge) dots on silicon system have been extremely studied recently. The Stranski–Krastanow growth method is usually used to obtain silicon substrates. After the wetting layer deposition, Ge dots are formed spontaneously to release misfit strain. Generally, these dots nucleate at random positions and exhibit rather large size dispersions. In Ge dots on silicon system for growth temperature around 600 °C, typically a bimodal distribution is found, including pyramid- and dome-shaped dots [1]. However, the size and position distribution of these dots is critical to their application in novel devices [2]. How to control the spatial distribution and uniformity of self-assembled dots consequently becomes an essential issue.

Some ideas have been addressed to improve the spatial aligning and the uniformity of Ge dots [3, 4]. Recently, major progress was reported by integrating lithography and self-assembly techniques for ordering of Ge dots on silicon substrates [5, 6]. We perform to fabricate Ge dots on the silicon substrates with trench-ridge nano-structures. Most surprisingly, well-aligned and mono-modal Ge dots are realized by 30nm-width trenches. Comparison with published papers, the ordering dots we proposed are grown on the silicon ridges (the area not exposed in lithography process) instead of grown in the trenches, pits, or holes (the area exposed in lithography process) [5-8]. This point is never discussed, emphasized or realized in previous published papers.

The whole process is briefly stated as below. At first, it is started with E-beam (ELS-7500EX) to make nano-patterns on the substrates. Next, RIE is exploited for trench-fabrication. After etching, the resist is removed. Subsequently, a wetting layer is deposited and Ge dots are grown by UHV-CVD (SIRIUS CVD 400). Finally, the surface morphology is investigated in air with atomic force microscope (AFM).

2 EXPERIMENTS

Two experiments have been shown. In the first experiment, Ge dots, as Fig. 1 shows its atomic force microscopy (AFM) images, are grown on the 100nm-100nm, 200nm-200nm, 300nm-300nm and 400nm-400nm trench-ridge nano-structures. Each patterned area consists of trenches with 50nm depth. 20nm silicon is deposited as wetting layer. Subsequently, 7 equivalent-monolayers (eq-MLs) Ge is deposited on the prepatterned silicon substrate at about 600 °C. It is found that Ge dots are grown easier in the trenches than on the ridges, as Fig. 1 shows, for there are more dots appeared in the trenches than on the ridges. The sample with 400nm-400nm trench-ridge nano-structures is observed that Ge dots are both grown in the trenches and on the ridges. The others are observed that Ge dots are grown almost in the trenches.

In the second experiment, 30nm-70nm, 30nm-120nm, and 30-170nm trench-ridge nano-structures are fabricated under the same process described in the first experiment. The depth of trenches is also 50nm depth. A 20nm silicon layer is deposited on the prepatterned areas following by a 10 nm SiGe layer with Ge content =20%. Subsequently, 7 eq-MLs Ge are deposited at 600 °C. Fig. 2 shows the AFM pictures of these three samples and the sample with dots grown on no-patterned area. From Fig. 2, dots are only grown on the ridges. It is never seen in published papers. In other words, nano-trenches are used to confine the position of Ge dots and make it grow well.

3 DISSCUSSION AND MODEL

An important result is that Ge dots grown on the prepatterned substrate are mono-modal. For the growth temperature around 600 °C, a bimodal distribution is found, including pyramid- and dome-shaped dots, in the Ge dots on silicon system. From the AFM data, the statistics of Ge-dots diameter and height are gathered. It is observed that the diameter-to-height ratio (D/*h*) of pyramid-shaped dots is about 10, and the D/*h* of dome-shaped dots is about 5. We classify Ge dots according to D/*h*. Fig. 3(a)-(c) show the percentage of Ge-dots D/*h*, which compare the area without nano-structures and the one with 400nm-400nm trench-ridge nano-structures. It is noticed that the dots grown on prepatterned area, compared to the ones grown on no-patterned area, have the bimodal tendency, both on the ridges and in the trenches. Fig. 3(d) and (e) show the percentage of Ge-dots D/*h*, which compare the area without nano-structures and the one with 30nm-120nm trench-ridge nano-structures. Obviously, the Ge dots are mostly pyramid-shape, for D/*h* is mostly about 10. The mono-modal inclination of Ge dots on 30nm-120nm trench-ridge nano-structures is much more conspicuous than the ones on no-patterned substrate. Thus, we conclude that the dots grown in prepatterned area are uniform than those in no-patterned area.

We also consider the relationship between the morphology of structures and the dots distribution. Fig. 4(a) demonstrates 400nm-400nm trench-ridge nano-structures morphology versus Ge distribution, for x-axis represents the 400nm-400nm trench-ridge nano-structure, and y-axis represents the percentage of Ge distribution, which is taken the volume into account. The figure shows that Ge is not appeared at the edge of the ridge but tends to congregate at the corner of the trench. According to this observation, a model is hypothetically set up, which is used to conceptually explain the position of Ge quantum dots grown on the silicon substrate with nano-structures. We define a parameter *surface mechanical potential energy* (E_{smp}), which strain and structure curvature is attributed to it and determines the position of the Ge-dots growth. Under our definition, dots are preferentially grown in the lower E_{smp} area and unlikely to form in the higher E_{smp} area. Fig. 4(b) illustrates our idea that how the nano-structures morphology affects E_{smp}. It is expected that Ge is preferred to form dots in low E_{smp} area than in high E_{smp} one, and the Fig. 4(a) shows that Ge are not appeared at the edge of the ridge but tends to congregate at the corner of the trench. Thus, we deduce that the ridge edge is the high E_{smp} area while the trench corner is the low E_{smp} one, as shown in Fig. 4(b). Based on the thoughts, we plot morphology versus E_{smp} for the sample with 30nm-120nm trench-ridge nano-structures, as shown in Fig. 5. Because of the width of the trenches reducing to 30nm and the narrowness between two ridge edges, the E_{smp} of the trench corners is higher than that of ridge center. This is why the Ge quantum dots are formed on the ridges instead of being grown in the trenches. In other words, we make use of the ridge-edge strain to restrain dots from being grown in the trenches and align them well. Furthermore, modulation of the 30nm trenches can be used to control the spatial position of dots.

4 CONCLUSIONS

We fabricate Ge dots on silicon substrate with nano-structures. A model is hypothetically set up to explain the phenomenon according to Ge-dots statistics. Moreover, we exploit nano-trenches to align Ge dots. We claim strongly that dots can not be grown in the 30nm-trench area due to the high E_{smp} of the ridge edge. Under this technique, fabrication of well-aligned and mono-modal Ge dots on the silicon substrate with nano-trenches is successfully realized. It offers the potential applications of array for the implementation of nano-devices.

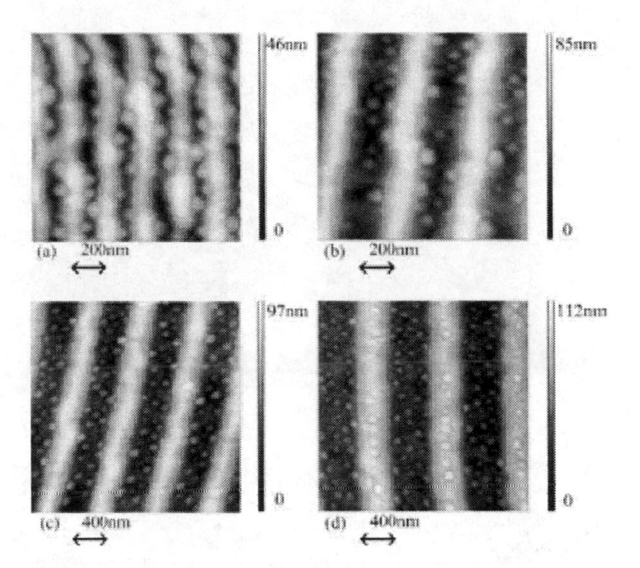

Fig. 1: AFM graphs of Ge quantum dots on the silicon substrate with (a) 100nm-100nm, (b) 200nm-200nm, (c) 300nm-300nm and (d) 400nm-400nm trench-ridge nano-structures

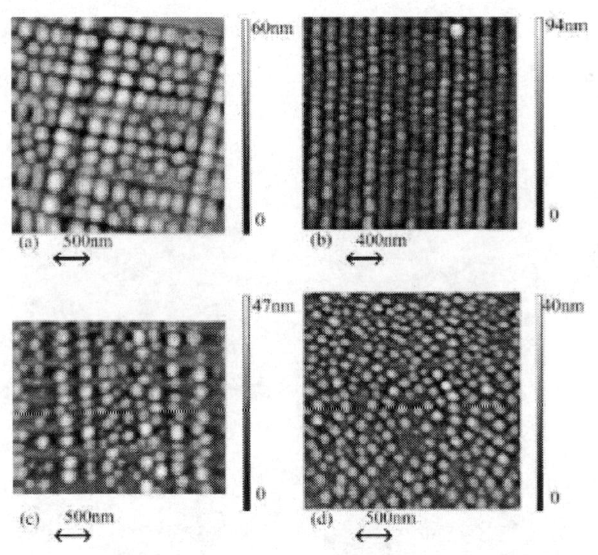

Fig. 2: AFM graphs of Ge quantum dots on the silicon substrate with (a) 30nm-70nm, (b) 30nm-120nm, (c) 30nm-170nm, and (d) no nano-structures

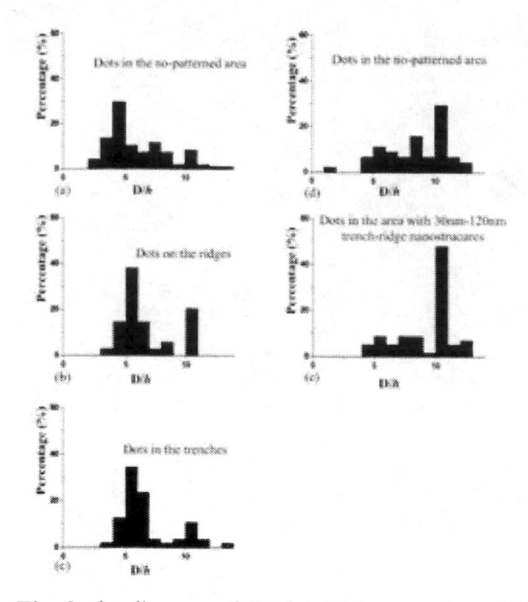

Fig. 3: the diagram of Ge-dots D/*h* percentage, which (a), (b), and (c) show respectively the area without nano-structures, the one with 400nm ridge, and the one with 400nm trench in the first experiment, and (d) and (e) show respectively the area without nano-structures and the one with 30nm-120nm trench-ridge nano-structures.

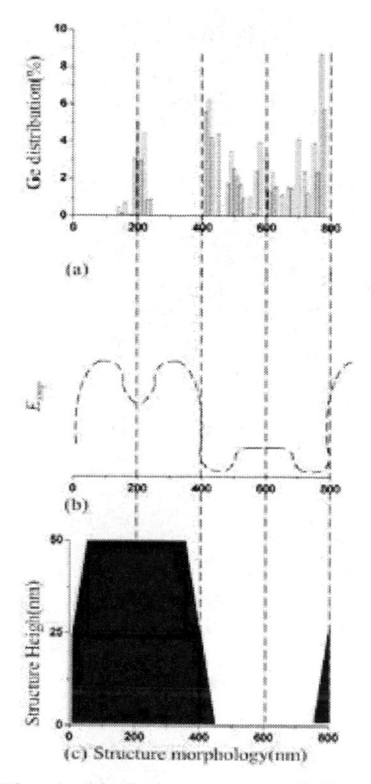

Fig. 4: (a) the percentage of Ge distribution, (b) the E_{smp} (hypothesized), and (c) the morphology versus 400nm-400nm trench-ridge nano-structures.

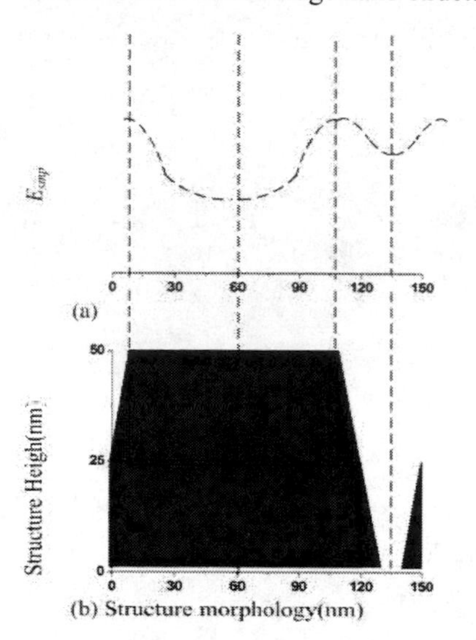

Fig. 5: (a) the E_{smp} (we hypothesize), and (b) the morphology versus 30nm-120nm trench-ridge nano-structures

REFERENCES

[1]Robert E. Rudd, G. A. D. Briggs, A. P. Sutton, G. Medeiros-Ribeiro, and R. Stanley Williams, Phys. Rev. Lett. **90**, 146101 (2003).

[2]Oliver G. Schmidt and Karl Eberl, IEEE Trans. Electron Devices **48**, 1175 (2001).

[3]I. Berbezier, A. Ronda, A. Portavoce, and N. Motta, Appl. Phys. Lett. **83**, 4833 (2003).

[4]Kunihiro Sakamoto, Hirofumi Matsuhata, Martin O. Tanner, Dawen Wang, and Kang L. Wang, Thin Solid Films **321**, 55 (1998).

[5]G. Jin, J. L. Liu, S. G. Thomas, Y. H. Luo, K. L. Wang, and Bich-Yen Nguyen, Appl. Phys. Lett. **75**, 2752 (1999).

[6]Zhenyang Zhong, A. Halilovic, M. Mühlberger, F. Schäffler, and G. Bauer, J. Appl. Phys. **93**, 6258 (2003).

[7]M. H. Baier, S. Watanabe, E. Pelucchi, and E. Kapon, Appl. Phys. Lett. **84**, 1943(2004).

[8]S. Watanabe, E. Pelucchi, B. Dwir, M. H. Baier, K. Leifer, and E. Kapon, Appl. Phys. Lett. **84**, 2907(2004).

Application of magnetic neutral loop discharge plasma in deep quartz and silicon etching process for MEMS/NEMS devices fabrication

Yasuhiro Morikawa, Toshio Hayashi, Koukou Suu, and Michio Ishikawa

Institute for Semiconductor Technologies, ULVAC, Inc.
1220-1 Suyama, Susono, Shizuoka 410-1231, Japan, yasuhiro_Morikawa@ulvac.com

Recently, deep quartz etching technologies to realize micro-opto-electromechanical systems (MOEMS) devices are required. And, silicon deep etching technology is also important. We report on the development of a novel etching system which micro and nano - electromechanical systems (MEMS/NEMS). Etching of quartz and silicon with one chamber using the newly NLD apparatus. The effective plasma production of the NLD method causes a low electron-temperature and high-density plasma in a low-pressure region below 1 Pa. This is one of the characteristics of the NLD plasma. The other important characteristic is controllability for uniform etching by changing the magnetic coil current radially and vertically [1,2]. A schematic of the NLD etching system is shown in Fig.1. Figure 2 shows the etching process flow, (a) KMPR (KAYAKU MICROCHEM Co.) or EB resist coating, (b) lithography, (c) NLD etching, respectively. KMPR has a same performance

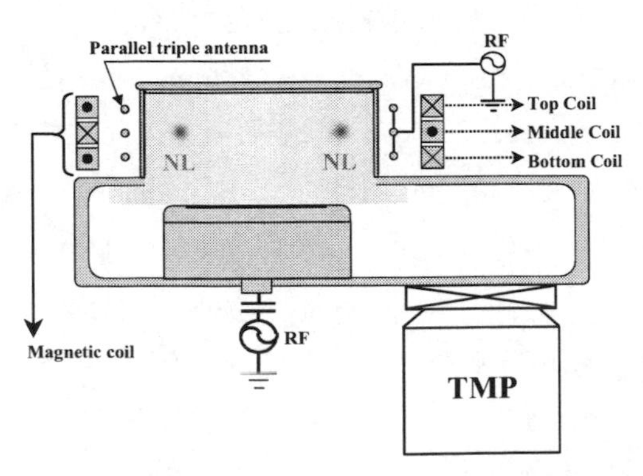

Fig. 1. NLD plasma etching apparatus.

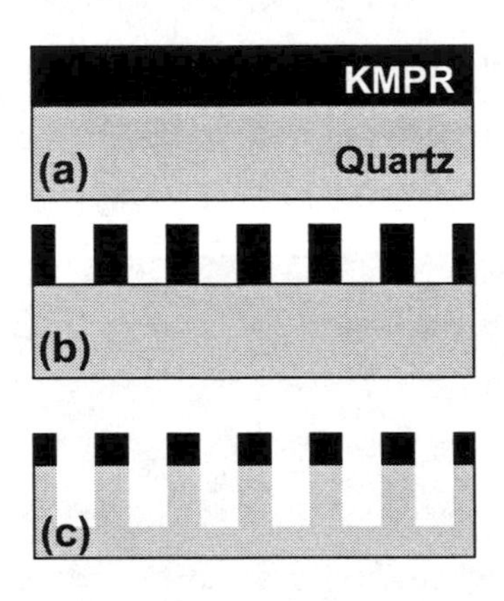

Fig.2. Fabrication process flow.

to a SU-8, and it is superior to SU-8 at the adhesion and crack on the silicon wafer, which were problems. Figure 3 shows KMPR pattern on the quartz substrate. The film thickness of KMPR was 50 um. Figure 4 shows SEM images of quartz deep etching profile after

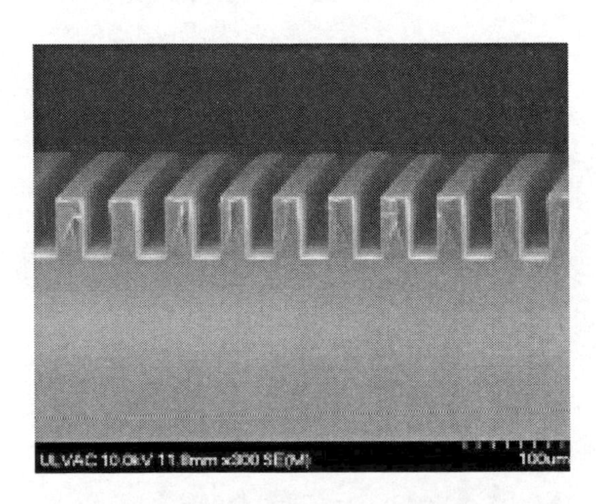

Fig.3. KMPR pattern on the quartz substrate.

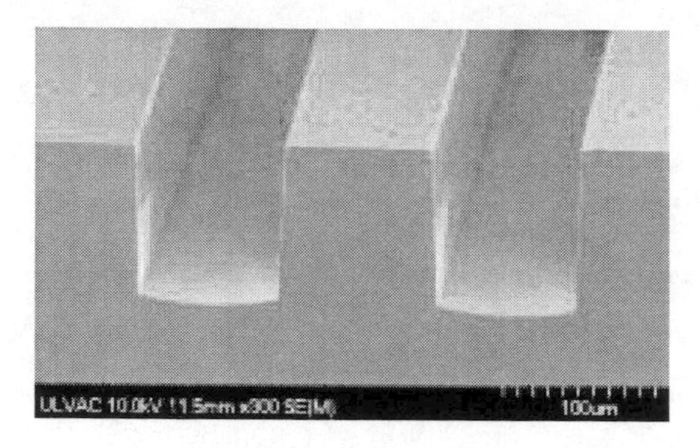

Fig. 4. Quartz deep trench etching

resist removal. We have succeeded in achieving the world's first smooth and vertical deep etching of crystal to a depth of 125 um and at a rate of 1.2 um/min., further enhancing the glass etching performance. And next, case of nano etching for nano-device fabrication, lithography target dimensions are below the 100 nm realm, ArF or EB (electron beam) lithography are considered to be a promising technique. However, these resists have poor etching resistance, which brings on low mask selectivity and results in striation or pitting by resist degradation. This is a serious problem in a nano device fabrication, for example DNA or proteins sorting chip [3]. Figure 5 shows a nano-hole structure made of a quartz plate employing an EB (ZEP-520) lithography and a NLD etching. The diameter, depth of the hole and the aspect ratio were 40 nm, 800 nm, and 20, respectively.

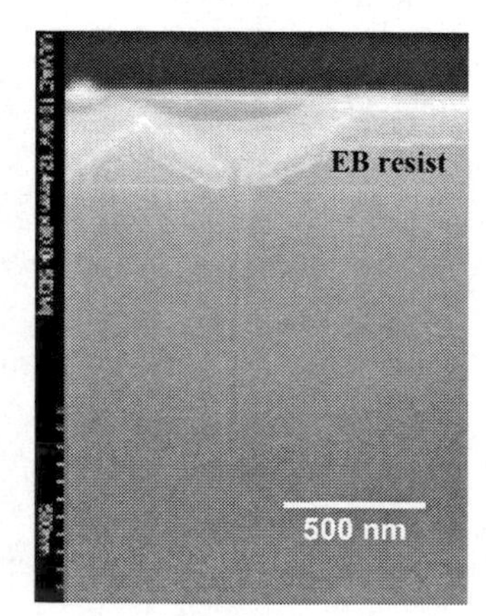

Fig.5. A nano-hole

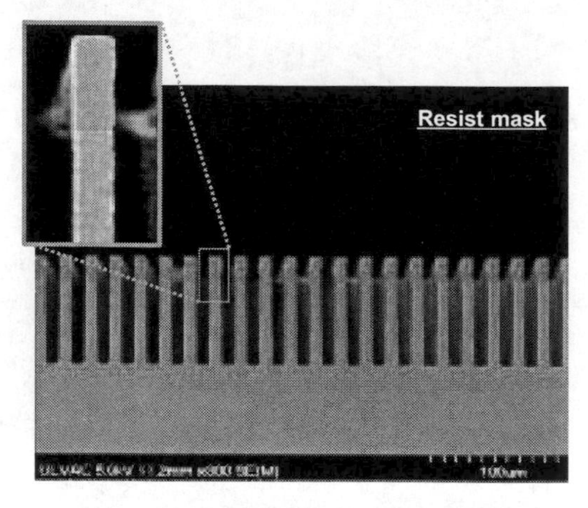

Fig. 6. Silicon deep trench etching

And finally, we have developed a new etching system for MEMS and NEMS application. This system provides combined plasma of NLD and a kind of capacitive coupled plasma (CCP), which is named as NLD-Si plasma. Using this system, deep silicon etching is capable even if fluorocarbon gases are not fed in the etching process. So this system is using our original technology. Silicon deep etching was carried out with resist mask in the NLD-Si plasma modulation, in which the electrodes were timely modulated for etching and deposition at a substrate temperature region above -10 ～ 30 ℃. The selectivity of 300 or more was obtained for resist mask. The etch rate was about 20 um/min when SF_6 / Ar was fed. Typically, the depth of 100 um with the line width of 10 um was anisotropically etched by the electrode modulation method in the NLD-Si plasma shown in fig.6, in which cleaning step was not necessitated. Thus, even if the fluorocarbon gas were not used, high selectivity and anisotropic deep silicon etching was achieved. Using NLD-Si system, vertical profile was successfully fabricated. This is the novelty silicon deep etching system for silicon-MEMS devices. The NLD-Si etching system is expected to be new de-facto-standard tool for R&D and mass production of current and next-generation MEMS/NEMS.

References

[1] W. Chen et al, J.Vac. Sci. Technol. A16(3) (1998) 1594

[2] Y. Morikawa et al., J.Vac. Sci. Technol. B21(4) (2003) 1344 .

[3] Y. Horiike, *Proc. 3rd International Symp. Dry Process*, (2003)

Microfabricated Silicon Apertures for Ion Channel Measurement

S. J. Wilk[*], M. Goryll[*], L. Petrossian[*], G. M. Laws[*], S. M. Goodnick[*],
T. J. Thornton[*], M. Saraniti[**], J. M. Tang[***] and R. S. Eisenberg[***]

[*]Arizona State University, Tempe, AZ, USA, seth.wilk@asu.edu
[**]Illinois Institute of Technology, Chicago, IL, USA
[***]Rush Medical College, Chicago, IL, USA

ABSTRACT

Well-known cleanroom microfabrication technologies have been used to etch an aperture in silicon, pattern a capacitance reducing SU-8 layer and passivate the surface with polytetrafluoroethylene to form an integrated support to study specific ion channel proteins. The silicon support has demonstrated successful lipid bilayer sealing measurements with repeatable seal resistances in the giga-ohm range. Characteristic measurements of OmpF porin ion channel proteins have been made.

Keywords: microfabrication, ion channel, sensor

1 INTRODUCTION

Cells are made up of phospholipid bilayer membranes that serve as high resistance impermeable barriers to the flow of charged ions between the intra and extra-cellular regions. Ion channels are proteins that form a pore across the cell membrane so that specific ions can pass more easily through the cell wall. The conductance of the channel towards the specific ions changes due to voltage or ligand gating events and local salt concentration. Recordings of ionic current are made by keeping the transmembrane voltage constant while observing step current fluctuations due to the gating mechanisms. Patch clamping is a technique that allows for physiological measurements of single ion channel proteins by sucking a cell into a glass pipette and forming a GΩ seal [1]. A stable high resistance GΩ seal is crucial to limit leakage current and enable low noise measurements of single proteins [2, 3].

Planar substrates with micrometer sized diameters have recently been used to span phospholipid bilayers so that ion channels can be measured in an artificial environment. This allows for the study of only the particular ion channels of interest added into the system. Glass [4, 5], plastics [6, 7] and Si/SiO_2 [8-12] have been used as the substrate to span lipid bilayers and record ion channel activity. Glass apertures have been fabricated with heavy ion irradiation and wet etching to form low capacitance devices suitable for low noise bilayer measurement [4, 5]. Small holes have also been punched in Teflon AF and polystyrene using a metal stylus so that a conical shaped aperture for planar bilayer experiments is formed [6, 7].

Current state of the art silicon technology has the advantage of precise micromachining capabilities with high throughput facilities. In previous articles we have demonstrated the fabrication of a silicon aperture that has been hydrophobically functionalized to allow for stable high resistance bilayer membranes [8]. In this article, similar techniques were used to fabricate an aperture of 150um diameter in a thinned silicon substrate with an additional layer of SU-8 added to reduce the capacitance. The surface was rendered hydrophobic with chemical vapor deposition of a polytetrafluoroethylene (PTFE, Teflon) surface layer. The samples were then tested in a bi-chambered Teflon cell using Montal Mueller techniques [13] to form stable bilayers and insert OmpF porin ion channel proteins.

2 EXPERIMENTAL

Samples were prepared using 4", double-sided polished Si (100) wafers having a thickness of 440 microns. The aperture was designed to have a 150um diameter similar to that currently used for Teflon devices [7]. An aspect ratio of 1:1 of the diameter to height of the aperture is desirable for planar lipid bilayer formation [14] so a central region of 1mm diameter was thinned to a final thickness of 150um. The substrates were patterned using photolithography and standard AZ4330 resist and then etched in a deep silicon reactive ion etcher (STS Advanced Silicon Etcher) using the Bosch process. After etching of the aperture, a thermal oxidation of 200nm followed to produce an electrically insulating layer on the surface. The device was then coated with 75um of SU-8 and patterned with conventional photolithography so resist entered the thinned region and decreased the overall capacitance of the device (Figure 1). Finally, a Teflon layer was chemically vapor deposited using the deep etcher and C_4F_8 as the gas source and measured with a Woolam ellipsometer.

Lipid bilayer experiments were performed using a Teflon bilayer chamber with a 5 mm diameter opening in between two baths of electrolyte solution. Both baths were filled with 3 ml of 1 M KCl solution, buffered with 20 mM N-(2-Hydroxyethyl) piperazine-N'- (2-ethanesulfonic acid) (HEPES) at pH 7.4. The device was sandwiched between the baths with the aperture in the center of the opening. Lipids (1,2-Dioleoyl-sn-Glycero-3-Phosphoethanolamine and 1,2-Dioleoyl-sn-Glycero-3-Phosphocholine) (DOPE: DOPC, 4:1) were dissolved in n-Decane (10mg/ml) and

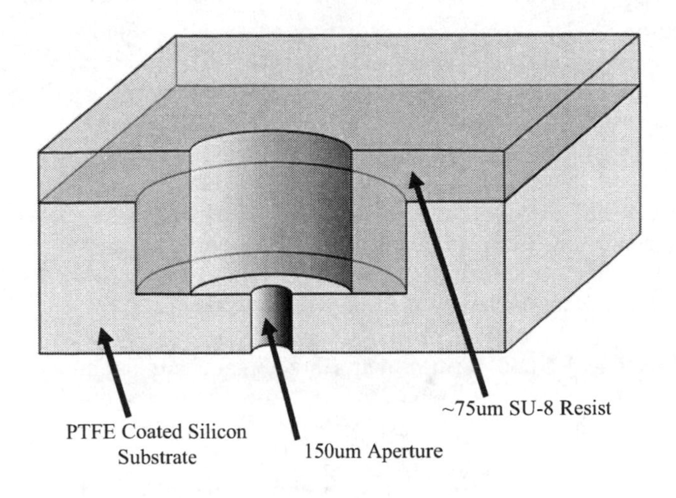

Figure 1: Conceptual drawing of silicon substrate device for transmembrane protein characterization where 75um of SU-8 has been patterned into a thinned aperture and then coated with Teflon.

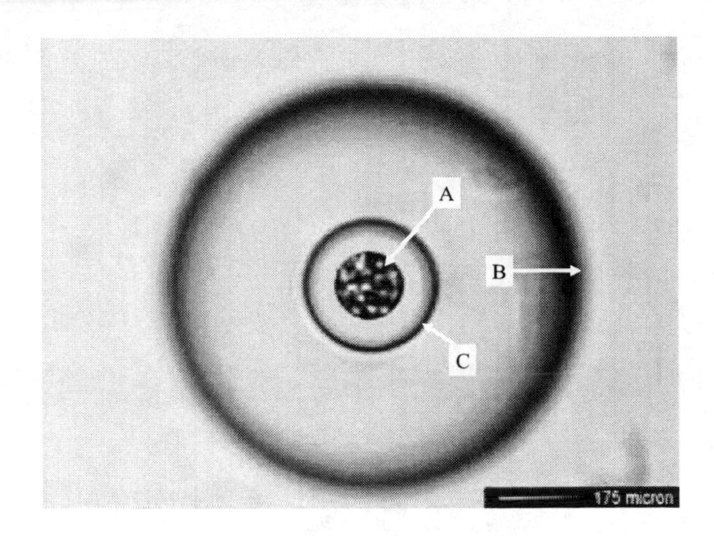

Figure 2: Optical micrograph of device showing (A) the inner aperture of 150um, (B) the outer thinned region and (C) the SU-8 layer for capacitance reduction.

used to form a bilayer with the techniques of Montal and Mueller [13]. Current and bilayer capacitance were measured using an Axon Instruments Axopatch amplifier [15], a Standford Research Systems SRS 830 lock-in amplifier and a National Instruments DAQ PCI card programmed with Labview software. Recordings were performed at a sampling rate of 5kHz and filtered with a four-pole low pass filter. The bilayer resistance was derived from the slope of the current trace. Additionally, it was checked if the layer formed could be broken by the application of a short voltage pulse with Vpulse > 0.5 V. Ion channels were inserted into the membrane by adding OmpF porin to the trans (ground side) bath.

3 DISCUSSION AND RESULTS

Devices were fabricated such that the aperture closely resembles those found in current planar device geometries with a 1:1 aspect ratio. The aspect ratio is important because it determines the shape of the torus and affects the formation and stability of the final bilayer membrane. A 75um layer of SU-8 resist, dielectric constant of ~3, was then patterned over the thinning recess to reduce the capacitance of the device. SU-8 is a chemically amplified epoxy based negative resist that allows for high aspect ratios and smooth sidewall features. Figure 2 shows the final device with SU-8 patterned in the etched recess surrounding the 150um aperture.

It is important to reduce the capacitance of the device in order to decrease the noise of the recordings and increase the possible recording bandwidth. The interaction of the input voltage noise of the amplifier headstage with the input capacitance of the device (septum capacitance coupled with the electrode and membrane capacitance) as well as the dielectric noise due to thermal fluctuations in lossy dielectric materials limits the minimum total rms noise of the measurements. A more in-depth discussion of the noise factors in such measurements can be found in references [2, 6, 7]. The recording bandwidth is proportional to the inverse of the input capacitance, so a decrease in device capacitance will increase the overall bandwidth of the device. High bandwidth is desirable because it enables the recording of fast channel gating events. The capacitance of the devices was found to be 20±5pF using the lock-in amplifier or by applying a triangular waveform and measuring the current response[15].

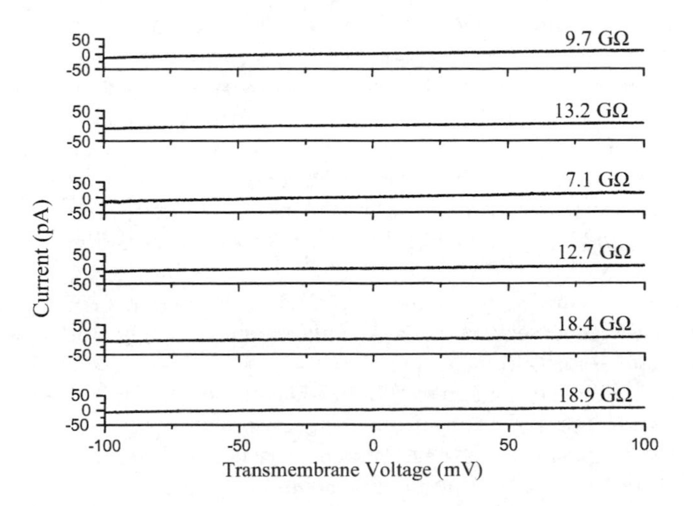

Figure 3: Current-voltage traces of bilayers formed and reformed on silicon substrate device with capacitance reducing SU-8 layer. Initial formation on a fresh sample (bottom plot) followed by bilayers reformation after rupturing using Montal Mueller techniques.

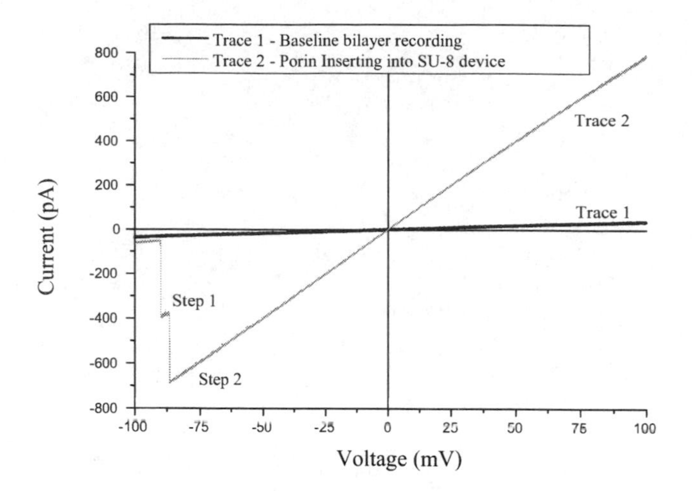

Figure 4: Current-voltage plots of the insertion of two trimers of OmpF porin into lipid bilayer membranes. Trace one shows the baseline before protein insertion with a seal resistance of 2.7 GΩ. Trace two shows the insertion of a single OmpF porin ion channel protein (step 1) followed by a subsequent insertion of a second protein (step 2).

In order to form high resistance stable bilayer membranes the surface of the device must be rendered hydrophobic. A hydrophobic surface lowers the surface energy helping to increase contact with the lipid hydrocarbon chains [13] and allow formation of a high resistance seal. Recently, apertures have been functionalized using self-assembled monolayers to enhance attraction between the substrate and n-decane lipid solvent [10].

After patterning the SU-8 layer, Teflon was vapor deposited on the surface of the device. Teflon has been chemically vapor deposited on substrates [16-18] and has been shown to have better adhesion then spin coated or evaporated films [17]. It serves as a hydrophobic passivation layer with contact angles of 108° [17]. The hydrophobic properties of Teflon make it ideal for lipid bilayer experiments because it enhances the attraction between the lipid tails and substrate and enables formation of a GΩ seal. The additional Teflon layer has previously been demonstrated to enable reproducible formation of a high resistance seal between the bilayer and substrate for ion channel measurements [8].

After device fabrication lipid bilayers were formed across the aperture using Montal Mueller techniques and the OmpF porin ion channel protein was inserted into the membrane by adding it to the trans bath. Multiple bilayers were formed and ruptured on one device with a mean resistance of 9.6 GΩ before addition of the protein. Reformation of the bilayer after rupturing with a short voltage pulse helps ensure phospholipids have not blocked the aperture and have formed a true lipid bilayer. Figure 3 shows the initial bilayer followed by bilayers formed after rupturing. These bilayers were formed over the span of several hours and showed reproducible GΩ seal formations.

After formation of several bilayers, OmpF porin was added to the trans bath of the experimental setup and the bath was stirred until protein insertion occurred. Figure 4 shows the baseline current recording of a bilayer with resistance of 2.7 GΩ (Trace 1) and then the insertion of two OmpF porin proteins (Trace 2). OmpF porin ion channel proteins are trimers with three channels and in 1M KCl solution the conductance of a single channel is about 1.2nS [19]. The first step in the graph corresponds to the conductance of one porin protein (three open channels) and the second step corresponds to the insertion of a second protein. The lower seal resistance of the bilayer can be attributed to the addition of detergent during protein insertion. The detergent changes the molecular orientation of the lipids in the bilayer and helps to facilitate protein insertion.

4 CONCLUSION

Silicon apertures have been fabricated using well-known cleanroom technologies for the measurement of ion channel proteins. Microfabrication of the aperture offers the advantage of precise control of the diameter and high volume throughput over the common method of drilling or burning a hole in a thin sheet of Teflon. After etching the aperture a capacitance reducing layer of SU-8 was patterned and the surface was hydrophobically modified with Telfon. The ion channel protein porin OmpF was inserted into lipid bilayer membranes formed across the aperture and characteristic current voltage measurements of the protein were made. Using microfabrication techniques and silicon as a substrate offers the advantage of integrating microelectronics onto the device and the fabrication of parallel apertures for high throughput screening methodologies.

This work was supported by the Defense Advanced Research Projects Agency as part of the MOLDICE program.

REFERENCES

[1] Hamill, O.P., A. Marty, E. Neher, B. Sakmann, and F.J. Sigworth, Improved Patch-Clamp Techniques for High-Resolution Current Recording from Cells and Cell-Free Membrane Patches. Pflugers Archiv-European Journal of Physiology, 1981. 391(2): p. 85-100.

[2] Levis, R.A. and J.L. Rae, Constructing a Patch Clamp Setup. Methods in Enzymology, 1992. 207: p. 14-66.

[3] Levis, R.A. and J.L. Rae, Patch-Clamp Applications and Protocols, in Technology of Patch Clamp Recording Electrodes, A.B.W. Walz and G. Baker, Editors. 1995, Humana Press: Totowa, NJ. p. 1-36.

[4] Fertig, N., R.H. Blick, and J.C. Behrends, Whole Cell Patch Clamp Recording Performed on a Planar Glass Chip. Biophysical Journal, 2002. 82: p. 3056-3062.
[5] Fertig, N., M. Klau, M. George, R.H. Blick, and J.C. Behrends, Activity of single ion channel proteins detected with a planar microstructure. Applied Physics Letters, 2002. 81(25): p. 4865-4867.
[6] Mayer, M., J.K. Kriebel, M.T. Tosteson, and G.M. Whitesides, Microfabricated Teflon Membranes for Low-Noise Recordings of Ion Channels in Planar Lipid Bilayers. Biophysical Journal, 2003. 85: p. 2684-2695.
[7] Wonderlin, W.F., A. Finkel, and R.J. French, Optimizing Planar Lipid Bilayer Single-Channel Recordings for High-Resolution with Rapid Voltage Steps. Biophysical Journal, 1990. 58(2): p. 289-297.
[8] Wilk, S.J., M. Goryll, G.M. Laws, S.M. Goodnick, T.J. Thornton, M. Saraniti, J. Tang, and R.S. Eisenberg, Teflon (TM)-coated silicon apertures for supported lipid bilayer membranes. Applied Physics Letters, 2004. 85(15): p. 3307-3309.
[9] Goryll, M., S. Wilk, G.M. Laws, T. Thornton, S. Goodnick, M. Saraniti, J. Tang, and R.S. Eisenberg, Silicon-based ion channel sensor. Superlattices and Microstructures, 2003. 34(3-6): p. 451-457.
[10] Pantoja, R., D. Sigg, R. Blunck, F. Bezanilla, and J.R. Heath, Bilayer Reconstruction of Voltage-Dependent Ion Channels Using a Microfabricated Silicon Chip. Biophysical Journal, 2001. 81: p. 2389-2394.
[11] Peterman, M.C., J.M. Ziebarth, O. Braha, H. Bayley, H.A. Fishman, and D.M. Bloom, Ion Channels and Lipid Bilayer Membranes Under High Potentials Using Microfabricated Apertures. Biomedical Microdevices, 2002. 4: p. 231-236.
[12] Pantoja, R., J. Nagarah, D. Starace, N. Melosh, R. Blunck, F. Bezanilla, and J.R. Heath, Silicon chip-based patch-clamp electrodes integrated with PDMS microfluidics. Biosensors and Bioelectronics, 2004. 20: p. 509-517.
[13] Montal, M. and P. Mueller, Formation of Bimolecular Membranes from Lipid Monolayers and a Study of Their Electrical Properties. Proceedings of the National Academy of Sciences of the United States of America, 1972. 69(12): p. 3561-3566.
[14] White, S.H., Analysis of the Torus Surrounding Planar Lipid Bilayers. Biophysical Journal, 1972. 12: p. 432-445.
[15] Ho, D., B. Chu, J.J. Schmidt, E.K. Brooks, and C.D. Montemagno, Hybrid protein-polymer biomimetic membranes. Ieee Transactions on Nanotechnology, 2004. 3(2): p. 256-263.
[16] Coburn, J.W. and H.F. Winters, Plasma-Etching - Discussion of Mechanisms. Journal of Vacuum Science & Technology, 1979. 16(2): p. 391-403.
[17] Jansen, H.V., J.G.E. Gardeniers, J. Elders, H.A.C. Tilmans, and M. Elwenspoek, Applications of Fluorocarbon Polymers in Micromechanics and Micromachining. Sensors and Actuators a-Physical, 1994. 41(1-3): p. 136-140.
[18] Okane, D.F. and D.W. Rice, Preparation and Characterization of Glow-Discharge Fluorocarbon-Type Polymers. Journal of Macromolecular Science-Chemistry, 1976. A 10(3): p. 567-577.
[19] van der Straaten, T.A., J.M. Tang, U. Ravaioli, R.S. Eisenberg, and N.R. Aluru, Simulating Ion Permeation Through the ompF Porin Ion Channel Using Three-Dimensional Drift-Diffusion Theory. Journal of Computational Electronics, 2003. 2(1): p. 29-47.

A Novel Method of Fabricating Optical Gratings Using the One Step DRIE Process on SOI Wafers

A. Cooper[*], P. T. Docker[**] and M. C. Ward[***]

[*] *Mircro and Nanotechnology Group, School of Engineering, Mechanical and Manufacturing Engineering, University of Birmingham, Edgbaston, Birmingham B15 2TT,*
Email: awc@bham.ac.uk , [**] p.t.docker@bham.ac.uk , [***] m.c.ward@bham.ac.uk

ABSTRACT

This paper describes a novel technique for manufacturing optical gratings using the one step DRIE (Deep Reactive Ion Etching) process. Utilising the notching effect documented in previous work when working with silicon on insulator (SOI) wafers, fully released, intact gratings have now been produced without the requirement for additional releasing processes. The one step process eliminates the possibility of stiction which can occur when "wet" processing chemistry is used in the release of microstructures. It should be noted that all DRIE etching in this work was performed using a Surface Technology Systems DRIE which uses a process developed from the Bosch process (R B Bosch Gmbh 1994 US Patent Specification 4855017 and German Patent Specification 4241045CI) termed 'time multiplexed deep etching' (TMDE) [1].

Keywords: gratings, notching, soi, dry-release, optical

1 INTRODUCTION

Deep reaction ion etching (DRIE) and silicon on insulator (SOI) wafers are well established technologies for the manufacture of released MEMS structures. These structures are normally released through the removal of the buried insulator layer, which is removed with an etchant [2]. Typically the etchant is then dispersed from the structure by rinsing with deionised (DI) water which is then dried through evaporation. Unfortunately the flexible microstructures can be pulled together or down onto the substrate at the last moment of drying by the capillary pressure induced by the droplet in the gap. These structures can remain stuck together even after completely dried, if the adhesion force between the contacted areas is larger than the elastic restoring force of the deformed structure. This phenomenon is more commonly called "stiction" in the field of microelectromechanical systems (MEMS).

When fabricating optical gratings, "stiction" is particularly undesirable because the beams of the gratings can stick to one another resulting in an imperfect grating.

In recent years, to try and eliminate the need for this wet processing, a number of workers [2-4] have developed a method of dry releasing microstructures on SOI wafers using DRIE. This method involves using the undercutting or notching phenomenon [5,6] that is observed when the buried oxide is exposed during the etching of high aspect ratio trenches.

Additional work was carried out by Docker *et al* [7] to further enhance this technology through the use of sacrificial 'waffle' type structures that allow large areas of silicon to be cleared from SOI wafers without the formation of silicon grass. These 'waffle' type structures were designed to simply float free from the surface of the SOI wafer when they had been under-etched and released. The waffles were originally used when etching device layers measuring 20 microns thick and failed catastrophically when they were released.

The new work demonstrates how modified waffle structures, when etched into a 300 micron device layer of a SOI wafer, can form intact released optical gratings. Unlike the standard waffle structures, these new gratings are made up of long trenches dimensionally wider by an order of magnitude. Their depth is also much greater than the standard waffle structures, measuring 300 microns as opposed to just 20 microns.

2 GRATING DEVELOPMENT

It is believed that the larger feature size of the gratings is the key feature responsible for the devices not being eradicated by the etching process. The height of the damage caused by the notching effect which releases the structures is limited, leaving the gratings with deep, parallel wall profiles. When released by the notching phenomenon, the gratings do not fail like the shallower waffles and they remain present on the wafer until it is removed from the etching chamber. The gratings can then be removed from the wafer by simply tipping the wafer through 180 degrees and letting the gratings fall onto a soft surface.

One of the key features in being able to etch the gratings to the 300 micron depth had been in the utilisation of a thermal oxide mask instead of the normal photo-resist mask for the etching process. This has enabled a very accurate feature size of the gratings to be maintained which would not have been possible using one very thick or multiple layers of photo resist.

The oxide mask is fabricated using conventional photolithography on top of the oxide layer which is then subjected to a plasma oxide etch to create the oxide mask. A one micron thick layer of oxide proved sufficient to withstand the DRIE process for the 130 minutes required to etch the 300 micron depth of the device layer and undercut the gratings allowing them to be released. The etch time required for release of the gratings was found simply by visual inspection.

3 FABRICATED GRATINGS

Figure 1 shows a grating with a 40 micron beam width and 40 micron spacing "in situ" on a 300 micron thick SOI wafer after etching.

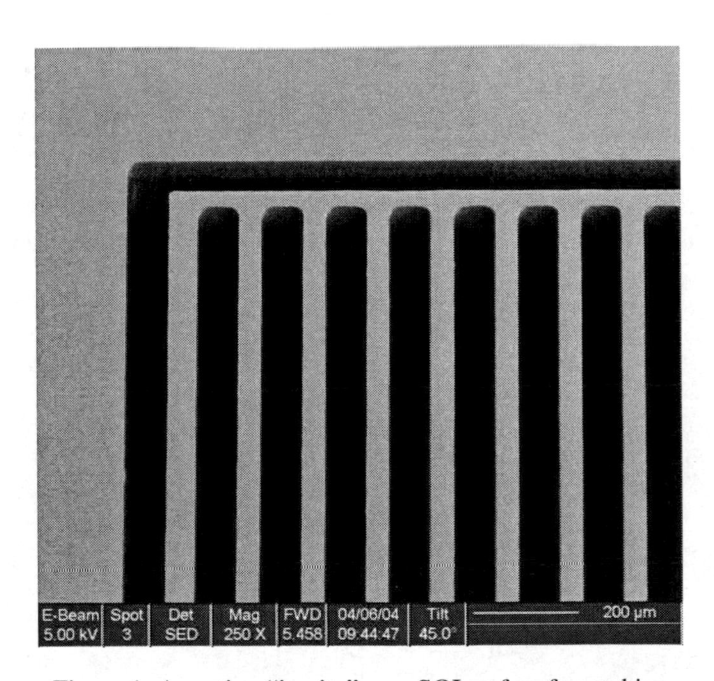

Figure 1: A grating "in-situ" on a SOI wafer after etching

Figure 2 shows the same grating after the under-etching process was completed and the grating was removed from the SOI wafer.

The photolithography performed to produce the gratings shown in figures 1 and 2 was done using acetate masks as opposed to conventional chrome on glass masks. This was done to reduce development costs due to the large number of grating masks that were needed for the project. For this reason the surface of the walls of the gratings is not as smooth as it is believed possible to achieve using conventional chrome on glass masks. This is due to the larger tolerance and minimum feature size that is a characteristic of the acetate masks.

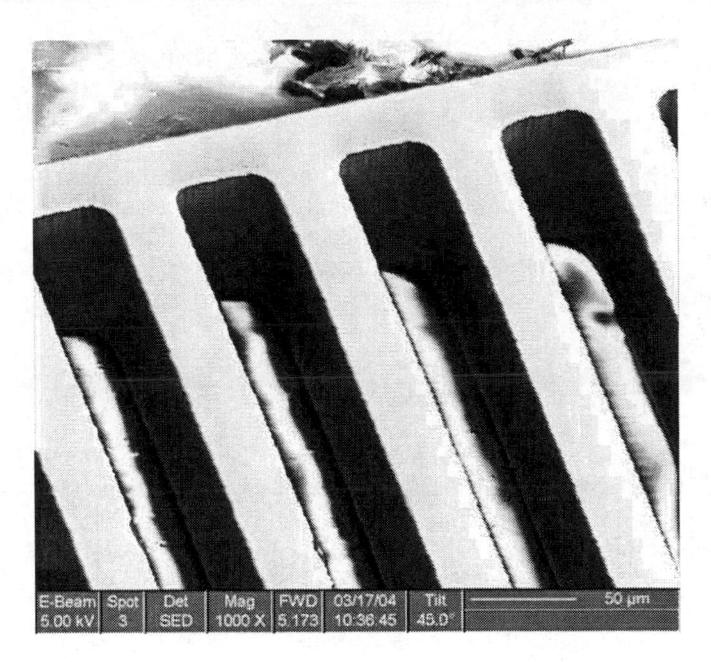

Figure 2: A grating removed intact from the SOI wafer

4 POTENTIAL APPLICATIONS FOR HIGH ASPECT RATIO GRATINGS

One of the most interesting features of the gratings is the high aspect ratio of the microstructures. With an aspect ratio of 7.5:1 the gratings lend themselves to the possibility of being used as optical switches that work by means of a tilting or rotating action. The gratings reported in this paper would require a tilting angle of only 7.6 degrees out of plane of the light source to stop light from passing through them.

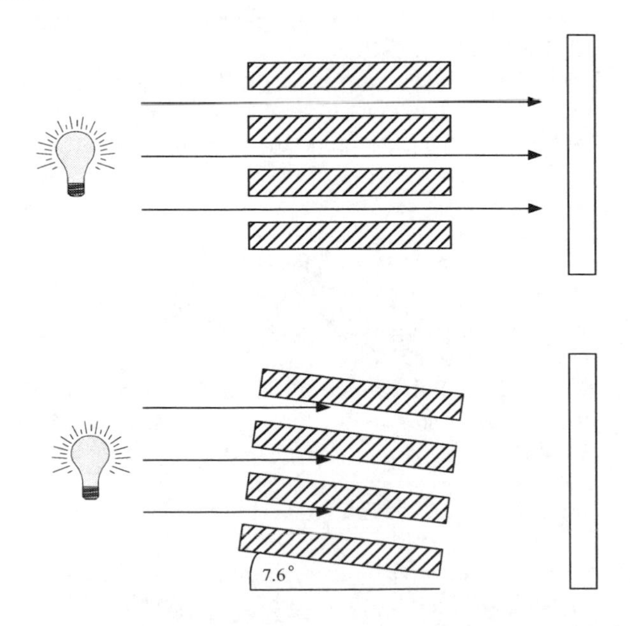

Figure 3: Schematic diagram of how the gratings might be used in switching applications

This is illustrated in figure 3 and it is believed that gratings with an aspect ratio of over 20:1 could be produced, reducing the tilting angle required to less than 3 degrees. In the field of microelectromechanical systems (MEMS) where the performance of devices such as switches is often governed by the displacement available from low voltage actuators, a device requiring such a small displacement could prove to be a valuable tool for use in optical switching technology

5 CONCLUSION

This paper describes a novel method of fabricating fully released optical gratings from SOI wafers using the one step DRIE process. Gratings with a 40 micron beam width and 40 micron gap width were fabricated from 300 micron thick device layer SOI wafers. The gratings were successfully dry released and removed from the wafer whilst maintaining their structural integrity.

The successful fabrication of these high aspect ratio gratings shows great promise for applications such as optical switching. The large aspect ratio of the grating means that they would only have to be tilted a few degrees out of plain with a light source to obstruct light from passing, effectively performing the role of an optical switch. These filters could prove to be a timely discovery with the current shift towards optical signal processing.

REFERENCES

[1] Bosch R B Gmbh 1994 US Patent Specification 48855017, and German patent Specification 4241045CI.

[2] Volland B E, Heerlein H, Kostic I and Rangelow I W, "The application of secondary effects in high aspect ratio dry etching for the fabrication of MEMS" Microelectron. Eng. 57-58, pp 641-650, 2001.

[3] Ayon A A, Ishihara K, Braff R A, Sawin H H and Schmidt M A, "Microfabrication and testing of suspended structures compatible with silicon on insulator technology" J. Vac. Sci. Technol. B 17, pp 339-349, 1999.

[4] Docker P T and Kinnell P K, "A dry single step process for the manufacture of released MEMS structures", J. Micromech. Microeng. 13, pp790-794, 2003.

[5] McAuley S A, Asraf H, Atabo L, Chambers A, Hall S, Hopkins J and Nicolls G, "Silicon micromachining using a high density plasma source", J. Phys. D: Appl. Phys. 34, pp2769-2774, 2001.

[6] Gormley C, Yallop K, Nevin W A, Bhardwaj J, Asraf H, Huggegett P and Blackstone S, "State of the art deep silicon anisotropic etching on SOI bonded substrates for dielectric isolation and MEMS applications", Fifth Symp. Semiconductor Wafer Bonding Svience, Technology and Application, The Fall Meeting of Electrochemical Society (Hawaii, USA) Oct 17-22, 1999.

[7] Docker P T, Kinnell P K and Ward M C L, "Development of the one-step DRIE dry release process for unconstrained fabrication of released MEMS devices", J. Micromech. Microeng. 14, pp941-944, 2004.

Absolute Pressure Measurement using 3D-MEMS Technology

Rick Russell* and Ulf Meriheinä**

*VTI Technologies Oy, Dearborn, MI, USA, rick.russell@vtitechnologies.com
**VTI Technologies Oy, Vantaa, Finland, ulf.meriheina@vti.fi

ABSTRACT

The use of 3D-MEMS processes has created yet another new breakthrough. Extremely small silicon capacitive absolute pressure sensors can now be realized by utilizing the same MEMS technology that has been in production for many years. The applications where absolute pressure measurement has not been possible due to size or current limitation can now be solved effectively.

The new breakthrough has enabled VTI Technologies to offer extremely robust capacitive pressure sensing elements with low current consumption and excellent performance. The hermetically sealed elements are made of single crystal silicon, which is close to perfect elastic material. These features along with its compact size make it possible to extend absolute pressure measurement to completely new areas such as wellness and safety. The same capacitive sensing element can also be used for battery-less measuring systems like tire pressure monitoring.

The development of a complete sensor component is underway and will only be 6 mm in diameter and 1.7 mm high. This device will be capable of detecting the smallest barometric pressure change, which enables extremely accurate weather forecasting and altitude measurements. The supply voltage (2.4 – 3.3V) and current consumption (0.5 - 50μA) make it easy to integrate into products like outdoor equipment and sports watches.

The excellent accuracy and resolution enables algorithms not seen in consumer weather stations or outdoor equipment before. In a sports watch application, the new sensor component can provide a resolution better than 0.2m (2PA) and consume less than 30μA when operated continuously at a rate of 2 times per second. When increasing the update interval to 2 seconds, the current consumption can be reduced to less than 10μA.

The absolute pressure sensor elements are available today and can be soldered directly onto a PCB as a SMD component. The development of a complete sensor component (SCP1000) will be launched in 2005.

Keywords: MEMS, capacitive, absolute, pressure sensor, barometric, altimeter.

1 3D-MEMS TECHNOLOGY USED FOR PRESSURE SENSORS

3D-MEMS is an optimized combination of technologies to achieve the best performance at the smallest size and lowest power. The technologies include wet and dry etching, capping with wafer bonding and glass insulation, electrode feed through structures and contacting for wire bonding and soldering. Unlike some other MEMS (Micro Electro Mechanical System) technologies, VTI offers real 3-dimensional structures (Figure 1), not just thin films on top of a silicon wafer. This gives flexibility in optimizing electrode insulation, stress minimization and capacitance dynamics for performance. Unlike piezoresistive pressure sensors, the relative capacitive change over the measuring range is typically 30-50% making it easy to measure, thus enabling high signal to noise ratio and accuracy even at low current levels.

Figure 1: VTI's 3D-MEMS absolute pressure sensor element manufactured on 150mm wafers

The glass insulation in use is relatively thick and gives high isolation resistance (low leakage current) and low stray capacitance. The two silicon wafers are anodically bonded together resulting in a hermetically sealed structure, where no chemicals or particles can intrude into the space between the capacitive electrodes. The mechanical material in use is single crystal silicon giving hysteresis free operation (no plastic deformation) and excellent over-range and shock performance. The sensor can withstand a pressure significantly higher than 10 times the measuring range. The excellent long-term stability of pressure sensors made in this technology was already proven in older sensor designs in the 1980's.

2 MANUFACTURING OF 3D-MEMS

To define the optimum shape and size of the APS4 pressure element, VTI had to perform many simulations in order to satisfy all the design parameters.

While the footprint of 1.4 x 1.4 mm^2 offers the best optimization for the length and width, the height has been optimized for different applications. The height of 0.85mm was chosen for high accuracy wire bondable applications. Taller devices are required to compensate for any mechanical stress related to direct soldering. The height of 1.25mm was chosen for both the 8 and 25 bar elements.

2.1 Transfer Function

The size of the element is the most important factor to keep a high dynamic measurement range. Hence the size was optimized to keep the capacitance range as high as 7pF for 0bar and 14pF for full scale pressure. This was achieved by an extremely small distance between the two capacitor plates with a nominal gap of 1µm. The passive parallel capacitance was kept small by using a thick glass layer between the upper and lower silicon layers. Silicon feed-throughs are used to make the necessary contacts. This technique provides a good active to passive capacitance ratio. The transfer function follows the equation:

$$C = C_{00} + \frac{C_0}{1 - p/p_0}$$

whereas C_0 is large compared to C_{00}. The realized values are in the range of: C_{00} = 1pF and C_0 = 5 pF. Figure 2 shows the measurement results of the first 1.2 bar prototypes.

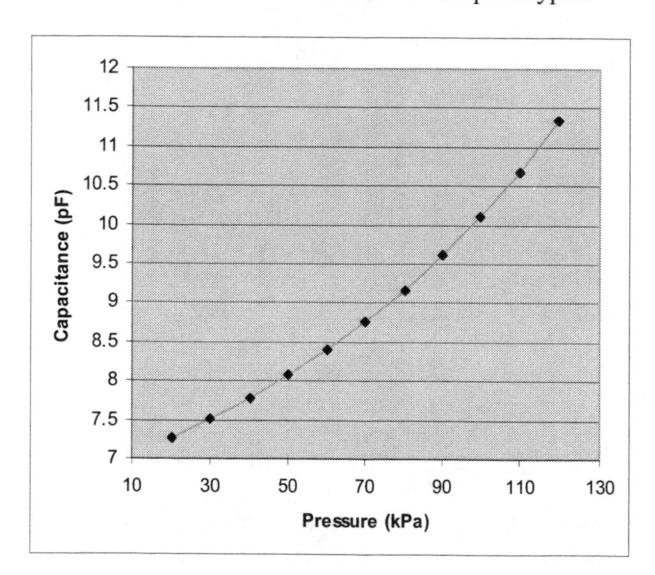

Figure 2: Transfer function of APS4 of first prototypes (measured results)

Reflecting the transfer curve over temperature provides the perspective that temperature compensation wouldn't be necessary for this type of element. The main reasons for this positive behavior are the small C_{00} impact and the fact that silicon is used as a conductor. This advantage needs to be preserved by using appropriate signal conditioning and mounting techniques to avoid thermal related stress to the sensing element.

2.2 Linearization

The linear behavior of the APS4 element is in line with the parallel plate model according to the following equation:

$$p = p_0 * (1 - \frac{C_0}{C - C_{00}})$$

Figure 3 shows that there is only a small deviation between the transfer curve of first sensor prototypes and the ideal parallel plate model. The maximum deviation is less than 20Pa (0.2mbar).

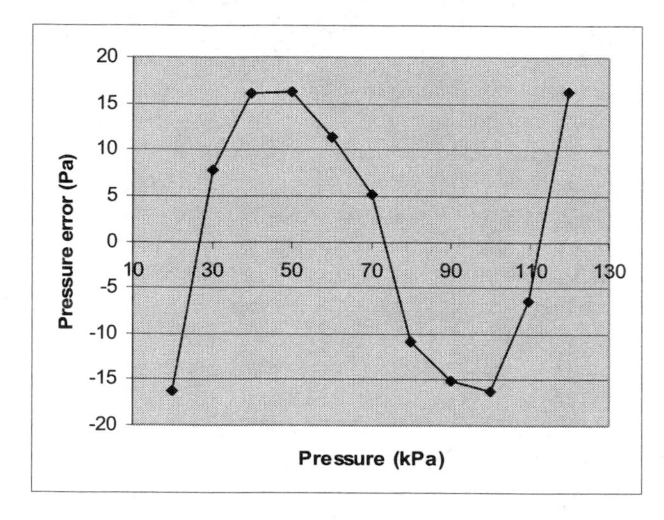

Figure 3: Difference between parallel plate model and Sensor Output

The APS4 sensor element gives engineers the ability to design ultra low current consumption circuits due to the high parallel resistance, which is greater 10 Gohm.

2.3 Optimizing the Diaphragm

The diaphragm can be dimensioned in such a way that sensor elements from 1bar (20µm) to 30bar (80µm) can be realized. By using wet etching, the height of the diaphragm can be processed quite accurate. The height or thickness of the diaphragm is determined by the full-scale pressure range. The gap is always kept at 1µm to remain with the same dynamic range for all the potential variants of this concept.

Underlying the design of the diaphragm was a requirement to avoid any signal changes caused by centrifugal force as well as any mechanical shock caused by rough disturbances. For that reason, the diaphragm has to be optimized in order to make a good compromise for the transfer function.

2.4 Designing for Long-Term Stability

The long-term stability of an absolute pressure sensor depends on the seal of the anodic glass bond, which can only come from highly controlled production processes. VTI has more than 20 years of experience in glass bonding technology. Glass is a well known material but it has some specific characteristics like outgasing and absorption of residual gas, which could change the internal pressure of the sensor.

To ensure good long-term stability, a wet etched gas reservoir was connected to the reference pressure gap. The difficulty in the given design was to find any space in the optimised element structure. The positioning of the vacuum reservoir is shown in figure 4 schematically and in figure 5 as a cross section in silicon.

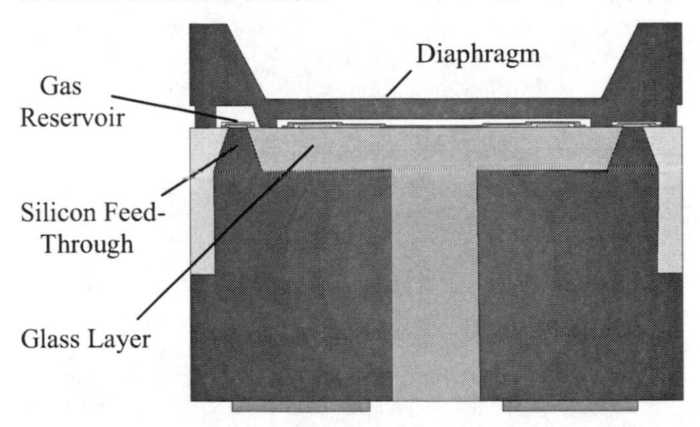

Figure 4: Schematic cross section APS4

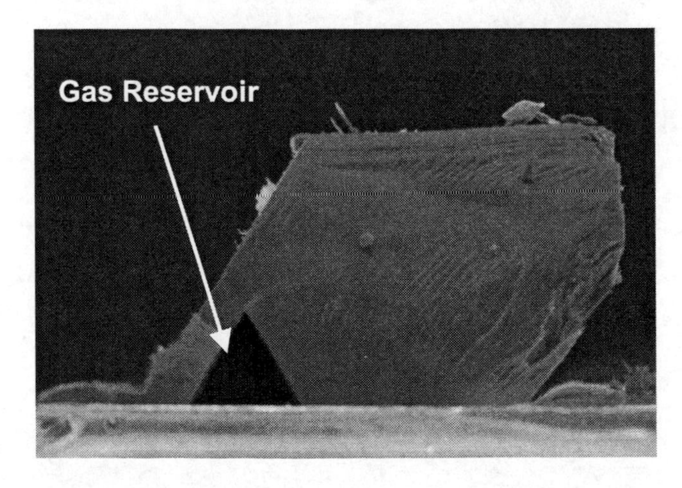

Figure 5: Vacuum reservoir for long term stability

2.5 Pad Location

VTI developed their first pressure sensors with the contact pads co-located on the same surface as the diaphragm. The contact pads were consuming space on the surface and it was difficult to protect them against aggressive media. To reduce the size of the element, the pads were moved to the bottom by adding a glass isolator in the middle of the element as shown in figure 6. The glass is structured in such a way that the offset capacity value is kept as small as possible.

Figure 6: VTI's APS4 pressure element showing contacts

3 PRICIPAL OF OPERATION

The principle of operation of VTI's pressure sensors can be seen in Figure 7. The outer pressure relative to the reference pressure inside the cavity between the silicon wafers causes a force on the membrane of the top wafer, bending it towards the bottom electrode. The elastic membrane acts as force gauge. The displacement of the membrane is detected as a change in the value of the capacitor (C) between the membrane and the counter electrode. The function 1/C is a fairly linear measure of the force acting on the membrane and it can easily be linearized with a simple polynomial for highest linearity.

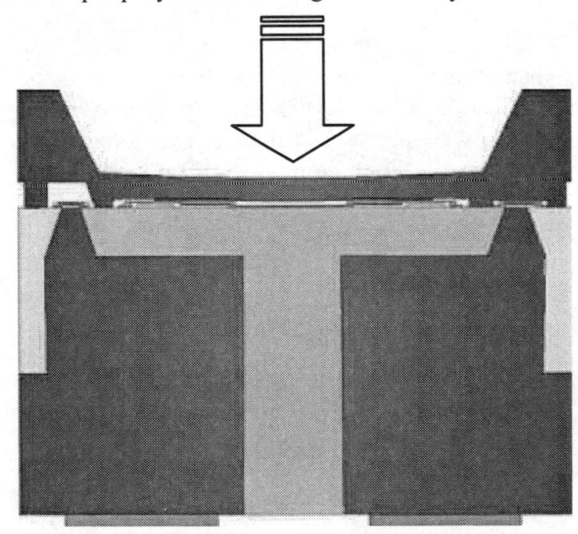

Figure 7: APS4 diaphragm under pressure

4 FEATURES OF VTI'S PRESSURE SENSORS

VTI is introducing a low-power absolute pressure sensor concept mainly intended for battery operated handheld devices. Measuring ranges extend from barometric (100kPa) to 25 bar (2500kPa). The output is digital over SPI (Serial Peripheral Interface). The SMD (Surface Mount Device) component SCP1000 is as such intended for non-corrosive environments, but being only 6.1mm in diameter and 1.7mm in height (Fig. 8) is easy to protect from aggressive media. The sensor component has a circular vertical wall for O-ring sealing, a feature important in a humid environment and when measuring in liquids like water.

Figure 8: VTI's low power pressure sensor component SCP1000 (ϕ = 6.1mm, h = 1.7mm).

The more detailed description of the features below is much related to the barometric sensor for the range 30…120kPa, corresponding roughly to -1000…+9000m in altitude. For other ranges the accuracy parameters roughly scale with the range.

SCP1000 has different measurement modes for optimum performance in different applications that require different resolution, speed and power consumption. All modes operate with at least 2.4…3.3V power supply. In the high-resolution mode the barometric sensor has a resolution of better than 2Pa, corresponding to about 16cm in an altimeter application at sea level. This resolution can be achieved with a twice per second update rate and at 20μA power consumption in continuous operation. In the high-speed mode the resolution is about 5 times worse and the speed 5 times higher.

To save power there are two options. One of them is to use the low power mode, where after each measurement the sensor goes into stand-by and consumes 3 - 4μA. In this mode the sensor stays ready for the next measurement command over its digital SPI interface. Switching off power between measurements can even save more power. Hereby the wake-up time being around 10ms is important.

SCP1000 has excellent accuracy and linearity. Under normal outdoor conditions for hand-held devices, 10…40°C, an accuracy of better than 2m can be reached at sea level. In a wider altitude and temperature range, i.e. 0…3000m and -20… +70°C, an accuracy of 30m or better can be reached. Hereby the non-linearity, being less than 10cm, is insignificant.

The SCP1000 includes a temperature sensor with 0.1°C resolution and ±1°C accuracy.

5 APPLICATIONS

There are numerous applications for the SCP1000 product family but the main focus area is wellness, sports and outdoor applications. In particular, handheld devices such as sport watches and diving computers, where a combination of small cost effective designs, low power, reasonable resolution and accuracy are important. The SCP1000 works well as an altimeter and is a perfect compliment to any GPS (Global Positioning System) signal for navigation systems.

The sub-meter altimeter resolution and 1m accuracy (under normal conditions) provides sports and fitness users a precise calculation of the accumulated risen height for a particular training session. On the other hand the high-speed mode and good performance in a wide altitude and temperature range give new opportunities in skydiving, paragliding and similar applications.

In sports and outdoor activities the barometric sensor can be used to forecast local weather and measure the oxygen pressure through the total air pressure.

Small and accurate low power pressure sensors are also finding their way to the medical field. Blood and brain pressure sensing are obvious applications. The barometric pressure sensor through its high resolution is an excellent fluid level. Its resolution is about 0.1mm water.

The 16 bar version of SCP1000 is intended for diving computers and water depth sensing. Of course it can be used in industrial fluid level applications as well. Another application area is pressure measurements in pneumatic systems, where the sensor through its low power consumption could be wireless or even be built into a batteryless transponder.

REFERENCES

[1] Jens Thurau, Dr. Jaakko Ruohio, Silicon Capacitive Absolute Pressure Sensor Elements for Battery-less and Low Power Tire Pressure Monitoring, AMAA 2004

Fabrication of Silicon Nanowires Using Atomic Layer Deposition

Deepthi Gopireddy*, Christos. G. Takoudis*, Dan Gamota**, Jie Zhang**, Paul W. Brazis**

* Advanced Materials Research Laboratory, Department of Chemical Engineering
University of Illinois at Chicago, Chicago IL 60607
** Motorola Advanced Technology Center, Schaumburg IL 60196

ABSTRACT

Growth of silicon nanodots at temperatures 550-625°C by thermal decomposition of silane in a low pressure chemical vapor deposition reactor is reported here. This procedure uses short periods of deposition time in the early stages of polysilicon growth. Atomic Force Microscopy was used a primary technique to investigate the growth of the silicon nanodots on three different substrates. The substrates used are clean silicon, silicon dioxide (SiO_2) substrate and dilute hydrofluoric acid treated silicon dioxide substrate. The analysis reveals that for a 1 min deposition time on a hydrofluoric acid treated substrate, the mean size of the nanodots obtained range between 5-10 nm in height and 20-30 nm in diameter. However, the size of the nanodots obtained is different for the three substrates used.

Keywords: silicon quantum dots, lpcvd, nanodots, silane decomposition

1 INTRODUCTION

The emerging field of nanotechnology offers possibilities for studying fundamental chemical and physical principles at nanoscale and provides avenues to new technologies such as nanodevices and nanosensors. Nanoscale structures have been extensively studied due to their unique properties that display important size-dependent quantum effects. These structures offer extremely attractive physical features and have a great potential for nano-scale applications since they can function both as active devices and interconnects. The quantum size effects and the coulombs blockade phenomenon make them extremely attractive to use in functional devices such as resonant tunneling device, single electron transistor and p-n junction transistors [1-3]. However, one current challenge is the fabrication and interconnection of these devices on semiconducting surfaces such as a Si wafer. We must precisely control the physical properties of these structures; thus, it is necessary to control with atomic precision their morphology, structure, composition and size. To realize silicon (Si) quantum effect devices, we desire to develop a simple technique to fabricate high-quality crystalline nanowires using Atomic Layer Deposition (ALD) technique. The method will enable control over several important aspects of the growth, including control of the nanowire diameter, length and structure. In this work, we present the first step to our approach – the formation of the Si nanodots that act as seeds to grow nanowires.

Low Pressure Chemical Vapor Deposition (LPCVD) has been widely used in the field of ultralarge-scale integrated circuits to form poly-crystalline Si films. Several groups have reported the formation of Si nanocrystals on insulators such as SiO_2 [4, 5], sapphire [6] and silicon nitride [7, 8] by controlling the early stages of LPCVD deposition. Here, we show that the Si dot density can be controlled by monitoring the surface properties and that the size can be controlled by varying the time and the temperature of deposition of silane.

2 EXPERIMENTAL

To study the nucleation process on Si, SiO_2 and hydrofluoric acid (HF) treated SiO_2, a clean hot-wall LPCVD system was used. The LPCVD system consisted of a quartz tube heated by a resistance furnace. The temperature was varied from 550 – 650° C and the partial pressure of silane was constant at 0.2 Torr. The p-type Si(100) substrates were chemical cleaned and some of the substrates were oxidized in dry oxygen at 1100° C to form an oxide layer of 15 nm thick. A few of these oxide layers were then treated with dilute HF solution (10%) for 60 s to form hydroxyl groups on the surface. Silicon nanodots were self-assembled on all the three substrates by LPCVD of pure silane. The deposition times for this study were short compared to the usual polysilicon deposition time.

In order to investigate the nucleation, growth and coalescence of the nanodots, the areal density and dimensions of the particles were probed and measured using AFM. The density of the particles was calculated using the "particle analysis" Nanoscope DI software.

3 RESULTS

Figures 1 (a), (b), and (c) show the top view image for SiO_2 substrates treated with HF with deposition times 1, 2 and 6 min, respectively, at 625°C and 0.25 Torr pressure. The hemispherical shaped nanodots vary in both diameter and height. Figure 1 also shows that several of the nanodots coalesce to form dots with larger diameter for longer deposition times. The mean size and size distribution of these nanodots with respect to both diameter and height on the three different substrates used is shown in figure 2. As the deposition time is increased not only do the diameter and the height of the dots increase but the dots tend to coalesce. Nakajima et.al. [4] speculate that this coalescence could be due to the Brownian migration of the particle. Si dots of about 10 nm are mobile on SiO_2 substrate. It should be noted from Figure 3 that though the size of the Si dots increases the density of the dots on the surface decreases with deposition time. The decrease in density is more significant on substrate treated with HF when compared to SiO_2, while the density of dots on clean Si substrate essentially remains constant. It is also evident from Figure 1 and that the dots are not only coalescing but also new Si dots are seen to be formed. The size distribution of the dots formed with deposition time (Figure 2) suggests the formation of new nucleation sites over a period of time. The lower limit of the size distribution essentially remains constant.

It has been suggested that a point imperfection is composed of silicon bonds [7]. In the case of HF treated SiO_2 substrate, the surface Si-O bonds dissociate creating dangling bonds [9]. At the thermal decomposition temperature of silane, the SiH_2 precursor reacts with the surface "OH" groups. Miyazaki et.al. have reported this reaction scheme based on variation of the concentration of surface "OH" groups achieved by annealing the surface for different times [9]. The Si dot distribution became significantly narrow with increasing concentration of the "OH" groups on the surface. In the case of SiO_2 substrate, the strain induced in the Si-O-Si bond due to oxidation is responsible for the Si dot nucleation sites. However, due to the stronger bond energy of Si-H (299 kJ mol^{-1}) the nucleation density on Si substrate is much smaller than that on SiO_2 and HF-treated SiO_2 substrates.

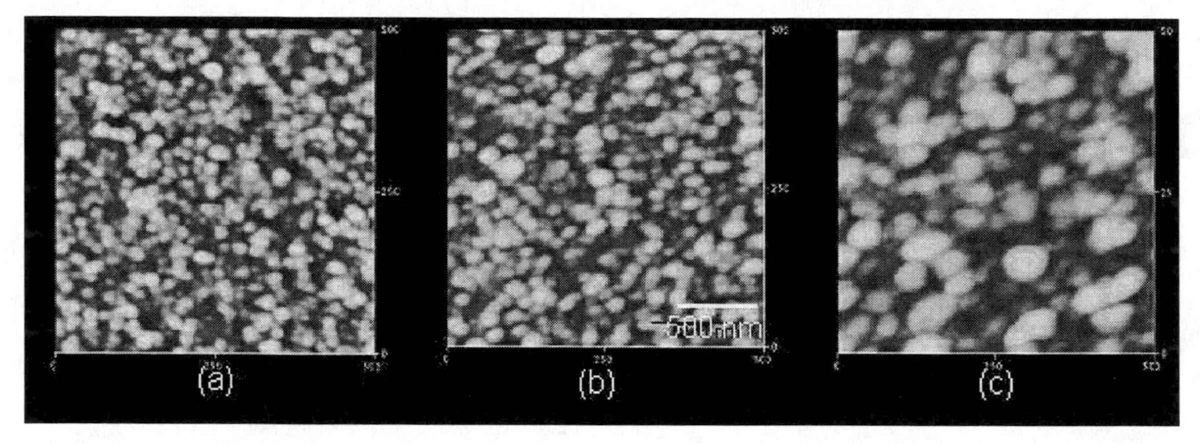

Figure 1: The evolution of Si dot density and size after a deposition time of (a) 1 min, (b) 2 min and (c) 6 min.

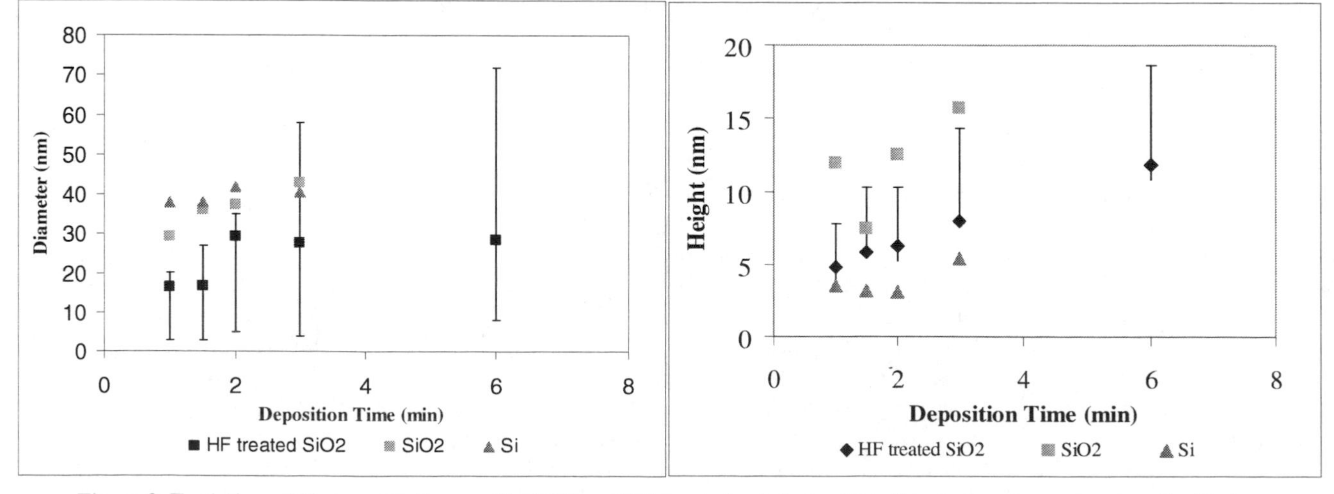

Figure 2. Evolution of Si size with time on Si, SiO2 and HF-treated substrates. This figure also shows the size distribution of the Si dots on HF treated SiO2 substrate.

The size and the density of the Si dots formed on Si substrates remain nearly the same. This suggests that the growth of the silicon dots is very slow and there is no evidence of Brownian motion on of Si dots on the clean Si surface.

The evolution of Si dot density and size with temperature was also explored. Figure 4 shows top view images for HF treated SiO_2 substrate. The deposition temperature affects the size of the Si nanodot formed. On the contrary, the deposition temperature does not seem to influence the density of nucleation of the nanodots. It appears that the size of the nanodots increases with temperature while the nucleation density remains constant. As the temperature increases, silane decomposes faster into SiH_2 precursor thus increasing the both the diameter and the height of the nanodots formed.

4 SUMMARY

In this study, we investigated the formation of Silicon nanodots on three different substrates. The size and the nucleation density of of Si dots formed were found to be dependent on the nature of substrates used. The HF-treated SiO_2 substrate resulted in higher nucleation rate than that of SiO_2 substrate. On a Si substrate, the nucleation rate did not depend on deposition time. To obtain a high nucleation density and narrow size distribution, high concentration of Si-OH group on the surface maybe required.

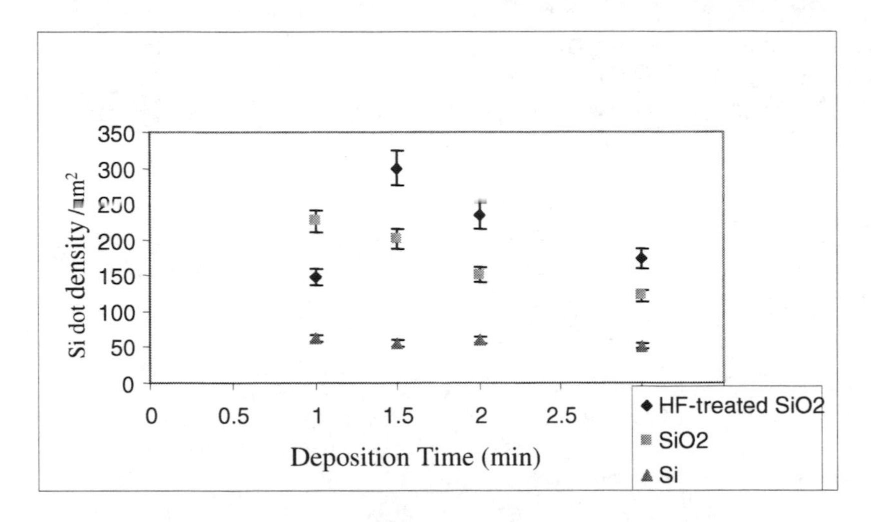

Figure 3. Variation of Si dot density for three different substrates - Si, SiO_2 and HF-treated SiO_2.

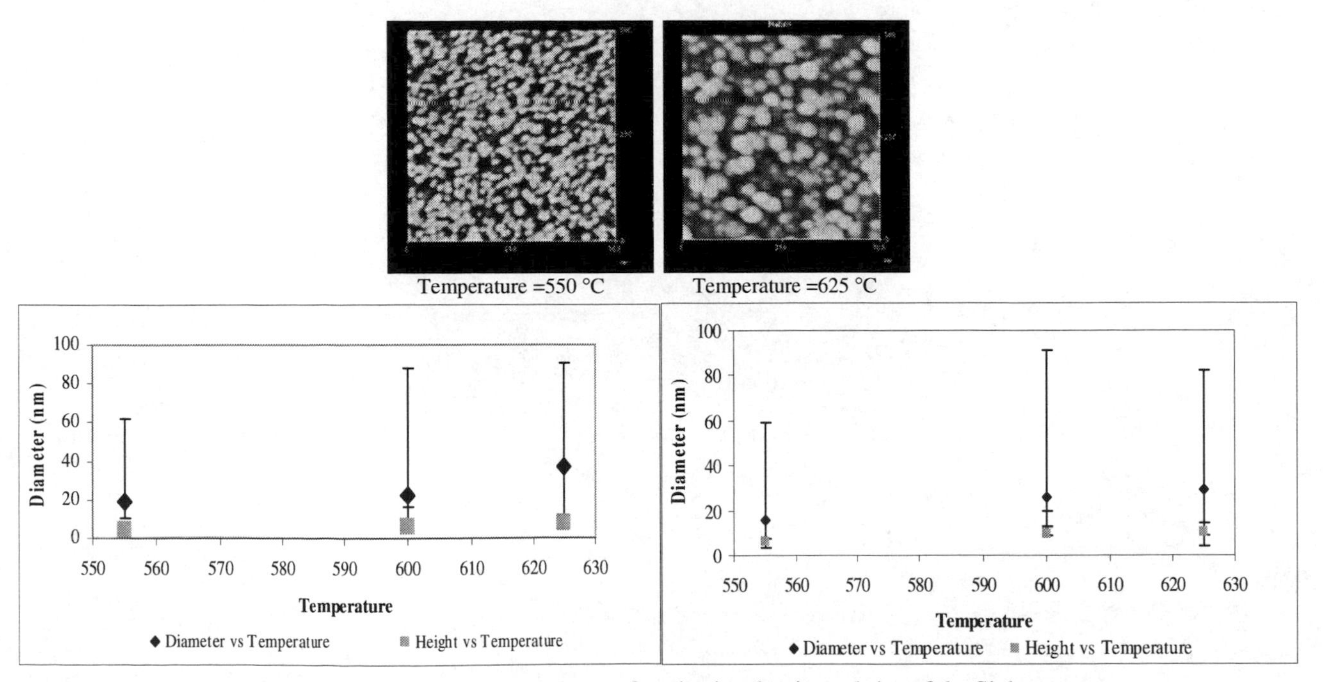

Figure 4. Temperature dependence of nucleation density and size of the Si dots

5 ACKNOWLEDGEMENTS

The authors would like to thank the Microfabrication Material Laboratory (MAL) at University of Illinois – Chicago and the Center for Microanalysis of Materials (CMM) at University of Illinois – Urbana Champaign. The AFM measurements were carried out in the Center for Microanalysis of Materials, University of Illinois, which is partially supported by the U.S. Department of Energy under grant DEFG02-91-ER45439. The LPCVD depositions were done in the MAL facility. We also thank the Motorola Advanced Technology center for funding the project.

6 REFERENCES

1. M. Fakuda, K. Nakagawa, K. Miyazaki, M hirose, Appl. Phys. Lett. 70 (1997) 2291
2. Y. Takahashi, A. Fujiwara, K. Yamazaki, H. Namatsu, K. Kurihara, K. Murase, Abs. Int. Conf. Solid Stae Devices Mater. (1998) 176
3. S. Tiwari, E. Rana, H. Hanafi, A. Hartstein. E.E. Crabb, K Chan, Appl. Phys. Lett. 68(1997) 1377
4. A. Nakajima, Y. Sugita, K. Kawamura, H. Tomita, N. Yokoyama, Jpn. J. Appl. Phys. 35 (1996) L189.
5. K. Nakagawa, M. Fukuda, S. Miyazaki, M. Hirose, MRS Symp. Proc. 452 (1997) 243.
6. M.S. Abrahams, C. J. Buiochhi, R. T. Smith, J.F. Corboy, Jr., J. Blanc and G. W. Cullen, J. App. Phy. $7 (1976) 5139.
7. W. A. P. Claassen, J. Bloem, J. Electrochem. Soc. 128 (1981) 1353.
8. W. A. P. Claassen, J. Bloem, J. Electrochem. Soc. 127 (1980) 194.
S. Mayazaki, Y. Hamamoto, E. Yoshida, M. Ikeda, M. Hirose, Thin Solid Films (2000) 55

Metal-oxide Nanowires for Toxic Gas Detection

D. P. Devineni[1], S. Y. Stormo, W. A. Kempf, J. Schenkel, R. Behanan, A. S. Lea[2] and D. W. Galipeau
Electrical Engineering Department, South Dakota State University, Brookings, SD 57007
605-688-4618, Fax: 605-688-5880, david.galipeau@sdstate.edu
[1]Current address: MtronPTI, Yankton, SD 57078
[2] EMSL, Pacific Northwest National Laboratory, Richland, WA 99352

ABSTRACT

The feasibility of using Electric field enhanced oxidation (EFEO) to fabricate metal-oxide nanowires for sensing toxic gases was investigated. The effects of fabrication parameters such as film thickness, ambient relative humidity, atomic force microscope (AFM) tip bias voltage, force, scan speed and number of scans on the growth of nanowires were determined. The chemical composition of indium-oxide (In_2O_3) nanowires was verified using Auger electron spectroscopy. It was found that oxygen to indium ratio was 1.69, 1.72, 1.71 and 1.84 at depths of 0, 1.3, 2.5, and 3.8 nm, which was near the 1.5:1 expected for stoichiometric In_2O_3 film. Future work will include characterizing the electrical and gas sensing properties of the metal-oxide nanowires.

Keywords— Electric field enhanced oxidation, nanosensors.

1. INTRODUCTION

Toxic gas emissions from automobiles and industry are a major source of hazardous gases such as nitric oxides (NO_x), ozone (O_3) and carbon monoxide (CO). Studies have shown that human health could benefit from even a slight reduction (in range of ppb) in the concentration of these pollutants. The ability to detect toxic gases may allow automobile manufacturers and industry to re-design air-fuel mixture systems to control toxic gas emissions and smog levels [1]. Automobile manufacturers and other industries use sensors interfaced with computer controlled fuel systems to increase fuel economy and to control toxic emissions to meet regulations set by federal agencies [2].

A wide variety of metal oxide materials have been investigated for toxic gas sensing in the past five decades. The first commercially available sensors for the detection of combustible gases were developed by Taguchi in 1962. The advantages of these sensors are low cost and sensitivity to hydrogen (H_2) and hydrocarbons but their drawbacks includes poor selectivity, stability and high power consumption (0.8W) since they must be heated [3, 4]. Since the late 1980's there has been a considerable effort to make use of microelectronic device fabrication techniques to manufacture low cost and low power microsensors such as those based on micro hotplates [5, 6]. Demarne and Grisel [5], and later, Corcoran et al [6] reported low cost, low power silicon-based tin oxide gas microsensors. In the mid-1990's Motorola Inc developed a low cost and low power commercial sensor based on the microhotplate technology for the detection of CO for the automobile industry [6]. In 2001, Mitzner et al [7] studied a microhotplate-based gas sensor array for detecting toxic gases such as hydrogen sulphide (H_2S), ammonia (NH_3) and methane (CH_4) in agricultural animal confinement facilities. It was reported that tin oxide/platinum (SnO_2/Pt), tungsten oxide/gold (WO_3/Au), and zinc oxide (ZnO) sensing films were sensitive to the target gases NH_3, H_2S and CH_4 respectively, but had some cross sensitivity. The limitations of this sensor array were baseline drift, high resistance, and long recovery time. In 2002, Sternhagen et al [8] studied a microhotplate-based gas sensor array for detecting automotive emissions and reported that SnO_2/aluminum (Al), SnO_2/Pt and undoped SnO_2 sensing films were sensitive to NO_2, CO and H_2O respectively. A limitation of this array was sensitivity to interferent gases such as NH_3, H_2S and CH_4.

Coles et al [9] investigated the advantages of using nanosize powders to produce metal oxide gas sensors. It was found that the sensitivity of nanosize alumina (Al_2O_3), zirconia (ZrO_2) and SnO_2 powders for detecting gases like CO, H_2 and CH_4 respectively was greater than corresponding microcrystalline powders. In particular, the sensitivity increased by a factor of five for SnO_2 powders. Gurlo et al [10] investigated the operation of nanocrystalline indium oxide (In_2O_3) and molybdenum oxide (MoO_3) - In_2O_3 thin films as toxic gas sensing materials. In their work, thin film layers were prepared using a sol-gel method. It was found that In_2O_3 and MoO_3-In_2O_3 thin films were very sensitive to low concentrations (100-200 ppb) of O_3 and NO_2, and the response of In_2O_3 films to NO_2 increased significantly when the grain size of the film was less than 50 nm. The operating temperature of these In_2O_3 based sensors was between 25° C and 250° C. The major disadvantage of the In_2O_3 films was that its performance characteristics were highly influenced by fabrication process temperatures.

Nanostructure-based metal oxide sensors appear to be promising candidates for toxic gas sensing since they offer a large surface area to bulk ratio and may have unique properties that could increase the sensitivity, selectivity and response time. Control of nanostructure growth is expected to increase the sensor stability and reproducibility while small size should reduce power consumption since there is less thermal mass and surface area.

Several investigators have used electrical field enhanced oxidation (EFEO) to grow nano-oxide wires from silicon, GaAs and Ti [11-14]. In the EFEO process, an atomic force microscope (AFM) tip is brought in contact with a metal or semiconductor film surface. This results in the formation of a water meniscus between the AFM tip and film due to humidity in the ambient air, as shown in Figure 1. When a negative bias voltage is applied to the AFM tip, the electric field dissociates the water meniscus into hydrogen and hydroxyl ions. The negatively charged hydroxyl ions react with the film surface to form the metal-oxide nanowires as the AFM tip is moved across the surface [11-14]. The feasibility of using EFEO to create indium, tin and tungsten (In, Sn and W) oxide nanowires however has not been demonstrated.

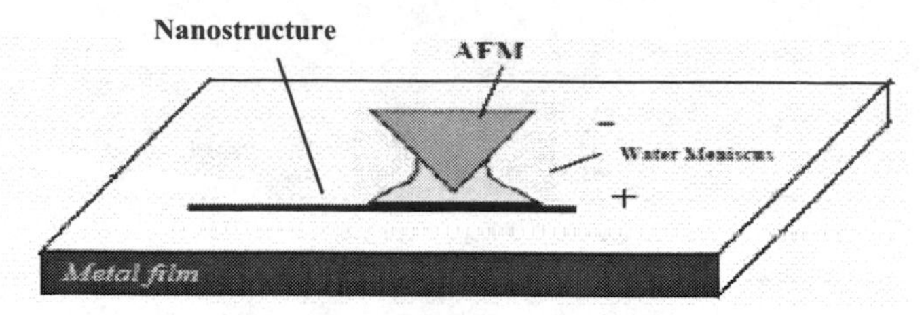

Figure.1. Electric field enhanced oxidation process.

The objectives of this work were to determine the feasibility of using EFEO to fabricate In_2O_3, SnO_2 and WO_3 nanowires that could potentially be used for sensing NO_2, O_3 and CO; and examine the effects of fabrication parameters such as film thickness, ambient relative humidity and AFM tip voltage, force and scan rate on the growth of nanowires.

2. EXPERIMENTAL PROCEDURES

Two photo masks were used in fabricating the metal oxide nanowires. Film patterning mask (mask 1) shown in Figure 2.1a was used in a liftoff process to pattern the metal films. The mask consisted of sixteen 100 µm X 100 µm squares for growing the nanowires. Each square was connected to the vertically adjacent square(s) with a 20 µm line for the electrical connection required to grow the metal oxide structures. Alignment marks were used on mask 1 to locate the nanowires after release and for bonding pad placement on the released nanowires to characterize their electrical properties. An alignment-mark cover mask (mask 2) shown in Figure 2.1b was used to protect the alignment marks on the glass substrate from being etched away during the release of the metal oxide nanowires.

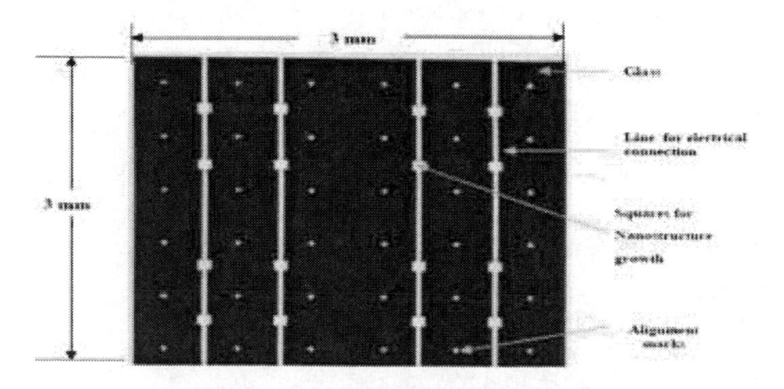

Figure 2.1a. Film patterning mask (mask 1) showing 100 µm X 100 µm squares for nanostructure growth and 20 µm wide lines for electrical connection.

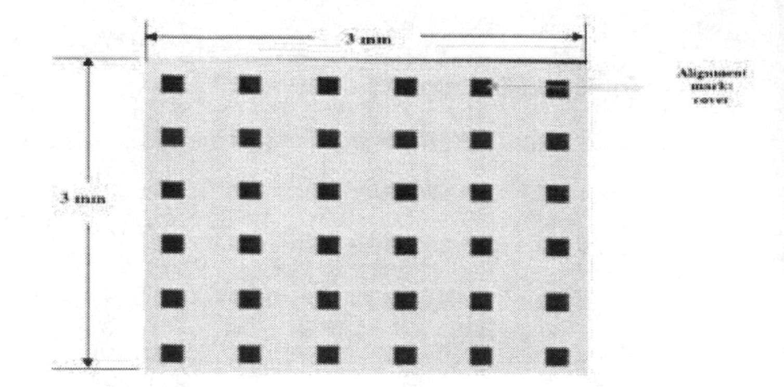

Figure 2.1b. Mask 2 for protecting alignment marks.

Micro glass slides (Corning) diced to 0.75 x 0.75 inch were used as substrates. Five to 20 nm thick metal films (In, Sn, W) were sputter deposited and patterned using the lift-off process. Uncoated silicon ultralever AFM tips were used to grow the SnO_2 and WO_3 nanowires. The number of scans per nanowire was from 1 to 1000. The silicon ultralever AFM tips used to grow the In_2O_3 nanowires were sputter coated with 10 nm thick titanium (Ti) and 20 nm thick Au layers to increase the tip conductivity and lifetime.

The experimental set up used for the EFEO process is shown in Figure 2.3. Silver paint was applied between magnetic mount and the metal film on the glass substrates to provide an electrical connection. The magnetic mount was placed on the sample holder of a Park Scientific AFM operated in the constant force mode. A negative bias voltage between –8 V and –17 V was applied to the AFM tip using an HP6216B power supply. Relative humidity was maintained at 50%. The parameters used for the growth of In_2O_3, SnO_2 and WO_3 nanowires are given in Table 1.

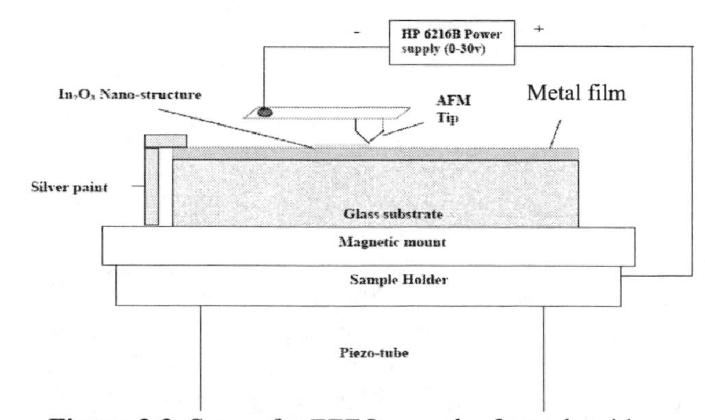

Figure 2.3. Set up for EFEO growth of metal-oxide nanowires.

After growing the nanowires, the alignment marks on the glass substrate were covered with photoresist using mask 2 to avoid chemical etch of the marks. The un-oxidized metal (100 µm x 100 µm squares), and 20 µm lines used for electrical connection on the surface of the glass substrates were etched using the parameters give in Table 2.

Table 1. Parameters for nanowire growth

Metal (Thickness)	Voltage (V)	Force (nN)	Frequency (Hz)	Scans
In (10 nm)	-12 to -17	28 - 32	0.01	3 - 10
Sn (10 nm)	-10 to -12	2 - 6	0.5 - 1.5	850
W (20 nm)	-8 to -15	2 - 6	0.04 - 2.1	1-1050

Table 2. Etch parameters for nanowire release

Metal	Etchant	Ratio	Time (sec)
In	HNO_3/H_2O	1: 99	90
Sn	HCL/ H_2O	1: 100	120
W	H_2O_2/ H_2O	1: 39	120

The chemical composition of the In_2O_3 nanowires was analyzed with Auger electron spectroscopy (AES). The Auger measurements were conducted on a Physical Electronics 680 Nanoprobe spectrometer equipped with a field-emitter electron beam source, cylindrical mirror analyzer, and multichannel plate detector. The measurements were taken using a 10 nA, 10 kV beam with a measured diameter of 26 nm. Due to potential charging of the glass substrate, a molybdenum mask was mounted on the sample to dissipate any charge build-up that might have occurred otherwise. Sample cleaning and depth profiling were performed by argon ion sputtering using sputtering rates of 2 nm/min (calibrated against SiO2). Quantization of the data was performed using pure element sensitivity factors (Handbook of Auger Electron Spectroscopy, Physical Electronics, 3rd edition) after the data had undergone a Savitsky-Golay 9-point smoothing and 5-point derivatization.

3. RESULS AND ANALYSIS

The liftoff process used to pattern the metal films was successful for 10 nm and 20 nm thick films but not successful below 10 nm, since the metal films lifted off when soaked in acetone.

Approximately 120 metal-oxide nanowires were fabricated with an average height of 10 nm, an average width of 600 nm and lengths ranging from 20 to 75 µm. Ambient relative humidity played a critical role in the growth of nanowires. When the relative humidity dropped below 45% there was either no growth or discontinuous growth. This was attributed to an insufficient water meniscus between the tip and the metal film surface. Fewer scans were required to grow In_2O_3 versus SnO_2 and WO_3 nanowires. This was attributed to higher AFM tip force and coated tips, which enhanced the electric field strength.

Figure 3.1 shows a 3D AFM image of an In_2O_3 nanowire that is representative of what was grown using the EFEO technique. This nanostructure had a height of 13 nm, width of 700 nm and length of 20 µm respectively. Most nanowires fabricated at the optimal parameters listed in Table 3 appeared continuous and free of defects.

Figure 3.2 shows similar oxide structure grown to illustrate x and y axis control using the EFEO technique on a 10 nm In film. This nano-oxide structure had an average height and width of 10 nm and 400 nm respectively while the length of nanowires varied from 2.5 to 8 µm.

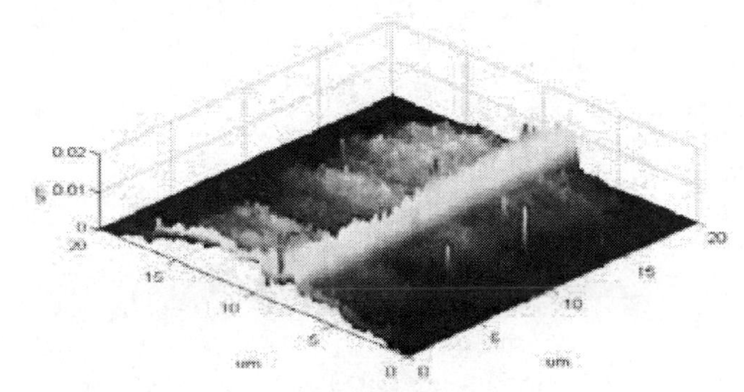

Figure 3.1. Three-dimensional AFM image of representative In_2O_3 nanostructure fabricated by EFEO.

Table 3. Optimal parameters for nanowire growth

Parameter	In	Sn	W
Film thickness (nm)	10	10	20
Relative humidity (%)	50	50	50
Applied voltage (V)	-15	-12	-15
AFM tip force (nN)	28	4	6
Scan speed (Hz)	0.01	0.5	0.04
Scans	10	850	1000

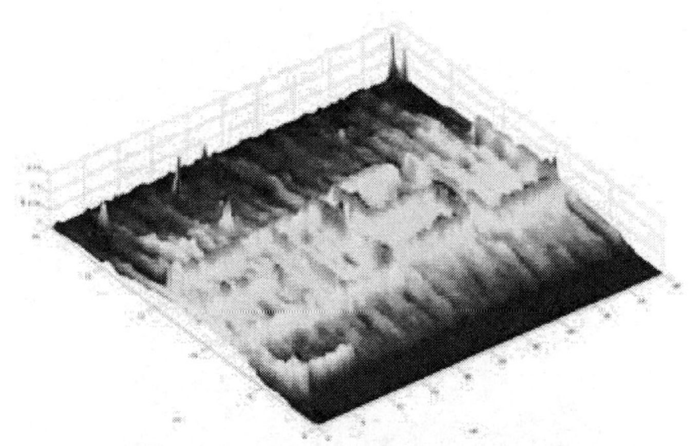

Figure 3.2. Three-dimensional AFM image of In_2O_3 "SDSU" nanostructure fabricated by EFEO.

About 90 nanowires grown on 10 nm thick In films were successfully released. Figure 3.3 shows optical microscope images of a representative 10 nm thick In film on a glass substrate before (left) and after (right) release. The nanowires are too small to be seen, however, the alignment marks protected with photoresist during the release process can be seen in both images. Figure 3.4 shows a 3D view of the In_2O_3 nanostructure after release. The released structure's average height and width were 12 nm and 700 nm respectively.

A Scanning electron microscopy (SEM) image of the nanostructure analyzed with AES is shown in Figure 3.5. The oxygen to indium ratio was found to be 1.69, 1.72, 1.71 and

1.84 at depths of 0, 1.3, 2.5, and 3.8 nm, which was near the 1.5:1 expected for stoichiometric In_2O_3 film.

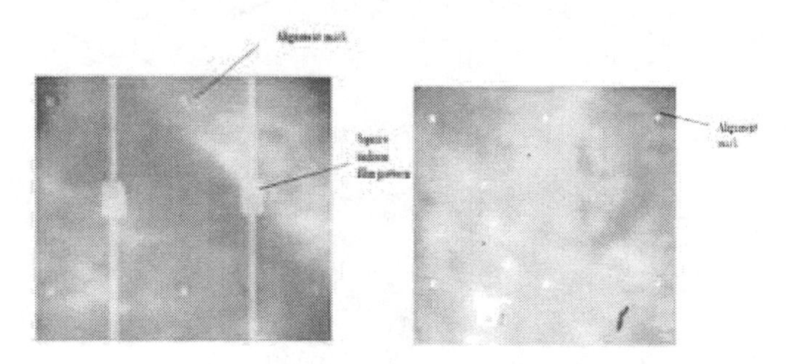

Figure 3.3. Optical image (250X) of a sample before (left) and after (right) In_2O_3 nanostructure release.

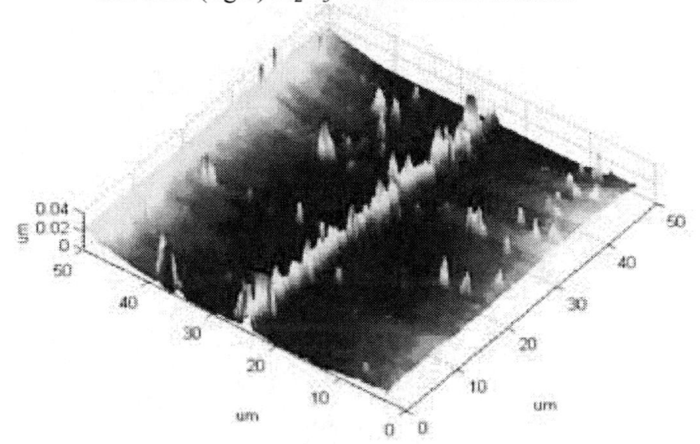

Figure 3.4. AFM image: 3D view of a representative In_2O_3 nanostructure after release.

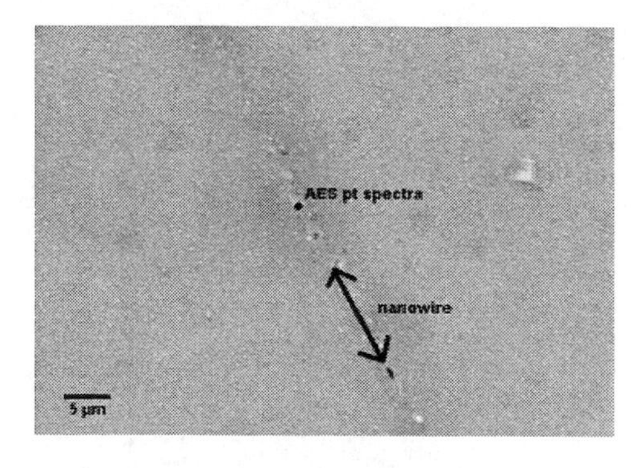

Figure 3.5. SEM image of In_2O_3 nanostructure on which AES analysis was performed.

4. CONCLUSIONS

Nanowires were successfully fabricated on indium, tin and tungsten films using the electric field enhance oxidation technique (EFEO) and released by chemical etching. The field enhanced oxidation process was strongly dependant on film thickness, relative humidity, AFM tip bias (electric field strength), force, scan speed and number of scans. The optimal values for these parameters to obtain consistent growth and release of the indium-oxide (In_2O_3), tin-oxide (SnO_2) and tungsten-oxide (WO_3) nanowires were determined.

5. FUTURE WORK

Future work should include fabrication of additional nanowires to measure their electrical and gas sensing properties. Methods to enhance the selectivity of indium-oxide, tin-oxide and tungsten-oxide nanowires to nitrogen dioxide, ozone and carbon monoxide should also be examined.

REFERENCES

[1] *Measure of Contaminants, Body Burdens, and Illnesses America's Children and The Environment*. US-EPA, EPA 240-R-03-001, February 2003, pp. 21-23.

[2] T.S.Kim, Y.B.Kim, K.S.Yoo, G.S.Sung and H.J.Jung, "Sensing characteristics of dc reactive sputtered WO3 thin films as an NOx gas sensor," *Sensors and Actuators B*, 62, 2000, pp. 102-108.

[3] N.Taguchi, "A Metal Oxide Gas Sensor," Japanese Patent No.45-38200, 1962.

[4] S. R. Morrison, Chemical Sensors, chpt. 8: Semiconductor Sensors edited by S.M.Sze, John Wiley and Sons, NY, 1994, pp. 283-410.

[5] V. Demarne and A. Grisel, "An integrated low power thin film CO gas sensor on silicon," *Sensors and Actuators B*, 13, 1988, pp. 301-313.

[6] P. Corcoran, H.V. Shurmer and J.W. Gardener, "Integrated tin oxide sensors of low power consumption for use in gas and odour sensing," *Sensors and Actuators B,* 15-16, 1993, pp. 32-37.

[7] K. Mitzner, J. Sternhagen, and D.W. Galipeau, Development of a Micromachined Hazardous Gas Sensor Array," *Sensors and Actuators B*, 93, 2002, pp. 92-99.

[8] J.D.Sternhagen, K. Mitzner, E.J. Berkempas and D.W.Galipeau, "A Micromachined Gas Sensor Array for Automotive Emissions," in *Proc. 2002 International MicroelectronicsSymposium,* Sept 2002, pp. 506-511.

[9] G.Coles, G.Williams, "Nanocrystalline Materials as Potential Gas Sensing Elements," *Mat. Res. Soc. Symposium*, December 1997, pp. 33-40.

[10] A.Gurlo, N.Barsan, M.Ivanovskaya, U.Weimar, W.Gopel, "In_2O_3 and MoO_3- In_2O_3 Thin Film Semiconductor Sensors: Interaction with NO2 and O3," *Sensors and Actuators B*, 47, 1998, pp. 92-99.

[11] Ph. Avouris, R. Martel, T. Hertel and R. Sandstrom, "AFM-Tip-Induced and Current-Induced Local Oxidation of Silicon and Metals," *Applied Physics,* A66, 1998, p 659- 667.

[12] J.T.Sheu, S.Yeh, C.Wu and K.You, "Nanometer-Scale Pattering on Titanium Thin Film with Local oxidation of Scanning Probe microscope," *Symposium on Nano Device Technology*, May 2002, pp. 363-367.

[13] Y.Okada, S.Amano, M.Kawabe and J.S.Harris, Jr., "Basic Mechanisms of an Atomic Force Microscope Tip-Induced Nano-Oxidation Process of GaAs," *Journal of Applied Physics,* Volume 83, June 1998, pp. 7998-8001.

[14] K.Matsumoto, M.Ishii, K.Segawa, Y.Oka, "Room Temperature Operation of a single electron transistor made by the STM Nanooxidation process for TiOx/Ti system," *Applied Physics Letters,* 68, January 1996, pp. 34-36.

Microfabrication of 3D Structures Using Novel Thermoplastic Elastomers

A.P. Sudarsan*, J. Wang** and V.M. Ugaz***

Department of Chemical Engineering, Texas A&M University, College Station, TX, USA
*arjun@tamu.edu, **jian.wang@chemail.tamu.edu, ***ugaz@tamu.edu

ABSTRACT

The use of thermoplastic elastomer gels as advanced substrates for construction of complex microfluidic systems is shown. These gels are synthesized by combining inexpensive polystyrene – (polyethylene / polybutylene) – polystyrene triblock copolymers with a hydrocarbon oil for which the ethylene/butylene midblocks are selectively miscible. The insoluble styrene endblocks phase separate into localized domains to form an optically transparent, viscoelastic, and biocompatible 3-D network (similar to PDMS), with the further advantage of melt-processability in the vicinity of 100 °C. Microfluidic devices can be fabricated in under 5 minutes by making an impression of the negative relief structures on a heated master mold. Multiple impressions can be made against different masters to construct geometries incorporating variable-height features, as well as intricate 3-D multilayered structures. Thermal and mechanical properties are tunable over a wide range through proper selection of gel composition.

Keywords: microfluidics, PDMS, thermoplastic elastomer, rapid prototyping, soft lithography

1 INTRODUCTION

The development of increasingly sophisticated chemical and biochemical assays, combined with the need to incorporate these processes within a compact device footprint suitable for massively parallel operation requires the construction of correspondingly complex microfluidic structures [1,2]. This ongoing drive toward increased device complexity requires corresponding advances in fabrication materials and technologies. For example, although a number of multilayer PDMS-based systems have been successfully constructed, the resulting fluidic networks are effectively 2-dimensional owing to the planar nature of the fabrication process. It is possible in principle to employ an arbitrary number of layers, however the entire device structure must be assembled at once due to the irreversibility associated with the curing process. This irreversibility can be advantageous in terms of ensuring excellent mechanical stability, however it also imposes limitations because the molded structures cannot be further modified after curing. Consequently, there is no straightforward process to fabricate structures incorporating features of variable height because only a single impression from a single master can be used. Finally, the range of viscoelastic properties available for design of fluidic components that operate by inducing deformations in the substrate material (e.g. valves, pumps) is somewhat limited.

2 THEORY

Novel thermoplastic elastomer gel substrates offer the capability to provide a greatly enhanced level of flexibility in microfluidic device design and construction. These gels are easily synthesized using a combination of inexpensive polystyrene – (polyethylene / polybutylene) – polystyrene (SEBS) triblock copolymers in hydrocarbon oils for which the ethylene/butylene midblocks are selectively miscible. The thermodynamic incompatibility between blocks induces microphase separation and self assembly of the insoluble polystyrene endblocks into distinct domains with characteristic size scales on the order of 10-20 nm [3,4]. The soluble midblocks emanating from these nanodomains penetrate into the solvent creating arrays of loops (beginning and terminating within a single nanodomain) and bridges (joining adjacent nanodomains) resulting in the formation of a 3-D viscoelastic gel network in which the polystyrene domains act as physical crosslink junctions. Like PDMS, this gel network is optically transparent, viscoelastic, and biocompatible, but also possesses the further advantage of melt-processability at temperatures in the vicinity of 100 °C.

3 EXPERIMENTAL

A series of thermoplastic elastomer gels were synthesized by combining commercially available SEBS copolymer resin (e.g. CP-9000, Kraton-G series) in mineral oil. The reisn and mineral oil were mixed and placed under vacuum overnight at room temperature in order to allow the oil to evenly wet the resin surface. The mixture was then heated to 120-170 °C (higher temperatures are required with increasing copolymer fraction) under vacuum for 2–4 hours to allow the resin and oil to intermix and to remove any residual air bubbles. Finally, the mixture was cooled to room temperature and the solidified gel was cut into smaller pieces used for molding devices. Gel compositions ranging from 9 to 41 wt% copolymer were studied.

4 RESULTS AND DISCUSSION

We investigated thermal transitions associated with these SEBS-mineral oil gels using small amplitude oscillatory shear experiments (Fig 1). A measure of the transition to liquid-like behavior can be inferred from the temperature T* at which the value of the loss modulus G″ exceeds that of the storage modulus G′. The range of gel compositions studied here allow the location of this transition to be varied over a range of approximately 50 °C. Moreover, the room temperature (plateau) value of the elastic modulus can be varied over an order of magnitude.

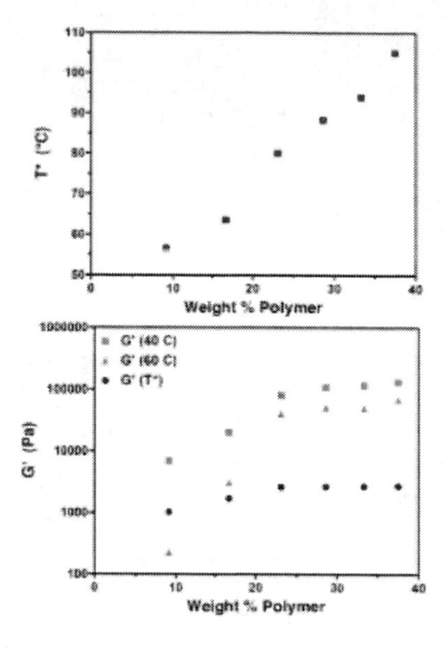

Figure 1: Variation of melt transition temperature (top) and plateau elastic modulus (bottom) with gel composition.

Fabrication of microfluidic devices is accomplished by placing a slab of elastomer on top of a master mold that has been preheated to 120 °C on a hot plate. Within seconds the elastomer begins to soften, after which a glass plate is placed on top of the slab and gentle pressure is applied by hand to ensure complete contact with the structures on the mold. After cooling and release, the solidified gel incorporates the shape of the structures on the master (Fig 2). Strong uniform bonds can be easily achieved, either with a glass or elastomer surfaces, by briefly heating the material to a temperature just below its softening point either on a hot plate or using a handheld heat gun. The entire fabrication process can be completed in about 5 minutes. We are also able to easily assemble a variety of complex multilayered structures in only a few minutes (Fig 2).

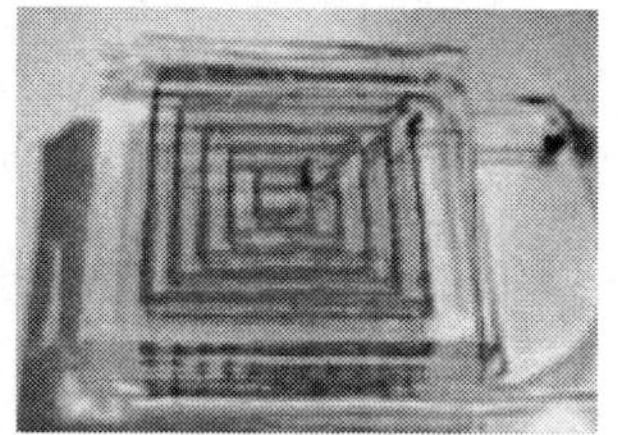

Figure 2: Planar (left) and interconnected multilayer (right) microfluidic channels constructed using elastomer gels (400 x 30 μm cross-section).

Individual layers are repositionable, thereby allowing precise alignment to be achieved prior to thermal bonding. More complex 3-D structures can be fabricated by direct casting (Fig 3), and interfaces with external fluidic supply lines can be readily sealed by locally heating the gel to melt it at the point where the lines are inserted into the substrate. An alternative fabrication approach involves directly embedding an intertwined bundle of fluidic channels within a bulk elastomer slab by dipping the channels into molten elastomer. Upon cooling, the channels become seamlessly embedded in the bulk gel, allowing the fabrication of braided or knotted flow geometries (Fig 4). Choosing a lower wt% gel formulation for the bulk material ensures the channels will remain intact during the embedding process.

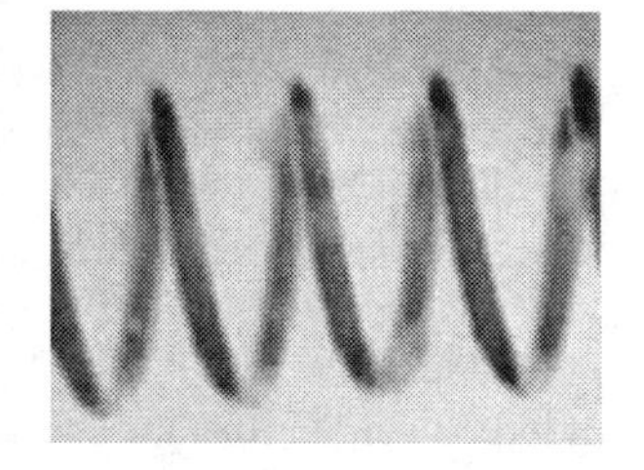

Figure 3: Helical channel (580 μm-diameter) made by direct casting of a spring. The spring was inserted into the molten elastomer, and then released after cooling by unscrewing from the solidified gel.

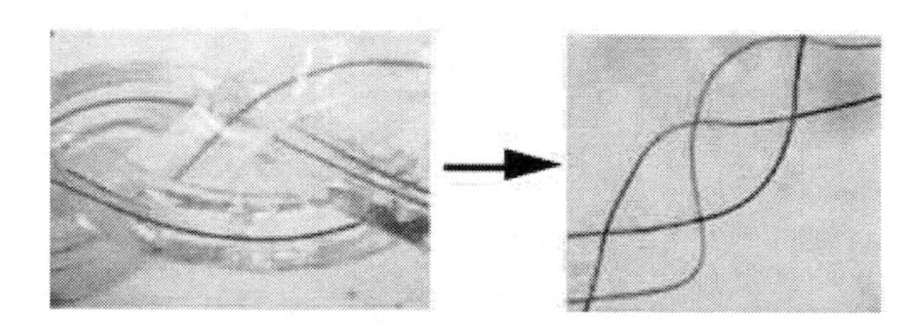

Figure 4: Braided microfluidic structure constructed by direct casting of 200 μm-diameter wires in 33 wt % elastomer gel. Individual channels are cut out, filled with dye, and knotted together (left) and embedded in a 9 wt % elastomer slab (right).

Perhaps the most compelling advantage of elastomer gel substrates is that, unlike PDMS, multiple impressions can be made against different masters to easily construct complex geometries incorporating variable-height features within the same channel network. After an initial impression is made using the first master, any number of subsequent impressions can be made in the same elastomer slab using different masters containing features with distinct heights and/or shapes (Fig 5).

NSTI-Nanotech 2005, www.nsti.org, ISBN 0-9767985-1-4 Vol. 2, 2005

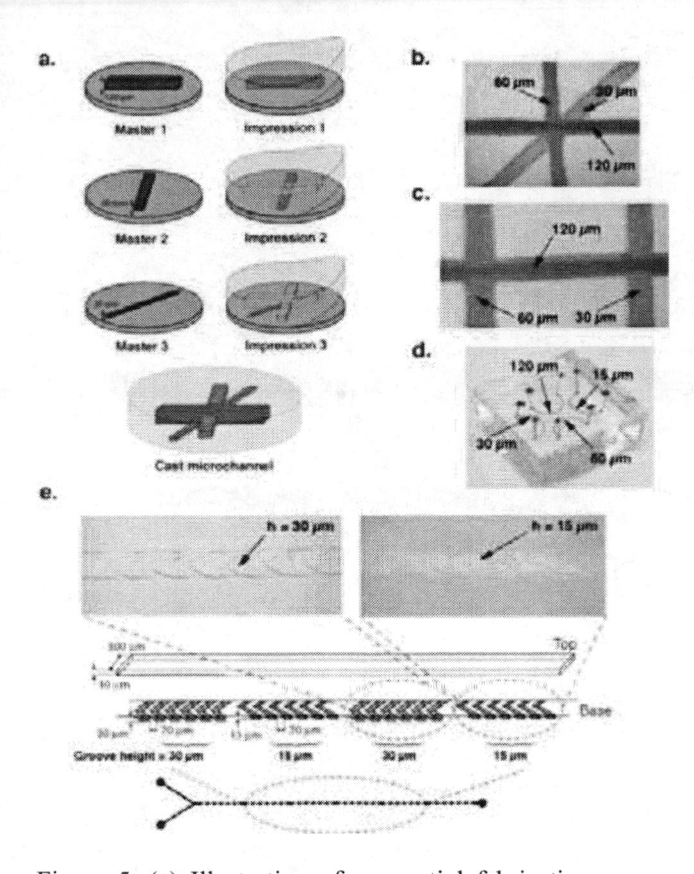

Figure 5: (a) Illustration of sequential fabrication process used to construct multi-height channel structures (fabrication of microchannel shown in (b) is depicted). (b-d) Examples of multi-height microchannels. Channels are filled with red dye to enable visualization of the flow network (note that even though all channels are filled with the same dye, the taller channels appear darker in color due to the increased optical path length). (b) Microfluidic junction formed by intersecting channels of three different heights (30, 60 and 120 µm; as indicated on figure). Channel cross sectional dimensions are as follows: horizontal = 300 x 120 µm; vertical = 200 x 60 µm; diagonal = 400 x 30 µm. (c) Horizontal 120 µm-high channel intersected at different locations by channels of heights 30 and 60 µm (as indicated on figure). Channel cross sectional dimensions are as follows: horizontal = 300 x 120 µm; vertical (left) = 500 x 60 µm; vertical (right) = 400 x 30 µm. (d) Four non-intersecting semi-circular 250 µm-wide channels of different heights (15, 30, 60 and 120 µm; as indicated on figure). (e) Close-up view of a multi-height herringbone channel structure incorporating a series of alternating 15 and 30 µm tall grooved structures bonded with a 300 µm-wide by 30 µm-tall top channel.

We have demonstrated the suitability of these elastomers as substrates for microfluidic applications by constructing devices for DNA electrophoresis and diffusive transport studies [5]. We further investigated biocompatibility of these elastomer materials by performing a *Eco* RI restriction digestion reaction on a lambda phage DNA target. Parallel reactions were performed using both untreated 0.2 mL reaction tubes, and tubes whose inner surfaces were coated with 23 wt% elastomer gel. All five digested bands were visible after 1 hour of incubation at 37 °C in both the untreated and elastomer coated tubes, thereby demonstrating a high level of compatibility with standard biochemical reagents and protocols (Fig 6).

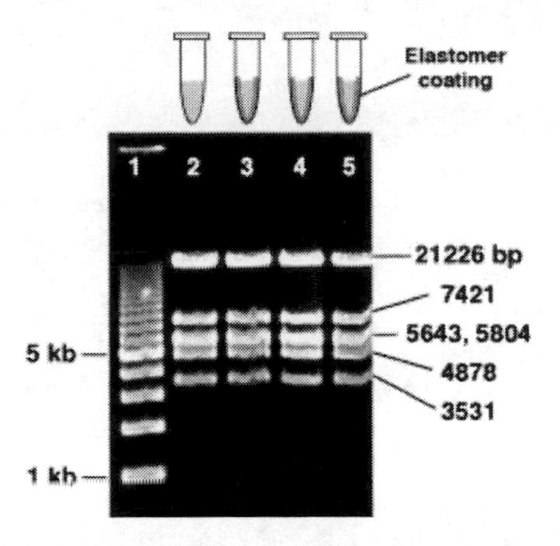

Figure 6: Products of *Eco* RI restriction enzyme digestion of lambda DNA performed in reaction tubes coated with 23 wt% elastomer gel to demonstrate biocompatibility. Lane 1: 1 kb sizing ladder, lane 2: control reaction performed in uncoated reaction tube, lanes 3-5: reactions performed in elastomer coated tubes. The expected fragment sizes are produced in all cases.

The high degree of versatility and fabrication ease, combined with their inherently inexpensive nature, make thermoplastic elastomer gels ideal substrates for many microfluidic applications.

REFERENCES

[1] T. Thorsen, S.J. Maerkl, and S.R. Quake, Science, 298, 580-584 (2002).

[2] S. Sia and G.M. Whitesides, Electrophoresis, 24, 3563- 3576 (2003).

[3] J.H. Laurer, J.F. Mulling, S.A. Khan, R.J. Spontak, and R. Bukovnik, J. Polym. Sci.: Part B: Polym. Phys., 36, 2379-2391 (1998).

[4] R. Kleppinger, N. Mischenko, H.L. Reynaers, and M.H.J. Koch, J. Polym. Sci.: Part B: Polym. Phys., 37, 1833-1840 (1999).

[5] A.P. Sudarsan and V.M. Ugaz, Anal. Chem., 76, 3229-3235 (2004).

Recent Advances in the Characterization of Free Volume in Model Fluids and Polymers: Shape and Connectivity

F. T. Willmore*, I. C. Sanchez**
Department of Chemical Engineering
University of Texas, Austin, TX 78712, USA
*willmore@che.utexas.edu
**sanchez@che.utexas.edu

ABSTRACT

The solubility and diffusivity of a penetrant species in an amorphous material are determined by the nanoscale properties of its free volume. Positron Annihilation Lifetime Spectroscopy,[i,ii,iii] mechanical measurements of equation of state behavior,[iv] and Voronoi tessellations[v,vi,vii,viii] have been applied to study free volume properties. A recent theoretical technique for sizing cavities in model fluids and polymers[ix] has been extended to characterize free volume in terms of shape and connectivity. A set of shape parameters is introduced, characterizing nanopores in terms of surface area, volume, radius of gyration, and span. Results are presented for a Lennard-Jones fluid, a hard sphere fluid, water, and for two high free volume polymers of interest to membrane scientists.

Keywords: nanopore, cavity, morphology, free volume, permeability

1 SPHERICAL VS NONSPHERICAL CAVITIES

As a first approximation to characterizing free volume, the technique developed by In't Veld et al. (termed Cavity Energetic Sizing Algorithm, or CESA) describes free volume in terms of spherical cavities. In this method, cavity centers are located by finding a minimum point of repulsion from other atomic centers, and are then sized by estimating how large of a particle might be inserted into the material at that center. The technique captures a fair amount (25-50%) of the free volume of the system, and provides a distribution of cavity sizes, which are correlated to permeability properties. It explains the differences observed in CO2 diffusivities between a random copolymer of tetrafluoroethylene with 2,2-bis(trifluoromethyl)-4,5-difluoro-1,3-dioxole (PTFE/BDD) and (poly-trimethyl-silyl-propane) poly(1-trimethylsilyl-1-propyne) (PTMSP). PTMSP and PTFE/BDD have a similar free volume (measured using the Bondi method) but the observed permeability of PTMSP with respect to CO2 is markedly higher. We have extended this technique in an effort to account for non-spherical cavities by means of overlap of the spherical cavities originally obtained.

1.1 Methodology: overlapping spherical cavities form cluster objects

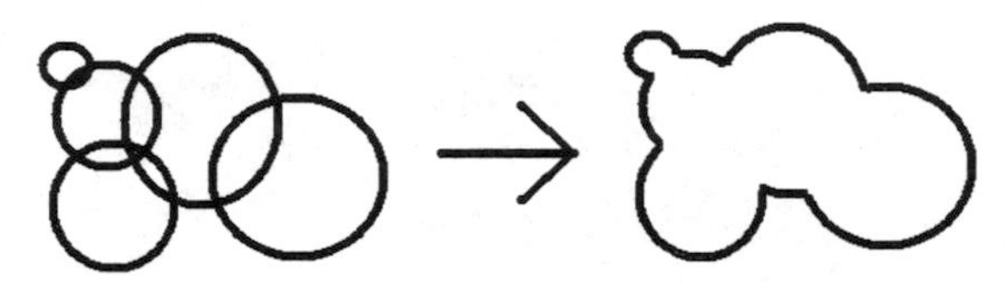

By considering the overlap of spherical cavities obtained using the CESA method, we attempt to understand the connectivity of the void spaces. Overlapping cavities are grouped into a new entity, called a cluster. Each cavity belongs to only one cluster, and each cluster contains at least one cavity. The shapes of these cluster objects is then characterized according to a set of 'shape parameters'.

1.2 The Shape parameters

Volume

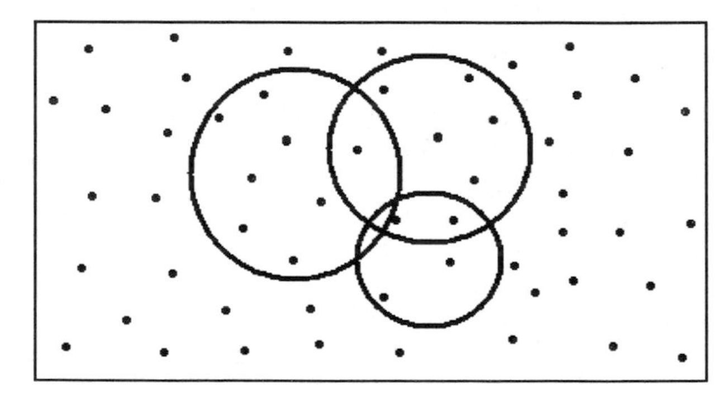

Volume of the nanopore is calculated using a straightforward monte carlo integration. Points are selected at random from within the simulation box, and are determined to lie within at least one of the component cavities of the cluster or not. This assures volume is not double-counted. The cluster volume is thus the ratio of points inside the cluster to total number of points selected, times the box volume.

Surface Area

Cluster surface area of the nanopore is determined by sampling points uniformly on all of the surfaces of the component cavities which comprise the cluster. Points which do not lie inside of another sphere in the cluster are tallied, and the surface area determined by multiplying this tally by the total suface area of all spheres and dividing by total number of points sampled.

Span

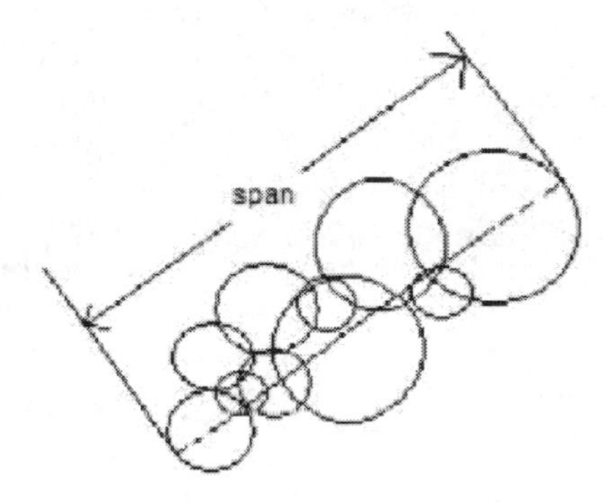

The span is defined as the farthest distance between any two points that lie within the nanopore.

Radius of Gyration

Radius of gyration is defined for our nanopore objects by using the same monte carlo sampling scheme used to determine volume, but in this instance all internal points are assigned an equal weight and used to determine a center of mass and radius of gyration for the cluster object.

1.3 Some Results Using this Technique

HS LJ and water

We compare three simple fluids: a hard sphere fluid, a Lennard Jones fluid (at liquid-vapor coexistence) and water (also at coexistence), all at the same reduced density of .75.

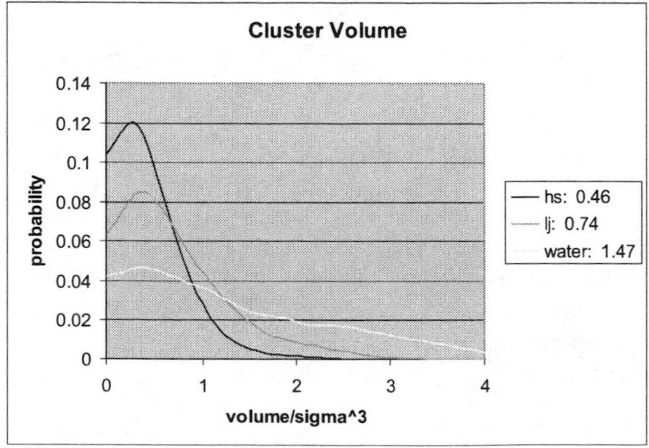

Hard spheres, with no attractive forces, have the smallest overall cluster volumes, as the spheres float freely and tend to 'jam up' what would otherwise be connected void space.

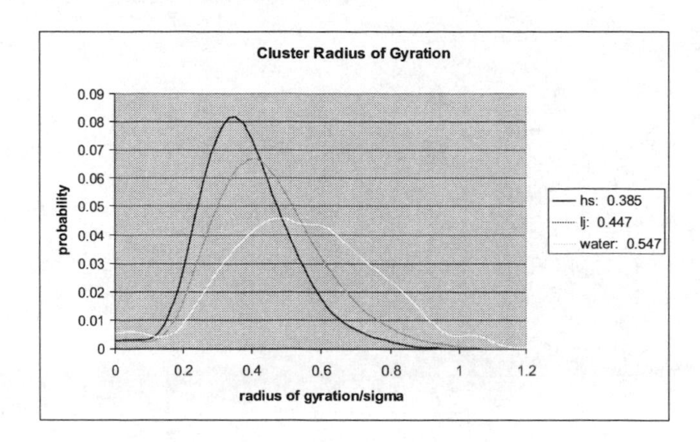

Water, with the highest level of intermolecular attraction shows free space which is most connected, as illustrated by having larger overall free volume cluster sizes, clusters of larger span, and clusters of larger radii of gyration.

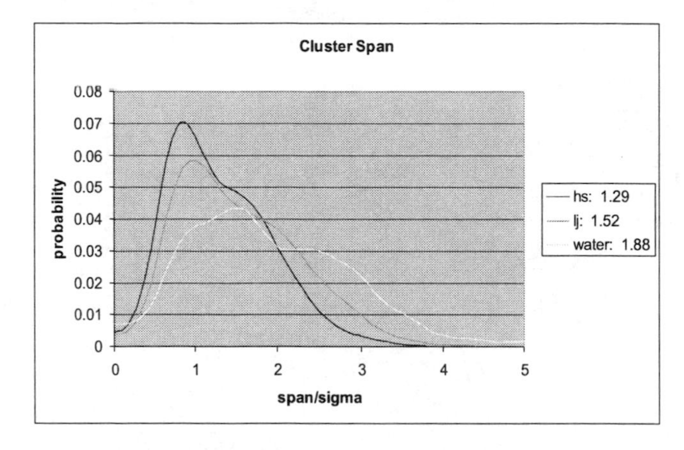

Note that the distributions generated have been weighted by cluster volume, that is, if you were to pick a point at random from all of the volume captured by all of the cavitites, it would lie in a cluster with the given value of surface area, volume, etc. Average values of each parameter are shown in the legend.

High Free Volume Polymers

We also applied this technique to our high free volume polymers PTMSP and PTFE/BDD. While the PTMSP shows the trend toward clusters of larger volume, the trends toward larger span and larger radius of gyration are more pronounced.

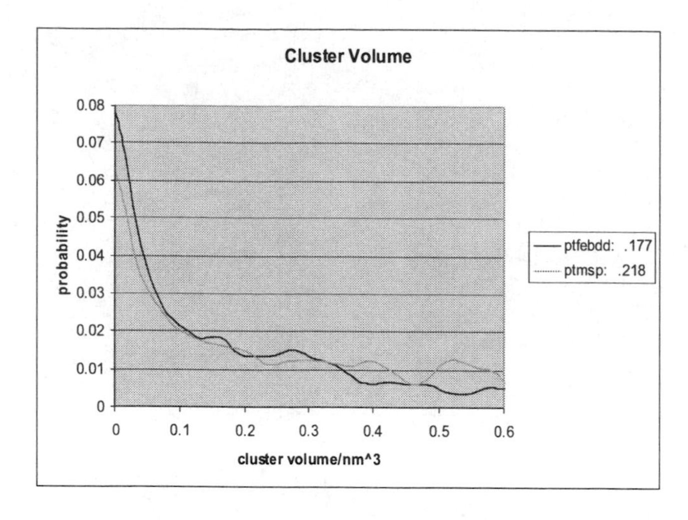

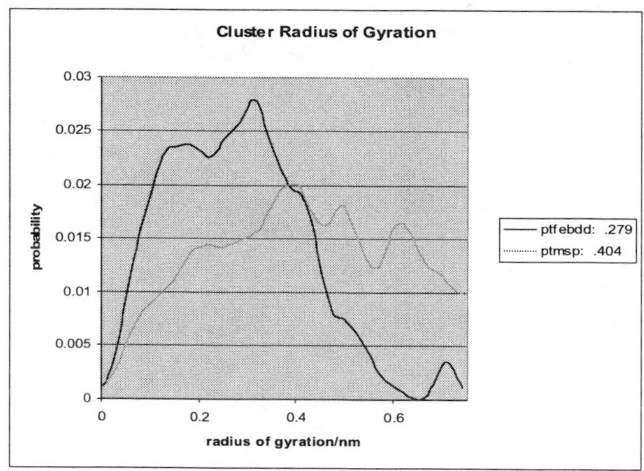

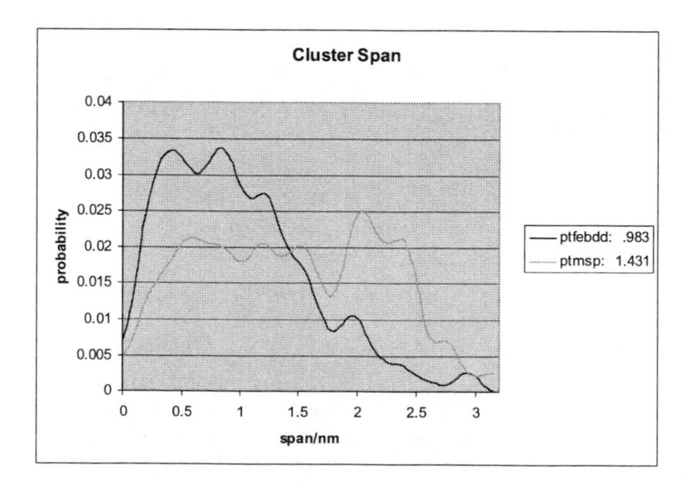

2 LAND AND SEA

One shortcoming of the cavity overlap technique is that the cavities are found in terms of an energetic repulsive minimum. This means that large amounts of free space are forsaken, in favor of choosing instead the local or not-so-local repulsive minimum at the center of free space. A new technique is introduced in an attempt to capture more of this free space.

2.1 Introducing the Method

This technique is dubbed the 'land and sea' model, in that it separates regions of space into two regions, one where a penetrant is not likely to be found ever (the land) and a region where our penetrant is very likely to be found floating about (the sea) based on the insertion energy for the penetrant into the configuration. On this energy landscape, 'sea level' is an arbitrarily chosen energy level (3/2 kT is used here, as the average thermal energy of penetrant species) on the inserertion energy landscape which distinguishes what regions are land (energy above) and sea (energy below).

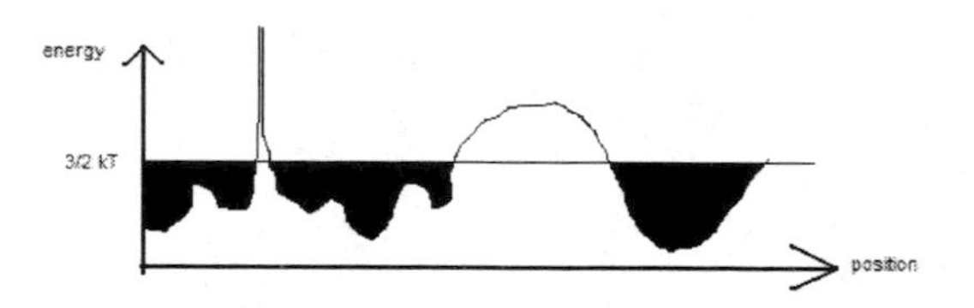

The technique is implemented by making trial insertions of a test particle at regular intervals, measuring insertion energy at each point, systematically mapping out the free space with respect to the test particle.

This model can be employed to generate sets of overlapping spheres (centered about these insertion points) similar to the sets of overlapping spherical cavities above, and can be analyzed using the same technique, with the drawback that high free volume systems evade analysis as the cluster objects have greater continuity and often percolate. In fact, in the limit of shrinking penetrant size, all structures percolate.

2.2 Measuring Fractional Cavity Volume

This technique was used to measure the fractional free volume of our two test polymers. Configurations of each polymer were probed using a penetrant particle roughly the size of CO_2 and with only a LJ interaction potential, epsilon = kT and sigma = .32 nm. The volume measured for PTMSP is in good agreement with that measured by the Bondi method, but PTFE-BDD showed a considerably lower 'available' free volume. This result is consistent with the lower fractional cavity volume observed using CESA.

Polymer	Fractional Free Volume (Bondi)	Fractional Cavity Volume (CESA)	FFV by Land and Sea w/CO_2
PTFE-BDD	.32	.132	.24
PTMSP	.29	.156	.34

3 CONCLUSIONS

CESA provides an effective means of distinguishing cavity size 'signatures' of materials in a way that predicts relative permeability. Extending the method for overlap opens the door to understanding the connectivity of the void space.

Free volumes predicted using the Bondi method may overpredict the 'available' free volume, the fraction of space where a penetrant particle is likely to be found. Both CESA and the land and sea model predict lower available free volumes for PTFE-BDD, consistent with the observed difference in diffusivity.

REFERENCES

[i] Marcello Sega; Pal Jedlovszky; Nikolai N Medvedev; Renzo Vallauri; 2004. **Free volume properties of a linear soft polymer: A computer simulation study** JOURNAL OF CHEMICAL PHYSICS 121, 2422-2427.

[ii] Shantarovich, VP; Azamatova, ZK; Novikov, YA; Yampolskii, YP. 1998. Free-volume distribution of high permeability membrane materials probed by positron annihilation. *MACROMOLECULES* 31 (12): 3963-3966.

[iii] Wang, CL; Hirade, T; Maurer, FHJ; Eldrup, M; Pedersen, NJ. 1998. Free-volume distribution and positronium formation in amorphous polymers: Temperature and positron-irradiation-time dependence. *JOURNAL OF CHEMICAL PHYSICS* 108 (11): 4654-4661

[iv] Jordan, SS; Koros, WJ. 1995. A free-volume distribution model of gas sorption and dilation in glassy polymers. MACROMOLECULES 28 (7): 2228-2235.

[v] Sega, M; Jedlovszky, P; Medvedev, NN; Vallauri, R. 2004. Free volume properties of a linear soft polymer: A computer simulation study. *JOURNAL OF CHEMICAL PHYSICS* 121 (5): 2422-2427.

[vi] Sastry, S; Debenedetti, PG; Stillinger, FH. 1997. Statistical geometry of particle packings .2. "Weak spots" in liquids. *PHYSICAL REVIEW E* 56 (5): 5533-5543, Part A.f

[vii] Lee, S; Mattice, WL. 1999. A "phantom bubble" model for the distribution of free volume in polymers. *COMPUTATIONAL AND THEORETICAL POLYMER SCIENCE* 9 (1): 57-61.

[viii] Corti, DS; Debenedetti, PG; Sastry, S; Stillinger, FH. 1997. Constraints, metastability, and inherent structures in liquids. *PHYSICAL REVIEW E* 55 (5): 5522-5534, Part A.

[ix] In 't Veld, PJ; Stone, MT; Truskett, TM; Sanchez, IC. 2000. Liquid structure via cavity size distributions. *JOURNAL OF PHYSICAL CHEMISTRY B* 104 (50): 12028-12034.

Simplified Crossover Droplet Model for Adsorption of Critical and Supercritical Fluids in Slit Nano-Pores

S. B. Kiselev and J. F. Ely

Chemical Engineering Department, Colorado School of Mines, Golden, Colorado 80401-1887, U.S.A.

ABSTRACT

In this work, we present a simplified crossover droplet (SCD) model for of pure fluids at a flat solid-liquid interface and in slit pores. The SCD contains size pore, L, as a parameter and at $L \gg \xi_b$ (where ξ_b is a bulk correlation length) is transformed into the generalized crossover model for the excess adsorption in semi-infinite systems. With $L=50$ nm, the SCD model reproduces all available experimental data for SF_6/graphite, including the critical isochore data at $\tau \to 0$, within experimental accuracy. The SCD model also yields excellent description of critical adsorption data for CO_2/silica gel system. Application of the SCD model to the description of experimental adsorption data for N_2O/silicia gel system is also discussed.

1 THEORETICAL BACKGROUND

The surface excess adsorption of pure fluids on a planar interface surface is defined as

$$\Gamma_\infty = \int_0^\infty (\rho(z) - \rho_b)\,dz \qquad (1)$$

where $\rho(z)$ is density of fluid at a distance z from the surface, and $\rho_b = \rho(\infty)$ is the bulk density of the fluid. The density profile $\rho(z)$ can be found from the optimization of the functional

$$\mathcal{F}[\rho(z)] = \int \left[\hat{A}(\rho) + c_0 (\nabla\rho)^2 + W_s(\rho)\right] dV \qquad (2)$$

where $\hat{A}(\rho) = \rho A(T,\rho)$ is a Helmholtz free-energy density of the bulk fluid and $W_s(\rho)$ is the surface contribution into the free energy density. Finally, the excess adsorption in a semi-infinite system is given by [1, 2]

$$\Gamma_\infty = \int_{\rho_1}^{\rho_b} \frac{(\rho - \rho_b)}{\left[\Delta\hat{A}(\rho)\right]^{1/2}} d\rho\,, \qquad (3)$$

where $\Delta\hat{A}(\rho) = \hat{A}(\rho) - \hat{A}(\rho_b) - (\rho - \rho_b)\mu(T)$ is the excess part of the free-energy density, and $\mu(T) = \mu(T,\rho_b)$ is the chemical potential of the bulk fluid. An equation of state for the surface order parameter, $m_1 = \rho_1/\rho_c - 1$, which can be written in the form

$$4c_0\Delta\hat{A}(\rho_1) = (2b_1 m_1 - h_1)^2 \rho_c^{-2}\,, \qquad (4)$$

provides a relationship between the surface ordering field h_1 and surface density ρ_1 at any fixed values of the temperature T and bulk density ρ_b. The temperature dependence appears in Eq. (4) through the excess free-energy density $\Delta\hat{A}(\rho_1)$, surface ordering field h_1, and parameters b_1 and c_0. In Eq. (4) the parameter $b_1 = b_{10}\sqrt{c_0 k_B T}$, while the surface field is represented by truncated Taylor expansion

$$h_1(\tau)/\sqrt{c_0 k_B T_c} = h_{10} + \sum_{i=1} h_{1i}\tau^i\,, \qquad (5)$$

where b_{10}, c_0, and h_{1i} are the system dependent coefficients.

2 DROPLET MODEL

In the crossover droplet model, a fluid near the critical point is considered as an "ideal gas" of homogeneous liquid droplets with the droplet radius equal to the correlation length of a bulk fluid at a temperature T and density ρ, $r = \xi_b(T,\rho)$. According to this picture, Eq. (3) in the droplet model for the excess adsorption in a slit pore should be replaced by

$$\Gamma_L = \int_{\rho_1}^{\rho_{L/2}} \frac{(\rho - \rho_{L/2})}{\left[\Delta\hat{A}(\rho)\right]^{1/2}} d\rho\,, \qquad (6)$$

where $\rho_{L/2}$ is the density at the center of the pore at $z = L/2$. In the case, when the correlation length is much smaller than the distance between walls in the pore, or the same $\xi_b \ll L/4$, the density $\rho_{L/2} = \rho_b(T,P)$, equation (6) is transformed into Eq. (3), and a slit pore becomes physically equivalent to a semi-infinite system,

 NSTI-Nanotech 2005, www.nsti.org, ISBN 0-9767985-1-4 Vol. 2, 2005

$\Gamma_L \equiv \Gamma_\infty$. In the opposite case, at $\xi_b \geq L/4$, the density $\rho_{L/2}$ becomes very close to the surface density ρ_1 ($\rho_{L/2} \to \rho_1$), and, as a consequence, $\Gamma_L \to 0$. In the intermediate region, when the correlation length is still smaller than L ($\xi_0 \ll \xi_b < L/4$), but the density at the center of the pore is not equal to the bulk density ρ_b ($\rho_1 > \rho_{L/2} > \rho_b$) the excess adsorption exhibits a crossover between these two regimes, $0 < \Gamma_L < \Gamma_\infty$, with the maximum at $\xi_b \cong L/4$.

All thermodynamic properties for a bulk fluid in the simplified crossover droplet model are calculated with a generalized corresponding sates crossover model [3], while an explicit dependence of the density $\rho_{L/2}$ on L and ξ_b is given by [2]

$$\rho_{L/2} = \rho_b - (\rho_b - \rho_1)\tanh\left(\frac{x^2}{1+x}\right), x = 4\xi_b/L \quad (7)$$

In Eq. (7) for the correlation length ξ_b we use the Ornstein-Zernike approximation

$$\xi_b = \sqrt{c_0\bar{\chi}_T/\chi_0^+}, \quad (8)$$

where χ_0^+ is the asymptotic amplitude in the power law $\bar{\chi}_T|_{\rho=\rho_c} = \chi_0^+\tau^{-\gamma}$ (where γ=1.24 is a universal critical exponent) for the isothermal compressibility $\bar{\chi}_T = \rho T(\partial\rho/\partial P)_T P_c\rho_c^{-2}T_c^{-1}$ along the critical isochore $\rho = \rho_c$ of a bulk fluid at $\tau = T/T_c - 1 \to +0$.

3 RESULTS AND DISCUSSION

The first system, which we considered in this work, is adsorption of hexafluoride on graphitized carbon black measured by Thommes *et al.* [4]. The theory predicts monotonically increasing behavior of the excess adsorption in a semi-infinite system along the critical isochore at $T \to T_c$, which diverges as $\Gamma \propto \tau^{-\nu+\beta} \propto \tau^{-0.6}$ at $\tau \to 0$ [1]. However, the excess adsorption in this system increases only down to a reduced temperature of $\tau \simeq 0.01$, but then Γ decreases sharply on approaching T_c. The "critical depletion" observed in experiment [4] is treated in the SCD model, as a density profile deformation, which appears in a slit pore when the size of the pore becomes comparable with the correlation length of a bulk fluid. At fixed temperature T, the condition $\xi_b \cong L/4$ along the critical isochore $\rho_b = \rho_c$ is achieved at in a slit pore with size $L \cong 4\xi_0\tau^{-\nu}$ (where $\xi_0 = \sqrt{c_0/\chi_0^+}$ is an asymptotic critical amplitude and the critical exponent ν=γ/2=0.62), or in the pore with a fixed size L at the reduced temperatures $\tau \cong (4\xi_0/L)^{1/\nu}$. Estimation of the characteristic size of the pore with equation

$$L_c = 4\xi_0\tau_1^{-\nu}, \quad (9)$$

where $\tau_1 = 0.01$ is the reduced reentrant temperature observed in the experiment [4] and $\xi_0 = 0.15 - 0.2$ nm is a reasonable estimate for pure SF_6. yields $L_c = 50 - 70$ nm, that is close to the value $L_c = 31$ nm reported by Thommes and Findenegg [5]. Taking into account the uncertainty in determination of the parameters τ_1, ξ_0, and the characteristic size L_c itself, we contend that this prediction is very good. Comparison of the predictions of the SCD model with excess adsorption data obtained by Thommes *et al.* [4] is shown in Fig. 1.

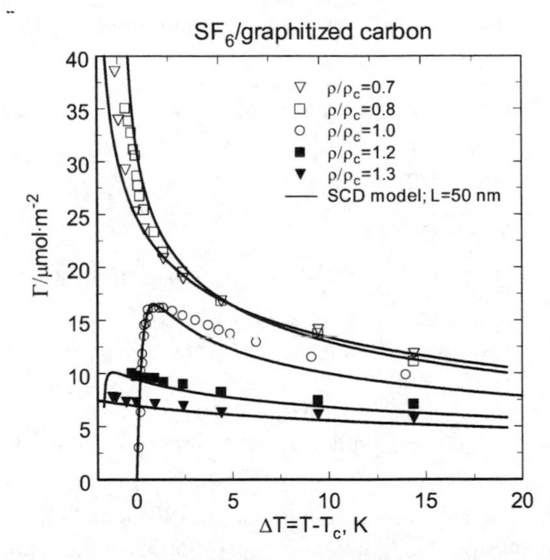

Figure 1: Surface excess adsorption isochores [4] of SF_6 on graphitized carbon black as functions of temperature.

The surface constants b_{10} and h_{1i} in Eq. (4) for SF_6/graphitized carbon system have been found from the optimization of the SCD model to experimental data with the fixed values of ξ_0 and L_c. As one can see from Fig. 1, along the critical isochore the excess adsorption calculated with the SCD model passes a maximum and, in agreement with experimental data, goes to zero as $T \to T_c$. At densities $\rho/\rho_c = 1.2$ and 1.3 the excess adsorption calculated with the SCD model increases only slightly as the saturated temperature $T_s(\rho)$ is approached, while at sub-critical densities, at $\rho/\rho_c = 0.7$ and 0.8, Γ increases sharply as $T \to T_s(\rho)$.

Another confined system considered here, is adsorption of carbon dioxide on silica gel, which is an adsorbent with a broad pore size distribution ranging from micropores of 0.8 nm to 16 nm [6, 7]. The quantity measured in the experiment [6] was the excess adsorption n^{ex} defined as

$$n^{ex} = \frac{v_{tot}}{V_{tot}} \int_{V_{tot}} \left[\rho(\vec{r}) - \rho_b\right] dV , \qquad (10)$$

where is specific pore volume $v_{tot} = V_{tot}/m_{sorb}$ and m_{sorb} is the mass of the adsorbent particle with the pore volume V_{tot}. In order to apply the SCD model for calculation of this quantity, one needs to specify the geometry and size distribution of the pores. In this work, silica gel was described as a porous media with one-dimensional slit pores of three different widths, L_1 with volume fraction x_{1v}, L_2 with volume fraction x_{2v}, and L_3 with the volume fraction

$$x_{3v} = 1 - x_{1v} - x_{2v} . \qquad (11)$$

In this case, the excess adsorption n^{ex} can be written in the form

$$n^{ex} = \left[\sum_{i=1}^{3} \left(\frac{2\Gamma_i}{L_i} + \rho_{li} \right) x_{iv} - \rho_b \right] v_{tot} , \qquad (12)$$

where Γ_i is the surface excess adsorption in the pore of size L_i calculated with Eq. (6). For the specific pore volume v_{tot} and merging pore sizes $L_1 = L_{min}$ and $L_2 = L_{max}$ we adopted the values reported in Ref. [7], while the parameters x_{1v}, x_{2v}, and L_3, as well as the surface constants $b_{10}, c_0, h_{10}, h_{11}$, and h_{12} in Eq. (4), have been found from optimization of the SCD model to experimental data [6]. Comparison of the predictions of the SCD model with experimental data for CO_2/silica gel is shown in Fig. 2. As one can see, in the entire density range $0 < \rho \le 2\rho_c$ and temperatures up to $1.53T_c$ excellent agreement between experimental data and predicted values of the excess adsorption for CO_2/silica gel [6] system is observed.

The last system considered in this work, is adsorption of N_2O on silica gel. Similar to the CO_2/silica gel system, the gel parameters x_{1v}, x_{2v}, L_3, and the surface constants b_{10}, c_0, and h_{1i} (i=0-2) have been found from optimization of the SCD model to the excess adsorption experimental data obtained for this system by Di Giovanni *et al.* [6]. The results of our calculations in comparison with experimental data for N_2O/silica gel are shown in Fig. 3.

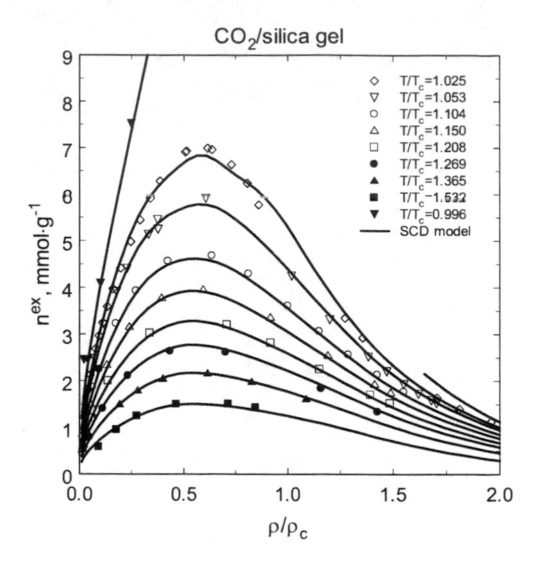

Figure 2: Excess adsorption isitherms [6] of CO_2 on silica gel as functions of density.

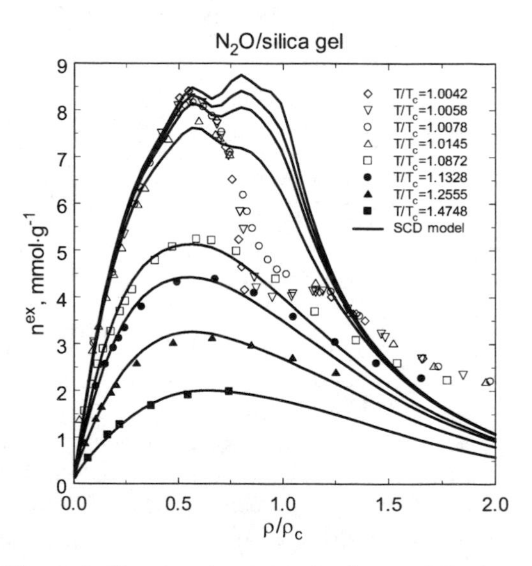

Figure 3: Excess adsorption isotherms [6] of N_2O on silica gel as functions of density.

As one can see, at low densities ($0<\rho<0.7\rho_c$) the SCD model yields very good representation of the excess adsorption data for N_2O/silica gel system in the entire temperature range $T_c<T<1.5T_c$. However, at near-critical isotherms large systematic deviations between experimental data and predicted values are observed at densities $0.8\rho_c < \rho < 1.2\rho_c$. The SCD model also fails to reproduce the excess adsorption data at high densities, $\rho > 1.5\rho_c$. As one can see, similar to CO_2/silica gel, for liquid N_2O at densities the SCD model predicts the values of the excess adsorption of about 0.8-1.1 mmol/g, while the experimental values for N_2O/silica gel system at near-critical isotherms are almost two times higher, $n^{ex} \approx 2\ \mathrm{mmol/g}$.

At present time we cannot say for sure what causes this discrepancy between experimental data and theoretical predictions for the excess adsorption in N_2O/silica gel system at high densities. We can only assume that except the density profile deformation, some other physical effects, which have not been considered in the SCD model, should be taken into account. One of them can be appearance of the orientational order parameter near the wall, which in principle can increase the excess adsorption of liquid N_2O on silica gel. In order to answer this question, additional theoretical and experimental studies are needed.

ACKNOWLEDGMENT

The research was supported by the U.S. Department of Energy, Office of Basic Energy Sciences, under the Grant No. DE-FG03-95ER14568.

REFERENCES

1. Kiselev, S.B. and J.F. Ely, *Adsorption of critical and supercritical fluids.* J. Chem. Phys., 2000. **112**(7): p. 3370-3383.
2. Kiselev, S.B. and J.F. Ely, *Simplified crossover droplet model for adsorption of pure fluids in slit pore.* J. Chem. Phys., 2004. **120**(17): p. 8241-51.
3. Kiselev, S.B. and J.F. Ely, *Generalized corresponding states model for bulk and interfacial propertiesof pure fluids and fluid mixtures.* J. Chem. Phys., 2003. **119**(16): p. 8645-62.
4. Thommes, M., G.H. Findenegg, and H. Lewandowski, *Critical Adsorption of SF_6 on a Finely Divided Graphite Substrate.* Ber. Bunsenges. Phys. Chem., 1994. **98**(3): p. 477-81.
5. Thommes, M., G.H. Findenegg, and M. Schoen, *Critical Depletion of a Pure Fluid in Controlled-Pore Glass. Experimental Results and Grand Canonical Ensemble Monte Carlo Simulation.* Langmuir, 1995. **11**(6): p. 2137-42.
6. Di Giovanni, O., et al., *Adsorption of Supercritical Carbon Dioxide on Silica.* Langmuir, 2001. **17**(14): p. 4316-31.
7. Hocker, T., A. Rajendran, and M. Mazzotti, *Measuring and Modeling Supercritical Adsorption in Porous Solids. Carbon Dioxide an 13X Zeolite and on Silica Gel.* Langmuir, 2003. **19**(4): p. 1254-67.

Atomistic Simulations in Nanostructures Composed of Tens of Millions of Atoms: Importance of long-range Strain Effects in Quantum Dots

M. Korkusinski[1], G. Klimeck[1,2], H. Xu[1], S. Lee[2], S. Goasguen[1], and F. Saied[1],

[1]Purdue University, West Lafayette, IN, 47907, USA, marekk@purdue.edu
[2]Jet Propulsion Laboratory, California Institute of Technology, Pasadena, CA, 91109, USA

ABSTRACT

Strain in self-assembled quantum dots is a long-range phenomenon, and its realistic determination requires a large computational domain. To tackle this problem for an embedded InAs quantum dot NEMO-3D uses the atomistic VFF Keating model containing up to 64 million atoms. Interatomic distance changes obtained are used to influence the sp3d5s* tight-binding electronic Hamiltonian defined in a subdomain containing up to 21 million atoms (matrix size of order of $4x10^8$). Targeted eigenstates with correct symmetry are found reliably even in such large systems. NEMO-3D is used to analyze the dependence of the dot states on the size of the strain domain and the boundary conditions. The energies of a deeply embedded dot depend dramatically on the strain domain size. For dots buried under a thin capping layer, on the other hand, the existence of a free surface at the top of the sample allows for an effective relaxation of atoms, and the penetration of strain into the barrier is small.

Keywords: nanoelectronics, quantum dots, atomistic simulations, strain, electronic structure, parallel computing, tight binding

1 INTRODUCTION

Quantum dots (QDs) are solid-state structures capable of trapping charged carriers so that their wave functions become fully spatially localized, and their energy spectra consist of well-separated, discrete levels. Existing nanofabrication techniques make it possible to manufacture QDs in a variety of types and sizes [1]. Among them, semiconductor QDs grown by self-assembly (SADs), trapping both electrons and holes, are of particular importance in quantum optics, since they can be used as detectors of infrared radiation [2], optical memories [3], single photon sources [4]. Arrays of quantum-mechanically coupled SADs can also be used as optically active regions in high-efficiency, room-temperature lasers [5].

The self-assembly of SADs is achieved in the Stranski-Krastanow growth mode [6] as a result of the mismatch of lattice constants of the material of the dot (e.g., InAs) and that of the substrate semiconductor (e.g., GaAs). This mismatch leads to the appearance of a long-range strain field, which strongly modifies the energy diagram of the system [7]. Therefore, device simulations must include the fundamental quantum character of charge carriers and the classical, long-distance strain effects on equal footing. The Nanoelectronic Modeling tool NEMO-3D [8] meets these requirements by modeling the strain and electronic structure of extended nanosystems (on the length scale of tens of nanometers, containing tens of millions of atoms) fully on the atomistic level.

This work reports on the application of NEMO-3D to study the impact of the strain on the electronic structure of a SAD. The first part of the paper gives a brief summary of NEMO-3D and its performance on parallel computing platforms. In the second part, NEMO-3D is used to study the sensitivity of the single SAD's energy spectrum to the size and boundary conditions of the strain domain, as well as to the geometry of the sample, the thickness of the capping layer and the distance between neighboring SADs in lateral quantum-dot arrays.

2 SYSTEM AND METHOD

2.1 Physical Structure

InAs dome-shaped QDs with diameter and height respectively of 18.09 nm and 1.7 nm (dot A) and 27.13 nm and 2.54 nm (dot B), positioned on a 0.6-nm-thick wetting layer, and embedded in the GaAs barrier material (Fig. 1) are considered. The computational strain domain denoted by lateral size d and vertical size h, can be much larger than the electronic structure domain (Fig. 1).

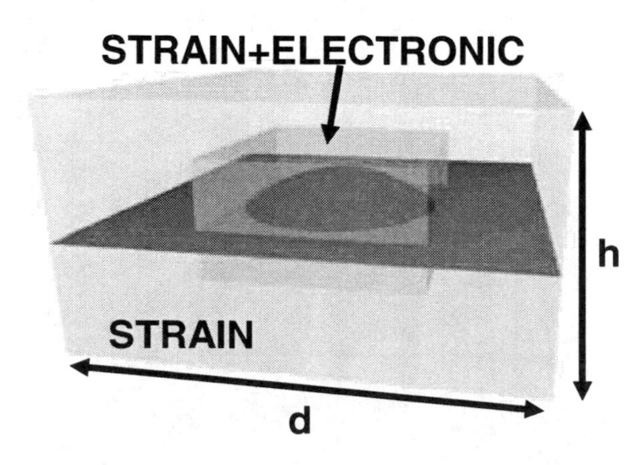

Fig. 1. Schematic view of the QD nanostructure, with two simulation domains: central for electronic structure, and larger for strain calculations

In the electronic calculation, only the confined states are targeted. Since their wave functions decay exponentially in the barrier, it is sufficient to consider a smaller computational box, as shown in Fig. 1.

2.2 NEMO-3D: Overview and Performance

NEMO-3D is an atomistic simulation tool designed to provide *quantitative* predictions of the electronic structure of nanodevices [8, 9]. The computation consists of two major steps. First, the strain distribution in the device is found by computing the positions of atoms yielding the minimal total elastic energy. Contributions to that energy from each distorted atomic bond are obtained in the frame of the valence-force-field method with Keating potentials, and the minimization procedure employs the conjugate gradient technique. In the second step, an atomistic tight-binding Hamiltonian is used to find the electron and hole energy levels and wave functions of the system. This Hamiltonian is created in the basis of 20 orbitals per atom (spin-degenerate s, p, d, and s*). Interactions between the orbitals within each atom and between the nearest neighbors are empirical fitting parameters, chosen so that the model reproduces the experimentally measured band structure of bulk semiconductors. Further, the tight-binding parameters are functions of interatomic bond length and angle distortions due to strain. The resulting Hamiltonian matrix is sparse; targeted eigenvalues and eigenvectors are found using the Lanczos algorithm.

Proper treatment of strain requires computational domains of tens of millions atoms. The electronic domains are usually smaller (up to 20 million atoms), but the sp3d5s* basis set results in Hamiltonian matrices of order of 20 times that number. Therefore, parallelization of NEMO-3D is a key consideration. The core computational engine of NEMO-3D was written in C with MPI used for message passing, which ensures its portability across major parallel computing platforms. Figure 2 shows the memory usage per CPU in the strain (a) and electronic part (b) as a function of the number of CPUs and the system's size. Both phases scale well for a small number of CPUs, but the strain part exhibits a saturation tendency for larger number of CPUs. This is due to a relatively large contribution of non-parallelizable data structures containing system information to the total memory occupied in this phase. In the electronic computation the memory usage is dominated by parallelizable data structures of the Lanczos algorithm, which results in a better scaling behavior.

The Lanczos procedure is particularly computationally challenging because (i) the spectrum of the Hamiltonian has a gap in its interior, and the sought eigenvalues lie above and below this gap; (ii) the eigenvalues are typically repeated, since at zero magnetic field the spin-up and spin-down electronic states are degenerate; (ii) the corresponding eigenvectors are effectively nonzero only on

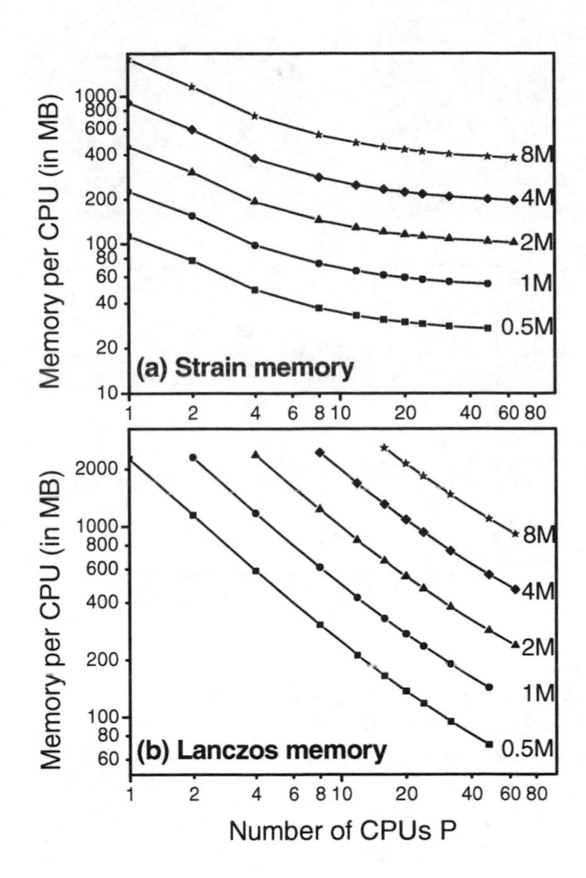

Fig. 2. Memory per CPU in the strain (a) and electronic (b) part of NEMO simulation as a function of the number of CPUs (P) for systems with 0.5 - 8 million atoms.

a small subset (a few tens of thousands of atoms) of the multimillion-atom domain. Therefore it is essential to prove that unique and targeted eigenstates can be extracted from the spectrum. Figure 3 shows a 2D slice through the electronic ground-state probability density for the same QD (dot A), but with the electronic domain containing 300 000 (a) and 21 million atoms (b). As can be seen, the orbital retains its unique, correct symmetry even for the largest electronic computational domain. The computational time in the electronic phase scales nearly linearly with the number of atoms, as demonstrated in Fig. 3 (c). A computation of four electronic and four hole eigenstates of a 21-million-atom domain takes about 12 hours on 64 Itanium 2 CPUs (1.5 GHz).

3 STUDY OF STRAIN IN SADS

3.1 Sensitivity to Size of the Strain Domain

The first NEMO-3D simulation of the SAD examines the sensitivity of its electronic levels on the size and boundary conditions of the strain domain. This first study

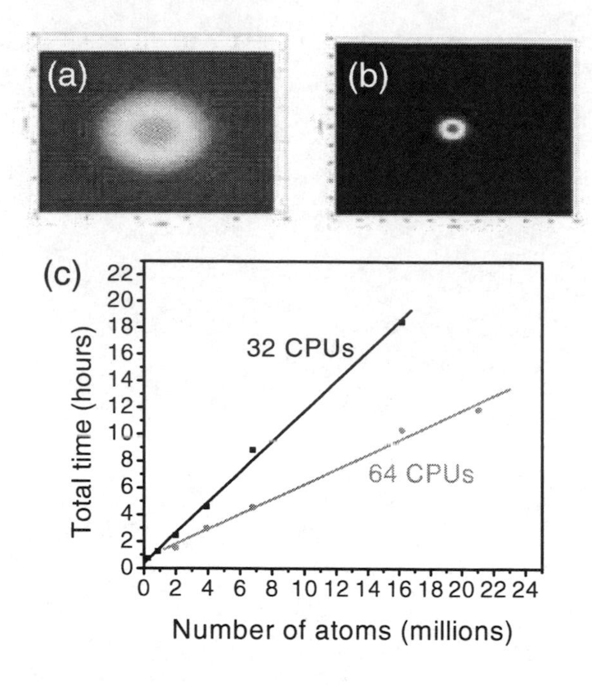

Fig. 3. Probability density for the electronic ground state computed with the computational domain containing (a) 300 000 and (b) 21 million atoms. (c) Total time of the Hamiltonian diagonalization versus system size on 32 and 64 CPUs of an Itanium 2 cluster.

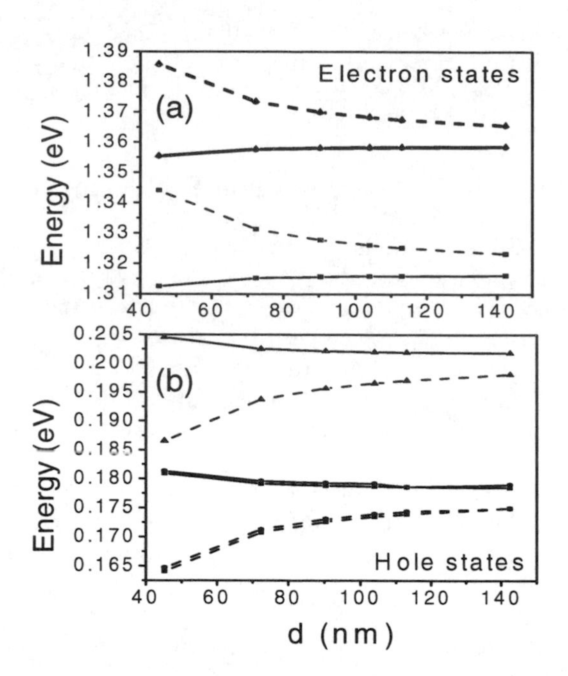

Fig. 4. Electronic (a) and hole (b) ground and two excited energy levels of dot A as a function of the lateral size d of the strain domain with mixed (solid lines) and fixed (dashed lines) BCs. In this calculation the ratio d/h=2.

treats a deeply embedded dot A, where the size of the strain domain around it is increased in all directions. Its volume is taken to contain from 2 to 64 million atoms, and its aspect ratio is fixed so that $d / h = 2$. Also, two sets of boundary conditions are used. In the first set the positions of all the atoms on the domain boundary are fixed to GaAs bulk positions (fixed BCs). In the second set only the bottom boundary is fixed; periodic conditions are assumed in the XY plane, and the top is free (mixed BCs). Once the strain calculation is completed, the electronic calculation is carried out on a 2-million-atom domain with fixed boundary conditions. Figure 4 shows the ground and two excited states for the electron and the hole as a function of the size d. The size of the strain domain influences the electronic levels dramatically, particularly for fixed BCs, where the convergence is not achieved even for the largest domain size. Much faster convergence is obtained for mixed BCs, which better reflect the actual physical setup of SAD arrays (the underlying substrate fixes the bottom of the domain and the SADs are covered with a capping layer of finite thickness, so that the elastic energy can be minimized by distorting the top surface). The mixed BCs are employed in the remainder of this work.

3.2 Strain Sensitivity to QD Shape

The height of the SAD is much smaller than its diameter. One might thus expect that the lateral extent of the strain field caused the QD might be different than its vertical extent. To verify this, a study similar to that presented in the previous Section was conducted with the strain domain aspect ratio set to $d / h = 0.5$, i.e., the domain is now a pillar, whose height is twice larger than its base length. Figure 5 shows a comparison of the electron (a) and hole (b) energies as a function of the size d for the two aspect ratios. The convergence is much faster for the pillar-shaped strain domain, which indicates that the strain field caused by the SAD has a dominant vertical component, whose penetration depth into the barrier is much larger than that in the XY plane. This observation is in agreement with the experimentally observed strain-induced stacking of SADs grown in vertical layers [10].

In the mixed BCs the top boundary of the strain domain is not fixed, and periodic boundary conditions are assumed on its side walls, which corresponds well to the experimental conditions. Care must be taken, however, with the bottom boundary of the strain domain, because positions of atoms placed on it are fixed to the bulk substrate positions. To justify this assumption, the strain domain must extend sufficiently far below the SAD structure. A convergence study of the electronic levels (not shown here) indicates that the distance from the bottom boundary to the bottom of the wetting layer has to be at least 55.4 nm for dot A, and at least 78.01 nm for dot B.

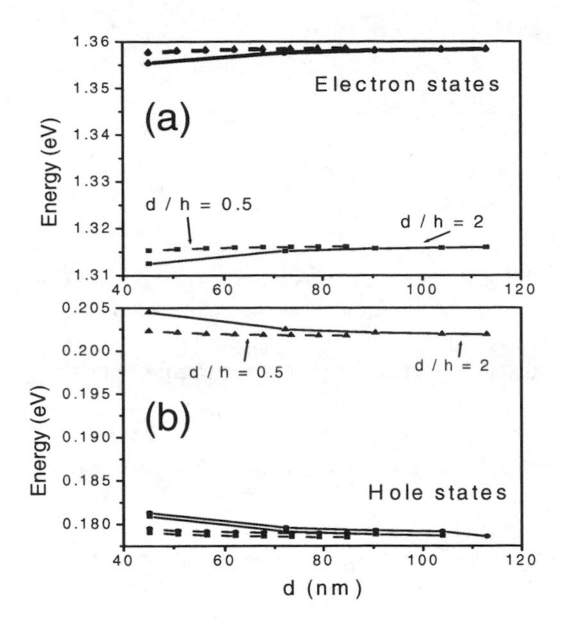

Fig. 5. Electronic (a) and hole (b) ground and two excited energy levels of the dot A as a function of the lateral size d of the strain domain for the aspect ratio d/h=2 (solid lines) and d/h=0.5 (dashed lines). Mixed BCs are used.

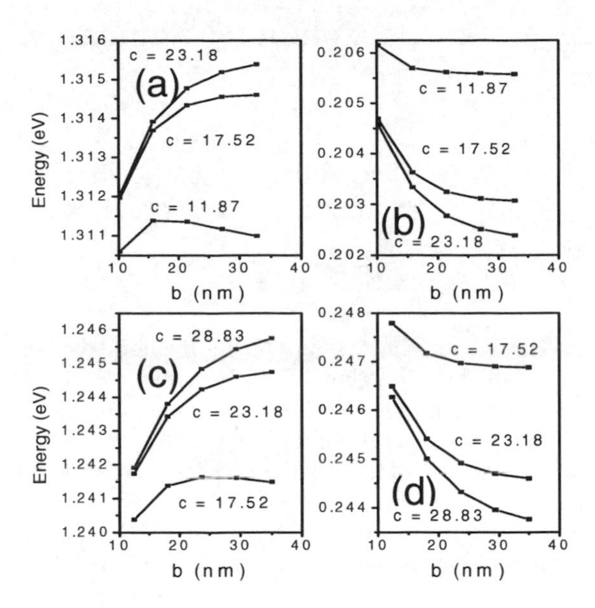

Fig. 6. Electronic (a, c) and hole (b, d) ground-state energies for dot arrays composed of dots A (top) and dots B (bottom) as a function of the distance b between SADs for different capping layer thicknesses c (given in nm).

3.3 Effect of Finite Capping Layer

Experiments probe SADs buried under a finite capping layer. To reproduce these conditions, in this study the capping layer thickness c, measured from the top of the wetting layer to the top boundary of the strain domain, is treated as a parameter and fixed. Periodic BCs are taken in the XY plane for both strain and electronic calculations (the side domain boundaries in both of them coincide), so that arrays of SADs, laterally coupled both via the classical strain field and quantum-mechanically, are studied.

Fig. 6 shows the electronic and hole ground states of an infinite SAD array as functions of the interdot distance b (measured from dot edge to edge) for several cap thicknesses c and for dots A (top panels) and B (bottom panels). In both cases the energies clearly depend on the interdot distance. For thicker cap layers the effective semiconductor gap increases with the increase of b, to reach saturation when SADs are far apart. For the thinnest caps, however, a nonmonotonic behavior is observed for the electronic ground state (it is seen particularly clearly for dots A in Fig. 6 (a)). For fixed b the effective gap increases as the cap thickness c increases, to reach saturation for thick capping layers (in accordance with Fig. 5). For b = 30 nm and c changing by about 10 nm, the total effective gap increases by about 7.5 meV.

Acknowledgements

The work described in this publication was carried out in part at the Jet Propulsion Laboratory, California Institute of Technology under a contract with the National Aeronautics and Space Administration. Funding at Purdue was provided by NSF under Grant No. EEC-0228390 and by the Indiana 21st Century Fund. At JPL funding was provided under grants from ONR, ARDA, and JPL.

REFERENCES

[1] For reviews and references see, e.g., Jacak, L., Hawrylak, P., and Wojs, A, "Quantum dots", Springer-Verlag, Berlin, 1998.
[2] Aslan, B., Liu, H.C., Korkusinski, M., Cheng, S.-J., and Hawrylak, P., Appl. Phys. Lett., 82, 630, 2003.
[3] Petroff, P.M., in "Single Quantum Dots: Fundamentals, Applications, and New Concepts", Peter Michler, Ed., Springer, Berlin, 2003.
[4] Michler, P., et al., Science, 290, 2282, 2000; Moreau, E., et al., Phys. Rev. Lett., 87, 183601, 2001.
[5] Arakawa, Y., and Sasaki, H., Appl. Phys. Lett., 40, 939, 1982; Fafard, S., et al., Science, 22, 1350, 1996; Maximov, M.V., et al., J. Appl. Phys., 83, 5561, 1998.
[6] Petroff, P.M. and DenBaars, S.P., Superlatt. Microstruct. 15, 15, 1994.
[7] For a review and references see, e.g., Tadic, M., et al., J. Appl. Phys. 92, 5819, 2002.
[8] Klimeck, G., et al., Computer Modeling in Engineering and Science, 3, 601, 2002.
[9] Oyafuso, F., et al., Journal of Computational Electronics, 1, 317, 2002.
[10] Wasilewski, Z.R., Fafard, S., and McCaffrey, J.P., J. Cryst. Growth, 201, 1131, 1999.

Computer Simulations of Nanoparticle Sintering

V. N. Koparde[*] and P. T. Cummings[**]

[*]Vanderbilt University, Nashville, TN, vishal.koparde@vanderbilt.edu
[**]Vanderbilt University, Nashville, TN, peter.cummings@vanderbilt.edu

ABSTRACT

During the vapor-phase synthesis of titanium dioxide (TiO_2) nanoparticles, sintering of the nanoparticles is an important aspect of their behavior and an understanding of this phenomenon is therefore important. In this work, molecular dynamics (MD) simulations of the coalescence of TiO_2 nanoparticles have been carried out. The driving force for sintering of nanoparticles is the reduction in potential energy due to the decrease in surface area. The loss of potential energy manifests itself as an increase in the temperature of the sintering particles. This work concentrates on 3 and 4nm anatase and rutile nanoparticles. Dependence of particle orientation on sintering is reported along with ion mobility studies in the core and neck regions.

Keywords: TiO_2, nanoparticles, sintering, molecular dynamics, simulation, anatase, rutile

1 INTRODUCTION

Nanoparticles have been the area of active research in recent years as they have novel and unique properties, which distinguish them from the bulk phase[1]. Considering the difficulties associated with experimental analysis at the nanoscale, these systems are good candidates for study using MD simulation. Titania nanoparticles are synthesized by oxidation of $TiCl_4$ in flame reactors at high temperatures and moderate-to-low pressures[2]. Nanoparticle sintering occurs in these reactors and is suitable for study at the molecular level. The stability between rutile and anatase has been reported to reverse with decrease in particle size[3]. So far, very few simulation studies have been reported on the coalescence of metal-oxide nanoparticles and their phase transformations. MD simulations will help to shed light on the fundamental transport mechanisms and the kinetics that govern the particle coagulation and to observe, at the atomistic level, any phase transformation.

Titania has three polymorphs found abundantly in nature, namely, anatase, rutile and brookite[4]. In bulk phase, anatase and brookite irreversibly and exothermally transform to rutile upon heating[5]. But at nanoscale, anatase is found to be the most stable phase[3]. In our studies, we concentrate on rutile and anatase, as they are the most important from applications point of view. Electronic and photocatalytic properties of these polymorphs tend to enhance considerably with decreasing particle size[3]. Recently, TiO_2 has been used as a photocatalyst for alcohol dehydration[6], in oxidation of aromatic compounds[7], degradation of paint pigments[8], removing contaminants from wastewater[9], and nitrogen oxide reduction[10].

Nanosized TiO_2 is manufactured by the "chloride" process[2], which involves vapor phase oxidation of $TiCl_4$ leading to the formation of amorphous nano-titania, which can then be annealed to the desired phase. The temperatures range from 973 to 1873K in this process[11]. During the thermal annealing following the chloride process, the titania nanoparticles collide with each other and sintering may occur to form larger particles. If the sintering time is less than collision time, then larger particles are produced; otherwise agglomerates are formed[12]. Various available sintering theories are not applicable to particles in the nanometer range as they cannot be treated as continuum elastic bodies but need to be subjected to more rigorous atomic treatment. In nanoparticles, the sintering mechanism is influenced by particle curvature and interatomic forces. Also, surface and grain boundary diffusion are found to be the most important transport processes in nanoparticle sintering. Some researchers also suggested that the high internal pressure of smaller particles could be responsible for the inapplicability of the available models to smaller particles. For example, the pressure inside a 10 nm TiO_2 particle is reported to be around 2000 atmospheres[13]. Such high internal pressures affect diffusion coefficients, which in turn affect the rates of particle sintering.

The impetus behind sintering of two TiO_2 nanoparticles is the decrease in the free energy due to a reduction in the surface area[14-16], which under the adiabatic environment causes a temperature rise. Formation of new chemical bonds may explain the rise in temperature. As solid-state diffusion, which is dependent on temperature, is the major transport mechanism in nanoparticle sintering, we can expect temperature to play a key role in process. Earlier studies found that the heat release due to particle coalescence may reduce the coalescence time by as much as a few orders of magnitude as in the case of silicon nanoparticles[16].

This work concentrates on the initial stages of TiO_2 sintering and the effect of phase, size and temperature on the sintering process. The systems under consideration contain two identical nanoparticles 3 or 4 nm in diameter. Rotating one of the particles in the system while keeping the other unchanged allows to look at the orientational dependence on sintering. The relative mobilities of ions in the neck and core regions of the sintering nanoparticles have also been reported.

2 SIMULATION DETAILS

A report[17] on the available force fields for TiO_2 suggests that the Matsui-Akaogi (MA)[18] is the most suitable for use in classical MD. This force field not only reproduces the structures of TiO_2 polymorphs but also outperforms more complex variable charge force fields. The MA force field describes the interaction between Ti and O ions as a sum of exponential-6 non-columbic terms and columbic interactions. It regards Ti and O ions to be rigid with +2.196 and -1.098 partial charges, respectively.

Spherical nanoparticles of the 3 or 4nm were cut out of larger anatase or rutile lattice[19]. Excess Ti or O ions were removed to obtain neutral particles. The sphere is duplicated and translated along x-axis to obtain a system containing two identical nanoparticles, which are separated by 1 nm. DLPOLY version 2.13[20,21] was used to carry out all the MD simulations with a timestep of 0.5 fs for a total time of 0.5 ns. The simulations were carried out in NVE ensemble in vacuum with no periodic boundary conditions, as this was the best way to imitate the low pressure conditions prevailing in the flame reactors. Simulations were repeated at different starting temperatures of 573, 973 and 1473K, which could be considered to represent the different temperature zones in the flame reactors.

3 RESULTS AND DISCUSSION

Snapshots of a typical TiO_2 nanoparticle sintering are shown in figure 1. The figure indicates that the particles are mutually attracting each other, before touching and forming a neck. It should be noted that no initial velocity is applied on the particles to induce collision. The non-elastic collision occurs in about 20-30 ps. The neck diameter gradually grows initially but then reaches a constant value.

The potential energy of the system drops as the nanoparticles come in contact as indicated in figure 2. But as the simulation is constant energy, the temperature simultaneously increases. For 3nm anatase or rutile particles the temperature rise is about 45K and for 4nm anatase or rutile particles it is about 30K. This rise is found to be independent of the initial temperature of the simulation as the amount of decrease in surface area is approximately identical in all cases, for a particular size.

To analyze the results in further detail a term called shrinkage is defined as follows,

$$shrinkage = \frac{\left(\frac{d_1 + d_2}{2}\right) - d_{COM}}{\left(\frac{d_1 + d_2}{2}\right)} \qquad (1)$$

where d_1 and d_2 are the diameters of the sintering nanoparticles while d_{COM} is the distance between their centers of mass at that instance.

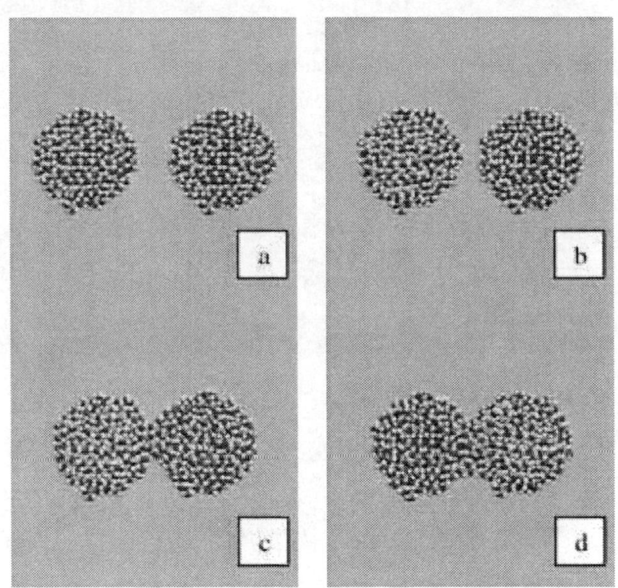

Figure 1. Snapshots of a typical sintering simulation

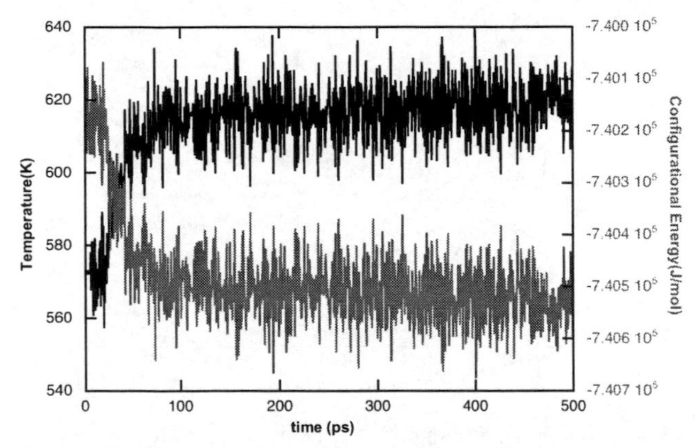

Figure 2. Time evolution of configurational energy and temperature of a typical sintering simulation (Two anatase 3nm particles with starting temperature of 573K)

Figure 3. shows the shrinkage variation with time for 3 and 4 nm anatase and rutile particle with a variety of different starting temperatures (mentioned in parenthesis). A negative shrinkage means that the particles are still separated while a positive shrinkage indicates particle interpenetration. Shrinkage seems to reach a constant value at the end of most of the simulations. This may be so because 0.5 ns is not long enough time to observe complete coagulation, which is reported to occur over a few microseconds[22]. If the simulations were run long enough, we would expect them to form a larger particle.

Neck diameter (not shown here) increased with temperature for 3nm anatase sintering particles but showed no such temperature dependence in case of 3nm rutile particles. For 3nm anatase case, it was ~17Å for a starting temperature of 573K while ~22Å for a starting temperature of 1473K.

Simulated x-ray diffraction patterns were calculated for the final agglomerates and they indicated no phase transformation over the time scale of the simulation. Examination of the neck regions explicitly showed that the neck region between two anatase particles was amorphous, while that between two rutile particles was rutile.

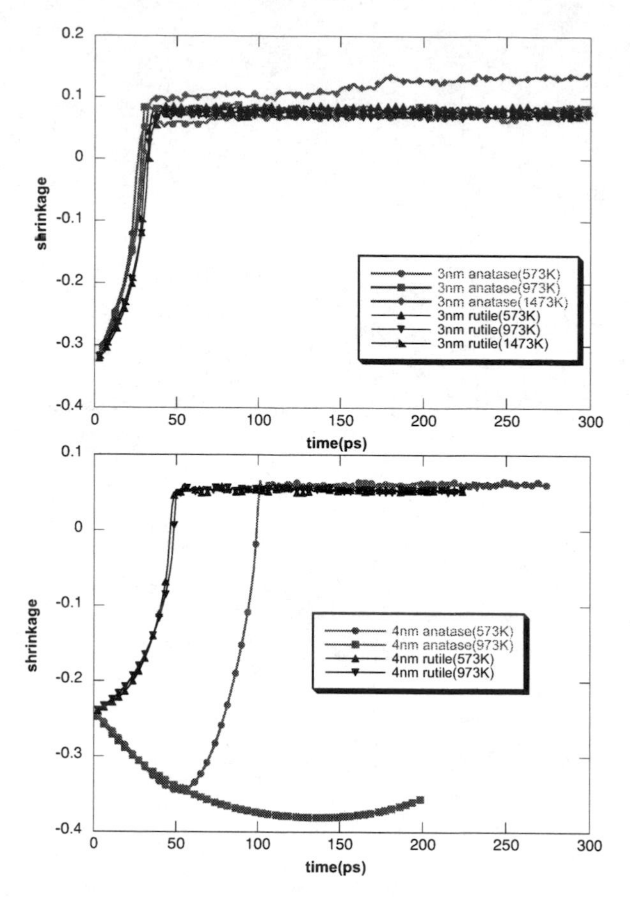

Figure 3. Shrinkage versus time for various simulations

3.1 Orientational Dependence

Crystallographic orientation is believed to play an important role in nanoparticle interaction. To study the effect of particle orientation the case of 3nm anatase sintering with the initial temperature of 573K was chosen and simulations were repeated after rotating the duplicated particle in the system through 20, 45, 90 and 180 degrees about z-axis at the beginning of the simulation. In all cases, except the 180-degree case, we observe neck formation. But in the 180-degree case the nanoparticles actually repel each other thus indicating that particle orientation is indeed important to the process of sintering. Also figure 4 shows that 90-degree orientation leads to maximum particle interpenetration. Such densification was experimentally observed and linked to anatase to rutile phase transformation[23] but the simulated x-ray diffraction patterns show no such phase transformation during the simulation.

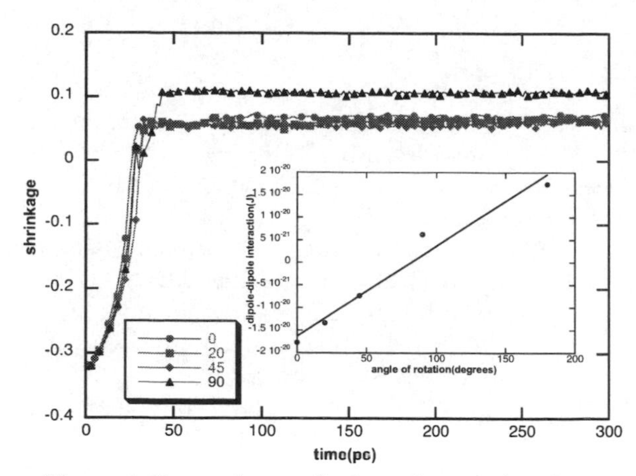

Figure 4. Dependence of orientation on sintering

To understand the orientational dependence in further detail, the dipole-dipole interactions between the nanoparticles at the beginning of the simulation were computed using the following formula[24],

$$W_{d-d} = \frac{\bar{\mu}_1.\bar{\mu}_2 - 3(\hat{n}.\bar{\mu}_1)(\hat{n}.\bar{\mu}_2)}{|\bar{r}_1 - \bar{r}_2|^3} \qquad (2)$$

where $\bar{\mu}_1, \bar{\mu}_2$ are the dipole moments of the particles, $\bar{r}_1, \bar{r}_2$ position vectors of the particles and $\hat{n}$ is the unit vector along $\bar{r}_1 - \bar{r}_2$. The dipole-dipole interaction becomes more repulsive with increase in orientational angle as indicated in figure 4 and at 180 degrees it is repulsive enough for the particle to not undergo neck formation. In all other cases, dipole-dipole interaction becomes attractive gradually as the simulation proceeds until the particles form a neck. Hence, it may be possible to model TiO_2 nanoparticles as spheres with fluctuating dipoles, before they collide.

3.2 Relative Ionic Mobilities

For further investigation of the sintering process, the relative ionic mobilities in the neck and core regions were determined. The relative displacement of the ith ion from time t to t' is[25],

$$\Delta r_i(t,t') = \left[\left(\left(r_i(t) - r_i(t')\right) - \frac{1}{m_{Ti}n'_{Ti} + m_O n'_O} X \left(\sum_{j=1}^{n'_{Ti}} m_{Ti}\left(r_j(t) - r_j(t')\right) + \sum_{k=1}^{n'_O} m_O\left(r_k(t) - r_k(t')\right)\right)\right)^2\right]^{1/2} \qquad (3)$$

Figure 5 shows the relative mobilities of Ti ions in the neck and core regions for the various simulations performed. Relative mobilities in the neck region are at least an order of magnitude greater than the core mobilities. There is no observable dependence of the neck ion relative mobilities on particle size or phase or initial temperature. At lower temperatures, core ion mobilities of anatase nanoparticles are measured to be lower than those of rutile nanoparticles.

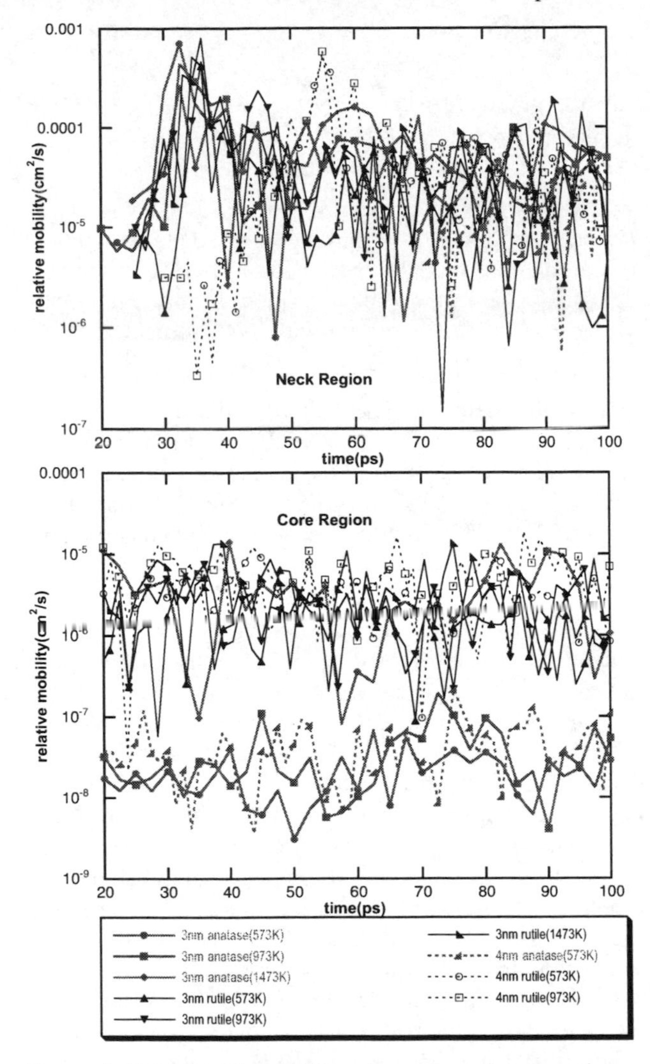

Figure 5. Relative Ti mobilities in neck and core regions

4 CONCLUSION

Initial stages of sintering of TiO_2 (anatase and rutile) nanoparticles have been studied using MD simulations. The sintering is very rapid and takes place in a few picoseconds. There is a temperature rise upon neck formation owing to the decrease in surface area. The temperature rise decreases with particle size but is independent of phase and initial temperature. Initial particle orientation is an important factor in nanoparticle sintering and its influence can be accounted for by considering the dipole-dipole interactions between the particles. Ions in the neck region are more mobile than those in the core. More simulations to study orientational dependence at higher temperatures are planned for the future. Further efforts will be made to correlate the simulation results and develop scaling laws for nanoscale sintering.

REFERENCES

[1] X.Z. Ding, X.H. Liu, Journal of Alloys and Compounds, 248, 143-145, 1997.

[2] M.K. Akhtar, X.O. Yun, S.E. Pratsinis, Aiche Journal, 37, 1561-1570, 1991.

[3] H. Zhang, J.F. Banfield, Journal of Materials Chemistry, 8, 2073-2076, 1998.

[4] C.M. Freeman, J.M. Newsam, S.W. Levine, C.R.A. Catlow, Journal of Materials Chemistry, 3, 531, 1993.

[5] M.M. Wu, J.B. Long, A.H. Huang, Y.J. Luo, S.H. Feng, R.R. Xu, Langmuir, 15, 8822-8825, 1999.

[6] M.A. Fox, M.T. Dulay, Chemical Reviews, 93, 341-57, 1993.

[7] M. Fujihira, Y. Satoh, T. Osa, Nature, 293, 206, 1981.

[8] P.A.M. Hotsenpiller, J.D. Bolt, W.E. Farneth, J.B. Lowekamp, G.S. Rohrer, Journal of Physical Chemistry B, 102, 3216-26, 1998.

[9] S.T. Martin, A.T. Lee, M.R. Hoffmann, Environmental Science & Technology, 29, 2567-2573, 1995.

[10] F. Gruy, M. Pijolat, Journal of American Ceramic Society, 75, 657, 1992.

[11] H.D. Jang, J. Jeong, Aerosol Science and Technology, 23, 553-560, 1995.

[12] M.R. Zachariah, M.J. Carrier, Journal of Aerosol Science, 30, 1139-1151, 1999.

[13] S.H. Ehrman, S.K. Friedlander, M.R. Zachariah, Journal of Aerosol Science, 29, 687-706, 1998.

[14] J.G. Gay, B.J. Berne, Journal of Colloid and Interface Science, 109, 90-100, 1986.

[15] K.E.J. Lehtinen, M.R. Zachariah, Physical Review B, 6320, art. no.-205402, 2001.

[16] K.E.J. Lehtinen, M.R. Zachariah, Journal of Aerosol Science, 33, 357-368, 2002.

[17] D.R. Collins, W. Smith, Council for the Central Laboratory of Research Councils, Daresbury, Research Report DL-TR-96-001, 1996.

[18] M. Matsui, M. Akaogi, Molecular Simulation, 6, 239-244, 1991.

[19] S.C. Abrahams, J.L. Bernstein, The Journal of Chemical Physics, 55, 3206-3211, 1971.

[20] W. Smith, Journal of Molecular Graphics, 5, 71-4, 1987.

[21] W. Smith, T.R. Forester, Journal of Molecular Graphics, 14, 136-41, 1996.

[22] A. Kobata, K. Kusakabe, S. Morooka, Aiche Journal, 37, 347-359, 1991.

[23] X.Z. Ding, X.H. Liu, Journal of Materials Research, 13, 2556, 1998.

[24] J.D. Jackson, Classical Electrodynamics, John Wiley & Sons Inc., 1975.

[25] H.L. Zhu, R.S. Averback, Philosophical Magazine Letters, 73, 27, 1996.

A Molecular Dynamics Simulation of Multi-wall Platinum Nanowires

Q.H. Cheng[1], H.P. Lee, C. Lu and S.J. Koh

Institute of High Performance Computing, Singapore
1 Science Park Road, #01-01 The Capricorn, Singapore 117528

ABSTRACT

Structural formation of platinum (Pt) nanowire (NW) is investigated using the classical molecular dynamics (MD) simulation method. A type of multi-shell Pt NWs is obtained from the simulations. These NWs consist of multi walls formed by rolling fcc (111) triangular network sheets. Experimental evidence of existence of such multi-wall Pt NWs has been reported (Kondo and Takayanagi, Science 289, 606, 2000). The simulations begin from initially random atom configurations. The initial configuration is minimized by the steepest descent method, and assigned a temperature of 601 K with a Maxwell-Boltzmann random distribution. Then simulated annealing is applied such that the temperature of the system is reduced gradually to 1 K and a stable NW structure is obtained. Structural characteristics of these Pt NWs, *i.e.* the lattice parameters, are further examined and presented in this paper.

Keywords: nanowire, molecular dynamics simulation, steepest descent method, periodic boundary condition, velocity verlet algorithm, Berendsen thermostat

1 INTRODUCTION

Metallic nanowire (NW) has been intensively studied in the past decade because of its importance in both fundamental low-dimensional physical theories and technological applications such as nanoelectronic devices. Investigation of configurations and structural properties of the NW is vital for broadening its applications.

Molecular dynamics (MD) simulations have been extensively employed to study metallic NWs. Interest has been paid to several fcc metals such as Au, Cu and Al [1-5]. However, simulations on another important fcc metal - platinum (Pt) - were scarcely reported.

This paper aims to study structural formation of Pt NWs using the classical MD simulation method. A type of multi-shell Pt NWs is obtained from the simulations. These NWs consist of walls formed by rolling an fcc (111) triangular network sheet, featuring different structures than those hexagonal solid Pt NWs which has also been observed in classical MD simulations [6].

Experimental evidence of existence of multi-wall Au NWs and single-wall Pt nanotube has been reported [7-8]. Theoretical investigations have been conducted on such multi-shell NWs of Au [9]. This type of structure was also referred to as helical multi-shell (HMS) NW [7] or cylindrical multi-shell (CMS) NW [10].

Each wall of these multi-shell NWs could be considered as a circular tube formed by folding an fcc (111) triangular network around a chord *OA* and a generatrix *OB*, as shown in Figure 1 where the thicker lines form a unit of unwrapped wire wall. We use the index *IJK* respectively to describe the atom strands in the $[\bar{1}10]$, $[01\bar{1}]$ and $[\bar{1}01]$ directions of the network. The formed cylindrical wall can be denoted by $n(m)$ in that the wall consists of n atom columns in *K*-direction, the left-handed spiral of the wall consists of m atom rows in *I*-direction and right-handed spiral of n-m rows in *J*-direction. Based on this notation, all tubes other than $n(0)$ and $n(n/2)$ with a even number of n are chiral, and $n(m)$ and $n(n\text{-}m)$ are mirror images of one another [9]. We further use $\alpha\beta\gamma$ for the angles between pairs of the three atom strands. It is obvious that, for an unstrained fcc (111) network, $\alpha = \beta = \gamma = 60^{\circ}$.

Referring to the notations introduced by Kondo and Takayanagi [7] and Tosatti [9], we use the label n_0-$n_1(m_1)$-$n_2(m_2)$-... to describe the multi-wall NWs consisting of coaxial shells. Besides using n_1, n_2 *etc.* to represent helical atom columns, m_1, m_2 *etc.* are also used to denote the chiral order of each helix wall. $n_0 = 1$ if there is a central strand at the wire axis; otherwise $n_0 = 0$ for a tunnel at the wire axis. For the special case of an ideal 1-$N(0)$ Pt NW in which the radius of the outer wall equals g_0 and the atom distance in the wall is d_{nn}, we can derive $N = 5$ approximately according to the relationship that the perimeter = $Nd_{nn} = 2\pi g_0$. g_0 is the distance between neighboring (111) layers and $g_0 = \sqrt{2/3}\, d_{nn}$.

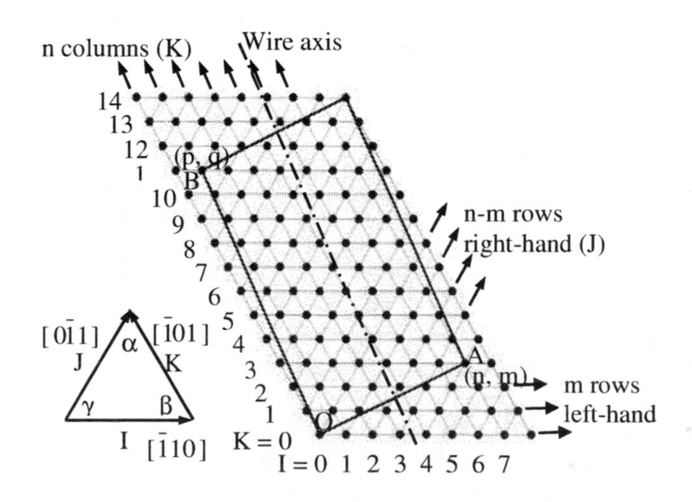

Figure 1: Folding of a triangular network.

2 SIMULATION METHOD

In this work, we investigated structural formation of thin and ultra-thin Pt NWs using a classical MD simulation. Interactions between Pt atoms were described by the Sutton-Chen many body potential [11], which was a development of one first proposed by Finnis and Sinclair [12]. The potential energy of an atomic system is written as:

$$V = \varepsilon\left[\sum_{i>j}\left(a_0/r_{ij}\right)^n - c\sum_{i}\rho_i^{1/2}\right] \tag{1}$$

where a_0 is the lattice parameter, r_{ij} the distance between atoms i and j and ρ_i local particle density, given by

$$\rho_i = \sum_{i>j}\left(a/r_{ij}\right)^m \tag{2}$$

The exponents n and m determine the range and shape of the potential and the parameters ε and a_0 determine the scales of energy and length respectively. The constants in the potential for Pt taken from [11] are $\varepsilon = 19.83$ meV, $c = 34.408$, $m = 8$, $n = 10$ and the lattice parameter $a_0 = 3.92$ Å. The interaction between atoms was truncated in the simulation at $2.7d_{nn}$ where d_{nn} is the nearest-neighbor distance and $d_{nn} = a_0/\sqrt{2} = 2.772$ Å. Another important parameter is the distance g_0 between neighboring (111) layers and $g_0 = a_0/\sqrt{3} = 2.263$ Å for Pt.

The Newton equations of the MD were integrated using the velocity Verlet algorithm [13] with a timestep $\Delta t = 2$ fs. To control temperature, the Berendsen thermostat [14] was applied with a characteristic relaxation time of 2.5 ps.

In the simulations, a periodic boundary condition (PBC) was considered in the axial direction of the NWs to simulate wires with infinite length. The simulations begin with initial configurations consisting of random distributions of atomic positions. Initial configurations of the Pt NWs were obtained by randomly arranged positions of atoms with a criterion that the nearest atom distance is 2.3 Å. This distance criterion is a compromise between feasibility of randomly generating the positions and a reasonably confined volume of the initial configuration. As this distance is much smaller than the lattice nearest-neighbor distance d_{nn}, a minimization procedures was first carried out by the steepest descendent method (SDM) to relax the initial configurations.

After minimization, simulated annealing was performed. The NWs were first assigned a temperature of 601 K and velocities of atoms were specified with a Maxwell-Boltzmann random distribution. Then the MD simulation runs 200 cycles. In each cycle, the temperature was reduced 3 K in the first step followed by 9999 steps with temperature control. Therefore the simulation time for each cycle is 20 ps and this simulated annealing phase lasted 4 ns. The initial temperature of 601 K is chosen so that the final temperature is approximately 1 K instead of 0 K whereby temperature control would encounter numerical singularity. When the phase finished, configurations of the NW became stable.

However, force in the NW along the axial direction may not equal to zero. So an MD simulation phase was further executed with 50 cycles to achieve a zero-stress state in the NWs. At the first step in each cycle, the residual force in the NW was measured from results of previous cycle and the length of the NW was accordingly adjusted. This was followed by 9999 relaxation steps with fixed NW length and temperature control. In all cases, the simulations converged to a stable NW length before 50 cycles finished.

Structural characteristics of these Pt NWs, *i.e.* lattice parameters, are then examined and presented in this paper.

3 RESULTS AND DISCUSSION

Figures 2-4 illustrates three single-wall NWs consisting of 69, 77 and 84 atoms, denoted as 1-6(0), 1-7(4) and 1-6(1) respectively. The final lengths of the NWs are always shorter than the initial ones. The atom distances in the central strands are smaller than d_{nn}, indicating the central strands are in a compressed state. With over five atom columns in the NWs, the radii of the outer walls are all greater than g_0, which should be in a 1-5(0) Pt NW as discussed in Section 2. The atom distances in the outer walls except d_I of the 84-atom NW are smaller than d_{nn}. Deviation of the atom bond angles from 60° shows the outer walls being distorted and excessive distortion of the 84-atom NW leads to $d_I > d_{nn}$.

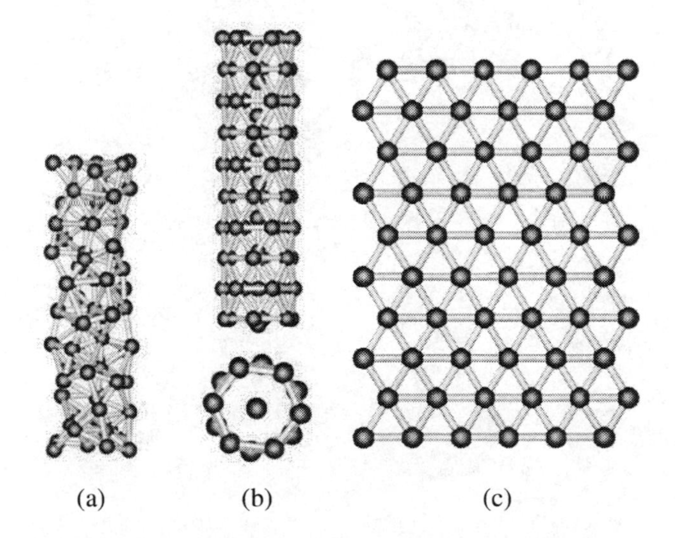

Figure 2: Configuration of a 1-6(0) Pt NW with 69 atoms: (a) initial; (b) final side and top views; (c) unwrapped outer wall. The NW has initially a length of 23.2 Å and finally 23.09 Å. The average neighboring atom distance in the central strand $d_0 = 2.564$ Å. The outer wall has a radius of 2.617 Å and the neighboring atom distance $d_I = 2.741$ Å and $d_J = d_K = 2.687$ Å. The atom bond angle $\alpha = 61.34°$ and $\beta = \gamma = 59.33°$.

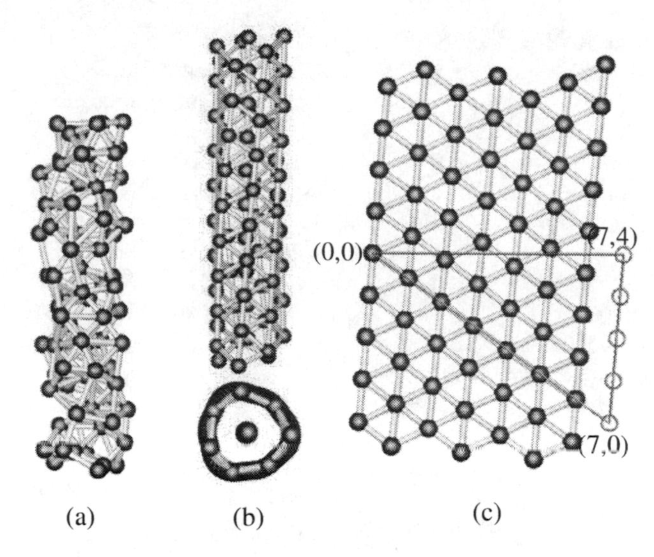

Figure 3: Configuration of a 1-7(4) Pt NW with 77 atoms: (a) initial; (b) final side and top views; (c) unwrapped outer wall. The NW has initially a length of 26.4 Å and finally 25.72 Å. The average neighboring atom distance in the central strand d_0 = 2.572 Å. The outer wall has a radius of 2.616 Å and the neighboring atom distance d_I = 2.688 Å, d_J = 2.711 Å and d_K = 2.699 Å. The atom bond angle α = 59.65°, β = 60.39° and γ = 59.96°. The open circles represent mirror images of atoms for assisting observation of the parameters $n_1(m_1)$.

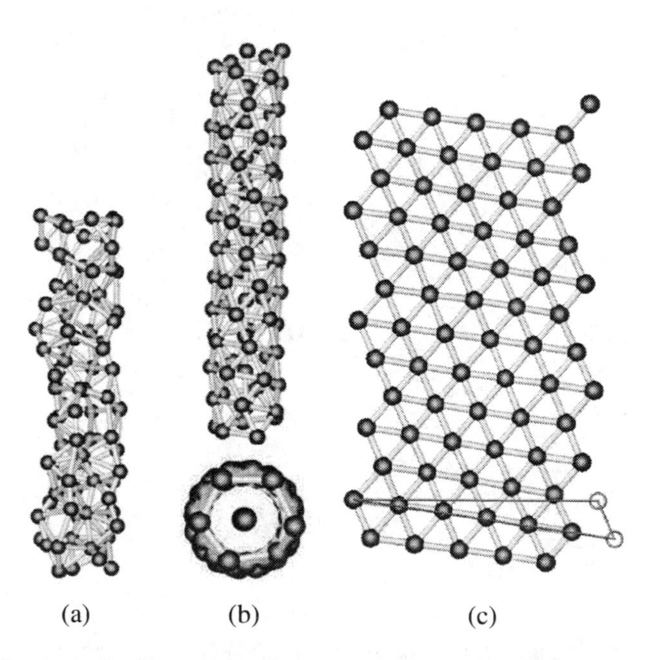

Figure 4: Configuration of a 1-6(1) Pt NW with 84 atoms: (a) initial; (b) final side and top views; (c) unwrapped outer wall. The NW has initially a length of 30 Å and finally 29.41 Å. The average neighboring atom distance in the central strand d_0 = 2.674 Å. The outer wall has a radius of 2.541 Å and the neighboring atom distance d_I = 2.872 Å, d_J = 2. 666 Å and d_K = 2.653 Å. The atom bond angle α = 65.34°, β = 57.59° and γ = 57.07°.

Figure 5 shows a double-wall 1-6(2)-12(2) Pt NW with 177 atoms. Again, the lengths of the final configurations are smaller than the initial and the atom distances in the central strands are smaller than d_{nn}. Though the radius of the inner wall is greater than g_0, the distance between the inner and outer walls is quite close to g_0. The inner wall is in excessive distortion leading to both d_I and d_J greater than d_{nn}. Distortion of the outer wall is not so intensive and only d_K is greater than d_{nn}.

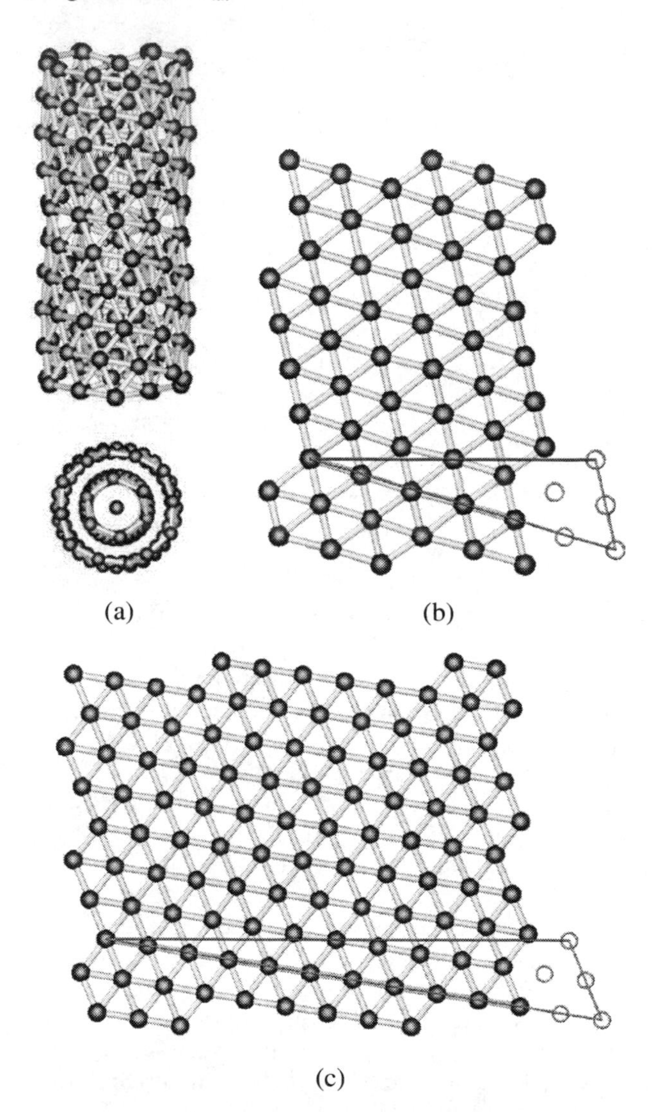

Figure 5: Configuration of a 1-6(2)-12(2) Pt NW with 177 atoms: (a) final side and top views; (b) unwrapped inner wall; (c) unwrapped outer wall. The NW has initially a length of 24.6 Å and finally 23.94 Å. The average neighboring atom distance in the central strand d_0 = 2.661 Å. The inner wall has a radius of 2.572 Å and outer wall 4.764 Å, thus a distance of 2.192 Å between them. The neighboring atom distance of the inner wall d_I = 3.011 Å, d_J = 2.870 Å and d_K = 2.630 Å and the atom bond angle α = 66.26°, β = 60.88° and γ = 52.86°; and outer wall d_I = 2.706 Å, d_J = 2.672 Å and d_K = 2.779 Å and the atom bond angle α = 59.51°, β = 58.30° and γ = 62.18°.

Figure 6 shows a triple-wall 1-6(4)-12(8)-18(14) Pt NW with 333 atoms. The atom distance in the central strand of this NW mostly approaches d_{nn} among the five simulated NWs, though still smaller than d_{nn}. The radius of the inner wall is still greater than g_0, but the distances between the inner and middle walls and between the middle and outer walls are quite close to g_0. Some faults exist in the NW. The inner wall has two atom columns in dislocation as indicated by two dashed lines; it is therefore not a pure (111) triangular network. That is why the average atom distances at all the three *IJK* directions in this wall are greater than d_{nn}. There are cavities in the middle and outer walls. The middle wall is intermediately distorted so that d_I and d_K are smaller than but $d_I > d_{nn}$. The outer wall is only slightly distorted and all the atom distances are smaller than d_{nn}.

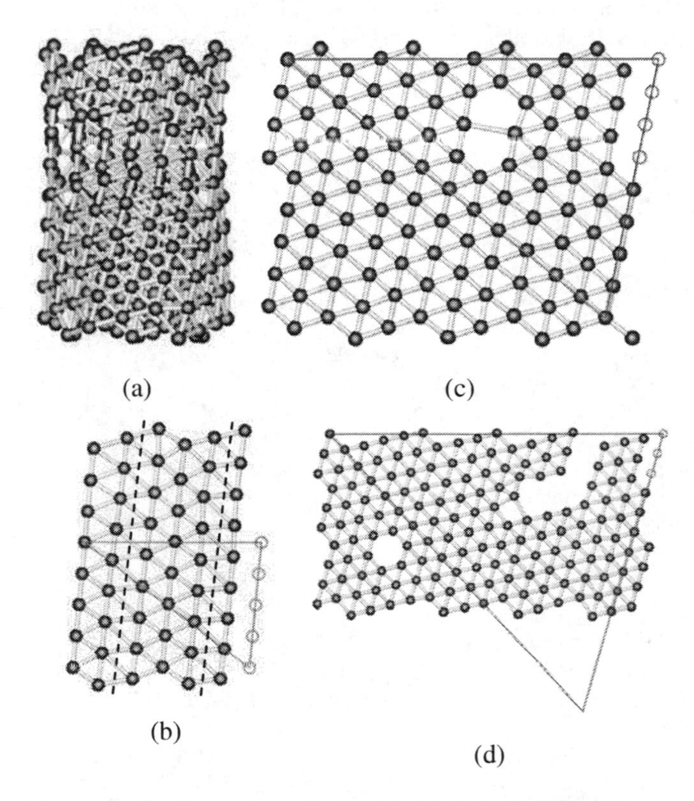

Figure 6: Configuration of a 1-6(4)-12(8)-18(14) Pt NW with 333 atoms: (a) initial; (b) unwrapped inner wall; (c) unwrapped middle wall; (d) unwrapped outer wall. The NW has initially a length of 25 Å and finally 24.28 Å. The average neighboring atom distance in the central strand d_0 = 2.708 Å. The inner wall has a radius of 2.502 Å, middle wall 4.739 Å and outer wall 6.962 Å, thus distances of 2.237 Å and 2.223 Å between them. The neighboring atom distance of the inner wall d_I = 2.814 Å, d_J = 2.907 Å and d_K = 2.811 Å and the atom bond angle $\alpha = 65.23^\circ$, $\beta = 60.97^\circ$ and $\gamma = 53.80^\circ$; middle wall d_I = 2.753 Å, d_J = 2.804 Å and d_K = 2.656 Å and the atom bond angle $\alpha = 60.45^\circ$, $\beta = 62.44^\circ$ and $\gamma = 57.11^\circ$; and outer wall d_I = 2.717 Å, d_J = 2.680 Å and d_K = 2.696 Å and the atom bond angle α = 60.48, $\beta = 59.44^\circ$ and $\gamma = 60.08^\circ$.

4 CONCLUSIONS

An MD study of the Pt NW was conducted. Beginning from random configurations of NWs, the simulations included a minimization procedure followed by simulated annealing with a thermostat algorithm. Five simulated Pt NWs are reported in this paper. Though geometric consideration of a Pt (111) triangular sheet dictates five atom columns for a single wall Pt NW, our simulations obtained NWs with six or seven columns in the single wall or inner wall, and a difference of six atom columns between each pair of walls in multi-wall NWs. By unwrapping the walls, structural parameters of the NWs are measured. Radii of the inner wall are always greater than the distance between (111) layers in bulk Pt, but the distance between the walls are quite close to that parameter. Atom distances in central strands and walls are smaller than the bulk lattice parameter except in cases where the walls are excessively distorted.

REFERENCES

[1] G. Bilalbegovic, Comp. Mater. Sci. 18, 333, 2000.

[2] G. Bilalbegovic, J. Phys. Condens. Mat. 13, 11531, 2001.

[3] J.W. Kang and H.J. Hwang, Nanotechnology 12, 295, 2001.

[4] J.W. Kang and H.J. Hwang, Mol. Simul. 28, 1021, 2002.

[5] A. Hasmy, P.A. Serena and E.Medina, Mol. Simul. 29, 427, 2003.

[6] Q.H. Cheng, C. Lu and H.P. Lee, Mol. Simul., in press, 2005.

[7] Y. Kondo and K. Takayanagi, Science 289, 606, 2000.

[8] Y. Oshima, H. Koizumi, K. Mouri, H. Hirayama, K. Takayanagi and Y. Kondo, Phys. Rev. B 65, 121401, 2002.

[9] E. Tosatti, S. Prestipino, S. Kostlmeier, A. Dal Corso and F.D. Di tolla, Science 291, 288, 2001.

[10] J.W. Kang and H.J. Hwang, Comput. Mater. Sci. **27**, 305, 2003.

[11] A.P. Sutton and J. Chen, Phil. Mag. 61, 139, 1990.

[12] M.W. Finnis and J.E. Sinclair, Phil. Mag. 50, 45, 1984.

[13] M.P. Allen and D.J. Tildesley, "Computer Simulation of Liquids," Clarendon Press, Oxford 1987.

[14] H.J.C. Berendsen, J.P.M. Postma, W.F. van Gunsteren, A. DiNola and J.R. Haak, J. Chem. Phys. 81, 3684, 1984.

[1] Tel: +65-6419-1547, Fax: +65-6419-1280, chengqh@ihpc.a-star.edu.sg

Molecular Dynamics (MD) Simulation on the collision of a nano-sized particle onto another nano-sized particle adhered on a flat substrate

Min-young Yi* and Jin-Won Lee**
Department of Mechanical Engineering,
Pohang University of Science and Technology, Pohang, 790-784 South Korea
*tao@postech.ac.kr, **jwlee@postech.ac.kr

ABSTRACT

Adhesion of a nano particle on a flat substrate both with and without deformation and also the behaviors of bullet and target particles after collision are simulated using the MD technique. The bullet particle, a low temperature solid Argon, is modeled by a Lennard-Jones(LJ) potential, and the target particle is modeled by a strong LJ potential. Parameters varied are the size of bullet and contaminant particles, adhesion force between the target particle and the substrate, and the velocity and collision angle of the bullet particle. Removal characteristics are different between weakly adhered and strongly adhered particles. For soft target particles high velocity at small angle favors removal. For hard particles the particle-substrate adhesion is the determining factor, and the ineffective angle is observed

Keywords: nano particle, nano particle adhesion, nano particle removal, molecular dynamics

1 INTRODUCTION

Particulate contamination seriously affects the manufacturing yield of micron or submicron scale devices, and semiconductor device feature is expected to decrease continuously, reaching 0.75μm by 2007 and 0.05μm by 2011[1]. Since the adhesion force is proportional to the size and drag force to the area of a particle, the use of drag force for cleaning becomes less efficient as the contaminant size is decreased, and it is generally agreed that these techniques work poorly for submicron particles [2,3] and other disadvantages with wet chemistry cleaning become more apparent at smaller scales [4].

Aerosol cleaning is a promising alternative to the classical cleaning methods. This technique has matured in industry for large particles in the micron range, but not for submicron or nano particles. Furthermore, the mechanism of contaminant removal by aerosol bombardment is not completely understood yet [5].

The objective of this study is to simulate the collision of a nano-sized particle with a rigid surface or a hard particle on a surface, in order to see the effect of various factors on the particle removal characteristics. The adhesion force between a particle and a flat substrate is dominated by the electrostatic force for the separation distance z_0>30nm but by the Van der Waals force for $z_0 \leq 30$nm, only the Van der Waals interaction is considered in this study [6].

2 MOLECULAR DYNAMICS

Techniques used for simulation in this study is the standard MD algorithm in which molecules interact via a pairwise Lennard-Jones potential: $V(r)=4\varepsilon\ [(\sigma/r)^{12}-(\sigma/r)^{6}]$, characterized by energy and distance scales ε and σ, respectively. All parameters are nondimensionalized with ε, σ and molecular mass *m*. The natural time unit is then $\tau = \sigma\sqrt{m/\varepsilon}$. Common values of the parameters for Argon are σ =3.405Å, ε/k_B=120°K, and mass *m*=40amu, where k_B is the Boltzmann constant [7]. Argon particle which is usually used as the bullet(cleanser) particle was simulated with the basic LJ potential. The contaminant particle, which may be a solid particle, an organic droplet, a fiber, or a metal ion and the like was simulated in three different forms: an LJ particle with number density(ρ) 1.0 and binding energy ε=10.0, an LJ particle with ρ=1.0 and binding energy ε=30.0, or an LJ particle with ρ=2.0 and binding energy ε=30.0. The substrate is assumed a rigid solid, considering the rather low level of kinetic energy and fragility of the bullet argon particles.

3 RESULTS AND DISCUSSION

When a particle on a substrate is hit by another particle, it can be detached from the surface, move along the surface, or stay at the same spot with some degree of vibration. The head-on collision of a soft nano particle onto a hard nano particle on a surface is simulated, and the post-collision motion of the surface particle is monitored (Figure 1).

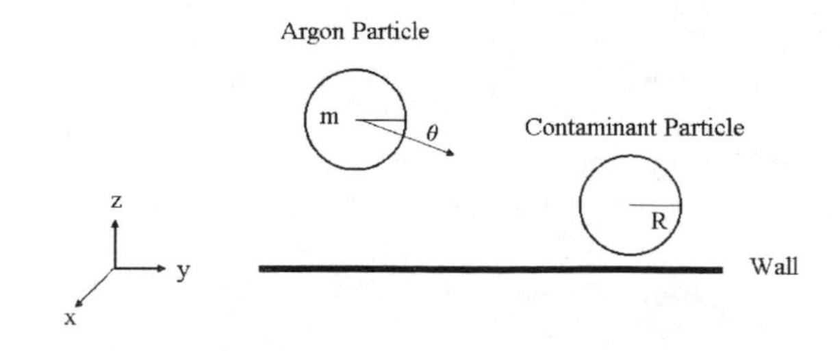

Figure 1: Schematic representation of collision of a bullet particle on a target particle on a substrate

The size of the computational domain is 60σ x 31σ x 71σ, and periodic boundary conditions are imposed in the x- and y- direction. The reflective boundary condition is used for the upper z boundary, and the domain size is set a little larger in the z-direction at 71σ(24.2nm) so that the effect of the reflective boundary condition can be reduced. Time step(Δt) was 0.002~0.005τ, and the total duration of the collision process was 30000~40000Δt. Cut-off radius was 2.5σ.

Initially, the system was equilibrated for 5000Δt, and then the contaminant particle is moved toward the substrate until the distance between the particle and substrate is smaller than one lattice parameter, 1.587. Then the particle is attracted to the substrate by the attractive Van der Waals force. After the contact is settled, additional 10000Δt is allowed for equilibration, and then the argon particle is shot toward the contaminant particle. The dimensionless bullet velocity is $v(\varepsilon/m)^{1/2}$, with v=1.0 corresponding to v=158 m/s in real dimension.

3.1 LJ Solid Particle of Medium Density and Binding Energy

The LJ solid particle with ε=10.0 is a little softer than Al or Cu particle, and sample results for a shooting angle of θ=45° are shown in Figure 2. The left figures show the states after contact equilibration. When $V_x = V_z = V_0 = 2.0(\varepsilon/m)^{1/2}$ (316m/s) with total kinetic energy of 1.38×10^{-19}J as in Figure 2(a), the argon particle is partially broken up, but the contaminant particle just slides and rolls to the +x direction, without departing from the surface. When the particle velocity is increased by 50% with total kinetic energy of 3.10×10^{-19}J as in Figure 2(b), the argon particle becomes totally broken up, and the contaminant particle becomes detached from the surface after collision.

It seems clear that a higher kinetic energy of the bullet particle will be effective in removing the target particle, but it is not clear whether the determining factor for particle removal is the kinetic energy or the momentum. In order to clarify this question, simulations are done with twice the kinetic energy as in Figure 2(a) but with different mass and velocity combinations: (a) 2m and V_0; (b) m and $\sqrt{2}\,V_0$; (c) m/2 and $2V_0$. At the large mass and low velocity condition, particle breakup is not complete, and the contaminant particle moves with sliding and rolling but stops after some distance. At the intermediate mass and velocity condition, particle breakup is almost complete, and the contaminant particle becomes detached after moving some distance with sliding and rolling. Small amount of contaminant molecules get disintegrated. At the small mass and high velocity condition, particle breakup is complete, and the contaminant particle gets detached from the surface with a high velocity soon after collision. Of the three cases (a) has twice as high and case (b) √2 times as high a momentum as case (c). From the simulation results it can be concluded that momentum(~ mv) cannot be an indicator for particle removal, and kinetic energy(~ mv^2) is not a proper indicator,

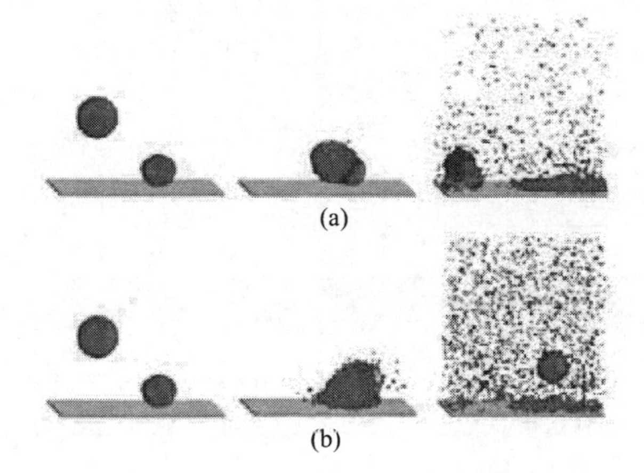

Figire 2: Snapshots at 25τ, 65τ, and 100τ for LJ particle: (a) $V_x=V_z=2.0(\varepsilon/m)^{1/2}$(316m/s);(b) $V_x=V_z=3.0(\varepsilon/m)^{1/2}$ (474m/s).

either. Particle removal seems to be determined by a new parameter with a much stronger dependence on velocity than mass like mv^3. It follows that an increased velocity is much more effective for particle removal than an increased mass.

The behavior of contaminant particle after collision can be classified into three modes: 1) just oscillates about a fixed point but does not move at all; 2) keeps moving with rolling and sliding but does not detach from the surface; 3) gets detached. The post-collision behaviors observed at various velocities and shooting angles are summarized in Table 1, where '×' means no motion, '□' moving without detachment, '•' detached, and '□×' stopping after sliding and rolling. It is clear from the table that a higher velocity gives a better removal, and shooting angles between 15° and 45° gives optimum performance. Particles in the cases marked □ or □× are thought to be removed in real situations due to the thermophoretic effect or by the carrier gas flow.

The argon particle breaks up when it collides with the substrate or the contaminant particle. To distinguish the effect of center-of mass kinetic energy from that of the burst induced by breakup, one artificial case is tried where the argon particle is shot normal to the substrate just ahead of the contaminant particle. In this case, the contaminant particle is affected solely by the breakup, not by the overall kinetic energy, of the bullet particle. This case is included as 'Burst only' in Table 1. When the velocity of the argon particle is high enough, the contaminant particle can be removed by the burst alone, but using the burst effect alone is not an efficient way of particle removal.

3.2 Strong Attraction : Deformation

Two different levels of deformation are considered. In weak deformation with ε=2.5, the adhesion energy is increased by a factor of 1.18, and in strong deformation with ε=5.0. the adhesion energy is increased by a factor of 2.28.

	Burst only	15°	30°	45°	60°	75°	K.E. (10^{-17}J)
V=67(m/s)	×	×	×	×	×	×	0.0313
V=223	□×	□	□	□×	□×	×	0.347
V=316	□	□	□	□	□	×	0.696
V=447	□	□	□	□	□	×	1.39
V=632	□	●	●	●	□	□	2.78
V=670	□	●	●	●	●	●	3.13
V=893	●	●	●	●	●	●	5.56

Table 1:Post-collision behavior of an LJ particle of medium binding energy at various velocities and angles of collision

Post-collision behavior of the strongly-bound target particle is much different from that of weakly-bound particle without deformation. When the bullet velocity is low and at low shooting angle, the contaminant particle starts to move but stops very soon, but at high shooting angles it does not move but gets deformed by collision. When the bullet velocity is high, the weakly-deformed particle gets detached if shot at low angle, but breaks up without moving if shot at high angle. The strongly-deformed particle breaks up, partially or fully, irrespective of the shooting angle, because the strong adhesion force prevents the contaminant particle from moving.

In generating Argon particles through supersonic expansion, the velocity of the argon particle cannot become much higher than 1000m/s. Then the results of Table 2 imply that particle bombardment may not be an effective means of removing deformed particles.

3.3 Effects of Higher Density and Binding Energy

In order to elucidate the behavior of a contaminant particle with a higher density and binding energy, simulations are tried with another artificial LJ particle with number density = 2.0, binding energy = 30.0 and σ = 0.80. Because Hamaker constant is proportional to density, the adhesion energy is increased due to the higher density, and a higher energy will be required to dislodge the target particle. Resultant behaviors of the new contaminant particle are summarized in Table 3.

When the results of Table 3 are compared with those of Table 1, it's evident that the removal efficiency for particles of higher density and binding energy is generally lower, particularly for the conditions of low velocities and small collision angles. At an intermediate collision angle of 45° the removal performance stays the same, and at higher angles of 60° and 75° removal even gets enhanced at high velocity conditions. This removal enhancement at high collisions angles and velocities can be attributed to the

Weakly deformed	15°	30°	45°	60°	75°	K.E. (10^{-16}J)
V=894(m/s)	□×	□×	×	×	×	0.551
V=1117	□	□×	×	×	×	0.861
V=1340	□×	□×	×	×	×	1.24
V=1564	□	□	□	□	□	1.71
V=1787	□	●	●	□	▲	2.23
V=1974	●	●	●	▲	▲	2.72
V=2070	●	●	●	▲	▲	3.00

strongly deformed	15°	30°	45°	60°	75°	K.E. (10^{-16}J)
V=1117	×	□×	×	×	×	0.861
V=1340	□×	□×	×	×	×	1.24
V=1564	□×	□▲	□▲	×	×	1.71
V=1787	□×	□▲	□▲	△	△	2.23
V=1974	□▲	□▲	□▲	▲	▲	2.72
V=2070	●▲	▲	▲	▲	▲	3.00

Table 2: Summary of removal characteristics for weakly (ε=2.5) and strongly (ε=5.0) deformed particles. '▲' denotes particle breakup. '△' denotes partial break up

elastic repercussion of the elastic target particle against the hard substrate.

In order to discriminate the effect of the target particle hardness, a little softer target particle with number density = 2.0 and binding energy = 10.0 is simulated. Then the general behavior after collision of this softer particle is seen nearly the same as that of a harder particle with number density = 2.0 and binding energy = 30.0. It implies that the removal characteristics of hard particles are dominated by the particle-substrate adhesion energy, not by the intra-particle binding energy.

3.4 Effects of Contaminant Particle Size

As a final example of particle removal problem the case of a larger contaminant particle size is simulated. The new target particle has a diameter of 16.2(5.5nm) instead of 12.7(4.6nm), and the density and binding energy is 1.0 and 30.0, respectively. Under these conditions the number of molecules and the total adhesion energy are the same as in the previous section 3.3 for a hard target particle, but only the diameter is changed. Then even at the same kinetic energy (or linear momentum) and collision angle of the

	Burst only	15°	30°	45°	60°	75°	K.E. (10^{-17}J)
V=67m/s	×	×	×	×	×	×	0.0313
V=223	×	□×	□×	□×	×	×	0.347
V=316	□×	□	□	□	×	×	0.696
V=447	□	□	□	□	×	×	1.39
V=632	□	□	□	●	●	●	2.78
V=670	●	□	□	●	●	●	3.13
V=893	●	●	●	●	●	●	5.56

Table 3: Removal behavior for a higher binding energy and density particle. ε = 30.0, σ = 0.80 and density = 2.0. Shaded and underlined symbol denotes the removal is deteriorated and enhanced compared with Table 1, respectively.

	Burst only	15°	30°	45°	60°	75°	K.E. (10^{-17}J)
V=67m/s	×	×	×	×	×	×	0.0313
V=223	×	□×	□×	□×	×	×	0.347
V=316	□×	□	□	□	×	×	0.696
V=447	□	●	□	□	□●	×	1.39
V=632	●	●	□	●	●	●	2.78
V=670	●	●	□●	●	●	●	3.13
V=893	●	●	●	●	●	●	5.56

Table 4: Removal behavior for a large contaminant particle whose radius is 8.1(2.75nm). ε= 30.0, σ = 1.0 and density = 1.0

bullet particle, the forcing effect on the target particle may change due to the difference in radius of rotation.

The change of removal behavior with shooting velocity and angle shows some anomalous pattern at an intermediate velocity and angle (Table 4). Under these conditions, the particle keeps in motion with sliding and rolling but does not get detached. This seemingly anomalous behavior has a theoretical ground. As Wang [8] formulated for a perfectly rigid particle, a particle adhered to a solid surface comes to move, upon impaction by another particle, in any of the three modes - lift-off, sliding, and rolling -, and among the three modes the onset force for rotation is usually the smallest. However, under some conditions, there appears a range of forcing angle over which particle rotation is very difficult to induce. Then particle removal becomes greatly reduced when the forcing angle is in this range, and he found that this ineffective zone for rotation extends from 42.1° to 47.9° of forcing angle. The seemingly ineffective zone observed in this study is about 30-45°, which is quite close to the theoretical prediction.

And the difference in forcing – this study considers only the head-on collision – may be the reason for the difference in the observed ineffective zone. Such a behavior was not observed with soft particles as in Table 1, and the reason for that is not clear so far. It may be that the behavior is predicted only for a perfectly rigid particle as is assumed in the theory, or that such a behavior is expected even for a soft particle but the small number of angles studied missed the narrow zone.

4 CONCLUSIONS

Higher kinetic energy is favored in particle removal, but from the view point of removal effectiveness an increased particle velocity is more efficient than an increased particle mass. Smaller particle with higher velocity is favored than large particle with low velocity. For particles adhered without strong deformation small shooting angles facilitate removal, and optimum performance is observed for shooting angles in the range of 30-45°. For strongly bound particles as in the presence of deformation, the target particle does not get detached but gets broken-up frequently. Then particle removal by particle bombardment is not effective. The effect of bullet particle breakup should be considered when the shooting velocity is high. Ineffective zone for rotation was observed for hard and large target particles, in close agreement with classical theory. The removal characteristics of hard particles are dominated by the particle-substrate adhesion energy, not by the intra-particle binding energy.

REFERENCES

[1] ITRS(International Technology Roadmap for Semiconductors), 2003

[2] Rimai, D.S. and Quesnel, D.J. Fundamentals of particle adhesion, Global Press, 2001

[3] Zapka W., Ziemlich W. and Tam A.C., Appl. Phys. Lett. 58, 2217, 1991

[4] Mahoney, J. F., Julius, P. and Carl, S. Removal of particulate and film contaminants by impacting surfaces with microcluster beams. In Particles on Surfaces 5&6 (Edited by Mittal, K.L.), pp. 311-325. VSP, 1999

[5] McDermott, W. and Sferlazzo, P. Argon Aerosol Surface Cleaning: An Overview, In Particles on Surfaces 5 & 6 (Edited by Mittal, K.L.), pp. 239-249. VSP, 1999

[6] Gady, B., Schleef, D., Rreifenberger, Rimai, D., and DeMejo, L.P. Proceedings of the Adhesion Society, 1996

[7] J.M. Haile, Molecular dynamics simulation, John & Wiley, 1997

[8] Wang, H.C., Aerosol Sci. and Technol. 13, 386, 1990

Numerical Analysis of the Effect of Diffusion and Creep Flow on Cavity Growth at the Nano-scale Level

F. W. Brust and J. Oh

Battelle Memorial Institute
Columbus, Ohio 43214, brust@battelle.org

ABSTRACT

In this study, a unified numerical method, which combines finite element analysis and finite difference analysis, for inter-granular cavity growth analysis is proposed. The numerical method includes three important mechanisms of cavity growth, which are grain boundary diffusion, surface diffusion along the cavity surface, and material creep flow, in simulating cavity growth at high temperature. An incremental cavity shape evolution scheme is used to simulate cavity shape transition from a spherical-shaped cavity to a crack-like cavity. The proposed numerical method was verified against two extreme cases; grain boundary diffusion controlled cavity growth and surface diffusion controlled cavity growth. When extremely fast surface diffusivity is dominant, the current numerical method predicts that the cavity maintains an initial spherical cavity shape. At the opposite extreme condition, that is surface diffusion controlled cavity growth, the current method successfully describes cavity shape change from initial spherical-shape to crack-like shape. Between these extremes, the cavity growth shape may evolve into different shapes including a diamond.

Keywords: creep, cavity growth, surface diffusion.

1 INTRODUCTION

Because of the complex physical phenomena, in most cases of numerical analysis for cavitation by diffusion, one of the two extreme cases, fast grain boundary diffusion or fast surface diffusion, is assumed. When grain boundary diffusivity is much faster than surface diffusivity (surface diffusion controlled process), the cavity shape will be similar to a crack because the atomic flow rate along the cavity surface is not fast enough to reduce surface curvature at the cavity tip. On the other hand, when surface diffusivity is much faster than grain boundary diffusivity (grain boundary diffusion controlled process), the cavity shape will be a spherical cap shape. Spherical cap refers to a sphere cut by a plane centered above or on the sphere center.

Despite its apparent importance, little analysis has been performed on continuous cavity growth analysis including grain boundary and surface diffusion mechanisms and viscoplastic deformation of the grain material. The numerical studies of these combined effects on the cavity growth will provide basic understanding of these synergies, and they will be useful in identifying the critical conditions where the combined effects become important. In this study, the effect of the ratio of the surface and grain boundary diffusivity and material viscoplastic deformation on the cavity growth rate is examined numerically. More details can be found in reference [1].

2 NUMERICAL MODEL

The atomic flow rates are proportional to driving forces, which are chemical potential gradients of the atom. When tensile stress is applied on the grain boundary, atoms diffuse from the cavity surface to the grain boundary due to the chemical potential gradient. As atoms diffuse from the cavity wall to the grain boundary, the grain boundary should accommodate the diffused atoms.

The chemical potential of the atom on the cavity surface 'μ_s' and that on the grain boundary 'μ_{gb}' respectively, are given by

$$\mu_s = -\gamma_s(\kappa_1 + \kappa_2)\Omega$$

$$\mu_{gb} = -\sigma_n \Omega .$$

The atomic volume, surface energy at cavity surface, principal curvatures of the cavity surface, and normal stress along the grain boundary are respectively represented by 'Ω', 'γ', 'κ_1', 'κ_2', and 'σ_n'. The driving force of the surface diffusion denoted by 'F_s' and that of the grain boundary diffusion denoted by 'F_{gb}', respectively, are

$$F_s = -\frac{\partial \mu_s}{\partial S}$$

$$F_{gb} = -\frac{\partial \mu_{gb}}{\partial r}$$

where 'S' is the curvilinear coordinate along the cavity surface and 'r' is the radial coordinate from the center of the cavity. Assuming the linear kinetic law, the surface and grain boundary atomic flow rates, denoted by 'j_s' and 'j_{gb}', respectively, are

$$j_s = M_s F_s$$

$$j_{gb} = M_{gb} F_{gb}$$

where 'M_s' and 'M_{gb}' are given by

NSTI-Nanotech 2005, www.nsti.org, ISBN 0-9767985-1-4 Vol. 2, 2005

$$M_s = \frac{D_s}{kT},$$

$$\text{where } D_s = D_{so}\delta_s \exp(-\frac{Q_s}{RT}),$$

$$M_{gb} = \frac{D_{gb}}{kT},$$

$$\text{where } D_{gb} = D_{gbo}\delta_{gb} \exp(-\frac{Q_{gb}}{RT}),$$

and "D_s", "D_{gb}", "k", "T", "$D_{so}\delta_s$", "$D_{gbo}\delta_{gb}$", "Q_s", "Q_{gb}", and "R" respectively, are the surface diffusivity, grain boundary diffusivity, Boltzman constant, absolute temperature, surface diffusion coefficient, grain boundary diffusion coefficient, activation energy for surface diffusion, activation energy for grain boundary diffusion, and gas constant.

In this analysis, cavity growth rates and cavity shape evolution are calculated by combining finite element and finite difference methods for a given time step. First, an extension of the finite element method by Needleman and Rice [3] is used to calculate the atomic flux (j_{gb} at cavity tip $= j_o$) by grain material deformation assisted grain boundary diffusion. In addition, the cavity shape change by grain material deformation, "jacking" effect for a given cavity geometry, and cavity tip stress, are included. When the "jacking" is not included material flux at the cavity tip is expressed as $2j_s = j_{gb}$, satisfying matter conservation law. However, material flow into the grain boundary contributes to the additional cavity volumetric growth. Therefore, this additional diffusion flux is included in the proposed numerical method (it has been neglected in many prior studies). A form of the open ended finite difference method is used to update cavity shape for a given time step.

After the cavity shape evolves for the current time step, the chemical potential of the atom at the cavity tip is calculated approximately from the value of principal curvatures of the node next to the cavity tip. At the next time step, the cavity tip stress, which is used for the FEM portion of the analysis, is calculated from the cavity tip curvatures obtained from previous FDM analysis and the same procedure is repeated until cavity coalescence occurs. Elastic-power law type material constitutive equations are assumed for creep flow. Figure 1 shows the structure of numerical calculation procedure (see [1]).

3 DIFFUSION GROWTH RESULTS

First results for grain boundary diffusion controlled growth are provided followed by combined growth.

2.1 Grain Boundary Diffusion

In this study, numerical analysis of an initial spherical-shaped cavity was carried out for $D_s/D_{gb} = 171$ (fast surface diffusion), $a_I/b = 0.1$ and $a_I/L=0.316$. The equilibrium

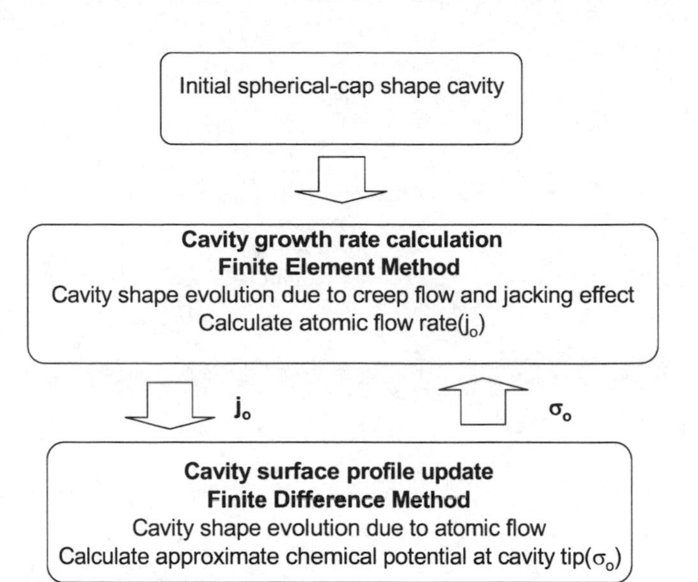

Figure 1. Illustration of numerical calculation structure for single cavity growth model. The unified numerical method, which combines finite element method and finite difference method, starts with the known spherical-cap shape cavity geometry. For the given time step, which is chosen to be sufficiently short, finite element method and finite difference method are employed to simulate cavity shape evolution (see [1] for details).

dihedral angle was chosen to be 70° and the creep exponent was taken to be 4.5. 'a' is cavity radius and 'b' is cavity spacing. Figure 2 shows the cavity radius increase with non-dimensional time, '$t\dot{\varepsilon}_e^{cr}$', where '$\dot{\varepsilon}_e^{cr}$' is the equivalent creep strain rate. From Fig. 2, it is clear that the proposed numerical method can reproduce Needleman and Rice [3] result exactly without the assumption of fast surface diffusivity. When the 'jacking' effect is not included, the present model overestimates cavity coalescence time. Therefore, it is also clear that the 'jacking' effect and creep flow of the surrounding grain material can cause significant difference in calculating cavity growth rates and cavity shape evolution especially when 'a+L' is smaller than 'b'. Chen and Argon [4] assumed that the grain boundary displacement ('jacking' effect) due to matter flow is constant over diffusion distance 'L'. However, Needleman and Rice [3] pointed out that material accommodation occurs mainly at cavity tip when creep deformation occurs. Also, when the diffusivity ratio is between the two extreme values (i.e. combined grain boundary diffusion/surface diffusion dominant) the cavity shape can evolve into more a complicated one, not spherical or crack-like one, due to the 'jacking effect'. This effect will be more important as 'a/L' increases since more severe matter accommodation occurs at the cavity tip. This is further explained in the next section. Analytical predictions are similar to the current numerical result up to a/b = 0.5. This gives us confidence in the model. Many other cases and results are presented in reference [1].

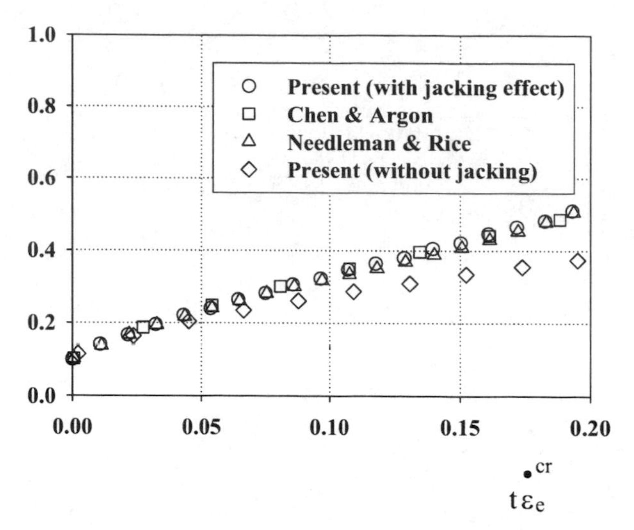

Figure 2. Normalized cavity major radius with time for $a_I/L=0.316$, $a_I/b_I=0.1$, and $D_s/D_{gb}=171$ (fast surface diffusion). When a/b is 0~0.5, the present numerical results (with jacking effect), which simulate cavity shape evolution, reproduce the Needleman and Rice results. In this range, the Chen and Argon analytical results also matches with the Needleman and Rice results. When jacking effect is not considered, the present result deviates with the other three results. That implies jacking effect is significant in this a/L range. Also, when a/b is 0.5~1.0, the present results still matches well with the Needleman and Rice results. However, Chen and Argon result starts to deviate with the Needleman Rice results.

2.2 Surface Diffusion

When surface diffusion controls cavity growth, the cavity elongates in the direction of the normal to the applied stress and the cavity growth kinetics change. In presenting the results, time is expressed in units of

$$t_s = \frac{a_I^4}{M_s \Omega \gamma_s}$$

and stress is expressed in non-dimensional form as

$$\Sigma = \frac{\sigma a_I}{\gamma_s}.$$

The diffusivity ratio is expressed as

$$f = \frac{D_{gb}}{D_s}.$$

The ratio of the applied normal stress to Young's modulus, σ_∞ / E, is 10^{-4} and initial "a_I/b_I" is 0.1. The capillarity angle at cavity tip is assumed to be 70°.

Fig.3 shows the cavity shape when a = 0.1, 0.3, 0.5, and 0.7 for f = 1 and f =10, respectively. For the results in Fig. 3, the initial cavity size divided by cavity spacing (a_I/b) was equal to 0.1. The values of a/b = 0.3, 0.5, and 0.7 occur after cavity growth is dominated by surface diffusion. Since a cavity becomes crack-like as it grows, as shown in Fig. 3, the first primary curvature increases at cavity tip. The nodes on the cavity surface were marked along the cavity surface in Fig. 3. Initially, 15 nodes existed along cavity surface with the same separation distance. As the cavity shape becomes more crack-like, more nodes exist along the high curvature area according to the node 'removal-creation' rule explained earlier. Nodes on the flat surface are removed when $\Delta\beta$ is less than 2° and moved to higher curvature area. This node 'removal-creation' procedure is accurate enough to simulate crack-like cavity evolution. Results not shown here show that the cavity becomes more crack-like when f = 10 than when f = 1. Therefore, the normal stress at the cavity tip (sintering stress) for f = 10 becomes higher than that for f = 1.

Based on the results from section 2.1 and 2.2, current numerical model is validated against two extreme cases; grain boundary and surface diffusion controlled cavity growth cases. With confidence in the accuracy of the model, we perform analyses in regimes where surface and grain boundary diffusion are both important.

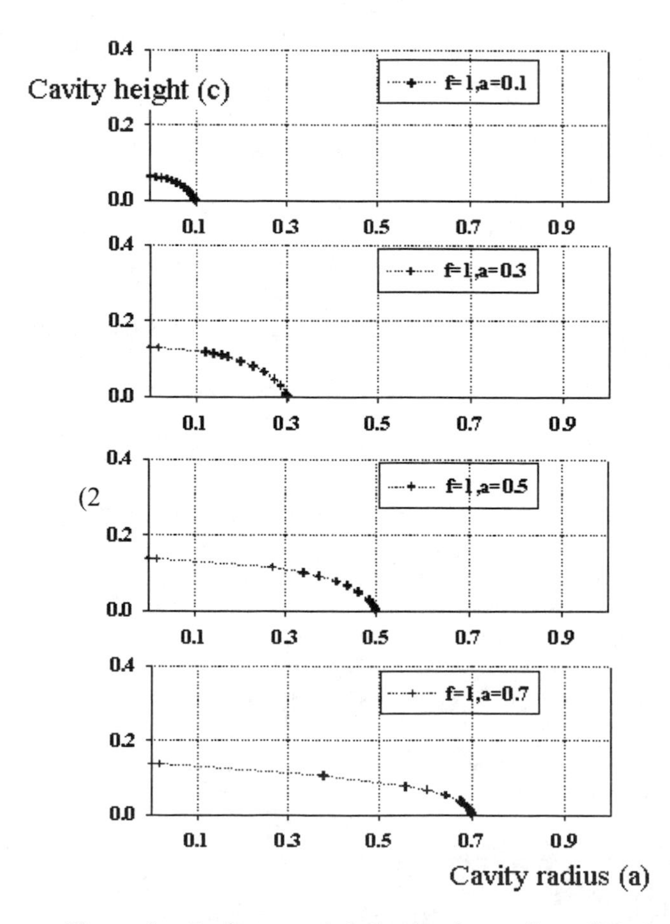

Figure 3. Cavity growth behavior for surface diffusion dominated cavity growth. Note the crack like nature.

NSTI-Nanotech 2005, www.nsti.org, ISBN 0-9767985-1-4 Vol. 2, 2005

2.3 Transition From Equilibrium Mode to Crack-like Mode

The grain material deformation not only assists the grain boundary diffusion but also changes the cavity shape. Therefore the cavity shape evolution can not be simply predicted based on simple analytical results. Fig. 4 shows cavity shapes in the case of $(a_I/L) = 1.0$ and $\alpha = 1.0$ when a/b = 0.3, 0.5, 0.7, and 0.9. The cavity shape is completed by reflecting nodes on the symmetric part of axis-symmetric geometry. When a/b = 0.3, the initial spherical-shaped cavity maintains its original shape with slightly increased aspect ratio. When a/b = 0.5 and 0.7, some nodes have negative curvature value because of cavity elongation at the cavity tip in the r-direction due to material diffusion and cavity elongation at the cavity top in z-direction due to the 'jacking' effect. When a/b = 0.9, the node with negative curvature fades away due to small '$\Delta\beta$' and cavity shape become V-shaped. These results are based on the assumption that the sintering stress is zero. When the sintering stress is considered, it is expected that the current numerical method predicts the cavity shape evolution and the final cavity coalescence time for general case in a physically more realistic manner. These 'diamond' shaped cavities have been observed in experiments.

3 CONCLUSIONS

In this study, a unified numerical method, which combines finite element analysis and finite difference analysis, for inter-granular cavity growth analysis was proposed. The numerical method includes three important mechanisms of cavity growth, which are grain boundary diffusion, surface diffusion along the cavity surface, and material creep flow, in simulating cavity growth at high temperature. The proposed numerical method was verified against two extreme cases; grain boundary diffusion controlled cavity growth and surface diffusion controlled cavity growth. When extremely fast surface diffusivity is dominant, the current numerical method predicts that the cavity maintains an initial spherical cavity shape, which was assumed by Needleman and Rice[19].

The present cavity growth rate prediction differs from the prediction by Chen and Argon[5] especially when 'a/b' is larger than 0.5. It is thought that this discrepancy is due to neglect of the proper 'jacking' effect in Chen and Argon model. At the opposite extreme condition, that is surface diffusion controlled cavity growth, the current method successfully describes cavity shape change from initial spherical-shape to crack-like shape. Although, in this method, the node next to cavity tip was used to calculate cavity tip normal stress, it was good approximation under physically reasonable diffusivity ratios.

It was shown that the cavity shape change is a complicated function of material diffusivity ratio, 'a/L', and cavity tip geometry when creep flow effects on cavity growth are important. It was shown that the cavity shape transition occurs gradually and the cavity growth rate during the transition phase is much higher than the prediction by Chen and Argon. As a/L increases, local accommodation at the cavity tip affects not only cavity shape but the overall cavity growth rate. This grain boundary displacement, which causes 'jacking' at the cavity tip, has been assumed to have a constant value in most research studies to date. Other results, including those compared with experiment, are presented in reference [1].

4 REFERENCES

[1] Oh, J., Katsube, N., and Brust, "Numerical Analysis of the Effect of Diffusion and Creep Flow on Cavity Growth", submitted to Journal Mech. Phys. Solids, Dec., 2004.

[2] Oh, J., Brust, F. W., "*Studies on Effect of Cyclic Loading on Grain Boundary Rupture Time*", Journal of Engineering Materials and Technology, reviews complete, to appear 2005.

[3] Needleman A. and Rice J.R. (1980). Plastic Creep Flow Effects in the Diffusive Cavitation of Grain-Boundaries. Acta Metallurgica, 28,1315. 1980.

[4] Chen I.-W. and Argon A.S. (1981). Diffusive Growth of Grain-Boundary Cavities. Acta Metallurgica, 29, pp 17-59.

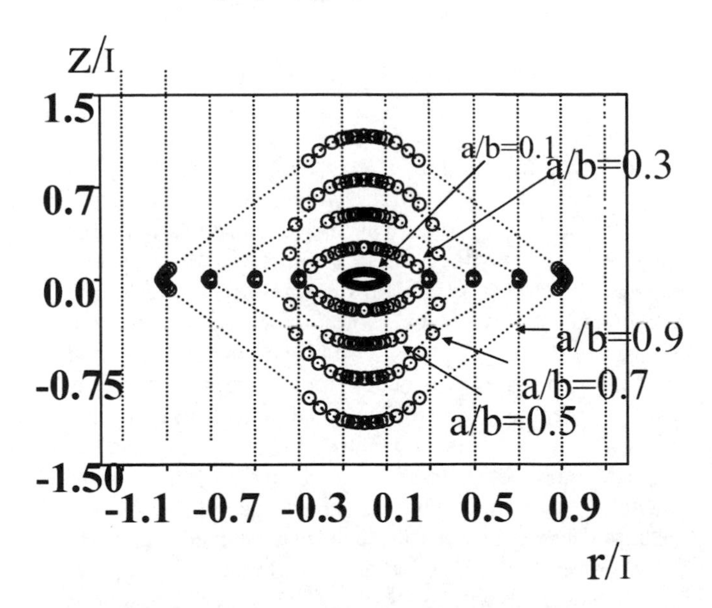

Figure 4. Cavity evolution from spherical-cap shape to V-shape for $a_I/L=1$, $\alpha=1$. Initial spherical-cap shape cavity maintains its original shape until a/L=0.3. When a/L reaches to 0.5, cavity shape changes to V-shape, since surface diffusivity is slow and material creep flow effect is significant.

Scaling relation for the size dependence of acoustic, dielectric and photoelectronic behaviour of nanosolid silicon

C. Q. Sun,* L. K. Pan and C. M. Li

School of Electrical and Electronic Engineering, Nanyang Technological University
Singapore 639798

ABSTRACT

Structural miniaturization provides us with a new freedom that is indeed fascinating. The new freedom of size not only allows us to tune the physical and chemical properties of a specimen by simply adjusting the shape and size but also enables us to gain information that is beyond the scope of conventional approaches. Here we show that a recent bond order-length-strength (BOLS) correlation could reconcile the size effect on nanosolid silicon with elucidation of information such as the single energy level of an isolated Si atom, the frequency of Si-Si dimer vibration, the upper limit of photoabsorption/emission, and dielectric suppression.

Keywords: Photoemission, photoabsorption, porous silicon, dielectrics, acoustic phonons and optical phonons

1 INTRODUCTION

Considerable efforts have been made in recent years to the understanding of the size effect on the acoustic, dielectric, and photoelectronic properties of nanosolid silicon because of the significance of nanosemiconductors in nanoelectronic devices. Among numerous theories that have been developed to explain the size effect such as the size induced blue shift in photoluminescence (PL) and photoabsorption (PA) as well as the phonon softening and phonon hardening, the "quantum confinement" theory is elegantly accepted [1]. However, without triggering the dominating factors in quantum confinement theory, electron-phonon (e-p) interaction or electron-hole (e-h) production, scanning tunneling spectroscopy/microscopy measurement revealed that the band gap expands from 1.1 to 3.5 eV when the Si nanorod diameter reduces from 7.0 to 1.3 nm associated with Si-Si bond length contraction of 10% [2]. This finding challenges the validity of the quantum confinement. Furthermore, all the size induced property changes arise from the same origin, therefore a model that unifies all the available data and provides consistent insight into the effect of size is desirable. Recently, we developed a bond order-length-strength (BOLS) correlation mechanism that is able to fulfill this task.

2 PRINCIPLE

The BOLS correlation indicates that the coordination (CN) imperfection of an atom at site surrounding a defect or near the surface edge causes the remaining bonds of the lower-coordinated atom to contract spontaneously. The spontaneous bond contraction is associated with bond-strength gain or atomic potential well deepening, which localize electrons and enhance the density of charge, mass, and energy in the relaxed region. The enhancement of energy density in the relaxed region perturbs the Hamiltonian and the associated properties such as the band-gap width [3], core-level energy [4], Stokes shift (electron-phonon interaction), and dielectric susceptibility [5]. On the other hand, bond-order loss lowers generally the cohesive energy of the lower-coordinated atom from the value of an atom with full CN, which dictates the thermodynamic process such as self-assembly growth, atomic vibration, thermal stability, and activation energies for atomic dislocation and diffusion [6].

Generally, the mean relative change of a measurable quantity of a nanosolid containing N_j atoms, with the dimensionless form of size K_j, being the number of atoms lined along the radius of a sphere or cross a thin plate, can be expressed as $Q(K_j)$; and as $Q(\infty)$ for the same solid without CN-imperfection contribution. The $Q(K_j)$ relates to $Q(\infty) = Nq_0$ as follows:

$$Q(K_j) = (N_j - N_s)q_0 + N_s q_s = N_j q_0 + N_s(q_s - q_0) \qquad (1)$$

The q_0 and q_s correspond to the local density of Q inside the bulk and in the surface region, respectively. $N_s = \sum N_i$ is the number of atoms in the surface atomic shells. Eq (1) leads to the immediate scaling relation:

$$\frac{\Delta Q(K_j)}{Q(\infty)} = \frac{Q(K_j) - Q(\infty)}{Q(\infty)} = \frac{N_s}{N_j}\left(\frac{q_s}{q_0} - 1\right) = \sum_{i \le 3} \gamma_{ij}(\Delta q_i / q_0) = \Delta_{qj} \qquad (2)$$

* E-mail: ecqsun@ntu.edu.sg; Fax: 65 6792 0415; www.ntu.edu.sg/home/ecqsun/

NSTI-Nanotech 2005, www.nsti.org, ISBN 0-9767985-1-4 Vol. 2, 2005

The weighting factor, γ_{ij}, represents the geometrical contributions from the size and dimensionality of the solid, which determines the magnitude of change. The quantity $\Delta q_i/q_0$ originates the change. The $\sum_{i\le 3}\gamma_{ij}$ drops in a K_j^{-1} fashion from unity to infinitely small when the solid dimension grows from atomic level to infinitely large. For a spherical dot at the lower end of the size limit, $\gamma_{1j} = 1$, $\gamma_{2j} = \gamma_{3j} = 0$, and $z_1 = 2$.

It is interesting to note that measurement of size dependence often follows a scaling law:

$$Q(K_j)-Q(\infty)=\begin{cases} bK_j^{-1} & (measured) \\ Q(\infty)\times\Delta_{qj} & (theory) \end{cases} \quad \textbf{(3)}$$

Given a functional dependence of the known quantities on bond length, bond strength, and bond nature, it is possible to predict the size dependence of the quantities.

3 EXPERIMENT

We prepared the p-Si samples and measured the size dependence of the PL, PA and E_{2p} peak energies. The particle size was controlled by varying the current density and estimated by matching the measured E_{PL} to the theory curve that has matched numerous sets of PL data of p-Si, CdS and CdSe nanosolids. XPS, impedance, and Raman measurements were conducted at room temperature.

4 RESULTS AND DISCUSSIONS

Calculations were performed with the following relations [7,8]: (4)

$$\left.\begin{array}{l} \dfrac{\Delta E_{PL}(K_j)}{E_{PL}(\infty)} \\ \dfrac{\Delta E_{PA}(K_j)}{E_{PA}(\infty)} \end{array}\right\} = \frac{\Delta E_G(K_j)\mp\Delta W(K_j)}{E_G(\infty)\mp W(\infty)}$$

$$\cong \sum_{i\le 3}\gamma_i\left[(c_i^{-m}-1)\mp B(c_i^{-2}-1)\right]$$

$$\left(B=\frac{A}{E_G(\infty)d^2};\quad \frac{W(\infty)}{E_G(\infty)}\approx\frac{0.007}{1.12}\approx 0\right)$$

$$\Delta_H = \sum_{i\le 3}\gamma_{ij}(c_i^{-m}-1)$$

$$\Delta_{e-p} = \sum_{i\le 3}\gamma_{ij}(c_i^{-2}-1)$$

$$\begin{cases} \dfrac{\Delta E_G(K_j)}{E_G(\infty)} = \dfrac{\Delta[\Delta E_v(K_j)]}{E_v(\infty)-E_v(1)} = \Delta_H \\ \dfrac{\Delta\chi(K_j)}{\chi(\infty)} = -(\Delta_H - B\Delta_{e-p}) + \sum_{i\le 3}\gamma_{ij}(c_i-1) \\ \dfrac{\Delta[\Delta\omega(K_j)]}{\omega(\infty)-\omega(1)} = \sum_{i\le 3}\gamma_{ij}\left[\dfrac{z_i}{z_b}c_i^{-\left(\frac{m}{2}+1\right)}-1\right] \end{cases}$$

where Δ_H, Δ_{e-p}, and Δ_{ph} correspond to the perturbation of Hamiltonain, electron-phonon coupling and phonon frequency due to the BOLS correlation.

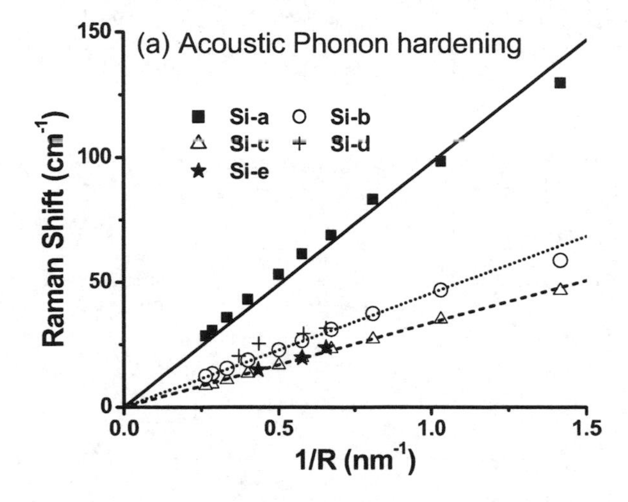

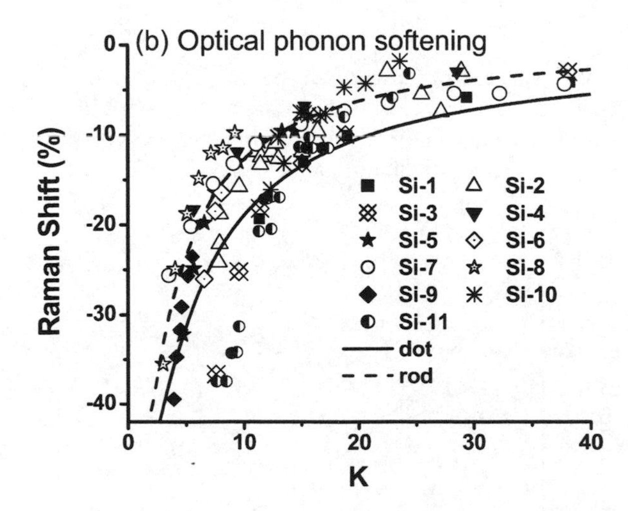

Figure 1 Agreement between BOLS predictions and observation on (a) acoustic phonon hardening, (b) optical phonon softening of nanosolid Si.

Figs 1-3 compare the BOLS predictions with the currently measured data and literature resourced data.

Agreement has been reached for the considered properties, indicating that the size induced property change arise from the lower-coordinated atoms in the surface skin.

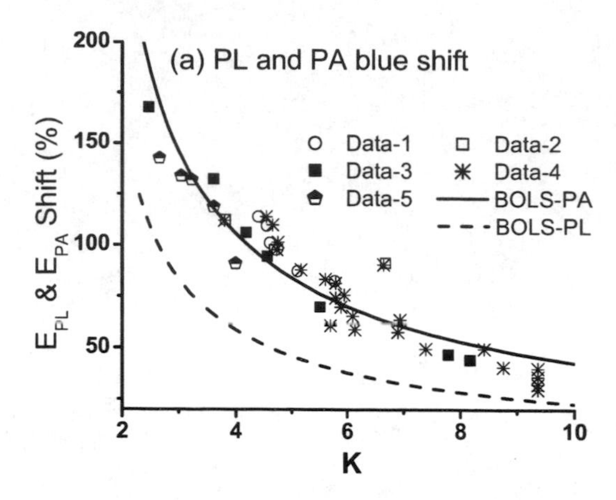

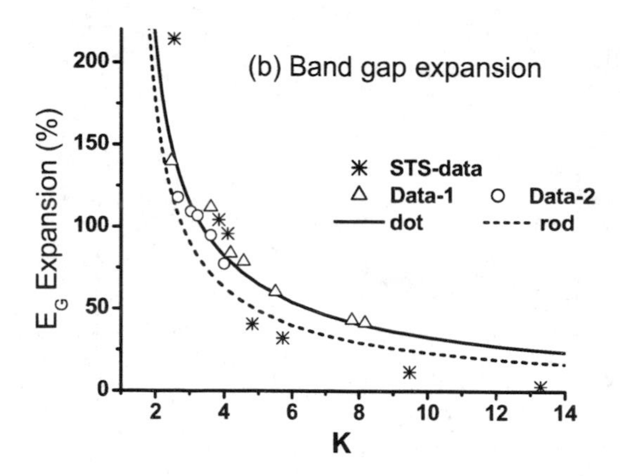

Figure 2 Agreement between BOLS predictions and observation on the size dependence of (a) blue shift in photoluminescence and photoabsorption, (b) band gap expansion of nano-Si

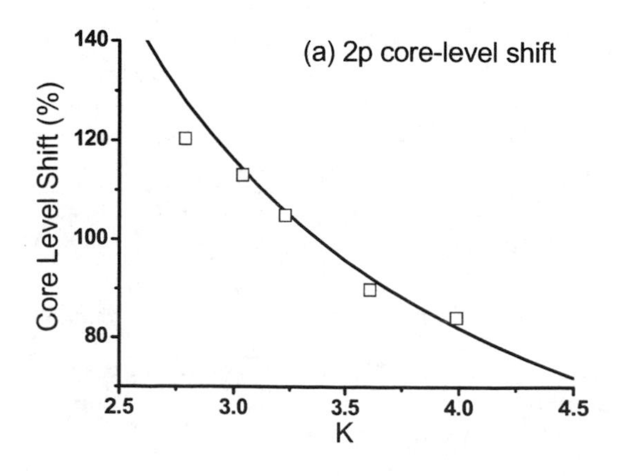

i

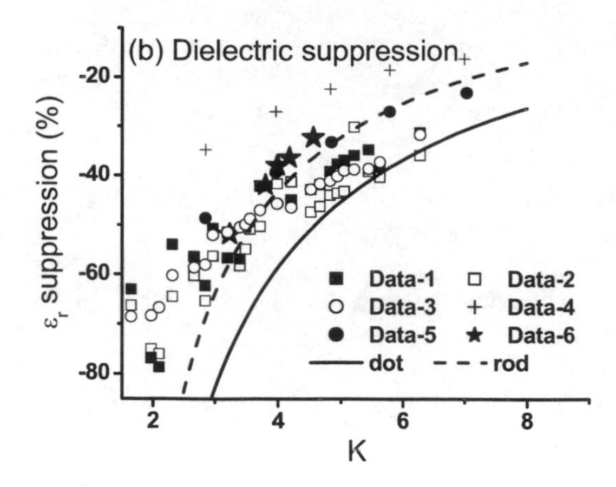

Figure 3 Agreement between BOLS predictions and observation on (a) 2p core-level shift and, (b) dielectric suppression of nansolid silicon. The index m = 4.88 represents the covalent nature of Si-Si bond.

5 CONCLUSION

Consistency between predictions and observations evidences the enormous impact of atomic CN imperfection to the low-dimensional systems and the validity, universality, and essentiality of the original BOLS correlation.

REFERENCES

[1] C. Q. Sun, T. P. Chen, B. K. Tay, “An extended 'quantum confinement' theory: surface-coordination imperfection modifies the entire band structure of a nanosolid,” J. Phys. D 34, 3470,2001.

[2] D.D.D. Ma, C. .S Lee, F. C. K. Au, S. Y. Tong, and S. T. Lee, “Small-diameter silicon nanowire surfaces,” Science 2003, 299, 1874.

[3] L. K. Pan, C. Q. Sun, B. K. Tay, “Photoluminescence of Si nanosolid near the lower end of the size limit,” J. Phys. Chem. B106, 11725, 2002.

[4] C. Q. Sun, Surface and nanosolid core-level shift: impact of atomic CN imperfection, Phys. Rev. B69, 045105, 2004.

[5] L. K. Pan, C. Q. Sun, T. P. Chen, C. M. Li, S. Li, B. K. Tay, “Dielectric suppression of nanosolid silicon,” Nanotechnology, 15, 1802, 2004.

[6] C. Q. Sun, W. H. Zhong, S. Li, “Atomic CN imperfection suppressed phase stability of ferromagnetic, ferroelectric, and superconductive nanosolids,” J Phys. Chem. B108, 1080, 2004.

[7] L. K. Pan, C. Q. Sun, C. M. Li, “Elucidating information of Si-Si dimer vibration from size dependence of Raman shift,” J. Phys. Chem. B108, L3819, 2004.

[8] C. Q. Sun, L. K. Pan, Y. Q. Fu, “Size dependent 2p-level shift of nanosolid silicon.” J. Phys. Chem. B107, 5113, 2003.

Elastic Network Model of Polymer Nanocomposites

Yunho Jang*, Shantikumar Nair** and Moon K. Kim***

*University of Massachusetts, Amherst, MA, USA, yujang@ecs.umass.edu
** University of Massachusetts, Amherst, MA, USA, nairsv@ecs.umass.edu
*** University of Massachusetts, Amherst, MA, USA, mkkim@ecs.umass.edu

ABSTRACT

To advance current understanding of nanoscale structure design for maximum toughness in polymer composites, a computational mechanics model called elastic network model (ENM) is addressed. In this method the potential energy of a system is approximated as a harmonic function and then the generalized eigenvalue problem is solved resulting in eigenvalues and eigenvectors, which are related to the natural frequencies and directions of corresponding motions, respectively. Using ENM, the physics of the deformation and the fracture processes at crack tips can be investigated. The computational cost of this method is significantly lower (e.g., a few hours in a desktop PC) than that of conventional molecular dynamics (MD) simulations performed by the aid of a supercomputer. The simulation results of a 2-D planar crack-tip topology and 3-D elastic network models are presented in this paper.

Keywords: elastic network model, toughness, polymer composite, eigenvalues, molecular dynamics

1 INTRODUCTION

Both micron-scale and nanoscale polymer composites have been investigated extensively not simply for the promise of a stiffer composite but for the promise of a composite that is significantly tougher as well. While work on conventional micron-scale composites goes back several decades, the work involving nanoscale polymer composites was greatly energized in the late 1990s following Toyota's demonstration that very low levels of a nanoscale rigid reinforcement can result in substantial modulus increases in a nylon 6 polymer. Considerable amount of effort has gone into the development of new nanoscale reinforcements, new processing routes to achieve such reinforcements in composites and the mechanical behavior of such nanoreinforced polymer composites.

Nair et al. showed indirect evidence that nanoscale reinforcements appeared to assist in shear plastic deformation at crack tips [1]. A high toughness was associated with a high crack-tip plastic zone size in the nanoclay-reinforced composites. However, the increase in toughness by the enhanced plasticity was offset by a decrease in toughness due to embrittling damage. From this study, it was proposed that toughening requires a proper balance between damage formation and enhanced plasticity in the crack-tip zone.

For modeling the structural response of materials to applied loads, finite element methods (FEM) have become widely accepted as a standard technique. However, FEM is a continuum method that cannot directly model atomistic or non-continuum effects, such as the interaction between polymer chains and nanoparticles. Thus, separate calculations, such as molecular dynamics (MD) simulations, must be employed to account for nanoscale effects [2].

In this study, we employ a novel concurrent multiscale approach based on Normal Mode Analysis (NMA), rather than MD, for the nanoscale model. One key advantage of NMA is that its resolution is adjustable from a true atomic scale to a scale of tens of nanometers, with the concomitant reduction in computational burden.

This macroscale tool was recently addressed to the study of micro and nanoscale systems, especially macromolecules such as proteins and nucleic acids. In both MD simulation and NMA, atomic trajectories are calculated by the classical Newtonian dynamics in which forces are derived from the empirical potential functions [3]. This potential is defined based on the structural data obtained from experiments such as X-ray crystallography and NMR. MD simulation can provide realistic molecular motions including the effects of surrounding solvent and ions, but the computational cost is tremendously high and it is also difficult to obtain long-time-scale collective motions from even millions of MD snapshots because MD results resemble Brownian motion, in which a time evolving conformation fluctuates rapidly.

As discussed above, MD simulation and NMA using all-atom empirical potentials follow the dynamics of macromolecules to the atomic level. However, the use of atomic approaches becomes computationally inefficient with the increased size of a system. To reduce a computational burden, an elastic network model was, alternatively, proposed in which a system is represented as a network of linear springs [4]. Sophisticated empirical potential models are replaced by a single-parameter Hookean potential and it was further simplified by reducing the number of degree-of-freedom (DOF) by coarse graining [5]. For example, only alpha-carbon atoms in a protein are treated as point masses and spatially proximal points are assumed to be linked with linear springs. Only structural (i.e., geometric) information is used to define a simple harmonic potential function. Such a coarse-grained elastic network model is suitable to describe the global behaviors of large macromolecules within reasonable time in a personal computer [6-7].

In NMA, the slow (i.e., low-frequency) modes, referred as the global modes, are usually insensitive to atomic details. In other words, the globally collective motions of the system are dominantly ruled by a few of the slowest modes. Statistical mechanics also predicts that the contribution to the corresponding eigenvalue [8]. That means that the low-frequency modes are naturally favorable to occur. Although a single normal mode cannot accurately represent the conformational changes during the deformation, the slowest normal mode indicate the most probable direction of the transformation [9].

Motivated by this situation, we have conducted a preliminary analysis of the molecular mechanisms of nanoscale polymer composites by performing computer simulations which iteratively generate the conformational changes from the results of NMA.

2 METHOD

2.1. Coarse-grained Elastic Network Modeling

The first step is defining a set of n representative atoms (e.g., C_α atoms in the case of proteins). We label the mass of the ith residue in the protein chain as m_i, and model the interaction between residues i and j with a linear spring having stiffness k_{ij}. The position of the ith atom at time t is denoted

$$\vec{x}_i(t) = [x_i(t), y_i(t), z_i(t)]^T \in R^3 . \quad (1)$$

If we define $\vec{\delta}_i(t)$ as a vector of small displacements such that

$$\vec{x}_i(t) = \vec{x}_i(0) + \vec{\delta}_i(t), \quad (2)$$

The total kinetic energy in a network of n point masses has the form

$$T = \frac{1}{2}\sum_{i=1}^{n} m_i \left\|\dot{\vec{x}}_i(t)\right\|^2 = \frac{1}{2}\dot{\vec{\delta}}^T M \dot{\vec{\delta}} , \quad (3)$$

where

$$\vec{\delta} = [\vec{\delta}_1^T, \cdots \vec{\delta}_n^T]^T \in R^{3n} . \quad (4)$$

In the present case, the global mass matrix M is diagonal. Likewise, the global stiffness matrix can be obtained from the total potential energy in a network of connected spring such that

$$V = \frac{1}{2}\sum_{i=1}^{n-1}\sum_{j=i+1}^{n} k_{ij}\left\{\left\|\vec{x}_i(t) - \vec{x}_j(t)\right\| - \left\|\vec{x}_i(0) - \vec{x}_j(0)\right\|\right\}^2 \quad (5)$$

and

$$k_{ij} = \begin{cases} 1 & if \left\|\vec{x}_i - \vec{x}_j\right\| \le d \\ 0 & if \left\|\vec{x}_i - \vec{x}_j\right\| > d \end{cases}, \quad (6)$$

where d is a cutoff distance between atoms at equilibrium and k_{ij} is the (i,j) element of k (called the "linking matrix" or "contact matrix"), which is assumed to be unity for all contacting pairs and zero for pairs not in contact. In this simple model, springs represent close residues, all interacting in identical ways, and the elastic potential energy follows a harmonic potential, appropriate for small deviations from equilibrium.

In general, Eq. 5 is a nonlinear function of the deformations even though each spring is linear. However, when we assume that the deformations are small, V becomes a classical quadratic potential energy function such that

$$V = \frac{1}{2}\vec{\delta}^T K \vec{\delta}. \quad (7)$$

The stiffness matrix K consists of an $n \times n$ array of 3×3 symmetric blocks denoted by K_{ij} [6-7]. If mass i is not connected to mass j, then the corresponding symmetric block is a 3×3 zero matrix. Generally,

$$K_{ij} = \begin{cases} -G_{ij} & if\ i \ne j \\ \sum_{k \ne j} G_{kj} & if\ i = j \end{cases}, \quad (8)$$

where

$$G_{ij} = k_{ij}\frac{(\vec{x}_i(0) - \vec{x}_j(0))(\vec{x}_i(0) - \vec{x}_j(0))^T}{\left\|\vec{x}_i(0) - \vec{x}_j(0)\right\|^2}. \quad (9)$$

If we set m_i to be a constant in a coarse-grained model, then normal modes are simply the eigenvectors of K. The physical meaning of each eigenvalue of the stiffness matrix is a scaled frequency of the harmonic motions and the corresponding eigenvectors indicate basic motions of the given conservative system. NMA can be used to evaluate potential motions about a single equilibrium conformation of a large molecule with relatively little computational cost.

2.2. Iterative NMA

The proposed procedure to simulate the dynamics of polymer nanocomposite is as follows: i) build the initial elastic network model at an equilibrium state, ii) perform NMA to find a few of the slowest modes, iii) perturb the initial conformation by adding a scaled slowest mode, iv) evaluate the plastic flow and the damage zone by measuring the change in spring lengths, v) rebuild the elastic network for the deformed conformation, and vi) back to step ii) and repeat the preceding procedure.

3 RESULTS

3.1 2-D Planar Crack-tip

The simulation results of a 2-D planar crack-tip topology are illustrated in Figure 1 for the case of a neat polymer. Each point represents a polymer chain uniformly distributed on the pure polymer matrix without nanoparticles. The secondary (i.e., hydrogen) bonds between chains are modeled as the linkages between points. Setting two threshold values of strain each of which, respectively, correspond to the yield point and the failure point, the growth of the crack-tip plastic zone and the growth of damage as crack opening displacement increases can be illustrated. The crack-tip deformation behavior of a polymer composite reinforced by nanoparticles is also simulated. They are represented as black dots in the polymer matrix (Figure 2). The strength of interface between the filler and the polymer is assumed to be 10 times stronger than the secondary bonds between polymer chains. Compared to the pure polymer in Figure 1, nanoscale reinforcement increases the stiffness of the polymer composite (represented by sparser plastic flow in Figure 2), but what is notable is that the radial extent of plastic flow is increased because of localization of flow between the nanoparticles. In this model the toughness actually decreased with failure occurring at the locations of plastic instability as would be expected from the various experimental observations in the literature.

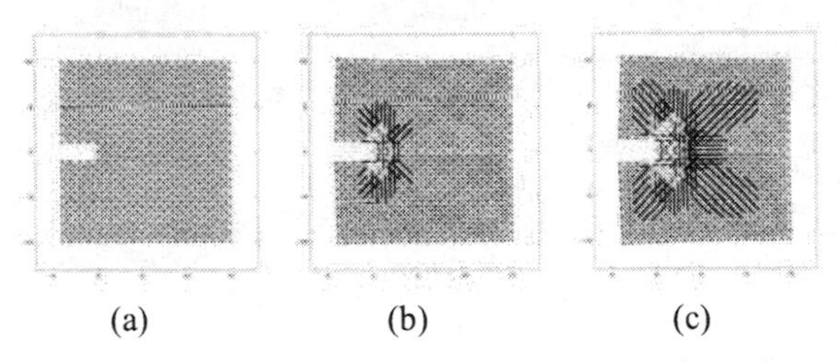

(a) (b) (c)

Figure 1 Plastic flow (dark line) and damage zone (light line) of a pure polymer matrix. (a) Initial conformation (b) 5th iteration (c) 10th iteration.

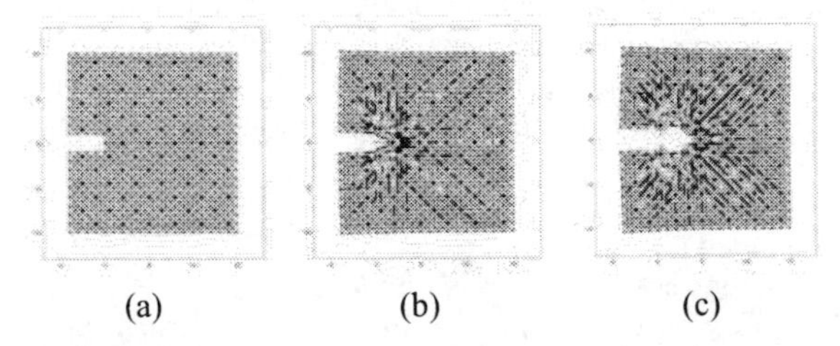

(a) (b) (c)

Figure 2 Plastic flow and damage zone of a polymer nanocomposite with strong reinforcement. (a) Initial conformation (b) 5th iteration (c) 10th iteration.

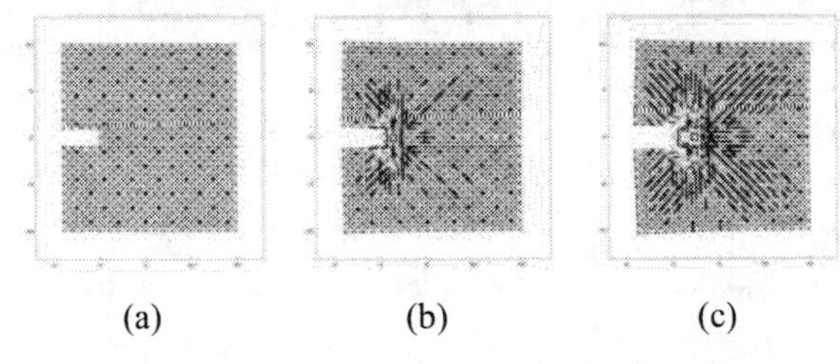

(a) (b) (c)

Figure 3 Plastic flow and damage zone of a polymer nanocomposite with weak reinforcement. (a) Initial conformation (b) 5th iteration (c) 10th iteration.

In Figure 3, the interface strength is reversely set to be 10 times weaker than that of the polymer matrix itself. Shear band propagation is more intense in this case, but the damage propagation is less than in Figure 2 for the same damage threshold. This suggests that a weaker interface may favor toughening in nanoparticle polymer composites.

3.2 3-D Models

To obtain more conformational changes of the polymer composite rather than those resulting from 2D approximation, 3D elastic network models are also simulated in this study. Using the fact that a polymer matrix is formed by entanglement of numerous numbers of polymer chain segments, a 3D elastic network model is illustrated in Figure 4.

In this simulation, the backbone of each polymer segment 3 μm long is modeled as an ideal chain and represented by 31 pointmasses evenly spaced and strongly linked to each other along the backbone. For each chain, an initial point is randomly chosen inside a cubic box 0.5μm wide, and then the next point is also randomly picked up from the surface of a sphere. The center of this sphere is located at the previous point and its radius is set to be 0.1 μm. This process is continued until we get a perfect random chain model which consists of 31 points and each point lies inside the cubic box used. In this context, 100 different random chains are created in the same manner. This sampling number is entirely subjected to the limitation of data storage of the personal computer used.

The density of polymer samples tends to decrease as it moves away from the center of the box. To restrict our attention to only the dense region enough to represent a randomly entangled polymer matrix, the box size is reduced by 0.3x0.3x0.3 μm^3 and the side of the resized box is cut off for a crack tip topology. Each direction is bounded by the latticed wall as illustrated in Figure 4.

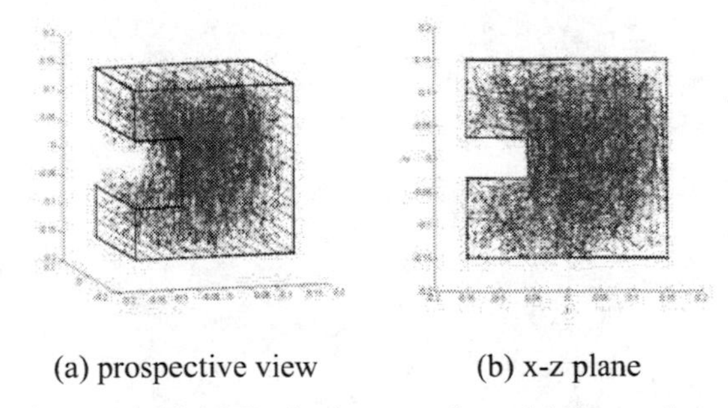

(a) prospective view (b) x-z plane

Figure 4 Initial 3-D elastic network model for a polymer matrix. Each polymer chain is represented by sampled points linked with line segments.

The simulation results of plastic zone of 3-D elastic network model are shown in Figure 5 as increasing iteration this model. Each line represents a plastic zone of elastic network model, and in order to provide a nicer view, the polymer chains (shown in Figure 4) and secondary bonds are not shown in this figure. The plastic flow starts from the crack-tip (5th), and it grows as crack opening displacement increases (10th). Figure 6 shows the damage (fracture) zone of 3-D elastic network model in the 5th and 10th iterations. The damage zone, represented by dark circles, also starts from the crack-tip and propagate asymmetrically as iteration progresses. Since the polymer chains are randomly generated, the local density of polymer matrix varies. It causes the plastic and damage zones not to increase in uniform way, but the pattern of growth in both plastic and damage zone looks similar to that of 2-D cases.

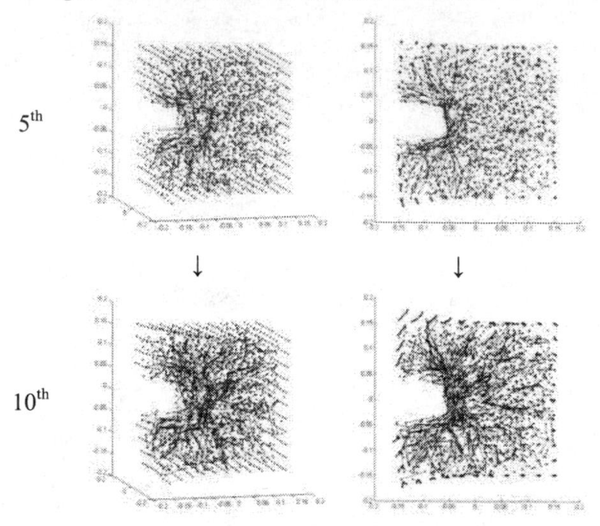

(a) Prospective view (b) x-z plane
Figure 5 Plastic zone of 3-D elastic network model for a polymer matrix in the 5th and the 10th iterations.

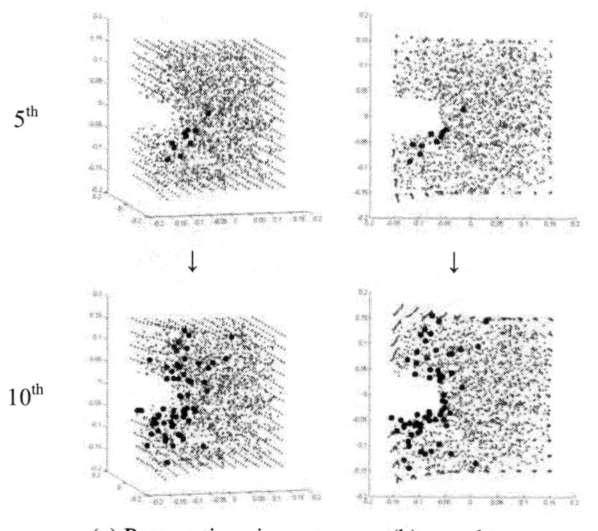

(a) Prospective view (b) x-z plane
Figure 6 Damage zone of 3-D elastic network model for a polymer matrix in the 5th and the 10th iterations.

4 CONCLUSION

We have conducted a preliminary computational study of the physics of deformation and fracture at crack tips of polymers and polymer nanocomposites, helping to define the fundamental questions involved.

Compared to the pure polymer (Figure 1), nanoscale reinforcement model (Figure 2) increases the stiffness of composite. However, when the polymer-particle interface is strong, the toughness decreases with failure occurring at the locations of plastic instability as would be expected from the various experimental observations. In the weaker interface case (Figure 3), the damage propagation is less for the same damage threshold, suggesting a toughness increse. In addition, 3-D random chain models are addressed for more realistic simulation of nanocomposite behaviors.

Iterative NMA is orders of magnitude more efficient computationally than MD while still providing the resolution required to model the interactions between polymer chains and nanoparticles. This computational and theoretical approach will not only elucidate the molecular mechanisms of deformation in polymer nanocomposites, but also suggest the strategy to improve both stiffness and toughness. The movie clips generated in this study are available at the following web site:

http://biomechanics.ecs.umass.edu/composites.html.

REFERENCES

[1] Nair, S. V., Goettler, L. A., and Lysek, B. A., "Toughness of nanoscale and multiscale polyamide-6,6 composites," Polym. Eng. Sci., 42, 1872-1882, 2002.

[2] Gersappe, D., "Molecular mechanisms of failure in polymer nanocomposites," Phys. Rev. Lett., 89, 2002.

[3] Brooks, B. and Karplus, M., "Harmonic dynamics of proteins: Normal modes and fluctuations in bovine pancreatic typsin inhibitor," Proc. Natl. Acad. Sci. USA, 80, 6571-6575, 1983.

[4] Tirion, M. M., "Large amplitude elastic motions in proteins from a single-parameter atomic analysis," Phys. Rev. Lett., 77, 1905-1908, 1996.

[5] Atilgan, A. R., Durell, S. R., Gernigan, R. L., Demirel, M. C., Keskin, O., and Bahar, I., "Anisotropy of fluctuation dynamics of proteins with an elastic network model," Biophys. J., 80, 505-515, 2001.

[6] Kim, M. K., Jernigan, R. L., and Chirikjian, G. S., "Efficient generation of feasible pathways for protein conformational transitions," Biophys. J., 83, 1620-1630, 2002.

[7] Kim, M. K., Chirikjian, G. S., and Jernigan, R. L., "Elastic model of conformational transitions in macromolecules," J. Mol. Graph. Model., 21, 151-160, 2002.

[8] Chirikjian, G. S., "A methodology for determining mechanical properties of macromolecules from ensemble motion data," Trac-trend. Anal. Chem., 22, 549-553, 2003.

[9] Miyashita, O., Onuchic, J. N., and Wolynes, P. G., "Nonlinear elasticity, proteinquakes, and the energy landscapes of functional transitions in proteins," Proc. Natl. Acad. Sci. USA 100, 12570-12575, 2003.

Investigation of the Sequence-Dependent dsDNA Mechanical Behavior using Clustered Atomistic-Continuum Method

Chang-An Yuan*, Cheng-Nan Han* and Kou-Ning Chiang**

*Research Assistant, Department of Power Mechanical Engineering, National Tsing Hua University, 101, Sec. 2, Kuang-Fu Rd., HsinChu Taiwan 300

**Professor, Department of Power Mechanical Engineering, National Tsing Hua University, 101, Sec. 2, Kuang-Fu Rd., HsinChu Taiwan 300, knchiang@pme.nthu.edu.tw

ABSTRACT

A novel clustered atomistic-continuum method (CACM) based on the transient finite element theory is proposed herein to simulate the dynamic structural transitions of the double strand DNA (dsDNA) under external loading. Moreover, the meso-mechanics of dsDNA molecules is then studied via the CACM model, including the base-stacking interaction between DNA adjacent nucleotide base pairs, the hydrogen bond of complementary base-pairs and electrostatic interactions along DNA backbones. Additionally, the mechanics of the dsDNA unzip is studied by the verified dsDNA CACM model.

Keywords: Single molecule simulation, DNA, Finite element method, Equivalent theory, Meso-mechanics

1 INTRODUCTION

The single molecule manipulation technique has been developed for years to measure the basic physical properties of double-stranded DNA (dsDNA) and to discover the interaction between dsDNA and proteins/enzymes [1]. Moreover, the results of the freely-untwisting dsDNA stretching experiment have indicated that a sharp structural transition occurs under roughly 65pN of tension, and that the classical B-form DNA structure dramatically transits to a S-form DNA structure [2-4]. However, the resolution of the single molecule measurement technique currently available restricts the researchers to completely clarify the mechanical behavior of stretching dsDNA as well as the continuous geometrical deformation of the sugar-phosphate chain during stretching. To overcome the resolution limitation of the single molecule manipulation technique, the molecular biology researcher essentially requires an accurate theoretical model to represent the dsDNA mechanical characteristics under specific external loading and boundary condition. However, a feasible numerical model to describe the dsDNA mechanics is difficult to achieve, because the meso-mechanics of single-molecule dsDNA include both quantum mechanics and continuum mechanics. Benham [5] have derived the analytical wormlike rod chain model (WLRC model), and Marko et al. [6] have improved the accuracy of Benham's WLRC model. These WLRC models could predict the DNA mechanical response under low level stretching. However, the WLRC model could not accurately describe the P-form and S-form DNA under high level stretching force and twisting torque. Zhou et al. [7] have proposed the unique Zhou, Zhang and Ou-Yang model (ZZO model), which considers the bending energy and the base pairs staking energy of dsDNA. The ZZO model could successfully describe the S-type DNA under high level stretching, but it could hardly represent the structural transition from the B-form DNA to the P-form DNA due to its limitation of geometric assumption. Additionally, these theoretical models mentioned above could not provide the dynamic dsDNA structural transition in virtuality. Therefore, a novel clustered atomistic-continuum method based on the transient finite element method with material/geometrical nonlinear properties is applied to illustrate the mechanical behavior of dsDNA under external loading.

The CACM comprises the clustered atomistic method and the atomistic-continuum method. In order to reduce the computational time efficiently, the clustered atomistic method treats the specific clustered atom groups as the clustered elements in the modeling (e.g. the sugar-phosphate backbone of dsDNA). Moreover, the virtual elements describe the interactions between atoms or clustered atomistic groups via the atomistic-continuum method. Therefore, a CACM model of the freely-untwisting dsDNA, which only single strand is fixed at each end, could be then conducted, as indicated in Figure 1 (Fig. 1). Notably, the numerical result could be obtained by the transient finite element method (LS-Dyna ®, version 970).

Although the CACM models have reduced the degree-of-freedoms (DOFs) of the freely-untwisting dsDNA, good agreement is achieved between the numerical simulation and experimental result. Moreover, we would apply the proposed dsDNA CACM model in the analysis of the dsDNA unzipping loading. In the preliminary result of the unzipping dsDNA, the mechanical mechanism of the hydrogen bond failure could be described by the numerical simulation. Furthermore, based on the robust model validated by the experimental results, one would further study the sequence-depended mechanism of the unzipping dsDNA. .

2 THEORY

2.1 Finite Element Theory

The finite element method considers the minimization of the total potential energy, which includes internal energy, bending energy, twisting energy, the contact energy and the external energy [8]. Moreover, the complex geometry of the double helix DNA can be described by discrete finite element with few geometrical limitations.

Using the principle of minimum potential energy, one can generate the equations for a constant-strain finite element. For each specific time ($t = t_i$), the total potential energy is a function of the nodal displacements $X(x,y,z)$ such that $\pi_p = \pi_p(X)$. Here the total potential energy is given by

$$\pi_B\big|_{t=t_n} = \left(P^{int} + P^{kin} - P^{ext}\right)\Big|_{t=t_n} \tag{1}$$

where P^{int}, P^{kin} and P^{ext} represent internal energy, kinematical energy and energy of external loading. The above equation can be rewritten as a finite element integrated form [8]. Therefore, the minimal potential energy with respect to each nodal displacement requires that:

$$\begin{aligned}\frac{\partial \pi_p}{\partial\{d\}}\bigg|_{t=t_i} &= \left(\iiint_V [\rho][N]^T[N]\{\ddot{d}\} + [B]^T[D][B]dV\right)\{d\}\bigg|_{t=t_i} \\ &- \left[\{P\} + \iint_S [N_S]^T[T_S]dS\right]_{t=t_i} = 0\end{aligned} \tag{2}$$

where $\{d\}$ represents the nodal vector, $\{\ddot{d}\}$ represents the nodal acceleration, ρ represents the density. $[B]$ is the strain-displacement matrix, $[D]$ is modulus of elasticity matrix, $[N]$ is the shape function matrix, $\{P\}$ is the external load vector and $[T_S]$ is the traction force matrix. Finally, solving the linear system shown in Eq.(2) at each specific time, one can obtain the {d} and the global nodal vector can be revealed.

2.2 Bonding Energies of the dsDNA

In the dsDNA modeling, two kinds of the bonding energies are considered herein. First, we consider the base-stacking interactions, which originate from the weak van der Waals attraction between the polar groups in the adjacent base pairs. Moreover, the hydrogen bond forces between the adjacent base-pairs are considered.

The base stacking energy is a short range interation, and their total effect could be described by the Lennard-Jones potential from (6-12 potential form [9]). Base-stacking interactions play a significant role in the stabilization of the DNA double helix. By the Crotti-Engesser theorem, one can obtain the L-J potential force versus displacement relationship:

$$\begin{aligned}f_{LJ} &= \frac{12AU_0}{l_0}\left(\frac{h_0 + \Delta l\cos\varphi_0}{h_0 + \Delta l}\right)^7\left[1 - \left(\frac{h_0 + \Delta l\cos\varphi_0}{h_0 + \Delta l}\right)^6\right] \\ &\square\left[\frac{h_0(1-\cos\varphi_0)}{h_0\tan\varphi_0(h_0 + \Delta l\cos\varphi_0)}\right]\end{aligned} \tag{3}$$

where f_{LJ} represents the stacking force, U_0 represents the base stacking intensity and Δl represents the distance between the adjacent base pairs. l_0, h_0 and φ_0 represent the initial specific length, base pair height and folding angle of the dsDNA, respectively [10].

The hydrogen bond (H-bond) force is the interaction between complementary bases. Moreover, the GC base-pair has 3 H-bonds and AT has 2. These bonding energies can transverse all of the bending moments and force, because both the distance (R_i) and the angle (θ_{iHj}) between the donor and the acceptor affect the hydrogen bond energy. The single H-bond energy E of base-pairs could be expressed as [11]:

$$E(R_i,\theta) = \sum_{R_{ij}} AD_0\left[5\left(\frac{R_0}{R_i}\right)^{12} - 6\left(\frac{R_0}{R_i}\right)^{10}\right]\cos^4\theta_{iHj} \tag{4}$$

where D_0 represents the hydrogen bond energies intensity. R_0 and R_i is the initial and recent distances of the H-bond. Moreover, we assume that the distances of H-bonds are the same along the dsDNA, the B-form DNA have the lowest H-bond potential and D-H-A at same line, and the H atom always at the center of H-bond at initial state. Figure 1 (Fig. 1) illustrates the hydrogen bond model we used, including both GC and AT cases.

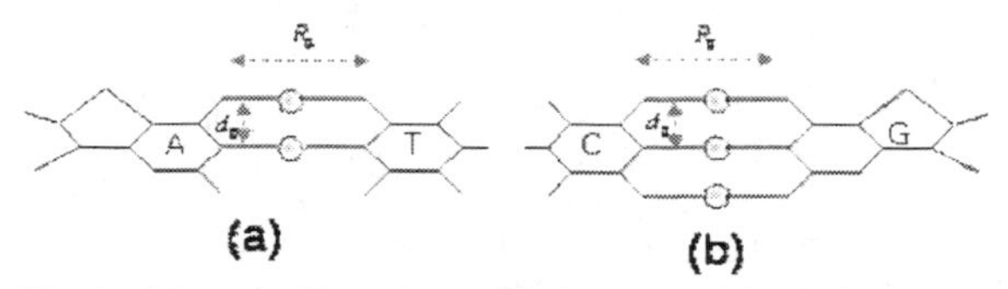

Fig. 1. Schematic illustration of hydrogen model. (a) is the AT hydrogen bond model, and (b) is the GC hydrogen bond model.

Furthermore, we assume that the hydrogen bound model in Fig. 1 would own two degree of freedoms, and the Fig. 2 illustrated the GC-based deformation scheme, and the AT-based deformation behaves similar to the GC one. By the Crotti-Engesser theorem, we could obtain the hydrogen bond reaction axial force by introducing the axial displacement vector Δr:

$$_{Axial}F_{HB}(\Delta r) = \frac{j}{R_0}D_0\left[\left(\frac{R_0}{R_0 + \Delta r}\right)^{11} - \left(\frac{R_0}{R_0 + \Delta r}\right)^{13}\right] \tag{5}$$

where j =180 for GC case and j=120 for AT case. The two direction shearing reaction forces could be expressed by the shearing displacement Δy :

$$\begin{aligned}{}_{s\text{-}shear}f_{GC} &= \frac{\partial}{\partial\Delta y}\left\{E\left(\sqrt{R_0^{\,2} + \Delta y^2} + \frac{d_0\Delta y}{R_0}, \pi - 2\frac{\Delta y}{R_0}\right) + E\left(\sqrt{R_0^{\,2} + \Delta y^2}, \pi - 2\frac{\Delta y}{R_0}\right)\right. \\ &\left. + E\left(\sqrt{R_0^{\,2} + \Delta y^2} - \frac{d_0\Delta y}{R_0}, \pi - 2\frac{\Delta y}{R_0}\right)\right\} \\ {}_{t\text{-}shear}f_{GC} &= 3\frac{\partial}{\partial\Delta y}E\left(\sqrt{R_0^{\,2} + \Delta y^2}, \pi - 2\frac{\Delta y}{R_0}\right)\end{aligned} \tag{6}$$

$$
\begin{aligned}
{}_{s\text{-}shear}f_{AT} &= \frac{\partial}{\partial \Delta y}\left\{E\left(\sqrt{R_0^{\,2}+\Delta y^2}+\frac{d_0\Delta y}{R_0},\pi-2\frac{\Delta y}{R_0}\right)\right. \\
&\qquad \left.+E\left(\sqrt{R_0^{\,2}+\Delta y^2},\pi-2\frac{\Delta y}{R_0}\right)\right\} \\
{}_{t\text{-}shear}f_{AT} &= 2\frac{\partial}{\partial \Delta y}E\left(\sqrt{R_0^{\,2}+\Delta y^2},\pi-2\frac{\Delta y}{R_0}\right)
\end{aligned}
\qquad (7)
$$

The two direction bending moment relationships could be expressed by the shearing displacement θ:

$$
\begin{aligned}
{}_{s\text{-}bending}M_{GC} &= \frac{\partial}{\partial \theta}\left\{E(R_1,2\theta)+E(R_2,2\theta)+E(R_3,2\theta)\right\} \\
{}_{t\text{-}bending}M_{GC} &= 3\frac{\partial}{\partial \theta}E(R_2,2\theta)
\end{aligned}
\qquad (8)
$$

$$
\begin{aligned}
{}_{s\text{-}bending}M_{GC} &= \frac{\partial}{\partial \theta}\left\{E(R_1,2\theta)+E(R_2,2\theta)+E(R_3,2\theta)\right\} \\
{}_{t\text{-}bending}M_{GC} &= 3\frac{\partial}{\partial \theta}E(R_2,2\theta)
\end{aligned}
\qquad (9)
$$

where R_1, R_2 and R_3 equals $\sqrt{(R_0+d_0\theta)^2+R_0^{\,2}\theta^2}$, $\sqrt{R_0^{\,2}+R_0^{\,2}\theta^2}$ and $\sqrt{(R_0-d_0\theta)^2+R_0^{\,2}\theta^2}$. At last, the torque reaction could be expressed as:

$$
\begin{aligned}
T_{GC} &= 2\square 60\square D_0\left(d_{0,GC}^4\right)\frac{R_0^{10}}{\left[R_0^{\,2}+d_{0,GC}^{\;2}\theta^2\right]^7}\theta^3\cos^4 2\frac{d_{0,GC}}{R_0}\theta \\
&\quad +8D_0\frac{R_0^{11}d_{0,GC}+6R_0^9 d_{0,GC}^3\theta^2}{\left[R_0^{\,2}+d_{0,GC}^{\;2}\theta^2\right]^6}\cos^3 2\frac{d_{0,GC}}{R_0}\theta\sin 2\frac{d_{0,GC}}{R_0}\theta
\end{aligned}
\qquad (10)
$$

$$
\begin{aligned}
T_{AT} &= 2\square 60\square D_{0,AT}\left(d_0^4\right)\frac{R_0^{10}}{\left[R_0^{\,2}+d_{0,AT}^{\;2}\theta^2\right]^7}\theta^3\cos^4 2\frac{d_{0,AT}}{R_0}\theta \\
&\quad +8D_0\frac{R_0^{11}d_{0,AT}+6R_0^9 d_{0,AT}^3\theta^2}{\left[R_0^{\,2}+d_{0,AT}^{\;2}\theta^2\right]^6}\cos^3 2\frac{d_{0,AT}}{R_0}\theta\sin 2\frac{d_{0,AT}}{R_0}\theta
\end{aligned}
\qquad (11)
$$

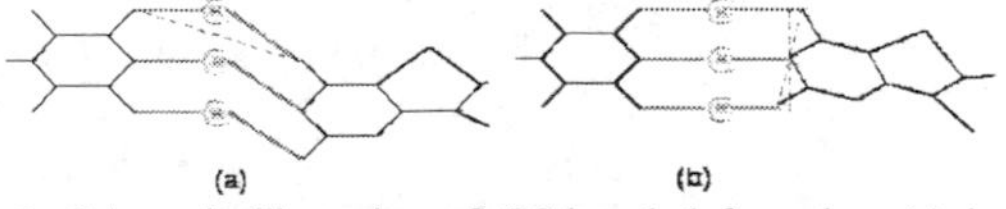

Fig. 2. Schematic illustration of GC-based deformation. (a) is the transverse type deformation and (b) is the rotation type deformation.

2.3 dsDNA CACM numerical modeling

In this section, we will construct the CACM model to simulate the dsDNA mechanical behavior. Due to that the classic B-DNA is stable in physiological aqueous solution, its geometrical structure has been chosen as the initial state of the said model, and the Writhe number of proposed dsDNA model has been assumed as zero. Moreover, both the major/minor groove and the sequence of the dsDNA were neglected for the sake of simplifying the proposed CACM model. Moreover, the CACM comprises both the clustered atomistic and atomistic-continuum methods.

The numerical modeling of the dsDNA would be divided into two parts. One aspect is the description of the mechanics of the double-helix structure of the nucleotide chain (backbones and base-pairs), where the atoms are bound by the covalent bonds. These atomic groups are treated as the individual clustered elements by the clustered atomistic mechanics method. However, before constructing the dsDNA numerical model, we must acquire the clustered atomistic properties of the double-helix sugar-phosphate backbone. Therefore, a feasible ssDNA transient finite element model was constructed, based on the continuum beam type elements.

The other aspect lies in the simulation of the binding energies which provide the stability of the dsDNA, such as the van der Waals forces (Eq. 3) between the adjacent base pairs and the hydrogen bonds between the complementary base pairs (Eq. 4-Eq. 11). Therefore, the virtual elements would be established between the clustered elements to represent the binding energies, where the physical properties of the virtual elements could be transferred by the atomistic-continuum method. Fig. 3 indicates concepts of the CACM modeling.

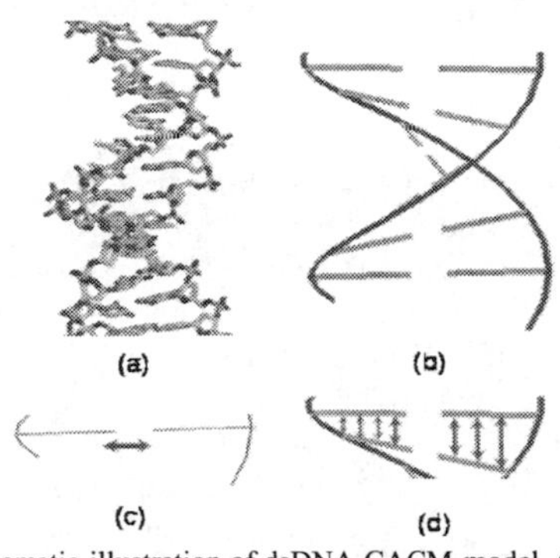

Fig. 3. Schematic illustration of dsDNA CACM model. (a) represents the chemical structure of dsDNA. (b) is the CACM model, including the backbone (red) and base-pairs (blue). The bonding energies are considered in (3) and (4), which are the Hydrogen bond energy and the stacking energies.

This model comprises 147 base pairs, with the initial length of the dsDNA equaling approximately 50nm. In the freely-untwisting dsDNA simulation, one end of the backbone was mechanically fixed. Besides, an external prescribed displacement was applied at the other end, which the prescribed displacment was strictly proportional to the simulation time.

3 SIMULATION RESULTS

The dsDNA CACM model consisted of 3,095 clustered elements and 584 virtual elements. The transient finite element model was solved by LS-DYNA3D® with a CPU time of 20,652 seconds on an IBM® SP2 SMP distributed computer.

The simulation results of the reacted forces, sensed by the bottom fixed point versus the external applied displacement, are shown in FIG. 4, where it shows that we achieved good agreement between the numerical and the experimental results. Moreover, the finite element simulation results of stretching the freely-untwisting dsDNA revealed a continuously structural transition, which may be characterized as three main stages in the stretching process [12].

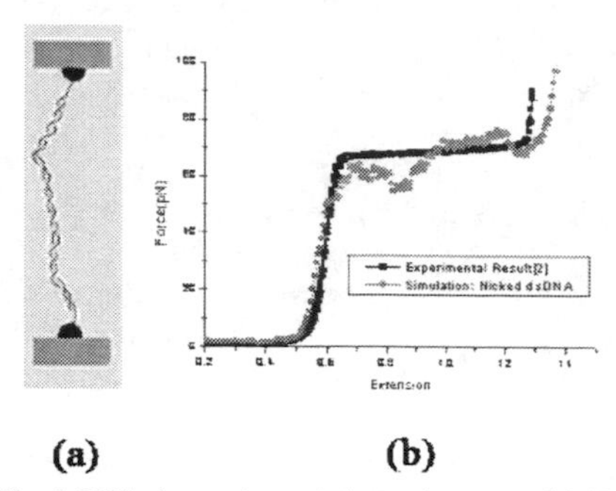

Fig. 4. The dsDNA clustered atomistic-continuum model (CACM) and simulation result. (a) is the schematic illustration of the nicked dsDNA boundary conditions. (b) is the simulation result based on the CACM, where the plateau is occurred at about 65 pN at experimental[2] and simulation results (on average).

Based on the validated dsDNA CACM model, the unzipping loading is applied on the model to simulate the dsDNA unzip mechanics. One strand of the dsDNA end is fixed and the other strand is applied to a prescribed motion. The simulation result is shown in Fig. 5. The reaction force plot in Fig. 6 indicated that the saw-tooth pattern of the hydrogen bond opening, and the a plateau would occur due to the rotation of the backbone, where no hydrogen bond breaks.

4 CONCLUSION

In this paper, a novel clustered atomistic-continuum method, based on a transient finite element method was proposed to simulate the dsDNA mechanics. To completely consider the hydrogen bond effects between base pairs, the detail theoretical derivation would be emphasized. Good agreement was achieved between the numerical simulation and experimental results in both freely-untwisting dsDNA. Furthermore, based on this robust model, the mechanics of the dsDNA unzipping could be studied.

ACKNOWLEDGEMENT

We are grateful to Dr. J. Marko, Dr. S. B. Smith and Dr. C. Bustemante. Also, we thank Dr. A. Sarkar for valuable discussions on single molucular experiment. C. A. thanks Dr. J. Day and Dr. C. Tsay for discussions on numerical simulation technique.

REFERENCES

[1] C. Bustamante, Z. Bryant and S. B. Smith, "Ten years of tension: single-molucule DNA mechanics," Nature, 421, 423-427, 2003.

[2] J. F. Leger, G. Romans, A. Sarkar, J. Robert, L. Bourdieu, D. Chatenay and J. F. Marko, "Structural transitions of a twisted and stretched DNA module," Phys. Rev. Letters, 83, 1066-1069, 1999.

[3] A. Sarkar, J. F. Leger, D. Charenay and J. F. Marko, "Structural transitions in DNA driven by external force and torque", Phys. Rev. E., 63, 51903, 2001

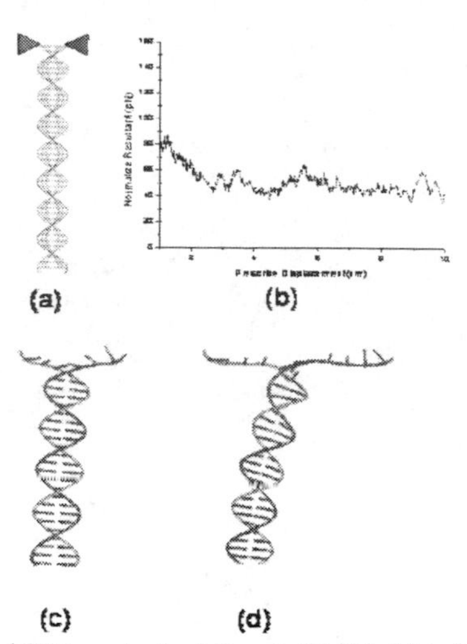

Fig. 5. The dsDNA unzip simulation via CACM. (a) is the schematic illustration of the unzip dsDNA boundary conditions, where red triangle means the fixed end and the blue one is applied a prescribed motion (b) is the reaction forces sensed by the fixed point. (c) and (d) is the continuous deformation of unzipping dsDNA.

[4] C. G. Baumann, V. A. Bloomfield, S. B. Smith, C. Bustamante and M. D. Wang, "Stretching of single collapsed DNA molecules", Biophys. J., 78, 1965-1978, 2000.

[5] C. Benham, "Onset of writhing in circular elastic polymers," Phys. Rev. A, 39, 2582-2586, 1989.

[6] J. F. Marko and E. D. Siggia, "Statistical mechanics of supercoiled DNA," Phys. Rev. E, 52, 2912-2938, 1995.

[7] H. Zhou, Y. Zhang and Z. Ou-Yang, "Elastic property of single double stranded DNA molecules: theoretical study and comparison with experiments," Phys. Rev. E, 62, 1045-1058, 2000.

[8] R. D. Cook, D. S. Malkus and M. E. Plesha, "Concepts and Application of Finite Element Analysis", Wiley, 367-428, 1989.

[9] W. Saenger, "Principles of Nucleic Acid Structure," Springer-Verlag, 1984

[10] C. A. Yuan and K. N. Chiang, "Investigation of dsDNA stretching meso-mechanics using finite Element Method", 2004 Nanotechnology Conference, March 7-11, 2004, Boston, Massachusetts, U.S.A.

[11]G. A. Jeffery, W. Saneger, "Hydrogen Bonding in Biological Structures", 1st ed., Springer-Verlag, Germay, 1994

[12] C. A. Yuan, C. N. Han and K. N. Chiang, "Atomistic to Continuum Mechanical Investigation of ssDNA and dsDNA using Transient Finite Element Method," Inter-Pacific Workshop on Nanoscience and Nanotechnology, Nov. 22-Nov. 24, 2004, City University of Hong Kong, Hong Kong SAR

Nano-scale Material by Design

D. Bleau*, E. Eisenbraun** and J. Raynolds**

*Benet Laboratories, Watervliet Arsenal
**College of Nanoscale Science and Engineering, University at Albany, State University of New York, Albany, NY 12203

ABSTRACT

This research is aimed at developing a fundamental understanding of how atomic scale processing and nano-scale film performance characteristics can be predictably correlated by the use of kinetic Monte Carlo and phase field simulations of PACVD processes. Our modeling builds on previous approaches in the use of first-principles density functional calculations to provide important thermodynamic and kinetic information to Monte Carlo and phase field simulations. By combining these approaches we will be utilizing a multi-scale modeling paradigm to bridge the length and time scales from the nanometer/nanosecond regime up to the macro (reactor-scale) realm. This work ultimately has direct applications toward cross-disciplinary functional capabilities in areas including propulsion, energy conservation, medical components, and nanoelectronics.

Keywords: CVD modeling, phase field, density functional theory, SiCN

1 INTRODUCTION

This report presents an overview of a joint theoretical/experimental collaboration aimed at the development of high performance thin-film coatings for a wide variety of protective coating applications. The desired properties of such coatings include high hardness, high thermal conductivity, high melting temperature as well as good mechanical stability. We have chosen to focus our efforts on the development of SiCN thin-films grown using plasma-assisted chemical vapor deposition (PACVD) processing. We seek to develop a detailed fundamental understanding of the connection between processing parameters (composition, temperature, etc) and the properties of the resulting thin film. The first part of this report is an overview of efforts aimed at the development of a predictive, fundamental theoretical understanding followed by a discussion of our experimental approach and recent progress.

2 THEORETICAL APPROACH

2.1 Modeling the growing surface

Two promising approaches have been introduced in recent years to model the detailed interplay between thermodynamic and kinetic influences occurring in thin-film growth: kinetic Monte Carlo (KMC) [1] and Phase-Field techniques [2].

2.2 Kinetic Monte Carlo vs. molecular dynamics

The KMC approach is an atomistic approach in which all relevant local chemical reactions at the growing surfaces are simulated to determine the composition and morphology of the growing surface vs. processing parameters. This approach has recently been successfully applied to the problem of CVD growth of diamond thin films [3].

Another approach that attempts to accurately treat the detailed chemical interactions at the growing surface employs the use of molecular dynamics (MD) simulations. In a molecular dynamics simulation, Newton's equations of motion are solved at each time step for an ensemble of atoms characterizing the growing surface. Unfortunately, however, in this approach one is drastically limited to very short time scales for the overall simulation. This is because of the fact that in a molecular dynamics simulation, the time step must be kept very small in order to correctly integrate the equations of motion governing very fast reactions. In the problem that we face (CVD deposition of thin films) there are chemical reactions occurring on varying time scales from very short to very long (reaction rates differ by orders of magnitude). The MD approach is therefore unsuitable for our purposes.

The KMC approach has the advantage that chemical reactions occurring at all time scales are correctly incorporated. This happens because, unlike in a MD simulation, the time step in a KMC calculation is not limited by the fastest occurring reaction. At each time step, a random number is generated, and based on this number, a reaction is chosen according to it probability of occurring during that time step. This probability is proportional to the reaction frequency. Thus over many times steps, fast reactions will have dominated the growth process (as is expected in the actual process) however slow reactions also take place.

One key drawback of the KMC approach, however, is the need to characterize all possible chemical reactions that can occur at the growing interface. In the case of diamond growth, the surface chemistry was well characterized. The input to these simulations was a list of dozens of possible reactions and corresponding kinetic data (free energy differences and rate constants for each reaction) [3]. For the case of SiCN, unfortunately, the details of the surface chemistry considerably less well known. Therefore, one of the goals of the present work, is the study of the detailed chemical reactions occurring at the surface using quantum chemical techniques as will be described in a following subsection.

2.3 Phase-field modeling

In contrast to KMC, phase-field modeling is a continuum approach that describes the morphology and composition profiles of the growing surface in terms of continuous fields. In this approach, one begins with a free-energy functional that, in the simplest case, depends on two fields, the composition u and the phase field ϕ. For a two-component system, for example solid and liquid, the phase field satisfies $-1 \leq \phi \leq 1$, taking on one value (e.g. -1 in the solid) in the one phase and the other (e.g 1 in the liquid). The interface between the two phases corresponds to localized regions in which the phases field varies between these two values.

As a simple example, we consider the generic free energy functional for a two-component system, which is written [4] as:

$$F[\phi,u] = F_1[u] + F_2[\phi] + F_{\text{int}}[\phi,u], \qquad (1)$$

where, the first and second terms depend on the respective fields u and ϕ separately, and the third term accounts or the mutual interaction of these fields. The third term would naturally account for the possibility that the phase field could depend on the composition (i.e. composition dependence of phase transitions).

Equations of motion for the composition and phase fields, u and ϕ, are obtained through minimization of the free-energy functional given in Eq. 1. These are:

$$\frac{\partial u}{\partial t} = \nabla \bullet \left[M(\phi,u) \nabla \frac{\delta F}{\delta u} \right], \qquad (2)$$

and,

$$\frac{\partial \phi}{\partial t} = -\gamma(\phi,u) \frac{\delta F}{\delta \phi} = -\gamma(\phi,u) \left[\frac{\delta F_2}{\delta \phi} + \frac{\delta F_{\text{int}}}{\delta \phi} \right]. \qquad (3)$$

Equation (2) is of the Cahn-Hilliard form for the conserved composition field u, while Eq. (3), for the non-conserved phase field ϕ is of the Allen-Cahn form [4]. The mobility coefficients M and γ, in general, depend on the phase field and the composition. The analytic forms for M and γ depend on the specific form chosen for the free energy functional $F[\phi,u]$ as discussed in Ref. [4].

Notable recent applications of phase-field techniques to surface growth include: (1) a study of spiral surface growth (spiral steps nucleating at a dislocation) [5] and, (2) the development of free-energy functionals giving rise to atomic scale density modulations including effects of strain energies associated with grain boundaries [6].

2.4 Solution of the phase field equations: finite elements

We are developing a phase field approach to CVD growth of SiCN thin films. The equations analogous to Eqs. (2) and (3) will be solved for two and three dimensions using finite-element techniques embodied in the commercial software package FEMLAB. The results of such calculations will be compared with experimental results to assess the accuracy of the model parameters. Through such comparisons and further input from quantum chemical calculations, the model will be refined.

2.5 Fundamental input from quantum calculations: density functional theory

An important component of this research is the use of first-principles calculations, based on the Density Functional Theory (DFT) to provide input for KMC and phase field models. As discussed previously, KMC requires kinetic data regarding chemical reactions occurring at the growing surface. In particular we need (free-) energy differences as well as reaction barriers. Such information is being determined from density functional calculations. In particular we employ the Generalized Gradient Approximation (GGA), which is known to improve the description of reaction barriers.

Two different methods are being employed for DFT calculations: (1) VASP [7] and (2) Abinit [8]. Both computer codes are implementations of DFT within a plane-wave/pseudopotential basis and they have complementary capabilities. VASP for example, is extremely computationally efficient but has relatively limited capabilities as compared to Abinit. Abinit, on the other hand, has an extraordinary array of properties that can be calculated at the price of being far less efficient.

In addition to providing chemical reaction data for KMC, first-principles calculations are being employed to determine free-energy functionals for use in the phase-field method. The current state of the art allows the calculation of compositional phase diagrams based solely on input from first-principles calculations [8]. Composition and thermal free energies can be calculated as well as strain-dependent free energies. Other important information includes: (1) lattice and elastic constants, (2) thermal expansion coefficients, (3) interfacial free energies, etc.

As an example, we consider the calculation of thermal free energies for a specified composition. The primary

NSTI-Nanotech 2005, www.nsti.org, ISBN 0-9767985-1-4 Vol. 2, 2005

ingredient for such a calculation is the detailed spectrum of lattice vibrations (phonons). Using self-consistent Density Functional Perturbation Theory (DFPT) it is now possible to calculate the full phonon spectrum of relatively simple (i.e. fairly small unit cells) materials [9]. Such a calculation, obtained from Abinit, is shown in Fig. 1 for Si.

The temperature-dependent thermal free energy is obtained by carrying out a certain integral over the allowed phonon states. Also, by considering strain-induced changes in the phonon spectrum we obtain thermal expansion coefficients. Thermal conductivities are also obtained by considering non-linear interactions of phonons that are also obtainable through the use of DFPT as implemented in Abinit. In addition to the phonon spectrum we also obtain, as a byproduct, the electron-phonon interaction matrix elements. Such information is important to understanding electronic transport properties.

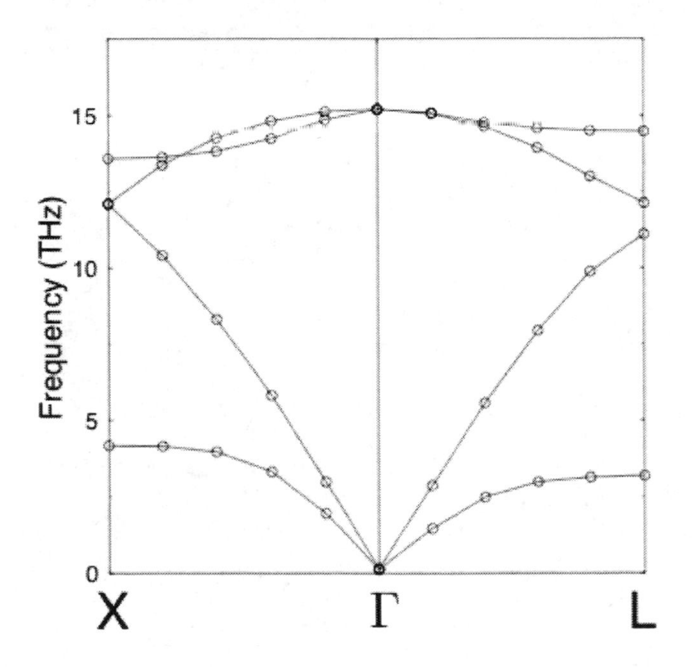

Figure 1: Calculated phonon band structure for Si. This calculation was carried out using the Abinit code and is based on DFPT.

To emphasize the importance of first-principles calculations of free-energies, we note that such calculations are completely parameter free and require no experimental input. The only input is description of the system in terms of the location of a collection of atoms in a representative unit cell along with the corresponding atomic numbers. Using such techniques allows us to make predictions of free energies vs. composition, temperature, and strain as is needed as input for the phase-field simulations.

The downside of using first-principles methods is the extreme demand on computational resources that is typically required. This results from the fact that such methods typically scale as the cube of the number of atoms in the system. Therefore we are forced to make approximations for large systems based on results obtained for small systems.

As an example, consider the computation of thermal expansion coefficients. For an ordered, periodic solid such as cubic SiC we can obtain accurate results from calculations of the phonon band structure (similar to that shown in Fig. 1). However, the material we wish to consider, SiCN is amorphous and requires a large unit cell for its proper description. For such a material, a complete calculation of the phonon band structure from DFPT would be prohibitively expensive. We can, however, obtain a reasonable approximation based on the non-linearity of the energy vs. volume relation.

The energy vs. volume relation for cubic SiC, based on a VASP calculation is illustrated in Fig. 2. From this curve we obtain the lattice constant, bulk modulus, and thermal expansion coefficient. This simple technique will be employed for large unit cells required to correctly describe amorphous structures.

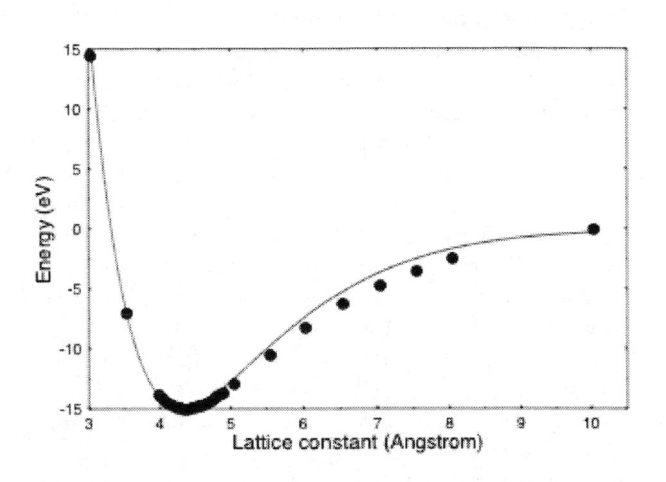

Figure 2: Plot of total energy vs. lattice constant for cubic SiC (calculated with VASP). Such curves will be calculated for large unit cells (describing amorphous structures) and used to obtain the lattice constant, bulk modulus, and thermal expansion coefficients.

2.6 Summary of theoretical approaches

We are pursuing a multi-scale modeling approach combining the use of Density Functional Theory, Kinetic Monte Carlo and phase field techniques to describe the CVD growth of SiCN thin films. The two modeling approaches (KMC) and phase-field methods yield similar information: the composition profile and morphology of the growing surface vs. processing parameters. The KMC approach, however, requires detailed information regarding plausible chemical reactions and their corresponding kinetic data. The resulting surface profile gives atomistic-level resolution.

The phase-field method is a continuum approach based on a free-energy functional. The phase-field model approach has the advantage that it is more easily implemented and solved using finite-element techniques. It also requires less information as input to the construction of the free-energy functional.

Calculations based on DFT are being carried out to provide input to both modeling approaches.

3 EXPERIMENTAL APPROACH AND PRELIMINARY RESULTS

In order to develop a complete solution for material development by design, a plasma-assisted chemical vapor deposition (PACVD) process was developed to enable the growth of SiCN nanoscale coatings with modulated composition. In particular, this process was developed so as to allow controllable, repeatable Si:C:N ratios via simple, reversible changes to the processing conditions. This work was carried out using a modified Varian MB^2 200-mm wafer cluster tool. The PACVD process employs 2,4,6-trimethyl-2,4,6-trisila-heptane as the Si and C source, and a mixture of H_2 and N_2 as co-reactants in a 13.56MHz plasma environment, to both bind up residual precursor fragments and to provide a source of nitrogen to the system. Precursor flow was fixed at 50sccm employing an MKS Instruments type 1153 vapor source mass flow controller and H_2/N_2 flows were varied between 50 and 1000 sccm employing MKS type 1179 mass flow controllers. The wafer temperature was 300°C, and the plasma power was held at 100W. Process pressure was 1 torr.

It was found that film composition could be easily modulated by varying the process parameters as described above. Figure 3 plots the composition as a function of H_2/N_2 flow ratio in the PACVD process as measure by Auger electron spectroscopy (AES). As can be seen, as the relative partial pressure of N in the process increases, the concentration of N in the films increases, at the expense of Si and C concentration, which generally decrease over the range investigated.

It can also be observed that the oxygen incorporation in the films varies over the processing regime investigated. Specifically, the amount of O in the films increases at both low and high H/N ratios. It is suggested that this is due to a lower specific film density or higher porosity at these extreme conditions, thus allowing O from ambient exposure to penetrate into the film matrix (since there is no O in the Si-C precursor material). It is expected that the degree of O incorporation will be related to the mechanical properties of the SiCN film.

These samples are currently undergoing mechanical performance testing, and the results will be compared with those predicted by our modeling and simulation approaches. Based on these results, a model will be developed to capture the relationship between processing conditions, film composition, and subsequent mechanical properties.

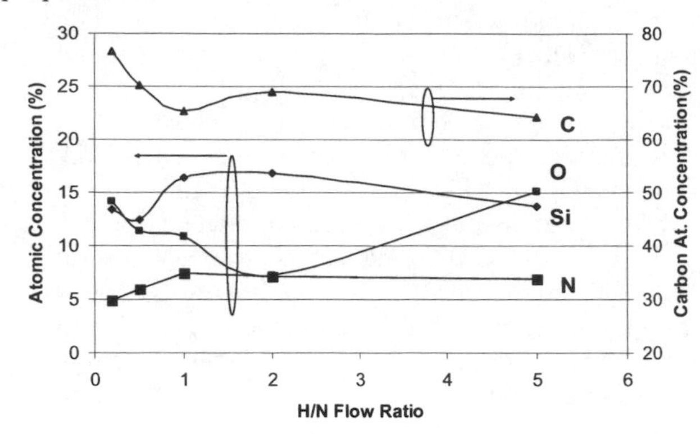

Figure 3: PACVD SiCN composition as a function of gas flow ratio as measured by Auger electron spectroscopy. N concentration can be controlled by altering the N flow. Note that carbon composition is plotted on the right axis for clarity, and the trend lines are included as a guide.

REFERENCES

[1] C. C. Battaile, and D. J. Srolovitz, Annu. Rev. Mater. Res. **32**, 297 (2002)

[2] W. Boettinger, J. A. Warren, C. Beckermann, and A. Karma. Ann. Rev. Mater. Res. **32**, 163 (2002).

[3] C. C. Battaile, D. J. Srolovitz, and J. E. Butler, J. Crystal Growth, **194**, 353 (1998).

[4] J. W. Cahn, P. Fife, and O. Penrose, Acta. mater. **45**, 4397 (1997).

[5] A. Karma and M. Plapp, Phys. Rev. Lett. **81**, 4444 (1998).

[6] K. R. Elder, M. Katakowski, M. Haataja, and M. Grant, Phys. Rev. Lett. **88**, 245701 (2002).

[7] G. Kresse, and J. Furthmueller, Phys. Rev. B **54**, 11169 (1996).

[8] *The abinit software project:* www.abinit.org.

[9] S. Baroni, S. de Gironcoli, A. Dal Corso, and P. Giannozzi, Rev. Mod. Phys. **73**, 515 (2001). (see also references within).

Effects of Temperature and Shape on the Tensile Behavior of Platinum Nanowires

S. J. A. Koh*, H. P. Lee, C. Lu and Q. H. Cheng

*Institute of High Performance Computing
1 Science Park Road, #01-01, The Capricorn, Singapore Science Park II, Singapore 117528, Singapore.
kohsj@ihpc.a-star.edu.sg

ABSTRACT

Molecular dynamics (MD) simulation was performed on an fcc solid platinum nanowire in this study. Infinite one-dimensional nanowires of circular and rectilinear cross sectional shapes were subjected to uniaxial tensile strain, and simulated at temperatures of 50K and 300K. The constitutive stress-strain response of the nanowire was investigated, the changes in crystal structure during axial deformation was studied and mechanical properties like first yield stress and strain, rupture strain, ductility and Young's modulus were deduced from the simulation statistics. It was observed that the nanowire with a rectilinear cross section generally exhibited greater ductility, with a correspondingly lower first yield stress and strain. The nanowire with a circular cross section displayed a higher tendency for the formation of a helical substructure during severe necking. These observations were more significant at the higher temperature of 300K. Young's modulus at nanoscale was also investigated.

Keywords: platinum nanowire, shape effect, temperature effect, molecular dynamics simulation, helical substructure.

1 OVERVIEW

Nanoscale research has flourished over the past decade. This was largely due to the overall enhanced mechanical, chemical, optical, electrical and magnetic properties of materials at characteristic sizes smaller than 100 nm [1]. The fabrication, characterization, experimentation and simulation of metallic nanomaterials had found much favor in nanoscale research during recent years. This could be attributed to its exceptionally high electrical conductivity, super-paramagnetic properties, superplasticity, high material strength, enhanced catalytic properties and high degree of biocompatibility at nanoscale.

Platinum was shown to have high resistance to corrosion, high ductility, low electrical resistance, high reactivity and is in relative abundance, as compared to other rare metals [2]. As such, platinum nanomaterials have been commonly used for catalysis, for opto-electronic purposes, for STM tips [3], as a superconductor, and as actuators [4] and very-high frequency (VHF) oscillators [5]. While ultrathin, single-walled platinum nanowires had been fabricated by Oshima *et. al.* [6], the mechanical behavior of such nanowires were scarcely reported [7,8]. Since platinum nanowires were currently used as AFM tips [3], actuators and VHF oscillators, detailed knowledge of its mechanical properties is essential.

This paper presents the MD simulation of a platinum nanowire under uniaxial tensile strain. A solid nanowire with an fcc crystal lattice atomic arrangement was simulated, using a stable crystal configuration which consists of 392 atoms. The crystal deformation behavior, constitutive response and mechanical properties were studied for infinite nanowires of circular and rectilinear cross sections. Thermal effects on the mechanical behavior were also investigated by performing the simulations at temperatures of 50K (the melting point of common gases like N_2 and O_2) and 300K (the standard laboratory temperature).

2 SIMULATION BRIEF

Classical MD simulation was performed on a 392-atom nanowire, with rectilinear and circular cross sections.

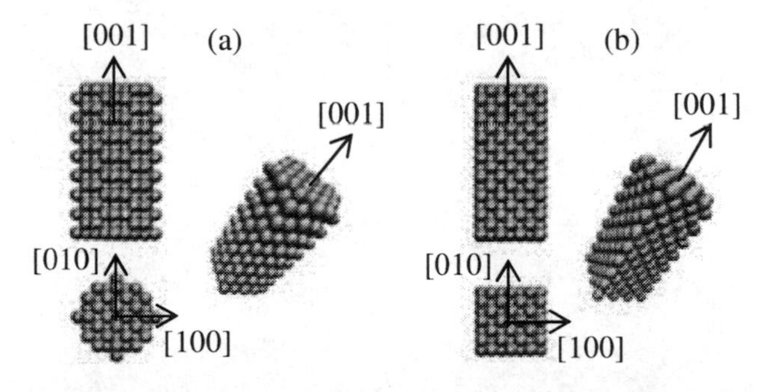

Figure 1: Platinum nanowire at initial stable configuration (a) circular cross section; (b) rectilinear cross section.

The circular cross section was obtained by building layers of atoms in an fcc lattice about the central [001] axis of the nanowire, which was terminated at the 8th nearest neighbor, giving a diameter of about 1.6 nm. The rectilinear section was obtained by repeated stacking of fcc unit cells in the [100] and [010] direction, terminated at the 6th unit cell in each direction, giving a square section of about 1.2 nm x 1.2 nm. 16 atomic layers in the [001] direction were simulated for both sections. Periodic boundary conditions were applied in the [001] direction only, in order to simulate an infinite one-dimensional nanowire. Figure 1 shows the initial stable configuration of the platinum nanowires. The nanowire was stretched at a constant strain rate of 0.04% ps^{-1} in the [001] direction. The Berendsen thermostat [9] was used for temperature control by coupling the simulated system to an external heat bath, with a time constant of $\tau = 2.5$ ps. The simulation time-step was fixed at 1.0 fs (10^{-15} s). The nanowire was initially allowed to stabilize and relax to its equilibrium configuration at the specified temperature over 10^5 time-steps (100 ps). Subsequently, the nanowire was strained at a fixed strain increment in accordance to the specified strain rate. Each strain step was followed by 9999 relaxation time-steps, giving each strain interval as 10 ps. System properties were obtained from the statistics of the final 2000 time-steps. The strain rate was therefore given as:

$$\dot{\varepsilon}_{zz} = \frac{\Delta\varepsilon_{zz}}{S \cdot \Delta t} \tag{1}$$

where $\varepsilon_{zz} = \Delta L_z / L_z$, refers to the nominal strain and $\Delta\varepsilon$ is the strain increment. S refers to the total time-steps during each strain increment, which was fixed at 10^4, and $\Delta t = 1.0$ fs. Hence, the strain increment of $\Delta\varepsilon = 0.4\%$ will give the specified strain rate of 0.04% ps^{-1}.

The Finnis and Sinclair [10] pair functional interatomic potential was commonly used for the simulation of metallic bonding. In this simulation, the interatomic potential parametrized by Sutton and Chen [11] for fcc metals were used for platinum. The potential in terms of interatomic separation r_{ij}, is as follows:

$$E_{SC} = \varepsilon\left[\frac{1}{2}\sum_{i=1}^{N}\sum_{\substack{j=1\\(j\neq i)}}^{N}\phi(r_{ij}) - c\sum_{i=1}^{N}\sqrt{s_i}\right] \tag{2a}$$

and $$\phi(r_{ij}) = \left(\frac{a}{r_{ij}}\right)^n, \qquad s_i = \sum_{\substack{j=1\\(j\neq i)}}^{N}\left(\frac{a}{r_{ij}}\right)^m \tag{2b}$$

The constant a in equation (2b) is fixed to the crystal lattice parameter, and ε, c, m and n are optimized against the equilibrium crystal configuration, cohesive energy per atom, bulk modulus and elastic constants C_{11}, C_{12} and C_{44}. Table I shows the optimized parameters for platinum. It should be noted here that ε in equation (2) refers to the energy constant of the interatomic potential, which must not be confused with ε_{zz} (strain in the [001] direction).

Finally, the axial stress on the nanowire at strain state ε, was given as follows [12]:

$$\sigma_{zz}(\varepsilon) = \frac{1}{N}\sum_{i=1}^{N}\eta_{zz}^{i}(\varepsilon) \tag{3a}$$

where,

$$\eta_{zz}^{i}(\varepsilon) = \frac{1}{\Omega^i}\sum_{\substack{j=1\\(j\neq i)}}^{N}F_z^{ij}(\varepsilon)r_z^{ij}(\varepsilon) \tag{3b}$$

F_z^{ij} refers to the [001] vectoral component of the pair-wise interatomic force between atoms i and j, obtained from differentiating equation (2) with respect to r_{ij}. r_z^{ij} is the interatomic distance in the [001] direction between an atomic pair. Ω^i refers to the volume of atom i, which was assumed as a hard sphere in a closely-packed undeformed crystal structure. From equation (3), the stress-strain response of the nanowire could be obtained from the simulation statistics and mechanical properties could be deduced.

Functional Parameter	Optimized value
a (Å)	3.92
ε (meV)	19.833
c	34.408
m	8
n	10

Table 1: Optimized parameters for platinum.

3 SIMULATION RESULTS

A virtual experiment was conducted to study the deformation characteristics and mechanical properties of the platinum nanowire. In this experiment, each cross section simulated at the specified temperature was assigned 8 different potential cutoff values. This gave 8 samples for each section at each temperature, with a total of 32 simulations conducted for this investigation. The potential cutoffs were varied uniformly between 1.91a and 2.50a, where a is the equilibrium lattice parameter. The former was adopted by Finbow *et. al.* [7] in their simulations of a 360-atom platinum nanowire, and the latter is the conventional value used for Lennard Jones pair potential [13]. Visualization of trajectory files was done using the Visual Molecular Dynamics (VMD) software, which was developed by the Theoretical Physics Group of University of Illinois at Urbana Champaign (USA).

Figures 2 to 5 present the stress-strain response of both circular and rectilinear nanowires simulated at T = 50K and T = 300K. Figure 2 shows the circular nanowire simulated at T = 50K. Figure 3 shows the rectilinear nanowire simulated at T = 50K. Figures 4 and 5 show the simulation at T = 300K for the circular and rectilinear nanowires respectively. The atomic configuration for the solid stress-strain curves (with substructures, which will be mentioned later) at various strain states – represented by markers {1} to {4} on the stress-strain plot – were shown alongside with the plots.

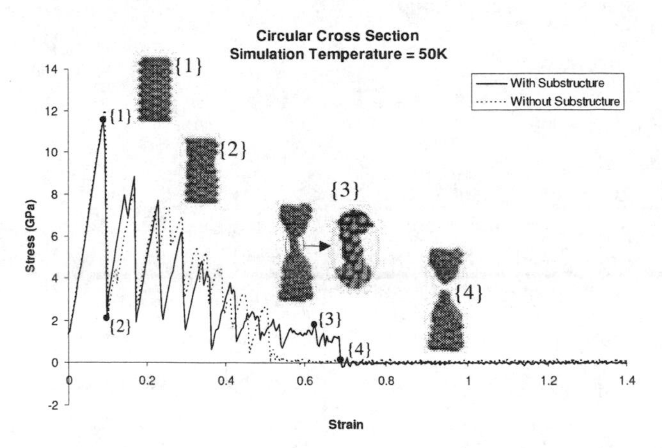

Figure 2: Stress-strain response of circular nanowire at T = 50K.

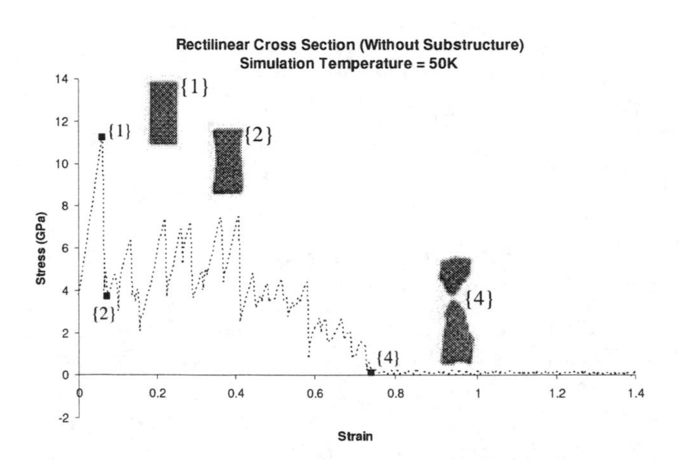

Figure 3: Stress-strain response of rectilinear nanowire at T = 50K.

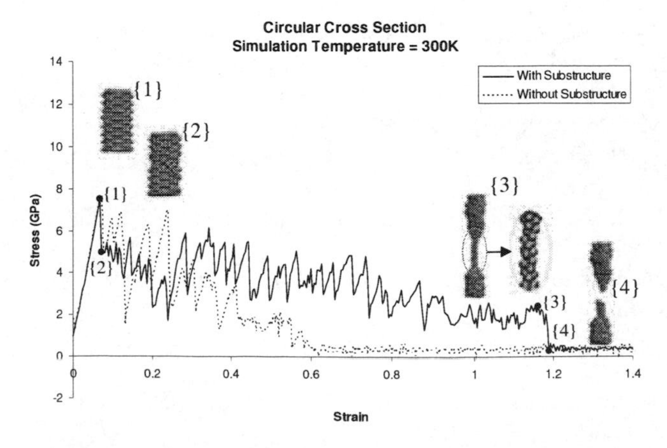

Figure 4: Stress-strain response of circular nanowire at T = 300K.

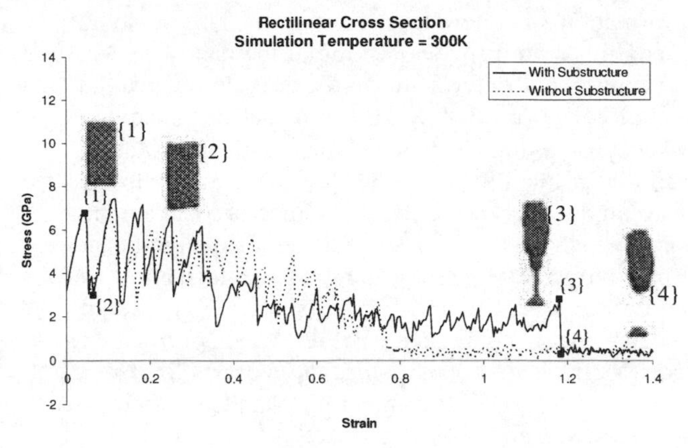

Figure 5: Stress-strain response of rectilinear nanowire at T = 300K.

From comparisons between Figures 2 and 4, and Figures 3 and 5, it was observed that first yield {1} occurred at a lower first yield stress (σ_{fy}) and first yield strain (ε_{fy}) for the nanowire at the higher temperature of T = 300K. At this strain state, there was an abrupt dislocation of the atomic positions from their equilibrium lattice positions, followed by an overall slippage along the (111) plane {2}. Due to the smaller Burger's vector for fcc crystals along the (111) plane, it was energetically more favorable for slippage to occur along this plane. From Figures 2 to 5, it was found that the first yield stress for the nanowire at T = 300K was about 35% to 40% lower than that at T = 50K, while the first yield strain was about 25% smaller. This was due to the higher atomic oscillation amplitude about their equilibrium lattice positions. This results in the destabilization of the crystal structure from its equilibrium configuration and encourages atoms to displace from its original lattice positions, and form slip planes. Furthermore, it was also observed that the rectilinear nanowire displayed marginally lower stress (3% - 10%) and a very much lower strain magnitude (30% - 35%) at first yield. This was a consequence of the shape effect whereby the rectilinear section has a higher degree of crystal order to be preserved as compared to that for the circular section. Due to the low coordination number for atoms located at the corners of the rectilinear section, the interatomic forces for these atoms was much larger than those located at the outermost layer of the circular section. This forces the corner line of atoms to collapse inwards into the internal bulk and therefore, resulting in slippage at a smaller first yield strain. Following the same argument, the large interatomic forces for the corner atoms of the rectilinear section would result in a slightly stiffer nanowire as compared to that of the circular section.

After the first yield, the nanowire undergoes a periodic recrystallization-yielding-relaxation process. This is a unique characteristic of crystalline materials at nanoscale. At this characteristic length scale, local atomic vibrations play a critical role in rearranging the crystal structure to a new equilibrium configuration after every planar slippage, by the formation of a new layer of atoms. This periodic process continues until severe necking of the nanowire sets in, where amorphous deformation at the neck dominated the deformation behavior. It was observed that the nanowire ruptured {4} at a higher rupture strain (ε_{ru}) and hence, displayed higher ductility at T = 300K. This was due to the higher crystal disorder and higher amplitude of local atomic vibrations, which result in the preservation of significant interatomic attractive forces, leading to the formation of new equilibrium configurations even at high strain magnitudes. As a consequence of the higher vibration amplitude, the crystal structure slips and recrystallizes at a faster rate at T = 300K and therefore, smaller period of yielding was observed.

An interesting phenomenon was observed at strain state {3}, where formation of substructures was noted at the neck area. The formation of a double helical substructure was observed for the circular nanowire at both simulation temperatures and a single wire of platinum atoms evolved for the rectilinear nanowire at T = 300K. The presence of stable, single-walled helical platinum nanowires was discovered and fabricated by Oshima *et. al.* [6] at temperatures of 680K, and Finbow *et. al.* [7] reported the formation of a single wire of platinum atoms in their simulation. In our virtual experiment, substructure formation was only observed for cutoff values of 2.10a and 2.20a for the circular nanowire at T = 50K and T = 300K respectively, and 2.50a, for the rectilinear nanowire at T = 300K. None was observed for the rectilinear nanowire at T = 50K. The remaining cutoff values displayed necking and rupture of nanowire without formation of substructures. From the comparison of the stress-strain response of the nanowire with and without substructure formation, it was observed from Figures 2 to 5 that the former contributed significantly to the ductility of the nanowire. It was further observed that, the amount of increase in ductility was dependent on the type of substructure formed and total development length of the substructure before complete rupture occurs. From Figure 2, the approximate development length of the helical substructure was 7 Å and from Figure 4, the approximate development length was 20 Å, and Figure 5 showed the development of a 19 Å single wire of platinum atoms. From Figure 2, the formation of the 7 Å helical substructure resulted in a 36% increase in rupture strain magnitude. This was much smaller than the magnitude of increase observed in Figure 4, where the formation of a 20 Å helical substructure resulted in a 92% larger rupture strain. On the other hand, Figure 5 indicated that the formation of the single wire substructure resulted in a 54% increase in the rupture strain magnitude. This implies that the single wire substructure resulted in a much smaller increase in ductility as compared to the double-wire helical substructure of an approximately similar development length.

Young's modulus (E_{pt}) was obtained from the gradient of the stress-strain response before the first yield occurs. It was could be seen from Table 2 that the average Young's modulus from 8 simulated samples was about 5% higher for the rectilinear nanowire at T = 50K, and no significant difference was observed at T = 300K.

4 CONCLUSION

[a]Units in GPa
[b]Rupture strain with substructure formation

	Circular Section		**Rectilinear Section**	
	***T*=50K**	***T*=300K**	***T*=50K**	***T*=300K**
ε_{fy}	0.088	0.068	0.060	0.044
[a] σ_{fy}	11.53	7.52	11.23	6.77
ε_{ru}	0.512	0.620	0.740	0.772
[b] ε^{s}_{ru}	0.688	1.188	No Data	1.188
[a] E_{pt}	121.67	97.16	128.03	96.44

Table 2: Summary of mechanical properties.

Table 2 shows a summary of the mechanical properties obtained for the platinum nanowire in this investigation. These properties, together with the stress-strain response of the nanowire, could aid in future studies for the mechanical behavior of platinum nanowires.

REFERENCES

[1] J. L. Beeby, *Condensed Systems of Low Dimension, NATO Advanced Study Institute, Series B* **253**: Physics (New York: Plenum Press, 1991).

[2] R. B. Ross, *Metallic Materials Specification Handbook* (London: Chapman & Hall, 1992).

[3] N. Agraït, G. Rubion and S. Vieira, Phys. Rev. Lett. **74**, 3995 (1995).

[4] S. X. Lu and B. Panchapakesan, *IEEE, ICMENS 2004*, edited by IEEE Computer Society (Los Alamitos, California, 2004), p. 36.

[5] A. Husain *et. al.*, Appl. Phys. Lett. **83**, 1240 (2003).

[6] Y. Oshima *et. al.*, Phys. Rev. B **65**, 121401 (2002).

[7] G. M. Finbow, R. M. Lynden-Bell and I. R. McDonald, Mol. Phys. **92**, 705 (1997).

[8] Q. H. Cheng, H. P. Lee and C. Lu, Mol. Sim. (to be published).

[9] H. J. C. Berendsen *et. al.*, J. Chem. Phys. **81**, 3684 (1984).

[10] M. W. Finnis and J. E. Sinclair, Philo. Mag. A. **50**, 45 (1984).

[11] A. P. Sutton and J. Chen, Philo. Mag. Lett. **61**, 139 (1990).

[12] M. F. Horstemeyer, M. I. Baskes and S. J. Plimpton, Acta. Mater. **49**, 4363 (2001).

[13] J. M. Haile, *Molecular Dynamics Simulation, Elementary Methods* (Singapore: John Wiley & Sons, Inc., 1992).

Theory and experiments on periodic lattice distortions that explain 1D conductivity along the CuO_2 plane ab and a-b diagonals in $YBa_2Cu_3O_7$ (YBCO) 50 nm film with a 24 DEG grain boundary (GB)

H.S. Sahibudeen, M.A. Navacerrada and J.V. Acrivos
San José State University, San José CA95192-0101 TEL: *408 924 4987/4972, FAX 408 924 4945*
Email: hizamss@yahoo.com; m_navacerrada@hotmail.com; jacrivos@athens.sjsu.edu

ABSTRACT

Self consistent field (SCF) calculations carried out using MOLECOLE Linux codes for a lamella L in the nano-particle $[T'\text{-}Nd_2CuO_4]_{18}$ explain the d_{xy} symmetry of the periodic lattice distortions (PLD) observed in X-ray diffraction (XRD) along preferred superconductivity direction, ab and a-b diagonals in the material CuO_2 plane.

Keywords: cuprate superconductor, XRD, XAS, SCF

INTRODUCTION

The continuous overlap along the O:2p_{xy} sigma orbital in the highest/lowest occupied molecular orbital (HOMO/LUMO) for the CuO_2 plane of cuprate superconductors showed that continuous 1D electron density, ρ_e appeared only along the ab (a-b) diagonals [1], in agreement with the preferred direction of superconductivity found later [2]. The continuous ρ_e in the material also governs the direction of 1D charge density waves (CDW) causing the PLD, observed in X-ray diffraction (XRD) [2]. This work describes the effects of a PLD by SCF calculations.

EXPERIMENTAL

XRD and X-ray absorption spectra (XAS) in fluorescence (F) were measured at the Stanford Synchrotron Radiation Laboratory (SSRL) and at the Lawrence Berkeley National Laboratory (LBNL-station 6.3.1 of the ALS) on 50 nm c-axis oriented YBCO films deposited on single crystal $SrTiO_3$ and bi-crystals with a 24° grain boundary (GB) at the Complutense University [4] (FIG. 1, 2).

DISCUSSION

The **(HKL)** = (1±12),(114),(221),(2±22),(224) planes (FIG. 1) intersect the PLD, $\mathbf{q} = (q_xq_yq_z)$ to produce XRD sidebands of intensity $A_{\pm1}$ relative to the center band A_0, independent of L and E = 8048 to 7040 eV (FIG. 1). Rocking curves plotted versus sa^2ds/H ($s = 2E\sin(\theta)/hc$ = scattering vector, h = Planck constant, c =velocity of light, θ =Bragg angle, χ,ϕ constant, and dH/H = dK/K = dL/L) obtain $2q_x = 2dH_{XRD,SB} \sim 1/12$ [2]. H, K scans with dL=0 give sidebands at $-dH_{XRD,SB} = dK_{XRD,SB} \Rightarrow q_x = -q_y$ when K=H>> L≠0, which identify **q** as a transverse wave along the ab diagonal in the CuO_2 plane. Thus the distortion at site **R** is caused by the transverse wave **u** $\exp(i(2\pi\mathbf{R}.\mathbf{q}+\phi))$, **u**/a = ($u_x$ $-u_xq_x/q_y$ 0) when a is the unit axis. The relative sideband intensities [2]:

$$A_0/A_{\pm1}=|J_0(z_{H\pm H\,L})/J_{\pm1}(z_{H\pm H\,L})|^2,\quad z_{H\pm H\,L}= 4\pi u_xH$$

where the J_n, Bessel functions of the first order obtain u_x =0.1 ±0.01. The effect of the PLD is simply to produce a twist of the 4 O relative to the center Cu atom (FIG. 3).

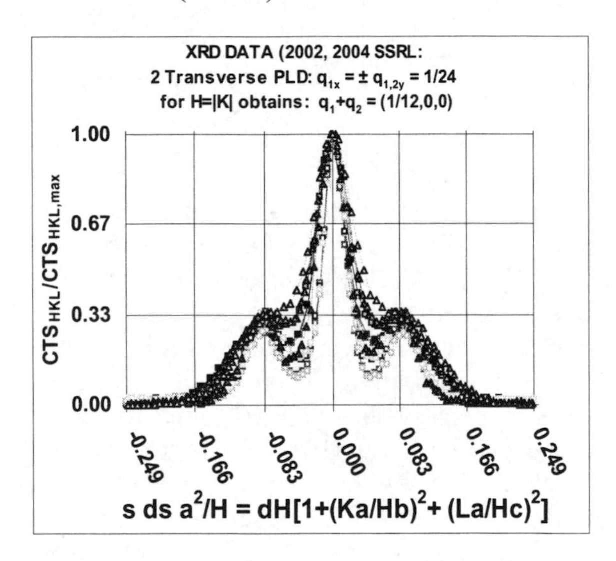

FIG. 1: *Film rocking curves for* ***(H K=±H L≠0).***

The main effect of the PLD is to focus the transport in 1D. A unique uniform 1D electron density along the ab, a-b diagonals of the CuO_2 plane for the Cu_4O_4 lamella HOMO/LUMO (FIG. 4) coincides with the directions of preferred superconductivity [3], normal to the 1D distortion CDW. The PLD are evident in XRD only when planes along the diagonal in reciprocal space (K = ± H) intersect **q** (FIG. 1).

SCF calculations can determine how the total energy and especially how the Mulliken overlap population (M-OP) ∝ ρ_e along a given bond depend on **q** as the CDW moves in the lattice.

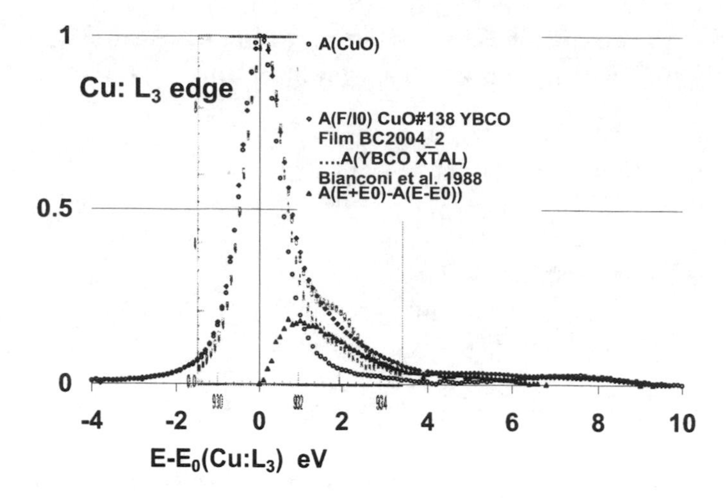

FIG. 2: Cu L_3 edge Absorbance, A for $YBa_2Cu_3O_{7-\delta}$ film compared to CuO reference and single crystal with different oxygen content, ▫: δ=0, o,x: δ>0.15 [ref. 5].

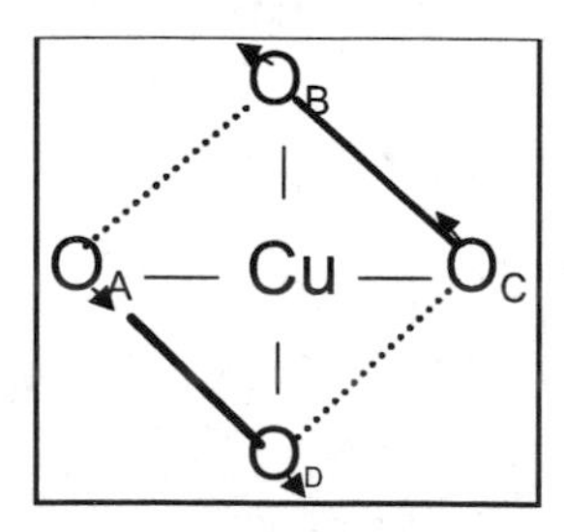

FIG. 3: Relative O atom motion with respect to Cu produced by transverse wave PLD.

The OP depends on the bond distance, but in the presence of a PLD, only O_A-O_D=O_B-O_C=$a/2^{1/2}$ remains constant (FIG. 3). All the other bond distances vary with sinusoidal amplitude less than $u_x\sin(2\pi q_x/2)$ ~0.013. This means that 1D electron conductivity due to MO overlap is favored along the direction of constant bond distance in the presence of a PLD. Since the HOMO are doubly degenerate, with O:$2p_{xy}$-O:$2p_{xy}$ overlap preferred along either ab and/or a-b diagonals (FIG. 4), two PLD normal to the diagonals are present in separate domains, and/or twin crystals. The 1D uniform ρ_e for N atom, n_a,n_b chains (FIG. 4) are described by Tight Binding, Alternant, LCAO-MO:

$$\chi_{O,\,chain\,na,nb}(\mathbf{r},\pi,\pi,k_z) = (2N)^{-1/2}\, i\cos(\pi(n_a+n_b))\,\Sigma_{chain\,O}[\psi_O(\mathbf{r}-\mathbf{R}_{O3a}) - \psi_O(\mathbf{r}-\mathbf{R}_{O3b})],$$

$$\chi_{Cu,\,na,nb}(\mathbf{r},\pi,\pi,0) = N^{-1/2}\cos(\pi(n_a+n_b))\,\Sigma_{chain\,Cu}\,\psi_{Cu}(\mathbf{r}-\mathbf{R}_{Cu}),$$

$$\chi_{donor\,D,na,nb}(\mathbf{r},\pi,\pi,k_z) = N^{-1/2}\cos(\pi(n_a+n_b))\Sigma_{chain\,D}\psi_D(\mathbf{r}-\mathbf{R}_D).$$

ψ_M = M atomic orbital at $\mathbf{R}$, $k_x = k_y = \pi$, k_z=0.

The predictive power of a continuous electron density in real space HOMO, shows the cause of PLD: oscillations in localized energy and charge at individual lamellae that is compensated in the 24 by 24 cell, and by a variation in the Mulliken overlap population of one percent in other directions is sufficient to focus 1D conductivity along a unique diagonal (FIG. 5a-d). Only a unique constant bond distance, e.g., O_A-O_D = O_B-O_C (FIG. 3) can give a constant overlap population: OP(O_A-O_D) = OP(O_B-O_C) = -0.79 that gives rise to continuous overlap conductivity.

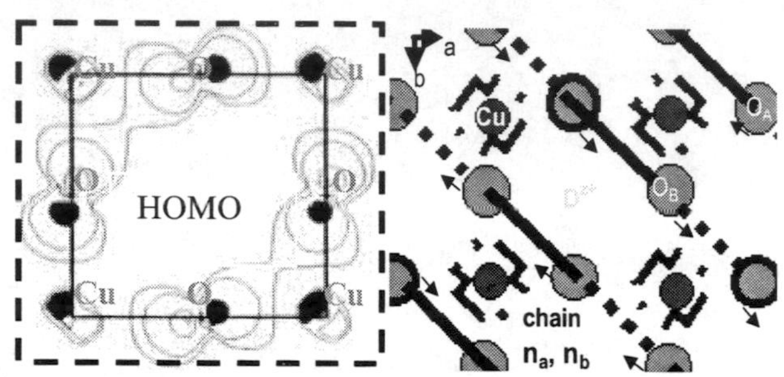

FIG. 4: HOMO electron density in CuO_2 plane, ρ_e>10^{-3}bohr^{-3} along ab and a-b diagonals [ref. 1] indicate the shape of the ball and stick model near the Fermi level: MO<0:... MO>0: -.

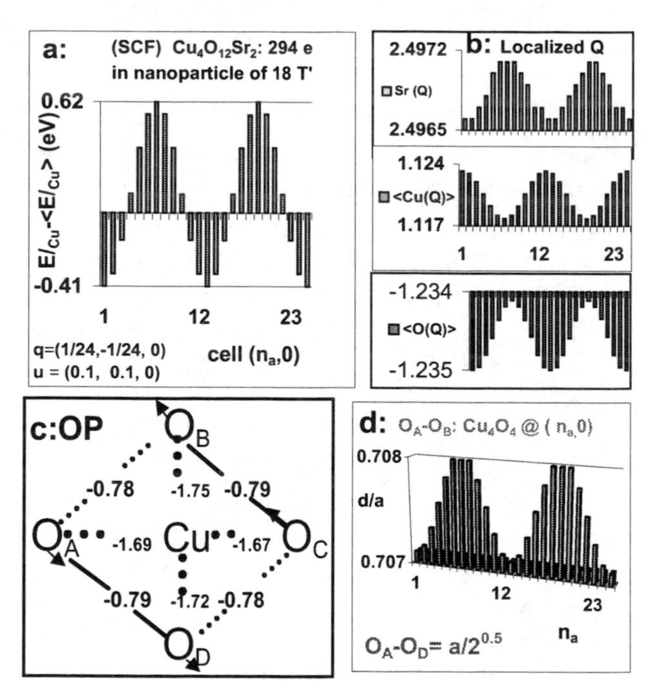

*FIG. 5: SCF calculations as PLD wave moves through origin of $Cu_4O_{12}Sr_2$**[T'Nd_2CuO_4]$_{18}$**: **a**: Total electronic energy deviation from <E> per Cu atom **b:** Localized charge Q at atoms in each of 24 lamella. **c**: Mulliken OP ∝ ρ_e in given bonds. **d**: O_A-O_B bond distance along a-axis [ref. 2].*

Charge transfer to lamella L^{+z}=$Cu_4O_{12}Sr_2^{+z}$ within **[T'-Nd_2CuO_4]$_{18}$** contributes to the transport properties of the material. The ab-initio SCF calculation shows how the stability E_{total}–E_{atoms} and charge, Q(M) at atom M in L changes and as the CDW passes through the center (Table I): Q(O) remains almost constant with charge transfer of $\pm$ 2 e to 14 unit cells, indicating that the $O{:}2p_{xy}$-$O{:}2p_{xy}$ form covalent bonds. Q(Cu) varies slightly more but the Sr atom ionizes according to the amount of charge transfer, indicating that the ionic layers act as a buffer. A neutral lamella L^0 and the positively charged L^{+2} obtain a SCF ground state with Q(Sr) = 2.5. Donors that produce L^{-2} obtain Q(Sr) = 0.6 in the SCF ground state, indicating that the optimum number of electrons in the CuO_2 covalent layer is controlled by the alkaline earth atom which can take on multiple valence. The positively charged L^{+2} is most unstable, with the largest CDW variation. The optimum SCF ground state charge Q(Cu) and Q(O) indicate that the total electronic charge is not localized as in an ionic lattice but distributed in the covalent bonds.

L in $[T'\text{-}Nd_2CuO_4]_{18}$	E-Ea eV	<Q(Sr)>	<Q(Cu)>	<Q(O)>
L^{-2}	-38.7	0.6	1.3	-1.2
L^{-2} CDW % DEV	0.3	0.1	0.3	0.1
L^{+0}	-38.7	2.5	1.1	-1.3
L^{+0} CDW % DEV	0.3	0.01	0.3	0.1
L^{+2}	-37.4	2.5	1.6	-1.3
L^{+2} CDW % DEV	0.3	1	0.5	0.1

***Table I**: SCF ground state energy stability, $E - E_a = (E_{total} - E_{atoms})/18$, average charges <Q(M)> at atoms, M in L^{+z} and percent deviation produced by CDW for different amount of charge transfer $\pm z$.*

The YBCO film Cu L_3 edge white line absorbance, $A = F/I_0$ ($c\wedge\varepsilon_{X\text{-rays}}=\pi/4$) shows a Gaussian shape for $E < E_0$, the edge threshold where it is comparable to A for CuO and YBCO single crystal and powders with a maximum oxygen content [5]. Many electron interactions in bulk metals may be responsible for enhanced absorption due to interactions near the Fermi energy when $E > E_0$ [6] but there is evidence for an X-ray exciton peak, 0.6 eV above E_0 when the absorption for the Gaussian shaped reference, A(CuO) is subtracted from A(F/I_0 YBCO film) (FIG. 2) [7 - 9]. The fully oxygenated samples [5] also show deviation from a Gaussian shape, 0.6 eV above E_0, but, the peak associated by Bianconi et al. with superconductivity in fully oxygenated YBCO powders and single crystals is 2 eV above E_0 and appears to be also present in the film though weaker than the exciton peak. Both CuO and YBCO film show an absorption peak 7 eV above E_0 which can not be due to interactions near the Fermi level because CuO is not a metal.

A valid question is how does a PLD favor two electron Bose pairing for superconductivity? A possible answer is that a change in direction from an ab to a-b chain needs to break an $O{:}2p_{xy}$ bond (FIG. 6), producing free electrons that interact with cation chain states, Cu or D = Ba, Y (FIG. 4).

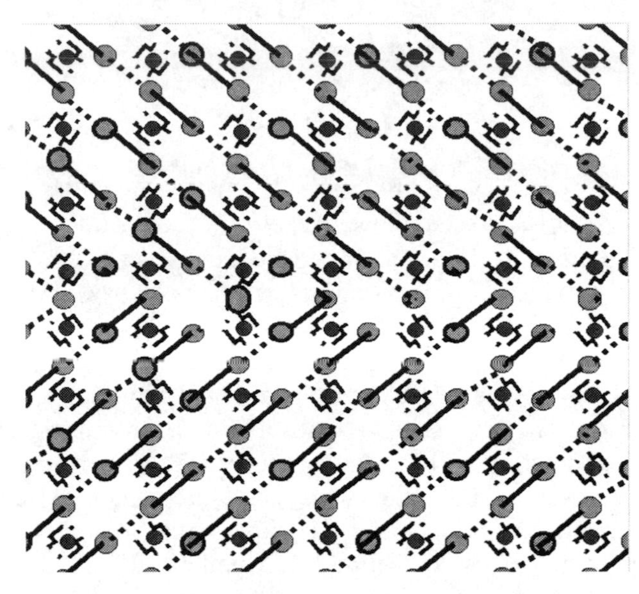

***FIG. 6:** Ball and stick model of PLD formation by breaking $O{:}2p_{xy}$ bonds along a chain. A 24 by 24 PLD arises when one bond in a 24 atom chain is broken, at an energy cost of one 24th of the neighbor resonant $O{:}2p_{xy}$-$O2p_{xy}$ integral [ref. 9].*

An electron pair obeying Bose statistics may be formed by the coupling of angular momenta in different chains, say an $O{:}2p_{3/2}$ chain defect electron with another in say, a $Cu{:}3d_{3/2}$ and/or $Ba{:}5d_{3/2}$ chain (FIG. 6) such that the sum $J = J_1 + J_2 = J_z = 0$. Eigenfunctions of J and J_z but not S_z of even parity for exchange, $\mathbf{P}_{12}$ of electrons 1 and 2 in $e_1e_2^{=}$ obtain product states arising from $d_{3/2}$(Cu or Ba) and $p_{3/2}$(O) chains: |J,0>, J = 3, 1, 0 with parity P= 2π:

$\|J,J_z\rangle$	P=		Matrix Coefficients		$d_{3/2}$ $p_{3/2}$
$\|3,0\rangle$	2π		$40^{-1/2}$[3 3 1 1]		$(I+e^{iP}\mathbf{P}_{12})$ $\|3/2\ 1/2\rangle_{d1} * \|3/2\ -1/2\rangle_{p2}$
$\|2,0\rangle$	π	=	$8^{-1/2}$[-1 -1 1 1]	*	$(I+e^{iP}\mathbf{P}_{12})$ $\|3/2\ -1/2\rangle_{d1} * \|3/2\ 1/2\rangle_{p2}$
$\|1,0\rangle$	2π		$40^{-1/2}$[1 1 -3 -3]		$(I+e^{iP}\mathbf{P}_{12})$ $\|3/2\ 3/2\rangle_{d1} * \|3/2\ -3/2\rangle_{p2}$
$\|0,0\rangle$	2π		$8^{-1/2}$[1 -1 1 -1]		$(I+e^{iP}\mathbf{P}_{12})$ $\|3/2\ -3/2\rangle_{d1} * \|3/2\ 3/2\rangle_{p2}$

where the matrix coefficients are obtained to satisfy the eigenvalues of $J^2 = J(J+1)$, J_z, and the raising and lowering angular momentum textbook operations [10]. **I** is the identity operator and electrons 1 and 2 are identified by the position in the product. When the principal axis of quantization is chosen parallel to a diagonal, $\mathbf{k}= \pm(\pi \pm\pi\ 0)$, the $e_1e_2^{=}$ product state |0,0> of even exchange parity is written:

$$|0,0>_{\text{chain jj couping}} = 5^{-1/2}[\mathbf{I}+ \mathbf{P}_{12}]/N$$
$$\{\Sigma_M\ Y_{1,0\uparrow}\ [R_{O:2p}(\mathbf{r}-\mathbf{R}_{O3a})\ -R_{O:2p}(\mathbf{r}-\mathbf{R}_{O3b})]_1$$
$$*\ \Sigma_M\ [Y_{2,0\downarrow}\ R_{Cu:3d}(\mathbf{r}-\mathbf{R}_{Cu})]_2$$
$$-\ \ \Sigma_M\ Y_{1,0\downarrow}\ [R_{O:2p}(\mathbf{r}-\mathbf{R}_{O3a})\ -R_{O:2p}(\mathbf{r}-\mathbf{R}_{O3b})]_1$$
$$*\Sigma_M\ [Y_{2,0\uparrow}\ R_{Cu:3d}(\mathbf{r}-\mathbf{R}_{Cu})]_2\}.$$

where $Y_{l,m}(\theta_M,\varphi_M)$ are spherical harmonics, the sub indexes $\uparrow\downarrow$ represent the spin $s_z = \pm1/2$ states and $R_{n,l}(r_M)$ is the atomic orbital radial dependence when (r_M,θ_M,φ_M) are the spherical polar coordinates of an electron relative to atom M = 1 to N in chain.

The elegant work of S.C. Zhang et al. [11] helps to unravel the $YBCO_7$ superconductivity-magnetism relation. They have shown that spin polarized transport is non-dissipating and that transport by a Bose condensate in the superconducting state is a sub-group of spin polarized transport. Many questions remain but a really important one is to distinguish whether spin polarized $Cu{:}3d_{3/2}$ states populated over the $3d_{5/2}$ band states interact with $O{:}2p_{3/2}$ chain defects to obtain Bose condensation in superconducting $YBCO_7$ and/or can an uneven spin population in 1D conjugate orbitals involving the $O{:}2p_{3/2}$ chains interacting with Ba and Y $nd_{3/2}$ and $4f_{3/2}$ chains achieve Bose condensation?

CONCLUSION

The experimental study of PLD was inspired by SCF calculations which first indicated the unique direction of continuous 1 D electron density along O:2p sigma orbitals parallel to the CuO_2 plane diagonals in superconducting cuprates. The formation of Bose pairs between free electrons in PLD related oxygen chain defects with cation chains in YBCO is not limited to the CuO_2 plane. Ba/Sr/Y/La atoms can form J_1, J_2 coupled chains as long as even parity product states $J = J_1+J_2 = 0= J_z$ are formed.

ACKNOWLEDGEMENT

Support for this work was given by 2002 and 2004 Dreyfus Senior mentor awards and NSF/DMR grant 9612873 at SJSU, and DOE at SSRL and LBNL-ALS. We are very grateful to Prof. Gina Corongiu and E. Clementi, who developed the MOLECOLE and the original METTECC codes [12] that allow large SCF and in particular, molecular orbital calculations for a nano particle. Professor Yao W. Liang, Cavendish Laboratory, Cambridge University is thanked for discussion and for moral support. We thank Drs. A. Mehta and P. Nachimuthu who gave us very valuable help in the data acquisition at SSRL and LBNL-ALS respectively.

REFERENCES

1. J.V. Acrivos and O. Stradella, *International Journal of Quantum Chemistry*, **46**, 55(1993)
2. M. A. Navacerrada and J.V. Acrivos, *NanoTech 2003*, **1**, 751 (2003) and ref. therein.
3. Z.-X. Shen, W.E. Spicer, D.M. King, D.S. Dessau, B.O.Wells, *Science* **267**, 343 (1995)
4. M. A. Navacerrada, M. L. Lucía and F. Sánchez-Quesada, *Europhys. Lett* **54**, 387, 2001 and references therein.
5. A. Bianconi et al, *Phys. Rev.* **B38,** 7196 (1988)
6. M. Hentschel, D. Ullmo and H. Baranger, *Phys. Rev. Lett.* 93, 176807 (2004)
7. J.V. Acrivos et al. to be published
8. G. Koster, T.H. Geballe and B.Moyzhes, *Phys. Rev.* **B66**, 085109 (2004)
9. J.V. Acrivos, Microchemical Journal (2005) in press
10. L.I. Sciff, “Quantum Mechanics”, McGraw Hill Book Co. Inc, 1949
11. S. Murakami, N. Nagaosa and S. C. Zhang, *Science* **301**, 1348 (2003)
12. E. Clementi, ed., *“Methods and Techniques in Computational Chemistry”, “METTECC-94 A, B and C”*, STEF, Cagliari (1993)

Electronic Structure and Transport in Biological Inorganic Hybrid Materials

R. R. Pandey, N. Bruque and R. K. Lake

LAboratory for Terascale and Terahertz Electronics
Department of Electrical Engineering
Univ. of California, Riverside, CA, rlake@ee.ucr.edu

ABSTRACT

We present ab-initio and semi-empirical calculations of hybrid bio-inorganic structures and interfaces of carbon nanotubes, DNA, and Peptide Nucleic Acid (PNA) with either amide or glutamate linkers. Such structures are studied for bio-assembled nanoelectronic applications. Simulation results of single strand PNA are presented and compared to those of single-strand DNA. Calculations of self-energies for the non-equilibrium Green function formalism (NEGF) within the ab-initio FIREBALL code are described and demonstrated with a calculation of the transmission through a (10,0) CNT.

Keywords: ab-initio, semi-empirical, self-energy, NEGF, electron-transport, PNA, DNA, CNT

1 INTRODUCTION

Biologically self-assembled, hybrid, organic-inorganic nanostructures have a wide range of applications in engineering and biotechnology. Self-assembly methods include chemical self-assembly based on peptides, DNA and PNA (peptide nucleic acid) based self-assembly, antigen-antibody self-assembly, and virus based self-assembly. Materials assembled are metals, semiconductor nanocrystals and nanowires, and carbon nanotubes (CNTs). There is currently an intense interest in bio-assembled nanosystems bringing together disparate materials such as metals, semiconductor nanocrystals, DNA, PNA, proteins, peptides, and CNTs [1–3]. This interest is not only academic, but also shared and funded by the semiconductor industry through the MARCO Focus Center Research Program FENA [4]. Electrochemical, optical, and photochemical biosensor systems have resulted from the integration of metallic nanoparticles and semiconductor nanoparticles with enzymes [5,6], nucleic acids [7–9], and antigens/antibodies [10]. Biomaterials have the features of specificity and recognition in their binding that make them promising for directed assembly of nanoscale hybrid materials. Examples include the complementary base pairing of DNA [11–13], antigen-antibody binding [13], and the material specific binding of virus bound peptides [14].

One of the most advanced examples of bio-assembly of an electronic device is the bio-assembled carbon nanotube field effect transistor (CNTFET) demonstrated by Keren et al. [13]. This example uses both antigen-antibody binding and DNA recognition. The assembly of a CNTFET with metalized DNA leads is particularly auspicious. Metallization of DNA has been demonstrated with Pd, Pt, Ag, Au, Cu, and Ni [15], with workfunctions ranging from 4.1 to 5.6 eV compatible with the range found in CNTs [16]. This is synergistic with the discovery and advancement of ohmic contact technology for CNTFETs [17–20]. Most recently, metal patterning of DNA in a sequence specific manner while still retaining biological functionality opens the possibilities for complex, DNA templated, self-assembled circuits of CNTFETs [21].

2 PEPTIDE NUCLEIC ACID

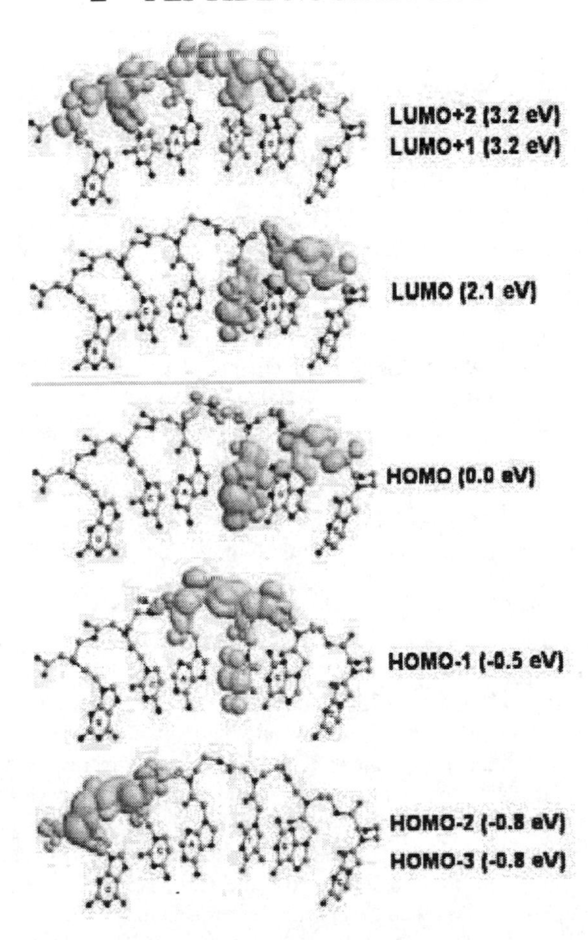

Figure 1: Nature of orbitals near the highest occupied molecular orbital (HOMO) and lowest unoccupied molecular orbital (LUMO) of a six base sequence (GCATGC) single strand PNA [22].

Fig. 1 shows the nature of orbitals near the highest occupied molecular orbital (HOMO) and lowest unoccupied molecular orbital (LUMO) of a six base sequence (GCATGC) single strand PNA. The calculations were done using the PM3 semi-empirical package [23] implemented in Gaussian03 program suite [24]. The HOMO-LUMO bandgap in the given PNA sequence is 2.1 eV. The HOMO and LUMO orbitals are relatively localized, confined on the thymine base, its backbone, and the backbone of its two nearest neighbors. Comparing with similar calculations for ss-DNA, we find for states close to the HOMO and LUMO, a significantly larger fraction of the molecular orbitals are on the peptide backbone of the PNA than on the sugar backbone of the DNA. This suggests that ss-PNA may be more suitable for metallization of the backbone compared to ss DNA [25].

3 HYBRID BIO-INORGANIC SYSTEMS

The structure in Fig. 2 consists of one unit of back-bone plus base (Guanine) segment of a single-strand PNA connected to a (10,0) CNT with a Glutamate (Glu) linker. This is the standard experimental procedure for functionalizing a CNT with PNA [26, 27]. We modeled the structure using GAUSSIAN03 to determine the energy levels and orbital locations as shown in Fig. 2. The hybrid density functional theory method B3LYP [28] was used to carry out a full geometry optimization of the single unit PNA attached to glutamate using the 6-31G* basis set. This optimized geometry was linked to an optimal 2 unit cell (10,0) zig-zag CNT and this assembly was used as a computationally cost effective model to investigate the PNA-Glu-CNT system. The molecular energies and orbitals are calculated using the PM3 method [23].

Fig. 2 shows that the HOMO of the PNA-Glu-CNT model is located on the Glu bridge between the CNT and the PNA, and the LUMO is located on the CNT. The HOMO-LUMO gap is 3.6 eV. For comparison, the HOMO-LUMO gap of the bare CNT is 3.2 eV. The large gap is the result of the short length (2 unit cells) of the modeled CNT. For an extended (10,0) CNT, the bandgap is 0.98 eV, and for a (13,0) CNT, the bandgap is 0.7 eV. The HOMO on the Glu sits in the energy gap of the CNT. The HOMO-1 orbital at -1.1eV (with respect to the HOMO) is also on the Glu. The HOMO-2 through HOMO-4 orbitals, lie on the CNT. They are closely spaced in energy 200-300 meV and begin at -1.7eV.

The relative alignment of the Glu states with respect to the the CNT states suggests that for a longer CNT for which the gap is close to its extended value of under 1eV, the Glu state will align closely with the valence band of the CNT enabling good hole transfer across the Glu bridge for p-type CNTs. The large band-gap seen here is due to the truncation of the CNT, that artificially increased the HOMO-LUMO gap by over a factor of 3 from 1 eV to 3.2 eV.

A second, similar example is the (10,0) CNT connected to a single unit of a guanine base, single-strand DNA with an amide linker shown in Fig. 3. The HOMO-LUMO gap of 3.1 eV is slightly less than the 3.2 eV gap of the bare 2 unit cell CNT. The HOMO orbital is confined on the CNT and pushed away from the amide linker at the interface. In contrast, the LUMO orbital extends across the linker suggesting good electron transfer across the amide bridge for n-type CNTs.

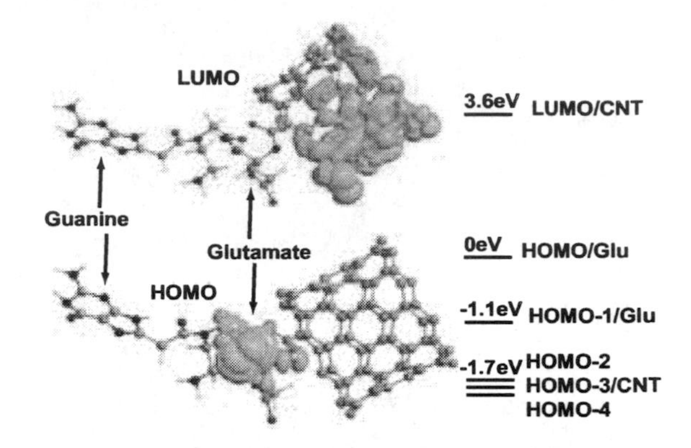

Figure 2: Highest occupied molecular orbital (HOMO) and lowest unoccupied molecular orbital (LUMO) for the CNT-Glu-PNA assembly [29].

The large HOMO-LUMO gap seen in both of these simulations is due to the truncation of the CNT. The system of real interest has an effectively semi-infinite CNT with a bulk band-gap of 1 eV. One can simulate the system with longer and longer sections of CNT so that the CNT states begin to more closely resemble the continuum of band states; however this approach uses more and more cpu and memory resources to model the large CNT.

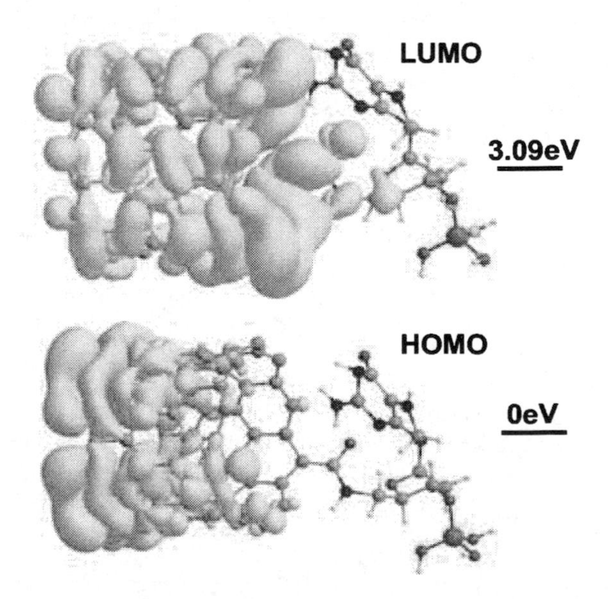

Figure 3: Highest occupied molecular orbital (HOMO) and lowest unoccupied molecular orbital (LUMO) for the model DNA-amide-CNT assembly [25].

A second approach is to terminate the CNT with a surface-self-energy matrix which exactly models the effect of the semi-infinite CNT extending to the left in Fig. 3. This is the NEGF approach to including the effect of semi-infinite regions via self-energies constructed from the surface Green functions. In this approach, the matrix size representing the system of Fig. 3 does not increase. It is simply modified at the end by the self-energy matrix representing the semi-infinite CNT. This modified matrix will then allow us to calculate the energy line-up of the molecular states with the band-states of the CNT.

4 NON-EQUILIBRIUM GREEN FUNCTIONS

We are implementing the NEGF approach within the FIREBALL code [30, 31]. The device Hamiltonian, the overlap matrix and the device-to-contact coupling matrices for an infinite (10,0) CNT are formed using the FIREBALL Hamiltonian matrix elements. The spatial extent of the non-zero matrix elements is determined by the pseudopotential cutoff limits and the FIREBALL orbital radii. The matrix elements extend 4 atomic layers which is exactly one unit cell of the (10,0) CNT. We label the unit cells such that cells $\{-\infty, \ldots, 0\}$ lie in the left contact, cells $\{1, \ldots, N\}$ lie in the "device," and cells $\{(N+1), \ldots, \infty\}$ lie in the right contact. The matrix elements of the Hamiltonian group into intra-cell subblocks $\mathbf{D}_{i,i}$ and inter-cell subblocks $\mathbf{t}_{i,i\pm1}$. We define effective off-diagonal matrix elements $\tilde{t}$ as $\tilde{t}_{i,j} = t_{i,j} - (E + i\eta)S_{i,j}$ where $S_{i,j}$ is the overlap matrix element between non-orthogonal orbitals i and j, E is the energy, and η is a convergence factor that is non-zero only in the leads.

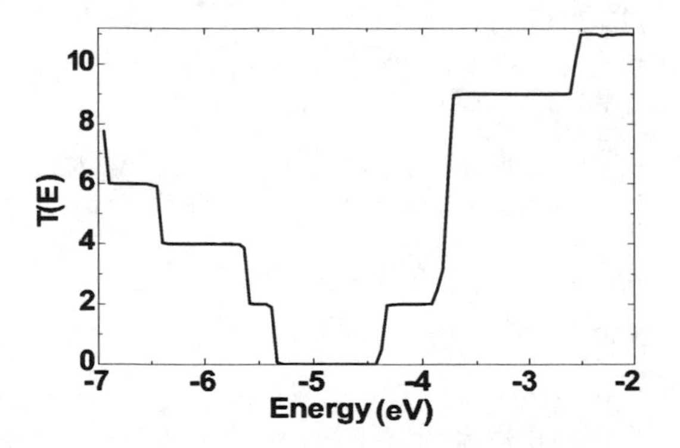

Figure 4: Transmission of a (10,0) CNT calculated with the matrix elements extracted from FIREBALL.

The calculation of the self-energy representing the semi-infinite CNT, begins with the calculation of the surface Green function. The surface Green function of the semi-infinite lead on the left $\mathbf{g}^R_{0,0}$ is calculated from the self-consistent expression:

$$\mathbf{g}^R_{0,0} = \left[(E + i\eta)\mathbf{S}_{0,0} - \mathbf{D}_{0,0} - \tilde{\mathbf{t}}_{0,-1}\mathbf{g}^R_{0,0}\tilde{\mathbf{t}}_{-1,0}\right]^{-1} \quad (1)$$

The convergence of this self-consistent calculation depends on the energy E. When the energy lies within a bandgap of the lead material, Eq. 1 converges quickly within 5 iterations. When the energy E lies in a band, convergence can be quite slow, taking up to 500 iterations with $\eta = 10$ meV and a convergence criteria of a percent difference change less than 0.0001. We update by taking the average of the old and new value of $\mathbf{g}^R_{0,0}$.

The non-zero blocks of the self-energy matrices are calculated using $\mathbf{\Sigma}^R_{1,1} = \tilde{\mathbf{t}}_{1,0}\mathbf{g}^R_{0,0}\tilde{\mathbf{t}}_{0,1}$ and $\mathbf{\Sigma}^R_{N,N} = \tilde{\mathbf{t}}_{N,N+1}\mathbf{g}^R_{N+1,N+1}\tilde{\mathbf{t}}_{N+1,N}$, where $\mathbf{g}$'s are the surface green functions calculated using iterative approach and $\tilde{\mathbf{t}}_{1,0} = t_{1,0} - (E+i\eta)S_{1,0}$ and $\tilde{\mathbf{t}}_{N,N+1} = t_{N,N+1} - (E+i\eta)S_{N,N+1}$.

One of the best tests of the self-energy matrices is to calculate the transmission through an ideal periodic structure of the same material, in this case the (10,0) CNT. The transmission should turn on in integer steps at the valence band maximums and conduction band minimums.

The transmission coefficient shown in Fig. 4 was calculated from [32]

$$T(E) = \mathrm{tr}\left\{\mathbf{\Gamma}^B_{1,1}\left[\mathbf{A}_{1,1} - \mathbf{G}^R_{1,1}\mathbf{\Gamma}^B_{1,1}\mathbf{G}^A_{1,1}\right]\right\} \quad (2)$$

where $\mathbf{\Gamma}^B_{1,1} = i\left(\mathbf{\Sigma}^{RB}_{1,1} - \mathbf{\Sigma}^{AB}_{1,1}\right)$, $\mathbf{A}_{1,1} = i\left(\mathbf{G}^R_{1,1} - \mathbf{G}^A_{1,1}\right)$, $\mathbf{\Sigma}^A = \left[\mathbf{\Sigma}^R\right]^\dagger$, and $\mathbf{G}^A = \left[\mathbf{G}^R\right]^\dagger$.

The integer turn-ons of the transmission matched the band maxima and minima of the E-k band diagram provided by FIREBALL as they should. We are now in the process of applying the NEGF/FIREBALL code to simulate the line-up of the CNT band states with the molecular states of the DNA, PNA, and linkers shown in Figs. 2 and 3.

5 SUMMARY

Ab-initio and semi-empirical calculations of hybrid bio-inorganic clusters have been presented. The results for the PNA-Glu-CNT cluster indicate that the Glu state will align closely with the valence band of the CNT enabling hole transfer across the bridge for p-type CNTs. For the DNA-amide-CNT cluster, the HOMO is localized away from the interface on the CNT, but the LUMO is extended across the interface suggesting good electron transfer across the amide bridge for n-type CNTs. An ab-initio, DFT, NEGF approach has been described to model the semi-infinite CNT - biomolecule electronic structure.

This work is supported by the NSF (DMR-0103248), DOD/DARPA/DMEA (DMEA90-02-2-0216), and the Microelectronics Advanced Research Corporation Focus Center on Nano Materials (FENA).

References

[1] E. Katz and I. Willner, Angew. Chem. Int. Ed., 43, 6042 – 6108, 2004.

[2] E. Katz and I. Willner, Chem. Phys. Chem, 5, 1085 – 1104, 2004.

[3] W. Fritzsche, ed., DNA-Based Molecular Electronics, vol. 725. AIP, New York, 2004.

[4] http://www.fena.org/.

[5] Y. Xiao, F. Patolsky, E. Katz, J. F. Hainfeld, and I. Willner, Science, 299, 1877 – 1881, 2003.

[6] V. Pardo-Yissar, E. Katz, J. Wasserman, and I. Willner, J. Am. Chem. Soc., 125, 622 – 623, 2003.

[7] R. Elghanian, J. J. Storhoff, R. C. Mucic, R. L. Letsinger, and C. A. Mirkin, Science, 277, 1078 – 1081, 1997.

[8] I. Willner, F. Patolsky, and J. Wasserman, Angewandte Chemie-International Edition, 40, 1861 – 1864, 2001.

[9] F. Patolsky, R. Gill, Y. Weizmann, T. Mokari, U. Banin, and I. Willner, J. Am. Chem. Soc., 125, 13918–13919, 2003.

[10] E. R. Goldman, E. D. Balighian, H. Mattoussi, M. K. Kuno, J. M. Mauro, P. T. Tran, and G. P. Anderson, J. Am. Chem. Soc., 124, 6378 – 6382, 2002.

[11] C. A. Mirkin, R. L. Letsinger, R. C. Mucic, and J. J. Storhoff, Nature, 382, 607 – 609, 1996.

[12] A. P. Alivisatos, K. P. Johnsson, X. G. Peng, T. E. Wilson, C. J. Loweth, M. P. Bruchez, and P. G. Schultz, Nature, 382, 609 – 611, 1996.

[13] K. Keren, R. S. Berman, E. Buchstab, U. Sivan, and E. Braun, Science, 302, 1380 – 1382, 2003.

[14] S. R. Whaley, D. S. English, E. L. Hu, P. F. Barbara, and A. M. Belcher, Nature, 405, 665 – 668, 2000.

[15] R. M. Stoltenberg and A. T. Woolley, Biomedical Microdevices, 6, 105 – 111, 2004.

[16] J. Zhao, J. Han, and J. P. Lu, Phys. Rev. B, 65, 193401 – 193404, 2002.

[17] A. Javey, J. Guo, Q. Wang, M. Lundstrom, and H. Dai, Nature, 424, 654–657, 2003.

[18] A. Javey, J. Guo, M. Paulsson, Q. Wang, D. Mann, M. Lundstrom, and H. Dai, Phys. Rev. Lett., 92, 106804, 2004.

[19] A. Javey, J. Guo, D. B. Farmer, Q. Wang, D. Wang, R. G. Gordon, M. Lundstrom, and H. Dai, Nano Lett., 4, 447–450, 2004.

[20] A. Javey, J. Guo, D. B. Farmer, Q. Wang, E. Yenilmez, R. G. Gordon, M. Lundstrom, and H. Dai, Nano Lett., 4, 1319–1322, 2004.

[21] K. Keren, R. S. Berman, and E. Braun, Nano Lett., 4, 323 – 326, 2004.

[22] X. Wang, K. V. Singh, R. R. Pandey, R. Lake, C. Tsai, A. Balandin, M. Ozkan, and C. S. Ozkan, submitted.

[23] J. Stewart, J. Comp. Chem., 10, 209.

[24] M. J. Frisch, G. W. Trucks, H. B. Schlegel, G. E. Scuseria, M. A. Robb, J. R. Cheeseman, J. A. Montgomery, Jr., T. Vreven, K. N. Kudin, J. C. Burant, J. M. Millam, S. S. Iyengar, J. Tomasi, V. Barone, B. Mennucci, M. Cossi, G. Scalmani, N. Rega, G. A. Petersson, H. Nakatsuji, M. Hada, M. Ehara, K. Toyota, R. Fukuda, J. Hasegawa, M. Ishida, T. Nakajima, Y. Honda, O. Kitao, H. Nakai, M. Klene, X. Li, J. E. Knox, H. P. Hratchian, J. B. Cross, V. Bakken, C. Adamo, J. Jaramillo, R. Gomperts, R. E. Stratmann, O. Yazyev, A. J. Austin, R. Cammi, C. Pomelli, J. W. Ochterski, P. Y. Ayala, K. Morokuma, G. A. Voth, P. Salvador, J. J. Dannenberg, V. G. Zakrzewski, S. Dapprich, A. D. Daniels, M. C. Strain, O. Farkas, D. K. Malick, A. D. Rabuck, K. Raghavachari, J. B. Foresman, J. V. Ortiz, Q. Cui, A. G. Baboul, S. Clifford, J. Cioslowski, B. B. Stefanov, G. Liu, A. Liashenko, P. Piskorz, I. Komaromi, R. L. Martin, D. J. Fox, T. Keith, M. A. Al-Laham, C. Y. Peng, A. Nanayakkara, M. Challacombe, P. M. W. Gill, B. Johnson, W. Chen, M. W. Wong, C. Gonzalez, and J. A. Pople, Gaussian 03, Revision C.02. Gaussian, Inc., Wallingford, CT, 2004.

[25] X. Wang, R. R. Pandey, M. Yang, K. Singh, K. Alim, R. Lake, A. Balandin, M. Ozkan, and C. S. Ozkan, submitted.

[26] K. A. Williams, P. T. M. Veenhuizen, B. G. de la Torre, R. Eritja, and C. Dekker, Nature, 420, 761, 2002.

[27] R. den Dulk, K. A. Williams, P. T. M. Veenhuizen, M. C. de Konig, M. Overhand, and C. Dekker, In W. Fritzsche, ed., DNA-Based Molecular Electronics, vol. 725, 25 – 31. AIP, New York, 2004.

[28] A. D. Becke, J. Chem. Phys., 98, 5648.

[29] K. V. Singh, X. Wang, R. R. Pandey, A. Martinez, K. Wang, R. Lake, A. Balandin, C. S. Ozkan, and M. Ozkan, submitted.

[30] O. F. Sankey and D. J. Niklewski, Phys. Rev. B, 40, 3979 – 3995, 1989.

[31] J. P. Lewis, K. R. Glaesemann, G. A. Voth, J. Fritsch, A. A. Demkov, J. Ortega, and O. F. Sankey, Phys. Rev. B, 64, 195103/1 – 10, 2001.

[32] R. Lake, G. Klimeck, R. C. Bowen, and D. Jovanovic, J. Appl. Phys., 81, 7845–7869, 1997.

Effects of the Addition of TiO_2 Nanoparticles for Polymer Electrolytes Based on Porous Poly(vinylidene fluoride-*co*-hexafluoropropylene)/ Poly(ethylene oxide-*co*-ethylene carbonate) Membranes

Jae–Deok Jeon*, Myung–Jin Kim**, Jae Woo Chung*** and Seung–Yeop Kwak****

School of Materials Science and Engineering, Seoul National University, San 56-1, Sillim-dong, Gwanak-gu, Seoul 151-744, Korea,
*jdjun74@snu.ac.kr, **odelly98@daum.net, ***cwfrank5@snu.ac.kr and ****sykwak@snu.ac.kr

ABSTRACT

We have attempted to combine a typical gel and a typical solvent-free polymer electrolyte through the conception of a pore-filling polymer electrolyte; instead of organic solvents, a viscous poly(ethylene oxide-*co*-ethylene carbonate) (P(EO-EC))/$LiCF_3SO_3$/TiO_2 electrolyte mixture is filled into the pores of a porous poly(vinylidene fluoride-*co*-hexafluoropropylene) (P(VdF-HFP))/P(EO-EC)/TiO_2 membrane. The porous membrane with 10 wt% TiO_2 showed better performance (*e.g.*, high uptake and good mechanical strength) than the others and was therefore selected as an optimal matrix to prepare a polymer electrolyte. All polymer electrolytes obeyed Arrhenius behavior and showed linear enhancement of ionic conductivity with increasing the temperature. In addition, maximum conductivity of 4.7×10^{-5} S cm^{-1} at 25 °C was obtained for the polymer electrolyte containing 1.5 wt% TiO_2 in electrolyte mixture.

Keywords: porous membrane, polymer electrolyte, TiO_2 nanoparticle, ionic conductivity, rechargeable lithium battery

1 INTRODUCTION

During the last two decades, ion-conducting polymer electrolytes have continued to attract great interest from both scientific and industrial perspectives due to their growing application potentials in electronic devices such as rechargeable batteries, displays, fuel cells, and so forth [1]. Among these, solvent-free polymer electrolytes (SPEs) formed by complexes of a lithium salt (LiX) with a polyether such as poly(ethylene oxide) (PEO) have received considerable attention due to their advantages in terms of the ease of fabrication, flexibility in dimensions, good mechanical properties, safety features, and excellent stability at the lithium interface. However, their low ionic conductivity has been the reason for them not being used in practical applications for rechargeable lithium batteries that require a value of above 10^{-4} S cm^{-1} at room temperature.

The ion transport of SPEs is induced via their rapid segmental motion combined with strong Lewis acid-base interactions between the cation and the donor atom present in the amorphous phase. Therefore, most approaches for enhancing the conductivity of PEO-based systems have been focused on lowering the degree of PEO crystallinity or reducing the glass transition temperature, T_g, through the modification of the polymer structures and the incorporation of plasticizers or ceramic fillers [2,3].

In this study, in order to ameliorate the performance of SPEs, we report on a pore-filling polymer electrolyte filled with a viscous poly(ethylene oxide-*co*-ethylene carbonate) (P(EO-EC)) complexed with $LiCF_3SO_3$ into a porous poly(vinylidene fluoride-*co*-hexafluoropropylene) (P(VdF-HFP))/P(EO-EC) membrane. However, the pore-filling polymer electrolyte requires the modification to enhance its mechanical strength and ionic conductivity. In order to satisfy these requirements, TiO_2 nanoparticles are added to both a porous membrane and a viscous electrolyte mixture. Generally, it is reported [4] that there are weak coordination interactions between alkylene oxide segments and TiO_2, resulting in the increase of mechanical strength. Also, hydroxyl groups on the surface of TiO_2 and carbonyl groups in the polymer have interactions by the formation of hydrogen bonds [5]. From these reasons, it is expected that TiO_2 would be dispersed well in both a porous membrane and an electrolyte mixture due to the use of P(EO-EC), which contains carbonate and ether groups, causing the mechanical strength to be increased. On the other hand, the addition of TiO_2 induces improvement of ionic conductivity of polymer electrolytes. In general, it is widely accepted that the addition of TiO_2 into semi-crystalline polymers suppresses the crystallization and enlarges the amorphous phase, resulting in the enhancement of conductivity. In this study, in order to improve the conductivity of polymer electrolytes, TiO_2 nanoparticles are added into the amorphous polymer. Although there is not enough research concerned with the addition of TiO_2 on polymer electrolytes consisted of amorphous polymers, it is possible to enhance their ionic conductivity because the free volume, which is produced by the hydrogen bonds between hydroxyl groups on the surface of inorganic nanoparticles and ether groups in polymers is increased. Therefore, we focus on the improvement of mechanical strength and ionic conductivity of pore-filling polymer electrolytes by the addition of TiO_2 nanoparticles with various contents into both a porous membrane and an electrolyte mixture.

2 EXPERIMENTAL

2.1 Preparation of Porous Membranes and Polymer Electrolytes

Porous membranes were prepared by a phase inversion method. To prepare porous membranes with TiO_2 (ST-01, Ishihara Sangyo Kaisha), the weight ratio of polymers with TiO_2, a solvent, and a nonsolvent was optimized at 1:8:1. The proportion of P(VdF-HFP) (M_w = 4.6 × 10^5 g mol^{-1}, Aldrich Chemicals) and P(EO-EC) (M_n = 1800 g mol^{-1}, synthesized in the laboratory) was chosen as 60 to 40 wt% in this study. TiO_2 content was varied as 0, 10, 20, 30, and 40 wt% in P(VdF-HFP)/P(EO-EC)/TiO_2. Firstly, a homogenous mixture was prepared in advance by dissolving P(VdF-HFP) and P(EO-EC) in acetone. Then, a measured amount of TiO_2 was added into TEA (nonsolvent) and acetone (solvent) mixture and dispersed by stirring and ultrasonication. Finally, the prepared TiO_2/TEA/acetone mixture was added into a homogenous P(VdF-HFP)/P(EO-EC) mixture. After complete dissolution, the solution degassed at 50 °C for 2 h was cast on a cleaned glass plate and desired thickness was made with a doctor blade. After a solution was stored at room temperature, the glass plate was put into steam atmosphere until the volatile acetone was evaporated, and then phase inversion occurred. TEA was removed by washing with methanol. After the residual solvent was allowed to slowly evaporate at room temperature, it was completely removed under vacuum at 50 °C for 24 h, thereby resulting in porous membranes. The resultant membranes had a thickness of 140–180 μm.

In order to prepare pore-filling polymer electrolytes, P(EO-EC) was first dissolved in acetone. After a homogenous solution was obtained, a measured amount of $LiCF_3SO_3$ (Aldrich Chemicals) and TiO_2 were added. TiO_2 content was varied as 0, 0.5, 1.0, 1.5, 2.0, 5.0, and 10.0 wt% in the P(EO-EC)/$LiCF_3SO_3$/TiO_2 mixture. Then, the solution was stirred until the $LiCF_3SO_3$ was completely dissolved. The resulting viscous solution was left to allow the acetone to evaporate under vacuum at room temperature. After perfect evaporation of acetone, porous membranes were filled for several times with the heated P(EO-EC)/$LiCF_3SO_3$/TiO_2 mixture by using a vacuum filter equipment, thereby producing pore-filling polymer electrolytes. The mixture remained on the surface was wiped with a filter paper. All procedures for preparing polymer electrolytes were carried out in a dry room.

2.2 Nomenclature

MTx denotes a porous Membrane with blend composition of P(VdF-HFP)/P(EO-EC) (the weight ratio of 6 to 4) containing x wt% TiO_2. Furthermore, ETx-y means a polymer Electrolyte filled with the P(EO-EC)/Li-salt mixture containing y wt% TiO_2 inside the pores of a MTx.

2.3 Characterization

The crystalline property of porous membranes was observed via a Perkin-Elmer GX IR spectrometer in the range of 400–4000 cm^{-1}. The uptake of P(EO-EC) in porous membranes was determined as follows:

$$\text{Uptake (\%)} = \frac{(W - W_o)}{W} \times 100 \qquad (1)$$

where W and W_o are the weights of the wet and dry membranes, respectively. Morphologies of porous membranes involving TiO_2 were investigated with a field emission scanning electron microscopy (FE-SEM) using JSM-6330F (JEOL, Japan), operating 5 kV, with the working distance of 15 mm. The mechanical properties of porous membranes were measured from stress-strain tests using a LLOYD LR10K universal testing instrument. Tensile load was measured by 100 N load cell, while air grips held the specimens. A standard ASTM D412-92T dumbbell was used with a 15.5 mm gauge length and a cross head speed of 20 mm min^{-1}.

The glass transition temperatures of electrolyte mixtures were determined by differential scanning calorimetry (DSC) with a TA instruments DSC 2920 at a heating rate of 10 °C min^{-1} under nitrogen atmosphere. The ionic conductivity of polymer electrolytes was measured using an ac impedance analyzer (Solartron 1260) within a temperature range of 5 to 95 °C and frequency range of 0.1 Hz to 1 MHz. The specimens were prepared by sandwiching the polymer electrolyte between two stainless steel (SS) electrodes. Each sample was allowed to equilibrate for 30 min at each temperature before the measurement was taken. All assemblies and testing operations of samples were carried out in a dry room.

3 RESULTS AND DISCUSSION

3.1 Porous Membranes

The Fourier transform infrared (FT-IR) spectra of pure TiO_2 and porous membranes including various contents of TiO_2 are shown in Fig. 1. The characteristic peaks of α-phase PVdF crystal were revealed at 531, 766, and 976 cm^{-1}. Also, the peaks at 484 and 840 cm^{-1} responded to γ- and β-phase of PVdF crystal. The stretching mode vibration of Ti-O band at 1640, 800, 450 cm^{-1} was occurred. The peak intensities corresponding to PVdF crystals were disappeared with the increase of TiO_2 content in porous membranes, except for the small remainder peak owing to α-phase of PVdF crystal at 976 cm^{-1}. The decrease of peak intensities was considered as the result from the fact that TiO_2 was blended within polymers very well, thereby bringing to a conclusion that TiO_2 dispersion in porous membranes was well established.

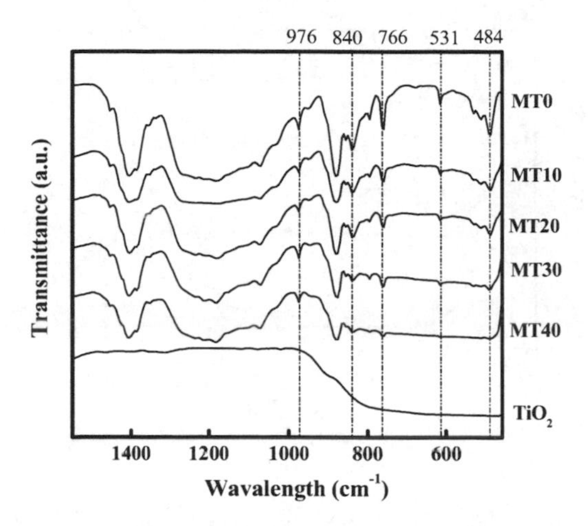

Figure 1: Infrared spectra of pure TiO_2 and porous membranes in the frequency of 1550–420 cm^{-1}.

The uptake of P(EO-EC) for porous membranes was measured. From the results, it was confirmed that the uptake of porous membranes was abruptly decreased above 30 wt% TiO_2. Consequently, the appropriate amount of TiO_2 in membranes was considered as 10 and 20 wt% (*i.e.*, uptake of MT10 and MT20 was 62.3 and 62.2%, respectively).

The morphologies of porous membranes loading various contents of TiO_2 were investigated with the field emission scanning electron microscopy (FE-SEM). Figure 2 shows the images of selected samples. The porous membrane (Fig. 2(a)) is the matrix of a polymer electrolyte (Fig. 2(d)) and the viscous electrolyte mixture confined into the pores of a membrane plays a role as an ion conductor. On the other hand, the surfaces of membranes became rougher with the increase of TiO_2 content. Furthermore, it was definitely observed that the size of TiO_2 agglomerate was up to a few micrometers because TiO_2 in inorganic/organic composites prepared by mechanically mixing easily became agglomerates due to their low compatibility with the polymer matrix. The bigger size of TiO_2 was occurred in MT30 in contrary to the case of MT20. These agglomerates may reduce the efficiency of TiO_2 to enhance mechanical properties of porous membranes. From these results, 10 and 20 wt% TiO_2 were regarded as the appropriate amounts to obtain the porous membranes containing relatively well-dispersed TiO_2. To confirm the presence and distribution of titanium in porous membranes, energy dispersive spectrometer (EDS) analysis was performed on the surface (Fig. 2(b)) and cross-section (Fig. 2(c)) of membranes. These images showed that TiO_2 in MT20 membrane had a homogenous distribution, implying that TiO_2 was well dispersed throughout the porous membrane.

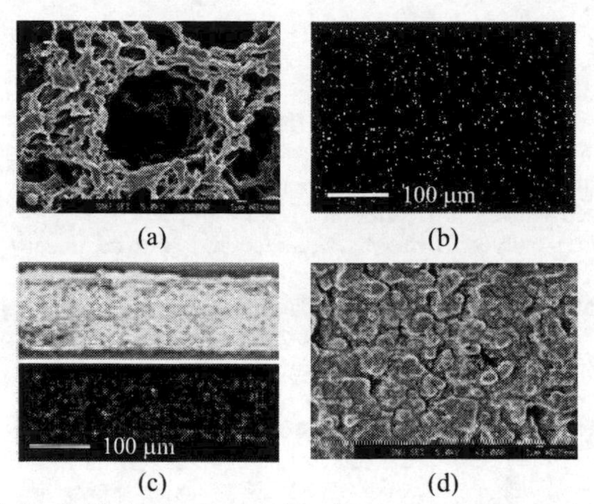

Figure 2: Typical micrographs of porous membranes and a polymer electrolyte: (a) MT0; (b) surface and (c) cross-section of MT20 measured using FE-SEM with EDS; (d) ET0.

The stress-strain data of porous membranes are described in Fig. 3. The maximum stress and tensile modulus values of porous membranes were increased up to 2.0 and 41.0 MPa until using 30 wt% TiO_2, respectively. This is due to the fact that TiO_2 nanoparticles play a role as a crosslinker, thereby resulting in the increase of mechanical strength. Then, they were decreased up to 1.2 and 34.5 MPa, respectively, when adding 40 wt% TiO_2. The strain values became decreased more and more as the increase of TiO_2 content in porous membranes. These results indicate that the addition of TiO_2 up to 20 wt% has an effect on the enhancement of mechanical properties of porous membranes as a result from the formation of interaction between TiO_2 nanoparticles and polymers. Overall, MT10 was chosen as the best one when considering the way to minimize the preparing cost.

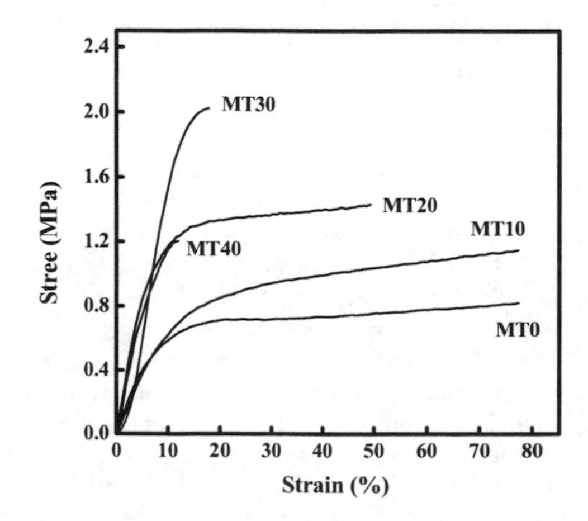

Figure 3: Stress-strain curves of porous membranes with different TiO_2 content.

3.2 Polymer Electrolytes

Ionic conductivity, σ of polymer electrolytes prepared by filling a P(EO-EC)/$LiCF_3SO_3$/TiO_2 mixture into the pores of a porous membrane was measured using ac impedance analyzer. All polymer electrolytes with the optimized Li-salt concentration of 1.5 mmol $LiCF_3SO_3$/g-P(EO-EC) obeyed Arrhenius behavior and showed linear enhancement of ionic conductivity with increasing the temperature. In addition, maximum conductivity of 4.7×10^{-5} S cm^{-1} at 25 °C was obtained for ET10-1.5 containing 1.5 wt% TiO_2 in electrolyte mixtures (Fig. 4). Generally, the increase of ionic conductivity for PEO-based polymer electrolytes is attributed to the role of ceramic filler in inhibition of crystallization of polymers from the amorphous state. However, P(EO-EC), which plays an important role as an ion conductor in this study, is so amorphous that the theory mentioned above is not suitable. Another role of ceramic fillers in the increase of ionic conductivity is the formation of Lewis acid-base interaction between oxide groups of polymers and hydroxyl groups of ceramic fillers such as TiO_2 [6]. To better investigate the role of TiO_2, thermal analysis using differential scanning calorimetry (DSC) was conducted.

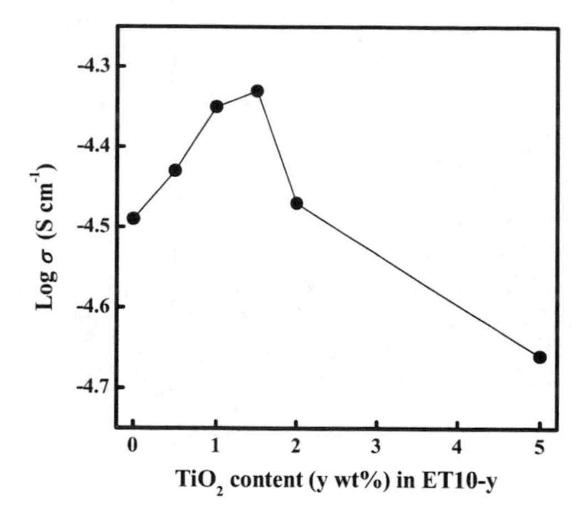

Figure 4: TiO_2 content dependence of ionic conductivity for polymer electrolytes (ET10-y).

The value of T_g was increased markedly by $LiCF_3SO_3$, *i.e.*, from *ca.* -42 to -23 °C. However, the value was lowered by the addition of TiO_2 (up to 1.5 wt%, see Fig. 5) due to the presence of the Lewis acid-base interaction between P(EO-EC) and TiO_2, which leads to weaker interaction between ether groups of the polymer and Li-ions. The decrease of T_g is contributed to the enhancement of conductivity through the weakened interaction between Li-ion and oxide groups in polymers as well as the increase of free volume among polymer segments. However, further addition of TiO_2 (above 1.5 wt%) caused the T_g to increase. This is supposed as a result of the TiO_2 agglomerates.

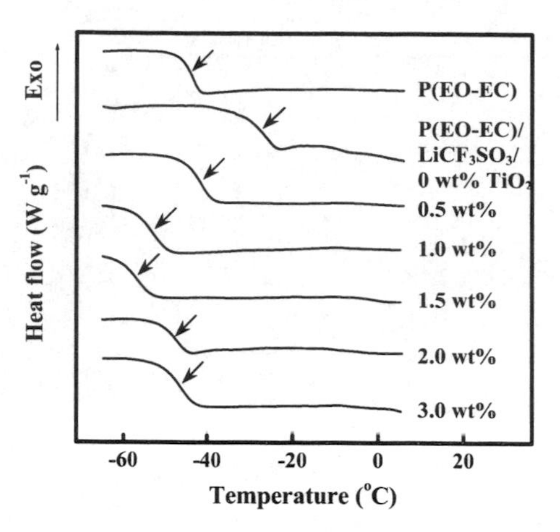

Figure 5: Changes in T_g with TiO_2 addition to P(EO-EC)/$LiCF_3SO_3$ complex.

4 CONCLUSIONS

This study has been focused on the characterization of the pore-filling polymer electrolyte with TiO_2 nanoparticles. From FT-IR results, it was confirmed that TiO_2 was well blended with polymers in porous membranes as the peak intensity of polymers became smaller with the increasing the TiO_2 content. The MT10 membrane showed better performance (*e.g.*, high uptake and good mechanical strength) than the others and was therefore selected as an optimal matrix to prepare the polymer electrolyte. The maximum conductivity of 4.7×10^{-5} S cm^{-1} at 25 °C was obtained for ET10-1.5. The T_g was increased markedly by $LiCF_3SO_3$ and then lowered by the addition of TiO_2 (up to 1.5 wt%), which was in good agreement with the ionic conductivity data.

Overall, these results indicate that pore-filling polymer electrolytes involving TiO_2 nanoparticles can be potentially useful as a new class of SPEs for rechargeable lithium batteries.

REFERENCES

[1] S. Rajendran, R. Kannan and O. Mahendran, Materials Letters 48, 331, 2001.

[2] Y. Liu, J. Y. Lee and L. Hong, J. Power Sources 129, 303, 2004.

[3] J.-H. Ahn, G. X. Wang, H. K. Liu and S. X. Dou, J. Power Sources 119-121, 422, 2003.

[4] P. Yang, D. Zhao, D. I. Margolese and B. F. Chmelka, Nature 396, 152, 1998.

[5] A. M. Rocco, C. P. Fonseca and R. P. Pereira, Chem. Mater. 15, 2936, 2003.

[6] M. Forsyth, D. R. MacFarlane, A. Best, J. Adebahr, P. Jacobsson and A. J. Hill, Solid State Ionics 147, 203, 2002.

Supported Mixed Metal Nanoparticles and PFA-Nafion Nanocomposite Membrane for Low Temperature Fuel Cells

Kwong-Yu Chan[a], Jiawen Ren[a], Jin Liu[b], Huanting Wang[c], Jie Ding[a], Kwok-Ying Tsang[a], and Tung-Chun Lee[a]

[a]Department of Chemistry, The University of Hong Kong
Pokfulam Road, Hong Kong.

[b]Department of Material Science and Engineering
Anhui Institute of Architectue and Industry
Hefei, Anhui, China 230022.

[c]Australian Key Centre for Microscopy & Microanalysis
Electron Microscope Unit
The University of Sydney
NSW, 2006, Australia

ABSTRACT

Performance of low temperature fuel cells depends critically on the nanostructures of the material components in the electrodes and membranes. Some studies are reported here for 1) mixed metal nanoparticles supported on mesoporous carbon and 2) modification of nanopores of Nafion via in-situ polymerization of furfuryl alcohol.

The anodic oxidation of small organic molecules such as alcohols in a low temperature fuel cell requires platinum based mixed metal electrocatalysts such as platinum-ruthenium. A number of techniques are now available for synthesizing mixed metal nanoparticles [1-3]. The challenge is to control both the size and composition of the mixed nanoparticles. The electrocatalysts are normally supported on activated carbon, such as Vulcan XC 72. A further challenge is to control the structure and properties of the support material. A number of novel carbon materials with well-defined nanostructures have been reported in the literature recently. We report here the synthesis of platinum and platinum-ruthenium nanoparticles supported on ordered mesoporous carbon CMK-3 [4]. The CMK-3 is synthesized from a template material SBA-15, an ordered mesoporous silica with uniform pores of a few nanometers in diameter. The mesoporous carbon supported Pt and Pt-Ru nanoparticles are characterized by HRTEM (Fig. 2) and EDX. The supported electrocatalysts are evaluated for electrochemical reduction of oxygen and electrocatalytic oxidation of methanol. A Pt-CMK-3 catalyst synthesized outperformed the commercial catalyst for oxygen reduction in a large area gas diffusion electrode. The Pt-Ru-CMK-3 catalyst did not perform as well, likely due to transport limitation in the long and narrow mesopores.

Methanol crossing over from anode to cathode has been a major barrier to the development of direct methanol fuel cells. Studies have been made to modify the polymer electrolyte membrane, Nafion. A novel approach is reported here using in-situ polymerisation of furfuryl alchol [5,6]. The monomer is hydrophilic and penetrates the nanopores of Nafion. Upon polymerisation catalysed by sulphuric acid, the polymer becomes hydrophobic. This increase in hydrophobicity increases the barrier to methanol. Improvement of the membrane performance in a methanol fuel cell is reported here. The methanol permeability and proton conductivity are measured as a function of methanol concentration. The performance of a membrane-electrolyte assembly at room temperature shows marked increase in current and power compared to the unmodified membrane (Fig. 5).

Keywords: mixed metal nanoparticles, mesoporous carbon support, PFA modified Nafion, fuel cell catalysts.

1 MIXED METAL NANOPARTICLES SUPPORTED ON MESOPOROUS CARON

Fuel cells using a safe liquid fuel and operating at temperature below 60°C are desirable for powering portable consumer products such as notebook computers and mobile phones. The performance of electrocatalysts, however, still falls below expectations for these applications in terms of power density and price per kW. Advances made in the nanostructuring of fuel cell catalysts are being made, most importantly in preparing mixed platinum based metal nanoparticles of uniform size and uniform composition and uniformly dispersed into a conducting carbon support. A number of synthetic methods are used to prepare uniform sized metal particles such as the colloidal method, the microemulsion method, and the glycol method[3]. While nanosize control has been made readily, simultaneous control of uniform composition at the nanoparticle level has not been generally achieved. On the other hand, advances have been made in controlling the structure of carbon support, mainly through the ordered mesoporous silica templating technique [4]. A notable example is the CMK type carbon templated from SBA-15 silica with uniform pores of 2-15 nm. Typical structures of these carbons are shown in Fig. 1. The challenge is to prepare mixed metal nanoparticles and have them loaded uniformly into carbon support with control nanostructures. The details of the synthetic procedure are given in [4]. We achived the loading of Pt-Ru mixed metal nanoparticles into CMK-3 mesoporous carbon, as shown in Fig. 2.

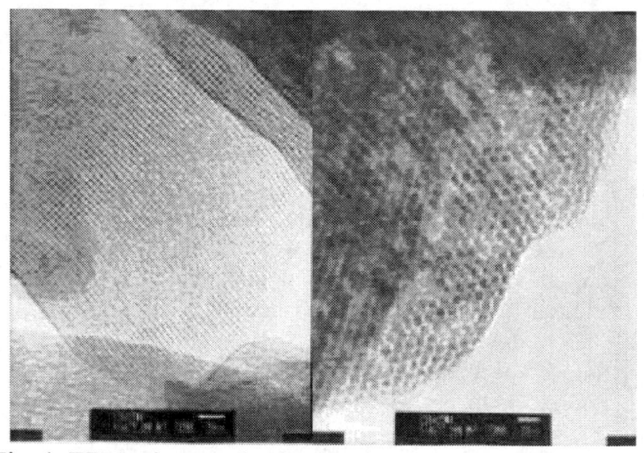

Fig. 1. TEM micrograph of ordered mesoporous carbon templated from SBA-15 silica.

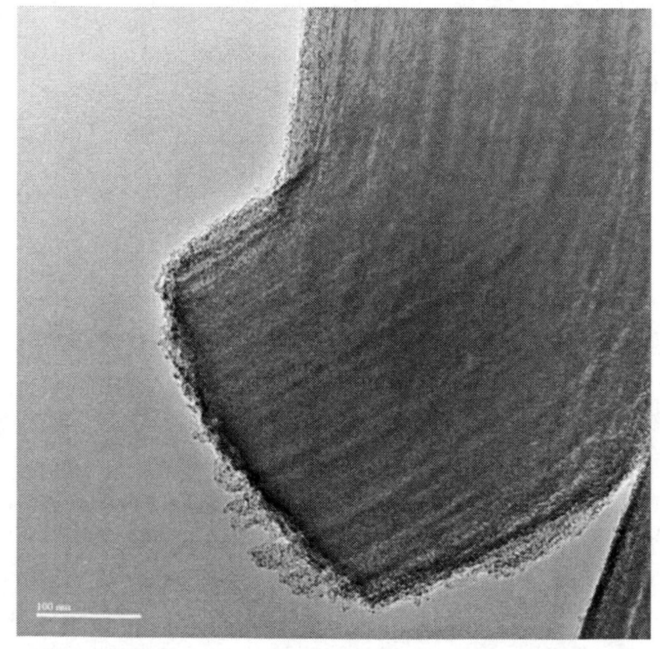

Fig. 2(a). TEM micrograph of Pt-Ru nanoparticles supported on mesoporous carbon templated from SBA-15 silica (CMK-3). The scale bar is 100 nm.

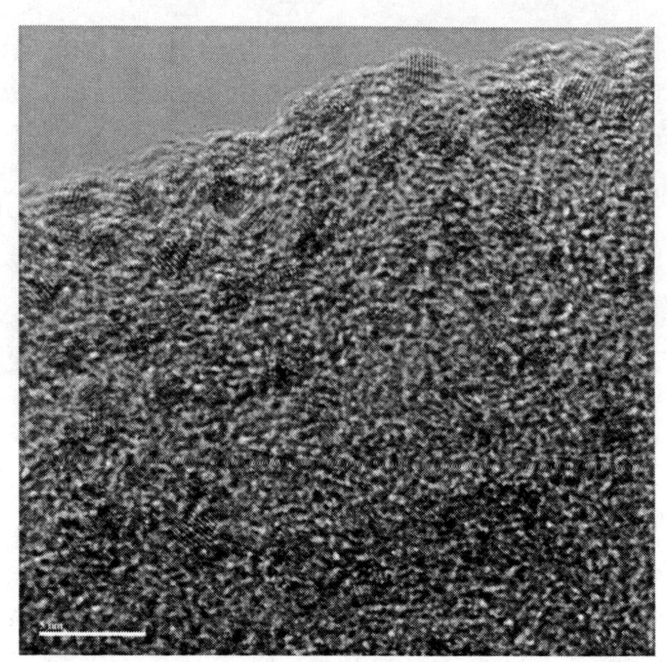

Fig. 2(b). HRTEM micrograph of Pt-Ru nanoparticles supported on mesoporous carbon templated from SBA-15 silica (CMK-3). The scale bar is 5 nm.

The performance of these Pt-Ru-CMK3 catalysts was tested for anodic oxidation of methanol in sulphuric acid at room temperature (20°C). As shown in Fig. 3, low activation polarization is indicated at low current. At higher current, however, polarization increases steadily, indicating ohmic and transport limitations. The interesting feature is some oscillatory V-I behaviour at high current. This may be due to the unsteady mass transfer in the long and narrow channels of CMK-3 carbon. Typical sizes of these CMK-3 particles are over a few hundred nanometers. One possible improvement is to produce smaller particles of these CMK-3 powders to about 50 nm of less in diameter.

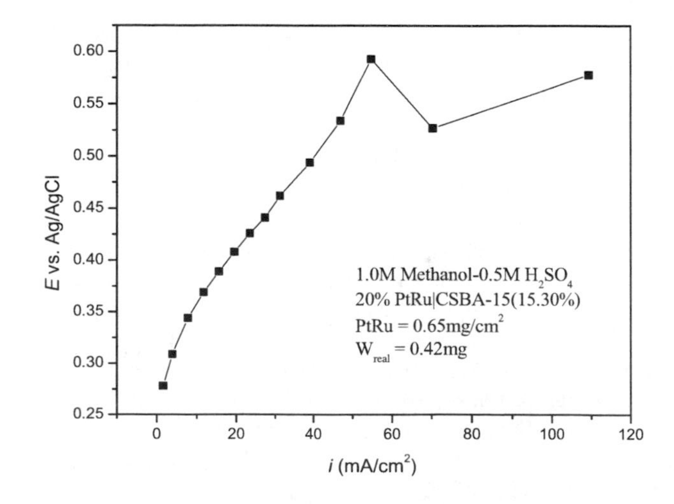

Fig. 3. Polarization curve of methanol oxidation on Pt-Ru/CMK-3 carbon electrode obtained from data of steady-state current discharged.

2. MODIFICATION OF NANOPORES OF NAFION VIA IN-SITU POLYMERIZATION OF FURFURYL ALCOHOL

The direct methanol fuel cell is a promising device to replace existing batteries for high energy density operation of portable electronic products. Current technology of direct methanol fuel cells uses a polymer electrolyte membrane between the anode and cathode to separate the anode and cathode contents, to prevent internal electronic current between the two electrodes, and most importantly to provide ionic conductivity within the fuel cell. Nafion© , a perfluorosulfonic polymers is the best proton electrolyte membrane developed for fuel cells because of its outstanding chemical, mechanical and thermal stability and high proton conductivity. In the operation of direct methanol fuel cells (DMFCs), however, high methanol permeation through Nafion membranes significantly lowers fuel efficiency and cell performance, and thus impedes the commercial development of DMFCs. Significant efforts have been made to modify Nafion by inserting a variety of materials such as palladium and its alloys, polyvinyl alcohol, polypyrrole, zeolite, silica, molybdophosphoric acid, and montmorillonite. The strategy of forming Nafion composite membranes has sometimes proven to be effective for reducing methanol permeation. However, either proton conductivity is undesirably affected, or additional costs and instability prevent their practical applications in DMFC.

Here, we describe a novel approach to the modification of solid polymer electrolyte membrane by in situ polymerization of an initially hydrophillic monomer, furfuryl alcohol to form the more hydrophobic polyfurfuryl alcohol (PFA) within the solid polymer electrolyte membrane porous domains. Furfuryl alcohol monomer is miscible with water and alcohols. It can be readily incorporated into hydrophilic zones of the internal pores of Nafion. After acid catalysed polymerization, PFA becomes more hydrophobic. The polymerization and change of hydrophobicity alter the microstructures and nanostructures of the polymer electrolyte and make it more impermeable to methanol. A homogenous PFA-Nafion nanocomposite membrane is demonstrated to be highly impermeable to methanol. In addition, PFA is chemically stable and relatively inexpensive. Conventional polymeric modifications to Nafion usually involve blend of polymers into the Nafion matrix and in effect filling either the hydrophillic or hydrophobic domains of the microstructures.

We found that the PFA-Nafion nanocomposite membrane outperforms the conventional Nafion membrane by more than a factor of two in the presence of methanol. Table 1 shows the reduced methanol permeability in several PFA-Nafion composite membranes with the appropriate weight percentages of PFA. The actual performance in an anode-membrane-cathode assembly (MEA) with 10% (vol) methanol in water as fuel is shown in Figures 3-4 under ambient air at room temperature and at an elevated temperature. The peak power density in the PFA modified MEA is more than two times higher than the MEA using an unmodified Nafion membrane.

A commercial Nafion 115 membrane was first sodium exchanged by boiling in 0.5 M NaOH solution and rinsed with deionized water. The dried membrane with an area of 2 cm x 2 cm was immersed in a mixture of 6 g furfuryl alcohol (FA, 98%, Lancaster), 12 g isopropanol, and 6 g deionized water. The fully swollen and saturated Nafion membrane was transferred into 1.0M sulphuric acid at room temperature for 2 min. Subsequently, the polymerization of furfuryl alcohol was carried out at 80 °C in an oven. The membrane was washed with ethanol, and boiled with 1.0M sulphuric acid to obtain the acidic form. The dark brown membrane was finally treated at 140° C for 10 min allowing for microstructural rearrangements of the membranes. The resulting PFA-Nafion membrane contains approximately 2.0 % by weight of PFA. Repeating the monomer loading and polymerization procedure can increase the final weight percentage of PFA to the desirable level.

Table 1 Properties of PFA-Nafion Nanocomposite

Sample	Nafion 115	Nafion-3.9PFA	Nafion-8.0PFA	Nafion-12.4PFA
Proton conductivity (S/ cm)	0.095	0.089	0.070	0.056
Methanol permeability ($\mu mol/cm^2 min$)	4.66	2.16	1.72	4.35

The microstructures of Nafion 115 membrane and Nafion-PFA nanocomposite membrane were examined with a scanning electron microscope (SEM, Philips Cambridge S360). The membranes were freeze-fractured in liquid N2 for SEM observations. A low voltage (5 kV) was operated to lower electron beam energy and avoid damage to the membranes. The representative cross-sectional images of the samples are shown in Figure 4. A protruded or rugged texture is observed from the plain Nafion 115 (Fig.4a). Flake-interlaced compact morphology develops when PFA is incorporated, and uniformly dispersed in Nafion structures (Fig.4b). The dimension of cluster aggregates evidently increases, and the thickness of interstitial regions in Nafion significantly decreases as the amount of PFA increases.

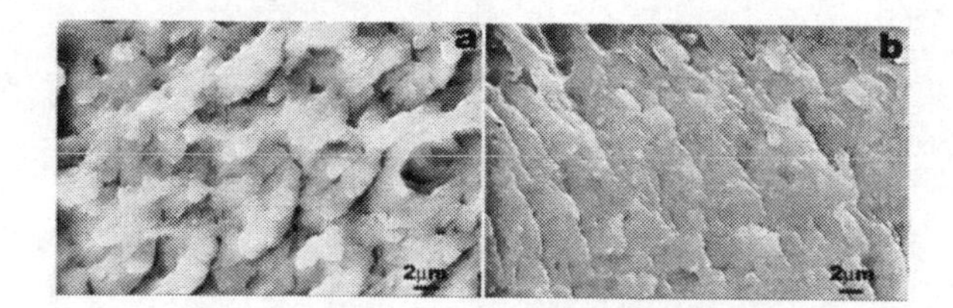

Fig. 4. SEM micrographs of cross-sections of fractioned unmodified Nafion membrane (a) and PFA-Nafion membrane (b)

The cathode electrode for the membrane –electrode assembly (MEA) was prepared by pasting a mixture of the required amount of the carbon-supported 60 wt% Pt catalyst (E-TEK) and 5 wt% Nafion solution (Aldrich) onto a teflonized carbon cloth (E-TEK) and drying in air at 80°C for 1 h. The anode was prepared by pasting a mixture of the required amount of the carbon-supported 60 wt% $Pt_{50}Ru_{50}$ onto a non-wet-proofing carbon cloth (E-TEK) and drying in air at 80°C for 1 h. Both electrodes prepared were then coated with Nafion solution (5wt% Nafion solution, Aldrich) and dry at 80°C. The metal loadings in cathode and anode were 2.5 mg cm^{-2} and 2 mg cm^{-2}, respectively. MEAs were fabricated by hot-pressing the made anode and cathode onto a pretreated Nafion or a polyfurfuryl alcohol modified Nafion 115 membrane at 140°C for 3 min. The performances of MEAs were evaluated with a self-made single cell test system with 4 cm^2 active geometrical area using galvanostatic technique. A 10 vol % methanol solution was supplied to the anode at 10 ml/min and an atmospheric dry air was flowed to the cathode. The performance at room temperature and 60° C of the MEAs is shown in Figs. 5.

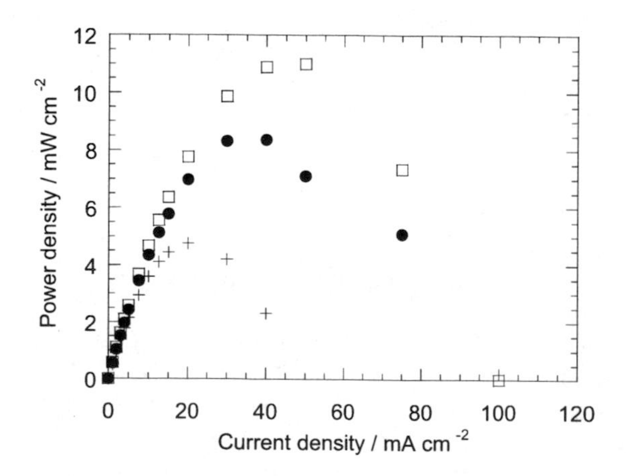

Fig.5. Cell performance of DMFC single cell made with unmodified Nafion 115 (crosses) membrane and two Nafion-PFA nanocomposite membranes operating at 60°C and ambient air.

ACKNOWLEDGEMENT

Support from Research Grants Council (HKU 7005/03P, HKU 7072/01P) is acknowledged.

REFERENCES

[1] 1. X. Zhang and K.-Y. Chan, "Microemulsion Synthesis and Electrocatalytic Properties of Platinum-Ruthenium Nanoparticles", Chemistry of Materials, 15(2) (2003) 451.

[2]. X. Zhang, K.Y. Tsang, and K.Y. Chan, "Electrocatalytic Properties of Supported Platinum-Cobalt Nanoparticles with Uniform and Controlled Composition", J. Electroan. Chem., 573 (2004) 1.

[3]. K.Y. Chan, J. Ding, Jiawen Ren, S.A. Cheng, and K.Y. Tsang, "Supported Mixed Metal Nanoparticles for Fuel Cell Electrode", J. Mater. Chem, 14 (2004) 505.

[4]. Jie Ding, Kwong-Yu Chan*, Jiawen Rena, and Feng-shou Xiao, "Platinum and Platinum-Ruthenium Nanoparticles Supported on Ordered Mesoporous Carbon and their Electrocatalytic Performance for Fuel Cell Reactions", Electrochimica Acta, to appear in 2005.

[5]. J. Liu, H.T. Wang, S.A. Cheng, and K.Y. Chan, "Nafion-polyfurfuryl alcohol nanocomposite membranes with low methanol permeation", Chem. Comm., 2004, 728.

[6]. J. Liu, H. Wang, S. Cheng, and K.Y. Chan, "Homogenous Nafion-polyfurfuryl alcohol nanocomposite membranes for direct methanol fuel cells", J. Membr. Sci., 246 (2005) 95.

Nano architecture MEA for next generation fuel cell

J. Schneider, G. Picard and M.F. Seye
Nanometrix Inc., 329 West, rue de la Commune, suite 200, Montréal (Québec), Canada H2Y 2E1, juanschneider@nanometrix.com

ABSTRACT

A new method for mass production of Membrane Electrode Assembly Fuel Cell (MEAFC) is presented. The Schneider-Picard (SP) method for industrial monolayer and ultra-thin film deposition has shown a great effectiveness for the production with nano scale architecture for fuel cells. Today, MEA's are built in such a way that no control is possible at the nano scale. Nanometrix process opens a new way showing the importance on the PEM performance of the effective triple contact line made by the contact between the colloidal Nafion, the platinum nano particles and the hydrogen gas access for high efficient catalytic reaction with ultra-low platinum loading.

Nanometrix industrial monolayer allows designing the catalytic reaction surfaces. Similar performance was obtained even after reducing the platinum loading by 100X. Furthermore, platinum-carbon chunks were substituted by the Carbon Nanotubes coated with platinum. Doing so, we were able to verify that the catalytic performance was extremely dependent on the spatial organization of the nano reactors. Platinum-PEM triple contact line and gas access are all important factors that are adjusted with the SP method to web process fuel cell MEA's with innovative nano materials and architecture, deployable now.

Keywords: nanotechnology, fuel cells, web, monolayer, platinum-CNT, catalyst, membrane electrode assembly

1 INTRODUCTION

Nanometrix was founded based on a new process for surface architecture at the nanoscale and up. This novel approach was a cherished idea since 1930's when another great inventor, physicist and Nobel Prize winner, developed, at the General Electric's facilities, the first monolayer characterisation tool, the Langmuir method. Mrs. Blodgett added shortly later the transferring principle, forming together the well known Langmuir-Blodgett technique. But, this process never achieved its most important goal: bring monolayers and their properties to the industry. This situation seemed to us like a dam without a turbine.

The SP method was invented to respond to this basic problem. Doing so, many sectors may now benefit from 50 years of fundamental research in the field of monolayers, worldwide.

Energy is one of the most important fields of application. Fuel cells are interfacial devices for catalytic reactions and monolayers are 2D interfacial architectures as well. It was natural for us to see the potential breakthrough.

A monolayer is an assembly of elements stacked side-by-side in an organized manner. So far, four sectors are already in development for specific applications. One of these sectors is the fuel cell electrode. The SP method was proof tested for making continuous high speed production of Membrane Electrode Assembly (MEA) and is described here.

Today, MEA's are built in such a way that no control is possible at the nano scale [1]. Slurries containing colloidal Nafion and carbon chunks (Figure 1) supporting platinum nano dots attached to their surfaces are mixed and "printed" on both sides of a proton exchange membrane (PEM). Performance and costs are the main drivers in MEA's business. Unfortunately, this way of handling nano dots hasn't shown any major improvement.

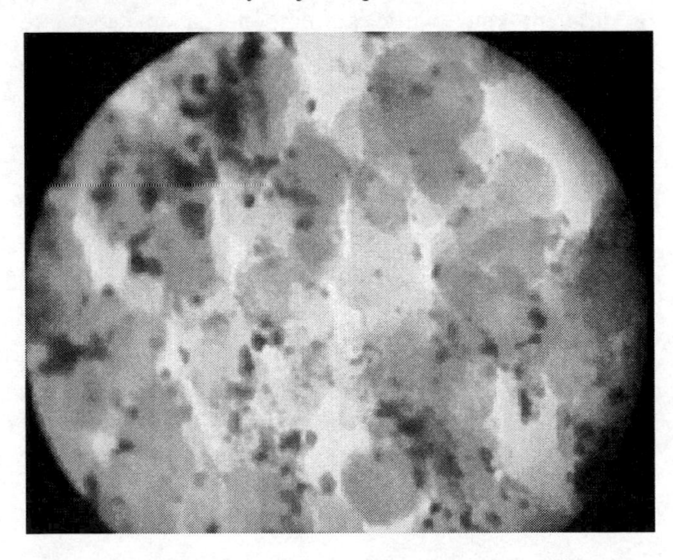

Figure 1: TEM image of carbon-platinum colloidal chunks.

Nanometrix MEA assemblies were bench tested and compared to commercially available MEA's with 0.1 mg/cm^2 as well as with a 0.3 mg/cm^2 of platinum load on each side of the proton exchange membrane.

Our coatings have shown significant catalytic performance. Similar power densities have been obtained with less than 10X platinum than a commercially available membrane with 0.3 mg/cm^2 of platinum loading.

Therefore, we first explain the basic ideas behind the SP method. This is followed by examples of MEAFC and their behavior. A discussion concludes this document with the future trends and new architecture design for catalytic surfaces.

2 NANOMETRIX FUNDAMENTALS

Nanometrix presents in this document a breakthrough technology, a new thin film generator concept to make monolayers and engineered surfaces at the nanometric scale, suitable for a vast myriad of applications, and more specifically to the fuel cell energy production. Energy is an important human concern and our approach is showing preliminary encouraging results for the industrial fabrication of Polymer Electrolyte Fuel Cells (PEFC).

In order to avoid confusion with the terminology, a monolayer is defined as a one layer thick assembly of elements. More than 20 different materials in nature and properties have been tested in our present industrial production tool. From nano diamonds to polymers, ceramics and CNT's have been tested successfully obtaining amorphous as well as tightly packed polycrystalline monolayers and ultra thin films. Multilayers have also been fabricated with similar and heterogeneous colloids, clusters, molecules and polymers of different kind. Particles from one nanometer up to colloids of a tenth of a micron have been assembled on flexible and rigid substrates.

Certainly that MEA's are not a straightforward and simple surface structure. This is a functional structure where, spatial organization, environment, humidity, pressure, temperature, gas flow and conductivity have to work simultaneously in a proper manner. This is a complex assembly of heterogeneous materials.

Structural-functional relations and interdependency have been studied by many R&D groups, testing different recipes, most of them from a macroscopic empirical approach. Others have been working through a more theoretical view combining, surface architecture and composition to maximize performance.

The SP method has shown great flexibility in handling the platinum, carbon, Nafion and membrane at once. The process has achieved single pass multilayer as well as the possibility to use a transfer membrane before application of the electrode on both sides of the proton exchange membrane. Multilayer structures have been also produced. Up to three complex assembly layers having a combination of Nafion colloids, platinum and carbon chunks for each layer, one on top of the other, have been produced showing no technical limitations for complex multilayer assembly.

3 THE TRIPLE CONTACT LINE: "THE THIN LINE OF EFFICIENCY"

Hydrogen gas is made of a pairs of atoms. The diatomic molecule is separated in two atoms at the surface of the platinum dot. Afterward, the atoms are separated into elementary charges at the triple contact line made with platinum, Nafion and hydrogen gas. Positive charges migrate through Nafion leaving the negative charges to find their path in a conductor, the carbon chunks so far.

With today's fuel cell configuration, a simple mathematical model applied to calculate for a 0.1 mg/cm^2 platinum loading, we obtain a total contact line length in the range of about 50 linear km/cm^2! This could mean that the catalytic loading ($grams/cm^2$) should be refined to be the parameter to characterize MEA performance. In fact, for a monolayer of nano platinum particles with for instance 10 nm diameter, arranged side by side in contact with a PEM, we obtain roughly the same linear contact length for the same unit surface. An experiment performed with cm^2 platinum vacuum deposition [2] validates our approach. Indeed, with the SP method, no vacuum is needed and the process runs at a speed of m^2 per minute with our present machine.

As mentioned before, Nanometrix technology is capable of producing MEA's with a performance matching commercially available membranes.

Our first target was to produce a MEA monolayer assembly. This was achieved with different carbon-platinum particles. Secondly, carbon platinum chunks where treated to reduce their average size. After treatment, the particles showed an average diameter of 100 nm. Nafion colloids and platinum nano dots were observed by TEM. Their diameter had a polydisperse size distribution with a mean value of 20 nm in diameter for the first one and 5 nm for the platinum.

It is interesting to point out that at this coating thickness, Nanometrix MEA's looked transparent enough to be able to read through it showing the small amount of platinum used. Figure 2 shows a commercially available 0.3 mg/cm^2 platinum load membrane compare to Nanometrix MEA's. Commercially available coated membranes have no light coming through the coated membrane.

Nafion colloids were entrapped in between the carbon chunks carrying the platinum nano dots and the Nafion membrane. The amount of Nafion colloids was tailored for better performance and minimum material needed. We wanted to expose as much as possible the catalytic reaction sites to the Nafion barrier with less intermediate material.

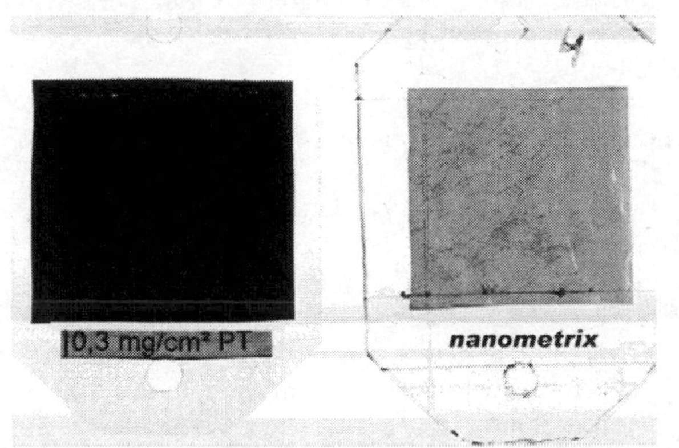

Figure 2: Nanometrix light transmission photography comparison to a commercially available MEA.

Nafion colloids were entrapped in between the carbon chunks carrying the platinum nano dots and the Nafion membrane. The amount of Nafion colloids was tailored for better performance and minimum material needed. We wanted to expose as much as possible the catalytic reaction sites to the Nafion barrier with less intermediate material. Performance was still there. Indeed, compared to a commercially available membrane with a loading of 0.1 mg/cm^2 of platinum, our assembly showed similar performance with almost 100X less platinum.

In the market, integrators are looking for membranes with higher energy density. Then, our next milestone was to produce a membrane with power performance comparable to a commercially available membrane, in the range of 0.3 mg/cm^2 of platinum. The results are showing almost the same behaviour but with substantially better performance. Indeed, 10X less platinum were needed showing great potential and performance with Nanometrix approach. Figure 3 present these results.

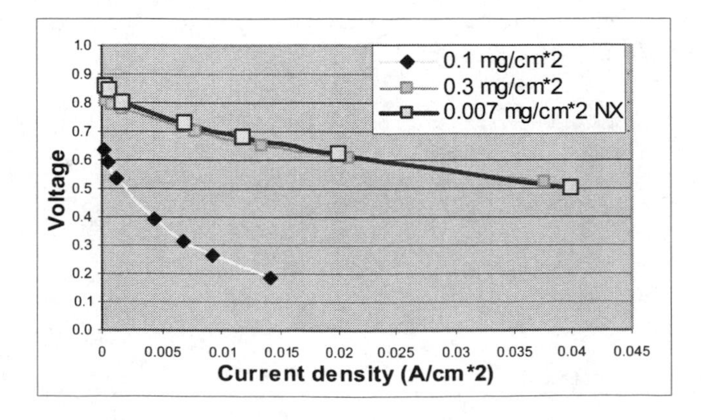

Figure 3: Polarization curves of three different platinum loadings. Nanometrix membrane load is 0.007 mg/ cm^2. It matches the 0.3 mg/cm^2 commercially available MEA's under the same conditions in our PEM test cell.

3 MONOLAYERS VS. SLURRY FILM

The slurry film presents several problems:

Diffusion: from one side, hydrogen and oxygen have to diffuse through fine pores. Diffusion being proportional to the fourth power of pore size, the **bigger the particles** are, the better the diffusion.

Electric contact: electrons must travel from the platinum to carbon chunks, and then from chunks to chunks up to the electrode. The **smaller the particles** are, the better the contacts.

The proton must travel from the platinum-Nafion-hydrogen triple contact line to Nafion colloid ionic channels to other ionic channels up to the Nafion membrane. The **smaller the particles** are, the better the proton conduction.

Flooding: water vapour produced should evacuate at the cathode against in coming oxygen and nitrogen. To avoid flooding, the **bigger the particles** are, the better the gas diffusion.

Fluid dynamics supports these points. Most of the concepts presented above are related to practical facts in many domains of physics and chemistry, and not only on the fuel cell technology.

It is evident that the requests go in opposite directions forcing the industry to fabricate electrodes through the difficult task of well balancing conditions using macroscopic tools. Therefore, all efforts of the present fuel cell industry are concentrated to find the optimum combination. It is possible to foresee that a monolayer of pure platinum colloids would have no problems with the hydrodynamic flow, more particularly with:

- Diffusion
- Electric contact
- Ionic channels
- Water flooding

The platinum surface and triple point are access free. The efficiency is therefore improved since the in and out flows do not have major restriction. It is expected that the yield with monolayer of platinum colloids will be at least an order of magnitude higher than the slurry film. The total mass of platinum onto the membrane should be in the order of the micrograms per square centimetre.

Due to the near molecular size of platinum nanodots, it is probable that such a monolayer on the Nafion membrane would need a performing larger structure like Carbon Nanodots. Its surface to volume ratio interfaces well Nafion and electrodes.

4 CARBON NANO TUBES

Carbon Nano Tubes (CNT) are of great performance. Extremely strong, flexible and with high conductivity, it seems to be a material of great potential for the fuel cell application. Several research groups have shown interest on the possibilities with carbon nano tubes (CNT) as platinum colloids support and charge carriers [3-5].

Further research guided us to see how nanotubes could be handled by our process. The SP Method showed easiness in handling this novel material. CNT's were pre treated by a new method for platinum dots adsorption on the surface of CNT's. This new catalytic material is under study and performance tests. Further data will be presented on the poster about recent results with platinum-CNT's.

Figure 4 is a TEM picture of CNT's mat with platinum nano dots on their surface. Our first approach was to deposit the CNT-platinum assembly in the same way that carbon chunks were deposited on Nafion membrane.

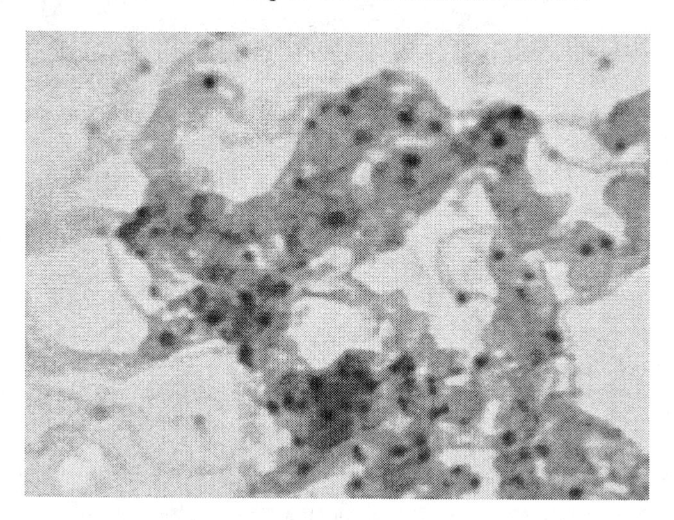

Figure 4: TEM image of platinum-CNT's mat.

5 COATED GDL's

Further development has been done with the gas diffusion layer (GDL) [6]. GDL can be also treated to be the receiving substrate where the catalytic reaction will take place. This eliminates the need of carbon chunks or other electric contact material. Platinum monolayers, combined with Nafion colloids are laid down and attached to the GDL closest fibers to the Nafion membrane. This approach is also currently tested in our facilities.

6 CONCLUSION

The Schneider-Picard method is a breakthrough technology. It is showing encouraging results in different domains of application. This technology can handle commercially available material used by the present fuel cell industry. New materials like CNT's have been successfully pre treated to carry platinum colloids becoming a possible substitution for the carbon chunks. Further development is in progress to improve efficiency and doing so, improving performance at the lowest cost possible.

The method has shown great versatility as well as ease of use, making the near future promising for significant advancements in the fuel cell industry.

REFERENCES

[1] V. Mehta and J. S. Cooper, Review and analysis of PEM fuel cell design and manufacturing, Journal of Power Sources 114, 2003, pages 32-53

[2] R. O'Hayre, S.-J. Lee, S.-W. Cha and F. B. Prinz, A sharp peak in the performance of sputtered platinum fuel cells at ultra-low platinum loading, Journal of Power Sources 109, 2002, pages 483-493

[3] Z. Liu, X. Lin, J. Y. Lee, W. Zhang, M. Han and L. M. Gan, Preparation and Characterisation of Platinum-Based Electrocatalysts on Multiwalled Carbon Nanotubes for Proton Exchange Membrane Fuel Cells, Langmuir 18, 2002, pages 4054-4060

[4] C. Wang, M. Waje, X. Wang, J. M. Tang, R. C. Haddon, and Y. Yan, Proton Exchange Membrane Fuel Cells with Carbon Nanotube Based Electrodes, Nano Letters Vol.4, No.2, 2004, pages 345-348

[5] W. Li, C. Liang, W. Zhou, J. Qiu, Z. Zhou, G. Sun and Q. Xin, Preparation and Characterization of Multiwalled Carbon Nanotube-Supported Platinum for Cathode Catalysts of Direct Methanol Fuel Cells, J. Phys. Chem. B 107, 2003, pages 6292-6299

[6] E. Antolini, Recent developments in polymer electrolyte fuel cell electrodes, Journal of Applied Electrochemistry 34, 2004, pages 563-576

Viscoelastic Relaxation and Molecular Mobility of Hyperbranched Poly(ε-caprolactone)s in Their Melt State

Hee Jae Song*, Jeongsoo Choi**, and Seung-Yeop Kwak***

School of Materials Science and Engineering, Seoul National University, San 56-1, Shinlim-dong, Gwanak-gu, Seoul 151-744, Korea,
*shjae01@snu.ac.kr, **jeongsu2@snu.ac.kr, ***sykwak@snu.ac.kr

ABSTRACT

The dynamic viscoelastic relaxation behavior and the molecular mobility of a series of hyperbranched poly(ε-caprolactone)s (HPCLs) possessing the molecular architectural variation and their linear counterpart (LPCL), were characterized and evaluated in conjunction with the different lengths of the linear backbone segments, and the different relative degrees of branching (DBs). The relative DBs, determined by the branching ratio values, decreased with increasing the linear backbone lengths. Dynamic viscoelastic relaxation measurements exhibited unentangled behavior for HPCLs compared to the apparently entangled linear, and the parallel $G'(\omega)$ and $G''(\omega)$ curves were observed for the HPCLs, while the LPCL exhibited a typical curve. The molecular mobility of three HPCLs, determined by the correlation time, τ_c, and the activation energy, E_a, was found to be higher than that of LPCL, and was observed to enhance with decreasing lengths of oligo(ε-caprolactone) segments and increasing relative DB.

Keywords: hyperbranched poly(ε-caprolactone), dynamic viscoelastic relaxation, molecular mobility; correlation time, activation energy

1 INTRODUCTION

Recently, the molecular motion in polymers has received considerable attention as a bridge to connect the structure-property relationships since the aspect of molecular behavior is inevitably related to the chemical structure and molecular environment and seems to exert the greatest influence on the relevant physical and mechanical properties in the end use.[1-3] Hyperbranched polymers have been considered to be developed into tailor-made materials with high performance and/or novel functionality at a reasonable cost due to the advantageous synthetic simplicity together with the unique mechanical, rheological, and compatibility properties.[2-6] In order to correlate the specific molecular structure and architecture with the unique properties in hyperbranched polymers, it is prerequisite to elucidate the molecular motion, which is systematically characterizing and explaining the correlation.

In the present paper, hyperbranched poly(ε-caprolactone)s (HPCLs), designed to incorporate the different lengths of the linear homologous oligo(ε-caprolactone) backbone segments consisting of 5, 10, and 20 ε-caprolactone monomer units (thereby referred to as HPCL–5, –10, and –20, respectively), were synthesized, and their dynamic viscoelastic relaxation behavior and the molecular mobility were compared to their linear counterpart (LPCL) whose chemical structure are identical and molecular weights are similar to those of HPCLs. The relative degrees of branching (DBs) for the HPCLs were determined by the branching ratio values calculated from the ratio of mean-square radius of gyration of each HPCL to that of LPCL. The focal point of this study is to elucidate and compare the dynamic viscoelastic relaxation behavior and the molecular mobility of the HPCLs and the LPCL in their melt state in conjunction with the molecular architectural difference which is the different length of the linear backbone segments and the different relative DB.

2 EXPERIMENTAL

2.1 Materials

Hyperbranched poly(ε-caprolactone)s (HPCLs), designed to incorporate the different lengths of the linear homologous oligo(ε-caprolactone) backbone segments consisting of 5, 10, and 20 ε-caprolactone monomer, were synthesized. HPCLs were synthesized according to a reaction developed by Trollsås ea al.[7] linear poly((ε-caprolactone) (LPCL), whose chemical structure and molecular weight is similar to those of the HPCLs, was commercially purchased and used as a linear counterpart to the HPCLs.

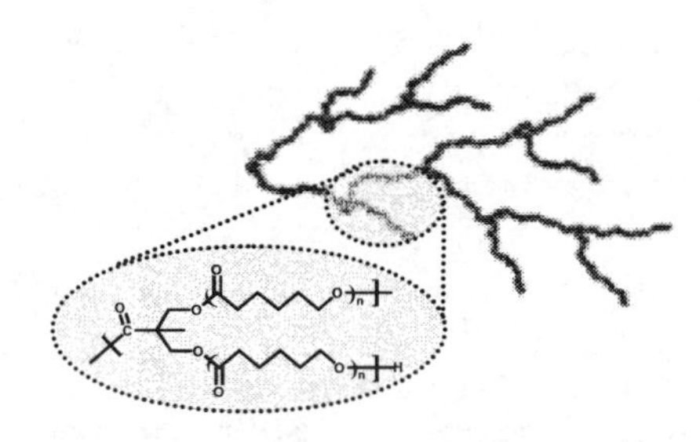

Figure 1: Schematic draw of hyperbranched poly(ε-caprolactone).

2.2 General Characterizations

Measurements of ^{1}H NMR and SEC-MALLS were performed to confirm molecular structures and weights of samples. The radii of gyration of samples were measured by small angle X-ray scattering (SAXS). The melting temperatures were determined by differential scanning calorimetry (TGA).

2.3 Dynamic Viscoelastic Relaxation Measurements

The dynamic viscoelastic relaxation properties of three HPCLs and an LPCL were measured with a Paar Physica UDS-200 mechanical spectrometer. The measurements were performed in dynamic shear oscillatory mode using 25 mm diameter parallel disc geometry with a gap setting of ca. 2 mm. The range of angular frequency was from 0.1 to 100 rad/s and the range of temperature was selected from 55 to 175 °C for three HPCLs, and 60 to 180 °C for an LPCL at an interval of 20 °C. The lowest temperature of the measurements was limited above the melting temperature, i.e., T_m+ 5 ~10 °C, depending on the samples, in which the slippage between the sample and the disk did not occur. The strain amplitude of 20% was selected to be large enough for accurate torque signals and small enough to keep the material response in the linear region.

3 RESULT AND DISCUSSION

3.1 Synthesis and Characterizations

The HPCLs were prepared to have the intrinsically different lengths of backbone segments by the use of the different AB_2 macromonomers, and referred to as HPCL–5, –10, and –20, respectively. Listed data in Table 1 are the general characteristics of three HPCLs and their linear counterpart, LPCL. The molecular weights were not significantly varied among the HPCLs and also similar to that of the LPCL.

Table 1: General characteristics of hyperbranched poly (ε-caprolactone)s and their linear counterpart.

Sample	$\langle N_{AB2}\rangle^{a}$	M_n, NMR^{b}	M_n^{c}	M_w/M_n^{d}
HPCL-5	8.1	11,510	11,800	1.8
HPCL-10	5.1	12,700	12,600	1.6
HPCL-20	3.3	15,100	15,700	1.5
LPCL			10,700	1.4

[a] Average number of the AB_2 macromonomer units incorporated in the HPCLs determined by ^{1}H NMR
[b] Number average molecular weights determined by ^{1}H NMR
[c] Obtained from SEC-MALLS

The information about DB of the HPCLs were alternatively characterized through the comparison of the molecular dimensions of hyperbranched molecules against their linear counterpart, taking into consideration the fact that as more highly branched structures are imparted, the molecular dimensions get decreased. The information about DB obtained in this way is a relative value. It is noteworthy that the branching ratio, g, which is the ratio of mean-square radius of gyration given hyperbranched polymer to that of its linear counterpart of the same chemical structure and the analogous molecular weights, provides an alternative means to characterize the branching structure and hence obtain information about relative DB. [8]

$$g = \frac{\langle R_g^{2}\rangle_{branched}}{\langle R_g^{2}\rangle_{linear}} \tag{1}$$

The parameter g is always less than 1 and decreases with the increase in DB because the branched molecules become more compact compared with the less branched molecules or the corresponding linear counterpart of similar molecular weight. The root-mean-square radii $\langle R^2\rangle^{1/2}$ (equivalently called as radii of gyration R_g) of three HPCLs and an LPCL were determined by SAXS by performing weighted nonlinear, least-square fits to the scattering curves over the accessible q range with the Zimm particle scattering function, $P_{Zimm}(R_g,q)=1/[1+q^2R_g^2/3]$, which was reported to be the most suitable for determining R_g for hyperbranched polymers.[9] Listed in Table 2 are the radii of gyration for all HPCLs and an LPCL and calculated values of branching ratio. The relative DB increased in the order of HPCL-20<HPCL-10<HPCL-5, indicating that the relative DB increased with the decrease in the length of linear oligo(ε-caprolactone) segments in the HPCLs.

Table 2: Radius of gyration and branching ratio of hyperbranched poly(ε-caprolactone)s and their counterpart.

Sample	R_g (nm)^{a}	g^{b}
HPCL-5	4.57	0.76
HPCL-10	4.77	0.83
HPCL-20	5.18	0.98
LPCL	5.23	1

[a] Determined from SAXS curves fit by Zimm scattering function.
[b] Branching ratio.

3.2 Dynamic Viscoelastic Relaxation Behavior and Molecular Mobility

From the master curves, all HPCLs and the LPCL exhibited the $G''(\omega)$ over $G'(\omega)$ with an increase in both $G'(\omega)$ and $G''(\omega)$ along shear rate within the experimental

NSTI-Nanotech 2005, www.nsti.org, ISBN 0-9767985-1-4 Vol. 2, 2005

temperature range, which is consistent with the terminal flow behavior for a non-entangled polymer. Also observed curve pattern was that the slopes of the logarithmic plots of $G'(\omega)$ versus ω and $G''(\omega)$ versus ω for the LPCL had values near 2 and 1, respectively, which are in good agreement with the recognized fact that $G'(\omega)$ is proportional to ω^2 and $G''(\omega)$ to ω in their terminal flow region.[10, 11] On the other hand, three HPCLs exhibited parallel $G'(\omega)$ and $G''(\omega)$ within the experimental temperature range, even though these polymers were expected to be in their terminal region from the measured temperature range ($T \approx T_m$+5 °C – T_m +125 °C). The parallel $G'(\omega)$ and $G''(\omega)$ have been reported in the previous works on dynamic viscoelastic behaviors of star polymer melts[12] and branched polymer melts having long chain branching and/or short chain branching.[13,14] The junction point of $G'(\omega)$ and $G''(\omega)$, occurring between the rubbery plateau and terminal flow, was observed only in the LPCL in the high frequency limit, while no crossovers were present for all HPCLs within the whole frequency region. (Figure 2)

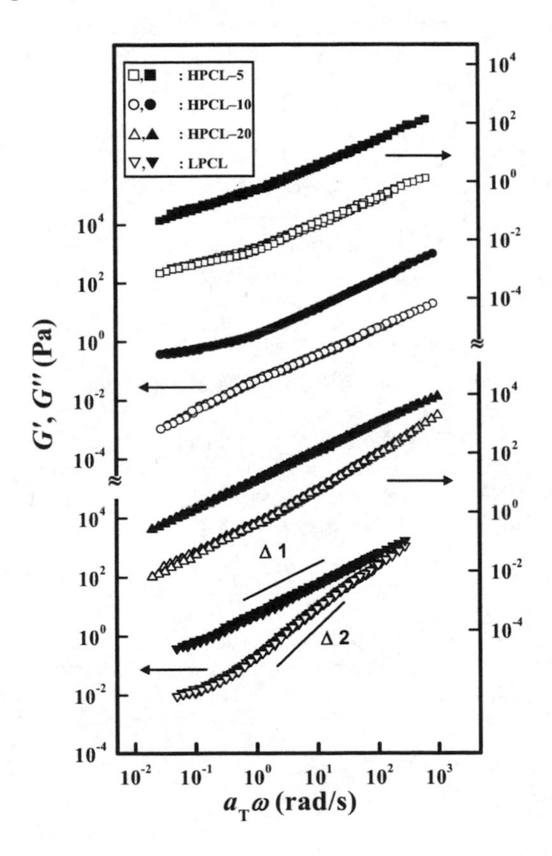

Figure 2: Master curves of the dynamic moduli $G'(\omega)$ (blank) and $G''(\omega)$ (filled), representing the behavior over an extended frequency scale (T_0=115 °C for HPCLs and 120 °C for LPCL). The moduli data were shifted vertically for clarity.

The dynamic mechanical data must be treated with empirical fitting functions in order to represent the relaxation behavior and determine the single average relaxation time. One of such empirical fitting functions with widespread acceptance is the Havriliak-Negami (H-N) distribution function of relaxation times. The empirical H-N equation is given by [15]

$$G^* = G' + iG'' = G_\infty + \frac{(G_0 - G_\infty)}{[1 + (i\omega\tau_{HN})^\alpha]^\gamma} \tag{2}$$

with $0 < \alpha, \gamma < 1$ where α is a parameter characterizing a symmetric broadening of the distribution of relaxation times and γ characterizes an asymmetrical broadening. G^* is the complex shear modulus, G_0 and G_∞ are the relaxed and unrelaxed modulus that can be estimated from the values of G' at low- and high-frequency, respectively, and τ_{HN} (τ_c) is the characteristic relaxation time. The correlation times of three HPCLs and an LPCL as a function of temperature are shown in Figure 3. Through the whole range of experimental temperatures, all HPCLs exhibited shorter τ_c values than those of the LPCL, indicating the HPCLs possess faster molecular motion than that of their linear counterpart. Moreover, among three HPCLs, as the length of oligo(ε-caprolactone) segments decreased and as the relative DB increased, the τ_c values decreased, and hence the molecular motion was fastened.

The molecular mobility evaluation are further confirmed by analyzing the apparent activation energy, E_a, determined by a fitting of the τ_{HN}'s to the following Arrhenius equation.[16]

$$\tau_{HN} = \tau_{HN,0} \exp\left(E_{a,\tau_{HN}} / RT\right) \tag{3}$$

where, $\tau_{HN,0}$ is the pre-exponential factor and R is the gas constant. Recognizing that the apparent activation energy corresponds to the barrier height for the potential hindering motion, the molecular mobility of three HPCLs was higher than that of their linear counterpart. In addition, the molecular mobility of each HPCL was enhanced with the decrease in length of oligo(ε-caprolactone) segments and the increase in the relative DB, which are in good agreement with the τ_{HN} results and can be explained by following aspects. The higher relative DB made the branches in the final HPCLs shorter, and intrinsically shorter branches were incorporated in the final HPCLs by using the AB_2 macromonomers having shorter oligo(ε-caprolactone) segments. Therefore, molecular motion in three HPCLs was generated more easily with the lower activation energy because of the existence of more mobile shorter branches, which were comparable to the LPCL possessing no such branched structure.

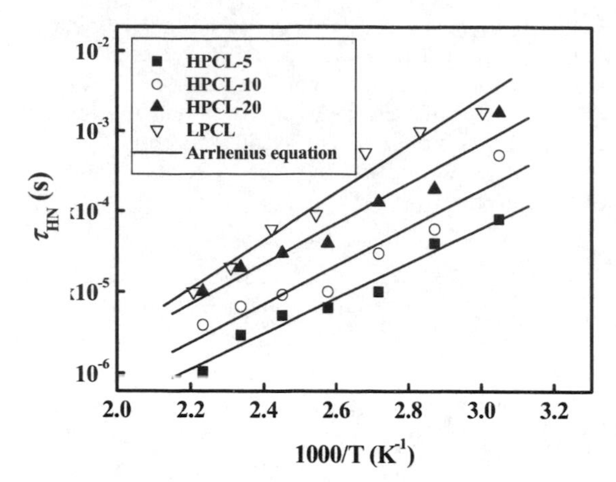

Figure 3: Correlation times, τ_c's, for HPCL–5 (■), HPCL–10 (○), HPCL–20 (▲), and LPCL (▽) as a function of temperature. The solid lines are the fits to the Arrhenius equation.

4 CONCLUSION

In the present paper, three hyperbranched poly(ε-caprolactone)s (HPCL–5, HPCL–10, HPCL–20) were prepared with the structural variation in the backbone chains, which was the different lengths of homologous oligo(ε-caprolactone) segments, and compared with respect to their linear counterpart (LPCL) whose chemical structure and molecular weight were similar to those of the HPCLs.

1. The molecular weights of the resulting HPCLs were not significantly varied. The ratio of mean-square radius of gyration of each HPCL to that of the LPCL, termed as branching ratio, resulted in the relative degree of branching (DB) for individual HPCL, which was found to be considerably influenced by the length of linear oligo(ε-caprolactone) segments; the shorter the oligo(ε-caprolactone) segments, the higher relative DB in the order HPCL–5 > HPCL–10 > HPCL–20.

2. The master curves, where three HPCLs and an LPCL exhibited the $G''(\omega)$ over $G'(\omega)$ with an increase in both $G'(\omega)$ and $G''(\omega)$ along shear rate within the experimental temperature range, indicate terminal flow behavior of non-entangled polymers. The HPCLs exhibited the parallel $G'(\omega)$ and $G''(\omega)$, of which origin is not clear but have been reported for the star polymers and branched molecules, while typical $G'(\omega)$ and $G''(\omega)$ versus ω curve was observed for the LPCL.

3. The correlation time, τ_c, was determined from the dynamic shear loss moduli and the empirical Havriliak-Negami (HN) equation, which provided a unique means to evaluate the molecular mobility. From the τ_c measurements, it was concluded that the molecular mobility of the HPCLs was higher than that of their linear counterpart, LPCL, and was enhanced with the increase in the relative DB of the HPCLs through the entire range of experimental temperatures.

4. From the curve fittings of the correlation times with the Arrhenius equation, novel information about the temperature-dependence of the molecular mobility and the apparent activation energy were provided. The lower apparent activation energy was resulted for the hyperbranched molecules compared to that of their linear counterpart, and the activation energy of HPCLs was decreased with increasing relative DB, indicating the higher molecular mobility and hence the easier molecular motion in the HPCL molecules as the hyperbranched structures were endowed with the shorter intrinsic branches and with the increase in the relative DB.

REFERENCES

[1] R. T. Bailey, A. M. North and R. A. Pethrick, "Molecular Motion in High Polymers," Oxford University Press, New York.
[2] S.-Y. Kwak and H. Y. Lee, Macromolecules, 33, 5536, 2000.
[3] S.-Y. Kwak and D. U. Ahn, Macromolecules, 33, 7557, 2000.
[4] J. M. Frechet, Science, 251, 887, 1994.
[5] O. W. Webster, Science, 251, 887, 1991.
[6] Y. H. Kim and O. W. Webster, J. Am. Chem. Soc., 112, 4592, 1990.
[7] M. Trollsås and J. L. Hedrick, Macromolecules, 31, 4390, 1998.
[8] S. Podzimek, J. Appl. Polym. Sci., 54. 91. 1994.
[9] T. J. Prosa and R. Scherrenberg, J. Polym. Sci. B: Polym. Phys., 35, 2913, 1997.
[10] C. D. Han and M. S. Jhon, J. Appl. Polym. Sci., 32, 3809, 1986.
[11] C. D. Han and J. K. Kim, Macromolecules, 22, 4292, 1989.
[12] D. Vlassopoulos, M. Pitsikalis and N. Hadiichristidis, Macromolecules, 33, 9740, 2000.
[13] J. R. Dorgan, D. M. Knauss, H. A. Al-Muallem, T. Huang and D. Vlassopoulos, Macromolecules, 36, 380, 2003.
[14] D. Yan, W.-J. Wang and S. Zhu, Polymer, 40, 1737, 1999.
[15] S. Havriliak and S. Negami, Polymer, 8, 161, 1967.
[16] J. D. Ferry, "Viscoelastic Properties of Polymer," Wiley, New York, 3rd ed., 1980.

Synthesis and Characterization of a Series of Star-Branched Poly(ε-caprolactone)s with Nanoscale Molecular Architectural Variation

Jeongsoo Choi[*], Hee Jae Song[**] and Seung-Yeop Kwak[***]

School of Materials Science and Engineering, Seoul National University,
San 56-1, Shinlim-dong, Gwanak-gu, Seoul 151-744, Korea
[*]jeongsu2@snu.ac.kr, [**]shjae01@snu.ac.kr, [***]sykwak@snu.ac.kr

ABSTRACT

A series of star-branched poly(ε-caprolactone)s (SPCLs) was synthesized with nanoscale molecular architectural variation in arm numbers and lengths by ring-opening polymerization under bulk condition. Arm numbers were varied to be 3, 4, and 6 by the use of multifunctional initiating cores while arm lengths were varied by controlling molar ratio of monomer to initiating hydroxyl group molar ratio. Molecular weights were determined by both ^{1}H NMR end-group analysis and MALDI-TOF mass spectrometry. The branching architecture of SPCLs was evaluated by the branching ratio, *g*, which is the ratio of the mean-square radius of SPCL to that of liner counterpart, linear poly(ε-caprolactone) (LPCL). The radii of gyration of SPCLs and LPCLs were determined using small-angle X-ray scattering (SAXS) from the initial slopes of Zimm plots. Branching ratios as well as thermal properties and crystallinity were found to be also dependent on structural variations.

Keywords: star-branched polymer, poly(ε-caprolactone), branching ratio, thermal property, crystallinity

1 INTRODUCTION

Macromolecules with complex molecular architectures have drawn great attention among many scientists and engineers in relations to the functional materials for nanoscale application as new and/or improved material properties can be achieved by altering their specific molecular architecture.[1] One such example in this context is the dendritic macromolecules, which are largely classified into dendrimers, hyperbranched, and star-branched polymers.[2] Star-branched polymers, which can be distinguished by the structure containing several linear arms of similar molecular weight that emanate from a central core, represent a special case of branched polymers. Like dendrimers, they can possess a globular architecture and defined inner and peripheral groups, which imparts unique properties to the molecules.[3] At the same time, similarly to hyperbranched polymers, their synthesis can be accomplished expeditiously, which makes them promising candidates for practical application.[4] Moreover, owing to the progress in controlled polymerization techniques, star-branched polymers can be easily prepared with narrow molecular weight distribution and with predictable lengths and numbers of arms under modest condition.[5]

Recently, the ring-opening polymerization (ROP) methods have been increasingly adopted for preparing dendritic macromolecules with well-defined building blocks, controlled molecular weights, and narrow molecular weight distributions. Hedrick et al. developed a novel synthetic approach to hyperbranched as well as dendrimer like star-branched aliphatic polyesters with a range of molecular architectural variation such as different lengths of linear backbone segments in each building block, the size of macromolecules, and the total molecular weights by means of ROP of ε-caprolactone.[6] However, the architectural characteristics as well as the correlation between architecture and properties of these hyper- and/or star-branched poly(ε-caprolactone)s have been much less pervasive than other types of dendritic macromolecules. Moreover, previous researches had focused mostly on synthetic variation of the composition of building blocks.

In this paper, we will describe the syntheses of a series of star-branched poly(ε-caprolactone)s (SPCLs) possessing architectural variations on arm numbers and lengths by means of ROP of ε-caprolactone. Chemical structures of synthesized SPCLs were confirmed by both ^{1}H and ^{13}C NMR spectroscopy, and the absolute molecular weights were determined by end-group analysis and matrix-assisted laser desorption/ionization time-of-flight (MALDI-TOF) mass spectrometry. The branching architectures were compared by the branching ratio estimated from the ratio of the mean-square radius of gyration of a given SPCL to that of its linear counterpart. Here, small-angle X-ray scattering (SAXS) was utilized for determining the radii of gyration. The thermal properties and the degree of crystallinity were evaluated using differential scanning calorimetry (DSC) and by performing a thermogravimetric analysis (TGA).

2 EXPERIMENTAL

2.1 Synthesis

Star-branched poly(ε-caprolactone)s (SPCLs) were prepared by ring-opening polymerization of ε-caprolactone (CL, Aldrich) which was initiated with multifunctional initiating cores in the presence of catalytic amount of tin

(II) 2-ethylhexanoate, $Sn(Oct)_2$. Variation in the number of arms was accomplished by using different initiating core molecules having 3 (trimethylol propane, TMP), 4 (pentaerythritol, PTO), and 6 (dipentaerythritol, DPTOL) hydroxyl groups (Scheme 1). On the other hand, the arm lengths were varied by controlling monomer-to-hydroxyl group molar ratio, which was $[CL]_0/[-OH]_0 = 5$, 10, and 15. The average degrees of polymerization of CL onto each hydroxyl group, n in Scheme 1, were found to be perfectly controlled by this method. The basic reaction procedure of the SPCLs is as follows. 500 mmol of CL and a certain amount of multifunctional core were put into the reaction flask, which was followed by the three repeated session of evacuation and argon purging processes. The flask was then immersed into an oil bath stabilized at 110 °C with vigorous stirring to form a homogeneous mixture, to which the catalytic amount of $Sn(Oct)_2$ was added. The evacuating and argon purging processes was repeated again and the polymerization was allowed to proceed under dry argon atmosphere for 24 h. On the basis of the different arm numbers (m) and arm lengths (n), the resulting SPCLs of the present study were accordingly designated as mSPCL–n where $m = 3, 4, 6$, and $n = 5, 10, 15$, respectively.

Linear poly(ε-caprolactone)s (LPCLs) were prepared by the same procedure as that of the SPCLs, except that EtOH which was used as an initiator and $[CL]_0/[-OH]_0$ was varied to be 17, 32, 42, and 63, respectively, in order to control the molecular weights of the resulting LPCLs.

Scheme 1: Synthesis of star-branched poly(ε-caprolactone)s

TMP → $Sn(Oct)_2$ → 3SPCLs

PTOL → $Sn(Oct)_2$ → 4SPCLs

DPTOL → $Sn(Oct)_2$ → 6SPCLs

$n = 5, 10, 15$

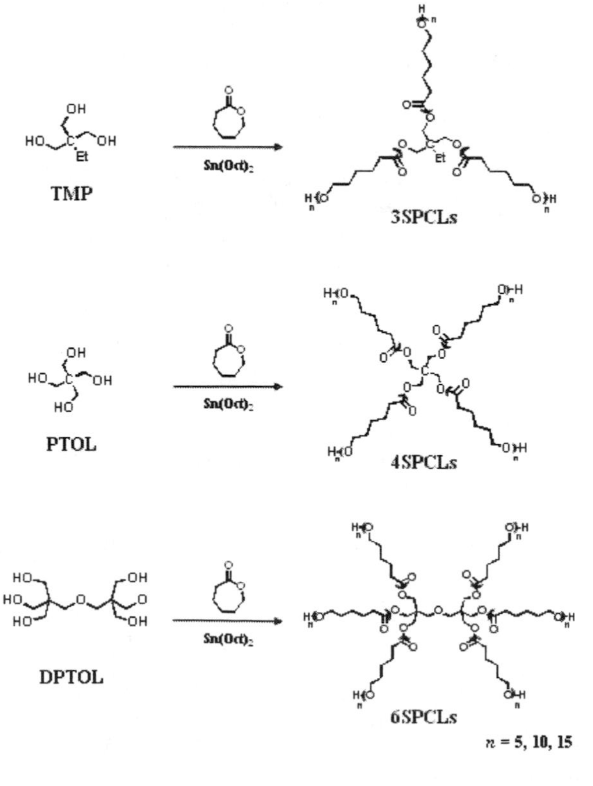

2.2 Characterization

Chemical structures of synthesized SPCLs as well as LPCLs were characterized via ^1H and ^{13}C nuclear magnetic resonance (NMR) spectroscopy. Both ^1H and ^{13}C NMR spectra were acquired on a Bruker Avance DPX–300 (300 MHz for ^1H and 75 MHz for ^{13}C) spectrometer using $CDCl_3$ as solvent and tetramethylsilane (TMS) as an internal standard. Matrix-assisted laser desorption/ ionization time-of-flight (MALDI-TOF) mass spectrometric measurements carried out on a PerSeptive Biosystems-Voyager-DE STR spectrometer equipped with a nitrogen laser (337 nm, 3 ns pulse width) in deflected mode. The accelerating voltage was fixed at 20 kV, the grid voltage 80 %, and the guide wire voltage 0.1 %. The radii of gyration, R_g's, of the SPCLs and their linear counterparts, LPCLs, were determined by small-angle X-ray scattering (SAXS). The SAXS intensity distribution, $I(q)$, was measured as a function of scattering vector, q, with a rotating-anode X-ray generator (Bruker AXS Nanostar) operated at 40 kV and 35 mA, of which X-ray source was a monochromatized CuKα (λ = 1.54 Å) radiation. Melting temperature, T_m, and the heat of fusion, ΔH_m, for the SPCLs were measured by differential scanning calorimetry (DSC) with a TA Instruments DSC 2920 at a heating rate of 20 °C/min under a nitrogen atmosphere. Thermogravimetric analysis (TGA) was carried out on a TA Instruments TGA 2050 thermal analyzer at a heating rate of 20 °C/min under a nitrogen atmosphere and the masses of samples were approximately 5-10 mg.

3 RESULTS AND DISCUSSION

3.1 Synthesis of SPCLs

Synthetic routes to the star-branched poly(ε-caprolactone)s (SPCLs) with the variation in both arm numbers and lengths are well represented in Scheme 1. The arm number variation was realized by using different multifunctional cores having 3, 4, and 6 hydroxyl groups which were TMP, PTOL, and DPTOL, respectively. ROP of CL with variable molar ratios of CL-to-initiation hydroxyl group in multifunctional cores ($[CL]_0/[-OH]_0 = 5$, 10, and 15) enabled a controlled variation of arm length of the resulting SPCLs. Table 1 summarizes the conditions for preparing SPCLs and LPCLs. The chemical structures of the resulting SPCLs were confirmed from both ^1H NMR and ^{13}C NMR spectra. In addition, ^1H NMR spectra of SPCLs and LPCLs were also analyzed to determine the average degree of polymerization of arm segments of SPCLs, $\langle DP_{CL} \rangle$, and the number-average molecular weights, $M_{n,NMR}$. In ^1H NMR spectra of SPCLs, the peak assigned to the chain ends ($-CH_2OH$, δ 3.65) and the peak assigned to the repeating methylene units ($-COCH_2$, δ 2.31) in poly(ε-caprolactone) segments were quite distinguishable, therefore, the $\langle DP_{CL} \rangle$ values could be easily calculated from the ratios of the integrated area of these peaks.

sample	Initiator entry	Number of OH	$[CL]_0$ /$[-OH]_0$	1H NMR		MALDI-TOF		DSC		DSC
				$<DP_{CL}>$	$M_{n,NMR}$	$M_{n,MALDI}$	M_w/M_n	T_m (°C)	X_c (%)	T_{d10} (°C)
3SPCL–5	TMP	3	5	5.6	2,100	2,000	1.08	37.9	52.9	319
3SPCL–10	TMP	3	10	10.4	3,700	3,600	1.17	45.5	54.6	322
3SPCL–15	TMP	3	15	15.4	5,400	5,300	1.22	50.2	56.1	326
4SPCL–5	PTOL	4	5	5.5	2,800	2,800	1.09	40.0	52.2	320
4SPCL–10	PTOL	4	10	10.6	5,100	5,100	1.17	45.7	54.1	323
4SPCL–15	PTOL	4	15	15.5	7,400	7,200	1.18	50.5	55.5	326
6SPCL–5	DPTOL	6	5	5.3	3,900	3,800	1.12	40.7	50.5	320
6SPCL–10	DPTOL	6	10	10.4	7,400	7,400	1.17	45.3	53.6	322
6SPCL–15	DPTOL	6	15	15.2	10,700	N/A	N/A	51.2	55.1	327

Table 1: Synthetic conditions and general characteristics for SPCLs and LPCLs.

As listed in Table 1, the calculated $<DP_{CL}>$ values for SPCLs as well as those for LPCLs were found to be in good agreement with those of the targeted values, which are solely predictable from the $[CL]_0/[-OH]_0$ ratios. The number-average molecular weights of the SPCLs and LPCLs, $M_{n,NMR}$, were calculated from the following equation (See Table 1):

$$M_{n,NMR} = MW_{Ini} + MW_{CL} \times \langle DP_{CL} \rangle \quad (1)$$

where MW_{Ini} is the molecular weight of initiating species and MW_{CL} is that of CL.

Figure 1 depicts the MALDI-TOF mass spectra for 4SPCLs. The mass difference between each adjacent peak, Δm, is 114 Da, which corresponds well with the mass of the CL repeating unit of the chain segments in SPCLs. Arm length variations made by controlled increase in the $[CL]_0/[-OH]_0$ shifted the whole mass distribution curves to higher molecular weights while the distribution shape showed no significant changes. The $M_{n,MALDI}$ and molecular weight distributions for SPCLs were derived from the mass spectra and listed in Table 1. For the SPCLs, $M_{n,MALDI}$ was found to be in good agreement with $M_{n,NMR}$.

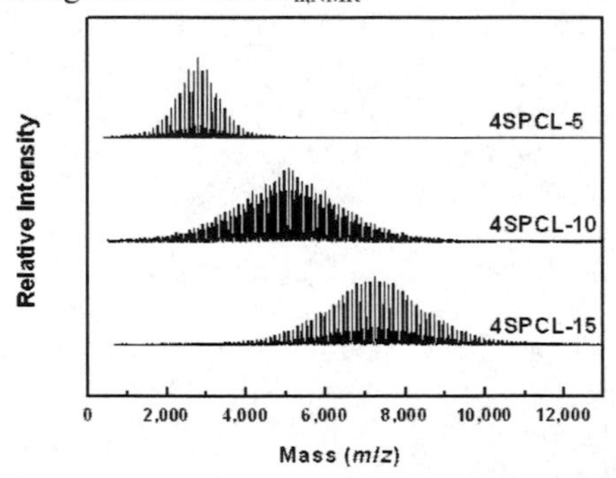

Figure 1: MALDI-TOF mass spectra of 4SPCLs.

3.2 Branching Ratio

In order to characterize the effect of the branching architecture on molecular dimension, the branching ratio, g, was estimated from the ratio of the mean-square radii of gyration between a given star-branched molecule and its linear counterpart with the same chemistry and similar molecular weight as follows:[7]

$$g = \frac{< R_g^2 >_{\text{branched}}}{< R_g^2 >_{\text{linear}}} \quad (2)$$

where $<R_g^2>_{branched}$ and $<R_g^2>_{branched}$ denote the mean-square radius of gyration of star-branched molecule and that of linear counterpart, respectively. It should be noted that the parameter g is always < 1, and decreases with increase in degree of branching, because branched molecules occupy less volume in solution than linear polymers of the same molecular weight. The radii of gyration for SPCLs as well as those for their linear counterpart were determined by small-angle X-ray scattering (SAXS). Assuming that solutions are very dilute and in a non-interacting system, the scattering curves obtained from polymer solutions can be simplified as a function of only model particle scattering function, $P(R_g,q)$, which enables us to extract the R_g values from the scattering curves by performing a weighted nonlinear, least-square that fit the scattering curves over the q range, $[q_{min}, q_{max}]$:[8]

$$I(q) = A \times P(R_g, q) + B, \ [q_{min}, q_{max}] \quad (3)$$

The most common model particle scattering function is one suggested by Zimm[7] which is shown below:

$$P_{Zimm}(R_g, q) = 1 / \left(1 + \frac{q^2 R_g^2}{3} \right) \quad (4)$$

Figure 2 shows the typical scattering curves for the SPCLs observed by SAXS. It should be noted that Zimm plot of 6SPCL–10 appeared linear over a relatively large q range.

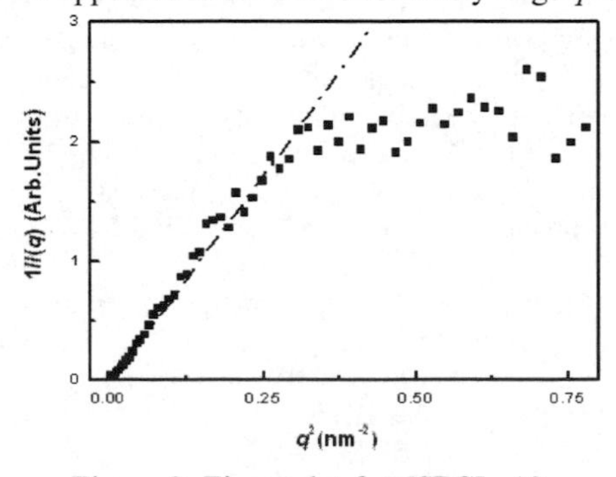

Figure 2: Zimm plot for 6SPCL–10.

In order to elucidate the effects of arm numbers and lengths on molecular dimension, the R_g values for 4SPCLs, SPCL–10's, and their counterparts were determined.

As illustrated in Figure 3, for the SPCLs with the same arm number, the g values remained almost constant (g = ca.0.8). This indicates that the segment density was not significantly varied when the arm length was increased. However, the g values were observed to decrease for SPCLs with increasing arm number having a constant arm length being constant. This indicates that more branching points resulted in more compactly packed volume in the solution.

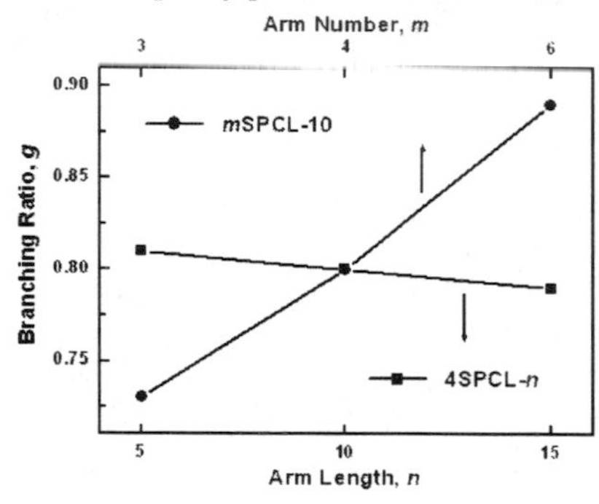

Figure 3: Dependence of g on arm lengths and numbers.

3.3 Thermal Properties

Thermal properties of SPCLs were characterized using both DSC and TGA. As listed in Table 1, the melting temperatures, T_m, for SPCLs of the same arm numbers are found to be fairly affected by arm length variation, which is assumed to be due to the presence of longer linear chains possibly facilitating the ordered chain packing to crystallize more efficiently.[9] The degree of crystallinity, X_c, of SPCLs were calculated as follows:

$$X_c(\%) = \frac{\Delta H_m}{\Delta H_m^{\ 0}} \times 100 \qquad (5)$$

where ΔH_m is the apparent enthalpy of melting and $\Delta H_m^{\ 0}$ is the extrapolated value of enthalpy corresponding to the melting of a 100% crystalline poly(ε-caprolactone), which was previously reported as 136.4 J/g.[10] As listed in Table 1, the X_c for SPCLs of the same arm number was increased with increasing arm length. However, the X_c for SPCLs of the same arm length was only minimal with increased arm numbers, which is regarded to be due to the competing effects of increased molecular weight and branching.

Thermal degradation stability of the SPCLs was evaluated using the temperature of 10% weight loss of the polymers, T_{d10}. As indicated in Table 1, the T_{d10} of 4SPCLs are seen to shift toward a higher temperature region as the arm length was increased. Considering that thermal degradation of poly(ε-caprolactone) involves an unzipping depolymerization from the hydroxyl end of the polymer chain.[11] increase in T_{d10} of SPCLs with longer arms are believed to be attributed to the lower number of polymer chain ends. On the contrary, T_{d10} for SPCLs of similar molecular weight were found to decrease as arm numbers increased, which may result in the higher number of chain ends and consequently resulting in the acceleration of unzipping degradation.

4 CONCLUSIONS

A series of star-branched poly(ε-caprolactone)s (SPCLs) was successfully synthesized with structural variation of arm numbers and length. Structural variations among SPCLs were verified to be precisely controlled as evidenced by ^{1}H and ^{13}C NMR spectra. The absolute values of molecular weight were obtained by both ^{1}H NMR end-group analysis and MALDI-TOF mass spectrometry, of which results were reasonably consistent. The molecular dimension of SPCLs was estimated using the radius of gyration, R_g, determined by SAXS curve fits to the Zimm scattering function. The branching ratio, g, was calculated from the ratio of the mean-square radius of gyration of each SPCL to that of its linear counterpart. In addition, g was observed to be fairly affected by arm number variation; the higher arm numbers, the smaller g values, and thus more compact molecular structure in order of 3SPCL–10 > 4SPCL–10 > 6SPCL–10, while the effect of arm length variation on g of SPCLs was found to be only minimal. We also found the correlative effect of structural variation on the thermal transitions as well as the degree of crystallinity. The T_m was found to increase with increasing arm length. Similarly, the T_{d10} was observed to be an increasing function of arm length. On the contrary, we did not observe any significant changes in both the T_m and T_{d10} for the SPCLs of the same arm length but with arm number variation. However, for the SPCLs of the equivalent molecular weight, the degree of crystallinity was found to decrease with increasing arm numbers.

REFERENCES

[1] J. M. J. Fréchet, Science 263, 1710, 1994.
[2] C. Gao, D.Yan, Prog. Polym. Sci. 29, 183, 2004.
[3] M. K. Mishra, S. Kobayash, "Star and Hyper-branched polymers," Dekker, 1999.
[4] D. R. Robello, A. André, T.A. McCovick, A. Kraus, T. H. Mourey, Macromolecules 35, 9334, 2002.
[5] C. J. Hawker, Acc. Chem. Res.30, 373, 1997.
[6] M. Trollsas,; J. L. Hedrick, J. Am. Chem. Soc. 120, 4644, 1998.
[7] B. H. Zimm, W. H. Stockmayer, J. Chem. Phys. 17, 1301, 1949.
[8] T. J. Prosa, B. J. Bauer, E. J. Amis, D. A. Tomalia, R. Scherrenberg, J. Polym. Sci. B: Polym. Phys. 35, 2913, 1997.
[9] H. Magnusson, E. Malmström, A. Hult, M. Johansson, Polymer 43, 301, 2002.
[10] V. Crescenzi, G. Manzini, G. Calzolari, C. Borri, Eur. Polym. J. 8, 449, 1972.
[11] Y. Aoyagi,; K. Yamshita, Y. Doi, Polym. Degrad. Stab. 76, 53, 2002

Synthesis and Characterization of Nanostructured Undoped/Doped CuO Films and their Application in Photoelectrochemical Water Splitting

Yatendra S. Chaudhary*, Rohit Shrivastav*, Vibha R. Satsangi** and Sahab Dass*

*Department of Chemistry, **Department of Physics & Computer Science, Faculty of Science, Dayalbagh Educational Institute, Dayalbagh, Agra-282 005, INDIA, sahabdass@yahoo.com

ABSTRACT

Metal oxide semiconductors are one of the most promising materials for photoelectrochemical production of hydrogen on account of a) easy availability, b) low cost, c) simple preparation methods and d) stability. The samples of undoped and (Cr/Fe) doped CuO were prepared using spray-pyrolysis technique. XRD analysis revealed the exclusive formation of CuO phase. Good adherence of films with substrate was confirmed by scotch tape method. The SEM analysis confirmed the granular surface of the film. Scherrer's calculation indicated the average grain size of the order of ~87 nm. Film samples were then converted into photoelectrodes by generating ohmic contact and photoelectrochemical behaviour was studied at pH 11 and 13 in NaOH. Enhanced photocurrent generation was observed in the sintered samples and also with the increase of electrolyte pH from 11 to 13, irrespective of nature and amount of dopant. However the variation in dopant concentration revealed a variable trend in the photocurrent generation.

Keywords: Photoelectrochemical properties, Cupric oxide, Spray-pyrolysis

1. INTRODUCTION

Versatile applications of metal oxides in technology development have infatuated the scientific community. Besides the advent of nanotechnology has given the new direction to tailor make the properties of materials for their potential high-tech applications due to high surface to volume ratio and enhanced surface effects. To overcome the problem of energy scarcity and global warming different ways of renewable energy production are being discovered. The pioneering work of Fujishima & Honda [1], followed by the development of efficient and cost effective semiconductor metal oxides for their application in the production of hydrogen by scientific community across the world.

In PEC cells, semiconductor photocatalysts are used as electrode to split water to produce H_2 – a valuable fuel, by utilizing vastly available sun light [2,3]. However, several oxides [3-13] have been investigated as privilege semiconductor materials, for this purpose, but on account of inappropriate band gaps and / or the instability of the semiconductor in the electrolyte has largely eluded the desired success so far. CuO on account of favorable bandgap (1.4 eV) absorbs throughout the visible region and is considered to be a material of choice as photoelectrode in PEC cell.

In the present work synthesis of nanostructured undoped/doped CuO films and their photoelectrochemcial behaviour has been investigated. The characterization of films by XRD and SEM has also been presented.

2. EXPERIMENTAL

2.1 Sample Preparation

Copper oxide films were deposited by spray-pyrolysis technique on conducting glass substrate (TCO). 0.1 M precursor solution was prepared by dissolving $Cu(NO_3)_2$ $2.5H_2O$ (99.99+%, Aldrich) and the calculated amount of dopant nitrate was added in double distilled water. The solution was sprayed onto heated (temperature $350^0C \pm 5^0C$) substrates of dimensions 4 cm. X 2.5 cm. Before spraying, nearly one-third length of the substrate was covered by an aluminium foil. Carrier gas used was compressed air with 3 Kg cm^{-2} pressure and spray rate was maintained at 4.4 ± 0.1 ml min^{-1}. The detailed experimental conditions and experimental set-up, used, are described elsewhere [6,14].

2.2 Characterization

The undoped/doped copper oxide films were characterized using glancing angle X-ray Diffraction Philips PW 3020 thin film diffractometer. The incident beam was slit collimated while the diffracted beam optics consisted of a thin film parallel plate Collimator and a flat graphite monochromator. Cu-Kα was used as the radiation source.

The average particle/grain size was calculated by Scherrer's equation (1).

$$B = \frac{0.9\lambda}{tCos\theta} \quad (1)$$

Where B is FWHM (full width at half maximum) of the broadened diffraction line on the 2θ scale (radians) and t is average diameter of particle/grain size. The B is given, by equation (2)

$$B^2 = B_M{}^2 - B_S{}^2 \quad (2)$$

Here B_S the measured breadth, at half maximum intensity, of the line from the standard while B_M is measured breadth of the diffraction line of sample.

In the present study surface morphology of films was performed for some arbitrarily chosen samples using JEOL JSM5800LV scanning electron microscope. The sputter coated samples with gold, were initially observed at low magnifications (a few 100 X) to get an overall picture of the features presented by the samples. Subsequently, the same features were observed under higher magnifications (up to 16000 X) to get the micro details of these features.

2.3 Photoelectrochemical Measurements

The film samples were converted into electrodes by making ohmic contact with copper wire using gallium-indium eutectic and silver paint. The exposed ohmic contact back side and edges were perfectly sealed with non-transparent and non-conducting epoxy resin, Hysol (Singapore).

The photoelectrochemical measurements involved monitoring current-voltage (I-V) characteristics, both under darkness (when the PEC cell was covered from outside using a thick black cloth cover) and under illumination (when the semiconductor electrode was illuminated using the light source) at different experimental conditions. The thin film electrode (working electrode, surface area of $1 cm^2$) was used in conjunction with platinum and saturated calomel electrodes (SCE), which were used as counter and reference electrodes, respectively. The potentiostat (Model ECDA-001, Con-Serv Enterprises) attached with PEC cell was used in this study. The light source for simulated solar radiation was a 250 W tungston-xenon lamp (Bentham) as light source. The electrolyte used was 1M NaOH (pH 13).

3 RESULTS AND DISCUSSION

The pyrolytic decomposition of copper nitrate formed its oxide. The films synthesized by this method were smooth, properly adhering with the substrate and were uniform. The film thickness of undped/doped film samples, determined by telestep was found of the order of 1.5 μm.

XRD analysis of film samples indicated the exclusive formation of CuO (Fig. 1). The formation of

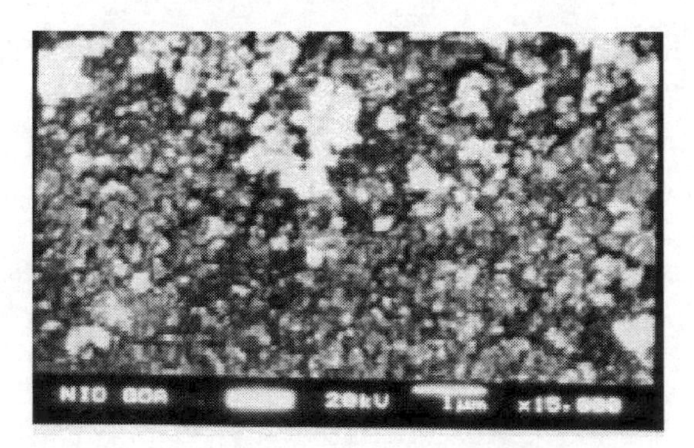

Fig. 2: Observed SEM image of CuO film (unsintered)

nanostructructred undoped/doped CuO films is clearly evident from Scherrer's calculation, which indicates the average particle/grain size of the order of 87 nm. SEM image also indicates the formation of crystalline structure. However, the average particle/grain size obtained from the SEM analysis is significantly higher then the values obtained from Scherrer's calculations. This disparity may be due the formation of particles/grains in cluster, which were infact observed by SEM analysis.

The current-voltage (I-V) characteristics of undoped CuO film electrodes before and after sintering are shown in Fig. 3(a & b). Similarly the I-V curves were obtained for other samples. The photocurrent density for all the samples is summarized in Table 1. It is evident that PEC response of doped samples was significantly higher, in most of the cases, than that of undoped samples, effect of variation in dopant concentration from 0.5% to 2% on the PEC response of the material has a varied trend. The sintered film electrodes showed enhanced photocurrent than the unsintered film electrodes. The effect may be understood in the light of the expected, and also observed through Scherrer's calculations, exhibiting clearly the increase in particles/grains size on sintering. With the increase in particles/grains size, the number of grain boundaries will decrease and this might be the reason for the observed higher photocurrent values upon sintering. The photocurrent density decreased with the increase in doping level in the case unsintered samples doped with Fe (at pH 11).

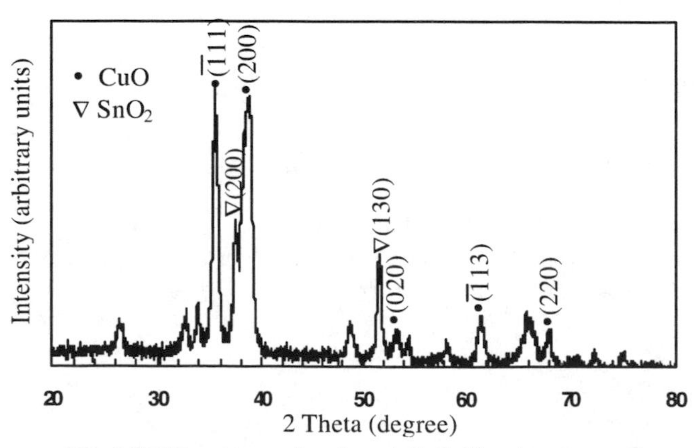

Fig.1 XRD pattern of undoped CuO film (unsintered)

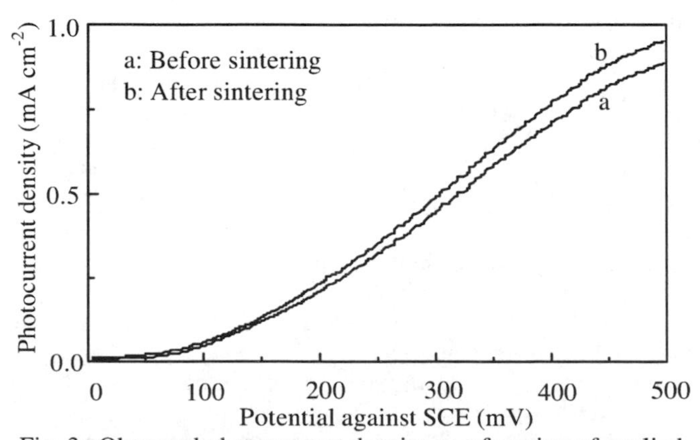

Fig. 3a Observed photocurrent density as a function of applied potential with CuO film (undoped) electrodes at pH-11

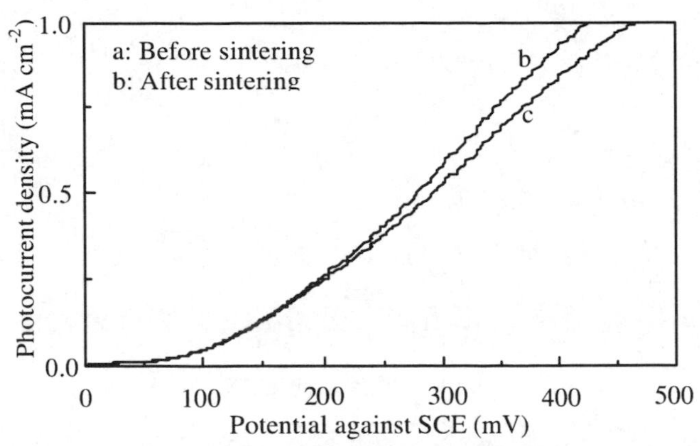

Fig. 3b Observed photocurrent density as a function of applied potential with CuO film (undoped) electrodes at pH-13

This can be explained by the increased electron concentration within quasi-neutral part of the electrode, where the probability of electron hole recombination increases. In case of unsintered Cr doped samples (at pH 11 and 13) and sintered samples doped with Cr (at pH 11) an increase in the photocurrent density with the increase in the doping concentration was seen. This might be due to the incorporation of the Cr in the CuO matrix, which increases the rate of charge transfer and hence the observed improvement in the electrochemical performance of the cell. Sintered samples doped with Cr (at pH 13) exhibit the increase in photocurrent density upto 1% doping, which decreases on increase. The decrease in photocurrent density at higher doping levels may be on account of greater rate of recombination, which is probably due to increased electron concentration within quasi-neutral part of the electrode. An unambiguous rise in photocurrent densities at higher pH may be attributed to the effect of increased ionic strength, which significantly reduces the internal resistance of the PEC cell.

4 CONCLUSION

This study reveals that nanostructured CuO film electrodes could be a potential candidate for PEC application. Better photocurrent generation was observed in the sintered samples and also with the increase in electrolyte pH from 11 to 13, irrespective of nature and amount of dopant. Thus sintering of samples and electrolytic pH also govern the PEC behaviour of material.

ACKNOWLEDGEMENT

The first author is thankful to Council of Scientific and Industrial Research, Govt. of India, New Delhi under Senior Research Fellowship award [ref. no. 9/607(20)/2003-EMR-I] for financial assistance. The authors also thank Dr. M. Shyamprasad, National Institute of Oceanography, Goa for SEM analysis.

REFERENCES

[1] A. Fujishima and K.K. Honda, Nature 238, 37, 1972.

[2] J.A. Turner, Science, 285, 687, 1999.

[3] P. Salvador, C. Guterrez, G. Campet and P. Hagenmeller, J. Electrochem. Soc. 131, 550, 1984.

[4] K.H. Yoon and K.S. Chung, J. Appl. Phys. 72, 5743, 1992.

[5] D.E. Stilwell and S.M. Park, J. Electrochem. Soc. 129, 1501, 1982.

[6] Y.S. Chaudhary, A. Agrawal, R. Shrivastav, V.R. Satsangi and S. Dass, Int. J. of Hydrogen Energy 29 (2), 131, 2004.

[7] Y. S. Chaudhary, M. Agrawal, Saroj, A. Shrivastav, R. Shrivastav and V. R. Satsangi, Ind. J. of Physics, 78A (2), 229, 2004.

[8] S.U.M. Khan and J. Akikusa, J. Phys. Chem. B 103, 7184, 1999.

[9] A. Agrawal, Y.S. Chaudhary, V.R. Satsangi, S. Dass and R. Shrivastav, Current Science 85, 101, 2003.

[10] C. Levy-Clement, A. Lagooubi, M. Newmann-Spullart, M. Robot and R. Tenne, J. Electrochem. Soc. L 69, 13, 1991.

[11] F.F. Fu-Ren and A.J. Bard, J. Am. Chem. Soc. 102, 3677, 1980.

[12] N. Chandra, B.L. Wheeler, A.J. Bard, J. Phys. Chem 89, 5037, 1985.

[13] S.M. Herman G. Franz, S.M. Mark, B. Slomn, S.U. Robert, D.N. Furlong, J. Soc. Faraday Trans. 91, 665, 1995.

[14] B. Joseph, K.G. Gopchandran, P.V. Thomas, P. Koshy, V.K. Vaidyan, Mater. Chem. Phys. 5, 71, 1999.

Table 1

Sintering Condition	pH	Photocurrent density (μA cm^{-2})						
		Undoped	Cr-doped			Fe-doped		
			0.5%	1.0%	2.0%	0.5%	1.0%	2.0%
Before sintering	**11**	440	395	546	758	850	430	360
	13	485	385	818	1087	576	516	1474
After sintering	**11**	526	709	1788	964	1235	566	155
	13	578	1451	2029	1246	1561	1475	1721

Color-masked and nano-size catalysts embedded in polymers for reducing dioxin emission

Jinseong Choi[a], Kyung-Shik Yang[b], Yong-Shik Jeong[c] and Jong-Shik Chung[d,*]

[a,b,d]*Department of Chemical Engineering,* [c,d]*School of Environmental Science and Engineering, Pohang University of Science and Technology, Pohang, 790-784, Korea.*

Email: [a]xksdir@postech.ac.kr, [b]ksyang@postech.ac.kr, [c]jys@postech.ac.kr, [d]jsc@postech.ac.kr

[*]Corresponding author: Phone: +82-54-279-2267 FAX: +82-54-279-5528.

Abstract

Instead of removing dioxin by the end-of-pipe technologies, there is an attempt currently to reduce the dioxin emission during the incineration of waste. In this concept, dioxin suppressing catalysts are pre-mixed in polymer materials during the manufacturing stage. During the calcinations of the wastes, the catalysts in the polymer can prevent from the formation of dioxin.

In this work, we have attempted to prepare polymer embedded catalysts for dioxin decomposition. Especially, nano-sized and color-masked Fe/Ti catalysts have pale or white color and showed better activity comparing individual Fe or TiO_2, and these materials were embedded inside polymer. During incineration of catalyst–embedded polymer, most of noxious gas was successfully removed. We can also control properties such as transmittance, tensile strength, etc. by changing catalytic materials embedded inside polymer.

Keywords: dioxin, Fe-based catalysts, color-masked catalysts, functional polymer, incineration

1 INTRODUCTION

Dioxin is released from stationary or mobile sources. Thermal waste incinerators are the most typical example among stationary sources. The flue gas from incinerator is composed of SO_x, NO_x, particulate matter, chlorinated volatile organic compounds (VOCs), etc.. Especially, polychlorinated dibenzodioxins(PCDDs) and dibenzofurans (PCDFs), so called dioxins, well known as a carcinogenic material is included that flue gas [1-2]. So many countries in the world have regulated those materials [3-5] due to their toxicities.

Many research groups have focused on catalytic decomposition of dioxin with end-of-pipe converter of incineration system. There are many catalysts for dioxin decomposition. Noble metals such as Pt, Pd and Ir were reported catalysts for dioxin decomposition [6-7]. But, since they are very expensive, several transition metal oxides such as CrO_x, V_2O_5-WO_3 have been reported catalysts for dioxin removal. However, CrO_x [8] and V_2O_5-WO_3 [9] based catalysts could act as toxic materials and pollutants themselves. Fe-based catalysts have alternated transition metal-based catalysts due to their environmental safety.

Instead of removing dioxin by the end-of-pipe technologies, there is an attempt currently to reduce the dioxin emission during the incineration of waste. In this concept, dioxin suppressing catalysts are pre-mixed in polymer materials during the manufacturing stage. During the incinerations of the wastes, the catalysts in the polymer can prevent form the formation of dioxin. For practical purpose, one may consider criterions when they produce the catalyst/polymer composites: these catalysts should be colorless for dyeing, highly active for dioxin decomposition and the catalyst particles are very small for homogeneous dispersion in the polymer. It was reported that nano-sized particles of α-FeOOH dispersed in polyethylene were active for reducing dioxin, but its brown color limited its usage [10].

In this work, bulk Fe-based catalysts are coated with white nano particles by zeta potential difference and Fe precursor are impregnated on pore of white nano particles in order to mask color and keep on small size. Therefore, our goals are to synthesize functional polymer embedded catalysts for dioxin decomposition and to remove dioxin and other pollutants during incineration of these polymers.

2 EXPERIMENTAL

2.1 Preparation of nano-sized Fe_3O_4 catalysts

It is important to prepare Fe_3O_4 nano particles in an aqueous solution with a molar ratio of Fe(II)/Fe(III) = 0.5. 10.285mL of 1N HCl and 25mL of purified, deoxygenated water were mixed. 8.65g of $FeCl_3 \cdot 6H_2O$ and 3.18g of $FeCl_2 \cdot 4H_2O$ were dissolved in the solution with vigorous stirring. The mother solution was dropped into 250mL of 1.5M NaOH solution under vigorous stirring. Generated precipitate was repeated centrifuging at 4000rpms and washing with purified deoxygenated water [11] until precipitate and water were not separated. The last step, particles were filtered by ultra filtration membrane (pore size 25nm) and dried at room temperature.

2.2 Preparation of color-masked catalysts

Prepared nano-sized Fe_3O_4 particles were coated by nano-sized TiO_2 using zeta potential difference under conditions of pH 6 and TiO_2/ Fe_3O_4 weight ratio 19/1.

Nano-sized FeO_x/TiO_2 catalysts were prepared by incipient wetness method. $FeCl_3$ $6H_2O$ acted as precursor and nano-sized TiO_2 (HOMBIKAT UV100, 100nm) was used as support. 5wt.% and 10wt.% Fe was impregnated on support. After impregnation, the catalyst was dried at 100℃ for 12hrs and then calcined at 500℃ for 5hrs.

2.3 Activity tests of catalysts

In order to test activity of the catalysts in a packed-bed reactor, we utilized 1,2-dichlorobenzene (*o*-DCB) instead of dioxin having similar structure and toxicity under reaction conditions of inlet concentration of 1,000ppm, weight hourly space velocity (WHSV) of 6,000 L/kg-cat/hr, and temperature of 473-973K.

2.4 Preparation of functional polymer

1wt.%, 5wt.% and 10wt.% of prepared catalysts were added into Poly Ethylene (PE) by electrically heated measuring mixer (Brabender). Procedure was like these.

① Heat the mixer to 200℃ and keep on.
② Put PE chips and catalysts
③ Mix PE and catalysts by Brabender for 7min.
④ Clean the mixer with PE until screw become clear.

In order to test tensile strength and transmittance, small film was prepared by MiniMax molder.

2.5 polymer incineration tests

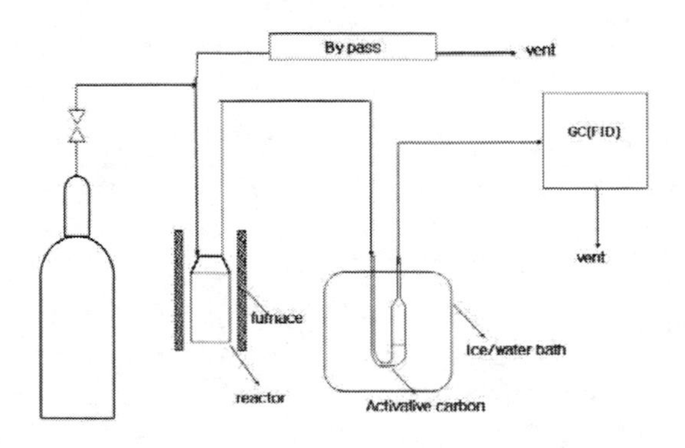

Figure 1: Apparatus for polymer incineration

Apparatus for incineration tests of polymer and functional polymer is shown in Figure 1. The flow rate of carrier gas was 100cc/min. Polymer was incinerated at 400℃. The noxious gases adsorbed on activated carbon at 0℃. And then gases desorbed from activated carbon at 400℃.

3 RESULTS AND DISCUSSION

3.1 Fe-based catalysts

Nano-sized Fe_3O_4 catalysts had black color and magnetic property. XRD pattern of Fe_3O_4 was shown in Figure 2. The size of particle was about 30nm by Scherrer equation. (Inverse triangle means typical peaks of Fe_3O_4)

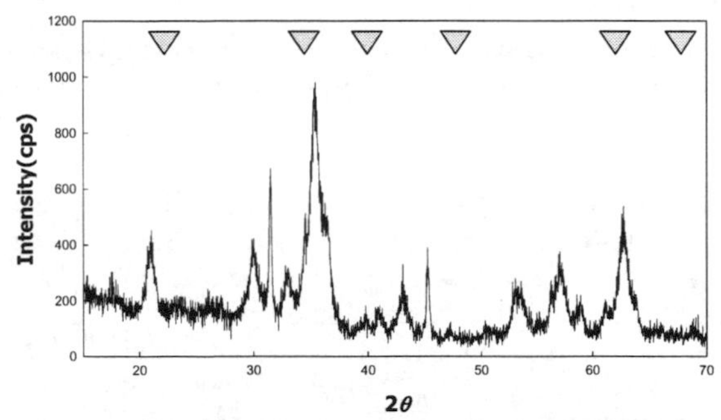

Figure 2: XRD pattern of Fe_3O_4 nano particles.

Figure 3: The SEM image of Fe_3O_4 particles.

Figure 3 shows the SEM image of the Fe_3O_4 particles. The shape of Fe_3O_4 was spherical and we could confirm its size was around 30nm. Zeta potential of Fe_3O_4 and TiO_2 particles are shown in Figure 4. Surface charge of Fe_3O_4 and TiO_2 are (+) and (-) respectively around pH 6. We, therefore, could obtain the optimal condition of zeta potential coating was pH 6.

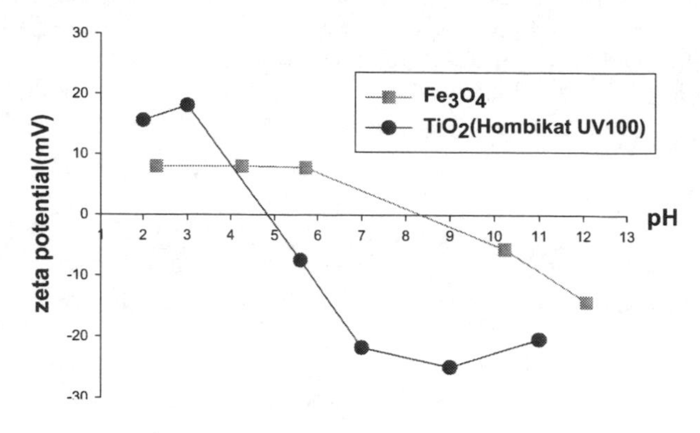

Figure 4: Zeta potential of particles.

3.2 Activity tests of catalysts

We tested activity of catalysts through *o*-DCB oxidation. As shown in Figure 5, Fe-based catalysts had conversion above 95% at about 500℃. Especially, FeO_x/ TiO_2 and TiO_2 coated Fe_3O_4 (Ti/Fe) catalysts have shown more active for *o*-DCB oxidation than bulk Fe_3O_4 due to Ti/Fe synergistic effect. In case of bulk Fe_3O_4, as temperature increases, crystal grows up and area of active site decreases. Since Fe-based catalysts are very cheap and eco-friendly, they can be used as alternative of V-based oxidation catalysts. Moreover, since their size is below 200nm and their color is pale or white, they can be applied to synthesize functional polymer.

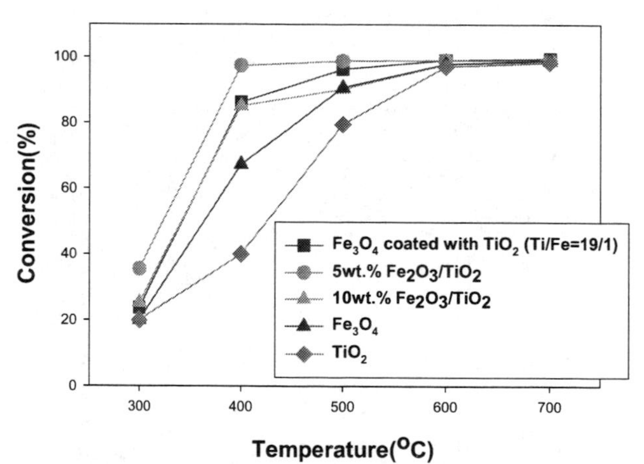

Figure 5: Activity test of catalysts by *o*-DCB oxidation.

3.3 Properties of functional polymer

PE embedded Fe_3O_4 and Ti/Fe particles were prepared and the weight ratio of particle/PE was 1/99. The transmittance of PE embedded Ti/Fe catalysts is better than in case of Fe_3O_4 particles as shown in Figure 6. As particles were added into polymer, the modulus increased. In case of blank PE, the inflection point was shown near 105℃. That point is not T_g. Commercial polymers have individual thermal history according to synthesis condition and that point is just caused by thermal history. Generally, the glass transition temperature (T_g) of polymer embedded inorganic particles increase because inorganic particles raise crystallinity between polymer chains. However T_g of PE embedded Ti/Fe particles is similar regardless of additive as shown in Figure 7. In other words, polymeric property is almost never affected by Ti/Fe catalysts. Therefore, functional polymer embedded Ti/Fe catalysts will use instead of blank PE in various applications and we will be control mechanical property according to changing amount of catalysts. So we will synthesize functional polymer with various specification according to our intention.

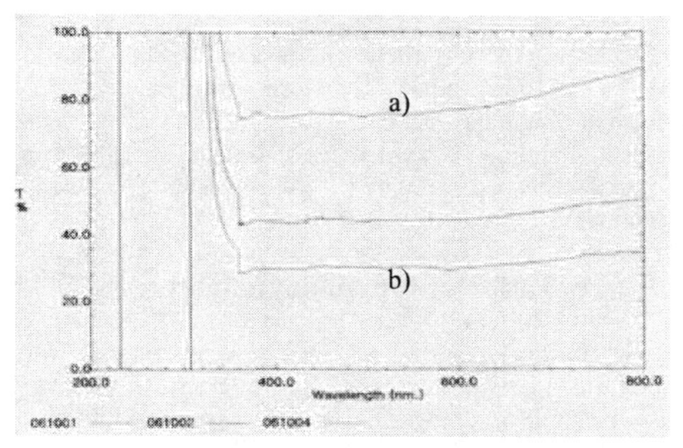

Figure 6: The transmittance comparison.
a) PE embedded Ti/Fe particles
b) PE embedded Fe_3O_4 particles

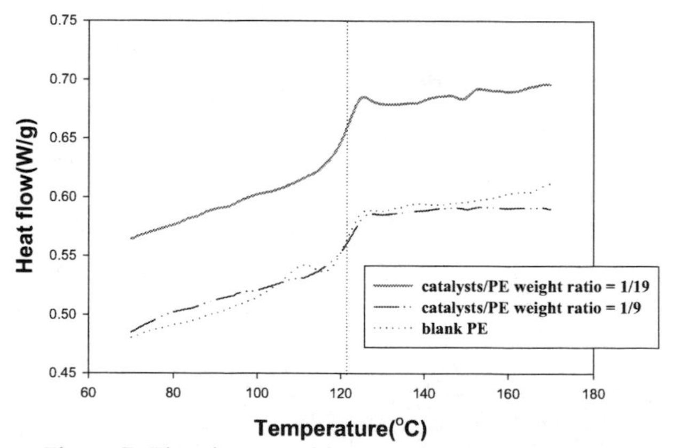

Figure 7: The glass transition temperature of polymer.

We could see that noxious gases were removed by catalysts during polymer incineration as shown in Figure 8. In case of blank PE, hydrocarbons from desorbed gases were detected. However, in case of PE embedded catalysts, hydrocarbons were removed. Especially, in case of PE embedded Ti/Fe particles, noxious hydrocarbons were almost removed.

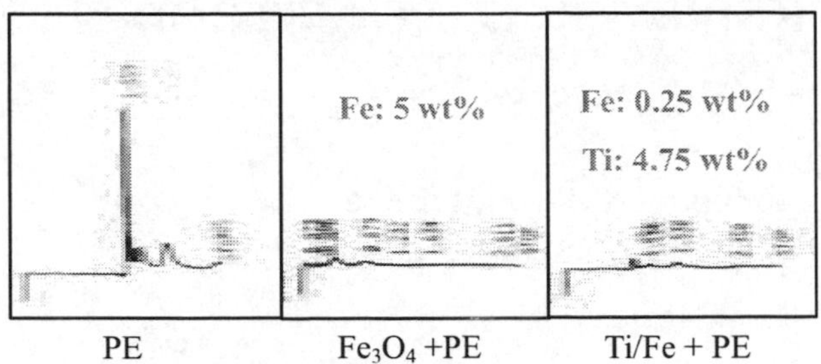

PE Fe_3O_4 +PE   Ti/Fe + PE

Figure 8: Polymer incineration test.

4 CONCLUSIONS

According to above discussion, Ti/Fe particles have better activity for *o*-DCB oxidation than bulk Fe-based particles. Fe-based catalysts are less harmful than other transition metal-based catalysts and have similar activity above 400℃. Moreover, Ti/Fe catalysts can be used as catalysts for synthesis of functional polymer embedded catalysts for dioxin removal during incineration. We have prepared PE embedded 1wt.%, 5wt% and 10wt.% of Ti/Fe particles. According to amount of particles, properties of PE were varied. However, T_g almost never changes regardless of additives. That is, polymeric property difference between blank PE and functional PE is negligible. So, functional PE can be used as alternative of PE. Through the incineration test, functional polymer embedded catalysts generate less noxious gases than blank polymer during incineration. Therefore functional polymer reduces chance for formation of dioxin.

ACKNOWLEDGEMENTS

The authors gratefully acknowledge the support from the Brain Korea 21 project and Nano Basic Research project.

REFERENCES

[1] D.J. Paustenbach, Regulatory Toxicology and Pharmacology, 36, 211-219, 2002.

[2] K. Olie, P.L. Vermeulen, O. Hutzinger, Chemosphere, 6, 455-459 1977.

[3] 17. BImSchVA vom 23.11.1990, BGB1 Ⅰ, 2545.

[4] Guideline for controlling PCDDs/DFs in MSW Management, the Advisory Committee for Controlling PCDDs/PCDFs in MSW Management, 1997.

[5] M. Hiroka, S. Sakai, T. Sakagawa and Y. Heta, Organohalogen Compounds, 31, 446, 1997.

[6] M. Hiraoka, N. Takeda and S. Okajima, Chemosphere, 19, 361-366, 1989.

[7] S. Toshihiko, I Toshihiko and S. Eiichi, European Patent 0 645 172 A1, 1995.

[8] S. Yim, D. Koh and I. Nam, Catalysis Today, 75, 269-276, 2002.

[9] R. Weber, T. Sakurai and H. Hagenmaier, Applied Catalysis B, 20, 249-256, 1999.

[10] T. Imai, T. Matsui, Y. Fujii, T. Nakai and Suminori Tanaka, Journal of Material Cycles and Waste Management, 3, 103-109, 2001.

[11] Y. Kang, S. Risbud, J.F. Rabolt and P. Stroeve, Chemistry of Materials, 8, 2209-2211, 1996.

MICROCHANNEL REACTOR STACKED BY BRAZING METHOD

S.P. Yu, W.Y. Lee, C.S. Kim, S.J Lee* and K.S. Song*

Fuel Cell Research Center, KIER, South Korea, spyu@kier.re.kr
*Catalyst Application Research Center, KIER, South Korea, kssong@kier.re.kr

ABSTRACT

Each stack of microstructured stainless steel and aluminum plates was coated and brazed in vacuum. As a supporter of catalyst, the coating layer was formed by sol-gel method and anodizing respectively. Though the number of brazed plates extended to a hundred, the thickness of coating layer on the respectively coated plate was relatively uniform and the leakage of assembly was effectively minimized. The critical variables for brazing were both thickness and shape of filler metal. Consequently, brazing method had good resolution for 200 m three dimensional multi layer structures. Through these results, coating to support catalysts in highly stacked layers by brazing can be applied to micro catalytic heat exchanger where the reaction and heat transfer occur simultaneously.

Keywords: microchannel, catalytic reaction, brazing, heat exchanger

1 INTRODUCTION

As the micro systems get more wireless and higher performed, their energy sources need to get denser. Recently, catalytic reaction which could be used to produces hydrogen is noticed as one of the issues of micro power systems. If we use chemical reaction energy directly in micro power systems, the density of that is as hundreds times as that of the latest lithium-ion battery system.

Catalytic combustion has three characteristics; the first is that the temperature of it can be controlled below 500°C or much lower, the second is that it has no limit of room for combustion such as quenching distance, and the last is that it can produce various desired chemicals in partial oxidation. Through these, catalytic reaction has been accepted to be very suitable to following micro systems ; the pre-processor of micro fuel cell system, micro chemical plant, micro propulsion system.

The microchannel reactors and heat exchangers have been made of mechanically microstructured metal foils of stainless steel, hastelloy, aluminum, copper, palladium, silver, and others [1]. A reactor is assembled from individual metal plates with micro channels having cross sections in the range of 50–300 μm. These metal plates are assembled by bolting, gaskets such as graphite or diffusion bonding. Brazing is employed to connect the parent metals : a filler metal, such as Nickel based alloys, is placed, for example in a foil form, between the parts to be brazed and heated upon its melting point. It allows to create tight chemical bonds between parts. Contrary to welding, parent parts are not melted. Below 450°C, brazing is called soldering. After brazing, there are little deformation and residual stresses in the parent metal which is not melted. Moreover, using metal as a bond, joints are tight and heat conductivity between the parent metals is higher than using graphite or polymer as a bond. On the other hand, for a wider application with chemical reactions it is necessary to provide oxide coatings as carriers for metal catalysts to increase the overall inner surface area of the microchannels [1]. Catalytic coatings are important in microstructured reactors, since they can minimize mass transfer resistance and pressure drop, and improve heat conduction for catalytic reactions. Several techniques have been utilized for coating microchannels with porous oxides: anodic oxidation which has been used to provide a porous layer on aluminum [2,3]; chemical vapor deposition [4]; deposition of nanoparticles [5]; and sol–gel process [1]. In these coating processes, sol-gel is relatively simple to use and compatible to various kind of metals, but the uniformity and adhesion of coating layers are weak. Anodizing leads to formation of unbranched, regular and nearly concentric pores, which is advantageous for catalytic reactions [6]. Moreover, the strength of adhesion of the anodic oxide layer to the support is strong. But it can only be applied to aluminum and not stainless steel.

In applying catalytic reaction to micro system, fabrication and coating to support catalysts are correlated with each other because the manufacturing process is dependent mutually. This study introduces vacuum brazing technique for assembly of microchanneled stainless steel and aluminum plates. It is also attempted to braze microchanneled stainless steel plates coated with alumina. Two cases were studied : alumina coatings were applied both before and after brazing. In the case of aluminum plates, anodizing is performed instead of the sol-gel method.

2 EXPERIMENTAL

The diagram of manufacturing process is shown in Figure 1. In the design of the micro heat exchanger, uniformity in fluid velocity distribution was considered, referring to Commenge et al. [7]. Main design factors were curvature ratio at the end of microchannels, ratio of depth to width in microchannels, a number of microchannels, length of microchannels and area ratio of wall to microchannels. Top and bottom covers were prepared by machining in 4 ×

4 cm2 of size. The name of stainless steel plate was SUS304 and that of aluminum plate was AL1050.

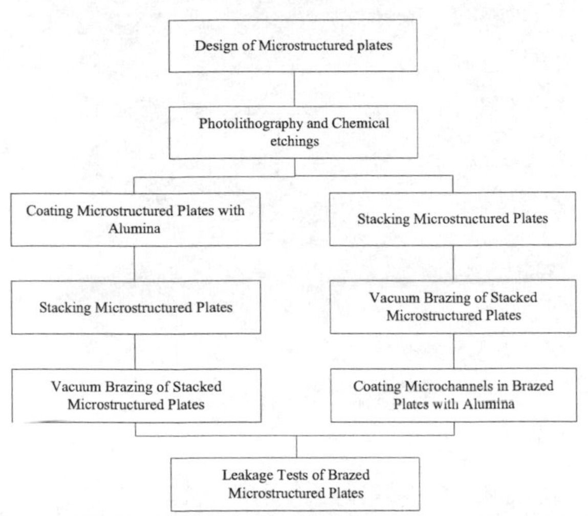

Figure 1. Diagram of manufacturing process for catalytic micro heat exchanger.

Microchannels were created by photolithography and chemical etchings and the details for microchanneled plates are summarized in Table 1.

Material	Stainless Steel	Aluminum
Plate geometry	$40 \times 40 \times 0.3$ mm^3	$40 \times 40 \times 0.3$ mm^3
Microchannel cross-sectional area	20×20mm^2	20×20 mm^2
Microchannel cross-sectional area	300 μm	500 μm
Microchannel cross-sectional area	200 μm	200 μm
Number of channels on one foil	34	21

Table 1. Details of microchanneled plates

For assembly of the microchanneled plates, they were stacked up as shown in Figure 2 and pure nickel foils were inserted between the microchanneled plates for vacuum brazing. The shapes of filler metal sheets are shown in Figure 3. The thickness of nickel foils was 38μm , and the width of them were ranged from 2.5mm to plate edge. At each case, the room between layers of brazed stack was different. If there are impurities on the surface of the parent metal, adhesion force of brazing can be diminished. Therefore, the surface of the parent metal was degreased by alcohol before brazing. Vacuum brazing lasted for 15minutes at 1000 °C, below 10^{-5} torr. The assembled microchanneled plates were tested for leakage with a leak detector (ASM 142, Alcatel). Leakage tests were performed in ambient air firstly, and then in latex globe filled with helium gas.

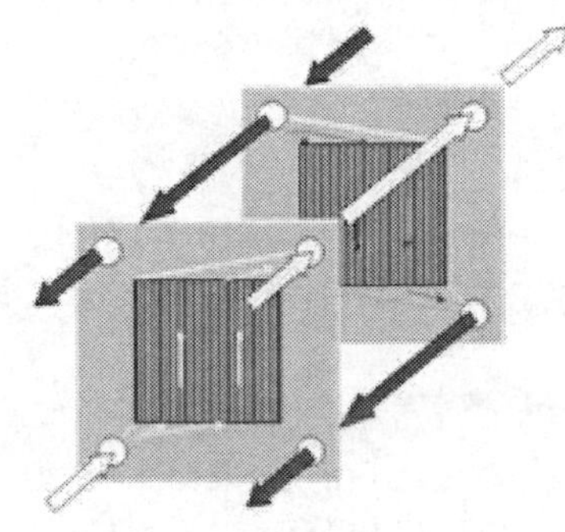

Figure 2. Schematics of stacking microstructured plates.

For incorporation of catalysts, a microchanneled stainless steel plate was coated with aluminum oxide films by dip-coating with alumina sol, followed by drying in air at 120 °C overnight. Alumina coated stainless steel plates were brazed under vacuum with nickel filler metal. On the other hand, microchannels in assembled stainless steel plates were coated by flowing alumina sol followed by blowing excess alumina sol with compressed air and then drying in air at 120 °C overnight.

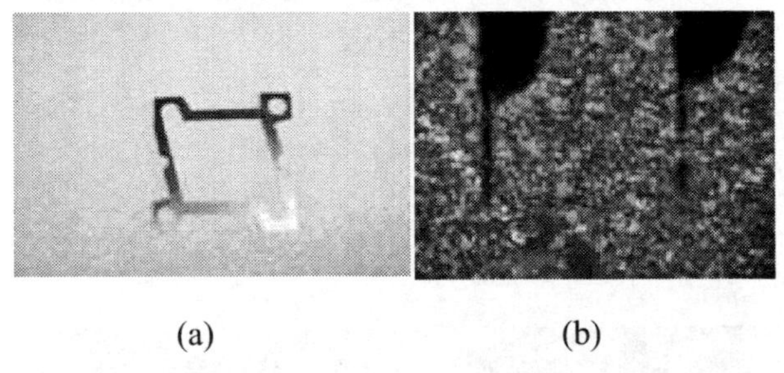

(a) (b)

Figure 3. The shapes of filler metal sheets and cross section at each brazing case. (a) nickel foil, thickness: 38μm, width 3.5mm (b) Cross section of brazed stack (×100)

The microchanneled aluminum foils were rinsed with ethanol and deionized water and then dried with compressed air. The electrode was immersed in 165 g/l H2SO4 solution which was circulated with a pump to remove heat generated during anodizing. For constant current operation, the current desired was set on a power supply. Anodizing time was varied from 10 to 30 min. The anodized aluminum foils were washed with deionized water followed by drying in air at room temperature. In aluminum stack, copper was used as filler metal and vacuum brazing lasted for 15 minutes at 400 °C, below 10-5 torr.

3 RESULT AND DISCUSSION

Microchanneled stainless steel plate and assembly of microchanneled stainless steel plates are shown in Figure 4. The assembly of microchanneled stainless steel plates was tested for leakage in air and helium environments. Leaking rates are 3.3×10^{-10} torr l/sec and 1.5×10^{-6} torr l/sec in air and helium environments, respectively. Considering that

Swagelok fittings are utilized below 1×10^{-6} torr l/sec, our results might be caused by leaking from the fittings rather than from gaps between brazed plates.

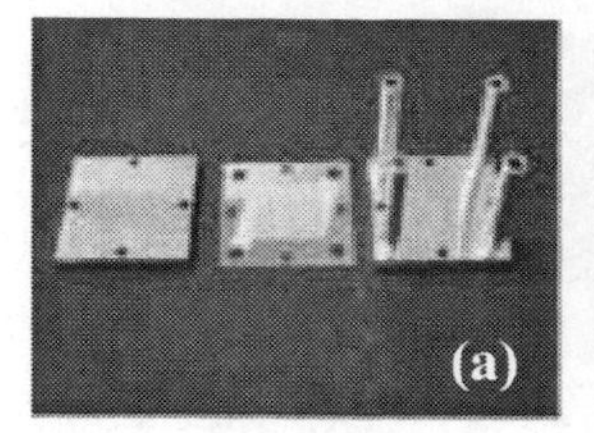

Figure 4. Photographs of (a) a microchanneled stainless steel plate and top and bottom covers and (b) assembly of microchanneled stainless steel plates by vacuum brazing.

In order to examine brazed plates, the assembly of microchanneled stainless steel plates was cut by wire-cutting method. Figure 5 shows both the cross sections and the brazed part of the assembly of microchanneled stainless steel plates. The microchannel shape is not rectangular but hemi-circular due to using photo-etchings. It is known that the amount of filler metals is important at brazing technology, as excess filler metals could block microchannels and lack filler metal could cause leaking. In our case, appropriate amount of filler metals was found by varying thickness and width of filler metals. Thus, no gaps are observed between brazed stainless steel plates (see Figure 5(b)). In the microchanneled stainless steel plates brazed and then coated with alumina, SEM picture shows that 3 μm of alumina film is successfully incorporated onto the microchannels (see Figure 6).

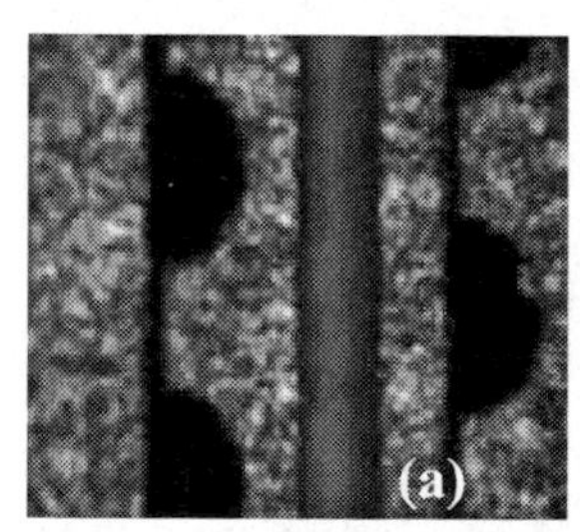

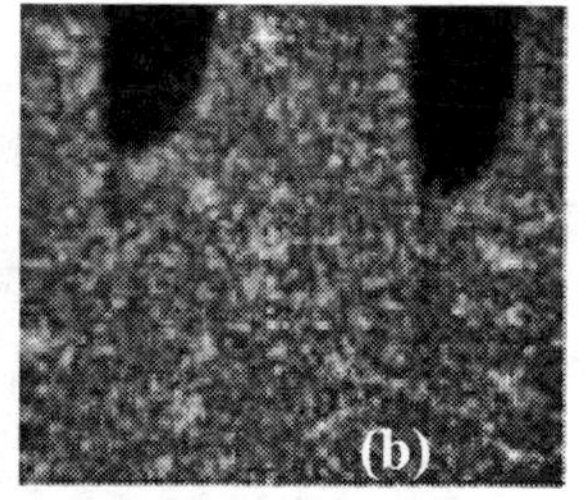

Figure 5. Photographs of (a) cross section and (b) brazed part of assembly of microchanneled stainless steel plates.

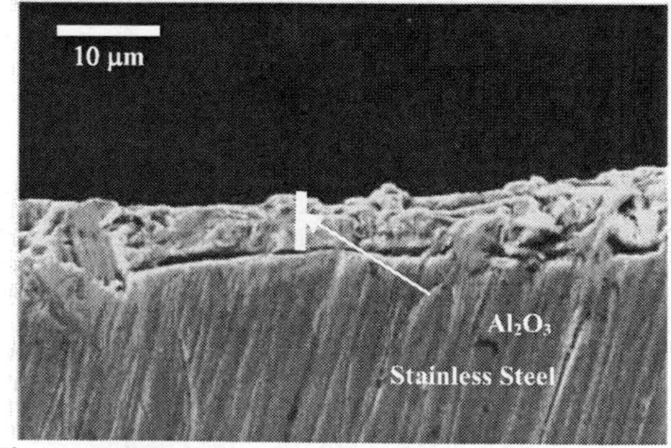

Figure 6. SEM picture of microchanneled stainless steel plates brazed and then coated with alumina.

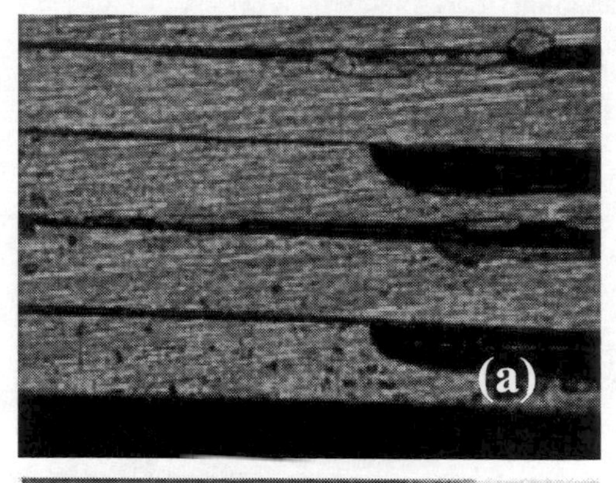

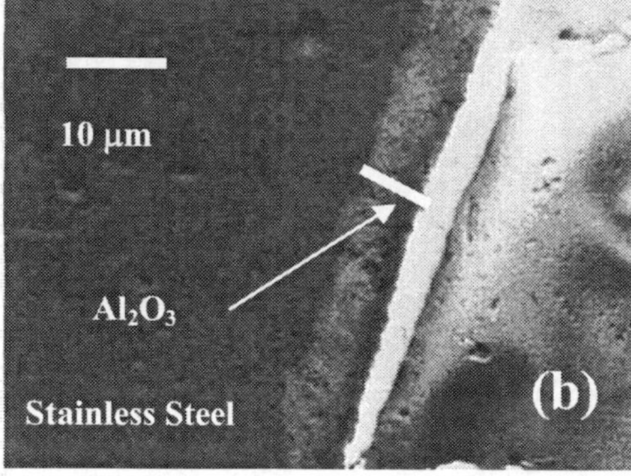

Figure 7. Photographs of (a) brazed part and (b) cross section of microchanneled stainless steel plates coated with alumina and then brazed.

In microstructured stainless steel plates coated with alumina and then brazed, a photograph of cross section of microchannels reveals that thickness of coated alumina film is about 2.8 μm (see Figure 7(b)). However, it is observed that large gaps between brazed plates are present, as vacuum brazing does not work with ceramics such as alumina (see Figure 7(a)).

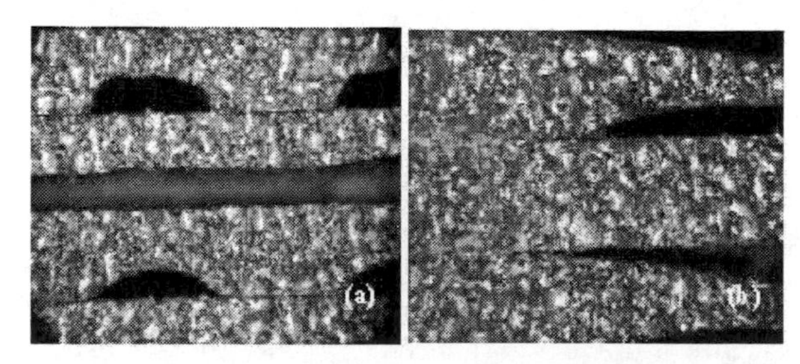

Figure 8. Photographs of (a) cross section and (b) brazed part of assembly of microchanneled aluminum plates by vacuum brazing.

As done in brazing stainless steel plates, proper amount of filler metals for aluminum brazing was investigated by varying thickness and width of filler metals. Figure 8 exhibits that cross section of brazed assembly of microstructured aluminum plates. The shape of microchannels on aluminum plates is irregular, since it is difficult to control etching rate with soft metals such as aluminum (see Figure 8 (a)). However, Figure 8(b) represents that microstructured aluminum plates can be successfully brazed as shown in brazing stainless steel plates.

4 CONCLUSION

Microstructured stainless steel plates were successfully brazed with nickel filler under vacuum. However, vacuum brazing did not work properly with the microstructured stainless steel plates coated with alumina. Thus, it was necessary to clean the surface for brazing or to coat only microchannels with alumina. It was also demonstrated that microchannels could be coated with alumina after brazing stainless steel plates. Microstructured aluminum plates were also brazed successfully. But in anodized aluminum plates brazing does not work effectively. Coating to support catalysts in highly stacked layers by brazing can be applied to micro catalytic heat exchanger where the reaction and heat transfer occur simultaneously.

REFERENCES

[1] K. Haas-Santo, O. Gorke, P. Pfeifer, K. Schubert, Chimia, 56 (11), pp.605-610, 2002.

[2] G. Wießmeier, K. Schubert, D. Hönicke, in Proceedings of the 1st International Conference on Microreaction Technology, pp. 20–26, Springer, Berlin, 1997.

[3] R. Wunsch, M. Fichtner, K. Schubert, in Proceedings of the 4th International Conference on Microreaction Technology, pp. 481–487, Atlanta, 6–10 March 2000.

[4] M. Janicke, H. Kestenbaum, U. Hagendorf, F. Schüth, M. Fichtner, K. Schubert, J. Catal. 192, pp.282–293, 2000.

[5] P. Pfeifer, M. Fichtner, K. Schubert, M. Liauw, G. Emig, in: Proceedings of the 3rd International Conference on Microreaction Technology, pp. 372–382, Frankfurt, 19–21 April 1999.

[6] G. Wießmeier, D. Hönicke, J. Micromech. Microeng, 6, pp.285. 1996.

[7] J. M. Commenge, L. Falk, J. P. Corriou and M. Matlosz, AIChE Journal, vol.48, pp.345-358, 2002.

Diamond Growth on An Array of Seeds: The Revolution of Diamond Production

James Chine-Min Sung[*,1,2,3], Ming-Fong Tai[4]

Address: KINIK Company, 64, Chung-San Rd., Ying-Kuo, Taipei Hsien 239, Taiwan R.O.C.
Tel: 886-2-2677-5490 ext.1150
Fax: 886-2-8677-2171
e-mail: sung@kinik.com.tw

[1] KINIK Company, 64, Chung-San Rd., Ying-Kuo, Taipei Hsien 239, Taiwan R.O.C.
[2] National Taiwan University, Taipei 106, Taiwan, R.O.C.
[3] National Taipei University of Technology, Taipei 106, Taiwan, R.O.C.
[4] Department of Electronic Engineering, Wufeng Institute of Technology, Chia-yi 621, Taiwan, R.O.C.

ABSTRACT

Industrial diamond production was pioneered by General Electric Company of U. S. in 1957 and followed by De Beers of S. Africa in 1961. The two companies optimized the production of diamond grits by heating alternative layers of graphite disks and metal catalyst (Fe, Ni, Co) under high pressure. Although there have been 20 countries attempting to duplicate the success of GE and DB, only Iljin Diamond of S. Korea is able to join the lucrative business starting with a technology provided by the author in 1989. In 1970s Winter of Germany developed a powdered mixture of graphite and metal and demonstrated that this assembly can greatly improve the diamond yield and quality. This technology has since been applied to make high grade saw diamond by the above "big three."

China start commercial production of diamond grits in 1966, since then, their diamond output has been growing at a much faster pace than that of the "big three." Today, Chinese makes 2/3 of diamond grits of the world consumption of about 600 tons. However, Chinese cannot make saw diamond larger than about 45 mesh (0.36 mm). Even so, the "big three" has felt the pressure of "yellow fever." As a consequence, De Beers reorganized their industrial diamond group in 2000 and renamed it Element Six, GE Superabrasives was sold to Little John in 2003, and Iljin Diamond suffered the first loss in 2003.

The world high pressure diamond synthesis is due for another technology advancement, this time, the random nucleation of diamond in the reaction cell is planted with a matrix of diamond seeds that will do away with the erratic growth of diamond. The result would be doubling of the diamond yield and quadrupling of the sales value.

Keywords: diamond synthesis, high pressure, diamond seed, diamond grit, De Beers

POSITIVE NUCLEATION CONTROL BY SEEDING

More than 700 tons of diamond were produced as superabrasives by subjecting graphite and a metal catalyst (Fe, Ni, Co or its alloy) under pressure of >5 GPa and heated to melt the catalyst. Graphite will transform into diamond by the catalytic action of the molten metal. Graphite and metal may form alternative layers in a high pressure cell, or they can be mixed as powder.

The layered structure of the cell design has many intrinsic advantages, such as it is simple to handle and easy to control. Moreover, a layered structure is particularly suited for high pressure belt apparatus of uniaxial design. In this case, the pressure is compressed and the current is transmitted in perpendicular to stacked layers. This uniaxial symmetry of cell design can allow a more uniform field of pressure and temperature, two critical factors for diamond growth. The problem of layered structure is the nucleation control and the interface uniformity. As the result of random nucleation and erratic growth, diamond yield and quality suffer.

The nucleation can be controlled by seeding of smaller diamond particles (e.g., 50 microns in size), for example, by fitting these diamond particles in holes drilled on catalyst (Showa-Denko 1984). A better way of controlled seeding is to guide diamond particles by a template. The seeded crystals can then be pressed directly into either catalyst or graphite (Sung, 2000). The diamond seed can be buried half way to allow for the protrusion into graphite volume when the stack of graphite and catalyst layers is assembled (Figure 1). In such a configuration, the molten catalyst can easily follow the diamond profile to form the metal envelope. This metal strait can facilitate the transport of graphite flakes across so the diamond can grow.

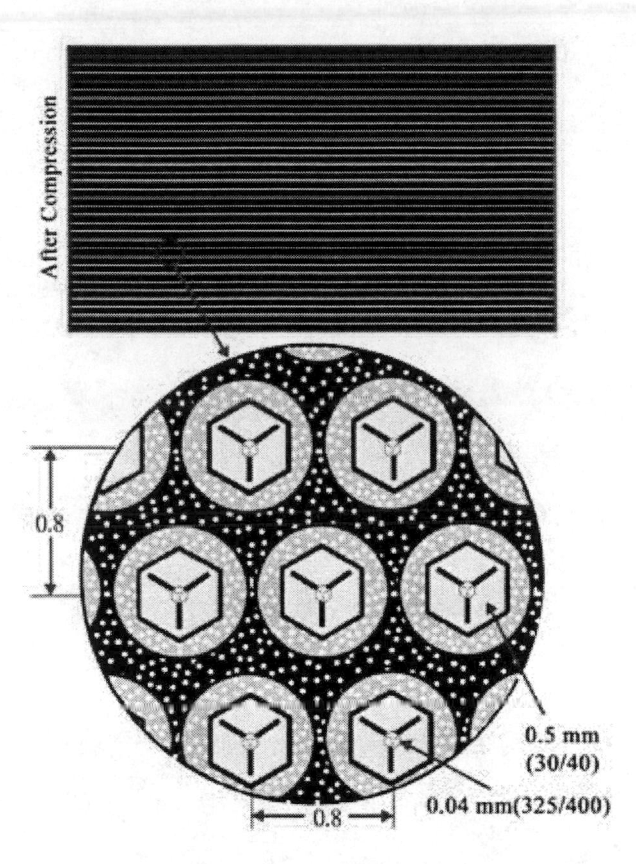

Figure 1: The design of a high pressure cell that shows the growth of diamond seeds to about 2000 times in volume. The typical growth conditions are about 5.2 GPa in pressure, about 1300 ℃ in temperature, and about 50 minutes in time. The diagram shows diamond is separated by the molten alloy (e.g. Invar or $Fe_{65}Ni_{35}$) during its growth. Graphite from the matrix will disperse into the molten alloy and rearrange to form diamond. The matrix contains the mixture of graphite and catalyst powder.

The seeded diamond may be overcoated with a thin layer of catalyst metal e.g., by electroplating Ni. This catalyst envelope can draw other catalyst particles to grow with the diamond. The catalyst envelope in the molten state is impregnated with suspended graphite flakes that feed to the growing diamond.

Assuming the saw grits to be grown are in 40 to 50 mesh (0.3 mm to 0. 4 mm), the most popular size for commercial applications, then the largest crystals (40 mesh) has a size of 0.4 mm and a surface area of about 0.5 mm2. The thickness of a catalyst envelope is less than 0.1 mm, hence after the synthesis, the volume of the catalyst covered each crystal is 0.05 mm3, or less than 0.5 mg. Such an amount of catalyst can form a sphere of about 0.4 mm in diameter

Assuming the diamond-to-diamond distance is two times of the largest diamond size, hence the growing area is 4 times of that covered by the diamond, or 0.7 mm^2. The thickness of the catalyst can be 0.15 mm, more than twice needed to envelop the crystal.

The graphite thickness can be 0.3 mm, twice of that for the catalyst. However, the mass of the former is only about 40% of the latter.

In the above example of cell design, the diamond grown will span the total thickness of graphite and catalyst. Each crystal will have a growing space of about 0.3 mm3. Hence, each cubic centimeter of the reaction cell will produce more than 3000 crystals that are more than 4 carats for 40-mesh diamond (770 pieces per carat). In this case, the graphite to diamond conversion rate is nearly 50%, a very high value for growing good quality diamond. However, because the distribution of diamond crystals is regular, each diamond occupies less than 15% of the reaction volume, so there are ample rooms for uninterrupted growth that are critical for the development of euhedral crystal shape with a minimized amount of inclusions.

The above cell design assumes that the graphite layer and catalyst layer is uniform in structure and composition. However, the reality is that these materials are quite heterogeneous. Hence, a better design is to use powdered graphite and catalyst with more uniform properties. But instead of mixing them and load into the cell, the mixture is first pressed to form disks, and then these disks planted with diamond seeds that are coated with catalyst metal. The powder layers with uniformly distributed diamond seeds are then loaded into the reaction cell for the diamond synthesis. Such a design has the advantages of both layered and powdered cell. Moreover, the diamond seeds would guarantee that the grown diamond crystals are in the same size and quality. Furthermore, because the elimination of the catalyst layer, the efficient of cell volume utilization is further improved. In such a cell, it is possible to achieve a yield of 5 carats per cubic centimeter with at least 4 carats in the desirable size and quality.

The diamond yield for this optimized cell design is compared with that for current arts of diamond synthesis as shown in Table 1.

Cell Size	Charge Type	Nucleation Uniformity	Seeding		Yield Good
			Used	Gross	
Small	Layer	Random	No	1	0.3
Large	Layer	Random	No	2	1
Large	Powder	Random	No	3	2
Large	Layer	Regular	Yes	4	3
Large	Powder	Regular	Yes	5	4

Table 1: Comparison of Diamond Yields.

The positive control of diamond nuclei was demonstrated in growing 30/40 mesh diamond with high grade with a cubic press (Fig. 2). The result indicates that Chinese cubic presses are capable to grow such high valued diamond crystals in competing with that produced in expensive belt apparatus.

Fig. 2: High pressure cells at diamond synthesis conditions heated for the same period of time. The left diagrams show diamond distributions in the original cell. The right diagrams show recovered diamond WITHOUT SIZING & SORTING. Note that Sung's cells produced larger diamond. Moreover, Sung's cells could be "dialed" for a tight range of size, shape, and quality. The above numbers in parenthesis are diamond yields in carats per cubic centimeter.

Novel Magnetic Separable Nano-Carriers For Chemical Catalysis & Bio-Catalysis

S.C. Tsang[a]*, K. Tam[b], X. Gao[a], C.H. Yu[a], C.M.Y. Yeung[a] and K.K.M. Yu[a]

[a]Surface and Catalysis Centre, School of Chemistry, The University of Reading, UK
[b]AstraZeneca, Mereside, Alderley Park, Macclesfield, Cheshire, SK10 4TG, UK

*Correspondence: Prof. S.C. (Edman) Tsang; e-mail: s.c.e.tsang@reading.ac.uk; fax: +44 118 378 6332

ABSTRACT

Separation of homogeneous catalyst species from product mixture in liquid phase has been a major problem in industry. Thus, the facilitated separation of nanosize magnetic body carrying catalytically active species is of a tremendous interest. The major advantage is that the magnetic nano-catalysts display an excellent mass transfer coefficient (high surface area to volume ratio) comparable to soluble species but can still be easily separated from liquid using magnetic interaction with an external applied inhomogeneous magnetic field. However, their stability in reactive environment remains to be the key issue. Here, we show that S-capping species can isolate and protect the nano-size FePt magnetic core from environment. The functionalized surfaces can also offer anchoring sites for immobilization of catalytically active species (nano-metal particulates, homogeneous catalysts and enzymes). With such small magnetic catalyst bodies, the merits of homogeneous and heterogeneous catalysis can then be combined.

Keywords: magnetic separation, nanocatalyst, sulfur capping reagent, surface passivation, FePt

1 INTRODUCTION

Magnetic separation involves (i) tagging or labeling of desired biological/chemical entity on nanosize magnetic particle for recognition of complementary species in solution, and (ii) separation of the resulting solid entities via a fluid-based magnetic separation followed by regeneration of the species from the particle [1-3]. This technique has now been widely adopted in protein purification, immunoassays [4], pre-processing in polymerase chain reactions [5] and pre-concentration of biological entities [2]. Recently, applications of the magnetic separation to catalysis and bio-catalysis areas in order to regenerate expensive catalyst species [6] or enzymes [7,8] from reaction mixture have been particularly denoted. Referring to these particular applications, two main problems arise. First, an undesirable catalysis could also be introduced if reactive surface of the magnetic core is exposed to substrate molecules. Secondly, many magnetic materials with a strong magnetization such as those of magnetic alloys are chemically susceptible for attacks (such as oxidation, hydrolysis) during use and thus deemed unsuitable for this purpose. Availability of reported chemical methods to modify surface properties and to add coating onto colloid nano-magnetic particles in solution is limited since the particles tend to agglomerate very easily during subsequent treatment(s).

In this paper, we show that the surface of new synthesized FePt magnetic nanoparticle can be capped or passivated by two different types of sulfur species, namely cystamine and 3-mercaptopropyl)-trimethoxysilane (MTPS). The former surface capped species could offer surface tethered amine end groups and the latter could give a porous silica overlayer with surface –OH for catalyst (enzyme) immobilization.

2 EXPERIMENTAL

2.1 Synthesis

Monodispersed FePt magnetic nanoparticle is synthesized using modified polyol process [9], followed by sulfur capping of the alloy surface as follows: Analytical Graded $Pt(acac)_2$ (0.1972 g, 0.5 mmol.), $FeCl_2{\cdot}4H_2O$ (0.0994 g, 0.5mmol.), 1,2-hexadecanediol (0.52 g, 2.0 mmol.) and octyl ether (30 mL) were added into a flask under N_2 atmosphere. The mixture was heated up to 95 ºC for 10 min. Then, oleic acid (0.16 mL, 0.5 mmol.) and oleyl amine (0.17 mL, 0.5 mmol.) were added upon before the mixture was continuously heated up to 200 ºC. After that $LiBEt_3H$ (1 M THF solution, 2.5 mL) was slowly added to the mixture *via* a syringe. The mixture was then heated to reflux at 263 ºC for 3 hours (formation of black FePt colloid was observed) before it was cooled to room temperature. 50 mL ethanol was added into the black colloid to induce precipitation. The black precipitate was separated by centrifugation (1,200 rpm, 2 hours) and ethanol washed for at least 3 times (reached 81.93% theoretical yields of FePt) prior to their re-dispersion using 30 mL mixture of hexane and ethanol (1 : 1 v/v).

Synthesis of cystamine-stabilized FePt: Excess cystamine dihydrochloride (0.52 g, 0.0023 mol.) was added into the mixture with stirring. The system was continuously stirred for overnight to allow the cystamine replace the oleic acid and oleyl amine adsorbed on the nano-magnet surface (no precipitation). The magnetic nanoparticles were then separated from the mixture by an external magnet (BHmax = 38MGOe). The supernatant was discarded. The precipitates were washed by both ethanol and the buffer solution three times to remove the excess cystamine and other organic species.

Synthesis of MPTS-stabilized FePt: Excess (3-mercaptopropyl)-trimethoxysilane (MPTS) (0.25 mL, 1.15 mmol.) was allowed to add into the hexane/ethanol FePt colloids slowly. The mixture was stirred overnight to allow the MPTS to displace the oleic acid and oleyl amine stabilizers on the FePt surface in order to produce MPTS modified FePt nanoparticle. It is particularly noted that the surface modified magnetic particles, similar to the above case, still remained in solution as a colloid. A black precipitate was then separated from solution by using an external magnetic field. It was found that the precipitate can be re-dispersed in 20 mL ethanol in a 250 mL round beaker under stirring.

Synthesis of S-capped FePt in silica: The ethanol solution containing MPTS-stabilized FePt was then hydrolyzed and condensed with added silica precursors to form S-capped FePt in silica. These were carried out by first adding 3 mL of sodium silicate solution (1.5 wt%) into the ethanol mixture. After that, another 200 mL ethanol was added before 0.2 mL TEOS (98%) was allowed to add into the mixture. Finally 1.0 mL ammonia solution (35 w/w% in water) was added gradually. The mixture was aged for 3 days with stirring where hydrolysis & condensation of the terminal surface attached silicon hydroxyl species (the hydrolytic product of MPTS) with added silica precursors (silicate and TEOS) to form silica overlayer. To ensure no precipitation during the synthesis we found that the procedure of adding ammonia solution was absolutely essential. It is believed each composite nanoparticles will carry the same but opposite charge on its surface (SiO^-) in alkaline solution hence avoiding them from aggregation. Notice that a black precipitate could only be collected from solution by centrifugation at 1,200 rpm for 2 hours. After washing the collected precipitate with ethanol and acetone, the final solid product was separated using the external magnetic field under nitrogen.

2.2 Characterization

Figure 1 shows a typical XRD pattern of for sulfur capped iron platinum alloy nanoparticle (1:1) in silica. Identical spectra were obtained over cystamine-stabilized FePt and MPTS-stabilized FePt samples (but no amorphous silica peak). The two strong peaks observed correspond to the d_{111} and d_{200} diffractions of a fcc iron-platinum alloy quoted in the International Centre For Diffraction Data, JCPDS-290718. From the pattern below, it also shows a very broad diffraction peak at lower angles corresponding to the non-crystalline amorphous silica. Using Debye-Scherrer equation (with taking into account of instrumental broadening) to apply onto (111) plane for each samples, the average diameter of FePt nanoparticle in each samples was calculated to be 3.11 ± 0.5 nm.

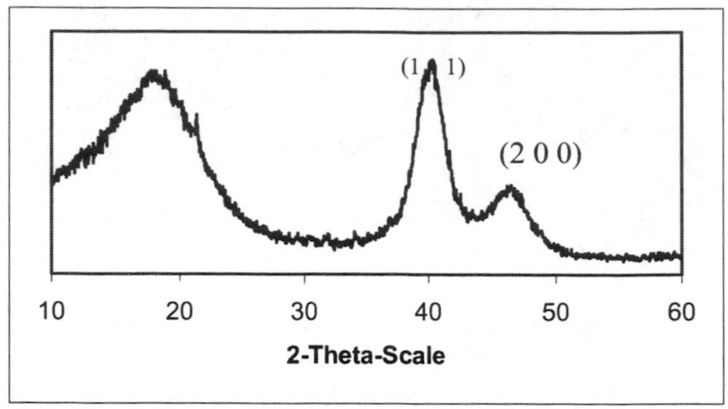

Figure 1: A X-ray diffraction spectrum of S capped iron platinum magnetic nanoparticle (1:1) in silica

Figure 2 shows a typical vibration saturation magnetization (VSM) of the silica encapsulated particles in a solid form (similar VSM responses were found over cystamine-stabilized FePt and MPTS-stabilized FePt samples). These materials exhibit a high saturation magnetization at > 1000 Oersted and thus can undergo magnetic induced precipitation upon exposure to external magnetic field (i.e. 38MGOe). One key challenge in developing material for magnetic separation is that any long range inter-particle magnetic interactions in the absence of magnetic field should be avoided. Otherwise, magnetic particles may not necessarily return to colloidal dispersed state upon removal of the field. It is noted that a long-range chemical ordered FePt alloy nanoparticles forming $L1_0$ phase has been reported when the sample is heat activated [9,15]. In contrast, all our data show extremely low coercivity and remanence values with no hysteresis (chemical disordered structure). Thus, these observations clearly suggest that all our coated magnetic nanoparticles (cystamine, MPTS and silica) synthesized at this low temperature will behave as little magnets in solution with respect to a strong external magnetic field leading to reversible precipitation (exhibit superparamagnetic behavior) as desired.

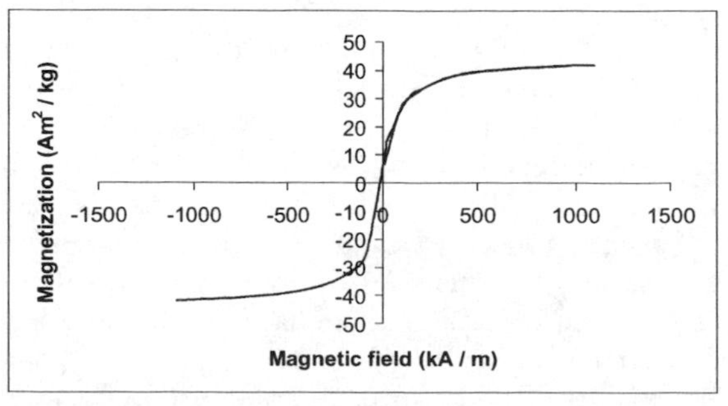

Figure 2: A VSM plot of S-capped FePt nanoparticle in silica (metal precursor molar ratio of Fe: Pt = 1 : 1).

Figure 3 shows the FT-IR spectra of FePt nanoparticles (from polyol) before and after addition of ***cystamine***. The appearance of weak bands at 2605 cm^{-1} and 2523 cm^{-1} after cystamine addition corresponds to the S-H stretching. It can also be seen that the bands at 2994 cm^{-1} and 1458 cm^{-1} corresponding to C-H stretching and bending modes become stronger after the sulfur species immobilization. The band around 1429 cm^{-1} corresponds to the C-N stretching and the band at 1544 cm^{-1} corresponds to the N-H bend. The broad weak band at the region between 700 cm^{-1} and 600 cm^{-1} corresponds to the N-H out-or-plane bend. All of these new bands arisen clearly suggest that the immobilization of cystamine onto the nanoparticle is evident.

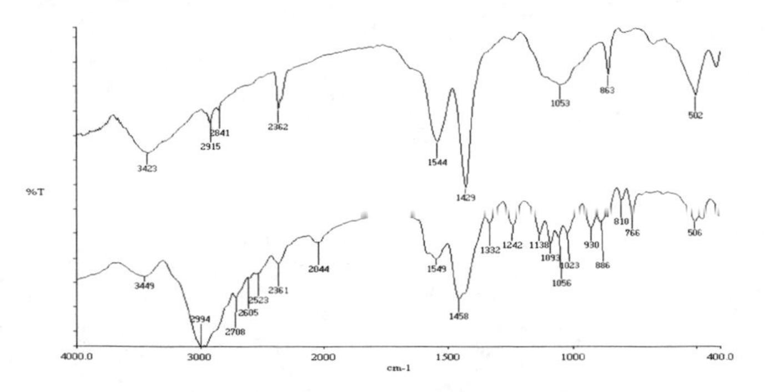

Figure 3. FTIR spectra of FePt nanoparticles before (upper curve) and after cystamine addition

Figure 4 shows the sulfur-stabilized FePt nanoparticles before (MPTS) and after the formation silica. Although no identification of Pt-S or Fe-S bonds is evident because of their weak absorptions at low wave-numbers, the C-Si [11] (~800cm^{-1}) and Si-O [12] (~1100cm^{-1}) linkages are clearly visibly. For the MPTS-stabilized FePt nanoparticle a weak but distinctive peak at 1019 cm^{-1} indicative of the Si-O-C stretching is clearly evident. After hydrolysis/condensation the characteristic broad Si-O-Si asymmetric stretching (~1117 cm^{-1}) together with a large shoulder of Si-O(H) stretching band (~ 980cm^{-1}) are present suggesting the formation of the immobilization of ***MPTS*** onto the surface followed by hydrolysis and condensation to form the silica over-layer as described in the Synthesis section.

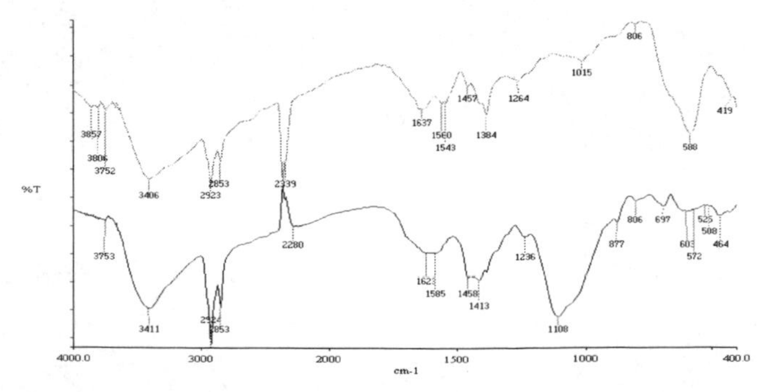

Figure 4: FTIR spectra of MPTS capped FePt nanoparticle before (-----) and after (----) hydrolysis & condensation to

Figure 5 shows the elemental mapping of isolated nanoparticle of ***S-capped FePt in silica*** by Energy Dispersive X-rays (EDX) randomly selected from the sample. After taking the correction of response factor for each element into account, the atomic ratios of the particle were found to be Fe : Pt : Si : O : S = 12.43 : 12.79 : 24.90 : 42.99 : 7.01 with a maximum standard deviation of ± 1.01 from these data. (excluding carbon analysis because the use of carbon filmed holder). Taking the particle size of 3.1 nm (XRD) into consideration the estimated dispersion of the metal cluster (surface atoms/total atoms) is of about 0.284. This matches excellently with the obtained sulfur to metal ratio of 0.278. This value implies that the sulfur atoms (excess MPTS was used) are indeed covering the entire surface metal sites on alloy particle by forming strong M-S linkages [10]. In addition, simple experiments of mixing the all the three S-treated FePt nanoparticles (***cystamine-stabilized FePt***, ***MPTS-stabilized FePt*** and **S-capped FePt in silica**) with hydrazine (64 wt%, Sigma) clearly demonstrated its total inactivity towards hydrogen formation whereas the freshly polyol prepared FePt nanoparticles liberated hydrogen from this reagent at room temperature (also the S-capped materials showed no hydrogen chemisorption value). Thus, the passivation of the exposed active metal surface *via* capping with these sulfur species is clearly established. It is interesting to find that the Si/S ratio of the particle indeed exceeded unity (3.55) with respect to the MPTS molecule. Together with the reduction of oxygen content approaching O/Si =2, the data undoubtedly support the fact that the terminal methoxyl groups of the capped MPTS had indeed undergone hydrolysis and condensation with the added silica precursors, giving the thin silica over-layer. The BET specific surface area and pore volume measurements of the S-capped FePt nanoparticle in silica in its solid condensed form using N_2 physisorption technique were carried out. The high surface area (223 m^2g^{-1}), highly porous nature (mean pore size of 0.8 nm) and high pore volume (0.89mlg^{-1} approaching to silica-gel value of 1.0 m^2g^{-1}) of the sample agree with the formation of porous silica overlayer on these magnetic particles.

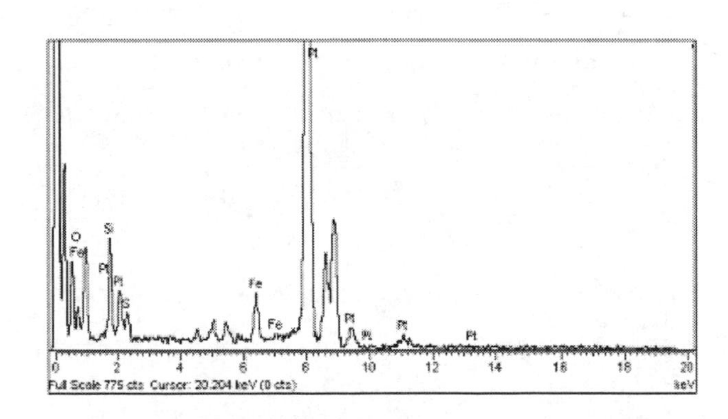

Figure 5: EDX spectrum of the sulfur capped FePt nanoparticle in silica (Fe : Pt = 1: 1).

As a result, we show in this paper that the presence of S species on the three samples passivate the exposure of metal surfaces (forming a strong M-S). In the case of cystamine capped FePt the terminal group on surface is $-NH_2$ and in the cases of MPTS and silica coated FePt the surfaces are covered Si-OH. Many well developed methodologies on making uses of these surface active groups to carry catalyst or enzyme species have been reported. In addition, the high porosity of the silica overlayer in the case of S-capped FePt in silica allows immobilization as well as physical storage of chemical or biological species in its pores as a magnetic recoverable nano-vehicle [6,7,14].

3 CONCLUSION

To conclude, magnetic recoverable nano-catalysts require stable magnetic nanoparticle to carry chemical or biological catalytically active species and the particles should reversibly be precipitated from and to solution upon application or removal of external magnetic field. The described simple synthetic methodologies of using sulfur compounds can incorporate both magnetic and chemical storage functions into a single nano-magnetic nanoparticle, which could extend the application of strong but reactive magnetic alloys such as FePt, FePd, FeCoPt, FePt-Ag for magnetic separation in catalysis. It is also believed that the ability to assemble functional magnetic porous nano-carriers from bottom-up construction approach through nano-chemistry synthesis as presently demonstrated may enable exploitation of these materials for other exciting applications such as magnetic switch, magnetic drug delivery, optical magnet and magnetic recording, etc..

REFERENCES

[1] J.M. Nam, C.S. Thaxton, C.A. Mirkin. Science, 301, 1884-1886, 2003.

[2] X.J. Zhao, R. Tapec-Dytioco, K.M. Wang, W.H. Tan. Anal. Chem. 75, 3476-3483, 2003.

[3] L.R. Moore, S. Milliron, P.S. Williams, J.J. Chalmers, S. Margel, M. Zborowski. Anal. Chem. 76, 3899-3907, 2004.

[4] M. Shinkai. J. Biosci. Bioeng. 94, 606-613, 2002.

[5] A. Spanova, B. Rittich, D. Horak, J. Lenfeld, J. Prodelalova, J. Sucikova, S. Strumcova. J. Chromatogr. A, 1009, 215-221, 2003.

[6] S.C. Tsang, V. Caps, I. Paraskevas, D. Chadwick, D. Thompsett. Angew.Chem.-Int. Edit. 43, 5645-5649, 2004.

[7] X. Gao, K.M.K. Yu, K.Y. Tam, S.C. Tsang. Chem. Commun. 2998-2999, 2003.

[8] S.C.Tsang, K.Y. Tam, X. Gao. International Patent Application No. PCT/GB2004/003103 filed on 16th July 2004.

[9] S.H. Sun, S. Anders, T. Thomson, J.E.E. Baglin, M.F. Toney, H.F. Hamann, C.B. Murray, B.D. Terris. J. Phys. Chem. B, 107, 5419-5425, 2003.

[10] R. Brito, V.A. Rodriguez, J. Figueroa, C.R. Cabrera. J. Electroanal. Chem. 520, 47-52, 2002.

[11] R. Voicu, R. Boukherroub, V. Bartzoka, T. Ward, J.T. C. Wojtyk, D.D.M. Wayner. Langmuir, 20, 11713-11720, 2004.

[12] V.E. Vamvakas, D. Davazoglou. J. Electrochem. Soc. 151, F93-F97, 2004.

[13] K.M.K. Yu, C.M.Y. Yeung, D. Thompset, S.C. Tsang. J. Phys. Chem. B, 107, 4515-4526, 2003.

[14] S.S. Prakash, C.J. Brinker, A.J. Hurd, S.M. Rao. Nature, 374, 439-443, 1995.

[15] S.H. Sun, C.B. Murray, D. Weller, L. Folks, A. Moser. Science, 287, 1989-1992, 2000.

Wire-Streaming Processors on 2-D Nanowire Fabrics

Teng Wang, Mahmoud Ben-Naser, Yao Guo, Csaba Andras Moritz

Electrical and Computer Engineering Department
University of Massachusetts Amherst, MA, USA
{twang, mbennase, yaoguo, andras}@ecs.umass.edu

ABSTRACT

Most of the research in the field of nanoelectronics has been focused on nanodevices and fabrication aspects and as a result a variety of nanodevice technologies have been demonstrated. By contrast, very little work has been reported on the design and evaluation of circuits and computational architectures using nanodevices. There is similarly not much work on the impact of device and fabric (e.g., the 2-D nanowire array) properties on computing. In this paper, we focus on computing architectures based on silicon nanowires. We explore a simple stream processor developed on 2-D nanowire fabrics and compare its density to a 30nm CMOS implementation. We also identify techniques to work around fabric-specific constraints. Our initial evaluation shows that this stream processor has great density advantage compared to CMOS technology.

Keywords: nanoscale circuits, architecture, NASIC, silicon nanowires, carbon nanotubes

1 INTRODUCTION

Nanotechnology is one of the most promising replacements for CMOS technology. Perhaps the most exciting nanodevices today for nanoscale integrated circuits are semiconductor nanowires (NWs) and arrays of crossed carbon nanotubes (CNTs). Researchers have already built FETs and diodes out of NWs [4]–[6] and CNTs [8]. While there are many practical challenges still remaining, it seems that it will soon be possible to build regular nanoarrays from uniform-length CNTs or NWs [7]. By contrast, we have seen very little work on the design and evaluation of circuits and computational architectures using nanodevices. Similarly, little work on the impact of device and fabric properties on computing has been made so far.

Our previous work [1]–[3] has addressed some of the challenges and technical constraints when building computing systems on 2-D nanowire fabrics. These challenges include:

- The 2-D regular NW array, where doping is fixed in each direction, significantly impacts the density of circuits due to the diagonal problem, when the logic is cascaded, only the diagonal portion of the nanotile is utilized [1].
- Having 2-level logic instead of multi-level on 2-D nanoarrays reduces the density further.
- Control circuits and bypass networks are difficult to implement. This is because building feedback paths on 2-D nanoarray is challenging.
- Similarly, it is hard to design high-density sequential circuits using traditional MOS-like approaches, because they require feedback paths.

In this paper, we present a complete design of a simple stream processor developed on 2-D nanowire fabrics, called WISP-0. WISP-0 is the first version of our *Wire-Streaming Processors* (WISP). In WISP, in order to preserve the density advantages of nanodevices, data are streamed through the fabric with minimal control/feedback paths. Intermediate values produced during processing are often stored on the nanowires without requiring explicit latching.

We also compare the density of a fully implemented CMOS design and the NW-based WISP-0. Our evaluation shows that it is possible to preserve the density advantages of nanodevices in WISP despite the fabric constraints.

2 OVERVIEW OF NASICS

The WISP architecture proposed here is a key part of our effort to build NASICs: *Nanoscale Application-Specific Integrated Circuits* [1]–[3]. NASICs are based on extensive research on understanding emerging device and fabrication constraints. A NASIC design has a hierarchical and tiled architecture and it is optimized to deal with various manufacturing, fabric and device constraints. Next, We briefly describe the key components of NASICs. More details can be found in our previous papers [1]–[3].

2.1 Nanotiles

Nanotiles are the basic building blocks of NASICs. Figure 1 shows a typical nanotile. The orthogonally crossed nanowires form a nanoarray. The junctions can be programmed as FETs or detached [7]. Nanoarrays

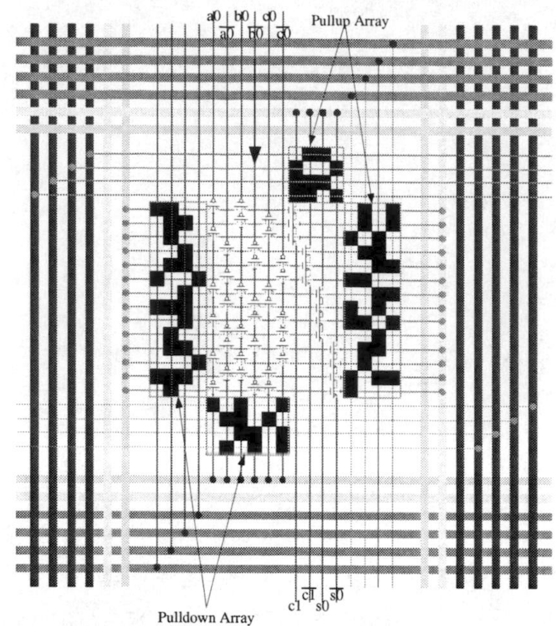

Figure 1: A nanotile for a 1-bit full adder. The thicker wires are MWs and the thinner wires are NWs. NWs in different directions have different doping types.

are surrounded by microwires (MWs). Each signal is expressed in its both original and complementary forms. Microwires provide signals and power supply. Pull-up/down arrays act as the interface between MWs and NWs. To program each NW junction, the number of MWs must be at least logarithmic to the number of NWs [7].

2.2 Dynamic Nanotiles and Pipeline

One of the most common components in any processor design is the datapath. Registers or latches are required to pipeline the data flow. Due to topological and doping constraints, latch circuits are however very difficult to implement on nanoarrays. In NASICs, we use a new dynamic circuit style to achieve temporary storage in stead of using explicit flip-flops. A pipelined NASIC circuit can be built by cascading dynamic nanotiles without explicit latching of signals. Use of latch circuits would have affected density considerably due to the diagonal problem [1]–[3].

2.3 Interconnect and Multi-tile Designs

A nanoscale processor may have thousands of nanotiles. Achieving efficient communication between nanotiles is a critical design aspect. In NASICs, local communication between adjacent nanotiles is provided by NWs for area efficiency. MWs are used for global communications only.

3 WISP-0

WISP-0 is an initial version of our wire-streaming processors. It exercises many of the design strategies

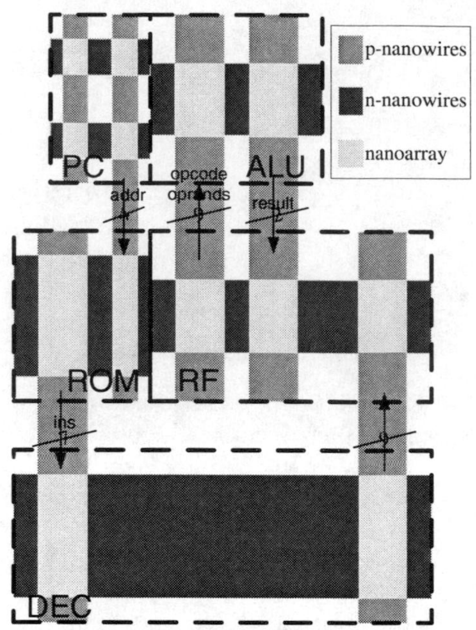

Figure 2: The floorplan of WISP-0

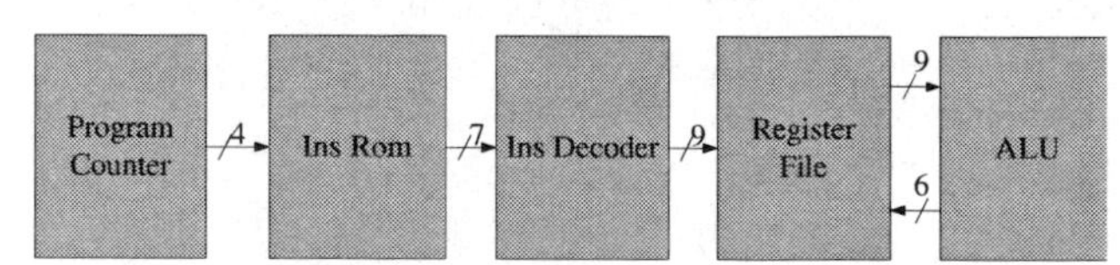

Figure 3: The schematic of WISP-0

and optimizations in NASICs. WISP-0 contains a 2-bit datapath and a 3-bit opcode. Figure 2 shows the overall layout of WISP-0. In this figure, each box surrounded by dashed lines represents a nanotile. All adjacent nanotiles are connected by a set of NWs. These nanotiles are all designed in dynamic style and are cascaded together "on the wire" to form a 5-stage pipeline: fetch, decode, register file, execute and write back. No explicit latches are used. Figure 3 is the schematic of WISP-0.

The *PC* block implements the program counter. It generates a 4-bit address for each cycle. *ROM* is the instruction ROM. It fetches one instruction according to the address from *PC*. The instruction goes to *DEC* (decoder) and is decoded into opcode and operands. Next, the opcode and operands enter *RF* (register file) stage and read the value of operands from the registers. The instruction will be executed in *ALU* and the result will be written back to the register file.

Control logics and bypass networks are difficult to implement on 2-D fabrics. Carefully selected ISA minimizes these circuits. Currently WISP-0 supports the following instructions: *nop*, *mov*, *movi*, *add* and *mult*. All fields in these instructions have fixed lengths (like RISC) in order to simplify the design of the decoder and ALU.

Due to limited space, we only include three key blocks: *PC*, *RF* and *ALU*. MWs are not shown in the following

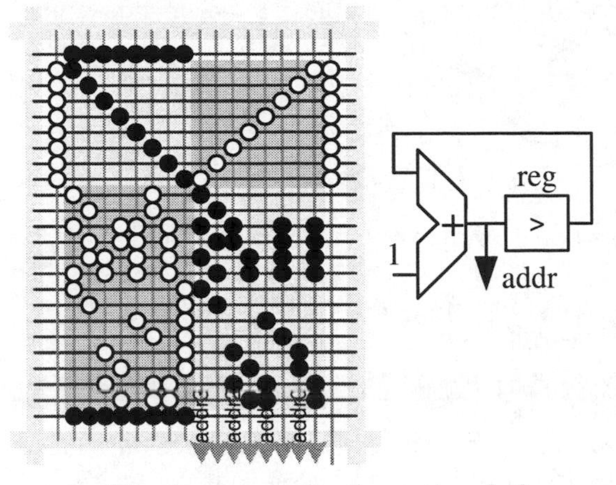

Figure 4: The layout and schematic of the program counter in WISP-0. Black dots represent p-FET and white dots represent n-FET.

figures to improve the readability. For simplicity, the pull-up/down networks are also omitted.

Figure 4 is the layout and schematic of the program counter. This block implements a 4-bit accumulator in a nanotile. It has two components: a 4-bit incrementer (bottom half in the layout) and a 4-bit latch (top half in the layout). In each cycle, the output of the incrementer is delayed and fed back to the input. The address is therefore increased by one in each cycle.

The design of the register file is shown in Figure 5 and Figure 6 is its schematic. In this block, data are stored on the 16 horizontal NWs at the bottom. They are selected by the 2-bit 4-to-1 multiplexer (*2-bit MUX41* in the figure) by $operand_1$ and $operand_2$ respectively. The data ($operand_a$) read out by $operand_1$ is sent directly to "ALU". The data read out by $operand_2$ is sent to another multiplexer (*2-bit MUX21*). This data and $operand_2$ are selected by *opcode* to produce $operand_b$. The reason is that some instructions (e.g., *movi*) will use the immediate data provided by the instruction instead of the values from registers. At the same time, *opcode* and *dest* (destination register address) are pipelined to "ALU". If "ALU" needs to write results back to the register file, the data and control signals will enter from the top right corner of the tile and update the values on the bottom 16 horizontal NWs.

Figure 7 shows the layout of ALU in WISP-0. This block executes the instructions and generates the result (*result* in the figure) and control signals (*rf3˜0*). The top part (*2-4 decoder*) is a 2-4 decoder. It selects the register to be written back according to the destination address (*dest*). The bottom part (*adder/multiplier*) represents an arithmetic unit. It calculates the summation or product (decided by *opcode*) of $operand_a$ and $operand_b$.

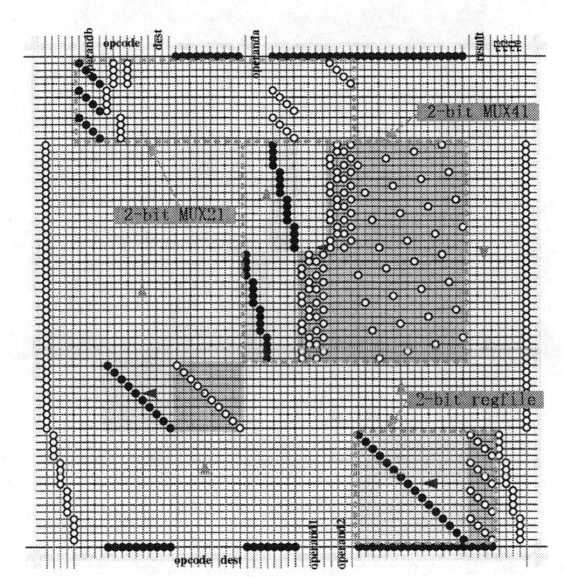

Figure 5: Register file in WISP-0. It has four 2-bit registers.

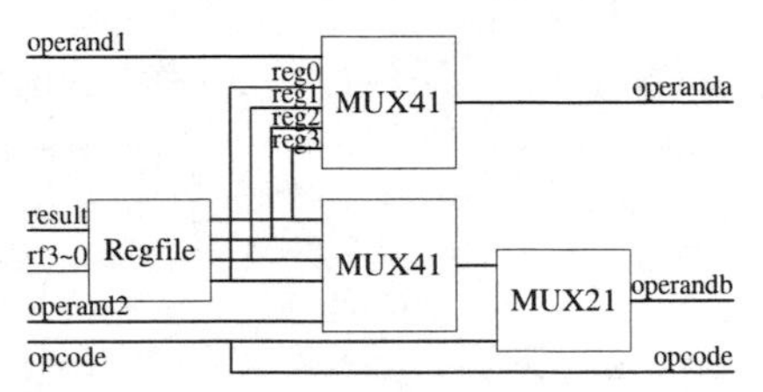

Figure 6: The schematic of register file.

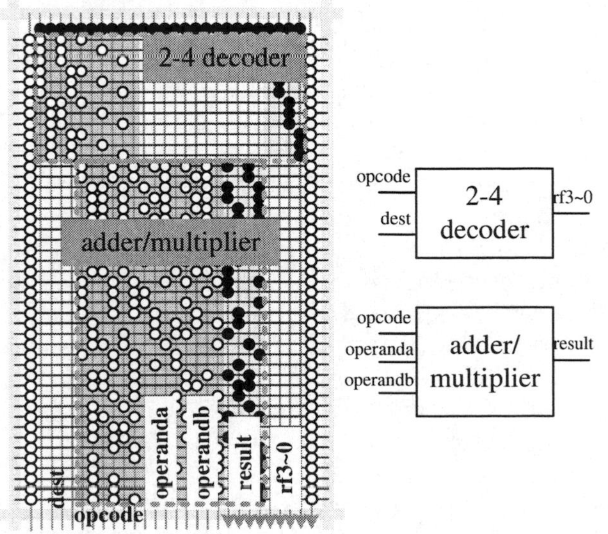

Figure 7: The layout and schematic of the ALU in WISP-0.

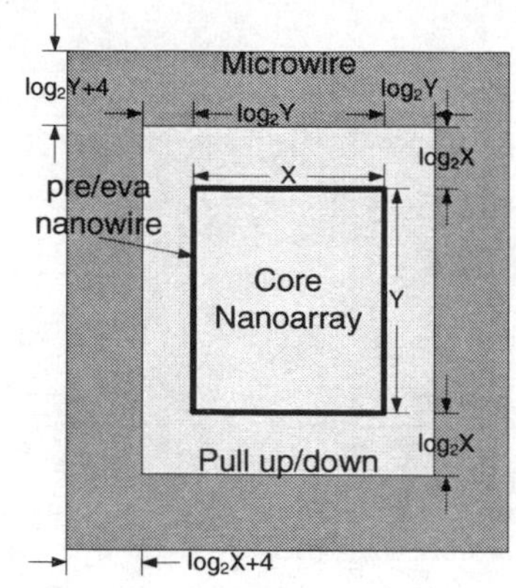

Figure 8: Area breakdown of a typical nanotile. This figure shows the minimum size of pull up/down networks and MWs.

4 DENSITY EVALUATION

The key advantage of nanoscale devices is their density. However, we have found that without new circuit and architecture approaches, this density advantage could be lost due to manufacturing, fabric and device constraints when building nanoscale systems [1]–[3]. In this section, we make an initial density evaluation on WISP-0.

First let us analyze a typical nanotile in NASIC. Figure 8 shows the area breakdown. Assuming that the size of a nanoarray is $X * Y$, we need at least $2log_2Y$ vertical and $2log_2X$ horizontal NWs as pull-up/down networks plus $2log_2X+4$ vertical and $2log_2Y+4$ horizontal MWs (4 MWs as power supply and ground). The total area becomes: $[X+2log_2Y+s*(log_2X+4)]*[Y+2log_2X+s*(log_2Y+4)]$.

In this expression, s is the pitch ratio of MWs to NWs. We assume that the pitch of NWs is 10nm and $s = 9$. This is a relatively conservative assumption for the rapidly improving nanodevice technology. We use this expression to calculate all blocks in WISP-0. The total area is $9.04\mu m^2$. Among this area, nanoarrays take only $0.77\mu m^2$ while the overhead of MWs dominates. The area efficiency $(Area_{nanoarray}/Area_{total})$ is only 8.5%. With larger nanotiles, however, this efficiency can be improved significantly as will be shown next.

To compare with CMOS technology, we implemented a CMOS prototype in Verilog. We use the Design Compiler from Synopsys to synthesize it with 180nm technology. The total area is $4{,}098\mu m^2$. We scale this value down to 30nm (might be expected in 10 years based on the semiconductor industry roadmap) to make a fair comparison. The area is $113\mu m^2$ in 30nm CMOS technology.

The density ratio between WISP-0 and the CMOS prototype is $113/9.04 = 12.5$. But, as we mentioned before, MWs take up most area in WISP-0. Because the number of MWs is logarithmic to the number of NWs, when the design is larger, the area percentage of MWs goes down. In practice, a typical nanotile would have around 1,000×1,000 NWs inside, and the area efficiency would go up to 76% and the corresponding density ratio to 115.

5 CONCLUSION

In this paper, we have shown a complete design of a simple stream processor and compared its density with an implementation in 30nm CMOS technology. Our research indicates that we successfully preserve the density advantage of nanodevices in WISP despite fabric constraints.

Our future work includes fault tolerance on nanoarrays. This is especially important since nanodevices will likely have a higher defect rate than CMOS devices. Currently we are focusing on built-in fault tolerance at circuit and architecture levels. In WISP, since every signal is generated in both original and complementary forms, the system already provides some redundancy.

REFERENCES

[1] T. Wang, Z. Qi, and C. A. Moritz, "Opportunities and challenges in application-tuned circuits and architectures based on nanodevices", in First ACM International Conference On Computing Frontiers, pp.503-511, Apr 2004.

[2] C. A. Moritz and T. Wang, "Latching on the Wire and Pipelining in Nanoscale Designs", in Non-Silicon Computing Workshop (NSC-3), ISCA-31, pp.39-45, Jun 2004.

[3] T. Wang and C. A. Moritz, "NASIC: Nanoscale Application-Specific ICs and Architectures", Boston Area Architecture Workshop, BARC'04, Boston, MA, Jan 2004

[4] Y. Huang, X. Duan, Y. Cui, L.J. Lauhon, K-Y. Kim, and C.M. Lieber, "Logic Gates and Computation from Assembled Nanowire Building Blocks" *Science* 294, 1313 (2001).

[5] Y. Cui, Z. Zhong, D. Wang, WU Wang, and CM Lieber, "High Performance Silicon Nanowire Field Effect Transistors" *Nano Letters*, Vol.3, pp.149-152, 2003.

[6] Z. Zhong, D. Wang, Y. Cui, M. M. W. Bockrath, and C. M. Lieber, "Nanowire Crossbar Arrays as Address Decoders for Integrated Nanosystems" *Science*, Vol. 302, 2003, p. 1377.

[7] A. DeHon, "Array-Based Architecture for FET-Based, Nanoscale Electronics" IEEE Transactions on Nanotechnology, Vol. 2, No. 1, pp.23-32, Mar 2003.

[8] R. Martel, V. Derycke, J. Appenzeller, S. Wind, and Ph. Avouris, "Carbon Nanotube Field-Effect Transistors and Logic Circuits" *DAC 2002* New Orleans, ACM (2002).

IrO_2 Nano Structures by Metal Organic Chemical Vapor Deposition

Fengyan Zhang*, Robert Barrowcliff*, Greg Stecker*, Deli Wang**, Sheng Teng Hsu*

* Sharp Laboratories of America, Inc.
5700 NW Pacific Rim Blvd, Camas, Washington 98607, USA, Fzhang@sharplabs.com
** Department of Electrical & Computer Engineering
University of California, San Diego
9500 Gilman Drive, Mail Code 0407, La Jolla, CA 92093-0407, USA, dwang@ece.ucsd.edu

ABSTRACT

This paper will discuss the fabrication of three different shapes of IrO_2 nano structures, nanorods, nanotubes and nanowires, using (methylcyclopentadienyl) (1, 5-cyclooctadiene) iridium (I) as the precursor and using Metal Organic Chemical Vapor Deposition (MOCVD) method with no template. Vertically aligned dense array of IrO_2 nanorods were fabricated with 100nm in diameter and 500nm-2μm long and with sharp tips at the end. Square shape hollowed IrO_2 nanotubes have also been fabricated by tuning the growth conditions. The wall of the nanotubes is only 10-20 nm thick. Also we have successfully grown high aspect ratio single crystal IrO_2 nanowires at about 10-50nm in diameter and 1-2 um long. The sharp tip at the end of the nanowire is about 5nm. Detailed experiment conditions for growing these different nano structures will be discussed. Furthermore, single IrO_2 nanowire test structure has been fabricated and the electrical properties of IrO_2 nanowires were studied.

Keywords: IrO_2, nanorods, nanotubes, nanowires, MOCVD

1 INTRODUCTION

Nano structures, such as semiconductor nano rods and nano wires, have draw more and more attention due to their potential application for nanoelectronics [1-7], optoelectronics [6, 8-11], and sensorics [12-14]. On the other hand, metallic nano structures can be used in nanoscale interconnects [15] and field emission devices [16]. Nano structures can be grown with or without templates using CVD or PVD methods [17-22], the shapes, compositions and crystal structures strongly depend on the processing conditions, the substrates and the catalysts on the substrate surface. A subtle change in any of the above conditions can results in dramatic changes on the physical forms of the nano structures. It will also affect the electrical, mechanical, chemical, and thermal properties of the nano structures that are important properties for integrating them into future nano devices and systems.

IrO_2 is a conductive metal oxide that has stable electrical and chemical properties even at high temperature in O_2 ambient [23, 24]. IrO_2 can also be used as pH sensor material [25]. This paper will discuss the deposition of varies shapes of IrO_2 nano structures using (methylcyclopentadienyl) (1, 5-cyclooctadiene) iridium (I) as the precursor and using Metal Organic Chemical Vapor Deposition (MOCVD) method with no template. Three different shapes of IrO_2 nano structures have been successfully synthesized: they are IrO_2 nanorods [26], IrO_2 nanotubes and high aspect ratio IrO_2 nanowires [27]. The IrO_2 nanorods and nanowires are good candidates for field emission and sensor applications. IrO_2 nanotubes also have the potential to carry chemicals and medicines in bio and medical fields.

2 EXPERIMENT

IrO_2 nano structures are fabricated using an in house cold wall MOCVD system. (methylcyclopentadienyl) (1, 5-cyclooctadiene) iridium (I) was used as the precursor. The growths were conducted at different pressures and temperatures that are in the range of 10-50 torr and 200-500°C. Oxygen gas was used directly as the carrier gas. The precursor and delivery line were maintained at 80°C. Ti has been used as buffer metal on Si substrate. The IrO_2 nano structures were imaged using JEOL JSM 6400F field emission scanning electron microscopy (SEM) with an accelerating voltage of 5 kV. The phase and composition of the nano structure were analyzed by Philips X'pert XRD. For transmission electron microscopy (TEM) studies, the NWs were suspended in ethanol, dispersed onto a lacey carbon film substrate (Ted Pella), and imaged with a JEOL 2010F TEM with an accelerating voltage of 200 kV. The electrical properties of IrO_2 nanowires and its temperature dependency were researched by fabricating IrO_2 single nanowire test structures. To fabricate the devices, the IrO_2 nanowires were dispersed on the SiO_2 wafer, followed by Ti metal deposition at 300°C and patterning by photolithography using a regular metal pad arrays. The I-V characteristics were collected using a HP 4156A precision semiconductor parameter analyzer.

3 RESULTS AND DISCUSSION

We found that the shapes of IrO_2 nano structures are strongly depend on the processing conditions and substrates. In order to grow IrO_2 nanorods, the pressure needs to be in the range of 30-50 torr and the growth temperature in the range of 300-400°C. If the pressure is lower than 10 torr or the temperature below 250°C or higher than 450°C, Ir or IrO_2 films will be formed. Also sufficient supply of the precursor is critical in order to obtain large array of vertically aligned IrO_2 nanorods. Figure 1 shows the SEM images of the IrO_2 nanorods. They are aligned vertically and densely with about 100nm in diameter and 500nm-2um long. There are naturally formed sharp tips that make them more ideal for field emission application. The XRD spectrum confirmed that the nanorods are polycrystalline IrO_2 with a preferred orientation of (101).

(a)

(b)

(c)

Figure 1. IrO_2 nanrods grown by MOCVD
(a) vertically aligned IrO_2 nanorods
(b) intentionally bended IrO_2 nanorods at the edge to show the sharp tips
(c) XRD spectrum of IrO_2 nano rods

We also found that by increasing the initial chamber pressure, hollowed square shape IrO_2 nano tubes can be formed. By carefully studying the forming process, we found that these hollowed nanotubes actually started by four nanorods to grow in clusters as shown on Figure 2(a). More interestingly, individual IrO_2 nanotubes can be formed on Si substrate with a tube wall as thin as 10-20 nm as shown on Figure 2(c).

(a)

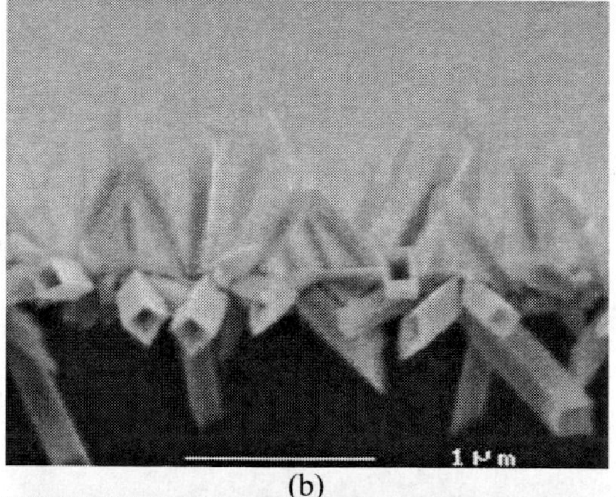

(b)

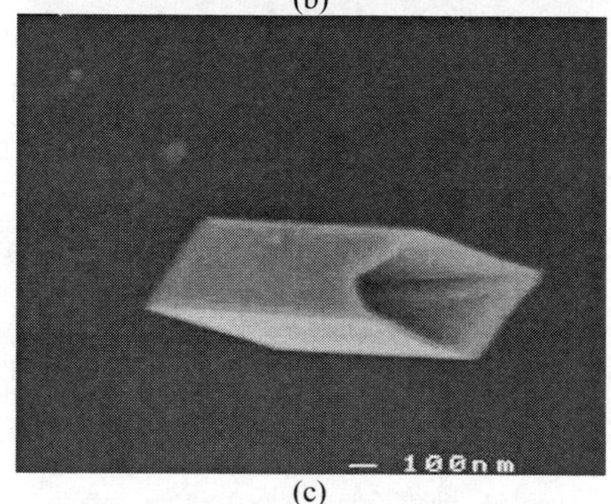

(c)

Figure 2. IrO_2 nano tubes grown by MOCVD
(a) cluster growth of IrO_2 nanorods
(b) IrO_2 nanotubes grown at the wafer edge
(c) Individual IrO_2 nanotube grown on Si substrate

Furthermore, by increasing the deposition temperature to 400°C and using Si substrate with a very thin Ti buffer layer, we have obtained very high aspect ratio IrO_2

NSTI-Nanotech 2005, www.nsti.org, ISBN 0-9767985-1-4 Vol. 2, 2005

nanowires with naturally formed sharp tips. The nanowires are about 10-50nm in diameter and 1-2 um long. The sharp tip is about 5nm under the HRTEM analysis. Selective growth of IrO_2 nano wires has also been achieved by patterning the Ti buffer layer. Unlike the IrO_2 nanorods and nanotubes that are polycrystalline structure with some dislocations and defects, HRTEM show that the IrO_2 nano wires are single crystals structure with [001] orientation.

(a)

(b)

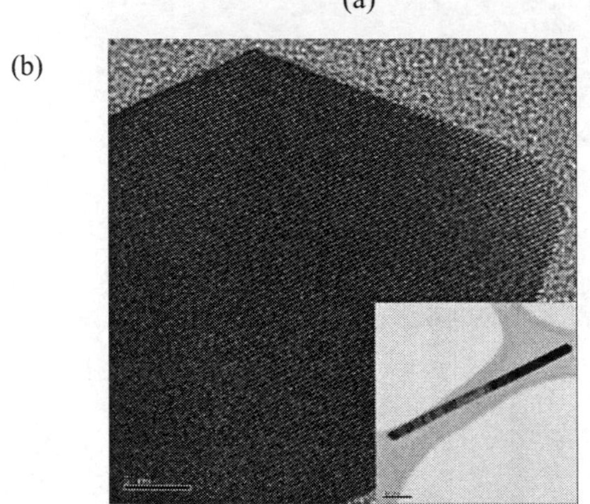

(c)

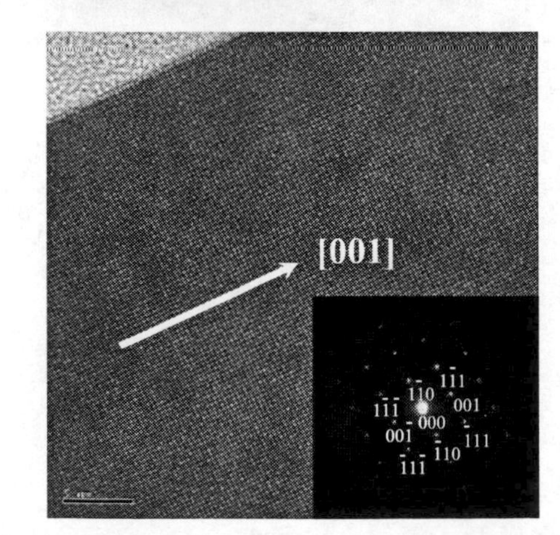

Figure 3. IrO_2 nanowires and the TEM analysis
(a) Selective growth of IrO_2 nanowires
(b) LRTEM and HRTEM of the tip region of IrO_2 nanowires
(c) HRTEM and the diffraction pattern of the IrO_2 nanowires

Single IrO_2 nanowire test structure has been successfully fabricated. Electrical measurements on IrO_2 single nanowire show ohmic contact were achieved between IrO_2 nanowire and Ti metal pads with no annealing. The IrO_2 nanowires are metallically conducting with a resistance of 2-10 Kohm for a nanowire at about 1 um long and 30-40 nm in diameter. The resistivity of IrO_2 nanowire is around 300-400 μohm•cm, which is higher compared to that of bulk IrO_2 (32-51 μohm•cm) and IrO_2 thin film (80 μohm•cm), most likely due to the surface scattering effect. We also measured the temperature coefficient of resistivity from -40°C to 60°C. As shown on Figure 3(c), the temperature coefficient of resistivity of IrO_2 nanowire is about 3 x 10^{-3} K^{-1}.

(a)

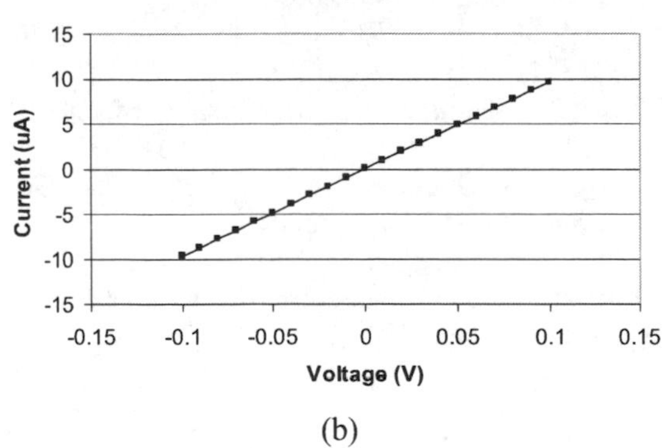

(b)

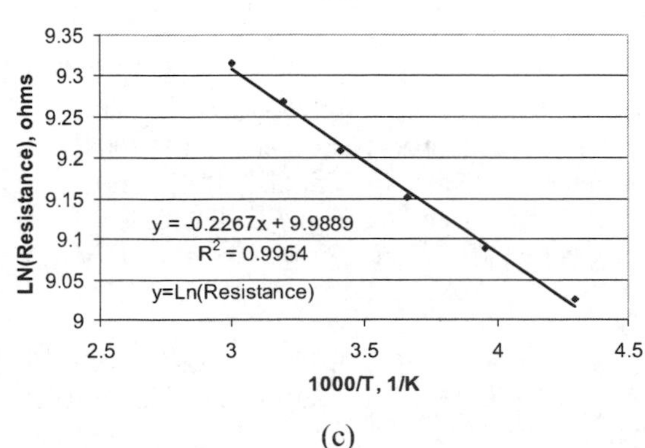

(c)

Figure 3. Electrical characterization of IrO_2 nanowires
(a) test structure of single IrO_2 nanowire
(b) IV characteristics of IrO_2 nanowire
(c) temperature coefficient of resistivity of IrO_2 nanowire

4 CONCLUSION

We have successfully fabricated three different shapes of IrO_2 nanostructures using MOCVD methods without any template. The IrO_2 nanorods are vertical aligned dense array with sharp tips at the end that make it a good candidate for field emission applications. The interesting hollowed IrO_2 nanotubes has a square shape and a thin wall at 10-20 nm that have a potential application in bio and medical field. High aspect ratio and single crystal IrO_2 nanowires can be selective deposited on a patterned Ti buffer layer on Si. It will also have a wide potential application in nanoelectronics, optoelectronics, sensorics and field emission devices.

REFERENCES

[1] Y. Cui, X. Duan, J. Hu, C. M. Lieber, J. Phys. Chem. B 104, 5213, 2000

[2] J. Y. Yu, S. W. Chung, J. R. Heath, J. Phys. Chem. B 104, 11864, 2000

[3] Y. Cui, C. M. Lieber, Science, 291, 851, 2001

[4] Y. Huang, X. Duan, Y. Cui, L. J. Lauhon, K.-H. Kim, C. M. Lieber, Science, 294, 1313, 2001

[5] Z. Zhong, F. Qian, D. Wang, C.M. Lieber, Nano Lett. 3, 343, 2003

[6] Y. Cui, Z. Zhong, D. Wang, W. U. Wang, C. M. Lieber, Nano Lett. 3, 149, 2003

[7] H. T. Ng, J. Han, T. Yamada, P. Nguyen, Y. P. Chen, M. Meyyappan, Nano Lett., 4, 1247, 2004

[8] X. Duan, Y. Huang, Y. Cui, J. Wang, C. M. Lieber, Nature, 409, 66, 2001

[9] J. Wang, M.S. Gudiksen, X. Duan, Y. Cui, C. M. Lieber, Science, 293, 1455, 2001

[10] H.J. Choi, J. C. Johnson, R. He, S. K. Lee, F. Kim, P. Pauzauskie, J. Goldberger, R. J. Saykally, P. Yang, J. Phys. Chem. B107, 8721, 2003

[11] X. Duan, Y. Huang, R. Agarwai. C. M. Lieber, Nature, 421, 241, 2003

[12] Y. Cui, Q. Wei, H. Park, C. M. Lieber, Science, 293, 1289, 2001

[13] M. Law, H. Kind, B. Messer, F. Kim, P. Yang, Angew. Chem. Int. Ed. 41, 2405, 2002

[14] J. Hahm, C. M. Lieber, Nano Lett., 4, 51, 2004

[15] N. Gothard, C. Daraio, J. Jaillard, R. Zidan, S. Jin, and A. M. Rao, Nano Letters, 4 (2004), 213

[16] R.-S. Chen, Y.-S. Huang, Y.-M. Liang, C.-S. Hsieh, D.-S. Tsai, and K.-K. Tiong, Appl. Phys. Lett. 84, 1552, 2004

[17] M. Vazquez, K. Pirota, M. Hernandez-Velez, V.M. Prida, D. Navas, R. Sanz, and F. Batallan, J. Appl. Phys. 95, 6642, 2004

[18] B. Erdem Alaca, Huseyin Sehitoglu, and Taher Saif, Appl. Phys. Lett. 84, 4669, 2004

[19] C.A. Decker, R. Solanki, J.L. Freeouf, and J.R. Carruthers, D.R. Evans, Appl. Phys. Lett. 84, 1389, 2004

[20] T.C. Wong, C.P. Li, R.Q. Zhang, and S.T. Lee, Appl. Phys. Lett. 84, 407, 2004

[21] O. Gurlu, O. A. O. Adam, H. J. W. Zandvliet, and B. Poelsema, Appl. Phys. Lett, 83, 4610, 2003

[22] A. M. Morales, C. M. Lieber, Science, 279, 208, 1998

[23] B. Jiang, V. Balu, T.-S. Chen, S.-H. Kuah, J. C. Lee, P. Y. Chu, R. E. Jones, P. Zurcher, D. J. Taylor, M. l. Kottke, and S. J. Gillespie, IEEE, symposium on VLSI Technology, 26, 1996

[24] F. Zhang, J.-S. Maa, S. T. Hsu, S. Ohnishi and W. Zhen, Jpn. J. Appl. Phys. 38, L 1447, 1999

[25] S. Yao, M. Wang and M. Madou, J. Electrochem. Soc, 148, H29, 2001

[26] R.-S. Chen, Y.-S. Chen, Y.-S. Huang, Y.-L. Chen, Y. Chi, C.-S. Liu, K.-K. Tiong, and A. J. Carty, Chem. Vap. Deposition, 9, 301, 2003

[27] F. Zhang, R. Barrowcliff, G. Stecker, W. Pan, D. Wang, S-T Hsu, "Synthesis of Metallic Iridium Oxide Nanowires via Metal Organic Chemical Vapor Deposition", submitted for publication on JJAP, Part 2

Conductive Films of Ordered High-Density Nanowire Arrays

Jaideep S. Kulkarni*, Brian Daly*, Kirk J. Ziegler*, Timothy Crowley*,
Donats Erts**, Boris Polyakov**, Michael A. Morris and Justin D. Holmes*

*Materials Section and Supercritical Fluid Centre, Department of Chemistry, University College Cork, Cork, Ireland
**Institute of Chemical Physics, University of Latvia, LV-1586 Riga, Latvia

ABSTRACT

Encapsulation of nanowires within an ordered template offers the possibility of manipulating nanowires into useful configurations and allows their aspect ratios and, hence, their physical properties to be tailored. In our laboratories, we have prepared high-density, ordered arrays of semiconductor nanowires within the pores of mesoporous thin films (MTFs) and anodized aluminum oxide (AAO) matrices using a supercritical fluid solution-phase inclusion technique. Conductive atomic force microscopy (C-AFM) was utilized to study the electrical properties of the nanowires within these arrays. Nearly all of the semiconductor nanowires contained within the AAO substrates were found to possess similar electrical properties demonstrating that the nanowires are continuous and reproducible within each pore. The ability to synthesize ordered arrays of semiconducting nanowires is a key step in future 'bottom-up' fabrication of multi-layered device architectures for potential nanoelectronic and optoelectronic devices.

Keywords: germanium nanowires, supercritical fluids, conductive AFM, anodic aluminium oxide

1. Introduction

The ability to pack high densities of memory storage and processing circuitry into specific nanoscale arrays, and utilize the unique transport properties associated with these architectures, is expected to lead to future generations of computer processors with device sizes many times smaller and faster than current silicon based processors [1,2]. However, both physical and economic constraints are expected to limit continued miniaturization of electronic and optical devices using current lithography based methods. Consequently, alternative nonlithographic methodologies for constructing the smallest features of an integrated circuit will soon be required.

Nanowires are expected to play a role in future integrated circuits as both devices and interconnects [3]. One of the most successful approaches for producing nanowires is based on the vapour–liquid–solid (VLS) growth process [4,5]. One inherent problem with these approaches however, is the formation of entangled meshes of nanowires. While significant progress has been made in manipulating these nanowire meshes into useful configurations for potential electronic devices [6] some researchers have focused on forming nanowires in predefined architectures to allow easier processing and integration of the nanostructures into functioning devices [7-11].

Encapsulation of nanowires within an ordered template offers the possibility of manipulating nanowires into useful configurations and allows their aspect ratios and, hence, their physical properties to be tailored. Nanomaterials have been synthesized within the pores of mesoporous solids [7,8] having uni-directional arrays of pores using chemical vapour deposition (CVD) [9], electrodeposition [10] and incipient wetness techniques [11]. But these approaches often lead to incomplete pore inclusion due to pore plugging where the surface tension of the liquid solvent prevents precursor penetration into the pores. CVD approaches are less prone to pore plugging but can undergo capillary condensation resulting in liquid phases within the pores.

In our laboratories, we have developed a supercritical fluid (SCF) solution-phase method for forming metal, metal oxide, and semiconductor nanowires within mesoporous and anodized aluminium oxide (AAO) membranes [12]. The high-diffusivity, high precursor solubility, and reduced surface tension of the SCF results in rapid nucleation and growth of the nanowires within the pores reducing the reaction time for pore inclusion by at least an order of magnitude compared to CVD. Additionally, SCFs cannot be condensed to a liquid phase reducing the problems of pore plugging and incomplete inclusion seen with CVD, electrodeposition, and incipient wetness techniques.

Here we describe the preparation of germanium nanowire arrays within AAO membranes (GeNW-AAO). Conductive atomic force microscopy (C-AFM) was utilized to investigate the electrical properties of GeNW-AAO. Additionally, the nanowires formed within each pore of the AAO membranes were observed to have similar electrical characteristics. These results suggest that each nanowire is continuous throughout the length of the substrate demonstrating the reproducibility of nanowire synthesis within the pores. Furthermore, nearly all of the nanowires formed within the AAO membranes are conducting suggesting nearly complete inclusion of nanowires within the matrix.

2. Experimental

AAO membranes of appropriate sizes were fabricated using methods reported in literature. Germanium nanowires were synthesized within the pores AAO membranes (GeNW-AAO) by the degradation of diphenylgermane in supercritical CO2. Briefly, the AAO membranes were placed inside a 25 mL high-pressure reaction cell with diphenylgermane placed inside an open top quartz glass boat adjacent to the membranes under an inert atmosphere. The reaction cell was attached via a three-way valve to a stainless steel reservoir (48 mL). A high-pressure ISCO pump was used to pump CO_2 through the reservoir into the reaction cell. The reaction cell was placed in a tube furnace and heated to 873 K using a platinum resistance thermometer and temperature controller. The pressure was simultaneously ramped to 37.5 MPa and the reaction proceeded at these conditions for 30 min.

To remove contaminants and possible oxides present at the end of the nanowires within the AAO membranes, the surface was cleaned by RF Ar plasma. This plasma cleaning removed approximately 30 nm of the surface layer. The AAO membranes were also mechanically polished to remove any bundles of nanowires. GeNW-AAOs were placed inside the chamber of a sputter coater and a gold film was then deposited onto the film surface 200 nm thick. A custom-built conductive atomic force microscope was utilized for simultaneous measurement of the surface topography and conductance mapping as shown in Fig. 1.

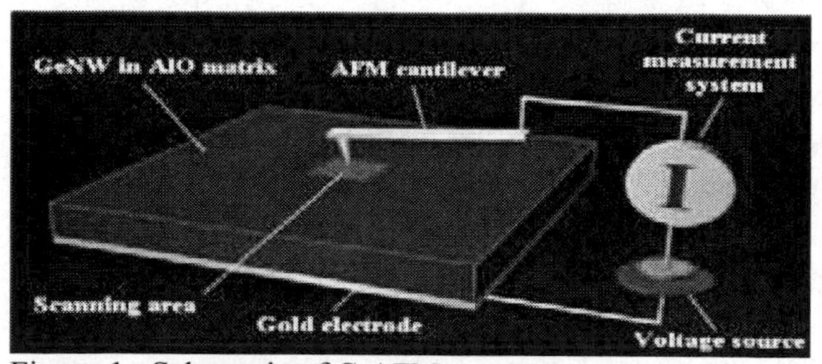

Figure 1. Schematic of C-AFM setup.

3. Results and Discussion

The inclusion of nanowires in AAO membranes combined with C-AFM yields the electrical properties of thousands of individual nanowires that are aligned in ordered architectures.

AAO membranes have pores running perpendicular through the substrate hence all the pores should be accessible to the C-AFM tip. This would rule out the problems associated with the MTF preparation. The C-AFM results in Fig. 3 show that a large portion of the substrate is conductive. Furthermore, the areas of conductivity correlate strongly with the pore diameter seen in the SEM and the nanowire diameter seen in the topography map. The uniformity of the conductance suggests that there is little difference between the nanowires within each pore. This is an important result and suggests that the substrate consists of highly reproducible continuous nanowires within each pore that are uniform and running throughout the substrate.

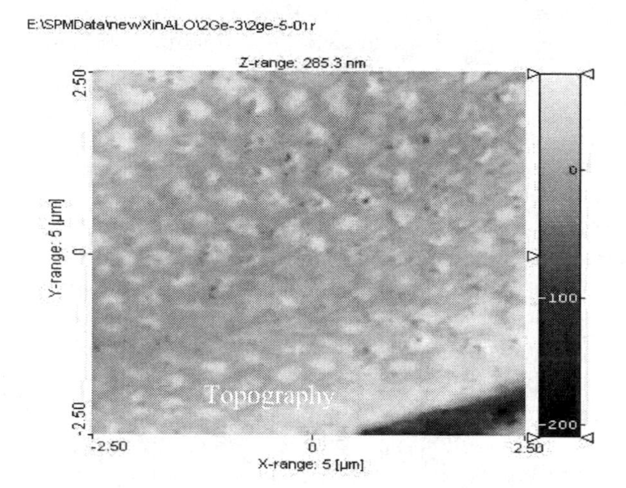

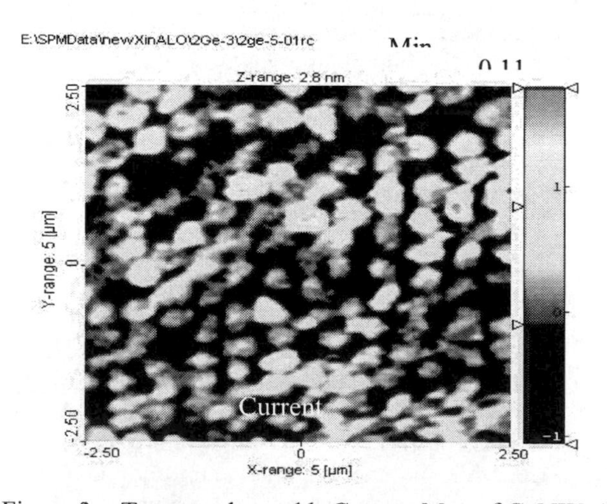

Figure 3 a-Topography and b-Current Map of GeNW-AAO

However, the uniformity of the conductance through each pore contained within the substrate clearly suggests complete pore inclusion of continuous nanowires from end-to-end. Comparison of the average density of pores within the AAO substrate to the average density of conducting

GeNWs within the AAO substrate shows that approximately 90% of the pores contain conducting nanowires.

To obtain a measure of the quality of the contact between the gold coating and the AAO membrane, a gold electrode 2 mm in diameter was deposited onto the top surface of the GeNW-AAO membranes in a similar approach to the bottom contact and is shown in Figure 4. The resistance of an individual GeNW measured above through C-AFM was approximately R_i = 450 GΩ. The number of GeNWs contained within the gold contact area was determined to be approx. 1.9×10^7 nanowires (N). The expected resistance between the electrodes would be $R_{calc} = R_i/N = 0.02$ MΩ; however, the measured resistance between the two contacts is often close to 5 MΩ. The high measured resistance suggests that better contacts are required to further study the electrical properties of individual nanowires.

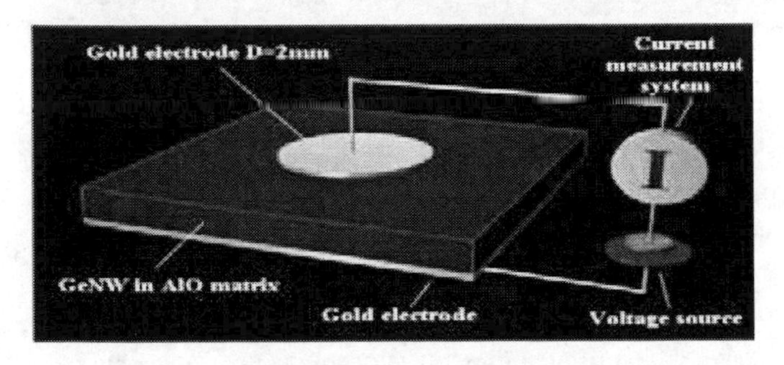

Figure 4: Schematic of a macrocontact measurement setup.

4. Conclusions

The preparation of GeNW-AAO substrates shows that a significant portion of the pores are filled with nanowires and that nearly all of the nanowires formed are conducting. Furthermore, the consistency of the conductivity of the nanowires within these GeNW-AAO substrates suggests that the nanowires are continuous throughout the length of the substrate and do not vary from pore to pore. The preparation of better contacts will also be required before the electrical properties of individual nanowires within the arrays can be studied.

5. References

1. Xia, Y.; Yang, P.; Sun, Y.; Wu, Y.; Mayers, B.; Gates, B.; Yin, Y.; Kim, F.; Yan, H., *Adv. Mater.* **2003,** *15*, 353.
2. Trindade, T.; O Brien, P.; Pickett, N. L., *Chem. Mater* **2001,** *13*, 3843.
3. Zhang, X.; Zhang, L.; Meng, G.; Li, G.; Jin, N., *Adv. Mater.* **2001,** *13*, 1238.
4. Wagner, R. S.; Ellis, W. C.,*Appl. Phys. Lett.* **1964,** *4,* 89
5. Wagner, R. S.; Ellis, W. C.; Jackson, K. A.; Arnold, S. M.,*J. Appl. Phys.* **1964,** *35*, 2993.
6. J. Wang, M. S. Gudiksen, X. Duan, Y. Cui and C. M. Lieber, *Science,* 2001, ***293***, 1455
7. Schult, F.; *Chem. Mater.* **2001,** *13,* 3184.
8. Attard, G.S.; Glyce, J. C.; Goltner, C. G.; *Nature,* **1995,** *378,* 366.
9. Dag, O.; Ozin, G. A.; Yang, H.; Reber, C.; Bussiere, G.; *Adv. Mater.;* **1999,** *11,* 474.
10. Banerjee, S.; Dan, A.; Chakrovorty, D.; *J. Mater. Sci.,* **2002,** *37,* 4261
11. Zhang, Z.; Dia, S.; Blom, D. A.; Shen, J.; *Chem. Mater.,* **2002,** *14,* 965
12. Holmes, J. D.; Lyons, D. M.; Ziegler, K. J.,*Chem. Eur. J.* **2003,** *9*, 2144.

Novel Method for the Electrochemical Synthesis of Nickel-Rich Oxide Nanowires

S.A. Thorne, Y.V. Bhargava, T.S. Mintz, V. Radmilovic[2], Y. Suzuki, and T.M. Devine
Department of Materials Science and Engineering
210 Hearst Mining Building, University of California, Berkeley, CA
sthorne@berkeley.edu
(510) 642-3956
2. National Center for Electron Microscopy
Lawrence Berkeley National Laboratory
Berkeley, CA

ABSTRACT

Using a novel electrochemical process, functionalized and self-assembled Ni-rich oxide nanowires were grown from an Alloy 600 (Ni – 15.5a/o Cr, 8a/o Fe) substrate. The formation of the nanowires occurs in a high-temperature (238-288°C), high-pressure (1500psi) aqueous solution by subjecting Alloy 600 to a specific progression of reducing and oxidizing conditions. The nanowires have an average diameter of 25nm with a range from 3nm to 100nm with a maximum observed length of 20μm. Energy dispersive spectra of individual nanowires have indicated a composition of 47a/o Ni, 4a/o Fe, 49a/o O. Their composition, in conjunction with the oxidizing growth environment, suggests that the nanowires are corrosion resistant, an important property for biotechnology applications. SQUID magnetometry shows the nanowires to be ferromagnetic, with a coercivity of approximately 80Oe and a remnant field of 0.027emu/g.

1 BACKGROUND

In recent years, the field of nano-systems, and in particular nanowires, has received much attention for its potential application ranging from arrays of sensors for detection of particular biological agents [1-3] to the building blocks for logic circuits [4-5]. However, the progression of the field has been retarded by the limited number of synthesis techniques and subsequent lack of control of material properties needed to functionalize the nanowires. Thus, any novel nanowire synthesis process that can produce new chemical compositions of nanowires with functional properties represents a step forward for the field.

We have discovered a novel electrochemical method for the synthesis of nickel-rich oxide nanowires formed by self-assembly. The advantage of self-assembly is that growth occurs directly on the substrate, without the use of a template or catalyst [6-9] (which requires later processing steps to remove). Preliminary analysis has shown the nanowires to already have a desirable set of functionalizable properties including electrical conductivity (as demonstrated by their appearance in electron microscopes) and ferromagnetism. Additionally, because many metal oxides have low solubilities over a wide range of pH [10], our nanowires should be viable for application in aqueous solutions, especially in highly oxidizing environments, such as the conditions found in the human body.

2 EXPERIMENTAL METHOD

Synthesis takes place in an autoclave designed to simulate the conditions found inside of a pressurized water reactor (PWR) of a nuclear power plant. The autoclave and attached piping are constructed out of titanium to prevent corrosion, and are built to withstand operating temperatures of 288°C and pressures of 1500psi (necessary to prevent the aqueous solution from boiling off at operating temperatures). The inner chamber, (shown in Figure 1) where the samples are held, is approximately 50cc and has interlocks for a thermocouple, two Cu/CuO reference electrodes, and for wires connected to the sample.

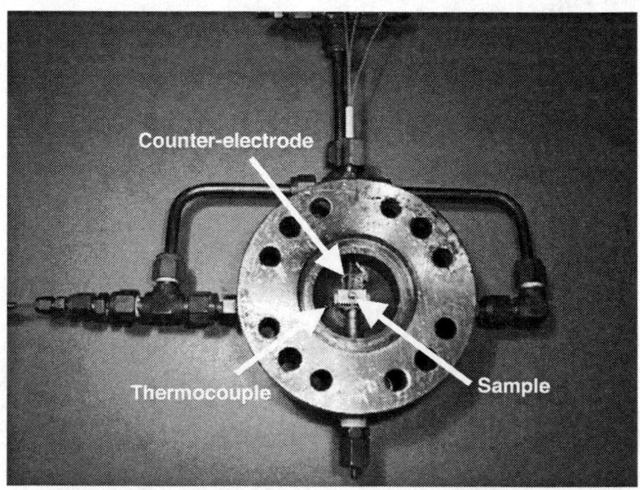

Figure 1: Open autoclave with mounted sample.

Each substrate is prepared using the same procedure. Using a low-speed saw, samples are cut into approximately 1/8 x 1/8 x 3/8 inch pieces, which is the maximum size than can be accommodated by the autoclave sample holder. The samples are then notched using a file, such that a Teflon® pin holds the sample in place in the autoclave's sample carrier.

Each substrate is then polished using a progression of finer grit polishing paper, concluding with 600 grit SiC, in order to remove any thick layer of surface oxide. The samples are then cleaned in a sonication bath using twice deionized water. Finally using a spot welder, a Teflon®-sleeve protected stainless steel wire is attached so that electrical connection can be made to the sample and the substrate is mounted in the sample carrier of the autoclave. Two samples of the same type can be mounted and simultaneously tested for each trial.

Once the samples are mounted, the chamber is sealed. Gold is used for the body and window seals and Teflon® for the wire seals to prevent the possibility of contamination due to the oxidation of other species during the experimental process. The chamber is connected to the pump and heating system and filled with an aqueous solution composed of deionized water containing 2ppm Li and 1200ppm B. The normal flow rate of the system is approximately 50cc/min, such that the experimental chamber is refilled with fresh solution once a minute. Both samples, the gold counter electrode, and the Cu/CuO reference electrode are connected to a Solartron SI 1287 potentiostat. The autoclave is heated to 288°C and concurrently pressured to 1500psi to prevent the solution from boiling. The potentiostat is activated after the autoclave has exceeded 170°C because the electrical conductivity of the Zirconia tube that houses the Cu/CuO is too low at temperatures below this value.

The samples are then taken through a polarization progression which can be broken into 3 basic polarization categories. The first step is cathodic polarization (-1600mV with respect to the standard hydrogen electrode, SHE), to reduce any oxide films formed with exposure to the air before the experiment or during the course of the heating of the water. Second, the substrate is polarized to -850mV in order to grow a Cr_2O_3 film, whose identity has been verified by previous surface enhanced Raman spectroscopy (SERS) studies [11]. Finally, the sample is polarized between -250mV to -25mV vs. SHE, where it has been shown that nanowire growth occurs.

Analysis of each substrate begins with an examination of the surface with a field-emission scanning electron microscope (FESEM) in order to determine the area, extent, and morphology of nanowire growth. Once this is catalogued, one of the two samples is immersed in an isopropyl-alcohol (IPA) bath and sonicated, effectively shaking wires off the surface of the sample into the solution. A drop of that solution is placed on a holey carbon transmission electron microscope (TEM) grid and the IPA is allowed to evaporate, leaving the nanowires (and whatever other oxides that were also stripped off in the process). Diffraction analysis is performed using a JEOL 200CX and compositional analysis is performed using electron dispersive spectroscopy (EDS) and verified using electron energy loss spectroscopy (EELS) on a Philips CM-200.

Samples for magnetic analysis are prepared by first selecting an area of the sample completely covered by nanowires (as shown by FESEM). Once a suitable area is identified, the nanowires are torn off the surface using a piece of household Scotch® tape. The mass of the nanowires removed is determined by measuring the weight of the tape before and after the nanowires are applied. The substrate is subsequently examined on the FESEM to verify that only a portion of the nanowire layer was removed (i.e. other surface oxide layers were not torn off as well). The tape is then placed in an inert gel capsule and inserted into a SQUID magnetometer where two magnetic tests are performed. The first examines the hysterisis of the nanowires by measuring magnetization as a function of applied field for a given temperature, in order to determine remnant field and coercivity. Second, magnetization is measured with changing temperature in order to determine if the Curie temperature of the nanowires lies within the temperature range of the magnetometer.

3 RESULTS / DISCUSSION

3.1 Processing

Approximately 15 tests with various values of process parameters have been conducted in which nanowires have been successfully synthesized.

The first parameter to be investigated is the role of substrate. Nanowires (shown in Figure 2) were first discovered on Alloy 600 (Ni – 15.5a/o Cr, 8a/o Fe), while conducting stress corrosion cracking investigations of the alloy. The majority of subsequent tests have also been conducted on this alloy however a variety of other samples have also been tested using the same processing parameters that produced nanowires on Alloy 600. Nanowires were grown on a Ni – 30a/o Cr binary alloy (Figure 3a) as well as a Ni – 20a/o Cr – 10a/o Mo ternary (Figure3b), while pure Cr produced nanoparticles (aspect ratio about 4:1), and Ni and Fe did not produce nanowires.

Figure 2: Nanowires grown on Alloy 600 at 288°C. Substrate held at -1600mV for 24 hours, -850mV for 1 hour, and -25mV for 40 hours.

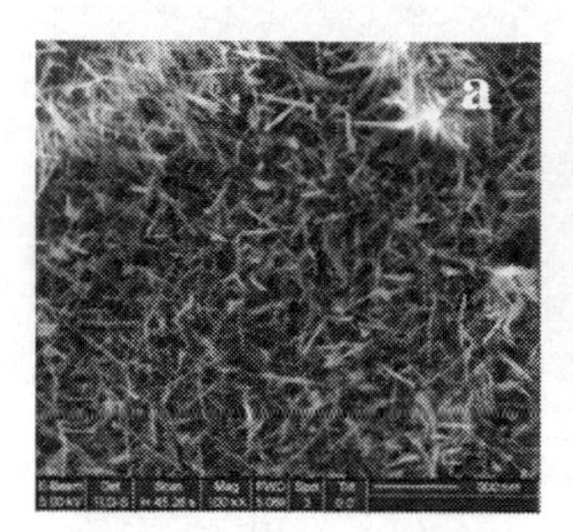

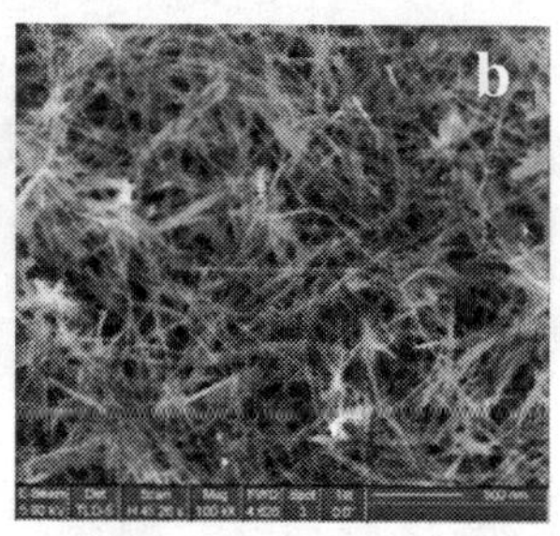

Figure 3: Nanowires grown on (a) Ni – 30a/o Cr binary alloy and (b) Ni – 20a/o Cr – 10a/o Mo ternary using same hydrogen- oxygen cycling.

The next major processing parameter found to be of importance to the synthesis process is the composition of the aqueous solution. Alloy 600 anodically polarized in an aqueous solution of deionized, deoxygenated water, with 2ppm Li, and 1200ppm B grew nanowires while the same test conducted in deionized, deoxygenated water without any Li and B did not. An important difference between the two cases is the pH of the solution at 288°C. With the electrolyte, the calculated pH is 6.75 [12], while the pH of pure water is 5.5 at 288°C. If the edge of the thermodynamic stability of the nanowires lies somewhere in between these two values, this would explain the observation.

Finally, growth of nanowires was found to be highly dependent on the potentials applied and the duration of the applications. Tests have shown that it is necessary to first bring the substrate to a reducing potential (in-situ surface enhanced Raman spectrometry has shown -1600mV to be very effective at this), perhaps to reduce any surface oxides that may be residing on the surface to clear a way for the nucleation of the nanowires. It may also be the case that in order for nanowires to grow, a Cr_2O_3 layer needs to be grown first, which has been found to form at -850mV [13]. Finally, tests show that potentials above -250mV need to be applied for the growth of nanowires. Typically, -25mV is utilized because these tests produce full surface coverage of nanowires, but potentials above -25mV have not been attempted yet. We have also shown that reducing potentials need to be applied for as little as an hour and other tests have shown that nanowire nucleation begins in less than 30 minutes of -25mV being applied (see Figure 4). However, for complete surface coverage, -25mV needs to be applied for over 4 hours (most tests usually lasted about 2 days at this potential).

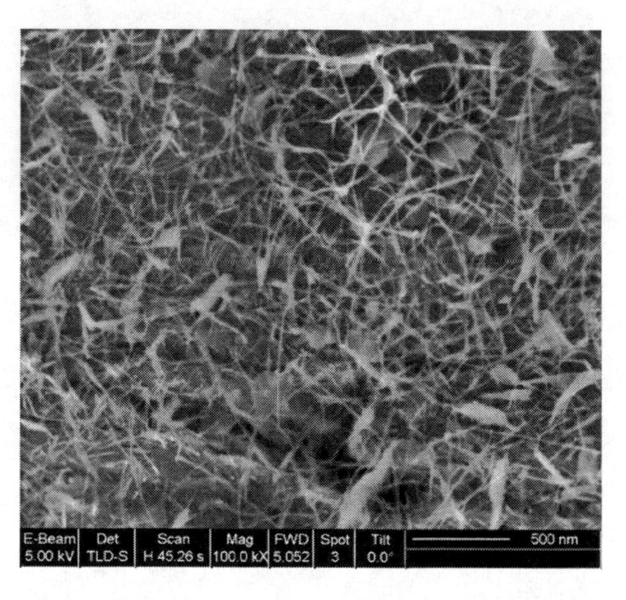

Figure 4: Nanowires beginning to form after being held at -1600mV for 24 hours, -850 mV for 1 hour, and -25mV for 30 minutes.

3.2 Properties

Scanning electron micrographs reveal nanowires as long as 20 microns with uniform thicknesses along the length of single nanowires, although a distribution of thicknesses exists between different wires. TEM analysis of over 300 wires shows the average diameter of the wires to be 25nm, with a standard deviation of 17nm. The maximum thickness was 90nm with the majority of the wires between 20-30nm. The nanowires have aspect ratios of 100:1 to 1000:1.

EDS analysis conducted on a TEM (and verified by EELS) shows the wires grown on Alloy 600 to be 47a/o Ni, 4a/o Fe, 49a/o O. This coupled with diffraction analysis suggests a crystal structure very similar to bulk NiO, however, it is unclear whether the role of Fe is substitutional or if it exists as a separate phase. Diffraction does show the nanowires to be single crystalline.

SQUID magnetometry indicates the nanowires grown on Alloy 600 are ferromagnetic as shown by the hysterisis loop in Figure 5. At 5K, the nanowires produce hysteretic loops with coercive fields of 150Oe and remanence of 0.039emu/g. Magnetization as function of temperature with an applied field of 250Oe shows that the Curie temperature lies above 400K (the temperature limit of the magnetometer).

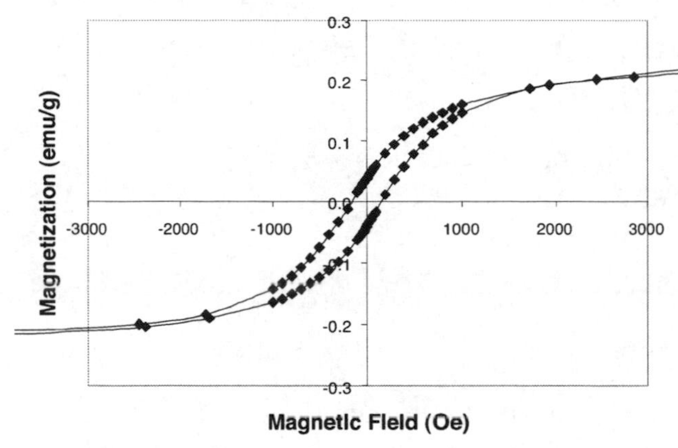

Figure 5: Magnetization versus applied field at 5K for a 0.3mg sample of nanowires grown from Alloy 600.

4 CONCLUSION

We have demonstrated that self-assembled nickel-rich oxide nanowires can be grown using a novel electrochemical process on a variety of nickel-based alloys. Nanowires grown on Alloy 600 have an average diameter of around 25nm and aspect ratios up to 1:1000, a composition of 47a/o Ni, 4a/o Fe, 49a/o O, and ferromagnetic behavior under applied magnetic fields. Future research will be focused on identifying and understanding the mechanism of growth as well as determining electrical, optical, and structural properties.

REFERENCES

[1] Y. Cui, Q. Wei, H. Park, C. M. Lieber, "Nanowire nanosensors for highly sensitive and selective detection of biological and chemical species," *Science* **293**, 1289-1292, (2001).

[2] M. Yun, N.V. Myung, R.P. Vasquez, W. Jianjun, H. Monbouquette, "Nanowire growth for sensor arrays," *SPIE-Int. Soc. Opt. Eng. Proceedings of Spie - the International Society for Optical Engineering* **5220**, 37-45 (2003).

[3] J. Hahm, C.M. Lieber, "Direct ultrasensitive eletrical detection of DNA and DNA sequence variations using nanowire nanosensors," *Nano Letters* **4**, 51-54, (2004).

[4] S.J. Tans et al., "Individual single-wall carbon nanotubes as quantum wires," *Nature*, **386**, 474-477 (1997).

[5] S.J. Tans, R.M. Verschueren, C. Dekker, "Room-temperature transistor based on a single carbon nanotube," *Nature*, **393**, 49-52, (1998).

[6] T.A. Crowley, K.J. Ziegler, D.M. Lyons, D. Erts, H. Olin, M.A. Morris, J.D. Holmes, "Synthesis of metal and metal oxide nanowire and nanotube arrays within a mesoporous silica template," *Chem. Mater.* **15**, 3518-3522 (2003).

[7] X. Peng, A. Chen, "Electrochemical fabrication of novel nanostructures based on anodic alumina," *Nanotechnology* **15**, 743-749 (2004).

[8] B. Tian, X. Liu, H. Yang, S. Xie, C. Yu, B. Tu, D. Zhao, "General synthesis of ordered crystallized metal oxide nanoarrays replicated by microwave-digested mesoporous silica," *Advanced Materials* **15**, 1370-1374 (2003).

[9] M, Yazawa, M. Koguchi, A. Muto, M. Ozawa, K. Hiruma, "Effect of one monolayer of surface gold atoms on the epitaxial growth of InAs nanowhiskers," *Applied Physics Letters* **61**, 2051-2053 (1992).

[10] M. Pourbaix, "Atlas of Electrochemical Equilibria in Aqueous Solutions," Pergamon Press, New York (1966).

[11] A. Machet, A. Galtayries, S. Zanna, L. Klein, V. Maurice, P. Jolivet, M. Foccault, P. Combrade, P. Scott, P. Marcus, "XPS and STM study of the growth and structure of passive films in high temperature water on a nickel-based alloy," *Electrochimica Acta*, **49**, 3957-3964 (2004).

[12] B. Beverskog and I. Puigdomenech, "Revised Pourbaix diagrams for nickel at 25-300°C," *Corrosion Science*, **39**, 969-980 (1997).

[13] T. Mintz, *Doctoral Thesis, UC Berkeley (2003).*

Synthesis of nickel nanoparticles by reduction of an organometallic precursor with hydrogen

Tibisay del C. Golindano M., Susana I. Martínez M., Omayra Z. Delgado G., y Guaicaipuro P. Rivas R.

Urb. Santa Rosa, Los Teques 1201- Miranda, Venezuela.
PDVSA-INTEVEP.
Tel.(0058)2123306228. Fax:(0058)-2123307139.
martinezsi@pdvsa.com

ABSTRACT

Nanometer sized clusters have received considerable attention in recent decades, because their unique properties, such as quantum size effects, surface and interfacial effects, and other novel phenomena, besides their potential application in industry, for example, catalysis, high-performance ceramics, magnetic recording. In this work, a new method of synthesis of nickel nanoparticles is reported. It has been achieved in an organic solution at the presence of non-ionic surfactant using hydrogen as reductor agent. Nickel nanoparticles of black colour were formed after 8 h. TEM shows a particle size distribution between 1 and 4 nm, with an average value of 2,1 nm. The electron diffraction pattern of nickel evidenced nickel in metallic state with fcc structure and very weak ring, which indicate nickel oxide reflections of cubic face centred (fcc). XPS results showed that the phase layer on Ni nanoparticles is composed of $Ni(OH)_2$.

Keywords: synthesis, nanoparticles, nickel and characterization.

1 INTRODUCTION

With increased interest in fabricating catalysts with nanometric size, much attention has been focused on exploiting a general route to control nanoscale materials size and morphology in synthetic route is indispensable to exploit the properties of this materials [1-7]. In recent years, nanoscale catalytic materials have attracted much interest due to the potential application in the industry. Supported nickel catalysts, comprise one of the most important kinds of heterogeneous catalysts, due to the widespread applications of these systems in a variety of applications, like methanation [8-10], partial oxidation[11,12], and steam reforming [13,14]. A flexible synthetic route is indispensable to exploit catalytic materials. So far, a number of physical and chemical routes have also been applied to produce nanoscale catalytic materials, including as photolytic reduction [15], radiolytic reduction [16], sonochemical method [17], solvent extraction reduction [18], microemulsion technique [19], polyol process [20], and alcohol reduction [21] have been developed for the preparation of metal nanoparticles. For the synthesis of various kinds of metal nanoparticles, some metals such as nickel, copper, and iron are relatively difficult because they are easily oxidized. Nickel nanoparticles have important applications in catalysts and conducting and magnetic materials.. However, the size distribution of the products is not ideal; it appears very difficult to produce particles with sizes in the range of 1-10 nm with relatively good monodispersity. Also, only a few works on the preparation of nickel nanoparticles have been reported up to now. Recent developments of the organometallic route to produce high quality nanocrystals included the thermal decomposition and reduction of organometallic compounds with pure hexadecylamine (HDA) [22] or polyvinylpyrrolidone (PVP) [23] as the stabilizing agents. Within this context, this study presents a new synthesis for to produce nanostructured nickel material with a simple method of preparation.

2 EXPERIMENTAL PROCEDURES

2.1 Preparation of nickel nanoparticles

Typically, an appropriate amount a nickel organometallic precursor (0.1–1 M) and no-ionic surfactant agent were added directly in n-nonane. Then, this mixture was introduced in 300 ml stainless steel autoclave with stirring to appropriate temperature (100-200°C) and hydrogen pressure between 200 and 500 psi. Nickel nanoparticles of black colour were formed after 8 h. The solvent was evaporated under reduced pressure and the solid was washed several times and stored in n-nonane solution, in order to avoid the oxidation. The reduction reaction could be expressed as:

$$(NiR_2) + H_2 \rightarrow Ni^0 + 2RH$$

2.2 Characterization

Transmission Electronic Microscopy (TEM) images were recorded using a Jeol JEM 1200 microscope operated at 100 kV. It provides information from a typical nanometer scale area of the specimen which can moreover be

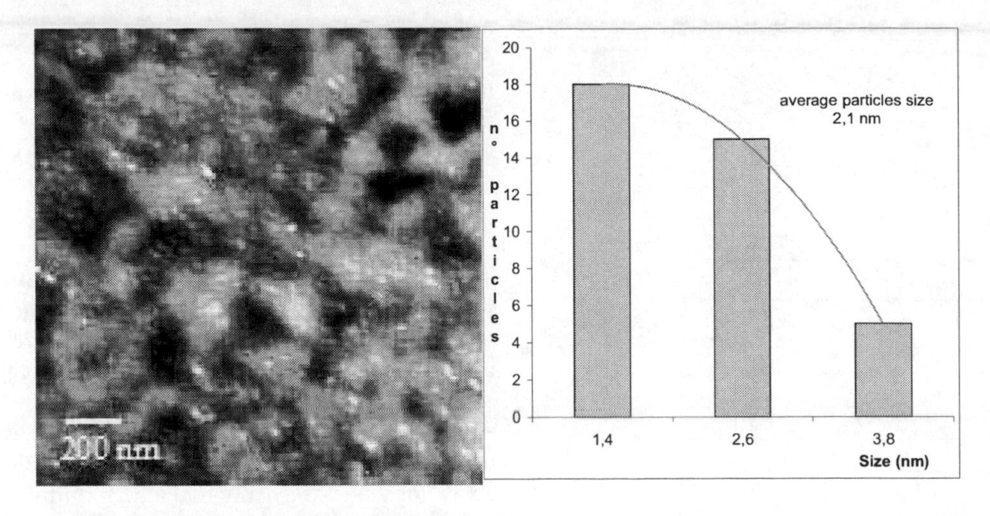

Figure 1: Micrographic and histogram of nickel nanoparticles stabilized with no-ionic surfactan

identified by imaging diffraction modes. Dry Specimens were prepared by depositing one droplet of the particle suspension onto glow-discharged carbon-coated copper grids. After 1 min, the liquid in excess was blotted with filter paper and the remaining film allowed drying. Approximately 100 objects were measured in order to estimate the mean size. Structural information was obtained using selected-area electron diffraction (SAED) patterns.

X-ray photoelectron spectroscopy (XPS) of nanostructure material was performed with a Leybold–Heraeus commercial surface analysis apparatus (LHS 11), equipped with a single channel detector, and employing AlKα radiation (1486.6 eV) at 360W (power settings: 12 kVand 30 mA). The 100 mm radius hemispherical analyzer was set in the constant pass energy mode (pass energy = 200 eV).The normal operating pressure inside the turbo-pumped analysis chamber was kept below 5×10^{-8} Torr during data collection.
Each spectral region was signal-averaged for a given number of scans to obtain good signal-to-noise ratios. Surface charging was observed on most samples, and accurate binding energies (BE) were determined by charge referencing by means of either adventitious carbon at 284.6 eV or with the reported value for metallic nickel component of 852.3 eV [24]. The nickel particles used for XPS spectral determination were de-agglomerated in n-nonane. Thin layers of particles were added onto a gold plated copper disk. Each added pipet of suspension was given time for the n-nonane to evaporated, until gold substrate was completely coverted.

3 RESULTS AND DISCUSSION

3.1 Microstructural characterization

Figure 1 shows typical dark-field transmission electron micrographs with the corresponding distributions of grain size of the nickel particles. These analysis showed a particle size distribution between 1 and 4 nm, with an average value of 2,1 nm. No coalescence can be detected for these particles. It can be suggested that the no-ionic surfactant might form a protective layer around the particle surface, preventing the aggregation or agglomeration.
The typical electron diffraction pattern of nickel nanoparticles is shown in Figure 2. The pattern evidence that corresponds to nickel in metallic state with fcc structure (d_{111} = 2.02 Å, d_{200} = 1.76 Å, d_{311} = 1.06 Å, d_{222} = 1.02 Å) and very weak ring, which indicate that nickel oxide reflections of cubic face centred (fcc) (d_{111} = 2.41 Å, d_{200}= 2.08 Å, d_{220}= 1.47 Å, d_{311}= 1.26 Å, and d_{222} =1.20 Å) are present.

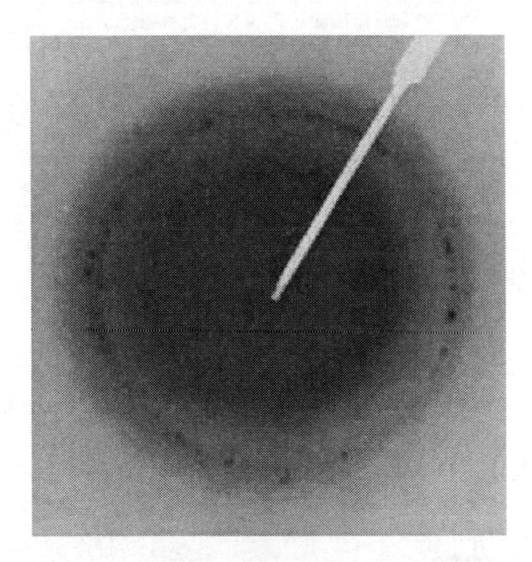

Figure 2: Electron Diffraction Patterns of nickel nanoparticles

Table 1 shows the theoretical and experimental interplanar diameter, being observed correspondence between them; all Miller Index was corroborated with the literature. The electron diffraction pattern of nickel nanoparticles did not reveals the presence another crystalline phase (carbide or hydroxide). These results indicate that metallic nickel nanoparticles were obtained, but these are not 100 % pure.
The presence of nickel oxidize is probably due to the fact that during the sample preparation, it was exposed to the atmosphere. Due to the presence of this phase the study must be continue with these materials, mainly

improving the conditions of the sample preparation for microscopy.

d_{int} exp. (Å)	d_{int} teoric Ni fcc (Å)	d_{int} teoric NiO fcc (Å)
2.43		2.41
2.09		2.08
2.02	2.02	
1.46		1.47
1.74	1.76	
1.26		1.26
1.20	1.20	1.20
1.06	1.06	
1.02	1.02	

Table 1: Experimental and theory interplanar diameter of metallic and nickel oxide.

3.2 Superficial characterization

Table 2 shows the XPS Ni2p data of the fine nickel particles. Clear Ni2+ peaks (the spectra not shows) are located at binding energies 851 and 873 eV. These peaks corresponds to binding energy measurements in literature for hydroxide nickel[24].

Element	Energía de Enlace (eV)
Ni 2p	855.7
O 1s	532.1
N 1s	400.1
C 1s	284.6
Si 2p	102.3

Table 2: XPS Ni2p data of Ni nanoparticles.

Possibly, the layer consists of innermost crystalline NiO, and outermost amorphous $Ni(OH)_2$. XPS indicate that the surface of the particles were completely covered with $Ni(OH)_2$ thin layer. The electron diffraction did not reveal the presence of hydroxide nickel; perhaps due, both the signal overlaps with the nickel oxide signal or this phase of nickel hydroxide is amorphous. But the first reason was discarded, because there isn't correspondence between the Miller Index here reported and the Miller Index of crystalline nickel hydroxide reported in the literature. The carbon presence (table 2) was found to be a contamination possibly from the n-nonane solvent or the XPS high vacuum chamber, and not a carbonaceous product on the Ni particle surface.
In this moment was not possible to determine the degree of oxidation of the nickel nanoparticles.
The structure of the surface layer may be attributed to oxidation and hydration phenomena during the particle exposure to air during material preparation or storage. Due to this, the study must be continue, mainly improving both the conditions of the sample preparation for microscopy and in the synthesis procedure.

4 CONCLUSIONS

The reduction technique of an organometallic precursor with hydrogen was used to synthetisize nickel nanoparticles, which have been characterized to be nickel crystalline of fcc structure converted by both, nickel oxide and hydroxide by TEM and XPS.
The analysis of the sample has showed that this material consists of very small nickel metallic particles, which have an average size of 2.1 nm and a distribution between 1 and 4 nm.
These particles are surrounded by a crystalline surface oxide layer and other phase of $Ni(OH)_2$. XPS showed that there were a superficial $Ni(OH)_2$ layer on the nickel metallic nanoparticles. Was not possible to detect this phase by electron diffraction, because it is amorphous. These results indicate that metallic nanoparticles of nickel were obtained, but these are not 100 % pure.

The quantitative and reproducible chemical approach in organic solvent allows the control of the mean particle size, morphology and stability in organic solvent. Using a relatively simple procedure, it can be applied at low temperature to the synthesis of numerous finely and stable oxides simple or mixed of transition metals which are crucial for industrial applications.

5 ACKNOWLEDGEMENTS

The autors thank the Electron Microscopy service of Universidad Simón Bolívar team, specially Lic. Edgar Cañizales for his support in performing the TEM analysis. Also, we wish to thank specially PhD Gema Gonzalez for making the crystallography and Lic. Juan Carlos de Jesús for his support in performing the XPS.

6 REFERENCES

1. V.F. Puntes, K.M. Krishnan, A.P. Alivisatos, Science 291 (2001) 2115.
2. Pillai, V., Kumar, P., Hou, M. J., Ayyub, P., *Adv. Colloid Interface Sci.* 55, 241 (1995).
3. Chahavorty, D., and Giri, A. K., *in* "Chemistry of Advanced Materials" (C. N. R. Rao, Ed.), p. 217. Blackwell Scientific, London, 1993.
4. Henglein, A., *J. Phys. Chem.* 97, 5457 (1993).
5. Benedetti, A., and Fagherazzi, G., *J. Mater. Sci.* 25, 1473 (1990).
6. B. Nagy, J., *Colloids Surf.* 35, 201 (1989).
7. Awschalom, D. D., DiVincenzo, D. P., and Smyth, J. F., *Science* 258, 414
8. M. Agnelli, H.M. Swaan, C. Marquez-Alvarez, G.A. Martin, C. Mirodatos, J. Catal. 175 (1998) 117.
9. A. E. Aksoylu, Z. Onsan, Appl. Catal. A Gen. 164

(1997)1.
10. I. Alstrup, J. Catal. 151 (1995) 216.
11. R. Jin, Y. Chen, W. Li , W. Cui, Y. Ji, C. Yu, Y. Jiang, Appl. Catal.. A Gen. 201 (2000) 71.
12. Z. Liu, K. Jun, H. Roh, S. Baek, S. Park, J. Mol. Catal. A: Chem. 189 (2002) 283.
13. H.S. Bengaard, J.K. Norskov, J. Sehested, B.S. Clauser, L.P. Nielsen, A.M. Molenbroek, J.R. Rostrup-Nielsen, J. Catal. 209 (2002) 365.
14. T. Borowiecki, A. Golebiowski, B. Stasinska, Appl. Catal. A. Gen. 153 (1997) 141.
15. S. Remita, M. Mostafavi, M.O. Delcourt, Radiat. Phys. Chem. 47 (1996) 275.
16. J.H. Hodak, A. Henglein, M. Giersig, G.V. Hartland, J. Phys. Chem. B 104 (2000) 11708.
17. Y. Mizukoshi, K. Okitsu, Y. Maeda, T.A. Yamamoto, R. Oshima, Y. Nagata, J. Phys. Chem. B 101 (1997) 7033.
18. M. Brust, M. Walker, D. Bethell, D.J. Schiffrin, R. Whyman, J. Chem. Soc. Chem. Commun. 801 (1994).
19. K. Osseo-Asare, F.J. Arriagada, Ceram. Trans. 12 (1990) 3.
20 L.K. Kurihara, G.M. Chow, P.E. Schoen, Nanostruct. Mater. 5 (1995) 607.
21. H.H. Huang, X.P. Ni, G.L. Loy, C.H. Chew, K.L. Tan, F.C. Loh, J.F. Deng, G.Q. Xu, Langmuir 12 (1996) 909.16. F.
22. Dumestre, S. Martínez, D. Zitoun, M-C, Fromen, M-J. Casanove, C. Amiens, and B.Chaudret. Faraday Discuss., 125 (2004) 265-278.
23. Dongliang Tao and Fei Wei,. Materials Letters 58(2004)3226-3228.
24. Handbook of X-Ray Photoelectron Spectroscopy, Perkin–Elmer, Eden Prairie, 1978.

Fabrication of Metal Oxide Coaxial Nanotubes Using Atomic Layer Deposition

Daekyun Jeong[1], Taekwan Oh[1], Hyunjung Shin[1], Woo-Gwang Jung[1], Jaegab Lee[1], Myung Mo Sung[2] and Jiyoung Kim[1,*]

[1.] School of Advanced Materials Engineering, Kookmin University,
[2.] Department of Chemistry, Kookmin University
Seoul, 136-702, Korea,
[*]jiyoung@kookmin.ac.kr

ABSTRACT

We fabricated nanotubules of TiO_2 and ZrO_2 using atomic layer deposition (ALD) technique with polycarbonate (PC) nanoporous filters as a template. Alkylsiloxane monolayers on the both sides of PCs were formed by blanket type contact printing in order to achieve one-step process of the freestanding oxide nanotubes. TiO_2 and ZrO_2 nanotubes with 30 ~ 200 nm of diameter were successfully fabricated by ALD at 140ºC and subsequent chemical etching of the PC. Very high aspect ratio of 160:1 was achieved in both oxide nanotubes. Growth rates of the wall thickness in oxides nanotubes were 0.5 and 0.6 Å/cycles for 200 and 50 nm pore sizes of PC templates, respectively, showing ultra-precise control of the wall thickness, so as to inner diameter of the tubes. Prepared oxide nanotubes were characterized by high-resolution transmission electron microscopy (HR-TEM), field emission scanning electron microscopy (FE-SEM), and atomic force microscopy (AFM). Oxide nanotubes were filled with CdS using chemical bath deposition (CBD). We successfully demonstrated formation of coaxial TiO_2/CdS nanocables. Further we developed MO-CVD processes for Cu layer. As a result, coaxial nanotubes of TiO_2/Cu and ZrO_2/Cu were successfully fabricated. Combination of the ultra-precise wall thickness control of oxide nanotube with high-aspect ratio filled Cu layer provides us a possible quantum coaxial cable for nanoelectronic applications.

Keywords: Nanotube, Self-Assembled Monolayers, Atomic Layer Deposition, Template, Wall thickness

1 INTRODUCTION

Metal oxide nanotubes as well as CNTs [1] are expected to be useful for various applications such as nanoelectronics, biological sensors, nano-fluidic devices, nano-electromechanical system (NEMS) and energy storage, etc [2]. Particularly, one-dimensional structures of inorganic materials have attracted a lot of attention because of their unique properties (e.g., electronic, magnetic, optical, etc.) and potential applications in the near future

In this study, we reported that more controllable and feasible process for nanotube of TiO_2 as well as ZrO_2 using atomic layer deposition (ALD) and nanoporous poly-carbonate (PC) membrane as a template.[3] We reported that the formation of TiO_2 and ZrO_2 nanotubes by ALD oxide thin film[4] coating on the nanoporous membranes (commercially available polycarbonate filters) as templates and subsequently removing the templates.[5] Applying this method, we successfully developed novel fabrication routes for coaxial nanotubes of metal oxides with metal oxide nanotubes with semiconducting CdS core as well as metallic Cu layer.

2 EXPERIMENTAL

Fig. 1 shows the schematic flow of the fabrication process of the oxide nanotubes in this study. Membranes used in this study as nano-templates are hydrophilic poly-carbonate (PC) filters. The PC filters are commercially available (Whatman Co., UK). PC filters used in this study have nano-pores of diameter of 200 nm, 50 nm, or 30 nm. The pores in these filters are randomly distributed across the filter, and pore densities are approximately 3×10^8 pores/cm^2.

Coatings on inner wall of PC filters were carried out using ALD up to 140 °C. Self-limiting surface reaction in our home-made ALD processes for thin films of TiO_2 and ZrO_2 were confirmed up to 200 °C. Above 140 °C, the PC filters were distorted due to the thermal stress in ALD reaction chamber. Therefore process temperatures for oxide nanotube fabrication were kept lower than 140°C. Zirconium *t*-butoxide and Titanium *iso*-propoxide were used as sources of Ti and Zr. Water vapor with the flow rate of 5 sccm was used as an oxidant. N_2 was used as a carrier gas and Ar gas was used for purging. A cycle of ALD includes Zr and/or Ti source/purging/H_2O(g)/purging (e.g. 2/80/2/240 sec., respectively, in case of ZrO_2 coating). Growth of TiO_2 thin films with 0.4 Å/cycle at 150 °C can be achieved. In a typical experiment, 100 ~ 800 cycles of ALD were performed to synthesis nanotubes of TiO_2 and ZrO_2. After the desired cyclic ALD processes, the PC filters were etched away using the solution of chloroform at 60 °C.

NSTI-Nanotech 2005, www.nsti.org, ISBN 0-9767985-1-4 Vol. 2, 2005

To avoid any subsequent processing steps – e.g. etching process of the unwanted films deposited onto both sides of PC templates – to fabricate free-standing oxide nanotubes, selective area depositions of ALD has been adopted using contact printed OTS-SAMs as passivation layers. OTS-SAMs are deposited onto the both sides of PC filter by contact printing as shown in Fig. 1(b). Micro-contact printing of OTS resulted in transition of surface characteristics of the template from hydrophilic to hydrophobic [6], which eventually caused selective deposition in hole of nanotemplate without deposition on the surface of the filters. Subsequently, ALD processes make conformal coatings of TiO_2 or ZrO_2 only onto the inner-wall of PC template. Then 2nd layer was formed either CdS using either chemical bath deposition (CBD) or Cu deposited by CVD. PC template was etched by an organic chemical etching solvent such as chloroform. Nanotubes were remained selectively without significant damaged.

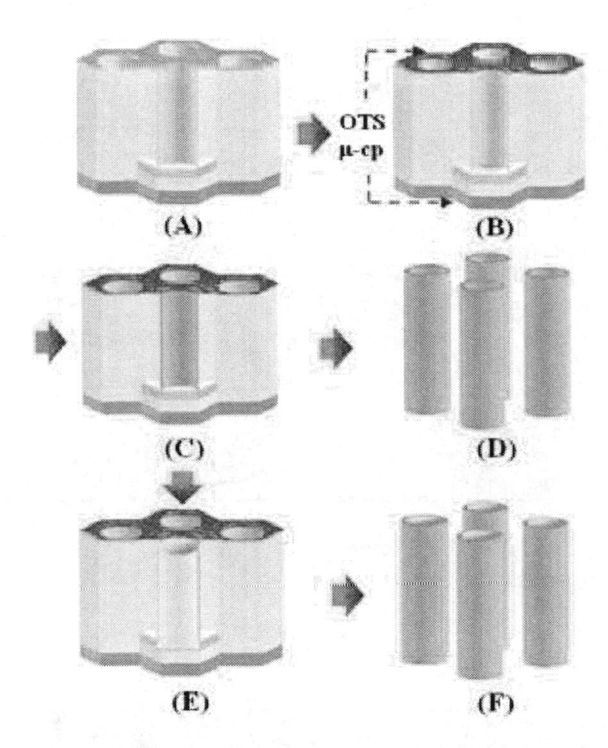

Figure 1: The fabrication processes for the oxide/semiconducting coaxial nanotubes in schematic sequence. a) PC template which has various pore size and thickness. b) Micro-contact printing with SAMs such octadesyl- tetrachlorosilane c) Selective deposition of the metal oxide layers on inner hole of nanotemplate d) Etched the PC template. Remained the metal oxide nanotubes e)Before etching the PC template, Cu was deposited by pulsed MOCVD while CdS was deposited by CBD. f) Etched the PC template. Nanocables were fabricated by the same method

As-prepared oxide nanotubular structures were characterized by field emission scanning electron micro-scopy (FE-SEM, Hitachi 4500, Japan) and atomic force microscopy (AFM, SPA-400, Seiko Instruments, Japan). High resolution transmission electron microscopy (HR-TEM) was used to analyze the nanotube crystal structures and wall thicknesses. And SAED patterns were used for evaluation of crystallization of the nanotubes.

3 RESULTS & DISCUSSION

3.1 Oxide Nanotubes

Fig. 2 shows the FE-SEM image of TiO_2 nanotubes with OTS treatment after etching the polycarbonate nanotemplate in aqueous solvent. In this figure, the average diameters of the nanotubes are about 50 nm and lengths are 8 μm, respectively. They formed a highly ordered array over all area of nanotemplate (that is, PC filters). OTS treatment prevents the connecting the nanotubes by unnecessary deposition on nanotemplate surface. Because CH_3^- which is the head group of OTS has hydrophobic property, OTS changes surface properties from hydrophilic to hydrophobic. This results in prohibition of nucleation of TiO_2 layer because ALD mechanism is based on surface reaction on OH^- bonding layers.

By changing the precursors and reactants, various nanotube materials were made. Zr-t-butoxide, Ti-isopropoxide precursors and H_2O oxidizing agent made ZrO_2 and TiO_2 nanotubes with wall thickness of 10nm. Especially, in Fig. 3(a), SAED pattern of TiO_2 nanotube shows amorphous nanostructure, otherwise ZrO_2 nanotubes, in Fig. 3(b), have a tetragonal phase. One significant advantage of this technique is easiness of geometric of nanotubes simply choosing adequate nano-template. Through variation of pore sizes of the nanotemplates, hole sizes of nanotubes are easily controlled. Using different thicknesses of nanotemplates, lengths of nanotubes are easily varied, too.

Figure 2 : FE-SEM image of TiO_2 nanotube which has the hole size of 50nm and length of 8 μm with 1:160 aspect ratio

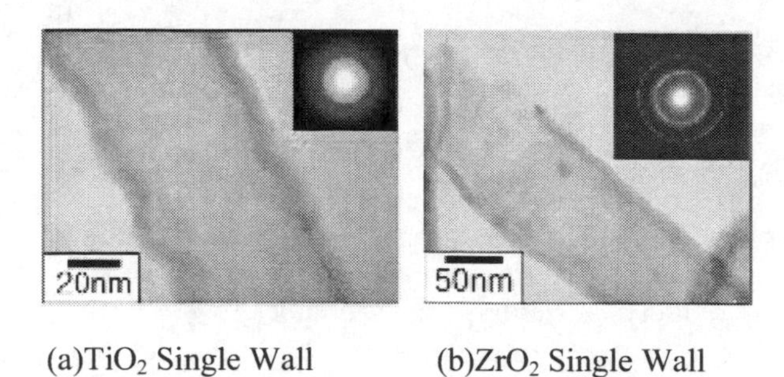

(a)TiO_2 Single Wall (b)ZrO_2 Single Wall

Figure 3: HR-TEM images of the oxide single wall nanotubes

3.2 Oxide and Semiconductor Layers Nanotubes

Cadmium sulfide (CdS) is an important II-VI semiconductor material compound which a good transparency in region of visible light and excellent photoconductive property.[7] CdS has been used in opto-electronic devices, window layer of thin film solar cells based on CdTe or $CuInSe_2$ or Cu_2S, SO_x gas sensor, optical memories and photoconductive cell. It can be applied widely in the development of other functional devices. In this fabrication, CdS was filled into metal oxide nanotubes using chemical bath deposition(CBD).

To fabricate TiO_2/CdS nanocables, nanotemplate surface was treated with SAM(OTS) by micro contact printing as we mentioned above. After oxide deposition by ALD, CdS was deposited by ammonia free chemical bath deposition technique. In ammonia free CBD condition, KOH, NTA acid, $CdSO_4$ and Thiourea are mixed in DI water and stirred. CdS was deposited about 100nm at pH 9 in 20 minutes. Unnecessary CdS removed by sonification. Then PC deposited TiO_2 / CdS was etched with Chloroform heated 60°C about 30minutes. Nanocables were chracterized by FE-SEM, HR-TEM and an energy dispersive spectroscopy (EDS) analysis. Fig.4 (a) shows a FE-SEM image and fig.4 (b) shows a HR-TEM micrograph. They show that outlet hole is TiO_2 nanotube and inlet hole is CdS nanowire, which is deposited uniformly on TiO_2 nanotube wall. Fig. 4 (c) shows EDS data of the TiO_2/CdS nanocable confirming chemical nature of the cable.

As increasing the deposition times, thickness of CdS was increased monotonously. Using this result, we could control the inner wall thickness easily. Fig. 5(a) is the TiO_2/CdS double wall nanotube that CdS was deposited for 20 minutes. It has that the outer TiO_2 thickness is 15nm, inner CdS thickness is 30nm and inner hole diameter is about 50nm. Otherwise, deposition times were increased for 40 minutes, CdS was jammed in TiO_2 nanotubes. As this result Fig. 5(b), TiO_2/CdS formed the nanocables. In the TiO_2/CdS nanocable, SAED pattern in Figs.5 show the inner CdS tube was polycrystal tetragonal structure.

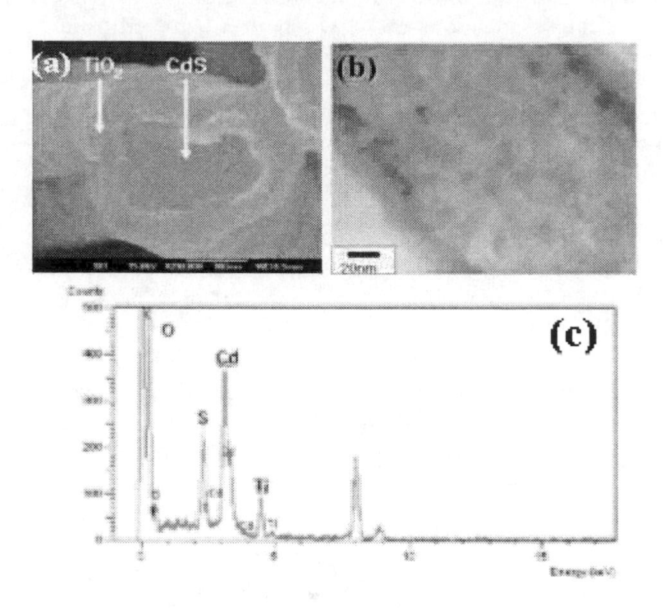

Figure 4: (a) FE-SEM and (b) HR-TEM images (c) EDS analysis of TiO_2/CdS nanotube

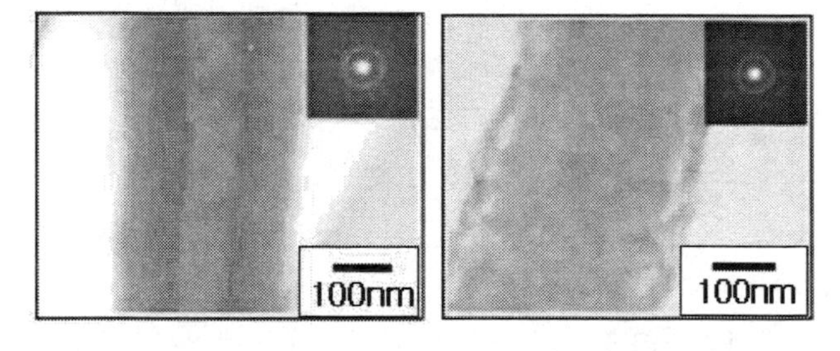

Figure 5: TiO_2/CdS nanotubular structures (a) Double wall nanotubes (b) Nanocable

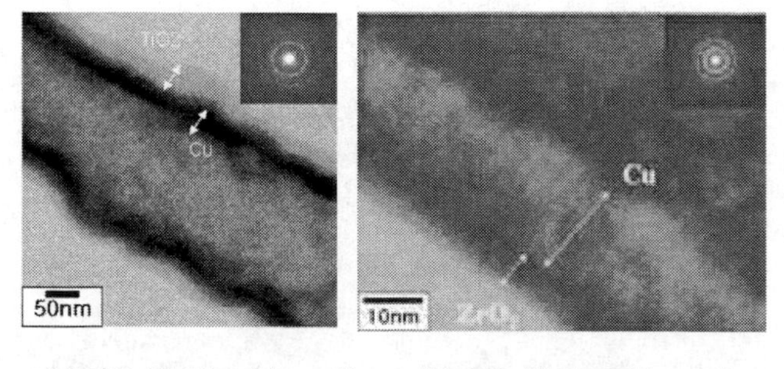

(a)TiO_2/Cu Double wall (b)ZrO_2/Cu Double wall

Figure 6: HR-TEM images of the copper metal and oxide double wall nanotubes

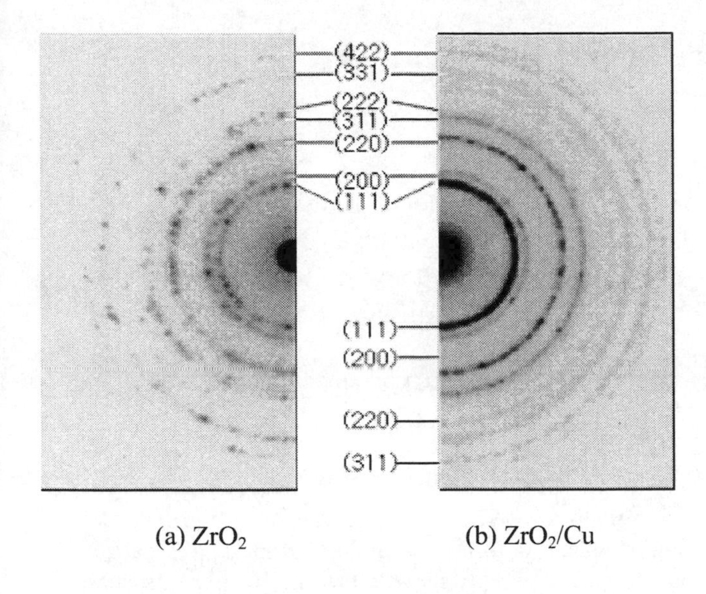

(a) ZrO_2 (b) ZrO_2/Cu

Figure 7: Diffraction pattern of the ZrO_2/Cu nanotubes

3.3 Oxide and Metal Layers Nanotubes

After TiO_2 layer was deposited on PC template, Cu layer was deposited by pulsed MOCVD by alternating the supply of the (hfac)Cu(DMB) and H_2 reactant. A typical pulsed deposition cycle consisted of 3 sec of (hfac)Cu(DMB) cycle, then a 6 sec of Ar purging time, followed by 3 sec long H_2 reactant pulse, and finally, a 6 sec of Ar purge cycle. The Cu source was heated at 30 °C, with the gas delivery line maintained at 40 °C. Then PC deposited TiO_2/ Cu was etched with Chloroform as the same method.

Fig 6. shows the metal copper and oxide double wall nanotubes. Cu layers were deposited on both oxide tube walls. And figs. 7 are the selective area electronic diffraction patterns of the ZrO_2 single wall and ZrO_2/Cu double wall nanotubes. In the case of ZrO_2/Cu double wall nanotubes, we clearly observed (111), (200), (220), (311) patterns of Cu as well as ZrO_2 patterns. In Fig. 7, SAED pattern of ZrO_2 nanotubes shows polycrystalline nanostructure which has tetragonal structure. Even though equilibrium phase of zirconia is monoclinic, tetragonal phase may results from nano-scale geometric effects, which cause high material pressure at tube-wall because of extremely high curvature.

4 SUMMARY

In this study, nanotubes which have various hole sizes, lengths and wall thicknesses were fabricated on PC nanotemplates by atomic layer deposition. SAMs treatment on the PC filters effectively reduced cross-link between oxide nanotubes because the treatment inhibited nucleation of oxide films. 30 ~ 200 nm in diameter of TiO_2 and ZrO_2 nanotubes were successfully fabricated by ALD at 140 °C and subsequent chemical etching of the PC. Very high aspect ratio of 160:1 was achieved in both oxide nanotubes. Growth rates of the wall thickness in oxides nanotubes were 0.5 and 0.6 Å/cycles for 200 and 50 nm pore sizes of PC templates, respectively, showing ultra-precise control of the wall thickness, so as to inner diameter of the tubes. After metal oxide nanotube was made, CdS could be filled into metal oxide nanotube by method of CBD. It was confirmed by FE-SEM and HR-TEM. In addition, we successfully fabricated nanotubes with double walls of metal and oxide layers.

ACKNOWLEDGEMENT

The authors gratefully acknowledge the financial support through Center for Nanostructured Materials Technology by Korean Ministry of Science and Technology (03K1501-02410).

REFERENCES

[1] S. Iijima. “Helical microtube of graphitic carbon,” Nature, vol.354, pp.56, 1991
[2] Hari Singh Nalwa, "Encyclopedia of Nanoscience and Nanotechnology" ASP, No.1, pp.547-634, 2004
[3] H. Shin, et. al "Formation of TiO_2 and ZrO_2 nanotubes using atomic layer deposition with ultra-precise control of the wall thickness" Wiley-Vch, Advanced materials, vol.16, pp.1197, 2004.
[4] D. Jeong, et. al, “Effects of various oxidizers on the ZrO_2 thin films deposited by atomic layer deposition” Taylor & Francis Ltd., Integrated Ferroelectrics, vol. 67 pp.41. 2004
[5] D. Gong, et. al “Titanium oxide nanotube arrays prepared by anodic oxidation” Mater. Res. Soc., J. Mater. Res., vol.16, pp.3331, 2001
[6] H. K. Kim, et. al "Thermal Decomposition of Alkylsiloxane Self-Assembled Monolayers in Air" ACS, J. Phys. Chem. B., vol.107, pp.4348. 2003.
[7] J.M. Dona, et. al. “Dependence of electro-optical properties on the deposition conditions of chemical bath deposited CdS thin films” ECS, J. Electrochem. Soc. vol. 144, pp.4091, 1997

Bulk Nanostructured Metals by Severe Plastic Deformation (SPD)

M. Zehetbauer*, M. Krystian**, W. Lacom***

*Institute of Materials Physics, Vienna University, A-1090 Austria, zehet@ap.univie.ac.at
**Institute of Materials Physics, Vienna University, A-1090 Austria,
and ARC Seibersdorf GmbH, Seibersdorf, A-2444 Austria, krystian@ap.univie.ac.at
***Austrian Research Center Seibersdorf GmbH, Seibersdorf, 2444 Austria, Wolfgang.Lacom@arcs.ac.at

ABSTRACT

The methods of manufacturing bulk nanostructured metals (BNMs) by means of SPD are concerned. The principle and merits of the two most promising methods are outlined out. New results of properties of SPD-processed BNMs are elucidated, especially in view of commercial applications. The benefits of stronger metals and alloys mainly for automotive industry, higher fatigue resistance of materials for aerospace industry, improved hydrogen storage kinetics in Mg-alloys for fuel cell technology, improved magnets for electronic industry, stronger biocompatible Ti- and Mg-alloys for prostheses, implants and stents are also pointed out.

Furthermore, a model based on lattice defect kinetics is presented. It provides quantitatively reliable results for the hardening behavior and microstructural evolution during SPD and post-SPD deformation.

Keywords: SPD, bulk nanostructured materials, mechanical properties and modeling, hydrogen storage

1 INTRODUCTION

Manufacturing of BNMs, i.e. bulk metals with a mean grain size of some hundreds of nm and resulting in advanced physical properties, requires very large plastic deformations at relatively low temperatures. Traditional methods, such as rolling, forging, drawing or extrusion cannot meet these requirements without crack formation. First attempts to produce BNMs followed a variety of methods like (i) inert gas condensation, (ii) high-energy ball milling with subsequent consolidation, (iii) electro-deposition, and (iv) crystallization from an amorphous state. However, these so-called bottom-up methods cannot accomplish sample dimensions of more than several mm. Moreover, a number of difficulties i.e. residual porosity (i), impurities (ii, iii), and exposure to dangerous nanopowders (ii), have prevented these techniques from reaching larger practical applications, in addition to the fact that they are not suited for fabrication on an industrial scale.

In recent years, however, as a versatile alternative a top-down method called Severe Plastic Deformation (SPD) has gained importance because of the *direct* conversion of *bulk* metals and alloys with conventional grain sizes to nano-scaled materials with outstanding new properties [1].

2 METHODS FOR BNMs BY SPD

2.1 General principle

Nano-scaling by SPD is considered to be the result of very high plastic deformation at low temperatures carried out under special conditions with the aim of microstructure refinement. That means the formation of an extensive dislocation cell network with *small angle* boundaries, developing to a microstructure with mainly *high angle* grain boundaries within the whole volume. The direct refinement of conventional grain size into nano-scaled bulk materials demands appropriate conditions: essential for SPD are mainly the *multidirectionality* of deformation, the enhanced *hydrostatic* pressure and the preservation of the original shape of samples [2].

During the last decade many special methods of deformation by mechanical means have been developed and tested. Although all of the techniques can lead to the BNMs of pure metals, alloys, steels and intermetallics, some technological difficulties have so far hampered up-scaling and technical utilization. For the moment, the two most well-known and promising SPD methods are considered to be the Equal Channel Angular Pressing (ECAP) and the High Pressure Torsion (HPT).

2.2 Equal Channel Angular Pressing (ECAP)

During ECAP a billet of at least 70 mm in length and a cross section diameter of up to 60 mm is multiple pressed through a special die with two *equal* channels (Fig. 1a). The deformation occurs via *pure shear* without changing the cross section area of the ingot. That fact enables accumulative deformation up to very high strains by repeated pressing. The angle of intersection of the two channels **Φ** is usually 90° but higher values, e.g. 120°, can be chosen to reduce the load or, together with elevated temperature and/or back-pressure, to engineer the structure. Between passes the billet can be rotated around its longitudinal axis through the angle of 0° (route A), 90° (route B) or 180° (route C) which, together with the number of passes, strongly determines the final microstructure (for additional information see [1] and [3]).

2.3 High Pressure Torsion (HPT)

The HPT-method, shown schematically in Fig. 1b and described in detail in [1] and [3], is suited for disks of 10-20 mm in diameter and up to 1 mm in thickness. A sample is strained in torsion at room temperature – or at any other temperature in the range from -200 to 500°C – many times between plungers on which an outer pressure is applied. The latter is of highest importance because it ensures the material to be deformed under an enhanced hydrostatic pressure of several GPa. These conditions enable to process hard-to-deform or brittle materials up to very large strains without failure or cracks. High friction forces between the rough dies and the ingot ensure deformation by *shear* during rotation of the plunger. Provided that the number of rotations applied is high enough, an almost homogeneous nanostructure can be achieved.

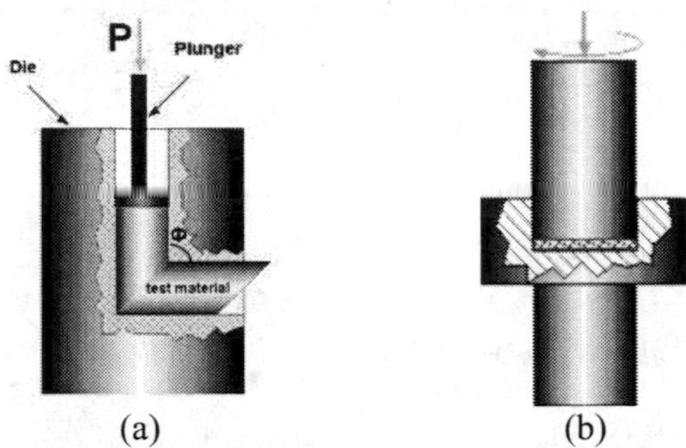

Figure 1: Sketch of SPD techniques: (a) ECAP, (b) HPT

2.4 Further SPD methods

SPD techniques with more industrial potential are shown in Fig. 2. Accumulative Roll-Bonding (ARB) (Fig. 2a, [4]) will enable the production of BNM sheets. Cyclic Extrusion-Compression (CEC) (Fig. 2b, [5]) and Twist Extrusion (TE) (Fig. 2c [6]) are suitable for manufacturing rods. Other SPD techniques are e.g. shock wave loading, multiple forging (see e.g. [1] and [3]) or a commercial version of ECAP, the Continuous Confined Strip Shearing (CCSS, Fig. 2d, [7]). Undoubtedly a lot of work still needs to be done to finally develop these methods by improving efficiency and overcoming some technical drawbacks so that a mature SPD technology is made available for industry.

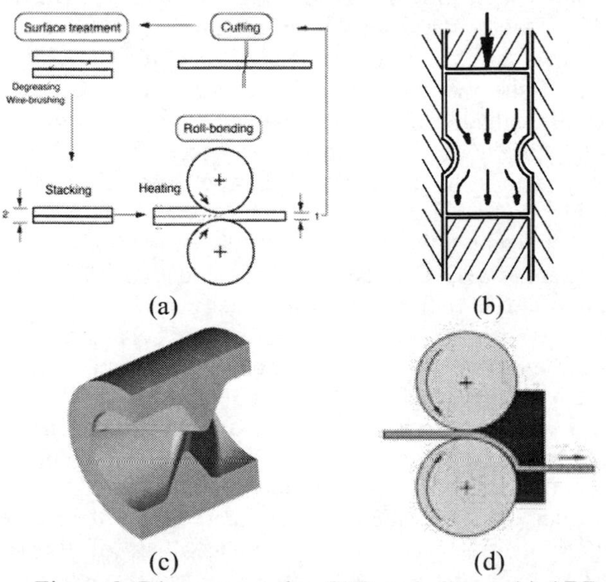

Figure 2: Diagram another SPD techniques: (a) ARB, (b) CEC, (c) TE, (d) CCSS.

3 ADVANTAGES OF SPD - BNMs

3.1 New properties of BNMs

As a result of the fairly unusual microstructure of BNMs new outstanding physical and mechanical properties occur. Some of them like changes in the Curie and Debye temperatures, atomic disordering, anomalies in the internal friction and optical properties of semiconductors or even elastic moduli are fundamental physical phenomena and surely of high scientific interest. Others possess the potential for practical applications as well. Let us just mention: saturation magnetization and magnetic hysteresis, enhanced kinetics of diffusion, improved corrosion behavior, low temperature and/or high strain rate superplasticity, enhanced high-cycle fatigue life time, formation of a supersaturated solution and metastable states in high carbon steels and – last but probably most important – enhanced strength at still high ductility. Plastic deformation induced by conventional methods can significantly increase strength. This increase, however, is accompanied by a loss of ductility. So it is possible to obtain e.g. high strength Cu, but only with significantly reduced ductility (see the Cu line in Fig. 3. It has been reported that pure Cu after 16 ECAP passes reveals even higher strength with much higher ductility in comparison to coarse-grained metals [8]. Also in pure Ti after 5 HPT revolutions a very strong increase in strength together with an only slight decrease of ductility was observed. This extraordinary mechanical behavior is now called paradoxon for SPD materials.

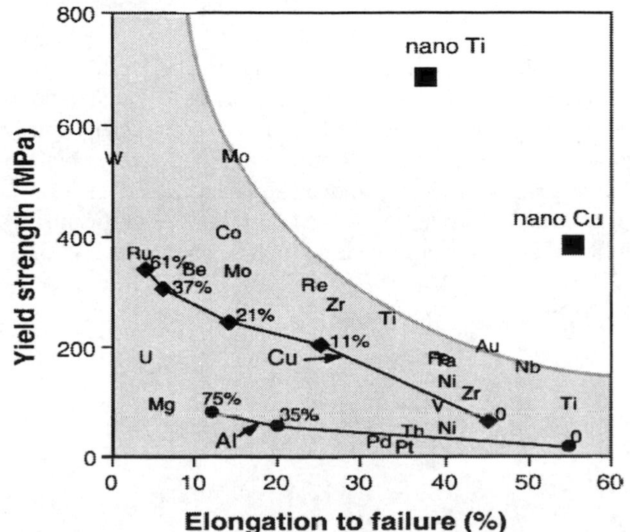

Figure 3: High strength and ductility of SPD pure Cu and Ti compared with conventional coarse-grained metals [8].

3.2 Applications of SPD produced Nanometals

The spectacular mechanical properties recommend SPD materials for the automotive and aerospace industry, as tools for metals machining, and for advanced sputtering targets. NanoSPD Fe-alloys can be part of highly effective magnets for computer hardware, video heads, generators, electric motors, transformers, mechatronic systems, MEMS etc. Many branches of the industry can benefit from SPD-processed NMs. Stronger Fe-, Al- and Mg-alloys permit weight saving and therefore less fuel consumption for cars and aircrafts. The high superplastic formability of SPD materials will speed up the production processes of deep-drawn products (bottles, vessels) from Al- and Mg-alloys by at least a factor of 10 and thus cut costs for machining. Aerospace industry will demand light nano-SPD materials with higher fatigue resistance than conventional ones. BNM out of CP- Ti as well as high-strength and flexible Ti-alloys will have a great impact on medical applications. The use of pure but nevertheless strong SPD-Ti is promising due to its enhanced biocompatibility compared to that of traditionally used medical alloys. The application of prostheses, implants and stents made of long-life BNMs will be essential to reduce the need for intermediate reoperations [9].

An example of applied ECAP is the forcefill technique used in the field of micro-electro-mechanical-systems for metallization and interconnection of conductors on different layers [10]. As shown in Fig. 4, metal from blanket film is forced by pressure into via holes which act as small equal channels (compare Fig 1a). By this way connections between integrated circuits not only can be achieved cheaply but also reveal improved mechanical and magnetic properties.

In developing alternatives for automotive fuels, hydrogen in combination with a fuel cell is considered to be the most promising solution. The crucial problem, however, is to achieve a safe and low-cost hydrogen storage technology. Since both liquid and high pressure hydrogen storage requires serious security precautions, the acceptance by customers may be restricted. Mg-alloys can solve this problem by providing hydrogen storage in the solid Mg-MgH_2 system. However, the main difficulties still are (i) the low ab-/desorption kinetics and (ii) the high desorption temperature. Nanocrystalline pure Mg with very high density of grain boundaries and thus enhanced diffusion of hydrogen (and formation of MgH_2) can significantly accelerate the ab-/desorption kinetics. As shown in Fig. 5,

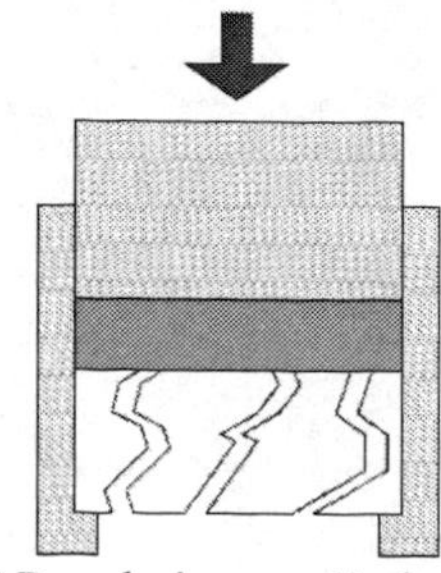

Figure 4: ECAP as being applied to forcefills [10]

in nano-Mg the hydride develops within only a few *minutes*. This kinetics can be further enhanced by catalysts. Although the presented results were achieved with powders from ball milling [11], first tests on ECAP Mg show a similar behavior [12].

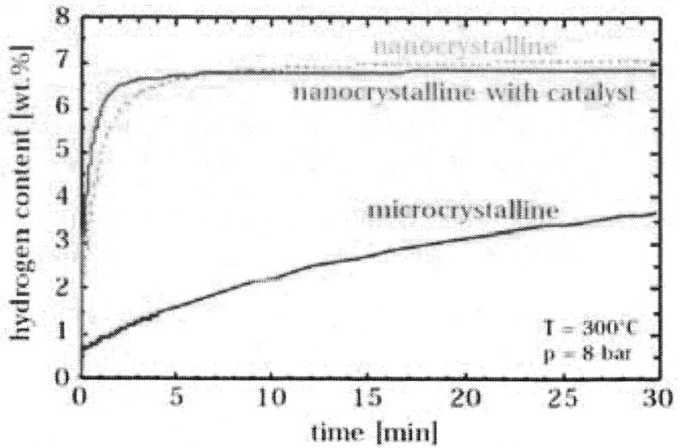

Figure 5: Kinetics of hydrogen absorption at 300°C and 8 bar for nano-Mg compared with conventional coarse-grained pure Mg [11].

4 MODELING AND SIMULATION OF SPD AND POST-SPD DEFORMATION

From a scientific point of view as well as for an optimized technical performance a deep understanding of the structural mechanisms during formation of the nano-structure by SPD is of great importance. On the other hand, efforts must be made to describe quantitatively the mechanical properties of the *as-produced* BNM. This means that one has to carefully distinguish between "in-situ" and "post SPD" modeling. In the recent literature the model by Zehetbauer et al. was shown to be capable of both, simulation of strengthening in BNMs *during* SPD [2] and simulation *after* processing [13]. The model is based on lattice defect kinetics and deals with five fitting parameters which contain basic physical properties. One of the big successes of the model is the proper simulation of the influence of the hydrostatic pressure p being present during processing [2]. It is shown that with increasing p the diffusion and thus the annihilation of edge dislocations is increasingly restricted so that an enhanced dislocation density is available providing more grain boundaries and/or smaller grain sizes. Concerning the simulation of the hardening of a BSM after SPD, it has been recently shown that the model can also correctly describe this behavior, as well as the evolution of the microstructure, when passing through the three post-SPD deformation stages II, III and IV. Fig. 6 clearly demonstrates the good agreement of the simulation by Zehetbauer's model with the experimental hardening data of ECAP-predeformed CP-titanium [13]. The evolution of certain microstructural parameters like the dislocation density and, in particular, that of the deformation induced vacancy concentration is reflected in quantitative agreement with the measurements (Fig. 7).

Moreover, the model accurately predicts the evolution of the cell size within the given limits of experimental evaluation (Fig. 8, [13]).

At present, efforts are made to quantify the enhanced ductility of SPD metals which, according to hitherto performed experiments, arises from structural recovery taking place between subsequent ECAP passes and/or even during the SPD process itself.

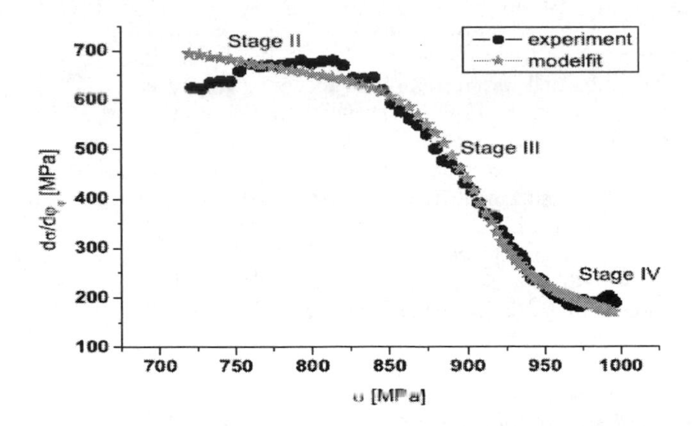

Figure 6: Experimental data of hardeing during post-deformation of nanoSPD CP-Ti after 8 ECAP passes compared with the prediction by Zehetbauer's model.

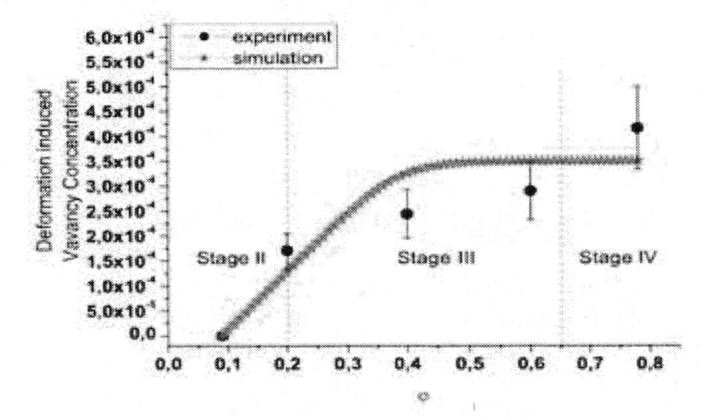

Figure 7: Comparison between the deformation induced vacancy concentration determined by residual electrical resistivity and the simulation by Zehetbauer's model.

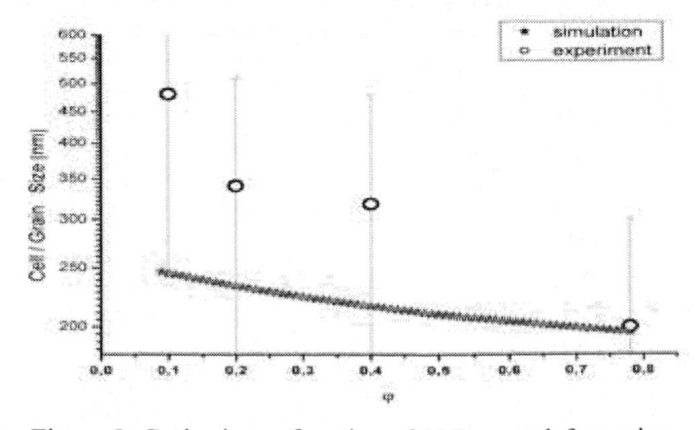

Figure 8: Grain size as function of SPD-post deformation estimated by TEM and simulated by Zehetbauer's model.

Recently, the first attempts to simulate the ECAP-process by neuronal networks have been made showing, however, that a large amount of data for training runs is necessary for the simulation to be successful.

5 SUMMARY AND OUTLOOK

It was shown that today's SPD techniques are able to produce full-dense bulk materials with enhanced strength (1), but highly ductile or even superplastic properties (2), with high fracture toughness (3), high fatigue life (4), and enhanced hydrogen diffusion (5). Some SPD nanomaterials exhibit highly advanced magnetic properties (6). Such outstanding properties suggest a number of applications in the automotive and aerospace industries (1, 4, 5), in final forming and shaping (2, 3), in medical fields (1, 3, 4), and as highly effective bulk magnets (6). At present, several attempts to combine the basic SPD techniques with new ones in order to fulfill the conditions of a low-cost and fast-manufacturing method (e.g. continuous operation).

The physics of the SPD-process can be understood by the constitutive model of Zehetbauer [2,13], which involves concrete physical parameters – such as dislocation density, vacancy concentration, grain size etc. – and is thus suitable for simulation of the grain refinement during and after SPD.

REFERENCES

[1] R.Z.Valiev, I.V.Islamgaliev, I.V.Alexandrov, Progr. Mater. Sci. 45, 103-189, 2000.

[2] M.Zehetbauer, H.Stüwe, A.Vorhauer, E.Schafler, J.Kohout, Adv. Eng. Mater., 5, 330-337, 2003.

[3] eds. M.Zehetbauer & R.Z.Valiev, "Nanomaterials by Severe Plastic Deformation", Proc."NANOSPD2", Dec. 9-13 (2002), Wien, Austria, Wiley VCH, Weinheim, Germany, 2004.

[4] N.Tsuji, Y.Saito, S.H.Lee, Y.Minamino, Adv. Eng. Mater. 5, 5, 338-344, 2003.

[5] J.Richert, M.Richert, Aluminium, 62, 604-607, 1986.

[6] Y.Beygelzimmer, V.Varyukhin, D.Orlov, B.Efros, V.Stolyarov, H.Salimgareyev, in: Proc. 2nd Int. Symp. UFG Materials, 2002 TMS Annual Meeting, Seattle, USA, 43-46, 2002.

[7] J.C.Lee, H.K.Seok, J.Y.Shu, Acta Mater., 50, 4005-4019, 2002.

[8] R.Z. Valiev, I.V. Alexandrov, Y.T. Zhu, T.C. Lowe, J.Mater. Res., 17, 1 , 5-8, 2002.

[9] L.Zeipper, M.Zehetbauer, B.Mingler, E.Schafler, G.Korb, H.P.Karnthaler, p. 810-816, in [3].

[10] Y.Estrin, H.S.Kim, M.Kovler, G.Berlar, R.Shaviv, E.Rabkin, J. Appl. Phys, 93, 5812-5815, 2003.

[11] T.Klassen, R.Bohn, G.Fanta, W.Oelerich, N.Eigen, F.Gärtner, E.Aust, R.Bormann, H.Kreye, Z. Metellkd., 94, 610-614, 2003.

[12] M.Skipnyuk, E.Rabkin, Y.Estrin, R.Lapovok, Acta Mater., 52, 405-414, 2004.

[13] L.F.Zeipper, M.J.Zehetbauer, Ch.Holzleithner, Mater. Sci. Eng. A, submitted for publication.

Surface plasmon absorption of metallic nanospheres in an amplifying medium

Andrei Y. Smuk and Nabil M. Lawandy
Solaris Nanosciences
46 Amaral St., East Providence, RI 02915, asmuk@solarisnano.com

ABSTRACT

Numerous applications that rely on the local field enhancement associated with the Surface Plasmon Resonance (SPR) have been suggested. With host material being an integral part in forming SPR, systems with optically transparent host have been studied exhaustively, while the case of an absorbing host still generates discussions in the literature. Amplifying host in the context of SPR has until now received no attention in the literature.

This paper considers the problem of resonant absorption of electromagnetic radiation by small metallic particles embedded in an amplifying medium. A recent theoretical analysis showed for the first time that presence of gain in the host medium could result in a considerable enhancement of the SPR of colloidal metals in that host. Local field intensities can be enhanced by as much as two orders of magnitude, compared with those obtained near surface plasmon resonance of metal nanoparticles in non-amplifying media. We further develop these ideas in a rigorous manner as a generalized Mie solution. Possible applications for these systems include rapid single molecule detection for gas and biological sensors and ultra-low intensity optical tweezers for manipulation of nanostructures.

Keywords: plasmon resonance, gold nanoparticles, Mie scattering, amplifying medium, plasmon enhancement

1 INTRODUCTION

The classic phenomenon of surface plasmon resonance (SPR) of small metallic particles has recently become the focus of renewed interest and intensive exploration due to the development of generally attainable techniques of synthesis of these materials. Numerous applications that rely on the local field enhancement associated with SPR have been suggested, including surface-enhanced Raman scattering [1] and harmonic generation [2]. The magnitude of the local electromagnetic field is, therefore, the key to efficiency in any of the SPR applications. With host material being an integral part in forming SPR, systems with optically transparent host have been studied exhaustively [3], while the case of an absorbing host still generates discussions in the literature [4]. Amplifying host in the context of SPR has until now received no attention in the literature.

This paper considers the problem of resonant absorption of electromagnetic radiation by small metallic particles embedded in an amplifying medium. In a recent theoretical analysis [5], performed in electrostatic approximation, it was shown for the first time that the presence of gain in the host medium could result in a considerable enhancement of the SPR of colloidal metals suspended in that host. This effect is the result of the mutual cancellation of the imaginary parts of dielectric functions of the metal sphere and host medium in the denominator of the first-order partial electric wave of the sphere. Complex dielectric function of the particle's surrounding, obtained by means of introduction of gain, transfers the normally complex natural frequencies of the sphere into the real domain, and thus makes it possible to increase local field intensities by as much as an order of magnitude, compared with those obtained near surface plasmon resonance of metal nanoparticles in non-amplifying media. We further develop these ideas in a rigorous manner as a generalized Mie solution for absorption of a coated gold nanosphere, utilizing numerical algorithms for evaluation of Bessel-Riccati functions and their derivatives.

2 COMPUTATION OF EFFICIENCY FACTORS FOR COATED SPHERE

We follow the formalism developed by Aden and Kerker [6] for scattering of light by stratified spheres in notations of Kerker [7]. In this theory, the exact solution of wave equation is obtained for scattering of electromagnetic radiation by a particle whose complex index of refraction varies radially from the center to the outer surface. We limit our treatment by the case of two-layer particle, i.e. a coated sphere. The inner layer (core) has radius a and refractive index n_1, and the outer layer (shell) has radius b and refractive index n_2. The refractive index of the host medium is n_3.

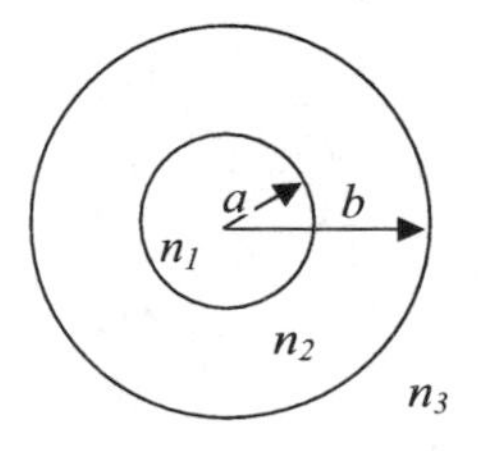

Figure 1. Geometry of a scattering coated sphere.

As usually, the calculation is performed in terms of dimensionless size parameters:

$$\alpha = \frac{2\pi a n_3}{\lambda},$$
$$v = \frac{2\pi b n_3}{\lambda}$$

where λ is the wavelength of light in vacuum, as well as relative indices of refraction, which are introduced as follows:

$$m_1 = \frac{n_1}{n_3},$$
$$m_2 = \frac{n_1}{n_3}$$

Kerker [7] gives the solution for the scattering coefficients in terms of Riccati-Bessel functions $\psi_n(z)$, $\chi_n(z)$, and their linear combination $\zeta_n(z)=\psi_n(z)+i\cdot\chi_n(z)$, which relate to the spherical Bessel functions of order n of the first kind $j_n(z)$ and of the second kind $y_n(z)$:

$$\psi_n(z) = z \cdot j_n(z),$$
$$\chi_n(z) = -z \cdot y_n(z)$$

The scattering coefficients are:

$$a_n = \frac{\begin{vmatrix} \psi_n'(m_2\alpha) & \chi_n'(m_2\alpha) & \psi_n'(m_1\alpha) & 0 \\ m_2\psi_n(m_2\alpha) & m_2\chi_n(m_2\alpha) & m_1\psi_n(m_1\alpha) & 0 \\ \psi_n'(m_2v) & \chi_n'(m_2v) & 0 & \psi_n'(v) \\ m_2\psi_n(m_2v) & m_2\chi_n(m_2v) & 0 & \psi_n(v) \end{vmatrix}}{\begin{vmatrix} \psi_n'(m_2\alpha) & \chi_n'(m_2\alpha) & \psi_n'(m_1\alpha) & 0 \\ m_2\psi_n(m_2\alpha) & m_2\chi_n(m_2\alpha) & m_1\psi_n(m_1\alpha) & 0 \\ \psi_n'(m_2v) & \chi_n'(m_2v) & 0 & \zeta_n'(v) \\ m_2\psi_n(m_2v) & m_2\chi_n(m_2v) & 0 & \zeta_n(v) \end{vmatrix}},$$

$$b_n = \frac{\begin{vmatrix} m_2\psi_n'(m_2\alpha) & m_2\chi_n'(m_2\alpha) & m_1\psi_n'(m_1\alpha) & 0 \\ \psi_n(m_2\alpha) & \chi_n(m_2\alpha) & \psi_n(m_1\alpha) & 0 \\ m_2\psi_n'(m_2v) & m_2\chi_n'(m_2v) & 0 & \psi_n'(v) \\ \psi_n(m_2v) & \chi_n(m_2v) & 0 & \psi_n(v) \end{vmatrix}}{\begin{vmatrix} m_2\psi_n'(m_2\alpha) & m_2\chi_n'(m_2\alpha) & m_1\psi_n'(m_1\alpha) & 0 \\ \psi_n(m_2\alpha) & \chi_n(m_2\alpha) & \psi_n(m_1\alpha) & 0 \\ m_2\psi_n'(m_2v) & m_2\chi_n'(m_2v) & 0 & \zeta_n'(v) \\ \psi_n(m_2v) & \chi_n(m_2v) & 0 & \zeta_n(v) \end{vmatrix}}$$

Efficiency factors for extinction and scattering are obtained from the scattering coefficients as follows:

$$Q_{sca} = \frac{2}{v^2} \cdot \sum_{m=1}^{\infty} (2m+1) \cdot \mathrm{Re}(a_m + b_m),$$

$$Q_{ext} = \frac{2}{v^2} \cdot \sum_{m=1}^{\infty} (2m+1) \cdot \left(|a_m|^2 + |b_m|^2 \right)$$

The efficiency factor for absorption can be obtained as $Q_{abs} = Q_{ext} - Q_{sca}$. We truncated the series when relative increment decreased below 10^{-6}.

Our model system consists of a gold core and an amplifying shell immersed in water. Refractive index of gold was taken from experimental data [8]. Idealized shell material was considered to model an aqueous solution of amplifying "dye" whose gain line shape is described by the Lorenzian:

$$g(\omega) = \frac{\Delta_\omega}{(\omega - \omega_0)^2 + \left(\frac{\Delta_\omega}{2}\right)^2}$$

where Δ_ω is the linewidth, and ω_0 is the center frequency of the gain line. Dielectric function of such a dye follows as:

$$\varepsilon_2(\omega) = n_3^2 + 4\pi A g(\omega) \left(\frac{\omega_0 - \omega}{\Delta_{a\omega}} + i \cdot \frac{1}{2} \right),$$

where $n_3 = 1.33$ is the refractive index of water and A is gain factor, proportional to the concentration of the dye. We choose our dye to have peak gain at 520 nm, and linewidth $\Delta_\omega = 4\cdot10^{14}$ s^{-1}. Plotted in Figures 2-4 is absorption efficiency of the core, defined as the total absorption cross-section divided by the geometrical cross-section of the gold core:

$$\alpha_Q^{abs} = \frac{C_{abs}}{\pi a^2} = Q_{abs} \cdot \frac{b^2}{a^2}$$

This quantity is proportional to the absorption coefficient of a suspension of corresponding particles in water, and allows direct comparison of Figures 2-4.

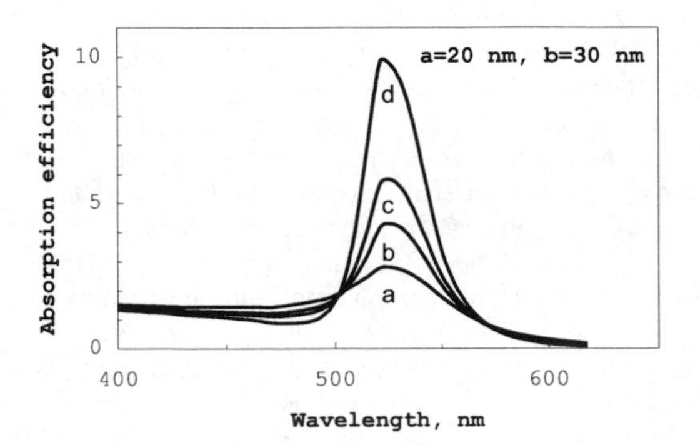

Figure 2. Absorption efficiency for a 20 nm core, 30 nm shell. Gain coefficient is (in cm^{-1}): a-0, b- $2.8\cdot10^4$, c-$3.9\cdot10^4$, d-$5.4\cdot10^4$.

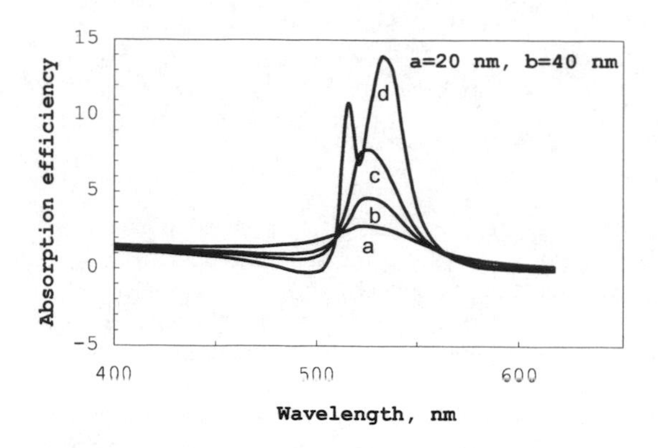

Figure 3. Absorption efficiency for a 20 nm core, 40 nm shell. Gain coefficient is (in cm^{-1}): a-0, b- $2.8 \cdot 10^4$, c-$3.9 \cdot 10^4$, d-$5.4 \cdot 10^4$.

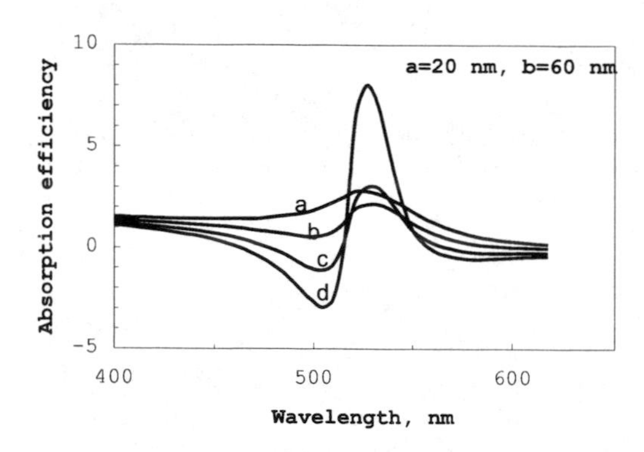

Figure 4. Absorption efficiency for a 20 nm core, 60 nm shell. Gain coefficient is (in cm^{-1}): a-0, b- $1.4 \cdot 10^4$, c-$2.8 \cdot 10^4$, d-$3.9 \cdot 10^4$.

Thus introducing amplification into a shell around a gold particle can lead to increase in absorption cross-section, in our case by a factor of 7 or more. This enhancement of absorption is necessarily accompanied by comparable increase of optical field in the vicinity of the particle, i.e the effects of the SPR are more pronounced.. Further shell amplification increase leads to absorption efficiency becoming negative in parts of the spectrum, which means overall amplification at the appropriate wavelength. However, the local field calculation, which has more direct and immediate relation to phenomena such as surface-enhanced Raman scattering and ultimately to the applications is the subject of another paper, which will be published elsewhere. Among suggested applications for the systems of metal nanostructures in amplifying host is rapid single molecule detection for gas and biological sensors and ultra-low intensity optical tweezers for manipulation of nanostructures.

REFERENCES

[1] K. Kneipp, H. Kneipp, I. Itzkan, R.R. Dasari, and M.S. Feld, J. Phys. **14**, R597-R624 (2002).

[2] P.M. Hui, C. Xu, and D. Stroud, Phys. Rev. B **69**, 014203 (2004).

[3] C.F. Bohren and D.R. Huffman, *Absorption and Scattering of Light by Small Particles* (Wiley, New York, 1998)

[4] A.N. Lebedev and O. Stenzel, Eur. Phys. J. D **7**, 83 (1999), I.W. Sudiarta and P. Chylek, Appl. Opt. **41**, 3545 (2002), G. Videen and W. Sun, Appl. Opt. **42**, 6724 (2003).

[5] N.M. Lawandy, Appl. Phys. Lett. **85**, 5040 (2004).

[6] A.L. Aden, and M. Kerker, J. Appl. Phys. **22**, 1242 (1951)

[7] M. Kerker, *The Scattering of Light and Other Electromagnetic Radiation*, chapter 5 (Academic Press, New York, 1969)

[8] P.B. Johnson and R.W. Christy, Phys. Rev. B **6**, 4370 (1972).

First-Principles Studies of SnS_2 Nanotubes

Hyunju Chang, Eunjeong In, Ki-jeong Kong, Jeong-O Lee,Youngmin Choi, Byung-Hwan Ryu

Korea Research Institute of Chemical Technology, Daejeon, 305-600, KOREA, hjchang@krict.re.kr

ABSTRACT

First principles calculations are used to predict the stability and electronic structures of SnS_2 nanotubes. Optimization of several structures and their corresponding strain energies confirms the stability of SnS_2 nanotube structures. Elelctronic structure calculations show that SnS_2 nanotubes could have moderate band gaps regardless of their chirality. It suggests that SnS_2 nanotubes would be well suited to use as semiconductor wires in nanoelectronic devices if they are synthesized. Adsorption of NH_3 onto SnS_2 is also investigated and discussed with regard to potential sensor application.

Keywords: SnS_2, metal-chalcogenide, nanotubes, first principles, gas sensors

1 INTRODUCTION

Since carbon nanotube(CNT) structure was found[1], there have been extensive investigations on applications of CNTs. CNTs are fascinating because of their diversities of electronic structure, from metal to semiconductor, depending on their chiralities. However, these diversities became a major disadvantage in CNT's real applications for electronic devices. Besides CNTs, syntheses of new nanotubes from various inorganic materials were reported recently. Especially inorganic nanotubes from metal chalcogenide, MoS_2 and WS_2, have received a great deal of attention, because they were found to be semiconductors regardless of their chiralities[2]. Since Tenne et al.[3] had found fullerene-like nanoparticles derived from a layered structure of WS_2, syntheses of nanotubes of WS_2 and MoS_2 were reported by the same group[2] and others[4]. Very recently, Tenne's group also reported synthesis of fullerene-like nanoparticles of SnS_2 [5]. SnS_2 is very similar to MoS_2 and WS_2, since it has a layered structure and it shows semiconductor properties. SnS_2 based materials show semoconducting properties, even when it has nanoporous structures [6]. Even though SnS_2 nanotubes are not synthesized experimentally yet, it is worthwhile to investigate them theoretically for future applications. In this paper, we report a possibility of SnS_2 nanotubes and their electronic structures investigated in the framework of density functional theory. The adsorption of NH_3 on SnS_2 nanotubes are also discussed for sensor application.

2 COMPUTATIONAL METHODOLOGY

The most stable structure of layered SnS_2 consists of a triple layer of S-Sn-S, similarity to MoS_2. These triple layers are stacked together by van der Waals interactions separated 5.94 Å. Since 2-D projected structure of the triple layers is similar to a graphene sheet, the SnS_2 nanotubes can be classified to "armchair" (n,n) nanotubes and "zig-zag" (n,0) nanotubes, depending on rolling directions, as in CNT's[7].

We have performed first-principles calculations of several hypothetical SnS_2 nanotubes using atomic orbital based density functional theory (DFT) with the generalized gradient approximation (GGA) of Perdew and Wang, implanted in $Dmol^3$[8]. The details of calculations can be found in the previous report [9].

3 RESULTS AND DISCUSSIONS

The optimized structures of armchair and zigzag nanotubes are shown in Fig. 1, and the optimized radiuses are listed in Table 1. These optimization of the several nanotubes confirm the stability of nanotube (NT) structures of SnS_2. We have also calculated the strain energies of SnS_2 NTs. The strain energy per atom is defined as $E_S = E_{NT} - E_{Layer}$, where E_{NT} is the energy per atom of the SnS_2 NT and E_{Layer} is the energy per atom of the SnS_2 layer. The strain energies per atom increase with $1/R^2$ following the classical $1/R^2$ strain law, as shown in Fig. 2, albeit we have calculated only four different NTs. In Fig. 2, we have compared the calculated strain energies of SnS_2 NTs with the reported values of MoS_2 NTs and CNTs [11,12]. From the $1/R^2$ strain energy of SnS_2, one can expect very large strain energy for small diameter nanotubes, since SnS_2 NTs are constructed by rolling a triple layer of S-Sn-S.

Table 1. The radiuses and HOMO-LUMO gaps of optimized nanotube structure of SnS_2

		radius (Å)	HOMO-LUMO gap (eV)
zigzag	(10,0)	6.65	0.64
	(12,0)	7.75	0.84
armchair	(8,8)	8.80	0.85
	(10,10)	11.3	1.00

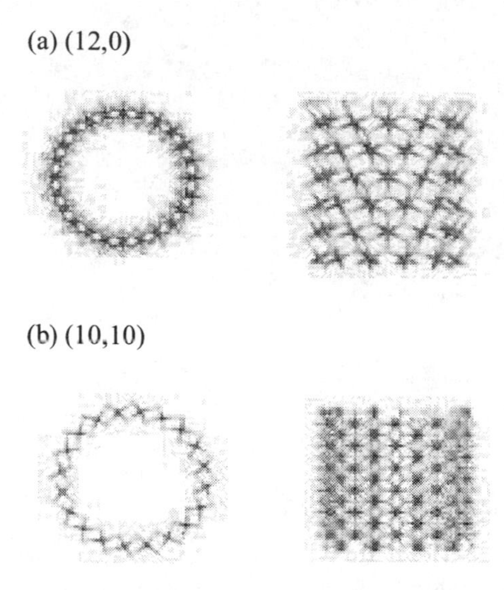

Fig. 1 Optimized nanotube structures; (a) (12,0) zigzag nanotube , (b) (10,10) armchair nanotube

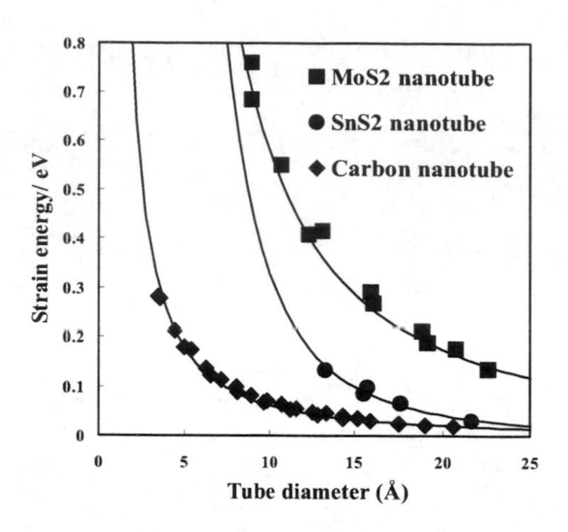

Fig. 2 Strain energies vs. tube diameters of SnS_2 nanotubes compared with those of MoS_2 nanotubes[11] and CNTs[12].

It is reasonable for SnS_2 NTs to require much more energies than CNTs for narrow tubes. However, the strain energies of SnS_2 NTs are comparable to those of CNTs for larger diameter than 20 Å. It is noted that the strain energies of SnS_2 are much smaller than those of MoS_2, even both of them are constructed from triple layer of chalcogenides. It infers that SnS_2 NTs can be easily synthesized considering that MoS_2 NTs were already synthesized even though they require more strain energies than SnS_2 NTs. The calculated energy gaps between the highest occupied molecular orbital (HOMO) and the lowest unoccupied molecular orbital (LUMO) are listed in Table 1. The gap values are in the range from 0.6 eV to 1.0 eV regardless of chiralities of SnS_2 NTs, which are comparable to the band gaps of semiconducting CNTs.

In order to model NH_3 adsorption on the theoretically obtained SnS_2 NT, a NH_3 molecule is attached to a (10,0) NT unit cell as shown in Fig. 3. To find a bound configuration, we calculated binding energy as a function of the distance between N and the nearest S atom of the NT. The binding energy, E_b, is defined as $E_b = E_t(NT+NH_3) - E_t(NT) - E_t(NH_3)$, where $E_t(NT+NH_3)$ and $E_t(NT)$ are total energies of the NT with and without a NH_3 molecule, respectively, and $E_t(NH_3)$ is the total energy of an isolated NH_3 molecule. After finding the minimum E_b varying the N-S distance, the structure of NT with NH_3 are reoptimized allowing the atoms of NH_3 and nearby 4 S atoms and 3 Sn atoms to relax, while other atoms in the NT are constrained to the their initial positions.

The calculated binding energy is -0.19 eV for NH_3 molecule on the SnS_2 NT. The charge transfer is found to be 0.04 electrons from a NH_3 to SnS_2 NT. These values are very close to those of CNT (-0.18 eV for binding energy and 0.04 e for charge transfer) in our previous calculations [13]. The adsorption seems physisorption rather than chemisorption, that is plausible for sensor application. The electron density of HOMO is shown in Fig. 3-b. The most of HOMO charge density is located near NH_3 molecule and some charges are on the SnS_2 NT which seems transferred from NH_3 molecule.

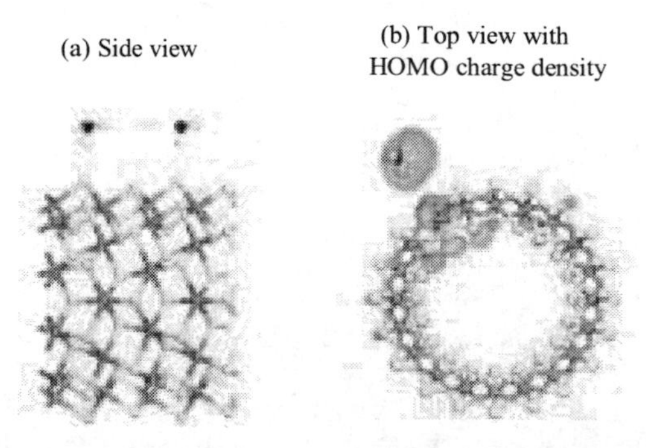

Fig. 3 Optimized structure of NH_3 molecules attached to (10,0) SnS_2 nanotube; (a) side view where NH_3 molecules are separated by 6.30 Å, (b) top view of the same system with HOMO charge density

4 CONCLUSION

In conclusion, optimizations of the several SnS_2 NTs and their strain energies confirm the stability of NT structures of SnS_2, even they are not experimentally synthesized yet. The calculated HOMO-LUMO gaps clearly show that SnS_2 NTs could be semiconductors with moderate band gaps regardless of their chiralities. It suggests that SnS_2 NTs can be easily used as semiconductor wires in nano-electronic devices, that is a great advantage in real applications, compared to CNTs and MoS_2 NTs. In order to see the possibility of gas sensor application of this hypothetical SnS_2 NT, we have shown that adsorption of NH_3 on SnS_2 NT is physisorption and charge transfer occurs from NH_3 to SnS_2 NT. If SnS_2 NTs is synthesized, semiconducting SnS_2 NTs could be a very plausible 1-dimensional sensor material for NH_3 detection.

ACKNOWLEGEMENTS

This work was supported by MOST of Korea through the National R & D Project for Nano Science and Technology.

REFERENCES

[1] S. Ijima, Nature, **354**, 56 (1991),

[2] R. Tenne, *Colloids and Surfaces* A, **208**, 83 (2002)

[3] R. Tenne, L. Margulis, M. Genut, and G. Hodes, *Nature*, **360**, 444 (1992).

[4] M. Remskar et al., *Science,* **292**, 479 (2001)

[5] S. Y. Hong, R. Popovitz-Biro, Y. Prior, and R. Tenne, *J. Am. Chem. Soc.*, **125**, 10470 (2003)

[6] T. Jiang, G. A. Ozin, A. Verman and R. L. Bedard, *J. Mater. Chem.*, **8**(7), 1649 (1998)

[7] M. S. Dresselhaus, G. Dresselhaus and P. C. Eklund, "*Science of Fullerenes and Carbon Nanotubes",* Academic,New York (1996)

[8] G. Kresse, J. Hafner, Phys. Rev. **B 47**, C558 (1993)

[9] DMol^3 is a registered software product of Accelrys Inc.; B. Delley, J. Chem. Phys., **92**, 508 (1990)

[10] H. Chang, E. In, K. Kong, J.-O Lee, Y. Choi, and B.-H. Ryu, J. Chem. Phys. B, **109**, 30 (2005)

[11] G. Seifert, H. Terrones, M. Terrones, G. Jungnickel, and T. Frauenheim, Phys. Rev. Lett., **85**, 146 (2000)

[12] H. Hernández, C. Goze, P. Bernier, and A. Bubio, Phys. Rev. Lett., **80**, 4502 (1998)

[13] H. Chang, J. D. Lee, S. M. Lee and Y. H. Lee, App. Phys. Lett., **79**, 3863 (2001)

Growth of 3C-SiC nanowires on nickel coated Si(100) substrate using dichloromethylvinylsilane and diethylmethylsilane by MOCVD method

B.-C. Kang, J. S. Hyun and J.-H. Boo

Department of Chemistry, Sungkyunkwan University, Suwon 440-746 South Korea
chemcvd@skku.edu, memoriesed@skku.edu, jhboo@skku.edu

ABSTRACT

We have grown 3C-SiC nanowires on nickel coated Si(100) substrates using single source precursors by thermal metal-organic chemical vapor deposition (MOCVD) method. Dichloromethylvinylsilane ($CH_2CHSi(CH_3)Cl_2$) and diethylmethylsilane ($CH_3SiH(C_2H_5)_2$) were used as a single precursor without any carrier and bubbler gas. 3C-SiC nanowires with 40 ~ 100 nm diameter could grow on substrates at temperature as low as 900 °C. XRD pattern showed that SiC nanowires were cubic silicon carbide. TEM analysis showed that an amorphous carbon layer surrounds the as-deposited SiC nanowires, and the 3C-SiC nanowire has [111] growth direction with well-crystallized structure. XPS and EDX analyses showed that the as-obtained SiC nanowire has an atomic Si and C composition of about 1.0:1.2, suggesting possible applications for both electronic devices and field emitters.

Keywords: 3C-SiC, nanowire, MOCVD, Ni catalyst

1 INTRODUCTION

3C-SiC is an important material for the fabrication of electronic devices because it has a high thermal conductivity, high hardness, high wear resistance and excellent chemical resistance. Therefore it can operate at high power, high temperature and in harsh conditions. In the meantime, the miniaturization of devices is an irresistible trend for both industrial manufacture and academic research. Nanowires are interesting building blocks for the fabrication of various devices on nanometer scale. Therefore, the fabrication and understanding of the properties of SiC nanowires are thus decisive for the development of SiC based nanodevices. Various methods have been used to grow SiC nanowires. Among them, the easiest way is a chemical vapor deposition (CVD) method. However, conventional SiC CVD process requires high deposition temperature. Therefore, many research groups have tried to find suitable metal-organic (MO) precursors for growing SiC nano materials [1-7].

In this study, therefore, we have deposited 3C-SiC nanowires on nickel thin film deposited Si(100) substrates at as low as 900 °C by thermal MOCVD method using two kinds of single molecular precursors ; dichloromethylvinylsilane ($CH_2CHSi(CH_3)Cl_2$) and diethylmethylsilane ($CH_3SiH(C_2H_5)_2$). This paper will be present that the structural characteristics, compositions and surface morphologies of deposited 3C-SiC nanowires on nickel coated Si(100) substrates.

2 EXPERIMENTAL

Growth of 3C-SiC nanowires were performed in homemade vertical metal-organic chemical vapor deposition (MOCVD) system. We expected that nickel thin film plays an important role in growing 3C-SiC nanowires, so we prepared the nickel thin film as a catalyst of about 20 nm on Si(100) substrate using rf magnetron sputtering method. Si(100) substrate was pretreated in an ultrasonic cleaner in order of methanol, DI water, and acetone. Dichloromethylvinylsilane ($CH_2CHSi(CH_3)Cl_2$) and diethylmethylsilane ($CH_3SiH(C_2H_5)_2$) were chosen as a single molecular precursor because they have already Si-C bonds in them and are very volatile at low temperature. We can suggest that these two kinds of single molecules are very suitable for growing 3C-SiC nanowires at low temperature. The base pressure of the MOCVD system was 5×10^{-5} Torr, and the deposition pressure was kept at 50 mTorr. The deposition was carried out at 900 °C for 1 ~ 3 h. The as-grown nanowires were firstly characterized by x-ray diffraction (XRD) and scanning electron microscopy (SEM). In order to further define the structure of the 3C-SiC nanowires, transmission electron microscope (TEM) images and transmission diffraction (TED) patterns were obtained. To identify the composition of the 3C-SiC nanowires, x-ray photoelectron spectroscopy (XPS) and energy dispersive x-ray (EDX) analysis were also carried out.

3 RESULTS AND DISCUSSION

Fig. 1 shows the XRD patterns of the 3C-SiC nanowires deposited on nickel deposited Si(100) substrates under 50 mTorr, 900 °C for 3 h. using (a) dichloromethylvinylsilane and (b) diethylmethylsilane. They show the crystalline peaks at 2θ = 35.6°, 41.4° and 60.0°, which are good attributed to diffraction of the 3C-SiC(111), (200) and (220) planes, respectively. Peaks from either the other phases of SiC or the nickel catalyst are not appeared. The SiC nanowires grown by two different kinds of precursors show the same diffraction pattern and tendency. However, the 3C-SiC nanowire grown by dichloromethylvinylsilane shows a good crystallinity than that of the SiC nanowire deposited by diethylmethylsilane. Conclusively XRD analysis show that the deposited 3C-SiC nanowire on nickel deposited Si(100) substrate has a poly crystalline zinc-blende structure at 900 °C, and the 3C-SiC nanowire from dichloromethylvinylsilane as a precursor has good crystallinity than that from diethylmethylsilane precursor.

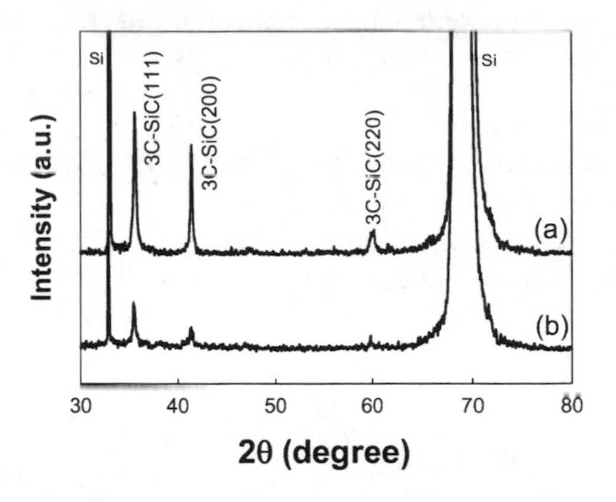

Fig. 1. XRD patterns of the 3C-SiC nanowires deposited on nickel deposited Si(100) substrates under 50 mTorr, 900 °C for 3 hr. using (a) dichloromethylvinylsilane and (b) diethylmethylsilane.

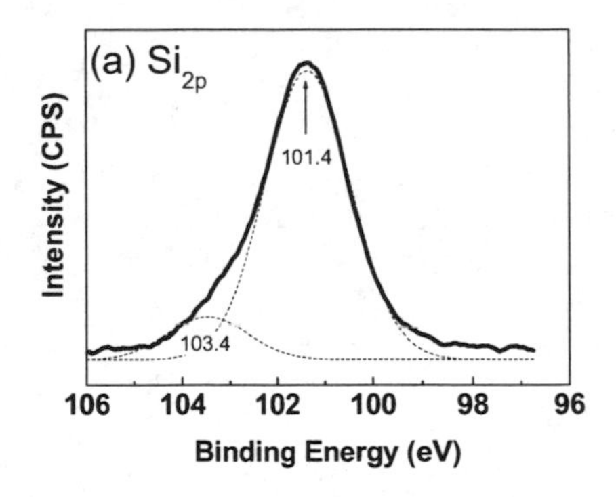

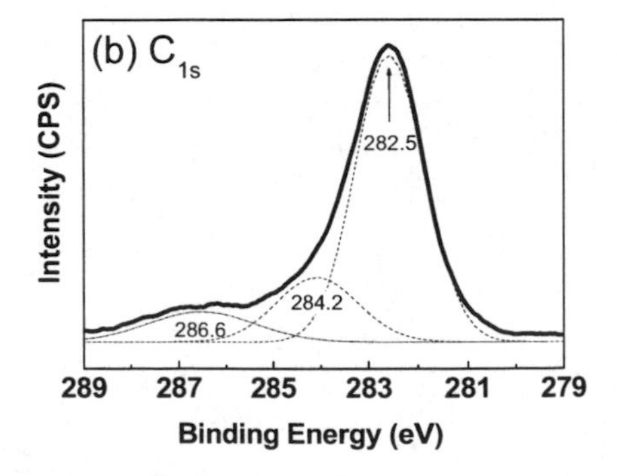

Fig. 2. Typical XPS high resolution spectra of the 3C-SiC nanowires grown on nickel covered Si(100) at 900 °C : (a) and (b) show Si_{2p} and C_{1s} peaks, respectively.

Further evidence for the formation of 3C-SiC can be obtained through the XPS analysis. The typical XP survey spectrum (not shown here) of an as-deposited 3C-SiC nanowire at 900 °C shows the strong XP peaks of Si_{2p} and C_{1s} as well as O_{1s}. We infer that the O_{1s} peak at binding energy of 531 eV is mainly attributed to the SiO_2 and/or CO_2 that formed by both oxidation reaction and adsorption at air condition. Fig. 2(a) and 2(b) show the Si_{2p} and C_{1s} high-resolution XP spectra of obtained 3C-SiC nanowires. For the case of Si_{2p} (Fig. 2(a)), the spectrum can be decomposed to two Gaussian components located at 101.4 eV and 103.4 eV. These two Si states are attributed to SiC and SiO_2. And the spectrum of C_{1s} (Fig. 2(b)) consists of three components centered at 282.5 eV, 284.2 eV, and 286.6 eV. The peaks at 282.5 eV correspond to C_{1s} for SiC, and the other two C_{1s} peaks can be attributed to a small amount of both the residual carbon originated from the metalorganic precursor and adsorbed CO_2 on the SiC nanowire surface, respectively. With these XPS quantification analysis, we thus obtained that 3C-SiC nanowire has an atomic ratio of Si and C to be about 1.0:1.2.

Fig. 3 shows the typical SEM images of the obtained 3C-SiC nanowires grown by (a) dichloromethylvinylsilane and (b) diethylmethylsilane at 900 °C. It can be seen clearly that straight nanowires are randomly grown on the substrate with a high density. The differences of two SEM images are the thickness and length of grown SiC nanowires. In the case of dichloromethylvinylsilane as a precursor (Fig. 3(a)), about 40 ~ 100 nm diameter and over 10 μm nanowires can grow easily. However, in the case of diethylmethylsilane (Fig. 3(b)), deposited SiC nanowires show that the thicker and shorter SiC nanowires, and the surface roughness looks very high than that do Fig. 3 (a).

Fig. 3. Typical SEM images of the obtained 3C-SiC nanowires grown by (a) dichloromethylvinylsilane and (b) diethylmethylsilane at 900 °C.

To investigate more detail structure of as-deposited β-SiC nanowire, TEM combined with EDX analyses were performed. For TEM experiments, the 3C-SiC nanowires grown at 900 °C were prepared. Fig. 4 shows a typical TEM image obtained from a 3C-SiC single nanowire. TEM analysis results exhibit the grown 3C-SiC nanowires with a diameter of about 100 nm, and these are wrapped with an amorphous layer with the thickness of about 2 nm. We can confirm that this layer is amorphous carbon by EDX data. Because the metal-organic sources used this experiment contains lots of carbon contents, we could guess that this amorphous carbon layer was originated from the precursor. Also, TEM image of this nanowire shows that they are crystalline, but showing that they have defects, including numerous stacking faults, etc. We speculate that the nanowire was grown via the vapor-liquid-solid (VLS) process from a nickel-containing catalytic droplet, although the catalytic droplet was not observed at the ends of the nanowire because the nanowire was broken and the droplet was removed during the ultrasonic treatment. Based on this speculation, we can thus guess that changes in diameter and well-defined facets can be recognized from the 3C-SiC nanowire grown via the VLS process from catalytic droplet. Theses two features indicate that stick-slip motion occurred during the growth; when sticking, the diameter, which was defined by the area of the LS interface, increased to form energetically favorable facets. When growing along the facets, the diameter became larger, then the wetting angle became smaller and the component of VL interface tension in the LS interface became larger. Eventually the droplet slipped driven by LS and VL interface tensions resulting in an abrupt decrease in diameter [8]. The inset image (left) of Fig. 4 shows the corresponding selected area electron diffraction (ED) pattern obtained from the same sample. It shows the bright spots corresponding to 3C-SiC(200) and (111), and streaks, which are perpendicular to the stacking faults. These stacking faults are generally thought to originate from thermal stress during growth process [7]. EDX spectrum corresponding to the stem of a 3C-SiC nanowire was also investigated. There are only three peaks due to Si, C, and Ni. Through quantification analyses of the stem of a 3C-SiC nanowire, an average atomic % ratio of Si and C is estimated about 1.0:1.2, while that of surface regime wrapped with carbon is observed about 1.0:11.5, respectively.

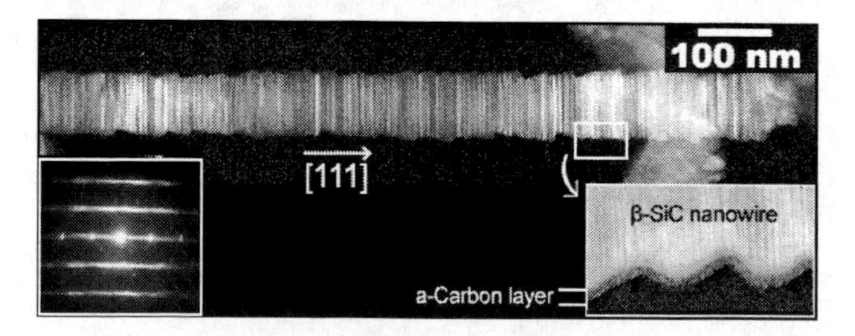

Fig. 4. Typical TEM image of single 3C-SiC nanowire grown at 900 °C. Inset (left) shows the electron diffraction pattern of the same sample, and insert (right) shows a zoom-image of 3C-SiC nanowire.

4 CONCLUTIONS

Cubic silicon carbide (3C-SiC) nanowires have been deposited on nickel thin film deposited Si(100) using a single molecular precursor at 900 °C by the metal-organic chemical vapor deposition (MOCVD) method. Dichloromethylvinylsilane ($CH_2CHSi(CH_3)Cl_2$) and diethylmethylsilane ($CH_3SiH(C_2H_5)_2$) were used as a single precursor without any carrier and bubbler gas to increase mass transportation or to remove contaminants in the nanowires.

XRD and SEM analyses show that the deposited 3C-SiC nanowire on nickel deposited Si(100) substrate has a poly crystalline zinc-blende structure at 900 °C, and the 3C-SiC nanowires from dichloromethylvinlysilane as a precursor have good crystallinity and narrow nanowires with smooth surface than that from diethylmethylsilane. High-resolution TEM analysis showed a detailed structure that amorphous carbon layer surrounds the as-deposited 3C-SiC nanowires, and the 3C-SiC nanowire has [111] growth direction with well-crystallized structure. Based on the XPS and EDX analyses, near same atomic composition ratio of Si to C (1.0:1.2) was obtained, indicating carbon rich species on surface regions, suggesting possible applications to both electronic devices and field emitter.

REFERENCES

[1] K. C. Kim, C. I. Park, J. I. Roh, K. S. Nahm, Y. B. Hahn, Y.-S. Lee, and K.Y. Lim, J. Electrochem. Soc. 148 (2001) C383.
[2] S. Madapura, A.J. Steckl, and M. Lobada, J. Elecrochem. Soc. 146 (1999) 1197.
[3] Y. Narita, T. Inubushi, M. Harashima, K. Yasui, and T. Akahane, Appl. Surf. Sci. 216 (2003) 575.
[4] Th. Kunstmann, and S. Veprek, Appl. Phys. Lett. 67 (1995) 3126.
[5] J.-H. Boo, S.-B. Lee, K.-S. Yu, M.M. Sung, and Y. Kim, Surf. Coat. Tech. 131 (2000) 147.
[6] B.-C. Kang, S.-B. Lee, and J.-H. Boo, Thin Solid Films 464-465 (2004) 215.
[7] H. J. Li, Z. J. Li, A. L. Meng, K, Z. Li, X. N. Zhang, and Y. P. Xu, J. of Alloys and Compounds 352 (2003) 279.
[8] H. Kohno, H. Yoshida, Solid State Communications 132 (2004) 59.

Acid transformation of TiO_2 nanotubes to nanoparticles

Dmitry V. Bavykin*, Alexei A. Lapkin*, Jens M. Friedrich**, Frank C. Walsh**

*Catalysis and Reaction Engineering Group, Department of Chemical Engineering, University of Bath, Bath BA2 7AY, UK, tel: +44 1225 384483, fax: +44 1225 385713, e-mail: D.V.Bavykin@bath.ac.uk
**Electrochemical Engineering Group, School of Engineering Sciences, University of Southampton, Highfield, Southampton SO17 1BJ, UK, e-mail: F.C.Walsh@soton.ac.uk

ABSTRACT

The long-term stability of TiO_2 nanotubes in acid, neutral and basic water suspensions was studied at room temperature. In neutral and basic (0.1 M NaOH) solutions, the TiO_2 nanotubes are stable and undergo minimal morphological changes. In dilute solutions of 0.1M H_2SO_4, suspended TiO_2 nanotubes slowly transform to rutile nanoparticles having an ellipsoid shape. The porosity and crystal structure of TiO_2 are also changed during the transformation. The rate of conversion of nanotubes to nanoparticles depends on the nature of the acid and is high fin the case of sulfuric acid. Thermodynamic and kinetic aspects of the acid transformation are discussed.

Keywords: TiO_2 nanostructures, dissolution, crystallisation

1 INTRODUCTION

Nanotubular titanium dioxide, produced by alkali hydrothermal treatment, is a relatively novel [1] and intensively studied material [2,3] which is characterized by nano-sized diameter morphology and unique physico-chemical properties. The material shows promise for applications including hydrogen sensors [4], photocatalysts [5], catalyst supports [6], solar cell photosensitisers [7], ion-exchange materials [8,9] and a new generation of lithium electrodes [10,11]. In all of these applications, except opto-electronic devices, nanotubes of TiO_2 can be exposed to chemically aggressive media. Titanium dioxide nanotubes are relatively fragile and can be easily be broken under ultrasonic treatment [12] resulting in shorter nanotubes. During hydrothermal treatment in 0.1M HNO_3 at 100ºC for 7 h TiO_2, the nanotubes completely transform to anatase [8]. Annealing of the protonated form of TiO_2 nanotubes in air at temperatures higher than 500ºC results in transformation of the nanotubes and formation of anatase [13]. Sodium saturated form of TiO_2 nanotubes is stable up to 850ºC [8].

For the successful application of TiO_2 nanotubes it is necessary to determine the range of operational conditions under which nanotubes are stable. In this work, the long-term stability of multilayered wall TiO_2 nanotubes in the presence of bases and acid was studied. During the acid transformation of nanotubes all intermediate states were characterised.

2 EXPERIMENTAL DETAILS

The alkali hydrothermal method of preparing TiO_2 nanotubes was based on studies by Kasuga et al.. 9 g of titanium dioxide (anatase) was added to 300 mL of 10M NaOH solution in a PTFE (Teflon) beaker under vigorous stirring. The solution was placed in a PTFE-lined autoclave and heated for 22 hours at 140°C. The white, powdery TiO_2 produced was thoroughly washed with water until the washing solution achieved pH 7, then it was washed with 0.05M H_2SO_4 on a glass filter for over 30 min then washed with water to pH 7. The sample was dried in vacuum at 50°C.

The surface area and BJH pore distribution of the synthesised TiO_2 were measured using a Micromeritics ASAP 2010 instrument. SEM images were obtained with a JOEL 6500 FEG-SEM scanning electron microscope. XRD patterns were recorded using a Bruker AXS D8 Discover x-ray diffractometer, with Cu-K_α radiation $\lambda = 0.154$ nm and a graphite monochromator in the 2θ range of 20° - 50°.

For the tong-term TiO_2 nanotube suspension ageing studies, 0.2 g of TiO_2 nanotubes were placed to the 9 vials and 10 mL of decimolar solution of NaOH, HCl and HNO_3 was added. 10 mL of 0.1M sulfuric acid was added to the five other vials. One vial was filled with 10 mL of water. All vials were closed, ultrasonicated for 10 min and kept at room temperature (22ºC). The samples were washed with water and dried in vacuum at 50ºC

3 RESULTS AND DISCUSSION

Acids and bases typically react with the surface of metal oxide powders resulting in their modification. For example, in the case of photoactive anatase, TiO_2 particles treatment with sulfuric acid improves the activity of this photocatalyst [14] without changing the crystal structure of particles and only modifies the surface properties. In contrast, long-term treatment of TiO_2 nanotubes in diluted sulfuric acid results in a complete change of morphology of particles. In Figure 1, pore size distributions produced from a N_2 desorption curve using the BJH algorithm are shown for samples of nanotubular TiO_2 treated with 0.1 M H_2SO_4 for controlled times. The average pore size increases at longer there is a significant fall of BJH pore volume (see Table 1), but the BET surface area remains unchanged.

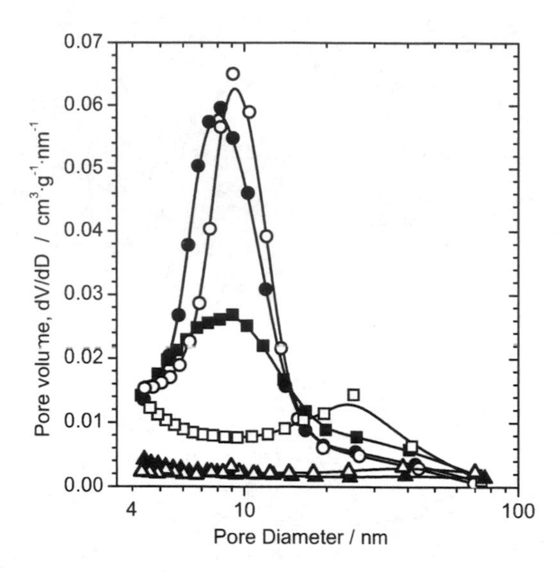

Figure 1. Pore volume distribution (BJH desorption) of TiO_2 nanotubes soaked in 0.1M H_2SO_4 for controlled time at room temperature: (■) – initial powder, (●) – 5 days, (○) – 15 days, (□) – 32 days, (▲) – 2 months, (△) – 5 months

In Figure 2, high resolution SEM micrographs of the initial TiO_2 nanotubes and rutile produced by ageing of nanotubes in 0.1M H_2SO_4 for 5 months are presented. Initial nanotubes having an internal tube diameter *ca.* 3-5 nm, external diameter *ca.* 7-10 nm, and length more than several hundreds of nanometers slowly transform to ellipsoid particles having typical dimensions of 50x200 nm. These particles consist of smaller particles of rutile nanocrystallites.

Despite some debate over the crystal structure of TiO_2 nanotubes [15, 16] we denote the apparent pattern of initial nanotubes (see Figure 3a) to the trititanic acid $H_2Ti_3O_7$ having followed characteristic reflections of 24.38, 28.98, 37.95 and 48.4 degree. Long-term washing of the nanotubes with sulfuric acid results in complete disappearance of the trititanic acid phase accompanied by appearance of the rutile phase having characteristic reflections 27.46, 36.10, 41.25 and 44.07 degree. The coherence area for the rutile crystals was calculated from the half-width at half-height of the (110) plane diffraction peak ($2\theta = 27.65^0$) using the Scherrer equation and was found to be approximately 3.9 nm for samples e) and f) in Figure 3. The sample produced after 32 days of washing in sulfuric acid is an approximately equimolar mixture of nanotubular $H_2Ti_3O_7$ and rutile. In all of the samples, the amount of anatase phase is negligible. Previously, it has been reported [17] that SO_4^{-2} ions in solution stimulate the preferential formation of the anatase phase of TiO_2 during hydrolysis reactions. In our case, the presence of sulfate ions during the transformation of nanotubes does not promote the anatase phase. Generally speaking, the formation of the rutile phase at room temperatures is relatively rare. Adding relatively concentrated 1M HCl or HNO_3 can promote recrystallization of amorphous TiO_2 to rutile at room temperature [18]. Addition of sulfuric acid, however, results in formation of anatase. Phase transformation of protonic layered titanates $H_2Ti_4O_9$ to the anatase TiO_2 in water suspension begins at 225°C without the formation of a rutile phase [19].

Figure 2. HRSEM pictures of TiO_2 a) initial nanotubes, b) nanotubes transformed to rutile nanoparticles after 5 months ageing in 0.1M H_2SO_4 at room temperature

Data on the stability of a suspension of TiO_2 nanotubes in 0.1M solutions of NaOH, HCl, or HNO_3 are presented in Figure 4 and Table 1. There is no significant change in pore size distribution, surface area or pore volume for samples aged for 5 months in NaOH solution. Probably, the high pH and high sodium ion level in the solution stabilise the nanotubular TiO_2. Ageing of TiO_2 nanotubes for 2 months in HCl and HNO_3 results in the partial transformation of nanotubes.

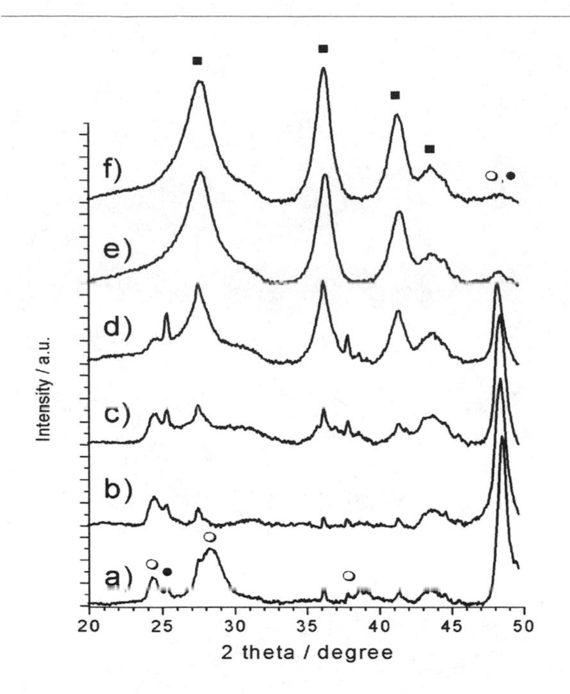

Figure 3. XRD patterns of TiO_2 nanotubes and acid washed samples for different time at room temperature: a) initial sample; b) 5 days; c) 15 days; d) 32 days; e) 2 months; f) 5 months. (■) rutile, (●) anatase and (○) $H_2Ti_3O_7$ reflections

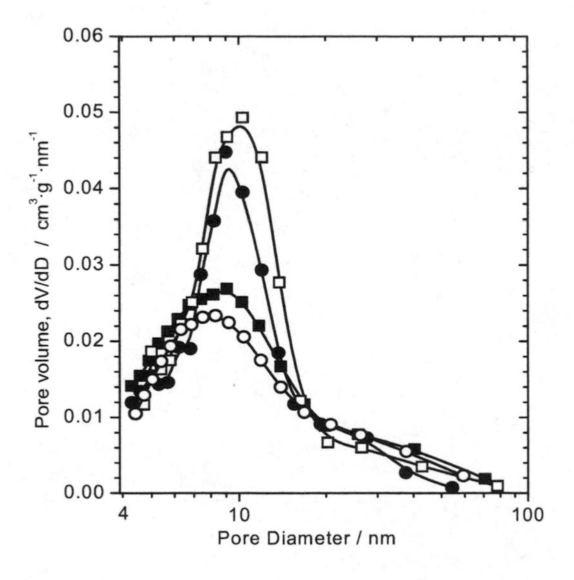

Figure 4. Pore volume distribution (BJH desorption) of TiO_2 nanotubes soaked in 0.1M solutions at room temperature: (■) – initial powder, (●) – 2 months in HNO_3, (○) – 5 months in NaOH, (□) – 2 months in HCl.

The pore size distribution of nanotubes aged for 2 months in HNO_3 and HCl is similar to the pore size distribution of nanotubes aged in H_2SO_4 for 5 and 15 days, respectively.. Thus, it can be estimated that the rate of TiO_2 nanotube transformation in HCl is 4 times (and in HNO_3 12 times) less than the rate of transformation in H_2SO_4. Such a difference in reaction rate may be attributable to the differences in TiO_2 solubility in different acids which decreases in the order: $H_2SO_4 > HCl > HNO_3$. The value of pH could also be a key factor that initiates the process of transformation. Since trititanic acid is a weak acid in water, the following dissociation occurs:

$$(H_2Ti_3O_7)_s \leftrightarrow (HTi_3O_7^-)_s + H^+ \qquad (1)$$

This results in a negative zeta potential in water (see Table 1). Addition of NaOH slightly increases the potential due to the adsorption of sodium ions [8,9]. In contrast, addition of 0.1M sulfuric acid significantly increases the zeta potential, resulting in a change of sign for the surface charge, the zeta potential having a small positive value. Probably, excess adsorption of protons from solution onto the surface of nanotubular TiO_2 destabilises the crystal structure of the multilayered trititanic acid, resulting in leaching of the surface TiO_6 octahedrons. A very slow, recrystallization of titania over several months results in formation of the most stable (rutile) phase of TiO_2.

Solution	Time of treatment	S_{BET} / $m^2 g^{-1}$	V_{BJH} / $cm^3 g^{-1}$	Zeta potential / mV
H_2O	0 day	199	0.70	-42.69
0.1M H_2SO_4	5 day	240	0.69	6.59
0.1M H_2SO_4	15 day	251	0.69	6.59
0.1M H_2SO_4	32 day	235	0.64	n/a
0.1M H_2SO_4	2 month	184	0.25	n/a
0.1 MH_2SO_4	5 month	246	0.30	n/a
0.1M NaOH	5 month	161	0.56	-36.71
0.1M HCl	2 month	247	0.73	n/a
0.1M HNO_3	2 month	199	0.55	n/a

Table 1 Change of TiO_2 nanotubes properties during slow transformation in acids and alkali.

4 CONCLUSIONS

The long-term stability of TiO_2 nanotubes in acid, neutral and basic water suspensions was studied. It was found in water and basic (0.1M NaOH) solutions that the TiO_2 nanotubes are stable and undergo minimal morphological changes. In dilute sulfuric acid (0.1M H_2SO_4) suspended TiO_2 nanotubes slowly transform to rutile nanoparticles having an ellipsoid morphology. In 0.1M solutions of HCl and HNO_3 the process of transformation is several times slower due to the lower solubility of TiO_2 in these acids. Slow recrystallization of TiO_2 nanotubes in acids results in the formation of a rutile phase having a very high surface area (BET N_2 adsorption) of 246 $m^2 g^{-1}$.

REFERENCES

1 T. Kasuga, M. Hiramatsu, A. Hoson, T. Sekino, K. Niihara, *Langmuir* **1998,** 14, 3160-3163

2 C.C. Tsai, H. Teng, *Chem. Mater.* **2004,** 16, 4352-4358

3 Z.Y. Yuan, B.L. Su, *Colloids and Surfaces A: Physicochem. Eng. Aspects*, **2004**, 241, 173–183

4 O.K. Varghese, D.Gong, M. Paulose, K.G. Ong, C.A. Grimes, *Sensors and Actuators B*, **2003**, 93, 338–344

5 C.H. Lin, C.H. Lee, J.H. Chao, C.Y. Kuo, Y.C. Cheng, W.N. Huang, H.W. Chang, Y.M. Huang and M.K. Shih, *Catalysis Letters,* **2004**, 98, 61-66

6 J. Cao, J.-Z. Sun, H.-Y. Li, J. Hong, M. Wang, *J. Mater. Chem,* **2004**, 14**,** 1203-1206

7 M. Adachi, Y. Murata, I. Okada, S. Yoshikawa, *J. Electrochemical Society,* **2003**, 150 (8), G488-G493

8 X. Sun, Y. Li, *Chem. Eur. J.,* **2003**, 9, 2229-2238

9 J.J. Yang, Z.S. Jin, X.D. Wang, W. Li, J.W. Zhang, S.L. Zhang, X.Y. Guo, Z.J. Zhang, *Dalton Transactions,* **2003**, 20, 3898-3901

10 J. Li, Z. Tang, Z. Zhang, *Electrochemistry Communications*, **2005**, 7, 62–67

11 L. Kavan, M. Kalbac, M. Zukalova, I. Exnar, V. Lorenzen, R. Nesper, M. Graetzel, *Chem. Mater.* **2004,** 16, 477-485

12 D.V. Bavykin, V.N. Parmon, A.A. Lapkin, F.C. Walsh, *J. Mater. Chem.*, **2004**, 14, 3370

13 M. Zhang, Z. Jin, J. Zhang, X. Guo, J. Yang, W. Li, X. Wang, Z. Zhang, *Journal of Molecular Catalysis A: Chemical*, **2004**, 217, 203–210

14 D.V. Kozlov, D.V. Bavykin, E.N. Savinov, *Catalysis Letters*, **2003,** Vol. 86, No. 4, 169-172.

15 Q. Chen, G.H. Du, S. L.-M. Zhang Peng, *Acta Cryst. B*, **2002**, 58, 587-593

16 J.J. Yang, Z.S. Jin, X.D. Wang, W. Li, J.W. Zhang, S.L. Zhang, X.Y. Guo, Z.J. Zhang, *Dalton Transactions,* **2003**, 20, 3898-3901

17 M. Yan, F. Chen, J. Zhang, *Chemistry Letters*, **2004**, Vol.33, No.10, 1352-1353

18 S. Yin, R. Li, Q. He, T. Sato, *Materials Chemistry and Physics*, **2002**, 75, 76–80

19 S. Yin, S. Uchida, Y. Fujishiro, M. Akib, T. Sato, *J. Mater. Chem.*, **1999**, 9, 1191–1195

Conductivity of the crystalline Boron nanowires measured in TEM.

Oleg Lourie and Mike Kundmann,
GATAN Inc., 5933 Coronado Lane, Pleasanton, CA 94588, USA

In this experiment we report on using in-situ STM-TEM system and EELS to characterize the electronic properties of the boron nanowires. We use low loss EELS approach to study the band structure and the dielectric response of these nanostructures. The STM-TEM system is used for a contact measurement of the conductivity of the individual Boron nanowire imaged in Transmission Electron Microscope (TEM). The in-situ EELS analysis has been completed with the structural information provided with HRTEM.

Recently synthesized crystalline boron nanowires [1] are predicted to have unique combination of electronic and mechanical properties compare to other popular nanostructures such as carbon and BN nanotubes. Boron fibers have been successfully used in high performance composites, where high strength, stiffness, and temperature resistance are required. Boron nanowires are expected to have similar mechanical properties. Based on the recent preliminary studies [1], their electronic properties have been reported to be close to those of the semiconductor [2].

The Boron nanowires have been synthesized using simple CVD method [1]. The reactive precursor gas was passed over the catalyst NiB powder in a hot furnace at 1100 C. After the reaction, the catalyst powder with grown nanostructures has been removed from the alumina substrate and transferred onto the gold wire used as a substrate.

The EELS study of the Boron nanowires has been done with GIF Tridiem (Gatan Inc.) installed on JEOL 2010 FasTEM microscope. Typically, a STEM probe of 3nm of effective size has been used to image and excite the EELS spectrum. The EELS spectrometer was utilized in STEM SI mode with dispersion 0.1eV/channel to acquire 1D spectrum image. The analysis and processing of acquired data have been performed with AutoFilter package software (Gatan Inc.). The software includes the Kramers-Kronig analysis of the low loss EELS data. It has been used to obtain the dielectric function of the analyzed nanostructure. The conductivity measurements have been done using STM-TEM system from Nanofactory AB.

Fig. 1(a) shows typical Boron nanostructure studied in the experiment. HRTEM has been used in conjunction with STEM (Fig. 2(a)) to add nanowires structural details into the analysis. Using Kramers- Kronig analysis of the spectra we have been able to evaluate the refractive index, band gap, inelastic mean free path and dielectric constant for both type of the nanostructures. The energies of the bulk plasmon excitations are also provided. In particular, we found the band gap of 3eV and the refractive index of 2.5 for Boron nanowires. The conductivity measurement is shown in Fig. 1(b)-2(b). Further EFTEM analysis of the Boron data revealed that the nanowires might have a thin oxidation layer on the surface, which possibly contributes into the measured constants. The effect and possible contribution from the amorphous carbon as a result of surface contamination are discussed further in detail.

References:
[1] Jones Otten C., Lourie O.R., Yu M-F., Cowley J. M., Dyer M.J., Ruoff R.S., Buhro W.E., "Crystalline Boron nanowires", J. Am. Chem. Society 124, 464-4565, (2002)
[2] Wang, D.; Lu, J.G.; Otten, C.J.; Buhro, W.E. *Appl. Phys. Lett.* 2003 , *83* , 5280-5282

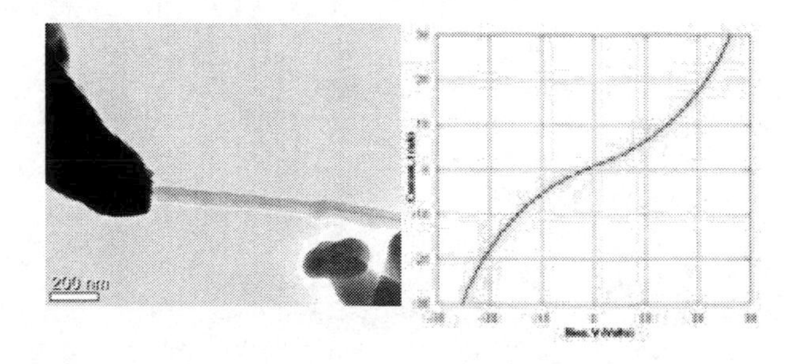
Fig. 1(left) STM tip in contact with Boron nanowire; (right) acquired I-V curve;

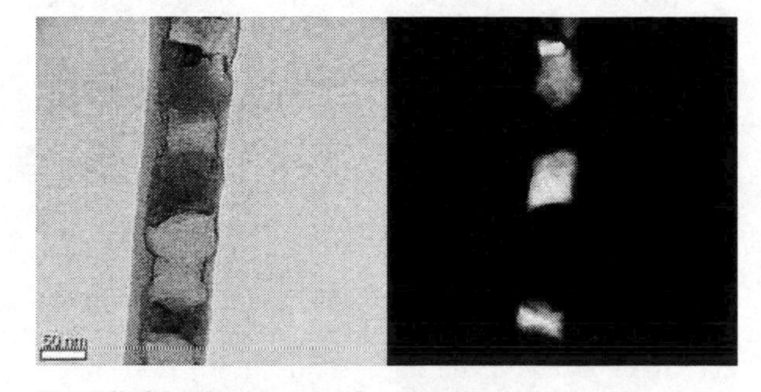
Fig. 2(left) TEM image of melted nanowire; (right) Energy filtered Boron map

Growth of Si Nanowires on Nano Catalyst Corners

C. Wang*, K. Malladi**, and M. Madou***
Department of Mechanical and Aerospace Engineering,
University Of California, Irvine, CA92697, USA
* chunleiw@uci.edu , ** kmalladi@uci.edu , ***mmadou@uci.edu

ABSTRACT

One dimensional structures with nanometer diameters, such as Si nanowires, have great potential for understanding and testing fundamental concepts about the roles of dimensionality in optical, electrical and mechanical properties and for applications ranging from nanoelectronics to biosensing. Here we successfully grow high aspect ratio Si nanowires using Ni catalyst. At high temperature, we found that Ni catalyst layer changed to irregular starfish like structure. Detailed microscopy results showed that catalyst particle size has no relation with the diameter of Si nanowires. Most Si wires grow from the irregular catalyst tentacles. Furthermore, to understand how catalyst size and shape affect Si wires formation, we use EB writer and lift-off process to make various types of Ni nano catalyst corners.

Keywords: Si nanowires, Ni catalyst, EB lithography, SLS, lift off.

1 INTRODUCTION

Semiconductor nanowires, such as: Si nanowires, have attracted tremendous interest recently because of their potential applications as building blocks in future large scale nanoelectronics, optoelectronics and biological devices [1-3]. Si nanowires can be produced by the well known vapor-liquid-solid (VLS) mechanism [4] and solid-liquid-solid (SLS) mechanism [5]. In VLS mechanism, Si atoms in the vapor phase, supplied by gas decomposition, are absorbed on the surface of liquid droplets forming at high temperature and are solved into them, causing the super-saturation of Si in the droplets and growth of Si nanowires. In the SLS mechanism, at a very high temperature the liquid droplet of metal-Si is formed and then the Si from the substrate solves into the droplet continuously making it supersaturated with Si because of continuous diffusion. This initiates the growth of SiNW's resulting from the precipitation of Si atoms from the droplets.

Here in this work, we take advantage of SLS mechanism to grow high aspect ratio Si nanowires. EB lithography was used to pattern Ni nano catalyst. We investigated and compared the growth behavior of Si nanowires from both nano catalyst particles and nano catalyst corners.

2 FABRICATION & PROCESSING

In this work, the substrates we used were Si (100) surfaces. Ni (100Å) was evaporated on cleaned Si substrates by electron beam (EB) evaporation method. Some of samples were directly used to grow Si nanowires. In other samples, Ni catalyst layer were patterned to various nano shapes including squares and corners. The typical process flow is listed in figure 1.

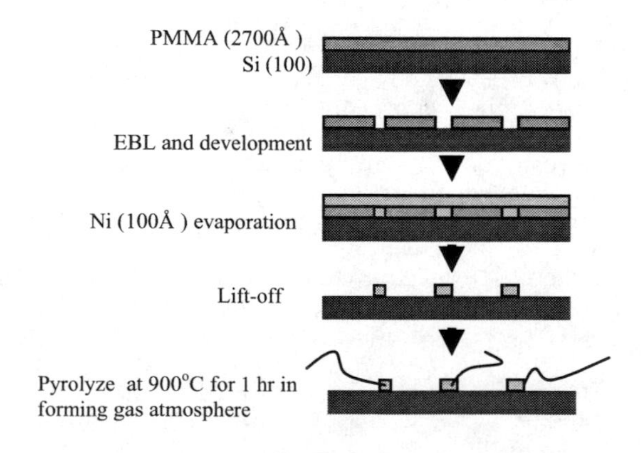

Figure 1: Fabrication details for Si nanowires grown on catalyst corners. (a) PMMA film of 0.27μm thickness was spin-coated on Si; (b) patterning the PMMA by electron beam, then developing it; (c) evaporation of Ni (100 Å) by EB evaporation; (d) making Ni nano patterns by lift-off method; and (e) pyrolyzing the samples at 900°C in forming gas environment.

3 RESULTS & DISCUSSION

Figure 2 (a) and (b) shows typical SEM images of as-grown nanowires from Ni (100 Å) layer. EDX investigation confirmed that the nanowires were composed of amorphous Si. In Figure 2(a) it can be seen that the Si nanowires originate/terminate with the star-fish shaped Ni catalyst islands. The Ni film appears to break down into several catalyst islands of Ni of around 0.1-0.5μm. These islands act as a starting point for the growth of the Si nanowires following the SLS mechanism since Si substrate is the only available source for the continuous diffusion of Si . The wires were found to be

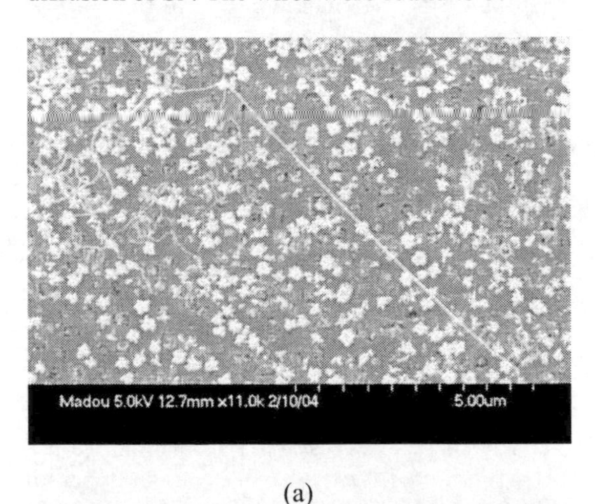

(a)

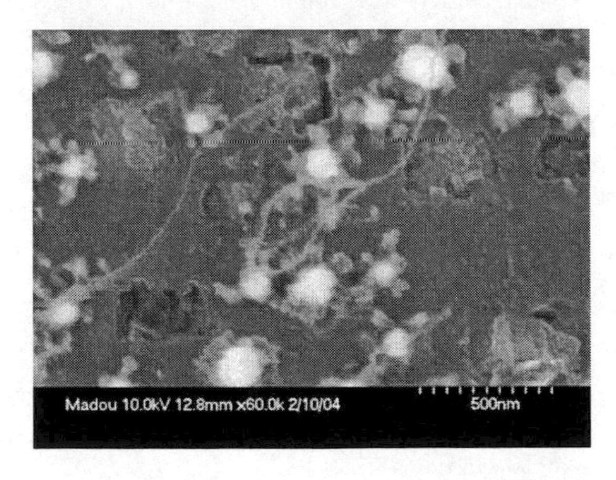

(b)

Figure2. Typical SEM images of Si nanowires grown on Si (100) with Ni (100 Å) catalyst layer.

about 20-50nm in diameter and extending upto more than 5μm. Figure 2(b) shows a higher magnification SEM of the Si nanowires. The catalyst islands can be seen clearly to form a star-fish shaped structure. It can also be seen that three separate nanowires are interlocked to one wire. There are a lot of etch pits on Si substrates.

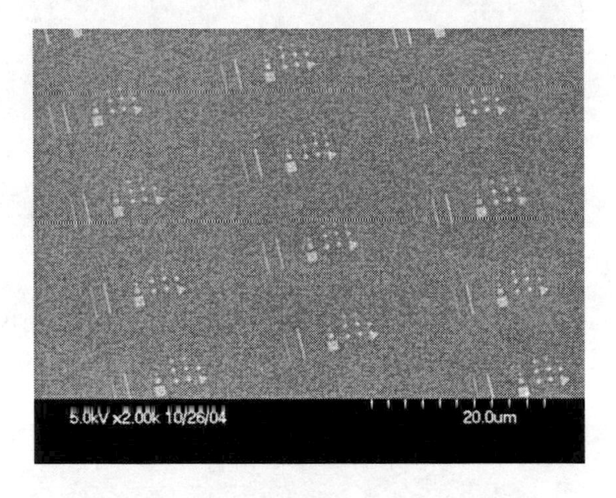

(a)

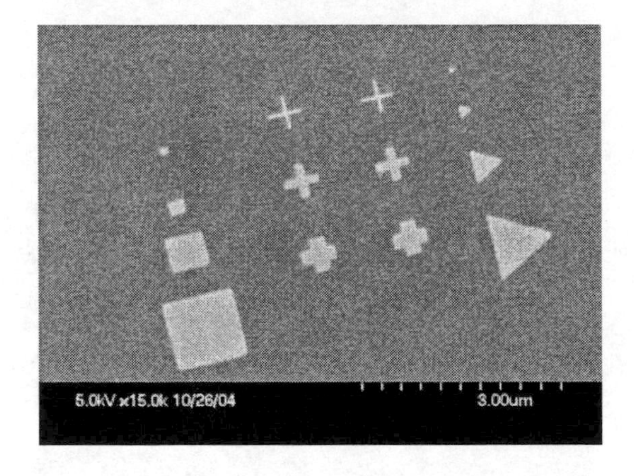

(b)

Figure 3. Typical SEM images of Ni nano structures (100 Å thick) fabricated by EBL as catalyst on Si substrate (100).

Next in order to see if the nucleation sites of Si nanowires happens on nano corners or not, we made catalyst nano corners by EBL before Si growth. Figure 3 (a) shows the SEM image of various nano-corner pattern arrays which were patterned with the EB lithography. Figure 3(b) shows a high magnification SEM image, in each array, the patterns had various nano shapes like triangles, squares, cross, lines. Their sizes ranged from 100nm to 1μm.

(a)

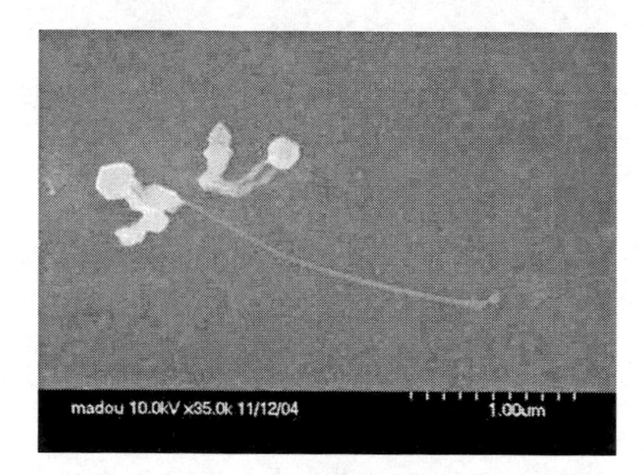

(b)

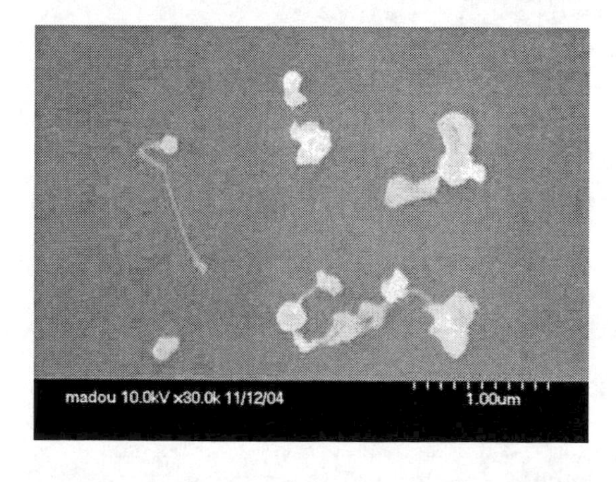

(c)

Figure4. Typical SEM images of Si nanowire grown from nucleated Ni catalyst nano-corners.

Figure 4 (a) shows the SEM image of the pyrolysed sample at 900°C in forming gas environment. We find that the catalyst pattern broke down into much smaller Ni irregular domains at the size of typically around 10-400nm. One high aspect ratio nanowire grows from one nano-domain and connects several irregular nano domains. From the high magnification SEM images as shown in Figure 4 (b) and 4 (c), it is observed that nanowires in different length grow from the broken nano domains without any relation to the original Ni nano corners. In a word, unlike the wire growth from starfish type tentacles shown in Figure 2, here artificially made nano tentacles didn't contribute to the nanowires growth. Further detailed studies will be done.

4 CONCLUSIONS

In conclusion, we successfully grow high aspect ratio Si nanowires using Ni catalyst and SLS mechanism. Detailed microscopy results showed that in our case catalyst particle size has no relation with the diameter of Si nanowires. At high temperature, we found that Ni catalyst layer changed to irregular starfish like structure. Most Si nanowires grew form the irregular catalyst tentacles. In order to understand how catalyst size and shape affect Si wires formation, we use EB writer and lift-off to make various types of Ni nano corners. It is found that artificially made nano tentacles didn't contribute to the nanowires growth.

ACKNOWLEDGMENTS

This work is supported by the NSF grant, DMI-0428958. The authors would thank Dr.Quinzhou Xu, INRF, UCI, for his assistance in EBL operation and useful discussions.

REFERENCES

[1] Xiangfeng Duan, Yu Huang, Yi Cui, Jianfang Wang& Charles M. Lieber, Nature,Vol 409, ,66(2001).

[2] Michael H. Huang,et al, Science, Vol 292, 5523, 1897(2001) .

[3] Y. Cui, Q. Wei, H. Park, C. M. Lieber, Science 293, 1289(2001).

[4] Alfredo M. Morales, Charles M. Lieber, Science, Vol 279, 208(1998).

[5] Xing et al, Chinese physics letters, Vol 19, No.2, 240(2002)

Mechanical nonlinear generation with coupled torsional harmonic cantilevers for sensitive and quantitative atomic force microscopy of material characteristics

O. Sahin, C. F. Quate and O. Solgaard
E. L. Ginzton Laboratory, Stanford University, Stanford, CA 94305

ABSTRACT

Tapping-mode has been the most widely used mode of operation in atomic force microscopy. Recent studies have shown that higher harmonics of the cantilever vibrations carries information about material properties such as stiffness, viscoelasticity or capillary forces. A major problem with higher harmonic imaging is the low signal to noise ratio. Here we present a micromachined cantilever that enhances the signal levels at a particular higher harmonic by 40 dB. We demonstrate that the higher harmonic signal is sensitive to the thickness of an oxide film thermally grown on silicon, and enable mapping of chemical composition variations across a polymer surface.

Keywords: atomic force microscopy, cantilever, higher harmonics, tapping mode, material properties

1 INTRODUCTION

Atomic force microscopy (AFM) is used to image materials with nanoscale lateral resolution. Although imaging with atomic resolution has been demonstrated, new methods that simultaneously map various material properties and topography are needed. Tapping-mode atomic force microscopy has been the most widely used mode of operation [1]. In this technique the AFM cantilever is vibrated near its fundamental resonance frequency in the vicinity of the sample so that the tip of the cantilever periodically contacts the sample. Due to non-linear tip-sample forces, the dynamics of the cantilever motion carries extensive information about the sample. The two measured quantities, amplitude and phase of the cantilever vibration, only relate to the average values of interaction forces however [2,3]. Recent studies have shown that higher harmonics of the cantilever vibrations has the potential to image material properties such as stiffness, viscoelasticity or capillary forces [3-8]. The signal to noise ratio at the higher harmonics is however not sufficient for practical measurements [9]. Previously we have demonstrated 20 dB enhancement of the higher harmonic response in a specially micromachined cantilever that uses a higher order vibration mode to resonantly enhance a particular harmonic [10].

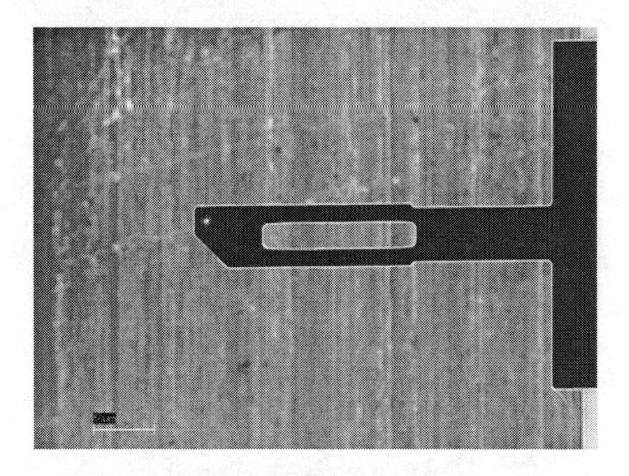

Fig. 1 SEM image of a torsionally coupled harmonic cantilever. The cantilever is 300 um long and 3 um thick. The width near the base is 40 um. The width of the arms is 15 um and the dimensions of the rectangular opening are 20 um by 130 um. The tip is located 12 um away from the longitudinal axis. The rectangular opening reduces the torsional bending stiffness and the torsional resonance frequency.

2 ENHANCING HIGHER HARMONIC GENERATION

Here we present recent advancements we made in higher harmonic imaging. We have designed, fabricated and experimentally demonstrated the use of a new class of micromachined cantilevers called coupled torsional harmonic cantilevers. These cantilevers have an asymmetric shape with an offset tip (see Fig. 1). When used in tapping-mode, the tip-sample forces excite torsional modes through higher harmonic generation.

In addition to the resonant enhancement, the torsional modes have a geometrical advantage over the flexural modes traditionally used to detect cantilever motion. Because the width of the cantilever is much smaller than its length, a small tip displacement due to torsional bending will generate a larger angular displacement compared to flexural modes. The torsional vibrations can therefore be effectively measured with regular four quadrant position sensitive photo-detectors that are commonly used in atomic force microscopes.

When a higher harmonic matches a torsional resonance frequency, the vibrations at that harmonic are resonantly enhanced because of the large quality factors of the torsional modes (~1000). This operation principle is similar to the resonant enhancement demonstrated in ref. 10. Because of the geometrical advantage explained above, measuring torsional-mode deflection is a much more efficient method for detecting higher-harmonics.

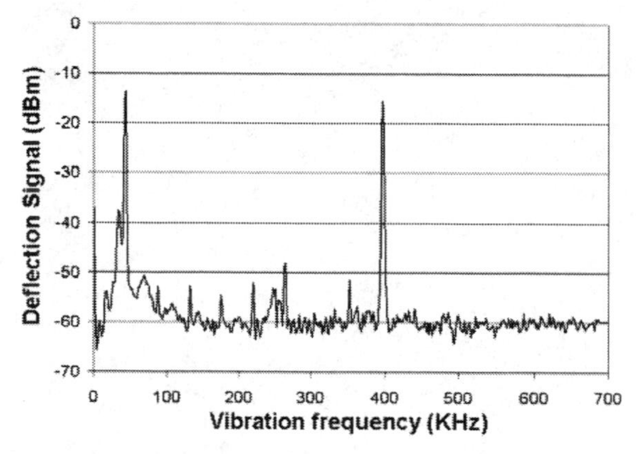

Fig. 2 Torsional vibration spectrum of a torsionally coupled harmonic cantilever. The torsional resonance frequency of this cantilever is at the exact 9th integer multiple of the fundamental flexural resonance frequency. The signal at the 9th harmonic is resonantly enhanced and has 45 dB signal to noise ratio in 1 KHz bandwidth.

In Fig. 2 we show the torsional vibration spectrum of a coupled torsional cantilever used in tapping-mode. The cantilever is driven at its first flexural resonance frequency of 44.14 KHz. It is the tip-sample forces that generate the torsional vibration spectrum shown in Fig. 2. The first peak in this spectrum is also at 44.14 KHz. Although torsional vibrations have a component at this frequency, the signal measured at this frequency is mainly due to the cross talk from the large flexural vibration. A small misalignment of the cantilever and photo detector can result in significant cross talk. There are other peaks in the vibration spectrum that appear at the integer multiples of the driving frequency. These are the higher harmonics. The torsional resonance frequency of this cantilever is at 397.26 KHz, which is also the 9th harmonic of the drive frequency. Therefore, the signal at this harmonic is much stronger. It has a signal-to-noise ratio of 45 dB in 1 KHz bandwidth.

There are a number of strategies available to tune the ratio of the torsional resonance frequency to the fundamental flexural resonance frequency to a desired integer. The cantilever in Fig. 1 has a rectangular opening that reduces the torsional stiffness of the cantilever. Reduced torsional stiffness results in a reduction in the torsional resonance frequency. On the other hand, the flexural modes of the cantilever are symmetric to the longitudinal axis and the presence of the hole does not affect the flexural stiffness as long as the effective width is the same. To match the integer ratio precisely, we have fabricated cantilevers with slight differences in the widths of the rectangular openings and parallel arms. Fabrication of the cantilevers was based on a silicon on insulator process described in detail previously [10].

2.1 Nanomechanical Measurements

Theoretical modeling of higher-harmonic generation has shown that the amplitude of a higher harmonic strongly depends on the stiffness of the sample under test [6,11]. In order to observe this experimentally, we have prepared a thin film of thermal oxide with a thickness gradient grown on a silicon wafer. The thickness gradient is obtained by polishing the wafer at a shallow angle after a uniform oxidation step. The polished wafer has a bare silicon surface on one side of the wafer, which develops an extremely thin native oxide layer after the polishing step. The thickness of the thermal oxide layer increases linearly beyond the native oxide region. A schematic of the cross section of the silicon wafer is given in Fig. 3. The Young's modulus of silicon dioxide is less than the Young's modulus of silicon. Therefore, the effective stiffness of the surface is that of the silicon where the film is negligibly thin, and then it is monotonously reduced to the stiffness of silicon dioxide as the thickness of the thermal oxide increases.

We have scanned the surface of the wafer along the thickness gradient in tapping mode with a coupled torsional harmonic cantilever. The amplitude of the enhanced higher harmonic is given in Fig. 3. The schematic of the wafer cross-section is drawn to relate the measured amplitudes and the position on the wafer. On the native oxide, the signal level has a constant value, while the harmonic amplitude drops rapidly to a minimum, as the film gets thicker. Beyond a certain thickness, the amplitude signal is slightly decreased as the thickness of the oxide is further increased. This oscillatory response of the harmonic amplitude with respect a monotonous change in effective Young's modulus of the surface has been predicted previously [11]. The response of the harmonic amplitude as well as the location of the first minimum depends on the spring constant of the cantilever, the free-oscillating and set point amplitude of the vibrations, and the geometry of the tip. A careful choice of the cantilever design and tapping conditions will optimize the sensitivity over a specific stiffness range. The measurements in Fig. 3 are done with a cantilever spring constant of 1 N/m. Silicon and silicon dioxide are relatively stiff materials, so the sensitivity to variations in the thickness of the thermal oxide can be improved by using a cantilever with a larger spring constant.

The measurements show that for the first 300 nm of the oxide film, the sensitivity to the film thickness is approximately 0.1 nm/√Hz. This calibration is done by slowing the scan speed and reducing the scan size around the linear region of the harmonic amplitude response, and then measuring the minimum thickness variation that results in a variation in the harmonic amplitude above the noise level.

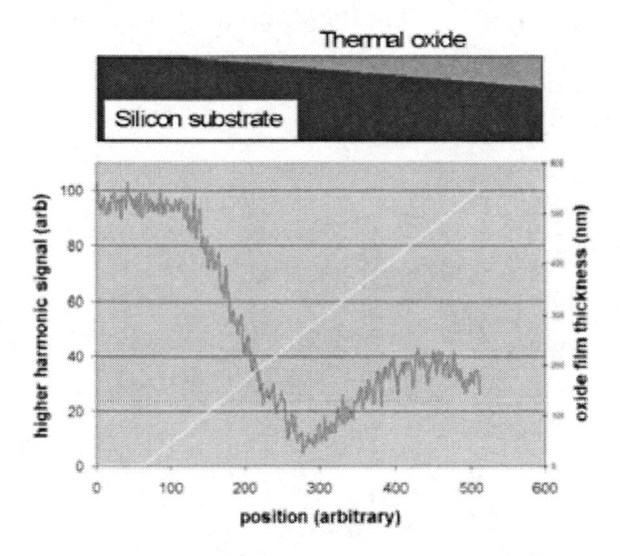

Fig. 3 Schematic of the thermally oxidized wafer polished at an angle and the measured higher harmonic amplitude across the wafer. The schematic cross section is shown to illustrate the location of the signals measured.

2.2 Mapping Chemical Compositions

Mechanical properties of materials highly depend on their chemical components. This relation allows us to map chemical composition variations across surfaces with the harmonic imaging technique. The imaging is done in tapping mode while the amplitude of the enhanced higher harmonic is monitored. Therefore the topography and higher harmonic image is generated simultaneously and they are perfectly registered. In Fig. 4, the topography and higher-harmonic image of a polymer surface are shown. The surface is composed of two types of polyethylene. One is amorphous and the other is ordered to some degree, resulting in a difference in hardness. The topography image on the right barely shows the features on the surface, whereas the harmonic image has three distinct colors, dark brown, light brown, and white. These levels correspond to the two types of polyethylene and the silicon substrate. Silicon is the stiffest of all and appears as white (the highest signal). Amorphous regions are relatively softer and they appear as dark brown.

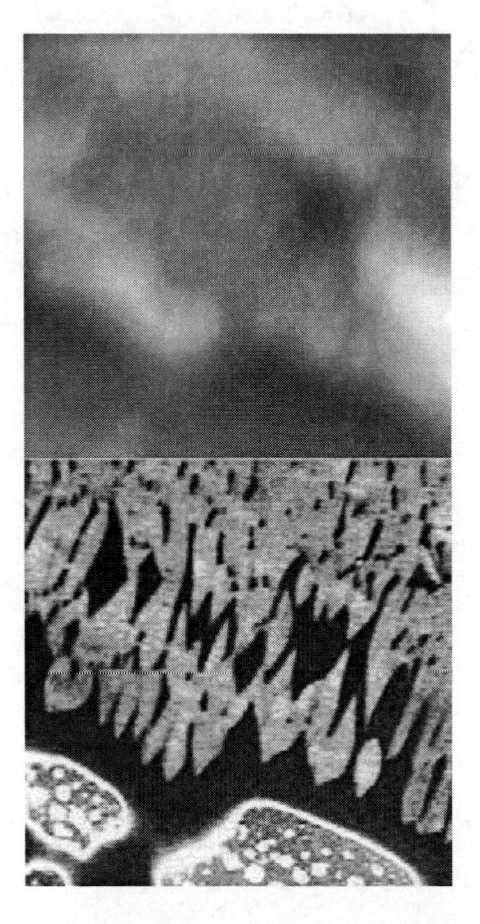

Fig. 4 Topography (top) and higher harmonic amplitude (bottom) image of a polymer surface on a silicon substrate. The harmonic image shows three color levels, white, light brown, and dark brown, corresponding to the three materials on the surface; silicon, amorphous polyethylene, and ordered polyethylene. Scan area is 5.6 microns

3 CONCLUSIONS

The enhanced higher harmonic generation achieved with the coupled torsional harmonic cantilever presented in this paper enables sensitive measurement of mechanical properties with nanoscale lateral resolution. The applications of harmonic imaging demonstrated here show the potential of nanomechanical measurements for the study and design of materials engineered at the nanoscale.

References

[1] Q. Zhong, D. Inniss, K. Kjoller, and V. B. Elings, "Fractured polymer/silica fiber surface studied by tapping mode atomic force microscopy," *Surf. Sci.* **280**, L688 (1993).

[2] J. P. Cleveland, B. Anczykowski, A. E. Schmid, and V. B. Elings, "Energy dissipation in tapping-mode atomic force microscopy" *Appl. Phys. Lett.* **72**, 2613 (1998).

[3] A. S. Paulo and R. Garcia, "Unifying theory of tapping-mode atomic force microscope" *Phys. Rev B.* **66,** 041406(R) (2002).

[4] R. Hillenbrand, M. Stark, and R. Guckenberger, "Higher-harmonics generation in tapping-mode atomic force microscopy: Insights into tip-sample interaction" *Appl. Phys. Lett.* **76**, 3478 (2000).

[5] W. Stark and W. M. Heckl, "Fourier transformed atomic force microscopy: tapping mode atomic force microscopy beyond the Hookian approximation" *Surf. Sci.* **457**, 219 (2000).

[6] O. Sahin and A. Atalar, "Simulation of higher harmonics generation in tapping-mode atomic force microscopy" *Appl. Phys. Lett.* **79**, 4455 (2001)

[7] M. Stark, R. W. Stark, W. M. Heckl, and R. Guckenberger, "Inverting dynamic force microscopy: from signals to time resolved forces" *PNAS* **99**, 8473 (2002).

[8] R. W. Stark and W. M. Heckl, *Rev. Sci. Instrum.* **74**, 5111 (2003).

[9] T. R. Rodriguez and R. Garcia, "Tip motion in amplitude modulation (tapping mode) atomic force microscopy: Comparison between continuous and point-mass models" *Appl. Phys. Lett.* **80**, 1646 (2002)

[10] O. Sahin, G. Yaralioglu, R. Grow, S. F. Zappe, A. Atalar, C. F. Quate, and O. Solgaard, "High resolution imaging of elastic properties using harmonic cantilevers" *Sensors and Actuators A,* **114,** 183 (2004)

[11] O. Sahin, A. Atalar, C. F. Quate, and O. Solgaard, "Resonant harmonic response in tapping-mode atomic force microscopy," *Phys. Rev. B.* **69** 165416 (2004)

Visualizing Nano-Electromechanics
by Vector Piezoresponse Force Microscopy

B.J. Rodriguez,[1] S. Jesse,[2] A.P. Baddorf,[2] S.V. Kalinin,[2] A. Gruverman[1]

[1]Department of Materials Science and Engineering, North Carolina State University, Raleigh, NC 27695
[2]Condensed Matter Sciences Division, Oak Ridge National Laboratory, Oak Ridge, TN 37831

ABSTRACT

A novel approach for nanoscale imaging and presentation of the orientation dependence of electromechanical properties, vector piezoresponse force microscopy (Vector-PFM), is proposed. The relationship between local polarization, piezoelectric constants and crystallographic orientation is described. The image formation mechanism in vector PFM and conditions for complete three-dimensional (3D) reconstruction of the electromechanical response vector and evaluation of the piezoelectric constants from PFM data are discussed. The developed approach can be applied to crystallographic orientation imaging in piezoelectric materials with a spatial resolution below 10 nm. Several approaches for data representation in two-dimensional (2D)-PFM and 3D-PFM are described.

Keywords: scanning probe microscopy, piezo response force microscopy, vector PFM,

1 INTRODUCTION

In the last decade, Piezoresponse Force Microscopy (PFM) has been established as a primary technique for imaging and non-destructive characterization of piezoelectric and ferroelectric materials on the nanometer scale.[1,2,3,4] The term piezoresponse, introduced in Ref. [4] for the description of this voltage-modulated contact mode of scanning probe microscopy, comes from the fact that the measured signal is dominated by the piezoelectric deformation of the ferroelectric sample. In the original papers on PFM, only the normal component of the tip displacement related to the out-of-plane component of polarization vector has been measured, an approach further referred to as vertical PFM (VPFM).[5] In 1998, Eng et al.[6,7] proposed a lateral PFM (LPFM) imaging method for measuring the in-plane component of polarization by monitoring the angular torsion of the cantilever. However, it is recognized that the electromechanical response is a vector having three independent components, while vertical and lateral PFM provide only two independent components. To address this problem, an approach for three-dimensional (3D) reconstruction of polarization using a combination of the VPFM data with two LPFM data sets obtained at different scanning directions has been developed.[6,8] However, in these experiments the absolute sensitivities in vertical and lateral directions were not calibrated, precluding the unambiguous reconstruction of the piezoresponse vector. The analysis by Kalinin et al.[9] has shown that the measured PFM signal also depends on local elastic properties, necessitating simultaneous measurement of the latter. Here, we discuss the information on local materials properties that can be obtained from quantitative measurements of full electromechanical response vector, an approach further referred to as Vector PFM, principles of the technique and its resolution limits. Vector PFM is shown to be a powerful tool for local orientation imaging in piezoelectrically active materials, such as micro- and nanocrystalline ferroelectric thin films and biological systems.

2 PRINCIPLES OF PFM

Piezoresponse force microscopy is based on the detection of the bias-induced piezoelectric surface deformation. The tip is brought into contact with the surface, and the piezoelectric response of the surface is detected as the first harmonic component, $A_{1\omega}$, of the tip deflection, $A = A_0 + A_{1\omega}\cos(\omega t + \varphi)$, induced by the application of the periodic bias $V_{tip} = V_{dc} + V_{ac}\cos(\omega t)$ to the tip. Here, the deflection amplitude, $A_{1\omega}$, is assumed to be calibrated and given in the units of length. When applied to the pyroelectric or ferroelectric materials, the phase of the electromechanical response of the surface, φ, yields information on the polarization direction below the tip.

Application of the bias to the tip results in the surface displacement, **w**, with both normal and in-plane components, $\mathbf{w} = (w_1, w_2, w_3)$. The usual assumption in the interpretation of PFM data is that the displacement of the tip apex in contact with the surface is equal to the surface displacement.[10] This is reasonable since the effective spring constant of the tip-surface junction is typically 2-3 orders of magnitude higher than the cantilever spring constant.

In addition to the vertical component of tip deflection, the use of a four-quadrant photodetector allows the lateral piezoresponse component in the direction normal to the cantilever axis (lateral transversal displacement) to be determined as torque of the cantilever. The fundamental difference between VPFM and LPFM is that in the latter case the displacement of the tip apex can be significantly smaller than that of the surface, e.g. due to the onset of sliding friction. Therefore, while in VPFM the response amplitude is expected to scale linearly with the modulation amplitude, in LPFM the response amplitude will eventually saturate. This sliding friction also minimizes the contribution from the piezoresponse component along the cantilever axis (lateral longitudinal displacement), as discussed in detail elsewhere.[11]

Information on the response component along the cantilever is usually difficult to obtain and in most cases only a 2D image can be acquired. For 3D imaging the second lateral component has to be measured by physically rotating the sample with respect to the cantilever as a simple change of the scan angle will not produce this result. This requirement necessitates locating the same microscale region on the surface after sample rotation, a task which is possible only for the samples with clear microscopic topographic markers.

The next step in the interpretation of the PFM data is detailed analysis of materials properties that can be extracted from the displacement vector. Assuming the scanning probe microscope is properly calibrated, the set of the piezoresponse data can be converted into the full electromechanical response vector, $\mathbf{w} = (w_1, w_2, w_3)$. The piezoelectric properties of materials are described by the third-order piezoelectric constant tensor d_{ij}, where $i = 1,..,3$, $j = 1, .., 6$, that defines the relationship between the strain tensor and the electric field: $X_j = d_{ij} E_i$.[12] Here, the components of d_{ij} are given in the laboratory coordinate system, in which axis 3 is normal to the surface and axis 1 is oriented along the long cantilever axis and the reduced Voigt notation[13] is used. The piezoresponse signal measured in the PFM experiment has the same dimensionality as the piezoelectric constants, suggesting the close relationship between d_{ij} and components of the electromechanical surface response vector. This is remarkable, since PFM is thus an SPM technique, which is sensitive to the tensorial properties of materials. In the local excitation case, the PFM signal can be semi-quantitatively approximated as $vPR_l = d_{33}$, $xPR_l = d_{35}$ and $yPR_l = d_{34}$, where the coordinate system for d_{ij} is laboratory coordinate system related to the cantilever orientation. For materials with known crystallographic orientation, the elements of the d_{ij}^0 tensor can be semiquantitatively determined by VPFM, 2D- or 3D-PFM measurements performed on crystals with different orientations. For materials with known piezoelectric constants, d_{ij}^0, the local crystallographic orientation in each point (orientation imaging) can be determined from vector 3D-PFM data. For materials systems with known constraints on possible crystallographic orientation (small number of domains), domain structure reconstruction can be obtained from partial VPFM or 2D-PFM data.

3 MODELLING SIGNAL GENERATION

One of the outstanding questions in PFM and mechanical modulation techniques such as Atomic Force Acoustic Microscopy is the signal generation volume that determines both lateral resolution and depth sensitivity. Both of these techniques are ultimately sensitive to bias and displacement-induced changes in indentation depth. The complete description of the nanoelectromechanics of piezoelectric indentation for a special case of transversally isotropic material as applied to PFM and AFAM was given by Kalinin and Karapetian.[9,10]

In particular, from the known displacement field below the tip, the signal generation volume in PFM can be found as $\delta u_z / \delta V$, whereas in AFAM the signal generation volume is $\delta u_z / \delta d$, where $u_z = u_z(\rho, z)$ is the normal displacement field below the tip and δu_z is its change in response to the change in tip potential, δV, or sample base position, δd. For piezoelectric indentation, the normal displacement field could be represented as a linear superposition of the fields $u_z = u_{z,m}(a) + u_{z,e}^0(a)\psi_0$. Here $u_{z,m}$ is the solution of the indentation problem with purely mechanical boundary conditions and zero potential and $u_{z,e}^0(a, R)\psi_0$ is the solution of the purely electrical problem. Shown in Figure 1a,b is the vector map of electrical field inside material induced by tip bias for $a = 5$ nm and $R = 50$ nm for PZT6b and $BaTiO_3$ respectively. The field distribution and penetration depth is determined primarily by anisotropy of the dielectric constant tensor of the material. Figure 1c,d illustrate the strain field induced by the mechanical load applied to the tip. In both cases, the electromechanical coupling affects the corresponding field distributions only weakly, as can be shown by comparison with similar maps calculated for zero electromechanical coupling, $e_{ij} = 0$. Finally, shown in

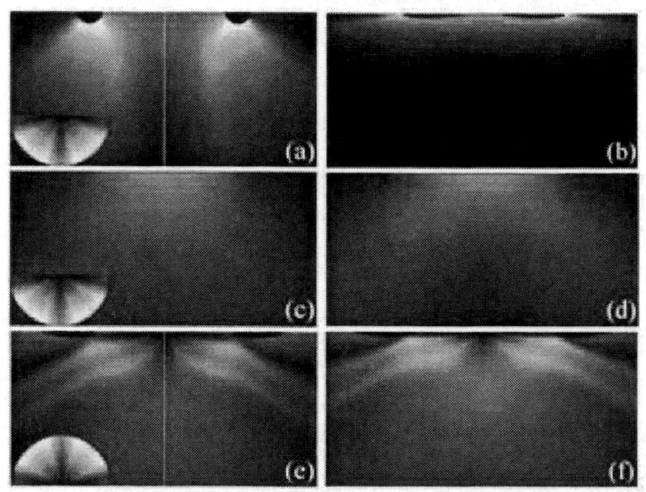

Figure 1:Vector map of electric field magnitude and direction induced by tip bias for PZT6b (a) and $BaTiO_3$ (b). Brightness indicates the magnitude of electric field, whereas color indicates direction. Shown is map for $(\ln(E_\rho), \ln(E_z))$, where E_ρ and E_z is lateral and vertical components of electric field. Vector maps for indentation-force induced strain field below the tip for PZT6b (c) and $BaTiO_3$ (d). Shown is map for (u_ρ, u_z), where u_ρ and u_z are lateral and vertical strain components. Vector map for bias induced strain field below the tip for PZT6b (e) and $BaTiO_3$ (f). Shown is map for (u_ρ, u_z), where u_ρ and u_z are lateral and vertical strain components.

Figure 1e,f is the strain field induced by the bias applied to the tip, which originates directly from electromechanical coupling.

This analysis suggests that the signal generation volume, and hence resolution, is limited by the contact area between the tip and the surface, typically of order of 3-5 nanometers. This is in a close agreement with experimentally observed resolution in PFM of order of 5-10 nm, suggesting the potential of Vector PFM to image local electromechanical properties with sub-10nm resolution.

4 EXPERIMENTAL RESULTS

An outstanding issue for the Vector PFM is image representation. While for most SPMs the use of pseudocolor images is common, the color is generally used for the enhancement of the contrast of the scalar data. Here we demonstrate two approaches to represent 2D and 3D Vector PFM data. Shown in Figure 2a,b are vertical and lateral PFM, $A_{1\omega}\cos(\varphi)$ signals, from $PbTiO_3$ thin film. In the VPFM image, high intensity corresponds to the regions with a strong vertical component of electromechanical response in a positive *z*-direction, while low intensity corresponds to a strong response in the negative *z*-direction. Gray areas of intermediate intensity correspond to a weak out-of-plane response component. Similarly, the LPFM image provides information on the in-plane component perpendicular to the cantilever axis. Upon close examination, the PFM images in Figure 2a,b show a decrease of the effective PFM signal in the center of the grain. Such behavior can be attributed to a change in magnitude of the electro-mechanical response vector, either due to an intrinsic change of the material's properties or tip-surface contact. However, interpretation of separate LPFM and VPFM data is not straightforward.

To address this problem, we employ vector representation for PFM data similar to an approach well-known in electron microscopy. The VPFM and LPFM images are normalized with respect to the maximum and minimum values of the signal amplitude so that the intensity changes between -1 and 1, i.e. $vpr, lpr \in (-1,1)$. While not strictly rigorous, this procedure is expected to provide the correct answer for systems where grains with all possible orientations of the response vector are present. Using commercial software,[14] these 2D vector data (vpr, lpr) are converted to the amplitude/angle pair, $A_{2D} = \mathrm{Abs}(vpr + I\ lpr)$, $\theta_{2D} = \mathrm{Arg}(vpr + I\ lpr)$. These data then used to generate a pseudocolor map where the color corresponds to the orientation, while color intensity corresponds to the magnitude of the response vector, as illustrated in Figure 2c. We refer to this method for PFM data representation as a 2D vector PFM image. Here both color and intensity convey information and the "color wheel" legend illustrates the direction and magnitude of the response vector.

Notably, the vector-PFM image illustrates that color is virtually uniform inside the grains, while the intensity varies between the central part of the grain and the circumference. This difference illustrates that only the magnitude, but not the orientation, of the piezoresponse vector changes. This information can be represented in the scalar form by plotting separately phase, θ_{2D}, and magnitude, A_{2D}, data, as illustrated in Figure 2d,e, respectively. In the phase image (Figure 2d), the phase of the 2D response vector is clearly uniform within individual grains, whereas magnitude (Figure 2e) changes from the grain center to its circumference.

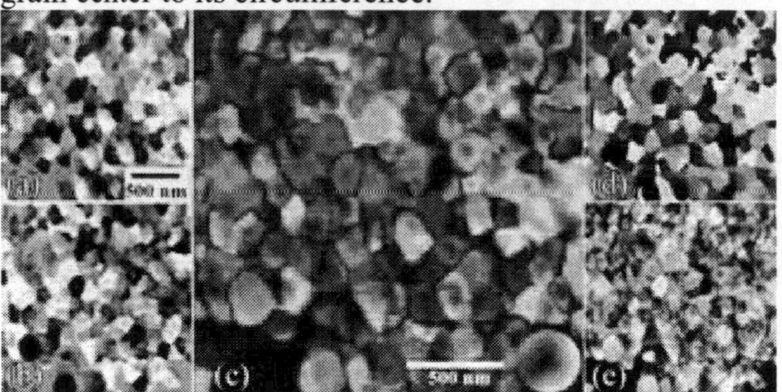

Figure 2. Vertical (a) and lateral (b) PFM image of PZT thin film. (c) Vector representation of 2D PFM data. (d) Angle and (e) amplitude images.

An example of 3D-PFM imaging is illustrated in Figure 3 using an etched $LaBGeO_5$ glass ceramic. Figure 3a is an image showing the 3D vector piezoresponse of the ferroelectric grain. Because of the relatively large grain size (~50 μm), the same region could be imaged several times with different orientations of the cantilever relative to the sample, thus allowing 3D PFM data to be collected. Here, VPFM, *x*-LPFM and *y*-LPFM data sets containing information on all three components of the electro-mechanical response vector were collected.

The representation of complete 3D vector field, as opposed to a 2D subset, represents a more challenging problem. The solution requires using the entire color palette available including hue, intensity, and lightness. Here, the VPFM and *x,y* LPFM images are normalized with respect to the maximum and minimum values of the signal amplitude so that the intensity changes between -1 and 1, i.e. $vpr, xlpr, ylpr \in (-1,1)$. These 3D vector data $(vpr, xlpr, ylpr)$ are mapped on the red, green, blue color scale, represented as vector (R, G, B), where *R*, *G* and *B* are mutually orthogonal and vary from 0 to 1. The magnitude of the *z*-component is represented by lightness/darkness, variation in direction in the *x,y*-plane is given by hue, and the magnitude of the vector is represent by color saturation (note that black and white are colors).

The transformation involves rotating the (R, G, B) coordinate system and shifting it so that $(R, G, B) = (0.5,\ 0.5,\ 0.5)$ corresponds to zero in the PFM coordinate system, $(vpr, xlpr, ylpr) = (0,0,0)$. This transformation is expressed in the equations:

$$(R \quad G \quad B) = \frac{1}{2}\left[\frac{1}{\sqrt{3}}\cdot\left(R_X(\theta)\cdot R_Z(\phi)\right)^T\cdot(xlpr \quad ylpr \quad vpr) + (1 \quad 1 \quad 1)\right]$$

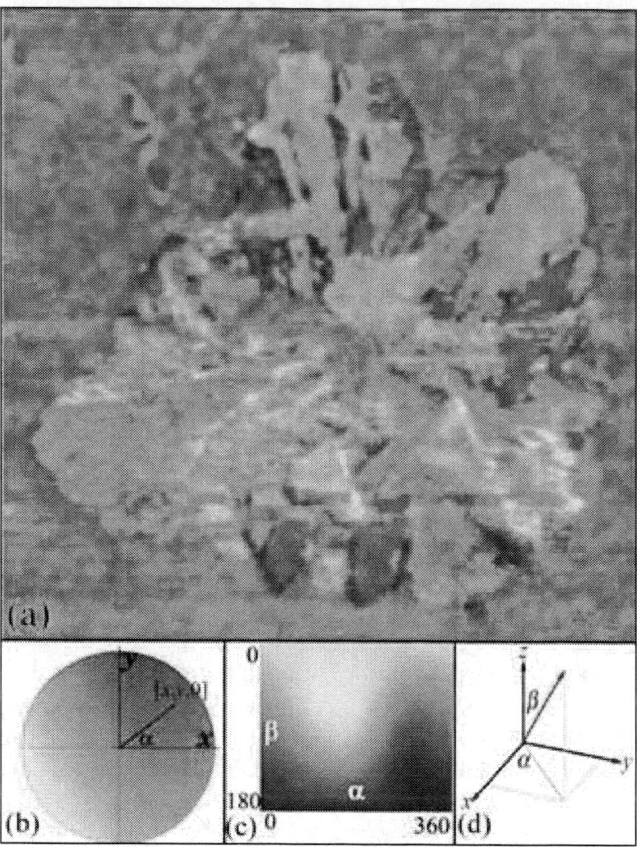

Figure 3: (a) 3D vector piezoresponse representation of an etched $LaBGeO_5$ glass ceramic. (b) color wheel map relating hue to lateral piezoresponse. (c) map relating vector direction to color for a unit vector. (d) relationship between angles in (c) and laboratory coordinate system.

where $R_x(\theta_r)$ and $R_z(\phi_r)$ are rotation matrices and $\theta_r = \tan^{-1}\sqrt{2}$ and $\phi_r = \pi/4$ are Euler angles.

Note that light shading indicates the vector pointing out of the page and dark shading indicates a vector pointing into the page. This relationship is shown in the color key in Figure 3c in accordance with the coordinate axis in Figure 3d. Gray areas indicate regions where the magnitude of the response vector is relatively small. Intense or saturated hues indicate a strong lateral response with small a vertical component. This is represented in the color wheel in Figure 3b which shows the color coding for purely lateral displacement vectors (no vertical component).

5 CONCLUSIONS

An approach for vector electromechanical imaging by SPM, referred to as vector piezoresponse force microscopy (Vector PFM), is proposed. The relationship between detected vertical and lateral signal components and local electromechanical response vector is discussed. The relationship between 3D-PFM data and local materials' properties is established and it is shown that 3D PFM can be used as a powerful tool for (a) local electromechanical property measurements on the nanoscale or (b) local orientation imaging on the sub-10 nm level. Finally, several approaches for data representation in 2D-PFM and 3D-PFM are presented. The developed approach can be applied for nanoscale electromechanical characterization of a broad range of material systems including polymers, composites, and biomaterials.

ACKNOWLEDGEMENTS

Research performed in part as a Eugene P. Wigner Fellow and staff member at the Oak Ridge National Laboratory, managed by UT-Battelle, LLC, for the U.S. Department of Energy under Contract DE-AC05-00OR22725 (SVK). Support from ORNL SEED funding is acknowledged (ABP and SVK). AG acknowledges financial support of the National Science Foundation (Grant No. DMR02-35632). $LaBGeO_5$ sample is courtesy of P. Gupta, H. Jain, and D.B. Williams (Lehigh).

[1] *Ferroelectrics at Nanoscale: Scanning Probe Microscopy Approach*, edited by M. Alexe and A. Gruverman (Springer Verlag, 2004).

[2] *NanoScale Phenomena in Ferroelectric Thin Films,* edited by S. Hong (Kluwer Academic Publishers, Boston, 2004).

[3] P. Gunther and K. Dransfeld, Appl. Phys. Lett. **61**, 1137 (1992).

[4] A.Gruverman, O.Auciello and H.Tokumoto, J. Vac. Sci. Technol. **B 14**, 602-605 (1996).

[5] A. Gruverman, O. Auciello, and H. Tokumoto, Annu. Rev. Mat. Sci. **28**, 101-123 (1998).

[6] L.M. Eng, H.-J. Guntherodt, G.A. Schneider, U. Kopke and J.M. Saldana, Appl. Phys. Lett. **74**, 233 (1999)

[7] L. M. Eng, H.-J. Güntherodt, G. Rosenman, A. Skliar, M. Oron, M. Katz, and D. Eger, , J. Appl. Phys. **83**, 5973-5977 (1998).

[8] B.J. Rodriguez, A. Gruverman, A.I. Kingon, R.J. Nemanich, and J.S. Cross, J. Appl. Phys. **95**, 1958 (2004).

[9] Sergei V. Kalinin, E. Karapetian, and M. Kachanov, Phys. Rev. B 70, 184101 (2004)

[10] E. Karapetian, M. Kachanov, and S.V. Kalinin, , Phil. Mag., in print

[11] Sergei V. Kalinin, B.J. Rodriguez, S. Jesse, J. Shin, A.P. Baddorf, P. Gupta, H. Jain, D.B. Williams, and A. Gruverman, Microscopy and Microanalysis, submitted

[12] *Physical Properties of Crystals,* J.F. Nye, (Oxford University Press, 1985).

[13] Due to symmetry, the indices i,j in full piezoelectric coefficient tensor, d_{ijk}, and strain tensor, X_{ij}, are substituted as 11→1, 22→2, 33→3, 12→ 6, 13→ 5, 23→ 4.

[14] Mathematika 5.0, Wolfram Research

Quantification of Properties of Ferroelectric Thin Film Using Piezoresponse Force Microscopy

M. G. Cain*, M. J. Lowe, A. Cuenat, M. Stewart and J. Blackburn

National Physical Laboratory
Hampton Road, Teddington, Middlesex, TW11 0LW, UK.
* Markys.Cain@npl.co.uk

ABSTRACT

Piezoresponse force microscopy has been employed to monitor the effects of surface contamination and tip-sample contact on the piezoelectric coefficients of PZT thin films. Calibration has been achieved through the use of an interferometer, offering sub-picometer resolution. Surface contamination leads to the piezoelectric properties being reduced by an order of magnitude. Cleaning can remove the surface contamination leading to a restoration of the piezoelectric properties. In addition, modelling of the tip-sample interaction has been performed using finite difference methods. Incorporating non-ferroelectric layers between tip and sample leads to large variations in the electric field across the ferroelectric film. The use of such models provides a better understanding of the tip-sample interactions, allowing for more robust measurements of piezoelectric properties using PFM.

Keywords: piezoresponse force microscopy, lead zirconate titanate, ferroelectric, thin films, characterisation

1 INTRODUCTION

Ferroelectric thin films represent an important class of materials for use in applications such as non-volatile memories [1] and micro-electro-mechanical (MEMS) devices [2]. At these reduced dimensions, device performance may be reduced due to effects such as residual stresses during the growth process [3], although island patterning can reduce clamping [4]. Such factors may result in a degradation of the piezoelectric coefficients, namely d_{33} and d_{31}, or changes to the remnant polarisation and coercive voltage of the ferroelectric material. In order to monitor these effects, experimental techniques are necessary which can offer nanometer resolution and below. Piezoresponse force microscopy (PFM) has been developed as a variation of the atomic force microscopy [5], which can provided both quantitative and qualitative information. Ferroelectric domains can be observed with a resolution below 10 nm [6], whilst on application of a known voltage, sample motion can be detected down in the picometer range. Despite the range of work performed using PFM, quantification is still limited due to a lack of complete understanding of the tip-sample contact. Various theoretical work has been performed to address this [7,8]. However, as most PFM work is performed under ambient conditions, a surface contaminant layer must be accounted for. Previous work has speculated the presence of atmospheric oxide and carbonate layers at PZT surfaces [9]. Thermodynamic analysis using MTDATA software from NPL show that lead carbonate may form in the presence of trace amounts of carbon dioxide [10]. In this work, a quantitative PFM study has been performed to provide a better understanding of the tip-sample interactions. In order to provide a comparison with PFM experiments, finite difference (FD) modelling has been performed. In addition, results have been compared with a compact differential interferometer system, designed at NPL, which offers sub-picometer resolution through the use of lock-in techniques.

2 EXPERIMENTAL PROCEDURE

PFM experiments were performed using a modified Dimension 3000 AFM. Two tip types were used in this work. For quantitative work, Pt-coated and Pt/Ir-coated Si tips ($f_0 \sim 75$ kHz, $k \sim 3$ N/m) were employed, whilst for domain imaging, softer Pt/Ir-coated Si tips ($f_0 \sim 13$ kHz, $k \sim 0.1$ N/m) were utilised to reduce the possibility of damage to the thin films. Imaging was achieved using a frequency close to the resonant tip frequency and a sinusoidal voltage (1 V amplitude). Quantitative measurements were obtained at a tip frequency of approximately 20 kHz, well below the resonant frequency. Measurements were obtained using two methods, either by applying the voltage directly to the tip (PFM mode) or to a top electrode (AFM mode). In each case, the bottom electrode was grounded.

The NPL interferometer, shown in figure 1, was used for calibrating PFM measurements. The system makes use of common path optics to direct a measuring and reference laser beam onto the sample under investigation. A differential set up is utilised to remove errors due to sample bending or external vibrations. Through the use of lock-in techniques, a voltage is applied to the sample, giving rise to displacements that may be measured with sub-picometer accuracy. In order to model the tip-surface interactions, finite difference modelling was employed. A 2-D static model was employed, with the AFM tip represented by a trapezium with semi-circular apex held at a voltage of 1 V. The bottom edge of the PZT film was clamped and fixed at 0 V. Successive overrelaxation (SOR) and conjugate gradient (CG) solutions were utilised in the modelling.

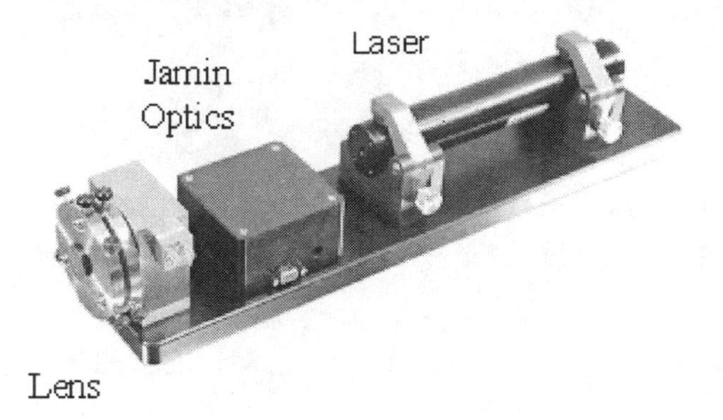

Figure 1: A photograph of one arm of the NPL interferometer, showing the major features.

3 EXPERIMENTAL RESULTS

In order to provide calibration for the PFM, d_{33} was determined for a bulk quartz sample, which should yield a value of 2.3 pm/V [11]. In addition, this was compared with that obtained by the interferometer. Figure 2, shows the results of the calibration process. Using the interferometer yielded a value for d_{33} of 2.07 pm/V, whilst PFM gave 2.06 pm/V. Both of these numbers agree with each other but fall below the expected value by approximately 10 %. This is believed to be due to the sample mounting in each case, with sample motion reduced by the holding adhesive.

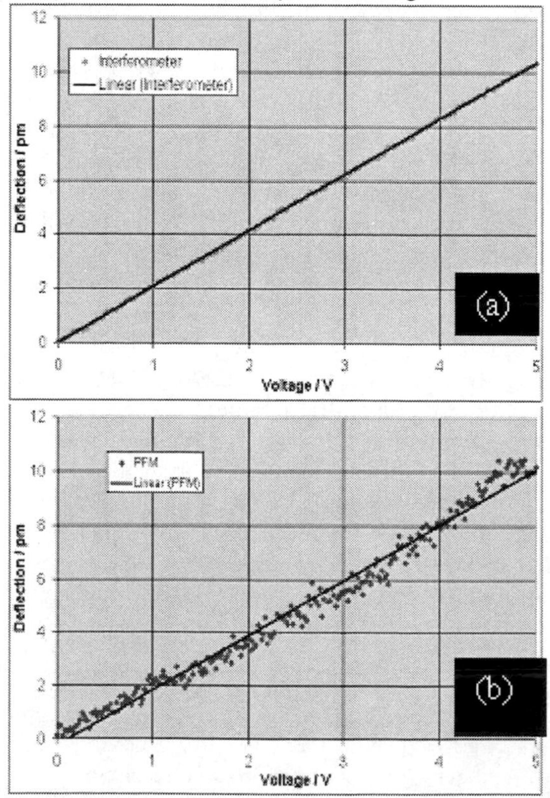

Figure 2: Plots of deflection against voltage for a quartz calibration sample using (a) NPL interferometer; and (b) PFM.

On performing the calibration procedure, results were then obtained from a variety of PZT samples. An early test was to measure d_{33} from a bulk PZT cubed sample using PFM and compare with the value found using the interferometer. For these measurements, the top surface of the PZT cube was electroded, but the experiment was performed in PFM mode. Results, not shown here, indicate good linearity and agreement between the two measurement techniques. Interferometry yielded a value of 263 pm/V, whilst in PFM mode, a value of 259 pm/V was achieved. The slight discrepancy in these results may be put down to the use of a reflector on the top electrode surface of the cube to enhance the laser signal.

Despite the good agreement shown between the two experimental techniques when using bulk samples, when attention was turned to thin films, discrepancies arose which could be explained by poor tip-sample contact and the presence of a non-ferroelectric surface layer. Initial studies were performed on a 500 nm PZT(30/70) thin film. On application of an increasing ac voltage to the film, a value for d_{zz} of 19.0 pm/V was measured. However, when measurements were attempted using PFM, d_{zz} was much smaller than expected. Figure 3 shows a frequency sweep against deflection when measurements were attempted in both AFM and PFM mode. In AFM mode, the deflection corresponds to a value for d_{zz} of approximately 22 pm/V, whereas in PFM mode, the value was reduced to below 5 pm/V. It should be noted that these experiments were recorded from the same region of the sample.

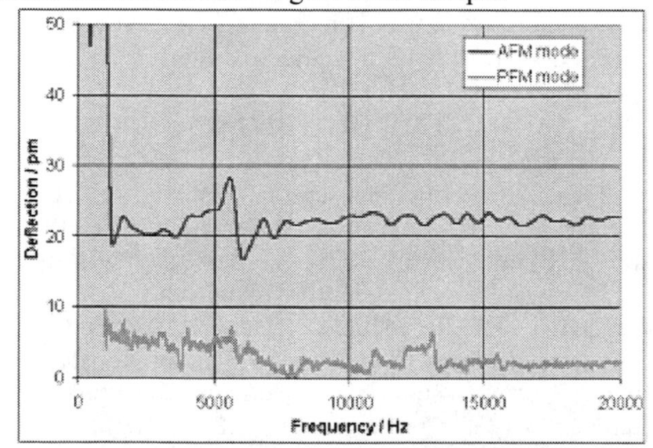

Figure 3: Plots of deflection against frequency for a 500 nm PZT(30/70) thin film when an ac voltage of 1V is applied to the film in (a) AFM mode; and (b) PFM mode.

Measurements were also obtained from two samples of similar thickness. A 210 nm PZT(30/70) film grown on an ITO/glass substrate (sample P1) gave a value for d_{zz} of 34.9 pm/V, whereas for a 200 nm PZT(30/70) film grown on a $Ti/Pt/SiO_2/Si$ substrate (sample P2), d_{zz} was as low as 2 pm/V. This difference also shows up in PFM domain images from the two samples. Figure 4 shows PFM images from a 1×1 μm^2 area of each sample. Both images are shown with the same piezoresponse scales. It is apparent that the image from sample P1 yielded greater contrast, with features as small as 15 nm being resolved. Sample P2 shows a larger minimum feature size, ascribed to the different grain size, but the image contrast is reduced due to the presence of a low permittivity surface layer producing reduced signal to noise levels.

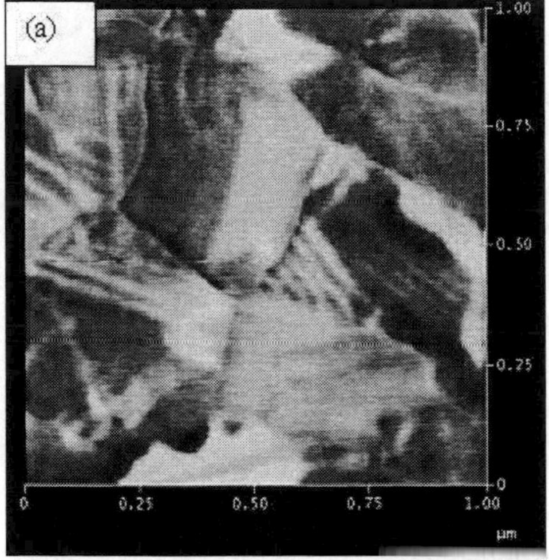

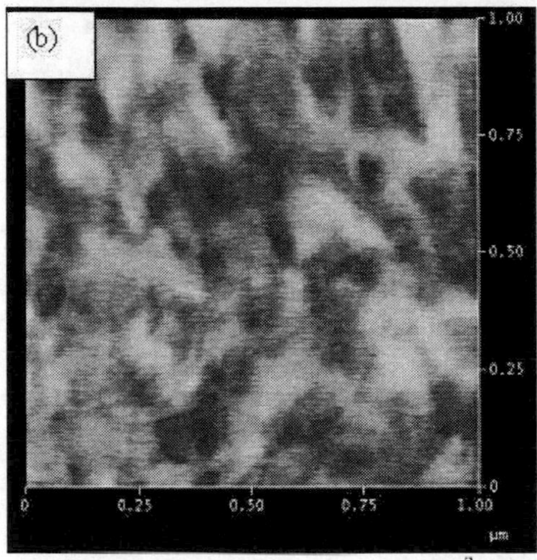

Figure 4: PFM domain images from 1×1 μm^2 areas of (a) sample P1; and (b) sample P2. In each case, the piezoresponse scale is the same to provide a comparison of the contrast.

To test the hypothesis of a contamination layer at the surface, attempts were made to clean the PZT thin film surfaces. For this work, a 400 nm thick PZT(30/70) film was selected. The PZT film was cleaned in a plasma barrel etcher using CF_4 + 5 % O_2 as the etchant. The sample film temperature was approximately 100 °C during etching, with no post clean annealing being performed. Figure 5 shows deflection vs frequency plots for the sample before and after etching. It is clear that the cleaning process removes some surface contamination, although the value for d_{zz} of 14.3 pm/V is lower than would be expected for these thin films. It is believed that either some surface recontamination occurred after the cleaning process, or some contamination was unable to be removed. Further work involving X-ray photoelectron spectroscopy is planned to resolve some of these cleaning issues.

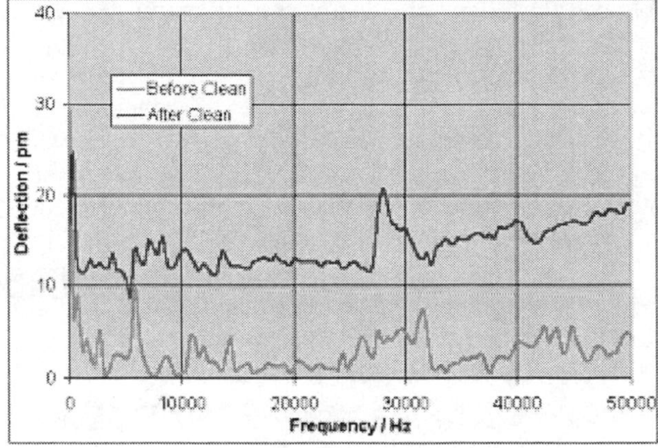

Figure 5: Plots of deflection against frequency for a 400nm thick PZT(30/70) film before and after cleaning. IT is clear that cleaning has a beneficial effect on the piezoelectric response.

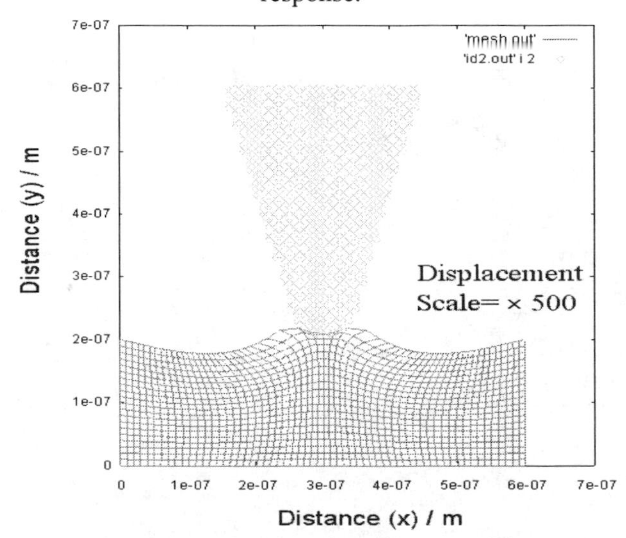

Figure 6: Result from the FD modelling, showing the film displacement from a 200 nm thick single domain PZT film. Displacements are magnified by a factor of 500 for clarity.

To investigate the tip sample interactions occurring during PFM, FD modelling was performed. The properties of the PZT layer were assumed to be as for a PZT-5H bulk sample [12]. Figure 6 shows the results of calculations on a 200 nm thick single domain PZT film when a voltage of 1V is applied to the conducting tip. The tip height of approximately 400 nm is much smaller than that used in experiments, but it is expected that this yields errors less than 5 % in the film displacement. In figure 6, the film displacement, magnified by a factor of 500, is found to be approximately 50 pm. The difference between experiment and theory is put down in part to the different piezoelectric properties of the experimental system, compared to the ideal bulk PZT properties used in the FD model.

The case of a single domain can be unrealistic when studying thin films experimentally. In this way, calculations were obtained for the more realistic cases of a multi-domain PZT thin film (100 nm thick) with and without a 20 nm non-ferroelectric dead layer, exhibiting a permitivity of 80 and a stiffness equal to PZT. In both cases, domains of width 100 nm were used, exhibiting periodic up and down polarisations. Figure 7 shows the displacement plots for each sample geometry. Again for the multi-domain film with no dead layer, the film displacement is approximately 50 pm. However, when the dead layer was incorporated, the displacement was reduced by approximately an order of magnitude, which agrees with the experimental findings from the 400 nm thick PZT(30/70) film.

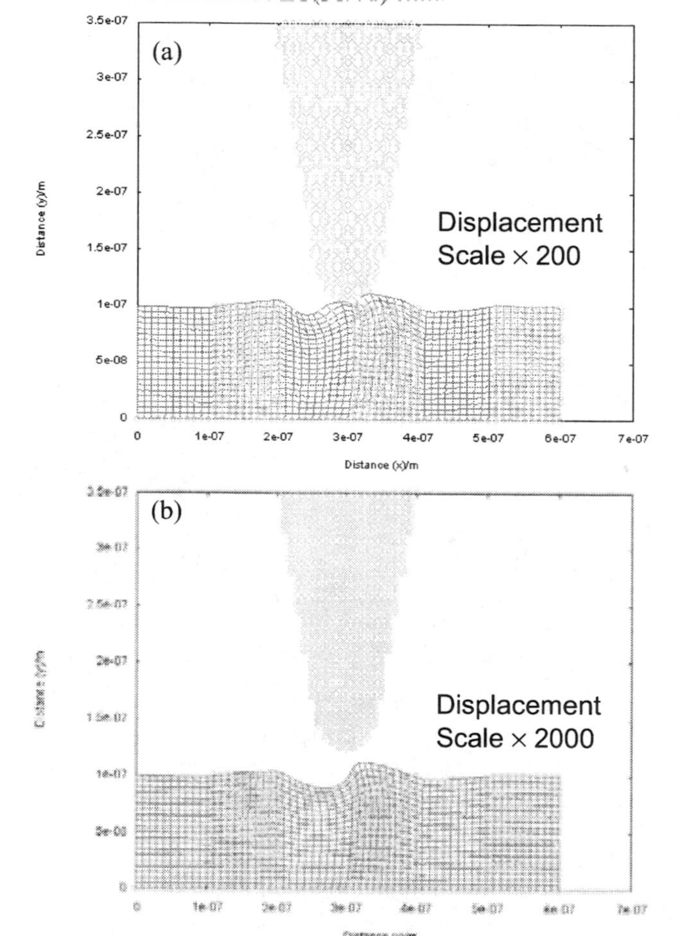

Figure 7: Result from the FD modelling, showing the film displacement from a 100 nm thick multi domain PZT film (a) without; and (b) with a non-ferroelectric dead layer. Displacements are magnified for clarity. Green symbols show undisplaced locations of –y oriented PZT domains.

The presence of the dead layer at the surface leads to a drop in the electric field across the ferroelectric thin film by an order of magnitude. This produces an effect which is similar to that found if an air gap exists between the tip and sample, but the magnitudes of the separations differ due to the difference in permittivity of the two insulating layers. A similar effect has also been found for $BaTiO_3$ using Green's functions and incorporating a variable thickness water layer [7]. At present further experimental work is necessary to characterize the chemical state of the PZT surface, although some work is now becoming available [13].

4 CONCLUSIONS

PFM studies have been conducted on various PZT sample and compared with an interferometer system which offers traceable calibration. It is shown that for good tip-sample contact, results obtained using PFM agree with those from the interferometer. However, good tip-sample contact is essential for quantitative PFM. Cleaning the thin films can remove some contaminants and yield piezoelectric properties that agree with expectations. Finite difference modelling shows that the presence of a non-ferroelectric dead layer can result in the voltage dropped across the thin film being reduced, leading to reduced piezoelectric properties. It is therefore essential that ferroelectric surfaces are free from contamination to enable accurate determination of the piezoelectric coefficients of such materials.

REFERENCES

1 O. Auciello, J. F. Scott and R. Ramesh, Phys. Today **51,** 22 – 27, 1998.

2 P. Muralt, J. Micromech. Microeng. **10**, 136 – 146, 2000.

3 T. M. Shaw, Z. Suo, M. Huang, E. Liniger, R. B. Laibowitz and J. D. Baniecki, Appl. Phys. Lett. **75**, 2129 – 2131, 1999.

4 V. Nagarajan, A. Roytburd, A. Sanishevsky, S. Prasertchoung, T. Zhao, L. Chen, J. Melngailis, O. Auciello and R. Ramesh, Nature Materials **2**, 43 – 47, 2003.

5 A. Gruverman, O. Auciello and H. Tokumoto, Annu. Rev. Mater. Sci. **28**, 101 – 123, 1998.

6 S. Dunn, C. P. Shaw, Z. Huang and R. W. Whatmore, Nanotechnology **13**, 456 – 459, 2002.

7 F. Felten, G. A. Scheider, J. Munoz Saldana and S. V. Kalinin, J. Appl. Phys. **96** 563 – 568, 2004.

8 S. V. Kalinin and D. A. Bonnell, Phys. Rev. B **65** 125408-1 – 125408-11, 2002.

9 M. G. Cain, S. Dunn and P. Jones, Technical Proceedings of the 2004 Nanotechnology Conference and Trade Show, Volume **3**, 362 – 365, 2004.

10 Analysis using MTDATA: software copywrite NPL.

11 V. E. Bottom, J. Appl. Phys. **41**, 3941 – 3944, 1970.

12 Vernitron Bulletin, "Five Modern Piezoelectric Ceramics", Bulletin 66011/F, Jan 1976.

13 F. Peter, K. Szot, R. Waser, B. Reichenberg, S. Tiedke and J. Szade, Appl. Phys. Lett. **85**, 2896 – 2898, 2004.

Nonlinear Response in Nanoshear Modulation Force Microscopy: Fourier Analysis

G. Haugstad
University of Minnesota, Minneapolis, MN, USA, haugs001@umn.edu

ABSTRACT

Fourier analysis of oscillating forces at a shear-modulated tip provides new insight into static-to-kinetic friction transitions (stiction). In addition to contrast in conventional friction force microscopy, layers of autophobically dewetted PVA films exhibit remarkable differences in stiction. These differences relate to strong adsorption of first layer to mica substrate and concomitant conformational arrest, as compared to bulk-like behavior in the second layer. The third Fourier harmonic is found to be a sensitive gauge to variable degrees of sliding as a function of both drive amplitude and normal load (tensile to compressive). For a nanoscale drive, it is discovered that a largely static contact at compressive loads becomes a largely sliding contact at tensile loads. This finding has implications for the analysis of shear modulation force microscopy of polymers in the context of contact mechanics models, and for studies under variable sample compliance (temperature or solvent induced).

Keywords: shear modulation, confinement, AFM, stiction, polyvinyl alcohol

1 INTRODUCTION

Oscillatory methods in scanning probe microscopy have enabled nanoscale tribology and rheology investigations on polymeric systems. Among these methods, shear modulation force microscopy has been used to probe phenomena such as the stick-to-slide transition [1], time-delayed viscoelastic response [2], and near-surface glass transition [3], all at the nanoscale. Advantages of shear over normal modulation include a much stiffer torsional spring constant (comparable to that of the analyzed material), a simplified analytical expression for contact stiffness (independent of contact mechanics model), and minimal normal motion under shear drive (whereas parasitic shear motion is unavoidable under normal drive because of cantilever tilt). The commonly adopted approach is to measure the dynamic torsional forces on the tip via amplitude and phase outputs of a lock-in amplifier [1-3], to quantify the response to sinusoidal, nanometer-scale shear oscillations (of sample or tip). Missing in this approach are the details of real-time response that may reveal nonlinear interactions: strongly anharmonic distortions due to sliding, plastic yield or intrinsically nonlinear responses (e.g., nonlinear viscoelasticity, shear thinning). Given that fundamental polymeric responses are being studied, it is imperative to understand shear modulation phenomenology at a fundamental level. To this end we have examined nonlinear dynamic behavior under shear modulation via Fourier analysis of the real-time response. The methodology exploits the third harmonic of response as a gauge of nonlinearity due to partial sliding [4].

2 EXPERIMENTAL DETAILS

2.1 Materials.

Atactic polyvinyl alcohol (Aldrich, 99% hydrolyzed, Mw=85,000-146,000) was dissolved in distilled/deionized water at 1 wt% concentration by heating to 90° C for two hours under stirring, then diluted to 10^{-3} wt% and cast onto freshly cleaved, muscovite mica from a droplet as the water slowly evaporated over several hours. High-force scanning abrasion revealed a variable film thickness of approximately 1-5 nm depending on surface location.

2.2 Scanning probe microcopy.

Nanoscope III/Multimode (Digital Instruments) PicoScan/PicoSPM (Molecular Imaging) scanning probe microscopes (SPMs) were operated in contact mode in ambient conditions for general purpose imaging, and the latter was employed for shear-modulation protocols. The X-modulation signal (100-500 Hz, 5-500 mV amplitude sinusoidal corresponding to 0.25-25 nm X movement) was generated by a NI-6110E I/O card (National Instruments) controlled by LabView, or by a Hewlett Packard 33120A, then added to the X scanning signal from the PicoScan with a home-built circuit. During some measurements, scanning across the surface was one-dimensional (Y) and the "scan frequency" setting for X determined the sampling frequency; data were acquired such that the "X" dimension of "images" was instead the time dimension. In other measurements during 1-second approach-retract Z cycles ("force curves") at point locations, both normal cantilever deflection and torsion (shear force) in raw units of volts were measured and the data processed in two ways: (1) calculating the root-mean-square deviation of torsion per period of cycling and tabulating against the average normal deflection; (2) parsing the torsional data into five-period subintervals and fast Fourier transforming each subinterval. Shear modulation measurements utilized a single diving-board shaped, uncoated silicon cantilever with nominal spring constant of 3 N/m and nominal tip radius of curvature of 10 nm (Nanosensors). Approach-retract curves extended to a maximum load of $\approx$10 nN. Y-scanned measurements were performed under constant applied loads of 10-20 nN.

3 RESULTS AND DISCUSSION

PVA films prepared as described above contained a complex of holes, exemplified in the 20x20 μm height image of Figure 1a. The holes were roughly one to several microns across, and 3-4 nm deep. Corresponding strong contrast in sliding friction force was observed, with lower friction force inside of holes (lower surface elevations). This was quantitatively probed by acquiring lateral force images during both trace (left-to-right) and retrace (right-to-left), then subtracting the two images following a several-pixel offset of X to account for piezoscanner hysteresis. The resulting "friction loop" image is shown in Figure 1b. The friction image is dominated by two principal levels of friction force, within holes (lower friction force) and outside of holes (higher friction force). The ratio of mean friction force is 2.0, meaning the higher topographic regions are twice as dissipative as the lower regions within the holes.

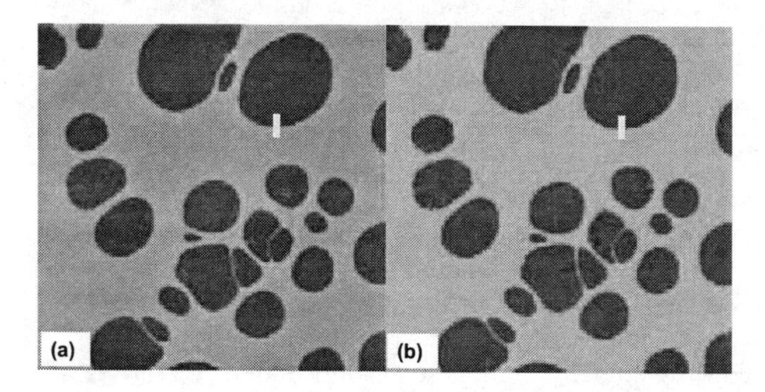

Figure 1. Topography (a) and friction-loop magnitude (b) images (20x20 μm) on an autophobically dewetted PVA film. Small vertical bars in (a) and (b) depict a typical domain for Y-scanned shear-modulation measurements.

High-force SPM scratching procedures at increasing loads sequentially revealed the selective removal of the higher surface regions, and ultimately the removal of a 1-nm thick first layer completely covering the substrate. These results indicate that *autophobic* dewetting, the phenomenon of a (usually polar) liquid not spreading on its own monolayer [5], occurred during film preparation. The reduction in sliding friction on the substrate-adsorbed layer ("Layer #1") is not as significant as found on highly crystalline microdomains (where chain packing strongly reduces dissipative motions) as characterized in other studies [6], but nonetheless implies reduced molecular freedom due to strong adsorption to substrate. The ultrathin geometry of this layer relative to the radius of gyration in the bulk (16.5 ± 2.5 nm) implies a high molecular asphericity, consistent with the strong polar interaction with mica. It is known that under strong substrate interaction, adsorbed chains have flattened, almost two-dimensional configurations. It is also theoretically expected that the substrate-adsorbed chains form a repulsive surface for chains not adsorbed to substrate, depleting chain density immediately atop the substrate-adsorbed layer (bright regions in Figure 1, "Layer 2"). These structural differences strongly impact dynamics, manifest here as differences in frictional response [7].

Time-dependent shear-modulated force response on Layers #1-2 was measured during Y-scanning of a small region as depicted in Figure 1, and over a set of drive amplitudes spanning from 0.25 nm to 25 nm. The results were Fourier analyzed and ratio of third to first harmonic calculated, as a gauge of nonlinearity due to partial sliding [4]. This ratio is plotted in Figure 2 as measured on Layer #1 and Layer #2. Each value derives from ≈10,000 analyzed cycles. An inflection point in A_3/A_1 is reached at drive amplitude of 1-2 nm on Layer #1 and 3-4 nm on Layer #2. It is within these intervals of drive amplitude that a characteristic transition from static- to kinetic-dominated friction takes place (static friction being significantly higher than kinetic friction). Above about 7-nm drive amplitude the ratio A_3/A_1 is surprisingly greater on Layer #2, albeit large in both cases signifying dominant sliding.

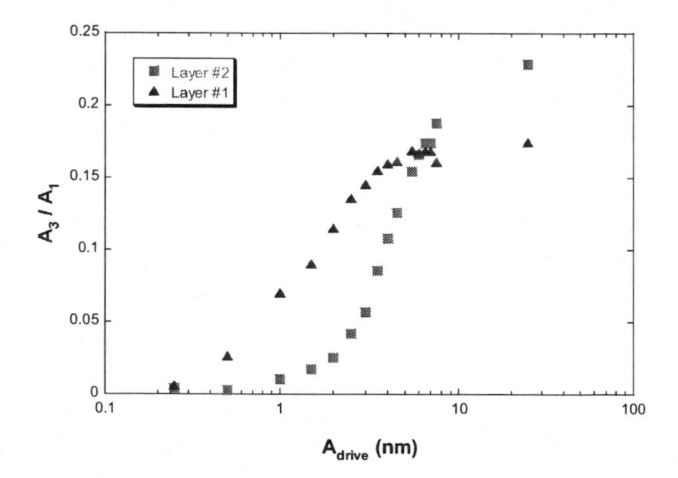

Figure 2. Fourier analysis of shear modulation measurements acquired over a set of drive amplitudes A_{drive} spanning from 0.25 nm to 25 nm. Plotted values are the ratio of the amplitudes of third to first harmonic, A_3/A_1.

That the transition from static to kinetic friction takes place at approximately 2:1 values of drive amplitude is highly suggestive. This finding implies that the differences of conformational mobility underlying a 2:1 ratio of static friction, or resistance to the onset of sliding, also result in a 2:1 ratio of kinetic or sliding friction. In turn this suggests that the ensemble of viscoelastic relaxations that comprise the sliding friction force [6] is similar to the ensemble of irreversible activation barriers that in aggregate must be overcome to initiate sliding.

Other measurements that help to assess stick-to-slide transitions, including the role of load, were obtained during approach-retract cycles. We present these results in Figure 3 in part to illustrate how our results relate to shear-modulation point measurements reported by others. The time dependence of shear and normal forces during an approach-retract cycle, as implemented by Wahl et al. [8], is shown.

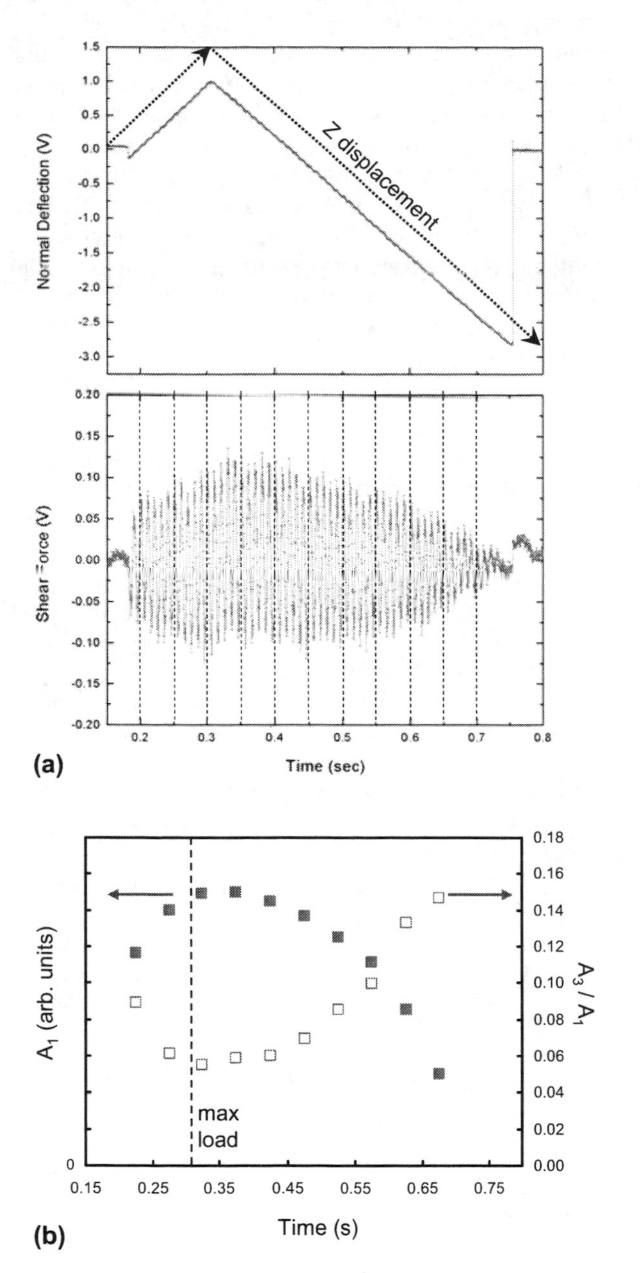

Figure 3. Time-domain response during an approach-retract cycle at a point on Layer #2. (a) Top: Vertical cantilever deflection (solid curve) with Z piezoscanner ramping motion depicted by dotted arrows. Bottom: Corresponding shear force response to sinusoidal X-displacement, subdivided into five-cycle time intervals for Fourier analysis. (b) Amplitude of first harmonic A_1 (closed squares) and ratio of amplitudes of third to first harmonic A_3/A_1 (open squares) determined from fast Fourier transforms of shear force response within each time interval denoted in bottom of (a).

Figure 3a contains representative raw shear force and normal (vertical) deflection signals acquired on Layer #2 under drive amplitude of 2.5 nm, spanning from compressive loads of ≈10 nN to tensile loads of ≈-30 nN just prior to pull-off. Figure 3b contains Fourier analysis of the shear force following the parsing of 50 oscillation periods into 10 subintervals as demarcated by vertical dashed lines in the lower graph of Figure 3a. The increase and subsequent decrease of the shear force amplitude that is apparent in Figure 3a (bottom) is more clearly represented by the plot of A_1 in Figure 3b. One finding is that the maximum in A_1 is reached later in time than the maximum in normal deflection (vertical dashed line) during Z ramping, as reported by Wahl et al. on polymeric systems [8] and attributed to the viscoelastic nature of tip-sample contact. A new finding in the present analysis is the corresponding evolution in A_3/A_1 seen during the approach-retract cycle: A_3/A_1 decreases to a minimum that coincides with the maximum in A_1, then increases up to break of adhesive contact in correspondence with a decrease in A_1. This result demonstrates, for the first time to our knowledge, that contact mechanics models may be naïve representations of what is actually occurring in many shear modulation studies on polymers (even if including viscoelasticity): a change from predominant stick to significant sliding depending on loading. Our results suggest that for many polymers, shear modulation measurements may be well-described by contact mechanics models only if the drive amplitude is in the sub-nanometer drive range, to study fixed contacts, or the tens-of-nanometer drive range to study sliding contacts.

Apart from the obvious difference of compression and tension, one also anticipates a change from weak stick under slight compression at low load (or low sample compliance) to strong stick under large compression at high load (or high sample compliance). Compliance-derived transitions may be induced for example by the glass-to-rubber transition whether temperature [3] or solvent (plasticization) induced.

Given that shear modulation microscopy studies generally have utilized the output of a lock-in amplifier, it is instructive to examine the load dependence of root-mean-squared (RMS) shear force response under 2.5-nm drive amplitude, Figure 4. Here each datum was computed from one cycle of real-time response, and is plotted versus the average value of vertical cantilever deflection (normal force, i.e. load) during the cycle. The data were collected during retraction from three point locations on each layer (triangles, Layer #1; squares, Layer #2). What is particularly notable on Layer #2 is the strong curvature of the data trends, compared to a nearly linear trend on Layer #1. Given the above identification of significant sliding under tension, we interpret the steep slope well into the tension regime (negative vertical deflections) as more characteristic of the coefficient of kinetic friction. The near-parallel nature of the data trends into the compressive regime (positive normal deflections) on both layers are misleading, because the mechanisms of tip-sample interaction are very different on the two layers: mainly stick on Layer #2, whereas substantial sliding on Layer #1. The plot of RMS amplitude versus load on Layer 2 actually exhibits an approximate 1/3 power law as predicted by static contact mechanics models, even though Fourier analysis clearly indicates variable degrees of sliding. This suggests that the load dependence of RMS amplitude as conventionally measured with a lock-in amplifier may be misleading: one must definitively establish a stick condition before interpreting (fractional) exponents in the load dependence of shear modulation amplitude, or fitting this load dependence with functional forms predicted by contact mechanics models.

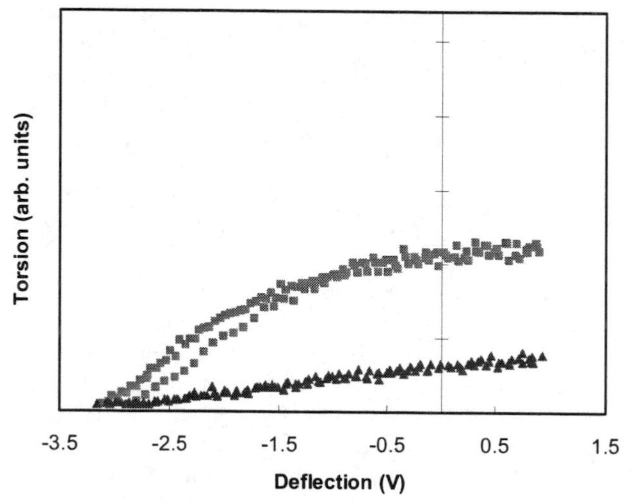

Figure 4. Root-mean-squared shear force response to sinusoidal shear displacement, versus vertical cantilever deflection during retraction from three locations on Layer #1 (triangles) and Layer #2 (squares) at 2.5-nm drive amplitude.

4 SUMMARY/CONCLUSIONS

Fourier analysis of spatially resolved, shear-modulated forces sheds new light on the transition from static to kinetic friction on polymers. The ratio of the amplitudes of third and first harmonic response is a sensitive gauge of the degree of sliding as a function of shear modulation amplitude. On autophobically dewetted polyvinyl alcohol film layers, this gauge identifies an approximate 2:1 ratio in the shear force needed to initiate sliding on second and first layers respectively, in correspondence with an approximate 2:1 ratio of sliding friction force as seen in conventional friction force microscopy. These differences apparently derive from conformational arrest in the strongly adsorbed first layer, as compared to more bulk-like mobility in the second layer.

Fourier analyzed response to a 2.5-nm shear modulated tip during approach-retract cycles reveals a change from a regime of primarily static friction at compressive loads of $\approx$10 nN to significant sliding at tensile loads of $\approx$-30 nN. This change of behavior impacts the functional relationship between RMS shear force response (as typically obtained with a lock-in amplifier) and applied load, rendering conventional contact mechanics models invalid. In using shear-modulated modes for mechanical and/or tribological studies, care should be exercised in choosing operating parameters so as to place the system in a well-defined regime (dominant stick or sliding), rather than a transitional regime where contrast interpretations and analytical modeling are problematic. More generally the interpretation of amplitude and/or phase images under variants of shear modulation including the recent "torsional resonance mode" [9] should be similarly impacted by the regime of operation.

REFERENCES

[1] H. –U. Krotil, E. Weilandt, Th. Stifter, O. Marti and S. Hild, Surf. Interface Anal. 27, 341, 1999.
[2] K. J. Wahl, S.V. Stepnow and W.N. Unertl, Trib. Lett. 5, 103, 1998.
[3] S. Ge, Y. Pu, W. Zhang, M. Rafailovich, J. Sokolov, C. Buenviaje, R. Buckmaster and R. M. Overney, Phys. Rev. Lett. 85, 2340, 2000.
[4] G. Haugstad, Trib. Lett. (in press).
[5] G. Reiter and R. Khanna, Langmuir 16, 6351, 2000.
[6] G. Haugstad, J.A. Hammerschmidt and W.L. Gladfelter, in Interfacial Properties on the Submicrometer Scale, J. Frommer and R.M. Overney, Editors. Oxford University Press: Washington, D.C (2001).
[7] J. Baschnagel, K. Binder, and A. Milchev, in Polymer Surfaces, Interfaces and Thin Films, A. Karim and S. Kumar, Editors. World Scientific Publishing: Singapore (2000).
[8] K. J. Wahl, S.V. Stepnowski and W.N. Unertl, Trib. Lett. 5, 103, 1998.
[9] T. Kasai et al., Nanotechnology 15, 731, 2004.

Simultaneous *T*opography and *REC*ognition Mapping with PicoTREC™: A Powerful New Technology That Can Be Used To Map Nanometer-Scale Molecular Binding Sites On A Variety Of Surfaces

W. T. Johnson*, G. Kada*, C. Stroh**, H. Gruber**, H. Wang***, F. Kienberger**, A. Ebner**, S. Lindsay***, and P. Hinterdorfer**

*Molecular Imaging, Tempe, AZ, USA, www.molec.com
**Johannes Kepler University of Linz, Linz, Austria, www.jku.at
***University of Arizona, Tempe, AZ, USA, www.asu.edu

ABSTRACT

PicoTREC™ is a powerful new chemical/biological detection technology that combines nanoscale *T*opographic imaging the AFM with single molecule *REC*ognition mapping. The technique has been used to map interactions between, for example charged species, antibodies/antigens, and ligands/receptors in order to analyze the chemical composition of a variety of samples. PicoTREC combines *in-situ* nanometer scale, resolution imaging with the selectivity and sensitivity of single molecule, piconewton sensitivity to resolve and detect recognition events. It is label-less, so it is not dependent on fluorescence, radioactivity, or enzyme-linked detection schemes. When a ligand for a particular receptor is attached to an AFM tip, the AFM tip becomes a chemically selective sensor so that the forces required to break chemical bonds with the specific binding epitopes on target molecules can be detected and resolved. Consequently, maps of binding sites across a variety surfaces can quickly and easily be obtained.

Keywords: AFM, biosensor, antibody-antigen, ligand-receptor, nanobiotechnology

1 INTRODUCTION

AFM offers a unique solution to study biological process at the nanometer scale. The development of the atomic force microscope (AFM) has allowed scientists and engineers to visualize, probe, and analyze the molecular structure of biological molecules and other substrates in their native environments with unprecedented resolution and without the need for rigorous sample preparation or labeling. AFM can be used to study the nanomechanical properties of a wide variety of biological, organic and inorganic samples, for example, adhesion, hardness, and elasticity of sample surfaces [1]. Nanoscale biological adhesion events affect a variety of important physiological phenomena, for example, DNA replication, RNA transcription, cellular growth and differentiation, tissue growth, the action of drugs, hormones, and toxic substances, to the performance of the immune system.

AFM is unique in its ability to detect piconewton single molecule interaction forces with nanoscale resolution [2]. This has opened the possibility of measuring inter- and intra-molecular forces of biomolecules on the single molecule level. TREC (*T*opography and *REC*ognition imaging) combines the high resolution imaging capability of AFM, with its ability to detect single molecule binding events on the piconewton scale [3,4]. Consequently, it is now possible to investigate interactions of single molecules with their specific receptors while simultaneously recording a high-resolution topography image. Here we will review the principles of TREC together with some applications to various biological systems.

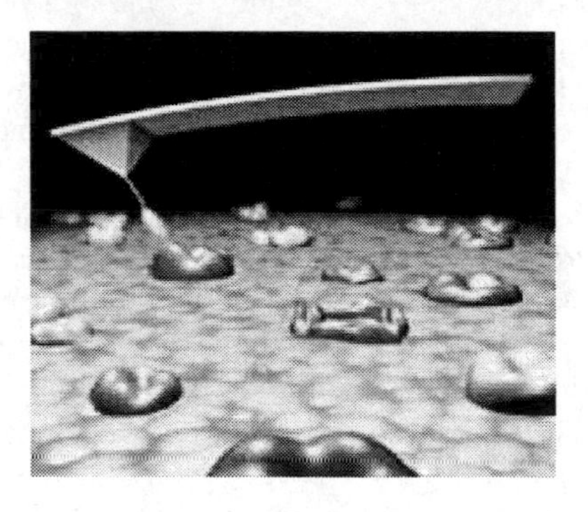

Figure 1: A specific ligand, attached to an AFM tip via a flexible PEG tether, binds to a target molecule.

2 WHAT IS PICOTREC?

PicoTREC is new technique in which sensor molecules are attached to the AFM tip on the end of an elastic polyethylene glycol (PEG) tether. This PEG tether gives the sensor molecule the freedom to reorient itself properly to bind to its target on the surface. When a recognition event occurs, the AFM detects a variety of signals. Included are the signals that correspond to the forces required to break hydrogen bonds and the weaker short range interactions that are involved in specific molecular binding events. PicoTREC resolves these signals and plots them separately. This technique is being detect the binding forces between antibodies and antigens, drug or hormones and their receptors, and protein-DNA complexes. When combined with MAC Mode™ AFM (Molecular Imaging, Tempe, AZ), entire maps of specific binding sites can be obtained in real time. PicoTREC records two separate images by scanning the tethered sensor molecule across the surface to detect binding events. One image provides the topography of the target molecules on the surface. The second image displays a map of specifically recognized target molecules on the surface. Using MAC Mode, PicoTREC, and optimized cantilevers, a ligand can be kept in close proximity to the surface, allowing efficient recognition and gentle interaction between tip and sample during scanning. PicoTREC resolves molecular recognition during the lateral scan by processing the asymmetric reduction of the oscillation amplitude. In this way, the locations of the target molecules are easily determined from their coordinates on the recognition image.

3 SINGLE-MOLECULE RECOGNITION STUDIES WITH AFM

Force-spectroscopy is a type of molecular recognition experiment that is often used to study and quantify binding interactions between single molecule on an AFM tip and molecules immobilized on a surface [5]. In this technique, an AFM tip, to which ligands are immobilized, is brought into contact with a surface that contains complimentary binding sites or receptors to the ligands on the AFM tip. As the molecules on the AFM and those on the surface come into contact, a ligand-receptor bond is formed (figure 1). Upon retraction of the AFM tip away from the surface, the ligand-receptor binding complex deforms until the hydrogen bonds and other attractive interactions which hold the complex together reaches a critical force and the complex dissociates or unbinds. By performing numerous unbinding experiments with the AFM, rate constants and affinity of binding for the recognition and binding events can be calculated [6]. Structural data of the binding pocket can also be inferred from force-spectroscopy studies. AFM tip sensitivity, tip-sensor design, and attachment chemistry are all critical factors in recognition experiments.

4 ATTACHMENT CHEMISTRY FOR RECOGNITION STUDIES

Various strategies have been used, with varying success, to bind ligands such antibodies, drugs, vitamins, hormones, and nucleic acids to AFM tips for recognition studies [7]. In most of the early AFM-spectroscopy studies, ligands were directly bound to the AFM tip surface. However, direct attachment of a ligand to this solid surface does not always allow the ligand to orient itself freely away from the relatively massive AFM tip, so direct attachment can severely constrain the ligand and, more importantly interfere with its ability to bind correctly to the site of interest. Consequently, direct attachment of ligands to AFM tips may not provide the most favorable environment for specific recognition events to take place. Due to the limitations of direct attachment of ligands to AFM tips in recognition studies, flexible polymer linkers, which provide space between the ligand and the surface of the tip have been developed and utilized with great success. In the unbound state, flexible linkers allow an unbound ligand to freely orient and diffuse within a certain volume. Therefore, the ligand is allowed to diffuse about much like it would if it were in solution and less constrained binding of a ligand to its receptor can be achieved [8]. Furthermore, flexible PEG linkers can stretch, allowing movement of the AFM tip to some extent without severely disrupting the binding interaction (figure 2).

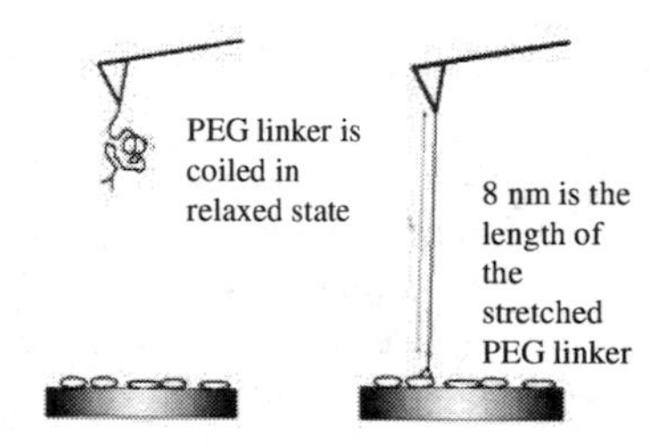

Figure 2: Flexible, heterobifunctional polyethylene glycol (PEG) linkers provide a means to attach ligands to AFM tips while permitting unconstrained binding to target molecules on the surface. The length of the linker is a critical parameter in TREC imaging as discussed below.

The PEG tether is a critical component for PicoTREC and recognition studies in general. In order to optimize the interactions between the ligand and it target, we use an elastic heterobifunctional polyethylene glycol (PEG) linker of ~8 nm length to attach the ligands to the AFM tip. Relatively short PEG linkers permit ligands on the AFM tip to diffuse within a defined volume. They impart additional degrees of freedom that are necessary so that the ligand can reorient itself to bind with an immobilized complimentary binding epitope in an optimal conformation.

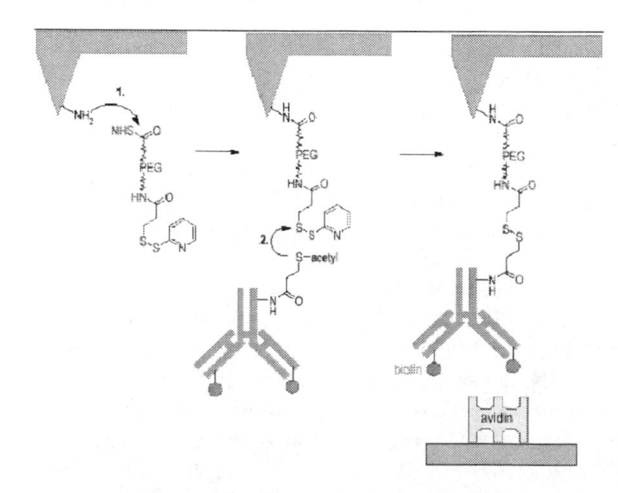

Figure 3: Immobilizing a specific ligand (biotinylated IgG) to an AFM tip and attaching a positively charged target (avidin) to negatively charged mica.

For AFM imaging, the target molecules also must be bound to a proper substrate. The substrate must be as flat and as smooth as possible so that the AFM tip that can resolve sample features over any features that are present on the substrate. The chemistry of the substrate must permit firm attachment of the target in order to prevent the target molecules from being pushed away by the AFM tip as it moves across the surface, but the attachment must not perturb the integrity and morphology of target. Also, in recognition studies, the target molecules must be adhered to the substrate more firmly than the specific bonds between the ligand and the target. If the target molecules are too loosely bound, they will simply be pulled off of the surface as the AFM tip is retracted. There are as many attachment chemistries are there are classes of samples. Mica is a negatively charged substrate, so if the target has sufficient positive charge (e.g., avidin, lysozyme, ferritin), then the target may be sufficiently immobilized by electrostatic binding using appropriate the buffers. Some molecules of interest in imaging and recognition studies are negatively charged (e.g., DNA, RNA, and certain proteins) and these molecules can often be immobilized to mica and imaged by adding, for example, positively charged Magnesium, Nickel, or ammonium ions to the attachment and imaging buffers [9]. Other biological molecules lack the ionic properties that are necessary for electrostatic immobilization and, therefore, they need to be covalently immobilized to the substrate. If the target is a protein with lysine residues on the surface, such as is the case for the histone proteins found in chromatin, the sample can be bound to mica that has been pretreated with APTES (aminopropyltriethoxysilane) and glutaraldehyde [10].

5 TOPOGRAPHY AND RECOGNITION IMAGING WITH PICOTREC

As noted previously, force-spectroscopy experiments can unveil important information about specific interactions between molecules, such as rate constants and affinity of binding, and it these experiments are particularly useful on pure samples of known identity. Unfortunately, most biological samples are composed of more than one component. Important data such as the sample's size, shape, and nanoscale morphology is also lacking in force-spectroscopy recognition experiments. Furthermore, force-spectroscopy experiments tend to be lengthy (generally taking hours to complete), which is a negative factor when dealing with biological samples, which quickly and easily denature and degrade. Simultaneous size and morphology information is also needed to localize the ligand-receptor interactions to particular areas. It is critical to image and characterize the components of biological mixtures before the samples have a chance to denature or degrade. Therefore, it is apparent that high speed topographical and recognition results are especially useful or even necessary in AFM studies on biological samples that are composed of mixtures of discrete components [12]. For mixed samples, protein-protein or DNA-protein complexes, and biological surfaces such as cells or membrane fragments, high-resolution topography imaging should be combined with high speed chemical recognition mapping, so that binding sites can properly be assigned to topographical features.

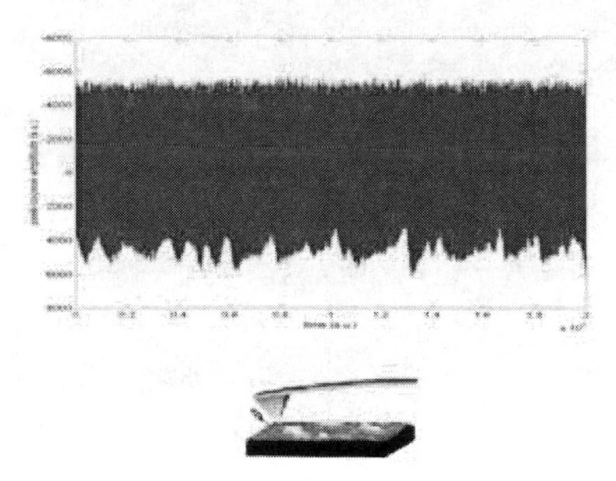

Figure 4A: MAC Mode signal in the absence of a recognition event

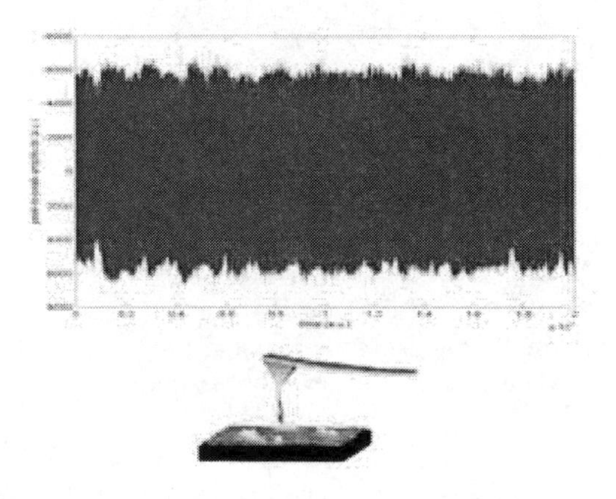

Figure 4B: MAC Mode signal in the presence of a recognition event

PicoTREC combines high resolution, fast, topographical information with specific recognition information. In PicoTREC, a magnetically coated AFM tip (MAC Mode Lever) is oscillated by an alternating magnetic field while being rastered over the sample. The magnetic field is generated by a magnetic coil that is placed either above or below the sample. The magnetically coated AFM tip comes in gentle contact with the sample, but only at the end of its downward movement, which reduces the contact time between the tip and the sample, and minimizes friction forces. As shown in figure 4A, this tip-surface interaction causes the amplitude of the oscillation to be reduced. The signal that arises from the oscillation is then processed into an image of the sample. Since the tip contacts the surface only intermittently, MAC Mode is an extremely gentle imaging tool that is useful for obtaining the highest resolution images possible of soft biological samples in their native environments (e.g., physiological pH) including individual protein molecules and membranes [11]. When specific ligands or antibodies, that are specific for surface-immobilized receptors, are attached to the AFM tip and the tip is oscillated and rastered across the sample surface, the amplitude of the oscillation is further modified. PicoTREC resolves the topography signal from the recognition signal to provide simultaneous Topography and RECognition (TREC) imaging. Figure 4B shows what happens when a MAC Mode AFM tip that has been derivatized with a ligand binds to a receptor immobilized on the surface. As the specific ligand molecules on the AFM tip bind to and then unbind from their targets on the surface, the unbinding force is detected in the top portion of the oscillation.

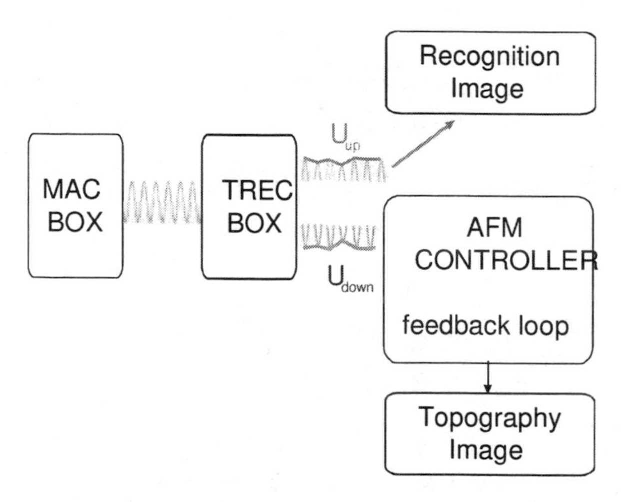

Figure 5: PicoTREC simultaneously records topography and recognition images using the maxima and minima of each sinusoidal cantilever oscillation

As shown in figure 5, the information from the unbinding event is resolved from the topographic information by PicoTREC. In this manner, molecules that bind to ligands on the AFM tip can be differentiated from molecules that do not bind.

The length of the PEG linker and the amplitude of MAC Mode oscillation (~5 nm) used with PicoTREC have been optimized and are extremely valuable components of the system. The amplitude is slightly smaller than the extended PEG tether length of ~8 nm (see figure 2) so that the ligand molecules on the tip can remain bound as the tip oscillates over the receptor-binding site. This results in high binding efficiencies and greater pixel densities in the recognition image. Figure 6 shows a TREC image of chromatin which was imaged with a MAC Mode AFM tip that was modified with a PEG linker and antihistone H3.

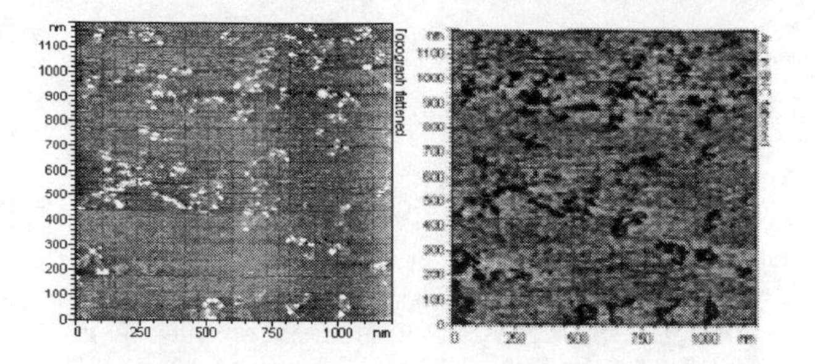

Figure 6: Simultaneous Topography (left) and RECognition (right) images of chromatin (histone-DNA complexes) using antihistone H3 modified AFM tips. Recognition sites are indicated by dark areas in the RECognition image.

6 SUMMARY

Simultaneous Topography and RECognition imaging has been demonstrated and proven to be a useful tool to identify molecules based on nanoscale topography and chemical composition. The technique has expanded AFM beyond basic imaging studies and quantifying molecular recognition events on mono-component samples to include pharmacology, toxicology, immunology, and the study of biological structure-function relationships on complex samples and mixtures of samples in real time. PicoTREC gives the atomic force microscope the ability to chemically distinguish between discrete molecular entities at the single-molecule level along with nanometer scale topographic imaging. The technique can be especially useful to analyze the components of heterogeneous samples and to resolve chemical information from the samples that can not be resolved from topographic images alone. PicoTREC has been demonstrated to be applicable to a variety of biological systems in their native environments (including antilysozyme-lysozyme, a variety of antihistone-chromatin interactions, and biotin-avidin).

7 REFERENCES

[1] Liu, Y. Z., *et al*: Langmuir 8547-8548 (1999)
[2] Hinterdorfer, P.: Handbook of Nanotechnology (Ed.: B. Bushan), Springer Verlag, Heidelberg, 475–494 (2004)
[3] Stroh C. M., *et al*: PNAS 101, (2004)
[4] Stroh C. M., *et al*: Biophys. J. (2004)
[5] Noy, A., *et al*: Annu. Rev. Mater. Sci. **27,** 381–421 (1997).
[6] Allison D. P., *et al*: Curr. Opinion Biotechn. 13, 47–51 (2002)
[7] Riener C., *et al*: Anal. Chim. Acta 497, 101–114 (2003)
[8] Kienberger F., *et al*: Single Mol. 1, 123–128 (2000)
[9] Han, W., *et al*: Nature. 563 (1997)
[10] Wang, H., *et al*: *Biophys. J.* **83,** 3619–3625 (2002)
[11] Kienberger F., *et al*: Biol. Proc. Online 6, 120–128 (2004)
[12] Raab A., *et al*: Nat. Biotechnol. 17, 901–905

Optimization of Nano-Machining With Focused Ion Beams

Lucille A. Giannuzzi, Paul Anzalone, and Daniel Phifer
FEI Company, 5350 NE Dawson Creek Drive, Hillsboro, OR 97124

ABSTRACT

FIB milled lines can be optimized for nano-machining with an understanding of ion-solid interactions. In particular, the incident angle of the ion beam with respect to the sample surface as well as the direction of milling can affect the shape and aspect ratio of the FIB milled cut.

Keywords: FIB, nano-machining, ion milling

1 INTRODUCTION

A DualBeam (DB) instrument consists of a focused ion beam (FIB) and scanning electron microscope (SEM) on the same platform. A 4k x 4k digital pattern generator can be utilized to automate precise movements of either the ion beam or the electron beam. FIB milling can be used for site-specific nano-machining of surfaces. In addition, by introducing a suitable precursor (e.g., organometallic gas) into the DB chamber, site specific metal lines can be produced via ion beam or electron beam assisted chemical vapor deposition processes. Many investigators have used DBs or FIB instruments to create nano-structures by either FIB milling or FIB/SEM deposition [1]. As will be shown below, the aspect ratio of a FIB milled line can be controlled by altering the incidence angle of the ion beam with respect to the sample surface through an understanding of ion-solid interactions.

2 EXPERIMENTAL TECHNIQUES

Lines of 1 μm in length were FIB milled into Si using 30 keV Ga^+ ions with a beam current of 100 pA at a nominal depth of 250 nm at either 52° or 0° incidence angle using an FEI Strata 400S DualBeam instrument. The lines were first filled in with electron beam deposited Pt first to protect the line, followed by FIB deposited Pt to facilitate cross-sectioning. The lines were cross-sectioned using conventional FIB methods [1]. The geometry of the lines were analyzed by SEM and compared with theoretical ion-solid interactions as shown below.

FIB milling is flexible enough such that either the ion beam dwell time, the depth of mill desired (i.e., time that the beam is on), and/or the number of beam passes can be defined. These parameters are interrelated and thus, changing one variable affects the others. In the next series of experiments, a 1000 pA ion beam was milled into Si at various incident angles at a set time (i.e., depth), using just one beam pass, but altering the start and end direction of the ion beam. These lines were FIB milled into Si using 30 keV Ga^+ ions with a beam current of 1000 pA at either 52° or 0° incidence angle using an FEI Quanta 200 3D DualBeam instrument using only a single beam pass.

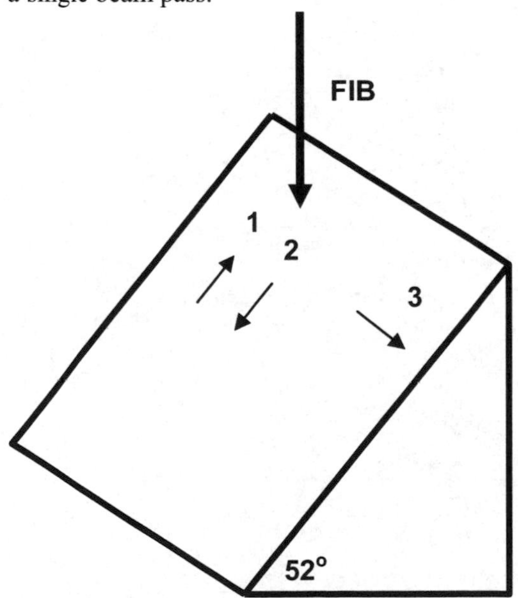

Figure 1: A schematic diagram of single pass lines FIB milled into Si.

A schematic diagram of the single ion beam passes performed at an incident ion angle of 52° is shown in figure 1. Each line was FIB milled either up the slope as shown by arrow 1, down the slope as shown by arrow 2, or across the slope, as indicated by arrow 3. The lines were then filled in with electron beam deposited Pt to protect the line, followed by FIB deposited Pt.

The lines were cross-sectioned using conventional FIB methods [1]. The geometry of each line was analyzed using an FEI Quanta 600 FEG SEM and compared with theoretical ion-solid interactions as presented below.

3 RESULTS AND DISCUSSION

Figure 2 shows results from TRIM simulations of ion trajectories of 30 keV Ga^+ ions into Si [2]. The images in figure 2 show the ion trajectories at 52° (top) and 0° (bottom) incidence angle. Note that the ion trajectories at 52° incidence angle are closer to the sample surface, resulting in a higher sputter yield (9 atoms/ion for 52° vs. 3 atoms/ion for 0°) and less damage to the surface. Since the sputter yield is higher at higher incidence angles, one would expect that, for all other FIB parameters kept constant, the ion beam would create a deeper cut, thereby creating a higher aspect ratio FIB milled line.

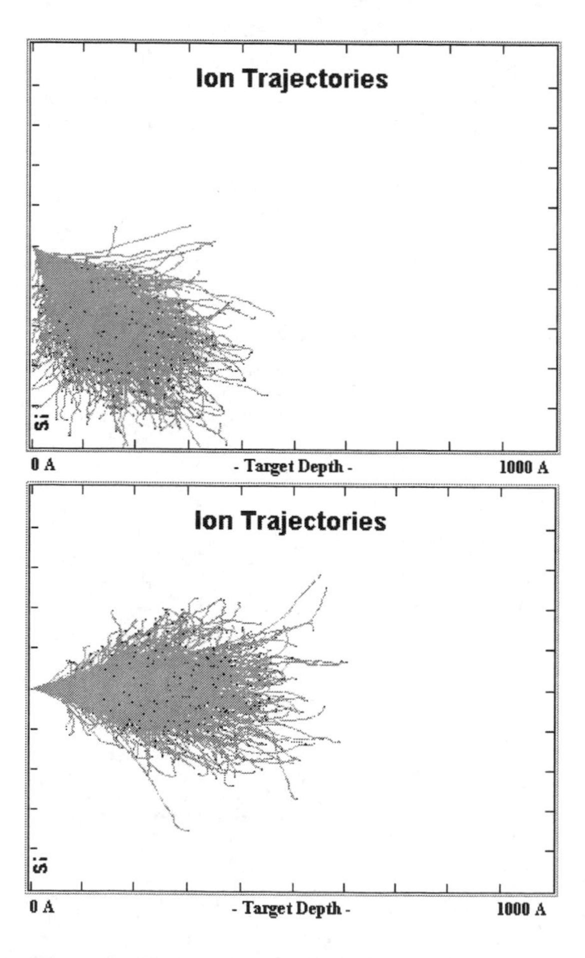

Figure 2: TRIM simulations of ion trajectories of 30 keV Ga into Si at (top) 52° incidence angle and (bottom) 0° incidence angle.

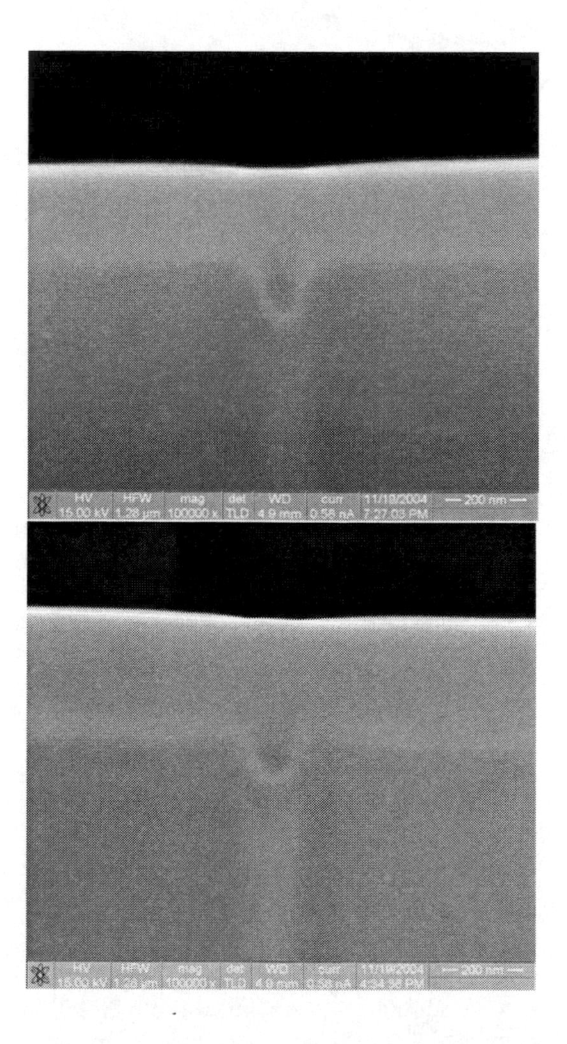

Figure 3: (top) FIB line milled at 52° incidence angle (DB stage tilt of 0°). (bottom) FIB line milled at 0° incidence angle (DB stage tilt of 52°).

NSTI-Nanotech 2005, www.nsti.org, ISBN 0-9767985-1-4 Vol. 2, 2005

Figure 3 shows SEM images of FIB milled cross-sections of FIB milled lines performed at 52° incidence angle (top) and 0° incidence angle (bottom). The differences in the aspect ratios of the FIB milled lines are evident in the SEM images in figure 3, where the 52° incidence angle cut shows a deeper cut with an overall improvement in the aspect ratio of the cut from ~ 2:1 to 3:1. Since different materials exhibit different collision cascade characteristics which also vary with incidence angle [3], it is expected that different FIB milled aspect ratios will vary with material as well as incidence angle that are consistent with ion-solid interaction theory.

A cross-section SEM image of a FIB line milled in a single line pass up a 52° slope is shown in figure 4 (top). A cross-section SEM image of a FIB line milled down a 52° slope is shown in figure 4 (middle), and a cross-section SEM image of a FIB line milled across the slope is shown in figure 4 (bottom). The dimensions shown in figure 4 have been corrected for SEM observation tilt angle.

Note that the FIB line milled up the slope (figure 4 top) is much shallower than the FIB line milled down the slope (figure 4 middle). This is consistent with the shape of the ion trajectories shown in figure 2. When the beam traverses up the slope, a new collision cascade is generated with each new position of the beam, and thus, the sputtered material is defined by the new position of each collision cascade.

However, when the beam traverses down the slope, each position of the beam overlaps a region devoid of material formerly removed by the previous collision cascade. Thus, milling down the slope occurs on an open sidewall, and the overall effect is that a deeper cut is produced when milling is performed down the slope of the inclined surface. In addition, the aspect ratio of the milled cut performed down the slope is greater than the aspect ratio of the milled cut performed up the slope. When a single beam traverses across the slope as shown by the SEM image in figure 4 (bottom), its leading edge angle is consistent with the incident angle defined by the ion-solid interactions. Milling on the trailing edge of the beam is diffuse since the side of the beam can interact with the sample surface, creating a larger hole opening in the top portion of the cut. Note also that the depth of the cut obtained across the slope (figure 4 bottom) is consistent with the depth of cut performed down the slope (figure 4 middle).

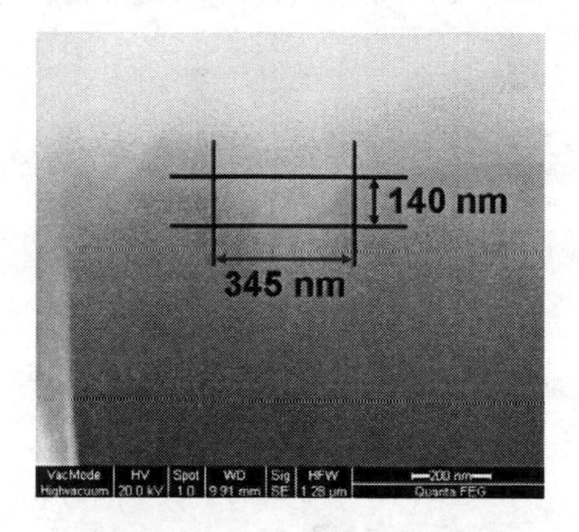

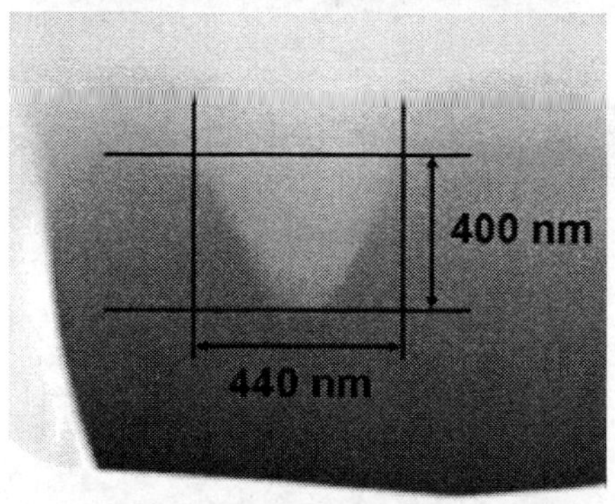

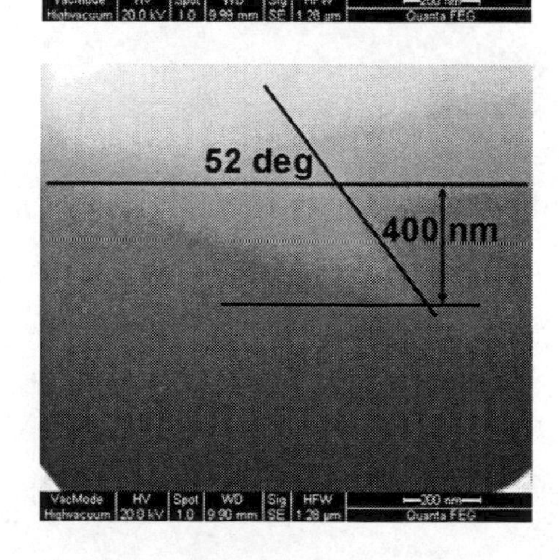

Figure 4: Lines FIB milled into Si at 52° incidence angle either, (top) up the slope, (middle) down the slope, or (bottom) across the slope.

4 CONCLUSIONS

Nano-machining can be accomplished using focused ion beams. The geometry of the FIB milled cuts can be altered consistently with ion-solid interaction theory.

REFERENCES

[1] "Introduction to Focused Ion Beams," eds. L.A. Giannuzzi and F.A. Stevie, Springer, NY, (2005).

[2] J.F. Ziegler, SRIM 2003, www.srim.com.

[3] B.I. Prenitzer et al., *Micros. Microanal.*, 9, 216-236, (2003).

Advanced Particle Beam Technologies for Nano Characterization and Fabrication

J.J.L. Mulders[*], J. Greiser[**]

[*]FEI Electron Optics, Building AAE, Achtseweg Noord 5
PO Box 80066, 5600 KA Eindhoven, The Netherlands, jjm@nl.feico.com
[**]FEI Electron Optics, Building AAE, Achtseweg Noord 5
PO Box 80066, 5600 KA Eindhoven, The Netherlands, jgreiser@nl.feico.com

ABSTRACT

Nanotechnology developments focus on materials and life science functionality at the very small scale, created by the local structure and interactions of the materials involved. Essential for understanding this local functionality is the visualization and the material-analysis of the nano-structures. In particular the 3-dimensional information at the nano and micro scale is an important way to characterize the material. This can be achieved by using high-resolution TEM tomography and by 3D slice and view applications of a DualBeam system, where its FIB column is used for milling and its SEM column for viewing and analysis. The DualBeam tool can also be used for the actual fabrication of nano scale structures by patterned ion beam milling, electron and ion beam induced depositions and electron beam lithography. The deposition technique offers direct deposition by control of a materials growth process from a gas phase and offers both high resolution and control of the growth in the third dimension.

Keywords: FIB, SEM, TEM, nano-charcterization, nano-scale production

1 INTRODUCTION

Since many years Transmission Electron Microscopes (TEM) and Scanning Electron Microscopes (SEM) have been used for high-resolution imaging and analysis of prepared, thin samples and of unprepared bulk samples respectively. Resolution improvements, low-dose techniques, adaptations to the sample's environment and addition of analytical capabilities have been recent changes for those instruments. As a result sample preparation for SEM has been minimized, the imaging capability for TEM has recently been extended down to the sub 1 Ångström level, and the availability of FIB capability has resulted in an new instrument where both the SEM and the FIB beam can be used in a combined way, allowing the full split between milling for sample modification on the one hand and non-destructive imaging and analysis on the other hand. Although many of the changes have been made due to the needs in the semi-conductor industry, they are generally applicable and fit very well into today's and near-future demands of research for nano technology. An interesting aspect of nano technology is the need for a multi-discipline approach having aspects of chemistry and physics and biology in addition to electronics that allows implementation of system control and measurement at the very small scale. An example can be found in the recent development of sensors with the direct aim to combine a minimum amount of material, to be analyzed with a maximum amount of accuracy or - in short - a higher specificity combined with a higher sensitivity. These objectives can only be achieved by manipulating material and building measurement systems at the nano scale.

2 LIFE-SCIENCE

2.1 Structural information

Over many years the structures in life-science have been studied with high resolution TEM, for a long time being the only technique allowing to visualize the high details of cell-based structures and mechanisms. As life-science materials are aqueous-based structures and electron microscopy is a high vacuum technique, sample preparation has been a key, for the application of the TEM on life-science materials. In addition, at very high magnification the electron dose applied to the sample is so large, that the structures can be destroyed by the imaging and analysis, and therefore special precautions have to be taken.

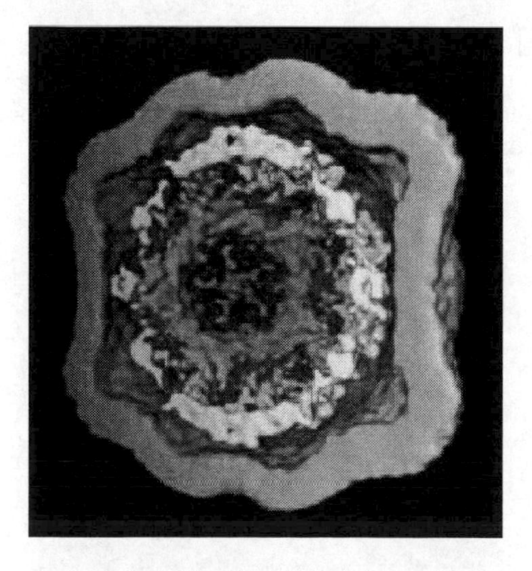

Figure 1: Cowpea mosaic virus: image obtained at low dose and at liquid He temperature

In practice this has been realized by keeping the sample at very low temperature (liquid Helium) with the electron density generally below 10 electrons / $Å^2$. An example of an image made under these conditions where the structure remains in-tact is given in figure 1.

As structures represent 3 dimensional information, the traditional 2D projection image form a TEM has been expanded to muti-angle recordings of digital images (typically 140), that are processed by Software to obtain a real, 3 Dimensional reconstruction of a volume of interest. Automation of this procedure on the TEM has allowed this technique to be applied routinely. In this way structures can be positioned in space and this helps for a better understanding of their actual function. An example of such a 3D reconstruction is given in Figure2.

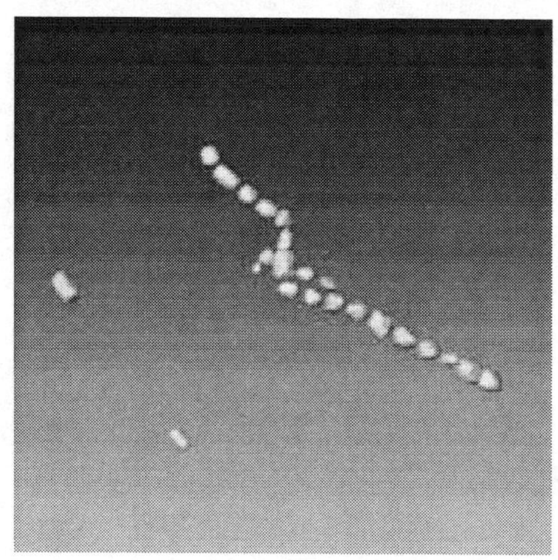

Figure 2 [1]: Visualization of small magnetite crystals in magnetotactic bacteria. This image is a part of a 3D dataset with the reconstructed volume, allowing any viewing angle.

2.2 Imaging in the natural state

The SEM technique has expanded to view wet samples in their natural state by adapting the vacuum conditions, to allow saturated water vapor in the chamber creating an equilibrium between the sample and its environment and hence preventing the sample from drying out.

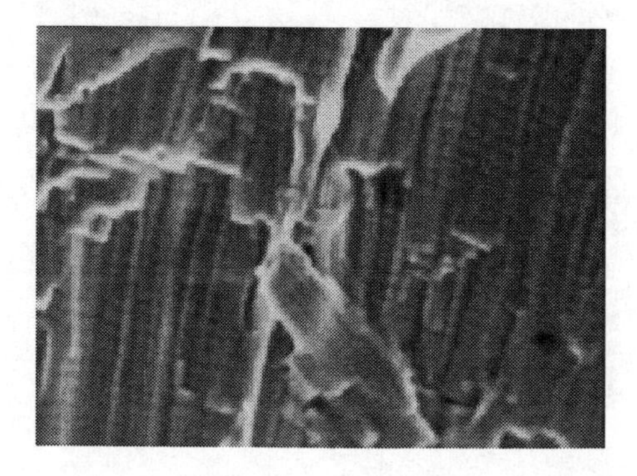

Figure 3: Muscle tissue in fully hydrated, natural state

Corresponding imaging techniques using dedicated detectors still allow the imaging of the sample surface under these conditions, using secondary electrons, hence showing topography. This technique is referred to as ESEM and has been implemented in most of today's FEI SEM instrumentation. An example of imaging under these sample environment conditions is given in figure 3.

2.3 Opening the 3D dimension of a bulk sample such as tissue or single particle

In addition to natural direct imaging using the ESEM technique, another way of preserving water containing samples is by quickly freezing them and, while frozen at a very low temperature of < 130 °C, create a fracture to open up the sample and view the inside of it. This is technique is known as cryo-SEM and it is applied to many life-science samples, but also to related materials, such as food and cosmetic products. An example of this is given in figure 4.

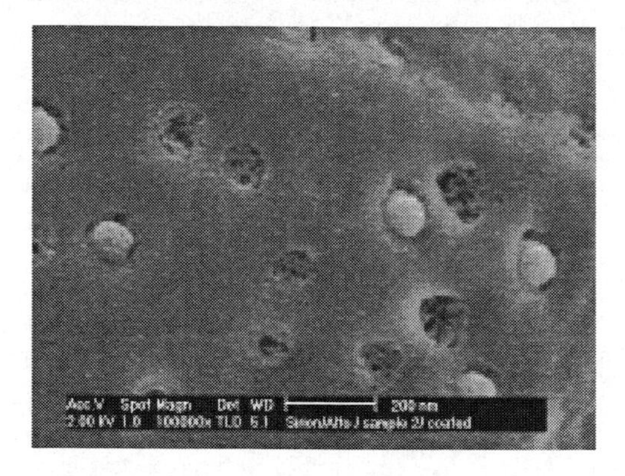

Figure 4: Cell membrane structure, generated by cryo fracture of the plunge frozen sample

By applying a FIB column from a DualBeam instrument, for the creation of a cross-section, it is even possible to cut through a very specific region of interest and hence reveal its internal structure, precisely at the area of interest. An example of a cut through a single Lacto Bacillus bacterium is given in Figure 5.

Figure 5: FIB cross-section through a selected site of a single, flash-frozen bacterium

3 MATERIAL SCIENCE

The electron and ion beam based techniques offer suitable analytical capabilities for materials at the micro and nano scale. Not only topographical information but also elemental distributions, crystallographic information and even magnetic information can be obtained with these instruments. In addition to 2 dimensional distributions, are replaced by three-dimensional information for a more fundamental description of the material.

3.1 Structural information

The latest developments in TEM have resulted in a break-through in resolution, where the 1 Ångström information limit has been exceeded. Modern electron optic lens design, correcting for higher order aberrations in combination with low beam energy spread, have allowed to pass this barrier and as a consequence atomic stacking in crystals can be shown with excellent clarity. Grain borders and dislocations can now be studied in more detail and with low uncertainty. An example of this achievement is given in figure 6, where the technique is applied to show the grain boundary of a gold crystal.

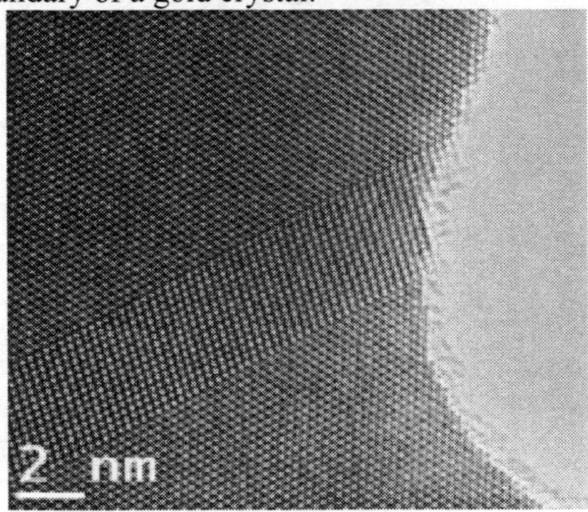

Figure 6: Image of a gold crystal showing the edge and the very clear separation on the twin grain boundary. Atomic structures shown at the boundary (some Moiree effect is introduced by the .pdf conversion).

3.2 Three dimensional image and analysis

The FIB capability to mill away material can be expanded to successive slicing in the material and subsequent imaging of the newly revealed surface. This automated process, referred to as slice an view will then allow the creation of a three dimensional volume of the internal structure of the material. This be done with images, but also with analysis data of the material and hence a 3D data set of a volume is created. For example, a materials crystallography can be determined using EBSD mapping on each, newly revealed slice, create by FIB. In this way a three dimensional crystallographic volume is visualized, and available for statistical analysis of the voxels.

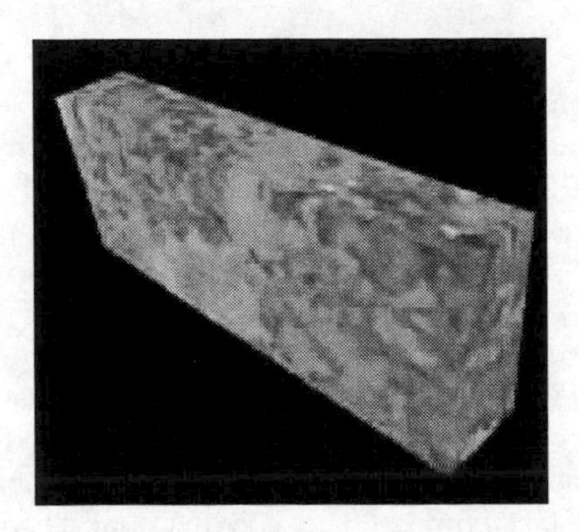

Figure 7: DualBeam 3D analysis of a 3 x 1 um volume of steel. Each color represents an orientation of a steel grain. Data suited for statistical crystallographic analysis.

4 CREATION OF STRUCTURES

Electron and ion beam equipment cannot only be applied for imaging and analysis at nano and micro-scale, but they can also be used to create structures at this level.

4.1 Pattern transfer with FIB

As the focused ion beam has enough energy and momentum transfer to remove material, by adding a patterning capability, it is possible to remove material in a very well controlled way and with a controllable depth. This process can be purified and / or made more efficient, but the local addition of etching gases close to the sample. Accuracy of this process and its material independency have allowed to apply this technique over a wide range of materials. Patterning can be applied to simple, repetitive structures or to more complex patterns such as controlled by a color-coded bit map, where intensities and pixels are translated into dwell times and actual beam positions. The technique can be applied for example in the creation process of a photonic array, where the regularity and the small size of holes milled in the material, are crucial for the final functionality.

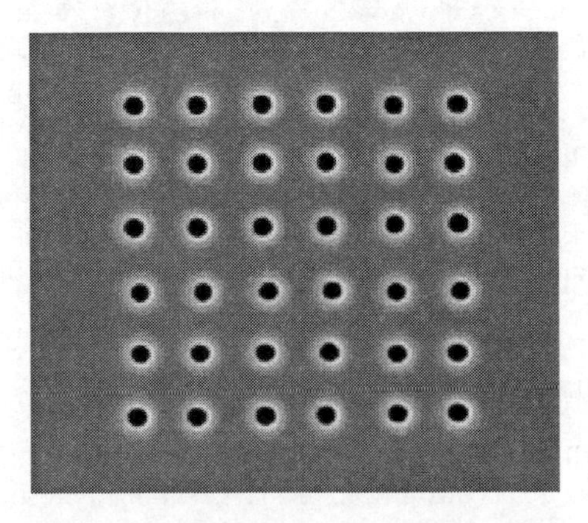

Figure 8: Matrix of 150 nm holes in InP, directly created by milling with the ion beam.

4.2 TEM sample preparation

As the ion beam can be so well controlled, it is also ideal for generating a very small, thin slice of material - thin enough to make it electron transparent and this at a location that is of interest to the researcher. In other words, the ion beam can be applied to create a site-specific TEM lamella that can be studied in a high resolution TEM. Because the ion beam milling process is quite independent of the material type, it is ideal to make TEM samples of interfaces between hard and soft material, such as for metal / polymer interfaces. The process is highly automated, accurate and fast compared to conventional TEM sample preparation techniques, that are not site specific and may suffer from a mix-up of the materials during the preparation and hence destruction of the actual interface of interest.

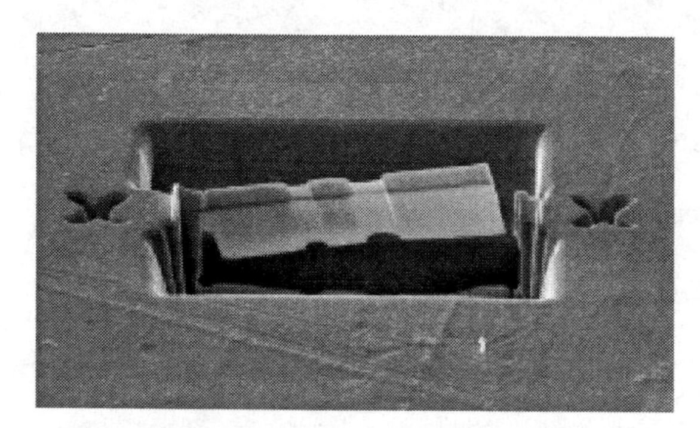

Figure 9: Ion beam created TEM sample in a coated polymer, ready for lift-out and transfer to the TEM

With this technique it now becomes possible to generate many TEM samples from a small area of the bulk sample, and hence ensure statistical validity of the analysis.

4.3 Patterning with the electron beam

Electron beam lithography using the electron beam from the SEM is a well-established technique to transfer a (CAD) pattern into resist and later into for example in a metal film. Also the electron column from a DualBeam is suited to do this and structures down to 20 nm can easily be created this way. Of course, resist processing is an essential part of this technique but there are many capabilities using high resolution or duo-layer resist that is thin enough to create very small structures. The stability of the FEG based equipment allows exposure over a longer period of time and using back-ground patterned (gold) structures, multi-field stitching and layer to layer alignment become possible as well.

4.4 Direct deposition techniques

With the ion beam or the electron beam it is possible to decompose a gaseous precursor that is supplied close to the beam impact area. This technique referred to as Electron Beam (or Ion Beam) Induced deposition allows the creation of very small structures by patterning and direct deposition of for example a metal, that is released from the gaseous organo-metallic vapor. Indeed small structures can be created with this as is shown in figure 10 and 11.

Figure 10: A 30 nm diameter Cobalt tip grown with electron beam induced deposition, on top of a diamond needle, to create a magnetic tip with very small apex radius for high resolution MFM measurements.

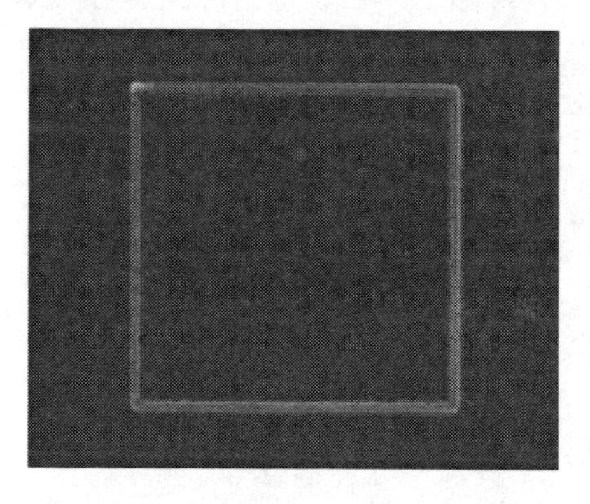

Figure 11: Rectangular frame of directly deposited gold. The frame is 500 nm wide. Line width and height 15 nm

5 ENABLING TECHNOLOGY

The capabilities of modern TEM, SEM, FIB and DualBeam instrumentation is still expanding and focus is given on both technological improvements, but clearly also on making this new technology accessible to the user by very extensive software control still condensed into an easy-to-use graphical user interface. Current limits will be pushed, new ideas implemented and working together with researchers in nano technology will help to implement the most relevant new functions in a user friendly way. In this way FEI's technology is now and will stay an enabling technology for the researchers in nano technology.

IMAGE COURTESEY

[1] M. Weyland, R. Dunin-Borkowsk, P. Midgley - Cambridge University and M. Otten, FEI Company

An Evaluation Of A Scanning Mobility Particles Sizer With NIST-Traceable Particle Size Standards

J. Vasiliou

Duke Scientific Corporation, Palo Alto, CA USA, vasiliou@dukesci.com

ABSTRACT

A scanning mobility particle sizer (SMPS -- TSI Model 3936-Series) was evaluated using Duke Scientific NIST-traceable particle size standards and Standard Reference Materials from the National Institute of Standards and Technology (NIST SRM's). The importance of instrument setup, electrospray operation and sample preparation for polystyrene spheres are discussed as well as the results from 14 different size reference standards. Correlations between the SMPS system and established electron microscopy and dynamic light scatting methods are also shown in tabular and graphical forms. Results show that with proper operation, the SMPS results fall within the uncertainty of the NIST traceable diameters in the range that was evaluated — 20 to 100 nanometers.

Keywords: particle, standard, size, nist, smps

1 INTRODUCTION

A Scanning Mobility Particle Sizer (SMPS) manufactured by TSI incorporated is used for sizing particles from 5 nanometers (nm) to 1 micron (μm) in size. The entire sizing system can be made up from many different components depending on the end-user's requirements. The three basic parts to the system include an aerosol nebulizer, a differential mobility analyzer (DMA) and a condensation particle counter (CPC).

Duke Scientific's interest lies in the smaller size ranges, therefore, we have selected the components that optimize the precision and accuracy of the measurement data in the smallest range. Following is a short description of the components used in this evaluation:

- TSI Model 3480 Electrospray Aerosol Generator
  2-100nm particle size range
- TSI Model 3085 Nano DMA
  2-150nm particle size range
- TSI Model 3081 Long DMA
  10-1000nm particle size range
- TSI Model 3025 Ultrafine CPC
  Concentration range from 0 to ~$1x10^5$#/mL

This report describes some of the instrument operating conditions that optimize the performance of the SMPS system when measuring polystyrene microspheres. It discusses sample preparation and data collection. The SMPS system was also evaluated using NIST traceable particle size standards and the results are compared to other methods including transmission electron microscopy (TEM) and Dynamic Light Scattering.

2 EXPERIMENTAL SETUP

2.1 Electrospray

Almost all commercial particles below one micron are packaged in an aqueous suspension. To suspend these particles into air requires some type of nebulizer. The electrospray generates an aerosol through a combination of a pressure differential and an electric field. The instrument runs on a mixture of CO_2 and air. The gases are filtered and dried. A small (1.5 mL) vial of a mixture of the suspension and an electrolyte is placed into the electrospray chamber and pressurized. The liquid travels through a capillary under approximately 3 PSIG of positive pressure and an electric field. At the other end of the capillary the droplets form a cone-jet that creates a uniform distribution of fine liquid droplets. The cone-jet is controlled by varying the strength of the electric field. The droplets evaporate almost immediately and the non-volatile material inside the droplets remain. These particles are highly charged and must be neutralized before they can be used by the DMA. A radioactive source of ions (Polonium-210) is used to bring the particles to a neutral state called Boltzman Equilibrium. By the time the aerosol has left the electrospray, it is dry and neutrally charged.

2.2 Sample Preparation

Small polystyrene latex microspheres can be easily nebulized in the electrospray. The typical sizes range from 150 nanometers and below. It is possible to generate droplets as large as 500 nanometers in diameter including the latex particle. However, as the particle size increases, so does the chance of forming partial or complete blocks of the capillary tube and, in most instances, ruining the capillary.

In general, from 20 μl up to one or two drops of a standard 1% suspension of particles in the standard buffer solution (1.5 mL of 20 mM ammonium acetate) will suffice for running on the SMPS system. However, there are some added steps to the sample preparation that may give better results.

2.3 Surfactant Removal From The Sample

Surfactant exists in almost all general purpose, aqueous-suspended, polystyrene products. The surfactant is usually negatively charged, and is added to stabilize the small particles and keep them from agglomerating. The amount of surfactant added varies quite a bit and is usually also a function of particle size. The smaller particles require more surfactant to remain stable.

The presence of surfactant is generally not a concern for most applications, however, in aerosol applications it can cause problems. The surfactant is non-volatile, so when the droplets exiting the electrospray evaporate, there will be two possible results. First, any liquid droplets that do not contain polystyrene particles will evaporate, leaving a small surfactant particle. Second, any liquid droplets that do contain a polystyrene particle will evaporate and leave a surfactant shell around the PSL particles, thus, increasing their size. Whether the surfactant will actually affect the particle size measurement is a function of the surfactant concentration, particle size, sample concentration, and electrospray droplet output.

Anionic surfactant can be removed with a procedure called ion-exchange. The method is simple and effective and can be conducted on small samples. Our procedure is outlined below:

1. *Obtain some ion-exchange resin (we have used Bio-Rad AG501-X8)*
2. *For a 15 mL bottle of particles at 1% solids use 3 to 4 grams of resin*
3. *Wash the resin thoroughly to remove potential contaminants*
    a. *Wash resin with five portions of 200 mL DI water*
    b. *Allow the resin to settle, and pour off the water*
4. *Add the particle suspension to the resin in a small bottle. You can add extra water if needed.*
5. *Roll the mixture for 4 to 6 hours and filter through washed glass wool to remove the resin*
6. *Alternatively you can let the resin settle and you can pour off the suspension into another clean bottle.*
7. *The suspension should be surfactant free and ready to use.*

For applications involving the electrospray, just 1 mL of a suspension can be ion-exchanged and diluted with water to form 15 mL's of liquid. Generally, a 1% solution of particles below 100 nm is too concentrated, so a 15:1 dilution is reasonable and saves the other 14 mL of particles in a more stable form. Ion-exchanged particles can be unstable and shouldn't be stored for more than a few days.

It is also important to note that the surfactant shell around small particles can artificially increase the apparent size of the PSL particles. Usually this contribution is insignificant, but if the surfactant concentration is high or the PSL size is very small, it may be worth taking the time to ion-exchange the sample to avoid this problem.

One last note about sample preparation concerns particle stability. Generally, PSL particles have been found to be stable in a 20 mM ammonium acetate buffer solution, but there have been instances where the particles immediately flocculate to the point of settling out of suspension. It is important to test a small amount of the sample in the standard 1.5 mL container with the buffer solution. After shaking the suspension, there should not be any visible inhomogeneity in the solution. This simple test can avoid having to replace the capillary due to clogging. Flocculated material will immediately, and in almost all cases, permanently clog the capillary.

2.4 DMA And CPC Instruments

Both the 3080 Electrostatic Classifier and the 3025 Ultrafine CPC were operated according to the instrument manual. In all particle analysis below 60 nanometers, the nanoDMA was used in conjunction with the 3080 classifier. For particles larger than 60 nanometers, the long DMA was used. The flow rates on the 3080 were verified using a Gilibrator bubble flow calibrator. The voltage regulator was assumed to be correct. In all scans, the sheath flow rate was as high as allowable—generally between a 10:1 and 12:1 sheath/aerosol flow ratios.

The CPC was operated in its high flow mode of 1.5 Lpm of aerosol. With the electrospray output of approximately 1.1 Lpm, 0.4 Lpm of make-up air was provided upstream of the DMA. This corresponds to running the SMPS system in a slight underpressure mode as outlined in the instrument manuals.

3 DATA COLLECTION

Outside of the Electrospray stability tests, all data was collected using TSI's Aerosol Instrument Manager (AIM) version 4.3. This software computes a particle's mean size and geometric distribution by rapidly stepping up the voltage across the DMA from 1 to 10,000 volts. The software uses algorithms to convert from a voltage to a particle size taking into account all of the variables associated with the DMA transfer function and operating parameters.

Scan times ranged from 120 seconds to 300 seconds for an upscan from 1 to 10,000 volts. A 15 second downscan (from 10,000 to 1 volt) is also performed with this software, but the data is not used. The data is corrected for multiple charged particles and the information is presented graphically and in table format.

Tabular data including particle diameter and number % were extracted from the AIM program and copied into an Excel worksheet. This data was incorporated into a statistics worksheet to allow comparisons between different sizing methods that measured different moment weighted diameters.

4 RESULTS AND DISCUSSION

Thirteen NIST traceable particle size standards and one NIST SRM were evaluated on the SMPS system. The results were in good agreement with the TEM measurements as well as measurements made with Dynamic Light Scattering (DLS) Instruments. DLS instruments report an intensity weighted mean diameter, so the SMPS number mean values were converted to intensity weighted values for comparison purposes.

Figure 1 (on the following page) shows excellent agreement between the SMPS measured diameter and the reference standards. Horizontal error bars indicate the uncertainty of the reference standard measurement. In all cases, the SMPS mean diameter (or calculated intensity weighted mean diameter) fell within the uncertainty of the reference standard.

An analysis of the width of the distribution for each particle size has been omitted from this paper due to complications arising from the different measurement techniques. Actual mean diameters calculated from 100% of the SMPS data will not correctly correlate with TEM data for the following reasons: TEM data is difficult to obtain below 40 nanometers, so a low-end tail of a distribution will not be seen with TEM data; surfactant crystals and multiply-charged doublets can cause extra counts on the low side of a distribution in the SMPS analysis. For the SMPS values listed above, the mean diameter was calculated using as range similar to, or slightly larger than, the TEM measurements. Obvious outliers and multiply charged particles were also excluded. In general, it seems that the SMPS system tends to broaden the distribution very slightly, although that may have more to do with the limitations of the software rather than the physical instrument.

5 CONCLUSION

The SMPS system can give very accurate and repeatable results if it is operating correctly. In our setup, the Electrospray was the most critical component for achieving accurate results. A newer and cheaper capillary design from the manufacturer has allowed the electrospray to provide a constant aerosol output over most ordinary scan times. Clean, unobstructed capillaries can deliver consistent concentrations of polystyrene particles for an accurate measurement. In addition, difficulties due to surfactant and additives can be eliminated through the use of dilution or ion-exchange methods.

Dynamic Light Scattering, Transmission Electron Microscopy and the SMPS methods all correlate very well. The standard AIM software with a 300 second scan was used for the SMPS method. Better resolution can be obtained using a manual scan, but that is not within the scope of this paper. It is clear from the results in this paper that a properly operated and maintained SMPS instrument can accurately and precisely measure small particles with a reliability similar to other established methods.

REFERENCES

[1] Aerosol Instrument Manager® Software Ver. 4.3, TSI Incorporated, St. Paul, MN.

[2] Hinds, W. C. (1982). *Aerosol Technology: Properties, Behaviour, and Measurement of Airborne Particles.* John Wiley & Sons, inc., New York.

[3] TSI Incorporated. (2000). *Model 3080 electrostatic classifier: instruction manual (revision C).* St. Paul, MN: TSI, Inc.

[4] TSI Incorporated. (2001). *Model 3936 SMPS (scanning mobility particle sizer: instruction manual (revision F).* St. Paul, MN: TSI, Inc.

[5] TSI Incorporated. (2000). *Model 3025A ultrafine condensation particle counter: instruction manual (revision H).* St. Paul, MN: TSI, Inc.

[6] TSI Incorporated. (2001). *Aerosol instrument manager software: instruction manual (revision A).* St. Paul, MN: TSI, Inc.

[7] TSI Incorporated. (2000). *Model 3480 electrospray aerosol generator: instruction manual (revision A).* St. Paul, MN: TSI, Inc.

Catalog PN Tested	TEM Diameter* [nm]	SMPS Diameter [nm]	SMPS Intensity** Weighted [nm]	DLS/PCS Diameter [nm]
3020A	N/A	14.5 nm	21 nm	21 nm ± 1.5 nm
3030A	N/A	23.0 nm	33nm	33 nm ± 1.4 nm
3040A	N/A	30.2 nm	41 nm	41 nm ± 1.8 nm
PD-047	47 nm ± 2 nm	46 nm	49 nm	50 nm
3050A	50 nm ± 2.0 nm	51 nm	54 nm	54 nm
3060A	60 nm ± 2.5 nm	58 nm	62 nm	64 nm
PD-064	64 nm ± 2 nm	63 nm	65 nm	64 nm
3070A	73 nm ± 2.6 nm	73 nm	76 nm	76 nm
PD-080	80 nm ± 5 nm	82 nm	85 nm	83 nm
3080A	83 nm ± 2.7 nm	81 nm	83 nm	86 nm
PD-083	83 nm ± 2 nm	83 nm	85 nm	84 nm
3090A	96 nm ± 3.1 nm	97 nm	100 nm	97 nm
PD-100	100 nm ± 5 nm	100 nm	101 nm	102 nm
NIST1963	100.7 nm ± 1 nm	100.2 nm	101 nm	101 nm

Table 1: SMPS Mean Diameters vs. Reference Standards

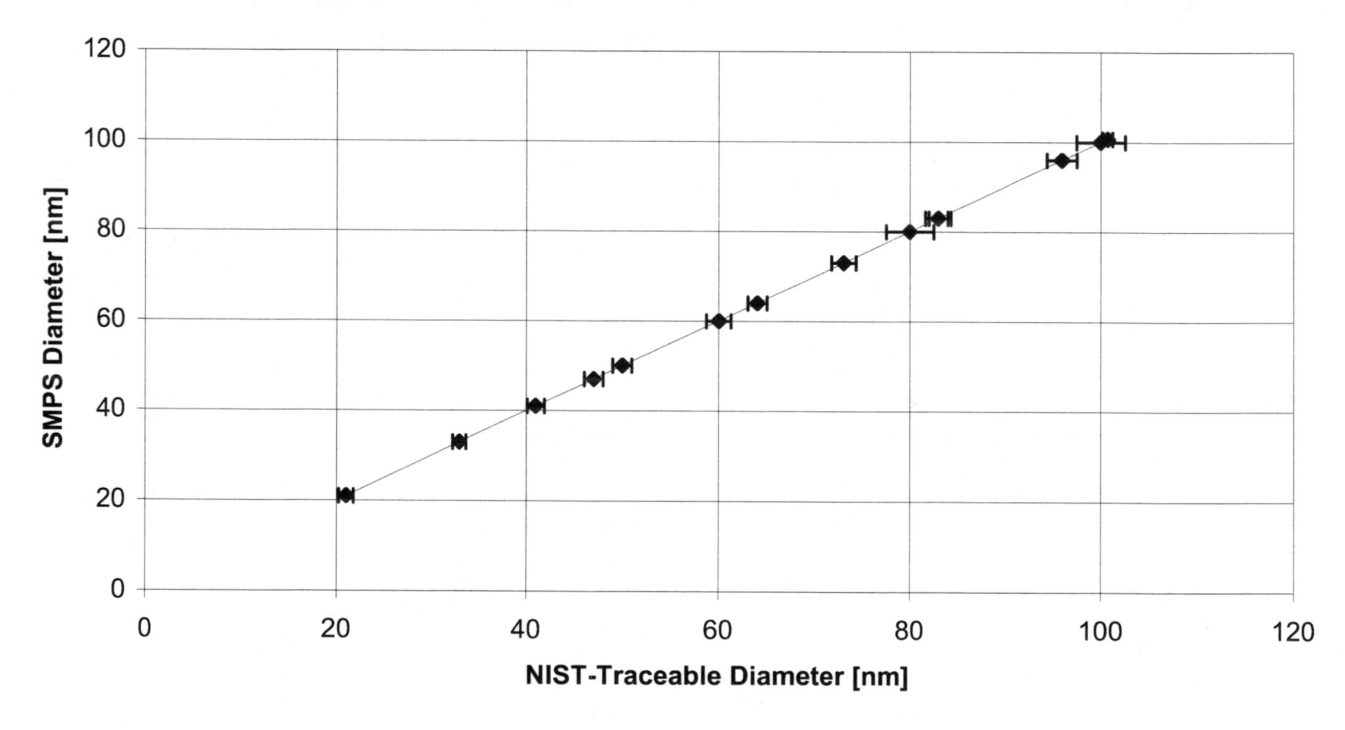

Figure 1: SMPS Mean Diameters vs. Reference Standards

NSTI-Nanotech 2005, www.nsti.org, ISBN 0-9767985-1-4 Vol. 2, 2005

Integrated optical profiler and AFM : a 3D metrology system for nanotechnology

M. Jobin[1,2], A. Du [1] and R. Foschia[2]

[1]Geneva School of Engineering (EIG), University of Applied Sciences
4 rue de la Prairie, 1202 Genève, Switzerland, jobin@eig.unige.ch
[2] NanoFeel, 4 Tir-au-Canon, 1227 Carouge, Switzerland rfoschia@nanofeel.com

ABSTRACT

A white light interference optical profiler (WL-IOP) operating in the vertical scanning interferometry (VSI) mode has been home-made developed and integrated into a commercial Atomic Force Microscope (AFM). The result is a 3D metrological tool operating with sub-nanometer resolution vertically over a wide range laterally: from 1mm down to 10nm. This is a faster and in many case a higher performance alternative to the more standard stylus profiler / AFM combination.

Keywords: 3D Metrology, White-light interference Microscopy, Atomic Force Microscopy

1 INTRODUCTION

The ability to perform accurate metrology measurement at the nanometer scale is a requirement in many nanotechnology applications, particularly in semiconductors and MEMS industries. Common 3D metrology tools widely used in those industries includes stylus profilometry, optical profilometry and closed-loop atomic force microscopy (AFM). The later has the advantage of a very good resolution, both laterally (2nm) and vertically (0.1nm), which makes AFM the choice instrument for analyzing defects or structures at high resolution. A drawback of AFM is its speed of operation, in terms of analyzed sample surface per analysis time. In addition, the field of view per image is limited, usually 100um or below. Therefore, it is natural to combine AFM with a larger field of view metrology system in order to first quickly localize the area of interest (e.g. defects) and then to measure it with AFM.

The most common choice along this line is to combine AFM with stylus profiler, probably because the two techniques compares easily (an AFM can be viewed as a high resolution mechanical profiler) and also because the mechanical integration is quite straightforward. However, the combination with optical profilometry might be a better solution in most applications, since it is non-contact (completely non-invasive) and faster compared to stylus profilometry.

White-light interference microscopy [1] (also referred to as low coherence microscopy) is a technique using an interference objective, a white-light illumination and a piezoelectric positioning mechanism allowing to modify the sample /objective distance with nanometer accuracy. The technique has a vertical range in principle unlimited (in practice it is limited by the vertical scanning stage on which the objective or the sample is mounted) and a sub-nanometer resolution in the vertical axis.

2 EXPERIMENTAL SETUP

2.1 Mechanical and Optical setup

We chose to design our own white light interference profiler and to integrate it into a commercial AFM [2]. The general set-up of the instrument is shown in Fig 1.

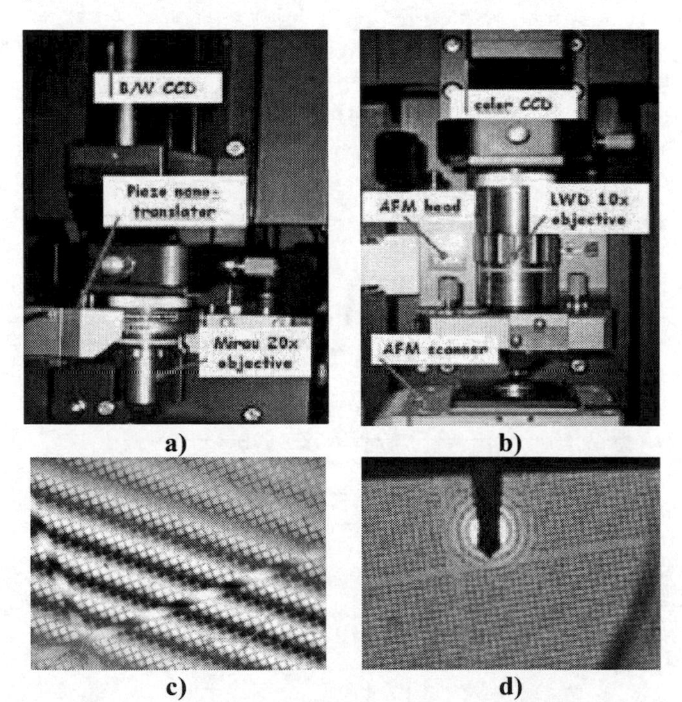

Figure 1: Setup of the instrument in a) WL-IOP mode b) AFM mode, with the corresponding optical images in c) and d).

To switch from optical profilometry (WL-IOP) to AFM, both the AFM head and the microscope objective are slide laterally. Mirau-type interferometric objectives (10x, 20x and 50x) are used in WL-IOP mode and a long working distance 10x objective in AFM mode. The white-light illumination system (standard halogen bulb) is used for both modes. The interferometric objective is mounted on a closed-loop piezoelectric translator with 1nm resolution and 200um range.

A two-axis tilt stage (+/- 2°) is mounted in between the AFM scanner and the mechanical XY positioning stage. Due to the thickness of the scanner, the sample is not in a perfect gimbal configuration. However, this gives only minor annoyance for the alignment.

2.2 Reconstruction Algorithm

A typical WL-IOP acquisition consists of recording images 20nm apart over the whole range of focus of the specimen under inspection. Our setup allows CCD frame acquisition at a rate of 30 fps, which corresponds to an acquisition time of 1.6 sec per micron of sample corrugation.

The optical intensity at a given pixel (x,y) as a function of the z position (i.e optical path difference) is called a correlogram. Due to the slight tilt given to the sample relative to the Mirau objective, the interference fringes as seen on Fig 1b) will slowly move across the image as the objective is scanned vertically. The correlogram $I_{xy}(z)$T has a maximum modulation at zero optical path difference (OPD), corresponding approximately to the focus point.

A typical untreated correlogram $I_{xy}(z)$ is shown in Fig 2a). It has the expected features for a low coherence interference signal, i.e fringes are localized over a short distance around the zero OPD.

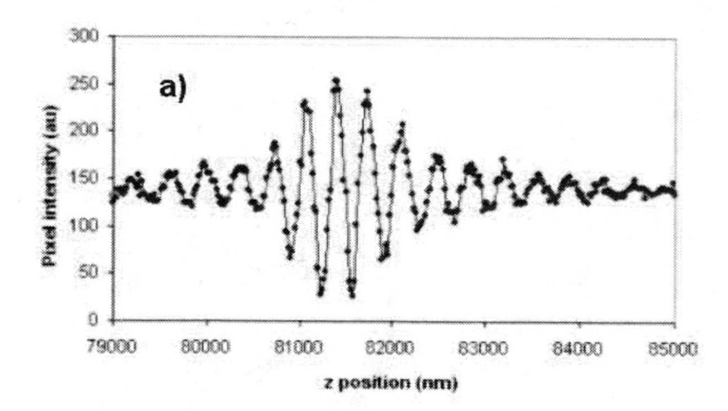

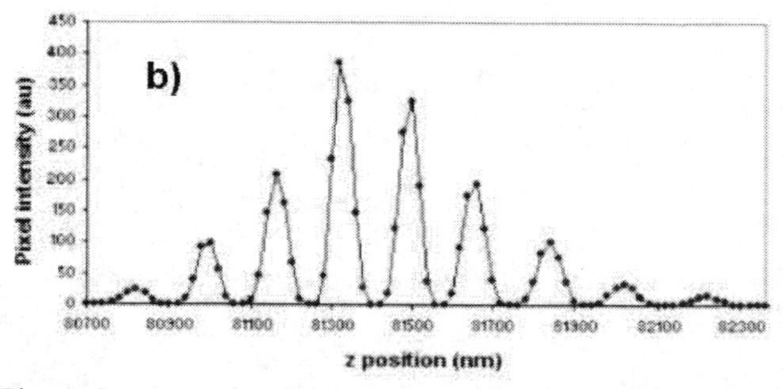

Figure 2 : a) untreated correlogram, b) processed correlogram for peak detection

To reconstruct the true z-position from the correlogram on each pixel, an algorithm has to be used to detect the peak position of the envelope of the correlogram. Many algorithms have been proposed for this purpose [3] [4] [5]. In our case, an important criterion for the algorithm is the computational time as we want to use the WL-IOP as a fast defects detection tool before AFM investigation.

We developed our own algorithm which mix Fourier transform and centroid calculation. The processed correlogram used for the final determination of the zero OPD is shown in Fig 2b. Each correlogram take an average of 0.29 ms to reconstruct. For an actual 640 x 480 image reconstruction (which is standard, but we also used a 1024 x 784 CCD occasionally), this corresponds to 90 seconds reconstruction time.

The actual calculated profile on a mirror (commercial grade) is shown in Fig 3. Sub-nanometer resolution is clearly achieved and the groove indicated by the arrow is 1.4 nm deep.

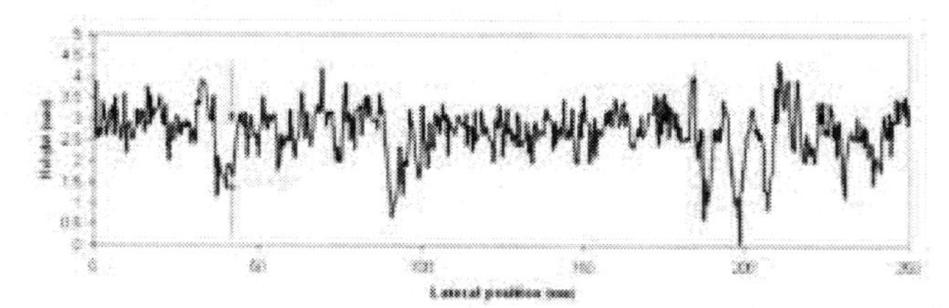

Figure 3 : Cross section of a polished miror in the WL-IOP mode

3 RESULTS AND PERFORMANCES

Fig 3 is an example of the integrated WL-IOP/AFM operation. The sample we used was a standard calibration grating (pitch 9.9um) made of silicon. No significant oxide thickness variation is expected for this sample, excluding then multiple interference artifacts [6] in white-light interferometry.

The optical interference profiler is used at first to image a large field of view (FOV) of the sample. Fig 4a) is a 3D reconstruction view [7] of the WL-IOP data for a field of view of 350um x 270 um. Many defects can be seen on this reconstruction, defects that cannot be detected with standard bright field microscopy, for example groves, pinholes and small dust particles. For a careful inspection of the surface, we found often advantageous to digitally zoom into the surface, as shown in Fig 4b). At this stage, we choose the area to be scanned with the AFM, represented in Fig 4b with a white square. An example of cross section analysis on the WL-IOP is shown in Fig 4c). The coordinate of the area of interest (AOI) is recorded and transferred to the AFM in order for the AFM probe to be directly positioned correctly when the system will be switched to AFM mode. Fig 4d) shows a high resolution AFM image over the AOI.

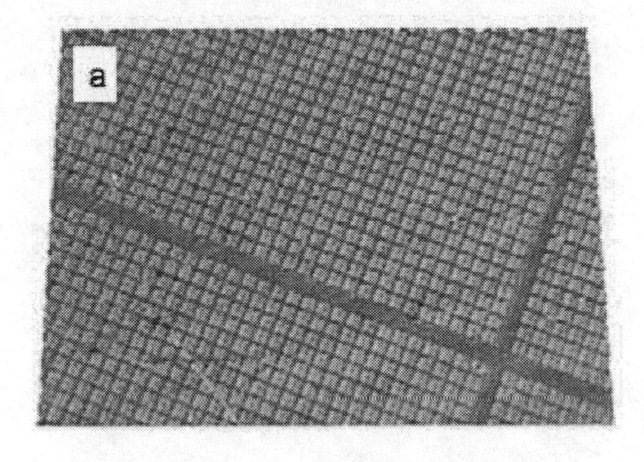

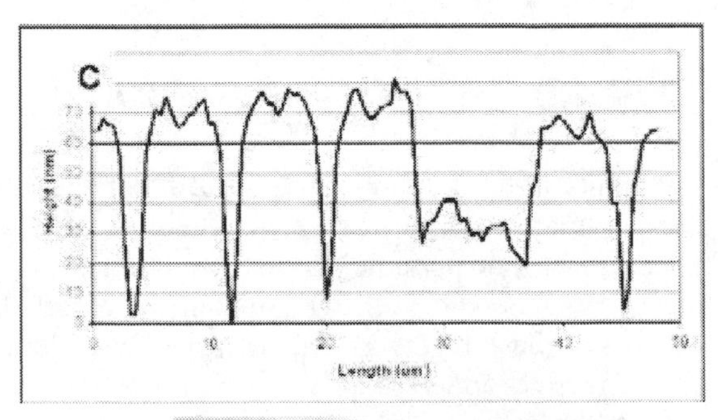

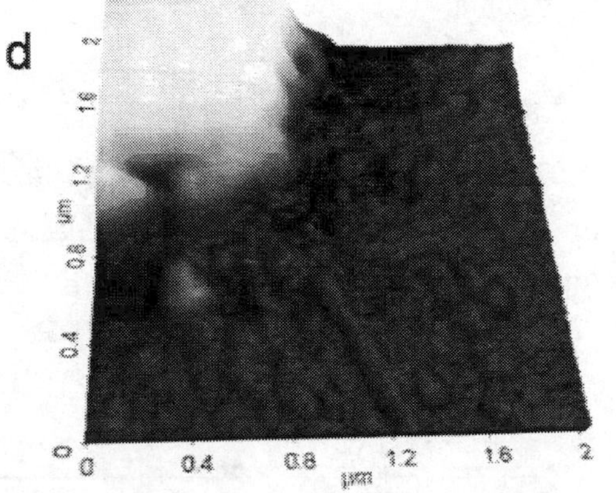

Fig 4 : Example of intergrated WL-IOP /AFM operation. a) Large field of view (350um x 270um) WL-IOP image in 3D view b)Gigital zooming and cross section selection of the area of interest c) Cross section in the W: -IOP mode d) AFM image (2um x 2um) over the selected area in b).

4 CONCLUSIONS

Our combined white-light interference optical profiler (WL-IOP) and AFM provides a fast, reliable and efficient 3D metrology system operable from 10nm to 1mm laterally with sub-nanometer vertical resolution over the whole lateral range. Such a combination should be particularly useful for quality control application in MEMS industry.

Further development of our system includes the phase-shift interfermometry (PSI) mode and improved reconstruction algorithms for batwings effects [8] cancellation.

REFERENCES

[1] See for example: P.J. Caber, *Appl. Opt.* **32**, 3438 (1993)

[2] Model XE 100 from PSIA, Seoul, Korea (www.psia.co.kr). We also integrate it to a Nano-R from Pacific Nanotechnology, Santa Clara, CA (www.pacificnanotech.com)

[3] L. Deck and P. de Groot, *Appl. Opt.*, **33**, 7334 (1994)

[4] G.G Larkin, *J. Opt. Soc. Am.*, **13**, 832 (1996)

[5] P. Sandoz, *Opt. Lett.*, **22**, 1065 (1997)

[6] M. Roy, I. Cooper, P. Moore, C. Sheppard and P. Hariaran, *Optics Express*, **13**, 164 (2005)

[7] we used SPIP software from ImageMetrology, Copenhagen, Denmark (www.imagemet.com)

[8] A. Harasaki and J.C Wyant, *Appl. Opt.*, **13**, 2101 (2000)

New developments in Spectroscopic Ellipsometry for Nano Sciences

JP.PIEL[1], A.DARRAGON[1], C.DEFRANOUX[1].

[1] SOPRA-SA. 26 rue Pierre Joigneaux. F-92270 Bois Colombes, FRANCE.
Tel. +33 1 46 49 67 00 – Fax + 33 1 42 42 29 34
jeanphilippe.piel@sopra-sa.com

ABSTRACT

Spectroscopic Ellipsometry (SE) is an optical technique, non destructive, to characterize the complex reflectivity of surfaces and layers from the deep UV (190nm) to the mid InfraRed (20µ). SE does not need any reference surface like interferometry nor reference materials or reference beam like in absorption spectroscopy (IRRAS): this is an absolute technique often called first principal technique.
A new optical configuration to avoid the beam reflected from the back face of the sample has been developed particularly for Infra Red application, the patented idea will be disclosed.
The layer thicknesses become thinner in the nano-sciences and in the case of surface preparation.
Spectroscopic Ellipsometry is the technique of choice to characterize not only the thickness with a fraction of angstrom sensitivity and the roughness but also the molecular composition of layers at the nano-scale or single mono molecular.
A new application of SE is in imaging of coated planar surfaces. The weakness of the technique is in its spatial resolution (several µ) particularly because of the incidence angle and limited divergence.
The kinetics can be measured for the full spectrum in real time with resolution of 0.1sec in the case of In-situ real time deposition or growth.
Several examples of applications will be given using different wavelengths and different configurations of SE.

Keyword : Spectroscopic Ellipsometry, Infra-Red Ellipsometry, Imaging Ellipsometry.

1 INTRODUCTION

Spectroscopic Ellipsometry has long been recognized as a powerful technique to characterize thin films and complex multilayers. The first spectroscopic ellipsometer was commercially available in 1983 [1] .

In the present paper, we use the last generation of SOPRA ellipsometers to show the potentialities of this technique for the characterization of materials and structures [2], [3].

The first application presented in this paper is the characterization of an OLED structure, which is composed of a multiplayer stack. Multi-angle ellipsometry measurements are used to extract accurately optical indices and thicknesses.

Infrared ellipsometry allows molecular bound detection but also resistivity measurements of metallic films [4]. Complete structure can be control with production oriented instrument.

Spectroscopic Ellipsometry Imager (SEI) is a also a very powerful approach to perform measurments with a good lateral resolution down to 10 µm on the sample. One example taken obtained on a pattern wafer will be presented.

2 OLED CHARACTERISATION

Organic light-emitting diodes (OLEDs) based on conjugated polymers have attracted much attention during the last decade, due to their possible application in large area flat panel display A schematic OLED structure is reported in Figure 1. Each of these layers can be characterized in term of thickness, refractive index and extinction coefficient by ellipsometry technique.

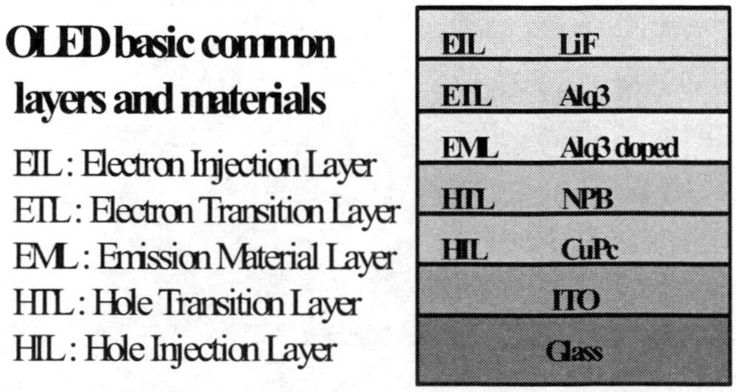

OLED basic common layers and materials

EIL : Electron Injection Layer
ETL : Electron Transition Layer
EML : Emission Material Layer
HTL : Hole Transition Layer
HIL : Hole Injection Layer

EIL	LiF
ETL	Alq3
EML	Alq3 doped
HTL	NPB
HIL	CuPc
	ITO
	Glass

Figure 1 : Schematic diagram an standard OLED structure

2.1 Characterization of ITO layers

The bottom ITO layer is used as conductive and transparent electrode for the OLED structure. For this layer, infrared spectroscopic ellipsometry is the technique of choice to measure at the same time the layer thickness and its conductivity in a non-contact, non-destructive way. Indeed, the Drude law can be used to describe the free-carrier absorption effects that can be detected on the absorption coefficient in the infrared region. The free carrier concentration can be deduced [5]. In Figure 2 we have reported the experimental and simulated ellipsometric curves obtained on an ITO electrode layer on glass. The adjustment is made at the same time on the layer thickness and the optical index of the ITO layer using a Drude model. In this way the refractive index and the absorption coefficient are deduced independently.

In Figure 3, the extinction coefficients measured on two ITO layers with and without thermal annealing are reported. The annealing effect at 450°C is easily detected. The different physical parameters of the ITO layers can be easily deduced as reported in Table I.

Sample	Thickness (nm)	Conductivity (S/m)	Resistivity (mΩ/cm)
As deposited	162.2	1.67E+5	0.60
Annealed 450°C	169.1	3.96E+5	0.25

Table I: Physical parameters deduced from Fig 3

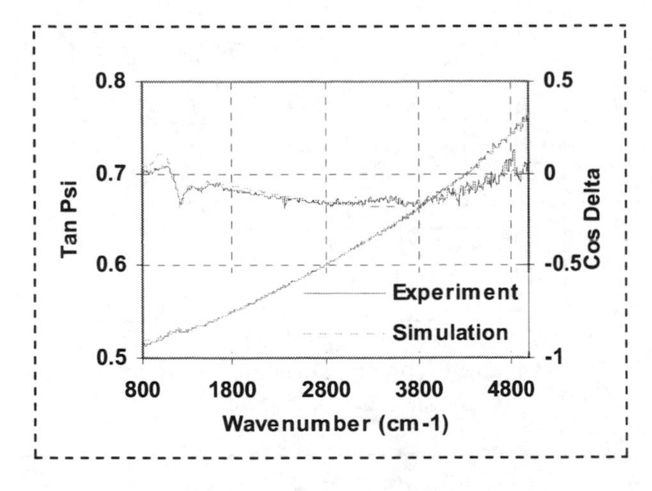

Figure 2 : Measurement of a ITO layer on glass in the infrared range. The layer thickness is 162nm

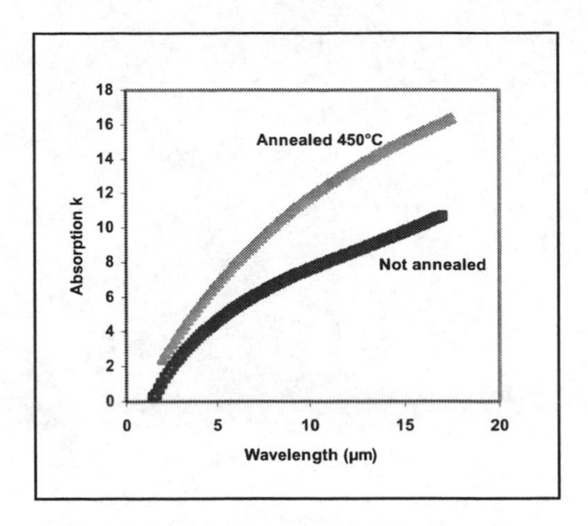

Figure 3 : Measured extinction coefficient of two ITO layer with and without thermal annealing

2.2 Characterization of CuPc layers

Absorption bands in complex organic materials, like CuPc need to be carefully characterized in the visible range to predict the emission properties of the display. In this respect, variable angle spectroscopic ellipsometry is the technique of choice to measure independently the refractive index and the absorption coefficient of these films. In figure 4 some experimental curves obtained on a hole injection layer (CuPc on glass) are reported with the simulations. Three incidence angles have been used at 65, 70 and 75°. From these data, the optical indices of the CuPc film and its thickness are deduced for each wavelength independently has shown in Fig. 5, and the accuracy of this extraction can be estimated (generally lower than ±0.02 on n and k).

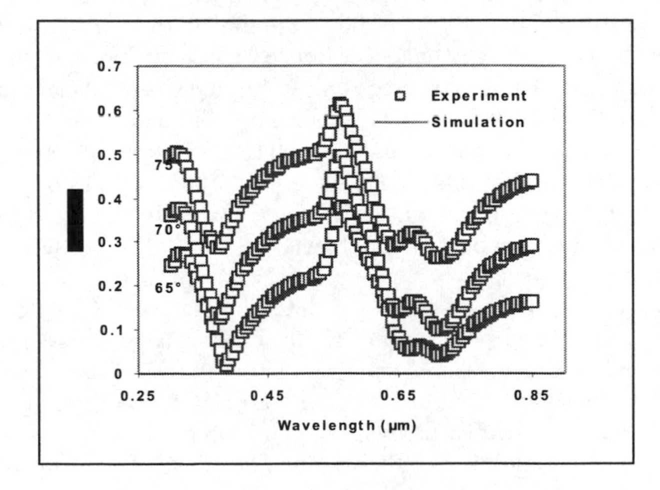

Figure 4 : Variable angle SE measurement on CuPc/Glass sample

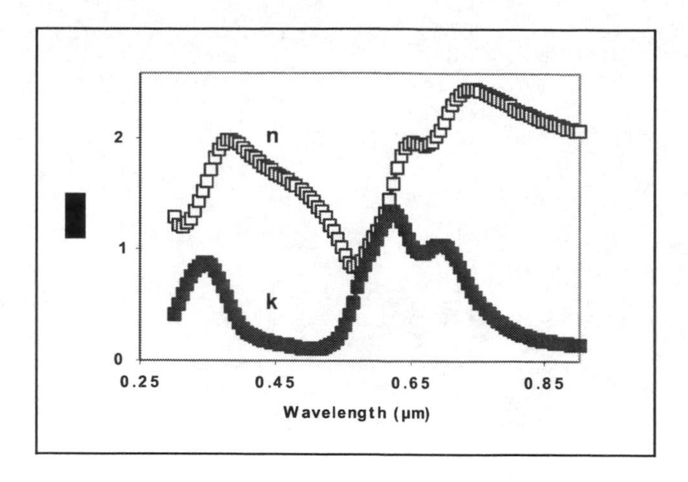

Figure 5 : Optical indices of the CuPc layer of Figure 4.

When all the materials have been carefully characterized, SE can control the complete OLED structure and the thickness of the different layer can be deduced independently. The same equipment and analysis method can be used to characterize polymer films.

3 LOW K DIELECTRIC MATERIALS

Reduction of the CD dimensions in the IC's induces an increase of the circuit density and consequently of the RC parameters between isolated contacts. To reduce this effect, SiO_2 is replaced by new low K dielectric materials.. Si-O-C-H films deposited by plasma enhanced chemical vapor deposition (PECVD) have shown the formation of nano-sized voids due to $Si\text{-}CH_3$ and OH related bonds included in the film [6]. These features are controlling the dielectric properties of the films and must then be measured accurately. Infrared ellipsometry controls thickness and chemical properties of thin films at the same time. The properties of the substrates, roughness and interfaces are taken into account in the analysis model. As shown in Figure 7, different absorption peaks can be detected in the infrared region in addition to the O-Si-O bounds classically detected on silica. A well defined peak around $1270cm^{-1}$ associated to $Si\text{-}CH_3$ bounds is also measured and two other peaks at $2350cm^{-1}$ and 3000cm-1 associated to the water absorbed by the films. We have used these measurements to extract the optical indices of different samples as shown in Figure 9. The relative concentration of carbon incorporated in the film can be calculated normalizing to the peak area of the O-Si-O stretch using:

$$Relative\_carbon\_concentration(\%)=\frac{A_c}{A_o+A_c}$$

where A_o and A_c are the peak areas of the Si-O stretching vibration mode at $1040cm^{-1}$ and the $Si\text{-}CH_3$ vibration mode at $1270cm^{-1}$ respectively. A summary of the results is reported in Table II.

These characterizations are obtained on the SOPRA IRSE 3000 which is an automatic spectroscopic Ellipsometer covers the mid Infrared range from 600 cm-1 to 7000 cm-1 (1.4 µm to 16.6 µm).

Main hardware components are reported on the following schematic (Fig. 6.):

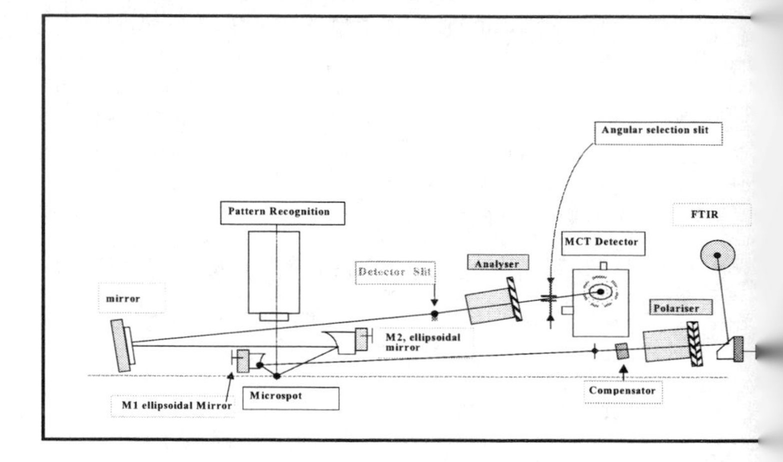

Fig 6: Mounting Schematic

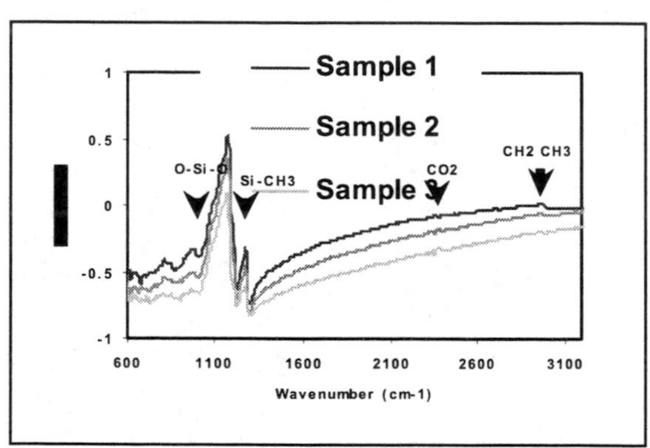

Fig 7: SE measurements on three Si-O-H-C low K dielectric with variable C content. Different absorption bounds are easily detected.

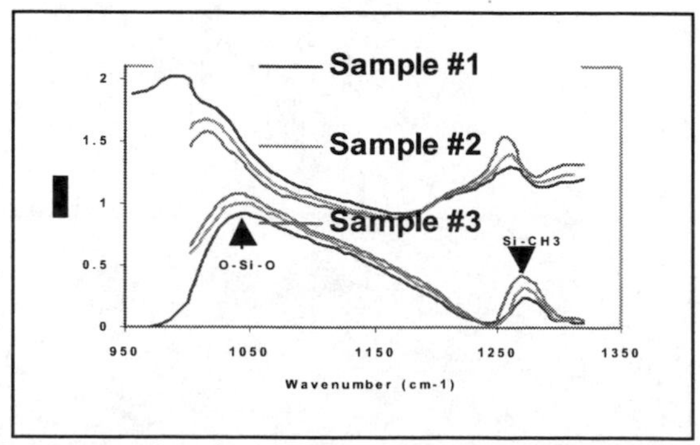

Fig 8: Measured optical indices of the Si-O-C-H films of Fig 7. The thickness of the films is between 300 and 500nm

	A_0	A_c	Carbon (%)
Sample 1	119.4	8.3	6.5
Sample 2	132.5	10.4	7.3
Sample 3	150.7	15.6	9.4

Table II: Carbon content of the Si-O-C-H low K dielectrics of Fig.7 & 8

4 IMAGING ELLIPSOMETRY

4.1 Introduction

In conventional ellipsometers the light beam is typically a few mm in diameter but can be reduced down to 20µm using a microspot by focusing the measuring beam [7]. This kind of system is used for in line analysis in semiconductor production. The time to obtain an image of the surface of the sample is nevertheless too long to study rapidly varying phenomena. It is the reason why a Spectroscopic Ellipsometer Imager (SEI) was developed. With this instrument we measure a full image of the surface of the samples with a spatial resolution of 10µm, or better, at a rate of a few wavelengths per image. The total image covers an area of 0.6x2mm with 480x780 pixels.

4.2 Experimental results

The spatial resolution and anamorphous effect of our new instrument have been evaluated by imaging a 200x200µm target (Fig 9). The raw gray scale image obtained at 577nm and 45° of polariser angle of this target shows an important (but expected) anamorphous effect. Taking into account the angle of incidence of the imaging measurement we can expect a reduction of Y scale of a factor of 3. An intensity profile along the line A of Figure 9.a shows the spatial resolution obtained with our instrument (cf. Figure 9.b). It is found around 10µm with some diffraction effects detectable on sharp boundaries. The edges of the patterns diffract the light, which cannot be collected in the 2° limited aperture angle of collection. In addition, due to the diffraction limit of 15µm, the pixels at edges of image are not correctly illuminated.

On Figure 10 we have reported the first SEI observation on a patterned SiO2/Si sample. As shown in the figure, each pixel contains a complete spectroscopic spectrum of the tanψ and cosΔ parameter. Zones with large SiO2 thickness around 105nm can be easily detected (red areas of tan ψ parameter at 436nm for example). Analysis treatment of such a set of images can be made as follows: First, different kinds of areas can be detected using pattern recognition or colour filtering on one well contrasted image. Then one structural model is attributed to each type of area. Standard regressions using the same model can then be applied to all the pixels of one area assuming that the structural parameters are continuous inside the area. Images of the different structural parameters can finally be deduced with the appropriate models. This technique or fuzzy logic or neural network can be applied for determination of complex multilayer structures corresponding to various different zones and models. Multilayer model can be applied and at the difference of single wavelength ellipsometer, no shadow zone nor cycle problem occur.

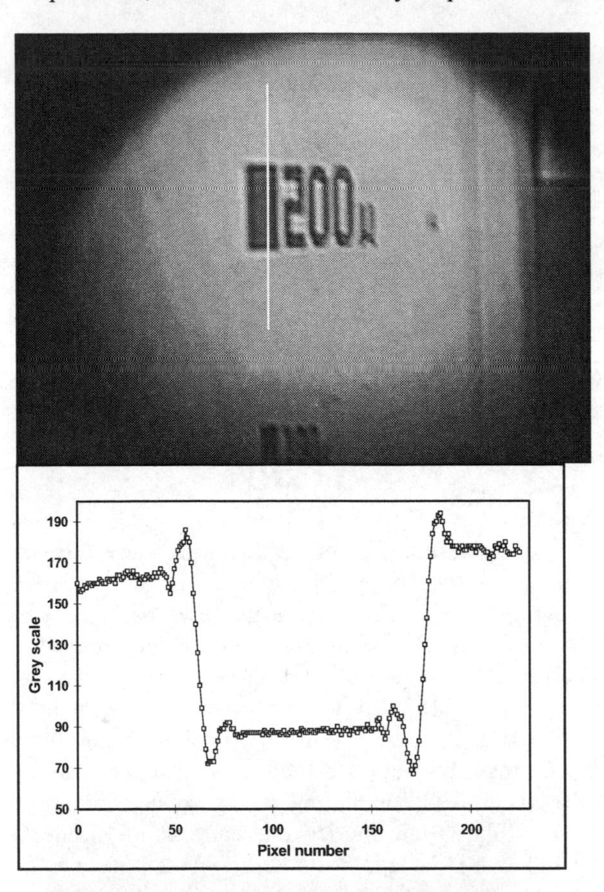

Fig 9: Image of 200 µm squared box on silicon substrate at $\lambda = 577$ nm

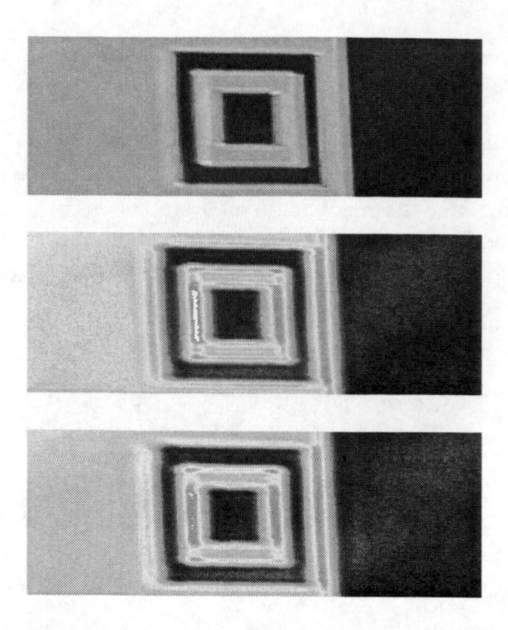

Fig 10: Cos Delta values obtained for 4 differents wavelengths

5 CONCLUSION

These different applications show clearly that spectroscopic ellipsometry is a powerful technique to characterize extremely thin layers in term of thickness but also indices. The possiblility to perform measurements on a very broad spectral range from 137 up to 20 µm is also extremely powerful, depending of the application. In mid Infra Red range, molecules signatures will be seen and if the material is conducting, using some appropriate model like Drude model, one can extract electrical parameters like mobility and conductivity of the layers.

Imaging ellipsometry is a very promising technique to overpass the limit of lateral resolution to reach 10 µm resolution and in a next future up to 1 µm. We can see applications of this technique not only in microelectronic but also in biotechnology field.

REFERENCES

[1] F. Ferrieu, F. Bernoux, J.L. Stehle, Le Vide les couches Minces, suppl. 233, 17 (1986).

[2] D. Zahorski, J.L. Mariani, L. Escadafals, J. Gilles, Thin Solid Films, 234, 412-415 (1997)

[3] P. Boher, J.P.Piel, C.Defranoux, J.L.Stehle, L.Hennet, SPIE's 1996 International Symposium on Microlithography, Santa Clara, March 10-15 (1996)

[4] C. Pickering, W. Leong, J. Glasper, P. Boher, J.Piel, EMRS Spring meeting, Strasbourg (2001)

[5] P. Boher, M. Bucchia, J.P. Piel, C. Defranoux, J.L. Stehle, C. Pickering, SPIE Conference, 29 july-3 august, vol. 444 (2001)

[6] Y.H. Kim, S.K. Lee, H.J.Kim, J. Vac. Sci. Technol., A18, 4, 1216 (2000)

[7] Piwonka-Corle, et al. « Focused beam spectroscopic ellipsometry method and system", US patent n°5608526, March 1999

Novel Ion Beam Tools for Nanofabrication

Q. Ji[a, 1], X. Jiang[a, b], L. Ji[a, b], Y. Chen[a, b], B. v. d. Akker[c], K.-N. Leung[a, b]

[a] Lawrence Berkeley National Laboratory, University of California, Berkeley, CA 94720
[b] Department of Nuclear Engineering, University of California, Berkeley, CA 94720
[c] Department of Physics, San Francisco State University, San Francisco, CA 94132

ABSTRACT

The drive towards controlling materials properties at nanometer length scales relies on the availability of efficient tools. Currently, high resolution ion beam processing is limited to direct-write techniques with mostly gallium ion beams, or the use of pre-structured masks. In this paper, several novel ion beam tools that have been developed at Lawrence Berkeley National Laboratory (LBNL) are reviewed: Maskless Micro-Ion-Beam Reduction Lithography system, Multiple Focused Ion Beam (FIB) system and ion beam imprinter, and a FIB/SEM dual beam system. With the availability of multicusp plasma ion sources, these ion beam tools can provide versatile ion beams in nanometer scale for future integrated circuit manufacturing, thin film media patterning, and micromachining

Keywords: ion projection lithography, focused ion beam, ion beam imprinter, maskless lithography

1 INTRODUCTION

Current research to understand the science associated with nanoscale structures has shown the need for new approaches to the fabrication of future small-scale devices. For the last several decades, ion beams have played a significant role both in material modification (e. g., ion implantation, and micromachining) and analysis (e. g., secondary ion mass spectrometry). Conventional FIB systems, which utilize liquid gallium ion sources to achieve nanometer resolution, severely limit some applications and fundamental studies due to gallium contamination. Demands have been continuously increasing for ion beam tools with a variety of ion species in selectively doping or surface modifications in nanoscale device structures.

In this paper, several novel ion beam tools that have been developed at Lawrence Berkeley National Laboratory (LBNL) are reviewed: Maskless Micro-Ion-Beam Reduction Lithography system, Multiple Focused Ion Beam (FIB) system and ion beam imprinter, and a FIB/SEM dual beam system. With the availability of multicusp plasma ion sources, these ion beam tools can provide versatile ion beams in nanometer scale for future integrated circuit manufacturing, thin film media patterning, and micro-machining.

2 MULTICUSP PLASMA ION SOURCES

Multicusp ion sources have been used for many applications, such as neutral-beam injectors for fusion devices, particle accelerators, ion implantation systems, compact neutron tubes, and proton therapy machines.[1, 2] Ions of virtually all elements in the periodic table can be generated by either filament dc discharge or RF induction discharge. Compact RF-driven ion sources have been developed for various ion species production, such as H^+, He^+, Ar^+, O^+, B^+, P^+ etc. [3]. Besides gaseous elements, they can also be employed to produce metallic ions, e.g. Cu^+, Ni^+, Cr^+, Pd^+,[4] and molecular ions, e.g. C_{60}^+. A large area and uniform plasma is achievable with the multicusp ion source by arranging the magnets around the source chamber in such a way as to generate line-cusp magnetic fields. By optimizing the source configuration and extractor geometry, beam brightness as high as 440 A/cm^{2}Sr has been measured, which represents a 30 times improvement over prior work.[5] The ion beam systems described here are all developed based on this type of ion sources.

3 NOVEL ION BEAM TOOLS FOR NANOFABRICATION

3.1 Maskless Micro-Ion-Beam Reduction Lithography (MMRL) system

The manufacture of CMOS integrated circuits will eventually require techniques for patterning sub-10 nm features, with sub-25 nm half-pitch. Mask costs for deep-UV (eventually EUV) lithography will continue to escalate with each new generation of technology, and will even become prohibitive for low-volume IC products. Maskless patterning techniques are desirable in order to circumvent these issues.

A proof-of-concept maskless ion beam lithography machine called Maskless Micro-beam Reduction Lithography system has been developed in Lawrence Berkeley National Laboratory.[6-8] The MMRL system (Figure 1) uses inductively coupled rf multicusp plasma source to generate helium or hydrogen plasma, which has uniform density over a large volume. A mask or a pattern generator is directly placed on the plasma electrode. Due to the low ion energy inside the plasma source, less than 3V biasing voltage can switch ion beamlets on and off. The prototype MMRL system has eight electrostatic electrodes to accelerate ion beams to 75 keV and to reduce the beam size by 10X. A limiting aperture (Figure 2) is placed on the beam cross-over position to limit ion beam half angle and reduce the geometrical and first-order chromatic aberrations. Because of the small wavelength of ion beams, much smaller NA can be used in ion beam lithography machines, which gives much larger exposure field size than direct-write e-beam systems.

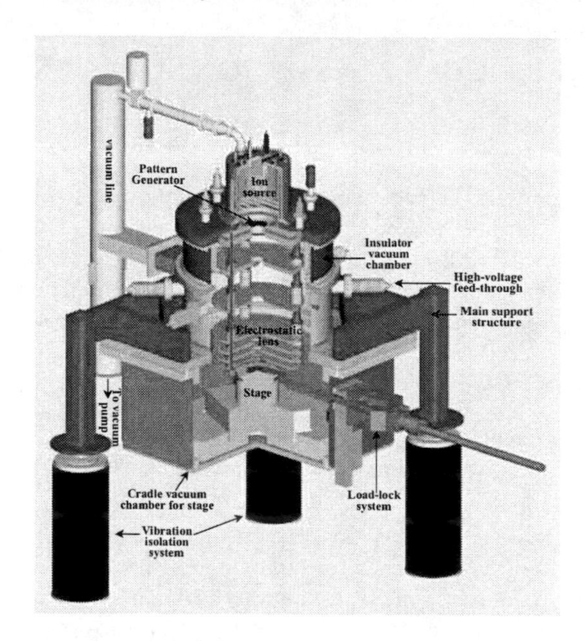

Figure 1: Maskless Micro-Ion Beam Reduction Lithography System uses a universal pattern generator (beam-forming electrode) to form lithographic patterns on wafers.

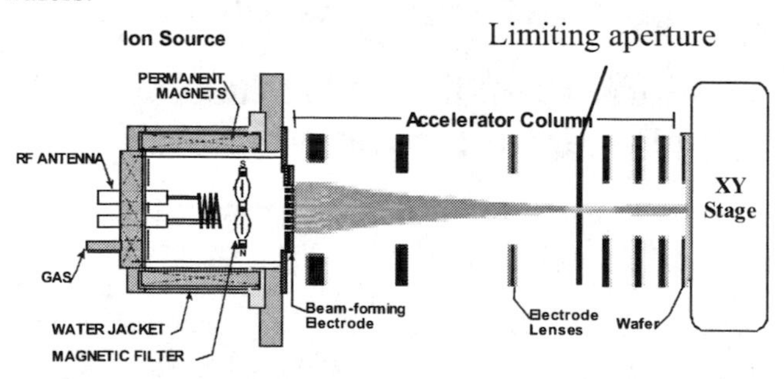

Figure 2: In the MMRL system, a limiting aperture is inserted at the beam cross-over plane for aberration reduction.

The MMRL tool can work in projection and maskless mode. Sub-100 nm resolution (Figure 3) has been demonstrated on the prototype MMRL system in projection mode. Employing the switchable pattern generator on the beam-forming electrode plane and combining the scanning of the high precision stages, the MMRL can perform multiple-beam maskless lithography and offer a possible solution for low-cost advanced ASIC fabrication

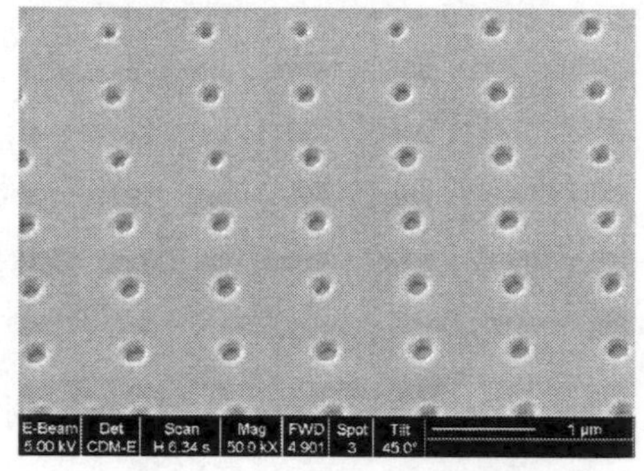

Figure 3: Ion beam exposure results on PMMA resist.

A pattern generator, which can individually switch ion beamlets on and off, is being fabricated in the microfabrication laboratory of UC-Berkeley. Figure 4 shows a schematic diagram of the pattern generator. The heavily boron doped silicon layer faces the plasma side. Under the conductive P^+ layer, there is a thermally grown SiO_2 insulation layer. A metal layer is then deposited under the insulation layer. The metal layer is patterned into lines and pads so that biasing voltages can be individually applied to each beam-forming aperture.

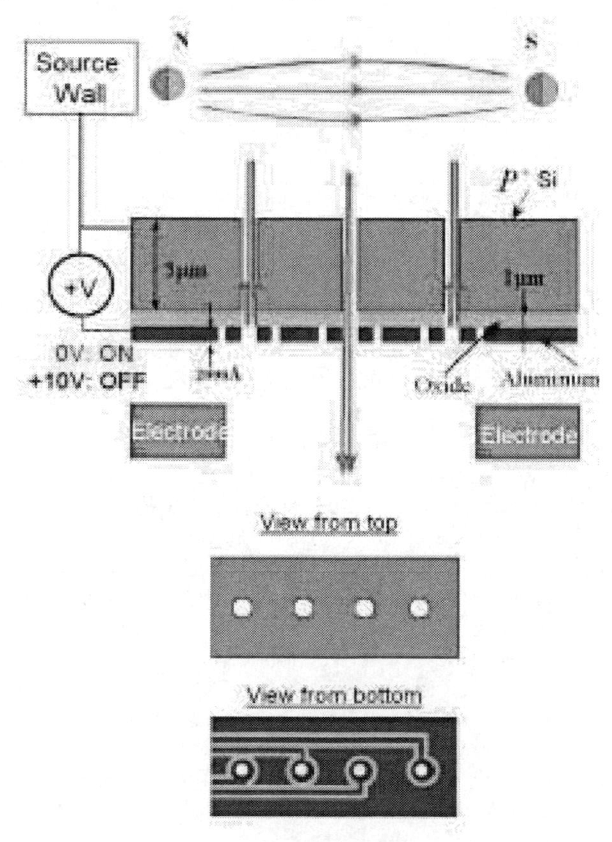

Figure 4: Switchable pattern generator for MMRL system.

3.2 Multiple Focused Ion Beam (FIB) system and Ion Beam Imprinter

Throughput is always an issue for focused ion beam system. In order to achieve higher throughput, parallel processes using multiple beamlets system is one of the promising solutions. Figure 5 shows the schematic diagram of a multiple focused ion beam system which employs a multicusp plasma ion source. As mentioned above, the ion source can generate uniform and large area plasma, therefore, only one source is needed to generate multiple beams. The pattern generator used in the MMRL system can also be employed as extraction electrode. Individual beamlet can be switched off simply by applying +10V on the extraction electrode relative to the source. A stack of electrodes with multiple apertures, which act as lens element, can further accelerate the ions to the desired energy, and then focus to a small beam spot at the target. Maskless lithography will be realized by switching the ion beamlets on/off and scanning the substrates in x and y directions.

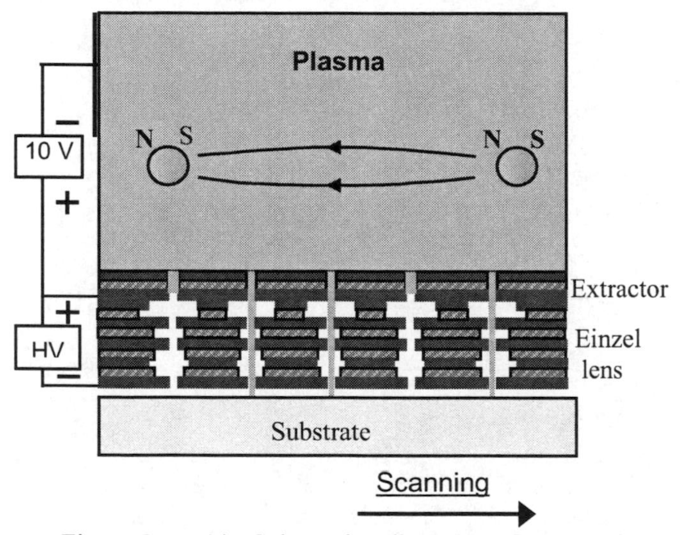

. Figure 5: (a) Schematic diagram of a multiple focused ion beam system. (b) Schematic diagram of the ion beam imprinter.

As illustrated in Figure 6, a stencil mask that consists of different features, such as lines, arcs, round holes, and other arbitrary shapes can also be used as a plasma electrode. In this case, ions extracted through the aperture reach the sample with the same pattern as those on the mask. This ion-beam imprint technique can transfer patterns not only onto a planar target, but also onto non-planar surfaces, for example the outer and inner surfaces of cylinders, which have numerous applications in micromachining and surface topology modification.[9]

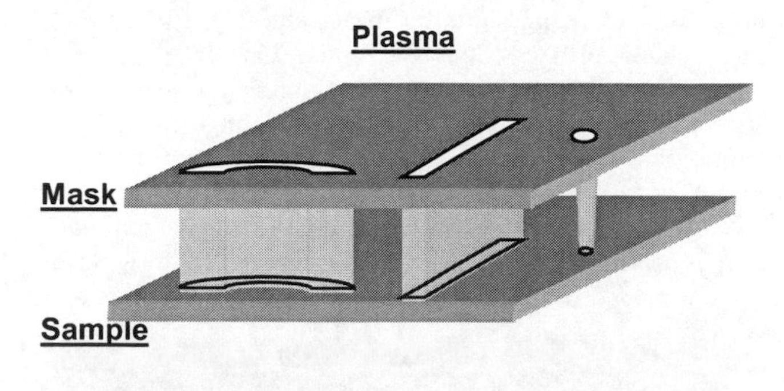

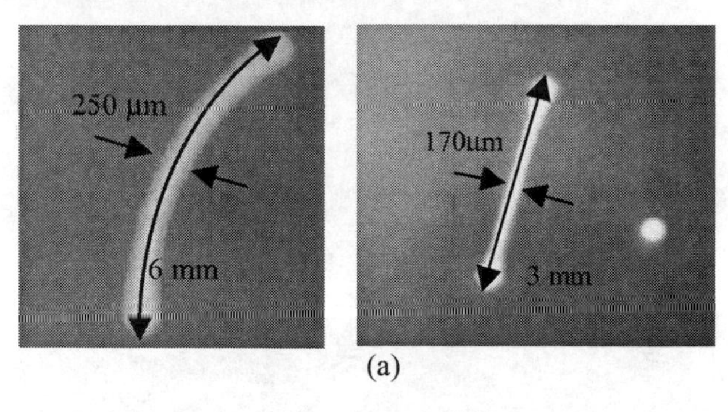

(a)

(b)

Figure 6: (a) Schematic diagram of the ion beam imprinter. (b) Microscopic image of the arc (left) and rectangular (right) shape feature milled on the sample using ion beam imprinting.

3.3 A FIB/SEM dual beam system

Lack of integrated diagnostics in conventional systems also limits the information available from nanofabrication experiments. An integrated FIB/SEM dual-beam system will not only improve the accuracy, resolution and reproducibility when performing ion beam sculpting[10] and direct implantation processes, but also enable researchers to perform cross-sectioning, imaging, and analysis with the same tool.[11,12]

As shown in Figure 7, the FIB/SEM dual system developed between Harvard University and LBNLfor direct doping or surface modification employs a mini-RF driven plasma source to generate focused ion beam with various ion species, a FEI two-lens electron (2LE) column for SEM imaging, and a five-axis manipulator system.[13] The mini-RF plasma source consists of a 1.5 cm inner diameter ceramic chamber and two layers copper wire as external antenna. Ion beams are extracted through a 50-µm-diameter extraction aperture. Ar^+ ion current density as high as 100 mA/cm^2 has been obtained with only 150 W of input RF power. An all-electrostatic two-lens column has been designed to focus the ion beam extracted from the source. Based on the ion optics simulation, beam spot sizes as small as 100 nm can be achieved at beam energies between 5 to 35 keV if a 5-µm-diameter extraction aperture is used.

Smaller beam spot sizes can be obtained with smaller apertures by sacrificing some beam current. The FEI 2LE column, which utilizes Schottky emission, electrostatic focusing optics, and stacked-disk column construction, can provide high-resolution (as small as 20 nm) imaging capability, with fairly long working distance (25 mm) at 25 keV beam voltage. The picture of the FIB/SEM dual beam system after assembly is shown in Figure 8. A major advantage of this tool is the ability to produce a wide variety of ion species tailored to the application.

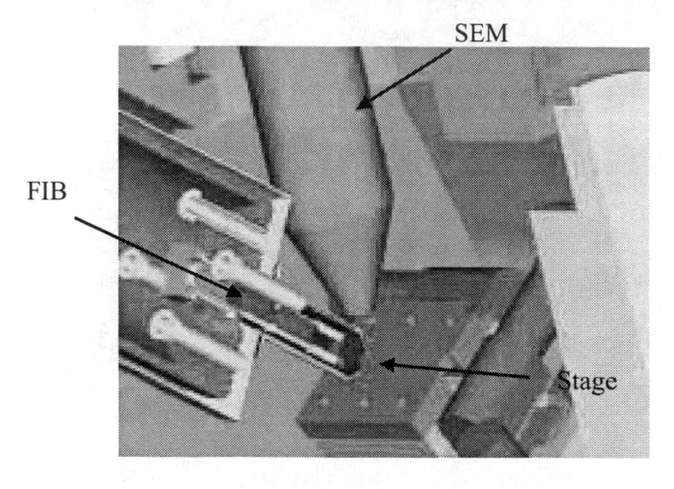

Figure 7: Schematic diagram of the FIB/SEM dual beam system.

Figure 8: A picture of the FIB/SEM dual beam system.

4 SUMMARY

Several ion beam tools have been developed to provide ion beams at nanometer scale with choices of a variety of ion species. They not only can be applied to many research areas including nanostructure patterning, selectively doping, and surface modification, but also will open up new opportunities in nanofabrication and analysis.

5 ACKNOWLEDGEMENTS

The work is supported work is supported by DARPA and the U.S. Dept. of Energy under Contract No. DE-AC03-76SF00098 and National Science Foundation under contract No. DMR-0216297. Support from staff members of the Plasma and Ion Source Technology Group in Lawrence Berkeley National Laboratory is also gratefully acknowledged.

REFERENCES

[1] K. N. Leung, J. Vac. Sci. Technol. B17, 2776, 1999.
[2] K. N. Leung, Rev. Sci. Instrum. 69, 998, 1998.
[3] Q. Ji, T.-J. King, K.-N. Leung, and S. Wilde, Rev. Sci. Instrum., vol.**73**, 822, 2002.
[4] L. Ji et al, submitted to EIPBN 2005.
[5] Q. Ji, X. Jiang, T.-J. King, K.-N. Leung, K. Standiford, S. B. Wilde, *J. Vac. Sci. Technol.* **B20**, 2717(2002).
[6] V. V. Ngo, W. Barletta, R. Gough, Y. Lee, K. N. Leung, N. Zahir, and D. patterson, *J. Vac. Sci. Technol.* **B17**, 2783(1999).
[7] V. V. Ngo, B. Akker, K. N. Leung, I. Noh, K. L. Scott, and S. Wilde, *J. Vac. Sci. Technol.* **B21**, 2293(2003).
[8] X. Jiang, Q. Ji, L. Ji, A. Chang, and K. N. Leung, *J. Vac. Sci. Technol.* **B21**, 2724(2003).
[9] Y. Chen et al, accepted by SPIE Microlithography meeting 2005.
[10] J. Li, D. Stein, C. McMullan, D. Branton, M. J. Aziz, J. A. Golovchenko, *Nature*, Vol. **412**, 166(2001).
[11] S. D. Berger, D. Desloge, R. J. Virgalla, T. Davis, T. A. Paxton, D. Witko, *Proceeding of SPIE* Vol. **4344**, (2001) 423.
[12] D. Sidawi, *R&D Magazine*, **45**, no. 11, (2003) 33.
[13] Q. Ji et al, to be published the proceeding of 18th International conference on the application of accelerators in research & industry, October 10-15, 2004, Texas, USA.

---

[1] Email: Qji@lbl.gov. PHONE: 510-486-4802, FAX: 510-486-5101.

MEMS/NEMS Dynamics Measurement Tool Using The Stroboscopic Principle

Hsin-Hung Liao and Yao-Joe Yang

Department of Mechanical Engineering,
National Taiwan University
No. 1 Roosevelt Rd. Sec. 4, Taipei, Taiwan, ROC
FAX: 886-2-23631755 TEL: 886-2-23646491 EMAIL: yjy@ntu.edu.tw

ABSTRACT

High resolution and noncontacted measurement tools for the dynamic characterization of micro devices are necessary to develop high performance/precision and reliable microelectromechanical systems (MEMS). In this paper, a three-dimensional dynamic measurement system based on stroboscopic and interferometric methods is presented. This system can capture bright-fielded and interferometric images at the same time and will not have any phase error when constructing three-dimensional solid-models. The accuracy of in-plane motion measurement using this system has been calibrated by using piezoelectric nano-platform. The measurement of in-plane motions of the comb-drive actuator using this developed tool have been demonstrated.

Keywords: MEMS, characterization, machine vision, stroboscopic, interferometer

1 INTRODUCTION

Recently, the rapid growth of micro- and nano-electromechanical systems (MEMS/NEMS) boosts the demands of advanced diagnostic tools and characterization techniques. One of the most critical demands is the high-resolution tool for measuring MEMS/NEMS dynamic behaviors, such as displacement and velocity. In [1], the laser Doppler vibrometer is used to characterize a micro-mirror. In [2], a white-light interferometry surface profiling technique is used for material characterization and device inspection. In [3,4], in-plane motions of MEMS devices can be measured up to nano-meters resolution using the stroboscopic principle as well as gradient-based algorithm. Also, with a focus controller, the out-of-plane motion can be measured [5]. In [6], the focus controller was replaced with a splitter and vibrating mirrors that can be integrated inside a microscope, and the out-of-plane resolution was significantly increased. The combination of the optical interferometry and the stroboscopy enables 3D measurement of MEMS devices [7-11]. The hybrid laser Doppler vibrometer/strobe video system can be used to measure periodic in-plane motions and real-time out-of-plane motions [12]. In this work, we develop a MEMS/NEMS dynamics measurement tool which employs the stroboscopic principle.

2 METHOD

2.1 Stroboscopic System

Typically, the maximum frame rate of a traditional CCD camera is on the order of tens of Hz. Therefore, stroboscopic illumination is widely used for measuring the high frequency devices with a low-frame-rate CCD. Figure 1 shows the principle of a stroboscopic measurement system. Here we use the arbitrary waveform generator (Tektronix AWG2005) to generate the driving signals for actuating devices to be measured, and the pulses for activating LED strobe light. The strobe light is used to freeze the motion of a periodic moving microstructure for CCD image capturing. The time resolution is given by the duration of the stroboscope flash. At shot 1, the LED flashes several times at the same phase (e.g., phase I) of different cycles of the driving signal. Similarly, at shot 2, the LED exposes at another phase (e.g., phase II) in each cycle. Therefore, we can snap the images at different phases of the device moving in periodicity.

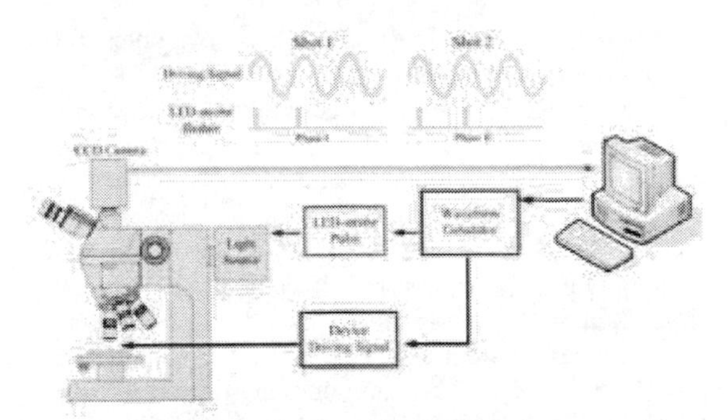

Figure 1: The principle of the stroboscopic system.

2.2 Setup of the Measurement System

The system for measuring the dynamics of microstructures uses strobe-bright-field for the measurement of in-plane motions and stroboscopic interferometric methods for the measurement of out-of-plane motions. Figure 2 is the proposed schematic of the strobe-bright-field/interferometric microscope. One objective lens is focused on the measured device and the other is focused on the reference mirror. When the LED flashes, CCD1 snaps the bright-field images of the device and CCD2 acquires the interference images at the same times. Therefore, the images of the bright field can be totally matched with the interference images and the measurement of three-dimensional motions can be completed in a single experiment. The quarter-wave-plate between P1 and NBS2 ensures that the light is directed onto a CCD2 instead of returning to the CCD1. A piezoelectric actuator that actuates the reference mirror is used to change the phase in the interferogram by adjusting the optical path-length difference.

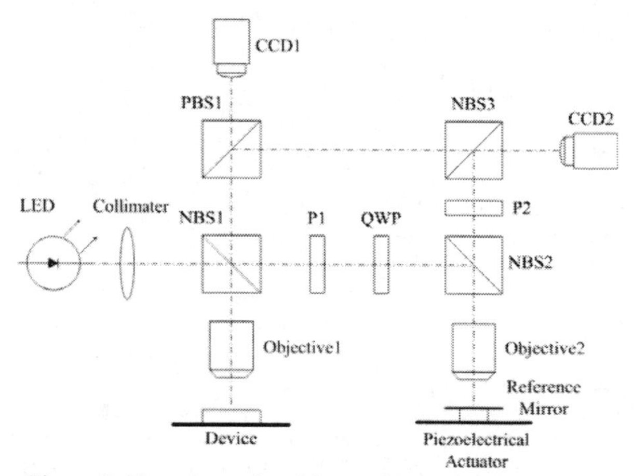

Figure 2: The schematic of the stroboscopic system. QWP is the quarter-wave-plate, PBS is polarization beam splitter, NBS is nonpolarization beam splitter, P is linear polarizer.

2.3 Procedure of Image Processing

The structure of the algorithm, which computes full three-dimensional motion, is implemented in MATLAB®. Figure 3 is the procedure of the three dimensional dynamic measurement and analysis. First, we set the region of interest (ROI) in the reference image (usually we do not actuate the device when define the ROI). The least-square optimization algorithm is used to analyze the bright-field images to obtain the in-plane displacement in one-pixel resolution. If the ROI is too small, it will discern in error and get the wrong results of the displacement. However, if the ROI is too big, the computation time increases unnecessarily. Furthermore, the gradient-based algorithm [4] is used to analyze the bright-field images. The interference images are also analyzed in frequency domain. Then the in-plane displacement of sub-pixel resolution and out-of-plane displacement are obtained separately. The three-dimensional dynamic motions of the devices can be rendered by combining the results of the in-plane and out-of-plane motions.

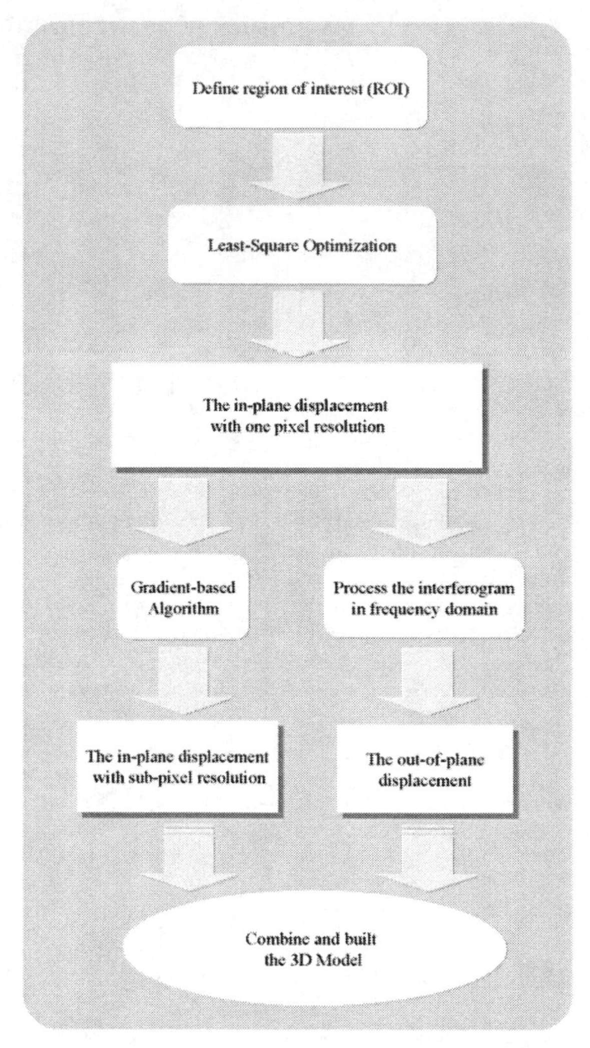

Figure 3: Procedure of estimating the 3D dynamic motion for MEMS device.

3 EXPERIMENTS

3.1 Piezoelectric Nano-platform

In order to verify the results of in-plane motion measurement, a piezoelectric nano-platform (PI P-621.20L) is used. To supply the accurate voltage to the piezoelectric nano-platform, we use the Keithley 2400 SourceMeter® to drive the nano-platform for a one-axial motion. The results of the measurement are shown in figure 4. It is shown that the maximum difference between the measurement value and the theoretic value of the piezoelectric platform is 10 nanometers.

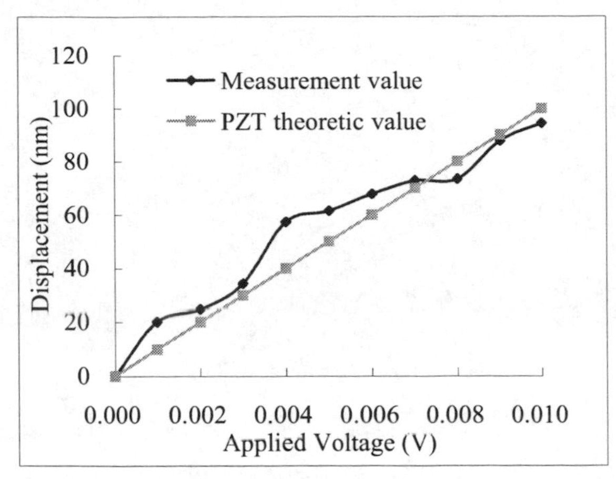

Figure 4: The displacement versus applied voltage for the piezoelectric nano-platform.

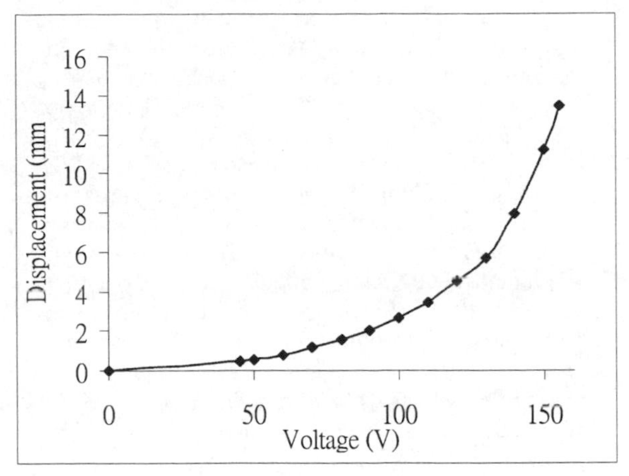

Figure 6: The displacement versus applied voltage for the comb-drive device.

3.2 Comb-drive device

The test subjects for this study is a comb-drive actuator fabricated by MUMPs® process which is shown in Figure 5. The comb-drive device is actuated by applying voltage to either of two comb electrodes. The driving signal is applied to pad 2, and pad 1 is grounded. Two sets of folded springs provide a mechanical restoring force that pulls the shuttle back to its rest position when the voltage is removed. At the right side of the figure, there are pictures captured when we applied the driving signal.

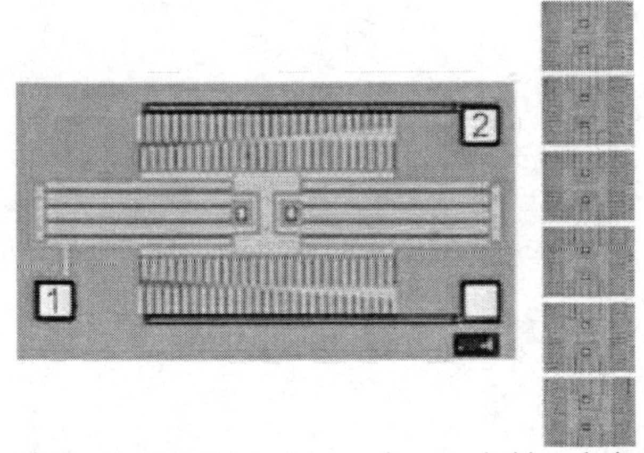

Figure 5: The CCD picture of a comb-drive device measured in this work.

The images at different applying voltages are taken to analyze displacement. The experiment results of displace vs. applied voltages for the comb-drive device is shown in Figure 6. The displacements are proportional to voltage-squared as governed by the electrostatic force.

4 CONCLUSIONS

This paper presents a three-dimensional dynamic measurement tool that can capture bright-field images and interferometric images at the same time. We have constructed a motion measurement system based on stroboscope. Measurements of in-plane motion implemented on a piezoelectric nano-platform and a comb-drive device have demonstrated that the in-plane resolution is about 10 nm. At present, we are trying to do the compensation of the gradient-based algorithm to improve the in-plane resolution to the level of less than 5nm. The construction of the section of interferometric system to measurement the out-of-plane motion is under way.

ACKNOWLEDGEMENT

This work is partially supported by the NSC (National Science Council, Taiwan, ROC) through the Grant contact No: NSC-93-2212-E-002-023.

REFERENCES

[1] Eric M. Lawrence, Kevin E. Speller, Duli Yu, "MEMS Characterization using Laser Doppler Vibrometry," *Reliability, Testing and Characterization of* MEMS/MOEMS II, SPIE Proceedings Vol. 4980, January 2003

[2] C. O'Mahony, M. Hill, M. Brunet, R. Duane and A. Mathewson, "Characterization of micromechanical structures using white-light interferometry," *Measurement Science and Technology*, Vol. 14, Issue 10, October 2003, pp. 1807-1814

[3] C. Quentin Davis, Dennis M. Freeman, "Using a light microscope to measure motions with nanometer accuracy," *Optical Engineering*, Vol. 37, No. 4, April 1998, pp. 1299~1304

[4] C. Quentin Davis, Dennis M. Freeman, "Statistics of subpixel registration algorithms based on spatiotemporal gradients or block matching," *Optical Engineering*, Vol. 37, No. 4, April 1998, pp. 1290-1298

[5] D. M. Freeman, A. J. Aranyosi, M. J. Gordon and S. S. Hong, "Multidimensional motion analysis of MEMS using computer microvision," *Solid-State Sensor and Actuator Workshop*, Hilton-Head Island SC, June 8-11, 1998, pp. 150-155

[6] W. Hemmert, M. S. Mermelstein, D. M. Freeman, " Nanometer Resolution of Three-Dimensional Motions Using Video Interference Microscopy," in Proc. of *12th IEEE International Conference on Micro Electro Mechanical Systems (MEMS '99)*, 17-21 Jan. 1999, pp. 302-308

[7] C. Rembe, H. Aschemann, S. aus der Wiesche, E. P. Hofer, H. Debéda, J. Mohr and U. Wallrabe, "Testing and improvement of micro-optical-switch dynamics," *Microelectronics Reliability*, Vol. 41, March 2001, pp. 471-480

[8] C. Rembe, B. Tibken and E. P. Hofer, "Analysis of the Dynamics in Microactuators Using High-Speed Cine Photomicrography," *Journal of Microelectromechanical Systems*, Vol. 10, No. 1, March 2001, pp. 137-145

[9] C. Rembe, P. Caton, R. M. White, R. S. Muller, "Stroboscopic Interferometry for Characterization and Improvement of Flexural Plate-Wave Transducers," *Reliability, Testing and Characterization of MEMS/MOEMS II*, SPIE Proceedings Vol. 4558, 2001

[10] C. Rembe, L. Muller, R. S. Muller, R. T. Howe, "Full Three-Dimensional Motion Characterization of a Gimballed Electrostatic Actuator," in *Proc. of IEEE International Reliability Symposium*, Orlando, Florida, April 30~May 3, 2001

[11] C. Rembe and R. S. Muller, "Measurement System for Full Three-Dimensional Motion Characterization of MEMS," *Journal of Microelectromechanical Systems*, Vol. 11, No. 5, Oct. 2002

[12] Eric M. Lawrence, Christian Rembe, "MEMS Characterization using New Hybrid Laser Doppler Vibrometer / Strobe Video System," *Reliability, Testing and Characterization of* MEMS/MOEMS III, SPIE Proceedings Vol. 5343, January 2004

Comparative Study of fabrication patterns of a ferroelectric polymer P(VDF-TrFE) on gold thin film and gold ball via Dip-pen Nanolithography

Q. Tang, S. Q. Shi*, L. M. Zhou and C. H. Xu

Department of Mechanical Engineering, The Hong Kong Polytechnic University.
Hung Hom, Kowloon, Hong Kong, mmsqshi@polyu.edu.hk

ABSTRACT

A Gold ball with many atomically flat facets was prepared using melting method, and a gold thin film was prepared by sputtering. A comparative study of fabrication of a ferroelectric polymer poly(vinylidene fluoride-trifluorethylene) [P(VDF-TrFE)] patterns on the gold ball and gold thin film via dip-pen nanolithography (DPN) was presented in this paper. The transport rate of P(VDF-TrFE) to the gold ball was greater than that to the gold thin film due to less friction force.

Key Words: dip-pen nanolithography, gold thin film, gold ball, atomic force microscopy, [P(VDF-TrFE)]

1 INTRODUCTION

Atomic force microscopy (AFM) was invented in 1986 by Binnig et al. At that time the main function of AFM was to visualize surface morphology at high spatial resolution [1]. In the following years, it was found that AFM tip could change the surface property mechanically or chemically. With the development of cantilevers, several AFM-based nanotechnologies, such as nanooxidation, nanomanipulation, nanografting, force nanolithography, nanocatalysis and dip-pen nanolithography, were developed rapidly, making AFM a promising technique for nanofabrication [2].

Dip-pen nanolithography [3] has drawn great attention and obtained rapid development since its invention due to its simplicity and high resolution. Arbitrary nanostructures composed of chemical(s) were fabricated at ambient condition by the deposition of chemical(s) from a coated tip to the substrate of interest during the contact between the tip and the substrate. The line width constructed via DPN was reported to be as narrow as 15 nm. Various materials such as thiols [3], biomaterials [4], inorganic materials or sol-based inks [5-6], and organic materials [7] were used as DPN inks. Nanostructures of some conducting polymers were also created by direct surface patterning or in situ polymerizing the generated patterns composed of monomer ink [8-9]. The substrate was extended from gold to Si, SiOx, glass and mica, and the tip was developed from a single tip to tip array [10], thus making DPN a versatile and efficient nanolithography.

Ferroelectric polymer is an important class of functional materials, which has potential applications in electrical, optical, biomedical, robotic, and sensing devices [11]. P(VDF-TrFE) was reported to show the highest ferroelectric polarization and electromechanical response among the known ferroelectric polymers [12], therefore, it was widely used as acoustic sensors and transducers [13-14]. Micro- and nanostructures of P(VDF-TrFE) were fabricated onto polycrystalline gold film via DPN. P(VDF-TrFE) molecules in these patterns are highly oriented and hold ferroelectric property [15].

The quality of DPN can be affected by many parameters, such as temperature, relative humidity, chemical affinity between the ink and the substrate, and surface property of the substrate. Temperature and relative humidity effect on DPN have been extensively studied in the thiol (ink) - gold (substrate) system [16-18].

In present study, we studied the effect of surface property on DPN by comparing the results of P(VDF-TrFE) transferred to a gold ball and a polycrystalline gold thin film. It is found that surface property plays important role in DPN process.

2 EXPERIMENTAL

2.1 Materials

P(VDF-TrFE) with a mole ratio of VDF/TrFE as 4/1 (Piezotech S. A), 99.5% acetone (Ajax Finechem), 30% hydrogen peroxide solution (BDH Laboratory), concentrated sulfuric acid (Aldrich) were used as received. 99.99% gold wire was supplied by TSL Jewelry Shop.

2.2 Preparation of Polycrystalline Gold Thin Film

Initially 10 nm Ti and then 30 nm gold were coated on *n*-Si (100) wafer by sputtering with ExplorerTM 14 Denton Vacuum.

2.3 Preparation of the Gold Ball

The preparation process of the gold ball is referred to literatures [19-20] with minor modification. A gold wire with diameter of 0.15 mm was rinsed with acetone and dipped in piranha solution (V/V=7/3 concentrated sulfuric acid / 30% H_2O_2. **Note**: piranha solution reacts violently with organic solvents) for 15 min to remove organic materials on gold surface. The wire was melted in CH_4 flame, and a small ball about 3 mm diameter was formed when the melted gold left the flame and re-solidified in the atmosphere.

2.4 DPN with P(VDF-TrFE)

A solution of 3.6 mg of P(VDF-TrFE) dissolved in 5 mL of acetone was used as the ink. The AFM tip was coated by dipping the cantilever in the above solution for 1 min, and blown dry with nitrogen, a thin film of P(VDF-TrFE) was coated on the cantilever surface after the volatilization of acetone, as visualized with a field emission scanning electron microscopy. DPN patterning and AFM imaging were performed with the coated tip and a commercial AFM (Solver P47H, NT-MDT, Russia). The silicon cantilever (contact "golden" silicon cantilever, CSG11) used for DPN was purchased from NT-MDT company, its typical spring constant, tip curvature radius and resonance frequency were about 0.03 N/m, 10 nm and 10 kHz, respectively. DPN patterning was operated in contact mode with a contact force of 2.0 nN. The AFM system was placed in a chamber. The relative humidity (RH) was controlled by dry N_2 or humid N_2 purged into the chamber at a rate of 0.2 L/min and measured with a humidity and temperature sensor with an accuracy of ±2% (Cole-Parmer Instrument Company). Fig. 1 is the schematic diagram of the DPN experimental system. All the experiments were conducted at room temperature (about 22°C) unless otherwise stated. Each data was averaged from five independent measurements.

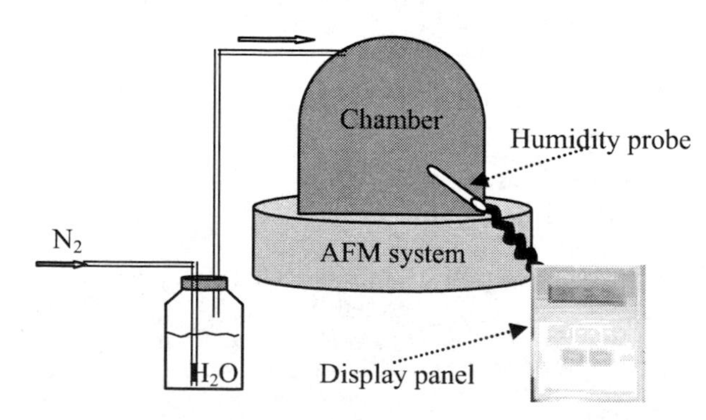

Figure 1: Schematic diagram of the AFM experimental system for DPN. AFM was placed in a chamber. Purge dry N_2 in to the system to obtain low relative humidity, whilelet N_2 pass through a water bottle to obtain high relative humidity.

3 RESULTS AND DISCUSSION

The surface topology of the flat facets on the gold ball was characterized using AFM, as shown in Fig. 2, some surface steps can be seen clearly, and the step height is ranged from 0.25 nm to 0.30 nm, close to the reported theoretical inter-atomic distance of gold, 0.288 nm [21]. Within each

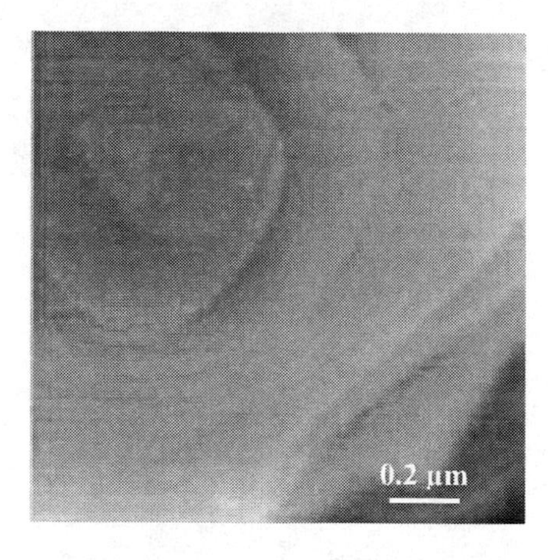

Figure 2: topography of flat facets on a gold ball

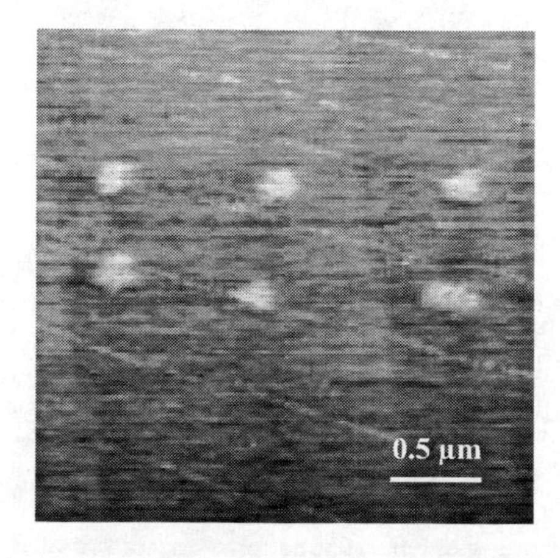

Figure 3: topography of a P(VDF-TrFE) point array on gold ball via DPN (contact time: 60s, 80% RH, 22°C).

NSTI-Nanotech 2005, www.nsti.org, ISBN 0-9767985-1-4 Vol. 2, 2005

step the surface is atomically flat with a roughness of less than 1 Å, while the roughness of the gold thin film prepared by sputtering is about 0.6 nm.

The topography and lateral force microscopy (LFM) images of the P(VDF-TrFE) patterns on a gold thin film fabricated via DPN may be referred to literature [15]. P(VDF-TrFE) was deposited on flat facets of the gold ball during the contact between the coated tip and the gold substrate and formed desired nano- to microstructures, which could be recognized in topography and LFM images. Fig. 3 gives a clear topography of a P(VDF-TrFE) point array on gold ball. The contrast is better and the pattern is more uniform than the patterns on a gold thin film due to less roughness, and the thickness of P(VDF-TrFE) in the patterns is about 1 nm, the same as that on the gold thin film. This indicates that P(VDF-TrFE) patterns on gold ball fabricated via DPN contain two layers of molecules [22].

Like on the gold thin film [15], hydrophobic polymer P(VDF-TrFE) on the gold ball exhibited darker contrast, i. e., smaller friction force between the tip and the substrate was produced, as illustrated in Fig. 4, the LFM image of a point array of P(VDF-TrFE) with nearly identical diameter shows darker contrast compared with gold substrate.

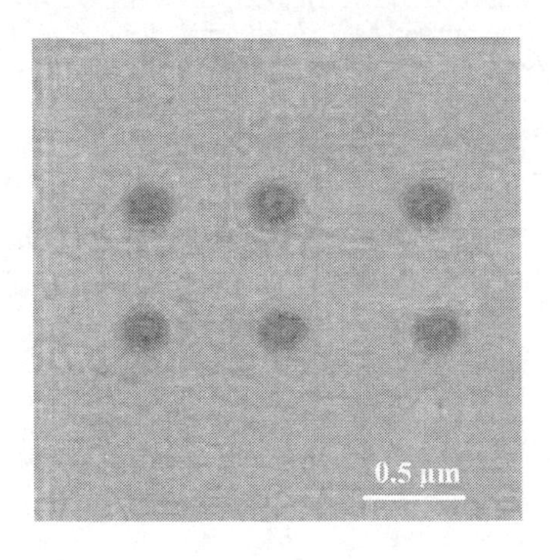

Figure 4: LFM images of a P(VDF-TrFE) point array fabricated via DPN on gold ball (contact time:60s, 80% RH, 22°C).

Fig. 5 is the curve of dot area (A) as a function of contact time (t) for the P(VDF-TrFE) patterns on the gold ball and gold thin film. The area was measured with an error of less than 10%. It can be found that A changes linearly with the contact time for both the gold ball and the gold thin film but with different rates. The slope of $A-t$ curve on the gold ball, 380 ± 34 nm^2 / s, is greater than the slope of $A-t$ curve on gold thin film, 150±13 nm^2 /s. (Here we defined the slope of $A-t$ curve as the transport rate of ink, v, and $v=\frac{dA}{dt}$). Therefore, at the same condition, the transport rate of P(VDF-TrFE) to gold ball is greater than that to gold thin film. The reason caused the difference is believed to be the surface topography of substrate. As described above, the roughness of the polycrystalline gold film (about 0.673 nm) is greater than the roughness of the facets on the gold ball (less than 0.1 nm). Kasupke and Henzler have proved that roughness strongly influences the sticking coefficient (S_0) of molecule on solid surface, and S_0 increases greatly with increasing surface roughness [23], Daikhin and Urbakh also reported that roughness increases the frictional force in confining system [24]. The increased sticking coefficient or friction force on the rough surface make the diffusion of the ink molecules more difficult, resulting in slower transport rate compared with less rough surface.

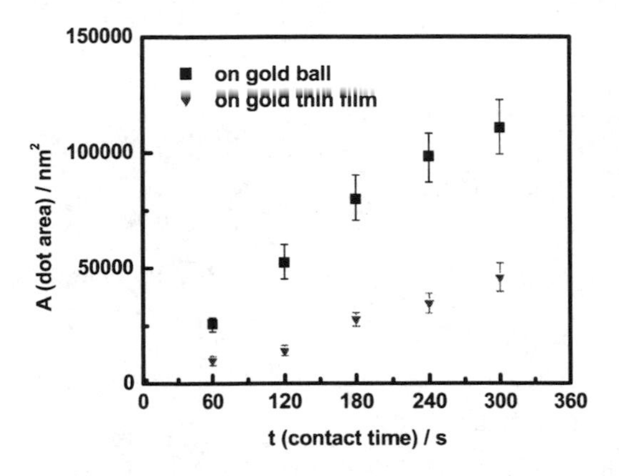

Figure 5: Plot of point area (A) as a function of contact time (t) at 80% RH, 22°C.

In addition, the gold ball can be reused for two more times for DPN after cleaned using a simple procedure. In this procedure, the gold ball was copiously rinsed with acetone, subsequently immersed in piranha solution for 15min, then copiously rinsed with deionized water, followed blown dry with N_2, finally cauterized on CH_4 flame to orange color. There is little difference between the patterns transferred to a fresh gold ball, and a gold ball cleaned once or twice. However, surface roughness increases with cleaning times. A gold ball is not suitable for DPN after being cleaned above three times.

In present work, the methods to prepare the gold film and gold ball were different, which results in different surface property. In order to study the quantitative relationship between surface roughness and the transport rate of ink, we will prepare a series of gold substrates with various surface roughness using the same method in future.

4 CONCLUSIONS

The paper reported the effect of substrate surface on DPN using a ferroelectric polymer P(VDF-TrFE) as ink. Two kinds of gold surfaces, polycrystalline thin film and atomically flat facets were prepared by sputtering and melting methods, respectively. A comparative study of P(VDF-TrFE) patterns fabricated onto a gold ball and gold thin film via dip-pen nanolithography was presented. The surface roughness plays important role in DPN. The transport rate of ink to a rough substrate is slower than that to a relative flat substrate due to the increased friction force or sticking coefficient of molecules.

5 ACKNOWLEDGEMENT

This work was funded by a research grant from the Hong Kong Polytechnic University (G-T687).

REFERENCES

[1] G. Binnig, C. F. Quate and Ch. Gerber, Phys. Rev. Lett. 56, 930,1986.

[2] Q. Tang, S. Q. Shi and L. M. Zhou, J. Nanosci. Nanotech. 4, 948, 2004.

[3] R. D. Piner, J. Zhu, F. Xu and C. A. Mirkin, Science 283, 661, 1999.

[4] J. Hyurk, D. S. Ginger and K. B. Lee, Angew. Chem. Int. Ed. 42, 2309, 2003.

[5] L. M. Demers and C. A. Mirkin, Angew. Chem. Int. Ed. 40, 3069, 2001.

[6] L. Fu, X.G. Liu, Y. Zhang, V. P. Dravid and C. A. Mirkin, Nano Letters 3, 757, 2003.

[7] P. Manandhar, J. Jang, G. C. Schatz, M. A. Ratner and S. Hong, Phys. Rev. Lett. 90, 115505-1, 2003.

[8] P. Xu and D. L. Kaplan, Adv. Mater. 16, 628, 2004.

[9] M. Su, M. Aslam, L. Fu, N. Q. Wu and V. P. Dravid, Appl. Phys. Lett. 84, 4200, 2004.

[10] M. Zhang, D. Bullen, S. W. Chung, S. Hong, K. S. Ryu, Z. F. Fan, C. A. Mirkin and C. Liu, Nanotechnology 3, 212, 2002.

[11] H. S. Nalwa, Ferroelectric Polymers: chemistry, physics, and applications (New York: Marcell Dekker, Inc), 1995.

[12] H. Ohigashi and K. Koga, Jpn. J. Appl. Phys. 21, L455, 1982.

[13] P. Lum, M. Greenstein, C. Grossman and T. L. Szabo, IEEE Transactions on Ultrasonics Ferroelectrics and Frequency Control 43, 536, 1996.

[14] M. Toda, IEEE Transactions on Ultrasonics Ferroelectrics and Frequency Control 49, 299, 2002.

[15] Q. Tang, S. Q. Shi, H. T. Huang and L. M. Zhou, Superlattices and Mircostructures 36, 21, 2004.

[16] P. V. Schwartz, Langmuir 18, 4041, 2002.

[17] S. Rozhok, R. D. Piner and C. A. Mirkin, J. Phys. Chem. B 107, 751, 2003.

[18] S. Rozhok, P. Sun, R. D. Piner, M. Lieberman and C. A. Mirkin, J. Phys. Chem. B 108, 7814, 2004.

[19] T. Hsu and J. M. Cowley, Ultramicroscopy 11, 239, 1983.

[20] C. B. Ross, L. Sun and R. M. Crooks, Langmuir 9, 632, 1993.

[21] V. M. Hallmark, S. Chiang, J. F. Rabolt, J. D. Swalen and R. J. Wilson, Phys. Rev. Lett. 59, 2879, 1987.

[22] A. V. Bune, V. M. Fridkin, S. Ducharme, L. M. Blinov, S. P. Palto, A. V. Sorokin, S. G. Yudin and A. Zlatkin, Nature 391, 874, 1998.

[23] N. Kasupke and M. Henzler, Surf. Sci. 92, 40, 1980.

[24] L. Daikhin and M. Urbakh, Phys. Rev. E 49, 1424, 1994.

Theoretical and Experimental Study of Synthetic MFM Tips

Yihong Wu*, Gang Han* and Yuankai Zheng**

*Department of Electrical and Computer Engineering, National University of Singapore
4 Engineering Drive 3, Singapore 117576, elewuyh@nus.edu.sg
**Data Storage Institute, 5 Engineering Drive 1, Singapore 117608

ABSTRACT

One of the extensively investigated issues for magnetic force microscopy is its low resolution as compared to other scanning probe techniques such as atomic force microscopy and scanning tunneling microscopy. This is mainly caused by the fact that the magnetically responsive coating covers the entire body of the tip as well as the relatively large sample-tip spacing. Recently, we have developed a new type of tip which effectively has a magnetic coating only at the tip apex and at the same time can be used at a smaller sample-tip distance due to its low moment. In this paper, the superior performance of the new tip is studied theoretically using an analytical model. The analytical results are supported by the experimental data.

Keywords: MFM, point-dipole, synthetic, high-resolution, analytical model

1 INTRODUCTION

Magnetic force microscopy (MFM) has been used extensively to study the magnetic properties of materials. The resolution of MFM is determined by many factors including the geometric shape of the tip, the materials and structures used to form the coating, the tip-sample distance, and in some cases the interaction between the tip and the sample. Generally speaking, a high-resolution MFM requires tips with small lateral dimensions and stable magnetic coatings with a highly localized stray field. The latter allows the scanning to be performed with a small tip-sample distance which gives a high resolution without disturbing the sample domain structures. Therefore, there are two approaches to improve the resolution of MFM. The first approach is to engineer the tip shape to achieve a smaller lateral dimension and the second to use different magnetic coating materials, such as superparamagnetic and antiferromagnetic materials to reduce the tip stray field. So far, many efforts had been made to improve the resolution of MFM through sharpening the tips using different approaches such as attaching carbon nanotubes to the original tips,[1,2] trimming the tips by focus ion beam (FIB),[3-6] electron beam lithography,[5,7,8] and ion beam etching,[9] selective deposition by self-field emission,[10] electron beam irradiation,[8,10,13,14] and focused electron beam decomposition and deposition.[16,17] Although all these techniques, to a certain extent, can improve the resolution, the primary drawback is that the tips must be processed one-by-one instead of using a batch process.

Based on this background, recently we have developed two different types of MFM tips.[18,19] In one of the designs, an exchange biased bilayer is used as the magnetic coating which significantly improves the stability of weak moment tips. In the second design, the stability is improved further using a ferromagnetic FM/Ru/FM trilayer. Furthermore, the tip functions effectively as a point-dipole due to the formation of a flux-closed structure at the tip apex when the two FM layers are coupled antiferromagnetically. In this paper, we analyze the tip using an analytical model and compare the results with experimental data.

2 THEORETICAL CONSIDERATION

Fig.1 shows the schematic of the synthetic MFM tip. The magnetic coating consists of two FM layers separated by a thin Ru layer. It is well known from exchange coupling studies that the FM layers can couple either ferromagnetically or anti-ferromagnetically, depending on the Ru thickness. The first antiferromagnetic coupling appears when the Ru layer is about 0.6-0.8 nm. In this case, as shown in Fig.1, an effective point-dipole will be formed at the tip apex due to the flux closure requirement. Intuitively, one can understand that the resolution of this kind of tip will be better than that of the single layer tip.

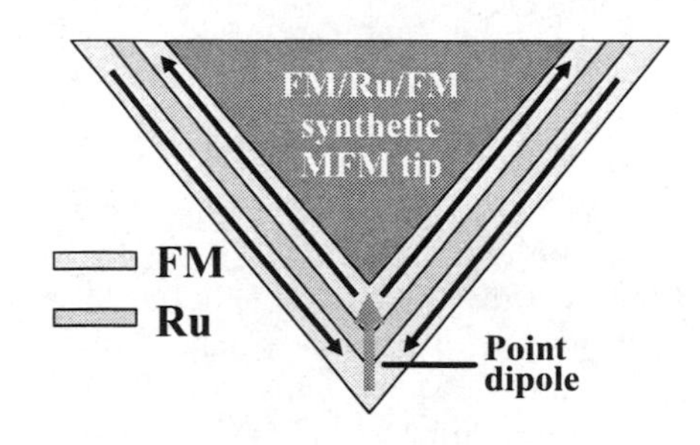

Fig.1. Schematic of the synthetic tip.

The MFM tip, as a magnetized body, when brought into the stray field of a sample, will have a magnetostatic potential energy E given by

$$E = -\mu_0 \int (\vec{M}_{tip} \cdot \vec{H}_{sample}) dV_{tip} \qquad (1)$$

Then the force acting on the MFM tip is given by

$$\vec{F} = -\nabla E = -\mu_0 \iiint \nabla(\vec{M}_{tip} \cdot \vec{H}_{sample}) dV_{tip} \quad (2)$$

Generally, we use the dynamic detecting mode, in which the signal detected is proportional to the force gradient in the linear range. For the tip with a uniform magnetization distribution in z direction, the force gradient in z direction is given by

$$F_z' = -\mu_0 \iiint M_z(\vec{r}') \frac{\partial^2 H_{sample}}{\partial z^2} dV_{tip}(\vec{r}' + \vec{r}) \quad (3)$$

Here, $\vec{r}$ is the position vector of the tip apex and $\vec{r}'$ is the position vector of a point inside the tip with respect to the tip apex. The above equation implies that the MFM response is given by a three dimensional convolution between the magnetization distribution on the entire tip body and the fringe field from the samples. Although the integration in real space can generate the MFM signal at each point, it is more instructive to deal with this kind of the problem in the spatial frequency domain in which the tip magnetization and sample fringe field will be decoupled in the linear-response regime. The latter is generally referred to as frequency response analysis of MFM.[20] In order to simplify the problem and at the same time to have a clear physics picture, S. Porthun *et al.*[21] has developed a one-dimensional model for MFM imaging of a periodical magnetization pattern given by

$$M_z(k, x) = M_s \cos(kx), \quad (4)$$

For a bar-shaped tip with a uniform magnetization M_t, depth D in y-direction, height H in z-direction and width t in x-direction (Fig.2), the frequency spectrum of the force gradient is given by

$$F_z'(k,d) = -\mu_0 \frac{M_s M_t}{2}(D \cdot kt) e^{-kd}(1 - e^{-k\delta})(1 - e^{-kH}) \frac{\sin(kt/2)}{kt/2} \quad (5)$$

Here, μ_0 is the permeability of vacuum, M_s is the sample saturation magnetization and M_t is the tip magnetization, d is the tip-sample distance, δ is the thickness of the media, and k is the spatial frequency. In the above equation, the last term sets a cut-off frequency beyond which the MFM tip cannot produce truthful information of the sample fringe field. The term e^{-kd}, originating from the tip-sample distance, simply reduces the high-frequency components against the low-frequency ones. It is interesting to note that the media thickness and tip height actually enhance the high-frequency response, particularly when they are comparable with the feature sizes of the magnetic patterns on the sample. The latter can be satisfied easily for the sample thickness, but it is not true for the tip height because it is normally in the micrometer range for commercial tips.

However, if one has two identical tips with their magnetization being oriented in opposite directions and their physical position being shifted against each other by a small distance Δd in the z direction, the net frequency response will be given by

$$F_z'(k,d) = -\mu_0 \frac{M_s M_t}{2}(D \cdot kt) e^{-kd}(1 - e^{-k\delta})(1 - e^{-kH})(1 - e^{-k\Delta d}) \frac{\sin(kt/2)}{kt/2} \quad (6)$$

The multiplication factor $(1 - e^{-k\Delta d})$ will boost the high-frequency components because $(1 - e^{-k\Delta d}) \approx k\Delta d$ for small Δd. When H >> Δd, the tip function effectively as a point-dipole tip.

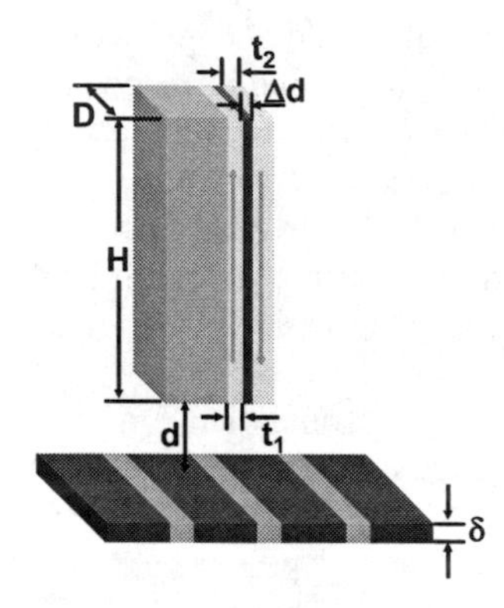

Fig.2. Schematic of simplified MFM tip.

The synthetic tip which we have reported in our earlier work consists of two ferromagnetic layers antiferromagnetically coupled via a thin non-magnetic layer. Although the inner and outer layers have a slightly different shape, they are more or less equivalent to the two identical tips that we have just discussed above. This explains why the synthetic tip has a higher resolution than the conventional single layer tip. Of course, it should be noted that the absolute value of the MFM signal from the synthetic tip weakens due to the partial cancellation of the signal from the two layers with opposite magnetization directions. It is also obvious that the multiplication factor $(1 - e^{-k\Delta d})$ will become $(1 + e^{-k\Delta d})$ when the two magnetic layers are coupled in parallel. This will result in the degradation of resolution, though the absolute strength of the signal will be enhanced.

The above discussion suggests that it is important to control the Ru thickness precisely so as to obtain a higher resolution in the synthetic tip. The resolution enhancement of synthetic tip has been demonstrated in our previous work. In this paper, we report a systematic study of the effect of the Ru thickness on the performance of the FM/Ru/FM synthetic tips. Fig.3 shows typical calculated frequency response of both a single-layer tip and synthetic tips with different interlayer spacing. Other parameters used for the simulation are shown in the figure. As compared to the single-layer tip, the synthetic tips have a suppressed response at low frequency, which contributes to the improvement of resolution, though the cutoff frequency is still the same. In practice, the synthetic tip is expected to have a much better performance when the FM layers are very thin because it is thermally more stable than the single-layer tip.

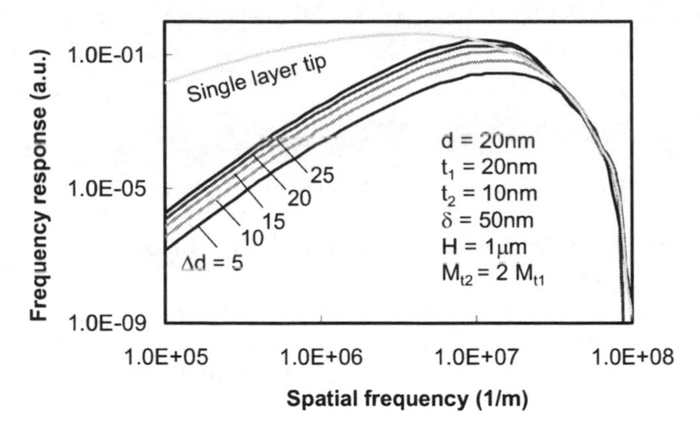

Fig.3. Calculated frequency response of single layer and synthetic tips.

3 RESULTS AND DISCUSSION

A series of MFM tips with the structure Cr(3 nm)/CoCrPt (20 nm)/Ru /CoCrPt (20 nm)/Cr(3 nm) (CCP is Co74Cr13Pt13) were fabricated by sputtering the multiple layers on the bare FMR tips from Nanosensor. The bottom Cr layer was used as a seed and adhesion layer. The sputtering conditions were optimized in our work on spin-valves. All the tips were designed to have the same layer configurations except that the nominal thickness of the Ru layer is varied from 0.2 nm to 1.2 nm with a step of 0.2 nm, and at 1.6 nm, 2.4 nm, and 3.2 nm, respectively. The nominal values were obtained from the growth rate on flat substrates; the actual thickness on the tips may differ from the nominal values due to the topographical effect near the tip apex. Nevertheless, the comparison among different tips is still valid and meaningful because care has been taken to make sure that all other parameters are the same and the only changing parameter is the thickness of the Ru layer. All the MFM imaging experiments were performed at room temperature in air using a commercial scanning probe microscopy (Digital Instrument Dimension 3100), operated at the tapping/lift mode with a lift-height of 30 nm. As it is expected, the fabricated synthetic tip shows oscillator behavior with respect to the Ru layer thickness. Detailed comparison of the performance of tips with different Ru thicknesses will be discussed elsewhere. Here we only focus on those with a nominal thickness of 0.6-0.8 nm which leads to an anti-parallel coupling between the two FM layers.

The MFM images obtained from a synthetic tip and a conventional MESP tip are shown in Figs. 4(a) and 4(b), respectively, together with the averaged down track profile. The full width at half maximum of the transition decreases from 130 nm for the MESP tip to 110 nm for the synthetic tip for the specific patterns which we have measured. It is obvious that not only the zigzag transition but also the structure of the grains can be seen clearly in the images obtained from the synthetic tip. The superior performance of the synthetic tip compared to the conventional tip is attributed to the stable magnetic configuration and the concentration of magnetic charges at the tip apex to form the point dipole. The latter makes the effective volume of the tip much smaller than that of the conventional tip. The point-dipole response can be seen from the overshot in the down track profile of the synthetic tip, but not in that of the conventional tip.

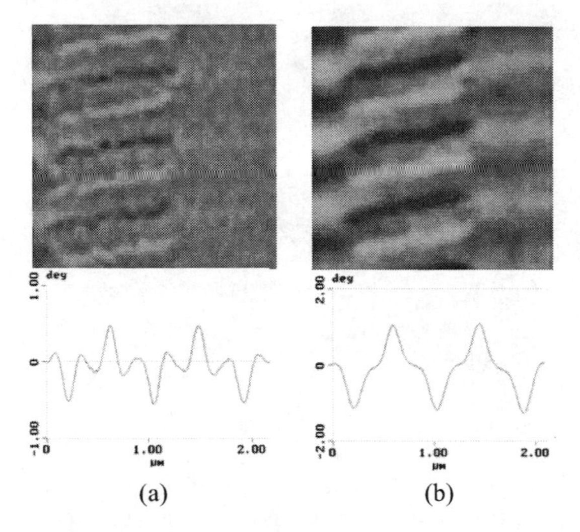

Fig.4. MFM images and averaged down-track profile of the magnetic response. (a) synthetic tip and (b) conventional tip.

The superior performance of the synthetic tip is also reflected in the imaging of Neel domain walls as shown in Fig.5. The cross-wall profile obtained from the synthetic tip is almost the same as the theoretical calculations.

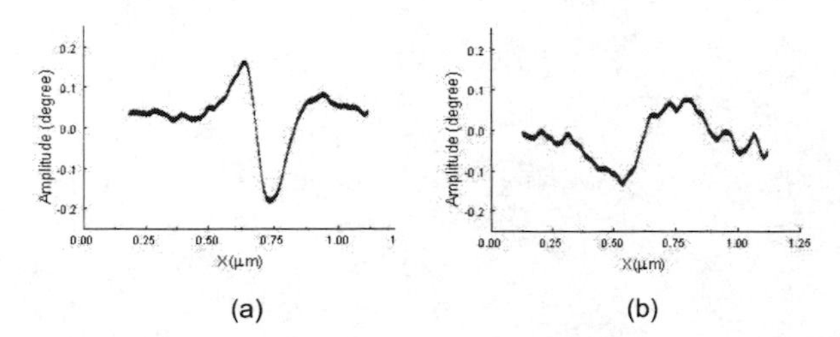

Fig.5. Cross-wall profile of Neel domain walls: (a) synthetic tip and (b) conventional tip.

The combination of the synthetic structure and point-dipole greatly reduces the spreading of the stray field of the tip. The anti-parallel coupling between the two FM layers also greatly reduces the interactions between the tip and the sample. This can be seen in the results shown in Fig.5 for imaging of cross tie 90^0 domain walls. This type of domain walls is very sensitive to external field and, therefore requires the use of a very low moment tip to image them. As it is shown in Fig.5, a very clear image of the cross tie 90^0 domain wall was obtained using a synthetic tip, but it could not be imaged using the conventional tip.

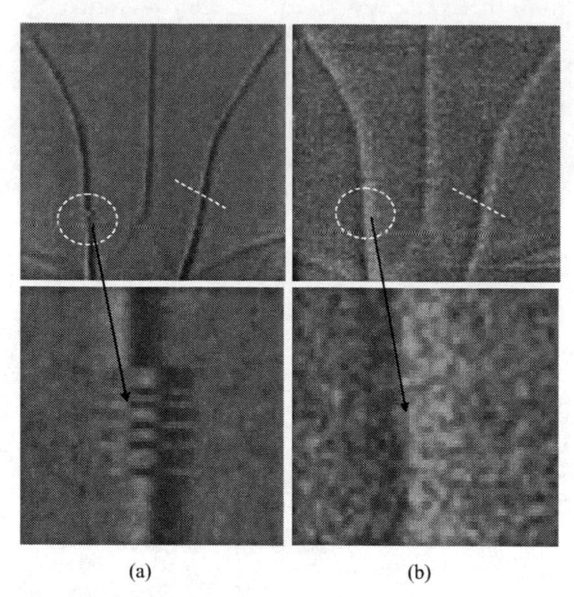

(a) (b)

Fig.5. MFM images of cross tie domain walls. (a) synthetic tip; (b) conventional tip.

The reduced tip-sample interaction can also be seen in the results shown in Fig.6 which are MFM images of a soft magnet nanostructure obtained at different lift heights. The influence of the tip-sample interaction in conventional tips can be seen from the movement of the boundary lines of the domain structures in different scans, which is not so obvious in images obtained from synthetic tips.

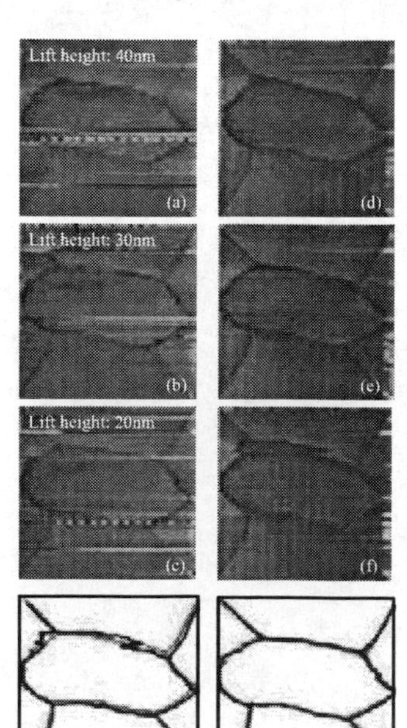

Fig.6. MFM images of a soft magnet. (a)-(c), conventional tip at a lift height of 40 nm, 30 nm and 20 nm, respectively; (d)-(f), synthetic tip at a lift height of 40 nm, 30 nm and 20 nm, respectively. (g) boundary lines of the domain walls for (a)-(c); (h) boundary lines of the domain walls for (d)-(f).

4 CONCLUSIONS

We have developed a novel type of MFM tips which does not only have a high resolution but also exhibits reduced sample-tip interactions. The superior performance of the synthetic tip has been studied both theoretically and verified experimentally. Further work is being done to improve the coating process so as to fabricate tips in a well-controlled fashion.

REFERENCES

[1] Takayuki Arie, Hidehiro Nishijima, Seiji Akita, and Yoshikazu Nakayama, J. Vac. Sci. Technol B **18**,1, 2000.
[2] E. S. Snow, P. M. Campbell, and J. P. Novak, Appl. Phys. Lett. **80**, 2002, 2002.
[3] G.N.Phillips, M. Siekman, L. Abelmann, and J .C .Lodder, Appl. Phys. Lett. **81**, 865, 2002.
[4] Dmitri Litvinov and Sakhrat Khizroev, Appl. Phys. Lett. **81**, 1878, 2002.
[5] S.E.Huq, L. Chen, and P. D. Prewett, J. Vac. Sci. Technol. B **13**, 6, 1995.
[6] L.Folks, M. E. Best, P. M. Rice, B. D. Terris, and D. Weller, Appl. Phys. Lett. **76**, 909, 2002.
[7] Paul B.Fischer, Mark S. Wei, and Stephen Y. Chou, J. Vac. Sci. Technol. B **11**, 6, 1993.
[8] Stephen Y. Chou, mark S. Wei, and Paul B. Fischer, IEEE Tran. Magn, **30**, 4485, 1994.
[9] P.Leinenbach, U. Memmert, J. Schelten, and U. Hartmann, Appl. Surf. Sci. **145**, 492, 1999.
[10] C.H.Oon, J.T.L.Thong, Y. Lei, and W. K. Chim, Appl. Phys. Lett. **81**, 3037, 2002.
[11] M.R.Koblischka, U. Hartmann, and T. Sulzbach, Thin Solid Film 428, 93, 2003.
[12] M. Ruhrig, S. Porthun, and J. C. Lodder, Rev. Sci. Instrum. **65**, 3224, 1994.
[13] R. Jumpertz, P. Leinenbach, A. W. A. Van der Hart, J. Schelten, Micorelectronic Engineering, **35**, 325, 1997.
[14] George D. Skidmore and E. Dan Dahlberg, Appl. Phys. Lett. **71**, 3293, 1997.
[15] I. Utke, P. Hoffmann, R. Berger, L. Scandella, Appl. Phys. Lett. **80**, 4792, 2002.
[16] A. Wadas and H. J. Hug, J. Appl. Phys. 72, 203, 1992.
[17] J. Lohau, S. Kirsch, A. Carl, G. Dumpich, and E. F. Wassermann, J. Appl. Phys. 86, 3410, 1999.
[18] Z. Y. Liu, D. You, J. J. Qiu, and Y. H. Wu, J. Appl. Phys. **91**, 8843, 2002.
[19] Y. H. Wu, Y. T. Shen, Z. Y. Liu, K. B. Li, and J. J. Qiu, Appl. Phys. Lett. **82**, 1748, 2003.
[20] Hans J. Hug, B. Stiefel, P. J. A. Van Schendel, A. Moser, R. Hofer, S. Martin, and H. –J. Guntherodt, J. Appl. Phys. **83**, 5609 (1998).
[21] S. Porthun, L. Abelmann, and C. Lodder, J. Magn. Magn. Mater. **182**, 238 (1998).

Atomic Force Microscope As a Tool for Nanometer Scale Surface Patterning

S. Lemeshko*, S. Saunin* and V. Shevyakov**

* Nanotechnology Instruments Europe B. V. Arnhemshweg 34 D
7331-BL, Apeldoorn, The Netherlands, lemeshko@ntmdt.com
** Moscow Institute of Electronic Engineering, Zelenograd, Russia, 124498

ABSTRACT

Progress in scanning probe microscopy (SPM) has transformed scanning tunneling microscopes (STMs) and atomic force microscopes (AFMs) from measuring devices into technological tools. Local surface modification by scanning probe allows us to treat materials with resolution from the angstrom to micron level [1-3]. The demonstration of single electron transistor creation by LAO [4] opens the way to the development of industrial nanolithography processing.

However, an insufficient knowledge in detailed understanding of the LAO mechanism limits the integration of this process into bulk production. The first experiments in STM tip oxidation of the hydrogen passivated Si surface was demonstrated in [5]. It was shown that electrical and structural properties of the positive-biased surface are changed irreversibly in air at room temperature under the tip effect. The common explanation of these changes in surface topography is formation of the oxide under the tip. The tendencies obtained in various works of the oxide pattern shape and its growth kinetics on the conditions of tip-induced treatment allows us to propose an electrochemical mechanism of LAO. Significant effect of the humidity on oxidation velocity is observed in [6]. This fact confirms the necessity of adsorbed electrolyte layer presence to produce oxide that is in good agreement with the electrochemical model. Moreover, there are other technological parameters that affect the oxidation kinetics. For example, the conductivity of an oxidized material affects the oxidation rate [7].

Keywords: nanolithography , AFM, anodic, oxidation.

1 EXPERIMENT

Nanopatterning was performed on thin titanium layers (2 ± 1 and 8 ± 1 nm) evaporated by a cathode arc deposition technique [8] on thermally-oxidized silicon substrate. Such evaporation method allows to deposit of ultrathin continuous amorphous films with a surface roughness about 0.1 nm. LAO cam be done on Si or GaAs wavers as well. Low value roughness is dramatically important because the surface morphology is changed within few nanometers during oxidation.

A commercial scanning probe laboratory Ntegra developed by NT-MDT Co. has used for LAO process. As probes for nanooxidation and further topography measurements silicon cantilevers with hard and high-conductive coatings based on W_2C have used [9]. Experimental scheme of local anodic oxidation process is shown (see Fig. 1). Several force constant lever types have been chosen, generally in the range of 10–20 N/m. The stiff silicon cantilevers had resonant frequencies in the range of 100–200 kHz respectively. Experiments in LAO were carried out in 10-50% of free cantilever oscillation amplitude. Free cantilever amplitude range about 60-500 Å. This mode allows one to increase the lifetime of the conductive cantilevers and thus to raise the nanolithography process reproductivity. Using environmental control in SPM chamber we were able to vary a relative humidity during LAO within range 10-90%. By the software it was possible to apply bias voltage on tip or on sample within ± 10V with accuracy 0,001V and voltage impulse duration 0,01-1000ms.

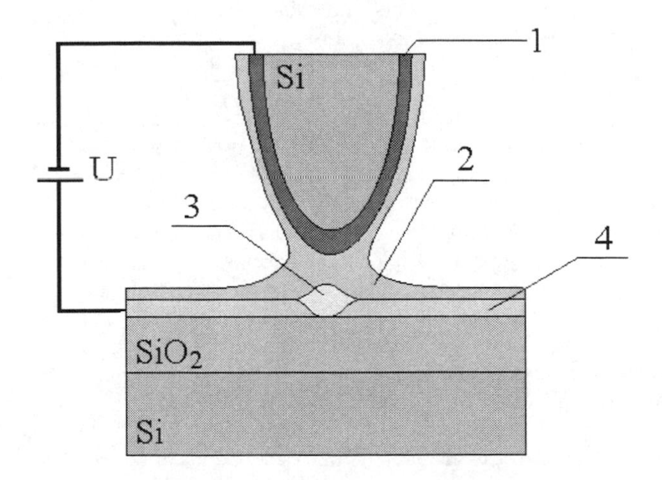

Figure 1: General scheme of LAO process. -1 Conductive coating, -2 Adsorption bridge, -3 Anodic oxide, - 4 Metal film.

2 RESULTS AND DISCUSSION

The main idea of this article is to show possibilities in development of local anodic oxidation for surface patterning on nanoscale with reproducible and linear patterns even on relatively large areas. Physical-chemical model of local anodic oxidation of thin titanium films were described in our previous article [10] and here we would like to focus on its applications.

It is well know n that conventional piezotube scanners are characterized by creep, drift and nonlinearity effects. These disturbance causes to unparallel oxide lines formed by LAO (see Fig. 2).

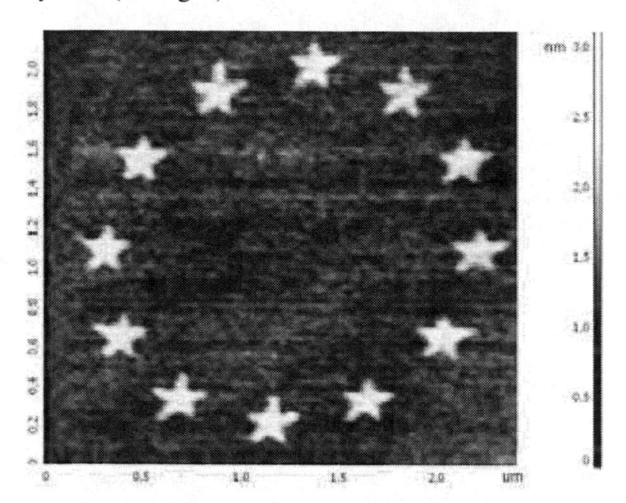

Figure 2: AFM image of Ti film after LAO process. Patterning and measurements were done with open-loop scanner. The small distortion of a ring stars is clearly visible.

In our experimental set up we have used scanning probe microscope with integrated capacitive sensors for closed-loop control scanner movement. Capacitance sensors are characterized by low noises for instance about 10ppm. Another important value of using capacitance sensors integrated in a piezotube scanner is improved linearity, thus scanners equipped with such sensors residual nonlinearity 0,1% in all three axis can be achieved. That high linearity ambles for SPM to compete in metrological field for example to create by LAO method nanometer scale test gratings for AFM or SNOM calibration.

We made an attend to make a patterning on large areas by using linearised scanner. As a sample we used intrinsic type of GaAs wafer in order to extend patterning time. On a figure 3 is demonstrated AFM image of GaAs surface after LAO patterning during 10minutes. Oxidizing started ftom left bottom corner of a square and then carried out continuously till tip returned back to the same position. On figure 4 shown cross sections over x and y directions. The distance between the left and right lines comparing with top and bottom lines within one pixel on AFM image. Image quality is 512x512 data points over 25x25 micrometer are, so scan step is 48 nm.

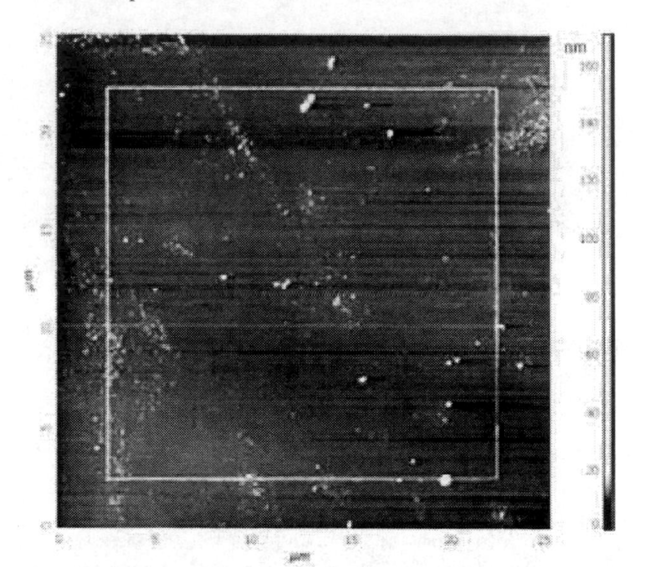

Figure 3: AFM image of GaAs surface after LAO process. In order to make this pattern probe started form left bottom corner made a square path and returned back within 10minutes.

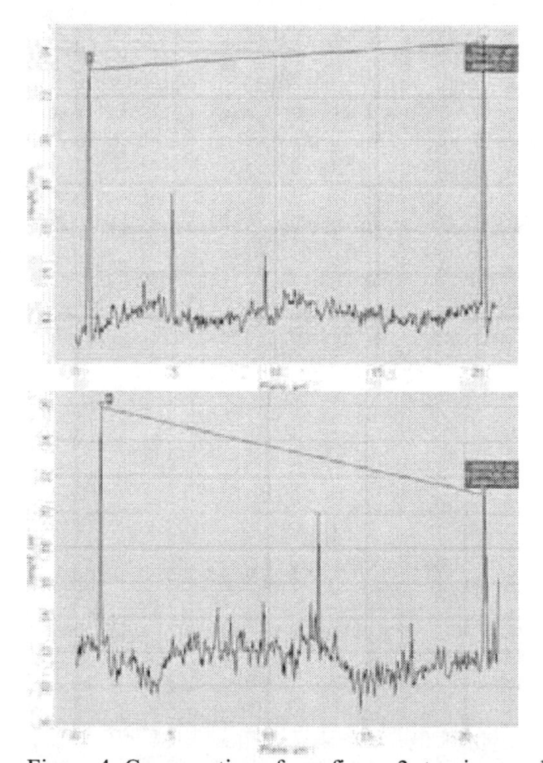

Figure 4: Cross-sections from figure 3, top image along X axis, bottom image along Y axis. The difference in distance between top-bottom and right-left lines is 50 nm, that determined AFM pixels resolution.

In order to check how good a tip returns back to starting point we have made a zoon in left bottom corner of square patters with 512x512 data point over 5x5 micrometers. On a figure 5 a close look on lines intersection is shown. The overlapping is 54nm and 29nm over 20micrometers lines in X and Y directions respectively.

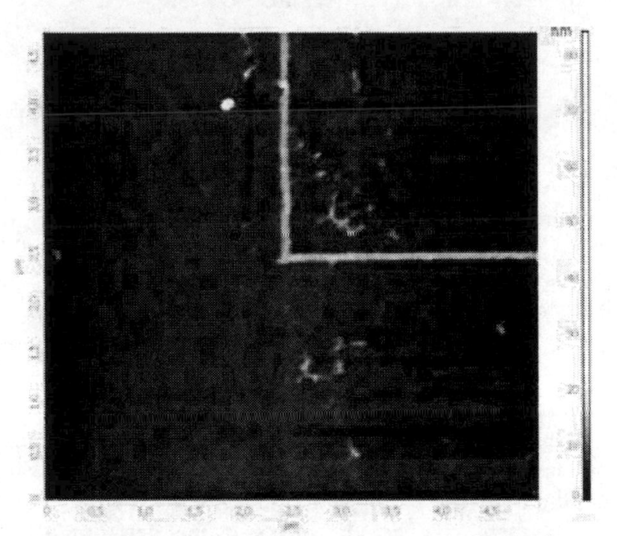

Figure 5: Zoom in to left bottom corner of AFM image shown in figure 3. Line width on half of height is 50nm.

Another interesting application of LAO process can be 2D nanopatterning. Looking for applications of the LAO process we are offering the method for ultrathing dielectric films formation. If we move the tip with step less then diameter of the oxide spot and apply variable voltage or impulse duration in each point according to some low, then within scan area we can modify the surface. For incarnation of the described idea the SPM software has been modified by such way that it is possible to load a mask as bitmap grayscale format. A loading mask may contain any information which we would like to transfer on surface by LAO, it can be drawing of nano-integrated circuit or just a portrait. In first step software is estimating a histogram of point's distribution over a grayscale image and associates the brightness in to percentage within 0-100% range. Then during LAO in every point of the raster we apply constant voltage but vary the voltage impulse duration according to certain brightness level of the particular point of the loaded mask. After performing such scanning with simultaneous LAO process it is necessary to check what's happened with the surface, for that we scan the same area in resonant mode AFM with same tip. On the figure 6 is shown AFM image of the modified surface of titanium film. Such image creation takes about 5-10min.

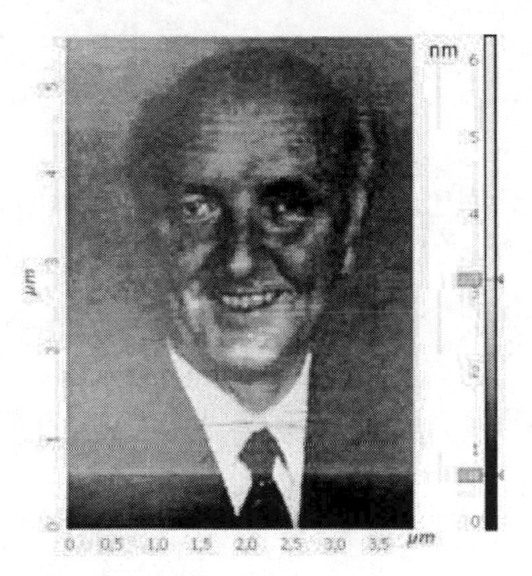

Figure 6: AFM image of Ti film after raster mode LAO.

3 CONCLUSIONS

In this work the possibility of LAO process accuracy improving is demonstrated. The main factor that can improve patterning process reproducibility are reduction of piezoscanner creep effect by integrating closed loop operation sensors for scanning control. Another important thing is a thermal drift that can also dramatically reduced buy making closed temperature stabilized chamber around a scanner, tip and sample.

REFERENCES

[1] E. Dubois and J. L. Bubbendorff, Solid-State Electron. 43, 1085, 1999.

[2] D. Saluel, J. Daval, B. Bechevet, C. Germain and B. Valon, J. Magn. Mater. 193, 488, 1999.

[3] R. Workman, C. Peterson and D. Sarid, Surf. Sci. 423, L277, 1999.

[4] K. Matsumoto, Physica B 227, 92, 1996.

[5] J. Dagata, J. Schneir, H. Harray, C. Evans, M. Postek and J. Bennett, Appl. Phys. Lett. 56, 2001, 1990.

[6] R. Held, T. Heinzel, P. Studerus and K. Ensslin, Physica E 2, 748, 1998.

[7] J. Dagata, F. Perez-Murano, G. Abadal, K. Morimoto, T. Inoue, J. Itoh and H. Yokoyama, Appl. Phys. Lett. 76, 2710, 2000.

[8] R. Boxman and S. Goldsmith, IEEE Trans. Plasma Sci.17, 705, 1989.

[9] V. Shevyakov, S. Lemeshko and V. Roschin, Nanotechnology 9, 352, 1998.

[10] S Lemeshko, S Gavrilov, V Shevyakov, V Roschin and R Solomatenko, Nanotechnology 12, 273, 2001.

High Resolution Mapping of Compositional Differences at Electrode Interfaces by Electric Force Microscopy

G. A. Edwards, J. D. Driskell, A. J. Bergren, R. J. Lipert, and M. D. Porter*

Ames Laboratory-U.S.D.O.E. and the Institute for Combinatorial Chemistry
Iowa State University, Ames, Iowa 50011, mporter@porter1.ameslab.gov

ABSTRACT

This presentation examines the mechanistic basis for the ability of electric force microscopy (EFM) to map terminal group differences of spatially patterned organic monolayers. It compares the experimentally observed response to that from modeling calculations of the dipole moments of gold-bound adlayers prepared from a series of benzyl mercaptans, and serves as a starting point for gaining insight into the contrast mechanism. While preliminary, the results show that the dipole moment of the adlayer plays an important role in the contrast mechanism.

Keywords: electric force microscopy, monolayer, monomolecular film, dipole, modeling

1 INTRODUCTION

Scanning probe microscopic characterizations of organic thin films are widely utilized to investigate a range of interfacial processes (e.g., electrocatalysis, corrosion inhibition, conductivity of organic electronic devices, and biocompatibility).[1] Recent reports have described the ability of electric force microscopy (EFM) [2], an offshoot of atomic force microscopy, to map the terminal group differences of patterned organic monolayers that are buried under a thick (~500 nm) film of an organic polymer.[3] This presentation describes preliminary findings from experiments that seek to unravel the mechanistic basis of the contrast mechanism.

There are two steps in an EFM experiment. The first is a line-scan characterization of sample topography, which is usually collected by tapping mode. The electronic properties of the sample are interrogated on a second path across the sample in lift-mode (i.e., the tip is rastered across the same line at a constant tip-sample separation as determined by the previous topographic line-scan). In the second scan, a dc bias voltage is applied across a gold-coated tip and a grounded conductive substrate. The electrical forces (F_{elec}) interacting with the oscillating tip can be qualitatively represented in Eqn. 1 [4], where C is the capacitance of the media between the tip and sample, V_{sample} and V_{tip} are the voltages of the sample and tip, respectively, and z is the tip-sample separation. This formulation neglects contributions from the pyramidal-shape of the tip as well as the extended structure of the cantilever by treating the system as a parallel plate capacitor.

$$F_{elec} = \frac{\partial C}{\partial z}\frac{(V_{tip} - V_{sample})^2}{2} \quad (1)$$

Force measurements can be accomplished by monitoring the amplitude or phase shift of the tip oscillation, with phase shift being the more sensitive of the two approaches. The phase shift ($\Delta\Phi$) is proportional to the force gradient between the tip and sample, as shown in Eqn. 2. This equation indicates that a plot of $\Delta\Phi$ with respect to V_{tip} will have: 1) a parabolic shape with respect to a change in the tip-sample voltage; 2) a parabolic shape (e.g., the length of the *latus rectum*, *LR*) dependant on the capacitance of the media between the tip and sample; and 3) a minimum (i.e., the parabolic vertex) when V_{tip} and V_{sample} are identical. The adlayer results in a variation of the capacitance and voltage gradient between the tip and substrate, which gives rise to the image contrast mechanism.

$$\Delta\Phi \propto \frac{\partial F_{elec}}{\partial z} \propto \frac{\partial^2 C}{\partial z^2}\frac{(V_{tip} - V_{sample})^2}{2} \quad (2)$$

This paper utilizes a set of gold substrates, which are modified with five different *para*-substituted benzyl thiolates as a test system to begin to elucidate the interfacial properties that contribute to image contrast. Molecular modeling is utilized to predict how these monomolecular films will modify the electronic properties of the gold-air interface, such that the expected surface potential (ΔU_{model}) can be calculated. Electrochemical measurements are presented to confirm adlayer deposition, determine an experimental double layer capacitance (C_{dl}), and estimate adlayer surface coverage (Γ). This system of films is experimentally interrogated utilizing EFM to measure interfacial electronic properties. The model developed to predict EFM contrast is then compared to the experimentally observed values.

2 EXPERIMENTAL

2.1 Chemicals

Absolute ethanol was purchased from Aaper Alcohol. The *para*-substituted benzyl mercaptans (H, *t*-butyl, Cl, F, and Br) and NaOH (semiconductor grade) were obtained from Aldrich. All chemicals were used as received.

2.2 Substrate and Monolayer Preparation

Template-stripped gold (TSG) was prepared as previously reported.[5,6] In brief, 300 nm of gold was resistively evaporated at a pressure of 7.5×10^{-7} Torr onto silicon(111) wafers (University Wafer) at $0.2\ nm\ s^{-1}$. Microscope slides (Fisher, 1 × 1 cm) were epoxied to the gold surface using Epo-Tek 377 (Epoxy Technology), and cured at 150 °C for 105 min. Samples were removed from the silicon wafer, and immediately immersed for 18 h in 1.0 mM ethanolic thiol solutions, removed from solution, rinsed copiously with ethanol, and dried. The resulting gold surface has an electrochemical roughness factor of 1.3.[6]

2.3 Molecular Modeling

Molecular models of the aromatic thiols were created by energy minimization using Chem3D Ultra 8.0.[7] The dipole moments were calculated utilizing the CS MOPAC Pro package. The MOPAC software employs the Parameterized Model (revision 3, PM3) to generate the potential energy function with a closed shell wave function. Dipole moments were calculated using the Milliken charge approximation. The three-dimensional molecular coordinates, in conjunction with the dipole vectors, were then utilized to determine the thickness and magnitude of the dipole moment (parallel to the alignment of the S-H bond) of the adlayers.

2.4 Electrochemistry

Electrochemical measurements were carried out using a CH Instruments model 600A potentiostat and a three-electrode cell with a platinum wire auxiliary electrode and a silver/silver chloride (saturated sodium chloride) reference electrode. All potentials are reported with respect to this reference. A solution of 0.5 M NaOH in high purity water was utilized for all experiments. Solutions were purged with nitrogen for 10 min prior to electrochemical measurements, and a blanket of nitrogen was held over the solution throughout each experiment.

The interfacial capacitance (C_{dl}) of the samples were determined utilizing cyclic voltammetry (CV).[8] CVs were collected at three sweep rates (50, 75, and 100 mV s^{-1}) by scanning between -0.3 V and +0.1 V. The double layer charging current at 0.0 V was used to calculate C_{dl}, which was independent of scan rate. Electrochemical measurements were also utilized to determine the surface concentration (Γ) of each adlayer by integrating the charge under the one-electrode desorption wave for the thiolate-based coating and accounting for the roughness factor of TSG.[8] Desorption voltammograms were collected by scanning cathodically from 0.0 V at 100 mV s^{-1}.

2.5 Electric Force Microscopy (EFM)

Data were collected with a Nanoscope 3A Multimode AFM, equipped with a signal access module for external control of V_{tip}. $\Delta\Phi$ was measured while changing V_{tip} at a tip-sample separation of 100 nm, and an amplitude for the tip oscillation of ~15 nm. Cantilevers were purchased from MikroMasch (resonance frequency 265-400 kHz, force constant 20-75 N/m). These tips are coated with a 20-nm Cr film and then a 20-nm Au film, which resulted in a tip radius of ~50 nm. The sample compartment was purged with argon for 40 min prior to imaging, which was maintained for the duration of the experiments.

3 RESULTS

3.1 Molecular Modeling

Fig. 1 depicts the molecular model of benzyl mercaptan. In this visualization, the gold surface would be positioned on the right side of the image, aligned normal to the H-S bond. This architecture results in the aromatic ring aligned close to the surface normal, with the H-S bond then utilized as an internal marker for adlayer orientation.

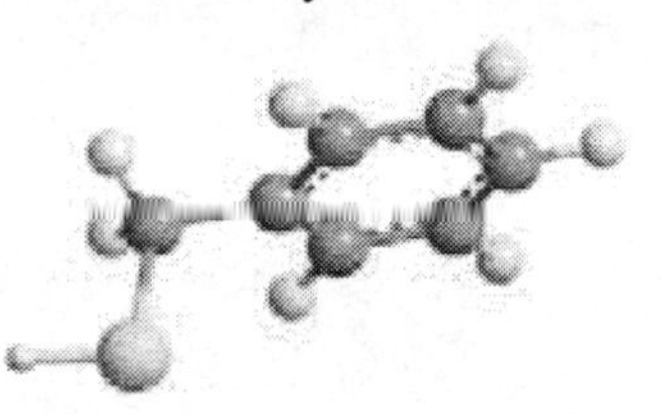

Fig. 1: Molecular model of the energy-minimized benzyl mercaptan molecule. The orientation and confirmation of all the *p*-substituted benzyl mercaptans were consistent with this depiction.

The molecular coordinates of these molecules, after minimization, were employed to calculate the thickness of the adlayer. The thickness is given as the distance from the sulfur to the substituent (X) in the direction parallel to the H-S bond; it neglects the length of the gold-sulfur bond. These results are presented in Fig. 2A, and are consistent with expectations. That is, the monolayer with H in the *para* position has the lowest thickness, whereas the *t*-butyl containing adlayer has the largest thickness. In all cases, the results yield an adlayer in which the plane of the aromatic ring is tilted $11.6° \pm 0.5°$ from the surface normal. These data are in general agreement with thicknesses determined by optical ellipsometry [9] and with the ring orientations found by molecular dynamics calculations [10] in similar systems. The molecular coordinates were also utilized to calculate both the closest packed area per molecule (A) and Γ for each of these adlayers, and are listed in Table 1. This analysis utilized the van der Waals radius of each atom, coupled with the projected ring orientation. The trend in molecular area follows intuition, such that the larger substituents (e.g., *t*-butyl) exhibit a larger surface area.

The most important results from these analyses are the dipole moments. Although the MOPAC software modeled the thiol precursors, the calculated dipole moment should be directly proportional to that of the corresponding adlayer because of the strong similarity in the thiolate linkages to the gold substrate. Data from X-ray photoelectron spectroscopy support this argument in that the positions of the S(2p) couplet are virtually identical for similar systems.[11] This claim also applies to contributions from image dipoles created in the substrate by the adlayers. The results of the dipole moment calculations are shown in Fig. 2B and tabulated in Table 1. As evident, the fluorinated adlayer has the largest dipole moment, followed by the brominated, chlorinated, and hydrogenated adlayers; the *t*-butyl derivative has the lowest value. This trend follows insights based on electronegativity and charge separation considerations. It suggests that the fluorinated monolayer should induce the largest modulation of the voltage gradient between the tip and substrate.

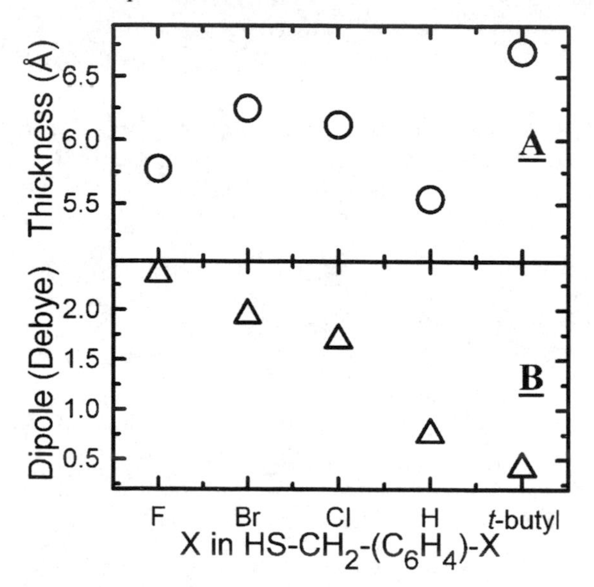

Fig. 2: (A) Calculated adlayer thickness (○) and (B) dipole moment (Δ) with respect to the surface normal.

The modeled data are utilized to calculate the theoretically predicted surface potential (ΔU_{model}) using Eqn. 3 [12], where A is in units of m^2, ε is the estimated relative permittivity of the monolayer (4.705) [13], ε_0 is the permittivity of free space (8.854×10^{-12} C m), $\mu_\perp$ is the dipole moment perpendicular to the surface (Debye), and ΔU is the expected change in surface potential with respect to the unmodified surface (V). The results of these calculations are presented in Table 1, and follow the trend based on the values of $\mu_\perp$.

$$\Delta U_{model} = 3.34 \times 10^{-30} \frac{\mu_\perp}{\varepsilon \varepsilon_0 A} \quad (3)$$

3.2 Electrochemistry

The C_{dl} and Γ of the adlayers were determined by electrochemistry. These results are presented in Table 1. As is evident, the values of C_{dl} are much lower than that of an uncoated gold electrode (~20 μF cm^{-2}).[8] Although nearly masked by the uncertainty of these measurements, the correlation of the capacitance values and substituent identity is close to that observed for the calculated dipole moments and surface potentials. That is, the C_{dl} for the adlayer with the *t*-butyl and hydrogen substituents are lower than those with the halogens in the *para* position. There is, however, one notable difference: the C_{dl}-value for the brominated adlayer is marginally less than that for the two remaining halogenated systems as well as that for the hydrogenated adlayer.

The values of Γ follow the same general trend as the theoretical values from the modeling calculations, with one exception: the brominated adlayer resulted in a higher coverage than the other adlayers. This discrepancy potentially stems from two factors: 1) the presence of faradic process upon adlayer desorption, and/or 2) incomplete electrochemical background subtraction. Experiments are being designed to examine the possibilities for the discrepancies in the C_{dl} and Γ findings.

3.3 EFM

Values of $\Delta\Phi$ were measured as a function of V_{tip} at a constant lift height (100 nm). These results are plotted in Fig. 3, and summarized in Table 1 in terms of the experimentally determined surface potential (ΔU_{exp}) and the shape of the resulting profile (i.e., the length of the *LR*). In each case, the dependence of $\Delta\Phi$ on V_{tip} exhibits a parabolic shape as predicted by Eqn. 2. The plots also show that the *t*-butyl monolayer has the most negative value of ΔU_{exp}, whereas the Cl-terminated monolayer has the most positive value.

Table 1: Summary of theoretical and experimental values related to EFM contrast mechanism for a series of *para*-substituted benzyl thiolate monolayers on gold.

	Modeling Results					Experimental Results			
Substituent	Thickness[a]	A[b]	Γ[c]	$\mu_\perp$[d]	ΔU_{model}[e,f]	C_{dl}[g]	Γ[c]	ΔU_{exp}[e]	*Latus Rectum*[e]
F	5.77	27.0	6.15	2.357	0.699	4.4 (0.6)	10.4 (0.5)	1.94	12.17
Br	6.25	30.2	5.49	1.947	0.516	3.8 (1.1)	13.9 (8.4)	1.23	11.35
Cl	6.12	28.8	5.76	1.700	0.472	4.1 (0.8)	10.2 (0.7)	2.18	11.08
H	5.54	24.8	6.69	0.755	0.243	4.0 (0.8)	11.5 (1.1)	-0.08	11.27
t-butyl	6.70	40.0	4.16	0.416	0.084	3.2 (0.7)	9.1 (1.4)	-0.21	12.24

[a] Å; [b] $Å^2$ molecule^{-1}; [c] mole cm^{-2}; [d] Debye; [e] V; [f] calculated using ε for all adlayers equal to 4.705 [13]; [g] μF cm^{-2}; error presented as one standard deviation for at least 8 samples.

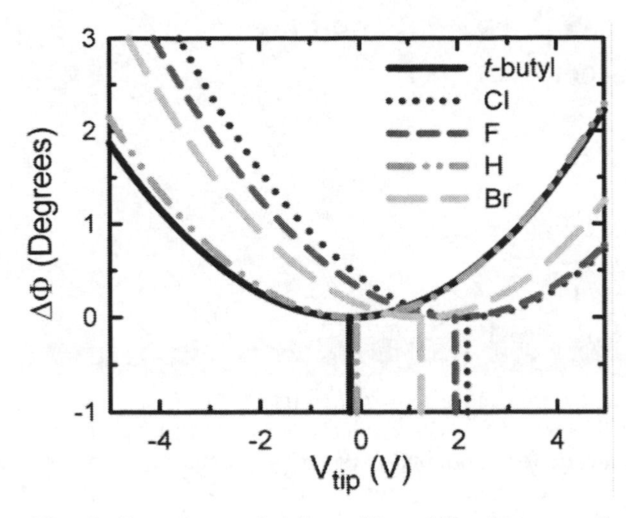

Fig. 3: Experimental $\Delta\Phi$ *vs*. V_{tip} with substrates of five different *para*-substituted benzyl mercaptan monolayers.

4 DISCUSSION

The curves in Fig. 3 yield several important observations. First, the V_{tip} at the parabolic vertex (ΔU_{exp}) of the plot for each adlayer is clearly different. Moreover, the trend in ΔU_{exp} (*t*-butyl < H < Br < F < Cl) closely, but not fully, matches with that for ΔU_{model} (*t*-butyl < H < Cl < Br < F). This agreement supports the possibility that the dipole moment is the major contributor to image contrast in EFM.

However, the lack of complete agreement points to several issues that require further investigation. From the prospective of modeling, the key limitation in the calculation of ΔU_{model} rests with the values for ε. Our analysis is presently limited by the lack of literature data for all the test systems. While ε for benzyl mercaptan has been reported [13], we have not been able to locate values for the remaining precursors. The literature for structurally similar compounds (e.g., benzenes and *para*-substituted toluenes) indicate that the value of ε may differ by up to 60%, indicating that only a qualitative comparison between ΔU_{model} and ΔU_{exp} can be reliably made at this time.

There are also several important refinements to address in the experimental area. First, the $\Delta\Phi$ curves of Fig. 3 have a parabolic shape that is dependant upon the capacitance of the media between the tip and substrate. However, the shapes, as judged from the *LR* values, do not track with the values of C_{dl}. We suspect that this situation reflects the presence of adventitious impurities adsorbed on the tip and sample. To address this issue, we are examining issues related to possible surface contaminates on the adlayer and tip as well as approaches to improve determination of C_{dl} and Γ, part of which will address methods to more reproducibly prepare the adlayers.

5 CONCLUSION

This paper has presented preliminary results that tested the effect of electronegative substituents on benzyl mercaptan-based monolayers assembled on gold surfaces. Electric force microscopy was utilized to interrogate the surface potential and capacitance of these adlayers. A theoretical model was developed to predict the expected trends in the EFM response. While preliminary, the results show that the dipole moment of the adlayer plays an important role in the contrast mechanism. Experiments to extend these first findings are planned.

ACKNOWLEDGEMENTS

This work was supported by the Basic Energy Sciences program of the U.S. Department of Energy. G.A.E. gratefully acknowledges a Conoco-Phillips graduate research fellowship. Valuable discussions with M. Gordon, I. Adamovic, and D. Zorn are also acknowledged. The Ames Laboratory is operated by Iowa State University for the U.S. Department of Energy under Contract W-7405-ENG-82.

REFERENCES

[1] Takano, H.; Kenseth, J. R.; Wong, S.-S.; O'Brien, J. C.; Porter, M. D. *Chem. Rev.* **1999**, 99, 2845-90.

[2] Fujihira, M. Annu. *Rev. Mater. Sci.* **1999**, 29, 353-80.

[3] a) Takano, H.; Porter, M. D. *J. Am. Chem. Soc.* **2001**, 123, 8412-3; b) Takano, H.; Wong, S.-S.; Harnisch, J. A.; Porter, M. D. *Langmuir* **2000**, 16, 5231-3.

[4] Sarid, D. *Scanning force microscopy: with applications to electric, magnetic, and atomic forces*; Oxford University Press: New York, 1994.

[5] Stamou, D.; Gourdon, D.; Liley, M.; Burnham, N. A.; Kulik, A.; Vogel, H.; Duschl, C. *Langmuir* **1997**, 13, 2425-8.

[6] Wong, S.-S.; Porter, M. D. *J. Electroanal. Chem.* **2000**, 485, 135-43.

[7] Chem3D, Ultra 8.0, CambridgeSoft: Cambridge MA, 2004.

[8] Widrig, C. A.; Chung, C.; Porter, M. D. *J. Electroanal. Chem.* **1991**, 310, 335-59.

[9] Jung, H. H.; Won, Y. D.; Shin, S.; Kim, K. *Langmuir* **1999**, 15, 1147-54.

[10] a) Tao, Y.-T.; Wu, C.-C.; Eu, J.-Y.; Lin, W.-L. *Langmuir* **1997**, 13, 4018-23; b) Howell, S.; Kuila, D.; Kasibhatla, B.; Kubiak, C. P.; Janes, D.; Reifenberger, R. *Langmuir* **2002**, 18, 5120-5.

[11] Zhong, C.-J.; Brush, R. C.; Anderegg, J.; Porter, M. D. *Langmuir* **1999**, 15, 518-25.

[12] Evans, S. D.; Urankar, E.; Ulman, A.; Ferris, N. *J. Am. Chem. Soc.* **1991**, 113, 4121-31.

[13] Lide, D. R., Ed. *Handbook of Chemistry and Physics*, 76 ed.; CRC Press: Boca Raton, 1995.

A Novel Battery Architecture Based on Superhydrophobic Nanostructured Materials

V. A. Lifton, S. Simon

mPhase Technologies, Inc.
587 Connecticut Avenue
Norwalk, CT 06854-1711
vlifton@mphasetech.com, ssimon@mphasetech.com

ABSTRACT

The paper presents the details of a novel battery architecture based on the superhydrophobic nanostructured materials. In a proposed configuration both electrodes of a battery are formed on the nanostructured silicon surfaces that are subsequently treated to render them superhydrophobic, effectively separating liquid electrolyte from the active electrode material. When battery is activated to provide power, a phenomenon called electrowetting promotes electrolyte penetration into the electrode space to initiate electrochemical reaction. Such architecture provides for an extremely long shelf life, instantaneous ramp up to full power, and chemistry independent functionality.

Keywords: battery, nanostructured, superhydrophobic, electrowetting, shelf life.

1. INTRODUCTION

For decades semiconductor technologies enabling many portable electronic devices have been rapidly advancing at very high rates, at times doubling every twelve to eighteen months. In the meantime, battery technology has been advancing at only 3% to 5% per year and reserve battery capabilities at an even slower rate. Devices containing semiconductor components requiring primary battery or battery back up needs have rapidly proliferated throughout every major national and global economic sector imaginable: transportation, health, defense, security, energy, environmental, and many others. These devices take on forms ranging in physical size from large scale computing devices, personal computers, handheld devices, embedded devices, remote sensors, and even into newly emerging nano-sized devices. Furthermore, the applications of these proliferated devices have become more and more decision critical therefore there is a strong need for breakthrough technology to occur within battery technology.

mPhase Technologies along with its partner Lucent/Bell Labs, has been jointly conducting research over the past year that demonstrates control and manipulation of fluids on superhydrophobic surfaces to create power cells by controlling wetting behavior of electrolyte on nanostructured electrode surfaces [1]. The scientific research we have conducted this year has set the groundwork for continued exploration in the development of intelligent nanotechnology power cells (nano-batteries), and forms a path to commercialization of the technology for a broad range of market opportunities.

In this market environment, the proposed nanobattery would be considered a disruptive technology, a technology that provides a fundamental paradigm shift in battery and power technology. Size, weight and shelf life characteristics could be dramatically enhanced, and chemistries could be chosen to be appropriate for designs that that would be used for powering numerous portable electronic devices such as sensors and transmitters for sensor networks as just one example. Packaging could be more flexible with the potential of shaped or conformal batteries, but still could be designed to work within the requirements of the physical dimensions of existing electronic devices requiring power. Batteries would have intelligence with the potential of being able to activate cells as needed to the power output needed, thus extending the useful duty cycle of the battery. In a reserve configuration, power could always be available as a backup to the primary source because of the negligible capacity loss in such a battery in a reserved state due to its unique architecture. Shelf life would be increased dramatically because the electrolyte would not come in contact with the electrode until activation. Because of the very small distances of the nanostructures, response time to full activation power would be enhanced, creating a power source for high performance electronics needing these characteristics. For the Military, Homeland Security, and First Responders, this feature would be highly desirable because batteries could be stored, for decades, without fear of dissipation. Conventional batteries in storage dissipate as much as 10% per year before use, while the nanobattery is projected to last 15-20 years.

In advanced configurations the silicon based architecture of the battery that is currently being developed has the potential for integration of electronic components directly into the manufacturing process of the battery, to create new classes of tightly integrated devices such as integrated active RFID tags and lab-on-the-chip applications for both defense and commercial markets.

2 Principle of Operation

The proposed battery architecture places our battery into a so-called reserve battery class [2]. The main function of these batteries is to provide power when required while enduring prolonged storage periods (essentially, a battery may sit its entire life in reserve to be activated only for a brief period of time as a back-up power source).

In order to be able to provide such a long shelf life, a traditional reserve battery normally has a mechanical separator to keep electrolyte away from the active electrode materials. Clearly this leads to reduced power density because inert filler materials, actuation mechanism and separators occupy significant part of the battery volume.

Our approach is to employ novel nanostructured superhydrophobic materials coupled with electrowetting phenomenon to create an architecture that allows to keep electrolyte and electrode separate from each other and yet to provide significant reduction in a dead volume in the battery. In the following discussion each component will be individually addressed.

2.1 Nanostructured superhydrophobic surfaces.

Our development capitalizes on the fundamental work performed at Lucent Bell Labs on the dynamic tuning and control of fluids on nanostructured surfaces [1]. The battery utilizes surfaces with the regularly spaced nanostructures etched by a suitable semiconductor fabrication method. The surface represents an array of cylindrical posts, 1 to 5 microns apart, 5-10 microns tall, 100 to 300 nanometers in diameter. When treated with an appropriate fluorocarbon polymer such a surface demonstrates superhydrophobic behavior that distinguishes itself from a regular surface by substantially higher contact angle of a liquid on such a surface (e.g. 120 degrees vs. <90 degrees). A droplet of electrolyte when placed on this surface does not stick to it but rather remains highly mobile as a consequence of small solid-liquid interface. In essence, the droplet is only supported by the very tips of the nanoposts and does not penetrate into the space between them.

Electrowetting gives one the ability to change the contact angle of the solid-liquid interface by applying voltage to the liquid. It has been successfully applied to create a variety of optical devices such as lenses, diffraction gratings and is now combines with nanostructures to create novel batteries and battery architecture with unique characteristics [3-6].

In a situation when a pool of electrolyte sits on top of the nanoposts of a nanostructured surface, it can be triggered to penetrate the nanoposts space and initiate the electrochemical reaction with active electrode material deposited on the bottom.

To fully characterize operation principle, characteristics and proof of concept of our novel concept we decided to start with a well known and well documented battery chemistry to make quick assessment and comparison possible. Therefore, Zn/MnO_2 electrode pair in $ZnCl_2/NH_4Cl$ electrolyte has been chosen to be the first proof of concept system. The next section deals with the advantages that such a battery will have as well as technical challenges that need to be overcome for successful commercialization.

Figure 1 gives a schematic depiction of such a battery showing two electrodes and details of the nanostructure.

3 Nanobattery Advantages

The advantages to using this nanostructured approach gives the battery the following characteristics:

- Improved power densities compared to other reserve battery technologies for its size, due to better utilization of internal surface area of the batteries design that does not require the ancillary structures to create physical separators between the electrolyte and electrodes (Our energy and power density calculations are early extrapolations based on our initial proof of concept feasibility prototype of reserved nano cell based on Zn/MnO_2 chemistries, and we expect similar improvements with other chemistries).
- Flexible architecture allows for wide range of aqueous and non-aqueous chemistries adapted to fit the design.
- Long shelf life, predicted to last over 15-20 years.
- Unique architecture design that allows for individual addressability of nano cells to be selectively activated as required.
- Applicability for primary and reserve battery applications.
- Readily scalable, easy to miniaturize battery.
- Fast ramp up to full power ~ 1 ms.
- Compatible with semiconductor processing and inexpensive to mass produce using microelectronic manufacturing techniques.
- Can be integrated into package with devices it is powering.

4 Technical challenges and demonstration.

The manipulation and control of the electrolytic solution on the electrodes of our battery is the result of our undertaking into the study of organic coatings on the nanostructures such that a superhydrophobic state is maintained on the portions of the nanostructures to repel the aqueous electrolyte, while other potions of the nano structures are kept in hydrophilic state. This superhydrophobic/hydrophilic transition provides the underlying physics of our design that allow the contact angle of the electrolyte solution to change, from having no physical contact with the electrode, in an inactive state, to an active state, where the electrolyte comes into contact with the electrodes, thus producing voltage in the battery.

In Figure 2 we show a typical image of the Zn electrode deposited in the nanopost space using a modified electroplating process. It gives reliable and controllable way of producing metal deposits on a highly conjugated superhydrophobic surface.

A variety of coatings have been evaluated to render surfaces hydrophobic. The material of choice so far remains a coating of conformal low-surface energy fluorocarbon polymer conveniently deposited in a commercial system.

A counter electrode made out of MnO_2 can be created on similar nanostructured surfaces or planar substrate as well. The following Figure 3 presents an actual voltage trace taken from the battery in its inactive state, followed by the triggering pulse with voltage being generated. One can clearly see all the features of the proposed system: dormant state of zero voltage, trigger by a voltage pulse, and voltage generated in accordance with thermodynamic predictions.

In conclusion, we have demonstrated a working reserve battery prototype based on superhydrophobic nanostructured surfaces. It has three distinct features: first, inactive state in which electrolyte is completely separated from the electrodes by the nanostructure; second, battery actuation by a voltage pulse, and third, stable voltage generation. We are now in the stage to refine and fully characterize battery parameters to compare them with the conventional battery structures. The architecture proposed is a disruptive technology, a technology that provides a fundamental paradigm shift in battery and power technology.

REFERENCES

[1] T.N.Krupenkin, J.A.Taylor, T.M.Schneider, S.Yang, Langmuir, 20, 3824, 2004.

[2] D.Linden, T.B.Reddy, eds., "Handbook of Batteries," McGraw-Hill, 2002.

[3] S.Yang, T.N. Krupenkin, P.Mach, E.Chandross, Adv. Mater., 15 (11), 940, 2003.

[4] T.N.Krupenkin, S.Yang, P.Mach, Appl. Phys. Lett., *82*, 316, 2003.

[5] F.Cattaneo, K.Baldwin, S.Yang, T.N.Krupenkin, J.A.Rogers, J. Microelectromech. Sys., *12,* 907, 2003.

[6] J.Hsieh, P.Mach, F.Cattaneo, Y.Yang, T.N. Krupenkin, K.Baldwin, J.A.Rogers, IEEE Photon. Techn. Lett., 15 , 81, 2003.

ACKNOWLEDGEMENT

Present research is being performed in collaboration with Bell Labs/Lucent Technologies and New Jersey Nanotechnology Center.

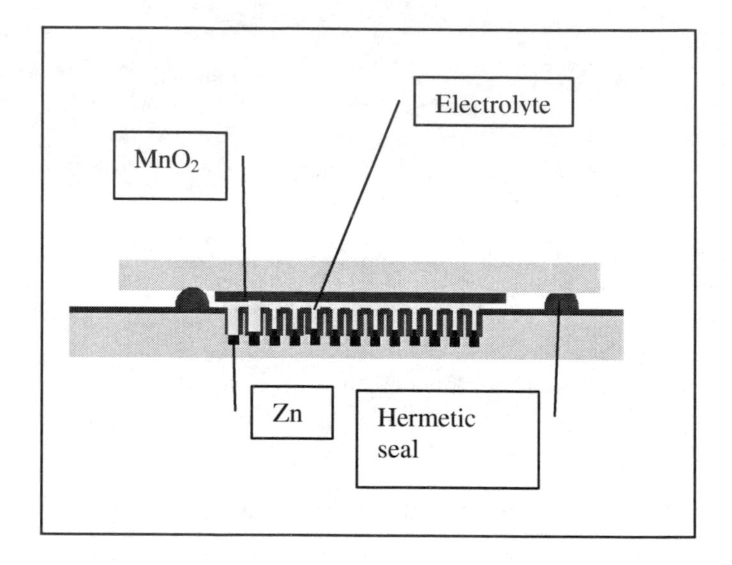

Figure 1. Cross-section of nanobattery in active state. Only essential parts are shown, such as nanoposts, electrolyte penetrating the nanopost space, Zn plated into the nanostructure and MnO_2 deposited onto a planar substrate.

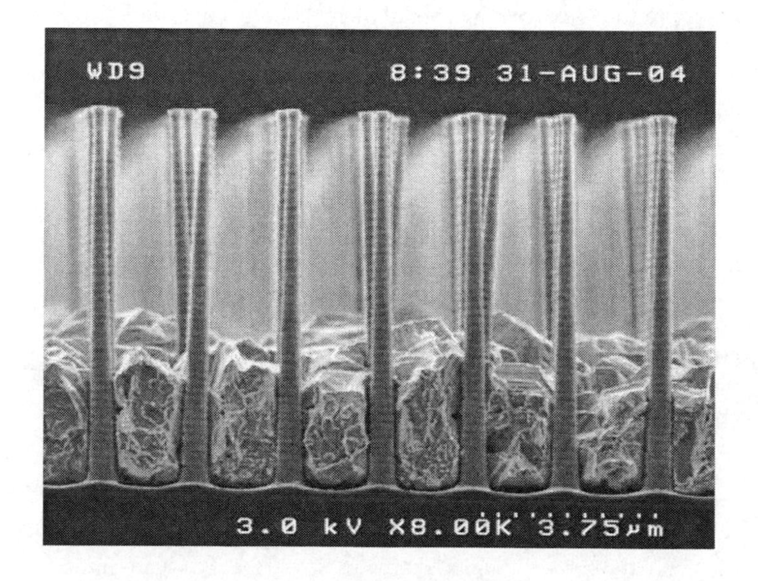

Figure 2. SEM image of Zn deposit on nanostuctructure.

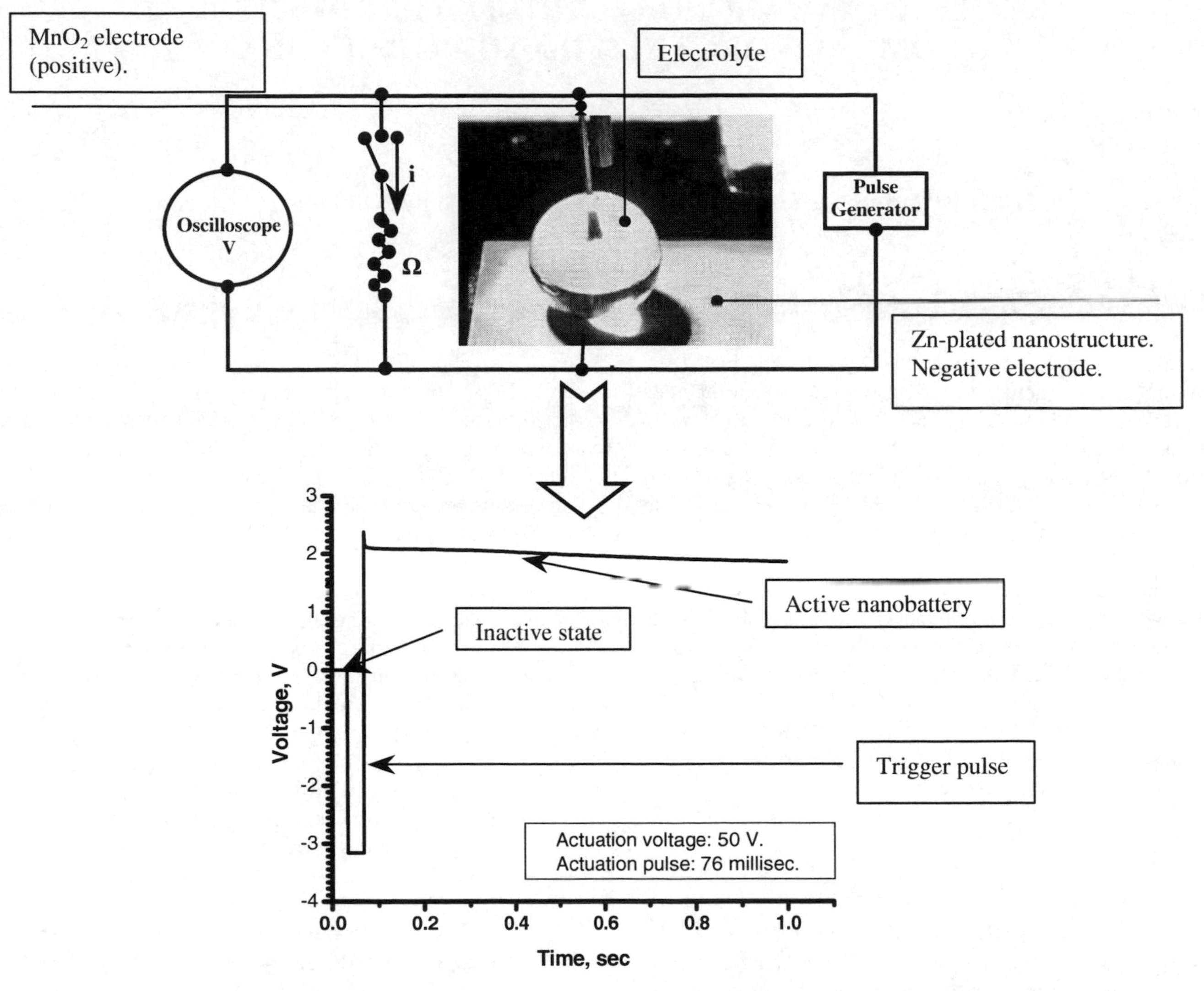

Figure 3. Schematic of battery triggering test and the actual graph of the battery output, showing 3 regions: 1) inactive battery, 2) actuation pulse, and 3) active power generation.

NANO- AND MICROFABRICATED PRODUCTS FROM SELF-ORGANIZED ANODIC ALUMINA

Dmitri Routkevitch

Synkera Technologies Inc.
2021 Miller Drive, Suite B, Longmont, CO, USA, droutkevitch@synkera.com

ABSTRACT

Synkera's product development efforts are based on nano- and microstructured materials and advanced methods for their integration into functional devices (www.synkera.com). We offer two main product groups. The first is a broad group of chemical sensor products and related devices. Synkera's chemical sensors are bridging the gap between the low cost of traditional solid-state sensors and the performance of liquid electrolyte electrochemical sensors. The other group of products, which will be reviewed in this presentation, originates from a unique nano/microfabrication technology platform, which affords multiscale engineering of advanced materials and their device integration.

Keywords: nanoporous anodic alumina, templated nanofabrication, ceramic MEMS

1 TECHNOLOGY

At the core of Synkera's technology platform is the synergy of two capabilities, *templated nanofabrication* and *micromachining*, both based on nanostructured *anodic aluminum oxide (AAO)*. AAO is self-organized into a nanoscale "honeycomb", formed by uniform and parallel nanopores that are aligned perpendicular to the surface of the material (Figure 1).

Synkera's well-established capabilities to engineer nanoscale dimensions and morphology of this material span pore diameter from below 5 to over 200 nanometers and pore length from 0.1 up to 300 microns, covering the size domain of interest to nanotechnology.

Self-organized anodic alumina is an attractive host for templated deposition [1, 2], and is widely used in fabrication of high-density arrays of prepackaged nanostructures (Figure 1, left). In addition, Intrinsic anisotropic morphology and chemistry of anodic alumina enables its micromachining via a process that combines features of both bulk and surface MEMS (Figure 1, center) and offers extensive opportunities for the development of Ceramic MEMS devices.

2 PRODUCTS

This technology was successfully used to develop and produce a number of products. Unique features and performance of these products are derived from nanoscale engineering of desired materials' properties and architecture. On the other hand, our Ceramic MEMS processing provides a convenient route for integrating these materials into practical devices. Synkera's current product development efforts include:

- Nanostructured *microsensor platform* [3] for solid-state gas sensing, including metal oxide conductimetric, catalytic combustible and electrochemical types of sensors (Figure 2, details described in a related talk [4]).
- Three types of *nanotemplates* for nanomaterials research that share unique morphology of anodic alumina (pore diameter 10-100 nm), but have distinctly different pore termination (Figure 3). These nanotemplates are currently offered in both standard and custom specifications to R&D market.
- Monodisperse ceramic *membranes* for separation and filtration applications, including membrane reactors for hydrogen generation and composite membranes for hydrogen separation (Figure 4, Figure 5).
- Other *Ceramic MEMS* components and devices (Figure 6) that utilize flexible design and fabrication capabilities of the process as well as robustness of alumina ceramic in harsh operating environments.
- *BioMEMS* components, such as high surface area substrates for bioanalytical applications and biocompatible chips for tissues culturing (Figure 6).
- Integrated and non-integrated metal *nanowire arrays* for energy conversion, bioengineering, bioanalysis and other applications.

We continue to develop and expand our product portfolio, and welcome inquiries on establishing collaborative partnerships for joint application and market development.

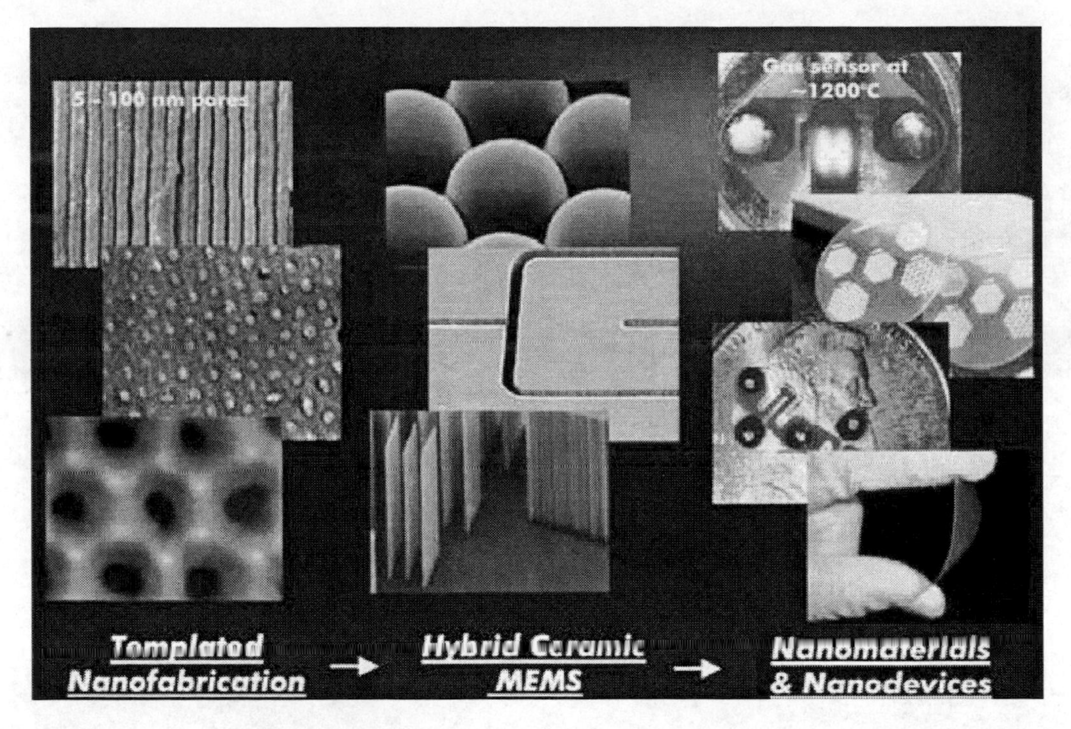

Figure 1: Synkera's nano- and microfabrication product development sequence (www.synkera.com)

Figure 2: Synkera's packaged microsensors and microheater operating in air at ~1100°C.

AAO on Al foil

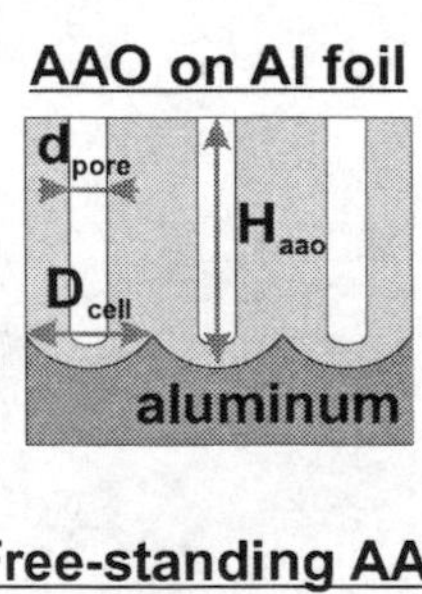

Free-standing AAO

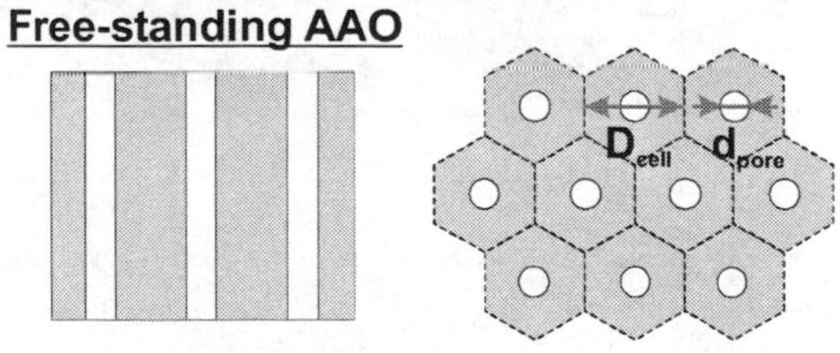

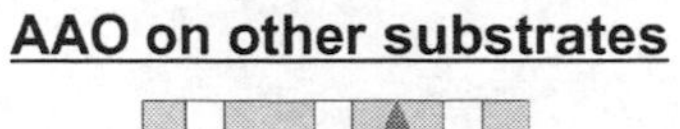

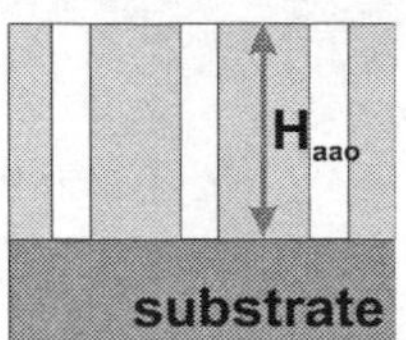

Figure 3: Three types of anodic aluminum oxide (AAO) nanotemplates offered by Synkera.

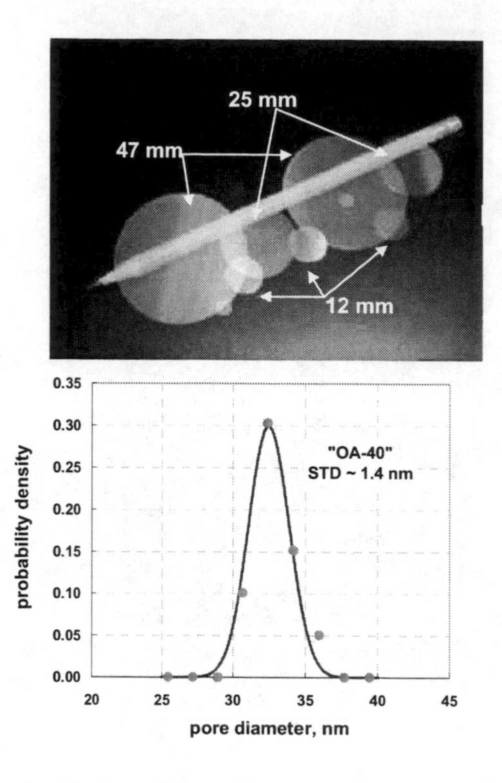

Figure 4: Synkera's monodisperse nanoporous ceramic membranes and membrane pore size distribution

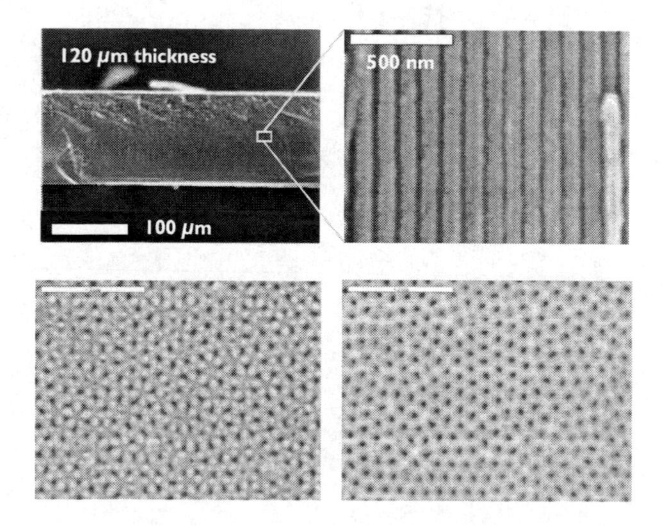

Figure 5: Scanning electron microscopy images of the cross-section and the plane view of the opposite faces of anodic alumina membranes produced at Synkera.

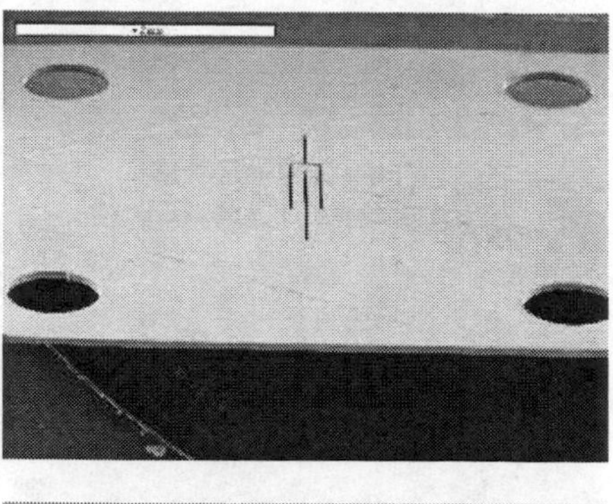

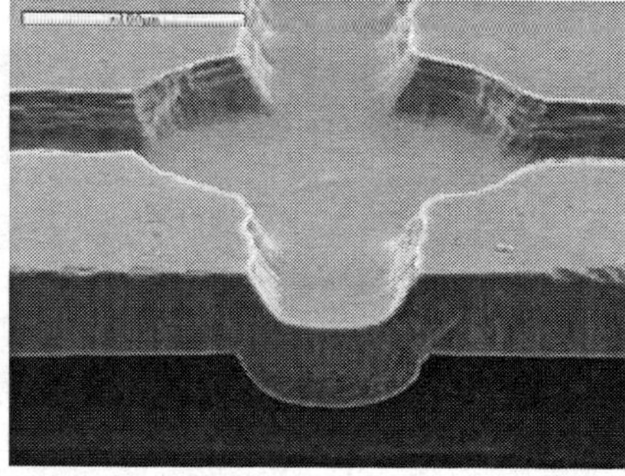

Figure 6: Examples of Synkera's Ceramic MEMS components. Top: gas microsensor substrate with thermally isolated sensing element. Bottom: micromachined chip for guided neuronal networks.

REFERENCES

[1] H. Masuda, M. Satoh, *Jpn. J. Appl. Phys.*, 35, L126 1996.

[2] D. Routkevitch, T. Bigioni, M. Moskovits, J. M. Xu, *J. Phys. Chem.*, 100, 14037, 1996.

[3] D. Routkevitch, P. Mardilovich, A. Govyadinov, S. Hooker, S. Williams, *US patent* No 6,705,152 B2, filed 05/07/1999; granted 03/16/04.

[4] D. Routkevitch, O. Polyakov, D. Deininger and C. Kostelecky, Nanotech 2005, Abstract Number 823, "Nanostructured Gas Microsensor Platform", 2005.

Nanotechnology Platforms for Commercialization

Alan Rae, NanoDynamics Inc., Buffalo NY USA, arae@nanodynamics.com

ABSTRACT

Much of the publicity surrounding nanotechnology covers leading-edge applications with the potential to create tremendous disruptive changes in key industries such as semiconductors. Whilst the intensive support of these projects is critical, nearer-term opportunities must be grasped across a range of technology platforms and markets in order to gain experience in the production of new nano materials and systems in more immediate opportunities. These product enhancements - more significant than evolution, less dramatic than disruption - can generate significant volumes and opportunities. Examples of "Enhanced" products and systems developed by NanoDynamics Inc. in the processing of metals and non-metals will be reviewed and conclusions drawn.

Keywords: nanotechnology, silver, copper, atomic cluster, self-assembly, golf ball.

HISTORY

Many disruptive technologies have foundered because the compelling reason to go to market was compromised. Many inventors do not have strong business backgrounds; many business people have a very conservative approach to new products removed from their existing product line; and while a product languishes in no-man's land the existing technology has time to make up ground by increasing performance or lowering cost… or the market may transform due to external factors.

One example comes from the ceramics industry. Ceramic turbochargers - replacing metal turbochargers - were touted in the 1980s as the answer to increasing output and efficiency of automotive engines by harnessing high-temperature exhaust gas energy while maintaining performance and response time. The impetus came from the first "oil shock" of the 1970's and much of the materials technology dated from the 1960's. Many major corporations and startups focused on advanced ceramics worldwide. A great deal of energy and money was spent on ceramic materials and process technology but metals engineering fought back – and four valves per cylinder and variable valve timing using metals with no turbocharger gave the performance upgrade that was needed

A second example comes from the electronics industry where advanced optoelectronics were promoted as the way to cope with increased bandwidth demand. Industry roadmaps claimed that copper circuits could not exceed 2 Gbps whereas optical systems were capable of 40 Gbps. The optoelectronics technology (much of it developed in the 1970's for voice communication) spawned a huge expansion in existing companies and a slew of startups – but the telecom bubble burst in 2001 and left us with a customer base that would only use existing technology (at dramatically reduced prices) and a copper supply base that fought back with a five-fold improvement in the data transmission rate of copper.

BRIDGING THE GREAT DIVIDE

Marketing professionals will always tell you that "market pull" is better than "technology push". In general that's true, but always relying on market pull from customers will limit companies to evolutionary developments. Truly revolutionary developments that can obsolete a business come from a different technology base – think of the replacement of video tapes by DVDs and subsequently by on-demand video.

The "great divide" (sometimes referred to as "The Valley of Death) comes from two incompatible positions.

The Inventor

- I have an idea.
- I have made a prototype and it works.
- I believe there is a market for this invention.

The Producer

- I have an unmet market need.
- I need a fully characterized product that is scalable, complies with all applicable regulations (TSCA, OSHA, NIOSH, shipping regulations), has a full MSDS and has a process capability proven through statistics. And it must give a total lower cost than previous solutions.

For the Inventor the cost of scaling a process to pilot or production stage can be extreme and not possible to finance readily e.g. in the case of a new or modified polymer. Even in more modest examples the inventor may be forced to go into debt or lose significant equity in the invention in order to raise capital.

The barriers on the Producer's side may also be high. Mature companies are generally risk-averse and are uncomfortable moving away from what is often a very narrow definition of core competency. They also have a pace of operation and legal mindset that can prove frustrating to the Inventor (who's looking at short-term survival) and the Producer (who finds the Inventor to be an impatient nuisance).

THE NANODYNAMICS MODEL

NanoDynamics was formed in 2002 to produce products enabled by nanotechnology. The company brought together seasoned professionals, enthusiastic research teams and a process development group and developed a number of platform products through licensing, sponsored research or internal development. Much of the technology has been developed by universities or individuals.

In one example, NanoDynamics has licensed novel metal precipitation technology from Clarkson University and continues to support researchers there. Taking the technology from gram to kilogram to tonne has taken less than 2 years. The market pull was from the electronics industry, which needs finer metal powders to produce smaller devices.

The technology push was from a new precipitation process yielding truly monosize particles that were crystalline and protected with a polymeric coat which prevents agglomeration and cold welding. The process yields 100% fine particles so there are no yield issues as with many other processes. From the customer's point of view the powder is easy to process and control and there are applications now in industries where NanoDynamics and Clarkson team members have experience and personal connections.

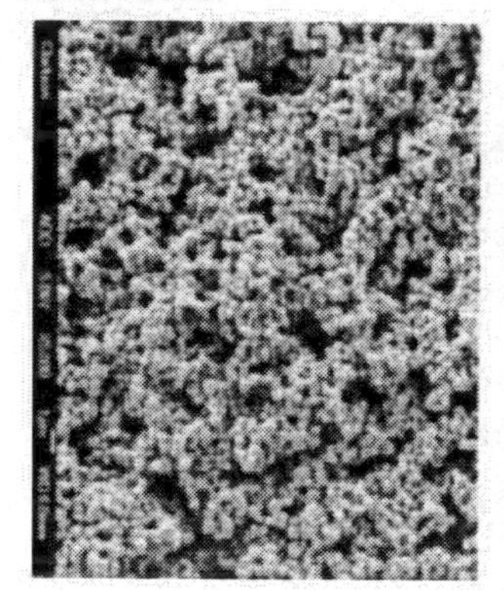

Figure 1. 200nm and 2 micron copper produced by the same precipitation process.

A second example is NanoDynamics' program with Nano Cluster Devices Ltd., a spin-off company based on an invention from the University of Canterbury in Christchurch, New Zealand. This technology combines the lithography techniques used in the semiconductor industry with selective atom cluster deposition to self-assemble conducting or semiconducting wires 1/00 of the size achievable by conventional lithography. The application to sensors and semiconductors is obvious, and although the commercialization stage is earlier than that of metal powders, the team again has the experience and contacts in the electronics industry that will allow this technology to become widely adopted.

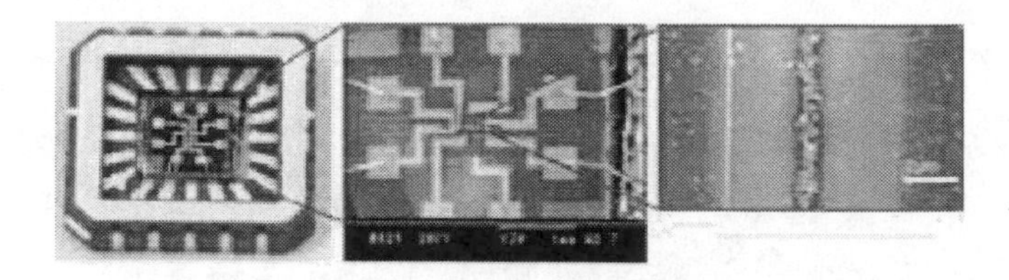

Figure 2. Directed assembly of wires and their incorporation into sensors (Nano Cluster Devices)

A third example is a consumer application, the NDMX ™ golf ball! This uses a nanostructured hollow titanium sphere in place of the rubber core in conventional balls. The titanium sphere, strengthened and recrystallized by controlled deformation, moves the mass to the surface of the ball improving directional stability during flight and rollout. It is also a surprisingly economic solution and allows the ball to be priced competitively with existing high-performance balls. Working with the inventors, Noonan LLC., the development of the ball to meet USGA guidelines has taken approximately a year. In this case NanoDynamics has assembled a team of partners to ensure a robust supply chain, necessary approvals and marketing channels.

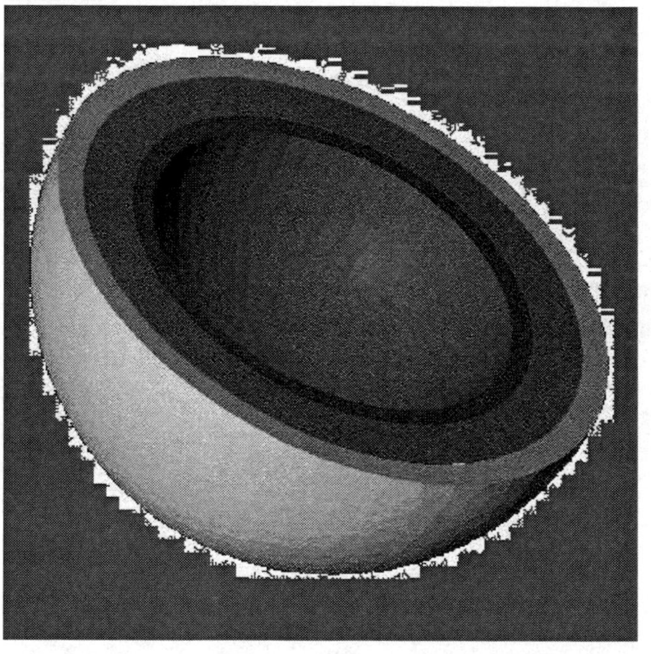

Figure 3. the NDMX ™ golf ball utilizing a nanostructured titanium shell.

Other product platforms include carbon, mineral and polymer nanotubes, Quasam ™ quasi-amorphous diamond coatings, single and mixed oxides and novel gel-casting technology. Many of these are incorporated in a novel solid oxide fuel cell system to be launched this year.

Figure 4. The Revolution 50 ™ solid oxide fuel cell system using nanomaterials in its construction to reduce size and increase performance.

SUMMARY

Nanotechnology is proposed as the answer to materials issues as diverse as the impending "silicon crunch" in 2015 when silicon integrated circuits can be made no smaller, to biocidal clothing, implants, cancer therapy, biological analysis, super-strong composites, novel conductive materials etc.

Most startup companies are looking at the revolutionary changes – the major steps where commercialization and revenue is some years out. These companies are supported by SBIR and venture funds but have to cross the chasm of commercialization. Many don't have the commercial experience or financial backing.

Many established companies are looking at the evolutionary changes – the "quick fix"- where nanomaterials can be plugged into an existing material or system that they are comfortable with. They rapidly come to the conclusion that the incremental benefits do not justify the increased cost of production of nanomaterials.

One way to bridge the gap is to create an agile organization run by individuals with industry and commercial experience who have an open mind and attitude to risk, and to allow them to identify technology platforms that can be married to a demonstrated industrial need. The technologies can be situated with individuals, with startup companies, with universities or can be orphan technologies neglected by other companies.

So what have we learned from previous technology waves and what lessons have we applied in commercializing these technologies?

- Choose your team carefully. Balance experience and innovation!
- Choose your backers carefully. Make sure you know their preferences and sensitivities.
- Don't rely on a single market.
- Don't rely on a single technology – "the one trick pony".
- Make sure that your IP is free and clear.
- Be realistic in your valuation. Just because $10 million was spent in development doesn't mean the technology is worth $10 million…or that it won't need another $10 million.
- Assume the existing technology will catch up when prodded by competitive pressure. Your performance advantage may and probably will erode in the time it takes to reach market. Watch the rear-view mirror!
- Develop a portfolio with managed time-to-market and managed market size.
- Apply appropriate management techniques to ensure that development targets are met.
- Integrate the supply chain as appropriate.
- Ensure that all the processes chosen will make you a low-cost supplier.
- Partner as appropriate to commercialize the materials (production partnership, sales partners, logistics partners, research partners). You can't do it all yourself!

Application-Driven Fine Particle Solutions

M. Oljaca, P. Atanassova, J. P. Shen, S. Haubrich, M. Hampden Smith, and T. Kodas
Cabot Corporation, Albuquerque, NM, USA, miki_oljaca@cabot-corp.com

ABSTRACT

Working closely with end-users and partners, Cabot SMP is developing solutions and manufacturing fine particle products for energy materials, electronics, displays, and other applications. In this manuscript, we address the importance of understanding the applications requirements and provide selected examples of application-driven particle manufacturing and processing. The implementation of fine particle solution for catalysis and development of dispersions and formulations for electronics and other applications will be discussed. This paper will focus on development of revolutionary sorbent materials for absorption-enhanced natural gas reforming for production of high purity hydrogen. The advantages of CSMP spray-based manufacturing method to produce advanced CaO-based reversible CO_2 sorbent powders will be described. The sorbent materials produced using CMSP's manufacturing method have been formed as extrudates and shown greatly improved durability, retaining high CO_2 absorption capacity and carbonation/de-carbonation kinetics through multiple cycles required for a commercial application. The presentation will also discuss the manufacturing challenges associated with production and processing of engineered ultra-fine particles for applications in electronics, lighting, and displays. In particular, special attention to the development of conductor type materials for inkjet processes for these applications will be addressed.

Keywords: engineered particles, spray, aerosol, sorbents, catalysts.

1 INTRODUCTION

The continuous push for performance enhancement in important existing and emerging applications has generated substantial interest in engineered fine particles. There are many examples of the improved optical, electronic, and physical properties achieved by controlling the morphology, composition and surface functionality of fine particles. To realize fully the broad benefits of fine particle solutions, it is necessary to implement an application-based approach to solving R&D problems and implementing commercially viable solutions. Some of the conventional particles manufacturing methods often have limited capability to provide control over various particle properties which are needed to satisfy increasingly stringent application requirements. For example, the conventional manufacturing methods for producing sorbent powders are liquid precipitation and impregnation, followed by forming of the powders into extrudates or monoliths. Typically targeted properties for these materials are high surface area and desired pore size distribution with precise control over the porosity, crystalline phases, surface composition, impurities and dispersion of the active phase. Such control is often difficult to achieve simultaneously for various properties due to limitations in the liquid precipitation processes. CSMP has developed a patented spray-based powder-manufacturing platform that is capable of producing a wide variety of sorbent materials with unique microstructures combined with economic suitability for high volume manufacturing. The uniqueness of the spray-based method to construct specific microstructures and compositions derives from the sequential application of liquid phase and solid-state chemistries that can be resolved by both temperature and time. A key feature of the process is that the physical (surface area. porosity, dispersion) and/or chemical (composition, phase) evolution of the particles can be arrested at any stage by quenching the reaction media. The fact that this process can involve a relatively high processing temperature for a relatively short amount of time while maintaining control over the particle size is valuable for the formation of complex composition materials such as mixed metal oxide sorbents.

2 REVOLUTIONARY SORBENT MATERIALS FOR NATURAL GAS REFORMING

The conventional methods for natural gas reforming for this application, steam methane reforming (SMR) and autothermal reforming (ATR) lead to relatively low hydrogen content gas streams. The fuel feeds are highly contaminated by CO and CO_2 and require extensive purification prior the delivery to a fuel cell stack.[1-3] Therefore, it is highly desirable to develop a method of natural gas reforming in which a high concentration, high purity H_2 stream is produced. This can be achieved by an absorption enhanced reforming (AER) process (Reaction 4) that combines SMR (Reaction 1), the WGS (Reaction 2), and CO_2 sorption (Reaction 3) to produce a synthesis gas with relatively high hydrogen purity and low CO_2 and CO content. The potential benefits are well known and have been the subject of a number of studies.[4-6]

(Reaction 1) $CH_4 + H_2O \rightarrow 3H_2 + CO$
(Reaction 2) $CO + H_2O \rightarrow H_2 + CO_2$
(Reaction 3) $CaO + CO_2 \rightarrow CaCO_3$
(Reaction 4) $CaO + CH_4 + 2H_2O \rightarrow 4H_2 + CaCO_3$

Figure 1 illustrates the degree to which the CO_2 sorption shifts the chemical equilibrium to the product side. At 600°C in the presence of a CO_2 sorbent, AER can achieve at least 98% conversion to H_2 as compared to only 75 % conversion according to thermodynamic calculations under normal conditions of SMR or ATR. To achieve a similar

 NSTI-Nanotech 2005, www.nsti.org, ISBN 0-9767985-1-4 Vol. 2, 2005

equilibrium conversion for SMR, it is necessary to operate at 720°C. For this reason it is desirable to remove the CO_2 during the reforming step as opposed to post WGS CO_2 removal.

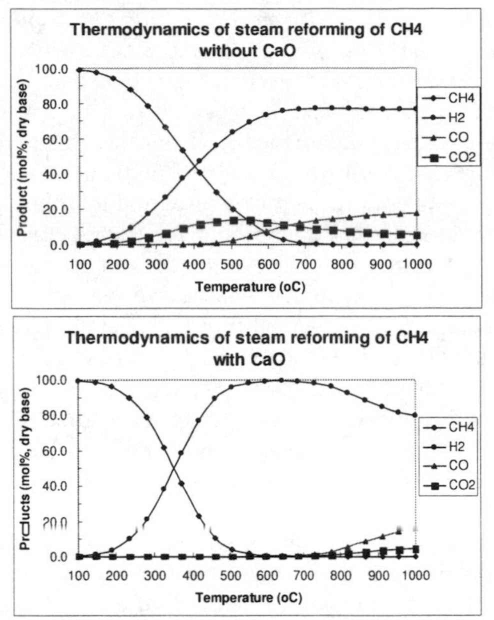

Figure 1. *Reformate composition (dry basis) predicted by Gibbs free energy minimization.*

The major barrier to the implementation of this approach has been the development of a recyclable CO_2 sorbents with a performance (CO_2 absorption capacity) that does not degrade significantly over the number of cycles required for a commercial product.[4-6, 8, 9] The problem of reduced absorption capacity for conventional sorbent materials stems from the fact that the product of the carbonation reaction, $CaCO_3$ has a much lower density than the reagent CaO. Therefore, for a fixed mass of CO_2 sorbent in a reactor bed, there is a large increase in volume as CaO is converted to $CaCO_3$. Due to the high temperature typically required for decarbonation, particle sintering occurs leading to a massively reduced CO_2 absorption capacity on the second cycle. Therefore a strategy that achieves a high absorption capacity through the production of a high surface area CaO powder based on small primary particle CaO alone will not lead to retention of a high capacity in subsequent cycles.

CSMP has addressed this problem by applying a spray-based powder manufacturing to design and produce materials with the necessary microstructure and composition to achieve the required high capacity and performance over multiple cycles. In the current work we use a unique spray-based powder manufacturing approach to develop and produce powders that have been specifically designed for absorption enhanced reforming using logical materials design concepts.[12,13] The materials are specifically designed for pelletization with tailored and optimized catalyst/sorbent proximity in the mm to nm length scale and with additional compositional and microstructural features that lead to enhanced carbonation/decarbonation kinetics and WGS activity.

CSMP has developed this patented spray-based powder manufacturing platform and has demonstrated that this powder production method is capable of producing a wide variety of materials and microstructures.[13] The major attributes of this approach are the extreme flexibility to achieve combinations of compositions and microstructures that cannot be achieved by other powder manufacturing methods, combined with economic suitability for high volume manufacturing. The flexibility to achieve unique combinations of compositions and microstructures comes from the fact that the spray process combines chemistries from both liquid phase and solid state processing.

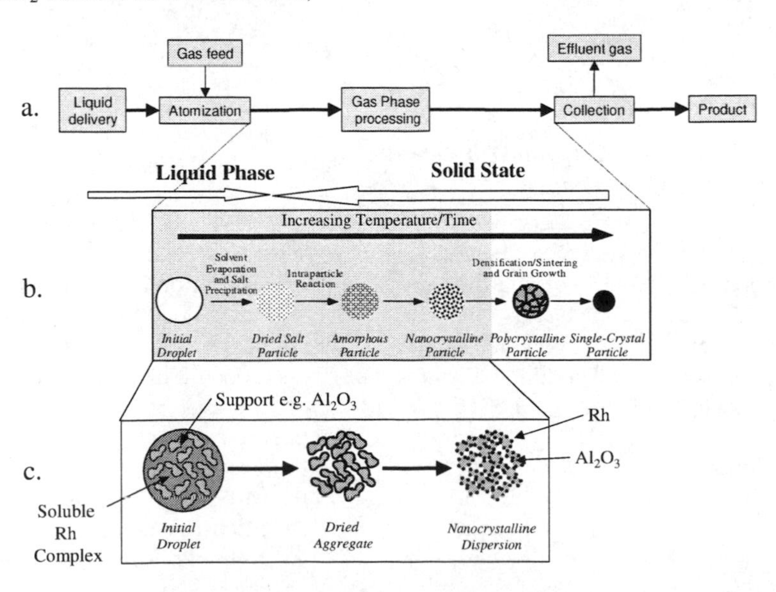

Figure 2. *a. Schematic representation of spray-pyrolysis process flow: b, Schematic representation of processes occurring during spray-based production of unsupported materials: c, Schematic representation of processes during the spray-based production of supported catalysts.*

A schematic representation of the spray-based process is illustrated in Figure 2 and has been described elsewhere in detail.[12,14] It starts with the formulation of a liquid that contains either dissolved or suspended reagents which act as precursors to the final product. The liquid, together with a gas, is then fed to an atomization unit where the liquid is converted into an aerosol. The size and size distribution of the droplets that comprise the aerosol are carefully controlled because each droplet becomes a particle (or aggregate in the case of supported catalyst powders) after gas phase processing. Many methods of droplet formation exist providing a broad spectrum of droplet sizes, atomization rates (measured by the amount of liquid phase atomized per unit time) and droplet size and size distribution. The gas stream containing the aerosol is then heated in a gas phase processing unit to effect the physical and chemical conversion of the droplets to the final powder. The final powder is separated from the gas stream using conventional powder collection methods, leaving only a gaseous effluent (no liquid effluent to be disposed). The final powder microstructure and composition depends on the residence time, temperature, the reactive nature of droplet components and the composition of the gas. A schematic representation of these processes is provided in Figure 2b and typically involves solvent evaporation, thermally or chemically induced liquid phase and/or solid state reactions, crystal nucleation and crystal growth. *The uniqueness in the ability to construct specific microstructures and compositions is derived from the sequential application of liquid phase and solid state chemistries that can be resolved by both temperature and time.* The physical and/or chemical evolution of the particles can be arrested at any stage by quenching of the reaction media. Therefore this process can be used to produce a wide variety of materials compositions combined with unique microstructures and morphologies.

A number of CaO-based powders have been synthesized and their CO_2 sorption capacity evaluated. Figure 3 shows an examples of two samples with large variation in their sorption capacity and stability as measured by Thermo-gravimetric Analysis (TGA) under the following conditions: carbonation: 600°C; decarbonation: 750°C. These data show that the reactivity of the sorbents (in terms of CaO reaction fraction) of a properly designed material can be extremely high (over 70 % for sample A, Figure 3) and that it can be retained after multiple cycles of carbonation and decarbonation. To our knowledge this is the best performance observed for these type of materials and while this is only a relatively small number of cycles (over 140) it demonstrates the feasibility of the proposed approach.

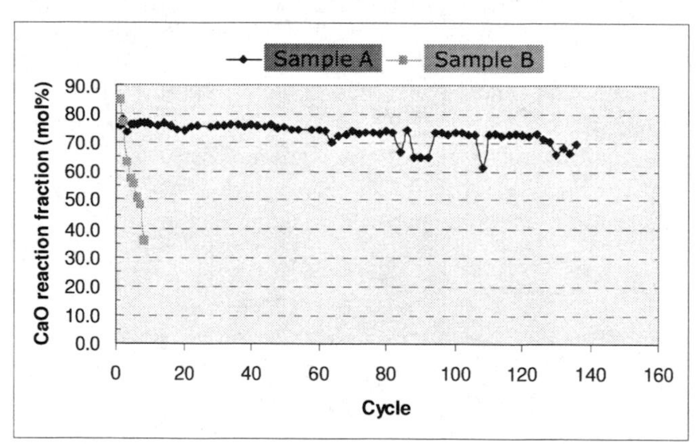

Figure 3: *Comparison of reversible CO_2 uptake in commercial in two CSMP sorbents*

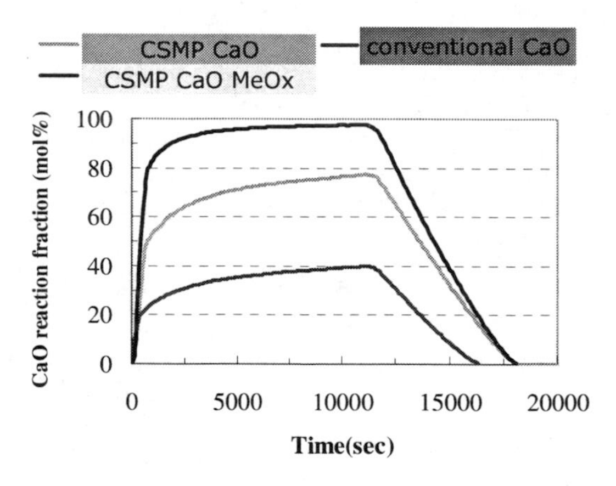

Figure 4: *Comparison of the rates of carbonation and decarbonation in CSMP spray-produced sorbents*

The recycle time in the fuel processing system is dictated by the size of the CO_2 sorbent beds, the number of beds present and the rate at which the CO_2 is absorbed and removed. High and reproducible rates of carbonation and decarbonation are desirable. Based on the initial studies of the materials produced to date and the nature of the equilibrium between CaO + CO_2 and $CaCO_3$, it can be concluded that the carbonation reaction has a relatively high rate at a lower temperature compared to the rate of decarbonation during the calcination step. This assumes that by controlling the microstructure, we can avoid the formation of an impervious $CaCO_3$ layer, which has been shown in the literature to limit the rate of CO_2 uptake after all the exposed surfaces of CaO have reacted.

The more difficult problem to solve is achieving decarbonation kinetics that is sufficiently rapid to ensure that this step does not limit the recycle time. Increasing the temperature can increase the rate constant for the decarbonation reaction, but this may lead to some sintering of the material and reduce the cycle life. Therefore, alternative strategies need to be employed to minimize the time required for decarbonation. Figure 4 shows a plot of the carbonation and decarbonation rates under the conditions specified for a spray-based material compared to a standard CaO sample. In this experiments, the CO_2 is present in large excess.

REFERENCES

1. "Fuel Cells and Their Applications", K. Kordesch and G. Simader, VCH, Weinheim, 1996.
2. "Fuel Cell Systems Explained", J. Laramie and A. Dicks, Wiley, Chichester, 2000.
3. "Fuel Cell Handbook", 6th Edition, DOE/NETL-2002/1179, EG&G Technical Services, USDOE, November 2002.
4. Squires A.M. Fuel Gasification, p205, 1996.
5. Balasubramanian, B.; Lopez Ortiz, A.; KaytaKoglu, S.; and Harrison, D.P., Chemical Engineering Science, 54, 3534, 1999.
6. Han, C., and Harrison, D.P., Chemical Engineering Science, 49, 5875, 1994.
7. "Sorption-Enhanced Reaction process….." DOE report # DE-FC36-95G010059, Mayorga, S.G.; Hufton, J.R.; Sircar, S.; Gaffney, T.R., Air Products and Chemicals.
8. "A Novel Steam Reforming Reactor for Fuel Cell Distributed Power", California Energy Commission, May 2000 Contract # 500-97-038.
9. Samtani, M., D. Dollimore, and K.S. Alexander, THERMOCHIMICA ACTA., 135, 2000.
10. Ahmed, S, and Krumpelt, M., International Journal of Hydrogen Energy, Elsevier, 26, P291, 2001.
11. Stevens, J.F., US Patent .
12. Kodas, T.T. and M.J. Hampden-Smith, Aerosol Processing Of Materials. 1999, New York: Wiley-VCH.
13. Hampden-Smith, M.J., T.T. Kodas, Q.P. Powell, D.J. Skamser, J. Caruso, and C.D. Chandler, US 6,338,809.
14. Hampden-Smith, M.J., Atanassova, P. Atanassov. P., and Kodas, T., The Handbook of Fuel Cell Technology, Volume 3, Chapter 3.3.4, Ed.s Gasteiger et al., Wiley, in Press, 2003.
15. "Structural Inorganic Chemistry", A.F. Wells, 5th Edition, Oxford Science, 1991.
16. Lvov, B.V., Mechanism and Kinetics of Thermal-Decomposition of Carbonates. THERMOCHIMICA ACTA, 2002. 386(1): p. 1-16.
17. Lvov, B.V., Mechanism of Thermal-Decomposition of Alkaline-Earth Carbonates. THERMOCHIMICA ACTA, 1997. 303(2): p. 161-170.
18. Shui, M., L.H. Yue, Y.M. Hua, and Z.D. Xu, The Decomposition Kinetics of the SiO2 Coated Nano-Scale Calcium-Carbonate. THERMOCHIMICA ACTA, 2002. 386(1): p. 43-49.
19. Muehling, J.K., H.R. Arnold, and J.E. House, Effects of Particle-Size on the Decomposition of Ammonium Carbonate. THERMOCHIMICA ACTA, 1995. 255(MAY): p. 347-353.

New Product Development for Nanomaterials Systems and Solutions

G. Varga, M. Kröll;
Degussa AG, Advanced Nanomaterials
Rodenbacher Chaussee 4, Geb.1040-004, Hanau, Germany

ABSTRACT

In the past several years there has been a tremendous increase of activitiy in the field of Nanomaterials Technology. This recent increase is in part a result of improved analytical methods which allow scientists and engineers to custom tailor nanoscale features into products which result in completely new functionalities. Realizing the intended functionalities of these materials most often requires that they undergo post-synthesis treatment and modification, and integration into dispersions and formulations. These developments will allow for the intended benefits to be realized in many different areas of application including energy, electronics, information technology, and optics. In this paper we present the synthesis techniques of our new nanoscaled products, their physical and chemical properties, and their further processing into systems which are tailored for use in specific applications.

Keywords: nanomaterials, synthesis, modeling, modification, dispersions

1 R&D APPROACH

Degussa's core competence of gas-phase synthesis of Nanomaterials was extended in a three-year strategic R&D project known as the Projechouse Nanomaterials. The goal of the project was to understand and control the important characteristics of these materials starting with particle morphology, which refers to the primary particle size and shape and also the aggregate structure of the materials, and also to understand and control other important characteristics which are established during synthesis, such as crystallinity, porosity, surface features and surface chemistry. It is necessary to control morphology with processing conditions which are optimiZed with regard to economics.

2 THE SCIENCE OF PARTICLE FORMATION

Synthesis processes involving gas-to-particle conversion allow for the production of particles having controlled nanometer-scale dimensions, which are built up from atomic or molecular size in the gas phase to the desired particle size. This general approach has been used for decades by Degussa and other companies to produce various nanoscale materials such as aluminum oxide, titanium dioxide, fumed silica and carbon black on a commercial scale. Although a number of variations exist for gas-phase processes, they all have in common the fundamental processes of particle dynamics which occur once the product species is generated [1]. As the superheated vapor cools, the saturation ratio of the vapor builds up and leads to the nucleation of thermodynamically stable clusters. These clusters can grow by collision and coalescence (collision-controlled growth), by vapor condensation (condensation-evaporation-controlled growth) or by surface reaction, in which the precursor reacts on the cluster surface. The clusters can grow further to a size where they are considered particles by collisions with other product particles, which leads to coagulation, or with product molecules, which results in a condensation, or by reaction of the precursor on the particle surface. Collisions between small particles at high temperatures lead to complete coalescence to form spherical particles. As the temperature decreases and particle coagulation continues, further collisions result in hard or soft agglomerates.

The most common reactor design used for the manufacturing of commercial quantities of nanoparticles such as SiO_2, TiO_2 or Al_2O_3 is the flame reactor. Flame reactor technology allows for current production rates typically in the range of several hundred to several thousand tons per year for each individual reactor, depending on the nanoscale oxide being produced. Such high temperature gas-phase synthesis typically leads to a product powder consisting of aggregates composed of up to several hundred primary particles. The structure of the aggregates and the product particle size distribution are determined by the particle formation processes occurring in the reactor. Three mechanisms dominate particle formation in flame reactors:

- Chemical reaction of the precursor: This leads to formation of product monomers (clusters) by *nucleation* and to the growth of particles by reaction of precursor molecules on the surface of newly formed particles, which is called *surface growth* [2].
- Coagulation: A critical level of particle concentration in the reactor leads to collisions of the product particles with other particles or with product monomers, which results in coagulation if the particles stick to each other [3]. Coagulation may occur in the *free-molecular regime* (particle diameter much smaller than the mean free path of the surrounding gas) or in the *continuum regime*, where the particles are larger than the mean free path.
- Sintering: In the high temperature zone of the flame reactor, coalescence and fusion are sufficiently fast which results in a reduction in the level of aggregation or the formation of spherical particles due to *sintering* processes [4].

NSTI-Nanotech 2005, www.nsti.org, ISBN 0-9767985-1-4 Vol. 2, 2005

2.1 Modeling and Simulation of Gas-Phase Synthesis Processes

Mathematical modeling and simulation of these production processes can improve product quality and performance characteristics of nanoparticles, since these often depend on the particle size distribution, morphology, and the degree of aggregation of the particles. The ultimate characteristics of the powder are determined by fluid mechanics and particle dynamics within a few milliseconds at the very early stages of flame synthesis. Therefore the product quality can be influenced by an intelligent selection of the process parameters. Process simulations can improve the understanding of the physical and chemical fundamentals of gas phase synthesis by connecting process parameters, such as temperature, reactant state or reactor geometry, to particle characteristics. The calculated particle size evolution is compared to experimental data obtained by thermophoretic sampling. Particle sizes and morphology can be measured directly in the flame at various distances (residence times) from the burner by shooting a probe into the flame, which collects particles by thermophoresis on a TEM-grid [5]. The resulting understanding and control of particle formation enables the production of tailor-made morphology and composition including nanocomposite particle structures. One example is a nanocomposite particle structure consisting of superparamagnetic iron oxide domains encased within an SiO_2 particle matrix.

3 ORGANIZATIONAL APPROACH TO NEW BUSINESS DEVELOPMENT

In order to generate new business we must make the transition from supplying simple nano powders to suppling highly functional specialized Nanomaterials and solutions. This includes tailoring materials with respect to composition, crystallinity and morphology, but also applying post-processing, surface modification, and forward integration of these materials into rheological or pH-stable dispersions or coating formulations in order to match a customer system.

Our organizational approach is to address the opportunities as an internal corporate venture, or an internal Start-up. The goal of this approach is to combine the agility, the flexibility and the innovative atmosphere of a Start-up together with the advantages of our more established internal business unit partners. We seek to leverage our technical innovation together with Degussa business units having the required production infrastructure for fast, effective, low-cost scale up of processes; the relevant applied tech expertise needed to thoroughly understand the customers' needs; or most importantly market access- preferably higher up in the value chain.

4 TAILOR-MADE SOLUTIONS

4.1 Custom Synthesis

The example of custom-tailored nanocomposite particles made up of superparamagnetic iron oxide in a silica matrix enables customers in the adhesives industry to remotely heat their formulations containing these particles using an AC magnetic field. Using this approach, parts can be bonded and de-bonded quickly, without heating the parts that are to be joined. In this case, the silica matrix provides rheological functionalities while the superparamagnetic domains contained within the silica provide the further functionality of remote heating, enabling efficient bonding and de-bonding processes. The ability to produce such multi-functional composite particles is a result of a thorough understanding of the science of particle formation.

4.2 Post-processing and Modification

Often times, materials are made more suitable for specific applications by realizing new characteristics and functionalities of standard products. These are achieved through post-processing or modification of standard, as-synthesized nanomaterials. One example is the thermal reduction of indium-tin-oxide nanoparticles in order to increase the density of highly mobile charge carriers. This post processing results in a significant increase in both the electrical conductivity and the IR absorption coefficient of the material, leading to higher effectiveness in anti-static and IR shielding applications in coating formulations containing the nanoscale particulate indium-tin-oxide. Another example is the granulation of nanoscale silica, which imparts free-flowing behavior into the product while maintaining high specific surface area, making it compatible with the material handling processes of certain customers. A third and very important example of particle modification is the hydrophobization of hydrophilic oxide nanomaterials, in order to improve dispersion characteristics, compatibility, and performance in various formulations and systems. In the case of sunscreen applications, hydrophobization of Zinc Oxide and Titania improves skin feel and suppresses photo catalytic activity.

4.3 Dispersions and Formulations

Often times, it is desirable to forward integrate these custom-synthesized and/or modified nanomaterials into dispersions and formulations. In some cases, customers prefer to utilize their know-how in these areas and profit from these value-added steps themselves. However, in other cases this opportunity can be realized by the nanomaterial supplier in order to increase margins by providing ready-to-go systems for customers in the fields of coatings, CMP, polymers, and printed films. Even in cases where the customer is willing to prepare their own dispersions and formulations, the material supplier often requires these competencies in order to show a proof-of-concept or to provide application guidance. Such competencies can also assist the particle design efforts in the synthesis of materials which are compatible with a given dispersion or formulation system. In these cases, a thorough understanding of the challenges facing customers one or more steps up the value chain can assist in the design of nanomaterials which are most suitable for a given system, providing an advantage over competing suppliers. In the case of inorganic nanomaterials for use in sunscreens, the desired transparency to visible wavelengths not only requires the appropriate particle morphology, but also the effective dispersion of the particles so that visible light is not scattered. Without the proper dispersion methods, transparency cannot be achieved. This also holds true for UV-resistant coatings on other surfaces such as wood or in polymer films. In the case of CMP, particle design is the first critical step, but the incorporation of the particles into a suitable and stable dispersion is just as critical to achieving the requirements of the IC wafer polishing process. The combination of high removal rate, a low defect or scratch rate, high degree of planarity, favorable selectivity between polishing different materials, and dispersion stability requires an in-depth knowledge of dispersions and formulations.

5 SUMMARY

In summary, developing new business in the field of nanomaterials technology requires a broad range of technical competencies. This starts with the understanding and ability to control a cost-effective, scaleable synthesis technology. Furthermore, winning new business often requires the synthesis of materials which are tailored, modified, or incorporated into a system intended for a specific application field. As a raw material supplier, significant opportunities for new business and higher profit margins can be realized through vertical integration up the value chain. But even to succeed as a material supplier, it is important to have a high degree of understanding of the entire value chain so that new products can be properly designed for specific applications which deliver the advantages and benefits that customers demand.

REFERENCES

[1] Kodas, T.T.; Hamden-Smith, M.: Aerosol Processing of Materials, Wiley-VCH, New York, 1999.

[2] Pratsinis, S.E.; Spicer, P.T.: "Competition Between Gas Phase and Surface Oxidation of $TiCl_4$ During Synthesis of TiO_2 Particles", Chemical Engineering Science, 53, 1861, 1998.

[3] Hinds, W.C.: Aerosol Technology- Properties, Behavior, and Measurement of Airborne Particles, John Wiley & Sons, New York, 1999.

[4] Koch, W.; Friedlander, S.K.: "The Effect of Particle Coalescence on the Surface Area of a Coagulating Aerosol", Journal Colloid Interface Science, 140,419,1990.

[5] Arabi-Katbi, O.I.; Pratsinis, S.E.; Morrison, P.W.; Megaridis, C.M.: "Monitoring Flame Synthesis of TiO_2 Particles by In-situ FTIR Spectroscopy and Thermophoretic Sampling", Combustion and Flame, 124, 560, 2001.

Converting of nanoparticles in industrial product formulations: Unfolding the innovation potential

S. Schaer, G. Arnosti, S. Pilotek, F. Tabellion and H. Naef

Buhler AG, Division Engineered Products, Uzwil 9240, Switzerland, samuel.schaer@buhlergroup.com

ABSTRACT

Potential users of nanoparticles seek for particles in a ready-to-use form. The market currently offers practically only raw materials and standard nanoparticles not tailored to a specific application. A newly developed chemo-mechanical converting process for ready-to-use nanoparticles and formulation is presented. This technology is based on the combination of the mechanical comminution in agitator bead mills and the small molecule surface modification concept. Tailored nanoparticle formulations are developed in close cooperation with the client, scaled to the required amounts and subsequently are produced. Our business model is focused on tailor-made solutions for our client's product innovation allowing the flexible and in-time implementation as well as an easy access to nanotechnology at a low risk.

Keywords: nanoparticle dispersion, converting technology, surface modification, agitator bead mill, tailored processing

1 NANOPARTICLE MARKET

Market experts estimate the potential of nanoparticles to be overwhelming. For example the US-NSF (National Science Foundation) projected an expected market size for nanomaterials alone of USD 340 billion by 2010-2015. Product developers around the world agree on the large potential linked to the use of nanoparticles in products. Actual and future application areas of nanoparticles will include the sectors of coatings, inks, plastics, resins, sealants, adhesives, technical pastes, ceramics as well as consumer products, catalytic applications ranging from chemistry to electronics, pharmaceutical and cosmetic products. Furthermore the use of nanomaterials will not only enhance the performance of already existing products, but also lead to entire new products and production processes. From this point of view, it can be assumed that these materials will be ubiquitous in our near future.

However, sales in ceramic nanoparticles only totaled an estimated USD 150 million in the US for 2003. If one accepts the NSF figures, that market should be growing at incredibly high double-digit percentage rates, which it currently does not. This is even more astonishing, since in the last decade a tremendous amount of research was focused on the production of these fine particles. The production technologies for various materials including SiO_2, Al_2O_3, TiO_2, ZrO_2, ITO etc. are well known and the necessary industrial production capacities are established.

But why are the sales and consequently the utilization of nanomaterials still on such low levels and far away from the predictions?

1.1 The Gap in the value chain

One of the main reasons for this is the fact that industrial converting techniques suitable for the user, meaning the manufacturer of the final products, are not available or not developed yet. This includes the whole processing route, starting from the particle handling, particle processing and converting up to the final material components and application. Nanopowders suffer from agglomeration due to the enormous particle-to-particle interaction and chemical forces, which cannot be resolved by traditional dispersion technology. Using a nanoparticle to unfold its inherent functionality in an application means incorporating or technically speaking converting it into a product on the nanoscale. This is where the difficulty starts: Nanopowders currently available on the market are not readily dispersible and therefore not ready-to-use.

In other words, today we see a massive gap in the value chain of nanomaterials.

On the one side of the value chain are more than 200 producers of nanoparticles selling raw materials, not customized to the costumer's prerequisite. On the other side are thousands of potential users of nanoparticles representing various industry braches.

1.2 Standard vs. tailored product approach

Chemical companies as well as start-ups have introduced standard nanoproducts in the form of dispersions into the market. It is self-evident though, that such standard materials in most cases cannot fulfill the different requirements of the diverse industrial applications. Typically product developers of companies interested in using nanoparticles test these materials on a trial and error basis in their formulation. As those standard nanoparticle dispersions are not tailored to the end product formulation, dispersion and stability does not work out. In consequence the nanoparticles agglomerate. In addition, the nanoparticles are typically not functionalized for the end application. Thus developers are not obtaining the improvement results they are looking for through the use of nanoparticles. Scratch-proof coatings end up not being scratch-proof enough or too expensive, nanocomposites do not show the desired enhancement of the mechanical properties, transparent UV protection using nanoparticles ends up not being transparent enough just to name a few.

To bridge the gap in the value chain, Buhler is perusing a different approach to overcome the massive disadvantages of standard products. This approach is based on tailor made nanoparticle formulations specifically customized for the users needs and requirements. Nanoparticle formulations are developed in close cooperation with the client, scaled to the required amounts and subsequently produced. The business model allows the flexible and in-time implementation and enables access to nanotechnology at a low risk. The underlying technology, the business model including the products and services will discussed in the following.

2 TECHNOLOGY

2.1 Chemomechanical processing approach

Conventional industrial powder converting tasks involve processes including three-roller-mills, mixers, dissolvers, extruders and agitator bead mills. Whereas this technology is sufficient for micron- and sub-micron-sized particles, a new approach is mandatory for the processing of nanoparticles.

We propose a technique for converting nanoparticles in agitator bead mills comprising the simultaneous utilization of chemical and mechanical competences called chemomechanical processing.

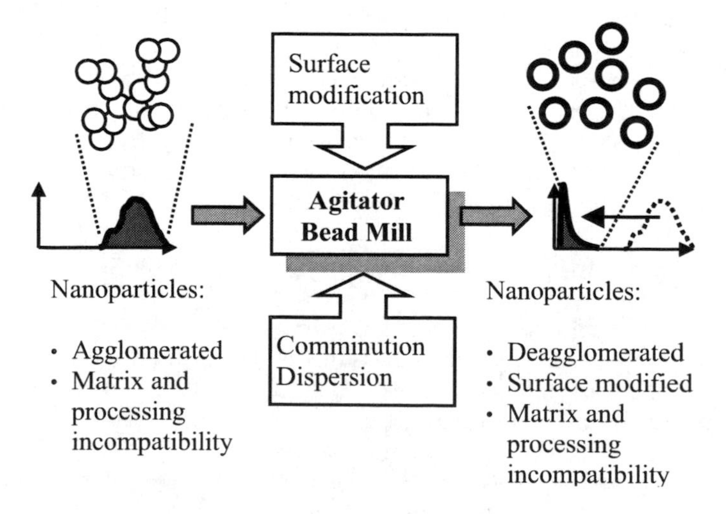

Figure 1: Chemomechanical processing

This technique is based on the combination of a) the mechanical comminution of agglomerated nanopowders in agitator bead mills and b) the rules of colloidal chemistry utilizing surface modification in order to tailor the particle surface chemistry. Using this technology agglomerated nanoparticles, which are incompatible for the further processing techniques will be converted into tailor made deagglomerated and functionlized dispersions ready for the later processing (fig.1). Furthermore this technology has the potential for the production of nanoparticles and nanodispersions by real comminution of larger particles. This applies also for application areas, where to-date no production technology exists.

2.2 Agitator bead mill reactor

Agitator bead mills provide a number of advantages in the processing of nanoparticles. Agitator bead mills are devices to process suspensions with high energy at controlled conditions. Solid state nanoparticles are processed in the liquid phase, which facilitates their easy further processing into the end products. The method itself is established, known and used for a long time. Therefore, a lot of experience is available about principles, constructive possibilities as well as about the processing details. Grinding in agitator bead mill is a scaleable process and thus industrially relevant. The large variety of available constructive details provides many possibilities for optimising processes, e.g. different grinding chamber geometries, grinding media separation, and cladding materials. For nanoparticle processing the milling chamber has to be understood as a reactor where complex chemical reactions can be conducted under well-defined mechanical conditions in the wet phase. Agitator bead mills are advantageous in this respect, as they provide a defined volume and a readily controllable energy input. Because of the large tendency of nanoparticles to form conglomerates, the challenge of separating the particles to obtain a highly dispersed state is comparable to a real comminution rather than a dispersion process. Therefore, the mill will be operated usually in the recirculation mode, as this set-up allows for a higher energy input than passage-wise operation. (fig.2).

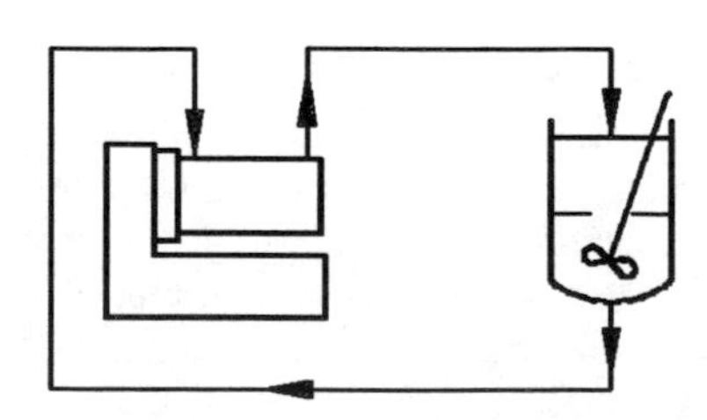

Figure. 2: Flow sheet of passage-wise operation of agitator bead mills

Experience has shown, that for traditional wet grinding processes a particle size of ca. 100-200 nm seems to constitute a major barrier in comminution. In many cases, increasing the energy input does not lead to finer particles. This has to be interpreted in terms of specific surface area and Van der Waals forces, which increases rapidly by reducing the particle size and approaching the barrier of 200 nm. If no precaution for the stabilization is carried out, a balance between deagglomeration and agglomeration leads to a stagnation of the comminution process and the resulting particle size distribution.

2.3 Chemical surface modification

In principle suspensions can be stabilized by making use of electrostatic, steric and electrosteric stabilization techniques. The purpose of these techniques is to prevent the re-agglomeration, meaning the formation of large agglomerates and the subsequently occurring destabilization of the suspension.

The simplst way to stabilize nanoparticles is to make use of the electric repulsion by generating an electrical charge at the particle surface due to a pH value change.

Unfortunately, the usability of this approach in product formulations is heavily restricted and permits its broad utilization. The second principle of steric stabilization refers to geometric arguments of holding particles apart, whereas electrosteric stabilization describes a combination of both means. Such sterical/electrosterical stabilizers are mostly polymeric or oligomeric compounds with high molecular weight. These large molecules provide several bonding sites to occupy the surface groups of the nanoparticles. The utilization of such large molecules for the stabilization of nanoparticles, especially particles in the size range between 10-100 nm, suffers from different drawbacks. As a first geometric argument, large molecules are able to hide nanoparticles inside of their large volume. This phenomenon might be relevant for certain applications, where particle-medium interaction is important, e.g. scratch resistance. At the same time, the solid content in terms of volume % is rather limited. Albeit for functional nanoadditives small degrees of filling are preferred, the freedom of formulation is reduced, if the dispersant constitutes a major part in the functional nanoadditive. Importantly, with nanoparticles, the many binding sites of a polymeric dispersant may lead to the generation of a new type of agglomerates, when the sites do not bind to the same but to adjacent particles. In dependence of the molecular size of the dispersant, these conglomerates may even encompass several particles. The resulting conglomerates will not show the wanted properties of nanodispersions.

Based on these considerations the herein presented chemomechanical processing technology utilizes the so called small molecule surface modification concept. This concept relies on the chemical bond principles known from solution chemistry and employs them for the anchoring of the surface modifiers (small molecules) on the nanoparticle surface. The basic principle of the surface modification, using organic acids as an example, is shown in figure 3.

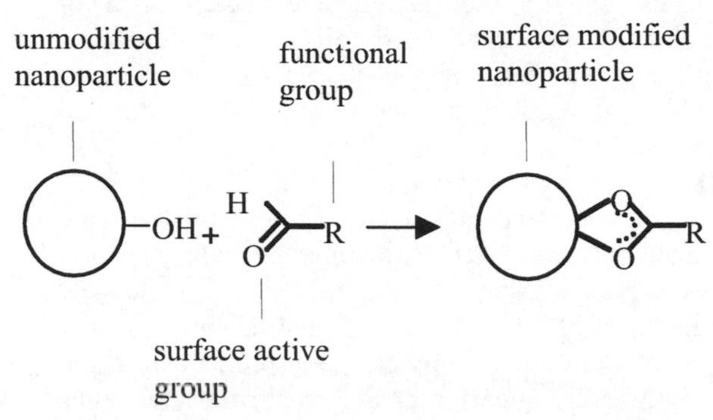

Figure. 3: Basic principle of a surface modification

As surface modifiers for nanoparticles bifunctional molecules are employed. One of the groups is reactive towards the surface of the particles to form a stable connection between particle and surface modifier, whereas the other group is to be chosen in terms of the later application or the required functionality. For the stabilization of nanoscaled suspensions the surface modifier needs to be compatible with the solvent, i.e. the respective part of the surface modifier needs to be soluble. Those small molecules still provide sufficient barrier for holding nanoparticles apart, but reduce the organic content to a minium. Moreover, molecular surface modifiers are paradigm for nanoparticles, as they allow the specific tailoring of the particles' surface by introducing a functinallity. This in turn is indispensable for manufacturing processable nanoparticle dispersions, as nanoparticles are highly responsive to their environment.

3 BUSINESS MODEL, PRODUCTS AND SERVICES

Buhler is a global technology, know-how and project partner for efficient production systems, engineering solutions and related services. As the world market leader in grinding & dispersion equipment we consequently own massive know-how in comminution down to the nano range and are powerfully present in the vital markets. From this point of view, it was a natural evolution to move into the nanoparticle technology market making the chemomechnical processing technology utilizable for the benefit of our customers. For this reason, three years ago Buhler has formed a strategic alliance with the INM (Institute for New Materials), which is a world wide leading institute for nanotechnology with more than ten years experience in this area. This alliance comprises the compulsory competence, know how and IP in mechanical comminution and surface modification. In combination with Buhler's competences in engineering and logistic as well as its strong market position all the vital components for a successful market implementation are given.

The Buhler nanotechnology business model is particular focused on tailor-made solutions for our clients product innovation by closing the gap in the value chain of nanomaterials. This will be accomplished by an overall processing solution for the production of customized nanoparticle formulation as well as customized toll production.

From the company perspective the mission of nanotechnology is clear: a) to foster and stimulate the existing machinery business, b) to open up larger market shares in the area of process sales and c) to create new business in the area of products (Fig 4.). This strategic direction towards a more process and product oriented business will subsequently create recurring revenues compared to the traditional machinery business.

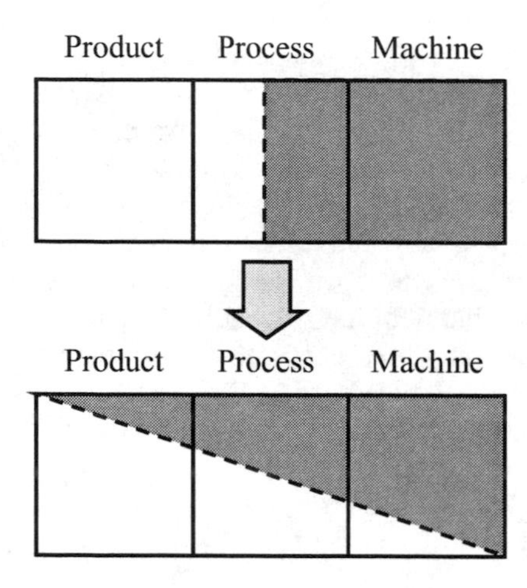

Figure 4: Mission of Nanotechnology

Beside the existing markets, where Buhler is already strong, our material independent technology allows to target various additional markets ranging form currently hot areas in the coating and lacquer industry, the electronic or cosmetic industry to markets where the use of nanoparticles is on a low level or even nanoparticles are not introduced yet.

From the customer's perspective Buhler`s strategy offers easy excess to nanoparticle technology, especially to a tailored one. Furthermore this approach permits the customers to protrude into hithero unanticipated markets, or to distinguish themselves from their competitors. Because customized solutions are specifically developed for the ennoblement of the customer's product, the time to market will be short. In contrast to classical contract research, carried out by institutes, the development and adaptation necessary to achieve a tailored solution will be carried out by Buhler, on our own expense, tremendously reducing the risk and capital expenditure for the customer. Additionally, the required development goes far beyond the horizon of the capability of classical contract research covering all steps from the laboratory, to the pilot plant till to the final industrial scale.

Foreseeing the diverse customers requirements, especially regarding the production place, Buhler has launched the product Buhler-Tailored Nanoprocess® and the service Buhler-Tailored Nanobatch®. Both offers are focused on the production of the tailored nanoparticle formulations. In the case of the Nanoprocess, the customer will be put into the position to carry out the production at its own facility, whereas in the case of a Nanobatch® Buhler is the toll producer and supplies the ready to use formulation. Apart from a small symbolic retention free paid by the customer Buhler is covering the whole expenses for the development. The retention free is later refunded by a discount on the purchase of materials from Buhler.

4 CONCLUSION

Buhler offers, for the first time, with its product Buhler-Tailored Nanoprocess® and the service Buhler-Tailored Nanobatch® an easy access to tailored nanoparticle formulations and production processes. It can be anticipated that such tailored solutions will close the currently existing gap in the value chain of nanomaterials and sustainably shape the landscape of this new emerging technology by unfolding the potential of the inherent particle functionality in real-life applications.

NILCom® – Commercialization of Nanoimprint Lithography

H. Luesebrink*, T. Glinsner*, Ch. Moormann**, M. Bender**, A. Fuchs**

* EV Group, Erich Thallner GmbH, Erich Thallner Strasse 1, A-4780 Schaerding, h.luesebrink@evgroup.com

**AMO GmbH, Huyskensweg 25, 52074 Aachen, Germany, moormann@amo.de

ABSTRACT:

NILCom® is a consortium and technology platform for Nanoimprint Lithography (NIL) processes (Fig 1.)[1]. NILCom® focuses on commercially available infrastructure for nanoimprint applications in nano electronics, life sciences, data storage, and opto electronics, whereas previous work in NIL examines specific phenomena of imprint technologies and feasibility studies in feature size resolution and its specific process challenges[2,3]. Today NIL is considered a member of next generation lithography (NGL) and with that, it enjoys its consideration and application in product development and manufacturing if compared to other patterning technologies.

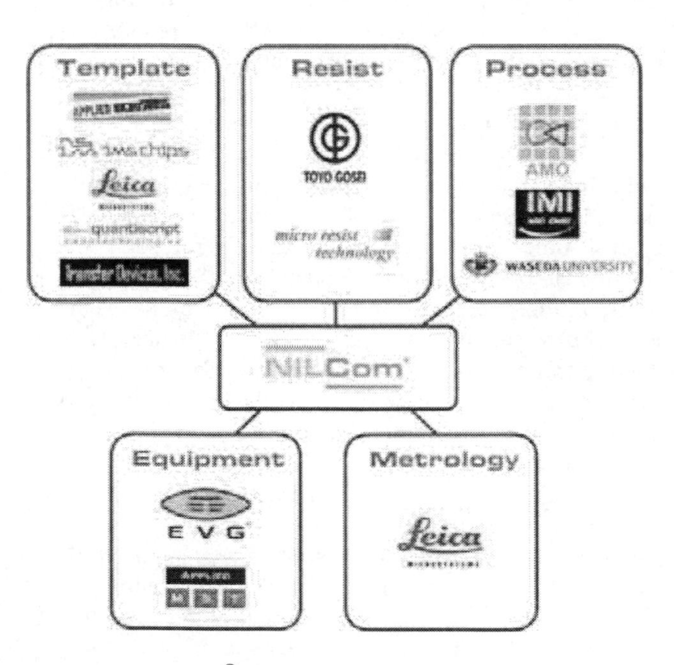

Figure 1 NILCom® Consortium Members

Keywords: NILCom®, nanoimprint lithography (NIL), UV-NIL, μTAS, patterned media, biosensors, commercialization

1 NILCom® IN THE INDUSTRY:

The industry looks at NIL as an enabling technology for novel devices and to replace high-end substrate materials with lower cost polymers [4]. In addition it offers opportunities to apply NIL technology as a promising and cost efficient way for advanced pattern replication amongst which the most prominent and promising is in nano electronics. Although the 65nm node is already destined for high volume manufacturing [5] there are challenges to provide a cost efficient continuing path with current NGL technologies, especially with Extrem UV (EUV).

Status	**NGL (EUV, EPL, ML2)**	**NIL**
Exposure equipment	Feature size depends on wavelength of light source	Existing lithography light source
Resist Material	New resist sensitivity development	Existing polymers available
Mask Making	New resist sensitivity needs	Existing technology can be used (LIGA, E-beam, etc..)
Metrology	Similar Requirements	

Table 1: EUV and NGL status.

Other successful application areas are in data storage with an opportunity and a roadmap to continually increase the data storage capacity for the growing demand on information mobility in consumer electronics. Advances in nanoimprint lithography (NIL) are being employed to introduce new manufacturing processes for next-generation, thin-film disks, which use patterned substrates for land-and-groove structures that magnetically isolate individual data tracks from each other. The concept, called discrete track recording (DTR), promises to significantly increase data storage in hard-disk drives because recording heads will no longer be used to define data tracks. Physical tolerances for recording heads will also be relaxed, and other performance improvements are expected from DTR [6].

Further the broad field of life sciences, its bio-sensors and fluidic devices, as well as in opto electronics are on the roadmap for NIL manufacturing technologies. NIL has been identified to combine multiple process steps in one, such as being promoted by Sematech [7] for dual damascene processes in semiconductor manufacturing. Its capability to generate three-dimensional patterns in a single step make it an ideal process technology for building structures on existing devices in surface or bulk patterning methods. It offers process simplification in particular with multi material systems prone to etch rate differences.

NILCom® serves application developers in the above-mentioned market segments as well as material suppliers alike to develop qualified NIL manufacturing methods. The combination of polymeric material development and availability of multiple centers of excellence provide processing expertise to accelerate prototyping and transfer from R&D to high volume manufacturing.

The goal of NILCom® is to leverage NIL synergies and support market segment acceleration for processes and infrastructure in its main areas of interest: nano electronics, data storage, life science, and opto electronics. In doing so NILCom® promotes process standardization and equipment automation for top down nano patterning technologies. The technology platform offers qualification opportunities for commercially available infrastructure (Fig. 2) such as templates, stamps, resists and substrates, processing equipment and metrology continuously taking advantage of developments along the International Technology Roadmap for Semiconductors in following ongoing scaling requirements in the nanometer regime driven by Moor's Law.

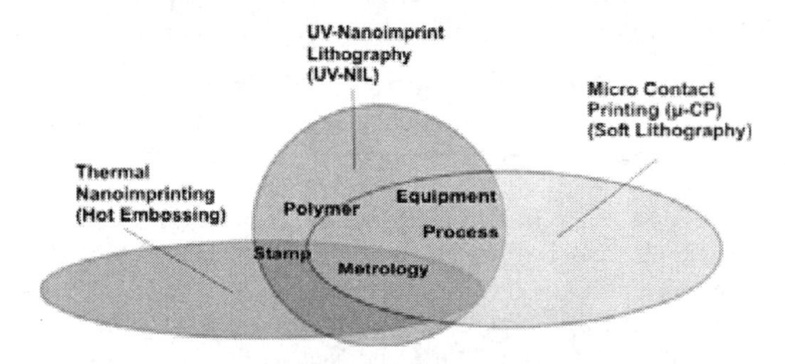

Figure 2 Synergies of NIL Technologies

2 CONNECTION WITH ITRS:

The difference between standard optical photo lithography (OPL), and NIL based pattern replication is that standard OPL depends on the control and range of its emitting wavelength. NIL technology mechanically replicates the pattern from a template (or stamp) either in a printing mode, under UV exposure or via a thermal imprint. As such high-end OPL systems are used today for template manufacturing only. High volume manufacturing can be done on a more cost efficient NIL system (Fig. 3) with reduced process steps due its intrinsic 3D capability [8].

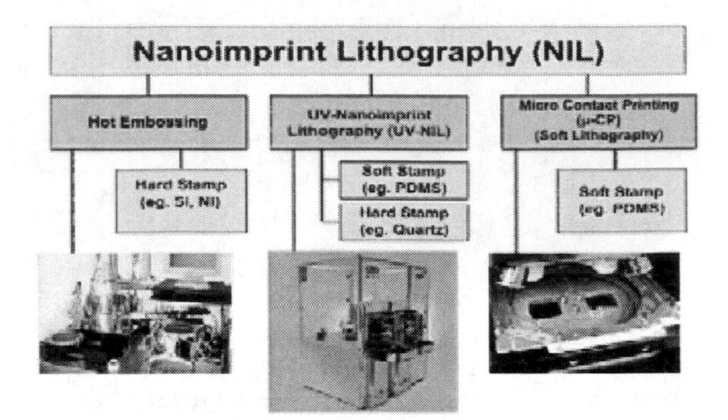

Figure 3 NIL Manufacturing Systems

Applications and components developed with NILCom® consortium partner's technology can take full advantage of the most advanced and cost efficient pattern replication methods promoting an accelerated product commercialization path. This becomes apparent when combining the high yield and high volume manufacturing specifications for semiconductors with low cost application requirements in life sciences for consumable products.

An area of particular interest is in the broad field of life sciences. Since NIL technology offers the possibility to work with biocompatible materials at room temperature it can take full advantage of critical processing environments and process time constraints. An example for such an application can be a cell culture platform that employs thermal imprinting for replicating larger patterns in sheet polymer (Polycarbonate (PC), Polymethylmethacrylate (PMMA)). Further it may use UV-NIL for nano structuring dedicated areas and finally applies solutions in dedicated zone providing functionality, fully automated and precision aligned. The same approach can be employed for manufacturing flexible displays or solar cells.

Of particular interest is the demonstration of nanoimprint technologies as contenders to 193nm optical lithography [9] with detailed results and demonstrations in defect control (Fig. 4). Since in all cases the template is in contact with the substrate, success of nanoimprint is strongly coupled to the absence of particles between the template and the substrate. While particles on a photomask in OPL lead to a local defect in the feature, a particle on the template acts as a bumper between template and substrate during the imprint procedure. Depending on its size areas up to the whole imprint area can be affected due to the non-parallelism between both surfaces.

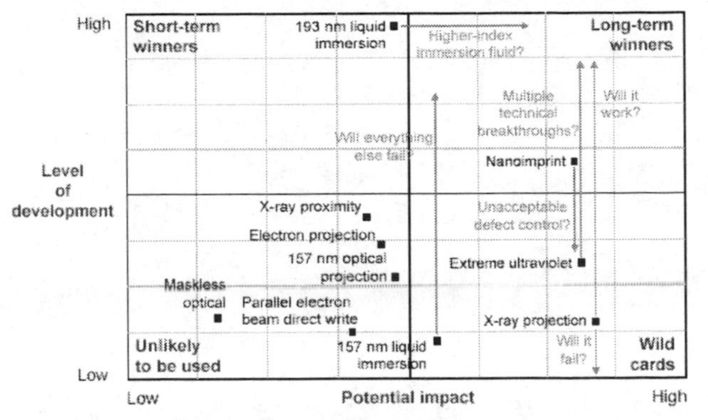

Figure 4 Nanoimprint Opportunities

However reapplying the same template in several consecutive imprint cycles can minimize the effect of small particles inside a template feature. It has been observed that particles tend to stick rather to the imprint resist than to the template resulting in some kind of "self-cleaning" effect of the template. This phenomena is observed were a defect caused by a particle on a template after 10 consecutive imprint steps. It can clearly be seen, that the defect size is dramatically reduced during the first five imprint cycles and nearly disappears after ten cycles [10].

Special nano-scale coatings can be used as "anti-stiction" or "release" layers in nano-imprint lithography. The MVD™ coatings can be applied to virtually any material using special adhesion layers [11]. Studies with release promoters are currently ongoing to satisfy the industry's demand for high repeatability and pattern fidelity.

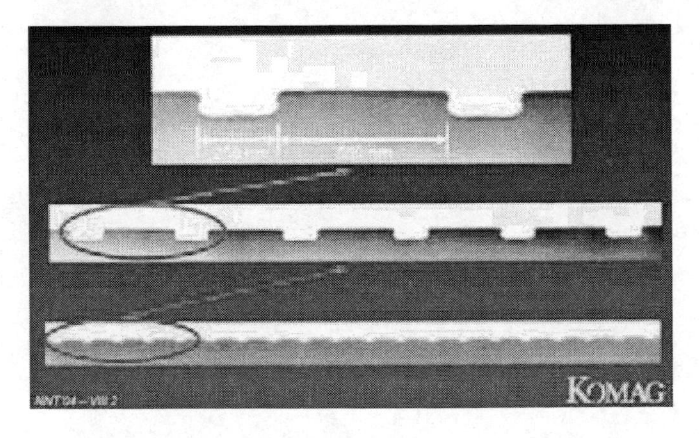

Figure 5 NIL on 95mm patterned media

Applied NIL technology in hard disk manufacturing for data storage aims at zero defect to avoid contact between the read and write head and the disc platter. Since the glide height over the platter during operation is in the range of 200nm, any contact between the head and the platter may cause a fatal error. Joint collaboration with individual NILCom® members and Komag, Inc. a hard disk manufacturer headquartered in California, USA, demonstrated the current state of the art in discrete track recording (DTR), where NIL is used as a cost efficient technology to create a land and grove structure on the platter to mechanically isolate the magnetic information on the disc and hence increase the data storage capability (Fig. 5).

The progress report (fig. 6) illustrates the developments in defect control on the final platter caused by the imprint process as well as subsequent processing steps. The current status lists 120wph throughput for this type of application and additional challenges remain to transfer this product successfully to commercialization [12].

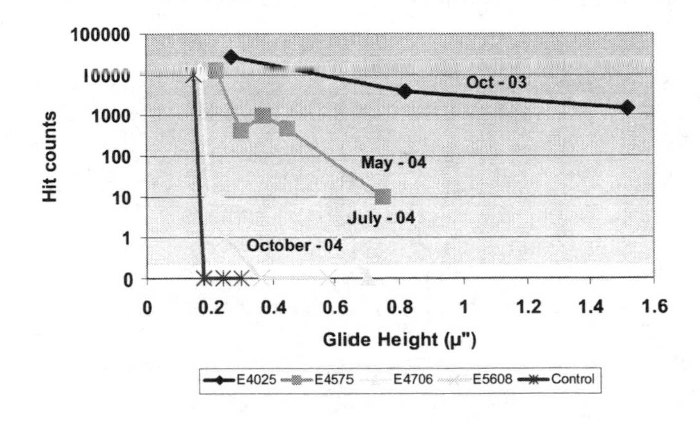

Figure 6 Defect Control

3 SUMMARY:

The NILCom® consortium has access to three centers of excellence located in Europe, Japan and North America for NIL infrastructure qualification and process development in nanoelectronics, data storage, life sciences and opto electronics. The qualified processes build upon results in large scale ultraviolet based nanoimprint lithography and hot embossing developments aiming to provide an off-the-shelf manufacturing platform for top-down nano patterning.

Main industrial opportunities for NIL relate to:

1) Enabling technology for novel devices
2) Manufacturing on low cost materials
3) Low cost manufacturing through process simplification
4) ITRS: 32nm node for nano electronics

REFERENCES

[1] www.NILCom.org

[2] J. Palensky, et al., 2nd *New England International Nanomanufacturing Workshop,* "UV-based Nanoimprint Lithography Stamp Application in Nanomanufacturing", June 28-30, 2004

[3] B. Vratzov et al, *J. Vac. Sci. Technol.*, B, Vol. 21, No. 6, Nov/Dec 2003

[4] H. Luesebrink, et al., *Journal of Nanoscience and Nanotechnology*, "Transition of MEMS Technology to Nanofabrication", Vol. X, 1-5, 2005

[5]http://www.intel.com/pressroom/archive/releases/20040830net.htm

[6] D. Wachenschwanz et al., *Solid State Technology Magazine*, „Nanoimprint lithography enables patterned tracks for high-capacity hard-disks", March, 2005

[7] W. Trybulla, *Nanoprint and Nanoimprint Technology Conference,* "Sematech, AMRC, and nano", December 1-3, 2004

[8] J. Beauvais, et al, *Nanoimprint and Nanoprint Technology Conference,* "Three Dimensional Imprint Template Produced using evaporated e-beam resist and reactive ion etching", December 1-3, 2004

[9] M. Nordan, Semi *NanoForum*, "Which Nanolithography platforms will succeed 193nm OPL" November 17, 2004

[10] M. Bender, et al., *Microelectronic Engineering,* "Multiple Imprinting in UV-based Nanoimprint Lithography: Related Material Issues", page 61-62, 407 (2002)

[11] www.appliedmst.com

[12] P. Dorsey et al., *Nanoimprint and Nanoprint Technology Conference,* "Diskrete Track Recording (DTR) Media Fabricated using Nanoimprint Lithography", December 1-3, 2004

PatGen DB – a consolidated genetic patent database implementing standard data mining resources

R. Rouse*

*HTS < Resources/PatentInformatics
P.O. Box 948586, La Jolla, CA, USA, rjdrouse@htsresources.com

ABSTRACT

Compared to the wealth of online resources covering genomic, proteomic and derived data, the bioinformatics community is rather under served when it comes to genetic patent information. This paper describes how PatGen DB has been compiled. This is a case study demonstrating how integrated theme based patent databases can be developed without resorting to expensive commercial data providers

Keywords: open patent services, XML, patent information, relational database, open source

1 THE STATE OF PATENT INFORMATION

Patent information is voluminous. According to the 2003 United States Patent and Trademark Office (USPTO) annual report, the office received 333,452 applications; this accounts for 913 applications a day (1). There is approximately 12,000 new patents issued every month by the USPTO. Given this scenario, effectively accessing patent information is important and yet a challenging task for a non patent professional.

Despite the presence of many commercial patent databases, public patent resources such as the European Patent Office Espacenet and INPADOC are excellent resources for accessing comprehensive patent information. Although these databases contain a wider collection of patent information than the commercial counterparts, effectively accessing this data has been difficult for a professional patent searcher.

This report demonstrates how freely available open source tools can be used to generate high quality patent information that can be reliably used when defining intellectual property space. As an example, a brief overview of how PatGen DB, our integrated open patent database has been compiled.

2 COMPOSITION OF PATGEN DB

Essentially PatGen DB is a consolidated database containing non-redundant data from public resources. Data from the EBI (2), DDBJ (3) and the NCBI (4) was collected from GenBank, EMBL and fasta formatted batch files vial ftp. Sequence data was also retrieved from the World Intellectual Property Organization (WIPO) (5). These sequences are derived from world patent applications in accordance with the Patent Cooperation Treaty (PCT). These files are in the required patent sequence-listing format (6).

As well as containing sequence data, PatGen DB also contains bibliographic data obtained from the European Patent Office (OPS). This is a web service where one can access current bibliographic, family and legal information in XML format (7). This can encompass multiple patent offices. This type of information enables the user to determine the legal status of both patent applications and issued patents. The data is accessed in real-time via a SOAP-based web service, delivering up-to-date information in a seamless and completely transparent manner. Full text patent information (i.e., background, specification and claims) can also be accessed.

The PatGen DB software architecture is entirely based on open source tools and deployed on a Suse 9.0 Linux server (8). Data acquisition and parsing of the flat files from the various sources is implemented using Perl (9) standard libraries. The consolidated dataset is stored and served from a MySQL relational database management system (10) using Bioperl-DB schema and parsers (11). The software interface that accesses the data is written in PHP (12). The OPS web service is being accessed using SOAP through PEAR – the PHP extension and application library (13). Full text is accessed using Perl libwww-perl library (14).

3 FEATURES OF PATGEN DB

PatGen DB is significantly more comprehensive than any public repositories. This database currently contains almost twice as many nucleic acids as well as significantly more amino acids. To remain current, PatGen DB is updated monthly. In order to keep PatGen DB non redundant we add patent documents that are not already in the database.

In PatGen both issued patents and pending patent applications can be searched via fulltext queries against the bibliographic data to retrieve disclosed genetic sequences. The simple query form provides fields for searches based on title, abstract, inventors, applicants (i.e., the inventors' assignees) and date of publication. The interface also provides direct access to patent-related sequences via the patent publication code. Each search displays the retrieved sequences as a tabulated list with links to detailed sequence information such as sequence taxonomy, genetic code and a brief description. Alternatively the entire list of sequences can be accessed in fasta format for bioinformatic analysis. Sequence taxonomy searching is available as well as a sequence search feature using BLAST.

PatGen DB is a resource where one can perform both patent and bioinformatic analysis. Patent analysis is used to determine whether to enter into licensing agreements and an essential component in profiling the technology of a given industry and thus relevant in many business activities. Bioinformatics analysis creates opportunities to develop new types of patent strategies, in particular now that annotation of newly sequenced genomes and comparing sequences across organisms have become straightforward and commonplace in biological laboratories.

In establishing a consolidated patent genetic database using open source tools, our intent is to integrate genetic sequence data with patent information. As a result we have established a modular system that can be used to compile customized theme based patent information databases upon request.

REFERENCES

[1] USPTO 2003 Performance and Accountability Report, http://www.uspto.gov/web/offices/com/annual/2003/index.html

[2] European Bioinformatics Institute, ftp://ftp.ebi.ac.uk/pub/databases/embl/patent

[3] DNA Database of Japan, ftp://ftp.ddbj.nig.ac.jp/database/ddbj

[4] National Center for Biotechnology Information, for nucleic acids - ftp://ftp.ncbi.nlm.nih.gov/genbank for amino acids - ftp://ftp.ncbi.nlm.nih.gov/blast/db/FASTA

[5] World Intellectual Property Organization, http://www.wipo.int/pct/en/sequences/listing.htm

[6] Information Concerning the Filing of International Applications Containing Large Nucleotide/Amino Acid Sequence Listings and/or Tables in the United States Receiving Office, http://www.uspto.gov/web/offices/pac/dapps/pct/aipart8.html

[7] European Patent Office Open Patent Service, http://ops.espacenet.com/

[8] Suse Linux, http://www.suse.com/us/index.html

[9] Comprehensive Perl Archive Network, http://www.cpan.org/

[10] MySQL, http://www.mysql.com/

[11] BioPerl, http://www.bioperl.org/

[12] PHP, http://www.php.net/

[13] PEAR::SOAP, http://pear.php.net/packages/SOAP

Virtual Probe Microscope

M. Heying*, J. Oliver*, S. Sundararajan**, P. Shrotyria**, Q. Zou** and A. Sannier*

*Virtual Reality Applications Center,
Human Computer Interaction Program
1620 Howe Hall, Iowa State University,
Ames, IA 50011, USA, mjheying@gmail.com
** Mechanical Engineering Department,
2025 H. M. Black Engineering Building,
Iowa State University, Ames, IA, 50011, USA

ABSTRACT

Virtual Probe Microscope (VPM) is a tool that has been developed to train users on Atomic Force Microscope (AFM) operation. The benefits from training with VPM include: reduced cost of training and increased transfer of training. The graphical user interface of VPM is laid out similar to common commercial AFM software packages. Along with standard AFM controls, users are given an additional graphical 3D window to view the probe traversing across a surface. Users are also allowed to manipulate probe geometry variable to increase understanding of AFM operation. VPM will be used in a graduate level scanning probe microscopy class in the spring of 2005 at Iowa State University to supplement traditional lab and classroom instruction.

Keywords: atomic force microscope, scanning probe microscope, simulator, simulation

1 INTRODUCTION

With the emergence of every new technology, a demand arises for training scientist and engineers the tools of the new technology. The dawning era of Nanotechnology has created a demand for scientist and engineers to learn many new tools. One of these tools is the Atomic Force Microscope (AFM). An AFM is a type of Scanning Probe Microscope (SPM) primarily used for imaging micro to nano-sized objects. The role of the AFM instrument will become even more important in the decades to come with the shift towards miniaturization.

Training large groups of users to operate an AFM has become a standard procedure with the creation of university courses and industry training sessions. Needless to say, teaching a large group of users can be a very daunting task. The equipment is expensive, the controls are many, and the training is tedious. One of the biggest challenges instructors face is training a large group of novice users on basic AFM operation. Novice users are unfamiliar with equipment and therefore require more attention in a hands-on learning environment. Any instructor will tell you that the combination of hands-on training and repetition is the best way to learn a new skill. However, receiving repetitious hands-on training on an AFM is costly, in terms of time and money.

Parallels of AFM training can be made to aircraft pilot training. Novice pilots do not simply strap into an airplane and take off the run way. Pilots must follow an extensive training program that incorporates both ground-based instruction as well as flight time with an instructor. However, flight time with an instructor is extremely expensive and an instructor can only effectively teach one trainee at a time. To solve these problems, the aviation industry developed flight simulators to provide low cost, multiple user training in a consequence-free environment. Flight simulators decrease the amount of actual flight time that pilots need and increase the amount of positive learned behavior gained from training. Teaching users how to operate an AFM is no different than teaching a pilot how to fly.

A fully interactive AFM training simulator named VPM has been developed to alleviate the problems of training large groups of users on basic AFM operation. VPM is a windows-based simulator that can simultaneously train a room full of users without the need of an actual AFM. Instructors can use this tool to demonstrate the exact same instruction that a trainee would receive in an AFM lab within the confines of a classroom or computer lab. The general mechanics and applications of VPM will be discussed.

2 SIMULATOR COMPONENTS

2.1 Probe-Sample Interaction

An AFM creates an image of a sample by dragging an extremely sharp probe in a raster scanned pattern across a surface. The probe height is sampled at a set interval during the scan creating a 2D array of height values also known as a height map. The height map is then used to create the image of scanned surface.

Theoretically, a perfect AFM image could be achieved using an infinitely sharp probe. In reality, the probe has a measurable geometry. The geometry of the probe will affect the image of the scan as shown in Figure 1. The probe-sample interaction is an important feature for an AFM simulator. Implementing this interaction provides users a useful lesson that a user's scanned images are dependent upon probe geometry.

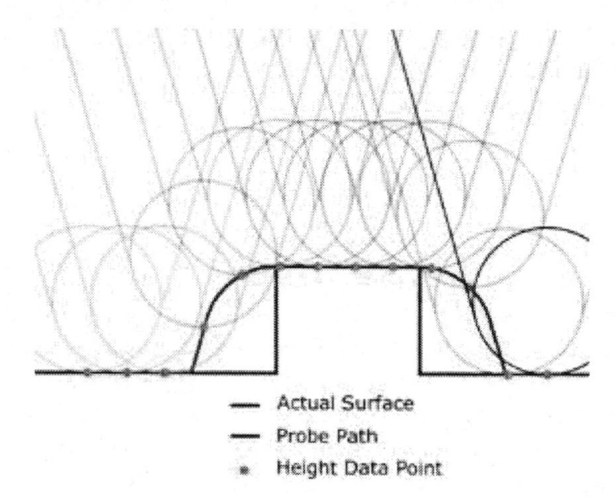

Figure 1: Effect of Probe Geometry on AFM Imaging

Previous AFM simulations have used Mathematical Morphology to simulate the probe-sample interaction [1, 2]. Mathematical morphology uses geometric set theory of shapes to calculate the interaction between two geometries. Using this technique, a surface can be mathematically determined given a probe shape and a surface shape.

VPM takes a different approach to determine the interaction between a probe and the surface, VPM uses an open source collision detection library named OPtimized COllision DEtection (OPCODE) [3]. Collision detection has been primarily used for video game development and virtual assembly simulations, but has been used increasingly more in simulator applications due to advancements in accuracy [4]. Collision detection can be just as accurate as mathematical morphology if used properly. A collision detection algorithm runs on a loop that checks whether two geometries have penetrated at a set interval. If two geometries inner-penetrate further than allowed in between a loop step, the algorithm will attempt to correct the mistake on the next loop forcing the geometries to separate. This separation can cause a jittering effect of the geometries if the interval between two consecutive collision detection loops is not small enough. A comparison study was conducted between the collision detection method and the mathematical morphology method. In the study, a sphere shaped probe scanned a cylindrical feature on a flat surface using the collision detection method [Figure 2]. The results of the study concluded that a collision detection algorithm could be used to obtain a comparable scan to the mathematical morphology method.

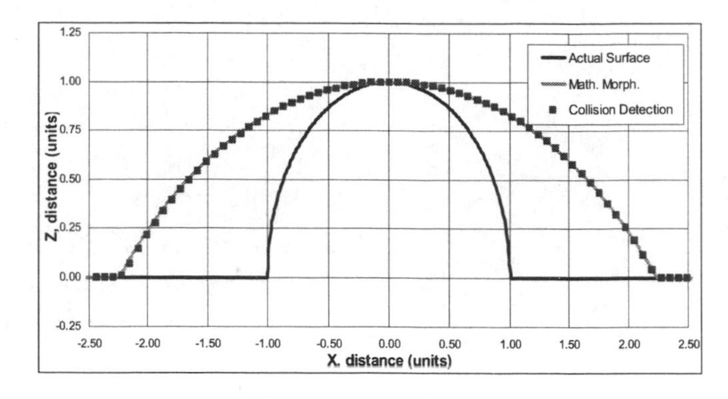

Figure 2: Probe-Sample Interaction Comparison Study

3 SYSTEM DYNAMICS

Tracking the change in height of a probe in a simulated environment is simple. In realty, tracking the height of the probe is extremely complex. An actual AFM uses a combination of a probe mounted cantilever, a piezoelectric crystal, a laser and a photo detector. To accurately replicate a real AFM, a dynamic model must be utilized. The model shown in Figure 3 is used to simulate "contact mode" or "constant force" operation. The VPM user sets a desired force to be applied to the sample, a PID controller adjusts the force applied to the piezoelectric crystal to achieve the desired applied load. When the probe comes across a change in height of the sample, the piezo must react with a change in height. The change in height of the piezo is tracked and stored as the value used to create the image height map.

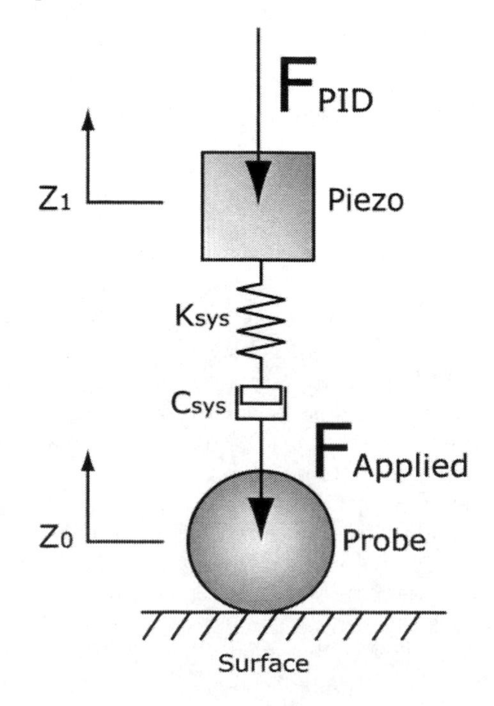

Figure 3: VPM Contact Mode Model

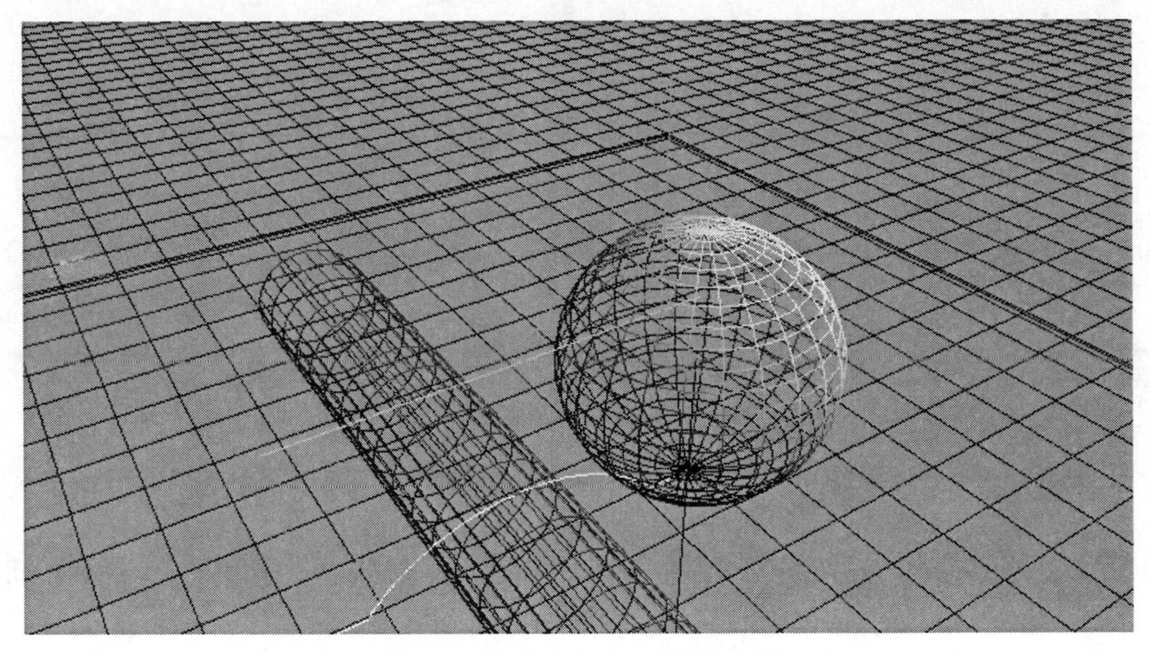

Figure 4: Real time 3D graphical view of a probe and sample

VPM utilizes an open source physics engine named Open Dynamics Engine (ODE) for the dynamic simulation [5]. ODE is a high performance library for simulating rigid body dynamics used in computer games, 3D authoring tools and simulation tools. ODE in conjunction with OPCODE can be used to create systems of joints and geometries to simulate the desired dynamic system.

4 SIMULATOR INTERFACE

The graphical user interface (GUI) for VPM has been created using Microsoft's .NET framework which allows the VPM software to be run on any Windows-capable machine. Furthermore, the Visual Studio development environment allows the simulator to be quickly reconfigured to replicate any desired commercial AFM interface. Replicating the interface that users train for increases the physical fidelity of the simulator. The physical fidelity of a simulator is the extent to which the simulator "looks like" the operational environment [6]. Users who train on a simulator interface similar in layout to the actual interface will generally perform better than if they had trained with a generic interface.

Along with standard AFM controls and windows, several other features were added to the interface to enhance user comprehension and control of the simulator. These features can be found in the following control panels:

- **Simulation Parameters.** This control panel allows users to adjust various parameters of the collision detection and simulation such as simulation time step.
- **Probe Geometry.** Changing probes on an actual AFM involves a steady hand with tweezers. In VPM, users can simply go to the probe geometry control panel and change various dimensions to get achieve a desired probe tip radius.
- **3D Window and Graphics Control.** A real time three-dimensional view has been implemented in addition to the standard 2D top view and line trace view [Figure 4]. The addition of this view is designed to visually aid comprehension of how the AFM system behaves in real time. The controls for this new view are related to view angle and graphics performance.
- **Sample Surface Control.** This control panel allows users to load a desired height map into VPM to be used as a sample surface.

4 VPM IN THE CLASSROOM

In the spring of 2005, a new experimental course entitled ME 561x: Scanning Probe Microscopy will be offered at Iowa State University. The course provides students an introduction to SPM techniques intended for novice users. The format of the course is an 80 minute lecture followed by 50 minutes of laboratory. The lab session will allow students to get a hands-on experience on an AFM. Each lab section has approximately three students sharing time on the atomic force microscope.

To test the usability of VPM, users enrolled in the class will be able to use VPM to assist them in lab activities that include the following:

- Gain optimization
- Contact Mode Operation
- Tip shape effects
- Force curves

At the end of the course, users will be surveyed to assess the student's knowledge of AFM operation. These results will then be compared to data gathered from users who are exclusively trained on the VPM alone. The performance of both groups will be compared to determine the overall effectiveness of the simulation. The results from this study can be used to determine the optimal ratio of time spent on lecture, lab and simulator for AFM training.

5 FUTURE WORK

Virtual Probe Microscope is still in its infancy. Further work will involve adding complexity to the simulator, increasing VPM's functional fidelity. Functional fidelity is similar to the concept of Physical Fidelity discussed in the previous section. Functional fidelity is the degree to which the simulator "acts like" the operational environment - does the simulator produce the same response as the actual instrument when given the same input [6]? Increasing the functional fidelity of a simulator will generally increase the amount of positive learned behavior that can be transferred from the training simulator to the operational environment. Areas of future work include the following:

- **Contact Mechanics.** To accurately simulate force curves on an AFM, adhesive forces must be employed. Users will be allowed to select which contact theory to use in the simulation: Hertzian, DMT or JKR.

- **Advanced Probe Geometry.** Users will be able to load probe geometry from file, or construct non sphere-swept cone geometry. Non-symmetrical geometry can be used to demonstrate tip characterization.

- **Deformable Surfaces**. Because VPM uses a rigid body dynamics engine, the tool is limited to scanning materials with an infinitely hard surface. A deformable surface could be implemented to simulate the scanning of soft or biological materials.

- **Lateral Force Mode**. This AFM operation mode allows users to calculate friction on surfaces independent of variation of sample topology. Many courses cover this mode and it would benefit VPM users to be able to operate in this mode.

6 CONCLUSION

Virtual Probe Microscope has been successfully developed as an AFM simulator to train a large group of users on basic AFM operation. Advantages of using the simulator for training include a reduction in cost of training, reduction in time of training, and an increase in hands-on learning. Additional feature not found on standard AFM interfaces have been implemented to increases transfer of learning from simulator to operational environment. Future usability studies on VPM will provide insight into the best possible use of the simulator.

REFERENCES

1. J.S. Villarubia, "Morphological Estimation of Tip Geometry for Scanned Probe Microscopy," Surface Science, 321 (3), 287, 1994
2. G.Varadhan, W. Robinett, D. Erie, R.M. Taylor II, "Fast Simulation of Atomic-Force-Microscope Imaging of Atomic and Polygonal Surfaces Using Graphics Hardware," Proceedings of SPIE Conference on Visualization and Data Analysis, 2002
3. OPCODE, www.codercorner.com/Opcode.htm
4. C. Wagner, M.A. Schill, and R. Männer, "Collision Detection and Tissue Modeling in a VR-Simulator for Eye Surgery," Eigth Eurographics Workshop on Virtual Environments, 27 – 36, 2002.
5. ODE, www.ode.org
6. Baum, D.R., Riedel, S., Hays, R.T., and Mirabella, A., 1982, "Training effectiveness as a function of training device fidelity" (ARI Technical Report 593). Alexandria, VA: U.S. Army Research Institute.

NANOSPRINT: An Infrastructure for Nanotechnology Foresight

Florin Ciontu

NanoSPRINT, France, florin.ciontu@nanosprint.com

ABSTRACT

NanoSPRINT exploits advances in knowledge engineering to provide a new generation of tools for nanotechnology foresight. Today, the foresight paradigm must cope with new challenges including an increased magnitude of the endeavor undertaken by the scientific community and increasingly complex interactions between the actors influencing the evolution of a scientific field. Nanotechnology accentuates these challenges through some of its specific characteristics: disruptive character and interdisciplinary character. NanoSPRINT's foresight program started as a "from within" effort of the scientific community. NanoSPRINT aims at building a knowledge base incorporating inference capabilities that would empower the researcher in his day-to day endeavor.

Keywords: technological foresight, nanotechnology, knowledge engineering

1 INTRODUCTION

The relationship between society and research is fundamentally grounded in economic reasoning – society funds research and in turn, research provides the means for social progress. Although always existent, this "deal" has become more explicit in recent years through a stronger conditioning of funding on research with a clear if not immediate feed-back to society. Besides becoming more explicit, the "deal" is now implemented through a complex interweaving of interests of practically all the players of the social system: government, industry and civil society.

In light of these changes, the research community is confronted with the challenge of "dreaming pragmatically". Initially, this issue has been tackled through a schism between (1) those at the front-line of technology, with research roadmaps largely determined by the interest of major industrial actors and (2) those preferring to retire to the "ivory tower" of fundamental science. Over the past decade or more, the system has undergone a reform destined to diffuse more homogenously the responsibility of a clear formulation of the feed-back to society.

However, even this status quo is being challenged by fields like nanotechnology whose dynamics are shaped by antagonistic forces: tackling fundamental scientific issues like new molecular structures or manufacturing paradigms under the pressure for results resulting in the spectacular increase in funding of recent years.

In reaction to these types of challenges, the past decades have witnessed the crystallization of the concept of technological foresight and its adoption at the national and corporation-level of decision-making.

The context briefly sketched here outlined the main drivers for the systematic incorporation of foresight at the most basic level of research environment. However, there is a real challenge in adopting a holistic view of scientific, technological and social factors without hampering the creativity of the research process. It is essential that a foresight-centered paradigm transposed at the individual level does not act as a filter but rather as an extension of the scientist's grasp of a cognitive universe of increasing complexity.

This remainder of the paper focuses mainly on the motivational aspects of the NanoSPRINT project and its approach in defining such a paradigm. In Section 2 we discuss few general characteristics of the foresight process from the perspective of emerging challenges. Section 3 focuses on challenges specific to nanotechnology. In Section 4 we sketch a few approaches used to cope with the aforementioned challenges. Section 5 focuses more specifically on the approach adopted in NanoSPRINT's foresight program. Section 6 is dedicated to the final conclusions and the envisioned development of the project.

2 FORESIGHT IN THE KNOWLEDGE ERA

The initial crystallization of the concept of foresight took the form of forecast – a rather simplistic vision of a "unique", predetermined future and a similarly simple goal – predicting that future. The first forecast programs originated in the '50s in the US defense sector. Later on, the concept integrated the notion of forecast into a more complex vision – several scenarios are possible and the evolution according to these scenarios can be influenced by today's decisions.

The champion of integrating forecast/foresight in the technological policy decision making was Japan starting with the '70s. Arguably, this approach contributed to the transition of Japan's position from a producer of cheap, low-quality goods to a technological leader. Over the next two decades, technological foresight has become a crucial component of technological policy decision-making at the national and corporate level. Most "big countries" (US,

France, UK, Germany) followed the Japanese model based on extensive Delphi forecasts. Smaller countries like the Netherlands implemented programs relying rather on expert panels. As for the today's vision on foresight, Ben Martin [1] identifies and specifically formulates the continuous, systematic, process-like character (as opposed to a technique-like view) of foresight.

At the corporate level, ~65% of large American companies and more than 82% of very large companies with more than $10bn in sales had forecast programs [2]. Still, at the most basic level, the research group, the process is very informal and relies mostly on the individual's "entrepreneurial" spirit.

The common denominator of today's foresight programs is the overwhelming reliance on communication and networking [2]. The advantage of this approach is that it makes easier to extract undocumented tacit knowledge. Its cons, poor reuse of the effort, poor scalability with respect to the increasing complexity and poor standardization of the process at the individual level, are an obvious consequence of the "handcraft" character.

Moreover, significant challenges emerge in today's environment, ranging from the magnitude of the technological endeavors undertaken by the scientific community to the increasingly complex interactions between the players active around a technological field.

The accelerated rate of disruptive technologies is the biggest contributor to the first complexity source. This challenges the grounds of one of the previously well-established hypotheses of foresight: products within a given industry depend on a finite number of technologies and components and the foresight problem can be shifted at the level of each of these technologies. The second factor is the obvious increase in R&D cycles in both time and money.

The players' level is currently undergoing a major shift towards the triple-helix model of interaction between government-research-industry introduced in [3]. The model could be furthermore extended to include other components like the civil society (e.g. the role of non-governmental agencies in the generically modified organisms).

2. CHALLENGES FOR NANOTECHNOLOGY FORESIGHT

All the aforementioned challenges find a direct instantiation in the case of nanotechnology and are reinforced by the considerable magnitude of the two forces shaping the dynamics of the field. First, nanotechnology involves a considerable amount of tackling fundamental scientific issues like new molecular structures or manufacturing paradigms (e.g. self-assembly). Despite an undisputable disruptive potential these issues have natural fuzziness in terms of evolution and outcomes. Secondly, as of 2004, it has become a $3bn public funding issue – more than enough to require a level of pragmatism from which research dealing with fundamental issues was traditionally spared. These two forces have antagonistic components that are intrinsically difficult to reconcile.

Moreover, the fact that the research spectrum of nanotechnology spans from fundamental research to applications leads to a layered evolution of different trends. This evolution can be structured on three successive waves: instrumentation, nanostructures and applications. Instrumentation companies selling tools to pursue nanotech R&D (e.g. selling Atomic Force Microscopes) are entering currently a low growth phase as most labs have been equipped. The big question is whether this industry will evolve into an industry similar to the semiconductor equipment one but this in turn depends on the emergence of large-volume applications. Companies focusing on the fabrication of nanostructures (raw materials, the basic bricks) are booming with one common objective: delaying mass production solutions. With costs scaling down two orders/year in some cases (carbon nanotubes), the common target is to make costs competitive within 5 years. Finally, the momentum for applications is slowly developing with an increasing number of start-ups appearing each year. At the same time, more and more big companies commit R&D budgets to nanotech. However, the applications segment will enter the high-growth slope of the S-curve in 5-10 years. The central message of this picture is the correlation between the evolutions of different areas of nanotechnology which renders imprecise predictions based on data from focused segments.

An anatomy of a nanotech subfield, carbon nanotubes, is particularly relevant for this discussion. Discovered in 1991 "by accident", carbon nanotubes [7] are molecular level nanostructures with remarkable properties. They are 100 times stiffer than steel for a sixth of the weight, they are very good electrical and thermal conductors and can be used as field emitters. However, interest in the field literally exploded only at the end of the '90 when synthesis methods lowered the production costs of carbon nanotubes enough to allow experiments in laboratories without any experience in their synthesis. At this point the maturity (due to development in the '80s) and large scale diffusion of instrumentation with atomic resolution facilitated the explorations of the remarkable properties mentioned above. With around 44 commercial providers currently focusing on scalable production methods and costs going down at an exponential rate, carbon nanotubes are one of the components with the strongest potential for real-life applications. Although extensive research on applications of nanotubes is on going, it will probably take longer than 5 years to see the cost of nanotubes scaled down to economically competitive levels and incorporated in commercial applications. This example illustrates the fact that in addition to the inter-layer dependency mentioned above, each vertical segment has its own S-curve.

Among the extra-technological factors, the presence of government is particularly strong in nanotechnology and not only though the traditional vehicle of funding research. Governmental support targets are not only universities but also venture capital and small firms. Additionally governmental organizations are particularly active in intermediating interactions between the universities, industry and the general public.

Finally, non-governmental organizations are expected to play a particularly important role. Although nanotechnology is in an incipient phase there are already ethical and environmental concerns. And even if, for instance, the potential toxicity of carbon nanotubes is considered very seriously through studies of specific problems [8],[9] various factors like the popular press [10],[11] can bring in a serious amount of unpredictability .

3 COPING WITH CHALLENGES

A generic solution to tackling complexity consists of shifting the decision process at the level of patterns. However, a practical foresight approach based on this principle requires being able to derive these patterns automatically. Moreover, the content from which the patterns would be extracted should cover technological as well as socio-economic data and be obtainable through concurrent gathering of disparate information units. Finally, irrespective of the degree of automation, the workflow should include verification and checking mechanisms.

Whether these desiderata are fully attainable or not could be a matter of interminable debates. Instead, we prefer to focus on a few instances where at least partial success has been demonstrated.

A very illustrative example for pattern extraction from technological data is TRIZ [11] (a Russian acronym for a Theory of Inventive Problem Solving). This methodology evolved from extensive studies of technical and patent information and consists of a set of principles that are applied in different forms in the innovation process. Although the penetration of TRIZ has been relatively slow, more and more success stories have been documented in recent years including Samsung hiring six TRIZ specialists from the former URSS for 18 months and accounting for $91 million in direct benefits. As of today, the TRIZ users-list includes companies like General Motors, Ford, IBM, Samsung, Motorola, Boeing etc. An example of foresight approach based on TRIZ is provided in [12]. Besides methodologies like TRIZ, some tools try to tackle (e.g. [13]) the complexity issue by extracting statistical correlations from technical data like patents. However, their impact is relatively limited and they have little predictive value.

As mentioned in the our previous brief analysis of the challenges of the foresight process, no pertinent foresight program can be implemented without taking into account the interactions between technological and non-technological factors (sociological, political etc.). From the perspective of pattern extraction, Eleonor Glor's work [14] is a good example (though many others could be listed here), where sociological factors ranging from individual motivation to organizational culture and the challenge environment are systematically taken into account in order to derive conclusions on creativity, implementation and the outcome of innovation.

A common characteristic of the cases mentioned here is their lack of formal character. By consequence, their results range from qualitative observations to at best an empirical approach. TRIZ for instance is far from a formal science and integrating it is a very subjective experience which translates into some of the trainees loving it while others find it sterile. A step forward is made by products [14] which combine syntactic and text search with elements of TRIZ.

Perhaps not surprisingly, the most audacious approach in coping with the challenges of an increased complexity through a **systematic approach** comes from the US defense sector. DARPA's High Performance Knowledge Bases (HPKB) [4] project was the first program destined to build on recent advances in Knowledge Representation and Reasoning [5] in order to create a new generation of tools for foresight. Test problems included crisis management and battle space related problems. The final conclusion of this $36 million effort was that KRR are mature enough to be considered for applications but the cost of encoding knowledge was prohibitive. The Rapid Knowledge Formation (RFK) [6] program focused on developing methodologies and tools designed to scale down the cost of encoding knowledge in databases. More precisely, the objective of this program was to empower subject matter knowledge experts without prior training in AI to encode information. The results of the project confirm the ability of subject matter experts with minimal AI training (2-3 weeks) to encode knowledge with results comparable to the ones obtained by Knowledge Engineers.

The use of knowledge engineering techniques in the two DARPA projects mentioned above, with the specific goal of coping with complexity in the foresight process and the results are particularly encouraging for the approach we envision for the NanoSPRINT project.

4 NANOSPRINT

NanoSPRINT's foresight program builds on the experience accumulated by TIMA Labs (http://tima.imag.fr) in dealing with successive technological waves in a strong interaction with industry, other laboratories or governmental agencies. Our approach tackles the challenges discussed at the beginning of this paper through a "from within" effort of the research community to systematically integrate foresight into the researcher's work approach. Simultaneously, the project aims at providing the tools to bridge the knowledge gap between players involved in the development of nanotechnology and players from outside the nanotech world.

The latter category covers (1) decision makers from traditional industries (2) research & development engineers (3) students and (4) the general public. In order to bridge this gap, NanoSPRINT provides the tools to store knowledge on nanotechnology and to extract custom information on an on-demand basis.

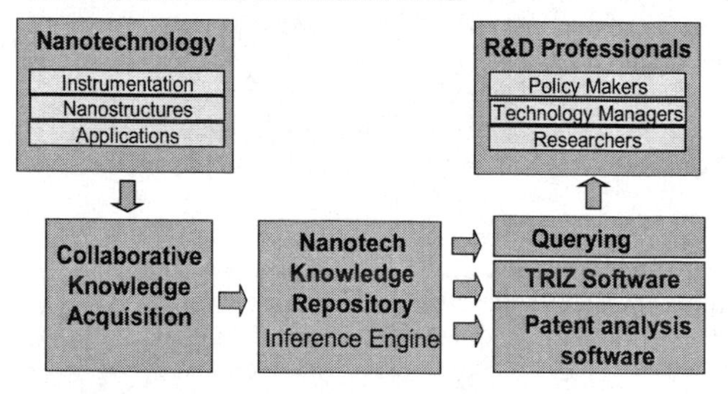

Figure 1: NanoSPRINT bridges the knowledge gap between nanotech players and players from traditional industries. Using a knowledge engineering centered approach.

NanoSPRINT's approach is centered around a Nanotechnology Knowledge Repository that (1) stores facts and meta-knowledge on nanotechnology and (2) (semi)-automatically derives answers to new questions based on stored facts.

The server is integrated in the following flow:

1. Knowledge is acquired from multiple sources by scientific analysts and encoded into a knowledge database. For example, the formalisms used to encode knowledge could use notions like: (1) concepts, (2) instances of the concepts, (3) rules and (4) axioms (which are equivalent to assertions or real-world facts). At this point, knowledge is expressed in a formal manner.

2. A reasoning server uses this "formalized" knowledge to infer answers to new questions that require knowledge which is not available immediately in the encoded texts. Using these mechanisms, knowledge is generated based on tractable implementations of subsets of the first-order logic.

3. Users access this knowledge by (1) browsing a content structure or, more importantly, (2) through specific questions. As the current state of Natural Language Processing does not allow free querying, current efforts are directed towards the development of graphical query interfaces.

The workflow presented here is oriented towards automation at each of its stages: from data gathering to decision-making. However, inherent problems arising from fundamental limitations (e.g. determining if a query results from a knowledge base is not decidable) will require extensive testing and human supervision before application to real-life issues.

5 CONCLUSIONS

This quick review of foresight from the perspective of emerging challenges was designed as a motivational paper for the need for a new approach making use of recent advances in knowledge representation and reasoning. The flow sketched here delineates some clear stages: knowledge acquisition, reasoning paradigm as well as issues related to the interface with the end-user. At each stage there are still many issues that remain to be solved in order to make economically feasible the application of this paradigm. Ongoing work at TIMA Laboratory, tackles this issue through a Pilot Project focusing on carbon nanotubes. Aside from presenting our motivations, this paper constitutes an open invitation to the carbon-nanotubes community to participate in this endeavor.

6. REFERENCES

[1] B.R. Martin, Regional Conference on Technology Foresight for Central and Eastern Europe and the Newly Independent States, 4-5 April 2001, Vienna

[2] G. Reger, Technology Analysis & Strategic Management, Vol. 13, No. 4, pp.533–553, 2001

[3] L. Leydesdorff, uPUBLISH.com: Universal Publishers, 2003

[4] http://reliant.teknowledge.com/HPKB/

[5] Ronald Brachman et *al*, "Knowledge Representation and Reasoning", Morgan Kaufmann, 2004

[6] http://reliant.teknowledge.com/RKF/

[7] M S Dresselhaus et *al*, , "Science of Fullerenes and Carbon Nanotubes" (Academic Press, New York)

[8] D. B. Warheit et *al*, Toxicological Sciences **77**, 117-125 (2004)

[9] Lam CW, Toxicol Sci. 2004 Jan;77(1):126-34. Epub 2003 Sep 26.

[10] ETC Group, "The big down", www.etcgroup.org/documents/TheBigDown.pdf

[11]Genrikh S. Altshuller, Creativity As an Exact Science, CRC Press January 1, 1984

[12] M.G. Moehrle , Int. J. Technology Intelligence and Planning, Vol. 1, No. 1, pp.87–99, 2004

[13] http://www.thevantagepoint.com/

[14] E. Glor, The Innovation Journal: The Public Sector Innovation Journal. Vol. 6(3), 2001

[15] https://gfi.goldfire.com/

Nanotechnology and the Environment: Will Emerging Environmental Regulations Stifle the Promise?

Mark C. Kalpin* and Melissa Hoffer**

Wilmer Cutler Pickering Hale and Dorr LLP
60 State Street, Boston, Massachusetts 02109, USA
*mark.kalpin@wilmerhale.com
**melissa.hoffer@wilmerhale.com

INTRODUCTION

The United States and the European Union, in a manner that is consistent with the approach taken by each in the past with respect to emerging high technologies, are again taking divergent approaches to developing environmental regulatory programs for nanomaterials and nanoproducts. Because the market for nanomaterials and nanoproducts is global, it is not clear whether the potential inconsistencies that exist in these approaches could stifle the development and distribution of those materials and products. At the same time, however, independent voluntary standards-setting entities, such as the American National Standards Institute (ANSI) and the International Standards Organization (ISO), propose to develop standards for measuring and evaluating toxicity effects, environmental impact, risk assessment, metrology, and methods of analysis for nanomaterials. Because these "voluntary" standards will form the basis of any regulatory programs that are developed for nanomaterials, they could significantly influence the environmental regulatory programs that ultimately are developed in the United States and the European Union.

Keywords: environment, health, safety, TSCA, REACH, ANSI

1. ENVIRONMENTAL REGULATION OF NANOMATERIALS IN THE UNITED STATES

In the United States, the manufacture, use, transport and disposal of engineered nanomaterials currently is unregulated. Despite the novelty and unique features of these materials, the development of new environmental regulatory programs specifically tailored to nanomaterials is unlikely. Instead, existing regulatory programs such as the Toxic Substances Control Act (TSCA) and the Occupational Health and Safety Act (OSHA) likely will be applied, or adapted to apply, to the regulation of nanomaterials. However, environmental and health-based regulation of nanomaterials is unlikely to progress significantly until a standardized nomenclature for nanomaterials is developed.

1.1 The ANSI Nanotechnology Standards Panel (NSP)

At the end of last year, the ANSI NSP released its "priority recommendations" with respect to the need for nanotechnology standardization. Among the "broad standardization topics" identified by the ANSI NSP, the need for standardization was classified as "most urgent" in the following four areas: (i) general terminology for nanoscience and technology; (ii) systematic terminology for materials composition and features; (iii) toxicity effects/environmental impact/risk assessment; and (iv) metrology/methods of analysis/standard test methods. The ANSI NSP has recommended that standards be achieved in these areas within the next year.

Standards developed by voluntary standards-developing entities such as ANSI and the American Society for Testing and Materials International (ASTM) are quasi-regulatory in nature, and readily are incorporated into regulatory programs. This pattern is not uncommon in the United States, especially in the context of environmental regulatory programs. For example, after the Comprehensive Environmental Response, Compensation, and Liability Act (CERCLA) was passed in 1980, ASTM developed standards for performing environmental site assessments to assess the potential presence and level of contaminants at commercial properties. Until recently, those standards provided the basis at both the federal and state level for determining whether a prospective purchaser of property had performed all appropriate inquiry in order to qualify as an innocent purchaser. ASTM, which is accredited by ANSI, is a member of the ANSI NSP, and announced in January of this year its likely involvement in developing standardized nomenclature for nanomaterials. It should be assumed that any ANSI NSP and ASTM standards that are developed for the purpose of measuring the toxicity effects of nanomaterials, as well as assessing the potential environmental impacts and risks associated with the manufacture, use, and disposal of nanomaterials, will be used in a similar manner.

ANSI also is the only U.S. representative to the ISO and the International Electrotechnical Commission (IEC), key international standards-developing bodies. ANSI promotes the use of U.S. standards internationally, and advocates for the adoption of those standards at the international level via ISO and IEC. Similarly, the British Standards Organization already has proposed that ISO develop nanotechnology standards that would provide suitable instruments for the evaluation of risk and the protection of health and the environment. Once an ISO standard is developed (as evidenced by the prior issuance of the ISO 14000 International Environmental Management Standards), it frequently takes on a quasi-regulatory effect and forms the basis for determining the adequacy of environmental-related activities on a world-wide basis.

The standards being developed by ANSI NSP, and ultimately by ISO, will have a tremendous impact on the future direction of nanotechnology development both in the US and internationally, especially in the area of environmental regulation. For this reason, entities interested in the development of nanomaterials should actively monitor and participate in the standards-development process for nanomaterials that is currently underway.

1.2 TSCA: New or SNU?

One of the most intriguing legal questions concerning the environmental regulation of nanomaterials in the United States is whether the Environmental Protection Agency (EPA) will treat nanomaterials as "new" chemical substances under existing environmental laws that regulate the manufacture and use of chemical substances, such as TSCA. The two key aims of TSCA are to ensure that adequate risk assessment data are developed with respect to a given substance, and appropriate actions are taken, using regulations based on those data, to mitigate human and environmental exposure to any unreasonable risk. Prior to manufacturing or using a "new chemical substance," TSCA requires chemical manufacturers and importers to submit a pre-manufacture notification (PMN) and risk assessment information to the EPA.

Subject to certain exemptions, manufacturers of nanomaterials will be required to comply with TSCA's PMN requirement in the event that EPA determines that nanomaterials constitute "new chemical substances." Alternatively, EPA may determine that certain nanomaterials are Significant New Uses (SNU) of existing chemical substances. For example, the same Material Safety Data Sheet (MSDS) for graphite potentially could be used for carbon nanotubes (CNTs), fullerenes, and carbon black. In fact, a review of MSDSs for various CNTs currently available for sale indicates that manufacturers have used a range of Chemical Abstract Registry (CAS) numbers, including those of graphite and carbon. However, because common elements like carbon behave very differently at the nano-scale, EPA could find that the manufacture of CNTs, for example, constitutes a SNU; and last year EPA staff indicated that the Agency is likely to do so. Such a determination would require manufacturers to file, subject to certain exemptions, a PMN with the EPA.

Recently, a CNT manufacturer applied to EPA for a low-volume exemption (LVE) to manufacture single walled CNTs under TSCA. If EPA determines that the CNTs are covered by an existing TSCA Inventory listing (*e.g.*, for carbon), the LVE application would be denied as unnecessary. If, however, EPA grants the application, its decision may signal the Agency's determination that such CNTs do not fall within the definition of Inventory-listed substances, or constitute a significant new use of listed substance(s). EPA's decision on that filing is expected soon.

Under TSCA, the manufacturer or importer of a substance that has been determined to be a SNU must file a PMN at least 90 days before it plans to begin manufacturing or importation. Along with the PMN, the entity must provide data that it believes will show that the intended significant new use of the chemical substance will not present an unreasonable risk of injury to health or the environment. There are very far-reaching requirements concerning the test data that must be provided. In essence, all data related to health and environmental effects in the possession, custody or control—defined broadly to include several categories of related corporate entities if they are associated with the research, development, and marketing of the substance—of the company must be submitted to EPA. Additionally, any other data known to the Company concerning health and environmental effects must be described in the PMN, as far as "reasonably ascertainable." Further, EPA may require testing where it determines unreasonable risk may exist and existing data on health and environmental effects are insufficient. These data submission requirements highlight the need for businesses to think carefully about the designs for any studies, engage qualified experts to assist, consult with counsel early on, and maintain records of data developed to ensure full TSCA compliance.

2. ENVIRONMENTAL REGULATION OF NANOMATERIALS IN THE EUROPEAN UNION

Similar to the situation in the United States, the manufacture, use, transport and disposal of engineered nanomaterials currently is unregulated in the European Union. Unlike the United States, however, the novelty and unique features of these materials make it likely that the European Union will develop new environmental regulatory directives that are specifically tailored to nanomaterials. In the meantime, member states with existing environmental regulatory programs that apply to chemical substances, such as the United Kingdom, likely will apply those programs (either directly or after adaptation) to regulate nanomaterials as new chemical substances.

2.1 The Precautionary Principle

The European Union (EU) was created in its current form in 1993, and currently has 25 member states (with the applications of four additional states pending). Within the EU, the European Commission (EC) is responsible for proposing legislation (in the form of "directives"), as well as the administration and enforcement of enacted legislation.

The so-called "precautionary principle" forms the basis for all environmental directives that are under consideration or have been issued by the EC. The approach envisioned by the precautionary principle stands in direct contrast to that under TSCA, in that under the precautionary principle a manufacturer may be restrained from manufacturing and distributing a material unless it can conclusively demonstrate that the material is "safe."

According to the EC, the precautionary principle acts to ensure a high level of protection of the environment and of human, animal and plant health whenever the available scientific data do not permit a complete evaluation of the potential risk. The precautionary principle may be invoked when the potentially dangerous effects of a new product or process – such as those related to the development of nanomaterials and nanoproducts – have been identified through a scientific evaluation that does not allow the level of risk to be determined with a sufficient degree of certainty. In such a case, the burden of proving that a material is "safe" is shifted to the manufacturer, and distribution of the material can be halted unless the manufacturer can "prove the negative" and meet this burden of proof.

2.2 REACH

In 2003, the EC proposed a comprehensive system, known as REACH (for Registration, Evaluation, and Authorisation of Chemicals), that would impose greater responsibility on industry to identify and manage the risks associated with chemical substances and ensure that safety information concerning these substances is provided to regulatory agencies and the public. At its core, REACH would reverse the burden of proof from public agencies to industry for ensuring the safety of any chemicals that are introduced into the EU market. As such, REACH would replace over 40 existing Directives and Regulations, and would fully implement the precautionary principle with respect to both existing and new chemical substances. Direct costs to industry that are associated with the implementation of REACH have been estimated to total 2.3 billion euros over an 11 year initial period. Over a 50 year period, it has been estimated that the costs of REACH to both industry and end-users could total 50 billion euros over a 30-year period.

Numerous concerns have been raised – by EU member states, industry, and even the United States government – concerning the high regulatory costs and burdens associated with implementing the REACH proposal. In response to those concerns, the EC has signaled that it is prepared to make major changes in the proposed legislation. Key changes under consideration include prioritizing the testing process for more than 50,000 chemicals, based on volume and toxicity considerations, as well as implementing a "one substance, one registration" system for all chemicals. At this time, it is not known whether formal amendments to the REACH proposal to implement these revisions will be made after the first reading of the legislation in the second half of 2005, or if the first reading will be postponed so that the revisions can first be made.

2.3 The EU's Proposed Environmental Approach to Nanotechnology

Environmental issues concerning the EU fall within the jurisdiction of the European Environmental Agency (EEA). The EEA recently developed its Sixth Action Programme for the Environment, which establishes the environmental priorities for the EU for the period from 2001 through 2010. These priorities include the management of natural resources and waste, and the development of an integrated product policy. In the context of nanomaterials and nanoproducts, the management of waste materials generated and the development of "end-of-life" vehicles is of primary importance.

In December 2004, the EC released a formal communication entitled "Towards a European Strategy for Nanotechnology." In that communication, the EC stressed that nanotechnology must be developed in a safe and responsible manner. As such, the EC urged that any potential public health, safety, environmental and consumer risks be addressed up front by generating the data needed for risk assessment, integrating risk assessment into every step of the lifecycle of nanotechnology-based products, and adapting existing methodologies (and, as necessary, developing new ones) for the regulation of nanomaterials and nanoproducts. Until that strategy is implemented, the EC has advised member states to make maximum use of their existing regulatory programs to address public health, worker and consumer safety, and environmental protection issues that may arise as a result of the manufacture and use of nanomaterials and nanoproducts.

2.4 Independent Observations

Concerns regarding potential environmental and health-related impacts associated with the widespread manufacturing and use of nanomaterials is not limited to the EC or its member states.

In July 2004, The Royal Society and The Royal Academy of Engineering released a comprehensive report on the opportunities and uncertainties associated with nanoscience and nanotechnologies (the "Royal Report"). The Royal Report noted that there is virtually no information available about: the effect of nanoparticles on species other than humans; how nanoparticles would behave in the air, water, or soil; or their ability to bioaccumulate in the food chain. As a result, the Royal Report recommended that, as a precautionary measure, entities producing nanoparticles and nanotubes should treat those materials as if they were hazardous and remove them from their waste streams. In addition, the Report recommended that the use of free nanoparticles in environmental applications (such as the remediation of groundwater) be prohibited. Finally, the Royal

Report concluded that chemicals produced in the form of nanoparticles and nanotubes should be classified as new chemical substances under both the existing and proposed regulatory frameworks present in the UK and the EU.

In 2004, Swiss Re, the world's second largest insurer, released a report that highlighted numerous questions related to the opportunities and potential hazards associated with nanotechnology. While Swiss Re was careful not to sound an alarmist cry, it nevertheless questioned whether nanotechnology would, despite its potential to be used in innovative and beneficial applications, suffer the same fate as did asbestos. As a result, Swiss Re concluded that, in the absence of an existing regulatory scheme that applied to nanotechnology, a new nano-specific framework of regulation is needed, along with an internationally valid system of standardization of nanomaterials.

3. CONCLUSION

A large degree of uncertainty remains regarding the shape of the environmental regulatory programs that will be developed and implemented in the United States and the European Union with respect to nanotechnology. Whether those programs will restrict, either by design or through inconsistencies, the manufacture and use of nanomaterials on a global basis is an open question. Current events indicate, however, that the development or refinement of those programs will be based, in substantial part, on the nomenclature and risk assessment standards that independently are being developed by organizations such as ANSI, ASTM, and ISO. At a minimum, those standards could lead to nanomaterials being regulated as either new chemical substances or significant new uses of existing chemical substances.

REFERENCES

[1] ANSI-NSP Priority Recommendations Related to Nanotechnology Standardization Needs (Nov. 2004).

[2] Toxic Substances Control Act, 15 U.S.C. §§ 2601-2692; 40 CFR Parts 700-789.

[3] www.ansi.org

[4] The European Commission: A Guide for Americans. European Commission (2004).

[5] Towards A European Strategy for Nanotechnology. Communication from the Commission of European Communities (Dec. 2004).

[6] Nanosciences and Nanotechnology: Opportunities and Uncertainties. The Royal Society and the Royal Academy of Engineering (July 2004).

[7] Nanotechnology: Small Matter, Many Unknowns. Swiss Reinsurance Company (2004).

[8] Industrial Application of Nanomaterials Chances and Risks. German Association of Engineers (2004).

[9] Analyzing the European Approach to Nanotechnology. Evan Michelson, Woodrow Wilson International Center for Scholars (Nov. 2004).

Nanobiotechnology: Responsible Action on Issues in Society and Ethics

D. J. Bennett* and D. Schuurbiers**

*European Federation of Biotechnology Task Group on Public Perceptions of Biotechnology, Oude Delft 60, 2611 CD Delft, The Netherlands, david.bennett@efbpublic.org

** Kluyver Centre for Genomics of Industrial Fermentation, Delft University of Technology, Julianalaan 67, 2628 BC Delft, The Netherlands, daan.schuurbiers@efbpublic.org

ABSTRACT

If claims about the major long-term impact of nanotechnologies prove true, then the implications for societies and economies around the globe will be profound. The contrast between expected impact on the one hand, and very low awareness of nanotechnology by the public on the other, has led many to consider nanotechnology, and especially its nanobio- applications, to be the next major public, NGO, media and political issue after nuclear energy and GM food and agriculture.

As yet few researchers have begun to explore seriously the ethical and societal issues in nanobiotechnology. The overall aim of this initiative therefore is, firstly, to anticipate the societal and ethical issues likely to arise as nanobiotechnologies develop and, secondly, to use the lessons from the European GM debate [1] to respond pro-actively and responsibly to the probable public, media and political concerns.

Keywords: nanobiotechnology, ethics, society, public, media

1 CURRENT STATE OF AFFAIRS

There are many repeated claims about the major long-term impacts of nanotechnologies upon global society: for example, that it will provide radical advances in medical diagnosis and treatment, cheap sustainable energy, environmental remediation, more powerful IT capabilities and improved consumer products. If even only some of these predictions prove true then the implications for global society and the economies of many nations are profound. The general experience with successful new technologies is that the time needed for their development and introduction is optimistically underestimated but that the magnitude of their long-term impact is grossly underestimated too.

Many industries welcome the opportunities which are seen in nanotechnology and this is reflected in the rapid increase in research funding by industry and governments. The European Commission has allocated €1,300 million to nanotechnology as one of the thematic priorities in its 2003-6 Sixth Framework Programme. According to estimates for 2003, public research funding alone totalled $3 billion worldwide and is still increasing. Sales revenues from products manufactured using nanotechnology have already reached 11-digit figures and are predicted to generate 12-digit $ sums by 2010.

Following a relatively short phase of research and development a number of new products have already been launched onto the market including cosmetics, sunscreen lotions and water-repellent and self-cleaning coatings. It is possible, but yet unproven, that materials involved, while not normally toxic to humans or the environment, may be so as nano-sized particles. Hence a quite different approach to detection of possible hazard, risk assessment and regulatory control is required. Concerns arise because of the potential nanotoxicity or pollution associated with certain nanomaterials and the likely widespread presence of nanoproducts in the near future across industry sectors, companies and countries throughout the world.

As the UK Royal Society and Royal Academy of Engineering report published in July 2004 states: *"Nanoscience and nanotechnologies are widely seen as having huge potential to bring benefits to many areas of research and application, and are attracting rapidly increasing investments from Governments and from businesses in many parts of the world. At the same time, it is recognised that their application may raise new challenges in the safety, regulatory or ethical domains that will require societal debate."* [2]

Current awareness of nanotechnology by the public is very low. In the 2002 Eurobarometer survey: *"Europeans and Biotechnology in 2002"* [3], over 50% of the sample answered 'don't know' when asked whether they thought that nanotechnologies would improve or make worse their way of life over the next 20 years. The very low awareness of nanotechnology by the public in Europe and the very high percentage of "don't knows" provides the opportunity for improving public understanding and ensuring effective public communication from the outset. This is clearly necessary to help achieve public acceptance of nanotechnology and not to leave a perception 'vacuum' to be occupied by activist NGOs as happened with GM food and agriculture.

Some already consider nanotechology, and especially its nanobio- applications, to be the next major public, NGO, media and political issue after nuclear energy and GM food and agriculture. The so-called "grey-goo problem" became a media topic and as "*global ecophagy by biovorous self-replicating nanorobots*" was made the subject of study [4]. Michael Crighton published a best-selling science fiction novel "*Prey*" on the theme.

The majority of the scientists in the academic and industrial nanotechnology communities tend to approach their science with what has been characterised in the epistemology of science as "logical positivism or empiricism, Popperian falsifiability and Kuhnian paradigmatic" views. This is not true of all and for many there has been a discernible, and continuing, shift towards more "relativistic" and "social constructionist" appreciations. Many nanobiotechnologists say that they have considered, or are willing to consider, the ethical and social impacts of their work. Currently, though, many still to tend to dismiss the concerns which people have (and probably will have increasingly) about the nanotechnologies in terms of being "just another new technology" and the solution being in the so-called "deficit model" of science communication by "providing the public with information and they will accept it".

A study by the Joint Centre for Bioethics at the University of Toronto, Canada, "*Mind the Gap: Science and Ethics in Nanotechnology*" [5], published in the journal *Nanotechnology* reportedly said: *"There is a danger of derailing nanotechnology if serious study of its ethical, environmental, economic, legal and social implications does not reach the speed of progress in the science"*. It further said they fear *"a showdown of the type we saw with genetically-modified crops"*. Dr Peter Singer, one of the study's authors, is quoted as saying: *"I don't want the science to slow down. I want the ethics to catch up."*

As yet few ethicists have begun to explore seriously the ethical issues in the nanosciences, nanotechnology and nanobiotechnology. Doubtless, as happened for example in the GM agriculture and food, and embryo stem cell research fields, progressively more ethicists will take up such studies as they become aware of the issues involved in these areas and of their importance.

2 NANOBIO-RAISE

It is for these reasons that the Nanobio-RAISE project is being established by the European Commission Research Driectorate-General under its "Science and Society" programme. The overall aim of this initiative is both, firstly, to anticipate the societal and ethical issues likely to arise as nanobiotechnologies develop and, secondly, to use the lessons from the European GM debate to respond pro-actively and responsibly to the probable public, media and political concerns. The project aims to bring ethicists and nanobiotechnologists together with the aim of helping the former to help the latter, and vice versa, by enabling scientists to better understand and respond to the concerns of the wider society which the nanobiotechnologies will undoubtedly benefit.

The main objectives therefore are to:

- bring together the key relevant players in the field including committed ethicists, European Commission Nano2Life Network of Excellence, European Federation of Biotechnology's Task Group on Public Perceptions of Biotechnology, EuropaBio, DECHEMA (Fachsektion Nanotechnologie), Royal Institute for Technology in Stockholm (Philosophy Unit), Church of Scotland Society, Religion and Technology Project, SMEs and major companies using nanobiotechnology,
- horizon-scan for the scientific and commercial developments which are likely to cause public and political concern,
- clarify the ethical issues and public concerns involved or as they arise, and recommend and carry out strategies for public communication to address the emerging questions,
- take on board the experiences and lessons learned from the European GM debate of the last decade and apply them with this project to the nanobiotechnology discussions,
- incorporate the recommendations of the European Commission's Communication "*Towards a European Strategy for Nanotechnology*" [6] and the results of its Nanoforum public consultation which surveys European public opinion on these issues [7],
- prepare for the relevant actions in the European Commission's Action Plan for Nanotechnology to be published in Spring 2005 and the Technology Platforms on Nanotechnology foreseen in its Seventh Framework Programme commencing in 2006 [8].

3 PROGRAMME

The project will realise these objectives by carrying out the following activities.

3.1 Expert Group

To help explore the potential questions a multi-disciplinary expert working group will be formed which brings together, on the one hand, experts from the relevant areas of scientific research, industry and regulation relating to nanobiotechnology and enhancement, and, on the other, specialists in ethics, social sciences, media and public attitudes. Expert groups have proved immensely valuable in elucidating and exploring issues in a non-aligned context. The group will identify both current and future issues, explore different value perspectives, disciplinary insights and societal contexts.

The expert group meetings will engage scientific practitioners with the ethical and social implications of their research and inform ethical and social science experts in what is and is not realistic in the science.

3.2 Horizon scanning workshops

Two workshops will bring together the key players in the nanobiotechnology scientific and commercial field together with the leading ethicists and public communication experts concerned with it. Their aim will be to forecast the serious societal and ethical issues likely to emerge and to recommend the responses which should be made.

The main conclusions from these workshops will be published as summary reports. These reports will be used in the training courses for nanobiotechnologists and will be widely disseminated to policy makers, scientists, companies, public interest organisations and journalists, and will be made available as PDF files via the project website

3.3 Public opinion focus group discussions

Four focused discussions will be conducted with small groups of lay people led by a professionally-trained ethicist moderator in west, north, south and east Europe for variation in opinions in the different regions.

The main conclusions from these focus group discussions will published as a summary report which will be used in the training courses and widely disseminated in the same ways as described above for the horizon-scanning workshop reports.

3.4 Ethics & Public Communication Courses for Nano-biotechnologists

Four 5-day residential courses for research and company scientists will be held to increase awareness of the issues, to encourage, support and train the participants in communicating with the public, the media and politicians about the ethically-related topics in their scientific fields and to discuss models for institutionalisation.

The training courses will enable the participants to carry out a wide variety of public communication activities discussing the ethical implications of their work with confidence. From the experience of previous similar courses the participants take what they have learned enthusiastically back to their colleagues and institutions, acting as "amplifiers", undertaking and organising further outreach and representational activities and working to establish these approaches in the courses and activities of their own institutions.

3. 5 Information & Dissemination

A series of activities will be carried out designed to raise awareness, augment skills and enhance attitudes to nanobiotechnology.

Ethics lecturers will make presentations during scientific conferences and will be available for media interviews, discussions with politicians and similar occasions.

Three types of publications designed for their respective audiences will be produced and widely disseminated. These will be in the form of:
– issue papers for government officials and politicians
– briefing papers for consumers, teachers, media, etc.
– ethical briefings for scientists.

A public relations professional will assist with government and media-related activities in order to achieve the maximum possible impact for the project's activities and publications to EU and national politicians and policy makers, the media, public interest organisations and to the public at large.

A regularly updated and user-friendly website will make all of the project's quality controlled publications and activities available.

3.6 On-line Forum and Database

An on-line forum linking the members of the main partner organisations and the other participants will enable them to communicate easily and efficiently by email about all of the project's activities.

An annotated bibliographic database of nanobiotechnology-related ethics research publications, public opinion surveys, researchers, organisations, conferences, reports, "grey literature", etc. will provide a selected and quality-controlled collection of key reference and resource material for the use of the project's participants and others interested in the topics covered by the Nanobio-RAISE project.

3.7 Support to EC Nanotechnology Action Plan and FP7 Technology Platform activities

The project has been designed in so far as is possible to anticipate the actions recommended in the European Commission's Action Plan for Nanotechnology to be published in Spring 2005 and for the Nanotechnology Technology Platform foreseen in the Seventh Framework Programme. Activities not covered in the presently proposed project will be the subjects of (a) further proposal(s) when the Action Plan is published and the Technology Platform is (being) established.

4 PARTICIPANTS

Dr David Bennett, **Coordinator**, Secretary, European Federation of Biotechnology Task Group on Public Perceptions of Biotechnology, The Netherlands
Dr Patrick Boisseau, Coordinator, Nano2Life Network of Excellence, France
Henriette Bout, Free University, Amsterdam, The Netherlands
Prof Richard Braun, former Chairman, European Federation of Biotechnology Task Group on Public Perceptions of Biotechnology, Switzerland
Dr Donald Bruce, Director, Church of Scotland, Society, Religion and Technology Project, United Kingdom
Dr Emilio Rodriguez Cerezo, Action Leader, Sustainability in Agriculture, Food and Health Unit, Institute for Prospective Technological Studies, European Commission-Joint Research Centre, Spain
Prof Sven Ove Hansson, Philosophy Unit, Royal Institute for Technology in Stockholm, Sweden
Prof Sir Brian Heap, Fellow, St Edmund's College, Cambridge and Royal Society, United Kingdom
Dr Johan Vanhemelrijck, Secretary General, EuropaBio, Belgium
Prof George Khushf, Department of Philosophy, Center for Bioethics, the University of South Carolina, United States of America
Prof Julian Kinderlerer, Chairman, European Federation of Biotechnology Task Group on Public Perceptions of Biotechnology and Sheffield Institute for Biotechnology Law and Ethics, United Kingdom
Dr Beatrix Rubin Lucht, University of Basel, Switzerland
Dr Onora O'Neill, Principal, Newnham College, Cambridge, United Kingdom
Lino Paula, Institute for Biology, Leiden University, The Netherlands
Prof Arie Rip, Centre for Studies of Science, Technology and Society, Twente University, The Netherlands
Dr Cristina Román Vas, Executive Director, European NanoBusiness Association, Belgium
Daan Schuurbiers, Project Manager, Kluyver Centre for Genomics of Industrial Fermentation, Delft University of Technology, The Netherlands
Prof Ludwig Siep, Department of Philosophy, Center for Bioethics, Westfälische Wilhelms-Universität, Germany
Dr Cristoph Steinbach, Secretary, Working Group on Nanotechnology, DECHEMA, Germany
Dr Klaus-Michael Weltring, Ethics Working Group, Nano2Life Network of Excellence, Germany

REFERENCES

[1] "The Great GM Food Debate", UK Parliamentary Office of Science & Technology (POST) Report 138, 2000.

[2] "Nanoscience and nanotechnologies: opportunities and uncertainties", London: The Royal Society & The Royal Academy of Engineering, 2004.

[3] G. Gaskell, N. Allum and S. Stares, "Europeans and Biotechnology in 2002, Eurobarometer 58.0" 2nd Edition: March 21st, 2003.

[4] R. A. Freitas Jr, "Some Limits to Global Ecophagy by Biovorous Nanoreplicators with Public Policy Recommendations" at: http://www.foresight.org/NanoRev/Ecophagy.html.

[5] A. Mnyusiwalla, A. Daar and P.A. Singer, "'Mind the gap': science and ethics in nanotechnology", Nanotechnology 14, R9–R13, 2003.

[6] "Towards a European Strategy for Nanotechnology", Luxembourg: Office for Official Publications of the European Communities, 2004.

[7] I. Malsch and M. Oud, "Outcome of the Open Consultation on the European Strategy for Nanotechnology", Nanoforum, 2004.

[8] The European NANOMEDICINE Technology Platform: Nanobiotechnologies for Medical Applications ftp://ftp.cordis.lu/pub/era/docs/3_nanomedicine_tp_en.pdf

European Nanoelectronics Initiative Advisory Council (ENIAC): ftp://ftp.cordis.lu/pub/era/docs/2_nanoelectronics_tp_en.pdf

The Austrian NANO Initiative
Small Country with Strong Expertise

M. Haas and E. Glenck

Austrian Research Promotion Agency (FFG)
Canovagasse 7, A-1010 Vienna, Austria
margit.haas@ffg.at, emmanuel.glenck@ffg.at
www.nanoinitiative.at or www.nanoforum.at

ABSTRACT

The Austrian NANO Initiative is the multi-annual public funding programme for nanoscale sciences and nanotechnology. It coordinates different action lines at national and regional levels. The Austrian Research Promotion Agency (FFG) runs the programme on behalf of the Ministry for Transport, Innovation and Technology (BMVIT). This initiative is supported by further Ministries, States and funding agencies. The public annual budget is EURO 10-15 m. In 2004, five outstanding clusters consisting of 39 projects were selected after an international evaluation procedure. 11 Universities, 12 companies and 2 Centres of Competence are leading the clusters.
The topics are:
- diamonds for biotechnological applications
- new materials for drug targeting, release and imaging
- integrated organic sensors and optoelectronics
- coatings for multifunctional surface design
- nanostructured surfaces and interfaces

Keywords: Austrian NANO Initiative, public funding programme, biotech, organic sensors, coatings, surfaces

1 AUSTRIAN NANO INITIATIVE

Nanotechnology is a strategic focus for Austrian research and industry. The strategic objectives are to strengthen and network the Austrian players in science and industry and to build critical masses by clustering competencies in large scale R&D Projects. Feasibility studies enable the early identification of R&D-project ideas with high potential. Training and education measures will support scientific and industrial staff.
The Austrian NANO Forum is the national information and communication platform open to all players and networks in Austria. A strong financial support for networking activities is therefore provided both in a national and an international context.
Relevant information on the Austrian NANO Initiative is provided under:
www.nanoinitiative.at and www.nanoforum.at

1.1 Strategic Goals

The strategic goals are:
- Mid/long term cooperation and networking between scientists and companies to develop new technologies and systems.
- Building critical masses for positioning the Austrian NANO players in international competition.
- Strengthening the national and international RTD cooperation.
- Increase the know-how through education and training.
- Building and expanding infrastructure as well as building centres in basic research and in application-oriented special fields.

1.2 Measures

The Austrian NANO Initiative encompasses 4 programme action lines:

1. Research and Technology Development Project Clusters
2. Networks and Confidence Building
3. Training and Education Measures
4. Accompanying Measures

The NANO Initiative addresses all NANO players from university and research institutions, as well as enterprises located in Austria. International partners are welcome to participate in all Programme Action Lines.

The objective of the project clusters is to develop new processes and applications, building on nanoscience findings. In a project cluster, several research institutions and enterprises cooperate in projects of several years, ranging from basic research to industrial research and technology development. The five outstanding clusters are introduces in chapter 3.

Confidence building activities are relying on new industrial requirements or the valorisation of new R&D results in companies. This exploring activity aims at investigating the possibilities of developing new functionalities and/or new specifications with a high risk and a high economic potential in a feasibility study.

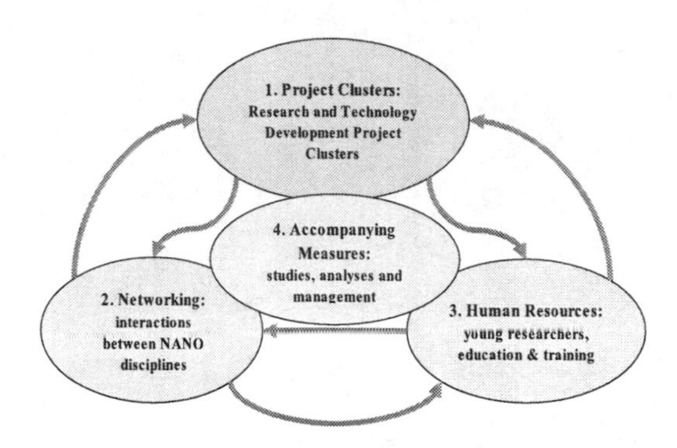

Figure 1: Four Programme Action Lines

Training and Education Measures aim at building up and increasing the human resources required for ensuring the qualitative and quantitative growth of NANO in Austria whereas Accompanying Measures support the strategic orientation of the NANO Initiative.

2 STRONG AUSTRIAN EXPERTISE

One of the main goals of the NANO Initiative is to initiate long lasting co-operations between industry and science. Therefore the first call focused on both the scientific-technological excellence of the projects and on their economic potential to successfully develop technologies or new functionalities. For the period 2004 - 2006 altogether 35 millions EURO are provided for all programme action lines.

3 OUTSTANDING NANO CLUSTERS

The first five successful Austrian NANO project clusters have started their ambitious research and technological development work.

3.1 NaDiNe

The project NaDiNe "Nano Diamond Network" unites seven projects and fifteen partners. The ultra-nano-crystalline diamond (UNCD) has excellent propertie to be used as material for biotech applications. The main topics are cell engineering, nano-biosensors, MEMS and NEMS. The Competence Centre Medicine in Tyrol (KMT) coordinates the project cluster together with the company Rho-BeSt coating [1].

3.2 Nano-Health

NANO-HEALTH "Nanostructured Material for Drug Targeting, Release and Imaging" is coordinated by JOANNEUM RESEARCH and focuses on the development of new therapy solutions for diseases such as diabetes and Alzheimer applying innovative nano-particles. The cluster deals with new strategies using nano-techniques to deliver active substances to specific targets and organs [2], [3]. This will open new opportunities in invasive therapy forms, but also will drastically improve the quality of life of the patients.

3.3 Isotec

ISOTEC "Integrated Organic Sensor and Optoelectronics Technologies" – lead by JOANNEUM RESEARCH and the University of Technology in Graz – will open new opportunities in the combination of new organic semi-conductors as well as structuring and production methods applicable for sensors and optoelectronics [4], [5].

3.4 Nanocoat

NANOCOAT "Development of Nanostructured Coatings for the Design of Multifunctional Surfaces" is coordinated by the Materials Competence Centre in Leoben (MCL) and focuses on the development of concepts for suitable manufacture technologies to design multi-functional surfaces and interfaces. The focus is put on friction layers with additional functions like auto-lubrication or anti-adhesive effects. The application fields are tool and manufacturing units coatings as well as functional coatings (e.g. consumer goods) [6], [7].

3.5 NSI

NSI "Nanostructured Surfaces and Interfaces" is lead by the NanoScience & Technology Centre in Linz (NSTL). The main topics are biocompatible nano-structures, polymers and nanocomposites as well as metal surfaces and interfaces [8], [9].

REFERENCES

[1] www.nanoinitiative.at or www.nanoforum.at
[2] A. Bernkop-Schnürch, AH. Krauland, VM. Leitner, T. Palmberger, Thiomers, „Potential excipients for non-invasive peptide delivery systems", Eur. J. Pharm. Biopharm., Vol. 58, 253-63, 2004.
[3] M. Junghans, J. Kreuter, and A. Zimmer, "Antisense delivery using Protamine - oligonucleotide – particles", Nucleic Acids Res. Vol. 28, E45, 2000.
[4] C. Suess, F. P. Wenzl, G. Jakopic, M. Wuchse, S. Muellegger, N. Koch, A. Haase, K. Lamprecht, M. Schatzmayr, C. Mitterbauer, F. Hofer, G. Leising,

„Combined XPS, AFM, TEM and ellipsometric studies on nanoscale layers in organic light emitting diodes", Surface-Science, Vol. 507-510, 473-9, 2002.

[5] M. Gaal, Ch. Gadermaier, H. Plank, E. Moderegger, A. Pogantsch, G. Leising, E.J.W. List, „Imprinted Conjugated Polymer Laser", Advanced Materials, Vol. 15, 1165-67, 2003.

[6] P.H. Mayrhofer, P.Eh. Hovsepian, C. Mitterer, W.-D. Münz; "Calorimetric evidence for frictional self-adaptation of TiAlN/VN superlattice coatings", Surf. Coat. Technol. Vol. 177-178, 341, 2004.

[7] J. M. Lackner, W. Waldhauser, R. Ebner, B. Major, T. Schöberl, "Pulsed laser deposition of titanium oxide coatings at room temperature - structural, mechanical and tribological properties", Surf. Coat. Technol. Vol. 180-181, 585, 2004.

[8] J. Hesse, M. Sonnleitner, A. Sonnleitner, G. Freudenthaler, J. Jacak, O. Höglinger, H. Schindler, G.J. Schütz, "Single molecule reader for high-throughput bioanalysis", Anal.Chem, Vol.76, 5960-5964, 2004.

[9] L. D. Sun, M. Hohage, P. Zeppenfeld, R. E. Balderas-Navarro, and K. Hingerl, "Enhanced Optical Sensitivity to Adsorption due to Depolarization of Anisotropic Surface States", Phys. Rev. Lett. Vol. 90, 106104, 2003.

Through The Nano-Needle's Eye: Models, Examples, And A Template For The Successful Transition From R&D To Commercialization

Michael O'Halloran

Industrial Design & Construction (IDC)
2020 SW Fourth Avenue, Suite 350, Portland, Oregon 97201, michael.ohalloran@idc-ch2m.com

ABSTRACT

As small-scale technologies emerge and are slated for commercialization, the critical challenge is scaling up prototypes into viable commercial applications. Many new technologies have faced this challenge, and many lessons learned from those experiences can be applied to the diverse fields comprising nanotechnology.

This paper details issues and strategies developed to improve the success of shifting from promising nanoscale research into profitable commercial production.

A key theme in this paper, based on findings from specific case histories involving nanotech research and manufacturing entities, is the contention that advancing nanotechnology research into commercial viability requires a clear understanding of the entire life cycle of technology development. A key target audience for this paper is senior management representatives.

Keywords: nano-technology, R&D, pilot, commercialization.

1 INTRODUCTION

Small-scale technologies (micro and nano) include a wide variety of products and applications. They include new materials or modified materials that may be of use on their own or to modify the bulk properties of existing materials. These include developments such as the use of carbon nano-tubes to build composite "super strength' materials. In other cases, small scale technologies are used to develop new solutions or build up products related to MEMS (or NEMS) devices, sensors, displays, bio-chips, lab-on-chip, microelectronics and so on. Indeed, Business Week recently identified these later applications as those expected to have first significant commercial applications for nanotechnologies.

It is our experience that the "eye of the needle" as it relates to the commercialization of technology is a business problem. If the technology is based on good science, it will be funding, organizational issues and schedule issues that kill the development.

1.1 Types of Commercialization

For the moment I would like to segregate the scope of the topic of transition from R&D to commercialization into two types. Type "A" will be those technologies that transfer cleanly into the commercial world. This is because frequently, once proven, these technologies become integrated into the manufacturing of existing products. The movement of semiconductor manufacturing technology into nano-scale is an example. Newly developed technology, once proven, is inserted into manufacturing processes and the technology steps forward into commercialization. Type "B" will be those technologies that result in the development of new and potentially disruptive products.

This paper is primarily focused on Type "B" technologies. The distinction may not always be clear, but for the moment we will examine the case where new technology is being developed that will result in new products which must be built in new factories. Generally, the nature of such technology-based products is that the associated manufacturing factories are relatively unique, and they are also complex and expensive.

2 THE COMMERCIALIZATION PROCESS FOR NEW PRODUCTS

As nanotechnology development moves forward with new products, experience indicates that the product development will go through a lifecycle sequence moving from R&D through Pilot/Development and into Mass Production (this is true of most complex technology-based development processes).

Figure 1: Formatting dimensions for manuscripts

It is the complexity and expense of the technology and manufacturing process that causes the need for this development model. The significant issues of funding, organizational structure and schedule must be addressed at each phase.

2.1 Funding

Let us begin at the end of the process and work our way backward. The end goal is a viable commercial product being sold in the marketplace for a profit. Millions of dollars, perhaps hundreds of millions of dollars will be invested in the factory. As the technology moves into production we will need additional investment above the pilot level. Our product must be protected by patents, manufacturing know how and/or other barriers to entry such as investment requirements or marketing and distribution channels. Investors like to be aware of as many of these barriers as possible. Prior to investing in the manufacturing plant, investors want to know:

- What is the cost of the product?
- What is the output of the factory?
- What will be the plant cost and cash flow requirements?
- What is the schedule?
- How big is the market?
- What is the selling price?
- Who or what is the competition and how do we keep them away?
- How many staff do you need and where do they come from?

The investors want quantitative (not qualitative) answers to these questions. With an accuracy to within +/-10% to at most +/-25%, they want a business plan. And they want the data to back up the answers. This information normally comes from the Pilot/Development phase of the program.

The Pilot phase is not as expensive as the production phase, but costs can be substantial. Investors ask similar questions when we move to production, but expect less from the answer. They will also be asking questions related to the science; specifically, is it sound?

The important issue to the R&D organization is that the better and more quantitative the answers to investors at both the Pilot and Manufacturing phases, the less risk investors will perceive. The less risk, the faster they will invest, the more they will invest and the less they will seek in return. It is our experience that failure to properly execute the Pilot Phase results in significant problems with obtaining investment, or (worse) significant financial and schedule problems during production.

2.2 Objectives of the Pilot/Development Phase

The pilot phase should use real (commercial) manufacturing equipment in a simulated manufacturing environment to produce commercial product. This will prove the technology within a "product" orientation. It will develop manufacturing run rates and produce product for the development of the marketplace. The objectives of pilot production are:

- Processes are optimized and brought into statistical control.
- Costs are developed.
- Suppliers are developed and pricing negotiated.
- Trial product is produced to seed distribution channels and test markets.

2.3 Organizational Issues

The following model (figures 2, 3 and 4) show the characteristics of the organization as it migrates from R&D to Production. A study of the characteristics indicates the significant scope of change an organization must accept as transitions are made. A stable R&D organization morphs into a pilot organization. Then, after just a short time, the organization morphs again into a production mode. This is exceptionally difficult. Think about the model as a business manager.

Again, management must answer questions:

- How to hire personnel for a pilot phase and fire them in 2 years?
- Are researchers disciplined enough to work in manufacturing? Will they lose interest?
- Will R&D people or manufacturing people be able to deal with the chaos and urgency of the pilot phase?

Staff	•Define and demonstrate theoretical concepts in a lab-scale industrial environment. •Technical staff hard science- and research-oriented (80% technologists, 20% engineers). •High degree of individual contribution. •Primary compensation based 90% individual, 10% team. •Focused on discrete events and intradiscipline interactions.
Equipment/ Facilities	•Provide a lab-scale industrial environment for development and prototyping activities. •Uncharacterized tools utilized with non-optimized equipment recipes. •Tool set flexible, portable, and multifunctional. •Work areas decentralized with layout optimized for intradiscipline research and development.
Process, Product, Procedures	•Define individual process steps and confirm initial sequence of operations. •Process variables understood through modeling/simulation and individual step sensitivity studies. •Chemical reactions and scaling parameters understood. •Produce a fully featured and functional prototype. •Initial prototype produced and characterized with respect to key performance variables. •Provide a flexible framework for the coordination of diverse development activities.
Materials	•Provide a basic set of materials specifications including initial sensitivity analysis with respect to intramaterial variation. •Initial experiments done with lab purity materials to obtain highest theoretical properties. •Material alternatives and substitutions freely examined with "decision to use" based on first-order impacts.

Figure 2: R&D Phase

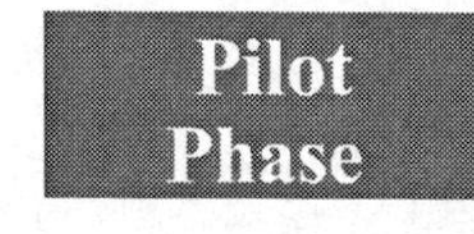

Staff	•Integration of developed concepts in a prototype manufacturing environment. •Technical staff balanced (50% hard science- and research-oriented and 50% engineering). •High degree of intrafunctional teams. •Focused on system-level events with balance between intra- and interdiscipline interaction.
Equipment/ Facilities	•Provide a manufacturing-scale environment for initial production equipment burn-in and pilot production. •One-of-each fully sized tools with optimized equipment recipes. •Equipment set user-friendly, repeatable, and functionally optimized. •Layout optimized for efficient manufacturing flow and support area centralization. •Involve vendors in partnership relationships.
Process, Product, Procedures	•Integrate process steps and define manufacturing flow. •Process integration variables understood through sensitivity analysis. •Manufacturing process model defined and characterized. •Produce a fully featured and functional production-worthy product in limited volumes. •Final production product defined framework supporting manufacturing requirements. •Provide a flexible but defined framework supporting manufacturing requirements. •Production support infrastructure defined and implemented.
Materials	•Freeze production bill-of-materials and provide initial intermodule/intermaterial sensitivity analysis. •Define final material purity and compositional requirements. •Determine proper cost vs. performance material trade-offs with 'decision to use' based on first-order. •Develop qualification requirements for vendors and materials. •Involve vendors in partnership relationships.

Figure 3: Pilot Phase

Staff	•Sustain and continually improve the ongoing production operations. •Technical staff 90% engineering and 10% hard science- and research- •High degree of interfunctional teams. •Primary compensation based 40% individual, 60% team. •Focused on system-level events and interdiscipline interaction.
Equipment/ Facilities	•Provide a fully operational manufacturing environment for high-volume •Multiplexed, fully characterized production tool set running stable, frozen equipment •Equipment set fully instrumented, in-situ monitored, and optimally •Layout optimized for maximum output, minimal cycle time, and lowest manufacturing
Process, Product, Procedures	•Running a frozen manufacturing process flow. •Process driven by statistical controls. •Manufacturing process model only changed through continued characterization in incremental steps market-driven demand changes. •Fully characterized products running in high volumes. •Final production product specifications frozen. •Provide a stable, defined framework preventing variation. •Production support infrastructure optimized.
Materials	•Bill-of-materials components optimized for cost reduction and supply •Low cost materials substitutes investigated and qualified. •Cost vs. performance trade-offs controlled tightly. •Vendors become full partners and part of the manufacturing flow.

Figure 4: Production Phase

2.4 Schedule Issues

The R&D to commercialization transition includes many activities. The following time frames are intended to be representative and can vary significantly.

- Negotiation with Pilot investors takes 2 to 6 months.
- Site selection/negotiations for a building or land takes 2 to 3 months (starting from when you have money).
- Conceptual design and permits for a pilot or production facility takes 2 to 4 months (this depends on the facility complexity and location. Code processes vary by state and country.)
- Negotiations with pilot/manufacturing equipment suppliers take 2 to 4 months.
- Delivery of manufacturing equipment takes 6 to 9 months.
- Remodel of a building to accept a pilot or production operation takes 3 to 6 months.
- New construction takes 12 to 15 months.
- Installation and turn of manufacturing equipment takes 1 month.
- Start up and shakedown of "the process" takes 3 to 12 months (this is heavily dependent on preplanning and complexity).
- Negotiations and development of distribution channels takes 6 to 18 months.
- Market development takes 12 months (but this can vary widely).

3 SUMMARY

Technology developments do not "speak for themselves." They require extensive effort, time and money to get to the point where there is an impact on society. The process can seem overwhelming. But, if we focus on the Pilot Phase (the eye of the needle), we can handle the process in manageable parts.

NSTI-Nanotech 2005, www.nsti.org, ISBN 0-9767985-1-4 Vol. 2, 2005

Index of Authors

Index of Keywords

 NSTI-Nanotech 2005, www.nsti.org, ISBN 0-9767985-1-4 Vol. 2, 2005